W9-CSF-025

HANDBOOK OF ACOUSTICS

EDITORIAL BOARD

HANDBOOK OF ACOUSTICS

MALCOLM J. CROCKER, *Editor-in-Chief*
Auburn University

A Wiley-Interscience Publication
JOHN WILEY & SONS, INC.
New York • Chichester • Weinheim • Brisbane • Singapore • Toronto

This text is printed on acid-free paper. 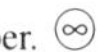

Library of Congress Cataloging in Publication Data:

Crocker, Malcolm J.
Handbook of acoustics / Malcolm Crocker.
p. cm.
"A Wiley-Interscience Publication"
Includes bibliographical references and index.
ISBN 0-471-25293-X (cloth : alk. paper)
1. Acoustics—Handbooks, manuals, etc. I. Title.
QC228.8.C76 1998
534—dc21 97-41670

Printed in the United States of America.

10 9 8 7 6 5 4 3 2 1

CONTENTS

PREFACE

This book has been designed to fill the need for a comprehensive source book on acoustics. Although there are several excellent textbooks on different aspects of acoustics theory or practice and some handbooks on specialized topics in acoustics and vibration, no single volume gives a comprehensive treatment. Up-to-date discussions on all topics of acoustics are certainly to be found in journals, periodicals, books, and conference proceedings. But the serious reader who wants a good understanding of acoustics needs to consult at least fifteen or twenty journals and five to ten books. Considering the continuing rapid expansion of scientific knowledge and the fact that many journal articles are written at a very high level for just a few specialists, it is very difficult for a reader to obtain even an elementary grasp of many acoustics topics without considerable effort. For these same reasons, it became apparent to me some years ago that no one author could really do justice to the whole field of acoustics and that a multiple-author book would be needed to introduce the reader to important topics in the whole field of acoustics.

Discussions with colleagues in the 1980s soon convinced me that there was a need for a comprehensive treatment of acoustics, and I began to plan a detailed outline of a book with 160 chapters. At that time, I was fortunate to obtain the assistance of additional colleagues and form the Editorial Board shown at the beginning of this book. The Editorial Board members provided invaluable help in suggesting the best possible author to write each chapter.

Before the authors began to write their chapters in the early 1990s, I prepared a detailed contents of the whole book with the help of the Editorial Board and some authors; this included a brief outline of each chapter. The tentative book contents and chapter outlines were given to each author before they began to write their chapters. In this way, it has been possible to reduce extensive overlap in treatment of topics and to ensure that there is adequate cross-referencing.

Each chapter was evaluated by several reviewers, and their anonymous comments were shared with the authors who updated their chapters and provided their final versions late in 1995 and early 1996. Each author was asked to write at a level accessible to general readers, not just specialists. At the same time, they were asked to write chapters that were at a reasonably high scientific level and to provide suitable, up-to-date, readily accessible references for the reader who would wish to study a topic in more depth. To prevent the book becoming too long, each author was given a page allocation. Although it may have been unreasonable to expect authors to meet all of these objectives in a limited space, my impression is that most authors responded admirably. The result was the four-volume, 2000-page *Encyclopedia of Acoustics* published by John Wiley & Sons in 1997. It contains 166 chapters. This book is hardly portable, however, and is mostly of use to readers in libraries as a reference tool.

Many readers have requested a more concise single-volume version of the *Encyclopedia* be made available to a larger group of readers. The *Handbook of Acoustics* is the result. Unfortunately, to make it more concise it has been necessary to omit two parts of the *Encyclopedia* (those on bioacoustics and animal bioacoustics) and about 20% of the remaining chapters, although one new chapter on Analyzers has been included. However, cross-referencing is maintained to the chapters that are omitted here but appear in the *Encyclopedia* should readers need to refer to those chapters.

The *Handbook* is divided into 16 main parts and contains a total of 113 chapters. A member of the Editorial Board has been associated with each part and has normally written the first chapter to serve as an introduction to that part. Coverage includes linear and nonlinear acoustics and cavitation, aeroacoustics and atmospheric sound propagation, underwater acoustics, ultrasonics, mechanical vibration and shock, statistical methods, noise control, architectural acoustics, signal processing, physiological acoustics, psychological acoustics, speech communication, musical acoustics, measurements and instrumentation, and transducer design. With coverage of so many topics, it is hoped that the *Handbook of Acoustics* will be of interest to engineers, scientists, architects, musicians, physicians, psychologists, and all those whose life acoustics touches in one way or another.

I should like to thank all of the authors who worked so hard to make this book possible. In addition, I am very much indebted to the 400 reviewers who donated their time to read first drafts of the chapters and to make helpful comments to the authors. The members of the Editorial Board also provided considerable assistance for which I am grateful. I was supported with a very able staff at my university including Yana Sokolova and Lisa Beckett who provided really splendid assistance. I would also like to thank Xinche Yan and Avinash Patil for their help with proofreading. The staff at Wiley must also be thanked including Bob Argentieri and Bob Hilbert, who guided both this *Handbook* and the earlier *Encyclopedia* to successful conclusions. Last and not least I want to thank my wife Ruth and daughters Anne and Elizabeth for their support, patience, and understanding and my mother Alice for her continual encouragement with this and all of my projects.

MALCOLM J. CROCKER

November, 1997

CONTRIBUTORS

Robert E. Apfel, Department of Mechanical Engineering, Yale University, New Haven, CT 06520-8286

Bishnu S. Atal, Bell Laboratories, Lucent Technologies, 600 Mountain Ave., Murray Hill, NJ 07974

James E. Barger, BBN Systems and Technologies, 70 Fawcett St., Cambridge, MA 01238

Henry E. Bass, National Center for Physical Acoustics, University of Mississippi, University, MS 38677

James W. Beauchamp, University of Illinois at Urbana-Champaign, Urbana, IL 61801

Elliott H. Berger, E-A-R/Aearo Company, 7911 Zionsville Rd., Indianapolis, IN 46268-1657

A. J. Berkhout, Laboratory of Seismics and Acoustics, Delft University of Technology, The Netherlands

David A. Bies, Department of Mechanical Engineering, University of Adelaide, South Australia, 5005, Australia

David T. Blackstock, Mechanical Engineering Department and Applied Research Laboratories, The University of Texas at Austin, Austin, TX 78713-8029

William K. Blake, David Taylor Model Basin, Bethesda, MD 20084-5000

J. M. Bowsher, Formerly with Department of Physics, University of Surrey, Guildford, Surrey GU2 5XH U.K.

Mack A. Breazeale, National Center for Physical Acoustics, University of Mississippi, University, MS 38677

Robert D. Bruce, Collaboration in Science and Technology Inc., 15835 Park Ten Place, Houston, TX 77084-5131

P. V. Brüel, Brüel Acoustics, Holte, Denmark

John C. Burgess, Department of Mechanical Engineering, University of Hawaii, Honolulu, HI 96822

Ilene J. Busch-Vishniac, Mechanical Engineering Department, The University of Texas at Austin, Austin, TX 78712

Søren Buus, Communication and Digital Signal Processing Center, Department of Electrical and Computer Engineering, Northeastern University, Boston, MA 02115

Marvin Camras, Illinois Institute of Technology, Chicago, IL 60616

John H. Cantrell, NASA Langley Research Center, Hampton, VA 23681-0001

John G. Casali, Industrial and Systems Engineering, Virginia Tech, Blacksburg, VA 24061

Josko A. Catipovic, Woods Hole Oceanographic Institution, Woods Hole, MA 02543

William J. Cavanaugh, Cavanaugh Tocci Associates, Inc., Sudbury, MA 01776

Robert D. Collier, Consultant in Acoustics, rcollier@tpk.net

A. Craggs, Department of Mechanical Engineering, University of Alberta, Edmonton, Alberta, Canada

D. G. Crighton, Department of Applied Mathematics & Theoretical Physics, University of Cambridge, Cambridge CB3 9EW U.K.

Malcolm J. Crocker, Department of Mechanical Engineering, Auburn University, Auburn, AL 36849

Lawrence A. Crum, Applied Physics Laboratory, University of Washington, Seattle, WA 98105

Gilles A. Daigle, National Research Council, Ottawa, Ontario K1A 0R6 Canada

P. O. A. L. Davies, Institute of Sound and Vibration Research, The University of Southampton, Southampton SO17 1BJ U.K.

Earl H. Dowell, School of Engineering, Duke University, Durham, NC 27708-0271

A. P. Dowling, Engineering Department, Cambridge University, Cambridge, U.K.

Louis R. Dragonette, Applied Acoustics Section, Naval Research Laboratory, Washington, DC 20375

Ira Dyer, Massachusetts Institute of Technology, Cambridge, MA 02139

Kenneth McK. Eldred, Ken Eldred Engineering, East Boothbay, ME 04544

S. J. Elliott, Institute of Sound and Vibration Research, University of Southampton, Southampton SO17 1BJ U.K.

E. Carr Everbach, Department of Engineering, Swarthmore College, Swarthmore, PA 19081-1397

F. J. Fahy, Institute of Sound and Vibration Research, University of Southampton, Southampton SO17 1BJ U.K.

Gunnar Fant, Department of Speech Music and Hearing, KTH, Box 70014, Stockholm S-10044 Sweden

Wolfgang Fasold, Friedenstrasse 2, D-15566 Schöneiche, Germany

David Feit, David Taylor Model Basin, Bethesda, MD 20084-5000

J. E. Ffowcs Williams, Department of Engineering, University of Cambridge, Cambridge CB2 1PZ U.K.

Sanford Fidell, BBN Systems and Technologies, 21120 Vanowen Street, Canoga Park, CA 91303-2853

J. L. Flanagan, Rutgers University, Piscataway, NJ 08855-1390

Neville H. Fletcher, Research School of Physical Sciences and Engineering, Australian National University, Canberra 0200 Australia

C. R. Fuller, Vibration and Acoustics Laboratories, Virginia Polytechnic Institute and State University, Blacksburg, VA 24061

Thomas B. Gabrielson, Applied Research Laboratory, The Pennsylvania State University, State College, PA 16804

Charles F. Gaumond, Naval Research Laboratory, Washington, DC 20375

Thomas L. Geers, Center for Acoustics, Mechanics and Materials, Department of Mechanical Engineering, University of Colorado, Boulder, CO 80309-0427

David M. Green, 399 Federal Point Road, East Palatka, FL 32131

Mark F. Hamilton, Department of Mechanical Engineering, The University of Texas at Austin, Austin, Texas 78712-1063

Colin H. Hansen, Department of Mechanical Engineering, University of Adelaide, South Australia 5005, Australia

Manfred A. Heckl, Institut für Technische Akustik, Technische Universität, Berlin, Germany

Robert Hickling, Sonometrics Inc., 8306 Huntington Rd., Huntington Woods, MI 48070

Joseph E. Hind, University of Wisconsin, Madison, WI 53706

Elmer L. Hixson, Electrical and Computer Department, The University of Texas at Austin, Austin, TX 78712

R. M. Hoover, Hoover & Keith, Inc., 11381 Meadowglen, Houston, TX 77082

Larry E. Humes, Department of Speech and Hearing Sciences, Indiana University, Bloomington, IN 47405

Finn Jacobsen, Department of Acoustic Technology, Technical University of Denmark, Lyngby DK-2800 Denmark

R. H. Keith, Hoover & Keith, Inc., 11381 Meadowglen, Houston, TX 77082

Howard F. Kingsbury, Acoustical Consultant

Govindappa Krishnappa, National Research Council, Vancouver, British Columbia, V6T 1W5 Canada

Robert W. Krug, Cirrus Research, Inc., Waukausha, WI 53188

William A. Kuperman, Marine Physical Laboratory, University of California, San Diego, La Jolla, CA 92037

K. Heinrich Kuttruff, Aachen University of Technology, D–52056 Aachen, Germany

Werner Lauterborn, Drittes Physikalisches Institut, Universität Göttingen, D-37073 Göttingen, Germany

Moisés Levy, Department of Physics, University of Wisconsin–Milwaukee, Milwaukee, WI 53201

James Lighthill, Department of Mathematics, University College London, London WC1F 6BT U.K.

Richard H. Lyon, RH Lyon Corp, 691 Concord Ave., Cambridge, MA 02138

Robert C. Maher, University of Nebraska at Lincoln, Lincoln, NE 68588

Jerome E. Manning, Cambridge Collaborative, Inc., 689 Concord Avenue, Cambridge, MA 02138

Philip L. Marston, Department of Physics, Washington State University, Pullman, WA 99164-2814

A. Harold Marshall, Marshall Day Associates, Wellesley St., Auckland 01 Australia

Julian D. Maynard, Department of Physics, The Pennsylvania State University, University Park, PA 16802

Dennis McFadden, Institute for Neuroscience and Department of Psychology, The University of Texas at Austin, Austin, TX 78712

Harry B. Miller, Naval Undersea Warfare Center, New London, CT

Brian C. J. Moore, Department of Experimental Psychology, University of Cambridge, Cambridge CB2 3EB U.K.

Charles T. Moritz, Collaboration in Science and Technology Inc., 15835 Park Ten Place, Houston, TX 77084-5131

Michael Möser, Institut für Technische Akustik, Technische Universität, Berlin, Germany

Victor Nedzelnitsky, National Institute of Standards and Technology, Gaithersburg, MD 20899

P. A. Nelson, Institute of Sound and Vibration Research, University of Southampton, Southampton SO17 1BJ U.K.

David E. Newland, Department of Engineering, University of Cambridge, Cambridge CB2 1PZ U.K.

Kanji Ono, Department of Materials Science and Engineering, University of California, Los Angeles, CA 90095-1595

Emmanuel P. Papadakis, Quality Systems Concepts, Inc., 379 Diem Woods Drive, New Holland, PA 17557-8800

G. Pavic, Acoustics Department, C.E.T.I.M., Senlis 60300 France

William T. Peake, Research Laboratory of Electronics, Massachusetts Institute of Technology, Cambridge, MA 02139 and Eaton-Peabody Laboratory, Massachusetts Eye and Ear Infirmary, Boston, MA 02114

Karl S. Pearsons, BBN Systems and Technologies, 21120 Vanowen Street, Canoga Park, CA 91303-2853

Allan D. Pierce, Department of Aerospace and Mechanical Engineering, Boston University, Boston, MA 02215

Allan G. Piersol, Piersol Engineering Company, 23021 Brenford Street, Woodland Hills, CA 91364

J. Pope, Pope Engineering Company, Newton Centre, MA 02159

Alan Powell, Department of Mechanical Engineering, University of Houston, Houston, TX 77204-4792

M. G. Prasad, Noise and Vibration Control Laboratory, Department of Mechanical Engineering, Stevens Institute of Technology, Hoboken, New Jersey 07030

Lawrence R. Rabiner, AT&T Labs-Research, 600 Mountain Ave., Murray Hill, NJ 07974

Richard Raspet, Department of Physics and Astronomy, University of Mississippi, University, MS 38677

Paul J. Remington, BBN Systems and Technologies, Cambridge, MA 02138

John J. Rosowski, Department of Otology and Laryngology, Harvard Medical School and Eaton-Peabody Laboratory, Massachusetts Eye and Ear Infirmary, Boston, MA 02114

Thomas D. Rossing, Physics Department, Northern Illinois University, Dekalb, IL 60115

Mario A. Ruggero, Department of Communication Sciences and Disorders, Northwestern University, Evanston, IL 60208-3550

Joseph Santos-Sacchi, Department of Surgery/Section of Otolaryngology, School of Medicine, Yale University, New Haven, CT 06510-2757

Bertram Scharf, Northeastern University, Boston, MA 02115 and CNRS Centre de Recherche en Neurosciences Cognitives, Marseille, France

Susan C. Schneider, Department of Electrical and Computer Engineering, Marquette University, Milwaukee, WI 53233

A. F. Seybert, Department of Mechanical Engineering, University of Kentucky, Lexington, KY 40506

Edgar A. G. Shaw, Institute for Microstructural Sciences, National Research Council, Ottawa, Ontario K1A OR6 Canada

F. Douglas Shields, National Center for Physical Acoustics, University of Mississippi, University, MS 38677

U. S. Shirahatti, University of Chicago, Chicago, IL 60637

Norma B. Slepecky, Department of Bioengineering and Neuroscience, Institute for Sensory Research, Syracuse University, Syracuse, NY 13244-5290

Robert C. Spindel, Applied Physics Laboratory, University of Washington, Seattle, WA 98105

Bradley M. Starobin, Polk Audio, 5601 Metro Drive, Baltimore, MD 21215

Peter Stepanishen, Department of Ocean Engineering, University of Rhode Island, Narragansett, RI 02882-1197

Kenneth N. Stevens, Research Laboratory of Electronics and Department of Electrical Engineering and Computer Science, Massachusetts Institute of Technology, Cambridge, MA 02139

Kenneth S. Suslick, Department of Chemistry, University of Illinois, Urbana, IL 61801

Louis C. Sutherland, Consultant in Acoustics, 27803 Longhill Dr., Rancho Palos Verdes, CA 90275

Gregory W. Swift, Condensed Matter and Thermal Physics Group, Los Alamos National Laboratory, Los Alamos, NM 87545

Gregory C. Tocci, Cavanaugh Tocci Associates Inc., Sudbury, MA 01776

Mikio Tohyama, Kogakuin University, Hachioji-shi, Tokyo 192 Japan

Herbert Überall, Department of Physics, Catholic University of America, Washington, DC 20064

Eric E. Ungar, BBN Corporation and Acentech Incorporated, 125 Cambridge Park Drive, Cambridge, MA 02140

Lawrence N. Virgin, School of Engineering, Duke University, Durham, NC 27708-0300

W. Dixon Ward, Department of Otolaryngology, University of Minnesota, Minneapolis, MN 55455

A. C. C. Warnock, Institute for Research in Construction, National Research Council, Ottawa Ontario K1A OR6 Canada

Gabriel Weinreich, University of Michigan, Ann Arbor, MI 48109-1120

D. E. Weston, BAeSEMA Ltd., Apex Tower, 7 High Street, New Malden, Surrey KT3 4LH U.K.

Ewart A. Wetherill, Paoletti Associates, 40 Gold Street, San Francisco, CA 94133

J. Woodhouse, Engineering Department, Cambridge University, Cambridge, U.K.

T. W. Wu, Department of Mechanical Engineering, University of Kentucky, Lexington, KY 40506

William T. Yost, NASA Langley Research Center, Hampton, VA 23681-0001

H. K. Zaveri, Brüel & Kjøer, Nøerum, Denmark

William E. Zorumski, NASA Langley Research Center, Hampton, VA 23681

Allan J. Zuckerwar, NASA Langley Research Center, Hampton, VA 23681

PART I

GENERAL LINEAR ACOUSTICS

1

INTRODUCTION

MALCOLM J. CROCKER

1 INTRODUCTION

The fluid mechanics equations, from which the acoustics equations and results may be derived, are quite complicated. However, because most acoustic phenomena involve very small perturbations, it is possible to make significant simplifications to these fluid equations and to linearize them. The results are the equations of linear acoustics. The most important equation, the wave equation, is presented in this chapter together with some of its solutions. Such solutions give the sound pressure explicitly as functions of time and space, and the general approach may be termed the wave acoustics approach. This chapter presents some of the useful results of this approach but also briefly discusses some of the other alternative approaches, sometimes termed ray acoustics and energy acoustics, that are used when the wave acoustics approach becomes too complicated.

The first purpose of this chapter is to present some of the most important acoustics formulas and definitions, without derivation, which are used in the chapters following in Part I and in many of the other chapers. The second purpose is to make some helpful comments about the chapters that follow in Part I and about other chapters as it seems appropriate.

2 WAVE MOTION

Some of the basic concepts of acoustics and sound wave propagation used in Part I and also throughout the rest of this book are discussed here. For a more advanced mathematical treatment the reader is referred to Chapter 2 and later chapters.

Wave motion is easily observed in the waves on stretched strings and as the ripples on the surface of water. Waves on strings and surface water waves are very similar to sound waves in air (which we cannot see), but there are some differences that are useful to discuss.

If we throw a stone into a calm lake, we observe that the water waves (*ripples*) travel out from the point where the stone enters the water. The ripples spread out circularly from the source at the wave speed which is independent of the wave height. Somewhat like the water ripples, sound waves in air travel at a constant speed, almost independent of their strength. Like the water ripples, sound waves in air propagate by transferring momentum and energy between air particles. There is no net flow of air away from a source of sound, just as there is no net flow of water away from the source of water waves. Of course, the waves on the surface of a lake are circular or two dimensional, while in air, sound waves in general are spherical or three dimensional.

As water waves move away from a source, their curvature decreases, and the *wavefronts* may be regarded almost as straight lines. Such waves are observed in practice as *breakers* on the seashore. A similar situation occurs with sound waves in the atmosphere. At large distances from a source of sound, the spherical wavefront curvature decreases, and the wavefronts may be regarded almost as plane surfaces.

Plane sound waves may be defined as waves that have the same acoustic properties at any position on a plane surface drawn perpendicular to the direction of propagation of the wave. Such plane sound waves can exist and propagate along a long straight tube or duct (such as an air conditioning duct). In such a case, the waves propagate in a direction along the duct and the plane

Note: References to chapters appearing only in the *Encyclopedia* are preceded by *Enc.*

waves are perpendicular to this direction (and are represented by duct cross sections). Such waves in a duct are one dimensional, like the waves on a long string or rope under tension (or like the ocean breakers described above).

Although there are many similarities between one-dimensional sound waves in air, waves on strings, and surface water waves, there are some minor differences. In a fluid such as air, the fluid particles vibrate back and forth in the same direction as the direction of wave propagation; such waves are known as *longitudinal*, *compressional*, or *sound waves*. On a stretched string, the particles vibrate at right angles to the direction of wave propagation; such waves are usually known as *transverse waves*. The surface water waves described are also partly transverse partly longitudinal waves, with the complication that the water particles move up and down and back and forth horizontally. (This movement describes elliptical paths in shallow water and circular paths in deep water. The vertical particle motion is much greater than the horizontal motion for shallow water, but the two motions are equal for deep water.) The water wave direction is, of course, horizontal.

Surface water waves are not compressional (like sound waves) and are normally termed *surface gravity waves*. Unlike sound waves, where the wave speed is independent of frequency, long wavelength surface water waves travel faster than short wavelength waves, and thus water wave motion is said to be *dispersive*.[1] Bending waves on beams, plates, cylinders, and other engineering structures are also dispersive (see Chapter 10). There are several other types of waves that can be of interest in acoustics: shear waves, torsional waves, and boundary waves (see *Enc.* Ch. 12), but the discussion here will concentrate on sound wave propagation in fluids.

3 PLANE SOUND WAVES

If a disturbance in a thin element of fluid in a duct is considered, a mathematical description of the motion may be obtained by assuming that (1) the amount of fluid in the element is conserved, (2) the net longitudinal force is balanced by the inertia of the fluid in the element, (3) the process in the element is adiabatic (i.e., there is no flow of heat in or out of the element), and (4) the undisturbed fluid is stationary (there is no fluid flow). Then the following equation of motion may be derived.[1–6]

$$\frac{\partial^2 p}{\partial x^2} - \frac{1}{c^2}\frac{\partial^2 p}{\partial t^2} = 0. \tag{1}$$

This equation is known as the one-dimensional equation of motion, or *acoustic wave equation*, and it relates the second rate of change of the sound pressure with the coordinate x with the second rate of change of the sound pressure with time t through the square of the speed of sound c. Identical wave equations may be written if the sound pressure p in Eq. (1) is replaced with the particle displacement ξ, the particle velocity u, fluctuating density ρ, or the fluctuating temperature T. However, the wave equation in terms of the sound pressure in Eq. (1) is perhaps most useful, since the sound pressure is the easiest acoustic quantity to measure (using a microphone) and is the acoustic perturbation we sense with our ears. The sound pressure p is the acoustic pressure perturbation or fluctuation about the time-averaged, or undisturbed, pressure p_0.

The *speed of sound waves* c is given for a perfect gas by

$$c = (\gamma RT)^{1/2}. \tag{2}$$

The speed of sound is proportional to the square root of the absolute temperature T. The ratio of specific heats γ and the gas constant R are constants for any particular gas. Thus Eq. (2) may be written as

$$c = c_0 + 0.6T_C, \tag{3}$$

where, for air, $c_0 = 331.6$ m/s, the speed of sound at 0°C, and T_C is the temperature in degrees Celsius. Note that Eq. (3) is an approximate formula valid for T_C near room temperature. The speed of sound in air does not depend on the atmospheric pressure. For a complete discussion of the speed of sound in fluids, see Chapter 5.

A solution to (1) is

$$p = f_1(ct - x) + f_2(ct + x), \tag{4}$$

where f_1 and f_2 are arbitrary functions such as sine, cosine, exponential, log, and so on. It is easy to show that Eq. (4) is a solution to the wave equation (1) by differentiation and substitution into Eq. (1). Varying x and t in Eq. (4) demonstrates that $f_1(ct - x)$ represents a wave traveling in the positive x-direction with wave speed c, while $f_2(ct + x)$ represents a wave traveling in the negative x-direction with wave speed c (see Fig. 1).

The solution given in Eq. (4) is usually known as the *general solution*, since, in principle, any type of sound wave form is possible. In practice, sound waves are usually classified as impulsive or steady in time. One particular case of a steady wave is of considerable importance. Waves created by sources vibrating sinusoidally in

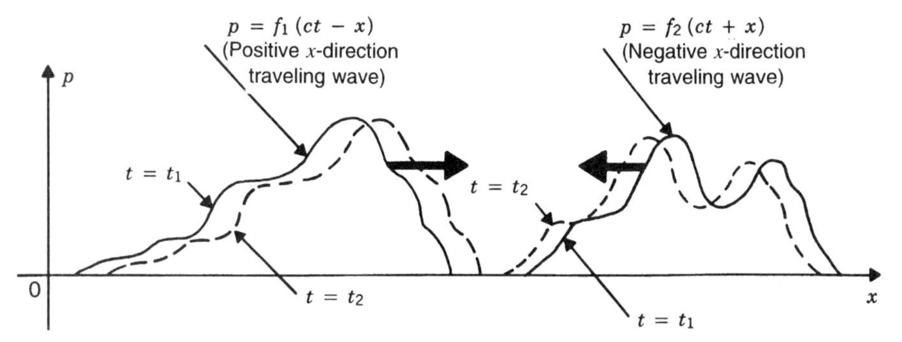

Fig. 1 Plane waves of arbitrary wave form.

time (e.g., a loudspeaker, a piston, or a more complicated structure vibrating with a discrete angular frequency ω) vary both in time t and space x in a sinusoidal manner (see Fig. 2):

$$p = p_1 \sin(\omega t - kx + \phi_1) + p_2 \sin(\omega t + kx + \phi_2). \quad (5)$$

At any point in space, x, the sound pressure p is simple harmonic in time. The first expression on the right of Eq. (5) represents a wave of amplitude p_1 traveling in the positive x-direction with speed c, while the second expression represents a wave of amplitude p_2 traveling in the negative x-direction. The symbols ϕ_1 and ϕ_2 are *phase angles*, and k is the *acoustic wavenumber*. It is observed that the wavenumber $k = \omega/c$ by studying the ratio of x and t in Eqs. (4) and (5). At some instant t the sound pressure pattern is sinusoidal in space, and it repeats itself each time kx is increased by 2π. Such a repetition is called a wavelength λ. Hence, $k\lambda = 2\pi$ or $k = 2\pi/\lambda$. This gives $\omega/c = 2\pi f/c = 2\pi/\lambda$, or

$$\lambda = \frac{c}{f}. \quad (6)$$

The wavelength of sound becomes smaller as the frequency is increased. At 100 Hz, $\lambda \approx 3.5$ m ≈ 10 ft. At 1000 Hz, $\lambda \approx 0.35$ m ≈ 1 ft. At 10,000 Hz, $\lambda \approx 0.035$ m ≈ 0.1 ft ≈ 1 in.

At some point x in space, the sound pressure is sinusoidal in time and goes through one complete cycle when ω increases by 2π. The time for a cycle is called the period T. Thus, $\omega T = 2\pi$, $T = 2\pi/\omega$, and

$$T = \frac{1}{f}. \quad (7)$$

4 IMPEDANCE AND SOUND INTENSITY

We see that for the one-dimensional propagation considered the sound wave disturbances travel with a constant

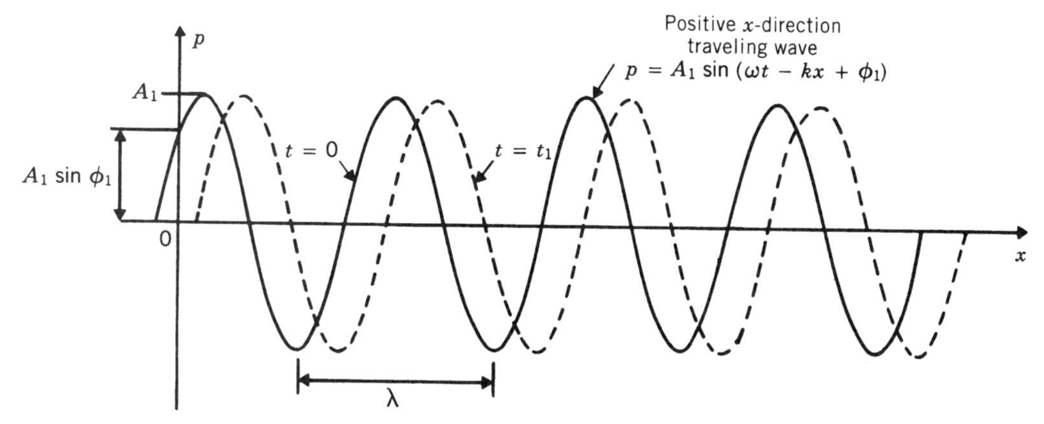

Fig. 2 Simple harmonic plane waves.

wave speed c, although there is no net, time-averaged movement of the air particles. The air particles oscillate back and forth in the direction of wave propagation (x-axis) with velocity u. We may show that for any plane wave traveling in the positive x-direction at any instant

$$\frac{p}{u} = \rho c, \tag{8}$$

and for any plane wave traveling in the negative x-direction

$$\frac{p}{u} = -\rho c. \tag{9}$$

The quantity ρc is called the characteristic impedance of the fluid, and for air, $\rho c = 428\ \mathrm{kg/m^2 \cdot s}$ at 0°C and 415 $\mathrm{kg/m^2 \cdot s}$ at 20°C.

The *sound intensity* is the rate at which the sound wave does work on an imaginary surface of unit area in a direction perpendicular to the surface. Thus, it can be shown that the *instantaneous* sound intensity in the x-direction, I, is obtained by multiplying the *instantaneous* sound pressure p by the *instantaneous* particle velocity in the x-direction, u. Therefore

$$I = pu, \tag{10}$$

and for a plane wave traveling in the positive x-direction this becomes

$$I = \frac{p^2}{\rho c}. \tag{11}$$

The time-averaged sound intensity for a plane wave traveling in the positive x-direction, $\langle I \rangle_t$, is given as

$$\langle I \rangle_t = \frac{\langle p^2 \rangle_t}{\rho c}, \tag{12}$$

and for the special case of a sinusoidal (pure-tone) wave

$$\langle I \rangle_t = \frac{\langle p^2 \rangle_t}{\rho c} = \frac{p_1^2}{2\rho c}, \tag{13}$$

where p_1 is the pressure amplitude, and the mean-square pressure is thus $\langle p^2 \rangle_t = p_{\mathrm{rms}}^2 = \frac{1}{2}p_1^2$.

We note, in general, for sound propagation in three dimensions, that the instantaneous sound intensity $\mathbf{I}$ is a vector quantity equal to the product of the sound pressure and the instantaneous particle velocity $\mathbf{u}$. Thus $\mathbf{I}$ has magnitude and direction. The vector intensity $\mathbf{I}$ may be resolved into components $\mathbf{I}_x$, $\mathbf{I}_y$, and $\mathbf{I}_z$. For a more complete discussion of sound intensity and its measurement see Chapter 106 and Ref. 7.

5 THREE-DIMENSIONAL WAVE EQUATION

In most sound fields, sound propagation occurs in two or three dimensions. The three-dimensional version of Eq. (1) in Cartesian coordinates is

$$\frac{\partial^2 p}{\partial x^2} + \frac{\partial^2 p}{\partial y^2} + \frac{\partial^2 p}{\partial z^2} - \frac{1}{c^2}\frac{\partial^2 p}{\partial t^2} = 0. \tag{14}$$

This equation is useful if sound wave propagation in rectangular spaces such as rooms is being considered. However, it is helpful to recast Eq. (14) in spherical coordinates if sound propagation from sources of sound in free space is being considered. It is a simple mathematical procedure to transform Eq. (14) into spherical coordinates, although the resulting equation is quite complicated. However, for propagation of sound waves from a spherically symmetric source (such as the idealized case of a pulsating spherical balloon known as an omnidirectional or monopole source) (Table 1), the equation becomes quite simple (since there is no angular dependence):

$$\frac{\partial^2 (rp)}{\partial r^2} - \frac{1}{c^2}\frac{\partial^2 (rp)}{\partial t^2} = 0. \tag{15}$$

Here, r is the distance from the origin and p is the sound pressure at that distance.

Equation (15) is identical in form to Eq. (1) with p replaced by rp and x by r. The general and simple harmonic solutions to Eq. (15) are thus the same as Eqs. (4) and (5) with p replaced by rp and x with r. The general solution is

$$rp = f_1(ct - r) + f_2(ct + r), \tag{16}$$

or

$$p = \frac{1}{r} f_1(ct - r) + \frac{1}{r} f_2(ct + r), \tag{17}$$

where f_1 and f_2 are arbitrary functions. The first term on the right of Eq. (17) represents a wave traveling outward from the origin; the sound pressure p is seen to be inversely proportional to the distance r. If the distance r is doubled, the sound pressure level [Eq. (29)] decreases by $20 \log_{10}(2) = 20(0.301) = 6$ dB. This is known as the *inverse-square law*. The second term in Eq.

TABLE 1 Models of Idealized Spherical Sources: Monopole, Dipole, and Quadrupole[a]

Monopole distribution representation	Velocity distribution on spherical surface	Oscillating sphere representation	Oscillating force model
Monopole 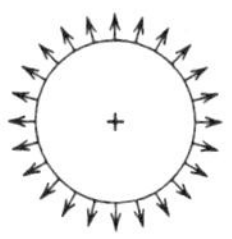			
Dipole 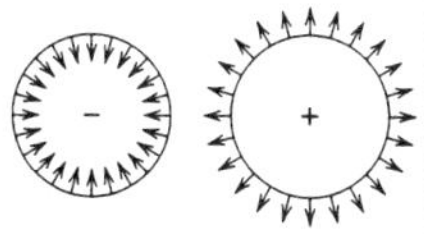	Dipole 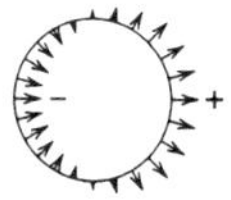	Dipole 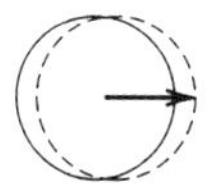	Dipole 
Quadrupole (Lateral quadrupole shown) 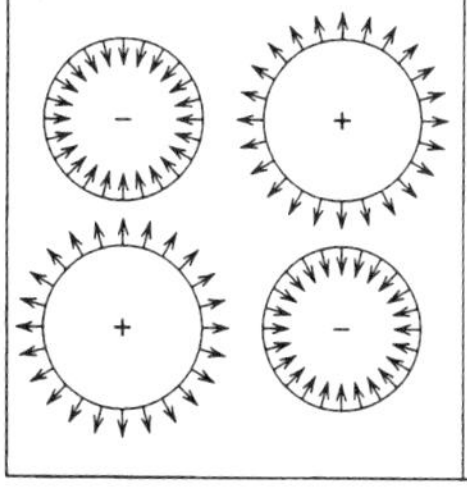	Quadrupole (Lateral quadrupole shown) 	Quadrupole (Lateral quadrupole shown) 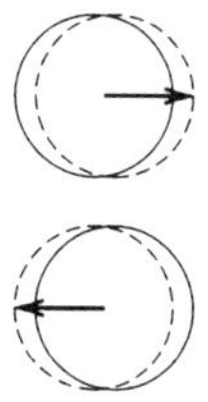	Quadrupole (Lateral quadrupole) (Longitudinal quadrupole) 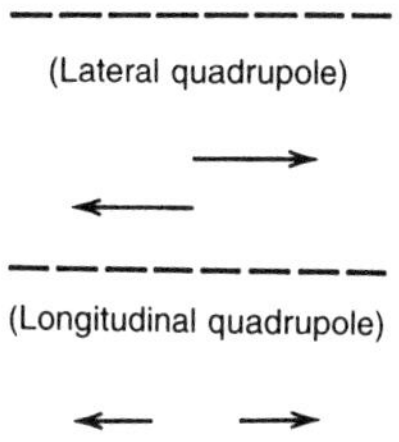

[a]For simple harmonic sources, after one half-period the velocity changes direction; positive sources become negative and vice versa, and forces reverse direction with dipole and quadrupole force models.

(17) represents sound waves traveling inward toward the origin, and in most practical cases these can be ignored (if reflecting surfaces are absent).

The simple harmonic (pure-tone) solution of Eq. (15) is

$$p = \frac{p_1}{r}\sin(\omega t - kr + \phi_1) + \frac{p_2}{r}\sin(\omega t + kr + \phi_2), \quad (18)$$

where p_1 and p_2 are the sound pressure amplitudes.

6 SOURCES OF SOUND

The second term on the right of Eq. (18), as before, represents sound waves traveling inward to the origin and is of little practical interest. However, the first term represents simple harmonic waves of angular frequency ω traveling outward from the origin, and this may be rewritten as[5]

$$p = \frac{\rho c k Q}{4\pi r}\sin(\omega t - kr + \phi_1), \quad (19)$$

where Q is termed the *strength of an omnidirectional (monopole) source* situated at the origin, and $Q = 4\pi p_1/\rho c k$. The mean-square pressure p_{rms}^2 may be found[5] by time-averaging the square of Eq. (19) over a period T:

$$p_{\mathrm{rms}}^2 = \frac{(\rho c k)^2 Q^2}{32\pi^2 r^2}. \quad (20)$$

From Eq. (20), the mean-square pressure is seen to vary with the inverse square of the distance r from the origin of the source for such an idealized omnidirectional point source everywhere in the sound field. Again, this is known as the inverse-square law. If the source is idealized as a sphere of radius a pulsating with a simple harmonic velocity amplitude U, we may show that Q has units of volume flow rate (cubic metres per second). If the source radius is small in wavelengths so that $a \ll \lambda$ or $ka \ll 1$, then we can show that the strength $Q = 4\pi a^2 U$.

Many sources of sound are not like the simple omnidirectional monopole source just described. For example, an unbaffled loudspeaker produces sound both from the back and front of the loudspeaker. The sound from the front and the back can be considered as two sources that are 180° out of phase with each other. This system can be modeled[5,8] as two out-of-phase monopoles of source strength Q separated by a distance l. The sound pressure produced by such a dipole system is

$$p = \frac{\rho c k Q l \cos\theta}{4\pi r} \cdot \left[\frac{1}{r}\sin(\omega t - kr + \phi) + k\cos(\omega t - kr + \phi)\right], \quad (21)$$

where θ is the angle measured from the axis joining the two sources (the loudspeaker axis in the practical case). Unlike the monopole, the dipole field is not omnidirectional. The sound pressure field is directional. It is, however, symmetric and shaped like a figure-eight with its lobes on the dipole axis.

For a dipole source the sound pressure has a near-field and a far-field behavior similar to the particle velocity of a monopole. Close to the source (the near field), for some fixed angle θ, the sound pressure falls off rapidly, $p \propto 1/r^2$, while far from the source (the far field $kr \gg 1$), the pressure falls off more slowly, $p \propto 1/r$. In the near field the sound pressure level decreases by 12 dB for each doubling of distance r. In the far field the decrease in sound pressure level is only 6 dB for doubling of r (like a monopole). The phase of the sound pressure also changes with distance r, since close to the source the sine term dominates and far from the source the cosine term dominates. The particle velocity may be obtained from the sound pressure [Eq. (21)] and use of Euler's equation [Eq. (22)]. It has an even more complicated behavior with distance r than the sound pressure, having three distinct regions.

As discussed in Chapter 8, an oscillating force applied at a point in space gives rise to results identical to Eq. (21), and hence there are many real sources of sound that behave like the idealized dipole source described above, for example, pure-tone fan noise, vibrating beams, unbaffled loudspeakers, and even wires and branches (which sing in the wind due to alternate vortex shedding).

The next higher order source is the quadrupole. It is thought that the sound produced by the mixing process in an air jet gives rise to stresses that are quadrupole in nature. See Chapters 8, 24, and 25. Quadrupoles may be considered to consist of two opposing point forces (two opposing dipoles) or equivalently four monopoles. (See Table 1.) We note that some authors use slightly different but equivalent definitions for the source strength of monopoles, dipoles, and quadrupoles. The definitions used in Sections 6 and 8 of this chapter are the same as in Refs. 5 and 8 and result in expressions for sound pressure, sound intensity, and sound power, which although equivalent are different in form from those in Chapter 8, for example.

The expression for the sound pressure for a quadrupole is even more complicated than for a dipole. Close to the source, in the near field, the sound pressure $p \propto 1/r^3$. Farther from the source, $p \propto 1/r^2$; while in the far field, $p \propto 1/r$.

Sound sources experienced in practice are normally even more complicated than dipoles or quadrupoles. The sound radiation from a vibrating piston is described in Chapters 8 and 10, the sound radiation from vibrating cylinders in Chapter 10.

The discussion in Chapter 8 considers steady-state radiation. However, there are many sources in nature and created by people that are transient. As shown in Chapter 9, harmonic analysis of these cases is often not suitable and time-domain methods have given better results and understanding of the phenomena. These are the approaches adopted in Chapter 9.

7 SOUND INTENSITY AND DIRECTIVITY

The radial particle velocity in a spherically spreading sound field is given by Euler's equation as

$$u = -\frac{1}{\rho}\int \frac{\partial p}{\partial r}\, dt, \quad (22)$$

and substituting Eqs. (19) and (22) into (10) then using Eq. (20) and time-averaging gives the magnitude of the radial intensity in such a field as

$$\langle I \rangle_t = \frac{p_{\mathrm{rms}}^2}{\rho c}, \quad (23)$$

the same result as for a plane wave [see Eq. (12)].

The sound intensity decreases with the inverse square of the distance r. Simple omnidirectional monopole sources radiate equally well in all directions. More complicated idealized sources such as dipoles, quadrupoles, and vibrating piston sources create sound fields that are directional (see Fig. 3). Of course, real sources such as machines produce even more complicated sound fields than these idealized sources. (For a more complete discussion of the sound fields created by idealized sources, see Chapter 8.) However, the same result as Eq. (23) is found to be true for any source of sound as long as the measurements are made sufficiently far from the source. The intensity is not given by the simple result of Eq. (23) close to sources such as dipoles, quadrupoles, or more complicated sources of sound. Close to such sources Eq. (10) must be used for the instantaneous radial intensity, or

$$\langle I \rangle_t = \langle pu \rangle_t \tag{24}$$

for the time-averaged radial intensity.

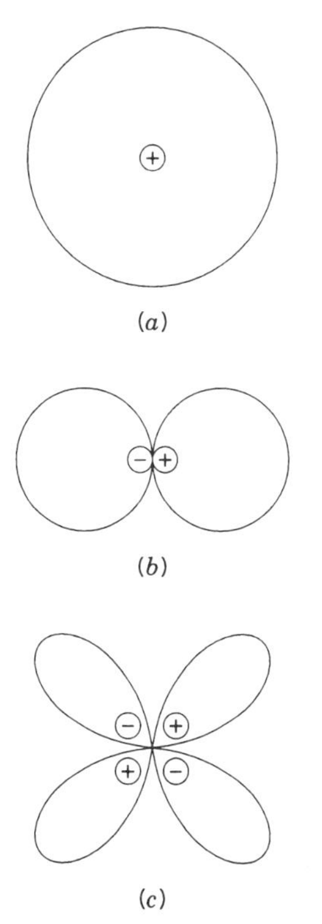

Fig. 3 Polar directivity plots for the radial sound intensity in the far field of (*a*) monopole, (*b*) dipole, and (*c*) (lateral) quadrupole.

The time-averaged radial sound intensity for a dipole is given by[5,8]

$$\langle I \rangle_t = \frac{\rho c k^4 (Ql)^2 \cos^2\theta}{32\pi^2 r^2}. \tag{25}$$

The sound intensity radiated by a dipole is seen to depend on $\cos^2\theta$. In general, a directivity factor $D_{\theta,\phi}$ may be defined as the ratio of the radial intensity $\langle I_{\theta,\phi}\rangle_t$ (at angles θ and ϕ and distance r from the source) to the radial intensity $\langle I_s\rangle_t$ at the same distance r from an omnidirectional source of the same total power. Thus

$$D_{\theta,\phi} = \frac{\langle I_{\theta,\phi}\rangle_t}{\langle I_s\rangle_t}. \tag{26}$$

A directivity index $\mathrm{DI}_{\theta,\phi}$ may be defined, where

$$\mathrm{DI}_{\theta,\phi} = 10\log D_{\theta,\phi}. \tag{27}$$

8 SOUND POWER

The *sound power* P of a source is given by integrating the intensity over any closed surface S around the source (see Fig. 4).

$$P = \int_s \langle I_n\rangle_t \, dS. \tag{28}$$

The normal component of the intensity I_n must be measured in a direction perpendicular to the elemental area dS. If a spherical surface is chosen, then the sound power of an omnidirectional (monopole) source is

$$P_m = \langle I_r\rangle_t 4\pi r^2, \tag{29}$$

$$P_m = \frac{p_{\mathrm{rms}}^2}{\rho c} 4\pi r^2, \tag{30}$$

and from Eq. (20) the sound power of a monopole is[5,8]

$$P_m = \frac{\rho c k^2 Q^2}{8\pi}. \tag{31}$$

It is apparent from Eq. (31) that the sound power of an idealized (monopole) source is independent of the distance r from the origin, where the source is located. This is the result required by conservation of energy and also to be expected for all sound sources.

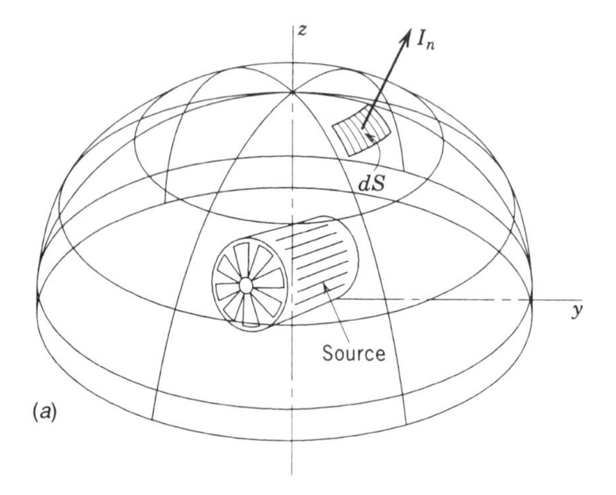

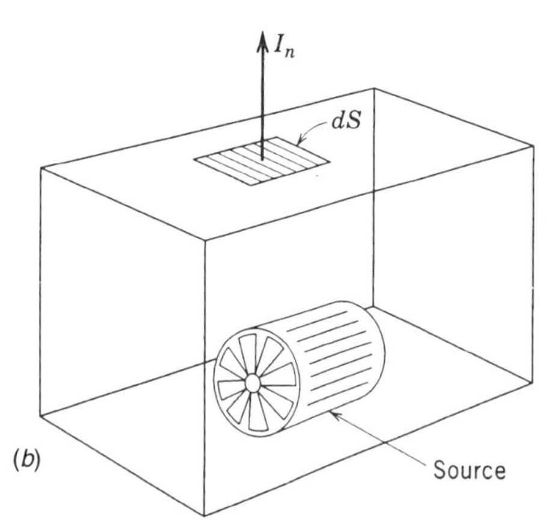

Fig. 4 Normal component of sound intensity I_n being measured on (*a*) a segment dS of a hemispherical enclosure surface, and (*b*) an elemental area dS of a rectangular enclosure surface surrounding a source having a sound power P.

Equation (30) shows that for an omnidirectional source (in the absence of reflections) the sound power can be determined from measurements of the mean-square pressure made with a single microphone. Of course, for such a source, measurements should really be made with a *reflection-free (anechoic) environment* or very close to the source where reflections are presumably less important.

The sound power of a dipole source is obtained by integrating the intensity given by Eq. (25) over a sphere around the source. The result for the sound power is[5,8]

$$P_d = \frac{\rho c k^4 (Ql)^2}{24\pi}. \qquad (32)$$

The dipole is obviously a much less efficient radiator than a monopole, particularly at low frequency.

In practical situations with real directional sound sources and where reflections are important, use of Eq. (30) becomes difficult and less accurate, and then the sound power is more conveniently determined from Eq. (28) with a sound intensity measurement system. See Chapter 106.

It is important to note that the sound power radiated by a source can be significantly affected by its environment. For example, if a monopole source with a high internal impedance (whose strength Q will be unaffected by the environment) is placed on a floor, its sound power will be doubled (and its sound power level increased by 3 dB). If it is placed at a floor–wall intersection, its sound power will be increased by four times (6 dB); and if it is placed in a room corner, its power is increased by eight times (9 dB).

9 DECIBELS AND LEVELS

The range of sound pressure magnitude and sound power experienced in practice is very large (see Figs. 5 and 6). Thus, logarighmic rather than linear measures are often used for sound pressure and power. The most common is the *decibel*. The decibel represents a relative measurement or ratio. Each quantity in decibels is expressed as a ratio relative to a *reference sound pressure*, *power*, or *intensity*. Whenever a quantity is expressed in decibels, the result is known as a *level*.

The decibel (dB) is the ratio R_1 given by

$$\log_{10} R_1 = 0.1, \qquad 10 \log_{10} R_1 = 1 \text{ dB}. \qquad (33)$$

Thus, $R_1 = 10^{0.1} = 1.26$. The decibel is seen to represent the ratio 1.26. A larger ratio, the *bel* is sometimes used. The bel is the ratio R_2 given by $\log_{10} R_2 = 1$. Thus, $R_2 = 10^1 = 10$. The bel represents the ratio 10.

The *sound pressure level* L_p is given by

$$L_p = 10 \log_{10}\left(\frac{\langle p^2\rangle_t}{p_{\text{ref}}^2}\right) = 10 \log_{10}\left(\frac{p_{\text{rms}}^2}{p_{\text{ref}}^2}\right) = 20 \log_{10}\left(\frac{p_{\text{rms}}}{p_{\text{ref}}}\right) \quad \text{dB}, \qquad (34)$$

where p_{ref} is the reference pressure $p_{\text{ref}} = 20\ \mu\text{Pa} = 0.00002\ \text{N/m}^2$ (= 0.0002 μbar). This reference pressure was originally chosen to correspond to the quietest sound (at 1000 Hz) that the average young person can hear.

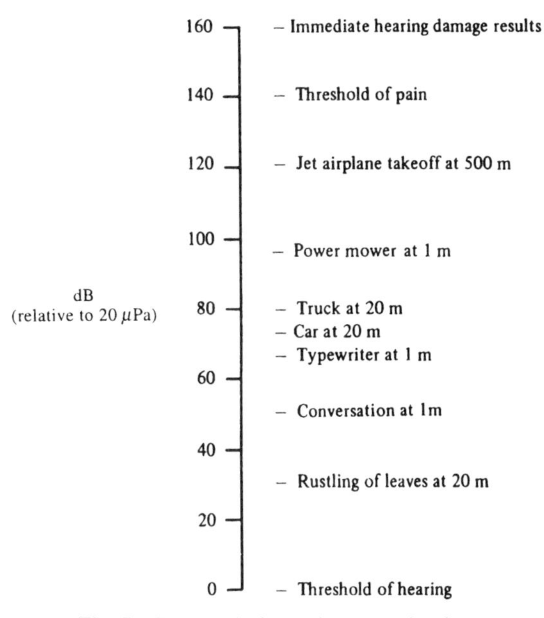

Fig. 5 Some typical sound pressure levels, L_p.

The sound power level L_P is given by

$$L_P = 10 \log_{10} \left(\frac{P}{P_{\text{ref}}} \right) \qquad \text{dB}, \tag{35}$$

where P is the sound power of a source and $P_{\text{ref}} = 10^{-12}$ W is the reference sound power.

The sound intensity level L_i is given by

$$L_i = 10 \log_{10} \left(\frac{I}{I_{\text{ref}}} \right) \qquad \text{dB}, \tag{36}$$

where I is the component of the sound intensity in a given direction and $I_{\text{ref}} = 10^{-12}$ W/m^2 is the reference sound intensity.

Some typical sound pressure and sound power levels are given in Figs. 5 and 6.

If two sound sources radiate independently (uncorrelated sources), then because the interfering effects of the source pressures time-average to zero, the mean

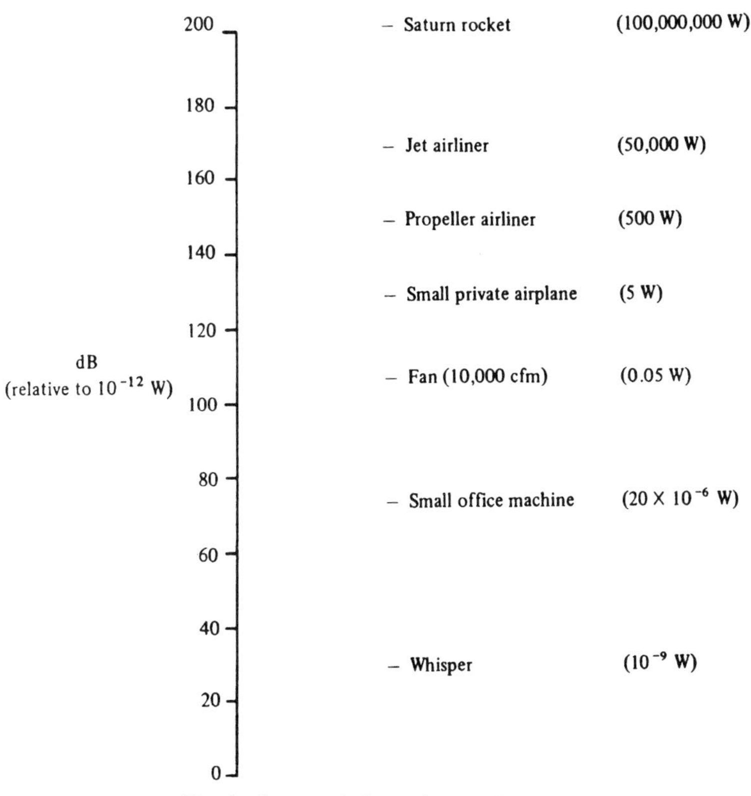

Fig. 6 Some typical sound power levels, L_P.

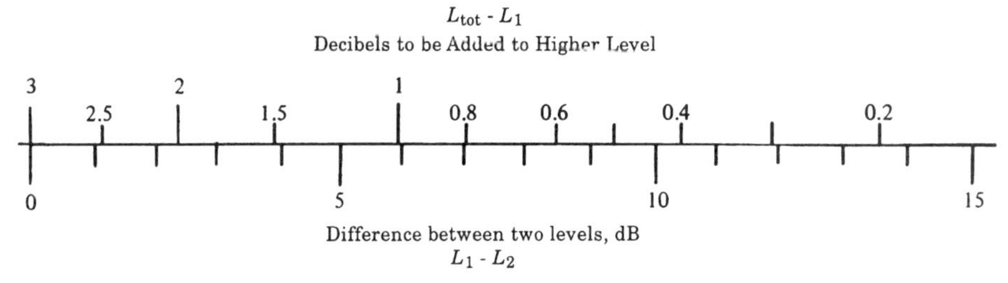

Fig. 7 Chart for combination levels of decibels (for uncorrelated sources).

square pressures are additive and the total sound pressure level at some point in space, or the total sound power level, may be determined using Fig. 7. *First example:* If two independent sound sources each create sound pressure levels operating on their own of 80 dB, at a certain point, what is the total level? *Answer:* The difference in levels is 0 dB; thus the total sound pressure level is 80 + 3 = 83 dB. *Second example:* If two independent sound sources have sound power levels of 70 and 73 dB, what is the total level? *Answer:* The difference in levels is 3 dB; thus the total sound power level is 73 + 1.8 = 74.8 dB.

Figure 7 and these two examples do *not* apply to the case of two pure tones of the same frequency. For such a case in the first example instead of 83 dB, the total sound pressure level can range anywhere between 86 dB (for in-phase sound pressures) and $-\infty$ dB (for out-of-phase sound pressures). For the second example the total sound power radiated by the two pure-tone sources depends on the phasing and separation distance. As discussed in Section 7, two out-of-phase closely spaced pure-tone sound sources are termed a *dipole*.

10 REFLECTION, REFRACTION, SCATTERING, AND DIFFRACTION

For a homogeneous plane sound wave at normal incidence on a fluid medium of different characteristic impedance ρc, both reflected and transmitted waves are formed (see Fig 8).

From energy considerations (provided no losses occur at the boundary) the sum of the reflected intensity I_r and transmitted intensity I_t equals the incident intensity I_i,

$$I_i = I_r + I_t, \tag{37}$$

and dividing throughout by I_i,

$$\frac{I_r}{I_i} + \frac{I_t}{I_i} = R + T = 1, \tag{38}$$

where R is the *reflection coefficient* and T is the *transmission coefficient*. For plane waves at normal incidence on a plane boundary between two fluids (see Fig. 8)

$$R = \frac{(\rho_1 c_1 - \rho_2 c_2)^2}{(\rho_1 c_1 + \rho_2 c_2)^2}, \tag{39}$$

and

$$T = \frac{4\rho_1 c_1 \rho_2 c_2}{(\rho_1 c_1 + \rho_2 c_2)^2}. \tag{40}$$

Some interesting facts can be deduced from Eqs. (39) and (40). Both the reflection and transmission coefficients are independent of the direction of the wave, since interchanging $\rho_1 c_1$ and $\rho_2 c_2$ does not affect the values of R and T. For example, for sound waves traveling from air to water or water to air, almost complete reflection occurs, independent of direction, and the reflection coefficients are the same and the transmission coefficients are the same for the two different directions.

As discussed before, when the characteristic impedance ρc of a fluid medium changes, incident sound is

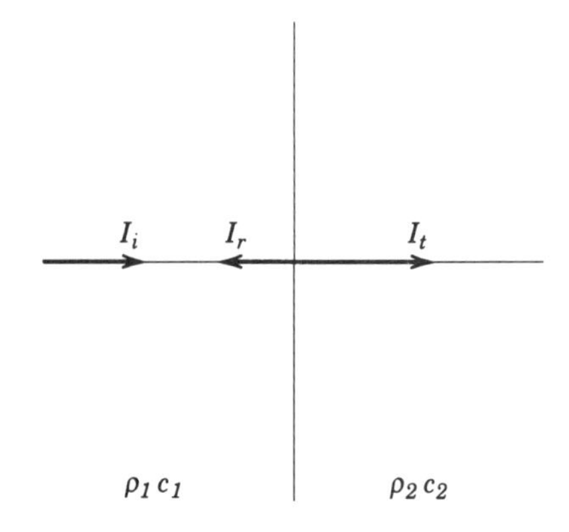

Fig. 8 Incident intensity I_i, reflected intensity I_r, and transmitted intensity I_t in a homogeneous plane sound wave at normal incidence on a plane boundary between two fluid media of different characteristic impedances $\rho_1 c_1$ and $\rho_2 c_2$.

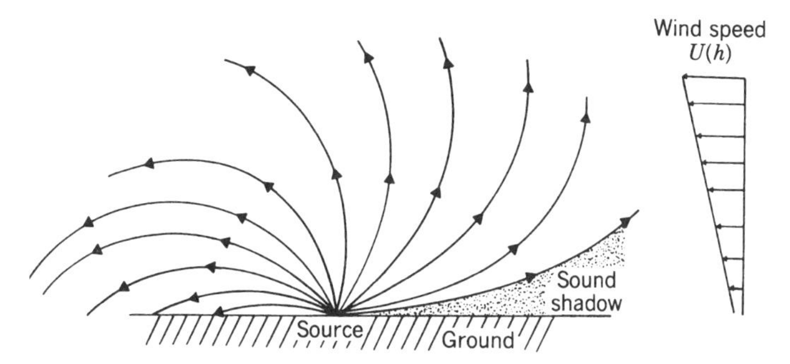

Fig. 9 Refraction of sound in air with wind speed $U(h)$ increasing with altitude h.

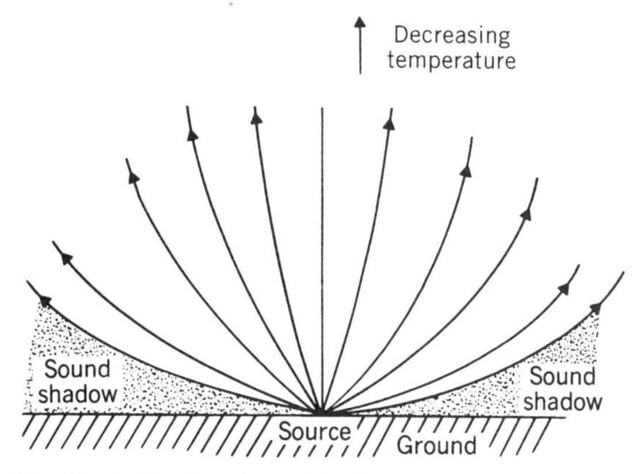

Fig. 10 Refraction of sound in air with normal temperature lapse (temperature decreases with altitude).

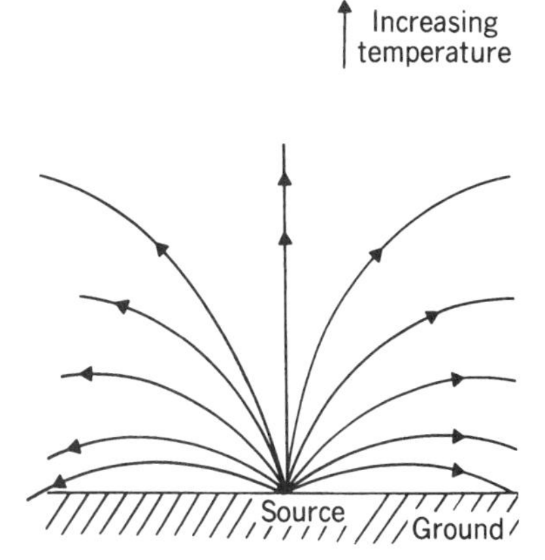

Fig. 11 Refraction of sound in air with temperature inversion.

both reflected and transmitted. It can be shown that if a plane sound wave is incident at an oblique angle on a plane boundary between two fluids then the wave transmitted into the changed medium changes direction. This effect is called *refraction*. Temperature changes and wind speed changes in the atmosphere are important causes of refraction.

Wind speed normally increases with altitude, and Fig. 9 shows the refraction effects to be expected for an idealized wind speed profile. Atmospheric temperature changes alter the speed of sound c, and temperature gradients can also produce sound shadow and focusing effects, as seen in Figs. 10 and 11.

When a sound wave meets an obstacle, some of the sound wave is deflected. The *scattered* wave is defined to be the difference between the resulting wave with the obstacle and the undisturbed wave without the presence of the obstacle. The scattered wave spreads out in all directions interfering with the undisturbed wave. If the obstacle is very small compared with the wavelength, no sharp-edged sound shadow is created behind the obstacle. If the obstacle is large compared with the wavelength, it is normal to say that the sound wave is reflected (in front) and *diffracted* (behind) the obstacle (rather than *scattered*). In this case a strong sound shadow is caused in which the wave pressure amplitude is very small. In the zone between the sound shadow and the region fully "illuminated" by the source the sound wave pressure amplitude oscillates. These oscillations are maximum near the shadow boundary and minimum well inside the shadow. These oscillations in amplitude are normally termed *diffraction bands*. One of the most common examples of diffraction caused by a body is the diffraction of sound over the sharp edge of a *barrier* or *screen*. For a plane homogeneous sound wave it is found that a strong shadow is caused by high-frequency waves where $h/\lambda \gg 1$ and a weak shadow where $h/\lambda \ll 1$, where h is the barrier height and λ is the wavelength. For intermediate cases where $h/\lambda \approx 1$ a variety of interference and diffraction effects are caused by the barrier. Scattering is caused not only by obstacles placed in the wave field, but also by fluid regions where the properties of the medium such as its density or compressibility change their values from the rest of the medium. Scattering is also caused by turbulence (see Chapter 28) and from rain or fog particles in the atmosphere and bubbles in water and by rough or absorbent areas on wall surfaces.

11 RAY ACOUSTICS

There are three main modeling approaches in acoustics, which may be termed wave acoustics, ray acoustics, and energy acoustics. So far in this chapter we have mostly used the wave acoustics approach in which the acoustic quantities are completely defined as functions of space and time. This approach is practical in certain cases where the fluid medium is bounded and in cases where the fluid is unbounded as long as the fluid is homogenous. However, if the fluid properties vary in space due to variations in temperature or due to wind gradients, then the wave approach becomes more difficult and other simplified approaches such as the ray acoustics approach described here and in Chapter 3 are useful. This approach can also be extended to propagation in fluid-submerged elastic structures, as described in *Enc.* Ch. 4. The energy approach is described in Section 12.

In the ray acoustics approach, rays are obtained that are solutions to the simplified eikonal equation [Eq. (41)].

$$\left(\frac{\partial S}{\partial x}\right)^2 + \left(\frac{\partial S}{\partial y}\right)^2 + \left(\frac{\partial S}{\partial z}\right)^2 - \frac{1}{c^2} = 0. \quad (41)$$

The ray solutions can provide good approximations to more exact acoustic solutions. In certain cases they also satisfy the wave equation.[6] The eikonal $S(x, y, z)$ represents a surface of constant phase (or wavefront) that propagates at the speed of sound c. It can be shown that Eq. (41) is consistent with the wave equation only in the case when the frequency is very high.[6] However, in practice, it is useful, provided the changes in the speed of sound c are small when measured over distances comparable with the wavelength. In the case where the fluid is homogeneous (constant sound speed c_0 and density ρ throughout) S is a constant and represents a plane surface given by $S = (\alpha x + \beta y + \gamma z)/c_0$, where α, β, and γ are the direction cosines of a straight line (a ray) that is perpendicular to the wavefront (surface S). If the fluid can no longer be assumed to be homogeneous and the speed of sound $c(x, y, z)$ varies with position, the approach becomes approximate only. In this case some parts of the wavefront move faster than others, and the rays bend and are no longer straight lines. In cases where the fluid is not stationary, the rays are no longer quite parallel to the normal to the wavefront. This ray approach is described in more detail in several books and in Chapter 3 (where in this chapter the main example is from underwater acoustics). The ray approach is also useful for the study of propagation in the atmosphere and is a method to obtain the results given in Figs. 9–11. It is observed in these figures that the rays always bend in a direction toward the region where the sound speed is less. The effects of wind gradients are somewhat different since in that case the refraction of the sound rays depends on the relative directions of the sound rays and the wind in each fluid region. *Enc.* Ch. 4 presents an extension of the ray approach to fluid-submerged structures such as elastic shells. The same high-frequency assumption is made that λ/L must be small, where L is a characteristic dimension in the fluid field (environment or structural dimension).

12 ENERGY ACOUSTICS

In enclosed spaces the wave acoustics approach is useful, particularly if the enclosed volume is small and simple in shape and the boundary conditions are well-defined. In the case of rigid walls of simple geometry, the wave equation is used, and after the applicable boundary conditions are applied, the solutions for the natural (eigen) frequencies for the modes (standing waves) are found. See Chapters 6 and 75 for more details. However, for large rooms with irregular shape and absorbing boundaries, the wave approach becomes impracticable and other approaches must be sought. The ray acoustics approach together with the multiple-image-source concept is useful in some room problems, particularly in auditorium design or in factory spaces where barriers are involved. However, in many cases a statistical approach where the energy in the sound field is considered is the most useful. See Chapters 60–62 and 74 and 75 for more detailed discussion of this approach. Some of the fundamental concepts are briefly described here.

For a plane wave progressing in one direction in a duct of unit cross section, all of the sound energy in a column of fluid c metres in length must pass through the cross section in 1 s. Since the intensity $\langle I \rangle_t$ is given by $p_{\text{rms}}^2/\rho c$, then the total sound energy in the fluid column c metres long must also be equal to $\langle I \rangle_t$. The energy per unit volume ϵ (joules per cubic metre) is thus

$$\epsilon = \frac{\langle I \rangle_t}{c} \quad (42)$$

or

$$\epsilon = \frac{p_{\text{rms}}^2}{\rho c^2} \quad (43)$$

The energy density ϵ may be derived by alternative

means and is found to be the same as that given in Eq. (42) in most acoustic fields, except very close to sources of sound and in standing-wave fields. In a room with negligibly small absorption in the air or at the boundaries, the sound field created by a source producing broadband sound will become very reverberant (the sound waves will reach a point with equal probability from any direction). In addition, for such a case the sound energy may be said to be diffuse if the energy density is the same anywhere in the room. For these conditions the time-averaged intensity incident on the walls (or on an imaginary surface from one side) is

$$\langle I \rangle_t = \tfrac{1}{4}\epsilon c, \tag{44}$$

or

$$\langle I \rangle_t = \frac{p_{\text{rms}}^2}{4\rho c}. \tag{45}$$

In any real room the walls will absorb some sound energy (and convert it into heat). The *absorption coefficient* $\alpha(f)$ of the wall material may be defined as the fraction of the incident sound intensity that is absorbed by the wall surface material:

$$\alpha(f) = \frac{\text{sound intensity absorbed}}{\text{sound intensity incident}}. \tag{46}$$

The absorption coefficient is a function of frequency and can have a value between 0 and 1. The *noise reduction coefficient* (NRC) is found by averaging the absorption coefficient of the material at the frequencies 250, 500, 1000, and 2000 Hz (and rounding off the result to the nearest multiple of 0.05). See Chapter 75 for more detailed discussion on the absorption of sound in enclosures.

If we consider the sound field in a room with a uniform energy density ϵ created by a sound source that is suddenly stopped, then the sound pressure level in the room will decrease. We define a reverberation time in such a room as the time that the sound pressure level takes to drop by 60 dB. We may show that the reverberation time T_R is given as

$$T_R = \frac{0.161V}{S\overline{\alpha}}, \tag{47}$$

where V is the room volume in cubic metres, S is the wall surface area in square metres, and $\overline{\alpha}$ is the average absorption coefficient of the wall surfaces.

By considering the sound energy radiated into a room by a broadband noise source of sound power W, we may sum together the mean squares of the sound pressure contributions caused by the direct and reverberant fields and after taking logarithms obtain the sound pressure level in the room:

$$L_p = L_w + 10\log\left(\frac{D_{\theta,\phi}}{4\pi r^2} + \frac{4}{R}\right), \tag{48}$$

where $D_{\theta,\phi}$ is the directivity factor of the source (see Section 7) and R is the so-called room constant

$$R = \frac{S\overline{\alpha}}{1-\overline{\alpha}}. \tag{49}$$

A plot of the sound pressure level against distance from the source is given for various room constants in Fig. 12. It is seen that there are several different regions. The near and far fields depend on the type of source (see Section 11 and Chapter 8) and the free field and reverberant field. The free field is the region where the direct term $D/4\pi r^2$ dominates, and the reverberant field is the region where the reverberant term $4/R$ in Eq. (48) dominates. The so-called critical distance $r_c = (D_{\theta,\phi}R/16\pi)^{1/2}$ occurs where the two terms are equal.

13 SOUND RADIATION FROM IDEALIZED STRUCTURES

The sound radiation from plates and cylinders in bending (flexural) vibration is discussed in Chapter 10. There are interesting phenomena observed with free-bending waves. Unlike sound waves, these are dispersive and travel faster at higher frequency. The bending-wave speed is $c_b = (\omega\kappa c_l)^{1/2}$, where κ is the radius of gyration $h/(12)^{1/2}$, h is the thickness, and c_l is the longitudinal wave speed $\{E/[\rho(1-\sigma^2)]\}^{1/2}$, where E is Young's modulus of elasticity. When the bending-wave speed equals the speed of sound in air, the frequency is called the critical frequency (see Fig. 13). The critical frequency is

$$f_c = \frac{c^2}{2\pi\kappa c_l}. \tag{50}$$

Above this frequency f_c the coincidence effect is observed because the bending wavelength λ_b is greater than the wavelength in air λ (Fig. 14) and trace wave matching always occurs for the sound waves in air at some angle of incidence. See Fig. 15. This has important consequences for the sound radiation from structures

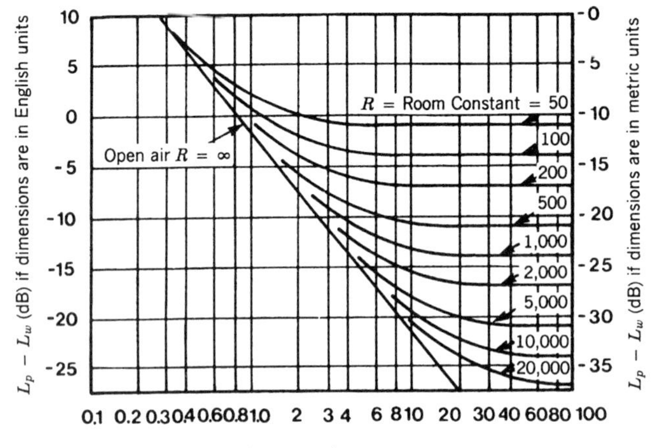

Fig. 12 Sound pressure level in a room (relative to sound power level) as a function of distance (r).

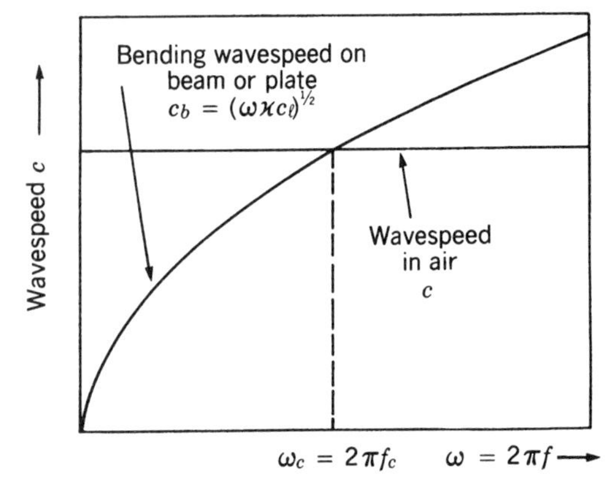

Fig. 13 Variation of frequency of bending wave speed c_b on a beam or panel and wave speed in air c.

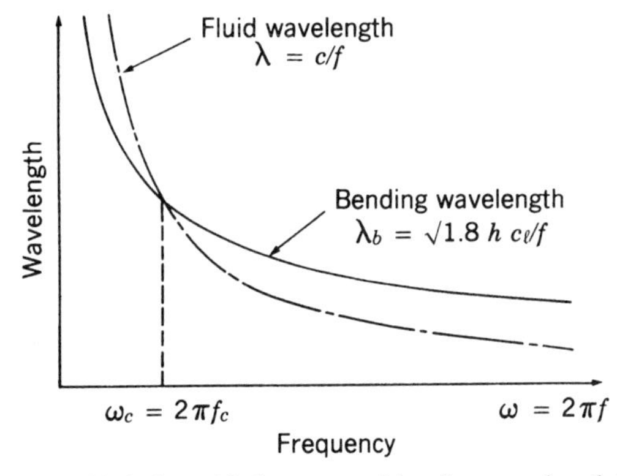

Fig. 14 Variation with frequency of bending wavelength λ_b on a beam or panel and wavelength in air λ.

and also for the sound transmitted through the structures from one air space to the other.

For free-bending waves on infinite plates above the critical frequency the plate radiates efficiently, while below this frequency (theoretically) the plate cannot radiate any sound energy at all. (See Chapter 10.) For finite plates, reflection of the bending waves at the edges of the plates causes standing waves that allow radiation (although inefficient) from the plate corners or edges even below the critical frequency. In the plate center, radiation from adjacent quarter-wave areas cancels. But radiation from the plate corners and edges, which are normally separated sufficiently in acoustic wavelengths, does not cancel. At very low frequency, sound is radiated mostly by corner modes, then up to the critical frequency mostly by edge modes. Above the critical fre-

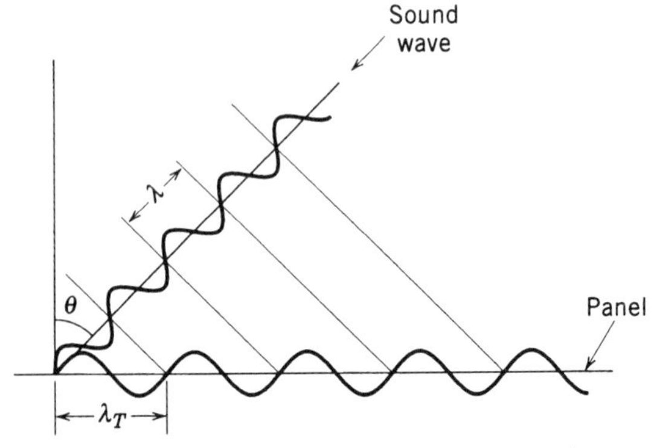

Fig. 15 Diagram showing trace wave matching between waves in air of wavelength λ and waves in panel of trace wavelength λ_T.

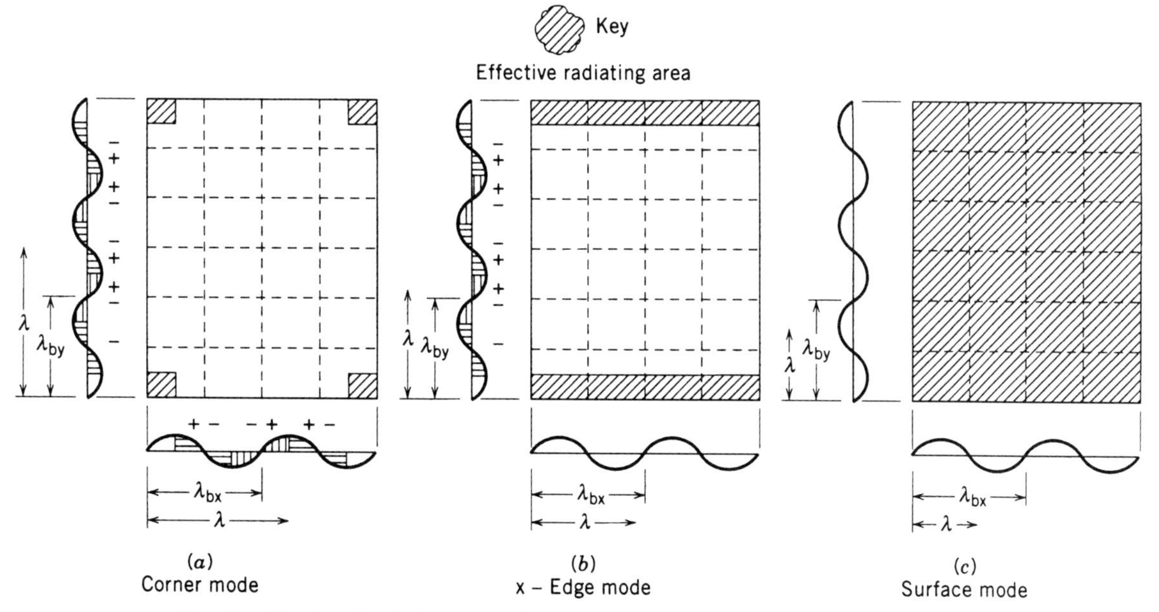

Fig. 16 Wavelength relations and effective radiating areas for corner, edge, and surface modes. The acoustic wavelength is λ, and λ_{bx} and λ_{by} are the bending wavelengths in the x- and y-directions, respectively.

quency the radiation is caused by surface modes with which the whole plate radiates efficiently (see Fig. 16). Radiation from bending waves in plates and cylinders is discussed in detail in Chapter 10. Sound transmission through structures is discussed in Chapters 66, 76, and 77.

14 BOUNDARY WAVES IN ACOUSTICS

Boundary waves in acoustic systems and elastic structures are observed on all scales—from the submillimetre (acoustic delay lines) to planetary size (Rayleigh modes in seismology). As their existence is essentially independent of that of the "body" or "volume" waves of acoustics or elasticity, they constitute a distinct and separate class of modes. The essential mechanism is one of storage at a boundary; the role of the boundary is *intrinsic*. In contrast, body or volume waves are in no way governed by the boundary, whose role is thus *extrinsic*. The energy density of boundary waves decreases exponentially or at a higher rate with increasing distance from the boundary. Reciprocally the excitation of such modes decreases exponentially with the distance of the source from the boundary. Homogeneous plane waves incident on a plane boundary do not generate true surface modes. The theory and practice of acoustics and elasticity offer many examples of such modes: surface waves traveling along a rigid periodic structure of corrugations on a hard wall bounding a fluid half space, Stoneley waves along plane elastic–elastic interfaces, Scholte waves along plane fluid–elastic interfaces, boundary roughness modes, Rayleigh waves on the surface of an elastic solid, and so on. Boundary waves exhibit the remarkable property of channelling spherically propagating energy produced by a point source into two-dimensional propagating configurations, that is, from an r^{-2} geometric energy spreading loss to an r^{-1} law. That is why Rayleigh waves of shallow-focus earthquakes dominate seismograms at middle or long ranges—an effect also observed in the laboratory with boundary roughness modes in acoustics. Boundary waves are discussed in detail in *Enc.* Ch. 12.

15 STANDING WAVES

Standing-wave phenomena are observed in many situations in acoustics and the vibration of strings and elastic structures. Thus they are of interest with almost all musical instruments (both wind and stringed) (see Chapters 100, 101, 102, 103, see also *Enc.* Ch. 132, 135, 137); in architectural spaces such as auditoria and reverberation rooms; in volumes such as automobile and aircraft cabins; and in numerous cases of vibrating structures, from tuning forks, xylophone bars, bells and cymbals, to windows, wall panels and innumerable other engineering systems including aircraft, vehicle, and ship structural members. With each standing wave is associated an eigen (or natural) frequency and a mode shape (or shape of vibration). Some of these systems can be idealized to simple one-, two-, or three-dimensional systems. For example with a simple wind instrument such as a whistle, Eq. (1) above together with the appropriate spatial boundary conditions can be used to predict the predominant frequency of the sound produced. Similarly, the vibration of a string on a violin can be predicted with an equation identical to Eq. (1) but with the variable p replaced by the lateral string displacement. With such a string, solutions can be obtained for the fundamental and higher natural frequencies (overtones) and the associated standing-wave mode shapes (normally sine shapes). In such a case for a string with fixed ends, the so-called overtones are just integer multiples (2, 3, 4, 5, ...) of the fundamental frequency. The standing wave with the whistle and string can be considered mathematically to be composed of two waves of equal amplitude traveling in opposite directions.

A similar situation occurs for bending waves on bars, but because the equation of motion is different (dispersive), the higher natural frequencies are not related by simple integers. However, for the case of a beam with simply supported ends, the higher natural frequencies are given by $2^2, 3^2, 4^2, 5^2, \ldots$ or 4, 9, 16, 25, ... and the mode shapes are sine shapes again.

The standing waves on two-dimensional systems (such as bending vibrations of plates) may be considered mathematically to be composed of four opposite traveling waves. For simply supported rectangular plates the mode shapes are sine shapes in each direction. For three-dimensional systems such as the air volumes of rectangular rooms, the standing waves may be considered to be made up of eight traveling waves. For a hard-walled room, the sound pressure has a cosine mode shape with the maximum pressure at the walls, and the particle velocity has a sine mode shape with zero normal particle velocity at the walls. See Chapter 6 for the natural frequencies and mode shapes for a large number of acoustic and structural systems.

16 WAVEGUIDES

Waveguides can occur naturally where sound waves are channeled by reflections at boundaries and by refraction. The ocean can be considered to be an acoustic waveguide that is bounded above by the air–sea interface and below by the ocean bottom (see Chapter 31). Similar channeling effects are also sometimes observed in the atmosphere. (See Chapters 3 and 28.) Waveguides are also encountered in musical instruments and engineering applications. Wind instruments may be regarded as waveguides. In addition, waveguides comprised of pipes, tubes, and ducts are frequently used in engineering systems, for example, air conditioning ducts and the ductwork in turbines and turbofan engines. The sound propagation in such waveguides is similar to the three-dimensional situation discussed in Section 15 but with some differences. Although rectangular ducts are used in air conditioning systems, circular ducts are also frequently used, and theory for these must be considered as well. In real waveguides, air flow is often present and complications due to a mean fluid flow must be included in the theory.

For low-frequency excitation only plane waves can propagate along the waveguide (in which the sound pressure is uniform across the duct cross section). However, as the frequency is increased, the so-called first cut-on frequency is reached above which there is a standing wave across the duct cross section caused by the first higher mode of propagation.

For excitation just above this cut-on frequency, besides the plane-wave propagation, propagation in this higher order mode can also exist. The higher mode propagation in each direction in a rectangular duct can be considered to be composed of four traveling waves each with a vector (ray) almost perpendicular to the duct walls and with a phase speed along the duct that is almost infinite. As the frequency is increased, these vectors move increasingly toward the duct axis, and the phase speed along the duct decreases until at very high frequency it is only just above the speed of sound c. However, for this mode, the sound pressure distribution across the duct cross section remains unchanged. As the frequency increases above the first cut-on frequency, the cut-on frequency for the second higher order mode is reached, and so on. For rectangular ducts, the solution for the sound pressure distribution for the higher modes in the duct consists of cosine terms with a pressure maximum at the duct walls, while for circular ducts, the solution involves Bessel functions. Chapter 7 explains how sound is propagated in both rectangular and circular waveguides and includes discussion on the complications created by a mean flow, dissipation, discontinuities, and terminations. Chapter 112 discusses the propagation of sound in another class of waveguides: horns.

17 ACOUSTIC LUMPED ELEMENTS

When the wavelength of sound is large compared with the physical dimensions of the acoustic system under consideration, then the lumped-element approach is useful. In this approach it is assumed that the fluid mass, stiffness, and dissipation distributions can be "lumped" together to act at a point, significantly simplifying the analysis of the problem. The most common example of this approach is its use with the well-known Helmholtz resonator in which the mass of air in the neck of the resonator vibrates at its natural frequency against the stiffness of its volume. A similar approach can be used in the design of loudspeaker enclosures and the concentric resonators in automobile mufflers in which the mass of the gas in the resonator louvres (orifices) vibrates against the stiffness of the resonator (which may not necessarily be regarded completely as a lumped element). Dissipation in the resonator louvers may also be taken into account. Chapter 11 reviews the lumped-element approach in some detail.

18 NUMERICAL APPROACHES: FINITE ELEMENTS AND BOUNDARY ELEMENTS

In cases where the geometry of the acoustic space is complicated and where the lumped-element approach cannot be used, then it is necessary to use numerical approaches. In the late 1960s, with the advent of powerful computers, the acoustic finite element method (FEM) became feasible. In this approach the fluid volume is divided into a number of small fluid elements (usually rectangular or triangular), and the equations of motion are solved for the elements, ensuring that the sound pressure and volume velocity are continuous at the node points where the elements are joined. The FEM has been widely used to study the acoustic performance of elements in automobile mufflers and cabins.

The boundary element method (BEM) was developed a little later than the FEM. In the BEM approach the elements are described on the boundary surface only, which reduces the computational dimension of the problem by 1. This correspondingly produces a smaller system of equations than the FEM and thus saves computational time considerably because of the use of a surface mesh rather than a volume mesh. For sound propagation problems involving the radiation of sound to infinity, the BEM is more suitable because the radiation condition at infinity can be easily satisfied with the BEM, unlike with the FEM. However, the FEM is better suited than the BEM for the determination of the natural frequencies and mode shapes of cavities.

Recently FEM and BEM commercial software has become widely available. The FEM and BEM are described in Chapters 12 and 13.

19 ACOUSTIC MODELING USING EQUIVALENT CIRCUITS

Electrical analogies have often been found useful in the modeling of acoustic systems. There are two alternatives. The sound pressure can be represented by voltage and the volume velocity by current, or alternatively the sound pressure is replaced by current and the volume velocity by voltage. Use of electrical analogies is discussed in Chapter 11. They have been widely used in loudspeaker design and are in fact perhaps most useful in the understanding and design of transducers such as microphones where acoustic, mechanical, and electrical systems are present together and where an overall equivalent electrical circuit can be formulated (see Chapters 112 and 113, see also *Enc.* Ch. 161). Beranek makes considerable use of electrical analogies in his books.[9,10] In Chapter 14, their use in the design of automobile mufflers is described.

REFERENCES

1. M. J. Lighthill, *Waves in Fluids*, Cambridge University Press, Cambridge, 1978.
2. P. M. Morse and K. U. Ingard, *Theoretical Acoustics*, Princeton University Press, Princeton, NJ, 1987.
3. L. E. Kinsler, A. R. Frey, A. B. Coppens, and J. V. Sanders, *Fundamentals of Acoustics*, Wiley, New York, 1982.
4. A. D. Pierce, *Acoustics: An Introduction to Its Physical Principles and Applications*, McGraw-Hill, New York, 1981.
5. M. J. Crocker and A. J. Price, *Noise and Noise Control*, Vol. 1, CRC Press, Cleveland, 1975.
6. R. G. White and J. G. Walker (Eds.), *Noise and Vibration*, Halstead Press, Wiley, New York, 1982.
7. F. J. Fahy, *Sound Intensity*, Second Edition, E&FN Spon, Chapman & Hall, London, 1995.
8. E. Skudrzyk, *The Foundations of Acoustics*, Springer Verlag, New York, 1971.
9. L. L. Beranek, *Acoustical Measurements*, rev. ed., Acoustical Society of America, New York, 1988.
10. L. L. Beranek, *Acoustics*, Acoustical Society of America, New York, 1986 (reprinted with changes).

2

MATHEMATICAL THEORY OF WAVE PROPAGATION

ALLAN D. PIERCE

1 INTRODUCTION

This chapter gives a more detailed discussion of some of the basic concepts referred to in Chapter 1. Of particular concern here is the mathematical embodiment of those concepts that underly the wave description of acoustics. This mathematical theory[1] began with Mersenne, Galileo, and Newton and developed into its more familiar form during the time[2,3] of Euler and Lagrange. Prominent contributors during the nineteenth century include Poisson, Laplace, Cauchy, Green, Stokes, Helmholtz, Kirchhoff, and Rayleigh.

The basic mathematical principles underlying sound propagation, whether through fluids or solids, include the principles of continuum mechanics, which include a law accounting for the conservation of mass, a law accounting for the changes in momenta brought about by forces, and a law accounting for changes in energy brought about by work and the transfer of heat. The subject also draws upon thermodynamics, upon symmetry considerations, and upon various known properties of the substances through which sound can propagate. The theory is inherently approximate, and there are many different versions, which differ slightly or greatly from each other in just what idealizations are made at the outset. The concern here is primarily with the simpler mathematical idealizations that have proven useful in acoustics, and in particular with those associated with linear acoustics, but the discussion begins with the more nearly exact nonlinear equations and explains how the linear equations of acoustics follow from them.

Note: References to chapters appearing only in the *Encyclopedia* are preceded by *Enc.*

2 CONSERVATION OF MASS

The matter through which an acoustic wave travels is characterized by its density ρ, which represents the local spatial average of the mass per unit volume in a macroscopically small volume. This density varies in general with position, generically denoted here by the vector $\mathbf{x}$, and with time t. The material velocity $\mathbf{v}$ with which the matter moves is defined so that $\rho\mathbf{v}$ is the mass flux vector within the fluid. The significance of the latter is that, were one to conceive of a hypothetical stationary surface within the material with a local unit-normal vector $\mathbf{n}(\mathbf{x})$ pointing from one side to the other, then $\rho\mathbf{v}\cdot\mathbf{n}$ gives the mass crossing this surface per unit time and per unit area of the surface. Here, also, a local average is understood; the material velocity $\mathbf{v}(\mathbf{x}, t)$ associated with a point and time may be considerably different from the instantaneous velocity of a molecule in the vicinity of that point.

The concept of a mass flux vector leads to the prediction that the net rate of flow of mass out through any hypothetical closed surface S within the material is the area integral of $\rho\mathbf{v}\cdot\mathbf{n}$. (The volume V enclosed by this surface is sometimes referred to as a control volume.[4]) The conservation of mass requires that this area integral be the same as the rate at which the volume integral of ρ decreases with time, so that

$$\int_S \rho\mathbf{v}\cdot\mathbf{n}\,dS = -\frac{d}{dt}\int_V \rho\,dV. \tag{1}$$

The partial differential equation expressing the conservation of mass results, after application of Gauss's theorem[5] to the surface integral, with the recognition that the volume of integration is arbitrary, so that

$$\frac{\partial \rho}{\partial t} + \nabla \cdot (\rho \mathbf{v}) = 0, \tag{2}$$

where the partial derivative with respect to time implies that the position $\mathbf{x}$ is held fixed in the differentiation.

An alternate form of the above partial differential equation is

$$\frac{D\rho}{Dt} + \rho \nabla \cdot \mathbf{v} = 0 \tag{3}$$

with the Stokes abbreviation[6]

$$\frac{D}{Dt} = \frac{\partial}{\partial t} + \mathbf{v} \cdot \nabla \tag{4}$$

for the total time derivative operator. The two terms in the latter correspond to (i) the time derivative as would be seen by an observer at rest and (ii) the convective time derivative.

If the material is incompressible, then the density viewed by someone moving with the flow should remain constant, so $D\rho/Dt$ should be zero, and consequently the conservation of mass requires that $\nabla \cdot \mathbf{v} = 0$. Although much of the analysis of fluid mechanics is based on this idealization, it is ordinarily inappropriate for acoustics, because the compressibility of the material plays a major role in the propagation of sound.

3 THE EULER EQUATION

The local rate of material acceleration is identified as the total time derivative $D\mathbf{v}/Dt$ of the material velocity. Consequently, the generalization of Newton's second law to a continuum requires that $\rho D\mathbf{v}/Dt$ equal the force per unit mass exerted on the material. The force on any small identifiable aggregate of material consists of a sum of a body force $\mathbf{F}_B$ and a surface force $\mathbf{F}_S$. Possible body forces are gravitational or electromagnetic; the former is customarily expressed as

$$\mathbf{F}_B = \int_V \rho \mathbf{g} \, dV, \tag{5}$$

where $\mathbf{g}$ is the acceleration associated with gravity. The surface force is expressible as

$$\mathbf{F}_S = \int_S \sum_{i,j=1}^{3} \sigma_{ij} \mathbf{e}_i n_j \, dS, \tag{6}$$

where the σ_{ij} are the Cartesian components of the stress tensor.[7] Here the quantities $\mathbf{e}_i$ are the unit vectors appropriate for the corresponding Cartesian coordinate system, and the n_j are the Cartesian components of the unit vector pointing outward to the enclosing surface S. The stress tensor's Cartesian component σ_{ij} can be regarded as the ith component of the surface force per unit area on a segment of the surface of a small element of the continuum, when the unit outward normal to the surface is in the jth direction. This surface force is caused by interactions with the neighboring particles just outside the surface[8] or by the transfer of momentum due to diffusion of molecules across the surface. The condition that the net force per unit volume be finite in the limit of macroscopically infinitesimal volumes requires[9] that the stress components σ_{ij} be independent of the shape and orientation of the hypothetical surface S, so the stress is a tensor field associated with the material that depends in general on position and time. Considerations of the net torque[10] that such surface forces would exert on a small element and the requirement that the angular acceleration of the element be finite in the limit of very small element dimensions lead to the conclusion $\sigma_{ij} = \sigma_{ji}$, so the stress tensor is symmetric.

With the above identifications for the surface and body forces and a subsequent application of Gauss's theorem to transform the surface integral to a volume integral, the application of Newton's second law yields Cauchy's equation of motion, which is written in Cartesian coordinates as

$$\rho \frac{D\mathbf{v}}{Dt} = \sum_{ij} \mathbf{e}_i \frac{\partial \sigma_{ij}}{\partial x_j} + \mathbf{g}\rho. \tag{7}$$

Here the Eulerian description is used, with each field variable regarded as a function of actual spatial position coordinates and time.

For disturbances in fluids, such as air and water, and with the neglect of viscous shear forces, the surface force associated with stress can only be normal[11] to the surface. Since such would have to hold for all possible orientations of the surface, the stress tensor has to have the form

$$\sigma_{ij} = -p\delta_{ij}, \tag{8}$$

where p is identified as the pressure and where δ_{ij}, equal to unity when the indices are equal and zero otherwise, is the Kronecker delta. The pressure p is understood to be such that $-p\mathbf{n}$ is the force per unit area exerted on the material on the interior side of a surface when the outward unit-normal vector is $\mathbf{n}$. The insertion of the above idealization for the stress into Cauchy's equation of motion (7) yields Euler's equation of motion for a fluid:

$$\rho \frac{D\mathbf{v}}{Dt} = -\nabla p + \mathbf{g}\rho. \qquad (9)$$

Although viscosity is often important adjacent to solid boundaries[12] and can lead to the attenuation of sound for propagation over very large distances, it is often an excellent first approximation to use the Euler equation as the embodiment of Newton's second law for a fluid in the analysis of acoustic processes. Also, gravity[13] has a minor influence on sound and can often be ignored in analytical studies.

The Euler equation can be alternately written, via a vector identity, as

$$\frac{\partial \mathbf{v}}{\partial t} - \mathbf{v} \times (\nabla \times \mathbf{v}) + \nabla\left(\frac{\mathbf{v}^2}{2}\right) = -\frac{1}{\rho}\nabla p + \mathbf{g}, \qquad (10)$$

and the curl of this yields the corollary

$$\frac{D}{Dt}(\nabla \times \mathbf{v}) + (\nabla \times \mathbf{v})\nabla \cdot \mathbf{v} - [(\nabla \times \mathbf{v}) \cdot \nabla]\mathbf{v} = \frac{1}{\rho^2}\nabla\rho \times \nabla p. \qquad (11)$$

This equation governs the dynamics of the vorticity ($\nabla \times \mathbf{v}$) and provides a means for assessing the common idealization that the vorticity is identically zero.

4 THERMODYNAMIC PRINCIPLES IN ACOUSTICS

The quasi-static theory of thermodynamics[14] presumes that the local state of the material at any given time can be completely described in terms of a relatively small number of variables. One such state variable is the density ρ; another is the internal energy u per unit mass (specific internal energy). For a fluid in equilibrium, these two variables are sufficient to specify the state. Other quantities such as the pressure p and the temperature T can be determined from equations of state that give these in terms of ρ and u. An important additional state variable, although one not easily measurable, is the equilibrium entropy $s(\mathbf{x}, t)$ per unit mass (specific entropy), the changes of which can be related to heat transfer, but which also can be regarded as a function of u and ρ. The physical interpretation of s requires that there be no heat transfer when s is constant, and conservation of energy requires that work done by pressure forces on a surface of a fluid element during a quasi-static process result in a corresponding increase in internal energy $\rho V\,du$. For a sufficiently small element, this work is p times the net decrease, $-dV$, in volume. Since mass is conserved, $\rho\,dV = -V\,d\rho$, and the work is consequently $p\rho^{-1}V\,d\rho$. This yields $\rho V\,du = p\rho^{-1}V\,d\rho$ when $ds = 0$, and leads to the prediction that, to first order in differentials, ds must be proportional to $du - p\rho^{-2}\,d\rho$, where the proportionality factor must be a function of u and ρ. There is an inherent arbitrariness in the actual form of the function $s(u, \rho)$, but the theory of thermodynamics requires that it can be chosen so that the reciprocal of the proportionality factor has all the properties commonly associated with absolute temperature. Such a choice yields

$$T\,ds = du - p\rho^{-2}\,d\rho, \qquad (12)$$

with the identification of $T\,ds$ as the increment of "heat energy" that has been transferred per unit mass during a quasi-static process.

For an ideal gas with temperature-independent specific heats, which is a common idealization for air, the function $s(u, \rho)$ is given by

$$s = \frac{R_0}{M}\ln(u^{1/(\gamma-1)}\rho^{-1}) + s_0. \qquad (13)$$

Here the constant s_0 is independent of u and ρ^{-1}, while M is the average molecular weight (average molecular mass in atomic mass units), and R_0 is the universal gas constant (equal to Boltzmann's constant divided by the mass in an atomic mass unit), equal to 8314 J/kg·K. The quantity γ is the specific heat ratio, equal to approximately $\frac{7}{5}$ for diatomic gases, $\frac{5}{3}$ for monatomic gases, and $\frac{9}{7}$ for polyatomic gases whose molecules are not collinear. For air, γ is 1.4 and M is 29.0. (The expression given here for entropy neglects the contribution of internal vibrations of diatomic molecules, which cause the specific heats and γ to depend slightly on temperature. In the explanation of the absorption of sound in air, a nonequilibrium entropy[15] is used that depends on the fractions of O_2 and N_2 molecules that are in their first excited vibrational states in addition to the quantities u and ρ^{-1}.) With the aid of the differential relation (14), the corresponding expressions for temperature and pressure appear as

$$T = (\gamma - 1)\frac{M}{R_0}u, \qquad p = (\gamma - 1)\rho u, \tag{14}$$

and from these two relations emerges the ideal-gas equation

$$p = \rho \frac{R_0}{M} T. \tag{15}$$

For other substances, a knowledge of the first and second derivatives of s, evaluated at some representative thermodynamic state, is sufficient for most linear acoustic applications.

An immediate consequence of the above line of reasoning is that, for a fluid in equilibrium, any two state variables are sufficient to determine the others. The pressure, for example, can be expressed as

$$p = p(\rho, s), \tag{16}$$

where the actual dependence on ρ and s is intrinsic to the material. For an ideal gas, for example, this function has the form

$$p = K(s)\rho^{\gamma}, \tag{17}$$

where

$$K(s) = (\gamma - 1)p_0\rho_0^{-\gamma}\exp\left((\gamma - 1)\frac{M}{R_0}(s - s_0)\right) \tag{18}$$

is a function of specific entropy only.

The formulation of acoustics that is most frequently used, and with which the present chapter is concerned, assumes that the processes of sound propagation are quasi-static in the thermodynamic sense. This implies that the p given by the equation of state (16) is the same as that which appears in the Euler equation (9). Also, Eq. (12) must hold for any given possibly moving element of fluid, so that the differential relation yields the partial differential equation

$$T\frac{Ds}{Dt} = \frac{Du}{Dt} - p\rho^{-2}\frac{D\rho}{Dt}. \tag{19}$$

With the neglect of heat transfer mechanisms within the fluid, the conservation of energy[16] requires that

$$\rho\frac{D}{Dt}\left\{\frac{v^2}{2} + u\right\} = -\nabla\cdot\{p\mathbf{v}\} + \rho\mathbf{g}\cdot\mathbf{v}, \tag{20}$$

where the quantity within braces on the left is the energy (kinetic plus potential) per unit mass and the terms on the right correspond to the work done per unit volume by pressure and gravitational forces, respectively. The Euler equation (9) yields the derivable relation

$$\rho\frac{D}{Dt}\left\{\frac{v^2}{2}\right\} = -\mathbf{v}\cdot\nabla p + \rho\mathbf{g}\cdot\mathbf{v}, \tag{21}$$

and the subtraction of this from Eq. (20) yields

$$\rho\frac{Du}{Dt} = -p\nabla\cdot\mathbf{v} = p\rho^{-1}\frac{D\rho}{Dt}. \tag{22}$$

The latter equality results from the version (3) of the conservation-of-mass equation. Therefore, the conservation-of-energy relation (20) requires that the right side of Eq. (19) be identically zero, so

$$\frac{Ds}{Dt} = 0, \tag{23}$$

which implies that the entropy per unit mass of any given fluid particle remains constant. A disturbance satisfying this relation is said to be isentropic.

The total time derivative of Eq. (16) with the replacement of Ds/Dt by 0 yields the equation

$$\frac{Dp}{Dt} = c^2\frac{D\rho}{Dt}, \tag{24}$$

where c^2 abbreviates the quantity

$$c^2 = \left(\frac{\partial p}{\partial \rho}\right)_s. \tag{25}$$

Here the subscript s implies that the differentiation is carried out at constant entropy. Thermodynamic considerations based on the requirement that the total entropy increases during an irreversible process requires in turn that this derivative be positive. Its square root c is identified as the speed of sound, for reasons discussed in a subsequent portion of this chapter.

The relation between the total time derivatives of pressure and density that appears in Eq. (24) can be combined with the conservation-of-mass relation (3) to yield

$$\frac{Dp}{Dt} + \rho c^2\nabla\cdot\mathbf{v} = 0, \tag{26}$$

which is often of greater convenience in acoustics than

the mass conservation relation because the interest is typically in pressure fluctuations rather than in density fluctuations.

For an ideal gas, the differentiation of the expression $p(\rho, s)$ in Eq. (17) yields

$$c^2 = \gamma K(s)\rho^{\gamma - 1}, \quad (27)$$

which can equivalently be written[17]

$$c^2 = \frac{\gamma p}{\rho} = \gamma \frac{R_0}{M} T. \quad (28)$$

For air, R_0/M is 287 J/kg·K and γ is 1.4, so a temperature of 293.16 K (20°C) and a pressure of 10^5 Pa yield a sound speed of 343 m/s and a density ρ of 1.19 kg/m^3.

Acoustic properties of materials are discussed in depth elsewhere in this book, but some simplified formulas for water are given here as an illustration of the thermodynamic dependences of such properties. For pure water, the sound speed[18] is approximately given in metres-kilograms-seconds (MKS) units by

$$c = 1447 + 4.0\Delta T + 1.6 \times 10^{-6} p. \quad (29)$$

Here c is in meters per second, ΔT is the temperature relative to 283.16 K (10°C), and p is the absolute pressure in pascals. The pressure and temperature dependence of the density is approximately given by

$$\rho \approx 999.7 + 0.048 \times 10^{-5} p - 0.088\Delta T - 0.007(\Delta T)^2. \quad (30)$$

The values for sea water[19] are somewhat different because of the presence of dissolved salts. An approximate expression for the speed of sound in sea water is given by

$$c \approx 1490 + 3.6\Delta T + 1.6 \times 10^{-6} p + 1.3\Delta S. \quad (31)$$

Here ΔS is the deviation of the salinity in parts per thousand from a nominal value of 35.

5 THE LINEARIZATION PROCESS

Sound results from a time-varying perturbation of the dynamic and thermodynamic variables that describe the medium. The quantities appropriate to the ambient medium are customarily represented[20] by the subscript 0, and the perturbations are represented by a prime on the corresponding symbol. The total pressure is $p = p_0 + p'$, and there are corresponding expressions for fluctuations in specific entropy, fluid velocity, and density. The linear equations that govern acoustic disturbances are then determined by the first-order terms in the expansion of the governing nonlinear equations in the primed variables. The zeroth-order terms cancel out completely because the ambient variables should themselves correspond to a valid state of motion of the medium. Thus, for example, the linearized version of the conservation-of-mass relation in Eq. (2) is

$$\frac{\partial \rho'}{\partial t} + \mathbf{v}_0 \cdot \nabla \rho' + \rho' \nabla \cdot \mathbf{v}_0 + \mathbf{v}' \cdot \nabla \rho_0 + \rho_0 \nabla \cdot \mathbf{v}' = 0. \quad (32)$$

The possible forms of the linear equations differ in complexity according to what is assumed about the medium's ambient state and according to what dissipative terms are included in the original governing equations. In the establishment of a rationale for using simplified models, it is helpful[21] to think in terms of the order of magnitudes and characteristic scales of the coefficients in more nearly comprehensive models. If the spatial region of interest has bounding dimensions that are smaller than any scale length over which the ambient variables vary by a respectable fraction, then it may be appropriate to idealize the coefficients in the governing equations as if the ambient medium were spatially uniform or homogeneous. Similarly, if the characteristic wave periods or propagation times are much less than characteristic time scales for the ambient medium, then it may be appropriate to idealize the coefficients as being time independent. Examination of orders of magnitudes of terms suggests that the ambient velocity may be neglected if it is much less than the sound speed c. The first-order perturbation to the gravitational force term can be ordinarily neglected if the previously stated conditions are met and if the quantity g/c is sufficiently less than any characteristic frequency of the disturbance.

A discussion of the restrictions on using linear equations and of neglecting second- and higher order terms in the primed variables is outside the scope of the present chapter, but it should be noted that one regards p' as small if it is substantially less than $\rho_0 c^2$ and $|\mathbf{v}'|$ as small if it is much less than c, where c is the sound speed defined via Eq. (25). It is not necessary that p' be much less than p_0, and it is certainly not necessary that $|\mathbf{v}'|$ be less than $|\mathbf{v}_0|$.

6 ACOUSTIC FIELD EQUATIONS

The customary equations for linear acoustics for a medium where there is no ambient flow can be derived from the Euler equation (9) with the neglect of the gravity term and from Eq. (26), which combines the conservation of mass with thermodynamic principles. The ambient medium is taken as time independent, so the ambient quantities depend at most only on position $\mathbf{x}$. Since the zeroth-order equations must hold when there is no disturbance, the Euler equation requires that the ambient pressure be independent of position. It is not precluded, however, that the ambient density ρ_0 and sound speed c_0 vary with position. Thus, the linearized version of Eq. (26) becomes

$$\frac{\partial p'}{\partial t} = c_0^2 \left(\frac{\partial \rho'}{\partial t} + \mathbf{v}' \cdot \nabla \rho_0 \right). \tag{33}$$

The inference that

$$p' = c_0^2 \rho' \tag{34}$$

applies only if the ambient entropy is constant, which for the present circumstances (no ambient flow) requires ρ_0 to be constant. If $\nabla \rho_0$ were nonzero, the appropriate alternate would be

$$p' = c_0^2 (\rho' + \xi \cdot \nabla \rho_0), \tag{35}$$

where ξ is the displacement vector (position shift) of the fluid particle nominally at $\mathbf{x}$ in its ambient position.

To allow for the possibility that the ambient medium is inhomogeneous, it is more convenient to deal with Eq. (26), and thereby take advantage of $\nabla p_0 = 0$, rather than to deal with the mass conservation equation. Equation (26) for the circumstances just described leads after linearization to

$$\frac{\partial p'}{\partial t} + \rho_0 c_0^2 \nabla \cdot \mathbf{v}' = 0, \tag{36}$$

and the Euler equation of Eq. (9) leads to

$$\rho_0 \frac{\partial \mathbf{v}'}{\partial t} = -\nabla p'. \tag{37}$$

The two equations just stated are a complete set of linear acoustic equations, insofar as there are as many partial differential equations as there are unknowns.

An alternate set of equations of comparable simplicity results when there is an ambient irrotational homoentropic flow. The term *homoentropic* means that the entropy is uniform (homogeneous) throughout the ambient medium. Because the total time derivative of the entropy is constant for the idealization considered here in which there is no heat transfer, a fluid that is homoentropic at any given instant will remain so for all time, so s will be independent of both position and time. This means that the pressure can be regarded as a function of density only, so that

$$\frac{\partial p}{\partial t} = c^2 \frac{\partial \rho}{\partial t}, \qquad \nabla p = c^2 \nabla \rho, \qquad p' = c_0^2 \rho', \tag{38}$$

where the latter results from the linearization of $p = p(\rho, s_0)$, even though c_0^2 is not necessarily constant.

When the fluid is homoentropic, it follows from Eq. (38) that $\nabla \rho \times \nabla p = 0$, so the vorticity dynamics equation (11) reduces to

$$\frac{D}{Dt}(\nabla \times \mathbf{v}) + (\nabla \times \mathbf{v}) \nabla \cdot \mathbf{v} - [(\nabla \times \mathbf{v}) \cdot \nabla] \mathbf{v} = 0. \tag{39}$$

This has the implication that, if the vorticity $(\nabla \times \mathbf{v})$ is initially everywhere zero, then it will stay everywhere zero for all time. This allows one to conceive of an ambient medium that is both homoentropic and irrotational, so that $\nabla \times \mathbf{v}_0 = 0$.

For these circumstances, with a homoentropic and irrotational ambient medium, the linearized equations, after a modest amount of manipulation, take the form

$$\rho_0 \left[\frac{\partial}{\partial t} + \mathbf{v}_0 \cdot \nabla \right] \left(\frac{p'}{\rho_0 c_0^2} \right) + \nabla \cdot (\mathbf{v}' \rho_0) = 0, \tag{40}$$

$$\frac{\partial \mathbf{v}'}{\partial t} - \mathbf{v}_0 \times (\nabla \times \mathbf{v}') = -\nabla \left[\frac{p'}{\rho_0} + \mathbf{v}' \cdot \mathbf{v}_0 \right]. \tag{41}$$

In these equations, the ambient quantities $\mathbf{v}_0$, ρ_0, and c_0 may depend on both time and position, although they are constrained by the zeroth-order fluid dynamic equations. A lengthier[22] analysis suggests that these equations remain a good approximation for arbitrary inhomogeneous time-dependent media with ambient flow, provided that the characteristic wavelengths of the sound disturbance are sufficiently short. Furthermore, they remain good approximations for all but extremely low infrasonic frequencies when one incorporates gravity into the derivation, with account taken of the associated variation of density with height (as for the air in the atmosphere).

In the limit when $\mathbf{v}_0$ is zero and when ρ_0 and c_0 are constant, Eqs. (36) and (40) are equivalent, while Eqs.

(37) and (41) are also equivalent. Because neither of the two sets of equations just given involve the ambient pressure p_0 and the density perturbation ρ', it is customary to delete the prime on p' and the subscript 0 on ρ_0 and c_0 in discussions of linear acoustics. A further simplification is to delete the prime on $\mathbf{v}'$ but to retain the subscript on $\mathbf{v}_0$, so that the total fluid velocity is $\mathbf{v}_0 + \mathbf{v}$. In the remainder of this chapter, p is the acoustic part of the total pressure, $\mathbf{v}$ is the acoustic part of the fluid velocity, and ρ is the ambient density, unless stated otherwise.

7 WAVE EQUATIONS

For a medium that is at rest but possibly inhomogeneous, the field equations (36) and (37), upon elimination of $\mathbf{v}$, lead to the partial differential equation

$$\rho c^2 \nabla \cdot \left(\frac{1}{\rho} \nabla p \right) - \frac{\partial^2 p}{\partial t^2} = 0. \tag{42}$$

A very good approximation to this when the ambient density ρ is slowly varying with position (but not necessarily constant) is

$$c^2 \nabla^2 \left(\frac{p}{\sqrt{\rho}} \right) - \frac{\partial^2}{\partial t^2} \left(\frac{p}{\sqrt{\rho}} \right) = 0, \tag{43}$$

but when the ambient density ρ is constant, one can use the ordinary wave equation

$$\nabla^2 p - \frac{1}{c^2} \frac{\partial^2 p}{\partial t^2} = 0, \tag{44}$$

where the dependent field variable is simply p.

The wave equation is sometimes rewritten in a more compact form,

$$\Box^2 p = 0, \tag{45}$$

where the operator

$$\Box^2 = \nabla^2 - c^{-2} \frac{\partial^2}{\partial t^2} \tag{46}$$

is called the d'Alembertian, because d'Alembert was the first (in 1747) to derive the one-dimensional version of Eq. (44), although his derivation was for the case of a vibrating string. The discovery of the wave equation for sound in fluids is due primarily to Lagrange and Euler.

For the linear acoustic equations (40) and (41), which apply when there is an ambient inhomogeneous flow, an analogous wave equation[23] also results. Equation (41) admits the possibility that $\nabla \times \mathbf{v} = 0$. Acoustic disturbances in fluids typically adhere to this and consequently can be described in terms of a velocity potential Φ such that

$$\mathbf{v} = \nabla \Phi. \tag{47}$$

Equation (41) is then satisfied identically with

$$p = -\rho D_t \Phi, \qquad D_t = \frac{\partial}{\partial t} + \mathbf{v}_0 \cdot \nabla, \tag{48}$$

where the operator D_t represents the total time derivative operator following the ambient flow. The substitution of Eqs. (47) and (48) into Eq. (40) then yields the wave equation

$$\frac{1}{\rho} \nabla \cdot (\rho \nabla \Phi) - D_t \left(\frac{1}{c^2} D_t \Phi \right) = 0. \tag{49}$$

This reduces to the wave equation of Eq. (44) when ρ and c are constant and there is no ambient flow.

8 ENERGY CONSERVATION COROLLARY

A consequence of the linear acoustic equations (36) and (37) (no ambient flow, medium possibly inhomogeneous, but time independent) is the sound energy conservation corollary[24]

$$\frac{\partial w}{\partial t} + \nabla \cdot \mathbf{I} = 0, \tag{50}$$

with

$$w = \frac{1}{2} \rho v^2 + \frac{1}{2} \frac{1}{\rho c^2} p^2, \tag{51}$$

$$\mathbf{I} = p\mathbf{v} \tag{52}$$

identified as the sound energy density and sound intensity (energy flux vector), respectively. The two terms in the energy density can be interpreted as the kinetic energy per unit volume and the potential energy per unit volume, respectively. The intensity corresponds to sound power flow per unit area.

This interpretation of an equation such as Eq. (50) as a conservation law follows after integration of both sides over an arbitrary fixed volume V within the fluid and after replacement of the volume integral of the divergence of **I** with a surface integral by means of Gauss's theorem. Doing this yields

$$\frac{d}{dt}\int_V w\,dV + \int_S \mathbf{I}\cdot\mathbf{n}\,dS = 0, \qquad (53)$$

where **n** is the unit-normal vector pointing out of the surface S enclosing V. This relation states that the net rate of increase of "sound energy" within the volume must equal the "sound power" flowing into the volume across its confining surface. If dissipation terms such as those associated with viscosity or relaxation processes are included within the linear acoustic equations, then the zero on the right side is replaced[25] by a term equal to the negative of the energy that is being dissipated per unit time within the volume.

No equation of simplicity comparable to Eq. (50) applies with generality to a medium with inhomogeneous ambient flow. However, if c, ρ, and $\mathbf{v}_0$ are constant, then Eqs. (40) and (41) yield

$$\frac{\partial w}{\partial t} + \nabla\cdot(\mathbf{I} + \mathbf{v}_0 w) = 0, \qquad (54)$$

where the quantities w and **I** are the same as given above. The interpretation that results from this is that the w given by Eq. (51) is the sound energy density regardless of whether there is an ambient flow. The quantity **I** is the energy flux vector relative to the medium itself. If the medium is flowing with an ambient velocity, then the energy flux vector seen by someone in a stationary coordinate system must also include the energy convected by the flow, which is just $\mathbf{v}_0 w$, so the energy flux vector perceived by someone at rest is $\mathbf{I} + \mathbf{v}_0 w$.

For acoustic disturbances in inhomogeneous moving flows, it is customary to use w and $\mathbf{I} + \mathbf{v}_0 w$ as identification for energy density and intensity, but without the expectation that sound energy is conserved, as there is a possibility that energy can be interchanged between the ambient flow and the acoustic field. (In the geometric acoustics[26] limit, where the disturbance everywhere locally resembles a plane wave, a conservation law does apply, but it is not interpretable as the conservation of sound energy. Instead, what is conserved is a quantity termed *wave action*.)

9 EQUATIONS OF ELASTICITY

Sound in solids is governed by analogous equations to those that govern sound in fluids. Cauchy's equation of motion [Eq. (7)] applies to solids as well as to fluids, but because a solid can support shear stresses, it is inappropriate to take the stress tensor as diagonal, with diagonal components equal to the negative of a pressure, as is done in Eq. (8). Instead, the idealization[27] of a linear elastic homogeneous solid is often used, and the Cauchy equation is linearized at the outset, with ρ set to the ambient density and with $D\mathbf{v}/Dt$ taken as the second derivative with respect to time of a particle displacement vector, with components (ξ_1, ξ_2, ξ_3), where $\xi_i(\mathbf{x}, t)$ is the ith Cartesian component of the displacement of the particle nominally at $\mathbf{x}$ from its ambient position. For any given Cartesian coordinate system, the Cauchy equation of motion then reduces to

$$\rho\frac{\partial^2 \xi_i}{\partial t^2} = \sum_{j=1}^{3}\frac{\partial \sigma_{ij}}{\partial x_j}, \qquad (55)$$

with the normally insignificant gravitational term deleted. The generalization of Hooke's law to a linear elastic solid causes the stress tensor components σ_{ij} to be linearly related to the strain tensor components

$$\epsilon_{ij} = \frac{1}{2}\left(\frac{\partial \xi_i}{\partial x_j} + \frac{\partial \xi_j}{\partial x_i}\right). \qquad (56)$$

Given the usual idealization that the material is isotropic, the stress–strain relations take the form

$$\sigma_{ij} = 2\mu\epsilon_{ij} + \lambda\delta_{ij}\sum_{k=1}^{3}\epsilon_{kk}. \qquad (57)$$

The Lamé constants λ and μ (the latter being the same as the shear modulus G) are related to the elastic modulus E and Poisson's ratio ν by the relations

$$\lambda = \frac{\nu E}{(1+\nu)(1-2\nu)}, \qquad \mu = G = \frac{E}{2(1+\nu)}. \qquad (58)$$

Alternative quantities that are convenient to use are

$$c_1^2 = \frac{\lambda + 2\mu}{\rho}, \qquad c_2^2 = \frac{\mu}{\rho}, \qquad (59)$$

The quantities c_1 and c_2 are referred to as the dilatational and shear wave speeds, respectively.

When λ and μ are both constant, the substitution of the stress–strain relation into Cauchy's equation of motion yields

$$\frac{\partial^2 \xi}{\partial t^2} = (c_1^2 - c_2^2)\nabla(\nabla \cdot \xi) + c_2^2 \nabla^2 \xi. \qquad (60)$$

This leads to the ordinary wave equation (44) for two special circumstances. If the displacement field is irrotational such that $\nabla \times \xi = 0$ (as is so for sound in fluids), then one can set $\xi = \nabla\Phi$, where the displacement scalar potential Φ is constant outside the acoustically perturbed region. In this circumstance, Φ and the components ξ_i satisfy

$$\nabla^2\Phi - \frac{1}{c_1^2}\frac{\partial^2\Phi}{\partial t^2} = 0. \qquad (61)$$

The other circumstance is when the displacement field is solenoidal, such that $\nabla \cdot \xi = 0$. Then it is possible to set $\xi = \nabla \times \Psi$, where the components of the vector potential Ψ satisfy the wave equation

$$\nabla^2\Psi - \frac{1}{c_2^2}\frac{\partial^2\Psi}{\partial t^2} = 0. \qquad (62)$$

For relatively general circumstances, any displacement field governed by the elastodynamic equations is decomposable to

$$\xi = \nabla\Phi + \nabla \times \Psi, \qquad (63)$$

where Φ satisfies Eq. (61) and the components of Ψ satisfy Eq. (62).

An energy conservation corollary of the form (5) also holds for sound in solids. The appropriate identifications for the energy density w and the components I_i of the intensity are

$$w = \frac{1}{2}\rho\sum_i\left(\frac{\partial\xi_i}{\partial t}\right)^2 + \frac{1}{2}\sum_{i,j}\epsilon_{ij}\sigma_{ij}, \qquad (64)$$

$$I_i = -\sum_j \sigma_{ij}\frac{\partial\xi_j}{\partial t}. \qquad (65)$$

10 PLANE WAVES IN FLUIDS

A solution of the wave equation that plays a central role in many acoustic concepts is that of a plane wave, which is such that all acoustic field quantities vary with time and with one Cartesian coordinate, taken here as x, but are independent of y and z. The Laplacian ∇^2 reduces thus to $\partial^2/\partial x^2$, and the d'Alembertian can be expressed as the product of two first-order operators, so that the wave equation takes the form

$$\left(\frac{\partial}{\partial x} + \frac{1}{c}\frac{\partial}{\partial t}\right)\left(\frac{\partial}{\partial x} - \frac{1}{c}\frac{\partial}{\partial t}\right)p = 0. \qquad (66)$$

The general solution of

$$\left(\frac{\partial}{\partial x} + \frac{1}{c}\frac{\partial}{\partial t}\right)f = 0 \qquad (67)$$

is any function $f(x - ct)$ that depends on x and t only in the combination $x - ct$. Similarly, the general solution of

$$\left(\frac{\partial}{\partial x} - \frac{1}{c}\frac{\partial}{\partial t}\right)g = 0 \qquad (68)$$

is any function $g(x + ct)$ that depends on x and t only in the combination $x + ct$. Consequently, the general solution of Eq. (66) is given by

$$p(x, t) = f(x - ct) + g(x + ct), \qquad (69)$$

where f and g are two arbitrary functions.

The quantity $f(x - ct)$ represents a plane wave traveling forward in the $+x$-direction at a speed c, while $g(x + ct)$ represents a plane wave traveling backward in the $-x$-direction, also at a speed c. The appropriateness of these identifications is established by the setting of x to the position $x_P(t)$ of a moving sensor. The function $f(x_P - ct)$ will appear to this sensor to be constant in time if $dx_P/dt = c$. Thus, crests, troughs, zero crossings, and other characteristic waveform features, at which the wave amplitude is some fixed value, appear to be moving with the speed c. An individual term, such as $f(x - ct)$ or $g(x + ct)$, which represents a wave traveling in just one direction, is referred to as a traveling wave. For a traveling plane wave, in contrast to traveling spherical waves (discussed in a subsequent section), not only the shape but also the amplitude is conserved during propagation.

The above description can be generalized to propagation in an arbitrary direction $\mathbf{n}$, the acoustic part of the pressure then being given by

$$p = f(\mathbf{n} \cdot \mathbf{x} - ct) \qquad (70)$$

for some generic function $f(\psi)$. Furthermore, p is the solution of the progressive wave equation

$$\frac{\partial p}{\partial t} + c\mathbf{n} \cdot \nabla p = 0, \qquad (71)$$

which is the generalization of Eq. (67).

The expression (70) for a plane traveling wave may be regarded as also being of the form

$$p = F(t - \tau), \tag{72}$$

where τ, referred to as the eikonal, is a function of position $\mathbf{x}$ only. If the traveling wave is propagating in the direction $\mathbf{n}$, then

$$\tau = \frac{1}{c}\,\mathbf{n}\cdot\mathbf{x}, \qquad \nabla\tau = \frac{\mathbf{n}}{c}, \tag{73}$$

and this establishes that, regardless of the direction of $\mathbf{n}$, the function τ must be a solution of the eikonal equation

$$(\nabla\tau)^2 = \frac{1}{c^2}. \tag{74}$$

This particular equation[28] is used to describe the long-term evolution of waves that locally resemble plane waves but for which the wavefronts are possibly curved and for circumstances where the ambient sound speed may possibly vary with position. (Wavefronts are surfaces in space at which similar waveform features, such as crests, are being simultaneously received.)

If there is an ambient flow, then the idealization that $\mathbf{v}_0$, ρ, and c are constant allows Eq. (49) to apply directly to the pressure, so that

$$\nabla^2 p - \frac{1}{c^2}\,D_t^2 p = 0. \tag{75}$$

Plane traveling waves for such circumstances exist of the generic form

$$p = F(t-\tau) = F\left(t - \frac{1}{v_{\text{ph}}}\,\mathbf{n}\cdot\mathbf{x}\right), \tag{76}$$

where the phase speed v_{ph} is the solution of

$$\left(\frac{1}{v_{\text{ph}}}\right)^2 = \frac{1}{c^2}\left(1 - \frac{\mathbf{v}_0\cdot\mathbf{n}}{v_{\text{ph}}}\right)^2, \tag{77}$$

or, equivalently, where the eikonal τ is the solution of the eikonal equation

$$(\nabla\tau)^2 = \frac{1}{c^2}\,(1 - \mathbf{v}_0\cdot\nabla\tau)^2, \tag{78}$$

with the correspondence $\nabla\tau = \mathbf{n}/v_{\text{ph}}$. The latter is consequently referred to as the wave slowness vector.

The appropriate solution of Eq. (77) for the phase speed, being that which reduces to the sound speed c when there is no ambient flow, is

$$v_{\text{ph}} = c + \mathbf{v}_0\cdot\mathbf{n}, \tag{79}$$

which indicates the sound speed is augmented by the ambient velocity component in the direction normal to the wavefront.

The wave equation (75) for homogeneous moving media can also be obtained from the ordinary wave equation (44) for homogeneous nonmoving media by means of a Galilean transformation, to a new coordinate system where the coordinate axes are moving with velocity $-\mathbf{v}_0$ relative to the original coordinate axes. Such causes the substitutions

$$t \to t, \qquad \mathbf{x} \to \mathbf{x} - \mathbf{v}_0 t, \tag{80}$$

$$\frac{\partial}{\partial t} \to \frac{\partial}{\partial t} + \mathbf{v}_0\cdot\nabla, \qquad \nabla \to \nabla. \tag{81}$$

The plane-wave solution (70) consequently transforms to

$$p = f(\mathbf{n}\cdot\mathbf{x} - [\mathbf{v}_0\cdot\mathbf{n} + c]t). \tag{82}$$

which is of the general form of Eq. (76), with the identification (79) for the phase speed v_{ph}. This can be alternately written as

$$p = f(\mathbf{n}\cdot[\mathbf{x} - \mathbf{v}_{\text{gr}}t]), \tag{83}$$

where the group velocity[29] $\mathbf{v}_{\text{gr}}$ is given by

$$\mathbf{v}_{\text{gr}} = c\mathbf{n} + \mathbf{v}_0. \tag{84}$$

The term *group velocity* is used here because this is the velocity with which a wave disturbance of finite spatial extent, although locally resembling a plane wave, would appear to be propagating through space. The general rule represented by these above relations is that acoustic waves that locally resemble plane waves always propagate relative to the ambient fluid itself in directions perpendicular to their wavefronts with the sound speed. If the ambient medium is not moving, then the disturbance has a velocity $c\mathbf{n}$. However, if the ambient medium is moving with a velocity $\mathbf{v}_0$, then an observer at rest sees a velocity that is the vector sum of the velocity of the fluid and the velocity of the wave relative to the fluid.

Unlike the case for electromagnetic waves, the wave equation for sound does not appear the same for two different coordinate systems, one of which is moving at constant velocity relative to the other. (That such is the case for the wave equation governing electromagnetic waves in a vacuum is a principal tenet of Einstein's theory of relativity.)

11 RELATIONS BETWEEN ACOUSTIC VARIABLES IN A PLANE SOUND WAVE

For the special case of a plane sound wave moving in the $+x$-direction through a nonmoving homogeneous medium, the acoustic part of the pressure is given by Eq. (85), and Euler's equation yields

$$v_x = \frac{1}{\rho c} f(x - ct) - \frac{1}{\rho c} g(x + ct). \tag{85}$$

Thus, in the expression for the acoustic part of the pressure, the term $f(x - ct)$, which corresponds to propagation in the same direction as that in which the fluid flow component v_x is directed, is multiplied by a factor $1/\rho c$, with a positive sign. The other term $g(x + ct)$, which corresponds to propagation in the direction opposite to that in which the fluid flow component v_x is directed, is multiplied by a factor $-1/\rho c$, with a negative sign.

The general rule that emerges from the one-dimensional example just cited is that, for a traveling plane wave propagating in the direction corresponding to unit vector $\mathbf{n}$, where, as in Eq. (70), the acoustic part of the pressure is given by $p = f(\mathbf{n}\cdot\mathbf{x} - ct)$ for some generic function $f(\psi)$, the acoustically induced fluid velocity is

$$\mathbf{v} = \frac{\mathbf{n}}{\rho c} p. \tag{86}$$

Because the fluid velocity is in the same direction as that of the wave propagation, such waves are said to be longitudinal. (Electromagnetic plane waves in free space, on the other hand, are transverse. Shear waves in solids, as are discussed further below, are also transverse.)

The relation (86) also holds for a plane wave propagating through a medium with an ambient fluid velocity $\mathbf{v}_0$. This can be derived from the linear acoustic equations (40) and (41), but it also follows directly because the above relation is invariant under Galilean transformations. The unit vector $\mathbf{n}$ appears the same to an observer moving with speed $-\mathbf{v}_0$ as it does to one at rest. The acoustic portion $\mathbf{v}$ of the total fluid velocity is the fluid velocity relative to the ambient flow velocity. Although the ambient flow velocity may appear different to observers moving relative to each other, the acoustic portion of the fluid velocity appears the same. The pressure and the density both appear unchanged when viewed by observers moving at constant speed. The sound speed c, because it is a thermodynamic property of the medium, as stated in Eq. (25), is also a Galilean invariant.

Both the density and the temperature fluctuate in an acoustic disturbance. Given that the medium is homogeneous, with possibly a uniform flow, and regardless of whether or not the disturbance is a plane wave, the density fluctuation is related to the pressure fluctuation by Eq. (34), $p' = c'\rho'$. The temperature fluctuation T' can be derived from the equation of state that expresses absolute temperature in terms of total pressure and total density, so that

$$T' = \left\{ \left(\frac{\partial T}{\partial p}\right)_{\rho,0} + \frac{1}{c^2}\left(\frac{\partial T}{\partial \rho}\right)_{p,0} \right\} p', \tag{87}$$

which by thermodynamic identities reduces to

$$T' = \left(\frac{T\beta}{\rho c_p}\right)_0 p', \tag{88}$$

where

$$\beta = \rho\left(\frac{\partial(1/\rho)}{\partial T}\right)_{p,0}, \qquad c_p = T_0\left(\frac{\partial s}{\partial T}\right)_{p,0} \tag{89}$$

denote the coefficient of thermal expansion and the specific heat at constant volume. If the fluid is an ideal gas, where $\beta = 1/T_0$ and $c_p = (R_0/M)\gamma/(\gamma - 1)$, the temperature fluctuation reduces to

$$T' = (\gamma - 1)\left(\frac{T}{\rho c^2}\right)_0 p'. \tag{90}$$

The prediction that the temperature fluctuation is nonzero is in accord with the requirement that sound propagation is more appropriately idealized as an adiabatic (no entropy fluctuation) process rather than as an isothermal process.

Relations such as Eqs. (34), (86), and (88), which relate different field components in a traveling wave, are categorically referred to as polarization relations, in analogy with those relations with the same name that characterize the ratios of the electric and magnetic field components in a propagating electromagnetic wave.

For a traveling plane acoustic wave in a homogeneous nonmoving medium, the energy conservation corollary of Eq. (50) applies. It follows from Eqs. (51) and (86) that the kinetic and potential energies are the same (Rayleigh's principle[30] for progressive waves) and that the energy density is given by

$$w = \frac{1}{\rho c^2} p^2. \tag{91}$$

The intensity becomes

$$\mathbf{I} = \mathbf{n} \frac{p^2}{\rho c}. \tag{92}$$

In regard to propagation through a moving medium, the intensity (perceived by an observer at rest when a plane wave is passing through) is, in accord with Eq. (54),

$$\mathbf{I} + \mathbf{v}_0 w = (c\mathbf{n} + \mathbf{v}_0)w. \tag{93}$$

This yields the interpretation that the energy in a sound wave is moving relative to the fluid with the sound speed in the direction normal to the wavefront. The velocity with which the energy is moving as seen by an observer at rest is consequently $c\mathbf{n} + \mathbf{v}_0$ and includes as an additive term the ambient velocity of the fluid. The appearance of the factor $c\mathbf{n} + \mathbf{v}_0$ in Eq. (93) supports the identification given in the previous section of this chapter that such a quantity is a group velocity (energy velocity).

12 PLANE WAVES IN SOLIDS

Plane acoustic waves in isotropic elastic solids have properties similar to those of waves in fluids. Dilatational (or longitudinal) plane waves are such that the curl of the displacement field vanishes, so the displacement vector must be parallel to the direction of propagation. A comparison of Eq. (61) with the wave equation of Eq. (44) indicates that such a wave must propagate with the speed c_1 determined by Eq. (59). Thus, a wave propagating in the $+x$-direction has no y- and z-components of displacement and has an x-component described by

$$\xi_x = F(x - c_1 t), \tag{94}$$

where F is an arbitrary function. The stress components can be deduced from Eqs. (56)–(59). These equations as well as symmetry considerations require, for a dilatational wave propagating in the x-direction, that the off-diagonal elements of the stress tensor vanish. The diagonal elements are given by

$$\sigma_{xx} = \rho c_1^2 F'(x - c_1 t), \tag{95}$$

$$\sigma_{yy} = \sigma_{zz} = \rho(c_1^2 - 2c_2^2)F'(x - c_1 t). \tag{96}$$

(Here the prime denotes a derivative with respect to the total argument.)

The divergence of the displacement field in a shear wave is zero, so a plane shear wave must cause a displacement perpendicular to the direction of propagation. Shear waves are therefore transverse waves. Equation (62), when considered in a manner similar to that described above for the wave equation for waves in fluids, leads to the conclusion that plane shear waves must propagate with speed c_2. A plane shear wave polarized in the y-direction and propagating in the x-direction will have only a y-component of displacement, given by

$$\xi_y = F(x - c_2 t). \tag{97}$$

The only nonzero stress components are the shear stresses

$$\sigma_{yx} = \rho c_2^2 F'(x - c_2 t) = \sigma_{xy}. \tag{98}$$

13 EQUATIONS GOVERNING WAVES OF CONSTANT FREQUENCY

Insofar as the governing equations are linear with coefficients independent of time, disturbances that vary sinusoidally with time can propagate without change of frequency. Such sinusoidally varying disturbances of constant frequency have the same repetition period T (reciprocal of frequency f) at every point, but the phase will in general vary from point to point. In mathematical descriptions of waves with fixed frequency, it is customary to use an angular frequency $\omega = 2\pi f$, which has units of radians per second.

For a plane wave of fixed angular frequency ω traveling in the $+x$-direction at the sound speed c, such that the sound pressure is a function of $t - (x/c)$, as in the first term of Eq. (69), one can in general write

$$p = |P| \cos\left[\omega\left(t - \frac{x}{c}\right) + \phi_0\right] = |P| \cos(\omega t - kx + \phi_0), \tag{99}$$

where $|P|$ is the amplitude of the disturbance, ϕ_0 is a phase constant, and $k = \omega/c$ is termed the wavenumber. The wavelength λ is the increment in propagation distance x required to change the argument of the trigonometric function by 2π radians, so $k = 2\pi/\lambda$. Also, the increment in t required to change the argument by 2π is the period T, which is the reciprocal of the frequency f; this observation yields the simple rule $\lambda = c/f$, relating wavelength, sound speed, and frequency.

In general, for a disturbance of fixed frequency or for one frequency component of a multifrequency disturbance, it is convenient to use a complex-number representation, such that each field amplitude is written[31]

$$p = \text{Re}\{\hat{p}e^{-i\omega t}\}. \tag{100}$$

Here $\hat{p}$ is called the complex amplitude of the sound pressure and in general varies with position. Two conventions are in common use for such a complex-number representation; the second uses $e^{+i\omega t}$ instead of $e^{-i\omega t}$ in expressions such as that of Eq. (100). In the latter case the identification of $\hat{p}$ would be the complex conjugate of what is used here. The $e^{-i\omega t}$ convention is predominant in literature on wave propagation as such, while the latter is predominant in literature on vibrations. If the reader wishes to translate from the first convention to the latter in the equations that appear in the remainder of this chapter, it is only necessary to replace i by $-i$.

Insertion of an expression such as that of Eq. (100) into a homogeneous linear ordinary or partial differential equation with real time-independent coefficients yields a result that can always be written in the form $\text{Re}\{\Phi e^{-i\omega t}\} = 0$, where the quantity Φ is an expression depending on the complex amplitudes and their spatial derivatives, but not depending on time. The requirement that the real part of $\Phi e^{-i\omega t}$ should be zero for all values of time consequently can be satisfied if and only if $\Phi = 0$. Moreover, the form of the expression Φ can be readily obtained from the original equation with a simple prescription: Replace all field variables by their complex amplitudes and replace all time derivatives using the substitution

$$\frac{\partial}{\partial t} \rightarrow -i\omega. \tag{101}$$

Thus, for example, the linear acoustic equations given by Eqs. (36) and (37) reduce to

$$-i\omega\hat{p} + \rho c^2 \nabla \cdot \hat{\mathbf{v}} = 0, \tag{102}$$

$$-i\omega\rho\hat{\mathbf{v}} = -\nabla\hat{p}. \tag{103}$$

The wave equation in Eq. (44) reduces to

$$\nabla^2\hat{p} + k^2\hat{p} = 0, \tag{104}$$

which is the Helmholtz equation[32] for the complex pressure amplitude.

14 SPHERICAL WAVES

Another wave type of fundamental importance is that of a spherically symmetric wave spreading out radially from a source in an unbounded medium. The symmetry implies that the acoustic field variables be functions of only the radial coordinate r and time t. The Laplacian reduces then to

$$\nabla^2 p = \frac{\partial^2 p}{\partial r^2} + \frac{2}{r}\frac{\partial p}{\partial r} = \frac{1}{r}\frac{\partial^2 (rp)}{\partial r^2}, \tag{105}$$

so the wave equation of Eq. (44) becomes

$$\frac{\partial^2 (rp)}{\partial r^2} - \frac{1}{c^2}\frac{\partial^2 (rp)}{\partial t^2} = 0, \tag{106}$$

with the general solution

$$p(r,t) = \frac{f(r-ct)}{r} + \frac{g(r+ct)}{r}. \tag{107}$$

Causality considerations (no sound before source is turned on) lead to the conclusion that the second term on the right side of Eq. (107) is not an appropriate solution of the wave equation when the source is concentrated near the origin. The expression

$$p(r,t) = \frac{f(r-ct)}{r}, \tag{108}$$

which describes the sound pressure in an outgoing spherically symmetric wave, has the property that listeners at different radii will receive (with a time shift corresponding to the propagation time) waveforms of the same shape but of different amplitudes. The factor of $1/r$ is characteristic of spherical spreading and implies that the peak waveform amplitudes in a spherical wave decrease with radial distance as $1/r$.

The fluid velocity associated with an outgoing spherical wave is purely radial and has the form

$$v_r = \frac{1}{\rho c}[-r^{-2}F(r-ct) + r^{-1}f(r-ct)]. \tag{109}$$

Here the function F is such that its derivative is the function f that appears in Eq. (108). Because the first term (a near-field term) decreases as the square rather than the first power of the reciprocal of the radial distance, the fluid velocity v_r asymptotically approaches $p/\rho c$, which is the same as the plane-wave relation of Eq. (86).

For outgoing spherical waves of constant frequency,

the complex amplitudes of the pressure and fluid velocity are

$$\hat{p} = A\,\frac{e^{ikr}}{r}, \qquad \hat{v}_r = \frac{1}{\rho c}\left[1 - \frac{1}{ikr}\right]\hat{p}, \tag{110}$$

where A is a constant. The expression for $\hat{p}$ here is a solution of the Helmholtz equation (104), with the Laplacian given by Eq. (105).

15 CYLINDRICAL WAVES

For cylindrically symmetric waves, there is no dependence on the azimuthal angle or on the axial coordinate, so the Laplacian in cylindrical coordinates reduces to

$$\nabla^2 = \frac{1}{r}\frac{\partial}{\partial r}\left(r\frac{\partial}{\partial r}\right), \tag{111}$$

where r is here the radial distance from the symmetry axis. Consequently, the wave equation of Eq. (44) takes the form

$$\frac{\partial^2(\sqrt{r}p)}{\partial r^2} - \frac{1}{c^2}\frac{\partial^2(\sqrt{r}p)}{\partial t^2} + \frac{\sqrt{r}p}{4r^2} = 0. \tag{112}$$

and the Helmholtz equation of Eq. (104) can be written in either of the forms

$$\frac{d^2\hat{p}}{dr^2} + \frac{1}{r}\frac{d\hat{p}}{dr} + k^2\hat{p} = 0,$$

$$\frac{d^2(\sqrt{r}p)}{dr^2} + \left[k^2 + \frac{1}{4r^2}\right]\sqrt{r}p = 0. \tag{113}$$

The solution of the latter that corresponds to an outgoing wave is

$$\hat{p} = AH_0^{(1)}(kr), \tag{114}$$

where the indicated function is a Hankel function[33] of the first kind, which asymptotically approaches the limit

$$\lim_{kr\to\infty} H_0^{(1)}(kr) = \left(\frac{2}{\pi kr}\right)^{1/2} e^{-i\pi/4}e^{ikr}. \tag{115}$$

For cylindrical waves that are not of constant frequency, an outgoing solution can be taken as

$$p = \int_{-\infty}^{\infty} R^{-1}F(t - c^{-1}R)\,dz_0, \tag{116}$$

where $R = [r^2 + z_0^2]^{1/2}$ and F is an arbitrary function. Waveform shapes of outward propagating cylindrical waves tend to distort with increasing propagation distance, especially so at small r. However, at larger values of r, it is often a good approximation to neglect the last term in Eq. (112), resulting in an approximate solution of the generic form, $p(r,t) \approx f(r - ct)/\sqrt{r}$, which is similar to the expression of Eq. (108) for an outgoing spherical wave, only here the amplitude drops off with r as $1/\sqrt{r}$. The latter approximate expression is consistent with the constant-frequency solution given by Eq. (114) when the Hankel function is replaced by its asymptotic limit (115).

The fluid velocity induced by outgoing cylindrical waves is not as simply related to the corresponding sound pressure as that induced by a plane wave, although symmetry directs that the velocity must be in the appropriate radial direction when the propagation is cylindrically symmetric. For the constant-frequency case, an expression may be determined from the radial component of (103) such that the velocity is expressed in terms of the Hankel function of first order. However, a simple approximate result emerges in the limit of large radial distance r, this being the plane-wave relation of Eq. (86). (Here large r implies that it is large compared to a characteristic wavelength, or that it is large compared to c divided by a characteristic angular frequency.)

16 SOUND SOURCES AND SOUND POWER

Many sound fields can be idealized as being steady, such that long-term time averages are insensitive to the duration and the center time of the averaging interval. Constant-frequency sounds and continuous noises fall into this category. For such sounds, the time derivative of the sound energy density will average out to zero over a sufficiently long time period, so the acoustic energy corollary of Eq. (50), in the absence of dissipation, yields the time-averaged relation

$$\nabla \cdot \mathbf{I}_{\text{av}} = 0. \tag{117}$$

This implies that the time-averaged vector intensity field is solenoidal in regions that do not contain sound sources. This same relation holds for any frequency component of the acoustic field or for the net acoustic contribution to the field from any given frequency band. The above yields the integral relation

$$\int_S \mathbf{I}_{\text{av}} \cdot \mathbf{n}\, dS = 0, \tag{118}$$

which is interpreted as a statement that the net sound power flowing out of any region not containing sources must be zero on the time average and for any given frequency band.

For a closed surface that encloses one or more sources such that the governing linear acoustic equations do not apply at every point within the volume, the reasoning above allows definition of the time-averaged net sound power of these sources as

$$\mathcal{P}_{\text{av}} = \int_S \mathbf{I}_{\text{av}} \cdot \mathbf{n}\, dS, \tag{119}$$

where the surface S encloses the sources. If follows from Eq. (118) that the sound power of a source computed in such a manner will be the same for any two choices for the surface S, provided that both surfaces enclose the same source and no other sources. The value of the integral is independent of the size and of the shape of S. This result is of great practical importance, as it allows considerable lattitude in the measurement of source power.

In the computation of the time-averaged intensity for a field of fixed frequency, a useful mathematical relation is

$$\{a(t)b(t)\}_{\text{av}} = \tfrac{1}{2}\,\text{Re}\{\hat{a}\hat{b}^*\}, \tag{120}$$

where $a(t)$ and $b(t)$ are any two quantities oscillating with the same angular frequency and $\hat{a}$ and $\hat{b}$ are their complex amplitudes, in the sense of Eq. (100). Thus the time-averaged intensity becomes

$$\mathbf{I}_{\text{av}} = \tfrac{1}{2}\,\text{Re}\{\hat{\mathbf{v}}\hat{p}^*\}. \tag{121}$$

17 BOUNDARY CONDITIONS AT INTERFACES

For the model of a fluid without viscosity or thermal conduction, such as is governed by Eqs. (36) and (37), the appropriate boundary conditions at an interface are that the normal component of the fluid velocity be continuous and that the pressure be continuous. At a rigid nonmoving surface, the normal component must consequently vanish, but no restrictions are placed on the tangential component of the velocity. The model also places no requirements on the value of the temperature at a solid surface.

In many cases involving boundary surfaces, it is helpful to make use of the concept of specific acoustic impedance[34] or unit area acoustic impedance $Z_S(\omega)$, which is defined as

$$Z_S(\omega) = \frac{\hat{p}}{\hat{v}_{\text{in}}}, \tag{122}$$

where $\hat{v}_{\text{in}}$ is the component of the fluid velocity directed into the surface under consideration. Typically, the specific acoustic impedance, often referred to briefly as impedance without any adjective, is used to describe the acoustic properties of materials. In many cases, surfaces of materials abutting fluids can be characterized as locally reacting, so that Z_S is independent of the detailed nature of the acoustic pressure field. In particular, the locally reacting hypothesis implies that the velocity of the material at the surface is unaffected by pressures other than in the immediate vicinity of the point of interest. At a nominally motionless and passively responding surface, and when the hypothesis is valid, the appropriate boundary condition on the complex amplitude $\hat{p}$ that satisfies the Helmholtz equation (104) is

$$i\omega\rho\hat{p} = -Z_S\,\nabla\hat{p}\cdot\mathbf{n}, \tag{123}$$

where $\mathbf{n}$ is the unit-normal vector pointing out of the material into the fluid. A surface that is perfectly rigid has $|Z_S| = \infty$. The other extreme, where $Z_S = 0$, corresponds to the ideal case of a pressure-release surface. This is, for example, what is normally assumed for the upper surface of the ocean in underwater sound. Since a passive surface absorbs energy from the sound field, the time-averaged intensity component into the surface should be positive or zero. This observation leads to the requirement that the real part (specific acoustic resistance) of the impedance should always be nonnegative. The imaginary part (specific acoustic reactance) may be either positive or negative.

18 REFLECTION AT PLANE SURFACES AND INTERFACES

When a plane wave reflects at a surface with finite specific acoustic impedance Z_S, a reflected wave is formed such that the angle of incidence θ_I equals the angle of reflection (law of mirrors). Here both angles are reckoned from the line normal to the surface and correspond to the directions of the two waves. If one takes the y-axis as pointing out of the surface and the surface as coinciding with the $y = 0$ plane, then an incident plane wave propagating obliquely in the $+x$-direction will have a complex pressure amplitude

$$\hat{p}_{\text{in}} = \hat{f} e^{ik_x x} e^{-ik_y y}, \tag{124}$$

where $\hat{f}$ is a constant. (For transient reflection, the quantity $\hat{f}$ can be taken as the Fourier transform of the incident pressure pulse at the origin.) The two indicated wavenumber components are $k_x = k \sin \theta_{\text{I}}$ and $k_y = k \cos \theta_{\text{I}}$. The reflected wave has a complex pressure amplitude given by

$$\hat{p}_{\text{refl}} = \mathcal{R}(\theta_{\text{I}}, \omega) \hat{f} e^{ik_y y} e^{ik_y y}, \tag{125}$$

where the quantity $\mathcal{R}(\theta_{\text{I}}, \omega)$ is the pressure amplitude reflection coefficient.

Analysis that makes use of the boundary condition of Eq. (123) leads to the identification

$$\mathcal{R}(\theta_{\text{I}}, \omega) = \frac{\xi(\omega) \cos \theta_{\text{I}} - 1}{\xi(\omega) \cos \theta_{\text{I}} + 1} \tag{126}$$

for the reflection coefficient, with the abbreviation $\xi(\omega) = Z_S/\rho c$, which represents the ratio of the specific acoustic impedance of the surface to the characteristic impedance of the medium.

The above relations also apply, with an appropriate identification of the quantity Z_S, to sound reflection[35] at an interface between two fluids with different sound speeds and densities. Translational symmetry requires that the disturbance in the second fluid have the same apparent phase velocity (ω/k_x) (trace velocity) along the x-axis as does the disturbance in the first fluid. This requirement is known as the trace velocity matching principle and leads to the observation that k_x is the same in both fluids.

If the trace velocity is higher than the sound speed c_2, then $c_2 < c_1/\sin \theta_{\text{I}}$ and a propagating plane wave (transmitted wave) is excited in the second fluid, with complex pressure amplitude

$$\hat{p}_{\text{trans}} = \mathcal{T}(\omega, \theta_{\text{I}}) \hat{f} e^{ik_x x} e^{ik_2 y \cos \theta_{\text{II}}}, \tag{127}$$

where $k_2 = \omega/c_2$ is the wavenumber in the second fluid and θ_{II} (angle of refraction) is the angle at which the transmitted wave is propagating. The trace velocity matching principle leads to Snell's law[36]:

$$\frac{\sin \theta_{\text{I}}}{c_1} = \frac{\sin \theta_{\text{II}}}{c_2}. \tag{128}$$

The change in propagation direction from θ_{I} to θ_{II} is the phenomenon of refraction.

The requirements that the pressure and normal component of the fluid velocity be continuous across the interface yield the relation

$$1 + \mathcal{R} = \mathcal{T}, \tag{129}$$

and this in conjunction with the continuity of the normal component of the fluid velocity yields the reflection coefficient

$$\mathcal{R} = \frac{Z_{\text{II}} - Z_{\text{I}}}{Z_{\text{II}} + Z_{\text{I}}}, \tag{130}$$

which involves the two impedances $Z_{\text{I}} = \rho_1 c_1/\cos \theta_{\text{I}}$ and $Z_{\text{II}} = \rho_2 c_2/\cos \theta_{\text{II}}$.

The other possibility, that the trace velocity is lower than the sound speed c_2, can only occur when $c_2 > c_1$ and, moreover, only if θ_{I} is greater than the critical angle $\theta_{\text{cr}} = \arcsin(c_1/c_2)$. In this circumstance, an inhomogeneous plane wave propagating in the x-direction but dying out exponentially in the $+y$-direction is excited in the second medium. Instead of Eq. (140), the transmitted pressure is given by

$$\hat{p}_{\text{trans}} = \mathcal{T}(\omega, \theta_{\text{I}}) \hat{f} e^{ik_x x} e^{-\beta k_2 y}, \tag{131}$$

with

$$\beta = \left[\left(\frac{c_2}{c_1} \right)^2 \sin^2 \theta_{\text{I}} - 1 \right]^{1/2}. \tag{132}$$

The previously stated equations governing the reflection and transmission coefficients are still applicable, but with the replacement of $\cos \theta_{\text{II}}$ by $i\beta$. This causes the magnitude of the reflection coefficient $\mathcal{R}$ to become unity, so the time-averaged incident energy is totally reflected. Sound energy is present in the second fluid, but its time average over a wave period stays constant once the steady state is reached.

REFERENCES

1. R. B. Lindsay, *Acoustics: Historical and Philosophical Development*, Dowden, Hutchinson, and Ross, Stroudsburg, 1972. (Many of the older articles cited in this chapter are reprinted in English translation in this book.)
2. C. Truesdell, "The Theory of Aerial Sound, 1687–1788," *Leonhardi Euleri Opera Omnia*, Ser. 2, Vol. 13, Orell Füssli, Lausanne, 1955, pp. 19–72.

3. C. Truesdell, "Rational Fluid Mechanics, 1687–1765," *Leonhardi Euleri Opera Omnia*, Ser. 2, Vol. 12, Orell Füssli, Lausanne, 1954, pp. 9–125.
4. G Batchelor, *An Introduction to Fluid Dynamics*, Cambridge University Press, Cambridge, 1967.
5. O. D. Kellogg, *Foundations of Potential Theory*, Dover, New York, 1953.
6. G. G. Stokes, "On the Theories of the Internal Friction of Fluids in Motion, and of the Equilibrium and Motion of Elastic Fluids," *Trans. Camb. Phil. Soc.*, Vol. 8, 1845, pp. 74–102.
7. P. A. Thompson, *Compressible-Fluid Dynamics*, McGraw-Hill, New York, 1972.
8. Kirkwood, J. G., "The Statistical Mechanical Theory of Transport Processes, I: General Theory," *J. Chem. Phys.*, Vol. 14, 1946, pp. 180–201.
9. Y. C. Fung, *Foundations of Solid Mechanics*, Prentice-Hall, Englewood Cliffs, NJ, 1965.
10. C. S. Yih, *Fluid Mechanics*, McGraw-Hill, New York, 1969.
11. H. Lamb, *Hydrodynamics*, Dover, New York, 1945.
12. L. Cremer, "On the Acoustic Boundary Layer Outside a Rigid Wall," *Arch. Elektr. Übertrag.*, Vol. 2, 1948, pp. 136–139.
13. P. G. Bergmann, "The Wave Equation in a Medium with a Variable Index of Refraction," *J. Acoust. Soc. Am.*, Vol. 17, 1946, pp. 329–333.
14. A. H. Wilson, *Thermodynamics and Statistical Mechanics*, Cambridge University Press, London, 1957.
15. A. D. Pierce, "Aeroacoustic Fluid Dynamic Equations and Their Energy Corollary with O_2 and N_2 Relaxation Effects Included," *J. Sound Vibr.*, Vol. 58, 1978, pp. 189–200.
16. G. R. Kirchhoff, "On the Influence of Heat Conduction in a Gas on Sound Propagation," *Ann. Phys. Chem.*, Vol. 134, 1868, pp. 177–193.
17. P. S. Laplace, "On the Velocity of Sound through Air and through Water," *Ann. Chim. Phys.*, Ser. 2, Vol. 3, 1816, pp. 238–241.
18. W. D. Wilson, "Speed of Sound in Distilled Water as a Function of Temperature and Pressure," *J. Acoust. Soc. Am.*, Vol. 31, 1959, pp. 1067–1072.
19. W. D. Wilson, "Equation for the Speed of Sound in Sea Water," *J. Acoust. Soc. Am.*, Vol. 32, 1960, pp. 641–644, 1357.
20. A. D. Pierce, *Acoustics: An Introduction to Its Physical Principles and Applications*, McGraw-Hill, New York, 1981.
21. C. Eckart, "Vortices and Streams Caused by Sound Waves," *Phys. Rev.*, Vol. 7, 1948, pp. 68–76.
22. A. D. Pierce, "Wave Equation for Sound in Fluids with Unsteady Inhomogeneous Flow," *J. Acoust. Soc. Am.*, Vol. 87, 1990, pp. 2292–2299.
23. D. I. Blokhintzev, "The Propagation of Sound in an Inhomogeneous and Moving Medium I," *J. Acoust. Soc. Am.*, Vol. 18, 1946, pp. 322–328.
24. G. R. Kirchhoff, *Vorlesungen über mathematische Physik: Mechanik*, Teubner, Leipzig, 1877.
25. C. Eckart, "The Thermodynamics of Irreversible Processes," *Phys. Rev.*, Vol. 58, 1940, pp. 267–269.
26. W. D. Hayes, "Energy Invariant for Geometric Acoustics in a Moving Medium," *Phys. Fluids*, Vol. 11, 1968, pp. 1654–1656.
27. S. H. Crandall, N. C. Dahl, and T. J. Lardner, *An Introduction to the Mechanics of Solids*, McGraw-Hill, New York, 1978.
28. P. G. Frank, P. G. Bergmann, and A. Yaspan, "Ray Acoustics," in *Physics of Sound in the Sea*, Department of the Navy, Headquarters Naval Material Command, Publication NAVMAT P-9675, 1969, pp. 41–68.
29. M. J. Lighthill, *Waves in Fluids*, Cambridge University Press, Cambridge, 1978.
30. J. W. S. Rayleigh, "On Progressive Waves," *Proc. Lond. Math. Soc.*, Vol. 9, 1877, pp. 21–26.
31. C. J. Bouwkamp, "Contributions to the Theory of Acoustical Radiation," *Phillips Res. Rep.*, Vol. 1, 1946, pp. 251–277.
32. H. Helmholtz, "Theory of Air Oscillations in Tubes with Open Ends," *J. Reine Angew. Math.*, Vol. 57, 1860, pp. 1–72.
33. M. Abramowitz and I. A. Stegun, *Handbook of Mathematical Functions*, Dover, New York, 1965.
34. P. M. Morse and R. H. Bolt, "Sound Waves in Rooms," *Rev. Modern Phys.*, Vol. 16, 1944, pp. 69–150.
35. G. Green, "On the Reflexion and Refraction of Sound," *Trans. Camb. Phil. Soc.*, Vol. 6, 1838, pp. 403–412.
36. W. B. Joyce and A. Joyce, "Descartes, Newton, and Snell's Law," *J. Opt. Soc. Am.*, Vol. 66, 1976, pp. 1–8.

3

RAY ACOUSTICS FOR FLUIDS

D. E. WESTON

1 INTRODUCTION

Ray acoustics is also called geometric acoustics and is probably the easiest and best known way of thinking about sound propagation problems. In a homogeneous medium the sound may be pictured as traveling in a straight line directly from the source to the receiver, and at high frequencies this is virtually what does happen, just as in ray optics.

In general, there is a three-dimensional distribution of sound speed c, and the ray approach proceeds in two stages. First, it finds the ray path or paths by considering the wavefront or the equivalent acoustic phase, which may vary quite rapidly with position. Second, it models the acoustic amplitude, which is assumed to vary only slowly. Reflections from the continuous changes in speed and density are ignored, so the result is usually an approximation, though often a good one.

Technically the approach stems from Huygens' principle, in which each point on a wavefront is imagined to be the source of a spherically spreading wavelet, as shown in Fig. 1. After a short time interval the envelope of the wavelets is taken to define the new wavefront, and then the process repeats indefinitely. The direction of ray travel is normal to the wavefront, except that there can be complications for moving fluids and anisotropic solids.

This chapter presents the basic equations, discusses applications (one is dramatized in Fig. 2, to be examined more fully later), gives a very brief account of expansions and extensions to ray theory, and finally comments on validity.

Note: References to chapters appearing only in the *Encyclopedia* are preceded by *Enc.*

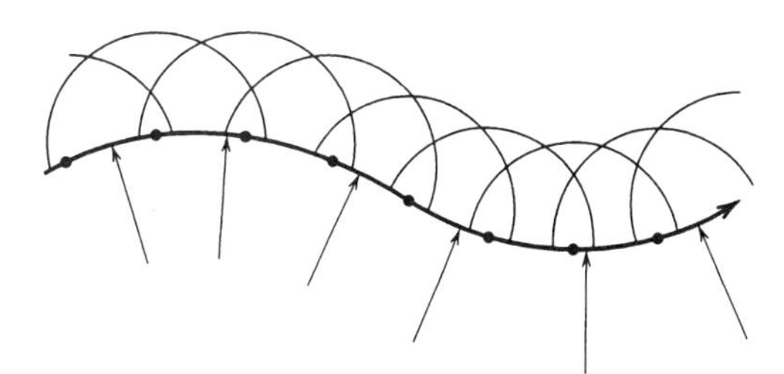

Fig. 1 Wavefront, Huygens wavelets, and ray directions.

2 BASIC EQUATIONS

It is necessary to start by introducing the eikonal $\tau(x, y, z)$, named from the Greek for image but actually specifying the travel time to a point on the acoustic wavefront. Thus any constant value of the eikonal τ defines a surface that is a wavefront. For example, with a point source in a homogeneous medium each τ value corresponds to a different spherical surface. For an extended source or an inhomogeneous medium the τ value may depend upon position in a complicated manner, with wavefronts that are also complicated, as in Fig. 1.

Defining the eikonal as a time has advantages both in simplicity and in reaching an equation that does not involve frequency, since the ray concept is independent of frequency. But it has sometimes been defined as the length $c_0\tau$, where c_0 is a reference sound speed, and sometimes as the phase $\omega\tau$, where ω is the angular frequency. Huygens' principle leads to the eikonal equation,[2–4] expressed in Cartesian coordinates as

$$\left(\frac{\partial\tau}{\partial x}\right)^2 + \left(\frac{\partial\tau}{\partial y}\right)^2 + \left(\frac{\partial\tau}{\partial z}\right)^2 = \frac{1}{c^2}. \qquad (1)$$

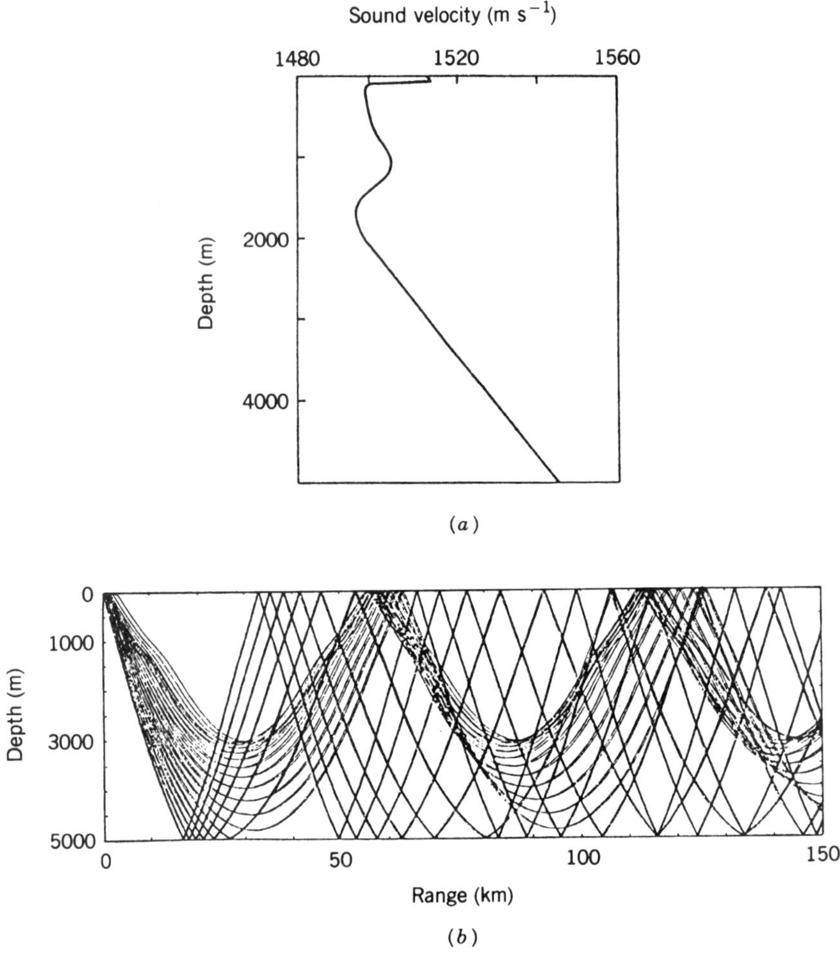

Fig. 2 (*a*) Typical deep-water profile of sound velocity vs. depth, Northeast Atlantic, October. (*b*) Computer-generated ray trace for profile (*a*), source at 38 m depth, ray angles at source every 1° between ±17°. Vertical scale is exaggerated by a factor of 10. From Ref. 1, © Crown Copyright, reproduced with the permission of the Controller of HMSO.

The point of Eq. (1) is that it enables the development in time of the wavefront to be followed. If the fluid is allowed to move, there are necessary modifications, as discussed by Pierce.[2]

Another form of Eq. (1) uses the local wavenumber k, really a propagation vector, together with the wavenumber components k_x, k_y, k_z resolved along the Cartesian axes,

$$k_x^2 + k_y^2 + k_z^2 = k^2. \tag{2}$$

The ray direction normal to the wavefront is specified by the local direction cosines k_x/k, k_y/k, and k_z/k, so that Eq. (1) or (2) also allows the tracing of rays.

In the most straightforward case k_x, k_y, and k_z are all real, and we have an ordinary ray or wave in which the amplitude variations are very slow. But if, for example, k_x^2 is negative and k_x imaginary, it is still possible to find a physical interpretation. It is that of a so-called inhomogeneous plane wave[5] in which the amplitude changes exponentially across the wavefront as $\exp(ik_x x)$: a good example is its occurrence in a second medium when the waves from the first suffer total internal reflection.

The eikonal equation is the basic one in ray acoustics but takes us on to another important result, Fermat's principle. This states that the travel time along a ray path from a to b is at a local extremum with respect to any small change δ in the path s,

$$\delta\left(\int_a^b c^{-1}\,ds\right) = 0. \tag{3}$$

In practice this usually turns out to be a path of least time.

To model amplitudes or intensities, narrow ray tubes or beams must be considered. Energy or power is conserved along the tube, and a transport equation or flux approach shows that intensity must vary *inversely as the tube cross-sectional area*. Thus, in a homogeneous medium the intensity will vary as the familiar r^{-2} law, where r is the range or distance from a point source.

3 APPLICATIONS

For any acoustic propagation problem a ray theory treatment should at least be considered. Problems run from those with solid structures in the laboratory and elsewhere; to those with fluids in the laboratory, industry, and even physiology; to noise in factories and music in concert halls (Chapters 74 and 75)[6]; and to long-range noise in the atmosphere (Chapter 28),[7] in the ocean (Chapter 31),[1,4] and in the earth's crust. This chapter concentrates on fluids; however, in scattering from targets and from irregularities (*Enc.* Chs. 4 and 43) ray theory may be applied not only within structures but also for surface-guided or creeping waves on the outside. Here the applications are considered less for their own sake, since they are covered elsewhere, and more as illustrations of particular ray effects.

Perhaps the classic ray acoustic effects are those for high frequencies in the laboratory, where many ray optic phenomena may be reproduced, including that of focusing with a suitable lens.

Figure 3 introduces multiple paths and images for a homogeneous medium with range-independent geometry. This could represent a laboratory arrangement, a first approximation to room acoustics, or a coastal water experiment. In fact, the acoustics of rooms, or of concert halls, involves the extreme form of this type of complication since the enclosures are three dimensional and the multiple arrivals determine the critical time history on reception. Pure ray theory breaks down on each reflection at a sharp boundary, and the connection to a continuing ray has to be made by solving a boundary condition problem (Chapter 2). In a room most boundaries are lossy and are not simple plane surfaces, so the process

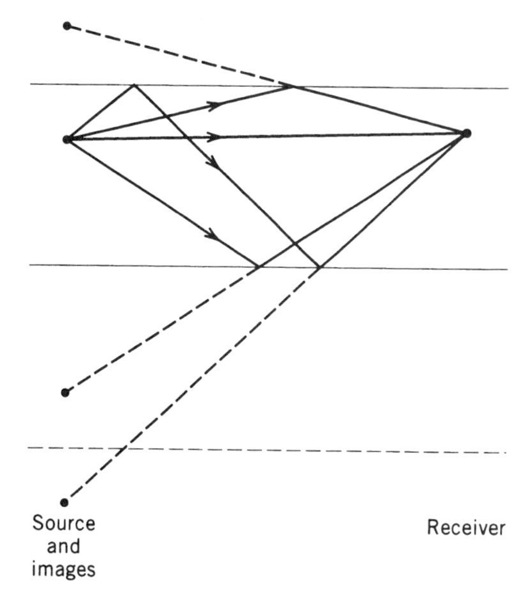

Fig. 3 Some multiple ray paths due to boundary reflections, illustrating the concept of image sources.

is one of scattering or diffraction. In practice, for room acoustics the geometry is usually so complicated that ray acoustics is to be preferred over wave acoustics, detailed modeling with the latter being too difficult.

In any application with multiple paths it is necessary to combine the various contributions. With continuous sources the simplest way is to add intensities, or, strictly, since intensity is a vector, to add the mean-square values of one of the field quantities, such as acoustic pressure. This incoherent addition may be satisfactory if source and receiver are well away from boundaries and if an average result is acceptable. Otherwise a great improvement is possible by coherent addition of amplitudes, that is, taking account of phase. In all these propagation complexities the principle of reciprocity tells us that the transmission is unchanged if the source and receiver are interchanged, however complicated the environment, provided the transducers and the medium remain still.

There remain the major applications to atmospheric acoustics and to underwater acoustics, and the discussions on these will be intertwined since each involves a layered medium and there are basic similarities. In effect, long-range underwater acoustics is an upside-down version of long-range atmospheric acoustics. The height of the isothermal atmosphere is of the same order as the depth of the deep ocean. But sound speed in water is more than 4 times that in air, and comparisons are better made for the same wavelength rather than the same frequency. Also, the absorption coefficients for air are about two orders of magnitude greater than those for water at the same frequency. Thus at around 100 Hz any acoustic signal in air disappears beyond a few 100 km, whereas in water it can extend halfway round the world. (An alternative basis for comparison of air and water is to choose the respective frequencies so that the absorptions are the same.) For both media the ranges can greatly exceed the heights or depths, and it is not surprising that the grazing angles for the rays are typically very small. Also, for both media, ray methods have to compete with a variety of wave approaches.

The scene is really set by the profiles of sound speed c, whereas the density variation, especially in a ray treatment, is unimportant. The sound speed (Chapter 5) in air is mainly a function of the temperature and the wind, although gravity plays an indirect part at very low frequencies. Sound speed in water depends mainly on temperature, salinity, and ambient pressure or depth, with current as a minor influence. A common assumption in modeling is that the environment, specifically the sound speed profile, is independent of range.

In a general range-independent layered medium the connection between the grazing angle θ of a ray (i.e., measured with respect to the horizontal) at one height or one depth and that at another is given by Snell's law, expressed as the invariant

$$c \sec \theta = \text{const.} \tag{4}$$

For example, if the velocity gradient c' is constant, this produces ray paths that are arcs of circles of radius c/c', c being measured where the ray is horizontal.

For layering with central or circular symmetry there is Bouguer's law:

$$R^{-1} c \sec \theta = \text{const.} \tag{5}$$

where R is the range from the center of symmetry and θ is measured with respect to the local layering. This may be used to allow for the sphericity of the earth, by equating R with the earth radius, but the resulting correction is small.

One special aspect of atmospheric acoustics is the magnitude of the refraction differences between day and night. In the daytime the warmer air, with higher sound speed, usually lies near the ground and tends to curve the rays upward, away from the surface. At night the cooler air is usually near the ground and tends to curve the rays downward. Another aspect is the magnitude of the dependence on wind direction; for sound propagating upwind the rays tend to curve upward, and for propagation downwind the rays tend to curve downward. Further atmospheric ray problems concern the influence of topography and propagation within city streets or within forests.

In underwater acoustics the propagation literature is very extensive, reflecting the great variability in the environmental conditions. There is the deep ocean, the Continental Slope of intermediate depth, and the shallow coastal waters, all with a variety of sound speed profiles and ray diagrams. In addition, underwater landscape is as varied as that above water.

A detailed example of ray behavior is shown in Figure 2, with a fan of rays from the source. Although the illustration could have come from either air or water, this diagram is for range-independent layered water. There are shadow zones where no rays penetrate; note especially the clear near-surface region around the 20-km range. The closer range boundary of a shadow is typically formed by one ray, the limit ray. The longer range boundary of a shadow is typically formed by the envelope of a series of rays that form a caustic, nominally with an infinite concentration of rays. Of course, the presence of an infinity indicates that ray theory is not working properly at these positions. But high levels are certainly encountered in the neighborhood of a caustic; note in particular the high concentration of rays at the convergence zones near the surface, around both 60 and 120 km in range. Convergence zones can occur at comparable ranges in the atmosphere.

The profile in Fig. 2*a* shows sound speed minima at three depths that identify very efficient sound channels; the surface sound channel, the shallow sound channel, and the main, or SOFAR, channel. For a source within a channel, some downgoing rays are refracted upward above the channel bottom, then either refracted shallow or reflected at the surface, and continue cycling along in this manner. Channeling also occurs in atmospheric acoustics.

It is possible to cope with range-dependent profiles, but it is common to assume cylindrical symmetry about the source. An intensity formula follows from the ratio of ray tube areas that also covers the range-independent case:

$$\frac{I_b}{I_a} = \left| \frac{\cos \theta_a}{x \sin \theta_b} \frac{\partial \theta_a}{\partial x} \right|_{\text{const } z} = \left| \frac{\cos \theta_a}{x \cos \theta_b} \frac{\partial \theta_a}{\partial z} \right|_{\text{const } x} . \tag{6}$$

Here a and b refer to source and receiver, I_b is the desired intensity, I_a is the intensity at unit range (1 m), x is horizontal range, and z is measured vertically. The result might be quoted as a transmission loss $10 \log(I_a/I_b)$ in decibels. In practice, I_b should be corrected for the attenuation along the path as well as for losses on reflection, where appropriate.

One special application of Eq. (6) is to the location of caustics, actually surfaces in three dimensions, found by putting $\partial\theta_a/\partial x$ or $\partial\theta_a/\partial z$ equal to infinity. Quite complicated caustic formations can occur.

4 WENTZEL–KRAMERS–BRILLOUIN APPROXIMATION FOR MODES

The ray concept comes in three varieties.[8] First, there are the rays that join a specific source and a specific receiver, as in Figure 3. Second are the rays broadcast in a fan from a source, defining a ray field as in Fig. 2*b*. Third is the infinite-distance concept with neither specific source nor specific receiver, the so-called plane-wave ray, for which in a homogeneous medium the wavefront is planar. This last type brings us toward the concept of normal modes (Chapter 31).

Consider a layered medium in which refraction or reflection at both upper and lower points leads to channeling. The rate of phase change can be integrated over a complete loop or cycle of the plane-wave ray, with the phase changes ϕ_1 and ϕ_2 at top and bottom included. The condition for reinforcement provides an estimate of the mode eigenvalues, that is,

$$2\int k_z\,dz + \phi_1 + \phi_2 = 2n\pi. \tag{7}$$

Reinforcement occurs if n is integral, so that after one loop the ray has the same phase as when it started, and n is then known as the mode number. This is the Wentzel–Kramers–Brillouin (WKB) approximation.[4] By adding together the downgoing and upgoing rays, it is possible to model the model intensity as a function of depth. Although based on ray theory the WKB approach avoids problems such as those with caustics.

There is in general a dichotomy between the description of propagation using rays and using modes,[4] either exact modes or WKB modes. Ray descriptions can be transformed into WKB descriptions and vice versa, whether the angular interval considered is complete or restricted. For example, a single ray may be replaced by a summation over modes for which the corresponding plane-wave angle is close to the ray angle or even by an approximate integration over these modes. This last idea may also be used in propagation modeling, especially as in the RAYMODE program.[9]

5 IMPROVEMENTS

Much work has gone into modifying or improving ray theory, often by patching in a local wave theory solution in the neighborhood where ray theory breaks down. The prime example is due to Brekhovskikh,[5] who fits an Airy function to the wave field near a caustic. Among other things this shows that the geometric calculation of phase for a ray that has passed through a simple caustic must be corrected by a $\frac{1}{2}\pi$ phase advance; physically the ray may be thought of as sensing the region of higher speed beyond the caustic. The same phenomenon occurs in WKB theory at refraction turning points, where the ray is horizontal.

Another improvement concerns reflection at a real medium, as opposed to an idealized sound-hard or sound-soft boundary, and this medium will be assumed to have a higher sound speed. The rays are assumed to be relatively shallow in angle and of the plane-wave type. They will undergo total internal reflection, nominally without loss in amplitude but with a phase change ϕ that depends on the angle. This phase change is equivalent to the reflection having taken place at an imaginary sound-soft boundary (i.e., a free or pressure-release surface) situated a distance δ_v beyond the real boundary, equivalent again to a "horizontal" displacement δ_h along the boundary. These wave shifts or displacements are given as

$$\delta_v = \frac{\pi + \phi}{2k_z}, \qquad \delta_h = \frac{k_x(\pi + \phi)}{k_z^2}. \tag{8}$$

They can be useful in the calculation of modal phase velocities and eigenvalues; for example, simplified version of Eq. (7) may be possible.

However, if the concern is with the amplitude distribution for a beam having a small range of angles, a different pair of displacements must be used:[5,8]

$$\Delta_v = \frac{1}{2}\frac{\partial\phi}{\partial k_z}, \qquad \Delta_h = -\frac{\partial\phi}{\partial k_x}. \tag{9}$$

These are the better known beam displacements, and Δ_h is the same as the Goos–Hänchen shift in optics. It can be used to improve ray theory calculations of group or energy velocity and of modal cycle distance, the latter being illustrated in Fig. 4.

Note that as grazing angle θ approaches zero, δ_v and

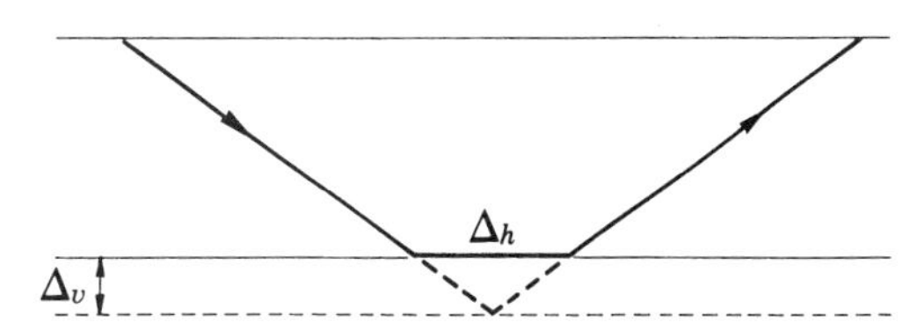

Fig. 4 Beam displacement on reflection, with effect on cycle distance.

Δ_v approach the same constant value. The beam displacement idea is related to that of the lateral ray,[5] and in underwater refraction shooting this is the ground wave arrival. In effect, it travels along the upper boundary of the seabed at a speed equal to that of the seabed material and for an indefinite distance. It is not a trapped or channeled interface wave. Note that Fermat's principle still applies to special ray types such as this one.

6 RANGE-DEPENDENT ENVIRONMENT

The general case is a three-dimensional medium with substantial and complicated sound speed changes as a function of all three coordinates. Ray acoustics may work in theory, but in practice, any calculations will be difficult to organize and perhaps also difficult to interpret. However, a conventional ray approach can cope well with a range-dependent layered medium, especially if the layering changes only slowly and, as already noted, there is cylindrical symmetry about the source that obviates cross-track complications. For sufficiently slow changes there is a ray invariant[10] given by an integration of ray angle with respect to a vertical line, so that the ray paths at one locality can be found from those at another locality without the need to know what has happened in between:

$$\int \frac{\sin\theta\, dz}{c} = \frac{1}{\omega}\int k_z\, dz = \text{const.} \tag{10}$$

This can be useful for long ranges in both atmospheric and underwater acoustics. For a homogeneous ("isospeed") medium this result states that $H\sin\theta$ is constant, where H is channel width. For example, for shallow coastal water, with propagation downslope, H increases, and the successive reflections at the sloping bottom cause $\sin\theta$ to decrease in compensation.

The result in Eq. (10) is obviously closely related to the WKB eigenvalue equation (7). It is equivalent to assuming that all the energy in a given normal mode at one locality will transfer into the corresponding normal mode at another locality, where the layering and mode characteristics may be quite different, provided again that the layering changes sufficiently slowly. This assumption, known as the adiabatic approximation, ignores coupling between different modes.

Development of the idea of either ray invariants or adiabatic modes gives a measure of the average intensity as a function of range r and of vertical coordinate (height or depth).[10] In effect, this is a range averaging over a distance interval comparable to cycle distance X; it gets rid of fluctuations due to interference, and the result $\overline{I_b}$ is a useful quantity. Keeping the values of grazing angle small yields

$$\frac{\overline{I_b}}{I_a} = \frac{4}{r}\int \frac{d\theta_a}{\theta_b X_b}. \tag{11}$$

For the isospeed constant-depth case this becomes simply $2\theta_m/rH$. Here θ_m is the maximum allowed angle, perhaps set by the critical angle for total internal reflection.

If the environment does vary across the track, the rays or modes may be deflected sideways, and this horizontal refraction of the projected ray path can be treated using another application of ray theory, in two dimensions.[1,8] The effect is usually small—but not always!

7 VALIDITY OF RAY APPROACH

The derivation of the eikonal equation (1) using Huygens' principle already suggests the approximate nature of the result. An alternative derivation from the more firmly based wave equation (Chapter 2) gives the validity condition[3]

$$\left|\frac{A''}{A(\omega\tau')^2}\right| \ll 1. \tag{12}$$

Here A is the amplitude of a field quantity such as acoustic pressure, the prime refers to spatial differentiation, and the inequality must hold for any direction. One consequence of this is the condition[4]

$$\left|\frac{1}{k_z}\frac{d}{dz}(\ln k_z)\right| \ll 1, \tag{13}$$

where the resulting inequality again must hold for all directions, the choice of z being merely illustrative. This result is more useful than inequality (12) since it controls reflectivity. Thus a ray normally incident on a discontinuity with a sharp change in sound speed will have a significant reflected component, whereas it will pass virtually straight through if there is a gradual change of the same magnitude. Inequality (13) is also applicable and also important for the validity of the WKB mode calculation. But it does not cover changes in density.

A weaker and simpler version of inequality (13) is given below followed by two equivalents of this weaker version, these three calling attention to different aspects of validity:

$|\lambda c'/c| \ll 1$, or the fractional velocity change over a wavelength should be very small.

$|\lambda A'/A| \ll 1$, or the fractional amplitude change over a wavelength should be very small.

$|\lambda c\tau''| \ll 1$, or the ray radius of curvature should be much larger than a wavelength.

Generally it appears that the wavelength λ should be smaller than any other pertinent length, such as shadow edge thickness (including caustics), focal size, obstacle size, roughness scale, range from source, and channel width H.

Concerning height of atmosphere or water depth, a different type of criterion is obtained by calculating the approximate ratio of the number of effective ray arrivals to effective mode arrivals. Conceptually and for convenience it can pay to use the ray approach if this ratio is less than unity. For source and receiver both well away from the boundaries of a channel, this produces[1]

$$\frac{r\lambda}{2H^2} < 1. \qquad (14)$$

Comprehensive advice on when to use ray theory is difficult to formulate, as it depends on which version of the formulas is to be used, and it is often best just to try ray acoustics. It is likely to do well in comparison with other methods at high frequency, at short range, for great channel width, and for range-dependent environments. It can give exact results in some simple situations.

8 CONCLUDING REMARKS

Ray acoustics is a very useful tool for a wide variety of applications; it should not be misused, but it can sometimes be rewarding to walk the tightrope at the limit between ray and wave approaches.

REFERENCES

1. D. E. Weston and P. B. Rowlands, "Guided Acoustic Waves in the Ocean," *Rep. Prog. Phys.*, Vol. 42, 1979, pp. 347–387.
2. A. D. Pierce, *Acoustics: An Introduction to Its Physical Principles and Applications*, McGraw-Hill, New York, 1981; revised edition, Acoustical Society of America, New York, 1989.
3. P. G. Frank and A. Yaspan, "Ray Acoustics," in *Physics of Sound in the Sea*, Chapter 3, Department of the Navy, Washington, DC, 1969; originally issued 1946.
4. I. Tolstoy and C. S. Clay, *Ocean Acoustics: Theory and Experiment in Underwater Sound*, McGraw-Hill, New York, 1966; paperback edition, Acoustical Society of America, New York, 1987.
5. L. M. Brekhovskikh, *Waves in Layered Media*, Academic, New York, 1960; second edition, 1980.
6. H. Kuttruff, *Room Acoustics*, 3rd ed., Applied Science, London, 1991.
7. J. E. Piercy, T. F. W. Embleton, and L. C. Sutherland, "Review of Noise Propagation in the Atmosphere," *J. Acoust. Soc. Am.*, Vol. 61, 1977, pp. 1403–1418.
8. D. E. Weston, "Rays, Modes and Flux, in L. B. Felsen (Ed.), *Hybrid Formulation of Wave Propagation and Scattering*, Nijhoff, Dordrecht, 1984, pp. 47–60.
9. P. C. Etter, *Underwater Acoustic Modeling*, Elsevier, New York, 1991.
10. D. E. Weston, "Acoustic Flux Formulas for Range-dependent Ocean Ducts," *J. Acoust. Soc. Am.*, Vol. 68, 1980, pp. 269–281.

4

INTERFERENCE AND STEADY-STATE SCATTERING OF SOUND WAVES

HERBERT ÜBERALL

1 INTRODUCTION

Although the science of acoustics reaches back into antiquity, the father of modern acoustics is without any doubt the English physicist John William Strutt, Lord Rayleigh (1842–1919), whose textbook *Theory of Sound* is still relevant today.[1] Among his many achievements is an elegant mathematical expression describing the scattering of sound waves by material objects that is termed the *Rayleigh series*.

From these beginnings, acoustics has seen a rapid and extensive development that today covers many different applications such as nondestructive materials testing, medical acoustics, ocean acoustics and the detection and recognition of submerged targets, and so on. Apart from the understanding of acoustic propagation required here, many of these applications are based on the subject of acoustic *scattering*, which represents an important ingredient in their design and interpretation. In the following, acoustic scattering will be discussed regarding its history and modern developments. These, starting from Rayleigh's normal-mode series, have proceeded to a separation of the scattering amplitude into "geometric" parts (specular reflection, transmission, and internal reflection) and parts that correspond to diffraction via the effects of circumferential (surface) waves on the scattering object. The latter are shown to give rise to prominent resonance effects for submerged elastic scatterers, which were clarified theoretically by the *resonance scattering theory* (RST) of Flax, Dragonette, and Überall and experimentally by the *method of isolation and identification of resonance* (MIIR) of Maze and Ripoche. The modifications of these phenomena for objects in air and for multiple scatterers are also discussed.

Note: References to chapters appearing only in the *Encyclopedia* are preceded by *Enc.*

2 RECENT HISTORY: THE CONCEPT OF NORMAL MODES AND SURFACE WAVES

The World War II effort gave a great impetus to acoustics research. The subsequent fundamental work of Faran[2] should be stressed. He derived the acoustic field scattered from submersed elastic objects having the "canonical" shapes of cylinders and spheres, in response to an incident plane wave, mathematically representing the fields in the form of Rayleigh series. He evaluated the field expressions numerically and carried out experiments that verified the correctness of the theory. More interesting from the physics viewpoint, he showed that resonances of the scattering objects were excited when the frequency of the incident sound wave coincided with one of the "natural" or eigenfrequencies of vibration of the submerged object. Using incident sound pulses, he observed that a "ringing" of the resonances (i.e., a continuing vibration of the object after the traveling pulse had passed on) took place when the carrier frequency of the pulse was equal to the eigenfrequency.

Pulse scattering was subsequently studied experimentally by Barnard and McKinney,[3] who observed periodic echoes returned by finite-size smooth metallic objects in water in response to a single incident pulse. Diercks et al.[4] interpreted this physically as the effects of circumferential pulses generated by the incident pulse that encircle the scatterer and radiate off shock waves as they propagate along, the shock waves from successive revolutions of the circumnavigating pulse reaching the observer in a sequence. This picture is based on the "creeping waves" discussed earlier by Franz[5] for the case of impenetrable objects. The difference is, however, that on impenetrable (rigid or soft) objects the circumferential (peripheral, or surface) waves propagate outside the object in the ambient fluid with a speed below the sound speed in the fluid, while for a (penetrable) elastic body the strongly dominant peripheral waves are of an elastic nature, propagating on the interior side of the scatterer's surface with speeds that are close to the higher wave speeds in the elastic material (although the slower Franz waves are present here also, but with smaller amplitudes). Both wave types radiate into the fluid: the fast elastic-type surface waves emit "shock" waves at a grazing angle ϑ with the scatterer's surface, given by

$$\cos \vartheta = \frac{c}{c_l}, \tag{1}$$

where c is the speed of sound in the ambient fluid and c_l the phase velocity of the lth-type surface wave. For the Franz-type (creeping) waves where $c_l < c$, the radiated wave is emitted tangentially and may be called a "slip" wave.[6] This situation is shown in Fig. 1. Note that the locus on the scatterer at which the surface waves are generated is also determined by Eq. (1). The field radiated by the creeping waves may be identified with the phenomenon of diffraction.

For the case of steady-state scattering, where one does not have an incident pulse but has a "long" pulse that has been approaching for an infinite time and continues to do so, all these wave types are superimposed on each other, together with a geometrically (or specularly) reflected wave as well as "transmitted" waves refracted into the inside of the elastic scattering object. (For an impenetrable, i.e., rigid or soft, object, which is a mathematical idealization not found in nature, only the reflected and the creeping waves would be present).

As to the internally transmitted waves, Fig. 1 only shows their first refraction. They continue being retransmitted to the exterior of the scatterer or reflected back into it (Fig. 2). At every interaction with the boundary, the interior waves, which are of the two types that an elastic medium can support (compressional, or P, and shear, or S), further split into these two types again (*mode conversion*). These retransmitted waves, which are of dominant importance at very large frequencies ($ka >>> 1$, where $k \equiv \omega/c = 2\pi/\lambda$ is the wavenumber in the fluid and a is a typical dimension of the scatterer) have been experimentally demonstrated by Quentin et al.[7] The scattering at lower frequencies ($ka \sim 1$, but mainly $10 \lesssim ka \lesssim 100$) is dominated by the effects of the elastic surface waves (at least for the common case of metal objects in water), where the circumferential (surface) waves at certain frequencies can match their phases when closing into themselves after repeated circumnavigations, forming standing waves around the object and thus generating highly noticeable resonance effects in the scattered amplitude when plotted versus frequency. By themselves, the surface waves would thus give rise to a series of isolated peaks on such a plot, which can be referred to as the *resonance spectrum* of a scattering object characteristic for its shape and composition. In practice, however, one always has the reflected wave present in addi-

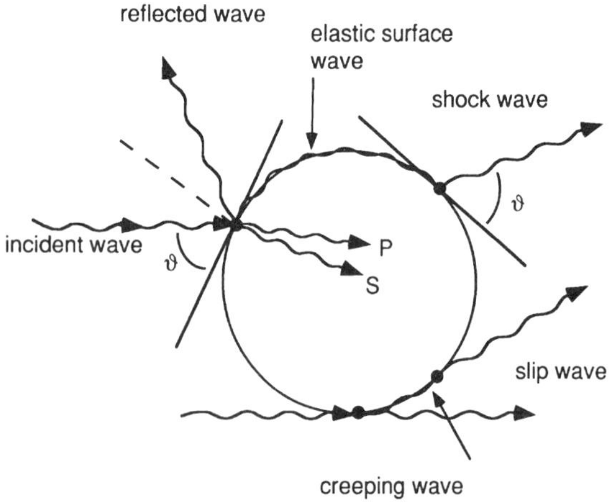

Fig. 1 Peripheral wave types and the waves radiated from them; also, reflected and transmitted (refracted) waves of compressional (P) and shear type (S).

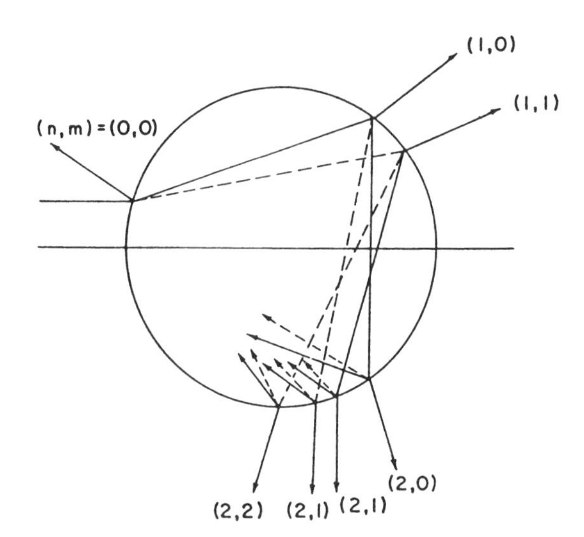

Fig. 2 Geometric-optical picture of refracted waves inside an elastic object; n indicates the total number of internal segments of a ray and m the number of segments of shear (S) type.

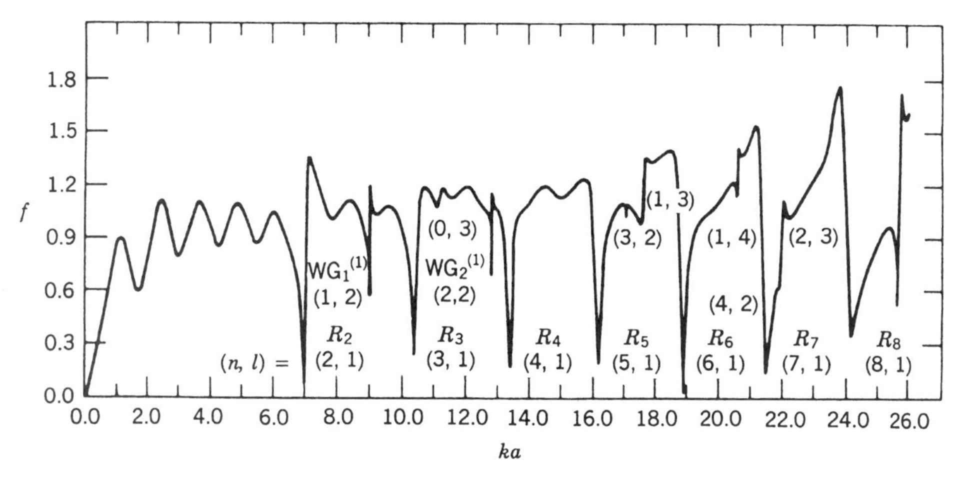

Fig. 3 Modulus of the form function f (backscattering amplitude less geometric spreading factor) plotted vs. ka for a WC sphere in water. (From Ref. 8.)

tion (as well as the retransmitted waves wherever they are of importance), and when the scattering amplitude is plotted versus frequency, the reflected (plus transmitted) wave(s) give rise to a smooth background spectrum with which the isolated resonances interfere, for the case of metal objects in water mainly destructively, so that they manifest themselves as a series of isolated dips. This is shown in Fig. 3 for a solid tungsten carbide (WC) sphere in water.[8] The resonances are labeled by two numbers (n, l), where l is the type of circumferential wave that causes them (on a solid body, $l = 1$ is referred to as a *Rayleigh wave*, causing the succession of deep dips labeled R_n, and $l \geq 2$ as "*Whispering Gallery*" *waves*, causing narrow dips; on an air-filled shell, the various wave types l are referred to as *Lamb waves*) and n is the number of wavelengths of the resonant standing wave closing into itself over the circumference of the scatterer. The quantity plotted here is the modulus of the *form function*, defined as the far-field backscattering amplitude with its geometric spreading factor ($1/r^{1/2}$ for two-dimensional scattering, e.g., from cylinders, $1/r$ for three-dimensional scattering) removed.

If the "background" of the reflected wave is subtracted out from the total scattering amplitude, then indeed only the pure resonances remain, as shown in Fig. 4 for the resonance spectrum of a WC sphere in water,[8] indicat-

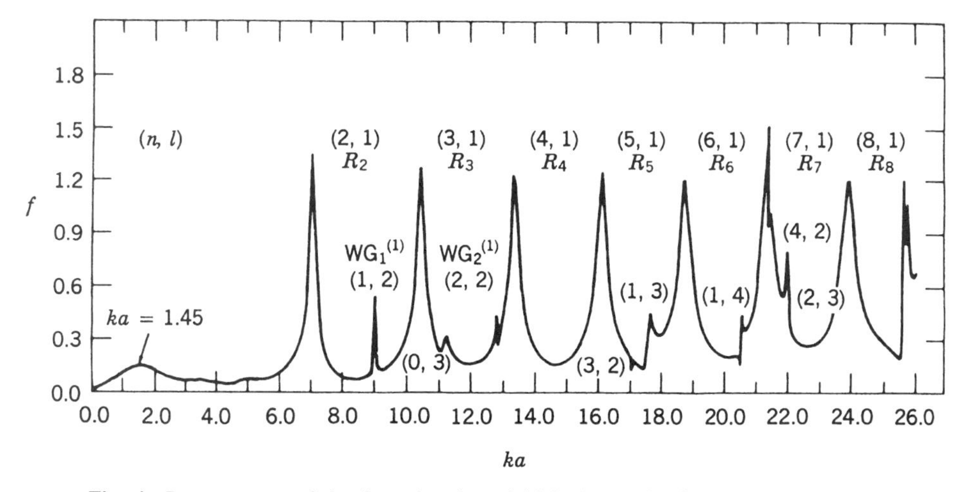

Fig. 4 Resonant part of the form function (rigid background subtracted), or resonance spectrum, of a WC sphere plotted vs. ka. (From Ref. 8.)

ing broad Rayleigh wave and narrow Whispering Gallery wave resonances. The RST of Flax et al.,[9,10] to be discussed below, describes the isolated resonances by a resonance formula, and a prerequisite for its application is the separation of the "geometric" (reflected and refracted/transmitted) background amplitude (known in nuclear scattering theory as the "*potential scattering term*[11]) from the resonance terms in the scattered amplitude. Experimentally, this separation has been achieved by the long-pulse MIIR of Maze and Ripoche,[12] and a long series of investigations has been performed by these acousticians at the University of Le Havre, France, on all aspects of resonance scattering.[13–15] However, a short-pulse method by Numrich et al.[16] and de Billy[17] has been devised for the same purpose. On the theoretical side, scattering from objects of noncanonical shape (mainly spheroidal) has been calculated by Werby[18,19] employing the so-called T-matrix method of Waterman[20]; he also has derived an accurate background formula for the specular reflection from air-filled shells[21] that correctly goes over into a soft background for very thin shells and into a rigid background for thick shells (including solid bodies). Related work is given in Refs. 22 and 23.

The form function, or rather the normalized far-field scattering amplitude at an observer's angular position ϑ, for a sphere can be written in the form of a Rayleigh series as[9]

$$f(\vartheta) = \frac{2}{ka} \sum_{n=0}^{\infty} (2n+1) e^{i\delta_n} \sin \delta_n P_n(\cos \vartheta). \tag{2}$$

Here, $P_n(\cos \vartheta)$ is the Legendre polynomial. The individual terms in the series are referred to as *normal modes* or *partial waves*. We introduced the scattering function

$$S_n = e^{2i\delta_n}, \tag{3}$$

where δ_n is the scattering phase shift. For a rigid sphere, for example, one has

$$S_n^{(r)} = -\frac{h_n^{(2)\prime}(x)}{h_n^{(1)\prime}(x)} \equiv e^{2i\xi_n}, \tag{4}$$

where $x = ka$ is a normalized frequency variable, a is the sphere radius, k the wavenumber in the ambient fluid, and $h_n^{(1)\prime}(x)$ the derivative of the spherical Hankel function. For elastic spheres, S_n is known from the boundary conditions. As in Eq. (4), it has a quotient form with denominator $D_n(x)$, which is a function of two variables: the frequency x and the mode number n.

The equation

$$D_n(x) = 0 \tag{5}$$

can be solved in either variable if the other one is held (real and) constant. Keeping the frequency x real and constant leads to a solution

$$n \rightarrow \nu_l(x), \qquad l = 1, 2, \ldots, \tag{6}$$

which introduces poles of the scattering amplitude in the complex mode number plane (known as *Regge poles* in quantum physics[24]). They lead to the dominant contributions (as shown by the so-called Watson transformation[5]) expressed via the asymptotic forms of P_n:

$$P_{\nu_l}(\cos \vartheta) \cong \frac{2}{\pi(\nu_l + \frac{1}{2}) \sin \vartheta} \frac{1}{2} \sum_{\epsilon = \pm 1} e^{i\epsilon(\nu_l + 1/2)\vartheta - i\epsilon\pi/4}, \tag{7}$$

which quite obviously describe circumferential waves that encircle the sphere in the angular direction ϑ with propagation constant $\nu_l + \frac{1}{2}$ in both senses ($\epsilon = \pm 1$).

Alternately, keeping n constant and equal to real integers, Eq. (5) has the solution

$$x = x_{nl}, \qquad l = 1, 2, \ldots, \tag{8}$$

where x_{nl} is a complex-pole position in the complex-frequency plane, as shown for a WC sphere[19] in Fig. 5. In this representation, the poles are called the *singularity expansion poles*, from a corresponding electromagnetic scattering theory.[25] Figure 5 shows that the poles fall into distinct families, corresponding to the Rayleigh wave ($l = 1$) and to the Whispering Gallery waves ($l = 2, 3, \ldots$).

Scattering theory may be extended to noncanonical objects (e.g., spheroids) by applying the T-matrix (null field, or extended boundary condition) method of Waterman.[20] In acoustic scattering, this has been extensively carried out by Werby.[18,19,26] (The method requires time-consuming computation at higher frequencies and may even show stability and convergence problems there.) One may introduce a spherical basis

$$\begin{aligned} \psi_n(\mathbf{r}) &= \gamma_{mn}^{1/2} h_n^{(1)}(kr) Y_{mn}^{\sigma}(\hat{r}), \\ \hat{\psi}_n(\mathbf{r}) &= \gamma_{mn}^{1/2} j_n(kr) Y_{mn}^{\sigma}(\hat{r}), \end{aligned} \tag{9}$$

with γ_{mn} the normalization constants and Y_{mn}^{σ} the spherical harmonics. The incident wave is expanded as[8]

$$p_{\text{inc}} = \sum_n a_n \hat{\psi}_n(\mathbf{r}), \tag{10}$$

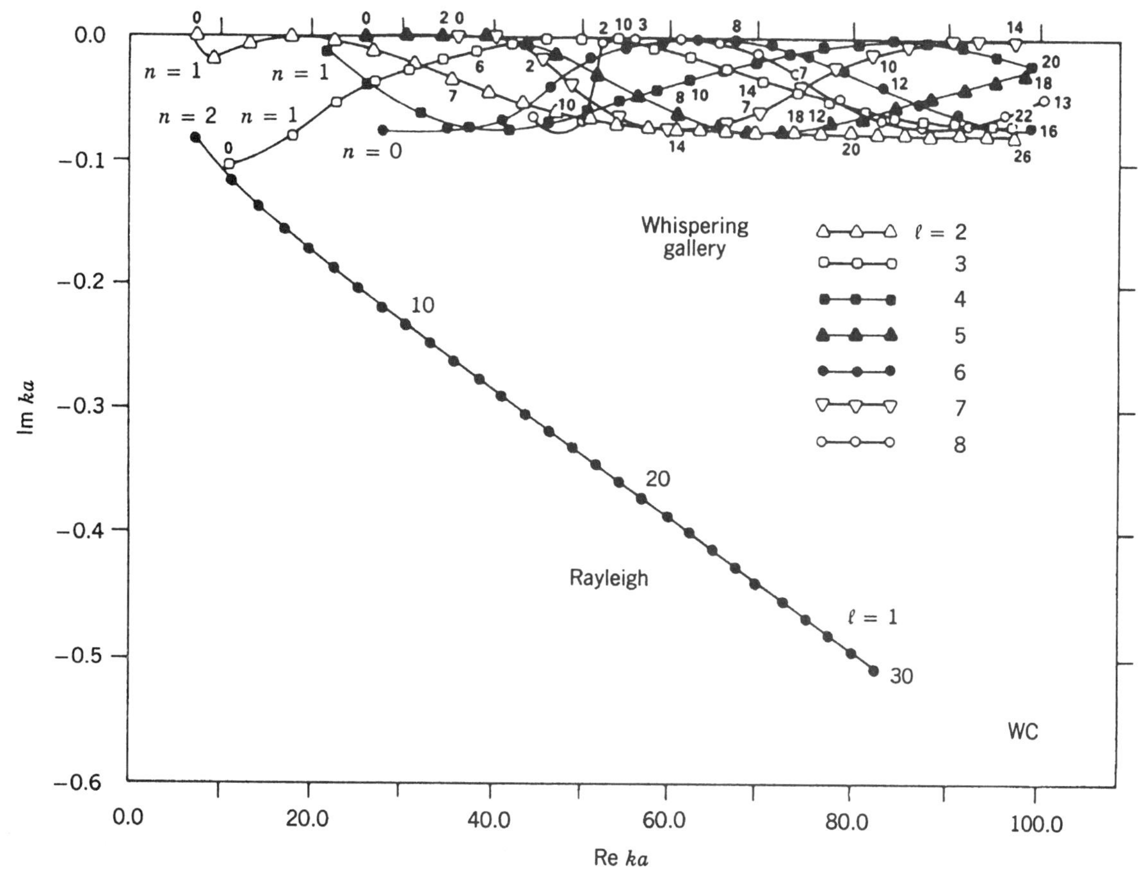

Fig. 5 Complex-frequency poles of the scattering amplitude for a WC sphere of radius a immersed in water (units ka, where $k = \omega/c$, c = sound speed in water). (From Ref. 19.)

with known coefficients a_n, the scattered wave as

$$p_{\text{sc}} = \sum_n c_n \psi_n(\mathbf{r}), \tag{11}$$

and the interior field (assuming a fluid body) as

$$p_{\text{int}} = \sum_n d_n \hat{\psi}_n^{(0)}(\mathbf{r}), \tag{12}$$

where, in $\hat{\psi}_n^{(0)}$, $k \equiv \omega/c$ is replaced by $k_0 \equiv \omega/c_0$ (c_0 being the sound velocity in the interior, assumed fluid). The unknown expansion coefficients c_n and d_n are expressed by a_n by satisfying the appropriate boundary condition on the scatterer's surface S. The resulting relation

$$c_i = T_{ij} a_j \tag{13}$$

determines the scattered wave via the T matrix, which is found as

$$T = -Q^{-1}\hat{Q} \tag{14}$$

(a generalization of the mentioned former quotient), where, for example,

$$Q_{pq} = k \int_S \left(\frac{\rho_0}{\rho} \hat{\psi}_p^{(0)} \frac{\partial \psi_q}{\partial n} - \psi_q \frac{\partial \hat{\psi}_p^{(0)}}{\partial n} \right) dS, \tag{15}$$

ρ_0 being the density of the interior fluid.

Calculating form functions and resonance spectra in

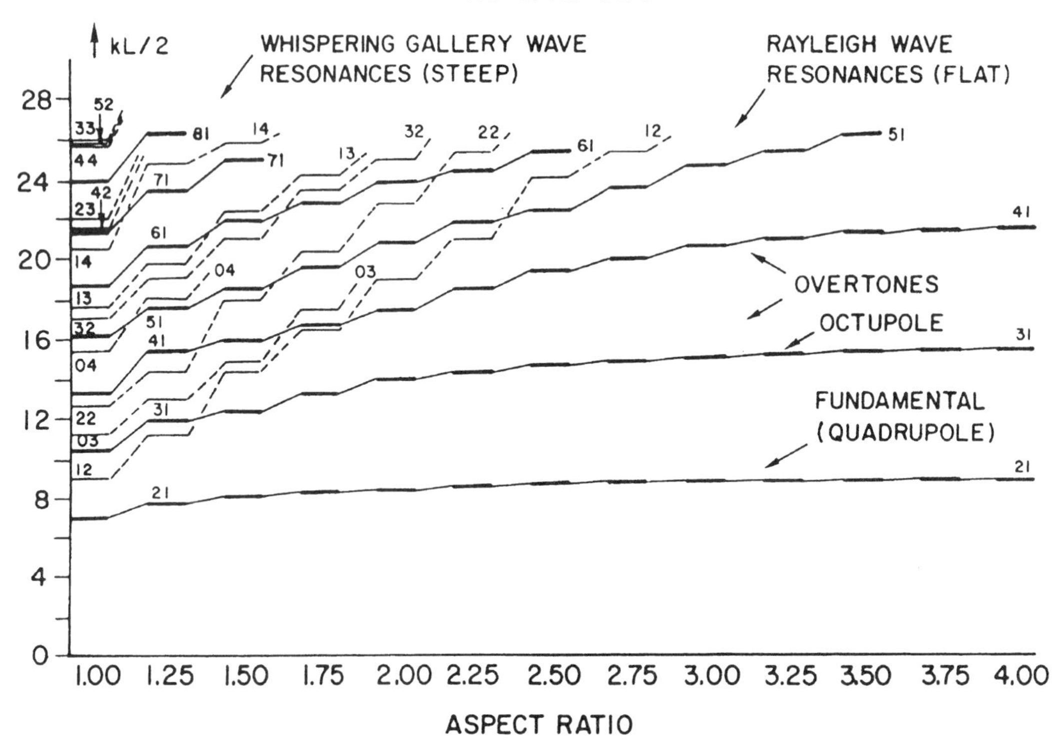

Fig. 6 Spectroscopic level diagram of the real resonance frequencies of prolate WC spheroids, as obtained from a T-matrix calculation for excitation by an axially incident plane wave as a function of aspect ratio. (For aspect ratio 1, levels are the real parts of pole positions in Fig. 5). (From Ref. 18.)

this way, the resonance frequencies have been obtained for a series of WC spheroids of various aspect ratios (up to 4 : 1) in water, subject to an axially incident plane wave. This is presented in Fig. 6, showing that the Rayleigh wave resonances move slowly upward with increasing aspect ratio, while the Whispering Gallery wave resonances move rapidly and cross over the Rayleigh resonances, resembling the "level crossing" phenomenon of deformed nuclei in nuclear physics.

3 RESONANCE SCATTERING THEORY AND PHASE MATCHING OF SURFACE WAVES

Resonance scattering theory was formulated in acoustics by Flax et al.[9,10] in 1977–1978. Patterned after quantum mechanical resonance scattering, it showed that the form function of Eq. (2) could be written as

$$f(\vartheta) = \frac{2}{ka} \sum_n (2n+1) e^{2i\xi_n} \cdot \left[e^{-i\xi_n} \sin \xi_n + \sum_l \frac{\frac{1}{2}\Gamma_{nl}}{x_{nl} - x - \frac{1}{2} i\Gamma_{nl}} \right] \cdot P_n(\cos \vartheta), \tag{16}$$

which makes obvious its superposition character of a smooth specular background (first term in braces) and a series of resonances (second term) with width Γ_{nl}. This expression demonstrates that the singularity expansion poles are located at

$$x = x_{nl} - (i/2)\Gamma_{nl}, \tag{17}$$

that is, in the lower complex x-plane, as shown in Fig. 5. These x-poles may be obtained from the Regge poles by solving Eq. (6):

$$\nu_l(x_{nl}) = n, \tag{18}$$

which is essentially the resonance condition.

This discussion has so far referred to spheres, and it may similarly be applied to cylinders. For finite objects of more general shape, it can be shown that Eq. (18) can be written in the form[27,28] (surface wave phase-matching condition)

$$\oint \frac{ds}{\lambda_l} = n + \tfrac{1}{2}, \tag{19}$$

the wavelength of the lth surface wave, of speed $c_l(k)$, being

$$\lambda_l(k) = \frac{(2\pi/k)c_l(k)}{c}. \tag{20}$$

Equation (19) is integrated over the shortest path (i.e., a geodesic) over the object's surface that closes into itself, following Fermat's principle. The term $\frac{1}{2}$ in Eq. (19) stems from a $\lambda/4$ phase jump at each of two caustics that the geodesics touch. It is absent for cases with no caustics. If the λ_l or $c_l(k)$ are known or are modeled, Eq. (19) may serve to obtain the complex resonance frequencies in this case. At the resonance frequencies, the phase and group velocities may alternately be found as

$$\frac{c_l(k)}{c} = \left.\frac{ka}{n + \frac{1}{2}}\right|_{\text{at res}}, \tag{21}$$

$$\frac{c_l^{gp}(k)}{c} = \left.\frac{d(ka)}{dn}\right|_{\text{at res}}. \tag{22}$$

In this way, the resonance frequencies, for example, for spheroids, similar to those given in Fig. 5 for a sphere, may be obtained from the phase-matching principle. The latter can be applied even for more complex objects where the T-matrix method fails. Numerous examples of resonance scattering have been discussed in Ref. 29 and of surface waves in Ref. 30.

4 METHOD OF ISOLATION AND IDENTIFICATION OF RESONANCES

This experimental method was devised by Maze and Ripoche[12] and is based on the scattering of long pulses of a spatial extent much larger than the size of the scatterer. A modulated square pulse is incident, as in Fig. 7*a*. If its carrier frequency does not coincide with a resonance frequency, its shape will be unchanged after scattering. At the resonance frequency, an initial transient appears, as well as a ringing tail, as shown in Fig. 7*b* for an aluminum cylinder in water. The ringing amplitude (free transient) is a sensitive function of frequency, and when plotted versus ka (bottom part of Fig. 8), it perfectly furnishes the resonance spectrum of the cylinder.[31] Measuring the amplitude of Fig. 7*b* a little ahead of the start of the ringing (where the incident pulse has not yet cut off, i.e., a quasi-stationary "forced" regime has become established) furnishes the form function, however (top part of Fig. 8; a detector bias here reduced its value at both ends).

The shape of Fig. 7*b* is explained[6,32] as shown in Fig. 9. The observed specular wave train and the successive circumferential wave trains overlap (in the case of Fig. 7, the wave train wraps around the cylinder 60 times!). Off resonance, the phases are random and the circumferential waves annihilate each other, leaving only the specular return. At resonance, they are in phase through Eq. (18) or (19); all overlapping waves add up, but in the example of Fig. 7, they interfere destructively with the specular wave, thus causing the central constriction in Fig. 7*b*.

The MIIR also identifies the resonances by assigning their order number n. The cosine of $(n + \frac{1}{2})\vartheta$, combining the two exponentials in Eq. (7) at resonance, shows that the number of lobes in the angular diagram of $f(\vartheta)$ for a sphere is determined by the value of n. A similar expression holds for a cylinder (with n instead of $n + \frac{1}{2}$ since there is no caustic), or approximately also for more generally shaped bodies.[8] The measured "daisy pattern" of Fig. 10 for the scattered-wave angular distribution[33] shows eight lobes (note that the incision at $\vartheta = 180°$, i.e., backscattering, stems from the source and receiver blocking each other), thus indicating $n = 4$ for the 330-kHz resonance of the object (a hemispherically capped cylindrical brass shell in water).

5 SCATTERING FROM OBJECTS IN AIR

When metallic scattering objects are contained in air, rather than submersed in water as was considered above, their large impedance contrast with air will lead to a negligible excitation of elastic transmitted waves as well as of (subsurface) elastic-type circumferential waves by an incident airborne acoustic signal; that is, they will appear as if they were rigid objects. Consequently, only the airborne incident and the Franz-type exterior circumferential ("creeping") waves will be present. Their complex resonance frequencies (poles of the scattering amplitude)

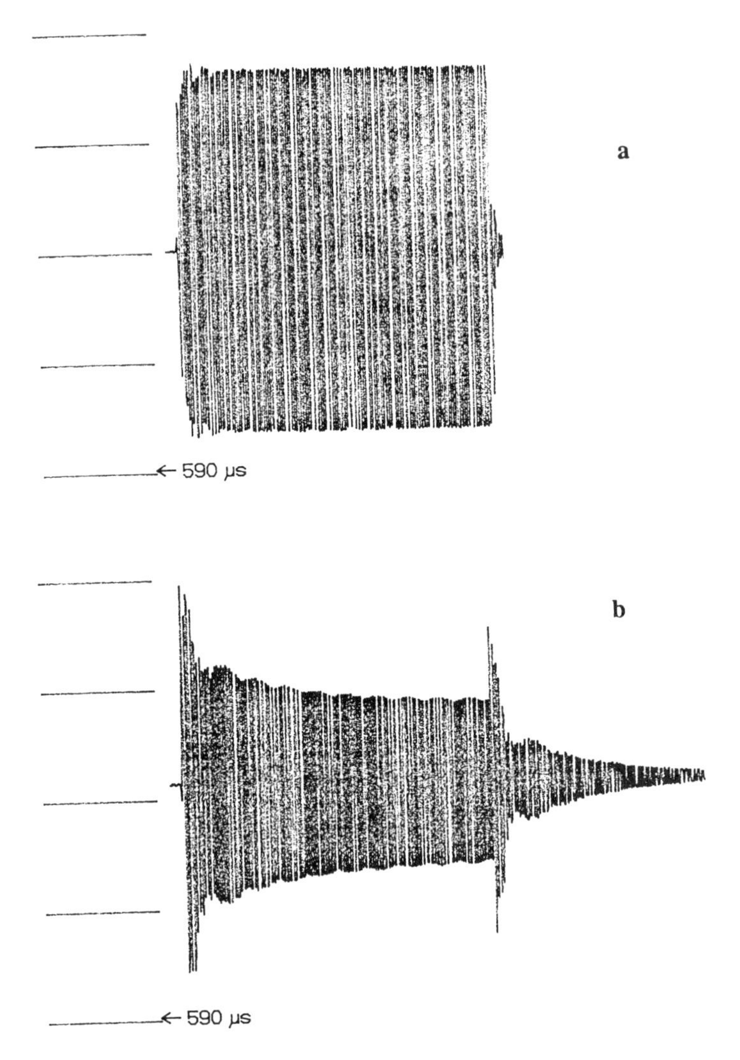

Fig. 7 Reflected pulse for an aluminum cylinder in water (*a*) off resonance and (*b*) at resonance. (From Refs. 12 and 14.)

would appear in the WC sphere plot of Fig. 5 considerably below[34] those of the Rayleigh poles, due to their larger imaginary parts (attenuations), but close to those of the Franz poles of the actual WC sphere (not shown in Fig. 5) and with a downward-sloping pattern similar to that of the Rayleigh poles. Measurements of sound scattering by duraluminium cylinders and spheres in air[35] could therefore be interpreted on the basis of creeping waves on a rigid cylinder alone. These measurements included a verification of their theoretical dispersions and absorptions as a function of frequency. Figure 11 shows the normalized *target strength* σ (the absolute-squared backscattering amplitude with spreading factor removed) versus frequency for a rigid cylinder of radius a and a normally incident plane wave with the following theoretical results:

1. *Dashed Curve.* Exact rigid-cylinder scattering theory based on the cylinder analog of Eq. (2)
2. *Solid Curve.* Creeping-wave theory based on the Watson transformation and retention of only the lowest order creeping wave

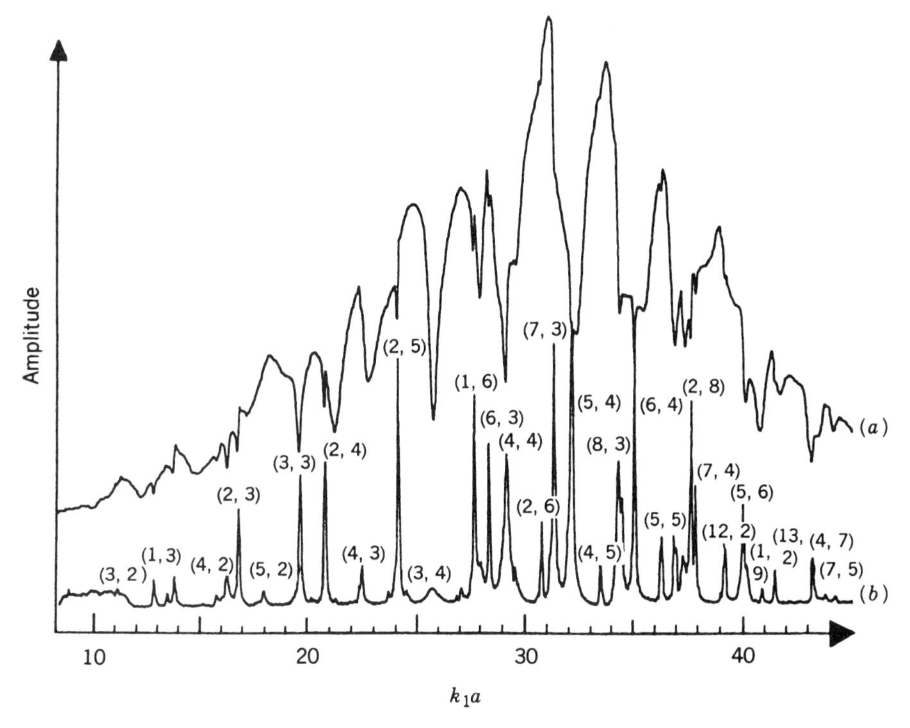

Fig. 8 Backscattering response for an aluminum cylinder in water: (*a*) form function (forced regime); (*b*) resonance spectrum (free transient). (From Refs. 14 and 31.)

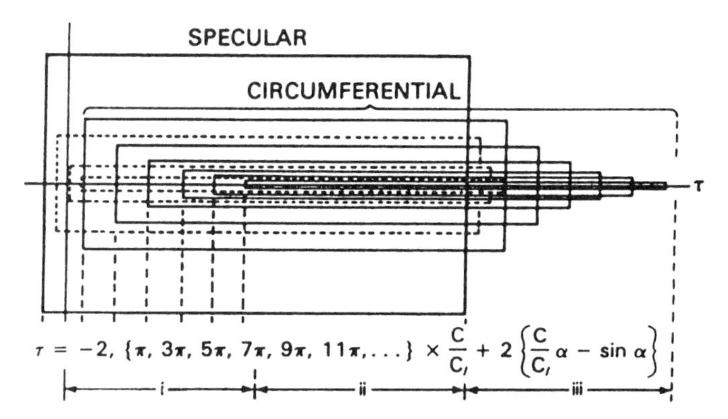

Fig. 9 Schematic view of superposition (without any coherent or incoherent additions) of specular (large solid rectangle), penetrating and multiply internally reflected (dotted rectangles), and circumferential wave trains (small solid rectangles), showing initial transient region (i), quasi-steady-state region (ii), and final transient region, or ringing (iii).

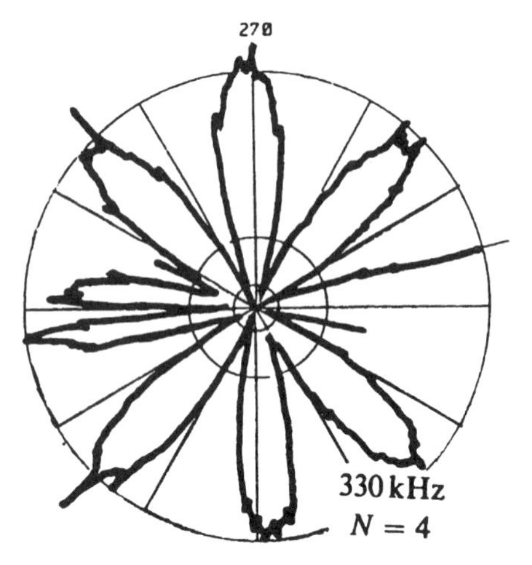

Fig. 10 Angular scattering diagram for a hemispherically capped cylindrical brass shell in water. (From Ref. 33.)

3. *Dot–Dashed Curve.* Use of the "physical optics" or Kirchhoff approximation[36] (which is obviously inapplicable here)

The dots are the experimental results, confirming the physical interpretation based on creeping waves. The undulatory character of the curves is due to the interference of the creeping wave with the (flat) specular background. The wide peaks are not interpretable as creeping-wave resonances since the large imaginary parts of the creeping-wave poles render these excessively broad. Rather, the interference minima can be shown to arise mathematically from zeros in the resonant amplitude that lie near the real axis.

For scattering objects of a density much below that of metals, appreciable penetration of sound into their interior will occur. Experimental results showing this, employing a spherical scatterer of polyurethane foam in air, are available together with their theoretical interpretation.[37] This analysis showed that it was not possible to model the scatterer as an absorbing fluid sphere. Rather, the foam material had to be described by the theory of Biot[38] for porous solids, in which in addition to the shear wave a fast and a slow dilatational wave are present, coupled by the entrained fluid mass in the pores and its friction. The experiment employed an incident one-cycle sine pulse; it measured both transmitted fields inside the sphere and scattered fields in the exte-

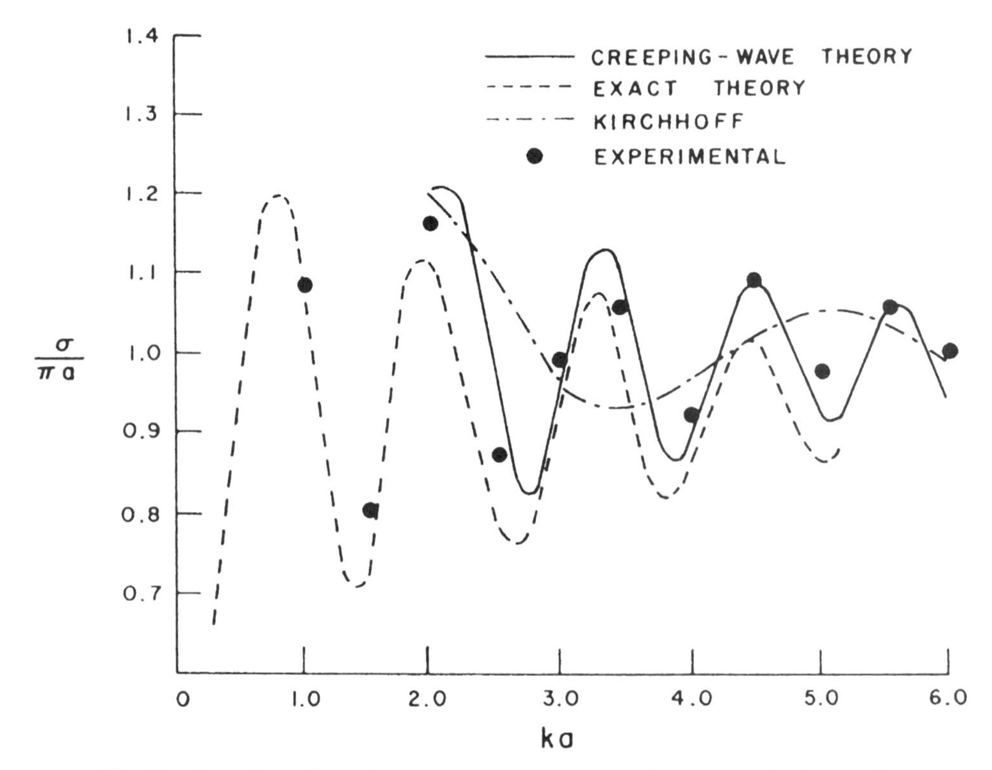

Fig. 11 Comparison of experimental data with calculations for the normalized scattering intensity of a rigid circular cylinder as a function of frequency. (From Ref. 35)

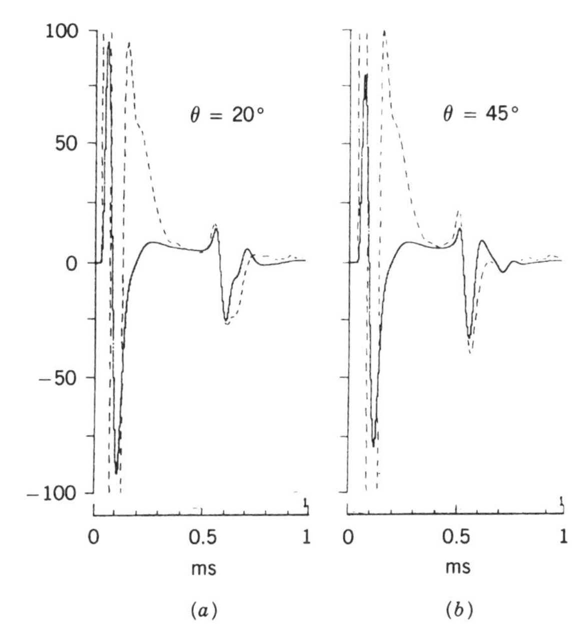

Fig. 12 Pulse scattering amplitude vs. time, at 20° or 45° from the backscattering direction, for a viscoelastic polyurethane foam sphere in air. (From Ref. 37.)

rior. The transmitted pulses exhibit a time delay due to the dispersion of the interior waves. The calculated scattered pulses, which are shown as solid curves in Fig. 12 as observed at $\theta = 20^\circ$ and 45° from the backscattering direction and which closely agree with the measured pulses, consist of the specularly reflected pulse and the later arriving first creeping pulse. They are compared with calculated pulse shapes for a rigid sphere (dashed curves). The difference is due to the absorptivity of the porous material, leading to an attenuation of the observed specular (and to a lesser degree, creeping) pulse that increases with θ.

6 SCATTERING FROM MULTIPLE OBJECTS

The theory of scattering from a manifold of objects goes back to Foldy, Lax, and later Twersky,[39] who assumed the objects to be an arbitrary configuration of parallel cylinders. This assumption is sufficient to highlight the physics involved in this process: The resulting field consists of the incident wave plus a sum of various orders of scattering. The first order, which may be considered as the *single-scattering approximation* results from the excitation of each object by only the incident wave, or primary excitation. The second order of scattering results from the excitation of each object by the first order of scattering from the remaining objects, and so on, to an infinite order of scattering. The first order thus consists of waves scattered by one object, the second order of waves scattered by two objects, and so on. Mathematically, this is treated by a technique of *multiple scattering* in which a hierarchy of coupled multiple-scattering equations is developed, usually to be truncated at a given stage in order to facilitate the solution of the problem.

This approach was applied to randomly distributed cylinders of arbitrary cross section[40] and to finite arbitrary gratings[41] and periodic gratings[42] of compliant tubes. Scattering from two parallel rigid cylinders[43] or two cylindrical shells[44] was considered as a special case.

As an illustration, Fig. 13 shows the insertion loss (defined as $20 \log_{10} p_i/p_t$, where p_i is the incident pressure and p_t the pressure transmitted through a grating), both measured (solid or dash–dotted curves) and calculated (dashed or dotted curves), for two different gratings of parallel steel tubes of highly eccentric rectangular cross section in water. The geometry of the two gratings is indicated in the figure. The first dip position is normalized to unity by dividing by f_1, the single-tube (without water loading) fundamental resonance frequency, and the resonance frequency of the tubes in the second grating, f_2, is indicated. In the case of randomly distributed scatterers,[40] one may consider the imbedding medium (the "matrix") together with the included scatterers as a composite "effective medium," for which the wave phase velocities and attenuations may be obtained from the multiple-scattering theory.

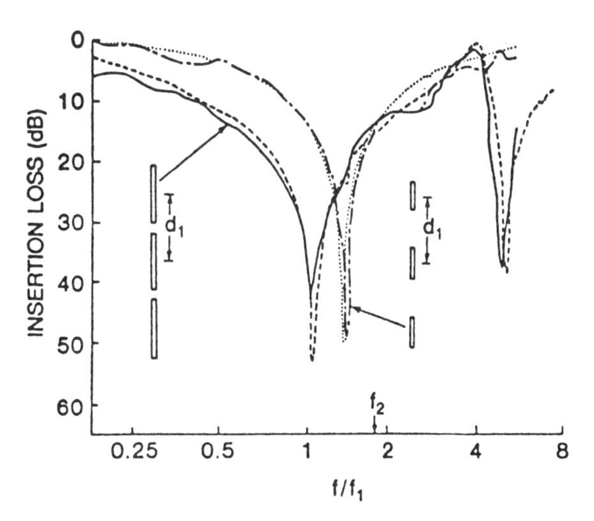

Fig. 13 Measured and calculated insertion loss for two single gratings of cylindrical steel tubes with highly eccentric rectangular cross sections in water. (From Ref. 42.)

It is hoped that this chapter has conveyed to the reader a picture of the present status of acoustic scattering theory and experiment, showing that modern methods of both measurement and analysis have brought us a long way in understanding the scattering phenomenon since the pioneering days of Faran.[2]

REFERENCES

1. Rayleigh, Lord (John William Strutt), *Theory of Sound*, 1877; reprinted by Dover, New York, 1937.
2. J. J. Faran, "Sound Scattering by Solid Cylinders and Spheres," *J. Acoust. Soc. Am.*, Vol. 23, 1951, p. 405.
3. G. R. Barnard and C. M. McKinney, "Scattering of Acoustic Energy by Solid and Air-filled Cylinders in Water," *J. Acoust. Soc. Am.*, Vol. 33, 1961, p. 226.
4. K. J. Diercks, T. G. Goldsberry, and W. W. Horton, "Circumferential Waves in Thin-walled Air-filled Cylinders in Water," *J. Acoust. Soc. Am.*, Vol. 35, 1963, p. 59.
5. W. Franz, "Über die Greenschen Funktionen des Zylinders and der Kugel," *Z. Naturf.*, Vol. A9, 1954, p. 705.
6. N. Veksler, *Acoustic Resonance Spectroscopy*, Springer, Berlin and Heidelberg, 1993.
7. G. Quentin, M. de Billy, and A. Hayman, "Comparison of Backscattering of Short Pulses by Solid Spheres and Cylinders at Large *ka*," *J. Acoust. Soc. Am.*, Vol. 70, 1981, p. 870.
8. M. F. Werby, H. Überall, A. Nagl, S. H. Brown, and J. W. Dickey, "Bistatic Scattering and Identification of the Resonances of Elastic Spheroids," *J. Acoust. Soc. Am.*, Vol. 84, 1988, p. 1425.
9. L. Flax, L. R. Dragonette, and H. Überall, "Theory of Elastic Resonance Excitation by Sound Scattering," *J. Acoust. Soc. Am.*, Vol. 63, 1978, p. 723.
10. H. Überall, "Modal and Surface Waves Resonances in Acoustic-Wave Scattering from Elastic Objects and in Elastic-Wave Scattering from Cavities," in J. Miklowitz and J. Achenbach, Eds., *Proceedings of the International Union Theoretical and Applied Mechanics (IUTAM) Symposium: Modern problems in elastic wave propagation*, Wiley, New York, 1978, pp. 239–263.
11. A. M. Lane and R. G. Thomas, "*R*-Matrix Theory of Nuclear Reactions," *Rev. Mod. Phys.*, Vol. 30, 1958, p. 257.
12. G. Maze and J. Ripoche, "Méthode d'isolement et d'identification des resonances (MIIR) de cylindres et de tubes soumis à une onde acoustique plane dans l'eau," *Rev. Phys. Appl.*, Vol. 18, 1983, p. 319.
13. B. Poirée, Ed., N. GESPA, *La Diffusion Acoustique*, CEDOCAR, Paris, 1987.
14. H. Überall, Ed., *Acoustic Resonance Scattering*, Gordon and Breach, New York, 1991.
15. H. Überall, G. Maze, and J. Ripoche, "Experiments and Analysis of Sound Induced Structural Vibrations," in A. Guran and D. J. Inman, Eds., *Stability, Vibration, and Control of Structures*, World Scientific, River Edge, NY, and Singapore, 1994.
16. S. K. Numrich, N. H. Dale, and L. R. Dragonette, "Generation of Plate Waves in Submerged Air-filled Shells," in G. C. Everstine and M. K. Au-Yang, Eds., *Advances in Fluid–Structure Interaction 1984*, American Society of Mechanical Engineers, New York, PVP vol. 78/AMD vol. 64, 1984, pp. 59–74.
17. M. de Billy, "Determination of the Resonance Spectrum of Elastic Bodies via the Use of Short Pulses and Fourier Transform Theory," *J. Acoust. Soc. Am.*, Vol. 79, 1986, p. 219.
18. M. F. Werby, J. J. Castillo, A. Nagl, R. D. Miller, J. M. D'Archangelo, J. W. Dickey, and H. Überall, "Acoustic Resonance Spectroscopy for Elastic Spheroids of Varying Aspect Ratios, and the Level Crossing Phenomenon," *J. Acoust. Soc. Am.*, Vol. 88, 1990, p. 2822.
19. M. F. Werby, Y. J. Stoyanov, J. W. Dickey, M. Keskin, J. M. D'Archangelo, A. Nagl, N. J. Stoyanov, J. J. Castillo, and H. Überall, "Resonant Acoustic Scattering from Elastic Spheroids," *J. d'Acoust.*, Vol. 3, 1990, p. 201.
20. P. C. Waterman, "New Formulation of Acoustic Scattering," *J. Acoust. Soc. Am.*, Vol. 45, 1969, p. 1417.
21. M. F. Werby, "Recent Developments in Scattering from Submerged Elastic and Rigid Targets," in H. Überall, Ed., *Acoustic Resonance Scattering*, Gordon and Breach, New York, 1991; "The Acoustical Background for Submerged Elastic Shells," *J. Acoust. Soc. Am.* Vol. 90, 1991, p. 3279; "The Isolation of Resonances and the Ideal Acoustical Background for Submerged Elastic Shells," *Acoust. Lett.*, Vol. 15, 1991, p. 65.
22. S. G. Kargl and P. L. Marston, "Longitudinal Resonances in the Form Function for Backscattering from a Spherical Shell: Fluid Shell Case," *J. Acoust. Soc. Am.*, Vol. 88, 1990, p. 1114.
23. N. D. Veksler, "Intermediate Background in Problems of Sound Wave Scattering by Elastic Shells," *Acustica*, Vol. 76, 1992, p. 1.
24. S. C. Frautschi, *Regge Poles and S-Matrix Theory*, W. A. Benjamin, New York, 1963.
25. C. E. Baum, "The Singularity Expansion Method," in L. B. Felsen, Ed., *Transient Electromagnetic Fields*, Springer, Berlin and Heidelberg, 1976, pp. 129–179.
26. M. F. Werby and G. J. Tango, "Numerical Study of Material Properties of Submerged Elastic Objects Using Resonance Response," *J. Acoust. Soc. Am.*, Vol. 79, 1986, p. 1260.
27. H. Überall, Y. J. Stoyanov, A. Nagl, M. F. Werby, S. H. Brown, J. W. Dickey, S. K. Numrich, and J. M. D'Archangelo, "Resonance Spectra of Elongated Objects," *J. Acoust. Soc. Am.*, Vol. 81, 1987, p. 312.
28. D. G. Vasil'ev, "The Frequency Distribution Function of a Shell of Revolution Immersed in a Liquid," *Dokl. Akad.*

Nauk SSSR, Vol. 248, 1979, p. 325 (in Russian); English translation *Sov. Phys. Dokl.*, Vol. 24, 1979, p. 720.

29. L. Flax, G. C. Gaunaurd, and H. Überall, "Theory of Resonance Scattering," in W. P. Mason and R. N. Thurston, Eds., *Physical Acoustics*, Vol. 15, Academic, New York, 1981, pp. 191–294.
30. H. Überall, "Surface Waves in Acoustics," in W. P. Mason and R. N. Thurston, Eds., *Physical Acoustics*, Vol. 10, Academic, New York, 1973, pp. 1–60.
31. G. Maze, "Diffusion d'une onde acoustique plane par des cylindres et des tubes immergés dans l'eau: Isolement et identification des résonances," Ph.D. Dissertation, University of Rouen, 1984.
32. W. E. Howell, S. K. Numrich, and H. Überall, "Complex Frequency Poles of the Acoustic Scattering Amplitude and Their Ringing," *IEEE Trans. Ultrason. Ferroelec. Freq. Control*, Vol. UFFC-34, 1987, p. 22.
33. S. K. Numrich and H. Überall, "Scattering of Sound Pulses and the Ringing of Target Resonances," in A. D. Pierce, Ed., *Physical Acoustics*, Vol. 21, Academic, New York, 1991, pp. 235–318.
34. H. Überall, G. C. Gaunaurd, and J. D. Murphy, "Acoustic Surface Wave Pulses and the Ringing of Resonances," *J. Acoust. Soc. Am.*, Vol. 72, 1982, p. 1014.
35. M. L. Harbold and D. N. Steinberg, "Direct Experimental Verification of Creeping Waves," *J. Acoust. Soc. Am.*, Vol. 45, 1969, p. 592.
36. P. M. Morse and H. Feshbach, *Methods of Theoretical Physics*, McGraw-Hill, New York, 1953.
37. G. Deprez, "Sphère absorbante: diffraction aérienne en régime impulsionnel," in B. Poirée, Ed., N. GESPA, *La Diffusion Acoustique*, CEDOCAR, Paris, 1987.
38. M. A. Biot, "Theory of Propagation of Elastic Waves in Fluid-saturated Porous Solids," *J. Acoust. Soc. Am.*, Vol. 28, 1956, p. 168.
39. V. Twersky, "Multiple Scattering of Radiation by an Arbitrary Configuration of Parallel Cylinders," *J. Acoust. Soc. Am.*, Vol. 42, 1952, p. 42.
40. V. K. Varadan, V. V. Varadan, and Y. H. Pao, "Multiple Scattering of Elastic Waves by Cylinders of Arbitrary Cross Section," *J. Acoust. Soc. Am.*, Vol. 63, 1978, p. 1310.
41. C. Audoly and G. Dumery, "Modeling of Compliant Tube Underwater Reflectors," *J. Acoust. Soc. Am.*, Vol. 87, 1990, p. 1841.
42. R. P. Radlinski, "Scattering by Multiple Gratings of Compliant Tubes," *J. Acoust. Soc. Am.*, Vol. 72, 1982, p. 607.
43. J. W. Young and J. C. Bertrand, "Multiple Scattering by Two Cylinders," *J. Acoust. Soc. Am.*, Vol. 58, 1975, p. 1190.
44. J. P. Sessarego and J. Sageloli, "Diffusion acoustique par deux coques cylindriques parallèles," *J. d'Acoust.*, Vol. 5, 1992, p. 463.

5

SPEED OF SOUND IN FLUIDS

ALLAN J. ZUCKERWAR

1 INTRODUCTION

Sound is a physical phenomenon identified with the propagation of a mechanical disturbance through a medium. The *speed of sound* is defined as the distance traversed per unit time by a given point on the disturbance, provided the disturbance does not change its shape (as would be the case in an ideal medium). If the disturbance is resolved into spatially harmonic components, an observed point on any one component may be designated by a phase angle, and if directional information is included, then the distance traversed per unit time is called the *phase velocity*. If the phase velocity is the same at all frequencies, the medium is said to be *nondispersive*; otherwise it is *dispersive*.

The phase velocity appears in the familiar wave equation, which governs the propagation of sound through the medium. The wave equation is derived from an equation of motion, a constitutive equation or equation of state, and a continuity equation. In an unbounded homogeneous medium the magnitude of the phase velocity, that is, the speed of sound, depends on the adiabatic bulk modulus and the density of the undisturbed medium. The speed of sound is sensitive to such properties as temperature, pressure, composition, and absorption. These are considered in this chapter for gases and liquids. Additional effects as the sensitivity to sound pressure amplitude, to specific absorption processes, and to medium boundaries are discussed in Chapters 6–8 and 42.

Note: References to chapters appearing only in the *Encyclopedia* are preceded by *Enc.*

2 SPEED OF SOUND IN GASES

2.1 Molecular Foundation

Ideal Gas Consider a gas having a pressure P, volume V, density ρ, temperature T, entropy S, and molecular mass M. Kinetic analysis yields, for the speed of sound,

$$c = \sqrt{\frac{M_s}{\rho_0}}, \tag{1}$$

where ρ_0 is density of the undisturbed medium and M_s the adiabatic bulk modulus defined by

$$M_s = \rho\left(\frac{\partial P}{\partial \rho}\right)_s. \tag{2}$$

The equation of state of an ideal gas,

$$P = \frac{\rho R T}{M}, \tag{3}$$

and the first two laws of thermodynamics,

$$T\,dS = C_v\,dT - \frac{P}{\rho^2}\,d\rho = 0, \tag{4}$$

yield the adiabatic bulk modulus $M_s = \gamma P$ and, from this, the speed of sound in terms of the properties of the gas:

$$c = \sqrt{\frac{\gamma P}{\rho_0}} \tag{5}$$

$$= \sqrt{\frac{\gamma RT}{M}}, \tag{6}$$

where γ is the specific heat ratio and R the ideal-gas constant. Equation (6), based on the ideal-gas law, is slightly less accurate than Eq. (5), which is amenable to real-gas values. The speed of sound increases with temperature, decreases with molecular mass, and for a gas obeying Eq. (3) precisely is independent of pressure.

The specific heat ratio $\gamma = C_p/C_v$ can be written as

$$\gamma = \frac{C_v + R}{C_v} \tag{7}$$

since the specific heat at constant pressure $C_p = C_v + R$ for an ideal gas. The specific heat C_v at constant volume contains contributions from the molecular translational, vibrational, and rotational degrees of freedom (electronic contribution being negligible):

$$C_v = \frac{3R}{2} + R\sum_i \Phi_v\left(\frac{\Theta_{vi}}{T}\right) + \frac{R}{2}\sum_i \Phi_R\left(\frac{\Theta_{Ri}}{T}\right). \tag{8}$$

The translational contribution is simply $\frac{1}{2}R$ for each of the three degrees of freedom, as given by first term on the right.

The vibrational contribution depends upon the Planck–Einstein function

$$\Phi_v(x) = \frac{x^2 e^x}{(e^x - 1)^2}, \tag{9}$$

where $x = \Theta_{vi}/T$ and

$$\Theta_{vi} = \frac{h\nu_i}{k_B}. \tag{10}$$

In Eq. (10), h is Planck's constant, k_B Boltzmann's constant, Θ_{vi} the *characteristic temperature*, and ν_i the frequency of the ith vibrational mode. If the characteristic temperature is low ($x \ll 1$), then $\Phi_v(x) \to 1$, the classical value; if it is high ($x \gg 1$), then $\Phi_v(x) \approx x^2 e^{-x}$ is small and the mode is said to be "frozen in." In a molecule consisting of N atoms, the number of vibrational modes is $3N - 5$ if the molecule is structurally linear and $3N - 6$ if it is nonlinear. Each vibrational mode, containing both kinetic and potential energy, is responsible for two degrees of freedom. The characteristic vibrational temperatures of most molecules are so high that at standard temperature (273.15 K) the vibrational degrees of freedom make only a small contribution to the specific heat.

The characteristic temperatures Θ_{Ri} for rotation, in contrast to those for vibration, are generally so small that the rotational degrees of freedom are fully excited at room temperature and the rotational function $\Phi_R \approx 1$ (H_2 being a notable exception). Since there are three axes of rotation, the rotational contribution to the specific heat is generally $\frac{3}{2}R$; but if a molecule is linear, rotation about the collinear axis is frozen in and makes no measurable contribution. Therefore, for nearly all gases at standard temperature Eq. (8) can be written as

$$C_v = \frac{R}{2}(3 + F) + R\sum_i \Phi_v\left(\frac{\Theta_{vi}}{T}\right) \tag{11}$$

where F is 0 for a monatomic molecule, 2 for a diatomic or linear molecule, and 3 for a nonlinear molecule; the sum is taken over all the vibrational modes.

If the vibrational contribution is neglected, then Eqs. (7) and (11) give the following values of γ:

$$\gamma = \begin{cases} \frac{5}{3} & \text{for a monatomic gas} \\ \frac{7}{5} & \text{for a diatomic gas or linear polyatomic gas,} \\ \frac{4}{3} & \text{for a nonlinear polyatomic gas.} \end{cases}$$

Example Compute the speed of sound in CO_2 at standard temperature and pressure (STP) (T = 273.15 K and P = 1 atm = 1.0133×10^5 Pa) from molecular data: M = 0.0441 kg/mol, Θ_{v1} = 1997.5 K (symmetric stretch), Θ_{v2} = 960.1 K (doubly degenerate bending), Θ_{v3} = 3380.2 K (antisymmetric stretch), F = 2 (linear).

From Eqs. (11), (9), (7), and (6) one finds

$$\Phi_v\left(\frac{\Theta_{v1}}{T}\right) = 0.0357, \qquad \Phi_v\left(\frac{\Theta_{v2}}{T}\right) = 0.390,$$

$$\Phi_v\left(\frac{\Theta_{v3}}{T}\right) = 0.00065,$$

$$C_v = (3 + 2)(\tfrac{1}{2}R) + (0.0357 + 2 \times 0.390 + 0.00065)R = 3.316R,$$

$$\gamma = \frac{(3.316 + 1)R}{3.316R} = 1.302,$$

$$c = \sqrt{\frac{1.302 \times 8.3145 \times 273.15}{0.0441}} = 258.9 \text{ m/s}$$

[from Eq. (6)],

which is somewhat higher than the experimental determination[1] $c = 257.5$ m/s. Equation (6) yields too high a sound speed because Eq. (3) underestimates the density.

Knowledge of the sound speed of a gas permits one to infer an important microscopic property. Comparison of Eq. (6) with the well-known expression for the root-mean-square (rms) thermal speed of the gas molecules $v_{\text{rms}} = \sqrt{(3RT/M)}$ yields the relationship

$$v_{\text{rms}} = \sqrt{\frac{3}{\gamma}}\, c. \qquad (12)$$

Thus CO_2 molecules move at an rms speed of $\sqrt{(3/1.302)} \times 257.5 = 390.8$ m/s at STP.

The validity of Eqs. (5) and (6) is based on three fundamental assumptions: namely, that the medium is isentropic, linear, and continuous. The consequences of departure from these assumptions are considered in Sections 2.2–2.5.

2.2 Isothermal Sound Speed

In situations where sound propagation is confined to a narrow duct or other small enclosure, rapid heat exchange between the solid boundaries and the gas may dictate the use of the isothermal sound speed,

$$c_T = \frac{c}{\sqrt{\gamma}}, \qquad (13)$$

rather than the adiabatic sound speed. Quantitatively, the use of Eq. (13) is appropriate when the thickness of the thermal boundary layer

$$d_h = \sqrt{\frac{2K}{\rho\omega C_p}} \qquad (14)$$

exceeds the lateral dimensions of the enclosure.[2] In Eq. (14), K is the thermal conductivity of the gas and ω the acoustic angular frequency, related to frequency f by $\omega = 2\pi f$.

Example The air gap between the membrane and backplate of a condenser microphone is typically $h = 25$ μm. Does sound propagate within at the adiabatic or isothermal sound speed?

Since there is a boundary layer at each boundary, the criterion for choosing the isothermal sound speed becomes $h < 2d_h$. Substitution of known values for air at 20°C into Eq. (14) leads to the result

$$f < (2 \times 0.0025/25 \times 10^{-6})^2 = 40 \text{ kHz}$$

Consequently sound in the air gap propagates at the isothermal sound speed at frequencies of up to 40 kHz.

2.3 Speed of Sound at High Gas Pressures

With increasing pressure, deviations from the ideal-gas law become more pronounced and affect the sound speed through modifications to Eqs. (3) and (7). Based on the virial equation of state,[3] these equations are replaced by

$$P = \frac{\rho RT}{M}(1 + B_p P + \cdots), \qquad (15)$$

$$\gamma = \gamma^0\left\{1 + \frac{PR}{C_v^0}\left[\frac{RT}{C_p^0}\frac{\partial^2(TB_p)}{\partial T^2} + 2\frac{\partial(TB_p)}{\partial T} + \cdots\right]\right\}. \qquad (16)$$

Substitution of Eqs. (15) and (16) into Eq. (5) leads to the sound speed of a real gas:

$$c = c^0\left\{1 + 2P\left[B_P + \frac{R}{C_v^0}\frac{\partial(TB_p)}{\partial T} + \frac{1}{2}\frac{R^2T}{C_v^0C_p^0}\frac{\partial^2(TB_p)}{\partial T^2}\right] + \cdots\right\}^{1/2}, \qquad (17)$$

where the superscript 0 designates ideal-gas values [Eqs. (3), (6), and (7)]. Here B_p is the second virial coefficient for a power series in pressure; it is related to B_v, the corresponding coefficient for a power series in volume, by $B_p = B_v/RT$.

Example Correct the sound speed of CO_2 at STP for the nonideality of the gas. Repeat for $P = 10$ atm.

Reference 3 gives the second virial coefficient of CO_2 as a function of temperature (converted here to TB_p):

$$TB_p = 0.335 - 8092T^{-1.5} - 3.489 \times 10^{10}T^{-4.5}\text{atm}^{-1}.$$

After inserting into Eq. (17) one obtains

$$\begin{aligned} c &= 258.9[1 + 2 \times 1 \times (-0.00672 + 0.00485 - 0.00206)]^{1/2} \\ &= 257.9 \text{ m/s}, \end{aligned}$$

which now brings the sound speed close to the experimental value. Note that the magnitudes of all the correction terms in Eq. (17) are comparable. Thus the effect of pressure on the specific heat ratio [Eq. (16)] is substantial, and it is not sufficient to account for the bulk modulus [Eq. (15)] alone. At $P = 10$ atm, Eq. (17) yields $c = 248.5$ m/s, but no precise measured value is available for comparison.

2.4 Speed of Sound at Low Gas Pressures

A decrease in gas pressure to sufficiently low levels introduces some interesting physical effects, which have a profound impact upon the sound speed. The prominence of any of these effects depends upon the relative magnitudes of three characteristic lengths: the propagation path s (i.e., enclosure dimension), the acoustic wavelength λ, and the molecular mean free path l. Accordingly, these effects are organized into the following dispersion regimes: *classical*, $s \gg l$, $\lambda \gg l$; *Burnett*, $s \gg l$, $\lambda \approx l$; *transition*, $s \approx l$; and *near*, $s \ll l$.

In the classical regime collisions among molecules are far more frequent than their collisions with the walls of the enclosure; furthermore, the transport properties (like viscosity and thermal conductivity) attendant to such collisions respond with such rapidity, relative to the period of the perturbing sound wave, that sound propagation remains nearly adiabatic and conforms to the relationships given in the previous sections. In the Burnett regime the intermolecular collisions are still dominating, but the lengthened response time of the transport properties, now comparable to the period of the sound wave, leads to significant "classical absorption" and accompanying dispersion of the sound wave. In the near regime intermolecular collisions concede to molecule–wall collisions; the constitutive relationships of continuum mechanics are no longer valid. Both the transport properties and sound speed are functions of frequency and geometry of the enclosure. In the transition regime the intermolecular collision rate is comparable to that with the walls.

Classical theory is based on linear relationships among the transport properties. In a nonrelaxing gas (see Chapter 42) the conservation laws and constitutive equations yield a quadratic equation for the square of the complex wavenumber. Greenspan[4] has shown that the solution for a propagating wave can be approximated by a remarkably simple equation:

$$\left(\frac{\alpha\lambda}{2\pi} + i\,\frac{c^0}{c}\right)^2 = -\left(1 + i\,\frac{\gamma_F}{r}\right)^{-1}, \tag{18}$$

where γ_F is an effective specific heat ratio ($\frac{7}{5}$ for a monatomic gas, $\frac{4}{3}$ for a polyatomic gas having a Prandtl number equal to $\frac{3}{4}$, α the attenuation due to viscous and thermal transport, c^0 the sound speed in an ideal gas, and r a viscous Reynolds number,

$$r = \frac{P}{2\pi f\mu}, \tag{19}$$

dependent upon μ the absolute viscosity. Note that the γ_F do not correspond to the γ given in Section 2.1. Equation (18) can be used to find the sound speed over the classical and Burnett regimes.

The physical idea behind the theory of sound propagation in the near regime is that the distribution function of molecular velocities is dominated by the velocity imparted by the sound transmitter rather than by the thermal velocities. The resulting expression for the sound speed involves an integral that cannot be evaluated in closed form; but it may be concluded that the sound speed is independent of pressure and falls very slowly with the product $2\pi f s$.

The plot in Fig. 1 of normalized sound speed versus the frequency–pressure ratio shows typical boundaries of demarcation for the classical, Burnett, transition, and near regimes in air. The first two are based on the real part of Eq. (18), which shows good agreement with experiment. The last is estimated from the data of Ref. 4 and corresponds to specific experimental conditions; scaling laws for other experimental conditions are cited in this reference.

2.5 Sound Dispersion in a Relaxing Gas

The theory of molecular relaxation in a gas appears in a later chapter (Chapter 42). For the present purpose it is important only to note that a relaxation process is characterized by two parameters: a relaxation time τ and a

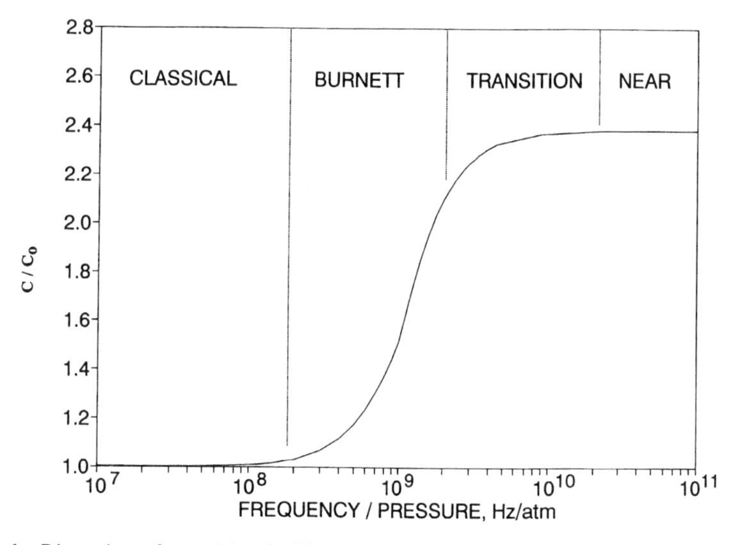

Fig. 1 Dispersion of sound in air. The sound speed c, normalized to the low-frequency limiting value c_0, is plotted against frequency–pressure ratio. The boundaries represent a gradual transition from one regime to another.

relaxation strength ϵ. The phenomenologic impact of a relaxation process on sound propagation is that it causes sound absorption and dispersion, revealed as a *relaxation peak* and *dispersion step*, respectively. The relationship between the squared sound speed and relaxation parameters is given in the equation[5]

$$\frac{c^2}{c_0^2} = \left(1 + \frac{\epsilon}{1-\epsilon}\frac{\omega^2\tau^2}{1+\omega^2\tau^2}\right). \tag{20}$$

This function is plotted in Fig. 2. Here it is seen that the "center" of the dispersion step occurs at a frequency

$$f_R = (2\pi\tau)^{-1}, \tag{21}$$

the *relaxation frequency*, and that its "height" has a magnitude

$$\frac{c_\infty^2 - c_0^2}{c_0^2} = \frac{\epsilon}{1-\epsilon}. \tag{22}$$

At low frequencies ($\omega\tau \ll 1$) $c = c_0$, the "relaxed" sound speed, while at high frequencies ($\omega\tau \gg 1$) $c = c_\infty$, the "unrelaxed" sound speed. The relaxation strength of a relaxing degree of freedom in an ideal gas is found to be

$$\epsilon = \frac{RC_i}{C_v(C_p - C_i)}, \tag{23}$$

where C_v and C_p are low-frequency specific heats and C_i the specific heat of the relaxing degree of freedom, taken from Eq. (8).

Example For CO_2 at STP find the relaxation strength ϵ, the unrelaxed sound speed c_∞, and, given the vibrational relaxation time $\tau = 4.7 \times 10^{-6}$ s, the relaxation frequency f_R and the sound speed at $f = 1$ kHz and at $f = 100$ kHz.

Assume that the relaxation time is that of the doubly degenerate bending mode. The data of Section 2.1 shows $C_v = 3.316R$, $C_p = 4.316R$, and $C_i = 0.780R$. Equations (20), (22), and (23) yield

$$\epsilon = \frac{R \times 0.708R}{3.316R(4.316R - 0.780R)} = 0.0645,$$

$$c_\infty = c_0(1-\epsilon)^{-1/2} = 257.5(1 - 0.0645)^{-1/2}$$

$$= 266.2 \text{ m/s},$$

$$f_R = (2\pi\tau)^{-1} = (2\pi \times 4.7 \times 10^{-6})^{-1} = 34 \text{ kHz},$$

$$c(1 \text{ kHz}) = 257.5\left(1 + \frac{0.0645}{1 - 0.0645}\frac{0.0295^2}{1 + 0.0295^2}\right)^{1/2}$$

$$= 257.5 \text{ m/s},$$

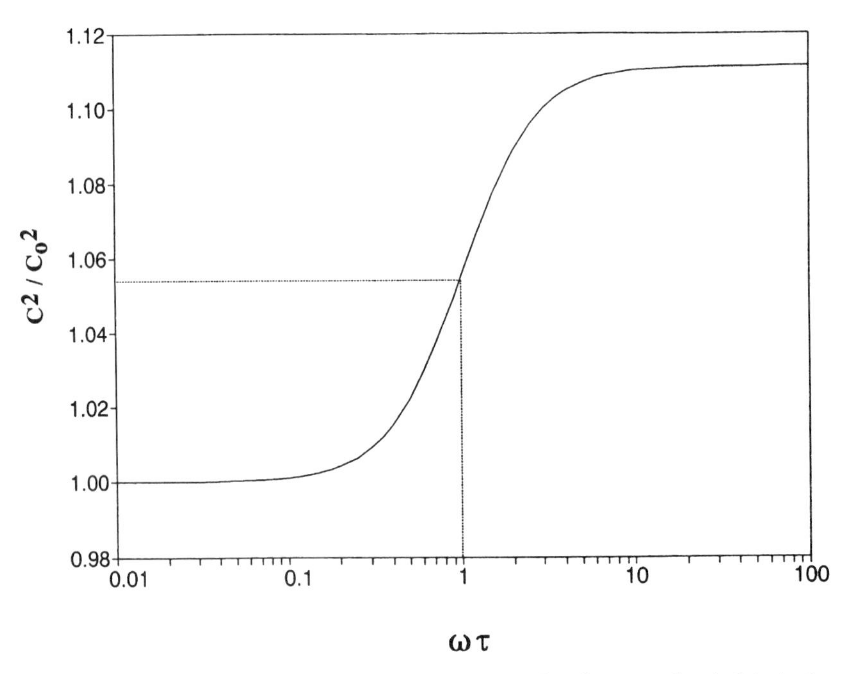

Fig. 2 Dispersion of sound in a relaxing gas. Here the relaxation strength ϵ is 0.1. At the relaxation frequency, corresponding to the condition $\omega\tau = 1$, the square of the sound speed is halfway between the low-frequency and high-frequency limits.

$$c(100\text{ kHz}) = 257.5\left(1 + \frac{0.0645}{1 - 0.0645}\frac{2.95^2}{1 + 2.95^2}\right)^{1/2}$$
$$= 265.4\text{ m/s},$$

since $\omega\tau = 0.0295$ at $f = 1$ kHz ($0.0294 f_R$) and 2.95 at $f = 100$ kHz ($2.94 f_R$). This example illustrates the effect of frequency on the sound speed of a relaxing gas.

2.6 Speed of Sound in Gas Mixtures

Equation (6) holds for a mixture of ideal gases having mole fractions $X_1, X_2, \ldots$ if one substitutes

$$\gamma = 1 + \frac{R}{C_v} = 1 + R(X_1 C_{v1} + X_2 C_{v2} + \cdots)^{-1}, \quad (24)$$

$$M = X_1 M_1 + X_2 M_2 + \cdots. \quad (25)$$

Example Find the speed of sound in air at STP for relative humidities of 0 and 100%.

First consider dry air as an ideal gas. The data are compiled in Table 1. Equations (24), (25), and (8) yield $M = 28.9644$ kg/kmol, $\gamma^0 = 1.4005$, and $c^0 = 331.39$ m/s, all of which agree with Greenspan.[6] Now consider dry air as a real gas. Greenspan cites data for B_v, converted here to data for B_p:

$$B_p = -6.023 \times 10^{-4}\text{ atm}^{-1},$$

$$\frac{d}{dT}(TB_p) = 2.9446 \times 10^{-3}\text{ atm}^{-1},$$

$$\frac{d^2}{dT^2}(TB_p) = -2.532 \times 10^{-5}\text{ atm}^{-1}\cdot\text{K}^{-1}.$$

Then Eq. (17) yields

$$c = 331.39\left[1 + 2 \times 1\left(-6.023 + \frac{29.446}{2.4967} - \frac{273.15 \times 0.2532}{2 \times 3.4967 \times 2.4967}\right) \times 10^{-4}\right]^{1/2} = 331.45\text{ m/s}.$$

In humid air containing a mole fraction X_h of water vapor, the mole fractions of all the other constituents are

TABLE 1 Properties of the Constituents of Air at $T = 0°C$

Constituent	M_j, kg/kmol[a]	X_j[a]	C°_{vj}/R[b]
N_2	28.0134	0.78084	2.5019
O_2	31.9988	0.209476	2.5208
Ar	39.948	0.00934	1.5001
CO_2	44.00995	0.000314	3.3255
CO	28.01	0.19×10^{-6}	2.503
H_2	2.01594	0.5×10^{-6}	2.4418
Ne	20.183	18.18×10^{-6}	1.5000
Kr	83.80	1.14×10^{-6}	1.4995
CH_4	16.04303	2×10^{-6}	3.2822
He	4.0026	5.24×10^{-6}	1.5000
N_2O	44.0128	0.27×10^{-6}	3.6165
Xe	131.30	0.087×10^{-6}	1.4998
Air (dry)	28.9644	—	2.4967
H_2O (100% RH[c])	18.01534	0.006026	3.01584[d]
Air (100% RH)	28.8984	—	2.4998

[a]From Ref. 7.
[b]For N_2–H_2, $C^\circ_v = C^\circ_p - R$, where C°_p is taken from Ref. 8 at 0.01 atm. For Ne–Xe, $C^\circ_v \approx C'_p - R$, where C'_p is the real gas value taken from Ref. 7; the contributions from these gases scarcely affect the fifth significant figure.
[c]Relative humidity.
[d]Based on Eq. (11) using data from Ref. 9.

reduced to $X'_j = X_j(1 - X_h)$. At 273.15 K the vapor pressure of water is 4.58 Torr, corresponding to a mole fraction $X_h = 4.58/760 = 0.006026$ (100% relative humidity). When the X_j are substituted into Eqs. (24) and (25), B_p is assumed unchanged, and H_2O is included, then Eq. (17) yields $c^0 = 331.71$ m/s and $c = 331.77$ m/s.

3 SPEED OF SOUND IN LIQUIDS

3.1 Liquid Structure and Its Relation to Sound Speed

From an acoustical point of view the liquid is a state intermediate between a gas and a solid. It is gaslike in that, in the absence of losses, it does not offer resistance to a shear stress and solidlike in that its bulk modulus is determined by intermolecular bonding forces and not by external forces (e.g., gravity) or constraints (enclosure walls), as in a gas.

This state of affairs is best explained by a modern picture of liquid structure.[10] The radial distribution function (molecular number density vs. radial distance) obtained from x-ray and neutron diffraction patterns clearly shows regions of highly ordered structure as found in a crystalline solid. These regions, called *aggregates* or sometimes *clusters*, are of limited spatial extent, ranging from a few molecular diameters to macroscopic dimensions. Liquids showing little tendency toward aggregation are called *unassociated*; those containing a substantial distribution of small-sized aggregates are called *associated*; and those containing long chains of aggregates are called *polymerized*. The prevailing amorphous matter contains an abundance of microcavities, called *holes*, of varying size and distribution, which are responsible for the fluidity of the liquid. The major impact of the holes is an enhancement of the *free volume*, which plays a key role in the emergence of many equilibrium and transport properties of the liquid.

The derivation of the sound speed of a liquid from its microscopic properties requires realistic expressions for the equation of state and the internal energy, leading in turn to the bulk modulus and specific heat. Successful theoretical developments in these areas have been limited. Among the difficulties are the incorporation of the free volume into the equation of state and the tabulation of contributions to the free energy, which is not as straightforward as in a gas. The most successful efforts are, for the most part, at least partially empirical.

3.2 The Equation of State

The general form of an equation of state

$$P = \rho f(\rho, T, \ldots) \tag{26}$$

yields a bulk modulus

$$M_S = \gamma\rho\left(\frac{\partial P}{\partial \rho}\right)_T = \gamma\left[P + \rho\left(\frac{\partial f}{\partial \rho}\right)_T\right]. \tag{27}$$

As a gas condenses to a liquid, the second term becomes decisive owing to the large increase in density.

The development of a successful equation of state of a liquid must address the issues of molecular bonding, association and polymerization, and free volume in both the ordered and disordered structure. The principal approaches are as follows:

1. *Semirigorous Derivations Based on a Liquid Model.* These include the celebrated Lennard-Jones and Devonshire cell model, the Eyring hole model, and correlation models to determine the radial distribution function. Although these models have had some success in predicting many physical properties of liquids, they have been less successful in predicting sound speed.

2. *Semiempirical Modifications to Existing Equations of State*. A noteworthy example is the theory of Schaaffs,[11] who modified the van der Waals equation to include an improved expression for molecular volume as well as the specifics of bonds between various types of atoms or molecular complexes. Schaaffs's treatment has enjoyed remarkable success in predicting the sound speeds of unassociated organic liquids.
3. *Empirical Expressions Fitted to Experimental Data*. A common procedure is to expand one state variable in a series (not necessarily a power series) of the others. In applications to sound speed, however, the specific heat at one reference point remains an unknown parameter. Therefore it is just as useful to expand the sound speed itself in terms of state variables, an approach taken by van Itterbeek and co-workers[12] for cryogenic liquids.

3.3 Speed of Sound in Unassociated Liquids: Schaaffs's Method

Schaaffs's formulas[11] to compute the sound speed are

$$c = c_R \left(\frac{\rho B}{M} - \frac{\beta}{\beta + B} \right), \tag{28}$$

$$M = \sum_i z_i M_i, \tag{29a}$$

$$B = \sum_i z_i B_i, \tag{29b}$$

$$\beta = \sum_i z_i \beta_i, \tag{29c}$$

where M is the molecular mass, B the *external addendum* (molecular volume per mole), β the *internal addendum* (heavy atom corrective volume), z_i the number and M_i the mass of the ith species, $c_R = 4450$ m/s, and B_i and β_i contributions from the ith species obtained from a table. Some excerpts are listed in Table 2.

TABLE 2 Parameters Entering Schaaffs' Formulas (28)–(29) for Selected Atomic and Molecular Bonds

Bond Type	External Addendum, B_i
H (all bonds)	1.06
C (4 monovalent)	3.06
C (1 divalent)	3.36
C–Cl (monovalent)	6.92
Bond Type	**Internal Addendum, β_i**
C–Cl (monovalent)	0.25, 0.52, 0.75, 0.87 for 1, 2, 3, 4 atoms present
C–OH (monovalent)	0.6

Example Compute the sound speed of carbon tetrachloride (CCl_4), chloroform ($CHCl_3$), benzene (C_6H_6), and methanol (CH_3OH) at 20°C. Values of the respective densities ρ are 1.5939, 1.487, 0.878, and 0.7913 g/cm^3. The atomic weights of H, C, Cl, and O are 1.008, 12.01, 35.457, and 16.00 g/mol.

Inserting the data into Eqs. (28)–(29) yields, for CCl_4,

$$M = 1 \times 12.01 + 4 \times 35.457 = 153.838 \text{ g/mol},$$

$$B = 1 \times 3.06 + 4 \times 6.92 = 30.74 \text{ cm}^3,$$

$$\beta = 4 \times 0.87 = 3.48 \text{ cm}^3,$$

$$c = 4450\left(1.5939 \times \frac{30.74}{153.838} - \frac{3.48}{3.48 + 30.74}\right)$$

$$= 964.8 \text{ m/s}.$$

The results for the above liquids are summarized in Table 3. The agreement between theory and experiment is remarkable, even for CH_3OH, which is considered an associated liquid.

3.4 Speed of Sound in Cryogenic Liquids

Despite the simplicity of cryogenic liquid structure, which consists of monatomic molecules bonded by central forces, Schaaffs-like expressions for the sound speed are not available. Over the years van Itterbeek and coworkers have presented expressions for the sound speed of cryogenics versus temperature and pressure based on polynomial fits to a wide range of data. For example, their expressions for liquid nitrogen (LN_2) and liquid oxygen (LO_2) at the saturation temperatures (77.395 K and 90.19 K at 1 atm), after interpolation and unit conversion, are

$$c(LN_2) = 854.1 + 0.8370P - 0.9072 \times 10^{-3}P^2 + 0.9697 \times 10^{-6}P^3 - 0.4904 \times 10^{-9}P^4, \tag{30a}$$

$$c(LO_2) = 907.7 + 0.5641P - 0.4389 \times 10^{-3}P^2 + 0.3632 \times 10^{-6}P^3 - 0.1429 \times 10^{-9}P^4, \tag{30b}$$

where c is in metres per second and P is in atmospheres. At $P = 1$ atm, Eqs. (30a) and (30b) yield 854.9 and 908.3 m/s, compared to experimental values[13] of 852.9 and 907.0 m/s.

TABLE 3 Sound Speed in Organic Liquids after Schaaffs

Liquid	CCl_4	CH_3Cl	C_6H_6	CH_3OH
M (g/mol)	153.838	119.389	78.108	32.042
B (cm^3)	30.74	24.88	26.52	10.77
β (cm^3)	3.48	2.25	0	0.16
Theoretical c (m/s)	964.8	1009.9	1326.5	1118.4
Experimental c (m/s)	935	1001	1326	1121

3.5 Speed of Sound in Associated Liquids (Water)

A successful treatment of many physical properties of an associated liquid must account for the role of structural complexes. For illustration, Hall's model of water[3]—considered an exemplary associated liquid—assumes the liquid composition to be a mix of the open *tridymite* (icelike) structure and a more closely packed quartzlike structure. Eucken's model[3] assumes the structure to contain a distribution of polymeric units, namely the monomer, dimer, tetramer, and octamer, the last of which has the open tridymite structure. As the temperature increases toward the boiling point, the closely packed constituents gain at the expense of the vanishing tridymite. The competition between structural redistribution and thermal expansion explains such exceptional properties of water as the density maximum at 4°C and the sound speed maximum at 74°C.

A precise expression for the sound speed in saturated water versus temperature, promising an error not to exceed 0.05%, is given by Chavez, Sosa, and Tsumura[14]:

$$c = \left(1 - \frac{T}{T_c}\right)^a \sum_{k=0}^{5} b_k T^k \qquad \text{m/s}, \tag{31}$$

where T_c, a, and the coefficients are given in Table 4. Equation (31) is plotted in Fig. 3. At 20°C, Eq. (31) yields $c = 1481.8$ m/s. An increase in pressure increases the sound speed.[15]

3.6 Speed of Sound in Electrolytic Solutions

The presence of ions in a polar solvent such as water completely reorganizes the structure due to the orientation of solvent dipoles about the ions. The considerable electric fields of the ions compress the volume of the solvent, thereby increasing the bulk modulus and speed of sound.

Although theoretical treatments based on microscopic properties are lacking, Weissler and Del Grosso[16] made an interesting observation concerning the contribution of the solutes to the sound speed of sea water. If the ith solute at a molarity X_i individually produces a sound speed $c_i(X_i)$, then the sound speed of the composite solution is

$$c = c_0 + \sum_i [c_i(X_i) - c_0], \tag{32}$$

where c_0 is the sound speed of pure water (implying negligible ionic interaction). Values in metres per second for the seven leading salts in sea water at 30°C are listed in Table 5. When the speed of pure water $c_0 = 1510.0$ m/s and total increment of 36.2 m/s are inserted into Eq. (32), the resulting sound speed is 1546.2 m/s, compared to a measured value of 1545.8 m/s.

3.7 Temperature Dependence of the Sound Speed in Liquids: Rao's Constant

Rao[17] established an empirical relation between the temperature coefficient of the sound speed and the coefficient of thermal expansion for an unassociated liquid:

$$\frac{1}{c}\left(\frac{\partial c}{\partial T}\right) \Big/ \frac{1}{V}\left(\frac{\partial V}{\partial T}\right)_P = -k, \tag{33}$$

TABLE 4 Parameters Entering Chavez, Sosa, and Tsumura's[a] formula (31) for Temperature Dependence of Sound Speed in Pure Water

k	b_k
0	−19214.88484
1	230.1609318
2	−1.028803876
3	0.002414336487
4	$-2.902395566 \times 10^{-6}$
5	$1.430493449 \times 10^{-9}$
a	0.75
T_c (K)	647.067

[a]Reference 14.

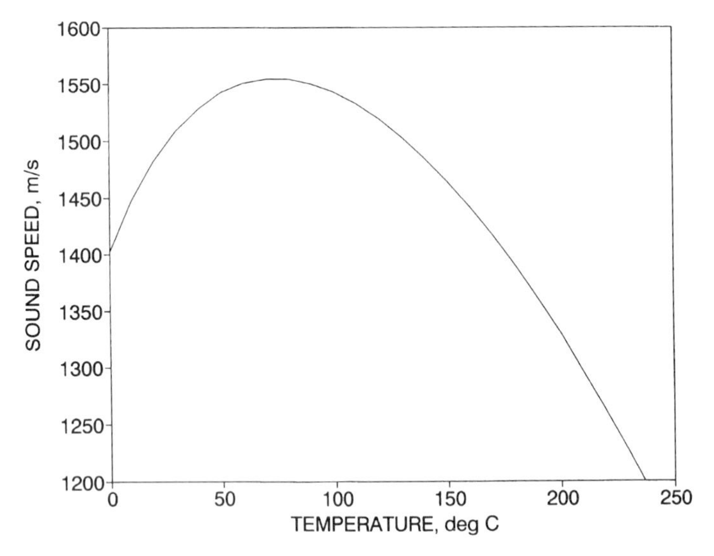

Fig. 3 Speed of sound in saturated water vs. temperature.

where the constant k is nearly independent of temperature and has a nominal value of 3. Although Rao proposed Eq. (33) as a strictly empirical relation, Schaaffs provides a derivation based on the semiempirical concepts leading to Eq. (28). Deviations from Eq. (33) are presumed to indicate association of the liquid. Integration of Eq. (33) yields

$$Vc^{1/k} = \mathcal{R}, \tag{34}$$

where the constant $\mathcal{R}$ is of theoretical interest because it is related to the repulsive exponent in the Lennard-Jones potential, and it serves as the basis for an additivity relationship in liquid mixtures. The value of Rao's constant k is listed in Table 6 for several liquids.

4 SPEED OF SOUND IN TWO-PHASE FLUIDS

The speed of sound in two-phase fluid mixtures, such as bubbly liquids or foggy gases, is strongly influenced by interfacial heat, mass, and momentum transfer. In the idealized mixture (absence of temperature differentials, phase transitions, and slip between phases), the density ρ and sound speed c are weighted combinations of those of the individual phases[18]:

$$\rho = \rho_1(1 - \phi) + \rho_2\phi, \tag{35}$$

$$\frac{1}{c^2} = \rho\left(\frac{1-\phi}{\rho_1 c_1^2} + \frac{\phi}{\rho_2 c_2^2}\right), \tag{36}$$

where subscripts 1 and 2 refer to the liquid and gas,

TABLE 5 Effect of Solutes on Sound Speed in Sea Water at 30°C

Solute	Molarity, X_i	Sound Speed, $c_i(X_i)$ (m/s)	Increment, $c_i(X_i) - c_0$ (m/s)
NaCl	0.4649	1538.2	28.2
$MgSO_4$	0.0281	1513.4	3.4
$MgCl_2$	0.0263	1512.9	2.9
$CaCl_2$	0.0105	1510.9	0.9
KCl	0.00997	1510.6	0.6
$NaCHO_3$	0.00246	1510.2	0.2
NaBr	0.00083	1510.0	0.0
Total			36.2

TABLE 6 Rao's Constant *k* for Selected Liquids

Liquid	Temperature Range, K	*k*
Nitrogen	70–76	2.4
Oxygen	80–90	2.3
Potassium	337–433	0.97
Tin	505–653	3.10
Glycerine	293–313	0.78
CCl_4	288–318	2.8
Octane	298–318	3.0
Acetone	293–313	2.6
Benzene	288–318	3.0
Toluol	293–313	3.1

respectively. The quantity ϕ is the *void fraction*, defined as

$$\phi = v_2 \rho, \tag{37}$$

where v_2 is the volume of the gas per unit mass of the mixture. According to Eq. (36), the sound speed of the mixture is less than that of the individual phases.

If ϕ is small but sufficiently large that $\phi \rho_1 c_1^2 \gg (1 - \phi)\rho_2 c_2^2$, as in a moderately bubbly liquid, then the density is essentially that of the liquid, but the compressibility is dominated by the gas. The sound speed increases strongly with pressure but is still bounded by that of the liquid.

If the gas is a condensing vapor on the gas–liquid phase boundary, as in the presence of fog in the atmosphere, then ϕ can vary due to a phase change induced by acoustic pressures. When the condensation/evaporation entropy is taken into account, then the sound speed can be approximated by

$$c^2 = \frac{P^2 L^2 M_2^2}{C_{p1} R^2 \rho_1^2 T^3}, \tag{38}$$

where L is the enthalpy of evaporation, M_2 the molecular weight of the vapor, and C_{p1} the specific heat at constant pressure of the liquid.

Wei and Wu[19] modeled sound propagation through atmospheric fog as a relaxation process by including the effects of droplet growth; their low-frequency limit of the sound speed is somewhat more complex than Eq. (38).

REFERENCES

1. F. W. Giacobbe, "Precision Measurement of Acoustic Velocities in Pure Gases and Gas Mixtures," *J. Acoust. Soc. Am.*, Vol. 94, 1993, pp. 1200–1210.
2. P. M. Morse and K. U. Ingard, *Theoretical Acoustics*, McGraw-Hill, New York, 1968, p. 290.
3. K. F. Herzfeld and T. A. Litovitz, *Absorption and Dispersion of Ultrasonic Waves*, Academic, New York and London, 1959.
4. M. Greenspan, "Sound Waves in Gases at Low Pressures," in W. P. Mason, Ed., *Physical Acoustics*, Vol. IIA, Academic, New York and London, 1965.
5. H. J. Bauer, "Theory of Relaxation Phenomena in Gases," in W. P. Mason, Ed., *Physical Acoustics*, Vol. IIA, Academic, New York and London, 1965.
6. M. Greenspan, "Comments on 'Speed of Sound in Standard Air,'" *J. Acoust. Soc. Am.*, Vol. 80, 1987, pp. 370–372.
7. G. S. K. Wong and T. F. W. Embleton, "Variation of Specific Heats and of Specific Heat Ratio in Air with Humidity," *J. Acoust. Soc. Am.*, Vol. 76, 1984, pp. 555–559.
8. J. Hilsenrath et al., *Tables of Thermal Properties of Gases*, National Bureau of Standards Circular 564, U.S. Department of Commerce, Washington, DC, 1955.
9. A. J. Zuckerwar and R. W. Meredith, "Low-Frequency Absorption of Sound in Air," *J. Acoust. Soc. Am.*, Vol. 78, 1985, pp. 946–955.
10. A. Muenster, "Theory of the Liquid State," in A. van Itterbeek, Ed., *Physics of High Pressures and the Condensed Phase*, North-Holland, Amsterdam, 1965.
11. W. Schaaffs, *Molekularakustik*, Springer-Verlag, Berlin, 1963.
12. A. Van Itterbeek and W. Van Dael, "Velocity of Sound in Liquid Oxygen and Liquid Nitrogen as a Function of Temperature and Pressure," *Physica*, Vol. 28, 1962, pp. 861–870.
13. A. J. Zuckerwar and D. S. Mazel, "Sound Speed Measurements in Liquid Oxygen–Liquid Nitrogen Mixtures," *NASA TP 2464*, National Technical Information Service, Springfield, VA, 1985.
14. M. Chavez, V. Sosa, and R. Tsumura, "Speed of Sound in Pure Water," *J. Acoust. Soc. Am.*, Vol. 77, 1985, pp. 420–423.
15. C. C. Chen and F. J. Millero, "Reevaluation of Wilson's Sound-Speed Measurements for Pure Water," *J. Acoust. Am.*, Vol. 60, 1976, pp. 1270–1273.
16. A. Weissler and V. A. Del Grosso, "The Velocity of Sound in Sea Water," *J. Acoust. Soc. Am.*, Vol. 23, 1951, pp. 219–223.
17. M. R. Rao, "A Relation between Velocity of Sound in Liquids and Molecular Volume," *Indian J. Physics*, Vol. 23, 1940, pp. 109–116.
18. V. E. Nakoryakov, B. G. Pokusaev, and I. R. Shreiber, *Wave Propagation in Gas–Liquid Media*, CRC Press, Boca Raton, FL, 1993.
19. R. Wei and J. Wu, "Absorption of Sound in Water Fog," *J. Acoust. Soc. Am.*, Vol. 70, 1981, pp. 1213–1219.

6

STANDING WAVES

U. S. SHIRAHATTI AND MALCOLM J. CROCKER

1 INTRODUCTION

The main focus of this chapter is on standing waves or modes of vibration. Engineers tend to think of vibrations in terms of modes and sound in terms of waves, and quite often it is forgotten that the two are simply different ways of looking at the same physical phenomenon. When considering the interactions between sound waves and the vibration of structures, it is important to have a working knowledge of both physical models.

A mode of vibration on a taut, fixed string can be interpreted as being composed of two waves of equal amplitude and wavelength traveling in opposite directions between the bounded ends. Alternatively, it can be interpreted as a "standing wave," that is, the string oscillates with a spatially varying amplitude within the confines of a specific stationary waveform.

Central to this concept of wave–mode duality is the finite or limited extent of the medium. In fact, it is not possible to have standing waves or modes in a medium of infinite or unlimited extent. A further point to be remembered is that an infinite or unlimited medium can vibrate freely at any frequency. In contrast, a finite- or limited-extent medium can vibrate freely only at specific frequencies known as the natural frequencies.[1]

2 ONE-DIMENSIONAL STANDING WAVES

In Chapter 2, the derivation of the wave equation for a fluid medium is presented [see Eq. (44)]. This form of wave equation can also be used to study standing waves or modes in strings, bars, and acoustic waveguides. The only thing that is different is the wave speed c, which depends on the nature of the medium. The wave equation may be interpreted as

Acceleration of any particle at any instant
= square of wave speed
× curvature of waveform at that point
and instant.

The expressions for the speeds of the several other important kinds of waves will not be derived here. The results of such derivations may be summarized by stating that the *wave speed c* is always given by some simple fraction under the square-root sign. The numerator of the fraction is always that particular elastic coefficient that defines the elastic property of the medium that is responsible for the wave motion. The denominator of the fraction is always a term denoting the mass, linear density, or total density (i.e., mass per unit length, area, or volume) of the medium transmitting the wave. In general,

$$c = \sqrt{\frac{E}{\rho}}. \qquad (1)$$

Table 1 lists expressions for wave speeds of several types of waves.

The speed of almost all waves that exist in mechanical media such as strings, plates, liquids, or solids is usually too great for the eye to be able to follow the waveform as it travels. For small wave speeds, the elastic constant E should be small and the inertia ρ large.

Note: References to chapters appearing only in the *Encyclopedia* are preceded by *Enc.*

TABLE 1 Values of Wave Speed $c = \sqrt{E/\rho}$

Type of Wave	E Represents	ρ Represents
Transverse wave in a string	Tension T	Linear density
Longitudinal (compressional) waves in a liquid	Bulk modulus K	Density
Longitudinal waves in a rigid rod	Young's modulus E	Linear density
Transverse waves in a rigid rod	Rigidity modulus G	Linear density
Torsional waves in a rigid rod	Moment of torsion J_0	Linear density
Compressional waves in a gas	Atmospheric pressure p_0	Density

2.1 Standing Waves in Strings

Free Vibrations Since all physical media that carry waves are bounded, that is, limited in extent, the simpler results that follow from this fact should be described, at least. In the case of waves on a real string, which is never very long, the idea of a wave that travels on and on indefinitely is certainly unrealistic.

D'Alembert has shown that the solution of the wave equation consists of the superposition of two terms, the functions $f_1(ct-x)$ and $f_2(ct+x)$. The function $f_1(ct-x)$ denotes a wave moving to the right (or positive x-direction) and the function $f_2(ct+x)$ denotes a wave moving to the left. There are several equivalent expressions for the string displacement $y = f(ct-x)$, such as

$$y = A \sin \frac{2\pi}{\lambda}(ct - x), \tag{2}$$

$$y = A \sin 2\pi\left(ft - \frac{x}{\lambda}\right), \tag{3}$$

$$y = A \sin 2\pi(\omega t - kx), \tag{4}$$

where $k = 2\pi/\lambda = \omega/c$ is the wavenumber, f is the frequency in hertz, and λ is the wavelength in metres. It is also possible to represent y using complex exponential notation as

$$y = Ae^{i(\omega t - kx)}, \qquad i = \sqrt{-1}. \tag{5}$$

For example, let us consider a string of length l with its ends rigidly clamped. The displacement y on the string at any point is given by

$$y = Ae^{i(\omega t - kx)} + Be^{i(\omega t + kx)}, \tag{6}$$

where A and B are complex constants to be determined from the boundary conditions at $x = 0$ and $x = l$. These boundary conditions for a clamped string are $y = 0$ at $x = 0$ and $x = l$. The condition $y = 0$ at $x = 0$ yields $A = -B$ from Eq. (5); physically this means that the wave is reflected at either end with a 180° phase change. Hence

$$y = Ae^{i\omega t}(e^{-ikx} - e^{ikx}) \tag{7}$$

$$= (-2i)Ae^{i\omega t} \sin kx. \tag{8}$$

The condition $y = 0$ at $x = l$ for all times t leads to $\sin kl = 0$, or $kl = n\pi$ ($n = 1, 2, 3$), or $k = n\pi/l$, which in terms of angular frequency becomes $\omega_n = n\pi c/l$.

These frequencies are called *natural frequencies* or *eigenfrequencies*. When the string is vibrating at a particular natural frequency, it has an associated deflected shape, known as its *mode of vibration*. It may be noticed that a string can theoretically vibrate in an infinite number of *modes*. Returning now to Eq. (5), we see that the *displacement* of the nth mode of vibration is given by

$$y_n = (A_n \cos \omega_n t + B_n \sin \omega_n t) \sin\left(\frac{\omega_n x}{c}\right), \tag{9}$$

where the *amplitude* of the nth mode is given by

$$\sqrt{A_n^2 + B_n^2} = 2A. \tag{10}$$

Forced Vibrations Consider a very long string such that when it is set into oscillation with an oscillator, waves can be seen moving away with a speed c, but no waves are seen returning. At $x = 0$ the force applied on the string by the oscillator is $F_y = F_0 e^{i\omega t}$ and it can be shown that, at $x = 0$, the ratio of the force to the velocity v becomes

$$\frac{F_0 e^{i\omega t}}{v} = \frac{T}{c} = Z_0, \tag{11}$$

where Z_0 is the characteristic mechanical impedance of the lossless infinite string. Since from Table 1 $c = \sqrt{T/\rho}$, we may also write the input impedance as $Z_0 = \rho c$, which is a real number and is not complex. Every lossless medium in which a wave propagates has a characteristic impedance of this type.

If the string is of finite length l and it is driven at $x = 0$ as before but is supported at $x = l$, the *input impedance is quite different* because of the reflected waves. The outgoing and the reflected waves add at all times to give a motion contained within the wave envelope. The string at any point x moves up and down through its equilibrium position $y = 0$ with a frequency ω. The amplitude of the so-called standing-wave pattern (mode of vibration) is different at various points x.

Energy of a Vibrating String A vibrating string possesses both kinetic and potential energy. The kinetic energy of an element of length dx and linear density ρ is given by $\frac{1}{2}\rho\, dx(\partial y/\partial t)^2$ and the total kinetic energy is the integral of this term along the length of the string. Hence

$$E_k = \frac{1}{2}\rho \int_0^L \left(\frac{\partial y}{\partial t}\right)^2 dx. \tag{12}$$

The potential energy is the work done by the tension T in extending an element dx to a new length ds when the string is vibrating. Hence

$$E_p = \int_0^L T(ds - dx)$$

$$= T\int_0^L \left\{\left[1 + \left(\frac{\partial y}{\partial x}\right)^2\right]^2 - 1\right\} dx, \tag{13}$$

or using Taylor's approximation and neglecting higher order terms

$$E_p = \frac{1}{2} T \int_0^L \left(\frac{\partial y}{\partial x}\right)^2 dx. \tag{14}$$

In a standing wave the total energy is a constant. However, energy is continually being exchanged between kinetic and potential forms. Further, when the displacement is maximum, all the energy is potential, while with zero displacement, the velocity is maximum and all the energy is kinetic. It may be noted also that maximum values of both potential and kinetic energies are identical.

TABLE 2 Standing-Wave Patterns for a Vibrating String[a]

Situation	Boundary Conditions	Natural Frequency Equation	Mode Shape
Unequal, spring loaded; k_1 and k_2 are spring stiffness, T is tension in string	$\left(\frac{\partial y}{\partial x}\right)_{x=0} = \frac{K_1}{T} y(0)$, $\left(\frac{\partial y}{\partial x}\right)_{x=l} = \frac{K_2}{T} y(l)$, $n = 1, 2, 3, \ldots$	$\omega_n^2 = \frac{\Omega_n^2 C^2}{l^2}$, $\tan \Omega_n = \frac{\Omega_n(\alpha_1 - \alpha_2)}{\alpha_1\alpha_2 + \Omega_n^2}$	For $\alpha_1 \neq \infty$: $y_n(x) = \frac{\alpha_1}{\Omega_n} \sin\left(\frac{\Omega_n x}{l}\right) + \cos\left(\frac{\Omega_n x}{l}\right)$,
		for $\alpha_1 \neq 0$: $\alpha_1 = \frac{lk_1}{\rho c^2}, \alpha_2 = \frac{lk_2}{\rho c^2}$	$y_n(x) = \sin\left(\frac{\Omega_n x}{l}\right) + \frac{\Omega_n}{\alpha_1} \cos\left(\frac{\Omega_n x}{l}\right)$
Free–free	$k_1 = k_2 = 0$	$\omega_n = \left(\frac{n\pi c}{l}\right)$	$y_n(x) = \cos\left(\frac{n\pi x}{l}\right)$
Clamped–clamped	$k_1 = k_2 = \infty$	$\omega_n = \left(\frac{n\pi c}{l}\right)$	$y_n(x) = \sin\left(\frac{n\pi x}{l}\right)$
Equal, spring loaded	$k_1 = k_2$	$\omega_n = \left(\frac{n\pi c}{l}\right)$, or	$y_n(x) = \frac{\alpha_1}{n\pi} \sin\left(\frac{n\pi x}{l}\right) + \cos\left(\frac{n\pi x}{l}\right)$
		$y_n(x) = \sin\left(\frac{n\pi x}{l}\right) + \frac{\alpha_1}{n\pi} \cos\left(\frac{n\pi x}{l}\right)$, $n = 1, 2, 3, \ldots$	

[a]From Ref. 2 with permission.

Natural Frequencies and Mode Shapes of a String for Different Boundary Conditions The natural frequencies and mode shapes or standing-wave patterns for a vibrating string with different boundary conditions are listed (see Table 2).[2]

2.2 Standing Waves in a Duct with Various Boundary Conditions

Free Vibrations Standing waves are readily formed in a tube of length l with various boundary conditions at the two ends. Table 3 lists some typical boundary conditions along with the resonance frequencies. Table 3 also gives expressions for sound pressure, particle velocity, and intensity inside the duct.

Forced Vibrations Consider plane sound waves traveling along a hard-walled tube such that only a single frequency is present. These plane waves are excited by the oscillations of a piston located at $x = 0$. Now, at $x = l$, it is possible to impose three types of terminations.

- *A Perfect Sound Absorber.* There is no reflection of sound at the termination of the tube; therefore no standing waves are formed.
- *A Perfectly Rigid Termination.* The waves are perfectly reflected at the termination. The reflected waves traveling back to the left have the same amplitude as incident waves traveling to the right. These two traveling waves combine to give rise to perfect standing waves.
- *A Partially Absorptive Termination.* There is a partial reflection at the termination so that a weaker wave returns from the right to the left. The two opposite traveling waves add together to give rise to a weaker form of standing waves.

In Table 4, expressions are given for sound pressure, particle velocity, intensity, and natural frequencies for various boundary conditions.

From the expressions in Table 4 note the following:

1. If $Z = \rho c$ (*anechoic termination*), then

TABLE 3 Standing Waves in a Duct: Free Vibrations

Case	Boundary Condition	Field Variables
Closed at both ends (x=0, x=l)	$\zeta(0) = 0$, $\zeta(l) = 0$	$\zeta = B \sin kx \sin \omega t$ $p = -\rho c \omega B \cos kx \sin \omega t$ $u = \omega B \sin kx \cos \omega t$ $I = -\frac{1}{4} \rho c \omega^2 B^2 \sin(2kx) \sin(2\omega t)$ $\omega_n = \left(\frac{n\pi c}{l}\right)$
Open at both ends (x=0, x=l)	$\left(\frac{\partial \zeta}{\partial x}\right)_{x=0} = 0$, $\left(\frac{\partial \zeta}{\partial x}\right)_{x=l} = 0$	$\zeta = A \sin kx \sin \omega t$ $p = \rho c \omega A \sin kx \sin \omega t$ $u = \omega A \cos kx \cos \omega t$ $I = \frac{1}{4} \rho c \omega^2 A^2 \sin(2kx) \sin(2\omega t)$ $\omega_n = \left(\frac{n\pi c}{l_a}\right)$, where $l_a = 0.6a + l + 0.6a$
Open at x = l and Closed at x = 0 (x=0, x=l)	$\zeta(0) = 0$, $\left(\frac{\partial \zeta}{\partial x}\right)_{x=l} = 0$	$\zeta = B \sin kx \sin \omega t$ $p = \rho c \omega B \cos kx \sin \omega t$ $u = \omega B \sin kx \cos \omega t$ $I = -\frac{1}{4} \rho c \omega^2 B^2 \sin(2kx) \sin(2\omega t)$ $\omega_n = \left(\frac{n\pi c}{l_a}\right)$, where $l_a = 0.6a + l + 0.6a$

Note: ζ = particle displacement, p = sound pressure, u = sound velocity, I = instantaneous intensity, a = radius of duct, l = length of duct; A and B are constants.

TABLE 4 Standing Waves in a Duct: Forced Vibrations

Case	Boundary Conditions	Field Variables
1. *Piston and closed end* (x=0, x=l)	$\zeta(0) = r \sin \omega t$ $\zeta(l) = 0$	$\zeta = r \dfrac{\sin[k(l-x)]}{\sin kl} \cos \omega t$ $u = \omega r \dfrac{\sin[k(l-x)]}{\sin kl} \cos \omega t$ $p = \rho c \omega r \dfrac{\cos[k(l-x)]}{\sin kl} \sin \omega t$ $I = \frac{1}{4} \rho c (\omega r)^2 \dfrac{\sin[2k(l-x)]}{\sin^2 kl} \sin(2\omega t)$ $\langle I \rangle = 0$, where $\langle \cdot \rangle$ signifies time average
2. *Piston and open end* (x=0, x=l)	$\zeta(0) = r \sin \omega t$ $\left(\dfrac{\partial \zeta}{\partial x}\right)_{x=l} = 0$	$\zeta = r \dfrac{\cos[k(l-x)]}{\cos kl} \sin \omega t$ $u = \omega r \dfrac{\cos[k(l-x)]}{\cos kl} \cos \omega t$ $p = -\rho c \omega r \dfrac{\sin[k(l-x)]}{\cos kl} \sin \omega t$ $I = -\frac{1}{4} \rho c (\omega r)^2 \dfrac{\sin[2k(l-x)]}{\cos^2 kl} \sin(2\omega t)$ $\langle I \rangle = 0$, where $\langle \cdot \rangle$ signifies time average
3. *Piston and Termination* Impedance $Z = R + jX$ (x=0, x=l)	$u(0) = \omega r e^{j\omega t}$ $\dfrac{p(l)}{u(l)} = Z$	$p = \rho c \omega r \left(\dfrac{\cos[k(l-x)] + j(\rho c/Z) \sin[k(l-x)]}{(\rho c/Z) \cos kl + j \sin kl} \right) e^{j\omega t}$ $u = \omega r \left(\dfrac{(\rho c/Z) \cos[k(l-x)] + j \sin[k(l-x)]}{(\rho c/Z) \cos kl + j \sin kl} \right) e^{j\omega t}$ $I = \rho c \omega r^2 \left(\dfrac{(\rho c/Z) \cos^2[k(l-x)] + j(\rho c/Z + 1) \cos[k(l-x)] \sin[k(l-x)]}{(\rho c/Z)^2 \cos^2(kl) - \sin^2 kl + 2j(\rho c/Z) \cos kl \sin kl} \right) e^{2j\omega t}$ $\langle I \rangle = \dfrac{(\frac{1}{2})(\rho c \omega r)^2 (2R/ZZ^*)}{(\rho c/ZZ^*)\rho c \cos^2 kl + \sin^2 kl - (\rho c/ZZ^*) \cos kl \sin kl(2X)}$

Note: Asterisk denotes complex conjugate

TABLE 5 Standing Waves in Membranes of Finite Area

Case	Boundary Conditions	Mode Shape	Natural Frequency
Clamped rectangular membrane	$W(x, 0, t) = 0$, $W(x, d, t) = 0$, $W(0, y, t) = 0$, $W(b, y, t) = 0$,	$W(x, y) = \sin\left(\dfrac{m\pi x}{b}\right) \sin\left(\dfrac{n\pi y}{d}\right)$, $m, n = 1, 2, 3, \ldots$	$\omega_{mn}^2 = (m\pi)^2 + \left(\dfrac{n\pi b}{d}\right)^2$ $m, n = 1, 2, 3, \ldots$
Clamped solid circular membrane	$W(b, \theta) = 0$	$W_{mn,j}(r, \theta) = J_m\left(\dfrac{\omega_{mn} r}{b}\right) \psi_j(m\theta)$, $j, m = 0, 1, 2, \ldots, n = 1, 2, \ldots$, $\psi_j(m\theta) = \sin(m\theta + j\pi/2)$, $m = 0, 1, 2, \ldots,\ j = 0, 1$	$J_m(\omega_{mn}) = 0$, $m = 0, 1, 2, \ldots$, $n = 1, 2, 3, \ldots$

$$ZZ^* = (\rho c)^2, \qquad R = \rho c, \tag{15}$$

and the time-averaged intensity at any point in the duct $\langle I \rangle = \frac{1}{2}\rho c(\omega r)^2$, which indicates that there are no standing waves.

2. If $Z \rightarrow \infty$ (*rigid termination*), then $\langle I \rangle = 0$, which indicates perfect standing waves.
3. If

$$Z = \frac{c}{\pi a^2}\left(\frac{k^2 a^2}{4} + j0.6ka\right),$$

TABLE 6 Roots of $J_m(\omega_{mn}) = 0$

m	$n = 1$	$n = 2$	$n = 3$	$n = 4$	$n = 5$
0	2.408	5.5201	8.6537	11.7915	14.9309
1	3.8317	7.0156	10.1735	13.3237	16.4706
2	5.1356	8.4172	11.6198	14.7960	17.9598
3	6.3802	9.7610	13.0152	16.2235	19.4094
4	7.5883	11.0647	14.3752	17.6610	20.8269
5	8.7714	12.3386	15.7002	18.9801	22.2178
6	9.9361	13.5893	17.0038	20.3208	23.5861
7	11.8064	14.8213	18.2876	21.6416	24.9349

TABLE 7 Acoustical Modes and Natural Frequencies

Description	Figure	Natural Frequency f_{ijk} (Hz)	Mode Shape Φ_{ijk}
Slender tube, both ends closed	D<<L, D, x, L	$\frac{ic}{2L}$ $D \ll \lambda$, where $\lambda = c/f$	$\cos\frac{i\pi x}{L}$ $\quad i = 0, 1, 2, \ldots$
Slender tube, one end closed, one end open	D<<L, D, x, L	$\frac{ic}{4L}$ $D \ll \lambda$, where $\lambda = c/f$	$\cos\frac{i\pi x}{2L}$ $\quad i = 1, 3, 5, \ldots$
Slender tube, both ends open	D<<L, D, x, L	$\frac{ic}{2L}$ $D \ll \lambda$, where $\lambda = c/f$	$\sin\frac{i\pi x}{L}$ $\quad i = 1, 2, 3, \ldots$
Closed rectangular volume	y, x, z, L_y, L_x, L_z	$\frac{c}{2}\left(\frac{i^2}{L_x^2} + \frac{j^2}{L_y^2} + \frac{k^2}{L_z^2}\right)^{1/2}$	$\cos\frac{i\pi x}{L_x}\cos\frac{j\pi y}{L_y}\cos\frac{k\pi x}{L_z}$ $\quad i = 0, 1, 2, \ldots$; $j = 0, 1, 2, \ldots$; $k = 0, 1, 2, \ldots$
Closed cylindrical volume	x, θ, r, 2R, L	$\frac{c}{2\pi}\left(\frac{\lambda_{jk}^2}{R^2} + \frac{i^2\pi^2}{L^2}\right)^{1/2}$ λ_{jk} from Table (a) below	$J_j\left(\lambda_{jk}\frac{r}{R}\right)\cos\frac{i\pi x}{L}\begin{cases}\sin j\theta_i \\ or; \\ \cos j\theta\end{cases}$ $\quad i = 0, 1, 2, \ldots$; $j = 0, 1, 2, \ldots$; $k = 0, 1, 2, \ldots$ $J_j = j$th-order Bessel function
Closed spherical volume	r, R	$\frac{c\lambda_i}{2\pi R}$ λ_i from Table (b) below	Modes symmetric about center $\frac{R}{\lambda_i r}\sin\frac{\lambda_i r}{R}$ $\quad i = 0, 1, 2, \ldots$
Arbitrary closed volume	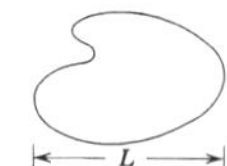	L—Maximum linear dimension fundamental natural frequency (approximate): $\frac{c}{2L}$	Finite element analysis

Table (a)

λ_{jk}	j						
k	0	1	2	3	4	5	6
0	0.	1.8412	3.0542	4.2012	5.3176	6.4156	7.5013
1	3. 8317	5.3314	6.7061	8.0152	9.2824	10.5199	11.7349
2	7. 0156	8.5363	9.9695	11.3459	12.6819	13.9872	15.2682
3	10.173	11.7060	13.1704	14.5859	15.9641	17.3128	18.6374

$\lambda_{j=0,k} = \pi(k + 1/4)$ for $k \geq 3$ $\quad (J_j'(\lambda_{jk}) = 0)$

Table (b)

i	0	1	2	3	4
λ_i	0	4.4934	7.7253	10.9041	14.0662

$\lambda_i = \pi(i + 1/2)$ for $i \geq 4$ $\quad \tan\lambda_i = \lambda_i$

Reprinted, with permission, from Ref. 3: R. D. Blevins, "Fluid Systems," in *Formulas for Natural Frequency and Mode Shape*, Van Nostrand Reinhold, New York, 1984 (and Ref. 4).

where $ka < 0.5$ is the radiation impedance for a pipe with an infinite flange radiating out into the atmosphere, the results from case 3 in Table 4 must be used and not those from case 2. In this case a weaker form of standing waves exists.

3 TWO-DIMENSIONAL STANDING WAVES

To illustrate two-dimensional standing waves, consider membranes. The two-dimensional wave equation is represented by[2]

$$\frac{\partial^2 W}{\partial t^2} = c^2 \nabla^2 W, \tag{16}$$

where ∇^2 is the two-dimensional Laplace operator and W is the membrane displacement. In Table 5, the boundary conditions, mode shapes, or standing-wave patterns along with the natural frequencies for rectangular and circular membranes are given.

TABLE 8 Summary of Equations for Lateral Vibration of Single-Span Uniform Beams

End Conditions	Equations[a]
0 l x	(1) $u(0,t) = u'(0,t) = u(l,t) = u'(l,t) = 0$ (2) $\cos \beta l \cosh \beta l = 1$ (3) $\phi(x) = A(\cos \beta x - \cosh \beta x) + (\sin \beta x - \sinh \beta x)$ $A = -\frac{\sin \beta l - \sinh \beta l}{\cos \beta l - \cosh \beta l} = \frac{\cos \beta l - \cosh \beta l}{\sin \beta l + \sinh \beta l}$
 0 l x	(1) $u''(0,t) = u'''(0,t) = u''(l,t) = u'''(l,t) = 0$ (2) $\cos \beta l \cosh \beta l = 1^*$ (3) $\phi(x) = A(\cos \beta x + \cosh \beta x) + (\sin \beta x + \sinh \beta x)$ $A = -\frac{\sin \beta l - \sinh \beta l}{\cos \beta l - \cosh \beta l} = \frac{\cos \beta l - \cosh \beta l}{\sin \beta l + \sinh \beta l}$
 0 l x	(1) $u(0,t) = u'(0,t) = u(l,t) = u''(l,t) = 0$ (2) $\tan \beta l = \tanh \beta l$ (3) $\phi(x) = A(\cos \beta x - \cosh \beta x) + (\sin \beta x - \sinh \beta x)$ $A = -\frac{\sin \beta l - \sinh \beta l}{\cos \beta l - \cosh \beta l} = -\frac{\sin \beta l + \sinh \beta l}{\cos \beta l + \cosh \beta l}$
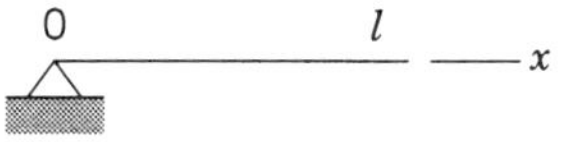 0 l x	(1) $u(0,t) = u''(0,t) = u''(l,t) = u'''(l,t) = 0$ (2) $\tan \beta l = \tanh \beta l^*$ (3) $\phi(x) = A \sin \beta x + \sinh \beta x$ $A = \frac{\sinh \beta l}{\sin \beta l} = \frac{\cosh \beta l}{\cos \beta l}$
 0 l x	(1) $u(0,t) = u''(0,t) = u(l,t) = u''(l,t) = 0$ (2) $\sin \beta l = 0$ (3) $\phi(x) = A \sin \beta x$
 0 l x	(1) $u(0,t) = u'(0,t) = u''(l,t) = u'''(l,t) = 0$ (2) $\cos \beta l \cosh \beta l = -1$ (3) $\phi(x) = A(\cos \beta x - \cosh \beta x) + (\sin \beta x - \sinh \beta x)$ $A = -\frac{\sin \beta l + \sinh \beta l}{\cos \beta l + \cosh \beta l} = \frac{\cos \beta l + \cosh \beta l}{\sin \beta l - \sinh \beta l}$

Lateral deflection $\sum \phi(x)q(t)$ $\quad m$ = length mass $\quad \omega$ = frequency (rad/s)

Beam equation: $\frac{\partial^2 u}{\partial t^2} + a^2 \frac{\partial^4 u}{\partial x^4} = 0 \qquad \frac{d^4 \phi}{dx^4} - \beta^4 \phi = 0 \qquad \frac{d^2 q}{dt^2} + \omega^2 q = 0$

$$\beta^4 = \frac{\omega^2}{a^2} = \frac{m\omega^2}{EI}$$

[a](1) End condition, (2) frequency equation; (3) eigenfunction, * semidefinite. Reprinted, with permission, from Ref. 5: F. S. Tse, I. E. Morse, and R. T. Hinkle, *Mechanical Vibrations*, Allyn and Bacon, Boston, 1978.

4 THREE-DIMENSIONAL STANDING WAVES

The analysis of standing waves in a tube can be readily extended to a three-dimensional cavity in an enclosure. For simple geometries and boundary conditions the acoustic modes can be expressed analytically as shown in Table 7.[3,4] If the geometry is complicated, it becomes essential to use a numerical method such as the finite element method, which is discussed in Chapter 12.

4.1 Standing Waves in Beams and Plates

Beams In Chapter 49, Section A.2.1, the equation of motion for bending waves in beams (the Bernoulli–Euler equation) has been given. We set here for convenience

TABLE 9 Frequency Coefficients and Different Mode Shapes for Plates

Boundary Conditions	Deflection Function or Mode Shape	N	K
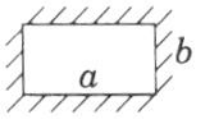 a, b	$\left(\cos\frac{2\pi x}{a}-1\right)\left(\cos\frac{2\pi y}{b}-1\right)$	2.25	$12+8\left(\frac{a}{b}\right)^2+12\left(\frac{a}{b}\right)^4$
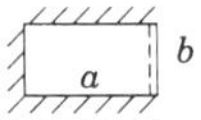 a, b	$\left(\cos\frac{3\pi x}{2a}-\cos\frac{\pi x}{2a}\right)\left(\cos\frac{2\pi y}{b}-1\right)$	1.50	$3.85+5\left(\frac{a}{b}\right)^2+8\left(\frac{a}{b}\right)^4$
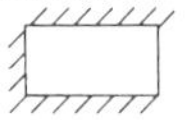	$\left(1-\cos\frac{\pi x}{2a}\right)\left(\cos\frac{2\pi y}{b}-1\right)$	0.340	$0.0468+0.340\left(\frac{a}{b}\right)^2+1.814\left(\frac{a}{b}\right)^4$
	$\left(\cos\frac{2\pi x}{a}-1\right)\sin\frac{\pi y}{b}$	0.75	$4+2\left(\frac{a}{b}\right)^2+0.75\left(\frac{a}{b}\right)^4$
	$\left(\cos\frac{2\pi x}{a}-1\right)\frac{y}{b}$	0.50	$2.67+0.304\left(\frac{a}{b}\right)^2$
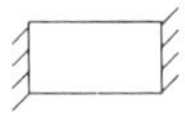	$\cos\frac{2\pi x}{a}-1$	1.50	8
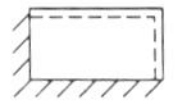	$\left(\cos\frac{3\pi x}{2a}-\cos\frac{\pi x}{2a}\right)\left(\cos\frac{3\pi y}{2b}-\cos\frac{\pi y}{2b}\right)$	1.00	$2.56+3.12\left(\frac{a}{b}\right)^2+2.56\left(\frac{a}{b}\right)^4$
	$\left(\cos\frac{3\pi x}{2a}-\cos\frac{\pi x}{2a}\right)\left(1-\cos\frac{\pi y}{2b}\right)$	0.227	$0.581+0.213\left(\frac{a}{b}\right)^2+0.031\left(\frac{a}{b}\right)^4$
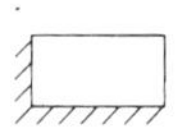	$\left(1-\cos\frac{\pi x}{2a}\right)\left(1-\cos\frac{\pi y}{2b}\right)$	0.0514	$0.0071+0.024\left(\frac{a}{b}\right)^2+0.0071\left(\frac{a}{b}\right)^4$
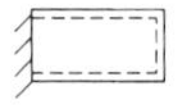	$\left(\cos\frac{3\pi x}{2a}-\cos\frac{\pi x}{2a}\right)\sin\frac{\pi y}{b}$	0.50	$1.28+1.25\left(\frac{a}{b}\right)^2+0.50\left(\frac{a}{b}\right)^4$
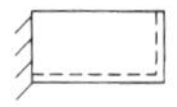	$\left(\cos\frac{3\pi x}{2a}-\cos\frac{\pi x}{2a}\right)\frac{y}{b}$	0.333	$0.853+0.190\left(\frac{a}{b}\right)^2$
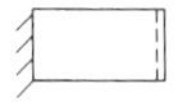	$\cos\frac{3\pi x}{2a}-\cos\frac{\pi x}{2a}$	1.00	2.56
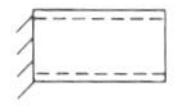	$\left(1-\cos\frac{\pi x}{2a}\right)\frac{\pi^2}{b^2}\sin\frac{\pi y}{b}$	0.1134	$0.0156+0.0852\left(\frac{a}{b}\right)^2+0.1134\left(\frac{a}{b}\right)^4$
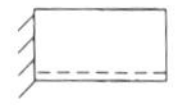	$\left(1-\cos\frac{\pi x}{2a}\right)\frac{y}{b}$	0.0756	$0.0104+0.0190\left(\frac{a}{b}\right)^2$
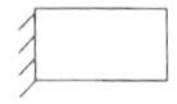	$1-\cos\frac{\pi x}{2a}$	0.2268	0.0313
	$\sin\frac{\pi x}{a}\sin\frac{\pi y}{b}$	0.25	$0.25+0.50\left(\frac{a}{b}\right)^2+0.25\left(\frac{a}{b}\right)^4$
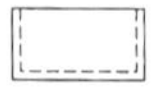	$\left(\sin\frac{\pi x}{a}\right)\frac{y}{b}$	0.1667	$0.1667+0.0760\left(\frac{a}{b}\right)^2$
	$\sin\frac{\pi x}{a}$	0.50	0.50

Reprinted, with permission, from Ref. 6: A. Leissa, *Vibration of Plates*, Acoustical Society of America, New York, 1993.
Key: clamped (C); simply-supported (SS); free (F).

$\kappa^2 S = I$, the moment of inertia of the beam, and $\rho = m/S$, is the mass per unit length of the beam.

In Table 8 the boundary conditions and the natural frequency equation along with the mode shape or eigenfunction are presented for several beam end conditions.

Plates The governing equation for bending waves in orthrotropic plates using the Euler–Bernoulli theory is given in Chapter 49, Section A.5. In the case of an isotropic plate

$$D_1 = D_3 = 2(D_\mu + 2D_G) = \frac{Eh^3}{12(1-\mu^2)} = D, \qquad (17)$$

and the governing differential equation becomes

$$D\,\nabla^4 \zeta_2 + \rho h \frac{\partial^2 \zeta_2}{\partial t^2} = 0. \qquad (18)$$

Note that E is Young's modulus, h is the thickness of the plate, μ is Poisson's ratio, ζ_2 is the displacement perpendicular to the plane of the plate, and ρ is the mass per unit volume of the plate. One can rewrite this equation in terms of x- and y-coordinates and $\tilde{\rho}$, the density per unit area, as

$$D\,\nabla^4 w + \tilde{\rho} \frac{\partial^2 w}{\partial t^2} = 0, \qquad (19)$$

where we have set $\zeta_2 = w$.

Altogether there are 21 combinations of simple boundary conditions [that is, clamped (C), simply supported (SS), or free (F)] for rectangular plates. In Table 8 the boundary conditions, mode shape, and factors for computing the natural frequency using $\omega^2 = (\pi^4 D/a^4 \rho)(K/N)$ are presented. The results presented here were computed using Rayleigh's method and hence yield only the upper bound on the fundamental frequencies.[6]

REFERENCES

1. M. P. Norton, *Fundamentals of Noise and Vibration Analysis for Engineers*, Cambridge University Press, Cambridge, 1989.
2. E. B. Magrab, *Vibrations of Elastic Structural Members*, Sijthoff & Nordhoff, MD, 1979.
3. R. D. Blevins, "Fluid Systems," in *Formulas for Natural Frequency and Mode Shape*, Van Nostrand Reinhold, New York, 1984.
4. L. Beranek and I. Ver, *Noise and Vibration Control Engineering*, Wiley, New York 1992.
5. F. S. Tse, I. E. Morse, and R. T. Hinkle, *Mechanical Vibrations*, Allyn and Bacon, Boston, 1978.
6. A. Leissa, *Vibration of Plates*, Acoustical Society of America, New York, 1993.

7

WAVEGUIDES

P. O. A. L. DAVIES

1 INTRODUCTION

This chapter is concerned with sound propagation in pipes and ducts, where, in contrast to free space, it is always strongly influenced or controlled by the presence of the confining boundaries. Pipe and duct networks feature widely in the engineering infrastructure of the process, power generation, automotive, and transport industries, of domestic services, and so on. In many such practical applications, the frequencies of interest remain sufficiently low, so that the wavelength remains a fraction of the transverse dimensions of the pipe or waveguide. Then the propagating waves remain plane, or one dimensional. When such is not the case, wave propagation may occur in one or more higher order modes. Both possibilities are addressed in this chapter.

This account of methods for the description and calculation of sound propagation in waveguides, or ducts, begins with an outline of the classical approach to establishing the guiding physical principles and analytical techniques. This is essentially a search for solutions of the appropriate form of the wave equation that satisfy the boundary conditions imposed by the duct walls.

2 CLASSICAL THEORY AND ASSUMPTIONS

The usual assumptions of classical theory are as follows:

1. The acoustic medium is a frictionless, homogeneous (ideal) fluid.
2. The processes associated with the wave motion are isentropic.
3. Fluctuating pressure amplitudes are sufficiently small that the linearizing acoustic assumptions remain valid.
4. The wave propagation remains wholly axial and directed along x.
5. The duct walls are rigid (acoustically hard), continuous, and of infinite axial extent.
6. The duct is rectangular or cylindrical with constant dimensions.

The classical account of waveguide acoustic behavior continues to the end of Section 5, with the application to lined ducts in Section 6.

The account continues with a sequence of selected examples, since in practical application due consideration must also be given to one or more of the particular features of the problem, such as the following:

1. Uniform sections of duct are of finite length L.
2. Boundary conditions will then include wave reflection and transmission at terminations and other discontinuities.
3. There are significant effects arising from the presence of a mean flow.
4. Thermal and viscous effects may introduce significant damping.
5. The temperature and flow conditions may vary axially or transversely within the duct.
6. Relevant geometric and other features such as axial taper or irregularities exist in the wall shape or properties.

Excluded here, however, is any discussion of the sources of excitation (see, e.g., Chapter 8). These selected examples are followed in Section 11 by a discussion of para-

Note: References to chapters appearing only in the *Encyclopedia* are preceded by *Enc.*

metric models of waveguide performance, and then a short summary in Section 12 completes the chapter.

2.1 The Wave Equation

The generation of propagating sound waves in a duct containing fluid by a source in the duct is subject to certain conditions. First the fluctuating acoustic pressure, density, and velocities must satisfy an appropriate wave equation, which itself must satisfy the conservation of momentum, mass, and energy in the bulk of the fluid. The second condition is associated with the physical boundary conditions at the duct walls, where the fluctuating pressure and the components of the displacement of the fluid near the duct wall must equal those on the duct wall. It is customary to apply the continuity of velocity rather than displacement, since for simple harmonic motion the two conditions are identical except when there is relative motion between the fluid close to the wall and the wall. The third condition is the boundary at the source, where the pressure and displacement fields on the source side and on the fluid side must be continuous. Waves reflecting from the walls interfere with each other, giving rise to interference patterns over the cross section of the duct. An infinite number of such modes is possible, but at a given frequency only a finite number can propagate down the duct.

The sound field in the waveguide must satisfy the wave equation

$$\frac{D^2p}{Dt^2} = c^2 \nabla^2 p, \tag{1}$$

where $p = p(x_i, t)$ is the fluctuating acoustic pressure at the coordinate position x_i, c is the adiabatic speed of sound, and the symbols D/Dt denote the material derivative. In classical acoustics it is normal to assume that there is no mean flow, so that Eq. (1) becomes

$$\frac{\partial^2 p}{\partial t^2} = c^2 \nabla^2 p, \tag{2}$$

when assumptions 1–6 above are included. With the existence of a mean flow having a uniform time-averaged axial velocity u_0 so that $Dp/Dt = \partial p/\partial t + u_0\, \partial p/\partial x$, while $D^2p/Dt^2 = (\partial/\partial t + u_0\, \partial/\partial x)\, Dp/Dt$, the term on the left side of Eq. (1) becomes

$$\frac{D^2p}{Dt^2} = \frac{\partial^2 p}{\partial t^2} + 2u_0 \frac{\partial^2 p}{\partial x\, \partial t} + u_0^2 \frac{\partial^2 p}{\partial x^2}, \tag{3}$$

which is relevant to situations of practical interest considered later. When significant additional mean velocity gradients are present, as may be the case with bends or area and other discontinuities, it will be necessary to include the associated factors in the expansion of D^2p/Dt^2; see, for example Section 16.5 of Ref. 1 and Ref. 2. Finally, it is worthwhile noting that a wave equation expressing the distribution in space and time of the scalar fluctuating velocity potential φ may be more readily derived and more easily and conveniently solved than the corresponding form with the acoustic pressure p. See, for example, Section 9.

2.2 Application of the Wave Equation

To apply the wave equation to a specific problem, it is appropriate to select the coordinate system that conveniently matches the wall boundaries when expressing the Laplacian $\nabla^2 p$ occurring on the right side of Eq. (1) or (2). Thus with rectangular ducts the rectangular coordinates x, y, z are appropriate, so that $\nabla^2 = \partial^2/\partial x^2 + \partial^2/\partial y^2 + \partial^2/\partial z^2$. Similarly, with straight cylindrical pipes the selection of cylindrical polar coordinates x, r, θ gives

$$\nabla^2 = \frac{\partial^2}{\partial x^2} + \frac{1}{r}\frac{\partial}{\partial r}\left\{r\frac{\partial}{\partial r}\right\} + \frac{1}{r^2}\frac{\partial^2}{\partial \theta^2}. \tag{4}$$

With conical expansions and contractions spherical polar coordinates are appropriate, and so on. Similar considerations apply to the selection of the appropriate finite elements when these are used to obtain numerical solutions, for example as explained in Chapter 16 of Ref. 1 and Chapter 7 of Ref. 3, both of which have extensive sets of relevant references, such as the sequence by Astley and Eversman. An alternative numerical scheme may be based on boundary elements.[4]

A review of appropriate mathematical methods of solution, with listed references illustrating their application, can be found in Section 5 of Ref. 5. The separation-of-variables method was chosen for what follows here, since insight is thus maintained into the distribution of the fluctuating pressure and velocity associated with the modes that can exist in any specific case.

3 MODES OF A RECTANGULAR WAVEGUIDE

Consider a duct extending to infinity on the x-axis but bounded at $y = 0$ and l_y and $z = 0$ and l_z by acoustically hard (rigid) duct walls. Note that it is convenient here to have the axis run along one corner. It is assumed that the medium inside the duct is homogeneous and has no time-averaged motion so that the sound field must satisfy Eq. (2). Since this equation is linear in $p(x, y, z, t)$

with coefficients that are functions only of the selected spatial coordinates x, y, z, no generality is lost if it is Fourier transformed on the time t. Then let p be represented by the product of four functions each depending on one variable only, so that

$$p(x, y, z, t) = X(x)Y(y)Z(z)\exp(i\omega t), \quad (5)$$

where ω is the radian frequency of excitation. From Eqs. (2) and (5) one obtains

$$\frac{1}{X}\frac{d^2X}{dx^2} + \frac{1}{Y}\frac{d^2Y}{dy^2} + \frac{1}{Z}\frac{d^2Z}{dz^2} + k^2 = 0, \quad (6)$$

where the first factor is a function of x only, the second only of y, and the third only of z, and $k^2 = (\omega/c)^2$. One can investigate each spatial dependency in turn, with the corresponding constants denoted by $-k_x^2$, $-k_y^2$, and $-k_z^2$. One then has $(1/X)(d^2X/dx^2) = -k_x^2$, and similarly, for the other two space functions, and so for Eq. (6) to be satisfied, one must have

$$k^2 = k_x^2 + k_y^2 + k_z^2. \quad (7a)$$

Corresponding to this, one can define a vector wavenumber $\mathbf{k}$ with the x, y, z components k_x, k_y, k_z.

As mentioned, with $X(x)$ now to be determined as the solution of the ordinary differential equation $d^2X/dx^2 + k_x^2X = 0$, one can take

$$X(x) = A^+\exp(-ik_xx) + A^-\exp(ik_xx), \quad (7b)$$

where A^+ and A^- correspond, respectively, to the complex amplitudes of the component waves traveling along the waveguide in the direction of positive and negative x, respectively.

In the same way, one finds that the spatial dependency in the y- and z-directions is described respectively by

$$Y(y) = C_1\cos(k_yy) + C_2\sin(k_yy), \quad (7c)$$
$$Z(y) = D_1\cos(k_zz) + D_2\sin(k_zz). \quad (7d)$$

The forms of solution in Eq. (7b) and in Eqs. (7c) and (7d) are equivalent, but the latter form is more convenient when acoustically hard boundary conditions at the wall are applied.

3.1 Boundary Conditions

The spatial distribution in the y- and z-directions must also satisfy the boundary conditions imposed by the walls. Thus with hard walls, the particle displacement and hence, for simple harmonic fluctuations, the fluctuating particle velocity normal to them must be zero. This can be related to the fluctuating pressure by satisfying dynamic equilibrium (Euler's equation) at the wall, which is expressed for the y-direction and associated particle velocity v by $\rho\ \partial v/\partial t = -\partial p/\partial y$, giving $v = (-1/i\omega\rho)(\partial p/\partial y)$, with a similar expression for the velocity w in the z-direction. The hard-wall boundary conditions are satisfied for $\partial p/\partial y = 0$ when $y = 0, l_y$ and also for $\partial p/\partial z = 0$ when $z = 0, l_z$. Substitution of these conditions into Eqs. (7c) and (7d) leads to $C_2 = D_2 = 0$, $k_y = m\pi/l_y$, $k_z = n\pi/l_z$, m, $n = 0, 1, 2, \ldots$. The integers m and n indicate the number of fluctuating pressure amplitude minima (nodes) in the spatial standing-wave distribution across the duct in the y- and z-directions, respectively. Thus $m = n = 0$ corresponds to plane-wave propagation with the fluctuating pressure in phase with the velocity and at constant phase over the whole cross section.

With the inclusion of the wall boundary conditions and substitution of all three spatial dependencies defined by Eqs. (7b)–(7d) into Eq. (5), acoustic propagation in a hard-walled rectangular waveguide in the direction of positive x is expressed by

$$p(x, y, z, t) = A_{mn}^+\cos\left(\frac{m\pi y}{l_y}\right)\cos\left(\frac{n\pi z}{l_z}\right) \cdot \exp[i(\omega t - k_{x,mn}x)], \quad (8)$$

with $k_{x,mn}^2 = k^2 - (m\pi/l_y)^2 - (n\pi/l_z)^2$ and A_{mn}^+ the amplitude distribution in the (m, n) duct mode. Wave propagation in the opposite direction (negative x) is described by replacing the factors $A^+\exp[i(\omega t - k_xx)]$ by $A^-\exp[i(\omega t + k_xx)]$ in Eq. (8). Both components normally occur together in ducts of finite length as sets of interfering axially propagating waves.

4 MODES OF A CYLINDRICAL WAVEGUIDE

Consider a hard-walled cylindrical duct of radius a with its axis aligned along x. Substituting the appropriate expression [Eq. (4)] for the Laplacian $\nabla^2 p$ into Eq. (3) leads to the wave equation

$$\frac{\partial^2 p}{\partial x^2} + \frac{1}{r}\frac{\partial p}{\partial r} + \frac{\partial^2 p}{\partial r^2} + \frac{1}{r^2}\frac{\partial^2 p}{\partial \theta^2} - \frac{1}{c^2}\frac{\partial^2 p}{\partial t^2} = 0. \quad (9)$$

Again using the method of separation of variables described in Section 3, one writes

$$p(x, r, \theta, t) = X(x)R(r)\Theta(\theta)\exp(i\omega t). \quad (10)$$

Substituting Eq. (10) into Eq. (9) leads to

$$\frac{1}{X}\frac{d^2X}{dx^2}+\frac{1}{R}\left(\frac{d^2R}{dr^2}+\frac{1}{r}\frac{dR}{dr}\right)+\frac{1}{r^2\Theta}\frac{d^2\Theta}{d\theta^2}+k^2=0, \qquad (11)$$

where $k^2=(\omega/c)^2$. Proceeding initially as before with rectangular ducts, letting $-k_x^2=(1/X)(d^2X/dx^2)$, one finds that the resulting axial distribution function $X(x)$ is again given by Eq. (7b). Selecting next the circumferential distribution function Θ, one notes that the periodicity in the coordinate θ together with the physical limitation on the sound field that it be single valued at any position θ imply that

$$\Theta(\theta)=\exp(\pm im\theta), \qquad m=0,1,2,\ldots, \qquad (12a)$$

so that $p(x,r,\theta,t)$ is represented by a Fourier series in θ, with coefficients depending on r, x, t. The integer m is the number of circumferential modes (or radial nodal lines) in the harmonic complex pressure amplitude. The plus-or-minus sign in Eq. (12a) corresponds to interference patterns rotating in the negative/positive θ directions, respectively.

The radial distribution function $R(r)$ must then satisfy

$$\frac{d^2R}{dr^2}+\frac{1}{r}\frac{dR}{dr}+\left\{k_r^2-\left\{\frac{m}{r}\right\}^2\right\}R=0, \qquad (11a)$$

where $k_r^2=k^2-k_x^2$, with the radial wavenumber k_r constant. A transformation of variable from r to $k_r r$ converts this to a standard form of Bessel's equation, for which the appropriate solution for a cylindrical duct is

$$R(k_r r)=J_m(k_r r), \qquad k_r \neq 0, \qquad (12b)$$

where $J_m(k_r r)$ is the Bessel function of the first kind of order m. The complete solution to Eq. (11a) may also include a corresponding Bessel function of the second kind, the Neumann functions N_m, which have values that are infinite for zero argument. Thus the latter set are omitted for cylindrical ducts but should be included[6] in the solution for annular ducts.

Some general features of Bessel functions $J_m(\psi)$, $\psi=k_r r$ in terms of their order m and argument ψ, are of practical interest. That of zero order ($m=0$), which is plotted in Fig. 1, differs from all those of higher order ($m>0$),

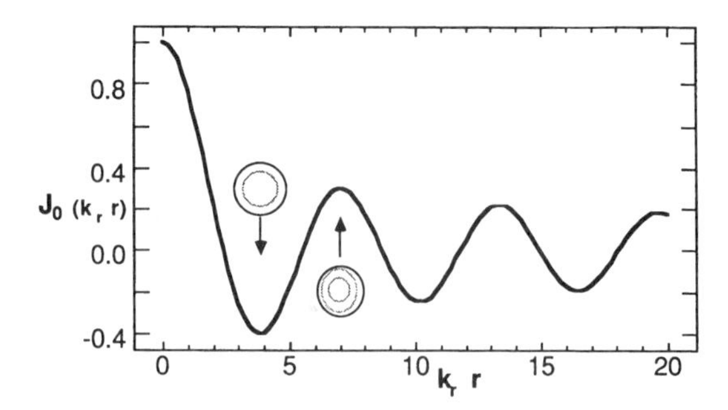

Fig. 1 $J_0(k_r r)$, defining radial pressure distribution in a cylindrical waveguide with no spinning modes ($m=0$). The two insets give the positions of the nodal lines for $n=1$ and $n=2$ with rigid walls.

being unity at $\psi=0$ and oscillating as ψ increases in the general manner of a damped cosine wave. Those with $m>0$ are all zero at $\psi=0$, increasing in value with ψ to a peak at approximately $\psi=m+1$ and then again continuing to oscillate in the manner of a damped cosine wave, as can be seen for $m=9$ in Fig. 2. Finally, the slope $J'_m(\psi)=d[J_m(\psi)]/d\psi$ is zero at $\psi=0$ in all cases except when $m=1$, when it is 0.5.

Finally, it remains to specify the values of $k_r r$ that satisfy the boundary conditions imposed by the duct walls. With hard walls, as for rectangular ducts, we require $\partial p/\partial r=0$ at the duct wall at radius a corresponding to the values of $k_r r$ where the gradient $d[J_m(\psi)]/d\psi$ is zero. These are designated as the physically tenable values $k_r=k_{mn}$, where $n=0, 1, 2, 3, \ldots$ and k_{mn} represents the nth root of the corresponding eigenequation. Collecting the solutions (7b), (12a), and (12b) and substituting them in Eq. (10) leads to the pressure pattern

$$p(x,r,\theta,t)=A_{mn}^{\pm}J_m(k_{mn}r)\exp\ i(\omega t\mp m\theta\mp k_x x), \qquad (13)$$

where $A^{\pm}$ and the subsequent minus-or-plus signs relate respectively to the positively and negatively spinning- and propagating-wave patterns [see also the discussion following Eqs. (8) and (12a)].

The influence of the presence of spinning modes on the radial distribution of fluctuating pressure is demonstrated by comparing Fig. 2, where there are nine present, with Fig. 1, where there are none. The relative positions of the nodal lines associated respectively with one and two radial modal circles in the two figures demonstrate the progressive concentration of acoustic energy toward the duct periphery that is found to occur as the number of spinning modes increases. A comprehensive study of the sound field in rigid-walled annular ducts can be found in Ref. 6.

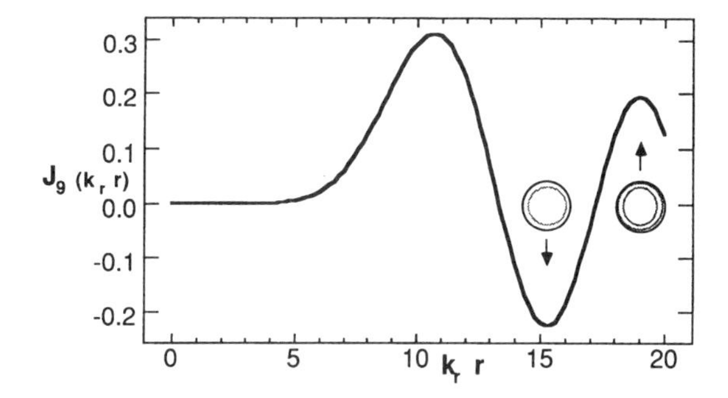

Fig. 2 $J_q(k_r r)$, defining radial pressure distribution in a cylindrical waveguide with nine spinning modes ($m = 0$). The two insets give the position of the nodal lines for $n = 1$ and $n = 2$ with rigid walls.

5 AXIALLY PROPAGATING MODES, CUTOFF MODES

Although an infinite number of modes are physically possible, only those whose axial wavenumbers k_x remain wholly real can propagate indefinitely, while the remainder will decay exponentially. The modes that propagate can be identified by reference to the appropriate condition for rigid-walled rectangular ducts, which is [see Eq. (8)]

$$k_{x,mn}^2 = \left(\frac{\omega}{c}\right)^2 - \left(\frac{m\pi}{l_y}\right)^2 - \left(\frac{n\pi}{l_z}\right)^2 > 0. \quad (14)$$

Similarly, for cylindrical ducts,

$$k_{x,mn}^2 = \left(\frac{\omega}{c}\right)^2 - (k_{r,mn})^2 > 0. \quad (15)$$

These equations demonstrate that the plane-wave mode ($m = n = 0$) always propagates, while higher order modes will not propagate unless condition (14) or (15) is satisfied.

At frequencies for which the condition is not satisfied, one has $k_{x,mn} = \mp i|k_{x,mn}|$, with minus or plus respectively, in the $\pm x$-direction. Thus the modes decay axially at an exponential rate with distance away from a source or reflection location. The axially decaying modes are called cutoff modes, with the frequency at which this occurs for each one called the cutoff frequency. Axial propagation will take place in modes with frequencies of excitation above the cutoff frequency, and these modes are then termed cut-on.

With rectangular ducts the condition for propagation becomes $\omega/c > \pi/l_y$ when l_y is the larger duct dimension, that is, when the free-field wavelength $\lambda_f < 2l_y$.

TABLE 1 Values of $k_{mn}a$ for $J'_m(k_{mn}a) = 0$ Corresponding to Cut-on of Axial Propagation in Rigid Cylindrical Waveguide of Radius a

	$n =$				
J_m	0	1	2	3	4
J_0	0	3.83	7.02	10.17	13.32
J_1	1.84	5.33	8.54	11.71	14.86
J_2	3.05	6.71	9.97	13.17	16.35
J_3	4.20	8.02	11.35	14.59	17.79
J_4	5.32	9.28	12.68	15.96	19.20
J_9	10.71	15.29	19.00	22.50	25.89

The corresponding conditions for cylindrical ducts, with radius a, can be established by reference to Table 1. This lists the first few values of $k_{mn}a$, for which $J'_m(k_{mn}a)$ is zero, thus defining the cut-on frequencies for the relevant modes. Noting that $k = \omega/c$, one sees from the table that the first purely spinning mode ($m = 1, n = 0$) propagates so long as the reduced frequency $\omega a/c > 1.84$, the second ($m = 2, n = 0$) when $\omega a/c > 3.05$, and so on.

Alternatively, inspection of Eq. (13) shows that at any fixed axial position x, the angular velocity of spin $\Omega_p = \partial\theta/\partial t = \omega/m$, while the corresponding peripheral velocity at the wall is $\omega a/m$ with Mach number $M_p = \omega a/mc$. Since such spinning modes propagate when $\omega a/c > \phi(m)$, where the values of $\phi(m)$ are given in the table, an alternative condition that such modes should propagate is $M_p > \phi(m)/m$. Noting too that $\phi(m) \approx (m+1)$ with $n = 0$, an approximate condition is that the spinning modes will propagate so long as $M_p > (m+1)/m$. In general, one can show that the cutoff condition represents a very rapid decay rate for such modes, which diminishes as $M_p^2 \to 1$.

6 LINED DUCTS

When the walls are not rigid and are lined with an absorbing material, the boundary conditions at the wall or lining surface require that the fluctuating pressures and normal fluid displacement (or velocity) at the boundary surface must equal those on the surface. The absorbent properties of the boundary are usually characterized by the normal acoustic impedance Z_b, which is defined as the ratio of acoustic pressure to normal velocity at the surface.

6.1 Lined Rectangular Waveguides with Absorbing Boundaries

If the walls are symmetrically lined with respect to the duct axis, it is more convenient to have a coordinate system based on the axis. The acoustic modes will be symmetric or antisymmetric about the axis, giving $\partial p/\partial y = 0$ and $\partial p/\partial z = 0$ for symmetric and $p = 0$ for antisymmetric modes for rectangular cross-sectional ducts at $y = z = 0$. The boundary conditions at $y = \pm\frac{1}{2}l_y$ become for

$$Z_{by} = \rho c \frac{k}{k_y} i \cot(k_y l_y/2), \quad \text{for symmetric modes,} \tag{16a}$$

$$-Z_{by} = \rho c \frac{k}{k_y} i \tan(k_y l_y/2) \quad \text{for antisymmetric ones,} \tag{16b}$$

with similar expressions for Z_{bz}. These transcendental equations are known as eigenequations, having an infinite number of solutions yielding k_{ym} and k_{zm}, which have complex values when the boundaries are absorbing. The axial wavenumber will then be

$$k_{mn}^2 = k^2 - k_{ym}^2 - k_{zn}^2 = (\beta_{mn} - i\alpha_{mn})^2, \tag{17}$$

which represents an axially decaying wave motion due to energy absorption at the walls. The pressure distribution can be obtained by the appropriate substitution in Eq. (8). A sharp cutoff frequency does not exist for this case as all modes decay to some extent.

6.2 Lined Cylindrical Waveguides

The corresponding eigenequation for a lined cylindrical duct with a boundary impedance Z_a, which is normally complex, is given by

$$ikJ_m(k_r a) + \frac{Z_a}{\rho c} k_r J_m'(k_r a) = 0, \tag{18}$$

which can be solved for k_r, which is in general complex. The prime denotes the derivative with respect to the argument. The axial wavenumber components k_{mn} of the wave vector then follow from Eq. (15), where these are in general also complex. Their imaginary parts then determine the axial decay rates for the cut-on modes, as with Eq. (17), with again no sharp cutoff frequency.

7 DUCTS WITH UNIFORM AXIAL FLOW

Sound propagation through flow ducts is described by the appropriate form of the convected wave equation (1). Solutions of the wave equation for a cylindrical waveguide with a steady uniform axial flow with velocity u_0 are presented here to illustrate the significant effects on axial wave propagation of the existence of a mean flow. The wave equation (1) is then expressed by

$$\left[\frac{D^2}{Dt^2} - c^2\left(\frac{\partial^2}{\partial x^2} + \frac{\partial^2}{\partial r^2} + \frac{1}{r}\frac{\partial}{\partial r} + \frac{1}{r^2}\frac{\partial^2}{\partial \theta^2}\right)\right] p = 0, \tag{19}$$

with D^2p/Dt^2 expressed by Eq. (3) where the velocity u_0 is explicitly included. Sound propagation in rectangular ducts is described by substituting the appropriate expression for the Laplacian $\nabla^2 p$ in Eq. (1), while solutions for more general cases can be found among the references cited.

7.1 Axial Wavenumber k_x

For rigid boundaries in a cylindrical duct, substitution of Eq. (10) into Eq. (19) followed by separation of variables leads to the general eigenequation for X,

$$(1 - M^2)\frac{\partial^2 X}{\partial x^2} - 2ikM\frac{\partial X}{\partial x} + (k^2 - k_r^2)X = 0,$$

where the flow Mach number $M = u_0/c$ and k_r^2 and k^2 are both constants. This has a solution $X(x) = A\exp(-ik_x x)$, with k_x expressed as

$$k_x = \frac{-kM \pm [k^2 - (1 - M^2)k_r^2]^{0.5}}{1 - M^2}, \tag{19a}$$

which, with $M = 0$, is the result previously expressed by Eq. (15) found with rigid boundaries. In this case the pressure pattern in the duct will be described by Eq. (13) with the appropriate substitution for k_x. We note, from Eq. (19a), that k_x is wholly real when $k^2 > (1 - M^2)k_r^2$. Thus the frequency at which modes become cut on is reduced by the factor $(1 - M^2)^{0.5}$. Also, by analogy with the results in Sections 2 and 3, the axial wavenumber k_x for a rigid rectangular duct is given by the substitution of $k_y^2 + k_z^2$ for k_r^2 in Eq. (19a), with a consequent reduction to the frequency of cut-on by the same factor for the propagating modes described by Eq. (14).

With plane-wave excitation and propagation one sets $k_r = 0$, so that $k_x = k(M \pm 1)/(1 - M^2)$. Thus with propagation in the direction of positive x, $k_x = k/(1 + M)$, and in the direction of negative x, $k_x = -k/(1 - M)$. Thus the wavelength is stretched relative to that in the free field for waves traveling in the direction of flow and contracted for waves traveling in the reverse direction, due to the convection by the mean flow.

7.2 Radial Wavenumber k_r in Lined Ducts with Mean Flow

To investigate the convective effects of the mean flow on the values of the radial wavenumber k_r (and thus on k_x), one must first adopt the appropriate matching conditions at the interface. One cannot match particle velocity in this case since, with particle displacement ζ one finds in general that $u = D\zeta/Dt = \partial\zeta/dt + u_0\ \partial\zeta/\partial x$ in the flow while in the lining u_b is simply $\partial\zeta/\partial t$, without a convective term. Expressing conservation of radial momentum (Euler's equation) in terms of ζ in the lining, one has $(-1/\rho)(\partial p/\partial r) = \partial u_0/\partial t = \partial^2\zeta/\partial t^2$, with $u_b = d\zeta/\partial t = i\omega\zeta$ and $\zeta = u_b/i\omega = p/i\omega Z_b$, where Z_b is the surface impedance of the lining. In the flow, conservation of radial momentum is expressed by $(-1/\rho)\,\partial p/\partial r = D^2\zeta/\partial t^2$. Hence, matching radial displacements ζ at the interface $r = a_i$, we find the boundary condition to be satisfied is expressed by $(1/\rho)(\partial p/\partial r)+(1/i\omega Z_b)\,D^2p/Dt^2 = 0$.

With this boundary condition one finds that the radial wavenumber k_r must satisfy

$$\frac{Z_b}{\rho c}\frac{\partial^2 p}{\partial r\,\partial t} + \frac{1}{c}\left(\frac{\partial}{\partial t} + u_0\frac{\partial}{\partial x}\right)^2 p = 0, \qquad r = a_i.$$

The corresponding values of k_r are then given by

$$\frac{Z_b}{\rho c}\,k_r J'_m(k_r a_i) + ik\left(\frac{1 - Mk_x}{k}\right)^2 J_m(k_r a_i) = 0. \qquad (20)$$

The implications of this result show that when the wall impedance is high, so that the radial displacements remain relatively insignificant at the wall, the modal patterns approach those for rigid boundary since $J'_m(k_r a_i) \to 0$. However, the radial wave numbers found with soft walls and hence the radial pressure patterns are modified by the convective effects of mean flow, while the circumferential modes are not. A similar result is found for lined rectangular ducts.[3] The combined effects of soft walls and sheared flow have been described by Ko.[7] Similar studies[8] have been made for annular flow ducts.

7.3 Sound Power Flux

The usefulness of sound power stems from the fact that the total acoustic energy, the sum of its kinetic and potential energy components, is conserved if the acoustic medium is lossless (i.e., for an inviscid non-heat-conducting fluid). In a stationary acoustic medium the sound power W is defined as $\int_S I_i\,dS_i$, where $I_i = \langle pu_i\rangle$ are the components of the sound intensity, which is the time-averaged sound energy flux, with $\langle\cdot\rangle$ denoting the time averaging. In a moving medium, however, the sound energy flux includes transport of the mean energy density and intensity by the mean flow. Provided viscous and thermal diffusion effects are negligible and the flow is steady, irrotational, and uniform, the sound power flux per unit area normal to the flow, or the sound intensity,[9] is given by

$$I = (1 + M^2)\langle pu\rangle + M\left(\frac{\langle p^2\rangle}{\rho c} + \rho c\langle u^2\rangle\right). \qquad (21)$$

With propagation in several modes simultaneously, care may be necessary when performing the averages or in interpreting overall sound power measurements. Intensity probes do not appear to give reliable measurements in this environment so that other methods of measurement should be adopted.[10]

7.4 Sound Propagation with Vibrating Walls

When the duct walls are not rigid but react elastically to the fluctuating internal pressure field, the cross section will vary with time according to $S(t) = \pi[a(t)]^2$ for a cylindrical tube with radius a. With mean flow neglected, the phase speed c_v will be defined[11] by

$$\left(\frac{1}{c_v}\right)^2 = \left(\frac{1}{c_0}\right)^2 + \left(\frac{1}{c_w}\right)^2 = \left(\frac{1}{S_0}\right)\left|\frac{d(\rho S)}{dp}\right|_{p=p_0},$$

where p_0 and S_0 are, respectively, the undisturbed or time-averaged pressure and area, c_0 is the usual isentropic sound speed, and c_w is a perturbation resulting from the wall vibration. Adopting the linear approximation $|dS/dp|_{p=p_0} = 2\pi a_0 |da/dp|_{p=p_0}$ and a locally reacting wall with (per unit area) mass m, resistance R, and stiffness K, the perturbation in phase velocity becomes

$$c_w^2 = \frac{a_0}{2\rho_0}\,[m(\omega_0^2 - \omega^2) + iR\omega],$$

where $\omega_0 = (K/m)^{0.5}$. The corresponding wavenumber modified by wall vibration is, to first order,

$$k_v = \left(\frac{\omega}{c_0}\right)\left(1 + \frac{\rho_0 c_0^2/a_0}{m(\omega_0^2 - \omega^2) + iR\omega}\right). \qquad (22)$$

One should note that this approximation is not strictly valid when the excitation frequency $\omega = \omega_0$, the resonance frequency of the wall, since radiation damping of the vibration may then make a significant contribution to R.

8 TERMINATIONS AND SOURCES

In practice, a waveguide comprises one or more sections of uniform or regularly shaped duct, each of length L_i, say, $i = 1, 2, \ldots$, but in general of different cross section, connected in sequence between the primary source of excitation and the final termination. Axial sound propagation in each individual section is described by relations such as Eqs. (8) and (13) once the modal distribution has been related to the shape and physical characteristics of the walls as well as to the frequency and to the properties of the acoustic medium and where relevant to the mean flow. The values of the two sets of arbitrary constants $A^{\pm}_{mn}$ that these equations include depend on the boundary conditions at the ends of each section. Thus they depend on the factors controlling excitation and wave reflection at the source end, those controlling reflection and wave transfer at the junctions between the component elements, and those controlling wave reflection and transmission at the final termination. The acoustic response of any uniform duct of finite length to any specific form of excitation is strongly influenced by the boundary conditions at this termination as well as its length. (See Section 11.) Thus it is logical to begin acoustic analysis of the system at its final termination,[10] where conditions are normally well defined and invariant, and then proceed along the system toward the source.

The termination may be characterized by its termination impedance $Z_{mn} = p_{mn}/u_{mn}$, by its radiation resistance, by its pressure reflection coefficient A^{-}_{mn}/A^{+}_{mn}, or by the corresponding power reflection coefficient for each mode. In the absence of reflections (i.e., $A^{-}_{mn} = 0$), the termination impedance is equal to $\rho ck/k_{mn}$. Specification of the reflection and transfer characteristics at the termination and at any junction between elements requires that the complete set of expressions describing the sound field are matched at the transfer plane for continuity of pressure and displacement[10,12–15] (or alternatively when appropriate, the velocity). Inclusion of the relevant boundary conditions demonstrates that, in general, the matching at such discontinuities involves the addition of further modes to the reflected and transmitted waves, though these may decay rapidly and thus not propagate. An interesting result[13] is that any well-cut-on mode of higher order suffers little reflection but carries on past the termination as if it were continuing down an infinite duct of the same cross section.

A primary sound source may be represented by a distribution of fluctuating velocity or pressure or of both together. Similar matching conditions apply at the source plane, so that, for example, the sum of the component velocities over all modes must equal the source velocity. The sound power transferred from the source to the waveguide in each mode will depend on the complex load impedance offered by the duct at the source plane with the corresponding internal impedance of the source (see Section 11.2). Sound may also be generated or amplified at other positions along flow ducts by various mechanisms,[16,17] as a result of the transfer of energy from the mean motion by the action of shear layers, with their associated vorticity. Those sources associated with turbulent mixing, such as those of boundary layer noise, tend to be broadband while those associated with resonant or reactive behavior are tonal in character. Such secondary sources may be situated at specific locations or distributed along elements of the system.

9 AXIAL SOUND PROPAGATION THROUGH PRACTICAL WAVEGUIDES

The classical approach to the analysis of sound propagation in realistic waveguides with complex geometry normally neglects mean flow and thus ignores features that may be significant in practice. In some instances, the boundary conditions have also been oversimplified, as for example with applications of an analogy with lumped electrical networks. A more physically valid approach is considered here, illustrated by reference to specific examples. For instance, flow duct elements, or the junctions between them, may concern rapid or distributed axial changes in the duct geometry or in the magnitude and direction of the wave propagation vector. Acoustic transfer can be described by satisfying conservation of mass, energy, and momentum both within and at the bounding surfaces of an appropriate control volume.[10] Successful predictions that correlate well with observation have generally been confined to plane waves or where significant energy transfer between the modes does not occur.[18] More recent examples are described in Ref. 19, which includes some 20 contributed papers, and others are summarized in a survey by Cummings[20] with many more relevant references.

As noted in Section 7, flow ducts may be represented as a sequence of uniform or geometrically regular elements with discontinuities at their junctions. For simplicity, the discussion presented here is restricted to axial acoustic propagation along the elements and across the junctions between them. The pressure density and velocity will all be functions of axial displacement $\mathbf{x}$ and time t and expressed as the sum of mean time-averaged and fluctuating parts, with pressure expressed as $p_0(\mathbf{x}) +$

$p(\mathbf{x}, t)$, and so on. For potential mean flow along the elements with velocity potential $\phi_0(\mathbf{x})+\phi(\mathbf{x}, t)$, one has $u_0(\mathbf{x}) + u(\mathbf{x}, t) = -\partial\phi_0(\mathbf{x})/\partial\mathbf{x} - \partial\phi(\mathbf{x}, t)/\partial\mathbf{x}$, the first term representing the irrotational mean flow velocity $u_0(\mathbf{x})$ and the second with zero time average, the fluctuating particle velocity. As before, the pressures, velocity, and so on, will be Fourier transformed on the time t and the discussion will refer to spectral components of frequency $\omega = 2\pi f$.

Doak[21] has shown that given any three-dimensional geometric region (regular element) in which there is a low-Mach-number mean potential flow for which the velocity potential is $\phi_0(\mathbf{x})$, then the Fourier spectral density $\Phi(\mathbf{x}, \omega)$ of any acoustic velocity potential $\phi(\mathbf{x}, t)$, due only to sources on the terminations of the region is given to first order in the mean flow Mach number $-\nabla\phi_0(\mathbf{x})/c_0$ by

$$\Phi(\mathbf{x}, \omega) = \exp\left[-i\left(\frac{k}{c_0}\right)\phi_0(\mathbf{x})\right]\varphi_0(\mathbf{x}, \omega), \tag{23}$$

where $\varphi_0(\mathbf{x}, \omega)$ satisfies the (no-flow) scalar Helmholtz equation $(\nabla^2 + k^2)\varphi_0(\mathbf{x}, \omega) = 0$. In addition, the approximate spectral density $\Phi(\mathbf{x}, \omega)$ of the acoustic velocity potential [and hence $\varphi_0(\mathbf{x}, \omega)$] must satisfy the appropriate boundary conditions at both end junctions of the element (region). For example, with a uniform flow duct of constant cross section, Eq. (23) becomes[10]

$$p(x, \omega) = p^+(0, \omega)\exp(ik^*Mx)[\exp(-ik^*x) + r_0 \cdot \exp(ik^*x)]\exp(i\omega t), \tag{24}$$

where $p^+(0, \omega)$ is the complex amplitude of the component wave incident at the downstream termination $x = 0$, where the reflection coefficient is r_0 and $k^* = (\omega/c)/(1 - M^2)$. This describes the superposition of a standing wave and a traveling wave.

9.1 Wave Transfer across Junctions and Other Discontinuities

Where the junctions between elements represent an area or any other discontinuity, a part of the incident-wave energy is reflected, with the remainder transmitted or dissipated, though generally the associated processes remain almost isentropic so that the resulting entropy fluctuations may be negligibly small. Following the procedure outlined above, one first defines an appropriate control volume V with surface S at the junction, including any relevant fixed boundaries. With axial wave motion, the integral relations describing conservation of mass and momentum over the surface S are expressed as

$$\int_S \rho' u'\, dS = 0, \tag{25a}$$

$$\int_S [\rho'(u')^2 + p']\, dS = \sum F, \tag{25b}$$

where $\rho' = \rho_0 + \rho$, $u' = u_0 + u$, $p' = p_0 + p$, and $\sum F$ represents the externally applied boundary forces, which are zero for rigid surfaces. With isentropic processes the acoustic pressure and density fluctuations are related by $c_0^2\rho = p$. When the entropy fluctuations are significant, though small, one can represent the fluctuating entropy[2] by $\varepsilon = (-1/\rho_0 T_0)\delta/(\gamma - 1)$ and then $c_0^2\rho = (p + \delta)$. Conservation of energy is expressed as

$$\int_S [h' + 0.5(u')^2]\, dS = 0, \tag{25c}$$

$$h' = h_0 + \frac{p}{\rho_0} + T_0\varepsilon, \tag{25d}$$

where T_0 is the ambient absolute temperature.

To evaluate the acoustic characteristics of the junction, one first subtracts the terms relating to mean motion from Eqs. (25a)–(25d) and then discards all fluctuating terms of order higher than the first (the acoustic approximation). Having expressed fluctuating velocity u, density ρ, and pressure p in terms of the equivalent progressive component wave amplitudes p^+ and p^-, one Fourier transforms the result on the time t. These three equations, with any necessary equations defining the boundary conditions, are then solved for each spectral component $p_1^\pm$ on one side in terms of $p_2^\pm$ on the other and, where relevant, δ. Where the shape of the wavefront changes across the junction, for example, from spherical to plane,[22] V should represent a finite volume within which the readjustment takes place. When the propagating waves remain plane on both sides, V may also be reduced to a plane, with an appropriate end correction included at the junction,[10] or alternatively, an appropriate set of higher order modes[18,23] may be introduced there. Such additions are necessary to account for the presence of evanescent waves required to satisfy the boundary conditions, with any corresponding adjustments to the wavefronts from one side to the other.

The transmission of plane waves at an abrupt area expansion with outflow provides a common example of such wave transfer. An analytic solution[18] that includes mean flow but with the restrictive assumption that reflected waves are absent beyond the junction shows fair agreement with corresponding observations.[24] Another analytic solution[23] that avoids this assumption but in which mean flow is neglected shows good agreement with measurements without flow. An alternative approach[10,24] that also provides predictions that agree well with all these and further sets of measurements, including those with mean flow, represents the evanescent waves and the associated fluctuating mass at the junction by an appropriate end correction. Similar comparisons[25,10] can be made concerning transfer at an expansion with reversal of the direction of propagation or with the addition of a tuned side branch[10,23] at the discontinuity.

Examples of extended control volumes include the ends of conical expansions with the associated transfers between spherical and plane-wave motion,[22] the analysis for which remains valid provided the flow does not separate from the walls. This has been shown to provide predictions that compare well with observation[19] for Mach numbers up to 0.2. An earlier analysis of sound propagation in a flow duct with perforated walls backed by an enclosed annular cavity provides predictions that also closely match observation.[26,27] Here the acoustic field within the lined duct has been matched to that within the surrounding annulus by specifying the impedance[26,28] of the perforated wall with grazing flow. Other examples (e.g., describing the acoustic performance with various arrangements of absorbing linings) can be found in Ref. 19 as contributed papers with extensive lists of relevant references.

10 VISCOUS AND THERMAL INFLUENCES

The propagation of sound in ducts subject to viscous and thermal effects was first described by Kirchhoff in 1868. A comprehensive review of the problem with a numerical solution for tubes of Kirchhoff's equation has been presented by Tijdeman[29] and supplemented by further comment by Kergomard.[30] The main parameters governing sound propagation in rigid cylindrical tubes are found to be the shear wave, or Stokes, number $S_n = a(\omega/\nu)^{0.5}$, where a is the tube radius and ν is the kinematic viscosity of the medium, as well as the reduced frequency, or Helmholtz, number $\omega a/c$, together with γ, the ratio of the specific heats and $\sigma = 1/\sqrt{P_r}$, with P_r the Prandtl number for the medium. The solution with Kirchhoff's wide-tube approximation,[31] which is shown to be appropriate for $S_n > 10$, replaces the free-space wave propagation constant $k = \omega/c$ by $\beta = k + \alpha(1 - i)$, where α is given by

$$\frac{\alpha}{k} = \frac{(1/S_n)[1 + (\gamma - 1)\sigma]}{\sqrt{2}}. \tag{26}$$

For air the factor $[1+(\gamma - 1)\sigma]/\sqrt{2}$ is close to unity. The resulting wave dispersion (i.e., the dependence of wave speed on frequency) represented by the factor α/k is inversely proportional to the Stokes number and thus to the square root of frequency. Thus these effects become relatively less significant as the frequency of excitation increases although α increases in proportion to the Stokes number (but see Ref. 30 for very high frequencies). An extension to tubes of arbitrary cross section is provided by Stinson.[32] With narrow tubes, the viscous effects predominate, and both dispersion and attenuation become dominant. In the presence of a turbulent mean flow, the damping of acoustic waves is influenced to some extent by the action of the turbulent stresses near the wall.[33] With smooth walls, so that a laminar sublayer is present, the influence depends on the value of the dimensionless acoustic boundary layer thickness $\delta^+ = (2u_\tau^2/\nu\omega)^{1/2}$, where u_τ is the wall friction velocity $(\tau_0/\rho_0)^{1/2}$ and τ_0 the wall shear stress. With $\delta^+ < 12.5$, say, observations indicate that the influence of wall shear stress may be neglected.[33] Thus with the exception of smooth walls with low frequencies of excitation and perhaps high-mean-flow Mach numbers or for very narrow tubes, the influence of mean pipe friction losses on sound propagation remains small. When such is the case, a useful first approximation to plane-wave propagation along flow ducts is expressed by Eq. (24) when k is replaced by β evaluated from Eq. (26).[10,19] Note that the Froude friction factor term given in Ref. 3, and first introduced in Eq. 1.99 there, is based on a quasi-static assumption that applies only at very low frequencies.

Flow ducts with axial temperature gradients provide a common practical problem in which both the flow and physical properties vary with gas temperature and hence with axial distance x along the duct. Thus, associated with the flow temperature $T(x)$ there will be a corresponding axial velocity $u(x,t)$, density $\rho(x,t)$, pressure $p(x,t)$, and entropy $s(x,t)$, with corresponding changes in gas properties. Relations expressing conservation of mass, momentum, and energy for a general case are respectively

$$\frac{1}{\rho}\frac{D\rho}{Dt} + \frac{\partial u_i}{\partial x_i} = 0, \tag{27a}$$

$$\frac{Du_i}{Dt} + \frac{1}{\rho}\frac{\partial p}{\partial x_i} - \frac{1}{\rho}\frac{\partial e_{ij}}{\partial x_j} = 0, \tag{27b}$$

$$\frac{\rho\, Ds}{Dt} - \frac{\partial}{\partial x_j}\left(\frac{k_T}{T}\frac{\partial T}{\partial x_j}\right) = \frac{1}{T}\left[\frac{k_T}{T}\left(\frac{\partial T}{\partial x_j}\right)^2 + e_{ij}\frac{\partial u_i}{\partial x_j}\right], \tag{27c}$$

$$\rho = \rho(s, p), \tag{27d}$$

where k_T is the thermal conductivity of the gas and e_{ij} is the viscous stress tensor. If one assumes small viscosity and neglects diffusion, conservation of energy simplifies to $\rho(Ds/Dt) = 0$, with corresponding changes to Eq. (27b). With one-dimensional flow in the x-direction $D/Dt = \partial/\partial t + u_0\,\partial/\partial x$, as for Eq. (3). Solutions[34] for this problem with plane waves and moderate temperature gradients, so that $T_x = T_0(1 - \alpha x)$, $\alpha x < 0.1$, justify subdividing the duct into appropriately short axial segments, with flow conditions and gas properties in each segment represented by their averaged properties. An approximate solution for plane waves with an exponential axial temperature distribution has been derived by Cummings.[35]

Wave propagation where the fluctuating pressure amplitude is sufficiently large to introduce significant changes in the gas properties and speed of sound represents a nonlinear problem that is best approached in the time domain. Though the general literature on this subject is extensive, successful predictions of continuous sound propagation under these conditions appear to be lacking so far, except for some rather special circumstances.[36, 37]

11 CALCULATION AND SPECIFICATION OF WAVEGUIDE PERFORMANCE

If is often convenient for acoustic performance assessment (e.g., during design optimization for noise control) to employ compact descriptions summarizing the acoustic behavior of the waveguide or each of its constituent elements. For simplicity, the discussion will be confined to plane-wave propagation, since the principles and approach can readily be extended to higher order modes. Acoustic plane-wave transmission across an individual element or sequence of elements is illustrated in Fig. 3. Here S represents the source of excitation for the element or sequence T, for which Z represents the acoustic impedance at its termination. With isentropic plane waves the complex amplitudes p^+ and p^- are related to the corresponding acoustic pressure p and particle velocity u by

$$p = p^+ + p^-, \tag{28a}$$

$$\rho c u = p^+ - p^-, \tag{28b}$$

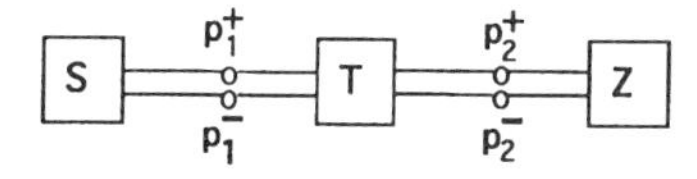

Fig. 3 Plane acoustic wave transfer across a waveguide or one of its elements.

Equation (28a) is valid[10] irrespective of the presence of a steady mean flow u_0, while Eq. (28b) is similarly valid so long as viscothermal effects remain negligible.

The wave components $p_1^\pm$ on the source side of T can be related to $p_2^\pm$ on the load side by

$$p_1^+ = T_{11}p_2^+ + T_{12}p_2^-, \tag{29a}$$

$$p_1^- = T_{21}p_2^+ + T_{22}p_2^-, \tag{29b}$$

where T_{11}, $T_{12}T_{21}$, and T_{22} are the four elements of the scattering transmission matrix $[T]$ defining the transfer. The complex values of these four elements are functions of geometry, of the undisturbed values of ρ and c, of the frequency of excitation, of the mean flow Mach number M, but remain independent of the value of Z. Obviously Eqs. (28a) and (28b) imply that the fluctuating acoustic pressure p_1 and velocity u_1 on the source side of T can be related to p_2 and u_2 on the load side in terms of the corresponding impedance transfer matrix $[t]$ by expressions analogous to Eqs. (29a) and (29b). The relative difficulty of measuring acoustic particle velocity u, compared with measurements of p combined with the restricted validity[10] of Eq. (28b), suggests that the scattering matrix $[T]$ provides a more robust description of the measured wave transfer than does the impedance matrix $[t]$.

To return to Fig. 3, we see that acoustic wave transmission across T can also be expressed in terms of the transmission coefficients

$$T_i = \frac{p_1^+}{p_2^+}, \tag{30a}$$

$$T_r = \frac{p_1^-}{p_2^-}, \tag{30b}$$

together with the reflection coefficient r, where

$$r = \frac{p_2^-}{p_2^+} \tag{31a}$$

or

$$r = \frac{\zeta - 1}{\zeta + 1}, \tag{31b}$$

and $\zeta = Z/\rho c$, also noting $p_1^-/p_1^+ = rT_r/T_i$. The transmission coefficients are related to the elements of the scattering matrix $[T]$ by

$$T_i = T_{11} + rT_{12}, \tag{32a}$$

$$T_r = \frac{T_{21}}{r} + T_{22}, \tag{32b}$$

and it is evident that T_i and T_r are both functions of r and hence of Z as well as of the element's geometry, ρ, c, f, and M. To evaluate the elements of $[T]$, one can easily show[38] that calculation or measurement of T_i and T_r with two different loads Za, Zb are necessary unless $[T]$ possesses reciprocal properties. This cannot be true when mean flow is present or when viscothermal or flow-associated losses are significant and, as it turns out, when the element is not geometrically reciprocal. Thus transfer coefficients might be preferred to transfer matrix representations, despite the apparent convenience of the latter's independence of the load Z. Finally, one should note that both representations remain independent of the source of excitation, so long as it remains external to the element.

11.1 Single-Parameter Descriptions of Acoustic Performance

Single-parameter descriptions have several practical uses. Normally they define the acoustic attenuation of the system or element T or of its transmission loss index TL or insertion loss index IL, all expressed in decibels. The transmission loss of an acoustic element is usually defined as the difference in power flux in free space between that incident on and that transmitted across it and is then an invariant property of the element. With reference to the element T in Fig. 3, after establishing anechoic conditions by setting $Z = \rho c$ or $r = 0$, then

$$\mathrm{TL} = 10 \log_{10} \frac{S_1 I_1}{S_2 I_2}, \tag{33}$$

where the intensity or power flux may be found from Eq. (21) and S_i is the cross-sectional area of the duct.

The insertion loss is the measured change in power flux at a specified receiver when the acoustic transmission path between it and the source is modified by the insertion of the element, provided that the impedances of both source and receiver remain invariant. If W_1 is the power at the receiver before the insertion of the element and W_2 that measured or calculated afterward, then the insertion loss index is defined by

$$\mathrm{IL} = 10 \log_{10} \frac{W_1}{W_2}. \tag{34}$$

In practice, it is normally necessary to provide a reference duct system of known performance to measure or calculate W_1 and then repeat this procedure with the element or system of interest. This also implies that the source impedance should be known and invariant; see Eqs. (39) and (40). Since the impedance presented at the source by the reference system will normally differ from that under test, due compensation to the results may be necessary to account for any corresponding changes to the source when its impedance is not effectively infinite.

A simple index of performance, which remains independent of an external source but is dependent only on the acoustic characteristics of the waveguide with its operating load Z, is the attenuation index AL, expressed as

$$\mathrm{AL} = 20 \log_{10} |T_i|, \tag{35}$$

which can provide a convenient guide when the source impedance is unknown.

11.2 Waveguide Excitation

The transfer of acoustic energy or sound power from the source to the system is strongly influenced by the acoustic impedance or load that the system presents to the source. One recalls too that the acoustic characteristics of both system and source are functions of the frequency, so the discussion that follows here applies strictly to each spectral component, although it also applies in general to all of them. The acoustic circuit corresponding to Fig. 3 has been extended to include two representations of the source in Fig. 4.

In the acoustic circuit in Fig. 4*a* the source is now represented by a fluctuating volume velocity with amplitude V_s and has an internal effective shunt impedance Z_e, while the system impedance at the source plane $Z_1 = p_1/u_1$. This representation corresponds to a fluctuating mass injection with an effective volume velocity $V_s = u_s S_s$. One notes that at the source plane the circuit model represents continuity of pressure with the source strength equated to a discontinuity of volume velocity. Thus this is equivalent to an acoustic monopole source in free space. The power output of this source is given by

$$W_r = 0.5\,\mathrm{Re}(p_1^* V_s) = \frac{0.5|V_s^2|\,\mathrm{Re}[Z_1/(1 + Z_1/Z_e)]}{S_s}, \tag{36a,b}$$

where the asterisk represents the complex conjugate. Alternatively, when the source is associated with fluctuating aerodynamic forces, the circuit model in Fig. 4*b* now represents continuity of velocity u_S across the source plane with the source strength f_s equated to the discontinuity of pressure $(P_s - p_1)$ acting over area S_s, or $(P_s - p_1)S_s$. This is equivalent to an acoustic dipole source in free space. The power output is given by

$$W_f = 0.5\,\mathrm{Re}(f_s^* u_s) = 0.5|f_s^2|\,\mathrm{Re}(M_l), \tag{37a,b}$$

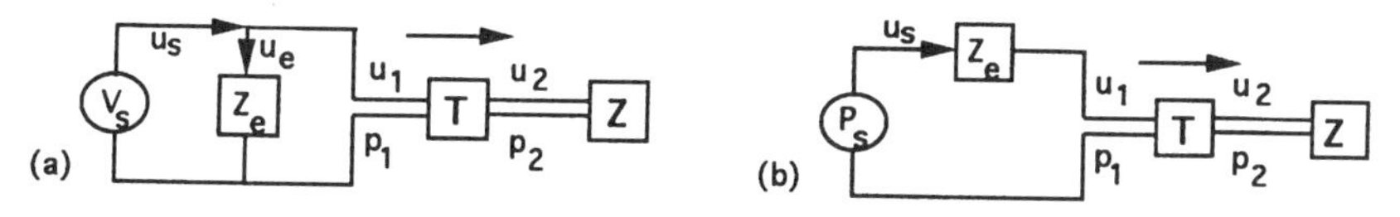

Fig. 4 Acoustic circuits for system excitation: (*a*) by fluctuating volume velocity; (*b*) by fluctuating aerodynamic forces.

where the *input mobility* $M_I = u_s/f_s$, which also equals $1/Z_eS_s$ for the circuit in Fig. 4*b*. Analogous expressions can be derived for the sources of excitation by mixing noise associated with turbulent shear layers.

In practice, the source of excitation may include a combination of factors related to fluctuating mass, fluctuating force, and flow turbulence. In all such cases the associated acoustic power delivered by the source to the system depends on the magnitude of the source impedance, among other factors, while the acoustic emission from the open termination to the surroundings depends also on the efficiency of power transfer through it. To illustrate the influence of the source and system impedance on such emissions, consider again the acoustic circuit in Fig. 4*a*. At the source plane the mass flux balance is given by

$$u_1 = u_s - \frac{p_1}{Z_e} \tag{38a}$$

or

$$Z_e u_s = p_1 + \rho_0 c_0 u_1 \zeta_e, \tag{38b}$$

where $\zeta_e = Z_e/\rho_0 c_0$. Substitution for p_1 and $\rho_0 c_0 u_1$ in terms of the corresponding wave component amplitudes $p_1^\pm$ from Eqs. (28a) and (28b) into Eq. (38b) and then making use of Eqs. (30a) and (30b) and Eq. (31a), one obtains

$$p_2^+ = \frac{Z_e u_s}{(1+\zeta_e)T_i + (1-\zeta_e)rT_r}. \tag{39}$$

If W_1 is the observed reference power and W_2 is the observed power with system T_2, when a reference system T_1 is replaced by another system T_2, provided the termination impedance Z and thus r remain constant and also similarly for Z_e and u_s, one finds that

$$\frac{W_1}{W_2} = \left(\frac{(p_2^+)_1}{(p_2^+)_2}\right)^2, \tag{40}$$

which can be substituted for W_1/W_2 in Eq. (34).

12 CONCLUSIONS

Any practical waveguide consists of a sequence of elements having uniform and regularly shaped boundaries connected by discontinuities with various configurations including expansions, branches, resonators, changes in bounding wall impedance, and so on. Wave energy is transmitted along the duct from a source or distributed sources of excitation to its final termination normally open to the surroundings. Provided that the fluctuating pressure amplitude remains a sufficiently small fraction of the ambient pressure in the medium, wave propagation is well described by adopting the acoustic approximation and formulated with linear models. Provided also that the transverse dimensions remain a small fraction of the acoustic wavelength, a further valid approximation is that wave propagation along each element remains effectively one dimensional.

A physically realistic analysis of sound propagation along waveguides must begin with a complete analytic model of the associated fluid motion, that is, the equations of mass, linear momentum, and energy transport and the fluid's constitutive equations, taken together with the geometric and surface properties of the boundaries. In many cases, useful and sufficiently valid descriptions are obtained by assuming that transfer processes remain isentropic and the system is time invariant. With the acoustic approximation, sound propagation along the regular elements can then be described for small mean flow Mach number by a linearized wave equation that may normally be Fourier transformed on the time t. The resulting solutions will then describe the spectral characteristics of the acoustic wave motion in such elements.

Similar spectral descriptions of wave transfer across discontinuities may also be derived by matching the associated fluid motion over appropriate surfaces on either side of the discontinuity, which together with any relevant fixed boundaries form a control volume. The equations expressing conservation of mass, energy, and momentum are then expressed in integral form. The processes associated with the boundary conditions may include losses as well as generation of nonpropagating (evanescent) higher order modes, whose influence, normally reactive, must be included. One classical method of doing so is by the addition of an appropriate end correction at such junctions. Many applications cited in the literature reduce the control volume to a plane, with continuity of pressure and particle velocity across it, the latter replacing continuity of particle displacement, though this simplification may not always be realistic.

ACKNOWLEDGMENT

The author expresses his thanks to M. J. Fisher for his helpful suggestions and discussion during the preparation of this chapter.

REFERENCES

1. R. G. White and J. G. Walker, *Noise and Vibration*, Ellis Horwood, Chichester, 1982.
2. P. Munger and G. M. L. Gladwell, "Acoustic Wave Propagation Is a Sheared Fluid Contained in a Duct," *J. Sound. Vib.*, Vol. 9, 1969, pp. 28–58.
3. M. L. Munjal, *Acoustics of Ducts and Mufflers*, Wiley, New York, 1987.
4. R. D. Ciskowski and C. A. Brebia (Eds.), *Boundary Element Methods in Acoustics*, Elsevier, London, 1991.
5. W. Rostafinski, "Monograph on Propagation of Sound Waves in Curved Ducts," NASA reference publication 1248, 1991.
6. C. L. Morfey, "Rotating Pressure Patterns in Ducts: Their Generation and Transmission," *J. Sound Vib.*, Vol. 1, 1964, pp. 60–87.
7. S. H. Ko, "Sound Attenuation in Acoustically Lined Circular Ducts in the Presence of Uniform Flow and Shear Flow," *J. Sound Vib.*, Vol. 22, 1972, pp. 193–210.
8. S. H. Ko, "Theoretical Prediction of Sound Attenuation in Acoustically Lined Annular Ducts in the Presence of Uniform Flow and Shear Flow," *J. Acoust. Soc. Am.*, Vol. 54, 1974, pp. 1592–1606.
9. C. L. Morfey, "Acoustic Energy in Non Uniform Flows," *J. Sound Vib.*, Vol. 14, 1971, pp. 159–170.
10. P. O. A. L. Davies, "Practical Flow Duct Acoustics," *J. Sound Vib.*, Vol. 124, 1988, pp. 91–115.
11. M. J. Lighthill, *Waves in Fluids*, Cambridge University Press, Cambridge, 1978.
12. P. E. Doak, "Excitation, Transmission and Radiation of Sound from Ducts," *J. Sound Vib.*, Vol. 31, 1973, I, pp. 1–72, II, pp. 137–174.
13. C. L. Morfey, "A Note on Radiation Efficiency of Acoustic Duct Modes," *J. Sound Vib.*, Vol. 9, 1969, pp. 367–372.
14. P. Munger, H. E. Plumblee, and P. E. Doak, "Analysis of Acoustic Radiation in a Jet Flow Environment," *J. Sound Vib.*, Vol. 36, 1974, pp. 21–52.
15. R. H. Munt, "Acoustic Transmission Properties of a Jet Pipe with Subsonic Flow. I. The Cold Jet Reflection Coefficient," *J. Sound Vib.*, Vol. 142, 1990, pp. 413–436.
16. E. A. Müller (Ed.), *The Mechanics of Sound Generation in Flows*, Springer, Berlin, 1979.
17. P. O. A. L. Davies, "Flow Acoustic Coupling in Ducts," *J. Sound Vib.*, Vol. 77, 1981, pp. 191–209.
18. A. Cummings, "Sound Transmission at a Sudden Area Expansion with Mean Flow," *J. Sound Vib.*, Vol. 38, 1975, pp. 149–155.
19. H. G. Jonasson (Ed.), *Science for Silence*, Vol. 1: *Proc Internoise 90*, Göteburg, Sweden, Noise Control Foundation, NY, 1990, pp. 517–702.
20. A. Cummings, "Prediction Methods for the Performance of Flow Duct Silencers," in H. G. Jonasson (Ed.), *Science for Silence*, Göteburg, Sweden, Noise Control Foundation, NY, 1990, pp. 17–38.
21. P. E. Doak, "Acoustic Wave Propagation in a Homentropic Irrotational Low Mach Number Mean Flow," *J. Sound Vib.*, Vol. 155, 1992, pp. 545–548.
22. P. O. A. L. Davies and P. E. Doak, "Wave Transfer to and from Conical Diffusers with Mean Flow," *J. Sound Vib.*, Vol. 138, 1990, pp. 345–350.
23. K. S. Peat, "The Acoustical Impedance at the Junction of an Extended Inlet or Outlet Duct," *J. Sound Vib.*, Vol. 150, 1991, pp. 101–110.
24. D. Ronneberger, "Experimentelle Untersuchungen zum akustiskhen Reflexionsfaktor von unstetigen Querschnittsänderungen in einem luftdurchströmten Rohr," *Acoustica*, Vol. 19, 1967/68, pp. 222–235.
25. A. Cummings, "Sound Transmission in a Folded Annular Duct," *J. Sound Vib.*, Vol. 41, 1975, pp. 375–379.
26. J. W. Sullivan, "A Method of Modelling Perforate Tube Muffler Components," *J. Acoust. Soc. Am.*, Vol. 66, 1979, I, pp. 772–778, II, pp. 779–788.
27. J. L. Bento Coelho, "Acoustic Characteristics of Perforate Liners in Expansion Chambers," Ph.D. Dissertation, University of Southampton, 1983.
28. J. L. Bento Coelho and P. O. A. L. Davies, "Prediction of Acoustical Performance of Cavity Backed Perforate Liners in Flow Ducts," *Proc 11 Institute of Acoustics Meeting, UK*, 1983, pp. 317–321.
29. H. Tijdemann, "On the Propagation of Sound Waves in Cylindrical Tubes," *J. Sound Vib.*, Vol. 39, 1975, pp. 1–33.
30. J. Kergomard, "Comments on Wall Effects on Sound Propagation in Tubes," *J. Sound Vib.*, Vol. 98, 1985, pp. 149–152.
31. Lord Rayleigh, *Theory of Sound*, Macmillan, London 1896.
32. M. Stinson, "The Propagation of Plane Sound Waves in Narrow and Wide Circular Tubes and Generalization to Uniform Tubes of Arbitrary Cross-Sectional Shape," *J. Acoust. Soc. Am.*, Vol. 89, 1991, pp. 550–558.
33. M. C. A. M. Peters, A. Hirshberg, and A. P. J. Wijnands, "The Aero-acoustic Behaviour of an Open Pipe Exit," *Proc. Noise-93*, Vol. 6, 1993, pp. 191–196.
34. P. O. A. L. Davies, "Plane Acoustic Wave Propagation

in Hot Gas Flows," *J. Sound Vib.*, Vol. 122, 1988, pp. 389–392.

35. A. Cummings, "Ducts with Axial Temperature Gradients," *J. Sound Vib.*, Vol. 51, 1977, pp. 55–68.
36. P. O. A. L. Davies and G. Jiajin, "Finite Amplitude Wave Reflection at an Open Exhaust," *J. Sound Vib.*, Vol. 141, 1990, pp. 165–166.
37. N. Sugimoto, "Burgers Equations with Fractional Derivatives; Hereditary Effects on Nonlinear Acoustic Waves," *J. Fluid Mech.*, Vol. 225, 1991, pp. 631–653.
38. P. O. A. L. Davies, "Transmission Matrix Representation of Exhaust System Acoustic Characteristics," *J. Sound Vib.*, Vol. 151, 1991, pp. 333–338.

8

STEADY-STATE RADIATION FROM SOURCES

A. P. DOWLING

1 INTRODUCTION

Insight into the behavior of many practical sound sources, such as vibrating surfaces, jet flows, and combustion, can be obtained by considering elementary sources. This chapter begins by discussing sources that are so small in comparison with the wavelength of the sound they produce that the sources can be considered as concentrated at a single point.

The simple monopole point source was introduced in Chapter 2. It is an acoustic source that leads to an omnidirectional sound field and causes an unsteady volume outflow of fluid from any surface enclosing the source. A small pulsating sphere is one practical realization of a point monopole source. The characteristics of the sound field generated by a monopole source are discussed in more detail in this chapter, with particular emphasis on the case where the source is periodic in time. Dipole and quadrupole point sources are introduced. These are less effective than monopoles at generating a distant sound field, a characteristic that has important implications in low-Mach-number aeroacoustics (see Part III).

Many sources of sound involve vibrating surfaces, and analytical results for some simple geometries are reviewed. These include pulsating and vibrating spheres and cylinders and baffled pistons. The different forms of the sound field radiated from these surfaces at low and at high frequencies provide physical insight into practical acoustic sources involving vibrating surfaces, such as loudspeakers and active sonar systems.

The chapter concludes with a discussion of the effects of source motion on the radiated sound field.

Note: References to chapters appearing only in the *Encyclopedia* are preceded by *Enc.*

2 MONOPOLE POINT SOURCE

2.1 Sound Pressure Field

A source that is concentrated at a point and produces an omnidirectional sound field is called a *simple source* or a *monopole point source*. It generates a pressure field consisting of spherical wavefronts traveling out from the source with the sound speed c, with amplitude decaying inversely with distance from the source. The pressure at position $\mathbf{x}$, a distance r away from a monopole point source of *strength* $Q(t)$, is[1–6] given as

$$p_M(\mathbf{x}, t) = \frac{Q(t - r/c)}{4\pi r} \tag{1}$$

at time t; the pressure perturbation a distance r away from the source is related to the source strength at the earlier time $t - r/c$.

If $Q(t)$ is simple harmonic with frequency ω, it is convenient to write the function $Q(t)$ in terms of its complex amplitude $\hat{Q}$. Then $Q(t) = \mathrm{Re}(\hat{Q}e^{i\omega t})$ and Eq. (1) reduces to

$$p_M(\mathbf{x}, t) = \frac{\hat{Q}}{4\pi r} e^{i\omega(t - r/c)}. \tag{2}$$

The real part of $\hat{Q}e^{i\omega t}$ represents the actual source strength, and the real part of the complex pressure in Eq. (2) describes the actual pressure.

2.2 Velocity Field

The particle velocity $\mathbf{u}_M(\mathbf{x}, t)$ produced by a monopole point source is radially outward from the source and can be readily calculated from the radial momentum equation $\rho\, \partial u_{Mr}/\partial t = -\partial p_M/\partial r$ to give[1–6]

$$u_{Mr}(\mathbf{x}, t) = \frac{1}{\rho c}\left(1 + \frac{1}{ikr}\right)\frac{\hat{Q}}{4\pi r}\, e^{i\omega(t - r/c)}$$
$$= \frac{1}{\rho c}\left(1 + \frac{1}{ikr}\right) p_M(\mathbf{x}, t), \qquad (3)$$

where $k = \omega/c$ is the wavenumber and ρ is the mean density.

In the *near field*, the region $kr \ll 1$, the velocity field varies with the inverse square of distance from the source and lags the pressure perturbation by 90°.

In the *far field*, $kr \gg 1$, $u_{Mr} = p_M/\rho c$, recovering the unidirectional plane-wave result.

2.3 Equivalence to Injection of Fluid

Integration of the radial velocity over the surface of a small sphere centered on the source shows that there is a net outward volume flow rate from the source of $\hat{Q}e^{i\omega t}/i\omega\rho$. A point monopole is equivalent to the injection of fluid at the point, the monopole source strength being equal to the product of the density and the rate of change of the volume flux.

This equivalence highlights the main characteristic of any physical source that approximates to a point monopole. It must produce an unsteady volume outflow from a region that is very small in comparison with the wavelength. Pulsating bubbles, sirens, and unsteady combustion are all examples of monopole sources, as are small bodies vibrating so that they undergo a change in volume.

2.4 Intensity[1–6]

The mean intensity I_M is entirely radial and decays with the inverse square of distance from the source. It is given by

$$I_{Mr} = \overline{p_M u_{Mr}} = \frac{\overline{p_M^2}}{\rho c} = \frac{|\hat{Q}|^2}{32\pi^2 \rho c r^2}. \qquad (4)$$

2.5 Sound Power[1–6]

Since the sound field is omnidirectional, the radiated sound power $P_M = 4\pi r^2 I_{Mr}$, that is,

$$P_M = \frac{|\hat{Q}|^2}{8\pi\rho c}. \qquad (5)$$

2.6 Inhomogeneous Wave Equation

The pressure field due to a monopole point source of strength $Q(t)$ at the origin satisfies the inhomogeneous wave equation[3–5]

$$\left(\frac{1}{c^2}\frac{\partial^2}{\partial t^2} - \nabla^2\right) p_M = Q(t)\delta(\mathbf{x}). \qquad (6)$$

In this equation ∇^2 is the Laplacian operator $\partial^2/\partial x^2 + \partial^2/\partial y^2 + \partial^2/\partial z^2$. The function $\delta(\mathbf{x})$ is zero for $\mathbf{x} \neq \mathbf{0}$, indicating that the source is concentrated at $\mathbf{x} = \mathbf{0}$. Away from the origin the pressure perturbations satisfy the usual homogeneous acoustic wave equation (see Chapters 1 and 2).

3 SUPERPOSITION OF MONOPOLE SOURCES

Since sound is a linear motion, the sound pressure due to two sources of strengths $Q_1(t)$ and $Q_2(t)$ at $\mathbf{y}_1$ and $\mathbf{y}_2$, respectively, can be found by adding their individual pressure fields:

$$p(\mathbf{x}, t) = \frac{Q_1(t - r_1/c)}{4\pi r_1} + \frac{Q_2(t - r_2/c)}{r\pi r_2}, \qquad (7)$$

where r_1 is the distance of $\mathbf{x}$ from the source at $\mathbf{y}_1$, $r_1 = |\mathbf{x} - \mathbf{y}_1|$, and r_2 is the distance of $\mathbf{x}$ from $\mathbf{y}_2$, $r_2 = |\mathbf{x} - \mathbf{y}_2|$.

Similarly the pressure field due to a distributed source can be determined by integration of the field of a single source.

3.1 Line Source

A line source concentrated on the 3-axis of uniform strength $q_l(t)$ per unit length produces a pressure perturbation

$$p_l(\mathbf{x}, t) = \int_{-\infty}^{\infty} \frac{q_l(t - |\mathbf{x} - \mathbf{y}|/c)}{4\pi|\mathbf{x} - \mathbf{y}|}\, dy_3, \qquad (8)$$

where $\mathbf{y} = (0, 0, y_3)$.

If $q_l(t)$ is simple harmonic, $q_l(t) = \hat{q}_l e^{i\omega t}$, Eq. (8) simplifies to

$$p_l(\mathbf{x}, t) = \frac{\hat{q}_l}{4i} H_0^{(2)}(k\sigma) e^{i\omega t}. \qquad (9)$$

Here, $H_n^{(2)}(z)$ is a Hankel function[7] of order n and σ is the radial distance of $\mathbf{x}$ from the 3-axis, $\sigma = (x_1^2 + x_2^2)^{1/2}$. The line source produces a volume outflow rate of $\hat{q}_l e^{i\omega t}/i\omega\rho$ per unit length of 3-axis.

In the near field, $k\sigma \ll 1$,

$$p_l(\mathbf{x}, t) \simeq -\frac{\hat{q}_l}{2\pi}\ln(k\sigma)e^{i\omega t}. \quad (10)$$

In the far field, $k\sigma \gg 1$.

$$p_l(\mathbf{x}, t) \simeq \sqrt{\frac{1}{8\pi k\sigma}}\,\hat{q}_l e^{i\omega(t-\sigma/c)-i\pi/4}. \quad (11)$$

Waves travel radially outward with speed c, their amplitude decaying with the inverse square root of distance from the line source.

Intensity and Power The mean intensity I_l is entirely radial and decays with the inverse of the radial distance from the 3-axis:

$$I_l = \frac{|\hat{q}_l|^2}{16\pi\rho\omega\sigma}. \quad (12)$$

The radiated sound power per unit length of the 3-axis is

$$P_l = \frac{|\hat{q}_l|^2}{8\rho\omega}. \quad (13)$$

A small length l of the line source radiates a sound power $l|\hat{q}_l|^2/8\rho\omega$. In contrast, Eq. (5) shows that the power radiated by a point source with the same volume outflow rate, $l\hat{q}_l e^{i\omega t}/i\omega\rho$, is $l^2|\hat{q}_l|^2/8\pi\rho c$. The power output of a small source of length l is enhanced by the factor $\pi(kl)^{-1}$ when it acts alongside other sources to produce a two-dimensional sound field.

3.2 Surface Source

A surface source concentrated in the $x_1 = 0$ plane of uniform strength $q_a(t)$ per unit area produces a pressure perturbation

$$p_a(\mathbf{x}, t) = \int_{-\infty}^{\infty}\int_{-\infty}^{\infty} \frac{q_a(t - |\mathbf{x} - \mathbf{y}|/c)}{4\pi|\mathbf{x} - \mathbf{y}|}\, dy_2\, dy_3, \quad (14)$$

where $\mathbf{y} = (0, y_2, y_3)$.

If $q_a(t)$ is simple harmonic, $q_a(t) = \hat{q}_a e^{i\omega t}$, Eq. (14) simplifies to

$$p_a(\mathbf{x}, t) = \frac{c\hat{q}_a}{2i\omega} e^{i\omega(t-|x_1|/c)}; \quad (15)$$

one-dimensional waves travel away from the $x_1 = 0$ plane with speed c and unchanging amplitude. This source produces a volume outflow rate of $\hat{q}_a e^{i\omega t}/i\omega\rho$ per unit area.

Intensity and Power[2] The mean intensity I_a is in the 1-direction and its amplitude is independent of x_1:

$$I_a = \operatorname{sgn}(x_1)\frac{c|\hat{q}_a|^2}{8\rho\omega^2}. \quad (16)$$

The radiated sound power per unit area is given by

$$P_a = \frac{c|\hat{q}_a|^2}{4\rho\omega^2}. \quad (17)$$

A small area A of the surface source radiates a sound power $Ac|\hat{q}_a|^2/4\rho\omega^2$. A point source with the same volume outflow rate of $A\hat{q}_a e^{i\omega t}/i\omega\rho$ only generates a sound $A^2|\hat{q}_a|^2/8\pi\rho c$. The power output of the source of small area A is enhanced by the factor $2\pi/k^2A$ when it acts alongside other surface sources to produce a one-dimensional sound field. The power output of a small loudspeaker is enhanced by a horn, which constrains the radiation to be one dimensional near the source.

3.3 Two Monopole Point Sources with Different Phase

Consider two monopole point sources of frequency ω and amplitude $\hat{Q}$ positioned a distance l apart, as shown in Fig. 1. One source leads the other by β, that is, $Q_1(t) = \hat{Q}e^{i(\omega t+\beta)}$ and $\hat{Q}_2(t) = \hat{Q}e^{i\omega t}$. These sources generate the pressure field

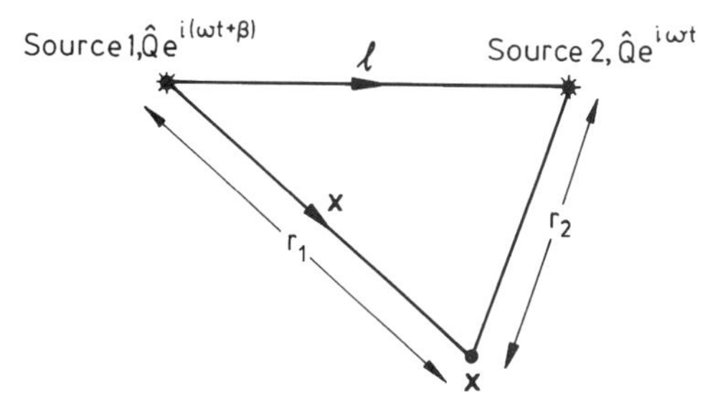

Fig. 1 Two monopole point sources a distance l apart.

$$p(\mathbf{x}, t) = \frac{\hat{Q}}{4\pi r_1} e^{i[\omega(t - r_1/c) + \beta]} + \frac{\hat{Q}}{4\pi r_2} e^{i\omega(t - r_2/c)}, \quad (18)$$

where r_1 is the distance from $\mathbf{x}$ to the source at the origin, $r_1 = |\mathbf{x}|$, and r_2 is the distance from $\mathbf{x}$ to the source at l, $r_2 = |\mathbf{x} - \mathbf{l}|$.

The sound power emitted by each individual source can be found by integrating the intensity over a small sphere centered on the source. This gives

$$P_1 = \text{sound power emitted by source 1}$$
$$= \frac{|\hat{Q}|^2}{8\pi\rho c}\left(1 + \frac{1}{kl}\sin(kl + \beta)\right), \quad (19a)$$

$$P_2 = \text{sound power emitted by source 2}$$
$$= \frac{|\hat{Q}|^2}{8\pi\rho c}\left(1 + \frac{1}{kl}\sin(kl - \beta)\right), \quad (19b)$$

$$P_\infty = \text{sound power radiated to infinity}$$
$$= P_1 + P_2 = \frac{|\hat{Q}|^2}{4\pi\rho c}\left(1 + \frac{1}{kl}\sin(kl)\cos\beta\right). \quad (19c)$$

Comparison with the single-source result in Eq. (5) shows that the sound power emitted by a source is altered when there is a coherent source nearby.

Special Cases

$$\text{(i)}\quad \beta = 0, \quad P_1 = P_2 = \frac{P_\infty}{2} = \frac{|\hat{Q}|^2}{8\pi\rho c}\left(1 + \frac{1}{kl}\sin(kl)\right). \quad (20)$$

The power output of a monopole source is increased when there is a second source of the same phase within half a wavelength, that is when $l < \frac{1}{2}\lambda$, where the wavelength $\lambda = 2\pi/k$:

$$\text{(ii)}\quad \beta = \tfrac{1}{2}\pi, \quad P_1 = \frac{|\hat{Q}|^2}{8\pi\rho c}\left(1 + \frac{1}{kl}\cos(kl)\right), \quad (21a)$$

$$P_2 = \frac{|\hat{Q}|^2}{8\pi\rho c}\left(1 - \frac{1}{kl}\cos(kl)\right), \quad (21b)$$

$$P_\infty = \frac{|\hat{Q}|^2}{4\pi\rho c}. \quad (21c)$$

There is now an exchange of energy between the two sources. Both sources emit power $|\hat{Q}|^2/8\pi\rho c$, which is radiated to infinity, but source 1 emits $\cos(kl)|\hat{Q}|^2/8\pi\rho ckl$ additional power, which is absorbed by source 2:

$$\text{(iii)}\quad \beta = \pi, \quad P_1 = P_2 = \frac{P_\infty}{2} = \frac{|\hat{Q}|^2}{8\pi\rho c}\left(1 - \frac{1}{kl}\sin(kl)\right). \quad (22)$$

The power output of a monopole source is decreased when there is a second source out of phase within half a wavelength, that is when $l < \lambda/2$. In particular, for $kl \ll 1$, Eq. (22) becomes

$$P_1 = P_2 = \frac{P_\infty}{2} = \frac{(|\hat{Q}|kl)^2}{48\pi\rho c}. \quad (23)$$

3.4 Line Array of Monopole Sources[3]

Consider a line array consisting of N harmonic monopole point sources equally spaced along the 1-axis at distance d apart, as shown in Fig. 2.

If all these sources have the same monopole strength $\hat{Q}e^{i\omega t}$, they generate the pressure field

$$p(\mathbf{x}, t) = \sum_{n=0}^{N-1} \frac{\hat{Q}}{4\pi r_n} e^{i\omega(t - r_n/c)}, \quad (24)$$

where $r_n = |\mathbf{x} - nd\mathbf{e}_1|$. When $\mathbf{x}$ is sufficiently far from the line array so that $r = |\mathbf{x}| \gg L$, where $L = (N-1)d$ is the length of the array, $r_n \simeq r - nd\cos\theta$, with $\cos\theta = x_1/r$. The sum in Eq. (24) can then be evaluated to give

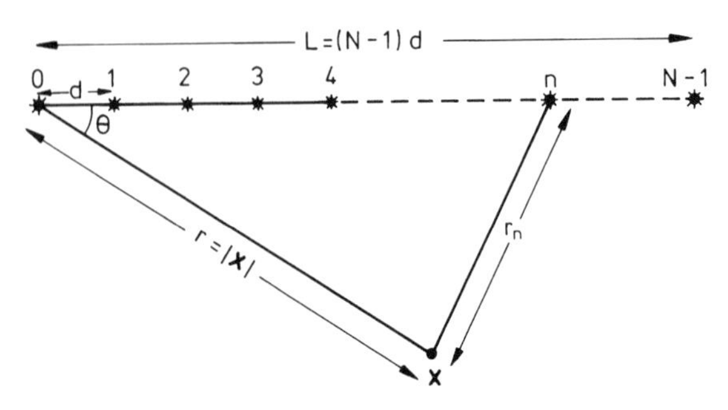

Fig. 2 The N sources forming a line array.

$$p(\mathbf{x}, t) = \frac{N\hat{Q}}{4\pi r} \frac{\sin(\frac{1}{2}Nkd\cos\theta)}{N\sin(\frac{1}{2}kd\cos\theta)} \times e^{i[\omega t - kr - (k/2)L\cos\theta]}. \quad (25)$$

At any angle θ, the mean intensity vector is radial, with

$$I_r(r, \theta) = \frac{N^2|\hat{Q}|^2}{32\pi^2\rho c r^2} \left| \frac{\sin(\frac{1}{2}Nkd\cos\theta)}{N\sin(\frac{1}{2}kd\cos\theta)} \right|^2. \quad (26)$$

In a direction normal to the array ($\theta = 90°$)

$$I_r(r, \tfrac{1}{2}\pi) = \frac{N^2|\hat{Q}|^2}{32\pi^2\rho c r^2}. \quad (27)$$

Hence at a general angle θ to the array,

$$I_r(r, \theta) = I_r(r, \tfrac{1}{2}\pi) \left| \frac{\sin(\frac{1}{2}Nkd\cos\theta)}{N\sin(\frac{1}{2}kd\cos\theta)} \right|^2. \quad (28)$$

The function $D(\theta) = |\sin(\frac{1}{2}Nkd\cos\theta)/[N\sin(\frac{1}{2}kd\cos\theta)]|^2$ describes how the acoustic intensity varies with angle. It is usually called the *directional factor* and is plotted as a function of polar angle θ in Fig. 3 for $N = 5$ and various values of kd.

Low-Frequency Limit A source whose size L is small in comparison with k^{-1} is said to be *compact*. For $Nkd \ll 1$, the array length is compact, and $D(\theta)$ is approximately equal to unity for all values of θ. The pressure field is omnidirectional (see Fig. 3*a*) and its strength is N times that produced by a single monopole point source.

High-Frequency Limit The term $D(\theta)$ has a maximum value of unity that it attains in directions normal to the array ($\theta = 90°$). The pressure field vanishes whenever

$$\cos\theta = \frac{2\pi n}{Nkd} \quad \text{for integers } n \text{ that are not multiples of } N \quad (29)$$

The region $\pi - \cos^{-1}(2\pi/Nkd) > \theta > \cos^{-1}(2\pi/Nkd)$ is called the *major lobe*.

If kd is large enough, the pressure field can have a

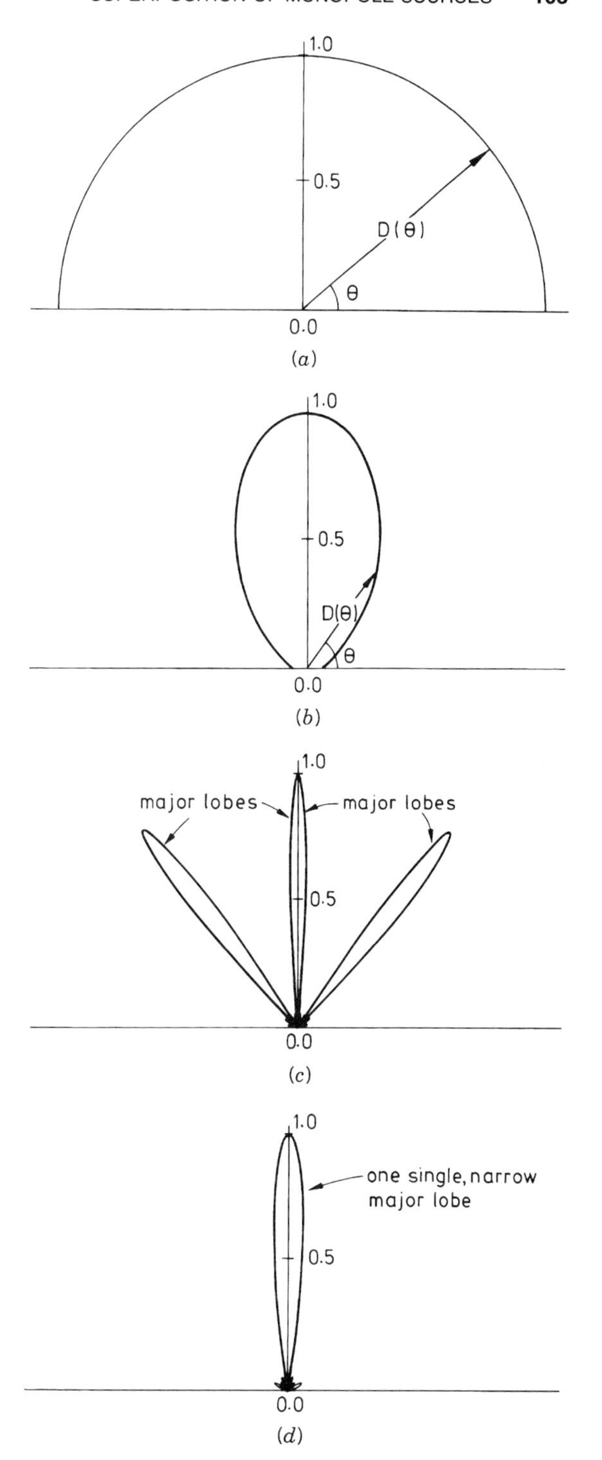

Fig. 3 Directional factor $D(\theta)$ for a line array with $N = 5$: (*a*) $kd = 0.1$; (*b*) $kd = 1.0$; (*c*) $kd = 10.0$; (*d*) $kd = 2\pi(N-1)/N = 1.6\pi$.

second major lobe, since $D(\theta)$ is also equal to unity at $\theta = \cos^{-1}(2\pi/kd)$ for $kd > 2\pi$. In general, a major lobe is obtained whenever $\theta = \cos^{-1}(2\pi M/kd)$ for some integer $M < kd/2\pi$.

In some applications a single major lobe as narrow as possible is required. A good approximation to this can be achieved by choosing $kd = 2\pi(N-1)/N$. This ensures that there is no second major lobe (see Fig. 3*d*). The primary major lobe is confined to angles $\pi - \cos^{-1}[1/(N-1)] > \theta > \cos^{-1}[1/(N-1)]$, and the width of this lobe can be decreased by increasing the number of elements N.

Steering Sometimes it is required to transmit a beam at a specific angle to the array. This can be achieved by inserting a time delay $n\tau$ in the source strength of the nth element. Then

$$p(\mathbf{x}, t) = \sum_{n=0}^{N-1} \frac{\hat{Q}}{4\pi r_n} e^{i\omega(t - n\tau - r_n/c)}. \tag{30}$$

This procedure is called *steering* and, for $r \gg L$, leads to the directional factor

$$D(\theta) = \left| \frac{\sin[\frac{1}{2}Nk(d\cos\theta - c\tau)]}{N\sin[\frac{1}{2}k(d\cos\theta - c\tau)]} \right|^2. \tag{31}$$

The primary major lobe is now transmitted in a direction $\theta_s = \cos^{-1}(c\tau/d)$ to the line array.

4 DIPOLE POINT SOURCE

4.1 Sound Pressure Field[1–6]

A *dipole point source* at the origin of *vector strength* $\mathbf{F}(t)$ produces a pressure field

$$p_D(\mathbf{x}, t) = -\operatorname{div}\left(\frac{\mathbf{F}(t - r/c)}{4\pi r}\right)$$

$$= -\sum_{i=1}^{3} \frac{\partial}{\partial x_i}\left(\frac{F_i(t - r/c)}{4\pi r}\right), \tag{32}$$

where $r = |\mathbf{x}|$. The derivative can be evaluated to show that

$$p_D(\mathbf{x}, t) = \frac{1}{4\pi}\left(\frac{\mathbf{x}\cdot\mathbf{F}(t - r/c)}{r^3} + \frac{\mathbf{x}}{r^2 c}\cdot\frac{\partial\mathbf{F}(t - r/c)}{\partial t}\right). \tag{33}$$

When the direction of the dipole is constant, this simplifies to

$$p_D(\mathbf{x}, t) = \frac{\cos\theta}{4\pi}\left(\frac{1}{cr}\frac{\partial F(t - r/c)}{\partial t} + \frac{F(t - r/c)}{r^2}\right), \tag{34}$$

where $F = |\mathbf{F}|$ and θ is the angle between $\mathbf{F}$, the direction of the dipole axis, and $\mathbf{x}$, the radius vector to the observation point.

If $F(t)$ is simple harmonic of frequency ω, $F(t) = \operatorname{Re}(\hat{F}e^{i\omega t})$, Eq. (34) reduces to

$$p_D(\mathbf{x}, t) = \frac{ik\cos\theta}{4\pi r}\left(1 + \frac{1}{ikr}\right)\hat{F}e^{i\omega(t - r/c)}. \tag{35}$$

The pressure field is directional and proportional to $\cos\theta$. It has maximum amplitude in line with the dipole axis and decreases to zero at $\theta = 90°$. In the far field, $kr \gg 1$, the pressure varies inversely with distance from the source, while in the near field, $kr \ll 1$, it is proportional to the inverse square of the distance from the source.

The ratio of the pressure perturbation at $r = R \gg k^{-1}$ to that at $r = \epsilon \ll k^{-1}$ in the same direction θ is $k\epsilon^2/R$. This is a factor $k\epsilon(\ll 1)$ smaller than the corresponding result for the pressure generated by a monopole point source. A point dipole is therefore less efficient than a monopole at converting the near-field pressures into far-field sound.

4.2 Velocity Field[1,2,5,6]

The particle velocity follows directly from Eq. (35) and the momentum equation. In terms of spherical polar coordinates (r, θ, ϕ), centered on the point dipole,

$$u_{Dr} = \frac{ik\cos\theta}{4\pi\rho c r}\left(1 + \frac{2}{ikr} - \frac{2}{k^2 r^2}\right)\hat{F}e^{i\omega(t - r/c)}, \tag{36a}$$

$$u_{D\theta} = \frac{\sin\theta}{4\pi\rho c r^2}\left(1 + \frac{1}{ikr}\right)\hat{F}e^{i\omega(t - r/c)}, \tag{36b}$$

$$u_{D\phi} = 0. \tag{36c}$$

In the far field the particle velocity is radial with $u_{Dr} = p_D/\rho c$, recovering the plane-wave result. The near acoustic field is more complicated than that generated by a monopole point source, and the particle velocity is not entirely radial.

4.3 Equivalence to a Point Force

The dipole source produces no net volume outflow. However, it exerts a force $\mathbf{F}(t)$ on the fluid. A vibrating body, which is very small in comparison with the wavelength and whose volume is fixed, approximates to a point dipole (apart from the special case when the body has the same density as the surrounding medium; then the net dipole strength is zero).[2]

4.4 Intensity[1,2,5,6]

The mean intensity vector $\mathbf{I}_D = \overline{p_D \mathbf{u}_D}$ is entirely radial and decays with the inverse square of the distance from the source:

$$I_{Dr} = \frac{k^2 \cos^2 \theta |\hat{F}|^2}{32\pi^2 \rho c r^2}. \qquad (37)$$

4.5 Sound Power[1–6]

The sound power radiated to infinity is given as

$$P_D = \frac{k^2 |\hat{F}|^2}{24\pi\rho c}. \qquad (38)$$

4.6 Inhomogeneous Wave Equation[4,5]

The pressure field due to a dipole point source at the origin of vector strength $\mathbf{F}(t)$ satisfies the inhomogeneous wave equation

$$\left(\frac{1}{c^2}\frac{\partial^2}{\partial t^2} - \nabla^2\right) p_D = -\mathrm{div}(\mathbf{F}(t)\delta(\mathbf{x})). \qquad (39)$$

4.7 Dipole as Superposition of Monopoles[1,6]

For two simple sources close together, equal in magnitude, and of opposite sign (as shown in Fig. 1 with $\beta = \pi$), Eq. (18) can be expanded as a Taylor series in l. This shows that

$$p(\mathbf{x}, t) = -\mathrm{div}\left(\frac{\hat{Q}\mathbf{l}}{4\pi r} e^{i\omega(t - r/c)}\right) \qquad (40)$$

for $l = |\mathbf{l}| \ll r$ and k^{-1}. This is equal to the dipole sound field described by Eq. (32) provided $\mathbf{F} = \hat{Q}\mathbf{l}e^{i\omega t}$. Therefore a point dipole is equivalent to two equal and opposite nearby point monopoles. The vector dipole strength is equal to the product of the monopole source strength and their distance apart in magnitude and is in the direction of the separation between the monopoles.

5 QUADRUPOLE POINT SOURCE

5.1 Sound Pressure Field[4,5]

A *quadrupole point source* at the origin of strength $T_{ij}(t)$, $i = 1, 2, 3$, $j = 1, 2, 3$, produces a pressure field

$$p_Q(\mathbf{x}, t) = \sum_{i=1}^{3} \sum_{j=1}^{3} \frac{\partial^2}{\partial x_i \, \partial x_j} \frac{T_{ij}(t - r/c)}{4\pi r}. \qquad (41)$$

The quadrupole source strength can be conveniently written as a matrix:

$$(T_{ij}) = \begin{pmatrix} T_{11} & T_{12} & T_{13} \\ T_{21} & T_{22} & T_{23} \\ T_{31} & T_{32} & T_{33} \end{pmatrix}. \qquad (42)$$

The strength of a point quadrupole source is specified by the functional form of its components.

A quadrupole point source can be considered as the superposition of two nearby dipoles of equal magnitude and opposite strengths (see Section 5.3). One practical example of a quadrupole source is a tuning fork, whose prongs vibrate in antiphase. Quadrupole sources are important because the momentum flux in air flows is a quadrupole source and is the main source mechanism in jet noise (see Chapters 24 and 25).

If $T_{ij}(t)$ is harmonic of frequency ω, $T_{ij}(t) = \mathrm{Re}(\hat{T}_{ij}e^{i\omega t})$, the derivatives in Eq. (41) can be evaluated to give

$$p_Q(\mathbf{x}, t) = -\sum_{i=1}^{3} \sum_{j=1}^{3} \left[\frac{x_i x_j}{r^2}\left(1 + \frac{3}{ikr} - \frac{3}{k^2 r^2}\right) - \delta_{ij}\left(\frac{1}{ikr} - \frac{1}{k^2 r^2}\right)\right] \frac{k^2 \hat{T}_{ij}}{4\pi r} e^{i\omega(t - r/c)}. \qquad (43)$$

In the far field this simplifies to

$$p_Q(\mathbf{x}, t) = -\frac{k^2 \hat{T}_{rr}}{4\pi r} e^{i\omega(t - r/c)}, \qquad (44)$$

where $\hat{T}_{rr} = (x_i x_j / r^2)\hat{T}_{ij}$ is the component of $\hat{T}_{ij}$ in the observer's direction.

Simple expressions for the pressure field can be determined if only one component of T_{ij} is nonzero. If this is one of the diagonal terms, the quadrupole is called *longitudinal*.[4] If T_{11} is the only nonzero component, then

$$p_Q(\mathbf{x}, t) = -\frac{k^2\hat{T}_{11}}{4\pi r}\left[\cos^2\theta\left(1 + \frac{3}{ikr} - \frac{3}{k^2r^2}\right) - \left(\frac{1}{ikr} - \frac{1}{k^2r^2}\right)\right]e^{i\omega(t-r/c)}. \quad (45)$$

where $\cos\theta = x_1/r$. The far-field pressure of a longitudinal quadrupole has directivity $\cos^2\theta$ and decays inversely with distance from the source. In the near field the pressure varies with the inverse cube of distance from the source and has directivity $1 - 3\cos^2\theta$.

If one of the off-diagonal terms of T_{ij} is the only nonzero component, the quadrupole is called *lateral*.[4] If T_{12} is the only nonzero component, then

$$p_Q(\mathbf{x}, t) = -\frac{k^2\hat{T}_{12}}{4\pi r}\cos\theta\sin\theta\cos\phi \cdot \left(1 + \frac{3}{ikr} - \frac{3}{k^2r^2}\right)e^{i\omega(t-r/c)}, \quad (46)$$

where (r, θ, ϕ) are spherical polar coordinates with $\cos\theta = x_1/r$, $\sin\theta\cos\phi = x_2/r$. A lateral quadrupole has the same directional dependence in the pressure field at all distances from the source. The far-field pressure decays inversely with distance from the source, while the near-field pressure decays with the inverse cube of distance from the source.

A quadrupole source that has $T_{11} = T_{22} = T_{33} \neq 0$ and all its off-diagonal components $T_{12} = T_{13} = T_{23} = 0$ is called an *isotropic quadrupole source*. It produces a pressure field

$$p_Q(\mathbf{x}, t) = -\frac{k^2\hat{T}_{11}}{4\pi r}e^{i\omega(t-r/c)}. \quad (47)$$

The pressure field of an isotropic quadrupole is identical to that of a monopole of strength $-k^2\hat{T}_{11}e^{i\omega t}$. It is omnidirectional and everywhere decays inversely with distance from the source.

5.2 Sound Power[4]

A longitudinal quadrupole $\hat{T}_{11}e^{i\omega t}$ emits a sound power $k^4|\hat{T}_{11}|^2/40\pi\rho c$. A lateral quadrupole $\hat{T}_{12}e^{i\omega t}$ emits a sound power $k^4|\hat{T}_{12}|^2/120\pi\rho c$.

5.3 Quadrupole as a Superposition of Dipoles[4]

A quadrupole point source can be considered as two nearby dipoles of equal but opposite strengths. Equation

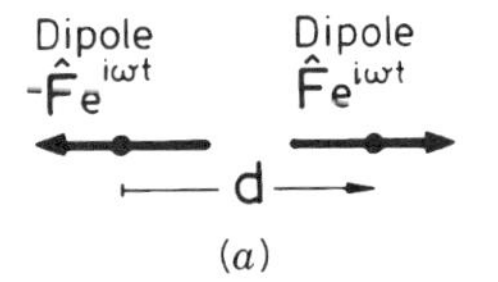

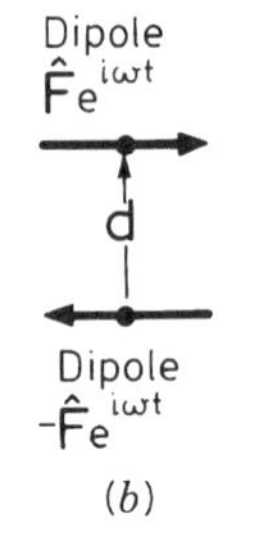

Fig. 4 Quadrupoles as a superposition of dipoles: (*a*) a longitudinal quadrupole; (*b*) a lateral quadrupole.

(32) shows that the pressure field generated by a point dipole of strength $\hat{\mathbf{F}}e^{i\omega t}$ can be written in the form

$$p_D(\mathbf{x}, t) = -\text{div}\left(\frac{\hat{\mathbf{F}}e^{i\omega(t-r/c)}}{4\pi r}\right).$$

The combined pressure field generated by a dipole of strength $\hat{\mathbf{F}}e^{i\omega t}$ at $\mathbf{x} = \mathbf{d}$ and $-\hat{\mathbf{F}}e^{i\omega t}$ at $\mathbf{x} = \mathbf{0}$ can be written as

$$p(\mathbf{x}, t) = \sum_{i=1}^{3}\sum_{j=1}^{3} d_i\hat{F}_j\frac{\partial^2}{\partial x_i\,\partial x_j}\left(\frac{e^{i\omega(t-r/c)}}{4\pi r}\right) \quad (48)$$

for $|\mathbf{d}| \ll k^{-1}$ and r. This is equal to the quadrupole sound field described by Eq. (41) provided $T_{ij}(t) = d_i\hat{F}_je^{i\omega t}$. A point quadrupole is equivalent to two equal and opposite nearby point dipoles.

For a longitudinal quadrupole the separation between the dipoles is in the direction of their axis, as shown in Fig. 4*a*. For a lateral quadrupole the separation is perpendicular to the dipole axis (Fig. 4*b*).

6 RECIPROCITY OF SOURCE AND FIELD POINTS

6.1 Reciprocity in an Unbounded Region

Denote the sound field generated at $\mathbf{x}$ by a harmonic monopole point source of unit strength at $\mathbf{x}_0$ by $G(\mathbf{x}|\mathbf{x}_0)$.

It is evident from Eq. (2) that

$$G(\mathbf{x}|\mathbf{x}_0) = \frac{e^{i\omega(t - |\mathbf{x} - \mathbf{x}_0|/c)}}{4\pi|\mathbf{x} - \mathbf{x}_0|}. \quad (49)$$

This is symmetric in $\mathbf{x}$ and $\mathbf{x}_0$, so that

$$G(\mathbf{x}_0|\mathbf{x}) = G(\mathbf{x}|\mathbf{x}_0); \quad (50)$$

from the definition of G. The function $G(\mathbf{x}_0|\mathbf{x})$ denotes the pressure at $\mathbf{x}_0$ due to a unit point monopole at $\mathbf{x}$. Equation (50) states that this is identical to the pressure at $\mathbf{x}$ due to a unit point monopole at $\mathbf{x}_0$. The pressure field is unchanged when the positions of the source and the listener are exchanged. This is referred to as *reciprocity* and can be generalized to other sources.

The pressure field $p_Q(\mathbf{x})$ due to a point monopole of strength $\hat{Q}e^{i\omega t}$ at $\mathbf{x}_0$ is given by

$$p_Q(\mathbf{x}) = \hat{Q}G(\mathbf{x}_0|\mathbf{x}). \quad (51)$$

That is, the pressure at $\mathbf{x}$ due to a monopole at $\mathbf{x}_0$ is equal to the product of the monopole source strength and the field at $\mathbf{x}_0$ due to a unit monopole at $\mathbf{x}$.

The pressure field $p_F(\mathbf{x})$ due to a point dipole of strength $\hat{\mathbf{F}}e^{i\omega t}$ at $\mathbf{x}_0$ is given by

$$p_F(\mathbf{x}) = \sum_{i=1}^{3} \hat{F}_i \frac{\partial}{\partial x_{0i}} G(\mathbf{x}_0|\mathbf{x}) = -\sum_{i=1}^{3} \hat{F}_i \frac{\partial}{\partial x_i} G(\mathbf{x}_0|\mathbf{x}). \quad (52)$$

That is, the pressure field at $\mathbf{x}$ due to a dipole at $\mathbf{x}_0$ is equal to the product of the dipole source strength and the gradient of the field at $\mathbf{x}_0$ due to a unit monopole at $\mathbf{x}$.

The pressure field $p_T(\mathbf{x})$ due to a point quadrupole of strength $\hat{T}_{ij}e^{i\omega t}$ at $\mathbf{x}_0$ is given by

$$p_T(\mathbf{x}) = \sum_{i=1}^{3}\sum_{j=1}^{3} \hat{T}_{ij} \frac{\partial^2}{\partial x_{0i}\,\partial x_{0j}} G(\mathbf{x}_0|\mathbf{x}) = \sum_{i=1}^{3}\sum_{j=1}^{3} \hat{T}_{ij} \frac{\partial^2}{\partial x_i\,\partial x_j} G(\mathbf{x}_0|\mathbf{x}). \quad (53)$$

That is, the pressure field at $\mathbf{x}$ due to a quadrupole at $\mathbf{x}_0$ is equal to the product of the quadrupole source strength and the double gradient of the field at $\mathbf{x}_0$ due to a unit monopole at $\mathbf{x}$.

6.2 Reciprocity in a Bounded Region

The reciprocal theorems expressed in Eqs. (50)–(53) were derived here for an unbounded region. They remain true[4] in the presence of a wide variety of different types of surfaces. These include hard and soft surfaces, elastic surfaces such as membranes, shells and plates,[8] and locally reacting surfaces,[5] whose boundary conditions can be expressed in the form

$$\frac{\partial p}{\partial n} = K(\mathbf{x}, \omega)p \quad \text{for some } K. \quad (54)$$

Here, $K(\mathbf{x}, \omega)$ may be a function of both position and frequency.

7 MULTIPOLE-SOURCE EXPANSION[2,4,5]

Consider a source distributed over a volume V with strength $\hat{q}(\mathbf{x})e^{i\omega t}$/unit volume. This source radiates a pressure field

$$p(\mathbf{x}, t) = \frac{e^{i\omega t}}{4\pi} \int \hat{q}(\mathbf{y}) \frac{e^{-i\omega|\mathbf{x}-\mathbf{y}|/c}}{|\mathbf{x} - \mathbf{y}|}\, dV. \quad (55)$$

Let the origin be within V and the maximum radial distance of points within V from the origin be l. If kl is small, the source region is said to be *compact*.

For a compact source region, when $r = |\mathbf{x}|$ is large in comparison with l, $e^{-i\omega|\mathbf{x}-\mathbf{y}|/c}/|\mathbf{x} - \mathbf{y}|$ can be expanded as a Taylor series in $\mathbf{y}$:

$$\frac{e^{-ik|\mathbf{x}-\mathbf{y}|}}{|\mathbf{x} - \mathbf{y}|} = \frac{e^{-ikr}}{r} - (\mathbf{y} \cdot \nabla)\left(\frac{e^{-ikr}}{r}\right) + \frac{1}{2}(\mathbf{y} \cdot \nabla)(\mathbf{y} \cdot \nabla)\left(\frac{e^{-ikr}}{r}\right) + \cdots. \quad (56)$$

where the operator $\nabla = (\partial/\partial x_1, \partial/\partial x_2, \partial/\partial x_3)$.

Substitution of this expansion into Eq. (55) leads to

$$p(\mathbf{x}, t) = \frac{\hat{Q}e^{i\omega(t-r/c)}}{4\pi r} - \hat{\mathbf{F}} \cdot \nabla\left(\frac{e^{i\omega(t-r/c)}}{4\pi r}\right) + \sum_{i=1}^{3}\sum_{j=1}^{3} \hat{T}_{ij} \frac{\partial^2}{\partial x_i\,\partial x_j}\left(\frac{e^{i\omega(t-r/c)}}{4\pi r}\right) + \cdots \quad \text{for } r \text{ and } k^{-1} \gg l, \quad (57)$$

where

$$\hat{Q} = \int_V \hat{q}(\mathbf{y})\, dV, \tag{58a}$$

$$\hat{\mathbf{F}} = \int_V \mathbf{y}\hat{q}(\mathbf{y})\, dV, \tag{58b}$$

$$\hat{T}_{ij} = \int_V \frac{1}{2} y_i y_j \hat{q}(\mathbf{y})\, dV. \tag{58c}$$

At distances large compared with the source size, the acoustic field generated by a source distributed over a compact region is equivalent to a multipole expansion of point sources at the origin.

7.1 Compact Source Region with Monopole Far Field

If $\hat{Q} \neq 0$, $\hat{F}_i \sim O(l\hat{Q})$, $\hat{T}_{ij} \sim O(l^2\hat{Q})$, and so on, and

$$p(\mathbf{x}, t) \simeq \frac{\hat{Q} e^{i\omega(t - r/c)}}{4\pi r} \quad \text{for} \quad r \text{ and } k^{-1} \gg l. \tag{59}$$

When the total monopole source strength is nonzero, a distributed source generates a distant sound field equivalent to that of a monopole point source[2] at the origin of strength $\int \hat{q}(\mathbf{y})\, dV$.

7.2 Compact Source Region with Dipole Far Field

If $\hat{Q} = 0$ and $\hat{\mathbf{F}} \neq \mathbf{0}$, then

$$p(\mathbf{x}, t) \simeq -\hat{\mathbf{F}} \cdot \nabla \left(\frac{e^{i\omega(t - r/c)}}{4\pi r} \right) \quad \text{for} \quad r \text{ and } k^{-1} \gg l. \tag{60}$$

The sound field is equivalent to that of a dipole point source at the origin,[2] the strength of the equivalent dipole being the first moment of $\hat{q}(\mathbf{y})$, and $\hat{\mathbf{F}} = \int \mathbf{y}\hat{q}(\mathbf{y})\, dV$.

7.3 Compact Source Region with Quadrupole Far Field

If $\hat{Q} = 0$, $\hat{\mathbf{F}} = \mathbf{0}$, and $\hat{T}_{ij} \neq 0$, then

$$p(\mathbf{x}, t) = \sum_{i=1}^{3} \sum_{j=1}^{3} \hat{T}_{ij} \frac{\partial^2}{\partial x_i\, \partial x_j} \left(\frac{e^{i\omega(t - r/c)}}{4\pi r} \right). \tag{61}$$

The sound field is equivalent to that of a quadrupole point source at the origin, the strength of the equivalent quadrupole being dependent on the second-order moments of $\hat{q}(\mathbf{y})$, $\hat{T}_{ij} = \int \frac{1}{2} y_i y_j \hat{q}(\mathbf{y})\, dV$.

8 HARMONICALLY PULSATING AND VIBRATING SPHERES

8.1 Pulsating Sphere

Consider an expanding and contracting sphere whose radius at time t is $a + \hat{\epsilon} e^{i\omega t}$, where $\hat{\epsilon}$ is small in comparison with both a and k^{-1}. Such a sphere generates a pressure field[2–6]

$$p(\mathbf{x}, t) = p(r, t) = \frac{i\omega a^2 \rho \hat{v}_s}{r(1 + ika)} e^{i\omega[t - (r - a)/c]}, \tag{62}$$

where $\hat{v}_s e^{i\omega t}$ is the surface velocity on $r = a$, $\hat{v}_s = i\omega\hat{\epsilon}$. This pressure field is omnidirectional and has the same dependence on radial distance r as that produced by a monopole point source [see Eq. (2)]. The *specific acoustic impedance* on $r = a$, Z_s, is defined by

$$Z_s = \frac{p(a, t)}{\hat{v}_s e^{i\omega t}} = \frac{ika}{1 + ika} \rho c. \tag{63}$$

The sound power radiated to infinity can be determined by integrating the energy flux through a large sphere,

$$P_\infty = 2\pi \frac{(ka)^2}{1 + (ka)^2} \rho c a^2 |\hat{v}_s|^2. \tag{64}$$

High-Frequency Limit[2,4,5]

$$p(\mathbf{x}, t) = \frac{a}{r} \rho c \hat{v}_s e^{i\omega[t - (r - a)/c]} \quad \text{for} \quad ka \gg 1. \tag{65}$$

The specific acoustic impedance on $r = a$ is ρc, recovering the plane-wave (or geometric acoustics) result. The factor a/r in Eq. (65) leads to a factor $(a/r)^2$ in intensity, and accounts for the fact that energy fed over the surface area $4\pi a^2$ of the sphere spreads out over the larger area $4\pi r^2$ at radius r. The time delay $(r - a)/c$ in Eq. (65) is the time taken for sound to reach $\mathbf{x}$ from the nearest point on the sphere. In the geometric acoustics limit, the pressure perturbation at $\mathbf{x}$ is produced by sound traveling along a ray, which takes the shortest path from the body surface to $\mathbf{x}$.

Since the surface pressure and velocity are virtually in phase, the sphere does a significant amount of work on the fluid, which is radiated away as sound energy, and

$$P_\infty = 2\pi\rho c a^2 |\hat{v}_s|^2 \quad \text{for} \quad ka \gg 1. \tag{66}$$

Low-Frequency Limit[2–6]

$$p(\mathbf{x}, t) = \frac{\hat{Q}}{4\pi r} e^{i\omega(t - r/c)} \quad \text{for} \quad ka \ll 1, \tag{67}$$

with $\hat{Q} = i\omega\rho 4\pi a^2 \hat{v}_s$. A compact pulsating sphere is equivalent to a monopole point source of strength $\hat{Q}e^{i\omega t}$, where $\hat{Q}e^{i\omega t}$ is the product of the density and the rate of change of volume enclosed by the sphere (cf. Section 2.1). This remains true even if the change in radius is not small in comparison with a, provided that $\hat{Q}e^{i\omega t}$ is the time derivative of the actual volume enclosed by the sphere, the sphere remains compact, and the speed of the surface motion is much smaller than the speed of sound.[4,9] On a compact sphere, the surface pressure and velocity are virtually 90° out of phase. This means that the sphere does little work on the surrounding fluid. The surface motion is therefore inefficient at radiating sound power. Indeed, $P_\infty = 2\pi(ka)^2\rho c a^2|\hat{v}_s|^2$, a factor $(ka)^2$ smaller than the power radiated by the large sphere in Eq. (66).

8.2 Vibrating Sphere

A rigid sphere vibrating with small amplitude about the origin in the 1-direction with speed $\hat{U}e^{i\omega t}$ has a normal surface velocity $\hat{U}\cos\theta e^{i\omega t}$, where $\cos\theta = x_1/r$. This leads to a pressure field[2,4–6]

$$p(\mathbf{x}, t) = p(r, \theta, t) = -\frac{(ka)^2 \rho c a \cos\theta}{2(1 + ika) - (ka)^2}\left(1 + \frac{1}{ikr}\right) \times \frac{\hat{U}}{r} e^{i\omega[t - (r-a)/c]}, \tag{68}$$

where $k = \omega/c$. The pressure field has the same directivity and dependence on r as that produced by a dipole point source with its axis in the 1-direction [see Eq. (35)].

The specific acoustic impedance on $r = a$ is given by

$$Z_s = \frac{ika(1 + ika)}{2(1 + ika) - (ka)^2} \rho c. \tag{69}$$

The force $\mathbf{f}(t)$ exerted on the fluid by the sphere is in the 1-direction, with

$$f_1(t) = \int_0^\pi p(a, \theta, t)\cos\theta \, 2\pi a^2 \sin\theta \, d\theta = \frac{2\pi}{3} \frac{ika(1 + ika)}{1 + ika - \frac{1}{2}(ka)^2} \rho c a^2 \hat{U} e^{i\omega t}. \tag{70}$$

The *mechanical impedance* of the surface of the sphere, Z_m, can be found by dividing $f_1(t)$ by the complex velocity in the 1-direction:

$$Z_m = \frac{2\pi}{3} \frac{ika(1 + ika)}{1 + ika - \frac{1}{2}(ka)^2} \rho c a^2. \tag{71}$$

The real part of Z_m is a measure of the work done by the sphere on the fluid and is commonly called the *resistance*. The imaginary part of Z_m is called the *reactance*.

The sound power radiated to infinity is equal to

$$P_\infty = \frac{1}{2} \operatorname{Re} Z_m |\hat{U}|^2 = \frac{\pi}{6} \frac{(ka)^4}{1 + \frac{1}{4}(ka)^4} \rho c a^2 |\hat{U}|^2. \tag{72}$$

High-Frequency Limit[2,4,5]

$$p(\mathbf{x}, t) = \frac{a}{r} \rho c \hat{U} \cos\theta e^{i\omega[t - (r-a)/c]} \quad \text{for} \quad ka \gg 1, \tag{73}$$

and the specific acoustic impedance is ρc, the geometric acoustics result. The force exerted on the fluid by the sphere is in phase with the velocity:

$$f_1(t) = \frac{4\pi}{3} \rho c a^2 \hat{U} e^{i\omega t} \quad \text{for} \quad ka \gg 1. \tag{74}$$

The mean rate at which this force does work on the fluid is $(2\pi/3)\rho c a^2|\hat{U}|^2$, and this power is radiated to infinity:

$$P_\infty = \frac{2\pi}{3} \rho c a^2 |\hat{U}|^2 \quad \text{for} \quad ka \gg 1. \tag{75}$$

Low-Frequency Limit[2,4,5]

$$p(\mathbf{x}, t) = -\frac{1}{2} (ka)^2 \rho c \hat{U} \cos\theta \frac{a}{r} \left(1 + \frac{1}{ikr}\right) e^{i\omega(t - r/c)} \quad \text{for} \quad ka \ll 1. \tag{76}$$

It is evident from a comparison with Eq. (35) that a compact vibrating sphere produces the same pressure field as a point dipole: The strength of the equivalent point dipole is $2\pi i\omega\rho a^3 \hat{U} e^{i\omega t}$.

On $r = a$, $Z_s \simeq \frac{1}{2} ika\rho c$. The surface pressure leads the normal velocity by approximately 90°. The largest term in the force **f** is therefore in phase with the acceleration and describes the force the sphere must exert on the fluid to overcome its inertia. There is also a very much smaller component of **f** in phase with the velocity. That in-phase component of **f** is $\frac{1}{3}\pi(ka)^4\rho ca^2\hat{U}e^{i\omega t}$. The rate at which the sphere does work on the fluid is therefore $\frac{1}{6}\pi(ka)^4\rho ca^2|\hat{U}|^2$. This power is radiated to infinity,

$$P_\infty = \frac{\pi}{6}(ka)^4\rho ca^2|\hat{U}|^2 \quad \text{for} \quad ka \ll 1. \tag{77}$$

This is a factor $(ka)^4/4$ smaller than the power radiated by the large vibrating sphere in Eq. (75) and is a measure of the very small amount of power that is radiated into the sound field by the rigid motion of a compact sphere.

9 HARMONICALLY PULSATING AND VIBRATING CYLINDERS

9.1 Pulsating Cylinder[1]

The sound field generated by an expanding and contracting cylinder whose radius at time t is $a + \hat{\epsilon}e^{i\omega t}$ (where $|\hat{\epsilon}| \ll a$ and k^{-1}) is given by

$$p(\mathbf{x}, t) = i\rho c \frac{H_0^{(2)}(k\sigma)}{H_1^{(2)}(ka)}\hat{v}_s e^{i\omega t}. \tag{78}$$

Here, σ is the radial distance of **x** from the cylinder axis and $H_n^{(2)}(z)$ is a Hankel function[7] of order n; $\hat{v}_s e^{i\omega t} = i\omega\hat{\epsilon}e^{i\omega t}$ is the radial surface velocity. This cylindrically symmetric pressure field has the same dependence on σ as that due to the line source in Section 3.1.

The specific acoustic impedance on $\sigma = a$,

$$Z_s = i\frac{H_0^{(2)}(ka)}{H_1^{(2)}(ka)}\rho c. \tag{79}$$

The sound power radiated to infinity per unit length of cylinder can be determined by integrating the energy flux through a large cylinder:

$$P_\infty = \frac{2\rho c}{k}\frac{|\hat{v}_s|^2}{|H_1^{(2)}(ka)|^2} \quad \text{per unit length of cylinder.} \tag{80}$$

High-Frequency Limit

$$p(\mathbf{x}, t) = \left(\frac{a}{\sigma}\right)^{1/2}\rho c\hat{v}_s e^{i\omega[t-(\sigma-a)/c]} \quad \text{for} \quad ka \gg 1. \tag{81}$$

Equation (81) describes the geometric acoustics result. The pressure and particle velocity are related by the plane-wave formula $p = \rho c u_\sigma$. The factor $(a/\sigma)^{1/2}$ in Eq. (81) leads to an intensity proportional to a/σ. This accounts for the fact that energy fed in over a surface area $2\pi al$ of the cylinder spreads out over the larger area $2\pi\sigma l$ at radius σ. Since the surface pressure and velocity are virtually in phase, the cylinder does a significant amount of work on the fluid, which is radiated away as sound energy, and

$$P_\infty = \pi\rho ca|\hat{v}_s|^2 \quad \text{per unit length of cylinder } ka \gg 1. \tag{82}$$

Low-Frequency Limit[4]

$$p(\mathbf{x}, t) = \frac{\hat{q}_l}{4i}H_0^{(2)}(k\sigma)e^{i\omega t} \quad \text{for} \quad ka \ll 1, \tag{83}$$

with $\hat{q}_l = i\omega\rho 2\pi a\hat{v}_s$. A pulsating cylinder of compact radius is equivalent to a line source (see Section 3.1), where the strength of the line source is the product of the density and the rate of change of volume enclosed by the cylinder. The pressure field has a different form in the near and far fields, as discussed in Section 3.1.

The specific acoustic impedance on $\sigma = a$ is given by

$$Z_s \simeq -ika\ln(ka)\rho c \quad \text{for} \quad ka \ll 1. \tag{84}$$

Since the surface pressure and velocity are virtually 90° out of phase, the cylinder does little work on the surrounding fluid. Indeed, $P_\infty = \frac{1}{2}\pi^2(ka)\rho ca|\hat{v}_s|^2$ per unit length of cylinder, a factor $\frac{1}{2}\pi ka$ smaller than the power radiated per unit length by a cylinder of large radius in Eq. (82).

9.2 Vibrating Cylinder[1]

A rigid cylinder with axis in the 3-direction and vibrating with small velocity $\hat{U}e^{i\omega t}$ in the 1-direction about the origin produces a pressure field

$$p(\mathbf{x}, t) = p(\sigma, \theta, t) = -i\rho c \frac{H_1^{(2)}(k\sigma)}{H_1^{(2)\prime}(ka)} \hat{U} \cos\theta e^{i\omega t}, \quad (85)$$

where $\sigma = (x_1^2 + x_2^2)^{1/2}$, $\cos\theta = x_1/\sigma$, and the prime denotes differentiation with respect to the argument.

The specific acoustic impedance on $\sigma = a$ is given by

$$Z_s = -i \frac{H_1^{(2)}(ka)}{H_1^{(2)\prime}(ka)} \rho c. \quad (86)$$

The force $\mathbf{f}(t)$ exerted on the fluid by unit length of the cylinder is in the 1-direction, with

$$f_1(t) = \int_0^{2\pi} p(a, \theta, t) \cos\theta \, a \, d\theta$$

$$= -i\pi \frac{H_1^{(2)}(ka)}{H_1^{(2)\prime}(ka)} \rho c a \hat{U} e^{i\omega t}. \quad (87)$$

The mechanical impedance per unit length of the cylinder is given as

$$Z_m = \frac{f_1(t)}{\hat{U} e^{i\omega t}} = -i\pi \frac{H_1^{(2)}(ka)}{H_1^{(2)\prime}(ka)} \rho c a. \quad (88)$$

The sound power radiated to infinity is calculated as

$$P_\infty = \frac{\rho c |\hat{U}|^2}{k |H_1^{(2)\prime}(ka)|^2} \quad \text{per unit length of cylinder.} \quad (89)$$

High-Frequency Limit

$$p(\mathbf{x}, t) = \left(\frac{a}{\sigma}\right)^{1/2} \rho c \hat{U} \cos\theta \, e^{i\omega[t - (\sigma - a)/c]} \quad \text{for} \quad ka \gg 1, \quad (90)$$

and the specific acoustic impedance is ρc, the geometric acoustics result. The force exerted on the fluid by the cylinder is in phase with the velocity:

$$f_1(t) = \pi \rho c a \hat{U} e^{i\omega t} \quad \text{for} \quad ka \gg 1. \quad (91)$$

The mean rate at which the cylinder does work on the fluid is $\frac{1}{2}\pi\rho c a |\hat{U}|^2$ per unit length of cylinder. This power is radiated to infinity:

$$P_\infty = \tfrac{1}{2}\pi\rho c a |\hat{U}|^2 \quad \text{per unit length of cylinder for } ka \gg 1. \quad (92)$$

Low-Frequency Limit[1]

$$p(\mathbf{x}, t) = \tfrac{1}{2}\pi (ka)^2 \rho c \hat{U} \cos\theta \, H_1^{(2)}(k\sigma) e^{i\omega t} \quad \text{for} \quad ka \ll 1. \quad (93)$$

In the near field the pressure is inversely proportional to σ, whereas in the far field it decays as $\sigma^{-1/2}$. On $\sigma = a$, $Z_s \simeq ika\rho c$. The surface pressure leads the normal velocity by approximately 90°. The largest term in $\mathbf{f}$, the force per unit length, is therefore in phase with the acceleration,

$$f_1(t) \simeq \pi i \omega \rho a^2 \hat{U} e^{i\omega t} \quad \text{for} \quad ka \ll 1. \quad (94)$$

There is also a very much smaller component of $\mathbf{f}$ in phase with the velocity. That in-phase component of $\mathbf{f}$ is $\frac{1}{2}\pi^2 (ka)^3 \rho c a \hat{U} e^{i\omega t}$. The rate at which the cylinder does work on the fluid is therefore

$$P_\infty = \frac{\pi^2}{4} (ka)^3 \rho c a |\hat{U}|^2 \quad \text{per unit length of cylinder.} \quad (95)$$

This is a factor $\frac{1}{2}\pi(ka)^3$ smaller than the power radiated by the unit length of a cylinder of large diameter in Eq. (92).

10 BAFFLED CIRCULAR PISTON[1–5]

A circular piston of radius a mounted flush in an infinite hard wall, $x_1 = 0$, and vibrating with normal velocity $\hat{U} e^{i\omega t}$ generates a pressure field

$$p(\mathbf{x}, t) = \frac{i\omega\rho \hat{U} e^{i\omega t}}{2\pi} \int_0^a \int_0^{2\pi} \frac{e^{-i\omega R/c}}{R} \sigma \, d\sigma \, d\mu, \quad (96)$$

where the origin has been chosen to be at the center of the piston. Here, R is the distance from the observer at

x to an element of the piston face with cylindrical polar coordinates $(0, \sigma, \mu)$:

$$R = [x_1^2 + (x_2 - \sigma \cos \mu)^2 + (x_3 - \sigma \sin \mu)^2]^{1/2}. \quad (97)$$

10.1 Distant Sound Field

For $r = |\mathbf{x}| \gg a^2k$ and a, the integrals in Eq. (96) may be evaluated analytically to give

$$p(\mathbf{x}, t) = p(r, \theta, t) = \frac{i\omega\rho a^2 \hat{U}}{2r} \frac{2J_1(ka \sin \theta)}{ka \sin \theta} e^{i\omega(t - r/c)}, \quad (98)$$

where $k = \omega/c$, $\cos \theta = x_1/r$, and $J_1(z)$ denotes a Bessel function.[7] The function $2J_1(z)/z$ is plotted in Fig. 5.

At any angle θ, the mean sound intensity vector is radial, with

$$I_r(r, \theta) = \frac{(ka)^2 \rho c a^2 |\hat{U}|^2}{8r^2} \left| \frac{2J_1(ka \sin \theta)}{ka \sin \theta} \right|^2. \quad (99)$$

On the axis

$$I_r(r, 0) = \frac{(ka)^2 \rho c a^2 |\hat{U}|^2}{8r^2}. \quad (100)$$

Hence, at an angle θ to the axis,

$$I_r(r, \theta) = I_r(r, 0) \left| \frac{2J_1(ka \sin \theta)}{ka \sin \theta} \right|^2. \quad (101)$$

The directional factor $D(\theta) = |2J_1(ka \sin \theta)/(ka \sin \theta)|^2$ describes how the sound intensity varies with angle and is plotted as a function of polar angle θ in Fig. 6 for two different values of ka.

Low-Frequency Limit When the piston is compact, $2J_1(ka/ \sin \theta)/(ka \sin \theta)$ is approximately equal to unity for all values of θ. The pressure field is therefore omnidirectional and can be written in the form

$$p(\mathbf{x}, t) = \frac{\hat{Q}}{2\pi r} e^{i\omega(t - r/c)} \quad \text{for} \quad ka \ll 1, \quad (102)$$

with $\hat{Q} = i\omega\rho\pi a^2 \hat{U}$. The amplitude of the pressure field is twice that produced in unbounded space by a monopole point source with a volume outflow of $\pi a^2 \hat{U} e^{i\omega t}$. The factor 2 describes the effect of the rigid baffle. An image of the piston in the baffle surface leads to a doubling of the pressure field.

High-Frequency Limit The term $J_1(z)$ has zeros at $z = 3.83, 7.02, 10.17, \ldots$. For $ka \geq 3.83$, $D(\theta)$ therefore vanishes at an angle θ_1 such that $\sin \theta_1 = 3.83/ka$. The major lobe occupies the region $|\theta| < \theta_1$. If the piston is sufficiently noncompact that $ka \geq 7.02$, $D(\theta)$ has a second zero at an angle θ_2 such that $\sin \theta_2 = 7.02/ka$. The secondary lobe is detected at angles between θ_1 and θ_2. It is clear from the shape of the $2J_1(z)/z$ plot in Fig. 5 that the maximum value of the intensity in the secondary lobe is smaller than that in the major lobe (see Fig. 6*b*). If the piston is grossly noncompact, a large number of side lobes may be present, each with a smaller peak intensity than the preceding lobe.

10.2 Pressure Field on the Symmetry Axis[3,4]

On the symmetry axis, $x_2 = x_3 = 0$, the integrals in Eq. (96) can be evaluated exactly for any value of x_1 to yield

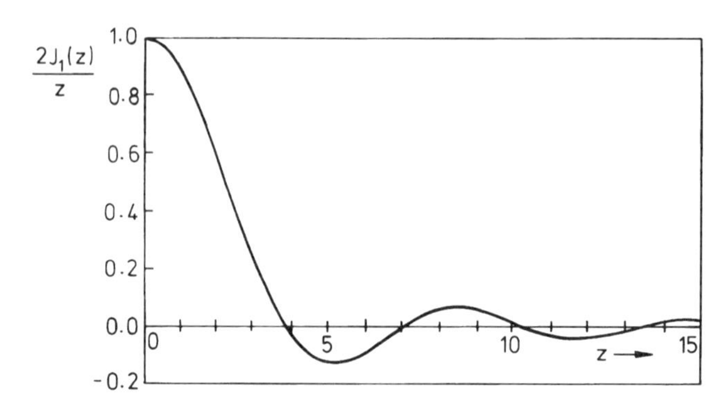

Fig. 5 Variation of $2J_1(z)/z$ with z.

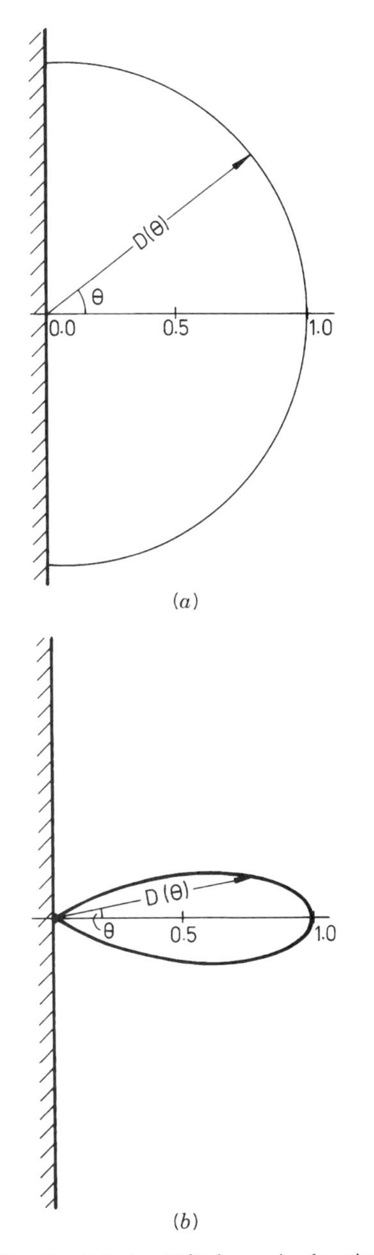

Fig. 6 Directional factor $D(\theta)$ for a circular piston in an otherwise rigid wall: (*a*) $ka = 0.5$; (*b*) $ka = 5.0$.

$$p(x_1, 0, t) = 2i\rho c\hat{U} \exp i\{\omega t - \tfrac{1}{2}k[x_1 + (x_1^2 + a^2)^{1/2}]\} \times \sin\{\tfrac{1}{2}k[(x_1^2 + a^2)^{1/2} - x_1]\}. \qquad (103)$$

The parameter $|\hat{p}|$ is plotted as a function of x_1 for $ka = 11\pi$ in Fig. 7.

The pressure vanishes whenever $k(x_1^2 + a^2)^{1/2} - kx_1$ is a multiple of 2π, that is, when

$$kx_1 = \frac{(ka)^2 - (2n\pi)^2}{4n\pi}, \qquad (104)$$

where n is a positive integer less than $ka/2\pi$. Thus when ka is 11π, as in Fig. 7, there are five pressure nodes along the x_1-axis.

In the distant field, $x_1 \gg ka^2$, $x_1 + (x_1^2 + a^2)^{1/2} \simeq 2x_1$ and $(x_1^2 + a^2)^{1/2} - x_1 \simeq \frac{1}{2}a^2/x_1$, and Eq. (103) reduces to

$$p(x_1, 0, t) = \frac{i(ka)\rho c a\hat{U}}{2x_1} e^{i\omega(t - x_1/c)}, \qquad (105)$$

in agreement with the form for the distant-field axial pressure obtained by putting $\theta = 0$ in Eq. (98).

10.3 Mechanical Impedance[3,4]

The force exerted by the piston on the fluid is in the 1-direction, and its strength is equal to the integral of the pressure over the surfaces of the piston:

$$f_1(t) = \int_S p(0, x_2, x_3, t)\, dx_2\, dx_3. \qquad (106)$$

After substitution for $p(\mathbf{x}, t)$ from Eq. (96), the integrals can be evaluated to give

$$f_1(t) = \pi\rho c a^2 \hat{U} e^{i\omega t}[R_1(2ka) + iX_1(2ka)], \qquad (107)$$

where $R_1(z) = 1 - 2J_1(z)/z$, $X_1(z) = 2\mathbf{H}_1(z)/z$, and $\mathbf{H}_1(z)$ is Struve's[7] function. The parameters $R_1(z)$ and $X_1(z)$ are plotted in Fig. 8.

The mechanical impedance Z_m is obtained by dividing this force by the complex velocity of the piston:

$$Z_m = \frac{f_1(t)}{\hat{U} e^{i\omega t}} = \pi\rho c a^2 [R_1(2ka) + iX_1(2ka)]. \qquad (108)$$

The resistance Re Z_m describes the rate at which work is done by the piston on the fluid. This power is radiated to infinity,

$$P_\infty = \tfrac{1}{2}\pi\rho c a^2 R_1(2ka)|\hat{U}|^2. \qquad (109)$$

High-Frequency Limit For $k_a \gg 1$ the functions $J_1(2ka)$ and $\mathbf{H}_1(2ka)$ can be approximated by their large-argument asymptotic forms[7]:

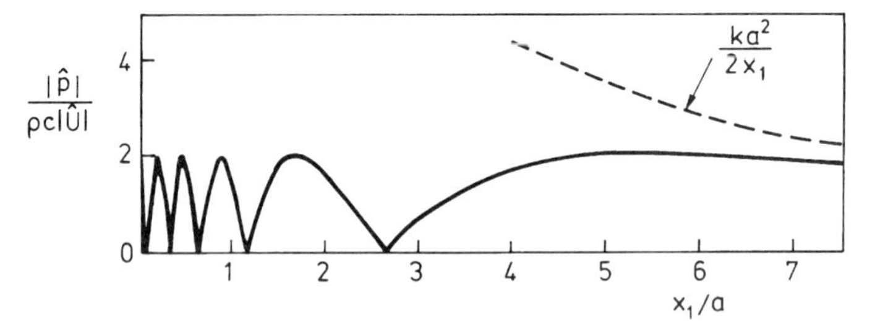

Fig. 7 Variation of $|\hat{p}|$ with x_1 on symmetry axis of a circular piston, $ka = 11\pi$. (Reprinted with permission, from Ref. 4, Fig. 5–10.)

$$R_1(2ka) \simeq 1 - \frac{\cos(2ka - 3\pi/4)}{\pi^{1/2}(ka)^{3/2}}, \quad (110a)$$

$$X_1(2ka) \simeq \frac{2}{\pi ka} + \frac{\sin(2ka - 3\pi/4)}{\pi^{1/2}(ka)^{3/2}}. \quad (110b)$$

The limiting values of 1 and $2/\pi ka$ are approached in an oscillatory manner, as shown in Fig. 8. To leading order,

$$Z_m = \rho c \pi a^2, \quad (111)$$

the geometric acoustics result. The force is in phase with the piston velocity and is equal to the product of the surface pressure $\rho c\hat{U}e^{i\omega t}$ and the piston surface area. The power radiated to infinity is simply given by the product of the mean intensity on the piston $\frac{1}{2}\rho c|\hat{U}|^2$ and the piston surface area:

$$P_\infty = \tfrac{1}{2}\pi\rho ca^2|\hat{U}|^2. \quad (112)$$

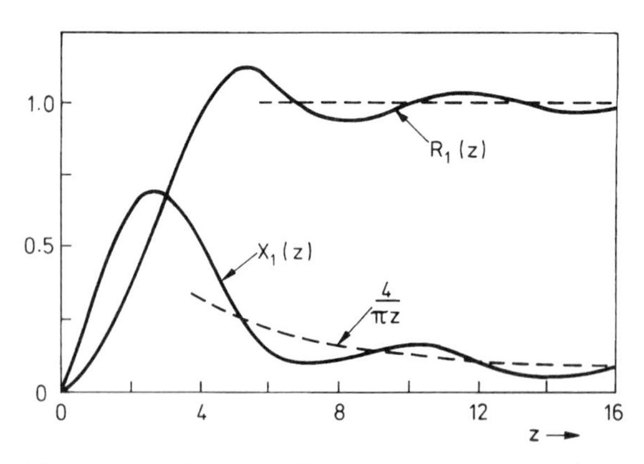

Fig. 8 Piston impedance functions $R_1(z)$ and $X_1(z)$ for a circular piston.

Low-Frequency Limit For small values of z, $J_1(z) \sim \frac{1}{2}z - \frac{1}{16}z^3$, and $\mathbf{H}_1(z) \simeq 2z^2/3\pi$, so that

$$Z_m = \pi\rho ca^2\left[\frac{(ka)^2}{2} + i\,\frac{8ka}{3\pi}\right]. \quad (113)$$

The largest term in Z_m is its imaginary part, the reactance. This describes the force the piston must exert on the fluid to overcome its inertia. That force is equal to $\frac{8}{3}i\omega\rho a^3\hat{U}e^{i\omega t}$, so that an effective volume $\frac{8}{3}a^3$ of fluid is accelerated.

The real part of Z_m describes the rate at which work is done on the fluid by the piston and hence the radiated sound power:

$$P_\infty = \tfrac{1}{2}\,\mathrm{Re}\,Z_m|\hat{U}|^2 = \tfrac{1}{4}\pi(ka)^2\rho ca^2|\hat{U}|^2, \quad (114)$$

which is a factor $\frac{1}{2}(ka)^2$ smaller than the high-frequency result in Eq. (112).

10.4 Generalization to Other Geometries

A baffled piston of arbitrary cross-sectional area mounted flush in an infinite hard wall, $x_1 = 0$, and vibrating with normal velocity $\hat{U}e^{i\omega t}$ generates a pressure field

$$p(\mathbf{x}, t) = \frac{i\omega\rho\hat{U}e^{i\omega t}}{2\pi}\int_S \frac{e^{-i\omega R/c}}{R}\,dy_2\,dy_3, \quad (115)$$

where $R = |\mathbf{x} - \mathbf{y}|$ and the integral is over the surface area S of the piston. At very large distances this simplifies to

$$p(\mathbf{x}, t) = \frac{i\omega\rho\hat{U}e^{i\omega(t - r/c)}}{2\pi r}\int_S e^{i\omega(x_2y_2 + x_3y_3)/rc}\,dy_2\,dy_3 \quad (116)$$

for $r \gg a$ and ka^2, where $r = |\mathbf{x}|$ and a denotes the piston size. In Section 9.1, this integral was evaluated for a circular piston. It can also be evaluated analytically when the piston is rectangular.[5] For an arbitrarily shaped piston, analytical results can only be derived in the limiting cases of low and high frequencies.

Low-Frequency Limit When the piston is compact, the pressure field does not depend on the details of the piston geometry. Then

$$p(\mathbf{x}, t) = \frac{\hat{Q}}{2\pi r} e^{i\omega(t - r/c)}, \tag{117}$$

where $\hat{Q} = i\omega\rho S\hat{U}$ and S is the surface area of the piston. This results in a radiated sound power

$$P_\infty = \frac{\rho(\omega S|\hat{U}|)^2}{4\pi c}. \tag{118}$$

High-Frequency Limit On the face of the piston, $p \simeq \rho c\hat{U}e^{i\omega t}$. Hence the force exerted on the fluid by the piston is approximately $\rho cS\hat{U}e^{i\omega t}$, and the radiated sound power is $\frac{1}{2}\rho cS|\hat{U}|^2$.

Here the baffle is an infinite-plane hard surface. Analytical results can also be obtained for a plane impedance surface and for a circular piston set in a rigid sphere.[1] Unbaffled radiators can be treated numerically by the boundary element method discussed in Chapter 13.

11 SOURCES IN UNIFORM MOTION

11.1 Monopole Point Source[1,5]

The sound field produced by a monopole point source of frequency ω in uniform motion with speed U_1 in the 1-direction satisfies the inhomogeneous wave equation

$$\left(\frac{1}{c^2}\frac{\partial^2}{\partial t^2} - \nabla^2\right) p = \frac{\partial}{\partial t}(\delta(\mathbf{x} - \mathbf{U}t)\hat{Q}e^{i\omega t}), \tag{119}$$

where $\mathbf{U} = (U_1, 0, 0)$ and $\mathbf{U}t$ is the source position at time t. This equation has solution

$$p(\mathbf{x}, t) = \frac{\partial}{\partial t}\left(\frac{\hat{Q}e^{i\omega\tau}}{4\pi R|1 - M\cos\theta|}\right). \tag{120}$$

The Mach number M is equal to U_1/c; R and θ are the distance and direction of the listener at $\mathbf{x}$ from the source at emission time τ, as illustrated in Fig. 9:

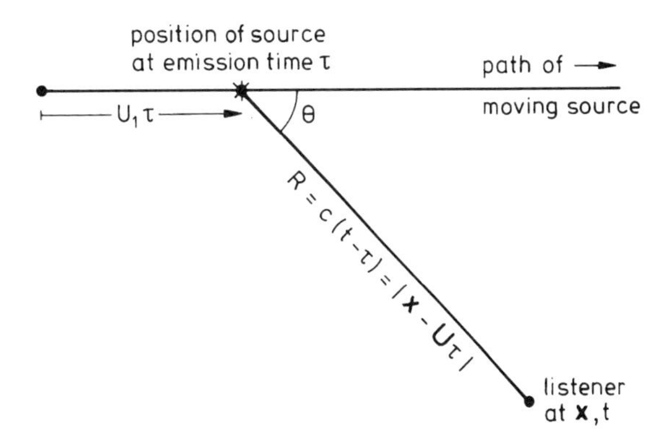

Fig. 9 Emission from a source with velocity **U**.

$$R = |\mathbf{x} - \mathbf{U}\tau| \quad \text{and} \quad \cos\theta = (x_1 - U_1\tau)/R. \tag{121}$$

The emission time τ satisfies

$$c(t - \tau) = |\mathbf{x} - \mathbf{U}\tau|. \tag{122}$$

If U_1 is subsonic, Eq. (122) has one root. When U_1 is supersonic, it has no real roots outside the Mach cone, $U_1 t - x_1 = (M^2 - 1)^{1/2}(x_2^2 + x_3^2)^{1/2}$, and two real roots within it. The pressure is then the sum of contributions from these two emission times.

Differentiation of Eq. (122) shows that

$$\left.\frac{\partial\tau}{\partial t}\right|_{\mathbf{x}} = \frac{1}{1 - M\cos\theta}, \tag{123}$$

which may be used to evaluate the derivative in Eq. (120):

$$p(\mathbf{x}, t) = \frac{i\omega\hat{Q}e^{i\omega\tau}}{4\pi R(1 - M\cos\theta)|1 - M\cos\theta|} + \frac{U_1(\cos\theta - M)\hat{Q}e^{i\omega\tau}}{4\pi R^2|1 - M\cos\theta|^3}. \tag{124}$$

In the far field the first term on the right-hand side of Eq. (124) is the largest. Then there is no effect of source motion in the direction $\theta = 90°$. The pressure directly in front of the moving source ($\theta = 0$) is larger than the pressure the same distance behind the source ($\theta = \pi$) by a factor $(1 + M)^2/(1 - M)^2$.

The frequency ω_l of the sound heard by a listener at $\mathbf{x}$ can be defined as the time derivative of the phase $\omega\tau$:

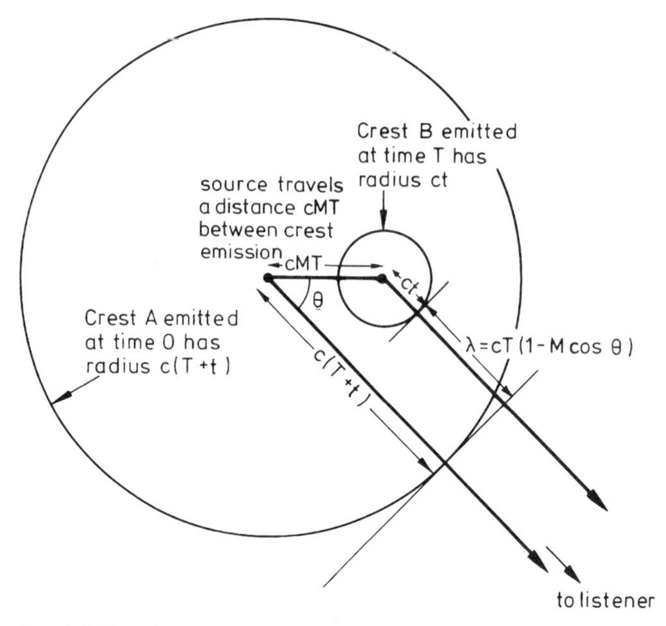

Fig. 10 Sound field at time $T + t$. Moving source emits crests at intervals $T = 2\pi/\omega$. In a direction θ, crest A leads crest B by a wavelength $cT(1 - M\cos\theta)$.

$$\omega_l = \omega \left.\frac{\partial\tau}{\partial t}\right|_{\mathbf{x}} = \frac{\omega}{1 - M\cos\theta}. \quad (125)$$

The sound radiated by a moving source of frequency ω is heard at $\mathbf{x}$ at the *Doppler-shifted* frequency $\omega/(1 - M\cos\theta)$. The frequency shift occurs because source motion causes a contraction $1 - M\cos\theta$ in the wavelength of sound traveling in the direction θ (see Fig. 10).

11.2 Dipole Point Source[1,5]

The distant sound field produced by a dipole point source of frequency ω in uniform motion with speed U_1 in the 1-direction satisfies the inhomogeneous wave equation

$$\left(\frac{1}{c^2}\frac{\partial^2}{\partial t^2} - \nabla^2\right) p = -\frac{\partial}{\partial x_i}(\delta(\mathbf{x} - \mathbf{U}t)\hat{F}_i e^{i\omega t}). \quad (126)$$

This has solution

$$p(\mathbf{x}, t) = -\frac{\partial}{\partial x_i}\left(\frac{\hat{F}_i e^{i\omega\tau}}{4\pi R|1 - M\cos\theta|}\right) \quad (127)$$

$$= \frac{i\omega\hat{F}_R e^{i\omega\tau}}{4\pi Rc(1 - M\cos\theta)|1 - M\cos\theta|} + \frac{[(1 - M^2)\hat{F}_R - M(1 - M\cos\theta)\hat{F}_1]e^{i\omega\tau}}{4\pi R^2|1 - M\cos\theta|^3}, \quad (128)$$

where $\hat{F}_R = \hat{\mathbf{F}}\cdot(\mathbf{x} - \mathbf{U}\tau)/R$ is the component of $\hat{\mathbf{F}}$ in the direction of the listener at emission time.

Dipole Axis in the Direction of Motion If $\hat{\mathbf{F}}$ is in the 1-direction, $\hat{\mathbf{F}} = (\hat{F}_1, 0, 0)$, Eq. (128) simplifies to

$$p(\mathbf{x}, t) = \frac{i\omega\cos\theta\hat{F}_1 e^{i\omega\tau}}{4\pi Rc(1 - M\cos\theta)|1 - M\cos\theta|} + \frac{(\cos\theta - M)\hat{F}_1 e^{i\omega\tau}}{4\pi R^2|1 - M\cos\theta|^3}. \quad (129)$$

Dipole Axis Normal to Direction of Motion If $\hat{\mathbf{F}}$ is in the 2-direction, $\hat{\mathbf{F}} = (0, \hat{F}_2, 0)$, Eq. (128) simplifies to

$$p(\mathbf{x}, t) = \frac{i\omega x_2 \hat{F}_2 e^{i\omega\tau}}{4\pi R^2 c(1 - M\cos\theta)|1 - M\cos\theta|} + \frac{x_2(1 - M^2)\hat{F}_2 e^{i\omega\tau}}{4\pi R^3 |1 - M\cos\theta|^3}. \qquad (130)$$

Far-Field Pressure When the listener is far from the source, the first term of the right-hand sides of Eqs. (128)–(130) describes the major contribution to the sound field.

11.3 Pulsating Sphere[5,10]

In Section 8.1, a stationary, compact pulsating sphere was found to produce the same sound field as a point monopole. That does not remain true when the sphere is in motion, even if the flow is considered to be irrotational. The far-field pressure generated by a compact sphere of variable radius $a + \hat{\epsilon} e^{i\omega t}$ and velocity $\mathbf{U} = (U_1, 0, 0)$ is given as

$$p(\mathbf{x}, t) = -\frac{\omega^2 \rho a^2 \hat{\epsilon} e^{i\omega\tau}}{R(1 - M\cos\theta)^{7/2}} \quad \text{for} \quad M^2 \ll 1. \qquad (131)$$

The effect of motion on a pulsating body is different from the effect of motion on a monopole point source. This is because a moving pulsating body producing a mass flux necessarily has a momentum flux associated with it. Hence the sound field produced is that of a moving monopole and a coupled dipole. Moreover, the pressure field due to the dipole is smaller by only a factor of the order of the Mach number than that due to the monopole and so affects the Doppler amplification factor.

11.4 Vibrating Sphere[10]

In Section 8.2, a stationary, compact vibrating sphere was found to produce the same sound field as a point dipole. However, its pressure field is affected differently by source motion from that of a point dipole. In irrotational flow, a rigid sphere with velocity $\mathbf{U} + \hat{\mathbf{U}} e^{i\omega t}$ generates a pressure field

$$p(\mathbf{x}, t) = -\frac{\omega^2 \rho a^3}{2Rc} \left(\frac{\hat{U}_R}{(1 - M\cos\theta)^4} - \tfrac{1}{3} M \hat{U}_1 \right) e^{i\omega\tau}, \qquad M^2 \ll 1, \qquad (132)$$

where $\mathbf{U} = (\mathbf{U}_1, 0, 0)$ and $M = U_1/c$; $\hat{U}_R$ is the component of $\hat{U}$ in the direction of the listener at emission time. If the sphere vibrates in the same direction as its mean velocity, the last term in Eq. (132) is nonzero. It represents an omnidirectional field, which produce amplification even at 90° to the motion. In practice, such a sphere would shed vorticity, leading to additional effects of flow.

REFERENCES

1. P. M. Morse and K. U. Ingard, *Theoretical Acoustics*, McGraw-Hill, New York, 1968; reissued Princeton University Press, 1987.
2. M. J. Lighthill, *Waves in Fluids*, Cambridge University Press, Cambridge, 1978.
3. L. E. Kinsler, A. R. Frey, A. B. Coppens, and J. V. Sanders, *Fundamentals of Acoustics*, Wiley, New York 1982.
4. A. D. Pierce, *Acoustics: An Introduction to Its Physical Principles and Applications*, McGraw-Hill, New York, 1981; revised edition, Acoustical Society of America, New York, 1989.
5. A. P. Dowling and J. E. Ffowcs Williams, *Sound and Sources of Sound*, Ellis Horwood, Chichester, 1983.
6. E. Skudrzyk, *Simple and Complex Vibratory Systems*, Pennsylvania State University Press, 1968.
7. M. Abramowitz and I. A. Stegun, *Handbook of Mathematical Functions*, Dover, New York, 1965.
8. L. M. Lyamshev, "A Question in Connection with the Principle of Reciprocity in Acoustics," *Sov. Phys. Dokl.*, Vol. 4, 1959, pp. 405–409.
9. P. A. Frost and E. Y. Harper, "Acoustic Radiation from Surfaces Oscillating at Large Amplitude and Small Mach Number," *J. Acoust. Soc. Am.*, Vol. 58, 1975, pp. 318–325.
10. A. Dowling, "Convective Amplification of Real Simple Sources," *J. Fluid Mech.*, Vol. 74, 1976, pp. 529–546.

9

TRANSIENT RADIATION

PETER STEPANISHEN

1 INTRODUCTION

Acoustic signals of interest are transient in nature since the signals exist over a limited time duration. Common acoustic transients include speech, impact noise such as a baseball striking a bat, musical sounds, and numerous other biologic and/or machinery-induced acoustic transients. In addition to these more well known sources of acoustic energy, acoustic transients are generated by a wide variety of devices or transducers for applications that include the remote location of underwater objects or subbottom oil and gas deposits and biomedical applications that involve noninvasive diagnostic testing and imaging using ultrasonic pulse echo systems.

This chapter briefly addresses the subject of linear acoustic transient radiation from a variety of sources, whereas the inclusion of nonlinear effects and the propagation of finite-amplitude waves in fluids that ,are of importance in transient explosive and large-amplitude problems are addressed in Chapter 17. The recent use of time-domain methods has led to an improved physical understanding of acoustic transient phenomena and will thus be emphasized here.

A statement of the general linear acoustic transient radiation problem is first presented. The solution of the initial-value problem in which the acoustic field is specified at a reference time in a fluid with no boundaries or sources is then presented, and the introduction of sources into the fluid is then addressed. Finally, acoustic transient radiation from sources on boundaries and impact noise are addressed.

Note: References to chapters appearing only in the *Encyclopedia* are preceded by *Enc.*

2 STATEMENT OF THE GENERAL PROBLEM

Consider now the acoustic transient radiation and scattering problem of interest as illustrated in Fig. 1. A specified acoustic field is assumed to exist in the fluid at time t_0. A general distribution of space- and time-dependent acoustic sources is also present in the fluid and is specified here to be zero for time $t < t_0$. In addition, an elastic structure of arbitrary shape is present in the fluid.

The basic problem of interest is to determine the acoustic pressure field in the fluid that results from the above-noted initial conditions, sources in the fluid, and/or the vibration of the structure. From a mathematical viewpoint the acoustic field can be represented as the solution of an initial-boundary-value problem.[1] For the sake of generality normalized space and time coordinates $(\mathbf{x}, t)$ are introduced here, as shown in Table 1.

The general initial boundary-value problem of interest can now be stated in normalized coordinates as

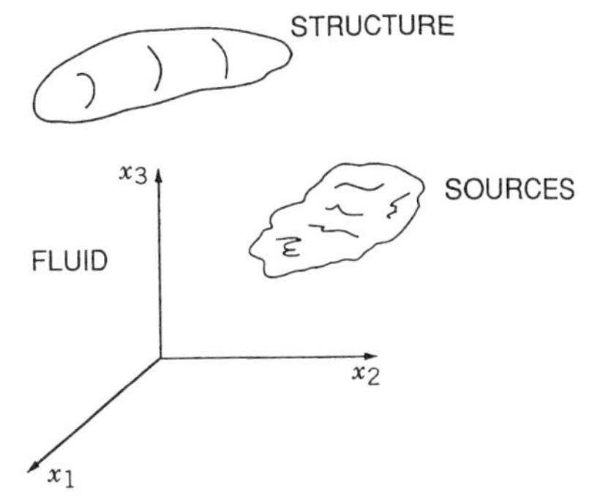

Fig. 1 Acoustic sources and a structure in a fluid.

TABLE 1 Normalization Factors

Variable	Normalization Factor
Length	L
Velocity	c
Density	ρ
Time	L/c
Pressure	ρc^2

Note: L = characteristic length; c = acoustic wave speed of fluid; ρ = density of fluid.

$$\left(\frac{\partial}{\partial t^2} - \nabla^2\right)\phi(\mathbf{x},t) = q(\mathbf{x},t) \quad \text{where } t_0 < t < \infty, \quad \mathbf{x} \text{ in } V;$$

$$\phi(\mathbf{x},t_0) \equiv f_1(\mathbf{x});$$

$$\frac{\partial\phi(\mathbf{x},t)}{\partial t} \equiv f_2(\mathbf{x}), \qquad t = t_0;$$

$$\hat{n}\cdot\nabla\phi = u(\mathbf{x},t), \quad \mathbf{x} \text{ on } \sigma;$$

$$p(\mathbf{x},t) = \frac{\partial\phi(\mathbf{x},t)}{\partial t}, \tag{1}$$

where $\phi(\mathbf{x},t)$ is the velocity potential, $p(\mathbf{x},t)$ is the pressure, $u(\mathbf{x},t)$ is the normal velocity of the structure in the normal direction $\hat{n}$, and $f_1(\mathbf{x})$ and $f_2(\mathbf{x})$ are initial conditions. The distribution $q(\mathbf{x},t)$ represents the sources in the fluid with volume V.

Rather than present a general solution to the above initial-boundary-value problem, a series of reduced acoustic transient problems will be addressed in the following sections. The solutions of these individual problems more clearly illustrate the general phenomena of interest associated with intitial conditions, sources in the fluid, and sources on structures. Furthermore, the individual solutions can be superimposed to address the more general problem since the basic problem as formulated in Eq. (1) is a linear one. The transient scattering and diffraction problem is addressed in Chapter 33.

3 INITIAL-VALUE PROBLEM

The simplified case of a source-free fluid with no structure and a specified set of finite-valued initial conditions in the fluid is first addressed using Eq. (1). After Laplace transforming Eq. (1) with respect to time and then solving the resulting inhomogeneous equation using a Green's function method, a straightforward inverse Laplace transform leads to the following solution[2] for the time-dependent velocity potential:

$$\phi(\mathbf{x},t) = \int_V dV_0\, g(\mathbf{x},t|\mathbf{x}_0,t_0) f_2(\mathbf{x}_0) + \frac{\partial}{\partial t}\left[\int_V dV_0\, g(\mathbf{x},t|\mathbf{x}_0,t_0) f_1(\mathbf{x}_0)\right], \tag{2}$$

where the indicated integrals are volume integrals and $g(\mathbf{x},t|\mathbf{x}_0,t_0)$ denotes the time-dependent free-space Green's function,[1,2] that is,

$$g(\mathbf{x},t|\mathbf{x}_0,t_0) = \frac{\delta(t - t_0 - |\mathbf{x}-\mathbf{x}_0|)}{4\pi|\mathbf{x}-\mathbf{x}_0|}, \qquad -\infty < t, t_0 < \infty, \qquad \mathbf{x},\mathbf{x}_0 \text{ in } V, \tag{3}$$

and δ denotes the Dirac delta function.

After positioning a spherical coordinate system at the spatial point of interest and using the sifting property of the Dirac delta function, it is easily shown that the volume integral in Eq. (2) can be reduced to the following surface integral solution, which was originally developed by Poisson (1819):

$$\phi(\mathbf{x},t) = \frac{1}{4\pi}(t-t_0)\int_\Omega d\Omega\, f_2(\mathbf{x} + (t-t_0)\hat{\mathbf{a}}_x) + \frac{1}{4\pi}\frac{\partial}{\partial t}(t-t_0)\int_\Omega d\Omega\, f_1(\mathbf{x} + (t-t_0)\hat{\mathbf{a}}_x) \quad \text{for } t \ge t_0, \tag{4}$$

where Ω indicates a solid angle and $\hat{\mathbf{a}}_x$ represents a unit vector in the radial direction centered at $\mathbf{x}$.

The solution of Eq. (4) has a simple and intuitively obvious interpretation. In brief, the two terms in Eq. (4) correspond to an initial acoustic pressure and a particle velocity field, respectively, that are distributed throughout space at $t = t_0$. As time progresses, the resultant field at any point is simply related to the spatial average of the initial field over a spherical surface centered at the spatial point of interest and with a radius equivalent to the elapsed time $t - t_0$. Such a wave interpretation is consistent with the use of Huygens's principle[3,4] to develop the solution to the initial-value problem.

4 SPACE- AND TIME-DEPENDENT VOLUMETRIC SOURCES

The next simplest type of transient field problem to address is the case of a field with a space- and time-dependent source distribution in the fluid but with no initial conditions or internal structure in the fluid. Mathematically the statement of the problem reduces to the solution of an inhomogeneous wave equation with a radiation condition on the field.[1,2] For a specified acoustic source distribution $q(\mathbf{x}, t)$ the use of a standard Green's function development leads to the following solution for the time-dependent velocity potential:

$$\phi(\mathbf{x}, t) = \int_{t_0}^{t} dt_0 \int_V d\mathbf{x}_0\, g(\mathbf{x}, t|\mathbf{x}_0, t_0) q(\mathbf{x}_0, t_0). \quad (5)$$

Transient radiation from point source time-dependent multipoles is of fundamental importance in understanding transient radiation from sources since they can be used as building blocks for addressing radiation from complex sources. Steady-state radiation from harmonic multipoles is addressed in Chapter 8. The most useful of these time-dependent multipoles for present purposes are the monopole and dipole. Monopole sources are associated with time-varying volume changes within a fluid and as such are of fundamental importance in sound generation by bubble oscillations,[5,6] combustion noise,[5] parametric sources,[7] and laser-induced sound.[8] Dipole sources are associated with forces acting on a fluid and are of fundamental importance in sound generation by structures or boundaries in a fluid.[9] Quadrupole sources are of fundamental importance in noise generation by turbulence, which is discussed in Chapter 24.

A simple illustration of a monopole source is a uniformly pulsating sphere with a volume flow rate denoted by $Q(t)$. Such a source generates an omnidirectional time-dependent field. In contrast to the monopole, the net volume flow rate for a dipole source is zero and the dipole source can be used to represent the radiation from a small translating sphere.[10] The pressure field for the monopole can be expressed in several forms as

$$\begin{aligned} p(\mathbf{x}, t) &= \frac{\partial}{\partial t} \frac{Q(\mathbf{x}_0, t - r)}{4\pi r} \\ &= \frac{\partial}{\partial t} [g(\mathbf{x}, t|\mathbf{x}_0, 0) \otimes Q(\mathbf{x}_0, t)], \end{aligned} \quad (6)$$

where $r = |\mathbf{x} - \mathbf{x}_0|$, $Q(\mathbf{x}_0, t)$ denotes the volume flow rate of the source at $\mathbf{x}_0$, and the $\otimes$ denotes the convolution operator in time,[11] that is,

$$a(\mathbf{x}, t) \otimes b(\mathbf{x}, t) = \int a(\mathbf{x}, t - \tau) b(\mathbf{x}, \tau)\, d\tau. \quad (7)$$

The pressure field for the dipole can also be expressed in several forms as

$$\begin{aligned} p(\mathbf{x}, t) &= \nabla_0 \cdot \frac{\mathbf{f}(\mathbf{x}_0, t - r)}{r} \\ &= g_d(r, t) \otimes f(t) \quad \text{where } \mathbf{f}(\mathbf{x}, t) = f(t)\hat{r}_f, \\ \cos\theta &= \hat{r}_f \cdot \hat{r}_r, \\ g_d(r, t) &= \cos\theta \frac{\partial g(r, t)}{\partial r} \\ &= \frac{\cos\theta}{4\pi} \left(\frac{\delta(t - r)}{r^2} + \frac{1}{r} \delta'(t - r) \right). \end{aligned} \quad (8)$$

In constrast to the field from the monopole source, which exhibits the same time history at all field points, apart from the inverse range dependence, the field from the dipole source exhibits a near-field r^{-2} and a far-field r^{-1} range dependence along with a directionality effect indicated by the cos θ dependence for the pressure field.

In general, the transient radiated pressure field from a monopole source distribution in a fluid can be determined via the principle of superposition, which leads to the equation

$$p(\mathbf{x}, t) = \frac{1}{4\pi} \frac{\partial}{\partial t} \int_V \frac{q(\mathbf{x}_0, t - |\mathbf{x} - \mathbf{x}_0|)}{|\mathbf{x} - \mathbf{x}_0|}\, dV_0, \quad (9a)$$

which in the far field can be expressed via the usual approximations as

$$p(\mathbf{x}, t) = \frac{1}{4\pi|\mathbf{x}|} \frac{\partial}{\partial t} \int_V q(\mathbf{x}_0, t - |\mathbf{x} - \mathbf{x}_0|)\, dV_0, \quad (9b)$$

where $q(\mathbf{x}, t)$ is used to denote a volumetric source strength density. In a similar manner the transient radiated pressure field from a dipole source distribution in a fluid can be determined via the principle of superposition, which leads to the equation

$$p(\mathbf{x}, t) = -\frac{1}{4\pi} \int_V \frac{\nabla_0 \cdot \mathbf{f}(\mathbf{x}_0, t - |\mathbf{x} - \mathbf{x}_0|)}{|\mathbf{x} - \mathbf{x}_0|}\, dV_0, \quad (10)$$

where $\mathbf{f}(\mathbf{x}, t)$ is used here to denote a force density. Equations (9) and (10) form the basis for evaluating the acoustic transient radiation from an elastic structure in a fluid, as noted in the following section.

To illustrate some interesting transient phenomena, consider a thin finite-length line source with a uniform source strength density $q(t)$. The volume integral in Eq. (9) can then be reduced to a line integral. In the far field it can be shown that the pressure resulting from an arbitrary pulsed excitation $q(t)$ can be represented as a sum of two space-dependent pulses where each pulse is a scaled and time-delayed replica of $q(t)$ which appears to originate at an end of the line distribution. If the pulse duration of $q(t)$ is less than the travel time over the source, then two distinct pulses will be observed in regions of the field.

As a second example, consider the case of a uniform monopole source density distribution distributed over a thin finite planar surface. Once again the pressure field can be simply obtained from Eq. (9), with the volume integral now reduced to a surface integral over the aperture. It can be shown that transient radiation from a finite-sized uniform planar source distribution can thus be expressed as a sum of a plane-wave contribution and an edge-wave contribution that originates from the boundary of the distribution. Such a result is to be expected in light of the importance of endpoint contributions for the line source problem as noted above.

5 TRANSIENT RADIATION FROM STRUCTURAL BOUNDARIES

Consider now the acoustic transient radiation from a structure in a fluid, as illustrated in Fig. 1. The initial conditions in the fluid are assumed to be zero and the medium is homogeneous and source free. It can be readily shown by a standard Green's function development[1,2] that the pressure field can be expressed as the following integral over the surface of the structure:

$$p(\mathbf{x},t) = \frac{1}{4\pi}\int_\sigma \left(\frac{1}{r}[a] + \frac{1}{r^2}\frac{\partial r}{\partial n}[p] + \frac{1}{r}\frac{\partial r}{\partial n}\left[\frac{\partial p}{\partial t}\right]\right) dS_0, \tag{11}$$

where $r = |\mathbf{x} - \mathbf{x}_0|$ and $[\cdot]$ denotes the retarded time of the quantity in the brackets, that is, $[a] = a(\mathbf{x}_0, t - r)$, where a is acceleration. It is thus apparent that the normal acceleration, pressure, and its time derivative over the surface are in general required to determine the pressure in the fluid via the use of Eq. (11), which is the well-known Kirchhoff retarded-potential solution of the acoustic transient-field problem.

An alternative form of the surface integral solution for the pressure field can be obtained via the sifting property of the Dirac delta function, which then leads to the expression

$$p(\mathbf{x},t) = \int_\sigma \left[\frac{\partial u(\mathbf{x}_0,t)}{\partial t} \otimes g(\mathbf{x},t|\mathbf{x}_0,0) + p(\mathbf{x}_0,t) \otimes g_d(\mathbf{x},t|\mathbf{x}_0,0)\right] dS_0, \tag{12}$$

which is a convolution integral form of the solution in which the dependence of the external field upon the surface acceleration and pressure is explicitly noted. It is immediately apparent from Eqs. (6) and (8) that the pressure field can now be interpreted as arising from a monopole source distribution associated with the surface acceleration and a dipole source distribution associated with the surface pressure.

If the structure of interest is compact for the time-dependent signals of interest, that is, if the smallest time scale of interest is much greater than the maximum acoustic travel time between any two points on the structure, Eq. (12) reduces to the field associated with a point monopole and a point dipole. The associated monopole and dipole strengths correspond to surface-integrated values. This low-frequency result is useful if the time-dependent velocity and pressure are known or can be estimated.

Acoustic transient radiation from a structure of arbitrary shape with a specified normal velocity or acceleration is now discussed as a result of its relative simplicity. Via a standard Green's function development[1,2] the field resulting from the specified normal velocity of the structural boundary surface shown in Fig. 1 can be expressed as

$$\phi(\mathbf{x},t) = \int_t \int_\sigma G(\mathbf{x},t|\mathbf{x}_0,t_0)u(\mathbf{x}_0,t_0)\,dS_0\,dt_0,$$

$$p(\mathbf{x},t) = \frac{\partial\phi(\mathbf{x},t)}{\partial t}, \tag{13}$$

where $G(\mathbf{x},t|\mathbf{x}_0,t_0)$ is the Green's function for the Neumann boundary-value problem of interest. Since the normal velocity of the structure, that is, $u(\mathbf{x},t)$, can be expressed as

$$u(\mathbf{x},t) = \sum_n u_n(t)\psi_n(\mathbf{x}) \tag{14}$$

where the $\psi_n(\mathbf{x})$ are a suitably chosen set of basis functions and $u_n(t)$ are the associated modal velocities, it follows after some simple algebra that the solution to the associated boundary-value problem can be expressed as

$$\phi(\mathbf{x}, t) = \sum_n \int_\sigma \int_t G(\mathbf{x}, t|\mathbf{x}_0, t_0) u_n(t_0) \psi_n(\mathbf{x}_0)\, dS_0\, dt. \tag{15}$$

An impulse response representation of the solution then results from integrating Eq. (15) over space to obtain

$$\phi(\mathbf{x}, t) = \sum_n h_n(\mathbf{x}, t) \otimes u_n(t), \tag{16}$$

where the $h_n(\mathbf{x}, t)$ are space- and mode-dependent impulse response functions defined as

$$h_n(\mathbf{x}, t) = \int_0 G(\mathbf{x}, t|\mathbf{x}_0, 0)\psi(\mathbf{x}_0)\, dS_0. \tag{17}$$

The case of a planar structure with a specified space- and time-dependent velocity is perhaps the simplest case. Although the Green's function for the planar problem is well known, that is, $G = 2g$, in cases of practical interest involving a finite-sized planar structure, the normal velocity is not known over the entire plane. It is a common assumption to assume that the structure is set in an infinite rigid planar baffle surrounding the radiator. The normal velocity over the entire plane of the structure reduces to that shown in Eq. (14), and the Green's function is known.

The impulse response method[12–16] for planar vibrators is well suited to investigate the spatial and temporal properties of transient acoustic fields from planar structures including ultrasonic transducers. General properties of the impulse response functions have been determined. For the case of planar radiators of finite size, the impulse response functions are nonzero only over a finite time duration, and the correspondence to finite impulse response filters[11] is noted here. Closed-form expressions for $h_n(\mathbf{x}, t)$ have been obtained for circular[13,14] and rectangular vibrators, and relatively simple expressions are available for $h_n(\mathbf{x}, t)$ with $\mathbf{x}$ in the far-field region of rectangular[15] and circular piston[16] sources.

To illustrate typical acoustic transient phenomena of interest for planar sources, consider a simple example of a circular piston or ultrasonic transducer with a specified gated sinusoidal velocity in which the characteristic length for the problem is the radius of the piston. The velocity of the piston is specified to consist of a single cycle with a normalized pulse duration T. In the far field, the on-axis pressure is simply a scaled replica of the acceleration, that is, a single-cycle cosine pulse, for all values of T. The on-axis far-field pressure is thus related to the volumetric acceleration of the source as previously indicated in Eqs. (6) and (9).

When $T \gg 1$, the pressure field from the piston source is essentially omnidirectional and the on-axis pressure is representative of the pressure field at all polar angles. This is to be expected since the piston is omnidirectional for the predominantly low frequency components in the energy spectrum of the velocity. As T decreases, there is an upward shift in the principal frequency components in the energy spectrum of the velocity, and the pressure field thus exhibits more directional effects. In addition a multipulse behavior occurs for $T \ll 1$. The multipulse structure is associated with the edge of the piston in a manner similar to that previously noted for the finite-line and planar source distributions. This behavior, which has been experimentally verified, is to be expected from the extensive work by Freedman[17–19] on related problems.

Acoustic transient radiation from a nonplanar structure of arbitrary shape can also be investigated using the impulse response method. It is noted that the normal velocity of such a structure can be represented as an eigenfunction expansion of the form shown in Eq. (14) in which the basis functions are of course dependent on the shape of the structure. Equations (16) and (17) are thus applicable to the more general nonplanar case when the appropriate Green's function G is used to evaluate the space- and mode-dependent impulse response functions. It is apparent from some related solutions of transient problems for specific geometries[20–22] that the impulse responses for nonplanar structures will in general exhibit decaying oscillatory responses characteristic of infinite impulse response filters, that is, the responses are not time limited or of finite time duration.

As a specific example of a nonplanar vibrator consider the case of a spherical vibrator.[20] Since the Green's function can be obtained in closed form using inverse Fourier transform methods, the impulse response functions can also be so determined. Hence, acoustic transient radiation from a spherical source with an arbitrary space- and time-dependent normal velocity can be determined using an impulse response method in the manner indicated earlier. Due to space constraints, the general case is not addressed here; however, some comments regarding the fields for sphere vibrating with a spherically symmetric time-dependent radial motion and a sphere vibrating as a rigid body with a rectilinear motion are noted here. The velocity of the sphere is again specified to consist of a single cycle of a sinusoidal signal with a normalized pulse duration T in which the characteristic length for the problem is now the radius of the sphere.

The pressure field resulting from a spherically symmetric time-dependent radial motion is first discussed. It can be shown that the pressure for $T \gg 1$ at each point is a cosine pulse that exhibits an inverse range dependence. The pressure in this regime is thus proportional to the surface acceleration, which is to be expected for this low-frequency case for the reasons noted earlier. The pressure in the high-frequency region ($T \ll 1$) corresponds to the velocity as would be expected. Results for the intermediate or resonance region T require more detailed study.

The pressure field for the case of a sphere vibrating as a rigid body with a rectilinear motion for a gated sinusoidal velocity is now discussed. The time history of the pressure field for this case changes with both range and pulse duration T. For the low-frequency case where $T \ll 1$ the spatial pressures exhibit a time dependence similar to that of the velocity. Spatial pressures in the high-frequency region ($T \ll 1$) generally exhibit an early time-impulsive-type behavior followed in time by an exponentially decaying negative tail. Such results illustrate the complex nature of the acoustic transient field from even a simple vibrator.

In light of the preceding example, it is apparent that rigid bodies in nonuniform motion can radiate acoustic transients. A relatively simple example is the pressure field generated by a rigid sphere with an impulsive acceleration[21–23]; that is, the velocity of the sphere is a step, or Heaviside, function. An exact solution for the pressure field can be obtained using either time-domain methods or the inverse Fourier transform method. The results for the pressure field indicate an exponentially decaying oscillatory pressure at each point in the field. This oscillatory field is associated with oscillatory surface pressure waves that traverse the surface of the sphere as a result of the spatially nonuniform normal velocity, which results in tangential pressure gradients and associated tangential accelerations. As such, the period of the oscillations is associated with the travel time for an acoustic disturbance to circumnavigate the sphere.

An interesting discussion of the energy exchange process between the source and the acoustic field[22] for the rigid-body sphere shows that acoustic energy is of course propagated to the far field during the acceleration of the body of a uniform velocity; however, an equivalent amount of energy is "entrained," or trapped, in the near field and is associated with the hydrodynamic mass. This equipartition of energy is clearly illustrated in the work of Junger,[21–23] who further illustrates that the total energy required to accelerate the body is radiated to the far field if the body is ultimately deaccelerated to rest or zero velocity. The kinetic energy associated with the acoustic near field is thus converted to the far field during the deacceleration process.

It is apparent from the preceding example that rigid bodies of arbitrary shape undergoing arbitrary transient motions will in general radiate oscillatory acoustic transients. As noted by Junger and Thompson,[22] these oscillations are to be expected as a result of the spatially nonuniform normal velocity and associated tangential accelerations that result in oscillatory surface pressure fields. In contrast to the case of the rigid sphere, more complex bodies undergoing spatially nonuniform time-dependent normal velocities can be expected to exhibit complex oscillatory pressure fields in which the periods of oscillations are linearly related to the size of the body. Although such a result can be inferred from the residue solution of the sphere, it is equally apparent from the solution of the Kirchhoff retarded-potential solution in Eq. (12). Such results are important in addressing the general subject of impact noise as noted in the following section.

6 IMPACT NOISE

Impact noise is an important topic involving acoustic transient phenomena. Such noise can arise in a vast number of different mechanical processes, as noted by Richards et al.[24–26] These processes include stamping, rivetting, forging, firing of reciprocating engines, and numerous other hammer-type operations. In addition to being of importance in the immediate workplace, the transmission of such impact noise through floors and other structures into other areas is also of concern. The subject of footfall noise is a specific case of interest, and related work has been recently summarized by Beranek and Ver,[27] who addressed periodic impact noise in buildings.

Impact noise can be broadly subdivided into two major areas: noise generated by the initially high surface accelerations and deaccelerations of impacting bodies during the time of contact and noise arising from the free vibration of the bodies following impact. The acceleration/deacceleration component of the process controls the early time response of the time-dependent field pressures, whereas the long time response is dominated by the free vibrations of the impacting bodies. Peak pressures are a function of the contact time of the excitation and the dynamic response of the impacting structures. The results of the preceding section clearly illustrate the complexity of the general problem.

The acceleration component of impact noise was addressed by Richards et al.[24] using a model of two impacting spheres. The importance of impact duration on the noise radiation process was clearly observed in an associated experimental study that led to curves for the peak sound pressure level as a function of contact time between the impacting bodies. For compact bodies incapable of flexural motion the study concludes that acceleration noise energy is of the same order of magnitude as that due to ringing and is less than 0.015% of the kinetic energy of the impacting bodies at the time of impact.

Subsequent studies by Richards and co-workers[25] focused on the importance of ringing noise following impact. These studies clearly illustrate the importance of flexural waves in the structures under impact in determining the relative contribution of acceleration versus ringing-noise energy. For noncompact bodies undergoing flexural motions following impact, the ringing-noise energy can be several orders of magnitude greater than the acceleration component. Although significant progress has been made in this area, it is clear that considerable work remains to be completed.

7 CONCLUDING COMMENTS

A brief overview of acoustic transient phenomena has been presented. Due to space limitations, a number of important related topics have been omitted. Some of these topics of interest are the following: acoustic transients in structures with internal fluid, for example, air conditioning and fluid piping systems exhibiting water hammer effects; acoustic transient loading effects in internally and externally fluid loaded structures; and acoustic transient radiation from fluid-loaded structures.

To address the acoustic transient radiation problem from structures subject to transient loading in air or a heavy fluid such as water, it is necessary to first address the transient structural vibration problem to determine the space- and time-dependent velocity or acceleration. Classical methods of addressing such transient vibration problems are well known.[28] A brief review of the vibration problem for bodies undergoing impact is also available.[29] Transient acoustic fluid-loaded vibration problems are a subject of present research interest. In this latter area the interested reader is directed to the texts of Junger and Feit[21] and Fahy,[30] which provide the basis for addressing the analogous harmonic problems. An understanding of the fundamentals related to the harmonic problem is considered essential prior to addressing the analogous transient problem.

REFERENCES

1. I. Stakgold, *Boundary Value Problems of Mathematical Physics*, Vol. 2, Macmillan, New York, 1968.
2. P. Morse and H. Feshbach, *Methods of Theoretical Physics*, McGraw-Hill, New York, 1953.
3. B. B. Baker and C. T. Copson, *The Mathematical Theory of Huygens Principle*, Clarendon, Oxford, 1953.
4. S. Hanish, "Review of World Contributions from 1945 to 1965 to the Theory of Acoustic Radiation," U.S. Naval Research Laboratory Mem. Rep. 1688, Washington, DC, 1966, Chapter 3.
5. A. P. Dowling and J. E. Ffowcs Williams, *Sound and Sources of Sound*, Halsted, Wiley, New York, 1983.
6. D. Ross, *Mechanics of Underwater Noise*, Pergamon, New York, 1976.
7. P. J. Westervelt, "Parametric Acoustic Array," *J. Acoust. Soc. Am.*, Vol. 35, 1963, pp. 535–537.
8. P. J. Westervelt and R. S. Larson, "Laser Excited Broadside Array," *J. Acoust. Soc. Am.*, Vol. 54, 1973, pp. 121–122.
9. S. Tempkin, *Elements of Acoustics*, Wiley, New York, 1981.
10. P. M. Morse and K. U. Ingard, *Theoretical Acoustics*, McGraw-Hill, New York, 1968.
11. A. Papoulis, *Signal Analysis*, McGraw-Hill, New York, 1977.
12. P. R. Stepanishen, "Transient Radiation from Pistons in an Infinite Planar Baffle," *J. Acoust. Soc. Am.*, Vol. 49, 1971, pp. 1629–1638.
13. P. R. Stepanishen, "Acoustic Transients from Planar Axisymmetric Vibrators Using the Impulse Response Approach," *J. Acoust. Soc. Am.*, Vol. 70, 1981, pp. 1176–1181.
14. P. R. Stepanishen, "Transient Radiation and Scattering from Fluid Loaded Oscillators, Membranes, and Plates," *J. Acoust. Soc. Am.*, Vol. 88, 1990, pp. 374–385.
15. P. R. Stepanishen, "Comments on Farfield of Pulsed Rectangular Radiator," *J. Acoust. Soc. Am.*, Vol. 52, 1972, p. 434.
16. P. R. Stepanishen, "Acoustic Transients in the Far Field of a Baffled Circular Piston Using the Impulse Response Approach," *J. Sound Vib.*, Vol. 32, 1974, pp. 295–310.
17. A. Freedman, "Transient Fields of Acoustic Radiators," *J. Acoust. Soc. Am.*, Vol. 48, 1970, pp. 135–138.
18. A. Freedman, "Farfield of Pulsed Rectangular Acoustic Radiator," *J. Acoust. Soc. Am.*, Vol. 49, 1971, pp. 738–748.
19. A. Freedman, "Sound Field of a Pulsed, Planar, Straight-Edged Radiator," *J. Acoust. Soc. Am.*, Vol. 51, 1972, pp. 1624–1639.
20. J. Brillouin, "Rayonnement transitoire des sources sonores et problemes conneses," *Ann. Telecommun.*, Vol. 5, 1950, pp. 160–172, 179–194.
21. M. C. Junger and D. Feit, *Sound Structures and Their Interaction*, MIT Press, Cambridge, MA, 1972.
22. M. C. Junger and W. Thompson, "Oscillatory Acoustic Transients Radiated by Impulsively Accelerated Bodies," *J. Acoust. Soc. Am.*, Vol. 38, 1965, pp. 978–986.
23. M. C. Junger, "Energy Exchange between Incompressible Near and Acoustic Far Field for Transient Sources," *J. Acoust. Soc. Am.*, Vol. 40, 1966, pp. 1025–1030.
24. E. J. Richards, M. E. Westcott, and R. K. Jeypalan, "On

the Prediction of Impact Noise, I: Acceleration Noise," *J. Sound Vib.*, Vol. 62, No. 4, 1979, pp. 547–575.

25. E. J. Richards, M. E. Westcott, and R. K. Jeypalan, "On the Prediction of Impact Noise, II: Ringing Noise," *J. Sound Vib.*, Vol. 65, No. 3, 1979, pp. 419–451.
26. J. M. Cushieri and E. J. Richards, "On the Prediction of Impact Noise IV: Estimation of Noise Energy Radiated by the Impact Excitation of a Structure," *J. Sound Vib.*, Vol. 86, No. 3, 1983, pp. 319–342.
27. I. L. Ver, "Interaction of Sound Waves with Solid Structures," in L. L. Beranek and I. L. Ver (Eds.), *Noise and Vibration Control Engineering*, Wiley, New York, 1992.
28. R. C. Ayre, in C. M. Harris and C. E. Crede (Eds.), *Shock and Vibration Handbook*, McGraw-Hill, New York, 1976.
29. W. C. Hoppmann, in C. M. Harris and C. E. Crede (Eds.), *Shock and Vibration Handbook*, McGraw-Hill, New York, 1976.
30. F. Fahy, *Sound and Structural Vibration: Radiation Transmission and Response*, Academic, New York, 1985.

10

ACOUSTIC INTERACTION BETWEEN STRUCTURES AND FLUIDS

F. J. FAHY

1 INTRODUCTION

Nearly all solid structures exist in surface contact with one or more fluid media, of which the most common are air and water. A few man-made exceptions now exist in space. Vibration generated in a solid structure is communicated to a fluid with which it is in contact via normal motion of the media interface. Everyday examples include the generation of audible sound in the air by vibrating structures such as machines, building components, and stringed instruments: This mechanism is one of the most common sources of acoustic noise. A related phenomenon of less obvious, but nonetheless considerable, practical importance is that of the excitation of vibration in solid structures by sound generated by sources in a contiguous fluid. In its more dramatic manifestations it can be responsible for severe damage to structures and connected components and even failure; a more commonly experienced example is the transmission of airborne sound through party walls between dwellings. Acoustic coupling between structural and fluid systems can significantly alter the free and forced vibration behavior of the coupled components from their uncoupled forms; as evidenced, for example, by the propagation of pressure waves in flexible pipes containing liquids, such as blood vessels: The speed of propagation is very much less than that in the fluid itself.

The general aims of this chapter are to explain the fundamental mechanism of the interaction phenomenon, to make the reader aware of situations and conditions in which fluid–structure interaction has to be considered as a major feature of system behavior, to present relevant and useful basic expressions and relationships, and to offer guidance on sources of more comprehensive information relating to the subject (including material contained in this book). The scope of the chapter does not extend to problems involving the interaction of vibrating structures with two-phase fluid media, such as the generation of surface waves in the sea by the motion of marine structures or the vibration of heat exchanger pipes filled with a mixture of vaporized and liquid water. The frequency range of concern is the so-called audio range of 20 Hz–20 kHz: Ultrasonic phenomena are excluded.

Note: References to chapters appearing only in the *Encyclopedia* are preceded by *Enc.*

2 NATURE OF FLUID–STRUCTURE INTERACTION

When a solid structure is caused to vibrate, it produces vibrational disturbances in any fluid with which it is in contact. Whatever the frequency of vibration, the resulting internal fluid forces (pressures) and motions are governed by the same equation, known as the *acoustic wave equation*: The disturbances constitute a *sound field*. This sound field is uniquely determined by (i) the properties of the fluid, (ii) the geometry of the vibrating surface(s), (iii) the acoustic properties and geometric distribution of any other "passive" surfaces bounding the fluid, and (iv) the spatial distribution of the component of vibrational acceleration normal to the surface of the vibrating structure(s). The agent that couples the fluid to the structure is the fluid pressure at the interface, and that which couples the structure to the fluid is the surface acceleration. The relationship is expressed mathematically by the Kirchhoff–Helmholtz integral equation (see Chapters 11 and 13).

It is useful to make a distinction between two basic geometric configurations: a structure may completely, or largely, enclose a fluid with which it interacts or it may be immersed in a fluid volume of which it does not itself form the outer boundaries and that may extend to very large distances in terms of structural dimensions. Examples are, respectively, the air within an aircraft fuselage and the sea in which a submarine is immersed. The reason for making this distinction is that the nature of the fluid reaction to boundary vibration takes significantly different forms in the two cases. An enclosed sound field exhibits the phenomenon of *standing waves* or *acoustic modes*, which possess characteristic frequencies; the fluid can consequently resonate and can also store vibrational energy. The result is that the fluid reaction to boundary motion can vary very strongly with frequency and energy can oscillate between the fluid and bounding elastic structures. An "unenclosed" fluid volume does not exhibit resonant behavior, and sound energy generated by boundary vibration flows away, not to return (although some energy flux oscillation does occur very close to a vibrating boundary). The fluid-loading effects in this latter case are generally less strong and less variable with frequency than those in the enclosed-fluid configuration.

Whether the independent sources of vibration operate directly on the fluid or on the structure, fluid–structure interaction couples the two components to form an integral vibrating system. The influence of such coupling on the vibrational behavior of structures and fluids varies greatly from system to system. The strength of fluid–structure interaction is proportional to mean fluid density; consequently, the fluid-loading effect of liquids and highly compressed gases is generally very much greater than that of gases at near atmospheric pressure. However, the high-frequency vibration of stiff, lightweight structures, such as aerospace honeycomb sandwich constructions, can also be significantly affected by air loading.

Vibrations of structural and fluid systems are wave phenomena. Consequently, fluid–structure acoustic interaction is, by definition, an "extended" reaction; this means that the field at any one point on the interface between the media is, in principle, influenced by conditions at all other points on the interface. This makes analysis and evaluation rather complicated in most practical situations for which simple analytical models are not generally available. In many engineering systems, the geometric configurations and material characteristics of the components are so complicated that only numerical solutions to the equations governing coupled fluid–structure vibration are viable (see Chapters 13 and 32). However, the general nature of the interaction process may be qualitatively revealed by idealized examples, as illustrated throughout this chapter.

3 INTERACTION OF PLANE STRUCTURES WITH SEMI-INFINITE FLUID VOLUMES

3.1 Simple Boundary Source Model: The Rayleigh Integral

Vibration of a surface structure displaces fluid volume at the interface. The pressure field generated in an infinitely extended fluid at a distance r from a small piston element of area δS set in an otherwise *rigid plane* surface, and in time-harmonic oscillation at circular frequency ω, is given by $p(r,t) = (i\omega\rho_0 V_n \delta S/2\pi r)\exp[i(\omega t - kr)]$, where V_n is the surface normal velocity, ρ_0 is the mean fluid density, and k is the wavenumber $(=\omega/c)$, where c is the fluid sound speed. The pressure is seen to decrease linearly with distance from the source. A continuous nonuniform distribution of surface normal velocity may be represented by a distribution of such piston elements, and according to the principle of linear superposition, the total sound pressure generated by such vibration may be represented by a summation, due account being taken of distance and phase through r and kr. The expression in integral form is

$$p(r,t) = \frac{i\omega\rho_0}{2\pi} e^{i\omega t} \int_S \frac{V_n(r_s)e^{-ikR}}{R} \, dS, \qquad (1)$$

where $R = |r - r_s|$ is the distance between the observation point and the position of the surface element at r_s.

This is known as Rayleigh's first integral. Note carefully that it strictly applies only to infinitely extended plane surfaces on which the surface normal velocity V_n is *known at all points* on the plane; consequently, it cannot be directly applied to a plane vibrating structure of limited extent, because the normal velocity on the plane beyond the limits of the structure is not known a priori. The fluid pressure (sound) field is determined by the distribution of *normal acceleration* $i\omega V_n$ over the plane surface. (Rayleigh's second integral expresses the radiated pressure in terms of the distribution of pressure over the whole surface.)

3.2 Fluid Loading on Oscillating Rigid Circular Piston

Application of the Rayleigh integral to a system consisting of a rigid, plane, circular piston of radius a that is surrounded by an in-plane infinite rigid baffle and vibrates harmonically in a direction normal to the plane yields the following expression for the ratio of the spatial average fluid pressure on the piston to the volume velocity $(V_n \pi a^2)$ of the piston, defined as the radiation impedance of the piston:

$$Z_{\text{rad}} = R_{\text{rad}} + iX_{\text{rad}}, \tag{2}$$

where

$$R_{\text{rad}} = \frac{\rho_0 c}{\pi a^2}\left[1 - \frac{2J_1(2ka)}{2ka}\right],$$

$$X_{\text{rad}} = \frac{\rho_0 c}{\pi a^2}\left[\frac{2H_1(2ka)}{2ka}\right],$$

where J_1 is the Bessel function of order 1 and H_1 is the Struve function of order 1. The real (in-phase) and the quadrature (phase $\pi/2$) components of the nondimensional impedance are known respectively as the acoustic resistance ratio and reactance ratio; they are plotted in Fig. 1. The piston circumference equals the acoustic wavelength when $ka = 1$. Note that the asymptotic value ($2ka \to \infty$) of the resistance ratio is unity (i.e., the same as a plane wave), whereas the reactance decreases to negligible values. By contrast, the reactance greatly exceeds the resistance when $ka < 2$. When $ka \ll 1$, the expressions for radiation resistance and reactance are well approximated by $R_{\text{rad}} \approx (\rho_0 c/\pi a^2)(ka)^2/2$ and $X_{\text{rad}} \approx (\rho_0 c/\pi a^2)(8/3\pi)(ka)$.

The resistive and reactive components of the radiation impedance have distinct physical interpretations. The former represents the ability of the piston to do net work on the fluid over one cycle of oscillation, that is, to radiate sound power into the fluid, given by $W = \frac{1}{2}\int_S \text{Re}\{pV_n^*\}\, dS$, where the quantities in brackets are complex amplitudes.

The latter represents the generation of reactive power by which kinetic energy is put into the fluid close to the piston during half a cycle, and equal energy is returned to the piston during the other half. Positive reactance indicates that the fluid applies an inertial-type load to the piston, which, when $ka \ll 1$, can be expressed in terms of an added mass of magnitude $M_a = \frac{8}{3}\rho_0 a^3$. The sound power radiation acts like a viscous damper of coefficient given by $B_a = \frac{1}{2}[\rho_0 c\pi a^2 (ka)^2]$. If such a piston of mass M is mounted on a damped spring suspension of stiffness S and damping coefficient B, the mechanical impedance of the coupled system will be given by $Z_c = i\omega(M + M_a - S/\omega^2) + (B + B_a)$, and the resonance frequency of the piston will be lowered by the fluid loading.

An analysis of *unbaffled* piston radiation is presented in Ref. 1. Figure 1 compares baffled and unbaffled radiation impedances.

The nondimensional radiation impedance $Z_{\text{rad}}(\pi a^2/\rho_0 c)$ of a rigid piston radiating into an anechoically terminated tube of the same radius, when $ka < 1$, is equal

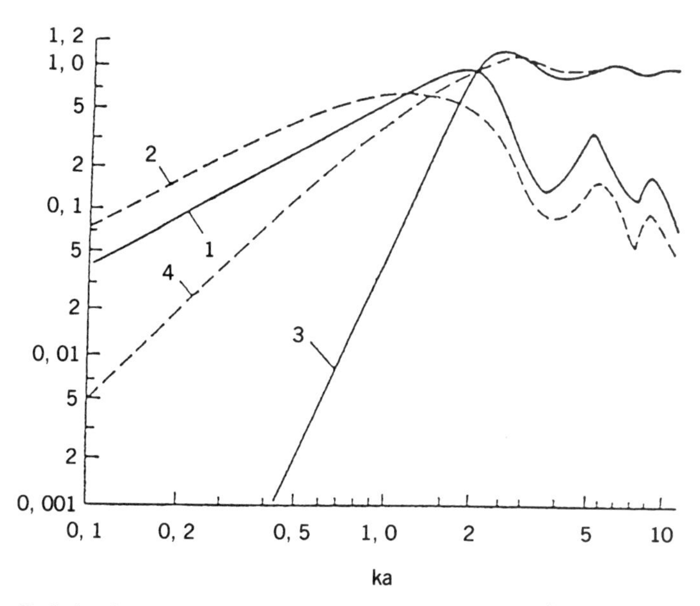

Fig. 1 Radiation impedance of an oscillating piston: (1) $-X_{\text{rad}}\pi a^2/\rho_0 c$ for an unbaffled piston; (2) $-X_{\text{rad}}\pi a^2/\rho_0 c$ for a baffled piston; (3) $R_{\text{rad}}\pi a^2/\rho_0 c$ for an unbaffled piston; (4) $R_{\text{rad}}\pi a^2/\rho_0 c$ for a baffled piston.[1]

to unity; the reactance is zero. The physical difference between these cases is that, close to a baffled piston, the fluid elements are free to move parallel to the piston surface and along elliptical trajectories, by which little strain is produced in the fluid at greater distances from the surface, whereas in a tube they are all forced to move in the direction normal to the piston surface, by which the strain is effectively passed on to fluid further along the tube.

Suppose a small ($ka \ll 1$) baffled piston is encircled by an annular piston of the same area that oscillates in opposite phase; the net volume velocity of the source is zero. The resistive component of pressure acting on each piston is negligible, but the reactive component acting on the central piston is altered only little from its value in the absence of the annulus. This simple example illustrates that the reactive component of fluid loading at a point on a vibrating surface depends mainly on the surface normal acceleration in the immediate vicinity of the point; however, the resistive component, and therefore the radiated power, generally depends upon the distribution of surface acceleration over a much larger area.

3.3 Sound Field Generated by Plane Surface Waves on Plane Boundary

Many structures of practical interest, such as room enclosures, machine casings, passenger compartment envelopes, and the hulls and bulkheads of ships, may, for the purpose of vibration and acoustic analysis, be represented by flat plates. Consequently, it is important to understand the nature of the acoustic interaction of vibrational waves traveling in plates and a contiguous fluid. Analysis of the sound field generated in a semi-infinite volume of fluid bounded by an *infinitely extended* plane surface carrying a harmonic plane transverse wave of wavenumber k_x yields the following expressions for the ratio of the complex amplitudes of surface pressure to surface normal velocity (specific acoustic impedance):

$$k_x < k: \qquad z = \frac{\rho_0 c}{[1 - (k_x/k)^2]^{1/2}}, \tag{3a}$$

$$k_x > k \qquad z = \frac{i(\rho_0 \omega / k_x)}{[1 - (k/k_x)^2]^{1/2}}. \tag{3b}$$

This form of impedance, which is associated with a given surface wavenumber, may be termed a *wave impedance*. The physical interpretation of these expressions is that when the surface wavenumber exceeds the acoustic wavenumber, the impedance presented by the fluid is purely reactive and inertial in character, and no sound energy is radiated away into the far field: The sound field decays exponentially with distance perpendicular to the plane. The inertial loading corresponds to that of a layer of fluid of thickness $(k_x^2 - k^2)^{-1/2}$ moving with the surface. When the surface wavenumber is less than the acoustic wavenumber, the impedance is purely resistive, and sound energy is radiated. In the limit $k/k_x \to 0$, the impedance approaches the characteristic impedance $\rho_0 c$ of the fluid. The fluid loading increases without limit as k_x approaches k. This is a feature unique to the infinitely extended surface wave field. The failure of the surface to radiate sound energy when $k_x > k$ may be qualitatively explained by analogy with the two-piston example described in Section 3.2 above: Regions of opposite phase motion are within a distance of less than half an acoustic wavelength of each other, and the resulting pressure phase is at 90° to that of the surface normal velocity. Clearly, k_x/k is a very important parameter, having a critical value of unity that separates two quite different regimes of fluid behavior.

If the plane-surface vibration is produced by *free* flexural propagation in a uniform plane elastic structure, the critical condition corresponds to equality of k and the free bending wavenumber $k_b = \omega^{1/2}(m/D)^{1/4}$ which occurs at a particular *critical frequency* given by $f_c = (c^2/2\pi)(m/D)^{1/2}$, where c is the sound speed in the fluid and D and m are, respectively, the bending stiffness per unit width and mass per unit area of the structure (see Fig. 2). For homogeneous isotropic plates of thickness h, f_c is uniquely related to h by a constant that depends only on the material properties and the fluid sound speed. Table 1 lists values for common materials in air; for water the values should be multiplied by 19. The acoustic wave impedance acts in series with the in vacuo plate wave impedance, which is given by $z_p = -(i/\omega)[Dk_x^4 - \omega^2 m]$. Free-wave propagation corre-

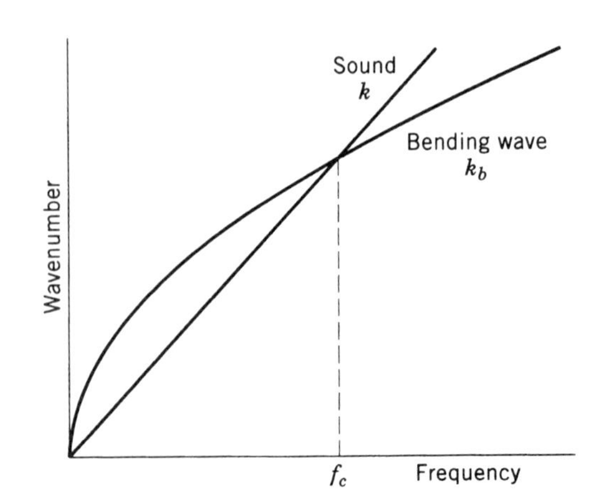

Fig. 2 Dispersion curves for sound waves and bending waves in a flat plate.

TABLE 1 Product of Thickness and Critical Frequency for Uniform Homogeneous Flat Plates in Air[a] at 20°C

Material	hf_c (m/s)
Steel	12.4
Aluminium	12.0
Brass	17.8
Copper	16.3
Glass	12.7
Perspex	27.7
Chipboard	23[b]
Plywood	20[b]
Asbestos cement	17[b]
Concrete	
Dense	19[b]
Porous	33[b]
Light	34[b]

[a]To obtain values in water, multiply by 18.9.
[b]Variations of up to ±10% possible.

sponds to a condition of zero external loading, and therefore, zero total wave impedance; free-wave propagation cannot occur at $k_b = k$, when the fluid wave impedance is infinite. The most dramatic effects of plate loading by an unenclosed fluid are produced by liquids at low frequencies. At frequencies for which $k_b \gg k$, the fluid-loaded free structural wavenumber is given approximately by $k'_b = (\omega^2\rho_0/D)^{1/5}$, which corresponds to a phase speed given by $c_{\text{ph}} = (\omega^3 D/\rho_0)^{1/5}$. At such frequencies the inertial loading by the fluid greatly outweighs the plate inertia, and the latter does not control the wave speed. At higher frequencies for which $k_b > k$, the inertial loading corresponds approximately to an additional mass per unit area given by $m' = \rho_0/k_b$. This loading can significantly lower structural natural frequencies in the ratio $[m/(m + m')]^{1/2}$. At frequencies close to and above the plate critical frequency, an undamped flexural wave can travel in a fluid-loaded plate at a phase speed slightly less than that of sound in the fluid, but this wave is very difficult to excite mechanically. In addition, above the plate critical frequency there exists a wave with a supersonic phase velocity. This is the so-called leaky wave, which radiates energy away at the Mach angle; the amplitude of the wave decays exponentially with distance along the plate as energy is lost. In cases of heavy fluid loading, this decay rate is very high, so that the wave is not present in the plate at large distances from the excitation.

3.4 Sound Radiation by Baffled Rectangular Flat Plates

As indicated above, the rectangular, uniform, flat-plate model is a useful approximation to many structures of interest. The vibration modes take a particularly simple form if simply supported boundary conditions are assumed. Many analyses of modal sound radiation have been published, for example, Refs. 2–7. Examples from Wallace are shown in Fig. 3, in which the radiation efficiency (ratio) is defined by $\sigma = W_{\text{rad}}/\rho_0 c S\langle v^2\rangle$, where W_{rad} is the radiated sound power and $\langle v^2\rangle$ is the space-averaged mean-square normal velocity of the surface of area S. The efficiency is asymptotic to unity when $k \gg k_b$. In most cases, the fundamental mode is the most efficient at frequencies below f_c. Notice that frequency does not appear in Fig. 3. Below f_c, adjacent cells of uniform vibration phase tend to short circuit each other, as in the two-piston case above. As a result, radiated sound originates principally from the uncanceled volume velocity at the boundaries; modes that radiate in this manner are termed *edge modes*. Consequently, the radiation efficiency of *unbaffled* panels is so low below f_c that, in practice, it can be neglected; above f_c it is unity. Perforated plates have, as expected, very low radiation efficiency.

A radiation efficiency curve for *resonant, multimodal* baffled plate radiation is presented in Fig. 4.[8] Note that low-frequency efficiency increases with decrease of plate size. (This curve is *not applicable* to acoustically excited plain panels or locally excited, nonuniform, nonreverberant panels.) Below f_c, incident plane waves couple most strongly with plate modes well above their resonance frequencies, and the associated transmission, which exceeds resonant transmission (except in cases of very lightly damped and/or small plates), is therefore mass controlled. The radiation loss factor is related to σ by $\eta_{\text{rad}} = (\rho_0/\rho_s)(1/kh)\sigma$ where ρ_s is the plate density.

Practical platelike structures are rarely uniform, which reduces the short-circuiting effect below f_c; σ may be considerably increased by the presence of irregularities such as localized stiffeners and concentrated masses. The addition of stiffeners or masses for the purposes of noise control influences both σ and the vibration level induced by mechanical sources. It is therefore impossible to present generalized data on this subject; however, it is advised that added stiffeners should also increase structural damping levels, wherever possible. Where plates are excited by localized force inputs, the region in the immediate vicinity of excitation radiates sound power, given by $W_f = \rho_0 c k^2 F^2/4\pi m^2\omega^2 = \rho_0 F^2/4\pi c m^2$, where F is the harmonic force amplitude and m is the plate mass per unit area. The ratio of sound powers radiated by the force and the surface is given by $W_f/W_s = (4\pi)(\omega/\omega_c)(\eta/\sigma)$. Excitation by a harmonic point velocity v_0 gives $W_v = 16\rho_0 v_0^2 D/\pi c m = 16\rho_0 v_0^2 c^3/\pi\omega_c^2$. The sound power generated per unit length of uniform line force excitation is $W'_f = \rho_0 c k \omega^2 F^2/4D^2 k_b^8 = \rho_0 F^2/4\omega m^2$, the ratio of pow-

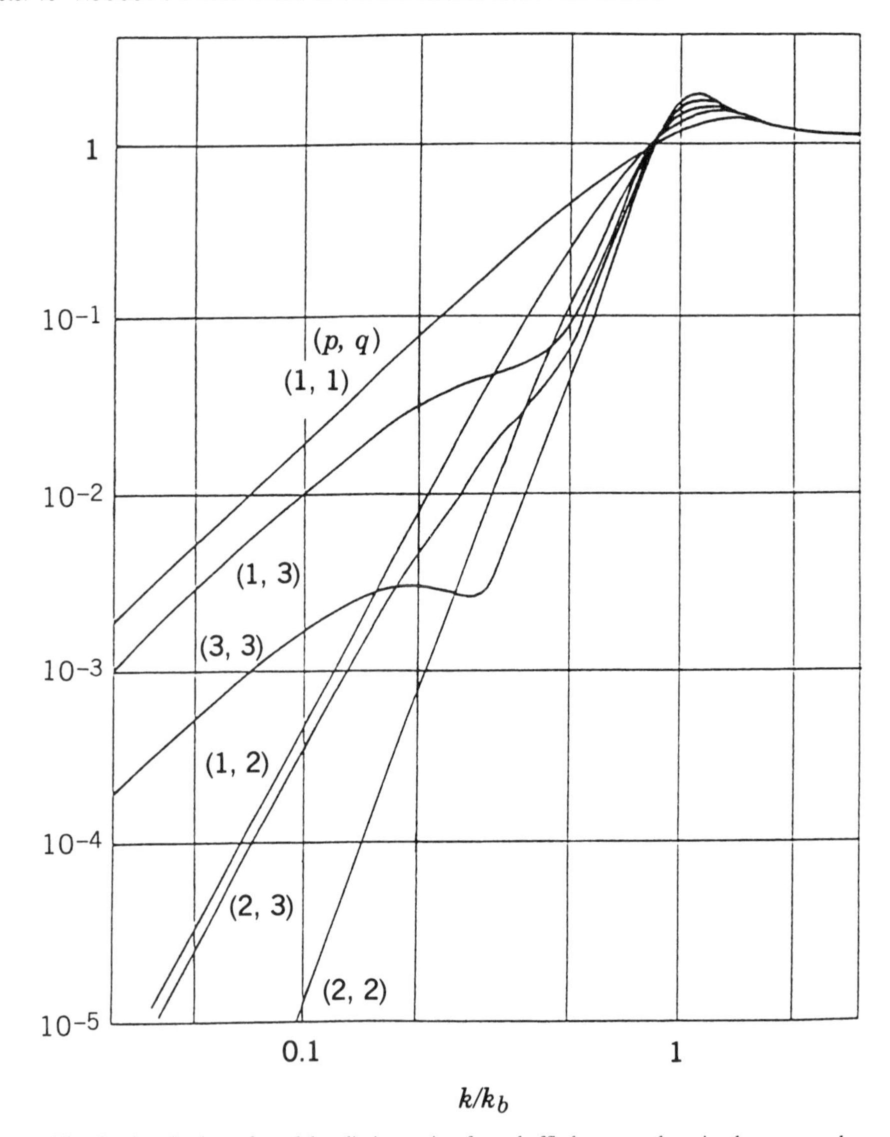

Fig. 3 A selection of modal radiation ratios for a baffled rectangular, simply supported, square plate (p, q = mode orders).[2]

ers radiated from the excitation region and the plate are given by $W_f'/W_s = 2(\omega/\omega_c)^{1/2}(\eta/\sigma)$, and the power radiated per unit length of line velocity excitation is given by $W_v' = 2\rho_0 D^{1/2} v_0^2/m^{1/2} = 2\rho_0 c^2 v_0^2/\omega_c$.

3.5 Fluid-Loading Effects on Flat Plates

The acoustic forces imposed on engineering structures by atmospheric air are generally rather small compared with the internal structural forces, although the stiffness and resonant reaction of air in a small enclosed cavity, such as that between wall partitions, can significantly influence structural motion.[9] Lightweight aerospace structures such as honeycomb panels are exceptions; acoustic damping often exceeds mechanical dissipation damping, especially in the frequency range close to f_c.[10] Loading by liquid significantly lowers the natural frequencies of flat plates, the effect decreasing with increasing mode

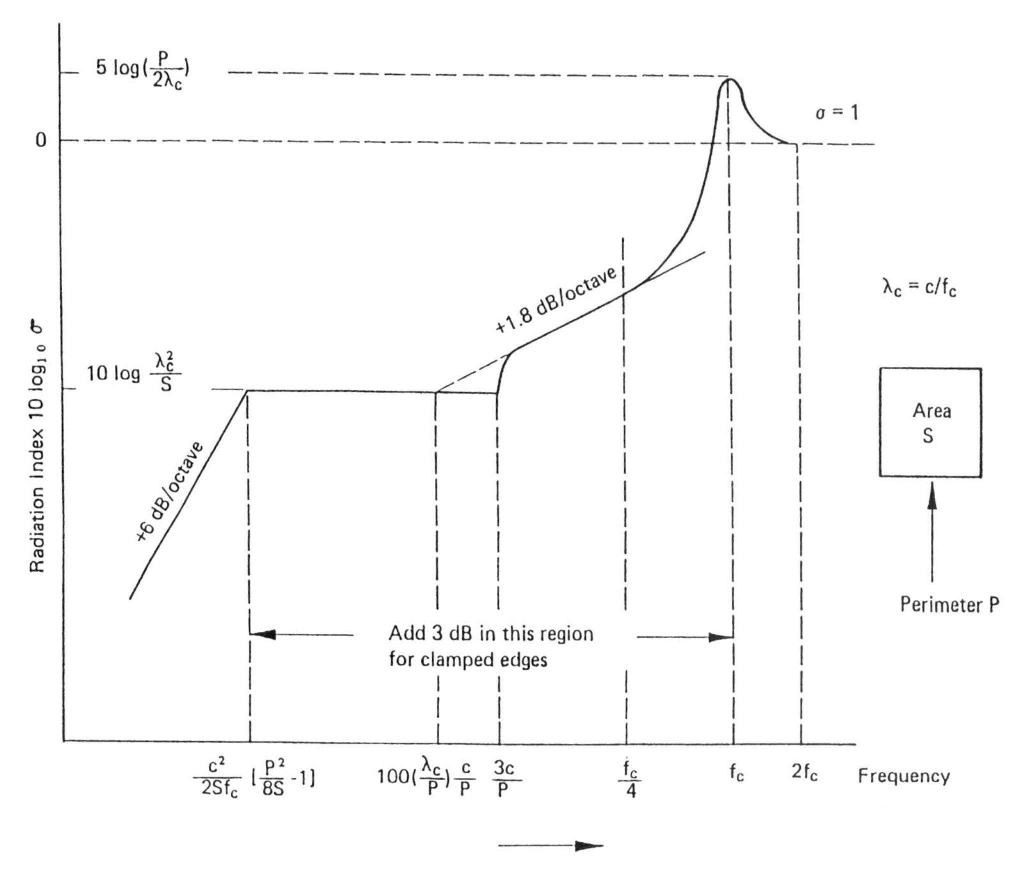

Fig. 4 Theoretical model-averaged radiation efficiency of a baffled rectangular panel.[8]

order. The ratio of fluid-loaded to in vacuo natural frequency is approximately given by $f_l/f \approx (1+\rho_0/mk_b)^{-1/2}$ where k_b is the free plate wavenumber.[11] Fluid loading also couples the in vacuo modes of structures,[12] but this coupling may often be neglected for practical purposes.

The critical frequencies for most plates in water are so high that uniform plate mode radiation efficiencies are very low; consequently, sound is radiated mainly from the sites of mechanical excitation and structural irregularities, such as stiffeners. For $\rho_0 c/\omega m \gg 1$ and at frequencies well below f_c, the sound radiated from plate structures underwater can be estimated rather accurately by assuming that twice the applied forces and plate–stiffener reaction forces are applied *directly* to the water. The expressions for sound powers radiated by point and line forces, respectively, are $W_f = k^2F^2/12\pi\rho_0 c$ and $W'_f = \pi k F^2/16\rho_0 c$. Note, these contain no plate parameters. A more detailed account of fluid-loading effects is contained in Chapter 32.

4 INTERACTION OF CIRCULAR CYLINDRICAL STRUCTURES WITH INFINITE FLUID VOLUMES

4.1 Vibration of Circular Cylindrical Shells

Many engineering structures have the basic form of circular cylindrical shells; examples include fluid-conducting pipes and tubes, aircraft fuselages, fluid storage tanks, and electric motor and generator casings. Vibrational waves in circular shells are very complex in nature because of coupling between the displacements in the axial, circumferential, and radial directions and because stresses arise from both flexural and median surface (membrane) strains.[13] Three "families" of wave types can propagate in the axial direction. Except at very low frequencies, each family contains waves in which displacements in one of the principal directions is dominant: They are consequently referred to as *longitudinal*, *torsional*, and *flexural* waves. The last mentioned is

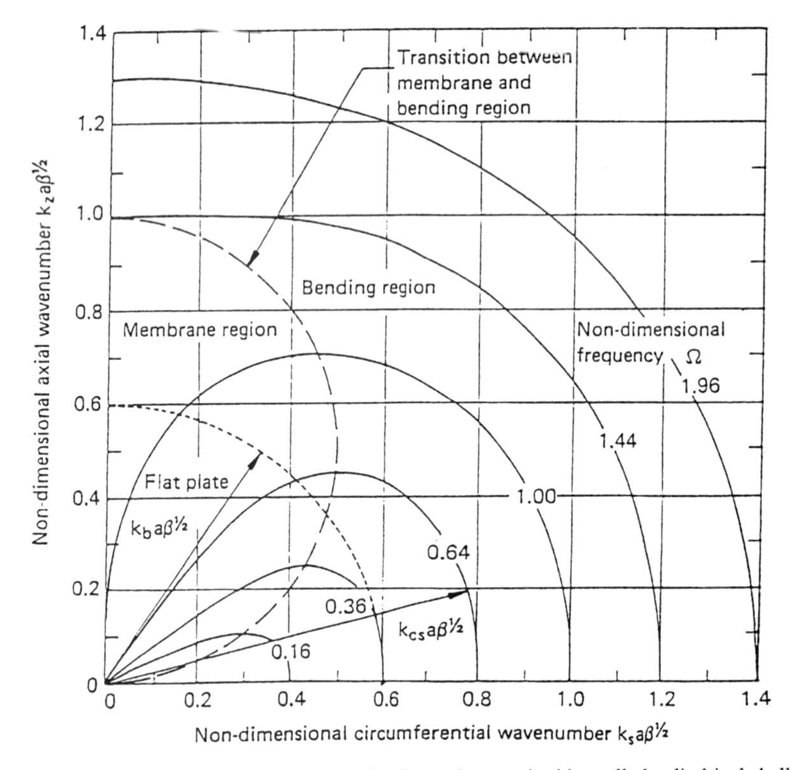

Fig. 5 Universal constant-frequency loci for flexural waves in thin-walled cylindrical shells.

the most involved in fluid–structure interaction because radial displacements are dominant.

4.2 Sound Radiation from Circular Cylindrical Shells

The acoustic interaction with surrounding fluids is considerably different in character from that of flat plates because shell curvature and the resulting membrane effects lead to the existence of surface vibrational wavenumbers less than that of sound at frequencies below the critical frequency of the flat plate of the same thickness as the shell wall. This behavior may be understood by reference to a shell wavenumber diagram, as shown in Fig. 5, using the coordinate system shown in Fig. 6. The constant-frequency loci for the equivalent flat

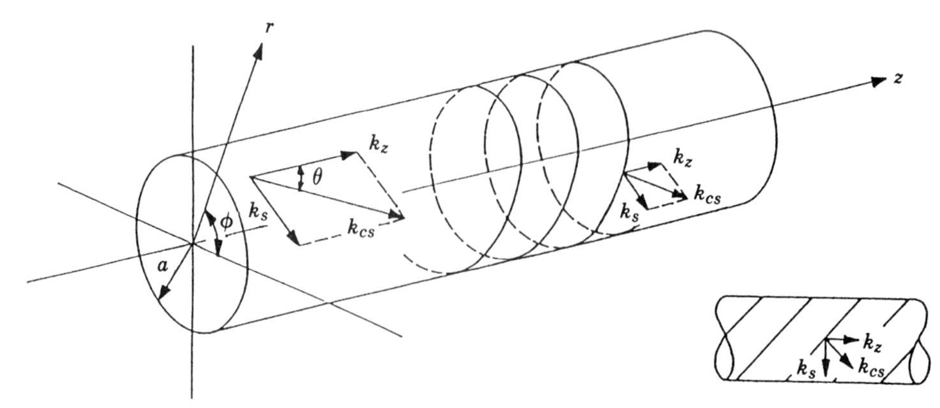

Fig. 6 Cylindrical shell coordinates and wavenumbers.

TABLE 2 Cutoff Frequencies of Flexural Modes of Thin-Walled Circular Cylindrical Shells[14]

Circumferential Mode, Order n	$\Omega_n/\beta n^2$	Ω_n/β
2	0.67	2.68
3	0.85	7.65
4	0.91	14.56
5	0.95	23.75
6	0.96	34.56
7	0.97	47.10

Note: Ω_n = frequency/ring frequency; β = thickness parameter = $h/(12)^{1/2}a$; h = shell wall thickness; a = shell radius.

plate are quarter circles of radius k_b as shown: membrane effects "pull" the loci toward the origin, reducing the shell wavenumbers and increasing the wave phase speed. The nondimensional frequency Ω is the ratio of the frequency to the ring frequency, which is given by $f_r = c_l'/2\pi a$, where a is the shell radius and c_l' is the phase speed of longitudinal waves in a plate. The wall thickness h is nondimensionalized in the parameter $\beta = h/\sqrt{12}a$. The circumferential wavenumber k_s can take only discrete values n/a (integer n). Flexural wave modes of circumferential order n may propagate axially only at frequencies greater than their cutoff frequencies. These are given in Table 2.

Analysis of the sound field radiated by a flexural shell wave of given circumferential wavenumber n/a traveling along an infinite uniform cylinder with axial wavenumber k_z produces resistive and reactive components of the nondimensional surface specific acoustic impedance of the form shown in Figs. 7*a, b*.[13] (*Note*: The curves in Fig. 7 relate to a specific ratio of axial wavelength to cylinder radius.) The impedance is purely reactive if $k_z > k$, irrespective of k_s: No sound power is radiated. The resistive impedance peaks when $k^2 = k_s^2 + k_z^2$, as expected and then asymptotes to unity when $k^2 \gg k_s^2 + k_z^2$. Some useful approximate expressions for the reactive and resistive components of cylinder radiation impedance are presented in Table 3.[13] The radiation characteristics of finite cylinders are difficult to evaluate and to describe in general terms because of the diffraction of the sound by the ends of the cylinder; numerical methods are generally employed for quantitative analysis. In some cases, it may be reasonable to assume that a cylinder is extended beyond its vibrating region by rigid cylindrical baffles. In this case the theoretical results for infinite cylinders may be applied by means of Fourier synthesis.

When the cylinder circumference considerably exceeds the acoustic wavelength at frequencies of interest, for example, in cases of large aircraft fuselages at frequencies in excess of 100 Hz, the curvature of the surface has little geometric influence on sound radiation. Consequently, the forms of analysis of modal radiation introduced in Section 3.4 for flat plates may be applied. Modes of which the principal axial wavenumber exceeds k and for which $n < ka$ exhibit the acoustic short-circuiting phenomenon along their lengths, leaving a ring of uncanceled sources at each end; these correspond to plate edge modes. Statistical analysis of resonant, multimode radiation by large-diameter, thin-walled shells yield the set of curves shown in Fig. 8, in which the parameter is the ratio of ring to critical frequencies.[15]

In many cases of practical interest, the acoustic wavelength greatly exceeds the cylinder circumference at frequencies of interest ($ka \ll 1$), for example, sound radiation from typical industrial pipes at frequencies below 300 Hz. Sound radiation is then dominated by the low-

TABLE 3 Asymptotic Values of Specific Radiation Impedance for Infinite Cylinders Carrying a Standing Wave of Axial Wavenumber k_z

$k_z > k$

$z_n = i\omega m_n$

where

$$m_n \approx \begin{cases} -\rho_0 a \ln(|k_z^2 - k^2|^{1/2} a) & n = 0 \quad (k_z^2 - k^2)a^2 \ll 2n + 1 \\ \rho_0 a/n & n \geq 1 \\ \rho_0/(k_z^2 - k^2)^{1/2} & \quad (k_z^2 - k^2)^{1/2}a \gg n^2 + 1 \end{cases}$$

$k_z < k$

$z_n = i\omega m_n + r_n$

where

$$m_n \approx \rho_0 a/2(k^2 - k_z^2)a^2 \qquad (k^2 - k_z^2)^{1/2}a \gg n^2 + 1$$

$$r_n \approx \begin{cases} \rho_0 c\pi ka(k^2 - k_z^2)^n a^{2n}/(n!)^2 2^{|2n-1|} & 0 < (k^2 - k_z^2)a^2 \ll 2n + 1 \\ \rho_0 c & (k^2 - k_z^2)^{1/2}a \gg n^2 + 1 \end{cases}$$

From Ref. 13.

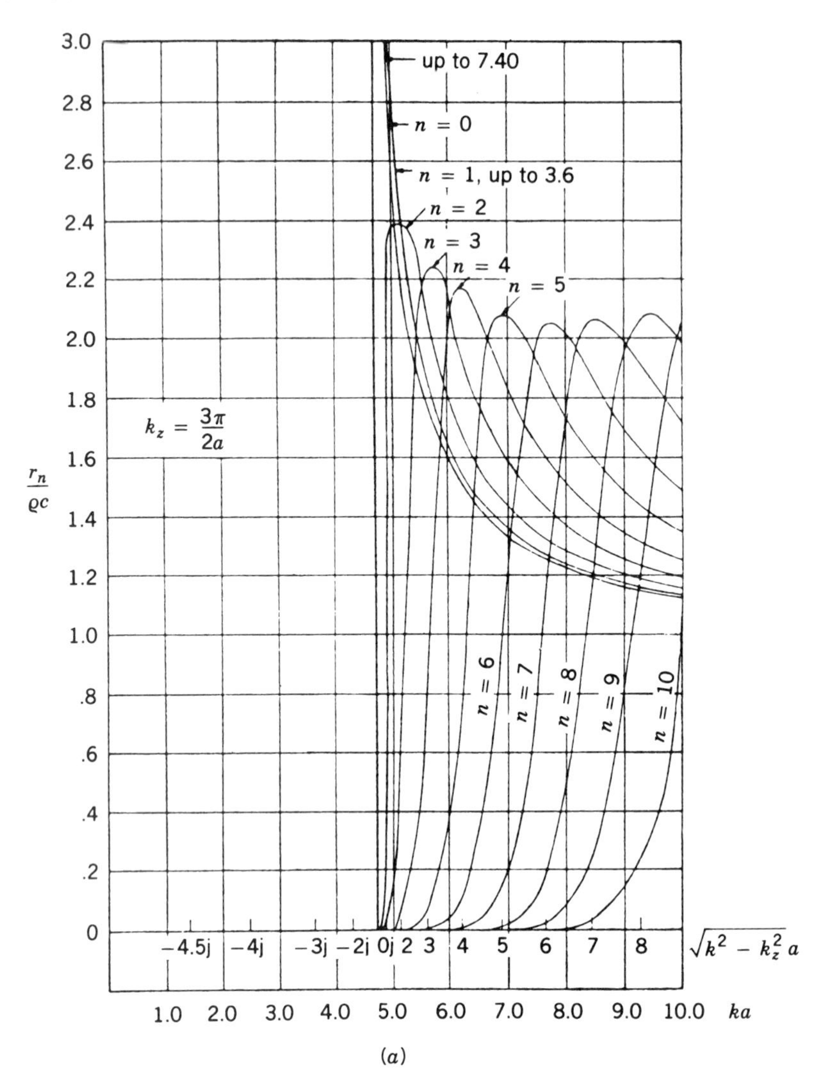

Fig. 7 (*a*) Specific acoustic resistance associated with cylindrical wave harmonics ($k_z = 3\pi/2a$).[13] (*b*) Specific acoustic reactance associated with cylindrical wave harmonics ($k_z = 3\pi/2a$).[13]

n modes (0, 1, 2). In practice, the "breathing" mode ($n = 0$) is not easily excited at low frequencies far below the ring frequency because it is very stiff. The "bending" mode ($n = 1$) is readily excited, even by plane sound waves in a contained fluid, through the presence of structural nonuniformities in the pipe, pipe bends, and pipe supports. The radiation efficiency of the bending mode is very low unless the axial structural wavelength exceeds the acoustic wavelength.[14] The $n = 2$ structural (ovalling) mode has a cutoff frequency given by $f/f_r = 2.68\beta$, above which it is often observed to dominate pipe vibration. In industrial pipes the axial wavelength associated with these modes usually greatly exceeds the acoustic wavelength, and the $n = 0, 1, 2$ modes radiate in a similar manner to line monopoles, dipoles, and quadrupoles, respectively. The radiation efficiency of a uniformly vibrating long, thin cylinder is shown in Fig. 9.[14] The addition of stiffening rings to thin cylindrical shells does not greatly alter the sound power generated by localized excitation of the shell wall because it is dominated by radiation from modes of low circumferential order, of which the strain energy is pre-

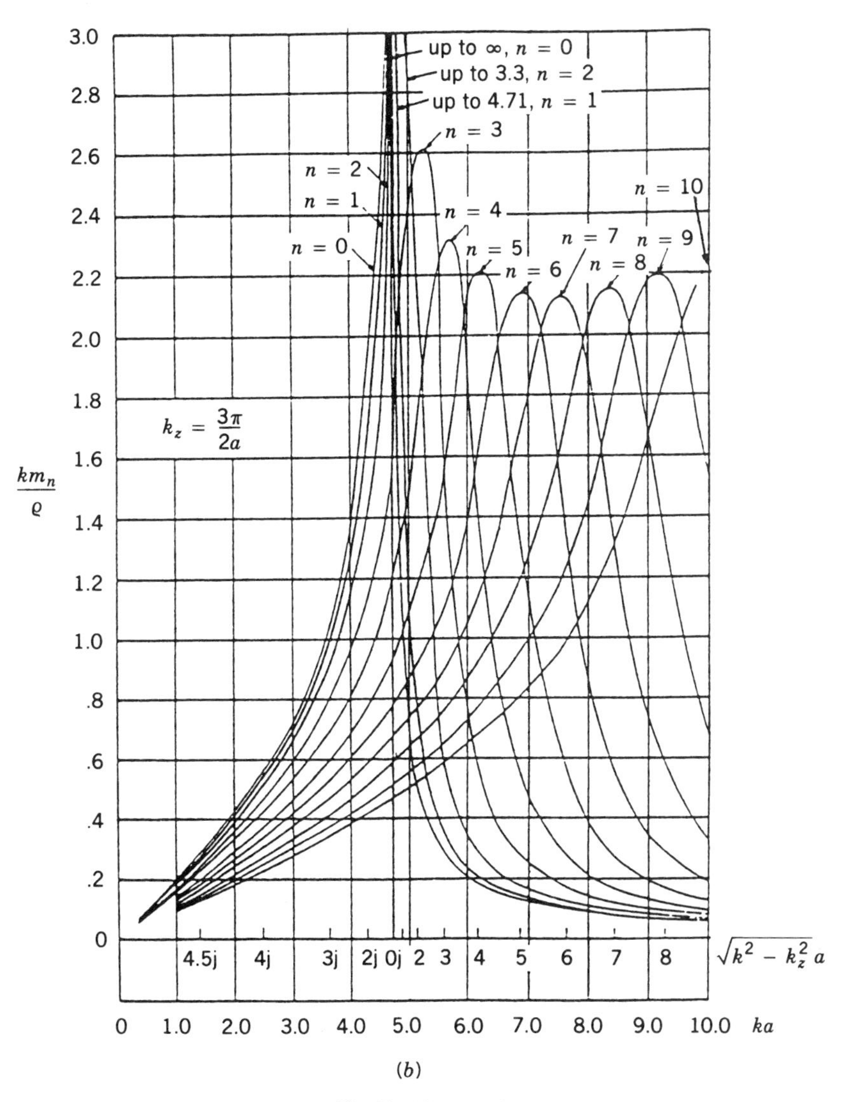

(*b*)

Fig. 7 (*Continued*)

dominantly associated with membrane (midplane) strain, and not with flexural deformation: Rings influence the former far less than the latter.

4.3 Fluid-Loading Effects on Circular Cylindrical Shells

Sound pressures on a cylinder vibrating in a surrounding fluid produce two distinct effects: (1) the component in phase with surface normal velocity drains away vibrational energy and therefore damps the motion and (2) the component at 90° to the velocity mass loads the structure, thereby reducing flexural wave speeds and modal natural frequencies. Air loading of cylinders generally has little effect, except in cases of large-diameter, honeycomb sandwich shells, in which the critical and ring frequencies may be comparable and the radiation loss factor may exceed the structural dissipation loss factor. One result is that damping treatment may be ineffective in reducing sound radiation. The most marked effects of fluid loading occur on structures submerged in a liquid, most commonly water.

Inertial loading can greatly reduce modal natural frequencies, especially those of low circumferential order,

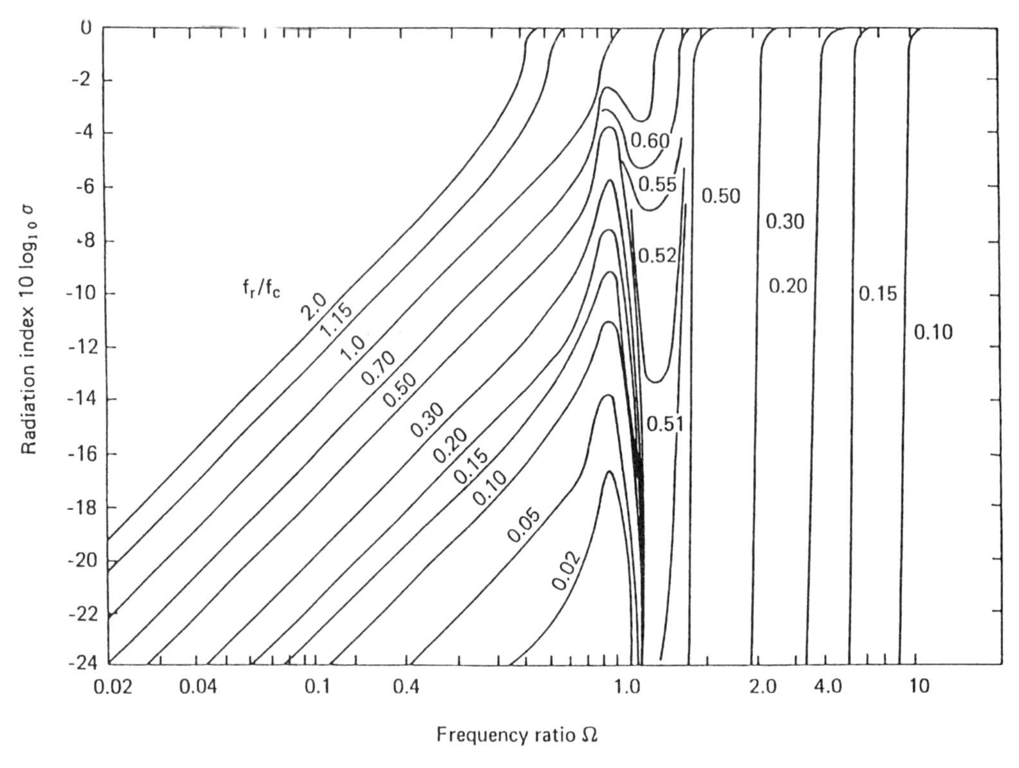

Fig. 8 Modal-averaged radiation efficiency of thin-walled, large-diameter cylindrical shells.[15]

as indicated in Table 3 and illustrated by Fig. 10. A method for estimating the reactive (inertial) effect of fluid loading on shell natural frequencies is presented in Ref. 16. The inertial loading experienced by thin-walled, pipelike structures vibrating in bending at frequencies for which $ka \ll 1$ is equivalent to an additional mass per unit length equal to that of a cylinder of fluid of radius a. Resistive fluid loading greatly influences the relative contributions to sound radiation by the various modes of submerged cylinders excited by vibrational forces. Those that have radiation loss factors close in value to their internal loss factors are the most effective. A remarkable result is that the sound power radiated by a point-force-driven cylindrical shell into water may be rather similar in magnitude to that which it radiates into air, although the modes principally responsible for power radiation are quite different.[17] Consequently, measures taken to increase mechanical damping can significantly alter the relative modal contributions to radiation.

Heavy fluid loading greatly affects the distribution of sound intensity radiated by a uniform plate or cylinder excited by a localized force or moment, tending to confine the principal region of radiation to the immediate vicinity of the point of excitation. As with the plate, the radiated sound power may be rather accurately estimated by assuming that twice the force (moment) is applied directly to the fluid.

5 INTERACTION OF STRUCTURES WITH CONTAINED FLUIDS

Many structures take the forms of fluid containment vessels; examples include steam pipes and car passenger compartments. Acoustic interaction is important in the first example because noise generated inside the pipe can cause structural damage as well as unacceptably high levels of radiated noise. In the second example, it is the noise levels created inside the compartment by various external noise and vibration sources that are of concern.

As explained in Section 2, interaction between fluids and the structures that contain them can be stronger and more frequency dependent than for surrounding fluids due to the resonant behavior of the fluid volume. In most cases of air containment, the effects of fluid loading on structural vibration are negligible, so that fully coupled vibration analysis may be avoided. However, if the resonance frequencies of a structural mode and an acoustic mode of the (assumed) rigidly bounded fluid volume are very close together, coupled modes having

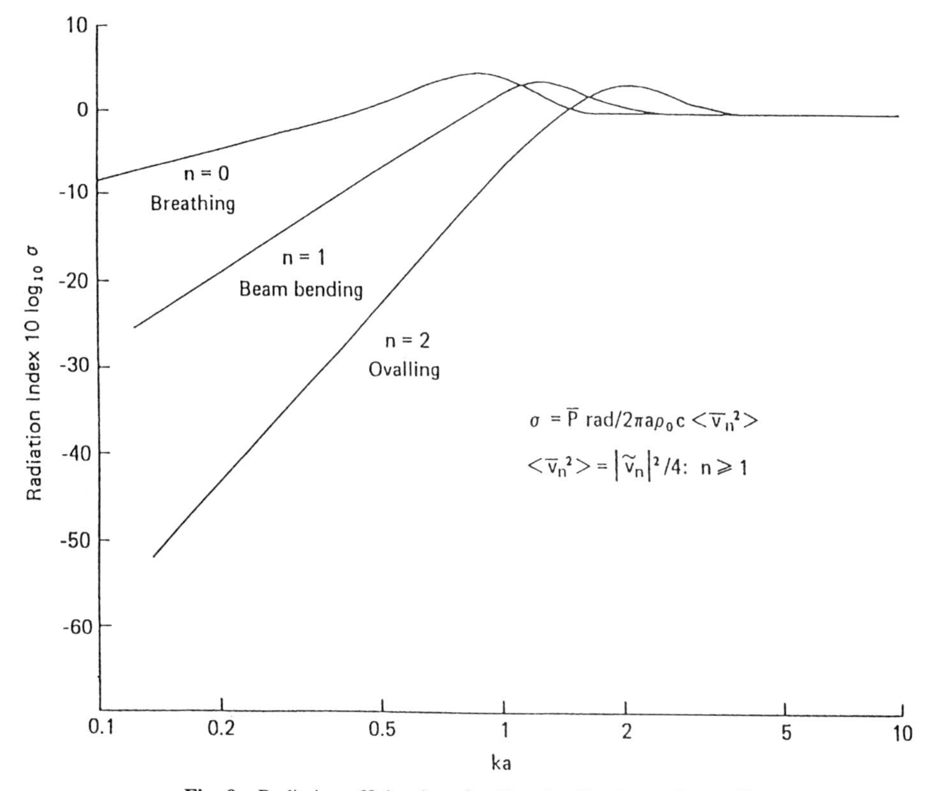

Fig. 9 Radiation efficiencies of uniformly vibrating cylinders.[14]

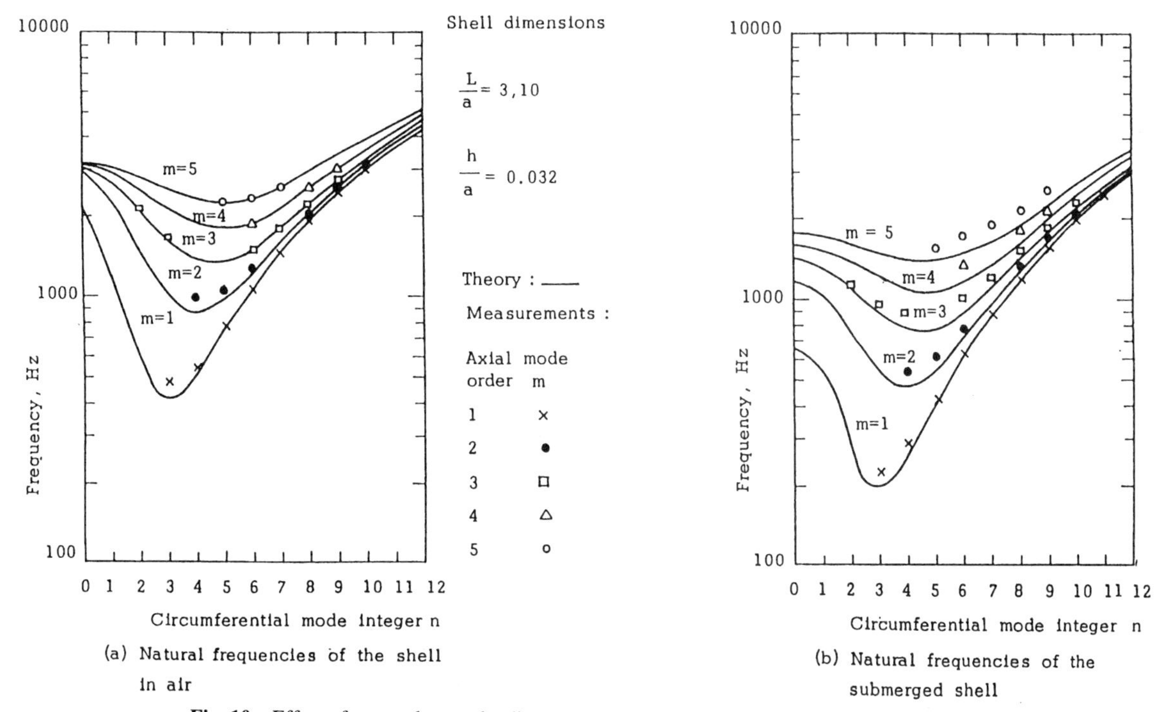

Fig. 10 Effect of external water loading on natural frequencies of aluminum cylindrical shell with closed ends.

frequencies slightly altered from the uncoupled values will occur.[14] Such behavior may be observed in vehicle compartments.

At low frequencies, a volume of fluid acts as an elastic spring, which can significantly affect the natural frequency of a coupled lightweight structure. For example, the air contained within a loudspeaker cabinet is usually stiffer than the mechanical suspension and controls the low-frequency performance. A shallow cavity can exert a strong effect on a large panel that bounds it, as exemplified by the mass–air–mass resonance in lightweight double partitions, which limits the maximum achievable low-frequency transmission loss of such structures (see Chapter 76).

Acoustic interaction between structures and the fluids they contain is most marked in liquid containment vessels, including pipes. Modes involve both fluid and structural motion and may not rigorously be described as "fluid modes" or "structural modes"; however, in most cases, the vibrational energy of a mode resides principally in either the fluid or solid component. One of the most common examples of this form of interaction is observed in water-filled hoses in which the flexibility of the walls significantly alters the speed of propagation of acoustic disturbances along the pipe. Well below the ring frequency of the pipe wall, the elastic reaction of the wall to expansion of the cross section greatly exceeds the inertial reaction. A simple model of the interaction, which accounts only for the elastic reaction of the pipe wall and neglects inertial effects, yields the following expression for wave speed[13]: $c_0/c = [1 + (2a\rho_0 c^2/h\rho_s c_p^2)]^{-1/2}$, where ρ_s and c_p are, respectively, the density and speed

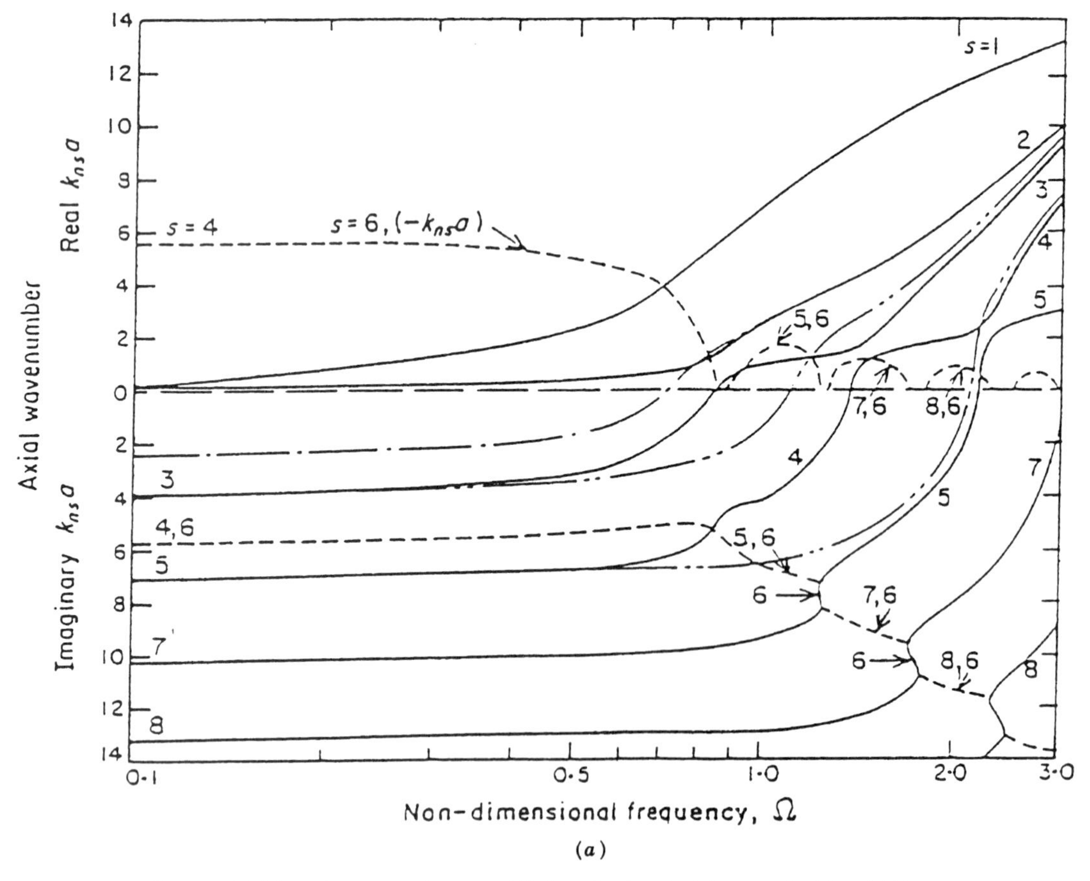

(*a*)

Fig. 11 (*a*) Dispersion curves for water-filled steel tube of thickness ratio $h/a = 0.05$, $n = 0$: (——) purely real and imaginary $k_{ns}a$; (---) real and imaginary parts of complex $k_{ns}a$; (— - —) pressure release duct solution; (— - - —) rigid walled duct solution. (*b*) Dispersion curves for water-filled steel tube of thickness ratio $h/a = 0.05$, $n = 1$: (——) purely real and imaginary $k_{ns}a$; (— - —) real and imaginary parts of complex $k_{ns}a$; (— - - —) rigid walled duct solution.

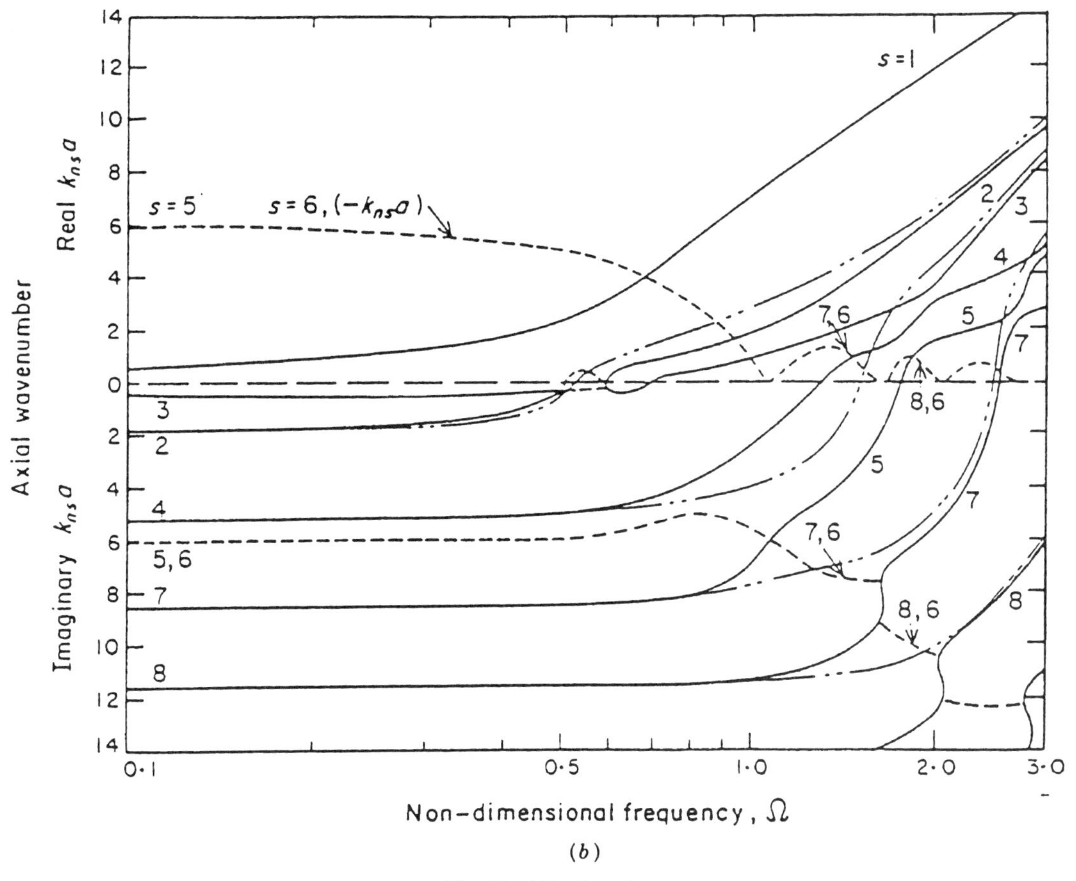

(*b*)

Fig. 11 (*Continued*)

of sound in the pipe material. Well above the pipe ring frequency the wall inertia dominates, and a corresponding model that neglects wall elasticity but accounts for wall mass yields the following expression for the axial wavenumber[13]: $\gamma = k[1 - (2\rho_0/\rho_s k^2 ha)]^{1/2}$ when $\rho_0 a/\rho_s h \ll 1$. At frequencies less than a cutoff frequency given by $f_c = (c/\pi)(\rho_0/2\rho_s ha)^{1/2}$, wave propagation is not possible in the absence of wall stiffness. At f_c, the fluid compressibility is balanced by the wall inertia. Blood vessels are so flexible that the speed of propagation of pressure disturbances is of the order of only 10–15 ms^{-1}. A general analysis of fluid-filled cylinders yields axial dispersion curves of which the form depends very much on the nondimensional parameter $K = \rho_0 a/\rho_s h$, where ρ_0/ρ_s is the ratio of fluid to solid densities and a/h is the ratio of pipe radius to wall thickness. When K is large (large-diameter, thin-walled cylinders containing liquid), one set of branches of the dispersion curves corresponds closely to the acoustic dispersion curve for a cylinder of fluid with a zero-pressure boundary. The "shell wave" branches are significantly altered from their in vacuo forms by fluid loading. The natural frequencies of purely circumferential, low-order (low-n) "shell" modes of cylinders are significantly reduced by fluid loading. The ratio of in vacuo (see Table 2) to fluid-filled natural frequencies for $n \geq 2$ is given approximately by $(1 + K/n)^{1/2}$, provided that the in vacuo frequency is below the ring frequency as well as below the lowest transverse acoustic mode cutoff frequency given by $f = 1.84c/2\pi a$.[18] Examples of the dispersion curves of a steel pipe filled with water for $n = 0$ and $n = 1$ are shown in Figs. 11*a*,*b*.[19]

6 ACOUSTICALLY INDUCED VIBRATION OF STRUCTURES

6.1 Practical Aspects

The process of airborne sound transmission through partitions involves the vibrational response of the structure to sound incident on one side, together with the consequent radiation from the other. The influence on partition motion of the sound pressures it generates is related to the nondimensional fluid-loading parameter $\rho_0 c/\omega m$. In most cases of practical interest in air, this parameter is much smaller than unity, in which case the influence of the air loading on the vibration of the partition is extremely small, except for very lightweight partitions such as plastic films and membranes and very low frequencies. In water, however, fluid-loading forces often dominate, and a partition has little influence on sound transmission. This topic is treated in Chapters 32 and 76.

High levels of sound (>140 dB) can cause fatigue damage to engineering structures and produce malfunctions in mechanical and electrical equipment, either by direct excitation or, more commonly, via vibration induced in supporting structures. Acoustically induced fatigue is of serious concern to aircraft designers, and the noise from the rockets of launch vehicles can adversely affect the operational integrity of electronic component packages onboard payloads. Noise has also been known to threaten industrial plant components such as steam and gas control valves and associated piping. Gas-cooled nuclear reactor structures may also be at risk from damage by gas circulator noise. The vibrational response of ship structures to incident sound and the consequent reradiation of sound constitute sources of interference to sonar receivers. The optimization of panel (membrane) sound absorbers for studio sound control requires understanding of the principle of matching of internal and radiation resistances.

6.2 General Principles

The sound field that results from the incidence of a sound wave upon a linear elastic structure may be expressed exactly as the sum of three components: (i) the unobstructed incident wave; (ii) the field produced by the interaction of the incident field with the body when considered to be completely rigid; and (iii) the field radiated by the vibrational surface motion produced by the incident sound. Superposition of components (i) and (ii) gives the "blocked" field. Field component (ii) may be thought of as the result of radiation by the body when vibrating in such a manner that the surface normal acceleration is equal in magnitude and opposite in sign to the corresponding component of the unobstructed incident field at the body surface. A structure vibrates in response to the total sound pressure on its surface, which is the sum of all three components. The structure is coupled to the fluid via the radiated field component. The general equation of structural normal displacement may be expressed as $L(w) + m\ddot{w} = p_{\text{bl}} + p_{\text{rad}}$, where L is a differential operator. Since p_{rad} is a function of $\ddot{w}$, through the Kirchhoff–Helmholtz integral equation (see Chapter 13), this relationship has the form of an integrodifferential equation. The physical effect of radiation loading is to increase the mass and damping (and occasionally stiffness) of the structure, and hence the term p_{rad} may be taken across to the left-hand side of the equation of motion, which in this modified form represents the response of the fluid-loaded structure to the blocked pressure. This form is particularly convenient when the in vacuo modes of a structure are known, since radiation loading generally alters the mode shapes rather little, and hence the generalized blocked force may be evaluated by rigid-body scattering analysis or modal reciprocity analysis, as indicated below.

6.3 Relationship between Acoustically Induced Structural Response and Sound Radiation

Point Reciprocity Lyamshev explicitly expresses Rayleigh's reciprocity relationship between sound pressure at a field point generated by a point-force-excited linear elastic structure and the response of that structure to sound generated by a point monopole source at the same field point[20] (Fig. 12). This relationship is of great practical value because it obviates the need to simulate vibrational forces in operating systems, for the purpose of determining the resulting sound, by replacing the direct measurement with an observation of the vibrational response to a point source placed at the sound field observation position.[21,22]

Modal Reciprocity The response of a structural mode to acoustic excitation by a point source in a free field may be expressed in terms of its sound radiation characteristics.[23,24] The ratio of the modal velocity amplitude V_m to the amplitude of the incident sound pressure at the position of, but in the absence of, the structure at the modal resonance frequency ω_m is given by

$$\frac{(V)_m^2}{P_0^2} = \frac{4\pi D(\theta,\phi)}{\rho_0 c k^2} \frac{R_{\text{rad}}^m}{(R_{\text{int}}^m + R_{\text{rad}}^m)^2}, \tag{4}$$

where $D(\theta,\phi)$ is the directivity factor of modal radiation in the direction of the position of the source and R_{rad}^m and R_{int}^m are the modal radiation and internal resistances, respectively. If the source is assumed to be very distant in terms of the maximum dimensions of the structure, this expression gives the response to plane-wave excitation from direction (θ,ϕ). Maximum response occurs when internal and radiation damping are equal. The expression for broadband excitation is

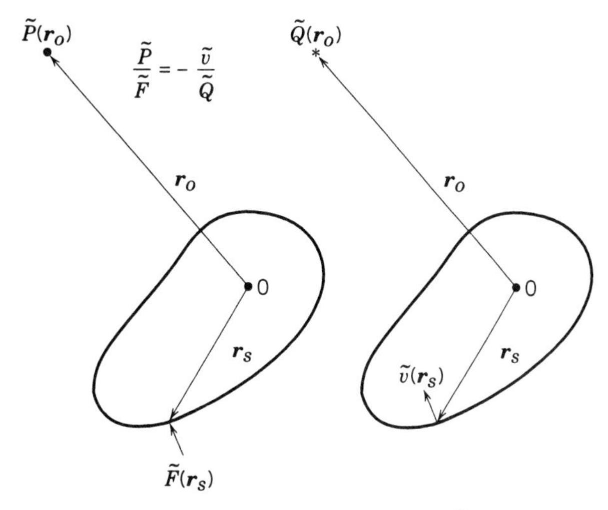

Fig. 12 Lyamshev's vibroacoustic reciprocity relationship: $\tilde{Q}$ = point monopole volume velocity (source strength); $\tilde{F}$ = point force; $\tilde{v}$ = vibration velocity; $\mathbf{r}$ = position vector; $\tilde{P}$ = sound pressure.

$$\frac{V_m^2}{G_{p0}(\omega)} = \frac{2\pi^2 cD(\theta,\phi)}{\rho_0\omega_m^2 M_m}\,\frac{R_{\text{rad}}^m}{R_{\text{int}}^m + R_{\text{rad}}^m}, \tag{5}$$

where $G_{p0}(\omega)$ is the uniform spectral density of the sound pressure and M_m is the modal mass. The corresponding expressions for diffuse field excitation are

$$\frac{V_m^2}{P_0^2} = \frac{4\pi}{\rho_0 ck^2}\,\frac{R_{\text{rad}}^m}{(R_{\text{int}}^m + R_{\text{rad}}^m)^2} \tag{6}$$

and

$$\frac{V_m^2}{G_{p0}(\omega)} = \frac{2\pi^2 c}{\rho_0\omega_m^2 M_m}\,\frac{R_{\text{rad}}^m}{R_{\text{int}}^m + R_{\text{rad}}^m}. \tag{7}$$

If a number of modes have close natural frequencies, the total energy E_s of vibrational response to broadband diffuse sound is given by

$$\frac{\overline{E_s}}{P_0^2} = \frac{2\pi^2 cn_s(\omega)}{\rho_0\omega_c^2}\left\langle \frac{R_{\text{rad}}^m}{R_{\text{int}}^m + R_{\text{rad}}^m}\right\rangle_m, \tag{8}$$

where $n_s(\omega)$ is the modal density (see Chapter 106) and the angle brackets denote modal average. The practical interpretation of these relationships is that the response is controlled by internal (dissipation) damping only if it considerably exceeds acoustic damping. When the latter greatly exceeds the former, the response to broadband excitation reaches an upper limit independent of damping, but the pure tone response *decreases* as R_{rad} increases.

7 NUMERICAL ANALYSES OF STRUCTURE–FLUID INTERACTION

The majority of systems of practical engineering interest exhibit such complexity of geometry and structural detail that their vibrational and acoustic behavior cannot be accurately predicted by analytical mathematical methods. Modern, computer-based techniques of numerical analysis are available; these are either finite element methods or boundary element methods or combinations thereof. These techniques are described in Chapter 13. The reader should be warned that the *broadband* noise radiation from complex vibrating structures it is generally rather difficult to predict with great precision for the following reasons: (i) the amplitude and phase distribution of surface vibration is not normally known with sufficient accuracy, or in sufficient detail, to provide a suitable input to the computational procedure; (ii) analysis is performed one frequency at a time, thereby requiring vast computing times for broadband noise problems (however, new techniques of frequency interpolation can significantly speed up the calculation); and (iii) most practical sources are highly irregular in their geometry, thereby requiring a large and complex array of surface elements. Another major impediment to the application of boundary element methods to practical problems at the present time is that they cannot routinely deal with the forms of sound insulation construction commonly installed in vehicles, pipe lagging, and machinery covers, where the sound-absorbent material may exhibit nonlocal reaction. Finite element analysis is best suited to the computation of sound fields in enclosed volumes, but external radiation fields may be represented by the *wave envelope* element method, which is available in several finite element software packages.

Alternative methods to the boundary element method for sound radiation calculation are currently under development, for example, the multipole representation by which an array of monopole sources is distributed within the source volume so as to produce a sound field that matches, as well as possible, the actual distribution of normal acceleration on the source surface.[25,26]

REFERENCES

1. E. L. Shenderov, "Sound Radiation by an Unbaffled Oscillating Disk (by a Disk in an Acoustically Compliant Baffle)," *Sov. Phys. Acoust.*, Vol. 34, 1988, pp. 191–198.
2. C. E. Wallace, "Radiation Resistance of a Rectangular Panel," *J. Acoust. Soc. Am.*, Vol. 51, 1972, pp. 946–952.
3. G. Maidanik, "Response of Ribbed Panels to Reverberant Sound Fields," *J. Acoust. Soc. Am.*, Vol. 34, 1962, pp. 809–826.
4. F. G. Leppington, E. G. Broadbent, and K. H. Heron, "The Acoustic Radiation Efficiency of Rectangular Panels," *Proc. Roy. Soc. Lond. Ser. A*, Vol. 382, 1982, pp. 245–271.
5. F. G. Leppington, E. G. Broadbent, and K. H. Heron, "Acoustic Radiation from Rectangular Panels with Constrained Edges," *Proc. Roy. Soc. Lond. Ser. A*, Vol. 392, 1984, pp. 67–84.
6. R. Timmel, "The Radiation Efficiency of Rectangular Thin Homogeneous Plates in an Infinite Baffle," *Acustica*, Vol. 73, 1991, pp. 1–11 (in German).
7. R. Timmel, "Investigations of the Effect of Edge Boundary Conditions for Flexurally Vibrating Rectangular Panels on the Radiation Efficiency as Exemplified by Clamped and Simply-Supported Panels," *Acustica*, Vol. 73, 1991, pp. 12–20 (in German).
8. L. L. Beranek and I. Vér (Eds.), *Noise and Vibration Control Engineering*, Wiley, New York, 1992.
9. N. Kiesewetter, "Impedance and Resonances of a Plate Before an Enclosed Volume of Air," *Acustica*, Vol. 61, 1986, pp. 213–217.
10. C. E. Wallace, "The Acoustic Radiation Damping of the Modes of a Rectangular Panel," *J. Acoust. Soc. Am.*, Vol. 81, 1987, pp. 1787–1794.
11. N. S. Lomas and S. I. Hayek, "Vibration and Acoustic Radiation of Elastically Supported Rectangular Plates," *J. Sound. Vib.*, Vol. 52, 1977, pp. 1–25.
12. H. G. Davies, "Low Frequency Random Excitation of Water-loaded Rectangular Plates," *J. Sound Vib.*, Vol. 15, 1971, pp. 107–126.
13. M. C. Junger and D. Feit, *Sound, Structures and Their Interaction*, 2nd ed., MIT Press, Cambridge, MA, 1986.
14. F. J. Fahy, *Sound and Structural Vibration*, Academic Press, London, 1987.
15. E. Szechenyi, "Modal Densities and Radiation Efficiencies of Unstiffenend Cylinders Using Statistical Methods," *J. Sound Vib.*, Vol. 19, 1971, pp. 65–82.
16. M. K. Au-Yang, "Natural Frequencies of Cylindrical Shells and Panels in Vacuum and in a Fluid, *J. Sound Vib.*, Vol. 57, 1978, pp. 341–355.
17. B. Laulagnet and J-L. Guyader, "Modal Analysis of a Shell's Acoustic Radiation in Light and Heavy Fluids," *J. Sound Vib.*, Vol. 131, 1989, pp. 397–416.
18. P. G. Bentley and D. Firth, "Acoustically Excited Vibrations in a Liquid-filled Cylindrical Tank," *J. Sound Vib.*, Vol. 19, 1971, pp. 179–191.
19. C. R. Fuller and F. J. Fahy, "Characteristics of Wave Propagation and Energy Distribution in Cylindrical Shells Filled with Fluid, *J. Sound Vib.*, Vol. 81, 1981, pp. 501–518.
20. L. M. Liamshev, "Theory of Sound Radiation by Thin Elastic Shells and Plates," *Sov. Phys. Acoust.*, Vol. 5, 1960, pp. 431–438.
21. T. Ten Wolde, "On the Validity and Application of Reciprocity in Acoustical, Mechano-acoustical and Other Dynamical Systems," *Acustica*, Vol. 28, 1973, pp. 23–32.
22. F. J. Fahy, "The Vibroacoustic Reciprocity Principle and Applications to Noise Control," *Acustica*, Vol. 81, 1995, pp. 544–558.
23. P. W. Smith, "Response and Radiation of Structural Modes Excited by Sound," *J. Acoust. Soc. Am.*, Vol. 34, 1962, pp. 640–647.
24. G. Chertock, "General Reciprocity Relation," *J. Acoust. Soc. Am.*, Vol. 34, 1962, p. 989.
25. G. H. Koopmann, L. Song, and J. B. Fahnline, "A Method of Computing Acoustic Fields Based on the Principle of Wave Superposition," *J. Acoust. Soc. Am.*, Vol. 86, 1989, pp. 2433–2438.
26. M. Ochmann and F. Wellner, "Calculation of the Three-dimensional Sound Radiation from a Vibrating Structure Using a Boundary Element–Multigrid Method," *Acustica*, Vol. 73, 1991, pp. 177–190 (in German).

11

ACOUSTIC LUMPED ELEMENTS FROM FIRST PRINCIPLES

ROBERT E. APFEL

1 INTRODUCTION

The term *acoustical lumped elements* refers to the significant simplification in the behavior of sound that occurs when sound interacts with a physical structure that is much smaller than an acoustic wavelength. These elements are analogous to electrical resistors (R), capacitors (C), and inductors (L); therefore, all the mathematical tools developed for these electrical elements can be adopted for acoustical and mechanical elements. This approach has found broad application in areas ranging from loudspeaker and muffler design,[1,2] to noise in the ocean from resonating bubbles,[1] to musical instruments based on Helmholtz resonators (e.g., when one blows over an opening of a wine bottle).

The direct analogy between electrical circuits and mechanical elements is made apparent by considering the equations for an R–L–C series circuit driven by an alternating voltage source and for a spring–mass–dashpot system driven by an alternating forcing function.

2 ACOUSTIC SYSTEM AND DEFINITION OF SYMBOLS

There is a fairly general familiarity among researchers in the electrical and mechanical sciences and engineering with lumped elements and their equivalent circuits. A comparison between electrical and mechanical, one-degree-of-freedom systems is given in Fig. 1.

$$L\ddot{Q} + R\dot{Q} + \frac{1}{C}Q = E_0 e^{i\omega t}, \qquad M\ddot{X} + c\dot{X} + kX = F_0 e^{i\omega t}.$$

Note: References to chapters appearing only in the *Encyclopedia* are preceded by *Enc.*

For simple harmonic motion:

$$Q = Q_0 e^{i\omega t}, \quad X = X_0 e^{i\omega t},$$

$$I = \dot{Q} = i\omega Q, \quad V = \dot{X} = i\omega X,$$

$$Z_E = \frac{E}{I} = R + i\omega L + \frac{1}{i\omega C}, \quad Z_M = \frac{F}{V} = c + i\omega M + \frac{k}{i\omega}.$$

The electrical impedance Z_E and mechanical impedance Z_M are defined roughly as the ratios of "cause" to "effect" [electomotive force (EMF) ÷ current for electrical impedance and force ÷ velocity for mechanical impedance].

For simple harmonic motion ($e^{i\omega t}$) in either of these systems, the differential equations can be simplified, leading to an impedance that has three terms: a real term (R or c), an imaginary term proportional to $i\omega$ (L or M), and an imaginary term proportional to $1/\omega$ ($1/C$ or k).

For distributed acoustic systems, the cause–effect relationship is between the change in pressure (disturbance) and the resulting particle velocity of the medium. The definitions of the relevant quantities are as follows:

Acoustic System Definitions

- p — Acoustic pressure ($= P - P_0$), instantaneous pressure change from its ambient value P_0 in a sound field
- v — Particle velocity; resulting from the passing of a sound wave, it is superimposed upon the thermally established and random molecular velocity
- C_s — Speed at which a sound wave propagates (not to be confused with v)

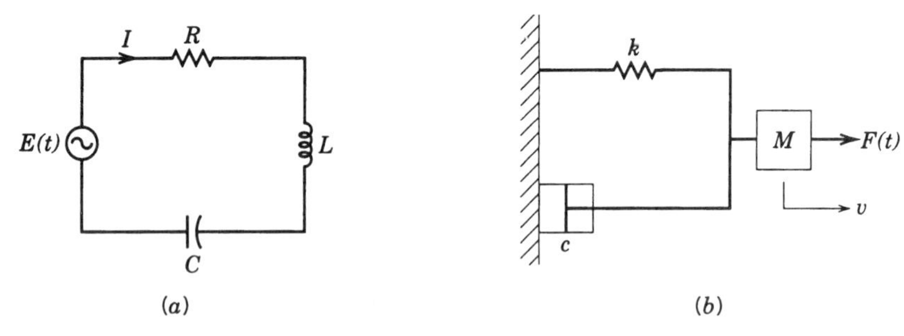

Fig. 1 (*a*) Electrical and (*b*) mechanical one-degree-of freedom systems: L = inductance, R = resistance, $1/C$ = inverse capacitance, E = exciting voltage, Q = charge, $I = dQ/dt$ = current, M = mass, c = damping constant, k = spring constant, F = exciting force, X = displacement, $V = dX/dt$ = velocity.

$p/v \equiv Z$ Specific acoustic impedance, a measure of how much fluid motion is produced for a given pressure disturbance

3 ACOUSTIC ELEMENTS

As a result of a mechanical disturbance in a medium, a sound wave is launched. The motions that arise behave Newton's laws of motion when the laws are applied to distributed (continuous) systems. A longitudinal pressure disturbance propagates with a wave velocity C_s. For a simple harmonic disturbance of frequency f, the wave will propagate a distance λ (the wavelength, = C_s/f) in one period of oscillation ($1/f$). If the wavelength of sound is long compared to the relevant physical length scales of the problem (as defined in the examples that follow), then acoustic lumped elements can be defined. It should be noted that this wavelength restriction reduces the problem from one of sound propagation to one of incompressible fluid hydrodynamics. Nevertheless, it should be realized that for the propagation of plane waves, the ratio of pressure to particle velocity is just the characteristic impedance $Z_c = \rho_0 C_s$, where ρ_0 is the fluid density (1.2 kg/m^3 for air at 20°C) and C_s is the sound velocity (344 m/s for air at 20°C), and Z equals 416 acoustic ohms (SI units). This impedance is real and thus represents a resistive impedance term, reflecting the fact that energy is lost to the source as acoustic radiation.

3.1 Inductance: Mass Element for a Tube

Consider a tube of length l and cross-sectional area A. A simple harmonic pressure disturbance across the ends caused by an acoustic wave will accelerate the mass of gas according to Newton's law, $F = Ap = M_{\text{gas}}\, dv/dt$, where dv/dt for harmonic motion can be written as $i\omega v$. The specific acoustic impedance $Z = p/v$ becomes, therefore, $Z = i\omega[M_{\text{gas}}/A]$.

The term in brackets is an inductance-like normalized mass of fluid in the tube. If this distributed mass is to move like a single slug, the pressure must be communicated throughout the tube in a time short compared to the acoustic period, τ, which is equivalent to requiring that the tube length be small compared to the acoustic wavelength. As an example, for a frequency of 100 Hz and an air-filled tube, the requirement for the lumped-element impedance expression to be valid would be that $l \ll 3.4$ m, or $l < 34$ cm.

3.2 Inductance: Mass Element for Vibrating Sphere

When a sphere vibrates, it moves the fluid surrounding it. By calculating the total kinetic energy of the surrounding fluid, with the assumption that the sphere is small compared to the acoustic wavelength, one finds

$$\text{KE} = \tfrac{1}{2} M_{\text{eff}} V_R^2,$$

where $M_{\text{eff}} = 4\pi R^3 \rho = 3 \times (\frac{4}{3}\pi R^3 \rho)$ and V_R is the velocity of the sphere at the outer diameter. The effective mass of fluid "felt" by the pulsating sphere is seen to be three times the mass of outer fluid that could fill the sphere. The force reaction on the sphere by the medium can, therefore, be written as

$$M_{\text{eff}} \frac{dV_R}{dt} = i\omega M_{\text{eff}} V_R \quad \text{(harmonic assumption)}.$$

The reaction impedance $Z = p/V_R = (F/A)/V_R = i\omega[M_{\text{eff}}/A]$, which is an inductance-type lumped element.

3.3 Inductance: Mass Element for Vibrating Piston

The mathematics is complex, but the results are simple for a uniformly vibrating circular piston of radius a:

$$M_{\text{eff}} = \rho \pi a^2 \Delta l,$$

where

$$\Delta l \doteq \begin{cases} \frac{8}{3}\pi a, & \text{piston in a large rigid baffle,} \\ 0.6a, & \text{no baffle.} \end{cases}$$

In both cases Δl represents the height of a "pill box" of fluid extending out from the vibrating piston. This pill box encloses the effective mass "felt" by the vibrating piston.

3.4 Resistive Elements

Viscous Resistive Elements When flow occurs through a slit of width w or a circular tube of radius a, energy is lost because of viscous dissipation (viscosity μ). When laminar flow conditions exist and the acoustic wavelength is long compared to the tube or slit length l, then the impedance is given by

$$Z_{\text{slit}} = \frac{12\mu l}{w^2}, \qquad Z_{\text{tube}} = \frac{8\mu l}{a^2}.$$

Radiative Acoustic Elements Whereas the resistance associated with plane sound waves is $\rho_0 C_s$, the acoustic resistance is less for low frequencies (where the waves are not planar). For a baffled circular tube of radius a, the radiation resistance in cases where the wavelength is much larger than the tube diameter is given by

$$Z_R = \rho_0 C_s \frac{(ka)^2}{2},$$

where k is the wavenumber ($= 2\pi f/C_s$).

3.5 Capacitance: Compliance Element

Consider a sealed loudspeaker box with a speaker diaphragm of area A and peak displacement d. By using the relationship between volume and pressure for an adiabatic change in a gas, one can write an expression relating the pressure change inside the loudspeaker to the diaphragm displacement d: $p = \gamma P_0 A d / V_0$, where γ is the ratio of the specific heats at constant pressure and volume, and it has been assumed that the volume change produced by the diaphragm displacement is small compared to the speaker's volume. For simple harmonic oscillatory motion with diaphragm velocity v, we find that $d = v/i\omega$, and the impedance can be computed as

$$Z = \frac{p}{v} = \frac{1}{i\omega C_M A},$$

where

$$C_M = \frac{V_0}{A^2 \gamma P_0} \quad \text{(mechanical compliance).}$$

Once again our analysis assumes that the wavelength of sound is large compared to the longest diagonal dimension of the loudspeaker (which assures that the pressure is essentially uniform within the box). Then the shape of the box is not a factor—only the volume.

4 RESONANCE BETWEEN MASS AND SPRING ELEMENTS

Consider now a sealed enclosure (as in Section 3.5) penetrated by a vibrating piston with a diaphragm of mass $M_{\text{diaphragm}}$ (Section 3.3) and area A. If resonance is defined by the vanishing of the imaginary part of the impedance, then for two acoustic elements in series (which implies that the same fluid volume oscillates in both elements), the resonance frequency f_0 is found by

$$Z = i\omega \frac{M_{\text{eff}}}{A} + \frac{1}{i\omega C_M A} = 0, \qquad \omega = \omega_0,$$

or

$$f_0 = \frac{\omega_0}{2\pi} = \frac{1}{2\pi}\sqrt{\frac{1}{M_{\text{eff}} C_M}}$$

$$= \frac{1}{2\pi}\sqrt{\frac{A\gamma P_0}{[M_{\text{diaphragm}} + \rho A(l + \Delta l)]}},$$

where Δl is given in Section 3.3 and is usually closer to $0.6a$ for most practical cases. This expression can be used for the *Helmholtz resonance*, as is found, for example, when one blows over the top of a partially filled, narrow-mouth bottle of neck length l. (Then, $M_{\text{diaphragm}} = 0$.)

The mechanical Q is a measure of the sharpness of the resonance as one varies the frequency, $Q = f_0/(f_2 - f_1)$, where f_2 and f_1 are the frequencies above and below the resonance frequency f_0 at which the resonator radiates half the power. For a diaphragm-free resonator with tube neck of length l and radius a,

$$Q \cong \frac{(l + 0.6a)C_s}{f_0 \pi a^2}.$$

REFERENCES

1. L. E. Kinsler, A. R. Frey, A. B. Coppens, and J. V. Sanders, *Fundamentals of Acoustics*, 3rd ed., Wiley, New York, 1982.
2. L. Beranek, *Acoustics*, American Institute of Physics, New York, 1986.

12

ACOUSTIC MODELING: FINITE ELEMENT METHOD

A. CRAGGS

1 INTRODUCTION

The problem being considered here is the following: Given an acoustic space with specified conditions on the bounding surfaces, either determine its response to a given source distribution or find its natural frequencies and normal modes. Now there are very few acoustic volumes amenable to an exact analytical solution as is possible with uniform tubes and rectangular and cylindrical cavities. An enclosure that has a complex geometry like the interior passenger space of an automobile or the piping configuration of the exhaust system has to be studied by an approximate numerical analysis, and the finite element method is one of several procedures available.

Essentially, a complex volume can be formed by assembling together a number of small elements that have a simpler geometry and smaller number of degrees of freedom. The more complex the geometry is, the greater the number of elements required. The number of elements is also influenced by the frequency range of interest, as usually the minimum dimension of an element should not be less than half a wavelength. If the frequency is very high, then perhaps the finite element procedure is not the best method to use.

There are a number of ways for formulating approximate numerical models for a continuous system. Some methods approximate the governing partial differential equation, which in this case is the wave equation, with finite differences; finite element equations can also be formed by the method of weighted residuals, the method of Galerkin being well established in this respect. Here, however, a variational procedure is outlined that is similar to that of the energy formulations used for elastic structures.

Note: References to chapters appearing only in the *Encyclopedia* are preceded by *Enc.*

2 THEORY

The procedure is outlined in many standard texts on finite element methods. It is based on a scalar functional that usually contains terms in the kinetic and potential energies and the boundary work.[1–5] The first variation of the functional leads to the exact governing differential equation and the boundary conditions. The belief is that the first variation of an approximate functional will give the optimum approximate equations.

2.1 One-Dimensional Elements

As they are more tractable, one-dimensional pipe finite elements will be considered first, and a further simplification is introduced by considering only harmonic motion that eliminates the time-dependent term in the wave equation. However, allowances are made for a varying pipe geometry. All elements are formulated in terms of the acoustic pressure.

The element is based on the scalar functional

$$\pi = \frac{1}{2}\int_0^l \left(\frac{dp}{dx}\right)^2 A\,dx - \left(\frac{\omega}{c}\right)^2 \frac{1}{2}\int_0^l p^2 A\,dx, \quad (1)$$

where p is the acoustic pressure, A is the cross-sectional area, c is the speed of sound, ω is the radian frequency, l is the length of the element, and x is the distance along the pipe. The first term in the functional is related to the kinetic energy and the second to the strain energy (but they are not equal). In this form the first variation leads directly to the classical form of the wave equation expressed in standard texts on acoustics and results in an eigenvalue problem in ω^2, rather than its reciprocal.

The stationary values of Eq. (1) lead to the well-known Webster horn equation

$$A\frac{d^2p}{dx^2}+\frac{dp}{dx}\frac{dA}{dx}+\left(\frac{\omega}{c}\right)^2 Ap=0 \qquad (2)$$

together with the boundary conditions $A(dp/dx)=0$ at $x=0$ and $x=l$. If a number of these elements are to be connected together, then at the element interfaces (see Fig. 2) the boundary conditions are

$$A\left(\frac{dp}{dx}\right)_2 - A\left(\frac{dp}{dx}\right)_3 = 0, \qquad (3)$$

$$p_2=p_3. \qquad (4)$$

Equation (3) is a form of the continuity equation, while Eq. (4) simply states that on connection the pressures are compatible.

If the problem is restricted to hard (i.e., immobile) boundaries, then the functional form (1) is all that is required. Certainly for many problems in room acoustics the hard boundary condition gives a very good approximation to the truth, and models based on this assumption give reliable predictions of the natural frequencies. However, if the boundary does have significant flexibility, extra terms are needed in the functional. For example, if one of the boundaries, say at $x=l$, has an impedance Z, then a suitable functional is

$$\pi=\frac{1}{2}\int_0^l\left(\frac{dp}{dx}\right)^2 A\,dx-\left(\frac{\omega}{c}\right)^2\frac{1}{2}\int_0^l p^2A\,dx - \frac{j}{2}\rho\omega\frac{p^2}{Z}A_l, \qquad (5)$$

where j is $\sqrt{-1}$, the complex operator.

This modified functional should only be applied to the terminal element or elements at the boundaries of the global system. Also, the form (5) is only suitable for conservative boundary conditions that involve no energy dissipation; the boundary is then either stiffness or mass controlled.

If the boundary has stiffness k per unit area, then $Z=k/j\omega$; if it is mass controlled with mass m per unit area, $Z=j\omega m$. In both of these simple cases the complex operator j disappears from the functional in Eq. (5) and retains a real form, with the extra term being added to either the first term of the functional (mass-controlled boundary) or the second term, the stiffness-controlled boundary. In each case, however, the terms disappear when either the stiffness or mass values become large.

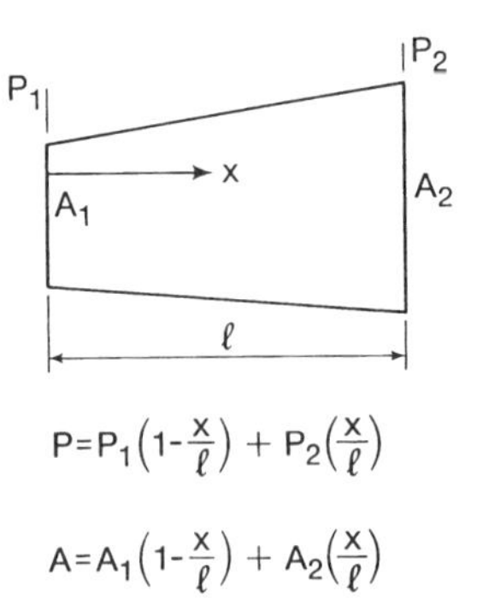

Fig. 1 Pipe acoustic finite element having linear variation of pressure and cross-sectional area.

When the boundaries are nonconservative, the functional has to be modified in the manner suggested by Morse and Ingard;[6] see also Craggs.[8]

The functional (1) will now be used to set up the approximate equations for a pipe element that allows a linear variation of acoustic pressure throughout the length. The cross-sectional area A will also be given a linear variation; refer to Fig. 1. Thus,

$$p=p_1\left(1-\frac{x}{l}\right)+p_2\left(\frac{x}{l}\right), \qquad (6)$$

and the cross-sectional area $A(x)$ can be written in terms of the end values A_1 and A_2:

$$A=A_1+(A_2-A_1)\left(\frac{x}{l}\right). \qquad (7)$$

To proceed further, Eq. (3) is written in a matrix form:

$$p=\{f(x)\}^{\mathrm{T}}\{P_e\}=\{P_e\}^{\mathrm{T}}\{f(x)\}, \qquad (8)$$

where $\{P_e\}$ is the vector $\{P_1/P_2\}$ containing the nodal acoustic pressures and $\{f(x)\}$ is a vector containing the linear polynomials $1-x/l$ and x/l. The superscript T denotes that the matrix has been transposed so that a column vector is transposed to a row matrix.

If Eq. (5) is differentiated with respect to x, then

$$\frac{dp}{dx}=\left\{\frac{df}{dx}\right\}^{\mathrm{T}}\{P_e\}=\{P_e\}^{\mathrm{T}}\left\{\frac{df}{dx}\right\}. \qquad (9)$$

Further progress in establishing the approximate equations is obtained by recognizing that

$$p^2 = \{P_e\}^{\mathrm{T}}\{f\}\{f\}^{\mathrm{T}}\{P_e\},$$

$$\left(\frac{dp}{dx}\right)^2 = \{P_e\}^{\mathrm{T}}\left\{\frac{df}{dx}\right\}\left\{\frac{df}{dx}\right\}^{\mathrm{T}}\{P_e\}; \qquad (10)$$

and after substitution, an approximate functional for the element is

$$\begin{aligned}\pi = {} & \frac{1}{2}\{P_e\}^{\mathrm{T}}\left(\int_0^l A_1\left\{\frac{df}{dx}\right\}\left\{\frac{df}{dx}\right\}^{\mathrm{T}} dx\right. \\ & \left. + \frac{A_2 - A_1}{l}\int_0^l x\left\{\frac{df}{dx}\right\}\left\{\frac{df}{dx}\right\}^{\mathrm{T}} dx\right)\{P_e\} \\ & - \frac{1}{2}\left(\frac{\omega}{c}\right)^2\{P_e\}^{\mathrm{T}}\left(\int_0^l A_1\{f\}\{f\}^{\mathrm{T}} dx\right. \\ & \left. + \frac{A_2 - A_1}{l}\int_0^l x\{f\}\{f\}^{\mathrm{T}} dx\right)\{P_e\}. \qquad (11)\end{aligned}$$

Carrying out the integrations, the result is

$$\pi = \frac{1}{2}\left(\{P_e\}^{\mathrm{T}}[S]\{P_e\} - \left(\frac{\omega}{c}\right)^2\{P_e\}^{\mathrm{T}}[R]\{P_e\}\right), \qquad (12)$$

where

$$[S] = \frac{A_1}{2l}\begin{bmatrix} 1 & 1 \\ -1 & 1 \end{bmatrix} + \frac{A_2}{2l}\begin{bmatrix} 1 & -1 \\ -1 & 1 \end{bmatrix} \qquad (13)$$

and

$$[R] = \frac{A_1 l}{12}\begin{bmatrix} 3 & 1 \\ 1 & 1 \end{bmatrix} + \frac{A_2 l}{12}\begin{bmatrix} 1 & 1 \\ 1 & 3 \end{bmatrix}, \qquad (14)$$

and setting the first variation to zero, that is, $\delta\pi = 0$, gives the approximate equation for one finite element,

$$[S]\{P_e\} - \frac{\omega^2}{c^2}[R]\{P_e\} = \{0\}. \qquad (15)$$

In the above it was assumed that there are no volume source terms, that is, the boundaries are immobile.

Equation (15) is for a single finite element; the assembly of two or more elements can be illustrated by considering the system shown in Fig. 2. Unconnected the elements have four unknowns, p_1, p_2, p_3, and p_4. If the elements are joined at node 2, then Eqs. (3) and (4) need

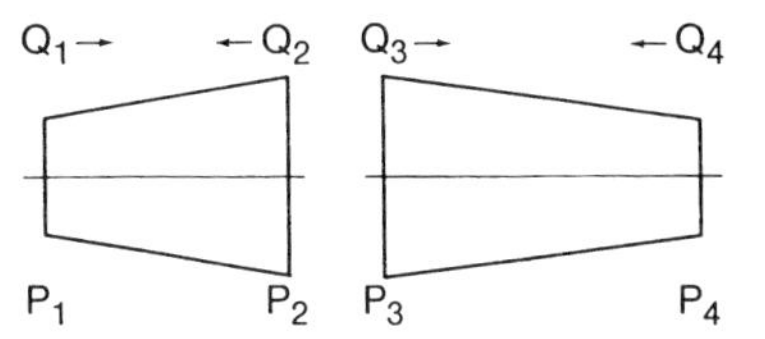

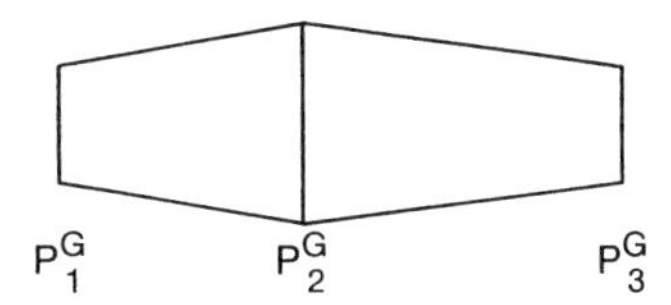

Fig. 2 Assembly of two linear pipe finite elements.

to be satisfied. As a result of these constraints, the four-degree-of-freedom system is reduced to a global system having three degrees of freedom:

$$\begin{bmatrix} S_{11}^A & S_{12}^A & 0 \\ S_{21}^A & S_{22}^A + S_{11}^B & S_{12}^B \\ 0 & S_{21}^B & S_{22}^B \end{bmatrix}\begin{Bmatrix} P_1^G \\ P_2^G \\ P_3^G \end{Bmatrix} - \frac{\omega^2}{c^2}\begin{bmatrix} R_{11}^A & R_{12}^A & 0 \\ R_{21}^A & R_{22}^A + R_{11}^B & R_{12}^B \\ 0 & R_{21}^B & R_{22}^B \end{bmatrix}\begin{Bmatrix} P_1^G \\ P_2^G \\ P_3^G \end{Bmatrix} = \begin{Bmatrix} 0 \\ 0 \\ 0 \end{Bmatrix}. \qquad (16)$$

The individual element matrices are then assembled by overlaying them at the connecting nodes. In the same manner many elements can be linked together. While it is common that they are connected in a chainlike fashion to form a one-dimensional system, it is also possible to form branched systems.

An element that is a little more accurate and allows a quadratic variation in the cross-sectional area and pressure has three degrees of freedom. Thus, if (refer to Fig. 3)

$$p = a_1 + b_1\left(\frac{x}{e}\right) + e_1\left(\frac{x}{e}\right)^2$$

and

$$A = a_2 + b_2\left(\frac{x}{e}\right) + c_2\left(\frac{x}{e}\right)^2,$$

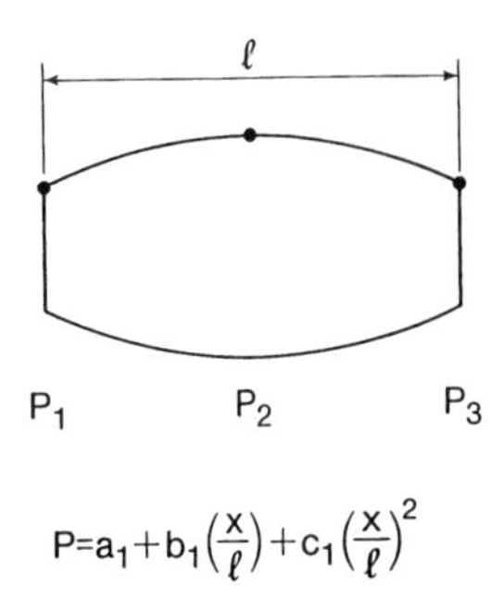

Fig. 3 The quadratic pipe element.

then the resulting matrices are

$$[S] = \alpha_1 \begin{bmatrix} 7 & -8 & 1 \\ -8 & 16 & -8 \\ 1 & -8 & 7 \end{bmatrix} + \alpha_2 \begin{bmatrix} 3 & -4 & 1 \\ -4 & 16 & -12 \\ 1 & -12 & 11 \end{bmatrix} + \alpha_3 \begin{bmatrix} 3 & -6 & 3 \\ -6 & 32 & -26 \\ 3 & -26 & 23 \end{bmatrix}, \quad (17)$$

where $\alpha_1 = A_1/3l$, $\alpha_2 = (4A_2 - 3A_1 - A_3)/6l$, and $\alpha_3 = (2A_1 - 4A_2 + 2A_3)/15l$, and

$$[R] = \beta_1 \begin{bmatrix} 4 & 2 & -1 \\ 2 & 16 & 2 \\ -1 & 2 & 4 \end{bmatrix} + \beta_2 \begin{bmatrix} 1 & 0 & -1 \\ 0 & 16 & 4 \\ -1 & 4 & 7 \end{bmatrix} + \beta_3 \begin{bmatrix} 2 & -4 & -5 \\ -4 & 64 & 24 \\ -5 & 24 & 44 \end{bmatrix}, \quad (18)$$

where $\beta_1 = A_1 l/30$, $\beta_2 = (4A_2 - 3A_1 - A_3)l/60$, and $\beta_3 = (2A_1 - 4A_2 + 2A_3)l/420$. If the cross section is uniform throughout, then $\alpha_2 = \alpha_3 = \beta_2 = \beta_3 = 0$ and only the first terms remain.

This latter element has been used to determine the natural frequencies of the wine bottle shown in Fig. 4; the interior dimensions being height 20.32 cm, base diameter 7.62 cm, and neck outlet diameter 1.52 cm. The model consists of eight of the quadratic finite elements that when assembled give a global matrix with 17 degrees of freedom. However, this was reduced by applying the boundary condition $p = 0$ at the open end. Figure 4 shows the four lowest modes and their natural frequencies.

The quadratic element was also used to model a one-dimensional version of the branched system shown in Fig. 5. Here the one-dimensional element results were compared with those from three-dimensional study (refer

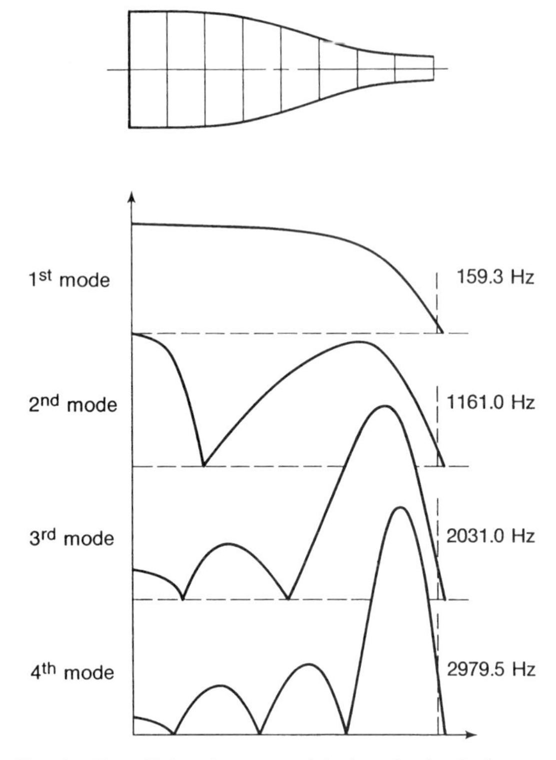

Fig. 4 (Top) Finite element model of a wine bottle from an assembly of quadratic elements. (Bottom) Lowest four normal modes and natural frequencies.

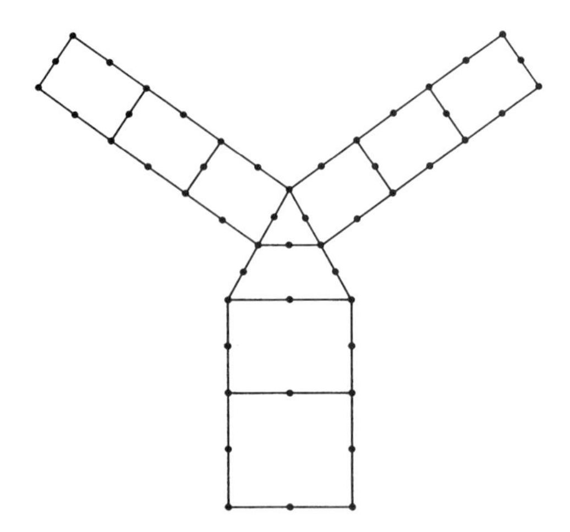

Fig. 5 Finite element model of a Y-branch. This was modeled with one-dimensional elements and three-dimensional elements (shown).

TABLE 1 Comparison of First 9 Modes of Y-System for Pipe Element and HX20

Mode	Pipe Element	HX20
1	6.854	6.772
2	9.874	10.03
3	39.60	40.78
4	62.15	61.50
5	90.16	93.97
6	164.4	164.7
7	179.4	177.4
8	300.0	305.6
9	403.5	400.2

to the next section). The results from the models are compared in Table 1, and there is close agreement for the first nine modes.

The pipe element is only valid as long as the acoustic wavelength is large compared with the cross-sectional dimensions. If this is not the case and the acoustic pressure is not uniform over the cross section, then it is necessary to allow for this with three-dimensional elements. These will be considered in the next section.

2.2 Three-Dimensional Elements

The steps in forming the finite element matrices [*S*] and [*R*] are the same as those given in the previous section. However, because of the extra two dimensions and the implied extra degrees of freedom, it is not a practical proposition to evaluate all of the terms analytically, as was done in the one-dimensional case. It is essential to use numerical procedures throughout, and this in turn results in a greater flexibility of approach.

The procedure is illustrated by developing a hexahedral element. In the first instance this will have a rectangular section, but later the geometry will be allowed to vary to produce elements with curved boundaries that will give the element a wider application:

The basic rectangular element is shown in Fig. 6, and neglecting the source effects, the functional has the form

$$\pi = \frac{1}{2}\iiint\left(\frac{\partial r}{\partial r}\right)^2 + \left(\frac{\partial p}{\partial s}\right)^2 + \left(\frac{\partial p}{\partial t}\right)^2 dr\,ds\,dt - \frac{\omega^2}{c^2}\frac{1}{2}\iiint p^2\,dr\,dr\,dt, \tag{19}$$

and the stationary values of this lead to the Helmholtz equation

$$\frac{\partial^2 p}{\partial r^2} + \frac{\partial^2 p}{\partial s^2} + \frac{\partial^2 p}{\partial t^2} + \frac{\omega^2}{c^2}p = 0, \tag{20}$$

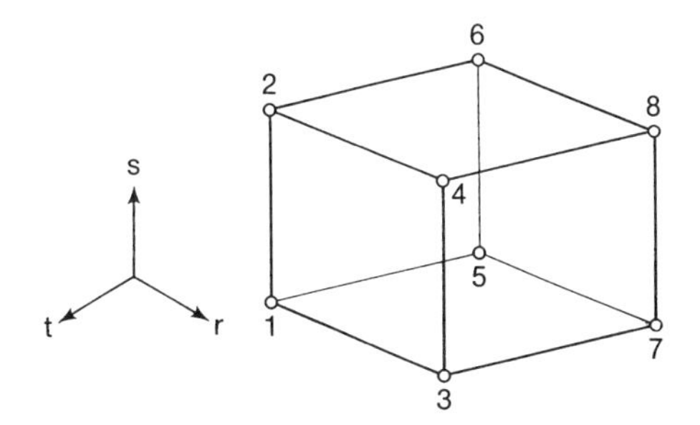

Fig. 6 Eight-node rectangular acoustic finite element.

with the boundary condition $\partial p/\partial \bar{n} = 0$ on the surface, where $\bar{\mathbf{n}}$ is the direction of the outgoing normal vector. The formation of the element matrices will be illustrated with the eight-node element, which has a node at each corner and allows a linear variation of the acoustic pressure:

$$p = a_1 + a_2 r + a_3 s + a_4 t + a_5 rs + a_6 rt + a_7 st + a_8 rst,$$
$$p = \{F\}^{\mathrm{T}}\{\alpha\},$$

where

$$\{F\} = \begin{Bmatrix} 1 \\ r \\ s \\ t \\ rs \\ rt \\ st \\ rst \end{Bmatrix}, \qquad \{\alpha\} = \begin{Bmatrix} a_1 \\ a_2 \\ a_3 \\ a_4 \\ a_5 \\ a_6 \\ a_7 \\ a_8 \end{Bmatrix}. \tag{21}$$

At this stage the pressure is expressed in terms of the generalized coefficients $a_1, \ldots, a_n$ that govern the contribution of each of the simple polynomial terms. It is more useful to relate these to the acoustic pressures at the individual node points. This can be achieved by placing the boundary values c_i, s_i, and t_i for each nodal pressure p_i, which in this case gives eight equations in eight unknowns that can be expressed in the matrix form

$$\{p\} = [T]\{\alpha\}, \tag{22}$$

$$\{\alpha\} = [M]\{p\}, \quad \text{where } [M] = [T^{-1}], \tag{23}$$

Using this result, the pressure at any point r, s, t within the element can be written in terms of the nodal pressures

$$p = \{F\}^{\mathrm{T}}[M]\{p\}. \tag{24}$$

Differentiating this equation, first with respect to r, then with respect to s and t, gives the derivative equations, which in matrix form are

$$\left\{\begin{matrix} \dfrac{\partial p}{\partial r} \\ \dfrac{\partial p}{\partial s} \\ \dfrac{\partial p}{\partial r} \end{matrix}\right\} = [G][M]\{P\} \tag{25}$$

The matrix $[G]$ is a 3×8 matrix, the rows containing the derivatives $\{\partial F/\partial r\}^{\mathrm{T}}$, $\{\partial F/\partial s\}^{\mathrm{T}}$, and $\{\partial F/\partial t\}^{\mathrm{T}}$. Substituting the results (2) and (3) into Eq. (1) gives the approximate functional for the element

$$\pi = \frac{1}{2}\{P\}^{\mathrm{T}}[S]\{P\} - \frac{1}{2}\left(\frac{\omega}{c}\right)^2 \{P\}^{\mathrm{T}}[R]\{P\}, \tag{26}$$

where

$$[S] = \iiint [M]^{\mathrm{T}}[G]^{\mathrm{T}}[G][M]\, dr\, ds\, dt, \tag{27}$$

$$[R] = \iiint [M]^{\mathrm{T}}\{F\}^{\mathrm{T}}\{F\}^{\mathrm{T}}[M]\, dr\, ds\, dt. \tag{28}$$

Setting the first variation $\delta\pi = 0$ in Eq. (4) gives the approximate form for the three-dimensional Helmholtz equation

$$\left([S] - \frac{\omega^2}{c^2}[R]\right)\{P\} = 0, \tag{29}$$

which is a linear eigenvalue problem. The limits of integration will depend upon the dimensions of the element. Since the volume is a regular parallelepiped, the integration is not difficult, but there are so many terms that it is best carried out numerically. For example, the eight-node element discussed above will require 64 integrations for each of the $[S]$ and $[R]$ matrices. The higher order element with 20 nodes will require 400 integrations and the 32-node element will require 1024.

The regular element discussed above will have a significant but limited application to rectangular sectioned rooms and ducts. The application can be greatly extended if the geometry also is allowed to be distorted, and it is possible to do this by using the same polynomials that govern the variation in the pressure to govern the variation in geometry. To achieve this, the regular geometry in the r, s, t system is transformed to an irregular geometry in x, y, z (refer to Fig. 7) by the transformations

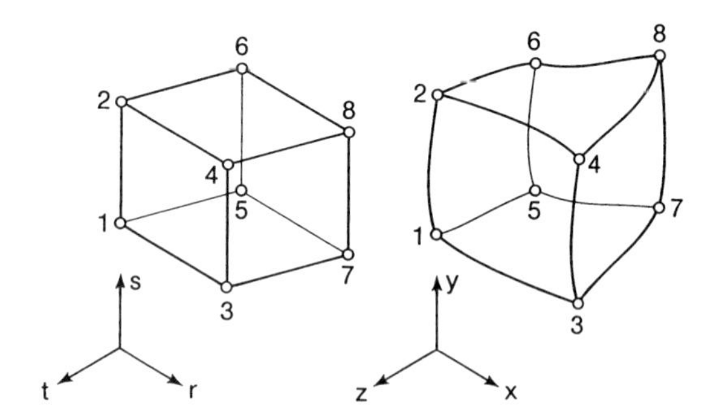

Fig. 7 Transformation from a regular rectangular element in r, s, t to an irregular element in x, y, z.

$$x = \{F\}^{\mathrm{T}}[M]\{X\}, \qquad y = \{F\}^{\mathrm{T}}[M]\{Y_i\},$$
$$z = \{F\}^{\mathrm{T}}[M]\{Z_i\}, \tag{30}$$

where $\{X\}$, $\{Y\}$, and $\{Z\}$ are the vectors containing the x-, y-, and z-coordinates at the node points. The functional for the element in x, y, z is

$$\pi = \frac{1}{2}\iiint\left(\frac{\partial p}{\partial x}\right)^2 + \left(\frac{\partial p}{\partial y}\right)^2 + \left(\frac{\partial p}{\partial z}\right)^2 dx\, dy\, dz$$
$$- \frac{\omega^2}{c^2}\frac{1}{2}\iiint p^2\, dx\, dy\, dz. \tag{31}$$

However, by making use of the results,

$$\left\{\begin{matrix} \dfrac{\partial p}{\partial x} \\ \dfrac{\partial p}{\partial y} \\ \dfrac{\partial p}{\partial z} \end{matrix}\right\} = [J]^{-1}\left\{\begin{matrix} \dfrac{\partial p}{\partial r} \\ \dfrac{\partial p}{\partial s} \\ \dfrac{\partial p}{\partial f} \end{matrix}\right\} \tag{32}$$

and $dx\, dy\, dz = \|J\|\, dr\, ds\, dt$, where $[J]$ is the Jacobian matrix and $\|\cdot\|$ denotes the absolute value of the determinant of $[J]$.

The matrices $[S]$ and $[R]$ are given by

$$[S] = \int_{-1}^{+1}\int_{-1}^{+1}\int_{-1}^{+1} [M]^{\mathrm{T}}[G]^{\mathrm{T}}[J^{-1}]^{\mathrm{T}}[J^{-1}][G][M]$$
$$\cdot \|J\|\, dr\, ds\, dt, \tag{33}$$

$$[R] = \int_{-1}^{+1}\int_{-1}^{+1}\int_{-1}^{+1} [M]^{\mathrm{T}}\{F\}\{F\}^{\mathrm{T}}[M]\|J\|\, dr\, ds\, dt. \tag{34}$$

It is worth noting that the integrals have been expressed in the r, s, and t domain where there is a regular geometry. Further, the limits of integration are -1 to $+1$, which conform with the standard limits of tabulated numerical integration schemes. Equations (24) and (25) are those used in several commercial codes, though some codes prefer using serendipity functions instead of the simple polynomials, as these allow the pressures and its derivatives to be written down directly in terms of the nodal pressures and avoid forming the inverse matrix $[M]$. However, this matrix need only be formed once and its elements stored in memory.

Although the results have been derived for the 8-node element, the same procedure can be used to form the element matrices with 20 or 32 nodes. These more powerful elements, however, need extra polynomials with higher order terms. While they give more accurate results on a degree-of-freedom basis, there is a penalty due to the greater number of integration points needed and extra complexity in the input data. They have their place in specialized programs. The 20-node element has probably the most common usage. An indication of the accuracy of the HEX8, HEX20, and HEX32 elements is given in Table 2, where the results for the lowest modes of a unit cube are given. The highest order element is seen to give the greatest accuracy; although it is not shown here, this gives a more rapid convergence to the solution as the number of elements is increased. A cavity representing a model of a car interior (Fig. 8) has been simulated with the 32-node element. Thirty of these elements were used, resulting in a system with 378 degrees of freedom. A comparison of the predicted and measured frequencies are given in Table 3.

While the three-dimensional isoparametric element is very versatile, there are situations in which the number of degrees of freedom can be reduced if the geometry has some form of symmetry. If there is axial symmetry, then a suitable finite element can be derived from the stationary values of the functional

$$\pi = \frac{1}{2} \iint \left(\frac{\partial p}{\partial r}\right)^2 + \left(\frac{\partial p}{\partial z}\right)^2 2\pi r \, dr \, dz - \frac{\omega^2}{c^2} \iint p^2 2\pi r \, dr \, dz, \tag{35}$$

TABLE 2 Frequency of Lowest Mode (0, 01), (0, 10) (1, 00) for a Unit Cube

Exact	HEX8	HEX20	HEX32
$\pi^2 = 9.8696$	12.000	11.595	9.8751

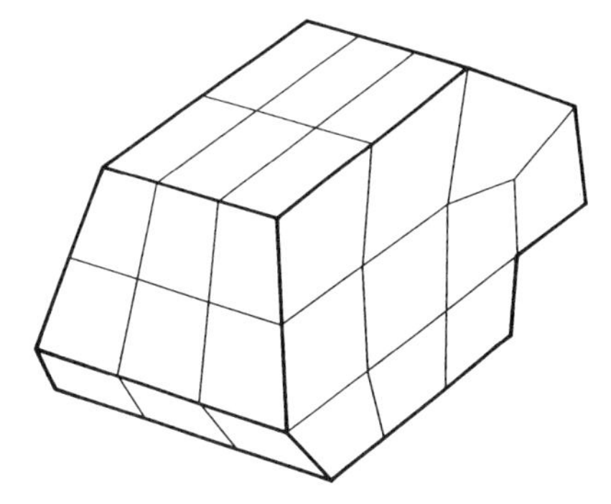

Fig. 8 Acoustic finite element grid for a model car enclosure. Assembly is of the 32-node isoparametric elements.

where r and z are the radial and axial dimensions, respectively. A very effective element is the four-node quadrilateral element. This element is very easy to form, it has only four degrees of freedom, it has pressures at each node and allows a linear variation in pressure and geometry, and yet it gives very acceptable results. It is particularly useful in evaluating axisymmetric mufflers.

2.3 Acoustostructural Problems

In an earlier section a very simple method for dealing with boundary flexibility was considered that is suitable for nondissipative reactive boundaries. However, if the bounding structure also has resonances in the same frequency range as the enclosed acoustic space, then a finite element model of the structure is also needed. The formulation of the mass matrix $[M]$ and the stiffness matrix $[K]$ for a structure is given in the text by Petyt.[10]

Recognizing that the structural equations are formu-

TABLE 3 Natural Frequencies of Irregular Cavity

Mode	Finite Element Simulation	Measured Frequency
100	140.8	154.6
001	160.3	168.2
010	208.9	220.2
200	243.3	251.0
101	267.8	281.8
201	288.8	301.8
110	299.9	312.2
110	299.9	312.2
002	315.4	323.8
102	353.2	355.4
111	347.0	366.6
020	371.3	378.2

lated in terms of displacements $\{w\}$ and the acoustics in terms of pressure $\{p\}$, the coupled equations governing harmonic motion have the form

$$\begin{bmatrix} S & 0 \\ \theta^{\mathrm{T}} & K \end{bmatrix} \begin{Bmatrix} p \\ w \end{Bmatrix} - \omega^2 \begin{bmatrix} \dfrac{1}{c^2} & \rho\theta \\ 0 & M \end{bmatrix} \begin{Bmatrix} p \\ w \end{Bmatrix} = \begin{Bmatrix} q \\ Q \end{Bmatrix}. \tag{36}$$

The matrix $[\theta]$ is a rectangular matrix found by integrating the product of the allowed pressure and displacement polynomials over the coupling surface area[7]; $\{q\}$ is an acoustic volume source term and $\{Q\}$ is a force vector representing distributed forces of excitation applied on the structure.

The technique given in Eq. (36) has formed the basis for determining the response of rooms to window excitation by sonic booms,[7] making sound transmission studies in cars,[8] and finding the effect of flexible walls on the transmission loss of silencers.[9]

3 FINAL COMMENTS

The finite element method is a good way to model the acoustics of an enclosed space, and even low-order elements with a linear variation of pressure can give acceptable results. In its simplest form, the resulting equations can involve huge matrices that are costly to solve, especially as they need to be solved over a spectrum of frequencies. Commercial computer codes can make work much easier in that they incorporate automatic grid generators and pre- and postprocessing facilities. Codes like ANSYS and NASTRAN have ready-made acoustic software built into a fully comprehensive engineering package. SYSNOISE is a program that deals solely with acoustic problems.

REFERENCES

1. G. M. L. Gladwell, "A Finite Element Method for Acoustics," paper presented at Fifth International Conference on Acoustics, Liege, 1965, paper L33.
2. V. Mason, "On the Use of Rectangular Finite Elements," Institute of Sound and Vibration Report No. 161, University of Southampton, 1967.
3. A. Craggs, "The Use of Simple Three Dimensional Acoustic Finite Elements for Determining the Natural Modes and Frequencies of Complex Shaped Enclosures," *J. Sound Vib.*, Vol. 23, 1972, pp. 331–339.
4. M. Petyt, J. Leas, and G. H. Koopman, "A Finite Element Method for Determining the Acoustic Modes of Irregular Shaped Cavities," *J. Sound Vib.*, Vol. 45, 1976, pp. 495–502.
5. D. Nefske and L. J. Howell, "Automobile Interior Noise Reduction Using Finite Element Methods," *Trans. SAE*, Vol. 87, 1978, pp. 1727–1737.
6. P. M. Morse and K. U. Ingard, *Theoretical Acoustics*, McGraw-Hill, New York, 1968, pp. 250–256.
7. A. Craggs, "The Transient Response of a Coupled Plate-Acoustic System Using Plate and Acoustic Finite Elements," *J. Sound Vib.*, Vol. 15, 1971, 509–529.
8. A. Craggs, "An Acoustic Finite Element Approach for Studying Boundary Flexibility and Sound Transmission between Irregular Enclosures," *J. Sound Vib.*, Vol. 30, No. 3, pp. 343–357.
9. C. I. J. Young and M. J. Crocker, "Finite Element Acoustical Analysis of Complex Muffler Systems with and without Wall Vibrations," *Noise Control Eng.*, Vol. 9, No. 2, 1977, pp. 86–93.
10. M. Petyt, *Introduction to Finite Element Vibration Analysis*, Cambridge University Press, 1990.

13

ACOUSTIC MODELING: BOUNDARY ELEMENT METHODS

A. F. SEYBERT AND T. W. WU

1 INTRODUCTION

The boundary element method (BEM) is a numerical technique for calculating the sound radiated by a vibrating body or for predicting the sound field inside of a cavity such as a vehicle interior. The BEM may also be used to determine the sound scattered by an object such as a microphone or for predicting the performance of silencers or mufflers. The BEM is becoming a popular numerical technique for acoustical modeling in industry. The major advantage of this method is that only the boundary surface (e.g., the exterior of the vibrating body) needs to be modeled with a mesh of elements. For infinite-domain problems, such as radiation from a vibrating structure, the so-called Sommerfield radiation condition is automatically fulfilled. In other words, there is no need to create a mesh to approximate this radiation condition. A typical BEM input file consists of a surface mesh, a normal velocity profile on the surface, the fluid density, speed of sound, and frequency. The output of the BEM includes the sound pressure distribution on the surface of the body and at other points in the field, the sound intensity, and the sound power. In this chapter, an overview of the BEM is presented with emphasis on its application to industrial noise control problems. Although the BEM may also be formulated in the time domain, the focus here will be on the frequency domain only.

1.1 The Boundary Element Mesh

Figure 1 shows a BEM for the calculation of the sound radiated by a tire. This is an example of an *exterior problem* in acoustics, so named because the acoustic domain is infinite or, in the case of Fig. 1, semi-infinite due to the reflecting ground surface. The geometry of the tire is represented by a boundary element mesh that is a series of points called *nodes* on the surface of the body that are connected together to form *elements* of either quadrilateral or triangular shape. Note in the case of the tire in Fig. 1 and many other structures there exists a symmetry plane so that only one-half of the body needs to be modeled.

Building the boundary element mesh or model is the first step in solving a problem with the BEM. The size of the elements must be chosen small enough to obtain an acceptable solution but not so small as to result in excessive computer time. In general, a boundary element mesh must meet three requirements to obtain an acceptable solution. First, it must model the geometry of the body accurately. This means that all major surfaces, as well as edges and corners, must be accurately represented. (However, it is not necessary to model regions of the body for which it can be concluded beforehand are not important acoustically, such as surface irregularities that are small compared to the acoustic wavelength.) Second, the mesh must be fine enough to represent the distribution of vibration on the surface of the body. This can be done by using a sufficient number of elements per *structural wavelength*. Finally, the mesh must be fine enough to represent the sound pressure distribution on the surface of the body. This can be done in most cases by selecting the element size to be no larger than a certain fraction of the *acoustic wavelength* for the highest frequency of interest. These last two requirements usually conflict, so one should select the largest element size that satisfies both requirements. The actual element size depends on the type of element used, as discussed in Section 2.3.

Note: References to chapters appearing only in the *Encyclopedia* are preceded by *Enc.*

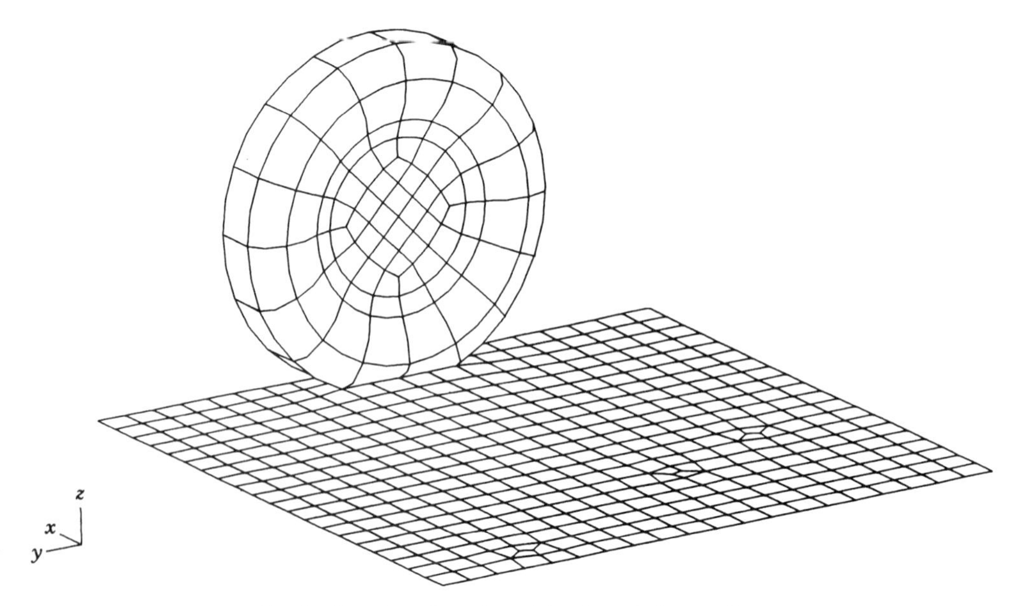

Fig. 1 Boundary element mesh for tire radiation problem.

Most commercially available pre- and postprocessing programs developed for the finite element method (FEM) may be used as well for constructing boundary element meshes. One uses shell or plate elements where the physical and material properties of the elements are immaterial since boundary elements have no thickness or properties. These same programs may be used to visualize the results of the BEM using their standard contouring capabilities.

1.2 Input to the Boundary Element Model

The boundary element mesh covers the entire surface of the radiating body. One needs to know some information (the so-called *boundary conditions*) about the problem at every node point of the mesh. Thus, for example, for most radiation problems one knows the vibration velocity normal to the surface at every point. Both the magnitude and phase of the velocity must be known. If a portion of the surface is covered by a sound-absorbing material, one must know the acoustic impedance of the material at each of the nodes that lie thereon.

1.3 Output of the Boundary Element Model

As described in Section 2, the BEM calculates the sound pressure distribution on the surface of the body from the geometry of the body and the distribution of vibration velocity or other boundary conditions provided by the user. First, once the sound pressure and the vibration velocity are known on the surface, the sound intensity, sound power, and sound radiation efficiency may be found. Second, the sound pressure, particle velocity, and sound intensity can be calculated at so-called *field points*, that is, points in the acoustic domain that are not on the surface of the body. These field points can even lie on a reflecting plane such as the ground plane shown in Fig. 1.

Figure 2 shows the contour of sound pressure level (SPL) radiated by the tire in Fig. 1 at 1000 Hz. Note that the BEM will determine the SPL at all points in the acoustic domain, both in the near and the far fields. The BEM will also calculate the SPL in the "shadow zone" behind a body due to radiation from the other sides. All of this is done without any additional approximations or restrictions than discussed previously, that is, the need for an adequate mesh and the requirement to know the boundary conditions on the surface of the body.

1.4 Application of the Boundary Element Mode to Interior Problems

A second class of problems in which the BEM may be used is the so-called *interior problem* in acoustics. In contrast to exterior problems in which the acoustic domain extends to infinity, the acoustic domain for interior problems is finite. Typical examples of interior problems include the prediction of the sound field inside a

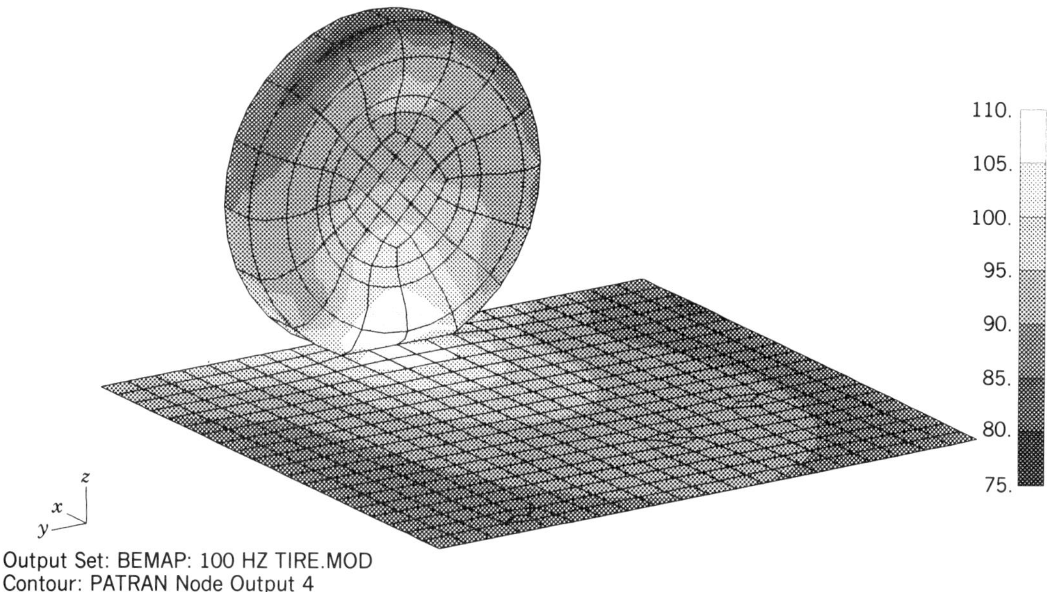

Fig. 2 The SPL contour plot of tire radiation at 1000 Hz.

vehicle or the calculation of the transmission loss of silencers.

From the user's standpoint, there is no difference between the application of the BEM to interior or exterior problems. First, the user creates a mesh with the same requirements discussed in Section 1.1. Second, boundary condition information at every node is supplied to the BEM. The BEM calculates the sound pressure on the inside surface of the cavity as well as the sound pressure level at field points inside the cavity.

Figure 3 shows the boundary element mesh for a simple expansion chamber muffler. Figure 4 shows the transmission loss (TL) of the muffler predicted by the BEM along with experimental results.[1] The BEM results were obtained by specifying the velocity on all nodes to be zero (since the muffler casing was assumed to be rigid) except those nodes at the inlet and exit of the muffler. At the inlet cross section, the velocity was specified as unity, and at the outlet the boundary condition was that the acoustic impedance was equal to the *characteristic impedance* of the medium.

The BEM can reveal a considerable amount of detail about the behavior of the system. For example, Fig. 5 shows the SPL contour inside the muffler at 2900 Hz, whereas the TL is approximately zero in Fig. 4. From Fig. 5 it may be seen that a cross mode in the chamber is excited at this frequency, resulting in a deterioration of the muffler's performance.

2 THEORY OF THE BOUNDARY ELEMENT METHOD

Although the mathematical foundation for the BEM is much older, it was not until the 1960s that the method could be used for problems of even modest size on the computers available at that time.[2–6] This early work provides a good basis for the BEM theory reviewed in this section.

2.1 The Boundary Element Method for Exterior Problems

Figure 6 shows a body denoted by B with surface S that radiates sound into an unbounded acoustic domain called B' having mean density ρ_0 and speed of sound c. The governing equation of linear acoustics is the well-known Helmholtz equation

$$\nabla^2 p + k^2 p = 0, \tag{1}$$

where p is the sound pressure at any point in the domain B' and $k = \omega/c$ is the wavenumber for harmonic waves of frequency ω. Boundary conditions on S are of the general form

$$\alpha p + \beta v_n = \gamma, \tag{2}$$

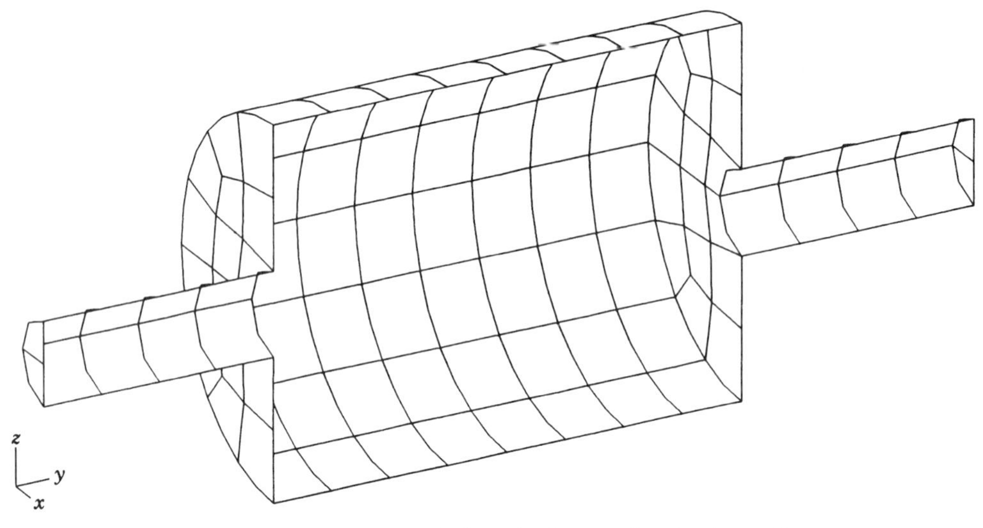

Fig. 3 Boundary element mesh for a simple expansion chamber muffler.

where v_n is the velocity (magnitude and phase) of any point on S in the normal direction, defined by the unit-normal vector n of S directed away from the acoustic domain (i.e., into the body B), and α, β, and γ are complex constants. A number of practical boundary conditions are included in Eq. (2). For example, Eq. (2) may be used to specify the normal velocity at a point by setting $\alpha = 0$, $\beta = 1$, and γ equal to the known velocity. Equation (2) can be used to represent a *locally reacting* normalized acoustic impedance z_n by setting $\alpha = 1$, $\beta = -z_n$, and $\gamma = 0$.

By using the direct formulation (via either the Green's second identity or the weighted residual formulation) and the Sommerfeld radiation condition, Eq. (1) is reformulated into a *boundary integral equation* as follows[7–11]:

$$C(P)p(P) = -\int_S [p(Q)G'(Q,P) + ikz_0 v_n(Q)G(Q,P)]\, dS(Q). \qquad (3)$$

In the above equation, $p(P)$ is the sound pressure at a

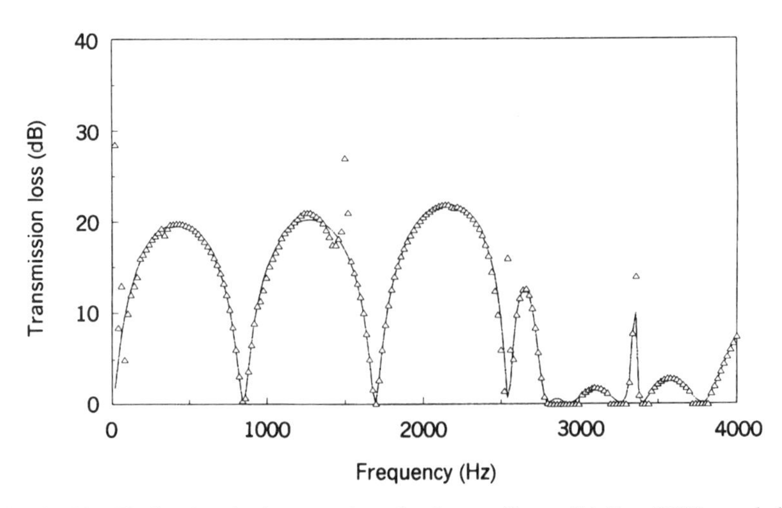

Fig. 4 The TL for the simple expansion chamber muffler: solid line, BEM; symbols, experiment.

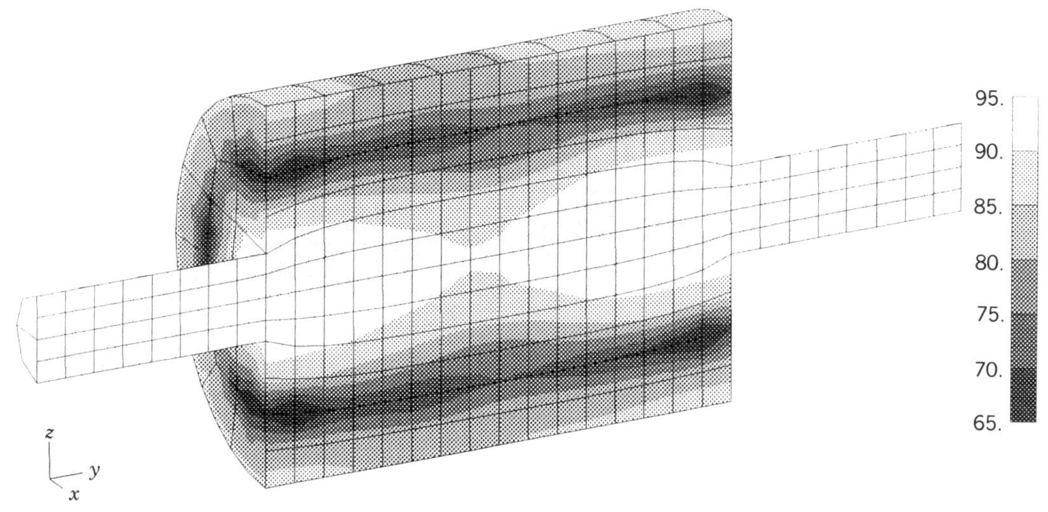

Fig. 5 The SPL contour plot for the expansion chamber muffler at 2900 Hz.

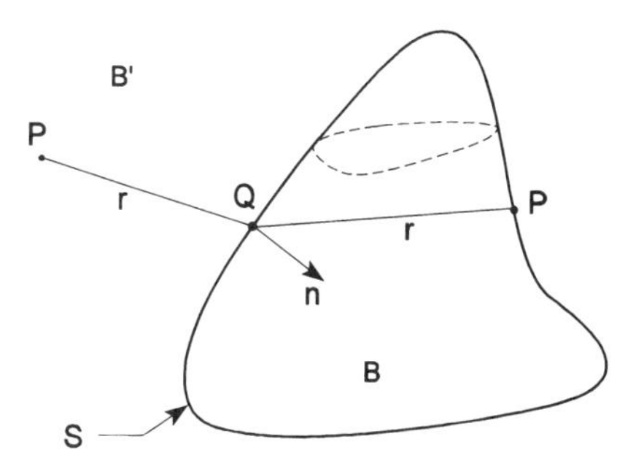

Fig. 6 Nomenclature for the BEM acoustic radiation problem.

point P, $p(Q)$ and $v_n(Q)$ are the sound pressure and normal velocity distributions on the surface of the body, $z_0 = \rho_0 c$ is the characteristic impedance of the medium, and $C(P)$ is a constant whose value depends on the location of the point P. Equation (3) is often referred to as the Helmholtz integral equation. In Eq. (3), G is the free-space Green's function

$$G(Q, P) = \frac{e^{-ikr}}{r}, \tag{4}$$

where $r = |Q - P|$ is the distance between points Q and P and G' is the derivative of G in the direction normal to the body. From a physical standpoint, Eq. (3) states that the sound pressure at any point P can be found by integrating (summing) the contributions of a dipole distribution [first term in Eq. (3)] and a monopole distribution [second term in Eq. (3)] over the boundary S to the point P in question.

The leading coefficient $C(P)$ in Eq. (3) is 4π for P in the acoustic domain B' and zero for P inside of the body B. When P is located on the surface S, the value of $C(P)$ is determined by[7]

$$C(P) = 4\pi - \int_S \frac{\partial}{\partial n}\left(\frac{1}{r}\right) dS(Q), \tag{5}$$

which has the physical interpretation of the exterior solid angle of S at point P. The value of $C(P)$ when P is on S will be 2π at points where the surface has a unique tangent plane. For corners and edges, $C(P)$ must be evaluated from Eq. (5).

It should be noted that when P coincides with Q on S, both integrals in Eq. (3) are singular because $r = 0$. The singularity of both terms is $O(1/r)$, which can be removed by a coordinate transformation before the integration is attempted.

2.2 The Boundary Element Method for Interior Problems

For interior problems such as the muffler shown in Fig. 3, the BEM formulation[12–14]

$$C^0(P)p(P) = -\int_S [p(Q)G'(Q,P) + ikz_0v_n(Q)G(Q,P)]\, dS(Q) \qquad (6)$$

is similar to that described in the previous section with several exceptions. First, the Sommerfeld radiation condition is not used to derive Eq. (6) because for interior problems the domain is finite. Second, the normal n that is directed away from the acoustic domain is opposite to the normal n for exterior problems. Finally, the value of the leading coefficient $C^0(P) = 4\pi - C(P)$, where $C(P)$ is given by Eq. (5).

2.3 Numerical Implementation of the Boundary Element Method

One (and only one) of the acoustic variables p or v_n may be specified at each point Q on the surface S. (If the impedance $z_n = p/v_n$ is known at a point Q, this is equivalent to knowing either p or v_n, although not explicitly.) A numerical solution to the boundary integral equation [Eq. (3) or (6)] is necessary to determine the unknown variable at each node. A numerical solution can be achieved by discretizing the boundary surface S into a number of surface elements and nodes. In most of the early work[2–6] only piecewise shape functions or planar element discretizations of the surface were used. As element technology developed, the BEM has been extended[7] to adapt *isoparametric elements* such as those shown in Figs. 1 and 3. With isoparametric elements, the same polynomials (*shape functions*) are used to approximate both the surface shape and the variation of the acoustic variables over the element.

Elements may be classified according to order as well as shape. *Linear* elements are elements in which the geometry and the acoustic variables (sound pressure and vibration on the surface of the body) are represented by a linear or first-order approximation. *Quadratic* elements provide a second-order or quadratic approximation (i.e., they have curvature). Compared to linear elements, fewer quadratic elements are needed to obtain a result of given accuracy. All of the elements in Fig. 1 are quadratic elements.

The arrangement of the nodes for quadratic and linear isoparametric elements is shown in Fig. 7. Each quadratic element, whether quadrilateral or triangular, consists of a node at each vertex and one on each side. It is the side nodes that permit the element to assume a curved shape. It is not necessary to locate the side nodes at the midpoint of the segment, but accuracy will decrease if the side nodes are located too close to the vertex nodes. Likewise, accuracy will decrease if the element is severely distorted. Good modeling technique is

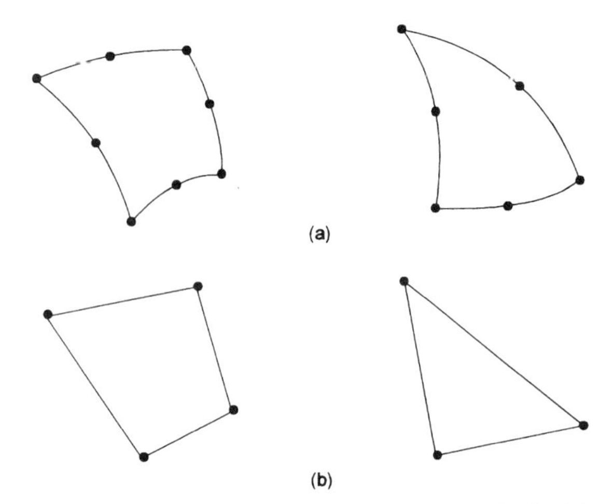

Fig. 7 Isoparametric boundary elements: (*a*) quadratic; (*b*) linear.

achieved by following simple guidelines and common sense with respect to element size and shape.

As mentioned in Section 1.1, the size of the boundary elements is determined by the geometry of the structure being modeled and the structural or acoustic wavelength, whichever is smaller. For linear elements, at least four elements per wavelength are required; for quadratic elements, at least two elements per wavelength are required.

A discretization of the surface S yields a discrete form of either Eq. (3) or (6). By placing a singular point P at each of the N nodes successively, N linearly independent equations are obtained. These equations can be arranged in matrix form to yield

$$[g(\omega)]\{p\} = [h(\omega)]\{v_n\}, \qquad (7)$$

where the quantities in curly brackets are $N \times N$ matrices formed by numerical integration of the G or G' and the element shape functions.[7] After inserting the boundary conditions, Eq. (7) may be solved to determine the unknown acoustic variable at each node. Once both p and v_n are known at every node, it is a straightforward calculation to determine the sound intensity, sound power, and radiation efficiency from these data. Then, using a discretized form of the boundary integral equation, one may determine the sound pressure, particle velocity, and sound intensity at any point P in the acoustic domain B' by simple numerical integration.

It may be seen from Eq. (7) that the frequency of vibration ω is contained in the g and h matrices. This means that these matrices must be recalculated for each frequency for which a solution is desired. It is sometimes advantageous to use the frequency interpolation technique[15,16] to reduce the central processing unit (CPU) time necessary for formation of the g and h matrices. However, this method does not substantially reduce *total* CPU time when the matrix is large because most of the CPU time is expended in solving rather than forming the matrix equation.

2.4 The Multidomain Boundary Element Method

As mentioned in Section 1, the traditional BEM can only be applied to homogeneous acoustic domains [i.e., as shown in Eq. (3) or (6), only one value of the characteristic impedance z_0 may be used]. However, by dividing the acoustic domain into several smaller domains in which z_0 may be different in each and applying the boundary integral equation to each smaller domain, it is possible to use the BEM to solve problems having a nonhomogeneous domain.[17,18] Where the various domains are joined, continuity of pressure and particle velocity, or other special boundary conditions, are enforced to effect a solution. If one of the domains is an interior one and one an exterior one, the multidomain BEM may be used to solve problems such as radiation of sound from openings in enclosures and sound radiated from open ducts.[19]

The muffler in Fig. 8 is an example where the multidomain BEM is ideally suited. Each of the volumes in the muffler has a different average temperature, which means that z_0 will be different. In addition, the first volume is filled with a *bulk-reacting* absorbing material that, in general, has a complex characteristic impedance (and cannot be modeled by a normal impedance z_n). With the multidomain BEM it is also possible to model perforations, as in the muffler inlet tube in Fig. 8.

Figure 9 shows the contour of the SPL inside the muffler in Fig. 8 at 700 Hz. For this problem the sound pressure at the inlet was selected to be approximately 85 dB. The progressive reduction of sound by each volume and tube in the muffler is clearly seen from the data in Fig. 9. The multidomain BEM also allows the BEM to be extended to thin bodies. The conventional single-domain BEM has numerical problems when the body is a thin structure, such as a plate or the tube connecting volumes 2 and 3 in Fig. 8.

2.5 Application of the Boundary Element Method to Half-Space Radiation Problems

The BEM may be applied to problems in which a body radiates sound near an infinite reflecting surface by modifying the Green's function G in Eq. (3). The modified Green's function

$$G_H = \frac{e^{-ikr}}{r} + R\,\frac{e^{-ikr_1}}{r_1}, \tag{8}$$

where $r_1 = |Q - P_1|$, the distance from a point on the body to an image point P_1 of the point P relative to the reflecting surface, and R is the reflection coefficient of the surface. The tire radiation problem in Figs. 1 and 2 shows an application of the BEM to half-space radiation problems.[20]

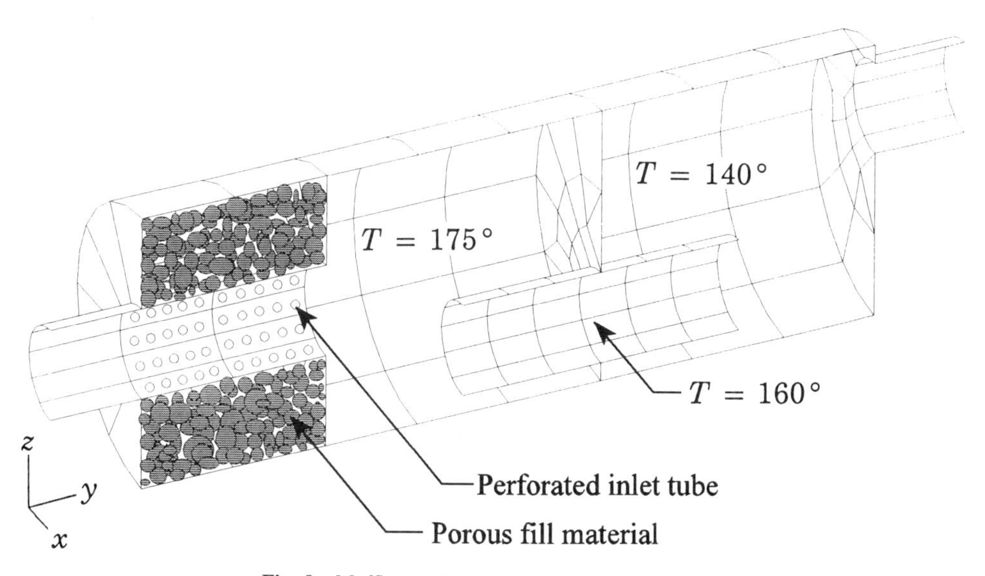

Fig. 8 Muffler model using the multidomain BEM.

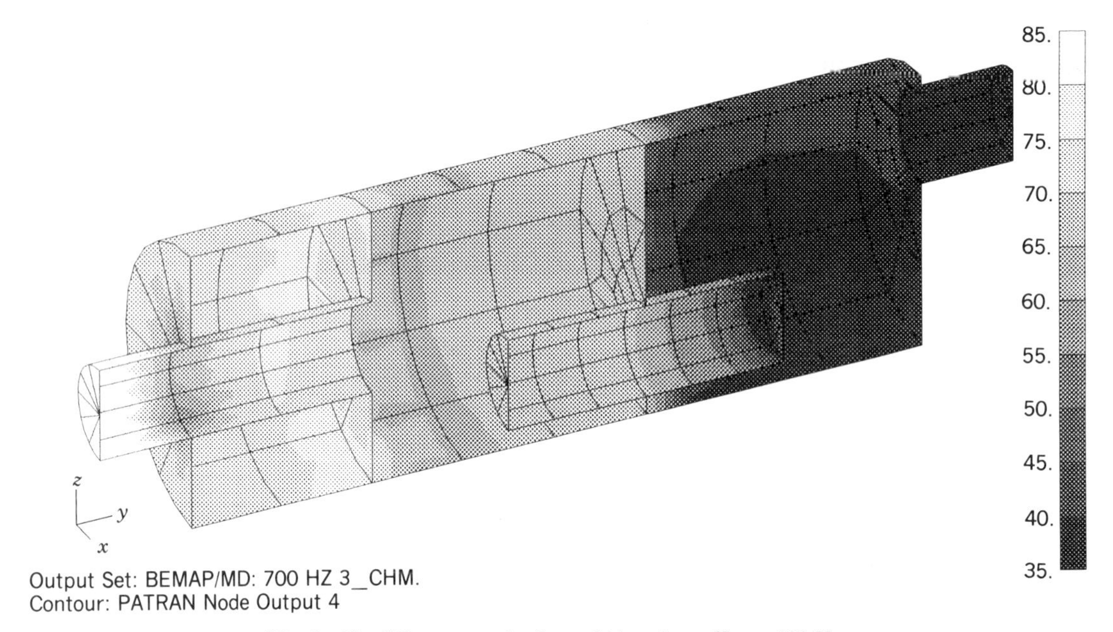

Fig. 9 The SPL contour plot for multidomain muffler at 700 Hz.

2.6 Nonuniqueness of the Boundary Element Method Solution

The exterior boundary integral equation [Eq. (3)] does not have a unique solution at certain characteristic frequencies.[21–23] This nonuniqueness is a purely mathematical problem arising from the integral formulation rather than from the nature of the physical problem. The interior boundary integral equation [Eq. (6)] does not have a nonuniqueness problem. Several methods have been developed to overcome the nonuniqueness problem, including CHIEF.[6] In the CHIEF method, Eq. (3) is collocated at one or more interior points of the body where $C(P) = 0$; these equations are combined with the original system of equations in Eq. (7) to obtain a solution by a least-squares approach. In the majority of radiation problems the CHIEF method provides a unique solution.[24] However, the selection of the interior points for collocation may not yield a unique solution if the points fall on the nodal surfaces of a related interior problem. More robust methods have been developed to remedy this problem.[25–31] In addition, the accuracy of the BEM solution may be confirmed by using a so-called interior point checking procedure[24] or an estimated matrix condition number.[32]

2.7 Other Boundary Element Method Formulations

In addition to the *direct* formulation [i.e., Eqs. (3) and (6)] there are various *indirect* integral equation formulations of the sound radiation problem.[33–35] The formulation is called indirect because an unknown density distribution on the boundary must be solved before any physical solutions can be found. Similar to the direct formulation, the indirect formulation also breaks down at a set of characteristic frequencies. The failure in fact is more catastrophic[21] than its counterpart in the direct formulation because it is a nonexistence problem instead of a nonuniqueness problem. Nevertheless, methods to overcome the nonexistence problem do exist.[21, 23]

When the body in Fig. 6 is an axisymmetric body, Eq. (3), or its interior counterpart, Eq. (6), may be separated into an integral over the generator of the body and a second integral over the angle of revolution.[36–39] The second integral is an elliptic integral that may be evaluated analytically. The integral over the generator may be evaluated using one-dimensional (i.e., line) boundary elements.

3 THE BOUNDARY ELEMENT METHOD AS AN ENGINEERING ANALYSIS TOOL

The BEM is a relatively new tool compared to other acoustic analysis techniques such as the FEM. Much progress has been made in enhancing and tailoring the BEM for acoustics. Similar to the FEM, however, it is computationally and memory intensive, perhaps more so for certain applications.

3.1 Comparison between the Boundary Element and the Finite Element Methods

The BEM and the more familiar FEM share a number of features such as common underlying assumptions, the use of element technology to effect a solution, and the use of pre- and postprocessing to manage the information obtained from each method. In some sense, however, the FEM is the more general of the two methods. The FEM may be used for nonlinear problems such as when the sound pressure is extremely high or when the acoustic domain is *nonhomogeneous*, as for example when there are temperature gradients. In its usual form, the BEM is restricted to linear, homogeneous problems. However, a modification of the BEM, referred to as the multidomain BEM, discussed in Section 2.4, overcomes this last limitation for many problems of practical interest.

For most radiation problems, the BEM is preferred over the FEM. With the FEM, one would need to extend the surface mesh into the acoustic domain using three-dimensional volume elements. This is not recommended, as it results in a large number of elements, but more important, it yields an error since no matter how far the mesh is extended, it is not possible to represent an infinite domain with a finite model using the FEM. This illustrates the fundamental difference between the BEM and the FEM: With the BEM one discretizes only the surface of the radiating body while with the FEM the entire acoustic domain must be discretized. Because with the BEM all numerical approximations are confined to the surface, a coarser mesh can be used as compared to the FEM for the same accuracy.

Most commercially available FEM programs are written for structural applications and, therefore, do not have many of the features that are important for acoustic modeling. For example, most FEM programs do not calculate sound intensity or sound power, nor is it possible with most FEM programs to use an impedance boundary condition (for modeling absorbing surfaces). However, it is sometimes useful to perform an eigenvalue search of an acoustic domain. For this objective, the FEM is ideally suited. However, it is possible with the BEM to use a forced-response calculation to determine the resonance frequencies of the acoustic domain.

A primary advantage of the BEM over the FEM is its simplicity, which allows the new user to become proficient in a much shorter time. Because the boundary element mesh is only a surface mesh, it is easy to construct and does not have as many potential variations as one has with the FEM.

3.2 Characteristics of Boundary Element Method Software

Regardless of the BEM formulation (direct or indirect), there are features that are important to most users for a wide variety of applications. The following list is not inclusive since there are certainly other features (such as structural/acoustic coupling or acoustic sensitivities) that are important in some problems. However, a software program that has the following features will be useful for a majority of problems in acoustics:

1. Capable of solving interior, exterior, and multidomain problems
2. Can handle a reflecting surface with an arbitrary reflection coefficient
3. Allows the definition of a plane of symmetry for modeling symmetric bodies
4. Accepts a range of boundary conditions, including the impedance boundary condition
5. Allows the input vibration to be discontinuous at nodes to permit relative motion between adjacent components and to model edges and corners without special attention by the user
6. Incorporates interfaces to popular commercial pre- and postprocessing programs
7. Calculates the sound radiation from surfaces using the Rayleigh integral (for preliminary predictions of sound radiation)
8. Calculates the sound intensity, sound power, and sound radiation efficiency
9. Can include perforates for modeling mufflers and silencers
10. Can include point sources for preliminary studies of cavities and resonance frequencies
11. Includes a robust method to overcome the nonuniqueness problem
12. Incorporates a frequency interpolation method to increase the speed of multifrequency runs
13. Automatically maintains the accuracy of a BEM calculation without increasing the computer time excessively
14. Has an element library that includes both linear and quadratic boundary elements

There are a number of commercially available BEM software programs that include most or all of these features as well as more specialized enhancements. Several such programs are listed in Table 1.

TABLE 1 Partial List of BEM Acoustics Software

Software Product Name	Company	Location
SYSNOISE	LMS	Belgium
COMET	Automated Analysis	Ann Arbor, MI
BEMAP	Spectronics	Lexington, KY

REFERENCES

1. A. R. Mohanty, "Experimental and Numerical Investigation of Reactive and Dissipative Mufflers," Ph.D. Dissertation, University of Kentucky, Lexington, KY, 1993.
2. L. H. Chen and D. G. Schweikert, "Sound Radiation from an Arbitrary Body," *J. Acoust. Soc. Am.*, Vol. 35, 1963, pp. 1626–1632.
3. G. Chertock, "Sound Radiation from Vibrating Surfaces," *J. Acoust. Soc. Am.*, Vol. 36, 1964, pp. 1305–1313.
4. L. G. Copley, "Integral Equation Method for Radiation from Vibrating Bodies," *J. Acoust. Soc. Am.*, Vol. 41, 1967, pp. 807–816.
5. L. G. Copley, "Fundamental Results Concerning Integral Representations in Acoustic Radiation," *J. Acoust. Soc. Am.*, Vol. 44, 1968, pp. 28–32.
6. H. A. Schenck, "Improved Integral Formulation for Acoustic Radiation Problems," *J. Acoust. Soc. Am.*, Vol. 44, 1968, pp. 41–58.
7. A. F. Seybert, B. Soenarko, F. J. Rizzo, and D. J. Shippy, "An Advanced Computational Method for Radiation and Scattering of Acoustic Waves in Three Dimensions," *J. Acoust. Soc. Am.*, Vol. 77, 1985, pp. 362–368.
8. W. Tobacman, "Calculation of Acoustic Wave Scattering by Means of the Helmholtz Integral Equation, I," *J. Acoust. Soc. Am.*, Vol. 76, 1984, pp. 599–607.
9. W. Tobacman, "Calculation of Acoustic Wave Scattering by Means of the Helmholtz Integral Equation, II," *J. Acoust. Soc. Am.*, Vol. 76, 1984, pp. 1549–1554.
10. K. A. Cunefare, G. H. Koopmann, and K. Brod, "A Boundary Element Method for Acoustic Radiation Valid at All Wavenumbers," *J. Acoust. Soc. Am.*, Vol. 85, 1989, pp. 39–48.
11. T. Terai, "On the Calculation of Sound Fields Around Three-Dimensional Objects by Integral Equation Methods," *J. Sound Vib.*, Vol. 69, 1980, pp. 71–100.
12. A. F. Seybert and C. Y. R. Cheng, "Applications of the Boundary Element Method to Acoustic Cavity Response and Muffler Analysis," *ASME Trans. J. Vib. Acoust. Stress Rel. Design*, Vol. 109, 1987, pp. 15–21.
13. R. J. Bernhard, B. K. Gardner, and C. G. Mollo, "Prediction of Sound Fields in Cavities Using Boundary Element Methods," *AIAA J.*, Vol. 25, 1987, pp. 1176–1183.
14. S. Suzuki, S. Maruyama, and H. Ido, "Boundary Element Analysis of Cavity Noise Problems with Complicated Boundary Conditions," *J. Sound Vib.*, Vol. 130, 1989, pp. 79–91.
15. H. A. Schenck and G. W. Benthien, "The Application of a Coupled Finite-Element Boundary-Element Technique to Large-Scale Structural Acoustic Problems," in C. A. Brebbia and J. J. Connor (Eds.), *Advances in Boundary Elements*, Vol. 2, Computational Mechanics Publications, Southampton, 1989, pp. 309–319.
16. A. F. Seybert and T. W. Wu, "Applications in Industrial Noise Control," in R. D. Ciskowski and C. A. Brebbia (Eds.), *Boundary Element Methods in Acoustics*, Computational Mechanics Publications, Southampton, 1991, Chapter 9.
17. C. Y. R. Cheng, A. F. Seybert, and T. W. Wu, "A Multidomain Boundary Element Solution for Silencer and Muffler Performance Prediction," *J. Sound Vib.*, Vol. 151, 1991, pp. 119–129.
18. T. Tanaka, T. Fujikawa, T. Abe, and H. Utsuno, "A Method for the Analytical Prediction of Insertion Loss of a Two-Dimensional Muffler Model Based on the Transfer Matrix Derived from the Boundary Element Method," *ASME Trans. J. Vib. Acoust. Stress Rel. Design*, Vol. 107, 1985, pp. 86–91.
19. A. F. Seybert, C. Y. R. Cheng, and T. W. Wu, "The Solution of Coupled Interior/Exterior Acoustic Problems Using the Boundary Element Method," *J. Acoust. Soc. Am.*, Vol. 88, 1990, pp. 1612–1618.
20. A. F. Seybert and B. Soenarko, "Radiation and Scattering of Acoustic Waves from Bodies of Arbitrary Shape in a Three-Dimensional Half Space," *ASME Trans. J. Vib. Acoust. Stress Rel. Design*, Vol. 110, 1988, pp. 112–117.
21. A. J. Burton and G. F. Miller, "The Application of Integral Equation Methods to the Numerical Solutions of Some Exterior Boundary Value Problems," *Proc. Roy. Soc. Lond. A* Vol. 323, 1971, pp. 201–210.
22. R. E. Kleinman and G. F. Roach, "Boundary Integral Equations for the Three-Dimensional Helmholtz Equation," *SIAM Rev.*, Vol. 16, 1974, pp. 214–236.
23. A. J. Burton, "The Solution of Helmholtz' Equation in Exterior Domains Using Integral Equations," National Physical Laboratory, Report No. NAC 30, Teddington, Middlesex, United Kingdom, 1973.
24. A. F. Seybert and T. K. Rengarajan, "The Use of CHIEF to Obtain Solutions for Acoustic Radiation Using Boundary Integral Equations," *J. Acoust. Soc. Am.*, Vol. 81, 1987, pp. 1299–1306.
25. T. W. Wu and A. F. Seybert, "Acoustic Radiation and Scattering," in R. D. Ciskowski and C. A. Brebbia (Eds.), *Boundary Element Methods in Acoustics*, Computational Mechanics Publications, Southampton, 1991, Chapter 3.
26. T. W. Wu and A. F. Seybert, "A Weighted Residual Formulation for the CHIEF Method in Acoustics," *J. Acoust. Soc. Am.*, Vol. 90, 1991, pp. 1608–1614.
27. G. Krishnasamy, L. W. Schmerr, T. J. Rudolphi, and F. J. Rizzo, "Hypersingular Boundary Integral Equations: Some Applications in Acoustic and Elastic Wave Scattering," *ASME J. Appl. Mech.*, Vol. 57, 1990, pp. 404–414.
28. T. W. Wu, A. F. Seybert, and G. C. Wan, "On the Numerical Implementation of a Cauchy Principal Value Integral to Insure a Unique Solution for Acoustic Radiation and Scattering," *J. Acoust. Soc. Am.*, Vol. 90, 1991, pp. 554–560.
29. D. J. Segalman and D. W. Lobitz, "A Method to Overcome Computational Difficulties in the Exterior Acoustic Problem," *J. Acoust. Soc. Am.*, Vol. 91, 1992, pp. 1855–61.

30. X. F. Wu, A. D. Pierce, and J. H. Ginsberg, "Variational Method for Computing Surface Acoustic Pressure on Vibrating Bodies, Applied to Transversely Oscillating Disks," *IEEE J. Oceanic Eng.*, Vol. OE-12, 1987, pp. 412–418.
31. J. G. Mariem and M. A. Hamdi, "A New Boundary Finite Element Method for Fluid–Structure Interaction Problems," *Intl. J. Numer. Meth. Engr.*, Vol. 24, 1987, pp. 1251–1267.
32. T. W. Wu, "On Computational Aspects of the Boundary Element Method for Acoustic Radiation and Scattering in a Perfect Waveguide," *J. Acoust. Soc. Am.*, Vol. 96, 1994, pp. 3733–3743.
33. C. R. Kipp and R. J. Bernhard, "Prediction of Acoustical Behavior in Cavities Using an Indirect Boundary Element Method," *ASME J. Vib. Acoust. Stress Reliabil. Design*, Vol. 109, 1987, pp. 15–21.
34. C. A. Brebbia and R. Butterfield, "The Formal Equivalence of the Direct and Indirect Boundary Element Methods," *Applied Mathematical Modeling*, Vol. 2, No. 2, 1978.
35. G. Chertock, "Integral Equation Methods in Sound Radiation and Scattering from Arbitrary Surfaces," Naval Ship Research and Development Center, Report 3538, June 1971.
36. A. F. Seybert, B. Soenarko, R. J. Rizzo, and D. J. Shippy, "A Special Integral Equation Formulation for Acoustic Radiation and Scattering for Axisymmetric Bodies and Boundary Conditions," *J. Acoust. Soc. Am.*, Vol. 80, 1986, pp. 1241–1247.
37. B. Soenarko, "A Boundary Element Formulation for Radiation of Acoustic Waves from Axisymmetric Bodies with Arbitrary Boundary Conditions," *J. Acoust. Soc. Am.*, Vol. 93, 1993, pp. 631–639.
38. P. Juhl, "An Axisymmetric Integral Equation Formulation for Free Space Non-Axisymmetric Radiation and Scattering of a Known Incident Wave," *J. Sound Vib.*, Vol. 163, 1993, pp. 397–406.
39. W. L. Meyer, W. A. Bell, M. P. Stallybrass, and B. T. Zinn, "Prediction of Sound Fields Radiated from Axisymmetric Surfaces," *J. Acoust. Soc. Am.*, Vol. 63, 1979, pp. 631–638.

14

ACOUSTIC MODELING (DUCTED-SOURCE SYSTEMS)

M. G. PRASAD AND MALCOLM J. CROCKER

1 INTRODUCTION

Ducted sources are commonly found in mechanical systems. Typical ducted-source systems include engines and mufflers and air-moving devices (flow ducts and fluid machines and associated piping). In these systems, the source is the active component and the load is the path, which consists of elements such as mufflers, ducts, and end terminations. The acoustic performance of the system depends on the source–load interactions. This chapter presents a system model based on electrical analogies that has been found useful in predicting the acoustic performance of systems. The various methods for determining the impedance of a ducted source are given. The acoustic performance of a system with a muffler as a path element is described in terms of the insertion loss and radiated sound pressure.

2 SYSTEM MODEL

The basic source–load representation of a ducted system is shown in Fig. 1.

The equations for the source–load system based on pressure and velocity (complex) source representations are given by

$$\tilde{p}_L = \frac{\tilde{p}_s \tilde{Z}_L}{\tilde{Z}_s + \tilde{Z}_L} \tag{1}$$

and

$$\tilde{V}_L = \frac{\tilde{V}_s \tilde{Z}_s}{\tilde{Z}_s + \tilde{Z}_L}, \tag{2}$$

Note: References to chapters appearing only in the *Encyclopedia* are preceded by *Enc.*

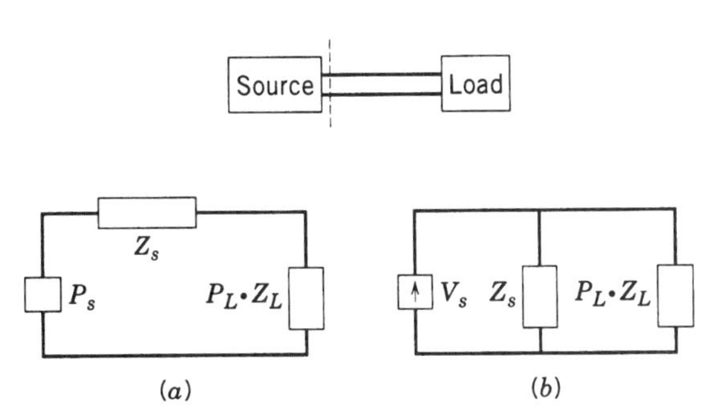

Fig. 1 Electrical analog of a ducted-source-load system: (*a*) pressure source; (*b*) volume velocity source.

where $\tilde{p}_s$ and $\tilde{V}_s$ are the source pressure and volume velocity, respectively; $\tilde{p}_L$ and $\tilde{V}_L$ are the pressure and volume velocity response of the source-load system, respectively; and $\tilde{Z}_s$ and $\tilde{Z}_L$ are the complex source and load impedances, respectively.[1,2]

3 SOURCE CHARACTERISTICS

The source at one end of a duct system represents one of the boundary conditions. It is more difficult to characterize the source compared to the termination because of the dynamic nature of the source.[3]

Although studies have been carried out both in the time and frequency domains, the transmission matrix approach in the frequency domain has been found to be effective in modeling and performance evaluations of ducted systems. Prior to the development of the direct and indirect methods for the measurement of source impedance, frequency-independent values of zero, characteristic (line) or infinite impedances were often assumed to characterize the source in the system model. However, such assumptions for source impedance do not normally yield good predictions.

3.1 Direct Methods

Direct methods for the measurement of source impedance are based on either[3] the standing-wave technique or the transfer function technique. These techniques have been successfully used for active test sources such as engines and pneumatic and electroacoustic drivers given a sufficient signal-to-noise ratio.[4,5] In this context, *signal* refers to the sound pressure level of the secondary measurement source and *noise* refers to the sound pressure level of the test source in operation. A minimum signal-to-noise ratio of 10 dB is required. In direct methods, the microphones are placed inside the duct. A modified transfer function method even for a low signal-to-noise ratio has been developed that only requires an additional measurement of a calibration transfer function.[6]

The complex measured transfer function $\tilde{H}_{12}$ is used to obtain the complex reflection $\tilde{R}$ from which the normalized complex impedance $\tilde{Z}_n$ looking at the test source is calculated:

$$\tilde{R} = e^{j(k_i + k_r)l}\left[\frac{\tilde{H}_{12} - e^{-jk_i s}}{e^{-jk_r s} - \tilde{H}_{12}}\right], \qquad (3)$$

$$\tilde{Z}_n = \frac{1 + \tilde{R}}{1 - \tilde{R}}, \qquad (4)$$

where s is the spacing between microphones, l is the distance between the test source and the microphone farthest from the source, and k_i and k_r are forward and backward wavenumbers in the presence of a mean flow. The transfer function method using white-noise excitation for the secondary source is more effective and also faster to use than the standing-wave method.

3.2 Indirect Methods

Indirect methods are based on the use of different loads and their corresponding responses.[7,8] In Eq. (1), the pressure response $\tilde{p}_L$ is measured for a known load of impedance $\tilde{Z}_L$. Assuming that the source pressure is invariant, the system of equations for two, three, four, or more load impedances is solved to calculate the source impedance. With two and three loads the complex pressure response needs to be used, whereas with four loads the sound pressure level response can be used.

The advantages of indirect methods are that no secondary source is required as the response is measured from the test source and the microphone can be placed outside the duct. However, the methods are sensitive even to slight errors in the measured response, which strongly influences the numerical aspects.

The direct and indirect methods are essentially experimental methods based on frequency-domain analysis. However, the analytical modeling efforts have been carried out mainly in the time domain based on the method of characteristics. There have been some studies based on the modeling of the geometry of sources.[3]

4 PATH ELEMENT AND FOUR-POLE MATRIX

A duct system can be seen as composed of three elements, namely, the source, path, and termination. The path element is the duct system between the predetermined source and termination junctions. The path element can be made up of several types of area discontinuities from a simple expansion to a complicated muffler geometry, as shown in Fig. 2.

The four-pole matrix approach is an effective way of cascading the different subelements of the path element, based on transmission line theory using variables, namely the sound pressure and volume velocity. Any two junctions 1 and 2 in the path can be related using

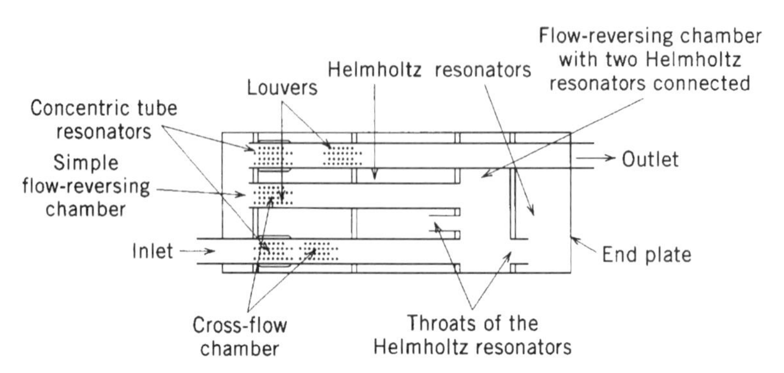

Fig. 2 Cross section of a common U.S. automobile muffler with different parts indicated.

$$\begin{bmatrix} \tilde{p}_1 \\ \tilde{V}_1 \end{bmatrix} = \begin{bmatrix} \tilde{A} & \tilde{B} \\ \tilde{C} & \tilde{D} \end{bmatrix} \begin{bmatrix} \tilde{p}_2 \\ \tilde{V}_2 \end{bmatrix}, \tag{5}$$

where $\tilde{p}$ and $\tilde{V}$ are the complex sound pressure and volume velocity and $\tilde{A}$, $\tilde{B}$, $\tilde{C}$, and $\tilde{D}$ are the complex four-pole parameters that describe the spectral response of the element. The four-pole parameters can be obtained for any geometric design and in the presence of mean flow and temperature gradients. The four-pole parameters can be obtained using both classical as well as numerical methods.[2] The four-pole parameters for a straight duct in the presence of a mean flow are given by

$$\begin{bmatrix} \tilde{p}_1 \\ \tilde{V}_1 \end{bmatrix} = e^{-Mk_cl} \begin{bmatrix} \cos k_cl & j\dfrac{\rho c}{S}\sin k_cl \\ j\dfrac{S}{\rho c}\sin k_cl & \cos k_cl \end{bmatrix} \begin{bmatrix} \tilde{p}_2 \\ \tilde{V}_2 \end{bmatrix}, \tag{6}$$

where M is the Mach number, $k_c = k/(1 - M^2)$ is the convected wavenumber, and l is the length of the duct element. The four-pole matrices for various types of duct elements are available in the literature.[2,10] It is seen that the four-pole matrix formulation is very convenient, particularly for use with a digital computer.

5 TERMINATION IMPEDANCE

The termination impedance $\tilde{Z}_r$ offered by the type of termination in a duct system influences the system performance. There are several types of terminations used, namely characteristic, flanged, and unflanged open end. The characteristic impedance refers to an infinitely long duct at the termination resulting in no reflections. An unflanged open end is commonly found in exhaust systems and is termed the radiation impedance $\tilde{Z}_r$, which can be obtained by Eq. (3) using the end reflection coefficient, $\tilde{R}$, given by[2]

$$\begin{aligned} \tilde{R} &= |\tilde{R}|e^{j(\pi - 2k_0\delta)}, \\ |\tilde{R}| &= 1 + 0.01336k_0r_0 - 0.59079(k_0r_0)^2 \\ &\quad + 0.33576(k_0r_0)^3 - 0.06432(k_0r_0)^4, \\ &\qquad 0 < k_0r_0 < 1.5, \\ \frac{\delta}{r_0} &= 0.6133 - 0.1168(k_0r_0)^2, \qquad k_0r_0 < 0.5, \\ \frac{\delta}{r_0} &= 0.6393 - 0.1104(k_0r_0), \qquad 0.5 < k_0r_0 < 2, \end{aligned} \tag{7}$$

where k_0 is the wavenumber, r_0 is the radius of the duct open end, and δ is the end correction factor.

6 SYSTEM PERFORMANCE

The acoustic performance of a ducted-source system depends on the impedances of the source and load and the four-pole parameters of the path. This complete description of the system performance is termed *insertion loss* (IL). The insertion loss is the difference between sound pressure levels measured at the same reference point (from the termination) without and with the path element, such as a muffler, in place. Figure 3 shows various descriptions:

$$\text{IL} = (L_{p2} - L_{p1}) \qquad \text{dB}. \tag{8}$$

Another useful description of the path element is given by the *transmission loss* (TL) as

$$\text{TL} = 10\log\frac{S_iI_i}{S_tI_t} \qquad \text{dB}, \tag{9}$$

where I_i is the intensity incident on the path element and I_t is the intensity transmitted through the path element; S_i and S_t are duct cross-sectional areas at incidence and transmission, respectively.

Noise reduction (NR) is another descriptor used to measure the effect of the path element and is given by

$$\text{NR} = (L_{p1} - L_{p2}) \qquad \text{dB}, \tag{10}$$

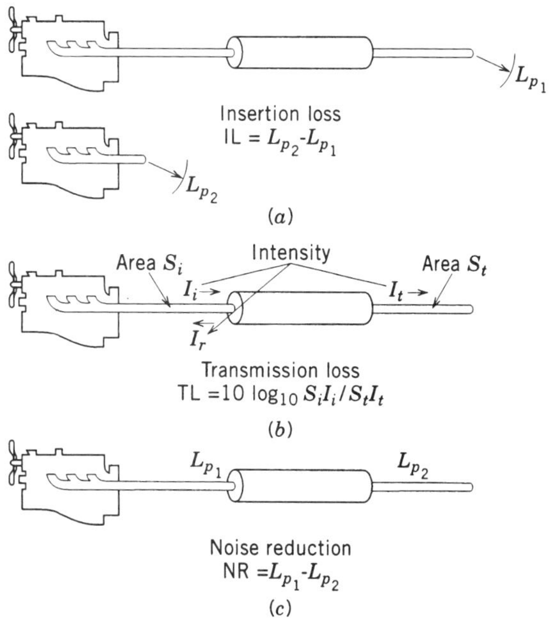

Fig. 3 Definitions of muffler performance.

where L_{p1} and L_{p2} are the measured sound pressure levels upstream and downstream of the path element (such as the muffler), respectively.

The three main descriptions of system performance are given by

$$\mathrm{IL} = 20\log_{10}\left|\frac{\tilde{A}\tilde{Z}_r + \tilde{B} + \tilde{C}\tilde{Z}_s\tilde{Z}_r + \tilde{D}\tilde{Z}_s}{\tilde{A}'\tilde{Z}_r + \tilde{B}' + \tilde{C}'\tilde{Z}_s\tilde{Z}_r + \tilde{D}'\tilde{Z}_s}\right| \quad \mathrm{dB}, \quad (11)$$

$$\mathrm{TL} = 20\log_{10}\left|\frac{1}{2}\left(\tilde{A} + \frac{\tilde{B}S}{\rho c} + \frac{\tilde{C}\rho c}{S} + \tilde{D}\right)\right| \quad \mathrm{dB}, \quad (12)$$

$$\mathrm{NR} = 20\log_{10}\left|\left(\tilde{A} + \frac{\tilde{B}}{\tilde{Z}_r}\right)\right| \quad \mathrm{dB}. \quad (13)$$

The primed four-pole parameters in Eq. (11) refer to the system without the noise-reducing path elements, such as mufflers, in place. Environmental factors such as mean flow and temperature gradient effects can be accounted for in the four-pole parameters.[4]

As seen in Eqs. (8), (9), and (10), the insertion loss is the most useful description for the user, as it gives the net performance of the path element (muffler), including the interaction of the source and termination impedances. Equation (11) shows that the insertion loss depends on the source impedance. It is easier to measure insertion loss than to predict it because the characteristics of most sources are not known. A difficult question to be addressed in system modeling is how the performance (insertion loss) of a muffler varies when used with different sources.

The transmission loss is easier to predict than to measure. The transmission loss is defined so that it depends only on the path geometry and not on the source and termination impedances. The transmission loss is a special case of insertion loss in which the complex source and termination impedances are both characteristic ($Z_s = Z_r = \rho c/S$).

The transmission loss is very useful for the acoustic design of the path geometry. However, it is difficult to measure as it requires two microphones to separate the incident and transmitted intensities. The description of the system performance in terms of noise reduction requires a knowledge of both the path element and the termination in terms of element four-pole matrix and termination impedance.

Although, the above descriptions provide system model performance, an even more important quantity is the sound pressure radiated from the system. This description requires a knowledge of the source strength in terms of the volume velocity $\tilde{V}_s$. The complex radiated sound pressure can be expressed as

$$|\tilde{p}_r| = |\tilde{Z}_{rs}||\tilde{V}_s|, \quad (14)$$

where $\tilde{Z}_{rs}$ is the transfer impedance of the source–path–termination system:

$$|\tilde{Z}_{rs}| = \frac{|\tilde{Z}_s|\,|\tilde{Z}_r + \rho_r c_r/S_r|}{2|\tilde{A}\tilde{Z}_r + \tilde{B} + \tilde{C}\tilde{Z}_s\tilde{Z}_r + \tilde{D}\tilde{Z}_s|} \cdot \sqrt{\frac{S_r}{4\pi r^2}\frac{\rho_0 c_0}{\rho_r c_r}[(1+M)^2 - (1-M)^2R^2]}, \quad (15)$$

$$|\tilde{V}_s| = \frac{|\tilde{p}'_r|}{|\tilde{Z}_s||\tilde{Z}_r|}|\tilde{A}'\tilde{Z}_r + \tilde{B}' + \tilde{C}'\tilde{Z}_s\tilde{Z}_r + \tilde{D}'\tilde{Z}_s| \quad (16)$$

where the primed quantities refer to a straight-duct system. The subscript "r" refers to tail pipe exit.

The radiated sound pressure usually refers to an open system with a finite-length tailpipe with an open end. However, the downstream sound pressure can be evaluated at a downstream field point inside the duct system.

7 APPLICATIONS

Although system modeling studies have been carried out on various ducted sources such as engines, fans, blowers, pumps, and so on, the most extensive work has been carried out on engine–muffler–tailpipe systems.[1–12] This is because unmuffled engine exhausts are usually the dominant noise sources in vehicles. System modeling studies are useful for application of the interaction of the source and load and the location of the muffler relative to the engine. Also, studies have useful application in the optimum design of the geometry of the source. Linear system modeling has been found to be effective and useful. Generally, nonlinear modeling would be challenging.[9]

Source impedances have been measured using the two microphone random-excitation method. In Fig. 4 the predicted insertion loss of a simple expansion chamber on an eight-cylinder engine is compared with that measured. The insertion loss of a simple expansion chamber is predicted using Eq. (11) using the measured engine (source) impedance. The system shown in Fig. 4 is an eight-cylinder engine with a simple expansion chamber. The studies have accounted both flow and temperature gradient effects.

Figure 5 shows comparisons of the sound pressure radiated made both from prediction and from measurements on the same engine–muffler system.[4] The applications here are presented only for an engine–muffler

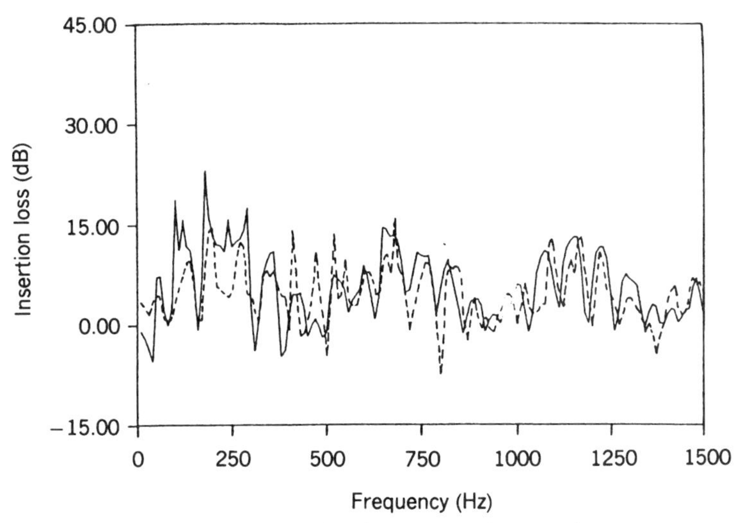

Fig. 4 Insertion loss of an expansion chamber on the engine operating at 2000 rpm, 10 in. Hg: (——) predicted (using measured engine impedance); (------) measured.

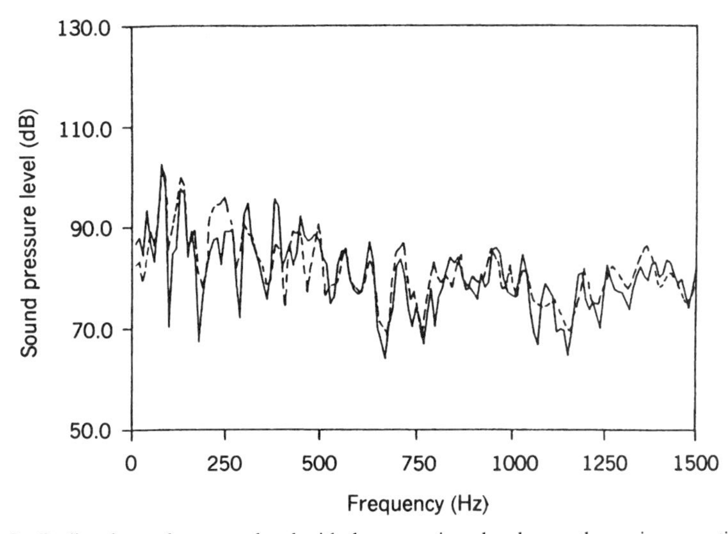

Fig. 5 Radiated sound pressure level with the expansion chamber on the engine operating at 2000 rpm, 10 in. Hg: (——) predicted (using measured engine impedance); (------) measured.

system. However, the modeling of ducted source–load systems is also important for many other ducted systems.[12–15]

REFERENCES

1. M. J. Crocker, "Internal Combustion Engine Exhaust Muffling," *Proceedings of NOISE-CON 77*, 1977, pp. 331–358.
2. M. L. Munjal, *Acoustics of Ducts and Mufflers*, Wiley, New York, 1987.
3. M. G. Prasad, "Characterization of Acoustical Sources in Duct Systems—Progress and Future Trends," in *Proceedings of Noise-Control*, Tarrytown, NY, 1991, pp. 213–220.
4. M. G. Prasad and M. J. Crocker, "Acoustical Source Characterization Studies on a Multi-Cylinder Engine Exhaust System" and "Studies of Acoustical Performance of a Multi-Cylinder Engine Exhaust Muffler System," *J. Sound Vib.*, Vol. 90(4), 1983, pp. 479–508.

5. A. G. Doige and H. S. Alves, "Experimental Characterization of Noise Sources for Duct Acoustics," *Trans. ASME: J. Vib. Acoust.*, Vol. 111, 1989, pp. 108–114.
6. W. Kim and M. G. Prasad, "A Modified Transfer Function Method for Acoustic Source Characterization Duct Systems," presented at the ASME Winter Annual Meeting, Paper 93-WA/NCA-5, 1993.
7. M. G. Prasad, "A Four Load Method for Evaluation of Acoustical Source Impedance in a Duct," *J. Sound Vib.*, Vol. 114, No. 2, 1987, pp. 347–355.
8. H. Boden, "On Multi-Load Methods for Determination of the Source Data of Acoustic One-Port Sources," *J. Sound Vib.*, Vol. 180, No. 5, 1995, pp. 725–743.
9. A. D. Jones, W. K. Van Moorhem, and R. T. Voland, "Is a Full Non-Linear Method Necessary for the Prediction of Radiated Exhaust Noise?" *Noise Control Eng. J.*, March-April 1986, pp. 74–80.
10. A. G. Galaitsis and I. L. Ver, "Passive Silencers and Lined Ducts," in L. L. Beranek and I. L. Ver (Eds.), *Noise and Vibration Control Engineering*, Wiley, New York, 1992, Chapter 10.
11. L. J. Eriksson, "Silencers," in D. E. Baxa (Ed.), *Noise in Internal Combustion Engines*, Wiley, New York, 1982.
12. S. Skaistis, *Noise Control of Hydraulic Machinery*, Marcel Dekker, New York, 1988, pp. 164–169.
13. J. Lavrentjev, M. Abom, and H. Boden, "A Measurement Method for Determining the Source Data of Acoustic Two-Port Sources," *J. Sound Vib.*, Vol. 183, No. 3, 1995, pp. 517–531.
14. M. Abom and H. Boden, "A Note on the Aeroacoustic Source Character of In-Duct Axial Fans," *J. Sound Vib.*, Vol. 186, No. 4, 1995, pp. 589–598.
15. H. Boden and M. Abom, "Maximum Sound Power from In-Duct Sources with Applications to Fans," *J. Sound Vib.*, Vol. 187, No. 3, 1995, pp. 543–550.

PART II

NONLINEAR ACOUSTICS AND CAVITATION

15

NONLINEAR ACOUSTICS AND CAVITATION

DAVID T. BLACKSTOCK AND MALCOLM J. CROCKER

1 INTRODUCTION

In Part I and most of the rest of this book, the wave behavior of sound, for example, propagation, reflection, transmission, refraction, and diffraction, is described in terms of the linear wave equation (see Chapters 1 and 2). If the wave amplitude becomes high enough, nonlinear effects occur and the linear wave equation becomes inadequate. In nonlinear acoustics, novel phenomena unknown in linear acoustics are observed, for example, waveform distortion, formation of shock waves, increased absorption, nonlinear interaction (as opposed to superposition) when two sound waves are mixed, amplitude-dependent directivity of acoustic beams, cavitation, and sonoluminescence. This chapter begins with a simple description of waveform deformation and then reviews, briefly, the chapters on nonlinear acoustics which follow in Part II.

2 WAVEFORM DISTORTION

This section is devoted to simple physical arguments to explain waveform distortion, which, more than any other phenomenon, defines nonlinear acoustics. Consider a plane wave propagating, say, in the x-direction in a lossless fluid. If it is a small signal (infinitesimal amplitude), the wave phenomena can be described in terms of the linear wave equation

$$\nabla^2\phi - \frac{1}{c_0^2}\frac{\partial^2\phi}{\partial t^2} = 0,$$

Note: References to chapters appearing only in the *Encyclopedia* are preceded by *Enc.*

where ϕ is the velocity potential, t is time, and c_0 is the small-signal sound speed (the value found in tables).

For such cases of small (infinitesimal amplitude) disturbances, the wave propagates without change of shape, i.e., without distortion because all points on the waveform travel with the same speed, namely, $dx/dt = c_0$. If the amplitude of the wave is finite, however, the propagation speed varies from point to point on the waveform. The variation has two causes. First, by its very existence, a propagating wave sets up a longitudinal velocity field u in the fluid through which it travels. Since the moving fluid helps carry, i.e., convects the sound wave along, the propagation speed with respect to a fixed observer is therefore

$$\frac{dx}{dt} = c + u, \tag{1}$$

where c is the sound speed with respect to the moving fluid. Second, c is not quite the same as c_0. To see this, let the fluid be a gas, for which the sound speed varies as $\sqrt{T}$, where T is the absolute temperature. The sound speed is a little higher where the acoustic pressure p is positive (because compression raises the temperature) and a little lower where the p is negative (expansion lowers the temperature). The mathematical description of this relationship is

$$c = c_0 + \frac{\gamma - 1}{2}u, \tag{2}$$

where γ is the ratio of specific heats of the gas. Although explained above in terms of the effect of temperature, the deviation of c from c_0 may be traced to nonlinearity of the pressure–density relation of the fluid. Combination of Eq. (2) with Eq. (1) yields

$$\frac{dx}{dt} = c_0 + \beta u, \tag{3}$$

where β, called the coefficient of nonlinearity, is given by

$$\beta = \frac{\gamma + 1}{2}. \tag{4}$$

The dependence of propagation speed on particle velocity, given by Eq. (3) and due physically to the convection and nonlinearity of the pressure–density relation, is what produces distortion of the traveling wave. A complete mathematical description of nonlinear acoustical phenomena requires a nonlinear wave equation. Model nonlinear equations for various propagation, standing wave, and diffraction problems are discussed in Chapter 16. Chapter 17 provides more detail, particularly on various one-dimensional propagation problems.

Examples of waveform distortion are shown in Chapter 17 (Fig. 2). Although even in nonlinear acoustics u is normally much smaller than c_0, the effect of the varying propagation speed is cumulative and eventually causes noticeable distortion. Moreover, the stronger the wave, the more quickly the distortion develops.

The example given above for propagation in a gas is easily extended to other media. Convection, which is responsible for the presence of the factor u in Eq. (1), is the same for compressional waves in all media. Nonlinearity of the pressure–density relation (or its equivalent for solids) is a great deal larger, however, for liquids and solids (compressional waves) than for gases. In the case of liquids, the (normalized) coefficient of the first nonlinear term in the pressure–density relation is known as $B/2A$, and B/A is called the parameter of nonlinearity. In this case Eq. (4) is replaced by

$$\beta = 1 + \frac{B}{2A}. \tag{5}$$

Chapter 18 is devoted to a discussion about B/A and a tabulation of its value for a large number of liquids. Nonlinear behavior of compressional waves in solids may be expressed in a similar fashion; see Chapter 19, which includes a table of values for β for a variety of solids.

3 REVIEW OF NONLINEAR ACOUSTICS TOPICS

There are two main classes of problems in nonlinear acoustics: (1) *source* problems, where the time variation of an acoustical field variable (e.g., pressure or particle velocity) is specified at a single spatial location and (2) *initial-value* problems, where the spatial variation of the field variable is specified at a particular time. Chapter 16 concentrates on the first class: *source* problems. Sound waves propagating from intense sound sources become distorted and almost always result in the formation of shock waves, which are discussed in Chapters 16, 20, and also in Chapter 27.

Chapter 17 is concerned with the second class: *initial-value* problems. Even when the particle velocity is much less than the speed of sound and the sound pressure is much smaller than $\rho_0 c_0^2$ (where ρ_0 is the static density), over large distances and long times waveform distortion occurs because of the cumulative effects of nonlinearity. The chapter begins with the Riemann analysis for plane waves without dissipation. Then diffusive mechanisms (viscosity and heat conduction) are included in the wave propagation formulation. The formation, propagation, interaction, and structure of shock waves are analyzed as well. Underwater acoustics and the effects of bubble concentrations are also discussed and analyzed in that chapter.

The nonlinearity in acoustic media contributes to the distortion in the wave shape during wave propagation. For fluids isentropic acoustic pressure fluctuations may be related to the density fluctuation through a Taylor series. The parameter of nonlinearity B/A is related to coefficients of the second and first terms in the Taylor series. The coefficient of nonlinearity is $\beta = 1 + B/2A$. Chapter 18 explains how these parameters can be derived from the equation of state of the material (if known) or else measured. Values of B/A are given in Tables 1 and 2 of that chapter for a variety of nonbiological and biological materials.

Chapter 19 discusses nonlinear propagation in solids and shows how the same basic approach can be used as for fluids. Tables 3–5 of that chapter provide values of B/A and β for gases, fluids, and some solids (crystals). The chapter also provides details of how these quantities are measured.

Finite-amplitude standing sound waves are experienced in a number of practical problems including mufflers for internal combustion engines, and, more recently, thermoacoustic heat engines and acoustic compressors. Several different approaches have been used to study such problems including Riemann-invariant analysis, the method of characteristics, and finite-difference analysis. Cavity walls cause phase speed dispersion and losses due to viscous and thermal boundary layers. The dispersion perturbs the resonance frequencies of rigid-wall cavities and influences the harmonic spectrum of the standing waves. In *Enc.* Ch. 22 a simplified nonlinear acoustic wave equation is solved. This approach accounts for shear viscosity and can be used to include additional losses. The equations are solved by a perturbation method.

Many applications of intense sound, such as those in aerospace rocket exhausts, sonar, medicine, and nondestructive evaluation, involve radiation from directive sources. Linear effects of diffraction determine length scales corresponding to nearfields, focal lengths, and beamwidths. However, nonlinear effects can be substantially different, and even dominant. The importance of the nonlinear effects depends on how the diffraction lengths compare with length scales corresponding to absorption and shock formation. Chapter 20 describes these competing effects in detail, beginning with analytic solutions for harmonic generation in weakly nonlinear Gaussian beams, both unfocused and focused. Asymptotic and numerical solutions are used to illustrate shock formation, nonlinear side lobe generation, and self-demodulation of pulses in sound beams radiated by circular pistons.

The lumped-element assumption, in which the sound wavelength is much greater than a typical element dimension, is discussed in Chapter 11. In many practical cases, such as in exhaust mufflers and in jets, intense sound waves exist, and the related nonlinear phenomena must be considered with lumped-element models. Elements in mufflers, for example, exhibit nonlinear behavior that must be included in the acoustic modeling. Chapter 21 discusses both how the nonlinear behavior may be modeled and measured experimentally.

Cavitation can occur in fluids because of a variety of causes, one of which is acoustic excitation. If sufficiently intense sound fields are created in a liquid, then cavitation can occur during the low-pressure phase of the pressure fluctuations when bubbles are produced. Strong noise occurs during acoustic cavitation. It is caused by the cavities or bubbles which are generated and set into oscillation in the sound field. The violent motion of the cavities in the sound field can cause destructive effects not only with soft materials but even with metallic materials and can lead to failure. Chapter 22 discusses all these phenomena in detail.

The last topics in Part II are sonochemistry and sonoluminescence. In the 1930s these phenomena were discovered in Germany when liquids containing bubbles were irradiated with intense sound waves. In such a case the bubbles oscillate, then collapse and omit a flash of light. Very high temperatures, high pressures, and high heating and cooling rates can occur. Chemical reactions can be speeded up immensely during this process. Some have even suggested that, in principle, temperatures within such microscopic bubbles might be raised high enough to cause cold fusion. Chapter 23 reviews all of the important phenomena associated with sonoluminescence.

16

SOME MODEL EQUATIONS OF NONLINEAR ACOUSTICS

DAVID T. BLACKSTOCK

1 PROPAGATION: MODEL EQUATIONS FOR PROGRESSIVE WAVES

Almost all model equations for propagation of finite-amplitude waves take advantage of the simplification that results when the fluid flow is so-called simple wave, that is, when the wave field is progressive. To illustrate, we consider small-signal plane waves. Let u stand for particle velocity, x distance, t time, and c_0 small-signal sound speed. If only forward-traveling waves are present, the (second-order) wave equation

$$\frac{\partial^2 u}{\partial x^2} - \frac{1}{c_0^2}\frac{\partial^2 u}{\partial t^2} = 0$$

may be integrated once to yield a first-order wave equation

$$\frac{\partial u}{\partial x} - \frac{1}{c_0}\frac{\partial u}{\partial t} = 0.$$

[*Proof:* The solution of the first-order equation is $u = f(t - x/c_0)$, where f is any function.] The first-order equation is further simplified by transforming from coordinates x, t to coordinates x, τ, where $\tau = t - x/c_0$ is retarded time. The result is

$$\frac{\partial u}{\partial x} = 0, \tag{1}$$

the solution of which is $u = f(\tau)$. Equation (1) is the building block on which most model equations for more complicated progressive waves are based.

In what follows we frequently distinguish between *source* problems [$u(t)$ specified at a single spatial location, often $x = 0$] and *initial-value* problems [$u(x)$ specified at a single time, usually $t = 0$]. Some of the literature of nonlinear acoustics is devoted to initial-value problems. Since the most common acoustic problems involve radiation, however, we stress source problems in this chapter.

1.1 Lossless Propagation

Plane Waves The model equation suitable for source-generated plane waves of finite amplitudes in a lossless fluid is[1]

$$\frac{\partial u}{\partial x} - \frac{\beta}{c_0^2}\, u\, \frac{\partial u}{\partial \tau} = 0 \tag{2}$$

[notice that linearization yields Eq. (1)], where β is the coefficient of nonlinearity (Chapter 15, Section 2). This equation, its sister equation appropriate for initial-value problems [$\partial u/\partial t + \beta u(\partial u/\partial x')$, where $x' = x - c_0 t$], and some solutions are discussed in Chapter 17, Sections 2–4. For sinusoidal source excitation, that is, $u = u_0 \sin \omega t$ at $x = 0$, Eq. (2) is satisfied by the Fubini solution[2]

$$u = u_0 \sum \frac{2}{n\sigma} J_n(n\sigma) \sin n\omega\tau, \tag{3}$$

where $\sigma = \beta\epsilon kx = x/\bar{x}$, $\bar{x} = (\beta\epsilon kx)^{-1}$ is the shock formation distance, $\epsilon = u_0/c_0$, and $k = \omega/c_0$ is the wavenumber. The size of σ, sometimes called the dimensionless

Note: References to chapters appearing only in the *Encyclopedia* are preceded by *Enc.*

distortion range variable, is a measure of the amount of distortion that has occurred. For example, $\sigma = 1$ indicates shock formation. Equation (15) in Chapter 17 gives a Fubini-like solution for the initial-value problem, $u = u_0 \sin kx$ at $t = 0$.

Other One-Dimensional Waves, Including Ray Theory For one-dimensional progressive waves that are not plane, geometric spreading, which is described mathematically by an extra term in the model equation, slows down the rate of distortion (see also Chapter 27, Section 4). If the waves are spherical or cylindrical and $kr \gg 1$, the model equation is[3]

$$\frac{\partial u}{\partial r} + \frac{a}{r} - \frac{\beta}{c_0^2} u \frac{\partial u}{\partial \tau} = 0, \tag{4}$$

where r is the radial distance coordinate, the retarded time is now $\tau = t - (r - r_0)/c_0$, r_0 is a reference distance (e.g., the radius of the source), and a has the value 1 for spherical waves, $\frac{1}{2}$ for cylindrical waves, and 0 for plane waves. Introduction of the coordinate stretching function

$$z = r_0 \ln \frac{r}{r_0} \qquad \text{(spherical waves)}, \tag{5}$$

$$= 2(\sqrt{r r_0} - r_0) \qquad \text{(cylindrical waves)} \tag{6}$$

and the spreading compensation function

$$w = \left(\frac{r}{r_0}\right)^a u \tag{7}$$

reduces Eq. (4) to plane-wave form,

$$\frac{\partial w}{\partial z} - \frac{\beta}{c_0^2} w \frac{\partial w}{\partial \tau} = 0. \tag{8}$$

Any plane-wave solution may therefore be extended to apply to spherical and cylindrical waves simply by replacing u and x with w and z, respectively. For example, the Fubini solution for spherical waves has the same form as Eq. (3). The expression for the distortion range variable for this case, $\sigma = \beta \epsilon k r_0 \ln r/r_0$, shows that the rate of distortion is much slower for spherical waves than for plane waves.

One-dimensional propagation also occurs in ducts of slowly varying cross section, such as horns or, in geometrical acoustics, ray tubes. Given the cross-sectional area $A(s)$ of the horn or ray tube, where s is the distance along the horn or ray tube axis, the coordinate stretching function is

$$z = \int_{s_0}^{s} \left(\frac{A_0}{A}\right)^{1/2} ds', \tag{9}$$

where s_0 is a reference distance similar to r_0 and $A_0 = A(s_0)$. The spreading compensation function is

$$w = \left(\frac{A}{A_0}\right)^{1/2} u. \tag{10}$$

In this way Eq. (8) is generalized to apply to finite-amplitude propagation in nonuniform ducts. The generalization may also be extended to cover cases in which the fluid is inhomogeneous.[4,5]

See Chapter 17, Section 10, for some applications of one-dimensional, nonplanar, finite-amplitude waves. Notice that although geometric spreading has been emphasized here, converging waves, which are often important in medical ultrasonics, may be treated in the same way.[1]

1.2 Dissipative Propagation

Although Eq. (8) is useful for a wide variety of finite-amplitude propagation problems, distortion almost always leads to formation of shocks, which are inherently dissipative. Losses must therefore be taken into account. Treated in this section are (1) the Burgers equation, the first really successful model that includes losses; (2) generalizations of the Burgers equation; and (3) weak-shock theory.

Classical Burgers Equation: Plane Waves in Thermoviscous Fluids Originally proposed as model for turbulence,[6] the Burgers equation was found to be an excellent approximation of the conservation equations for plane progressive waves of finite amplitude in a thermoviscous fluid.[7] In the form suitable for source problems, the Burgers equation is[8]

$$\frac{\partial u}{\partial x} - \frac{\beta}{c_0^2} u \frac{\partial u}{\partial \tau} = \frac{\delta}{2c_0^3} \frac{\partial^2 u}{\partial \tau^2}, \tag{11}$$

where $\delta = \nu[\mathcal{V} + (\gamma - 1)/\mathrm{Pr}]$ is the diffusivity of sound,[7] $\nu = \mu/\rho_0$ is the kinematic viscosity, μ is the shear viscosity coefficient, ρ_0 is the static density, $\mathcal{V} = \frac{4}{3} + \mu_B/\mu$ is the viscosity number, μ_B is the bulk viscosity coefficient, and Pr is the Prandtl number. Notice that the only difference between the Burgers equation and Eq. (2) is that the latter has no dissipation term. For this reason Eq. (2) is sometimes called the *lossless Burgers equation*.

The fact that the Burgers equation is exactly integrable, a property discovered by Cole[9] and Hopf,[10] makes it a very attractive model. In Chapter 17, Section 5, the form of the Burgers equation appropriate for initial-value problems is given, the Hopf–Cole transformation is presented, and several different applications are discussed.

The most popular acoustical problem to which the Burgers equation has been applied is that for sinusoidal source excitation. Even though exact, the solution is quite complicated.[8,11] For distances larger than $3\bar{x}$ ($\sigma > 3$), however, it reduces to the well-known Fay solution[12]

$$u = u_0 \frac{2}{\Gamma} \sum \frac{\sin n\omega\tau}{\sinh n(1+\sigma)/\Gamma}, \qquad (12)$$

where $\Gamma = \beta\epsilon k/\alpha = 1/\alpha\bar{x}$ characterizes the importance of nonlinear distortion relative to absorption, and $\alpha = \delta\omega^2/2c_0^3$ is the small-signal absorption coefficient at the source frequency.

Generalized Forms of the Burgers Equation The Burgers equation has been generalized in two ways. First, it has been extended to apply to other one-dimensional waves in thermoviscous fluids. For example, the Burgers equation for cylindrical and spherical waves is the same as Eq. (4) but with the zero on the right-hand side replaced by the right-hand side of Eq. (11). The equation for horns and ray tubes (stratified fluids) is similar. As shown in Chapter 17, Section 10, however, the generalized equation is not exactly integrable. The advantage of a known exact solution therefore stops with plane waves in homogeneous fluids.

In the second form of generalization,[13] plane waves are retained but other forms of dissipation are considered, for example, that due to relaxation effects[14–16] (Chapter 17, Section 7) or thermoviscous boundary layers[1] (Chapter 17, Section 8). Solutions of limited validity for special cases are known, but nothing comparable to the general solution of the classical Burgers equation.

Of course, the two forms of generalization may be combined, for instance in application to nonlinear geometric acoustics in the ocean or atmosphere.[17]

The chief merit of the various generalized forms of the Burgers equation is their relative simplicity (as compared to the full-fledged conservation equations for fluids). They offer a good starting point for numerical solution.[16,17]

Weak-Shock Theory Weak-shock theory[18–21] (see also Chapter 27, Section 4) is a very effective alternative to the Burgers equation for cases in which dissipation is primarily due to shocks in the traveling wave. Unlike the Burgers equation, weak-shock theory is as effective when generalized to apply to nonplanar waves and/or inhomogeneous media as it is for ordinary plane waves in homogeneous media. The lossless Burgers equation (8) is used for continuous sections of the waveform between shocks, and a low-amplitude approximation of the Rankine–Hugoniot shock relations, sometimes cast as the *equal-area rule*,[20] is used to calculate the position and amplitude of each shock. In this way the continuous sections of the waveform are connected to each other. Well-known applications are to N-waves and sawtooth waves[1,20,21]; see also Chapter 24 and Sections 6 and 7 in Chapter 27. Weak-shock theory has also been used to show the connection between the Fubini solution [Eq. (3)] and the Fay solution [Eq. (12)].[21]

Weak-shock theory loses its accuracy when the wave becomes so weak that its shocks are dispersed and no longer dominant centers of dissipation. At great distances, therefore, where the wave amplitude becomes very small, asymptotic results based on weak-shock theory may be unsatisfactory.[1]

A computer algorithm based on weak-shock theory, called the Pestorius algorithm,[22,23] has been developed to predict the propagation of complicated signals, such as finite-amplitude noise. The limitation described in the previous paragraph is overcome by periodically correcting the waveform for effects of ordinary dissipation.

2 STANDING WAVES, REFLECTION, AND REFRACTION

Standing waves, reflection, and refraction are problems of compound flow: The sound field is composed of overlapping forward- and backward-traveling waves. Since nonlinearity precludes superposition, compound flow cannot be described simply by adding the progressive wave expressions for the two waves. In other words the two waves interact with each other as well as propagate. One- and two-dimensional problems are considered separately here.

2.1 One-Dimensional Fields

The general nonlinear wave equation for planar flow in lossless gases is

$$c_0^2 \frac{\partial^2\phi}{\partial x^2} - \frac{\partial^2\phi}{\partial t^2} = \frac{\partial}{\partial t}\left(\frac{\partial\phi}{\partial x}\right)^2 + (\gamma-1)\frac{\partial\phi}{\partial t}\frac{\partial^2\phi}{\partial x^2} + \frac{\partial\phi}{\partial x}\left[\frac{\gamma-1}{2}\frac{\partial\phi}{\partial x}\frac{\partial^2\phi}{\partial x^2} + \frac{1}{2}\frac{\partial}{\partial x}\left(\frac{\partial\phi}{\partial x}\right)^2\right]. \qquad (13)$$

[When only forward-traveling waves are present, this equation reduces to simple-wave form, Eq. (6) in Chapter 17 or, with minor approximation, Eq. (2).] An alternative representation in terms of Riemann invariants is given in Chapter 17, Section 1. In general, compound-flow problems are much more difficult to solve than simple-wave problems.

The problem of standing waves in a closed-end resonance tube was solved by Chester[24]; for experimental verification, see Cruikshank.[25] Jimenez[26] (theory) and Sturtevant[27] (experiments) considered a range of different end conditions, from closed to open. *Enc.* Ch. 22 deals with closed-end tubes and also with the three-dimensional extension, rectangular cavities with rigid walls.

Reflection of a normally incident plane wave at an interface between two semi-infinite fluids is a related problem involving overlapping wave fields. First let the interface be a rigid wall. Pfriem[28] solved this problem for waves of continuous waveform, that is, a shock-free field. His result shows that pressure doubling occurs only for small signals.[1] Although the pressure amplification produced by the wall can be quite high for strong waves, the shock-free assumption severely limits the applicability of the result. Solution of the corresponding shock reflection problem[29] shows a much smaller wall amplification factor for shock waves, but still greater than 2. For a pressure-release surface the result for small signals, that the particle velocity doubles at the interface, continues to hold even for finite-amplitude waves.[1]

2.2 Two-Dimensional Fields

When plane waves are obliquely incident on the interface, refraction occurs as well as reflection. Equation (13) generalizes to

$$c_0^2\,\nabla^2\phi - \phi_{tt} = [(\nabla\phi)^2]_t + (\gamma-1)\phi_t\,\nabla^2\phi + \nabla\phi[\tfrac{1}{2}(\gamma-1)\nabla\phi\,\nabla^2\phi + \tfrac{1}{2}\nabla(\nabla\phi)^2]. \quad (14)$$

The reflection–refraction problem is very difficult, and only a few investigators have attempted it.[30,31] The question of whether the law of specular reflection and Snell's law hold for continuous finite-amplitude waves is still open. Some results are known for shock waves.[32]

3 MODEL EQUATIONS FOR TWO- AND THREE-DIMENSIONAL FIELDS

The model equations discussed in Section 1 are too simple for anything but one-dimensional propagation. The exact wave equations in Section 2 are more complicated than need be for most two- and three-dimensional problems in nonlinear acoustics (the third-order terms have little impact) and yet do not include the important effects of dissipation. Several more useful two- and three-dimensional model equations have been developed, usually in response to specific needs. For example, the Westervelt equation[33] was an almost incidental product of Westervelt's discovery of the parametric array (see *Enc.* Ch. 53); a desire to include the effect of diffraction in propagation of intense sound beams led Zabolotskaya and Khokhlov to what is now called the KZK equation.[34–36] Naze Tjøtta and Tjøtta and their co-workers have broadened the approach and obtained a family of very general model equations,[37–39] which may be used for a very wide variety of problems in nonlinear acoustics. All previously mentioned model equations are special cases. Here is one version of the Tjøttas equation for thermoviscous fluids:

$$\Box^2 p + \frac{\delta}{c_0^4}\frac{\partial^3 p}{\partial t^3} = -\frac{\beta}{\rho_0 c_0^4}\frac{\partial^2 p^2}{\partial t^2} - \left(\nabla^2 + \frac{1}{c_0^2}\frac{\partial^2}{\partial t^2}\right)\mathcal{L}, \quad (15)$$

where $\Box^2$ is the d'Alembertian operator

$$\Box^2 = \nabla^2 - \frac{1}{c_0^2}\frac{\partial^2}{\partial t^2} \quad (16)$$

and $\mathcal{L}$ is the Lagrangian density

$$\mathcal{L} = \frac{\rho_0 u^2}{2} - \frac{p^2}{2\rho_0 c_0^2}. \quad (17)$$

For example, for unidirectional sound beams $\mathcal{L} = 0$, and Eq. (15) reduces to the Westervelt equation (in Westervelt's original development the dissipation term was not included).

4 DIFFRACTION

Beginning with the pioneering work of Zabolotskaya and Khokhlov,[34] progress on effects of diffraction in nonlinear acoustics has been substantial. Most of the investigations have been on sound beams, and Chapter 20 is devoted to this topic. The applications are various, for example, parametric arrays (*Enc.* Ch. 53), harmonic distortion, self-demodulation, and focused beams. Pulses as well as time-harmonic signals have been studied, and the work has been both experimental and theoretical. The model equation used for almost all the analytical work in this area is the KZK equation, which, for an axisymmetric beam, is

$$\frac{\partial^2 p}{\partial z\,\partial \tau} = \frac{c_0}{2}\left(\frac{\partial^2 p}{\partial r^2} + \frac{1}{r}\frac{\partial p}{\partial r}\right) + \frac{\delta}{2c_0^3}\frac{\partial^3 p}{\partial \tau^3} + \frac{\beta}{2\rho_0 c_0^3}\frac{\partial^2 p^2}{\partial \tau^2}. \tag{18}$$

The coordinate system is cylindrical: z is the axial coordinate and r is the radial coordinate.

Much computational work has been done on nonlinear effects in beams, most based on the KZK equation (see Chapter 20), some on other models.[40,41]

REFERENCES

1. D. T. Blackstock, "Nonlinear Acoustics (Theoretical)," in D. E. Gray (Ed.), *American Institute of Physics Handbook*, McGraw-Hill, New York, 1972, pp. 3-183 to 3-205.
2. E. Fubini, "Anomalies in the Propagation of Acoustic Waves of Great Amplitude," *Alta Frequenza*, Vol. 4, 1935, pp. 530–581 (in Italian).
3. D. T. Blackstock, "On Plane, Spherical, and Cylindrical Sound Waves of Finite Amplitude in Lossless Fluids," *J. Acoust. Soc. Am.*, Vol. 36, 1964, pp. 217–219.
4. C. L. Morfey, "Nonlinear Propagation in a Depth-Dependent Ocean," Tech. Rep. ARL-TR-84-11, Applied Research Laboratories, University of Texas at Austin, May 1, 1984 (ADA 145 079).
5. W. D. Hayes, R. C. Haefeli, and H. E. Karlsrud, "Sonic Boom Propagation in a Stratified Atmosphere, with Computer Program," Aeronautical Research Associates of Princeton, NASA CR-1299, April 1969.
6. J. M. Burgers, "A Mathematical Model Illustrating the Theory of Turbulence," in R. von Mises and T. von Kármán (Eds.), *Advances in Applied Mechanics*, Vol. I, Academic, New York, 1948, pp. 171–191.
7. M. J. Lighthill, "Viscosity Effects in Sound Waves of Finite Amplitude," in G. K. Batchelor and R. M. Davies (Eds.), *Surveys in Mechanics*, Cambridge University Press, Cambridge, 1956, pp. 250–351.
8. D. T. Blackstock, "Thermoviscous Attenuation of Plane, Periodic, Finite-Amplitude Sound Waves," *J. Acoust. Soc. Am.*, Vol. 36, 1964, pp. 534–542.
9. J. D. Cole, "On a Quasi-Linear Parabolic Equation Occurring in Aerodynamics," *Q. Appl. Math.*, Vol. 9, 1951, pp. 225–236.
10. E. Hopf, "The Partial Differential Equation $u_t + uu_x = \mu u_{xx}$," *Commun. Pure Appl. Math.*, Vol. 3, 1950, pp. 201–230.
11. S. I. Soluyan and R. V. Khokhlov, "The Propagation of Acoustic Waves of Finite Amplitude in a Dissipative Medium," *Vestn. Mosk. Univ., Fiz. Astron.*, Ser. III, Vol. 3, 1961, pp. 52–61 (in Russian).
12. R. D. Fay, "Plane Sound Waves of Finite Amplitude," *J. Acoust. Soc. Am.*, Vol. 3, 1931, pp. 222–241.
13. D. T. Blackstock, "Generalized Burgers Equation for Plane Waves," *J. Acoust. Soc. Am.*, Vol. 77, 1985, pp. 2050–2053.
14. O. V. Rudenko and S. I. Soluyan, *Theoretical Foundations of Nonlinear Acoustics*, Consultants Bureau, Plenum, New York, 1977.
15. A. D. Pierce, *Acoustics: An Introduction to Its Physical Principles and Applications*, McGraw-Hill, New York, 1981, Chap. 11.
16. R. O. Cleveland, M. F. Hamilton, and D. T. Blackstock, "Time-Domain Modeling of Finite-Amplitude Sound in Relaxing Fluids," *J. Acoust. Soc. Am.*, Vol. 99, 1996, pp. 3312–3318.
17. R. O. Cleveland, J. P. Chambers, H. E. Bass, R. Raspet, D. T. Blackstock, and M. F. Hamilton, "Comparison of Computer Codes for the Propagation of Sonic Booms through the Atmosphere," *J. Acoust. Soc. Am.*, Vol. 100, 1996, pp. 3017–3027.
18. K. O. Friedrichs, "Formation and Decay of Shock Waves," *Commun. Pure Appl. Math.*, Vol. 1, 1948, pp. 211–245.
19. G. B. Whitham, "The Flow Pattern of a Supersonic Projectile," *Commun. Pure Appl. Math.*, Vol. 5, 1952, pp. 301–348.
20. L. D. Landau and E. M. Lifshitz, *Fluid Mechanics* (translated from the Russian by J. B. Sykes and W. H. Reid), Addison-Wesley, Reading, MA, 1959, Art. 95.
21. D. T. Blackstock, "Connection between the Fay and Fubini Solutions for Plane Sound Waves of Finite Amplitude," *J. Acoust. Soc. Am.*, Vol. 39, 1966, pp. 1019–1026.
22. F. M. Pestorius, "Propagation of Plane Acoustic Noise of Finite Amplitude," Tech. Rep. ARL-TR-73-23, Applied Research Laboratories, University of Texas at Austin, August 1973 (AD 778 868).
23. F. M. Pestorius and D. T. Blackstock, "Propagation of Finite-Amplitude Noise," in L. Bjørnø (Ed.), *Finite-Amplitude Wave Effects in Fluids*, Proceedings of 1973 Symposium in Copenhagen, IPC Press, Guildford, 1974, pp. 24–29.
24. W. Chester, "Resonant Oscillations in Closed Tubes," *J. Fluid Mech.*, Vol. 18, 1964, pp. 44–64.
25. D. B. Cruikshank, Jr., "Experimental Investigation of Finite-Amplitude Acoustic Oscillations in a Closed Tube," *J. Acoust. Soc. Am.*, Vol. 52, 1972, pp. 1024–1036.
26. J. Jimenez, "Nonlinear Gas Oscillations in Pipes. Part I. Theory," *J. Fluid Mech.*, Vol. 59, 1973, pp. 23–46.
27. B. Sturtevant, "Nonlinear Gas Oscillations in Pipes. Part II. Experiment," *J. Fluid Mech.*, Vol. 63, 1974, pp. 97–120.
28. H. Pfriem, "Reflexionsgesetze für ebene Druckwellen

grosser Schwingungsweite," *Forsch. Gebeite Ingenieurw.*, Vol. B12, 1941, pp. 244–256.

29. R. Courant and K. O. Friedrichs, *Supersonic Flow and Shock Waves*, Interscience, New York, 1948.
30. F. D. Cotaras, "Reflection and Refraction of Finite Amplitude Acoustic Waves at a Fluid–Fluid Interface," Tech. Rep. ARL-TR-89-1, Applied Research Laboratories, University of Texas at Austin, January 3, 1989 (ADA 209 800).
31. K. T. Shu and J. H. Ginsberg, "Oblique Reflection of a Nonlinear P Wave from the Boundary of an Elastic Half Space," *J. Acoust. Soc. Am.*, Vol. 89, 1991, pp. 2652–2662.
32. R. G. Jahn, "Transition Processes in Shock Wave Interactions," *J. Fluid Mech.*, Vol. 2, 1957, pp. 33–48.
33. P. J. Westervelt, "Parametric Acoustic Array," *J. Acoust. Soc. Am.*, Vol. 35, 1963, pp. 535–537.
34. E. A. Zabolotskaya and R. V. Khokhlov, "Quasi-Plane Waves in the Nonlinear Acoustics of Confined Beams," *Sov. Phys. Acoust.*, Vol. 15, 1969, pp. 35–40.
35. V. P. Kuznetsov, "Equations of Nonlinear Acoustics," *Sov. Phys. Acoust.*, Vol. 16, 1971, pp. 467–470.
36. N. S. Bakhvalov, Ya. M. Zhileikin, and E. A. Zabolotskaya, *Nonlinear Theory of Sound Beams*, American Institute of Physics, New York, 1987.
37. S. I. Aanonsen, T. Barkve, J. Naze Tjøtta, and S. Tjøtta, "Distortion and Harmonic Generation in the Nearfield of a Finite Amplitude Sound Beam," *J. Acoust. Soc. Am.*, Vol. 75, 1984, pp. 749–768.
38. J. Naze Tjøtta, and S. Tjøtta, "Interaction of Sound Waves. Part I: Basic Equations and Plane Waves," *J. Acoust. Soc. Am.*, Vol. 82, 1987, pp. 1425–1428.
39. J. Naze Tjøtta and S. Tjøtta, "Nonlinear Equations of Acoustics," in M. F. Hamilton and D. T. Blackstock (Eds.), *Frontiers of Nonlinear Acoustics—12th ISNA*, Elsevier Applied Science, London, 1990, pp. 80–97.
40. P. T. Christopher and K. J. Parker, "New Approaches to the Linear Propagation of Acoustic Fields," *J. Acoust. Soc. Am.*, Vol. 90, 1991, pp. 507–521.
41. P. T. Christopher and K. J. Parker, "New Approaches to Nonlinear Diffractive Field Propagation," *J. Acoust. Soc. Am.*, Vol. 90, 1991, pp. 488–499.

17

PROPAGATION OF FINITE-AMPLITUDE WAVES IN FLUIDS

D. G. CRIGHTON

1 INTRODUCTION

Sound waves propagate linearly when *both* their amplitude is very small *and* the times and distances over which they are observed are not too great. If either of these conditions is violated, one may have to take account of nonlinear effects. When disturbance amplitudes are large, as in explosions or close to a high-speed jet engine exhaust, then there are large local nonlinear effects, and the process cannot in any sense be thought of as acoustic. But even when disturbance amplitudes are locally small (i.e., particle velocity small compared with the speed of sound, pressure fluctuations small compared with the background pressure), a sound wave will still suffer severe waveform distortion, through the cumulative action of small nonlinearities, in its propagation over a sufficiently large time or distance. And no matter how small the initial disturbance, its long-time evolution will necessarily require nonlinear equations (unless the wave is heavily damped by dissipation).

We are concerned in this chapter with those weak nonlinear effects that become important in long-range propagation of a signal that locally satisfies the "infinitesimal-amplitude" criterion for linear acoustics; and if the amplitudes are small but larger than infinitesimal, then the important finite-amplitude effects will simply occur over shorter times and distances. These are the situations covered by the field of *nonlinear acoustics*.

The effects of nonlinearity (or finite amplitude) on sound waves are cumulative and lead eventually to very steep waves. As a consequence of the steep gradients induced in this way, various linear mechanisms become important, even though they were insignificant in the initial undistorted wave. In particular, diffusive mechanisms (viscosity, heat conduction) and frequency dispersion (minute in pure air or water but large in bubbly liquid, for example) are called increasingly into play as a sound wave steepens under its nonlinear self-distortion. What nonlinear acoustics is really about is this competition, first analyzed in these terms by Lighthill,[1] between nonlinear waveform-steepening mechanisms and linear waveform-easing mechanisms. In some cases these mechanisms come into balance, and then they determine waveforms of some characteristic shape—shock waves when diffusion or relaxation mechanisms balance nonlinearity, soliton pulses when (as in bubbly liquid) frequency dispersion balances nonlinearity. Recall, by contrast, that a linear wave can have any shape, and once that shape is set (by a source), then it remains unchanged in linear (nondissipative, nondispersive) propagation.

This chapter will give a summary of important results illustrating the basic issues in nonlinear acoustics. We start with the Riemann analysis for plane waves with no dissipation, which is important because it shows exactly what the effects of nonlinearity, however small or large, are. Then we show how to include diffusivity for weak waves and analyze the formation, propagation, interaction and structure of shock waves when steepening is resisted both by diffusion and by relaxation processes. In applications to underwater acoustics, the effects of small bubble concentrations are important for linear and nonlinear propagation of sound, and we discuss these effects along with, in later sections, those of divergence or convergence of the rays along which the waves can be regarded as propagating, of the effects of dissipation in tube wall boundary layers, and of wavefront turning and focusing associated with multidimensional nonlinear acoustics. The treatment is necessarily in mathematical

Note: References to chapters appearing only in the *Encyclopedia* are preceded by *Enc.*

terms, but the results are simply quoted, with emphasis throughout on interpretations in clear physical terms.

2 RIEMANN ANALYSIS FOR PLANE WAVES

For motions of an ideal fluid starting from a uniform rest state (suffix 0) the entropy S remains uniform, and this implies a definite functional relationship between pressure p and density ρ, $p = p(\rho)$. Then one can define the function of state, with dimensions of velocity,

$$P = \int_{\rho_0}^{\rho} \frac{c(\rho)}{\rho}\, d\rho = \int_{p_0}^{p} \frac{dp}{\rho c}, \tag{1}$$

where $c^2 = dp/d\rho$ defines the local sound speed c, with $P = 0$ in the undisturbed medium. Riemann rearranged the mass and momentum conservation equations in the form

$$\left(\frac{\partial}{\partial t} + (u \pm c)\frac{\partial}{\partial x}\right)(u \pm P) = 0, \tag{2}$$

where $u(x, t)$ is the fluid velocity at point x at time t in a one-dimensional (plane) wave. These can be regarded as ordinary differential equations along *characteristic curves* $C_{\pm}$ in the (x, t) plane:

$$\frac{dR_{\pm}}{dt} = 0 \quad \text{on } C_{\pm} : \frac{dx}{dt} = u \pm c, \qquad R_{\pm} = u \pm P, \tag{3}$$

so that the *Riemann invariant* $R_+ = u + P$ is constant along each C_+ and $R_- = u - P$ is constant along each C_-.[1-3]

If neither of $R_{\pm}$ has the same value everywhere, we have what is called a *compound-wave* motion, with waves traveling in both directions relative to the local flow. A flow with R_+ (or R_-) constant everywhere is called a left- (or right-) running *simple wave*.

For a perfect gas with specific heat ratio (adiabatic index) $\gamma = c_p/c_v$, we have $p/p_0 = (\rho/\rho_0)^{\gamma}$ and

$$c = \left(\frac{\gamma p}{\rho}\right)^{1/2}, \qquad R_{\pm} = u \pm \frac{2c}{\gamma - 1}. \tag{4}$$

3 CAUCHY PROBLEM

Figure 1 shows the (x, t) plane for a Cauchy (initial-value) problem in which $u(x, 0)$ and (say) $(p - p_0)(x, 0)$ are specified and vanish identically for $x < 0$, $x > L$. Imagine that the fluid occupies a long tube, along the x-axis, and is in a uniform state of rest at $t = 0$, except for the segment $0 < x < L$ in which a velocity and pressure perturbation are specified. Regions I, II, and III are undisturbed ($u = 0$, $c = c_0$); region IV is one of compound flow; regions V and VI are simple-wave regions with R_-, R_+ identically zero, respectively. The solution for compound flow must be obtained numerically through step-by-step construction of piecewise linear approximations to the $C_{\pm}$. For $t > t_1$ we have two independent simple-wave problems; we shall analyze here the right-running wave (region V). This has one initial condition, say for u, specified on the segment $O'L'$ and vanishing elsewhere, with time now measured from t_1. We write this initial condition as $u = f(x)$ at $t = 0$.

Riemann invariant $R_- = 0$ everywhere in V (and also in I and III) gives $c = c_0 + \frac{1}{2}(\gamma - 1)u$, and so the R_+ part of Eq. (3) gives

$$\frac{\partial u}{\partial t} + \left(c_0 + \frac{\gamma + 1}{2}u\right)\frac{\partial u}{\partial x} = 0, \tag{5}$$

or

$$\frac{du}{dt} = 0 \quad \text{on } C_+ : \quad \frac{dx}{dt} = c_0 + \frac{\gamma + 1}{2}u. \tag{6}$$

Thus the C_+ characteristics are straight in a simple wave with constant R_-, and if a C_+ is labeled by the point $x = \xi$ through which it passes at $t = 0$, then that C_+ carries the constant signal $u = f(\xi)$. The characteristic solution

$$u = f(\xi), \qquad x = \xi + \left[c_0 + \frac{\gamma + 1}{2} f(\xi)\right] t \tag{7}$$

is therefore an implicit solution for u, given (x, t), in terms of the distribution $u = f(x)$ at $t = 0$ ($t = 0$ is the start of the simple-wave flow). We shall explain what the physical content of Eq. (7) is in Section 5.

4 WEAK NONLINEARITY: MULTIPLE SCALES

The results above are exact and hold for waves of any amplitude. By specializing to the "weakly nonlinear" approximation that characterizes nonlinear acoustics, it is however possible to develop the theory greatly through the inclusion of many further effects and mechanisms that preclude exact analysis of the Riemann type. The idea is that *locally* (i.e., over a few wave periods or wavelengths) the nonlinear term in Eq. (5) is smaller

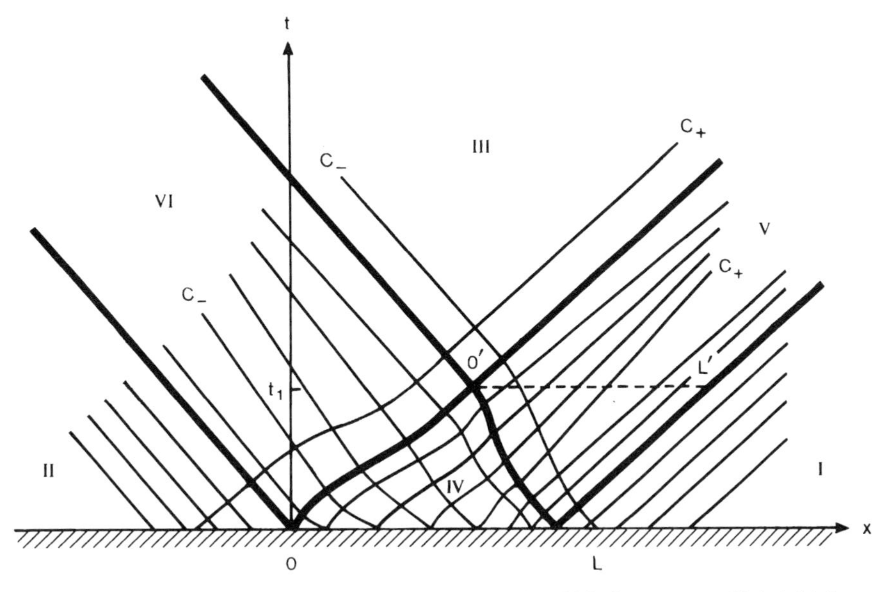

Fig. 1 The (x, t) plane for an initial-value problem in which the segment OL is initially disturbed. In regions I, II, and III the undisturbed state prevails. In region IV there is compound flow. Region V is one of simple-wave propagation (C_+ characteristics straight) to the right, region VI of simple-wave propagation (C_- characteristics straight) to the left. At time t_1 the simple-wave propagation begins with, for region V, an initially disturbed segment $O'L'$.

than the linear ones by a factor of order u/c_0, and therefore if $u/c_0 \ll 1$, the essential balance is between $\partial u/\partial t$ and $c_0\, \partial u/\partial x$. Accordingly, in any "small" term we may replace $\partial u/\partial t$ by $-c_0\, \partial u/\partial x$, or vice versa. Thus in nonlinear acoustics Eq. (5) may equally well be taken in the approximate form

$$\frac{\partial u}{\partial t} + c_0 \frac{\partial u}{\partial x} - \frac{\gamma + 1}{2c_0} u \frac{\partial u}{\partial t} = 0, \tag{8}$$

or, in terms of range x and retarded time (or phase) $\tau = t - x/c_0$ as independent variables, by

$$\frac{\partial u}{\partial x} - \frac{\gamma + 1}{2c_0^2} u \frac{\partial u}{\partial \tau} = 0. \tag{9}$$

This form is particularly suitable for *signaling problems*, which are more natural in acoustic contexts than Cauchy problems, and for such problems the simple-wave representation often applies throughout the flow with none of the complications of compound waves. If u is specified at $x = 0$, as $g(t)$ say, then we have a signaling problem, and the solution of Eq. (9) is

$$u = g(\phi), \qquad \tau = \phi - \frac{\gamma + 1}{2c_0^2} x g(\phi). \tag{10}$$

This would give the field u in a plane-wave mode propagating in a duct $0 < x < \infty$, with $g(t)$ representing measured data at $x = 0$ or the velocity of a piston, $g(t) = \dot{X}_p(t)$. [Strictly $u = g(t)$ should be imposed at the piston face $x = X_p(t)$, not at $x = 0$; however, this particular nonlinearity is a local one of absolutely no global consequence and can be ignored if $\dot{X}_p/c_0 \ll 1$.]

The nonlinear acoustics requirement $u/c_0 \ll 1$ is well satisfied even in very strong acoustic waves. A plane wave in a duct at a sound pressure level of 150 dB has $u/c_0 \simeq 10^{-2}$, and even in this rather extreme case the local linear balance $\partial u/\partial t + c_0\, \partial u/\partial x \simeq 0$ holds with a relative error of only about 1%.

It is natural to try to exploit the smallness of u/c_0 in perturbation theory.[4] A straightforward perturbation expansion applied to the mass and momentum equations, together with $(p/p_0) = (\rho/\rho_0)^\gamma$, is easily carried out to give

$$\frac{u(x,t)}{c_0} = \epsilon F(x - c_0 t) - \frac{\gamma + 1}{2} \epsilon^2 c_0 t F(x - c_0 t) F'(x - c_0 t) + \cdots, \tag{11}$$

where we consider only waves traveling to the right and

where ϵ = (velocity amplitude at $t = 0$)/(sound speed c_0), $u(x,0) = \epsilon c_0 F(x)$. We see that the weakly nonlinear corrections to the linear prediction contain *secular terms*, proportional to the elapsed time, which for large t make the expansion invalid (and which indicate, therefore, that *linear acoustics itself fails at large t*). The ratio of successive terms in Eq. (11) is of order $\epsilon\omega t$, where ω is a typical frequency, showing that the motion cannot be a weakly nonlinear perturbation of linear acoustics for $\omega t \sim \epsilon^{-1}$ and larger.

To find a description valid at such long times, we introduce *multiple scales*, the *fast scales* x, t and a *slow scale* $T = \epsilon t$, and seek a solution of the form

$$u = \epsilon u_0(x,t,T) + \epsilon^2 u_1(x,t,T) + \cdots,$$

with similar expansions of the other variables. At lowest order we find that u_0 depends on the "fast" variables only in the combination $x \pm c_0 t$ (local linear acoustics; the relations between the leading-order velocity, pressure, and density fluctuations are also simply those of linear theory). Then inspection of the equation for u_1 shows that secular terms will not arise—and therefore $u \simeq \epsilon u_0$ will remain a good approximation even at large times, $\omega t = O(\epsilon^{-1})$—provided the "slow," or long-time, evolution is governed by

$$\frac{\partial u_0}{\partial T} + \frac{\gamma+1}{2} u_0 \frac{\partial u_0}{\partial \chi} = 0, \tag{12}$$

where $\chi = x - c_0 t$ and we consider only waves running to the right. This is exactly Eq. (5) of the Riemann analysis. Alternatively, taking $X = \epsilon x$ as a slow space variable and setting $u = \epsilon u_0(x,t,X) + \epsilon^2 u_1(x,t,X) + \cdots$, we find that we can have (x,t) dependence only through $\tau = t - x/c_0$ (for right-running waves) and that

$$\frac{\partial u_0}{\partial X} - \frac{\gamma+1}{2c_0^2} u_0 \frac{\partial u_0}{\partial \tau} = 0, \tag{13}$$

which is precisely the (approximate) equation (9). All such equations are formally equivalent for $(x,t) = O(1)$ up to $O(\epsilon^{-1})$. They show that local linear acoustics holds over the fast scales but that *on the slow scales acoustic propagation is inherently nonlinear*.

Multiple-scales analysis of this kind has been used many times recently to derive certain canonical nonlinear wave equations (Burgers, Korteweg–de Vries, nonlinear Schrödinger, and many others) on a formal basis.[4] The underlying idea here is simply that there is *local* linear behavior, that is, $\partial u/\partial t \pm c_0\, \partial u/\partial x = 0$ and $p - p_0 = \pm\, \rho_0 c_0 u$ for right- (left-) running waves, together with some nonlinear modulation over the long or slow scales.

5 WAVE DISTORTION: SHOCK FORMATION

The process described by Eq. (5) is one of *pure distortion* of the initial waveform f (aside from trivial translation at speed c_0); the evolved waveform $u = f(\xi)$ contains precisely those "wavelets" (values of the signal u) that were initially present, but at time t a given wavelet has reached not the position $x = \xi + c_0 t$ of linear theory, but the position $x = \xi + c_0 t + \frac{1}{2}(\gamma+1)ut$. Thus the waveform is subject to uniform shearing[1] at a rate $\frac{1}{2}(\gamma+1)$ [i.e., the "excess wavelet speed" is $\frac{1}{2}(\gamma+1)u$] of which a contribution 1 comes from convective nonlinearity, and a contribution $\frac{1}{2}(\gamma-1)$ from the nonlinearity of the equation of state; the former dominates for air ($\gamma = 1.4$), the latter for water, where the constant equivalent to γ has a value around 7. The waveform distortion as described by Eqs. (12) and (13) is illustrated in Fig. 2. It is evident that the nonlinear processes are cumulative, and over long space and time intervals, $\omega x/c_0 = O(c_0/u_0), \omega t = O(c_0/u_0)$, there is substantial distortion of the wave profile, though there is no change of the values of the signal carried by the wave.

Such waveform distortion can also be seen as transfer of energy to the higher Fourier components produced as the wave steepens. A famous solution exhibiting this explicitly is that due to Fubini; if for Eq. (5) we have a pure sinusoid $u = u_0 \sin kx$ at $t = 0$, then[2,5]

$$u(x,t) = 2u_0 \sum_{n=1}^{\infty} (-1)^{n-1} \frac{J_n(n\beta k u_0 t)}{n\beta k u_0 t} \cdot \sin nk(x - c_0 t), \tag{14}$$

where $\beta = (\gamma+1)/2$. This comes from direct evaluation of the Fourier coefficients by transformation of the integration over x to one over ξ, a maneuver that requires $t < t_* = 1/\beta k u_0$. Figure 3 shows the initial variation of the Fourier amplitudes according to Eq. (14). Energy is pumped, at least initially, from the low harmonics to the higher ones. It can be shown from Eq. (14) that energy is conserved in this process, provided $t < t_*$, that is,

$$\sum_{n=1}^{\infty} \frac{J_n^2(n\beta k u_0 t)}{(n\beta k u_0 t)^2} = \text{const.}$$

This also follows directly in any continuous simple wave; thus from Eq. (5), for example, we have

$$\frac{d}{dt} \int_A^B u^2\, dx = 0,$$

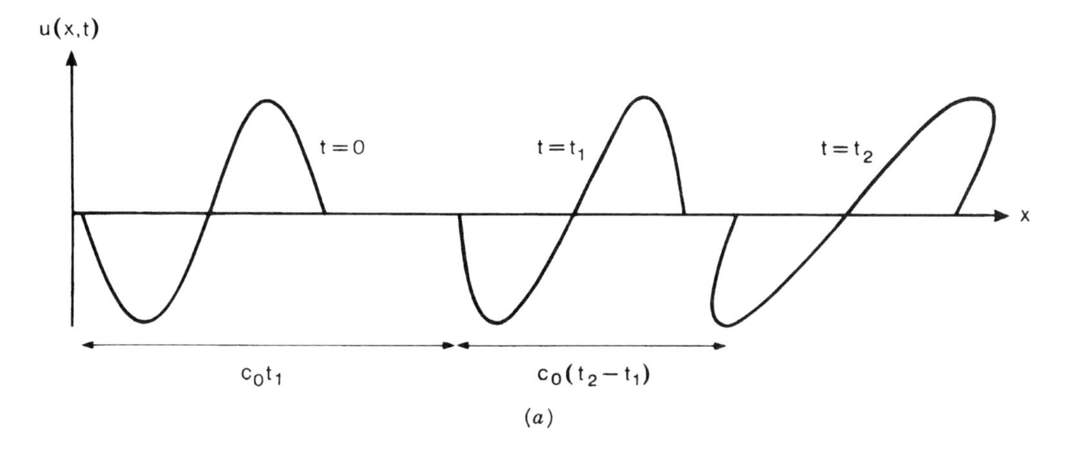

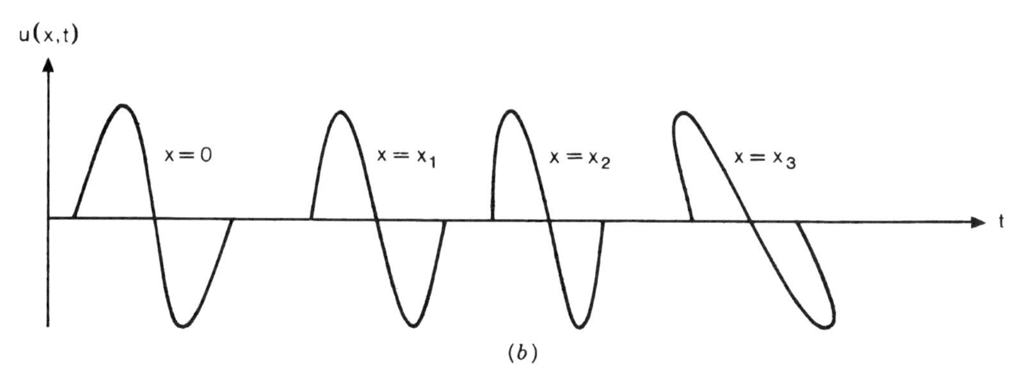

Fig. 2 Cumulative distortion by nonlinear convection, as seen in (*a*) initial-value problems, u as a function of x at increasing times t_i, and (*b*) signaling problems, u as a function of t at increasing ranges x_i from the signaling location.

where A, B correspond to points at which $u = 0$ or that differ by a wavelength of a periodic wave.

Now in all waves except those of pure expansion [pure expansion means $\partial u(x, 0)/\partial x > 0$] it is evident that the wave profile will develop an infinite gradient at a finite time $t = t_*$, and that, for $t > t_*$, $u(x, t)$ will be triple valued for some range of x. The "shock formation" time t_* can be found by integrating the equation $dS/dt + \beta S^2 = 0$ for $S = \partial u/\partial x$ along the characteristics; or by finding the time at which the Jacobian $\partial x/\partial\xi|_t$ of the transformation to characteristic coordinates first becomes singular; or directly by considering the waveform distortion; it is

$$t_* = 1/\max_{f'<0} |\beta f'(\xi)|. \tag{15}$$

The triple-valued waveform for $t > t_*$ is physically unacceptable and is prevented in reality by one or more linear mechanisms [the *nonlinear* theory of Eq. (5) is exact] that may have been initially negligible but increase in importance as the wave steepens under nonlinear effects. In a medium with significant dispersion, short-wave components generated by steepening may disperse away fast enough to prevent the catastrophe at $t = t_*$, but in acoustic media the dispersion is not normally sufficient, and instead it is dissipation that precludes wave overturning. This leads (see later) to a description in which, for $t > t_*$ and small dissipation, the wave is well described by branches of the "lossless" simple-wave solution, separated by thin regions that, on the overall wave scale, can be treated as *shock discontinuities*; see Fig. 4. To locate the discontinuity in the waveform, it can be argued[1–3] that the creation and propagation of it cannot change the mass or momentum perturbations carried by the wave, and since in the weakly

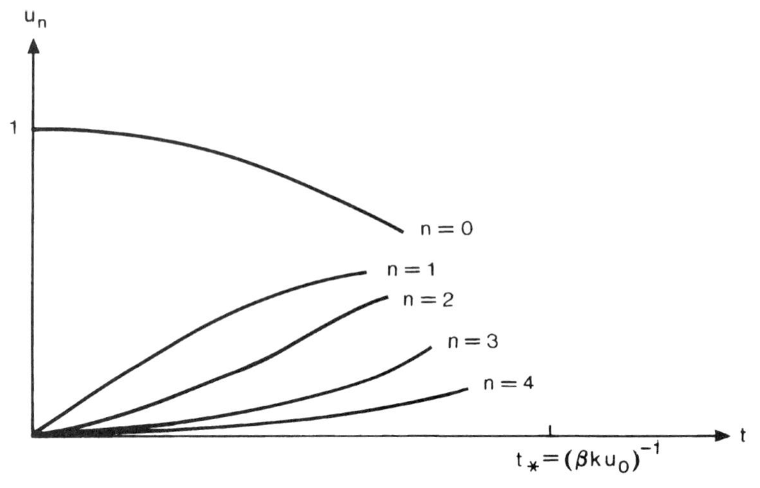

Fig. 3 Schematic to show energy pumping to higher frequencies according to the Fubini solution [Eq. (14)]. The ordinate is $u_n = 2J_n(nt/t_*)/(nt/t_*)$, with $t_* = (\beta k u_0)^{-1}$ the shock formation time.

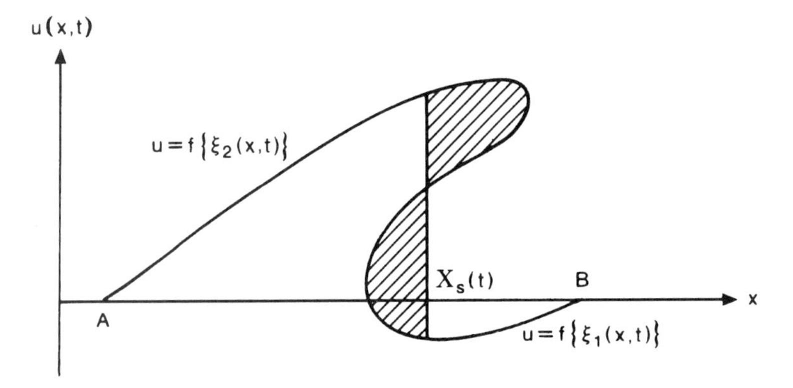

Fig. 4 Avoidance of triple-valued signals by insertion of a shock discontinuity satisfying the equal-areas rule; $u = f\{\xi_1(x,t)\}$, $u = f\{\xi_2(x,t)\}$ are branches of the simple-wave solution ahead of and behind the shock at $x = X_s(t)$.

nonlinear approximation each is proportional to $\int_A^B u\,dx$ (where $u = 0$ at A, B, or else $[A, B]$ contains an integral number of wavelengths of a periodic wave), the shock must cut off lobes of equal area[3] in the graph of u against x, as in Fig. 4 (*equal-areas* rule). The *energy* in segment $[A, B]$, proportional to $\int_A^B u^2\,dx$, is *not preserved* and in fact decreases at a rate that is then fixed regardless of the details of the dissipation mechanism. Those details simply provide the fine structure of the shock, though not all possible dissipation mechanisms are in fact able to provide a smooth transition between arbitrary signals, supplied by simple-wave theory, outside the shock (see Section 8).

Analytically, the equal-areas rule is expressed by

$$\dot{X}_s(t) = c_0 + \frac{\gamma + 1}{4}\,[f(\xi_{1s}) + f(\xi_{2s})], \tag{16}$$

that is, as the mean of the wavelet speeds $c_0 + \frac{1}{2}(\gamma + 1)u$ on the two sides of the shock. Here $x = X_s(t)$ is the shock path, $\xi_1(x,t)$, $\xi_2(x,t)$ are branches of the characteristic relation obtained by continuous variation from the single-valued regions ahead of and behind the shock, respectively, and the suffix s indicates their limiting values as $x \to X_s(t) \pm 0$.

In general, shock formation leads to wave reflection (and compound flow), and dissipation leads to entropy production. However, the entropy change can be found[1] as

$$S_2 - S_1 \simeq c_v \left(\frac{\gamma^2 - 1}{12\gamma^2} \right) \left(\frac{p_2 - p_1}{p_1} \right)^3, \qquad (17)$$

that is, of the third order in the shock strength $(p_2 - p_1)/p_1$, whereas changes in other linear mechanical and thermodynamic variables are of first order. Thus in nonlinear acoustics one can continue to use simple-wave theory and the uniform-entropy approximation even after shock formation.

Note that for normal fluids waves steepen forward, as in Fig. 2*a*, and any shock formed by dynamic evolution is necessarily *compressive* ($p_2 > p_1$); a rarefaction shock created somehow would immediately become continuous. There are, however, cases in gas dynamics in which the fundamental derivative $(\partial^2 \rho^{-1}/\partial p^2)_S$ [which is $(\gamma+1)/\gamma^2 p^2 \rho$ for a perfect gas] is negative; van der Waals gases at very high pressure have this property, and actual examples include certain hydro- and fluorocarbons. Then waves steepen *backward*, forming rarefaction shocks and compression expansions, though the entropy of course always increases through any shock [because the factor analogous to $\gamma + 1$ in Eq. (17) is then negative]. Fluids in which this occurs are often called "retrograde," with negative nonlinearity.[6] It is also possible for a gas to have a vanishing coefficient of quadratic nonlinearity on some manifold in the space of thermodynamic variables, and then in the neighborhood of that manifold one needs to account for quadratic and cubic nonlinearity together, with a model equation involving "mixed nonlinearity," of the form

$$\frac{\partial u}{\partial t} + (c_0 + \epsilon_m u + \alpha_m u^2)\, \frac{\partial u}{\partial x} = 0,$$

where ϵ_m is small but α_m is of order unity in an appropriate sense. Then one may have compression or expansion shocks, or, indeed, shocks that change their character from one to the other as the phase of the wave changes.[6]

The procedure—whatever the type of nonlinearity—in which one uses a combination of simple-wave theory together with shock discontinuities located according to an appropriate conservation principle is called *weak-shock theory*.[3] It holds for moderate times after shock formation, though not necessarily to very large times and distances (see Section 7 below).

6 BURGERS EQUATION

If one includes molecular diffusion of heat and momentum (*thermoviscous* diffusion) as a small effect, nominally of the same order of magnitude as nonlinearity over a few wave periods, then by a variety of approaches one comes for a simple wave to the Burgers equation[1–5]

$$\frac{\partial u}{\partial t} + \left(c_0 + \frac{\gamma + 1}{2}\, u \right) \frac{\partial u}{\partial x} = \frac{1}{2}\, \delta\, \frac{\partial^2 u}{\partial x^2}, \qquad (18)$$

where δ was called the "diffusivity of sound" by Lighthill[1]; $\delta = \nu[\frac{4}{3} + (\gamma - 1)/\sigma]$ in terms of kinematic viscosity ν and Prandtl number σ. This equation, proposed by Burgers as a model for turbulence, is a fundamental equation of wave theory. It is *exactly integrable* by the Hopf–Cole transformation of 1950–1951 (actually known at least as early as 1906 and now seen in a wider context as a Bäcklund transformation[7] between solutions of two different partial differential equations). In Eq. (18) put

$$u = -\delta \frac{\partial}{\partial X} \ln \psi, \qquad X = x - c_0 t, \qquad (19)$$

and then we find that ψ satisfies the linear diffusion equation

$$\frac{\partial \psi}{\partial t} = \frac{1}{2} \delta \frac{\partial^2 \psi}{\partial X^2}. \qquad (20)$$

Given $u(x, 0)$ we have

$$\psi(X, 0) = \exp\left(-\frac{1}{\delta} \int^X u(X'', 0)\, dX'' \right) = \exp\left(-\frac{1}{\delta} \int^X f(X'')\, dX'' \right),$$

in terms of which the solution of Eq. (20) is known to be

$$\psi(X, t) = \frac{1}{(2\pi\delta t)^{1/2}} \cdot \int_{-\infty}^{+\infty} \exp\left(-\frac{|X - X'|^2}{2\delta t} \right) \psi(X', 0)\, dX', \qquad (21)$$

and then $u(x, t)$ follows from Eq. (19) as single valued for all (x, t) for all $\delta > 0$.

The limit $\delta \to 0+$ can be examined quite generally[1–3] by standard techniques for asymptotics of integrals *pro-*

vided (x, t) are fixed as the limit is taken. Then one finds $u = f(\xi_1)$ or $u = f(\xi_2)$ [with branches of $\xi(x, t)$ defined from Eq. (7) by continuous variation from $t = 0$] with jumps between these branches at locations $X_s(t)$ satisfying the equal-areas rule, or the differential equation (16). Thus one recovers weak-shock theory [though not necessarily if x, t become large, e.g., $(x, t) = O(\delta^{-1})$ as $\delta \to 0$], and in addition one can obtain a description of the internal shock structure over a region of thickness $O(\delta)$ around $x = X_s(t)$. This can be written,[1] again with $\beta = \frac{1}{2}(\gamma + 1)$, as

$$u = \frac{U_2 + U_1}{2} - \frac{U_2 - U_1}{2} \tanh\left\{\frac{\beta}{2\delta}(U_2 - U_1)(x - X_s(t))\right\},$$

$$\dot{X}_s = c_0 + \beta\left(\frac{U_2 + U_1}{2}\right), \qquad (22)$$

and for normal fluids describes a smooth symmetric compressive transition from $U_1 = f(\xi_{1s})$ ahead to $U_2 = f(\xi_{2s})$ behind, with $U_2 > U_1$; note that U_1, U_2, $\dot{X}_s$ are not necessarily constant but can be taken so over the space and time scales $O(\delta)$ associated with passage of the shock. In the case where they *are* constant—a single shock wave separating semi-infinite media with uniform velocities U_1 and U_2—Eq. (22) is the exact traveling-wave solution of the Burgers equation first found by G. I. Taylor in 1910.

The solution, Eqs. (19)–(21), of the Burgers equation can also be used to examine shock–shock interaction.[1,3] In the Hopf–Cole variable ψ the N-shock solution is given by

$$\psi = \sum_{i=1}^{N+1} \exp\left\{-\frac{U_i}{\delta}(X - \overline{X}_i) + \frac{1}{2}\frac{U_i^2}{\delta}t\right\}, \qquad (23)$$

with the ordering $U_1 < U_2 < \cdots < U_{N+1}$. At $t = 0$ the signal $u(x, 0)$ comprises N shock transitions, the one from U_j to U_{j+1} being located at $(U_{j+1}\overline{X}_{j+1} - U_j\overline{X}_j)/(U_{j+1} - U_j)$, $j = 1, \ldots, N$, each of which then propagates at its individual speed $c_0 + \frac{1}{4}(\gamma + 1)(U_j + U_{j+1})$. If the jth and $(j + 1)$th shocks meet at t_j, then for $t > t_j$ there is [from Eq. (23)] no range of x for which u is close to U_{j+1}, and instead there is a single shock from U_j to U_{j+2}. After a finite time all possible *shock mergings* of this kind have taken place, and there remains only a single shock, from U_1 to U_{N+1}. All traces of the previous history of the wave are destroyed (with errors exponentially small, $O(e^{-1/\delta})$). As before, the levels U_j need not be constant, and if they are not, the shock paths are curved. The shock merging represents a spectral cascade both to lower wavenumbers (the waveform simplification means energy transfer to larger scales) and to higher ones (the few shocks remaining have increased strength and the smallest scale present is reduced to $O[\delta/(U_{N+1} - U_1)]$).

Propagation of the individual shocks and calculation of the various merging times and subsequent propagation of the new shocks can all be carried out in the (x, t) plane using just Eq. (16). If the shock structure is needed, it can be taken as given by Eq. (22) unless the times involved are very long, $O(\delta^{-1})$.

7 SIMPLE EXAMPLES: LONG-TIME BEHAVIOR

1. *Monopolar Pulse.* Suppose $f(x)$ is one signed, vanishing for $|x| > L$, and with $\int_{-\infty}^{+\infty} f(x)\,dx = UL$, say. The weak-shock solution depends on details of f but, for large t, has the same form as if f were initially triangular[3]; thus $u = 0$ for $x < c_0t - L$, $u = (x - c_0t + L)/\beta t$ for $c_0t - L < x < X_s(t)$, and $u = 0$ for $x > X_s(t)$, where $X_s(t) \sim c_0t + (2UL\beta t)^{1/2}$ and again $\beta = (\gamma + 1)/2$. Transition between $U_1 = 0$ and $U_2 = (2UL/\beta t)^{1/2}$ is given by Eq. (22), and examination of the exact Hopf–Cole solution shows that the weak-shock description continues to hold for arbitrarily large x, t. The reason is that the *Reynolds number* (ratio of nonlinear terms to diffusive terms on the overall wave scale) has a typical value $(\int_{-\infty}^{+\infty} u\,dx)/\delta = \text{const} = UL/\delta$, assumed large. Therefore the wave never becomes linear; the shock width is proportional to $\delta t^{1/2}$ and is small and fixed on the overall scale $t^{1/2}$. The ramp remains described by a nonlinear simple-wave solution, the shock by the (nonlinear and diffusive parts of) the Burgers equation.

2. *N-Wave.* The model case has the ideal form of Fig. 5. Weak-shock theory indicates[1] independent propagation of the positive and negative parts of the wave as in part 1 above, and the Hopf–Cole solution confirms this, for fixed (x, t) as $\delta \to 0$. However, the long-time behavior is very different. Diffusion of mass and momentum across the node $u = 0$ takes place, leading to a reduction of area of the positive or negative parts separately, hence of the wave Reynolds number and thus of the importance of nonlinearity. Ultimately all trace of a narrow shock disappears, and the wave becomes smooth and dies in *old age* under linear mechanisms,[1,8,9]

$$u \propto \left(\frac{X}{t^{1/2}}\right)\exp\left(-\frac{X^2}{2\delta t}\right), \qquad X = x - c_0t. \qquad (24)$$

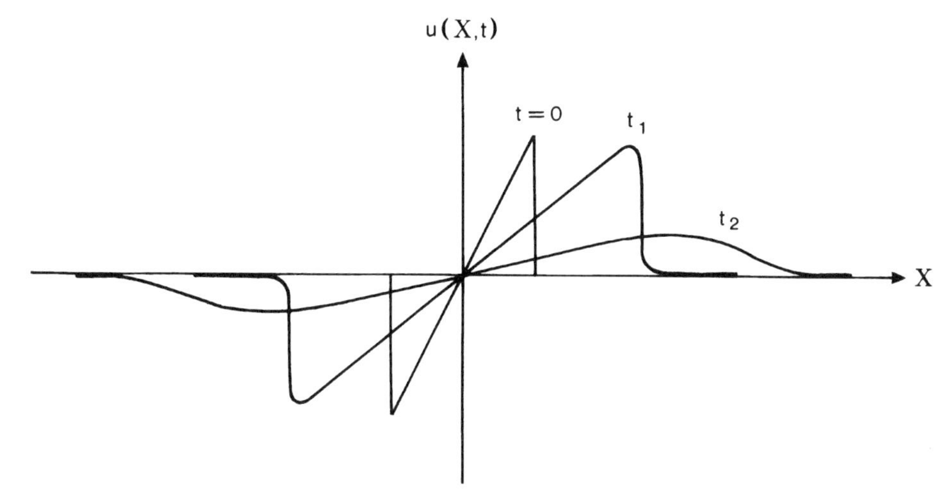

Fig. 5 Propagation of an N-wave, with $X = x - c_0 t$; at $t = t_1$, the wave is fully nonlinear, with overall length scale increasing as $t^{1/2}$ and with thin Taylor shocks of thickness proportional to $\delta t^{1/2}$; at $t = t_2$ the wave has decayed into old age, with propagation governed by linear mechanisms alone.

Weak-shock theory thus fails in this case for very large t (actually because the shocks get increasingly retarded by diffusive effects[1]—"shock displacement due to diffusion"—while keeping their weak-shock theory strength, thus reducing the Reynolds number as the propagation continues until it is so low that diffusion is important over the whole wave profile).

3. *Periodic Wave.* The familiar example is $u = u_0 \sin \omega x / c_0$ at $t = 0$. Shocks form at $\frac{1}{2}(\gamma + 1)\omega t_* = c_0/u_0$ and by symmetry are permanently centered on $\omega X / c_0 = (2n+1)\pi$. For $t_* \ll t \ll c_0^2/\omega^2\delta$ weak-shock holds, and the wave quickly develops a sawtooth form,[5] with Fourier representation

$$u = \frac{2c_0}{\beta\omega t} \sum_{n=1}^{\infty} \frac{(-1)^{n-1}}{n} \sin\left[n\frac{\omega}{c_0}(x - c_0 t)\right], \tag{25}$$

or

$$u = \frac{X}{\beta t} = \frac{2}{\gamma+1}\left(\frac{x}{t} - c_0\right), \qquad -\pi < \frac{\omega X}{c_0} < +\pi. \tag{26}$$

Significant in these is the fact that u is independent of u_0, a characteristic nonlinear effect known as *amplitude saturation.* The shocks have Taylor structure, with $\partial u/\partial x = O(\delta^{-1})$ in them; as a result the energy dissipation rate, proportional to $\delta(\partial u/\partial x)^2$ per unit length of $0x$, is concentrated at $O(\delta^{-1})$ in the shocks and when integrated over the shocks gives a total dissipation rate independent of δ and precisely equal to the rate of loss of energy implicit in the weak-shock solution. Analytically, if $E = \langle u^2 \rangle$, averaged over a cycle in x, then it follows from Eq. (18) that $dE/dt = -\delta \langle (\partial u/\partial x)^2 \rangle$. Evaluating this for $\delta \to 0$ from the shock contribution alone gives $dE/dt = -(2\pi^2/3\beta^2)(c_0^2/\omega^2 t^3)$, identical to what is obtained from the weak-shock theory, which from Eq. (26) gives $E = (\pi^2/3\beta^2)(c_0^2/\omega^2 t^2)$.

The shock thickness is $O(\delta/(U_2 - U_1)) = O(\delta t/u_0)$ and for $t = O(\delta^{-1})$ is comparable with the wavelength. At this phase of the evolution the effective Reynolds number is unity, and nonlinearity and diffusion are comparable everywhere (and comparable with $\partial u/\partial t$). From the Hopf–Cole solution one can then obtain an asymptotic description valid over the whole wave:

$$u(x,t) \sim \frac{\omega\delta}{\beta c_0} \sum_{n=1}^{\infty} \frac{(-1)^{n-1} \sin[n(\omega/c_0)(x - c_0 t)]}{\sinh(n\omega^2\delta t/2c_0^2)}. \tag{27}$$

This is associated with the name of Fay[5] (who worked directly with approximations to the governing equations and not with the Burgers equation itself); amazingly it turns out to be an exact solution of the Burgers equation, though it does not satisfy the initial condition posed here. For $\omega t \gg c_0^2/\omega\delta$ only the first harmonic survives, and

$$u \sim \frac{4}{\gamma+1}\left(\frac{\omega\delta}{c_0}\right)\exp\left(-\frac{\delta\omega^2}{2c_0^2}t\right)\sin\left[\frac{\omega}{c_0}(x-c_0t)\right], \tag{28}$$

showing linear old-age decay with amplitude saturation. Note that the time to old age is just the time at which diffusive decay would have taken over in the absence of nonlinear effects; under the conditions implied here, wave steepening, shock formation, and shock thickening all take place before that time and scramble all trace of the initial amplitude u_0. Weak-shock theory again fails in this case [now for $\omega t = O(c_0^2/\omega\delta)$], the essential mechanism being shock thickening.

8 RELAXATION AND DISPERSION

In many common situations, thermoviscous effects are often dominated over the initial wave scales by dissipation and dispersion associated with relaxation processes. In single-phase polyatomic gases the relaxation process may be the inability of one or more of the vibrational or rotational modes to instantaneously accept its full energy (as specified by the equipartition theorem); in a multiphase medium, such as a gas containing fine dust particles, the processes of relaxation involve the transfer of macroscopic heat and momentum between phases. Such a relaxing medium (with one relaxing mode for simplicity) may be characterized by a relaxation time T (with an assumed exponential relaxation rate $\exp(-t/T)$) and two sound speeds: c_0, the low-frequency (or equilibrium) speed, and $c_\infty > c_0$, the infinite-frequency (frozen) speed. Linear plane waves propagate in such a medium as $\exp[-\alpha x - i\omega(t - x/c)]$, where the attenuation coefficient $\alpha(\omega)$ and phase speed $c(\omega)$ are given by[2,9]

$$\alpha(\omega) = \frac{\omega^2 T(c_\infty - c_0)}{c_0^2 + c_\infty^2\omega^2T^2}, \qquad c(\omega) = \frac{c_0^2 + c_\infty^2\omega^2T^2}{c_0 + c_\infty\omega^2T^2}; \tag{29}$$

c increases monotonically from c_0 to c_∞, α increases monotonically from 0 to $(c_\infty - c_0)/c_\infty^2 T$, as ω increases from 0 to ∞.

The significant case for nonlinear acoustics is that in which the dispersion $(c_\infty - c_0)/c_0$ is small and comparable with the nonlinearity (Mach) parameter u_0/c_0; however, we want to deal with *finite rate* processes, so no restriction is placed on the typical frequency parameter ωT. Then with inclusion also of thermoviscosity, one can derive the model equation[2,9,10]

$$\left(1 + T\frac{\partial}{\partial t}\right)\left[\frac{\partial u}{\partial t} + \left(c_0 + \frac{\gamma+1}{2}u\right)\frac{\partial u}{\partial x} - \frac{1}{2}\delta\frac{\partial^2 u}{\partial x^2}\right] = (c_\infty - c_0)c_0T\frac{\partial^2 u}{\partial x^2}. \tag{30}$$

For waves with $\omega T \ll 1$ this reduces to the Burgers equation with a diffusivity of sound $\Delta = \delta + \delta_B$, $\delta_B = 2(c_\infty - c_0)c_0T$; relaxation processes on low-frequency waves are equivalent to a *bulk diffusivity* δ_B (often much larger than the molecular diffusivity δ). All the results for the Burgers equation then go through; in particular, relaxation processes prevent wave overturning, and the "shocks" produced have thickness $O(\delta_B/(U_2 - U_1))$, much larger than those structured by thermoviscous diffusion. This is important for sonic boom propagation in a humid atmosphere. For $\omega T \gg 1$ it can be shown from Eq. (30) that

$$\frac{\partial u}{\partial t} + \left(c_\infty + \frac{\gamma+1}{2}u\right)\frac{\partial u}{\partial x} + \lambda u - \frac{1}{2}\delta\frac{\partial^2 u}{\partial x^2} = 0, \tag{31}$$

with $\lambda = (c_\infty - c_0)/c_0T$. This equation, sometimes known as the Varley–Rogers equation when $\delta = 0$, is not integrable and no exact solutions are known except when $\delta = 0$. Then, by characteristics, one has

$$u = f(\xi)e^{-\lambda t}, \qquad x = \xi + c_\infty t + \frac{\gamma+1}{2}f(\xi)\left(\frac{1-e^{-\lambda t}}{\lambda}\right), \tag{32}$$

and in the absence of genuine molecular diffusion, wave overturning *will* occur in finite time—despite the relaxation processes—provided $|f'(\xi)| > 2\lambda/(\gamma+1)$ for some ξ with $f'(\xi) < 0$. Whether, for general ωT and $\delta = 0$, wave overturning will occur in finite time (necessitating the reintroduction of δ) can only be settled by numerical work, though that is frustrated in direct finite-difference or spectral schemes by "numerical viscosity." An intrinsic coordinate approach that gets around this has recently been developed.[11]

The steady traveling waves $u = F(\chi)$, $\chi = x - Vt$, are the only solutions to Eq. (30) that are known (even then we need $\delta = 0$). If we take $u = 0$ ahead of the wave, $u = U$ behind, we find $V = c_0 + \frac{1}{4}(\gamma+1)U$ (the shock speed for a medium in equilibrium). Then the integration gives

$$\frac{W}{U}\ln F + \left(1 - \frac{W}{U}\right)\ln(U - F) = \frac{\chi - \chi_0}{2VT}. \tag{33}$$

Here W is defined exactly as $\frac{1}{2}U - 2c_0(c_\infty - c_0)/(\gamma+1)V$, and for small dispersion we can put $V = c_0$ in the second term, so that $W = 2(V - c_\infty)/(\gamma + 1)$. If $V < c_\infty$, that is, $W < 0$, Eq. (33) gives a smooth transition between 0 and U, described as a *fully dispersed* relaxing shock. If $V \ll c_\infty$, the solution becomes the Taylor solution (22) with $U_1 = 0$, $U_2 = U$, and the diffusivity δ replaced by the bulk diffusivity δ_B. For stronger waves relative to the dispersion $(c_\infty - c_0)/c_0$ the wave profile is markedly asymmetric, and has a kink on the low-pressure side when $W = 0$. For $W > 0$ the curve starting from U doubles back on itself, there is no continuous transition structured by relaxation processes alone, and one must insert a discontinuity (or, equivalently, a thin Taylor shock structured by molecular diffusivity) within the outer relaxing shock.[1,2] Equation (33) is used from $u = U$ forward to the point X^* at which $V - u^* = c_\infty$, and at this point the discontinuity (or Taylor shock) takes the velocity down to zero. The Taylor shock speed of course turns out to be given by Eq. (16) again. In such a case we have a *partly dispersed* relaxing shock,[1] with an embedded diffusive shock; the reason for the discontinuity is that otherwise the lowest (and therefore linear) wavelets would have a propagation speed, into fluid at rest, greater than the highest available sound speed c_∞. Partly dispersed shocks, with a double structure, are common in air, though it is difficult to conduct repeatable experiments in atmospheric air because the relaxation time T (and even more the bulk diffusivity $\delta_B \sim T^2$) is extraordinarily sensitive to the humidity.

9 NONLINEAR WAVES IN TUBES

In tubes of finite length waves propagate in both directions, independently in the linear limit, except for coupling at the ends. For weakly nonlinear waves it is also found that the simple waves propagating in each direction do not interact in the body of the fluid, but only through the boundary condition coupling at the ends. It is the *self-distortion* of a simple wave that leads to cumulative long-time effects, not interaction with a wave traveling in the opposite direction. Reflection at a closed end just requires $u = 0$ there; at an open end one may assume $p = p_0$ (though there is often a need to model vortex-shedding processes that take place when shocks reflect from an open end[12]). Analytically these reflection processes are described by functional equations for the simple-wave shapes. There is an extensive literature on such functional mappings, and it is known, for example, that in some parameter ranges very complicated behavior can develop.

Dissipation by diffusive effects in the Stokes boundary layers on the tube walls is usually much greater than the thermoviscous dissipation in the uniform central core. The model equation for a right-running simple wave in a circular tube, with both effects included, was derived by Chester[13] in the form

$$\frac{\partial u}{\partial t} + \left(c_0 + \frac{\gamma+1}{2}u\right)\frac{\partial u}{\partial x} - \frac{1}{2}\delta\frac{\partial^2 u}{\partial x^2} = c_0\kappa\int_0^\infty \frac{\partial u}{\partial x}(x, t-\xi)\xi^{-1/2}\,d\xi, \qquad (34)$$

where u is velocity averaged over a section and $\kappa = (1/R)(\nu/\pi)^{1/2}[1 + (\gamma - 1)/\sigma^{1/2}]$ for a tube of radius R. As with relaxation, although the tube wall effects are generally larger than the mainstream diffusion, neglect of the latter often leads to partly dispersed shocks of the relaxing-media kind, a thin inner "mainstream diffusion" subshock of Taylor type being embedded in a highly asymmetric outer shock whose much larger thickness is controlled by κ. Limited analytical information is available on solutions of Eq. (34), together with asymptotic and numerical studies.[14,15]

Detailed confirmation of the applicability of Eq. (34) for waves in a very long tube was obtained by Pestorius and Blackstock.[16] They took a complicated signal u, as specified at one end in an experiment, and allowed it to propagate a small distance according to nonlinear lossless theory. The evolved signal was then Fourier analyzed, and phase shifts and amplitude decay factors were applied to the spectral components to account for linear attenuation and dispersion effects in the boundary layer over the small time step. Then the modified Fourier components were synthesized into a time waveform, which was then allowed to propagate a further time step with the same split procedure. Extremely close agreement with experiment was obtained, even over very long ranges. Such methods are probably the best that will ever be available. Equations such as (34) have been shown to be nonintegrable, and there is almost no chance of general analytical progress.

10 NONLINEAR PROPAGATION IN BUBBLY LIQUID

If α_0 is the volume concentration of small gas bubbles in liquid, then it is well known[17] that the low-frequency sound speed in the bubbly suspension is given by $c_0^2 = \gamma p_0/\rho_0\alpha_0(1 - \alpha_0)$ (except for α_0 very small or very close to 1), with p/ρ^γ = const the equation of state for the air in the bubbles and ρ_0 the mixture density. Even for quite small α_0 the speed c_0 may be not only less than the water sound speed c_w but also significantly less than the air sound speed; values as low as 30 ms^{-1} were measured and compared quite favorably with this prediction as long ago as 1954. This implies that a given velocity u_0, generated by a piston source say, is associated with nonlinear effects greater by a factor c_w/c_0 when the motion is generated in bubbly liquid than in pure water (provided frequencies are sufficiently low) and shocks form much earlier, at shorter ranges from a source. Detailed analysis[17] shows that the lossless simple-wave equation takes the form

$$\frac{\partial u}{\partial t} + \left(c_0 + \frac{1}{\alpha_0} u\right) \frac{\partial u}{\partial x} = 0. \tag{35}$$

Shock formation then takes place at time $t'_* = \beta\alpha_0 t_*$, where t_* is the pure-water time (for the same initial f) given in Eq. (15). (Again α_0 must not be too small.) For $t > t'_*$ the motion might be assumed to proceed as in a single-phase medium, with the emergence of thin shocks that might in the end thicken and lead to old-age decay.

The microstructure of the bubbly medium does, however, have a well-defined intrinsic time scale—the bubble volume oscillation period—when there is a sharp clustering of bubble sizes about a single value R_0. This leads to strong frequency dispersion. If $\omega_0 = (3p_0/\rho_w R_0^2)^{1/2}$ is the (Minnaert) isolated bubble resonance frequency, then the phase speed $c_p(\omega)$ of linear waves of frequency ω is given by[17]

$$\frac{1}{c_p^2(\omega)} = \frac{(1-\alpha_0)^2}{c_w^2} + \frac{1}{c_0^2[(1-\omega^2/\omega_0^2) - 2i\Delta(\omega/\omega_0)]}, \tag{36}$$

where we include an ad hoc damping factor Δ in the bubble response. Figure 6 shows Re c_p as a function of ω. If $\Delta = 0$ there is a forbidden frequency band in which the motion is purely reactive, with an infinite sound speed as one approaches the upper limit of this band from above. Dissipation (and variation of bubble size) blurs this picture, but all the essentials have been seen in experiment, including decrease of c_p from the (low) value c_0 almost to zero at $\omega = \omega_0$ and very large values of c_p(> 3000 ms^{-1}) at higher frequencies, with approach to $c_w/(1-\alpha_0)$ at the highest frequencies. In general, the dis-

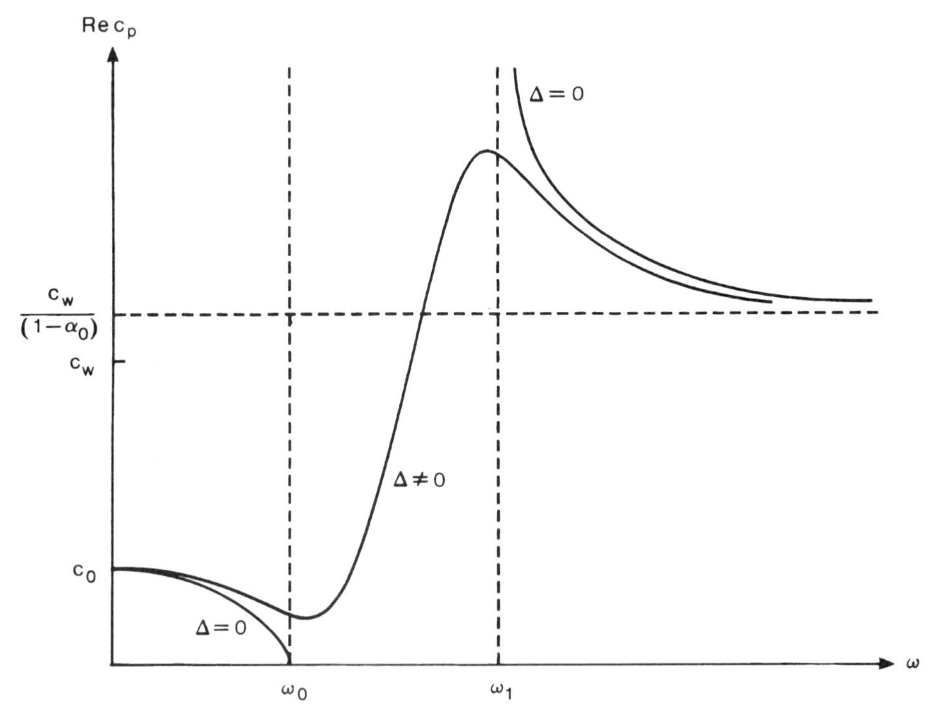

Fig. 6 Real part of acoustic phase speed $c_p(\omega)$ as function of frequency for propagation in a bubbly liquid (schematic only); Δ is the dissipation coefficient. If $\Delta = 0$, then Re $c_p = 0$ throughout the forbidden frequency band $\omega_0 < \omega < \omega_1$.

persion is only weak (which is necessary for strong nonlinear interaction between all the harmonics generated by nonlinearity) at low and high frequencies. At low frequencies the dispersion is cubic, $\omega = c_0 k - c_0^3 k^3/2\omega_0^2$, and the combined effects of nonlinearity and dispersion are represented by the Korteweg–de Vries (KdV) equation[17]

$$\frac{\partial u}{\partial t} + \left(c_0 + \frac{1}{\alpha_0} u\right) \frac{\partial u}{\partial x} + \frac{c_0^3}{2\omega_0^2} \frac{\partial^3 u}{\partial x^3} = 0. \tag{37}$$

This famous equation is the prototype for integrable equations with *soliton* solutions. These are traveling-wave solutions of sech^2 profile (bell shaped), necessarily representing positive-pressure pulses, traveling at an excess speed proportional to amplitude and with width inversely proportional to $(\text{amplitude})^{1/2}$. A new analytical method (inverse spectral, or scattering, transform) has been devised for such integrable equations and through a sequence of three linear problems provides the solution for $u(x, t)$ given $u(x, 0)$. Analysis shows that in general $u(x, t)$ will contain a definite number [related to properties of $u(x, 0)$] of solitons that propagate with fixed amplitude and shape, together with a decaying oscillatory wave train ("radiation"). The solitons preserve their identity after interaction with each other and with other waves and thus represent permanent large-scale features that dominate the wave field at large time. For a simple introduction to soliton theory, see Drazin and Johnson.[18]

Predictions of soliton theory have been compared quite favorably with experiments on pressure pulses in bubbly liquid. Dissipation cannot always be ignored and typically contributes a Burgers term $-\frac{1}{2}\delta\ \partial^2 u/\partial x^2$ to the left side of Eq. (37). Then one cannot have pulses in which the pressure returns behind the wave to its value ahead but rather shock waves in which the pressure is increased behind. If the dissipation coefficient δ is appropriately large relative to the dispersion coefficient $c_0^3/2\omega_0^2$, the shock transition is monotonic; if not, the signal approaches the value ahead monotonically but the value behind through a decaying oscillation.[19]

At frequencies far above the bubble resonance the dispersion takes the form $\omega = c_w k + (\omega_0^2 c_w/2c_0^2)k^{-1}$, and the appropriate nonlinear wave equation is the nonlinear Klein–Gordon equation

$$\frac{\partial}{\partial x}\left[\frac{\partial u}{\partial t} + \left(c_w + \frac{\Gamma + 1}{2} u \frac{\partial u}{\partial x}\right)\right] - \frac{\omega_0^2 c_w}{2c_0^2} u = 0, \tag{38}$$

where Γ plays the role for water of γ for air. Here the dispersion is not generally strong enough to prevent wave overturning, and shocks form in finite time (no wave overturning can occur for KdV). Little is known about the subsequent motion (except in the case of periodic waves[19]; the only bounded traveling-wave solutions of Eq. (38) are periodic), but the nonlinear and dispersive effects in this case are much less dramatic than in the low-frequency case.

11 GEOMETRIC AND MATERIAL NONUNIFORMITY

Consider now the propagation of quasi-plane waves in narrow horns (no higher order modes). It is natural in such cases (which include freely diverging or converging cylindrical and spherical waves) to formulate signaling, rather than initial-value, problems. Let s be a (range) coordinate along the horn, $\tau = t - (s - s_0)/c_0$ retarded time relative to a signaling location s_0 for a simple wave propagating "outward," $u(s, \tau)$ velocity averaged across a section of area $A(s)$. Then if the medium parameters ρ, c are uniform, energy conservation in linear nondissipative theory requires $\partial(A^{1/2}u)/\partial s = 0$, or $\partial u/\partial s + \frac{1}{2}u(d/ds)\ln A(s) = 0$. This is also the "transport equation" for high-frequency waves propagating along a ray tube; $\omega L/c_0 \gg 1$ is required, where L is the length scale for variations in $A(s)$. Now if nonlinearity and diffusion are included as of the same nominal magnitudes $[O(\omega L/c_0)^{-1}]$ as the geometric area variations, one has

$$\frac{\partial u}{\partial s} - \frac{\gamma + 1}{2c_0^2} u \frac{\partial u}{\partial \tau} + \frac{1}{2} u \frac{d}{ds} \ln A = \frac{\delta}{2c_0^3} \frac{\partial^2 u}{\partial \tau^2}, \tag{39}$$

called a *generalized Burgers* equation. It has been proved[8] that none of these can be exactly integrated, except when $\delta = 0$ or when $A'(s) = 0$ (Burgers equation), and few exact solutions are known. The transformations

$$v = \left[\frac{A(s)}{A(s_0)}\right]^{1/2} u, \qquad \zeta = \int_{s_0}^{s} \left[\frac{A(s_0)}{A(s)}\right]^{1/2} ds \tag{40}$$

put Eq. (39) in the form of a Burgers equation with range-dependent diffusivity,

$$\frac{\partial v}{\partial \zeta} - \frac{\gamma + 1}{2c_0^2} v \frac{\partial v}{\partial \tau} = \frac{\delta}{2c_0^3} G(\zeta) \frac{\partial^2 v}{\partial \tau^2}, \tag{41}$$

where $G(\zeta) = A^{1/2}[s(\zeta)]/A^{1/2}(s_0)$.

Included here are the cases of freely diverging cylindrical and spherical waves, for which $s = r$, and

$$A(r) \sim r, \qquad A(r) \sim r^2,$$
$$\zeta = 2r_0^{1/2}(r^{1/2} - r_0^{1/2}), \qquad \zeta = r_0 \ln \frac{r}{r_0}, \qquad (42)$$
$$G(\zeta) = 1 + \frac{\zeta}{2r_0}, \qquad \exp \frac{\zeta}{r_0},$$

respectively.

For $\delta = 0$, Eq. (41) is integrable by characteristics and the question of shock formation is covered by the plane-wave criteria and the range transformation. For example, if $u = u_0 \sin \omega t$ at $r = r_0$, then shocks form at ranges

$$r_* = r_0 + \frac{2c_0^2}{(\gamma + 1)u_0\omega}, \qquad r_0\left(1 + \frac{c_0^2}{(r+1)u_0\omega r_0}\right)^2,$$
$$r_0 \exp\left[\frac{2c_0^2}{(\gamma + 1)u_0\omega r_0}\right]$$

in the plane, cylindrical, and spherical cases, respectively.

The transformations also give well-known decay laws for practically important problems. For example, if the signal is an N-wave, then the velocity jump Δu at the shocks decays as $r^{-1/2}$ in plane flow (cf. Section 7), and therefore as $r^{-3/4}$ and $r^{-1}(\ln r)^{-1/2}$ for cylindrical and spherical N-waves, while for the time duration $T(r)$ of the whole N-wave at any range r one has $T(r) \sim r^{1/2}$, $r^{1/4}$, $(\ln r)^{1/2}$ for plane, cylindrical, and spherical N-waves. Cylindrical geometry is relevant to the sonic boom of a supersonic aircraft; the flight axis is timelike and the propagation is along ray paths on the surface of a cone, with $A(r) \sim r$.

In the weak-shock theory treatment of Eq. (39), shock discontinuities are inserted, as before, at equal-areas locations in the graph of u (or v) against τ, and the analysis of the lossless simple-wave parts is as for plane flow, after transformation via Eq. (40). For the diffusive structure of the shocks, this is effectively given by the Taylor solution [Eq. (22)] in which we simply replace δ by the equivalent diffusivity $\delta G(\zeta)$ at range r, at least if r is not large as $\delta \to 0$. As with plane waves, however, weak-shock theory often fails at large ranges; indeed, for the N-wave and sinusoidal signals, weak-shock theory holds indefinitely *only* for the exponentially converging horn.[8] For all others, one of a number of nonuniformities takes place at large range, indicating a change of the wave properties and dynamics. Shock thickening is one and is typical for horns diverging no more rapidly than cylindrically. For more rapid divergence, there is typically first a *local* nonuniformity, a breakdown in Taylor shock structure. Conditions outside the shocks are changing so rapidly (because of the rapid area variations) that the local nonlinear steepening–diffusion balance can no longer be preserved and the shocks become evolutionary [i.e., their structure can only be described by solutions of Eq. (39) in its full form, without neglect of $\partial u/\partial s$]. Then they decay to an error function form that thickens with increasing range (but more slowly than the Taylor shock would). Then a second *global* nonuniformity follows, in which the wave becomes smooth everywhere and dies in old age.[9] Spherically diverging waves typically follow this scenario, but there are several other possible scenarios[8] for the long-range behavior after weak-shock theory has ceased to be valid. The elucidation of these scenarios requires asymptotic and numerical study[20] in the absence of analytical solutions.

If the medium parameters are not uniform, then one must first solve the corresponding linear problem by ray methods to define the horn or ray tube geometry. Take, as a simple example, the case of an isothermal atmosphere, $c(\mathbf{x}) = c_0$, constant, but with density stratification $\rho(z) = \rho_0 \exp(-z/H)$ in the vertical direction z. Then the rays are straight and for a spherically symmetric excitation are just radial lines through the origin, $\theta = \text{const}$ (θ = angle from vertical). Energy conservation (in the absence of winds) implies $\partial(\rho^{1/2}A^{1/2}u)/\partial r = 0$, where $A \sim r^2$, $z = r\cos\theta$, and the retarded time is simply $\tau = t - (r - r_0)/c_0$. Generalizing this by the inclusion of nonlinear terms, we have, as the transport equation of *nonlinear geometrical acoustics*,

$$\frac{\partial u}{\partial r} - \frac{\gamma + 1}{2c_0^2}\, u\, \frac{\partial u}{\partial \tau} + \frac{u}{r} + \frac{u \cos\theta}{2H} = 0. \qquad (43)$$

There is a different nonlinear problem to be solved along each ray. Shock formation [easily examined by transformation of Eq. (43) to the form $\partial v/\partial\zeta - v\,\partial v/\partial\tau = 0$] is facilitated by the decreasing density for waves propagating along rays above the horizontal and hindered for waves propagating in directions below the horizontal to the extent that shocks may not form in finite range. If $c(\mathbf{x})$ is not uniform, an equation corresponding to Eq. (43) can be derived by consideration not of just the invariance of energy but also of Snell's law for the phase (see Rogers and Gardner[21] for an application to the propagation of N-waves in the lower atmosphere [$c(\mathbf{x})$ = const] and the upper atmosphere [$c(\mathbf{x}) \sim z$, giving rays that are arcs of circles]).

12 NONLINEARITY AND DIFFRACTION

The formal basis for nonlinear geometric acoustics (Section 11) rests on the idea that if the geometric (or material) nonuniformity parameter $(\omega L/c_0)^{-1}$ and the nonlinearity parameter u_0/c_0 are comparable, then the field has a phase structure given by local linear ray theory, with modulation along the rays expressed by a generalized Burgers equation. However, if a relatively rapid variation *along* the wavefronts is imposed at a boundary or allowed to develop by diffraction in two or three dimensions, then one-dimensional propagation along rays fails. In the simplest case, think of a beam propagating along the x-axis, with (modulation distance)/(wavelength) $= \epsilon^{-1}$, say, and suppose (scale for lateral y variations)/(wavelength) $= O(\epsilon^{-1/2})$. Here ϵ is a small parameter, comparable with the particle Mach number u_0/c_0. Then the balance between nonlinear modulation and lateral diffraction turns out to be governed by the dZK (dissipative Zabolotskaya–Khokhlov) equation[2,22]

$$\frac{\partial}{\partial x}\left[\frac{\partial u}{\partial t}+\left(c_0+\frac{\gamma+1}{2}u\right)\frac{\partial u}{\partial x}-\frac{1}{2}\delta\frac{\partial^2 u}{\partial x^2}\right] + \frac{c_0}{2}\frac{\partial^2 u}{\partial y^2}=0, \tag{44}$$

where we have two space dimensions and u is the x-velocity.

A great deal of work, mainly numerical, has been done on this equation, much of it by Russian scientists; see especially the monographs.[2,10,22,23] The essential feature of Eq. (44) is that its solutions display *nonlinear wavefront turning*, a property seen in numerical simulations but necessarily absent from nonlinear geometrical acoustics. A transformation between solutions of dZK has recently been found,[24] and the infinitesimal version of it can be used to demonstrate the wavefront turning due to amplitude variations along wavefronts.

A key question is whether, in the nondissipative limit $\delta = 0$, diffraction is able to prevent the caustic singularity formation produced in nonlinear geometrical acoustics by convergence of rays to a point. The answer is that it cannot in all cases. Classes of exact solutions to ZK [$\delta = 0$ in Eq. (44)] have now been found from the transformation mentioned above,[24] and these can be used to show that nonlinear caustic formation may indeed occur in ZK theory. The amplitude at the focus will actually be limited either by dissipative effects or by full-wave-mechanics effects (dZK itself may not be applicable in the caustic region) in a way not yet understood.

13 FURTHER COMMENTS

Several items should at least be mentioned here. First, an important practical and scientific field within nonlinear acoustics is concerned with *statistical problems*; for example, suppose the propagation is governed by a generalized Burgers equation, and the input signal $u(s = s_0, t) = u_0(t)$ can only be described statistically. Then one would like to know the statistics of the evolved signal [e.g., the mean-square value $\langle u^2(s)\rangle$ as a function of range s and the frequency spectrum $\Pi(\omega, s)$ of $\langle u^2\rangle$ at each s] in terms of the input statistics. Even in the weak-shock theory description this is a formidable problem, because one has to compute the statistics with allowance for shock formation and multiple-shock coalescence in each realization of the random field. Applications would include, for example, the propagation of high-intensity noise from jet aircraft engines. Other important problems include the propagation of an input pure-tone signal contaminated by broadband random noise and the possibility of enhancement or suppression of one of these components by the other. There is again much Russian work in this field, and the reader is referred to an important monograph[25] devoted entirely to statistical nonlinear acoustics.

Second, not all nonlinearity in acoustic media is of the quadratic-convective kind. For transverse and torsional waves in perfect elastic solids the nonlinearity is cubic-convective, $u^2\,\partial u/\partial x$, while for elastic media with microstructure (pores, voids, cracks) the stress–strain relation may be linear but with a different Young's modulus for compression and for expansion. For chemically active media with Arrhenius rate laws, transcendental nonlinearity $\exp u$ arises. Naturally, quite new phenomena arise in all cases. For cubic nonlinearity (the so-called modified Burgers equation) the striking feature is the formation in finite time, from general initial conditions, of *sonic shocks*.[26] These have a propagation speed precisely equal to the wavelet or characteristic speed ahead of the shock, and the signals on either side of the shock are related in the ratio 2 to -1. Such sonic shocks cannot occur in ordinary gas dynamics but are generic for cubic and all higher nonlinearity. For media with bilinear stress–strain laws, the weak-shock description can be analyzed and shows that a sinusoidal wave would develop shocks whose strength vanishes identically (and therefore the whole signal vanishes identically!) in finite time.[10]

For propagation of pressure waves in reacting media the dominant nonlinearity is in the chemistry initially and leads to explosive singularities in finite time unless fuel depletion is accounted for. Then a very complicated interaction takes place between the thermal field and the gas dynamic velocity field, leading to the launching of shock, deflagration, and ultimately detonation waves.

REFERENCES

1. M. J. Lighthill "Viscosity Effects in Sound Waves of Finite Amplitude," in G. K. Batchelor and R. M. Davies (Eds.), *Surveys in Mechanics*, Cambridge University Press, U.K., 1956, pp. 249–350.

2. O. V. Rudenko and S. I. Soluyan, *Theoretical Foundations of Nonlinear Acoustics*, Consultants Bureau, Plenum, New York, 1977.
3. G. B. Whitham, *Linear and Nonlinear Waves*, Wiley, New York, 1974.
4. J. K. Engelbrecht, V. E. Fridman, and E. N. Pelinovsky, *Nonlinear Evolution Equations*, Longman/Wiley, New York, 1988.
5. R. T. Beyer (Ed.), *Nonlinear Acoustics in Fluids*, Van Nostrand Reinhold, New York, 1984, (a collection of fundamental papers in nonlinear acoustics).
6. A. Kluwick, "Small-Amplitude Finite-Rate Waves in Fluids Having Both Positive and Negative Nonlinearity," in A. Kluwick (Ed.), *Nonlinear Waves in Real Fluids*, Springer-Verlag, Berlin, 1991, pp. 1–43.
7. C. Rogers and W. F. Shadwick, *Bäcklund Transformations and Their Applications*, Academic, New York, 1982.
8. J. J. C. Nimmo and D. G. Crighton, "Geometrical and Diffusive Effects in Nonlinear Acoustic Propagation over Long Ranges," *Phil. Trans. Roy. Soc. Lond.* Vol. A, 320, 1986, pp. 1–35.
9. D. G. Crighton and J. F. Scott, "Asymptotic Solutions of Model Equations in Nonlinear Acoustics," *Phil. Trans. Roy. Soc. Lond. A*, Vol. 292, 1979, pp. 101–134.
10. K. A. Naugol'nykh and L. A. Ostrovsky, *Nonlinear Wave Processes in Acoustics*, Cambridge University Press, U.K., 1996.
11. P. W. Hammerton and D. G. Crighton, "Overturning of Nonlinear Acoustic Waves. Part 1. A General Method. Part 2. Relaxing Gas Dynamics," *J. Fluid Mech.*, Vol. 252, 1979, pp. 585–599, 601–615.
12. J. H. M. Disselhorst and L. van Wijngaarden, "Flow in the Exit of Open Pipes During Acoustic Resonance," *J. Fluid Mech.*, Vol. 99, 1980, pp. 293–319.
13. W. Chester, "Resonant Oscillations in Closed Tubes," *J. Fluid Mech.*, Vol. 18, 1964, pp. 44–64.
14. N. Sugimoto, "Burgers Equation with a Fractional Derivative; Hereditary Effects on Nonlinear Acoustic Waves," *J. Fluid Mech.*, Vol. 225, 1991, pp. 631–653.
15. J. J. Keller, "Resonant Oscillations in Closed Tubes: The Solution of Chester's Equation," *J. Fluid Mech.*, Vol. 77, 1976, pp. 279–304.
16. F. M. Pestorius and D. T. Blackstock, "Propagation of Finite-Amplitude Noise," in L. Bjørnø (Ed.), *Finite-Amplitude Wave Effects in Fluids*, IPC Press, Guildford, 1974, pp. 24–29.
17. L. van Wijngaarden, "One-Dimensional Flow of Liquids Containing Small Gas Bubbles," *Ann. Rev. Fluid Mech.*, Vol. 4, 1972, pp. 369–396.
18. P. G. Drazin and R. S. Johnson, *Solitons: An Introduction*, Cambridge University Press, U.K. 1989.
19. D. G. Crighton, "Nonlinear Acoustics of Bubbly Liquids," in A. Kluwick (Ed.), *Nonlinear Waves in Real Fluids*, Springer-Verlag, Berlin, 1991, pp. 45–68.
20. P. W. Hammerton and D. G. Crighton, "Old-Age Behaviour of Cylindrical and Spherical Nonlinear Waves; Numerical and Asymptotic Results," *Proc. Roy. Soc. Lond. A*, Vol. 422, 1989, pp. 387–405.
21. P. H. Rogers and J. H. Gardner, "Propagation of Sonic Booms in the Thermosphere," *J. Acoust. Soc. Am.*, Vol. 67, 1980, pp. 78–91.
22. N. S. Bakhvalov, Ya. M. Zhileikin, and E. A. Zabolotskaya, *Nonlinear Theory of Sound Beams*, American Institute of Physics, New York, 1987.
23. B. K. Novikov, O. V. Rudenko, and V. I. Timoshenko, *Nonlinear Underwater Acoustics*, American Institute of Physics, Acoustical Society of America, New York, 1987.
24. A. T. Cates and D. G. Crighton, "Nonlinear Diffraction and Caustic Formation," *Proc. Roy. Soc. Lond. A*, Vol. 430, 1990, pp. 69–88.
25. S. N. Gurbatov, A. N. Malakhov, and A. I. Saichev, *Nonlinear Random Waves and Turbulence in Nondispersive Media: Waves, Rays, Particles*, Manchester University Press, U.K., 1991.
26. I. P. Lee-Bapty and D. G. Crighton, "Nonlinear Wave Motion Governed by the Modified Burgers Equation," *Phil. Trans. Roy. Soc. Lond. A*, Vol. 323, 1987, pp. 173–209.

18

PARAMETERS OF NONLINEARITY OF ACOUSTIC MEDIA

E. CARR EVERBACH

1 INTRODUCTION

Nonlinearity in acoustic media contributes to distortion of the wave shape during propagation and can be described by various parameters. Nonlinearity parameters include B/A, the ratio of the second to the first term in a Taylor series expansion of the pressure as a function of density, the coefficient of nonlinearity β, which takes self-convection into account, and the acoustic Mach number M, which differentiates the linear from the nonlinear regimes. During acoustic wave propagation, energy initially carried at the fundamental frequency is moved into the higher harmonics as the wave steepens due to the cumulative effects of nonlinearity. Additional heating of the medium and loss of amplitude at the fundamental frequency are negative consequences, but nonlinearity can also be used to infer properties of the medium or generate useful acoustic sources. Nonlinearity parameters may be measured by quantifying the relationship among thermodynamic properties of a medium or by measuring the waveform distortion and inferring the underlying nonlinearities. Mixture laws have been developed to aid in the calculation of nonlinearity parameters of mixtures of media whose component nonlinearities are known.

2 ACOUSTIC MACH NUMBER

The acoustic Mach number is defined via[1] $M = u_{\max}/c_0$, where $u_{\max}$ and c_0 are the maximum particle velocity and the small-signal acoustic phase speed in the propagation medium, respectively. For plane-wave propagation in a gas, $M = p_{\max}/\gamma P_0$, where $p_{\max}$ is the maximum deviation of pressure from the equilibrium pressure P_0 and γ is the ratio of specific heats C_p/C_v for the gas. For linear acoustics, $M \ll 1$.

3 ACOUSTIC NONLINEARITY PARAMETER *B/A*

For small deviations from P_0, the pressure P may be expanded in a Taylor series in terms of the density ρ and specific entropy s. For changes that induce temperature gradients small enough so that appreciable heat flow does not occur during a fraction of an acoustic period, the system is said to be adiabatic. With the further requirement that the changes are thermodynamically reversible, the entropy does not change from its equilibrium value s_0, the system is said to be isentropic, and the Taylor series becomes

$$p = P - P_0 = A\left[\frac{\rho - \rho_0}{\rho_0}\right] + \frac{B}{2!}\left[\frac{\rho - \rho_0}{\rho_0}\right]^2 + \frac{C}{3!}\left[\frac{\rho - \rho_0}{\rho_0}\right]^3 + \cdots, \tag{1}$$

where p is the excess or acoustic pressure and $(\rho - \rho_0)/\rho_0$ is the condensation or fractional change in density of the propagation medium. Also, for small deviations from equilibrium[2]

Note: References to chapters appearing only in the *Encyclopedia* are preceded by *Enc.*

$$A = \rho_0 \left[\left(\frac{\partial P}{\partial \rho} \right)_s \right]_{\rho = \rho_0} = \rho_0 c_0^2,$$

$$B = \rho_0^2 \left[\left(\frac{\partial^2 P}{\partial \rho^2} \right)_s \right]_{\rho = \rho_0},$$

$$C = \rho_0^3 \left[\left(\frac{\partial^3 P}{\partial \rho^3} \right)_s \right]_{\rho = \rho_0}, \quad (2)$$

and so on, to higher orders. When the condensation is infinitesimal, the higher order terms are negligible and the acoustic waves will propagate at the (constant) speed c_0. For finite-amplitude waves, the higher order terms B, C, ... become increasingly more important as the amplitude increases. Alternatively, for waves of a given amplitude, materials with larger values of B, C, ... will cause the local sound speed to differ increasingly from c_0. The acoustic nonlinearity parameter B/A is the ratio of the quadratic coefficient in this expression to the linear coefficient[3] and therefore provides a measure of the degree to which the local sound speed deviates from the small-signal (linear) case. Further manipulations yield equivalent expressions:

$$\frac{B}{A} = 2\rho_0 c_0 \left(\frac{\partial c}{\partial p} \right)_s \quad (3)$$

or

$$\frac{B}{A} = 2\rho_0 c_0 \left[\left(\frac{\partial c}{\partial p} \right)_T \right]_{\rho = \rho_0} + \frac{2 c_0 T \kappa_e}{C_p} \left[\left(\frac{\partial c}{\partial T} \right)_p \right]_{\rho = \rho_0} \quad (4)$$

or

$$\frac{B}{A} = \left[\frac{\partial (1/k)}{\partial P} \right]_s - 1, \quad (5)$$

where c is the local sound speed; T is the temperature in Kelvin; and κ_e, C_p, and k are the volume coefficient of thermal expansion, the specific heat at constant pressure, and the adiabatic compressibility (the reciprocal of stiffness) for the propagation medium, respectively.

Equation (3) shows that B/A is proportional to the change of sound speed for a change of pressure, provided that the pressure change is so rapid and smooth that isentropic conditions continue to hold. Equation (4) shows that B/A may be written as the sum of isothermal and isobaric components[4]; however, the isothermal component of B/A for most materials typically dominates.[5] Equation (5) suggests that materials whose stiffness changes greatly with changes in pressure, such as bubbly liquids,[6,7] will have a large value of the acoustic nonlinearity parameter. Other physical characteristics described by B/A include the geometric packing of molecules[8] and the form of the potential energy function that governs the forces between adjacent molecules in a material.[9] The connection between molecular structure and B/A has been exploited by several investigators[10,11] to determine the relative concentrations of "bound water" and "free water" in water–alcohol mixtures or the contribution of molecular subgroups to the overall nonlinearity of larger molecules in solution.[12] Other authors[13] have suggested a connection between the level of macroscopic structure in soft-organ tissue and its B/A value. Fluids in which B/A can be zero or negative are called "retrograde" fluids and have unusual thermodynamic properties.[14]

4 COEFFICIENT OF NONLINEARITY β

The acoustic nonlinearity parameter describes the steepening of acoustic waves as they propagate through a material.[15] A given point on a traveling acoustic waveform will propagate at the local sound speed dx/dt, given by

$$\frac{dx}{dt} = c_0 + \beta u, \quad (6)$$

where the so-called coefficient of nonlinearity $\beta \equiv 1 + B/2A$. Equation (6) shows that wave steepening has two distinct causes that are assumed to act independently: nonlinearities inherent in the material's properties, which are described by B/A, and those due to convection. The convective term arises from purely kinematic considerations[16] and would therefore exist even if no material nonlinearities were present, that is, for the case of $B/A = 0$. Equation (6) also shows that nonlinear effects are cumulative. The amount of distortion also depends upon the time or distance the wave travels.

5 SPECIFIC INCREMENT OF B/A

Because the increase in B/A value with solute concentration χ of solutions of biologic materials is often linear, the specific increment of B/A with concentration, $\Delta(B/A)/\chi$, can be used to determine the relative contribution of the solute to the total nonlinearity of the solution.[17] The specific increment is defined by differentiating Eq. (4):

$$\frac{\Delta(B/A)}{\chi} = 2\rho_0 c_0 \left\{ \frac{1}{\chi}\Delta\left(\frac{\partial c}{\partial P}\right)_{T_0} + ([c]+[\rho])\left(\frac{\partial c}{\partial P}\right)_{T_0} + \frac{\kappa_e T}{\rho_0 C_p}\left\{ \frac{1}{\chi}\Delta\left(\frac{\partial c}{\partial T}\right)_{P_0} + ([c]+[\kappa_e]-[C_p])\left(\frac{\partial c}{\partial T}\right)_{P_0} \right\}\right\}, \quad (7)$$

where $[c] = \Delta c/\chi c_0$, $[\rho] = \Delta\rho/\chi\rho_0$, $[\kappa_e] = \Delta\kappa_e/\chi\kappa_{e0}$, $[C_p] = C_{p0}/\chi C_p$, the subscript 0 denotes the solvent, and Δ means the difference between solution and solvent for the corresponding parameter.

6 HIGHER ORDER NONLINEARITY PARAMETERS

The higher order parameters C/A, $D/A, \ldots$ become important at large acoustic pressure amplitudes or in extremely nonlinear media.[18] From Eqs. (2) and (3) the third-order constant C/A may be written[19]

$$\frac{C}{A} = \frac{3}{2}\left(\frac{B}{A}\right)^2 + \frac{1}{k}\frac{\partial[B/A]}{\partial P}. \quad (8)$$

For most materials, the first term in Eq. (8) exceeds the second term by several orders of magnitude.[20,21]

7 MIXTURE LAWS FOR NONLINEARITY PARAMETERS

If the nonlinearity parameters of each of an n-component mixture of mutually immiscible materials are known, it is possible to derive[22] expressions for the corresponding nonlinear parameters of the mixture as a whole:

$$k^2\beta = \sum_{i=1}^{n} \frac{\rho Y_i k_i^2 \beta_i}{\rho_i} = \sum_{i=1}^{n} X_i k_i^2 \beta_i, \quad (9)$$

where Y_i and X_i are the mass and volume fractions of component i, respectively. In Eq. (9), k is the adiabatic compressibility and ρ the density of the mixture as a whole, given by

$$k = \sum_{i=1}^{n} k_i X_i \quad \text{and} \quad \rho = \sum_{i=1}^{n} \rho_i X_i,$$

respectively, where the subscripted quantities refer to the component properties. Similar expressions may be developed[23] for C/A, $D/A, \ldots$.

8 METHODS OF MEASURING NONLINEARITY PARAMETERS

Since the coefficients B and A are thermodynamic properties of the propagation medium, they can either be derived from the equation of state of the material or be measured empirically if the equation of state is not known. The B/A value for a gas obeying the perfect-gas law $P/P_0 = (\rho/\rho_0)^\gamma$ is $B/A = \gamma - 1$, and for a perfect gas, $C/A = (\gamma - 1)(\gamma - 2)$.

For materials for which no analytical equation of state exists, B/A must be measured.[24] The finite-amplitude method[25] is a technique in which the distortion of the wave as it propagates is measured via the growth of the second harmonic, and the B/A value is inferred from Eq. (6). Another technique is the thermodynamic method,[26] which relies upon changes in sound speed that accompany changes in ambient pressure and temperature via Eq. (4). The isentropic phase method makes use of Eq. (3): Sound speed is measured during a sufficiently rapid and smooth pressure change that the system is considered thermodynamically reversible.[27–31] This method has the advantage that a detailed knowledge of the thermodynamic properties of the propagation material and of the acoustic field of the source transducer is unnecessary. The precision with which B/A can be measured is about 10% for the finite-amplitude method and 5% for the thermodynamic method. Early isentropic phase methods were capable of measurement precisions of 4%, but methodological improvements[32,33] have increased these to within 1%. These more precise techniques have been necessitated by the use of the acoustic nonlinearity parameter in predictive models of tissue composition[34,35] and nonlinear acoustic propagation in biologic tissues. Other methods for measuring B/A include optical methods,[36] parametric arrays,[37,38] cavity resonance systems,[39–41] and methods involving the measurement of volumetric effects.[42]

Tables 1 and 2 show values of B/A for various materials reported in the literature.

TABLE 1 Published B/A Values of Nonbiologic Materials at 1 atm

Substance	T (°C)	B/A	Reference
Distilled water	0	4.2	43
	20	5.0	43
		4.985 ± 0.063	23
	25	5.11 ± 0.20	29
	26	5.1	31
	30	5.31	5
		5.18 ± 0.033	32
		5.280 ± 0.021	23
	40	5.4	43
	60	5.7	43
	80	6.1	43
	100 (liquid)	6.1	43
Sea water (3.5% NaCl)	20	5.25	43
Glycerol (4% in H_2O)	20	8.77	30
	25	8.58 ± 0.34	29
		8.84	30
	30	9.0	43
		9.4	5
		9.08	30
Methanol	20	9.42	19
	30	9.64	19
Ethanol	0	10.42	19
	20	10.52	19
	40	10.60	19
n-Propanol	0	10.47	19
	20	10.69	19
	40	10.73	19
n-Butanol	0	10.71	19
	20	10.69	19
	40	10.75	19
Benzyl alcohol	30	10.19	19
	50	9.97	19
Ethylene glycol	25	9.88 ± 0.40	29
	26	9.6	31
	30	9.7	43
		9.93	5
		9.88 ± 0.035	32
Acetone	20	9.23	19
	40	9.51	19
Benzene	20	9.0	43
		8.4	44
	25	6.5	3
	40	8.5	3
Chlorobenzene	30	9.33	19
Diethylamine	30	10.30	19
Ethyl formate	30	9.8	43
Heptane	30	10.0	43
	40	10.05	45
Hexane	25	9.81 ± 0.39	29
	30	9.9	43
	40	10.39	45
Methyl acetate	30	9.7	43
Cyclohexane	30	10.1	43
Nitrobenzene	30	9.9	43
1,2-DHCP	30	11.8	43

TABLE 1 (*Continued*)

Substance	T (°C)	B/A	Reference
Carbon bisulfide	10	6.4	3
	25	6.2	3
	40	6.1	3
Chloroform	25	8.2	3
Carbon tetrachloride	10	8.1	3
	25	8.7	3
		7.85 ± 0.31	29
	40	9.3	3
Toluene	20	5.6	3
	25	7.9	3
	30	8.929	46
Pentane	30	9.87	45
Octane	40	9.75	45
Aqueous *t*-butanol (60% by volume)	20	11.5	47
Bismuth	318	7.1	43
Indium	160	4.6	43
Mercury	30	7.8	43
Methyl iodide	30	8.2	43
Potassium	100	2.9	43
Sodium	110	2.7	43
Sulfur	121	9.5	43
Tin	240	4.4	43
Monatomic gas	20	0.67	43
Diatomic gas	20	0.40	43
Liquid argon	−183.16	5.67	48
Liquid nitrogen	−195.76	6.6	24
Liquid helium	−271.38	4.5	24
Saturated marine sediments	20	19	49
	20	11.78	50

TABLE 2 Published B/A Values of Biologic Materials

Substance	T (°C)	B/A	Reference
Isotonic saline	20	5.540 ± 0.032	23
	30	5.559 ± 0.018	23
Bovine serum albumin			
20 g/100 ml H_2O	25	6.23 ± 0.25	29
38.8 g/100 ml H_2O	30	6.68	5
Dextrose (25%)	30	5.96	5
		6.11 ± 0.4	32
Dextrose (30%)	26	5.9	31
Dextran T150 (24%)	30	6.05	5
Dextran T2000 (26%)	30	6.03	5
Sucrose (30%)	25	5.50	31
D-Glucose anhydrous	25	5.85	51
Bovine liver	23	7.5–8.0	5
	30	7.23–8.9	5
	30	6.88	52
Bovine brain	30	7.6	5
Bovine heart	30	6.8–7.4	5
Bovine milk	26	5.1	53
Bovine whole blood	26	5.5	53

TABLE 2 *(Continued)*

Substance	T (°C)	B/A	Reference
Chicken fat	30	11.270 ± 0.090	23
Porcine liver	26	6.9	31
Porcine heart	26	6.8	31
Porcine kidney	26	6.3	31
Porcine spleen	26	6.3	31
Porcine brain	26	6.7–7.0	51
Porcine muscle	26	6.5–6.6	51
	30	7.5–8.1	5
Porcine tongue	26	6.8	31
Porcine fat	26	9.5–10.9	51
	30	10.9–11.3	5
Porcine whole blood	26	5.8	53
Human liver	30	6.54	30
Human breast fat	22	9.206	30
	30	9.909	30
	37	9.633	30
Human multiple myeloma	22	5.603	30
	30	5.796	30
	37	6.178	30
Corn oil	20	10.666 ± 0.074	23
	30	10.574 ± 0.026	23
Castor oil	20	11.270 ± 0.044	23
	30	11.006 ± 0.051	23
Olive oil	20	11.136 ± 0.042	23
	30	11.066 ± 0.641	23
Peanut oil	20	10.911 ± 0.065	23
	30	10.680 ± 0.038	23
Safflower oil	20	11.610 ± 0.102	23
	30	11.161 ± 0.083	23
Cod liver oil	20	10.958 ± 0.022	23
	30	10.867 ± 0.029	23
Tung oil	20	11.278 ± 0.031	23
	30	11.064 ± 0.041	23
Lamp oil	20	11.156 ± 0.053	23
	30	10.918 ± 0.104	23
Mineral oil	20	11.331 ± 0.020	23
	30	11.497 ± 0.038	23
Pump oil	20	11.791 ± 0.026	23
	30	11.451 ± 0.018	23
Silicone oil	20	11.381 ± 0.052	23
	30	11.461 ± 0.017	23

REFERENCES

1. R. T. Beyer and S. V. Letcher, *Physical Ultrasonics*, Pure & Appl. Physics Ser., Academic, New York, 1969.
2. L. Bjørnø, "Nonlinear Acoustics," in R. W. B. Stephens and H. G. Leventhall (Eds.), *Acoustics and Vibration Progress*, Vol. 2, Chapman & Hall, London, 1976.
3. R. T. Beyer, "Parameter of Nonlinearity in Fluids," *J. Acoust. Soc. Am.*, Vol. 32, 1960, pp. 719–721.
4. I. Rudnick, "On the Attenuation of Finite Amplitude Waves in a Liquid," *J. Acoust. Soc. Am.*, Vol. 30, 1958, pp. 564–567.
5. W. K. Law, L. A. Frizzell, and F. Dunn, "Determination of the Nonlinearity Parameter B/A of Biological Media," *Ultrasound Med. Biol.*, Vol. 11, No. 2, 1985, pp. 307–318.
6. Y. A. Kobelev and L. A. Ostrovsky, "Nonlinear Acoustic Phenomena Due to Bubble Drift in a Gas–Liquid Mixture," *J. Acoust. Soc. Am.*, Vol. 85, No. 2, 1989, pp. 621–629.

7. J. Wu and Z. Zhu, "Measurements of the Effective Nonlinearity Parameter B/A of Water Containing Trapped Cylindrical Bubbles," *J. Acoust. Soc. Am.*, Vol. 89, No. 6, 1991, pp. 2634–2639.
8. H. Endo, "Prediction of the Nonlinearity Parameter of a Liquid from the Percus-Yevick Equation," *J. Acoust. Soc. Am.*, Vol. 83, No. 6, 1988, pp. 2043–2046.
9. B. Hartmann, "Potential Energy Effects on the Sound Speed in Liquids," *J. Acoust. Soc. Am.*, Vol. 65, No. 6, 1979, pp. 1392–1396.
10. K. Yoshizumi, T. Sato, and N. Ichida, "A Physiochemical Evaluation of the Nonlinear Parameter B/A for Media Predominantly Composed of Water," *J. Acoust. Soc. Am.*, Vol. 82, No. 1, 1987, pp. 302–305.
11. C. M. Sehgal, B. Porter, and J. F. Greenleaf, "Relationship between Acoustic Nonlinearity and the Bound and Unbound States of Water," *IEEE Ultrason. Symp.*, Vol. 2, 1985, pp. 883–886.
12. A. P. Sarvazyan, D. P. Kharakoz, and P. Hemmes, "Ultrasonic Investigation of the pH-Dependent Solute–Solvent Interactions in Aqueous Solutions of Amino Acids and Proteins," *J. Phys. Chem.*, Vol. 83, No. 13, 1979, pp. 1796–1799.
13. J. Zhang, M. S. Kuhlenschmidt, and F. Dunn, "Influences of Structural Factors of Biological Media on the Acoustic Nonlinearity Parameter B/A," *J. Acoust. Soc. Am.*, Vol. 89, No. 1, 1991, pp. 80–91.
14. M. S. Cramer, and R. Sen, "Shock Formation in Fluids Having Embedded Regions of Negative Nonlinearity," *Phys. Fluids*, Vol. 29, No. 7, 1986, pp. 2181–2191.
15. D. T. Blackstock, "Nonlinear Acoustics (Theoretical)," *AIP Handbook of Physics*, McGraw-Hill, New York, 1972.
16. M. F. Hamilton and D. T. Blackstock, "On the Coefficient of Nonlinearity β in Nonlinear Acoustics," *J. Acoust. Soc. Am.*, Vol. 83, No. 1, 1988, pp. 74–77.
17. A. P. Sarvazyan, T. V. Chalikian, and F. Dunn, "Acoustic Nonlinearity Parameter B/A of Aqueous Solutions of Some Amino Acids and Proteins," *J. Acoust. Soc. Am.*, Vol. 88, No. 3, 1990, pp. 1555–1561.
18. Y. A. Basin and V. M. Kryachko, "Experimental Study of the Propagation of Compression and Expansion Pulses in a Sound Beam in a Highly Nonlinear Medium," *Sov. Phys. Acoust.*, Vol. 31, No. 4, 1985, pp. 255–257.
19. A. B. Coppens, R. T. Beyer, M. B. Seiden, J. Donohue, F. Guepin, R. H. Hodson, and C. Townsend, "Parameter of Nonlinearity in Fluids. II," *J. Acoust. Soc. Am.*, Vol. 38, 1965, pp. 797–804. In Equation A4 there is an omission of the squared factor of $(\partial c/\partial p)$.
20. L. Bjørnø and K. Black, "Higher-order Acoustic Nonlinearity Parameters of Fluids," in U. Nigul and J. Engelbrecht (Eds.), *Nonlinear Deformation Waves*, Springer, Berlin, 1983, pp. 355–361.
21. K. P. Thakur, "Non-linearity Acoustic Parameter in Higher Alkanes," *Acustica*, Vol. 39, 1978, pp. 270–272.
22. E. C. Everbach, Z. Zhu, P. Jiang, B. T. Chu, and R. E. Apfel, "A Corrected Mixture Law for B/A," *J. Acoust. Soc. Am.*, Vol. 89, No. 1, 1991, pp. 446–447.
23. E. C. Everbach, Tissue Composition Determination via Measurement of the Acoustic Nonlinearity Parameter B/A, Ph.D. Dissertation, Yale University, New Haven, CT (1989), p. 66.
24. H. A. Kashkooli, P. J. Dolan, Jr., and C. W. Smith, "Measurement of the Acoustic Nonlinearity Parameter in Water, Methanol, Liquid Nitrogen, and Liquid Helium-II by Two Different Methods: A Comparison," *J. Acoust. Soc. Am.*, Vol. 82, No. 6, 1987, pp. 2086–2089.
25. See, for example, W. N. Cobb, "Measurement of the Acoustic Nonlinearity Parameter of Biological Media, Ph.D. Dissertation, Yale University, New Haven, CT (1982).
26. See, for example, W. K. Law, Measurement of the Nonlinearity Parameter B/A in Biological Materials Using the Finite Amplitude and Thermodynamic Method, Ph.D. Dissertation, University of Illinois at Urbana-Champaign (1984).
27. C. Kammoun, J. Emery, and P. Alias, Determination of the Acoustic Parameter of Nonlinearity in Liquids at Very Low Pressure, in the Seventh Intern. Symp. on Nonlinear Acoust., Virginia Polytechnics Inst. and State Univ., 1976, pp. 146–149.
28. J. Emery, S. Gasse, C. Dugué, "Coefficient de nonlinearité acoustique dans les melanges eau-methanol et eau-ethanol," *J. Phys.*, Vol. 11, No. 40, 1979, pp. 231–234.
29. Z. Zhu, M. S. Roos, W. N. Cobb, and K. Jensen, "Determination of the Acoustic nonlinearity Parameter B/A from Phase Measurements," *J. Acoust. Soc. Am.*, Vol. 74, No. 5, 1983, pp. 1518–1521.
30. C. M. Sehgal, R. C. Bahn, and J. F. Greenleaf, "Measurement of the Acoustic Nonlinearity Parameter B/A in Human Tissues by a Thermodynamic Method," *J. Acoust. Soc. Am.*, Vol. 76, No. 4, 1984, pp. 1023–1029.
31. X. Gong, Z. Zhu, T. Shi, and J. Huang, "Determination of the Acoustic Nonlinearity Parameter in Biological Media Using FAIS and ITD Methods," *J. Acoust. Soc. Am.*, Vol. 86, No. 1, 1989, pp. 1–5.
32. J. Zhang and F. Dunn, "A Small Volume Thermodynamic System for B/A Measurement," *J. Acoust. Soc. Am.*, Vol. 89, No. 1, 1991, pp. 73–79.
33. E. C. Everbach and R. E. Apfel, "An Interferometric Technique for B/A Measurement," *J. Acoust. Soc. Am.*, Vol. 98, No. 6, 1995, pp. 3428–3438.
34. R. E. Apfel, "Prediction of Tissue Composition from Ultrasonic Measurements and Mixture Rules," *J. Acoust. Soc. Am.*, Vol. 79, No. 1, 1986, pp. 148–152.
35. C. M. Sehgal, G. M. Brown, R. C. Bahn, and J. F. Greenleaf, "Measurement and Use of Acoustic Nonlinearity and Sound Speed to Estimate Composition of Excised Livers," *Ultrasound in Med. & Biol.*, Vol. 12, No. 11, 1986, pp. 865–874.
36. L. Adler and E. A. Hiedmann, "Determination of the

Nonlinearity Parameter B/A for Water and *m*-xylene," *J. Acoust. Soc. Am.*, Vol. 34, No. 4, 1962, pp. 410–412.

37. P. J. Westervelt, "Parametric Acoustic Array," *J. Acoust. Soc. Am.*, Vol. 35, 1963, pp. 535–537.
38. Y. Nakagawa, M. Nakagawa, M. Yoneyama, and M. Kikuchi, "Nonlinear Parameter Imaging Computed Tomography by Parametric Array," *Proc. IEEE Ultrasonics Symp.*, IEEE, New York, 1985, pp. 673–676.
39. D. T. Blackstock, "Finite-Amplitude Motion of a Piston in a Shallow, Fluid-filled Cavity," *J. Acoust. Soc. Am.*, Vol. 34, No. 6, 1962, pp. 792–802.
40. F. Eggers and T. Funck, "Ultrasonic Measurements with Milliliter Liquid Samples in the 0.5–100 MHz Range," *Rev. Sci. Instrum.*, Vol. 44, 1973, pp. 969–978.
41. A. P. Sarvazyan, "Development of Methods of Precise Ultrasonic Measurements in Small Volumes of Liquids," *Ultrasonics*, Vol. 20, 1982, pp. 151–154.
42. H. M. Merklinger, High Intensity Effects in the Nonlinear Acoustic Parametric End-Fire Array, Ph.D. Thesis, Appendix A, University of Birmingham, Birmingham, England (1971).
43. R. T. Beyer, *Nonlinear Acoustics*, Naval Ships Systems Command, Washington, D.C., 1974, Table 3–1.
44. O. Nomoto, "Nonlinear Parameter of the 'Rao Liquid'," *J. Phys. Soc. Jpn.*, Vol. 21, 1966, pp. 569–571.
45. K. L. Narayana and K. M. Swamy, "Acoustic Nonlinear Parameter (B/A) in *n*-pentane," *Acustica*, Vol. 49, 1981, pp. 336–339.
46. S. K. Kor and U. S. Tandon, "Scattering of Sound by Sound From Beyer's (B/A) Parameters," *Acustica*, Vol. 28, 1973, pp. 129–130.
47. C. M. Sehgal, B. R. Porter, and J. F. Greenleaf, "Ultrasonic Nonlinear Parameters and Sound Speed of Alcohol–Water Mixtures," *J. Acoust. Soc. Am.*, Vol. 79, No. 2, 1986, pp. 566–570.
48. K. M. Swamy, K. L. Narayana, and P. S. Swamy, "A Study of (B/A) in Liquefied Gases as a Function of Temperature and Pressure from Ultrasonic Velocity Measurements," *Acustica*, Vol. 28, 1973, pp. 129–130.
49. J. M. Hovem, "The Nonlinearity Parameter of Saturated Marine Sediments," *J. Acoust. Soc. Am.*, Vol. 66, No. 5, 1979, pp. 1463–1467.
50. L. Bjørnø, "Finite-Amplitude Wave Propagation through Water-saturated Marine Sediments," Acustica, Vol. 38, No. 4, 1977, pp. 196–200.
51. Z. Zhu, X. Gong, and J. Huang, Measurement of the Acoustic Nonlinearity Parameter B/A in Biological Media by Improved Thermodynamic Method, Proc. China-Japan Joint Conf. on Ultrasonics, May 11–14, Nanjing, 1987.
52. W. K. Law, L. A. Frizzell, and F. Dunn, "Comparison of Thermodynamic and Finite Amplitude Methods of B/A Measurement in Biological Materials," *J. Acoust. Soc. Am.*, Vol. 74, No. 4, 1983, pp. 1295–1297.
53. X. Gong, R. Feng, C. Zhu, and T. Shi, "Ultrasonic Investigation of the Nonlinearity Parameter B/A in Biological Media," *J. Acoust. Soc. Am.*, Vol. 76, No. 3, 1984, pp. 949–950.

19

FINITE-AMPLITUDE WAVES IN SOLIDS

MACK A. BREAZEALE

1 INTRODUCTION

The theory describing propagation of a finite-amplitude sound wave in air has been given.[1] An experimental proof of the validity of the approach also has been given.[2] The application of the theory to liquids has been made,[3] and the nonlinear equation for wave propagation in fluids has the same form as that for an ideal gas. Different thermodynamic quantities appear in the nonlinear equations, however. The same basic approach as used for fluids[4] has been used for solids.[5] The propagation of ultrasonic waves of finite amplitude in a crystal of cubic symmetry can be described by a nonlinear differential equation similar to that used for fluids. An approximation makes the equations identical. Using this approximation, one can show how nonlinear distortion of an initially sinusoidal ultrasonic wave takes place and how nonlinear distortion can be measured with the second-harmonic generation (SHG) technique. Subsequently it has been realized that the nonlinearity parameter is the controlling factor in nonlinear distortion of sound in air, fluids, and solids. We therefore describe the SHG technique and show how it is used to determine the nonlinearity parameter of crystalline solids.[6] Description of the evaluation of nonlinearity parameters of liquids and gases will be left to other authors, as will the evaluation of other third-order elastic constants by stress–velocity techniques.

2 THEORY OF SOLIDS

The propagation of a finite-amplitude ultrasonic wave in the principal directions [100], [110], or [111] in a cubic solid or in any direction in an isotropic solid is described by an equation of the form

$$\rho_0 \frac{\partial^2 \xi}{\partial t^2} = K_2 \frac{\partial^2 \xi}{\partial a^2} + (3K_2 + K_3) \frac{\partial^2 \xi}{\partial a^2} \frac{\partial \xi}{\partial a}, \qquad (1)$$

where ξ is the particle displacement and a is the distance measured in the direction of propagation. The coefficients K_2 and K_3 are to be interpreted as listed in Table 1, depending on the direction of propagation in the solid. For any direction in an isotropic solid one can use the K_2 and K_3 listed for the [100] direction.

3 THEORY OF FLUIDS

In an isentropic fluid the equation of state can be written in the form

$$P - P_0 = A \left[\frac{\rho - \rho_0}{\rho_0} \right] + \frac{B}{2!} \left[\frac{\rho - \rho_0^2}{\rho_0} \right]^2 + \cdots, \qquad (2)$$

where

$$A = \rho_0 \left[\frac{\partial P}{\partial \rho} \right]_{P=P_0} = \rho_0 c^2 \qquad (3)$$

and

$$B = \rho_0^2 \left[\frac{\partial^2 P}{\partial \rho^2} \right]_{P=P_0}, \qquad (4)$$

Note: References to chapters appearing only in the *Encyclopedia* are preceded by *Enc.*

TABLE 1 K_2 and K_3 for [100], [110], and [111] Directions in Cubic Lattice

Direction	K_2	K_3
[100]	C_{11}	C_{111}
[110]	$\frac{1}{2}(C_{11} + C_{12} + 2C_{44})$	$\frac{1}{4}(C_{111} + 3C_{112} + 12C_{166})$
[111]	$\frac{1}{3}(C_{11} + 2C_{12} + 4C_{44})$	$\frac{1}{9}(C_{111} + 6C_{112} + 12C_{144} + 24C_{166} + 2C_{123} + 16C_{456})$

Note: The coefficients C_{ij} are the ordinary elastic constants; the C_{ijk} are the third-order elastic constants.

where P is the pressure, ρ is the density, and c is the sound velocity. The subscript refers to reference values of the quantities. The quantity B/A has become an important measure of the nonlinearity of the fluid. For gases one can use an ideal-gas equation of state in the form

$$P = P_0\left(\frac{\rho}{\rho_0}\right)^{\gamma}, \tag{5}$$

where γ is the ratio of specific heats:

$$\gamma = \frac{C_p}{C_v}. \tag{6}$$

Substituting Eq. (5) into the general form Eq. (2) leads to the fact that, for an ideal gas,

$$\frac{B}{A} = \gamma - 1. \tag{7}$$

The nonlinear equation describing the propagation of a finite-amplitude wave in a fluid can be written in the form[5]

$$\frac{\partial^2 \xi}{\partial t^2} = c_0^2 \frac{\partial^2 \xi/\partial a^2}{(1 + \partial\xi/\partial a)^{B/A+2}}, \tag{8}$$

where c_0 is the velocity of sound measured with small amplitudes. The nonlinear equation for describing the propagation of a finite-amplitude wave in an ideal gas can be written in a similar form:

$$\frac{\partial^2 \xi}{\partial t^2} = c_0^2 \frac{\partial^2 \xi/\partial a^2}{(1 + \partial\xi/\partial a)^{\gamma+1}}, \tag{9}$$

where one recognizes the fact that $\gamma + 1$ plays the same role in an ideal gas as $B/A + 2$ does for any fluid. By expanding the denominator on the right of these equations in a binomial series, one finds that either Eq. (8) or Eq. (9) can be written in the same form as Eq. (1). These results can be summarized as in Table 2. Proper use of Table 2 allows one to make the appropriate substitutions such that Eq. (1) can be used to describe the propagation of finite-amplitude waves in solids, liquids, or gases.

4 SOLUTION OF THE NONLINEAR EQUATION

The solution of the nonlinear equation (1) appropriate to measurement is made by assuming that the wave is initially sinusoidal: At $a = 0$, $\xi = A_1 \sin(-\omega t)$. With this assumption, one can obtain a perturbation solution in the form

$$\xi = A_1 \sin(ka - \omega t) + \beta \frac{A_1^2 k^2 a}{4} \cos 2(ka - \omega t) + \cdots, \tag{10}$$

TABLE 2 Parameters Used in Describing Waves of Finite Amplitude in Gases, Liquids, and Solids

Wave	Velocity Squared, c_0^2	Nonlinearity Parameter, 2β	Discontinuity Distance, L
In ideal gas	$\frac{\gamma P_0}{\rho_0}$	$\gamma + 1$	$\frac{2c_0^2}{(\gamma + 1)\omega^2 A_1}$
In liquid	$\frac{A}{\rho_0}$	$\frac{B}{A} + 2$	$\frac{2c_0^2}{(B/A + 2)\omega^2 A_1}$
In solid (cubic or isotropic)	$\frac{K_2}{\rho_0}$	$-\left(\frac{K_3}{K_2} + 3\right)$	$\frac{2c_0^2}{-(K_3/K_2 + 3)\omega^2 A_1}$

where β is the nonlinearity parameter used by others in this book. The negative ratio of the coefficient of the nonlinear term to the linear term in Eq. (1),

$$2\beta = -\left(3 + \frac{K_3}{K_2}\right) \tag{11}$$

is the quantity to be determined from measurement. From 2β one can evaluate K_3, since $K_2 = \rho c_0^2$ is known. From values of K_3 in the three principal directions one can use Table 1 to determine combinations of third-order elastic constants from measured results.

Equation (10) shows that an initially sinusoidal ultrasonic wave generates a second harmonic as it propagates. To measure the nonlinearity parameter 2β, one needs to measure the absolute value of the fundamental displacement amplitude A_1 and that of the second harmonic,

$$A_2 = \beta \frac{A_1^2 k^2 a}{4}. \tag{12}$$

Since the propagation constant

$$k = \frac{2\pi}{\lambda} = \frac{\omega}{c_0} \tag{13}$$

is known and the sample length a can be measured, one can write

$$2\beta = \frac{8A_2 c_0^2}{A_1^2 \omega^2 a} \tag{14}$$

and determine 2β directly. It is common to plot A_2 as a function of A_1^2 and to evaluate the slope to determine 2β.

5 EXPERIMENTAL CONSIDERATIONS

5.1 Room Temperature Measurements

To determine the nonlinearity parameters, one needs only to measure the sample length a, the angular frequency ω from which one can calculate $k = \omega/c_0$, and the slope A_2/A_1^2. The problem is in measurement of A_2/A_1^2. This requires absolute measurement of amplitudes. At a fundamental frequency of 30 MHz the amplitude A_1 is of the order of 1 Å, and A_2 may be 1% of A_1. A very sensitive device is required.

Figure 1 is a block diagram of a system for measuring A_1 and A_2. A radio frequency (RF) tone burst of approximately 30 MHz is applied to an X-cut quartz transducer bonded to one end of the sample. The ultrasonic wave that propagates through the sample is detected at the other end by a capacitive receiver. The sample assembly is shown in Fig. 2. The capacitive receiver is a 1.016-cm-

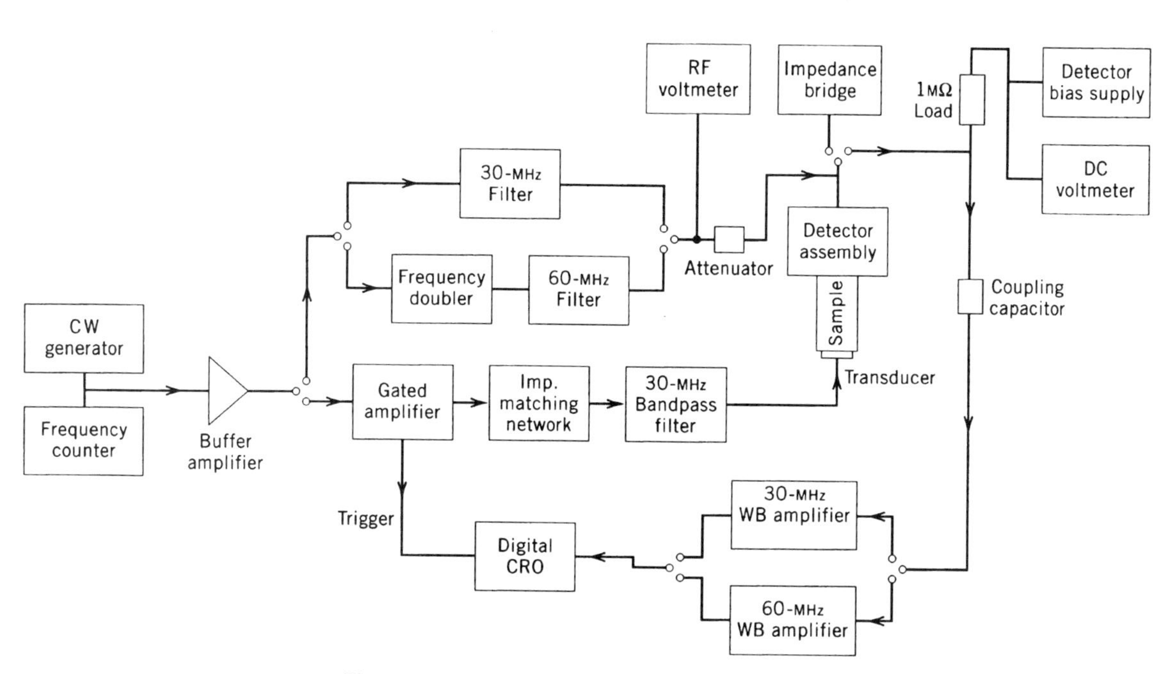

Fig. 1 Block diagram of room temperature apparatus.

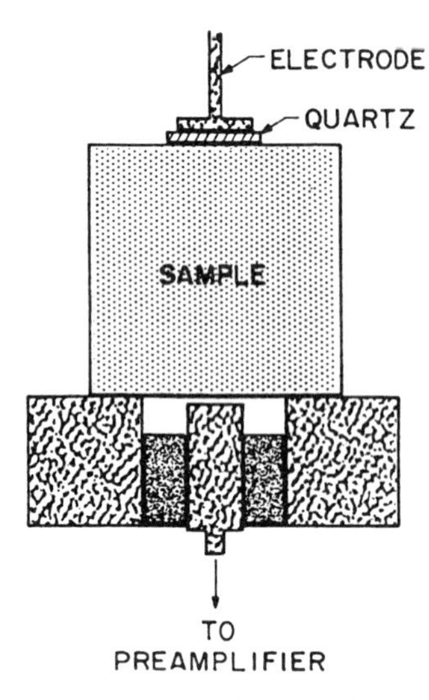

Fig. 2 Sample assembly.

diameter electrode placed approximately 5 μm from the conducting sample (both surfaces being optically flat), and a direct current (DC) bias on the order of 150 V is applied across the gap through a large resistor (approximately 1 MΩ). A substitutional signal giving the same output as the ultrasonic signal is introduced across the capacitive receiver in such a way that the current in the substitutional signal can be measured. This current i can be related to the amplitude A_1 of the ultrasonic wave by

$$i = 2A_1 V_b \omega \frac{\chi}{s},$$

where V_b is the DC bias on the receiver, ω is the angular frequency, χ is the capacitance of the receiver, and s is the spacing of the capacitive receiver plates.

The signal from the capacitive receiver is taken to either a 30- or a 60-MHz wide-bandpass amplifier. This amplified signal is detected and taken to the digital oscilloscope. The digital oscilloscope is used to select a portion of the first received echo and measure its amplitude.

After the fundamental and second-harmonic signals have been measured by the digital oscilloscope, a continuous-wave substitutional signal is introduced at the capacitive receiver. The 60-MHz signal is derived by doubling the 30-MHz signal with a ring bridge mixer. Both the 30- and the 60-MHz substitutional signals are filtered to ensure spectral purity. These two signals are adjusted to give the same output at the digital oscilloscope as the ultrasonic signals and are measured with an RF voltmeter. From these measurements and a knowledge of the circuit impedances, the current of the substitutional signal can be calculated that exactly matches the ultrasonic signal to be measured.

5.2 Samples

For precise measurements it is necessary to lap samples optically flat and parallel. Optical tolerance is usually one fringe over the sample surface and parallelism of the surfaces should be within 12 s of arc. Nonconductive samples must be coated with copper, silver, or gold to a thickness of approximately 1000 Å for electrical conductivity. Optimum sample size is 2.5 cm diameter and 2.5 cm long, the cylinder axis being one of the principal directions [100], [110], or [111]. The smallest practical sample size is a cube 0.5 cm on a side. The size of the ultrasonic transducer and the capacitive receiver are determined by the sample size. For smaller transducers it is necessary to correct the measurements for the effect of diffraction.[7]

5.3 Second-Order Elastic Constants

The second-order elastic constants are determined from measurement of velocity. Many sets of data, both at room temperature and lower temperatures, are readily available for comparison. Velocity measurements can be measured to an accuracy of 10^{-4}–10^{-2}, depending upon the sample. Relative measurements can be made to an accuracy of 10^{-6} with a stable frequency source. Such velocity measurements are described in Chapter 41.

5.4 Cryogenic Apparatus

The electronic apparatus for measurements at low temperatures is essentially the same as that for room temperature, with one exception. It is not necessary to calibrate at each temperature since relative measurements are adequate. A reference signal, then, is not necessary.

The sample holder is encased in a stainless steel can, as shown in Fig. 3. The can shown is surrounded by a second can, and the space between them is evacuated to provide an insulating jacket around the inner can shown. Radiation loss is reduced by polishing the cans to a high sheen. The cans are supported by three cupro-nickel tubes. Two of the tubes have a smaller cupro-nickel tube centered inside to provide a 50-Ω coaxial transmission line. The tubes simultaneously can be used as vacuum lines. The third tube houses leads for the temperature sensor and the heater and also is used to evacuate the inner can to a known pressure.

The entire apparatus can be suspended inside a standard helium research dewar. By sealing the dewar and pumping on the coolant, temperatures below the ambient pressure boiling point of the coolant (either liquid nitrogen or liquid helium) may be obtained.

The temperature in the inner can is controlled by an electric resistance heater connected to a commercial temperature controller. It can be varied continuously from

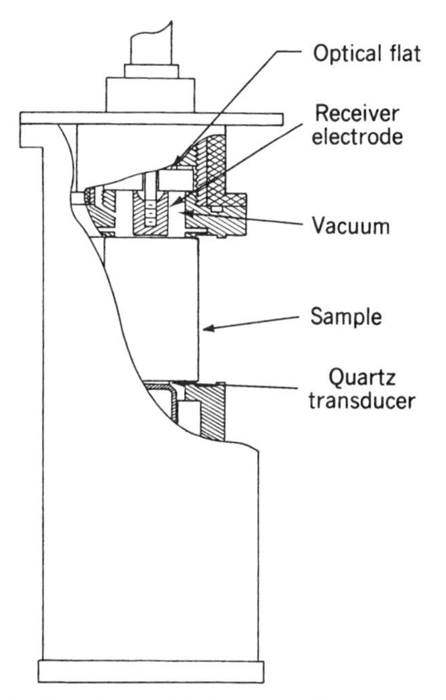

Fig. 3 Sample assembly for cryogenic measurements.

approximately 3 K to a few degrees above room temperature.

5.5 Cryogenic Nonlinearity Measurements

Only relative measurements need to be taken with the cryogenic apparatus. At each temperature one first adjusts the drive signal to the quartz transducer so that the fundamental ultrasonic wave received at the capacitive receiver has the same known value. The second harmonic then is measured by a slide-back technique: the DC bias voltage on the capacitive receiver is adjusted so that the electrical signal coming from the capacitive receiver is the same at the new temperature as it was at the previous temperature. The second-harmonic amplitude A_2 of the ultrasonic wave at the two different absolute temperatures (T_1 and T_2) are then related by

$$A_2(T_2) = \frac{A_2(T_1)V_b(T_1)}{V_b(T_2)}.$$

Use of this equation allows one to use experimental data to plot the temperature dependence of the second harmonic and ultimately to plot the nonlinearity parameter as a function of temperature. If desirable, one can interpret data on different crystalline orientations in terms of combinations of third-order elastic constants.

6 EXPERIMENTAL RESULTS

Many experimental results are available in the literature. The results for cubic crystals are summarized by Breazeale and Philip[6]; however, it may be useful to make a comparison of the magnitudes of the nonlinearity parameters for different substances. A few common gases are listed in Table 3. The nonlinearity parameters,

TABLE 3 Values of γ and 2β for Gases (15°C)

Gas	γ	$2\beta = \gamma + 1$
He	1.66	2.66
Ar	1.67	2.67
O_2	1.40	2.40
N_2	1.40	2.40
CO_2	1.30	2.30
H_2O(200°C)	1.31	2.31
CH_4	1.31	2.31

TABLE 4 Values of B/A and 2β for Fluids at Atmospheric Pressure

Liquid	Temperature (°C)	B/A	$2\beta = B/A + 2$
Water, distilled	0	4.16	6.16
	20	4.96	6.96
	40	5.38	7.38
	60	5.67	7.67
	80	5.96	7.96
	100	6.11	8.11
Acetone	30	9.44	11.44
Benzene	30	9.03	11.03
Benzyl alcohol	30	10.19	12.19
CCl_4	30	11.54	13.54

TABLE 5 Comparison of Structure, Bonding, and Acoustic Nonlinearity Parameters along the [100] Direction of Cubic Crystals

Structure	Bonding Type	2β Range
Zincblende	Covalent	1.8–3.2
Flourite	Ionic	3.4–4.6
FCC	Metallic	4.0–7.0
FCC (inert gas)	Van der Waals	5.8–7.0
BCC	Metallic	5.0–8.8
NaCl	Ionic	13.5–15.4

Abbreviations: FCC, face-centered-cubic; BCC, body-centered-cubic.

obtained simply by evaluating $\beta = \gamma + 1$, cover the range between 2.3 and 2.7. For comparison, Table 4 lists the nonlinearity parameters for common liquids. In liquids the nonlinearity parameters range between 6 and 14. In Table 5 are listed nonlinearity parameters for the [100] direction in cubic crystals having different structure and bonding. The solids listed cover the range of β between 2 and 15, a range that includes that of both gases and liquids. The relatively wide range of nonlinearity parameters of solids is one reason for interest in measuring this important physical property.

6.1 Evaluation of the Nonlinearity Parameters of Solids

Evaluation of specific nonlinearity parameters involves measurement of the amplitudes of the fundamental and the second harmonic at room temperature, then relative measurements at different temperatures. Recent data[8] on NaCl will serve as an example. The received second harmonic at room temperature as a function of the square of the fundamental is plotted in Fig. 4. The different

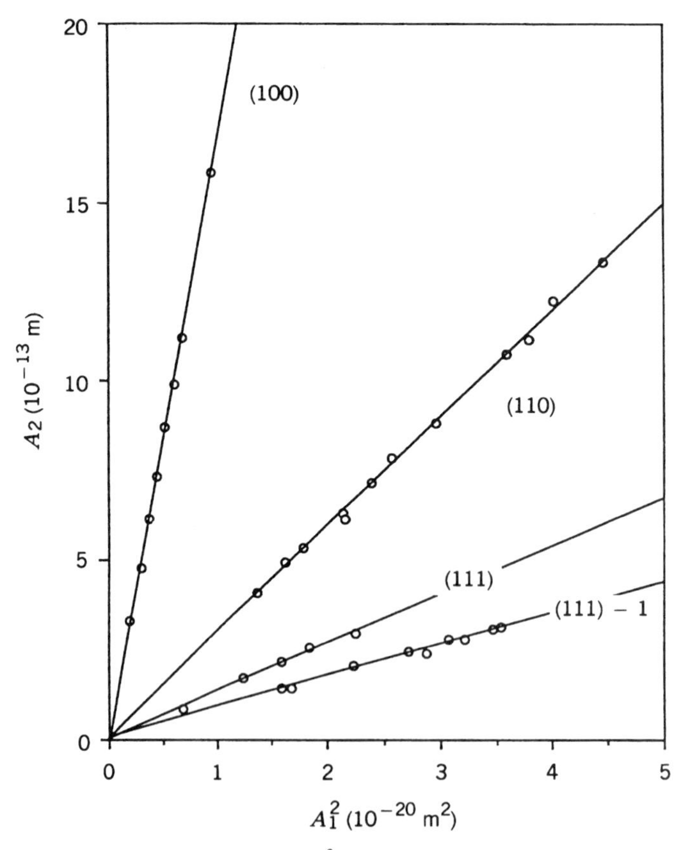

Fig. 4 Room temperature values of A_2 vs. A_1^2 for NaCl. (Two different samples gave different curves for the [111] direction.)

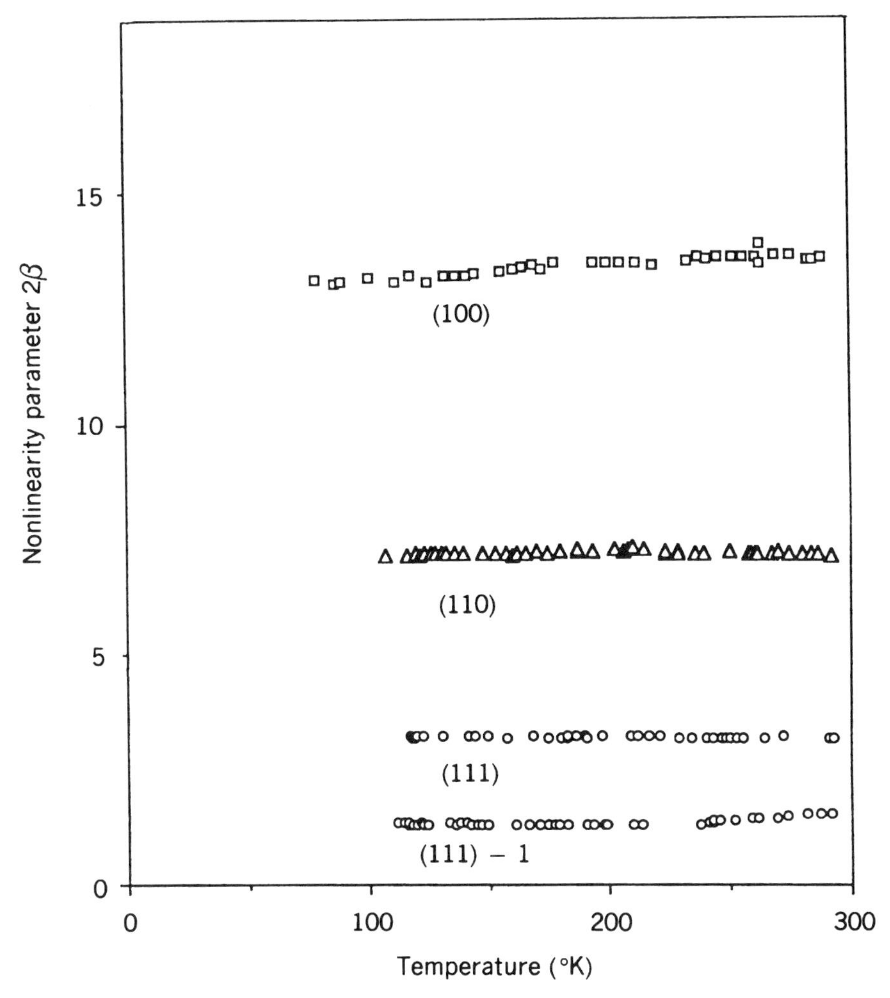

Fig. 5 Temperature dependence of the nonlinearity parameters of NaCl for different propagation directions.

slopes show that the nonlinearity parameter in the [100] direction is greater than that in other directions. Furthermore, in this case the nonlinearity parameter in the [111] direction was different for the two samples measured. By evaluating the slopes in Fig. 4, one can evaluate the nonlinearity parameters at room temperature; then relative measurement of the second harmonic at different temperatures allows one to evaluate the nonlinearity parameter as a function of temperature. A plot of the results for NaCl as a function of temperature is given in Fig. 5. Although the nonlinearity parameters typically are not strong functions of temperature, it is seldom that the linear behavior shown in Fig. 5 is observed for other cubic solids. A collection of data on a few other cubic and isotropic solids is shown in Figs. 6*a–c*. Liquid helium was used for the low-temperature measurements to show that there is not a wide variation in nonlinearity parameters with temperature. The range of nonlinearity parameters typically is between 0 and 16 for solids, although to date there is no specific reason to assume that nonlinearity parameters outside this range will not be observed.

Further data analysis is possible. In particular, third-order elastic constants calculated from these data and interpreted in terms of a model of the solid state is possible; however, such analysis is beyond the scope of the present work.

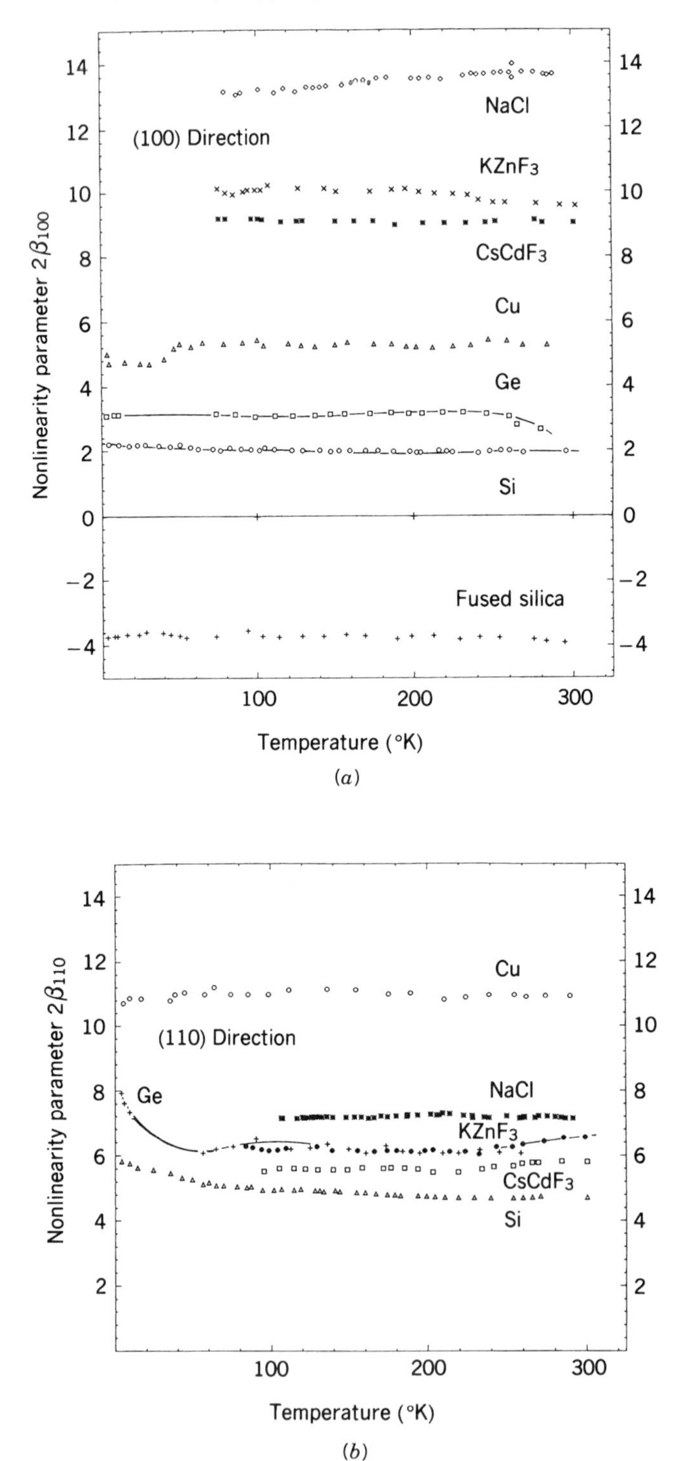

Fig. 6 Temperature dependence of the nonlinearity parameters of different cubic crystals for (*a*) [100]; (*b*) [110]; and (*c*) [111] directions.

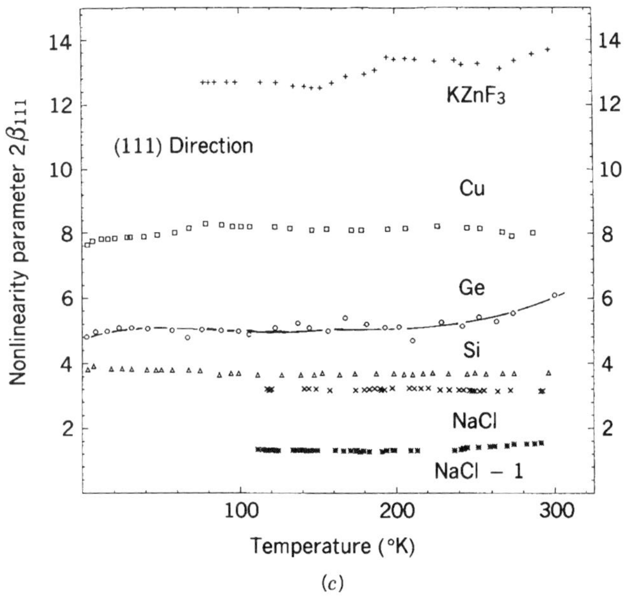

(*c*)

Fig. 6 (*Continued*)

REFERENCES

1. S. Earnshaw, "On the Mathematical Theory of Sound," *Phil. Trans. Roy. Soc. (Lond.)*, Vol. 150, 1860, p. 133.
2. A. L. Thuras, R. T. Jenkins, and H. T. O'Neil, "Extraneous Frequencies Generated in Air Carrying Intense Sound Wave," *J. Acoust. Soc. Am.*, Vol. 6, 1935, p. 173.
3. W. Keck and R. T. Beyer, *Phys. Fluids*, Vol. 3, 1960, p. 346.
4. R. T. Beyer, "Nonlinear Acoustics," in W. P. Mason (Ed.), *Physical Acoustics*, Vol. IIB, Academic, New York, 1965, pp. 231–264.
5. M. A. Breazeale and J. Ford, "Ultrasonic Studies of the Nonlinear Behavior of Solids," *J. Appl. Phys.*, Vol. 36, 1965, pp. 3486–3490.
6. M. A. Breazeale and J. Philip, "Determination of Third-Order Elastic Constants from Ultrasonic Harmonic Generation Measurements," in W. P. Mason and R. N. Thurston (Eds.), *Physical Acoustics*, Vol. XVII, Academic, New York, 1984, pp. 1–60.
7. B. D. Blackburn and M. A. Breazeale, "Nonlinear Distortion of Ultrasonic Waves in Small Crystalline Samples," *J. Acoust. Soc. Am.*, Vol. 76, 1984, pp. 1755–1760.
8. W. Jiang and M. A. Breazeale, "Temperature Variation of Elastic Nonlinearity of NaCl," *J. Appl. Phys.*, Vol. 68, 1990, pp. 5472–5477.

20

NONLINEAR EFFECTS IN SOUND BEAMS

MARK F. HAMILTON

1 INTRODUCTION

Many applications of intense sound, such as in sonar, medicine, and nondestructive evaluation, involve radiation from directional sources. Linear effects of diffraction determine length scales corresponding to near fields, focal lengths, and beamwidths. Depending on how these diffraction lengths compare with length scales corresponding to absorption and shock formation, the dominant nonlinear effects can be substantially different. This chapter describes these competing effects in detail, beginning with analytic solutions for harmonic generation in weakly nonlinear Gaussian beams, both unfocused and focused. Asymptotic and numerical solutions are used to illustrate shock formation, nonlinear side lobe generation, and self-demodulation of pulses in sound beams radiated by circular pistons. Hallmarks of nonlinear effects in sound beams include asymmetric waveform distortion and the increase in beamwidth and side lobe levels that accompanies shock formation along the axis of the beam. Another, in the case of bifrequency radiation, is the generation of a difference-frequency beam with high directivity and low side lobe levels.

2 PARABOLIC WAVE EQUATION

The combined effects of diffraction, absorption, and nonlinearity in *directional sound beams* (i.e., radiated from sources with dimensions that are large compared with the characteristic wavelength) are taken into account by the Khokhlov–Zabolotskaya–Kuznetsov (KZK) parabolic wave equation[1]:

$$\frac{\partial^2 p}{\partial z\,\partial \tau} = \frac{c_0}{2}\left(\frac{\partial^2 p}{\partial r^2} + \frac{1}{r}\frac{\partial p}{\partial r}\right) + \frac{\delta}{2c_0^3}\frac{\partial^3 p}{\partial \tau^3} + \frac{\beta}{2\rho_0 c_0^3}\frac{\partial^2 p^2}{\partial \tau^2}. \tag{1}$$

Here p is the sound pressure, z the coordinate along the axis of the sound beam (assumed to be axisymmetric and propagating in the $+z$-direction), $\tau = t - z/c_0$ a retarded time, c_0 the small-signal sound speed, r the distance from the z-axis, δ the diffusivity of sound,[2] which accounts for absorption due to both viscosity and heat conduction, β the coefficient of nonlinearity, and ρ_0 the ambient density. Equation (1) provides an accurate description of the sound field at distances beyond several source radii and in regions close to the z-axis (e.g., within about 20° from the z-axis in the far field). The KZK equation may thus be used to model most regions of interest in directional sound beams. To the order of accuracy inherent in the KZK equation, it is consistent to use the plane-wave impedance relation $p = \rho_0 c_0 u$, where u is the z-component of the particle velocity vector. Consequently, source conditions expressed in terms of velocity are easily transformed into source conditions for the pressure. In the absence of nonlinearity ($\beta = 0$), Eq. (1) models the small-signal propagation of a directional beam in a thermoviscous fluid. Alternatively, when the first term on the right-hand side is omitted, Eq. (1) reduces to the Burgers equation [Eq. (16) of Chapter 16].

3 QUASILINEAR SOLUTIONS FOR HARMONIC RADIATION

Quasilinear solutions are based on the assumption that linear theory provides a good description of the harmonic

Note: References to chapters appearing only in the *Encyclopedia* are preceded by *Enc.*

components radiated directly by the source. The linear pressure field p_1 is obtained by solving Eq. (1) with $\beta = 0$. The secondary pressure field p_2 may then be interpreted as radiation from a volume distribution of *virtual sources* with strengths proportional to p_1^2. Solutions for the secondary field are obtained by substituting the linear solution p_1 into the nonlinear term in Eq. (1) and solving the resulting linear, inhomogeneous differential equation for p_2. Quasilinear solutions ($p = p_1 + p_2$) obtained in this way are valid provided the nonlinear interactions are relatively weak, that is, for $|p_2| \ll |p_1|$ throughout the beam.

Consider a bifrequency source in the plane $z = 0$ that radiates simultaneously at the two angular frequencies ω_a and ω_b. The linear solution for the primary wave field p_1 may be written

$$p_1(r, z, \tau) = \frac{1}{2j}\,[q_{1a}(r, z)e^{j\omega_a\tau} + q_{1b}(r, z)e^{j\omega_b\tau}] + \text{c.c.}, \quad (2)$$

where q_{1a} and q_{1b} are the complex-pressure amplitudes of the individual primary beams, $|q_{1a}|$ and $|q_{1b}|$ are the corresponding peak physical pressures, and c.c. designates complex conjugates of the preceding terms. An equivalent statement of Eq. (2) is $p_1 = \text{Im}\{q_{1a}e^{j\omega_a\tau} + q_{1b}e^{j\omega_b\tau}\}$. The functions q_{1a} and q_{1b} are slowly varying in z because the rapid variations on the wavelength scale have been factored out. When q_{1a} and q_{1b} are real, Eq. (2) reduces to $p_1 = q_{1a}\sin\omega_a t + q_{1b}\sin\omega_b t$. The solutions for q_{1a} and q_{1b} are given by[3]

$$q_1(r, z) = \left(\frac{jk}{z}\right) e^{-\alpha_1 z - jkr^2/2z} \int_0^\infty q_1(r', 0) J_0\left(\frac{krr'}{z}\right) \cdot e^{-jkr'^2/2z} r'\,dr', \quad (3)$$

where the subscripts a and b have been suppressed, J_0 is the Bessel function of the first kind of order zero, $q_1(r, 0)$ is a source function, $k = \omega/c_0$ is the wavenumber, and $\alpha_1 = \delta\omega^2/2c_0^3$ is the thermoviscous attenuation coefficient. Since Eq. (3) is expressed in the frequency domain, it is not restricted to thermoviscous fluids. Simply replace the attenuation coefficients with suitable empirical relations or numerical values for the fluid under consideration. The same may be done in Eqs. (5) and (6) below.

The secondary wave field p_2 is composed of four components, the pressures at the second harmonics of the source frequencies ($2\omega_a$ and $2\omega_b$), the sum frequency ($\omega_+ = \omega_a + \omega_b$), and the difference frequency ($\omega_- = \omega_a - \omega_b$, where $\omega_a > \omega_b$ is assumed):

$$p_2(r, z, \tau) = \frac{1}{2j}\,[q_{2a}(r, z)e^{j2\omega_a\tau} + q_{2b}(r, z)e^{j2\omega_b\tau} + q_+(r, z)e^{j\omega_+\tau} + q_-(r, z)e^{j\omega_-\tau}] + \text{c.c.} \quad (4)$$

Quasilinear solutions for the individual spectral components are obtained by substituting the known solution for the primary field [Eqs. (2) and (3)] into the nonlinear term in Eq. (1), substituting the unknown expression for the secondary field [Eq. (4)] into the linear terms, and solving the resulting system of inhomogeneous wave equations for the complex pressures q_{2a}, q_{2b}, q_+, and q_-.[4-6] We assume that no secondary-frequency components are radiated directly by the source, that is, $p_2(r, 0, t) = 0$. The solution for the second-harmonic pressures is then

$$q_2(r, z) = \frac{j\beta k^2 e^{-\alpha_2 z}}{\rho_0 c_0^2} \int_0^z \int_0^\infty \frac{q_1^2(r', z')}{z - z'} J_0\left(\frac{2krr'}{z - z'}\right) \cdot e^{\alpha_2 z' - jk(r^2 + r'^2)/(z - z')} r'\,dr'\,dz'. \quad (5)$$

Equation (5) applies to either q_{2a} or q_{2b} once the appropriate subscript, a or b, is affixed to q_1, k, and α_2. The solution for the sum- and difference-frequency pressures is

$$q_\pm(r, z) = \pm\frac{j\beta k_\pm^2 e^{-\alpha_\pm z}}{2\rho_0 c_0^2} \cdot \int_0^z \int_0^\infty \frac{q_{1a}(r', z')q_{1b}^{(*)}(r', z')}{z - z'} J_0\left(\frac{k_\pm rr'}{z - z'}\right) \cdot e^{\alpha_\pm z' - jk_\pm(r^2 + r'^2)/2(z - z')} r'\,dr'\,dz', \quad (6)$$

where the upper sign of $\pm$ is used for the sum frequency, the lower sign for the difference frequency. The notation $q_{1b}^{(*)}$ indicates that the complex conjugate of q_{1b} must be used when evaluating the difference-frequency component, but not when evaluating the sum-frequency component.

4 GAUSSIAN SOURCES

Source distributions with Gaussian amplitude profiles generate sound fields for which the integrals in Section 3 can be solved analytically for a variety of cases. Let the source functions be given by

$$q_{1a}(r,0) = p_{ga} \exp\left[-\left(\frac{r}{\epsilon_a} \right)^2 \right], \tag{7a}$$

$$q_{1b}(r,0) = p_{gb} \exp\left[-\left(\frac{r}{\epsilon_b} \right)^2 \right], \tag{7b}$$

where p_g and ϵ (with the appropriate subscripts) are the peak pressure and characteristic radius, respectively, of each Gaussian source.

4.1 Primary Beams

The linear solution obtained by substituting either of Eqs. (7) into Eq. (3) is

$$q_1(r,z) = \frac{p_g e^{-\alpha_1 z}}{1 - jz/z_0} \exp\left(-\frac{(r/\epsilon)^2}{1 - jz/z_0} \right), \tag{8}$$

where the subscripts a and b are again suppressed. The quantity $z_0 = k\epsilon^2/2$ (which represents either $z_{0a} = k_a\epsilon_a^2/2$ or $z_{0b} = k_b\epsilon_b^2/2$) is referred to as the *Rayleigh distance*, and it marks the transition between the *near-field* and *far-field* regions of the primary beams. For $z \ll z_0$, Eq. (8) describes a collimated plane wave having a transverse amplitude distribution that matches that of the source. For $z \gg z_0$, Eq. (8) describes spherical waves having directivity

$$D_1(\theta) = \exp[-(\tfrac{1}{2}k\epsilon)^2 \tan^2\theta] \tag{9}$$

in terms of the angle $\theta = \arctan(r/z)$ with respect to the z-axis. The pressure amplitude decays as $z^{-1}e^{-\alpha_1 z}$ in the far field.

4.2 Second-Harmonic Generation

The quasilinear solution for the second-harmonic pressure is obtained by substituting Eq. (8) into Eq. (5):

$$q_2(r,z) = \frac{jP_g e^{-\alpha_2 z + j(2\alpha_1 - \alpha_2)z_0}}{1 - jz/z_0} \exp\left(-\frac{2(r/\epsilon)^2}{1 - jz/z_0} \right) \cdot \{E_1[j(2\alpha_1 - \alpha_2)z_0] - E_1[j(2\alpha_1 - \alpha_2)(z_0 - jz)]\}, \tag{10}$$

where $P_g = \beta p_g^2 k^2 \epsilon^2 / 4\rho_0 c_0^2$ has dimensions of pressure and $E_1(\sigma) = \int_\sigma^\infty u^{-1}e^{-u}\,du$ is the exponential integral, which can be evaluated with standard mathematical software packages and reduces to simple asymptotic forms for large and small arguments. Comparison of the exponentials in Eqs. (8) and (10) reveals $q_2(r) \propto q_1^2(r)$, and the second-harmonic beamwidth is narrower at all ranges z, by a factor of $1/\sqrt{2}$, than the width of the primary beam. For $\alpha_2 > 2\alpha_1$ (as is the case for thermoviscous fluids, for which $\alpha_2 = 4\alpha_1$) and in the far field $[z \gg \max\{z_0, (\alpha_2 - 2\alpha_1)^{-1}\}]$, Eq. (10) describes spherical waves that decay as $z^{-2}e^{-2\alpha_1 z}$ and have directivity $D_2(\theta) = D_1^2(\theta)$.

With no absorption ($\alpha_1 = \alpha_2 = 0$), Eq. (10) reduces to

$$q_2(r,z) = \frac{jP_g \ln(1 - jz/z_0)}{1 - jz/z_0} \exp\left(-\frac{2(r/\epsilon)^2}{1 - jz/z_0} \right), \tag{11}$$

which in the far field describes spherical waves that decay as $z^{-1}\ln(z/z_0)$, slightly slower than z^{-1} on account of energy transfer from the primary waves. Equation (11) reveals that the asymptotic decay rate $z^{-1}\ln(z/z_0)$ occurs not simply for $z/z_0 \gg 1$, which defines the far-field region in linear acoustics. Instead, the much stronger condition $\ln(z/z_0) \gg 1$ must be satisfied[5,6] [note that $\ln(-jz/z_0) = \ln(z/z_0) - j\pi/2$]. In general (not just for Gaussian beams), as long as effects of absorption are negligible, the far-field region in nonlinear acoustics may begin at distances that are orders of magnitude greater than in linear acoustics. Nonlinear near-field effects are particularly noticeable in harmonic beam patterns produced by piston sources (see Section 5.3).

4.3 Sum- and Difference-Frequency Generation

The general form of the quasilinear solution for the sum- and difference-frequency sound,[7] which is obtained from Eq. (6), requires numerical integration:

$$q_\pm(r,z) = \pm k_\pm P_\pm \exp\left(-\alpha_\pm z - \frac{k_\pm^2 r^2}{2f_\pm} \right) \cdot \int_0^z \exp\left(-\alpha_T^\pm z' - \frac{k_\pm^2 k_a k_b (z_{0a} \mp z_{0b})^2 r^2}{2f_\pm(g_\pm \mp jf_\pm k_\pm z')} \right) \cdot \frac{dz'}{g_\pm \mp jf_\pm k_\pm z'}, \tag{12}$$

where $P_\pm = \beta k_\pm^2 z_{0a} z_{0b} p_{ga} p_{gb} / 2\rho_0 c_0^2$, $\alpha_T^\pm = \alpha_a + \alpha_b - \alpha_\pm$, $f_\pm(z) = k_a z_{0a} + k_b z_{0b} - jk_\pm z$, and $g_\pm(z) = k_\pm^2 z_{0a} z_{0b} - j(k_b z_{0a} + k_a z_{0b})k_\pm z$. On axis, Eq. (12) reduces to

$$q_\pm(0,z) = \left(\frac{jP_\pm}{f_\pm}\right)\exp\left(-\alpha_\pm z \pm \frac{j\alpha_T^\mp g_\pm}{k_\pm f_\pm}\right)$$
$$\cdot\left\{E_1\left[\pm\frac{j\alpha_T^\mp g_\pm}{k_\pm f_\pm}\right]\right.$$
$$\left.-E_1\left[\pm\frac{j\alpha_T^\mp g_\pm}{k_\pm f_\pm}\left(1\mp\frac{jf_\pm k_\pm z}{g_\pm}\right)\right]\right\}, \quad (13)$$
$$= \left(\frac{jP_\pm}{f_\pm}\right)\ln\left(1\mp\frac{jf_\pm k_\pm z}{g_\pm}\right),$$
$$\alpha_a = \alpha_b = \alpha_\pm = 0. \quad (14)$$

When absorption is negligible, Eq. (12) reduces to

$$q_\pm(r,z) = \left(\frac{jP_\pm}{f_\pm}\right)\exp\left(-\frac{k_\pm^2 r^2}{2f_\pm}\right)$$
$$\cdot\left\{E_1\left[\frac{k_\pm^2 k_a k_b (z_{0a}\mp z_{0b})^2 r^2}{2f_\pm(g_\pm \mp jf_\pm k_\pm z)}\right]\right.$$
$$\left.-E_1\left[\frac{k_\pm^2 k_a k_b (z_{0a}\mp z_{0b})^2 r^2}{2f_\pm g_\pm}\right]\right\}. \quad (15)$$

Different asymptotic properties[8] are obtained for the sum- and difference-frequency fields according to the sign of the combined attenuation coefficient $\alpha_T^\pm$. Specifically, the relations $\alpha_T^+ < 0$ and $\alpha_T^- > 0$ apply to thermoviscous fluids (because $\alpha_\omega \propto \omega^2$), in which case the sum-frequency pressure decays in the far field as $z^{-2}e^{-(\alpha_a+\alpha_b)z}$ with directivity $D_+(\theta) = D_{1a}(\theta)D_{1b}(\theta)$.

The nonlinear interaction region where the difference-frequency sound is generated is often referred to as a *parametric array*.[9] An interesting and widely applied result is obtained with $\alpha_T^- > 0$, and for the case of equal source radii ($\epsilon_a = \epsilon_b = \epsilon$), neighboring primary frequencies ($\omega_a \approx \omega_b \gg \omega_-$) and strong absorption ($\alpha_T^- z_0 > 1$). The difference-frequency pressure then decays as $z^{-1}e^{-\alpha_- z}$ in the far field ($\alpha_T^- z \gg 1$), and the directivity is given by $D_-(\theta) = D_A(\theta)D_W(\theta)$, where[8]

$$D_A(\theta) = \exp\left[-\left(\frac{k_-\epsilon}{2\sqrt{2}}\right)^2\tan^2\theta\right], \quad (16a)$$

$$D_W(\theta) = \frac{1}{\sqrt{1+(k_-/2\alpha_T^-)^2\tan^4\theta}}. \quad (16b)$$

For small $k_-\epsilon$, as is often the case for $\omega_a \approx \omega_b \gg \omega_-$, the *aperture factor*[10] $D_A(\theta)$ is a relatively weak function of θ in comparison with the *Westervelt directivity*[9] $D_W(\theta)$, and the directivity of the difference-frequency radiation is therefore determined primarily by the latter. [The form of $D_W(\theta)$ given in Eq. (16b) is the parabolic (i.e., small-angle) approximation of the classical result obtained by Westervelt.[9]] Note that $D_W(\theta)$ does not depend on the source radius ϵ, only on the ratio k_-/α_T^-, and no side lobes are predicted. Consequently, the nonlinearly generated difference-frequency sound forms a more narrow beam than if it were radiated directly by the source of the primary beams. For sources other than Gaussian, the same far-field properties are obtained but with $D_A(\theta)$ replaced by the directivity function corresponding to direct, linear radiation at frequency ω_- by a source with amplitude distribution $q_{1a}(r,0)q_{1b}^*(r,0)$. See *Enc.* Ch. 53 for more extensive discussion of parametric arrays.

4.4 Focused Sources

When sources a and b are focused at distances d_a and d_b, respectively, Eqs. (7) are replaced by (with the subscripts a and b suppressed)

$$q_1(r,0) = p_g\exp\left[-\left(\frac{r}{\epsilon}\right)^2 + \frac{jkr^2}{2d}\right], \quad (17)$$

which reduces to Eqs. (7) when the sources are focused at infinity ($d = \infty$). Equations (7) are transformed into Eq. (17) if, in Eqs. (7), ϵ is replaced by $\tilde{\epsilon} = \epsilon(1-jG)^{-1/2}$, where $G = z_0/d$. Likewise, the effects of focusing on the radiated sound may be included simply by replacing ϵ by $\tilde{\epsilon}$ everywhere throughout the expressions for q_1, q_2, and q_+. In the expressions for q_-, however, ϵ_a is replaced by $\tilde{\epsilon}_a$ but ϵ_b is replaced by $\tilde{\epsilon}_b^*$. For example, Eq. (8) (with $\alpha_1 = 0$) and Eq. (11) become, respectively,

$$q_1(r,z) = \frac{p_g}{1-(1+jG^{-1})z/d}$$
$$\cdot\exp\left(-\frac{(1-jG)(r/\epsilon)^2}{1-(1+jG^{-1})z/d}\right), \quad (18)$$

$$q_2(r,z) = \frac{jP_g}{1-jG}\,\frac{\ln[1-(1+jG^{-1})z/d]}{1-(1+jG^{-1})z/d}$$
$$\cdot\exp\left(-\frac{2(1-jG)(r/\epsilon)^2}{1-(1+jG^{-1})z/d}\right). \quad (19)$$

Since $|q_1(0,d)| = Gp_g$, G is referred to as the linear *focusing gain*. The maximum value of $|q_1(0,z)|$ is located at $z = d(1+G^{-2})^{-1}$, and it approaches the geometric focus at $z = d$ with increasing G.

4.5 Approximation for Circular Piston Sources

Gaussian beam formulas can be used to approximate the field produced by a plane circular piston at locations beyond the near field and close to the axis (i.e., within the main radiation lobe). For a circular piston of radius a and with effective peak pressure $p_0 = \rho_0 c_0 u_0$, where u_0 is the velocity amplitude of the piston, Eq. (3) yields an axial pressure amplitude $(ka^2/2z)p_0 e^{-\alpha_1 z}$ and directivity $D_1(\theta) = 2J_1(ka \tan\theta)/ka \tan\theta$ for the far field. Comparison of these expressions with Eqs. (8) and (9) leads to the circular piston source transformation[11] $p_g = 2p_0$ and $\epsilon = a/\sqrt{2}$. Application of the transformation to Eq. (8) produces

$$q_1(r,z) = \frac{2p_0 e^{-\alpha_1 z}}{1 - j4z/ka^2} \exp\left(-\frac{2(r/a)^2}{1 - j4z/ka^2}\right). \tag{20}$$

Application of the same transformation to q_2 and $q_\pm$ yields good predictions for the secondary pressures beyond the near fields of the primary beams and close to the z-axis, particularly for weak attenuation (see Fig. A1 of Ref. 11).

5 PISTON SOURCES

Let the source condition be given by

$$p(r,0,t) = \begin{cases} p_0 E(t) \sin \omega_0 t, & r \le a, \\ 0, & r > a, \end{cases} \tag{21}$$

where the envelope function $E(t)$ modulates the carrier signal of frequency ω_0. Equation (21) is used to model the vibration of a circular piston having characteristic velocity amplitude $u_0 = p_0/\rho_0 c_0$. For $E(t) = 1$, that is, a monofrequency source, evaluation of Eq. (3) for the magnitude of the axial pressure field yields $|q_1(0,z)| = 2p_0 e^{-\alpha_0 z}|\sin(z_0/2z)|$, where $z_0 = k_0 a^2/2$, $k_0 = \omega_0/c_0$, and α_0 is the attenuation coefficient at frequency ω_0. The axial solution is in good agreement with the exact solution of the wave equation for a baffled circular piston (i.e., without the parabolic approximation; see Chapter 8) at distances $z/a \gtrsim (k_0 a)^{1/3}$. The corresponding far-field directivity function was given in Section 4.5.

Equations (5) and (6) are not available in closed form for either monofrequency or bifrequency piston sources, although asymptotic expressions have been obtained for the far field.[5] The main far-field properties are similar to those discussed in Section 4 for Gaussian beams. The quasilinear integral solutions have been evaluated numerically,[4–6] but sophisticated computational techniques are required because of the highly oscillatory nature of the integrands. An alternative approach is to solve the fully nonlinear KZK equation with finite-difference methods, either in the frequency domain[12] or in the time domain.[13]

5.1 Self-demodulation

Now let $E(t)$ vary slowly in comparison with $\sin \omega_0 t$. For strong absorption ($\alpha_0 z_0 > 1$), an approximate quasilinear solution ($p = p_1 + p_2$) of Eq. (1) for the axial pressure in the beam is given by[14]

$$p(0,z,\tau) = p_0 \left[f(\tau) - f\left(\tau - \frac{a^2}{2c_0 z}\right) + \frac{\beta p_0 a^2}{16 \rho_0 c_0^4 \alpha_0 z} \frac{d^2 E^2}{d\tau^2} \right] * \mathcal{D}(z,\tau), \tag{22}$$

where $\mathcal{D}(z,t) = (c_0^3/2\pi\delta z)^{1/2} \exp(-c_0^3 t^2/2\delta z)$ is the dissipation function for a thermoviscous fluid, the aster-

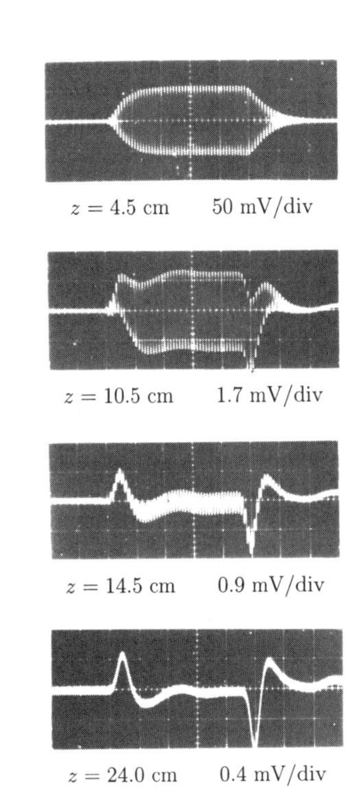

Fig. 1 Measured axial pressure waveforms that demonstrate the self-demodulation of a 10-MHz pulse in carbon tetrachloride.[16]

isk designates convolution with respect to time, and $f(t) = E(t)\sin\omega_0 t$. The high-frequency component (ω_0) is absorbed within the near field, and what remains in the far field is a distorted replica of the envelope $E(t)$. For $z \ll \alpha_E^{-1}$, where α_E is the attenuation coefficient at the characteristic frequency of the envelope, the far-field axial waveform is given simply by the Berktay result[15]

$$p(0,z,\tau) = \frac{\beta p_0^2 a^2}{16\rho_0 c_0^4 \alpha_0 z}\frac{d^2E^2}{d\tau^2}, \qquad \alpha_0^{-1} \ll z \ll \alpha_E^{-1}. \tag{23}$$

Experimental verification[16] of the *self-demodulation* phenomenon is shown in Fig. 1 for a 10-MHz pulse in carbon tetrachloride. At $z = 4.5$ cm the waveform is unaffected by nonlinearity. The self-demodulation process becomes visible near $z = 10.5$ cm, the high-frequency component is strongly attenuated in comparison with the envelope at $z = 14.5$ cm, and only the distorted (i.e., squared and twice differentiated) envelope remains at $z = 24.0$ cm. Equation (22) is in excellent agreement with measurements of this type.[14]

5.2 Waveform Distortion and Shock Formation

Predictions of waveform distortion and shock formation cannot be made on the basis of quasilinear solutions, and numerical computations are required. Shown in the left column of Fig. 2 are axial pressure waveforms, with the corresponding frequency spectra $S(\omega)$ in the right column (normalized by the peak spectral magnitude S_0 at $z = 0$), computed for a thermoviscous fluid with a finite-difference solution[13] of Eq. (1). The source waveform is a Gaussian tone burst with envelope $E(t) = \exp[-(\omega_0 t/3\pi)^2]$. The absorption is relatively weak ($\alpha_0 z_0 = 0.1$) and the source amplitude is relatively high [$p_0 = 4\rho_0 c_0^2/\beta(k_0 a)^2$, for which a plane wave would form a shock at distance $\frac{1}{2}z_0$ in a lossless fluid]. By the end of the (small-signal) near field ($z/z_0 = 1$), the combined effects of nonlinearity and diffraction on the waveform have produced sharpening of the positive cycles, rounding of the negative cycles, and the development of shock fronts. In addition, the peak positive pressures are approximately twice the peak negative pressures. The waveform distortion in the near field causes energy to be shifted primarily upward in the frequency spectrum. Farther away from the source, absorption filters out the nonlinearly generated high-frequency components and the shock fronts disappear ($z/z_0 = 10$), the relative importance of the low-frequency spectrum increases ($z/z_0 = 30$), and a self-demodulated, low-frequency signal is all that remains at $z/z_0 = 100$.

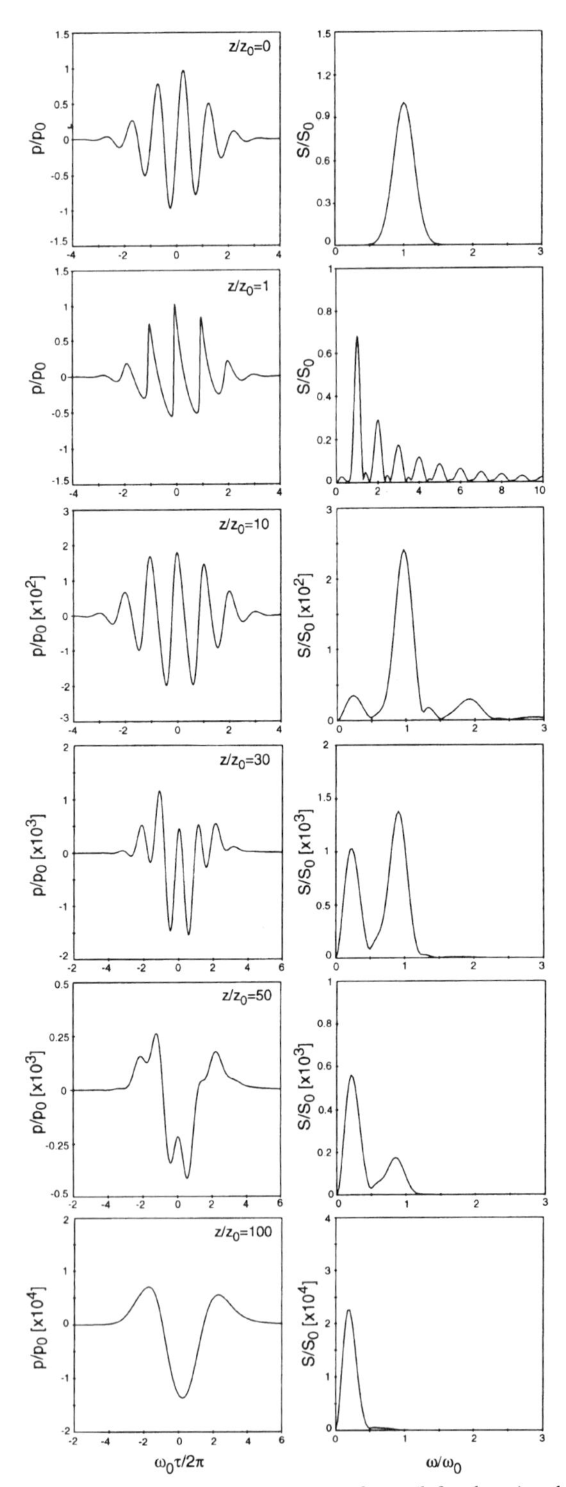

Fig. 2 Computed axial pressure waveforms (left column) and corresponding frequency spectra (right column) for a Gaussian tone burst radiated by a circular piston.[13]

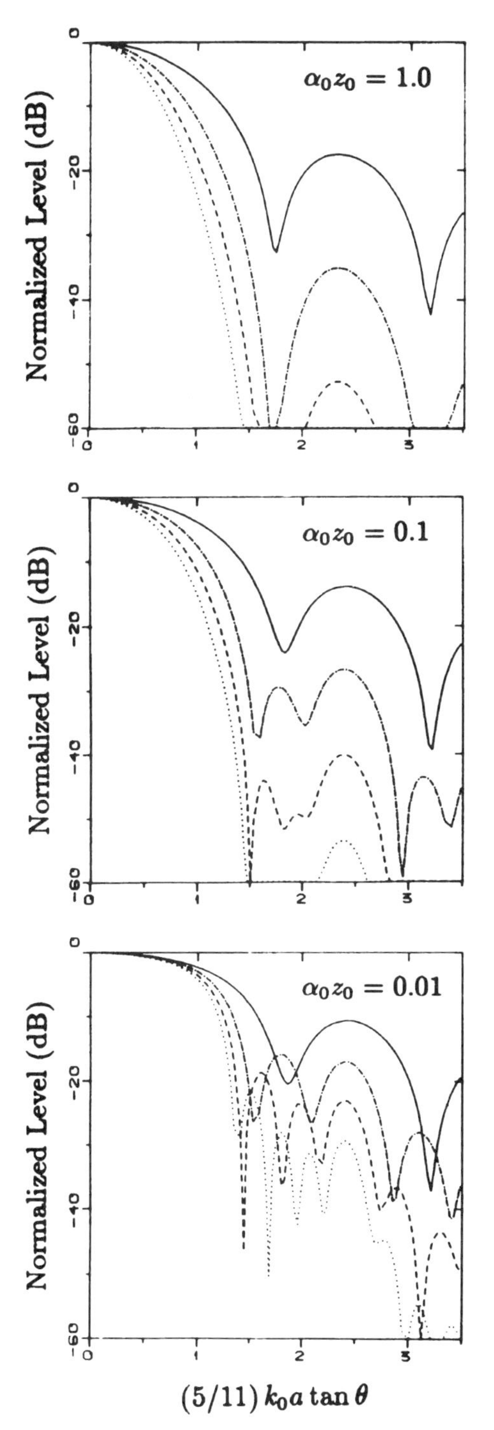

Fig. 3 Computed beam patterns at $z = 10z_0$ for the fundamental component $n = 1$ (—), second-harmonic component $n = 2$ (– · –), third-harmonic component $n = 3$ (– –), and fourth harmonic component $n = 4$ (· · ·) due to radiation of sound from a monofrequency circular piston.[17]

5.3 Harmonic Beam Patterns

In Fig. 3 are shown numerical calculations[17] of normalized beam patterns at range $z = 10z_0$ for $E(t) = 1$, that is, for a monofrequency source. The source pressure is $p_0 = 3\rho_0 c_0^2/\beta(k_0 a)^2$, for which the plane-wave shock formation distance is $\frac{2}{3}z_0$. Three levels of absorption are considered: $\alpha_0 z_0 = 1.0$, 0.1, and 0.01. In the case of strong absorption ($\alpha_0 z_0 = 1.0$), the far-field structure is fully established at $z = 10z_0$, and the directivity function $D_n(\theta)$ for the nth-harmonic component is given very nearly by $D_n(\theta) = D_1^n(\theta)$. With increasing n, the harmonic components become more directional and exhibit greater side lobe suppression. For weak absorption ($\alpha_0 z_0 = 0.01$), the nth-harmonic beam pattern possesses n times as many side lobes as are present in the beam pattern at the fundamental frequency. Similar phenomona also occur in focused sound beams.[18] The additional side lobes are due to *near-field effects*, and they decay faster with range than do side lobes in the fundamental beam pattern.[6] However, the decay rates of different side lobes may differ by only a factor of order $\ln(z/z_0)$, and consequently the nonlinear near-field effects may be significant out to hundreds of Rayleigh distances. Finally, it can be seen that decreasing the effect of absorption (or similarly, increasing the source level and therefore the relative effect of nonlinearity) produces flattening of the main lobes, which causes the relative levels of the side lobes to increase (compare results for $\alpha_0 z_0 = 1.0$ and $\alpha_0 z_0 = 0.01$). Ultimately, nonlinear effects in the main lobe can become sufficiently strong that *acoustic saturation*[19] occurs, and a further increase in source level produces no increase in the axial pressure at a given distance.

Acknowledgment

The Office of Naval Reearch is gratefully acknowledged for supporting much of the work that produced the results described in this chapter.

REFERENCES

1. N. S. Bakhvalov, Ya. M. Zhileikin, and E. A. Zabolotskaya, *Nonlinear Theory of Sound Beams*, American Institute of Physics, New York, 1987.
2. J. Lighthill, *Waves in Fluids*, Cambridge University Press, New York, 1980.
3. G. S. Garrett, J. Naze Tjøtta, and S. Tjøtta, "Nearfield of a Large Acoustic Transducer. Part I: Linear Radiation," *J. Acoust. Soc. Am.*, Vol. 72, 1982, pp. 1056–1061.
4. G. S. Garrett, J. Naze Tjøtta, and S. Tjøtta, "Nearfield of a Large Acoustic Transducer. Part II: Parametric Radiation," *J. Acoust. Soc. Am.*, Vol. 74, 1983, pp. 1013–1020.

5. G. S. Garrett, J. Naze Tjøtta, and S. Tjøtta, "Nearfield of a Large Acoustic Transducer. Part III: General Results," *J. Acoust. Soc. Am.*, Vol. 75, 1984, pp. 769–779.
6. J. Berntsen, J. Naze Tjøtta, and S. Tjøtta, "Nearfield of a Large Acoustic Transducer. Part IV: Second Harmonic and Sum Frequency Radiation," *J. Acoust. Soc. Am.*, Vol. 75, 1984, pp. 1383–1391.
7. C. M. Darvennes and M. F. Hamilton, "Scattering of Sound by Sound from Two Gaussian Beams," *J. Acoust. Soc. Am.*, Vol. 87, 1990, pp. 1955–1964.
8. C. M. Darvennes, M. F. Hamilton, J. Naze Tjøtta, and S. Tjøtta, "Effects of Absorption on the Nonlinear Interaction of Sound Beams," *J. Acoust. Soc. Am.*, Vol. 89, 1991, pp. 1028–1036.
9. P. J. Westervelt, "Parametric Acoustic Array," *J. Acoust. Soc. Am.*, Vol. 35, 1963, pp. 535–537.
10. J. Naze Tjøtta and S. Tjøtta, "Effects of Finite Aperture in a Parametric Acoustic Array," *J. Acoust. Soc. Am.*, Vol. 68, 1980, pp. 970–972.
11. M. F. Hamilton and F. H. Fenlon, "Parametric Acoustic Array Formation in Dispersive Fluids," *J. Acoust. Soc. Am.*, Vol. 76, 1984, pp. 1474–1492.
12. J. Naze Tjøtta, S. Tjøtta, and E. Vefring, "Propagation and Interaction of Two Collinear Finite Amplitude Sound Beams," *J. Acoust. Soc. Am.*, Vol. 88, 1990, pp. 2859–2870.
13. Y.-S. Lee and M. F. Hamilton, "Time-Domain Modeling of Pulsed Finite-Amplitude Sound Beams," *J. Acoust. Soc. Am.*, Vol. 97, 1995, pp. 906–917.
14. M. A. Averkiou, Y.-S. Lee, and M. F. Hamilton, "Self-Demodulation of Amplitude and Frequency Modulated Pulses in a Thermoviscous Fluid," *J. Acoust. Soc. Am.*, Vol. 94, 1993, pp. 2876–2883.
15. H. O. Berktay, "Possible Exploitation of Non-linear Acoustics in Underwater Transmitting Applications," *J. Sound Vib.*, Vol. 2, 1965, pp. 435–461.
16. M. B. Moffett, P. J. Westervelt, and R. T. Beyer, "Large-Amplitude Pulse Propagation—A Transient Effect," *J. Acoust. Soc. Am.*, Vol. 47, 1970, pp. 1473–1474.
17. M. F. Hamilton, J. Naze Tjøtta, and S. Tjøtta, "Nonlinear Effects in the Farfield of a Directive Sound Source," *J. Acoust. Soc. Am.*, Vol. 89, 1985, pp. 202–216.
18. M. A. Averkiou and M. F. Hamilton, "Measurements of Harmonic Generation in a Focused Finite-Amplitude Sound Beam," *J. Acoust. Soc. Am.*, Vol. 98, 1995, pp. 3439–3442.
19. J. A. Shooter, T. G. Muir, and D. T. Blackstock, "Acoustic Saturation of Spherical Waves in Water," *J. Acoust. Soc. Am.*, Vol. 55, 1974, pp. 54–62.

21

NONLINEAR LUMPED ELEMENTS

WILLIAM E. ZORUMSKI

1 INTRODUCTION

Acoustics is usually based on the assumption of small motions. This is a good approximation for sound in the tolerable range of hearing, but there are many instances, such as jet engine fans and internal combustion engine exhausts, where intense sound waves must be considered. The amplitudes of these waves are sufficiently large so that nonlinear effects are significant. Noise control is an area where this occurs often. The sound must be controlled because it is intense, and this leads to the consideration of nonlinear effects in the control process. Lumped elements are an approximation, as discussed in Chapter 11, used when the size of the element is small in comparison to a typical acoustic wavelength. This chapter will consider some representative causes of nonlinearity in lumped elements and then focus on some details of nonlinear resistance and reactance as is exhibited by some porous materials and perforated plates used in noise control. It will be shown how the effect of nonlinearity is to cause an interaction between sound at different frequencies, which is a considerable complication in understanding, analysis, measurement, and design.

2 SOURCES OF NONLINEARITY

There are many sources of nonlinearity, but the physical example of the interaction of a sound wave in air with a perforated plate illustrates most of the nonlinear effects. These effects are observed in irregular materials such as woven composites and fibermetals also. Consider a plate, as shown in Fig. 1, that has N holes of diameter d per unit face area of the plate and is mounted a fixed

Note: References to chapters appearing only in the *Encyclopedia* are preceded by *Enc.*

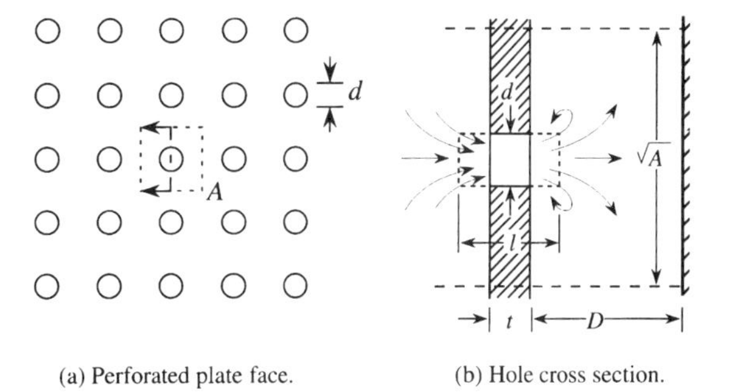

(a) Perforated plate face. (b) Hole cross section.

Fig. 1 Unsteady flow through a perforated plate.

distance D in from a rigid wall. The external area associated with a single hole is then $A = 1/N$, as denoted by the dashed square around the hole. Looking at the fine detail of a cross section through the plate, as shown on the right, we imagine air flow through the hole caused by a sound wave. The air stream must contract from a channel of cross section A and squeeze through the hole with area A_h. Energy is dissipated in this flow by viscous effects within the thickness t and by shed vortices as the flow exits the hole. As air flows into the space behind the plate, the pressure increases behind the plate, and as air flows out, the pressure decreases.

Now construct a harmonic oscillator model for this interaction. A "slug" of air of length l within the hole forms an effective mass $m = \rho A_h l$, but this mass is not constant. As velocity through the hole increases, pressure decreases due to the Bernoulli effect and density decreases. Also, a jet forms on the outflow side so that the length of the slug increases. Consequently, the inertial force is not simply proportional to acceleration $\ddot{x}$ but may depend as well on velocity $\dot{x}$ and displacement x. The damping force may be simply proportional to velocity $\dot{x}$ when velocity is small but can increase quickly when the exiting jet begins to shed vortices. The effective spring for the oscillator is the air between the plate and the wall, but the air pressure, assuming adiabatic compression and expansion, increases faster than the density so that the spring force is nonlinear also. The equation for the motion of the oscillator shown in Fig. 2 is

$$m\ddot{x} + c\dot{x} + kx = f \tag{1}$$

where $f = pA_h$ is the exciting force due to the pressure in front of the plate. This would be a standard equation, but the mass m, damping coefficient c, and spring coefficient k are not constants, as depicted by the straight dashed lines in Fig. 2, but instead are the slopes of the solid curves, which may all depend on x, $\dot{x}$, and $\ddot{x}$. These effects are all horribly complicated, even for the simple example of a perforated plate. A comprehensive discussion of the impedance of perforates is given by Melling.[1]

An alternative to modeling the flow details within the porous material is to adopt a phenomenologic approach, relying on measurements of certain assumed functions, to incorporate nonlinear effects into acoustic interactions. One such approach is given in the following section.

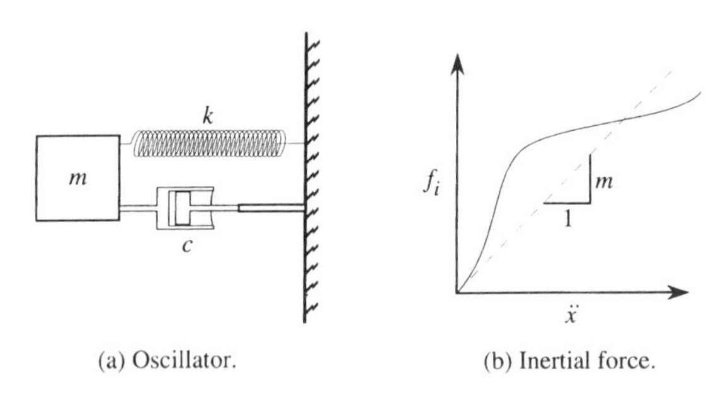

(a) Oscillator. (b) Inertial force.

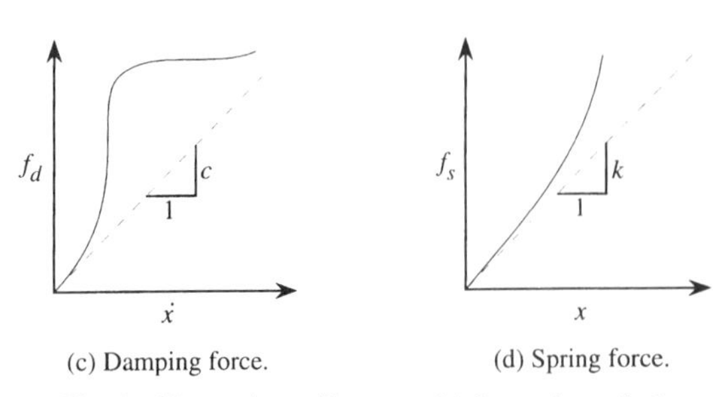

(c) Damping force. (d) Spring force.

Fig. 2 Harmonic oscillator model for perforated plate.

3 NONLINEAR MATERIALS

3.1 Spectral Impedance

Impedance is the parameter used in linear acoustics to relate acoustic pressure to acoustic velocity. An impedance equation is derived readily from Eq. (1) by assuming complex harmonic motions of the form $e^{i\omega t}$, where ω is circular frequency and t is time. Let the velocity $\dot{x} = v$ so that acceleration is $\ddot{x} = i\omega v$ and displacement is $x = -iv/\omega$. Equation (1) is then

$$\begin{aligned} p &= \left[\frac{c}{A_h} + i\left(\omega \frac{m}{A_h} - \frac{1}{\omega}\frac{k}{A_h} \right) \right] v \\ &= [R + iX]v \\ &= Zv. \end{aligned} \tag{2}$$

Equation (2) is still the equation of motion of a harmonic oscillator, but expressed in terms of variables common to acoustics. Resistance R is clearly proportional to the damping coefficient of the oscillator. Reactance X has two parts: The first, proportional to mass, increases with frequency, while the second, proportional to the spring constant, decreases. Acoustic impedance Z is the phased sum of the two parameters, $Z = R + iX$, and is the constant of proportionality relating acoustic pressure and velocity. More precisely, it is the constant of proportionality for complex pressure and velocity at the same frequency, and hence the qualifying adjective *spectral* is introduced to make this distinction from what follows.

3.2 Temporal Impedance

Temporal impedance is the phenomenologic operator introduced by Zorumski and Parrott[2] to relate instantaneous acoustic pressure and velocity when both are arbitrary functions of time:

$$Z_t = R_t + X_t \frac{\partial}{\partial t}. \tag{3}$$

Any attempt to derive this expression reduces eventually to so much hand waving. It is true if it works and only because it works. It is a postulate. The construction of the operator has X_t cast in the role of a masslike effect. An operator could be introduced with a springlike effect by adding an integral, with respect to time, to the right-hand side of Eq. (3). This term would complete the analogy with the spectral impedance operator. The more limited form shown in Eq. (3) was used in the instance where a porous plate or thin sheet of porous material was to be represented by an equation relating pressure difference across the sheet to velocity through the sheet:

$$\Delta p = Z_t v. \tag{4}$$

Here, Δp would be the pressure difference between the front and back of the plate or sheet. Since the backing space is not included, there is less need for an integral in the operator.

3.3 Measurement of Nonlinear Impedance

Resistance The postulate represented by Eqs. (3) and (4) has important and immediate consequences for the measurement of temporal impedance. Imagine a sample of material placed in a tube, as shown in Fig. 3. The instantaneous pressure drop across the material is Δp, and the velocity through the material is v. Before measurements may be made, some assumptions must be made about the nature of R_t and X_t. If they are constants, the problem is linear. Nonlinearity is represented by the dependence of R_t and X_t on (perhaps) particle displacement, velocity, or acceleration. Whatever the dependence, it must be valid for all possible time functions $\Delta p(t)$ and $v(t)$, and one of these possibilities is the case of steady flow. In this case the temporal impedance is identical to R_t, and this term can be at most a function of the velocity v. That is, R_t is the steady-flow resistance. This function is generated experimentally by adjusting the flow v through the material, measuring the pressure difference for each flow rate, and plotting $\Delta p/v$ versus v, as shown in Fig. 3.

Reactance Measurement of reactance is more challenging because it necessarily involves a dynamic test. The method of Ref. 2 was to expose the test specimen to a periodic (but not necessarily harmonic) signal and plot $\Delta p(t)$ versus $v(t)$, as shown in Fig. 4. Since the signal is periodic, the plot must form a closed trajectory with time being a parameter representing position along the trajectory. The trajectory on the left of Fig. 4 is a typical result of a low-intensity test where the sample is behaving linearly. The observed trajectory is an inclined ellipse. The slope of the major axis is proportional to the resistance, and the ratio of the minor axis to the major axis is proportional to the reactance. When time increases in a clockwise sense moving along the trajectory, reactance is positive. On the right of Fig. 4 is a typical trajectory formed by a high-intensity test where the sample exhibits nonlinear behavior. Note the loops that have formed at the "ends" of the trajectory. Time is increasing while moving in a counterclockwise sense around these loops, which is an indication of negative reactance during this portion of the cycle.

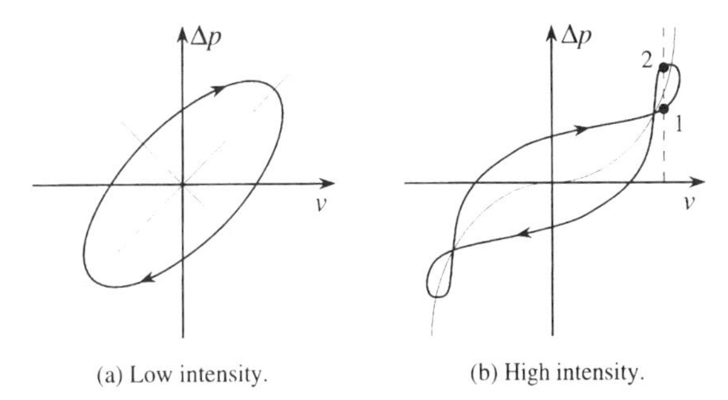

(a) Low intensity. (b) High intensity.

Fig. 4 Procedure for measuring temporal impedance.

Both resistance and reactance may be derived from a test of this sort if it is assumed that each is a function of velocity only. Simply select two times t_1 and t_2 on the trajectory where the velocities are equal by drawing a vertical line as shown on the right of Fig. 4. Accelerations $a = \dot{v}$ are not equal at these different times so that Eqs. (3) and (4) produce two simultaneous equations:

$$R_t[v] + a_1 X_t[v] = \Delta p(t_1), \qquad R_t[v] + a_2 X_t[v] = \Delta p(t_2). \tag{5}$$

Solving these equations gives both resistance and reactance for the selected velocity. Repeating the process for a sequence of velocities generates both functions.

It was verified in Ref. 2 that dynamically measured resistance was equal to flow resistance for fundamental test frequencies up to 4000 Hz and sound pressure levels up to 157 dB. This result supported the fundamental assumption of temporal impedance, that resistance is a function of velocity only, at least for that one material sample.

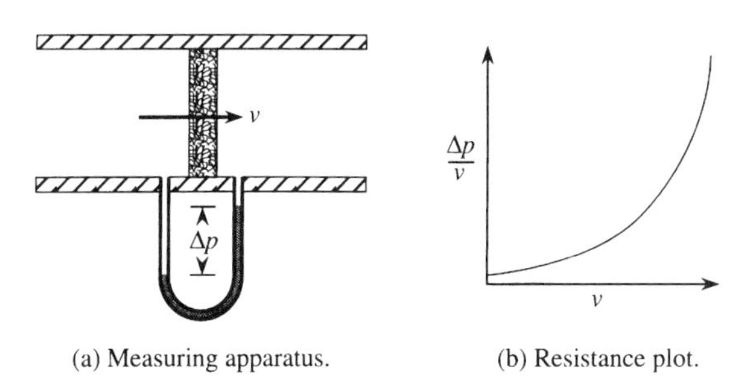

(a) Measuring apparatus. (b) Resistance plot.

Fig. 3 Flow resistance measurement.

4 EFFECTS OF NONLINEARITY

There are two cases where exact solutions are known[3] for the interaction of acoustic waves with a nonlinear lumped element. In the first case, an incident wave is transmitted through the element into a nonreflecting region behind the element. In the second case, the incident wave is periodic, and the space behind the element is one-quarter wavelength (of the fundamental) deep and terminates at a perfectly reflecting, or hard, wall. In both cases, a linear wave equation is assumed for the waves, but the lumped element is nonlinear. The waves thus give pressure and velocity in the forms

$$p(x,t) = f(x-t) + g(x+t), \qquad v(x,t) = f(x-t) - g(x+t). \tag{6}$$

All variables in Eq. (6) represent suitable nondimensional groups.

4.1 Nonreflecting Termination

Let the nonlinear element be placed at $x = 0$ and use subscripts 1 to denote waves where $x < 0$ and 2 to denote $x > 0$. Further, assume that the reactance is negligible so that the element is represented by

$$\Delta p(t) = R[v(t)]v(t), \qquad x = 0, \tag{7}$$

where v is the velocity through the element. The nonreflecting condition gives $g_2(x+t) = 0$. Expressions for velocity through and pressure drop across the element are then

$$v(t) = f_1(-t) - g_1(t) = f_2(-t),$$
$$\Delta p(t) = f_1(-t) + g_1(t) - f_2(-t). \tag{8}$$

Given the incident wave f_1, there results a single equation for the velocity $v(t)$:

$$(R[v(t)] + 2)v(t) - 2f_1(-t) = 0. \tag{9}$$

Since the resistance must be positive, the function on the left of Eq. (9) is a monotonically increasing function of $v(t)$ and has a single real-valued solution. The solution for the case where $R[v] = 1 + v^2$ and $f_1(x-t) = \cos 2\pi(x-t)$ is

$$v(t) = -\left(\cos 2\pi t + \sqrt{1 + \cos^2 2\pi t}\right)^{-1/3} + \left(\cos 2\pi t + \sqrt{1 + \cos^2 2\pi t}\right). \tag{10}$$

Clearly, a simple harmonic wave will be transmitted through a nonlinear element as a spectrum of harmonics of the incident wave.

4.2 Quarter-Wave Backing

Now consider the interaction of waves with the element at $x = 0$ but with a hard wall at $x = \frac{1}{4}$. Assume periodic waves such that $f(x - t + n) = f(x - t)$ and $g(x + t + n) = g(x + t)$ for all integers n. The condition of perfect reflection at $x = \frac{1}{4}$ gives $v(t + \frac{1}{2}) = -v(t)$, and if $R[v]$ is an even function, the following algebraic equation for the velocity through the element results:

$$(R[v(t)] + 1)v(t) + [f_1(\tfrac{1}{2} - t) - f_1(-t)] = 0. \tag{11}$$

Using the same resistance and incident wave as before, the velocity through the element is

$$v(t) = -\frac{2}{3}\left(\cos 2\pi t + \sqrt{\frac{8}{27} + \cos^2 2\pi t}\right)^{-1/3} + \left(\cos 2\pi t + \sqrt{\frac{8}{27} + \cos^2 2\pi t}\right). \tag{12}$$

5 CONCLUDING REMARKS

Lumped nonlinear acoustic elements cause interactions of waves with different frequencies. The effect of the element is influenced by the complete spectrum of the flow velocity through the element. The description of this complex process is simplified by using the time domain. In the time domain, acoustic impedance is replaced by a temporal impedance operator. The acoustic resistance that appears in this operator is identical to the steady-flow resistance for a broad range of frequencies so that the flow resistance is the primary descriptor of the nonlinear behavior of the element. The nature of the temporal reactance is not so clear. While it may be measured, it seems to depend on more than just the instantaneous flow velocity. Kuntz[4] has developed models of these functions for porous materials. This reference is recommended for the study of distributed nonlinear acoustic elements.

Grazing flow over porous materials has an important effect on their impedance. Rao and Munjal[5] have evaluated these effects for perforates. Further information on these effects and on the application of lumped-parameter methods to the analysis and design of internal engine mufflers is given by Sullivan and Crocker.[6–9]

REFERENCES

1. T. H. Melling, "The Acoustic Impedance of Perforates at Medium and High Sound Pressure Levels" *J. Sound Vib.*, Vol. 29, No. 1, 1973, pp. 1–65.
2. W. E. Zorumski and T. L. Parrott, "Nonlinear Acoustic Theory for Rigid Porous Materials," NASA TN D-6196, June 1971.
3. W. E. Zorumski, "Acoustic Scattering by a Porous Elliptic Cylinder with Nonlinear Resistance," Ph.D. Dissertation, Virginia Polytechnic Institute, March 1970.
4. H. L. Kuntz, II, "High Intensity Sound in Air Saturated Fibrous Bulk Porous Materials," ARL-TR-82-54, Applied Research Laboratories, University of Texas at Austin, September 1982.
5. K. N. Rao and M. L. Munjal, "Experimental Evaluation of Impedance of Perforates with Grazing Flows," *J. Sound Vib.*, Vol. 108, No. 2, July, 1986, pp. 283–295.
6. J. W. Sullivan and M. J. Crocker, "Analysis of Concentric-Tube Resonators Having Unpartitioned Cavities," *J. Acoust. Soc. Am.*, Vol. 64, No. 1, 1978, pp. 207–215.
7. J. W. Sullivan, "A Method for Modelling Perforated Tube Muffler Components. I. Theory," *J. Acoust. Soc. Am.*, Vol. 66, No. 3, 1979, pp. 772–778.
8. J. W. Sullivan, "A Method for Modelling Perforated Tube Muffler Components. II. Application," *J. Acoust. Soc. Am.*, Vol. 66, No. 3, 1979, pp. 779–788.
9. J. W. Sullivan, "Some Gas Flow and Acoustic Pressure Measurements Inside a Concentric-Tube Resonator," *J. Acoust. Soc. Am.*, Vol. 76, No. 2, 1984, pp. 479–484.

22

CAVITATION

WERNER LAUTERBORN

1 INTRODUCTION

Cavitation, the rupture of liquids including the effects connected with it, may be classified according to the scheme given in Fig. 1. Cavitation is brought about by either tension in the liquid or a deposit of energy. The main technical areas where tension in a liquid plays a role are hydrodynamics and acoustics. Hydrodynamic cavitation occurs with ship propellers, turbines, pumps, and hydrofoils, devices that all introduce a pressure reduction in the liquid via Bernoulli pressure forces and eventually generate strong tension. Acoustic cavitation occurs with underwater sound projectors and vibrating containers as, for example, used for cleaning. In sufficiently strong sound fields the tension generated during the pressure reduction phase of the sound wave may produce cavitation. Energy deposit can be brought about in various ways. When light of high intensity is propagating through a liquid, for example, laser light is focused into it, cavities may form by heating or dielectric breakdown. When, instead of photons, elementary particles are propagating (e.g., protons), cavities may form along their paths (bubble chamber). Other forms are local heat addition (boiling) and strong (static) electric fields (underwater spark). Acoustic cavitation has been emphasized in Fig. 1 because it is the topic of the present chapter. A selection of useful books and survey articles in descending order of their date of issue is given in Refs. 1–14.

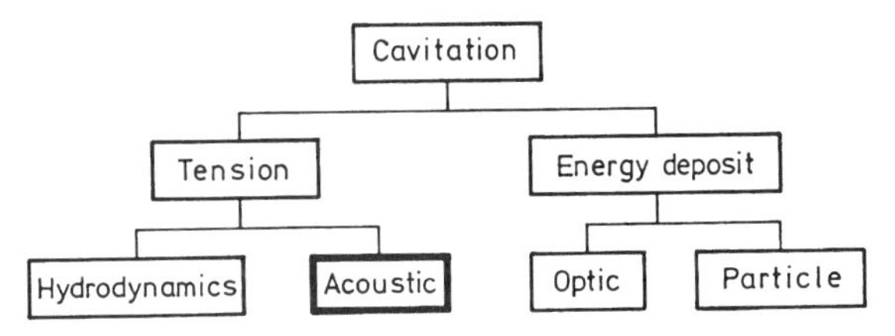

Fig. 1 Classification scheme for the different types of cavitation.

2 DEVICES

There exists a variety of devices to produce acoustic cavitation. The "Mason horn" (Fig. 2) is mainly used for cavitation erosion tests (besides in drilling). The probe to be tested forms the tip of an amplitude transformer (the horn) dipped into the liquid or is put at some distance opposite to the tip to get rid of the large accelerations encountered on the tip. The transducer driving the horn is usually made up of a sandwich of nickel plates (making use of magnetostriction) or plates of piezoelectric mate-

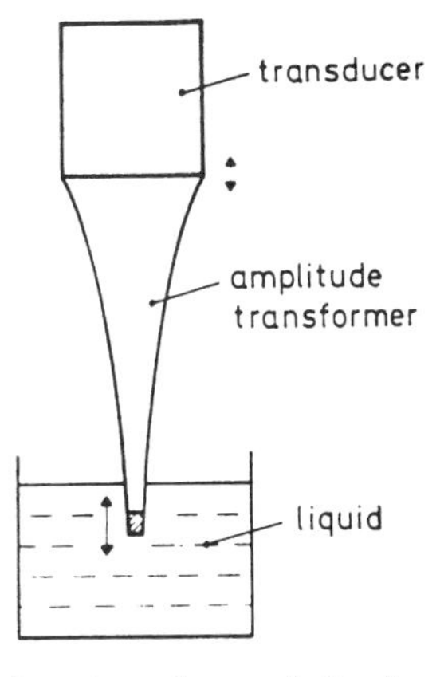

Fig. 2 Mason horn to produce cavitation for erosion studies.

Note: References to chapters appearing only in the *Encyclopedia* are preceded by *Enc.*

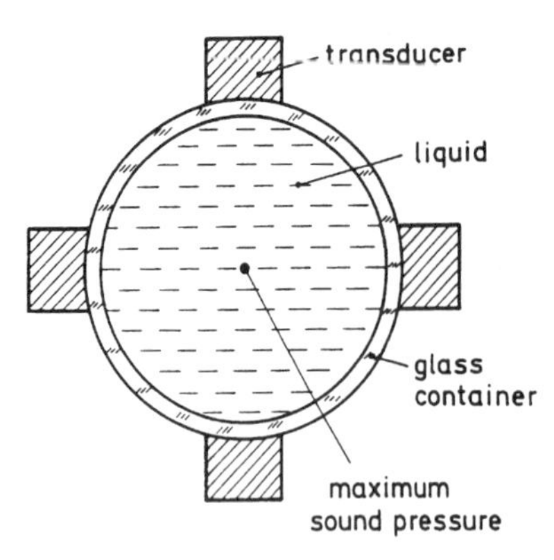

Fig. 3 Spherical or cylindrical container filled with liquid for cavitation threshold, bubble oscillation, and sonoluminescence studies.

rials ($BaTiO_4$ or PZT ceramics). Figure 3 shows a device to investigate cavitation thresholds, bubble oscillations, and sonoluminescence.[15,16] A glass container, a sphere or cylinder, is driven by several transducers to generate a standing-wave pattern in the liquid at the fundamental (or some upper resonance) of the whole system. The fundamental resonance has the advantage that only one maximum of the sound pressure (and of the tension) occurs in the liquid. Also bubbles placed in the liquid can be "levitated," thus allowing extended study of their vibrations. Figure 4 shows another simple device to study cavitation thresholds, the dynamics of cavitation bubble clouds, and cavitation noise emission (acoustic chaos[17,18]). It consists of a hollow cylinder of piezoelectric material submerged in the liquid to be cavitated. When the system is driven sinusoidally to excite the fundamental resonance, maximum sound pressure and tension occurs at the center of the cylinder.

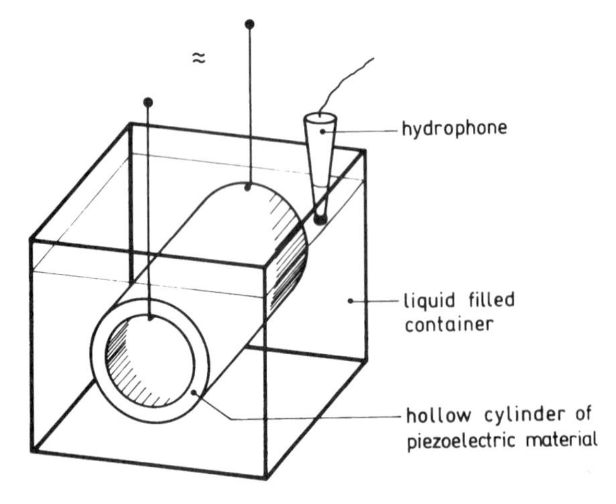

Fig. 4 Cylindrical transducer submerged in liquid for cavitation noise measurements (acoustic chaos).

3 CAVITATION THRESHOLD

Sound waves in liquids need a certain intensity for cavitation to appear. Cavitation inception thus is a threshold process. It is commonly agreed that nuclei must mediate the inception, that is, tiny bubbles being stabilized by some mechanism against dissolution. There is the crevice model, where gas is trapped in conical pits of solid impurities,[19] and the skin model, with organic or surface-active molecules occupying the bubble wall.[20] When these nuclei encounter a sound field, they are set into oscillation and may grow by a process called rectified diffusion.[21] Rectified diffusion itself is a threshold

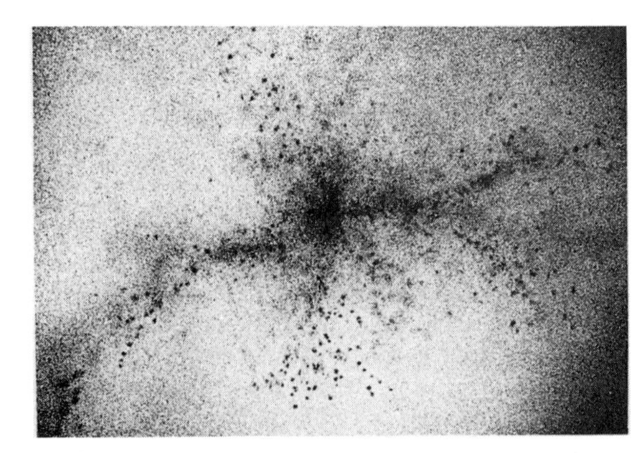

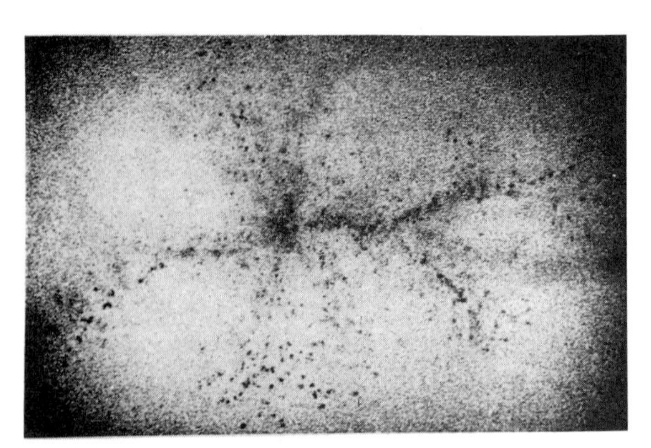

Fig. 5 Cavitation bubble cloud obtained inside a cylindrical transducer. Two planes from a holographic image. (Courtesy of F. Bader.)

process. It is the net effect of mass diffusion across the bubble wall due to varying bubble wall area and concentration gradients during oscillation. When a nucleus starts to grow, it eventually reaches a size where the oscillations turn into large excursions of the bubble radius, with strong and fast bubble collapse shattering the bubble into tiny fractions. This then marks the cavitation threshold, and acoustic cavitation sets in.

4 OBSERVED BUBBLE DYNAMICS AND PATTERN FORMATION

Substantial noise emission occurs upon acoustic cavitation. It is brought about by the action of the cavities or bubbles generated and set into oscillation in the sound field. Usually many bubbles are involved, forming a complex dynamic pattern (Fig. 5). The individu-

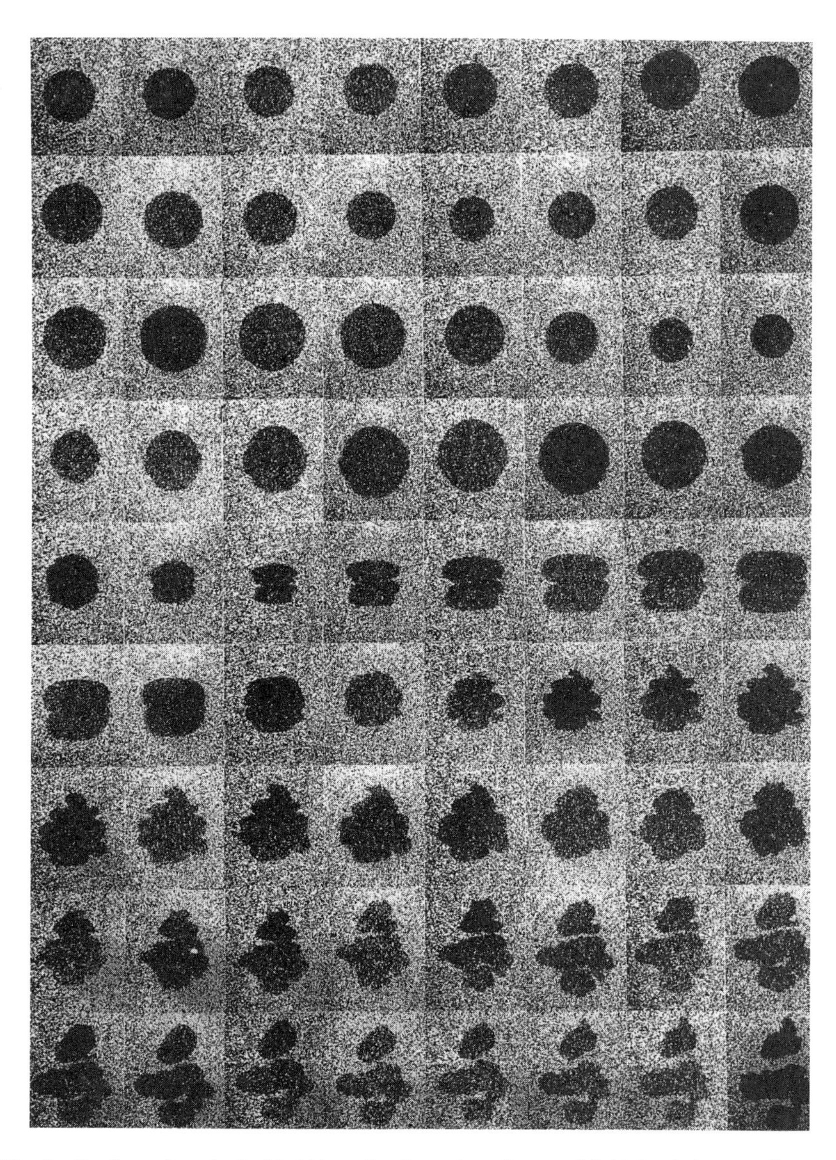

Fig. 6 Breakup of a spherical bubble under the action of a sound field that induces surface oscillations. Reconstructed images from a series of holograms taken at a rate of 66,700 holograms per second. Individual picture size is 2.4×2 mm. (Courtesy of W. Hentschel.)

ally moving and oscillating bubbles arrange themselves along a branchlike, filamentary structure, the filaments being called streamers. The pattern is steadily rearranging but is stable in gross appearance over hundreds of cycles of the driving sound field and resembles a Lichtenberg figure.

The processes leading to this type of structured bubble ensemble are very complex due to an interplay between competing mechanisms based on instabilities and cooperation. There are at least two time scales involved, a fast one connected with bubble oscillation and a slow one connected with bubble migration and spatial pattern formation. The spherical shape of a bubble is due to the surface tension of the liquid. This shape becomes unstable at higher oscillation amplitudes whereby surface waves[22,23] are set up on the bubble wall, eventually leading to a breakup of the bubble (Fig. 6). This shattering yields a fast proliferation of small bubbles that is stopped when the bubbles become too small to be excited to large amplitudes.

In addition to the appearance of surface waves destroying a bubble, there is a related mechanism operating in aspherical geometries surrounding a cavitation bubble. The collapse of a bubble, that is, its fast approach to minimum size, may proceed with the formation of a liquid jet piercing the bubble.[5,24,25] The jet leads to a long protrusion sticking out of the bubble and to the formation of a vortex ring decaying into a host of microbubbles.

There is also the competing opposite phenomenon of coalescence in which, upon expansion, bubbles come into contact with each other. These effects surely are not sufficient to describe the pattern formation, as some attracting and guiding forces are needed. These are the Bjerknes forces,[26] which are attracting when two bubbles are oscillating in phase, and the radiation pressure forces, which drive the bubbles toward the pressure antinodes in a standing acoustic field when they are smaller than resonant size. Moreover, the bubbles themselves radiate sound and are thus interacting additionally. A first approach to account for the observed pattern formation in acoustic cavitation is given elsewhere.[27]

5 SINGLE-BUBBLE DYNAMICS

Cavitation bubble clouds are difficult to investigate both experimentally and theoretically. Therefore much effort is spent on single-bubble dynamics. Laser-produced bubbles are used to investigate the involved dynamics of single bubbles near boundaries with jet formation (Fig. 7), shock wave radiation, and vortex ring formation.[25,28] Bubble collapse is accompanied by strong shock wave emission[28,29] (Fig. 8). It comes about

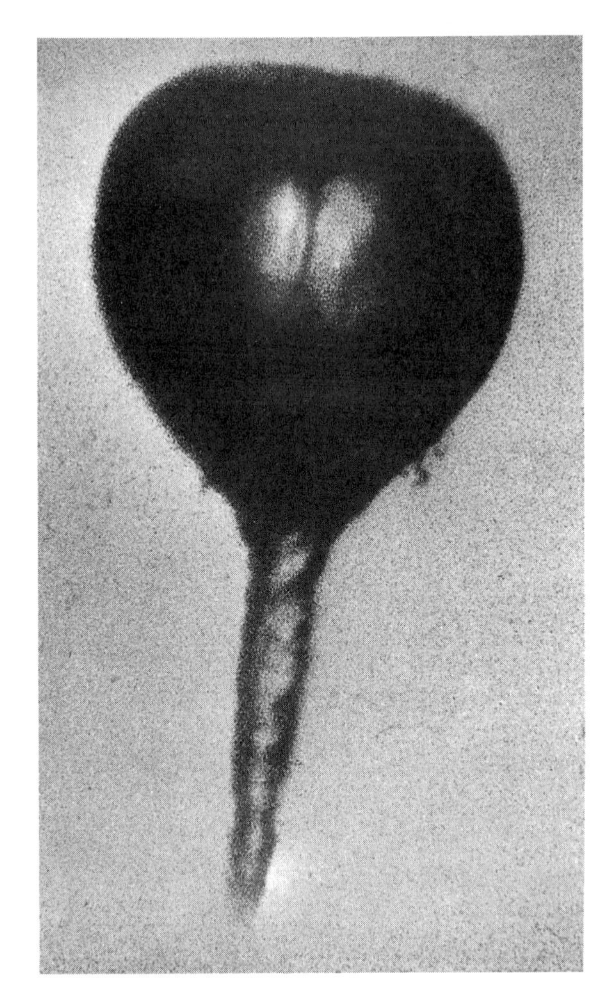

Fig. 7 Jet formation in the case of a bubble collapsing in the neighborhood of a solid wall (below the lower margin of the picture). The jet pierces the bubble from above and leads to the long protrusion.

through the sudden stop of the inflowing liquid when the bubble approaches a tiny fraction of its initial volume with a correspondingly high internal pressure. Simultaneously, high temperatures are reached inside the bubble, as indicated by the emission of a short flash of light (sonoluminescence).[15,16]

Theory is not yet developed sufficiently to cope with the experimental dynamic system completely. Just the single spherical bubble can be treated fairly well.[11,30] The simplest bubble model is called the Rayleigh–Plesset model:

$$\rho R\ddot{R} + \tfrac{3}{2}\rho\dot{R}^2 = P_i - P_e, \quad (1)$$

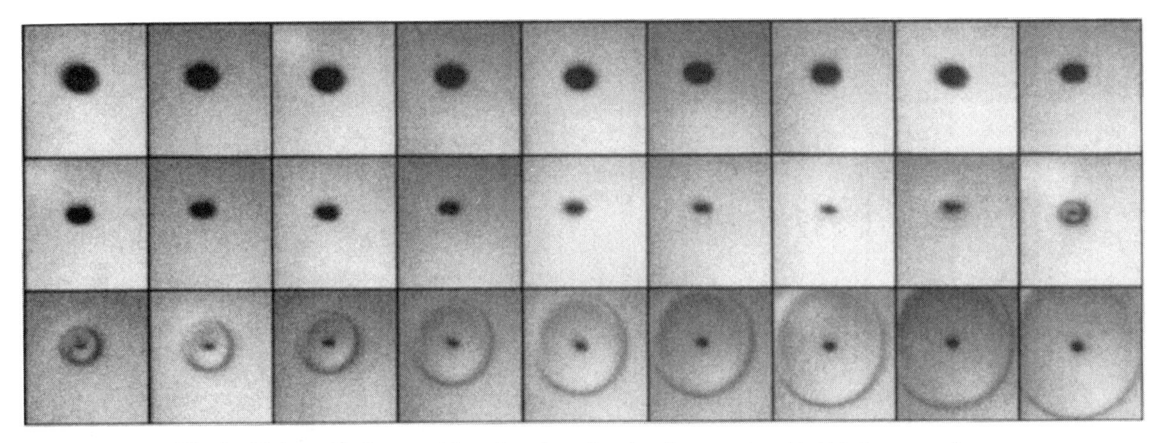

Fig. 8 High-speed photographic series of a collapsing, laser-produced bubble in water taken at 20.8 million frames per second. Picture size is 1.5 × 1.8 mm. (Courtesy of C. D. Ohl.)

where R is the bubble radius, an overdot means differentiation with respect to time, ρ is the density of the liquid, and P_i and P_e are internal (in the bubble) pressure and external (in the liquid including the surface to the bubble) pressure, respectively. The difference in pressure drives the bubble motion. The form of the inertial terms on the left-hand side is the result of the condensation of the spherical three-dimensional geometry to one dimension in the differential equation. Both P_i and P_e are functions of the radius R and time t, when gas and vapor fill the bubble, and surface tension σ, (kinematic) liquid viscosity μ, and a sound field are taken into account. Then the Rayleigh–Plesset model takes the form[31]

$$\rho R\ddot{R} + \frac{3}{2}\rho\dot{R}^2 = P_{gn}\left(\frac{R_n}{R}\right)^{3\kappa} + P_v - P_{\text{stat}} - \frac{2\sigma}{R} - \frac{4\mu}{R}\dot{R} - P(t), \quad (2)$$

with

$$P_{gn} = \frac{2\sigma}{R_n} + P_{\text{stat}} - P_v, \quad (3)$$

the equilibrium gas pressure inside the bubble at equilibrium radius R_n, and

$$P(t) = -P_A \sin \omega t, \quad (4)$$

the acoustic pressure of angular frequency ω and pressure amplitude P_A.

It is assumed that P_v, the vapor pressure in the bubble, remains constant during the motion. In the case of water at 20°C, P_v will be small and will change the value of P_{stat}, the static pressure in the liquid at infinity corrected by the hydrostatic pressure at the location of the bubble, only slightly. The gas in the bubble is assumed to be compressed according to a polytropic gas law with the polytropic exponent κ. Usually κ is assumed constant during the motion.

In the bubble model of Eq. (2) viscosity is the only damping mechanism. A model taking the damping by sound radiation into account is the Gilmore model[12,13,32]:

$$\left(1 - \frac{\dot{R}}{C}\right)R\ddot{R} + \frac{3}{2}\left(1 - \frac{1}{3}\frac{\dot{R}}{C}\right)\dot{R}^2 = \left(1 + \frac{\dot{R}}{C}\right)H + \left(1 - \frac{\dot{R}}{C}\right)\frac{R}{C}\dot{H}, \quad (5)$$

where C is the liquid sound velocity at the bubble interface and H the enthalpy evaluated at the bubble wall:

$$H = \int_{P\infty}^{P(R)} \frac{dp}{\rho}. \quad (6)$$

Even more advanced bubble models have been developed to better incorporate a time-varying pressure as given by an external sound field.[33] These models have been investigated for the response of a bubble in an acoustic field.[34,35] A behavior typical for nonlinear oscillators is found.[36] Period-doubling cascades to chaos (aperiodic oscillations with broadband spectra) abound in

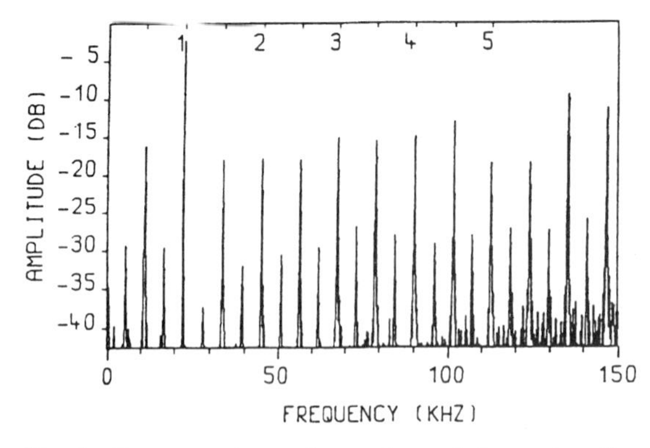

Fig. 9 Power spectrum of acoustic cavitation noise after second period doubling. One-fourth of the driving frequency (23.56 kHz) and its harmonics are present.[39]

all models, providing a clue to why the first subharmonic in the spectrum of the cavitation noise is so pronounced.

6 CAVITATION NOISE MEASUREMENTS

Measurements of the noise emission have been done with subsequent application of the new methods from nonlinear dynamics to the sound output from the liquid. These are phase space analysis, dimension analysis, and Lyapunov analysis.[37] A period-doubling route to chaos has been found.[38] Figure 9 gives a power spectrum of the noise output after two period doublings have taken place.[39] After a cascade of period doublings a chaotic noise attractor appears (Fig. 10). It is possible to determine the dimension of the noise attractors. Surprisingly small fractal dimensions between 2 and 3 are found. The appearance of fractal attractors as well as period-doubling sequences suggests that the system of oscillating bubbles in an acoustic field is a chaotic system. The definition of a chaotic system is that at least one of the Lyapunov exponents of the system making up the Lyapunov spectrum should be positive. A Lyapunov exponent is a measure of how fast two neighboring trajectories in phase space separate. The calculation of the Lyapunov spectrum for a (chaotic) noise attractor indeed yields a positive Lyapunov exponent.[40] Thus acoustic cavitation noise has been proven to be a chaotic system with only a small number of (nonlinear) degrees of freedom. According to the fractal dimension, only three variables should be sufficient to describe the dynamics of the system. This finding suggests that a high degree of cooperation must take place among the bubbles. The highly structured bubble ensemble confirms this view.

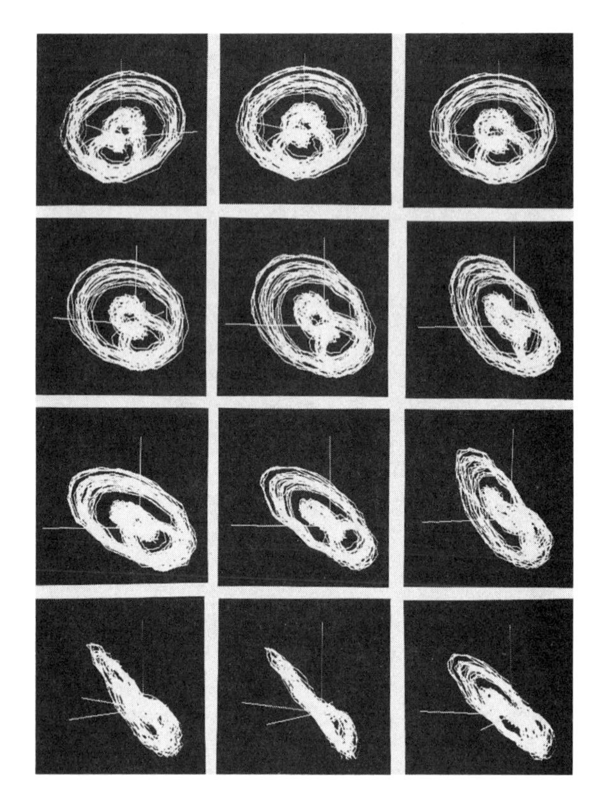

Fig. 10 Chaotic noise attractor as obtained by embedding the noise data into a three-dimensional phase space. (Courtesy of J. Holzfuss.)

7 CAVITATION EFFECTS

The violent motion of cavities in a sound field not only is the source of the peculiar noise emission discussed but also gives rise to other phenomena, notably the destructive action[41] on all kinds of material from soft tissue to hard steel. This action appears quite natural in view of the fact that high pressures and temperatures are reached in collapsing bubbles. This aggressive action may be put to good use for cleaning, for example, for removing grinding material from lenses. In the majority of cases, however, the destructive action is unwanted and may lead to failure of the parts suffering cavitation by progressive removal of material. This process is called cavitation erosion and was first observed with ship propellers. With acoustic cavitation it has been used for accelerated cavitation damage tests. Aluminum bronze and titanium proved to be quite cavitation resistant.[13,42] Figure 11 shows a damage pit from one single cavitation bubble collapsing on an aluminum specimen. The dam-

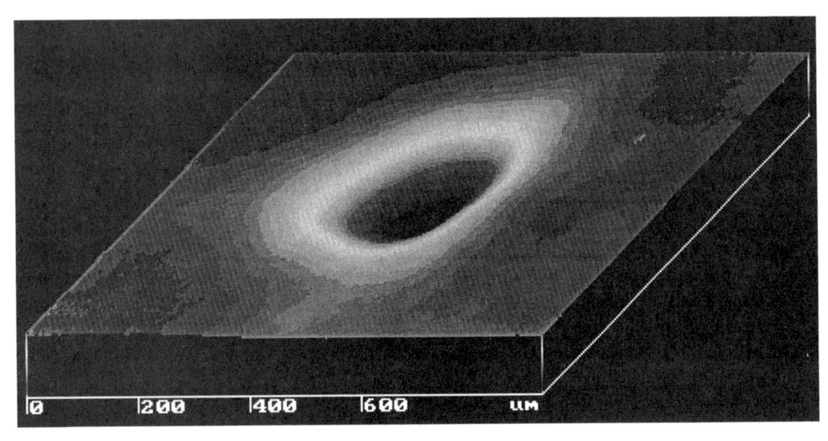

Fig. 11 Damage pit from a single laser-produced bubble collapsing in the neighborhood of a solid boundary (aluminum specimen). Pit depth 1.0 μm, pit diameter 560 μm. (Courtesy of A. Philipp.)

aging process is very involved and of a statistical nature even under highly controlled conditions. It seems that only a bubble collapsing in contact with the boundary leads to pit formation.

It has not yet been possible to study the dynamics of pit formation and its relation to bubble dynamics. There are several candidates presumably being involved; these are shock waves, the liquid jet, and the high pressure and temperature in the bubble that act on the surface when the bubble touches it in the final collapse phase.

Sound waves, in particular those in the megahertz range, are used in medicine for diagnosis and therapy. As cavitation is connected with high pressures and temperatures and the possibility of damage, it normally has to be avoided for reasons of safety. Therefore quite a number of experiments have been conducted to learn about the damage potential to biologic material (tissue, blooc cells, etc.). It has been found that, whenever gas bubbles are present, the damage potential of ultrasonic waves is high.[43] Even a few 10 mW/cm^2 peak intensities suffice to induce damage. Damage may be reduced by pulsed ultrasound with pulse lengths of a few microseconds only. Unless the duty cycle is also low, however, damage may nevertheless occur, and even in one acoustic cycle preexisting cavities may be set into violent motion if they fit the appropriate conditions.

8 APOLOGY

Some basic topics of acoustic cavitation have been reviewed, essentially those pertaining to the process itself. Space limitations have not allowed us to include the plethora of effects of ultrasound in liquids that have connections to cavitation, as there are lithotripsy, chemical effects,[44] sonoluminescence, biological effects, and so on. Also, the author apologizes to all those having contributed to acoustic cavitation but that could not be mentioned here because of space constraints.

REFERENCES

1. J. R. Blake, J. M. Boulton-Stone, and N. H. Thomas (Eds.), *Bubble Dynamics and Interface Phenomena*, Kluwer, Dordrecht, 1994.

1a. T. G. Leighton, *The Acoustic Bubble*, Academic Press, London, 1994.

2. L. M. Lyamshev, "Radiation Acoustics," *Colloque de Physique C2*, Vol. 51, 1990, pp. C2-1 to C2-7.

3. F. R. Young, *Cavitation*, McGraw-Hill, London, 1989.

4. A. A. Atchley and L. A. Crum, "Acoustic Cavitation and Bubble Dynamics," in K. S. Suslick (Ed.), *Ultrasound: Its Chemical, Physical and Biological Effects*, VCH Publishers, New York, 1988, pp. 1–64.

5. J. R. Blake and D. C. Gibson, "Cavitation Bubbles Near boundaries," *Ann. Rev. Fluid Mech.*, Vol. 19, 1987, pp. 99–123.

6. L. A. Crum, "Acoustic Cavitation," in B. R. Avoy (Ed.), *Proceedings of the 1982 Ultrasonics Symposium*, IEEE Press, New York, 1983, pp. 1–12.

7. L. van Wijngaarden (Ed.), *Mechanics and Physics of Bubbles in Liquids*, Martinus Nijhoff Publishers, The Hague, 1982.

8. R. E. Apfel, "Acoustic Cavitation," in P. D. Edmonds (Ed.), *Methods of Experimental Physics*, Vol. 19, Academic Press, New York, 1981, pp. 355–411.

9. W. Lauterborn (Ed.), *Cavitation and Inhomogeneities in Underwater Acoustics*, Springer, Berlin, 1980.
10. E. A. Neppiras, "Acoustic Cavitation," *Phys. Rep.*, Vol. 61, 1980, pp. 159–251.
11. M. S. Plesset and A. Prosperetti, "Bubble Dynamics and Cavitation," *Ann. Rev. Fluid Mech.*, Vol. 9, 1977, pp. 145–185.
12. L. D. Rozenberg (Ed.), *High Intensity Ultrasonic Fields*, Plenum, New York, 1971.
13. R. T. Knapp, J. W. Daily, and F. G. Hammitt, *Cavitation*, McGraw-Hill, London, 1970.
14. H. G. Flynn, "Physics of Acoustic Cavitation in Liquids," in W. P. Mason (Ed.), *Physical Acoustics*, Vol. 1, Part B, Academic, New York, 1964, pp. 57–112.
15. L. A. Crum, "Sonoluminescence," *Physics Today*, September 1994, pp. 22–29.
16. A. J. Walton and G. T. Reynolds, "Sonoluminescence," *Adv. Phys.*, Vol. 33, 1984, pp. 595–660.
17. W. Lauterborn and J. Holzfuss, "Acoustic Chaos," *Int. J. Bifurcation Chaos*, Vol. 1, 1991, pp. 13–26.
18. W. Lauterborn, J. Holzfuss, and A. Billo, "Chaotic Behavior in Acoustic Cavitation," in M. Levy, S. C. Schneider, and B. R. Avoy (Eds.), *Proceedings of the 1994 Ultrasonics Symposium*, IEEE Press, New York 1994, pp. 801–810.
19. L. A. Crum, "Nucleation and Stabilization of Microbubbles in Liquids," *Appl. Sci. Res.*, Vol. 38, 1982, pp. 101–115.
20. D. E. Yount, E. W. Gillary, and D. C. Hoffmann, "A Microscopic Investigation of Bubble Formation Nuclei," *J. Acoust. Soc. Am.*, Vol. 76, 1984, pp. 1511–1521.
21. L. A. Crum, "Rectified Diffusion," *Ultrasonics*, Vol. 22, 1984, pp. 215–223.
22. M. S. Plesset and T. P. Mitchell, "On the Stability of the Spherical Shape of a Vapor Cavity in a Liquid," *Q. Appl. Math.*, Vol. 13, 1956, pp. 419–430.
23. H. W. Strube, "Numerische Untersuchungen zur Stabilität nichtsphärisch schwingender Blasen," *Acustica*, Vol. 25, 1971, pp. 289–330.
24. T. B. Benjamin and A. T. Ellis, "The Collapse of Cavitation Bubbles and the Pressure Thereby Produced Against Solid Boundaries," *Phil. Trans. Roy. Soc. Lond.*, Vol. A260, 1966, pp. 221–240.
25. A. Vogel, W. Lauterborn, and R. Timm, "Optical and Acoustic Investigations of the Dynamics of Laser-Produced Cavitation Bubbles Near a Solid Boundary," *J. Fluid Mech.*, Vol. 206, 1989, pp. 299–338.
26. L. A. Crum, "Bjerknes Forces on Bubbles in a Stationary Sound Field," *J. Acoust. Soc. Am.*, Vol. 57, 1975, pp. 1363–1370.
27. I. Akhatov, U. Parlitz, and W. Lauterborn, "Pattern Formation in Acoustic Cavitation," *J. Acoust. Soc. Am.*, Vol. 96, 1994, pp. 3627–3635.
28. C. D. Ohl, A. Phillipp, and W. Lauterborn, "Cavitation Bubble Collapse Studied at 20 Million Frames per Second," *Ann. Phys.*, Vol. 4, 1995, pp. 26–34.
29. R. Hickling and M. S. Plesset, "Collapse and Rebound of a Spherical Bubble in Water," *Phys. Fluids*, Vol. 7, 1964, pp. 7–14.
30. A. Prosperetti, "Physics of Acoustic Cavitation," in D. Sette (Ed.), *Frontiers in Physical Acoustics*, North-Holland, Amsterdam, 1986, pp. 145–188.
31. B. E. Noltingk and E. A. Neppiras, "Cavitation Produced by Ultrasonics," *Proc. Phys. Soc. Lond.*, Vol. B63, 1950, pp. 674–685.
32. F. R. Gilmore, "The Growth or Collapse of a Spherical Bubble in a Viscous Compressible Liquid," California Institute of Technology Report No 26-4, Pasadena, CA, 1952, pp. 1–40.
33. J. B. Keller and M. Miksis, "Bubble Oscillations of Large Amplitude," *J. Acoust. Soc. Am.*, Vol. 68, 1980, pp. 628–633.
34. W. Lauterborn, "Numerical Investigation of Nonlinear Oscillations of Gas Bubbles in Liquids," *J. Acoust. Soc. Am.*, Vol. 59, 1976, pp. 283–293.
35. U. Parlitz, V. English, C. Scheffczyk, and W. Lauterborn, "Bifurcation Structure of Bubble Oscillators," *J. Acoust. Soc. Am.*, Vol. 88, 1990, pp. 1061–1077.
36. C. Scheffczyk, U. Parlitz, T. Kurz, W. Knop, and W. Lauterborn, "Comparison of Bifurcation Structures of Driven Dissipative Nonlinear Oscillators," *Phys. Rev. A*, Vol. 43, 1991, pp. 6495–6501.
37. W. Lauterborn and U. Parlitz, "Methods of Chaos Physics and Their Application to Acoustics," *J. Acoust. Soc. Am.*, Vol. 84, 1988, pp. 1975–1993.
38. W. Lauterborn and E. Cramer, "Subharmonic Route to Chaos Observed in Acoustics," *Phys. Rev. Lett.*, Vol. 47, 1981, pp. 1445–1448.
39. W. Lauterborn, "Acoustic Turbulence," in D. Sette (Ed.), *Frontiers in Physical Acoustics*, North-Holland, Amsterdam, 1986, pp. 124–144.
40. J. Holzfuss and W. Lauterborn, "Liapunov Exponents from Time Series of Acoustic Chaos," *Phys. Rev. A*, Vol. 39, 1989, pp. 2146–2152.
41. Y. Tomita and A. Shima, "Mechanisms of Impulsive Pressure Generation and Damage Pit Formation by Bubble Collapse," *J. Fluid Mech.*, Vol. 169, 1986, pp. 535–564.
42. I. S. Pearsall, *Cavitation*, Mills and Boon, London, 1972.
43. D. L. Miller, "Gas Body Activation," *Ultrasonics*, Vol. 22, 1984, pp. 261–271.
44. K. S. Suslick, "Sonochemistry," *Science*, Vol. 247, 1990, pp. 1439–1445.

23

SONOCHEMISTRY AND SONOLUMINESCENCE

KENNETH S. SUSLICK AND LAWRENCE A. CRUM

1 INTRODUCTION

High-energy chemical reactions occur during the ultrasonic irradiation of liquids.[1–6] The chemical effects of ultrasound, however, do not come from a direct interaction with molecular species. The velocity of sounds in liquids is typically about 1500 m/s; ultrasound spans the frequencies of roughly 15 kHz to 1 GHz, with associated acoustic wavelengths of 10–10^{-4} cm. These are not molecular dimensions. No direct coupling of the acoustic field with chemical species on a molecular level can account for sonochemistry or sonoluminescence. Instead, these phenomena derive principally from acoustic cavitation: the formation, growth, and implosive collapse of bubbles in a liquid. Cavitation serves as a means of concentrating the diffuse energy of sound. Bubble collapse induced by cavitation produces intense local heating, high pressures, and very short lifetimes. In clouds of cavitating bubbles, these hot spots[7,8] have equivalent temperatures of roughly 5000 K, pressures of about 2000 atm, and heating and cooling rates above 10^9 K/s.

Related phenomena occur with cavitation in liquid–solid systems. Near an extended solid surface, cavity collapse is nonspherical and drives high-speed jets of liquid into the surface.[9] This process can produce newly exposed, highly heated surfaces. Furthermore, during ultrasonic irradiation of liquid–powder slurries, cavitation and the shock waves it creates can accelerate solid particles to high velocities.[10] The resultant interparticle collisions are capable of inducing dramatic changes in surface morphology, composition, and reactivity.[11]

The chemical effects of ultrasound are diverse and include dramatic improvements in both stoichiometric and catalytic reactions.[12–15] In some cases, ultrasonic irradiation can increase reactivity by nearly a millionfold.[16] The chemical effects of ultrasound fall into three areas: homogeneous sonochemistry of liquids, heterogeneous sonochemistry of liquid–liquid or liquid–solid systems, and sonocatalysis (which overlaps the first two). Chemical reactions are not generally seen in the ultrasonic irradiation of solids or solid–gas systems.

Sonoluminescence may be considered a special case of homogeneous sonochemistry; however, recent discoveries in this field have heightened interest in the phenomenon in and by itself.[17,18] New data on the duration of the sonoluminescence flash suggest that under the conditions of single-bubble sonoluminescence (SBSL), a shock wave may be created within the collapsing bubble with the capacity to generate enormous temperatures and pressures within the gas.

Acoustic cavitation provides the potential for creating exciting new physical and chemical conditions in otherwise cold liquids and results in an enormous concentration of energy. If one considers the energy density in an acoustic field that produces cavitation and that in the collapsed cavitation bubble, there is an amplification factor of over 11 orders of magnitude. The enormous local temperatures and pressures so created result in phenomena such as sonochemistry and sonoluminescence and provide a unique means for fundamental studies of chemistry and physics under extreme conditions.

2 MECHANISTIC ORIGINS OF SONOCHEMISTRY AND SONOLUMINESCENCE

2.1 Hot-Spot Formation during Cavitation

The most important acoustic process for sonochemistry and sonoluminescence is cavitation. Compression of a

Note: References to chapters appearing only in the *Encyclopedia* are preceded by *Enc.*

gas generates heat. When the compression of cavities occurs in irradiated liquids, it is more rapid than thermal transport, which generates a short-lived, localized hot spot. There is a general consensus that this hot spot is the source of homogeneous sonochemistry. Rayleigh's early descriptions of a mathematical model for the collapse of cavities in incompressible liquids predicted enormous local temperatures and pressures.[19] Ten years later, Richards and Loomis reported the first chemical and biologic effects of ultrasound.[20] Alternative mechanisms involving electrical microdischarge have been occasionally proposed[21,22] but remain a minority viewpoint.

2.2 Two-Site Model of Sonochemical Reactivity

The transient nature of the cavitation event precludes conventional measurement of the conditions generated during bubble collapse. Chemical reactions themselves, however, can be used to probe reaction conditions. The effective temperature realized by the collapse of clouds of cavitating bubbles can be determined by the use of competing unimolecular reactions whose rate dependencies on temperature have already been measured. This technique of *comparative-rate chemical thermometry* was used by Suslick, Hammerton, and Cline to first determine the effective temperature reached during cavity collapse.[7] The sonochemical ligand substitutions of volatile metal carbonyls were used as these comparative-rate probes [Eq. (1), where the arrow with three small closing parentheses on top represents ultrasonic irradiation of a solution and L represents a substituting ligand]:

$$M(CO)_x \xrightarrow{)))} M(CO)_{x-n} + n\text{-}CO \xrightarrow{L} M(CO)_{x-n}(L)_n,$$
$$\text{where } M = \text{Fe, Cr, Mo, W.} \qquad (1)$$

These kinetic studies revealed that there were in fact *two* sonochemical reaction sites: the first (and dominant site) is the bubble's interior gas phase while the second is an *initially* liquid phase. The latter corresponds either to heating of a shell of liquid around the collapsing bubble or to droplets of liquid ejected into the hot spot by surface wave distortions of the collapsing bubble.

The effective local temperatures in both sites were determined. By combining the relative sonochemical reaction rates for Eq. (1) with the known temperature behavior of these reactions, the conditions present during cavity collapse could then be calculated. The effective temperature of these hot spots was measured at ≈ 5200 K in the gas-phase reaction zone and ≈ 1900 K in the initially liquid zone.[7] Of course, the comparative-rate data represent only a composite temperature: During the collapse, the temperature has a highly dynamic profile as well as a spatial temperature gradient. This two-site model has been confirmed with other reactions,[23,24] and alternative measurements of local temperatures by sonoluminescence are also consistent,[8] as discussed later.

2.3 Microjet Formation during Cavitation at Liquid–Solid Interfaces

Cavitation near extended liquid–solid interfaces is very different from cavitation in pure liquids. There are two proposed mechanisms for the effects of cavitation near surfaces: microjet impact and shock wave damage. Whenever a cavitation bubble is produced near a boundary, the asymmetry of the liquid particle motion during cavity collapse induces a deformation in the cavity.[9] The potential energy of the expanded bubble is converted into kinetic energy of a liquid jet that extends through the bubble's interior and penetrates the opposite bubble wall. Because most of the available energy is transferred to the accelerating jet, rather than the bubble wall itself, this jet can reach velocities of hundreds of metres per second. Because of the induced asymmetry, the jet often impacts the local boundary and can deposit enormous energy densities at the site of impact. Such energy concentration can result in severe damage to the boundary surface. Figure 1 shows a photograph of a jet developed in a collapsing cavity; Fig. 2 shows a micrograph of an impact site. The second mechanism of cavitation-induced surface damage invokes shock waves created by cavity col-

Fig. 1 Photograph of liquid jet produced during collapse of a cavitation bubble. The width of the bubble is about 1 mm.

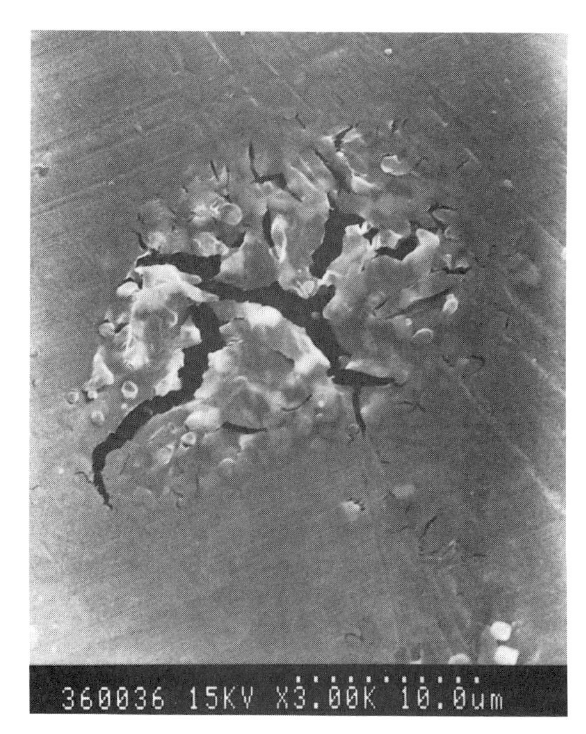

Fig. 2 Photomicrograph of a region of an aluminum foil (about 50 μm in thickness) exposed to collapsing cavitation bubbles produced by extracorporeal shock wave lithotriptor.

lapse in the liquid. The impingement of microjets and shock waves on the surface creates the localized erosion responsible for ultrasonic cleaning and many of the sonochemical effects on heterogeneous reactions. The erosion of metals by cavitation generates newly exposed, highly heated surfaces. Further details of jet and shock wave production and associated effects are presented elsewhere.[25–27]

Distortions of bubble collapse require a surface several times larger than the resonance bubble size. Thus, for solid particles smaller than $\approx$200 μm, damage associated with jet formation cannot occur with ultrasonic frequencies of $\approx$20 kHz. In these cases, however, the shock waves created by homogeneous cavitation can create high-velocity interparticle collisions.[10,11] Suslick and co-workers have found that the turbulent flow and shock waves produced by intense ultrasound can drive metal particles together at sufficiently high speeds to induce effective melting in direct collisions (Fig. 3) and the abrasion of surface crystallites in glancing impacts (Fig. 4). A series of transition-metal powders were used to probe the maximum temperatures and speeds reached during interparticle collisions. Using the irradiation of Cr, Mo, and W powders in decane at 20 kHz and 50

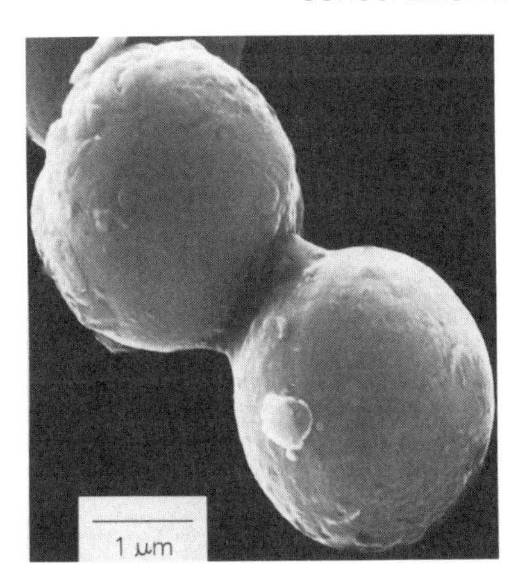

Fig. 3 Scanning electron micrograph of Zn powder after ultrasonic irradiation for 30 min at 288 K in decane under Ar at 20 kHz and $\approx$50 W/cm^2. (Reproduced from Ref. 10 with permission.)

W/cm^2, agglomeration and essentially a localized melting occur for the first two metals but not the third. On the basis of the melting points of these metals, the effective transient temperature reached at the point of impact during interparticle collisions is roughly 3000°C. From the volume of the melted region of impact, the amount of energy generated during collision was determined. From this, a lower estimate of the velocity of impact is roughly one-half the speed of sound.[10] These are precisely the effects expected on suspended particulates from cavitation-induced shock waves in the liquid.

3 SONOCHEMISTRY

High-intensity ultrasonic probes (50–500 W/cm^2) are the most reliable and effective source for laboratory-scale sonochemistry and are commercially available from several sources. Lower acoustic intensities can often be used

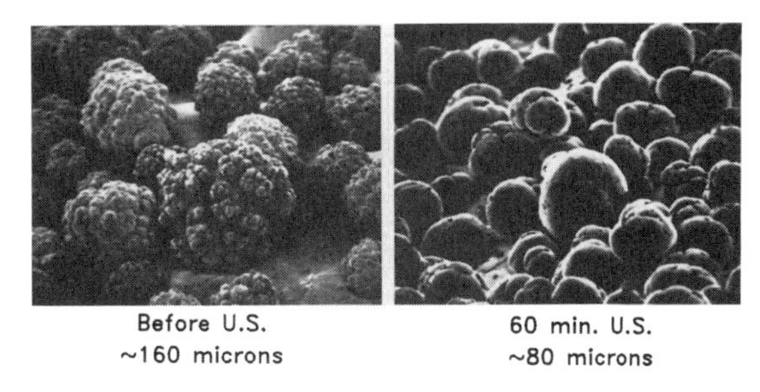

Fig. 4 Effect of ultrasonic irradiation on surface morphology of Ni powder. (Reproduced from Ref. 16 with permission.)

in liquid–solid heterogeneous systems, because of the reduced liquid tensile strength at the liquid–solid interface. For such reactions, a common ultrasonic cleaning bath will therefore often suffice. The low intensity available in these devices (≈ 1 W/cm^2), however, can prove limiting. In addition, the standing-wave patterns in ultrasonic cleaners require accurate positioning of the reaction vessel. On the other hand, ultrasonic cleaning baths are easily accessible, relatively inexpensive, and usable on a moderately large scale. Finally, for larger scale irradiation, flow reactors with high ultrasonic intensities are commercially available in modular units of ≈ 20 kW.

3.1 Comparison of Cavitation Conditions to Other Forms of Chemistry

Chemistry is the study of the interaction of energy and matter. Chemical reactions require energy, in one form or another, to proceed: Chemistry stops as the temperature approaches absolute zero. One has only limited control, however, over the nature of this interaction. In large part, the properties of a specific energy source determines the course of a chemical reaction. Ultrasonic irradiation differs from traditional energy sources (such as heat, light, or ionizing radiation) in duration, pressure, and energy per molecule. The immense local temperatures and pressures and the extraordinary heating and cooling rates generated by cavitation bubble collapse mean that ultrasound provides an unusual mechanism for generating high-energy chemistry. Similar to photochemistry, very large amounts of energy are introduced in a short period of time, but it is thermal, not electronic, excitation. As in flash pyrolysis, high thermal temperatures are reached, but the duration is very much shorter (by more than 10^4) and the temperatures are even higher (by 5- to 10-fold). Similar to shock-tube chemistry or multiphoton infrared laser photolysis, cavitation heating is very short lived but occurs within condensed phases. Furthermore, sonochemistry has a high-pressure component, which suggests that one might be able to produce on a microscopic scale the same macroscopic conditions of high temperature–pressure "bomb" reactions or explosive shock wave synthesis in solids.

Control of sonochemical reactions is subject to the same limitation that any thermal process has: The Boltzmann energy distribution means that the energy per individual molecule will vary widely. One does have easy control, however, over the intensity of heating generated by acoustic cavitation through the use of various physical parameters (including thermal conductivity of dissolved gases, solvent vapor pressure inside the bubble, and ambient pressure). In contrast, frequency appears to be less important, at least within the range where cavitation can occur (a few hertz to a few megahertz), although there have been few detailed studies of its role.

3.2 Homogeneous Sonochemistry: Bond Breaking and Radical Formation

The chemical effects of ultrasound on aqueous solutions have been studied for many years. The primary products are H_2 and H_2O_2; other high-energy intermediates have been suggested, including HO_2, $H^{\cdot}$, $OH^{\cdot}$, and perhaps $e^-_{(aq)}$. The elegant work of Riesz and collaborators used electron paramagnetic resonance with chemical spin-traps to demonstrate definitively the generation of $H^{\cdot}$ and $OH^{\cdot}$ during ultrasonic irradiation, even with clinical sources of ultrasound.[28] The extensive work in Henglein's laboratory involving aqueous sonochemistry of dissolved gases has established clear analogies to combustion processes.[23,24] As one would expect, the sonolysis of water, which produces both strong reductants and oxidants, is capable of causing secondary oxidation and reduction reactions, as often observed by Margulis and co-workers.[29]

In contrast, the ultrasonic irradiation of organic liquids has been less studied. Suslick and co-workers established that virtually all organic liquids will generate free radicals upon ultrasonic irradiation, as long as the total vapor pressure is low enough to allow effective bubble collapse.[30] The sonolysis of simple hydrocarbons (e.g., *n*-alkanes) creates the same kinds of products associated with very high temperature pyrolysis. Most of these products (H_2, CH_4, and the smaller *l*-alkenes) derive from a well-understood radical chain mechanism.

The sonochemistry of solutes dissolved in organic liquids also remains largely unexplored. The sonochemistry of metal carbonyl compounds is an exception.[31] Detailed studies of these systems led to important mechanistic understandings of the nature of sonochemistry. A variety of unusual reactivity patterns have been observed during ultrasonic irradiation, including multiple ligand dissociation, novel metal cluster formation, and the initiation of homogeneous catalysis (discussed later) at low ambient temperature, with rate enhancements greater than 100,000-fold.

Of special interest is the recent development of sonochemistry as a synthetic tool for the creation of unusual materials.[32] As one example is the recent discovery of a simple sonochemical synthesis of amorphous and nanostructured materials, including transition metals, alloys, carbides, and oxides.[33] A second example is the sonochemical preparation of protein microspheres,[34] which have applications for medical diagnostic imaging, drug delivery, and blood substitutes.

3.3 Heterogeneous Sonochemistry: Reactions of Solids with Liquids

The use of high-intensity ultrasound to enhance the reactivity of metals as stoichiometric reagents has become a routine synthetic technique for many heterogeneous organic and organometallic reactions,[12–15] especially those involving reactive metals, such as Mg, Li, or Zn. This development originated from the early work of Renaud and the more recent breakthroughs of Luche.[12] The effects are quite general and apply to reactive inorganic salts and to main-group reagents as well.[35] Less work has been done with unreactive metals (e.g., V, Nb, Mo, W), but results here are promising as well.[11] Rate enhancements of more than 10-fold are common, yields are often substantially improved, and byproducts are avoided. A few simple examples of synthetic applications of heterogeneous sonochemistry are shown in Eqs. (2)–(7), taken from the work of Ando, Boudjouk, Luche, Mason, and Suslick, among others:

$$C_6H_5Br + Li \xrightarrow{)))} C_6H_5Li + LiBr, \tag{2}$$

$$RBr + Li + R'_2NCHO \xrightarrow[2.\,H_2O]{1.\,)))} RCHO + R'_2NH, \tag{3}$$

$$2o - C_6H_4(NO_2)I + Cu \xrightarrow{)))} o\text{-}(O_2N)H_4C_6\text{—}C_6H_4(NO_2) + 2CuI, \tag{4}$$

$$RR'HC\text{—}OH + KMnO_{4(s)} \xrightarrow{)))} RR'C{=}O, \tag{5}$$

$$C_6H_5CH_2Br + KCN \xrightarrow[Al_2O_3]{)))} C_6H_5CH_2CN, \tag{6}$$

$$MCl_5 + Na + CO \xrightarrow{)))} M(CO)_6^- \quad (M = V, Nb, Ta). \tag{7}$$

The mechanism of the sonochemical rate enhancements in both stoichiometric and catalytic reactions of metals is associated with dramatic changes in morphology of both large extended surfaces and powders. As discussed earlier, these changes originate from microjet impact on large surfaces and high-velocity interparticle collisions in slurries. Surface composition studies by Auger electron spectroscopy and sputtered neutral mass spectrometry reveal that ultrasonic irradiation effectively removes surface oxide and other contaminating coatings.[11] The removal of such passivating coatings can dramatically improve reaction rates. The reactivity of clean metal surfaces also appears to be responsible for the greater tendency for heterogeneous sonochemical reactions to involve single electron transfer rather than acid–base chemistry.[36]

Green, Suslick, and co-workers examined another application of sonochemistry to difficult heterogeneous systems: the process of molecular intercalation.[37] The adsorption of organic or inorganic compounds as guest molecules between the atomic sheets of layered inorganic solid hosts permits the systematic change of optical, electronic, and catalytic properties for a variety of technologic applications (e.g., lithium batteries, hydrodesulfurization catalysts, and solid lubricants). The kinetics of intercalation, however, are generally extremely slow, and syntheses usually require high temperatures and very long reaction times. High-intensity ultrasound dramatically increases the rates of intercalation (by as much as 200-fold) of a wide range of compounds into various layered inorganic solids (such as ZrS_2, V_2O_5, TaS_2, MoS_2, and MoO_3). Scanning electron microscopy of the layered solids coupled to chemical kinetics studies demonstrated that the origin of the observed rate enhancements comes from particle fragmentation (which dramatically increases surface areas) and to a lesser extent from surface damage. The ability of high-intensity ultrasound to rapidly form uniform dispersions of micrometer-sized powders of brittle materials is often responsible for the activation of heterogeneous reagents, especially nonmetals.

3.4 Sonocatalysis

Catalytic reactions are of enormous importance in both laboratory and industrial applications. Catalysts are generally divided into two types. If the catalyst is a molecular or ionic species dissolved in a liquid, then the system is *homogeneous*; if the catalyst is a solid, with the reactants either in a percolating liquid or gas, then it is *heterogeneous*. In both cases, it is often a difficult problem either to activate the catalyst or to keep it active.

Ultrasound has potentially important applications in both homogeneous and heterogeneous catalytic systems. The inherent advantages of sonocatalysis include (1) the use of low ambient temperatures to preserve thermally sensitive substrates and to enhance selectivity, (2) the ability to generate high-energy species difficult to obtain from photolysis or simple pyrolysis, and (3) the mimicry of high-temperature and high-pressure conditions on a microscopic scale.

Homogeneous catalysis of various reactions often

uses organometallic compounds. The starting organometallic compound, however, is often catalytically inactive until the loss of metal-bonded ligands (such as carbon monoxide) from the metal. Having demonstrated that ultrasound can induce ligand dissociation, the initiation of homogeneous catalysis by ultrasound becomes practical. A variety of metal carbonyls upon sonication will catalyze the isomerization of *l*-alkenes to the internal alkenes,[31] through reversible hydrogen atom abstraction, with rate enhancements of as much as 10^5 over thermal controls.

Heterogeneous catalysis is generally more industrially important than homogeneous systems. For example, virtually all of the petroleum industry is based on a series of catalytic transformations. Heterogeneous catalysts often require rare and expensive metals. The use of ultrasound offers some hope of activating less reactive, but also less costly, metals. Such effects can occur in three distinct stages: (1) during the formation of supported catalysts, (2) activation of preformed catalysts, or (3) enhancement of catalytic behavior during a catalytic reaction. Some early investigations of the effects of ultrasound on heterogeneous catalysis can be found in the Soviet literature.[38] In this early work, increases in turnover rates were usually observed upon ultrasonic irradiation but were rarely more than 10-fold. In the cases of modest rate increases, it appears likely that the cause is increased effective surface area; this is especially important in the case of catalysts supported on brittle solids.[39] More impressive accelerations, however, have included hydrogenations and hydrosilations by Ni powder, Raney Ni, and Pd or Pt on carbon.[13] For example, Casadonte and Suslick discovered that hydrogenation of alkenes by Ni powder is enormously enhanced ($>10^5$-fold) by ultrasonic irradiation.[16] This dramatic increase in catalytic activity is due to the formation of uncontaminated metal surfaces from interparticle collisions caused by cavitation-induced shock waves.

4 SONOLUMINESCENCE

Ultrasonic irradiation of liquids can also produce light. This phenomenon, known as sonoluminescence, was first observed from water in 1934 by Frenzel and Schultes.[40] As with sonochemistry, sonoluminescence derives from acoustic cavitation. Although sonoluminescence from aqueous solutions has been studied in some detail, only recently has significant work on sonoluminescence from nonaqueous liquids been reported.

4.1 Types of Sonoluminescence

It is now generally thought that there are two separate forms of sonoluminescence: multiple-bubble sonoluminescence (MBSL) and single-bubble sonoluminescence (SBSL). When an acoustic field of sufficient intensity is propagated through a liquid, placing it under dynamic tensile stress, microscopic preexisting inhomogeneities act as nucleation sites for liquid rupture. This cavitation inception process results in many separate and individual cavitation events that would be distributed broadly throughout the acoustic field, especially if the liquid had a sufficiently large number of nuclei. Since most liquids such as water, have many thousands of potential nucleation sites per milliliter, the "cavitation field" generated by a propagating (or standing) acoustic wave typically consists of many bubbles, and is distributed over an extended region of space. If this cavitation is sufficiently intense to produce sonoluminescence, then we call this phenomenon multiple-bubble sonoluminescence.[17,41]

When an acoustic standing wave is excited within a liquid, pressure nodes and antinodes are generated. If a gas bubble is inserted into this standing wave, it will experience an acoustic force that will tend to force the bubble either to the node or antinode, depending upon whether it is respectively larger or smaller than its resonance size. Under the appropriate conditions, this acoustic force can balance the buoyancy force and the bubble is said to be "acoustically levitated." Such a bubble is typically quite small, compared to a wavelength (e.g., at 20 kHz, the resonance size is approximately 150 μm), and thus the dynamic characteristics of this bubble can often be examined in considerable detail, both from a theoretical and an experimental perspective.

It was recently discovered that under rather specialized but easily obtainable conditions a single, stable, oscillating gas bubble can be forced into such large-amplitude pulsations that it produces sonoluminescence emissions each (and every) acoustic cycle.[42,43] This phenomenon is called single-bubble sonoluminescence and has received considerable recent attention.[17,18,44,45]

4.2 Applications of MBSL to Measurements of Cavitation Thresholds

When the acoustic pressure amplitude of a propagating acoustic wave is relatively large (greater than ≈ 0.5 MPa) local inhomogeneities in the liquid often give rise to the explosive growth of a nucleation site to a cavity of macroscopic dimensions, primarily filled with vapor. Such a cavity is inherently unstable, and its subsequent collapse can result in an enormous concentration of energy. This violent cavitation event has been termed *transient* or *inertial cavitation* because the collapse of the cavity is primarily dominated by inertial forces.[46] A normal consequence of this rapid growth and violent collapse is that the cavitation bubble itself is destroyed. Although gas-filled residues from the collapse may give rise to reinitiation of the process, this type of cavitation is thought to be a temporally discrete phenomenon. When one examines the light emissions from transient inertial cavitation, one can see single isolated events associated (presumably) with individual collapses of imploding cavities.

Because acoustic cavitation is often associated with large energy concentrations and thus potentially damaging mechanical effects, its presence is often undesirable (e.g., in the use of diagnostic ultrasound for prenatal examinations). Since light can be detected at very low levels and with very high time resolution, sonoluminescence can be used as a sensitive indicator of violent acoustic cavitation.[47,48] This capability permits threshold determination of cavitation inception for acoustic pulses of short time duration, as often is the case for medical ultrasound.

The widespread use of medical ultrasound has made possible enormous advances in the noninvasive examination of internal organs and conditions. With these diagnostic devices, the acoustic pulses used to create images are often less than a microsecond in length and possess duty cycles on the order of 1 : 1000. Since the acoustic pressure amplitudes generated by these devices are relatively large, it has been possible to use sonoluminescence as a detection criterion for acoustic cavitation generated by microsecond-length pulses of ultrasound. These studies have indicated that cavitation can be generated by acoustic pulses similar to those used in diagnostic ultrasound instruments and that continued studies need to be undertaken to evaluate the potential risks of these devices.[49]

4.3 Origin of MBSL Emissions: Chemiluminescence

The spectrum of MBSL in water consists of a peak at 310 nm and a broad continuum throughout the visible region. An intensive study of aqueous MBSL was conducted by Verrall and Sehgal.[50] The emission at 310 nm is from excited-state $OH^{\bullet}$, but the continuum is difficult to interpret. The MBSL from aqueous and alcohol solutions of many metal salts have been reported and are characterized by emission from metal atom excited states.[51]

Flint and Suslick reported the first MBSL spectra of organic liquids.[52] With various hydrocarbons, the observed emission is from excited states of C_2 ($d^3\Pi_g$–$a^3\Pi_u$, the Swan lines), the same emission seen in flames. Furthermore, the ultrasonic irradiation of alkanes in the presence of N_2 (or NH_3 or amines) gives emission from CN excited states, but not from N_2 excited states. Emission from N_2 excited states would have been expected if the MBSL originated from microdischarge, whereas CN emission is typically observed from thermal sources. When oxygen is present, emission from excited states of CO_2, $CH^{\bullet}$, and $OH^{\bullet}$ is observed, again similar to flame emission.

For both aqueous and nonaqueous liquids, MBSL is caused by chemical reactions of high-energy species formed during cavitation by bubble collapse. Its principal source is most probably not blackbody radiation or electrical discharge. The MBSL is a form of chemiluminescence.

4.4 Origin of SBSL Emissions: Imploding Shock Waves

It is known that the spectra of MBSL and SBSL are measurably different. For example, an aqueous solution of NaCl shows evidence of excited states of both $OH^{\bullet}$ and Na in the MBSL spectrum; however, the SBSL spectrum of an identical solution shows no evidence of either of these peaks.[53] Similarly, the MBSL spectrum falls off at low wavelengths, while the SBSL spectrum continues to rise, at least for bubbles containing most noble gases.[54]

Perhaps the most intriguing aspect of SBSL is the extremely short duration of the sonoluminescence flash. Putterman and his colleagues, using the fastest photomultiplier tube (PMT) available, determined that this duration must be at least as short as 50 ps, perhaps even lower.[55] Moran et al., using a streak camera, presented evidence of a 12-ps flash duration.[56] Because these measurements test the limitations of current technology, an accurate assessment of this pulse duration cannot be reliably given at this time.

As described earlier, the most likely explanation for the origin of sonoluminescence is the hot-spot theory, in which the potential energy given the bubble as it expands to maximum size is concentrated into a heated gas core as the bubble implodes. To understand the origin of sonoluminescence emissions, it is necessary first to understand something about the bubble dynamics that describe the bubble's motion. The equation that describes the oscillations of a gas bubble driven by an acoustic field is known generally by the name *Rayleigh–Plesset*, one form of which, called the *Gilmore equation*, can be expressed as a second-order nonlinear differential equation given as

$$R\left(1-\frac{U}{C}\right)\frac{d^2R}{dt^2}+\frac{3}{2}\left(1-\frac{U}{3C}\right)\left(\frac{dR}{dt}\right)^2 - \left(1+\frac{U}{C}\right)H-\frac{R}{C}\left(1-\frac{U}{C}\right)\frac{dH}{dt}=0. \quad (8)$$

The radius and velocity of the bubble wall are given by R and U, respectively. The values for H, the enthalpy at the bubble wall, and C, the local sound speed, may be expressed as follows using the Tait equation of state for the liquid:

$$H = \frac{n}{n-1} \frac{A^{1/n}}{\rho_0} \left\{ (P(R) + B)^{(n-1)/n} - [P_\infty(t) + B]^{(n-1)/n} \right\}, \quad (9)$$

$$C = [c_0^2 + (n-1)H]. \quad (10)$$

The linear speed of sound in the liquid is c_0. The constants A, B, and n should be set to the appropriate values for water. Any acoustic forcing function is included in the pressure at infinity, $P_\infty(t)$. The pressure at the bubble wall, $P(R)$, is given by

$$P(R) = \left(P_0 + \frac{2\sigma}{R} \right) \left(\frac{R_0}{R} \right)^{3\gamma} - \frac{2\sigma}{R} - \frac{4\mu U}{R}, \quad (11)$$

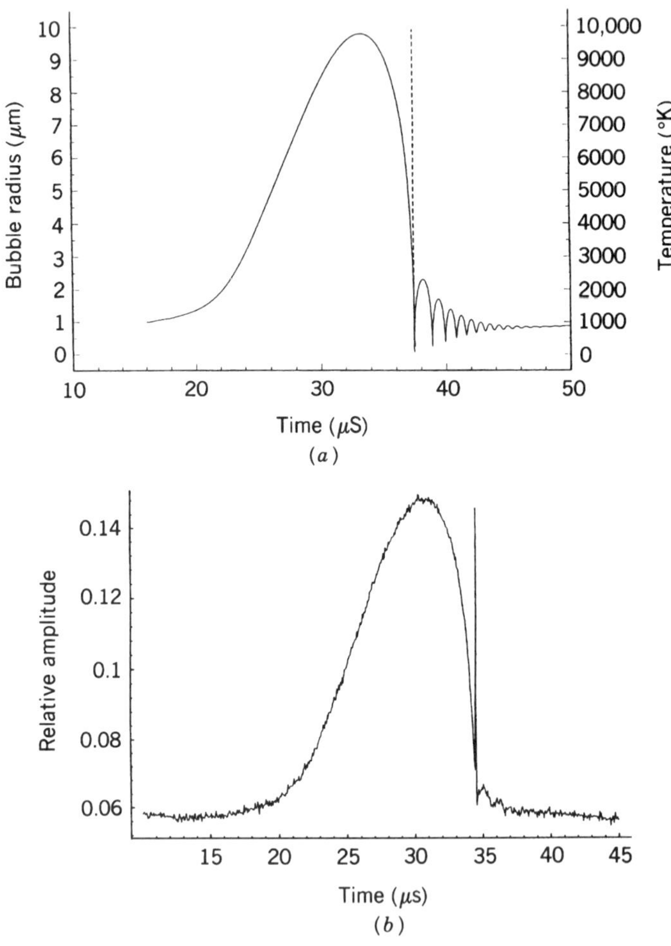

Fig. 5 (*a*) Theoretical response of a single gas bubble when driven under conditions of SBSL (solid line). Pressure amplitude 0.136 MPa, equilibrium radius 4.5 μm, and driving frequency 26.5 kHz. Bubble expands to several times its initial radius and then implosively collapses. Broken line: calculation of the interior bubble temperature assuming an adiabatic collapse of the cavity contents. (Courtesy of John Allen.) (*b*) Measured response of a gas bubble to conditions similar to those shown. Experimentally determined bubble radius reconstructed from the square root of the scattered laser light intensity. Intensity spike near bubble collapse is not due to the laser but results from the sonoluminescence emissions. With some minor adjustments, the theoretical and experimental curves for the bubble radius can be made to coincide. (Courtesy of Tom Matula.)

where the initial radius of the bubble at time zero is R_0. The ambient pressure of the liquid is P_0, the surface tension σ, the shear viscosity μ, and the polytropic exponent γ. The latter is set to 1.4 assuming the bubble behaves as an adiabatic system.

We can use the Gilmore equation to compute the behavior of a bubble undergoing SBSL for conditions similar to those in which this phenomenon is observed experimentally. Figure 5*a* shows an example of these computations. The solid line is the radius–time curve; the dashed line is the computed temperature in the interior of the bubble based upon the assumptions of this model.

It is possible to test experimentally certain aspects of these models. For example, using a light-scattering technique, various researchers have obtained measurements of the radius–time curve, simultaneous with the optical emissions,[41,57] as shown in Fig. 5*b*. Both the laser light scattered intensity and the SBSL can be acquired with a single photomultiplier tube. The SBSL emission is seen as the sharp spike in the figure, appearing at the final stages of bubble collapse. Note that these emissions occur at the point of minimum bubble size, as predicted by the hot-spot theory and that the general shape of the theoretical curve is reproduced. More quantitative assessment of these data has been made.[57,58]

The computations of single-bubble cavitation suggest that the temperature of the gas within the bubble would remain at elevated temperatures for times on the order of tens of nanoseconds; however, there is strong evidence that the pulse duration of the SBSL flash is at least three orders of magnitude shorter than this value. The most plausible explanation for this short flash interval and some of the observed spectra (see below) is that an imploding shock wave is created within the gas bubble during the final stages of collapse. If this shock wave does indeed exist, exciting possibilities can be inferred about the temperatures that could be attained within the bubble and the physics that might result. Indeed, speculations on the possibilities of inertial confinement fusion have been made.[59–61]

4.5 Spectroscopic Probes of Cavitation Conditions

Determination of the temperatures reached in a cavitating bubble has remained a difficult experimental problem. As a spectroscopic probe of the cavitation event, MBSL provides a solution. High-resolution MBSL spectra from silicone oil have been reported and analyzed.[8] The observed emission comes from excited state C_2 and has been modeled with synthetic spectra as a function of rotational and vibrational temperatures, as shown in

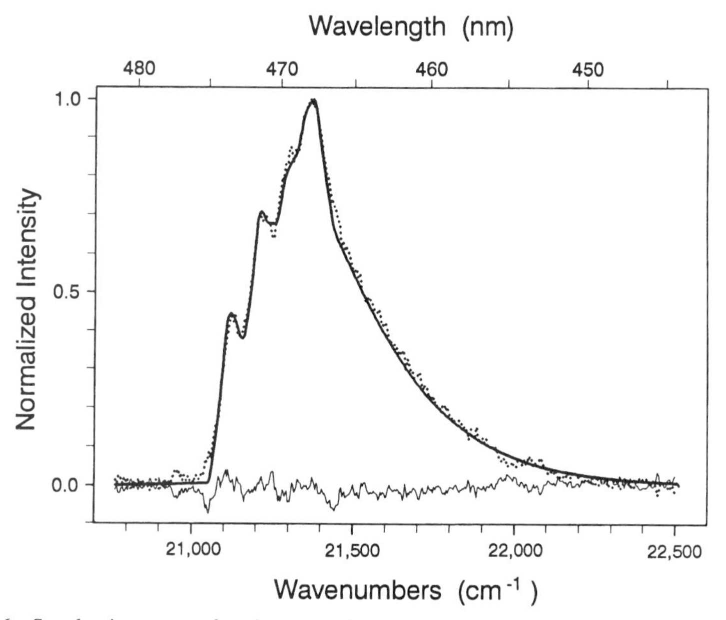

Fig. 6 Sonoluminescence of excited state C_2. Emission from the $\Delta\nu = +1$ manifold of the $d^3\Pi_g$–$a^3\Pi_u$ transition (Swan band) of C_2. (······) Observed sonoluminescence from silicone oil (polydimethylsiloxane, Dow 200 series, 50 cSt viscosity) under a continuous Ar sparge at 0°C. (——) Best fit synthetic spectrum, with $T_v = T_r = 4900$ K. (——) Difference spectrum.

Fig. 6. From comparison of synthetic to observed spectra, the effective cavitation temperature is 5050 ± 150 K. The excellence of the match between the observed MBSL and the synthetic spectra provides definitive proof that the sonoluminescence event is a thermal, chemiluminescence process. The agreement between this spectroscopic determination of the cavitation temperature and that made by comparative-rate thermometry of sonochemical reactions[7] is surprisingly close.

The interpretation of the spectroscopy of SBSL is much more unclear. Some very interesting effects are observed when the gas contents of the bubble are changed.[55] These results are shown in Fig. 7. Note that doping a nitrogen bubble with small quantities of noble gases drastically affects the emission intensity. Furthermore, the spectra show practically no evidence of OH emissions and, when He and Ar bubbles are considered, continue to increase in intensity for smaller and smaller wavelengths. These spectra suggest that temperatures considerably in excess of 5000 K may exist within the bubble and lend some support to the concept of an imploding shock wave. Several other alternative explanations for SBSL have been presented, and there exists considerable theoretical activity in this particular aspect of SBSL.[62–66]

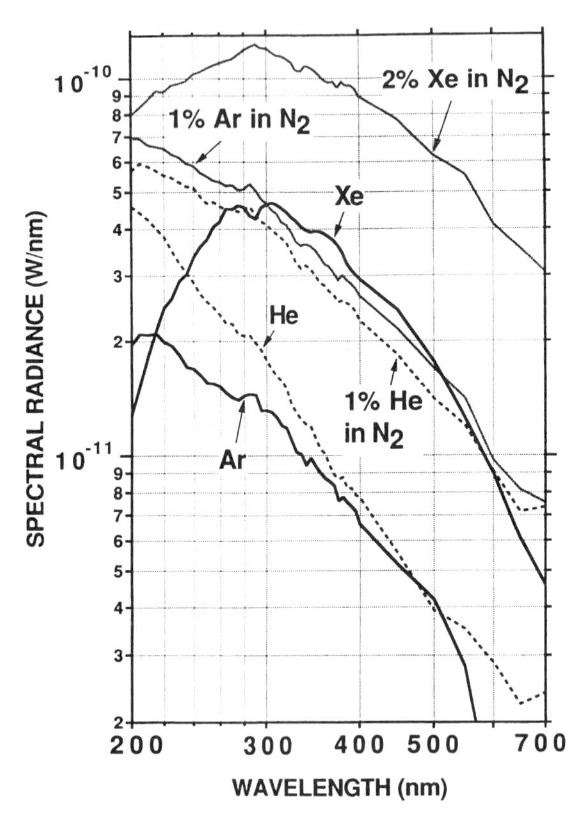

Fig. 7 Spectrum of SBSL for a variety of different gas mixtures in water. Pressure head 150 Torr. In contrast to spectra for MBSL, curves are relatively smooth and in some cases show a steady increase in intensity as one progresses to smaller wavelengths. Data have been corrected for the adsorption of water and quartz and for the quantum efficiency of the photodetector. (Reprinted with permission from Ref. 54. Copyright 1994 American Association for the Advancement of Science. Courtesy of Bob Hiller.)

Acknowledgments

We acknowledge the support of the Office of Naval Research (L. A. C.), the National Institutes of Health through grant numbers DK43881 (L. A. C.) and HL25934 (K. S. S.), and the National Science Foundation through grant numbers CHE-9420758 and DMR-89-20538 (K. S. S.) and PHY-9311108 (L. A. C.).

REFERENCES

1. K. S. Suslick (Ed.), *Ultrasound: Its Chemical, Physical, and Biological Effects*, VCH Publishers, New York, 1988.
2. K. S. Suslick, *Science*, Vol. 247, 1990, p. 1439.
3. T. J. Mason (Ed.), *Advances in Sonochemistry*, Vols. 1–3, JAI Press, New York, 1990, 1991, 1993.
4. T. J. Mason and J. P. Lorimer, *Sonochemistry: Theory, Applications and Uses of Ultrasound in Chemistry*, Ellis Horword, Chichester, United Kingdom, 1988.
5. G. J. Price (Ed.), *Current Trends in Sonochemistry*, Royal Society of Chemistry, Cambridge, 1992.
6. O. V. Abramov, *Ultrasound in Liquid and Solid Metals*, CRC Press, Boca Raton, FL, 1994.
7. K. S. Suslick, D. A. Hammerton, and R. E. Cline, Jr., *J. Am. Chem. Soc.*, Vol. 108, 1986, p. 5641.
8. E. B. Flint and K. S. Suslick, *Science*, Vol. 253, 1991, p. 1397.
9. T. G. Leighton, *The Acoustic Bubble*, Academic, London, 1994, pp. 531–551.
10. S. J. Doktycz and K. S. Suslick, *Science*, Vol. 247, 1990, pp. 1067.
11. K. S. Suslick and S. J. Doktycz, *Adv. Sonochem.*, Vol. 1, 1990, pp. 197–230.
12. C. Einhorn, J. Einhorn, and J.-L. Luche, *Synthesis*, Vol. 1989, 1989, p. 787.
13. P. Boudjouk, *Comments Inorg. Chem.*, Vol. 9, 1990, p. 123.
14. J. M. Pestman, J. B. F. N. Engberts, and F. de Jong, *Recl. Trav. Chim. Pays-Bas*, Vol. 113, 1994, p. 533.
15. K. S. Suslick, "Sonochemistry of Transition Metal Compounds," in R. B. King (Ed.), *Encyclopedia of Inorganic Chemistry*, Vol. 7, 1996, Wiley, New York, pp. 3890–3905.

16. K. S. Suslick and D. J. Casadonte, *J. Am. Chem. Soc.*, Vol. 109, 1987, p. 3459.
17. L. A. Crum, *Physics Today*, Vol. 47, 1994, p. 22.
18. S. J. Putterman, *Sci. Am.*, February 1995, p. 46.
19. Lord Rayleigh, *Philos. Mag.*, Vol. 34, 1917, p. 94.
20. W. T. Richards and A. L. Loomis, *J. Am. Chem. Soc.*, Vol. 49, 1927, p. 3086.
21. M. A. Margulis, *Ultrasonics*, Vol. 30, 1992, p. 152.
22. T. Lepoint and F. Mullie, *Ultrasonics Sonochem.*, Vol. 1, 1994, p. S13.
23. A. Henglein, *Ultrasonics*, Vol. 25, 1985, p. 6.
24. A. Henglein, *Adv. Sonochem.*, Vol. 3, 1993, p. 17.
25. C. M. Preece and I. L. Hansson, *Adv. Mech. Phys. Surf.*, Vol. 1, 1981, p. 199.
26. W. Lauterborn and A. Vogel, *Ann. Rev. Fluid Mech.*, Vol. 16, 1984, p. 223.
27. J. R. Blake and D. C. Gibson, *Ann. Rev. Fluid Mech.*, Vol. 19, 1987, p. 99.
28. P. Riesz, *Adv. Sonochem.*, Vol. 2, 1991, p. 23.
29. M. A. Margulis and N. A. Maximenko, *Adv. Sonochem.*, Vol. 2, 1991, p. 253.
30. K. S. Suslick, J. W. Gawienowski, P. F. Schubert, and H. H. Wang, *J. Phys. Chem.*, Vol. 87, 1983, p. 2299.
31. K. S. Suslick, J. W. Goodale, H. H. Wang, and P. F. Schubert, *J. Am. Chem. Soc.*, Vol. 105, 1983, p. 5781.
32. K. S. Suslick, *MRS Bulletin*, Vol. 20, 1995, pp. 29–34.
33. K. S. Suslick, S. B. Choe, A. A. Cichowlas, and M. W. Grinstaff, *Nature*, Vol. 353, 1991, p. 414.
34. K. S. Suslick and M. W. Grinstaff, *J. Am. Chem. Soc.*, Vol. 112, 1990, p. 7807.
35. T. Ando and T. Kimura, *Adv. Sonochem.*, Vol. 2, 1991, p. 211.
36. J.-L. Luche *Ultrasonics Sonochem.*, Vol. 1, 1994, p. S111.
37. K. Chatakondu, M. L. H. Green, M. E. Thompson, and K. S. Suslick, *J. Chem. Soc. Chem. Commun.*, 1987, p. 900.
38. A. N. Mal'tsev, *Zh. Fiz. Khim.*, Vol. 50, 1976, p. 1641.
39. B. H. Han and P. Boudjouk, *Organometallics*, Vol. 2, 1983, p. 769.
40. H. Frenzel and H. Schultes, *Z. Phys. Chem.*, Vol. 27b, 1934, p. 421.
41. L. A. Crum, *J. Acoust. Soc. Am.*, Vol. 95, 1994, p. 559.
42. D. F. Gaitan and L. A. Crum, in M. Hamilton and D. T. Blackstock (Eds.), *Frontiers of Nonlinear Acoustics*, 12th ISNA, Elsevier Applied Science, New York, 1990, pp. 459–463.
43. D. F. Gaitan, L. A. Crum, R. A. Roy, and C. C. Church, *J. Acoust. Soc. Am.*, Vol. 91, 1992, p. 3166.
44. B. P. Barber and S. J. Putterman, *Nature*, Vol. 352, 1991, p. 318.
45. L. A. Crum and R. A. Roy, *Science*, Vol. 266, 1994, p. 233.
46. H. G. Flynn, "Physics of Acoustic Cavitation in Liquids," in W. P. Mason (Ed.), *Physical Acoustics*, Vol. IB, Academic, New York, 1964, p. 157.
47. L. A. Crum and J. B. Fowlkes, *Nature*, Vol. 319, 1986, p. 52.
48. L. A. Crum and D. F. Gaitan, *Proc. Int. Soc. Opt. Eng.*, Vol. 1161, 1989, p. 125.
49. L. A. Crum, R. A. Roy, M. A. Dinno, C. C. Church, R. E. Apfel, C. K. Holland, and S. I. Madanshetty, *J. Acoust. Soc. Am.*, Vol. 91, 1992, p. 1113.
50. R. E. Verrall and C. Sehgal, in K. S. Suslick (Ed.) *Ultrasound: Its Chemical, Physical, and Biological Effects*, VCH Publishers, New York, 1988, pp. 227–287.
51. E. B. Flint and K. S. Suslick, *J. Phys. Chem.*, Vol. 95, 1991, p. 1484.
52. E. B. Flint and K. S. Suslick, *J. Am. Chem. Soc.*, Vol. 111, 1989, p. 6987.
53. T. J. Matula, R. A. Roy, P. D. Mourad, W. B. McNamara III, and K. S. Suslick, submitted.
54. R. Hiller, K. Weninger, S. J. Putterman, and B. P. Barber, *Science*, Vol. 266, 1994, p. 248.
55. B. P. Barber, R. Hiller, K. Arisaka, H. Fetterman, and S. J. Putterman, *J. Acoust. Soc. Am.*, Vol. 91, 1992, p. 3061.
56. M. J. Moran et al., "Direct Observations of Single Sonoluminescence Pulses," *UCRL-JC-118486* (Preprint), Lawrence Livermore National Lab, October 1994.
57. B. P. Barber and S. J. Putterman, *Phys. Rev. Lett.*, Vol. 69, 1992, p. 3839.
58. R. Lofstedt, B. P. Barber, and S. J. Putterman, *Phys. Fluids*, Vol. A5, 1993, p. 2911.
59. B. P. Barber, C. C. Wu, R. Lofstedt, P. H. Roberts, and S. J. Putterman, *Phys. Rev. Lett.*, Vol. 72, 1994, p. 1380.
60. C. C. Wu and P. H. Roberts, *Phys. Rev. Lett.*, Vol. 70, 1993, p. 3424.
61. W. C. Moss, D. B. Clarke, J. W. White, and D. A. Young, *Phys. Fluids*, Vol. 6, 1994, p. 2979.
62. L. Frommhold and A. A. Atchley, *Phys. Rev. Lett.*, Vol. 73, 1994, p. 2883.
63. R. G. Holt, D. F. Gaitan, A. A. Atchley, and J. Holzfuss, *Phys. Rev. Lett.*, Vol. 72, 1994, p. 1376.
64. C. C. Wu and P. H. Roberts, *Proc. R. Soc. Lond.* A, Vol. 445, 1994, p. 323.
65. J. Schwinger, *Proc. Natl. Acad. Sci.*, Vol. 89, 1992, p. 4091.
66. V. Kamath, A. Prosperetti, and F. N. Egolfopoulos, *J. Acoust. Soc. Am.*, Vol. 94, 1993, p. 248.

PART III

AEROACOUSTICS AND ATMOSPHERIC SOUND

24

INTRODUCTION

JAMES LIGHTHILL

1 BROAD OVERVIEW

1.1 Brief Survey of Part III

Part III, consisting of this introductory chapter and the six chapters following it, deals with sound in the air surrounding our planet. This nonhomogeneous fluid, usually in nonuniform motion, is the medium for much of the sound propagation that is of greatest human importance. Such propagation interacts with solid boundaries, including natural topography and man-made structures, while being strongly influenced also by distributions of wind and of atmospheric composition.

Sources of atmospheric sound long included a wide variety of natural and man-made phenomena at ground level, along with meteorologic processes ranging from lightning flashes to interactions between winds and structures. Moreover, in the twentieth century, exploitation of the air as a medium for transport introduced new sources of atmospheric sound, including aeroengine noise, airframe noise, and supersonic booms. Study of such sources forms a major part of aeroacoustics.

In addition to sound at frequencies to which the human ear is sensitive, atmospheric acoustics is also concerned with sources of sound at much lower frequencies (infrasound) and with its propagation, often over very long distances, through the atmosphere. By contrast, sound at ultrasonic frequencies achieves only modest penetration, being subject to a much higher level of atmospheric absorption; which, however, may be a source of interesting air motion known as acoustic streaming.

This chapter serves as a general introduction to Part III. First, some fundamental ideas that, historically, have helped to give physical understanding of the field are introduced. Then their power in this regard is illustrated through applications to aeroengine noise, airframe noise, and supersonic booms, to the propagation of sound through nonuniform winds, and to certain aspects of acoustic streaming. Next a more detailed modern account of aeroengine noise, with special emphasis on jet noise and on techniques for its suppression, is given in Chapter 25. Some subtle features of the interactions between fluid motion and sound are outlined in Chapter 26, and *Enc.* Ch. 26 describes the streaming effects generated by both standing and traveling sound waves. General accounts, both of blast waves generated by explosions or by lightning flashes and of booms originating in the flight of supersonic aircraft, are given in Chapter 27.

A wide-ranging survey of propagation of sound in the atmosphere follows in Chapter 28. This takes into account the different mechanisms of attenuation in the air itself and at the ground and some effects of these acting in combination with refraction due to nonuniformities of wind velocity and of atmospheric composition. It outlines diffraction and scattering processes, including scattering by atmospheric turbulence, and ends with an account of acoustic methods (echosound) for probing the atmosphere. Part III is then concluded with a comprehensive description (Chapter 29) of atmospheric infrasound, which includes accounts of sources, propagation, measurement methods, and perception by the human body.

So Part III relates mainly to interactions of sound with air, including transmission through the atmosphere and both generation of sound by, and propagation of sound in, airflows (e.g., man-made flows—around aircraft or air machinery—or natural winds) as affected by the air's boundaries and atmospheric composition, with (conversely) generation of airflows by sound (acoustic streaming).

Note: References to chapters appearing only in the *Encyclopedia* are preceded by *Enc.*

From linear theory in Part I we use the wave equation's physical basis (Chapter 2), the short-wavelength ray acoustics approximation (Chapter 3), acoustic attenuation (Chapters 5 and 7), waveguides (Chapter 7), and multipole sources with the long-wavelength compact-source approximation (Chapter 8). From nonlinear acoustics in Part II we use the physics of waveform shearing and shock formation (Chapter 17).

Historically, the key to understanding how airflows interact with sound has been a comparison between the dynamic equations for airflows and the wave equation approximations of the elementary theory of sound. Although a feature common to both theories is mass conservation (very briefly, mass flux is a vector field whose divergence gives the local rate of decrease of mass density), nonetheless fundamental differences emerge in their treatment of momentum. The following critique of these differences offers a basis for understanding the interactions.

1.2 Role of the Momentum Equation

The techniques presented in Part III are centered on the momentum equation for air. Differences between the momentum equation and a wave equation approximation include the following:

1. The *linear effects* of gravity acting on stratified air. These effects allow independent propagation of "internal" gravity waves and of sound, except at wavelengths of many kilometres, when the atmosphere becomes a waveguide for global propagation of interactive acoustic gravity waves[1] (Chapter 28).
2. More important are *nonlinear effects* of the momentum flux $\rho u_i u_j$; that is, the flux, or the rate of transport across a unit area, of any ρu_i momentum component by any u_j velocity component. This term, neglected in linear acoustics, acts as a stress (i.e., force per unit area, since the rate of change of momentum is the force):
   (i) An airflow's momentum flux $\rho u_i u_j$ generates sound as would a distribution of (time-varying) imposed stresses; thus not only do forces between the airflow and its boundary radiate sound as distributed dipoles, but also stresses (acting on fluid elements with equal and opposite dipole-like forces) radiate as distributed quadrupoles.[2,3]
   (ii) The *mean momentum flux* $\langle \rho u_i u_j \rangle$ in any sound waves propagating through a sheared flow (with shear $\partial V_i/\partial x_j$) is a stress on that flow[1,4]; the consequent *energy exchange* (from sound to flow when positive, vice versa when negative) is given as

$$\langle \rho u_i u_j \rangle \frac{\partial V_i}{\partial x_j}. \tag{1}$$

   (iii) Even without any preexisting flow, energy flux *attenuation* in a sound wave allows streaming to be generated by unbalanced stresses due to a corresponding attenuation in acoustic momentum flux; essentially, then, as acoustic energy flux is dissipated into heat, any associated acoustic momentum flux is transformed into a mean motion.[1,5]
3. Another (less crucial) momentum equation/wave equation difference is the *nonlinear deviation of pressure excess* $p - p_0$ from a constant multiple, $c_0^2(\rho - \rho_0)$ of density excess:
   (i) For sound generation by airflows, this adds an isotropic term to the quadrupole strength per unit volume:

$$T_{ij} = \rho u_i u_j + [(p - p_0) - c_0^2(\rho - \rho_0)]\delta_{ij}. \tag{2}$$

   The last term is considered important mainly for flows at above-ambient temperatures.[2,6]
   (ii) For propagation of sound with energy density E through flows of air with adiabatic index γ, the mean deviation is about $\frac{1}{2}(\gamma - 1)E$, and the *total radiation stress*[7]

$$\langle \rho u_i u_j \rangle + \tfrac{1}{2}(\gamma - 1)E\delta_{ij} \tag{3}$$

   adds an isotropic pressure excess to the mean momentum flux [although the energy exchange (1) is unchanged in typical cases with $\partial V_i/\partial x_i$ essentially zero].

[Note that the special case of sound waves interacting on themselves (see Part II) may be interpreted as a combined operation of the *self-convection effect* (difference 2 above) and the (smaller) *sound speed deviation* (difference 3).]

2 COMPACT-SOURCE REGIONS

2.1 Sound Generation by Low-Mach-Number Airflows

The main nondimensional parameters governing airflows of characteristic speed U and length-scale L are the Mach number $M = U/c$ (where, in aeroacoustics, c is taken as the sound speed in the atmosphere into which sound radiates) and the Reynolds number $R = UL/\nu$ (where ν is the kinematic viscosity). Low-Mach-number airflows are compact sources of sound, with frequencies either narrow banded at moderate R or broadbanded at high R when, respectively, flow instabilities lead to regular flow oscillations or extremely irregular turbulence. Since, in either case, a typical frequency ω scales as U/L (Strouhal scaling), the compactness condition that $\omega L/c$ be small (Chapter 8) is satisfied if $M = U/c$ is small.[3]

A solid body that, because of flow instability, is subjected to a fluctuating aerodynamic force F scaling as $\rho U^2 L^2$ (at frequencies scaling as U/L) radiates as an *acoustic dipole* of strength F, with (Chapter 8) mean radiated power $\langle \dot{F}^2 \rangle / 12\pi\rho c^3$.

This acoustic power scales as $\rho U^6 L^2/c^3$ (a sixth-power dependence on flow speed). Therefore *acoustic efficiency*, defined as the ratio of acoustic power to a rate of delivery (scaling as $\rho U^3 L^2$) of energy to the flow, scales as $(U/c)^3 = M^3$. [Exceptions to compactness include bodies of high aspect ratio; thus, a long wire in a wind (where the scale L determining frequency is its diameter) radiates as a lengthwise distribution of dipoles.]

Away from any solid body a compact flow (oscillating or turbulent, with frequencies scaling as U/L) leads to quadrupole radiation (see difference 2, part (i) above) with total quadrupole strength scaling as $\rho U^2 L^3$. Acoustic power then scales (Chapter 8) as $\rho U^8 L^2/c^5$: an eighth-power dependence[2,3] on flow speed. In this case acoustic efficiency (see above) scales as $(U/c)^5 = M^5$. Such quadrupole radiation, though often important, may become negligible near a solid body when dipole radiation due to fluctuating body force (with its sixth-power dependence) is also present.[8,9]

For bodies that are not necessarily compact there exists a more refined calculation, using Green's functions for internally bounded space rather than for free space, that leads in general to the same conclusion: Quadrupole radiation with its eighth-power dependence is negligible alongside the sixth-power dependence of dipole radiation due to fluctuating body forces. Important exceptions to this rule include sharp-edged bodies, where features of the relevant Green's function imply a fifth-power dependence on flow speed of acoustic radiation from turbulence.[10–13]

2.2 Sound Generation by Turbulence at Not-So-Low Mach Numbers

The chaotic character of turbulent-flow fields implies that velocity fluctuations at points P and Q, although they are well correlated when P and Q are very close, become almost uncorrelated when P and Q are not close to one another.

Statisticians define the correlation coefficient C for the velocities $\mathbf{u}_P$ and $\mathbf{u}_Q$ as $C = \langle \mathbf{v}_P \cdot \mathbf{v}_Q \rangle / \langle \mathbf{v}_P^2 \rangle^{1/2} \langle \mathbf{v}_Q^2 \rangle^{1/2}$ in terms of the deviations $\mathbf{v}_P = \mathbf{u}_P - \langle \mathbf{u}_P \rangle$ and $\mathbf{v}_Q = \mathbf{u}_Q - \langle \mathbf{u}_Q \rangle$ from their means. When two uncorrelated quantities are combined, their mean-square deviations are summed:

$$\begin{aligned}\langle \mathbf{v}_P + \mathbf{v}_Q \rangle^2 &= \langle \mathbf{v}_P^2 \rangle + 2\langle \mathbf{v}_P \cdot \mathbf{v}_Q \rangle + \langle \mathbf{v}_Q^2 \rangle \\ &= \langle \mathbf{v}_P^2 \rangle + \langle \mathbf{v}_Q^2 \rangle \quad \text{if } C = 0.\end{aligned}$$

Theories of turbulence define a correlation length l, with $\mathbf{u}_P$ and $\mathbf{u}_Q$ either well correlated or uncorrelated (C close to 1 or 0) when PQ is substantially $<l$ or $>l$. Roughly speaking, different regions of size l ("eddies") generate sound independently, and the mean-square radiated noise is the sum of the mean square outputs from all the regions.[14]

Typical frequencies in the turbulence are of order $\omega = v/l$, where v is a typical root-mean-square (rms) velocity deviation $\langle v^2 \rangle^{1/2}$, so that for each region the compactness condition $\omega l/c$ small is satisfied if v/c is small. Compactness, then, requires only that an rms velocity deviation v (rather than a characteristic mean velocity U) be small compared with c, which is less of a restriction on $M = U/c$ and can be satisfied at not-so-low Mach number.[6]

3 DOPPLER EFFECT

How is the radiation from such "eddies" modified by the fact that they are being *convected* at not-so-low Mach number? The term *Doppler effect*, covering all aspects of how the movement of sources of sound alters their radiation patterns, comprises (i) frequency changes,[1] (ii) volume changes,[2,3] and (iii) compactness changes.[6,15]

3.1 Frequency Changes

When a source of sound at frequency ω approaches an observer at velocity w, then in a single period $T = 2\pi/\omega$ sound emitted at the beginning travels a distance cT, while at the end of the period sound is being emitted from a source that is closer by a distance wT. The wavelength λ (distance between crests) is reduced to

$$\lambda = cT - wT = \frac{2\pi(c - w)}{\omega}, \qquad (4)$$

and the frequency heard by the observer (2π divided by the time λ/c between arrival of crests) is increased to the Doppler-shifted value[1]

$$\omega_r = \frac{\omega}{1-(w/c)} \quad \text{(relative frequency)} \tag{5}$$

that results from relative motion between source and observer. For an observer located on a line making an angle θ with a source's direction of motion at speed V, the source's velocity of approach toward the observer is $w = V\cos\theta$ and the relative frequency becomes

$$\omega_r = \frac{\omega}{1-(V/c)\cos\theta} \tag{6}$$

and is augmented or diminished, when θ is an acute or obtuse angle. Such Doppler shifts in frequency are familiar everyday experiences (Chapter 28).

3.2 Volume Changes

When an observer is approached at velocity w by a source whose dimension (in the direction of the observer) is l, sounds arriving simultaneously from the source's far or near sides have been emitted earlier or later by a time difference τ (say). In the time t for sound from the far side to reach the observer, after traveling a distance ct, the relative distance of the near side in the direction of the observer was increased from l to $l + w\tau$ before it emitted sound that then traveled a distance $c(t-\tau)$. Both sounds arrive simultaneously if

$$ct = l + w\tau + c(t-\tau), \tag{7a}$$

giving

$$\tau = \frac{l}{c-w} \quad \text{and} \quad l + w\tau = \frac{l}{1-(w/c)} = \frac{l\omega_r}{\omega}. \tag{7b}$$

The source's effective volume during emission is increased, then, by the Doppler factor ω_r/ω (since the dimension in the direction of the observer is so increased whereas other dimensions are unaltered).[2,3]

If turbulent "eddies" are effectively being convected, relative to the air into which they are radiating, at velocity V, then Eq. (6) gives, for radiation at angle θ, the Doppler factor ω_r/ω, which modifies both the frequencies at which they radiate and the effective volume occupied by a radiating eddy.

But Eq. (2) specifies the quadrupole strength T_{ij} per unit volume for such an eddy. Without convection the pattern of acoustic intensity around a compact eddy of volume l^3 and quadrupole strength $l^3 T_{ij}$ would be

$$\frac{\langle (l^3 \ddot{T}_{ij} x_i x_j r^{-2})^2 \rangle}{16\pi^2 r^2 \rho_0 c^5} \tag{8}$$

(Chapter 8); and since different eddies of volume l^3 radiate independently, we can simply sum mean squares in the corresponding expressions for their far-field intensities. This gives

$$\frac{l^3 \langle (\ddot{T}_{ij} x_i x_j r^{-2})^2 \rangle}{16\pi^2 r^2 \rho_0 c^5} \tag{9}$$

as the intensity pattern radiated by a unit volume of turbulence. The Doppler effect modifies this, when the compactness condition is satisfied, by five factors ω_r/ω (one for the change in source volume l^3 and four for the frequency change as it affects the mean square of a multiple of the second time derivative of T_{ij}), and this intensity modification brings about an important preference for forward emission[15] by a factor

$$\left[1 - \left(\frac{V}{c}\right)\cos\theta\right]^{-5}. \tag{10}$$

3.3 Compactness Changes

As V/c increases, however, the Doppler effect tends to degrade the compactness of aeroacoustic sources in relation to forward emission. Not only does $\omega l/c$ increase in proportion to Mach number, but an even greater value is taken by $\omega_r l/c$, a ratio that must be small if convected sources are to be compact. A restriction on the extent (10) of intensity enhancement for forward emission as V/c increases is placed by these tendencies.[6,15,16]

Indeed, they can develop to a point where the compact-source approximation (Chapter 8) may appropriately be replaced by the ray acoustics approximation (Chapter 3). Thus, for supersonic source convection ($V/c > 1$), the relative frequency (6) becomes infinite in the Mach direction

$$\theta = \cos^{-1}\frac{c}{V}, \tag{11}$$

and radiation from the source proceeds along rays emitted at this angle.[17]

Explanatory note: The source's velocity of approach w toward an observer positioned at an angle (11) to its direction of motion is the sound speed c; thus, not only is the generated wavelength (4) reduced indefinitely (the ray acoustics limit) but, essentially, different parts of a signal are observed simultaneously: the condition of stationary phase satisfied on rays (Chapter 3).

Further note: The influences placing a limit on the signal propagated along rays may include the duration δ of well-correlated emission from turbulent eddies and, in addition, nonlinear effects (see Section 4.2).

3.4 Uniformly Valid Doppler Effect Approximations

The correlation length l for turbulence was presented in Section 2.2; a correlation duration δ can be characterized by the requirement that moving eddies have, respectively, well-correlated or uncorrelated velocities at times differing by substantially $<\delta$ or $>\delta$. Combined use of correlation length l and duration δ affords an approximation to the radiation pattern from convected eddies that has some value at all Mach numbers, spanning the areas of applicability of the compact-source and ray acoustics approximations.

Figure 1 uses space–time diagrams in which the space coordinate (abscissa) is distance in the direction of the observer. Figure 1*a* for unconvected eddies approximates the region of good correlation as an ellipse with axes l (in the space direction) and δ (in the time direction). Figure 1*b* shows such a region for convected eddies whose velocity of approach toward the observer is w; thus, it is Fig. 1*a* sheared by distance w per unit time.

Signals from far points F and near points N, in either case, reach the observer simultaneously, as do signals from other points on the line FN, if this line slopes by distance c (the sound speed) per unit time:

(i) Compact-source case with w/c small: The space component of FN in Fig. 1*b* is $l[1-(w/c)]^{-1}$, just as in Eq. (7) for normal Doppler effects (neglecting finite δ).

(ii) Ray acoustics case with $w/c = 1$: The space component of FN is $c\delta$.

(iii) Intermediate case with w/c "moderately" <1: The space component of FN is l multiplied by an enhancement factor

$$\left[\left(1-\frac{w}{c}\right)^2+\left(\frac{l}{c\delta}\right)^2\right]^{-1/2}, \tag{12}$$

which represents the effective augmentation of source volume due to convection.[15]

Enhancement factor (12) is applied not only to the volume term l^3 in the quadrupole field (9) but also twice to each of the pair of twice-differentiated terms inside the mean square, essentially because time differentiations in quadrupole fields arise (Chapter 8) from differences in the time of emission by different parts of the quadrupole source region (and the time component of FN in Fig. 1*b* is simply the space component divided by c). As before, then, five separate factors (12) enhance the intensity field, and with w replaced by $V\cos\theta$, expression (10) for the overall intensity modification factor is replaced by

$$\left\{\left[1-\left(\frac{V}{c}\right)\cos\theta\right]^2+\left(\frac{l}{c\delta}\right)^2\right\}^{-5/2}. \tag{13}$$

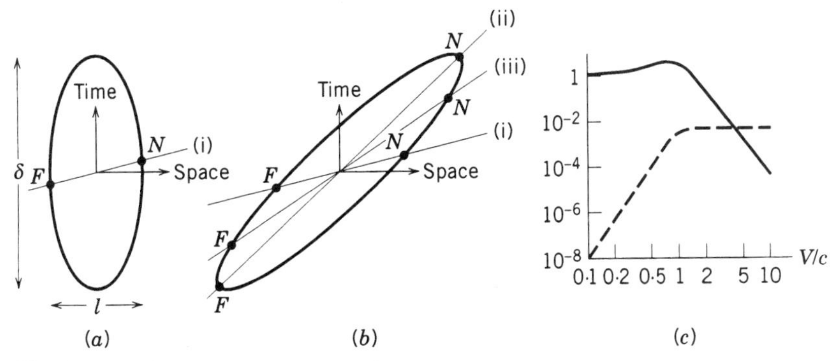

Fig. 1 A uniformly valid Doppler-effect approximation. (*a*) Space-time diagram for unconvected "eddies" of correlation length l and duration δ. (*b*) Case of "eddies" convected toward observer at velocity w; being Fig. 1*a* sheared by a distance w per unit time. Here, lines sloping by a distance c per unit time represent emissions received simultaneously by observer. Case (i): w/c small. Case (ii): $w/c = 1$. Case (iii): intermediate value of w/c. (*c*) The solid line is the average modification factor (13). The dashed line is the acoustic efficiency obtained by applying this factor to a low-Mach-number "quadrupole" efficiency of (say) $10^{-3}(V/c)^5$.

This modification factor affords us an improved description of the influence of the Doppler effect not only on the preference for forward emission but also on the overall acoustic power output from convected turbulence.[6,15] For example, Fig. 1*c* gives a (plain line) log-log plot of the average (spherical mean) of the factor (13) as a function of V/c on the reasonable assumption that $l = 0.6V\delta$. As V/c increases, this average modification factor rises a little at first but falls drastically as $5(V/c)^{-5}$ for $V/c \gg 1$.

Now low-Mach-number turbulence away from solid boundaries (Section 2.1) should radiate sound with an acoustic efficiency scaling as $(U/c)^5$, where U is a characteristic velocity in the flow. With V taken as that characteristic velocity (although in a jet a typical velocity V of eddy convection would be between 0.5 and 0.6 times the jet exit speed), the modification of (say) an acoustic efficiency of $10^{-3}(V/c)^5$ for low Mach number by the average modification factor would cause acoustic efficiency to follow the broken-line curve in Fig. 1*c*, tending asymptotically to a constant value 0.005 (aeroacoustic saturation) at high Mach number. Such a tendency is often observed for sound radiation from "properly expanded" supersonic jets (see below).

4 INTRODUCTION TO AIRCRAFT NOISE

4.1 Aeroengine and Airframe Noise

How are aeroacoustic principles applied to practical problems such as the reduction of aircraft noise?[18–20]

In any analysis of the generation of sound by airflows, we may first need to ask whether the geometry of the problem has features that tend to promote resonance. For example, a long wire in a wind (Section 2.1) generates most sound when vortex-shedding frequencies are fairly close (and so can "lock on") to the wire's lowest natural frequency of vibration; giving good correlation of side forces, and so also of dipole strengths, all along the wire.

Again, a jet emerging from a thin slit may interact with a downstream edge (parallel to the slit) in a resonant way,[21,22] with very small directional disturbances at the jet orifice amplified by flow instability as they move downstream to the edge, where they produce angle-of-attack variations. Dipole fields associated with the resulting side forces can at particular frequencies renew the directional disturbances at the orifice with the right phase to produce a resonant oscillation. Some musical wind instruments utilize such jet-edge resonances, reinforced by coincidence with standing-wave resonances (Chapter 6) in an adjacent pipe. However, in the absence of such resonances (leading to enhanced acoustic generation at fairly well defined frequencies), airflows tend to generate acoustic "noise" whose reaction on the flow instability phenomena themselves is negligible.

Resonances analogous to the above that need to be avoided in aircraft design include the following:

(a) Panel flutter, generated at a characteristic frequency as an unstable vibration of a structural panel in the presence of an adjacent airflow.[23]
(b) Screeching of supersonic jets from nozzles which, instead of being "properly expanded" so that an essentially parallel jet emerges, produce a jet in an initially nonparallel form followed by shock waves in the well-known recurrent "diamond" shock cell pattern. The first of these, replacing the edge in the above description, can, through a similar feedback of disturbances to the jet orifice, generate a powerful resonant oscillation.[22,24,25]

Undesirable resonances may also be associated with aeroengine combustion processes.[26] We now turn to the aircraft noise of a broadbanded character that remains even when resonances have been avoided.

Aeroengine jet noise proper[6] (i.e., the part unrelated to any interaction of jet turbulence with solid boundaries) tends to follow a broad trend similar to that in Fig. 1, where, however, because the eddy convection velocity V is between 0.5 and 0.6 times the jet exit speed U, the acoustic efficiency makes a transition between a value of around $10^{-4}M^5$ in order-of-magnitude terms for subsonic values of $M = U/c$ and an asymptotically constant value of 10^{-2} or a little less for M exceeding about 2.

The above tendency for $M < 1$ implies that noise emission from jet engines may be greatly diminished if a given engine power can be achieved with a substantially lower jet exit speed, requiring of course a correspondingly larger jet diameter L. Furthermore, with acoustic power output scaling as $\rho U^8 L^2/c^5$ (Section 2.1) and jet thrust as $\rho U^2 L^2$, noise emission for a given thrust can be greatly reduced if U can be decreased and L increased by comparable factors.

Trends (along these lines) in aeroengine design toward large turbofan engines with higher and higher bypass ratios, generating very wide jets at relatively modest mean Mach numbers, have massively contributed to jet noise suppression (while also reducing fuel consumption). On the other hand, such successes in suppressing jet noise proper (originally, the main component of noise from jet aircraft) led to needs for focusing attention upon parallel reductions of other aircraft noise sources[27]:

(a) Those associated with the interaction of jet turbulence with solid boundaries, where sharp-edged boundaries (Section 2.1) pose a particular threat
(b) Fan noise emerging from the front of the engine and turbine noise emerging from the rear

(c) Airframe noise, including acoustic radiation from boundary layer turbulence and from interaction of that turbulence with aerodynamic surfaces for control purposes or lift enhancement

Some key areas of modern research on aeroengine and airframe noise are as follows:

For jet noise, techniques that relate acoustic output to vorticity distributions,[28,29] and to any coherent structures,[30,31] in jet turbulence and that take into account (cf. Section 5.4) propagation through the sheared flow in a wide jet[32,33]

For noise from fans and propellers, mathematically sophisticated ways of reliably estimating the extent of cancellation of dipole radiation from different parts of a rotating-blade system (alongside a good independent estimate of quadrupole radiation)[34]

For airframe noise, a recognition[35,36] that massive cancellations act to minimize noise radiation from boundary layer turbulence on a flat surface of uniform compliance; and, therefore, that avoidance of sharp nonuniformities in airframe skin compliance may promote noise reduction

4.2 Supersonic Booms

In addition to aeroengine and airframe noise, any aircraft flying at a supersonic speed V emits a concentrated "boom"-like noise along rays in the Mach direction (11). We sketch the theory of supersonic booms with the atmosphere approximated as isothermal (so that the undisturbed sound speed takes a constant value c even though the undisturbed density ρ varies with altitude), a case permitting a simple extension of the nonlinear analysis (Chapter 17) of waveform shearing and shock formation.[37] Then the rays continue as straight lines at the Mach angle for reasons summarized in the explanatory note below expression (11). (Actually, the slight refraction of rays by temperature stratification in the atmosphere, when taken into account in a generalized version of the theory, produces only somewhat minor modifications of the results.)

As such straight rays stretch out from a straight flight path along cone-shaped surfaces with semi-angle (11), any narrow tube of rays has its cross-sectional area A increasing in proportion to distance r along the tube.[1,38] In linear theory (Chapter 3), acoustic energy flux $u^2\rho cA$ is propagated unchanged along such a ray tube [so that $u(\rho r)^{1/2}$ is unchanged], where u is air velocity along the ray tube. In nonlinear theory, $u(\rho r)^{1/2}$ is propagated unchanged but at a signal speed altered to

$$c + \frac{\gamma + 1}{2} u \tag{14}$$

by self-convection and excess-wave-speed effects (Chapter 17).

This property can be described[1] by

$$\left[\left\{\frac{1}{c} - \frac{\gamma + 1}{2}\frac{u}{c^2}\right\}\frac{\partial}{\partial t} + \frac{\partial}{\partial r}\right] u(\rho r)^{1/2} = 0, \tag{15}$$

where the quantity in braces is the altered value of the reciprocal of the signal speed (14). Now a simple transformation of variables,

$$x_1 = r - ct, \qquad t_1 = \int_0^r (\rho r)^{-1/2}\, dr,$$

$$u_1 = \frac{\gamma + 1}{2}\frac{u}{c}(\rho r)^{1/2}, \tag{16}$$

converts Eq. (15) into the familiar form (Chapter 17)

$$\frac{\partial u_1}{\partial t_1} + u_1 \frac{\partial u_1}{\partial x_1} = 0, \tag{17}$$

which describes the waveform shearing at a uniform rate associated with shock formation and propagation in nonlinear plane-wave acoustics.

From the physically relevant solutions of Eq. (17), namely, those with area-conserving discontinuities (representing shocks), the famous N-wave solution is the one produced by an initial signal (such as an aircraft's passage through the air) that is first compressive and then expansive. The rules governing N-wave solutions of Eq. (17) (Chapter 17) are that the discontinuity Δu_1 at each shock falls off as $t_1^{-1/2}$ while the space (change Δx_1 in x_1) between shocks increases as $t_1^{1/2}$. These rules for the transformed variables (16) have the following consequences for the true physical variables: At a large distance r from the flight path the velocity change Δu at each shock and the time interval Δt between the two shocks vary as

$$\Delta u \approx \left[(\rho r)\int_0^r (\rho r)^{-1/2}\, dr\right]^{-1/2} \quad \text{and}$$

$$\Delta t \approx \left[\int_0^r (\rho r)^{-1/2}\, dr\right]^{1/2}. \tag{18}$$

On horizontal rays (at the level where the aircraft is flying), ρ is independent of r and Eqs. (18) take the greatly simplified form

$$\Delta u \approx r^{-3/4} \quad \text{and} \quad \Delta t \approx r^{1/4}, \tag{19}$$

appropriate to conical N-waves in a homogeneous atmosphere. Actually, the rules (19) apply also to the propagation of cylindrical blast waves generated by an exploding wire, since, here also, ray tube areas increase in proportion to r. (For spherical blast waves, see Chapter 27.)

On downward-pointing rays in an isothermal atmosphere ρ increases exponentially in such a way that the time interval Δt between shocks approaches the constant value obtained in Eqs. (18) by making the integral's upper limit infinite.[1,38] On the other hand, the shock strength (proportional to the velocity change Δu) includes the factor $(\rho r)^{-1/2}$, where the large increase in ρ from the flight path to the ground (as well as in r) enormously attenuates the supersonic boom. Below Concorde cruising at Mach 2, for example, an observer on the ground hears two clear shocks with an interval of around 0.5 s between them and yet with strength $\Delta p/p$ of only about 0.001.

5 PROPAGATION OF SOUND THROUGH STEADY MEAN FLOWS

5.1 Adaptations of Ray Acoustics

Useful information on sound propagation through steady mean flows[4,39] can be obtained by adaptations of the ray acoustics approximation. We sketch these here before, first, applying them (in Section 5.3) to propagation through sheared stratified winds and, second, giving indications of how effects of such parallel mean flows are modified at wavelengths too large for the applicability of ray acoustics.

Sound propagation through a steady airflow represents an autonomous mechanical system, one governed by laws that do not change with time. Then small disturbances can be Fourier analyzed in the knowledge that propagation of signals with different frequencies ω must proceed without exchange of energy between them.

Such disturbances of frequency ω involve pressure changes in the form $P \cos \alpha$, where P varies with position and the phase α is a function of position and time, satisfying

$$\frac{\partial \alpha}{\partial t} = \omega \quad \text{(frequency) and}$$

$$-\frac{\partial \alpha}{\partial x_i} = k_i \quad \text{(wavenumber)}, \tag{20}$$

a vector with its direction normal to crests and its magnitude 2π divided by a local wavelength.

In ray theory for any wave system,[1,4] we assume that the wavelength is small enough (compared with distances over which the medium—and its motion, if any—change significantly) for a well-defined relationship

$$\omega = \Omega(k_i, x_i) \tag{21}$$

to link frequency with wavenumber at each position. Equations (20) and (21) require that

$$-\frac{\partial k_j}{\partial t} = \frac{\partial^2 \alpha}{\partial x_j\, \partial t} = \frac{\partial \omega}{\partial x_j} = \frac{\partial \Omega}{\partial k_i}\left(-\frac{\partial^2 \alpha}{\partial x_i\, \partial x_j}\right) + \frac{\partial \Omega}{\partial x_j}$$

$$= \frac{\partial \Omega}{\partial k_i}\frac{\partial k_j}{\partial x_i} + \frac{\partial \Omega}{\partial x_j}, \tag{22}$$

yielding the basic law (in Hamiltonian form) for any wave system: On rays satisfying

$$\frac{dx_i}{dt} = \frac{\partial \Omega}{\partial k_i} \tag{23a}$$

wavenumbers vary as

$$\frac{dk_j}{dt} = -\frac{\partial \Omega}{\partial x_j}. \tag{23b}$$

These are equations easy to solve numerically for given initial position and wavenumber. However, the variations (23) of wavenumber ("refraction") produce no change of frequency along rays:

$$\frac{d\omega}{dt} = \frac{\partial \Omega}{\partial k_i}\frac{dk_i}{dt} + \frac{\partial \Omega}{\partial x_i}\frac{dx_i}{dt}$$

$$= \frac{\partial \Omega}{\partial k_i}\left(-\frac{\partial \Omega}{\partial x_i}\right) + \frac{\partial \Omega}{\partial x_i}\frac{\partial \Omega}{\partial k_i} = 0, \tag{24}$$

so that rays are paths of propagation of the excess energy, at each frequency, associated with the waves' presence.

For sound waves we write k as the magnitude of the wavenumber vector, expecting that at any point the value of the relative frequency ω_r in a frame of reference moving at the local steady-flow velocity u_{fi} will be $c_f k$ (the local sound speed times k); this implies[1,4] that

$$\omega_r = \frac{\partial \alpha}{\partial t} + u_{fi}\frac{\partial \alpha}{\partial x_i} = \omega - u_{fi}k_i, \tag{25a}$$

giving

$$\omega = \omega_r + u_{fi}k_i = c_f k + u_{fi}k_i \tag{25b}$$

as the acoustic form of relationship (21). [*Note:* Rule (25) for relative frequency agrees with the Doppler rule (6), since the velocity of a source of frequency ω relative to stationary fluid into which it radiates is minus the velocity of the fluid relative to a frame in which the acoustic frequency is ω.]

Use of form (25b) of relationship (21) in the basic law (23) tells us that

$$\frac{dk_j}{dt} = -k\frac{\partial c_f}{\partial x_j} - k_i\frac{\partial u_{fi}}{\partial x_j} \tag{26a}$$

on rays with

$$\frac{dx_i}{dt} = c_f \frac{k_i}{k} + u_{fi}, \tag{26b}$$

where the last terms in these equations represent adaptations of ray acoustics associated with the mean flow. For example, the velocity of propagation along rays is the vector sum of the mean-flow velocity u_{fi} with a wave velocity of magnitude c_f and direction normal to crests.

5.2 Energy Exchange between Sound Waves and Mean Flow

The excess energy (say, E per unit volume) associated with the presence of sound waves is propagated along such rays; in particular, if attenuation of sound energy is negligible, then

$$\text{Flux of excess energy along a ray tube} = \text{constant.} \tag{27}$$

Note that this excess energy density E is by no means identical with the sound waves' energy density,

$$E_r = \langle \tfrac{1}{2}\rho_f u_{si}u_{si}\rangle + \langle \tfrac{1}{2}c_f^2\rho_f^{-1}\rho_s^2\rangle = c_f^2\rho_f^{-1}\langle\rho_s^2\rangle, \tag{28}$$

where the subscript s identifies changes due to the sound waves and the equality of the kinetic and potential energies makes E_r simply twice the latter in a frame of reference moving at the local flow velocity [compare definition (25) of ω_r]. The kinetic-energy part of the excess energy density E is

$$\langle \tfrac{1}{2}(\rho_f + \rho_s)(u_{fi} + u_{si})^2\rangle - \tfrac{1}{2}\rho_f u_{fi}u_{fi}, \tag{29}$$

which includes an extra term,

$$\langle \rho_s u_{fi} u_{si}\rangle = \left\langle \rho_s u_{fi} \frac{c_f}{\rho_f}\,\rho_s\,\frac{k_i}{k}\right\rangle = E_r \frac{u_{fi}k_i}{c_f k} = E_r\left(\frac{\omega}{\omega_r} - 1\right), \tag{30}$$

and E is the sum of expressions (28) and (30), giving

$$E = E_r \frac{\omega}{\omega_r}, \tag{31a}$$

or, equivalently,

$$E_r = E\,\frac{\omega_r}{\omega} \quad \text{or} \quad \frac{E_r}{\omega_r} = \frac{E}{\omega}. \tag{31b}$$

The quantity E/ω, called the *action density* in Hamiltonian mechanics, is identical in both frames of reference, and Eqs. (24) and (27) tell us that its flux along a ray tube is constant.[1,4]

However, Eq. (31) shows too that energy is exchanged between (i) the acoustic motions relative to the mean flow and (ii) the mean flow itself. For example, where sound waves of frequency ω enter a region of opposing flow (or leave a region where the mean flow is along their direction of propagation), the ratio ω_r/ω increases and so therefore does E_r/E: The sound waves gain energy at the expense of the mean flow.

The rate of exchange of energy takes the value (1) from Section 1. This is readily seen from the laws governing motion in an accelerating frame of reference, which is subject to an

$$\text{Inertial force} = -(\text{mass}) \times (\text{acceleration of frame}). \tag{32}$$

If at each point of space we use a local frame of reference moving with velocity u_{fi}, then fluid in that frame has velocity u_{si} but is subject to an additional force (32), where, per unit volume, mass is ρ_f and the frame's acceleration takes the form

$$u_{sj}\,\frac{\partial u_{fi}}{\partial x_j}$$

$$\text{Giving force} \quad -\rho_f u_{sj}\,\frac{\partial u_{fi}}{\partial x_j}$$

$$\text{Doing work} \quad -\rho_f\langle u_{si}u_{sj}\rangle\,\frac{\partial u_{fi}}{\partial x_j} \tag{33}$$

per unit time on the local relative motions. This rate of energy exchange (33) proves to be consistent with the fact that it is the flux, not of E_r but of action E_r/ω_r, that is conserved along ray tubes.

Energy can be extracted from a mean flow, then, not only by turbulence but also by sound waves; in both cases, the rate of extraction takes the same form (33) in terms of perturbation velocities u_{si}. It represents the effect (Section 1) of that

$$\text{Mean momentum flux} = \rho_f \langle u_{si} u_{sj} \rangle \qquad (34)$$

or Reynolds stress[40] with which either the sound waves or the turbulent motions act upon the mean flow. For sound waves, by Eq. (28) for E_r and by the substitution

$$u_{si} = \frac{c_f}{\rho_f}\,\rho_s\,\frac{k_i}{k}$$

we arrive at

$$\text{Mean momentum flux} = E_r\,\frac{k_i k_j}{k^2}, \qquad (35)$$

so that the Reynolds stress is a uniaxial stress in the direction of the wavenumber vector having magnitude E_r. Strictly speaking, the complete

$$\text{Radiation stress} = E_r\left(\frac{k_i k_j}{k^2} + \frac{\gamma - 1}{2}\,\delta_{ij}\right) \qquad (36)$$

for sound waves includes not only the momentum flux (35) but also the waves' mean pressure excess,

$$\frac{1}{2}\left(\frac{\partial^2 p}{\partial \rho^2}\right)_{\rho=\rho_f} \langle \rho_s^2 \rangle = \frac{\gamma - 1}{2}\,\frac{c_f^2}{\rho_f}\,\langle \rho_s^2 \rangle = \frac{\gamma - 1}{2}\,E_r, \qquad (37)$$

acting equally in all directions[7]; however (Section 1), this isotropic component produces no energy exchange with solenoidal mean flows.

5.3 Propagation through Sheared Stratified Winds

The extremely general ray acoustics treatment outlined above for sound propagation through fluids in motion has far-reaching applications (in environmental and, also, in engineering acoustics), which, however, are illustrated below only by cases of propagation through parallel flows, with stratification of velocity as well as of temperature.[1,4] The x_1-direction is taken as that of the mean-flow velocity $V(x_3)$, which, together with the sound speed $c(x_3)$, depends only on the coordinate x_3. Thus, V replaces u_{f1} in the general theory while c replaces c_f (and, for atmospheric propagation, x_3 is altitude). [*Note:* The analysis sketched here is readily extended to cases of winds veering with altitude, where u_{f2} as well as u_{f1} is nonzero.]

Either the basic law (23) or its ray acoustics form (26) provides, in general, "refraction" information in the form of three equations for change of wavenumber, while the single, far simpler, Eq. (24) is a consequence of, but by no means equivalent to, those three. By contrast, in the particular case when u_{f_1} and c_f are independent of x_1 and x_2, Eq. (24) and, additionally, Eqs. (26) for $j = 1, 2$ give three simple results along rays that may be shown fully equivalent to the basic law:

$$\omega = \text{const}, \qquad k_1 = \text{const}, \qquad k_2 = \text{const}. \qquad (38)$$

If now we write the wavenumber (a vector normal to crests) as

$$(k_1, k_2, k_3) = (\kappa \cos \psi, \kappa \sin \psi, \kappa \cot \theta), \qquad (39)$$

so that κ is its constant horizontal resultant, ψ its constant azimuthal angle to the wind direction, and θ its variable angle to the vertical, and use Eq. (25) in the form

$$\omega = c(x_3)k + V(x_3)k_1 = c(x_3)\kappa \operatorname{cosec} \theta + V(x_3)\kappa \cos \psi, \qquad (40)$$

we obtain an important extension to Snell's law from the classical case ($V = 0$) when the denominator is a constant:

$$\sin \theta = \frac{c(x_3)}{\omega\kappa^{-1} - V(x_3)\cos \psi}. \qquad (41)$$

This extended law (41) tells us how θ varies with x_3 along any ray, whose path we can then trace using Eq. (26) in the form

$$\frac{dx_1}{dt} = c(x_3)\cos \psi \sin \theta + V(x_3),$$

$$\frac{dx_2}{dt} = c(x_3)\sin \psi \sin \theta,$$

$$\frac{dx_3}{dt} = c(x_3)\cos \theta \qquad (42)$$

by simply integrating dx_1/dx_3 and dx_2/dx_3 with respect to x_3.

It follows that a ray tube covers the same horizontal area at each altitude, so that conservation of the flux of wave action E_r/ω_r along it implies that the vertical component

$$\frac{E_r}{\omega_r}\frac{dx_3}{dt} = E_r\kappa^{-1}\sin\theta\cos\theta \tag{43}$$

of wave action flux is constant along rays, from which, with Eq. (28), sound amplitudes are readily derived.

Wind shear is able to reproduce all the main types of ray bending (Chapter 3) associated with temperature stratification, often to an enhanced extent. Roughly, the downward curvature of near-horizontal rays in reciprocal kilometres comes to

$$3\{V'(x_3)\cos\psi + c'(x_3)\}, \tag{44}$$

where the velocity gradients are in reciprocal seconds and the factor 3 $(\text{km})^{-1}$s outside the braces is an approximate reciprocal of the sound speed.[1,4]

When (44) is negative, curvature is upward; its magnitude with zero wind is at most 0.018 km^{-1} (because temperature lapse rate in stable atmospheres cannot exceed 10°C/km, giving $c' = 0.006$ s^{-1}) but with strong wind shear can take much bigger values for upwind propagation ($\psi = \pi$). In either case Fig. 2*a* shows how the lowest ray emitted by a source "lifts off" from the ground, leaving below it a zone of silence (on ray theory; actually, a zone where amplitudes decrease exponentially with distance below that ray).

When (44) is positive, curvature is downward, as found with zero wind in temperature inversion conditions (e.g., over a calm, cold lake) and even more with strong wind shear for downwind propagation ($\psi = 0$). Figure 2*b* shows how this leads to signal enhancement through multiple-path communication.

In summary, then, the very familiar augmentation of sound levels downwind, and diminution upwind, of a source represent effects of the wind's shear (increase with altitude).

5.4 Wider Aspects of Parallel-Flow Acoustics

The propagation of sound through parallel flows at wavelengths too great for the applicability of ray acoustics can be analyzed by a second-order ordinary differential equation. Thus, a typical Fourier component of the sound pressure field takes the form

$$p_s(x_3)e^{i(\omega t - k_1 x_1 - k_2 x_2)}$$

with

$$\rho\frac{d}{dx_3}\left[\frac{1}{\rho(\omega - Vk_1)^2}\frac{dp_s}{dx_3}\right] + \left[\frac{1}{c^2} - \frac{k_1^2 + k_2^2}{(\omega - Vk_1)^2}\right]p_s = 0. \tag{45}$$

Equation (45) can be used to improve on ray acoustics:

(a) Near caustics (envelopes of rays), where it allows a uniformly valid representation of amplitude in terms of the famous Airy function, giving "beats" between superimposed waves on one side of the caustic and exponential decay on the other[1]

(b) At larger wavelengths by abandoning ray theory altogether in favor of extensive numerical solutions of Eq. (45)

(c) To obtain waveguide modes for sound propagation in a two-dimensional duct (between parallel planes)[41–43]

On the other hand, in the case of a three-dimensional duct carrying parallel flow $V(x_2, x_3)$ in the x_1-direction, Eq. (45) is converted to a partial differential equation

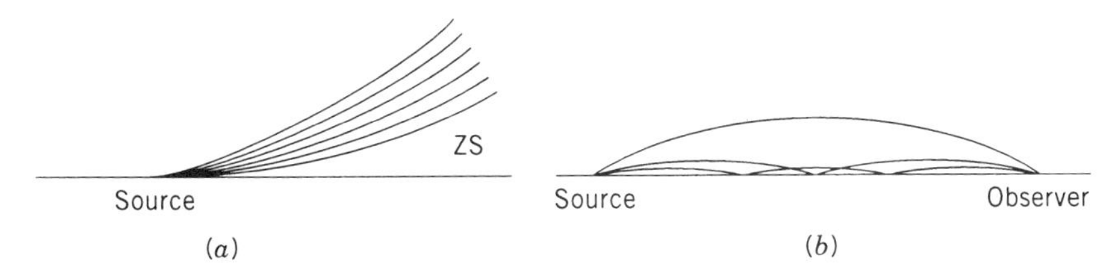

Fig. 2 Effects of ray curvature (44) on propagation from a source on horizontal ground. (*a*) Rays of given upward curvature (due to temperature lapse or upwind propagation) can leave a zone of silence (ZS) below the ray emitted horizontally. (*b*) Rays of given downward curvature (due to temperature inversion or downwind propagation) can enhance received signals through multiple-path communication.

(the first term being supplemented by another with d/dx_2 replacing d/dx_3, while k_2 is deleted) which is used for the following:

(d) To obtain waveguide modes in such ducts
(e) In calculations of propagation of sound through the wide jets, modeled as parallel flows, typical of modern aeroengines (Section 4.1)
(f) With aeroacoustic source terms included, in certain enterprising attempts at modeling jet noise generation and emission[32,33]

6 ACOUSTIC STREAMING

6.1 Streaming as a Result of Acoustic Attenuation

Sound waves act on the air with a Reynolds stress (34) even when mean flow is absent (so that subscript f becomes subscript zero). The j-component of force acting on a unit volume of air is then

$$F_j = -\frac{\partial}{\partial x_i}\langle \rho_0 u_{si} u_{sj}\rangle, \tag{46}$$

namely, the force generating acoustic streaming.[5]

However, the force given in Eq. (46) could not produce streaming for unattenuated sound waves; indeed, the linearized equations (Chapter 2) can be used to show that

$$\text{if:}\quad p^M = \langle \tfrac{1}{2}c_0^2\rho_0^{-1}\rho_s^2 - \tfrac{1}{2}\rho_0 u_{si}u_{si}\rangle$$

$$\text{then:}\quad F_j - \frac{\partial p^M}{\partial x_j} = \left\langle \frac{\partial}{\partial t}(\rho_s u_{sj})\right\rangle, \tag{47}$$

which is necessarily zero (as the mean value of the rate of change of a bounded quantity). Accordingly, the fluid must remain at rest, responding merely by setting up the distribution p^M of mean pressure whose gradient can balance the force. [*Note:* In the ray acoustic approximation (28), p^M is itself zero, but the above argument does not need to use this approximation.]

Attenuation of sound waves takes place as follows:

(a) In the bulk of the fluid through the action of viscosity, thermal conductivity, and lags in attaining thermodynamic equilibrium (Chapter 5)
(b) Near solid walls by viscous attenuation in Stokes boundary layers (Chapter 7)

All these effects produce forces [Eq. (46)] that act to generate acoustic streaming. It is important to note, furthermore, that even the forces due solely to viscous attenuation, being opposed just by the fluid's own viscous resistance, generate mean motions that do not disappear as the viscosity μ tends to zero.[44–47]

6.2 Jets Generated by Attenuated Acoustic Beams

Attenuation of type (a) produces a streaming motion u_{fj} satisfying

$$\rho u_{fi}\frac{\partial u_{fj}}{\partial x_i} = F_j - \frac{\partial p}{\partial x_j} + \mu\,\nabla^2 u_{fj}. \tag{48}$$

Substantial streaming motions can be calculated from this equation only with the left-hand side included,[48] although in the pre-1966 literature it was misleadingly regarded as "a fourth-order term" and so ignored, thus limiting all the theories to uninteresting cases when the streaming Reynolds number would be of order 1 or less.

We can use streaming generated by acoustic beams to illustrate the above principles. If acoustic energy is attenuated at a rate β per unit length, then a source at the origin that beams acoustic power P along the x_1-axis transmits the following distributions:

$$\text{Power } Pe^{-\beta x_1} \tag{49a}$$

and therefore

$$\text{Energy per unit length } c^{-1}Pe^{-\beta x_1}, \tag{49b}$$

which is necessarily the integral over the beam's cross section of energy density as well as of the uniaxial Reynolds stress (35). It follows that the force per unit volume (46) integrated over a cross section produces[5]

$$\text{Force } c^{-1}P\beta e^{-\beta x_1} \text{ per unit length in the } x_1\text{-direction.} \tag{50}$$

At high ultrasonic frequencies the force distribution (50) is rather concentrated, the distance of its center of application from the origin being just β^{-1} (which at 1 MHz, e.g., is 24 mm in air). Effectively, the beam applies at this center a total force $c^{-1}P$ [integral of distribution (50)].

The type[1,5] of streaming motion generated by this concentrated force $c^{-1}P$ depends critically on the value of $\rho c^{-1}P\mu^{-2}$: a sort of Reynolds number squared, which is about $10^7 P$ in atmospheric air (with P in watts). Streaming of the low-Reynolds-number "stokeslet" type predicted (for a concentrated force) by Eq. (48) with the left-hand side suppressed is a good approximation only for $P < 10^{-6}$ W.

For a source of 10^{-4} W power, by contrast, the force $c^{-1}P$ generates quite a narrow laminar jet with momentum transport $c^{-1}P$, and at powers exceeding 3×10^{-4} W this jet has become turbulent, spreading conically with a semiangle of about 15° and continuing to transport momentum at the rate $c^{-1}P$. Such turbulent jets generated by sound are strikingly reciprocal to a classical aeroacoustic theme!

At lower frequencies an acoustic beam of substantial power delivers a turbulent jet with a somewhat more variable angle of spread but one that at each point x_1 carries momentum transport

$$c^{-1}P(1 - e^{-\beta x_1}) \tag{51}$$

generated by the total force (50) acting up to that point. This momentum transport in the jet represents the source's original rate of momentum delivery minus the acoustic beam's own remaining momentum transport (49). In summary, as acoustic power is dissipated into heat, the associated acoustic momentum transport is converted into a mean motion (which, at higher Reynolds numbers, is turbulent).[5]

6.3 Streaming around Bodies Generated by Boundary Layer Attenuation

Sound waves of frequency ω well below high ultrasonic frequencies have their attenuation concentrated, if solid bodies are present (Chapter 7), in thin Stokes boundary layers attached to each body. Then the streaming generated near a particular point on a body surface is rather simply expressed by using local coordinates with that point as origin, with the z-axis normal to the body and the x-axis in the direction of the inviscid flow just outside the boundary layer—the exterior flow. The Stokes boundary layer for an exterior flow

$$(U(x, y), V(x, y))e^{i\omega t} \tag{52a}$$

has interior flow

$$(U(x, y), V(x, y))e^{i\omega t}[1 - e^{-z\sqrt{(i\omega\rho/\mu)}}]. \tag{52b}$$

[Note that our choice of coordinates makes $V(0, 0) = 0$ and that expressions (52) become identical outside the layer.] The streaming motion[1,5] is calculated from the equation

$$F_j^{\text{int}} - F_j^{\text{ext}} + \mu\frac{\partial^2 u_{fj}}{\partial z^2} = 0, \tag{53}$$

with certain differences from Eq. (48) explained as follows:

(a) The first term is the force (46) generating streaming within the boundary layer.

(b) We are free, however, to subtract the second, since (see Section 6.1) it can produce no streaming, and conveniently, the difference is zero outside the layer.

(c) Gradients in the z-direction are so steep that the third term dominates the viscous force and, indeed, in such a boundary layer, dominates also the left-hand side of Eq. (48).

The solution of Eq. (53), which vanishes at $z = 0$ and has zero gradient at the edge of the layer, is obtained by two integrations, and its exterior value is

$$u_{fj}^{\text{ext}} = \mu^{-1}\int_0^\infty (F_j^{\text{int}} - F_j^{\text{ext}})z\,dz, \tag{54}$$

where integration extends in practice, not to "infinity," but to the edge of the layer within which the integrand is nonzero. Expression (54) for the exterior streaming is yet again (see Section 6.1) independent of the viscosity μ, since Eq. (52) makes $z\,dz$ of order $\mu/\rho\omega$, and it is easily evaluated.

At $x = y = 0$ (in the coordinates specified earlier) the exterior streaming (54) has components

$$-U\frac{3\,\partial U/\partial x + 2\,\partial V/\partial y}{4\omega} \quad (x\text{-component}),$$

$$-U\frac{\partial V/\partial x}{4\omega} \quad (y\text{-component}) \tag{55}$$

with zero z-component. This is a generalized form of the century-old Rayleigh law of streaming (which covers cases when V is identically zero).

For the complete streaming pattern, expressions (55) are, effectively, boundary values for its tangential component at the body surface (because the Stokes boundary layer is so thin). Therefore, any simple solver for the steady-flow Navier–Stokes equations with specified tangential velocities on the boundary allows the pattern to be determined. It is important to remember that the inertia terms in the Navier–Stokes equations must *not* be neglected unless the Reynolds number R_s based on the streaming velocity (55) is of order 1 or less, when, however, the corresponding streaming motions would (as in Section 6.2) be uninterestingly small.

In the other extreme case when R_s is rather large (at least 10^3) the streaming motion remains quite close to the body[48] within a steady boundary layer whose dimension (relative to that of the body) is of order $R_s^{-1/2}$. This layer is by no means as thin as the Stokes boundary layer, but it does confine very considerably the acoustic streaming motion. Equations (55) direct this motion toward one of the exterior flow's stagnation points, whence the steady-boundary-layer flow emerges as a jet—yet another jet generated by sound.[49,50]

Acknowledgments

I am warmly grateful to D. G. Crighton and N. Riley for invaluable advice on the text as well as to the Leverhulme Trust for generous support.

REFERENCES

1. J. Lighthill, *Waves in Fluids*, Cambridge University Press, 1978.
2. M. J. Lighthill, "On Sound Generated Aerodynamically. I. General Theory," *Proc. Roy. Soc. A*, Vol. 211, 1952, pp. 564–587.
3. M. J. Lighthill, "Sound Generated Aerodynamically. The Bakerian Lecture," *Proc. Roy. Soc. A*, Vol. 267, 1962, pp. 147–182.
4. J. Lighthill, "The Propagation of Sound through Moving Fluids," *J. Sound Vib.*, Vol. 24, 1972, pp. 471–492.
5. J. Lighthill, "Acoustic Streaming," *J. Sound Vib.*, Vol. 61, 1978, pp. 391–418.
6. M. J. Lighthill, "Jet Noise. The Wright Brothers Lecture," *Am. Inst. Aeronaut. Astronaut. J.*, Vol. 1, 1963, pp. 1507–1517.
7. F. P. Bretherton and C. J. R. Garrett, "Wavetrains in Inhomogeneous Moving Media," *Proc. Roy. Soc. A*, Vol. 302, 1968, pp. 529–554.
8. N. Curle, "The Influence of Solid Boundaries on Aerodynamic Sound," *Proc. Roy. Soc. A*, Vol. 231, 1955, pp. 505–514.
9. J. E. Ffowcs Williams and D. L. Hawkins, "Sound Generation by Turbulence and Surfaces in Arbitrary Motion," *Phil. Trans. Roy. Soc. A*, Vol. 264, 1969, pp. 321–342.
10. J. E. Ffowcs Williams and L. H. Hall, "Aerodynamic Sound Generation by Turbulent Flow in the Vicinity of a Scattering Half Plane," *J. Fluid Mech.*, Vol. 40, 1970, pp. 657–670.
11. D. G. Crighton and F. G. Leppington, "Scattering of Aerodynamic Noise by a Semi-infinite Compliant Plate," *J. Fluid Mech.*, Vol. 43, 1970, pp. 721–736.
12. D. G. Crighton and F. G. Leppington, "On the Scattering of Aerodynamic Noise," *J. Fluid Mech.*, Vol. 46, 1971, pp. 577–597.
13. D. G. Crighton, "Acoustics as a Branch of Fluid Mechanics," *J. Fluid Mech.*, Vol. 106, 1981, pp. 261–298.
14. M. J. Lighthill, "On Sound Generated Aerodynamically. II. Turbulence as a Source of Sound," *Proc. Roy. Soc. A*, Vol. 222, 1954, pp. 1–32.
15. J. E. Ffowcs Williams, "The Noise from Turbulence Convected at High Speed," *Phil. Trans. Roy. Soc. A*, Vol. 255, 1963, pp. 469–503.
16. A. P. Dowling, J. E. Ffowcs Williams, and M. E. Goldstein, "Sound Production in a Moving Stream," *Phil. Trans. Roy. Soc. A*, Vol. 288, 1978, pp. 321–349.
17. J. E. Ffowcs Williams and G. Maidanik, "The Mach Wave Field Radiated by Supersonic Turbulent Shear Flows," *J. Fluid Mech.*, Vol. 21, 1965, pp. 641–657.
18. D. G. Crighton, "Basic Principles of Aerodynamic Noise Generation," *Prog. Aerospace Sci.*, Vol. 16, 1975, pp. 31–96.
19. M. E. Goldstein, *Aeroacoustics*, McGraw-Hill, New York, 1976.
20. M. E. Goldstein, "Aeroacoustics of Turbulent Shear Flows," *Ann. Rev. Fluid Mech.*, Vol. 16, 1984, pp. 263–285.
21. N. Curle, "The Mechanics of Edge-Tones," *Proc. Roy. Soc. A*, Vol. 216, 1953, pp. 412–424.
22. A. Powell, "On Edge-tones and Associated Phenomena," *Acustica*, Vol. 3, 1953, pp. 233–243.
23. E. H. Dowell, *Aeroelasticity of Plates and Shells*, Noordhoff, Leiden, 1975.
24. A. Powell, "The Noise of Choked Jets," *J. Acoust. Soc. Am.*, Vol. 25, 1953, pp. 385–389.
25. M. S. Howe and J. E. Ffowcs Williams, "On the Noise Generated by an Imperfectly Expanded Supersonic Jet," *Phil. Trans. Roy. Soc. A*, Vol. 289, 1978, pp. 271–314.
26. S. M. Candel and T. J. Poinsot, "Interactions between Acoustics and Combustions," *Proceed. Instit. Acoust.*, Vol. 10, 1988, pp. 103–153.
27. D. G. Crighton, "The Excess Noise Field of Subsonic Jets," *J. Fluid Mech.*, Vol. 56, 1972, pp. 683–694.
28. Powell, A. "Theory of Vortex Sound," *J. Acoust. Soc. Am.*, Vol. 36, 1964, pp. 177–195.
29. W. Möhring, "On Vortex Sound at Low Mach Number," *J. Fluid Mech.*, Vol. 85, 1978, pp. 685–691.
30. H. S. Ribner, "The Generation of Sound by Turbulent Jets," *Adv. Appl. Mech.*, Vol. 8, 1964, pp. 103–182.
31. J. E. Ffowcs Williams and A. J. Kempton, "The Noise from the Large-scale Structure of a Jet," *J. Fluid Mech.*, Vol. 84, 1978, pp. 673–694.
32. O. M. Phillips, "On the Generation of Sound by Supersonic Turbulent Shear Layers," *J. Fluid Mech.*, Vol. 9, 1960, pp. 1–28.
33. R. Mani, "The Influence of Jet Flow on Jet Noise, Parts 1 and 2," *J. Fluid Mech.*, Vol. 73, 1976, pp. 753–793.
34. A. B. Parry and D. G. Crighton, "Asymptotic Theory of Propeller Noise," *Am. Inst. Aeronaut. Astronau. J.*, Part I in Vol. 27, 1989, pp. 1184–1190 and Part II in Vol. 29, 1991, pp. 2031–2037.

35. A. Powell, "Aerodynamic Noise and the Plane Boundary," *J. Acoust. Soc. Am.*, Vol. 32, 1960, pp. 982–990.
36. D. G. Crighton, "Long Range Acoustic Scattering by Surface Inhomogeneities Beneath a Turbulent Boundary Layer," *J. Vibration, Stress and Reliability*, Vol. 106, 1984, pp. 376–382.
37. G. B. Whitham, "On the Propagation of Shock Waves through Regions of Non-uniform Area or Flow," *J. Fluid Mech.*, Vol. 4, 1958, pp. 337–360.
38. M. J. Lighthill, "Viscosity Effects in Sound Waves of Finite Amplitude," in G. K. Batchelor and R. M. Davies (Eds.), *Surveys in Mechanics*, Cambridge University Press, 1956, pp. 250–351.
39. D. I. Blokhintsev, *Acoustics of a Nonhomogeneous Moving Medium*, Moscow: Gostekhizdat, 1945. Also available in English translation as *National Advisory Committee for Aeronautics Technical Memorandum*, No. 1399, Washington, DC, 1956.
40. O. Reynolds, "On the Dynamical Theory of Incompressible Viscous Fluids and the Determination of the Criterion," *Phil. Trans. Roy. Soc. A*, Vol. 186, 1895, pp. 123–164.
41. D. C. Pridmore-Brown, "Sound Propagation in a Fluid Flowing through an Attenuating Duct," *J. Fluid Mech.*, Vol. 4, 1958, pp. 393–406.
42. P. Mungur and G. M. L. Gladwell, "Acoustic Wave Propagation in a Sheared Fluid Contained in a Duct," *J. Sound Vib.*, Vol. 9, 1969, pp. 28–48.
43. P. N. Shankar, "On Acoustic Refraction by Duct Shear Layers," *J. Fluid Mech.*, Vol. 47, 1971, pp. 81–91.
44. Lord Rayleigh, *The Theory of Sound*, 2nd ed., Vol. II, Macmillan, London, 1896.
45. W. L. Nyborg, "Acoustic Streaming Due to Attenuated Plane Waves," *J. Acoust. Soc. Am.*, Vol. 25, 1953, pp. 68–75.
46. P. J. Westervelt, "The Theory of Steady Rotational Flow Generated by a Sound Field," *J. Acoust. Soc. Am.*, Vol. 5, 1953, pp. 60–67.
47. W. L. Nyborg, "Acoustic Streaming," in W. P. Mason (Ed.), *Physical Acoustics, Principles and Methods*, Academic, New York, 1965.
48. J. T. Stuart, "Double Boundary Layers in Oscillatory Viscous Flow," *J. Fluid Mech.*, Vol. 24, 1966, pp. 673–687.
49. N. Riley, "Streaming from a Cylinder Due to an Acoustic Source," *J. Fluid Mech.*, Vol. 180, 187, pp. 319–326.
50. N. Amin and N. Riley, "Streaming from a Sphere Due to a Pulsating Acoustic Source," *J. Fluid Mech.*, Vol. 210, 1990, pp. 459–473.

25

AERODYNAMIC AND JET NOISE

ALAN POWELL

1 INTRODUCTION

The subject of aerodynamically generated noise became of considerable interest in about 1950 as the result of the appearance of the aircraft jet engine, soon recognized as a remarkably powerful sound source likely to cause environmental problems. Besides interest in measurements and empirical noise reduction techniques, attention focused on the *mechanism* by which the sound was generated, whereas the few precursor studies had mostly focused on the frequency of aerodynamic sound, as of the very familiar aeolian tone of the whistling of telephone wires in the wind.

This aeolian tone typifies aerodynamic sound in the presence of a *fixed* surface, since sound is still radiated even if the wire does not move at all. Instead it reacts to a fluctuating aerodynamic lift force and therefore is to be associated with sound radiation of *dipole* character (see Ref. 1 for an extended account).

The aeolian tone is often periodic, a nearly pure tone, due to the regular formation of vortices of alternating sign in its wake, forming what is often call a Kàrmàn vortex street. This situation is totally different to that of an organ pipe or whistle, both with rigid surfaces, where the side-to-side oscillations of a jet excites a resonator. The mass of air in the resonator fluctuates as it resonates, and therefore the rate at which the net air flow enters the atmosphere fluctuates sympathetically and oppositely, these pulsations generating sound of *monopole* (simple-source) character. Such systems are not discussed here (but see Ref. 2).

Sound may be generated even in the complex absence of a fixed surface. Jet noise typifies this situation, the turbulence of the jet generating noise long after it has left the jet nozzle. There is no surface to react to a net aerodynamic force, and so the dipole character is absent. Instead, it is of the next-order higher source type, namely the *quadrupole* (see Chapter 8, Section 5). Here one may visualize a spherical eddy. Its volume does not fluctuate if the flow speed is low enough, so there is no monopole sound. It cannot vibrate in position, for there is no surface to react to the force that would be needed to cause the momentum to fluctuate, and so there is no dipole radiation. But there is the possibility that the surface of the sphere vibrates like the four quarters of an orange skin, adjacent quarters moving in and out oppositely, there being no volume or momentum change. Such distortions can in fact be viewed as the source of aerodynamic sound in the absence of solid surfaces, though it is not the only physical interpretation.[3]

The following discussion is limited to the latter, that is, to the aerodynamic noise of free flows (no solid surfaces present). After consideration of the fundamentals of aerodynamic sound generation of very low speed flows, the more complex situations of subsonic and supersonic jets are briefly discussed.

There is now a very extensive literature covering practically all aspects of the subject. Generally the references are limited to a few key, historically significant or recent ones. For fuller treatments and more extensive bibliographies, covering closely related aspects of aeroacoustics, the reader may refer to some fairly recent reviews,[4–11] some books,[1,2,12] and a conference proceedings.[13] There is also a recent collection of comprehensive monographs covering aeronautic aeroacoustics in general.[14]

Note: References to chapters appearing only in the *Encyclopedia* are preceded by *Enc.*

2 AERODYNAMIC NOISE SOURCES

In this section, the size of the source region is assumed to be "compact," that is, very small compared to the acoustic wavelength, and the flow Mach number is assumed to be very small, $M \ll 1$.

2.1 Lighthill Theory

Lighthill[15] manipulated the exact equations of continuity with zero source strength,

$$\frac{\partial \rho}{\partial t} + \frac{\partial \rho v_j}{\partial y_j} = 0,$$

and of momentum with zero externally applied force,

$$\frac{\partial \rho v_i}{\partial t} + \frac{\partial \rho v_i v_j}{\partial y_j} + \frac{1}{\rho}\frac{\partial p_{ij}}{\partial y_j} = 0,$$

so as to obtain the inhomogeneous wave equation (see Chapter 8, Section 2.1)

$$\frac{\partial^2 \rho}{\partial y_i^2} - \frac{1}{c_0^2}\frac{\partial^2 \rho}{\partial t^2} = -\frac{1}{c_0^2}\frac{\partial^2 T_{ij}}{\partial y_i\, \partial y_j}, \quad \text{where } T_{ij} = \rho v_i v_j + p_{ij} - c_0^2 \rho \delta_{ij}, \tag{1}$$

in which the standard acoustic form appears on the left side and everything else is put on the right side. Here c_0 is the sound speed in the stationary uniform acoustic medium surrounding the volume V_0 of unsteady sound-generating fluid flow, v_i is the velocity in the y_i-direction, ρ is the fluid density, and p_{ij} is the stress tensor (the force per unit area in the i-direction on the surface element with inward normal in the j-direction). The suffices i and j take the directions 1, 2, and 3 in turn but are to be summed if either one is repeated in a single term (e.g., there are three momentum equations, one for each of $i =$ 1, 2, 3; but $\partial \rho v_i/\partial y_j \equiv \partial \rho v_1/\partial y_1 + \partial \rho v_2/\partial y_2 + \partial \rho v_3/\partial y_3$).

This is interpreted as *Lighthill's acoustic analogy*: The *stationary* uniform acoustic medium is now taken to be *everywhere*, including the volume corresponding to the flow region V_0 with a *monopole* source distribution, given by the right side, embedded in it. The source strength is zero outside volume V_0 of the sound-generating flow. The double space differential reflects that the monopole and dipole strengths are both zero (corresponding to the assumed zero source strength and zero externally applied force), so the result is a quadrupole source field of strength T_{ij} per unit volume, (see Chapter 8, Sections 4 and 5, where, however, T_{ij} is the total *point* source strength).

The equation is exact. As the density ρ appears on both sides, the solution is formally exact if the source strength T_{ij} is taken to be known *exactly*. Usually the source strength is approximated by use of the fluctuations in a similar incompressible flow; then terms representing the interaction of the sound with the flow itself (e.g., propagation and refraction of the sound through the moving fluid) are omitted. It is also commonly assumed that the extent of the source region is small compared to the acoustic wavelength λ of interest, so that the source arriving at the observation point $\mathbf{x}$ can be assumed to have been emitted simultaneously from throughout the source region, a great mathematical simplification. Then the sound pressure in the far field at distance x from the source region reduces to a point quadrupole:

$$p(\mathbf{x}, t) = \frac{1}{4\pi x c_0^2}\frac{\partial^2}{\partial t^2}\int_{V_0} T_{xx}\, dV(\mathbf{y})^*. \tag{2}$$

This corresponds to Eq. (40) in Chapter 8, where now the asterisk means that the retarded time $(t - x/c_0)$ is to be taken for the integral over the compact source volume V_0, and $(1/c_0^2)\, \partial t^2/\partial t^2$ replaces $-k^2$, so that all frequencies are represented. Here the $T_{xx} \equiv T_{ij}x_i x_j/x^2$ is summed over all the longitudinal and lateral quadrupole terms for which $i = j$ and $i \neq j$, respectively.

The effect of viscous stresses can be neglected at other than small Reynolds number.[15] Specifically, the effect of dissipation by viscous forces results in a monopole-like contraction as the flow loses kinetic energy (imagine the lessening pressure drop in the center of a decaying vortex) and an expansion due to the corresponding heating.[11] The net effect is rather small, being of the order of 1/(Reynolds number) compared to that due to the vorticity movement per se.

For adiabatic fluctuations of an unheated flow, $\delta p = c_0^2 \delta \rho$.

Then taking the density to have the unperturbed value, everywhere, $T_{ij} = \rho_0 v_i v_j$ and $T_{xx} = \rho_0 v_x^2$, where v_x is the velocity component in the $\mathbf{x}$-direction.

If the sound pressures at three orthogonally related equidistant points $\mathbf{x}$, $\mathbf{y}$, and $\mathbf{z}$ are summed, then the integrals sum to twice the kinetic energy of the flow, which may be taken to be constant. Only lateral quadrupoles satisfy the condition of the sound pressure at such three far-field points always summing to zero. This *three-sound-pressures theorem*[11] indicates that, under the stated conditions, the assembly of quadrupole sources must be reducible to lateral ones alone.

Whereas a monopole source can be considered due to a fluctuation in mass $m = \rho_0 Q$ and a dipole due to a fluctuating force, that is, changes in momentum $mv = \rho_0 Qv$, these quadrupoles are due to fluctuations in momentum flux mv^2. A spherical surface in the fluid undergoes no vibratory expansion, or overall displacement, but only the distortions of the lateral quadrupole. These are similar to the surface of the four quarters of an orange, adjacent quarters moving oppositely in or out.

The sound intensity $I = p^2/\rho c$, with $q = \partial^2 (v_r^2)/\partial t^2$, is given by

$$I(\mathbf{x}) = \frac{\rho_0}{16\pi^2 x^2 c_0^5} \left\langle \int_{V_0} q(y_1)\, dV(y_1) \times \int_{V_0} q(y_2)\, dV(y_2) \right\rangle, \quad (3)$$

where $\langle \cdots \rangle$ indicates that the time average of the quantity is to be taken. Now

$$\langle \cdots \rangle = \iint \langle q(y_1) q(y_2) \rangle\, dV(y_1)\, dV(y_2).$$

If homogeneous turbulence is considered, this time average depends only on the difference between y_2 and y_1, so put $y_2 = y_1 + y'$ and define the *correlation coefficient* R by

$$\langle q^2 \rangle R(y') = \langle q(y_1) q(y_1 + y') \rangle. \quad (4)$$

Compactness of the whole source region, assumed earlier for convenience, may be relaxed somewhat, since now it is only necessary that $y' \ll \lambda$ when $R(y') \to 0$. Then

$$\langle \cdots \rangle = \langle q^2 \rangle \int dV(y_1) \int R(y')\, dV(y') = \langle q^2 \rangle V_0 \mathcal{V},$$

where the *correlation volume* $\mathcal{V}$ is defined by the second integral.

This $\mathcal{V}$ may be considered to be the definition of the volume of a *noise-producing eddy* of strength $\langle q^2 \rangle$ per unit volume. There are $V_0/\mathcal{V}$ such eddies in the noise-generating volume V_0. Then

$$I(\mathbf{x}) = \frac{\rho_0 \langle q^2 \rangle \mathcal{V} V_0}{16\pi^2 x^2 c_0^5}. \quad (5)$$

2.2 Vortex Theory

It is shown in classical (incompressible) hydrodynamics that the fluid flow induced by a vortex ring (of circulation Γ and area A_i) is exactly the same as that induced by a sheet of dipoles (of strength Γ per unit area of the same area A_i).[11] The total dipole strength is ΓA_i and the momentum in both cases is $m_i = \rho_0 \Gamma A_i$; see Fig. 1. In this theory[16] the aeolian tone, of dipole nature, is generated by the fluctuating force $dF_i/dt = d^2 m_i/dt^2$ exerted by a cylinder on the surrounding fluid when the vortex ring formed by the "bound" circulation about the cylinder and the cast-off "starting" vortex are stretched, or rather, the rate of change of stretching, $d^2 A_i/dt^2$, that can be taken as the downstream acceleration of the vortex as it breaks loose from the cylinder. The line of action of the dipole is through the vortex and perpendicular both to it and to its acceleration (but may be considered to act at the cylinder with negligible quadrupole error). The maxima of the dipole sound field lie along the line of action, approximately normal to the flow direction. The far-field sound pressure is therefore given by

$$p(\mathbf{x}, t) = -\frac{\rho_0}{4\pi x c_0} \frac{\partial}{\partial t} \int_b L_x\, db^*, \qquad \mathbf{L} = \Gamma \times \boldsymbol{u}, \quad (6)$$

where the product of the cross-stream span b of the vortex and its downstream velocity u is $bu = dA_i/dt$. The simple similarity argument indicates that the sound power is proportional to the flow velocity to the sixth power.

In a free flow, no net force occurs between the flow and the surrounding fluid, so the force due to a given accelerating vortex *must* be exactly balanced by an equal and opposite action elsewhere in the flow. As each one acts as a dipole, together they must constitute a quadruple with strength proportional to the distance between them, the orientation determining the relative lateral and longitudinal components; see Chapter 8, Section 5.3, and

Fig. 1 Left: Streamlines of incompressible flow field due to vortex pair. Center: Streamlines due to a sheet of dipoles. Right: Streamlines due to bound vortex about cylinder and accelerating cast-off vortex (mean flow omitted).

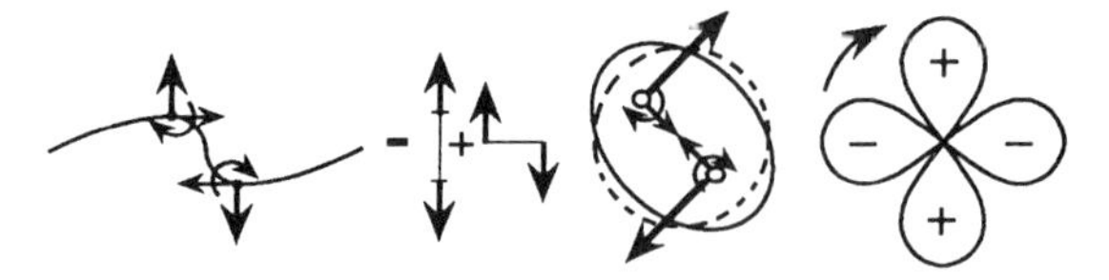

Fig. 2 Left: Horizontal acceleration of vorticity results in vertical dipoles, which form a longitudinal and a lateral quadrupole. Center: Spinning vortices accelerate toward their center, resulting in a rotating lateral quadrupole. Right: Rotating lateral quadrupole field due to spinning vortices.

Fig. 2. The far-field sound pressure is given by

$$p(\mathbf{x}, t) = -\frac{\rho_0}{4\pi x c_0^2} \int \frac{\partial^2}{\partial t^2} y_x \mathcal{L}_x \, dV^*, \qquad \mathcal{L} = \zeta \times \boldsymbol{u}. \tag{7}$$

The three-sound-pressure theorem and, for the sound power, the eighth-power law naturally apply.

Although their physical interpretations are very different, the vortex theory and Lighthill's are closely connected: since $\partial v_i v_j/\partial y_j = v_i \partial v_j/\partial y_j + v_j \partial v_i/\partial y_j$ (the first term vanishes because of continuity), which then equals $v_j(\partial v_j/\partial y_i - \partial v_i/\partial y_j) + \frac{1}{2}\partial v^2/\partial y_i$, that is, the velocity times the vorticity ζ (vortex strength per unit area), with the last term ultimately vanishing when integrated over the source volume.

Further developments include a recasting of the result into a source term depending on $\partial\zeta^3/dt^3$ rather than on L (possibly attractive for numerical methods) and the inclusion of the effect of entropy (temperature) variation in the flow.[11,17]

Apart from the useful physical interpretation, the vortex theory is very convenient when the vorticity can be specified, as for idealized vortex flows, for example, the spinning vortex pair (Fig. 2), vortex rings colliding head-on with each other, and simulated jet flows[13,18] when the vorticity can be viewed as inducing the local incompressible (hydrodynamic) flow field as well as the acoustic field.

2.3 Dilatation Theory

The dilatational theory is an alternative acoustic analogy that has a simple but unfortunately deceptive physical appeal. In an incompressible aerodynamic flow there are pressure fluctuations p_{inc}, and it may be imagined that if a small degree of compressibility is admitted then at some point where the pressure is positive, say, there would be a corresponding compression of the fluid. This compression then acts like a local monopole. For a free flow the pattern of positive and negative pressure volumes yields a quadrupole form, and simple similarity leads to the eighth-power law.

The theory may be developed *formally* for a free flow,[6] finding p_{inc} from the equation $\nabla^2 p_{\text{inc}} = \partial^2 T_{ij}/\partial y_i \, \partial y_j$, for example, so that

$$p(\mathbf{x}, t) = -\frac{1}{4\pi x c_0^2} \int \frac{\partial^2 p_{\text{inc}}}{\partial t^2} \, dV^*. \tag{8}$$

Unfortunately one cannot distinguish the pressure in the original T_{ij} source region from that in the surrounding resultant ("incompressible") near field, so as p_{inc} falls off only as r^{-3}, the region of integration must extend out to *at least* a wavelength: The source region is *essentially* noncompact.[11] While this spoils the simple appeal of the physical argument, the method is mathematically correct for some simple test cases[6].

2.4 Alternatives to Acoustic Analogies

Several alternatives to the acoustic analogy methods have been, or are being developed. Space permits only a brief mention of these newer important aspects of aeroacoustics.

Contiguous Method The imaginary rotating ellipsoid-like surface formed by the streamlines about the spinning vortices in Fig. 2 can be considered to drive the *contiguous* acoustic field external to them. In fact, if the pressure $p_a(\mathbf{a})$ on a surface of radius a can be estimated for some *incompressible* (hydrodynamic) flow contained within, then the pressure $p(\mathbf{x})$ in the acoustic field follows immediately; for the far field

$$p(\mathbf{x}) = \frac{1}{3} \frac{a^3}{x c_0^2} \frac{\partial^2 p_a^*}{\partial t^2}. \tag{9}$$

This simple formulation[11] [readily derived from Eq. (42) in Chapter 8] obviously depends on some knowledge of the source type, specifically lateral quadrupole for free flows.

Matched Asymptotic Expansions The preceding method has a mathematical counterpart in the method of asymptotic expansions (MAE), in which the *inner* hydrodynamic flow field, say of a vortex pair, is matched analytically to the *outer* external acoustic fields.[4] This method is more general and rigorous provided that corresponding care is taken, the inner flow not necessarily being incompressible or compact and the source type not needing to be known initially. An important aeroacoustic application to the instability waves of a jet is addressed later for supersonic jets.[19]

Computational Aeroacoustics The advent of very fast computers is making possible numerical solutions of the fundamental equations of motion of noise-producing flows, that is, direct numerical simulation without the introduction of "models" of any sort. For example, the results for the two-dimensional choked jet (neglecting viscosity and heat transfer) bear a satisfying likeness to photographic and other evidence.[20] Such methods may be used to confirm various models, for example, the vortex theory, that are very much more simple to use. In practice, though, it is more likely that the numerical model will be evaluated against a simple mode and, once validated, extended to conditions that the simpler models plainly cannot address. For example, for a pair of spinning vortices, direct computation yields results that agree precisely with the results of vortex theory (an acoustic model) at very small Mach number and extends the results to the compressible region (higher Mach numbers) that the acoustic model formulations are ill-equipped to handle.[11] Also possible are numerical simulation experiments to test simple physical hypotheses, as in the case of the edge tone, where the point dipole of theory can be inserted in the jet flow to represent the action of the edge[8] (see later).

3 JET NOISE

In this section some applications to the noise of turbulent jets are briefly considered. While the jet exhaust velocity may be supersonic, the convection velocity of the eddies in the flow is initially taken to be subsonic relative to the ambient sound speed.

3.1 Simple Similarity

Simple scaling based on Lighthill's theory can give useful indications, and its limitations gives some hint about important omitted effects.

By *simple similarity* is meant taking a basic equation, such as Eq. (5) or (7), and assuming that the turbulent velocities v_i are proportional to some characteristic velocity U (e.g., jet exit velocity) and the $\partial/\partial t$ are like frequency and so proportional to U/D and both V and V_0 are proportional to D^3, where D is a convenient characteristic dimension (e.g., jet nozzle diameter). Then it follows that $p(\mathbf{x}) \sim \rho_0 U^4 D/xc_0^2$ in any given direction and the corresponding intensity at a point is $I(\mathbf{x}) \sim \rho_0 U^8 D^2/x^2 c_0^5$. Lighthill's *eighth-power law*[15] for the sound power W follows:

$$W = \frac{K\rho_0 U^8 D^2}{c_0^5}, \tag{10}$$

where K is a constant and the factor $\rho_0 U^8 D^2/c_0^5$ is often called *Lighthill's parameter*. As the kinetic power of a jet, say, is proportional to $\frac{1}{2}\rho_0 U^2 \cdot UD^2$ (ignoring the density differences), the fraction of the power converted to sound energy is the *noise-generating efficiency* η:

$$\eta \sim M^5, \tag{11}$$

where $M = U/c_0$ is the Mach number of the flow referenced to the *ambient* speed of sound. No provision has been made for convection of the sources or for flow interaction effects.

The earliest measurements of jet noise showed that the intensity and noise power varied very closely with the eighth power of the jet exit velocity. In fact, it is now generally accepted that

$$P \approx \frac{(3\text{–}4) \times 10^{-5} \cdot \rho_0 U^8 D^2}{c_0^5}, \tag{12}$$

and in terms of the mechanical power P_M of the jet,

$$P \approx 10^{-4} M^5 P_M. \tag{13}$$

These apply quite well to unheated air jets (with low initial turbulence and noise entering the nozzle) from low speeds (say $U \approx 50$ ms^{-1}) to slightly above choking ($U \approx 310$ ms^{-1}) and to the jets of turbojets near full power ($U \approx 610$ ms^{-1}).[21,22] Note that it is the ambient density that occurs in the former equation, not that of the jet mixing region.

From the theoretical point of view, the interesting implication is that the sum effect of omitted considerations—such as source convection and refraction—are either not important or balance each other out, so far as *noise power* is concerned.

As this simply derived eighth-power law agrees rather well with experimental results, one may take the *empirical* approach that the similarity method can be applied to the most important noise-generating regions of a turbulent jet. There are two regions of the jet in which the turbulence is *self-preserving*, that is, follows a similarity relationship: first, the nearly two-dimensional thin shear layer in region A not too far from the nozzle exit and adjacent to the nonturbulent potential core (see Fig. 3) and, second, the axially symmetric region B relatively

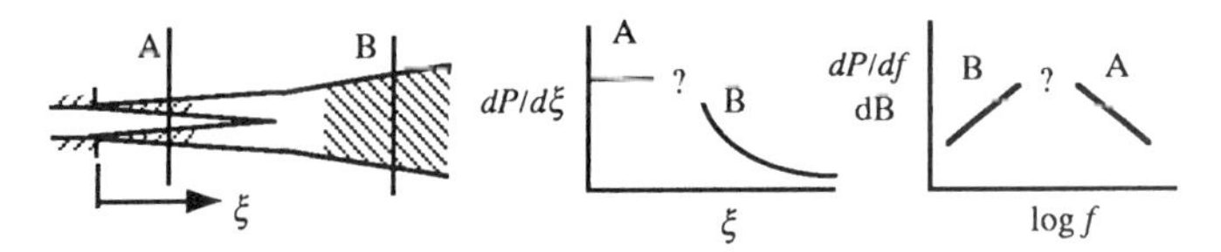

Fig. 3 Left: Similarity regions of a turbulent jet. Center: Axial source strength distribution. Right: Corresponding spectral shape.

far downstream. These are characterized by the lateral dimensions of the turbulent regions, growing very nearly proportionally to the distance ξ from the nozzle, and by the total shear velocity that equals the jet exit velocity U in region A but falls inversely with ξ in region B. Then the noise power produced per unit length is constant in region A but falls very rapidly, as ξ^{-7}, in region B, with an indeterminate transition between them.[5,21] The relative level for the two regions is indeterminate (see Fig. 3).

Early theoretical ideas suggested that high shear gradients "amplify" the turbulent sound generation; but according to similarlity considerations, high shear rates and the high frequency adjacent to the jet exit are just offset by the smallness of the eddy size and depth of the shear layer. If the unknown constants for the two regions are taken to be the same and the regions are assumed to join each other without a transitional region between them, then three-fourths of the noise is generated in region A.

Measurements in the very near field and the use of an acoustic focusing mirror do indicate that relatively little sound is generated in region B but that in region A it is weighted toward the transition region.[13]

The associated noise power spectrum $dP/df = dP/d\xi \cdot d\xi/df$, assuming similarity of local spectrum shapes, first increases as f^2 and then decreases as f^{-2}, the relative levels again being undefined[21] (see Fig. 3). The peak of the spectrum naturally follows some fixed value of Strouhal number fD/U. This result is independent of the spectral shape of the local source (except that it must have slopes steeper than the resultant ones).[5]

Experiments showed that the average spectra of noise power of unheated air jets, turbojets, and rockets do have a *shape* fairly close to that just discussed. However, these spectra mostly collapse onto the common shape using the parameter fD/c_j, where c_j is the speed of sound in the jet, *not* fD/U, as shown in Fig. 4. That the peak frequency does not scale proportionally to jet velocity was evident from the earliest measurements.[5]

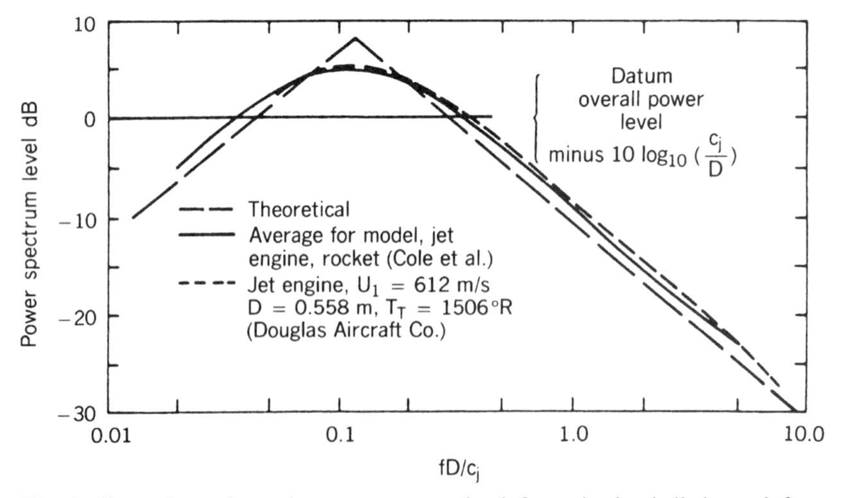

Fig. 4 Comparison of sound power spectrum level from simple similarity and from measurements of unheated air jets, jet engines and rockets. Note that the abcissa scale is fD/c_j, not fD/U. (From A. Powell, *Noise Control Eng. J.*, Vol. 8, 1977, pp. 69–80, 108–119. Copyright Institute of Noise Control Engineering of the USA, used with permission.)

3.2 Turbulent Self-Noise and Shear Noise

For jet noise, two classes of quadrupoles may be identified. First, *self-noise* is that due to the turbulence, assumed to act as if it were isotropic and locally homogeneous. These quadrupoles are lateral ones, oriented at random and therefore producing, on the average, a uniform directionality.[6]

The second is that due to the turbulence being sheared. In Lighthill's original theory the quantity $\partial/\partial t \cdot \rho_0 v_i v_j = pe_{ij}$, p being the local pressure and e_{ij} the local rate of strain tensor,[15] was interpreted as producing lateral quadrupole noise generators, with the maxima at 45° to the jet axis.

A more recent alternative[6] is to take from Eq. (1) the quadrupole source terms $2\rho_0\,\partial U_1/\partial y_2 \cdot \partial v_2/\partial y_1$, where U_1 is the mean velocity sheared in the direction y_2 (or, equivalent after integration, putting $v_1 = U_1 + v_1'$ in T_{ij}). This results in longitudinal quadrupoles with the maxima along the jet axis, having a $\cos^4\theta$ variation for their intensity. Assuming the turbulence to be isotropic and superimposed on the sheared flow, this *shear noise* power is estimated to be about equal to that of the self-noise.[6]

3.3 Convection and Refraction

The considerations leading to the eighth-power law, based on Eq. (5), have a major shortcoming: The observed strong downstream bias of jet noise (see Fig. 5) is completely absent. This bias has been attributed to the movement of the noise-producing eddies at about 0.6 of the jet exit velocity relative to the external medium into which they radiate. It may be incorporated by the method discussed in Chapter 8, Section 11, or by the use of moving axes[15] applied to the sources in the acoustic model. For quadrupoles, convection at Mach number $M_{con} = U_{con}/c_0$ introduces the directionality factor $(1 - M_{con}\cos\theta)^{-6}$ for the sound intensity, where θ is the angle measured from the downstream axis; however, the effective volume of the noise-producing eddies is reduced,[22] so the net effect becomes $(1 - M_{con}\cos\theta)^{-5}$. Thus there is a considerable amplification in the downstream direction, falling to none at right angles to the jet and with a reduction in the upstream direction. A corresponding factor applies to supersonic convection, $(M_{con}\cos\theta - 1)^{-5}$. These are both obviously unsatisfactory as $M_{con}\cos\theta \rightarrow 1$ when it might be considered that the degeneration of constituent monopoles into a quadrupole can no longer take place, and a proper combination is the factor

$$[(1 - M_{con}\cos\theta)^2 + \alpha^2 M_{con}^2]^{-5/2}, \tag{14}$$

where α depends on the length and time scale of the turbulence.[6,22] Plausible values of α result in the sound power in the region of $M_{con} = 1$ more modestly exceeding the extrapolation of the eighth-power law.

Thus the gross bias of the jet noise toward the downstream direction can be explained. Decreasing turbulence levels with jet velocity were suggested to partly compensate for the rise above the eighth-power law,[6,21] supported by some measurements of noise intensity at $\theta = 90°$ varying more slowly than with the eighth power of velocity.

However, there are two shortcomings. First, the necessary Doppler effect on frequency, $(1 - M_{con}\cos\theta)^{-1}$, worsens the aforementioned disparity with the observed spectra of noise power, and second, this directionality has the maximum in the jet direction instead of the observed sharp minimum.

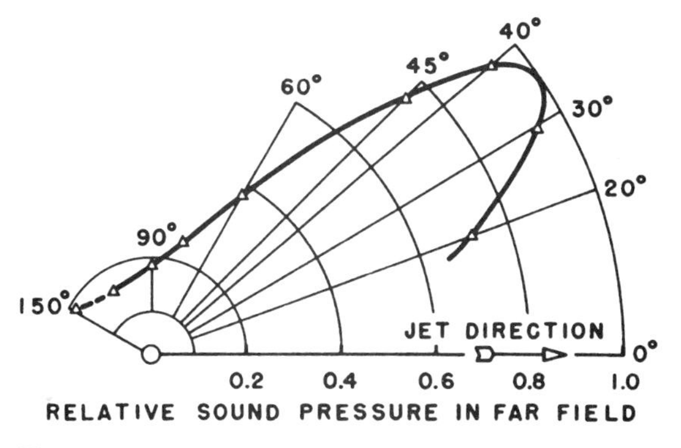

Fig. 5 Directivity of the sound pressure for a jet engine on a linear scale. (From A. Powell, "On the Generation of Noise by Turbulent Jets," ASME Paper 59-AV-53, 1958, with permission.)

Refraction The sound power of a source is not much affected by (say rigid) boundaries more than a quarter of a wavelength or so from it, that is, beyond some *region of influence*. It may be presumed that the same is true if there is a shear of the medium there. Thus the effects of convection will be limited to the differential velocities occurring within the region of influence, while mean velocity changes outside this region will cause *refraction*. For a given jet velocity, the shorter wavelengths of the local characteristic spectrum will scale with the local shear layer thickness in region *A* of Fig. 3, and so the convection and refraction effects will both tend to be constant. For increasing jet velocity, the position of the source of a given high frequency (relative to the local characteristic frequency) will move downstream and suffer constant convection effects but increasing refraction, away from the jet in the downstream quadrant but toward the jet in the upstream quadrant. The longest (local) wavelengths will undergo full convection effects and little refraction and the shortest little convection but strong refraction.[5] Experiments with a sound source introduced into a jet show the anticipated minimum along the downstream axis, while a very cold jet shows a local maximum there, the lower sound speed causing inward refraction.[6] Such considerations provide a plausible qualitative explanation of the directional characteristics of turbulent jets at subsonic Mach numbers, in particular, the fact that the peak frequency of the noise spectra in the important intense downstream direction varies very little with jet speed.

Flow Interaction The important effects of convection and refraction cannot be simply separated for moderate acoustic wavelengths in realistic flows that have flow shear; together they typify *flow interaction* of the mean flow with the acoustic radiation within it. The formal solution to Lighthill's equation is exact and includes all such effects, provided that the source terms are known *exactly*. The common use of incompressible approximations clearly eliminates all interaction between the acoustic perturbations and the mean flow, such as that due to refraction.

The alternative is to incorporate mean-flow interaction effects in the wave operator on the left side of the equation, so that the right (source) side can still be approximated in a fairly simple manner. This is a convective wave equation, with the convective velocity varying in a shear flow. Although such equations are difficult to solve, very significant progress has been made, providing much detailed insight.[14]

From the point of view of general fundamental theory, attention is drawn to developments describing flows in terms of mean, turbulent, acoustic, and thermal components.[23,24]

Turbulent Jet Noise Reduction The challenge in reducing the high noise levels of the jets of propulsive engines is to do so with the minimum impact on the thrust. By far the most progress has come about through "simply" reducing the jet velocity, maintaining the thrust constant, for then Eq. (10) gives $W \sim (\text{thrust})U^6/c_0^5$. The earliest emphasis was on altering the nozzle shape to variations on the "cooky-cutter" forms, some rather extreme, or of using multiple smaller nozzles. Some of these produced up to 10 dB noise reductions for a modest thrust loss. More recent by-pass and fan-jet engine designs involve much lower jet velocities, large streamlined center bodies, and annular jets that may be subdivided in various ways.[14]

4 SUPERSONIC JETS AND ROCKETS

While the noise of subsonic turbulent jets can be considered to be due to turbulent mixing at subsonic convection velocities, for supersonic jets the convection velocity is supersonic, and this brings about a fundamental change in the source mechanism. Moreover, there are shock waves present in the flow, and these introduce an entirely different source mechanism.

4.1 Shock-Free Jet Noise

Supersonic Convection The eighth-power law must ultimately give way to a power no greater than 3, or else the acoustic power would exceed the jet kinetic power. Actually, at the very high speeds of rockets the noise power does tend to the third power of the velocity,[5,21] the sound power then being about 0.5% of the jet power. The sound power for very large rockets (up to 10^{10} W) also levels out to about the same 0.5% of the mechanical power.[25] When the convection Mach number exceeds unity, the convection amplification factor discussed earlier, $(M_{\text{con}} \cos \theta - 1)^{-5}$, when combined with the U^8, does in fact tend to the U^3 that has been observed.[5,21] Of course, in the acoustic analogy, the source strength has to be assumed given, so to some extent such agreement might be fortuitous.

Instability Wave Radiation Close to the nozzle exit of supersonic jets, a pattern of almost straight sound waves of very high frequency, making an acute angle with the jet direction, has been long observed in schlieren and shadowgraph photographs, usually attributed simply to Mach waves generated by supersonic eddies of short lifetime. A more recent explanation is that the shear layer there is unstable and the instability waves radiate sound as their amplitude increases and then decreases. The acoustic analogy may be employed,[26] while alternatively the exact solution to the equations of motion (within plausible simplifications) also reveals these nearly plane sound waves, but propagating at a phase velocity somewhat less than the sound speed in the rather extensive near field.[27]

Most of the noise from a supersonic jet emanates from further downstream than for a subsonic jet and appears largely attributable to the large-scale eddy structures associated with instability waves there.[26–28]

4.2 Shock-Generated Noise

Vortex–Shock Interaction In the elementary one-dimensional case of sound transmission across a boundary, the incident wave gives rise to a transmitted and a reflected wave, these two waves being necessary to satisfy the two boundary conditions at the interface, namely the continuity of velocity and pressure. All the fluctuations occur in a single perturbation mode, namely that of acoustic waves. If the interface is a shock wave, with the fluid on the incident wave side approaching it at supersonic velocity, then no reflected wave can propagate upstream. In a sense, it cannot leave the shock wave interface and so causes fluctuations in its strength and therefore in the entropy change across it, these entropy changes being carried downstream at the subsonic flow velocity. Thus the two waves now necessary to satisfy the boundary conditions are a transmitted sound wave and a convected entropy (temperature) wave.[8] The sound wave is amplified, perhaps considerably, by this nonlinear interaction. The resultant perturbations are now of two modes: acoustic and entropy.

Similarly, an entropy wave convected into a shock wave results in the two modes of a convected entropy wave and a sound wave propagating downstream, the latter being relatively intense in acoustic measure.

In three dimensions, three perturbation modes occur, the additional one being vorticity. Thus a vortex swept into a shock wave results in a changed vortex being convected away from the shock wave, an entropy wave, and a sound wave.[29,30] In the same way the interaction of turbulence with a shock wave results the generation of sound.

Screech There are many situations in which *flow resonance* occurs. The flow—either a shear flow or of a jet—is inherently unstable over a certain frequency range, or more strictly, over a range of disturbance wavelengths in the flow. It is self-excited into resonance-like sound-generating oscillations in the region of maximum instability, all in the total absence of any physical resonator (such as an organ pipe).[9] The classic case is that of the edge tone, in which a laminar slit jet blows onto a wedge, the periodic interaction there resulting in the jet being disturbed as it leaves the orifice. The induced sinuosity of the jet grows very rapidly as it is convected toward the edge, taking energy from the mean jet flow, some of which is radiated as sound of dipole character associated with the fluctuating fluid force on the edge.

Two criteria, concerning the phase and loop gain, respectively, must be met if such oscillations are to be maintained.[8] First, one might think that there must be an integer number N of wavelengths in the feedback loop. But because of possible noncanceling phase changes at both the orifice and the edge, a noninteger fraction p must be introduced, so the possible oscillation frequencies must really be proportional to $N + p$. Second, in the limit cycle, the amplification in the jet—a factor of possibly hundreds or thousands—must result in the loop gain being unity. As the orifice-to-edge distance is increased, the sinuosities in the jet become longer and consequently less unstable. At some point the oscillation flips to a shorter, more unstable wavelength for a new limit cycle with the next greater integer value of N; the process may be repeated several times. Thus steady falls in frequency are interrupted by sudden jumps upward.

Supersonic jets are characterized by the formation of a stationary "cell" structure of nearly periodically varying pressure, with each cell terminating in shock waves across the jet, the nearly equal spacing between them increasing with jet pressure ratio. The characteristic "screech" noise of such jets is due to a feedback mechanism analogous to that of the classical edge tone, except that the acoustic source is attributed to the interaction of the jet sinuosities with the periodically spaced shock waves, the latter being virtually fixed in space. Together the acoustic sources—assumed to be monopoles—form a *stationary* phased array that is *not* compact, so unlike the simple dipole of the edge tone, the directionality of the sound is markedly frequency dependent, though of course the frequency still has to be such that the jet is excited in the region of maximum instability.[8,10,32]

As a simple example, assume that the sound is emitted equally from three shock waves spaced a distance s apart, then the sound pressure at angle β to the jet downstream direction is proportional to

$$\frac{1}{3} + \frac{2}{3} \cos\left(\frac{2\pi s}{\Lambda} (1 - M_{\text{con}} \cos \beta) \right), \tag{15}$$

where M_{con} is the convection Mach number and $\Lambda = M_{\text{con}}\lambda = M_{\text{con}}c/f$ is the wavelength of the jet instability.[32] This accounts for the relatively high sound levels passing upstream, past the nozzle exit. Moreover, if it is postulated that the feedback is maximized in the limit cycle, given that the oscillations are to be in the region of maximum jet instability, then the sound waves from all the sources arrive all in phase with each other at the nozzle; that is, there is constructive interference. With $\cos \beta = -1$, the resonant frequency is then given by the simple formula[8,10]

$$f = \frac{c}{s} \frac{M_{\text{con}}}{1 + M_{\text{con}}}. \tag{16}$$

This expression is independent of the number and strength of the assumed sources and of the distance of the source array from the nozzle. The same formula results for a continuously distributed source with its strength being proportional to the disturbance wave amplitude and the local pressure in the shock cells (the fundamental wavelength of which is the shock spacing).[10] It also applies to supersonic jets from convergent–divergent nozzles.[10] Thus, for screech the gain criterion fixes the frequency, whereas it was the phase criterion for the edge tone. The dipole of the edge tone is located at a definite point, that is, in the vicinity of the edge of the wedge, but for screech the distance to the effective source center is determined primarily by the limit cycle of the unstable waves in the jet, that is, by both the rate at which the instability grows and how rapidly it dissipates the shock structure.

The directionality of the screech tone can be estimated by formulas akin to Eq. (15). The agreement with experiment is very satisfactory, especially considering the simplicity of the formula.[33]

In the edge tone, frequency jumps occur as the nozzle–edge distance increases. For screech the shock wave spacing become greater as the pressure ratio increases: The instability mode for the screech of choked jets (from convergent nozzles) of high aspect ratio ("two dimensional") is (almost) always sinuous, and the acoustic wavelength is closely proportional to that spacing. But for round nozzles there is a change in the instability mode, taking the following form as the pressure ratio increases: first, one or two axially symmetric (varicose, toroidal); second, sinuous (flapping); third, a strong and very stable helical, (spinning); and sometimes fourth, the sinuous one reappears.[34] The corresponding jumps in frequency are all upward except for the last one.

For supersonic jets (from convergent–divergent nozzles) at Mach 1.41, the modes appear to be just axially symmetric (torroidal) and helical.[10]

Screech Reduction As screech involves feedback, the reduction of any of the factors in the gain criterion of the feedback loop may be hypothesized to reduce the amplitude of the limit cycle.[35] The principal means are as follows: First, the sound pressure perturbing the jet is lessened by removal of a flat face (thick lip) of the nozzle[33,35] its replacement by a sound absorbant face,[35] or shielding the jet at the exit.[35] Similarly a reflector at a variable axial distance upstream of the nozzle will accentuate the feedback at some distances and attenuate it at others.[36,37] Second, the growth of the initially miniscule instability waves in the jet may be disrupted by a rough edge to the nozzle[35] or the addition of cambered vanes protruding radially into the jet at the exit[35] or of small circumferential tabs pressing the jet boundary inward.[10] Third, the acoustic source vanishes when the shock cells are absent, as resulting from ventilating the nozzle of its excess pressure[35] or by the use of a shock-free divergent nozzle at its design pressure.[38]

Broadband Shock Noise Besides the discrete frequency instability modes of screech, any other large-scale structures (eddies) interacting with the shock waves will naturally result in sound emission. The everpresent more random eddy structure can be visualized as made up of a broadband of frequencies and spacewise Fourier components, each of which will radiate with the directionality given by the principles of Eq. (15) or refinements thereon.[10] Thus, in complete contrast to the directionality of turbulent mixing noise, the lowest frequencies will be emphasized by a tendency to constructive interference in the upstream direction and the highest frequencies in the downstream direction, according to the maxima of Eq. (15). Recasting Eq. (15) the angle of the directional maximum for a given frequency can be expected to be given by

$$\beta = \cos^{-1}\left(\frac{1}{M_{\text{con}}} - \frac{c}{sf}\right). \qquad (17)$$

Screech evidently may be viewed as a special case when feedback occurs, requiring that the maximum is close to the upstream direction.[10]

A detailed analysis of the interaction of the constituent instability waves with the nearly periodic structure of the jet results in good agreement with measurements of the noise field.[10] Measurements of the intensity of the broadband noise from choked jets (convergent nozzle) show a proportionality of the rms value to the shock strength for which there is a theoretical basis[10]; the same relationship holds for the supersonic jets from convergent–divergent nozzles.[10]

REFERENCES

1. W. K. Blake, *Mechanics of Flow-Induced Sound and Vibration*, Academic, 1986.
2. N. H. Fletcher and T. H. Rossing, *The Physics of Musical Instruments*, Pt. IV: *Wind Instruments*, Springer-Verlag, New York, 1990, pp. 345–491.
3. A. Powell, "Why Do Vortices Generate Sound?" *Trans. ASME*, Vol. 17, 1995, pp. 252–260.
4. D. G. Crighton, "Basic Principles of Aerodynamic Noise Generation," *Prog. Aerospace Sci.*, Vol. 16, 1975, pp. 31–96.
5. A. Powell, "Flow Noise: A Perspective on Some Aspects of Flow Noise, and of Jet Noise in Particular," *Noise Control Eng.*, Vol. 8, 1977, pp. 69–80, 108–119.

6. H. S. Ribner, "Perspectives on Jet Noise," American Institute of Aeronautics and Astronautics, Paper No. AIAA-81-0428R, 1981.
7. J. M. Seiner, "Advances in High Speed Jet Aeroacoustics," American Institute of Aeronautics and Astronautics, Paper No. AIAA-84-2275, 1984.
8. A. Powell, "Some Aspects of Aeroacoustics: from Rayleigh until Today," *J. Vibr. Acoust.*, Vol. 112, 1990, pp. 145–159.
9. D. Rockwell, "Oscillations of Impinging Shear Layers," *AIAA J.*, Vol. 21, No. 5, 1983, pp. 645–664.
10. C. K. W. Tam, "Supersonic Jet Noise," *Ann. Rev. Fluid Mech.*, Vol. 27, 1995, pp. 17–43.
11. A. Powell, "Why Do Vortices Generate Sound?" *Trans. ASME*, Vol. 117, 1995, pp. 252–260.
12. M. E. Goldstein, *Aeroacoustics*, McGraw-Hill, New York, 1976.
13. E.-A. Muller (Ed.), *Mechanics of Sound Generation in Flows*, Springer-Verlag, Berlin, 1979.
14. H. H. Hubbard, (Ed.) *Aerodynamics of Flight Vehicles: Theory and Practice*, Vol. 1: *Noise Sources*, Vol. 2: *Noise Control*. Acoustical Society of America, New York, 1994.
15. M. J. Lighthill, "On Sound Generated Aerodynamically, I General Theory," *Proc. Roy. Soc.*, Vol. A 211, 1952, pp. 564–587.
16. A. Powell, "Theory of Vortex Sound," *J. Acoust. Soc. Am.*, Vol. 36, 1964, pp. 177–195.
17. M. J. Howe, "Contributions to the Theory of Aerodynamic Sound, with Application to Excess Jet Noise and the Theory of the Flute," *J. Fluid Mech.*, Vol. 71, No. 4, 1975, pp. 625–673.
18. T. Kambe, "Acoustic Emissions by Vortex Motions," *J. Fluid Mech.*, Vol. 173, 1986, pp. 643–666.
19. C. K. W. Tam and D. E. Burton, "Sound Generation by Instability Waves of Supersonic Flows," *J. Fluid Mech.*, Vol. 138, No. 7, 1984, pp. 273–295.
20. Y. Umeda, R. Ishii, T. Matsuda, A. Yasuda, K. Sawada, and E. Shima, "Instability of Astrophysical Jets. II. Numerical Simulation of Two-Dimensional Choked Under-Expanded Slab Jets," *Prog. Theor. Phys.*, Vol. 4, No. 6, 1990, pp. 856–866.
21. A. Powell, "On the Generation of Noise by Turbulent Jets," ASME Paper 59-AV-53, 1958.
22. J. E. Ffowcs Williams, "The Noise from Turbulence Convected at High Speed," *Phil. Trans. Roy. Soc. (Lond.)*, Vol. A255, 1963, pp. 469–503.
23. P. E. Doak, "Momentum Potential Theory of Energy Flux Caused by Momentum Fluctuations," *J. Sound. Vib.*, Vol. 131, 1989, pp. 67–90.
24. P. L. Jenvey, "The Sound Power from Turbulence; A Theory of the Exchange of Energy between the Acoustic and Non-Acoustic Fields," *J. Sound. Vib.*, Vol. 131, 1989, pp. 37–66.
25. S. H. Guest, "Acoustic Efficiency Trends for High Thrust Boosters," NASA Tech. Note D-1999, 1964.
26. J. E. Ffowcs Williams and A. J. Kempton," The Noise from the Large-Scale Structure of a Jet," *J. Fluid Mech.*, Vol. 84, No. 4, 1978, pp. 673–694.
27. C. K. W. Tam, "Directional Acoustic Radiation from a Supersonic Jet Generated by Shear Layer Instability," *J. Fluid Mech.*, Vol. 46, No. 4, 1974, pp. 757–786.
28. J. M. Seiner, D. K. McLaughlin, and C. H. Liu, "Supersonic Jet Noise Generated by Large-Scale Instabilities," NASA Tech. Paper 2072, 1982.
29. H. S. Ribner, "Cylindrical Sound Waves Generated by Shock-Vortex Interaction," *AIAA J.*, Vol. 23, No. 11, 1985, pp. 1708–1715.
30. K. R. Meadows, A. Kumar, and M. Y. Hussain, "Computational Study on the Interaction between a Vortex and a Shock Wave," *AIAA J.*, Vol. 29, No. 2, 1991, pp. 171–179.
31. H. S. Ribner, "Spectra of Noise and Amplified Turbulence Emanating from Shock Turbulence Interaction," *AIAA J.*, Vol. 25, No. 3, 1987, pp. 436–442.
32. A. Powell, "On the Noise Emanating from a Two-Dimensional Jet Above the Critical Pressure," *Aero. Quart.*, Vol. 4, 1953, pp. 103–122.
33. T. D. Norum, "Screech Suppression in Supersonic Jets," *AIAA J.*, Vol. 21, No. 2, 1983, pp. 283–303.
34. A. Powell, Y. Umeda, and R. Ishii, "Observations of the Oscillating Modes of Choked Circular Jets," *J. Acoust. Soc. A.*, Vol. 92, 1992, pp. 2823–2836.
35. A. Powell, "The Reduction of Choked Jet Noise," *Proc. Phys. Soc. Series B*, Vol. 67, 1954, pp. 313–329.
36. T. D. Norum, "Control of Jet Shock Associated Noise by a Reflector," American Institute of Aeronautics and Astronautics, AIAA Paper No. 84-2279, 1984.
37. R. T. Nagel, J. W. Denham, and A. G. Papathanasiou, "Supersonic Jet Screech Tone Cancellation," *AIAA J.*, Vol. 21, No. 11, 1983, pp. 1541–1545.
38. C. K. W. Tam and H. K. Tanna, "Shock Associated Noise of Supersonic Jets from Convergent-Divergent Nozzles," *J. Sound. Vib.*, Vol. 81, No. 3, 1982, pp. 337–358.

26

INTERACTION OF FLUID MOTION AND SOUND

J. E. Ffowcs Williams

1 INTRODUCTION

Fluid motion and sound are not necessarily different things, and the cases in which they can be usefully separated and their interaction examined are rather special. If the waves are sufficiently short, their progress can be followed and their characteristics recognized to be essentially those of simple one-dimensional waves. The situation is different if the propagation properties vary more abruptly and the acoustic wavelength is not small. For example, sound is both reflected from and transmitted through a plane vortex sheet. The situation is complicated by the fact that the sheet is unstable; sound interacts best with flow when the flow is unstable. General flow/acoustic interaction phenomena are understood only in a qualitative way, and exact statements that bear on the problem are very rare. That is why this chapter is devoted to statements relating flow and sound and emphasizing the sources of sound in fast flow.

2 THE PHYSICAL PROBLEM

There is little risk of confusion in defining the small-amplitude vibrational field of a compressible homogeneous material as sound. Bulk motion convects the entire wave field, any particular wave crest moving at the speed of sound augmented by material drift. Inhomogeneous convective motion deforms the geometric arrangement of both the wave field and the material elements through which the waves propagate. Materials with low resistance to those staining motions are fluids, and the non-trivial cases of *fluid motions* all involve a substantial and continuous rearrangement of a particle's shape. Again it might be reasonable to define the small-amplitude vibration of this distorting material as sound, but to do so without recognizing that it might be quite unlike the vibrational field of a uniform material is to risk confusing the picture by oversimplification.

The simplest cases are those in which "packets" of unidirectional sound waves travel in virtually "uniform" material. If the waves are sufficiently short compared with the size of the uniform zone, their progress can be followed and their characteristics recognized to be identical with those of a simple one-dimensional wave; waves travel through matter in a direction normal to wavefronts, energy flowing with them. The existence of sound imparts the material with vibrational and elastic energy, and bulk motion augments both the wave speed and the rate at which energy flows. The packet of waves, identifiable by the crest structure spanning across the wave's path, moves through the material by radiation and additionally drifts with the flow. Gradual changes in flow conditions, those that cause the "port" and "starboard" sides of the packet to move at different speeds, will rotate the packet and bend the path it follows. These are the principles underlying the method of ray tracing, one of the most potent techniques for describing the progress made by sound propagating through slowly varying flows and through material of slowly varying composition. Short-wave sound moves along rays bent by gradients in either flow velocity or sound speed. The wave activity tends to grow where rays converge; speed of sound gradients do not destroy the sound's energy-conserving properties but wind gradients do. The general theory of ray tracing is clearly described by Lighthill,[1] who builds on the earlier foundations laid down by Brekovskikh[2] and Blokhintzev.[3] This is the best way of thinking about how sound travels through the strat-

Note: References to chapters appearing only in the *Encyclopedia* are preceded by *Enc.*

ified ocean or is refracted by wind gradients. Though the underlying theory exploits the fact that the propagation properties change very slowly, the effect of those changes need not be small. The wind can refract sound to create zones of effective shadow; it can also guide sound waves to concentrate and focus them. These effects are routinely accounted for in engineering practice, with the ray-tracing calculations being performed with the aid of widely available computational codes.

But the situation is different if the propagation properties vary more abruptly and the acoustic wavelength is not small compared to the scale of material change. Of course, ray theory may still provide a useful guide even though it has lost its formal validity. It is certainly interesting to note the qualitative similarity that exists between the refractive properties measured in a jet shear layer and those predicted by ray theory (though wave acoustics are much better).[4,5] It is also tempting to explain the apparent inability of sound to traverse high-speed boundary layers in terms of ray trajectories, even though the flow profile is not thick enough to justify that explanation.[6]

New effects enter with abrupt changes in flow, Miles[7] having been the first to show how sound was both reflected back from and transmitted through a plane vortex sheet, the acoustic energy changing because of work done by the flow during its interaction with sound. The situation is complicated by the fact that the sheet is unstable; local acoustic sources induce exponentially growing disturbances that soon render the linear calculations meaningless.[8] Despite that complication, linear perturbation methods have proved extremely impressive in describing how sound escapes from a pipe, interacting with and modifying the form of the early jet enclosed in the vortex sheet continuation of the pipe wall.[9] Acoustic seeding of emergent jet flow can provoke jets to modify their noise-producing behavior; the broadband turbulence-induced noise can be changed by harmonic jet excitation, but a proper explanation to that mechanism must await more inroads into the understanding of turbulence.[10,11]

Turbulence has known wave-scattering properties,[12] the modeling of which is one of the principal applications of new computational methods.[13,14] Ray techniques have pointed to the tendency for rays to cross, giving multiple paths and multiple arrival times for the receipt of signals. The atmospheric propagation of sound displays these features in a characteristic spasmodic fading of distant signals, with the waveform of the signals being grossly distorted because of its interaction with atmospheric turbulence. The boom from a high-flying supersonic aircraft evolves in a homogeneous atmosphere into a characteristic N-wave, but in a typical turbulent atmosphere both the starting and trailing shocks are commonly spread out over a couple of metres, with distinct individual spikes being formed by coalescing rays. Crow[15] has explained the origins of these changes to be in the unsteady Reynolds stresses that give a scattering source strength proportional to the product of the velocities in the sound and turbulence field.

Vorticity is an element of flow that is foreign to sound, but there is no doubt that unsteady vortex fields create sound as a by-product.[16] Vorticity is created in regions of high velocity gradient, at the lip of a jet nozzle for instance. Flow usually separates from a sharp edge rather than flowing tightly around it as would a potential flow, which sound is. But the usually noisy shedding of vorticity into a downstream vortical wake can, remarkably, act to reduce the scattered field. Howe[17] demonstrated that the sound caused by the vortex acted to oppose what would otherwise be scattered from the sharp edge. Another example of vorticity as a sound-absorbing feature was discovered by Howe[18] in his study of the acoustic properties of rigid but porous screens. The potential interaction of sound with rigid screens is conservative; they convert the energy of the incident sound into that of sound transmitted through and reflected from the screen. But flow through the screen tends to separate at the sharp edges of the perforations, and if the right balance of steady through-flow and porosity is selected, the incident sound can be completely absorbed in the flow,[19] an absorption dependent on the secondary field created by the unsteady vortex shedding.

The interaction between a line vortex and a sharp-fronted pressure wave has long been a source of fascination to aerodynamicists and acousticians, mainly because the flow can be realized in a simple shock tube[20] and because it might provide a tractable example of the otherwise far-too-complicated turbulence–acoustic interaction problem. Though little mystery now remains in that example,[21] the general interaction of sound with flow of nonnegligible Mach number has fallen into the hands of computational aeroacousticians, and their field is still in its infancy. What is sound and what is flow is hard to tell, with the subject's repertoire of worked examples corroborated by independent theoretical or experimental checks yet to be established. Brentner's[22] early calculation of the flow created by impulsive boundary motion and the evident interchangeability revealed by that calculation of flow and acoustic elements provide a fascinating preview of what might yet be to come.

Sound interacts best with flow if the flow is unstable or the sound is especially strong. In whistles, organ pipes, and other forms of wind-driven resonators the energizing feature of the flow is in the unsteady vortex elements shed into an unstable shear layer to grow as they travel toward an edge. There they are scattered into sound, a sound that reacts back to provoke the shedding of another vortex element, to travel and grow and repeat the cycle. The interaction is especially strong when the vortex-shedding frequency coincides with the resonance frequency.[23] The screech of an imperfectly expanded supersonic jet flow results from cyclic backscattering of instability waves from their interaction with compressive waves trapped in the jet.[24]

All these flow–acoustic interaction phenomena are understood in a qualitative way, but exact statements that bear on the problem are very rare. Remarkable in its ability to relate flow and sound is the acoustic analogy produced by Lighthill[25]; his is an exact statement that anchors the flow–acoustic interaction subject and yields definite relationships between flow and sound, though arguably they have only limited predictive force. At this stage in the subject where the most important acoustic-coupled flows are at nonnegligible Mach number, definite results can be invaluable, so little is certain from the inevitably noisy experimental data and intuitive theoretical models. That is the reason why the remaining part of this chapter is devoted to statements relating flow and sound, particularly the sources of sound in fast flow. They may be useful in drawing out effects that remain hidden from view until finite-Mach-number terms are emphasized.

3 SOUND SOURCES IN MOVING FLUID

Sound propagates through uniform still fluid according to the wave equation $\partial^2\phi/\partial t^2 - c^2\,\nabla^2\phi = 0$. If the fluid moves uniformly, it will convect the sound field with it and a Gallilean transformation equates the two cases.

We can regard $\partial^2\phi/\partial t^2 - c^2\,\nabla^2\phi$ as the source of sound; we call it q. Given q and a boundary condition, the sound field ϕ is determined. Should the source q not be specified independently of ϕ, any "solution" giving ϕ in terms of q would still be only a secondary equation for ϕ. Then, ϕ is not a sound field; it is something else, similar to sound only inasmuch as q is independent of ϕ.

When the fluid is moving and the sound is better specified relative to a moving reference frame, there are two distinct phenomena affecting the level and structure of the sound field. First, there is the Doppler effect, by which successive wave crests are closer together when laid down by a following source. Second, the waves can be stronger because they can spend longer within the source, accumulating more from the source all the time they remain in touch. A moving point source at $\mathbf{y} = \mathbf{y}_s(\tau)$ is modeled by

$$q(\mathbf{y}, \tau) = Q(\tau)\,\delta[\mathbf{y} - \mathbf{y}_s(\tau)] \tag{1}$$

and generates the outwardly propagating sound field

$$\phi(\mathbf{x}, t) = \frac{Q(t - r^*/c)}{4\pi c^2 r^* |1 - M_r|^*}, \tag{2}$$

$r = |\mathbf{x} - \mathbf{y}|$ signifying the distance separating the source from the observer and cM_r the speed at which the source approaches. The asterisk implies values at the time when the source emitted the sound destined to arrive at $(\mathbf{x}, t)$:

$$\tau^* = t - \frac{r^*}{c}, \qquad r^* = |\mathbf{x} - \mathbf{y}_s(\tau)^*|. \tag{3}$$

The source moves with velocity $d\mathbf{y}_s/d\tau$, so that it approaches the observer with speed

$$cM_r = \frac{d\mathbf{y}_s}{d\tau} \cdot \frac{(\mathbf{x} - \mathbf{y}_s)}{|\mathbf{x} - \mathbf{y}_s|}.$$

The source time τ and reception time t are connected for a particular source and a particular observation position $\mathbf{x}$. The sound reaching $\mathbf{x}$ at time t was emitted from the source at $\mathbf{y}_s(\tau)$ at time $\tau = \tau^*$, the emission time scale being stretched when the source approaches:

$$\frac{d\tau^*}{dt} = 1 - \frac{1}{c}\frac{dr^*}{dt} = 1 - \frac{d}{dt}\frac{|\mathbf{x} - \mathbf{y}_s(\tau^*)|}{c}$$

$$= 1 + \frac{(\mathbf{x} - \mathbf{y}_s)}{|\mathbf{x} - \mathbf{y}_s|} \cdot \frac{d\mathbf{y}_s}{d\tau}\frac{1}{c}\frac{d\tau^*}{dt},$$

$$\frac{d\tau^*}{dt} = \frac{1}{(1 - M_r)^*}. \tag{4}$$

The perturbations of a real fluid caused by unsteady vorticity, by inhomogeneous heating, by the motion of a flow structure, or by foreign bodies moving or deforming in flow, by anything at all in fact, can all be represented through Lighthill's acoustic analogy as being pure sound in a uniform medium at rest, the sources of that sound being prescribed through the analogy. The analogy is useful if the perturbations are clearly recognizable as sound over an extensive region, the sound sources being outside that region and specified independently. That is not generally the case, most fluid motion being different from sound. The analogy is then an exact but somewhat sterile prescription of what sources would be needed to drive a nonexistent acoustic medium to support the density perturbations of the real world. The analogy is designed for the sound radiated by powerful machinery into uniform flow, and for that it is uniquely useful, giving insight, approximation techniques, and an exact formal methodology; accounting for the interaction of fluid motion and sound is reduced to a matter of extracting useful results from the formal method, and the main difficulty of the subject is to avoid the confusion likely to follow the inevitable simplifying approximation.

Sound in a source-free region V can be calculated by Kirchhoff's theorem in terms of boundary conditions on S, a fixed control surface enclosing V with its unit normal $\mathbf{n}$ leading into V:

$$\phi(\mathbf{x}, t) = -\frac{1}{4\pi}\frac{\partial}{\partial x_i}\int_S \frac{\phi^* n_i}{r}\, dS(\mathbf{y}) - \frac{1}{4\pi}\int_S \frac{1}{r}\left[\frac{\partial \phi}{\partial n}\right]^* dS(y), \qquad (5)$$

the boundary values being integrated over all boundary positions $\mathbf{y}$ at the retarded time $\tau^* = t - r/c$. A difficulty of the general problem is illustrated by using Eq. (5) to estimate the field at $(\mathbf{x}, t)$ in terms of approximations to the boundary conditions. For example, when the boundary is hard and $\partial\phi/\partial n = 0$, the field is given unambiguously by the first of the two integrals, the dipole term. Even when the boundary is not hard, the field is still *determined* once the dipole term is known, so is the boundary value $\partial\phi/\partial n$, which cannot be arbitrarily specified on S. Approximating either of the two surface integrals introduces errors that are not easy to quantify.

A generalization of Eq. (5) to give the field in terms of conditions specified on a moving surface $S(\boldsymbol{\eta})$ that encloses the space surrounding the observation point has been given in Ref. 26:

$$\begin{aligned}\phi(\mathbf{x}, t) = &-\frac{1}{4\pi}\frac{\partial}{\partial x_t}\int_{S(\boldsymbol{\eta})}\left[\frac{\phi n_i A}{r|1-M_r|}\right]^* dS(\boldsymbol{\eta}) \\ &-\frac{1}{4\pi c^2}\frac{\partial}{\partial t}\int_{S(\boldsymbol{\eta})}\left[\frac{\phi y_n A}{r|1-M_r|}\right]^* dS(\boldsymbol{\eta}) \\ &-\frac{1}{4\pi}\int_{S(\boldsymbol{\eta})}\left[\left(\frac{\partial\phi}{\partial n} - \frac{v_n}{c^2}\frac{\partial\phi}{\partial\tau}\right)\frac{A}{r|1-M_r|}\right]^* \\ &\cdot dS(\boldsymbol{\eta}). \end{aligned} \qquad (6)$$

The surface element in the moving reference frame (fixed $\boldsymbol{\eta}$) moves with velocity $\mathbf{v}$ and passes the point $\mathbf{y} = \boldsymbol{\eta}$ at a fixed reference time, the unit area of the moving surface $S(\boldsymbol{\eta})$ corresponding to an area A of the instantaneous reference time surface $S(\mathbf{y})$.

Equations (5) and (6) are two exact equations, one giving the field in terms of conditions on a moving surface and the other in terms of conditions on a fixed control surface. They could, for example, both be used to specify sound fields generated by a motionless source outside both the fixed and moving boundary surfaces. The two equations appear completely different if casually regarded as the field of moving boundary sources without recognizing that the formal problem requires that the boundary values be the very special values determined from the solution itself. The sometimes singular values of the Doppler factor $|1-M_r|^{-1}$ in the high-Mach-number form of each term of Eq. (6) need not imply that the field shows any trace of this high-Mach-number effect; the Doppler factors arise because the control boundary surface is moving and need not imply the motion of any real source.

Both Eqs. (5) and (6) are true for any interior sound field generated by exterior sources, and it is obviously a hazardous matter to seek in the equation particular features of the sound without the knowledge of all the boundary conditions. The exact equations of motion are similar in this respect but different enough in general to allow some surprising features. Lighthill's analogy by which the real density fluctuations are regarded as linear waves driven by specific but wave-dependent sources takes a strikingly simple form,

$$\frac{\partial^2(\rho-\rho_0)}{\partial t^2} - c^2\,\nabla^2(\rho-\rho_0) = \frac{\partial^2 T_{ij}}{\partial x_i\,\partial x_j},$$

$$T_{ij} = \rho u_i u_j + p_{ij} - c^2(\rho-\rho_0)\delta_{ij}. \qquad (7)$$

Here, T_{ij} is Lighthill's stress tensor, ρ the mass density, u the fluid velocity, and p_{ij} the compressive viscous stress tensor. The constants ρ_0 and c represent the mean density and the speed of sound. Equation (7) is a statement of the Navier–Stokes equations manipulated to appear as an inhomogeneous wave equation. The quadrupole strength density T_{ij} is needed to drive an acoustic medium to mimic exactly the density perturbation of the real fluid. Reference 26 gives an integral form of this equation that gives the density field in terms of T_{ij} and conditions on a moving boundary, an equation that has become the starting point for many studies of the sound generated by bodies moving at high speed.

If a surface S marks the fluid particles enclosing a foreign body moving and deforming arbitrarily in flow, flow that is surrounded externally by weakly disturbed homogeneous material through which sound radiates out to infinity, then the first Ffowcs Williams–Hawkings representation of Eq. (7) is appropriate.

$$4\pi c^2(\rho - \rho_0)(\mathbf{x}, t) = \frac{\partial^2}{\partial x_i \, \partial x_j} \int_V \left[\frac{T_{ij} J}{r|1 - M_r|} \right]^* d^3\boldsymbol{\eta}$$
$$- \frac{\partial}{\partial x_i} \int_S \left[\frac{p_{ij} n_j A}{r|1 - M_r|} \right]^* dS(\boldsymbol{\eta})$$
$$+ \frac{\partial}{\partial t} \int_S \left[\frac{\rho_0 v_n A}{r|1 - M_r|} \right]^* dS(\boldsymbol{\eta}). \tag{8}$$

This is an exact result for the fully nonlinear equations and is surprisingly simple when compared with its linear form, which is effectively Eq. (6). At low Mach number the volume quadrupole term is usually negligible compared with the dipole and monopole surface integrals and tends often to be ignored.

At first sight the sound of propeller blades, in which the last (monopole) term is specified by kinematic conditions and the dipole term is known once the blade loading is determined, can be calculated by a straightforward evaluation of this equation. But in practice the blade loading is not easy to specify exactly even when the blade flow is steady and laminar, the turbulent surface loading is very problematic, and the influence of the volume quadrupole is unclear. It takes great skill to build this prescription of the sound field into a blade noise prediction scheme.[27]

The simplest case for illustrating the effect of flow on a boundary-induced source is when the motion is linear and T_{ij} is negligible. The relative flow at speed U is then very nearly uniform and parallel to the boundary surface, and the normal velocity v_n is equal to $\partial\xi/\partial t + (\partial\xi/\partial y_1)$, ξ being the surface elevation above the y_1-axis that is parallel to the uniform mean flow. When the boundary encloses a finite-sized body, it must be thin for this linearization to apply. In the particular case when the boundary surface is plane, Eq. (8) simplifies because S may then be taken as the $y_3 = 0$ surface; $A = 1$ and the monopole and dipole terms are equal [see Ref. 28 and Eq. (19) in Chapter 69]:

$$(\rho - \rho_0)(\mathbf{x}, t) = \frac{\rho_0}{2\pi c^2} \frac{\partial}{\partial t} \int_{\eta_3 = 0} \left[\frac{v_n}{r|1 - M_r|} \right]^* \cdot dS(\boldsymbol{\eta}) \tag{9a}$$
$$= -\frac{1}{2\pi c^2} \frac{\partial}{\partial x_3} \int_{\eta_3 = 0} \left[\frac{p}{r|1 - M_r|} \right]^* \cdot dS(\boldsymbol{\eta}). \tag{9b}$$

These are convenient forms for evaluating the sound of flow over a slightly deformed plane surface and are simplest when only a small section of the surface is displaced from its nominal position. Reference 29 gives the far field of such a planar compact source by evaluating Eq. (9a) and shows that

$$(\rho - \rho_0)(\mathbf{x}, t) = \frac{\rho_0 \ddot{Q}(\tau)^*}{2\pi r_0 |1 - M_{r0}|^3}, \qquad \text{where}$$
$$\ddot{Q}(\tau)^* = \int_{s_0} [\ddot{\xi}] \, dS(\boldsymbol{\eta}), \tag{10}$$

and r_0 is the distance of the observer at $\mathbf{x}$ from the vibrating part of the surface at the time when it launched its sound to the observer; the source region was then approaching at speed cM_{r0}. The effect of motion on this source is to modify the field strength at a large distance from the small wave *emission* zone by three powers of the Doppler factor and to alter the time scale of the sound from that of the source by the Doppler factor. The Doppler amplification is at first sight surprising, the three powers being a characteristic of quadrupole sources,[25] and warns that it is not always a trivial matter to deduce the influence of flow from exact representations of the field. It comes from Eq. (9a) only after recognizing that the part of v_n given by $U(\partial\xi/\partial y_1)$ tends to integrate to zero.

A second case in which the effect of low-speed motion can be worked out exactly is due to Dowling.[30] She considered a compact pulsating sphere of radius $a(\tau)$ moving at low Mach number through infinite homogeneous inviscid fluid at rest. She worked out the effect of slow flow by deriving an expression for the distant sound accurate to first order in M_{r0}, the Mach number at which the sphere was approaching the distant observer at the time it radiated the observed sound. The volume quadrupoles are negligible because of their low-Mach-number inefficiency, but a dipole arises as the sphere experiences an unsteady drag $2\pi\rho a^2 \dot{a} U$.[31] The linear element of the dipole term in Eq. (8) therefore contains a term arising from the component of drag directed toward the observer, $2\pi\rho a^2 \dot{a} c M_{r0}$, making

$$-\frac{\partial}{\partial x_i} \int_S \left[\frac{p_{ij} n_j A}{r|1 - M_r|} \right] dS(\boldsymbol{\eta}) = \frac{1}{r_0} \frac{\partial}{\partial t} \left[2\pi\rho_0 a^2 \dot{a} M_{r0} \right]^* \tag{11}$$

to first order in M_{r0}.

The monopole term of Eq. (8) can be evaluated once it is recognized that the contact between the fluid and surface must be continuous[32]:

$$\frac{\partial}{\partial t}\int_s \left[\frac{\rho_0 v_n A}{r|1-M_r|}\right] dS(\boldsymbol{\eta}) = \frac{\rho_0}{r_0}\frac{\partial}{\partial t}\left[\frac{4\pi\rho_0 a^2 \dot{a}}{(1-M_{r_0})^2}\right]^*. \quad (12)$$

The sum of Eqs. (11) and (12) give the first-order effects of source motion to be

$$4\pi c^2(\rho-\rho_0)(\mathbf{x},t) = \frac{4\pi\rho_0 a^2}{r_0}\frac{\partial}{\partial t}\left[\frac{\dot{a}(\tau)^*}{(1-M_{r_0})^{5/2}}\right]^*, \quad (13)$$

so that, bearing in mind the Doppler time contraction [Eq. (4)],

$$c^2(\rho-\rho_0)(\mathbf{x},t) = \frac{\rho_0 a^2 \ddot{a}(\tau)^*}{r_0(1-M_{r_0})^{7/2}}. \quad (14)$$

This $3\frac{1}{2}$ power of the Doppler factor amplification due to real source motion is more complicated than the effects of flow on hypothetical sources; flow affects both the coupling of a source to the sound and *also* the strength of the source. The small pulsating sphere, often regarded as the physical embodiment of a simple point source, is evidently quite different from it when flow is considered, Eqs. (14) and (2) representing their respective fields. The modification of sources by flow is just as important as the effect of flow on the radiation efficiency.

The distinction between sound and flow is difficult to maintain at high speed. Equation (8), though exact, is hard to interpret when T_{ij} depends on the sound; very few cases exist for evaluating the explicit form and importance of extensively distributed quadrupole terms. If the acoustic analogy is used to calculate known shock waves generated by impulsively started plane boundaries or attached to wedges in supersonic flow,[33] it is seen that the quadrupoles mainly account for the self-convective aspects of nonlinear waves and are weak provided the boundary-induced velocity is much smaller than the sound speed; when that is not so, the sources representing the waves in the exact analogy have a highly unphysical form; for example, they can arrive at the observer before their sound is heard. It seems likely that the general use of the analogy in high-speed flow is usefully confined to flows that are only weakly disturbed by boundary motions and where T_{ij} is negligible outside strictly limited source regions. Other fields, where a nonvanishing T_{ij} represents all nonlinear effects over extensive regions of space, are easier to approach through a modification of Lighthill's method. One such modification extends the analogy to sound generated near and interacting with the interface between two extensive regions of uniform fluid in relative motion, the sources of waves being confined to the proximity of the interface.

The interface matters a lot, even when it is only linearly disturbed by sound. Sound passing from one region to the other is partly reflected and partly refracted and exchanges energy with the moving fluid. The irradiated plane vortex sheet is one of the simplest idealizations of sound negotiating inhomogeneous flow, but even that is a far from straightforward problem. For a start the vortex sheet is highly unstable and if disturbed by a nearby source will soon degenerate as instability waves grow and destroy the otherwise calm state. This would be avoided if the source were anticipated by an exactly opposite instability wave to that produced by the source so that the two combine, by linear superposition, to cancel one another at large time. This noncausal response is the only way that the vortex sheet modeling is consistent with linear theory and is an essential part of the generalization of Lighthill's acoustic analogy to account for the presence of vortex layers.[34] The sound generated at the turbulent interface of moving streams is identical to the sound that would be generated by Lighthillian quadrupoles moving with the fluid but reflected in a hypothetical backing surface, which behaves exactly as if it were a linearly disturbed vortex sheet in otherwise uniform flow. Beyond the surface it is as if the moving quadrupoles' sound passed through a linearly disturbed vortex sheet into the uniformly moving fluid.

Instabilities are the cause of turbulence, but sound provoked the instability, to make turbulence, to make sound, and so on. In this view of a statistically steady turbulent flow finiteness of the linear response is a more important aspect to preserve in an analogy than is causality, which is patently inappropriate. This will be so whatever scheme is used to synthesize the actual sound by superposing a set of linear perturbations onto an unstable primary flow.

Even the planar shear layer displays considerable modifying influence on the sound of turbulence, Ref. 35 predicting that no sound can travel parallel to the layer and that the angular directivity of the sound is arranged in beams and shadows with tens of decibels contrast occurring within a few degrees variation of the angle at which sound leaves a highly supersonic flow. These aspects are not inconsistent with experimental jet noise studies, the vortex sheet modeling of the sound's interaction with jet flow bringing the acoustic analogy into far better accord with experiment than is possible when that interaction is ignored.[36]

Early jet noise modeling accounted for the convective effect of flow on the source but not its subsequent interaction with sound. Source motion through a still atmosphere amplifies the radiated sound and directs it preferentially ahead of the source. But if the source radiates its sound into uniform surrounding fluid that moves at the same speed, there is no such amplification, and the wrong impression is given by concentrating on the convective terms in the fixed-atmosphere form of the analogy. Jet noise displays both effects to varying degrees, the low-frequency noise being amplified by convection while the moving jet flow shields the high-frequency elements. Short waves are refracted away from the jet.[37] Ray theory is more appropriate to the short-wave case but is not easy to accommodate in the acoustic analogy. Equation (8) does give the sound once the flow is known, but T_{ij} is unlikely to be adequately specific other than for compact sources. There are hardly any examples simple enough to model exactly, and one should not expect more to come from the modeling than the useful qualitive deductions that can be made on dimensional grounds.

The sound is also given precisely in the vortex sheet extension of the acoustic analogy. In that, the solution of Lighthill's quadrupole-driven wave equation is applied in two stages. First, it is applied in the moving flow in a convected reference frame with the mean density and sound speed equal to those of the moving primary flow. Second, the Lighthill equation is applied in the fixed rest frame, with the density and sound speed of the surrounding fluid. The convective form is solved in the first region and the ordinary form in the other, the solution proceeding via a Green's function, which is the response of the homogeneous problem to an impulsive point source in the two regions connecting across a vortex sheet with the usual continuity of pressure and material position. This modified analogy[34] points to the field of a compact jet being closely related to that of a plane turbulent mixing layer, the density perturbation differing only because of a speed-sensitive directionality factor D_{ij} in the general expression

$$\rho'(\mathbf{x},t) = \frac{D_{ij}}{4\pi x c_0^4} \int \left[\frac{D}{Dt} \left\{ \frac{1}{1-N_r} \frac{D}{Dt} \cdot \left(\frac{\rho^* T_{ij}}{\rho(1-N_r)} \right) \right\} \frac{d^3\boldsymbol{\eta}}{1-N_r} \right], \quad (15)$$

where the square brackets indicate retarded time, N_r is the Mach number at which the mean flow approaches the observer at $\mathbf{x}$, ρ is the density at the emission time, and ρ^* is the density at the reference time when $\boldsymbol{\eta} = \mathbf{y}$. Mani's proof that the interaction of longitudinal quadrupoles with a jet flow results in an amplification term $(1-M_r)^{-2}(1-N_r)^{-3}$ is contained in this formula, M_r being based on the source convection speed and N_r on the jet speed. This is a good illustration of how regions of moving fluid affect both the generation and refraction of sound.

The interaction of sound with flow is known once the Reynolds stress terms in T_{ij} are known. That interaction is usually weak because of the inefficiency of compact quadrupoles but could be very strong should the flow contain material inhomogeneities. The isotropic element in T_{ij} contains the term $p - c^2\rho$, the unsteady parts of which depend on the difference between changes in $c^2\rho$ and change in p, a vanishing difference in a perfect acoustic medium. But inhomogeneous material has density gradients unrelated to pressure, and the acoustic balance fails, leaving a strong isotropic source in unsteady flow. In fact, Morfey[38] showed that this effect produces a linear source term even in isentropic fluid, the Lighthill equation then being effectively

$$\frac{\partial^2\rho}{\partial t^2} - c^2\,\nabla^2\rho = \frac{\partial}{\partial x_i}\left((\rho^* - \rho_0)\,\frac{\partial u_i}{\partial t}\right). \quad (16)$$

Each particle of mean density ρ^* accelerating in unsteady flow is being acted on by a force different to what would be needed to produce that acceleration in sound, $\rho_0(\partial u_i/\partial t)$, and that difference acts as a dipole source, fundamentally more efficient than the quadrupoles that account for the interactions between sound and homogeneous fluid. This effect is an important contributor to the noise of propulsive jets, which are inevitably hotter and lighter than their environment. The noise of many of those jets is actually dominated by this effect at subsonic speeds, increasing in proportion to M^6 rather than conforming with the familiar eighth-power law.

REFERENCES

1. M. J. Lighthill, *Waves in Fluids*, Cambridge University Press, Cambridge, 1978.
2. L. M. Brekovskikh, *Waves in Layered Media*, Academic, New York, 1960.
3. D. I. Blokhintzev, "Acoustics of a Nonhomogeneous Moving Medium," Tech. Mem. No. 1399, National Advisory Communittee on Aeronautics, Washington, DC, 1956.
4. L. K. Schubert, "Numerical Study of Sound Refraction by a Jet Flow I. Ray Acoustics," *J. Acoust. Soc. Am.*, Vol. 51, 1972, pp. 439–446.
5. L. K. Schubert, "Numerical Study of Sound Refraction by a Jet Flow II. Wave Acoustics," *J. Acoust. Soc. Am.*, Vol. 51, 1972, pp. 447–463.
6. D. B. Hanson, "Shielding of Prop-fan Cabin Noise by the Fuselage Boundary Layer," *J. Sound Vib.*, Vol. 92, 1984, pp. 591–598.

7. I. W. Miles, "On the Reflection of Sound at an Interface of Relative Motion," *J. Acoust. Soc. Am.*, Vol. 29, 1957, pp. 226–228.
8. J. D. Morgan, "The Interaction of Sound with a Semi-infinite Vortex Sheet," *Quart. J. Mech. Appl. Math.*, Vol. 27, 1974, pp. 465–487.
9. R. M. Munt, "The Interaction of Sound with a Subsonic Jet Issuing from a Semi-infinite Cylindrical Pipe," *J. Fluid Mech.*, Vol. 83, 1977, pp. 609–640.
10. C. J. Moore, "The Role of Shear-Layer Instability Waves in Jet Exhaust Noise," *J. Fluid Mech.*, Vol. 80, 1977, pp. 321–367.
11. D. Bechert and E. Pfizenmaier, "On the Amplification of Broadband Jet Noise by a Pure Tone Excitation," *J. Sound Vib.*, Vol. 43, 1975, pp. 581–587.
12. V. I. Tatarski, *Wave Propagation in a Turbulent Medium*, McGraw-Hill, New York, 1961.
13. C. R. Truman and M. J. Lee, "Effects of Organised Turbulence Structures on the Phase Distortion in a Coherent Optical Beam Propagating through a Turbulent Shear Flow," *Phys. Fluids A*, Vol. 5, No. 2, 1990, pp. 851–857.
14. P. H. Blanc-Benon, D. Juvé, and G. Comte-Bellot, "Occurrence of Caustics for High-Frequency Acoustic Waves Propagating through Turbulent Fields," *Theoret. Comput. Fluid Dynamics*, Vol. 2, 1991, pp. 271–278.
15. S. C. Crow, "Distortion of Sonic Bangs by Atmospheric Turbulence," *J. Fluid Mech.*, Vol. 37, 1969, pp. 529–563.
16. W. Möhring, "On Vortex Sound at Low Mach Number," *J. Fluid Mech.*, Vol. 85, 1978, pp. 685–691.
17. M. S. Howe, "The Influence of Vortex Shedding on the Generation of Sound by Convected Turbulence," *J. Fluid Mech.*, Vol. 76, 1976, pp. 711–740.
18. M. S. Howe, "The Influence of Vortex Shedding on the Diffraction of Sound by a Perforated Screen," *J. Fluid Mech.*, Vol. 97, 1980, pp. 641–653.
19. A. P. Dowling and I. J. Hughes, "The Absorption of Sound by Perforated Linings," *J. Fluid Mech.*, Vol. 218, 1990, pp. 299–335.
20. M. A. Hollingsworth and E. J. Richard, "A Schlieren Study of the Interaction between a Vortex and a Shock Wave in a Shock Tube," Aeronautical Research Council (report) 17,985, FM 2323, 1955, AD 140 845.
21. H. S. Ribner, "Cylindrical Sound Wave Generated by Shock-Vortex Interaction," *AIAA J.*, Vol. 23, 1985, pp. 2708–1715.
22. K. S. Brentner, "Direct Numerical Calculation of Acoustics: Solution Evaluation through Energy Analysis," *J. Fluid Mech.*, Vol. 254, 1993, pp. 267–281.
23. A. Powell, "On the Edgetone," *J. Acoust. Soc. Am.*, Vol. 33, 1961, pp. 395–409.
24. A. Powell, "The Noise of Choked Jets," *J. Acoust. Soc. Am.*, Vol. 25, 1953, pp. 385–389.
25. M. J. Lighthill, "On Sound Generated Aerodynamically, I General Theory," *Proc. Roy. Soc.*, Vol. 221A, 1952, pp. 564–587.
26. J. E. Ffowcs Williams and D. L. Hawkings, "Sound Generation by Turbulence and Surfaces in Arbitrary Motion," *Phil. Trans. Roy. Soc.*, Vol. A264, 1969, pp. 312–342.
27. F. Farasat and K. S. Brentner, "The Uses and Abuses of the Acoustic Analogy in Helicopter Rotor Noise Prediction," *J. Am. Helicopter Soc.*, Vol. 33, No. 1, 1988, pp. 29–36.
28. A. P. Dowling and J. E. Ffowcs Williams, *Sound and Sources of Sound*, Ellis-Horwood, Chichester, UK, 1983.
29. J. E. Ffowcs Williams and D. J. Lovely, "Sound Radiation into Uniformly Flowing Fluid by Compact Surface Vibration," *J. Fluid Mech.*, Vol. 71, 1978, pp. 689–700.
30. A. P. Dowling, "Convective Amplification of Real Simple Sources," *J. Fluid Mech.*, Vol. 74, 1976, pp. 529–546.
31. Milne-Thomson, *Theoretical Hydrodynamics*, 5th ed., Macmillan, New York, 1968.
32. D. G. Crighton, A. P. Dowling, J. E. Ffowcs Williams, M. Heckl, and F. G. Leppington, *Modern Methods in Analytical Acoustics Lecture Notes*, Springer-Verlag, 1992.
33. J. E. Ffowcs Williams, "On the Role of Quadrupole Source Terms Generated by Moving Bodies," AIAA 5th Aeroacoustics Conference, Paper 79-0567, 1979.
34. A. P. Dowling, J. E. Ffowcs Williams, and M. E. Goldstein, "Sound Production in a Moving Stream," *Phil. Trans. Roy. Soc.*, Vol. 288, No. 1353, 1978, pp. 321–349.
35. J. E. Ffowcs Williams, "Sound Production at the Edge of a Steady Flow," *J. Fluid Mech.*, Vol. 66, No. 4, 1974, pp. 791–816.
36. R. Mani, "The Influence of Jet Flow on Jet Noise Parts 1 and 2," *J. Fluid Mech.*, Vol. 73, No. 4, 1976, pp. 753–793.
37. P. A. Lush, "Measurement of Subsonic Jet Noise and Comparison with Theory," *J. Fluid Mech.*, Vol. 46, 1971, pp. 477–501.
38. C. L. Morfey, "Amplification of Aerodynamic Noise by Convected Flow Inhomogeneities," *J. Sound Vib.*, Vol. 31, 1973, pp. 391–397.

27

SHOCK WAVES, BLAST WAVES, AND SONIC BOOMS

RICHARD RASPET

1 INTRODUCTION

Many sounds in the atmosphere originate from strong impulsive sources. Blast waves from explosions, thunder from lightning, sonic booms from aircraft, ballistic waves from projectiles, and N-waves from spark sources all involve large-amplitude waves. This chapter serves as a guide to strong- and weak-shock theory as applied to impulses.

First the basic equations governing shock propagation and decay are examined. Next, the application of these equations to the prediction of source waveforms and decay with distance is described. Plots of the waveforms and calculated pressure decay with range and scaling laws to relate these plots to arbitrary sources are provided. These relations can be applied to spherical and cylindrical explosions, sparks, and exploding wires. Next, nonlinear propagation and weak-shock theory are reviewed, and their application to far-field predictions is described. More detailed information on finite-wave and weak-shock propagation is contained in Chapter 17.

2 BASIC EQUATIONS

The basic equations necessary to describe shock wave propagation express conservation of mass, Newton's second law for fluids, conservation of energy, and an equation of state relating pressure to density.[1] In regions away from shocks, the Eulerian form of these equations are

$$\frac{\partial \rho}{\partial t} + \nabla \cdot (\rho \boldsymbol{u}) = 0, \tag{1}$$

$$\rho \left(\frac{\partial \boldsymbol{u}}{\partial t} + \boldsymbol{u} \cdot \nabla \boldsymbol{u} \right) = -\nabla p, \tag{2}$$

$$\rho \frac{\partial}{\partial t} (\tfrac{1}{2} u^2 + e) + \boldsymbol{u} \cdot \nabla (\tfrac{1}{2} u^2 + e) = 0, \tag{3}$$

and

$$p = p(\rho, T), \tag{4}$$

where ρ is the fluid density, **u** is the particle velocity, T is the temperature, p is the pressure, and e is the internal energy per unit mass. These equations may be linearized in terms of the fluctuation densities and pressure to form the basic acoustic equations.

If Eqs. (1)–(4) are solved for wave propagation without any attenuation mechanism, they will result in multivalued pressure waveforms. In reality, attenuation mechanisms such as viscosity and heat conduction become important for large pressure gradients and lead to small but finite rise times of the pressure waveform.

For sufficiently strong waves, discontinuities form and the differential equations expressed in Eqs. (1)–(3) do not hold. The governing equations across the shocks are the Rankine–Hugoniot equations. These may be expressed as

$$[\rho(u - u_{\text{sh}})]_+ = [\rho(u - u_{\text{sh}})], \tag{5}$$

$$[\rho u(u - u_{\text{sh}}) + p]_+ = [\rho u(u - u_{\text{sh}}) + p]_-, \tag{6}$$

and

$$[\rho(\tfrac{1}{2}u^2 + e)(u - u_{\text{sh}}) + pu]_+ = [\rho(\tfrac{1}{2}u^2 + e)(u - u_{\text{sh}}) + pu]_-, \tag{7}$$

Note: References to chapters appearing only in the *Encyclopedia* are preceded by *Enc.*

where e is the internal energy per unit mass, the subscript minus signifies the variable just behind the shock, and the subscript plus signifies the variable just in front of the shock. Equations (5), (6), and (7) express conservation of mass, Newton's second law, and work–energy conservation, respectively.

In many cases the air or gas surrounding an explosion or energy release can be considered a polytropic gas, that is, a gas with a constant ratio of specific heats.[2] The shock relations for a polytropic gas are written conveniently in terms of shock strength $z = (p_- - p_+)/p_+$ and Mach number of the shock relative to the flow ahead, $M = (u_{\text{sh}} - u_+)/c_+$. The shock relations are then given by

$$\frac{u_- - u_+}{c_+} = \frac{z}{\gamma\{1 + [(\gamma + 1)/2\gamma]z\}^{1/2}}, \tag{8}$$

$$\frac{\rho_-}{\rho_+} = \frac{1 + [(\gamma + 1)/2\gamma]z}{1 + [(\gamma - 1)/2\gamma]z}, \tag{9}$$

and

$$\frac{c_-}{c_+} = \left(\frac{1 + z\{1 + [(\gamma - 1)/2\gamma]z\}}{1 + [(\gamma + 1)/2\gamma]z}\right)^{1/2}. \tag{10}$$

We examine shock theory as applied to explosion waves in the next section.

3 EXPLOSION WAVES

3.1 Spherical Explosions

The simplest model of an explosion is a point blast. Exact solutions for this problem have been developed by Von Neumann,[3] Taylor,[4] and Sedov.[5] The explosion is modeled as an energy release E concentrated at a point in space with ambient density ρ_0. Dimensional analysis can be used to derive relations between the position of the shock, $r(t)$, and the overpressure in terms of E, ρ_0, and t:

$$r(t) = k\left(\frac{E}{\rho_0}\right)^{1/5} t^{2/5} \tag{11}$$

and

$$p = \frac{8}{25}\,\frac{k^2\rho_0}{\gamma + 1}\left(\frac{E}{\rho_0}\right)\frac{2}{5}\,t^{-6/5}, \tag{12}$$

or, equivalently,

$$p = \frac{8}{25}\,\frac{k^5}{\gamma + 1}\,Er^{-3}. \tag{13}$$

The decay of a strong spherical shock as r^{-3} is a general characteristic of the strong-shock regime.

The dimensionless constant k is determined by requiring conservation of energy and is a function only of the ratio of specific heats. We require that

$$E = \int_0^{r(t)} \left(\frac{p}{\gamma - 1} + \frac{\rho u^2}{2}\right) 4\pi r^2\, dr \tag{14}$$

be conserved. This has been evaluated numerically by Taylor[4] and analytically by Sedov[5] and Von Neumann.[3] For $\gamma = 1.4$, at standard pressure, Eq. (13) becomes

$$p = \frac{0.155E}{r^3}. \tag{15}$$

Sedov presents a number of similarity solutions to more complicated problems, but the simple point explosion serves as a basis to understanding the numerical solutions for strong-shock development and for understanding scaling laws.

It is useful to express the variables in terms of dimensionless variables as used in similarity solutions. The development of explosion waves from realistic explosions or differing initial conditions is not expected to obey the scaling laws exactly since realistic initial conditions introduce a characteristic length. For strong enough sources, however, the influence of the initial geometry is reduced as the radius of the shock increases. The variables of overpressure, radius, and time can all be expressed in terms of dimensionless variables in terms of the ambient density and the total energy release[6]:

$$\frac{p}{p_0} \quad \text{(scaled pressure)}, \tag{16}$$

$$\frac{c_0 p_0^{1/3} t}{E^{1/3}} \quad \text{(scaled time)}, \tag{17}$$

$$\frac{p_0^{1/3} r}{E^{1/3}} \quad \text{(scaled distance)}, \tag{18}$$

where p_0 is the ambient pressure, c_0 is the speed of sound in air, E is the energy release, r is the radial distance, p is the shock pressure, and t is the time from the explosion. The model of a point energy release works well for nuclear explosions. For chemical explosions, the difference in the equation of state of the explosion products and the finite initial distribution of kinetic energy lead to lower initial shock pressures and higher efficiencies in the conversion of energy to the blast wave.

Brode numerically integrated the gas dynamic equations for a variety of spherical explosions with different initial conditions and equations of state. In his 1955 paper[7] he demonstrated that the blast wave from an isothermal sphere of 199 atm initial pressure "assumes the general shape and values of the point source solution (to within 10 percent) after the shock wave has engulfed a mass of air 10 times the initial mass of the sphere."

In a later paper, Brode[8] calculated the blast wave from the detonation of a sphere of TNT with a loading density of 1.5 g/cm^3. He used realistic equations of state for the TNT and for air. Figure 1 displays Brode's results for TNT in air compared to the point source in air and in an ideal gas. Also included are calculations from Berry, Butler, and Holt that used an ideal gas with $\gamma = 3$ for the explosion products and an ideal gas with $\gamma = 1.2$ for the surrounding gas and calculations by Wecken for RDX-TNT in an ideal gas. The similarity of the curves beyond a reduced radius of 0.2 shows that the blast wave from a strong explosion is only weakly dependent on initial conditions and that the scaling laws are a good approximation for relating different yield explosions. The shock decay follows a $1/r^3$ dependence down to about 10 atm. Brode[7] also suggests fits to the decay curve for the point source at lower pressures. At longer distances weak-shock theory holds and the decay approaches $1/r$. ANSI standard S2.20[9] approximates the long-range decay of explosion waves to be $r^{-1.1}$. This power law is an approximate fit to theoretical results prepared by the Air Force Weapons Laboratory.

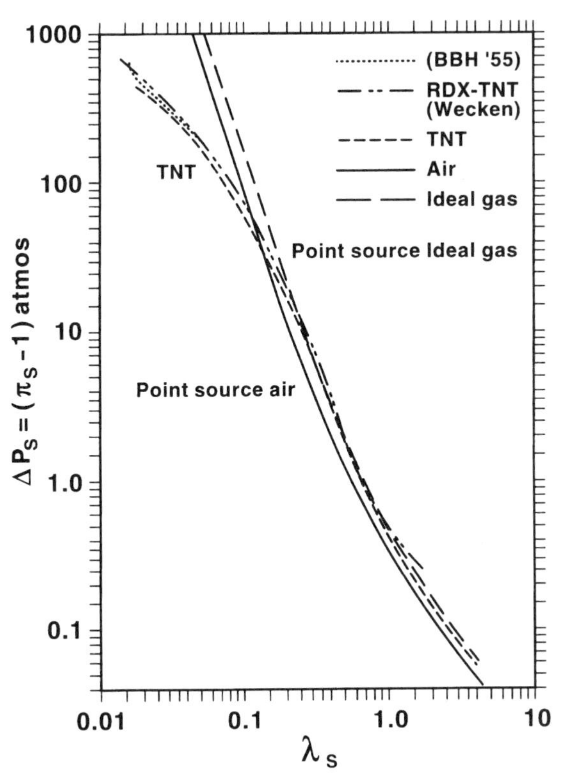

Fig. 1 Peak overpressure ($\Delta p_s = \pi_s - 1$) in atmospheres versus shock radius ($\lambda_s = R_s/\alpha$, $\alpha^3 = W_{tot}/p_0$) with comparisons between TNT and point source blasts in both real air and an ideal gas ($\gamma = 1.4$). The dash–double-dot curve is from the results of Wecken for 50% RDX, 50% TNT. The short dotted curve is from the Berry, Butler, and Holt calculations for PENT. (From H. L. Brode, "Blast Wave from a Spherical Charge," *Phys. Fluids*, Vol. 2, 1959, pp. 217–229.)

The principal difference in initial conditions is in the effective energy yield. Nuclear explosions approximate a point release of energy and leave a higher percentage of energy near the origin. The efficiencies of various explosives in terms of their weight are listed in Table 1 (taken from Ref. 9). Data from different types of explosions can be related by converting to the TNT equivalent weight.

Spherical blast waveforms for noise calculations can be developed from Brode's calculations. Figure 2 presents the pressure versus radius at fixed time based upon Brode's calculation for a TNT explosion. Extrapolations using finite-wave analysis and weak-shock theory may be used for lower amplitude waves. Figure 3 compares a measured waveform from a blasting cap to Brode's waveform extrapolated to larger distance with a finite-wave propagation algorithm.[10,11]

TABLE 1 Equivalent Explosive Weight Approximations Based on TNT

Explosive	Equivalent Weight
TNT	1.00
Tritonal	1.07
Composition B	1.11
HBX-1	1.17
HBX-3	1.14
TNETB	1.36
Composition C-4	1.37
H-6	1.38
Pentolite	1.42
PETN	1.27
Nitroglycerine	1.23
RDX-Cyclonite	1.17
Nitromethane	1.00
Ammonium nitrate	0.84
Black powder	0.46
Nuclear explosives	0.79

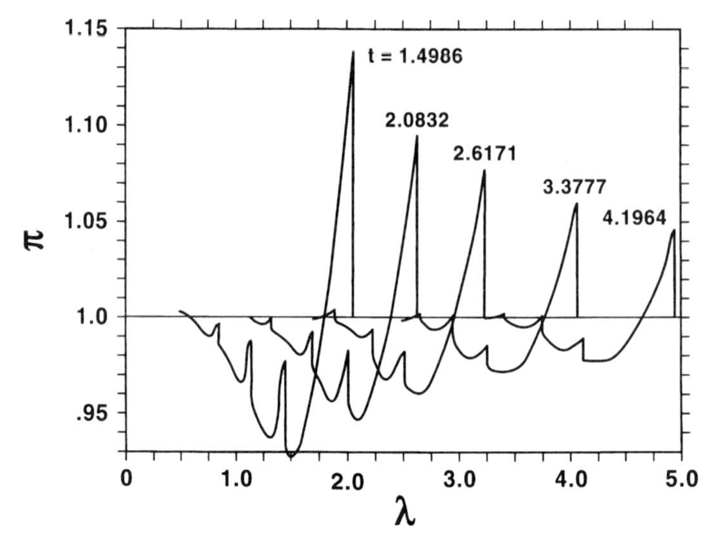

Fig. 2 Pressure ($\pi = p/p_0$) versus radius at indicated times for a TNT blast. (From H. L. Brode, "Blast Wave from a Spherical Charge," *Phys. Fluids*, Vol. 2, 1959, pp. 217–229.)

Measured and empirical waveforms can also be used in calculations. Figure 4 is a waveform for 0.57 kg of C-4 plastic explosive measured at 30 m over an acoustically hard surface.[12] The calibration constant is 3.17×10^5 Pa/V. Reed[13] has used the following waveform in blast noise attenuation calculations and it is a reasonable fit to measured waves:

$$p(t) = \Delta p\left(1 - \frac{t}{t_+}\right)\left(1 - \frac{t}{\tau}\right)\left[1 - \left(\frac{t}{\tau}\right)^2\right],\qquad 0 < t < \tau. \tag{19}$$

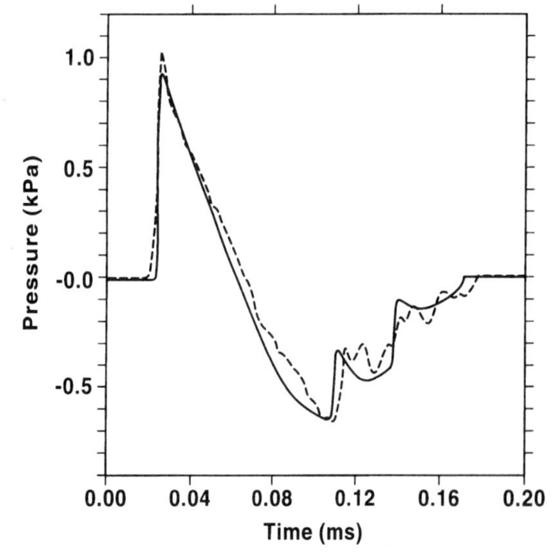

Fig. 3 Comparison of the measured waveform with the output of the finite-wave program. (From R. Raspet, J. Ezell, and S. V. Coggeshall, "Diffraction of an Explosive Transient," *J. Acoust. Soc. Am.*, Vol. 79, 1985, pp. 1326–1334.)

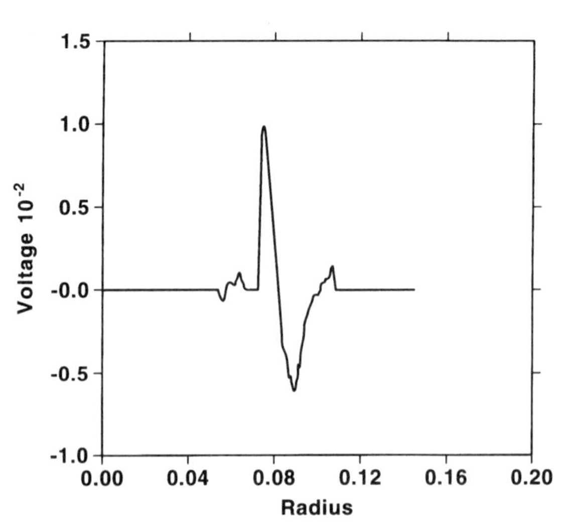

Fig. 4 Measured explosion pulse shape at 30 m for 0.57 kg of C-4 plastic explosive. (From R. Raspet, H. E. Bass, and J. Ezell, "Effect of Finite Ground Impedance on the Propagation of Acoustic Pulses," *J. Acoust. Soc. Am.*, Vol. 74, 1983, pp. 267–274.)

For a 1.0-kiloton nuclear explosion $R_0 = 2.7$ km, $t_+ = 0.375$ s, $\tau = 1.375$ s, and $\Delta p = 2.55$ kPa. The long negative duration of this pulse is typical of larger explosions. The waveforms illustrated may be scaled to different charge weights and explosive types by Eqs. (16)–(18) and Table 1. Charge weight or explosive mass may be substituted for the energy when scaling between different explosives.

3.2 Cylindrical Explosions

Cylindrical explosions and line source energy releases can be analyzed using appropriate scaling laws and solutions. The development of the wave from an instantaneous energy release along a line in an ideal gas has been solved by Lin[14] in connection with the problem of shock generation by meteors or missiles. Plooster has evaluated the development and decay of cylindrical sources with more realistic initial energy distributions and atmospheres.[15] Plooster applied his procedure to the prediction of the shock wave from long sparks[16] and from lightning.[17]

Plooster[15] and Lin[14] have used a scaled radius that involves an additional constant determined by the ratio of the specific heats of the fluid:

$$\frac{p}{p_0} \quad \text{(scaled pressure)}, \tag{20}$$

$$\frac{r}{r_0} \quad \text{(scaled distance)}, \tag{21}$$

$$\frac{c_0 t}{r_0}, \quad \text{(scaled time)}. \tag{22}$$

where $r_0 = (E_0/b\gamma p_0)^{1/2}$ and b is a constant dependent on γ, the ratio of the specific heats. For $\gamma = 1.4$, $b = 3.94$.

Figure 5 displays Plooster's result for A, the line source in an ideal gas; B, an isothermal cylinder, constant density, ideal gas; C, an isothermal cylinder, constant density, real gas; D, an isothermal cylinder, low density, ideal gas; and E, an isothermal cylinder, high density, ideal gas. As in the spherical case, the solutions differ mainly at short distances and in the efficiency of the energy release coupling into the explosion waves. Beyond a reduced radius of 0.09, the decay rates are similar. Figure 6 displays the pressure wave calculated by Plooster at scaled distances from 1.44 to 5.95.

Similar calculations for exploding wires were first carried out by Rouse[18] and by Sakurai.[19] Rouse found that the line source model produced good estimates of the shock strength and decay from exploding wires because the errors introduced by the presence of the copper vapor partially canceled the errors in the equation of state for air.

A strong-shock estimate of the decay of cylindrical waves from an instantaneous line source in air characterized by $\gamma = 1.4$ at standard pressure is given by

$$p = \frac{0.216E}{r^2}. \tag{23}$$

Decay proportional to r^{-2} is a general property of cylindrical strong shocks.

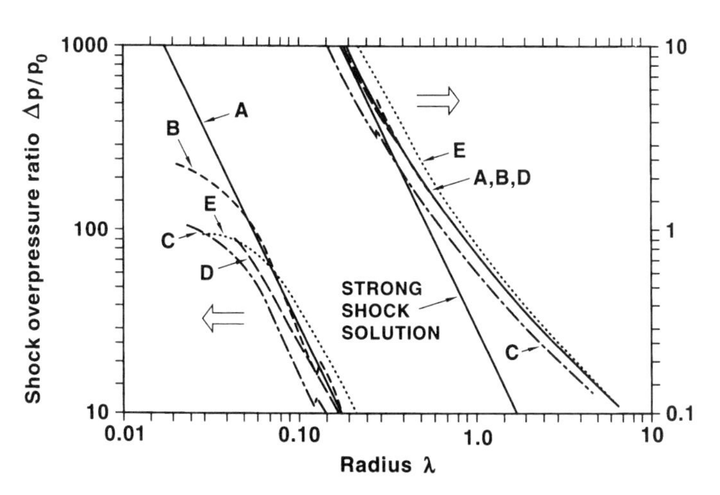

Fig. 5 Shock overpressure ratio $\Delta p/p_0$ versus dimensionless shock radius λ for initial conditions described in text. (From M. N. Plooster, "Shock Waves from Line Sources. Numerical Solutions and Experimental Measurements," *Phys. Fluids*, Vol. 13, 1970, pp. 2665–2675.)

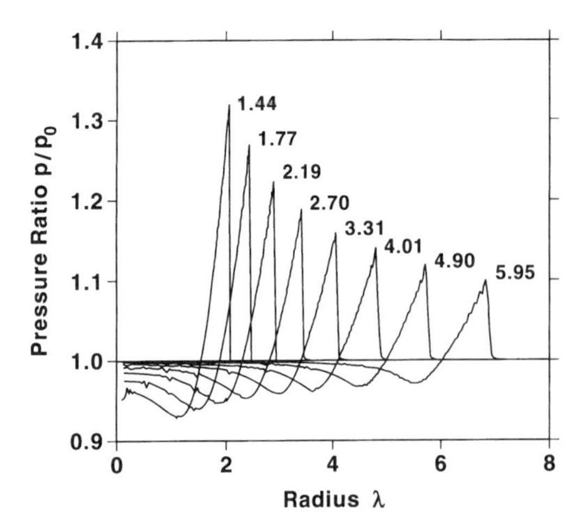

Fig. 6 Pressure ratio p/p_0 versus radius λ for a line source of energy in an ideal gas at indicated times after energy release. (From M. N. Plooster, "Shock Waves from Line Sources. Numerical Solutions and Experimental Measurements," *Phys. Fluids*, Vol. 13, 1970, pp. 2665–2675.)

4 NONLINEAR WAVE PROPAGATION OF PLANE, CYLINDRICAL, AND SPHERICAL WAVES

The development and propagation of sonic booms, ballistic waves from projectiles, and N-waves from weak sparks can usually be treated with nonlinear acoustic equations and with weak-shock theory.[1,2] The propagation of explosion waves can be treated in the same manner once the overpressure has decayed into the weak-shock region.

4.1 Nonlinear Propagation

If second-order terms are retained in the acoustic equations, the largest effect is the dependence of the sound velocity on amplitude, which may be expressed as

$$c \approx c_0 + \frac{\beta p'}{\rho_0 c_0} \tag{24}$$

for plane waves, where c_0 is the small-amplitude velocity. Here, $\beta = 1 + B/2A$, where

$$A = \left(\frac{\rho \, \partial p}{\partial \rho} \right)_0 \quad \text{and} \quad B = \left(\rho^2 \frac{\partial^2 p}{\partial \rho^2} \right)_0 .$$

For an ideal gas, $\beta = (\gamma + 1)/2$, where γ is the ratio of specific heats.

The dependence of sound speed on amplitude leads to the distortion of the waveform as it propagates. For plane waves, this distortion can be expressed parametrically by

$$p(x, t) = g(\psi), \tag{25}$$

$$t = \psi + \frac{x}{c_0} - \frac{x}{c_0^2} \frac{\beta g(\psi)}{\rho_0 c_0}, \tag{26}$$

where $g(t)$ is $p(0, t)$ and t is ψ when $x = 0$.

The distortion is illustrated in Fig. 2 of Chapter 17 for a single-cycle wave. The slope of a waveform dp/dx will become infinite at time

$$t = \frac{\rho_0 c_0^2}{\beta (dp/dt)_{\max}}, \tag{27}$$

where $(dp/dt)_{\max}$ is the largest negative slope of the initial waveform. At this time weak-shock theory must be employed to determine the subsequent propagation of the wave.

The nonlinear equations may be generalized to different geometries and to a refractive atmosphere through the definition of the age variable. In this construction it is assumed that nonlinear effects do not appreciably affect geometric acoustic ray directions and ray tube areas. Again, the principal effect is the distortion of the waveform. The strength of the wave and its distortion are inversely proportional to the square root of the ray tube area. The geometric decay of nonlinear waves in cylindrical and spherical geometries may be derived by combining the linear geometric decay with the change in travel times due to nonlinear effects.

The linear decay of a wave is expressed by

$$p = \left[\frac{A(r_0)}{A(r)} \right]^{1/2} g[t - \tau(r)], \tag{28}$$

where $A(r)$ is the ray tube area and $\tau(r)$ is the linear ray travel time from r_0 to r, and

$$c = c_0 + \frac{\beta p}{\rho_0 c_0} = \left(\frac{d\tau}{dr} \right)^{-1} + \left(\frac{d\mathcal{A}}{dr} \right) \left(\frac{d\tau}{dr} \right)^{-2} g(\psi), \tag{29}$$

where the age variable $\mathcal{A}(r)$ is given by

$$\mathcal{A} = \int_{r_0}^{r} \beta \left[\frac{A(r_0)}{A(r)} \right]^{1/2} \frac{\hat{e}_{\text{ray}} \cdot \hat{n}}{\rho_0 c_0 (c + \mathbf{v} \cdot \hat{n})^2} \, dr$$

$$= \frac{\beta}{\rho_0 c_0^3} \int_{r_0}^{r} \left[\frac{A(r_0)}{A(r)} \right] dr, \qquad (30)$$

where $\mathbf{v}$ is the wind velocity, $\hat{n}$ is the unit vector in the direction normal to the wavefront, and $\hat{e}_{\text{ray}}$ is the unit vector in the ray direction. The first integral in Eq. (30) is the general form for an inhomogeneous atmosphere with winds; the second applies only to a homogeneous quiescent atmosphere. If no shock forms, then $g(\psi)$ must be constant for an observer moving at the nonlinear velocity and the distortion can be determined from the parametric equation

$$t = \psi + \tau(r) - g(\psi)\mathcal{A}(l). \qquad (31)$$

This equation with Eq. (28) can be used to specify the distortion of cylindrical and spherical waves in a homogeneous atmosphere:

Cylindrical:

$$p = \left[\frac{r_0}{r} \right]^{1/2} g(\psi), \qquad (32)$$

$$t = \psi + \frac{r - r_0}{c_0} - 2g(\psi) \frac{\beta}{\rho_0 c_0^3} (\sqrt{r} - \sqrt{r_0}) \cdot \sqrt{r_0}. \qquad (33)$$

Spherical:

$$p = \frac{r_0}{r} g(\psi), \qquad (34)$$

$$t = \psi + \frac{r - r_0}{c_0} - g(\psi) \frac{\beta}{\rho_0 c_0^3} r_0 \ln\left(\frac{r}{r_0} \right). \qquad (35)$$

The distance required for an initial waveform to form a shock may also be calculated:

Cylindrical:

$$r = r_0 \left(1 + \frac{\rho_0 c_0^3 / \beta r_0}{(dp/dt)_{\max}} \right), \qquad (36)$$

Spherical:

$$r = r_0 \exp\left(\frac{\rho_0 c_0^3 / \beta r_0}{(dp/dt)_{\max}} \right). \qquad (37)$$

Weak-shock theory must be used once a shock forms.

4.2 Weak-Shock Theory

Nonlinear theory predicts the formation of multivalued waveforms. Since density and pressure are physical quantities, such waveforms are not physically possible. If the Rankine–Hugoniot equations are expanded for small density and pressure variations ($\Delta\rho/\rho < 0.1$), it can be shown that the entropy change across the shock is third order in the pressure fluctuation and the difference terms may be treated as isentropic derivatives yielding

$$u_{\text{sh}} - u_{\text{av}} = \pm c_{\text{av}} \qquad (38)$$

and

$$\Delta u = \pm \frac{\Delta p}{\rho_{\text{av}} c_{\text{av}}}. \qquad (39)$$

If we let $g(\psi_-)$ and $g(\psi_+)$ be the acoustic pressure behind and in front of the shock,

$$u_{\text{sh}} = c_0 + \tfrac{1}{2}\beta \, \frac{g(\psi_+) + g(\psi_-)}{\rho_0 c_0}, \qquad (40)$$

where $g(\psi)$ and ψ are defined in Eqs. (25) and (26). Comparison of this equation and Eq. (24) shows that the shock velocity is slower than the second-order acoustic velocity behind the shock and faster than the second-order velocity in front of the shock. The shock "takes in" points on the waveform in front of and behind the shock.

The position of the shock and behavior of the wave after a shock forms can be determined from the multivalued waveform by the equal-areas rule.[20] The equal-areas rule specifies that the area between the shock position and the multivalued predicted waveform must be zero:

$$A(t) = \int_{\psi_-}^{\psi_+} [x(\psi, t) - x_{\text{sh}}(t)] \frac{dg(\psi)}{d\psi} \, d\psi. \qquad (41)$$

See Fig. 4 of Chapter 17. This construction along with weak-shock theory in general is valid only until the shock becomes weak enough for dissipative effects to disperse the shock.

If weak-shock theory were uniformly valid, all impulses with a positive-overpressure front lobe and negative-overpressure rear lobe would form an N-shaped wave, that is, a front shock with positive overpressure $p_{\max}$ decaying linearly to a rear shock with negative pressure $-p_{\max}$. Sonic booms from aircraft, ballistic waves from bullets, and strong sparks often form N-waves.

Spherically spreading explosions and weak sparks decay too rapidly for the rear shock to form from the lower pressure and longer duration rear lobe of the pulse. The overpressure decay of the front lobe, however, can usually be treated by weak-shock decay of a triangular pulse.

We may use the equal-area rule to determine the decay and lengthening of an N-wave or a triangular pulse. The analysis of the corrected travel time along a path can be adapted for the analysis of N-waves by noting that the shock velocity is $c_0 + \beta p_{av}/\rho_0 c_0$.

An N-wave with initial overpressure $p(r_0)$ and duration $2T_0$ at r_0 will have duration $2T(r)$ and overpressure $p(r)$ at position r given by

$$T(r) = T_0 \left[1 + \frac{p(r_0)}{T_0} \mathcal{A}(r) \right]^{1/2}, \tag{42}$$

$$p(r) = \left[\frac{A(r_0)}{A(r)} \right]^{1/2} p(r_0) \left[1 + \frac{p(r_0)}{T_0} \mathcal{A}(r) \right]^{-1/2}. \tag{43}$$

The conserved quantity in the propagation of an N-wave is $P(r)T(r)[A(r)/(A(r_0))]^{1/2}$. The area under the N-wave is preserved as if the geometric spreading is taken into account. Specific forms are easily derived for plane, cylindrically, and spherically spreading waves:

Plane waves:

$$p(x) = p(x_0) \left[1 + \frac{\beta}{\rho_0 c_0^3} \frac{p(x_0)}{T_0} x \right]^{-1/2}, \tag{44}$$

$$T(x) = T_0 \left[1 + \frac{\beta}{\rho_0 c_0^3} \frac{p(x_0)}{T_0} x \right]^{1/2}. \tag{45}$$

Cylindrical waves:

$$p(r) = p(r_0) \sqrt{\frac{r_0}{r}} \cdot \left[1 + \frac{2\beta}{\rho_0 c_0^3} \frac{p(r_0)}{T_0} \sqrt{r_0}(\sqrt{r} - \sqrt{r_0}) \right]^{-1/2}, \tag{46}$$

$$T(r) = T_0 \left[1 + \frac{2\beta}{\rho_0 c_0^3} \frac{p(r_0)}{T_0} \sqrt{r_0}(\sqrt{r} - \sqrt{r_0}) \right]^{1/2}. \tag{47}$$

Spherical waves:

$$p(r) = \frac{r_0}{r} p(r_0) \left[1 + \frac{\beta}{\rho_0 c_0^3} \frac{p(r_0)}{T_0} r_0 \ln \frac{r}{r_0} \right]^{-1/2}, \tag{48}$$

$$T(r) = \frac{r_0}{r} T_0 \left[1 + \frac{\beta}{\rho_0 c_0^3} \frac{p(r_0)}{T_0} r_0 \ln \frac{r}{r_0} \right]^{1/2}. \tag{49}$$

Although weak-shock theory does not explicitly include any absorption mechanisms, these are included in the assumption that discontinuities are formed and that the Rankine–Hugoniot jump conditions hold.

The energy loss inherent in weak-shock propagation can be demonstrated using the N-wave propagation equations. The energy carried by the N-wave across a surface at distance r such that kr is large is given by

$$\begin{aligned} E(r) &= A(r) \int_{-\infty}^{\infty} \frac{p^2\, dt}{2\rho c} = \frac{A(r)}{3\rho c} p(r)^2 T(r) \\ &= E(r_0) \left[1 + \frac{p(r_0)}{T_0} \mathcal{A}(r) \right]^{-1/2}. \end{aligned} \tag{50}$$

The energy loss across the shock is not sensitive to the dissipative mechanism that produces the gradient. Only if the shock is dispersed are the details of the physical mechanism important.

The relationship between the duration and pressure of a spherically spreading N-wave [Eqs. (48) and (49)] has been exploited for the calibration of high-pressure wide-band capacitor microphones.[21,22] Wright has achieved calibration accurate to ±1.0 dB using 0.5–1.0 cm long sparks with 0.01–0.10 J energy discharge.

5 SOURCE MODELING OF BALLISTIC WAVES AND SONIC BOOMS

The shock wave from projectiles and sonic booms at moderate Mach numbers can be predicted using the non-linearization process described in the previous section. The generated shocks are sufficiently weak at low Mach numbers that the waveform can be calculated by the non-linear distortion of the wave predicted from linear theory.

An example of such a process is the prediction of the ballistic wave from supersonic thin cylindrical objects moving at velocity V.[1] The displacement of air by the body moving supersonically along the x-axis is equivalent to a distribution of monopole volume sources.

The result of this displacement is a wave with conical constant-phase surfaces. The ray paths are straight lines normal to the Mach cone with apex angle

$$\sin \theta_m = \frac{c_0}{V}. \tag{51}$$

Along the rays the linear solution can be put in the standard form for cylindrical propagation in terms of a reference distance s_0:

$$p = \left(\frac{s_0}{s}\right)^{1/2} g[t - \tau(l)], \tag{52}$$

where

$$\tau(l) = \frac{s}{c_0} + \frac{x_0}{V} \quad \text{and} \quad g(t) = \frac{\rho_0 V^2 \sqrt{m} F_w(Vt)}{\sqrt{m^2 - 1}\sqrt{2 s_0}}.$$

Here, F_w is the Whitham F-function

$$F_w(\xi) = \frac{1}{2\pi} \frac{d^2}{d\xi^2} \int_0^\infty \frac{A(\xi - \eta)}{\sqrt{\eta}}\, d\eta. \tag{53}$$

The nonlinear distortion of the wave can be calculated using Eqs. (28) and (30). The final waveform is calculated using the equal-areas rule.

The discussion above is restricted to steady supersonic motion of an axisymmetric body in a homogeneous quiescent atmosphere. Whitham[23] describes the extension of the F-function to include the effects of lift on a thin wing. Hayes[24,25] has incorporated the effect of an inhomogeneous atmosphere and winds on sonic booms into a computer program to predict sonic booms on the ground. If the ray tube area goes to zero, focusing occurs and diffraction corrections are necessary to predict the waveform. Recent work has applied improved numerical methods and computational fluid dynamics for the prediction of sonic booms for complicated aircraft shapes at high Mach numbers.

6 SPECIAL CASE: N-WAVE PROPAGATION FROM BALLISTIC WAVES AND SONIC BOOMS

If the attenuation mechanisms are weak so that the weak shocks are not dispersed, every waveform of finite duration and a positive pressure front lobe will asymptotically become an N-wave whose maximum pressure at cylindrical distance r is given as

$$p_{\max} = \frac{\rho_0 c_0^2 (M^2 - 1)^{1/2} S_{\max}^{1/2} K}{2^{1/4} \beta^{1/2} r^{3/4} L^{1/4}}, \tag{54}$$

where $S_{\max}$ is the maximum cross section of the body, L the length of the body, and

$$K^2 = \frac{\sqrt{L}}{S_{\max}} \max \int_{-\infty}^{\xi} F_w(\xi)\, d\xi, \tag{55}$$

where K depends only on the shape of the body.

The positive phase duration of the N-wave may be approximated as

$$T = \frac{2^{3/4} \sqrt{\beta} M S_{\max}^{1/2} K r^{1/4}}{L^{1/4} c_0 (M^2 - 1)^{3/8}}. \tag{56}$$

For these expressions to be valid, the weak shock must have formed an N-wave. Whether this occurs or not depends on the initial strength of the shock, the initial duration, and the spreading law. The predicted duration must be longer than the linear duration L/V for the N-wave to occur. Equations (54) and (56) may still be applied to a positive-pressure triangular pulse of a shock, even when the rear pulse has not formed a shock due to a longer initial duration.

DuMond, Cohen, Panofsky, and Deeds[26] present extensive measurements and weak-shock predictions for ballistic waves from a variety of small arms. The Army report, "Acoustical Considerations for a Silent Weapon's System: A Feasibility Study," contains a good review of the theory as well as measurements of the ballistic wave from a number of weapons.[27]

The shock wave from high-energy sparks and the ballistic wave from small arms typically form clean N-waves since they are relatively strong sources with small initial duration. Sonic booms from fighter aircraft usually form N-waves in spite of their long initial duration since the wave propagates a long distance while maintaining the weak-shock conditions. The sonic boom from large aircraft do not reach their asymptotic form by the time they propagate to the ground due to their large initial duration.

7 SOURCE MODELING OF FINITE-LENGTH SPARKS AND LIGHTNING

The strength and duration of a shock wave from a spherically symmetric or cylindrically symmetric source can be used to approximate the energy released. Detailed numerical calculations can be used to infer information about energy relations in the source region.[16]

For finite-length sparks, the shock development is primarily cylindrical close to the spark but becomes spherical at long distances from the spark. The study of thunder recorded at a distance as a diagnostic of energy released in lightning requires assumptions as to the nature of the propagation at long distances.

Wright and Medendorp[28] developed a model for a finite-length spark using a superposition of spherically propagating N-waves. They had good success reproducing the angular dependence of the waveshape on the propagation angle. They adjusted the duration of the spherical N-waves to best reproduce the measured data. The success of this linear model for shock propagation is due to the weak sparks used in the study (0.01–0.10 J).

Ribner and Roy[29] applied the Wright and Medendorp model for finite-length sparks to simulate thunder. They summed spherically spreading Wright–Medendorp waves emitted from zig-zag segments of a simulated lightning channel. These summations reproduce thunder that sounds realistic.

Few[30,31] modeled the shock development as cylindrical out to the relaxation length, then as spherical weak decay. The relaxation lengths that relate the spherical and cylindrical decay are defined as

$$r_0 = \left(\frac{E_t}{\frac{4\pi}{3} p_0} \right)^{1/3} \quad \text{(spherical)} \tag{57}$$

and

$$r_0 = \left(\frac{E_l}{\pi p_0} \right)^{1/2} \quad \text{(cylindrical)}, \tag{58}$$

where E_t is the total energy release and E_l is the energy release per unit length. Few argues that the tortuous nature of the lightning channel leads to spherical decay at ranges beyond one relaxation length and that the duration of the pulses are given by $2.6r_0$.

Plooster[17] and Bass[32] have modeled the propagation of thunder from lightning as finite-wave propagation from a cylindrical source. Since a cylindrical wave undergoes less decay with distance and greater finite-wave distortion, this approach leads to smaller estimates of the energy per unit length of the lightning. Few's estimates are usually on the order of 10^5 J/m while Bass and Plooster's estimates using cylindrical wave properties are two orders of magnitude smaller.

REFERENCES

1. A. D. Pierce, *Acoustics, an Introduction to Its Physical Principles and Applications*, Acoustical Society of America, Woodbury, NY, 1989.
2. G. B. Whitham, *Linear and Nonlinear Waves*, Wiley-Interscience, New York, 1974.
3. J. Von Neumann, "The Point Source Solution," in A. H. Taub (Ed.), *The Collected Works of John Von Neumann*, Vol. VI, Pergamon, New York, 1963, pp. 219–237.
4. G. I. Taylor, "The Formation of a Blast Wave by a Very Intense Explosion: I. Theoretical Discussion," *Proc. Roy Soc. A*, Vol. 201, 1950, pp. 159–174.
5. L. I. Sedov, *Similarity and Dimensional Methods in Mechanics*, Academic, New York, 1961.
6. W. E. Baker, *Explosions in Air*, University of Texas Press, Austin, TX, 1973.
7. H. L. Brode, "Numerical Solutions of Spherical Blast Waves," *J. Appl. Phys.*, Vol. 26, 1955, pp. 766–775.
8. H. L. Brode, "Blast Wave from a Spherical Charge," *Phys. Fluids*, Vol. 2, 1959, pp. 217–229.
9. "Estimating Air Blast Characteristics for Single Point Atmospheric Propagation and Effects," ANSI S2.20–1983, American National Standards Institute.
10. R. Raspet, J. Ezell, and S. V. Coggeshall, "Diffraction of an Explosive Transient," *J. Acoust. Soc. Am.*, Vol. 79, 1985, pp. 1326–1334.
11. H. E. Bass, J. Ezell, and R. Raspet, "Effect of Vibrational Relaxation on the Atmospheric Attenuation and Rise Time of Explosion Waves," *J. Acoust. Soc. Am.*, Vol. 74, 1983, pp. 1514–1517.
12. R. Raspet, H. E. Bass, and J. Ezell, "Effect on Finite Ground Impedance on the Propagation of Acoustic Pulses," *J. Acoust. Soc. Am.*, Vol. 74, 1983, pp. 267–274.
13. J. W. Reed, "Atmospheric Attenuation of Explosion Waves," *J. Acoust. Soc. Am.*, Vol. 61, 1977, pp. 39–47.
14. S-C. Lin, "Cylindrical Shock Waves Produced by Instantaneous Energy Release," *J. Appl. Phys.*, Vol. 25, 1954, pp. 54–57.
15. M. N. Plooster, "Shock Waves from Line Sources. Numerical Solutions and Experimental Measurements," *Phys. Fluids*, Vol. 13, 1970, pp. 2665–2675.
16. M. N. Plooster, "Numerical Simulation of Spark Discharges in Air," *Phys. Fluids*, Vol. 14, 1971, pp. 2111–2123.
17. M. N. Plooster, "Numerical Model of the Return Strokes of the Lightning Discharge," *Phys. Fluids*, Vol. 14, 1971, pp. 2124–2133.
18. C. A. Rouse, "Theoretical Analysis of the Hydrodynamic Flow in Exploding Wire Phenomena," in W. G. Chace and H. K. Moore (Eds.), *Exploding Wires*, Plenum, New York, 1959.
19. A. Sakurai, "On the Propagation of Cylindrical Shock Waves," in W. G. Chace and H. K. Moore (Eds.), *Exploding Wires*, Plenum, New York, 1959.

20. L. D. Landau, "On Shock Waves at Large Distances from the Place of Their Origin," *U.S.S.R. J. Phys.*, Vol. 9, 1945, pp. 496–503.

21. W. M. Wright, "Propagation in Air of N-Waves Produced by Sparks," *J. Acoust. Soc. Am.*, Vol. 73, 1983, pp. 1948–1955.

22. B. A. Davy and D. T. Blackstock, "Measurement of the Refraction and Diffraction of a Short N Wave by a Gas Filled Soap Bubble," *J. Acoust. Soc. Am.*, Vol. 49, 1971, pp. 732–737.

23. G. B. Whitham, "On the Propagation of Weak Shock Waves," *J. Fluid Mech.*, Vol. 1, 1956, pp. 291–318.

24. W. D. Hayes, R. C. Haefeli, and H. E. Kulsrud, "Sonic Boom Propagation in a Stratified Atmosphere with Computer Program," NASA CR-1299, 1969.

25. W. D. Hayes and H. L. Runyan, "Sonic Boom Propagation through a Stratified Atmosphere," *J. Acoust. Soc. Am.*, Vol. 51, 1972, pp. 695–701.

26. W. M. Dumond, E. R. Cohen, W. K. H. Panofsky, and E. Deeds, "A Determination of the Wave Forms and Laws of Propagation and Dissipation of Ballistic Shock Waves," *J. Acoust. Soc. Am.*, Vol. 18, 1946, pp. 97–118.

27. G. R. Garinther and J. B. Moreland, "Acoustical Consideration for a Silent Weapon System: A Feasibility Study," Human Engineering Laboratories, Technical Memorandum 10-66, Defense Documentation Center, Cameron Station, Alexandria, VA, AD521 727/H, 1966.

28. W. M. Wright and N. W. Medendorp, "Acoustic Radiation for a Finite Line Source with N-Wave Excitation," *J. Acoust. Soc. Am.*, Vol. 43, 1968, pp. 966–971.

29. H. S. Ribner and D. Roy, "Acoustics of Thunder: A Quasilinear Model for Tortuous Lightning," *J. Acoust. Soc. Am.*, Vol. 72, 1982, pp. 1911–1925.

30. A. A. Few, "Power Spectrum of Thunder," *J. Geophys. Res.*, Vol. 74, 1969, pp. 6926–6934.

31. A. A. Few, "Acoustic Radiation from Lightning," in H. Volland (Ed.), *CRC Handbook of Atmospherics*, CRC Press, Boca Raton, FL 1982, Vol. 2, pp. 257–290.

32. H. E. Bass, "The Propagation of Thunder Through the Atmosphere," *J. Acoust. Soc. Am.*, Vol. 67, 1980, pp. 1959–1966.

28

ATMOSPHERIC SOUND PROPAGATION

LOUIS C. SUTHERLAND AND GILLES A. DAIGLE

1 INTRODUCTION

This chapter summarizes the current state of knowledge in outdoor sound propagation, including geometric spreading and atmospheric absorption, propagation over ground, reflection and diffraction by obstacles, and refraction by a stationary but nonhomogeneous atmosphere. The important influence of turbulence on sound propagation in the atmosphere and the related topic of acoustic sounding of the atmosphere are treated at the end of the chapter. Propagation losses due to finite-amplitude effects are treated in Part II.

Recent reviews[1–3] reflect the great strides made in this field since the late 1960s, primarily in the area of atmospheric absorption, ground and terrain effects, numerical evaluation of propagation in nonhomogeneous media, and the development in 1968[1] of acoustic sounders (sodar) to probe the atmosphere acoustically.

2 SPREADING LOSSES IN OUTDOOR SOUND PROPAGATION

For this chapter, it will be convenient to describe the total attenuation A_T, in decibels, due to sound propagation as the sound level at a source minus the level at a receiver:

$$A_T = L_{ps} - L_{pr} = 20 \log \left[\frac{p(s)}{p(r)} \right], \qquad (1)$$

where L_{ps} is the sound pressure level with a root-mean-square (rms) sound pressure $p(s)$ at a distance s near the source and L_{pr} is the corresponding sound pressure level with an rms sound pressure $p(r)$ at the receiver a distance r from the source.

This attenuation, normally a positive quantity, will be expressed as the sum of three nominally independent terms:

$$A_T = A_s + A_a + A_e, \qquad (2)$$

where A_s is the attenuation due to geometric spreading, A_a is the attenuation due to atmospheric absorption, and A_e is the excess attenuation due to all other effects including attenuation A_g by the ground in a homogeneous atmosphere, refraction by a nonhomogeneous atmosphere, attenuation by diffraction and reflection by a barrier, and scattering or diffraction effects due to turbulence.

The "excess attenuation" term A_e is not subdivided into the individual elements suggested by its description since they may not all be present or may not necessarily act independently.

A general expression for the spreading loss A_s, in decibels, between any two positions at distances r_1, r_2 from an acoustic source can be given in the form

$$A_s = 20g \log \left(\frac{r_2}{r_1} \right), \qquad (3)$$

where r_2, r_1 are the distances between the acoustic center of the source and the farthest (i.e., receiver) and closest (i.e., source) positions, respectively, and $g = 0$ for plane-wave propagation such as within a uniform pipe (i.e., no spreading loss), $g = \frac{1}{2}$ for cylindrical propagation from a line source, and $g = 1$ for spherical wave propagation from a point source.

Note: References to chapters appearing only in the *Encyclopedia* are preceded by *Enc.*

The latter two conditions correspond to the commonly specified condition of 3 and 6 dB loss (respectively) per doubling of distance from the source.

Finite arrays of stationary, incoherent sources are commonly utilized to model community noise levels from large sources such as highways, industrial plants, or broad distributions of ambient noise sources. The spreading losses for these cases can be conveniently evaluated using linear or planar arrays of such incoherent sources in the form of finite line, circular, or rectangular source arrays[4] or infinite source arrays to model the horizontal[5] and vertical[6] distribution of ambient noise in a community. Prediction models have also been developed, empirically and experimentally, to describe sound propagation in urban, built-up areas from motor vehicles[7] and V/STOL aircraft such as helicopters.[8]

As discussed in Section 5, major deviations from these simple spreading loss models occur outdoors when nonuniformity of the atmosphere is considered. For example, downwind propagation from sources of very low-frequency (e.g., <25 Hz) energy, such as wind turbines,[9] can, under some conditions, exhibit a spreading loss corresponding to a line, instead of a point, source.

3 ATTENUATION OF OUTDOOR SOUND BY ATMOSPHERIC ABSORPTION

A sound wave traveling through air free of any particles is attenuated due to atmospheric absorption caused by (1) classical (heat conduction and shear viscosity) losses and (2) molecular relaxation losses associated with an exchange between molecular translational and molecular rotational or vibration energy. These loss components vary with temperature and atmospheric pressure and, for molecular vibrational relaxation, with humidity content.

The theoretical foundation[2] for atmospheric absorption by Knudsen and Kneser has been refined by recent experimental data to define two key humidity- and temperature-dependent parameters: the vibrational relaxation frequencies for oxygen and nitrogen. This combination of theory and experimentally based algorithms now provides a firm basis for standard expressions to accurately predict the attenuation coefficient for atmospheric absorption of pure-tone sounds.[10]

The attenuation A_a, in decibels, over a path length r, in meters, due only to atmospheric absorption can be expressed by

$$A_a = -20 \log \left[\frac{p(r)}{p(0)} \right] = -20 \log[\exp(-\alpha r)] = ar, \quad (4)$$

where $p(r)$ is the sound pressure after traveling the distance r, $p(0)$ is the initial sound pressure at $r = 0$, α is the attenuation coefficient in Nepers per meter, and a is the attenuation coefficient in decibels per meter ($= 8.686\alpha$).

While spreading losses are nominally independent of frequency and weather, atmospheric absorption losses are strongly dependent on these parameters. At long propagation distances and for high frequencies, atmospheric absorption is usually much greater than spreading losses for sound propagation outdoors.

The attenuation coefficient for pure tones is shown in Fig. 1 as a function of frequency, with relative humidity as a parameter, for a temperature of 20°C and a pressure of 1 standard atmosphere. Values for other atmospheric pressures can be obtained from this figure by employing the pressure scaling indicated by the abscissa and ordinate scales and the pressure-scaled relative humidity.[11] Values for this pure-tone attenuation coefficient, given in Table 1 for a limited range of frequencies, temperatures, and relative humidities, are readily computed for other conditions (i.e., <330 K and <2 atm) from the standard algorithms.[10]

The effective atmospheric attenuation of a constant-percentage band of a broadband noise is normally less than for a pure-tone sound due to the finite bandwidth and slope of the filter skirts. For a one-third octave band of noise with an exact midband frequency f_m, the atmospheric attenuation $A_{ab}(f_m)$ in decibels is closely approximated by the following expression. It is a theoretically based, empirically refined function of the total pure-tone atmospheric attenuation $A_{at}(f_m)$ at this same midband frequency. The expression is invalid if $A_{at}(f_m)$ is more than about 50 dB[12]:

$$A_{ab}(f_m) = A_{at}(f_m)[1 + 0.00533[1 - 0.2303A_{at}(f_m)]]^{1.6}. \quad (5)$$

Some atmospheric attenuation also occurs in fog,[3] in dust in the air,[13] and at frequencies below about 10 Hz, from absorption due to electromagnetic radiation of moist air molecules.[14]

4 ATTENUATION OF OUTDOOR SOUND PROPAGATION OVER THE GROUND

Sound propagation in a still, uniform atmosphere over a path near the ground surface, such as illustrated in Fig. 2, is changed by the interaction between the direct source-to-receiver sound over path r_1 and sound received from the ground surface. The latter consists primarily of ground-reflected sound over the path r_2. For a spheri-

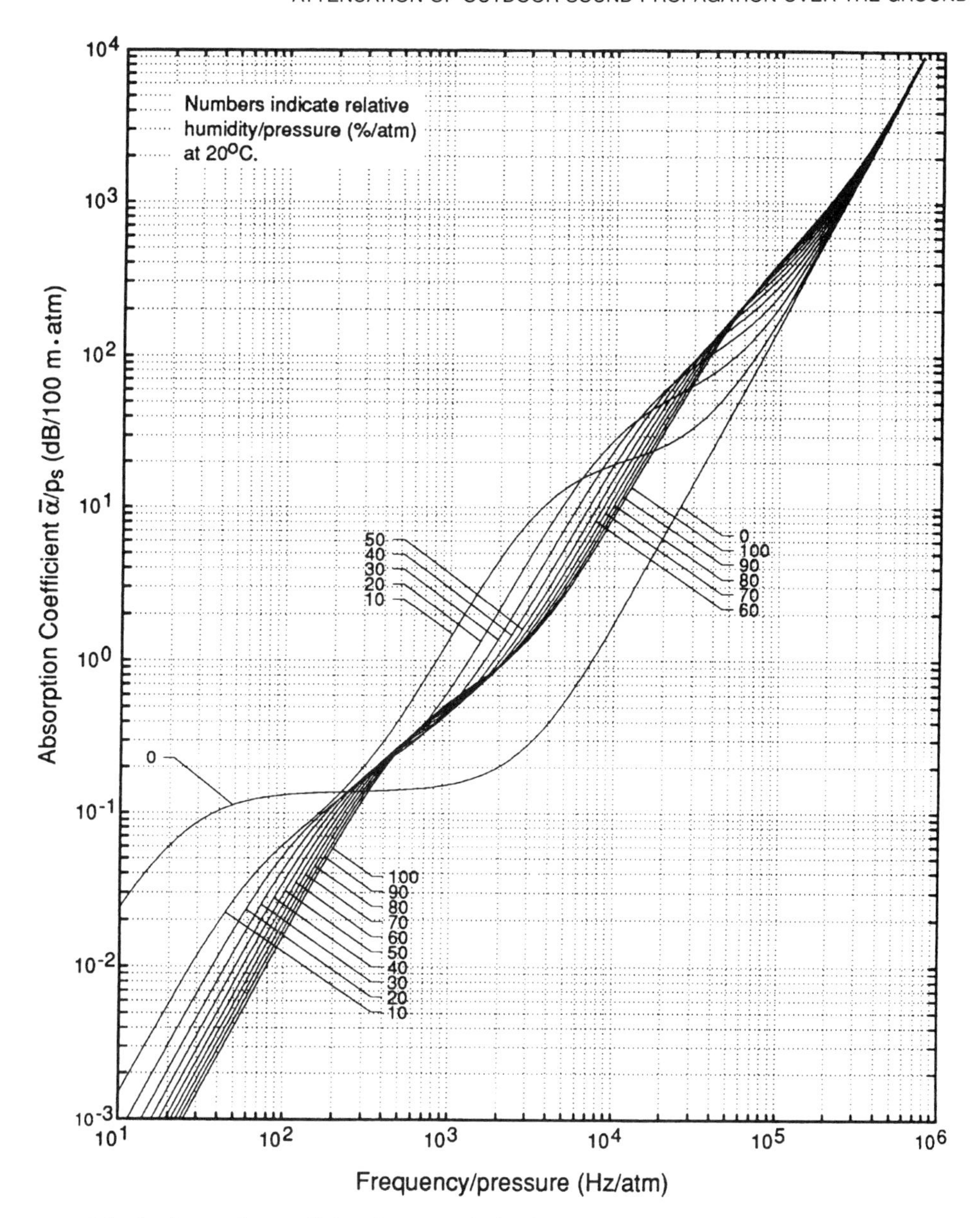

Fig. 1 Attenuation coefficient for atmospheric absorption as a function of frequency and relative humidity. All parameters are scaled by atmospheric pressure so the chart may be used for any pressure within linear limits of the perfect-gas law. (From Ref. 11 with permission.)

cal wavefront, this is augmented by a ground wave and, under some circumstances, by a surface wave. This section treats (1) acoustic impedance of the ground and (2) propagation of spherical sound waves over such ground.

4.1 Acoustic Impedance of Ground Surfaces

The specific acoustic impedance at a ground surface for normal incidence Z_2 is the complex ratio of the acoustic pressure at the surface and the resulting nor-

TABLE 1 Atmospheric Attenuation Coefficient *a* (dB/km) at Selected Preferred Frequencies[a]

Temperature	Relative Humidity (%)	62.5 Hz	125 Hz	250 Hz	500 Hz	1000 Hz	2000 Hz	4000 Hz	8000 Hz
30.0°C	10	0.362	0.958	1.82	3.40	8.67	28.5	96.0	260
(86 °F)	20	0.212	0.725	1.87	3.41	6.00	14.5	47.1	165
	30	0.147	0.543	1.68	3.67	6.15	11.8	32.7	113
	50	0.091	0.351	1.25	3.57	7.03	11.7	24.5	73.1
	70	0.065	0.256	0.963	3.14	7.41	12.7	23.1	59.3
	90	0.051	0.202	0.775	2.71	7.32	13.8	23.5	53.3
20.0°C	10	0.370	0.775	1.58	4.25	14.1	45.3	109	175
(68 °F)	20	0.260	0.712	1.39	2.60	6.53	21.5	74.1	215
	30	0.192	0.615	1.42	2.52	5.01	14.1	48.5	166
	50	0.123	0.445	1.32	2.73	4.66	9.86	29.4	104
	70	0.090	0.339	1.13	2.80	4.98	9.02	22.9	76.6
	90	0.071	0.272	0.966	2.71	5.30	9.06	20.2	62.6
10.0°C	10	0.342	0.788	2.29	7.52	21.6	42.3	57.3	69.4
(50 °F)	20	0.271	0.579	1.20	3.27	11.0	36.2	91.5	154
	30	0.225	0.551	1.05	2.28	6.77	23.5	76.6	187
	50	0.160	0.486	1.05	1.90	4.26	13.2	46.7	155
	70	0.122	0.411	1.04	1.93	3.66	9.66	32.8	117
	90	0.097	0.348	0.996	2.00	3.54	8.14	25.7	92.4
0.0°C	10	0.424	1.30	4.00	9.25	14.0	16.6	19.0	26.4
(32 °F)	20	0.256	0.614	1.85	6.16	17.7	34.6	47.0	58.1
	30	0.219	0.469	1.17	3.73	12.7	36.0	69.0	95.2
	50	0.181	0.411	0.821	2.08	6.83	23.8	71.0	147
	70	0.151	0.390	0.763	1.61	4.64	16.1	55.5	153
	90	0.127	0.367	0.760	1.45	3.66	12.1	43.2	138

[a]Values actually computed at the exact midband frequency $f_m = 1000 \times 10^{0.3n}$, where the integer n varies from -4 to $+3$.

mal component of particle velocity into the ground. For a semi-infinite media, this specific acoustic impedance for normal incidence is the same as the characteristic impedance Z_c throughout the medium and is expressed, here, in terms of its normalized value $Z_2/\rho_0 c_0$, where $\rho_0 c_0$ is the characteristic impedance of air. In this chapter, this normalized value will be referred to as simply the surface impedance Z_s.

The characteristic impedance for a medium with a

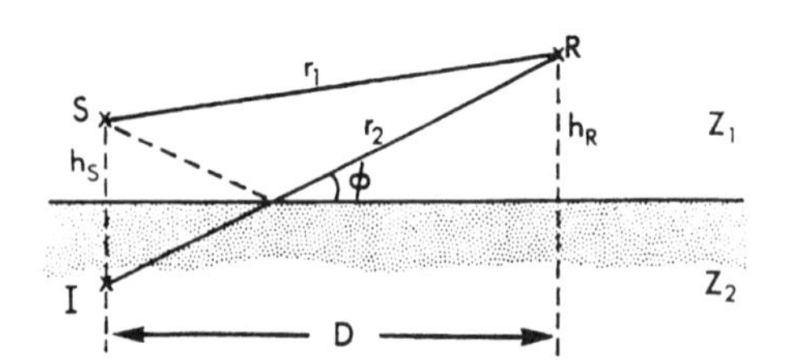

Fig. 2 Geometry of sound propagation over the ground illustrating the source (S), image source (I), and the direct (r_1) and reflected (r_2) path lengths to the receiver (R).

resistance to steady flow is a convenient first approximation to the surface impedance of a semi-infinite rigid-frame porous medium, a simple ground surface model. For such a medium with a density and speed of sound in air of ρ_0 and c_0 and flow resistivity σ in the medium at a frequency f, the characteristic impedance Z_c and acoustic propagation constant $k_b = \beta + i\alpha$ are given by[15]

$$Z_c = \rho_0 c_0 \left[1 + i \left(\frac{\sigma}{2\pi f \rho_0} \right) \right]^{1/2} \tag{6a}$$

and

$$k_b = \frac{2\pi f}{c} \left[1 + i \left(\frac{\sigma}{2\pi f \rho_0} \right) \right]^{1/2}. \tag{6b}$$

Transforming the complex roots, the characteristic impedance, normalized by the characteristic impedance of air, is

$$\frac{Z_c}{\rho_0 c_0} = [\tfrac{1}{2}(A+1)]^{1/2} + i[\tfrac{1}{2}(A-1)]^{1/2}, \qquad (7)$$

where $A = [1 + (2\pi f \rho_0/\sigma)^{-2}]^{1/2}$.

For time dependence in the form $\exp(-i2\pi f t)$, the reactive part of the impedance is positive, corresponding to a stiffness or spring reactance.[16] This "acoustic spring" represents the compressible air in ground surface pores.

A dimensionless frequency $f\rho_0/\sigma$ appears as the key scaling parameter in this approximation for the characteristic impedance and propagation constant for porous ground. While the general form of Eq. (7) is found in more complex models for propagation through porous media,[15] other parameters, as indicated later, besides flow resistivity are required to more accurately define surface impedance for many ground surfaces. However, flow resistivity retains its dominant influence on the frequency variation of impedance.

In a benchmark study, Delany and Bazley[17] found that the measured characteristic impedance Z_c and acoustic propagation constant k_b of a wide range of absorbent, porous materials could be defined in terms of the same dimensionless frequency $f\rho_0/\sigma$ found in Eq. (7). Using a standard value for ρ_0 of 1.205 kg/m^3, they developed the following expressions to define the measured characteristic impedance and propagation constant of these materials in terms of the ratio of frequency f, in hertz, to flow resistivity σ, in Pa · s/m^2 (MKS Rayls/m):

$$\frac{Z_c}{\rho_0 c_0} = \left[1 + 0.0511\left(\frac{f}{\sigma}\right)^{-0.75}\right] + i\left[0.0768\left(\frac{f}{\sigma}\right)^{-0.73}\right], \qquad (8a)$$

$$k_b = \frac{2\pi f}{c}\left\{\left[1 + 0.0858\left(\frac{f}{\sigma}\right)^{-0.70}\right] + i\left[0.175\left(\frac{f}{\sigma}\right)^{-0.59}\right]\right\}. \qquad (8b)$$

Although their experimental data fit these empirical expressions well for f/σ from 0.01 to 1.0 m^3/kg, Delany and Bazley cautioned that outside this range other power law relationships may be required. Nevertheless, Delany and Bazley as well as Chessell[18] subsequently found that Eq. (8a) provides a reasonable first approximation to surface impedance in their separate evaluations of ground effects on propagation of aircraft noise over grassy terrains. For such terrains, f/σ varies from approximately 10^{-4} to 0.1 m^3/kg.

Delany and Bazley assumed that the flow resistivity σ of their materials was actually an effective value σ_e equal to the "DC" value σ multiplied by the porosity Ω. The latter is typically about 0.4–0.7 for grass-covered ground surfaces.[19]

An alternative model by Miki[20] essentially eliminates one problem with the Delany–Bazley model, which predicts, at low frequencies, the physically unrealizable condition of a negative-resistance component for a thin, hard-backed layer for most ground surfaces.[21]

Subject to these limitations, the surface impedance for a semi-infinite ground can be estimated with Eq. (8a) given suitable values for the effective flow resistivity. Table 2 lists published values from a number of sources (e.g., Refs. 22–25) for empirically determined effective flow resistivity and measured or computed values of porosity for a full range of ground surfaces. Effective flow resistivities that best fit ground attenuation data are less than directly measured values for grassy surfaces, but the opposite is true for relatively homogeneous ground surfaces such as sand, silty soil, and snow. This supports the need for more complex models for surface impedance for the general case.[26]

Two examples of these more accurate models for the surface impedance of the ground are provided in the following.

Four-Parameter Model This general model has been employed in a benchmark paper presenting a comparison of sound propagation computation programs (see Ref. 51). With this model, the characteristic impedance Z_c (and hence the surface impedance for a homogeneous, locally reacting media) and acoustic propagation constant k_b can be expressed as

$$\frac{Z_c}{\rho_0 c_0} = \frac{q/\Omega}{B^{1/2}C^{1/2}} \quad \text{and} \quad k_b = \frac{2\pi f q}{c_0}\frac{C^{1/2}}{B^{1/2}}, \qquad (9)$$

where

$$B = \left[1 - \frac{1}{\epsilon\sqrt{i}}\,T(2\epsilon\sqrt{i})\right],$$

$$C = \left[1 + \frac{\gamma - 1}{\epsilon\sqrt{iN_{\text{pr}}}}\,T(2\epsilon\sqrt{iN_{\text{pr}}})\right]$$

and Ω is porosity, the ratio of air to total volume; $q^2 = \Omega^{-n'}$ a tortuosity parameter that accounts for the total

TABLE 2 Flow Resistivity and Porosity Data for Ground Surfaces[22–25]

Types of Ground	Effective Flow Resistivity,[a] kPa · s/m^2		
	Range	Average	Porosity
Upper limit[b]	2.5×10^5–25×10^5	800,000	
Concrete, painted	200,000	200,000	
Concrete, depends on finish	30,000–100,000	65,000	
Asphalt, old, sealed with dust	25,000–30,000	27,000	
Quarry dust, hard packed	5,000–20,000	12,500	
Asphalt, new, varies with particle size	5,000–15,000	10,000	
Dirt, exposed, rain-packed	4,000–8,000	6,000	
Dirt, old road, filled mesh	2,000–4,000	3,000	
Limestone chips, $\frac{1}{2}$–1-in. mesh	1,500–4,000	2,750	
Dirt, roadside with <4 in. rocks	300–800	550	
Sand, various types	40–906	317	0.35–0.47
Soil, various types	106–450	200	0.36–0.55
Grass lawn or grass field	125–300	200	0.48–0.55
Clay, dry (wheeled/unwheeled)	92–168	130	0.47–0.55
Grass field, 16.5% moisture content	75	75	
Forest floor (Pine/Hemlock)	20–80	50	
Grass field, 11.9% moisture content	41	41	
Snow, various types	1.3–50	29	0.56–0.76

[a]Effective resistivity value inferred from fitting measured ground attenuation data to predicted values based on Delany–Bazley impedance model and local reaction ground attenuation model (Ref. 22).
[b]Upper limit of effective flow resistivity due to thermal and viscous boundary layer, $= (1.16 \times 10^5) f^{1/3}$ (kPa · s/m^2) (an equivalent value, according to resistance portion of the Delany–Bazley impedance model; Ref. 22).

deviation of pore axes from a normal to the surface; $\epsilon = \sqrt{\pi/\Omega}(q/s_p)[f\rho_0/\sigma]^{1/2}$; $T(x) = J_1(x)/J_0(x)$ the ratio of Bessel functions of the first kind and order 1 and 0, respectively; n' a grain shape factor, typically taken to be 0.5; s_p a pore shape factor, typically 0.25 (these latter two parameters account for friction at the pore walls); N_{pr} the Prandtl number for air (0.71 at 15°C); and $\gamma = 1.4$ the ratio of specific heats for air.

Various approximations to Eq. (9) exist[26] for a restricted range of the same frequency parameter f/σ found in ϵ above and in Eq. (8).

Thin-Layer Model For a thin homogeneous layer (e.g., grassy sod) of thickness d over a rigid semi-infinite backing, the surface impedance Z_s is proportional to the characteristic impedance Z_c of the layer medium and is given by[15]

$$Z_s = Z_c \coth(-ik_b \cdot d) \tag{10}$$

where k_b is the acoustic propagation constant for the layer media.

Values for Z_c and k_b can be taken from Eq. (8) or (9).

In addition to applying prediction models, surface impedances for ground surfaces have also been obtained by (1) measurements with impedance tubes directed into the ground;[2] (2) calculation from standing-wave ratios,[2] Fourier spectra,[27] or phase gradients of interference patterns of ground-reflected sound;[25] or (3) trial-and-error calculation adjusting impedance parameters until predicted short-range ground attenuation matches measurements.[22–24] This latter semiempirical approach is considered a very practical way to determine surface impedance.

Published data on surface impedance for grassy surfaces obtained by the direct measurement methods are compared in Fig. 3 with predicted values for the characteristic impedance based on the Delany and Bazley[17] model. Predicted impedance values using Eq. (8a) for effective flow resistivities of 40 and 400 kPa · s/m^2 provide an approximate upper and lower bound for these measured surface impedance data from the different site–investigator data sets[27–31], identified in the figure.

4.2 Boundary Conditions at the Ground

Either local or extended reaction[15] conditions are normally assumed at a ground surface boundary. For a local reaction boundary, (1) the speed of sound in the ground

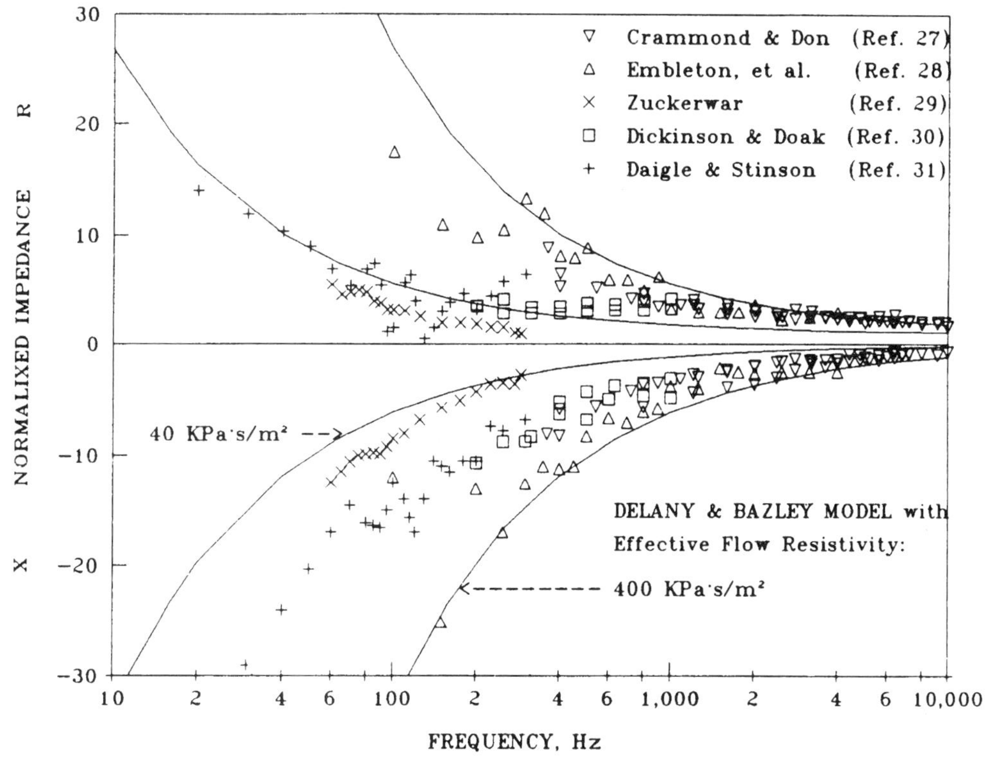

Fig. 3 Normal impedance measurements for grass surfaces compared to values predicted by the Delany–Bazley model. (Data from Refs. 27–31.)

surface is much less than the speed of sound in air, (2) the surface impedance is essentially independent of the incidence angle of any impinging sound wave, and (3) the dilatational sound waves transmitted into the ground travel normal to the surface. For an extended reaction boundary, the opposie is true and the refraction angle of the sound wave transmitted into the ground depends upon the incidence angle of the incident sound wave. While a locally reacting boundary model is usually suitable for predicting propagation over ground,[18] an extended reaction model may be more realistic when the acoustic surface properties vary significantly with depth or correspond to a medium with a relatively low flow resistance[32] or the "ground" is a body of water that has a much higher sound speed than air.

A third more complex impedance model treats the ground as an elastic medium capable of transmitting both acoustic dilatational and seismic shear waves from an acoustic wave impinging on the surface.[33] However, this complexity is normally not required for evaluation of sound propagation over the ground since the seismic impedance is 20–60 times greater than the acoustic impedance of the ground.

4.3 Attenuation of Spherical Acoustic Waves over Ground

According to Fig. 2, the sound pressure at the observer height h_r above the ground from a point source at a height h_s above the ground will be the sum of the direct sound pressure $p(r_1)[= P_d \exp(-kr_1)/r_1]$ from source to receiver over path r_1 and the indirect or ground-reflected sound pressure $p(r_2) = P_r \exp(-kr_2)/r_2$ nominally over the path r_2. A time variation $\exp(-i2\pi ft)$ is understood for both, and k is the wavenumber $2\pi f/c$. The indirect sound can be considered as originating from an image source located below the ground the same distance h_s as the actual source is above the ground. The ground attenuation A_g in decibels at the observer due solely to the presence of the ground surface can be expressed by[18]

$$A_g = 10\log\left[\frac{|P_d \exp(-kr_1)/r_1 + P_r \exp(-kr_2)/r_2|^2}{|P_d \exp(-kr_1)/r_1|^2}\right]$$

$$= 10\log\left[1 + \frac{r_1^2}{r_2^2}\,|Q_s|^2 + 2\,\frac{r_1}{r_2}\,|Q_s|C_c\right], \qquad (11)$$

where $Q_s = \overline{[p(r_2)/p(r_1)]} = |Q_s|e^{i\theta}$ is the spherical wave reflection factor with a magnitude $|Q_s|$ and phase angle θ, $C_c = \overline{p(r_2)p(r_1)}/[\overline{p^2(r_2)}\,\overline{p^2(r_1)}]^{1/2}$ is the correlation coefficient between the direct and ground "reflection," and the overbar denotes a time average.

For a wide-band random noise source with an approximately constant sound pressure spectrum level over a one-third octave frequency band with a bandwidth δf and midband frequency f_m, the correlation coefficient C_c is given by[18]

$$C_c = \frac{\sin[\mu\,\delta r\,f_m/c_0]}{[\mu\,\delta r\,f_m/c_0]}\cos\left[\tau\delta r\,\frac{f_m}{c_0} + \theta\right], \qquad (12)$$

where δr is the path length difference $(r_2 - r_1)$ between the direct and reflected wavefronts moving with the sound speed c_0. The phase shift in the latter, $[\tau\delta r(f_m/c_0) + \theta]$ is due to the path length difference δr and the phase angle θ of the spherical wave reflection factor. The parameter $\mu = \pi(\delta f)/f_m$ is 0.725 for a one-third octave band filter and 0 for a pure-tone source and $\tau = 2\pi[1 + (\delta f/2f_m)^2]^{1/2}$ is approximately 2π for a one-third octave band source and 2π for a pure tone.

Atmospheric turbulence can add an additional fluctuating random phase shift term to the argument of the cosine term in Eq. (12). For this condition, with large differences in path lengths, C_c can approach zero.

For a very hard ground, if the path difference δr is much less than the wavelength c_0/f, C_c approaches 1.0 and the excess ground attenuation A_g approaches −6 dB, equivalent to pressure doubling at a hard surface. If δr is much greater than a wavelength, C_c approaches 0 so the excess ground attenuation approaches −3 dB, equivalent to a simple energy summation of the two uncorrelated direct and reflected signals. As the incidence angle ϕ shown in Fig. 2 approaches 90°, the total phase shift between the direct and reflected signals [the argument of the cosine term in Eq. (12)] becomes dominated by the path length difference and is nearly independent of the surface impedance.

A "ground dip," or minimum, in excess ground attenuation occurs when the total phase shift between the direct and reflected signals (i.e., the argument of the cosine term in C_c) is equal to an odd multiple of π. As the incidence angle ϕ becomes very small, the path length difference δ_r also becomes small and the total phase shift is due almost entirely to the phase angle θ in the spherical wave reflection factor. The latter is dominated by the surface impedance, which therefore controls the frequency of the ground dip. This behavior is put to practical use to measure, indirectly, the surface impedance of the ground, given a model for the total ground attenuation pattern as

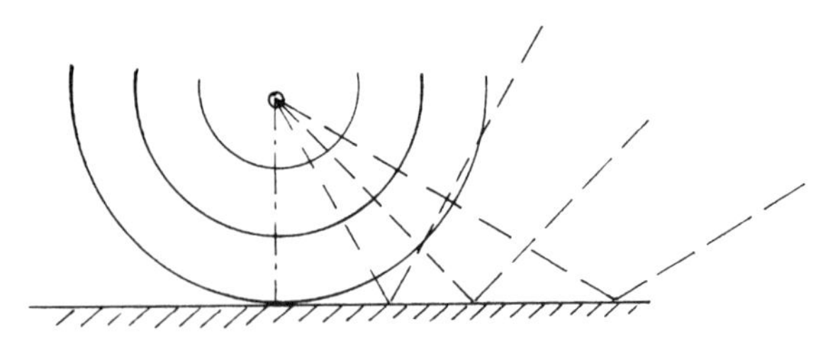

Fig. 4 Sketch illustrating varying incidence angle of spherical wavefront impinging on a plane surface.

a function of surface impedance parameters for a given source–receiver geometry.[22,34]

As shown in Fig. 4 by the straight sound ray paths from a point source in a uniform atmosphere, there is not a single incidence angle for all parts of the spherical wavefront striking the ground plane. The Weyl–Van der Pol solution provides a widely used approach to this classical problem of matching the boundary conditions for a spherical wavefront impinging on a plane, finite-impedance surface.[2,35] In this case, the spherical reflection factor Q_s is expressed as the sum of a plane-wave reflection factor R_p plus a boundary correction term $(1 - R_p)F(w)$. The added component provides the necessary correction to the reflected plane-wave component to match the boundary conditions of the spherical wavefront to the ground plane surface.[2] This component, called a ground wave, is governed by the same inverse-square spreading loss as the reflected wave but, its magnitude is proportional to the unreflected fraction $(1 - R_p)$ and, at large distances, has an additional inverse-square law loss of its own. Thus, the spherical reflection factor Q_s is given as

$$Q_s = R_p + (1 - R_p)F(w). \qquad (13)$$

The first application of this approach by Rudnick built on previous studies of the analogous electromagnetic field over a conducting surface.[2] The many subsequent solutions to the problem differ, for example, in how the spherical wavefront is synthesized, in the use of an extended or local reaction impedance model for the ground, or in the details of the complex contour integration involved.[2,35,36]

The plane-wave reflection coefficient term R_p is given by

$$R_p = \frac{Z_s \sin\phi - 1}{Z_s \sin\phi + 1}, \qquad (14)$$

where Z_s is the surface impedance defined earlier and ϕ is the incidence angle shown in Fig. 2.

The quantity $F(w)$, called the boundary loss factor by Rudnick,[2] is actually the first term of a more accurate asymptotic series solution[35] and is defined in terms of a complex complimentary error function $\text{erfc}(-iZ)$ by

$$F(w) = 1 + i[\pi w]^{1/2}e^{-w} \text{ erfc}(-iw^{1/2}), \tag{15}$$

where w is a numerical distance given, for a locally reacting medium, by[18]

$$w = i\,\frac{kr_2}{2}\,\frac{[\sin(\phi) + 1/Z_s]^2}{1 + (1/Z_s)\sin(\phi)}. \tag{16}$$

The denominator, in this expression $1 + (1/Z_s)\sin\phi$, is usually assumed equal to 1. This numerical distance, often expressed as $w^{1/2}$, can be considered a nondimensional distance r_2 in wavelengths modified by a function of the incidence angle ϕ and surface impedance Z_s. [Note that the positive square root of w is used in the argument $-iw^{1/2}$ for $\text{erfc}(-iw^{1/2})$]. Algorithms defined by Chien and Soroka[37] for computation of $F(w)$ are available in a more convenient form.[38] For an extended reaction surface, $F(w)$ must be multiplied by an additional correction term.[39]

The ground wave embodied in the boundary loss factor $F(w)$ is augmented by a surface wave that exists, as illustrated in Fig. 5, only between the flat ground and a conical surface (with an apex at the "image source" position) whose vertical projection defines the maximum value of the incidence angle for which a surface wave can occur. This angle is defined in terms of the following function of the surface impedance $|Z_s|e^{i\beta}$:

$$\sin(\phi) < \frac{[\text{Im}(Z_s) - \text{Re}(Z_s)]}{|Z_s|^2}. \tag{17}$$

For a grazing incidence wave with $\phi = 0$, this general condition requires only that the imaginary part, $\text{Im}(Z_s)$, of the surface impedance exceed the real part, $\text{Re}(Z_s)$. This criterion also applies for the existence of surface waves for propagation of near-grazing incidence plane waves traveling over an impedance ground.[40]

Over a range of distances (10–1000 m), frequencies (10–10,000 Hz), and flow resistivities (σ_e = 25–25,000 kPa · s/m^2) of practical interest, the magnitude of the numerical distance w will vary from less than 0.01 to greater than 100. Figure 6 shows the marked change in $20 \log|F(w)|$ over this range of $|w|$ with the surface impedance phase β as a parameter. For this plot, the values of $|w|$ were computed for the source and receiver on the surface of a local reaction ground.

The increase in the boundary loss factor $F(w)$ for $\beta > 45°$ is due to the addition of the surface wave component. For realistic values of β (< 90°), as $|w|$ becomes larger than about 10, this added surface wave decays exponentially along the ground to a value below the residual ground wave and the latter decreases by an additional 6 dB per doubling of $|w|$.

For very small values of $|w|$, $F(w)$ approaches 1.00, and the spherical reflection coefficient Q_s also approaches 1.0, so that the total sound pressure at a receiver is simply the direct signal plus a perfectly reflected signal modified in phase only by the path length difference δr and in magnitude by the different spreading loss (r_1/r_2). For nearly horizontal propagation, these quantities approach 0 and 1.0, respectively, resulting in the basic 6 dB pressure doubling as for a rigid plane. For $|w| \approx 1$, the ground wave and, when present, the surface wave component, will make a substantial difference in the total ground attenuation.

Surface wave propagation has been clearly demonstrated in laboratory studies and by a unique outdoor experiment with impulse sounds propagated over snow.[41] The important physical characteristics of surface waves observed in this study, confirming theory, are shown in Fig. 7 by the broad, delayed pulse following the initial impulse sound made up of the direct, reflected, and ground wave. This shows that the phase velocity of the

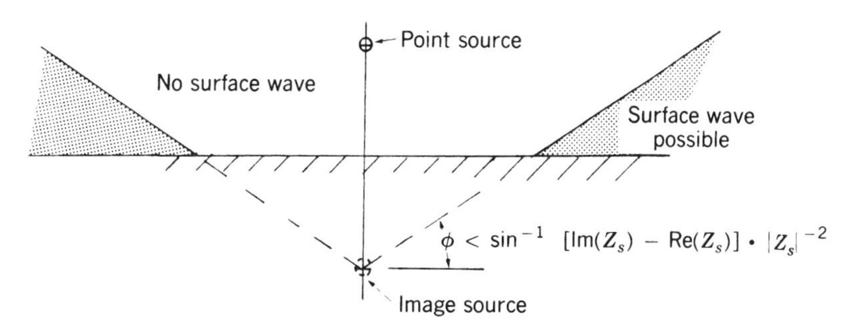

Fig. 5 Region where surface wave can exist. (Based on Ref. 40 with permission.)

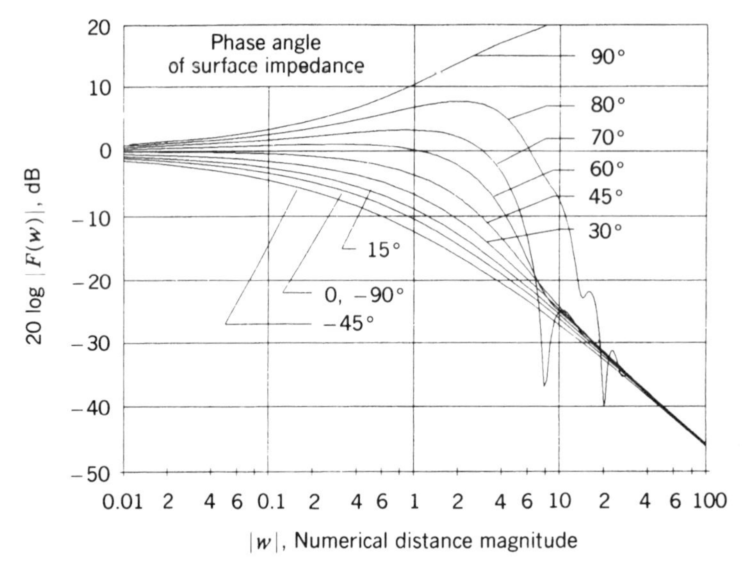

Fig. 6 Absolute magnitude of boundary loss factor $F(w)$ as a function of frequency with phase angle β of surface impedance as parameter. Source and receiver on surface of normally reacting ground.

surface wave is less than the speed of sound in air and the amplitude of the surface wave decreases exponentially[40] vertically and horizontally above the surface.

The separate wave components for propagation over the ground are illustrated by the classic long-range sound propagation data obtained by Parkin and Scholes.[42] Figure 8, from a previous review[2] of these data, breaks down the predicted contributions to the total ground attenuation by the direct (D) wave, the reflected (R) component, the ground (G) wave, and the surface (S) wave for three different distances representative of the range covered by the Parkin and Scholes data.

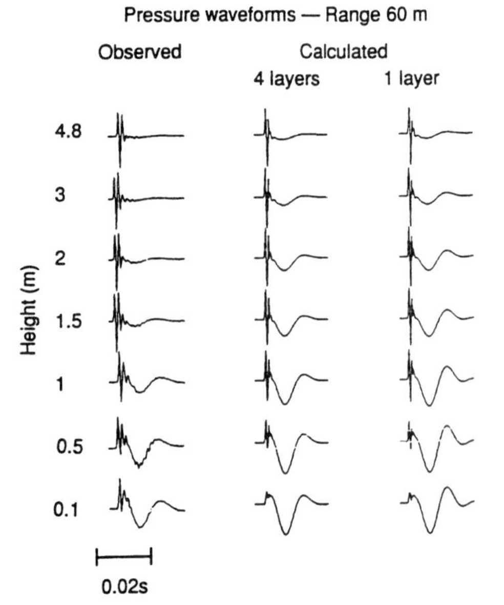

Fig. 7 Calculated and experimental measurements of surface waves over snow. (From Ref. 41.)

For a distance of 31.2 m from the source, there is no contribution by a surface (S) wave, and for frequencies greater than about 1 kHz, the minima are for path length differences that are approximately even multiples of one-half wavelength, indicating that the surface impedance is relatively low (e.g., soft surface) for this configuration and frequency range.

For a range of 125 m, there is a broad minimum (i.e., ground dip), just as for the shorter range, centered at about 500 Hz, which is also characteristic of the propagation for short distances over soft ground. It is the result of cancellation between direct and reflected waves caused primarily by the phase change on reflection.

The curves in Fig. 8 indicate that the ground dip broadens and deepens with increasing distance until most of the audible frequency range is included in this region of destructive interference (i.e., an acoustic "shadow"). The measurements are in reasonable agreement with theoretical predictions for 125 m using impedances obtained from a similar "soft" site.[2]

At 500 m, a further broadening of this ground dip or shadow zone extends to both higher and lower frequencies, which leaves most of the energy in the frequency

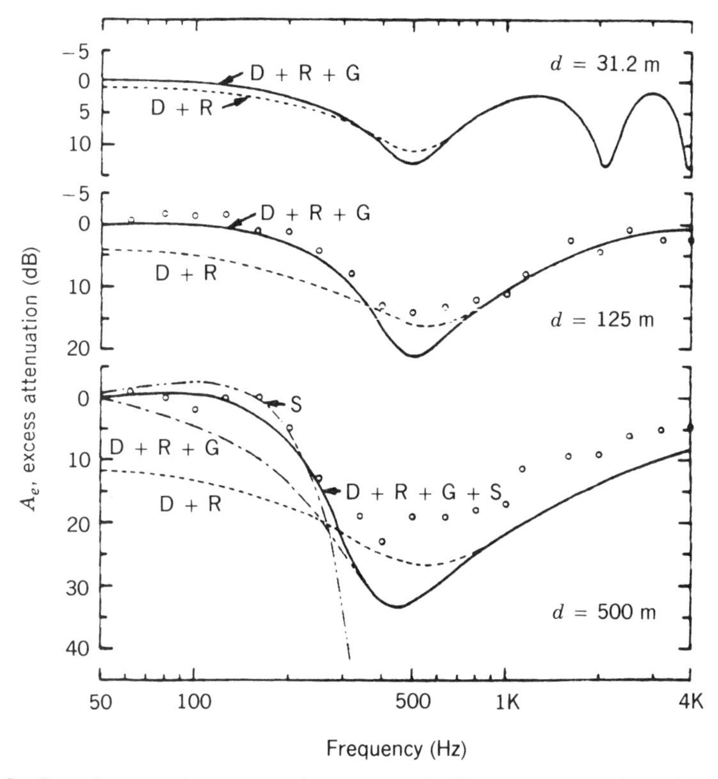

Fig. 8 Ground attenuation measured over several distances compared to theoretical prediction showing contributions by direct (D), plane-wave reflected (R), ground wave (G), and surface wave (S) components. (From Ref. 2 with permission.)

range 50–200 Hz in the surface wave. At high frequencies, the measured excess attenuation is consistently less than predicted due to turbulence effects discussed in Section 7.

4.4 Practical Examples of Ground Attenuation

To illustrate ground attenuation for a range of practical cases, values have been evaluated with Eqs. (8) and (10), assuming a homogeneous, locally reacting ground surface, for the following representative source heights h_s a receiver height h_r of 1.2 m and typical source–receiver distances encountered in the analysis of environmental sound:

Source	Source Height h_s, m	Receiver Height h_r, m
Tire noise	0.01	1.2
Passenger car engine exhaust noise	0.3	1.2
Truck exhaust noise (cab over engine) or aircraft engine exhaust (typical)	3.0	1.2

The resulting values for the ground attenuation are shown in Figs. 9*a*,*b* for distances from 15 to 1500 m for the first two of these three source heights for a typical grass surface with a flow resistivity of 250 kPa · s/m^2, approximately the same as for Fig. 8. The overall trend in Fig. 9 is similar to that of Fig. 8 but covers a wider range of source–receiver geometry. The ground attenuation for the higher source height of 3 m exhibited a pattern similar to that for the 0.3 m height but with less attenuation for the ground dip and a more complex pattern of constructive and destructive interference for the shortest distance of 15 m.

Significant departures from the curves in Fig. 9 can occur at high frequencies and long distances due to turbulence (see Section 7) when the excess attenuation due to interference between direct and reflected waves would otherwise be greater than about 20–25 dB, an approximate upper bound for total excess attenuation outdoors in a real atmosphere.

A limited examination of the effect of changing flow resistance is presented in Figs. 10*a*,*b* for a source height of 0.3 m and a range of 150 m. In Fig. 10*a* the ground is assumed to be homogeneous while in Fig. 10*b* a

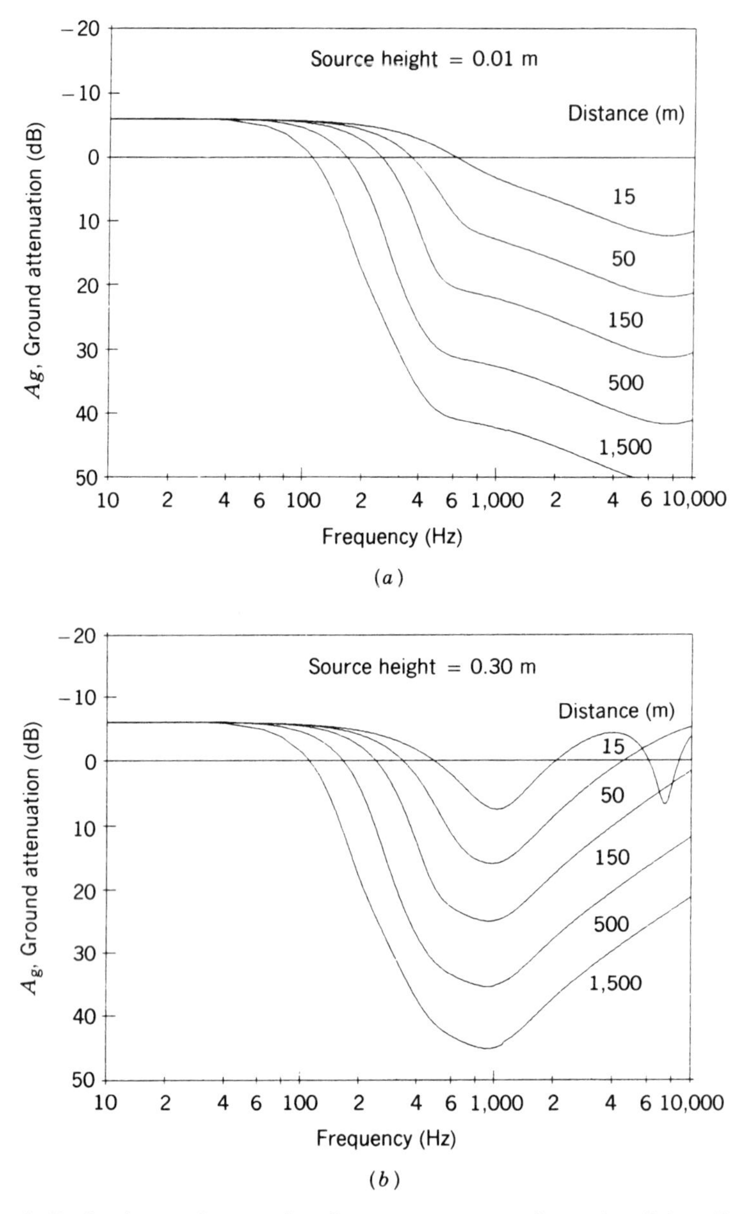

Fig. 9 Predicted ground attenuation for average grass surface using Delany–Bazley impedance model for receiver height h_r of 1.2 m, distances from 15 to 1500 m for source heights h_s of (*a*) 0.01 m (i.e., tire noise) and (*b*) 0.3 m (e.g., passenger car exhaust noise).

0.05-m-thick layer over a hard backing is assumed. For both parts, the range of values for σ_e (25–25,000 kPa $\cdot$ s/m^2) covers ground surfaces from soft snow to old, sealed asphalt. Figure 10 shows that the ground attenuation changes markedly as this key impedance parameter changes, showing, as expected, a maximum ground dip for snow and little ground attenuation for the hardest surface.

Figure 10*b* shows a significant effect of surface waves at low frequencies for the lowest flow resistivities due to the greater reactive component of surface impedance for this hard-backed layer model for the ground.

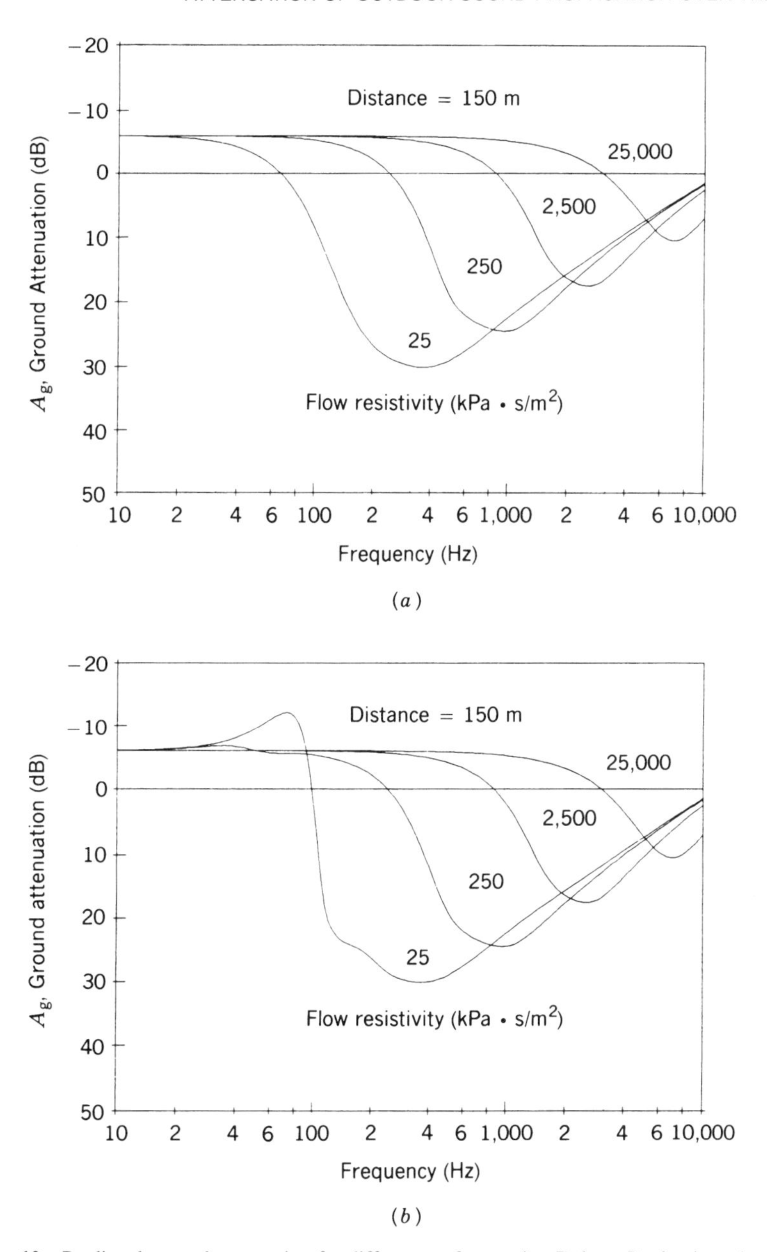

Fig. 10 Predicted ground attenuation for different surfaces using Delany–Bazley impedance model with effective resistivity σ_e of 25–25,000 kPa · s/m^2. Receiver height h_r = 1.2 m, distance 150 m and, source height h_s = 0.3 m. (*a*) Ground is semi-infinite; (*b*) 0.05 m thick layer over a hard backing.

When the surface impedance changes abruptly, such as at the edge of a roadway, analytical methods[43] are available as well as an approximate approach involving weighting the excess attenuation values according to the distance traveled over each surface.[25]

4.5 Attenuation through Foliage and Trees

For sound attenuation through foliage and trees, the main effect at low frequencies is to enhance ground attenuation, the roots making the ground more porous.[44] At high frequencies, where the dimensions of leaves become comparable with the wavelength, there is also a significant attenuation caused by scattering.[45]

In a forest, the vertical gradients of wind and temperature are reduced at elevations up to approximately the height of the trees, thus reducing the effects of refraction from such gradients as considered in Section 5.

4.6 Attenuation through Built-up Areas

As within forests, the effect of refraction on the propagation of sound in city streets is small, and for the same reason, the obstruction to the flow of air causes the viscous and thermal boundary layer to approach the height of buildings. The ground attenuation effect in the city is also much smaller than in the country, partly because of paving, but also because the buildings produce an interference pattern much different from that encountered over flat terrain. However, prediction of the propagation of noise from airports, freeways, and so on, out into the suburbs is based largely on empirical methods not well found on analytical models or extensive experimental data.[46]

5 REFRACTION IN OUTDOOR SOUND PROPAGATION

The preceding sections have discussed propagation in a still uniform atmosphere where sound follows straight ray paths. Although this is a reasonable assumption at shorter ranges, the simple description of the direct and ground-reflected path shown in Fig. 2 is no longer valid at longer ranges. Under most weather conditions both the temperature and the wind velocity vary with height above the ground. The speed of sound relative to the ground is a function of temperature and wind velocity, and hence it also varies with height, causing the sound waves to propagate along curved paths.[47]

During the day solar radiation heats the earth's surface, resulting in warmer air near the ground. This condition, called a temperature lapse, is most pronounced on sunny days but can also exist under overcast skies. A temperature lapse is the common daytime condition during most of the year and ray paths curve upward. After sunset there is often radiation cooling of the ground, which produces cooler air near the surface and forms a temperature inversion. Within the temperature inversion, the temperature increases with height and ray paths curve downward.

When there is wind, its speed decreases with decreasing height because of drag on the moving air at the ground. Therefore, the speed of sound relative to the ground increases with height during downwind propagation, and ray paths curve downward. For propagation upwind the sound speed decreases with height, and ray paths curve upward. There is no refraction in the vertical direction produced by wind when the sound propagates directly crosswind.

5.1 Downward Refraction

Downward-curving sound rays (see Chapters 1 and 3) are shown in Fig. 11. The effect of the downward-curving rays is to increase the grazing angle ϕ in Fig. 2 and hence modify the ground attenuation shown in Figs. 8 and 9. Further, under specific conditions that depend on source and receiver heights, horizontal range, and the strength of the refraction, additional ray paths are possible that involve one or more reflections at the ground. These additional rays can nullify the ground attenuation.

It is possible to modify[48] the Weyl–Van der Pol solution to account for refraction. However, the general solution[47] for the pressure $p(r)$ in Eq. (1) at a receiver a distance r from the source is expressed as a Hankel transform:

$$p(r) = \int_{\infty}^{\infty} H_0^1(Kr)P(K,z)K\,dK, \qquad (18)$$

where H_0^1 is the Henkel function of the first kind and of order 0. K is the horizontal wavenumber, and z is the vertical distance above the ground. In general, it is not possible to obtain closed-form solutions to Eq. (18), and numerical algorithms are required to compute Eq. (18). These will be discussed later, in Section 5.3. However, in the case of simple sound speed profiles, a residue series solution[49] to Eq. (18) can be obtained. The form of profile that is most convenient for physical interpretation and mathematical computation is one where the

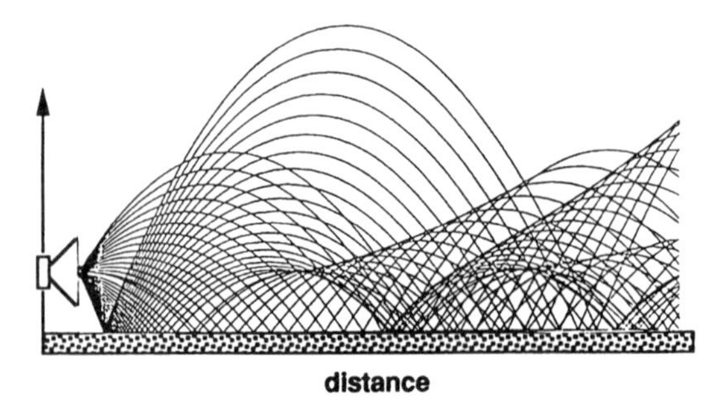

Fig. 11 Sound rays in the atmosphere during downward refraction. At the longer distance, additional ray paths are seen that involve more than one reflection at the ground. There are also regions of focusing called caustics.

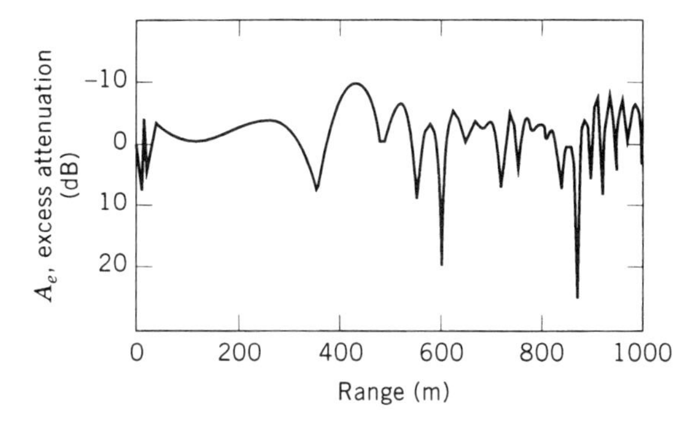

Fig. 12 Excess attenuation predicted for a frequency of 500 Hz for propagation during downward refraction. The ground impedance is characterized using the Delany–Bazley model [Eq. (8a)] with an effective flow resistivity σ_e of 200 kPa · s/m^2. The sound speed profile is given by $c(z) = 340(1 + 5.88 \times 10^{-4}Z)$ (m/s). Source height 2 m; receiver height 5 m.

sound velocity increases linearly with height z above the ground:

$$c(z) = c_0(1 + az). \qquad (19)$$

In Eq. (19), a is the coefficient of increase in velocity with height. The sound rays between source and receiver are circular arcs. We note that a linear variation with height is a good approximation for many cases, although it is not necessarily achieved in practice.

An example of the excess attenuation A_e, in decibels, obtained during downward refraction at 500 Hz is shown by the curve in Fig. 12. The excess attenuation is obtained using one of the numerical algorithms (see Section 5.3) to compute Eq. (18) and then using Eqs. (1)–(4). On average, the excess attenuation is constant with increasing distance. In simple terms, the attenuation due to the ground is nullified by the energy provided by the additional rays shown in Fig. 11. The sound pressure levels therefore increase to the levels predicted by geometric spreading and atmospheric absorption alone, but in general not above such levels. Increases above such levels are due to focusing of the various rays and are inevitably accompanied by decreases caused by defocusing elsewhere in the sound field, which leads to the various dips observed in Fig. 12.

5.2 Upward Refraction

When the sound speed decreases with height, the sound rays are bent upward, away from the ground. For realistic sound speed profiles, there is a shadow boundary formed by sound rays from various regions of the ensonified sound field and beyond which there is an acoustic shadow region. If the relation between sound speed and height is linear, as in Eq. (19), with a being negative, the rays are arcs of circle and the shadow boundary is a single limiting ray that just grazes the ground. A limiting ray with radius of curvature R is shown in Fig. 13 for a source located at a height h_s. Beyond the limiting ray no direct sound energy can penetrate causing the acoustic shadow region.

Equation (18) also applies in the case of upward refraction as the general solution for the pressure, and the numerical algorithms discussed later can be used for computation. There is also a residue series solution[50] to

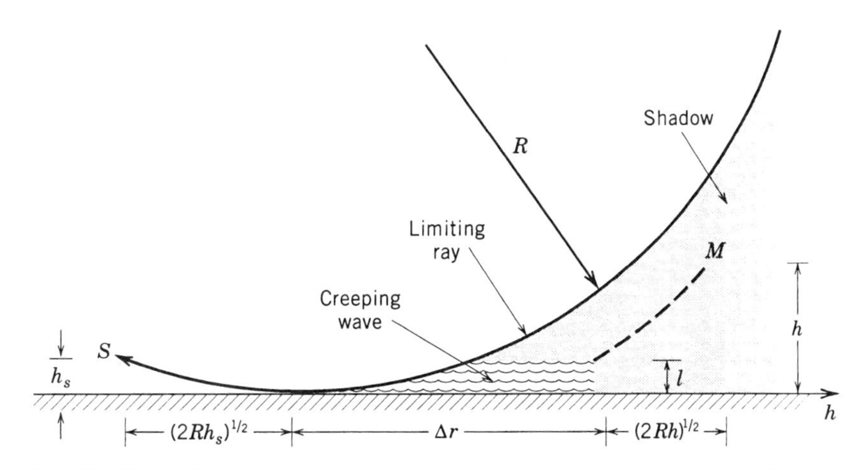

Fig. 13 Sketch illustrating the limiting ray during upward refraction in the case where the sound speed profile is a linear function of height. The limiting ray is an arc of a circle with radius of curvature R.

Eq. (18) in the case of a linear sound speed profile. The residue series provides a physical interpretation of the diffraction within the shadow region. When the receiver is deep within the shadow, but not close to the ground, the series can be approximated by the first term only and, further, the various functions in the solution can be approximated by their asymptotic behavior. The result is

$$p(r) = \left(\frac{1}{2k_0 r l^2}\right)^{1/2} \frac{1}{K}\left(\frac{l}{h_s}\right)^{1/4}\left(\frac{l}{z}\right)^{1/4} \exp(-\alpha\,\Delta r) \cdot \exp\left(i\omega\left[\tau(h_s) + \frac{\Delta r}{v} + \tau(z)\right]\right), \quad (20)$$

where $k_0 = \omega/c(0)$, $l = (R/2k_0^2)^{1/3}$, and τ corresponds to travel times. The function K essentially incorporates the impedance of the ground and the function α describes the attenuation of the wave along the portion of the path Δr with reduced phase velocity $v < c$. Because of the attenuation and reduced phase velocity, the term *creeping wave* is often used to describe the part of the sound field along the path Δr. The quantity l can then be identified as the thickness of the creeping wave.

Equation (20) suggests (see Fig. 13) that the energy received at M initially leaves the source and travels along the limiting ray to the ground with a travel time $\tau(h_s)$ along a distance $(2Rh_s)^{1/2}$. Then it propagates via the creeping wave in the air along the surface a distance Δr with reduced phase velocity v and weak attenuation. At the appropriate distance, the energy is then shed from the creeping wave and travels to M at a height $z = h$ along the ordinary geometric acoustic rays with travel time $\tau(h)$ along a distance $(2Rh)^{1/2}$.

Unfortunately, the simple picture provided by Eq. (20) has limited success in predicting the excess attenuation during propagation in an upward-refracting atmosphere. Beyond a few hundred meters Eq. (20) predicts large attenuation that is not supported by experimental data.[2,50] Atmospheric turbulence contributes significantly to the excess attenuation, and numerical methods that incorporate the effects of turbulence must be used to calculate the acoustic pressure (see Section 7).

5.3 Numerical Methods

Numerical techniques originally developed for applications in underwater acoustics have been adapted in recent years for atmospheric predictions in the presence of refraction. The numerical techniques can be broadly classified as variations of the fast-field program (FFP) or forms of the parabolic equation (PE). A review of the various FFP and PE techniques in common use, including a comprehensive list of references, can be found in Ref. 51.

Fast-Field Program Models Fast-field programs permit the prediction of sound pressure in a refracting atmosphere at an arbitrary receiver on or above a flat continuous ground from a point source somewhere above the ground. The sound speed can be specified from Eq. (19) or as an arbitrary function of height above the ground. The basis of the FFP method is to work numerically from exact integral representations of the sound field within the layered atmosphere in terms of coefficients that may be determined from the ground impedance. The method gets its name from the discrete Fourier transform used to evaluate these integrals.

The starting point is the Hankel transform in Eq. (18), from which it is straightforward to show that $P(K, z)$ satisfies

$$\frac{d^2P}{dz^2} + [k^2(z) - K^2]P = -2\delta(z - h_s). \quad (21)$$

The solution for $P(K, z)$ in Eq. (21) is found and the total field at frequency f is calculated at any range r by carrying out the inverse transform. The indefinite integral is replaced by a finite sum using discrete Fourier transforms. If the maximum value of wavenumber in the sum is $K_{\max}$ and N discrete values of K are introduced, then the wavenumber intervals are given by $\Delta K = K_{\max}/(N - 1)$ and correspond to range intervals $\Delta r = 2\pi/N\,\Delta k$, so, for example,

$$p(r_m) = 2(1 - i)\sqrt{\frac{\pi}{r_m}}\,\Delta K \sum_{n=0}^{N-1} P(K_n)\sqrt{K_n}\,e^{2i\pi nm/N}, \quad (22)$$

where $K_n = n\,\Delta K$ and $r_m = m\,\Delta r$ (or $r_0 + m\,\Delta r$, where r_0 is the desired starting range).

An implementation[51] of Eq. (22) was used to calculate the excess attenuation shown in Fig. 12 for the case of downward refraction. In the case of upward refraction, Eq. (22) agrees with the residue series solution described by Eq. (20) (see Section 7).

Parabolic Equation Models This technique assumes that wave motion for a particular problem is always directed away from the source or that there is very little backscattering. Making the change of variable $U = pr^{1/2}$, the far-field assumption ($kr \gg 1$) leads to the Helmholtz wave equation for the field U in two dimensions (r, z):

$$\frac{\partial^2 U}{\partial r^2} + \frac{\partial^2 U}{\partial z^2} + k^2 U = 0. \tag{23}$$

We define the operator $Q = \partial^2/\partial z^2 + k^2$, and if k is independent of range, Eq. (23) can be written as

$$\left(\frac{\partial}{\partial r} + i\sqrt{Q}\right)\left(\frac{\partial}{\partial r} - i\sqrt{Q}\right) U = 0. \tag{24}$$

The factors within the parentheses represent propagation of incoming and outgoing waves, respectively, if a time dependence $\exp(-i2\pi f t)$ is assumed. Considering only the outgoing wave, Eq. (24) reduces to

$$\frac{\partial U}{\partial r} = i\sqrt{Q}U. \tag{25}$$

Most implementations of the PE method can be traced back to Eq. (25).

It is convenient to define a new, more slowly varying wave u, where $u = U\exp(-ik_0 r)$. The equation for u is given by

$$\frac{\partial u}{\partial r} = i(\sqrt{Q} - k_0)u, \tag{26}$$

and the formal operator solution to Eq. (26) for advancing the field is

$$u(r + \Delta r) = e^{i\Delta r(\sqrt{Q} - k_0)}u(r). \tag{27}$$

That is, the field at a distance $r + \Delta r$ is calculated by multiplying the field previously calculated at r by the operator in Eq. (27). For example, the implementations[51] of Eq. (27) yield the same acoustics pressure, and hence the same excess attenuation in decibels, as Eq. (22).

6 DIFFRACTION IN OUTDOOR SOUND PROPAGATION

Far from any boundaries or above a simple infinite plane boundary, a sound field propagates in a relatively simple way, and as seen in the previous sections, this simplicity can be exploited by describing the propagation in terms of ray paths. However, if a large solid body blocks the sound field, the ray theory of sound propagation predicts a shadow region behind the body with sharply defined boundaries, so in principle, on one side of the boundary there is a sound field and close by on the other side of the boundary there is essentially silence. This does not happen in practice; as the waves propagate, sound "leaks" across this sharp boundary. Diffraction effects are most clearly evident in the vicinity of solid boundaries or along geometric ray boundaries such as the limiting ray discussed in Fig. 13. A more complete discussion of diffraction can be found in Ref. 47.

Acoustic diffraction occurs in conjunction with a wide range of solid bodies: Some such as thin solid barriers are erected alongside highways or are carefully located to shield residential communities from ground operations of aircraft; others such as buildings are often built for other purposes but fortuitously provide some beneficial shielding; yet others such as undulating ground or low hills occur naturally and provide shielding at much larger distances and bring forth other manifestations of diffraction such as the creeping waves referred to earlier.

The simplest and most widely used procedure for determining the reduction of sound pressure level due to diffraction around the edge of a barrier is described in terms of a Fresnel number.[47] This is simply the minimum increase in distance that the sound must travel around the edge of the barrier to go from source to receiver (Fig. 14) divided by the half-wavelength $\lambda/2$ at the frequency of interest. Thus, Fresnel number N is given as

$$N = \frac{2}{\lambda}(d_1 + d_2 - d). \tag{28}$$

The attenuation by diffraction A_d, in decibels, is then given as a function of Fresnel number by

$$A_d = 10\log(20N). \tag{29}$$

Equation (29) yields the curve in Fig. 14. This curve is obtained[52] from diffraction theory assuming a thin knife-edge barrier and no ground and then empirically allows for the presence of the ground by reducing the loss of sound level by about 2 dB. This prediction curve is not exact because the empirical correction does not account for the frequency dependence (here, the Fresnel number dependence) of the ground reflection interference in a specific configuration of source, barrier, and receiver heights and separation distances. In order to obtain a more exact prediction of the sound field behind the barrier, the complex interference spectrum resulting from the sum of four paths shown in Fig. 15 must be calculated. Nonetheless, the curve in Fig. 14 is correct to about ±5 dB in most cases and is the mean curve through the interference spectrum that would be measured, and can be predicted, in any specific circumstances.[53]

Rigorously, the attenuation provided by a barrier above a natural ground surface replaces the ground attenuation present before the construction of the barrier.

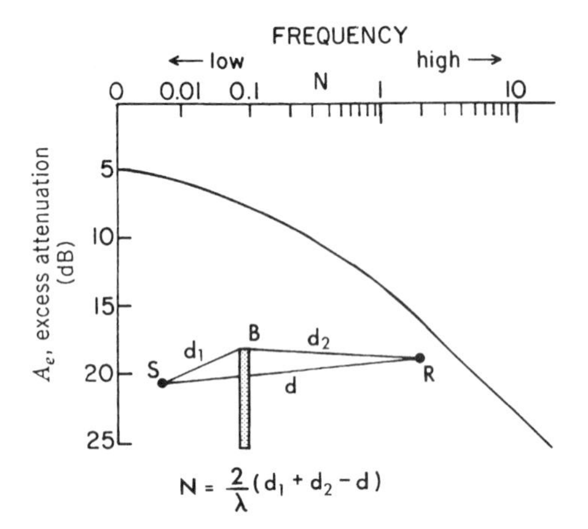

Fig. 14 Sketch defining the Fresnel number N in terms of the path difference and wavelength λ. The curve is the barrier attenuation as a function of Fresnel number.

Therefore, barrier performance is best measured by its insertion loss $\text{IL}_{\text{barrier}}$, in decibles, defined as the difference in sound pressure level before and after the barrier is constructed:

$$\text{IL}_{\text{barrier}} = L_p(\text{before}) - L_p(\text{after}). \tag{30}$$

Detailed studies of traffic noise reduction by barriers show average insertion loss of 5–8 dB but rarely exceeding 10 dB. In general, the insertion loss of a barrier is limited by the effects of atmospheric turbulence to about 15–25 dB. As discussed earlier (and see Section 7), turbulence scatters sound energy that penetrates the shadow behind the barrier, thus resulting in an upper limit to the insertion loss by a barrier.[54]

When barriers are used specifically to attenuate sound, it is good practice to locate them, when possible, as closely as possible to either the source or the receiver. A barrier of given height then results in a large value of the diffraction angle θ and a greater path difference $(d_1 + d_2 - d)$. At distances between source and receiver greater than a few hundred meters, it is difficult to provide man-made barriers large enough to provide any noticeable attenuation. Naturally occurring topographic features such as hills can often function as barriers, blocking the line of sight between source and receiver.

The use of absorbing materials can increase barrier performance,[55] and the general principles can be obtained from simplified calculations.[56] The calculations demonstrate that the effectiveness of absorbing material increases as either the source or receiver is moved closer to the barrier and decreases as the frequency decreases. Field studies also show that barrier effectiveness is reduced when the propagation over the barrier is downwind. The rays from the source reach the receiver behind the barrier by curving over the top of the barrier edge, thus providing direct ensonification. However, this does not mean that the barrier has completely lost its effectiveness. In this case, the "before" levels, L_p(before), in the calculation of the insertion loss correspond to the ray-tracing picture in Fig. 11. The multiple rays close to the ground responsible for the before levels in Eq. (30) are effectively blocked by the barrier. Model experiments and theory[57] suggest that the barrier still provides positive insertion loss during downwind propagation.

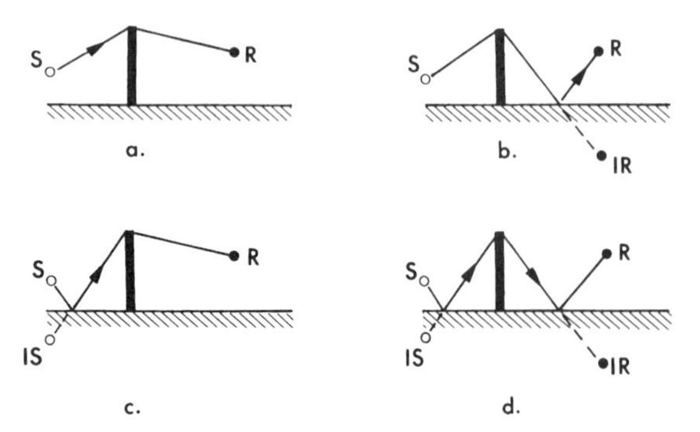

Fig. 15 Four paths contributing to the sound field behind a barrier above ground.

7 ATMOSPHERIC TURBULENCE

The atmosphere is an unsteady medium with random variations in temperature, wind velocity, pressure, and density. In practice, only the temperature and wind velocity variations significantly affect acoustic waves over a short time period. During the daytime these inhomogeneities are normally much larger than is generally appreciated. Fluctuations in temperature of 5°C that last several seconds are common and 10°C fluctuations not uncommon. The wind velocity fluctuates in a similar manner and has a standard deviation about its mean value that is commonly one-third of the average value. When sound waves propagate through the atmosphere, these random fluctuations scatter the sound energy. The total field is then the sum, in amplitude and phase, of these scattered waves and the direct line-of-sight wave, resulting in random fluctuations in amplitude and phase. The acoustic fluctuations are in some respects analogous to more familiar optical phenomena such as the twinkling of light from a star.

Turbulence can be visualized as a continuous spectrum of eddies.[58] For example, since the average horizontal wind velocity varies as a function of height (see Section 5), this variation creates turbulent motion or eddies of a size called the outer scale of turbulence. The size of the outer scale L_0, is typically of the order of metres. In the range of eddy size smaller than L_0, the kinetic energy is transferred to eddies of smaller size. This energy transfer can be visualized as a process of eddy fragmentation where large-scale eddies cascade into eddies of ever-decreasing size. The fluid motion is almost completely random and irregular, and its features can be described in statistical terms. This range of eddy sizes is called the inertial range. As the eddy size becomes smaller, virtually all the energy is dissipated into heat and almost no energy is left for eddies of size smaller than l_0. This size l_0 is called the inner scale of turbulence and is typically of the order of 1 mm. The three characteristic ranges of eddy sizes of the turbulent atmosphere are illustrated in Fig. 16. The points are an example of the power spectral density of the time-varying signal recorded by an anemometer of the wind velocity fluctuations. A similar power spectral density is obtained from the time-varying signal recorded by a fast-response thermometer.

The scattering of sound by turbulence produces fluctuations in the sound pressure level. The fluctuations initially increase with increasing distance of propagation, sound frequency, and strength of turbulence but quickly reach a limiting value.[59] This saturation minimizes the nuisance of coping with fluctuating levels during noise measurements from relatively distant sources. For example, when the noise from aircraft propagates under clearly line-of-sight conditions over distances of a few kilometers, the measured sound pressure levels fluctuate about their mean value with a standard deviation of no more than 6 dB.

Another effect of atmospheric turbulence that has traditionally been considered important is the direct attenuation of sound. In a highly directed beam, the turbulence attenuates the beam by scattering energy out of it. However, for a spherically expanding wave this attenuation is negligible. In a simpleminded way, the energy scattered out from the line of sight is replaced by energy scattered back to the receiver from adjacent regions. This implies that the energy level of the rms sound pressure in an unsteady medium is the same as the level in the absence of turbulence. The only mechanism by which turbulence could provide attenuation in a spherical wave field is backscattering. However, it seems that the attenuation provided by backscattering is much smaller than the attenuation due to molecular absorption. This is an important result for the use of the PE models.

An important effect of atmospheric turbulence is the degradation of the ground attenuation and the reduction of the deep shadows produced behind barriers or during propagation in upward refraction conditions. As dis-

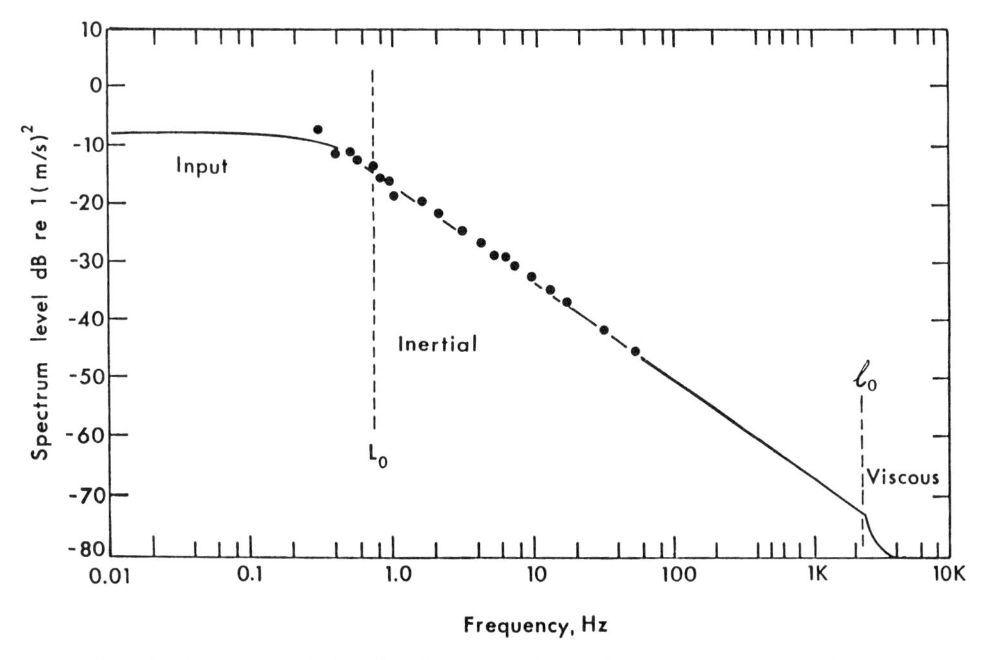

Fig. 16 Three ranges of eddy size of the atmospheric turbulence spectrum. The points are the result of a fast Fourier transform analysis of a wind velocity recording.

cussed in Section 4.3, the effects of atmospheric turbulence on the attenuation of sound over ground can be incorporated into the theory by adding an additional fluctuating random term in Eq. (12). The resulting modified expressions[60] provide an explanation for the discrepancy observed in Fig. 8 between the solid curve and the experimental points. The scattering of sound behind a noise barrier can be calculated [54] by using the standard cross section for scattering by atmospheric turbulence (see Section 8). In order to include the effects of turbulence in the case of propagation in upward refraction conditions, the numerical models discussed in Section 5.3 can be extended.[61,62]

For example, in the PE model, the wavenumber is separated into a part representing the variation of the sound speed with height z above the ground and a part that represents a small random perturbation $\mu \ll 1$,

$$k(r, z) = k_0 \left[\frac{c_0}{c(z)} + \mu(r, z) \right]. \tag{31}$$

The operator for advancing the field in Eq. (27) can then be approximated by the product

$$e^{i\Delta r(\sqrt{Q} - k_0)} = e^{i\langle\mu\rangle k_d \Delta r} e^{i\Delta r(\sqrt{Q_d} - k_0)}, \tag{32}$$

where $Q_d = \partial^2/\partial z^2 + k_d^2$ and $k_d = k_0 c_0/c(z)$. For computational purposes, the stochastic part of the index of refraction is assumed to have a simple correlation function approximated by a Gaussian distribution

$$\langle\mu_1\mu_2\rangle = \langle\mu^2\rangle \exp\left(\frac{-r^2}{L^2}\right). \tag{33}$$

The Fourier transform of the correlation function in Eq. (33) is then an approximation to the power spectral density shown in Fig. 16. For small-scale turbulence[63] near the ground, $L \approx$ 1–7 m and $\langle\mu^2\rangle \sim 10^{-6}$.

The curves in Fig. 17 were calculated[64] from the PE using the operator in Eq. (32) for a frequency of 500 Hz during upward refraction for $\langle\mu^2\rangle$ values of 0.5, 2, and 10×10^{-6} (lower, middle, and upper curves, respectively). Beyond 200 m, the calculation shows that the excess attenuation is between 20 and 30 dB, in agreement with results found in many field data sets.[2] The pictures in Fig. 18 illustrate more visually the effects of turbulence on the excess attenuation shown in Fig. 17. The top of Fig. 18 is a contour plot of sound levels when the atmosphere is assumed nonturbulent, that is, $\langle\mu^2\rangle = 0$. The same contour plot for $\langle\mu^2\rangle = 2 \times 10^{-6}$ is shown at the bottom of Fig. 18. Sound energy (white areas) scattered into the shadow region is clearly observed.

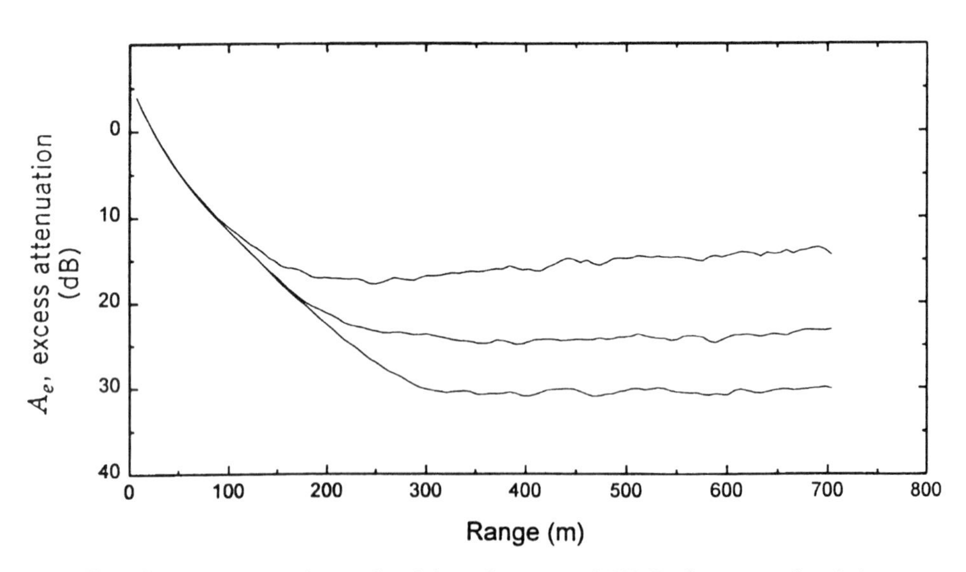

Fig. 17 Excess attenuation predicted for a frequency of 500 Hz for propagation during upward refraction in a turbulent atmosphere. The ground impedance is characterized using the Delany–Bazley model [Eq. (8a)], with an effective flow resistivity of 300 kPa · s/m^2. The sound speed profile is given by $c(z) = c(0) - 0.5 \ln(z)$ (m/s) and the turbulence strengths were $\langle\mu^2\rangle = 0.5$, 2, 10×10^{-6} (lower, middle, and upper curves, respectively). Source and receiver heights are 0.3 and 0.1 m, respectively.

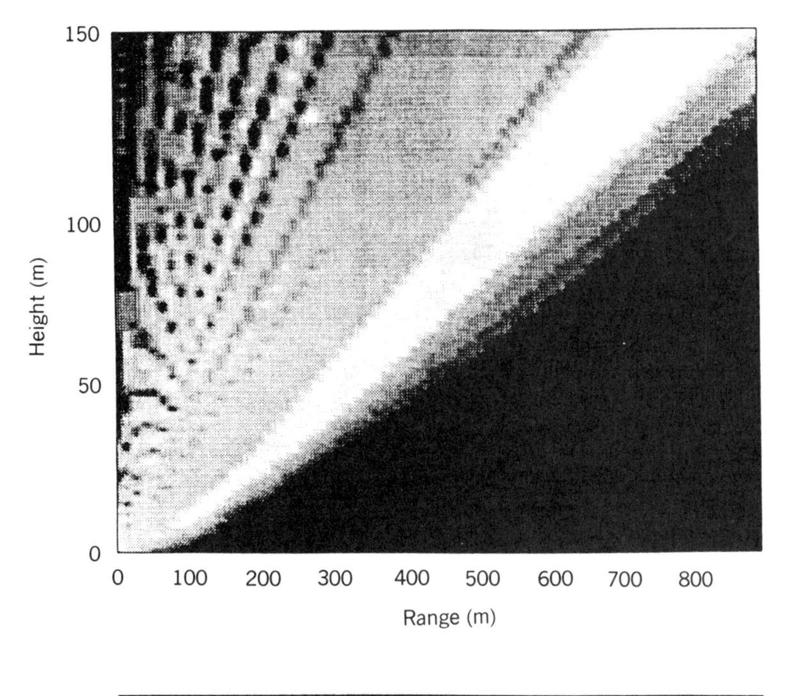

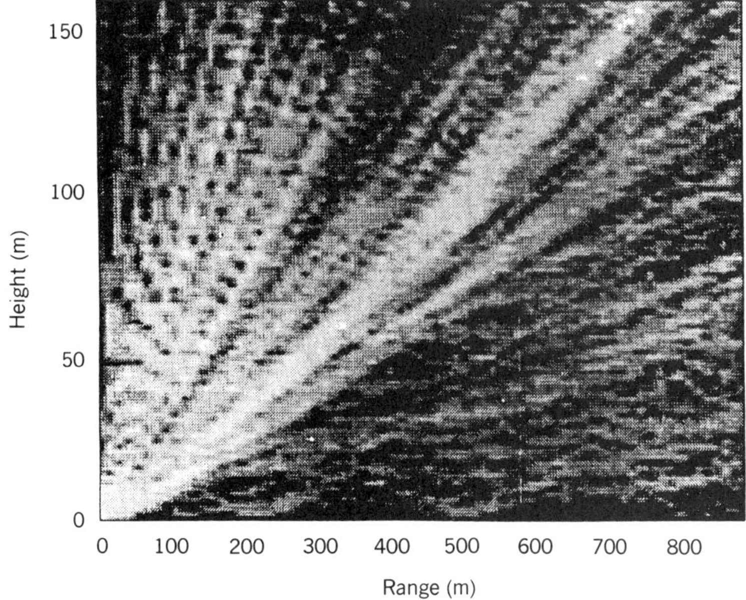

Fig. 18 Contour plots of sound levels in the presence of upward refraction. The white areas are regions of higher sound levels. (Top) No turbulence; (Bottom) in the presence of turbulence.

8 ACOUSTIC SOUNDING OF THE ATMOSPHERE

Meteorologists divide the structure of the atmosphere into a succession of layers, each having different characteristics. Beginning at the surface, the troposphere represents a region of, mainly, decreasing temperature with increasing altitude. The lowest part of the troposphere, the atmospheric boundary layer, which varies in thickness from 1 or 2 km during sunny daytime to a few hundred meters at night, departs from the general temperature decrease. An increasing temperature in the boundary layer is called an inversion and, for example, acts as a lid that prevents the dispersion of atmospheric pollutants. The stratosphere lies just above the troposphere at an altitude of between 10 and 20 km. The temperature in the upper level of the stratosphere increases with height. In the next layer, the mesosphere, at an altitude of about 50 km, temperatures again decrease with height, and the molecular composition of the atmosphere begins to change.

The invention of the echosonde (also called the sodar) added a powerful tool for atmospheric research and for applications to forecasting, especially in the atmospheric boundary layer. The use of the echosonde has become so widespread that results derived with it have become an ubiquitous part of the environmental impact studies required in many countries around the world. In its more complex form—the Doppler echosonde—the instrument can provide measurements of mean winds, sea breezes, and flows in and out of valleys. The monostatic echosonde relies on backscatter (scattering angle $\theta_s = \pi$) directly back along the ensonification path from a field of turbulence-driven temperature fluctuations along that path. The bistatic echosonde ($\theta_s < \pi$) uses a separated transmitter and receiver, with a signal that results from both temperature and velocity fluctuations. Figure 19 illustrates a bistatic configuration.

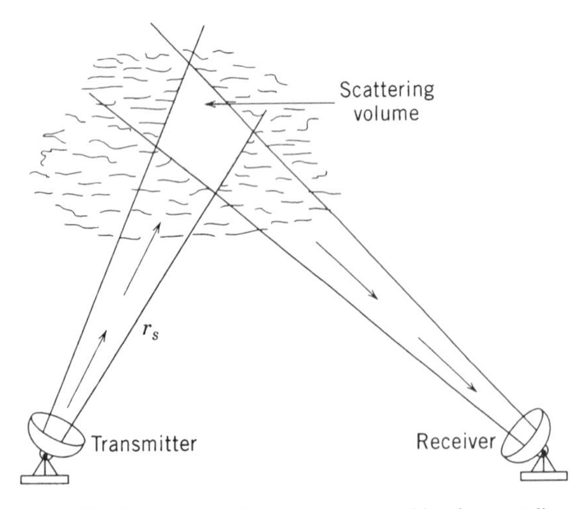

Fig. 19 Scattering volume used in a bistatic sounding measurement.

The equations of fluid mechanics, together with two assumptions valid in acoustic remote sensing (but invalid in aeroacoustics), that (1) that the turbulence remains a nonzero vorticity and incompressible field that produces no sound while (2) the acoustic wave remains a vorticity-free longitudinal compression field that produces no changes in the turbulence, lead to the governing acoustic remote-sensing equation. If it is further assumed that the turbulence "eddy" sizes lie within the inertial range (see Fig. 16), a good estimate of the scattering cross section (in reciprocal metres) as a function of scattering angle θ_s is obtained,[1]

$$\sigma_s = 1.52 k_0^{1/3} \cos^2\theta_s \left[0.13 C_n^2 + \left(\frac{C_v^2}{4c_0^2}\right)\cos^2\left(\frac{\theta_s}{2}\right)\right] \cdot \left[2\sin\left(\frac{\theta_s}{2}\right)\right]^{-11/3}. \qquad (34)$$

In Eq. (34), k_0 is the initial wavenumber, C_v^2 the structure parameter for the velocity fluctuations, and C_n^2 the structure parameter for refractive index fluctuations. There is no scattering at right angles, and Eq. (34) predicts the strong forward nature of the scattering by turbulence. In fact, for very small scattering angles Eq. (34) very quickly exceeds unity and becomes infinite at 0°. For this reason, although scattering can be invoked to explain the sound pressure level fluctuations discussed earlier, it is inappropriate to use Eq. (34) to quantify the fluctuations. Instead, other techniques such as perturbation methods are invoked. Equation (34) can be used, though, to calculate the scattered sound energy behind a barrier.[54]

For direct backscatter ($\theta_s = 180°$) velocity fluctuations do not contribute to the scattering because of the $\cos(\theta_s/2)$ multiplying the term containing the C_v^2. Therefore the intensity display for a monostatic echosonde system transmitter and receiver collocated reflects the temperature structure only. By monitoring the backscatter intensity as a function of time from transmission, the relative strength of temperature fluctuations as a function of altitude can be estimated.

The echosonde equation, like its radar and sonar counterparts, relates the received acoustic power P_r to the transmitted power P_t and frequency f and other atmospheric and system design variables. We define L as the propagation attenuation, S the pulse stretching factor ($S \sim 1$), $\alpha(\theta_s)$ a dimensionless aspect factor that accounts for the shape of the scattering volume, l_p the pulse length, and g the antenna directivity gain factor. Then the echosonde equation is

$$P_t = P_t L S \alpha(\theta_s) \frac{l_p}{2} \frac{g A_r}{r_s^2} \sigma_s(\theta_s, f), \qquad (35)$$

where A_r is the area of the receiver antenna aperture and r_s is the distance from the receiving antenna to the scattering volume. Equation (35) represents the key link between theory and experiments and provides the means for interpreting observation.

REFERENCES

1. E. H. Brown and F. F. Hall, Jr. "Advances in Atmospheric Acoustics," *Rev. Geophys. Space Phys.*, Vol. 16, 1978, pp. 47–110.
2. J. E. Piercy, T. F. W. Embleton, and L. C. Sutherland, "Review of Sound Propagation in the Atmosphere," *J. Acoust. Soc. Am.*, Vol. 61, 1977, pp. 1403–1418.
3. M. E. Delany, "Sound Propagation in the Atmosphere, A Historical Review," *Proc. Inst. Acoust.*, Vol. 1, 1978, pp. 32–72.
4. R. B. Tatge, "Noise Radiation by Plane Arrays of Incoherent Sources," *J. Acoust. Soc. Am.*, Vol. 52, 1972, pp. 732–736.
5. E. A. G. Shaw and N. Olson, "Theory of Steady-State Urban Noise for an Ideal Homogeneous City," *J. Acoust. Soc. Am.*, Vol. 51, 1972, pp. 1781–1793.
6. L. C. Sutherland, "Ambient Noise Level Above Plane with Continuous Distribution of Random Sources," *J. Acoust. Soc. Am.*, Vol. 57, 1975, pp. 1540–1542.
7. B. H. Sharp and P. R. Donavan, "Motor Vehicle Noise," in C. M. Harris (Ed.), *Handbook of Noise Control*, McGraw-Hill, New York, 1979, pp. 32–16.
8. P. R. Donavan and R. H. Lyon, "Model Study on the Propagation from V/STOL Aircraft into Urban Environs," *J. Acoust. Soc. Am.*," Vol. 55, 1974, p. 485(A).
9. W. E. Zorumski and W. L. Willshire, "The Acoustic Field of a Point Source in a Uniform Boundary Layer Over an Impedance Plane," Paper before AIAA Aeroacoustics Conference, July, 1986, Paper No. AIAA-86-1923.
10. American National Standards Institute, "Method for the Calculation of the Absorption of Sound by the Atmosphere," ANSI S1.26-1995 (Revision of ANSI S1.26-1978).
11. H. E. Bass, L. C. Sutherland, A. J. Zuckerwar, D. T. Blackstock, and D. M. Hester, "Atmospheric Absorption of Sound: Further Developments," *J. Acoust. Soc. Am.*, Vol. 97, 1995, pp. 680–683.
12. P. D. Joppa, L. C. Sutherland, and A. J. Zuckerwar, "A New Approach to the Effect of Bandpass Filters on Sound Absorption Calculations," *J. Acoust. Soc. Am.*, Vol. 88, S1, 1990, p. S73.
13. D. C. Henley and G. B. Hoidale, "Attenuation and Dispersion of Acoustic Energy by Dust," *J. Acoust. Soc. Am.*, Vol. 54, 1973, pp. 437–445.
14. J. B. Calvert, J. W. Coffman, and C. W. Querfeld, "Radiative Absorption of Sound by Water Vapour in the Atmosphere," *J. Acoust. Soc. Am.*, Vol. 39, 1966, pp. 532–536.
15. P. M. Morse and K. U. Ingard, *Theoretical Acoustics*, McGraw-Hill, New York, 1968.
16. G. A. Daigle, T. F. W. Embleton, and J. E. Piercy, "Some Comments on the Literature of Propagation Near Boundaries of Finite Acoustical Impedance," *J. Acoust. Soc. Am.*, Vol. 66, 1979, pp. 918–919.
17. M. E. Delany and E. N. Bazley, "Acoustical Properties of Fibrous Absorbent Materials," *Appl. Acoust.*, Vol. 3, 1970, pp. 105–116.
18. C. I. Chessell, "Propagation of Noise Along a Finite Impedance Boundary," *J. Acoust. Soc. Am.*, Vol. 62, 1977, pp. 825–834.
19. M. J. M. Martens, L. A. M. van der Heijden, H. H. J. Walthaus, and W. J. J. M. van Rens, "Classification of Soils Based on Acoustic Impedance, Air Flow Resistivity, and Other Physical Soil Properties," *J. Acoust. Soc. Am.*, Vol. 78, 1985, pp. 970–980.
20. Y. Miki, "Acoustical Properties of Porous Materials—Modifications of Delany–Bazley Models," *J. Acoust. Soc. Jpn. (E)*, Vol. 1, 1990, pp. 19–23.
21. L. C. Sutherland, "Review of Ground Impedance for Grass Surfaces—Delany and Bazley Revisited," *Proc. Sixth Int. Symp. on Long Range Sound Prop.*, Ottawa, Canada, 1994, pp. 460–479.
22. T. F. W. Embleton, J. E. Piercy, and G. A. Daigle, "Effective Flow Resistivity of Ground Surfaces Determined by Acoustical Measurements," *J. Acoust. Soc. Am.*, Vol. 74, 1983, pp. 1239–1244.
23. L. N. Bolen and H. E. Bass, "Effects of Ground Cover on the Propagation of Sound through the Atmosphere," *J. Acoust. Soc. Am.*, Vol. 69, 1981, pp. 950–954.
24. H. M. Hess, K. Attenborough, and N. W. Heap, "Ground Characterization by Short Range Propagation Measurements," *J. Acoust. Soc. Am.*, Vol. 87, 1990, pp. 1975–1986.
25. T. F. W. Embleton and G. A. Daigle, "Atmospheric Propagation," Chap. 12, in H. H. Hubbard (Ed.), NASA Ref. Pub. 1258, Vol. 2, WRDC Tech. Report 90–3052, Aug. 1991.
26. K. Attenborough, "Ground parameter Information for Propagation Modeling," *J. Acoust. Soc. Am.*, Vol. 91, 1992, pp. 418–427.
27. A. J. Crammond and C. G. Don, "Effects of Moisture Content on Soil Impedance," *J. Acoust. Soc. Am.*, Vol. 82, 1987, pp. 293–301.
28. T. F. W. Embleton, J. E. Piercy, and N. Olson, "Outdoor

Propagation Over Ground of Finite Impedance," *J. Acoust. Soc. Am.*, Vol. 59, 1976, pp. 267–277.

29. A. Zuckerwar, "Acoustical Ground Impedance Meter," *J. Acoust. Soc. Am.*, Vol. 73, 1983, pp. 2180–2186.
30. P. J. Dickinson and P. E. Doak, "Measurements of the Normal Acoustic Impedance of Ground Surfaces," *J. Sound Vib.*, Vol. 13, 1970, pp. 309–322.
31. G. A. Daigle and M. R. Stinson, "Impedance of Grass-Covered Ground at Low Frequencies Measured Using a Phase Difference Technique," *J. Acoust. Soc. Am.*, Vol. 81, 1987, pp. 62–68.
32. K. Attenborough, "Review of Ground Effects on Outdoor Sound Propagation from Continuous Broadband Sources," *Appl. Acoust.*, Vol. 24, 1988, pp. 289–319.
33. J. M. Sabatier, H. E. Bass, L. N. Bolen, and K. Attenborough, "Acoustically Induced Seismic Waves," *J. Acoust. Soc. Am.*, Vol. 80, 1986, pp. 646–649.
34. J. M. Sabatier, R. Raspet, and C. K. Fredericksen, "An Improved Procedure for the Determination of Ground Parameters Using Level Difference Measurements," *J. Acoust. Soc. Am.*, Vol. 94, 1993, pp. 396–399.
35. M. A. Nobile and S. I. Hayek, "Acoustic Propagation over an Impedance Plane," *J. Acoust. Soc. Am.*, Vol. 78, 1985, pp. 1325–1336.
36. Y. L. Li, M. J. White, and M. H. Hwang, "Green's Functions for Wave Propagation Above an Impedance Plane," *J. Acoust. Soc. Am.*, Vol. 96, 1994, pp. 2485–2490.
37. C. F. Chien and W. W. Soroka, "A Note on the Calculation of Sound Propagation Along an Impedance Surface," *J. Sound Vib.*, Vol. 69, 1980, pp. 340–343.
38. R. K. Pirinchieva, "Model Study of Sound Propagation Over Ground of Finite Impedance," *J. Acoust. Soc. Am.*, Vol. 90, 1990, pp. 2678–2682; see also Erratum in *J. Acoust. Soc. Am.*, Vol. 94, 1993, p. 1722.
39. K. B. Rasmussen, "Sound Propagation Over Grass Covered Ground," *J. Sound Vib.*, Vol. 78, 1981, pp. 247–255.
40. G. L. McAninch and M. K. Myers, "Propagation of Quasiplane Waves Along an Impedance Boundary," *Proc. 26th AIAA Aerospace Sciences. Mtg.*, Reno, NE, 1988, Paper No. AIAA-88-0179.
41. D. G. Albert, "Observations of Acoustic Surface Waves Propagating Above a Snow Cover," *Proc. Fifth Int. Symp. Long Range Sound Prop.*, 1992, pp. 10–16.
42. P. H. Parkin and W. E. Scholes, "The Horizontal Propagation of Sound from a Jet Engine Close to the Ground at Hatfield," *J. Sound Vib.*, Vol. 2, 1965, pp. 353–374.
43. G. A. Daigle, J. Nicolas, and J.-L. Berry, "Propagation of Noise Above Ground Having an Impedance Discontinuity," *J. Acoust. Soc. Am.*, Vol. 77, 1985, pp. 127–138.
44. D. Aylor, "Noise Reduction by Vegetation and Ground," *J. Acoust. Soc. Am.*, Vol. 51, 1972, pp. 201–209.
45. T. F. W. Embleton, "Sound Propagation in Homogeneous Deciduous and Evergreen Woods," *J. Acoust. Soc. Am.*, Vol. 35, 1963, pp. 1119–1125.
46. R. H. Lyon, "Role of Multiple Reflections and Reverberations in Urban Noise Propagation," *J. Acoust. Soc. Am.*, Vol. 55, 1974, pp. 493–503.
47. A. D. Pierce, *Acoustics—An Introduction to Its Physical Principles and Applications*, McGraw-Hill, New York, 1981.
48. K. M. Li, "A High-Frequency Approximation of Sound Propagation in a Stratified Moving Atmosphere Above a Porous Ground Surface," *J. Acoust. Soc. Am.*, Vol. 95, 1994, pp. 1840–1852.
49. R. Raspet, G. E. Baird, and W. Wu, "Normal Mode Solution for Low Frequency Sound Propagation in a Downward Refracting Atmosphere Above a Complex Impedance Plane," *J. Acoust. Soc. Am.*, Vol. 91, 1992, pp. 1341–1352.
50. A. Berry and G. A. Daigle, "Controlled Experiments on the Diffraction of Sound by a Curved Surface," *J. Acoust. Soc. Am.*, Vol. 83, 1988, pp. 2047–2058.
51. K. Attenborough, S. Taherzadeh, H. E. Bass, X. Di, R. Raspet, G. R. Becker, A. Güdesen, A. Chrestman, G. A. Daigle, A. L'Espérance, Y. Gabillet, K. Gilbert, Y. L. Li, M. J. White, P. Naz, J. M. Noble, and H. A. J. M. Van Hoof, "Benchmark Cases for Outdoor Sound Propagation Models," *J. Acoust. Soc. Am.*, Vol. 97, 1995, pp. 173–191.
52. Z. Maekawa, "Noise Reduction by Screens," *Appl. Acoust.*, Vol. 1, 1968, pp. 157–173.
53. T. Isei, T. F. W. Embleton, and J. E. Piercy, "Noise Reduction by Barriers on Finite Impedance Ground," *J. Acoust. Soc. Am.*, Vol. 67, 1980, pp. 46–58.
54. G. A. Daigle, "Diffraction of Sound by a Noise Barrier in the Presence of Atmospheric Turbulence," *J. Acoust. Soc. Am.*, Vol. 71, 1982, pp. 847–854.
55. S. I. Hayek, "Mathematical Modeling of Absorbent Highway Noise Barriers," *J. Acoust. Soc. Am.*, Vol. 31, 1990, pp. 77–100.
56. A. L'Espérance, J. Nicolas, and G. A. Daigle, "Insertion Loss of Absorbent Barriers on Ground," *J. Acoust. Soc. Am.*, Vol. 86, 1989, pp. 1060–1064.
57. Y. Gabillet, H. Schroeder, G. A. Daigle, and A. L'Espérance, "Application of the Gaussian Beam Approach to Sound Propagation in the Atmosphere," *J. Acoust. Soc. Am.*, Vol. 93, 1993, pp. 3105–3116.
58. V. I. Tatarskii, *Wave Propagation in a Turbulent Atmosphere*, McGraw-Hill, New York, 1961.
59. G. A. Daigle, J. E. Piercy, and T. F. W. Embleton, "Line-of-Sight Propagation through Atmospheric Turbulence near the Ground," *J. Acoust. Soc. Am.*, Vol. 74, 1983, pp. 1505–1513.
60. G. A. Daigle, "Effects of Atmospheric Turbulence on the Interference of Sound Waves Above a Finite Impedance Boundary," *J. Acoust. Soc. Am.*, Vol. 79, 1986, pp. 613–627.
61. K. E. Gilbert, R. Raspet, and X. Di, "Calculation of Turbulence Effects in an Upward Refracting Atmosphere," *J. Acoust. Soc. Am.*, Vol. 87, 1990, pp. 2428–2437.

62. R. Raspet and W. Wu, "Calculation of Average Turbulence Effects on Sound Propagation Based on the Fast Field Program Formulation," *J. Acoust. Soc. Am.*, Vol. 97, 1995, pp. 147–153.

63. M. A. Johnson, R. Raspet, and M. T. Bobak, "A Turbulence Model for Sound Propagation from an Elevated Source Above Level Ground," *J. Acoust. Soc. Am.*, Vol. 81, 1987, pp. 638–649.

64. X. Di and G. A. Daigle, "Prediction of Noise Propagation During Upward Refraction Above Ground," *Proc. INTER-NOISE 94*, Yokohama, Japan, 1994, pp. 563–566.

29

INFRASOUND

THOMAS B. GABRIELSON

1 INTRODUCTION

Ever present but generally inaudible, infrasound forms a continuum of acoustic radiation below the useful frequency range of human hearing. In contrast to the more familiar audible sound, natural infrasound often reflects global phenomena propagating over thousands of kilometres.

Infrasound covers that part of the acoustic spectrum below 20 Hz. Above about 0.003 Hz (300-s period), the restoring force for acoustic oscillations results from compressibility of the air; below this frequency in gravitationally stable air masses, transverse oscillations are possible in which buoyancy provides the restoring force. Waves having both compressional and transverse components are known as acoustic-gravity waves. At even lower frequencies (periods of hours), the transverse component overwhelms the compressional component and the waves are known as gravity waves.

Infrasound is used to monitor natural events such as volcanic eruptions, earthquakes, severe storms (including microbursts), oceanic disturbances ("microbaroms"), avalanches, meteors, and motion of the auroras; to monitor man-made events such as explosions or supersonic aircraft and rocket flight; and to probe the middle and upper atmosphere.

2 SOURCES OF INFRASOUND

Naturally occuring infrasound is produced continually by wind interaction with mountain ranges (mountain-associated infrasound) and intermittently from auroral motion, meteors, avalanches, storms, volcanoes, and earthquakes (Table 1). One of the most persistent source mechanisms is nonlinear interaction of ocean waves.[1] Particularly intense when storms are active at sea, these signals are known as microbaroms if the transmission path is through the atmosphere and microseisms if the transmission path is through the solid earth. The lack of correlation between microbaroms and microseisms is a reflection of the relatively rapid temporal changes in the atmospheric transmission path compared to the seismic transmission path.

Highly energetic infrasound is more commonly produced from man-made sources but does occur naturally on occasion. Two hours of infrasonic arrivals were recorded in Antarctica from the 1982 explosive eruption of El Chichon,[2] and infrasound from the 1980 eruption of Mt. St. Helens was recorded in China.[3] At least three man-made sources lead to long-distance propagation of infrasound: large explosions, rocket launches and reentry, and supersonic aircraft flight.

In any infrasonic measurement, there is a background of noise. Noise resulting from local nonacoustic pressure fluctuations (caused principally by wind at the sensor) can be substantially reduced by sensor construction (see Section 4) and electronic filtering. Acoustic noise generated by turbulence remote from the sensor is not so easily reduced. The common feature of this background is a power spectrum that decreases with increasing frequency. Power laws ranging from f^{-1} to f^{-3} (3–9 dB/octave) are evident in various data; however, f^{-2} (6 dB/octave) seems to be representative of this noise from about 0.1 to 10 Hz.[4] Theoretically, acoustic radiation from shear turbulence would give this same power law as would nonacoustic contributions from eddies and waves.

Note: References to chapters appearing only in the *Encyclopedia* are preceded by *Enc.*

TABLE 1 Natural Sources of Infrasonic Acoustic Waves

Source	Period (s)	Typical Levels	Characteristics	Mechanism	Altitude
Microbaroms	2–8	0.01–1 Pa	Diurnal, semidiurnal, seasonal variations	Nonlinear interaction of ocean waves	Sea level
Aurora	10–100	0.1–0.5 Pa	Not observed from 1200–1600 local time	Supersonic motion of auroral arc, joule heating or electromagnetic force	100+ km
Meteors	0.2–18	0.05–1 Pa at 200–1200 km	Often observe both stratospheric and thermospheric arrivals	Explosive interaction with the atmosphere	25 km and above
Avalanches	0.5–2	0.02–0.05 Pa at 100 km	Dominant spectral peak	Monochromatic structure may result from periodic leading-edge roll of avalanche	Surface
Volcanic eruptions	>100	15 Pa at 10,000 km from Mt. St. Helens	Long duration (hours) at long distances, multiple arrivals	High-energy explosive compression of atmosphere	Surface to 10+ km
Strong earthquakes	8–30	0.1–2 Pa at thousands of kilometres for large events	In addition to airborne arrivals, locally generated seismic-acoustic arrivals	Ground motion at epicenter and from intermediate regions, ground motion from seismic waves at receiving station	Surface
Mountain-associated waves	10–50	0.1–3 Pa	No diurnal variation, long duration, −10.5 dB/octave power spectrum	Associated with wake turbulence from wind flow over mountain ridges	Surface to several kilometres
Severe storms	0.1–50	0.05–0.3 Pa at 30–800 km	Not correlated with lightning, −3 to −6 dB/octave power spectrum from 2–16 Hz	Mechanism uncertain: vigorous convection, tornadoes, fronts?	Surface to tropopause

3 PROPAGATION CHANNELS

Infrasound propagates to extremely large distances in the atmosphere for two primary reasons: The absorption of infrasound is small, and strong refracting channels are formed naturally in the atmosphere. A secondary contributor is the high reflectivity of the ground at infrasonic frequencies. While its actual behavior has not been completely determined, below 20 Hz the ground is often adequately treated as a perfectly rigid reflector.

Atmospheric refracting channels are produced by vertical gradients in temperature and wind speed. Temperature in the "average" atmosphere first decreases with altitude (troposphere) to about 11 km and then is constant (tropopause) to about 20 km. Absorption of solar radiation by ozone produces a temperature increase (stratosphere) to about 45 km and a region of roughly constant temperature (stratopause) to about 55 km. Above that, the temperature decreases (mesosphere) to 85 km or so. The temperature then increases again (thermosphere) from solar heating of molecular oxygen. The sound speed profile based on the *1976 Standard Atmosphere*[5] temperature profile is shown in Fig. 1.

For ground-based observations of infrasound, the two principal propagation channels are the stratospheric duct, formed between the ground and the stratopause, and the thermospheric duct, formed between the ground and 100 km or so in the thermosphere. In addition, downwind propagation channels in the planetary boundary layer can form from vertical wind shear, and very strong, low-altitude channels are produced by strong temperature inversions common in the Arctic or over snow-covered surfaces in winter (Table 2). Vertical gradients in wind speed associated with the jet stream can also produce ducting.

High-altitude wind influences the stratospheric and thermospheric channels. There is a strong seasonal reversal of wind over a large region of altitude (tens of kilometres) in the vicinity of the stratopause.[6] In winter at midlatitudes, there is a strong west-to-east flow of up to 100 m/s at 70 km altitude; in summer, the dominant flow is east to west at up to 60 m/s and 50 km altitude. Since, without wind, the sound speed in the stratopause is approximately equal to that at the ground, the effectiveness of the stratospheric channel is strongly influenced by these winds.[7] For propagation paths from east to west, the winter wind reduces the effective sound speed in the stratopause, which allows much of the energy to escape into the thermospheric duct; the summer wind strengthens the stratospheric channel. For propagation from west to east, the channel is stronger in winter than summer.

Through the fall, winter, and spring, there is a strong semidiurnal tidal variation in the thermospheric wind[8] with an amplitude of up to 100 m/s above 80 km. This

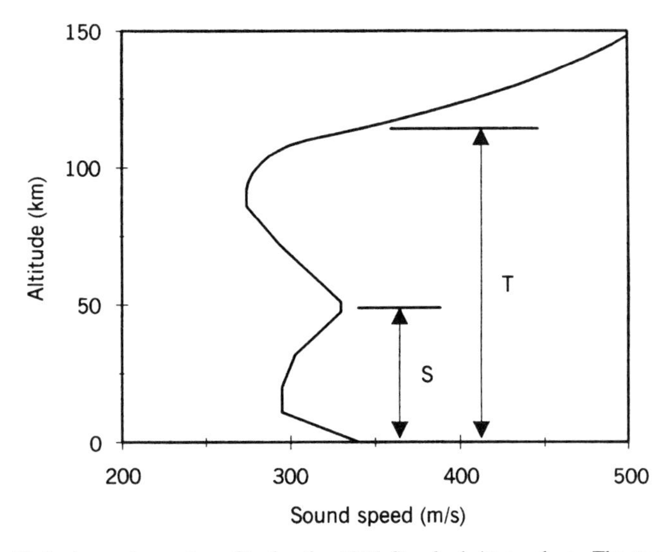

Fig. 1 Vertical sound speed profile for the 1976 Standard Atmosphere. The two primary propagation channels are the stratospheric channel (S) and the thermospheric channel (T).

introduces a semidiurnal fluctuation in infrasonic signals propagating through the thermosphere; this fluctuation can be used to identify this path.

Absorption of infrasound is so low in the troposphere and stratosphere that signals, even of a few hertz, can propagate to long distances. At these altitudes, the absorption can be determined using the procedures in Chapter 28. In the lower thermosphere, the pressure is low enough that the vibrational relaxation frequencies of oxygen and nitrogen are below the band of acoustic infrasonics. Over most of the band and at these altitudes, the ratio of frequency to ambient pressure is below 10 Hz/Pa, so that rotational relaxation need not be considered. Lacking detailed measurements of absorption at these low pressures, low frequencies, and very low humidity, the classical absorption coefficient is an adequate first estimate. For more details about the physics of absorption under these conditions, see Chapter 42.

The pressure at 80 km is about 1 Pa (or 10^{-5} atm). Assuming that the dry-air limits for relaxation frequency can be scaled by pressure, the oxygen and nitrogen relaxation frequencies should be 10^{-4}–10^{-5} Hz—below the acoustic region. For a bulk viscosity of 0.6 times the normal viscosity[9] η and a Prandtl number of 0.7, the classical absorption coefficient at the pressure p_s is

$$\alpha_{\text{cl}} = \frac{5\pi^2 \eta f^2}{\gamma p_s c}, \tag{1}$$

where γ is the ratio of specific heats and c is the sound speed. At 80 km altitude, this is $1.5 \times 10^{-6} f^2$ nepers per metre, or $13 f^2$ decibels per 1000 km. This would attenuate energy strongly above 1 Hz over the long paths typical of thermospheric propagation, but energy below a few tenths of hertz would propagate to long distances. This is consistent with the observed spectral content of infrasound refracted by the thermosphere.

Microbaroms (3–6-s periods) can be used to monitor conditions in the upper atmosphere by also monitoring microseisms to track changes in source strength.[8] Microbaroms recorded on the East coast show marked semidiurnal variation in the fall, winter, and early spring, indicating propagation through the thermospheric duct.

TABLE 2 Natural Propagation Channels

Channel	Altitude	Cycle Range	Frequency (Hz)
Thermosphere	100+ km	200–1000 km	<0.5
Stratosphere	50 km	10–300 km	<10
Planetary boundary layer	Several kilometres	1–10 km	
Arctic inversion	10–100 m	20–1000 m	

For propagation from the ocean to the coast (east to west), the dominant prevailing upper atmosphere wind reduces the effectiveness of the stratospheric duct during the winter. In the summer, when the stratospheric wind has reversed, propagation from east to west is enhanced in the stratospheric channel and the semidiurnal variation is not observed.

Systematic observations of long-range infrasound from transatlantic supersonic transport (SST) flights also reflect some of the features of high-altitude propagation channels.[10] In this case, the source (12–17 km altitude) is in the tropopause, a sound speed minimum. If the winds in the stratopause are in the same direction as the propagation and are of sufficient speed, energy refracted by the stratospheric channel reaches the ground. In this case, there are often two arrivals, one from stratospheric refraction and another from thermospheric refraction. Otherwise, the ground arrival is the thermospheric path. Most of the energy in the stratospheric arrival is between 0.5 and 2 Hz, while the thermospheric arrival is concentrated between 0.13 and 0.3 Hz.

At sufficiently low frequency, the effects of gravity become important in the propagation equation. The buoyant restoring force for vertical displacement becomes comparable to the restoring force from adiabatic compressibility. The resulting waves, which have both transverse and longitudinal components, are known as acoustic-gravity waves; their periods range from about 300 s to several hours.[11,12] For low enough frequency, the compressibility effects are negligible compared to buoyancy and the waves are called gravity waves.

Acoustic-gravity (and gravity) waves are supported only in those regions of the atmosphere that are convectively stable. That is, if a portion of air is displaced vertically and the buoyant force opposes that motion, then the region is stable. The natural frequency of the resulting free oscillation of such a displaced volume of air is the Brunt–Väisälä frequency. If air is treated as an ideal gas, this frequency is

$$f_{\mathrm{BV}}^2 = -\frac{g}{4\pi^2 T}\left[\frac{\rho g T(\gamma - 1)}{p_s \gamma} + \frac{dT}{dz}\right] \tag{2}$$

or, for $g = 9.8$ m/s^2, temperature T in kelvin, and height z in metres,

$$f_{\mathrm{BV}}^2 = \frac{0.0025}{T}\left[1 + 100\frac{dT}{dz}\right]. \tag{3}$$

If f_{BV}^2 is negative, the region is unstable for vertical displacements; a parcel of air displaced upward would continue to move upward. If f_{BV}^2 is positive, then the region is stable; f_{BV} is an upper limit for acoustic-gravity waves. The *1976 Standard Atmosphere* definition has no regions of instability, but in reality, the troposphere often has temperature gradients sufficiently negative for instability. Based on the *Standard Atmosphere*, the Brunt–Väisälä period in the tropopause, stratosphere, and stratopause is about 300 s. Long-range propagation of acoustic-gravity waves near this period are often considered diagnostic of source location in the tropopause or higher. Large storm systems that reach the tropopause can generate acoustic-gravity waves; the explosive eruption of El Chichon in 1982 penetrated the tropopause, and acoustic-gravity waves were received as far away as the antipodal point in Australia.[2]

Since the standard atmospheric sound speed profile supports refractive propagation to altitudes in excess of 100 km, not only are extremely long range paths possible, so also are extremely large amplitudes. Ignoring for the moment temperature differences, the density of the atmosphere decreases exponentially with altitude. Since the energy density in an acoustic wave is proportional to ρv^2, where v is the acoustic particle velocity, conservation of energy suggests that the velocity, and therefore the displacement, should increase exponentially (as $\rho^{-1/2}$). Consequently, displacement amplitudes of hundreds of metres (with a concomitant decrease in acoustic pressure) may be produced in the thermosphere.[12]

Not all infrasound is propagated entirely through the atmosphere. Strong earthquakes generate surface waves in the solid earth that have frequencies in the infrasound region. These waves travel with seismic speeds (kilometres per second) and so the coincidence angle for radiation into the atmosphere is almost vertical. Consequently, this component of the radiation is only detected acoustically when and where the seismic wave is present. The effective propagation speed from the epicenter is the seismic speed, and the acoustic amplitude is directly related to the amplitude of the ground motion. Strong infrasonic signals from the Alaskan earthquake of 1964 were received near Washington, DC, by this seismic-acoustic path.[13] About 5000 km from the epicenter, the Rayleigh wave produced ground motion with a 25-s period and 4-cm amplitude; this generated a local acoustic wave with an amplitude of 2 Pa. Other components of radiation, generated for example at the earthquake's epicenter or from regions with significant changes in terrain (mountain ranges), can also be detected under favorable propagation conditions in the atmosphere.

4 INSTRUMENTATION

Most systems for detecting infrasound below 0.1 Hz consist of a capacitor or moving-coil microphone element (or a pressure transducer) mounted inside a chamber. The chamber is divided into a front volume and a back volume with the microphone element mounted in the dividing wall. A capillary leak (the inlet) into the front volume determines the high-frequency cutoff, and another capillary leak between the front and back volumes determines the low-frequency cutoff.[14] In practice, the chamber volumes may be filled with steel wool and buried to reduce sensitivity to temperature fluctations. The inlet to the chamber is either exposed directly to the air or connected to a noise-reducing pipe. Above 1 Hz, capacitor microphone elements with only a back volume can be used in large wind screens. The back volume and capillary (back vent) roll the response off at lower frequencies. Some standard low-frequency capacitor microphones can be modified by reducing the back-vent aperture to increase the vent resistance, thereby lowering the low-frequency rolloff. In any infrasonic measurement system, it is particularly important to reduce the low-frequency response below the band of interest since the overall background spectrum goes roughly as f^{-2}. By doing this acoustically, the dynamic range requirement on the electronics is reduced.

Most infrasonic systems must be designed to reduce response to wind-generated local pressure fluctuations. Large wind screens are of some value, but these fluctuations can be reduced greatly with pipe arrays.[15,16] By effectively spreading the sampling region of a single sensor over a region much larger than the correlation distance of these pressure fluctuations but smaller than the acoustic wavelength of the intended signal, the signal-to-noise ratio can be improved substantially. This is achieved by connecting the sensor to one or more long pipes with capillary tubes spaced along the pipe to serve as sampling ports.

Based on observed correlation distances for nonacoustic pressure fluctuations, the minimum sampling port spacing is on the order of tens of centimetres for frequencies above 1 Hz, while below 0.01 Hz spacings of more than 5 m can be used. The acoustic wavelength sets an upper limit on the pipe length (and, therefore, the number of sampling ports). If the maximum dimension is greater than about 0.1 wavelength, the apparatus will start to exhibit noticeable directionality. This limits pipe array lengths to 30 or 40 m at 1 Hz, but at 0.01 Hz pipes of several kilometres may still be effective. Ideally, the signals should add coherently over all the inlets and the nonacoustic fluctuations should add incoherently. For N inlets, the signal power would then be proportional to N^2 and the noise power to N. Therefore, the theoretical gain in signal-to-noise power ratio would be N. In some circumstances, it may be more advantageous to distribute the sampling ports over an area with multiple pipes instead of over a line with a single pipe.

Frequently, several sensor systems are placed in a spatial array, and the outputs are processed for horizontal direction of arrival, horizontal phase speed ("trace" speed), and correlation. In this manner, further gain can be achieved against interfering signals and the origin of the desired signals can be determined. For example, a phase speed of the same order as the local sound speed indicates a nearly horizontal arrival as would be the case for a path through the stratospheric channel (since the sound speed in the stratopause is about equal to the sound speed at the ground). A phase speed of about twice the normal sound speed is often indicative of a thermospheric path. Phase speeds an order of magnitude higher than the local sound speed are observed for acoustic waves generated by local passage of seismic waves.

Suitable measures for infrasonic acoustic signals include frequency spectrum, time wave shape, spatial correlation, horizontal phase speed, and horizontal and vertical arrival direction. Local temperature and wind speed and direction must also be measured over the array site in order to relate measured phase speed to vertical direction of arrival. To locate unknown sources, the sound speed profile and the wind speed and direction profiles over the suspected propagation path must be known. For measurements undertaken in environments in which nonacoustic turbulence-generated signals are likely to interfere, the local wind speed and temperature profiles should be measured so that the turbulent stability of the flow can be determined.

This short introduction to infrasound can scarcely begin to cover the richness of the subject. Often ignored because it is not perceived, a continuous and highly variable background of infrasonic acoustic energy is present everywhere on earth and under the sea. In addition to the source mechanisms mentioned above, some machinery and transportation systems generate infrasound, which can excite very low frequency resonances in buildings and structures, which can in turn be a source of annoyance and fatigue to the occupants. A number of large mammals produce infrasonic sounds with sufficient variety and connection to behavior patterns that strongly suggest communication, perhaps over relatively long distances. Even a casual reading of the references will reveal the broad scope of existence and application of infrasound.

REFERENCES

1. E. Posmentier, "A Theory of Microbaroms," *Geophys. J. R. Astron. Soc.*, Vol. 13, 1967, pp. 487–501.
2. L. McClelland, T. Simkin, M. Summers, E. Nielsen, and T. Stein (Eds.), *Global Volcanism 1975–1985*, Prentice Hall, Englewood Cliffs, NJ, 1989, pp. 467–472.

3. L. Wen-Yi, "Detection and Analytical Results of the Infrasonic Signal from the Eruption of St. Helens Volcano," *J. Low Freq. Noise Vibr.*, Vol. 4, 1985, pp. 98–103.
4. E. Gossard, "Spectra of Atmospheric Scalars," *J. Geophys. Res.*, Vol. 65, 1960, pp. 3339–3351.
5. *U.S. Standard Atmosphere*, U.S. Government Printing Office, Washington, DC, 1976.
6. E. Batten, "Wind Systems in the Mesosphere and Lower Thermosphere," *J. Meteorol.*, Vol. 18, 1961, pp. 283–291.
7. T. Georges and W. Beasley, "Refraction of Infrasound by Upper-Atmosphere Winds," *J. Acoust. Soc. Am.*, Vol. 61, 1977, pp. 28–34.
8. W. Donn and D. Rind, "Microbaroms and the Temperature and Wind of the Upper Atmosphere," *J. Atmos. Sci.*, Vol. 29, 1972, pp. 156–172.
9. A. Pierce, *Acoustics*, Acoustical Society of America, New York, 1989, p. 553.
10. W. Donn, "Exploring the Atmosphere with Sonic Booms," *Am. Scientist*, Vol. 66, 1978, pp. 724–733.
11. T. Beer, *Atmospheric Waves*, Wiley, New York, 1974, pp. 29–34.
12. I. Tolstoy, *Wave Propagation*, McGraw-Hill, New York, 1973, pp. 63–66.
13. J. Young and G. Greene, "Anomalous Infrasound Generated by the Alaskan Earthquake of 28 March 1964," *J. Acoust. Soc. Am.*, Vol. 71, 1982, pp. 334–339.
14. J. Macdonald, E. Douze, and E. Herrin, "The Structure of Atmospheric Turbulence and Its Application on the Design of Pipe Arrays," *Geophys. J. R. Astron. Soc.*, Vol. 26, 1971, pp. 99-110.
15. F. Daniels, "Noise-Reducing Line Microphone for Frequencies Below 1 cps," *J. Acoust. Soc. Am.*, Vol. 31, 1959, pp. 529–531.
16. R. Burridge, "The Acoustics of Pipe Arrays," *Geophys. J. R. Astron. Soc.*, Vol. 26, 1971, pp. 53–69.

PART IV

UNDERWATER SOUND

30

INTRODUCTION

IRA DYER

In this introduction, I broadly survey the subject of underwater sound. My purpose is to help those wishing to glimpse its various facets, to better understand its at times confusing nomenclature, and to sense what is in the chapters of Part IV. I will set the stage by illuminating its application areas, describing its professional names and general outlooks, and discussing a few trends and scientific challenges that may help explain the formulation of some of its major topics. Many will find it appropriate to go directly to Chapters 31 and 39, see also *Enc.* Ch. 35, 37–40, 43, 44, 52, and 53, for they contain the central technical ideas of the subject.

Underwater sound involves a large range of distances and a wide spread of frequencies. Although no firm limits exist, present practice entails distances from about 1 m to 20,000 km and frequencies from about 1 Hz to 1 MHz. (Since the speed of sound in seawater is about 1.5 km/s, the latter correspond to wavelengths as large as 1.5 km and as small as 1.5 mm.) Within this broad sweep of parameters, a rich mixture of physical ideas, analytical and numerical methods, data, and empirical approaches reside, which unfold in Chapters 31 and 39, see also *Enc.* Ch. 35, 37–40, 43, 44, 52, and 53.

Sound waves are the principal means of long-distance wireless signaling in the ocean. While electromagnetic waves are carried by wires or fibers at the ocean bottom with high reliability, useful bandwidth, and low cost, in wireless use such waves cannot overcome the conductivity of seawater for the distances needed. On the other hand, sound waves propagating in seawater can provide long-distance signal links, although they do so with somewhat restricted bandwidths, imperfect reliability, and higher costs than we might like. But because sound can work, a wide range of applications are now to be found, from monitoring global climate change to detecting sunken ships on the bottom.

It might be tempting to think that sound waves in the ocean are direct analogs of electromagnetic waves in the air, since both can propagate in their respective media reasonably well. But acoustic wavelengths are relatively enormous (i.e., the sound propagation speed relatively tiny), so that the technologies are, with some exceptions, substantially different. For example, a side-scan sonar image of a familiar object on the sea bottom often requires an expert for its unequivocal identification, while a photographic image of the same object in air can be reliably identified without special training. The side-scan sonar image is simply a stack of *temporal* signals scattered by the object and taken in sequence as the sonar is towed by. The photographic image is, in contrast, a *spatial* display of signals scattered simultaneously by the object to a fixed receiver. While physically related, the two technologies, and their match to a human observer, are far different! One difference that often is crucial is the size–wavelength ratio for detecting and recognizing an object: The ratio is many orders of magnitude larger for the visible band of the electromagnetic spectrum, compared to the ratio for typical acoustic waves in seawater.

1 SOUND IN THE SEA

At the dawn of the sonar age, roughly from 1910 to 1940, sonars were high-frequency, short-range devices for which the classical three-dimensional inverse-square spreading law sufficed. (By short range I mean an observation distance comparable to or less than the ocean depth, which might be, as examples, less than 200 m

Note: References to chapters appearing only in the *Encyclopedia* are preceded by *Enc.*

on the continental shelf or less than 5000 m in the deep ocean.) With the growth of interest in longer ranges, the applicable spreading law changed from the three-dimensional model to a two-dimensional one; containment of sound at long ranges within the ocean's horizontal boundaries made a two-dimensional model essential. Since the ocean's gradients of ambient pressure, temperature, and salinity are larger in the vertical compared to the horizontal direction (*Enc.* Ch. 35), acousticians adopted range-independent water column approximations. These, then, when coupled to range-independent models for the bottom geology, completed the two-dimensional idea. A spreading law of inverse range, rather than inverse-*square* range, clearly means that larger sound power can be received in two-dimensional propagation and, therefore, that many applications at long ranges are feasible. In the main, this is the applications context today, as is reflected in most of the subsequent chapters of this Part.

Deterministic approaches to propagation prediction are relatively simple for short-range three-dimensional propagation. While more complicated for long-range two-dimensional cases, deterministic methods are nonetheless at hand and provide essential predictive and interpretive capabilities (see Chapter 31). Complexity in such methods arises precisely because the two-dimensional containment of sound produces multiple overlapped sound paths (multiple modes) with differing phases or time delays.

Statistical rather than deterministic methods that describe two-dimensional propagation are also in use.[1–6] These, in essence, treat the multipath phases as random, that is, the multipaths as incoherent. Statistical methods are easier to apply, implicitly average over the more troublesome complexities, agree well with the broad trend of measurements, but submerge details of the propagation process induced by coherence among the multipaths. To expand the latter point, the ocean is a dynamic medium with spatial and temporal fluctuation scales that readily perturb the phases of long-range sound paths. When these perturbations are not known or little understood, or when broad predictive/interpretive trends suffice, statistical methods are viable alternatives to the deterministic ones. Indeed, although powerful deterministic methods were already available, I estimate that up until about 1975, statistical methods were favored by practitioners.

About 1975, practitioners' choices for long-range two-dimensional propagation modeling began to shift strongly to determinism, and that trend continues to this day. Many factors account for the shift, the main one being that research and operational sonars increasingly include high-resolution capabilities. Large arrays and optimum array processing algorithms in such high-resolution systems refine the ocean's angular domain being measured or interrogated. Large bandwidths in such sonars enable coherent processing of signals in short well-resolved time windows and reduce the range-wise extent of the measurement/interrogation volume. With use of high-resolution sonars, the ocean thus can be viewed more deftly through a few or even one path, rather than through many overlapping paths. Consequently, sound is now a tool for studying detailed ocean processes or for more precisely detecting and tracking targets in the ocean. The limits of sonar in pursuit of such high-resolution applications are not yet reached, promising broader scientific challenges and further important applications for many years to come.

2 SUBFIELDS OF UNDERWATER SOUND

2.1 Ocean Acoustical Metrology

An early application of sound in the ocean (about 1925) was in depth measurement, a use that continues to this day, although often with much higher resolution of bottom topography than that needed for surface ship navigation (its original objective). Depth sounders are suggestive of a class of uses that can be called *ocean acoustical metrology*, in which one observes changes in sound waves as affected by one or more ocean properties (depth, sea height, water temperature, current, etc.). The observation is then inverted to reveal the ocean property. In many cases other sensors could do the same, but sound can measure remotely (as in depth), synoptically (as in eddy tomography), and rapidly (as in sea ice fracture), capabilities that are increasingly important for large-area monitoring, long-time sampling, or fast diagnosing.

Oceanographers, marine geologists/geophysicists, and marine biologists need acoustic measurement tools to understand basic oceanic processes (see Chapter 38), and many have adopted the name *acoustical oceanography* for this form of ocean acoustical metrology. Offshore petroleum engineers, marine soil engineers, naval architects, fishery resource managers, and ocean environmental engineers likewise need such tools (as in Chapters 38 and 39, see also *Enc.* Ch. 44). Because the problems of these user groups are different in scope and outlook, distinctions are likely to evolve, but for now the user groups are seen as one within the name acoustical oceanography.

Whatever distinctions might evolve, the disciplines in acoustical oceanography have invariably become joint ones, for example oceanography *and* acoustics. The implications of the interdisciplinary nature of acoustical oceanography are important and far-reaching: in the conduct of research and in the development of measurement systems, intersections of one or more relevant disciplines with acoustics are firmly established and highly productive. An observer of this subfield, or a newcomer to it, would better appreciate its activities with this in mind. But it must be said that in education such interdisciplinary intersections are at best still at the formative stages.

2.2 Ocean Acoustics

The design, development, and underlying science of acoustical systems for use in the oceans generally falls in a class of activities called *ocean acoustics*. Probably only those professionals close to acoustical oceanography or ocean acoustics care about the distinctions in the names, but the objectives of the two subfields are different. In the latter, one wants to detect an adversary's submarine, to navigate an underwater search vehicle with bottom-fixed beacons, to find harvestable schools of fish, and to receive and decipher mammalian vocalizations as just a few examples. In such applications the focus is on the task the system is intended to perform, as affected by the relevant ocean properties, and not on measurement of those properties per se (as in acoustical oceanography). There is, however, little or no difference between the fundamentals of the two subfields as they relate to ocean science, since issues such as propagation, scattering, and the like need to be addressed in each. Thus the chapters in Part IV are properly silent on the distinction between ocean acoustics and acoustical oceanography, although a sensitive observer will at times detect a leaning to engineering as well as to ocean science in ocean acoustics.

The bases of ocean acoustics are those interdisciplinary ones at the base of acoustical oceanography, almost always with additional intersections between acoustics and certain engineering sciences, such as signal processing, instrumentation, and system design and optimization. In its fullest form, therefore, ocean acoustics could be called "sonar engineering," but this name is less commonly used. Bonds in ocean acoustics at the various intersections of acoustics with its other disciplines are, for the most part, quite strong and well developed. This strength cuts across the avenues of research, development, design, and education.

I believe that two large challenges confront the science of ocean acoustics: (1) more detailed knowledge is required of the ocean related to the increasing resolution of sonars and (2) more powerful predictive/interpretive tools are needed where the two-dimensional environmental assumptions cannot be justified. Scattering from a rough ocean bottom, as an example under (1), begs for detailed knowledge of rock outcrops within *small* bottom footprints, each of which may be a sample from a spatially *nonstationary* random process. And, as an example under (2), propagation on the continental shelf requires acoustic models that can efficiently deal with anisotropic heterogeneities in the water column or the bottom.

Systems used in acoustical oceanography and ocean acoustics are of two kinds: *passive sonars* and *active sonars*; these are described in Chapter 37. A sonar that receives signals created by breaking surface gravity waves (to measure wind stress) is, for example, a passive sonar; one that radiates an acoustic signal *and* receives that signal scattered from an autonomous underwater vehicle (to track it) is an example of an active sonar. The fundamentals of passive and active sonars have much in common, and when they do, the following chapters need not distinguish between them. But the fundamentals can also differ. For an active sonar one needs to address interfering or unwanted acoustic scatter. This is scatter of the radiated acoustic signal from oceanic features, also called *reverberation*, rather than that from the target, and such cases are treated separately.

A major difference between passive and active sonars is in their signal processing design, implementation, and display. But there are also large areas of commonality, for example in array configuration and signal handling. All are topics of extraordinary interest, challenge, and importance and are covered in Chapter 37. These are to be coupled with other engineering topics such as transducers (which connect a sonar's shipboard subsystems to the ocean) and with topics in the acoustics of the oceans to form an overall view of ocean acoustics.

2.3 Marine Structural Acoustics

Another subfield of underwater sound is *marine structural acoustics*, which is covered in Chapters 32 and 33, see also *Enc.* Ch. 43 and (indirectly) 35. This subfield deals with elastic waves in floating or submerged structures and their coupling to acoustic waves in the water. Two factors account for its difficulty and importance: First, structures of practical interest are usually quite large and geometrically complicated and therefore can contain a variety of elastic wave types that interact at structural discontinuities to produce a total elastic wave field of high complexity. Second, since structure/water densities and wave speeds are not too dissimilar, this complicated structural wave field is strongly coupled to the acoustic field in the water. Thus, for example, sound scattered from a ship's hull is different from the incident signal, not only in its signal amplitude, but also in its time (or frequency) characteristics.

The underlying disciplines of marine structural acoustics are structural dynamics *and* acoustics, an intersection that is reasonably well established in research and development and nearly so in education. In theory and in practice, however, marine structural acoustics has yet to develop substantive intersections with basic disciplines in the strength of structures, perhaps because issues in the latter are largely static or quasi-static. The net result is that a marine structure is, with few exceptions, considered separately from the two perspectives, and thus not jointly optimized for strength and acoustics. I believe the challenge of joining marine structural acoustics with the strength-related disciplines is not only important but also achievable.

Other challenges in marine structural acoustics are also important. Structures of interest to the underwater sound community have size–wavelength ratios of about 1–10^3. The lower half of this regime (1–30) is one in which details of the acoustic–elastic wave coupling is crucial in high-resolution applications. But exact analytical or numerical methods cannot now account for the immense complexity in a practical structure. While the ultimate challenge for the lower half of the size regime is to develop such methods, an intermediate challenge is to develop hybridized partial models that can robustly predict/interpret acoustic–elastic coupling for practical structures. This is on the horizon and realistically achievable. Chapters 32 and 33 provide approaches to partial models that are not yet, however, in a hybrid amalgam.

In the upper half of the size–wavelength regime (30–10^3), acoustic–elastic coupling is dominated by the shape and material properties of the structure's outer skin. Recent progress in this size regime for scattering of underwater sound waves from elastic bodies is covered in *Enc.* Ch. 43. The methods described there are generally applicable to acoustic–elastic coupling, for example in radiation from structures as well as in scattering.

2.4 Hydroacoustics

Finally, *hydroacoustics* (also known as *hydrodynamic noise*) is a subfield of underwater sound that addresses noise from fluctuating forces on marine propellers, from turbulent boundary layers acting on moving vehicles, from vortices shed by flow over cavities, in fact from any portion of an unsteady-flow field. Chapter 34 covers this subfield. When the unsteadiness is strong, the noise can be detected at large distances. (Example: Propeller noise of a merchant ship traveling at high speed can be detected across an ocean!) But even when relatively weak, hydrodynamic noise can interfere with nearby active or passive sonars; such locally important noise is commonly termed *self-noise*. (Sidelight: Sonar domes or outer decouplers between sonar transducers and a flow field are used to reduce noise from turbulent boundary layers and, if properly configured, do not degrade acoustic transmission!)

Fluid dynamics *and* acoustics are the intersecting disciplines underlying hydroacoustics. In some important areas, marine structural acoustics is closely tied to hydroacoustics (e.g., in understanding and design of the aforementioned sonar domes); in such cases, exquisite attention needs to be given to the wavenumber content of both the acoustic and the fluctuating hydrodynamic fields as they may match/mismatch the wavenumber properties of the structure.

The greatest scientific challenges in hydroacoustics mirror those in fluid dynamics, at least for flows at low Mach and high Reynolds numbers. For example, the wavenumber content of anisotropic turbulence, up to and beyond the Kolmogorov scale, could well be important in propeller noise but is little understood.

3 MAJOR SCIENCE/ENGINEERING TOPICS OF UNDERWATER SOUND

As may be surmised from my emphasis on it in the previous sections, *propagation* commands a large share of attention among underwater sound professionals. It includes absorption of sound in sea water caused by chemical relaxation, notably the Epsom salt and the boric acid reactions, as described in *Enc.* Ch. 35. While present in quite small concentrations, these constituents are significant in setting the maximum practical range sound waves can reach at a set frequency (or the maximum practical frequency for a set range). Specifically, maximum range and maximum frequency are inversely related via chemical relaxation (or other loss mechanisms), so that long-range sonars operate at low frequencies, and vice versa.

Propagation also includes the effects of geometric spreading of sound, which, as discussed in Section 1, is, for short ranges, the classical three-dimensional inverse-square spreading law. For short ranges, one can usually disregard wave refraction, but not so for long ranges. Instead wave refraction, caused by spatial sound speed gradients (which originate from spatial gradients of temperature, salinity, and density, as detailed in *Enc.* Ch. 35), not only changes the spreading law but also modifies vital details of the multipaths (see Chapter 31). Specifically, to reach long ranges, the refracted sound in most cases interacts effectively with only *one* of the ocean's top and bottom interfaces and in a few cases with *neither* interface. Refraction in a range-independent ocean thus forms two-dimensional propagation at long ranges, but the containing horizontal interfaces need not be the physical top or bottom. With use of ray language (Chapter 31), refraction bends the sound rays away from one or both physical interfaces and does so periodically in range to form a *duct* or *channel* that contains the sound in two dimensions. Equivalently, with use of mode language (also Chapter 31), those modes whose amplitudes are small near one or both physical interfaces form the two-dimensional duct.

While sound will not transmit effectively from water

to air through the top interface, because of the large contrast in density and bulk modulus, it can do so effectively from water to the geological medium below the bottom interface. When a ray path strikes the physical bottom, or equivalently, when a mode's amplitude near the bottom is not small, the propagation parameters given in *Enc.* Ch. 37 need to be invoked. In this process, the propagation medium is greatly expanded via good coupling, since it then includes the geological medium, whose thickness is typically of the same order as or much thicker than the seawater itself. *Ocean seismic acoustics* is the name sometimes applied to these cases, to emphasize the physical importance of coupling between the two media.

Because they are usually rough, *scattering* can occur at the physical ocean interfaces provided, of course, that rays (modes) strike one or both of them. In propagation our concern is with *forward scatter*, for example, with that portion of the redirected sound field at the interface pointing around the forward propagation direction. *Enc.* Ch. 38 describes the main ideas and results in forward scatter. For active sonars, with the receiving sonar at or near the radiating sonar, our conern also is with *backscatter* at an interface, for this determines the interface *reverberation* (defined as backscatter reaching the receiving sonar).

Reverberation (see *Enc.* Chs. 40) can interfere with the signal and thereby limit sonar performance. For a given interface, the characteristics of forward scatter and backscatter are not in general the same, and consequently in the chapters just referred to they are covered separately.

Scattering also occurs in the ocean volume, with causes as diverse as fish schools or turbulence patches. Other scattering sources are found in the geological volume, such as shellfish debris buried in sediment. As in the interface case, volume scattering is divided into forward and back categories. *Enc.* Ch. 40 and 44 should be accessed for details, but the user of this textbook will surely note that less practical knowledge is available for volume than for interface scattering, a consequence of its lesser importance in the operation of most sonar systems.

Sometimes a given scattering mechanism degrades or interferes with a signal and sometimes it is the signal. Scattering from the rough underside of pack ice, for example, degrades active sonar tracking of an underwater vehicle in the Arctic Ocean or provides signals for the ice-keel avoidance sonar mounted on the vehicle.

In the context of active sonar, *target strength* deals with man-made structures (Chapter 33, see also *Enc.* Ch. 43) and fish and fish schools (*Enc.* Ch. 44) as sources of scatter. A sonar using scatter in the back direction (simply, "backscatter") is known as a *monostatic active* one. *Bistatic active* sonars that use an arbitrary scatter angle are technically feasible, although more costly because of the need for an additional sonar platform. (*Mono* in monostatic refers to the single location of the radiating and receiving subsystems of the active sonar, *bi* in bistatic to the two separate horizontal locations. *Multistatic active* sonars, with multiple horizontal locations, are also technically feasible. The origin of *static* carries no essential meaning as these names are used today.)

Another topic relating to man-made structures is *ship noise* (Chapter 35). Beyond the hydrodynamic and self-noise introduced previously, ships or ocean platforms typically contain noisy machinery. Machinery can be significant sources of self-noise and also of noise radiated to large ranges. As examples, machinery noise radiated by submarines, if uncontrolled, can lead to easy detection at long ranges, or machinery aboard offshore petroleum platforms is often of concern because of possible adverse effects on nearby sea life.

A ubiquitous feature of signals observed in the ocean is their *fluctuations* (see *Enc.* Ch. 39). Unsteadiness in source properties is not a factor; in most cases fluctuations are dominated by effects of ocean scattering and/or multipaths. (As described previously in the Introduction, multipaths are readily developed by refraction and concomitant trapping of sound within the ocean's top and bottom boundaries.) Fluctuations affect a sonar's ability to relay information, and thus their causes and characteristics are important in underwater sound. Whether the application entails a long-range sonar for detecting and tracking a target or a much shorter range sonar for *telemetry* of broadband data (see Chapter 39 on telemetry), fluctuations set performance bounds via time spreads, frequency spreads, short-term amplitude variations, and the like.

Performance also depends upon signal level relative to noise. Beyond backscatter (i.e., reverberation) for an active sonar, *ambient noise* in the ocean must be considered for both active and passive sonars as the ultimate limit to performance. Chapter 36 covers ambient noise. Such noise is created by various mechanisms (breaking surface waves, rain, ships distributed over a large area, etc.). As variable as the weather, ambient noise has its own spatial and temporal fluctuation scales that, besides its mean level, affect sonar performance.

Directional receivers discriminate against (spatially filter) ambient noise and therefore improve performance. In the main, directionality is achieved with use of an *array* of elemental receivers. To be useful, an array's size should be larger than about one wavelength; the larger it is compared to wavelength, the more directional it is. Simply put, an array filters noise by receiving it through the array's weakly responding side lobes, while the desired signal is received through the strongly responding main directional lobe. Chapter 37 should be read for details on receiving arrays.

Arrays are also used in sonars to radiate sound. (Again, see Chapter 37.) As in reception, an array's size compared to wavelength is important. A large array directs most of the sound power, through its main lobe, to the desired angular region, leaving only a small fraction of the total power to be radiated in undesired directions via the weak side lobes. In radiation, therefore, an array optimizes the use of sonar power.

Both receiving and transmitting arrays need to be steered, which is done by phase (or time) modulation among individual transducers forming an array. Early in sonar history, mechanical motion of the entire array accomplished steering, but except for a few cases, this older technology has been supplanted by electrical control of the signals among the array elements; see Chapter 37.

Parametric arrays (described in *Enc.* Ch. 53) make use of seawater nonlinearity in response to very high amplitude sound. Size for such an array is determined by the absorption length of high-amplitude sound at its two somewhat different "pump" frequencies, set much larger than the signal frequency (the difference in the pump frequencies). Thus the acoustical size of a parametric array, set by its absorption length, is considerably larger than its actual size, an advantage one must pay for with increased source power.

Sound waves in an ideal fluid are described by three scalar thermodynamic variables (acoustic perturbations in pressure, temperature, and density) and by one vector kinematic variable (the three components of particle velocity, say). Basic fluid equations connect these six variables so that the complete sound field can be uniquely described by one thermodynamic variable plus the one kinematic (vector) variable. Far from a discontinuity (sound source, scatterer, boundary), the list of variables required to describe the field can be reduced to one. For underwater sound, as in air acoustics, it is sound pressure. *Transducers* (*Enc.* Ch. 52) typically used in underwater sound produce an electrical output proportional to the sound pressure acting on it, and vice versa, and thereby connect the electrical domain of the sonar with the acoustical domain of the ocean. That sound pressure is the variable of underwater sound has no fundamental significance; its use reflects the simplicity of a scalar quantity and the accident of technological evolution centered on sound pressure transducers.

Close to a sound field discontinuity, more than one scalar variable is needed. This need arises mainly in research or testing contexts, and when it does, the sensors used are usually pressure and vector velocity transducers or vector intensity transducers.

Rather than electrical, some sources transduce chemical, hydaulic, or pneumatic energy to sound pressure (*Enc.* Ch. 47). Compared with electrically driven sources, the latter can produce very high source strengths. Thus, these are often used by petroleum exploration geophysicists to overcome high losses suffered by waves propagating in the sea bottom or by global climate monitoring scientists to sense sound signals propagated via very long (transoceanic) paths to give but two of many application examples.

The references that follow are particular to the development and fruition of statistical methods for prediction and interpretation of sound waves that propagate to long ranges in two-dimensional ocean ducts. These are included here because such methods still have practical use and are not treated elsewhere in Part IV.

Ample references to all other topics are included in Chapters 31 and 39, see also *Enc.* Ch. 35, 37–40, 43, 44, 52 and 53.

REFERENCES

1. L. M. Brekhovskikh, *Waves in Layered Media*, 1st ed. (trans. D. Lieberman), Academic Press, New York, 1960, pp. 415–426.
2. L. M. Brekhovskikh, "The Average Field in an Underwater Sound Channel," *Sov. Phys.–Acoust.*, Vol. 11, 1965, pp. 126–134.
3. P. W. Smith, Jr., "Averaged Impulse Response of a Shallow-Water Channel," *J. Acoust. Soc. Am.*, Vol. 50, 1971, pp. 332–336.
4. D. E. Weston, "Intensity-Range Relations in Oceanographic Acoustics," *J. Sound Vib.*, Vol. 18, 1971, pp. 271–287.
5. P. W. Smith, Jr., "Spatial Coherence in Multipath or Multimodal Channels," *J. Acoust. Soc. Am.*, Vol. 60, 1976, pp. 305–310.
6. L. M. Brekhovskikh and Yu. Lysanov, *Fundamentals of Ocean Acoustics*, Springer-Verlag, New York 1982, pp. 104–108.

31

PROPAGATION OF SOUND IN THE OCEAN

WILLIAM A. KUPERMAN

1 INTRODUCTION

The ocean is an acoustic waveguide bounded above by the air–sea interface and below by a viscoelastic layered structure, the latter commonly called the ocean bottom. The physical oceanographic parameters, as ultimately represented by the ocean sound speed structure, make up the index of refraction of the water column waveguide. The combination of water column and bottom properties leads to a set of generic sound propagation paths descriptive of most propagation phenomena in the ocean. This chapter first reviews the qualitative properties of the various propagation paths. Then the ocean acoustic wave equations with the appropriate coefficients and boundary conditions are presented. Sound propagation models are essentially algorithms for solving the equations; the word *model* is used because the technique used for realistic ocean scenarios typically is implemented as a computer model (program). Many models exist because it is not computationally efficient to use a single algorithm for all frequencies and ocean environments (e.g., water depth, variable bathymetry). Acoustic measurements at sea normally show significant fluctuations that are not predicted by the deterministic class of models considered here; an output from these models represents an average prediction. The models discussed are only as realistic as the input environment, and they do not include temporal ocean variability. A basic set of models are reviewed and results obtained from these models are presented to further elucidate the physics of sound propagation in the sea. Throughout this chapter, the depth coordinate z is positive in the downward direction. The appendix reviews some relevant units.

Note: References to chapters appearing only in the *Encyclopedia* are preceded by *Enc.*

2 QUALITATIVE DESCRIPTION OF OCEAN SOUND PROPAGATION PATHS

Enc. Ch. 35 summarizes essential details of physical oceanography vis à vis acoustics. Here we summarize aspects of oceanography that impact propagation paths before we go into more detail on the propagation itself.

2.1 Selective Review of the Ocean Environment

Figure 1 illustrates a typical set of sound speed profiles indicating greatest variability near the surface as a function of season and time of day. In a warmer season (or warmer part of the day), the temperature increases near the surface and hence the sound speed decreases with depth. In nonpolar regions, the oceanographic properties of the water near the surface result from mixing activity originating from the air–sea interface. This near-surface mixed layer has a constant temperature (except in calm, warm surface conditions as described above). In this isothermal mixed layer, the sound speed profile can increase with depth due to the pressure gradient effect. This is the "surface duct" region.

Below the mixed layer is the thermocline, in which the temperature decreases with depth and the sound speed decreases with depth. Below the thermocline, the temperature is constant (about 4°C, a thermodynamic property of salt water at high pressure) and the sound speed increases because of increasing pressure. Therefore, there exists a depth between the deep isothermal region and the mixed layer with a minimum in sound speed; this depth is often referred to as the axis of the deep sound channel. However, in polar regions, the water is coldest near the surface, and hence the minimum sound speed is at the ocean–air (or ice) interface, as indicated in Fig. 1. In continental shelf regions (shallow water) with

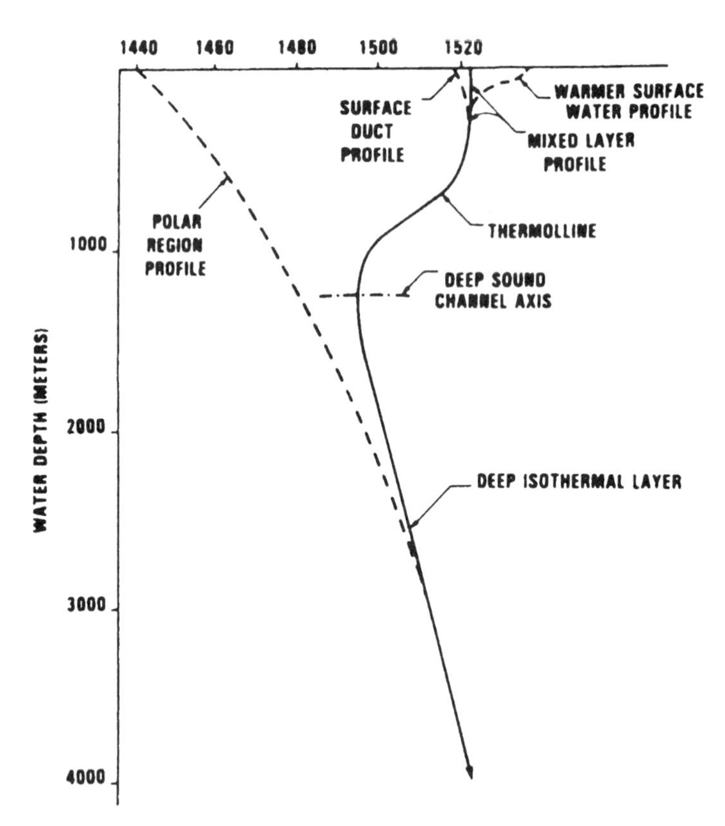

Fig. 1 Generic sound speed profiles.

water depths on the order of a few hundred metres, only the upper region of the sound speed profile in Fig. 1, which is dependent on season and time of day, affects sound propagation in the water column.

Figure 2 is a contour display of the sound speed structure of the North and South Atlantic,[1] with the deep sound channel indicated by the heavy dashed line. Note the geographic (and climatic) variability of the upper ocean sound speed structure and the stability of this structure in the deep isothermal layer. For example, as explained above, the axis of the deep sound channel becomes shallower toward both poles, eventually going to the surface.

2.2 Sound Propagation Paths in the Ocean

Figure 3 is a schematic of the basic types of propagation in the ocean resulting from the sound speed profiles (indicated by the dashed lines) discussed in the last section. These sound paths can be understood from a simplified statement of Snell's law: Sound bends locally toward regions of low sound speed (or sound is "trapped" in regions of low sound speed). Paths *A* and *B* correspond to surface duct propagation where the minimum sound speed is at the ocean surface (or at the bottom of the ice cover for the Arctic case). Path *C*, depicted by a ray leaving a deeper source at a shallow horizontal angle, propagates in the deep sound channel, whose axis is at the shown sound speed minimum. This local minimum tends to become more shallow toward polar latitudes converging to the Arctic surface minimum. Hence, for midlatitudes, sound in the deep channel can propagate long distances without interacting with lossy boundaries; propagation via this path has been observed over distances of thousands of kilometres. Also, from the above description of the geographical variation of the acoustic environment combined with Snell's law, we can expect that shallow sources coupling into the water column at polar latitudes will tend to propagate more horizontally around an axis that becomes deeper toward the midlatitudes. Path *D*, which is at slightly steeper angles than those associated with path *C*, is convergence zone propagation, a

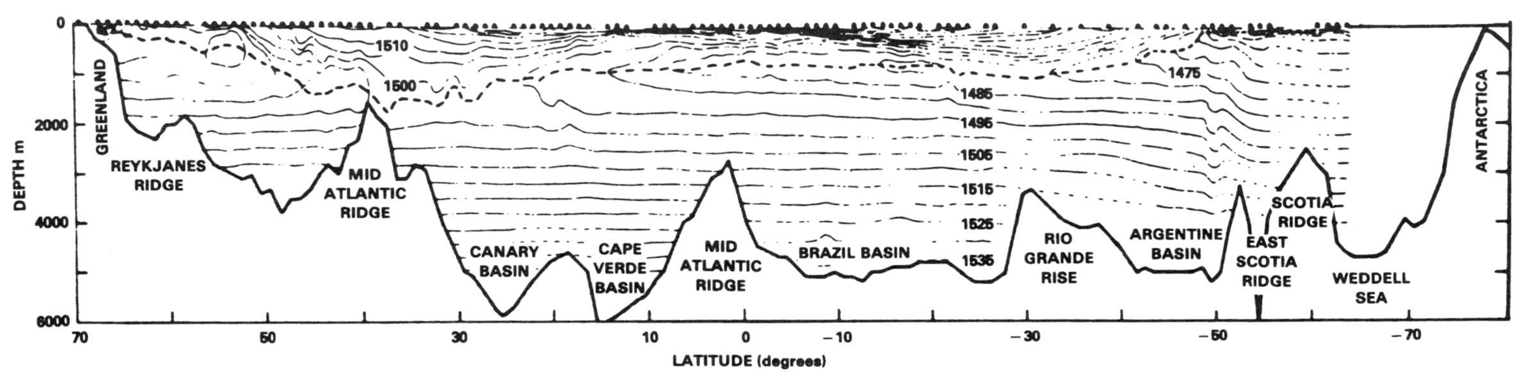

Fig. 2 Sound speed contours at 5-m/s intervals taken from North and South Atlantic along 30.50° W. Dashed line indicates axis of deep sound channel (from Ref. 1).

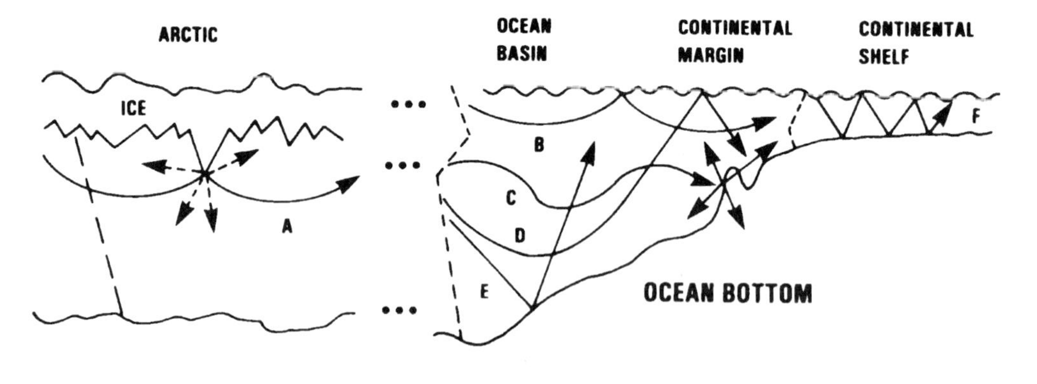

A. ARCTIC
B. SURFACE DUCT
C. DEEP SOUND CHANNEL
D. CONVERGENCE ZONE
E. BOTTOM BOUNCE
F. SHALLOW WATER

Fig. 3 Schematic representation of various types of sound propagation in the ocean.

spatially periodic (~35–65 km) refocusing phenomenon producing zones of high intensity near the surface due to the upward refracting nature of the deep sound speed profile. Referring back to Fig. 1, there may be a depth in the deep isothermal layer at which the sound speed is the same as it is at the surface. This depth is called the *critical depth* and, in effect, is the lower limit of the deep sound channel. A *positive* critical depth specifies that the environment supports long-distance propagation without bottom interaction, whereas a *negative* one implies that the bottom ocean boundary *is* the lower boundary of the deep sound channel. The bottom bounce path E, which interacts with the ocean bottom, is also a periodic phenomenon but with a shorter cycle distance and a shorter total propagation distance because of losses when sound is reflected from the ocean bottom. Finally, the right-hand side of Fig. 3 depicts propagation in a shallow-water region such as a continental shelf. Here sound is channeled in a waveguide bounded above by the ocean surface and below by the ocean bottom.

The modeling of sound propagation in the ocean is further complicated because the environment varies laterally (range dependence), and all environmental effects on sound propagation are dependent on acoustic frequency in a rather complicated way that often makes the ray-type schematic of Fig. 3 misleading, particularly at low frequencies. Finally, a quantitative understanding of acoustic loss mechanisms in the ocean is required for modeling sound propagation. These losses are, aside from geometric spreading, volume attenuation (*Enc.* Ch. 35 and 38), bottom loss (i.e., a smooth water–bottom interface is not a perfect reflector), and surface, volume (including fish), and bottom scattering loss (*Enc.* Ch. 38–40 and 44). Here we will only review those aspects of bottom loss that impact propagation structure.

2.3 Bottom Loss

Ocean bottom sediments are often modeled as fluids since the rigidity (and hence the shear speed) of the sediment is usually considerably less than that of a solid such as rock. In the latter case, which applies to the "ocean basement" or the case where there is no sediment overlying the basement, the medium must be modeled as an elastic solid, which means it supports both compressional and shear waves.

Reflectivity, the amplitude ratio of reflected and incident plane waves at an interface separating two media, is an important measure of the effect of the bottom on sound propagation. For an interface between two fluid semi-infinite half-spaces with density ρ_i and sound speed c_i, $i = 1, 2$, as shown in Fig. 4*a* [assuming a harmonic time dependence of $\exp(-i\omega t)$], the reflectivity is given by

$$\mathcal{R}(\theta) = \frac{\rho_2 k_{1z} - \rho_1 k_{2z}}{\rho_2 k_{1z} + \rho_1 k_{2z}}, \tag{1}$$

with

$$k_{iz} = \frac{\omega}{c_i} \sin\theta_i \equiv k_i \sin\theta_i, \qquad i = 1, 2. \tag{2}$$

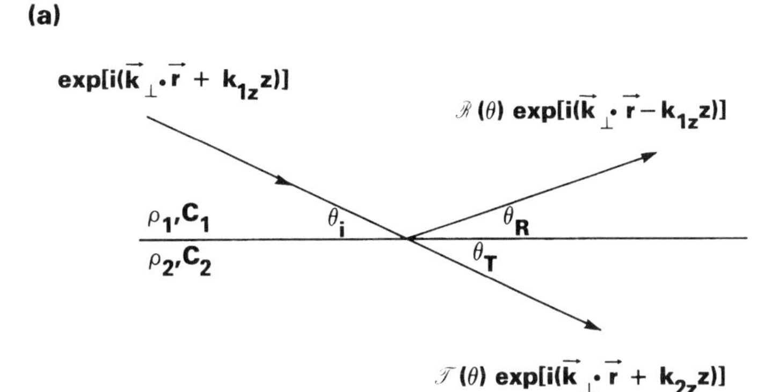

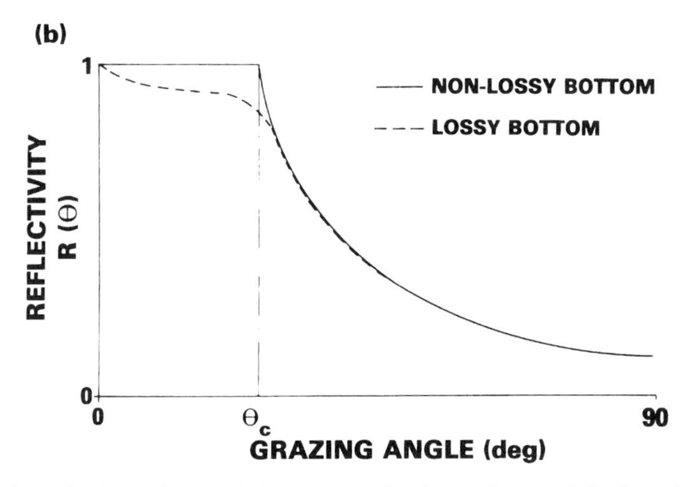

Fig. 4 Reflection and transmission process. Grazing angles are defined relative to the horizontal.

The incident and transmitted grazing angles are related by Snell's law,

$$k_\perp = k_1 \cos \theta_1 = k_2 \cos \theta_2, \tag{3}$$

where the incident grazing angle θ_1 is also equal to the angle of the reflected plane wave. The function $\mathcal{R}(\theta)$ is also referred to as the *Rayleigh reflection coefficient* and has unit magnitude (total internal reflection) when the numerator and denominator of Eq. (1) are complex conjugates. This occurs when k_{2z} is purely imaginary, and using Snell's law to determine θ_2 in terms of the incident grazing angle, we obtain the *critical grazing angle* below which there is perfect reflection,

$$\cos \theta_c = \frac{c_1}{c_2}, \tag{4}$$

so that a critical angle can exist only when the speed in the second medium is higher than that of the first. Using Eq. (2), Eq. (1) can be rewritten as

$$\mathcal{R}(\theta) = \frac{\rho_2 c_2/\sin \theta_2 - \rho_1 c_1/\sin \theta_1}{\rho_2 c_2/\sin \theta_2 + \rho_1 c_1/\sin \theta_1} \equiv \frac{Z_2 - Z_1}{Z_2 + Z_1}, \tag{5}$$

where the right-hand side is in the form of *impedances* $Z_i(\theta_i) = \rho_i c_i/\sin \theta_i$, which are the ratios of the pressure to the vertical particle velocity at the interface in the ith medium. Written in this form, more complicated reflection coefficients become intuitively plausible. Consider the case in Fig. 4*a* where the second medium is elastic and thus supports shear as well as compressional waves with sounds speeds c_{2s} and c_{2p}, respectively. The Rayleigh reflection coefficient is then given by

$$\mathcal{R}(\theta) = \frac{Z_{2,\text{tot}} - Z_1}{Z_{2,\text{tot}} + Z_1}, \tag{6}$$

with the total impedance of the second medium being

$$Z_{2,\text{tot}} \equiv Z_{2s} \sin^2 2\theta_{2s} + Z_{2p} \cos^2 2\theta_{2p}. \qquad (7)$$

Snell's law for this case is

$$k_1 \cos \theta_1 = k_{2s} \cos \theta_{2s} = k_{2p} \cos \theta_{2p}. \qquad (8)$$

In lossy media, attenuation can be included in the reflectivity formula by taking the sound speed as complex so that the wavenumbers are subsequently also complex, $k_i \to k_i + \alpha_i$.

Figure 4*b* depicts a simple bottom *loss* curve derived from the Rayleigh reflection coefficient formula where both the densities and sound speed of the second medium are larger than those in the first medium, with unit reflectivity indicating perfect reflection. For loss in decibels, 0 dB is perfect reflecting, 6 dB loss is an amplitude factor of $\frac{1}{2}$, 12 dB loss is of $\frac{1}{4}$, and so on. For a lossless bottom, severe loss occurs above the critical angle in the water column due to transmission into the bottom. For the lossy (more realistic) bottoms, only partial reflection occurs at all angles. With paths involving many bottom bounces (shallow-water propagation), bottom losses as small as a few-tenths of a decibel per bounce accumulate and become significant because the propagation path may involve many tens of bounces. Further information on the acoustic properties of the ocean bottom are given in *Enc.* Ch. 37.

Path E in Fig. 3, the bottom-bounce path, often involves paths that correspond to angles near or above the critical angle; therefore, after a few bounces, the sound level is highly attenuated. On the other hand, for shallow angles, many bounces are possible. Hence, in shallow water, path F, most of the energy that propagates is close to the horizontal and this type of propagation is most analogous to waveguide propagation. In fact, as shown in Fig. 5, there exists a small cone from which energy propagates long distances (θ_c is typically 10°–20°). Energy outside the cone is referred to as the near field (or continuous spectrum), which eventually escapes the waveguide. The trapped field originating from within the cone is referred to as the normal-mode field (or discrete spectrum) because there is a set of angles corresponding to discrete paths that constructively interfere and make up the normal (natural) modes of the shallow-water environment.

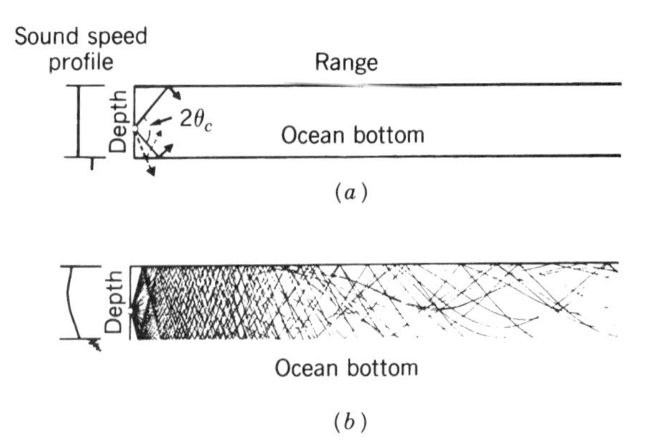

Fig. 5 Ocean waveguide propagation: (*a*) long-distance propagation occurs within a cone of $2\theta_c$; (*b*) same as (*a*) but with nonisovelocity water column showing refraction.

3 SOUND PROPAGATION MODELS

Sound propagation in the ocean is mathematically described by the wave equation, whose parameters and boundary conditions are descriptive of the ocean environment. There are essentially four types of models (computer solutions to the wave equation) to describe sound propagation in the sea: ray theory, the spectral method or fast-field program (FFP), the normal mode (NM), and the parabolic equation (PE). All of these models allow for the fact that the ocean environment varies with depth. A model that also takes into account horizontal variations in the environment (i.e., sloping bottom or spatially variable oceanography) is termed range dependent. For high frequencies (a few kilohertz or above), ray theory is the most practical. The other three model types are more applicable and usable at lower frequencies (below a kilohertz). The hierarchy of underwater acoustics models is shown in schematic form in Fig. 6. The models discussed here are essentially two-dimensional models since the index of refraction has much stronger dependence on depth than on horizontal distance. Nevertheless, bottom topography and strong ocean features can cause horizontal refraction (out of the range depth plane). Ray models are most easily extendable to include this added complexity; though fully three dimensional models are beginning to emerge, the latter are extremely computationally intensive. A compromise that often works for "weak" three-dimensional problems is the $N \times 2D$ approximation, which consists of combining a set of two-dimensional solutions along radials to produce a three-dimensional solution.[2]

3.1 Wave Equation and Boundary Conditions

The wave equation is typically written and solved in terms of pressure, displacement, or velocity potentials. For a velocity potential φ, the wave equation in cylindrical coordinates with the range coordinates denoted by $\mathbf{r} = (x, y)$ and the depth coordinate denoted by z (taken

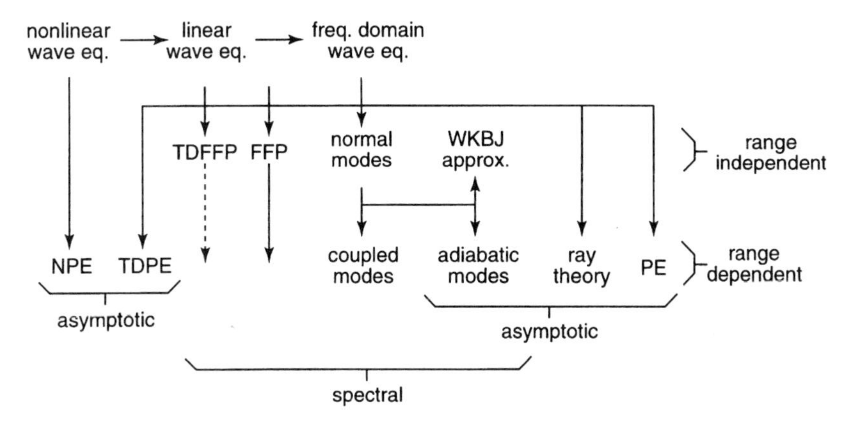

Fig. 6 Hierarchy of underwater acoustics models (TD, time domain).

positive downward) for a source-free region is

$$\nabla^2\varphi(\mathbf{r}, z, t) - \frac{1}{c^2}\frac{\partial^2\varphi(\mathbf{r}, z, t)}{\partial t^2} = 0, \tag{9}$$

where c is the sound speed in the wave-propagating medium. With respect to the velocity potential, the velocity $\mathbf{v}$ and pressure p are given by

$$\mathbf{v} = \nabla\varphi, \qquad p = -\rho\frac{\partial\varphi}{\partial t}, \tag{10}$$

where ρ is the density of the medium. The wave equation is most often solved in the frequency domain; that is, a frequency dependence of $\exp(-i\omega t)$ is assumed to obtain the Helmholtz equation ($K \equiv \omega/c$)

$$\nabla^2\varphi(\mathbf{r}, z) + K^2\varphi(\mathbf{r}, z) = 0. \tag{11}$$

In underwater acoustics both fluid and elastic (shear-supporting) media are of interest. In elastic media the field can be expressed in terms of three scalar displacement potentials $\Phi(\mathbf{r}, z) \equiv \{\phi(\mathbf{r}, z), \psi(\mathbf{r}, z), \Lambda(\mathbf{r}, z)\}$, corresponding to compressional (P), vertically polarized (SV), and horizontally polarized (SH) shear waves, respectively. In the limiting case of a fluid medium, where shear waves do not exist, $\Phi(\mathbf{r}, z)$ represents the compressional potential $\phi(\mathbf{r}, z)$. Most underwater acoustic applications involve only compressional sources that only excite the P and SV potentials, eliminating the SH potential $\Lambda(\mathbf{r}, z)$. The displacement potentials satisfy the Helmholtz equation with the appropriate compressional or shear sound speeds c_p or c_s, respectively,

$$c_p = \left[\frac{\lambda + 2\mu}{\rho}\right]^{1/2}, \qquad c_s = \left[\frac{\mu}{\rho}\right]^{1/2}, \tag{12}$$

where λ and μ are the Lamé constants.

The most common plane interface boundary conditions encountered in underwater acoustics are described below: For the ocean surface there is the pressure release condition where the pressure (normal stress) vanishes; for the appropriate solution of the Helmholtz equation, this condition is

$$p = 0, \qquad \varphi = 0 \quad \text{or} \quad \phi = 0. \tag{13}$$

The interface between the water column (layer 1) and an ocean bottom sediment (layer 2) is often characterized as a fluid–fluid interface. The continuity of pressure and vertical particle velocity at the interface yields the following boundary conditions in terms of pressure:

$$p_1 = p_2, \qquad \frac{1}{\rho_1}\frac{\partial p_1}{\partial z} = \frac{1}{\rho_2}\frac{\partial p_2}{\partial z}, \tag{14}$$

or velocity potential:

$$\rho_1\varphi_1 = \rho_2\varphi_2, \qquad \frac{\partial\varphi_1}{\partial z} = \frac{\partial\varphi_2}{\partial z}. \tag{15}$$

These boundary conditions applied to the plane-wave fields in Fig. 4*a* yield the Rayleigh reflection coefficient given by Eq. (1).

For an interface separating two solid layers, the boundary conditions are continuity of vertical displacement $w(\mathbf{r}, z_i)$, tangential displacements $\mathbf{u}(\mathbf{r}, z_i) = (u_x, u_y)$, normal stress $n = \sigma_{zz}$, and tangential stresses $\mathbf{t} =$

$(\sigma_{xx}, \sigma_{yz})$. In cyclindrical coordinates with azimuthal symmetry, the radial and vertical components of displacements (in homogeneous media) u and w, respectively, are

$$u(r,z) = \frac{\partial \phi}{\partial r} + \frac{\partial^2 \psi}{\partial z^2}, \tag{16}$$

$$w(r,z) = \frac{\partial \phi}{\partial z} - \frac{1}{r}\frac{\partial}{\partial r} r \frac{\partial \psi}{\partial r}, \tag{17}$$

and the normal and tangential stresses are

$$\sigma_{zz}(r,z) = (\lambda + 2\mu)\frac{\partial w}{\partial z} + \lambda \frac{\partial u}{\partial r}, \tag{18}$$

$$\sigma_{rz}(r,z) = \mu\left(\frac{\partial u}{\partial z} + \frac{\partial w}{\partial r}\right). \tag{19}$$

Continuity of these quantities at the interface between two solids are the boundary conditions. For a fluid–solid interface, the rigidity μ vanishes in the fluid layer and the tangential stress in the solid layer vanishes at the boundary. If at least one of the media is elastic, these boundary conditions permit the existence of interface or surface waves (see *Enc.* Ch. 37) such as Rayleigh waves at the interface between a solid and vacuum, Scholte waves at a fluid–solid interface, and Stoneley waves at a solid–solid interface. These waves are normally only excited when the source is acoustically close, in terms of wavelengths, to the interface.

The Helmholtz equation for an acoustic field from a point source with angular frequency ω is

$$\nabla^2 G(\mathbf{r}, z) + K^2(\mathbf{r}, z) G(\mathbf{r}, z) = -\delta^2(\mathbf{r} - \mathbf{r}_s)\delta(z - z_s),$$

$$K^2(\mathbf{r}, z) = \frac{\omega^2}{c^2(\mathbf{r}, z)}, \tag{20}$$

where the subscript s denotes the source coordinates. The range-dependent environment manifests itself as the coefficient $K^2(\mathbf{r}, z)$ of the partial differential equation for the appropriate sound speed profile. The range-dependent bottom type and topography appear as boundary conditions on scalars and tangential and normal quantities, as discussed above. The acoustic field from a point source $G(\mathbf{r})$ is obtained either by solving the boundary value problem of Eq. (20) (spectral method or normal modes) or by approximating Eq. (11) by an initial-value problem (ray theory, parabolic equation).

3.2 Ray Theory

Ray theory is a geometric, high-frequency approximate solution to Eq. (9) of the form

$$G(\mathbf{R}) = A(\mathbf{R}) \exp[iS(\mathbf{R})], \tag{21}$$

where the exponential term allows for rapid variations as a function of range and $A(\mathbf{R})$ is a more slowly varying "envelope" that incorporates both geometric spreading and loss mechanisms. The geometric approximation is that the amplitude varies slowly with range [i.e., $(1/A)\nabla^2 A \ll K^2$] so that Eq. (20) yields the eikonal equation

$$(\nabla S)^2 = K^2. \tag{22}$$

The ray trajectories are perpendicular to surfaces of constant phase (wavefronts) S and may be expressed mathematically as

$$\frac{d}{dl}\left(K \frac{d\mathbf{R}}{dl}\right) = \nabla K, \tag{23}$$

where l is the arc length along the direction of the ray and $\mathbf{R}$ is the displacement vector. The direction of average flux (energy) follows that of the trajectories, and the amplitude of the field at any point can be obtained from the density of rays.

The ray theory method is computationally rapid and extends to range-dependent problems. Furthermore, the ray traces give a physical picture of the acoustic paths. It is helpful in describing how noise redistributes itself when propagating long distances over paths that include shallow and deep environments and/or midlatitude to polar regions. The disadvantage of conventional ray theory is that it does not include diffraction and such effects that describe the low-frequency dependence (*degree of trapping*) of ducted propagation.

3.3 Wavenumber Representation or Spectral Solution

The wave equation can be solved efficiently with spectral methods when the ocean environment does not vary with range. The term *fast-field program (FFP)* had been used because the spectral methods became practical with the advent of the fast Fourier transform (FFT). Assume a solution of Eq. (20) of the form

$$G(\mathbf{r}, z) = \frac{1}{2\pi}\int_{-\infty}^{\infty} d^2\mathbf{k}\, g(\mathbf{k}, z, z_s) \exp[i\mathbf{k} \cdot (\mathbf{r} - \mathbf{r}_s)], \tag{24}$$

which then leads to the equation for the depth-dependent Green's function $g(\mathbf{k}, z, z_s)$,

$$\frac{d^2g}{dz^2} + [K^2(z) - k^2]g = \frac{1}{2\pi}\,\delta(z - z_s). \tag{25}$$

Furthermore, we assume azimuthal symmetry, $kr > 2\pi$ and $\mathbf{r}_s = 0$, so that Eq. (24) reduces to

$$G(r, z) = \frac{\exp(-i\pi/4)}{(2\pi r)^{1/2}} \int_{-\infty}^{\infty} dk\,(k)^{1/2} g(k, z, z_s)\exp(ikr). \tag{26}$$

We now convert the above integral to an FFT form by setting $k_m = k_0 + m\,\Delta k$; $r_n = r_0 + n\,\Delta r$, where $n, m = 0, 1, \ldots, N - 1$, with the additional condition $\Delta r\,\Delta k = 2\pi/N$, where N is an integral power of 2:

$$G(r_n, z) = \frac{\Delta k \exp[i(k_0 r_n - \pi/4)]}{(2\pi r)^{1/2}} \cdot \sum_{m=0}^{N-1} X_m \exp\left(\frac{2\pi imn}{N}\right),$$

$$X_m = (k_m)^{1/2} g(k_m, z, z_s)\exp(imr_0\,\Delta k). \tag{27}$$

Although the method was initially labeled "fast field," it is fairly slow because of the time required to calculate the Green's functions [solve Eq. (27)]. However, it has advantages when one wishes to calculate the "near-field" region or to include shear wave effects in elastic media[3]; it is also often used as a benchmark for other less exact techniques. Recently, a range-dependent spectral solution technique has been developed.[4]

3.4 Normal-Mode Model

Rather than solve Eq. (25) for each g for the complete set of k's (typically thousands of times), one can utilize a normal-mode expansion of the form

$$g(\mathbf{k}, z) = \sum a_n(\mathbf{k})u_n(z), \tag{28}$$

where the quantities u_n are eigenfunctions of the eigenvalue problem

$$\frac{d^2u_n}{dz^2} + [K^2(z) - k_n^2]u_n(z) = 0. \tag{29}$$

The eigenfunctions u_n are zero at $z = 0$, satisfy the local boundary conditions descriptive of the ocean bottom properties, and satisfy a radiation condition for $z \to \infty$. They form an orthonormal set in a Hilbert space with weighting function $\rho(z)$, the local density. The range of discrete eigenvalues corresponding to the poles in the integrand of Eq. (26) is given by the condition

$$\min[K(z)] < k_n < \max[K(z)]. \tag{30}$$

These discrete eigenvalues correspond to discrete angles within the critical angle cone in Fig. 5 such that specific waves constructively interfere. The eigenvalues k_n typically have a small imaginary part α_n, which serves as the modal attenuation representative of all the losses in the ocean environment (see Ref. 2 for the formulation of normal-mode attenuation coefficients). Solving Eq. (20) using the normal-mode expansion given by Eq. (28) yields (for the source at the origin)

$$G(r, z) = \frac{i}{4}\,\rho(z_s)\sum_n u_n(z_s)u_n(z)H_0^1(k_n r) \tag{31}$$

The asymptotic form of the Hankel function can be used in the above equation to obtain the well-known normal-mode representation of a cylindrical (axis is depth) waveguide:

$$G(r, z) = \frac{i\rho(z_s)}{(8\pi r)^{1/2}} \exp\left(\frac{-i\pi}{4}\right) \sum_n \frac{u_n(z_s)u_n(z)}{k_n^{1/2}} \cdot \exp(ik_n r). \tag{32}$$

Equation (32) is a far-field solution of the wave equation and neglects the continuous spectrum [$k_n < \min[K(z)]$ of inequality (30)] of modes. For purposes of illustrating the various portions of the acoustic field, we note that k_n is a horizontal wavenumber so that a "ray angle" associated with a mode with respect to the horizontal can be taken to be $\theta = \cos^{-1}[k_n/K(z)]$. For a simple waveguide the maximum sound speed is the bottom sound speed corresponding to $\min[K(z)]$. At this value of $K(z)$, we have, from Snell's law, $\theta = \theta_c$, the bottom critical angle. In effect, if we look at a ray picture of the modes, the continuous portion of the mode spectrum corresponds to rays with grazing angles greater than the bottom critical angle of Fig. 4*b* and therefore outside the cone of Fig. 5. This portion undergoes severe loss. Hence, we note that the continuous spectrum is the near (vertical) field and the discrete spectrum is the (more horizontal, profile-dependent) far field (falling within the cone in Fig. 5).

The advantages of the NM procedure are that the solution is available for all source and receiver configurations once the eigenvalue problem is solved; it is easily extended to moderately range dependent conditions using the adiabatic approximation; it can be applied (with more effort) to extremely range dependent environments using coupled-mode theory. However, it does not include a full representation of the near field.

3.5 Adiabatic Mode Theory

All of the range-independent normal-mode "machinery" developed for environmental ocean acoustic modeling applications can be adapted to mildly range dependent conditions using adiabatic mode theory. The underlying assumption is that individual propagating normal modes adapt (but do not scatter or "couple" into each other) to the local environment. The coefficients of the mode expansion, a_n in Eq. (28), now become mild functions of range, that is, $a_n(\mathbf{k}) \rightarrow a_n(\mathbf{k}, \mathbf{r})$. This modifies Eq. (32) as follows:

$$G(\mathbf{r}, z) = \frac{i\rho(z_s)}{(8\pi r)^{1/2}} \exp\left(\frac{-i\pi}{4}\right) \sum_n \frac{u_n(z_s)u_n(z)}{\overline{k}_n^{-1/2}} \cdot \exp(i\overline{k}_n r), \tag{33}$$

where the range-averaged wavenumber (eigenvalue) is

$$\overline{k_n} = \frac{1}{r}\int_0^r k_n(r')\,dr' \tag{34}$$

and the $k_n(r')$ are obtained at each range segment from the eigenvalue problem (29) evaluated at the environment at that particular range along the path. The quantities u_n and v_n are the sets of modes at the source and the field positions, respectively.

Simply stated, the adiabatic mode theory leads to a description of sound propagation such that the acoustic field is a function of the modal structure at both the source and the receiver and some average propagation conditions between the two. Thus, for example, when sound emanates from a shallow region where only two discrete modes exist and propagates into a deeper region with the same bottom (same critical angle), the two modes from the shallow region adapt to the form of the first two modes in the deep region. However, the deep region can support many more modes; intuitively, we therefore expect the resulting two modes in the deep region will take up a smaller more horizontal part of the cone of Fig. 5 than they take up in the shallow region. This means that sound rays going from shallow to deep tend to become more horizontal which is consistent with a ray picture of downslope propagation. Finally, fully coupled mode theory for range-dependent environments has been developed[5] but requires extremely intensive computation.

3.6 Parabolic Equation Model

The PE method was introduced into ocean acoustics and made viable with the development of the "split-step" algorithm, which utilized FFTs at each range step.[6] Subsequent numerical developments greatly expanded the applicability of the parabolic equation.

Standard PE–Split Step Algorithm The PE method is presently the most practical and encompassing wave-theoretic range-dependent propagation model. In its simplest form, it is a far-field narrow-angle (~ ±20° with respect to the horizontal, adequate for most underwater propagation problems) approximation to the wave equation. Assuming azimuthal symmetry about a source, we express the solution of Eq. (21) in cylindrical coordinates in a source-free region in the form

$$G(r, z) = \psi(r, z) \cdot J(r), \tag{35}$$

and we define $K^2(r, z) \equiv K_0^2 n^2$, n therefore being an *index of refraction* c_0/c, where c_0 is a reference sound speed. Substituting Eq. (35) into Eq. (11) in a source-free region and taking K_0^2 as the separation constant, J and ψ satisfy the equations

$$\frac{d^2J}{dr^2} + \frac{1}{r}\frac{dJ}{dr} + K_0^2 J = 0, \tag{36}$$

$$\frac{\partial^2\psi}{\partial r^2} + \frac{\partial^2\psi}{\partial z^2} + \left(\frac{1}{r} + \frac{2}{J}\frac{\partial J}{\partial r}\right) + \left(\frac{\partial\psi}{\partial r}\right) + K_0^2 n^2 \psi - K_0^2\psi = 0. \tag{37}$$

Equation (36) is a Bessel equation, and we take the outgoing solution, a Hankel function, $H_0^1(K_0 r)$, in its asymptotic form and substitute it into Eq. (37), together with the "paraxial" (narrow-angle) approximation

$$\frac{\partial^2\psi}{\partial r^2} \ll 2K_0 \frac{\partial\psi}{\partial r} \tag{38}$$

to obtain the parabolic equation (in r)

$$\frac{\partial^2\psi}{\partial z^2} + 2iK_0\frac{\partial\psi}{\partial r} + K_0^2(n^2 - 1)\psi - 0, \tag{39}$$

where we note that n is a function of range and depth. We use a marching solution to solve the parabolic equation. There has been an assortment of numerical solutions, but the one that still remains the standard is the so-called split-step algorithm.[6]

We take n to be a constant; the error this introduces can be made arbitrarily small by the appropriate numerical gridding. The Fourier transform of ψ can then be written as

$$\chi(r,s) = \frac{1}{2\pi}\int_{-\infty}^{\infty} \psi(r,z)\exp(-isz)dz, \tag{40}$$

which together with Eq. (39) gives

$$-s^2\chi + 2iK_0\frac{\partial\chi}{\partial r} + K_0^2(n^2-1)\chi = 0. \tag{41}$$

The solution of Eq. (41) is simply

$$\chi(r,s) = \chi(r_0,s)\exp\left[-\frac{K_0^2(n^2-1)-s^2}{2iK_0}(r-r_0)\right], \tag{42}$$

with specified initial condition at r_0. The inverse transform gives the field as a function of depth,

$$\psi(r,z) = \int_{-\infty}^{\infty}\chi(r_0,s)\exp\left[\frac{iK_0}{2}(n^2-1)\Delta r\right] \cdot \exp\left[-\frac{i\Delta r}{2K_0}s^2\right]\exp(isz)\,ds, \tag{43}$$

where $\Delta r = r - r_0$. Introducing the symbol $\mathcal{F}$ for the Fourier transform operation from the z-domain [as performed in Eq. (40)] and $\mathcal{F}^{-1}$ as the inverse transform, Eq. (4) can be summarized by the range-stepping algorithm

$$\psi(r+\Delta r,z) = \exp\left[\frac{iK_0}{2}(n^2-1)\Delta r\right] \cdot \mathcal{F}^{-1}\left[\left(\exp\left(-\frac{i\Delta r}{2K_0}s^2\right)\right)\mathcal{F}[\psi(r,z)]\right], \tag{44}$$

which is often referred to as the split-step marching solution to the PE. The Fourier transforms are performed using FFTs. Equation (44) is the solution for n constant, but the error introduced when n (profile or bathymetry) varies with range and depth can be made arbitrarily small by increasing the transform size and decreasing the range step size. It is possible to modify the split-step algorithm to increase its accuracy with respect to higher angle propagation.[7]

Generalized or Higher Order PE Methods Methods of solving the parabolic equation, including extensions to higher angle propagation, elastic media, and direct time-domain solutions including nonlinear effects, have recently appeared (see Refs. 8 and 9 for additional references). In particular, accurate high-angle solutions are important when the evironment supports acoustic paths that become more vertical, such as when the bottom has a very high speed and, hence, a large critical angle with respect to the horizontal. In addition, for elastic propagation, the compressional and shear waves span a wide-angle interval. Finally, Fourier synthesis for pulse modeling requires high accuracy in phase, and the high-angle PEs are more accurate in phase, even at the low angles.

Equation (39) with the second-order range derivative that was neglected because of inequality (38) can be written in operator notation as

$$[P^2 + 2iK_0P + K_0^2(Q^2-1)]\psi = 0, \tag{45}$$

where

$$P \equiv \frac{\partial}{\partial r}, \qquad Q \equiv \sqrt{n^2 + \frac{1}{K_0^2}\frac{\partial^2}{\partial z^2}}. \tag{46}$$

Factoring Eq. (45) assuming weak range dependence and retaining only the factor associated with outgoing propagation yields a one-way equation

$$P\psi = iK_0(Q-1)\psi, \tag{47}$$

which is a generalization of the parabolic equation beyond the narrow-angle approximation associated with inequality (38). If we define $Q = \sqrt{1+q}$ and expand Q in a Taylor series as a function of q, the standard PE method is recovered by $Q \approx 1 + 0.5q$. The wide-angle PE to arbitrary accuracy in angle, phase, and so on, can be obtained from a Padé series representation of the Q operator,[8]

$$Q \equiv \sqrt{1+q} = 1 + \sum_{j=1}^{n}\frac{a_{j,n}q}{1+b_{j,n}q} + O(q^{2n+1}), \tag{48}$$

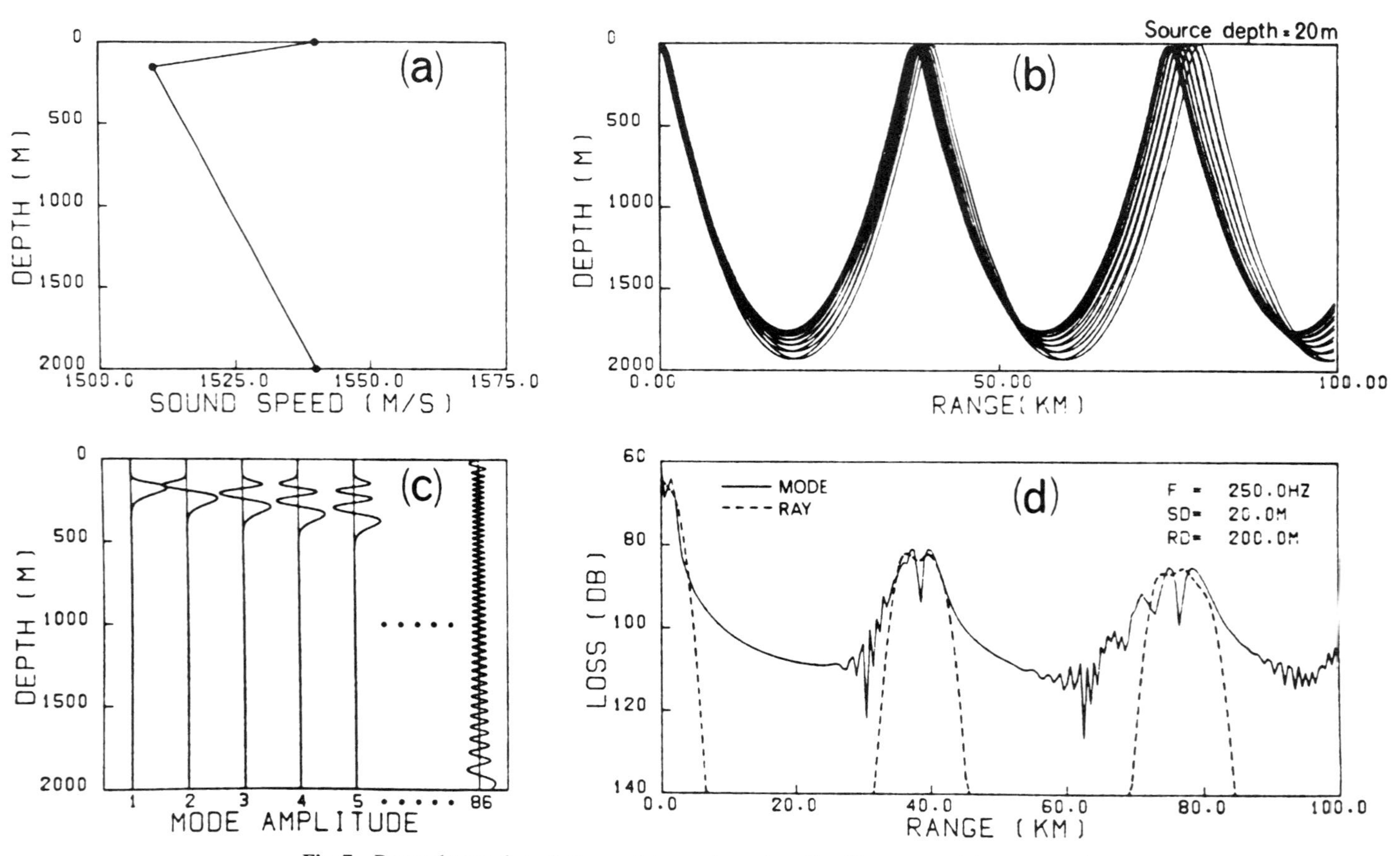

Fig. 7 Ray and normal-mode theory: (*a*) sound speed profile; (*b*) ray trace; (*c*) normal modes; (*d*) propagation calculations.

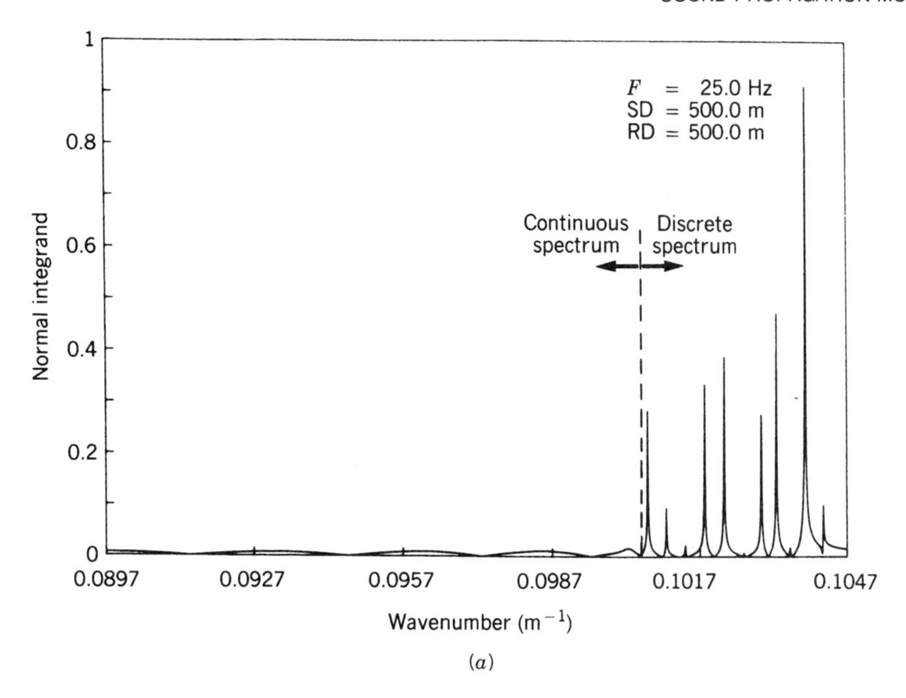

(*a*)

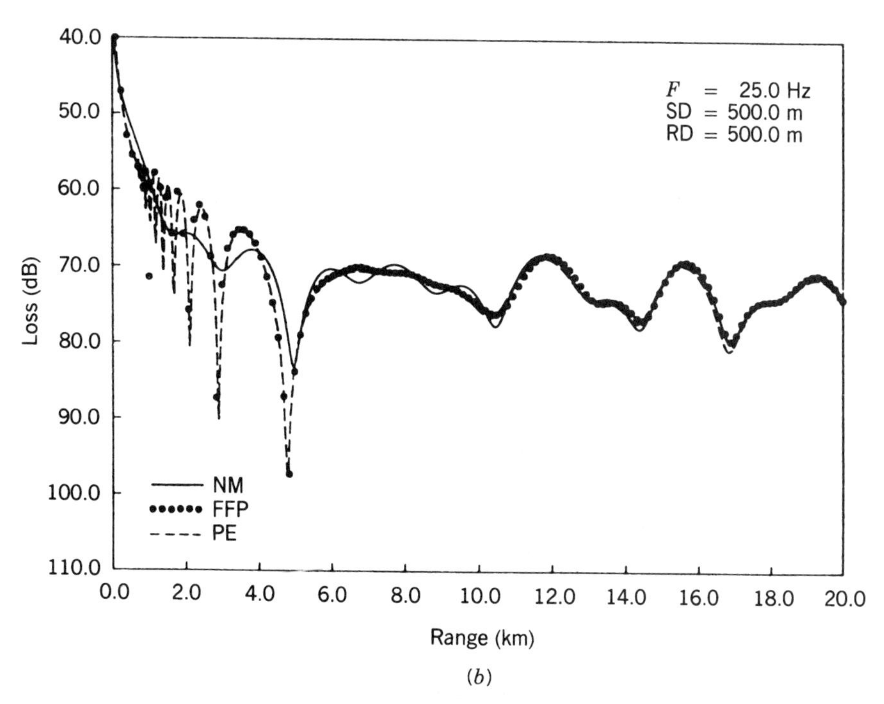

(*b*)

Fig. 8 Relationship between FFP, NM, and PE computations: (*a*) FFP Green's function from Eq. (25); (*b*) NM, FFP, and PE propagation results showing some agreement in near field complete agreement in far field.

where n is the number of terms in the Padé expansion and

$$a_{j,n} = \frac{2}{2n+1} \sin^2\left(\frac{j\pi}{2n+1}\right),$$

$$b_{j,n} = \cos^2\left(\frac{j\pi}{2n+1}\right). \tag{49}$$

The solution of Eq. (47) using Eq. (49) has been implemented using finite-difference techniques for fluid and elastic media.[8] A split-step Padé algorithm[10] has recently been developed that greatly enhances the numerical efficiency of this method.

4 QUANTITATIVE DESCRIPTION OF PROPAGATION

All of the models described above attempt to describe reality and to solve in one way or another the Helmholtz equation. They therefore should be consistent, and there is much insight to be gained from understanding this consistency. The models ultimately compute propagation loss, which is taken as the decibel ratio (see Appendix) of the pressure at the field point to a reference pressure, typically 1 m from the source.

Figure 7 shows convergence zone type propagation for a simplified profile. The ray trace in Fig. 7*b* shows the cyclic focusing discussed in Section 1.2. The same profile used to calculate normal modes is shown in Fig. 7*c*, which when summed according to Eq. (32) exhibit the same cyclic pattern as the ray picture. Figure 7*d* shows both the normal-mode (wave theory) and ray theory result. Ray theory exhibits sharply bounded shadow regions, as expected, whereas the normal-mode theory, which includes diffraction, shows that the acoustic field does exist in the shadow regions and the convergence zones have structure.

Normal-mode models sum the discrete modes, which roughly correspond to angles of propagation within the cone of Fig. 5. The spectral method can include the

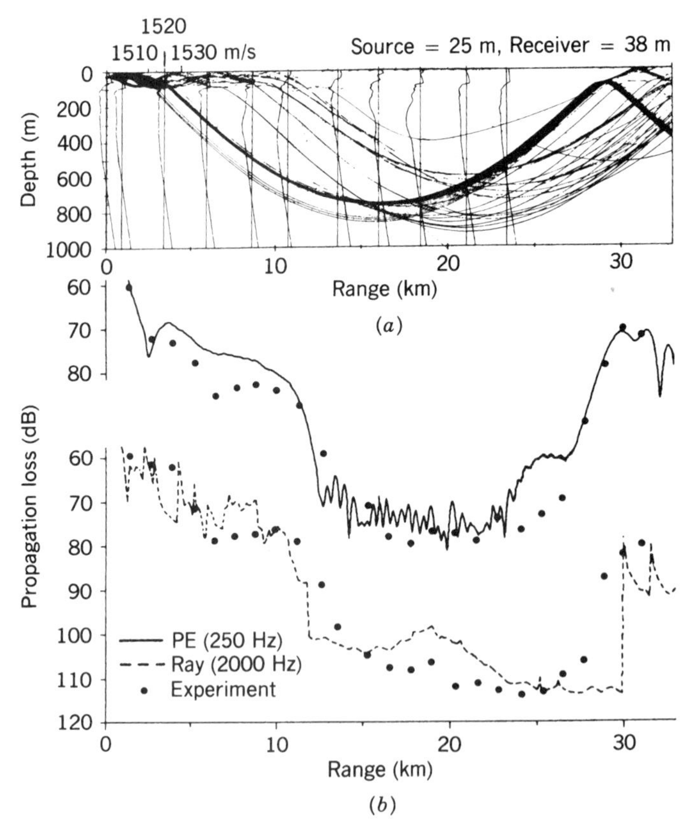

Fig. 9 Model and data comparison for a range-dependent case: (*a*) profiles and ray trace for a case of a surface duct disappearing; (*b*) 250 Hz PE and 2 kHz ray trace comparisons with data.

full field, discrete plus continuous, the latter corresponding to larger angles. The discussion below Eq. (32) defines these angles in terms of horizontal wavenumbers, and eigenvalues of the normal-mode problem are a discrete set of horizontal wavenumbers. Hence the integrand (Green's function) of the spectral method has peaks at the eigenvalues associated with the normal modes. These peaks are shown on the right of Fig. 8*a*. The smoother portion of the spectrum is the continuous part corresponding to the larger angles. Therefore, the consistency we expect between the normal mode and the spectral method and the physics of Fig. 5 is that the continuous portion of the spectral solution decays rapidly with range so that there should be complete agreement at long ranges between normal-mode and spectral solutions. The Lloyd's mirror effect, a near-field effect, should also be exhibited in the spectral solution but not the normal-mode solution. These aspects are apparent in Fig. 7*b*. The PE solution appears in Fig. 7*b* and is in good agreement with the other solutions, but with some phase error associated with the average wavenumber that must be chosen in the split-step method. The PE solution, which contains part of the continuous spectrum including the Lloyd mirror beams, is more accurate than the normal-mode solution at short range; more recent PE results[8] can be made arbitrarily accurate in the forward direction.

Range-dependent results[2] are shown in Fig. 9. A ray trace, a ray trace field result, a PE result, and data are plotted together for a range-dependent sound speed profile environment. The models agree with the data in general, with the exception that the ray results predict too sharp a leading edge of the convergence zone.

Upslope propagation is modeled with the PE in Fig. 10. As the field propagates upslope, sound is dumped into the bottom in what appears to be discrete beams.[11] The flat region has three modes, and each is cut off successively as sound propagates into shallower water. The ray picture also has a consistent explanation of this phenomenon. The rays for each mode become steeper as they propagate upslope. When the ray angle exceeds the critical angle, the sound is significantly transmitted into the bottom. The locations where this takes place for each of the modes are identified by the three arrows.

As a final example of how physical insight can be derived from models, we present a range-independent normal-mode[12] study of the optimum frequency of propagation in a shallow-water environment with a summer profile, as indicated in Fig. 11*a*, with the source (S) and receiver (R) also indicated. Frequency-versus-range contours of propagation loss obtained from a wideband experiment (analyzed in one-third-octave bands) and from an incoherent (no cross terms) sum of modes

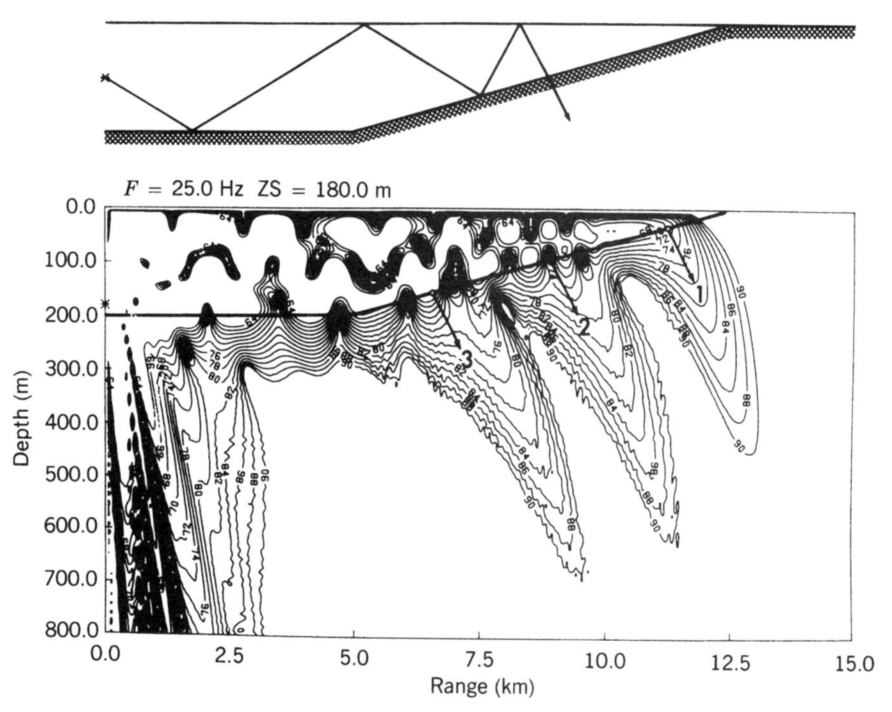

Fig. 10 Relation between up-slope propagation (from PE calculation) showing individual mode cutoff and energy dumping in the bottom and a corresponding ray schematic.

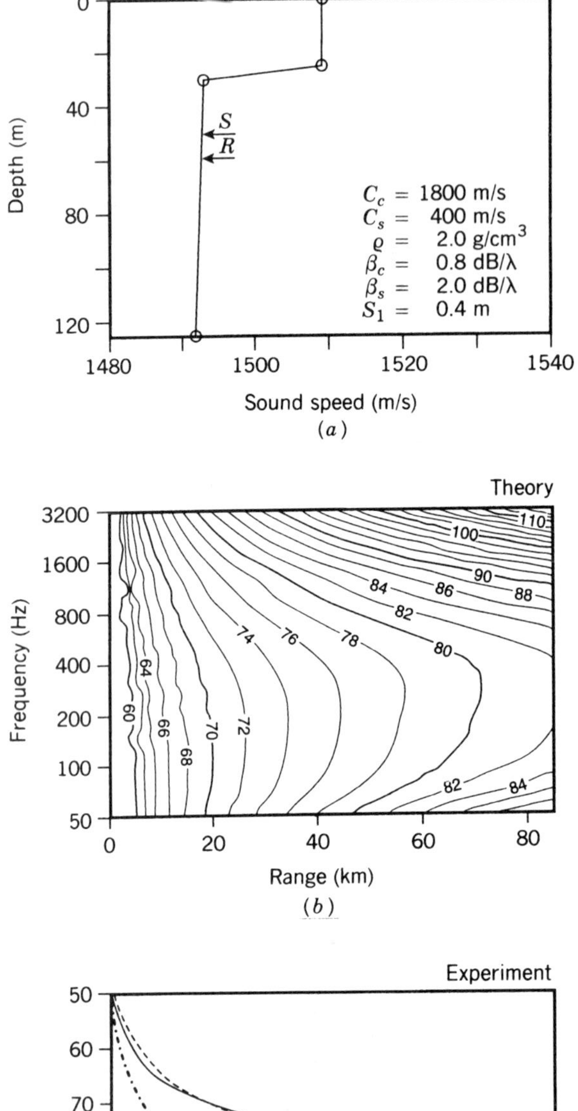

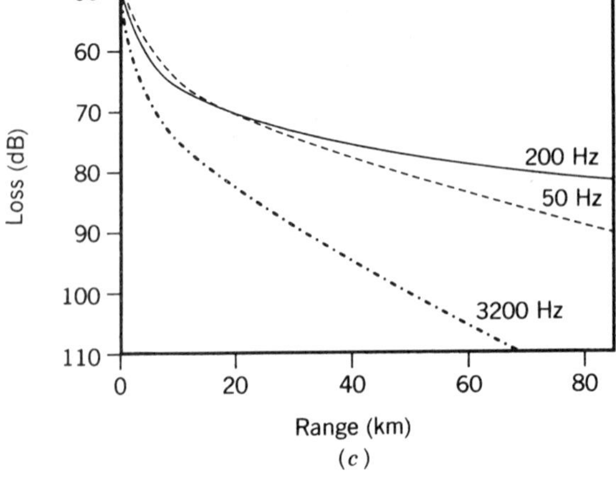

Fig. 11 (*a*) Shallow-water environment with summer and winter (isovelocity) profiles; (*b*) frequency vs. depth propagation loss contours; (*c*) third-octave propagation loss curves.

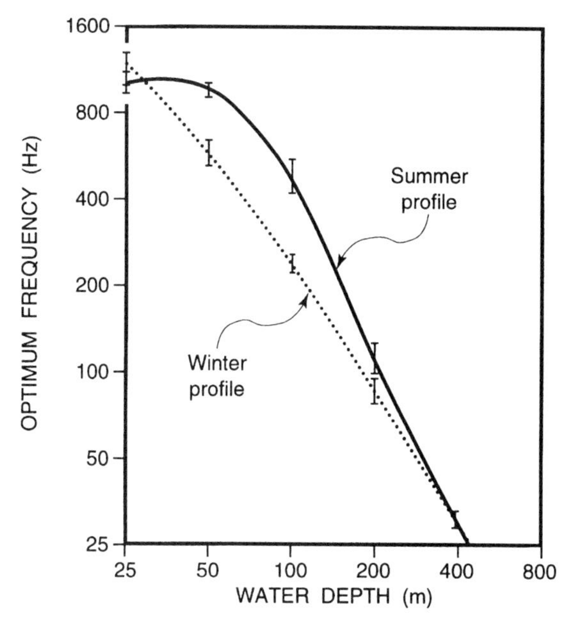

Fig. 12 Optimum frequency is strongly dependent on the depth of the ocean waveguide.

are shown in Figs. 11*b* and 11*c*. One obtains the conventional type of propagation loss curves by making a horizontal cut through the contour plots, as shown in Fig. 11*d*. We note here, as an aside, that in shallow-water environments propagation loss obtained by incoherently summing the modes is approximately equal to one-third-octave frequency averaging, which has the effect of averaging away model interference. The frequency-versus-range contours reveal an optimum frequency in the 200–400-Hz region. This can be seen by observing the 80-dB contour, which goes out to long ranges in the region, whereas other frequencies, at say a range of 70 km, have much higher losses.

The underlying physics of the optimum frequency is further elaborated in Fig. 12. For the sound speed profiles in Fig. 11*a*, the optimum frequency is shown to be a strong function of water depth. As a matter of fact, for the source receiver configuration under study it is actually a function of the dominant duct, which would be the whole water column for a isovelocity winter profile and the duct below the thermocline in the summer. Next, vis à vis the optimum frequency, after water depth in importance is bottom type, which also has a dominant effect on the actual propagation loss levels.

APPENDIX: UNITS

The decibel is the dominant unit in underwater acoustics and denotes a ratio of intensities (not pressures) expressed in terms of a logarithmic (base 10) scale. Two intensities I_1 and I_2 have a ratio I_1/I_2 in decibels of $10\log I_1/I_2$ decibels. Absolute intensities can therefore be expressed by using a reference intensity. The presently accepted reference intensity is based on a reference pressure of one micropascal: the intensity of a plane wave having an rms pressure equal to 10^{-5} dyn/cm^2. Therefore, taking 1 μPa as I_2, a sound wave having an intensity, of, say, one million times that of a plane wave of rms pressure 1 μPa has a level of $10\log(10^6/1) \equiv 60$ dB re 1 μPa. Pressure (p) ratios are expressed in dB re 1 μPa by taking $20\log p_1/p_2$, where it is understood that the reference originates from the intensity of a plane wave of pressure equal to 1 μPa.

The average intensity I of a plane wave with rms pressure p in a medium of density ρ and sound speed c is $I = p^2/\rho c$. In seawater, ρc is 1.5×10^5 g/cm · s, so that a plane wave of rms pressure 1 dyn/cm^2 has an intensity of 0.67×10^{-12} W/cm^2. Substituting the value of a micropascal for the rms pressure in the plane-wave intensity expression, we find that a plane-wave pressure of 1 μPa corresponds to an intensity of 0.67×10^{-22} W/cm^2 (i.e., 0 dB re 1 μPa).

REFERENCES

1. J. Northrup and J. G. Colborn, "Sofar Channel Axial Sound Speed and Depth in the Atlantic Ocean," *J. Geophys. Res.*, Vol. 79, 1974, p. 5633.
2. F. B. Jensen, W. A. Kuperman, M. B. Porter, and H. Schmidt, *Computational Ocean Acoustics*, AIP Press, Woodbury, NY, 1994.
3. H. Schmidt and F. B. Jensen, "A Full Wave Solution for Propagation in Multilayered Viscoelastic Media with Application to Gaussian Beam Reflection at Fluid-Solid Interfaces," *J. Acoust. Soc. Am.*, Vol. 77, 1985, p. 813.
4. H. Schmidt, W. Seong, and J. T. Goh, "Spectral Super-element Approach to Range-Dependent Ocean Acoustic Modeling," *J. Acoust. Soc. Am.*, Vol. 98, 1995, p. 465.
5. R. B. Evans, "A Coupled Mode Solution for Acoustic Propagation in a Waveguide with Stepwise Depth Variations of a Penetrable Bottom," *J. Acoust. Soc. Am.*, Vol. 74, 1983, p. 188.
6. F. D. Tappert, "The Parabolic Approximation Method," in J. B. Keller and J. S. Papadakis (Eds.), *Wave Propagation and Underwater Acoustics*, Springer Verlag, Berlin, 1977.
7. D. J. Thomson and N. R. Chapman, "A Wide-Angle Split-Step Algorithm for the Parabolic Equation," *J. Acoust. Soc. Am.*, Vol. 74, 1983, p. 1848.
8. M. D. Collins, "Higher-Order Padé Approximations for Accurate and Stable Elastic Parabolic Equations with Applications to Interface Wave Propagation," *J. Acoust. Soc. Am.*, Vol. 89, 1991, p. 1050.
9. B. E. McDonald and W. A. Kuperman, "Time Domain

Formulation for Pulse Propagation Including Nonlinear Behavior at a Caustic," *J. Acoust. Soc. Am.*, Vol. 81, 1987, p. 1406.

10. M. D. Collins "A Split-Step Padé Solution for the Parabolic Equation Method," *J. Acoust. Soc. Am.*, Vol. 93, 1993, p. 1736.

11. F. B. Jensen and W. A. Kuperman, "Sound Propagation in a Wedge Shaped Ocean with a Penetrable Bottom," *J. Acoust. Soc. Am.*, Vol. 67, 1980, p. 1564.

12. F. B. Jensen and W. A. Kuperman, "Optimum Frequency of Propagation in Shallow Water Environments," *J. Acoust. Soc. Am.*, Vol. 73, 1983, p. 813.

32

SOUND RADIATION FROM MARINE STRUCTURES

DAVID FEIT

1 INTRODUCTION

Marine structures radiate sound as a result of the time-dependent pressure fluctuations communicated to the surrounding water medium by the vibratory motions of their hull envelopes. These motions are in response to unsteady forces and moments generated within the hull by the many machines necessary to the ship's operation. The effects of the forces are transmitted to the hull plating via the structures supporting the machinery.

Ross (Ref. 1, pp. 326–347) characterizes the internal sources as being primarily due to (1) mechanical imbalances, (2) electromagnetic force fluctuations, and (3) friction between moving parts and impact sounds such as gear tooth impacts or piston slap. The present treatment is not concerned with the sources but examines the dynamic response of the hull and the sound radiation arising from the vibrations of the wetted hull surfaces due to the sources.

For present purposes water is assumed to be nonviscous and slightly compressible. Because of the inviscid nature of water, it is only the normal component of the hull-plating vibratory response that imparts its motion to the contiguous fluid particles, and this in turn generates unsteady pressure fluctuations in the water. These pressure fluctuations not only radiate sound to distant observers but also react back on the structure in the form of *radiation loading*. This latter effect significantly complicates the problem, altering the vibrations by effectively adding mass to the structure, thus reducing the frequency and magnitude of vibration. Furthermore, the fluid motions carry energy away from the structure, having the effect of an additional resistive force, acting to reduce the magnitude of vibration and dissipate the vibratory energy.

In addition to internal machinery sources, a ship's propeller also generates time-varying forces and moments as it rotates through the nonuniform flow that exists in the wake of the ship. These are transmitted to the hull via the propeller shaft and thrust bearing. At the same time, the propeller generates a rotating pressure field that acts as an unsteady force on nearby appendages or hull surfaces, creating vibrations and its accompanying radiation of sound.

In this chapter we assume that the unsteady forces and/or moments are prescribed and concern ourselves only with the vibrations and sound generated by the structure in response to these forces.

Note: References to chapters appearing only in the *Encyclopedia* are preceded by *Enc.*

2 VIBRATION AND RADIATION FROM SHIP HULLS

Ship hulls are very complex structures. Therefore, to gain an understanding of the mechanisms involved in ship sound radiation, the structures are highly idealized. This allows us to arrive at approximate explicit results to describe the fundamental phenomena that are involved, rather than give an exact one-to-one correspondence with realistic structural configurations. To effect the latter would require a detailed discussion of computational structural acoustics as applied to ship structures. The general subject of computational methods in acoustics is treated more thoroughly in Chapters 10, 12, and 13.

Besides the assumptions made with respect to ship structures, we deal with the radiation problem in as simple a way as possible. To achieve this, we assume that the structures representing the ships are completely submerged in an infinite body of water. Therefore, for surface ships the radiation results must be modified to account for the free surface. This may be accomplished by adding to the source distributions, determined by the ship's vibratory motions, image source distributions of equal and opposite strength located at image points with respect to the free surface. Such an assumption yields an overall dipolelike directivity pattern in the vertical plane for a surface ship far-field pressure.

Ship hulls are complicated structures, and the frequency range over which they vibrate and radiate sound is extremely wide. It is neither recommended nor possible to postulate a single all-encompassing model that can be used to investigate the problem over such a wide range. It is useful to divide the frequency range into three parts designated as low frequency (LF), mid frequency (MF), and high frequency (HF). Such a division is described by Ross (Ref. 1, pp. 100–102), among others.

Within each of these ranges we shall postulate appropriate modeling procedures that have proven to be useful. In practice, modeling of the structure used for estimating the hull vibrational response is more complicated than that required for estimation of the radiated sound. This is due to the wide variability and uniqueness of design for each type of ship structure. Here we use very simplified and idealized models for illustrative purposes only.

2.1 Low-Frequency Range

In the LF range, which extends from about 1 Hz up to the frequency at which an acoustic wavelength is on the order of half the ship's length (e.g., for a ship that is 250 ft long, this corresponds to 40 Hz), the hull vibrates as a nonuniform elastic beam. Because of the low frequencies involved and the stiffness of the structure, a localized excitation causes the entire structure to vibrate, and the vibration patterns are said to be *global*, that is, they extend over the entire length of the body.

In calculating the vibration response in this range the water is assumed to be incompressible. The water strongly affects the motion by the addition of *added mass*. This is the inertia effect of the surrounding medium. Sound radiation can then be estimated by assuming that the vibrating body is a distribution of acoustic volume sources whose strength is determined by the product of the normal acceleration and the surface area.

When considered as a rigid body, the hull has six degrees of freedom that are superimposed on its steady forward speed. The orientation of a ship relative to a typical set of coordinate axes is shown in Fig. 1. In this presentation we deal with the vibrations and sound radiation of ships that are completely submerged, and we therefore illustrate the ship as a prolate spheroid. We call the rigid-body translational motions in the x-, y-, and z-directions *surge, sway*, and *heave*, respectively, while the angular motions about the same axes are called *roll, pitch*, and *yaw*.

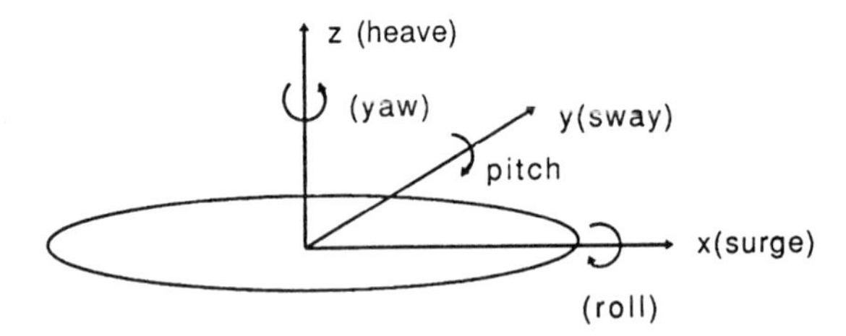

Fig. 1 Axes showing directions and names of a ship's rigid-body motions (both linear and angular).

The ship's elastic vibrations in which a cross section vibrates uniformly, back and forth, in the surge direction are referred to as *longitudinal* vibrations, while vibrations of the section in the two directions transverse to the longitudinal axis are known as *whipping, bending*, or *flexural* vibrations. The latter can occur in either the vertical or horizontal planes. As the hull stretches and shortens in longitudinal vibration, there are radial motions of the hull plating induced by the Poisson ratio effect. The motion normal to the structure's surface is the only vibratory component that is coupled to the water medium for a nonviscous fluid.

In modeling for vibration and sound radiation in the LF range, the ship structures are assumed to be beams, freely suspended in the water medium, with longitudinally varying cross-sectional properties. To handle the longitudinal variation, a *lumped-parameter* multi-degree-of-freedom model of the beams can be used in determining the ship's LF vibration response. For illustrative purposes only, we shall consider the axial variation of stiffness, both longitudinal and transverse bending, to be negligible. This allows us to introduce analytic expressions rather than numerical solutions for the vibration distributions.

2.2 Midfrequency Range

The MF range covers those frequencies between the LF and the HF range. Here the response of the hull to a localized force extends approximately over a compartment length. In this range the acoustic wavelength varies from being somewhat larger than a compartment length to a fraction of the circumference of a ship's cross section. As an example, we consider a submarine hull to be adequately modeled as a cylindrical shell separated into compartments by bulkheads. Thus for a submarine having a diameter of 30 ft and a length of 250 ft, the MF range would go from about 40 Hz to approximately 200 Hz where the cylinder's circumference is about four acoustic wavelengths.

Here we typically model the hull compartment as a finite-length elastic cylindrical shell terminated structurally by bulkheads, or end closures. Resonances of the shell section play a significant part in the sound radiation phenomena, and the entire length of the compartment is assumed to participate in the vibration and sound radiation. The resonance frequencies are sensitive to the boundary conditions assumed at the ends of the cylindrical segments as well as the dynamics of the structural systems internal to the hull.

To make the acoustics problem more amenable to analysis, the remaining portion of the hull is modeled as a rigid cylindrical baffle extending to infinity in both fore and aft directions. Any more refined investigation would necessitate an analysis that makes use of numerical methods.

2.3 High-Frequency Range

In this range the vibrating forces excite only a small portion of the hull. The lower boundary of this range is sometimes taken as the ring resonance frequency $f_r = c_p/2\pi a$, where c_p is the hull plating compressional wave velocity and a is the compartment cylinder radius. The effects of curvature of the hull plating and the importance of resonances diminish. Here details of the hull plating and the stiffening structures take on added significance to the radiation process. Analytical methods have been most successful in this range, and it is from such analyses that much of our intuitive knowledge about the radiation of elastic structures arises.

3 VIBRATION AND RADIATION IN LOW-FREQUENCY RANGE

3.1 Longitudinal Vibrations

We treat the hull as an equivalent beam in the LF range. The motions of a beamlike ship include longitudinal (when dealing with longitudinal motions beams are sometimes referred to as rods), flexural, and torsional vibrations. For the cases to be illustrated here, the torsional vibrations of the ship are of no significance to the acoustic radiation problem. For this treatment the internal mass distributions are assumed to be rigidly attached at discrete points along the length of the hull and for simplicity do not add any bending or longitudinal stiffness. They do, however, add distributed mass to the beam model. The internal mass is assumed to be distributed in such a way that there is no mass coupling of longitudinal and flexural vibrations.

With these assumptions, the equation of motion (assuming harmonic time variation of the form $e^{-i\omega t}$ for all time-dependent variables) that determines the longitudinal vibrations is given by

$$\frac{d}{dx}\left(EA\,\frac{dU}{dx}\right) + (\rho_s A + M)\omega^2 U = -F(x). \tag{1}$$

Here $U(x)$ is the longitudinal component of displacement, E is Young's modulus, $A(x)$ is the cross-sectional area, ρ_s is the density of the hull material, $M(x)$ is the internal mass distribution per unit length, and $F(x)$ is the externally applied longitudinal force distribution. The displacement $U(x)$ is subject to the free–free boundary conditions at the ends, given by

$$\frac{dU(0)}{dx} = \frac{dU(L)}{dx} = 0. \tag{1a}$$

Assuming that the cross-sectional area and the internal distribution are uniform, Eq. (1) can be solved in terms of the *eigenmodes* of the system $\cos(n\pi x/L), n = 0, 1, 2, \ldots, \infty$. For a compact force of magnitude F_0 applied at the longitudinal location $x = x_0$, the solution is

$$U(x) = \frac{F_0}{M_T L}\left[\frac{1}{\omega^2} + 2\sum_{n=1}^{\infty}\frac{\cos(n\pi x/L)\cos(n\pi x_0/L)}{\omega^2 - \omega_n^2}\right], \tag{2}$$

where $\omega_n = C_B\sqrt{M_P/M_T}(n\pi/L)$ is the nth *resonance* frequency of the hull vibration and M_P/M_T represents the ratio of pressure hull mass to the total mass of the ship.

Examination of Eq. (2) reveals that if the excitation frequency corresponds to any of the resonance frequencies, the solution becomes unbounded. This situation is circumvented by ascribing an ad hoc measure of damping to the structure. We achieve this by assuming that the Young's modulus is complex according to the relation $E^* = E(1 - i\eta_s)$, where η_s is the structural loss factor. The first term on the right-hand side of Eq. (3) represents the rigid-body surging motion of the body, while the terms appearing in the summation are the elastic vibration mode contributions. At each of the resonance frequencies of the hull the vibration displacement is inversely proportional to the structural loss factor.

In Fig. 2 we show the real and imaginary parts of the normalized response for two different excitation frequen-

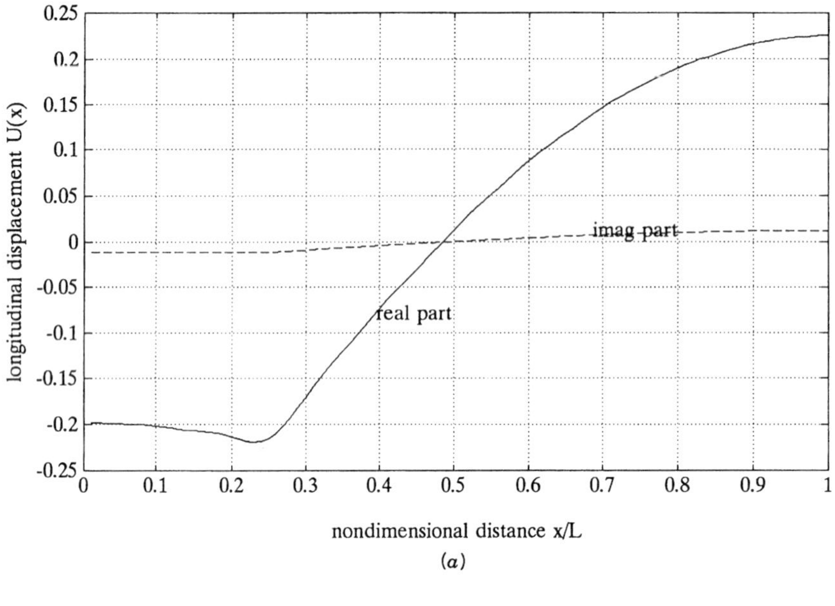

(*a*)

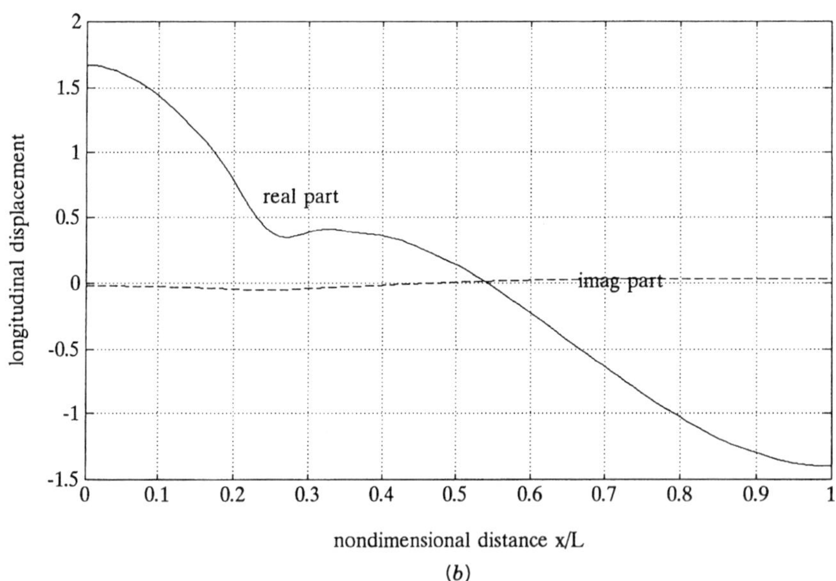

(*b*)

Fig. 2 Forced response: (*a*) nondimensional frequency = 0.5 and (*b*) nondimensional frequency = 1.5.

cies, one at a frequency below the first elastic mode and the other midway between the first and second resonance frequencies of the beam. The apparent discontinuity in the slope of the response at $x = L/4$ reflects the fact that the applied load is represented as a point load.

The far-field radiated pressure can be calculated in this LF range by using the response function (2) and calculating the equivalent volumetric source distribution on the surface of the cylinder. The pressure radiated by an idealized point source located at the origin of a Cartesian coordinate system is given by

$$p(R, \theta) = \frac{\rho \ddot{Q}}{4\pi R} e^{ikR}, \tag{3}$$

where $k = \omega/c$ is the acoustic wavenumber, c is the speed of sound in the acoustic medium, and $\ddot{Q}$ is the volumetric source strength. The parameter $R = \sqrt{x^2 + y^2 + z^2}$ measures the distance between the observation point and the source location.

The radiating sources are, in this case, the two end faces of the cylinder, with displacements $U(0)$, $U(L)$ and the cylindrical surface, which has a radial component of displacement due to Poisson's ratio effect given by $W(x) = -\nu a\ dU/dx$. The expression for the far-field pressure in terms of the hull response function, assuming a circular cross section of radius a, is then given by

$$p(R,\theta) = -\frac{\rho(\omega a)^2}{4\pi R_0}\, e^{ikR_0}\Bigg[U(L)e^{-i(kL/2)\cos\theta} - U(0)e^{i(kL/2)\cos\theta} - 2\nu\int\frac{dU(x)}{dx}\, e^{-ikx\cos\theta}\, dx\Bigg], \qquad (4)$$

where θ is the polar angle made by the observation point position vector and the centerline of the axis along which it vibrates. Here, R_0 is the distance to the origin. The first two terms inside the brackets on the right-hand side of Eq. (4) are the end-face contributions while the term under the integral represents the radiation from the cylindrical surface.

The far-field directivity patterns for the two cases illustrated in Fig. 2 are shown in Figs. 3 and 4. The higher frequency case shown in Fig. 4 reveals a more highly directive pattern. This is to be expected because the driving frequency is three times as high as that used in the first case, making the beam length three times longer in terms of acoustic wavelengths. In each pattern the levels are given in decibels normalized to the pressure at $\theta = 0$.

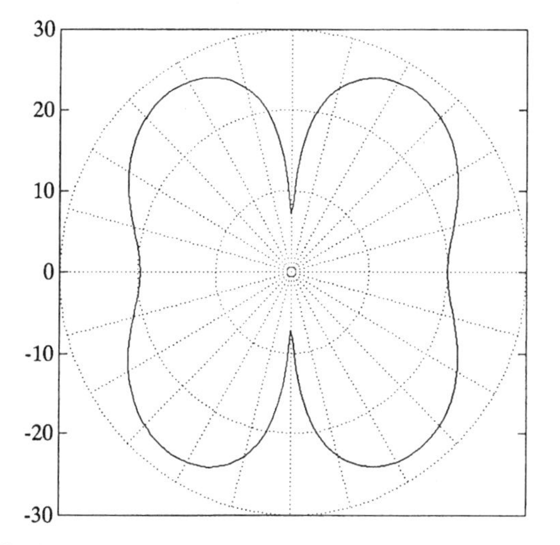

Fig. 3 Directivity pattern for nondimensional frequency = 0.5 (see Fig. 2*a*).

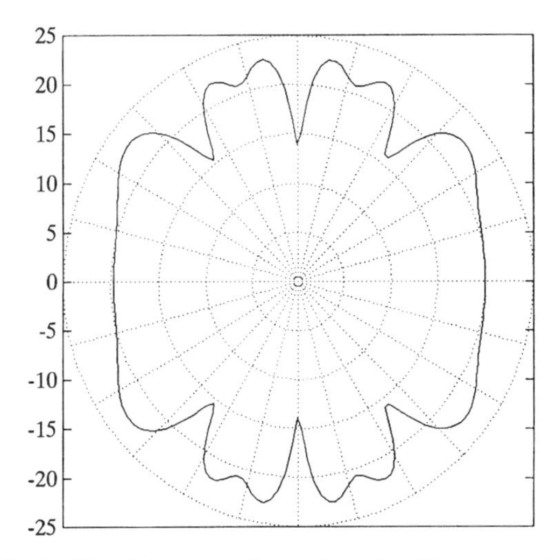

Fig. 4 Directivity pattern for nondimensional frequency = 1.5 (see Fig. 2*b*).

3.2 Flexural Vibrations

For forces that are applied in the vertical or horizontal direction, the beam representing the ship structure vibrates in flexure. The equation of motion (again assuming harmonic time variation) for such motions $V(x)$, using the theory of beam flexural vibrations, is given by

$$\frac{d^2}{dx^2}\left(EI\,\frac{d^2V}{dx^2}\right) - m(x)\omega^2 V = F_V, \qquad (5)$$

where I is the beam's cross-sectional moment of inertia and $m(x)$ is the total mass distribution per unit length, including both the beam's cross-sectional mass and its distributed internal mass.

A beam that is free at both ends must satisfy the condition that $d^2V/dx^2 = d^3V/dx^3 = 0$ at the ends $x = 0, L$. These conditions admit the two rigid-body modes, which in the vertical plane are a heaving and pitching motion and in the horizontal plane would be swaying and yawing motion. For this case as well as any other case of flexural vibrations, except that of a uniform beam simply supported at both ends, the non-rigid-body solutions to Eq. (5) are functions of eigenvalues that cannot be explicitly obtained. To solve such problems, especially in the realistic situation where the beam properties are a function of position along the length of the ship, numerical solutions are used.

For present purposes we assume that the flexural response has been determined by one of the numerical approaches. An approximate expression for the far-field radiated pressure assumes the radiation is from a distribution of dipoles oriented in the direction of the response along the beam axis. The result, first derived by Junger[2] and also by Chertock,[3] is

$$p(R,\theta,\phi) = \frac{i\rho\omega^2 k}{2\pi R} \sin\theta \cdot \cos\phi \int_0^L A(x)V(x)e^{-ikx\cos\theta}\,dx, \quad (5a)$$

where $A(x)$ is the cross-sectional area distribution, the angle ϕ is the angle between the plane of vibrations and the plane determined by the observation point position vector and the centerline of the ship, and θ is the angle in the vibration plane measured from the position vector projection to the beam axis perpendicular.

4 VIBRATION AND RADIATION IN MIDFREQUENCY REGION

In the MF range we assume that the ship's radiation can be modeled as if the radiation emanates from a limited region of the ship. Typically, the ship structure has a number of strong discontinuities such as bulkheads separating it into compartments. It then becomes appropriate in this MF region to model only that portion of the hull between major bulkheads.

An adequate representation of a ship's compartment in this frequency range is a ring-stiffened cylindrical shell simply supported by rigid and motionless end caps. The latter simulate the compartment bulkheads. The ring stiffeners are accounted for by assuming an equivalent orthotropic shell with a different bending stiffness in the circumferential and longitudinal directions. A cylindrical shell of length L_c, the compartment length, radius a, and thickness h is shown in Fig. 5.

The sound radiated by a shell, or for that matter by any structural model, depends only on the velocity component normal to the surface. We therefore focus our attention on w, the radial displacement (the radial velocity $\dot{w}$ in the case of periodic excitation is simply related to the displacement by the relation $\dot{w} = -i\omega w$, where ω is the circular frequency corresponding to the excitation frequency).

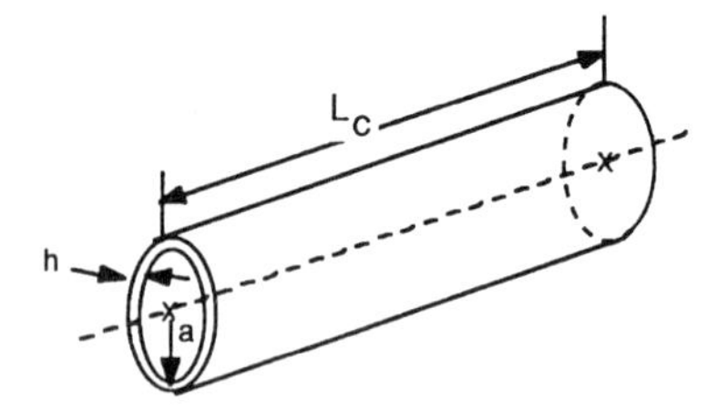

Fig. 5 Cylindrical shell.

The shell's motion is described as a superposition of responses in various modes. Each mode's response is determined by the generalized parameters such as modal mass, stiffness, and resistance. In this fashion, the radial response of the shell is written as a modal sum of the form

$$w(z,\phi) = \sum_{m=1}^{\infty}\sum_{n=1}^{\infty} W_{mn}\cos n\phi \sin k_m z, \quad (6)$$

where W_{mn} is the amplitude of the mode having $2n$ (circumferential mode number) nodes in the circumferential direction and $k_m = m\pi/L_c$ is the axial wavenumber, where m is the number of half wavelengths within the length L_c of the shell.

A highly exaggerated picture of the radial displacement around the shell circumference at a particular cross section is shown in Fig. 6 for $n = 1, 2, 3$. The mode $n = 1$ corresponds to the flexural vibration of a beam having a thin cylindrical shell cross section. The higher modes $n = 2, 3, \ldots$ are more characteristic of shells and are referred to as lobar modes. The $n = 0$ mode is axisymmetric (no ϕ-dependence) and is known as the breathing mode of the shell.

The modal amplitudes W_{mn} are determined by the generalized modal parameters, which are here explicitly listed:

- F_{mn} Modal force, or component of force that acts on a particular mode, a measure of how well the force distribution matches the displacement distribution in a given mode
- M_{mn} Modal generalized mass
- ω_{mn} Modal resonance frequency
- R_{mn} Total modal resistance, including the acoustic radiation resistance $R_{mn}^{(a)}$ and a structural modal resistance $R_{mn}^{(s)}$, related to the intrinsic loss factor discussed earlier

A detailed derivation of the above parameters, especially the modal resistance expressions, is beyond the

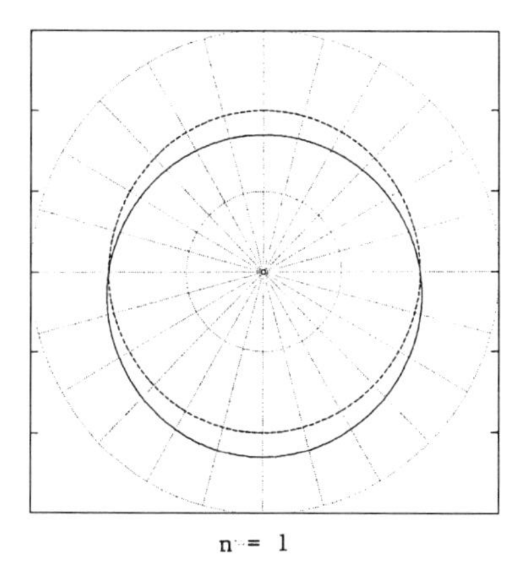

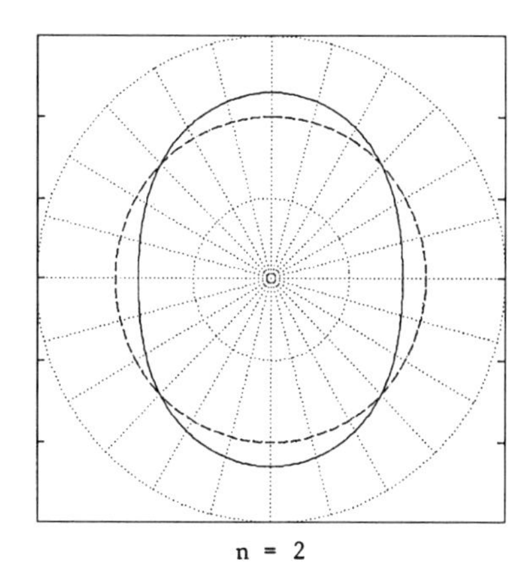

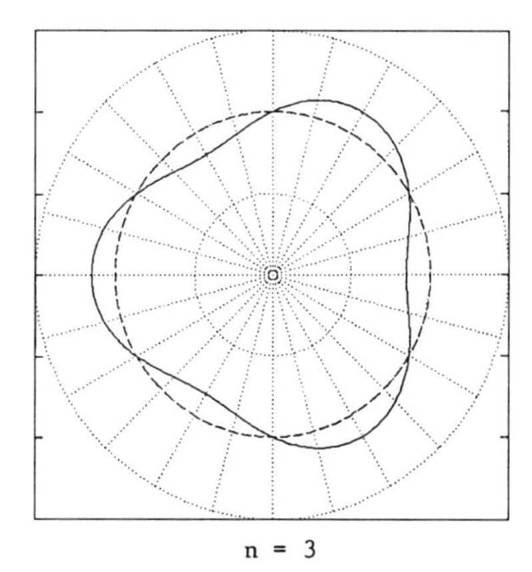

Fig. 6 Cylindrical shell model used in the midfrequency range.

scope of this presentation but can be found in Junger and Feit.[4] Here approximations to those parameters that are necessary for an estimation of the vibrational response are presented. We now consider the case of a cylindrical shell section of length L_c driven radially by a concentrated force of magnitude F at the midpoint of the shell section, $z = L_c/2, \phi = 0$; the modal amplitudes of the velocity response $\dot{W}_{mn}$ are given by

$$\dot{W}_{m0} = \frac{F}{\pi a L_c}\{\sin(m\pi/2)\}/R_{m0} - i\omega\rho_s h[1 - (\omega_0/\omega)^2] \cdot (1 + \rho L_c/\rho_s h_m m\pi), \quad (7a)$$

$$\dot{W}_{mn} = \frac{2F}{\pi a L_c}\{\sin(m\pi/2)\}/R_{mn} - i\omega\rho_s h(1 - (\omega_{mn}/\omega)^2) \cdot (1 + 1/n^2 + \rho/\rho_s h_m k_s). \quad (7b)$$

These expressions are in the form characteristic of the response of single-degree-of-freedom systems. The modal resonance frequencies can be approximated using the formulas

$$\omega_{m0} = \frac{c_p}{a}\frac{1}{(1 + \rho L_c/\rho_s h_m m\pi)^{1/2}}, \quad (8a)$$

$$\omega_{m1} = \frac{c_p}{a}\{[(1 - \nu^2)(m\pi/k_s L_c)^4 + (h^2/12a^2)(k_s a)^4]/(1 + \rho/2\rho_s k_s h_m)\}^{1/2}, \quad (8b)$$

$$\omega_{mn} = \frac{c_p}{a}\{[(1 - \nu^2)(m\pi/k_s L_c)^4 + [(m\pi/L)^2 + \sqrt{D_c/D_a}(n/a)^2]^2(h^3 a^2/12h_m)]/ [1 + \rho/\rho_s k_s h_m(n^2/n^2 + 1)]\}^{1/2}. \quad (8c)$$

The new quantities introduced in the above formulas, in addition to ν, Poisson's ratio (for steel $\nu = 0.29$), and ρ_s, the density of shell material, are $h_m = M_0/\rho_s A$, the effective shell thickness derived from the total shell mass M_0, the shell surface area A, the helical wavenumber given by the relation $k_s = [(m\pi/L_c)^2 + (n/a)^2]^{1/2}$, and D_c/D_a, the ratio of the circumferential bending stiffness to the axial bending stiffness calculated as for a flat orthotropic plate. In calculating the resonance frequencies of the modes given in Eq. (7), the added mass effects of the fluid loading have also been approximated using expressions derived in Junger and Feit.[4]

Some of the modal responses are critically dependent on the radiation resistance portion of the total modal resistance. To estimate this factor, we must consider the acoustic radiation properties of the modes. The modal radiation resistance is a measure of how efficiently the spatial characteristics of a particular mode give rise to a radiated sound pressure. The critical parameter of the vibrational pattern in this regard is the scale of the vibration pattern, or structural wavelength, relative to the acoustic wavelength at the frequency in question. The scale of the structural vibration is conveniently given by the axial wavelength $2L_c/m$ for any axial mode of order m and the circumferential wavelength $2\pi a/n$.

Using the results and experiences of previous studies, the modes can be organized into various types:

1. *Surface modes* are those whose helical wavelength is greater than the acoustic wavelength at a given frequency. The radiation resistance of such a mode is approximately

$$\frac{R_{mn}^{(a)}}{\rho c} \approx 1. \tag{9}$$

2. *Edge modes* are modes whose axial wavelength is less than the acoustic wavelength while the circumferential wavelength is larger than the acoustic wavelength. An estimate of the radiation resistance of such a mode is

$$\frac{R_{mn}^{(a)}}{\rho c} = 2\left(\frac{\omega L_c}{c}\right)\frac{1}{m^2\pi^2}, \qquad \frac{\omega L_c}{c} \gg 1, \tag{10a}$$

$$\frac{R_{mn}^{(a)}}{\rho c} = 4\left(\frac{\omega L_c}{c}\right)\frac{1}{m^2\pi^2}, \qquad \frac{\omega L_c}{c} \ll 1. \tag{10b}$$

3. *End modes* are a special case that arises for $n = 0$ modes when $\omega a/c < 1$, and an estimate of its radiation resistance is

$$\frac{R_{m0}^{(a)}}{\rho c} \approx 2\left(\frac{\omega L_c}{c}\right)\left(\frac{\omega a}{c}\right)\frac{1}{m^2\pi^2}, \qquad \frac{\omega L_c}{c} \gg 1, \tag{11a}$$

$$\frac{R_{m0}^{(a)}}{\rho c} = 4\left(\frac{\omega L_c}{c}\right)\left(\frac{\omega a}{c}\right)\frac{1}{m^2\pi^2}, \qquad \frac{\omega L_c}{c} \ll 1. \tag{11b}$$

4. For the remaining modes not satisfying the above conditions, the acoustic resistance is taken as zero.

The structural damping part of the modal resistance is given by

$$R_{m0}^{(s)} \approx \eta_s \rho_s h_m \left(\frac{\omega_{m0}}{\omega}\right)^2 \left(1 + \frac{\rho L_c}{\rho_s h_m m\pi}\right), \tag{12a}$$

$$R_{mn}^{(s)} \approx \eta_s \rho_s h_m \left(\frac{\omega_{mn}}{\omega}\right)^2 \left(1 + \frac{1}{n^2} + \frac{\rho}{\rho_s k_s h_m}\right), \qquad n \geq 1. \tag{12b}$$

The response of the shell and the near-field sound pressure are dominated by the resonant modes that radiate a negligible amount of sound power to the far field. Because they radiate little sound power, these modes are controlled by the structural damping, that is, $R_{mn}^{(s)} \gg R_{mn}^{(a)}$. When we combine Eqs. (7) and (12), we see that the modal amplitude and, therefore, the structural response can be effectively controlled by increased structural damping. Increased damping reduces the resonant response of the shell and hence the vibration level of the shell itself as well as its near-field sound pressures. It does not, however, significantly reduce the total radiated sound power that is dominated by a large number of nonresonant, efficiently radiating surface modes and a lesser number of resonant edge modes.

5 VIBRATION AND RADIATION IN HIGH-FREQUENCY REGION

As the excitation frequency increases, we reach the regime in which the effects of the radiating surface curvature become negligible. In this range circumferential stresses due to curvature become much less important than the out-of-plane bending stresses. We therefore assume that the behavior of the actual structure can be approximated by that corresponding to a flat elastic plate. The vibratory response of a plate is most easily evaluated using a structural theory rather than an elasticity theory when the thickness of the plate is small relative to a flexural wavelength. The flexural wavelength in inches of a steel plate is given by the formula $\lambda_f = 612\sqrt{h/f}$, where f is in hertz and h is in inches.

The parameters that determine the vibrational response of the plate are its mass density per unit area m and its bending stiffness $D = Eh^3/[12(1 - \nu^2)]$. In most applications the plates are stiffened by frames. If the distance between the frames is small compared to a flexural wavelength, then the stiffened plate can be conveniently idealized as an orthotropic plate with different bending stiffnesses for each of two orthogonal directions. We write D_x as the bending stiffness in the stiffest direction (for example, in the previous section we assumed the cylindrical shell was framed circumferentially and D_x would correspond to the circumferential bending stiffness) and D_y as the bending stiffness in the least stiff direction (for the previous case this corresponds to the axial direction, i.e., the direction perpendicular to the frames).

As in the previous discussions the motion of the plate is affected by the presence of the water. Depending on the excitation frequency, the water adds either additional mass to the plate or a resistive component dissipating the motion of the plate. We consider such a plate to be of infinite extent in all directions. Whether it be *isotropic* (plate of uniform thickness with no variation of bending stiffness in any direction) or *orthotropic* (bending stiffness different for different directions because of the framing), the vibrational response to an oscillating force consists of two parts. One part is an outwardly propagating wave that spreads out in all directions, analogous to the outwardly spreading wave caused by a stone dropped onto the quiet surface of a lake. This is referred to as the *propagating field*. The other part of the response consists of a pistonlike motion that is confined to the vicinity of the excitation point and is referred to as the *near field*.

The speed at which the wave crests of the propagating field spread out varies with frequency and the properties of the plate. This propagation speed c_f is called the flexural wave speed and for a steel plate is given by $c_f = 612\sqrt{fh}$ inches per second. From this formula we note that the flexural wave speed is proportional to the square root of the frequency. As the frequency of excitation increases, we find that there is a frequency at which the flexural wave speed coincides with the acoustic wave speed. This frequency is called the *coincidence frequency*.

If one were to calculate the flexural wavelength at the coincidence frequency or higher, we would find that it is equal to or greater than the acoustic wavelength at this same frequency. This gives rise to an enhanced radiation of sound in a specific direction known as the *coincidence direction*. The coincidence frequency for a steel plate in water is related to the plate thickness through the equation $f_c = 9300/h$, where frequency is given in hertz and h is measured in inches.

Consider an orthotropic plate of infinite extent lying in the (x, y) plane, driven by a normal point force $F_0 e^{-i\omega t}$ applied at the origin. The bending stiffness in the x-direction is D_x, and D_y is that in the y-direction. The radiated far-field pressure is then given by

$$p(R, \theta, \phi) = \frac{-ikF_0 e^{ikR}}{2\pi R} \left\{ \frac{\cos\theta}{1 - i(\omega m/\rho c)} \cdot \cos\theta [1 - (\omega/\omega_c)^2 \sin^4\theta (\cos^2\phi + \sqrt{D_y/D_x} \sin^2\phi)^2] \right\}, \tag{13}$$

where θ is the polar angle between the perpendicular to the plate and the observation point vector and ϕ is the circumferential angle between the x-axis and the projection of that same vector on the plane of the plate.

For very low frequencies such that $\omega m/\rho c \ll 1$, the pressure field is that of a force acting directly on an acoustic half-space,

$$p(R, \theta, \phi) = -\frac{-ikF_0 e^{ikR}}{2\pi R} \cos\theta. \tag{14}$$

From this expression we see that the plate presence has no effect on the radiated pressure.

At somewhat higher frequencies but still in the frequency range less than coincidence, the radiated pressure becomes

$$p(R, \theta, \phi) = \frac{-ikF_0 e^{ikR}}{2\pi R} \frac{\cos\theta}{1 - i(\omega m/\rho c)\cos\theta}. \tag{15}$$

Here it is only the mass per unit area of the plate that is significant, while the stiffness of the plate plays no role. The stiffness terms do not play a significant role until the frequency approaches coincidence and above.

In Fig. 7 we plot the polar angle dependence of the far-field radiated pressure for a number of frequencies. In each curve the pressures are plotted in decibels relative to the on-axis pressure, that is, on the normal to the plate. For illustrative purposes we have assumed that the plate is of uniform thickness, having no stiffeners and therefore isotropic. This accounts for the fact that the pressure is independent of the circumferential angle.

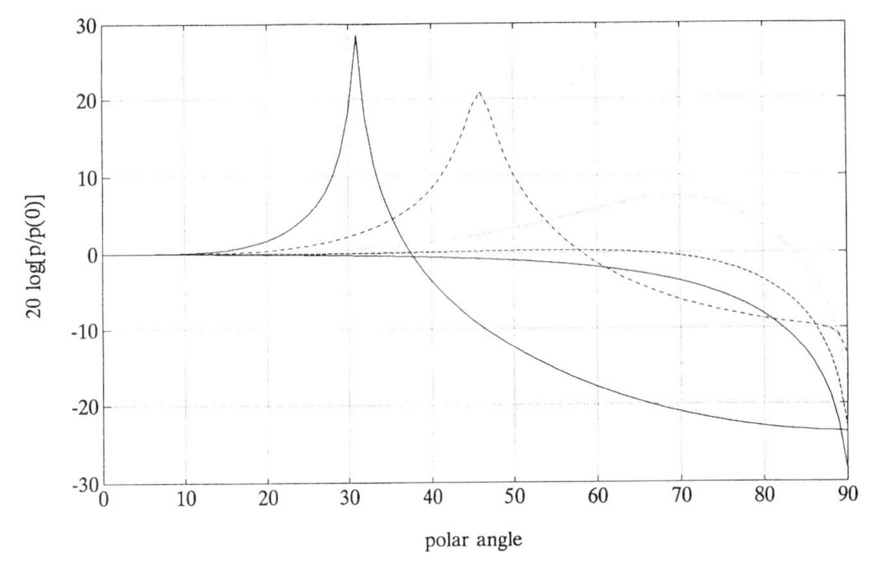

Fig. 7 Far-field directivity: polar angle plotted against far-field radiated pressure.

We see that in the frequency range above coincidence the pressure maximum occurs at a specific angle, the *coincidence angle*, which is off the normal and is given by the relation $\theta_c = \sin^{-1}\sqrt{f_c/f}$. As the frequency increases, the coincidence angle approaches the normal. In this extremely high frequency range, a higher order plate theory such as that of Timoshenko-Mindlin (Ref. 4, pp. 214–215) would be more appropriate. Had such a theory been used in the calculations, the coincidence angle would asymptotically approach a limiting angle $\theta_c = \sin^{-1}(c/c_R)$ where c_R is the Rayleigh wave speed for the plate material.

REFERENCES

1. D. Ross, *Mechanics of Underwater Sound*, Pergamon Press, New York, 1976, pp. 326–347.
2. M. C. Junger, "Sound Radiation by Resonances of Free-Free Beams," *J. Acoust. Soc. Am.*, Vol. 52, No. 1 (Part 2), pp. 332–334.
3. G. Chertock, "Sound Radiated by Low-Frequency Vibrations of Slender Bodies," Vol. 57, No. 5, May 1975, pp 1007–1016.
4. M. C. Junger and D. Feit, *Sound, Structures and Their Interaction*, 2nd ed., MIT Press, Cambridge, MA, 1986.

33

TRANSIENT AND STEADY-STATE SCATTERING AND DIFFRACTION FROM UNDERWATER TARGETS

LOUIS R. DRAGONETTE AND CHARLES F. GAUMOND

1 INTRODUCTION

Much of the understanding of the physics of underwater acoustic scattering has been derived from investigations of the scattering of plane waves by simple shapes, especially elastic spheres and spherical shells and the infinite-elastic cylinder and cylindrical shell. The spherical and cylindrical shapes lend themselves to exact solutions and thus to the very accurate calibration and verification of the experimental and numerical approaches required to deal with the more complicated problems of interest. The term *canonical* in this chapter will be limited to those two simple shapes.

Three formulations of the canonical problems are of particular interest.

The normal-mode series solution, or Rayleigh series, is the description best suited to accurate computation of scattering by the canonical shapes as a function of frequency.[1–12] The relative ease and accuracy of this technique over a broad bandwidth permits, among other uses, the simulation of transient responses[4,13] and the calibration of sophisticated transient and steady-state measurement techniques.

The "creeping-wave" formulation, a contour integral representation of the scattering problem,[14–19] led to the circumferential wave descriptions of scattering. This description has been especially valuable in the understanding of the contributions of diffraction (Franz waves), the "leaky-Rayleigh-surface" wave, and guided "leaky Lamb" waves and had a major influence in research on active classification approaches.[20,21]

Note: References to chapters appearing only in the *Encyclopedia* are preceded by *Enc.*

The formalization of a resonance theory of acoustic scattering,[22] analogous to nuclear reaction theory,[23] grew out of comparisons between the Rayleigh series and contour integral formulations. (See Chapter 4, Section 3.) Families of free-body resonances of an aluminum cylinder were identified with particular elastic surface waves.[24] A physically meaningful separation of the Rayleigh series solution into a geometric, rigid background term and a term containing the elastic resonance behavior was accomplished[22] and demonstrated.[25–27] The introduction of this resonance formalism[22] has been, subsequently, the source of a prolific area of published theoretical scattering research,[28–31] only a small sampling of which will be indicated here.

Scattering is quantitatively described in terms of the form function, which is related to the impulse response of a target, and target strength, which is the metric used in the active sonar equations. The techniques used to empirically determine these metrics will be outlined.

2 FORM FUNCTION

The form function is a dimensionless measure of the complex amplitude A of the pressure field scattered from a target. The form function requires the scattered pressure p_s to be normalized by the incident pressure p_0, the range r between the target and receiver, and a characteristic length dimension a of the target. The frequency is expressed by the dimensionless parameter $ka = 2\pi a/\lambda$, where λ is the wavelength of the incident sound in the surrounding water. The normalization factor for a given shape is, in general, chosen so that the magnitude of the backscattered form function is unity for perfectly rigid or soft boundary conditions in the high-frequency limit.

The scattering examples considered here will involve incident plane pressure waves, $p_0 = P_0 e^{-ikz}$, with harmonic time dependence $e^{+i\omega t}$; the scattered pressure field p_s approaches either $A(\mathbf{\Omega})e^{-ikr}/r$ in three dimensions or $A(\theta)e^{-ikr}/\sqrt{r}$ in two dimensions for the distances $r \gg a^2/\lambda$ large enough to be in the far-field or Fraunhofer zone.[32]

For a spherically shaped target the characteristic length dimension is the radius a and the far-field form function is defined as

$$f(\theta, ka) = \frac{2r}{a}\frac{p_s}{P_0}. \tag{1a}$$

The case most discussed in the literature is backscattering, $\theta = \pi$. The form function $f(\pi, ka)$ is then generally presented as a function of ka only; that is, when the angular dependence is subdued, backscattering $\theta = \pi$ is understood. Thus the backscattered form function is written as

$$f(ka) = \frac{2r}{a}\frac{p_r}{P_0}, \qquad \theta = \pi. \tag{1b}$$

For the infinite cylinder, the characteristic length dimension is again the radius a, but the far-field spreading of the scattered pressure has a $1/\sqrt{r}$ dependence. When the incident wave vector is normal to the cylinder axis, the far-field form function for backscattering is given as

$$f(ka) = \sqrt{\frac{2r}{a}}\frac{p_s}{P_0}. \tag{2}$$

3 NORMAL-MODE SOLUTIONS

Normal-mode series solutions for the far-field function of spherical[1–5,7,8,10–12] and cylindrical[1,2,6,9,12] shapes with a variety of boundary conditions are found in the literature. Consideration of these shapes, as evidenced by this large but not complete list, are predominate in the literature since these shapes lend themselves to solution by a separation-of-variables technique. The form function depends, of course, on the boundary conditions as well as the target shape. Results and interpretation for the case of solid elastic spheres and infinite cylinders are given below. The perfectly "rigid" (Neumann) and "soft" (Dirichlet) boundary conditions are considered here as special cases of the solid elastic target.

3.1 Spheres and Infinite Cylinders

For spherical geometries the far-field form function is given in Eq. (3) of Chapter 4 as

$$\sin \eta_n e^{i\eta_n} = \frac{j_n(ka)L_n - (ka)j_n'(ka)}{h_n^{(2)}(ka)L_n - (ka)h_n^{(2)\prime}(ka)}. \tag{3}$$

The functions j_n and h_n are spherical Bessel functions of the first and third kind, respectively, with primes denoting derivatives with respect to the argument. The factor L_n is a function of frequency and of the material properties of the sphere. For a rigid sphere $L_n = 0$, and for a pressure-release sphere $L_n = \infty$. The a_n are given by $j_n(ka)/h_n^{(2)}(ka)$ for the case of Dirichlet boundary conditions and $j_n'(ka)/h_n^{(2)\prime}(ka)$ for Neumann boundary conditions. The L_n for the solid elastic sphere with and without absorbtion are found in Ref. 10; for a spherical shell the L_n elements are found in Ref. 4. (Note that the complex conjugate of L_n is given in Ref. 4 where $e^{-i\omega t}$ is assumed.)

Normal-mode-series solutions in the infinite cylindrical geometry are found in many references[1,2,6,9,12] (Note that Refs. 1, 2, 9, and 12 use $e^{+i\omega t}$ while Ref. 6 uses $e^{-i\omega t}$.) In this geometry the far-field form functions have the form

$$f(ka) = \sqrt{\frac{2}{\pi ka}} \sum_{n=0}^{\infty} \epsilon_n C_n \cos(n\phi), \tag{4}$$

where ϵ_n is the Neumann factor ($\epsilon_n = 2, n = 0; \epsilon_n = 1, n > 0$) and C_n are given as

$$C_n = \frac{J_n(ka)L_n - kaJ_n'(ka)}{H_n^{(2)}(ka)L_n - kaH_n^{(2)\prime}(ka)}, \tag{5}$$

with L_n given as quotients of determinants B_n and D_n:

$$L_n = \frac{\rho}{\rho_1}\frac{B_n}{D_n}. \tag{6}$$

The rank and elements of the determinants depend on the boundary conditions; ρ and ρ_1 are the densities of water and the material, respectively. Flax[9] gives the matrix elements required to determine L_n for the case of a cylindrical shell comprised of two absorbing layers with a fluid-filled interior. Instructions are given in the same reference for the reduction of this broad solution to the simpler cases of the solid elastic cylinder and the air- and water-filled cylindrical shells using the appropriate subset of these matrix elements. The solutions of the Neumann and Dirichlet problems reduce, as in the case of the sphere, to simplified expressions for the L_n. These are

$$L_n = 0 \quad \text{and} \quad C_n = \frac{J_n'(ka)}{H_n^{(2)\prime}(ka)}$$

for the rigid cylinder and

$$L_n = \infty \quad \text{and} \quad C_n = \frac{J_n(ka)}{H_n^{(2)}(ka)}$$

for the soft cylinder.

The form functions for rigid and elastic spheres and infinite cylinders are given in Figs. 1*a* and 1*b*, respectively. The form functions contain contributions from both the geometric and elastic properties of the scatterer. Figure 1*a* shows $|f(ka)|$ for a rigid sphere and spheres of aluminum, tungsten carbide, and brass. The form function of the rigid sphere is determined by two physical mechanisms—specular reflection from the surface of the sphere and diffraction of energy around the sphere. These are sketched in Fig. 1 of Chapter 4. The diffracted component can be described mathematically as a dispersive surface wave traveling wholly in the surrounding fluid at a speed slower than the compressional wave speed in the fluid. Peaks in the form function of a rigid sphere (and cylinder) occur at ka values for which the specular reflection and the creeping wave add in phase; nulls when the waves destructively interfere. The initial peak occurs near the same ka value (ka equals 1.15 for the sphere and 0.85 for the cylinder) for both the rigid and for each metal scatterer, thus providing a shape clue, that is, one that is independent of material. All solid elastic spheres and cylinders fabricated from materials whose densities and shear speeds are greater than the density and the sound speed in water produce form functions that at low ka are similar to the rigid-body form function and may be said to deviate from a "rigid background" at mid and high ka values. The form function curves for liquid[33] and plastic[10] spheres have unique properties. The amplitude of the initial peaks in Figs. 1*a* and 1*b* are related to the densities of the material and approach $|f| = 1$ as the density becomes larger.

A more significant material property can be determined by the ka value at which the form function curves first deviate significantly from the rigid case. As predicted by Faran,[1] Vogt[11] demonstrated for elastic-solid spheres that substantial variations occur above a critical ka value that depends on the resonance frequencies of the free sphere. Figure 2, computed by Vogt,[11] shows the amplitudes of the resonance terms plotted versus ka for three metals. The modes are labeled by the integers n, l. The n's, which correspond to each term in the normal-mode series, can take on values from 0 to ∞; the l's represent the overtones beginning with the fundamental $l = 1$. Just as the creeping-wave interaction with the specular provides unique shape clues, the results given in Figs. 1 and 2 show that unique information about the material composition may be obtained directly from the form function. The lowest observable mode for an elastic sphere is the (2, 1) spheroidal mode. Vogt[11] demonstrated that the shear speeds of the materials considered in Fig. 2 are inversely related to the ka position of the (2, 1) mode, that is, for two materials

$$\frac{c_{s1}}{(ka)_1} \sim \frac{c_{s2}}{(ka)_2}, \tag{7}$$

where c_{s1} and c_{s2} are the shear velocities of materials 1 and 2 and $(ka)_m$ is the position of the (2, 1) mode for material m. This expression gives resonable accuracy for all homogeneous materials whose density and shear speeds are greater than the density and sound speed in water. Similar elastic phenomena are exhibited in cylinders.[22,25]

4 CREEPING-WAVE FORMULATION

Uberall and collaborators,[14–19] at the Catholic University, rewrote the Rayleigh series descriptions of the scattering by spheres and cylinders as contour integrals using the Watson transformation. This is discussed in Chapter 4. However, the growth of computer technology makes the Rayleigh series more computationally tractable for generating form functions of spheres and cylinders. The significant use of the Watson transform is the physical interpretation that follows. Residues of contributing poles (ν_l) are interpreted as attenuated circumferential waves having a phase velocity[24]

$$c_l(ka) = \frac{(ka)c}{\Re \nu_l}. \tag{8}$$

Franz poles were investigated first[14]; this work led to the creeping-wave description of the diffraction around a rigid cylinder or sphere. The term *creeping-wave series* is now commonly used as a descriptor of the Watson transform analysis, even though in elastic bodies important poles give rise to elastic circumferential waves with speeds generally much faster than the speed of

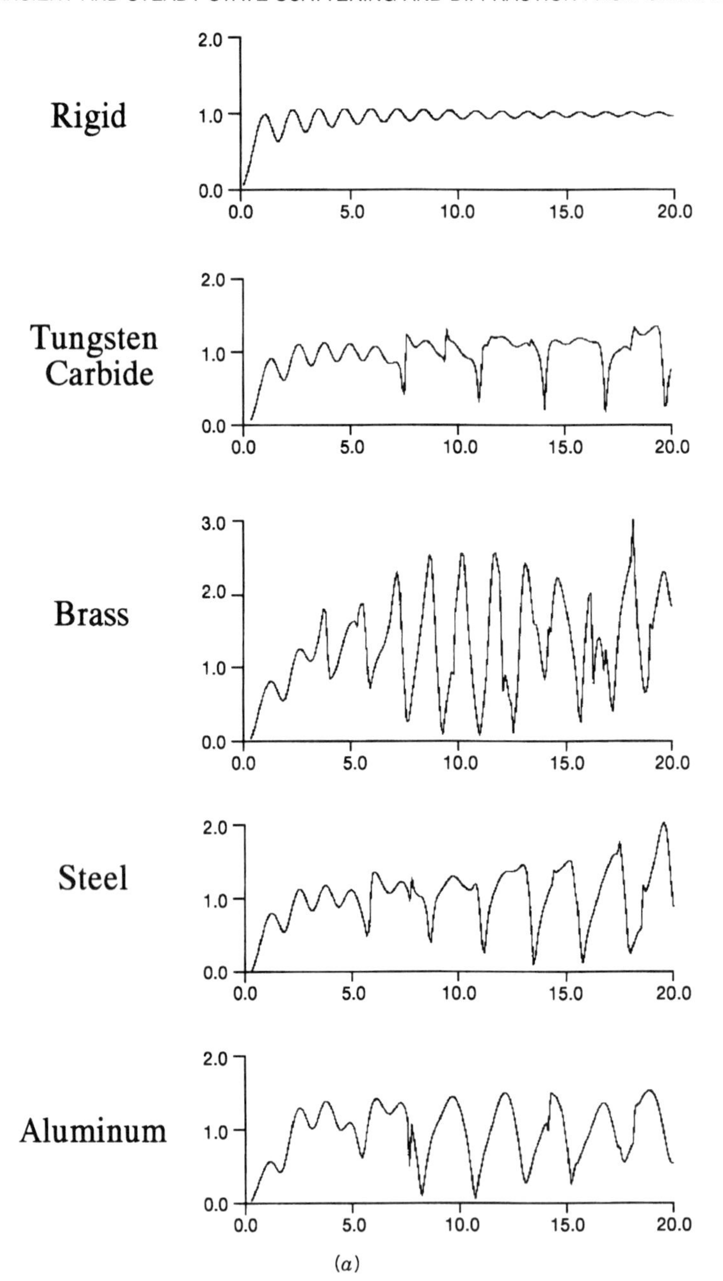

Fig. 1 (*a*) Form function for rigid, tungsten carbide, brass, steel, and aluminum spheres. (*b*) Form function for rigid, tungsten carbide, brass, steel, and aluminum cylinders. Computations require compressional speed, shear speed, and density as the independent variables. For the four materials these are: tungsten carbide (6860 m/s, 4185 m/s, 13,800 kg/m^3); brass (4700 m/s, 2110 m/s, 8600 kg/m^3); steel (5950 m/s, 3240 m/s, 7700 kg/m^3); aluminum (6376 m/s, 3120 m/s, 2710 kg/m^3).

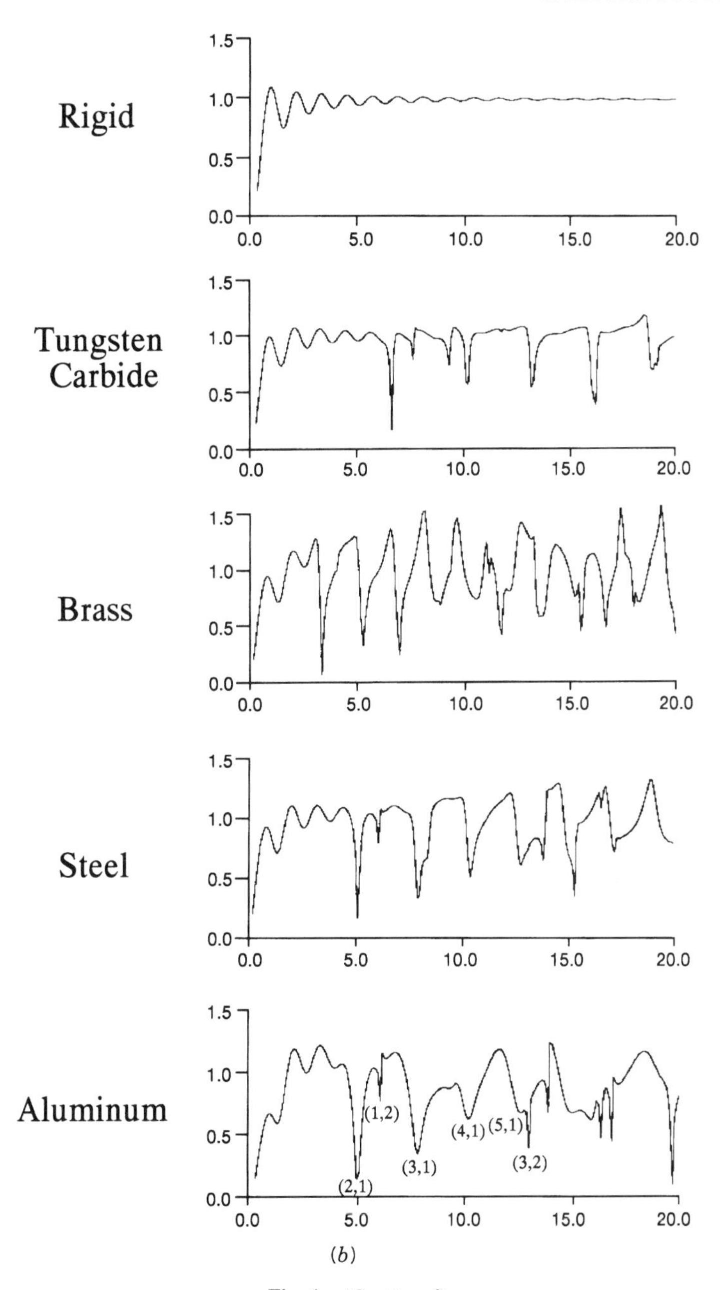

Fig. 1 (*Continued*)

sound in water. Poles corresponding to Rayleigh-like and Stonely-like waves (the Rayleigh and Stonely designations are defined for an infinite half-space) have been investigated[15]; whispering gallery waves,[19] which hug the inside surface, and general refraction mechanisms[18] have also been studied.

The physical relationship between the modal resonances observed in the form functions computed by the Rayleigh series and the circumferential wave descriptions obtained from the creeping-wave formulation have been established.[24–27] If we consider the form function for an aluminum cylinder given in Fig. 1*b*, the labeled

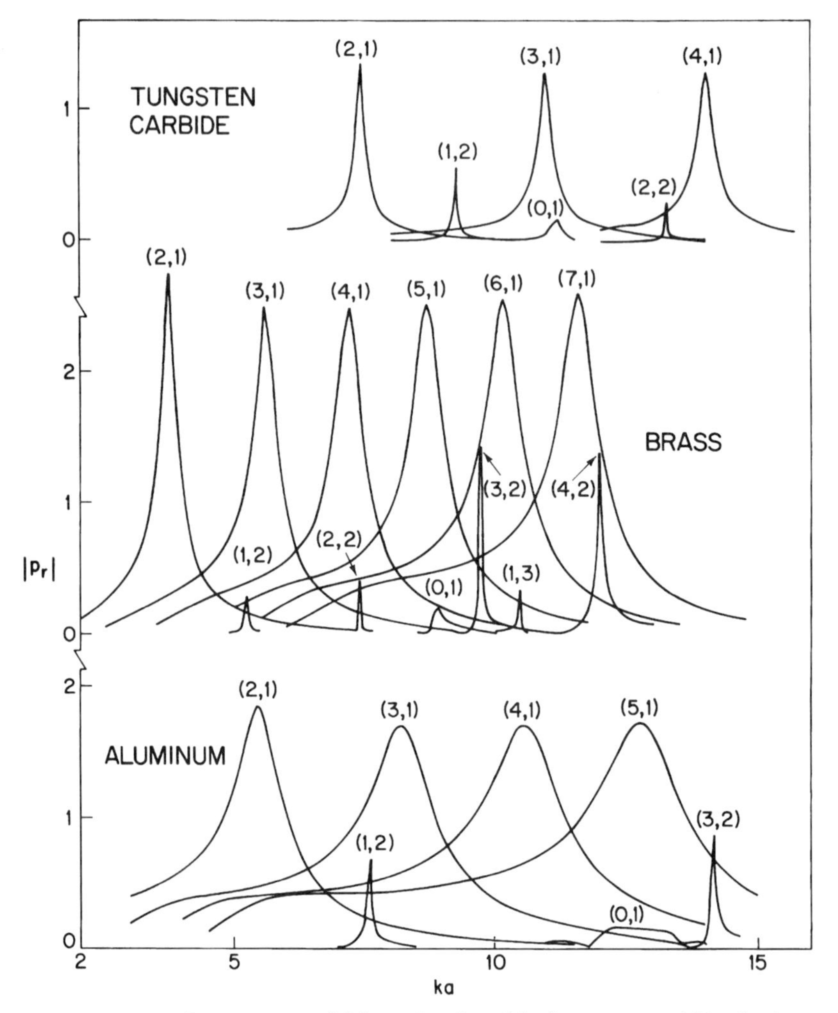

Fig. 2 Modulus of resonant terms $|f_n|$ as a function of ka for tungsten carbide, aluminum, and brass spheres.

resonance effects in $|f(ka)|$ occur near ka values that satisfy the in vacuo resonance condition.[24] Each term n in the normal-mode series is the nth multipole with $n = 0$ the breathing mode, $n = 1$ the dipole, $n = 2$ the quadrupole, and so on. These motions can be resolved into a pair of standing waves $e^{\pm in\phi}e^{+i\omega t}$ circumnavigating the target in opposite directions with a modal phase velocity

$$c_n(ka) = \frac{(ka)c}{n}. \tag{9}$$

A comparison of Eqs. (8) and (9) gives the physical relationship between the circumferential waves discussed in the Catholic University series of papers and the modal resonances.

When $\Re(\nu_l) = n$, Eqs. (8) and (9) become identical in form so that the modal velocity and the circumferential velocity become equal. For the case of the aluminum cylinder response given in Fig. 1*b*, the R_1 (Rayleigh-type) wave labeled by Doolittle[15] is related to the modes labeled $(n, 1)$ in Fig. 1*b*, the R_2 wave of Doolittle to the modes labeled $(n, 2)$, and so on. At the ka position of the (2, 1) resonance the circumference of the cylinder is exactly two wavelengths of the R_1 wave, the circumference is three wavelengths long at the position of the (3, 1) resonance, and so on. A particular circumferential wave R_l is related to all of the resonances labeled with the eigenvalue l.

5 RESONANCE THEORY OF SCATTERING FORMULATION

This topic is discussed in detail in Chapter 4. Briefly, clear resonance behavior is observed in the form function computations considered in Figs. 1*a* and 1*b*, but the exact Rayleigh series expressions [e.g., Eqs. (3) and (4)] do not explicitly show resonance terms. A resonance term would contain a denominator of the form $1|(k - k_n + i\Gamma_n)$, where k_n is the resonance wavenumber and Γ_n is the resonance width. The formalism of the classical resonance theory of nuclear reactions[23] was applied to the solid elastic infinite-cylinder and spherical geometries,[22] and a meaningful separation of the Rayleigh series into the form

$$f(ka) = \sum_{n=0}^{\infty} f_n(ka) = \sum_{n=0}^{\infty} f_n^R(ka) + \sum_{n=0}^{\infty} f_n^E(ka) \tag{10}$$

was demonstrated. In Eq. (10) each partial-wave term $f_n(ka)$ in the exact normal-mode series solution has been separated into a term $f_n^R(ka)$ from the rigid-body solution and a resonance term $f_n^E(ka)$ in the form $1/(k - k_n + i\Gamma_n)$. For descriptions and examples see Chapter 4. The utility of the resonance formalism in describing the physics of acoustic scattering has been demonstrated. Uberall and collaborators have applied the formalism to numerous problems with various background conditions.[28–31] References 28–31 are a small sample of the existing literature on this subject.

6 SHELLS

The three techniques of computation and analysis considered above have been applied to the shell problem.[4,6,16,29] A consideration of the properties of Lamb waves[34] on a flat plate provides a useful analysis tool for the high *ka* scattering by a curved shell. Figure 3 shows a computation of the phase velocity versus frequency–thickness product (fd) for the first asymmetric and first symmetric Lamb waves for a thin aluminum plate. The ordinate shows the phase velocity V_p; a surface wave can be generated by an incident sound wave in water at an angle of incidence $\sin\theta_i = c/V_p$. The symmetric wave can be generated by phase matching at any fd value represented on the plot; the asymmetric wave can only be generated above a critical fd value at which V_p is equal to the sound speed in water. This critical value is indicated by the arrow in Fig. 3. The Lamb wave curves presented are defined for a plate in vacuo. There are significant modifications caused by fluid loading[35,36] and curvature.[37] For an infinite cylin-

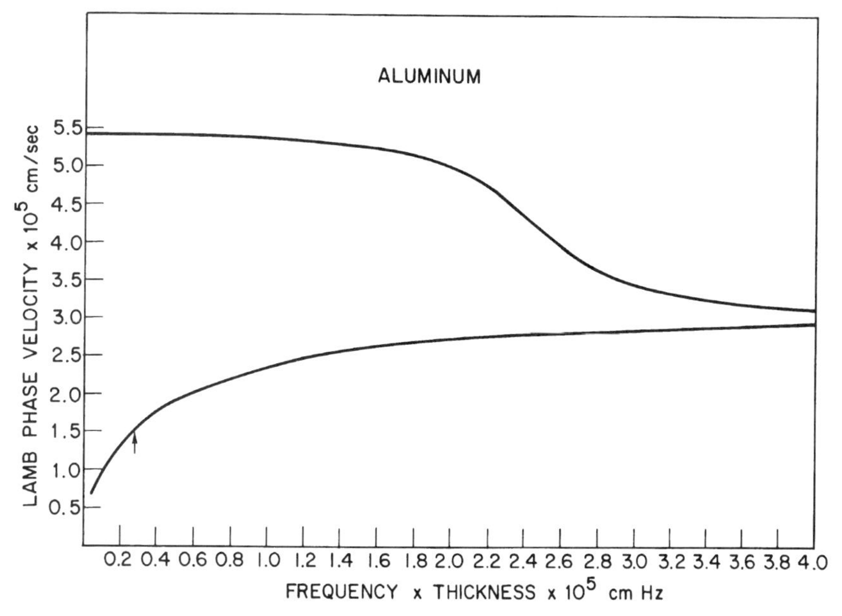

Fig. 3 Dispersion curves for the fundamental symmetric and asymmetric Lamb waves for an infinite aluminum plate in a vacuum.

der this simple model predicts that the high ka scattering below the coincidence frequency for asymmetric waves would consist of an interference between specular reflection and a symmetric leaky Lamb wave that follows the circumference of the cylinder. The backscattering from a steel shell with $d/a = 0.015$ (here d is the thickness) is given in Fig. 4, which plots the form function magnitude versus ka. Below $ka \sim 60$, the form function is well modeled by the interference of a specularly reflected wave and a symmetric Lamb wave. Above $ka \sim 60$ the coincidence frequency, the form function includes effects from the asymmetric wave.[36] The symmetric Lamb wave generated by phase matching is in phase with the incident acoustic wave. The specular reflection is out of phase with the incident wave at the lower ka values, and as frequency is increased, it becomes in phase. Thus the impedance boundary condition for a thin shell is frequency dependent. The shell presents a soft background at low ka and presents a rigid background in the high-frequency limit. The steady-state, low-frequency nulls in the form function occur at the ka values for which the cylinder circumference is an integral number of wavelengths of the symmetric Lamb wave. At these ka values the elastic response is a maximum due to constructive interference caused by the multiple circumnavigations of the Lamb wave. Nulls in the form function occur in the frequency band over which the specular response is 180° out of phase with the incident wave (soft background); peaks occur at high ka when the shell presents a rigid background and the specular is in phase. The absolute amplitudes of the interaction decrease with frequency because of the damping of the Lamb wave by reradiation into the water as it circumnavigates the cylinder. As frequency increases, the same circumferential path represents a larger number of wavelengths traversed. For other shapes similar plane-wave generation angles and paths can be determined to provide a qualitative description of the form of the expected response.

7 TARGET STRENGTH

Target strength is a descriptor of the scattering amplitude widely used in sonar[38]; it is defined as the logarithm of the ratio of the reflected and incident sound intensities weighted by the target-to-receiver range r:

$$\text{TS} = 10\log\left(\frac{I_s}{I_i}\right) + 20\log r. \tag{11a}$$

Since the steady-state intensity is $p^2(f)/\rho c$, this equation can be expressed as

$$\text{TS} = 20\log\left(\frac{p_r}{p_0}\right) + 20\log r, \tag{11b}$$

where p_r and p_0 are steady-state pressures. Note that the energy spectral density defined as $\tilde{p}(f) = \int pe^{-i2\pi ft}\,dt$ may be used instead of the steady-state pressure. Target strength is given as decibels relative to 1 m^2, since the unit of r currently in common use in the U.S. Navy is the metre. Previously the yard was used, which leads to a definition that is smaller by 0.78 dB. The use of standard units allows the use of TS in the sonar equation.[38] Table 1 gives the standard units of applicable terms in the sonar equation in the time domain and frequency domain. Using the terminology of the sonar equation, steady-state target strength can be expressed as

$$\text{TS} = \text{EL} - \text{SL} + \text{TL}_1 - \text{TL}_2, \tag{11c}$$

where $20\log(|\tilde{p}_0(f)|) = \text{SL} - \text{TL}_1$ and the other terms are defined in Table 1. Other definitions of TS that depend on the transient, incident pressure signal are integrated target strength (ITS) and peak target strength (PTS):

$$\text{ITS} = 10\log\frac{\int |p_s(t)|^2\,dt}{\int |p_i(t)|^2\,dt} + 20\log r, \tag{11d}$$

$$\text{PTS} = 10\log\frac{\max|p_s(t)|^2}{\max|p_i(t)|^2} + 20\log r. \tag{11e}$$

The target strength definitions implicitly assume that the receiver is located in the far field of the target where the effects of range can be compensated by the factor r. This "projection of the scattered amplitude back to a distance of 1 m from the target center" then removes the range dependence in the definition of TS.

Target strength can be found from expressions of form function. The backscattered TS of a spherically shaped scatterer can be obtained from Eqs. (1b) and (11b) as

$$\text{TS} = 20\log(a/2) + 20\log|f(ka)| \tag{12}$$

and for the infinite cylinder at normal incidence as

$$\text{TS} = 10\log(ar/2) + 20\log|f(ka)|. \tag{13}$$

An extensive collection of formulas for high-frequency limits for many shapes can be found in Urick.[38] These formulas assume that the targets are rigid; thus $|f(ka)| \approx 1$. An expression that approximates the target strength of a finite-length elastic cylinder at normal incidence is

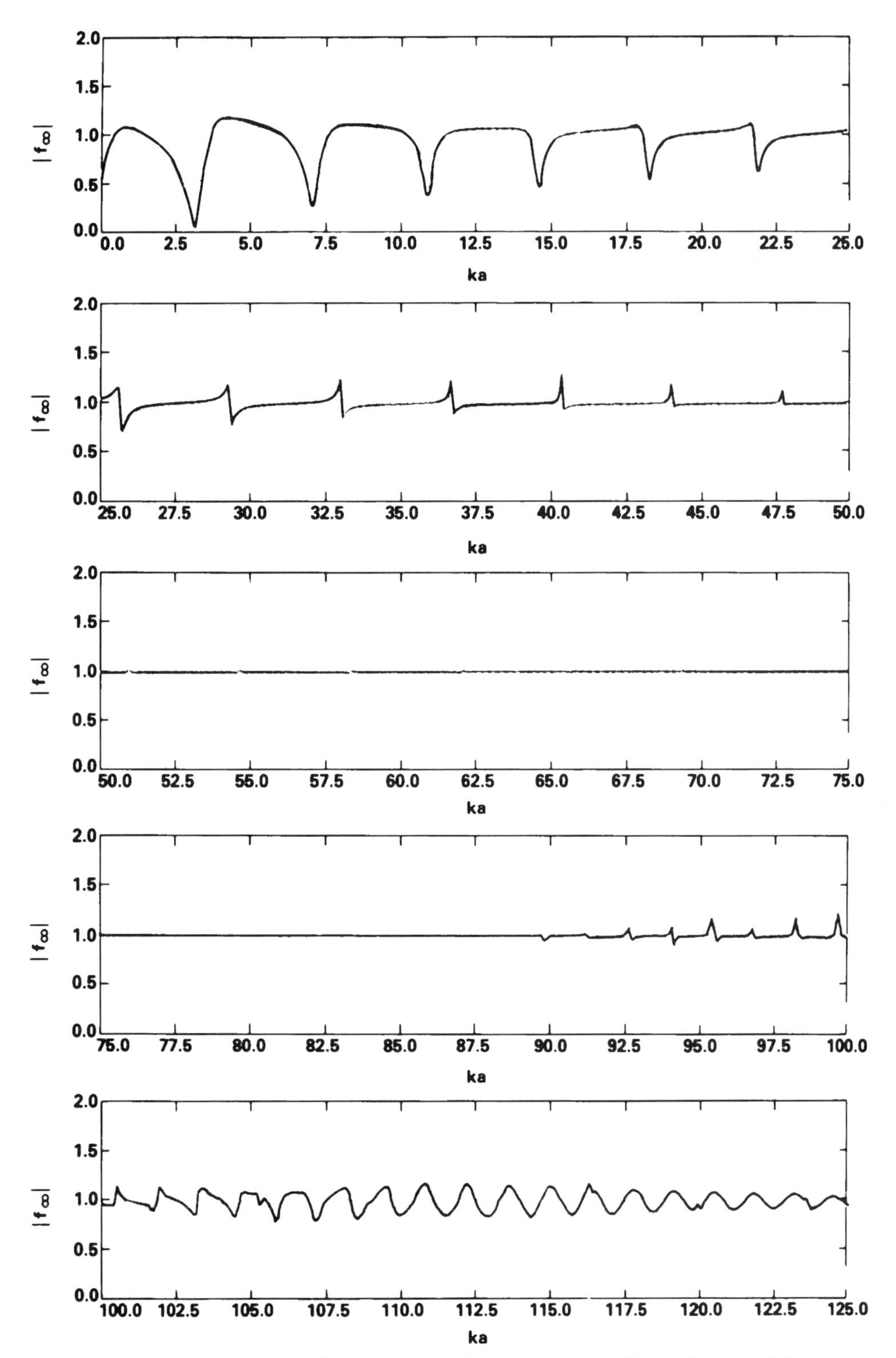

Fig. 4 Backscattered form function for an infinite cylindrical shell as a function of ka. The material is steel; the shell has an inner-to-outer-radius ratio b/a of 0.985. Indicated in parentheses on the ka scale is the equivalent fd value for a plate of the same thickness. The units of the fd values are $fd \times 10^{-5}$ cm·Hz.

TABLE 1 Standard Units

Quantity	Abbreviation	Expression	Unit
Pressure		$p(t)$	μPa
Signal level (time)	EL	$20 \log \lvert p(t) \rvert$	dB re μPa
Signal level (frequency)	EL	$20 \log \lvert \tilde{p}(f) \rvert^{a}$	dB re μPa · s
Source level (time)	SL	$20 \log \lvert A(t) \rvert^{b}$	dB re μPa · m
Source level (frequency)	SL	$20 \log \lvert \tilde{A}(f) \rvert^{a,b}$	dB re μPa · m · s
Transmission loss (time)	TL	$20 \log r^{b}$	dB re m
Transmission loss (time)	TL	$20 \log \lvert g(r, r', t - t') \rvert^{c}$	dB re m/ s
Transmission loss (frequency)	TL	$20 \log \lvert \tilde{g}(r, r', f) \rvert^{d}$	dB re m/ s
Noise level (time)	NL	$10 \log \left(\frac{1}{T} \int_0^T \lvert n(t) \rvert^2 dt \right)$	dB re μPa/Hz$^{1/2}$
Noise level (frequency)	NL	$10 \log \left(\frac{1}{T} \left\lvert \int_0^T n(t) e^{-i2\pi f t} dt \right\rvert^2 \right)$	dB re μPa/Hz$^{1/2}$
Form function (frequency)		$f(\theta, f)$	Dimensionless
Target strength (frequency)	TS	$10 \log \frac{a}{2} \lvert (f\theta, f) \rvert$	dB re m
Target strength (time)	TS	$10 \log \frac{a}{2} \lvert (\tilde{f}\theta, t) \rvert$	dB re m/s

[a] Sometimes 10 log 2 is added to these frequency-domain definitions in order to include the contributions of the negative frequencies to the energy density of the signal.
[b] Pressure is related to source level as $p(t) = A(t)/r$ or $\tilde{p}(f) = \tilde{A}(f)/r$ for an infinite, homogeneous, and isotropic medium.
[c] A time-invariant medium is assumed. In this case the pressure at time t and position r is related to a source at time t' and position r' by a temporal convolution with $g(r, r', t - t')$.
[d] A time-invariant medium is assumed and $\tilde{p}(f) = \tilde{A}(f)\tilde{g}(r, r', f)$.

derived from Eq. (11b) in this chapter and Section 10.6 in Junger and Feit[32] as

$$\mathrm{TS} = 10 \log \left(\frac{al^2}{2\lambda} \right) + 20 \log \lvert f(ka) \rvert, \tag{14}$$

where l is the length of the cylinder and λ is the wavelength of the incident sound wave. Note that the approximation uses the far-field expression for the form function of an infinite cylinder. Equation (14) is approximate since it does not include three-dimensional scattering effects such as the reflection and reradiation of axially traveling waves at the cylinder ends.

Table 2 gives several high-frequency, rigid-body target strengths that can be properly combined to form a more complicated scatterer. An extensive list is found in Urick.[38]

When specular scattering is the dominant echo mechanism, the bistatic TS may be estimated from the monostatic TS by a simple rotation of the monostatic beam pattern. Figure 5*a* shows the monostatic beam pattern computed for a rigid hemispherically end-capped cylinder from formulas given in Table 2. The pattern in Fig. 5*a* may be interpreted in two ways. The source/receiver can be fixed and the target rotated over 360° from the starting position shown, or the target can be considered fixed and the source/receiver rotated over the full 360°.

The bistatic beam pattern for the same target, with source and receiver separated by an angle θ_r, is estimated by rotating the monostatic pattern by the bistatic angle θ_b, defined as $\theta_b = \frac{1}{2}\theta_r$. In the example given in Fig. 5*b*, the source and receiver are separated by 60° and the monostatic beam pattern displayed in Fig. 5*a* is rotated through 30° to give the approximation to the bistatic pattern. The interpretation of Fig. 5*b* is similar to that of Fig. 5*a*; namely, the target may be considered fixed in the position shown while the source and receiver are rotated through 360°, maintaining a 60° separation, or the source and receiver are fixed and the target rotated through 360°. The source, receiver, and target positions giving the maximum response are indicated in Fig. 5*b*. The rotation technique for bistatic approximation will fail when elastic mechanisms are dominant and also when the receiver-to-source separation is much larger than 90°.

8 MEASUREMENT METHODS

A form function or target strength experiment requires measurement of the parameters given in Eq. (1) or (2),

TABLE 2 Target Strength of Various Shapes

Shape	Incident Angle	TS
Sphere	Any	$10 \log \frac{a^2}{4}$
General convex shape	Any	$10 \log \frac{r_1 r_2}{4}$
Infinite cylinder	Normal	$10 \log \frac{ar}{4}$
Finite cylinder	θ_i	$10 \log \frac{al^2}{2\lambda} \left(\frac{\sin \beta}{\beta}\right)^2 \cos^2 \theta_i$
Finite plate with area A and reflection coefficient α_r	Normal	$10 \log \left(\frac{A}{\lambda}\right)^2 + 10 \log \alpha^r$

Note: The symbols r_1 and r_2 are the radii of curvature at the point of specular reflection and $\beta = kl \sin \theta_i$.

namely the scattered pressure $\tilde{p}_s$, the incident pressure $\tilde{p}_0$, and the range r as functions of frequency. Obtaining $\tilde{p}_s$ and $\tilde{p}_0$ over a band of frequencies can be accomplished either by capturing a time-limited signal and Fourier transforming it or by selecting the steady-state portion of an echo generated by a narrow-band, continuous-wave pulse at each frequency. The scattered pressure should be measured at a range r sufficiently far from the scatterer so that an acceptably small phase error kl^2/r is incurred across the scatterer of lateral length l. A phase error of less than $\pi/8$ radians (45°) requires $r \geq 2l^2/\lambda$. The measurement of range r to sufficient accuracy is straightforward in a laboratory setting and needs no discussion. The measurement of range, or more properly transmission loss from the target to the receiver, in a natural environment can be quite problematical and is outside

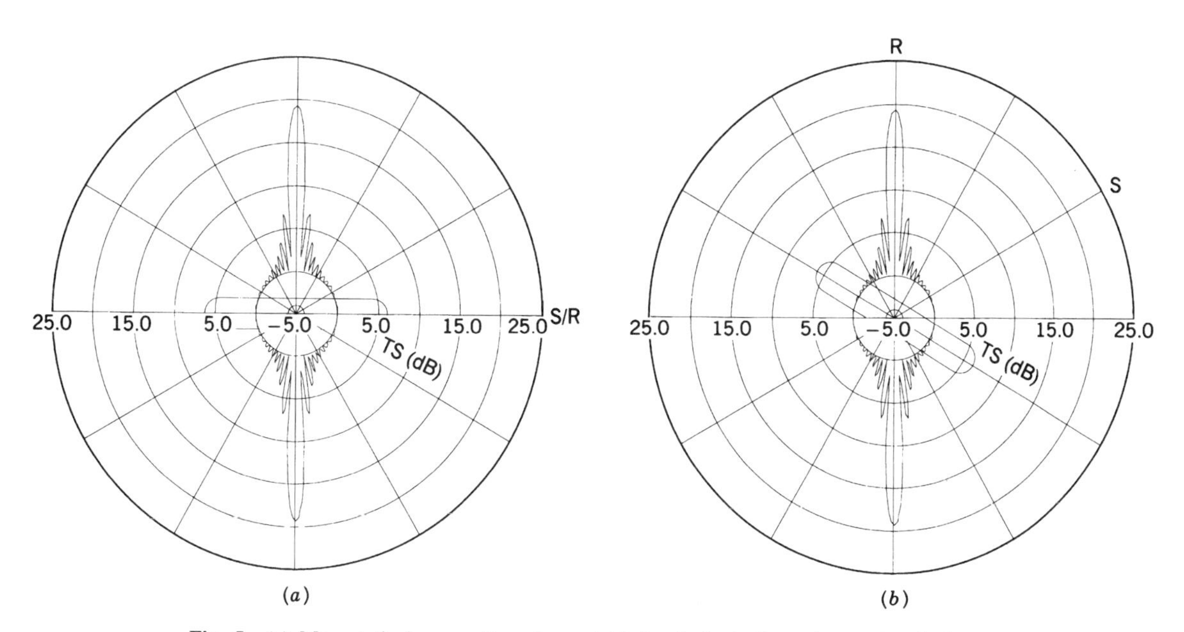

Fig. 5 (*a*) Monostatic beam pattern for a rigid, hemispherically end-capped cylinder as computed from the formula in the Table 2. (*b*) An approximation of the bistatic beam pattern that would be obtained with a source-to-receiver separation of 60°. The approximation is made by rotating the monostatic pattern through the bistatic angle of 30°.

the scope of this section. In controlled environments, pressure measurements can be made either directly or indirectly. The dynamic range between p_0 and p_s is usually sufficiently large to preclude simultaneous measurement with the same hydrophone. Often the measurement of p_0 is obtained by removal of the target and replacement with a desensitized hydrophone at the same position. Alternatively, the incident pressure can be obtained by measuring a reference pulse p_r with a desensitized hydrophone at a range r_{sr} from the source. Assuming spherical spreading from the source, the incident pressure p_0 can be inferred from p_r as

$$|\tilde{p}_0| = |\tilde{p}_r| \frac{r_{sr}}{r_{st}}, \qquad (15)$$

where r_{st} is the distance from the source to the target. In terms of measured quantities the target strength is

$$\mathrm{TS} = 20 \log \frac{|\tilde{p}_s|}{|\tilde{p}_r|} + 20 \log \frac{r_{st} r}{r_{sr}}. \qquad (16)$$

The dynamic range between p_s and p_r is still large enough to preclude using the same hydrophone system to measure both easily. One approach is to use a fast switch to increase the sensitivity of the hydrophone after the time when p_r is present at the hydrophone and before the arrival of p_s. This approach does not require a calibrated receiver. Another approach is to use two hydrophones of differing sensitivity to measure p_s and p_r. In this case each hydrophone must be adequately calibrated.

For controlled laboratory measurements the sphere calibration technique is often the easiest method for calibration of scattering measurements.[26] For $ka \gg 1$, the target strength of a rigid sphere is given by $20 \log(a/2)$. The target strength of a test object can in principle be obtained by comparison of the transient response of a rigid sphere, p_{sph}, with that of the test object to the same incident signal. If the sphere identically replaces the target, a simple modification of Eq. (16) gives

$$\mathrm{TS} = 20 \log \frac{|\tilde{p}_s(f)|}{|\tilde{p}_{\mathrm{sph}}(f)|} + 20 \log \frac{a}{2} + 20 \log \frac{r_{st} r}{r_{sr}}. \qquad (17)$$

If the sphere and test object are measured at different ranges, a simple modification to the range term is required. As indicated by the form functions discussed previously with reference to Fig. 1*a*, no sphere will act as a rigid reflector in water; however, for an impulsive incident signal there is a significant frequency band over which the elastic and geometric responses of a sphere can be temporally isolated.[26] Figure 6*a* shows the echo backscattered by a solid tungsten carbide sphere that was illuminated by a broadband signal whose useful band-

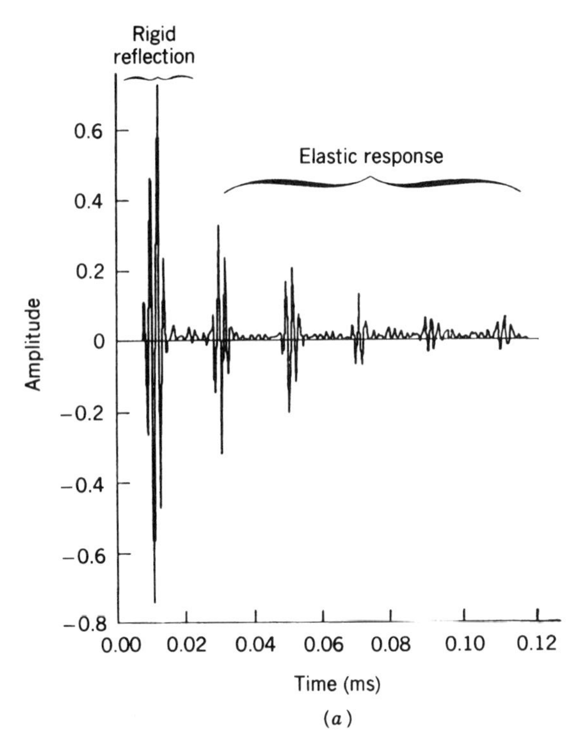

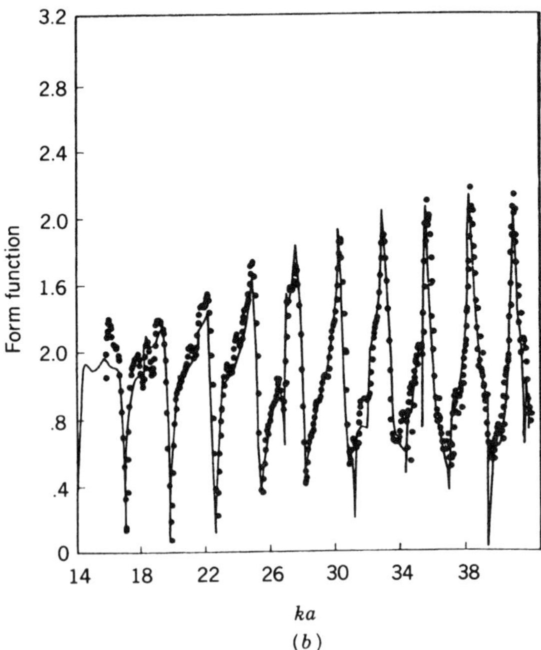

Fig. 6 (*a*) Backscattered echo from a tungsten carbide sphere, showing a clear separation of specular and elastic contributions. (*b*) Comparison of theory and experiment. The experiment was processed by using the rigid reflection identified in (*a*) as a calibration standard.

width covers the frequency range between $14 < ka < 40$. The form function of the sphere computed by a normal-mode series is given as the solid curve in Fig. 6*b*. In Fig. 6*a* the specular and elastic responses of the sphere are observed to be effectively isolated in the transient echo. The time-windowed Fourier transform of the initial specular echo is taken to produce $\tilde{p}_{\text{sph}}(f)$ in Eq. (17), and the Fourier transform of the entire response gives $\tilde{p}_s(f)$. The form function obtained using $\tilde{p}_{\text{sph}}(f)$ and $\tilde{p}_s(f)$ is given as the points in Fig. 6*b*.

High-frequency calibration methods that involve the use of liquid-filled spheres are found in Refs. 39 and 40. Descriptions of the techniques used for precise lake measurements are found in the work performed by the ARL University of Texas; Refs. 41 and 42 are recommended introductions to that body of work.

REFERENCES

1. J. J. Faran, "Sound Scattering by Solid Cylinders and Spheres," Office of Naval Research Technical Memorandum No. 22, DTIC No. AD827027, March 15, 1951.
2. J. J. Faran, "Sound Scattering by Solid Cylinders and Spheres," *J. Acoust. Soc. Am.*, Vol. 23, 1951, p. 405.
3. R. Hickling, "Analysis of Echoes from a Solid Elastic Sphere in Water," *J. Acoust. Soc. Am.*, Vol. 34, 1962, p. 1582.
4. R. Hickling, "Analysis of Echoes from a Hollow Metallic Sphere in Water," *J. Acoust. Soc. Am.*, Vol. 36, 1964, p. 1124.
5. R. R. Goodman and R. Stern, "Reflection and Transmission of Sound by Elastic Spherical Shells," *J. Acoust. Soc. Am.*, Vol. 34, 1962, p. 338.
6. R. D. Doolittle and H. Uberall, "Sound Scattering by Elastic Cylindrical Shells," *J. Acoust. Soc. Am.*, Vol. 39, 1966, p. 272.
7. Werner G. Neubauer, Richard H. Vogt, and Louis R. Dragonette, "Acoustic Reflection from Elastic Spheres, I. Steady-State Signals," *J. Acoust. Soc. Am.*, Vol. 55, 1974, p. 1123.
8. Louis R. Dragonette, Richard H. Vogt, Lawrence Flax, and Werner G. Neubauer, "Acoustic Reflections from Elastic Spheres and Rigid Spheres and Spheroids. II. Transient Analysis," *J. Acoust. Soc. Am.*, Vol. 55, 1974, p. 1130.
9. L. Flax and W. G. Neubauer, "Acoustic Reflection from Layered Elastic Absorptive Cylinders," *J. Acoust. Soc. Am.*, Vol. 61, 1977, p. 307.
10. Richard H. Vogt, Lawrence Flax, Louis R. Dragonette, and Werner G. Neubauer, "Monostatic Reflection of a Plane Wave from an Absorbing Sphere," *J. Acoust. Soc. Am.*, Vol. 57, 1975, p. 558.
11. R. Vogt and W. G. Neubauer, "Relationship between Acoustic Reflection and Vibrational Modes of Elastic Spheres," *J. Acoust. Soc. Am.*, Vol. 60, 1976, p. 15.
12. L. Flax, "High *ka* Scattering of Elastic Cylinders and Spheres," *J. Acoust. Soc. Am.*, Vol. 62, 1982, p. 1502.
13. A. J. Rudgers, "Acoustic Pulses Scattered by a Rigid Sphere Immersed in a Fluid," *J. Acoust. Soc. Am.*, Vol. 45, 1969, p. 900.
14. H. Uberall, R. D. Doolittle, and J. V. McNicholas, "Use of Sound Pulses for a Study of Circumferential Waves," *J. Acoust. Soc. Am.*, Vol. 39, 1966, p. 564.
15. R. D. Doolittle, H. Uberall, and P. Ugincius, "Sound Scattering by Elastic Cylinders," *J. Acoust. Soc. Am.*, Vol. 43, 1968, p. 1.
16. P. Ugincius and H. Uberall, "Creeping-Wave Analysis of Acoustic Scattering by Elastic Cylindrical Shells," *J. Acoust. Soc. Am.*, Vol. 43, 1968, p. 1025.
17. W. G. Neubauer, P. Ugincius, and H. Uberall, "Theory of Creeping Waves in Acoustics and Their Experimental Demonstration," *Zeitschrift Naturforschung*, Vol. 24, 1969, p. 693.
18. D. Brill, H. Uberall et al., "Acoustic Waves Transmitted through Solid Elastic Cylidners," *J. Acoust. Soc. Am.*, Vol. 50, 1970, p. 921.
19. J. W. Dickey, G. V. Frisk, and H. Uberall, "Whispering Gallery Wave Modes on Elastic Cylinders," *J. Acoust. Soc. Am.*, Vol. 59, 1976, p. 1339.
20. N. C. Yen, Louis R. Dragonette, and Susan K. Numrich, "Time-Frequency Analysis of Acoustic Scattering from Elastic Objects," *J. Acoust. Soc. Am.*, Vol. 87, 1990, p. 2359.
21. S. K. Numrich, L. R. Dragonette, and L. Flax, "Classification of Submerged Targets," in V. K. Varadan and V. V. Varadan (Eds.), *Elastic Wave Scattering and Propagation*, Ann Arbor Press, 1982, MI, p. 149–175.
22. L. Flax, L. R. Dragonette, and H. Uberall, "Theory of Elastic Resonance Excitation by Sound Scattering," *J. Acoust. Soc. Am.*, Vol. 63, 1978, p. 723.
23. G. Breit and E. P. Wigner, "Capture of Slow Neutrons," *Phys. Rev.*, Vol. 49, 1936, p. 519.
24. H. Uberall, L. R. Dragonette, and L. Flax, "Relation between Creeping Waves and Normal Modes of Vibration of a Curved Body," *J. Acoust. Soc. Am.*, Vol. 61, 1977, p. 711.
25. L. R. Dragonette, "Influence of the Rayleigh Surface Wave on the Backscattering by Submerged Aluminum Cylinders," *J. Acoust. Soc. Am.*, Vol. 65, 1979, p. 1570.
26. Louis R. Dragonette, S. K. Numrich, and Laurence J. Frank, "Calibration Technique for Acoustic Scattering Measurements," *J. Acoust. Soc. Am.*, Vol. 69, 1981, p. 1186.
27. J. W. Dickey and H. Uberall, "Surface Wave Resonances in Sound Scattering from Elastic Cylinders," *J. Acoust. Soc. Am.*, Vol. 63, 1978, p. 319.
28. L. Flax, G. C. Gaunaurd, and H. Uberall, "Theory of Resonance Scattering," in W. P. Mason and R. N. Thurston (Eds.), *Physical Acoustics*, Vol. 15, Academic Press, New York, 1981.
29. J. D. Murphy, J. George, A. Nagl, and H. Uberall, "Isola-

tion of the Resonant Component in Acoustic Scattering from Fluid-Loaded Elastic Spherical Shells," *J. Acoust. Soc. Am.*, Vol. 65, 1979, p. 368.

30. H. Uberall, Y. J. Stoyanov, A. Nagl, M. F. Werby, S. H. Brown, J. W. Dickey, S. K. Numrich, and J. M. D'Archangelo, "Resonance Spectra of Elongated Elastic-Objects," *J. Acoust. Soc. Am.*, Vol. 81, 1987, p. 312.
31. M. F. Werby, H. Uberall, A. Nagl, S. H. Brown, and J. W. Dickey, "Bistatic Scattering and Identification of the Resonances of Elastic Spheroids," *J. Acoust. Soc. Am.*, Vol. 84, 1988, p. 1425.
32. M. C. Junger and D. Feit, *Sound, Structures and Their Interaction*, 2nd ed., MIT Press, Cambridge, MA, 1986.
33. C. M. Davis, et al., "Acoustic Scattering from Silicon Rubber Cylinders and Spheres," *J. Acoust. Soc. Am.*, Vol. 63, 1978, p. 1694.
34. H. Lamb, "On Waves in an Elastic Plate," *Proc. Roy. Soc.*, Vol. A93, 1917, p. 114.
35. J. F. M. Scott, "The Free Modes of Propagation of an Infinite Fluid-Loaded Thin Cylindrical Shell," *J. Sound Vib.*, Vol. 125, 1988, p. 241.
36. M. Talmant and J. Ripoche, "Study of the Pseudo-Lamb Wave S_0 Generated in Thin Cylindrical Shells Insonified by Short Ultrasonic Pulses in Water," in H. M. Merklinger (Ed.), *Progress in Underwater Acoustics*, Plenum, New York, 1987, pp. 137–144.
37. P. L. Marston, "Phase Velocity of Lamb Waves on a Spherical Shell: Approximate Dependence on Curvature from Kinematics," *J. Acoust. Soc. Am.*, Vol. 85, 1989, p. 2663.
38. R. J. Urick, *Principles of Underwater Sound*, 3rd ed. McGraw-Hill, New York, 1983, p. 291.
39. B. M. Marks and E. E. Mikeska, "Reflections from Focused Liquid-Filled Spherical Reflectors," *J. Acoust. Soc. Am.*, Vol. 59, 1976, p. 813.
40. D. L. Folds and C. D. Loggins, "Target Strength of Liquid-Filled Spheres," *J. Acoust. Soc. Am.*, Vol. 73, 1983, p. 1147.
41. G. R. Barnard and C. M. McKinney, "Scattering of Acoustic Energy by Solid and Air-Filled Cylinders in Water," *J. Acoust. Soc. Am.*, Vol. 33, 1961, p. 226.
42. C. W. Horton and M. V. Mechler, "Circumferential Waves in a Thin-Walled Air-Filled Cylinder in a Water Medium," *J. Acoust. Soc. Am.*, Vol. 51, 1971, p. 295.

34

FUNDAMENTAL UNDERWATER NOISE SOURCES

WILLIAM K. BLAKE

1 INTRODUCTION

In this chapter we will examine the essential features of sound from underwater acoustic sources that typically emerge from hydrodynamic cavitation, bubbly flow, single-phase flow past elastic surfaces and bodies, and elastic surfaces that are driven by flow. Emphasis will be on fundamentals; interested readers will find more complete discussion in Refs. 1–3. The theme of this chapter is to bring out essential features of underwater noise source mechanisms in a unified way. The development of the fundamentals of these source types began in the 1950s, and extends to today. We shall survey the generic types of sources, examine the general acoustics of flow-driven surfaces, and finally summarize the characteristics of flow excitation forces.

Underwater acoustics is of obvious importance to underwater vehicles and ships, but it is also of importance to the noises radiated by piping systems and pumps in processing industries. Underwater acoustic sources also have competed with acoustic signals in mobile active acoustic devices used in Doppler positioning and oceanography as well as in towed streamers and fairings used in off-shore oil exploration.

2 PROPERTIES OF ELEMENTARY UNDERWATER SOUND SOURCES

2.1 Canonical Sources

The features of flow-induced sound from underwater sources differ somewhat from the features of aerodynamically generated sound. This is a consequence of the low Mach numbers of single-phase hydrodynamic flows, which make these canonical sources very inefficient. In practical instances, enhancements in the acoustic radiation efficiencies of hydrodynamic flow sources require the involvement of resonators, scattering sites in the form of spatially local discontinuities in the surface impedances of structures, bubbles, or cavitation. Bubbly flows and cavitation are among the most efficient of hydrodynamic sources because these are monopoles and have counterparts in subsonic combustion flow aeroacoustics. Underwater sources in the absence of cavitation are principally dipole and are dominated by vibration of contiguous surfaces as well as by localized flow dipoles. These sources also have aeroacoustic counterparts, although contribution from flow-induced vibration is often of lesser importance in aeroacoustics. Quadrupole noise is typically not of such importance in single-phase underwater acoustics as it is in aeroacoustics because of typically low Mach numbers of hydroacoustics. The presence of bubbles can, however, elevate the relevance of quadrupole sources. In the case of noncavitating flow, this different hierarchy of sources is determined by the generally long acoustic wavelengths of the sound in low-Mach-number liquid flow compared with the characteristic dimensions of the sources and the fact that fluid phase impedances may often be comparable to those of typical elastic boundary surfaces.

All hydroacoustic systems of sound sources may be regarded as a superposition of spatial distributions of three sources: monopoles, dipoles, and quadrupoles. These elementary classifications apply to both mechanically driven and flow-driven bodies. In the case of a flow-generated source system, a monopole source may be modeled as an unsteady mass injection $\dot{q}$, dipoles by a spatial distribution of force divergence $\nabla \mathbf{f}$, and

Note: References to chapters appearing only in the *Encyclopedia* are preceded by *Enc.*

quadrupoles by a spatial distribution of Reynolds stress $\partial^2 T_{ij}/\partial x_i\, \partial x_j$. The acoustic wave equation that expresses the field of these sources is

$$\nabla^2 p(\mathbf{x}, t) - c_0^2 p(\mathbf{x}, t) = -q(\mathbf{x}, t) + \nabla \mathbf{f}(\mathbf{x}, t) - \frac{\partial^2 T_{ij}(\mathbf{x}, t)}{\partial x_i\, \partial x_j}. \tag{1}$$

The source terms in this expression are per unit volume and they include all the stress distributions acting on the fluid, all the localized forces, and mass injection sites in the fluid. The effect of bounding surfaces that are adjacent to the fluid is simply to reflect and scatter the acoustic field pressures from these sources. These effects may be properly dealt with mathematically in the solution of the appropriate Green's function for the geometry involved.

In the case of mechanically driven plates and shells, the dipole classification applies to the mechanism of sound radiation from localized forces. Such forces may arise both from a mechanical drive (such as a machine attached to a plate) as well as from reaction forces that exist at stiffeners and attached masses on plates that are subjected to flow excitation. This latter mechanism results from local reactions at impedance discontinuities that can convert (scatter) plate vibration at one wavenumber into sound radiation to all directions. Without these wavenumber conversion (scattering) sites, the primary plate vibration field may only radiate sound weakly or into only preferred directions.

One physical system that allows us to conceptualize a source region such as that existing in the case of flow past bluff bodies is shown in Fig. 1. A blunt body is illustrated in a flow with downstream vortex shedding and associated unsteady lift fluctuations generated back on the shedding body. Due to low ambient static pressures, cavitation is shown to be developed at the low-pressure regions of the body surface and, in the more remote wake, a region of turbulent stresses is generated. The coexisting combination of unsteady cavitation, lift, and turbulent Reynolds stresses provides the required dynamics for sound generation by monopole, dipole, and quadrupole sources, respectively. Although the illustration is idealized, many real underwater flow sources are complex in the multiplicity of excitations and source types. In general, body vibration can make this picture even more complex by introducing additional dipoles.

Solutions to Eq. (1) for an unbounded fluid that is also free of reflecting boundaries apply to these sources as they are confined to specific regions. For the monopole sources

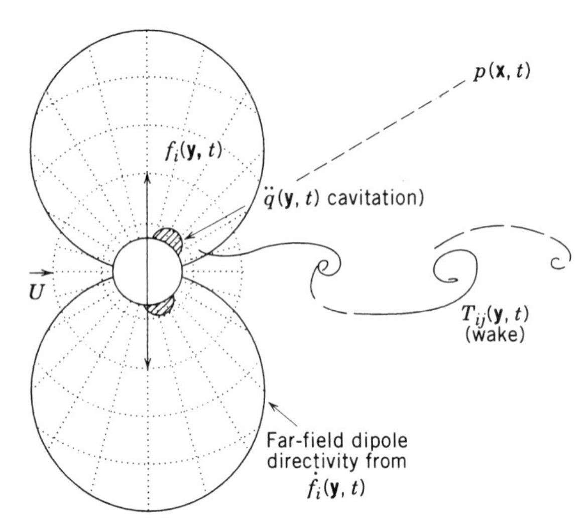

Fig. 1 Notional diagram of acoustic sources associated with flow–body interaction.

$$p(\mathbf{x}, t) = \frac{1}{4\pi} \iiint_v \frac{\ddot{q}(\mathbf{y}, t - r/c_0)}{r}\, dV(\mathbf{y}), \tag{2a}$$

which, in the case of a nonconvecting volume source, reduces to

$$p(\bar{x}, t) = \frac{\rho \ddot{Q}(\mathbf{y}_0, t - r/c_0)}{4\pi r}, \tag{2b}$$

where $\ddot{Q}$ represents the net volumetric acceleration in the case of acoustically compact monopole regions. Acoustically compact regions are those for which the characteristic spatial dimensions of the sources are small compared with both the acoustic range $r = |\mathbf{x} - \mathbf{y}|$ from the sources to the field observation point and an acoustic wavelength over the entire frequency range of interest. This statement invokes the notion of a characteristic frequency of the source dynamics in order to specify this wavelength. We continue to develop these notions in the context of Fig. 1. For the dipole sources

$$p(\mathbf{x}, t) = \iiint_v \frac{(\partial f_i/\partial y_i)(\mathbf{y}, t - r/c_0)}{4\pi r}\, dV(\mathbf{y}), \tag{3a}$$

which reduces to

$$p(x, t) = \frac{1}{4\pi c_0} \left(\frac{x_i}{r}\right) \frac{\partial F_i}{\partial t} \left(y_0, t - \frac{r}{c_0}\right) \tag{3b}$$

for compact dipole source regions, where F_i represents the resultant force on the fluid and x_i/r is the direction cosine ($\cos\phi$) from the force vector to the observer.

Finally, the quadrupole source field yields

$$p(x,t) = \iiint_v \frac{\partial^2 T_{ij}}{\partial y_i\, \partial y_j} \frac{1}{4\pi r}\, dV(\mathbf{y}), \tag{4a}$$

which reduces to

$$p(x,t) = \frac{1}{4\pi c_0^2} \frac{x_i x_j}{r^3} \iiint_v \frac{\partial^2 T_{ij}}{\partial t^2} \cdot \left(\mathbf{y}, t - \frac{r}{c_0}\right) dV(\mathbf{y}) \tag{4b}$$

for a compact (i,j) quadrupole source region. It is to be emphasized that Eqs. (2b), (3b), and (4b) apply to far-field acoustic pressures. The acoustic far field is loosely defined as r much larger than the largest spatial dimension of the source field and $k_0 r \gg 1$.

2.2 Speed Dependencies of Canonical Flow Sources

We still assume that the sources are compact, nonconvecting, and situated near a coordinate $\mathbf{y}_0$, but we now assume that the time variation of the sources is simple harmonic with a frequency ω_s. Source convection may be safely ignored in underwater acoustics since the convection Mach number is negligible with respect to unity. This allows the retardation $t - r/c_0$, where $r = |\mathbf{x} - \mathbf{y}_0|$, to be replaced by a simple phase representation. Then for monopoles

$$p(x,t) = -\rho\omega_s^2 Q \frac{e^{i(k_0 r - \omega_s t)}}{4\pi r}, \tag{2c}$$

where $Qe^{-i\omega_s t}$ represents the time-varying net volumetric acceleration. The dipoles give the field

$$p(x,t) = -i\left(\frac{\omega_s}{c_0}\right) F\cos\phi\left(\frac{e^{i(k_0 r - \omega_s t)}}{4\pi r}\right), \tag{3c}$$

where F represents the amplitude of the net harmonic force on the fluid and ϕ is the angle from the force vector to the observer position. A similar expression could be derived for quadrupoles, but it has limited practical utility.

Figure 2 illustrates these directional features for a force dipole induced by a propeller operating in nonuniform flow as measured in an anechoic wind tunnel.[4] Typically, the flow-generated forces have a resultant direction, arising from the orientation of lift or drag, thrust, or side-force fluctuations. Equation (3c) and Fig. 2 show that the sound field is directed along the axes of each of the resultant thrust and forces acting on the fluid by the propeller.

The quadrupole field is more amorphous, showing minimal directional preference because the random-turbulence quadrupole sources all contribute to the sound with a multitude of directivities, with mean convection being responsible for a slight directionality. This has been discussed elsewhere in this book in connection with aeroacoustic sources. Accordingly, quadrupole sources will receive little further consideration in this chapter.

These relationships, though approximate and strictly valid for acoustically compact sources, have valuable application to underwater acoustics. Typically due to the relatively low Mach numbers of underwater flows and the characteristic frequencies involved, the source regions are compact, so that Eqs. (2b), (2c), (3b), and (3c) actually become exact for certain underwater applications and have great utility in estimation and extrapolation.

It is a characteristic of flow-generated sources that the characteristic frequency is determined by the flow velocity U and a typical body dimension L (say, diameter in the case of a circular cylinder or boundary layer thickness in the case of extended flow surfaces) such that a dimensionless frequency (called a Strouhal number)

$$S_t = \frac{\omega_s L}{U} \tag{5}$$

may be defined. For dynamically similar flows, the Strouhal number is a constant and then necessarily determined by fluid dynamic considerations. Thus the wavenumber ω_s/c_0 appearing in Eqs. (3c) and (4c) may be written as

$$\frac{\omega_s}{c_0} = S_t\left(\frac{U}{c_0}\right) L^{-1}. \tag{6}$$

Substitution of Eq. (6) into Eqs. (2b) and (2c) yields the important dimensional relationships for the mean-square radiated sound pressure of the monopole

$$\overline{p^2}(\mathbf{r}) = \frac{\rho^2 U^4}{16\pi^2} \frac{L^2}{r^2} [S_t^4 \tilde{Q}^2]. \tag{7}$$

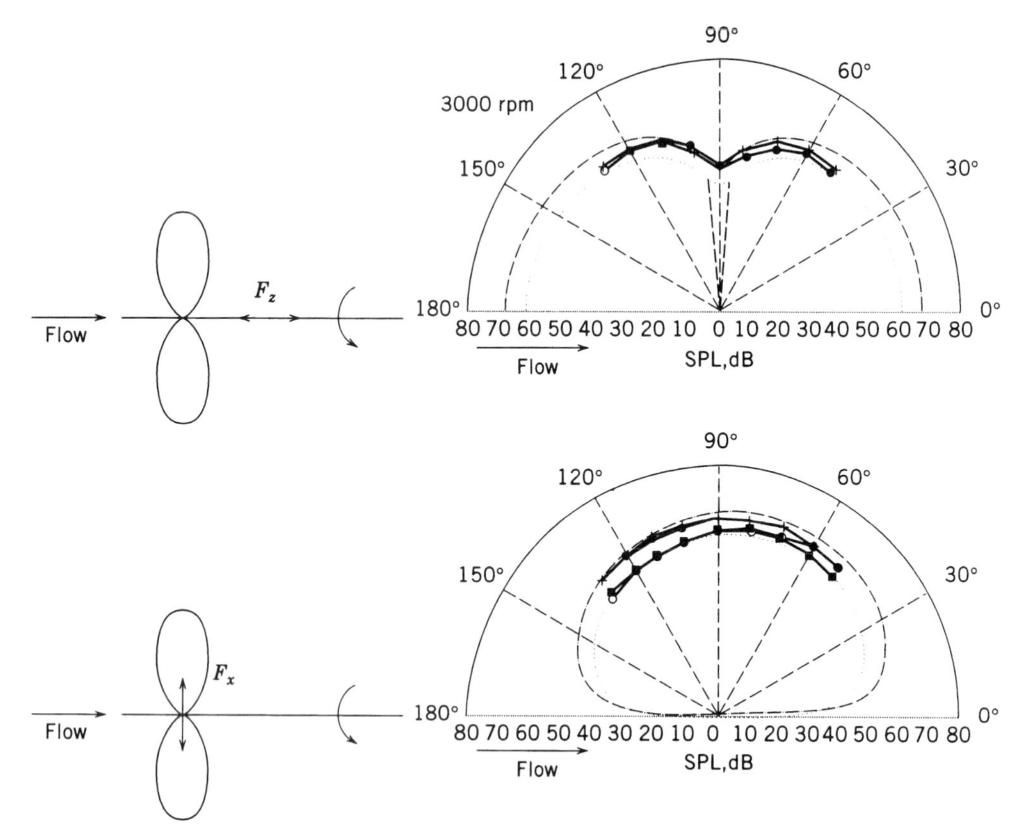

Fig. 2 Measured and predicted dipole sound patterns for a four-bladed propeller operating in three- and four-cycle wakes. Experiment: (○) 0° half-plane; (•) 90° half-plane; (×) 180° half-plane. Theory: (- - - -) low; (- · · -) high.

The term in square brackets is dimensionless, involving a dimensionless volume quantity $\tilde{Q}$, so that the leading terms carry all the dimensionality of the sound. No directivity functions appear in Eq. (7) because monopole sound (e.g., cavitation noise) is omnidirectional. For the dipole sound we have

$$\overline{p^2}(\mathbf{r}) = \frac{\rho^2 U^4}{16\pi^2} \frac{L^2}{r^2} \cos^2 \phi \left(\frac{U}{c_0} \right)^2 S_t^2 C_F^2, \quad (8)$$

where C_F represents a dimensionless force coefficient

$$C_F = \frac{F}{\rho U^2 L^2}.$$

Again, S_t and C_F are characteristics of the fluid dynamics and, particularly at low Mach numbers, they are strictly hydrodynamic parameters. The $\cos^2 \phi$ factor orients the directionality of the sound field with the direction of the resultant force. This directionality of the sound as well as the presence of the Mach number are direct consequences of the gradient of the forces that appears in Eqs. (1) and (3a). Equations (7) and (8) form the bases of many practical hydroacoustic scaling rules.

For the quadrupole sound, Eq. (4b) yields the familiar eighth-power rule for velocity dependence,

$$\overline{p^2} = \frac{\rho^2 U^4}{16\pi^2} \left(\frac{U}{c_0} \right)^4 \frac{L^2}{r^2} \tau(\theta, \phi) S_t^4, \quad (9)$$

where $\tau(\theta, \phi)$ represents a dimensionless resultant quadrupole strength term. This dimensionless term carries with it all the resultant directivity features of the quadrupole source field, as is discussed in Chapter 25.

These expressions show the hierarchy of increasing order of Mach number dependence and directivity of the sources as the geometric order of the sources increases. Thus, for the low Mach numbers of underwater acoustics, quadrupole sound is irrelevant in comparison to other dipoles and monopoles when they also exist. Accounting for the existence of monopoles would appear to be simple since these are typically due to volumetric pulsations of bubbles suspended in the flow and convected through a pressure field. In the linear acoustics limit (i.e., low Mach number of bubble wall motion), the sound is determined by the net value of $Q(\mathbf{y}, \omega_s)$. However, as we shall discuss in Section 3 on cavitating flows, such sources are hydrodynamically very complicated, and generalized descriptions of cavitation noise are only available by empirical means.

Flow dipoles may be generated by the resultant flow-induced forces on finite-extent blunt bodies and propellers, as illustrated in Fig. 2, as well as by turbulent flow over continuous but acoustically compact elastic surfaces with discontinuities such as stiffeners and trailing edges. In this case of localized unsteady flow sources, a commonly used expression for sound power radiated from a compact flow dipole is determined from Eq. (3c) by squaring and integrating over the polar angle to give the power spectral density

$$P_{\text{rad}}(\omega) = \frac{\omega^2 \overline{F^2}(\omega)}{12\pi\rho c_0^3}. \tag{10}$$

In the other extreme, as discussed in Section 4.3, flat rigid surfaces that support spatially homogeneous turbulent boundary layers generate quadrupole radiation because such surfaces serve only as reflectors. For rigid extended surfaces to generate dipole sound, they must have edges, curvature, bumps, or roughness since these elements represent wave conversion sites that scatter the relatively high flow-convected wavenumbers of turbulence (ω/U) to the lower wavenumbers of sound (ω/c_0). Such wavenumber conversion mechanisms will also be discussed in Section 4.3. All real surfaces are finite, curved, and stiffened elastic, so that such surfaces, when excited by nominally homogeneous turbulent boundary layers, emit dipole sound.

Underwater quadrupole sound as generated by jet flow is dominated by so-called lip dipoles and other possible coexisting sources so that quadrupoles per se generated by jet flows are also typically irrelevant. Lip dipoles are due to the interaction of the turbulence in the jet efflux grazing and interaction with the lip of the jet nozzle (which acts analogously to a trailing edge). These are dominant in low-Mach-number turbulent jet flows.

2.3 Low-Frequency Sound from Localized (Point) Forces on a Plate in Fluid

In this section we examine the necessary ingredient for the scattering of flow-driven vibration to sound at surface structural discontinuities. As described above, when a plate of mass impedances $(-i\omega m_s)$ contains localized attachments and stiffeners, a vibration incident on these stiffeners generates localized forces at these sites. When the incident vibration field is excited by flow, it is comprised of the length and time scales associated with the flow-induced surface pressure. Due to the interaction forces at the discontinuities of structure, the sound is locally radiated, and it may differ significantly in intensity and directionality from the sound that would result from the incident primary wave field on the plate without the stiffeners. The canonical radiation mechanism for this phenomenon is dipole and may be regarded as due to a point force applied to a plate that is immersed in fluid. A point force applied to a plate of area density m_s with dense fluid on one side and vacuum on the other, F, generates a sound pressure[5] that is given by the familiar dipole relationship

$$p(r, \phi, t) = k_0 F f(\phi) \frac{e^{i(k_0 r - \omega x)}}{2\pi r}, \tag{11}$$

where

$$f(\phi) = \left[\frac{\rho c_0/\cos\phi}{\rho c_0/\cos\phi + (-i m_s \omega)}\right](-i\cos\phi), \tag{12a}$$

$$f(\phi) = \begin{cases} \dfrac{\beta\cos\phi}{\cos\phi + i\beta}, & (12b) \\[2ex] \dfrac{Z_f}{Z_s + Z_f}(-i\cos\phi), & (12c) \end{cases}$$

where Z_f is the acoustic impedance of the fluid adjacent to the plate and Z_s is the inertial impedance of the plate. The factor

$$\beta = \frac{\rho c_0}{m_s \omega} \tag{13}$$

is a fluid-loading factor that expresses the ratio of fluid impedance to characteristic plate impedance $(i\omega m_s)$ and the angle is measured from the direction of the force. Figure 3 illustrates the dependence of this dipole on plate inertia and observation angle. We see that at low frequencies (large β) the directivity pattern of the sound is that of an isolated dipole and the dipole strength is double

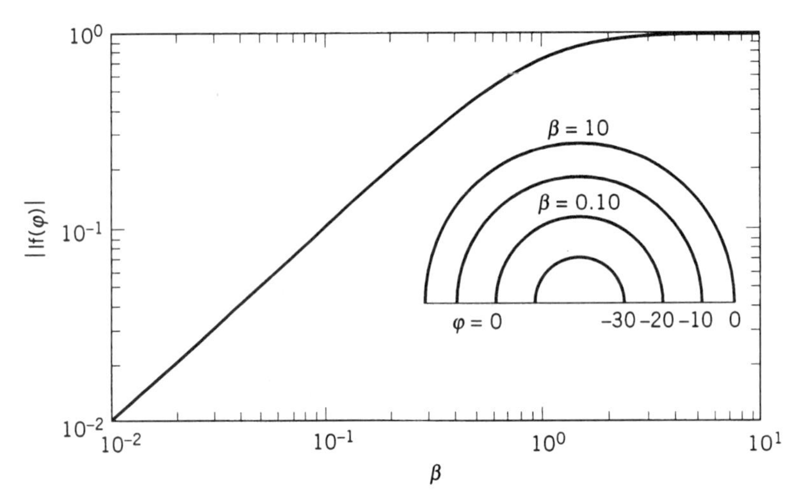

Fig. 3 Radiated sound pressure from a point force acting on an effectively infinite elastic plate with fluid on one side.

that of the original dipole due to reflection at the $y = 0$ surface. At higher frequencies (small β) the directivity appears omnidirectional (except near $\phi = \pm\pi/2$, where the fluid impedance exceeds the plate impedance near grazing incidence).

This equation applies remarkably well to practical underwater structures that are large enough that the radiated sound field is dominated by the drive point rather than by reflections from boundaries or effects of finite size of the structure. It shows that the sound is essentially that of a point dipole applied to the fluid whose strength is modified by the impedance of the plate, particularly at large β (low enough frequency that $m_s\omega$ is smaller than ρc_0).

It is well to remark that the intensity and directionality of the sound resulting from the primary plate vibration field is determined by the structural acoustics of the structure. Below the acoustic coincidence frequency of the plate, the above equation often accounts for the dominant contribution of the radiated sound from the point-driven structure. Analogous relationships apply for line force and moment-driven structures. We shall return to the acoustics of flow-driven surfaces in Section 4.

3 SOUND FROM CAVITATING FLOWS

Although the mechanism of sound generation from cavitation is fundamentally quite simple (i.e., monopole), hydrodynamic cavitation noise is an extremely complex subject about which little detail is known. A classical reference on hydrodynamic cavitation is the book by Knapp et al.[6] The adjective *hydrodynamic* is invoked here in order to distinguish flow-induced cavitation from spark-induced cavitation. The latter form of cavitation is used in the laboratory to study specific aspects of single-bubble-cavity dynamics. There has not been a substantive development in theoretical modeling of cavitation noise for engineering application since the work of Fitzpatrick and Strasberg in the 1950s.[7] Most follow-up approaches have been empirical and based on dimensional analysis of single-bubble analogies.

Theoretical work on cavity dynamics has focused on the dynamics of single bubbles and bubble clusters (e.g., Ref. 8; see also Ref. 2 for an extensive reference list). These researches have examined the influences of liquid phase compressibility, entrapped gas in bubbles, gas–liquid thermodynamics, and bubble interactions. Even for idealized bubble populations, the bubble mechanics are nonlinear and highly complex.[8,9] The resulting mathematics is accordingly involved. With regard to mathematical modeling of ship propeller cavitation, Baiter[10] has examined low-frequency sound and effects of modulation.

Practical description of hydrodynamic cavitation noise is made difficult because of the large number of flow parameters that can influence the acoustically relevant details of cavity dynamics. There is unfortunately little connection between cavitation noise theory and observed cavitation noise of ship propellers and turbines. Experimentally, it has been difficult to control (or even measure) some of the relevant parameters. These include Reynolds number, the concentration of dissolved and free gas in the liquid, the gas/liquid phase dynamic transfer properties of the liquid, particularly over short time scales, and distributions of cavitation nucleation sites. On the other end of the spectrum, analytical work has been largely devoted to single spherical bubbles and dynamics of bubble clouds. Controlled experimental studies of cavitation noise are also difficult. Some of the most recent are given in Refs. 11–15.

Notwithstanding these complexities, attempts have been made to develop scaling and prediction rules that have engineering use. These start with Eq. (2a) for linear acoustic monopole sources. Notions of single-bubble cavitation noise assist in the establishment of dimensional length and time scales for the monopole source; the resulting scaling then depends on the ambient hydrostatic pressure, the size of the body, and the vapor pressure of the fluid. Several rules for scaling and predicting cavitation noise from propellers have thus emerged over the past 20 years.[6,7,11–21]

Cavitation in propellers and turbomachine rotors occurs in the tip vortex or the gap between the end wall and the blade tip and on the blade surfaces. The blade surface cavitation takes two major forms: bubbly cavities occurring individually or in clusters on or near the blade surface on its suction side (or back) and sheet cavities that extend along the chord from the line of minimum pressure at the leading edge. Sheet cavitation can occur on either side of the rotor, depending on its load distribution or its operating point. The common names given to these forms of cavitation are tip vortex, gap, back bubble (BUBBLE), leading-edge pressure surface (LEPS), and leading-edge suction surface (LESS). Figure 4 illustrates the surface pressure characteristics on hydrofoils and turbomachinery blades, P_s which cause cavitation environments. Cavitation in the shear layers of free jets occurs as individual bubbles and bubble clusters which are convected with the turbulent vortices.

The principal parameter that determines the occurrence of cavitation and cavitation noise is the cavitation index σ, defined as[6]

$$\sigma = \frac{P - P_v}{(1/2)\rho U^2}, \tag{14}$$

where P is the ambient hydrostatic pressure, P_v is the vapor pressure and ρ the density of the fluid, and U is the relative velocity between the blade and the fluid. This dimensionless number is the single most relevant parameter in determining the practical scaling of cavitation onset and the occurrence of cavitation noise. The condition for cavitation is that $\sigma < \sigma_i$, where σ_i is the cavitation inception index for the blade. The cavitation inception index is a function of the hydrodynamic design of the blades, the disturbance field in which the blades must operate, a variety of viscous effects, and the amount of free and dissolved gas in the fluid. It is thus roughly predictable in the design stage of the propeller and turbine rotor, but it must eventually be measured on either a model or a prototype design. Simple dimensional analysis leads to a description of the spectrum of the sound pressure in proportional frequency bands ($\Delta f \propto f$, in this case we will consider one-third-octave bands) of the form that follows immediately from Eq. (7) by replacing U^2 by P and L by propeller or pump diameter D, that is,

$$\overline{p^2} \propto P^2 \left(\frac{D}{r}\right)^2 \tag{15}$$

with frequency scaled as

$$f \propto D^{-1}\sqrt{\frac{P}{\rho}}. \tag{16}$$

These relationships assume that strict geometric dynamic similitude for the cavitation occurs.[1,2,21] Due to several possible scale effects, strict similitude may not occur, particularly to the extent that $\sigma_i)_{\text{model}} = \sigma_i)_{\text{full}}$. This lack of similitude may result in differing speeds for a similar extent of cavitation at model and full scale. For example, the inception of tip vortex cavitation has a Reynolds number scale effect that delays cavitation inception on models, compared with full size. Thus, an attempt has been made to include the inception index in the semiempirical rule given below. This approach also borrows some notions of single-bubble cavitation dynamics and yields the form

$$\overline{p^2}(r, f, \Delta f) = p_m^2 G\left(\frac{f}{f_m}\right). \tag{17}$$

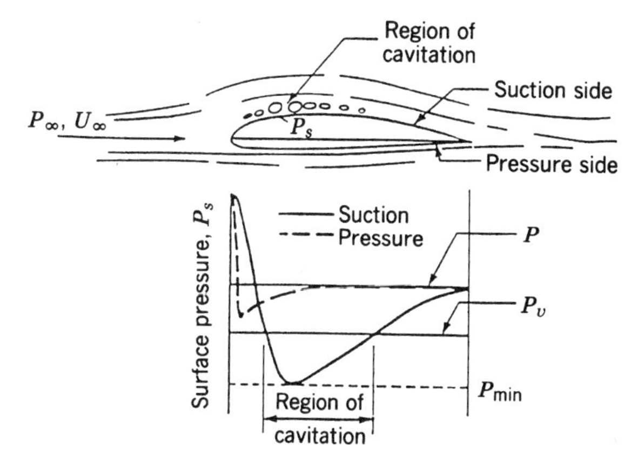

Fig. 4 Illustration of a cavitating hydrofoil, its surface pressure distribution, and its region of cavitation.

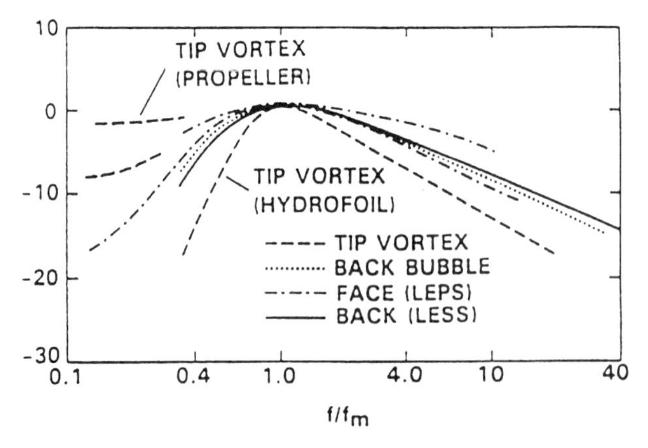

Fig. 5 Measured spectral forms for various types of hydrodynamic cavitation.

Figure 5, from Ref. 19, shows the characteristic spectrum forms for the four forms of cavitation. The tip vortex cavitation appears to have a variable level at low frequencies. This may occur because of the large-scale unsteadiness of propeller blade cavitation due to ship wake effects. Figure 6 shows an example of the collapse of the data for the maximum spectrum level, in this case the tip vortex cavitation, which has a roughly 10-dB spread. As an example of the utility of the cavitation noise prediction formulas for ship propellers, Fig. 6 shows a prediction for the twin propellers of the research vehicle *R/V Athena*. Here it is assumed that the dominant form of cavitation is the leading-edge suction side (LESS) type. The LESS type is the most common form of propeller surface cavitation and is noisier than tip vortex cavitation noise when it occurs. An example for axial flow turbomachinery is a little less reliable. Figure 7 shows measured and predicted noise levels for the drive pump of a water tunnel.[17] The measured sound levels are expressed as sound power due to the semireverberant nature of water tunnel enclosures.

4 GENERAL FEATURES OF COMPLEX HYDROSTRUCTURAL ACOUSTIC ELASTIC SYSTEMS

4.1 Description of the System

We now return the broad subject of the hydroacoustics of flow-excited structures in the absence of cavitation. We devote our attention to the role of structural surfaces in the conversion of flow energy to sound energy. The physical complexities of the hydroacoustic sources for

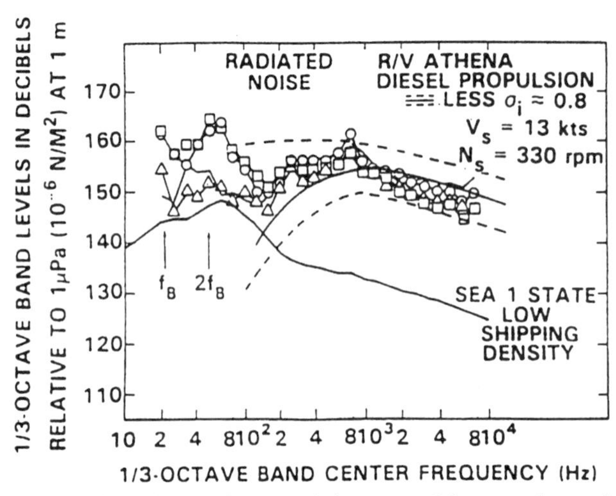

Fig. 6 Comparison of predicted and measured signatures of the research vessel *RV Athena*.

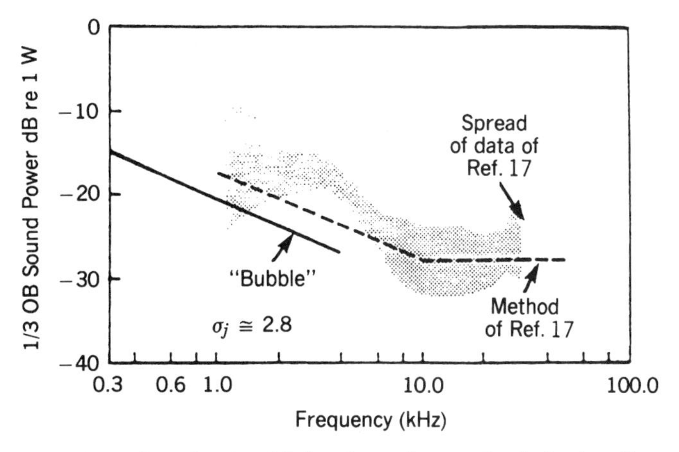

Fig. 7 Comparison of predicted and inferred sound power levels for impeller cavitation measured within the MIT water tunnel.

flow over real elastic surfaces can be viewed as a coupled system in which the flow–surface interaction leads to unsteady surface forces and pressure fluctuations in the fluid as well as to induced surface vibration. The fluid-borne pressures and forces and the structure-borne vibration combine to yield a sound field that is a superposition of both contributions. Thus the effect of surfaces in hydroacoustic source mechanisms is both hydrodynamic and structural acoustic. Hydrodynamic motion develops sources in the fluid and on the surface, while structural vibration induced by flow forcing introduces new surface sources. These new sources are due to impedance discontinuities occurring at localized structural stiffening and constraining boundaries as well as to surface finiteness and curvature.

Figure 8 illustrates these general interactions as a coupled hydrostructural acoustic system. The discipline of classical structural acoustics, the subject of Chapter 32, considers the coupling of vibration and sound as identified in the lowest three blocks in the figure. With the exception of cavitation, the discipline of aeroacoustics of rigid surfaces typically considers the upper portion of the figure. Thus, the science of underwater sound generation involves a combined discipline of hydrostructural acoustics in which are studied the sound, structure, and flow interactions. The characteristics of the sound generated by this system are determined by the relative acoustic length and time scales of the acoustic field, the hydrodynamic motion, and the flow-induced surface vibration. The sound field characteristics include the magnitude, the speed dependence, and the directivity.

Domination of the flow–surface interaction by both potential and viscous effects develops the flow unsteadiness, which establishes the spatial and temporal scales of the flow sources. Flows over bluff bodies thus often involve the development of viscous-dominated shear flows that accompany strong surface pressure gradients on the body (and thus body forces) and generate vortex street wakes. Thus trailing-edge flows, propeller unsteady forces, unsteady body forces of vortex street wakes, and turbulent boundary layer pressure fluctuations are all consequences of these flows. Cavitation occurs on surfaces when the surface pressure locally dips below a critical pressure (nearly the vapor pressure of the fluid). Inflow unsteadiness may cause the cavitation extent to become strongly time dependent if it results from variation of the surface pressure, as through angle-of-attack variations of lifting surfaces.

In the case of hydrodynamically generated sources due to flow–surface interaction, it is the nature of unsteady fluid mechanics that for certain flows there may be a coupling between the development of disturbances in the fluid and the flow-induced vibration of the surface. When surface vibration occurs, this motion can couple with the unsteady viscous-dominated shear flow at the surface and alter (typically strengthen) the generation of vortex sound sources (and forces). Such coupled flow–structure sources typically occur over a speed and frequency range for which the fundamental shear flow frequency becomes coincident with a frequency of a surface vibration mode and the amplitude of motion becomes comparable to a viscous disturbance scale of the shear flow. The essential behavior is a feedback mechanism that accounts for propeller singing, hole tones, and cavity tones; characteristics of many of these are discussed in Section 6. Damping and stiffening of the surface structure can significantly minimize the likelihood of these tones by adding dissipation (loop delay) and by detuning the characteristic frequencies of the structure and flow.

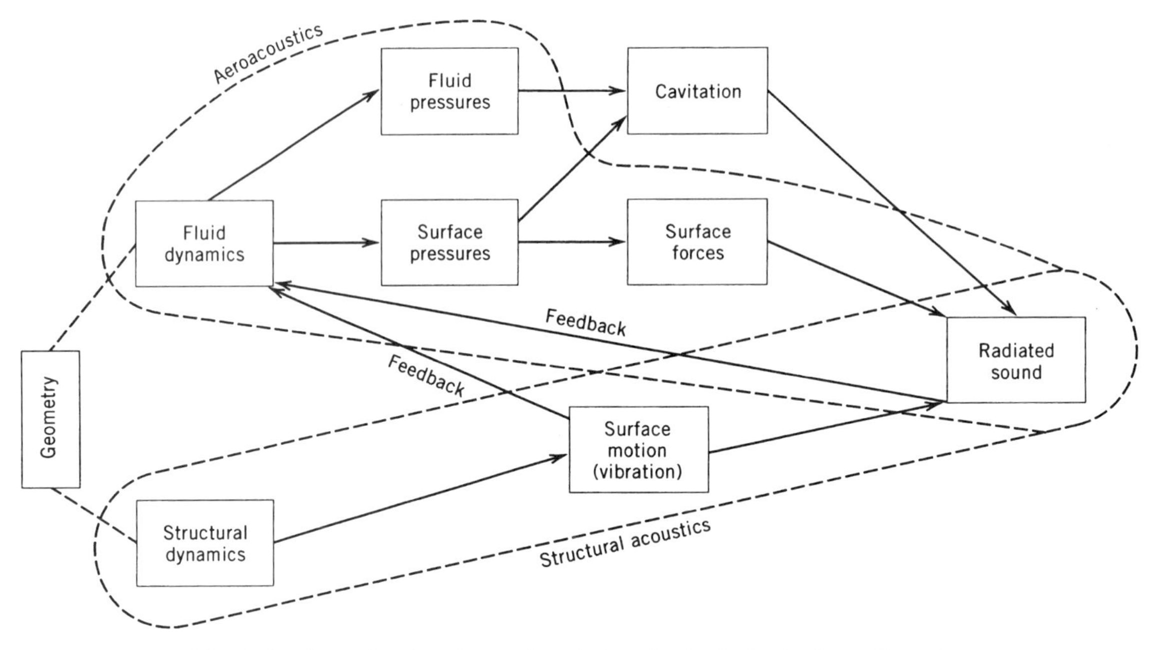

Fig. 8 Interfaces of various classes of mechanisms in a hydrostructural acoustic system.

Even when there is no flow–surface feedback mechanism, surface motion typically plays a significant role in hydroacoustics source mechanisms. This is because the motion of the surface can contribute to the sources. This is done in three fundamental ways: the surface becomes a finite-impedance reflector of the sources so that the radiated sound is simply altered by a reflection coefficient; the finite spatial extent of the surface or structural discontinuities or curvature of the surface may cause localized surface forces that contribute to the radiated field as dipoles; and trailing-edge flows of extended surfaces can generate additional wake sources (by dipole mechanisms).

The mathematical representation of the characteristics of the source complex are described with Helmholtz's integral equation, which gives the superposition of direct sound radiation and the contributions from surfaces. Thus

$$
\begin{aligned}
p(\mathbf{x}, \omega) = & \iiint_v \sigma(\mathbf{y}, \omega) g(\mathbf{x}, \mathbf{y}, \omega)\, dV(\mathbf{y}) \\
& + \iint_s \nabla_n p(\mathbf{y}, \omega) g(\mathbf{x}, \mathbf{y}, \omega)\, dS(\mathbf{y}) \\
& - \iint p(\mathbf{y}, \omega) \nabla_n(\mathbf{x}, \mathbf{y}, \omega)\, dS(\mathbf{y}), \qquad (18)
\end{aligned}
$$

where $\sigma(\mathbf{y}, \omega)$ represents the volume density of sources and $g(\mathbf{x}, \mathbf{y}, \omega)$ is the free-space Green function. On the surface, the normal gradient ∇_n of the pressure is related to the normal surface velocity $u_n(\mathbf{y}, \omega)$ by

$$
-\nabla_n p(\mathbf{y}, \omega) = \rho \dot{u}_n(\mathbf{y}, \omega). \qquad (19)
$$

The volume integral includes the effects of sources within a region of turbulent flow, and the two surface integrals include all effects of the surface motion $u_n(\mathbf{y}, \omega)$ and normal stresses $p(\mathbf{y}, \omega)$ on the surface–fluid interface. The effects of surfaces thus also may extend to the modification of the source density term $\sigma(\mathbf{y}, \omega)$ due to the generation of new turbulent sources. For a given turbulent volume source density, the surface integrals provide all the surface stresses, surface wave scattering, and geometric reflection and diffraction. For the remainder of this section we will assume that the fluid phase is devoid of bubble and cavitation motion.

The next two sections will examine the consequences of the limits of low- and high-frequency behavior of hydroacoustic sources that are made complex by the coexistence of turbulent sources and vibration response of adjacent surfaces. These sections are based on some of the earliest and fundamental work in flow-induced noise modeling.[22–24] Since then, there have been many extensions.[25–27]

We shall examine the behavior of the hydroacoustic system in two frequency limits in order to clearly see the parametric ranges for which important physical mechanisms are dominant. Understanding of these generalities is important in identifying generic control strategies for mitigating underwater sound. We shall see that the sound from flow-induced vibration is determined by the so-called blocked-flow forces on the rigid surface and the relative impedances of the fluid and surface. This finding is common to both acoustically compact and acoustically extended surfaces.

4.2 Sound from Flow around Acoustically Compact Elastic Bodies: Low-Frequency Limits

In the low-frequency limit, the radiating surface is compact compared with an acoustic wavelength. Figure 9 illustrates the case of a bluff body, shown here as a blunt lifting surface of chord C and span L in a uniform flow of mean velocity U. The surface is incompressible and is free to vibrate under the influence of flow-exciting pressure fluctuations on the upper and lower surfaces, designated as p^+ and p^-, where the superscripts plus and minus denote quantities evaluated on the "upper" and "lower" surfaces, respectively. We assume that the time dependence can be suppressed so that all disturbances may be considered in the frequency domain as Fourier transforms. The Helmholtz equation, Eq. (18), for this situation then gives the far-field sound pressure as a superposition of three contributions that involve the volume integration of the wake quadrupoles, surface pressure dipoles, and surface vibration dipoles. Values of the integrands are determined by the differentials between quantities on the upper and lower surfaces. Flow-induced vibration is constrained to be an oscillatory rigid-body translation so that the velocities on the upper and lower surfaces are equal and opposite,

$$u^+ = -u^- = u_n(\mathbf{y}_s, \omega), \tag{20}$$

because of the assumption of incompressibility of the surface. The compactness of the surface applies to low frequencies for which

$$\frac{\omega C}{c_0} \ll 1. \tag{21}$$

The compactness of the source allows simple expansions of the Green function with position in the source zone relative to the observer. In these expressions, the free-space Green function is

$$g(\mathbf{x}, \mathbf{y}, \omega) = \frac{e^{i(k_0 r - \omega t)}}{4\pi r}, \tag{22}$$

where $r = |\mathbf{x} - \mathbf{y}|$, so a simplified expression for the far-field pressure is

$$\begin{aligned} p(\mathbf{x}, \omega) = \frac{1}{4\pi} \iiint_v \frac{\sigma}{r} e^{ik_0 r}\, dV(\mathbf{y}) \\ + \iint_s [-i\omega u_n(\mathbf{y}, \omega)](g^+ - g^-)\, dS(\mathbf{y}) - \cdots \\ - \iint_s \nabla_n g(\mathbf{x}, \mathbf{y}_s, \omega) \\ \cdot [(p^+(\mathbf{y}, \omega) - p^-(\mathbf{y}, \omega)]\, dS(\mathbf{y}), \end{aligned} \tag{23}$$

where the Green functions on the upper and lower surfaces are given by

$$g^+ = g(\mathbf{x}, y^+, \omega), \qquad g^- = g(\mathbf{x}, y^-, \omega) \tag{24}$$

and

$$g = g(\mathbf{x}, \mathbf{y}_s, \omega),$$

respectively, and $\mathbf{y}_s$ is the coordinate of the median surface of the body with coordinate $\mathbf{y}_0$. Under these stipulations, Eq. (18) reduces to (see Refs. 2 and 28)

$$\begin{aligned} p(\mathbf{x}, \omega) \cong \iiint_v \sigma(\mathbf{y}, \omega) g(\mathbf{x}, \mathbf{y}_s)\, dV(\mathbf{y}) \\ + ik_0 \cos\phi \mathcal{L}\, g(\mathbf{x}, \mathbf{y}_s, \omega), \end{aligned} \tag{25}$$

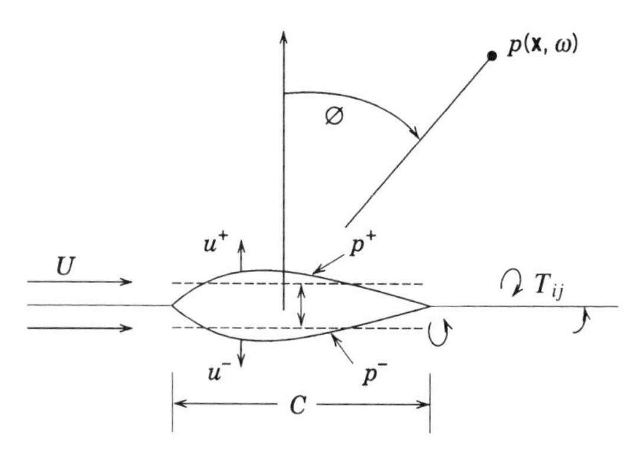

Fig. 9 Conceptual diagram of an acoustically compact elastic body interacting with flow.

where $\mathcal{L}$ denotes the spectrum of unsteady lift *on the surface* including the effect of a contribution to the pressure jump that is due to the fluid reaction to the motion of the surface, as discussed below.

The first term of this expression is the direct quadrupole sound emitted by the wake turbulence; it is of order $U/c_0 \ll 1$ relative to the remaining terms. The second term is the contribution from surface pressure dipoles.

Since the fluid reacts to surface motion, these dipoles are affected by the response of the surface. Equation (26) gives an approximate equation of motion of the surface,

$$\overline{Z_s \bar{u}_n} = \overline{(p^+ - p^-)} - \overline{\Delta p_{\text{fl}}} = (\mathcal{L}_h(\omega) + \frac{\mathcal{L}_h(\omega)}{\text{CL}}, \qquad (26)$$

where (Δp_{fl}) is the differential of the reaction pressure of the fluid on the vibrating surface, $\mathcal{L}_h(\omega)$ is the unsteady lift that would act on the surface if it was rigid, and Z_s represents the impedance of the surface to the differential pressure. In the acoustically compact limit, the pressure differential $\mathcal{L}_{\text{fl}}(\omega)$ is essentially reactive and determines the added mass.[29] Thus, approximately, for the inertial loading on the compact body

$$\frac{-\mathcal{L}_{\text{fl}}(\omega)}{CL} = -i\omega m_a \overline{U}_n. \qquad (27)$$

Upon substitution of these expressions into Eq. (25), we can isolate the dominant parametric behavior of the flow-excited vibrating body, particularly as it relates to the influence of body motion due to the relative impedances of the fluid and the body. Thus

$$p(\mathbf{x}, \omega) = \begin{cases} ik_0 \cos\phi \mathcal{L}_h(\omega) \left(\dfrac{e^{ik_0 r}}{4\pi r}\right) \left\{1 - \dfrac{Z_f}{Z_s + Z_f}\right\}, & (28a) \\[2ex] ik_0 \cos\phi \mathcal{L}_h(\omega) \left(\dfrac{e^{ik_0 r}}{4\pi r}\right) \left\{\dfrac{Z_f}{Z_s + Z_f}\right\}, & (28b) \end{cases}$$

where we have let $Z_f = -i\omega m_a$. The leading term is recognized as the basic dipole sound from the flow–body interaction if the body is not free to vibrate; as given by Eq. (3c), note that L_h is the negative of the force applied to the fluid. The second term in the curly brackets of Eq. (28a) gives an adjustment due to the contribution of surface dipoles caused by the forced motion of the surface. This term is recognized as the leading term of $f(\phi)$ in Eq. (12c). This contribution is determined by the radiation efficiency of the surface motion as expressed by the relative impedances of the body and fluid since these govern the level of vibration that responds to the unsteady lift on the body $\mathcal{L}_h(\omega)$. The effect of surface motion may destructively interfere with the direct dipole contribution if $Z_f \approx Z_s$, i.e. the sound from a compact, unrestrained rigid neutrally buoyant body vanishes. The second term in curly brackets we shall see is also analogous to a term appearing later in Eq. (36). For the many cases for which the excitation forces are due to turbulent flow, we must consider the sound in the statistical sense of a sound pressure spectral density. In this case the sound pressure spectrum is

$$\Phi_{pp}(\mathbf{x}, \omega) = \left[\frac{k_0^2 \langle \mathcal{L}_h^2(\omega) \rangle}{(4\pi r)^2}\right] \left|\frac{Z_s}{Z_s + Z_f}\right|^2. \qquad (28c)$$

4.3 Sound from Flow over Extended Surfaces: High-Frequency Limits

We will now consider the case of sound and vibration that are caused by turbulent flow past bodies extended in the flow direction, such as boundary layers on hulls (analogous to boundary layers on aircraft fuselages), internal flow in long pipes, and external flow along pipes. At high enough frequencies that the flow bounding surface is larger in the flow direction than an acoustic wavelength and all controlling length scales of the flow and flexural vibration wavelengths are also small, we may approximate the out-of-plane vibration of the surface by a flow-induced vibration of a flat surface. We will consider that the driving flow is turbulent and spatially homogeneous in the plane of the surface; this is a nonrestrictive provision that will greatly simplify analysis and permit ready conclusions. These surfaces will, in general, contain discontinuities such as stiffeners and other attached impedances; these features will determine the characteristics of the resulting sound and vibration fields. Under the driving of flow-induced surface pressures associated with the boundary layer, the plate will deform in flexure.

Figure 10 illustrates the model problem of this section, which was first solved by Ffowcs Williams,[22,23] with many extensions in the ensuing years.[25–27] This example problem will enable us to identify the controlling physical parameters that are relevant to the flow-induced vibration sound of extended elastic surfaces. Figure 10*a* shows the surface as a homogeneous one without stiffeners or impedance discontinuities; Fig. 10*b* shows the surface with such discontinuities. Both problems will be examined in order to show how such discontinuities become important to the sound field.

The mathematical treatment of these types of flows is

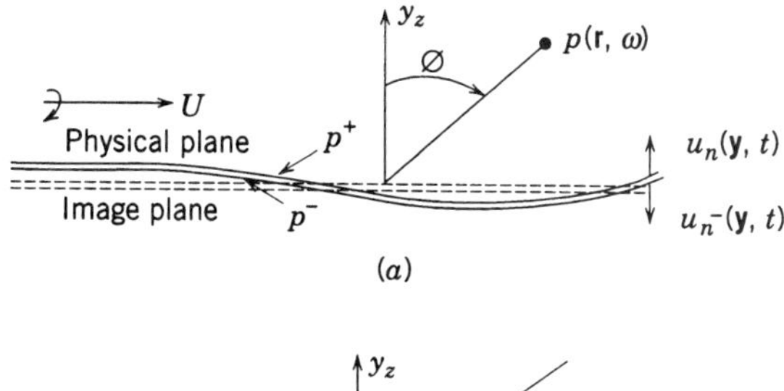

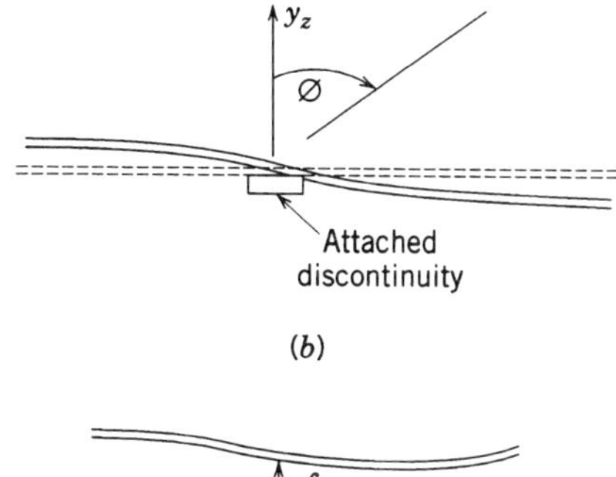

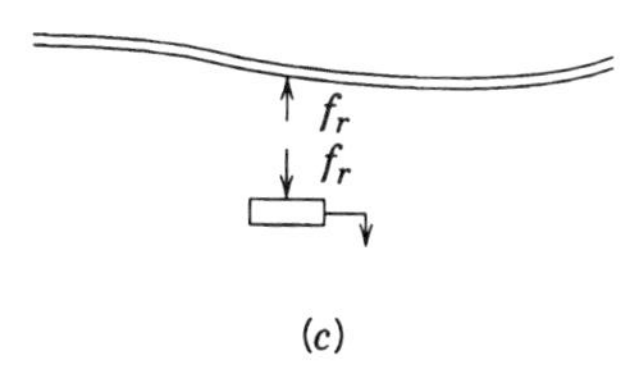

Fig. 10 Conceptual diagrams of flow-excited panels: (*a*) homogeneous flow and elastic surface; (*b*) homogeneous flow and impedance discontinuity; (*c*) free-body diagram of (*b*).

facilitated by considering the spatial Fourier transforms of pressure and flow-induced response. Thus, for example, we express the flexural surface velocity as

$$u(\mathbf{x}, t) = \iiint_{-\infty}^{\infty} U(\mathbf{k}, \omega) e^{i(k\mathbf{x} - \omega x)} \, d^2\mathbf{k} \, d\omega, \qquad (29)$$

where the magnitude squared of the transformed variable $U(\mathbf{k}, \omega)$ is the wavenumber of the frequency spectral density of the panel velocity. Similarly, we define a pressure transform

$$p(\mathbf{x}, t) = \iiint_{-\infty}^{\infty} P(\mathbf{k}, \omega) e^{i(k\mathbf{x} - \omega x)} \, d^2\mathbf{k} \, d\omega \qquad (30)$$

and a pressure spectral density that is the magnitude squared of $p(\mathbf{k}, \omega)$. The problem is solved by a method of images applied to the Helmholtz integral equation in order to eliminate one of the surface integrals. Following a method also used by Powell,[24] an image system is assumed in the lower half plane ($y_2 < 0$) (see also the chapters on statistical methods in acoustics).

We shall define the turbulent flow sources to be invariant upon reflection from the upper half plane to the lower half plane and occupy a thin region δ adjacent to the surface such that $k_0\delta \ll 1$. By invoking an impedance model for the surface velocity $U_n(\mathbf{k}, \omega)$, the transform of the surface pressure is

$$P(\mathbf{k}, \omega) = Z_s(\mathbf{k}, \omega) U_n(\mathbf{k}, \omega), \qquad (31)$$

where $Z_s(\mathbf{k}, \omega)$ is the impedance of the fluid-loaded surface that includes both the structural impedances and the additional impedance loading of the fluid, that is,

$$Z_s(\mathbf{k}, \omega) = Z_m(\mathbf{k}, \omega) + Z_f(\mathbf{k}, \omega), \qquad (32)$$

where m and f denote material and fluid properties, respectively, of the surface. Equation (31) is the analogy of Eqs. (26) and (27).

If the surface is perfectly rigid, the analysis gives only a perfect reflection (Powell's reflection principle[24]) and the familiar doubling of the surface pressure by the rigid wall. We denote this pressure the so-called blocked pressure $P_{\text{blocked}}(\mathbf{k}, \omega)$, and it is completely determined by the quadrupole sources in the turbulence above the plate. When the wall is elastic and spatially homogeneous, as currently modeled, the surface pressure becomes

$$P(\mathbf{k}, \omega) = P_{\text{blocked}}(\mathbf{k}, \omega) \left[\frac{Z_s(\mathbf{k}, \omega)}{Z_m(\mathbf{k}, \omega)} \right], \qquad (33)$$

and the surface pressure spectral density on the elastic surface is related to the blocked surface pressure spectral density by

$$\Phi(\mathbf{k}, \omega) = \left| \frac{Z_m(\mathbf{k}, \omega)}{Z_m(\mathbf{k}, \omega) + Z_f(\mathbf{k}, \omega)} \right|^2 \Phi_{\text{blocked}}(\mathbf{k}, \omega). \quad (34)$$

This surface pressure spectrum is still determined solely by the turbulent *quadrupole* field above the plate. Thus, the effect of a spatially homogeneous elastic plate is to reflect the sources; finite surface impedance modifies this reflection, but it does not introduce any additional sources.

The radiated sound pressure is given by an inverse transform

$$p_{\text{rad}}(\mathbf{x}, t) = \iiint_{-\infty}^{\infty} P(x_2 = 0, \mathbf{k}, \omega) \cdot \exp[i(\sqrt{k_0^2 - k^2}\, x_2 + \mathbf{k}_{1,3} \cdot \mathbf{x}_{1,3}) - i\omega t] \cdot d^2\mathbf{k} \, d\omega. \qquad (35)$$

Since the radical in the integrand becomes imaginary for $|k| > k_0$, pressure components at wavenumbers greater than acoustic are evanescent with distance above the plate. Thus, only surface pressure at wavenumbers that are less than or equal to k_0 are capable of contributing to the sound field. Since the dominant contribution of surface pressures is contained at convected wavenumbers

$$k_c = \frac{\omega}{U} \tag{36}$$

and since the acoustically relevant radiating pressures are determined by wavenumbers less than acoustic, that is,

$$k_c < k_0 = \frac{\omega}{c_0}, \tag{37}$$

the radiating pressures occupy only a fraction

$$\frac{k_c}{k_0} = \frac{U}{c_0} \tag{38}$$

of the total effective wavenumber spectral range of the pressure spectrum. Since the Mach numbers of underwater vehicles and flows are typically small, this fraction accounts for the weak radiation from the quadrupole sound that results from flow over geometrically and structurally homogeneous flat extended surfaces.

Fully analytical predictions that are based on the above relationships and methods are essentially limited by our knowledge of the spectrum of the blocked wall pressure, $\Phi_{\text{blocked}}(\mathbf{k}, \omega)$. The features of the wavenumber frequency spectrum of the blocked wall pressure are well known only for certain boundary layer flow types and only in a range of wavenumbers near ω/U. Figure 11 illustrates a typical spectrum that has been nondimensionalized on the mean wall shear stress τ_w and the boundary layer displacement thickness δ^*. The dominating behavior near the convective wavenumber ω/U is clear. The remainder of the spectrum is reliably known only down to wavenumbers of order ~0.03 ω/U.

When the surface has structural inhomogeneities, as depicted in Fig. 10*b*, then the sound field may be remarkably enhanced due to the creation of additional dipoles at the spatially localized stiffeners as discussed above. To see how the stiffener or attachment affects the vibration and to derive a simple solution that gives the relative importance of scattered to incident vibration sound, we consider the case of a single-point attachment to a plate that has a vertical translational input impedance Z_r, as illustrated in the free-body diagram in Fig. 10*c*. The attachment is situated at the origin of the plate coordinate system and has a velocity u_r and causes a reaction

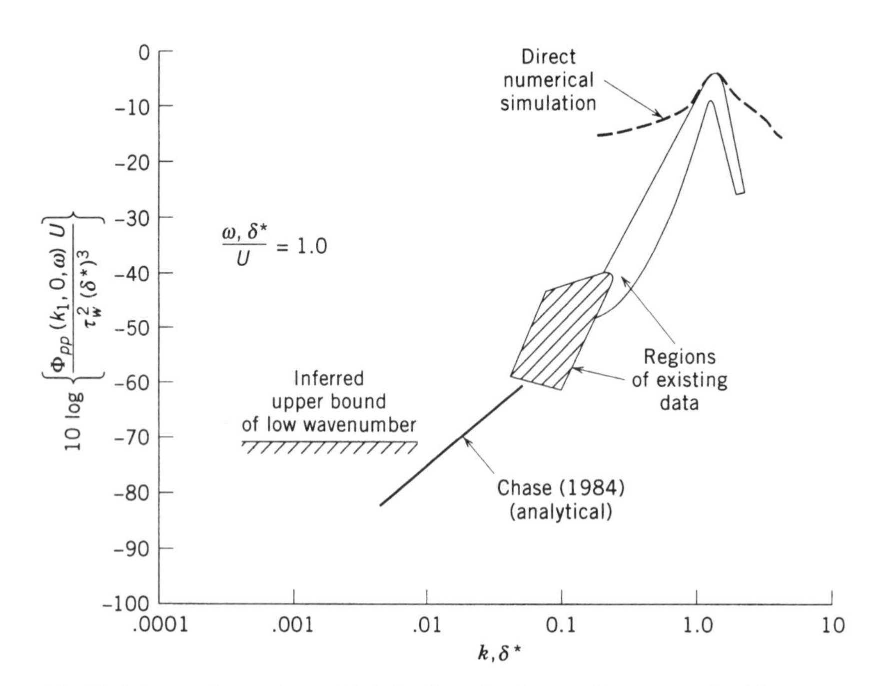

Fig. 11 Wavenumber spectrum of turbulent boundary layer wall pressure at fixed frequency.

force F_r on the plate. We shall consider the attachment at the plate to be a point in order to draw parallels with previous discussion in this chapter. This is not a limitation to the general conclusions, which also apply to other geometries of structural discontinuities. Incorporating this force on the fluid-loaded plate yields, instead of Eq. (31) (see, e.g., Ref. 23),

$$Z_s(\mathbf{k}, \omega)U_n(\mathbf{k}, \omega) = P(\mathbf{k}, \omega) + \frac{F_r}{(2\pi)^2}, \tag{39}$$

where the impedance $Z_s(\mathbf{k}, \omega)$ is given by Eq. (35).

The consequences of the discontinuity is to introduce a dipole of strength F_r to the sound field that would have existed without the presence of the discontinuity.

To bring out a general expression that discloses the effect of the discontinuity compared with the homogeneous plate, we apply the imaging approach to remove the explicit reference to the plate velocity to obtain Eq. (40) which is given in terms of the blocked pressure spectrum and the impedance variables:

$$P(\mathbf{k}, \omega) = P_{\text{blocked}}(\mathbf{k}, \omega)\left[\frac{Z_m(\mathbf{k}, \omega)}{Z_m(\mathbf{k}, \omega) + Z_f(\mathbf{k}, \omega)}\right] + \left\{1 - \frac{Z_m(\mathbf{k}, \omega) - Z_f(\mathbf{k}, \omega)}{Z_m(\mathbf{k}, \omega) + Z_f(\mathbf{k}, \omega)}\right\}\frac{F_r}{(2\pi)^2}k_0 \cdot \cos\phi\, G(x_2, \mathbf{k}, \omega), \tag{40}$$

where $G(x_2, \mathbf{k}, \omega)$ is the two-dimensional Fourier transform of the free-space Green function

$$G(x_2, \mathbf{k}, \omega) = \frac{i}{2}\frac{e^{i\sqrt{k_0^2 - k^2}|x_2|}}{\sqrt{k_0^2 - k^2}}. \tag{41}$$

The first term has already appeared [Eq. (33)] and the new term in curly brackets is identical to that given in Eqs. (11)–(13) for the impedance relationship of the point-driven fluid-loaded plate. Here, the driving force applied to the plate is F_r. To see this identity more clearly, we use inverse Fourier transformation of only the last term to recover the expression for the point dipole applied to the fluid,

$$p_{\substack{\text{rad}\\\text{dipole}}}(\mathbf{r}, \omega) = ik_0\left[\frac{Z_m - Z_f}{Z_m + Z_f}\right]F_r \cos\phi\,\frac{e^{i|k_0 r|}}{4\pi r} \tag{42a}$$

or

$$p_{\substack{\text{rad}\\\text{dipole}}}(\mathbf{r}, \omega) = ik_0\left[\frac{Z_f}{Z_m + Z_f}\right]F_r \cos\phi\,\frac{e^{i|k_0 r|}}{4\pi r}, \tag{42b}$$

where Z_m and Z_f represent values of impedance evaluated at wavenumbers $k = k_0 \sin\phi$, where ϕ is the angle between the observer vector and the plate normal. The force that is induced by the discontinuity depends on the relationship between the input impedance of the discontinuity and the point input impedance of the fluid-loaded plate. The point force on the plate (which is negative of that applied to the fluid) is

$$F_r = \frac{Z_r Z_{P_0}}{Z_r + Z_{P_0}}\iint \frac{P_{\text{blocked}}(\mathbf{k}, \omega)}{Z_m(\mathbf{k}, \omega) + Z_f(\mathbf{k}, \omega)}\, d^2\mathbf{k}, \tag{43}$$

where Z_{P_0} is the point impedance of the fluid-loaded plate. The term in brackets in Eq. (42b) will be recognized as analogous to the bracketed term of Eqs. (12a) and (28b) in expressing the essential dipole nature of the localized effects of real surface impedance and geometry nonuniformities in fluid-loaded elastic surfaces.

5 GENERALIZATIONS

The above three sections form the essential result of this section, and this result has general implications to the flow-forced vibration and sound of structures; see Fig. 8. The effect of the extended surface alone on the flow-induced sound from the turbulent sources is as a specular reflector; the sound field will be established by the identical wave vector spectral content in the vibration as the sources that produce the pressure field on the rigid surface; however, it is weighted by a function that expresses the effect of the plate impedance in altering the acoustic reflection coefficient. This is expressed by Eq. (34) for the extended surface and by Eq. (28c) for the compact surface. The surface which is acoustically compact in one dimension but not in the other dimension is just a one-dimensional extended surface in this context. The presence of structural discontinuities will introduce localized forces that cause nonspecular acoustic radiation. This nonspecular radiation results because the discontinuities convert the wave vectors of the vibrations that are the same as wave vectors of the pressure field into contributions at lower and higher wavenumbers by generating vibrations of the surface at these localized surface distortions. With the addition of the vibration, lower wavenumbers occur that are also less than or equal to the acoustic wavenumber, so that the radiated sound is enhanced. The magnitude of the interaction forces will be determined by the input impedances of the plate and attached structure; the acoustic directivity will be determined by the spatial geometry of the attachment (i.e., the dimensionality, the size and orientation, the type of constraint in translation or rotation). Regardless of detail, the overall effect will be to produce dominating dipole sound. These effects are the dipole source contribution adjustments that appear in Eq. (40) and are evaluated in Eqs. (11) and (42).

If there are distributions of multiple discontinuities (as rib arrays and the like), the above physics and solution approach still applies; however, the interactions between discontinuities through the plate and the adjacent fluid field make the solution procedure more complex due to the acoustical reinforcements and interferences associated with source arrays that are made coherent by the basic structure. The resulting sound field will still depend on the magnitude of the interaction forces between the plate and the discontinuities, but it will also depend on the phase and spatial separation scale of the array of discontinuities. In these cases, fluid–strength path interactions may introduce flow-excited surface resonances, and the array behavior could introduce preferred far-field acoustic directionality.

Since the dipole contribution to the flow-induced noise is likely to be dominant in most realistic situations, and since these situations are generally rather complex, it has been found useful to consider the effects of flow-induced plate vibration and sound in a product form that invokes the notion of the structural radiation efficiency σ_{rad}. This is done particularly for engineering analysis of one-third-octave-band levels. Thus, if the flow-induced mean-square vibration velocity of the ribbed plate averaged over its area is $\overline{V^2}(\omega)$, then a general expression for the acoustic power radiated is developed by methods given in Chapter 62 on statistical methods in acoustics and Refs. 2, 30, and 31. The acoustic power is

$$P_{\text{rad}}(\omega) = \rho_0 c_0 A \sigma_{\text{rad}} \overline{V^2}(\omega) \tag{44}$$

and the mean-square surface velocity is

$$\overline{V^2}(\omega) = \sum \int \int |Z_s(\mathbf{k}, \omega)|^{-2} \Phi_{\text{blocked}}(\mathbf{k}, \omega)\, d\mathbf{k}, \tag{45}$$

where the summation symbol represents summation over all modes of the flow-driven structure and the integration is over all wavenumbers such that $|k| < k_0$.

6 SURVEY OF FORCING CHARACTERISTICS OF SOME IMPORTANT FLOW-INDUCED SOUNDS

The previous sections discussed the essential acoustical features and radiation mechanism for various classes of flow sources and flow-driven surfaces. In this section, we briefly discuss the characteristics of various classes of flow-induced surface excitation forces and pressures for noncavitating flows. Cavitating flows have already been discussed in Section 3 as well as elsewhere in this volume. Specifically, in this section, we will discuss the behaviors of forcing function spectral quantities $\langle \mathcal{L}_h^2(\omega) \rangle$ and $\Phi_{\text{blocked}}(k, \omega)$ that appear in the preceding sections.

Thus, these flow-induced surface force distributions may do the following:

- Occupy one or two space dimensions [e.g., $\mathcal{L}_h(\omega)$ or $\Phi_{\text{blocked}}(\mathbf{k}, \omega)$]
- Be convected over the surface in a mean flow direction or spatially localized by separated flow [e.g., $\Phi_{\text{blocked}}(\mathbf{k}, \omega)$ of a turbulent boundary layer]
- Be broadband in frequency [e.g., $\Phi_{\text{blocked}}(\mathbf{k}, \omega)$ of a turbulent boundary layer] or be nearly pure tone [e.g., $\mathcal{L}_h(\omega)$ for vortex shedding from a bluff body]

The discussion of specific magnitudes and length–time scales for the relevant forcing function is beyond the scope of this chapter. They are fully surveyed and discussed in Refs. 1 and 2. We shall only describe the general features of these characteristics. The following is a summary of various major classes of forcing function by surface type.

6.1 Dipole Forcing Functions on Lifting Surfaces and Blunt Bodies

These are roughly one dimensional at the low Mach numbers of hydroacoustics since they are typically aligned with leading and trailing edges or the axis of a vortex-shedding body. Trailing-edge forcing may occupy a narrow frequency band when orderly vortex shedding occurs at bluff (blunt) trailing edges just as behind circular cylinders. Trailing-edge flow noise and forcing originate because of two classes of flows: turbulent flow across the edge of the surface and in the viscous wake of the trailing edge. Leading-edge forcing is caused by interaction of the lifting surface with incoming gusts. The gust occurs because of relative motion between the surface and its incoming mean flow. If the inflow contains a time mean distortion (as a pipe flow upstream of a

TABLE 1 Important Source Types and Their Parameters

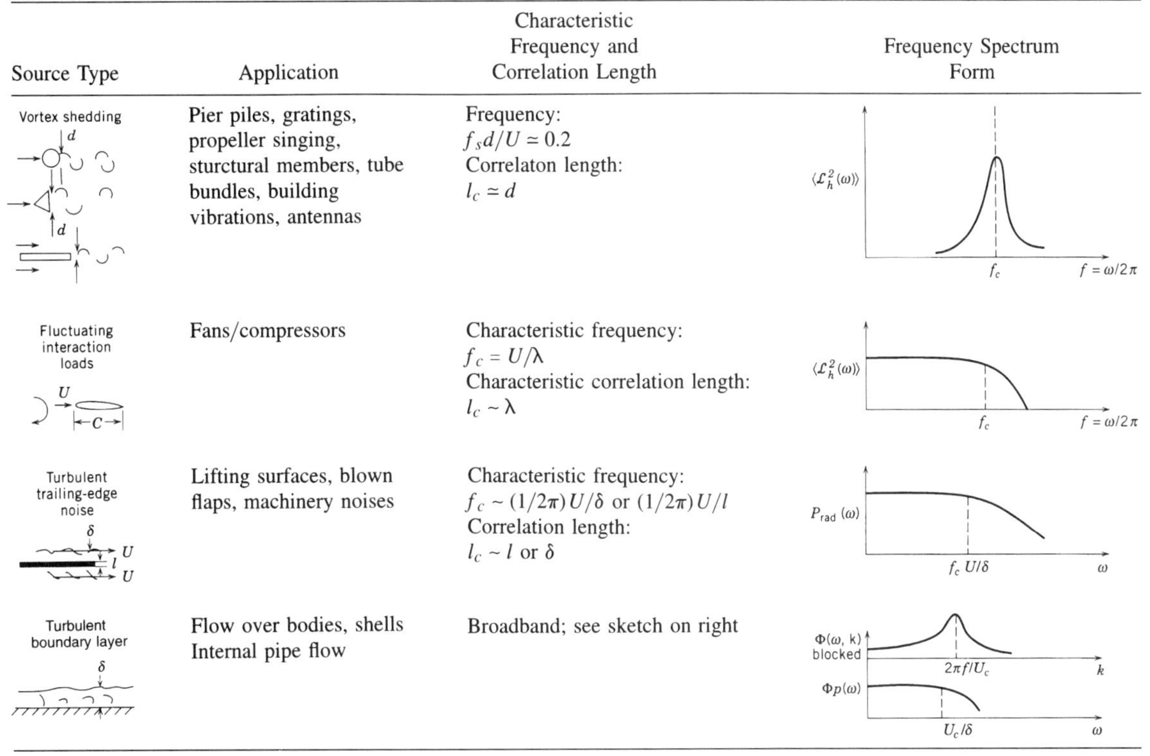

Source Type	Application	Characteristic Frequency and Correlation Length	Frequency Spectrum Form
Vortex shedding	Pier piles, gratings, propeller singing, sturctural members, tube bundles, building vibrations, antennas	Frequency: $f_s d/U \simeq 0.2$ Correlaton length: $l_c \simeq d$	
Fluctuating interaction loads	Fans/compressors	Characteristic frequency: $f_c = U/\lambda$ Characteristic correlation length: $l_c \sim \lambda$	
Turbulent trailing-edge noise	Lifting surfaces, blown flaps, machinery noises	Characteristic frequency: $f_c \sim (1/2\pi)U/\delta$ or $(1/2\pi)U/l$ Correlation length: $l_c \sim l$ or δ	
Turbulent boundary layer	Flow over bodies, shells Internal pipe flow	Broadband; see sketch on right	

Note: The symbols used are as follows:

f = frequency, $\omega/2\pi$
d = diameter
U = approach velocity
δ = boundary layer thickness
C = chord
l_c = correlation length
λ = characteristic eddy (gust) scale
l = thickness of blunt ($l > \delta$) trailing edge
$P_{\text{rad}}(f)$ = sound power in proportional frequency band $Df \propto f$
$\Phi(k, \omega)$ = wavenumber frequency spectral density of wall pressure

pump rotor), the interaction force on the rotor that passes through this flow distortion will be periodic. This periodicity is in accordance with the repetitive passage of the blades through the flow distortion. If the inflow is also turbulent, then the interaction forces will be broadband in frequency. Table 1 summarizes each of these three generic forcing function characteristics.

6.2 Spatially Distributed Turbulent Boundary Layer on Extended Surfaces

This class of excitation is distributed over a large area and is typically convected at a velocity U_c, where U_c is a substantial percentage of the mean advance velocity of the surface. This convection gives a spatial bias to the forcing so that the excitation is concentrated at spatial scales or order $U_c/2\pi f$ at each frequency. As previously described and illustrated in Fig. 11, the wave vector spectrum extends to low values of wave vector, and this is due to the nonfrozen characteristics of the turbulence. Thus in Table 1 we depict the surface pressure field as broadband both in frequency and wavenumber when the integral over all wavenumbers,

$$\Phi_p(\omega) = \iint_{-\infty}^{\infty} \Phi_p(\mathbf{k}, \omega)\, d^2\mathbf{k}, \tag{46}$$

expresses the overall frequency content of the boundary layer pressure. If flow separation occurs due to a localized geometric discontinuity on the surface, then the sound-producing character of surface pressure will more closely resemble that of an edge flow, as described above because of the stationary and localized (i.e., nonconvected) nature of the flow separation.

6.3 Flow Tones due to Feedback Mechanisms

In the cases of flow tones, all sources shown in Table 2 are dipoles, and the data derived for the table are essentially of aeroacoustic origin; see also Blake and Powell.[32] Little is known about the acoustic amplitudes because they depend on the details of each feedback mechanism that sustains the tone, particularly in underwater application. In aeroacoustic application, feedback is by a fluid path; in underwater acoustics the path may also be through the vibrating structure. The frequencies of the tones are often established by the matching of fundamental dispersion characteristics of allowed modes of flow instabilities and the resonance frequencies or geometry of the structure that is attached to the flow. Thus the frequencies shown are likely to have application to both airborne and waterborne flows.

The behavior of flow oscillator tones that occur with coupled flow elastic phenomena manifest tones that occur with $n = 1, 2, 3, \ldots$, resonance frequencies and have the frequency–speed relationship with jumps as illustrated in Fig. 12. In the case of edge, hole, and cavity tones a pronounced velocity gradient is produced by the jet or grazing flow across the opening. Such flows with strong velocity profiles are unstable and are very sound and vibration sensitive. These sensitive disturbances in the field have a characteristic wavelength in the flow direction, say $\lambda_c = U_c/f$, where U_c is the flow convection velocity of the sensitive disturbance and f is frequency. The sounds of such flows often exhibit a series of stages, or orders of tones. A downstream body situated a distance L away on which the flow impinges sends back a disturbance to the upstream point. When $\lambda_c = \alpha L/n$, where α is a factor that depends on the par-

TABLE 2 Types of Flow Tones

Source Type	Application	Characteristic or Fundamental Frequency
D-1 Edge tone 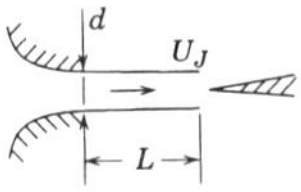	Laminar rectangular jet impinging on sharp edge, musical instruments	Tonal: $\frac{f_s L}{U_j} \cong \frac{1}{2}\left(n + \frac{1}{4}\right)$ $n = 1, 2, \ldots, 4 < L < 20$ s = jet width
D-2 Hole tone 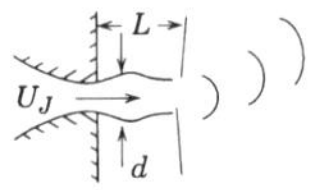	Staged throttling devices	Tonal: $0.011\sqrt{\mathcal{R}_d} > \frac{fb}{U_L} > 0.5$ $\mathcal{R}_d < 3000$ $\frac{L}{d} = 2, 2.5, 3$ d = jet diameter
D-3 Cavity tone 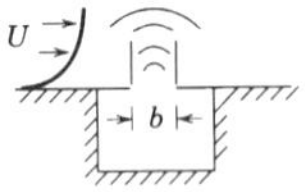	Cavities, bleed ports	Laminar boundary layer: $\frac{fb}{U} \cong 0.022\frac{b}{\theta}$ Turbulent boundary layer: $\frac{fb}{U} = 0.33\left(n - \frac{1}{4}\right)$ $n = 1, 2, \ldots$ b = opening width
D-4 Gaps 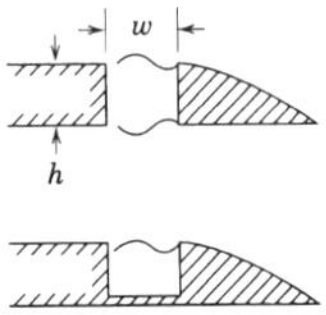	Flaps on lifting surfaces	$\frac{fw}{U} \sim (0.6, \ldots, 0.7)\left(n - \frac{1}{2}\right)$ $n = 1, 2, \ldots$ $\frac{fw}{U} \sim (0.5, \ldots, 0.6)\left(n - \frac{1}{4}\right)$ $n = 1, 2, \ldots$

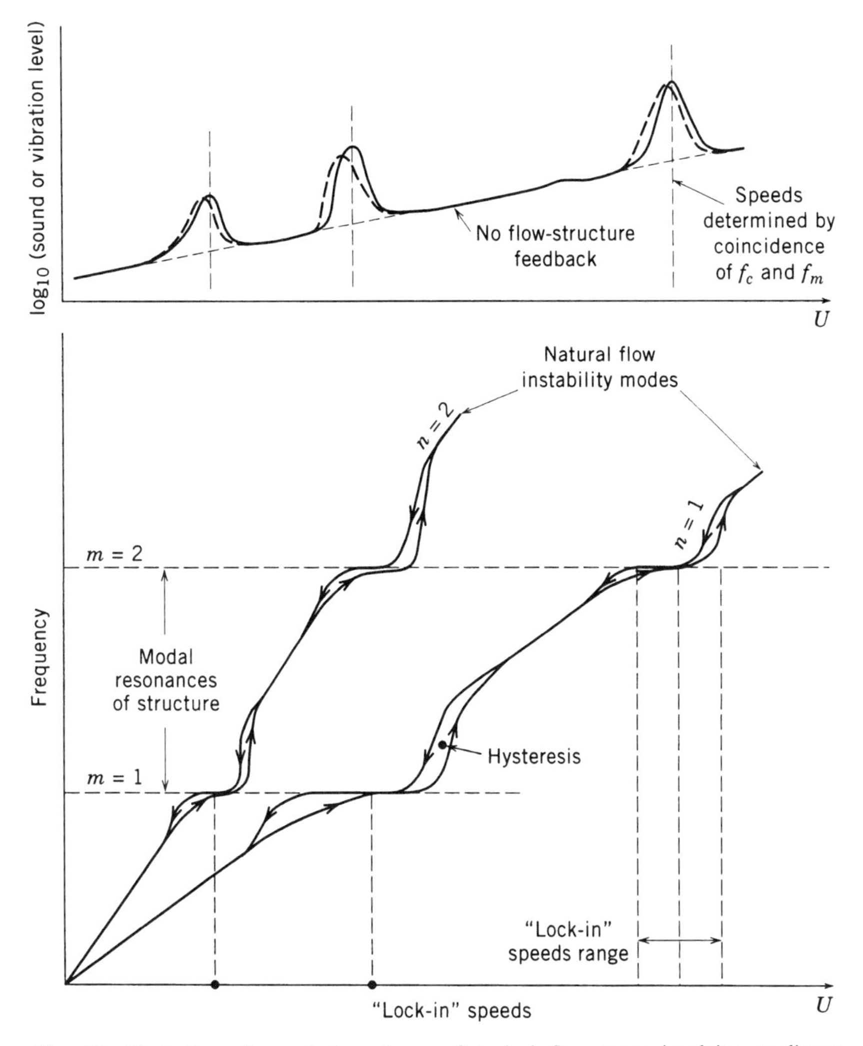

Fig. 12 Illustration of speed dependence of typical flow tones involving nonlinear flow–structure interaction. At "lock-in" very large amplitudes occur.

ticular flow geometry and n is the order of the tone, then the reinforced tone frequencies are

$$f_n = \frac{nU_c}{\alpha L}. \tag{47}$$

The flow presents multiple modes, each mode being roughly established by the number of half waves that "fit" between the leading and trailing edges of the opening. The exact end points of each stage usually depend on whether the mean flow velocity increases or decreases and the mobility characteristics of the structure. Each stage persists as long as the phase and disturbance growth may be sustained by the coupled flow acoustic (elastic) system.

In the case of singing hydrofoils and propellers, only a single fluid mode is typical and the tone is sustained by coupled modal vibration and vortex shedding. In this case, the structure has a series of resonance frequencies, say f_m, for which the fluid disturbance has a characteristic frequency, say f_c, which varies as, say, $f_c d/U$ = const when d is a transverse dimension of the body that generates the flow.

Thus the self-sustained frequency is always near both a resonance frequency of a flexural mode of the structure and an intrinsic instability frequency of the flow. Such instability frequencies are described by a Strouhal number, Eq. (5), where L is a characteristic dimension of the flow. Self-sustained vibration and sound at "lock-in" have a very much larger amplitude than at other speeds.

REFERENCES

1. D. Ross, *Mechanics of Underwater Noise*, Permagon Press, 1984.
2. W. K. Blake, "*Mechanics of Flow-Induced Sound and Vibration*," 2 Vols., Academic Press, New York, 1986.
3. D. G. Crighton, A. P. Dowling, J. E. Ffowcs Williams, M. Heckl, and F. G. Leppington, "*Modern Methods in Analytical Acoustics*," Springer Verlag, 1992.
4. S. Subramanian and T. J. Mueller, "An Experimental Study of Propeller Noise Due to Cyclic Flow Distortions," *J. Sound Vibration*, Vol. 183, 1995, pp. 907–923.
5. G. Maidanik and E. M. Kerwin, "Influence of Fluid Loading on the Radiation from Infinite Plates below the Critical Frequency," *J. Acoust. Soc. Am.*, Vol. 40, 1966, pp. 1034–1038.
6. R. T. Knapp, J. Daily, and F. G. Hammitt, *Cavitation*, McGraw-Hill, New York, 1970.
7. H. Fitzpatrick and M. Strasberg, "Hydrodynamic Sources of Sound," *Proceedings 1st Symposium of Naval Hydrodynamics*, Washington, D.C., 1956, pp. 241–280.
8. G. L. Chanine and C. R. Sirham, "Collapse of a Simulated Multibubble System," ASME Cavitation and Multiphase Flow Forum, Albuquerque, NM, 1985.
9. G. VanWijngaarden, "One Dimensional Flow of Liquids Containing Small Gas Bubbles," *Am. Rev. Fluid Mech.*, Vol. 4, 1970, pp. 369–396.
10. H. J. Baiter, "On Different Notions of Cavitation Noise and What They Imply," ASME International Symposium on Cavitation and Multiphase Flow Noise," Anaheim, CA, 1986.
11. W. K. Blake and M. M. Sevik, "Recent Developments in Cavitation Noise Research," ASME Symposium on Cavitation Noise, Phoenix, AZ, 1982.
12. W. K. Blake, "Propeller Cavitation Noise: The Problems of Scaling and Prediction," ASME Symposium on Cavitation and Multiphase Flow Noise, Anaheim, CA, 1986.
13. W. K. Blake, M. J. Wolpert, and F. E. Geib, "Cavitation Noise and Inception as Influenced by Boundary Layer Development on a Hydrofoil," *J. Mech.*, Vol. 80, 1974, pp. 617–640.
14. S. Ceccio and C. E. Brennan, "Observations of the Dynamics and Acoustics of Travelling Bubble Cavitation," *J. Fluid Mech.*, Vol. 233, 1991, pp. 633–660.
15. V. Arakeri and V. Shanmuganathan, "On the Evidence for the Effect of Bubble Interference in Cavitation Noise," *J. Fluid Mech.*, Vol. 159, 1985, pp. 131–150.
16. G. Bark, "Development of Distortions in Sheet Cavitation on Hydrofoils," ASME Jets and Cavities-International Symposium, Miami Beach, FL, 1985.
17. P. Abbot, D. S. Greeley, and N. A. Brown, "Water Tunnel Pump Cavitation Noise Investigations," ASME International Symposium on Cavitation Research Facilities and Techniques, Boston, MA, 1987.
18. N. A. Brown, "Cavitation Noise Problems and Solutions," in *International Symposium on Shipboard Acoustics*, Elsevier, 1977.
19. W. K. Blake, H. Hemingway, and T. C. Mathews, "Two Phase Flow Noise," Noise-Con 88, Purdue University, June 1988.
20. Y. L. Levkovskii, "Modelling of Cavitation Noise," *Sea Prop.-Acoust.* (Engl. Trans.), Vol. 13, 1968, pp. 337–339.
21. M. Strasberg, "Propeller Cavitation Noise after 25 Years of Study," *Proc. ASME Sym. Noise Fluids*, Atlanta, GA, 1977.
22. J. E. Ffowcs Williams, "Sound Radiation from Turbulent Boundary Layers Formed on Compliant Surfaces," *J. Fluid Mech.*, Vol. 22, 1965, pp. 347–358.
23. J. E. Ffowcs Williams, "The Influence of Simple Supports on the Radiation from Turbulent Flow Near a Plane Compliant Surface," *J. Fluid Mech.*, Vol. 26, 1966, pp. 641–649.
24. Powell, A., "Aerodynamic Noise and the Plane Boundary," *J. Acoust. Soc. Am.*, Vol. 32, 1960, pp. 982–990.
25. D. G. Crighton and J. E. Oswell, "Fluid Loading with Mean Flow. I. Response of an Elastic Plate to Localized Excitation," *Phil. Trans. R. Soc. Lon.*, Vol. A335, 1991, pp. 559–592.
26. D. G. Crighton and D. Innes, "The Modes and Forced Response of Elastic Structures under Heavy Fluid Loading," *Phil. Trans. R. Soc. Lon.*, Vol. A312, 1984, pp. 291–341.
27. M. S. Howe, "Sound Produced by an Aerodynamic Source Adjacent to a Partly Coated Finite Elastic Plate," *Phil. Trans. R. Soc. Lon.*, Vol. A436, 1992, pp. 351–372.
28. R. Martinez, "Thin Shape Breakdown (TSB) of the Helmholtz Integral Equation," *J. Acoust. Soc. Am.*, Vol. 90, 1991, pp. 2728–2738.
29. W. K. Blake, D. Noll, R. Martinez, and Y. T. Lee, "Dynamics of Fluid-Coupled Neighboring Substructures at Low Frequencies," *Int. J. Comp. Str.*, 1997 to be published.
30. R. H. Lyon, *Statistical Energy Analysis of Dynamical Systems*, MIT Press, Cambridge, MA, 1975.
31. F. Fahey, *Sound and Structural Vibration*, Academic Press, New York, 1985.
32. W. K. Blake and A. Powell, "The Development of Contemporary Views of Flow-Tone Generation," in *Recent Advances in Aeroacoustics*, A. Krothapalli and C. A. Smith (Eds.), Springer Verlag, New York, 1986.

35

SHIP AND PLATFORM NOISE, PROPELLER NOISE

ROBERT D. COLLIER

1 INTRODUCTION

Ship noise is a major part of the field of underwater acoustics. Ship noise reduction and control are important factors in the performance of underwater acoustic systems and in the habitability of the vessel for the crew and passengers. A singularly important text is *Mechanics of Underwater Noise*,[1] which provides detailed information on both mechanical and hydrodynamic noise sources and radiation. Chapter 32 summarizes the physics of sound radiation from ship structures and illustrates the basic mechanisms with simple mathematical models. The effects of fluid loading on acoustic radiation of plates and cylindrical structures are described in Chapter 10. In addition, detailed descriptions of hydroacoustic noise sources, including propellers, cavitation, vortex shedding, and turbulent boundary layer flow-induced noise, are covered in Chapter 34. These chapters provide valuable information directly related to the sources and characteristics of ship noise.

One of the principal objectives of this chapter is to provide engineering procedures for estimating ship machinery source levels and structural vibration transmission losses to arrive at hull vibration levels. The subsequent calculations for acoustic radiation are highly dependent on the details of the ship design and operational factors. Empirical data are presented to provide engineering guidance on radiated noise levels and, in particular, procedures for estimating propeller radiated noise are given, including estimates of cavitation noise. This chapter also discusses the underwater noise effects on ship sensors, that is, platform or self-noise, which has a significant impact on the performance of sonar systems (Chapter 37). Ship noise as it relates to human factors is a critical issue in ship design and operations. This chapter provides design guidance on criteria for environmental noise in interior ship compartments and discusses some of the material issues relating to noise control in the marine environment. Detailed information on noise control in interior spaces may be found in Parts VIII and IX.

Note: References to chapters appearing only in the *Encyclopedia* are preceded by *Enc.*

2 RADIATED NOISE

2.1 General Characteristics

In naval operations the noise radiated by a ship is a dominant source of information, that is, signal, for underwater sonar systems. Radiated noise from ships can be an important contributor to ocean ambient noise, as discussed in Chapter 36, and as a factor in oceanographic research and geophysical exploration and cruise ships operating in environmentally sensitive areas. Engineering estimates for noise predictions of specific classes of ship machinery are given in this chapter based on weight, power, and foundation types. The effectiveness of vibration isolation systems, including examples of two-stage systems with rafts, is presented in Chapter 55. The important role of structural transmission mechanisms and their interaction with fluid systems is dealt with in Chapter 10.

The four principal groups of radiated noise sources are (1) machinery vibration caused by propulsion machinery and ships' services and auxiliary machinery, including steam, water, and hydraulic piping systems; (2) propellers, jets, and other forms of in-water propulsion; (3) acoustic noise within compartments below the waterline; and (4) hydroacoustic noise generated external to the hull by flow interaction with appendages, cavities, and other

TABLE 1 Representative Ship-Radiated Noise Source Levels at 1 yd

	Source Noise Levels (dB re 1 μPa 1 Hz Band)						
Ship Class	0.1 kHz	0.3 kHz	1.0 kHz	3.0 kHz	5.0 kHz	10.0 kHz	25.0 kHz
Freighter, 10 knots	152	142	131	121	117	111	103
Passenger, 15 knots	162	152	141	131	127	121	113
Battleship, 20 knots	176	166	155	145	141	135	127
Cruiser, 20 knots	169	159	148	138	134	128	120
Destroyer, 20 knots	163	153	142	132	128	122	114

discontinuities. A summary of these noise sources and their radiated noise characteristics is discussed in Ref. 2, from which Table 1 gives representative source levels as a function of frequency for a range of surface ships. The data for ships operating between 10 and 20 knots illustrate the dominance of low-frequency noise.

The speed/power dependence of the radiated noise of surface ships is further illustrated by the measurements plotted in Fig. 1.[1] The 9-knot noise spectrum is governed by machinery sources while the significant increase in low-frequency noise as speed is increased is due to both propulsion machinery and the inception and development of propeller cavitation noise. The latter source of radiated noise is dealt with in Section 4. Ross[1] provides estimation formulas for broadband source levels as a function of size or displacement tonnage and speed or power. For example,

$$L_s = 134 + 60 \log \frac{U_a}{10\,\text{knots}} + 9 \log \text{DT} \qquad (1)$$

where U_a is ship speed in knots and DT is displacement tonnage. Ross states that this formula is applicable for frequencies above 100 Hz and ships weighing under 30,000 tons. The acoustic efficiencies of ships have been found to range from 0.3 to 5 W of acoustic power for ship mechanical propulsion power of 1 MW; Ross suggests acoustic conversion efficiencies of 1×10^{-6} for machinery sources and 1.5×10^{-6} for cavitating propellers.

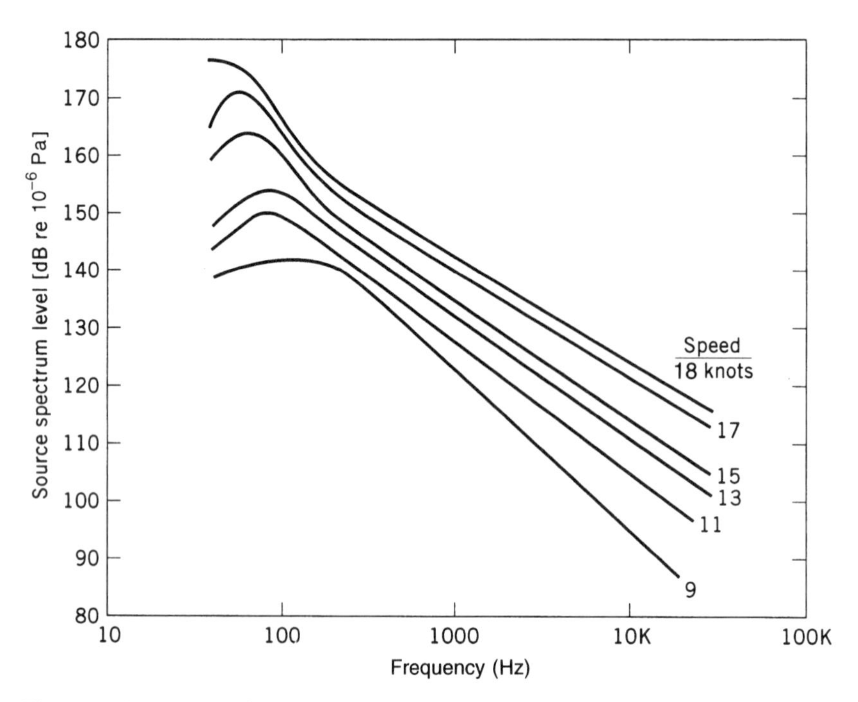

Fig. 1 Radiated noise of passenger ship *Astrid*, U.S. Office of Scientific R&D, published 1960.[1]

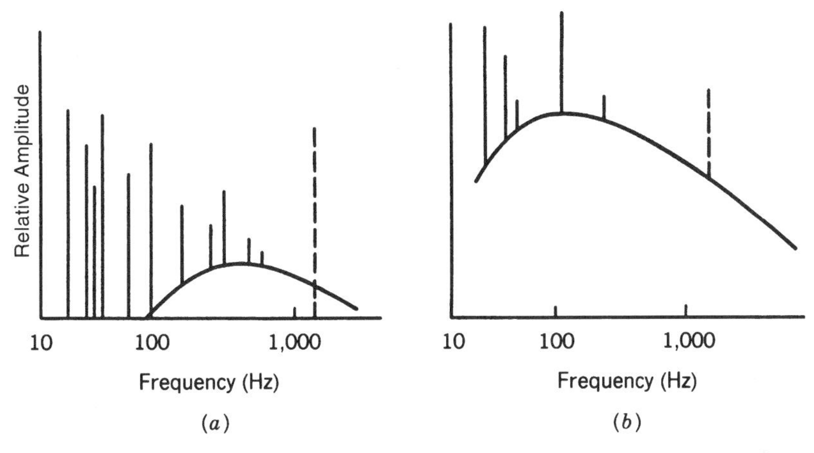

Fig. 2 Diagrammatic spectra of submarine noise at (*a*) low and (*b*) high speeds.[2]

2.2 Noise Spectra

Noise spectra are generally classified in two types: (a) broadband noise having a continuous spectrum such as that associated with cavitation and (b) tonal noise containing discrete frequency or line components related to machinery, gears, and modulation of broadband noise. In addition to steady-state noise, ship noise is also characterized by transient and intermittent noise caused by impacts, loose equipment, or unsteady flow that has particular spectral properties. Ship noise is generally a combination of continuous and tonal noise covering the audio spectrum and is usually concentrated in the low-frequency region. Figure 2 shows a diagrammatic comparison of two radiated noise spectra in which auxiliary machinery tonal noise governs the low-speed condition and propulsion system speed-related frequencies are superimposed on broadband propeller cavitation noise at high speeds. Figure 3 gives an overview of the frequency range of the major sources of radiated noise and

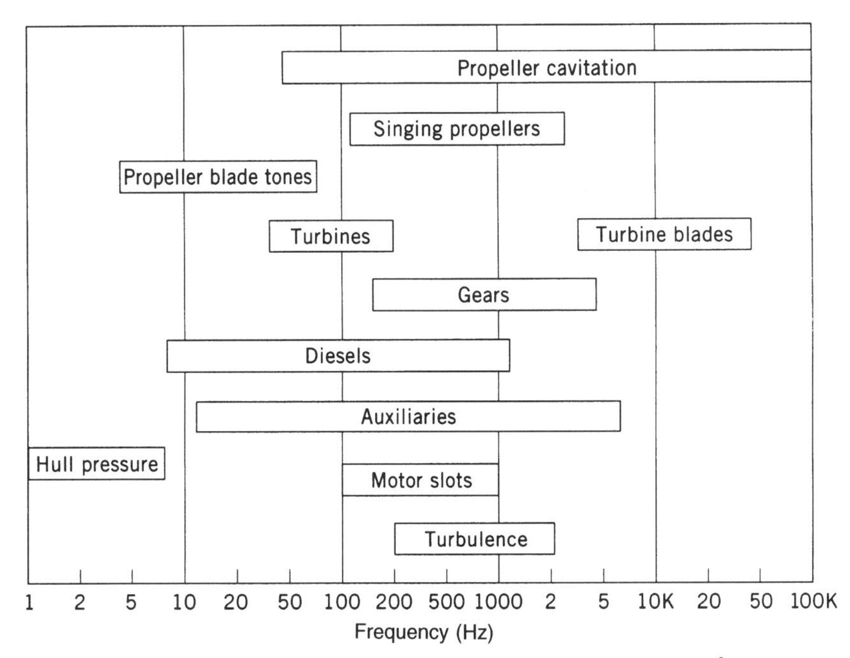

Fig. 3 Frequency ranges of noise radiated by ship noise sources.[3]

shows that the majority of machinery sources produce noise from 10 to 1 kHz and significant sources such as cavitation and turbine lines may extend into the 10-kHz region and above.[3]

2.3 Machinery Noise

The principal mechanisms that generate vibratory forces involve mechanical unbalance, electromagnetic force fluctuations, impact, friction, and pressure fluctuations. The classes of machinery that produce noise may be categorized according to their functions, such as (a) propulsion machinery (diesel engines, steam turbines, gas turbines, main motors, reduction gears, etc.) and (b) auxiliary machinery (pumps, compressors, generators, air conditioning equipment, hydraulic control systems, etc.).

Figure 4 is a schematic view of the machinery components of a diesel–electric propulsion system and their associated noise sources, which are described as follows:

1. Piston slap, which is a dominant noise mechanism of diesel engines and is caused by the impact of a piston against the cylinder wall, results in a spectrum made up of a large family of harmonically spaced tonals.
2. Mechanical imbalances of the generator and auxiliary machinery result in fluctuating forces and moments that are proportional to the square of the angular speed. Since the force is proportional to vibration velocity, the radiated power increases as the fourth power of rotational speed.
3. Electromagnetic force fluctuations of the main drive motor are related to changes in the flux density, which are a function of the number of poles and result in low-frequency line spectra.
4. Reduction gear noise is dominated by gear tooth impacts and results in tones at multiples of the tooth contact frequency. Helical gears are significantly quieter than spur gears.
5. Propeller noise, which is discussed in Section 3, consists of two major components: (a) direct radiation from the propeller blades and (b) low-frequency hull vibration modes induced by hydrodynamic fluctuating forces acting on the blades and transmitted through the propeller shaft and thrust bearings to the hull. The hull response is thus related to the shaft rpm and the number of propeller blades. The modes of hull vibration and resulting sound radiation are discussed in Chapter 32.

2.4 Machinery Vibration Levels

The prediction of radiated noise from machinery sources is based on the traditional noise model, which involves source levels, transmission path dynamics including vibration isolation and foundation transfer functions, and hull vibration and radiation. The role of foundation struc-

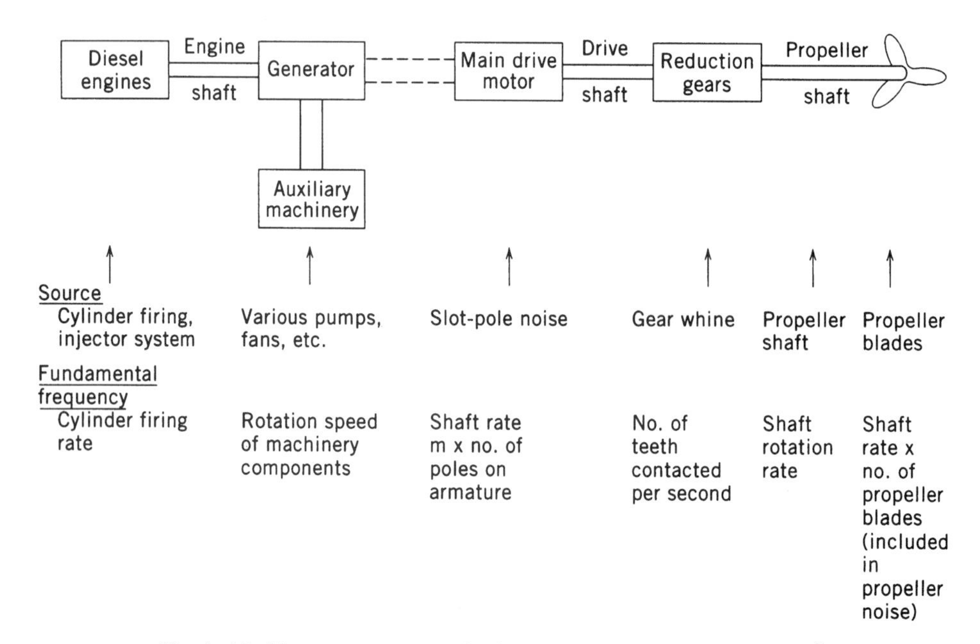

Fig. 4 Machinery components and noise sources on a diesel–electric ship.[2]

TABLE 2 Machinery Source Levels: Octave Band Baseline Vibration Levels

Machinery Type	Baseline Vibration Levels, L_{VB} (dB re 10^{-6} cm/s) — Center Frequency										
	16 Hz	31.5 Hz	63 Hz	125 Hz	250 Hz	500 Hz	1 kHz	2 kHz	4 kHz	8 kHz	16 kHz
Gas turbines											
Small high speed	80	85	95	100	100	90	110	100	100	95	95
Intermediate	110	104	106	103	98	93	86	80	73	67	57
Electric motors											
AC	102	96	90	84	78	72	66	60	54	48	42
DC	88	88	83	78	76	73	68	58	53	48	43

tural mobilities is a critical part of this system and directly determines the vibration levels of the source machinery, as discussed in Chapter 55. The purpose of this section is to provide engineering estimates for machinery vibration levels, based on an extensive data base, as an input to structural design and noise radiation calculations developed for SNAME.[4]

The source level algorithms are provided for engineering guidance. Reliable measured data should be used when available. Also, the parametric dependencies of the prediction algorithms can be used to scale from measured data for similar classes of machinery. It should be noted that the source vibration levels assume that the machine is mounted on low-frequency vibration isolation mounts. Thus the effect of rigid mounting is to decrease machinery source vibration levels due to the increased constraining effect of the foundation relative to that of resilient mounts.

Table 2 gives source vibration velocity levels in decibels relative to 10^{-6} cm/s as a function of octave bands for two types of gas turbines and for representative classes of electric motors.

Table 3 gives formulas for overall machinery source levels expressed as baseline vibration velocity in decibels relative to 10^{-6} cm/s. The machinery parameters include w = gross weight in kilograms, kW = rated power in kilowatts, and actual and rated rpm. The types of machinery include diesels, generators, pumps, and reduction gears. For each type, adjustments to the overall levels are given in Table 4 to obtain estimates of vibration levels for the designated octave bands. Alternative prediction procedures may be found elsewhere.[5–7]

2.5 Structural Vibration Transfer Functions

The effects of the transmission of vibratory energy (structure-borne noise) along and through different structural paths are expressed as transfer functions that relate the input (machinery) and output (foundation or hull) vibration levels. In this section an empirical procedure is given for estimating transfer functions for different transmission paths. The U.S. Navy has published several design handbooks to guide structural designers.[8–11] It should be noted that state-of-the-art PC-based noise models can provide greatly refined and more accurate transfer function estimates for specified structural configurations, and the accompanying estimates should only be used for guidance. Furthermore, several vibration trans-

TABLE 3 Machinery Vibration Overall Source Levels

Machinery Class	Overall Baseline Vibration Source Level, L_{VB} (dB re 10^{-6} cm/s)
Diesel	$-20\log(w) + 20\log(\text{kW}) + 30\log\left(\dfrac{\text{rpm}}{\text{rpm}_0}\right) + 136$
Reduction gears	$64 + 10\log(\text{kW})$
Generator	$53 + 10\log(\text{kW}) + 7\log(\text{rpm})$
Pumps	
Nonhydraulic	$65 + 10\log(\text{kW})$
Hydraulic	$63 + 10\log(\text{kW})$

Note: w = gross weight (kg), kW = rated power, rpm = given rotational speed, rpm_0 = rated rotational speed. See Table 4 for octave band adjustments.

TABLE 4 Machinery Vibration Source Levels[a]: Octave Band Adjustments to Overall Baseline Vibration Levels (Table 3)

Machinery Class	Vibration Source Levels										
	Center Frequency										
	16 Hz	31.5 Hz	63 Hz	125 Hz	250 Hz	500 Hz	1 kHz	2 kHz	4 kHz	8 kHz	16 kHz
Diesel	0	−3	−4	−4	5	−6	−6	−10	−18	−29	−44
Reduction gears	0	−2	1	−11	−12	−3	1	−5	−16	−32	−38
Generator	0	3	8	5	−1	−5	−10	−15	−21	−27	−35
Pumps											
Nonhydraulic	10	10	12	19	11	9	4	−6	−8	−15	−25
Hydraulic	10	20	27	32	33	36	30	25	20	5	−10

[a]See Table 3.

mission paths must generally be considered for ship noise predictions.

Vibration Isolation Mountings The transfer function or transmission loss is expressed as the log ratio of vibration velocity above the mounts at the attachment to the machine subbase to that on the foundation structure below the mounting system. In this design guide, sources of vibration are divided into three weight classes: class I, less than 450 kg; class II, 450–4500 kg; class III, over 4500 kg.

Table 5 gives representative transmission loss values for four types of mounting configurations: hard mounted (no vibration isolation), distributed isolation material (DIM), single-stage low-frequency isolation mounts, and

TABLE 5 Representative Transmission Loss Values versus Octave Band Center Frequency for Ship Machine Mounting Arrangements (dB)

Machinery Weight Class	Transmission Loss, dB								
	Center Frequency								
	31.5 Hz	63 Hz	125 Hz	250 Hz	500 Hz	1000 Hz	2000 Hz	4000 Hz	8000 Hz
	Hard Mounted								
I	13	10	8	5	3	2	1	0	0
II	9	7	5	3	1	0	0	0	0
III	5	3	2	1	0	0	0	0	0
	Distributed Isolation Material								
I	0	1	5	9	12	15	15	15	15
II	0	0	1	3	5	8	9	10	10
III	0	0	1	2	3	3	4	5	8
	Low-Frequency Isolation Mounts								
I	20	25	30	30	30	30	30	30	30
II	12	16	20	23	25	25	25	25	25
III	5	6	8	12	15	18	20	20	20
	Two-Stage Mounting System								
I	25	33	40	45	50	50	50	50	50
II	22	30	35	40	45	48	50	50	50
III	20	25	30	30	35	45	50	50	50

Note: Machinery weight classes: class I, under 450 kg; class II; 450–4500 kg; class III; over 4500 kg. Values based on relatively rigid high-impedance foundation structures.

two-stage isolation with intermediate rafts. The estimates are presented as average values for the designated octave bands. For the hard-mounted case lightweight machines can be expected to have a modest loss at low frequencies, while all classes of machinery have little or no loss above 250 Hz. The DIM installations are effective for lightweight machines above 250 Hz, but their performance decreases as the machinery weight increases. The losses of low-frequency mounts closely follow theoretical predictions for lightweight machines, that is, 20–30 dB transmission loss over the given frequency range. However, as the weight of a machine increases relative to that of the foundation structure, the degree of isolation decreases, reflecting the overall impedance relationships. Two-stage mounting systems have been implemented extensively in ship designs and have proved to be highly effective. Similar types of estimates of transfer functions for representative machinery foundation structures may be used to arrive at hull vibration levels.

2.6 Hull Vibration-Radiated Noise

The relationships between hull structural vibration levels and radiated noise are discussed in Chapters 10 and 32. Assuming that the radiation efficiency σ_r is known, a first-order estimate of the sound radiated from hull plating can be calculated by

$$L_s = L_v + 10 \log \sigma_r + 10 \log A_p + 10 \log n + 41 \qquad (2)$$

where L_s is the equivalent source level at 1 yd, L_v is the space average vibration velocity level of the hull plating, A_p is the area of a single hull plate in inches squared, and n is the number of radiating panels.

2.7 Hull Vibration Transmission

Hull vibration (calculated or measured) in the area of machinery foundation attachment locations (e.g., deep frames) is transmitted through the steel hull structure. Thus, larger areas of the hull may contribute to radiated noise. Also, the transmission of machinery excited hull vibration into sonar array structures can contribute to sonar self-noise (see Section 4). A useful guideline for broadband transmission loss in typical damped ship structures is 0.5–0.8 dB/ft for cases of free-layer damping and 1.7 dB/ft for structures, including wetted hulls, with constrained layer damping. Structural details are needed to establish frequency dependence.

2.8 Hull Grazing Sound Transmission Loss

Table 6 provides frequency-dependent expressions of propagation losses for sound traveling in the water along the length of the hull, that is, grazing sound. The hull is considered to be an air-backed baffle and the grazing sound transmission loss applies to a 1-yd source level for far-field radiation. The theoretical basis for these estimates involves energy transmission through both the hull structure and the water path.

TABLE 6 Hull Grazing Transmission Loss versus Frequency (dB)

Octave-Band Center Frequency (Hz)	Transmission Loss at Distance r (dB re $r_0 = 1$ yd)
250	$10 \log r/r_0$
500	$13 \log r/r_0$
1000	$17 \log r/r_0$
2000	$23 \log r/r_0$
4000	$27 \log r/r_0$
8000	$27 \log r/r_0$

3 PROPELLER NOISE

3.1 General Characteristics

Propulsion propellers constitute a major source of ship noise and, similar to fans, aircraft turbo-props, helicopters, and other devices with rotating blades operating in nonuniform flow fields, are the subject of continuing noise control efforts. The interaction of blade forms and hydrodynamic flows and the resulting dynamic response of the blades and the associated acoustic radiation are complex phenomena that depend on a wide range of design and operational variables that do not lend themselves to simple models and noise predictions. Hydrodynamic noise sources are discussed in Chapter 34; in particular, Section 3 in that chapter provides the theoretical background for cavitation inception, development, and associated noise.

The guidelines for estimating propeller noise included in this chapter are based on the original work of Ross[1] and the more recent engineering analyses presented in the report of the Nordic cooperative project.[6] Marine propeller noise reduction has benefited from parallel aerodynamic acoustic studies, which are the subject of other chapters. For example, significant reductions in both propeller and fan noise have been achieved through smoothing and control of inlet flows and the design of skewed blades.

3.2 Propeller Noncavitating Noise

There are three types of noncavitating propeller noise: (a) mechanical blade tonals related to propeller shaft speed

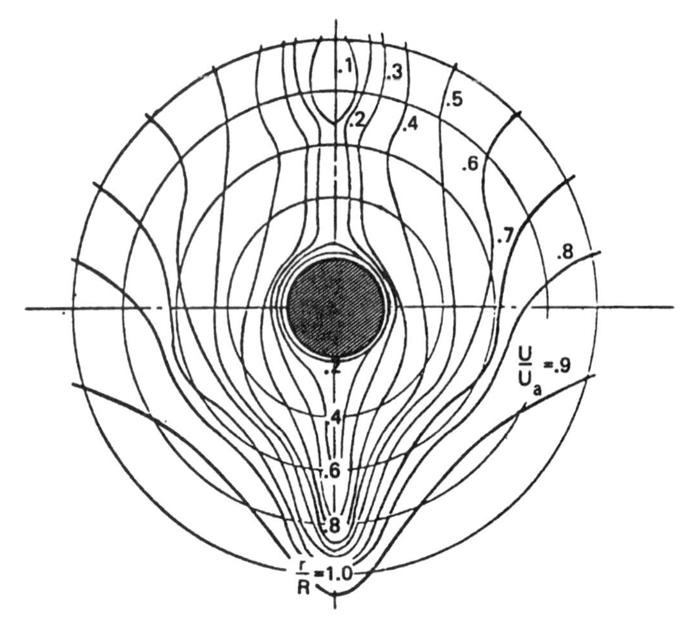

Fig. 5 Wake diagram for a single-screw merchant ship.[1]

and the number of blades; (b) propeller broadband noise related to blade vibratory response to turbulence ingestation and trailing-edge vortices; and (c) propeller singing dùe to coincidence of vortex shedding and blade resonant frequencies. Blade tonals and harmonics result from oscillating components of forces or propeller thrust variations caused by circumferential variations of the wake inflow velocity. Figure 5 illustrates, by use of equivelocity contours, the velocity variations in the plane of a single-propeller merchant ship. The flow speed varies from 10 to 90% of the forward speed of the propeller. These velocity differences cause large variations of the angle of attack and associated lift forces, which lead to significant fluctuations in thrust and torque during each revolution of the shaft and, in turn, to high-level, low-frequency hull vibration. Thus, the most important design consideration is the relationship between the harmonic structure of the wake and the number and blade form of the propellers.

The primary propeller design factors include diameter, shaft rpm, number of blades, expanded area ratio, blade load distribution, skew distribution, blade tip–hull clearance, and the spatial and temporal characteristics of the inflow field.

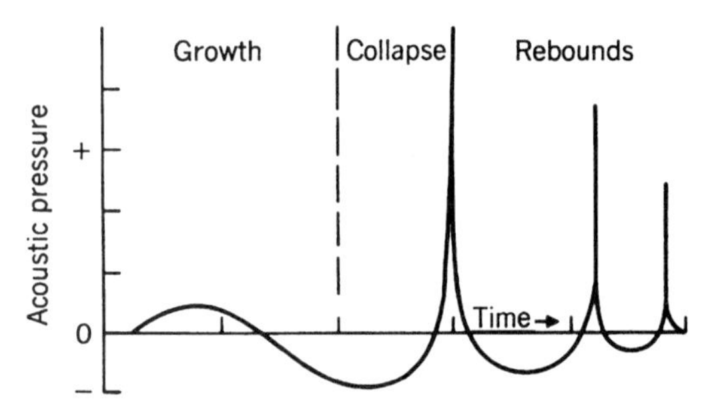

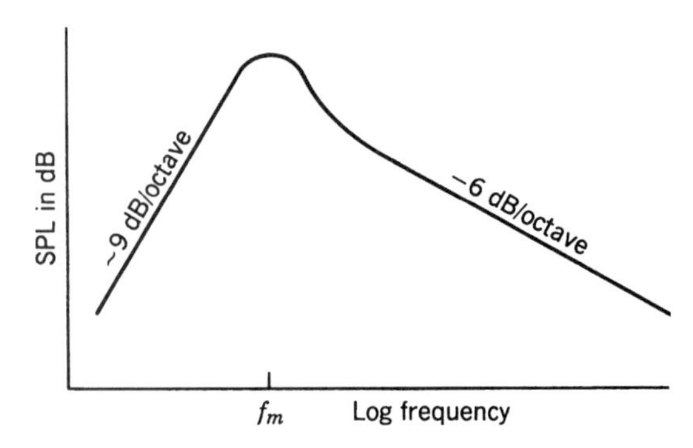

Fig. 6 Cavitation pressure pulses from collapsing cavity and idealized spectrum.[1]

3.3 Propeller Cavitation Noise

As stated above, a description of cavitation is provided in Chapter 34. There are four types of propeller cavitation: driving face, suction face, tip vortex, and hub vortex. The blade tip speed governs cavitation inception, as shown in Eq. 14 of that chapter. Broadband cavitation noise results from the growth and collapse of a sheet of bubbles occupying a volume on the individual blades. Figure 6 illustrates this process for a single cavitation bubble with the resultant idealized spectrum. The general noise spectrum of blade cavitation is shown in Figure 7 and has four principal spectral regions:

I. Low frequency: contains blade frequency ω and harmonics; mean power level increases as ω^4.

II. Midfrequency: starts at bubble frequency ω_b; mean power level increases as $\omega^{-5/2}$.

III. Intermediate frequency: transition region between regions II and IV.

IV. High frequency: starts at bubble frequency ω_c; mean power level decreases as ω^{-2}.

In regions I and II the fluctuations of the sheet cavitation volumes may be represented by a large bubble that acts as an acoustic monopole. In region IV, the power is caused by cavity collapse or by shock wave generation by nonlinear wave propagation. Region III contains a mixture of regions II and IV. Figure 8 is an example of a comparison between predicted and full-scale measured source levels for a 32,000-ton vessel.[6]

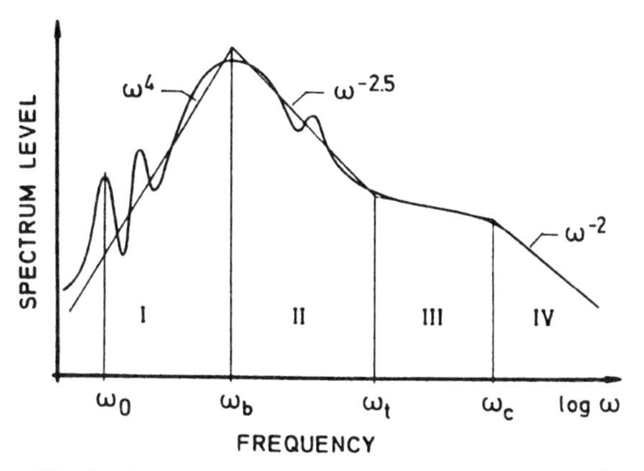

Fig. 7 General noise spectrum of a cavitating propeller.[6]

4 PLATFORM AND SONAR SELF-NOISE

4.1 General Characteristics

Conceptually, self-noise is that noise in a sonar system attributable to the presence of the platform, as illustrated

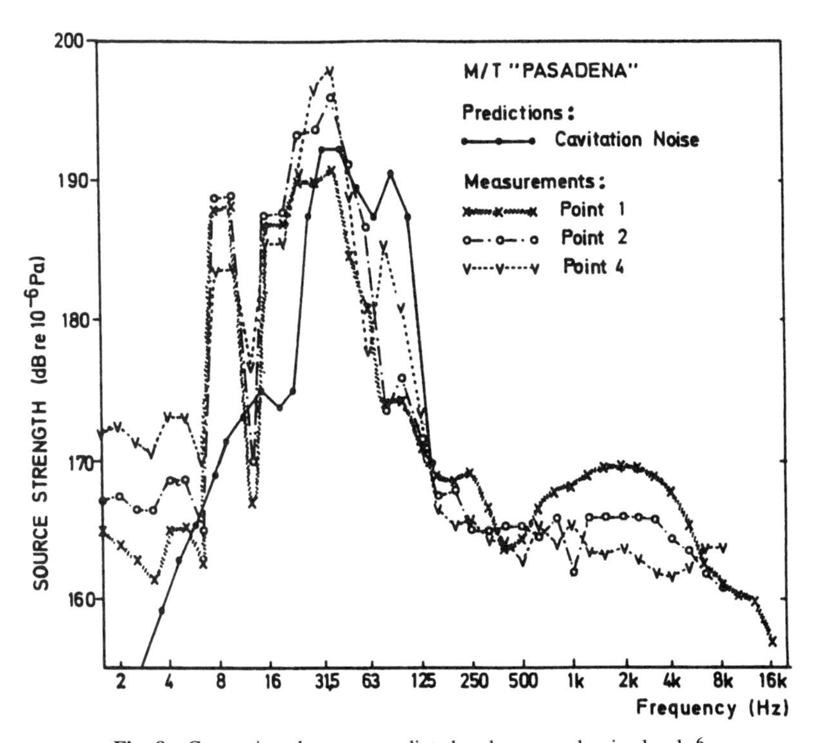

Fig. 8 Comparison between predicted and measured noise levels.[6]

in Figure 9. In practice, sonar self-noise measurements always include the contribution of the ambient noise. The self-noise term in the sonar equation set forth in Chapter 37 is also all inclusive. Thus, sonar self-noise is really the total noise level of a sonar when there is no target present.

Two kinds of self-noise measurements are made: "platform-noise" measurements are self-noise measurements made with omnidirectional hydrophones; "sonar self-noise" measurements are made at the output of the designated array. Platform noise is L_N in the sonar equation, while sonar self-noise L_e is the equivalent of $L_N - N_{DI}$, where N_{DI} is the measured array gain. It is important to measure the sonar self-noise directly because the array signal-to-noise gain depends upon the spatial properties of the noise field. The directivity index, which is based on isotropic noise, is a first-order approximation of the ability of an array to discriminate against noise, but it usually does not equal the actual array gain. In fact, coherent noise sources can appear as target signals on sonar displays.

In treating sonar self-noise, it is convenient to consider six dominant noise sources:

1. Ambient ocean noise
2. Local machinery sources
3. Remote machinery sources
4. Propeller noise
5. Local flow noise
6. Local cavitation and/or bubble sweepdown

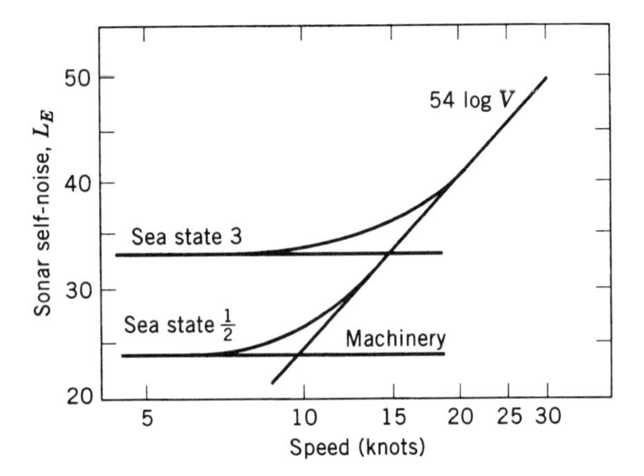

Fig. 10 Speed-dependent self-noise as a function of sea state, relative sonar noise level vs. speed, knots.[1]

Ambient noise is discussed in *Enc.* Ch. 47. It generally dominates only at slow speeds. Figure 10 illustrates how at low sea states (e.g., sea state $\frac{1}{2}$) ambient noise may control sonar self-noise only at low ship speeds before other speed-dependent noise sources begin to dominate. At sea state 3, ambient noise controls sonar self-noise up to higher ship speeds.

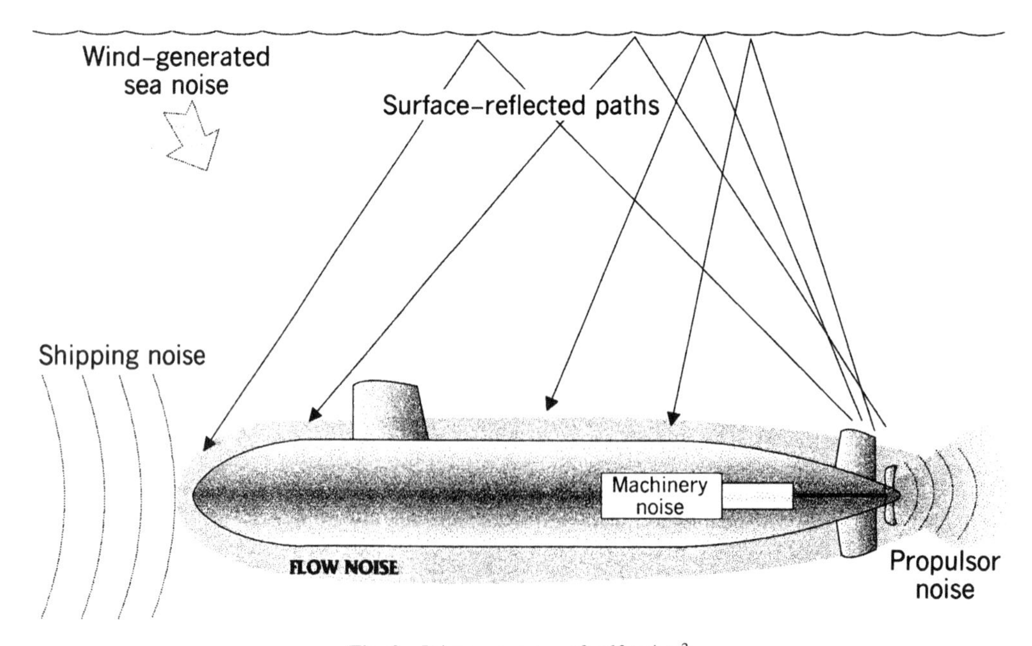

Fig. 9 Primary sources of self-noise.[3]

At the other extreme, *local cavitation* on the sonar dome is generally a problem only at the highest speeds of surface ships. In moderate to high sea states, bow wave splash is a significant noise source that extends along the line of the breaking bow wave. It is highly dependent on ship motion with respect to the seaway. Machinery can contribute noise into the sonar arrays by vibrational paths and these sources may be significant over a wide speed range. Propeller noise may be transmitted to sonar arrays by hull vibration, hull grazing (see Table 6), and by surface reflected paths. Each type of noise source has individual speed and power dependencies that contribute in varying degrees to the overall sonar self-noise level as a function of speed.

At higher speeds *local flow noise* is a dominant noise source. The mechanisms whereby turbulent boundary layer pressure fluctuations excite flush-mounted hydrophones, sonar domes, and structures local to the sonar are discussed in Chapter 34. The wavenumber of the convective turbulent excitation is much larger than that of the acoustic signal of the same frequency. It is thus possible to design hydrophones and arrays to discriminate against flow noise while maximizing signal gain.

The importance of remote machinery and propellers as contributors to self-noise is apparent when close correlations between radiated noise and sonar self-noise are recognized. The two principal paths are surface boundary reflection or scattering and hull vibration. Hull vibration is particularly important for low-frequency noise while the acoustic path surface ship sonar self-noise is often higher in shallow water than in deep water. Generally speaking, the reduction of radiated noise can also be important to improving sonar performance.

4.2 Dome Design

The term *dome* refers to a vaulted structure. It probably originated when rounded projections were first installed to protect protruding hydrophones. Today the term encompasses any structure housing arrays or hydrophones, whatever the shape, and sometimes even describes the supporting structure and array. Domes may now comprise the whole front portion of a submarine or the bulbous bow of a surface ship. They are either conformal domes (i.e., conforming to the general shape of the ship) or appendage domes (i.e., protruding into the water flow around the ship, as illustrated in Fig. 11).

The structural and acoustic factors involved in dome design and ship installations are dealt with in Ref. 12 and illustrated in Fig. 12. The acoustic factors in dome design, which are numbered in Fig. 12, are (1) transmission through material, (2) compressional and flexural coincidence angles, (3) flow excitation, (4) refraction, (5) internal reflection, (6) structure-borne noise, (7) waterborne noise, (8) reverberation, (9) cavitation, (10) bubbles, and (11) fouling.

When designing and installing sonar domes, the

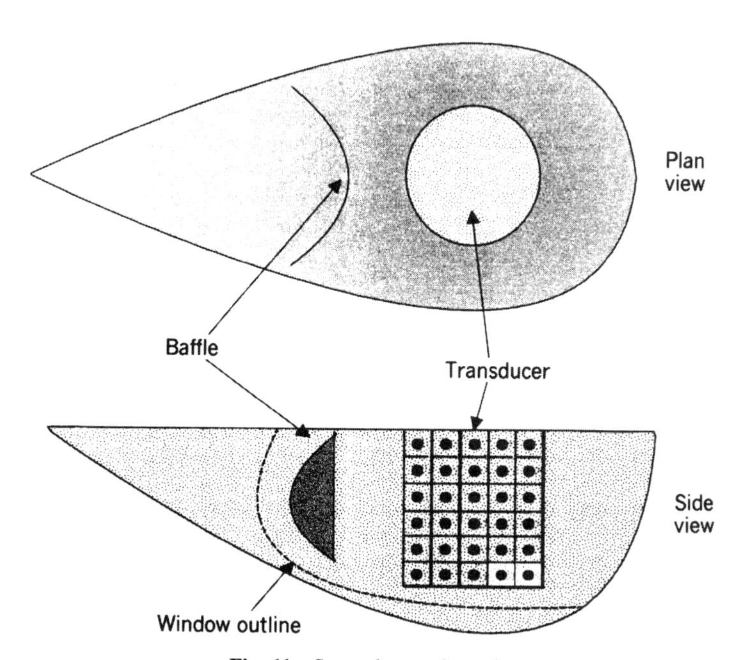

Fig. 11 Sonar dome schematic.

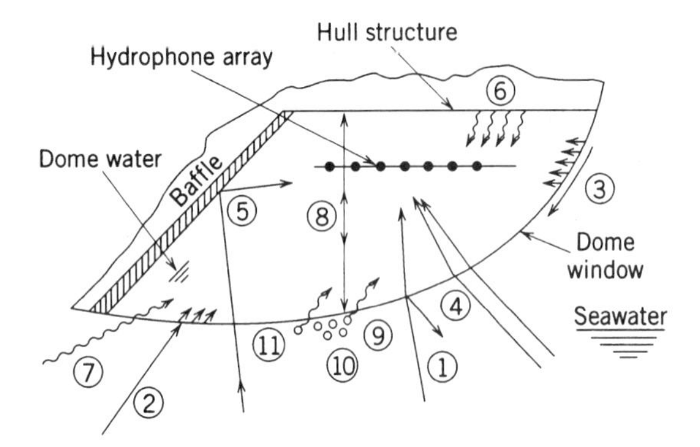

Fig. 12 Acoustic considerations in dome design.

emphasis is on protecting the tranducers, minimizing ship impact, and enhancing signal-to-noise ratio. The interposition of the dome, or window, reduces the signal level and, therefore, low-transmission-loss structures and materials are mandatory. At low frequencies, where large arrays and domes are required, the wavelengths are large and attenuation through the material is low. The thicker windows required for mechanical strength do not therefore cause unacceptable attenuation losses at these frequencies. Thus, design becomes a compromise between acoustic, hydrodynamic and structural requirements and the location, size, operating frequency, and shape of the array.

The dome or window should be made of tough, hard-to-damage material because hydroelastic excitation or flutter around a damaged portion can cause noise. It can also be excited by the flow of water past its surface and radiated to the hydrophone if it is constructed of an elastic material (e.g., steel or fiberglass). A highly damped, low-modulus material such as rubber reduces flow noise problems. But in every case, it is important to maintain both the fairness and smoothness of the dome and the adjacent hull surfaces. Step discontinuities must be avoided.

The fluid inside the dome may become warmer than the surrounding sea and, thus, may not have the same speed of sound as the external water. This results in refractions and focusing, which leads to degradation in beam former performance and gives erroneous bearings. Thus dome design should provide for flushing and replacement of the internal dome water.

The dome is usually equipped with baffles aft of the array to reduce the effect of ship machinery and propeller noise on the sonar array. However, these baffles can also mask targets in the stern sectors. Flat or curved surfaces inside the dome can reflect incoming signals, and non-reflecting surface materials may be needed to reduce the possibility of spurious signals appearing in the array output. Table 7 gives values for diffraction losses as a function of frequency for typical water and air-backed sonar baffles.

4.3 Flow-Induced Noise

Turbulence is generated along the hull of a ship as it passes through the water. The fluctuating pressures associated with a turbulent boundary layer radiate noise directly into the water (flow noise) and excite vibrations in ship structures because the mechanism of direct radiation of flow noise into the water is inefficient. Flow noise is usually not a significant source of ship noise. However, flow-induced noise transmitted through excitation and radiation of sonar domes and adjacent hull structures becomes a significant source of sonar self-noise at higher speeds. Chapters 10 and 34 provide more detailed information on hydroacoustic noise and structure–fluid interactions.

The estimates of platform self-noise given below are sound pressure levels at individual element locations within sonar domes. For flow-induced noise, platform noise levels inside sonar domes are estimated by adding the values given below to the baseline level:

$$L_N = 34 + 45\log(V) - 20\log(h) + 10\log(A) \tag{3}$$

where V is ship speed in knots, h is the thickness of the sonar dome in centimetres, and A is the surface area of the sonar dome in square metres.

Octave Band Adjustments to L_N (dB)

Center Frequency, Hz	31.5	63	125	250	500	1000	2000	4000	8000
Flow noise	39	28	19	16	13	10	7	4	1

TABLE 7 Hull-Mounted Sonar Array Baffle Diffraction versus Frequency Loss (dB)

	Diffraction Loss (dB)					
Type of Barrier/Baffle	250 Hz	500 Hz	1000 Hz	2000 Hz	4000 Hz	8000 Hz
Water backed	0	0	0	5	5	5
Air backed	8	11	14	17	20	23

TABLE 8 Noise Level Criteria for Interior Compartments for New Construction, U.S. Navy Ships

	Compartment Noise Level Criteria (dB re 20 μPa)								
Noise Category[a]	31.5 Hz	63 Hz	125 Hz	250 Hz	500 Hz	1000 Hz	2000 Hz	4000 Hz	8000 Hz
A-12	66	63	60	57	54	53	48	45	42
A-3	75	72	69	66	63	60	57	54	51
B	75	72	69	66	63	60	57	54	51
C	72	69	66	63	60	57	54	51	48
D	91	88	85	82	79	76	73	70	67
E	82	79	74	73	70	67	64	61	58

[a]Noise categories: (1) A-12, command and control centers, equivalent to large quiet offices to allow normal conversation at 6 ft; (2) A-3 and B, sonar rooms and crew's living quarters; (3) C, crew's birthing areas; (4) D, machinery rooms and working spaces; (5) D-1, engine rooms requiring ear protection; (6) E, compartments and spaces subjected to active sonar transmissions.

5 INTERIOR COMPARTMENT NOISE

5.1 General Characteristics

The airborne noise levels in ships constitute a major area of ship acoustics. The high airborne noise levels of propulsion machinery and reduction gears in confined machinery spaces constitute a serious hearing and communication problem. Living and working spaces must meet habitability criteria for the long-term health and comfort of ship's personnel. References 4, 5, and 7 provide applicable design guidelines.

The latest available (1991) Coast Guard and Navy airborne noise criteria are given by the noise category designations defined in Table 8:

- Category A: Spaces where direct speech communication must be understood with minimal error and without need for repetition. Acceptable noise level is based on a talker–listener distance of either 3 or 12 ft. Category A-3 applies when extreme talker-to-listener distance is less than 6 ft. Category A-12 applies when the extreme talker-to-listener distance is 6 ft or greater.
- Category B: Spaces where comfort of personnel in their quarters is the primary consideration and communication considerations secondary.
- Category C: Spaces where it is essential to maintain especially quiet conditions.
- Category D: High-noise-level areas where voice communication is not important, ear protection is not provided, and prevention of hearing loss is the primary consideration.
- Category E: High-noise-level areas where voice communication is at short distances and there is high vocal effort and where amplified speech and telephones are normally available.

TABLE 9 Machinery Noise: Sound Power Source Levels (dB re 10^{-12} W) Octave Band Adjustments

	Sound Power Source Levels (dB re 10^{-12} W)								
Machinery Class	31.5 Hz	63 Hz	125 Hz	250 Hz	500 Hz	1000 Hz	2000 Hz	4000 Hz	8000 Hz
				Equation (4)					
Diesel engines									
Intake	21	21	27	28	26	24	20	13	4
Exhaust	44	40	46	42	34	30	24	14	6
Casing	4	6	15	18	17	15	11	4	0
				Equation (5)					
Gas turbines									
Exhaust	22	22	22	22	22	20	16	14	4

Source: From Ref. 13.

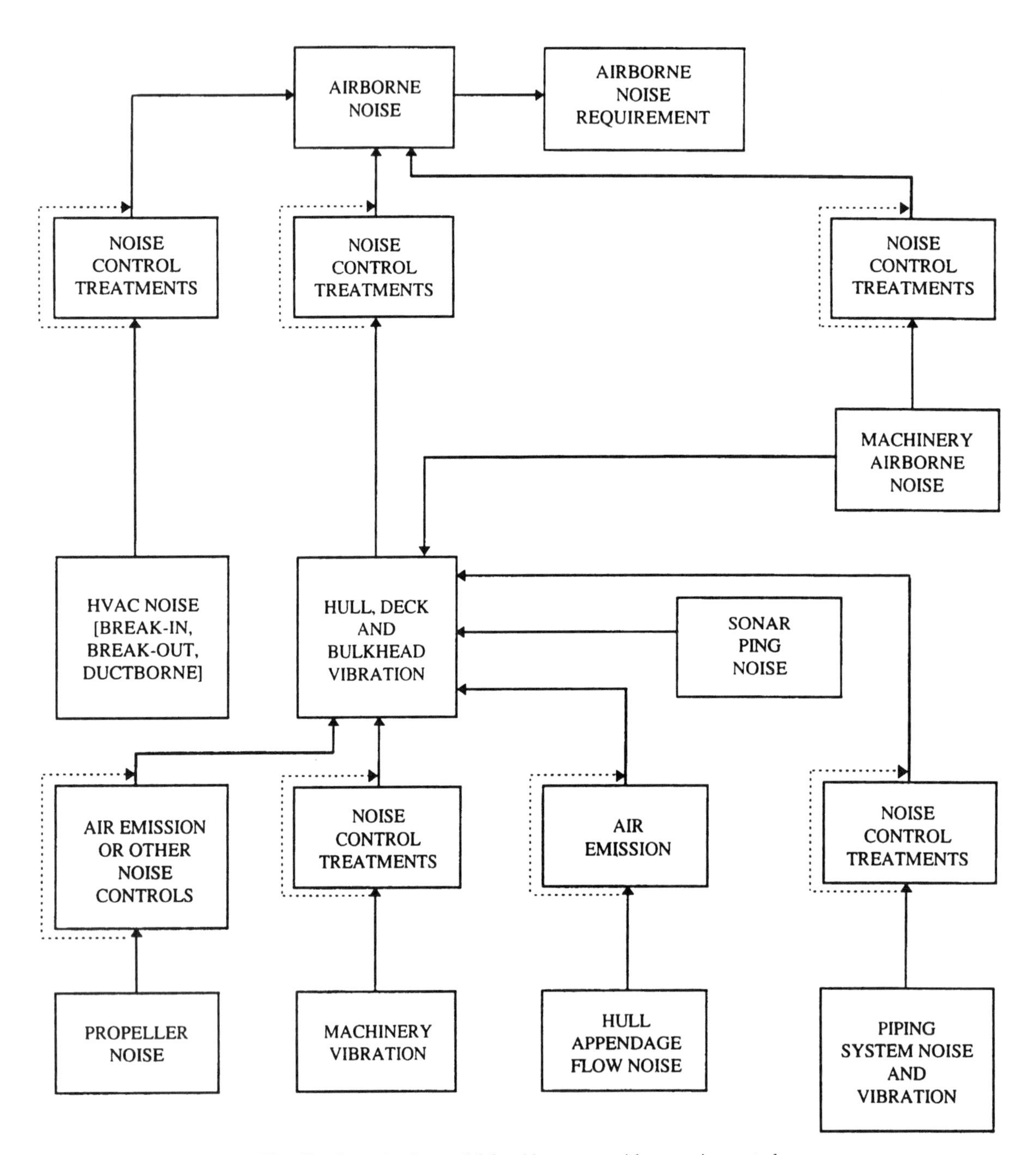

Fig. 13 General noise model for ship systems airborne noise control.

Associated with each noise category are sound pressure levels in octave bands and/or requirements on the speech interference level (SIL), where SIL is defined as the arithmetic average of the sound pressure levels in the 500-, 1000-, and 2000-Hz octave bands. For design purposes it is permissible to assign to each of these bands the value of the SIL requirement.

Additional noise categories are sometimes defined for spaces where noise levels are higher than the levels of noise category D and personnel are instructed to wear ear protectors, e.g., D-1 engine rooms.

5.2 Airborne Noise Sources

The acoustic source levels of representative machinery noise sources in ships are expressed as sound power in octave bands with adjustments given in Table 9[13]:

1. Diesel engines—intake, exhaust and casing radiation, baseline:

$$L_{\mathrm{WB}} = 58 + 10\log(\mathrm{kW})\ \mathrm{dB\ re}\ 10^{-12}\ \mathrm{W} \qquad (4)$$

2. Gas turbines, intermediate—exhaust:

$$L_{\mathrm{WB}} = 74 + 10\log(\mathrm{kW})\ \mathrm{dB\ re}\ 10^{-12}\ \mathrm{W} \qquad (5)$$

5.3 Acoustic Transmission Paths

Noise control measures in ships are similar to those in other architectural acoustic applications. An extensive and broad base of materials technology has been developed and applied by the U.S. Navy and industry to meet surface ship and submarine habitability and environmental requirements. This technology is largely available to nonmilitary ship designers. A general noise model is given in Fig. 13. In addition, Statistical Energy Analysis noise models are available to support alternative designs for noise control treatment combinations.

Examples of available noise control technology are as follows:

1. Internal high-transmission-loss "sandwich"-type composite treatments (e.g., fiberglass blanket/loaded vinyl septum/fiberglass blanket) or tuned

TABLE 10 Transmission Loss for Representative Ship Structure and Materials

Material No.	Panel Description, Type and Thickness	Transmission Loss (dB)						
		125 Hz	250 Hz	500 Hz	1000 Hz	2000 Hz	4000 Hz	STC
1	Bare aluminum bulkhead, $\frac{1}{4}$ in.	20	25	30	35	30	39	30
2	One layer, 2 in. MIL-A-23054	18	27	37	47	48	57	39
3	One layer, 2 in. MIL-A-23054 + 1 lb/ft^2 lead vinyl	14	37	51	56	59	69	38
4	Two layers, 2 in. MIL-A-23054 + 1 lb/ft^2 lead vinyl	23	44	57	65	68	75	38
5	One layer, 1 in. Mylar-faced fiberglass	17	23	33	44	45	59	35
6	One layer, 1 in. Mylar-faced fiberglass and W.R. 1 lb/ft^2 lead	21	34	49	60	63	75	40
7	Two layers, 1 in. Mylar-faced fiberglass and W.R. 1 lb/ft^2 lead	20	33	51	61	61	72	39
8	One layer, 2 in. Mylar-faced fiberglass	18	25	38	48	48	67	38
9	One layer, 2 in. Mylar-faced fiberglass and F.G.R. 1 lb/ft^2 LD	12	36	49	57	60	72	36
10	Two layers, 2 in. Mylar-faced fiberglass and F.G.R. 1 lb/ft^2 LD	17	38	54	57	60	78	41
11	Two layers, 2 in. Mylar-faced fiberglass (no septum)	16	33	47	56	55	73	40
12	One layer, 2 in. fiberglass with 14 oz/ft^2 lead face septum	14	35	45	57	58	67	38
13	Two layers, 2 in. fiberglass with 14 oz/ft^2 lead septum	17	38	52	58	55	65	41
14	Quilted blanket, two layers, 1 in. fiberglass with lead septum	14	28	40	47	48	53	37

Note: Tests conducted in accordance with ASTM E90.

TABLE 11 Coefficients of Sound Absorption for Representative Shipboard Materials

Material Type	Thickness (in.)	Coefficient of Sound Absorption					
		125 Hz	250 Hz	500 Hz	1000 Hz	2000 Hz	4000 Hz
Board	1.0	0.07	0.25	0.70	0.90	0.75	.70
2.5 ± 0.5							
lb/ft.^{2}	1.5	0.15	0.45	0.90	0.90	0.80	0.75
Navy II	2.0	0.25	0.70	0.90	0.85	0.75	0.75
	0.5	0.04	0.10	0.20	0.40	0.45	0.55
	1.0	0.06	0.20	0.45	0.65	0.65	0.65
	2.0	0.15	0.40	0.75	0.75	0.75	0.70
	3.0	0.20	0.60	0.90	0.80	0.80	0.75
	4.0	0.25	0.65	0.95	0.85	0.85	0.80
Navy III	2.0	0.43	0.96	1.0	1.0	0.70	0.35

double-wall constructions for reduction of airborne noise transmission through the hull or through partitions between compartments

2. Internal sound absorption materials for installation on bulkheads and overheads of ship compartments to reduce individual compartment noise levels
3. Vibration damping materials for application to internal ship structures to reduce transmission of structure-borne noise to interior compartments

Transmission loss data on a representative sample of commercial products that meet U.S. Navy requirements are given in Table 10.[14]

5.4 Acoustic Absorption Materials

There are two classes of acoustic materials for sound absorption applications in (1) "clean" spaces and (2) machinery and equipment spaces and a third class to supplement structural transmission loss of decks, bulkheads, and interior joiner work.

Type I treatments are simply designed to reduce the reverberant sound field by increasing the sound absorption characteristics of interior bulkheads and overheads. Typically, 2-in.-thick fiberglass blankets with perforated facings are used for this purpose. Such treatments typically provide on the order of 5 dB(A) noise reduction.

Type II treatments are similar in function to type I treatments but generally consist of a 2-in. layer of fibrous glass material faced with an impervious fabric for protection and to reduce the risk of degradation in absorption characteristics due to oil and water contamination. However, the facing material generally degrades the absorption performance of the overall treatment.

Type III treatments are designed to supplement the transmission loss afforded by baseline structural decks and bulkheads and interior joiner work. Typically, these treatments incorporate relatively high density fiberglass installed as two separate blankets separated by a thin septum of lead-loaded or barium-sulfate-loaded vinyl. The outer, exposed layer of fibrous glass material can have either a perforated facing to maximize sound absorption or a thin facing impervious to oil and water.

Table 11 provides data on ship acoustic materials.[15]

Acknowledgments

The author wishes to acknowledge the contributions of many colleagues to the database which underlies the information presented in this chapter. In particular, the selection and summarizing of this material have benefitted from discussions with Dr. Donald Ross, author of the text *Mechanics of Underwater Noise*, and Daniel Nelson, Bolt, Beranek and Newman, Inc., who originated a major part of the noise and vibration performance predictions for shipboard machinery systems.

REFERENCES

1. D. Ross, *Mechanics of Underwater Noise*, Peninsula Publishing, Los Altos, CA, 1987.
2. R. J. Urick, *Principles of Underwater Sound for Supervisors*, McGraw-Hill, New York, 1967.
3. D. Ross and R. D. Collier, *Mechanics of Underwater Noise*, Course Notes, Applied Technology Institute, Columbia, MD, 1985–94.
4. R. W. Fischer, C. B. Burroughs, and D. L. Nelson, "Design Guide for Shipboard Airborne Noise Control," *SNAME Tech. Res. Bull.*, 1983, pp. 3–37.
5. *Noise Control Handbook*, British Ship Research Association, Naval Architecture Department, February 1982.
6. S. Nilsson and N. P. Tyvand (Eds.), *Noise Sources in Ships: I Propellers, II Diesel Engines*, Nordic Cooperative

Project: Structure Borne Sound in Ships from Propellers and Diesel Engines, Nordforsk, Norway, 1981.

7. *Handbook for Shipboard Airborne Noise Control*, Bolt Beranek and Newman, Inc., Technical Publication 073-0100, U.S. Coast Guard and U.S. Naval Ship Engineering Center, February 1974.
8. *Design Handbook. Resilient Mounts*, U.S. Naval Sea Systems Command, NAVSEA 0900-LP-089-5010, 1977.
9. *Design Handbook. Distributed Isolation Material*, U.S. Naval Sea Systems Command, NAVSEA S9078-AA-HBK-010-DIM, 1982.
10. *Design Handbook. Piper Hangers*, U.S. Naval Sea Systems Command, NAVSEA S9073-A2 HBK-010, undated.
11. *Design Handbook of Vibration Damping*, Mare Island Naval Shipyard, Report No. 11-77, U.S. Naval Sea Systems Command, January 1979.
12. *Design Handbook for Sonar Installations*, U.S. Naval Underwater Systems Center, TD 6059, undated.
13. D. L. Nelson, *Main Propulsion Gas Turbine Exhaust Noise*, Bolt Beranek and Newman, Inc., TM-339, February 1977.
14. R. D. Collier, *Noise Control Materials and Application*, Bolt Beranek and Newman, Inc., Report No. 6637, 1987.
15. *Design Data Sheet. Ship Damping and Special Acoustic Materials*, Bolt Beranek and Newman, Inc., DDS-636-1, U.S. Naval Sea Systems Command, September 1980.

36

OCEAN AMBIENT NOISE

IRA DYER

1 INTRODUCTION AND DEFINITIONS

Ocean ambient noise is sound in the ocean, unwanted generally because it interferes with operation of sonar systems or other underwater sound devices. Except with use of frequency filters and/or receiving arrays, ambient noise typically cannot be controlled by system operators.

Ambient noise mechanisms are as diverse as sea surface agitation, mammalian vocalization, or distant shipping. It is usual to consider such noise as an oceanic property (in that it can be related, for the three examples cited, to breaking waves under action of wind stress, to food chains and water column parameters such as temperature, or to commercial ships that inject noise into the deep sound channel). Also since ambient noise in most cases propagates from its source to a sonar, the noise can be affected by sound propagation properties of the ocean.

Noise observed omnidirectionally is designated by its mean-square pressure $p_N^2(f)$, dependent upon frequency f. In general, the noise has a *spectral density* $S_n(f)$, so that the omnidirectional observation at f is

$$p_N^2(f) = \int_0^\infty S_n(f_0)F^2(f_0;f)\,df_0$$

$$\approx \int_{f-b/2}^{f+b/2} S_n(f_0)\,df_0 \approx S_n(f)b, \tag{1}$$

where F^2 is the mean-square response of the receiver's frequency filter (with normalization $F^2 = 1$ at its maximum) and b is the receiver bandwidth that, for the latter parts of Eq. (1), is assumed small enough to consider the noise spectral shape approximately constant within it and the filter response outside it to be essentially zero. Sonar bandwidth clearly is a spectral filter against the usually continuous and wide-frequency distribution of the noise. Note that p_N^2 is weighted by the electrical frequency filter and thus is considered an *effective* acoustical quantity.

Ambient noise is not spatially isotropic, and it is therefore necessary to consider its *spatial* or *directional* density. Hence spectral density is composed of

$$S_n(f) = \int_{4\pi} S(f,\Omega)\,d\Omega, \tag{2}$$

where the integrand is the spatial density at f and varies with the solid angle Ω. (In spherical coordinates, $d\Omega = \cos\theta\,d\theta\,d\phi$, with θ the vertical angle measured from the horizontal and ϕ the azimuthal angle.)

In log measure, the noise level L_N is given as

$$L_N = 10\log p_N^2(f) \qquad \text{dB re } 1\ \mu\text{Pa}, \tag{3}$$

and the spectrum level L_n is

$$L_n = 10\log S_n(f) \qquad \text{dB re } 1\ \mu\text{Pa and } 1\ \text{Hz}. \tag{4}$$

Also

$$L_N = L_n + 10\log b \qquad \text{dB re } 1\ \mu\text{PA} \tag{5}$$

in which b is in hertz.

When noise is observed with a receiving array, the performance of the array as a spatial filter against the generally wide spatial distribution of the noise is mea-

Note: References to chapters appearing only in the *Encyclopedia* are preceded by *Enc.*

sured by the array gain AG:

$$\mathrm{AG} = 10 \log \left[\frac{S_n(f)}{\int_{4\pi} S(f,\Omega) B^2(f,\Omega)\, d\Omega} \right] \quad \mathrm{dB}. \quad (6)$$

Here, B^2 is the array's mean-square beam pattern[1,2] (with normalization $B^2 = 1$ at the maximum of the array's main lobe) and is determined by the amplitude and phase response of the array's transducers, as distributed in the array. The equivalent (postarray) noise level L_{N_e}, or equivalent spectrum level L_{n_e}, is Eq. (3) or (4), respectively, less Eq. (6). (Alternatively, AG can be defined in terms of the noise spatial covariance[2]; both formulations will be used subsequently in this chapter.) The denominator of the argument of Eq. (6),

$$S_{n_e}(f) = \int_{4\pi} S(f,\Omega) B^2(f,\Omega) d\Omega, \quad (7)$$

is the effective (postarray) noise spectral density that, when compared with the in-water (prearray) density $S_n(f)$, determines AG.* It is clear that one needs both the spatial density $S(f,\Omega)$ and the mean-square beam pattern $B^2(f,\Omega)$ to determine spatial filtering of the noise by a receiving array. Good spatial filtering ultimately requires the array side lobes to have low sensitivity in the direction of the noise maximum. Note that if the array main lobe is directed toward the noise maximum, little or no spatial filtering occurs.

By comparison with Eq. (1), B^2 in the spatial domain is the analog of F^2 in the frequency domain. An approximation to Eq. (6) [or Eq. (7)] analogous to that of Eq. (1) can be reached with use of the array's main-lobe beamwidth Ω_e. But the side lobes of an array typically are much more energetic than the out-of-band response of a frequency filter. Thus the spatial maximum of the noise, when received via the side lobes, can be quite important in setting the total noise entering the sonar. An approximation such as in Eq. (1) is, therefore, not as useful, and the integral in Eq. (6) [or Eq. (7)] usually has to be evaluated.

Finally, ambient noise may be considered, in the time domain, as a statistically random process. In addition to frequency spectrum and spatial directionality, sonar performance depends directly on the statistical properties of ambient noise.[2] Its *probability* density, in general, is well established.[2] But its statistical stationarity, also crucial in determining performance,[2] is, in general, less so. Beyond ocean conditions, stationarity is affected by sonar design and operational parameters (bandwidth b, beamwidth Ω_e, sonar speed, etc.), which cannot be addressed in general in a chapter on ambient noise. Nevertheless, this chapter provides some information on stationarity, intended largely to direct the reader to its fuller treatment.

*Formulation of Eqs. (6) and (7) is based on the assumption of many incoherent noise components in bandwidth b and beamwidth Ω_e, a condition usually met in practice.

2 SPECTRAL DENSITY ESTIMATES

Some years ago Wenz published a compendium of ambient noise spectral levels for the open ocean,[2] shown in Fig. 1. Remarkably it has withstood the test of time and serves as the best single survey of such noise. He distinguished between prevailing and intermittent (or local) noise mechanisms, the former ever-present spatially and temporally, the latter observed sporadically. The prevailing noises are considered the more important and are described in Sections 2.1–2.4 in terms of their spectral density $S_n(f)$.

2.1 Turbulence-Related Pseudosound

At the lowest frequencies, turbulence in or graininess of the seawater can induce pressure fluctuations at a hydrophone. This noise is strictly pseudosound rather than a result of propagating sound waves. Its spectral level will depend on the turbulence or grain wavenumber spectrum as well as on the hydrophone size, shape, and motion. No general estimate can readily be given, since these details vary greatly from application to application. But this prevailing noise mechanism is rarely of practical importance, at least because most sonars operate at $f \geq 10$ Hz, for which noise from other sources is almost always more important.

2.2 Distant Shipping Noise

Noise radiated by a surface ship is distributed broadly in frequency,[4] but when propagated to long ranges, it is shaped by the low-pass filter of the ocean[5] to a peak in the range $10 \leq f \leq 10^2$ Hz. Distant shipping in most cases is the dominant noise in this frequency range. Busy shipping areas or lanes have $O(10^3)$ ships within them at any one time,[6] so that ship traffic noise is sensed as a virtually continuous rather than a discrete spectrum.

Wenz's summary for distant traffic noise in the deep ocean is roughly appropriate to all sensor depths except those near the ocean bottom, in which as much as 25 dB less can be observed below the critical depth[7] near the bottom. (At such depths, there are no strong long-range deep-channel paths, because sound speeds are higher there than those near the surface, where the noise is created.)

Distant shipping noise prevails over most of the ocean. Exceptions include the ice-covered Arctic Ocean and some seas sheltered from ocean traffic (provided their own traffic is very light).

2.3 Surface Agitation Noise

At still higher frequencies (about $10^2 \leq f \leq 10^4$ Hz) formation and oscillation of subsurface air bubbles created by breaking waves cause noise that dominates the spectrum. Only recently has this mechanism come under detailed study,[8–11] but it is highly probable that the essential ideas are now well in hand. Sea surface bubble noise is parameterized in Fig. 1 by wind force, and this remains

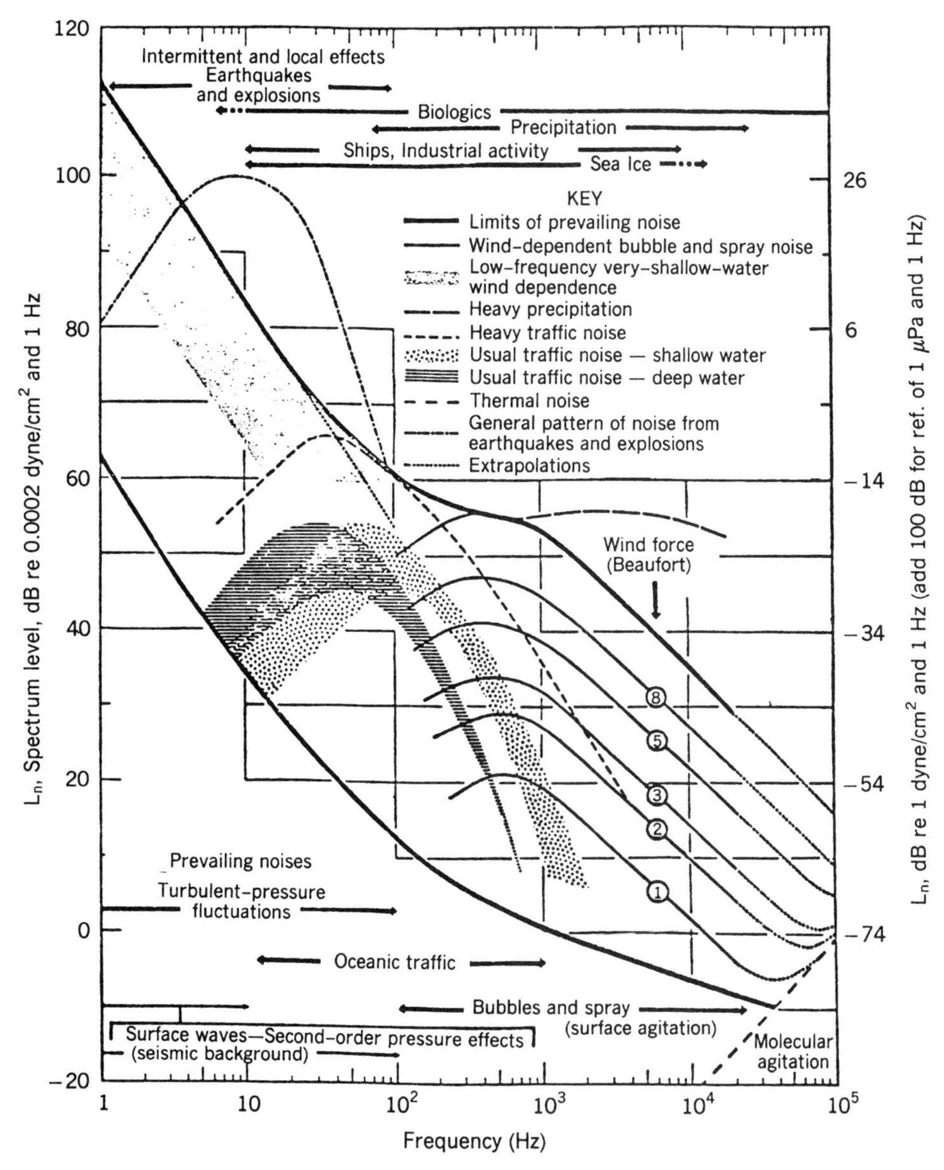

Fig. 1 Wenz's ambient noise spectra.[3] (Add 100 dB to the right-hand scale to obtain L_n in dB re 1 μPa and 1 Hz.)

the most practical way to describe noise generated by bubbles from breaking waves.[12,13] But it is also so that more fundamental parameters, such as wind stress and wave age, may ultimately succeed wind force for estimating the spectrum more precisely.

As stated, Wenz's summary of noise generated bubbles from breaking waves is for the deep open ocean. In shallow water no fully systematic estimates for bubble-caused noise are at hand, at least because shallow-water environments vary greatly in depth, sound speed profile, and bottom loss.[14] An estimate for a wide range of wind force, however, is in Fig. 2 and is useful for gaining at least a rough idea of such noise without detailed knowledge of the shallow-water environment.

As concluded in Section 2.2, noise from distant traffic typically dominates at low frequencies ($10 \le f \le 10^2$ Hz). But there are ocean areas that either are shielded or are remote from distant shipping. In such areas a wind-dependent noise component at low frequencies is observed instead;[12] see Fig. 3. The wind-related mechanism has not been firmly identified for $10 \le f \le 10^2$ Hz (but many possibilities have been suggested[13]).

2.4 Molecular Agitation Noise

At high frequencies ($f \ge 10^2$ kHz or so, dependent upon the wind agitation crossover), ambient noise is dominated by noise from molecular agitation. See Fig. 1. Dynamic forces from molecular momentum reversals at a pressure sensor cause this noise.[15] Molecules have kinetic energy set by the absolute temperature, but because absolute temperature varies only slightly even for the widest variations in ocean conditions, the single line in Fig. 1 can be considered broadly applicable and can be extrapolated as f^2 for $f \ge 10^2$ kHz. This noise is not strictly sound; its local molecular origin will be of importance later when its covariance properties are estimated.

2.5 Intermittent or Local Noise

In this chapter, I exclude details of several intermittent or local ambient noise sources, but they can be important and thus are included by reference. Earthquakes[16–18] and precipitation[19–21] have recently been studied as sources of noise and can dominate their respective spectral ranges for the duration of such events. Earthquakes cause noise below about 10 Hz, via induced motion of the bottom above the epicenter, which then creates noise in the sound channel by scattering from nearby rough boundaries. Because of its low frequencies, earthquake noise rarely is an issue in sonar performance. Precipitation noise, however, is usually observed for $f > 10$ kHz and can affect sonars operating in that frequency regime, of which there is a wide variety of designs for many different applications. Such noise is created by direct forcing

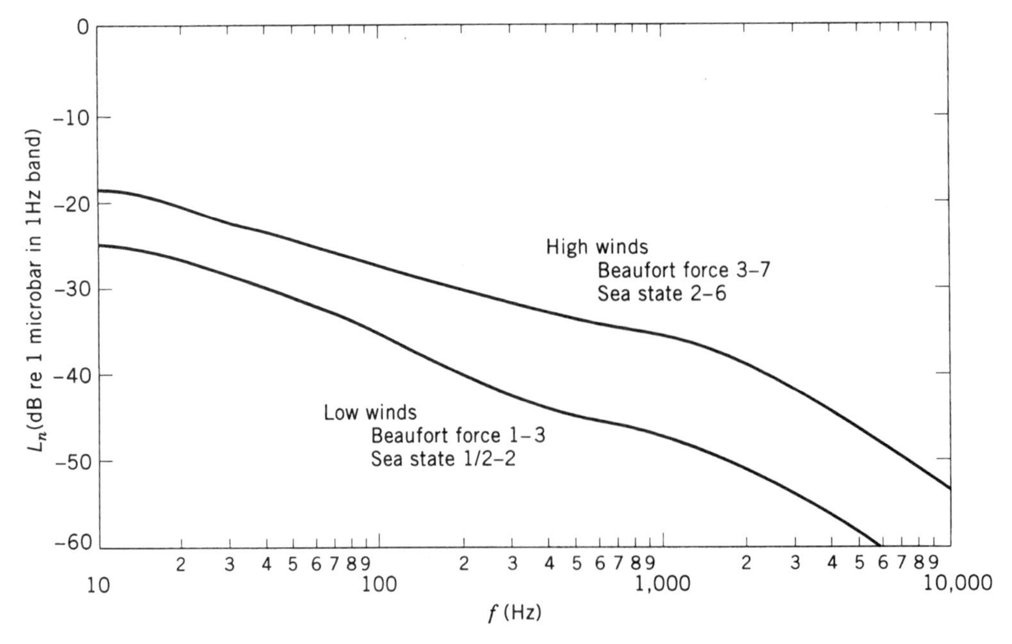

Fig. 2 Wind-driven sea surface noise in shallow water. (Add 100 dB to obtain L_n in dB re 1 μPa and 1 Hz.)

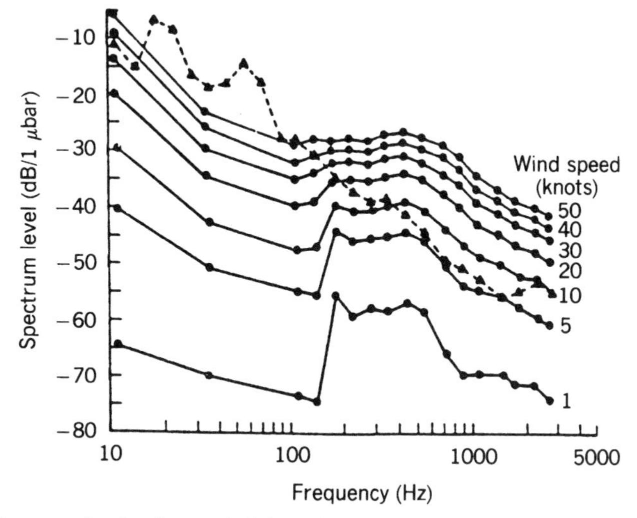

Fig. 3 Spectrum levels of non-wind-dependent (▲) and wind-dependent (●) portions of ambient noise, extracted from measured levels.[11] (Add 100 dB to obtain L_n in dB re 1 μPa and 1 Hz.)

of the water surface as the precipitates strike and by wake cavity formation and oscillation as the precipitates drive down into the water.

Noise near and under the pack ice of the Arctic also has been recently studied. It is dominated by various ice fracture processes,[22–26] is pervasive in the sonar frequency range, and is always present to a varying degree. I think of it as local or at least special in character, however, and exclude its details.

Noise from marine life is present to varying degrees

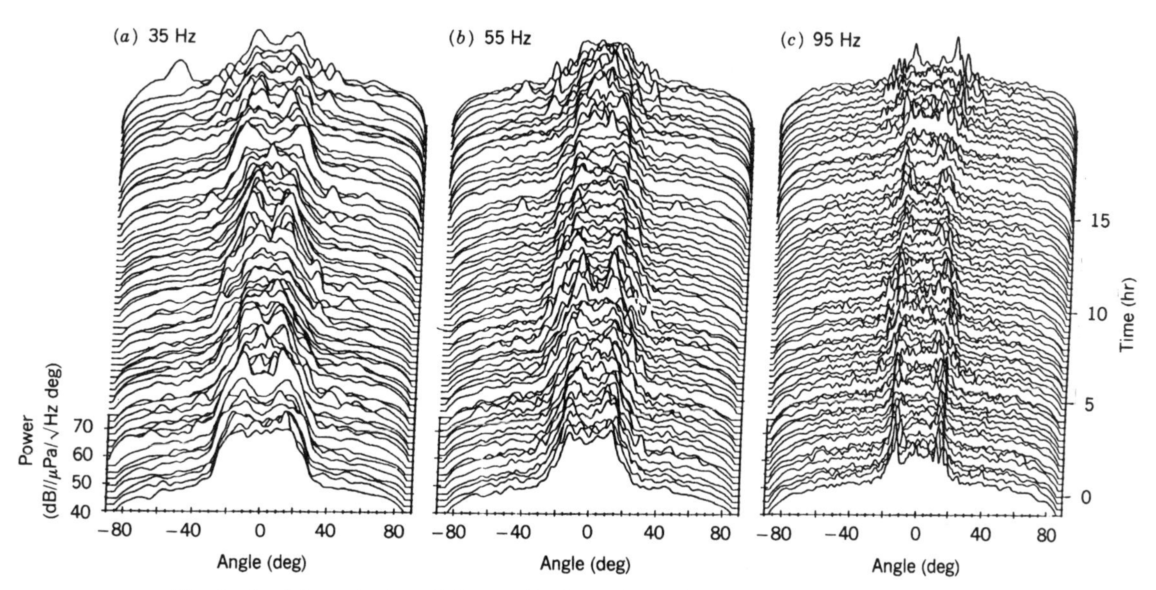

Fig. 4 Spatial distribution of distant shipping noise in deep water as a function of f, θ, and time.[28] Frequencies f are (*a*) 35, (*b*) 55, and (*c*) 95 Hz. Data were averaged in 0.5-Hz bandwidths. Time corresponds to wind increasing in speed from 2 to 12 m/s. The array was at a depth of 850 m, and the spatial density levels are in dB re 1 μPa, 1 Hz, and 1°.

in specific ocean areas of appropriate nutritional content. Marine life can create noise throughout the frequency range of interest, but its presence is temporally and spatially sporadic. While not all species create sound, enough do to provide even the occasional sonar operator with opportunities to marvel at its richness and to practice patience as its passage is awaited. Again the inclusion of such noise is by reference only.[27]

3 SPATIAL DENSITY ESTIMATES

3.1 Directionality of Distant Shipping Noise

In the deep ocean the distribution of distant shipping noise in vertical angle is closely symmetric about $\theta = 0$ (the horizontal) and has at most weak frequency and depth dependence. See Figs. 4 and 5.[28] Data in these figures show an angular "pedestal" standing 18 dB or so above a broad "platform" in vertical angle. At the lower and higher frequencies the pedestal width θ values are $\approx \pm 20°$ and $\approx \pm 15°$, respectively. These widths correspond to the sound field contained within the deep sound channel. The pedestal is more rounded at the lower frequencies, more peaked at the edges for the higher frequencies, the difference likely due at the lower frequencies to more intense scattering associated with distant ocean boundaries.[29] The platform level is wind speed dependent (an effect consistent with Fig. 3), and at the low wind speeds corresponding to distant shipping dominance, it is roughly estimated to be below the pedestal peak by $-18 + 15 \log|\cos\theta|$ decibels.

Synoptic measurements of the distribution of distant shipping noise in horizontal (azimuthal) angle ϕ are not available, but within a large shipping area one can surmise it to be approximately uniform in 2π, and well away from a shipping lane, to be roughly uniform in π.

Systematic data for the spatial distribution of shipping

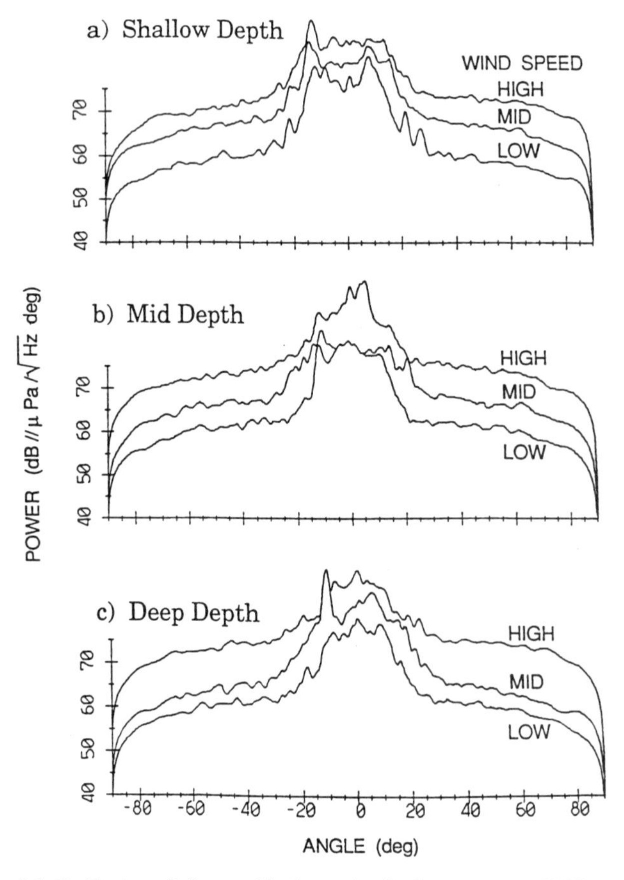

Fig. 5 Spatial distribution of distant shipping noise in deep water at 75 Hz versus depth (170, 850, and 2650 m) and wind speed (3, 7, and 11 m/s).[28]

noise in shallow water are also not available. Because propagation in shallow water is quite variable, each case would have to be estimated from knowledge of shipping activity and of sound speed profile and other ocean parameters. But it should be symmetric in θ, with a pedestal width determined by the low-order modes (which usually have little or no excess attenuation).

3.2 Directionality of Surface Bubble Noise

An estimate of the density in vertical angle θ for surface bubble noise is in Fig. 6, based on measurements for $0 \leq \theta \leq 90^\circ$ (the up-looking quadrant).[30] The data were acquired near the deep bottom, apparently below the critical depth. The density maximizes at $\theta = 90^\circ$ (vertically up), with a 3-dB-down width of about $35^\circ \pm 5^\circ$. Air bubble dynamics, the physical mechanism for surface noise, has dipolar radiation.[8] For a vertical dipole the directivity function is proportional to $\sin^2 \alpha$, where α is the vertical angle at the source measured from the horizontal. In the absence of refraction (and sea surface tilt), this transforms to a proportionality of $\sin \theta$ for the *spatial density* in solid angle Ω.* Thus the 3-dB-down width of the density would be expected to be 30°, reasonably close to the observed value of about 35°.

Similarly, in the nonrefracting limit, the density for a dipolar surface distribution near $\theta = 0^\circ$ (measured at the receiver) would have a deep minimum as a consequence of the deep minimum near $\alpha = 0^\circ$ (at the source). But refraction must be accounted for at such small angles. In cases of sound speed at depth larger than that at the surface, refraction has the effect that radiation is observed at depth only for $|\alpha| > 0^\circ$, based on Snell's law.[31] This accounts for the absence of a deep minimum in Fig. 6. For example, an *excess* of 1% of sound speed at depth over the surface sound speed would give a trough of only -9 dB at $\theta = 0^\circ$, rather than a deep minimum.

In cases of sound speed at observation depth smaller than that at the surface, the effect, while different from the foregoing one, is equally important. For example, a *decrement* of 1% of sound speed at depth would make the minimum centered on $\theta = 0^\circ$ wider by about 16°.

As shown in Fig. 6, there is evidence that a broad secondary maximum occurs around $\theta = -50^\circ$ in the downward-looking quadrant, a result of noise scattered from the ocean bottom.[32] Of course a secondary maximum cannot be observed too close to the bottom but is likely to be observed at midwater or higher depths.

The frequency dependence of the density shown in Fig. 6 is typically weak. Noise in the upper quadrant can become somewhat more directive (3-dB-down angle smaller than 35°) at high frequencies ($f > 10$ kHz) and high wind speeds (> 15 m/s). This is so largely because a quasistatic bubble layer exists just below the sea surface, the quiescent debris of bubbles created by earlier breaking waves. The layer acts to increase the attenuation for high-frequency and low-angle-noise radiation paths.[33]

*This assumes a uniform surface distribution of dipoles, takes account of inverse-square spreading, and is for a spatial density in constant increments of Ω.

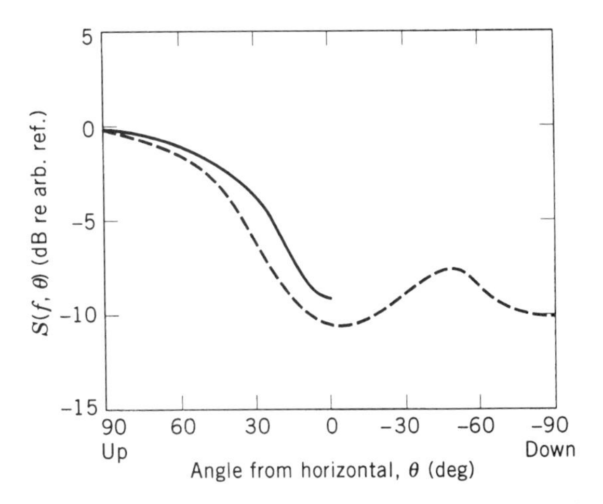

Fig. 6 Estimated spatial distribution of sea surface wind agitation noise in the deep ocean. [The arbitrary reference may be converted to an absolute one by setting the integral of $S(f, \Omega)$ from the figure to $S_n(f)$, as in Eq. (2).] The solid curve is based on data acquired deep in the water column for $200 < f < 1400$ Hz,[30] and the dash curve on data at a depth of 175 m for $f = 1000$ Hz.[31] For simplicity, both have been smoothed by the present author.

There is some indication from at-sea data that the spatial density is not azimuthally uniform, exhibiting small maxima for ϕ normal to the wave crests and shallow minima for ϕ parallel to them.[34]

Spatial density data in shallow water are not generally available, but measurements in a sloping basin of 500 m depth yield density estimates about the same as the foregoing.[35] The 3-dB-down angle, however, is somewhat broader (about 40° rather than 35°). The additional width of about 5° is perhaps within the uncertainty of the deep-water case or perhaps caused by more intense scattering from the nearby sloping bottom.

3.3 Postarray Noise from Molecular Motion

Noise from molecular agitation cannot be described by a spatial density originating from propagating sound waves.[15] But forces from the molecular motion acting on the sensor are spatially and temporally uncorrelated, the molecular scales for all practical purposes giving rise to a delta function space–time correlation. Consequently the postarray spectral density is precisely

$$S_{n_e}(f) = S_n(f), \tag{8}$$

where $S_n(f)$ is the quantity shown in Fig. 1 for molecular agitation. The postarray value is independent of array size and design, simply because the spectral density grows as $\sigma S_n(f)$, where σ is the total surface area of the sensor (or array) but the sensor (or array) has a gain of $1/\sigma$ against the noise, which is uncorrelated over it.[2]

4 ESTIMATES OF STATISTICAL STATIONARITY

4.1 Stationarity of Distant Shipping Noise

Distant shipping noise is at times statistically stationary, at times not. Its stationarity is affected by the bandwidth b and beamwidth Ω_e forming the noise time series, that is, by the number of distant ships at various ranges sensed simultaneously. Some examples help to provide an image of its wide variation. For omnidirectional data in a one-third-octave band centered on 63 Hz and for short observation periods, less than about 1 h, it is approximately statistically stationary. For longer periods, between about 4 and 24 h, it is nonstationary.[3] For very long observation periods, between about 1 and 10^2 days, it once again can be considered approximately stationary.[3] (One may surmise its return to a nonstationary process for periods longer than about 10^2 days but less than several years.) The major reason for these stationarity shifts is variability in long-range propagation, as affected by temporal fluctuation scales of the ocean.

Still other data show that distant shipping noise is stationary for observation periods less than about $\frac{1}{3}$ h when sensed through $b \approx 0.13$ Hz at 75 Hz, also omnidirectionally.[36] When sensed omnidirectionally in the horizontal but directionally in the vertical, in one-third-octave bands at several frequencies from 23 to 100 Hz, the noise is stationary for about 1 h.[37] These data indicate at least some of the sensitivity of random-process stationarity to observation conditions. Other factors are also important, such as wind-driven seas that can affect scattering in long-range propagation paths.

4.2 Stationarity of Surface Bubble Noise

Ambient noise due to bubbles can be taken as stationary, one might surmise, for observation periods less than the wind persistence time, typically about 4 h. Omnidirectional data, however, at several frequencies from 0.1 to 1.6 kHz in one-third-octave bands show it to be stationary for periods of at most $\frac{1}{20}$ h![38] Further, directional data in the octaves 0.8–1.6 and 1.6–3.2 kHz show stationarity for periods < 1 s in an upward-looking beam and < 10 s in a horizontally looking beam.[39,40]

Apparently, bubble creation by each breaking wave is temporally discrete (<1 s), so that an upward beam senses noise beyond it as a nonstationary eventlike process. A horizontal beam, in contrast, simultaneously senses a much larger ocean surface area and hence a superposition of many events; the typical wave period of about 10 s then sets the transition to nonstationarity.

Because its source is more local than that of distant shipping noise, one might also surmise that surface bubble noise does not get imprinted significantly with the fluctuation scales of long-range propagation paths.

4.3 Stationarity of Molecular Noise

Molecular noise in the ocean is a statistically stationary random process for indefinitely long observation periods.

REFERENCES

1. I. Dyer, *Fundamentals and Applications of Ocean Acoustics*, Cambridge University Press, New York, 1998 (in preparation).
2. J. E. Barger, "Sonar Systems," Chap. 37.
3. G. M. Wenz, "Acoustic Ambient Noise in the Ocean: Spectra and Sources," *J. Acoust. Soc. Am.*, Vol. 34, 1962, pp. 1936–1956.
4. R. J. Collier, "Ship Noise," Chap. 35.
5. F. H. Fisher and P. F. Worcester, "Essential Oceanography," Chap. 35, *Encyclopedia of Acoustics*, Wiley, New York, 1998.
6. I. Dyer, "Statistics of Distant Shipping Noise," *J. Acoust. Soc. Am.*, Vol. 53, 1973, pp. 564–570.
7. J. A. Shooter, T. E. Demary, and A. F. Wittenborn, "Depth Dependence of Noise Resulting from Ship Traffic and Wind," *J. Oceanic Eng.*, Vol. 15, 1990, pp. 292–298.
8. H. Medwin and M. M. Beaky, "Bubble Sources of the Knudsen Sea Noise Spectra," *J. Acoust. Soc. Am.*, Vol. 86, 1989, pp. 1124–1130.
9. A. Prosperetti, "Bubble Related Ambient Noise in the Ocean," *J. Acoust. Soc. Am.*, Vol. 84, 1988, pp. 1042–1054.
10. G. E. Updegraff and V. C. Anderson, "Bubble Noise and Wavelet Spills Recorded 1 m below the Ocean Surface," *J. Acoust. Soc. Am.*, Vol. 89, 1991, pp. 2264–2279.
11. H. C. Pumphrey and J. E. Ffowcs Willaims, "Bubbles as Sources of Ambient Noise," *J. Oceanic Eng.*, Vol. 15, 1990, pp. 268–274.
12. W. W. Crouch and P. J. Burt, "The Logarithmic Dependence of Surface-Generated Ambient Sea-Noise Spectrum Level on Wind Speed," *J. Acoust. Soc. Am.*, Vol. 51, 1972, pp. 1066–1072.
13. D. J. Kewley, D. G. Browning, and W. M. Carey, "Low-Frequency Wind-Generated Ambient Noise Source Levels," *J. Acoust. Soc. Am.*, Vol. 88, 1990, pp. 1894–1902.

14. F. Ingenito and S. N. Wolf, "Site Dependence of Wind-Dominated Ambient Noise in Shallow Water," *J. Acoust. Soc. Am.*, Vol. 85, 1989, pp. 141–145.
15. R. H. Mellen, "The Thermal Noise Limit in the Detection of Underwater Acoustic Signals," *J. Acoust. Soc. Am.*, Vol. 24, 1952, pp. 478–480.
16. R. H. Johnson, R. A. Norris, and F. K. Duennebier, "Abyssally Generated T-Phases," in L. Knopoff, C. L. Drake, and P. J. Hart, (Eds.), *The Crust and Upper Mantle of the Pacific Area*, Geophys. Mono. No. 12, Am. Geophys. Union, Washington D.C., 1968, pp. 70–78.
17. R. E. Keenan and I. Dyer, "Noise from Arctic Ocean earthquakes," *J. Acoust. Soc. Am.*, Vol. 75, 1984, pp. 819–825.
18. R. E. Keenan and L. R. L. Merriam, "Arctic Abyssal T Phases: Coupling Seismic Energy to the Ocean Sound Channel via Under-ice Scattering," *J. Acoust. Soc. Am.*, Vol. 89, 1991, pp. 1128–1133.
19. J. A. Scringer, D. J. Evans, and W. Yee, "Underwater Noise Due to Rain: Open Ocean Measurements," *J. Acoust. Soc. Am.*, Vol. 85, 1989, pp. 726–731.
20. H. C. Pumphrey, L. A. Crum, and L. Bjorno, "Underwater Sound Produced by Individual Drop Impacts and Rainfall," *J. Acoust. Soc. Am.*, Vol. 85, 1989, pp. 1518–1526.
21. D. G. Browning, D. G. Williams, and V. Sadowski, "Revised Standard Precipitation Noise Curves for Sonar Performance Modeling," NUWC-NL Tech. Memo. No. 931116, Naval Undersea Warfare Center, Division Newport, New London Detachment, New London, CT, 1993.
22. N. C. Makris and I. Dyer, "Environmental Correlates of Pack Ice Noise," *J. Acoust. Soc. Am.*, Vol. 79, 1986, pp. 1434–1440.
23. T. C. Yang, C. W. Votaw, G. R. Giellis, and O. I. Diachok, "Acoustic Properties of Ice Edge Noise in the Greenland Sea," *J. Acoust. Soc. Am.*, Vol. 82, 1987, pp. 1034–1038.
24. J. K. Lewis and W. W. Denner, "Higher Frequency Ambient Noise in the Arctic Ocean," *J. Acoust. Soc. Am.*, Vol. 84, 1988, pp. 1444–14552.
25. Y. Xie, and D. M. Farmer, "Acoustical Radiation from Thermally Stressed Sea Ice," *J. Acoust. Soc. Am.*, Vol. 89, 1991, pp. 2215–2231.
26. N. C. Makris and I. Dyer, "Environmental Correlates of Arctic Ice-Edge Noise," *J. Acoust. Soc. Am.*, Vol. 90, 1991, pp. 3288–3298.
27. R. J. Urick, *Ambient Noise in the Sea*, Peninsula Publishing, Los Altos, CA, 1986, pp. 7.1–7.19.
28. B. J. Sotirin and W. S. Hodgkiss, "Fine-Scale Measurements of the Vertical Ambient Noise Field," *J. Acoust. Soc. Am.*, Vol. 87, 1990, pp. 2052–2063.
29. W. M. Carey, R. B. Evans, J. A. Davis, and G. Botseas, "Deep-Ocean Vertical Noise Directionality," *J. Oceanic Eng.*, Vol. 15, 1990, pp. 324–334.
30. E. H. Axelrod, B. A. Schoomer, and W. A. Von Winkle, "Vertical Directionality of Ambient Noise in the Deep Ocean at a Site Near Bermuda," *J. Acoust. Soc. Am.*, Vol. 37, 1965, pp. 77–83.
31. W. A. Kuperman, "Propagation of Sound in the Ocean," Chap. 31.
32. B. A. Becken, "Sonar," in Ven Te Chow (Ed.), *Advances in Hydroscience*, Vol. 1, Academic Press, New York, 1964, p. 1.
33. D. M. Farmer and D. D. Lemon, "The Influence of Bubbles on Ambient Noise in the Ocean at High Wind Speeds," *J. Phys. Ocean.*, Vol. 14, 1984, pp. 1762–1777.
34. W. F. Hunter, "An Introduction to Acoustic Exploration," in R. W. B. Stephens (Ed.), *Underwater Acoustics*, Wiley-Interscience, London, 1970, pp. 91–127.
35. R. M. Kennedy and T. V. Goodnow, "Measuring the Vertical Directional Spectra Caused by Sea Surface Sound," *J. Oceanic Eng.*, Vol. 15, 1990, pp. 299–310.
36. W. J. Jobst and S. L. Adams, "Statistical Analysis of Ambient Noise," *J. Acoust. Soc. Am.*, Vol. 62, 1977, pp. 63–71.
37. V. C. Anderson, "Envelope Spectra for Signals and Noise in Vertically Directional Beams," *J. Acoust. Soc. Am.*, Vol. 65, 1979, pp. 1480–1487.
38. T. Arase and E. M. Arase, "Deep-Sea Ambient-Noise Statistics," *J. Acoust. Soc. Am.*, Vol. 44, 1968, pp. 1679–1684.
39. W. S. Hodgkiss and V. C. Anderson, "Detection of Sinusoids in Ocean Acoustic Background Noise," *J. Acoust. Soc. Am.*, Vol. 67, 1980, pp. 214–219.
40. V. C. Anderson, "Nonstationary and Nonuniform Oceanic Background in a High-Gain Acoustic Array," *J. Acoust. Soc. Am.*, Vol. 67, 1980, pp. 1170–1179.

37

SONAR SYSTEMS

JAMES E. BARGER

1 INTRODUCTION

The principal application of underwater acoustics is to sonar, the acoustical analog of radar. The variation among sonar systems is very great; long-range detection sonars may operate at frequencies lower than 50 Hz, while mine detection sonars may operate at frequencies higher than 50 kHz, a relative frequency range greater than three decades. The purpose of most sonar systems is to detect and localize a particular target: submarines, mines, fish, the ocean floor, surface ships. All of these systems have a common architecture, and this chapter deals with their common components. Another class of sonar systems are designed to measure some particular quantity: depth of the ocean, speed of a ship, or speed of ocean current. A final class seeks to image remote objects: side-scan sonars and sub-bottom profilers are examples.

The other chapters in this part deal with the individual physical processes that act on the performance of a sonar system. This chapter describes how the measures of these physical processes are combined to analyze the performance of sonar systems. The generalized active and passive sonar systems are separated into their components. First described are the functions of each component, then analytical performance models, and finally, system performance prediction equations.

Note: References to chapters appearing only in the *Encyclopedia* are preceded by *Enc.*

2 FUNCTIONS OF SONAR SYSTEM COMPONENTS

2.1 Generalized Active Sonar System

A schematic diagram of the generalized active sonar system is shown in Fig. 1, where the principal system components and their interconnections are shown. This section describes the function of each component in the order of signal flow.

The four system components shown in Fig. 1 that form the signal transmission portion represent the majority of sonar systems that use electroacoustic transducers as their sound source. These transducers are functionally linear, radiating sound pressure waveforms that are proportional to the electrical waveforms impressed upon them. Linear transducers are described in *Enc.* Ch. 52 and nonlinear transducers that are functionally linear are described in *Enc.* Ch. 53. The waveform generator initiates the sonar cycle by producing a pulse that can be a combination of up to four different types of waveforms. The first type of waveform is a constant-amplitude sinusoidal pulse that is useful to detect moving targets, because the Doppler frequency shift impressed by the target upon the reflected pulse enables the background of reverberation from fixed objects to be filtered out, unmasking the target echo. Moreover, the range rate of the target can be measured accurately from the magnitude of the Doppler frequency shift. The second type of waveform is a frequency shift pulse having sudden discrete changes in otherwise constant sinusoidal frequency used both to obtain the benefits of narrow-band

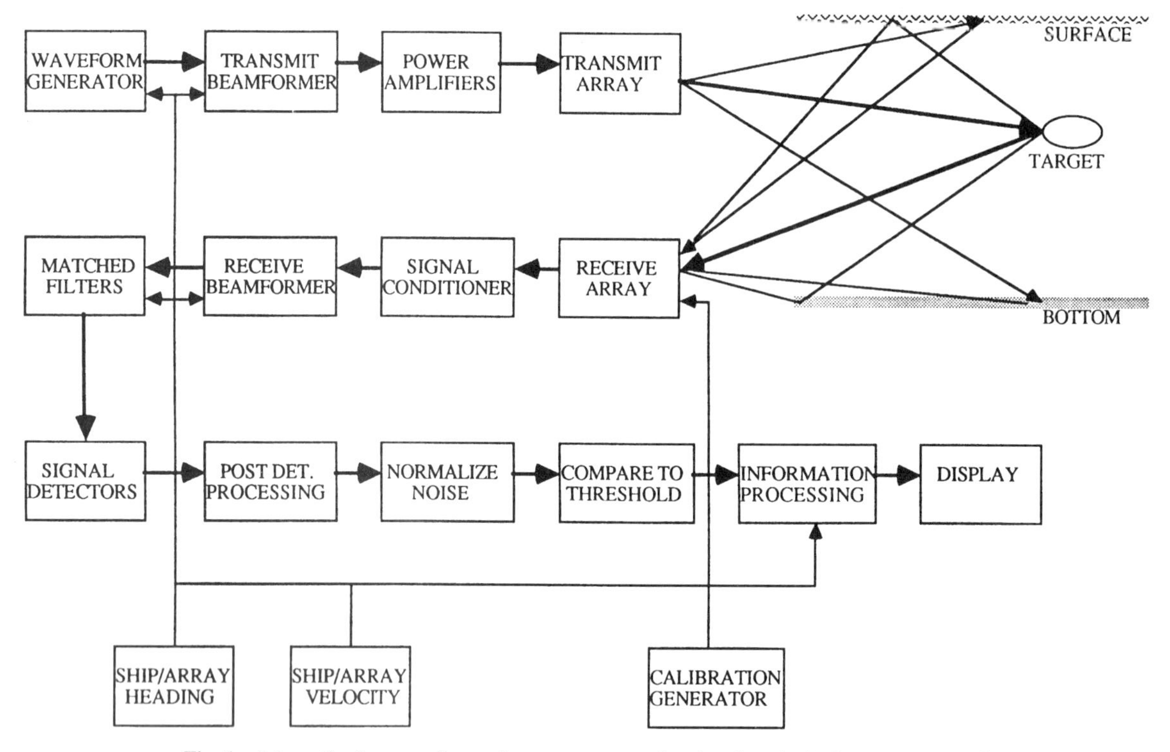

Fig. 1 Schematic diagram of an active sonar system showing the principal components and their interconnections.

pulses and to combat the propagation channel fading that inhibits sound propagation within bands at unpredictable frequencies. The third type of waveform is one having frequency modulation by one of several schemes. The increased bandwidth of these waveforms provides greater range resolution than from unmodulated narrowband pulses. The benefits are less reverberation power from extended scatterers and accurate target range measurement. The fourth type of waveform is pulse repitition of similar pulses. Pulse trains provide accurate measurement of target range rate and unmasking by reverberation and are the only type of waveform that can be used with systems that employ functionally nonlinear sound sources such as air guns, sparkers, explosives, and other impulsive sound sources. Impulsive sources are described in *Enc.* Ch. 47.

The transmit beamformer accepts waveforms from the generator and produces individual copies of them for each transducer in the transmit array. The function of the transmit beamformer and array together is to radiate beams of sound having power distributed narrowly in azimuth angle and vertical angle. There are two reasons for using relatively narrow transmit beams. First, many systems are input power limited, so that sufficient power can be radiated into only one narrow sector at one time. Second, reverberation power coming from any azimuth can be minimized with respect to the target echo by radiating preferentially into vertical angles that reach the target at the expense of angles that reach either the surface or the bottom. If directive transmit beams are used, it is often necessary to stabilize them in both vertical and azimuth angle against ship or array motion. Furthermore, it is necessary to transmit sequentially toward successive azimuth angles to ensonify all potential targets. This rotational directional transmission (RDT) may require that successive pulse waveforms contain mutually exclusive frequencies so that the direct path signal to the receiver from the nth + 1 pulse does not blank the receiver during reception of the nth pulse.

A bank of power amplifiers drive the transducers in the transmit array. These amplifiers have small gain amplitude and phase errors so that the beamformed signal is not distorted in a significant way. Typically, gain errors need to be less than ±0.5 dB and phase errors need to be less that ±10° to ensure that transmitted side-lobe amplitudes remain at least 30 dB below the main lobe.

The transmit array itself is composed either of electroacoustic transducers (the subject of *Enc.* Ch. 52 and 53) or of impulse sources (which are the subject of *Enc.* Ch. 47). Transmit arrays are subjected to several special requirements that are not discussed in the transducer chapters. It is necessary to avoid cavitation that occurs either at the face of the transducers or in the sound near field in front of the array whenever the peak sound pressure approaches the local ambient pressure. It is also necessary to avoid beam pattern distortion caused by mutual coupling between transducers. Whenever a transducer's radiation impedance is significantly modified by sound radiated from all other transducers in the array, the resulting transducer amplitude and phase are changed. Finally, the transmit array is often housed in a streamlined dome to facilitate flow around it. The dome must be acoustically transparent while not inducing cavitation on its surface in the intense sound field.

Sound from the transmitted beam can find its way to the receiver array by several processes, as shown in Fig. 1. Considering first the desired path, sound propagates to the target and then scatters from it, and the echo propagates back. Details of sound propagation are considered in Chapter 31, and details of sound scattering from targets are considered in Chapter 33, see also *Enc.* Ch. 43. There are four distinct features of sound propagation that are important to sonar systems. Sound attenuation, called transmission loss, is the loss of signal power in decibels between a point 1 m from the source and a distant point. Time spread is the time increase in length of a pulse propagating over the path. Frequency spread is the bandwidth increase of a pulse propagating over the path. Correlation time is the time interval over which the other propagation features remain essentially unchanged by the internal fluctuations within the ocean.

A portion of sound power that strikes the surface scatters back to the receiver as surface reverberation. The essential features of the surface backscattering of sound are covered in *Enc.* Chs. 39 and 40, where surface scattering strength is introduced as the measure of surface backscatter. Two different kinds of surface reverberation can occur. First, there is the continuous backscatter from the component of surface roughness that is uniformly distributed over the surface. Second, these occur at the surface when significant winds blow plumes of bubbles that are convected downward and act as discrete targets that can mimic submarines.

All sound paths produce continuous backscatter that originates at inhomogeneities of the ocean's mechanical properties (sound, speed, and density) or at marine animals. The backscattered sound is called volume reverberation, and its features are discussed in *Enc.* Ch. 39. Typically, volume reverberation is not important at frequencies below 1 kHz. The marine animals that cause the most intense backscatter are confined to a layer whose depth has diurnal migrations.

The final path of sound to the receiver shown in Fig. 1 represents bottom reverberation. The essential features of the bottom backscattering of sound are covered in *Enc.* Ch. 40, where the bottom scattering strength is introduced as the measure of bottom backscatter. Bottom backscatter has both continuous and discrete components, just as surface backscatter has. The discrete bottom targets that can mimic submarines are either concentrated bottom roughness or inhomogeneities of sub-bottom properties.

The receive array is often a spatially distributed array of individual hydrophones having the purpose of preferentially selecting incoming signals according to their arrival azimuth and vertical angles. In some systems, the receiving array function is performed by an acoustic lens or reflector. The subsequent signal processing equipment is separated into many parallel channels, called beams, each pointed at a different incoming azimuth and vertical angle. Two general purposes are served by the receiving array. First, the total noise in each beam is less than the total noise arriving at the receiving array because of the preferential array response in the direction of the beam. Second, the direction of the target is indicated by the direction of the beam that receives the target echo. The receive array takes many different forms in different applications. Towed line arrays are long, neutrally buoyant flexible hoses that contain hydrophones at regular intervals and are towed behind the sonar platform. Hull arrays are spherical, cylindrical, or conformal arrays attached below or on the platform hull and contained within a streamlined, transparent sonar dome or window. Variable-depth systems (VDSs) have arrays contained in tow bodies.

Each hydrophone in the receive array is connected to its own preamplifier and signal conditioning electronics. Generally, the preamplifier output is low-pass filtered at a frequency at or above the Nyquist sampling frequency and digitized. Many systems incorporate a test signal insertion point in series with each hydrophone. The array is calibrated by inserting the same signal simultaneously into each hydrophone. It is then convenient to measure the amplitude of all of the digitized signals and to multiply each one by a different constant that causes the resulting products for all hydrophones to be equal. This procedure corrects for all signal amplitude variations that may occur in the analog portion of each hydrophone signal path.

The receive beamformer accepts the conditioned signals from each hydrophone and combines them in many parallel channels to form a set of beams, usually about equal in number to the number of hydrophones in the array. Flush-mounted arrays that have flow over them are populated with many additional hydrophones to help average out the turbulent wall pressure fluctuations. Each beam is characterized by a maximum sensitivity to incoming signals on its own major lobe, or maximum response axis (MRA), with diminished sensitivity to incoming signals away from its MRA. Usually each MRA is defined by its relative azimuth and vertical angle, so that the MRA changes as the platform heading changes. To avoid this, stabilized beams are sometimes formed by considering the platform heading, so that the MRA is defined by its true heading. Forward platform motion induces frequency shifts in the echos coming from all targets except those that are exactly abeam. The beamformer can, by adjusting its clock rate, remove the Doppler shifts caused by platform motion by a process called own Doppler nullification (ODN).

The output of each beam is filtered in a way designed to maximize the signal-to-noise ratio at the filter output. The general name of this type of filter is *matched filter*. When the signal transmission sequence includes more that one kind of signal, there must be a matched filter for each signal type attached to each beam. When the transmitted signal is sensitive to Doppler distortion, there must be a bank of filters each of which is matched to a different resolvable Doppler shift.

Signal detectors follow the matched filters, because the filter outputs are zero-mean signals. Subsequent averaging to reduce the confounding effects of noise fluctuations would reduce the signals to zero, if not for the detectors. Any nonlinear process, such as half- or full-wave rectification or mth law exponentiation would do. The usual choice is square-law envelope detection.

Often the signal is spread out in time or frequency by the propagation path to and from the target. When this occurs, the matched filter performance is degraded somewhat, because the filter was matched to the transmitted waveform and not to the received waveform. The loss in matched filter performance can be substantially recovered by postdetection processing, which involves averaging the output of the detectors over the same frequency or time window that was introduced by the propagation paths. It is helpful to convert the data to its logarithm in certain cases of reverberation interference that contain a large number of false echoes that are louder than some target echoes.

The next step, the normalizer, would not be necessary if the noise could be characterized as a time-invariant random process. But it never can be, so that fluctuations in the noise can be confused with possible target echoes. Moreover, the reverberation component of interference dies out with increased range, so that it is not time invariant either. The normalizer essentially divides the output of each envelope detector by its mean value determined over a time longer that the signal duration but shorter that the characteristic time of noise power fluctuation. This normalization can also be done in azimuth by dividing the signal in one beam by the average signal in many adjacent beams at the same time (range). Signal normalization by its mean in time and bearing eliminates the effect of long-term changes in background interference, so that the nonsignal outputs always have the same mean.

The normalizer output is continuously compared to a fixed output value called the threshold value for detection. Each time the normalizer output exceeds this threshold value, a tentative detection (called a detection) is made and ascribed to the beam and Doppler filter to which that normalizer is attached. These detections are passed to the information processing subsystem for subsequent analysis.

The first purpose of the information processor, called classification, is to determine which of the detections are correct ones and which are false ones. Then the target position and its course made good are estimated by a process called localization and tracking. These functions rely upon several classes of logic and processing. One class is target connectivity. Successive detections are pieced together in tracks that are plausible ones in terms of target speed and maneuvers. Another class is echo parameter plausibility. The range and bearing extent of each detection is measured to see if it represents a target much larger that a submarine. Range rate (Doppler shift) is measured to see if it is consistent with the hypothetical track. Echo power level is compared with plausible echo power levels according to the estimated target range as a test that a submarine caused the echo. Sometimes various moments of the echo envelope are computed and compared to characteristic values for submarines. Localization parameters (range, relative bearing, and range rate) are estimated from the individual sonar detections. These values are used in algorithms called trackers to form geographical tracks of the target. Trackers begin new tracks when detections are inconsistent with previous tracks and prune tracks when no new detections are received for a long period that can extend them.

The final system block is the man–machine interface, or the display. Typically displays are formed on high-resolution cathode-ray tubes (CRTs). Multiple A-scan displays show signal-amplitude-versus-time plots of the outputs of several beam normalizers in a vertical stack. The common abscissa of each beam output is target range. If enough beams are included, the operator sees an intensity-modulated area in the azimuth–range plane. In some displays, the threshold crossing detections are overlayed on this display to call the operator's attention to the echo. A zoom feature allows any portion of the display to be greatly expanded to allow the individual time features of the echo to be seen and evaluated. A synoptic or geographical situation display is shown either in relative coordinates or geographical coordinates. In the first case, the sonar platform is shown at the center of polar coordinates range and true bearing. In the second case, the platform is shown at its geographical position. Detections and target tracks are shown on these displays, sometimes with bathymetric overlays to aid in the classification of bottom echoes.

2.2 Generalized Passive Sonar System

The generalized passive sonar system is a subset of the active system, because in the passive case the target itself radiates the signal that is to be detected by the sonar system. Figure 1 shows that there are three classes of sound propagation paths between the target and the receiver: direct, surface reflected, and bottom reflected. At ranges much greater than the water depth there is an ensemble of paths, most of which are reflected many times from both the bottom and the surface. As before, the ensemble of paths is characterized by propagation loss, called transmission loss. The time spread of the paths is of no direct concern, because target noise radiation is for the most part of a continuous nature. Frequency spread induced by the time-varying propagation path due to motion limits the narrowness of the processing bandwidth.

Receive arrays for passive systems do not differ in a fundamental way from those for active systems. Passive systems do not need to be colocated with a sound source but must be located in quiet places. Therefore, passive receive arrays are located on the ocean bottom, towed behind quiet ships, or mounted on hulls with elaborate sound attenuation systems between them and the hull. Since the hydrophones in the arrays do not need to serve the dual purpose of transmitting the sound signal, they are much smaller and lighter.

Signal conditioners and receive beamformers do not differ in a significant way from their active counterparts. Matched filters for passive systems do differ greatly, because the time-continuous noise emitted from targets is completely different than the pulse train sequences emitted from active sonar systems. There are two distinct types of passive sonar-matched filters. One type is designed to detect the discrete spectral line components in the target-radiated noise signature. The other type is designed to detect the continuous spectral background of the target-radiated noise signature. The former filter is used for narrow-band detection, while the latter filter is used for broadband detection.

Postdetection processing of both narrow-band and broadband filters consists of rather long time averaging. The longer the averaging, the weaker the signal that can be detected, within limits imposed by the noise and propagation path. In the active sonar case the averaging time is set equal to the time spread of the propagation path, for to average longer would reduce the signal amplitude. In the passive case the averaging time is also to be no longer than the duration of the signal. This time is determined by how long the target remains at a relatively low transmission loss range.

Normalizing is of crucial importance in passive sonar, because the signals last for relatively long times and therefore noise fluctuations over long times must be removed without removing the signals themselves. The principal interference in passive sonar systems is the noise radiated from individual merchant or other nontarget surface shipping, for these noise sources mimic the ships that are to be detected.

3 ANALYTICAL REPRESENTATION OF SONAR SYSTEM ELEMENTS

This section describes the performance of sonar system elements.

3.1 Arrays and Beamformers

A transmit or receive array and its beamformer functions as a unit either to radiate a pattern of sound or to receive a pattern of sound. The fundamental relationships between array parameters and array performance are described in this section. The parameters of beam patterns that affect signal reception are beamwidth, sidelobe levels, and directivity index. The parameter of beam patterns that affects noise reception is array gain.

Beamforming and Array Beam Patterns The beam pattern of a transmitting array describes its directional distribution of radiated sound power. In a similar way, the beam pattern of a receiving array describes its directional sensitivity to incoming sound power. A transmitting array consists of N discrete radiators and a receiving array consists of N discrete receiving hydrophones. When the radiators are linear, reciprocal transducers, they are often used also as the receiving hydrophones. A transmitting beamformer forms a time-delayed and amplitude-adjusted replica of the transmitted waveform for each transducer in the array to produce a desired transmission beam pattern. The principal consideration for transmit beam patterns is to maximize sound power in target directions by transmitting as little power as possible in other directions. A receiving beamformer obtains the outputs of each hydrophone and combines them all to produce a desired receiving beam pattern. The principal consideration for receive beam patterns is to minimize sensitivity in directions away from the target direction, thereby increasing the target signal-to-noise ratio. The beam pattern equations in this section are described as receiving array beam patterns, but they are equally valid as transmitting array beam patterns.

The sound pressure distribution at the array aperture caused by an incoming plane sound wave having sound pressure amplitude P_0, frequency f, and wave vector $\mathbf{k}$ is

$$p(\mathbf{x}, t) = P_0 e^{i(\mathbf{k} \cdot \mathbf{x} - 2\pi f t)}, \tag{1}$$

where the direction of $\mathbf{k}$ coincides with the direction of wave propagation and its magnitude is equal to $|k| = 2\pi f/c$. Beam patterns are defined for monochromatic waves, so that the sound pressure distribution is represented simply by its spatial factor:

$$P(\mathbf{x}) = P_0 e^{i\mathbf{k} \cdot \mathbf{x}}. \tag{2}$$

This sound pressure distribution is sampled spatially by an array of identical hydrophones. The distribution of hydrophone sensitivities over the array aperture is

$$s(\mathbf{x}) = \sum_{n=1}^{N} \sigma_n f(\mathbf{x} - \mathbf{x}_n) e^{i\mathbf{k}_0 \cdot \mathbf{x}}, \tag{3}$$

where $\mathbf{x}_n$ are the center locations of the hydrophones, σ_n are the receiving sensitivities of the hydrophones, $f(\mathbf{x})$ is the spatial distribution of hydrophone sensitivity over its face, and $\mathbf{k}_0$ is the wave vector to which the array beamformer is steered. The beamformer function is to add up all of the hydrophone outputs to obtain the beamformer output signal Y. The beam pattern D is equal to the normalized output signal power, $D = 10 \log |Y(k)|^2/|Y(k_0)|^2$:

$$Y = \int P(\mathbf{x}) s(\mathbf{x})\, d\mathbf{x} = P_0 F(\mathbf{k} - \mathbf{k}_0) \sum_{n=1}^{N} \sigma_n e^{i(\mathbf{k} - \mathbf{k}_0) \cdot \mathbf{x}_n}. \tag{4}$$

The function $F(\mathbf{k})$ is the Fourier transform of the hydrophone sensitivity function $f(\mathbf{x})$. Equation (4) is a product of two functions: $F(\mathbf{k})$ is the beam pattern of an individual hydrophone and the summation term is the beam pattern of an array of point hydrophones located at the centers of the actual ones. This result is known as the product theorem, or pattern multiplication theorem for arrays.

The product theorem is an important result, because it can be used to obtain the beam pattern of a planar array of hydrophones at the centers of a rectangular grid, a very common array arrangement. If the columns of hydrophones are considered to be line arrays with the beam pattern $F(\mathbf{k})$, and if the rows of hydrophones are considered to be line arrays with the beam pattern $G(\mathbf{k})$, then the beam pattern of the array is equal to $G(\mathbf{k})F(\mathbf{k})$. The beam patterns of linear point arrays are of fundamental importance.

The beam pattern of an unshaded linear point array of N hydrophones spaced d apart and steered in the direction θ_0 and measured from the array normal is given as

$$D(\theta) = \left(\frac{\sin NX}{N \sin X} \right)^2, \tag{5}$$

where

$$X = \frac{kd}{2} \sin(\theta - \theta_0).$$

The beam pattern is a periodic function of the sound arrival angle θ. The beam pattern has maximum values, or lobes, equal to 1 at values of $X = n\pi$. Only the beam pattern lobe centered at $X = 0$ is the desired lobe. Any other lobes that appear in the real sound angle range of $-90°$ to $+90°$ are unwanted "grating lobes." The condition for avoiding grating lobes for any steering angle is

$$d < \frac{\lambda}{2}. \tag{6}$$

The beamwidth Δ of the main lobe of the beam pattern is its angular width at the half-power points, or $\Delta = 2\theta'$, where $D(\theta') = 0.5$,

$$\Delta = \frac{\lambda}{L} \geq \frac{2}{N}. \tag{7}$$

the second equality holds at the grating lobe limit given by Eq. (6).

All other lobes in the range of sound angles θ besides the main lobe are called minor lobes, or sometimes side lobes. For the beam pattern described by Eq. (5), the minor lobes occur at values of X that satisfy the equation $\tan(NX) = N \tan(X)$. If the length of the array $L = d(N - 1)$ is many wavelengths, then the side-lobe arguments and levels are given in Table 1.

The near side-lobe levels for a linear unshaded array are not much less sensitive that the major lobe. Expected signal power levels from targets and interfering ships can range over values that differ by at least 30 dB. Therefore, a loud target in the direction of a side lobe can overwhelm a weaker target in the direction of the main lobe, causing it to be missed. The usual correction for this problem is to shade the amplitudes of the array elements to produce lower side-lobe levels.

TABLE 1 Side Lobes of Long Line Arrays

Side-Lobe Number	X	Side-Lobe Level (dB)
1	4.5	−13.3
2	7.7	−17.8
3	10.9	−20.8
4	14.1	−23.0
5	17.2	−24.7

TABLE 2 Amplitude Coefficients and Beamwidths of Four Chebyshev Arrays

Side-Lobe Level	A_0	A_1	A_2	Beamwidth (deg)
Unshaded	1	1	1	28
−20	4.32	3.35	2.33	33
−25	8.42	6.12	3.25	36
−30	15.98	10.92	4.27	38

Dolph–Chebyshev Arrays The shaded linear array that provides equal-level side lobes and at the same time the narrowest possible beam width is based on the Chebyshev polynomials.[1] The number of elements N is related to the order m of Chebyshev polynomial by

$$N = 2m + 1. \tag{8}$$

The ratio of main-lobe to side-lobe signal sensitivities is R ($R = 10$ for −20-dB side lobes). Two design parameters are a and b;

$$a = \frac{e + 1}{1 - \cos[kd(1 + \sin\theta_0)]},$$

$$b = \frac{e\cos[kd(1 + \sin\theta_0)] + 1}{\cos[kd(1 + \sin\theta_0)]}, \tag{9}$$

where

$$e = \cosh\left(\frac{1}{m}\cosh^{-1} R\right).$$

The mth-order Chebyshev polynomial can be derived from the following recursion relationship:

$$T_{m+1} = 2xT_m(x) - T_{m-1}(x), \tag{10}$$

where $T_0(x) = 1$ and $T_1(x) = x$.

The argument x is defined by the transformation $x = a\cos\psi + b$. The array shading coefficients A_n are obtained from

$$T_m(x) = A_0 + 2\sum_{n=1}^{(N-1)/2} A_n \cos n\psi. \tag{11}$$

The shading coefficients are determined by equating Eq. (11), term by term, in $\cos n\psi$. As an example, Bartberger[2] developed four six-element arrays, summarized in Table 2. Beam broadening is not great, and the side lobes are much reduced with respect to unshaded arrays, for which the first side-lobe level is only −13 dB. Typical arrays are shaded for −25- or −30-dB side lobes. For these arrays, an unwanted noise on a side lobe would have to be more than 30 dB louder than a weaker signal on the main lobe to mask it. Sometimes this difference is not great enough, and adaptive shading is used.

Side-lobe control can also be accomplished by spatial shading (as opposed to amplitude shading). The theory of aperiodic array design is described by Steinberg.[1] The objective of aperiodic array design is to obtain acceptable beam patterns with fewer array elements than are required to satisfy the grating lobe requirement [Eq. (6)].

Other shading schemes having particular uses are described by Steinberg[1] and Nielsen.[3]

Adaptive Beamforming The individual hydrophone gains are derived from within the beamformer, based on the nature of the hydrophone outputs, in adaptive beamforming. The most common goal of adaptive beamforming in sonar systems is called adaptive nulling. In this process, each beam is assigned an expected signal direction, and receiving sensitivity for that beam is maintained at a constant value in the assigned direction. When a strong signal is detected from another direction (outside the main lobe), a null sensitivity is generated in that direction. Several strong, out-of-main-lobe signals can be nulled at once. Several methods of adaptive nulling are described by Steinberg.[1]

Array Gain and Directivity Index Array gain (AG) is defined as the increase, in decibels, of the signal-to-noise ratio caused by a receiving array and its beamformer. Array gain depends both upon the geometry of the array and the nature of the noise field around the array. The directivity index (DI) of a transmitting array is defined as the ratio, in decibels, of the signal power in the transmitting direction to the signal power averaged over all directions. The directivity index of an array is numerically equal to its array gain in the special case of a three-dimensional isotropic noise field.

The signal and noise amplitudes at the output of each array element are given by $s(t)$ and $n(t)$. The signal from the beamformed beam pointing at the target is $y_s(t)$, and the noise in the same beam is $y_n(t)$. There are N elements in the array, and the sensitivity of the ith element is σ_i. The output signal and noise powers are

$$y_s^2(t) = \left(\sum_{i=1}^{N} \sigma_i S_i(t) \right)^2 ,$$

$$\overline{y_n^2} = \overline{\sum_{i=1}^{N} \sigma_i n_i(t) \sum_{j=1}^{N} \sigma_j n_j(t)}. \qquad (12)$$

The correlation coefficient of the noise at elements i and j is defined as the covariance of the noise amplitude at the two elements, normalized by the noise power. The noise power is assumed to have the same value in all elements. Moreover, the noise covariance C is a function only of the distance between the two elements x_{ij} if the noise field is statistically homogeneous over the array face:

$$C_{ij}(x_{ij}) = \frac{\overline{n_i n_j}}{\overline{n^2}} . \qquad (13)$$

The array gain of an N-element unshaded array ($\sigma_i = 1$) is equal to $10 \log(g)$, where

$$g = \frac{y_s^2/\overline{y_n^2}}{s^2/\overline{n^2}} = \frac{N^2}{N + 2\sum_{i=1}^{N} \sum_{j \neq i}^{N} C_{ij}(x_{ij})} . \qquad (14)$$

The gain ratio g can be written as the ratio of two terms: the ratio of beam signal power to element signal power and the ratio of beam noise power to element noise power. This leads to the connection between the logarithmic quantities array gain (AG), signal gain (SG), and noise gain (NG):

$$g = \frac{y_s^2}{s^2} \cdot \frac{\overline{n^2}}{\overline{y_n^2}} , \qquad \text{AG} = \text{SG} - \text{NG}. \qquad (15)$$

A common way of calibrating the beamformer is in terms of the incident signal sound power. This sound power is, of course, independent of the measurement array, so that the signal power at any element output should be the same as the signal power at the output of the beam that points at the incident wave. Unless the beamformer has loss due to incorrect element summation and delay, the signal gain ratio is 1, and SG = 0 dB. With this convention, the array gain is numerically equal to the negative of the noise gain. The noise power $\overline{y_n^2}$ is less than the element noise power $\overline{n^2}$.

Array Gain and Directivity Index of Unshaded Line Arrays In the important case of a line array of N elements spaced by distance a, the distance between the ith and jth elements is equal to $(j - i)a$. The double summation in Eq. (14) becomes a single summation over $m = j - i$:

$$g^{-1} = \frac{1}{N} + \frac{2}{N^2} \sum_{m=1}^{N-1} (N - m)C(am). \qquad (16)$$

When the noise field over the face of the array is uncorrelated from element to element, $C(x_{ij})$ is zero. From Eqs. (14) and (16), the array gain is in this case equal to $10 \log(N)$. The noise field correlation coefficients are tabulated in Table 3 for four different homogeneous noise fields.

The first tabulated noise field is composed of a superposition of plane sound waves of the same frequency having equal power in all propagation directions distributed throughout three dimensions. This is an idealized noise field that is seldom approximated in the ocean. This case corresponds to conditions for which the directivity index is defined, so that this case leads to the directivity index. The second noise field is a superposition of plane sound waves of the same frequency having equal

TABLE 3 Noise Correlation Coefficients for Four Noise Fields

Noise Field	Correlation Coefficient
Three-dimensional isotropic	$\sin(ka)/ka$
Two-dimensional isotropic, elements in propagation plane	$J_0(ka)$
Cosine directivity, elements parallel to null plane of cosine	$2J_1(ka)/ka$
Cosine directivity, elements parallel to peak of cosine	$[\cos(ka) + ka \sin(ka) - 1]/(ka)^2$

power in all propagation directions distributed over a single plane. This noise field is approximated in the ocean at low frequencies, where the noise is mostly generated by distant shipping and therefore arrives more or less horizontally from all azimuthal directions. The third noise field is a superposition of plane waves of the same frequency having power proportional to the cosine of their vertical angle. This noise field is approximated in the ocean at high frequencies where dipole noise generated at the surface by wind waves predominates. The correlation coefficient for the third tabulated case is for elements in a line parallel to the surface, so that it is appropriate for towed line arrays. The correlation coefficient for the fourth tabulated case is for elements in a line perpendicular to the surface, so that it is appropriate for vertical line arrays.

The array gain for the broadside beam of a 50-element line array is shown in Fig. 2 for each of the four different correlation coefficients listed in Table 3. All four cases asymptote to a value of $10\log(N)$ for large values of the argument ka. This is generally true, because the noise correlation functions all tend to zero at separations a that are large with respect to a wavelength. The element spacing a is equal to one-half wavelength at $ka = \pi$. This is the largest spacing that will not produce aliasing lobes when the array is steered toward endfire, and arrays are often not used at higher frequencies. But larger values of array gain occur up to $ka = 2\pi$. Values of array gain are about 2 dB larger than the DI against the two-dimensional noise field (representing low-frequency ambient noise). Values of array gain are about 1 dB smaller than the DI against the dipole noise field (representing high-frequency ambient noise). This illustrates why towed arrays provide higher gain at lower frequencies. The vertical line array achieves about 15 dB more gain than the others. This illustrates the beneficial effect of directing the main lobe in a direction free of the predominant noise.

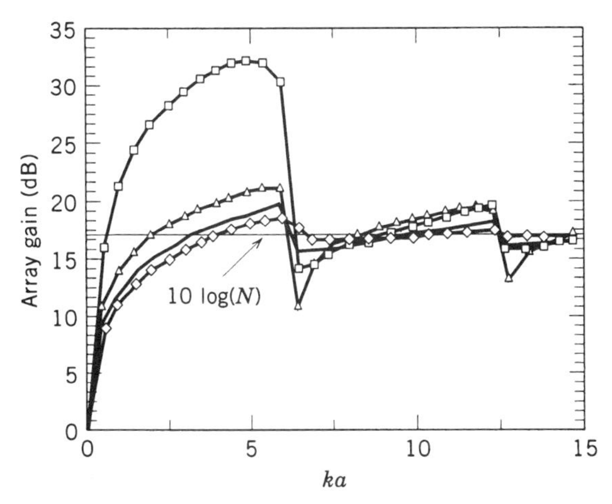

Fig. 2 Array gain for the broadside beam of 50-element line arrays as functions of normalized element spacing ka: (——) three-dimensional noise field; (△) two-dimensional omnidirectional noise field; (◇) line parallel to a surface of dipole sources; (□) line perpendicular to a surface of dipole sources.

The DI for the unshaded linear array can be written as $\mathrm{DI} = 10\log(2L/\lambda)$, expressing the fact that in the three-dimensional isotropic noise field the elements must be spaced at a distance of $\lambda/2$.

Directivity Index of Two-Dimensional Arrays

The DI (also the array gain in three-dimensional isotropic noise) of two-dimensional planar arrays is a function only of the area of the array A and the wavenumber k:

$$DI = 10\log\left(\frac{k^2A}{\pi}\right). \tag{17}$$

This equation is valid for an array of any shape, provided the smallest dimension is longer than about two wavelengths. This equation is also valid if the array is made up of N-point hydrophones if the spacing between adjacent hydrophones is less than one-half wavelength. For a uniformly spaced planar array, $A = Nsw$, where s and w are the hydrophone spacings in the length and width directions of the planar array.

Array Gain of Shaded Arrays The array gain of an array in isotropic noise is not sensibly changed by shading. An example of this is given by Bartberger.[2] He tabulated the array gain against a three-dimensional isotropic noise field for a six-element line array having Chebyshev polynomial shading. The results are listed in Table 4. Severe shading reduces the DI by only 0.7 dB. If a noise field is reliably nonisotropic, array gain will be increased by shading for low side-lobe levels in the noisy directions.

TABLE 4 Array Gain of Six-Element Chebyshev Shaded Line Array

Array	Gain (dB)
Unshaded	7.78
Shaded, 15-dB side lobes	7.75
Shaded, 20-dB side lobes	7.53
Shaded, 25-dB side lobes	7.26
Shaded, 30-dB side lobes	7.02

3.2 Signal Processing in Active Sonar Systems

The signal processing chain is a part of Fig. 1. In this section each component of the chain is described, more or less in the order that the signal proceeds through the system.

Matched Filters The purpose of the matched filters is to maximize the signal-to-noise ratio (SNR) in each beamformer output beam. Each different signal waveform requires a unique filter to achieve this purpose, thus the name "matched." The typical active sonar signal bandwidth is not so broad that the noise power spectral density over the band cannot be assumed flat. Moreover, the noise amplitude probability density function (PDF) is assumed to be Gaussian. The desired filter is the one that produces the maximum value of SNR when a signal $x(t)$, together with additive white Gaussian noise $u(t)$ having a power spectral density N_0, is passed through it. The filter is linear and characterized by the impulse response $h(t)$. The output of the filter caused by the signal input is $s(t)$, and the output caused by the noise input is $n(t)$:

$$s(t) = \int_{-\infty}^{\infty} x(t')h(t' - t)\,dt',$$

$$n(t) = \int_{-\infty}^{\infty} u(t')h(t' - t)\,dt'. \qquad (18)$$

The SNR is defined as the signal power at the filter output at time t_m that it reaches its maximum value divided by the noise power at the filter output:

$$\text{SNR} = \frac{s(t_m)^2}{n^2} = \frac{S(t_m)}{N}. \qquad (19)$$

The derivation of the matched filter impulse response $h(t)$ is explained in many books, for example, Davenport and Root[4] and Levanon.[5] The impulse response of the desired filter is $h(t) = Kx^*(t_m - t)$, where K is a constant. This result is interpreted as the complex conjugate of the input signal $x(t)$ that has been inverted in time and delayed by the time at which the output signal reaches its maximum value. It is necessary for t_m to be greater than the time span T of the signal $x(t)$ itself, so that the value of $h(t)$ is zero for negative times (causality). The value of SNR at the time t_m of maximum signal output, as defined in Eq. (19), is proportional to the signal energy:

$$\text{SNR} = \frac{E}{N_0}, \quad \text{where } E = \int_{-\infty}^{\infty} x(t)^2\,dt. \qquad (20)$$

The time duration τ of the matched filter output pulse can be deduced from the foregoing. Looking first at Eq. (20), we see the SNR is the ratio of two input energies. The filter gain factor K divides out, so that the ratio will apply also to the corresponding two output energies. Equating Eqs. (19) and (20) and denoting the signal energy by $S\tau$ and the noise power by N_0B lead to an important result,

$$\tau B = 1. \qquad (21)$$

The time duration τ of the output signal pulse is called the pulse resolution time. The pulse time span T can be much larger than the resolution time τ. The pulse compression ratio c_r is given by

$$c_r = \frac{T}{\tau} = TB. \qquad (22)$$

These important results mean that no parameter of the active sonar signal other than its energy and bandwidth has any bearing on the SNR or the range resolution at the output of the matched filter. It is shown in Section 3.3 that most of the system's detection performance depends only upon SNR and the bandwidth of the signal, so that only signal energy and bandwidth are prime system parameters.

Postfilter Envelope Detection Some form of nonlinear circuit is introduced following the matched filtering operation. The matched filters are linear and their inputs contain no DC component, so that their signal and noise outputs are both zero-mean time series. This is an inappropriate form for the signals to have for several reasons. First, the signal may have either positive or negative polarity. The difficulty of finding the signal in noise is increased if its larger portions can be either above or below the axis. Second, averaging or smoothing of the filter outputs is often helpful to supress noise fluctuations that tend to mimic targets. Averaging the signal in this zero-mean form will reduce its amplitude just as much as the noise with no net gain. Third, signals in this form display their entire waveforms. The only part of the waveform that is of interest in detection is its envelope. Therefore, each matched filter is followed immediately by an envelope detector.

There are several essentially equal methods of envelope detection, but the one most commonly used is an instantaneous square-law device followed by a low-pass filter. If the low-pass filter is "matched" to the signal at the squarer output, then it has a cutoff frequency equal to the bandwidth of the signal. It is equivalent to saying that the low-pass filter is an averager, which is matched to the squared signal when its averaging time equals the time span of the signal. It is common for the propagation path to be composed of several "multipaths." These multipaths are said to be irresolvable when the time delay between adjacent paths is less than the signal's resolution time. It is common in active sonar systems to "overaverage" with the low-pass filter, so that the averaging time is set to approximate the sum of the pulse resolution time and the multipath time spread. Doing this causes no loss in the amplitude of the filtered signal, but it reduces the fluctuations in the noise. Whether the low-pass filter overaverages or not, it will also eliminate the noise spectral components that are generated by the square-law device at twice the passband frequency of the matched filter. Failure to filter out these high-frequency noise components would unnecessarily double the noise power.

Signals for Active Sonar Systems The preceding section explained that sonar systems invariably use filters that are matched to the transmitted signals, for these filters maximize the peak SNR. It is intuitively obvious that maximum system detection performance is achieved by maximizing SNR. But there are other functions to be performed by the system. In particular, once the target is detected, it is necessary to measure accurately both the range to the target and its range rate. A different type of requirement is set by the maximum signal power output achievable from the transmitting array, so that some minimum pulse time span is required for the signal energy to produce a sufficiently large SNR.

Range to a target is determined by the time delay δ between the time the signal is transmitted and the time that its echo is received:

$$R = \frac{c\delta}{2} . \tag{23}$$

The effective sound speed c can be determined by calculating the travel time over the propagation path using methods described in Chapter 35. It is often sufficiently accurate to use a value of 0.8 nautical miles/s, which is found to match closely the effective sound speed of many different propagation paths in the ocean.

Measurement of target range can only be accomplished to an accuracy set by the time resolution of the pulse. The range resolution ΔR is given as

$$\Delta R = \frac{c\tau}{2} = \frac{c}{2B} . \tag{24}$$

Range rate is determined by measuring in some way the time scaling impressed upon an echo by a moving target. If the transmitted signal is $x(t)$, the echo from a point target moving with respect to the platform with range rate $\dot{R}$ will be scaled in time as $x(t')$, where t' is related to t by a function of Mach number M, where $M = \dot{R}/c$:

$$t' = t\,\frac{1+M}{1-M} . \tag{25}$$

The Mach number is always small, for even a 90-knot torpedo closes range at a Mach number of only 0.03. This leads to the common approximation of Eq. (25):

$$t' = t(1 + 2M). \tag{26}$$

For narrow-band signals, the motion-induced time scaling causes a frequency shift, called the Doppler shift. Denoting the transmitted frequency by f_t, the echo frequency by f_e, and the Doppler shift frequency by f_d, the following equation relates the relative Doppler frequency to the Mach number:

$$f_e = f_t + f_d = f_t(1 - 2M). \tag{27}$$

The negative sign means that for receding targets, for which range rate is positive, the Doppler shift is negative due to stretching in time and the echo has a lower frequency than has the transmitted signal.

Measurement of target range rate can only be accomplished to an accuracy set by the frequency resolution of the pulse. The frequency resolution $\Delta f = \nu$ is a fundamental relationship

$$\Delta f = \nu = \frac{1}{T} , \tag{28}$$

where T is the pulse time span.

The capacity for a sonar signal to measure target range (travel time) and target range rate (Doppler shift) is illustrated by its ambiguity function. The ambiguity function of a signal is defined to be its own matched filter's response to its time-delayed and Doppler-shifted self. In other words, the ambiguity function of a signal is a two-dimensional map of its own filter's output on the time delay, Doppler shift plane. This map shows how sharp the signal response is to changes in target range and speed. The output of the matched filter is given by Eq. (18). The signal waveform is expressed in the complex form $x(t) = u(t)e^{i2\pi f t}$, where $u(t)$ is the complex modulation function whose magnitude $|u(t)|$ is the envelope of the real signal and $2\pi f$ is the carrier frequency. The time-delayed and Doppler-shifted matched filter output is then[6]

$$\chi(\delta, f_d) = \int_{-\infty}^{\infty} u(t)u^*(t+\delta)e^{i2\pi f_d t}\, dt. \qquad (29)$$

The squared magnitude $|\chi(\delta, f_d)|^2$ is the ambiguity function.

The ambiguity function illustrates the pulse resolution time τ and resolution frequency ν by its central width in the δ, f_d coordinates. There are three general types of signal waveforms from the point of view of their ambiguity functions. First are single-frequency pulses, or "CW pulses," which are used to obtain a fine resolution of range rate or of echo frequency. The ambiguity function of such a pulse is shown in Fig. 3, both as a quasi-three-dimensional plot and as a contour plot. The time span T of the pulse is two delay units. The figure shows that the pulse resolution τ of the ambiguity function is relatively large, being equal to the pulse time span at the half-amplitude points. The frequency resolution of the pulse is relatively small, being equal approximately to the inverse of the pulse time span. The pulse compression ratio c_r is unity, so the time–bandwidth product BT for a sinusoidal pulse is unity. The CW pulses can be amplitude shaded to reduce side lobes in the Doppler dimension of the ambiguity surface. Long CW pulses are used to separate high Doppler targets from low Doppler reverberation.

A second general type of waveform is designed to provide both fine range and Doppler frequency resolution at the expense of range–frequency ambiguity. This type is illustrated by a linear frequency modulated pulse, in which the frequency changes by $f(t) = kt$, where k is a constant. The ambiguity function of such a pulse is shown in Fig. 4, both as a quasi-three-dimensional plot and as a contour plot. The time span of the pulse is $T = 1$ delay unit and $k = 10$, leading to a pulse bandwidth $B = 10$. The plot shows that the time delay width of the ambiguity function at zero Doppler (resolution time) is

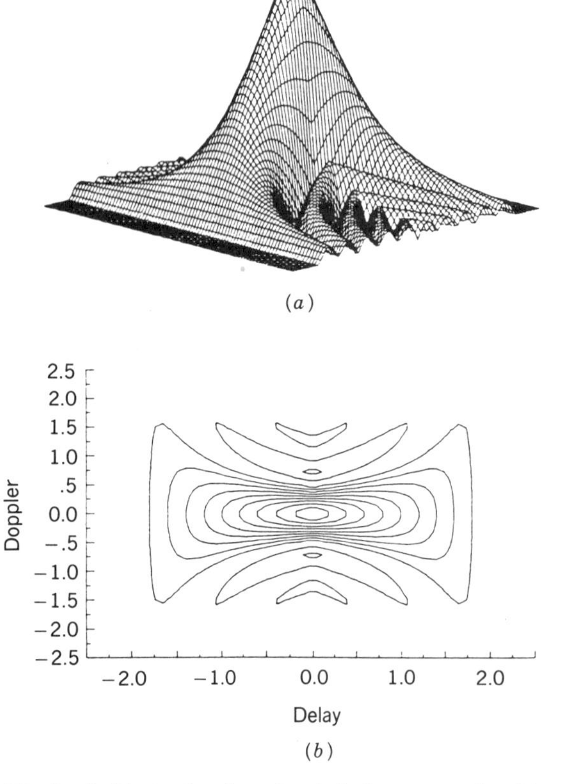

Fig. 3 Ambiguity function of a single-frequency pulse: (*a*) three-dimensional view; (*b*) contour plot. Pulse span $T = 2$. (From Ref. 5, Fig. 7.2.)

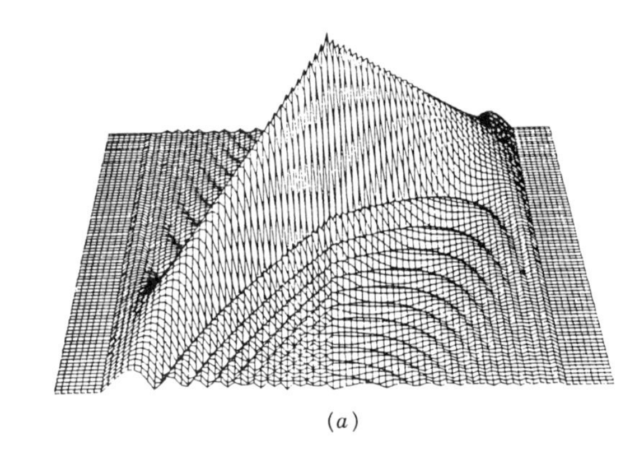

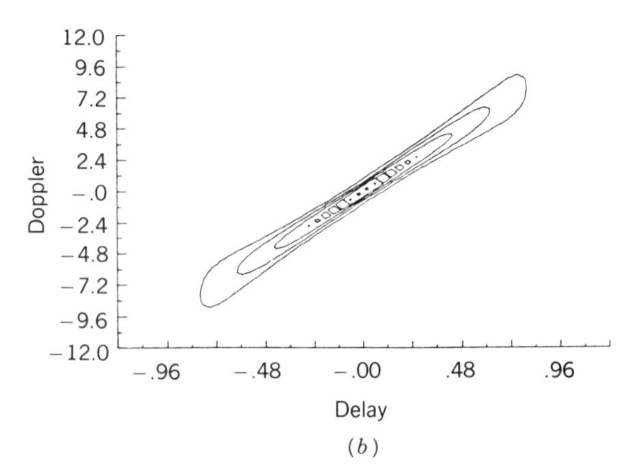

Fig. 4 Ambiguity function of a linear FM pulse ($t = 1$, $k = 10$): (*a*) three-dimensional view; (*b*) contour plot. (From Ref. 5, Fig. 7.5.)

only about equal to 0.1. The product TB is 10, and the pulse compression is equal to 0.1, in agreement with Eq. (22). The figure also illustrates the resolution frequency ν. At zero delay, the width of the ambiguity function is unity, in agreement with Eq. (28). If the pulse time span had been 10 units and $k = 1$, then the pulse resolution time would still have been equal to $1/B = 0.1$ but for a compression factor of 100. The resolution frequency would have been reduced to 0.1. This type of pulse is very useful if T must be large to obtain the necessary signal energy from the transmitting array. Even though T is large, it is possible still to resolve small targets because the pulse resolution time τ is small. There is a problem of range–Doppler ambiguity with this waveform, as shown in Fig. 4. Targets having nonzero Doppler appear to be at the wrong range. This ambiguity may be advantageous in reducing the number of matched filters needed to detect targets.

The third general type of waveform is designed to avoid the range–Doppler ambiguity of the linear frequency-modulated pulse but to retain both its pulse compression and Doppler resolution features. These waveforms are coded to have thumb tack ambiguity functions, meaning a significant response occurs only for targets jointly having a narrow dimension in both Doppler and range coordinates. An example of this type of coded pulse are Costas signals. Costas signals are sequences of equal-time-span single-frequency tones whose frequencies skip in a way that time and frequency shift overlaps are minimized. The ambiguity function of a seven-tone Costas signal is shown in Fig. 5. If the seven different frequencies are ranked by frequency, the signal sequence shown is 4, 7, 1, 6, 5, 2, 3. The pulse resolution τ of the sequence is equal to the inverse sequence bandwidth, in this case the frequency difference between tones 1 and 7. The frequency resolution of the function is equal to the inverse time span of the sequence, or 7 times the span of each tone. The particular choice of frequency sequence is made to keep all of the ambiguity surface away from the central peak flat, so that no target ambiguity occurs.

3.3 Target Detection Decisions by Active Sonar Systems

The outputs of the envelope detectors contain target echoes superimposed onto a background of noise. The echoes occur at unknown time delays δ and contain unknown Doppler shift frequencies f_d. Their pulse resolution times τ are known. The detection decision of whether an echo is present is to be made for each time cell of length τ in which an echo could occur. In general, this decision is made by comparing the envelope detector output in each time interval to a threshold value. The decision that a target is present is made when the detector output is larger than the threshold, and the target absent decision is made whenever the threshold is not exceeded. The value at which the threshold is set clearly has an important role in the validity of the detection decisions. If the detection threshold is set to a low value, even weak echoes will exceed it, and the probability of achieving detection is high. But the low threshold value will often be exceeded by large fluctuations of the noise, so the probability of a false detection (called a false alarm) will be high also. Conversely a high value of threshold will result in detecting only very large echoes, so that the probability of detection will be small. So also will the probability of false alarm be small. The proper adjustment of the threshold is a different matter for different statistical descriptions of both the signal echoes and the noise.

For each case of signal and noise statistics considered in this section, equations for both the probability of detection and the probability of false alarm are given as functions of SNR. The value SNR* needed to achieve jointly a particular set of probabilities of detection and false alarm is called the detection index, and when expressed in decibels, it is called the detection threshold DT (note that Urick[7] defines detection threshold in terms of a "signal to noise ratio referred to a 1-Hz band of noise," rather than the dimensionless ratio used in this book):

$$\mathrm{DT} = 10 \log \mathrm{SNR}^*. \tag{30}$$

We begin with single nonfluctuating pulse detection in stationary noise. But the requirements to measure range rate and to improve detection probability require that multiple pulses be used. Moreover, signal amplitudes fluctuate from pulse to pulse over real propagation paths. Several cases of increasing complexity and generality follow.

Single-Pulse Detection of Nonfluctuating Signals in Stationary Noise The simplest case is the detection of a single pulse in stationary noise having a Gaussian pdf. The probability of false alarm is determined by first calculating the pdf $p_n(r)$, which is the probability that the amplitude of the envelope detector output lies between r and $r + dr$, given that noise only is present. The probability of false alarm P_{FA} is determined by integration of $p_n(r)$ from the threshold value V_T to infinity. In other words, the probability of false alarm is equal to the probability that the noise envelope exceeds the threshold. The result is given by Levanon:[5]

$$P_{\mathrm{FA}} = e^{-V_T^2/2N}, \tag{31}$$

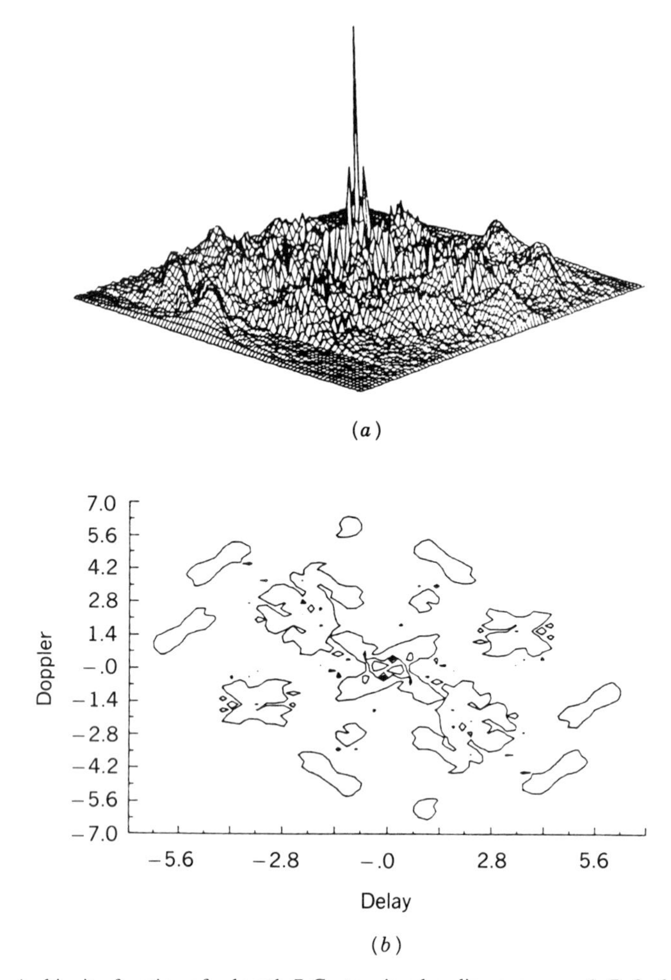

Fig. 5 Ambiguity function of a length 7 Costas signal-coding sequence 4, 7, 2, 6, 5, 2, 3: (*a*) three-dimensional view; (*b*) contour plot. (From Ref. 5, Fig. 8.3.)

where the noise power N is equal to N_0B. The probability of detection is determined by first calculating the pdf $p_{s+n}(r)$, which is the probability that the amplitude of the envelope detector output lies between r and $r+dr$, given that the signal is present in the noise. The probability of detection P_D is equal to the probability that the signal plus noise envelope exceed the threshold. An approximation that is valid for values of SNR $= E/N_0$ that are

$$P_D = \frac{1}{2}\left[1 - \text{erf}\left(\frac{V_T}{\sqrt{2N}} - \sqrt{\text{SNR}}\right)\right]. \tag{32}$$

The threshold value V_T can be eliminated from Eqs. (31) and (32), yielding a threefold relationship between P_D, P_{FA}, and SNR. This relationship is widely published (for example Fig. 12.6 in Urick[7]). The full curve, valid also for small SNR, can be found as Fig. 3.4 in Levanon.[5] An approximation due to Albersheim,[8] accurate for $10^{-7} < P_{\text{fa}} < 10^{-3}$ and $0.1 < P_D < 0.9$, is

$$\text{SNR} = A + 0.12AB + 1.7B, \tag{33}$$

where

$$A = \ln\frac{0.62}{P_{\text{FA}}}, \qquad B = \ln\frac{P_D}{1 - P_D}.$$

Detection of Multiple Nonfluctuating Pulses in Stationary Noise Single-pulse detection cannot achieve simultaneously large P_D and small P_{FA} without very large SNR, a luxury not often available. But the target will be in view for several consecutive pulses, and it is beneficial to defer the detection decision until after several pulses have been transmitted and received. There are two general ways to integrate consecutive pulses. One way is called coherent integration, a process that requires the propagation path to remain the same in all respects for the duration of the pulse ensemble. Due to motion, this typically does not happen in the ocean for usefully long time periods. Noncoherent integration requires only that the nominal sound speed of the propagation path remain the same over the pulse ensemble period, and this does obtain. In noncoherent integration, M pulses are combined after the envelope detectors by sampling the data stream simultaneously at the transmit pulse spacings and summing the samples obtained. The threefold relationship between P_{FA}, P_D, and SNR is given by Levanon[5] in terms of the integral $\Phi(T)$:

$$\Phi(T) = \frac{1}{\sqrt{2\pi}} \int_T^{\infty} e^{x^2/2} dx = \frac{1}{2} \operatorname{erfc}\left(\frac{T}{\sqrt{2}}\right). \quad (34)$$

Since the inverse of the complementary error function given by Eq. (34) appears in many of the SNR equations to follow, it is plotted in Fig. 6:

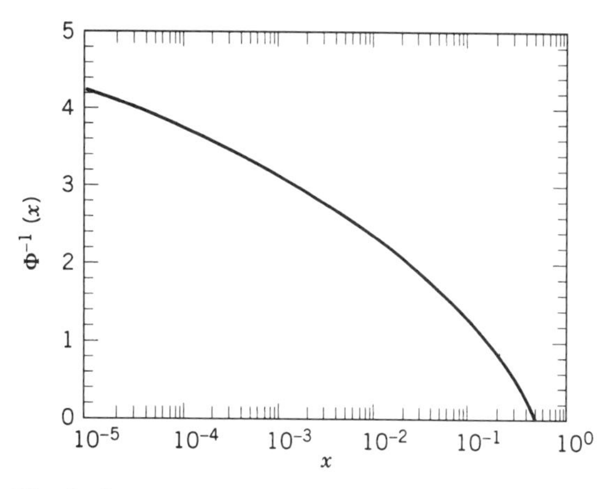

Fig. 6 Inverse complementary error function used in detection SNR equations.

$$\text{SNR} = \frac{\Phi^{-1}(P_{\text{FA}}) - \Phi^{-1}(P_D)}{\sqrt{M}}. \quad (35)$$

There is an Albersheim approximation to this threefold relationship that is analogous to Eq. (33):[8]

$$10 \log \text{SNR} = -5 \log M + \left[6.2 + \frac{4.54}{\sqrt{M + 0.44}}\right] \cdot x \log(A + 0.12AB + 1.7B). \quad (36)$$

Comparison of Eq. (36) with Eq. (33) shows that the noncoherent summation of M pulses requires a lower SNR to achieve a given P_D and P_{FA}. The required SNR is reduced by approximately $\sqrt{M}$. Therefore, noncoherent summation of pulses is very beneficial.

Detection of Single Fluctuating Pulses in Stationary Noise When considering pulse train to pulse train detection with fluctuations between trains, the simplest case is for a single pulse per train. The signal power of the echoes have a Rayleight pdf. For this case, the threefold relationship between P_D, P_{FA}, and SNR is particularly simple:

$$P_D = P_{\text{FA}} e^{1/(1+\text{SNR})}. \quad (37)$$

If SNR = 10, then the value of P_D required for $P_{\text{FA}} = 10^{-4}$ is 0.43. This compares to a value of 0.60 from Fig. 3.4 in Levanon.[5] Therefore, the Rayleigh fluctuations in echo amplitude cause a 28% decrease in the probability of detection at $P_{\text{FA}} = 10^{-4}$.

Detection of Multiple Fluctuating Pulses in Stationary Noise When the advantages of multiple pulse echo noncoherent integration are sought, it is usually found that the transmission loss over the propagation path changes between adjacent pulse trains and sometimes even within a pulse train. Four different cases are commonly differentiated. In Swerling cases I and III there is no amplitude change during the integration time, while in cases II and IV there is. In Swerling cases I and II the echo amplitudes are assigned a Raleigh pdf, while in Swerling cases III and IV the echo amplitudes are assigned a chi-square pdf. The Rayleigh pdf for echo amplitude represents physically the echoes from a large number of targets having similar target strength but randomly different ranges. This condition is met for large distributed targets such as surface or bottom reverberation. The chi-square pdf for echo amplitude represents physically the echoes from a dominant nonfluctuating target immersed in a sea of distributed targets. This condition is met for a point target surrounded in reverberation from an extended target at about the same range.

For the Swerling case I, the threefold relationship between P_D, P_{FA}, and SNR is given by Levanon.[5] The SNR is still defined as the ratio of signal energy to noise PSD [power spectral density; see Eq. (20)], but in this case of fluctuating signals, the signal energy is taken as the average of the Raleigh-distributed signal power times the pulse width:

$$\mathrm{SNR} = \frac{-\Phi^{-1}(P_{\mathrm{FA}})}{\ln(P_D)\sqrt{M}}. \tag{38}$$

Comparison of Eq. (38) with Eq. (35) shows that the SNR needed for specified values of P_{FA} and P_D is reduced by the square root of the number of pulses noncoherently summed both for fluctuating and for nonfluctuating targets.

For the Swerling II case the signal power of each pulse is an uncorrelated random variable, having Rayleigh pdf, within the pulse train. The threefold relationship between P_D, P_{FA}, and SNR is given by Levanon:[5]

$$\mathrm{SNR} = \frac{\Phi^{-1}(P_{\mathrm{FA}}) - \Phi^{-1}(P_D)}{\Phi^{-1}(P_D) + \sqrt{M}}. \tag{39}$$

Consideration of the different Swerling cases I and II provides an important result. For values of the operating parameters $10^{-7} < P_{\mathrm{FA}} < 10^{-3}$ and $0.1 < P_D < 0.9$ the SNR values required for detection are less for case II than for case I. The reason for this is that when no fluctuations take place within a pulse train, the whole train will be undetectable in a fade. When fluctuations do take place during the train, the fades are averaged out by the pulse train itself. This is an important consideration in designing active sonar pulse trains, because the fading time constant is a function of frequency, range, and range rate.

Swerling cases III and IV assume a chi-square distribution for the echo amplitudes. For the Swerling case II the signal power of each pulse is an uncorrelated random variable, having Rayleigh pdf, within the pulse train. Approximations to the threefold relationship between P_D, P_{FA}, and SNR are given by Levanon.[5] These relationships are accurate to a decibel for $M < 100$:

$$\mathrm{SNR} = \frac{2}{MQ}\left[\sqrt{M}\Phi^{-1}(P_{\mathrm{FA}}) + 2\right], \tag{40}$$

where

$$\ln(P_D) = \ln(1 + Q) - Q.$$

The threefold relationship between P_D, P_{FA}, and SNR for the Swerling case IV is also given by Levanon.[5] It is found to be identical to Eq. (35), which was given for noncoherent integration of nonfluctuating targets.

Binary Integration of Pulse Trains Binary integration is another way to implement noncoherent integration over the pulse echo train. After the thresholding operation, the data stream becomes a sequence of 0's and 1's, one digit for each time interval (range bin) in which an echo could occur, with 1's representing only the time intervals when the threshold was crossed. If N pulses were transmitted, the binary integrator samples the data stream at N range bins separated from each other by the time intervals between the N pulses. A detection decision is made whenever $M < N$ occurrences of 1's are found within the N range bins. Clearly the choice of M is similar to selecting a secondary threshold. If M is large or equal to N, then the SNR of all N echoes must be large enough to cross the primary threshold. Likewise, the chance that noise could exceed the threshold in all N range bins simultaneously is small. But if M were much smaller than N, then the probability that M echoes would cross the threshold is larger and the probability that M noise threshold crossings would occur is also larger.

The probability of detection for M of N binary integration is given by Levanon in terms of the probability p that an echo in a range cell will exceed the threshold value.[5] The probability of false alarm is given in terms of the probability P_F that noise will exceed the threshold in one range cell:

$$P_D = \sum_{i=M}^{N} \binom{N}{i} p^i (1 - P)^{N-i},$$

$$P_{\mathrm{FA}} = \binom{N-1}{M-1} P_F^M (1 - P_F)^{N-M+1}. \tag{41}$$

If the target range rate is nonzero, then the time spacing of the echoes will not be the same as the time spacing of the transmitted pulses. The time axis of the echoes will be scaled according to Eq. (26). In this event, no detections will ever occur because the binary integrator's range cells will not match more than one echo. The solution is to construct a family of binary integrators, each having its N range cells spaced by a differently scaled version of the transmission sequence. Each of these Doppler integrators is tuned to a different range rate. When detections occur, the target range is determined by the range cell having the detection, and the target range rate is determined by the binary integrator that made the detection.

When a region of intense backscatter, or clutter, is encountered, many range cells will contain echoes that cross the threshold. In this case, the M-of-N detectors will indicate detections at several contiguous range cells. Since a ship target could not return such a continuum of echoes, the fact of detection in contiguous range cells is sufficient to classify the target as clutter.

3.4 Signal Normalization for Constant False Alarm Rate

All the detection decision strategies described in the previous section assumed the noise to be stationary; that is, it can be characterized for all time by constant values of PSD and all moments of noise amplitude. This situation does not obtain in sonar practice. Ambient noise mean values fluctuate from time to time with a standard deviation of about 5 dB. Sonar self-noise can fluctuate by 10 dB or more as the platform speeds up or turns. Reverberation from continuous surfaces, such as the bottom, can easily increase the background noise by 10–20 dB. But the false alarm rate is extremely sensitive to changes in background noise. For the Gaussian noise, the probability of false alarm in a range cell is given by Eq. (31). If the desired $P_{\rm FA}$ is 10^{-4}, then the threshold must be set at a value equal to 4.29 times the rms noise amplitude. When the noise power increases by just 3 dB, then the $P_{\rm FA}$ increases by a factor of 100 to a value of 10^{-2} if the threshold is not changed.

There are many strategies for normalizing the outputs of the envelope detectors to achieve a constant false alarm rate (CFAR). The general idea is to form an estimate of the background noise in each range cell and then to divide the envelope amplitude in each cell by its own noise estimate. This normalized envelope will fluctuate, even in the absence of an echo, because the noise estimate for each cell will be only an estimate, not its true value. The detection threshold is set, often empirically, above the typical values achieved by the fluctuating normalized cell amplitudes, to achieve a specified false alarm rate. When the range cell contains an echo, its normalized envelope will cross the threshold if the SNR is large enough. The challenge is to avoid overestimating the noise, as will occur if false target or target echoes are present in the cells from which the noise estimates are made, for then the cell with the signal will be "normalized" too much and the detection will be missed. Equally important is to not underestimate the noise, for then the normalized outputs of cells containing only noise are increased, causing too many false alarms.

Cell-Averaging CFAR In a cell-averaging CFAR system the noise estimate is made from adjacent range cells, Doppler cells, azimuthal cells, or any combination of the three. The "split-window" normalizer uses two contiguous groups of $M/2$ cells on each side of the cell under test, separated from the cell under test by "guard bands" that are a few cells wide. The guard bands are to ensure that the echo from a spread target does not spill over into the cells from which the noise estimate is made. An analysis of the cell averaging is given by Levanon[5] for the case of a single pulse whose echo fluctuates with a Rayleigh pdf. The cell under test has been normalized by the average of M cells, half on each side of the cell under test.

$$\mathrm{SNR} = \frac{(P_D/P_{\rm FA})^{1/M} - 1}{1 - P_D^{1/M}}. \tag{42}$$

In the limit as M approaches infinity, the noise estimate will be perfect, and the SNR′ necessary for detection is given by

$$\mathrm{SNR}' = \frac{\log(P_{\rm FA}/P_D)}{\log P_D}. \tag{43}$$

The normalization loss associated with the CFAR process is defined as the increase in SNR required for CFAR detection as compared to the SNR′ needed for unnormalized detection:

$$\mathrm{Loss}_{\rm norm} = \frac{\mathrm{SNR}}{\mathrm{SNR}'}. \tag{44}$$

As a typical example, a system may be set to operate at a P_D of 0.5 and at a $P_{\rm FA}$ of 10^{-4}, with $M = 50$ noise estimation cells. For this example, the SNR for normalized operation is 13.5, and the SNR′ for unnormalized operation is 12.3. The normalization loss is 1.1, or 0.4 dB.

There are two situations that defeat the proper operation of this simple cell-averaging CFAR. When a rangewise region of clutter spans part but not all of the noise estimation cells, then the noise estimate can be too large or too small, depending on whether the cell under test is in the clear. Also, when one or more strong echoes lie in some of the noise estimation cells, then the noise estimate will be too large. The result can cause either too many false alarms or missed detections. The common way to mitigate these effects is to eliminate from the noise estimate all cells that have unusual amplitudes.

The method of censoring suspected target or false target echoes in the averaging cells is called multiple-pass outlier removal. In this method, all cells are used to form the first-pass estimate of the average noise amplitude. All cells are censored that have amplitudes that exceed the average estimate by a factor proportional to the standard deviation of the cell amplitudes. The remaining cells are used to form the second-pass estimate of the average noise amplitude, and again any cells used in the average are censored as before. The average formed from the remaining cells is then used to normalize the cell under test.

If the cell amplitudes were statistically homogeneous, then the problems leading to cell censoring would not occur. Whenever the cell amplitudes are so inhomogeneous that the censoring technique does not work well, then an alternative method called clutter map CFAR is available.

Clutter Map CFAR Clutter map CFAR uses averages of past values from the cell under test to normalize its present value. In order to work, clutter map CFAR requires that the cell amplitudes be independent from pulse train to pulse train. This requirement is often met because of platform motion and propagation fluctuations induced by ocean motion. Clutter map CFAR is memory intensive, for it requires that 50 or more past scans be stored for every range, Doppler, and azimuth cell. A common way to reduce the memory requirement is to form a moving average by adding a portion of each new cell amplitude to the current average value. This method is called exponential cell smoothing. If a_n is the nth value of the cell amplitude and z_n is the nth cell average estimate, then

$$z_n = (1 - W)z_{n-1} + Wa_n \qquad (45)$$

The SNR relationship with P_D and W is plotted for $P_{\mathrm{FA}} = 10^{-6}$ in Fig. 7. The $W = 0$ case represents no CFAR thresholding in a stationary noise field. The increased SNR required in a nonstationary field, dealt with by exponential cell smoothing, is the CFAR loss. When $W = 0.5$, the loss is 7 dB at $P_D = 0.5$.

3.5 Signal Processing by Passive Sonar Systems

The signals to be detected by passive systems are determined by the targets and are classified into two different types. The radiated-noise source PSDs of ships are described in Chapter 35. The first classification of signals to be detected is narrow-band spectral components, the so-called line components. The second classification of signals is the continuous-spectra component, the so-called broadband component. Different signal processing chains are needed for each type of signal.

Narrow-Band Passive Detection Processing The matched filter for a narrow-band signal is simply a filter with a passband that has the same bandwidth B as the signal. The SNR at the matched filter output is equal to the signal power S divided by the mean noise power N. It is convenient to write SNR in terms of the noise PSD:

$$\mathrm{SNR} = \frac{S}{N_0 B}. \qquad (46)$$

The outputs of the matched filters (a bank of narrow-band filters) are envelope detected in the same way as active signals. The postdetection smoothing is likewise similar to active signal processing, in that the signal is averaged for a time T that is equal to the time span of the signal. In the passive case, T might seem to extend for as long as the target is in view, but this is not the case. Signal fading of the kind discussed for active signals is impressed by motion through the medium, and T is limited to the characteristic time of fluctuations in propagation loss. The postdetection smoothing reduces the noise variance by a factor that is a function of the smoothing time–bandwidth product BT. This factor is plotted in Fig.

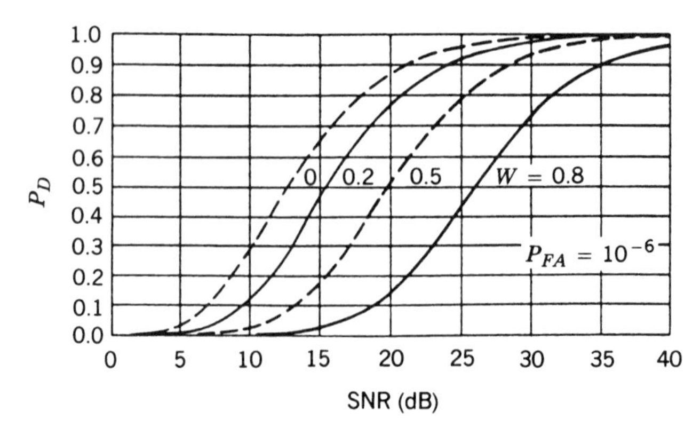

Fig. 7 Plot of P_D versus SNR for clutter map CFAR with exponential averaging weight W (Rayleigh fluctuating target). (From Ref. 5, Fig. 12.6.)

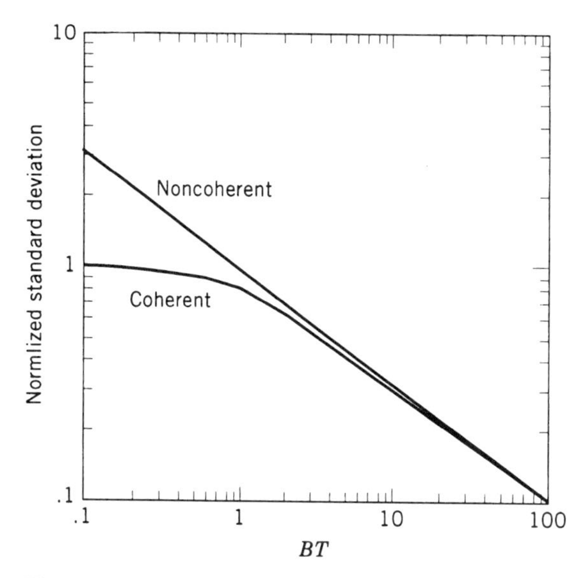

Fig. 8 Normalized noise standard deviation at the output of an envelope detector for white band-limited Gaussian noise versus bandwidth averaging–time product BT.

8, where it is seen to asymptote to $(BT)^{-0.5}$. Assuming that BT is large enough (larger than about 3), the required SNR* at the square-law detector and smoother output is

$$\text{SNR}^* = \left(\frac{S}{N}\right)\sqrt{BT} = \left(\frac{S}{N_0}\right)\sqrt{\frac{T}{B}}. \qquad (47)$$

The narrow-band detection threshold is the signal power to noise PSD ratio, in decibels, necessary to achieve the desired SNR*:

$$\text{DT}_0 = 10\log\left(\frac{S}{N_0}\right) = 10\log(\text{SNR}^*) + 5\log\left(\frac{B}{T}\right). \qquad (48)$$

The values of SNR necessary for specified detection and false alarm probabilities are given in Section 3.3. Equation (47) shows that DT is minimized (and therefore performance is maximized) by small bandwidths and large smoothing times as long as B is large enough to completely contain the signal.

Broadband Passive Detection Processing The matched filter for passive broadband signals is the Eckart filter. The Eckart filter has the spectral shape of the signal, weighted by the noise PSD.[9] Since neither the signal nor the noise spectral shapes are usually known beforehand, the total bandwidth of the broadband filter is simply kept small, so that both the signal and noise are likely to be white within the filter band. In this case, the SNR* at the output of the square-law detector and averager is, where again the asymptotic value of noise variance reduction is taken,

$$\text{SNR}^* = \left(\frac{S_0}{N_0}\right)^2 BT. \qquad (49)$$

The broadband detection threshold is the signal PSD to noise PSD level required for detection at the specified performance:

$$\text{DT} = 10\log\left(\frac{S_0}{N_0}\right) = 5\log\left(\frac{\text{SNR}^*}{BT}\right). \qquad (50)$$

Equation (50) shows that performance is maximized by large values of both bandwidth and smoothing time.

Normalization in Passive Detection Noise interference in passive receivers is the same as it is in active ones, apart from reverberation. The normalization process that removes the effect of changes in time, frequency, and bearing of the mean noise values is described in Section 3.4. In narrow-band passive detection, the usual detection decision is based upon signal power changes over time and frequency. Normalization of the envelope-detected matched filter outputs across the frequency coordinate is called *noise spectral equalization* (NSE), which is very similar to the cell-averaging CFAR described in Section 3.4, but in NSE the noise estimate for the cell under test is made from adjacent frequency cells. Whenever loud signals (outliers) from other targets or from other spectral components from the same target occur in the noise-estimating cells, they are usually censored by a two-pass outlier replacement scheme.

The first noise estimate is made as an average of all cells in the noise-estimating windows. Next, each cell value in the noise-estimating windows is compared to this initial average. Each cell value that exceeds the initial average value by a factor α is replaced by the initial average. A second noise estimate is made from the new values in the noise-estimating window. Each cell value that exceeds the second average by the factor α is replaced by the second average. Finally, the cell under test value is divided by the second noise estimate. If this ratio exceeds the detection ratio, a detection is called.

4 PERFORMANCE PREDICTION FOR ACTIVE SONAR SYSTEMS

The principal tools for performance prediction, called the sonar equations, relate the mean values of the principal sonar system and environmental parameters. The sonar equations are written in terms of detection threshold in this section. In practice, the required value of DT is determined as a function of the SNR needed for specified P_D and P_{FA} values. The sonar equations are then solved for transmission loss (TL). The value of TL thus determined is named the system figure of merit (FOM). The FOM is clearly a function of nonsystem parameters such as TS and N_0, so that FOM must be quoted together with the target and noise environment used. When system performance at ranges less that the FOM range are considered, the difference in TL between the FOM range and the reduced range is called the signal excess.

4.1 Performance of Active Sonar Systems

The fundamental parameter that determines the performance of an active sonar system is its SNR, defined by Eq. (20) as the ratio of signal energy to noise PSD at the processor input. The detection threshold is the required value of this ratio, expressed in decibels, to achieve a specified detection and false alarm probability, as given by Eq. (30). The sonar equation appropriate for noise-dominated interference is derived from the definition of detection threshold. (Another form of the sonar equation, written to describe the ratio of signal power to noise PSD, is given by Urick[7] and Tucker and Gazey[10]):

$$\mathrm{DT} = \mathrm{ESL} - \mathrm{TL}_1 + \mathrm{TS} - \mathrm{TL}_2 + \mathrm{SG} - L - N_0 - \mathrm{NG} \quad (51)$$

where DT is the detection threshold, the signal energy to noise power spectral density ratio, in decibels, needed for the specified detection and false alarm probability; ESL is the energy source level of the transmitted waveform, in dB re $\mu\mathrm{Pa}^2 \cdot \mathrm{s/m}$, directed at the propagation path leading to the target; TL_1 is the transmission loss from the transmitter to the target in decibels; TS is the target strength of the target in decibels; TL_2 is the transmission loss from the target to the receiver in decibels; SG is the signal gain of the receiving array, in decibels, in the beam directed at the propagation path from the target; L is the sum of all system processing losses in decibels; N_0 is the noise power spectral density at the receiving array in dB re $\mu\mathrm{Pa}^2/\mathrm{Hz}$; and NG is the noise gain of the receiving array in decibels. It is common to define the receiving array gain as AG = SG − NG in the noise-limited sonar equation.

A form of the sonar equation valid for distributed reverberation interference, applicable when the reverberation power in a receiving beam dominates the noise power, is

$$\begin{aligned}\mathrm{DT} = {} & \mathrm{TL}_1' - \mathrm{TL}_1 + \mathrm{TS} - \mathrm{SS} - 10\log A_s + \mathrm{TL}_2' - \mathrm{TL}_2 \\ & + \mathrm{SG} - \mathrm{SG}' - L + L', \end{aligned} \quad (52)$$

where primed symbols represent the reverberation paths for the same quantities unprimed; SS is the surface or bottom scattering strength in dB re m^2, or the volume scattering strength in dB re m^3; and A_s is the scattering area in square metres, or the scattering volume in cubic metres. The values of signal gain and system loss can be different for the reverberation signal than for the target signal because the two "targets" are not colocated and do not have the same size. The scattering area for bottom- and surface-distributed targets is equal to $\frac{1}{2}r\Psi ct$, where r is the range to the scattering area, Ψ is the effective azimuthal beamwidth of the receiver beam, c is the speed of sound, and t is the effective pulse length after matched filtering. The scattering volume for volume-distributed targets is equal to $\frac{1}{2}r^2\Psi\Theta ct$, where Θ is the vertical beamwidth of the receiver beam.

The sonar equations relate the mean values of the processes that comprise the signal chain. It is often necessary to estimate the variance of the SNR to determine the range of performance changes to expect. The symbol for the standard deviation of a process is S_a, where a designates the process

$$S_{\mathrm{DT}}^2 = S_{\mathrm{TL}_1}^2 + S_{\mathrm{TS}}^2 - 2CS_{\mathrm{TL}_1}S_{\mathrm{TS}} + S_{\mathrm{TL}_2}^2 + S_{\mathrm{AG}}^2 - S_{N_0}^2. \quad (53)$$

If all of the individual processes in the echo signal chain were statistically independent, then the variance of the detection threshold values would equal the sum of the variances of all of the other processes. If, on the other hand, two of the processes are correlated with each other, then a term containing the correction coefficient C appears. In Eq. (53) the outgoing transmission loss TL_1 is correlated with target strength. This kind of correlation can occur whenever the target is large with respect to the spatial scale of amplitude fluctuations in the transmitted sound field. In this case, the target averages out some of the fluctuations in the sound field. No term is included in Eq. (53) for variations in ESL, because all pulse trains are transmitted alike. An equation for the expected variance in DT values for reverberation-limited operation can be developed from Eq. (52). Correlation can occur between the outgoing transmission loss and the scattering strength term, because the scattering area is usually larger that the area over which variations of transmission loss occur.

4.2 Performance of Passive Sonar Systems

The passive sonar equations yield the expected value of the detection threshold in terms of the mean target radiated-noise source level, the mean transmission loss, the mean array gain, and the mean background noise. Two different detection thresholds are defined for passive sonar. The detection threshold for narrow-band detection,

DT_0, is defined in Eq. (48) as the required decibel ratio of the signal power and noise PSD at the matched filter input. The detection threshold for broadband detection, DT, is defined by Eq. (50) as the required decibel ratio of the signal PSD and noise PSD at the matched filter input.

Target-radiated noise source levels, called target signatures, are the subject of Chapter 94. The broadband components of target signatures are characterized by their PSD, referred to 1 m, and denoted by SL in dB re $\mu\text{Pa}^2/\text{Hz/m}$. The narrow-band components of target signatures are characterized by their power levels, referred to 1 m, and denoted by SL_0 in dB re $\mu\text{Pa}^2/\text{m}$.

The broadband passive sonar equation is

$$\text{DT} = \text{SL} - \text{TL} - N_0 + \text{AG} - L. \tag{54}$$

The narrow-band passive sonar equation is

$$\text{DT}_0 = \text{SL}_0 - \text{TL} - N_0 + \text{AG} - L. \tag{55}$$

These equations relate mean values of the processes to yield the expected value of the detection thresholds. Since none of the processes are correlated with each other, the variances of the detection thresholds are given by the sum of the variances of each process:

$$S^2_{\text{DT}} = S^2_{\text{TL}} + S^2_{\text{SL}} + S^2_{\text{AG}} + S^2_{N_0}. \tag{56}$$

The variances of source levels, SL, are small for submarines but can be large for surface ships. Surface ship source levels vary periodically with the ship's encounter rate with sea waves.

REFERENCES

1. B. D. Steinberg, *Principles of Aperture and Array System Design*, Wiley, New York, 1976.
2. C. L. Bartberger, *Lecture Notes on Underwater Acoustics*, Defense Documentation Center, Alexandria, Virginia, 1965.
3. R. O. Neilsen, *Sonar Signal Processing*, Artech House, Boston, 1985.
4. W. B. Davenport and W. L. Root, *An Introduction to the Theory of Random Signals and Noise*, McGraw-Hill, New York, 1958.
5. N. Levanon, *Radar Principles*, Wiley, New York, 1988.
6. M. I. Skolnik, *Introduction to Radar Systems*, 2nd ed., McGraw-Hill, New York, 1980.
7. R. J. Urick, *Principles of Underwater Sound*, McGraw-Hill, New York, 1975.
8. W. J. Albersheim, "A Closed-Form Approximation to Robertson's Detection Characteristics," *Proc. IEEE*, Vol. 69, 1981.
9. W. S. Burdic, *Underwater Acoustic System Analysis*, Prentice-Hall, Englewood Cliffs, NJ, 1984.
10. D. G. Tucker and B. K. Gazey, *Applied Underwater Acoustics*, Pergamon Press, New York, 1966.

38

OCEANOGRAPHIC AND NAVIGATIONAL INSTRUMENTS

ROBERT C. SPINDEL

1 INTRODUCTION

A wide variety of fundamental ocean measurements are made with underwater sound. The characteristics of sound propagation are affected by the properties of seawater itself, its temperature, salinity, and chemical composition, by objects within the ocean volume, and by interaction with surface and bottom boundaries. Thus, information about these parameters is embedded in propagating sound waves and can be extracted by appropriate techniques. Acoustic devices are used to measure water depth, geologic properties of marine sediments, currents, turbulence, internal waves, mesoscale variability, bubbles created by breaking waves, rainfall, particulate matter in the water column, and fish and plankton density, distribution, and type. Acoustic systems are also used to measure ship speed through the water and over the bottom and to provide high-accuracy positioning and navigation.

2 GEOLOGIC MEASUREMENTS

2.1 Echo Sounders

The most widely used acoustic instrument is the so-called fathometer, echo or depth sounder, which in its most common form measures the round-trip travel time of an acoustic pulse emitted by a transducer at the ocean surface, reflected from the ocean bottom and received back at the same transducer. Travel time is converted to range by multiplying half the round-trip time by the speed of sound c. Most echo sounders are calibrated for nominal sound speeds $c = 1500$ m/s or 800 fathoms/s (4800 ft/s). Accuracies achieved with no correction for departures from these nominal values are within a few percent of water depth. Where greater precision is required, corrections are made based on catalogued historical sound speed data or by measuring local sound speed directly. Carter has provided correction tables based on historical observations of worldwide sound speed profiles[1].

Factors affecting the performance of an echo sounder are its operating frequency, the characteristics of the emitted signal—its amplitude, duration, and modulation—the size of the transmitter and receiver aperture, and signal conditioning and processing algorithms. Depth accuracy is governed by the precision of the travel time measurement, σ_t, which is controlled by signal bandwidth BW and received signal-to-noise ratio SNR:

$$\sigma_t \propto \frac{1}{\mathrm{BW}\sqrt{\mathrm{SNR}}}. \tag{1}$$

Spatial resolution is governed by beamwidth ϕ, which is proportional to wavelength λ and inversely proportional to aperture size. For both a circular piston transducer of diameter d or continuous line array of length d, the full half-power beamwidth is approximately $\phi \approx \lambda/d$. The usual Rayleigh criterion for resolution implies that two features a distance R from the sonar will be resolved if their spatial separation exceeds $R\phi$. Since the earliest returned echo is the assumed bottom depth, wide beams that may span local depressions or elevations can result in erroneous readings.

Depth sounders intended for shallow, coastal waters operate in the 30–100-kHz range. Beams a few degrees wide can be produced with transducers only a few centimetres in diameter. The most common echo sounders for deep water operate near 12 kHz where a typical

Note: References to chapters appearing only in the *Encyclopedia* are preceded by *Enc.*

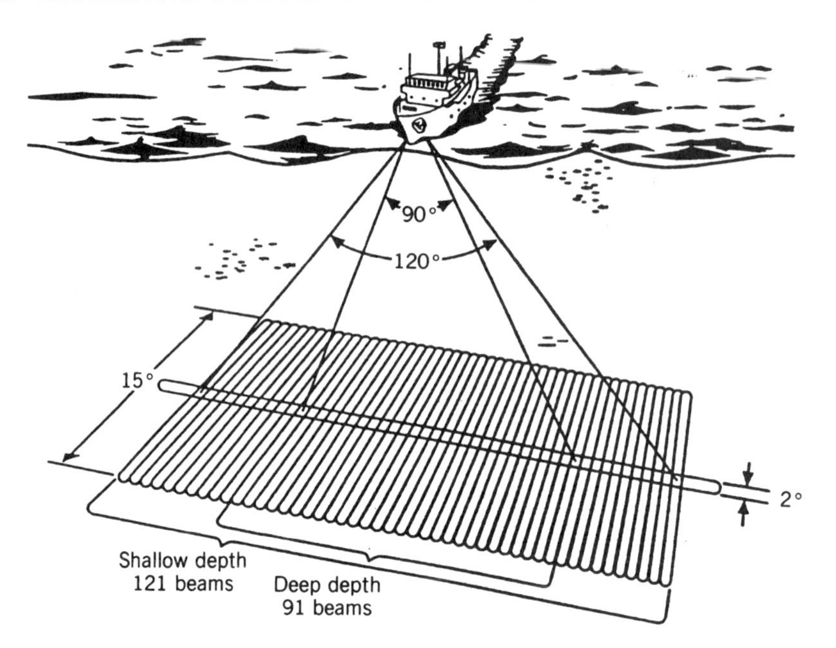

Fig. 1 Beam patterns of a typical multibeam, swath echo sounder. Swath width depends on water depth. Coverage in 3000 and 5000 m depth waters is about 120° and 90°, respectively. (Courtesy SeaBeam Instruments, Inc., East Walpole, MA.)

30-cm-diameter transducer produces an approximate 25° beam that resolves bottom features about 2 km apart in 5-km-deep waters. Advanced echo sounders use phased arrays with hull size apertures to transmit and receive simultaneously on many narrow beams, thereby providing high resolution and broad coverage. A typical multibeam swath echo sounder, as depicted in Fig. 1, has 1.5° beams, providing spatial resolution of about 150 m in 5-km-depth water. From 40 to 150 beams enable the system to sweep out a swath on the bottom up to 20 km wide. The beams are electronically stabilized to compensate for vessel pitch, heave, and roll. Automatic contour plotting algorithms convert measured travel times directly into bathymetric charts.

Precision echo sounders allow pulse duration to be varied from a fraction of a millisecond to tens of milliseconds. Shorter pulses provide better depth resolution but produce less intense echoes and therefore lower SNR. Pulse shapes are varied; the most common is rectangular. Some depth sounders transmit modulated signals, and returning echoes are processed by replica correlation. This results in SNR gain by deemphasizing noise that is uncorrelated with the signal.

2.2 Sub-bottom Profilers

Below about 3.5 kHz, energy penetrates readily to the bottom, where it is reflected from discontinuities and inhomogeneities in the sediments, thereby revealing sub-bottom stratigraphy and structures. Conventional piezoelectric and magnetic transducers are generally used at frequencies above 100 Hz; explosive materials, airguns, and electromechanical devices are used for lower frequencies. (See *Enc.* Ch. 37 and 47.)

2.3 Side-Scan Sonars

The side-scan sonar operates by transmitting a fan-shaped beam perpendicular to its direction of motion.[2] The beam, which is narrow in azimuth (the direction of motion) and wide in elevation, is produced by a linear array of transducers. Figure 2 is an example of the output of a typical side-scan sonar.

Reflections from the seafloor produce a shadowgraph of the swath swept out by the sonar. Some side-scan systems incorporate a second row of transducers parallel to the first to obtain differential phase or differential arrival time measurement of returning signals in order to estimate bottom relief. Such systems produce bathymetry as well as imagery. High-resolution side-scan sonars in the 100–500-kHz range are used for precision surveys for pipeline installation and detailed ocean bottom search. Their range is limited to several hundred metres due to the high acoustic attenuation at these frequencies, but the small acoustic wavelength allows resolution of 10 cm to 1 m with practical aperture sizes. Because of their limited

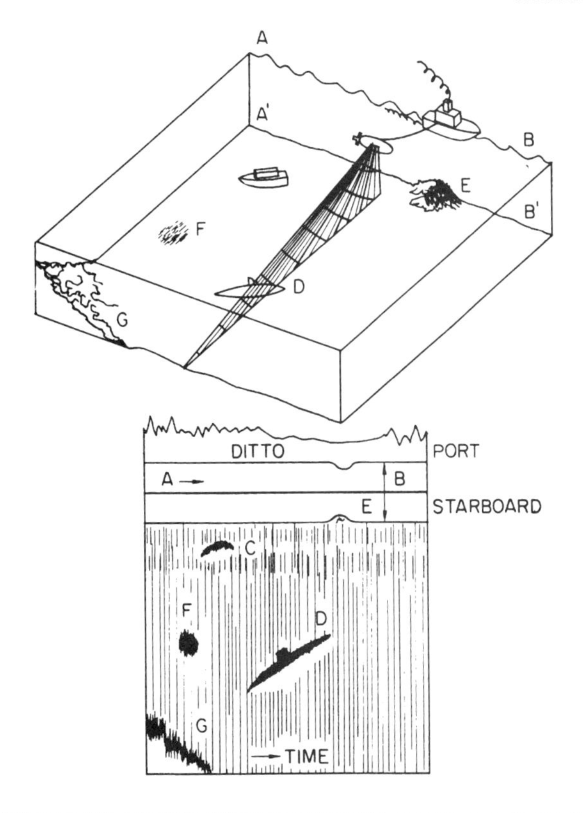

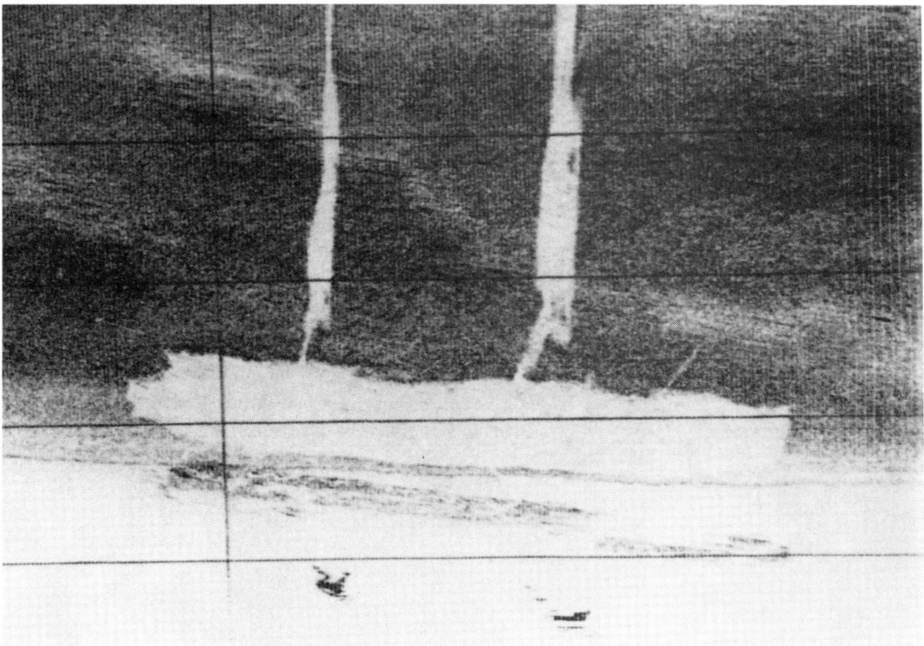

Fig. 2 Top drawings show how the side-scan sonar image is developed from a narrow beam along the direction of travel. (From H. Edgerton, *Sonar Images*, 1986, p. 15. Reprinted by permission of Prentice-Hall, Englewood Cliffs, NJ.) Bottom panel is a side-scan image of the *Breadalbane* on the seafloor in the High Arctic. (Courtesy Klein Associates, Inc., Salem, NH.)

range, in deep water they must be towed near the bottom. Low-frequency side-scan sonars operating at several kilohertz provide kilometer ranges and broad-area coverage but with less resolution because of the very large array needed to produce a narrow beam. They are used primarily for geologic surveying.

2.4 Synthetic Apertures for Echo Sounding

Synthetic aperture sonars, which operate on the same principle as synthetic aperture radars wherein an aperture of length L is synthesized from N subapertures of length $l = L/N$, have been tested but have not been widely used. The attractive feature of such systems is that synthetic apertures many wavelengths long can be formed with a smaller, real aperture system moving along a known (usually straight) path, thereby providing very narrow beams with consequently high spatial resolution mapping. Unfortunately, position of the subapertures must be known accurately, to within $\lambda/2$, to achieve full array gain, and this difficult requirement has limited the implementation of synthetic aperture sonars.

2.5 Seafloor Sediment Properties

Empirical geoacoustic models that relate acoustic properties of marine sediments, such as sound speed, impedance, and attenuation, to physical properties such as density, porosity, mean grain size, compressibility, and chemical composition are used to characterize sediments by acoustic measurements.[3] Signals reflected from the seafloor and sub-bottom are either fitted directly to empirical models to determine sediment type or inverted first to obtain impedance and attenuation coefficients, which are then related to sediment type. Wide-band sonars operating at 2–25 kHz have demonstrated most success.[4]

3 PHYSICAL OCEANOGRAPHY

3.1 Current Meters

Travel Time Instruments The difference in travel time, δt, of acoustic pulses traversing the same path r in opposite directions is proportional to the velocity of the current, u, along the path,

$$\delta t = \frac{2ru}{c^2}, \qquad (2)$$

where c is the average speed of sound along the path. Current meters based on this principle are in wide use. Typically, $r \sim 10$ cm, so $\delta t \sim 10^{-9}$ s for $u = 1$ cm/ s. Two orthogonal paths allow velocity to be resolved into vector components.

This technique also has been used to measure currents along $r \approx 1000$ km paths. (See Section 3.2.)

A related method for observing river and ocean flow depends on the advection by currents of particulate matter or turbulent cells that constitute acoustic impedance discontinuities. Scintillations in the forward scattered acoustic signal transmitted approximately orthogonal to the principal flow direction and received at two spatially separated points approximately parallel to the flow can be correlated, and the rate of advection can be calculated.[5]

Doppler Instruments Doppler shifts of acoustic pulses scattered from the ocean bottom or from scatterers within the water column are used to deduce water currents and ship speeds.[6] Acoustic Doppler current meters rely on scattering from particles or other acoustic impedance discontinuities in the water, such as bubbles or thermal microstructure that are swept along with the current. Instruments with multiple acoustic beams at various angles (usually orthogonal) allow computation of vector velocity. Estimates of the Doppler shift of short segments of the scattered return signal provide range resolution. The Doppler shift of a segment of the return signal t_0 seconds after transmission and τ seconds long gives an estimate of water velocity at range $ct_0/2$ with resolution $c\tau/2$. Typical Doppler current metres operate in the 50 kHz–200 kHz range, measure currents to centimetres per second accuracy, and when employed in the range-gated mode have ranges of several hundred metres and resolution of several metres.

Acoustic ship speed logs are described in Section 5.4.

3.2 Acoustic Tomography: Currents and Temperature

Ocean acoustic tomography is a technique for measuring the three-dimensional sound speed and current fields of an ocean volume. The travel time of acoustic signals transmitted between multiple points on the perimeter of a volume are related to the sound speed and current fields in the interior. Temperature change δT in degrees Celsius is related to sound speed change δc to first order; $\delta T \approx \delta c(m/s)/4.6$. Tomography consists of measuring the travel times and deducing the interior fields.[7]

Figure 3 shows a conceptual ocean tomography network sampling the ocean in both horizontal and vertical dimensions. Spatial resolution in the horizontal plane is determined by the number and placement of instruments. In deep-water applications, multipath propagation in the vertical plane arising as a result of the natural background ocean sound speed profile provides depth reso-

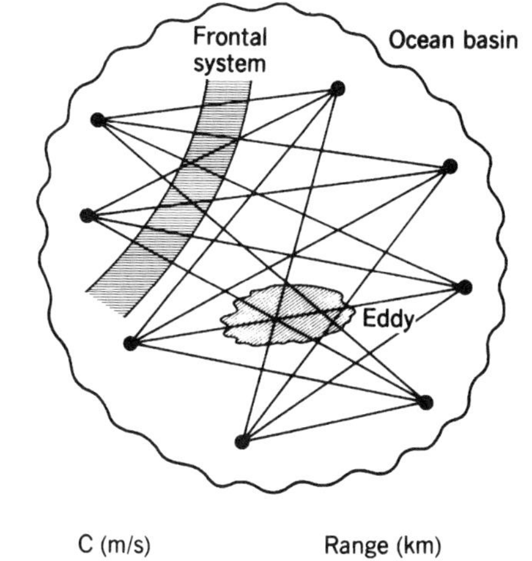

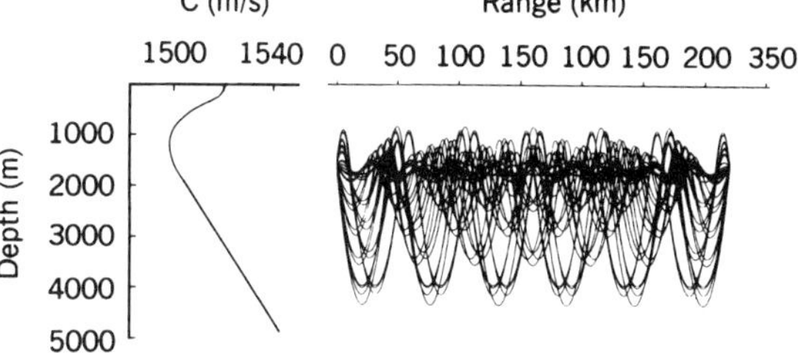

Fig. 3 In an ocean acoustic tomography system transceivers send and receive acoustic pulses along many paths, as shown in the upper panel plan view. Ocean features having different sound speed or current characteristics, such as eddies and frontal systems, alter the travel times of the pulses. The measured travel time changes are used to infer the sound speed and current fields interior to the tomographic array. The lower panel shows a typical natural background sound speed profile in the deep ocean (at left) and the multiple acoustic paths that exist between transceivers moored at 1200 m depth, 300 km apart. Acoustic energy directed downward is refracted upward, and vice versa, resulting in channeled, multipath propagation.

lution. Thus, a single instrument, usually at a depth near the axis of the deep sound channel, is sufficient at each measurement point. (See Chapter 31 for discussion of the deep sound channel.)

In the limit of geometric optics, the travel time along a path, i, in the presence of a current, $\overline{u}(\overline{x}, t)$, is given as

$$T_i(t) = \int_i \frac{ds}{c(\overline{x}, t) + \overline{u}(\overline{x}, t)}. \tag{3}$$

With a reference sound speed field $c_0(x)$ and a perturbation field $\delta c(x, t) \ll c_0(x)$, defined as

$$c(\overline{x}, t) = c_0(\overline{x}) + \delta c(\overline{x}, t) \tag{4}$$

and a reference travel time defined as

$$T_{0i} = \int_{0i} \frac{ds}{c_0(\overline{x})}, \tag{5}$$

the changes in travel time over the reference travel time, in opposite directions along the same path, due to the current and sound speed perturbation are

$$\delta T_i^+ = T_i^+ - T_{0i} = -\int_{0i} \frac{\delta c(\overline{x}, t) + \overline{u}(\overline{x}, t)}{c_0^2(\overline{x})}\, ds \tag{6}$$

and

$$\delta T_i^- = T_i^- - T_{0i} = -\int_{0i} \frac{\delta c(\overline{x}, t) - \overline{u}(\overline{x}, t)}{c_0^2(\overline{x})}\, ds, \tag{7}$$

respectively. Sums and differences of these two equations give

$$s_i = -2\int_{0i} \frac{\delta c(\overline{x}, t)}{c_0^2(\overline{x})}\, ds \tag{8}$$

and

$$d_i = -2\int_{0i} \frac{\overline{u}(\overline{x}, t)}{c_0^2(\overline{x})}\, ds. \tag{9}$$

Thus, the sum travel times are related to the sound speed perturbation; the difference travel times are related to the water velocity. The tomographic inverse problem determines the fields δc and u from measurement of the sum and difference travel times. The inversion is accomplished using standard linear inverse techniques in which the fields are parameterized with a finite number of discrete parameters derived from a model of the ocean.[8]

Tomography transceivers operating in the 200–500 Hz range have been used to measure ocean features over 1000-km basins.[9]

3.3 Inverted Echo Sounders

The round-trip travel times of acoustic pulses emitted by upward-looking echo sounders placed on moorings or on the ocean bottom are used to deduce changes in the depth of the main ocean thermocline, the region of rapidly decreasing temperature that separates warm surface waters from deep cold waters. The measurement is based on the fact that the integrated vertical sound speed is a function of vertical thermocline migration, increasing with decreasing thermocline depth.

Upward-looking sonars mounted on submarines and operating at high frequencies are used to monitor ice draft for submarine navigational purposes. Similar sonars operating near 300 kHz mounted on the seafloor or moorings are used to obtain time series of average ice thickness.

3.4 Wind and Rain

The impact of raindrops on the sea surface generates underwater ambient noise signatures in the frequency band 1–40 kHz depending on drop size, rainfall rate, and spatial density. Wind speed affects wave production and breaking, which in turn changes the background ambient noise.[10,11]. Thus, rainfall and wind speed can be deducted from ambient noise measurements (see Chapter 36).

3.5 Scatterometers

These are a class of high-frequency sonars used to measure a variety of phenomena based on backscattering from impedance discontinuities in the water column. They are used to study the intensity and distribution of material such as suspended particulate matter (pollutants, sediments), marine organisms, and bubbles.[12,13] These sonars operate in the range of several hundred kilohertz to several megahertz depending on the relative backscattering strength of the scatterers of interest. For a small nonresonant sphere of radius a, the backscattering cross section is

$$\frac{\sigma_s}{\pi a^2} = 4(ka)^4 \left[\left(\frac{e-1}{3e} \right)^2 + \frac{1}{3} \left(\frac{g^{-1}}{2g+1} \right)^2 \right], \quad ka \ll 1, \tag{10}$$

where e is the ratio of the elasticity of the sphere to water and g is the ratio of the density of the sphere to water. If the sphere is rigid, such as in the case of solid particulates, e, $g \gg 1$. Then, in the region $ka \ll 1$, known as the Rayleigh scattering region, scattering strength is proportional to f^4. Near $ka = 1$ there is a transition to a region of geometric scattering, where scattering strength no longer increases with frequency. If the scatterer is a gas bubble, such as in the case of fish swimbladders (see *Enc.* Ch. 44), or wave-generated bubbles, then e, $g \ll 1$, and scattering is large because the bracketed term dominates. At its resonant frequency f_r, where $a \approx \delta$, the scattering cross section of a bubble is larger still,

$$\frac{\sigma_s}{\pi a^2} = 4 \left(\frac{f_r}{\Delta f} \right)^2 = 4Q^2, \tag{11}$$

where Δf is the width of the resonance peak and Q varies from 10 to 100. Scatterer type and size are inferred from measurements of scattering strength and resonant frequency.[14]

4 BIOLOGICAL OCEANOGRAPHY

Sonars operating in the range of several kilohertz to 100 kHz are used routinely for fish finding and in some cases for estimating stock abundance (see *Enc.* Ch. 44). The fish swimbladder behaves roughly as a gas bubble. Bladder sizes, and therefore fish type and size, are differentiated by resonant frequency. At very high frequencies, several hundred kilohertz to several megahertz, scatterometers are used to study smaller species such as plankton, ctenophore, and some jellyfish.[13] (See Section 3.5.)

5 ACOUSTIC POSITIONING AND NAVIGATION

Acoustic systems are used for precise positioning and navigation of surface vessels and for underwater instruments and vehicles. They are used in oceanographic studies; geophysical prospecting; underwater exploration and surveying; subsea oil and gas production; vessel dynamic positioning; and test and performance evaluation of military systems such as sonars, torpedoes, and missile range and accuracy. These systems operate by measuring the time of arrival, phase, or Doppler shift of acoustic signals transmitted between a reference frame and the navigated point. In the case of missile tests, the origin of the underwater signature created by the impact is located by triangulation.

Acoustic positioning systems fall into three broad classes, long, short, and ultra-short baseline, distinguished by the separation between the reference elements. In a long-baseline system the reference elements are spaced about the same distance as the ranges from them to the navigated point, as shown in Fig. 4. Long baselines are commonly used when the reference elements are placed on the seafloor. In short-baseline systems the reference element spacing is much less than the ranges to the object. These are used in ship-mounted applications, where a long baseline, which gives greater precision, is not possible. Both the long- and short-baseline systems generally obtain ranges by measuring the transit time of acoustic pulses. The third class of system, the ultra-short-baseline system, has the reference element spacing ~λ. Azimuth and depression angles to the navigated object are obtained by measuring phase differences of a pulse emitted by the object and received at the reference elements (usually three). Ultra-short-baseline systems are compact but not as accurate as those with longer baselines.

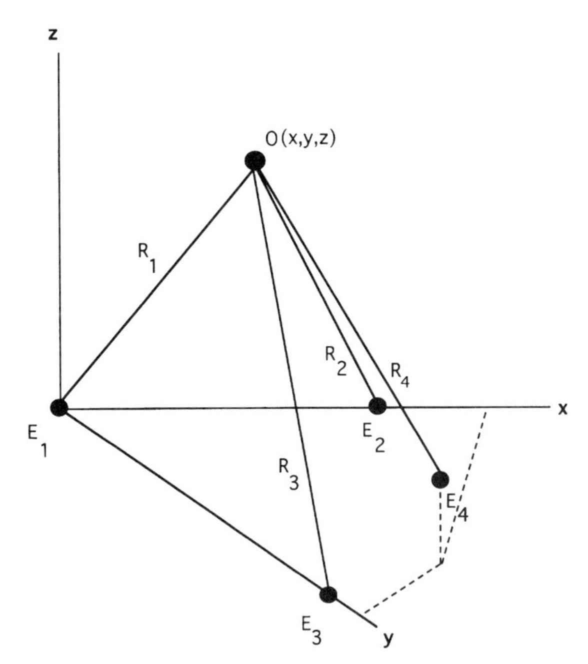

Fig. 4 Basic acoustic positioning system geometry. Four non-coplanar reference elements E_i locate the object $O(x, y, z)$ unambiguously. In most applications, three elements suffice, since it is known a priori where object lies with respect to reference element plane. Redundant ranges obtained from additional reference points beyond three or four can be used to reduce random errors due to system noise or fluctuations in acoustic travel times due to ocean inhomogeneities. Illustrated is a long-baseline system, where the ranges between the reference elements are roughly equal to the ranges to the navigated object. In a short-baseline system, distances between reference elements are much less than to the navigated object.

5.1 Travel Time Navigation

In a travel time system, the position of a point, $O(x, y, z)$, is computed from measurement of the ranges R_i obtained by determining the travel time of acoustic pulses transmitted between the reference elements and the point. Four non-coplanar reference elements locate $O(x, y, z)$ unambiguously; three locate it either above or below the plane defined by the three elements; and two locate it on a circle. All three implementations are used. When fewer than four elements are employed, the ambiguity is resolved by operational constraints. For example, if it is known that $O(x, y, z)$ is on the sea surface, three elements suffice; if it is known that $O(x, y, z)$ is both on the sea surface and on one side of the baseline defined by two reference elements, then two are sufficient.

To first order, the acoustic paths are assumed to be straight lines and the ranges are based on a nominal sound speed c or a computed average sound speed based on measurement of the local vertical sound speed profile $c(z)$. Thus, $R_i = ct_i$, where t_i are the measured travel times. More precise ranges are obtained by accounting for the actual path of the acoustic signal, which can depart significantly from a straight line under certain circumstances, for example, in deep water when the angle between the reference element plane and the range vector is small ($< 50°$).[15]

In one form of acoustic range employing travel time measurements, clocks in the reference elements and navigated point are synchronized. Signals are usually transmitted from $O(x, y, z)$ and received at the reference elements, but transmission in the opposite direction is also employed. In the latter case, signals from different reference elements are distinguished by modulation coding or frequency. A common variation replaces the reference elements with transponders, devices that respond to the reception of a signal by transmitting one of their own, thereby eliminating the need for synchronized clocks. The round-trip travel times are measured, and $R_i = ct_i/2$ is computed. Accuracy of pulse systems can be as high as a few centimetres, depending on the operating frequency and range. For ranges of kilometre order, frequencies in the 5–20-kHz range are used, and accuracies of less than a metre are easily obtained. For shorter range systems, greater acoustic attenuation can be tolerated, thus allowing the use of higher operating frequencies. These yield higher accuracies because shorter pulses with steeper rise times can be transmitted. Short-range systems operating in the several hundred kilohertz range give centimetre accuracy.

In short-baseline systems transmission is usually from $O(x, y, z)$ to a cluster of reference receivers. The reference element spacing is small compared to the range to $O(x, y, z)$, so small errors in travel time measurement result in relatively larger errors in position. Thus, overall accuracy is not as great as with a long baseline, but these systems have the advantage of being compact, and the reference elements are easily mounted on the hull of a vessel.

The ultra-short-baseline systems that measure signal phase rather than arrival time yield angles to the object; range is computed by triangulation. This is usually the least precise of the commonly used systems, but it is very compact and provides adequate positioning in many applications, particularly when short ranges are involved. To avoid ambiguities, receiving elements must be spaced less than $\lambda/2$. Some systems combine phase and travel time measurements, thus measuring both angle and range.[16]

5.2 Doppler Navigation

Acoustic navigation and positioning are also accomplished by measuring Doppler shifts induced by motion.

Doppler shift provides a direct measurement of velocity, and it also can be integrated to yield spatial translation. Either the reference elements, which are fixed in space, or the navigated point emit constant frequencies. In the former case, each reference element emits a different frequency f_i, and the Doppler-shifted receptions at the navigated object, f_{d_i}, are measured. In the latter case the navigated object emits a constant frequency, and the Doppler-shifted signals received by the fixed reference elements, f_{d_i}, are measured. Velocities along the paths from reference element to navigated point are

$$V_i = c\left(1 - \frac{f_{d_i}}{f_i}\right). \tag{12}$$

This technique is less common than transit time navigation, primarily because Doppler systems provide relative, rather than absolute, position. Also, because translation must be determined by integration, signal fades disrupt continuous tracking. However, under certain circumstances more precise relative positioning can be obtained than with a transit time system. Thus, the two systems are frequently combined.[17]

5.3 Reference Element Calibration

The position of the reference elements must be known. For short- and ultra-short-baseline systems element positions can often be determined easily at the time of manufacture or installation by direct measurement. For long-baseline systems, determining the position of the reference elements is more difficult, especially if the elements are placed on the inaccessible seafloor. In this case two methods are employed. In the first, a ship acoustically measures ranges to the bottom units at a sufficient number of points on the ocean surface to solve the set of resulting simultaneous equations that relate element positions, ship positions, and measured ranges. Six independent ship positions and ranges are sufficient to determine the relative position of three elements on the seafloor. The second method is self-calibrating. The reference elements transmit pulses between each other, thereby measuring their separations directly. This method is not always possible, especially in deep waters having negative sound speed gradients with consequent strong upward acoustic refraction that limits obtainable horizontal ranges. Optimized techniques for surveying reference element positions have been devised.[18]

5.4 Acoustic Speed Logs

Speed logs measure ship speed through the water. Two types are in general use. One, a high-frequency sonar in the 10–100-kHz range mounted on the hull of the vessel, relies on measuring the Doppler shift of signals scattered from particulate or biologic matter in the water. It is assumed that water motion, and hence scatterer motion, can be neglected, in comparison with ship speed. This assumption is not always valid, especially in high current regions such as the Gulf Stream, rivers, or tidal inlets where water currents can be as high as 2–4 m/s. Doppler-shifted returns of signals scattered from the ocean bottom are also used to estimate ship speed. Orthogonal beams allow forward and thwartship velocities to be computed.

The second type of log is the correlation log. An acoustic signal is projected vertically downward, and scattered signals from the bottom or from scatterers in the water column are received by transducers spaced several centimetres from the transmitter. The separation in time of the transmitted and received pulses, obtained by correlation, is a measure of ship speed. If the transmitter and received hydrophones are oriented along the bow to stern line, forward speed is measured; if oriented across the beam, thwartship speed is obtained. High frequencies (150–200 kHz) are used in shallow-water systems (300 m), and lower frequencies (10–25 kHz) are used in deep-bottom-tracking systems.[19]

REFERENCES

1. D. J. T. Carter, *Echo-Sounding Correction Tables*, 3rd ed., NP139, The Hydrographic Dept., Ministry of Defense, Taunton, England, 1980.
2. H. E. Edgerton, *Sonar Images*, Prentice-Hall, Englewood Cliffe, NJ, 1986.
3. R. D. Stoll, *Sediment Acoustics*, Springer Verlag, New York, 1989.
4. S. Panda, L. R. Leblanc, and S. G. Schock, "Sediment Classification Based on Impedance and Attenuation Estimation," *J. Acoust. Soc. Am.*, Vol. 95, No. 5, Pt. 1, 1994, pp. 3022–3035.
5. D. M. Farmer and G. B. Crawford, "Remote Sensing of Ocean Flows by Spatial Filtering of Acoustic Scintillations: Observations," *J. Acoust. Soc. Am.*, Vol. 90, No. 3, 1991, pp. 1582–1591.
6. G. F. Appell, T. N. Mero, R. Williams, and W. E. Woodward, "Remote Acoustic Doppler Sensing: Its Application to Environmental Measurements," in T. McGuinness and H. H. Shih (Eds.), *Current Practices and New Technology in Ocean Engineering*, Amer. Soc. Mech. Eng., New York, 1986.
7. W. Munk and C. Wunsch, "Ocean Acoustic Tomography: A Scheme for Large Scale Monitoring," *Deep-Sea Res.*, Vol. 26A, 1979, pp. 123–161.
8. B. Cornuelle, C. Wunsch, D. Behringer, T. Birdsall, M. Brown, R. Heinmiller, R. Knox, K. Metzger, W. Munk, J. Spiesberger, R. Spindel, D. Webb, and P. Worcester,

"Tomographic Maps of the Ocean Mesoscale. Part 1: Pure Acoustics," *J. Phys. Oceanogr.*, Vol. 15, 1985, pp. 133–152, WHOI. #5961.

9. P. F. Worcester, B. D. Cornuelle, and R. C. Spindel, "A Review of Ocean Acoustic Tomography: 1987–1990", in *U.S. National Report to the International Union of Geodosy and Geophysics (IUGG) 1987–1990, Contributions in Oceanography*, American Geophysical Union, 1991, pp. 557–570.
10. J. A. Nystuen, "Rainfall Measurements Using Underwater Ambient Noise," *J. Acoust. Soc. Am.*, Vol. 79, 1986, pp. 972–982.
11. J. A. Nystuen, C. C. McGlothin, and M. S. Cook, "The Underwater Sound Generated by Heavy Precipitation," *J. Acoust. Soc. Am.*, Vol. 93, 1993, pp. 3169–3177.
12. M. H. Orr and L. Baxter, "Dispersion of Particles after Disposal of Industrial and Sewage Wastes," in I. W. Duedall, B. H. Ketchum, P. K. Park, and O. R. Kester (Eds.), *Wastes in the Ocean*, Wiley, New York, 1983.
13. P. H. Wiebe, C. H. Greene, and T. K. Stanton, "Sound Scattering by Live Zooplankton and Micronekton: Empirical Studies with a Dual Beam Acoustical System," *J. Acoust. Soc. Am.*, Vol. 88, No. 5, 1990, pp. 2346–2360.
14. C. S. Clay and H. Medwin, *Acoustical Oceanography: Principals and Applications*, Wiley, New York, 1977.
15. M. M. Hunt, W. M. Marquet, D. A. Moller, K. R. Peal, W. K. Smith, and R. C. Spindel, "An Acoustic Navigation System," Rep. 74-6, Woods Hole Oceanographic Institution, Woods Hole, MA, 1974.
16. M. J. Morgan, *Dynamic Positioning of Offshore Vessels*, PPC Books, Petroleum Publishing, Tulsa, OK, 1978.
17. R. C. Spindel, R. P. Porter, W. M. Marquet, and J. L. Durham, "A High-Resolution Pulse-Doppler Underwater Acoustic Navigation System," *IEEE J. Oceanic Eng.*, Vol. 1, No. 1, 1976, pp. 6–13.
18. A. G. Mourad, D. M. Fubara, A. T. Hopper, and G. Y. Ruck, "Geodetic Location of Acoustic Ocean-Bottom Transponders from Surface Positions," *EOS, Trans. Amer. Geophysics Union*, Vol. 53, 1972, pp. 644–649.
19. B. Woodward, W. Forsythe, and S. K. Hole, "Estimating Backscattering Strength for a Correlation Log," *IEEE J. Oceanic Eng.*, Vol. 19, No. 3, 1994, pp. 476–483.

39

ACOUSTIC TELEMETRY

JOSKO A. CATIPOVIC

1 INTRODUCTION

The underwater acoustic channel permits data telemetry at modest rates. It has been exploited for simple voice communication as well as sophisticated digital telemetry of television images and command signals for control of subsea wellheads, underwater vehicles, and other instrumentation. Its capacity is limited primarily by bandwidth constraints and acoustic ambient noise and also by unique characteristics of the undersea channel, such as ducted, multipath propagation in the deep ocean, surface and bottom reflection and scattering, especially in shallow water, and random variability arising from time-varying surface changes and changes in the transmission medium itself. The channel transfer function is therefore stochastic at certain scales and is spread in both time and frequency. In this chapter we discuss the limitations imposed by these channel constraints and describe the implementation and performance of typical underwater telemetry systems.

2 CHANNEL CAPACITY

Transmission capacity constrained by the Shannon limit states that error-free data transmission is possible if and only if the data rate is lower than some maximum data rate, termed the channel capacity C[1]:

$$C = W \log_2 \left(1 + \frac{P}{W N_0} \right), \tag{1}$$

where P is the average received signal power, N_0 is the ambient noise level, and W is the bandwidth available for information transmission. Defining the signal-to-noise ratio (SNR) as

$$\text{SNR} = 10 \log_{10} \left(\frac{P}{W N_0} \right), \tag{2}$$

we note that for a SNR of 15 dB, a typical objective for relatively error free data transmission is $C/W \sim 5$ bits/s/Hz; that is, the maximum data rate is roughly 5 times the available bandwidth. A further consequence of Eq. (1) is the inverse relationship between received signal power and available bandwidth. As more bandwidth is available for transmission of each data bit, the power required to transmit that bit is reduced. Most systems operate either in the bandwidth-limited or energy-limited regimes; bandwidth-limited systems attempt to maximize the number of transmitted bits per second per hertz of available bandwidth and are able to operate at high SNR. An excellent example is telephone modems, some of which operate at 28,800 bps over the 2400-Hz telephone channels. Energy-limited systems are constrained by the amount of energy available for the transmission of a single data bit. An example of energy-constrained acoustic systems is deep ocean tomography transmitters, which typically send each waveform over many seconds and many hertz of bandwidth in an attempt to maximize energy efficiency. These energy-limited ocean basin transmitters attempt to approach the "infinite-bandwidth" channel capacity (in bits per second):

$$C = \frac{P}{N_0 \ln 2}. \tag{3}$$

Note: References to chapters appearing only in the *Encyclopedia* are preceded by *Enc.*

The system designer is constrained by both power and bandwidth. Typically, for short deployments or applications where ample power is available, the telemetry system attempts to maximize bandwidth efficiency. Conversely, for long, energy-limited deployments, the number of transmitted data bits per joule of battery power typically dominates, and bandwidth efficiency is sacrificed.

2.1 Bandwidth and Ambient Noise Constraints

Acoustic channel bandwidth is limited by sound absorption which roughly varies quadratically with frequency. The reader is referred to Ref. 2 and *Enc.* Ch. 109 for an in-depth discussion of acoustic attenuation. A useful rule of thumb is to select the frequency resulting in a 10-dB attenuation loss at a desired transmission range. For example, this results in an upper frequency limit of ~50 kHz at 1 km, 12 kHz at 10 km, and 1.5 kHz at 100 km.

Although the entire band between 0 Hz and the attenuation limit is available for use, the limited bandwidth of practical transducers often restricts systems to a single octave. Most existing acoustic telemetry systems operate within the octave of bandwidth upper limited by the attenuation limit in order to maximize available bandwidth and because ocean ambient noise levels usually decrease with frequency.

Ambient noise levels are readily available for a variety of underwater environments.[2] Published noise levels are generally adequate for predicting ambient conditions at the receiver. However, these generally refer to average conditions, whereas a robust system is primarily affected by extremal or infrequent high-noise events. Near man-made structures such as offshore drilling sites, peak noise levels are particularly bothersome. In many cases, the receiver platform noise dominates SNR considerations, particularly if the receiver is on a ship or is being moved through the water. Decreasing receiver acoustic self-noise is possibly the most important operational aspect of undersea telemetry.

2.2 Multipath and Fluctuation Limits

The relatively slow sound speed in the ocean gives rise to a very long multipath when compared to other channels characterized by speed-of-light signal propagation. The SOFAR waveguide and many coastal and harbor environments have characteristic reverberation times from tens of milliseconds to several seconds, and time-variant long-delay multipath must be recognized by the system designer as a basic channel characteristic present in all but a few propagation geometries.

The SOFAR channel at ranges over 1 convergence zone yields a number of distinct arrivals. Typical multipath duration is ~1 s, with individual path root-mean-square (rms) fluctuation of approximately 10 ms. The fluctuation statistics reflect the underlying time-variable environmental process, such as the surface elevation spectrum or internal wave spectrum. For example, for a surface-reflecting ray, the rms travel time fluctuation is given by:

$$\langle(\Delta t)^2\rangle \sim \frac{2\sigma_s\sqrt{M}}{c}\sin\theta, \tag{4}$$

where σ_s is the rms surface elevation, M is the number of surface reflections, and θ is the angle of incidence. The fluctuation spectrum is directly related to the surface elevation spectrum.[3] Surface-induced fluctuation is typically the most significant contribution to received signal dynamics in the deep ocean.

In shallow water, reflections from objects and channel boundaries dominate the multipath; the problem becomes geometry specific, and no generic solutions are available. It is important to note that the multipath itself is not a detrimental phenomenon. Rather, single-path temporal fluctuations and multipath time stability are primarily performance problems.

3 ACOUSTIC CHANNELS

The underwater acoustic channels are readily divided into four categories for purposes of data telemetry. Each category spans a range of propagation geometries with similar channel-induced performance limits. Each category has seen the development of distinct types of telemetry systems, ranging from simple "grafts" of telephone modems to elaborate joint channel–data estimators required for operation with highly dynamic multipaths.[4]

1. The simplest case is the near-vertical deep-water channel encountered, for example, when communicating between the ocean surface and ocean bottom instrumentation in the deep ocean. Transmission range is typically 3–10 km. Over this channel, often called the reliable acoustic path (RAP), the multipath is limited to discrete surface and bottom bounces and can be easily eliminated by appropriate transducer directionality or baffling. The direct path undergoes minimal fluctuation, and the received signal is well modeled as a combination of the transmitted signal and additive noise. Data rate is limited primarily by the available bandwidth and the ambient noise level. A number of communication systems developed for the radio frequency (RF) or telephone channel have been demonstrated successfully in deep water. The

15–30-kHz frequency band is typically utilized, yielding data rates of 10–50 kbits/s. Several systems for deep-water telemetry are currently commercially available.[3]

2. The very shallow water 2–10-km-range channel exhibits a complex and dynamic multipath structure, particularly in enclosed bodies such as harbors and bays. Significant acoustic interaction with the surface, midwater microstructure, and bottom produces a rapidly variant, extended multipath that forms the principal limitation to data telemetry in this environment. The upper frequency limit ranges from 10 to 100 kHz. To date, equalization techniques have been unable to track the multipath dynamics in this channel, although this is an area of active research. This has greatly compromised available data rates, and most current systems operate below 10 kbits/s, except at very short (<500-m) ranges.[3]

3. The continental shelf is amenable to telemetry at ranges up to ~100 km. Primary performance limitation in this case is the high spatial variability of the received acoustic pressure field. The spatial and frequency dependence of transmission loss makes it difficult to maintain a continuous telemetry channel, although data can be transmitted robustly for at least several hours per day. This channel also introduces complex bottom-interacting multipath, and multipath duration can be a performance constraint for higher data rates. Recent work in equalization of the continental shelf channel has resulted in 1–2 kbits/s data rate over ~80-km ranges.[4]

4. The long-range deep-water SOFAR channel supports telemetry at several convergence zones at data rates up to ~1 kbits/s. While significant multipath is present, it is relatively stable and hence can be equalized.[5] The primary performance limitation in this case is the ambient noise at the receiver; that is, this channel is essentially power limited, particularly since deep-ocean systems are difficult to deploy and long deployment times are required. Reducing noise levels at the receiver and placing transmitters and receivers deep in the SOFAR channel are key operational steps to long-range data telemetry. A prime example of ocean-basing scale telemetry systems are acoustic tomography transceivers, described in Chapter 94.

4 DATA TRANSMISSION SYSTEMS

This section describes acoustic telemetry systems in order of increased data rate and reliability. Possibly the first undersea communication system was the UQC-1 underwater telephone, which used analog modulation to transmit human voice. Soon after, the first Frequency Shift Keyed (FSK) systems were developed for simple control of undersea instrumentation.[6]

The FSK modulation consists of transmitting one of two possible frequency tones to signal the state of a single bit. The transmitter transmits at a frequency f_0 or f_1 for a duration Δt. Typically $\Delta t = 1/|f_1 - f_0|$. Note this technique requires a time–bandwidth product of 2 to transmit a single bit; that is, the required bandwidth in hertz is twice the data rate in bits/per second.

The time–frequency parameters are selected primarily with regard to multipath duration L. If $\Delta t \gg L$, the effect of channel multipath is negligible. However, the system may become Doppler sensitive if the tone frequency separation $1/\Delta t$ becomes smaller than a channel Doppler shift B. In practice, if $\Delta t < 10$ ms, the system becomes unacceptably sensitive to multipath in all but the vertical acoustic channel. Thus 100 bits/s has been a practical limitation to FSK telemetry.

Several techniques are used to increase FSK data rate. Frequency hopping the tones allows for a longer channel clearing time following each tone transmission, but at the expense of increased bandwidth per data bits/per second. While this method indeed increases data rate, it has been largely supplanted in favor of Multiple FSK, or MFSK, which implements a number of FSK systems in parallel. Its principal advantage is the ability to tailor the time–frequency cell to channel multipath and Doppler characteristics without compromising data rate, since the number of tones is adjustable to occupy the desired bandwidth. Typical tone spacings range from ~1 Hz for the long-range deep-water channel to ~80 Hz for the shallow-water channels where underwater vehicle motion and currents cause significant Doppler shifts.

The MFSK system found wide use in digital acoustic telemetry because it is well matched to typically encountered multipath durations and Doppler spreads while being insensitive to details of multipath fluctuation. A number of MFSK systems were constructed, and the method is widely used for underwater data transmission.[3]

4.1 MFSK Receivers

The MFSK receivers are essentially banks of narrowband filters that integrate and dump the received waveform.[7] The demodulation is typically performed with a fast Fourier transform (FFT) and the data decoding with a digital signal processing (DSP) chip. The FFT resolution and duration are matched to the tone frequency spacing and Δt. An external synchronization system is used to section data prior to FFT demodulation. Doppler corrections are typically performed by adjusting the digitizing frequency to match Doppler drift.[3]

For each FSK tone pair at the output of the demodulator FFT, the cell with larger received energy is declared as the received bit. In many underwater acoustic links, the received energy is χ^2 (chi-square) distributed; that is, the channel behaves as the Rayleigh fading channel. The received bit error probability for this case is[7]

$$P_{err} = \frac{1}{2 + \text{SNR}}, \quad (5)$$

where SNR is the received signal-to-noise ratio per bit. This results in generally mediocre performance. Error correction coding is used to improve bit error performance at the cost of increased bandwidth and receiver complexity. Error correction coding for a fading channel is outside the scope of this discussion, but an excellent source is Ref. 7. At the cost of increasing bandwidth utilization to 4 Hz/bit/s, and a modest increase in receiver complexity, bit SNR required for acceptable system performance is typically reduced to 15–20 dB. Additional techniques such as interleaving, spatial diversity, automatic repeat request, and code combining offer additional robustness in extremal channel conditions.

In summary, incoherently detected MFSK is a robust modulation method for underwater telemetry. Robust performance can generally be reached at 0.25 bit/s/Hz at 20 dB SNR. However, this performance point is well below the performance bound determined by channel capacity considerations [Eq. (1)], which bounds performance at ~6.5 bits/s/Hz at 20 dB SNR.

4.2 Coherent Modulation

Phase-coherent signal modulation can approach channel capacity much closer than incoherently detected MFSK. This method encodes information into the phase, as well as amplitude, of the transmitted signal and is therefore more bandwidth efficient.[7] The transmitted signal can be expressed as

$$x(t) = a(t)e^{[j\omega t + \Phi(t)]} \quad (6)$$

where $a(t)$ and $\Phi(t)$ are selected at each bit duration. Quadrature phase shift keying (QPSK) assigns four phase angles to each of four possible two-bit messages:

Data	00	01	10	11	
$\Phi(t)$	0	90	180	270	(degrees)
$a(t)$	1	1	1	1	

A more bandwidth-efficient quadrature amplitude modulation jointly modulates amplitude and phase:

Data	000	001	010	011	100	101	110	111
$\Phi(t)$	0	45	90	135	180	235	270	315
$a(t)$	$1+\sqrt{3}$	$\sqrt{2}$	$1+\sqrt{3}$	$\sqrt{2}$	$1+\sqrt{3}$	$\sqrt{2}$	$1+\sqrt{3}$	$\sqrt{2}$

A large number of modulation methods are useful in this context.[7] The receiver receives the waveform after it has propagated through a randomly time variant, multipath channel. Since received phase and amplitude are severely perturbed by the time-variant acoustic channel, phase-coherent receivers are required to track and equalize the channel multipath and phase fluctuations.[6] A phase-coherent receiver is typically based around a detrital channel estimator/equalizer. A possible receiver architecture is shown in Fig. 1. This receiver uses a feed-forward linear equalizer to correct for channel fluctuations and remove some intersymbol interference (ISI). It is followed by a Doppler correction stage implemented with a digital phase-locked loop (PLL) and finally by a data decoder/feedback equalizer, which removes residual ISI and recovers the data stream.[4] Other receiver realizations have also been used.[5,7]

Performance of phase-coherent systems is typically limited by channel variability. The shallow water and continental shelf channels, in particular, have significant surface interactions and exhibit high levels of dynamics,

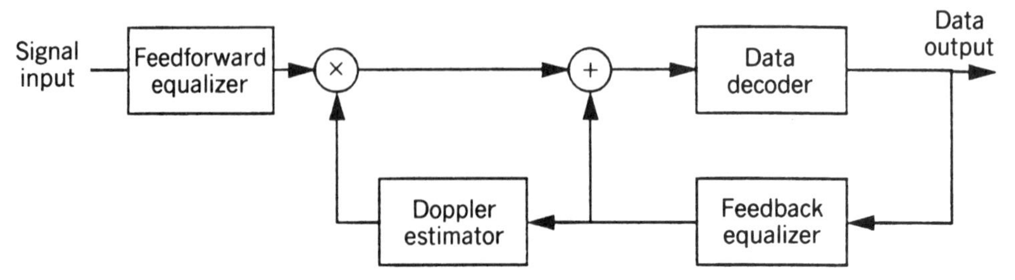

Fig. 1 Phase-coherent receiver structure.

TABLE 1 Phase-Coherent Telemetry Performance Milestones

Channel	Data Rate	Range (km)
Shallow water	20 kbits/s	8
Continental shelf	2000 bits/s	80
Deep water	1000 bits/s	200

as described by Eq. (4). The deep-ocean vertical channel is amenable to coherent telemetry. Receiver structures developed for RF and telephone channels operate reliably. A number of operational systems exist, with data rates from 5 bits/s to 30 kbits/s over the 5-km vertical channel. Recent improvements in equalizer performance have permitted coherent signaling over the other three types of underwater channels.[4] Some milestone results are summarized in Table 1.[4,5]

Phase-coherent acoustic data transmission is currently an active research topic, and further improvements in reliability, bandwidth, and power efficiency are likely. Performance improvement over incoherently demodulated MFSK telemetry is significant. For example, QAM bandwidth efficiency is 3 bits/s/Hz. It is decodable at 14 dB SNR over the long-range deep-water channel, offering a roughly order-of-magnitude improvement in bandwidth and power efficiency over MFSK.[5] However, phase-coherent methods are much more sensitive to environmental variability, and there is presently a large subset of acoustic channels where coherent methods do not perform well.

5 RELATIVE MERITS OF THE SIGNALING METHODS

It is a mistake to exclude FSK and MFSK signaling from consideration simply because phase-coherent methods outperform them on some occasions. For instance, FSK signaling is in wide use on a number of undersea acoustic products, such as transponders and releases. These systems are frequently built around a hard-limiting (clipper) detector or a phase-locked loop (PLL) tone detector. The receiver can be realized in simple hardware, and its performance is adequate for a large number of applications.[8] Its performance can be considerably enhanced by digitizing the acoustic signal and using an adaptive-threshold detector combined with a matched input filter and a forward error correction code.[9] Typically, data rates up to 80 bits/s can be achieved quite robustly at ranges up to 10 km, although higher data rate–range combinations have been reported.[9]

Increasing this data rate is easily done with a MFSK system, where multiple tones are transmitted on adjacent bands to fill up the available bandwidth. Tone "chip" duration of 80 s^{-1} appears to yield an efficient trade-off between Doppler sensitivity and robustness to multipath-induced fading. Systems with data rates of 5–10 kbits/s at ~10-km ranges have been implemented.[3] There are also several military systems that operate at much longer ranges.

The MFSK systems are typically implemented around a DSP FFT implementation. Incoming data are corrected for Doppler with an adaptive digital sampler, blocked, and match-filtered with an FFT. The FFT outputs are used by the decision device to recover the data stream. In strongly fading channels, receiver spatial diversity offers a large performance increase at some additional complexity.[10]

Phase-coherent systems offer the highest performance, but as yet, their robustness has not been demonstrated in long-term deployments. A number of performance-limiting issues are being studied but have not been definitely resolved to data:

1. Long multipath duration increases the number of equalizer taps required for channel compensation. Since the taps are adaptively updated, increasing their number augments receiver self-noise, which eventually dominates the system SNR. In particular, long-range SOFAR propagation is characterized by multipath arrivals spread over several seconds, and phase-coherent techniques have difficulties. Overall multipath duration can be decreased by combining an equalizer with a beamformer or by using narrow transmitter beams to constrain the channel to a narrow beam.[5,11] Both techniques have shown considerable promise but have yet to be implemented in practical systems
2. Doppler tracking, in particular tracking and compensating for Doppler shifts of several frequency-spread multipath arrivals, has not been solved. There are several applicable frequency-tracking algorithms available, but they all appear to significantly increase receiver self-noise.
3. System robustness to interference, whether broadband impulsive sources such as snapping shrimp or strong tonal interference, is inadequate in current systems. Again, several solutions are available but have not been sufficiently tested in operation.

The above examples should alert the reader that phase-coherent systems are not the solution for many applications, although a great deal of work in this area is underway and they are likely to mature in the near future.

Meanwhile, the simple incoherent systems are likely to stay with us for the forseeable future.

REFERENCES

1. R. Kennedy, *Fading Dispersive Communication Channels*, Wiley, New York; 1969. A. B. Baggeroer, "Acoustic Telemetry—An Overview," *IEEE J. Oceanic Eng.*, Vol. OE-9, October 1984, pp. 229–235.
2. R. J. Urick, *Principles of Underwater Sound for Engineers*, McGraw-Hill, New York, 1967.
3. J. Catipovic, "Performance Limitations in Underwater Acoustic Telemetry," *IEEE J. Oceanic Eng.*, Vol. 15, No. 3, July 1990, pp. 205–216.
4. M. Stojanovic, J. Catipovic, and J. Proakis: "Phase-Coherent Digital Communications for Underwater Acoustic Channels," *IEEE J. Oceanic Eng.*, Vol. 10, No. 1, 1994, pp. 100–111.
5. M. Stojanovic, J. Catipovic, and J. Proakis: "Adaptive Multichannel Combining and Equalization for Underwater Acoustic Communication," *J. Acoust. Soc. Am.*, Vol. 94, No. 3, pt. 1, September 1993, pp. 1621–1631.
6. *IEEE Journal of Oceanic Engineering*, Special Issue on Acoustic Telemetry, January 1991.
7. J. Proakis, *Digital Communications*, McGraw-Hill, New York, 1989.
8. P. Hearn: "Underwater Acoustic Telemetry," *IEEE Trans. on Comm. Tech.*, Vol. CT-14, December 1963, pp. 839–843.
9. R. S. Andrews and L. F. Turner, "On the Performance of Underwater Data Transmission Systems Using Amplitude Shift-Keying Techniques," *IEEE Trans. Sonics Ultrasonics*, Vol. SU-23, No. 1, January 1976, pp. 64–67.
10. J. Catipovic and L. Freitag, "Spatial Diversity Processing for Underwater Acoustic Telemetry," *IEEE J. Oceanic Eng.*, Vol. 16, No. 1, January 1991, pp. 86–87.
11. H. A. Quazi and W. L. Konrad, "Underwater Acoustic Communications," *IEEE Comm. Mag.*, 1982, pp. 24–30.

PART V

ULTRASONICS, QUANTUM ACOUSTICS, AND PHYSICAL EFFECTS OF SOUND

40

INTRODUCTION

PHILIP L. MARSTON

1 INTRODUCTION

Mechanical radiation and vibrations of matter are of fundamental importance to various areas of physics and instrumentation. An understanding of the coupling of ultrasound with molecular, thermal, rheological, and electronic properties can facilitate an analysis of attenuation mechanisms and, in some cases, facilitate the use of ultrasound as a probe of microscopic systems and processes. Quantum theory is relevant to microscopic processes affecting the propagation of sound and also to the detection of weak mechanical vibrations. Part V contains chapters on selected physical aspects of sound and vibrations of general and contemporary interest. The present chapter includes some introductory material as well as summary descriptions of relevant physics issues not described elsewhere in Part V or elsewhere in this encyclopedia.

The scientific literature pertaining to fundamental aspects of ultrasonics and physical effects of sound is sufficiently large that it would be impossible to give a uniform coverage of areas in Part V. Consequently some references will be included in the present chapter as entry points to the literature for certain specialized topics. Monographs that are especially useful are listed in Refs. 1–5.

Special attention will be given in the present chapter to certain topics of pedagogical value relating macroscopic and quantum descriptions of vibrations. The representation of classical sound waves as a coherent superposition of phonons is described in Section 2. This serves to introduce the discussion of the theory of phonons in Chapter 43. Section 2 is also related to discussions in Sections 3 and 4 of the detection of low-amplitude vibrations and the dissipation of sound in solids. These topics are relevant to the detection of vibrations with dimensionless strain amplitudes of approximately 10^{-19}, as reviewed in Ref. 6–9 where the applications include attempts to detect gravitational radiation. These developments have furthered the understanding of quantum limitations to the measurements of small-amplitude vibrations, and certain concepts of general value will be reviewed here.

2 CLASSICAL SOUND WAVES AS A COHERENT SUPERPOSITION OF PHONONS

To model the longitudinal vibrations of a crystalline solid, consider a chain of equal masses coupled only through nearest-neighbor harmonic potentials.[5] This is, of course, an approximation to the coupling between atoms for real solids; however, it is sufficient for introducing the relationship between phonons and classical sound waves. The procedure for calculating the normal modes and frequencies for such a chain is described in Chapter 43. Associated with the kth normal mode is a normal coordinate, designated here as ζ_k, that describes the displacement amplitudes for a given mode [see Eq. (1)]. Phonons are the elementary excitations of this chain given by quantizing the uncoupled harmonic oscillator corresponding to each normal coordinate. The quantization procedure, also reviewed in Chapter 43, is described in Refs. 10 and 11 for a broader class of harmonic excitations of many-particle systems. The quantum number of the oscillator corresponding to the kth mode is a nonnegative integer $n_k = 0, 1, 2, \ldots$ and is commonly referred to as the occupation number of that mode. Raising or lowering n_k by unity corresponds to adding or removing a

Note: References to chapters appearing only in the *Encyclopedia* are preceded by *Enc.*

phonon from the kth mode. The energy increment per phonon is $\hbar\omega_k$ where $\hbar = 1.054 \times 10^{-34}$ J · s and ω_k is the radian frequency of the mode.

A question of general interest is how to understand classical sound waves from the phonon (quantized strain field) point of view. Since we are considering a linearized description of vibrations, the problem is analogous to the relationship between classical electromagnetic fields and photons reviewed by Scully et al.[12] Consider a chain with $N + 1$ masses m with equilibrium locations $x_j = ja$, $j = -N/2, \ldots, N/2$ where N is even and a denotes the space between masses. For definiteness, attention is restricted to the case of free ends and to symmetric modes. The contributions to the displacement from equilibrium for the jth mass due to the kth normal coordinate ζ_k are related by[10]

$$u_j = \left[\frac{2}{m(N+1)}\right]^{1/2} \sum_{k=1}^{N} \zeta_k \sin\left[\frac{\pi(2k-1)j}{N}\right], \quad (1)$$

where the normalization is such that the dimensions of ζ_k are length·(mass)$^{1/2}$. (For the system under consideration ζ_k corresponds to a superposition of the amplitudes denoted by U_k and U_{-k} in Chapter 43.) The coordinate ζ_k is treated as the position coordinate q of a harmonic oscillator having a natural frequency ω_k corresponding to that of the kth classical mode. The conjugate momentum for the normalization used here becomes[10] $p_k = \partial L(\zeta_k, \dot{\zeta}_k)/\partial\dot{\zeta}_k = \dot{\zeta}_k \equiv d\zeta_k/dt$, where L denotes the Lagrangian for the kth mode. Recall that the wave function of a quantized harmonic oscillator[13] in the nth state is proportional to a product of the nth Hermite polynomial and a Gaussian function denoted here as $\phi_n(\zeta_k)$. The general wave function of the kth normal mode is the superposition

$$\Psi_k(\zeta_k, t) = \sum_{n=0}^{\infty} c_n \exp\left[-i\left(n + \frac{1}{2}\right)\omega_k t\right]\phi_n(\zeta_k), \quad (2)$$

where for convenience the index k has been omitted from the occupation number n_k. The question under consideration becomes how the coefficients c_n are to be chosen such that the expectation value for the normal coordinate

$$\langle\zeta_k\rangle = \int_{-\infty}^{\infty} \Psi_k^*(\zeta_k, t)\zeta_k\Psi_k(\zeta_k, t)\,d\zeta_k \quad (3)$$

reduces to that of a classical harmonic oscillator in the many-phonon limit. The coefficients are also to be chosen such that the wave packet described by $|\Psi_k|^2$ oscillates coherently in time without spreading and with a minimal uncertainty product $\Delta\zeta_k\,\Delta p_k$. From an analysis of the harmonic oscillator problem due to Schrödinger, it is known that the appropriate superposition is[12–15]

$$c_n = \frac{\alpha^n}{(n!)^{1/2}} \exp\left(-\frac{|\alpha|^2}{2}\right), \quad (4)$$

where $\alpha = \zeta_{k0}/\sigma_k$, ζ_{k0} is the expected magnitude, and $\sigma_k = (2\hbar/\omega_k)^{1/2}$ describes the width of the packet. The wave packet is Gaussian

$$|\Psi_k|^2 = \sigma_k^{-1}\pi^{-1/2} \exp\left[-\left(\frac{\zeta_k - \langle\zeta_k\rangle}{\sigma_k}\right)^2\right], \quad (5)$$

and Eq. (3) reduces to $\langle\zeta_k\rangle = \zeta_{k0}\cos\omega_k t$. See Fig. 1. The probability of n phonons in the kth mode becomes

$$|c_n|^2 = \frac{|\alpha|^{2n}}{n!}\exp(-|\alpha|^2), \quad (6)$$

which is a Poisson distribution.[12–16] The expectation value for the number of phonons in the kth mode becomes[15]

$$\langle N_k\rangle = \sum_n n|c_n|^2 = |\alpha|^2. \quad (7)$$

The superposition in Eqs. (2)–(6) is analogous to Glauber's description of a coherent superposition of photons giving Poisson statistics for photon counting.[15] It is noteworthy that the coherent distribution given by Eq. (6) is peaked at n near $\langle N_k\rangle$, which depends on the expected amplitude ζ_{k0}. The distribution differs greatly from the Planck distribution, which is descriptive of thermally excited photons and phonons.[12,15,16] Traveling classical waves in solids can be treated as an appropriately phased superposition of the standing waves considered here. For example, the traveling wave $\sin(bx - \omega t)$ may be written as $\sin(bx)\cos(\omega t) - \cos(bx)\sin(\omega t)$ where b is the wavenumber.

The uncertainty in the coordinate is usually too small to be significant in acoustics. The packet width σ_k is slightly larger than the uncertainty $\Delta\zeta_k$ and specifies where $|\Psi_k|^2$ has decreased by e^{-1}. Associated with σ_k is a coordinate spread at an antinode that from Eq. (1) becomes

$$\sigma_u = \left[\frac{2}{m(N+1)}\right]^{1/2}\sigma_k \approx 2\left(\frac{\hbar}{\omega_k M}\right)^{1/2}, \quad (8)$$

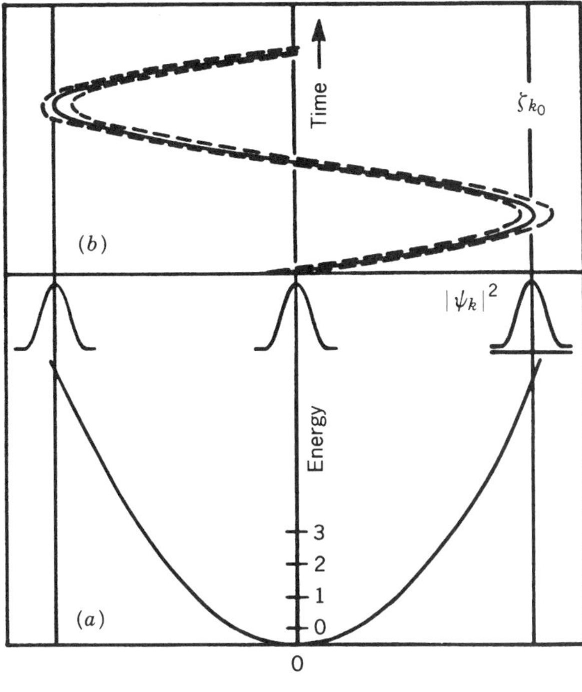

Fig. 1 Coherent state phonon description of harmonic sound waves in a crystal lattice: (*a*) shows the potential as a function of the normal coordinate ζ_k that specifies the strain amplitude for the *k*th normal mode. The lowest quantized energy levels are indicated corresponding to occupation numbers n of 0, 1, 2, and 3. Classical sound waves correspond to a coherent superposition of number states, specified by Eqs. (2) and (4), that give the harmonically oscillating Gaussian wave packet illustrated. The oscillating expectation value $\langle \zeta_k \rangle$ is indicated by the solid curve in (*b*). Associated with the width of that wave packet is a spread of ζ_k indicated by the dashed lines in (*b*).

where $m(N+1)$ is the mass M of the system. This has the same magnitude as the position uncertainty[6] for a quantized oscillator of mass M and frequency ω_k. For the purpose of illustration, consider the oscillations of a large crystal having $M = 1$ kg and a fundamental frequency $\omega_k/2\pi = 30$ kHz. These parameters are taken to correspond roughly to the sapphire rods described in Ref. 6 that are of interest because of their high mechanical quality factor Q (see Section 4). The coordinate spread from Eq. (8) is $\sigma_u \approx 5 \times 10^{-20}$ m. The estimated small magnitude of the uncertainty illustrates why the quantum uncertainty is generally of little importance in acoustics.

One essential character of a classical sound wave, the harmonic oscillations of $\langle \zeta_k \rangle$, can be preserved with a simpler superposition of phonon number states. Only two values of c_n having $n \gg 1$ need be nonvanishing, as described in Chapter 43. For that superposition, however, the width of the oscillating wave packet does not remain constant as in the case of the coherent state illustrated in Fig. 1.

3 DETECTION OF LOW-AMPLITUDE VIBRATIONS: SOME PHYSICAL LIMITATIONS

3.1 Thermal and Quantum Limitations

The physical limitations to the detection of the response of a mechanical resonator to weak applied forces will now be summarized. This will extend the discussion of uncertainties given in conjunction with Eq. (8) to other limitations. Figure 2 shows a diagram of the essential aspects of the detection problem. The displacement response of the resonator is converted to an electrical signal by a transducer and the signal is amplified and filtered. The electrical components are modeled as ideal

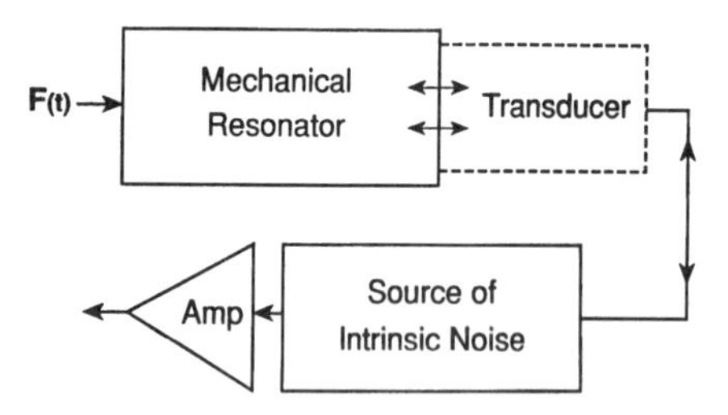

Fig. 2 System for detecting vibrations in response to a weak applied force that may be distributed over the resonator to couple into the mode of interest. With appropriate cooling and mechanical isolation of the resonator, vibration detection may be quantum mechanically limited. With a weakly coupled electrical transducer, modifications of this system may be used to study the intrinsic quality factor Q for the normal mode as described in Ref. 6.

with their inherent electrical noise introduced separately as indicated in the diagram. At low signal levels the flow of energy from the noise source back through the transducer and into the resonator becomes important. The applied force $\mathbf{F}(t)$ may represent, for example, the coupling to an external acoustic field or other influences such as gravitational radiation[6–9] or probe forces as in atomic force microscopy.[17] The summary below relies on discussions in Refs. 6–9 and on standard results for the Brownian motion of oscillators.

Consider the case of a resonator with free ends. A convenient measure of the mode amplitude is the surface displacement x measured by a transducer positioned at a displacement antinode. [For comparison with Section 2, x scales with the normal coordinate ζ_k by a factor of order $(2/M_k)^{1/2}$.] The equipartition theorem of statistical mechanics gives the mean-squared displacement from the thermal energy expression

$$\tfrac{1}{2} M_k \omega_k^2 \langle x^2 \rangle_{\text{th}} = \tfrac{1}{2} k_B T, \tag{9}$$

where k_B denotes Boltzman's constant, T the absolute temperature of the resonator, and M_k and ω_k are the effective mass and frequency of the resonator for the mode under consideration. Typically the mode with the lowest ω_k is the one of interest. In Eq. (9), $M_k \omega_k^2$ represents the effective stiffness of the resonator, and this result has been used to estimate the thermal noise limitation of fiberoptic sensors.[18] The root-mean-square (rms) noise displacement is seen to be proportional to $T^{1/2}$. The resonator is taken to have a quality factor $Q \gg 1$ for the mode under consideration; this is the number of radians of oscillation for the mode energy to be reduced by a factor e^{-1}. The amplitude decay time is $\tau_k = 2Q/\omega_k$. The time required for thermal motion to cause a change in displacement having the same magnitude as the rms thermal displacement is determined by τ_k. Suppose the transducer samples x with a sampling time $\tau_s \le \pi/\omega_k$. During an interval τ_s the thermal change in displacement Δx_{th} is limited in magnitude by approximately[7]

$$|\Delta x_{\text{th}}|^2 \approx 2 \frac{\tau_s}{\tau_k} \langle x^2 \rangle_{\text{th}} = \frac{\tau_s \omega_k}{Q} \langle x^2 \rangle_{\text{th}} \tag{10}$$

for $\tau_s \ll \tau_k$ and factors of order unity neglected. The important point is that if the displacements are only due to thermal motion, the successive displacement samples are correlated over a range that is much smaller than the rms displacement from Eq. (9). From Eq. (10), the magnitude of displacement changes in response to suitable externally applied forces (such as impulsive forces) is limited in principle by[6,7]

$$|\Delta x_{\text{th}}| \approx \left[\frac{k_B T \tau_s}{M_k \omega_k Q} \right]^{1/2} \tag{11}$$

instead of the rms displacement from Eq. (9). Comparison with Eq. (9) gives an effective noise temperature that can be much less than the actual temperature T when Q is large and narrow-band sensitivity is desired. In practice, the sensitivity may also be limited by amplifier noise and the coupling of that noise back to the resonator.[6–8]

The discussion above is sufficient to understand why resonators with large Q's have been developed for the purpose of detecting responses to weak applied forces. It is reasonable to examine how large Q needs to be before the limiting sensitivity is determined by quantum mechanical uncertainty instead of by thermal excitations. The estimated threshold value is[6] $Q_{\text{qm}} \approx k_B T \tau_s / \hbar$, which is in agreement with the magnitude estimated by comparing Eqs. (8) and (11). Cryogenic resonators have been constructed having quality factors in excess of Q_{qm} for sampling times $\approx 10^{-4}$–10^{-3} s, indicating that quantum mechanical considerations may be important for some mechanical resonators.[6] (See the discussion in Section 4 concerning limitations on the Q of resonators.)

3.2 Squeezed States of Mechanical Resonators and Phonons

Methods for extending the limits of detectability beyond those for an oscillating wave packet in the coherent state described by Eqs. (2)–(7) have been proposed by various authors. The experimental emphasis has been on improving the measurability of selected features of light waves by superposing photon states in a way that differs from the classical coherent state described in Section 2.[19] Suppose that it is desired to improve the quantum limitations on the measurement of amplitude beyond those implied by the optical version of Fig. 1 in which the normal coordinate ζ_k is replaced by the instantaneous electric field amplitude. The coefficients c_n of the superposition in Eq. (2) are changed to reduce the amplitude uncertainty with a corresponding increase in the momentum uncertainty Δp_k. The resulting nonclassical states of light are referred to as squeezed states.[19] The squeezing of the quantized strain of a mechanical resonator has been proposed as a method of extending the quantum limit of detectability of weak vibrations.[9] Theoretical aspects of squeezed phonon states have also been discussed. Squeezing is closely related to the concept of quantum nondemolition measurements for evading or altering the usual consequences of measurement for a quantized oscillator.[7,19] The interest in squeezing uncertainties is not limited to quantum noise. Thermal noise (Brownian motion) of a mechanical resonator has been reduced by a form of squeezing.[17] The resonator was a cantilever with a flexural normal mode.

4 INTRINSIC LOSS MECHANISMS AND THE ATTENUATION OF SOUND IN SOLIDS AND FLUIDS

While the attenuation of sound is a phenomenon of general interest, the mechanisms of attenuation also limit the quality factor Q of the mechanical resonators mentioned in Section 3. Generally for a perfect nonmetallic crystal, the most important attenuation mechanisms for longitudinal waves are thermoelastic dissipation and phonon–phonon interactions.[2,3,5,6] Thermoelastic dissipation is the consequence of heat flow associated with local temperature changes resulting from the oscillating compression and expansion of the materials. (The analogous attenuation mechanism for fluids is discussed in Chapter 42). Phonon–phonon interactions are a consequence of the anharmonicity of the crystal lattice that is not included in the phonon formulations of Section 2 and Chapter 43. In addition to the sound wave of interest, thermal phonons will be present with a Planck distribution of energies and typical frequencies of 10^{12} Hz. Assuming that the acoustic wavelength is much longer than the mean free path of the thermal phonons, the lattice nonlinearity results in a shift of the phonon frequencies. The return of the phonon distribution to thermal equilibrium contributes to the attenuation of sound as originally pointed out by Akhiezer.[2,3,6] The resulting attenuation generally increases in proportion to ω^2 where ω is the frequency of the sound wave. An important intrinsic mechanism for the attenuation of sound in metals is the interaction of electrons and phonons, which is discussed in Chapter 44. The aforementioned mechanisms are thought to be the important ones for limiting the Q of mechanical resonators at moderate to low temperatures. Generally Q tends to increase with decreasing temperature T. For frequencies on the order of 10^4 Hz and $T \approx 4$ K, Q values of 5×10^9 and 2×10^9 have been achieved for single crystal bars of sapphire and silicon.[6,7] Engineering solids exhibit a broad range of attenuation mechanisms. These include lattice defects and mechanisms resulting from the polycrystalline internal structure.[2,3,6] Various specialized interactions leading to attenuation of sound and transverse waves in solids have also been studied.[1–3]

The attenuation of sound in gases and liquids is generally dominated by viscous, thermal, and relaxation mechanisms as explained in Chapter 42. Ordinary sound waves in liquid helium, discussed in *Enc.* Ch. 59, are noteworthy because of the combination of low attenuation and low sound speed.

5 BALLISTIC PHONONS, PARTICLE DETECTION, AND PICOSECOND ULTRASONICS

In nonmetallic crystals, the principal source of thermal conduction is thermal phonons having typical frequencies of 10^{11}–10^{12} Hz. Measurements for crystals of the thermal conductivity tensor for temperatures T below about 20 K show that conduction tends to decrease as T is lowered. This indicates that the phonon mean free path has grown to the dimensions of the crystal and is limited by boundary scattering. Consequently for sufficiently low T, heat energy that ordinarily propagates diffusively as a consequence of internal phonon scattering can be carried ballistically across the crystal without scattering.[16] This phenomenon has a number of potential applications and has enabled studies of elastic properties at ultrahigh frequencies. Typically the ballistic phonons are produced on the surface of the crystal as a burst of nonequilibrium phonons excited by a focused laser beam. Time-resolved arrivals of groups of phonons associated with the different classes of elastic waves (longitudinal, fast transverse, and slow transverse) are detected with high-speed bolometers typically made of superconducting thin film. Since the phonons are localized in both space and time, they consist of a superposition of the occupation number states for several normal modes of the lattice.

An important property of ballistic phonons is that as a consequence of propagation in nonspecific directions of the anisotropic crystal, the phonon flux is focused along caustic directions. This follows from the usual anisotropic properties of crystals at ordinary ultrasonic frequencies.[20] The flux pattern at the surface may in principle be inverted to determine the location of the transient heat source. A promising application is the detection of nuclear radiation incident on silicon crystals.[21] Sensitivity has been demonstrated to α particles and x-rays.

A development that has extended the frequency range of room temperature ultrasonic measurements is linked with the development of high-power short-pulse lasers. The availability of intense laser sources having durations of 0.1–1 ps has facilitated the photoacoustic generation of transient stress waves having duration ~10 ps and frequencies in the 100-GHz range.[22] The duration of the waves is determined by the intrinsic material response rather than the duration of the optical pulse. The reflectivity of short lower power optical pulses, delayed relative to the initial pulse, are used to monitor the reflectivity and propagation of acoustic transients. At room temperature the technique is primarily useful for investigating the propagation across thin-layered structures.

6 RADIATION PRESSURE, ACOUSTIC LEVITATION, AND LOW-FREQUENCY OSCILLATIONS DRIVEN BY MODULATED ULTRASOUND

The radiation pressure on objects placed in propagating and standing acoustic waves in fluids is an effect of sound that has diverse applications both within and outside the field of acoustics. For example, the average force on liquid or elastic spheres can be used to measure the acoustic amplitude.[23,24] It is noteworthy, however, that the radiation pressure produced by a beam of sound is affected by the boundary conditions on the fluid.[24,25] Several applications of radiation pressure make use of the ability to levitate a fluid or solid object without mechanical contact. Containerless acoustic trapping of solid, liquid, and molten materials is carried out with ultrasonic standing waves in air (or other gases) where the weight of the object is supported by the radiation pressure.[26] The method has been used for measuring the properties of materials. For trapping objects acoustically in microgravity (such as in NASA's Space Shuttle) sound in the low kilohertz range has been used.[27] Acoustic trapping of tiny bubbles in water has been used to facilitate controlled studies of radial bubble pulsations (see Chapters 22 and 23). The radiation force on a bubble in a standing wave is sometimes termed the primary Bjerknes force. There can also be a radiation force due to the sound scattered by a neighboring bubble, which is usually attractive, and is termed the secondary Bjerknes force.[28]

Low-frequency oscillations of levitated fluid objects (drops or bubbles) can be excited by temporally modulating either the levitation standing wave or a second auxiliary standing wave introduced to control the shape of an object. This has been demonstrated for oil drops[27,29] and large bubbles[30] levitated in water and drops of water and other liquids levitated in air. The excitation of such low-frequency oscillations is noteworthy in that though the coupling is second order in the acoustic pressure amplitude as with other aspects of radiation pressure, the response of the object need not be small when the modulation is tuned to one of the resonances. Capillarity typically provides the restoring force. There is a low-frequency flow induced by the oscillations that is distinct from the acoustic streaming normally associated with attenuation as described in *Enc.* Ch. 30. Sound can also be used to apply controlled radiation torques on levitated objects.[27]

7 DIFFRACTION OF LIGHT BY SOUND: ACOUSTOOPTICS

Traveling or standing acoustical waves affect light because of the variations of the optical index of refraction that accompany the density variations in the sound wave. This acoustooptic effect has been used to measure the wavelength of sound and is briefly discussed in Chapter 41. Other important applications include the modulation and control of light[31,32] and the imaging of acoustic wavefields by schlieren methods.[33] A compilation of several of the original studies has been published,[31] and Ref. 1 gives reviews of the various applications. Strain-induced modulations of the refractive index are also relevant to fiberoptic hydrophones discussed in Refs. 18 and 32 and in Chapter 109.

8 GENERATION OF SOUND WITH MODULATED LASER LIGHT: PHOTOACOUSTICS

The absorption of modulated laser light by solids, liquids, and gases results in a time-dependent thermal expansion that is accompanied by the radiation of sound. The rate of thermalization and expansion is such that the acoustic frequencies attainable are limited to about 100 GHz as noted in Section 5. Most applications are at much lower frequencies and the local heating rate is simply proportional to $\alpha I(\mathbf{r}_s, t)$ where α is the absorption coefficient, I is the space- and time-dependent optical irradiance in watts/centimetre square, and $\mathbf{r}_s$ is a source point location under consideration. The temporal profile of the acoustic pressure depends on the spatial properties of the source and the position of the observer.[32] For example, consider an N-dimensional source with $N = 1, 2$, and 3 corresponding to absorption confined to a thin fluid layer, thin cylinder, and small droplet, respectively. In the absence of any boundary reflections, the corresponding pressures for the low-frequency spectral components $p(\mathbf{r}_s, t')$ evaluated at time t' retarded by the propagation delay, become proportional to $I(\mathbf{r}_s, t)$, $(d/dt)^{1/2}I(\mathbf{r}_s, t)$, and $dI(\mathbf{r}_s, t)/dt$, respectively, as reviewed by Diebold et al.[34] Here $(d/dt)^{1/2}$ is the half-order derivative operator. For the case of a distant observer and a compact source ($N = 3$), the singularity in p implied by dI/dt for a step change in I is softened, indicating that the relationships noted above need to be corrected for many real situations. Furthermore, for large I the generation process becomes nonlinear and can be complicated by boiling.[32]

Photoacoustic sources have numerous applications in materials characterization[35] and nondestructive evaluation[36] where they have the advantages of a noncontact source. See also reviews in Ref. 1. Photoacoustics is useful for measuring weak optical absorption coefficients, and the acoustical signal can be enhanced by acoustical resonances of the absorber. As noted in Chapter 42, photoacoustic sources can be useful for investigating relaxation processes affecting the attenuation of sound.

9 SURFACE ACOUSTIC WAVE DEVICES AND ULTRASONIC INSTRUMENTATION

Chapter 44 reviews the role of surface acoustic waves (SAW) in the investigation of various microscopic processes, though SAW devices have much broader applications. They are extensively used in real-time filtering and signal processing with typical operating frequencies in the range from 10 MHz to 1 GHz.[37] Applications include the sensing of a wide range of physical properties.[1] Chapter 45 concerns ultrasonic instrumentation for nondestructive testing and evaluation. Relevant principles are discussed in Ref. 37.

10 THERMOACOUSTIC ENGINES AND REFRIGERATORS

Chapter 46 describes a relatively recent development. Reliable refrigerators have been demonstrated that are driven by sound waves.[38] One of the advantages is that the working fluids can be environmentally safer than conventional refrigerants. A reciprocal process where sound is generated by temperature differences has also been developed.

Acknowledgment

Preparation of this chapter was supported by the U.S. Office of Naval Research.

REFERENCES

1. W. P. Mason and R. N. Thurston or R. N. Thurston and A. D. Pierce (Eds.), *Physical Acoustics*, Vols. 1–23, Academic, New York. (Note: Current volumes typically list the chapter titles of previous volumes.)
2. R. T. Beyer and S. V. Letcher, *Physical Ultrasonics*, Academic, New York, 1969.
3. R. Truell, C. Elbaum, and B. B. Chick, *Ultrasonic Methods in Solid State Physics*, Academic, New York, 1969.
4. P. D. Edmonds (Ed.), *Ultrasonics, Methods of Experimental Physics*, Vol. 19, Academic, New York, 1981.
5. C. Kittel, *Introduction to Solid State Physics*, 6th ed., Wiley, New York, 1986.
6. V. B. Braginsky, V. P. Mitrofanov, and V. I. Panov, *Systems with Small Dissipation*, University of Chicago Press, Chicago, 1985; M. F. Bocko and R. Onofrio, "On the Measurement of a Weak Classical Force Coupled to a Harmonic Oscillator: Experimental Progress," *Rev. Mod. Phys.*, Vol. 68, 1996, pp. 755–799.
7. K. S. Thorne, "Gravitational Radiation," in S. W. Hawking and W. Israel (Eds.), *Three Hundred Years of Gravitation*, Cambridge University Press, Cambridge, 1987, pp. 330–458.
8. P. F. Michelson, et al., "Near Zero: Toward a Quantum-limited Resonant-Mass Gravitational Radiation Detector," in J. D. Fairbank et al. (Eds.), *Near Zero: New Frontiers of Physics*, Freeman, New York, 1988, pp. 713–730.
9. J. N. Hollenhorst, "Quantum Limits on Resonant-Mass Gravitational-Radiation Detectors," *Phys. Rev. D*, Vol. 19, 1979, pp. 1669–1679.
10. A. L. Fetter and J. D. Walecka, *Theoretical Mechanics of Particles and Continua*, McGraw-Hill, New York, 1980, Chaps. 4 and 6 and Problem 4.15.
11. A. L. Fetter and J. D. Walecka, *Quantum Theory of Many-Particle Systems*, McGraw-Hill, New York, 1971.
12. M. O. Scully and S. F. Jacobs, "Coherence—A Sticky Subject," *Appl. Opt.*, Vol. 9, 1970, pp. 2414–2422; M. O. Scully and M. Sargent, "The Concept of the Photon," *Phys. Today*, Vol. 25, No. 3, 1972, pp. 38–47.
13. L. I. Schiff, *Quantum Mechanics*, 3rd ed., McGraw-Hill, New York, 1968, pp. 66–76.
14. E. Goldin, *Waves and Photons*, Wiley, New York, 1982, Chaps. 7 and 8.
15. R. Loudon, *The Quantum Theory of Light*, University Press, Oxford, 1983, Chap. 4.
16. J. P. Wolfe, "Ballistic Heat Pulses in Crystals," *Phys. Today*, Vol. 33, No. 12, 1980, pp. 44–50.

17. D. Rugar and P. Grutter, "Mechanical Parametric Amplification and Thermomechanical Noise Squeezing," *Phys. Rev. Lett.*, Vol. 67, 1991, pp. 699–702.
18. T. J. Hofler and S. L. Garrett, "Thermal Noise in a Fiber Optic Sensor," *J. Acoust. Soc. Am.*, Vol. 84, 1988, pp. 471–475.
19. M. C. Teich and B. E. A. Saleh, "Squeezed and Antibunched Light," *Phys. Today*, Vol. 43, No. 6, 1990, pp. 26–34; P. Meystre and D. F. Walls (Eds.), *Nonclassical Effects in Quantum Optics*, American Institue of Physics, New York, 1991.
20. M. J. P. Musgrave, *Crystal Acoustics*, Holden-Day, San Francisco, 1970.
21. B. A. Young, B. Cabrera, and A. T. Lee, "Observations of Ballistic Phonons in Silicon Crystals Induced by α Particles," *Phys. Rev. Lett.*, Vol. 64, 1990, pp. 2795–2798.
22. H. T. Grahn, H. J. Maris, and J. Tauc, "Picosecond Ultrasonics," *IEEE J. Quant. Electronics*, Vol. 25, 1989, pp. 2562–2569.
23. T. Hasegawa, "Comparison of Two Solutions for Acoustic Radiation Pressure on a Sphere," *J. Acoust. Soc. Am.*, Vol. 61, 1977, pp. 1445–1448.
24. R. T. Beyer, "Radiation Pressure—The History of Mislabeled Tensor," *J. Acoust. Soc. Am.*, Vol. 63, 1978, pp. 1025–1030.
25. C. P. Lee and T. G. Wang, "Acoustic Radiation Pressure," *J. Acoust. Soc. Am.*, Vol. 94, 1993, pp. 1099–1109.
26. E. H. Trinh, "Compact Acoustic Levitation Device for Studies in Fluid Dynamics and Material Science in the Laboratory and Microgravity," *Rev. Sci. Inst.*, Vol. 56, 1985, pp. 2059–2065.
27. T. G. Wang, "Equilibrium Shapes of Rotating Spheroids and Drop Shape Oscillations," *Adv. Appl. Mech.*, Vol. 26, 1988, pp. 1–26.
28. L. A. Crum, "Bjerknes Forces on Bubbles in a Stationary Sound Field," *J. Acoust. Soc. Am.*, Vol. 57, 1975, pp. 1363 1370.
29. P. L. Marston and R. E. Apfel, "Quadrupole Resonance of Drops Driven by Modulated Acoustic Radiation Pressure—Experimental Properties," *J. Acoust. Soc. Am.*, Vol. 67, 1980, pp. 27–37.
30. T. J. Asaki, P. L. Marston, and E. H. Trinh, "Shape Oscillations of Bubbles in Water Driven by Modulated Ultrasonic Radiation Pressure: Observations and Detection with Scattered Laser Light," *J. Acoust. Soc. Am.*, Vol. 93, 1993, pp. 706–713.
31. A. Korpel (Ed.), *Selected Papers on Acousto-Optics*, SPIE, Bellingham, WA, 1990.
32. L. M. Lyamshev, "Lasers in Acoustics," *Sov. Phys. Usp.*, Vol. 30, 1987, pp. 252–279.
33. S. Stanic, "Quantitative Schlieren Visualization," *Appl. Opt.*, Vol. 17, 1978, pp. 837–842.
34. G. J. Diebold, T. Sun, and M. I. Khan, "Photoacoustic Monopole Radiation in One, Two, and Three Dimensions," *Phys. Rev. Lett.*, Vol. 67, 1991, pp. 3384–3387.
35. D. A. Hutchins and A. C. Tam, "Pulsed Photoacoustic Materials Characterization," *IEEE Trans. Ultrasonics, Ferro., Freg. Cont.*, Vol. UFFC-30, 1986, pp. 429–449 (and other articles in this special issue).
36. F. A. McDonald, "Practical Quantitative Theory of Photoacoustic Pulse Generation," *Appl. Phys. Lett.*, Vol. 54, 1989, pp. 1504–1506.
37. G. S. Kino, *Acoustic Waves: Devices Imaging, and Analog Signal Processing*, Prentice-Hall, Englewood Cliffs, NJ, 1987.
38. G. W. Swift, "Thermoacoustic Engines and Refrigerators," *Phys. Today*, Vol. 48, No. 7, 1995, pp. 22–28.

41

ULTRASONIC VELOCITY

JOHN H. CANTRELL AND WILLIAM T. YOST

1 INTRODUCTION

Measurements of the sound velocity and attenuation can provide considerable information about the physical properties of solids, liquids, and gases. Measurement of the variation of the sound velocity as a function of some intrinsic material variable, such as temperature or pressure, provides additional information about basic (often nonlinear) material properties. It is important to distinguish in measurements of velocity the difference between group and phase velocities. This distinction is addressed below in the context of acoustic propagation in a dispersive medium. Similarly, it is important to distinguish between attenuation and absorption in acoustic measurements. Absorption refers to the loss of amplitude of an acoustic wave that results in an increase of temperature (however slight) in the propagation medium. Attenuation refers to the total change in the amplitude of an acoustic wave resulting from all mechanisms responsible for such changes, including but not limited to absorption, scattering, diffraction, and phase cancellation.

Numerous techniques have been reported in the literature for making velocity and attenuation measurements. A number of review articles have been published emphasizing different applications or aspects of the measurement techniques.[1–6] The most common techniques are based on optical, pulse, and continuous-wave methods or hybrids of these methods. The particular technique chosen is dictated in part by the availability of equipment and ease of set-up for the particular application, the specific material and ultrasonic properties of the material to be measured, and the accuracy and precision desired in the measurement.

Note: References to chapters appearing only in the *Encyclopedia* are preceded by *Enc.*

This chapter is mainly concerned with an examination of representation measurement techniques in each of the aforementioned categories. Pulse methods are slightly emphasized because of their widespread popularity and ease of application. The accuracy and precision of a given measurement are intrinsically related to error sources and are among the most important considerations in evaluating data from any technique. The chapter concludes with a general discussion of the most common, and often most correctable, sources of error.

2 PULSE SYSTEMS

2.1 Basic Pulse–Echo Method

An equipment arrangement for the basic pulse–echo system[7,8] that allows for both velocity and attenuation measurements is shown in Fig. 1. For velocity measurements all switches are set to the "out" position and a pulsed ultrasonic signal is transmitted into the sample by means of a transducer attached to the sample surface. The ultrasonic pulse travels through the sample and reflects between the sample boundaries, eventually decaying away because of sample attenuation. Each time the ultrasonic pulse strikes the end of the sample coupled to the transducer, an electrical signal is generated that is amplified and displayed on an oscilloscope. The received signals are commonly rectified and filtered (i.e., detected) before being displayed on the oscilloscope. The velocity of ultrasonic wave propagation is determined by measuring the transit time between corresponding reference points of the reflected pulses and the corresponding pulse propagation distance in the sample.

Adjusting the attenuator to maintain a constant receiver signal level for each displayed echo minimizes

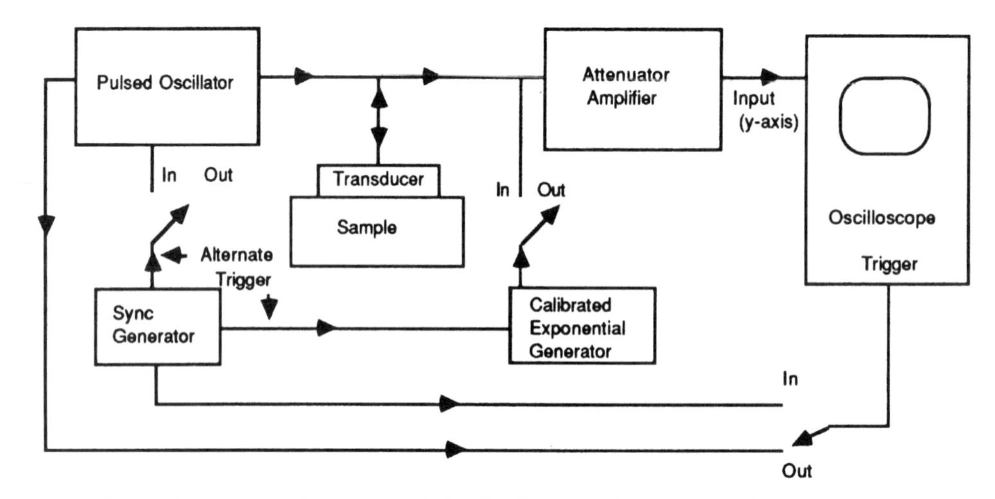

Fig. 1 Basic pulse–echo system (switches "out") for velocity and attenuation measurements and improvements (switches "in") for attenuation measurements.

the error due to amplifier nonlinearities and reduces the uncertainty in the signal reference points. The ultrasonic attenuation can also be read directly from the calibrated attenuator with this procedure.

A small transit time error is introduced into velocity measurements of solids using contact transducers in part because of the partial transmission and reflection of the ultrasonic pulse at the sample–transducer interface. This phenomenon changes the shape of the pulse with each successive echo and foils any attempt to find some characteristic "mark" on the pulses to use as an accurate reference point to measure transit time. Reference mark or echo-matching measurements,[9] while expedient, can lead to errors as large as a few parts in 10^3. Cycle-for-cycle matching of the echoes[10] eliminates this particular aspect of the problem and is discussed in Section 2.3. Ultrasonic phase shifts incurred upon pulse reflection at the sample–transducer interface are another source of error that also affect cycle-for-cycle matching and are addressed in Section 2.2.

In measurements of solids using contact transducers, the basic pulse–echo method using leading-edge echo matching is generally accurate to a few parts in 10^3. Attenuation measurements accurate to about 5 parts in 10^2 are possible. In liquids and gases the sound velocity can be determined by measuring the distance the transducer must be moved to delay the received echo by a specified amount. This variable-path-length procedure increases the accuracy of the basic pulse–echo method to about 5 parts in 10^4, since multiple reflections need not be considered.

The accuracy of attenuation measurements can be improved by replacing the calibrated attenuator with either a calibrated exponential generator or a pulser having a calibrated attenuator. This arrangement corresponds to setting all switches to the "in" position in Fig. 1. In the pulser arrangement,[11] measurements are made by matching the comparison pulse amplitude to the selected echo on alternate sweeps of the oscilloscope. Accuracies of the order 2 parts in 10^2 are possible with this set-up.[12] If the exponential generator is used,[13] the generated curve is matched to a selected pair of peaks of the decaying echo train. In both cases correction for diffraction losses is made for the echoes selected. Automated systems using peak detection of two selected echoes have been developed that achieve greater sensitivities than the manual systems.[5]

Finally, a modification useful for velocity measurements may be made in the basic pulse–echo method by placing a second transducer (used as a receiver) at the end of the sample opposite the transmitting transducer.[14] The received signal is used to retrigger the pulse generator, thereby generating a continuous succession of pulses. The repetition rate of the pulse sequence may be used to determine the ultrasonic velocity, since the repetition rate depends on the pulse travel time in the medium.

An inherent error in the pulse repetition rate (due to time delays in the electronic circuits, changes in pulse shape from attenuation, and presence of transducer–sample interfaces) limits the accuracy of this sing-around method for absolute velocity measurements. Accuracies of a few parts in 10^4 for absolute velocity measurements are possible with this method. Improved versions of the sing-around method[15] allow relative velocity measurements (i.e., changes in velocity) to be measured with accuracies to 1 part in 10^7.

2.2 Gated Double-Pulse Superposition Method

This method[16] utilizes a pulse generated by gating a continuously running oscillator. The gated, continuous, ultrasonic wave (toneburst) is transmitted into the sample followed by a second ultrasonic toneburst phase locked but delayed in time with respect to the first. The two tonebursts reflect between the ends of the sample, giving rise to two pulse–echo trains. The time delay is adjusted such that superposition of the desired echoes from the two pulse trains is achieved. The resulting signal is received by a transducer at the opposite end of the sample, amplified, and then displayed on an oscilloscope.

The superimposed signal may be made zero, independently of time, by adjusting the oscillator frequency $f = \omega/2\pi$ such that the time delay τ between the pulses and the phase shift γ due to reflection at the sample ends satisfy the condition[17] $\omega\tau - \gamma = (2n+1)\pi/2$, where n is a nonnegative integer. The pulse transit time is then calculated from this expression after n and γ have been determined. The phase angle γ is given in Ref. 17 in terms of the acoustical characteristic impedances Z_B, Z_T, and Z_S (i.e., the product of the mass density and the sound velocity) of the bonding material, transducer, and sample, respectively, and the thicknesses and velocities of sound of the transducer and the bonding material.

If the bond thickness can be neglected, the expression for γ is greatly simplified. For noncontact transducers, $Z_T = 0$ and the transit time is simply calculated as $\tau = 1/2(f_{n+1} - f_n)$. Accuracies of 1 part in 10^4 are possible with the gated double-pulse superposition method.[16]

2.3 Pulse Superposition Method

The equipment arrangement for the pulse superposition method[18,19] is shown in Fig. 2. A series of pulses (tonebursts) from a radio frequency (RF) pulse generator is introduced into the sample. The repetition rate of these pulses, controlled by the frequency of a continuous-wave (CW) low-frequency oscillator is adjusted to correspond approximately to some multiple p of an acoustic round-trip transit time δ $(=2\tau)$ in the sample. The oscillator frequency is adjusted to produce an "in-phase" superposition of pulses such that a maximum amplitude is achieved. The measured time delay T between superimposed pulses is the reciprocal of the CW oscillator frequency and is written as[19]

$$T = p\delta - \frac{p\gamma}{360f} + \frac{n}{f}, \tag{1}$$

where γ again is the phase angle associated with the pulse reflections, f is the ultrasonic frequency, and n is now a positive or negative integer that indicates the cyclic mismatch between pulses.

The pulse transit time τ is calculated from Eq. (1) once a value of T corresponding to a given n is determined. Procedures for obtaining the appropriate value for T is given in Ref. 19. For measurements in solids using bonded-contact transducers the unknown bond thickness introduces some uncertainty in the values of the γ's. In this case the phase angles are calculated as a function of estimated bond thickness, and these values are used to calculate the change in T (i.e., ΔT) as a function of the bond thickness. Such calculations establish that ΔT has a limited range for a given value of p and n. Thus all experimentally measured values of T and ΔT outside the limits calculated can be eliminated from consideration.[19]

Absolute velocity measurements having accuracies of 2 parts in 10^4 have been reported with the pulse superposition method.[18,19] Transit time measurements, hence relative velocity measurements, can be made to 1 part in 10^7.

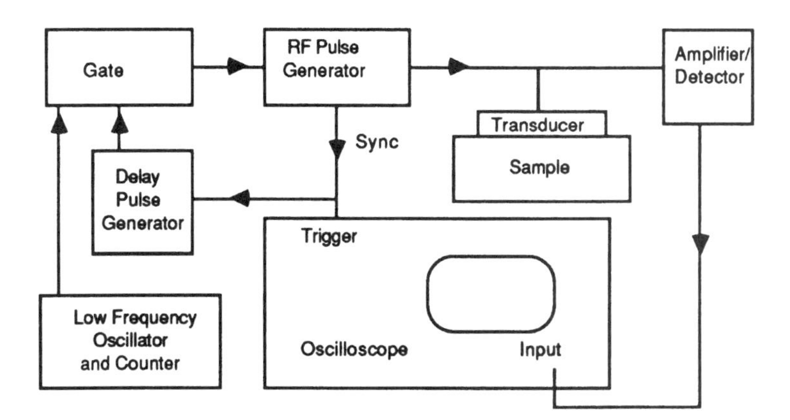

Fig. 2 Equipment arrangement for pulse superposition method.[18,19]

2.4 Echo-Overlap Method

A schematic of the equipment arrangement for the echo-overlap method[9,20,21] is presented in Fig. 3. A series of RF pulses from an RF-pulsed oscillator is transmitted into the sample. The trigger signal is derived by frequency dividing the output of a CW low-frequency (typically 100–1000 kHz) oscillator by a factor of 10^3. The trigger is used to control the pulse repetition rate and to trigger a double-time-delay (strobe) generator that actuates the Z-axis intensity gate on the oscilloscope. The delays are adjusted to intensify any chosen pair of displayed echoes. The oscilloscope is operated with a linear sweep (trigger-sweep generator position) during this adjustment so that many echoes appear on the screen.

During the acoustic transit time measurement the oscilloscope is switched to an X–Y mode of operation. In this mode the low-frequency CW oscillator provides the sweep without decade division and the oscilloscope intensity is reduced so that only the two intensified echoes of interest are visible. If the CW oscillator frequency is adjusted to correspond to the reciprocal of the travel time between the echoes, the echoes can be made to overlap cycle for cycle. Overlap also occurs for an integral multiple m of this frequency, since an echo then appears for every mth sweep. Centering of the overlap pattern can be accomplished with the phase shift network. The round-trip acoustic transit time δ ($=2\tau$) in the sample is obtained from Eq. (1), where the time delay actually measured between the echoes is the product of m and the reciprocal of the oscillator frequency. The procedure for obtaining the correct cycle-for-cycle matching of the chosen echo pair (i.e., $n = 0$ condition) and the calculation of the phase shift due to reflection are outlined in Refs. 18 and 19.

Absolute velocity measurements of a few parts in 10^5 have been reported with the echo-overlap method.[9]

2.5 Pulse Interferometer Methods

In the pulse interferometer methods an ultrasonic tone-burst is transmitted into the sample by gating a CW

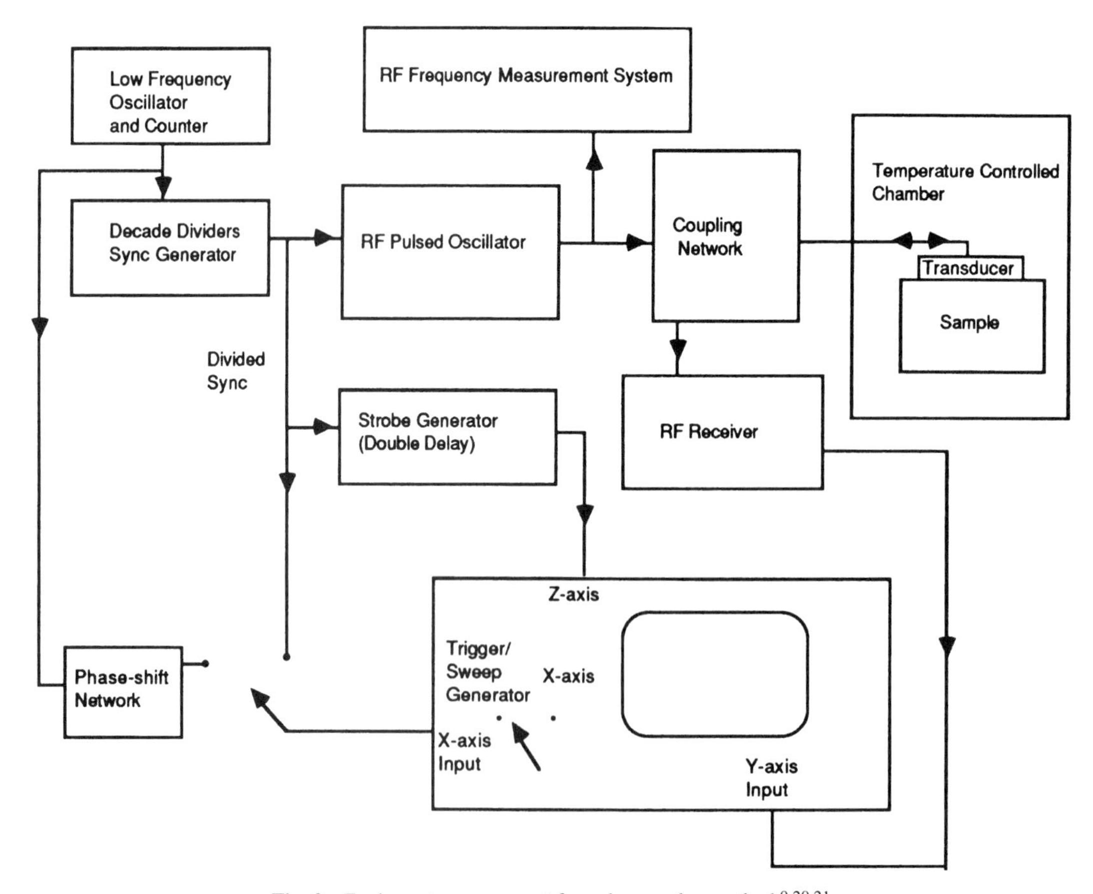

Fig. 3 Equipment arrangement for echo–overlap method.[9,20,21]

reference oscillator. The received pulse (an echo in a single-transducer arrangement) is combined with the CW reference signal in a phase-sensitive detector. If the relative phase between the received acoustic signal and the CW reference signal is then changed, the output of the phase-sensitive detector varies in response. The change in relative phase may result, for example, from a variation in the acoustic path or from a variation in the sound velocity as a function of temperature.

Two basic means are reported to measure the change in relative phase. One means employs a phase shift network that monitors the amount of phase added to the reference signal necessary to maintain a constant output from the phase-sensitive detector.[22,23] The frequency of the CW reference signal is held fixed during this process. The second means uses a change in the frequency of the CW reference signal to maintain a constant detector output.[24,25] In the pulse phase-locked loop system a voltage-controlled oscillator network replaces the phase shift network.[25]

In both the constant-frequency and variable-frequency techniques, the total phase seen by the phase detector includes acoustic contributions related to the acoustic wavelength in sample length as well as collective nonacoustic contributions. The frequency-dependent nonacoustic contributions include those from electrical signal propagation paths and phase perturbations associated with various electrical components. In measurements of phase velocity in fluids, for example, the change in phase corresponding to a given change in path length may be measured. For the constant-frequency technique both the wavelength and the nonacoustic phase term are constants. The phase velocity is obtained from a measurement of the slope of the linear curve giving the change in phase as a function of the change in path length. For the variable-frequency technique the wavelength and the nonacoustic phase term are not fixed and the velocity can be calculated only after correcting for the variations in the nonacoustic phase term.

Variations in the pulse transit time can be measured with a sensitivity of parts in 10^8 with pulse interferometer methods. Absolute velocity measurements can be made to accuracies that are comparable to the pulse superposition and echo-overlap methods.

3 CONTINUOUS-WAVE METHODS

Continuous-wave methods are among the most expedient for measuring sound velocity and attenuation of materials with appropriate geometry. Consider a sample of length $l/2$ with flat and parallel faces and assume that one face is driven continuously by a noncontacting transducer with particle velocity $v = v_0 \cos \omega t$. The multiple reflections of the wave from the parallel sample faces results in a steady-state particle velocity amplitude at the position of the driving transducer given by[6,26]

$$v = \frac{v_0 e^{\alpha l/2}}{\sqrt{2}(\cosh \alpha l - \cos kl)^{1/2}}, \tag{2}$$

where α is the attenuation coefficient in nepers per centimetres and k is the propagation constant. The amplitude response thus consists of a series of mechanical resonances (standing waves) whose frequencies correspond to an integral number of half wavelengths within the sample length. The condition for the nth resonance is $kl = 2\pi n(\alpha l \ll 1)$, where n is an integer.

For materials having a fixed length and sufficiently small dispersion the phase velocity c_p can be determined from the frequency difference $f_n - f_{n-1}$ between two adjacent resonances as $c_p = l(f_n - f_{n-1})$. The attenuation α can be calculated from a determination of the frequency width at half power, $\Delta\omega$, as $\alpha = \Delta\omega/2c_p$. These results are based on the assumption that the sample responds as a simple resonator driven by noncontacting transducers. If contacting transducers are used, the sample and transducers form a composite resonating system whose resonant frequencies are predicted by the roots of a transcendental equation. Approximate solutions to the transcendental equation have resulted in correction factors for the above formulas.[27] However, errors in velocity determination can exceed 10% under certain conditions.

In the method of ultrasonic resonance spectroscopy[28,29] measurements of the modal frequencies for a sample of given geometry are compared to predictive analytical models of the sample modal frequencies. Analysis of the amplitude–frequency spectrum allows an assessment of the elastic constants as well as the internal friction (attenuation) of the sample.

For gases, fixed-frequency CW methods such as the resonating-tube technique and Kundt's dust-tube method are useful for velocity measurements. In the latter method cork dust initially distributed uniformly through the tube piles up at the nodes of a standing wave set up in the tube. The distance between the nodes are used to determine the wavelength λ in a gaseous medium, and the sound velocity c_p is calculated from $c_p = f\lambda$. In the resonating-tube method, a tuning fork or other single-frequency source of sound held at the open end of a gaseous column produces sound reinforcement whenever the vibrational mode frequency of the column is equal to the vibrational frequency of the source. If the column length is then varied by a half wavelength, sound reinforcement again occurs. A high-frequency transducer mounted on a micrometer screw can yield an accurate determination of the speed of sound in relatively small amounts of gas.

A particularly useful aspect of CW methods is the determination of small changes in the attenuation or sound velocity. Specific frequencies can be chosen to isolate small changes in the propagation constant or attenuation resulting from a change in some intrinsic material variable.[26]

4 OPTICAL METHODS

Optical techniques can be used for sound velocity measurements in transparent media. The methods generally are based on the optical determination of acoustic wavelength either from diffraction or scattering of light by the sound waves. The wavelength together with the acoustic frequency allows determination of phase velocity. For techniques based on diffraction it is essential to know the type of diffraction experienced by the light.

4.1 Diffraction of Light by Sound (Fraunhofer Diffraction Conditions for Light)

Consider an ultrasonic transducer generating compressional waves in a transparent propagation medium. If the propagating sound wave encounters light traversing the same medium at some angle incident to the sound wave, the light is diffracted. Assuming that the Fraunhofer conditions hold for the light diffraction (i.e., parallel light rays are brought to a focus), two distinct physical processes can produce the diffraction effects. One involves the formation of a corrugation in the phase fronts of the light due to the acoustically induced spatial variation of the index of refraction. This phenomenon is called Raman–Nath diffraction.[30,31] The second process involves the reflection of light from the (evenly spaced) crests of the sound wave that occurs under conditions similar to that of x-ray diffraction. This phenomenon is called Bragg diffraction.[32] Both types of diffraction can be used to determine the sound velocity of the material.

A dimensionless parameter has been introduced to determine which type of diffraction predominates for given experimental conditions. The parameter Q is defined by[33]

$$Q = \frac{k^2 L}{\mu_0 k^*}, \tag{3}$$

where k is the ultrasonic propagation constant, L is the width of the ultrasonic beam, μ_0 is the index of refraction, and k^* is the propagation constant of the light in vacuo. If $Q > 9$, Bragg diffraction is predominate. If $Q < 1$, Raman–Nath diffraction occurs. If $1 < Q < 9$, the diffraction is mixed. A typical transparent solid has an acoustic phase velocity of the order 5×10^3 m/s and an index of refraction of order 1.5. Consider an ultrasonic beam 1.27×10^{-2} m in width and illuminated with light of wavelength 632.8 nm. If the ultrasonic frequency is approximately 27 MHz or less, the interaction satisfies Raman–Nath diffraction conditions, according to Eq. (3). If the frequency is greater than approximately 81 MHz, the interaction is governed by Bragg diffraction. In liquids the demarcation frequencies are often quite different from solids. In water for the same ultrasonic beam width and wavelength of light as the above example, Raman–Nath diffraction occurs for acoustic frequencies less than 7.7 MHz while Bragg diffraction occurs above 23 MHz.

Raman–Nath Diffraction For Raman–Nath diffraction with the light incident at an angle $\pi/2 - \phi$ to the sound beam direction, the light is diffracted into diffraction orders given by the expression[6]

$$\sin(\theta_n + \phi) - \sin(\phi) = \frac{n\lambda^*}{\lambda}, \tag{4}$$

where θ is the angle of diffraction with respect to the incident beam, λ^* is the wavelength of light in the medium, λ is the ultrasonic wavelength, and n is an integer denoting the diffraction order. The ultrasonic wavelength can be determined from Eq. (4).

Bragg Diffraction For Bragg diffraction, the angle of diffraction θ_n is set equal to twice the negative angle of incidence $(-\phi)$ such that

$$n\lambda^* = 2\lambda \sin \phi_B, \tag{5}$$

where ϕ_B is the Bragg angle. Bragg diffraction is used to measure wave velocities in the frequency range from approximately 100 MHz to several gigahertz. The technique has been used to measure wave velocity, including local velocities,[34] to an accuracy estimated to be better than 0.1%. Measurements with precision better than 0.01% in homogeneous samples have been reported.

4.2 Direct Measurements (Fresnel Diffraction Conditions for Light)

Consider a standing ultrasonic wave in a sample through which collimated light perpendicular to the sound beam is passed. A measuring microscope or similar optical device is focused so that the image of the sound wave can be viewed at the instrument's focal plane. The light that is diffracted by the sound wave is viewed under experimental conditions such that nonparallel light rays are focused to form the image (Fresnel diffraction). Because the images of the wavefronts are $\lambda/2$ apart under such conditions, it is possible to determine directly the wavelength in the medium. The technique[35] is sufficiently sensitive to detect local variations in phase velocity greater than 0.01%.

4.3 Scattering Techniques: Brillouin Scattering

Vibrational Modes in a Crystal Lattice Lattice vibrations are caused by thermal and other excitations of the lattice. These vibrations form vibrational modes called normal modes within the lattice and play a major role in determining the thermal properties (e.g., heat capacity) of the solid. The normal modes form two major branches. The acoustic branch includes normal-mode frequencies that vanish linearly as the propagation vector k approaches zero (long-wavelength limit). The optical branch includes normal-mode frequencies that do not vanish as the propagation vector approaches zero.

The quantum mechanical treatment of the lattice vibrations predicts that the normal modes gain and lose energy in quantized units just as in the case of the harmonic oscillator. Because of the similarity of the lattice vibrations to the electromagnetic radiation field in a cavity, a quantum of lattice vibrational energy called the phonon is defined, primarily for convenience, in analogy to the electromagnetic quantum of energy called the photon. As with photons, phonons are given the property that each phonon carries an energy $\hbar\omega$, where $\hbar$ is Planck's constant divided by 2π, and under certain conditions[36] behaves as if it carries a crystal momentum of value $\hbar k$. The reader is referred to Chapter 43 for detail.

Interaction of Light with Vibrational Modes Light interacting with the normal modes in the material can gain or lose energy and momentum. Brillouin scattering occurs when the modes of the acoustic branch interact with light. Consider the interaction of a photon with a material whose index of refraction is μ_0. Application of the laws of conservation of energy and momentum, respectively, to the scattering process gives

$$\hbar\omega^{*\prime} = \hbar\omega^{*} + \hbar\omega, \qquad \hbar\mu_0 k^{*\prime} = \hbar\mu_0 k^{*} + \hbar k + \hbar K, \quad (6)$$

where $\omega^{*\prime}$ and $k^{*\prime}$ are the frequency and propagation vectors, respectively, of the incident photon; ω^{*} and k^{*} are the frequency and propagation vectors of the scattered photon; ω and k are the frequency and crystal momentum assigned to the phonon; K is the reciprocal lattice vector (assumed to be zero in this case); and μ_0 is the index of refraction of the material. Since the photon frequency shifts are small for this process, the propagation vectors are very nearly the same length. The vector addition expressed by Eq. (6) gives an isosceles triangle whose apex angle is θ, the complement of the scattering angle. Equation (6) leads to the result[37]

$$\omega = \pm 2\omega^{*}\left(\frac{c}{C_l}\right)\sin\left(\frac{\theta}{2}\right), \quad (7)$$

where C_l is the speed of the photon in the medium, c is the speed of the phonon, and θ is the scattering angle. The frequency ω is equal to the difference in frequencies between the incident and scattered photons.

The technique produces a scattered photon spectra of three lines, one of which is unshifted. The two frequency-shifted lines are called the Stokes line (a phonon is created) and the anti-Stokes line (a phonon is annihilated). These lines can be separated and measured with optical devices such as the Fabry–Perot interferometer to obtain ω. If desirable, the Bragg condition can be used for constructive reinforcement[38] by adjusting θ. Equation (7) can be solved for the velocity c in terms of the other quantities measured.

Brillouin scattering is useful in the investigation of the sound velocity of solids and liquids, especially near a phase transition where attenuation can become very large.[39] The technique is also advantageous in that it is one of the few techniques capable of assessing sound velocity at frequencies of several gigahertz. The technique has been used for sound velocity measurements for both longitudinal and mixed (longitudinal–transverse) modes. Uncertainties in the index of refraction, the scattering angle, and line width of the Stokes and anti-Stokes lines influence the accuracy of the measurement. The precision is considerably better, however, with values of 0.1% possible.

5 SOURCES OF ERROR

A number of sources of error arise in the measurements of velocity and attenuation. Their effect on the measurements cannot be overstated. Errors arising from time or length measurements can be evaluated directly, but among the less easily evaluated errors are those arising from velocity dispersion, material inhomogeneity, phase cancellation resulting from lack of parallelism of the transducers, diffraction, and the effects of thickness of the transducer and transducer bond. Many aspects of the sources of error are best addressed in the context of the measurement technique used, but certain characteristics of the error sources have more generic implications that are considered here.

5.1 Dispersion

In a linear, dispersive, propagation medium the acoustic waveform $u(x,t)$ at position x and at time t in the medium is given by the expression[40]

$$u(x,t) = F(x,t)e^{i\Omega(x,t)} + \text{c.c.}, \qquad \Omega(x,t) = kx - \omega t, \quad (8)$$

providing

$$\frac{d\omega}{dk} - \frac{x}{t} = 0, \qquad \frac{d^2\omega}{dk^2} \neq 0, \quad (9)$$

where ω is the angular frequency and c.c. is the complex conjugate. The functional dependence of the angular frequency on the wavenumber k defines the dispersion relation. The condition $d^2\omega/dk^2 \neq 0$ is the defining condition for a dispersive propagation medium and $d\omega/dk = c_g$ is defined to be the group velocity. For a given choice of x and t, Eq. (9) may be solved for k if the dispersion function $\omega(k)$ is known. Thus, different parts of the waveform corresponding to different given pairs (x,t) propagate with different group velocities $c_g(k)$ corresponding to specific wavenumbers determined from Eq. (9).

Both the complex amplitude $F(x,t)$ and the phase function $\Omega(x,t)$ in Eq. (8) are functionally dependent on x and t. Equation (8) has the form of an elementary waveform, but $F(x,t)$ is actually a position- and time-varying envelope of the waveform with the position- and time-varying phase $\Omega(x,t)$ defining nonuniformly spaced maxima and minima of the wave. Equation (8) thus defines a disturbance that in general changes shape during its propagation through a dispersive medium.

The propagation of a given constant value of phase Ω_0 defined by $\Omega(x,t) = \Omega_0$ moves with the phase velocity $c_p(x,t) = dx/dt = \omega/k$. The phase velocity $c_p(x,t)$ is generally different for different values of Ω_0 and for different values of x and t. An observer following a particular phase (e.g., crest of a wave) moves with the local phase velocity having a locally varying frequency and wavenumber. Thus, the phase velocity of a wave of finite length propagating in a dispersive medium can only be defined locally (i.e., for given x and t) and for a given phase Ω_0. Hence, if a given pulse measurement technique follows a given phase Ω_0 of the waveform at different monotonically increasing or decreasing values of the space–time coordinates (x_i, t_i), then the local phase velocity at (x_α, t_α) for phase Ω_0 can be obtained by plotting x_i as a function of t_i and measuring the tangent to the curve at (x_α, t_α). For dispersive media the curve is generally nonlinear, but for nondispersive media the curve is linear and coincides with the curve for the group velocity $c_g = x/t$. Techniques employing a change in the operation frequency of the system cannot generally be used to accurately measure phase velocity in highly dispersive media, since the shift in operation frequency violates conditions under which the phase velocity is defined.

Measurements using gated continuous waves (i.e., tonebursts) of finite length can be made to approximate CW conditions if the pulse lengths are sufficiently long. The appropriate pulse length L is dependent on the intrinsic error M of the measurement technique and is given by the expression[22]

$$L \geq \frac{\pi(d^2\omega/dk^2)}{Mc_p}. \quad (10)$$

5.2 Diffraction

Because the transducer is of finite size, the acoustic wave generated in the sample spreads out into a diffraction field that can introduce error in both attenuation and velocity measurements. The effects of diffraction are usually treated[41–44] for the case of circular, axially concentric transmitting and receiving transducers of the same radius a. The transmitting transducer is treated as a finite piston source in a semi-infinite medium. The acoustic field is found at each point in the propagation medium, and an integration is performed over the area in the field presented by the receiving transducer. In one treatment the effects of diffraction can be represented for waves of wavelength λ propagating along the z-direction in an isotropic medium by the expression[43]

$$u(z,t) = 2(I_x^2 + I_y^2)^{1/2}\cos(kc_pt - kz - \phi), \qquad \phi = \tan(I_y/I_x), \quad (11)$$

where the quantities I_x and I_y are integral functions of the dimensionless parameter $S = z\lambda/a^2$.

These equations show that the diffraction error involves an intensity variation in the diffraction pattern and a distortion of the otherwise planar wavefront. Phase variations in the surface plane of the receiving transducer give rise to interference effects and a loss in signal when integrated over a phase-sensitive transducer surface. The factor ϕ in Eq. (11) produces a variation in the phase of the propagating wave that results in an error in the velocity measurements, if not properly corrected.

5.3 Phase Cancellation

A variation in the phase of the wave over the area presented by the surface of a phase-sensitive transducer of radius a can give rise to phase cancellation of different portions of the received signal and set limits on the measurement accuracy of both velocity and attenuation. Sources of such phase variation include diffraction, media inhomogeneity, temperature gradients, wedging, and nonparallelism of the sample surfaces. Nonparallelism of opposite surfaces of the sample by an angle θ is representative of the effects of phase cancellation. For example, in measurements employing acoustic pulses of wavenumber k reflecting from the ends of the sample the amplitude of the pulse received after n double reflections is diminished by the factor $J_1(2kan\theta)/kan\theta$ due to nonparallelism.[5] The use of phase-insensitive devices such as the "acoustoelectric transducer" can minimize phase cancellation error,[45] although such devices are generally less sensitive than the commonly used phase-sensitive piezoelectric elements.

5.4 Transducer Bond

Major sources of error in measurements of velocity are the effects of the transducer and the bond coupling the transducer to the sample. Both the transducer and bond introduce additional phase into the measurement system, which can lead to errors of the order of parts in 10^2 in the measurement of velocity. Methods for correcting such errors are well established and have been addressed as appropriate to the specific measurement techniques discussed. An alternative to such corrections is the use of noncontacting transducers such as capacitive[46] and electromagnetic transducers[47] and optical techniques.

6 CONCLUSION

The methods presented in this chapter for measuring velocity and attenuation in solids, liquids, and gases are basic and instructive. It is not possible to single out a best method to use in all measurement situations. Considerations of sample dimensions, environmental constraints, and specific physical and chemical properties of the material play a role in selecting the measurement method. Many variations of the basic methods delineated in this chapter have been developed, some of which are useful in specific applications.[1–6]

It must be emphasized that any method is only as good as the care taken by the investigator to correct for sources of error. For example, a critical analysis of the accuracies of the pulse superposition and the echo-overlap methods shows that absolute accuracies of the order of parts in 10^6 are possible for velocity measurements with either method if care is taken to correct for diffraction, wedging, transducer and bond thicknesses, and other sources of error. Similar accuracies are expected to be achievable with other methods such as pulse interferometry and some of the CW methods.

In addition to the error sources discussed here, a serious limitation on the accuracy of absolute velocity measurements is the determination of the length of the sample. Length considerations are not a factor in relative velocity measurements where sensitivities of parts in 10^8 are theoretically possible. For attenuation measurements error corrections are no less critical. The aforementioned error sources and the intrinsic limitations on instrument calibration for measuring wave amplitudes geneally limit the accuracy of attenuation measurements to parts in 10^3 even after corrections are applied.

Careful sample preparation and apparatus set-up together with the appropriate error corrections and efforts to control equipment instability and environmental factors (such as temperature and pressure) allow measurements of velocity, relative velocity, and attenuation that approach the theoretical limits for a given method.

REFERENCES

1. H. J. McSkimin, in W. P. Mason (Ed.), *Physical Acoustics*, Vol. IA, Academic, New York, 1964, pp. 271–417.
2. E. P. Papadakis, in W. P. Mason (Ed.), *Physical Acoustics*, Vol. IVB, Academic, New York, 1968, pp. 269–328.
3. E. P. Papadakis, in W. P. Mason and R. N. Thurston (Eds.), *Physical Acoustics*, Vol. XI, Academic, New York, 1975, pp. 151–211.
4. E. P. Papadakis, in W. P. Mason and R. N. Thurston (Eds.), *Physical Acoustics*, Vol. XII, Academic, New York, 1976, pp. 277–374.
5. R. Truell, C. Elbaum, and B. B. Chick, *Ultrasonic Methods in Solid State Physics*, Academic, New York, 1969.
6. M. A. Breazeale, J. H. Cantrell, Jr. and J. S. Heyman, in P. H. Edmonds (Ed.), *Methods of Experimental Physics, Ultrasonics*, Vol. 19, Academic, New York, 1981, pp. 67–135.
7. F. A. Firestone and J. R. Frederick, "Refinements in Supersonic Reflectoscopy. Polarized Sound," *J. Acoust. Soc. Am.*, Vol. 18, 1946, pp. 200–211.
8. J. R. Pellam and J. K. Galt, "Ultrasonic Propagation in Liquids: I. Application of Pulse Technique to Velocity and Absorption Measurements at 15 Megacycles," *J. Chem. Phys.*, Vol. 14, 1946, pp. 608–614.
9. E. P. Papadakis, "Ultrasonic Phase Velocity by the Pulse-Echo-Overlap Method Incorporating Diffraction Phase Corrections," *J. Acoust. Soc. Am.*, Vol. 42, 1967, pp. 1045–1051.
10. H. J. McSkimin, "Measurement of Ultrasonic Wave

Velocity for Solids in the Frequency Range 100–500 Mc," *J. Acoust. Soc. Am.*, Vol. 34, 1962, pp. 404–409.

11. R. L. Roderick and R. Truell, "The Measurement of Ultrasonic Attenuation in Solids by the Pulse Technique and Some Results in Steel," *J. Appl. Phys.*, Vol. 23, 1952, pp. 267–279.
12. J. H. Andreae, R. Bass, E. L. Heasell, and J. Lamb, "Pulse Techniques for Measuring Ultrasonic Absorption in Liquids," *Acustica*, Vol. 8, 1958, pp. 131–142.
13. B. Chick, G. Anderson, and R. Truell, "Ultrasonic Attenuation Unit and Its Use in Measuring Attenuation in Alkali Halides," *J. Acoust. Soc. Am.*, Vol. 32, 1960, pp. 1186–1193.
14. N. P. Cedrone and D. R. Curran, "Electronic Pulse Method for Measuring the Velocity of Sound in Liquids and Solids," *J. Acoust. Soc. Am.*, Vol. 26, 1954, pp. 963–966.
15. R. L. Forgacs, "An Improved Sing-Around System for Ultrasonic Velocity Measurements," *IRE Trans. Instrum.*, Vol. 9, 1960, pp. 359–367.
16. J. Williams and J. Lamb, "On the Measurement of Ultrasonic Velocity in Solids," *J. Acoust. Soc. Am.*, Vol. 30, 1958, pp. 308–313.
17. M. Redwood and J. Lamb, "On the Measurement of Attenuation in Ultrasonic Delay Lines," *Proc. Inst. Electr. Eng.*, Vol. 103B, 1956, pp. 773–780.
18. H. J. McSkimin, "Variations of the Ultrasonic Pulse-Superposition Method for Increasing the Sensitivity of Delay-Time Measurements," *J. Acoust. Soc. Am.*, Vol. 37, 1965, pp. 864–871.
19. H. J. McSkimin and P. Andreatch, Jr., "Measurement of Very Small Changes in the Velocity of Ultrasonic Waves in Solids," *J. Acoust. Soc. Am.*, Vol. 41, 1967, pp. 1052–1057.
20. J. E. May, Jr., "Precise Measurement of Time Delay," *IRE Nat. Conv. Rec.*, Vol. 6, 1958, pp. 134–142.
21. E. P. Papadakis, "Ultrasonic Attenuation and Velocity in Three Transformation Products in Steel," *J. Appl. Phys.*, Vol. 35, 1964, pp. 1474–1482.
22. W. T. Yost, J. H. Cantrell, and P. Kushnick, "Fundamental Aspects of Pulse Phase-Locked Loop Technology-Based Methods for Measurement of Ultrasonic Velocity," *J. Acoust. Soc. Am.*, Vol. 91, 1992, pp. 1456–1468.
23. W. M. Whitney and C. E. Chase, "Velocity of Sound in Liquid Helium at Low Temperatures," *Phys. Rev. Lett.*, Vol. 9, 1962, pp. 243–245.
24. R. J. Blume, "Instrument for Continuous High Resolution Measurement of Changes in the Velocity of Ultrasound," *Rev. Sci. Instrum.*, Vol. 34, 1963, pp. 1400–1407.
25. J. S. Heyman and E. J. Chern, "Ultrasonic Measurement of Axial Stress," *J. Testing Evaluation*, Vol. 10, 1982, pp. 202–211.
26. D. I. Bolef and J. G. Miller, in W. P. Mason and R. N. Thurston (Eds.), *Physical Acoustics*, Vol. VIII, Academic, New York, 1971, pp. 96–201.
27. E. J. Chern, J. H. Cantrell, Jr., and J. S. Heyman, "Improved Formula for Continuous-Wave Measurements of Ultrasonic Phase Velocity," *J. Appl. Phys.*, Vol. 52, 1981, pp. 3200–3204.
28. O. Anderson, "Rectangular Parallelepiped Resonance-A Technique of Resonance Ultrasound and Its Applications to the Determination of Elasticity at High Temperatures," *J. Acoust. Soc. Am.*, Vol. 91, 1992, pp. 2245–2253.
29. H. Ledbetter, C. Fortunko, and P. Heyliger, "Elastic Constants and Internal Friction of Polycrystalline Copper," *J. Mater. Res.*, Vol. 10, 1995, pp. 1352–1353.
30. C. V. Raman and N. S. N. Nath, "The Diffraction of Light by High Frequency Sound Waves: Part 1," *Proc. Indian Acad. Sci.*, Sect. A, Parts 1 and 2, Vol. 2, 1935, pp. 406–420.
31. C. V. Raman and N. S. N. Nath, "The Diffraction of Light by High Frequency Sound Waves: Part V. General Considerations—Oblique Incidence and Amplitude Changes," *Proc. Indian Acad. Sci.*, Sect. A, Part V, Vol. 3, 1936, pp. 459–465.
32. A. B. Bhatia and W. J. Noble, "Diffraction of Light by Ultrasonic Waves I. General Theory," *Proc. R. Soc. London*, Ser. A 220, 1953, pp. 356–368.
33. W. R. Klein, B. D. Cook, and W. G. Mayer, "Light Diffraction by Ultrasonic Gratings," *Acustica*, Vol. 15, 1965, pp. 67–74.
34. F. Michard and B. Perrin, "New Optical Probe for Local Elastic Study in Transport Media," *J. Acoust. Soc. Am.*, Vol. 64, 1978, pp. 1447–1456.
35. W. G. Mayer and E. A. Hiedemann, "Optical Methods for the Ultrasonic Determination of the Elastic Constants of Sapphire," *J. Acoust. Soc. Am.*, Vol. 30, 1958, pp. 756–760.
36. C. Kittel, *Introduction to Solid State Physics*, 3rd ed., Wiley, New York, 1966, pp. 134ff.
37. N. W. Ashcroft and N. D. Mermin, *Solid State Physics*, Saunders, Philadelphia, 1976, pp. 480–483.
38. H. F. Pollard, *Sound Waves in Solids*, Pion Limited, London, 1977, pp. 308–309.
39. P. A. Fleury, in W. P. Mason and R. N. Thurston (Eds.), *Physical Acoustics*, Vol. VI, Academic, New York, 1970, pp. 37–58.
40. G. B. Whitham, *Linear and Nonlinear Waves*, Wiley-Interscience, New York, 1974.
41. H. Seki, A. Granato, and R. Truell, "Diffraction Effects in the Ultrasonic Field of a Piston Source and Their Importance in the Accurate Measurement of Attenuation," *J. Acoust. Soc. Am.*, Vol. 28, 1956, pp. 230–238.
42. A. S. Khimunin, "Numerical Calculation of the Diffraction Corrections for the Precise Measurement of Ultrasound Absorption," *Acustica*, Vol. 27, 1972, pp. 173–181.
43. G. C. Benson and O. Kiyohara, "Tabulation of Some Integral Functions Describing Diffraction Effects in the Ultrasonic Field of a Circular Piston Source," *J. Acoust. Soc. Am.*, Vol. 55, 1974, pp. 184–185.

44. E. P. Papadakis, "Ultrasonic Diffraction Loss and Phase Change in Anisotropic Materials," *J. Acoust. Soc. Am.*, Vol. 40, 1966, pp. 863–876.
45. J. S. Heyman, "Phase Insensitive Acoustoelectric Transducer," *J. Acoust. Soc. Am.*, Vol. 64, 1978, pp. 243–249.
46. J. H. Cantrell, Jr., and M. A. Breazeale, "Elimination of Transducer Bond Corrections in Accurate Ultrasonic Wave Velocity Measurements by Use of Capacitive Transducers," *J. Acoust. Soc. Am.*, Vol. 61, 1977, pp. 403–406.
47. D. J. Meridith, R. J. Watts-Tobin, and E. R. Dobbs, "Electromagnetic Generation of Ultrasonic Waves in Metals," *J. Acoust. Soc. Am.*, Vol. 45, 1969, pp. 1393–1401.

42

ULTRASONIC RELAXATION PROCESSES

HENRY E. BASS AND F. DOUGLAS SHIELDS

1 INTRODUCTION

Studies of acoustic absorption and dispersion have proven to be an effective tool in developing a physical description of molecular interactions in gases and liquids. Experimentally, absorption or velocity dispersion is measured as a function of frequency (or, in gases, as a function of frequency and pressure). These same quantities are computed using the theory outlined below, which requires some insight into the particular physical processes that give rise to the absorption or dispersion (e.g., vibrational relaxation). Computed values are compared to experiment, and the assumed physical process or interaction parameters are varied until agreement is achieved. The theoretical model that gives rise to predicted absorption and dispersion that agree with experiment are assumed to correctly represent the interactions taking place at the microscopic level.

This chapter is concerned with ultrasonic relaxation processes. The relaxation contribution to absorption of sound in air at lower frequencies is addressed in Chapter 28. One source of relaxation absorption is that associated with chemical reactions, especially in fluids. That type of relaxation is not considered here. This chapter will be concerned, primarily, with the theory that relates the microscopic excitation/deexcitation interactions to acoustic absorption or dispersion.

2 PHENOMENOLOGICAL EXPLANATION OF RELAXATION

For the purpose of developing an understanding of relaxation processes, first consider an ideal gas made up of diatomic molecules. The individual molecules are free to move translationally in three directions, rotate about two perpendicular axes (actually three but the third has zero moment of inertia and so has no energy), and vibrate along the bond joining the atoms. Some energy is associated with each of these allowed motions.

Translational motion can be considered nonquantized. Any energy is allowed. As the molecules translate, they collide, exchanging energy with their collision partners. At atmospheric pressure molecular collisions take place at a rate of about 10^{11} s^{-1}. A single collision is typically sufficient to transfer translational energy from one molecule to another. However, a certain period of time is required to randomize energy associated with excess velocity in a particular direction. This time is often referred to as the translational relaxation time. Kohler, following Maxwell,[1] associates viscosity with this relaxation time writing $\tau_{tr} = \eta/p = 1.25\tau_c$, where p is gas pressure and τ_c is the time between collisions.

As the pressure is lowered, the rate of collisions decreases proportionately. At 1 Torr (1/760 atm), the translational relaxation time is about 10^{-8} s; at 1 mTorr it is 10^{-5} s. These time regimes can be effectively studied using ultrasonics.

Unlike translational motion, rotation and vibration are noticeably quantized. During a collision, a change in rotational or vibrational state can only occur if the change in energy of another state is sufficient to allow at least one quantum jump. For rotational energy transfer, the spacing between energy levels is given by $2(J+1)B$, where J is the rotational quantum number $B = \hbar^2/2I$ and I is the effective moment of inertia. If one assumes J is the most probable value (from a Boltzman distribution), the value of $2(J+1)B$ for a typical molecule (say N_2) in temperature units is about 1°. This means that in a gas above 1 K, essentially all collisions will have sufficient translational energy to cause multiple changes in joules.

Note: References to chapters appearing only in the *Encyclopedia* are preceded by *Enc.*

As a result, rotation rapidly equilibrates with translation.

An exception is hydrogen, which has much larger rotational energy level spacing due to the small moment of inertia. On the average, as many as 350 collisions[2] may be necessary to transfer a quantum of rotational energy in H_2. At a pressure of 1 atm, this gives a relaxation time of about 2×10^{-8} s. It should be noted that a given collision does or does not transfer a quantum of energy. The 350-collision average means that only 1 collision in 350 has the proper geometry and energy to cause a transfer of one quantum of rotational energy. The number of collisions necessary, on the average, to transfer one quantum of energy is referred to as the collision number Z. When rotational energy is involved, the subscript "rot" is typically added (Z_{rot}). The inverse of this dimensionless quantity is the probability of transferring a quantum in a collision (P_{rot}). Since rotational energy level spacings are unequal, a $1 \to 2$ transition should be more probable than a $2 \to 3$ transition. These events are distinguished by using the symbols $P_{\text{rot}}^{1\to 2}$ or $P_{\text{rot}}^{2\to 3}$.

Generally speaking, the probability for transferring a quantum of energy in a collision decreases rapidly with the size of the quantum transferred. Since vibrational levels are much more widely spaced than rotational energy levels, vibrational relaxation times are much longer than rotational. Vibrational levels in a single vibrational mode are approximately equally spaced. This means that energy can be exchanged between levels (i.e., the vibrational quantum number goes up in one molecule and down in the other) with very little energy exchanged between vibration and translation. The result is that such vibration-to-vibration exchanges take place very rapidly. The vibrational relaxation time is controlled by the time it takes energy to transfer between translation and the lowest lying vibrational level. Since this energy level varies greatly for different molecules, so do the probabilities of vibrational energy transfer during a collision. During N_2 collisions with N_2, Z_{10} is near 1.5×10^{11}, so the relaxation time is near 15 s[3] (Z_{10} is the number of collisions needed to transfer energy from the lowest vibrational level to translation). Large molecules have vibrational energy levels that are very close together. A molecule such a C_2H_6 requires only 100 collisions to transfer a quantum of vibrational energy from the first vibration level into translation.[4] This very wide range of relaxation times presents interesting experimental challenges.

To this point, there has been no attempt to rigorously define relaxation times in terms of energy transfer probabilities. In fact, such a relation is possible in a simple form for only the few cases where gases exhibit a single relaxation time. Nevertheless, the simple case provides valuable insight into the behavior of more complex systems and deserves a detailed description.

Consider the case where the population of a vibrational state is excited to an energy E_v that is greater than the energy $E_v(T_{\text{tr}})$, which it could have in Boltzmann equilibrium with translation. In this case, the excess vibrational energy will equilibrate with translation according to a standard relaxation equation,

$$-\frac{dE_v}{dt} = \frac{1}{\tau}[E_v - E_v(T_{\text{tr}})]. \tag{1}$$

The return to equilibrium occurs due to energy transfer during individual collisions.

The rate of energy transfer k_{10} is defined as the rate at which molecules go from the first excited state to the ground state due to collisions at a pressure of 1 atm. This rate is just the collision frequency M times the probability of energy transfer, $P^{1\to 0}$, times the mole fraction of molecules in the first excited state, x_1. During some collisions, the reverse process will occur, that is, some molecules in the ground state will become excited at a rate k_{01}. In equilibrium, equal numbers of molecules go in both directions, so

$$k_{10}x_1 - k_{01}x_0 = 0. \tag{2}$$

As explained above, energy is quickly shared from the first excited level of the vibrational mode to higher levels of the mode by vibration-to-vibration exchanges. Assuming quantum mechanical laws hold for probabilities of energy exchanges between vibrational levels of a harmonic oscillator, Landau and Teller showed[5] that

$$-\frac{dE_v}{dt} = k_{10}(1 - e^{-hv/kT})[E_v - E_v(T_{\text{tr}})], \tag{3}$$

where v is the vibrational frequency of the relaxing mode. By comparison with Eq. (1),

$$\tau = \frac{1}{k_{10}(1 - e^{-hv/kT})}.$$

The link between the relaxation time and ultrasonic absorption and dispersion is understood by noting that the relaxation process makes the specific heat of the gas time (or frequency) dependent. This time dependence can be obtained from the energy relaxation equation. Consider, as above, that the specific heat of a simple gas can be divided into translational, rotational, and vibrational contributions. For now, assume that the translation and rotational energy both equilibrate rapidly enough to follow any acoustically induced temperature variations. In this case, the effective specific heat can be written as

$$(C_v)_{\text{eff}} = C_v^{\infty} + C' \frac{dT'}{dT_{\text{tr}}}, \tag{4}$$

where C_v^{∞} is the sum of rotational and translational specific heats, C' is the relaxing specific heat, and T' is the instantaneous temperature of the relaxing mode (in this case, vibration). From the energy relaxation equation [Eq. (3)], for small periodic variations in T_{tr} and T_v about their equilibrium values,

$$(C_v)_{\text{eff}} = C_v^{\infty} + \frac{C'}{1 + i\omega\tau}, \tag{5}$$

where ω is the angular frequency of the acoustic wave.

The acoustic propagation constant can be written in the form

$$\frac{k^2}{\omega^2} = \left(\frac{1}{c} - \frac{i\alpha}{\omega}\right)^2 = \frac{\rho_0 \kappa_T}{\gamma_{\text{eff}}}, \tag{6}$$

where c is the acoustic velocity, α is the attenuation, ρ_0 is the equilibrium density, κ_T is the compressibility, and

$$\gamma_{\text{eff}} = \frac{(C_v)_{\text{eff}} + R}{(C_v)_{\text{eff}}} \tag{7}$$

with R the gas constant. For this simple single relaxation, assuming $\alpha/\omega \ll 1/c$,[6]

$$\alpha\lambda = \pi \left(\frac{c}{c_0}\right)^2 \epsilon \frac{\omega\tau_s}{1 + (\omega\tau_s)^2} \tag{8}$$

and

$$\left(\frac{c_0}{c}\right)^2 = 1 - \frac{\epsilon\omega^2\tau_s^2}{1 + \omega^2\tau_s^2}, \tag{9}$$

where $\epsilon = (c_\infty^2 - c_0^2)/c_\infty^2$, λ is the wavelength, c_0 is the speed of sound for $\omega\tau_s \ll 1$, and c_∞ is the speed of sound at frequencies much greater than the relaxation frequency. The adiabatic relaxation time τ_s is related to the isothermal relaxation time τ used earlier by

$$\tau_s = \frac{C_v + R}{(C_v^{\infty} + R)\tau}. \tag{10}$$

The relaxation frequency f_r, defined as the frequency at which the maximum absorption per wavelength occurs, is related to τ_s by

$$f_r = \frac{c_\infty}{c_0(1/2\pi\tau_s)}. \tag{11}$$

Figure 1 shows typical curves for absorption per wavelength and velocity dispersion due to a single relaxation process. The example here is Fl_2 at 102°C. The figure compares curves representing the above theory with measured values.[7]

In the case of polyatomic gases or mixtures of relaxing diatomic gases, the different relaxing modes can be coupled together by vibration-to-vibration exchanges. Such complex or multiple relaxation processes exhibit the general behavior given by Eqs. (8) and (9), but the magnitude of the absorption and dispersion and the relaxation frequencies can take on new meaning. In this case of multiple relaxing internal energy modes, Eqs. (8) and (9) take on the form[8]

$$\left(\frac{c}{c^{\infty}}\right)^2 = 1 + \sum_j \frac{\delta_j k_s/k_s^{\infty}}{1 + (\omega\tau_{s,j})^2}, \tag{12}$$

$$\alpha\lambda\left(\frac{c}{c^{\infty}}\right)^2 = -\pi \sum_j \frac{\delta_j k_s/k_s^{\infty}}{1 + (\omega\tau_{s,j})^2}\,\omega\tau_{s,j}, \tag{13}$$

where $\delta_j k_s/k_s^{\infty}$ is a relaxing adiabatic compressibility (negative) and j indicates that there might be more than one relaxation process. In these complex cases, $\tau_{s,j}$ can no longer be related to a single energy transfer reaction and $\delta_j k_s/k_s^{\infty}$ can no longer be related to relaxing energy of a specific mode. Instead, the various modes and reaction pathways are coupled. The sums in Eqs. (12) and (13) are over eigenvalues of the energy transfer matrix which simultaneously accounts for all reactions. Equations (12) and (13) have been used to calculate the sound absorption in moist air as a function of frequency and temperature. The standard for such calculations is now based upon these equations. The process is explained in Chapter 28.

Not only can Eqs. (12) and (13) be used to calculate sound absorption and velocity, but the reverse process is also possible, that is, the transition rates can be extracted from measured values of absorption and velocity. However, the number of possible relaxation paths multiplies rapidly with the increase in the number of relaxing modes, and the identification of specific rates becomes a tedious process and usually involves some assumptions. It has been done for only a few special cases. The next section discusses one example, that is, SO_2/O_2 mixtures.

So far, only relaxation processes that involve the exchange of energy in gases have been considered. Conceptually, relaxation processes in liquids result in similar relaxation equations but there are important differences in detail. These differences are brought about by the greater density of molecules that make multibody interactions typical. In this case, the idea of a rate equa-

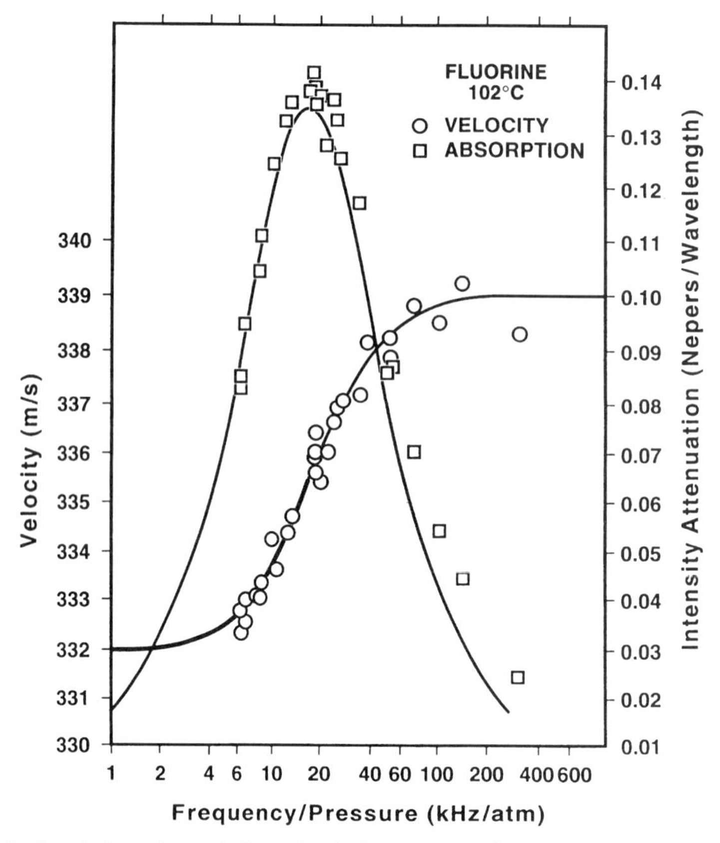

Fig. 1 Sound absorption and dispersion in fluorine at 102°C. The curves are calculated from specific heat data using a value of 17 kHz/atm for the frequency-to-pressure ratio when maximum absorption occurs.

tion is less precise, but the existence of a relaxation time as a measure of the time required for the system to return to equilibrium following a perturbation remains valid.

Various interesting relaxation processes occur in liquids. In some cases, such as CS_2 and a number of organic liquids, the relaxation mechanism seems to be the same as that described above for gases; that is, the internal energy of the individual molecules is being excited in "collisions." Such liquids generally have a positive temperature coefficient of absorption and are called Kneser liquids.[9,10] It is more common, however, that in liquids molecules temporarily bond to form large groups that may reconfigure themselves when an ultrasonic wave passes through. Such configuration relaxations tend to be very fast and the frequency dependence of the absorption and dispersion may indicate a distribution of relaxation times. Water is an exmaple of such a liquid, generally referred to as associated liquids.[11]

A further complication arises from chemical reactions. For a reversible chemical reaction with heat of reaction ΔH, ΔH enters into the relaxation equations in a manner similar to ΔE for vibrational relaxation. A major difference is that chemical reactions allow the possibility that the number density of molecules can change. Such changes bring about additional relaxation absorption and dispersion.

3 EXPERIMENTAL TECHNIQUES

Before modern techniques were available for measuring transients, most ultrasonic absorption and velocity measurements were made with ultrasonic interferometers.[12–15] This instrument continues to be used with refined precision and modern methods of control and measurement.[16] Both velocity and absorption can be measured with the interferometer. A column of gas or liquid of an adjustable length forms the load on a quartz crystal vibrating at its resonant frequency. The loading effect of the gas or liquid column increases whenever the length of the gas column is a whole multiple of half-wavelengths of sound. This loading effect is reflected in the driving circuit of the crystal. The separation between the peak values determines the sound wavelength and the variation of the magnitude of the peaks with length of the column allows the determination of the absorption coefficient of the sound in the test medium.

Another continuous-wave method for measuring velocity and absorption in gases uses a source and a receiver whose separation can be varied that are mounted in a tube so as to avoid standing waves.[17,18] This traveling-wave tube has been used to make measurements at audible frequencies and reduced pressures and therefore is capable of measurements over the f/p range where many of the interesting relaxation processes occur in gases and gas mixtures. The absorption is determined from the decrease in sound amplitude at the receiver as the source–receiver separation is increased. The wavelength of the sound is equal to the distance between points of equal phase in the received sound. This change in phase is easily observed by comparing the receiver and source signals. With the development of the capability for generating and measuring tone bursts, it has been possible to use the pulse–echo technique.[19,20] In this case the sound velocity is determined from the time of flight and the absorption from the variation of the tone burst amplitude with source and receiver separation.

At lower frequency-to-pressure ratios resonant tubes are used.[3] The velocity is determined from the value of the resonant frequency and the absorption from the width of the resonant peak.

In the case of liquids, special techniques have been developed to measure absorption and velocity at very high frequencies where many of the interesting relaxation effects occur.[21]

When relaxation effects are studied in gases the measurements of sound absorption are designed to provide absorption as a function of frequency divided by pressure over a range of values where the period of the sound wave is approximately equal to the relaxation time. When studying relaxation in molecular N_2, this requires measurements of very small absorption (less than 1 dB/m) at f/p values as low as a fraction of a hertz per atmosphere. On the other hand, measurements of absorption in UF_6 involve very large attenuation (~100 dB/m) at f/p values as high as 10 MHz/atm.[22]

To separate the effects due to vibrational relaxation, the measured absorption must be corrected for the following:

1. Viscous and thermal losses in the body of the gas
2. Viscous and thermal losses to the measuring chamber walls
3. Radiation or leakage losses from the chamber
4. Spreading losses
5. Losses due to rotational relaxation

These different effects are generally small enough to be additive. The quantity usually reported is the absorption coefficient α, in terms of which the amplitude of the plane wave is written as

$$A = A_0 e^{-\alpha(x_2 - x_1)}. \tag{14}$$

Plotting log A versus x gives α as a slope in nepers per metre. The measured α is due to a combination of the losses listed above. Careful design of the experimental apparatus is necessary if accurate corrections are to be made for wall and/or spreading losses.

A common geometry for sound absorption measurements is a cylindrical tube. This method generally avoids losses 3 and 4 (due to spreading and leakage) above and involves well-known corrections for 1 and 2. In this case, the total absorption is given by

$$\alpha_{tot} = \alpha_{cl} + \alpha_{rot} + \alpha_{tu} + \alpha_v, \tag{15}$$

where α_{cl} (classical absorption) is the absorption due to viscosity and thermal conductivity, α_{rot} is absorption due to relaxation processes not of interest to the study (rotational relaxation for example), α_{tu} is the tube absorption, and α_v is the absorption due to the relaxation process of interest. The term α_{cl} can be computed from known or measured values of viscosity and thermal conductivity[23]; it is proportional to f^2. The term α_{rot} generally is estimated based upon other studies. At frequencies well below the rotational relaxation frequency, it also is proportional to f^2. The tube absorption α_{tu} depends upon the viscosity, thermal conductivity, and tube radius (r). It varies approximately as[24] $[f/(pr^2)]^{1/2}$.

Only plane waves will be present in the tube if the sound frequency is maintained below the cutoff frequency for the first nonplane mode, that is, $f < 0.586c/d$, where c is the sound speed and d the tube diameter. Plane waves can be maintained for higher frequencies if the transducer generating the waves fills the tube and is carefully maintained perpendicular to the tube axis and its surface vibrates as a piston.[25] Interferometers used for high-frequency measurements in liquids and gases will generally satisfy these conditions. As the wavelength becomes small compared to the tube diameter, wall losses become negligible.

Low-frequency absorption measurements in liquids are difficult. Because of the high sound velocity and small absorption, large volumes are required. If the measurements are made in an unbounded liquid, spreading losses will likely dominate. If they are made in an enclosure, the small compressibility of the liquid makes the rigid-wall assumption invalid and it is difficult to correct for wall losses.

In the above discussion of relaxation in gases the relaxation absorption and dispersion of sound resulted when a finite time was required for the passage of the energy of translation into the vibrational energy of the gas molecules. Relaxation mechanisms can also be studied in special cases using the reverse processes. The laser light of a particular frequency is used to excite a vibrational mode. This excess vibrational energy then relaxes into translational energy generating an increase in translational temperature and pressure. If the laser light is chopped, a sound wave is generated. This phenomenon is referred to as the *optoacoustic effect* and the device as a *spectrophone*.[26] The relaxation time can be determined from the phase relations between modulated laser light and the resulting sound pressure.

The primary attraction of optoacoustic measurements is the ability to excite a specific internal mode and observe how energy from that mode makes its way to translation. Experimentally, only a limited number of internal modes can be excited with available lasers, which limits the list of systems that can be studied.

4 TYPICAL RESULTS

A comprehensive review of all studies of ultrasonic relaxation is far beyond the scope of this chapter. No such review has been published in recent years. The best available is that by Herzfeld and Litovitz, which is now 37 years old.[1] A large compilation of experimentally measured absorption and dispersion is provided in Ref. 27. For the purposes of this chapter specific studies will be selected that illustrate the physics involved. Three cases will be treated. The first is the halogen family of diatomic molecules, specifically F_2, Cl_2, Br_2, and I_2. This

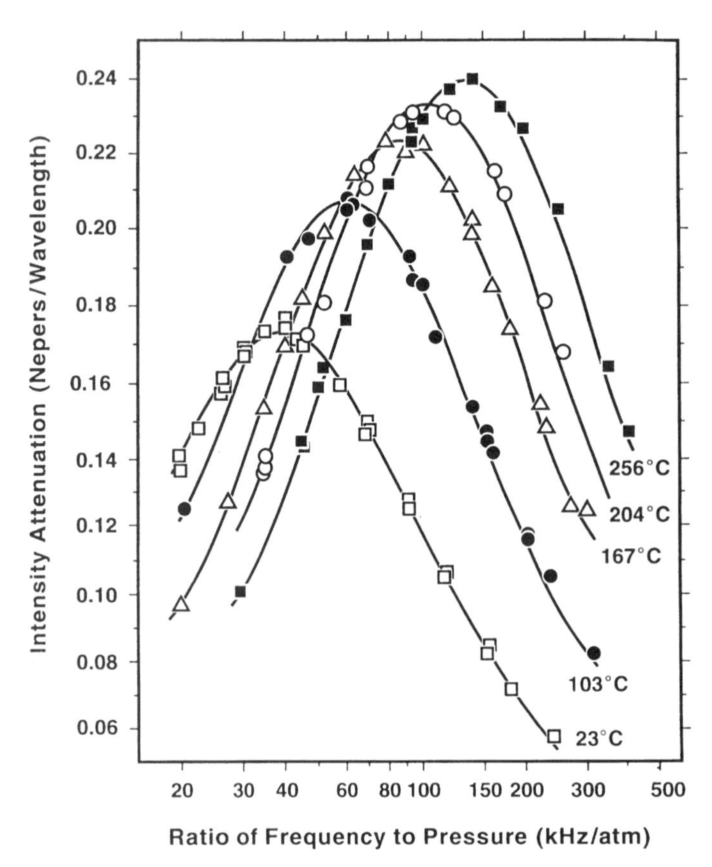

Fig. 2 Relaxation absorption coefficient per wavelength vs. log (frequency/pressure) for chlorine. The solid curves are the theoretical absorption with values of A_m and f_m adjusted to give the best fit of the experimental points.

example illustrates the functional forms presented earlier for absorption due to relaxation processes, gives typical values for probability of energy transfer, $P^{1\rightarrow 0}$, and provide some indication of how $P^{1\rightarrow 0}$ varies with temperature, molecular weight, and vibrational frequency. The second example is a more complex molecule, SO_2, in mixtures with Ar and O_2. This example illustrates the complexity of the relaxation process when different vibrational modes exchange energy. The final example will be for CS_2, which illustrates the case where relaxation has been studied in both the liquid and gas phases.

4.1 Simple Relaxation in Halogen Gases

Shields used an acoustic traveling-wave tube similar to that developed by Angona[17] to measure acoustic absorption in F_2, Cl_2, Br_2, and I_2 as a function of frequency, pressure, and temperature.[7,28] Results for Cl_2 at five different temperatures are plotted in Fig. 2. The absorption and velocity in F_2 at 102°C are shown in Fig. 1. The curves drawn through the experimental points were calculated using the theory discussed in Section 2.

Several features of these curves are interesting. First, when more than one point is plotted for the same value of f/p, it means that absorption was measured at two different pressures but at different frequencies so that the ratio was the same. The agreement between such measured values confirms that the classical and tube corrections were being made correctly and that the relaxation absorption varied as f/p. Second, note that the relaxation time decreases as temperature increases (the relaxation frequency increases). This means that although the density is lower for a given pressure, the probability of energy transfer increases more rapidly than the collision frequency decreases. Finally, note that the maximum absorption increases as the temperature increases. This is a result of increasing vibrational specific heat. The Planck–Einstein relation predicts that

$$C' = R\left(\frac{\theta}{T}\right)^2 \frac{e^{\theta/T}}{(e^{\theta/T}-1)^2}, \qquad (16)$$

where C' is the vibrational specific heat, θ is the vibrational temperature, and T is the gas temperature.

Results of Shields's study are summarized in Fig. 3, which shows the probability for deexciting the first excited vibrational level in a collision. The log of this probability is plotted versus $T^{-1/3}$. This type of plot was suggested by an early theory by Landau and Teller.[5] They predicted an approximately linear relationship between log P^{10} versus $T^{-1/3}$. From similar measurements on a great many other gases one can expect the following trends for vibration-to-translation transitions:

1. The log of the transition probability increases roughly linearly with $T^{-1/3}$ for a particular molecular collision pair.

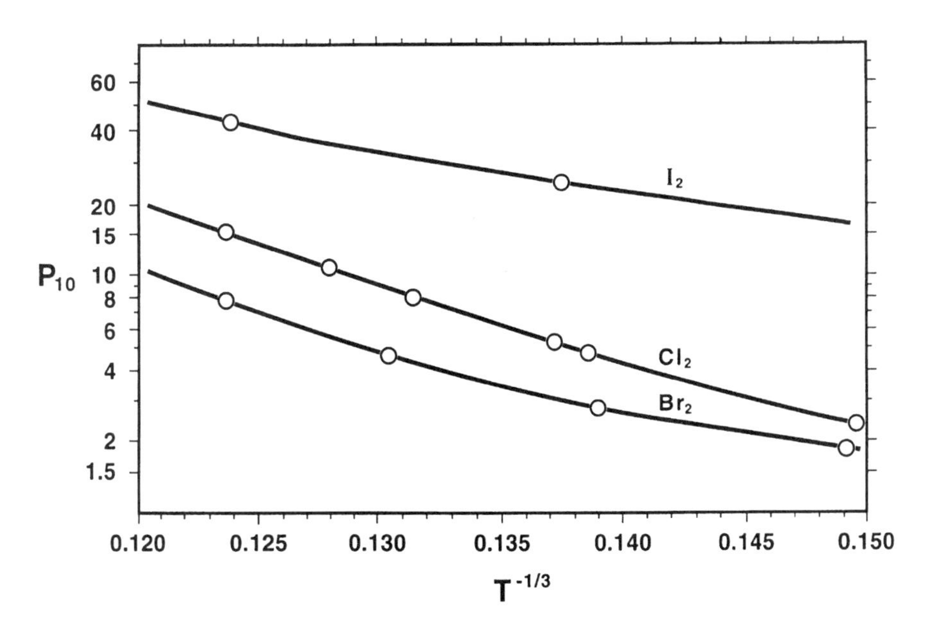

Fig. 3 Log of collision efficiency vs. absolute temperature to the minus-one-third power. The values of the ordinate should be multiplied by 10^{-5} for Cl_2 and by 10^{-4} for Br_2 and I_2.

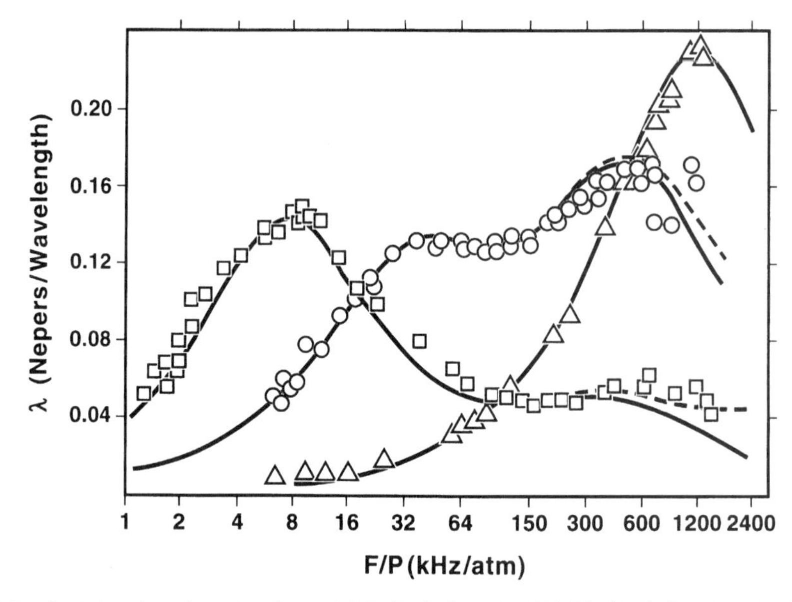

Fig. 4 Relaxation absorption in 75% SO_2/25% O_2 (Δ), 20% SO_2/80% O_2(○), and 5% SO_2/95% O_2 (□) at 500 K. The solid curve was computed assuming series relaxation and the transmission probabilities in Refs. 34–37. From Ref. 34.

2. The transition probability is very sensitive to the amount of energy that must be transferred between vibration and translation in the transition, increasing rapidly as this energy decreases.
3. The transition probability is very sensitive to the time involved in the molecular collision, increasing rapidly as this time decreases. What this means is that the transition probability increases with increasing temperature (see 1 above) and with decreasing molecular mass.
4. For polyatomic molecules, bending vibrations are more easily excited than stretching vibrations (as one might expect from the geometry of the collision process).
5. Water vapor and a few other molecules with low moments of inertia and, therefore, high rotational velocities are very efficient in shortening relaxation times when added to gases, even in small amounts. For such mixtures the relaxation time is much less sensitive to temperature change and can even increase with temperature. This effect is now attributed to the coupling between vibration and rotation in such collisions.[29–33]

4.2 Relaxation in a Polyatomic Molecule

As an example of the use of sound absorption measurements to determine the relaxation scheme in a polyatomic gas mixture, consider the case of SO_2–O_2 mixtures.[34] The SO_2 has three vibrational modes. These can be classified as a bending mode with a fundamental vibrational frequency of 518 cm^{-1}, a symmetrical stretching mode, and an asymmetrical stretching mode with frequencies of 1151 and 1361 cm^{-1}, respectively. Oxygen has a vibrational frequency of 1580 cm^{-1}. Sulfur dioxide was one of the first gases in which the sound absorption versus frequency curve evidenced more than one relaxation time. Figure 4 shows measurements in three different SO_2–O_2 mixtures at 500 K. The solid curves were calculated using the theory of Section 2. Twelve different transition rates were adjusted to make the theoretical curves simultaneously fit these data and the data for pure SO_2, SO_2–Ar mixtures,[35,36] and pure O^2.[37]

As an illustration of the kind of information that can be obtained from such studies, the following conclusions from the SO_2–O_2 and SO_2–Ar measurements are listed:

1. The relaxation in SO_2 is primarily a series process where translational energy flows into the bending mode and from there is shared with the symmetrical stretch with a two quantum for one exchange and from there with the asymmetrical stretch with a one-for-one exchange.
2. Argon is only about one-tenth and O_2 about one-

fourth as effective as another SO_2 molecule in deexciting the bending vibration in SO_2.

3. The first excited level of the O_2 vibration exchanges energy in vibration-to-vibration exchange primarily with the first excited level of the asymmetric stretch mode of SO_2.
4. The transition rate for the SO_2 vibration-to-vibration transfer increases with temperature faster than the vibration-to-translation rates. This is a peculiar property of SO_2 and has been attributed to its dipole moment.

4.3 Relaxation in CS_2 Gas and Liquid

Gaseous CS_2 has an unusually high vibrational specific heat at room temperature. From what has been said above, it is expected, therefore, to have a large relaxation absorption at moderate f/p values. Such was found early to be the case.[38,39] Only a single relaxation absorption peak was found, indicating a series relaxation process with energy passing from translation first into the low-energy bending vibration and from there shared quickly with the stretching modes. The frequency of maximum absorption increased from about 370 to 560 kHz/atm as the temperature increased from 0°C to 160°C. This corresponds to a transition probability that decreases relatively slowly (by about one-third) in this temperature interval.[39]

More interesting is the absorption of CS_2 as a liquid. It is an example of a "Kneser" liquid where the vibrational relaxation can be identified in the liquid state just shifted to a higher frequency due to the higher molecular collision rate in the liquid state.

The concept of a collision loses meaning when dealing with liquids since all molecules are interacting with one or more other molecules simultaneously. If one does assume that the liquid is a very dense gas, a collision frequency in the range of 10^{13} s^{-1} would be appropriate. This immediately suggests that relaxation processes in liquids will be fast. In fact, for most liquids up to a frequency of 10 MHz, no relaxation peak is observed so we must assume that $\omega\tau \ll 1$. Invoking this low-frequency limit, Eq. (8) gives

$$\alpha = bf^2, \tag{17}$$

where b depends upon the relaxation time and relaxation strength. This is the same frequency dependence exhibited by viscous losses. As a result, relaxation, viscous, and thermal conduction losses are difficult to sort out for most liquids.

Carbon disulfide has a maximum in the quantity α/f^2 in the range of 100 MHz.[40] Takagi has shown that the experimental data are best explained in terms of two relaxation processes. The strongest process is associated with relaxation of the two lowest vibrational modes (ν_2 and ν_1) and has a relaxation time of 2.2 ns. The weaker relaxation is associated with relaxation of ν_3 with a relaxation time of 30 ns.

The observation that relaxation in CS_2 behaves like thermal relaxation in a dense gas makes it a very interesting liquid. Carbon disulfide also strongly absorbs 337 nm UV, which makes it amenable to study with a pulsed N_2 laser.[41] Optoacoustic studies using a pulsed N_2 laser give relaxation times for the slow process of 90 ns. Given the difficulties in extracting the relaxation time for this weak process, a factor of 3 difference should not be surprising.

REFERENCES

1. K. F. Herzfeld and T. A. Litovitz, *Absorption and Dispersion of Ultrasonic Waves*, Academic, New York, 1959.
2. T. G. Winter and G. L. Hill, "High-Temperature Ultrasonic Measurements of Rotational Relaxation in Hydrogen, Deuterium, Nitrogen, and Oxygen," *J. Acoust. Soc. Am.*, Vol. 42, No. 4, 1967, p. 848.
3. A. J. Zuckerwar and W. A. Griffin, "Resonant Tube for Measurement of Sound Absorption in Gases at Low Frequency/Pressure Ratios," *J. Acoust. Soc. Am.*, Vol. 68, No. 1, 1980, p. 218.
4. G. L. Hill and T. G. Winter, "The Effect of Temperature on the Rotational and Vibrational Relaxation Times of Some Hydrocarbons," *J. Chem. Phys.*, Vol. 49, 1968, p. 440.
5. L. Landau and E. Teller, "Zur Theorie der Schalldispersion," *Phys. Z. Sowjet-Union*, Vol. 10, 1936, p. 34.
6. R. T. Beyer and S. V. Letcher, *Physical Ultrasonics*, Academic, New York, 1969, p. 103.
7. F. D. Shields, "Thermal Relaxation in Fluorine," *J. Acoust. Soc. Am.*, Vol. 34, No. 3, 1962, p. 271.
8. H. E. Bass, L. C. Sutherland, J. Piercy, and L. Evans, "Absorption of Sound by the Atmosphere," *Phys. Acoust.*, Vol. XVII, 1984, p. 161.
9. R. T. Beyer and S. V. Letcher, *Physical Ultrasonics*, Academic, New York, 1969, Chap. 5.
10. K. F. Herzfeld and T. A. Litovitz, *Absorption and Dispersion of Ultrasonic Waves*, Academic, New York, 1959, Chap. 8.
11. K. F. Herzfeld and T. A. Litovitz, *Absorption and Dispersion of Ultrasonic Waves*, Academic, New York, 1959, Chap. 12.
12. J. C. Hubbard, "The Acoustic Resonator Interferometer: I. The Acoustic System and Its Equivalent Electric Network," *Phys. Rev.*, Vol. 38, 1931, pp. 1011–1019.
13. J. V. Connter, "Ultrasonic Dispersion in Oxygen," *J. Acoust. Soc. Am.*, Vol. 30, 1958, p. 297.
14. E. S. Stewart and J. L. Stewart, "Rotational Dispersion

in the Velocity, Attenuation, and Reflection of Ultrasonic Waves in Hydrogen and Deuterium," *J. Acoust. Soc. Am.*, Vol. 24, 1952, p. 194.

15. R. T. Lagemann, "The Use of Glass Parts in Ultrasonic Interferometers," *J. Acoust. Soc. Am.*, Vol. 24, 1952, p. 86.
16. B. Jacobs, R. Cerami, J. R. Olson, and R. C. Amme, "Vibrational Relaxation in SiF_4 Using a Computer Controlled Ultrasonic Interferometer," *J. Acoust. Soc. Am.*, Vol. 88, 1990, pp. 2812–2815.
17. F. A. Angona, "Attenuation of Sound in a Tube," *J. Acoust. Soc. Am.*, Vol. 25, 1953, p. 336.
18. F. D. Shields, "Measurements of Thermal Relaxation in CO_2 Extended to 300°C," *J. Acoust. Soc. Am.*, Vol. 31, 1959, p. 248.
19. H. E. Bass, T. G. Winter, and L. B. Evans, "Vibrational and Rotational Relaxation in Sulfur Dioxide," *J. Chem. Phys.*, Vol. 54, 1971, p. 644.
20. F. D. Shields, H. E. Bass, and L. N. Bolen, "Tube Method of Sound-Absorption Measurement Extended to Frequencies Far above Cutoff," *J. Acoust. Soc. Am.*, Vol. 62, 1977, pp. 246–253.
21. K. F. Herzfeld and T. A. Litovitz, *Absorption and Dispersion of Ultrasonic Waves*, Academic, New York, 1959, Chap 9.
22. D. Cravens, F. D. Shields, H. E. Bass, and W. D. Breshears, "Vibrational Relaxation of UF_6: Ultrasonic Measurements in Mixtures with AR and N_2," *J. Chem. Phys.*, Vol. 71, 1979, pp. 2797–2802.
23. F. D. Shields and R. T. Lagemann, "Tube Corrections in the Study of Sound Absorption," *J. Acoust. Soc. Am.*, Vol. 29, No. 4, 1957, p. 470.
24. F. D. Shield and J. Faughn, "Sound Velocity and Absorption in Low-Pressure Gases Confined to Tubes of Circular Cross Section," *J. Acoust. Soc. Am.*, Vol. 46, No. 1, 1969, p. 158; F. D. Shields, "An Acoustical Method for Determining the Thermal and Momentum Accommodation Coefficients of Gases On Solids," *J. Chem. Phys.*, Vol. 62, No. 4, 1975, p. 1248.
25. F. D. Shields, H. E. Bass, and L. N. Bolen, "Tube Method of Sound-Absorption Measurement Extended to Frequencies Far above Cutoff," *J. Acoust. Soc. Am.*, Vol. 62, No. 2, 1977, p. 346.
26. H. E. Bass and H. X. Yan, "Pulsed Spectrophone Measurements of Vibrational Energy Transfer in CO_2," *J. Acoust. Soc. Am.*, Vol. 74, No. 6, 1983, p. 1817.
27. W. Schaafs, "Landolt–Bornstein Numerical Data and Functional Relationships in Science and Technology; Group II; Atomic and Molecular Physics," Vol. 5, *Molecular Acoustics*, Springer Verlag, Berlin, 1967.
28. F. D. Shields, "Sound Absorption in the Halogen Gases," *J. Acoust. Soc. Am.*, Vol. 32, No. 2, 1960, p. 271.
29. J. W. L. Lewis and K. P. Lee, "Vibrational Relaxation in Carbon Dioxide/Water Vapor Mixtures," *J. Acoust. Soc. Am.*, Vol. 38, No. 5, 1965, p. 813.
30. F. D. Shields and J. A. Burks, "Vibrational Relaxation in CO_2/D_2O Mixtures," *J. Acoust. Soc. Am.*, Vol. 43, No. 3, 1968, p. 510.
31. F. D. Shields, "Sound Absorption and Velocity in H_2S and CO_2/H_2S Mixtures," *J. Acoust. Soc. Am.*, Vol. 45, No. 2, 1969, p. 481.
32. F. D. Shields, and G. P. Carney, "Sound Absorption in D_2S and CO_2/D_2S Mixtures," *J. Acoust. Soc. Am.*, Vol. 47, No. 5, 1970, p. 1269.
33. H. E. Bass and F. D. Shields, "Vibrational Relaxation and Sound Absorption in O_2/H_2O Mixtures," *J. Acoust. Soc. Am.*, Vol. 56, No. 3, 1974, p. 856.
34. B. Anderson, F. D. Shields, and H. E. Bass, "Vibrational Relaxation in SO_2/O_2 Mixtures," *J. Chem. Phys.*, Vol. 56, No. 3, 1972, p. 1147.
35. F. D. Shields, "Vibrational Relaxation in SO_2 and SO_2/Ar Mixtures," *J. Chem. Phys.*, Vol. 46, No. 3, 1967, p. 1063.
36. F. D. Shields and B. Anderson, "More on Vibrational Relaxation in SO_2/Ar Mixtures," *J. Chem. Phys.*, Vol. 55, No. 6, 1971, p. 2636.
37. F. D. Shields and K. P. Lee, "Sound Absorption and Velocity Measurements in Oxygen," *J. Acoust. Soc. Am.*, Vol. 35, No. 2, 1963, p. 251.
38. W. T. Richards and J. A. Reid, "Acoustical Studies, IV. The Collison Efficiences of Various Molecules in Exciting Lower Vibrational States of Ethylene, Together with Observations Concerning the Excitation of the Rotational Energy in Hydrogen," *J. Chem. Phys.*, Vol. 2, 1934, p. 206.
39. J. C. Gravitt, "Sound Absorption in Carbon Disulfide Vapor as a Function of Temperature," *J. Acoust. Soc. Am.*, Vol. 32, No. 5, p. 560.
40. K. Takagi, "Vibrational Relaxation in Liquid Carbon Disulfide," *J. Acoust. Soc. Am.*, Vol. 71, 1982, pp. 74–77.
41. C. H. Thompson, S. A. Cheyne, H. E. Bass, and R. Raspet, "Optoacoustic Observation of Internal Relaxation in Liquid CS_2," *J. Acoust. Soc. Am.*, Vol. 85, No. 6, 1989, pp. 2405–2409.

43

PHONONS IN CRYSTALS, QUASICRYSTALS, AND ANDERSON LOCALIZATION

JULIAN D. MAYNARD

1 INTRODUCTION

The term *phonon* refers to the effects of a transition between eigenstates of a system of coupled quantum mechanical oscillators. The phonon is sometimes referred to as a "quantized sound wave," but phonons and sound waves differ in some ways: phonons usually apply to discrete, strictly linear systems, while sound waves are usually derived from continuous, intrinsically nonlinear systems in the limit of small amplitudes. The difference appears in the relation between energy and momentum (second-order quantities), but the difference is small in real systems. Phonons may occur in all phases of matter, but they are most readily apparent in crystalline solids. In order to understand the effects of disordered and quasiperiodic systems, it is necessary to know some details of the periodic system. In Section 3 we derive the formulas for a system of harmonically coupled classical particles in one dimension. Section 4 covers a one-dimensional system of harmonically coupled quantum particles. Section 5 discusses the effects of unit cells with more than one particle and the extension of the theory to two and three dimensions. Section 6 treats the effects of disorder in the system and explains Anderson localization. Section 7 discusses the effects of quasicrystalline symmetry.

2 BRIEF HISTORY

The first evidence for phonons in solids occurred in the observation by Dulong and Petit in 1819 that the specific heat of a solid is usually about twice that of the corresponding gas. This suggested that in addition to the kinetic energy that gas molecules possess, solids also have a way of storing potential energy. Calculations of wave motion in periodic lattices were done by Lord Kelvin in 1881, but the relevance to the properties of solids was not known then. An atomistic theory of the specific heat of solids was first proposed by Einstein in 1907. He assumed that the kinetic and potential energies arose from atoms oscillating about their equilibrium lattice positions, and using the newly formulated Planck quantum theory, he related the energy to the frequency of vibration. Einstein made the simplifying assumption that the atoms oscillated independently of each other, and his formula for the specific heat was consequently incorrect. The coupling of the atomic oscillators was properly taken into account by Peter Debye in 1912, and later in more detail by Max Born, Theodore van Karman, and Moses Blackman. Their theory correctly matched the experimental measurements of the temperature dependence of the specific heat of solids. In 1912 Debye recognized the role of phonons in explaining thermal conductivity, and in the same year Frederick Lindemann showed that the lattice vibrations were related to thermal expansion and the melting of solids. These last two effects are due to nonlinearities in the forces between the atoms in the solid.

3 HARMONICALLY COUPLED CLASSICAL PARTICLES

We consider a one-dimensional sequence of masses m_j ($j = 0, N + 1$) connected with ideal springs of stiffness s_j ($j = 0, N$). The spring s_j lies between mass m_j and

Note: References to chapters appearing only in the *Encyclopedia* are preceded by *Enc.*

mass m_{j+1}. The displacement of the mass m_j is denoted u_j; u_0 and u_{N+1} serve as boundary conditions at each end of the system. Inside the boundaries the motion of each mass is governed by Newton's law and Hooke's law:

$$m_j \frac{d^2 u_j}{dt^2} = -s_{j-1}(u_j - u_{j-1}) + s_j(u_{j+1} - u_j). \quad (1)$$

We assume that the motion can be Fourier analyzed in time, and adopt a time dependence $\exp(i\omega t)$. The first term in Eq. (1) becomes $-m_j\omega^2 u_j$.

At this point standard derivations make the sequence periodic and invoke Bloch's theorem, which provides the form for the normal modes of the coupled system. However, Bloch's theorem is usually derived for an infinitely periodic system or a system that has periodic boundary conditions. In the derivation presented here, a technique will be used that is valid for arbitrary boundary conditions and that will furthermore provide useful insight into the properties of a disordered (nonperiodic) system.

The system of equations represented by Eq. (1) may be written as 2×2 matrix equations:

$$\begin{pmatrix} u_{j+1} \\ u_j \end{pmatrix} = \begin{pmatrix} 2\alpha_j & -\beta_j \\ 1 & 0 \end{pmatrix} \begin{pmatrix} u_j \\ u_{j-1} \end{pmatrix}, \quad (2)$$

where $\omega_j = 2\sqrt{(s_j/m_j)}$, $\alpha_j = (1 + \beta_j)/2 - 2(\omega/\omega_j)^2$ and $\beta_j = s_{j-1}/s_j$.

To proceed, we look for the singular value decomposition of the matrix in Eq. (2); that is, we need to find matrices D_j and inverses D_j^{-1} such that

$$\begin{pmatrix} 2\alpha_j & -\beta_j \\ 1 & 0 \end{pmatrix} = D_j^{-1} \begin{pmatrix} \lambda_j^+ & 0 \\ 0 & \lambda_j^- \end{pmatrix} D_j. \quad (3)$$

The singular values $\lambda_j^\pm$ are found by solving the secular equation given by setting a determinant equal to zero, as in a standard matrix eigenvalue problem.

An equation, which will later be instructive, for periodic and disordered systems is one that provides the normal modes of the system, expressed as a product of the matrices in Eq. (3):

$$\begin{pmatrix} u_{j+1} \\ u_j \end{pmatrix} = D_j^{-1} \begin{pmatrix} \lambda_j^+ & 0 \\ 0 & \lambda_j^- \end{pmatrix} D_j D_{j-1}^{-1} \begin{pmatrix} \lambda_{j-1}^+ & 0 \\ 0 & \lambda_{j-1}^- \end{pmatrix} D_{j-1},$$
$$\ldots, D_1^{-1} \begin{pmatrix} \lambda_1^+ & 0 \\ 0 & \lambda_1^- \end{pmatrix} D_1 \begin{pmatrix} u_1 \\ u_0 \end{pmatrix}. \quad (4)$$

We now specialize to the case of a periodic system. We let $m_j = m$, $s_j = s$, $\omega_j = \omega_0 = 2\sqrt{(s/m)}$, $\beta_j = 1$, $D_j = D$, and $\alpha_j = \alpha = 1 - 2(\omega/\omega_0)^2$. Because of the cancellation of DD^{-1} in the product in Eq. (4), we have a simple expression for the displacements u_j:

$$\begin{pmatrix} u_{j+1} \\ u_j \end{pmatrix} = D^{-1} \begin{pmatrix} (\lambda^+)^j & 0 \\ 0 & (\lambda^-)^j \end{pmatrix} D \begin{pmatrix} u_1 \\ u_0 \end{pmatrix}. \quad (5)$$

The expression for $\lambda^\pm$ for the periodic system is

$$\lambda^\pm = \begin{pmatrix} \alpha \mp i(1-\alpha^2)^{1/2} & \text{if } 0 < \omega < \omega_0 \\ \alpha \mp (\alpha^2 - 1)^{1/2} & \text{if } \omega > \omega_0 \end{pmatrix}. \quad (6)$$

Note that if $\omega > \omega_0$, then $|\lambda^+| < 1$ and $|\lambda^-| > 1$. Because of the powers of j in Eq. (5), the larger singular value grows exponentially while the smaller singular value decays; these solutions correspond to evanescent waves, which are usually not regarded as solutions of interest, but they are nevertheless important; they account for the band gaps in periodic systems and for the exponentially localized states that appear in the gaps in disordered systems. If $0 < \omega < \omega_0$, then $\lambda^+ = (\lambda^-)^*$, $|\lambda^\pm| = 1$, and we can write $\lambda^\pm = e^{\pm ika}$. The choice of notation for the phase, ka, will be evident later. From the ω dependence in λ (through α), we can solve for ω in terms of k and obtain the dispersion relation

$$\omega_k = \omega_0 \sin(ka/2). \quad (7)$$

The subscript k has been added to emphasize the dispersion. Note that $0 < k < \pi/a$. This dispersion relation is illustrated in Fig. 1.

We now consider some limits relevant to a continuous system. We assume that the springs connecting the masses have been stretched to a periodic spacing a with the application of a static tension T, so that $T = sa$. We

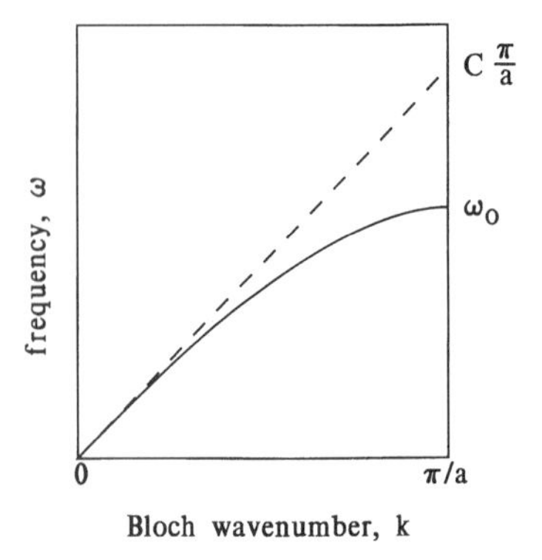

Fig. 1 Dispersion relation for a one-dimensional system of harmonically coupled particles.

now take the limit $m \rightarrow 0$ and $a \rightarrow 0$, while keeping the mass per unit length, $\mu = m/a$, constant. As $a \rightarrow 0$, then $\sin(ka/2) \rightarrow ka/2$, and the dispersion relation becomes $\omega_k = ck$, where $c = \sqrt{(T/\mu)}$ is the usual speed of sound in a string of mass per unit length μ and tension T. Thus the dispersion relation for the continuum limit is obtained.

If one considers the dispersion relation for the discrete system in the limit of small k, one also obtains $\omega = ck$. However, if this is extrapolated to the maximum $k = \pi/a$, one finds a frequency $\omega = c\pi/a$, whereas the actual frequency should be ω_0, which is a factor of $2/\pi$ smaller than the extrapolated value of $c\pi/a$. This dispersion is a fundamental difference between the discrete coupled oscillators appropriate for phonons and sound waves in a continuous system.

In addition to providing the singular values $\lambda^{\pm}$, Eq. (3) is solved for the D matrices. If these and $\lambda^{\pm}$ are inserted into Eq. (5), then the normal modes are obtained:

$$u_j = e^{ik(ja)} U_k + e^{-ik(ja)} U_{-k}, \tag{8}$$

where

$$U_k = \frac{u_1 - u_0 e^{-ika}}{(2i)\sin(ka)}. \tag{9}$$

The values of u_0, u_1, U_k, and restrictions on k (eigenvalues) are determined by the boundary conditions and initial conditions. For example, for a system clamped at the ends, $u_0 = 0$, $u_{N+1} = 0$:

$$u_j = \sin[k(ja)] \left[\frac{u_1}{\sin(ka)} \right], \tag{10}$$

and k is quantized with $k = [n/(N+1)]\pi/a$, $n = 1, 2, \ldots, N$. For an infinite system or a system with periodic boundary conditions one finds $U_k = 0$ or $U_{-k} = 0$. If one uses a coordinate system with the origin at the mass m_0, then the position of the jth mass is $x = ja$, and we can write for infinite or periodic boundary conditions

$$u_k(x) = e^{ikx} U_k, \tag{11}$$

which is the usual Bloch wave result. The subscript k has been added to the displacement field $u(x)$ as a label for the normal mode. The final solution will be a linear combination of the normal modes:

$$u_j = \sum_k (e^{ik(ja)} U_k + e^{-ik(ja)} U_{-k}) e^{i\omega_k t}. \tag{12}$$

For any boundary conditions, clamped, periodic, and so forth, one can show that the normal modes are orthogonal and obtain the normalization constants. With this one can derive the expression for the total energy of the system:

$$E = m \sum_k |U_k|^2 \omega_k^2. \tag{13}$$

4 HARMONICALLY COUPLED QUANTUM PARTICLES

For a system of harmonically coupled quantum mechanical particles, we use a Lagrangian formulation instead of the Newton's law formulation in Eq. (1). The Lagrangian is

$$L = \frac{1}{2} \sum_j m_j \left(\frac{du_j}{dt} \right)^2 - \frac{1}{2} \sum_j s_j (u_{j+1} - u_j)^2. \tag{14}$$

The canonical momenta are $p_j = m_j(du_j/dt)$, and the Hamiltonian is

$$H = \sum_j p_j \left(\frac{du_j}{dt} \right) - L. \tag{15}$$

The system is quantized with the commutation relations $[u_j, p_{j'}] = i\hbar\delta_{j,j'}$, where $\hbar$ is Planck's constant divided by 2π. We now assume a periodic system with $m_j = m$ and $s_j = s$. Guided by our results with the classical system, we let

$$u_j = \sum_k (e^{ikja} U_k + e^{-ikja} U_{-k}). \tag{16}$$

The Hamiltonian becomes

$$H = \frac{1}{2m} \sum_k P_k P_{-k} + \frac{1}{2} m \sum_k \omega_k^2 U_k U_{-k}, \tag{17}$$

where $P_k = m(dU_{-k}/dt)$ and ω_k is defined as before in Eq. (7). One could use this Hamiltonian and construct a Schrödinger wave equation for a field $\psi(U)$, where U is a point in the $2N$-dimensional U_k space, and $P_k\psi(U) = -i\hbar(\partial\psi/\partial U_k)$. However, a better approach is to use another transformation and construct the properties of the eigenfunctions and eigenvalues using the commutation relations for u_j and p_j. We define

$$a_k = \left(\frac{m}{2\hbar\omega_k} \right)^{1/2} \left(\omega_k U_k + \frac{i}{m} P_{-k} \right), \tag{18}$$

and $a_k^+ = a^*_{-k}$, with $\omega_{-k} = \omega_k$. The Hamiltonian becomes

$$H = \sum_k \hbar\omega_k \left(N_k + \frac{1}{2} \right), \tag{19}$$

where $N_k = a_k^+ a_k$. Using the commutation relations for u_j and p_j, one finds the commutation relation for a_k^+ and a_k as $[a_k^+, a_{k'}] = \delta_{k,k'}$.

We now write ψ_{kn} for an eigenfunction of the operator N_k with eigenvalue n_k. One can show that the eigenvalues must be positive. Using the commutation relations, one finds that $a_k\psi_{kn}$ is an eigenfunction with eigenvalue $(n_k - 1)$, and $a_k^+\psi_{kn}$ is an eigenfunction with eignevalue $(n_k + 1)$. We then consider the sequence of eigenfunctions given by $(a_k)^p\psi_{kn}$, with p a positive integer. The eigenvalues form the sequences $(n_k - 1), (n_k - 2), \ldots, (n_k - p)$, etc. If this sequence were allowed to continue indefinitely, then for $p > n_k$, the eigenvalues would become negative, which cannot happen. If n_k is such that $(n_k - p) = 0$, then the sequence will terminate without going negative. Thus we have $n_k = p$; that is, n_k is quantized to be a positive integer.

A general state of the system will be a superposition of eigenstates:

$$\psi = \sum_k \sum_n C_{kn}\psi_{kn}, \tag{20}$$

where $|C_{kn}|^2$ gives the probability that the system is in the state ψ_{kn}. The expectation value of the total energy is (using $E\psi = H\psi$)

$$\langle\psi|E|\psi\rangle = \sum_k \sum_n |C_{kn}|^2 \hbar\omega_k \left(n_k + \frac{1}{2} \right). \tag{21}$$

By examining the expectation value of the square of the momentum operator, one finds

$$\frac{1}{m} \langle\psi|P^2|\psi\rangle = m \sum_k \langle\psi|U_k|\psi\rangle\omega_k^2 \tag{22}$$

$$= \sum_n \sum_k |C_{nk}|^2 \hbar\omega_k n_k = \langle\psi|E|\psi\rangle - E_0, \tag{23}$$

where $E_0 = \sum_k |C_{nk}|^2\hbar\omega_k/2$ is the zero-point energy. If $|C_{nk}|$ is appreciable only for $n_k \gg \frac{1}{2}$, then $\langle\psi|E|\psi\rangle - E_0 \sim \langle\psi|E|\psi\rangle$, and we have from Eqs. (22) and (23)

$$\langle\psi|E|\psi\rangle = m \sum_k \langle\psi|U_k^2|\psi\rangle\omega_k^2, \tag{24}$$

which corresponds to the classical expression in Eq. (13).

If the expectation value of the position operator is examined, it is found that it couples a state ψ_{nk} only to the state $\psi_{(n+1)k}$ or $\psi_{(n-1)k}$. As a consequence, the time dependence will behave like

$$e^{i[(n+1)-n]\omega_k t} + e^{i[(n-1)-n]\omega_k t} = 2\cos(\omega_k t). \tag{25}$$

If $C_{nk} \sim C_{(n\pm1)k}$ and is appreciable for only one k and for $n_k \gg \frac{1}{2}$, then

$$\langle\psi|U_k|\psi\rangle = \left(\frac{2\langle\psi|E|\psi\rangle}{m\omega_k^2} \right)^{1/2} \cos(\omega_k t), \tag{26}$$

which corresponds to the normal mode oscillating between the classical turning points at the frequency ω_k. It is interesting to note that in order to observe a "classical sound wave," the quantum system must not only have a large n_k so that $n_k\hbar\omega_k$ corresponds to the macroscopic classical energy, the quantum state must contain a superposition of two eigenstates, ψ_{nk} and $\psi_{(n\pm1)k}$.

The term *phonon* can now be defined in the following sense: We can say that a phonon is emitted or absorbed when a system of harmonically coupled quantum particles makes a transition from a state ψ_{nk} to a state $\psi_{(n-1)k}$ or $\psi_{(n+1)k}$.

5 LARGER UNIT CELLS AND HIGHER DIMENSIONS

Instead of having a periodic system such that $m_j = m$, and so on, one may have different masses or spring constants but still have a periodic system. That is, several different masses and springs may be joined to form a unit cell, and this unit cell may be repeated periodically. For M different masses or springs, the matrix in Eq. (2) will enlarge from 2×2 to $2M \times 2M$, and instead of a single dispersion relation as in Eq. (7), there will be M dispersion relations and M "branches" in the eigenvalue spectrum. If the system is extended into two or three dimensions, the matrix representing the dynamics of the system is further enlarged, the displacement fields become vector fields, and the eigenfunctions may be characterized as transverse or longitudinal. A thorough discussion of the dynamics of a three-dimensional system of coupled oscillators is available in Ref. 1.

5.1 Phonon Density of States

In statistical physics, one calculates the properties of a system by taking a product of the quantity of interest with a distribution function and integrating over all possible states. The distribution functions are functions of the energy of the state, and for this reason it is usually easier to transform the integration variable from the state label, such as the Bloch wave vector k in our one-dimensional example, to the energy, which is related to ω in our example. In our one-dimensional example, the transformation is accomplished using

$$dk = \left(\frac{d\omega}{dk}\right)^{-1} d\omega = \frac{d\omega}{v_g}, \qquad (27)$$

where $v_g = d\omega/dk$ is the group velocity. For the general case in higher dimensions, the Jacobian of the transformation is the density of states, which also varies as $1/v_g$. Values of the state parameter (e.g., k) where the dispersion relation ω_k is flat are called critical points, and at these points the group velocity is zero and the density of states is singular. Such singularities are called van Hove singularities.[1,2] In our one-dimensional example,

$$\frac{d\omega}{dk} = \frac{2\omega_0}{a} \cos\frac{ka}{2} = \frac{2}{a}(\omega_0^2 - \omega^2)^{1/2} \qquad (28)$$

so that $d\omega/dk = 0$, and the density of states is singular, when $\omega = \omega_0$. In two and three dimensions the singularities may be classified according to the nature of $|\nabla_k \omega| = 0$, whether it is a maximum, minimum, or one of two types of saddlepoint. The van Hove singularities are sometimes important in determining the properties of a material relating to phonons.

6 EFFECTS OF DISORDER: ANDERSON LOCALIZATION

6.1 Introduction

If one takes the modulus of an eigenfunction for a periodic system [e.g., Eq. (11)], then one obtains a constant, independent of position in the system. That is, the eigenfunction has the same amplitude throughout the entire system. Such eigenfunctions are said to be extended. Extended eigenfunctions are very important in solid-state theory; for example, they explain why metallic crystals have high electrical conductivity. In a metal, an electron is very strongly scattered by the positive ions, and on this basis a metal should have a very low electrical conductivity. But because the ions in a crystal are arranged periodically, the quantum mechanical eigenfunctions for the electron are extended, allowing the electron to propagate freely and resulting in high electrical conductivity.

That the eigenfunctions for a periodic system are extended is guaranteed by Bloch's theorem. We now consider what happens if we take a periodic system and introduce some disorder throughout the system. One would suppose that the eigenfucntions, instead of having uniform amplitude everywhere in the system, would simply obtain some variations about the original constant value, but would otherwise remain extended. Surprisingly, this is not what happens; instead, the eigenfunctions become exponentially localized, having a maximum value at some site in the system and, at least in a probabilistic sense, decaying exponentially from that site. This phenomenon, referred to as Anderson localization, was used by P. W. Anderson and Sir Neville Mott to explain the electrical conductivity of disordered metals, and this accomplishment was cited in their Nobel prize in 1977.

While much of the research with Anderson localization has dealt with electron Schrödinger waves in disordered metals, the phenomenon is relevant for any form of wave propagation in a disordered system. Anderson localization is important for the understanding of phonon quantum wave propagation and its effects on specific heat and thermal conductivity,[3] and for the understanding of classical wave propagation in disordered systems, as in acoustic geophysical survey, oil exploration, medical imaging, the vibration of complex structures, sound propagation in the ocean, radar propagation in vegetation, and so forth. Anderson localization remains an active area of research for both quantum and classical systems in condensed matter physics.

Many of the advances in the theory of Anderson localization have occurred through the application of renormalization group methods and other modern techniques of statistical physics. As with critical phenomena in statistical physics, Anderson localization is strongly dependent on the dimensionality of the system. In one dimension it is found that any amount of disorder will result in some degree of localization; very weak disorder will result in localization, but the characteristic length for the exponential decay may be large. In three dimensions it is believed that a critical amount of disorder must be exceeded before all of the eigenstates become localized. Two dimensions is the "critical dimension," where theory becomes difficult; it is believed that in two dimensions all eigenstates are localized, as in one dimension, but the localization is much weaker. In one dimension there are rigorous theorems that describe Anderson localization, but in two and three dimensions there are no comprehensive rigorous theorems. In the paragraphs below Anderson localization in one dimension, based on the coupled oscillator model described in the first sections of this chapter, will be treated. More detailed reviews of Anderson localization, particularly in dimensions higher than one, may be found in Refs. 4–6.

6.2 Explanation of Anderson Localization

While it is easy to describe Anderson localization, it is very difficult to understand how it occurs; one would naively expect that disorder, especially weak disorder, would simply put variations in the amplitude of the otherwise extended eigenfunctions. Of the many studies treating Anderson localization, most simply assume the phenomenon as an established fact and few offer explanations. The rigorous theory for one dimension was found by Furstenburg in 1963[7]; Furstenburg's study is long and difficult, with references to earlier theorems, and so it offers little physical insight into the phenomenon. In this section, a simple but nonrigorous explanation of Anderson localization will be presented. An excellent, but longer, explanation may be found in a study by Luban and Luscombe.[6]

A careful statement of the Anderson localization effect reveals some of the difficulty. Anderson localization does not refer to the behavior of a single realization of a disordered system, but rather refers to the results of averaging over an ensemble of realizations, with the disorder of each realization randomly determined with some distribution function. Furthermore, the rigorous result occurs in the limit as the system becomes infinite in size. The actual eigenfunctions of a single realization do not have the form $\exp(-2|x-x_l|/\Lambda)$ (here x_l would be the site of localization and Λ would be the localization length); in an actual realization, an eigenfunction at some site far from x_l may fluctuate above the small exponential tail, requiring an unbounded prefactor to the exponential. When averages are taken, the fluctuations can be ignored, and the eigenfunctions, while not actually having the form $\exp(-2|x - x_l|\Lambda)$, behave as though they did, in some statistical sense.

In the treatment of the coupled oscillators presented in the first sections, it was found that the eigenfunctions could be expressed as products of matrices, as in Eq. (4). For an ensemble of disordered systems, we consider a product of random matrices for a single realization, which we write symbolically as

$$P_j = \prod_{j'=1}^{j} M_{j'}. \tag{29}$$

We are interested in the ensemble average $\overline{P}_j$ of the product of random matrices. While this is a difficult mathematical problem, we consider an analogy with a much more familiar problem, the ensemble average of a sum of random variables $X_{j'}$ with distribution function $F(X)$:

$$S_j = \sum_{j'=1}^{j} X_{j'}. \tag{30}$$

We expect from intuition, and know from rigorous theorems, that the average of the sum is just the average of the random variable, $\overline{X} = \int XF(X)\, dX$, times j, so that $S_j = j\overline{X}$ in the limit as j goes to infinity. Comparing Eq. (29) with Eq. (30) we imagine taking the log of the product of random matrices and obtain a sum of random variables:

$$\ln(P_j) = \sum_{j'=1}^{j} \ln(M_{j'}). \tag{31}$$

We might then expect the average of $\ln(P_j)$ to be j times some mean $\overline{M}$, so that $\overline{\ln(P_j)} = j\overline{M}$. Exponentiating both sides of this expression suggests that $P_j = \exp(j\overline{M})$. This indicates that the ensemble average of a product of random matrices may grow or decay exponentially as j goes to plus or minus infinity. This effect might also be seen in Eq. (4), showing a product of random matrices, including their singular value decomposition. If the product in Eq. (4) is ensemble averaged, one might expect that the combinations DD^{-1} would tend to cancel; the diagonal matrices cannot cancel, leaving a matrix of the form:

$$P_j = \begin{pmatrix} \prod_{j'=1}^{j} \lambda_{j'}^{+} & 0 \\ 0 & \prod_{j'=1}^{j} \lambda_{j'}^{-} \end{pmatrix} \tag{32}$$

For the periodic system we would have $|\lambda_{j'}^{\pm}| = 1$, resulting in extended eigenfunctions. However, for the disordered system we have $|\lambda_{j'}^{-}| > 1$ and $|\lambda_{j'}^{+}| < 1$, so that in Eq. (32) the singular values (and the eigenfunctions) grow or decay exponentially.

These discussions of the average of a product of random matrices are not rigorous, but they do suggest what Furstenburg proved rigorously. The statistical nature of the Anderson localization is actually stated somewhat differently using Furstenburg's theorem: For waves propagating in a disordered system, the eigenfunctions decay exponentially to zero at plus or minus infinity with probability one. This probabilitistic statement permits actual wave functions for a given realization to fluctuate above an exponential envelope.

The problem with the rigorous statement of Anderson localization is that it is difficult to see how to apply it to practical situations. That is, real experiments may have only one finite-size realization available; one cannot examine the system out to infinity: even if one could, it would not be evident if something were happening with "probability one." A treatment of Anderson localization that illustrates the localization and its probabilistic nature in a single finite-size realization is the one by Luban and Luscombe.[6] In this treatment one considers the effects of the rigorous theorems on a computer evaluation of a finite product of random matrices. For a complete understanding of Anderson localization, the reference by Luban and Luscombe[6] is highly recommended.

The most common description of Anderson localization, namely that "the eigenfunctions are exponentially localized" is not particularly useful for single realizations of finite-size systems. The eigenfunctions only behave as though they were exponentially localized if many systems are ensemble averaged. A better way of describing the effects of disorder is with a quantity called the participation ratio, $P(N)$ [or $P(L)$], which is defined with

$$P(N) = \frac{\left(\sum_0^N |y_j|^2\right)^2}{\left(\sum_0^N |y_j|^4\right)} \tag{33}$$

for eigenfunctions of discrete systems or

$$P(L) = \frac{\left(\int_{-L/2}^{L/2} |\psi(x)|^2 dx\right)^2}{\left(\int_{-L/2}^{L/2} |\psi(x)|^4 dx\right)} \tag{34}$$

for continuous systems. The participation ratio is a measure of the localization of a function: A function that is spread out has a larger value of $P(L)$ than a function that is more localized. If $|\psi(x)| = \psi_0$ then $P(L) = L$, that is, the entire system participates. If $|\psi(x)| = \psi_0 \exp(-2|x|/\Lambda)$, then $P(L) \to \Lambda$, the localization length, as $L \to \infty$. A practical method of determining whether a particular eigenfunction is localized or not is to monitor $P(L)$ as L is increased beyond an anticipated localization length Λ. If $P(L)$ keeps increasing as L is increased, then the state is not localized; if $P(L)$ levels off at a finite value as L is increased, then the state is localized. In either case $P(L)$ may fluctuate about L or Λ.

7 QUASICRYSTALLINE SYMMETRY

Until a few years ago it was believed that solid matter could exist in two basic forms: crystalline and amorphous (glassy). In an amorphous material the atoms are positioned at random positions, and a macroscopic sample would be homogeneous and isotropic. Crystals are formed by taking a unit cell and periodically repeating it to fill three-dimensional space. In order to fill all space with long-range periodic order, only unit cells of particular shapes are allowed; only 14 shapes fit together to fill space, and these form the bases for the 14 Bravis lattices of crystallography. Since there are only 14 possible unit cells, only certain rotational symmetries are allowed; in particular, fivefold rotational symmetry is forbidden. However, a few years ago Shechtman[8] made an aluminum alloy whose x-ray diffraction pattern indicated fivefold rotational symmetry. While it is possible to have atoms in clusters with fivefold (icosahedral) short-range order, the diffraction patterns had sharp spots, indicating long-range order. At first the existence of such materials seemed impossible, but theorists pointed out that while it is impossible to have fivefold rotational symmetry and long-range periodic order, it is possible to have fivefold rotational symmetry and another type of long-range order that is quasiperiodic. Thus atoms may minimize their energy by arranging themselves quasiperioidically, forming quasicrystals.

In one dimension the notion of quasiperiodicity is relatively easy to understand. For a line with periodic lattice sites at a spacing a, the Fourier transform (or diffraction pattern) has a sharp line at π/a. If a line has points positioned randomly, then the Fourier transform would show a broad spectrum. Now suppose one takes a line with periodic lattice spacing b, and superimposes it on the line with periodic spacing a. The result may appear similar to the line with random spacing. However, if the two lattice constants a and b are commensurate, that is, their ratio a/b is equal to the ratio of two integers, then the superposition will be periodic. The period may be much larger than a or b, but it will be exactly periodic, and the powerful theorems pertaining to periodic systems (Bloch's theorem, group theory, etc.) will apply, and the properties of the system that follow from symmetry may be calculated exactly. On the other hand, if the two lattice constants a and b are not commensurate, that is, their ratio a/b is equal to an irrational number, then the superposition is not periodic, and there is in general no easy way to calculate the properties. Nevertheless, the superposition does possess long-range order; it is constructed from simple rules, and its Fourier transform contains sharp peaks at π/a and π/b. The superposition of periodic lines with incommensurate lattice constants is an example of one-dimensional quasiperiodicity.

While quasiperiodicity in one dimension is fairly straightforward, in two and three dimensions the notion is considerably more interesting. For higher dimensions, another method of generating quasiperiodic patterns is used. One begins with a periodic lattice in a higher dimension (e.g., for a one-dimensional quasiperiodic lattice one uses a square lattice in two dimensions). The higher dimensional lattice is traversed with a lower dimensional surface (e.g., a one-dimensional straight line passing through the two-dimensional square lattice, making an angle θ with one of the lattice directions). Next a window width is defined, and lattice points falling within that window are projected onto the lower dimensional surface. If the direction cosines describing the orientation of the lower dimensional surface are irrational, then the sites projected onto the surface will form a quasiperiodic pattern. In the one-dimensional example a particularly interesting case occurs when the slope of the line is such that $\tan\theta = \tau = (\sqrt{5}+1)/2$, the golden mean (also known as the divine ratio), which is "the most irrational number." In this case the pattern of sites can be related to a Fibbonacci sequence, and rigorous theorems describing the properties of this special quasiperiodic symmetry can be derived. An algorithm that generates this sequence is

$$u_j = \tau - (\tau - 1)\left[\mathrm{Int}\left(\frac{j+1}{\tau}\right) - \mathrm{Int}\left(\frac{j}{\tau}\right)\right], \tag{35}$$

where $\mathrm{Int}(x)$ indicates the integer part, and $j = 0, 1, 2, 3, \ldots$. The u_j will take on the values τ and 1 in a quasiperiodic sequence.

General theories for one-dimensional systems have been reviewed by Simon[9]; more recently renormalization group and dynamic mapping techniques have been used.[10] Some special properties of one-dimensional quasiperiodic systems are that the eigenvalue spectrum is a Cantor set, and the eigenfunctions may be extended, localized, or critical.

If one begins with a six-dimensional periodic lattice (e.g., a hypercubic lattice) and intersects it with a three-dimensional surface, then one may obtain a three-dimensional quasiperiodic pattern; such patterns may describe the recently discovered aluminum alloy quasicrystals. If one begins with a five-dimensional periodic lattice and intersects it with a two-dimensional (plane) surface, then one may obtain a two-dimensional quasiperiodic pattern, or Penrose tilting[11,12]; one such pattern is illustrated in Fig. 2. If one tries to tile a plane with only one tile shape (unit cell), then as in three dimensions only certain shapes and rotational symmetries are allowed. However, if one is allowed to use tiles of two or more shapes, then fivefold and other symmetries are possible. Algorithms for generating two- and three-dimensional Penrose tilings are quite complicated.[13,14]

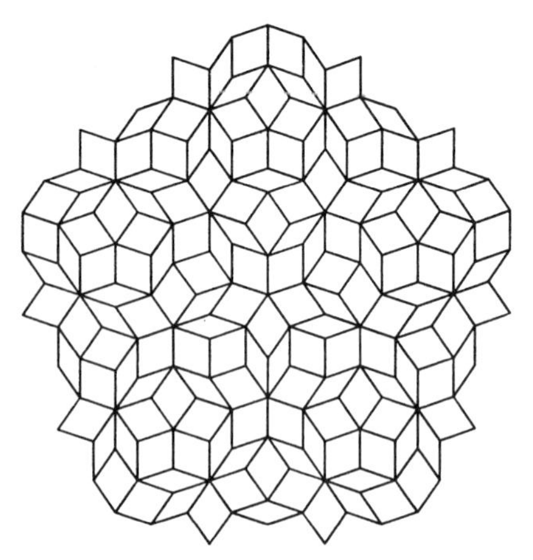

Fig. 2 Penrose tiling pattern consisting of two rhombuses whose areas are in the ratio of the golden mean, $(\sqrt{5}+1)/2$.

Although Penrose tile patterns are not periodic, they do have some symmetry properties.[11–15] Special patterns may have "inflation symmetry," so that a decoration of the tiles with certain lines will produce a replication of the original Penrose pattern, but with a reduced scale. There is also Conway's theorem, which states that given any local pattern (having some nominal diameter), an identical local pattern will be found within a distance of two diameters. The typical Penrose tiling shown in Fig. 2 is formed with two different rhombuses, a fat one and a skinny one; the ratio of the areas is the golden mean, $(\sqrt{5}+1)/2$.

Given a wave equation with a potential field having quasiperiodic symmetry, one can ask how the quasicrystalline symmetry affects the eigenvalues and eigenfunctions. Unlike the theorems for the quasiperiodic patterns in one dimension, theorems for two and three dimensions, if they exist, have not yet been discovered. For a pattern that is periodic, Bloch's theorem may be used to obtain exact solutions. For a pattern that is random, statistical methods may be used to make predictions about properties (e.g., Anderson localization). However a quasicrystalline pattern is not periodic, so that Bloch's cannot be used; and it is not random, so that statistical methods cannot be used. Experimental measurements[16] have been undertaken to determine the effects of quasicrystalline symmetry on eigenvalues and eigenfunctions, and it has been found that the eigenvalue spectrum occurs in bands and gaps whose widths are in the ratio of the golden mean. It will be interesting to see if a "Quasi-Bloch's" can be developed for two- and three-dimensional quasicrystals.

REFERENCES

1. A. A. Maradudin, E. W. Montroll, and G. H. Weiss, *Theory of Lattice Dynamics in the Harmonic Approximation*, Academic Press, New York, 1963.
2. J. M. Ziman, *Principles of the Theory of Solids*, Cambridge University Press, Cambridge, 1964.
3. J. E. Graebner, B. Golding, and L. C. Allen, "Phonon Localization in Glasses," *Phys. Rev.*, Vol. 34, 1986, pp. 5696–5701.
4. P. A. Lee and T. V. Ramakrishnan, "Disordered Electronic Systems," *Rev. Mod. Phys.*, Vol. 57, 1985, pp. 287–337.
5. B. L. Al'Tshuler and P. A. Lee, "Disordered Electronic Systems," *Phys. Today*, Vol. 41, December, 1988, pp. 36–44.
6. M. Luban and J. H. Luscombe, "Localized Eigenstates of One Dimensional Tight-binding Systems: A New Algorithm," *Phys. Rev.*, Vol. B35, 1987, pp. 9045–9055.
7. H. Furstenberg, "Noncommuting Random Products," *Trans. Am. Math Soc.*, Vol. 108, 1963, pp. 337–428.
8. D. Shechtman, I. Blech, D. Gratias, and J. W. Cahn, "Metallic Phase with Long Range Orientational Order and No Translational Symmetry," *Phys. Rev. Lett.*, Vol. 53, 1984, pp. 1951–1953.
9. B. Simon, "Almost Periodic Schrodinger Operators: A Review," *Adv. Appl. Math.*, Vol. 3, 1982, pp. 463–490.
10. M. Kohmoto and J. R. Banavar, "Quasiperiodic Lattice: Electronic Properties, Phonon Properties and Diffusion," *Phys. Rev.*, Vol. B34, 1986, pp. 563–566.
11. R. Penrose, *Bull. Instit. Math. Appl.*, Vol. 10, 1974, p. 266.
12. M. Gardner, "Mathematical Games: Extraordinary Nonperiodic Tiling That Enriches the Theory of Tiles," *Sci. Am.*, Vol. 236, 1977, pp. 110–115.
13. D. Levine and P. J. Steinhardt, "Quasicrystals. I. Definition and Structures," *Phys. Rev.*, Vol. B34, 1986, pp. 596–616.
14. J. E. S. Socolar and P. J. Steinhardt, "Quasicrystals. II. Unit-Cell Configurations," *Phys. Rev.*, Vol. B34, 1986, pp. 617–647.
15. P. J. Steinhardt, "Quasicrystals," *Am. Sci.*, Vol. 74, 1986, pp. 586–597.
16. S. He and J. D. Maynard, "Eigenvalue Spectrum, Density of States, and Eigenfunctions in a Two-dimensional Quasicrystal," *Phys. Rev. Lett.*, Vol. 62, 1989, pp. 1888–1891.

44

SURFACE WAVES IN SOLIDS AND ULTRASONIC PROPERTIES

MOISÉS LEVY AND SUSAN C. SCHNEIDER

1 INTRODUCTION

Surface acoustic wave (SAW) devices are used for many applications, particularly for high-frequency electronic signal processing. In order to optimize the SAW devices, the properties of the piezoelectric substrates, SAW interdigital electrode (IDT) optimization techniques, and SAW temperature compensation techniques have all been extensively studied.[1–4] When a SAW device is used in experiments to study thin films, however, the SAW device is used as both the vehicle to launch the SAW waves and as a support for the thin film. In such experiments, the SAW velocity and attenuation measured are determined by *both* the thin film and the SAW device. In this chapter, a brief review of the fundamentals of SAW propagation through metallic layers deposited on substrates and the SAW velocity of the layered system is given in Section 3. The attenuation of SAWs due to electron–phonon interaction in both the normal and superconducting state is described in Section 4.1; the attenuation of SAWs due to the acousto-electric effect in both normal and superconducting state for homogeneous and granular metallic films is discussed in Section 4.2. The quantum Hall effect as it is related to SAW attenuation is summarized in Section 4.3, and the attenuation of SAWs due to magnetoelastic interaction is described in Section 4.4.

2 SAW DEVICES

A drawing to represent a "typical" SAW device used in SAW attenuation experiments is shown in Fig. 1. The

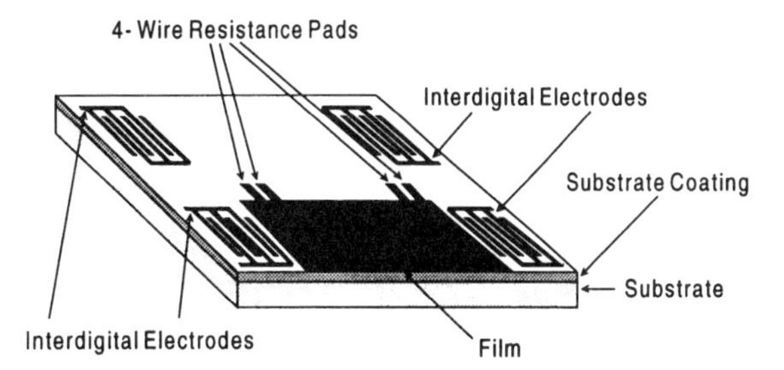

Fig. 1 Typical surface acoustic wave (SAW) device.

SAW device consists of a piezoelectrically active substrate, such as quartz or $LiNbO_3$, on which one or more sets of IDTs are deposited. In some cases, the SAW device is fabricated with a piezoelectrically active coating such as ZnO or AlN deposited on a base substrate material such as fused quartz or sapphire. The IDTs are fabricated with a microphotolithographic process. The fundamental frequency of operation of the SAW device is determined by the width of individual electrodes as well as the spacing between the electrodes and the SAW velocity of the substrate. In addition, the design of the IDTs can be optimized for the generation of overtone frequencies to allow the frequency dependence of an attenuation mechanism to be studied.

When a SAW device is used in experiments to study the properties of thin films, the material to be studied may be deposited on the substrate between one of the sets of IDTs, as shown in the figure. When available on the SAW device, a second set of IDTs is used to monitor the attenuation of a "clear path" on which no film has been deposited. The measurements taken on the clear path provide information about the background attenuation of the substrate. To simultaneously monitor the resistance of a

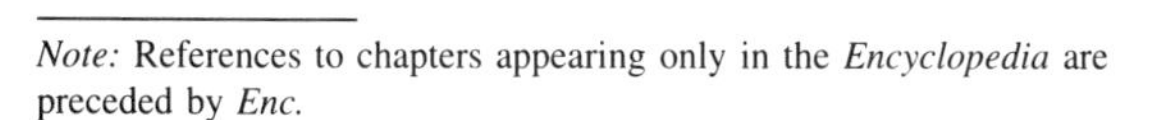
Note: References to chapters appearing only in the *Encyclopedia* are preceded by *Enc.*

thin film using four-wire resistance measurements, many devices have additional electrodes deposited directly on the film (or on the substrate) that do not interfere with the sound path in the film.

3 SURFACE ACOUSTIC WAVE VELOCITY IN FILM–SUBSTRATE GEOMETRIES

When a thin film is deposited on the surface of a SAW device, the SAW velocity of the film–substrate combination is determined by the elastic constants, density, and thickness of the thin film as well as the elastic constants and density of the (piezoelectric) substrate. In general, the SAW velocity for the film–substrate combination can be obtained by solving the equations of motion in the film and the substrate subject to the appropriate boundary conditions at the free surface of the film and the interface between the film and substrate.[5] The simplest case for this calculation is when both the substrate and the film are considered to be isotropic elastic bodies. The calculation method for the SAW velocity for this case is outlined here.

In the following it is assumed that the SAW with wave vector $\boldsymbol{q}$ propagates along the x-axis, and the z-axis is perpendicular to the film plane. The crests of the plane wave are along the y-axis. The film thickness is h. The film substrate interface is the $z = 0$ plane and the free surface of the film is the plane is $z = h$; the substrate lies in the region where $z < 0$. The surface wave is assumed to have the general form $\boldsymbol{\psi}(\boldsymbol{r}, t) = (\alpha, \beta, \gamma)e^{bqz}e^{i(qx-\omega t)}$, where α, β, and γ are the partial-wave coefficients of the wave and ω is the angular frequency of the wave. In terms of the displacement vector $\boldsymbol{\psi}$, the equation of motion has the form

$$\rho\ddot{\boldsymbol{\psi}} = \boldsymbol{\nabla}\cdot\boldsymbol{T}, \tag{1}$$

where the stress tensor $\boldsymbol{T}$ is defined as

$$\boldsymbol{T}_{ij} = \lambda\sum_k \epsilon_{kk}\delta_{ij} + 2\mu\epsilon_{ij}. \tag{2}$$

The strain is defined by $\epsilon_{ij} \equiv (\nabla_i\psi_j + \nabla_j\psi_i)/2; \lambda$ and μ are the isotropic Lamé constants, and ρ is the density of the material. Equation (1) may be rewritten in terms of the displacement to give the equation of motion

$$\rho\ddot{\boldsymbol{\psi}} = (\lambda+\mu)\boldsymbol{\nabla}(\boldsymbol{\nabla}\cdot\boldsymbol{\psi}) + \mu\boldsymbol{\nabla}^2\boldsymbol{\psi}. \tag{3}$$

In order to find $\boldsymbol{\psi}$ for the layered half-space, the equation of motion is solved separately for each medium. The resulting solutions must then satisfy the three boundary conditions, which may be summarized as follows:

1. Stress free surface: $T^F_{zz} = T^F_{zx} = 0$ at $z = h$.
2. Continuous stress: $T^F_{zz} = T^S_{zz}$ and $T^F_{zx} = T^S_{zx}$ at $z = 0$.
3. Continuous amplitude: $\psi^F = \psi^S$ at $z = 0$.

The superscripts F and S refer to the film and substrate, respectively.

The resulting six equations from the above boundary conditions can be written in matrix form as

$$\begin{pmatrix} a & 1 & -a_F & -1 & -a_F & -1 \\ -1 & -d & 1 & d_F & -1 & -d_F \\ A\dfrac{\mu}{\mu_F} & 2d\dfrac{\mu}{\mu_F} & -A_F & -2d_F & A_F & 2d_F \\ -2a\dfrac{\mu}{\mu_F} & -A\dfrac{\mu}{\mu_F} & 2a_F & A_F & 2a_F & A_F \\ 0 & 0 & A_Fe^{a_Fqh} & 2d_Fe^{d_Fqh} & -A_Fe^{-a_Fqh} & -2d_Fe^{-d_Fqh} \\ 0 & 0 & -2a_Fe^{a_Fqh} & -A_Fe^{d_Fqh} & -2a_Fe^{-a_Fqh} & -A_Fe^{-d_Fqh} \end{pmatrix} \cdot \begin{pmatrix} C_1 \\ C_2 \\ C_3 \\ C_4 \\ C_5 \\ C_6 \end{pmatrix} = 0, \tag{4}$$

where the subscript F refers to the film. The coefficients without subscripts refer to the substrate.

For the substrate

$$a = \left(1 - \frac{v^2}{v_t^2}\right)^{1/2}, \qquad d = \left(1 - \frac{v^2}{v_l^2}\right)^{1/2}, \tag{5}$$

where $v_l = [(\lambda+2\mu)/\rho]^{1/2}$ and $v_t = (\mu/\rho)^{1/2}$ are the bulk longitudinal and transverse velocities of the substrate, v is the SAW velocity, and $A = 1 + a^2$.

For the film

$$a_F = \left(1 - \frac{v^2}{v_{tF}^2}\right)^{1/2}, \qquad d_F = \left(1 - \frac{v^2}{v_{lF}^2}\right)^{1/2}, \tag{6}$$

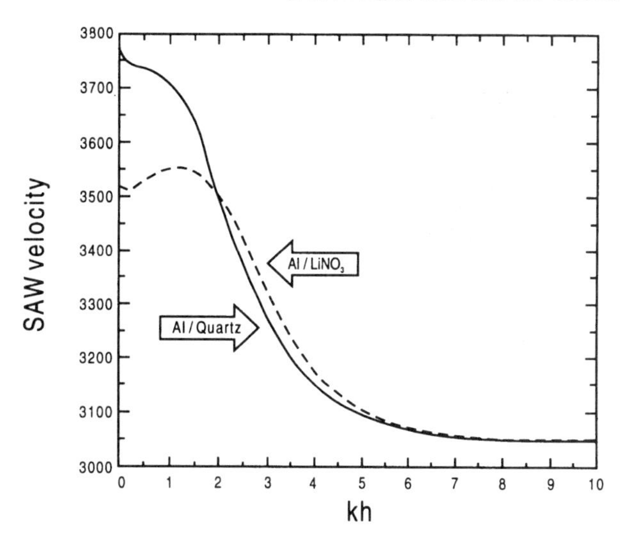

Fig. 2 SAW velocity calculation as a function of qh for an aluminum film on a quartz substrate (solid line) and a $LiNbO_3$ substrate (dashed line).

where $v_{lF} = [(\lambda_F + 2\mu_F)/\rho_F]^{1/2}$ and $v_{tF} = (\mu_F/\rho_F)^{1/2}$ are the longitudinal and transverse bulk velocities of the film and $A_F = 1 + a_F^2$. Equation (4) has a nontrivial solution only if the determinant of the coefficient matrix equals zero. The velocity of the surface wave is the only undetermined parameter in this determinant. A computer program is used to search for those values of the velocity that make the determinant vanish for a chosen frequency ω and particular value of h. The results of the SAW velocity calculation for an aluminum film on a $LiNbO_3$ substrate and a quartz substrate are shown in Fig. 2. The velocity of the SAW in the film–substrate combination is determined by the substrate Rayleigh velocity at zero film thickness, and as the film thickness increases, the SAW velocity approaches the Rayleigh velocity of the film material.

4 SAW ATTENUATION IN FILM–SUBSTRATE GEOMETRIES

Just as the presence of a film on a substrate results in a change in the SAW velocity, the presence of the film can also lead to a change in the attenuation of the SAW. The SAW devices are not lossless, but in general, the attenuation of the SAW due to the substrate alone is known or can be determined. Any additional attenuation of the SAW measured when a film has been deposited on the substrate can then be attributed directly to the presence of the film. The SAW attenuation mechanisms in layered film–substrate combinations can be placed into two broad classes. Surface acoustic wave attenuation produced by the electron–phonon interaction in a metallic film is an example of the "mechanical class" of SAW attenuation mechanisms. In the case of electron–phonon interaction, it is the film that is primarily responsible for the observed SAW attenuation, with the substrate serving to launch the SAW, support the film, and carry the mechanical disturbance. The other class of SAW attenuation mechanisms in layered film–substrate combinations can be characterized as "electrical" in nature, such as SAW attenuation produced through the acoustoelectric effect. In the acoustoelectric effect, a metallic film interacts with the piezoelectric substrate by electrically "shorting out" the surface of the substrate; in this case, the substrate launches the SAW and carries the electrical disturbance but does not necessarily serve as a support for the film. Regardless of the class of mechanism involved in producing the SAW attenuation in the film–substrate calculations, the SAW attenuation is proportional to the energy absorbed by the SAW in the material divided by the total mechanical energy of the SAW in both the film and substrate.

4.1 Attenuation of SAW due to Electron–Phonon Interaction

The first measurement of the attenuation of bulk waves due to electron–phonon interaction was reported by Bommel on a single crystal of lead in 1954.[6] Theoretical models to describe the attenuation of longitudinal and transverse bulk waves produced by electron–phonon interaction include models developed by Mason,[7] Morse,[8] and Pippard,[9] for example. The

SAW measurements in which the attenuation is due to electron–phonon interaction have been performed on a wide variety of films.[10–18] At present, there are no ab initio calculations for the electron–phonon interaction in thin films as there are for bulk waves. However, there are several models[19,20] for SAW attenuation that incorporate the results of the theoretical models developed for bulk wave attenuation produced by electron–phonon interaction. For this reason, the attenuation of sound waves in bulk materials due to electron–phonon interaction in both the normal and superconducting states will be summarized.

Electron–Phonon Interaction in Bulk Materials

Pippard[9] found that the relevant parameter for predicting the attenuation of bulk sound waves in the normal state for a metal is ql, where q is the phonon wave vector and l is the electron mean free path. When $\omega\tau \ll 1$ and $ql \ll 1$, where $\omega = qv_l$ is the angular frequency and $\tau = l/v_F$, the attenuation of longitudinal waves is given by

$$\alpha_L = \frac{4}{15}\,\frac{Nmv_F^2\omega^2\tau}{\rho v_l^3}. \tag{7}$$

The attenuation of transverse sound waves due to electron–phonon interaction in the same limit is given by

$$\alpha_T = \frac{1}{5}\,\frac{Nmv_F^2\omega^2\tau}{\rho v_t^3}. \tag{8}$$

In these equations, N is the number of electrons per unit volume, v_F is the Fermi velocity, v_l and v_t are the materials's longitudinal and transverse sound velocities, respectively, ρ is the mass density of the material, and τ is the relaxation time for electron–phonon interaction. Note that for both cases the attenuation is linearly dependent on the relaxation time and quadratically dependent on the frequency. Many other authors have derived similar expressions for the attenuation.[7,8,21–23]

In the superconducting state, the ultrasonic attenuation of bulk waves is given by the Bardeen, Cooper, and Schrieffer[24] (BCS) expression

$$\frac{\alpha_s}{\alpha_n} = \frac{2}{e^{\Delta/kT}+1}. \tag{9}$$

This result for the ratio of the attenuation coefficient in the superconducting state, α_s, to that in the normal state, α_n, was derived in 1957[24] for the ultrasonic attenuation coefficient for longitudinal waves in the limit $ql \gg 1$. Subsequently, it was shown that this result is also valid for both longitudinal and transverse waves in the limit $ql \ll 1$,[25] the limit in which the SAW attenuation data are generally acquired.

Electron–Phonon Interaction in Thin Films

The computation of the attenuation due to electron–phonon interaction produced by a thin metallic film deposited on a piezoelectric substrate is complicated by several factors. These factors are (a) the attenuation is only produced in the film; (b) the substrate has different elastic constants and density than the film; (c) the longitudinal SAW motion is *not purely compressive* and the transverse SAW motion is *not purely shear*; and (d) theoretical calculations are performed in the limit where the piezoelectric interaction is considered to be negligible.

In earlier derivations it was assumed that either the displacement[19] or the strains[26] were continuous. A later approach[20] satisfied both of these boundary conditions. This later approach to calculating the attenuation of SAWs due to electron–phonon interaction combines one of the techniques used to theoretically model the attenuation of bulk waves due to electron–phonon interaction[7] with the method described previously to calculate the velocity of SAW propagation in film–substrate geometries. The attenuation of SAW is calculated by including viscous loss terms in the macroscopic equation of motion. The viscous loss terms were proposed by Mason[7] to calculate the attenuation of bulk sound waves due to electron–phonon interaction in the limit $ql \ll 1$. In terms of the displacement vector $\boldsymbol{\psi}$ the equation of motion is given by Eq. (1), where the stress tensor $\boldsymbol{T}$ includes viscous damping terms and is defined as

$$T_{ij} = \lambda \sum_k \epsilon_{kk}\delta_{ij} + 2\mu\epsilon_{ij} + \chi \sum_k \dot{\epsilon}_{kk}\delta_{ij} + 2\eta\dot{\epsilon}_{ij}. \tag{10}$$

As before, the strain is defined by $\epsilon_{ij} \equiv \frac{1}{2}(\nabla_i\psi_j + \nabla_j\psi_i)$; λ and μ are the isotropic Lamé constants. The compressional and shear viscosity coefficients χ and η are now explicitly included in the equation of motion. The bulk modulus is related to λ and μ by $B = \lambda + 2\mu/3$; and one may relate the bulk coefficient of viscosity κ to the compressional viscosity χ and shear viscosity η through a similar equation, $\kappa = \chi + 2\eta/3$. The bulk coefficient of visocity is equal to zero for a monatomic gas, such as a free electron gas, since uniform compressions in such a system would produce no viscous losses; and therefore, $\chi = -2\eta/3$. The shear viscosity coefficient η was originally derived by Mason[7] in his calculation of the attenuation of bulk waves due to electron–phonon interaction, $\eta = Nmv_F^2\tau/5$. When these definitions of the shear and compressional viscosity, together with the time dependence of the SAW, $\boldsymbol{\psi}(\boldsymbol{r},t) = \boldsymbol{\psi}(\boldsymbol{r})e^{-i\omega t}$, are used, the following equation of motion for the SAW results:

$$-\rho\omega^2\boldsymbol{\psi}(\boldsymbol{r}) = (\lambda + \mu - \tfrac{1}{3}i\omega\eta)\nabla(\nabla\cdot\boldsymbol{\psi}(\boldsymbol{r})) + (\mu - i\omega\eta)\nabla^2\boldsymbol{\psi}(\boldsymbol{r}). \quad (11)$$

This equation is the same as that of a nonlossive medium where λ is replaced by $\lambda+\frac{2}{3}i\omega\eta$ and μ by $\mu - i\omega\eta$. Hence, the solutions for the displacement fields and the propagation velocity are the analytic continuation of the solutions to the equation of motion without explicit losses and can be found by an extension of the same methods described previously to calculate the velocity of SAW in layered film–substrate geometries when the longitudinal and transverse velocities of the film are modified to take into account the explicit loss produced by the viscous damping; $v_{lF} = [(\lambda_F + 2\mu_F - 4i\omega\eta/3)/\rho_F]^{1/2}$ and $v_{tF} = [(\mu_F - i\omega\eta)/\rho_F]^{1/2}$ are the longitudinal and transverse bulk velocities of the film. Thus the viscous terms make the coefficients in the matrix complex. The same procedure described previously to determine the SAW velocity is followed to determine the SAW attenuation; that is, the matrix has a nontrival solution only if the determinant of the coefficient matrix equals zero. The velocity of the surface wave is the only undetermined parameter in this determinant, but this time, the velocity is a complex quantity. The amplitude attenuation is then given by the imaginary part of the complex wave vector q, $\alpha_{\text{amp}} = \text{Im}(q) = \text{Im}(\omega/v)$. The amplitude attenuation of the SAW as a function of qh for an aluminum film on either a quartz or a $LiNbO_3$ substrate is shown in Fig. 3. The value of η for aluminum used in this calculation was $\eta = 2.19 \times 10^{-3}$ P. In the limit of $qh \ll 1$, the SAW attenuation calculated with this approach appears to be linear with respect to qh.

The derivation for the attenuation of SAW due to electron–phonon interaction in the superconducting state also exploits the results obtained for the attenuation of bulk waves due to electron–phonon interaction in bulk materials. Since the attenuation of a SAW can be ascribed to the shear viscosity of the electron gas, and since the shear viscosity of the electron gas follows the BCS region as derived for transverse bulk waves, which are purely shear, it should follow that the attenuation of a surface acoustic wave in the $ql \ll 1$ limit should follow the BCS relation

$$\left(\frac{\alpha_s}{\alpha_n}\right)_{\text{SAW}} = \frac{2}{e^{\Delta/kT} + 1}. \quad (12)$$

This same result is also obtained from models that ascribe the attenuation of the SAW to the sum of the compressive and shear losses, since both of these losses follow the BCS relation in the $ql \ll 1$ limit.

Magnetic Field Effects. Superconductors may be divided into two types, types I and II. This classification depends upon whether the surface energy required to produce a superconducting–normal state interface is positive or negative, which in turn is determined by the ratio of the London penetration length λ to the coherence length ξ.

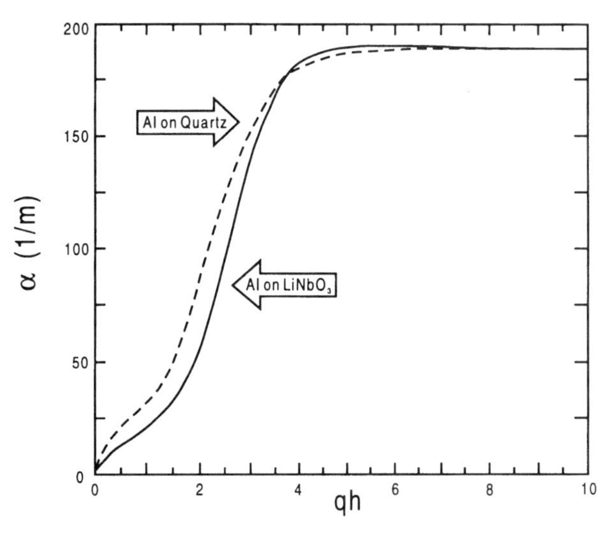

Fig. 3 Amplitude attenuation of the SAW as a function of qh for an aluminum film on either a quartz or a $LiNbO_3$ substrate.

A type I superconductor has $\lambda/\xi < 1$, since the super conducting condensation energy expended for producing a normal cylinder of radius ξ in a superconductor will be higher than the energy gained by allowing a magnetic field to occupy a cylinder of radius λ around the same axis. A type II superconductor has $\lambda/\xi > 1$ for the opposite argument. Therefore, a type I superconductor is perfectly diamagnetic up to the thermodynamic critical field H_c, whereupon the magnetic field totally penetrates the superconductor, destroying superconductivity. Sound waves traveling through the interior of the superconductor do not sense the presence of an external magnetic field and therefore are not affected; thus the attenuation remains constant up to H_c and regains the normal state value when superconductivity is quenched. A type II superconductor remains in the diamagnetic state up to a lower critical field H_{c1}, whereupon magnetic flux lines of value $\phi = h/2e$ and radius ξ start to penetrate (h is Planck's constant and e is the electronic charge). The superconductor remains in this vortex state, in which normal-state cylinders coexist in a superconducting matrix up to the upper critical field H_{c2} when the flux lines coalesce. The magnetization immediately below H_{c2} is linearly dependent upon the applied magnetic field. Type II superconductors should be further divided into dirty and clean type II behavior. In dirty type II superconductors, the electron mean free path is shorter than the coherence length. Electrons inside one of the normal cylinders suffer several collisions before escaping and therefore contribute the normal-state value to the attenuation. Since the fraction of volume occupied by flux lines is proportional to the quantity 1 minus the magnetization $(1-M)$, the attenuation should depend linearly[27] upon the applied magnetic field, $(\alpha_s-\alpha_n)/\alpha_n \simeq (H_{c2} - H)$. In clean type II superconductors, the electron mean free path is longer than the coherence length, and the electrons take a spatial average of the superconducting order parameter Δ, which is proportional to the square root of the magnetization. By placing such an average for Δ in the BCS expression for the attenuation [Eq. (12)] and making a small value expansion, it is found[28] that $(\alpha_s - \alpha_n)/\alpha_n \approx \sqrt{H_{c2} - H}$. Thus for dirty type II superconductors, the attenuation should depend linearly on the applied magnetic field, while for clean type II superconductors, it should depend upon the square root of the difference between the upper critical field H_{c2} and the applied field H. An example of the linear dependence may be found in measurements of Fredricksen et al.[29] when SAWs were used to measure the attenuation coefficient of a Nb_3Sn film near H_{c2} at 700 MHz.

Experimental Measurements of SAW Attenuation due to Electron–Phonon Interaction In 1969, Akao[10] and Krätzig et al.[11] reported the first SAW measurements done on superconducting films. Akao[10] measured attenuation as a function of temperature on In and Pb films. Krätzig et al.[11] reported the magnetic field dependence of the SAW attenuation in a Pb film with the magnetic field either parallel or perpendicular to the film surface. In 1974, Robinson et al.[12] reported on the temperature dependence of SAW attenuation in a granular Al film. In 1978, Park et al.[13] reported on SAW attenuation measurements as a function of temperature in PbCu and PbAg sandwiches. In 1979, Barley and Marshall[15] reported attenuation measurements as a function of temperature in Zn films. That same year, Fredricksen et al.[14] reported on magnetic field measurements on Nb_3Sn. In 1980, Salvo et al.[16] reported on SAW attenuation measurements as a function of temperature on Nb_3Ge and Fredricksen et al.[17] reported similar measurements on Nb_3Sn. In 1991, Baum et al.[18] reported attenuation measurements for a thin film of niobium. Many of these films appear to follow what could be called the classical BCS relation [Eq. (9) and (12)] exactly. The other films, such as Nb_3Ge, exhibit deviations from this relation. The results of several of these SAW attenuation measurements in superconducting thin films are summarized in Table 1. The parameters reported in this table include the superconducting transition temperature T_c reported for the film and the zero-temperature energy gap $2\Delta(0)$. Shown in Fig. 4 is an example of SAW attenuation data for a Nb_3Sn film on a $LiNbO_3$ substrate when compared to BCS theory [Eq. (12)]. The zero-temperature energy gap for this film is $2\Delta(0) = 3.5kT_c$.

In addition to determining the zero-temperature energy gap and the superconducting critical temperature, SAW attenuation measurements can also be used to determine a value for the electron mean free path in the material. Using the value of the normal-state attenuation obtained from experimental measurements, a value for the electron mean free path can be obtained, $l_{\alpha,\mathrm{SAW}}$, which can be compared to the value of the electron mean free path obtained from conductivity/resistivity measurements on the films, l_{elec}. The value for $l_{\alpha,\mathrm{SAW}}$ may be deduced from plots such as those shown in Fig. 3 by selecting the appropriate value for τ in η. An alternate and simpler approach, though approximate, would be to use the procedure outlined in Ref. 19. Values for the electron mean free path obtained from both SAW attenuation measurements and resistivity measurements are shown in Table 2 for three superconducting films that are granular in nature. In all cases shown in Table 2 the value of the mean free path obtained from SAW attenuation measurements is greater than that obtained from the electrical resistivity measurement. In the SAW attenuation measurements, the value of the mean free path obtained is consistent with one given by diffuse scattering over the surface of grains in the film. The SAWs appear to sample the granules in the film in parallel and

TABLE 1 Superconducting Energy Gap as Determined by SAW Attenuation Measurements on Thin Films

Film	Thickness, h (Å)	T_c (K)	$2\Delta(0)/kT_c$	Reference
Nb_3Sn	5000	18	3.5	17
Aluminum	300	1.7	3.6	12
Zinc	3000	1.5	4.2	15
Zinc	10,000	1.31	3.8	15
Indium	500, 1500	—	4.3	10
Lead	1000, 2500	—	3.9	10
Niobium	3000	8.5	3.5	18
Nb_3Ge	5000	21	3.5	16

are thus sensitive to the longest mean free path in the granules. In the resistivity measurements for l, the electrical current passes through the granules in series, and therefore, resistivity measurements are sensitive to the smallest mean free path in the granules. Thus, the two measurements for electron mean free path yield complementary information.[19]

4.2 Acoustoelectric Effect

The attenuation of SAW in piezoelectrics coated with semiconducting films has been described by Ingebrigtsen[30] and Adler,[31] for example. Adler used an intuitive approach to derive an expression for the attenuation (gain or loss) of an acoustic surface wave that was related to frequency, the effective electromechanical coupling constant K, and the semiconductor drift velocity. This same type of intuitive approach can be used to describe the acoustoelectric interaction of SAWs with superconducting films.

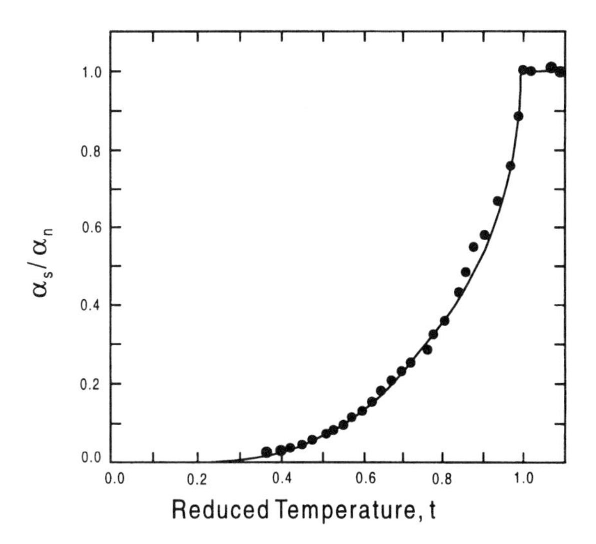

Fig. 4 SAW attenuation data as a function of reduced temperature, $t = T/T_c$, for a Nb_3Sn film on a $LiNbO_3$ substrate when compared to BCS theory [Eq. (9)].

The effect of acoustoelectric interaction with superconducting films and with a two-dimensional electron gas that is in close proximity to a piezoelectric surface carrying a SAW may be described as follows. A SAW propagating through a piezoelectric substrate produces surface electric polarization charges associated with the SAW strains. When a conducting film either is deposited on or is brought into close proximity to the piezoelectric substrate surface, the polarization charges in the substrate induce image charges in the conducting film. These image charges will move in the conducting film with the velocity of the SAW, producing energy dissipation that attenuates the SAW. If the conductivity of the film is infinite, there will be zero energy dissipation; if the conductivity is zero, there will be no image charges and the dissipation will again be zero. Only if the conductivity is finite will there be energy dissipation. Therefore, the attenuation process may be described by a debye relaxation expression, which for the case when the film is deposited directly onto the piezoelectric substrate is given by

$$\alpha = \frac{\kappa^2}{v_0} \frac{\omega^2 \tau}{1 + \omega^2 \tau^2}, \tag{13}$$

where α is the energy attenuation coefficient in nepers per centimetre, κ^2 is the electromechanical coupling con-

TABLE 2 Electron Mean Free Path Comparison

Film	$l_{\alpha,\mathrm{SAW}}$ (Å)	l_{elec} (Å)	Reference
Aluminum	33	1.3	19
Niobium	1050	450	18
Nb_3Ge	140	10	16

stant, v_0 is the velocity of the SAW, ω is the angular frequency of the SAW, and τ is a relaxation time. For very thin films, τ is given by $\omega\tau = v_0(\epsilon_0 + \epsilon)/\sigma_\square$, where ϵ_0 and ϵ are the dielectric permittivity of vacuum and the substrate, respectively, and $\sigma_\square$ is the sheet conductivity of the conducting film ($\sigma_\square = \sigma d$, where d is the thickness of the film and σ is the electrical conductivity of the film material). Very thin films are defined as having a thickness that is smaller than $\lambda/2$, where λ is the wavelength of the SAW. The attenuation α will have a maximum when $\omega\tau = 1$, which occurs for $\sigma_{2D} = v_0(\epsilon_0 + \epsilon)$, generally equal to about $10^{-6}\,\Omega^{-1}$; σ_{2D} is known as the characteristic conductivity of the system.

Since Eq. (13) is a typical relaxation expression for the attenuation, the SAW velocity will also change with $\omega\tau$ as follows[31]:

$$\frac{\Delta v}{v_0} = -\frac{\kappa^2}{2}\frac{1}{1+\omega^2\tau^2}, \tag{14}$$

so that, for $\omega\tau \to \infty$, $\Delta v/v \to 0$, and for $\omega\tau \to 0$, $\Delta v/v_0 = -\kappa^2/2$. This behavior should be expected since the case $\omega\tau \to \infty$ results when the piezoelectric surface is exposed to free space and then the normal SAW velocity should be obtained. The latter case, $\omega\tau \to 0$, results when the surface is covered by a perfectly conducting film. This perfectly conducting film shorts out the surface electric fields, thus quenching the electrical energy carried by the wave and leaving only the mechanical energy; since the elastic constants are given by the second partial derivative of the Gibbs free energy, they should also be reduced.

It is convenient to recast Eq. (13) in terms of σ_{2D} as

$$\alpha = \frac{\omega\kappa^2}{v_0}\frac{\sigma_\square/\sigma_{2D}}{1+(\sigma_\square/\sigma_{2D})^2}. \tag{15}$$

For $\sigma_\square \ll \sigma_{2D}$, the joule loss in the film is of the form of a constant voltage source, V^2/R, since the acoustoelectric currents through the film are very small. However, for $\sigma \gg \sigma_{2D}$, there is enough redistribution of charge in the conductor to almost completely screen the electric field and the loss is of the form of I^2R, such as for a constant current generator.

In some cases, it is not possible to deposit the film onto a piezoelectric substrate directly. Instead, the film is deposited on another substrate, which is then brought into close proximity to the piezoelectric substrate. Schenstrom[32] has derived an expression for proximity coupling between the surface of the SAW substrate and the conducting film.

Even superconducting films with the highest normal-state resistivities have values of $\sigma_\square$ that are much larger than σ_{2D}. Therefore, in such films, $\omega\tau < 1$, and it is possible to simplify Eq. (13) to

$$\alpha = \frac{\kappa^2}{v_0}\omega^2\tau = \frac{\omega\kappa^2}{v_0}(\epsilon_0+\epsilon)R_\square, \tag{16}$$

where $R_\square = 1/\sigma_\square$. Equation (16) implies that in the limit of $\omega\tau < 1$ the attenuation should be proportional to the sheet resistivity of the conducting film. If the conducting film undergoes a superconducting transition, then the SAW attenuation produced by the acoustoelectric interaction should vanish. Any change in attenuation should be proportional to the change in the resistivity of the film, since the dielectric constants and the electromechanical coupling constants should not be affected by the superconducting transition. Schmidt et al.[33] has shown that the change in attenuation produced by superconducting quenching of the acoustoelectric effect is proportional to the change of resistance in an amorphous indium film. The broadness of the transition in this film may be accounted for by superconducting fluctuations[34] above the midpoint of the transition and by the motion of unbound vortex pairs below the midpoint.[35] This justifies the deduction that the attenuation due to the acoustoelectric effect should be proportional to the sheet resistivity through the whole of a superconducting transition.

However, data obtained on a granular lead film[36] do not exhibit this proportionality in the superconducting state. Not only is the acoustoelectric attenuation larger than would be predicted by Eq. (16), it also has a finite value in the superconducting state even when the sheet resistivity has vanished. To account for both the shape of the attenuation curve in the superconducting state and its magnitude with respect to the sheet resistivity curve, a percolation model was developed. The percolation model for attenuation produced by the acoustoelectric effect is motivated by the granular nature of some of the films studied. It is assumed that the grains in the film are separated by oxide barrier junctions through which electrical currents may tunnel and, for simplicity, that these tunneling junctions are arranged in a two-dimensional square lattice network. The essence of the model is that measurements of DC electrical resistivity sample the percolation resistivity of such a network with infinite dimensions, or at least the dimensions of the film, while the acoustoelectric attenuation coefficient depends on the average resistivity of squares whose dimensions are comparable to the SAW wavelength,[37] specifically $\lambda/2\pi$.

The percolation resistance of such networks can be determined using a model due to Ambegaokar, Halperin, and Langer (AHL).[38] The tunneling junctions are replaced by resistors whose values correspond to those of the junctions. All these resistors are removed from the network and then returned to the network according to the following scheme: First, the lowest value resistors are returned at random, then the next lowest value resistors are returned, and so on. When the fraction of resistors that have been returned to the network is equal to the critical percolation fraction—$\frac{1}{2}$ for the case of an infinite square network—a percolation path will be established through the resistor. For a two-dimensional network, the sheet resistivity R_n is approximately equal to that value of the critical resistor r_c, which establishes this percolation path. This claim is plausible, since resistors with values larger than r_c do not contribute much to the sheet resistivity because they are shunted by resistors whose values are on the order of r_c; resistors with values less than r_c cannot by themselves provide current transport over macroscopic distances. The assumptions made by Levy et al.[37] yield a parabolic distribution function for resistors with a minimum resistance r_1 and a maximum resistance r_2, and $R_n = (r_1 r_2)^{1/2}$. As mentioned above, R_n is a measure of the sheet resistivity of a film with a DC potential or an AC potential with infinite wavelength. The acoustoelectric effect produces AC potentials with finite wavelengths, namely the SAW wavelength λ. Therefore, the AC potentials sample the resistance of small squares whose dimensions are comparable to λ. In fact, Levy et al.[37] postulate that this dimension should be $l = \lambda/2\pi$. It is easy to deduce the factor of 2 in this expression since the electric potential changes sign twice over a wavelength. The factor of 2π comes from the belief that since most acoustic effects have some sort of transition when $ql \simeq 1$, then ql would be the important parameter in any acoustic propagation equation. This dimension l can be conveniently expressed in terms of the average number of grains L contained in l. Since the granular film is inhomogeneous, the local sheet resistance of these small sections, $R(L)$, will vary from location to location. Therefore, the ultrasonic attenuation in Eq. (16) should be governed by the average sheet resistance $\boldsymbol{R}(L)$ of these small sections of the film, each of which can be represented by a finite resistor network, and not the sheet resistance of the entire film. It was found[33] that films with over 300 grains per $\lambda/2\pi$ may be considered homogeneous for the purpose of acoustoelectric attenuation measurements.

As the temperature is lowered, the metallic grains become superconducting at their transition temperature T_g, but the Josephson junctions between the grains remain resistive until a temperature is reached where the Josephson coupling energy E_j is equal to or higher than the thermal energy. Thus, as the temperature is lowered, more junctions become superconducting by passing Josephson currents, the lower resistance junctions becoming superconducting first. The percolation resistance of the film is then computed using an effective medium approximation[39] wherein the junction resistances have a binary distribution; the junctions are either superconducting with a zero resistance if the temperature is below their particular E_j or the junctions have a resistance value equal to R_n. Then the sheet resisivity of the film may be calculated by assuming a linear dependence of the overall resistance with respect to the fraction of R_n resistors that are present, the resistance going to zero when half of the junctions become superconducting. This binary distribution is adequate for describing the resistivity of the small sections sampled with SAW since there are only a few tunneling junctions per wavelength. However, for computing the DC resistivity of films that encompass a very large number of junctions per wavelength, it is necessary to take into account the full distribution of the junction resistances in the normal state.[40] By doing so, even when plotting the resistivity data logarithmically, Schmidt et al.[40] find that this approach fits the resistivity data for several granular films very well.

The same procedure may be used to obtain the sheet resistance of each individual finite resistor network sampled by the SAW. For this calculation, it is necessary to remember that the critical percolation fraction for each network is no longer $\frac{1}{2}$ but has a value given by a Gaussian distribution centered around $\frac{1}{2}$. In this case, Levy et al.[37] find reasonable agreement for granular Pb and In films.

Recently, acoustoelectric SAW attention measurements at 168 MHz were performed on an $YBa_2Cu_3O_7$ film deposited on a YZ cut $LiNbO_3$ substrate.[18] The attenuation curve deviates from the resistance curve; there is also a significant difference between the predicted and observed attenuation changes. A scanning electron micrograph of the film shows that the size of the grains is about 3000 Å and that therefore $L = 10$. By using Eqs. (15) and (20) in Ref. 31 and $R_n = (r_1 r_2)^{1/2}$, it is possible to find the minimum resistance of the network r_1 and from this the intrinsic resistivity of the individual grains in the $YBa_2Cu_3O_7$ film. If it is assumed that r_1 measures the resistance of a grain of pure material, the resistivity $\rho = r_1 t_{\text{film}} = 12.5\ \mu\Omega \cdot \text{cm}$, where t_{film} represents the thickness of the film. X-ray analysis of the film has shown that the individual grains are oriented with the c-axis perpendicular to the surface, and therefore both the electric and acoustoelectric resistance measurements are essentially confined to the better conducting ab planes. It is interesting to note that the value obtained for the intrinsic resistance of the grains along the ab planes is at least a factor of 2 lower than what has been measured electrically on the best single crystals of $YBa_2Cu_3O_7$. Therefore, it may be possible that acoustoelectric measurements circumvent any twinning imperfections that may be present in the single crystals which could contribute to the resistivity. Thus, this may be a technique for measuring the average intrinsic resistivity of grains whose dimensions are even less than 3000 Å.

4.3 Quantum Hall Effect

The quantum Hall effect (QHE) gets its name from the fact that the Hall resistance ρ_{xy} of some two-dimensional electron (2DE) systems appears to be quantized in steps to a very high degree of accuracy. The resistance ρ_{xx} of such a system exhibits peaks at the edge of the steps. Since the acoustoelectric attenuation depends on σ_{xx}/σ_{2D}, by evaluating $\sigma_{xx} = \rho_{xx}/(\rho_{xx}^2 + \rho_{xy}^2)$, associating σ_{xx} with $\sigma_\square$, and substituting the result into Eq. (15), one finds that there should be an attenuation minimum at the position of the peak in σ_{xx} with a maximum on either side of the minimum. This is because when $\sigma_{xx}(\text{peak}) > \sigma_{2D}$, the attenuation should go through a maximum on either side of the peaks at the fields at which $\sigma_{xx} = \sigma_{2D}$, resulting in a minimum in attenuation at the position of the peaks in σ_{xx}. This theoretical prediction agrees fairly well with experimental results obtained by Schenstrom et al.[41] using proximity coupling to a III–V semiconducting heterojunction. The observation that the experimental attenuation peaks are broader than the predicted ones provides evidence for delocalization of the quantized electronic states on the scale of the dimensions probed by the 210-MHz SAW, about 2.6 μm.

4.4 Magnetoelastic Interaction

Very large magnetic-field-dependent attenuation of 618 MHz Rayleigh mode SAWs propagating on a quartz substrate through evaporated 200-Å Ni films was discovered in 1976 by Krischer et al.[42] The impressive results described in this work are that a small change in magnetic field from 10 to 90 Oe can produce a change in attenuation of 30 dB/cm and that the maximum in attenuation is located at 10 Oe. The experimentally observed correlation between the attenuation and the in-plane magnetic susceptibility of the Ni films[43] led to a phenomenological model for the observed attenuation response.[44,45] The films are characterized as having in-plane uniaxial anisotropy with single-domain magnetization that changes in direction, through coherent spin rotation, when an external magnetic bias field is applied.

Energy dissipation is produced by magnetoelastic

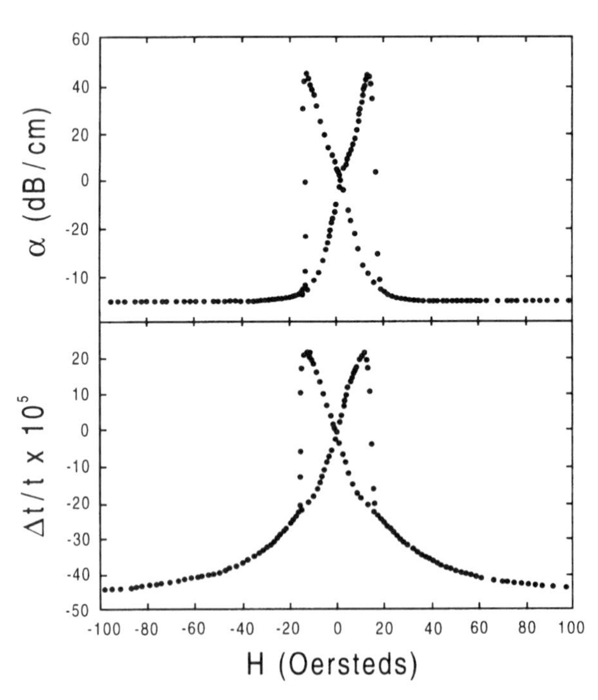

Fig. 5 SAW attenuation and transit time (velocity) as a function of applied magnetic field produced by a 372-Å nickel film deposited on a quartz substrate.[47] The SAW attenuation data are shown at the top and transit time data is displayed at the bottom.

interaction. The time-dependent strains ϵ of the SAW produce oscillating magnetic fields via magnetoelastic interaction. These small magnetic fields oscillate the film magnetization about its equilibrium orientation determined by the larger DC magnetic field. Since there is a finite relaxation time for oscillations of the magnetization, the magnetization does not follow the SAW in phase. This results in energy dissipation per unit volume per cycle given by[45]

$$E_d = \frac{\pi[(\Delta H_A/2)\sin(2\theta_0 - 2\phi)]^2[B^2 + H_s^2(1 + G^2)]H_s MG}{[AB - H_s^2(1 + G^2)]^2 + [H_s G(A + B)]^2}, \tag{17}$$

where H_A is the uniaxial anistropy field that acts along the film's easy axis of magnetization; M is the magnetization; H is the applied field; θ and ϕ are, respectively, the angular positions of M and H_A with respect to H; G is the Gilbert damping parameter; $A = H_A \cos(2\theta_0 - 2\phi) + H \cos\theta_0, B = H\cos\theta_0 + 4\pi M, H_s = \omega/\gamma$, where γ is the electronic magnetomechanical ratio; and $\Delta H_A = \eta_m \epsilon$, where η_m is the magnetoelastic coupling coefficient of the film. The term A represents the restoring force on $\boldsymbol{M}$ in the film plane, while B represents the restoring force on $\boldsymbol{M}$ perpendicular to the film plane. The change in velocity is given by[46] a similar expression.

The shapes of the curves obtained from Eq. (17) and the one for the velocity changes look remarkably close to those displayed in Fig. 5 for the attenuation and transit time (velocity) produced by a 372-Å nickel film deposited on a quartz substrate,[47] where the maximum attenuation change is almost 80 dB/cm. This attenuation change is large enough to be used to develop frequency tunable SAW dispersion lines and filters.[48]

In summary, SAW attenuation measurements provide an invaluable experimental tool with which to probe relevant electrical and physical properties of thin films. Surface acoustic wave attenuation in thin films is an exciting and fruitful area of research whose potential for device application is only beginning to be tapped.

REFERENCES

1. B. A. Auld, *Acoustic Fields and Waves in Solids, Vols. I and II*, 2nd ed., R. E. Krieger, Malabar, FL, 1990.
2. G. S. Kino, *Acoustic Waves: Devices, Imaging and Analog Signal Processing*, Prentice-Hall, Englewood Cliffs, NJ, 1987.
3. D. P. Morgan, *Surface Wave Devices for Signal Processing*, Elsevier, Amsterdam, 1985.
4. S. Datta, *Surface Acoustic Wave Devices*, Prentice-Hall, Englewood Cliffs, NJ, 1986.
5. G. W. Farnell and E. L. Adler, in W. P. Mason and R. N. Thurston (Eds.), *Physical Acoustics*, Vol. IX, Academic, New York, 1972, pp. 35–127.
6. H. E. Bommel, "Ultrasonic Attenuation in Superconducting Lead," *Phys. Rev.*, Vol. 96, 1954, p. 220.
7. W. P. Mason, "Ultrasonic Attenuation Due to Lattice-Electron Interaction in Normal Conducting Metals," *Phys. Rev.*, Vol. 97, 1955, p. 557; W. P. Mason, *Physical Acoustics and the Properties of Solids*, Van Nostrand, Princeton, NJ, 1958.
8. R. W. Morse, "Ultrasonic Attenuation in Metals by Electron Relaxation," *Phys. Rev.*, Vol. 97, 1955, p. 1716.
9. A. B. Pippard, "Ultrasonic Attenuation in Metals," *Phil. Mag., Ser. 7*, Vol. 46, 1955, p. 1104.
10. F. Akao, "Attenuation of Elastic Surface Waves in Thin Film Superconducting Pb and In at 316 MHz," *Phys. Lett.*, Vol. 30A, 1969, p. 409.
11. E. Krätzig, K. Walther, and W. Shilz, "Investigation of Superconducting Phase Transitions in Pb-films With Acoustic Surface Waves," *Phys. Lett.*, Vol. 30A, 1969, p. 411.
12. D. A. Robinson, K. Maki and M. Levy, "Effects of Superconducting Fluctuations on the Ultrasonic Attenuation in a Thin Aluminum Film," *Phys. Rev. Lett.*, Vol. 34, 1974, p. 709.
13. J. G. Park, P. Tai, and S. Frake, "Ultrasonic Surface Wave Attenuation by Induced Superconductivity in PbCu and PbAg 'Sandwiches'," *J. Phys.*, LT15, Vol. 39, No. C-6, Pt. 1, 1978, p. C6/681.
14. H. P. Fredricksen, M. Levy, J. R. Gavaler, and M. Ashkin, "The Ultrasonic Attenuation of 700 MHz Surface Acoustic Waves in Thin Films of Superconducting NbN," *Phys. Rev.*, Vol. B27, 1983, p. 3065.
15. W. E. Barley and B. J. Marshall, "Ultrasonic Attenuation of Surface Acoustic Waves in Superconducting Zinc," *Phys. Rev.*, Vol. B19, 1979, p. 3467.
16. H. L. Salvo, Jr., H. P. Fredricksen, M. Levy, and J. R. Gavaler, "Ultrasonic Attenuation by a Thin Film of High Transition Temperature Nb_3Ge," *Solid State Comm.*, Vol. 33, 1980, p. 781.
17. H. P. Fredricksen, H. L. Salvo, Jr., M. Levy, R. H. Hammond and T. H. Geballe, "Ultrasonic Attenuation of Surface Acoustic Waves in Thin Film of Superconducting Nb_3Sn," *Phys. Lett.*, Vol. 75A, 1980, p. 389.
18. H. P. Baum, B. K. Sarma, M. Levy, J. Gavaler, and A. Hohler, "SAW Measurements on a Nb Film and an $Y_1Ba_2Cu_3O_7$ Film," *IEEE Trans. Mag.*, Vol. 27, 1991, p. 1280.
19. M. Tachiki, H. Salvo, Jr., D. A. Robinson, and M. Levy, "Ultrasonic Mean Free Path in a Granular Aluminum Film," *Solid State Comm.*, Vol. 17, 1975, p. 635.
20. D. R. Snider, H. P. Fredricksen, and S. C. Schneider, "Surface Acoustic Wave Attenuation by a Thin Film," *J. Appl. Phys.*, Vol. 52, 1981, p. 3215.
21. T. Holstein, "Notes on Electroacoustical Phenomena," Westinghouse Research Memo 60-94698-3-MI7, 1956.

22. A. B. Pippard, "Theory of Ultrasonic Attenuation in Metals and Magneto-acoustic Oscillations," *Proc. Roy. Soc.*, Vol. A257, 1960, p. 165.
23. R. S. Sorbello, "Transport Theory of Ultrasonic Attenuation," *Phys. Stat. Sol. (B)*, Vol. 77, 1976, p. 141.
24. J. B. Bardeen, L. N. Cooper, and J. R. Schrieffer, "Theory of Superconductivity," *Phys. Rev.*, Vol. 108, 1957, p. 1175.
25. M. Levy, "Ultrasonic Attenuation in Superconductors for $ql < 1$," *Phys. Rev.*, Vol. 131, 1963, p. 1497.
26. M. Levy, H. Salvo, Jr., D. A. Robinson, K. Maki, and M. Tachiki, "Surface Acoustic Wave Investigation of Superconducting Films," in J. de Klerk (Ed.), *Proceedings 1976 Ultrasonics Symposium*, IEEE 76 CH 1120-5 SU, IEEE, New York, 1976, p. 633.
27. B. R. Tittmann, "Ultrasonic Attenuation in a Type II Superconducting Alloy," *Phys. Rev.*, Vol. B2, 1970, p. 625.
28. R. Kagiwada, M. Levy, I. Rudnick, H. Kagiwada and K. Maki, "Ultrasonic Attenuation in a Pure Type II Superconductor Near H_{c2}," *Phys. Rev. Lett.*, Vol. 18, 1967, p. 74; R. Kagiwada, M. Levy, and I. Rudnick, "Ultrasonic Attenuation in Type II Superconductors," in A. S. Borovik-Romanov and V. A. Tallin (Eds.), *Proceedings of the 10th International Conference on Low Temperature Physics*, Moscow, 1966.
29. H. P. Fredricksen, H. L. Salvo, Jr., M. Levy, R. H. Hammond and T. H. Geballe, "Ultrasonic Attenuation of Surface Acoustic Waves in a Superconducting Thin Film of Nb_3Sn in an Applied Magnetic Field," in J. de Klerk and B. R. McAvoy, (Eds.), *Proceedings 1979 Ultrasonics Symposium*, IEEE 79 CH 1482-9 SU, IEEE, New York, 1979, p. 435.
30. K. A. Ingebrigtsen, "Linear and Nonlinear Attenuation of Acoustic Surface Waves in a Piezoelectric Coated With a Semiconducting Film," *J. Appl. Phys.*, Vol. 41, 1970, pp. 454–459.
31. R. Adler, "Simple Theory of Acoustic Amplification," *IEEE Trans. Sonics Ultrasonics*, Vol. SU-18, 1971, p. 115.
32. A. H. Schenstrom, "Proximity Coupling of Surface Acoustic Waves to Quasi 2-Dimensional Systems," Ph.D. Dissertation, University of Wisconsin-Milwaukee, 1987.
33. J. Schmidt, M. Levy and A. F. Hebard, "Ultrasonic Investigation of Amorphous Superconducting Films," *Phys. Rev.*, Vol. B50, 1994, pp. 3988–3994.
34. L. G. Aslamazov and A. I. Larkin, "Effect of Fluctuations on the Properties of a Superconductor Above the Critical Temperature," *Fiz. Tverd. Tela*, Vol. 10, 1968, p. 1104 (*Sov. Phys. Solid State*, Vol. 10, 1968, p. 875).
35. J. M. Kosterlitz and D. J. Thouless, "Ordering, Metastability and Phase Transitions in Two Dimensional Systems," *J. Phys.*, Vol. C6, 1973, p. 1181.
36. H. Tejima, J. Schmidt, C. Figura, and M. Levy, "Surface Acoustic Wave Investigation of Granular Lead Films," in B. R. McAvoy (Ed.), *IEEE 1983 Ultrasonics Symposium Proceedings*, 83 CH 1947-1, IEEE, New York, 1983, p. 1100.
37. M. Levy, J. Schmidt, A. Schenstrom, M. Revzen, A. Ron, B. Shapiro, and C. G. Kuper, "Surface Acoustic Attenuation in Granular Lead Films," *Phys. Rev.*, Vol. B34, 1986, p. 1508; C. G. Kuper, M. Levy, M. Revzen, A. Ron, and B. Shapiro, "Surface Acoustic Attenuation in Granular Lead Films," in U. Eckern, A. Schmid, W. Weber, and H. Wahl (Eds.), *Proceedings of the 17th International Conference on Low Temperature Physics*, Elsevier Science, Amsterdam, 1984, p. 913.
38. V. Ambegaokar, B. I. Halperin and J. S. Langer, "Hopping Conductivity in Disordered Systems," *Phys. Rev.*, Vol. B4, 1971, p. 2612.
39. For a discussion of the effective medium approximation, see, S. Kirkpatrick, "Classical Transport in Disordered Media: Scaling and Effective-Medium Theories," *Phys. Rev. Lett.*, Vol. 27, 1971, p. 1722.
40. J. Schmidt, M. Levy, and A. F. Hebard, "Ultrasonic Investigation of Granular Superconducting Films," *Phys. Rev.*, Vol. B43, 1991, p. 505.
41. A. Schenstrom, B. K. Sarma, M. Levy, and H. Morkoc, "Frequency Dependent Breakdown of the Dissipationless State in the Quantum Hall Effect," *Solid State Comm.*, Vol. 68, 1988, p. 357.
42. C. Krischer, I. Feng, J. B. Block, and M. Levy, "Magnetic Field-Dependent Attenuation of Surface Waves by Nickel Thin Films," *Appl. Phys. Lett.*, Vol. 29, 1976, p. 76.
43. I. Feng, H. Fredricksen, C. Krischer, M. Tachiki and M. Levy, "Correlation between the Magnetic Susceptibility of Nickel Films and Their Effect on Surface Acoustic Waves," in J. de Klerk and B. R. McAvoy, (Eds.), *Proceedings 1977 IEEE Ultrasonics Symposium*, 77CH 1264-1 SU, 1977, p. 328.
44. I. Feng, "Interaction of Surface Waves with Thin Nickel Films," Ph.D. Dissertation, University of Wisconsin-Milwaukee, 1979.
45. I. Feng, M. Tachiki, C. Krischer, and M. Levy, "Mechanism for the Interaction of Surface Waves with 200 Å Nickel Films," *J. Appl. Phys.*, Vol. 53, 1982, p. 177.
46. D. R. Walikainen, "Ultrasonic Surface Acoustic Wave Investigation of Thin Magnetostrictive Films," Ph.D. Dissertation, University of Wisconsin-Milwaukee, 1990.
47. R. F. Wiegert and M. Levy, "Enhanced Magnetically Tunable Attenuation and Relative Velocity of 0.6 GHz Rayleigh Waves in Nickel Thin Films," *Appl. Phys. Lett.*, Vol. 54, 1989, p. 995.
48. M. Levy, R. G. Wiegert, B. R. McAvoy, H. L. Salvo, Jr., and D. Bailey, "Variable Bandwidth SAW Filter Using Magnetoelastic Coupling," in B. R. McAvoy (Ed.), *IEEE 1985 Ultrasonics Symposium Proceedings*, 85 CH 2209-S, IEEE, New York, 1985, p. 88.

45

ULTRASONIC INSTRUMENTS FOR NONDESTRUCTIVE TESTING

EMMANUEL P. PAPADAKIS

1 INTRODUCTION

This chapter deals with ultrasound as used in nondestructive testing (NDT). Ultrasonics as limited to NDT in this chapter can be defined as actively generated low-intensity mechanical radiation above human hearing that can be directed for interrogation purposes. As ultrasonic waves propagate in materials, they are influenced not only by the flaws[1] but also by the mechanical properties[2–5] and can be used to determine some of them nondestructively.[6] Since NDT is such a results-oriented and ad hoc interdisciplinary field, it is appropriate to focus on ultrasonic instruments. To conduct actual nondestructive testing one does not interlace a pulse generator, a piezoelectric plate, an amplifier, and an oscilloscope with BNC cables as in a laboratory; one buys a portable, battery-operated flaw detector with built-in pulser, amplifier, oscilloscope, touch-pad controls, and computerized functions that will fit down the conning tower of a submarine.[7] The only cable goes to the potted transducer. For a different kind of job, one may install a large system with manipulators and multiaxis computer control to move sensors (transducers) around in an immersion tank as large as two lanes of an Olympic swimming pool.[8] To get an idea of size and proportion, see Fig. 1. Succinctly, the result desired determines the system configuration, while the system is self-contained and specialized.

The self-contained and specialized instruments all operate to make measurements on the basis of a few fundamental principles. The fundamentals will be treated first in a generic way such that the reader can refer to other chapters in this book for details. Then certain special cases will be handled. Special-purpose and general-purpose instruments will be illustrated. A section will be dedicated to a buyers' guide for instruments and accessories by category and function.

Note: References to chapters appearing only in the *Encyclopedia* are preceded by *Enc.*

2 FUNDAMENTALS

2.1 General

Ultrasound propagates in materials.[9] The propagation is characterized by the mode of the wave,[10] by the variables velocity and attenuation,[11] by the phenomenon of refraction,[12] and by the reflectivity of a discontinuity the wavefront may meet.[13] To make a measurement, the ultrasound must be transmitted and received. The latter is the process of transduction in which a device called a transducer changes an electrical variable (voltage or current in some waveform) into a mechanical wave, and also does the reverse, outputting an electrical signal characteristic of the interaction of the mechanical radiation with the workpiece.[14] To perform an ultrasonic nondestructive measurement on a workpiece, it is necessary to assemble a proper combination of instrument, transducer, coupling (since the waves travel through materials), and possibly focusing, refraction, and mode conversion mechanisms to introduce a propagating wave of a desired mode into the workpiece of a particular geometry.

2.2 Modes

In the medium, the mechanical radiation will propagate in allowed modes.[10] In a bulk solid, the waves can be longitudinal (compressional) and transverse (shear).[2] If the bulk solid is isotropic (the same in all directions),

Fig. 1 Immersion tank with ultrasonic scanning equipment showing size and scale of commercial instrumentation. (Courtesy of Douglas Aircraft Company.)

there will be two velocities, one longitudinal and one transverse. If the bulk solid is anisotropic, having directional properties, then there are three velocities (one longitudinal and two shear) with directional variability. (Actually, they are quasi-longitudinal and quasi-shear.) The symmetries of materials can be treated as in crystallography with point groups and elastic moduli.[15] The longitudinal and shear modes (or quasi-longitudinal and quasi-transverse except along pure mode axes) can be used to characterize the workpiece in different ways.

On the surface of a solid, Rayleigh (surface) waves can propagate. Stonley waves run along interfaces.[16]

For plates, the modes are termed Lamb waves (longitudinal and flexural) and Love waves (horizontally polarized shear waves).[17] The Lamb waves have components of particle motion both normal to the plate and in the direction of propagation (in the plane of the plate), while the horizontally polarized shear waves have particle motion in the plate but normal to the direction of propagation. These plate waves are of different orders depending on the number of nodal planes there are in the plate.[18] The most useful wave mode for NDT is the zeroth-order Lamb mode, which can be visualized as a symmetric set of bulges and contractions propagating along the plate (the ostrich swallowing the orange). The useful frequency range for this mode is near the low-frequency limit where the wavelength is much longer than the plate thickness so that the mode is not dispersive. Modes with even numbers of nodal planes (including zero) are flexural.

Waves in wires are similar. There are extensional, flexural, and torsional modes.[19]

2.3 Velocity

The velocity of a wave is determined by the elastic moduli of the medium and the mode of a propagation.[10] As the mode depends on the boundary conditions, the moduli are combined differently to express the velocities of various modes.[20] Fundamentally, however, the dependence of the velocity of a mode upon the elastic moduli of the medium permits one to monitor some of the mechanical properties of the medium.[21] Processing steps that produce desired mechanical properties also change the moduli.[22] Velocity can be used to monitor the process,[21] perform quality assurance,[6] carry out statistical process control,[23] and sort with 100% testing (NDT).[24] If the velocity is known, then travel time measurements yield a thickness gauge with one-sided access.

2.4 Attenuation

The attenuation of a wave is determined by scattering[25] and absorption,[3] which are properties of the medium, and by beam spreading from a finite source.[26] As the scattering and absorption are frequently determined by processing steps, attenuation can be used in quality and process control in some cases.[27] In practice, the measurement of attenuation poses some problems, making its use somewhat limited.

2.5 Reflectivity

All wave modes are partially reflected by changes in the acoustic impedance along their path.[28] Except in some special applications, the reflectivity of air versus a solid can be considered 100%. Reflectivity makes ultrasound ideal for the detection and measurement of cracks and voids in solids.

2.6 Transduction

Transduction from electrical to mechanical energy and from mechanical to electrical energy[14] is performed by specialized, ruggedized devices known as transducers.[29] The most common form is based on an encapsulated longitudinal mode piezoelectric plate.[30] Other types are magnetostrictive, inductive (electromagnetic acoustic transducers), or capacitative. Special front members are used for focusing and refraction with emphasis on mode conversion to get shear waves into solids by means of longitudinal transducers.

2.7 Focusing, Refraction, and Mode Conversion

All three processes depend on refraction. Ultrasound is refracted by Snell's law just as in optics except that a solid medium can support two refracted waves from a single incident wave.[31] These are the longitudinal and shear waves with particle motion in the sagittal plane. A typical use would be to introduce both waves into a solid from a longitudinal wave in a liquid. Often the incident angle is specified so that the refracted angle of the longitudinal wave is beyond the critical angle in the solid, leaving only the vertically polarized shear wave in the solid. Refraction, of course, can be used with lenses for focusing. In going from a liquid into a solid, a wave is bent away from the surface normal. A plano-concave plastic lens accomplishes focusing from a solid transducer into water.[32]

3 ULTRASOUND, ELECTRONICS, AND READ-OUT

3.1 Generic Instrument

What might be termed a "generic ultrasonic instrument" is shown in Fig. 2. Under command of a synchronization generator (at, say, 1000 repetitions per second), the pulser sends electrical energy to the transducer through a pulse limiter circuit, which clips the voltage seen by the amplifier down to a threshold but lets the full voltage hit the transducer. The transducer transmits the ultrasonic wave and receives the echoes. The resulting voltage, not clipped by the pulse limiter because the echo voltage is below the clipping threshold, goes to the amplifier and then to the display. The computer may or may not be present and may simply represent some analog function.

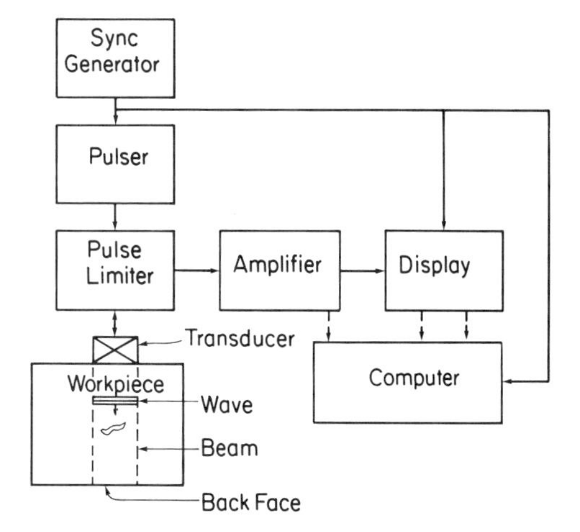

Fig. 2 Generic ultrasonic instrument in block diagram form. A synchronization generator establishes a repetition rate for a pulse input and a display module (with or without computer). The transducer sends ultrasound pulses into the workplace and receives echoes. The electronics allows the echoes to be displayed and manipulated.

3.2 The Generic Display: A-Scan

For the flaw detection instrument with no "intelligence," the display is an oscilloscope with limited adjustability. With well-damped transducers, the resulting display of amplitude versus time (and hence amplitude versus distance in the workpiece) is shown in Fig. 3*a*. With less damped or with electrically tuned transducers, the echoes contain more radio frequency (RF) cycles; usually the result is rectified and detected before display, as in Fig. 3*b*. These sketches could be termed the "generic ultrasonic display." From a technical point of view, they are termed an A-scan, which is simply amplitude versus time for the echoes within the beam of the transducer.

In the A-scan, one sees multiple echoes between the front and back faces of the workpiece plus the echoes from flaws. Beyond the first back face reflection, the echo pattern may become very garbled but may, on the other hand, yield especially interesting data in some cases.[7]

3.3 Analog Modifications to A-Scans for Amplitude

Commercial flaw detection instruments provide various analog processes to help analyze the data. The first, mentioned already, is to rectify and detect the echoes. A second is a baseline suppressor, which eliminates a few decibels of amplitude just above zero amplitude, thus making the more significant flaw echoes easier to discern. A third is a range gate, which picks out a time range of interest from t_1 to t_2 along the A-scan baseline. The signals within the range gate may be either (a) suppressed or (b) analyzed to the exclusion of other signals. The analysis may be by internal computer, external computer, internal alarm analog circuit, or external analog instrument. Note the C-scan discussion below. An alarm can be activated if echoes within the range gate exceed a certain amplitude or threshold.

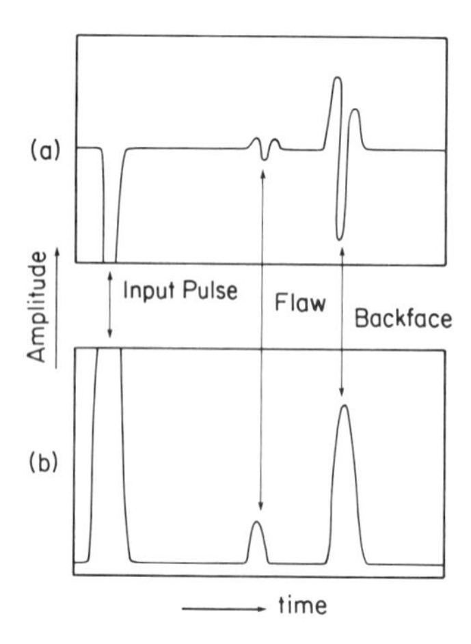

Fig. 3 Generic ultrasonic display for the instrument in Fig. 2. The amplitude is shown, either as RF in (*a*) or as rectified and detected in (*b*), versus time, which is equivalently distance into the workpiece. This display is known as an A-scan.

3.4 Analog Modifications to A-Scans for Time Measurement

The time from the electrical pulsing of the transducer to the arrival of the first back face echo represents either the thickness of the workpiece if the velocity is known or the velocity in the workpiece if the thickness is known. The ultrasonic thickness gauges, which do not require access to the back side of a part, work on the principle of knowing (or at least calibrating for) the velocity. The synchronizing timer starts the charging of a capacitor through a resistor in an *RC* network. The voltage on the capacitor increases as $V = V_0(1 - e^{-t/RC})$, which is close to linear for times such that $V \ll V_0$. The *RC* time constant is built such that this condition holds over the range of time to be measured in one round trip travel time of the wave in the thickest specimen for the given range of calibration, for example, 0.05–1.00 cm, then 1.00–20.00 cm, and so forth. The arrival of the first back-wall reflection triggers the cessation of the charging of the capacitor. The voltage then represents the travel time and hence can be read out on a light-emitting diode (LED) or liquid-crystal display (LCD) as thickness, given the velocity as a multiplier, as noted earlier. If the thickness is known, then the velocity can be measured by varying the velocity-setting dials until the read and known thicknesses match. Obviously, a flaw could vitiate both types of time-dependent measurements. Most thickness gauges are small and portable with no video displays except numerals.

3.5 Mechanical Scanning

General Since a picture is worth a thousand words, developments have been made to provide ultrasonic pictures of the interiors of workpieces. Some sort of mechanism to pick up signals from a number of areas to represent a slice through a part is required. Then a two-dimensional electronic or electromechanical display can provide a picture of the slice.

The most straightforward and simplistic methodology is to scan the transducer over the part or to move the part in front of the transducer with mechanical drives in some sort of scanning pattern. See Fig. 4 where the transducer is shown able to move in the x- and y-directions while propagation is in the z-direction (travel time t is along

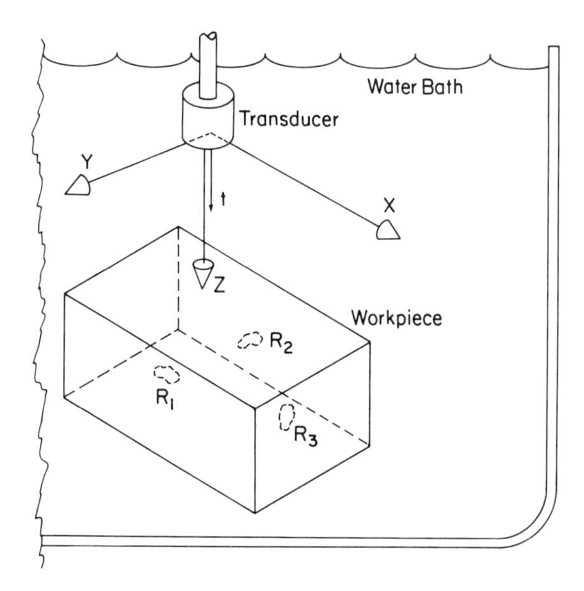

Fig. 4 Schematic diagram of a scanning mechanism in an immersion tank. The transducer can be translated and angulated to aim at a part appropriately. Some display and recording means is always also attached.

the z-direction). There is to be a one-to-one relationship between a point on a display and a region in the workpiece from which an echo comes. How is this achieved? There are two standard scans:

B-Scans A B-scan will portray an x–z slice of the workpiece at a set position Y_i, then make another x–z slice at Y_{i+1}, and so forth. The transducer is moved along x at a rate permitting the ultrasonic pulse to interrogate the workpiece many times over the x-dimension of the workpiece. The pulse repetition rate must be low enough to permit the previous pulse to be attenuated below the noise level before the next pulse is transmitted. The coordinates of the display are the position x_R where the transducer was at the time of arrival t_R of a reflection from a reflector R (R_1, R_2, or R_3 in the figure, depending on whether the beam in slice y passes over them). Then t_R is the other coordinate of the point (x_R, t_R). If the display means has a gray scale, then the brightness of the display corresponds to the amplitude of the reflection. For an on–off scale, a threshold is usually set so that only echoes above a certain amplitude are displayed. The B-scan is used extensively in medical ultrasonics, particularly obstetrics.[33]

C-Scans In the C-scan configuration, the transducer is scanned along x for consecutive steps y_i. However, the coordinate on the display is (x, y) and the display is nonzero only if a reflector R is within the beam along z centered at (x, y). In this configuration, the echoes from the front and back faces would swamp the display, so they are eliminated by setting a range gate to accept and transmit echoes to the display if and only if they arise from volumes below the top face and above the bottom face of the workpiece. Obviously, the range gate can be set narrower such that only signals from a slice thinner in the z-dimension are allowed. Also, a threshold can be set within the gate so that only signals above a certain amplitude will be displayed.[34] Alternately, a gray scale can be added.

For the B-scan as well as the C-scan, a computer can be appended for data storage and image enhancement. Pseudocolor gray scales on color monitors are useful.

3.6 Two Transducers and Other Items

Worth mentioning for completeness are methods where the receiver is a separate transducer. Two configurations are common: pitch and catch and through transmission.

Pitch-and-Catch Configuration Here the two transducers are on the same face of the workpiece, often close together.

Through-Transmission Configuration Here the two transducers are on opposite faces of the workpiece. Often they are in a water tank with the workpiece between them.

Squirters When a tank would be inconvenient, a squirter can be used to convey the ultrasound into and out of the workpiece. A squirter is a laminar flow water hose nozzle with a transducer mounted coaxially inside. Its flow can be aimed at the workpiece to allow normal incidence or refraction at angles.

4 EXAMPLES OF INSTRUMENTS

4.1 General

Following the strategy in Section 1, which delineated flaw detection and materials properties measurement as the two big areas of ultrasonic NDT, both types of instruments will be illustrated. Because flaw detection is the heart of NDT, flaw detection instruments will be treated first.

4.2 Flaw Detection

Flaw detection by ultrasound could be traced back to Langevin's work on piezoelectric crystals for sonar dur-

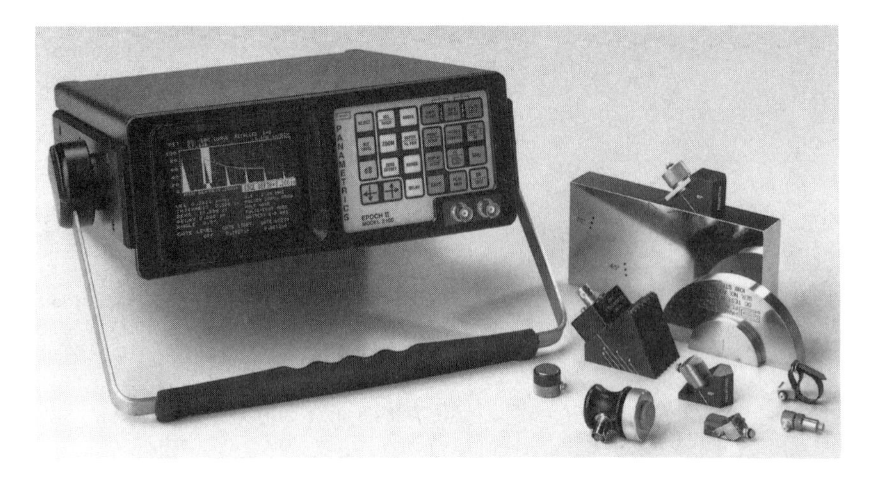

Fig. 5 Modern flaw detection instrument with CRT display. Along side are transducers and calibration blocks. (Courtesy of Panametrics, Inc.)

ing World War I. However, more modern pulse-echo radar when combined with piezoelectric crystals led to pulse–echo ultrasonics in the early 1940s. The first model supersonic reflectoscope by Firestone[35] was as large as a refrigerator. (One will note that the date is early enough to be prior to the division of nomenclature leaving "supersonic" to aerodynamics and "ultrasonic" to mechanical vibrations). A few of these were made and applied in war production. Interestingly enough, just such a flaw detection instrument with a shear wave transducer was instrumental in the first demonstration by Firestone and Fredericks of ultrasonic double refraction in worked metals,[36] a *mechanical property* having to do with differences in shear wave velocity as a function of polarization direction.

Figure 5 shows a modern flaw detection instrument with a cathode ray tube (CRT) as its display. Its pulser is a charged capacitor switched onto the transducer by a charge-conducting diode (CCD); the charge then drains away through a damping resistor. The frequency is broadband and is determined by the highly damped mechanical resonance[37] of the piezoelectric element in the transducer.[30]

The direction of future instruments may be defined by the instrument shown in Fig. 6. In this subcompact piece, the bulk and the weight have been drastically reduced by eliminating the CRT with its high-voltage transformer and substituting a digital LCD display. In the terminology of Fig. 2, all functions beyond the amplifier have been digitized and computerized. All the functions mentioned in Sections 3.3 and 3.4, in addition to rectification mentioned in Section 3.2, have been put into software to

Fig. 6 Subcompact flaw detection instrument with maximum possible digital processing and control and flat LCD display. (Courtesy of DuPont NDT Systems.)

miniaturize the system down to 3 lb (less than 1.5 kg). Controls are also computer operated. The analog section is relegated to the function of acquiring data and "talking" to the computer through an analog-to-digital converter, which takes a digital "snapshot" of what would be the display in Fig. 3*a*.

Deviating from the tendency to miniaturization is the type of installation that is dedicated to testing huge parts such as aircraft panels and structures. The flaw detection instruments are state-of-the-art, of course. One example of a system utilizing water jets (squirters) on large contoured honeycomb panels is shown in Fig. 7. Scanning motion in this photograph is across the width of the gantry. The jet holders may be rotated 90 degrees to allow scanning along the length of the gantry. Both upper and lower jet holders move in X, Y, Z, and two angular axes to maintain the jets normal to the part surface. The motion of the lower head is controlled by data downloaded from a computer-assisted design (CAD) system. The upper head motion is the mirror image of the lower head, while maintaining a constant spacing between the ends of the jets. Parts within a $20 \times 20 \times 6$ ft envelope ($6 \times 6 \times 2$ m) may be scanned with this system.[38]

Much more advanced signal processing methodologies can be called upon to interpret the echoes in terms of the size and shape of the reflector or scatterer. Among these methods are the synthetic aperture focusing technique[39] (SAFT), familiar from radio astronomy, and computed tomography[40] (CT), familiar from x-ray radiography.

4.3 Materials Properties

As soon as ultrasonic waves could be propagated and received in beams by instruments, it was realized that physical properties associated with elasticity could be measured by ultrasound. Leading these experimentalists were T. A. Litovitz,[3] W. P. Mason,[2] and R. Truell.[41] They tackled both ultrasonic attenuation and ultrasonic velocity, breaking new ground that led their disciples to develop measurement methods and instrumental embodiments to determine mechanical properties in practical applications. Their pioneering work led to other engineering devices and to elegant science, as well.

One instrument[42] for measuring ultrasonic velocity to high absolute accuracy[43] is shown in Fig. 8. It employs the pulse–echo overlap method for ultrasonic travel time measurements.[44,45] Under ideal conditions and using calibration of its time base against frequency standards, this instrument can make measurements to an absolute accuracy[43] (not just precision) of ± 5 ppm. For quality assurance of certain hi-tech materials, such as recoined

Fig. 7 Large steered scanning ultrasonic test system using squirters to inspect large contoured parts. (Courtesy Boeing Commercial Airplane Division.)

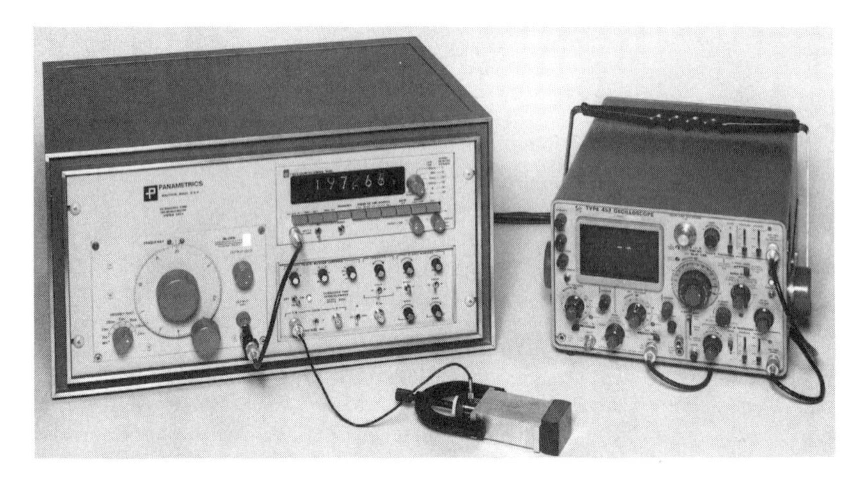

Fig. 8 Commercial pulse–echo overlap instrument for highly accurate ultrasonic velocity measurements. This instrument is used in quality control, scientific experimentation, and development of NDT procedures dependent on ultrasonic velocity. (Courtesy of Panametrics, Inc.)

Fig. 9 Ultrasonic system for measuring ultrasonic velocity in modular iron in a foundry for quality assurance. Parts are tested immersed in the tank. (From Ref. 21. Courtesy of Academic Press.)

sintered powder metal parts, high-density ceramics, and vitreous silica, this degree of accuracy is necessary. The laboratory prototype of this instrument was used in quality assurance efforts on vitreous silica for dispersive delay lines[46] in which a few degrees of phase error across the broad ultrasonic beam would cause degradation of performance.[47] The instrument in Fig. 8 is sold worldwide.

Instruments[21] like the ones shown in Figs. 9 and 10 are in actual factory operations throughout the world. This automatic system measures ultrasonic velocity in cast iron to assure quality of nodular cast iron.[21] Molten iron is treated with an additive to cause its carbon to nucleate and grow into microscopic spheres (nodules) when nodular cast iron is the desired product. Nodular iron is a substitute for forged steel in strength and is lower in cost, being cast near-net-shape. Having the 14% by volume of carbon in graphite spheres instead of in graphite flakes or other spread-out forms of graphite maximizes the connectedness of the iron and minimizes stress risers. Thus, the nodular iron has both the maximum strength and the maximum modulus of the possible resultant irons.[48] Maximum modulus implies maximum velocity, so ultrasonic velocity is a quantitative NDT test for the strength of nodular cast iron. Having this quality assurance tool available is necessary because failure modes and effects analysis[49] (FMEA) indicates that improper treatment in the molten state causes deviations from the nodular condition. Improprieties can be caused by "fading" over time before solidification in a batch

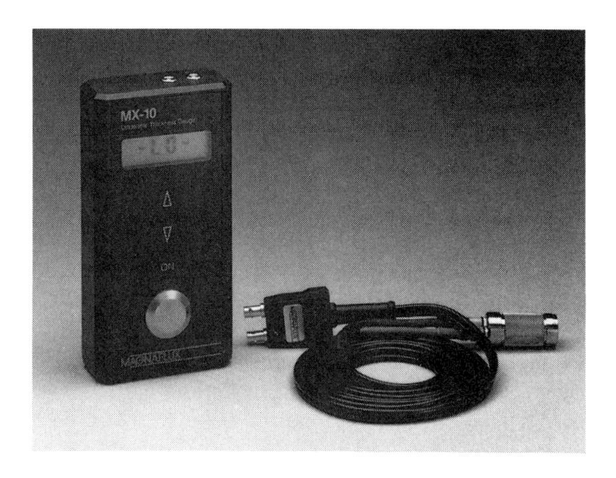

Fig. 11 An ultrasonic thickness gauge about the size of a five-pack of cigars. (Courtesy of Magnaflux, Inc.)

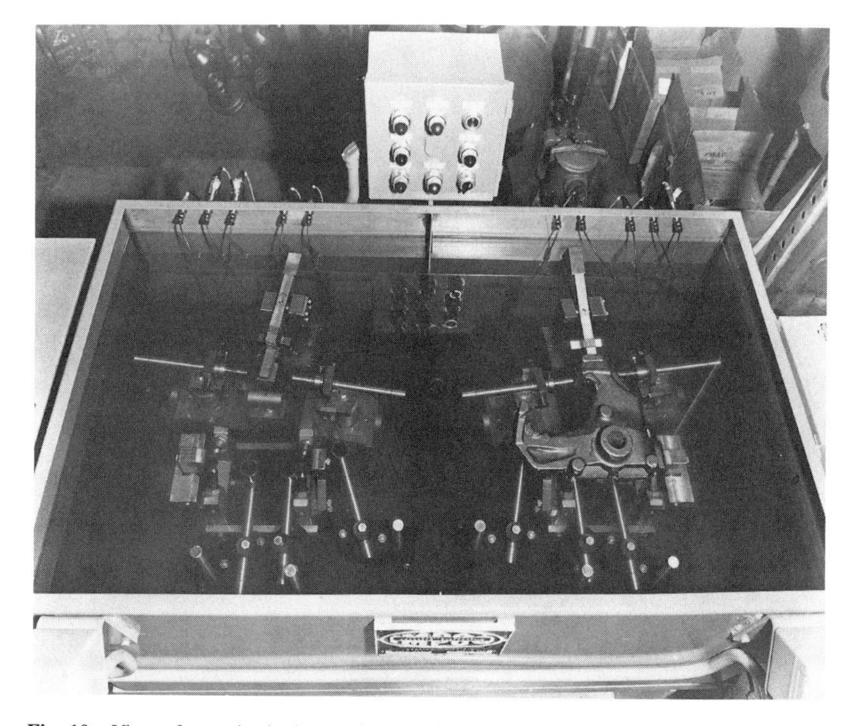

Fig. 10 View of parts in the immersion tank in Fig. 9. One can see the transducers aiming through the part. (From Ref. 21. Courtesy of Academic Press.)

treatment process or else too fast or too slow dissolving of the treatment agent (or too little of it) in an in-mold treatment process. Thus, ultrasound is paying its way in safe car and train parts, to name a few items.

The last item, a thickness gauge, is shown in Fig. 11. This palm-of-the-hand model is one among many that has evolved for practical use where thickness is to be measured from access to one side of the material only. Thickness gauges are placed here under materials properties since the ultrasonic velocity in the workpiece must be known so that the measured travel time can be translated into thickness.

5 BUYER'S GUIDE

The coordinating society for NDT in the United States is the American Society for Nondestructive Testing (ASNT). On a worldwide basis it is one of the top few and probably is the leader. Annually, the ASNT publishes a "Buyer's Guide" in one issue of *Materials Evaluation* (its journal) with a practical operational emphasis. As companies are founded, acquired, divested, and terminated at a rate too great to enter them into a handbook of enduring worth, this section is devoted to listing the categories under which buyers can find ultrasonic testing items in the ASNT "Buyer's Guide".[50] This list may change in future years with editorial policy and technological growth; one can get an idea of the subject from this listing:

Airborne ultrasound	Squirters
Blocks, test	Systems, computerized
Bond testers	Systems, custom
Bubblers	Systems, EMAT
Calculators, flaw	Systems, handling
Couplant, dry	Systems, imaging
Couplant, wet	Systems, piping tubing
Detectors, flaw	Systems, velocity/integrity
Extensometers (bolt gauges)	Tank, immersion
Flaw discriminators	Testers, concrete
Gages, thickness	Testers, hardness
Instruments, multichannel	Testers, wood
Manipulators	Transducer analyzers
Processors, signal	Transducers
Pulsers/receivers	Other ultrasonic products
Scanners, lab	

REFERENCES

1. F. A. Firestone, "The Supersonic Reflectoscope for Interior Inspection," *Metals Progress*, Vol. 505, September, 1945.
2. W. P. Mason, *Physical Acoustics and the Properties of Solids*, Van Nostrand, Princeton, 1958.
3. K. F. Herzfeld and T. A. Litovitz, *Absorption and Dispersion of Ultrasonic Waves*, Academic, New York, 1959.
4. D. Setti (Ed.), "*Dispersion and Absorption of Sound by Molecular Processes*," Course XXVII, in *Proceedings of the International School of Physics "Enrico Fermi*," Academic, New York, 1963.
5. R. T. Beyer and S. V. Letcher, *Physical Ultrasonics*, Academic, New York, 1969.
6. E. P. Papadakis, "Challenges and Opportunities for Nondestructive Inspection Technology in the High-Volume Durable Goods Industry," *Mat. Evaluation*, Vol. 39, No. 2, 1980, pp. 122–130.
7. T. M. Mansour, "Ultrasonic Inspection of Spot Welds in Thin-Gage Steel," *Mat. Evaluation*, Vol. 46, No. 5, 1988, pp. 640–658.
8. D. J. Hagemaier, "Ultrasonic Maintenance Testing of Aircraft Structures," in *ASNT Handbook*, Vol. 7, American Society for Nondestructive Testing, Columbus, OH, 1991.
9. R. B. Lindsay, *Mechanical Radiation*, McGraw-Hill, New York, 1960.
10. B. A. Auld, *Acoustic Fields and Waves in Solids*, Vols. I and II, Wiley-Interscience, New York, 1973.
11. E. P. Papadakis, "The Measurement of Ultrasonic Velocity" and "The Measurement of Ultrasonic Attenuation" in R. N. Thurston and A. D. Pierce (Eds.), *Physical Acoustics*, Vol. XIX, *Ultrasonic Measurement Methods*, Academic, Boston, 1990, pp. 81–106 and 107–155.
12. D. L. Arenburg, *J. Acoust. Soc. Am.*, Vol. 20, No. 1, 1948, pp. 1–26.
13. P. M. Morse and K. U. Ingard, *Theoretical Acoustics*, McGraw-Hill, New York, 1968, pp. 400–441.
14. O. E. Mattiat, *Ultrasonic Transducer Materials*, Plenum, New York, 1971.
15. G. Simmons and H. Wang, *Single Crystal Elastic Constants and Calculated Aggregate Properties: A Handbook*, MIT Press, Cambridge, MA, 1971.
16. L. M. Brekhovskikh, *Waves in Layered Media*, 2nd ed., Academic, New York, 1980, pp. 14, 38–41, 48–50, 53, 95.
17. L. M. Brekhovskikh, *Waves in Layered Media*, 2nd ed., Academic Press, New York, 1980, pp. 60–70. See also I. A. Vicktorov, *Rayleigh and Lamb Waves*, Plenum, New York, 1967.
18. T. R. Meeker and A. H. Meitzler, "Guided Wave Propagation in Elongated Cylinders and Plates," in W. P. Mason (Ed.), *Physical Acoustics: Principles and Methods*, Vol. I A, Academic, New York, 1964, pp. 111–167.
19. T. R. Meeker and A. H. Meitzler, "Guided Wave Propagation in Elongated Cylinders and Plates," in W. P. Mason (Ed.), *Physical Acoustics: Principles and Methods*, Academic, New York, 1964, especially pp. 134–140.
20. R. C. McMaster (Ed.), *Nondestructive Testing Handbook*, Vol. II, American Society for Nondestructive Testing; Roland, New York, 1959, pp. 43.8–43.11.

21. E. P. Papadakis, "Ultrasonic Velocity and Attenuation: Measurement Methods with Scientific and Industrial Applications," in W. P. Mason and R. N. Thurston (Eds.), *Physical Acoustics: Principles and Methods*, Vol. XII, Academic, New York, 1976, pp. 277–374.
22. E. P. Papadakis, "Physical Acoustics and the Microstructure of Iron Alloys," *Inter. Met. Rev.*, Vol. 29, No. 1, 1984, pp. 1–24.
23. E. P. Papadakis, "A Computer-Automated Statistical Process Control Method with Timely Response," *Eng. Costs Product Econ.*, Vol. 18, 1990, pp. 301–310.
24. E. P. Papadakis, "Sampling Plans and 100% Nondestructive Inspection Compared," *Quality Progress*, Vol. 15, No. 4, 1982, pp. 37–39.
25. E. P. Papadakis, "Ultrasonic Attenuation Caused by Scattering in Polycrystalline Media," in W. P. Mason (Ed.), *Physical Acoustics: Principles and Methods*, Vol. IVB, Academic, New York, 1968.
26. H. Seki, A. Granato, and R. Truell, "Diffraction Effects in the Ultrasonic Field of a Piston Source and Their Importance in the Accurate Measurement of Attenuation," *J. Acoust. Soc. Am.*, Vol. 28, No. 2, 1956, pp. 230–238.
27. E. P. Papadakis, "Ultrasonic Attenuation and Velocity in Three Transformation Products in Steel," *J. Appl. Phys.*, Vol. 35, 1964, pp. 1474–1482.
28. W. P. Mason, *Physical Acoustics and the Properties of Solids*, Van Nostrand, Princeton, 1958, p. 27.
29. M. G. Silk, *Ultrasonic Transducers for Nondestructive Testing*, Adam Hilger, Bristol, England, 1984.
30. E. P. Papadakis, "Theoretical and Experimental Methods to Evaluate Ultrasonic Transducers for Inspection and Diagnostic Applications," *IEEE Trans.*, Vol. SU-26, 1979, pp. 14–27.
31. W. P. Mason, *Physical Acoustics and the Properties of Solids*, Van Nostrand, Princeton, 1958, pp. 23–32.
32. E. P. Papadakis, "Lens Equation for Focused Transducers," *Int. J. Nondestruct. Test.*, Vol. 4, 1972, pp. 195–198.
33. R. R. Price and A. C. Fleischer, "Sonographic Instrumentation," in A. C. Fleischer et al., *Ultrasonography in Obstetrics and Gynecology*, 4th ed., Appleton & Lange, Norwalk, CT, 1991, pp. 25–38.
34. T. M. Mansour, "Evaluation of Ultrasonic Transducers by Cross-Sectional Mapping of the Near Field Using a Point Reflector," *Mat. Evaluation*, Vol. 37, No. 7, 1977, pp. 50–54.
35. R. Straw, "Do You Hear What I Hear?—The Early Days of Pulse-Echo Ultrasonics," *Mat. Evaluation*, Vol. 42, No. 1, 1984, pp. 24–28.
36. F. A. Firestone, and J. R. Fredericks, "Refinements in Supersonic Reflectoscopy. Polarized Sound," *J. Acoust. Soc. Am.*, Vol. 18, No. 1, 1945, pp. 200–211.
37. E. P. Papadakis, "Broadband Flaw Detection Transducers: Application to Acoustic Emission Pulse Shape and Spectrum Recording Based on Pulse Echo Response Spectrum Corrected for Beam Spreading," *Acustica*, Vol. 46, 1980, pp. 293–298.
38. W. Woodmansee and R. D. Whealy, Boeing Commercial Airplane Division, Seattle, Washington, private communication.
39. T. Sato and O. Ikeda, "Sequential Synthetic Aperture Sonar System: A Prototype of a Synthetic Aperture Sonar System," *IEEE Trans.*, Vol. SU-24, 1977, pp. 253–259.
40. R. K. Muller, M. Kaveh, and G. Wade, "Reconstructive Tomography and Applications to Ultrasound," *IEEE Proc.*, Vol. 67, 1979, pp. 567–587.
41. R. Truell, C. Elbaum, and B. B. Chick, *Ultrasonic Methods in Solid State Physics*, Academic Press, New York, 1969.
42. E. P. Papadakis, "New, Compact Instrument for Pulse-Echo-Overlap Measurements of Ultrasonic Wave Transit Times," *Rev. Sci. Instrum.*, Vol. 47, 1976, pp. 806–813.
43. E. P. Papadakis, "Absolute Accuracy of the Pulse-Echo-Overlap Method and the Pulse-Superposition Method for Ultrasonic Velocity," *J. Acoust. Soc. Am.*, Vol. 52 (Part 2), 1972, pp. 843–846.
44. J. E. May, Jr., "Precise Measurement of Time Delay," *IRE Ntl. Convection Rec.*, Vol. 6 (Part 2), 1958, pp. 134–142.
45. E. P. Papadakis, "Ultrasonic Phase Velocity by the Pulse-Echo-Overlap Method Incorporating Diffraction Phase Corrections," *J. Acoust. Soc. Am.*, Vol. 42, 1967, pp. 1045–1051.
46. E. P. Papadakis, "Variability of Ultrasonic Shear Wave Velocity in Vitreous Silica for Delay Lines," *IEEE Trans.*, Vol. SU-16, 1969, pp. 210–218.
47. G. A. Coquin and R. Tsu, "Theory and Performance of Perpendicular Diffraction Delay Lines," *Proc. IEEE*, Vol. 53, No. 6, 1965, pp. 581–591.
48. E. Plenard, "The Elastic Behavior of Cast Iron," National Metal Congress, Cleveland, 1964.
49. Engineering and Research Staff, *Potential Failure Mode and Effects Analysis: An Instruction Manual*, Ford Motor Company, Dearborn, MI, September, 1979.
50. "Eighth Annual NDT Buyers' Guide," *Mat. Evaluation*, Vol. 48, No. 6, 1990, pp. 725–784.

46

THERMOACOUSTIC ENGINES

GREGORY W. SWIFT

1 INTRODUCTION

Thermoacoustic engines harness the thermal effects present in sound waves to produce acoustic power from heat or to produce refrigeration from acoustic power. In most acoustic phenomena, the temperature oscillation accompanying the pressure oscillation in a sound wave is unimportant, except as it contributes to the attenuation of sound in bulk and at boundaries. However, under some circumstances the temperature oscillation and in particular its interaction with boundaries—thermoacoustic processes—can be harnessed to produce powerful, useful thermodynamic effects. Both basic types of heat engine can be realized thermoacoustically: In a thermoacoustic prime mover, heat flow from high temperature to lower temperature produces work in the form of acoustic power; in a thermoacoustic heat pump, acoustic power is used to pump heat from low temperature to higher temperature. Either a predominantly standing wave or a predominantly traveling wave can be used, although geometries for the two types of waves are substantially different. Qualitative understanding of thermoacoustic engines is easiest when considering the position, pressure, volume, and temperature oscillations of typical parcels of gas. Quantitative analysis is based on the acoustic approximation.

2 EXAMPLES

The earliest thermoacoustic engine was the Sondhauss tube, shown in Fig. 1*a*. Over 100 years ago, glassblowers found that when a small, hot glass bulb was being blown on a cool glass tubular stem, the stem tip sometimes radiated sound. Rayleigh[1] understood that the sound was generated by oscillatory thermal expansion and contraction of the air in the tube, which in turn was due to acoustic motion of the air toward and away from the hot end of the tube. The Sondhauss tube is a thermoacoustic prime mover, converting heat into mechanical work in the form of sound. It utilizes a predominantly standing acoustic wave, with the bulb and stem forming a resonator. A key geometric requirement for this and other standing-wave thermoacoustic engines is that, in the region of substantial temperature gradient, the smallest transverse dimension must be comparable to the thermal penetration depth in the air, which is roughly the distance that heat can diffuse in a time $1/\pi f$, where f is the frequency of the wave.

Hofler's thermoacoustic refrigerator[2] is another example of a standing-wave thermoacoustic engine and illustrates the separate functions of resonator, power transducer, and heat exchanger components. As shown in Fig. 1*b*, it consisted of a helium-filled resonator driven by a loudspeaker and containing a "stack" with one end thermally anchored at room temperature by a hot heat exchanger. With a static pressure of 1 MPa and driven at an amplitude of 30 kPa, its cold heat exchanger reached 200 K; at somewhat higher temperatures, it produced 3 W of cooling power with a coefficient of performance of 0.12 of the Carnot coefficient of performance.

The resonator, consisting of tubing segments and a spherical bulb, produced a resonance frequency of 500 Hz compactly and with minimal dissipation of acoustic power. The heart of this thermoacoustic refrigerator was the stack, a ribbon of plastic sheet spirally wound to make a 3.8-cm-diameter, 8-cm-long assembly. The spacing of 0.4 mm between turns of the spiral was maintained by plastic spacers; this spacing is several thermal penetration depths at the 500-Hz operating frequency and

Note: References to chapters appearing only in the *Encyclopedia* are preceded by *Enc.*

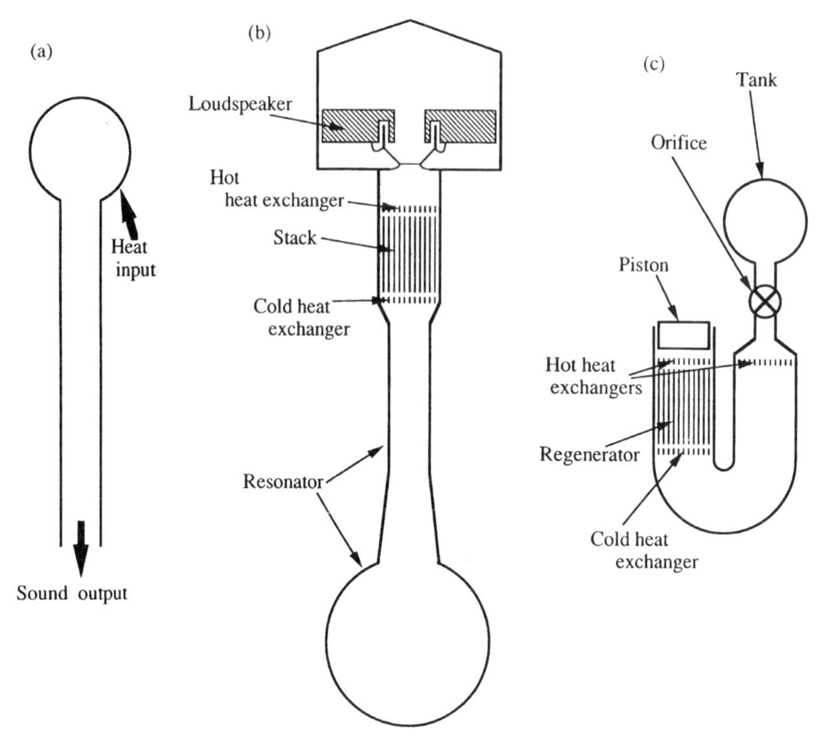

Fig. 1 Examples of thermoacoustic engines: (*a*) the Sondhauss tube, a standing-wave thermoacoustic prime mover; (*b*) Hofler's refrigerator, a standing-wave thermoacoustic heat pump; (*c*) the orifice pulse-tube refrigerator, a traveling-wave thermoacoustic heat pump.

enabled heat to be pumped thermoacoustically from the cold heat exchanger to the hot heat exchanger. These heat exchangers consisted of well-spaced copper strips extending across the resonator at either end of the stack to ensure thermal contact between the gas at the stack ends and the resonator walls nearby.

Ceperley[3] has discussed thermoacoustic engines using traveling waves. One example is the orifice pulse-tube refrigerator, reviewed by Radebaugh[4] and shown in Fig. 1*c*. A piston functions as a source of acoustic power at one end of the refrigerator, and an orifice and tank act as a sink of acoustic power at the other end. Between them, a regenerator, heat exchangers, and an open tube are filled with gas. The regenerator and adjacent heat exchangers function just like those of a Stirling cycle refrigerator and are the heart of this refrigerator; with the hot heat exchangers anchored at room temperature, the cold heat exchanger reaches typically 40 K, and refrigeration is produced with a coefficient of performance of typically one-fourth of Carnot's. The key geometric requirement is that the smallest transverse dimension in the regenerator be much smaller than the thermal penetration depth in the gas at the operating frequency; this requirement is characteristic of traveling-wave thermoacoustic engines, in contrast to standing-wave thermoacoustic engines. The open tube functions as a temperature buffer space between the cold heat exchanger and the warm orifice.

Many attempts to produce practical and commercially viable thermoacoustic devices are underway worldwide. Standing-wave refrigerators having cooling powers of tens or hundreds of watts at 0°C are under development for satellite and residential use. The technology is inexpensive, reliable, and environmentally benign; commercial success will require improvements in efficiency. Standing-wave prime movers driving orifice pulse tube refrigerators comprise the only cryogenic refrigeration technology having no moving parts; this combination is under development for cooling cryogenic electronics and sensors and for cryogen liquefaction. Again, low cost and reliability are the technology's chief virtues; developers are striving to increase efficiency.

3 LAGRANGIAN EXPLANATION OF PHENOMENA

Time phasing is an important factor in the operation of traditional heat engines: Pistons, valves, and sparks must have the correct relative timing in order to produce the desired thermodynamic cycle in the working gas. Thermoacoustic engines contain no obvious moving parts to perform these functions, yet the acoustic stimulation of heat flux and the generation (or absorption) of acoustic power point to some type of timed phasing of thermodynamic processes, here achieved in a remarkably simple way.

The key to phasing in thermoacoustic engines is the presence of two thermodynamic media: gas and solid. As the gas oscillates along the solid at the acoustic frequency, it experiences oscillations in temperature. Part of the temperature oscillations comes from compression and expansion of the gas by sound pressure, and the rest is a consequence of the local temperature of the solid itself.

To understand thermoacoustic phenomena qualitatively, it is best to adopt a Lagrangian point of view, following a typical parcel of gas as it moves. Real thermoacoustic engines operate sinusoidally, but it is simplest here to consider square-wave motion and pressure in the gas, with pressure and displacement in phase for standing waves and pressure and velocity in phase for traveling waves. Viscosity is ignored in this section, even though it has a significant, harmful effect on thermoacoustic efficiency. With these conceptual simplifications, the thermodynamic cycle for standing waves becomes identical to the Brayton cycle, with two reversible adiabats and two irreversible constant-pressure heat transfers, while the cycle for traveling waves becomes two isotherms and two reversible polytropic processes.

3.1 Standing-Wave Heat Pump

The standing-wave thermoacoustic heat pump can be understood with reference to Fig. 2, where the four steps of its cycle are displayed for one typical parcel of gas. The gas oscillates along the adjacent solid in steps 1 and 3 in response to the standing wave and is also adiabatically pressurized and depressurized in steps 1 and 3, respectively, absorbing and doing work dW''. Steps 2 and 4 allow thermal equilibration between gas and adjacent solid.

There is a critical longitudinal temperature gradient $\nabla T_{\text{crit}} = p/\xi\rho c_p$ at which no heat is transferred in steps 2 and 4. (Here, p is the acoustic pressure amplitude, ξ is the acoustic displacement amplitude, ρ is density, and c_p is isobaric heat capacity per unit mass.) This occurs when the temperature rise due to adiabatic compression in step 1, $2p/\rho c_p$, exactly matches the rise in adjacent wall temperature that the gas experiences because of its motion along the wall $2\xi\nabla T_{\text{crit}}$. For all temperature gradients less than ∇T_{crit}, heat transfer between gas and solid is as shown in Fig. 2, with heat flow dQ from gas to solid in step 2, at the higher temperature and position, and heat flow dQ' from solid to gas in step 4, at the lower temperature and position. Thus, the net effect of this parcel of gas is to move a little heat a short distance up the solid, from slightly lower temperature to slightly higher temperature, during each cycle of the sound wave. All the other parcels do the same thing; so the overall effect,

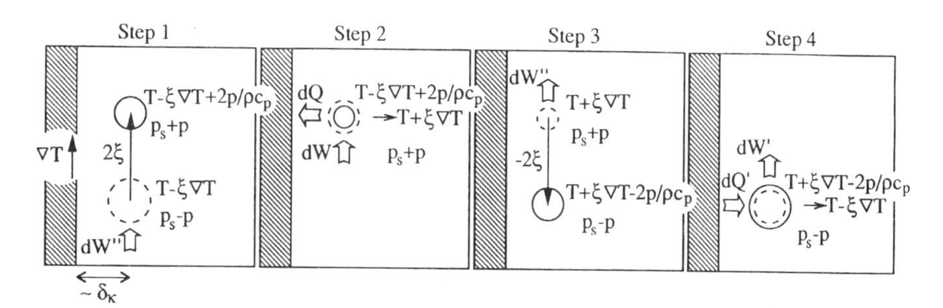

Fig. 2 Typical gas parcel executing the four steps of the cycle of a standing-wave thermoacoustic heat pump, assuming an inviscid gas and square-wave acoustic motion and pressure. In each step, the dashed and solid circles show respectively the initial and final positions and volumes of the parcel, while labels show the initial and final temperatures and pressures of that step, and broad arrows show directions of attendant heat flows (either from the parcel to the solid or from the solid to the parcel) and work flows (either into or out of the parcel). Averaged over one cycle, the parcel absorbs work $dW - dW'$ from the wave.

much as in a bucket brigade, is the net transport of heat from the cold end of the stack to its hot end.

Imperfect thermal contact between gas and solid is required for the cycle to operate as described. If thermal contact between gas and solid were perfect, the gas would arrive at its uppermost position at the end of step 1 already equilibrated at the local solid temperature, so that no heat would be transferred during step 2; and similarly for steps 3 and 4. At the other extreme, if thermal contact were nonexistent, no heat could be transferred in steps 2 and 4. Thus, intermediate, *imperfect* thermal contact between gas and solid is required for the cycle to operate as described. For the sinusoidal motion and pressure characteristic of real engines, this imperfect thermal contact is achieved for gas neither much closer nor much farther than a thermal penetration depth from the nearest solid. Thus, effective standing-wave engines fill the entire cross section of the resonator with a stack having spacings of a few thermal penetration depths, as shown in Fig. 1*b*.

3.2 Standing-Wave Prime Mover

Similarly, the standing-wave thermoacoustic prime mover is illustrated in Fig. 3. Here, the temperature gradient is greater than ∇T_{crit}, so the directions of heat transfer in steps 2 and 4 are reversed relative to the standing-wave heat pump. Thus, the net thermal effect of the parcel of gas shown is to move a little heat down the solid, from higher to lower temperature. Again, all parcels have a similar effect, so that overall a substantial amount of heat is carried down along the solid from the hot heat exchanger to the cold heat exchanger.

Since heat flows into the gas in step 2 and out of it in step 4, the gas expands at high pressure in step 2 and contracts at low pressure in step 4. Thus, the net effect in the thermoacoustic prime mover is the production of work at the acoustic frequency by this and all other gas parcels. One can imagine all the gas in the stack region of a thermoacoustic prime mover thermally expanding and contracting because of its displacement along the temperature gradient, with the correct phasing relative to the oscillatory standing-wave pressure to do net work. In the Sondhauss tube, for example, this net work maintains the standing wave against viscous and thermal losses in the resonator and provides power to radiate into the room.

3.3 Traveling-Wave Heat Pump

Traveling-wave thermoacoustic engines differ from standing-wave engines in two fundamental ways: phasing between pressure and motion and spacing between solid elements. These differences are illustrated for the traveling-wave heat pump in Fig. 4. Isothermal compression in step 2 transfers heat from the gas parcel to the solid at a higher temperature, and isothermal expansion at step 4 transfers heat from solid to gas at a lower temperature. The other two steps serve to preheat and precool the gas, reversibly storing and recovering heat in the solid. The net effect of one parcel over an entire cycle is, again, to move a little heat a short distance up the solid, up the temperature gradient. Excellent thermal contact between gas and solid is required in steps 1 and 3 for effective preheating and precooling, respectively; in realistic, sinusoidal traveling-wave engines, this is achieved by maintaining transverse dimensions much smaller than a thermal penetration depth.

Traveling-wave prime movers are also possible in principle,[3] although none has been built.

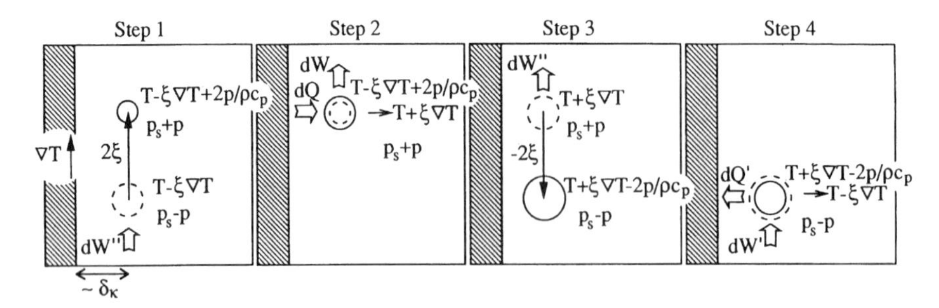

Fig. 3 Typical gas parcel executing the four steps of the cycle of a standing-wave thermoacoustic prime mover, assuming an inviscid gas and square-wave acoustic motion and pressure. Symbols have the same meaning as in Fig. 2. Averaged over one cycle, the parcel delivers work $dW - dW'$ to the wave.

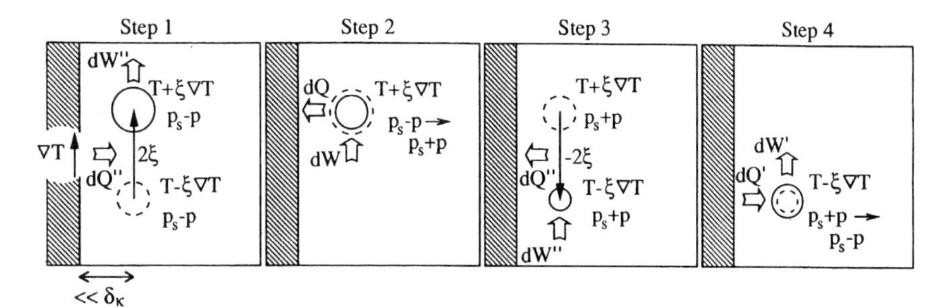

Fig. 4 Typical gas parcel executing the four steps of the cycle of a traveling-wave thermoacoustic heat pump, assuming an inviscid gas and square-wave acoustic motion and pressure. Symbols have the same meaning as in Fig. 2.

4 METHODS OF COMPUTATION

4.1 Accurate Method

Applicable to both standing- and traveling-wave thermoacoustic engines, the wave equation for the complex pressure amplitude p in a gas in a small channel with a temperature gradient is[5]

$$\frac{\omega^2}{c^2}\,[1+(\gamma-1)f_\kappa]p+\rho\,\frac{d}{dx}\left[\frac{(1-f_\nu)}{\rho}\,\frac{dp}{dx}\right]-\frac{1}{T}\,\frac{dT}{dx}\,\frac{f_\kappa-f_\nu}{1-\sigma}\,\frac{dp}{dx}=0, \tag{1}$$

where f_ν and f_κ are defined below. The acoustic power flowing through the channel is

$$W=\frac{A}{2\omega\rho}\,\mathrm{Im}\left[p\,\frac{d\tilde{p}}{dx}\,(1-\tilde{f}_\nu)\right] \tag{2}$$

and the enthalpy flow (total energy flow) through the channel is

$$H=\frac{A}{2\omega\rho}\,\mathrm{Im}\left[p\,\frac{d\tilde{p}}{dx}\left(1-\tilde{f}_\nu-\frac{f_\kappa-\tilde{f}_\nu}{1+\sigma}\right)\right]+\frac{Ac_p}{2\omega^3\rho(1-\sigma)}\,\frac{dT}{dx}\,\frac{dp}{dx}\,\frac{d\tilde{p}}{dx}\,\mathrm{Im}\left(\tilde{f}_\nu+\frac{f_\kappa-\tilde{f}_\nu}{1+\sigma}\right)-(AK+A_sK_s)\,\frac{dT}{dx}\,. \tag{3}$$

Here

$$f_\nu=\frac{\tanh[(1+i)y_0/\delta_\nu]}{(1+i)y_0/\delta_\nu},\qquad f_\kappa=\frac{\tanh[(1+i)y_0/\delta_\kappa]}{(1+i)y_0/\delta_\kappa} \tag{4}$$

for parallel-plate channels of separation $2y_0$; $\delta_\nu=\sqrt{2\nu/\omega}=\sqrt{2\mu/\omega\rho}$ and $\delta_\kappa=\sqrt{2\kappa/\omega}=\sqrt{2K/\omega\rho c_p}$ are the viscous and thermal penetration depths, where ν is kinematic viscosity, μ is dynamic viscosity, κ is thermal diffusivity, and $\omega=2\pi f$ is the radian frequency of the wave; σ is the Prandtl number, γ is the ratio of isobaric to isochoric specific heats, and T is the temperature; A and A_s and K and K_s are the areas and thermal conductivities of gas and solid, respectively; and $i=\sqrt{-1}$, Im signifies the imaginary part of what follows, and the tilde denotes complex conjugation. Results shown here are for ideal gases only; results for liquids can be found in Ref. 6. Expressions for f_ν and f_κ for circular channels, rectangular and triangular channels, and pin arrays can be found in Refs, 5, 7, and 8, respectively.

When numerically integrating these equations to predict the performance of thermoacoustic engines, recognize that Eqs. (1) and (3) comprise a set of three second-order coupled real differential equations in the variables $T(x)$, $\mathrm{Re}[p(x)]$, and $\mathrm{Im}[p(x)]$. With the addition of the definition

$$p_x=\frac{dp}{dx}, \tag{5}$$

Eqs. (1), (3), and (5) become a set of five first-order coupled real differential equations in x in the five variables T, Re[p], Im[p], Re[p_x], and Im[p_x]. The enthalpy flow H is a given constant, independent of x within the stack or regenerator because a steady state is assumed and no energy flows in or out in the transverse directions (except at the heat exchangers, where the value of H is established). The gas and solid thermophysical properties depend implicitly on x through their dependence on T.

A solution may be obtained numerically using a Runge–Kutta method starting at one end of the stack or regenerator and integrating to the other end. Required to start the integration is a choice of five attendant initial conditions: the temperature, complex pressure amplitude, and complex impedance at the starting position.

Such a Runge–Kutta thermoacoustic code has been imbedded as a subroutine in a more complete program[9] including resonator shape and loss, power transducer impedance, heat exchanger geometry, and the like. The larger program can iteratively call the thermoacoustic subroutine, so that the user can supply more convenient boundary conditions such as hot and cold temperatures, and the program can then find variables such as resonance frequency, acoustic power, and enthalpy flow. The results are generally in good agreement with experimental data.[10]

4.2 Approximate Method, Standing Waves

The energy flows in parallel-plate standing-wave thermoacoustic engines are given approximately[6] by

$$H \simeq \frac{A\delta_\kappa}{4y_0} \frac{p\langle u\rangle}{(1+\sigma)(1-\delta_\nu/y_0+\delta_\nu^2/2y_0^2)} \cdot \left(\Gamma \frac{1+\sqrt{\sigma}+\sigma}{1+\sqrt{\sigma}} - 1 - \sqrt{\sigma} + \frac{\delta_\nu}{y_0}\right) + (AK + A_s K_s)\frac{\Delta T}{\Delta x}, \tag{6}$$

$$W \simeq \frac{A\delta_\kappa\,\Delta x(\gamma-1)\omega p^2}{4y_0\rho c^2} \cdot \left[\frac{\Gamma}{(1+\sqrt{\sigma})(1-\delta_\nu/y_0+\delta_\nu^2/2y_0^2)} - 1\right] - \frac{A\delta_\nu\,\Delta x}{4y_0} \frac{\omega\rho\langle u\rangle^2}{(1-\delta_\nu/y_0+\delta_\nu^2/2y_0^2)}, \tag{7}$$

where

$$\Gamma = \frac{\rho c_p \langle u\rangle}{\omega p} \frac{\Delta T}{\Delta x}, \tag{8}$$

Δx is the length of the engine along the propagation direction, ΔT is the temperature difference spanned, and $\langle u\rangle$ is the transverse spatial average (perpendicular to the gas motion) of the velocity amplitude u. Here, the sign conventions are that ΔT, p, and $\langle u\rangle$ are all real and positive and H (the enthalpy flow) and W (the net acoustic power) are positive for a prime mover. It is assumed that the engine is short enough that thermophysical properties p and u are essentially independent of x and can be approximated by their values in the middle of the engine and further that the channels are large enough (typically $y_0 \geq 2\delta_\kappa$) that the hyperbolic tangents in Eqs. (4) are essentially unity. Equations (6) and (7) are generally accurate to within a factor of 2. They are easily computed with a programmable calculator or simple computer program and are useful in finding overall trends and approximate magnitudes.

For $\Gamma = 0$ and $\delta_\nu/y_0 \ll 1$, Eq. (7) reduces to the expression for thermoviscous wall losses in standing waves (cf. *Enc.* Ch. 22).

4.3 Approximate Method, Traveling Waves

The order of magnitude of the gross refrigeration power in well-designed traveling-wave thermoacoustic heat pumps is[4]

$$H = \frac{Ap\langle u\rangle T_C}{30T_H}, \tag{9}$$

where T_C and T_H are the cold and hot heat exchanger temperatures. Considering only regenerator effects, the efficiency is approximately half of Carnot's efficiency.

5 AUXILIARY COMPONENTS

Two heat exchangers are required in thermoacoustic engines to supply and extract heat at the two ends of the stack or regenerator. The length of each along the direction of the acoustic oscillation should be of the order of the local displacement amplitude ξ; transverse dimensions should be of the order of those of the adjacent stack or regenerator. Equation (7) with $\Gamma = 0$ can be used to estimate the acoustic power dissipation in the heat exchangers.

The heat exchangers for small engines can simply consist of multiple copper strips spanning the engine at the stack ends. Larger diameter engines may require integral heat pipes, pumped-fluid channels, or electric resistance heaters imbedded in the heat exchangers.

A resonator is required for standing-wave thermoacoustic engines. It should minimize unnecessary dissipation of acoustic power into heat. Unusual geometries[2] such as that of Fig. 1*b* can reduce resonator dissipation by almost a factor of 2 relative to uniform-diameter right-circular-cylinder resonators. Resonator dissipation can also be calculated using Eq. (7) with $\Gamma = 0$.

6 OVERALL PERFORMANCE

Calculations using Eqs. (1) and (3) generally show that standing-wave thermoacoustic engines have second-law efficiencies (i.e., efficiencies normalized by the Carnot efficiency) between 0.2 and 0.4, considering effects in the stack only. Including resonator, heat exchanger, and transducer losses reduces this efficiency to between 0.15 and 0.3 for large-cross-section engines, for which resonator losses are not a significant fraction of the total power; small-diameter engines have even lower efficiencies. Traveling-wave thermoacoustic engines can have higher efficiencies.

The power per unit of cross-sectional area in standing-wave thermoacoustic engines can be estimated crudely by $H/A \sim (p/p_s)^2 p_s c/10$. Thus, high amplitude, high static pressure, and high sound speed increase power density. Heat flows of the order of 1 MW/m^2 are easily achievable in high-pressure helium gas. The pressure amplitude is often 5–10% of static pressure; the gas displacement amplitude can exceed the length of heat exchangers and approach the length of the stack. At such high amplitudes, performance deviates significantly[10] from the predictions of Eqs. (1) and (3).

REFERENCES

1. Lord Rayleigh (J. W. Strutt), *The Theory of Sound*, 2nd ed., Vol. 2, Dover, New York, 1945, Sec. 322.
2. T. Hofler, "Thermoacoustic Refrigerator Design and Performance," Ph.D. Dissertation, University of California at San Diego, 1986; summarized in T. Hofler, "Concepts for Thermoacoustic Refrigeration and a Practical Device," in *Proc. 5th Int. Cryocoolers Conf.*, August 1988, Monterey CA, p. 93.
3. P. H. Ceperley, "Gain and Efficiency of a Short Traveling Wave Heat Engine," *J. Acoust. Soc. Am.*, Vol. 77, 1985, p. 1239.
4. R. Radebaugh, "A Review of Pulse Tube Refrigeration," in *Adv. Cryogenic Eng.*, Vol. 35, Plenum, New York, 1990, p. 1191.
5. N. Rott, "Damped and Thermally Driven Acoustic Oscillations in Wide and Narrow Tubes," *Z. Angew. Math. Phys.*, Vol. 20, 1969, p. 230; "Thermally Driven Acoustic Oscillations, Part III: Second-Order Heat Flux," *Z. Angew. Math. Phys.*, Vol. 26, 1975, p. 43.
6. G. W. Swift, "Thermoacoustic Engines," *J. Acoust. Soc. Am.*, Vol. 84, 1988, p. 1145.
7. W. P. Arnott, H. E. Bass, and R. Raspet, "General Formulation of Thermoacoustics for Stacks Having Arbitrarily Shaped Pore Cross Sections," *J. Acoust. Soc. Am.*, Vol. 90, 1991, p. 3228.
8. G. W. Swift and R. M. Keolian, "Thermoacoustics in Pin-Array Stacks," *J. Acoust. Soc. Am.*, Vol. 94, 1993, p. 941.
9. W. C. Ward and G. W. Swift, "Design Environment for Low Amplitude Thermoacoustic Engines," *J. Acoust. Soc. Am.*, Vol. 95, 1994, p. 3671. Fully tested software and users guide available from Energy Science and Technology Software Center, US Department of Energy, Oak Ridge TN. For a beta-test version, contact ww@lanl.gov (Bill Ward) via Internet.
10. J. R. Olson and G. W. Swift, "Similitude in Thermoacoustics," *J. Acoust. Soc. Am.*, Vol. 95, 1994, p. 1405.

PART VI

MECHANICAL VIBRATIONS AND SHOCK

47

INTRODUCTION

MANFRED A. HECKL

1 INTRODUCTION

Within the scope of acoustics mechanical vibrations are defined as motions of rigid or elastic solid bodies provided the spectrum of the motion contains frequencies within the acoustic range. Other names for this field are structure-borne sound or vibroacoustics. Typical topics of mechanical vibrations in acoustics are as follows:

- Structural vibrations caused by unbalanced forces, misalignments, rolling over rough surfaces, internal combustion, parametric excitation, stick-slip, and so on. Such phenomena occur in all kinds of machines and vehicles. They are of great importance for noise control problems (see also Chapter 55).
- Vibrations of space vehicles (acoustic fatigue), aircraft surfaces, windows, walls, and so on, caused by strong airborne sound sources or turbulent flows (see Chapter 10, see also *Enc.* Ch. 70).
- Sound transmission through walls, ceilings, enclosures.
- Shock and vibration isolation (see Chapter 52).
- Ground vibrations near factories, forge hammers, underground railways, and so on.
- Measurement of elastic moduli and loss factors of materials (mechanical spectroscopy) (see Chapter 56).
- High-frequency elastic waves in delay lines or in (ultrasonic) nondestructive testing equipment (see Chapters 44 and 54)
- Machine monitoring and machine diagnosis (see Chapters 53 and 57).

Note: References to chapters appearing only in the *Encyclopedia* are preceded by *Enc.*

Shocks constitute a subclass of mechanical vibrations. They are associated with rather short and sudden signals. Usually they are of high amplitude; therefore, nonlinear effects are of concern for shock absorbers. Consideration of shocks is important in packaging, in the field of machine reliability, and in military applications (see Chapter 52).

2 BASIC RELATIONS

2.1 Newton's Law and Hooke's Law

When the vibrations are small enough so that the principle of superposition (linearity) can be employed and when losses are negligible, all equations dealing with mechanical vibrations can be seen as a consequence of Newton's law and Hooke's law. For lumped-parameter systems consisting of masses m and massless springs of stiffness s, these laws are

$$m\frac{\partial^2 \xi_i}{\partial t^2} = m\frac{\partial v_i}{\partial t} = F_i \quad \text{(Newton)},$$

$$F_i = s_i u_i \quad \text{(Hooke)}. \qquad (1)$$

These two equations describe in most simple terms the inertia forces and the elastic forces necessary for the existence of elastic waves. The terms ξ_j and v_i are the components of displacement and velocity of the mass; F_i are the components of the force acting on m; and u_i is the length change of the spring due to the force F_i.

The equations contain two approximations. One of them consists in the replacement of a total derivative by a partial derivative; that is, convective terms and very high amplitudes are excluded. The second is that the spring behaves linearly.

In Chapter 48 Newton's and Hooke's laws are applied extensively to derive the equations of motion of single- and multi-degree-of-freedom mechanical systems. If the two laws are applied to continua, the mass has to be replaced by a mass per unit volume (i.e., density ρ), the forces become stresses σ_{ij}, and the relative length change has to be replaced by the strains. This way one finds (see, e.g., Refs. 1 and 2)

$$\rho \frac{\partial^2 \xi_i}{\partial t^2} = \frac{\partial \sigma_{ij}}{\partial x_j} + F_i'',$$

$$\sigma_{ij} = C_{ijkl} \frac{1}{2}\left(\frac{\partial \xi_k}{\partial x_l} + \frac{\partial \xi_l}{\partial x_k}\right). \quad (2)$$

Here F_i'' is the force (if any) per unit volume that acts from outside and C_{ijkl} is the elasticity tensor characteristic of the material. In these expressions the summation convention has been used; that is, one has to sum over identical indices.

For the most important case, a homogeneous, isotropic material with shear modulus G and Poisson number μ the stresses in Eq. (2) can be written as

$$\sigma_{ij} = G\left[\frac{2\mu}{1-2\mu}\left(\frac{\partial \xi_k}{\partial x_k}\right)\delta_{ij} + \left(\frac{\partial \xi_i}{\partial x_j} + \frac{\partial \xi_j}{\partial x_i}\right)\right]. \quad (3)$$

Instead of G and μ one might write Eq. (3) also in terms of G and Young's modulus E, or (as in many books on theoretical physics) in terms of the so-called Lamé constants λ and μ_L. These quantities are related to each other by

$$E = 2(1+\mu)G, \qquad \lambda = \frac{E\mu}{(1+\mu)(1-2\mu)}, \qquad \mu_L = G. \quad (4)$$

For short bars of cross section S and length l Young's modulus E and μ are defined in Fig. 1. As can be seen, μ is related to lateral contraction or extension when a bar is elongated or compressed.

The main difference between waves in gases or fluids and in elastic solids is that, in the absence of loss mechanisms, there are only compressional (longitudinal) waves in gases or fluids, whereas there can be compressional and shear waves in solids. Other wave types such as bending waves, torsional waves, and Rayleigh (surface) waves can be understood as superpositions of compressional and shear waves.

2.2 Loss Mechanisms

In the absence of loss mechanisms or damping the vibratory energy in finite systems would remain constant if after some sort of excitation it is left free. As a consequence, a system would never "forget" its past history; for example, a simple resonator that is driven since $t = 0$ with a frequency close to its resonance frequency would exhibit the beating frequency forever. Since such a behavior does not agree with reality, the basic equations have to be modified to account for the effect of losses.

The usual way to include linear damping in the basic equations is to modify the force–displacement or

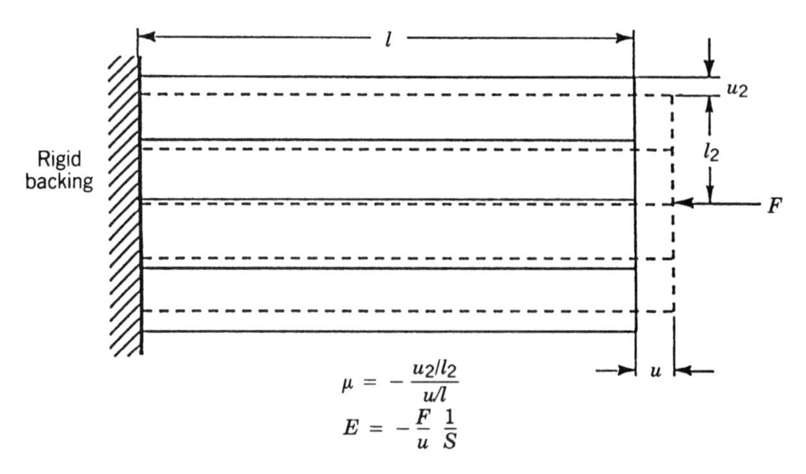

Fig. 1 Definition of Poisson's ratio and Young's modulus for static loads: (----) undeflected state; (——) compressed state.

stress–strain relation in Eqs. (1) and (2). In the simplest case a velocity-proportional, viscous force is added so that the second part of Eqs. (1) becomes

$$F = su + b\,\frac{\partial u}{\partial t} = su + bv. \tag{5}$$

Here b is the damping constant and v the relative velocity of the two sides of the spring. For elastic continua a viscous term has to be added to Eq. (2). Normally this is not done in acoustics. The concept of complex moduli [see Fig. (8)] is employed instead.

Equation (5) is suitable to represent dampers that are in parallel to springs, but they do not correctly describe the behavior of materials with inner losses. Other and, if the parameters are chosen properly, better relations for lossy, elastic structural elements are[3]

$$F + \frac{\partial F}{\partial t} = su + \beta\,\frac{\partial u}{\partial t} \quad \text{or}$$

$$F(t) = s_0 u(t) - \int_0^{\infty} u(t-\tau)\varphi(\tau)\,d\tau. \tag{6}$$

In these expressions α and β are material constants and $\varphi(\tau)$ is the so-called relaxation function. Usually $\varphi(\tau)$ consists of a sum of terms of the type $(D/\tau_R)\exp(-\tau/\tau_R)$, where D is a constant and τ_R the relaxation time, which may be anywhere between nanoseconds and days. The second version of Eq. (6) goes back to Boltzmann. As well as the first expression it represents a material that has memory; that is, the instantaneous value of $F(t)$ depends not only on the instantaneous value of the displacement $u(t)$ but also on the previous history $u(t-\tau)$.

Other loss mechanisms are dry (Coulomb) friction in the interface between two structural elements and radiation damping, which is caused by waves that propagate into the surrounding medium (e.g., air, water, sand, ground, etc.). See also Sections 3.5 and 4.3 in Chapter 10.

Equations (6) lead to lengthy expressions when they are used to derive equations of motion, but they still give linear expressions because no products or powers of the field quantities appear.

Mechanical vibrations are very often expressed as spectra; that is, the time dependence is written as a sum of terms of the type

$$\xi(t) = \mathrm{Re}\{\underline{A}_n c^{j\omega_0 nt}\}. \tag{7}$$

Here $\underline{A}_n$ is the complex amplitude (spectral value) and ω_0 the fundamental frequency. With this type of representation it is very convenient to introduce the concept of complex moduli, that is,

$$\underline{s} = s(1+j\eta), \qquad \underline{E} = E(1+j\eta), \qquad \underline{G} = G(1+j\eta), \tag{8}$$

where the underbars indicate that the quantity is complex ($j = \sqrt{-1}$). The term η is the loss factor. For many applications it can be assumed to be independent of frequency.

The great advantage of the loss factor is that loss mechanisms can be taken into account simply by replacing the real-valued elastic moduli by complex ones. It should be kept in mind, however, that complex moduli are an artefact, which works only for harmonic motions or sums of harmonic motions. The transformation of frequency spectra to the time domain that may be desirable in shock applications is problematic. The loss factor can also be expressed in terms of energies, as shown in Section 1 in Chapter 48. For more on the loss factor and its relation to other quantities, see Chapter 57 and Table 3 in Chapter 56.

2.3 Range of Validity

The relations given in Sections 2.1 and 2.2 can be used to find the equations of motion of mechanical structures, and if the appropriate boundary and initial conditions are used, they can be applied to practical situations. But even though many solutions of interest can be found this way, there remain also open questions. Usually they have to do with nonlinear effects or with time-varying parameters (e.g., stiffness). The causes for nonlinearities are (see also Chapter 50) as follows:

- Material nonlinearities; that is, the quantities s, E, G, b, α, ... in Eqs. (5) and (6) depend on amplitude, as may happen for very large vibrations or for strong shocks.
- Frictional forces that increase very strongly with amplitude; this effect is very often purposely applied in shock absorbers.
- Geometric nonlinearities, which occur when there is no longer a linear relation between the length change of a spring and the displacements of the adjacent bodies. Hertzian contact,[4] where the contact area changes with the displacement, and large-amplitude vibrations of thin plates,[5] where the length variations of the neutral plane can no longer be neglected, are examples of this type.
- Constraints at boundaries that cannot be expressed by linear equations; an example of this type is dry Coulomb friction, or motion between free clearances.

- Convective forces that are neglected in the inertia terms of Eqs. (1) and (2).

The consequence of slight nonlinearities is the existence of higher harmonics that are not contained in the excitation. When the nonlinearity is very pronounced, chaotic behavior may occur (see also Chapter 50). Examples for parameter changes (see also Section 3 in Chapter 50) are as follows:

- Length change of a pendulum
- Stiffness change due to rapid variations of the point of contact (e.g., teeth in a gearwheel, which can be considered as short cantilever beams that are loaded at different points)
- Impedance variations "seen" by a body that moves with constant speed over periodic supports (e.g., wheel on rail)

3 ENERGY RELATIONS AND VARIATIONAL PRINCIPLES

3.1 Hamilton's Principle and Rayleigh's Principle

In the previous section the motion of mechanical systems was expressed in terms of forces and displacements or similar dynamical and kinetic quantities. An alternative approach is to describe such systems by their energies and their displacements. The most powerful method in this type of calculation is Hamilton's principle,

$$\delta \int_{t_1}^{t_2} (W_{\text{kin}} - W_{\text{pot}})\, dt = 0, \tag{9}$$

which states that "for an actual motion of the system, under the influence of the conservative forces, when it is started according to any reasonable initial conditions, the system will move so that the time average of the difference between kinetic and potential energies will be a minimum (or in a few cases, a maximum)" (Ref. 6, p. 281). In Eq. (9) W_{kin} and W_{pot} are the total kinetic and potential energies of a system. The symbol δ indicates that the variation has to be taken. For a few cases the potential and kinetic energies for surface or line elements are given in Table 1 in Chapter 49. If loss mechanisms have to be included, one can again employ the concept of the loss factor. As an alternative one can extend Eq. (9) by a dissipation function (see Ref. 7, Sections 81 and 82), which is a quadratic expression involving the relative velocities. Another extension of Eq. (9) can be made by introducing the work done by outside sources; this way all types of linear, forced vibrations can be treated.

Hamilton's principle is an extremely useful relation in vibroacoustics. It can be used as a starting point for the finite element method (FEM).[8,9] By employing the calculus of variations, it is also possible to derive equations of motion from Eq. (9),[10] to calculate the vibrations of coupled systems, and to find the phase speeds of different wave types.[11] (See also Refs. 12 and 13.)

A special solution of Eq. (9) is $\overline{W}_{\text{kin}} = \overline{W}_{\text{pot}}$, where the overbar denotes the time average. This relation is known as Rayleigh's principle[13] (see also Ref. 7, Section 88). It states that at resonance the time averages of kinetic and potential energies are equal. Since usually

$$\overline{W}_{\text{kin}} \sim \left(\frac{\partial \xi_i}{\partial t} \right)^2 \sim \omega^2 \overline{\xi_i^2}, \tag{10}$$

Rayleigh's principle is a very convenient way of finding resonances. Its main advantage, however, is that a first-order error in the assumed spatial distribution of the displacements ξ_i (which are needed to find W_{kin} and W_{pot}) leads only to a second-order error in the resonance frequencies, provided that the boundary conditions are fulfilled. It also is known that Rayleigh's principle gives always an upper bound for the resonance frequency. Finding approximations for the lowest resonance frequency is just one application of Rayleigh's principle. It also can be used to find higher resonances and mode shapes (Rayleigh–Ritz method).[12,13]

As an example one may make the (wrong) assumption that the displacement of a simply supported beam in bending motion is given by $\xi_2 = 1 - 4x_1^2/l^2$. If one then uses the formulas in the appendix to Chapter 49 for a Bernoulli–Euler beam, Rayleigh's principle yields an approximate resonance frequency that is only a few percent higher than the correct value.

For the application of Rayleigh's principle to find the first stopping and passing bands of a periodic system, the reader is referred to Ref. 14.

3.2 Lagrange's Equation

Applying the calculus of variations to Hamilton's principle yields Lagrange's equation of the second type:

$$\frac{d}{dt}\frac{\partial L}{\partial \dot{\xi}_n} - \frac{\partial L}{\partial \xi_n} = 0, \qquad L = W_{\text{kin}} - W_{\text{pot}}. \tag{11}$$

Here ξ_n is the nth displacement coordinate and $\dot{\xi}_n$ the nth velocity coordinate. If forces F_n are driving the system from outside, the zero in the above equation has to be

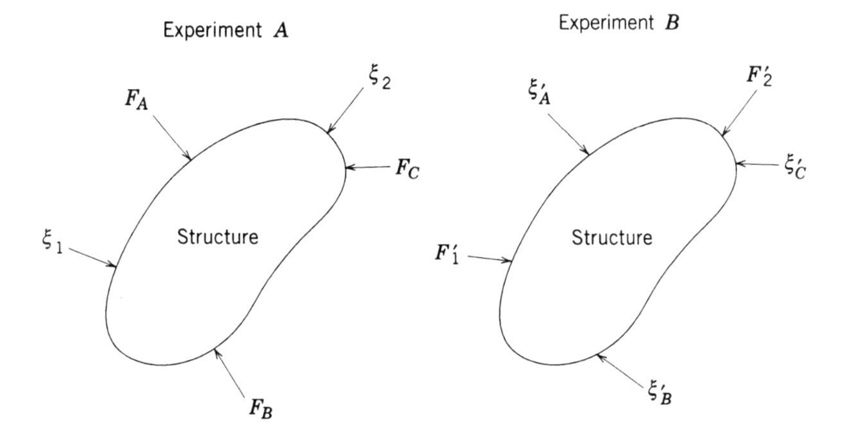

Fig. 2 Law of conjugate energies.

replaced by F_n. The main application of Eq. (11) is in the field of vibrations of multi-degree-of-freedom systems (see Chapter 63) where it can be used to calculate resonance frequencies, the shapes of eigenvectors, impulse responses, frequency responses, and so on.

3.3 Reciprocity

If the motion of the different parts of a mechanical system consists of small deflections about the equilibrium points, kinetic and potential energies are quadratic forms of these deflections. As a consequence, Hamilton's principle leads to symmetric differential equations and Eq. (11) is a set of symmetric, linear differential equations.* For such so-called self-adjoint expressions it is known that the principle of reciprocity holds (see, e.g., Ref. 15). In its simplest form this principle relates a force F_A acting at position x_A and generating a displacement ξ_1 at a point x_1 to a force F'_1 acting at x_1 and generating a displacement ξ'_A at x_A. It states

$$F_A \xi'_A = F'_1 \xi_1 \quad \text{or} \quad \frac{F_A}{\xi_1} = \frac{F'_1}{\xi'_A}. \tag{12}$$

Instead of ξ_1, ξ'_A one might also have used the velocities or accelerations. Care has to be taken that the directions of F_A and ξ_1 as well as F'_1 and ξ'_A are equal. The principle of reciprocity holds also for other pairs of field quantities, provided their product gives a power or energy, for example, angular velocity and moment (torque). In Section 6.3 in Chapter 10, the principle of reciprocity is used to relate the sound radiation from vibrating surfaces to the excitation of the same structure by a sound field.

An extension of the principle of reciprocity is Heaviside's law of conjugate energies[16] that can already be found in Ref. 7, Section 72. It is shown in Fig. 2. There F_A, F_B, ... are the forces acting at positions x_A, x_B, ... in the first experiment; ξ_1, ξ_2, ... are the displacements generated by them at the positions $x_1, x_2, \ldots; F'_1, F'_2, \ldots$ are the forces acting at x_1, x_2, ... in the second experiment; and ξ'_A, ξ'_B, ... are the displacements generated by them at x_A, x_B, ... The above equations hold for the complex amplitudes when the $\exp(j\omega t)$ notation is used.

Reciprocity is a very useful property. It can be used to calibrate transducers[17] and to identify unknown forces by making some auxiliary measurements.[18–20] For example, when F_A is unknown and unaccessible, one can measure ξ_1 and then make a measurement with an auxiliary force F'_1. When conjugate energies are used, more auxiliary measurements are necessary to get a set of linear equations for the unknown quantities.

4 GENERAL RESULTS

4.1 Frequency Spectra

The behavior of vibrating mechanical structures is quite often characterized by frequency spectra that determine

*In the literature one can also find nonsymmetric linear differential operators. They are the consequence of inconsistent approximations that are made in the derivation. Usually they give results that are very close to the correct ones, but the apparent violation of the principle of reciprocity is a certain drawback.

the response for purely harmonic excitation of the type $\exp(j\omega t)$. Usually such spectra are given as the ratio of two (complex) amplitudes.

They are either measured by slowly sweeping through the frequency range of interest or are obtained as a result of a two-channel fast Fourier transform (FFT) measurement (transfer function). Typical examples are as follows:

$$\begin{array}{ll} \text{Input mobility (admittance)} & \underline{A}_F = \mathbf{y}(x_0)/\mathbf{F}(\mathbf{x}_0) \\ \text{Input impedance} & \underline{Z}_F = \mathbf{F}(x_0)/\mathbf{v}(x_0) \\ \text{Apparent mass} & \mathbf{F}(x_0)/\mathbf{a}(x_0) \\ \text{Dynamic stiffness} & \mathbf{F}(x_0)/\boldsymbol{\xi}(x_0) \\ \text{Transfer mobility} & \underline{A}_T = \mathbf{v}(x_1)/\mathbf{F}(x_0) \\ \text{Velocity ratio} & \mathbf{v}(x_1)/\mathbf{v}(x_2) \\ \text{Moment mobility} & \underline{A}_M = \mathbf{w}(x_0)/\mathbf{M}(x_0) \\ \text{Reflection coefficient} & \mathbf{v}_r/\mathbf{v}_j \end{array} \tag{13}$$

Obviously many other combinations are possible.

Some of the above quantities are of great importance because they can be related to the energy flow, as is shown by Eq. (25). Examples are

$$\begin{aligned} P &= \frac{1}{2}\,|\mathbf{F}(x_0)|^2\,\mathrm{Re}\{\mathbf{A}_F\}, \\ P &= \frac{1}{2}\,|\mathbf{v}(x_0)|^2\,\mathrm{Re}\{\mathbf{Z}_F\}, \\ P &= \frac{1}{2}\,|\mathbf{M}(x_0)|^2\,\mathrm{Re}\{\mathbf{A}_M\}, \\ \alpha &= 1 - \left|\left|\frac{\mathbf{v}_r}{\mathbf{v}_i}\right|\right|^2. \end{aligned} \tag{14}$$

Here, $\mathbf{F}(x_0)$, $\mathbf{M}(x_0)$ are the driving force or moment at the excitation point; $\mathbf{v}$, $\mathbf{a}$, $\boldsymbol{\xi}$ represent velocity, acceleration, and displacement either at the excitation point x_0 or at some other point x_1 or x_2; $\mathbf{w}(x_0)$ is the angular velocity at the excitation point; $\mathbf{v}_i$, $\mathbf{v}_r$ are the amplitudes of the waves that are incoming to or reflected from a discontinuity; P is the mechanical power that is transmitted into a structure; and α is the absorption coefficient. To emphasise their complex nature, the quantities are underlined. Examples of, for example, spectra and mobilities can be found in the following chapters.

When the spectral behavior of a mechanical system is known over the whole range of interest, the above quantities can be used to find time histories. As an example, we consider a force $\mathbf{F}_{x0}(t)$ with period T and fundamental frequency $\omega_0 = 2\pi/T$. If this force is acting at x_0 and if the spectrum of the transfer mobility $\mathbf{A}_T$ is known, the time dependence of the velocity $v_{x1}(t)$ at point x_1 is given by

$$v_{x1}(t) = \frac{1}{2\pi} \sum_n \mathbf{F}^v_{x0}(n\omega_0) c^{jn\omega_0 t} / \mathbf{A}_T(n\omega_0) \tag{15}$$

with

$$\mathbf{F}^v_{x0}(n\omega_0) = \frac{1}{T} \int_0^T \mathbf{F}_{x0}(t) c^{-jn\omega_0 t}\, dt.$$

The extension to nonperiodic signals follows the same lines as in other areas of acoustics; see Chapters 80–82.

4.2 Green's Function: Impulse Response

The basic idea of the application of Fourier analysis is to express all quantities as a sum of purely harmonic signals, or when a space dependence is also involved as a sum of plane; monochromatic, progressive waves. Thus these simple motions are used as "building blocks" to generate more complicated solutions. The great advantage of this approach is that, because the response of a system to purely harmonic time dependence or to plane-wave excitation can quite often be found, it takes only a summation to get the general solution (for the plane-wave case see wave impedances or transmission impedances and wave number spectra in Section 2.5 in Chapter 49).

An alternative to this "Fourier approach" is the use of Green's functions. In this case the response of a system to an idealized pulse is used as a building block for more complicated problems.

A Green's function $g(x, t|x_0, t_0)$ or impulse response function is defined by

$$L\{g(x, t|x_0, t_0)\} = A\delta(x - x_0)\delta(t - t_0). \tag{16}$$

Here $L\{\cdot\}$ is the linear differential equation (or system of such equations) that describes the mechanical structure that is tested; $\delta(\cdot)$ is Dirac's delta function; and A is a constant that usually has a dimension (e.g., momentum transmitted by the pulse). Equation (16) shows that the Green's function gives the response (velocity, displacement, etc.) of a structure at point x and time t when it is excited by a pulse at point x_0 and time t_0.

Equation (16) is the general definition of Green's functions. A special case is when excitation point x_0 and observation point x are identical. In this case x and x_0 do not appear explicitly. For example, in other problems one is looking only for the response of structures when they are excited by a purely harmonic point force, and so forth. In this case all quantities are proportional to $\exp(j\omega t)$ and the time t does not appear explicitly. For examples, see Eqs. (27a)–(27d) in Chapter 49. The Green's function approach has great advantages because any excitation function [e.g., the outside pressure $p(x, t)$ or a shock] can be expressed as

$$p(x,t) = \int p(x_0,t_0)\delta(x-x_0)\delta(t-t_0)\,dx_0\,dt_0. \tag{17}$$

In physical terms this means that any pressure distribution can be written as a sum of short, localized pulses of magnitude $p(x_0, t_0)\,dx_0\,dt_0$. Because of linearity, one can conclude that the response to any excitation distribution consists of the sum of the responses (Green's functions) of the short localized pulses; that is,

$$v(x,t) = \frac{1}{A}\int p(x_0,t_0)g(x,t|x_0,t_0)\,dx_0\,dt_0. \tag{18}$$

In this equation the response is expressed as a velocity, but it might also be another field quantity provided the proper Green's function is used. The constant A, which in the literature very often is set equal to unity, is retained here, because it makes it easier to keep track of the dimensions. The operation defined by Eq. (18) is also known as convolution, or *Faltung*.

Impulse response functions are very helpful when the response to transient excitation (e.g., shocks) is of interest. Shock design to a large degree consists in finding useful approximations for impulse response functions of complicated systems. See also Chapter 52.

Since Green's functions (or sets of Green's functions for different x and x_0) are solutions of the equation of motion of a structure, they contain all the information that is relevant for the vibration of the structure. That makes them very suitable for numerical problems when time histories of a motion have to be calculated. Similarly, they are very useful for problems where a linear structure is coupled to a localized nonlinear device (see e.g., Ref. 21).

The Green's function approach and Fourier transforms are just two sides of the same coin. A strong connection between them exists because the Fourier transform of a transfer function for normalized excitation is equal to the Green's function and vice versa. This fact is often used to find Green's functions. See Eq. (20).

4.3 Normal Modes

It is known (see, e.g., Refs. 12, 13, and 22 for applications or 6 and 23 for the mathematical foundations) that the vibrations of lightly damped mechanical structures can be expressed as

$$v(x) = \sum_{n=1}^{\infty} \frac{A_n\varphi_n(x)}{\omega_n^2(1+j\eta_n)-\omega^2}. \tag{19}$$

Here $v(x)$ is the velocity (it might also be another quantity), A_n are constants that depend on the excitation and on mechanical properties, ω_n are the eigenfrequencies, η_n the (modal) loss factors, and $\varphi_n(x)$ the eigenfunctions (modes). Equation (19) holds for sinusoidal excitation. By applying a Fourier transform, it can be shown that the modal approach can also be used to describe the motion after the excitation has stopped, for example, after a short pulse or a shock. The result is

$$v_1(x,t) = \sum_n \frac{I_n}{\omega_n}\varphi_n(x)e^{-\eta_n\omega_n t/2}\sin\omega_n t. \tag{20}$$

Here I_n depends on the location and the type of excitation. Equation (20) shows that the free motion consists of several ringing modes that decay according to their modal loss factors. When the system is excited by a very short pulse, the total vibration consists of many modes. For longer pulses and for most shock applications, the ringing is concentrated to a few (or only one) low-frequency modes.

The eigenfunctions $\varphi_n(x)$ are solutions of the equation of motion for vanishing excitation. They have to fulfil the boundary conditions for the problem to be investigated. Equation (19) holds if the internal damping is small and if the energy flow through the boundaries can be neglected. As mentioned before, ω_n are the resonance frequencies. They determine the ringing sound after a structure has been excited (e.g., by a pulse). The $\varphi_n(x)$ are the vibration patterns or mode shapes that belong to these resonances. For many purposes it is very helpful to know that the eigenfunctions have the orthogonality property:

$$\int m''(x)\varphi_n(x)\varphi_m(x)\,dx = \begin{cases} 0 & \text{for } n \neq m, \\ A_n & \text{for } n = m. \end{cases} \tag{21}$$

Here the integration has to be taken over the whole structure. The term $m''(x)$ is the mass density. It is known[22] that at high frequencies the number ΔN of modes in a frequency band $\Delta\omega$ is given by

$$\frac{\Delta N}{\Delta\omega} = \frac{2m}{\pi}\,\mathrm{Re}\left\{\frac{1}{\mathbf{Z}_F}\right\}. \tag{22}$$

Here m is the total mass of the system and $\mathbf{Z}_F$ its input impedance for point force excitation when there would be no reflection from boundaries. Thus $\mathbf{Z}_F$ is the impedance of the corresponding infinite system (see also Section 2.7 in Chapter 49). Apart from a few exceptions (see Section 2.9 in Chapter 49), eigenfunctions and eigenfrequencies cannot be represented by simple analytical functions. For most applications they have to be found numerically, usually by the FEM. (See, e.g., Refs. 8 and 9.) The great advantage is that once the φ_n and ω_n are known, the vibrations of a structure can in general be expressed in a very concise way by Eq. (19) because for many applications a few dozen modes (or less) give a rather accurate description (although thousands of elements may have been used to calculate them originally). Another advantage of the modal approach comes up in the high-frequency range where Eq. (21) is very helpful to obtain average values. This topic is treated more thoroughly in connection with statistical energy analysis (SEA) in Chapter 78.

TABLE 1 Material Properties[a]

Material	ρ (kg/m^3)	E ($\times 10^9$ N/m^2)	G ($\times 10^9$ N/m^2)	μ	c_L (m/s)	c_T (m/s)
Aluminum	2,700	72	27	0.34	5,160	3,160
Brass	8,400	100	38	0.36	3,450	2,100
Copper	8,900	125	48	0.35	3,750	2,300
Gold	19,300	80	28	0.43	2,030	1,200
Iron	7,800	210	80	0.31	5,180	3,210
Lead	11,300	16	5.6	0.44	1,190	700
Magnesium	1,740	45	17	0.33	5,080	3,100
Nickel	8,860	200	76	0.31	4,800	2,960
Silver	10,500	80	29	0.38	2,760	1,660
Steel	7,800	200	77	0.3	5,060	3,170
Tin	7,290	54	20	0.33	2,720	1,670
Zinc	7,140	100	41	0.25	3,700	2,400
Perspex	1,150	5.6	—	—	2,200	—
Polyamid	1,100–1,200	3.6–4.8	—	—	1,800–2,000	—
Polyethylene	≈900	3.3	—	—	1,900	540
Rubber[b]						
30 shore	1,010	0.0017	—	≈0.5	41	
50 shore	1,110	0.0023	—	≈0.5	46	
70 shore	1,250	0.0046	—	≈0.5	61	
Polyvinylchloride	1,050	0.005–3.0	—	—	69–1,650	
Asphalt	1,800–2,300	7–21	—	—	1,900–3,100	
Brick						
Solid	1,800–2,000	≈16	—	—	2,600–3,100	
Hollow	700–1,000	3.1–7.8	—	—	2,100–2,800	
Concrete						
Dense	2,200–2,400	≈26	—	—	3,400	
Poured						
Concrete						
Light	1,300–1,600	≈4.0	—	—	1,700	
Porous	600–800	≈2.0	—	—	1,700	
Cork	120–250	0.02–0.05	—	—	430	
Fir	440–700	1.0–4.0	—	—	1,400–3,500	
Glass	2,500	≈4.4	—	—	4,200	
Gypsum board	700–950	4.1	—	—	2,000–2,500	
Oak	700–950	1.8–7.2	—	—	1,500–3,000	
Chipboard	600–750	≈4.9	—	—	≈2,700	
Plaster	≈1,700	≈4.3	—	—	≈1,600	
Plywood	600	5.4	—	—	3,000	

[a]For loss factors see Fig. 8, Chapter 55.
[b]Dimension of test sample smaller than wavelength. Large rubber samples have c_L = 1500 m/s.

5 LIST OF MATERIAL PROPERTIES

For mechanical vibrations the most important material properties are density, elastic moduli, and the loss factors. In Table 1 some values of moduli are given. Since for many applications it is useful to know the wave speeds, the shear wave speed $c_T = \sqrt{G/\rho}$, and the longitudinal wave speed in rods, $c_L = \sqrt{E/\rho}$ is also given. Data on loss factors are shown in Fig. 8 in Chapter 55. Since loss factors vary with load, temperature, preconditioning, (sometimes) frequency-only ranges are given in the figure.

REFERENCES

1. L. Brekhovskikh and V. Goncharov, *Mechanics of Continua and Wave Dynamics*, Springer, Berlin, 1985, Chapters 1–4.
2. J. D. Achenbach, *Wave Propagation in Elastic Solids*, North-Holland, Amsterdam, 1973.
3. C. Zener, *Elasticity and Anelasticity of Metals*, University of Chicago Press, Chicago, 1948.
4. K. L. Johnson, *Contact Mechanics*, Cambridge University Press, Cambridge, 1985, Chapters 4 and 7.
5. S. P. Timoshenko and S. Woinowsky-Krieger, *Theory of Plates and Shells*, McGraw-Hill, New York, 1959, Chapter 13.
6. P. M. Morse and H. Feshbach, *Methods of Theoretical Physics*, Vol. 1, McGraw-Hill, New York, 1953, Chapter 3.2.
7. Lord J. W. Rayleigh, *The Theory of Sound*, Vol. I, Macmillan, London, 1894, or Dover, New York, 1945.
8. M. Petyt, "Finite Element Techniques for Structural Vibration," and "Finite Element Technique for Acoustics," in R. G. White and J. G. Walker (Eds.), *Noise and Vibration*, Wiley, New York, 1982, Chapters 15 and 16.
9. O. Zienkiewicz, *The Finite Element Method*, McGraw Hill, London, 1977.
10. M. C. Junger and D. Feit, Sound, Structures and Their Interaction, MIT Press, Cambridge, 1972, Chapter 9.
11. M. Heckl, "Einfache Anwendung des Hamilton'schen Prinzips bei Kö rperschallproblemen," *Acustica*, Vol. 72, 1990, pp. 189–196.
12. L. Meirowitch, *Elements of Vibration Analysis*, McGraw-Hill, New York, 1986.
13. R. E. D. Bishop and D. C. Johnson, *The Mechanics of Vibration*, Cambridge University Press, London, 1979, Chapters 3.9, 5.3, 7.10.
14. D. J. Mead, "Response of Periodic Structures to Noise Fields," in *Noise and Vibration*, R. G. White and J. G. Walker (Eds.), Wiley, New York, 1982, Chapter 13.
15. L. M. Lyamshev, "A Question in Connection with the Principle of Reciprocity in Acoustics," *Sov. Phys. Doklady*, Vol. 4, 1959, pp. 406–409.
16. O. Heaviside, *Electrical Papers*, Vol. I, p. 520 and Vol. II, p. 202.
17. L. L. Beranek, *Acoustic Measurements*, Wiley, New York, 1949, Chapter 4.
18. T. ten Wolde and G. Gadefelt, "Development of Standard Measurements for Structure-Borne Sound," *Noise Control Engr.*, Vol. 28, 1987, pp. 5–15.
19. M. Heckl, "Anwendungen des Satzes von der wechselseitigen Energie," *Acustica*, Vol. 58, 1985, pp. 111–117.
20. I. L. Vier, "Use of Reciprocity and Superposition in Predicting Power Input to Structures Excited by Complex Sources," *Proceedings INTER-NOISE*, Newport Beach, California, 1989, Vol. 89, pp. 543–546.
21. M. E. McIntyre, R. T. Schumacher, and J. Woodhouse, "On the Oscillations of Musical Instruments," *J. Acoust. Soc. Am.*, Vol. 74, 1983, pp. 1325–1345.
22. L. Cremer, M. Heckl, and E. E. Ungar, *Structure-Borne Sound*, 2nd ed., Springer, Berlin, 1988.
23. R. Courant and D. Hilbert, *Methoden der mathematischen Physik I*, Springer, Berlin, 1931, Chapters V, VI.

48

EXPERIMENTAL AND THEORETICAL STUDIES OF VIBRATING SYSTEMS

PAUL J. REMINGTON

1 INTRODUCTION

This chapter provides an overview and introduction to analytical and experimental methods applied to vibrating systems. The initial focus is on the single-degree-of-freedom oscillator, not only because it is simple and easy to understand but also because it is a useful idealized model of many complicated vibrating systems. The simple oscillator serves as a vehicle for the discussion of free and forced response, Fourier transform methods, impulse response functions, energy balance, and equivalent circuits. The focus then shifts to more complicated multi-degree-of-freedom lumped-parameter systems where the concepts of mode shape functions and natural frequencies are introduced. The chapter concludes with a discussion of experimental methods. Similar to the discussion of analytical methods, preceding it, the discussion of experimental methods begins with the single degree-of-freedom oscillator and its use in the determination of material properties. The chapter concludes with a discussion of the extension of these techniques to multi-degree-of-freedom systems, commonly referred to as experimental modal analysis.

2 SINGLE-DEGREE-OF-FREEDOM SYSTEMS

2.1 Free Vibration

Equations of Motion While very simple, the linear, single-degree-of-freedom oscillator depicted in Fig. 1 is a very useful idealized model of many real-world vibrating systems. A simple example is an engine on elastic mounts where the engine represents the mass and the elastic mounts the spring and damper. Another example is a piezoelectric accelerometer in which a small seismic mass is mounted on a slightly resilient piezoelectric crystal. The main reason, however, for treating single-degree-of-freedom oscillators rather extensively here is the fact that in the neighborhood of resonances many systems, especially when they are lightly damped, behave very much like single-degree-of-freedom systems even though they may have many resonances.

In Fig. 1 the displacement of the mass m is denoted by u. This motion is assumed to be in a straight line with no rotation. Thus the inertial force is

$$m\,\frac{d^2u}{dt^2}.$$

The force acting on the damper element, which is assumed to be viscous in nature, is given by

$$b\,\frac{du}{dt},$$

where b is the damping coefficient and the spring force is ku, where k is the spring constant. When the three relations are combined and an outside force $f(t)$ is added as shown in Fig. 1, the equation of motion

$$m\,\frac{d^2u}{dt^2} + b\,\frac{du}{dt} + ku = f(t) \tag{1}$$

of a single-degree-of-freedom system is obtained.

Note: References to chapters appearing only in the *Encyclopedia* are preceded by *Enc.*

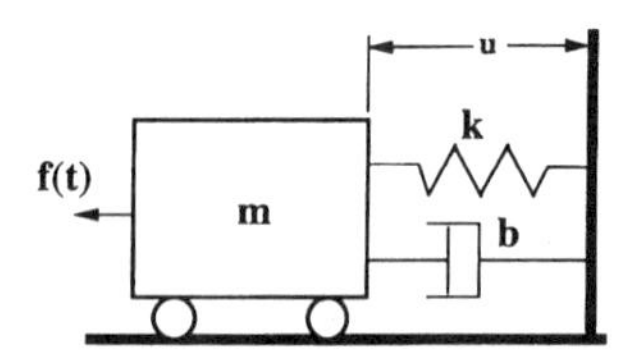

Fig. 1 Single-degree-of-freedom oscillator.

Response Two distinct solutions can be found to Eq. (1), the free response and the forced response. The free response (transient response) of the system is the response that occurs in the absence of any external force. In this section the free response will be considered by setting the external force $f(t)$ in Eq. (1) to zero and computing the resulting motion of the mass when it is given an initial displacement and velocity and then released. To keep things simple, the undamped case when $b = 0$ will be considered first. Although there is some damping in all real physical systems, the undamped case is a reasonable approximation to very lightly damped systems. If the mass is displaced a distance u_0 from its equilibrium position and is released with an initial velocity du_0/dt, then the motion of the mass as a function of time becomes

$$u = A_0 \cos(\omega_0 t + \phi_0), \tag{2}$$

where

$$A_0 = \left[\left(\frac{du_0/dt}{\omega_0}\right)^2 + u_0^2\right]^{1/2},$$

$$\phi_0 = \tan^{-1}\left(\frac{-du_0/dt}{\omega_0 u_0}\right),$$

$$\omega_0 = \left(\frac{k}{m}\right)^{1/2}.$$

As Eq. (2) shows, the motion of the mass is a pure sinusoid of period $2\pi/\omega_0$. The quantity ω_0 is called the natural frequency of the system and in the undamped case is given by the very simple formula shown above.

If a small amount of damping is now added to the system, the free response of the mass will change to

$$u = A_d e^{-\zeta\omega_0 t} \cos(\omega_d t + \phi_d), \tag{3}$$

where

$$A_d = \left[u_0^2 + \frac{(du_0/dt + \zeta\omega_0 u_0)^2}{\omega_d^2}\right]^{1/2},$$

$$\phi_d = \tan^{-1}\left[\frac{-(du_0/dt + \zeta\omega_0 u_0)}{u_0\omega_d}\right],$$

$$\omega_d = \omega_0(1 - \zeta^2)^{1/2},$$

$$\zeta = \frac{1}{2}\frac{b}{(mk)^{1/2}} < 1,$$

and u_0 and du_0/dt, as before, are the initial displacement and velocity, respectively. As the equation indicates, when the mass is displaced from its equilibrium position, it will oscillate at frequency ω_d (slightly lower than ω_0) and the oscillations will decay exponentially with time. The rate of decay depends on the damping ratio ζ and the undamped natural frequency ω_0. In many practical instances the shift in natural frequency due to damping is very small and the motion of the mass is a very slowly decaying, nearly pure harmonic oscillation at the undamped frequency ω_0. Figure 2 illustrates the resulting motion when the mass is displaced u_0 from its equilibrium position and released with no initial velocity when the damping ratio ζ is 0.01.

Equation (3) is correct as long as ζ, the damping ratio, is less than 1. If ζ exceeds 1, no oscillations will occur, and when the mass is displaced from its equilibrium position, it will simply decay back to its equilibrium position with no oscillation according to the equation

$$u = -\frac{du_0/dt + \omega_0 u_0 b_2}{2\omega_d'} e^{-b_1\omega_0 t} + \frac{du_0/dt + \omega_0 u_0 b_1}{2\omega_d'} e^{-b_2\omega_0 t}, \tag{4}$$

where

$$b_1 = (\zeta^2 - 1)^{1/2} + \zeta,$$
$$b_2 = \zeta - (\zeta^2 - 1)^{1/2},$$
$$\omega_d' = \omega_0(\zeta^2 - 1)^{1/2}.$$

Figure 3 illustrates the response of an "overdamped" system ($\zeta = 1.05$) when, as in the previous example, the mass is displaced u_0 from its equilibrium position and released with no initial velocity. As the figure shows, no oscillations of the mass occur in this case, and the mass

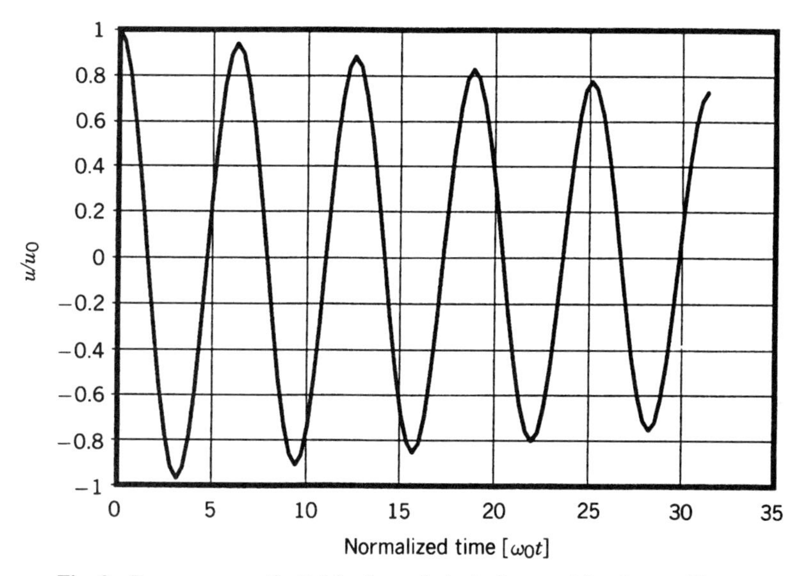

Fig. 2 Free response of a lightly damped single-degree-of-freedom oscillator.

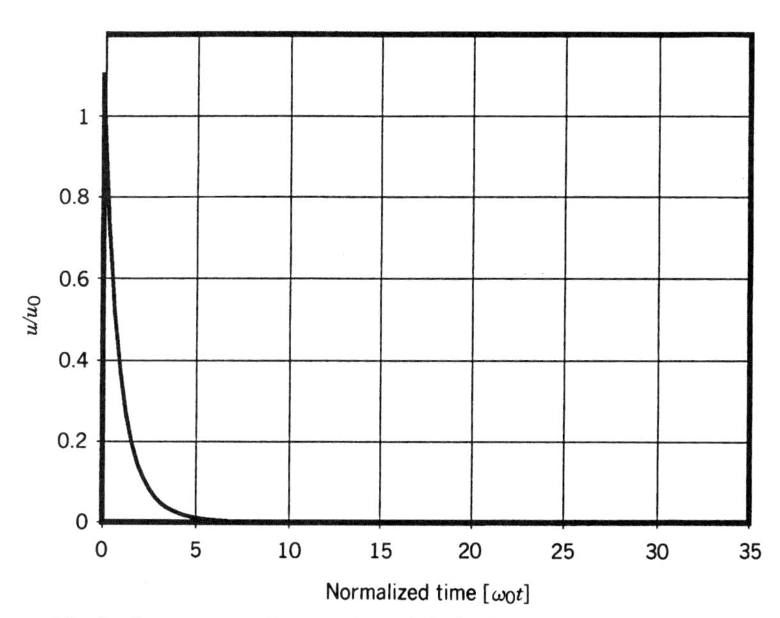

Fig. 3 Free response of an overdamped single-degree-of-freedom oscillator.

simply settles back to its equilibrium position. The *in-between* case in which $\zeta = 1$ or $b/2m = \omega_0$ is called critical damping. For such a system the motion of a mass initially displaced by u_0 and set into motion with velocity du_0/dt is given by

$$u = \left[u_0 + \left(\frac{du_0}{dt} + \omega_0 u_0 \right) t \right] e^{-\omega_0 t}. \tag{5}$$

Systems with critical damping and those that are overdamped do not occur very often in real life. However, in the field of servomechanisms and electromechanical positioning systems, they are of importance.

Energy The total energy associated with the oscillator is given by the sum of the kinetic energy in the mass plus the potential energy in the spring:

$$W = \tfrac{1}{2}m\left(\frac{du}{dt}\right)^2 + \tfrac{1}{2}ku^2, \qquad (6)$$

which for the undamped system yields

$$W = \tfrac{1}{2}m\omega_0^2 A^2 \sin^2(\omega_0 t + \phi_0) + \tfrac{1}{2}kA^2 \cos^2(\omega_0 t + \phi_0),$$

or on simplifying the equation with $\omega_0^2 = k/m$,

$$W = \tfrac{1}{2}kA^2 = \tfrac{1}{2}m(\omega_0 A)^2. \qquad (7)$$

As the equation shows the total energy in the system is equal to the kinetic energy when the mass attains its greatest speed or to the potential energy when the spring attains its greatest deflection.

If we multiply Eq. (1) by du/dt, we find that

$$m\,\frac{d^2u}{dt^2}\,\frac{du}{dt} + ku\,\frac{du}{dt} + b\left(\frac{du}{dt}\right)^2 = f(t)\,\frac{du}{dt},$$

which is identical to

$$\frac{d}{dt}\left[\frac{1}{2}\,m\left(\frac{du}{dt}\right)^2 + \frac{1}{2}\,ku^2\right] + b\left(\frac{du}{dt}\right)^2 = f(t)\,\frac{du}{dt}.$$

Recognizing the quantity in brackets as W, we can write

$$\frac{dW}{dt} + P_d = P_i, \qquad (8)$$

where P_d is the dissipated power and P_i is the power injected by the external force. For a lightly damped system we can use Eq. (7) to approximate the average energy $\overline{W}$ as

$$\overline{W} = \tfrac{1}{2}m\left(\frac{du}{dt}\right)^2_{\max} = m\overline{(du/dt)^2}, \qquad (9)$$

where the overbar means the time average over one cycle. The average dissipated power can be written similarly as

$$\overline{P}_d = b\overline{(du/dt)^2} = \frac{b}{m}\,\overline{W}. \qquad (10)$$

In the steady state where $dW/dt = 0$, Eq. (8) gives a simple energy balance that states that $P_d = P_i$ or that the dissipated power equals the injected power. As a result the time average dissipated power and injected power are related in the steady state by

$$\overline{P}_d = b\overline{(du/dt)^2} = \frac{b}{m}\,\overline{W} = \overline{P}_i. \qquad (11a)$$

In the transient state when $\overline{P}_i = 0$, the power balance in Eq. (8) gives

$$\frac{d\overline{W}}{dt} + \frac{b}{m}\,\overline{W} = \overline{P}_i \quad \text{or} \quad \overline{W} = W_0 e^{-(b/m)t} = W_0 e^{-2\zeta\omega_0 t}. \qquad (11b)$$

In this equation $\overline{W}$ is a slowly varying function because it has been averaged over one (or more) cycles.

Instead of the damping coefficient b or the damping ratio ζ, the loss factor η is often used. It is defined by

$$\eta = \frac{1}{2\pi}\left(\frac{W_{\text{cycle}}}{W_{\text{rem}}}\right), \qquad (12)$$

where W_{cycle} is the energy lost per cycle and W_{rem} is the energy remaining after the cycle. For lightly damped systems $W_{\text{rem}} \approx \overline{W}$ and $W_{\text{cycle}} = \overline{P}_d T$, where T is the time duration of one cycle of the system or $T = 2\pi/\omega_0$. Using these quantities in Eq. (12) we obtain the definition of the loss factor in terms of the damping ratio

$$\eta = \frac{b}{m\omega_0} = \frac{b}{(mk)^{1/2}} = 2\zeta. \qquad (13)$$

More information about damping can be found in Chapter 55.

Example As a simple example consider the point mass suspended at the end of a cantilever beam as shown in Fig. 4. It is assumed that the mass of the beam is much less than the suspended mass and that the rotary inertia of the suspended mass is too small to affect the motion of the beam. The force deflection relationship for the cantilever beam is given by

$$f = \left(\frac{3EI}{L^3}\right) u \quad \text{or} \quad k = \frac{3EI}{L^3},$$

where EI is the bending stiffness of the beam and L is the length, which gives, for the natural frequency of the mass m on the cantilever,

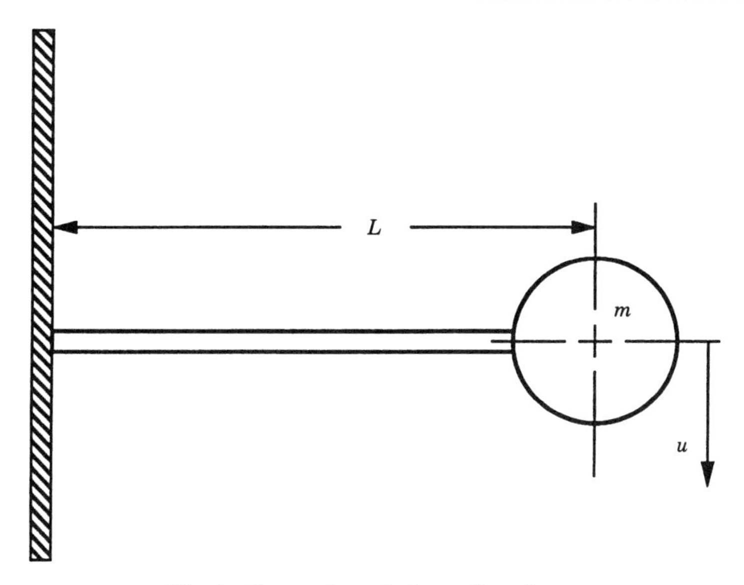

Fig. 4 Mass on the end of a cantilever beam.

$$\omega_0 = \left(\frac{k}{m}\right)^{1/2} = \left(\frac{3EI}{mL^3}\right)^{1/2}.$$

For a lightly damped system if the mass is displaced a distance u_0 from its equilibrium position and if it is released with no initial velocity, the position of the end of the beam as a function of time will be [see Eq. (3)]

$$u = \frac{u_0}{(1-\zeta^2)^{1/2}} e^{-\zeta\omega_0 t} \cos(\omega_0\sqrt{1-\zeta^2}t + \phi_d)$$

$$\approx u_0 e^{-\zeta\omega_0 t} \cos \omega_0 t, \qquad \zeta < 1,$$

$$\phi_d = \tan^{-1}\left(\frac{-\zeta}{(1-\zeta^2)^{1/2}}\right). \tag{14}$$

As the equation shows, the free response of the beam will depend on the magnitude of the damping ratio ζ. Unfortunately, the analytical determination of damping in dynamic systems is usually very difficult. Consequently, it is usually estimated based on measurements. In fact, as will be described below, an experiment involving the natural decay of an oscillator, the same as that described by Eq. (14), can be used to estimate the damping.

2.2 Forced Response

Harmonic Forcing If the external force in Eq. (1) is not zero, but is suddenly applied to the mass, then the response will be a linear combination of the free or transient response, just described, and the forced response. In real dynamic systems the transient response quickly dies out due to damping, and one is left with only the forced response.[1] If a harmonic force of the form $\text{Re}\{F_0 e^{j\omega_f t}\}$ (where $\text{Re}\{\cdot\}$ means the real part of the quantity in brackets) is applied to the mass of Fig. 1, the forced displacement of the mass relative to its equilibrium point will be of the form

$$u(t) = \text{Re}\{U e^{j\omega_f t}\},$$

where ω_f is the forcing frequency. In what follows we will use phasor notation and omit the symbol $\text{Re}\{\cdot\}$.

The ratio of the complex displacement amplitude to the force amplitude can be calculated as[1,2]

$$\frac{U}{F_0} = \left\{k\left[1 - \left(\frac{\omega_f}{\omega_0}\right)^2 + \frac{2j\zeta\omega_f}{\omega_0}\right]\right\}^{-1}. \tag{15}$$

The ratio of system response amplitude to harmonic force amplitude as a function of frequency is often referred to as the frequency response function of the dynamic system. Both velocity and acceleration amplitude are commonly used in place of the displacement. A variety of different terminologies have been used over the years to name these various ratios and their inverses.

TABLE 1 Terminology for Frequency Response Functions

Response Measurement	Response/Force	Force/Response
Displacement	Receptance Compliance	Dynamic stiffness
Velocity	Admittance Mobility	Impedance
Acceleration	Accelerance Inertance	Apparent mass

Table 1 lists the ratios and some of the names associated with each. The reactance given in Eq. (15) will be used in sections to follow as will the admittance or mobility, which is given by

$$Y(\omega_f) = \frac{j\omega U}{F_0} = \left\{ \frac{k}{j\omega_f} \left[1 - \left(\frac{\omega_f}{\omega_0} \right)^2 + 2j\zeta \frac{\omega_f}{\omega_0} \right] \right\}^{-1},$$

or in terms of amplitude and phase

$$Y(\omega) = |Y(\omega)| e^{j\phi},$$

$$|Y(\omega_f)| = \left\{ \frac{k}{\omega_f} \right\}^{-1} \cdot \left\{ \left[1 - \left(\frac{\omega_f}{\omega_0} \right)^2 \right]^2 + \left[2\zeta \left(\frac{\omega_f}{\omega_0} \right) \right]^2 \right\}^{-1/2},$$

$$\phi = \tan^{-1} \frac{1 - (\omega_f/\omega_0)^2}{2\zeta(\omega_f/\omega_0)}. \tag{16}$$

Figure 5 illustrates the magnitude and phase of the admittance as a function of the forcing frequency ω_f. Figure 5 shows a peak in the admittance that occurs for very light damping at $\omega_f = \omega_0$, the natural frequency of the oscillator. The bandwidth of the resonance is also shown in Figure 5. It is measured at the point where the admittance drops by 3 dB and for a lightly damped oscillator is given approximately by

$$\delta\omega = 2\zeta\omega_0 = \eta\omega_0.$$

At the undamped natural frequency of a lightly damped oscillator, the admittance is given by

$$Y(\omega_f) = \{2\zeta\omega_0 m\}^{-1}, \qquad \omega_f = \omega_0.$$

The admittance is real and controlled by the magnitude of the damping. For forcing frequencies well below ω_0 the admittance is given approximately by

$$Y(\omega_f) \approx \frac{j\omega_f}{k}, \qquad \omega_f \ll \omega_0.$$

The admittance is now complex with a phase shift between velocity and force of 90° and appears to be approximately a pure stiffness. Finally for frequencies well above the natural frequency the admittance becomes

$$Y(\omega_f) \sim \{j\omega_f m\}^{-1}, \qquad \omega_f \gg \omega_0,$$

and is once again purely imaginary with a phase shift now of −90°, and the admittance approximates a pure mass.

In the above the frequency response functions have all been given as if the damping were viscous, that is, assuming that the damping coefficient b is constant with frequency. In many practical cases the damping is better modeled as $b = c/\omega$, where c is a constant independent of frequency. This form of damping is referred to as *hysteretic* or *structural* damping. It is equivalent to assuming that the stiffness is not real but is complex and of the form $k = k_0(1 + j\eta)$.

For structural damping the admittance takes the form

$$Y(\omega_f) = \left\{ \frac{k}{j\omega_f} \left[1 - \left(\frac{\omega_f}{\omega_0} \right)^2 + j\eta \right] \right\}^{-1}.$$

Here η is the loss factor defined in Eq. (12) and described more fully in Chapter 55.

Equivalent Circuit It is often useful when visualizing complex arrangements of springs, masses, and dampers to utilize an equivalent electrical circuit representation of the system analogous to the equivalent electrical circuit discussed in Chapter 11 for acoustical systems. If the analogy is made that force is voltage and velocity is current, one obtains the representation shown in Fig. 6 for the oscillator of Fig. 1. The external force becomes a voltage source, the mass is equivalent to an inductor, the spring is analogous to a capacitor, and the damper is a resistor. While this approach offers little additional insight for the simple oscillator, in truly complex systems the equivalent circuit can be of significant help in visualizing the interactions of the various elements. In addition the equivalent circuit approach is not limited to lumped-parameter systems but can be extended to visualize the interactions between continuous systems (beams, plates,

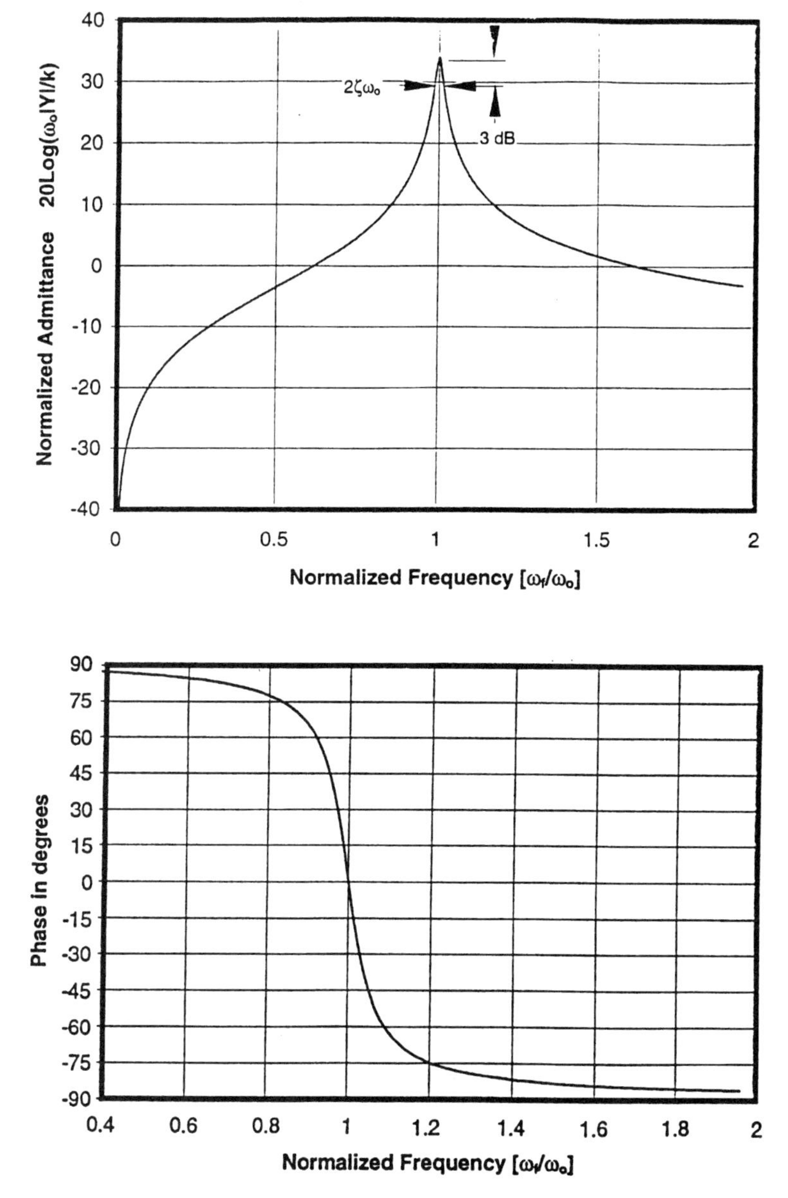

Fig. 5 Admittance of a single-degree-of-freedom oscillator with a damping ratio of 0.01.

acoustic volumes, etc.) by introducing frequency-dependent impedances. Figure 7 illustrates the equivalent circuit for a slightly more complex dynamic system. The simple oscillator is now attached to a finite beam, and the influence of the beam is included through the use of the point impedance of the beam at the point of attachment of the oscillator. It is interesting to note that the series arrangement of the spring–damper with the beam in the

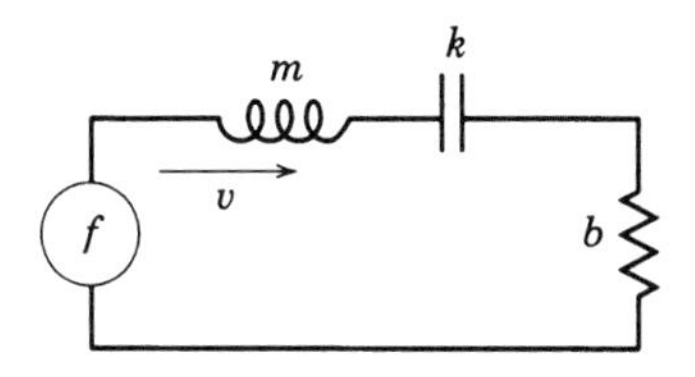

Fig. 6 Equivalent circuit for a spring–mass system.

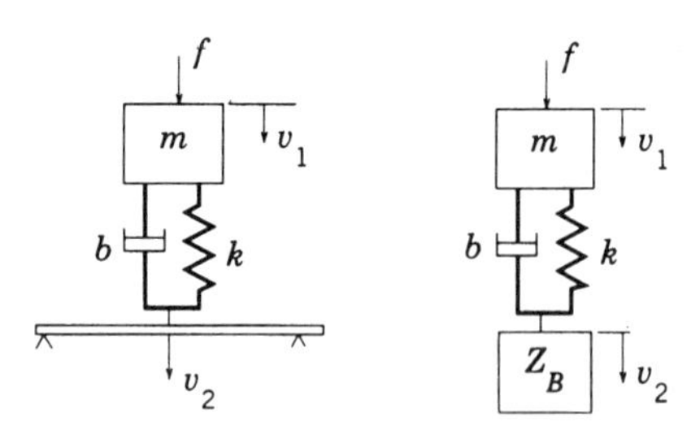

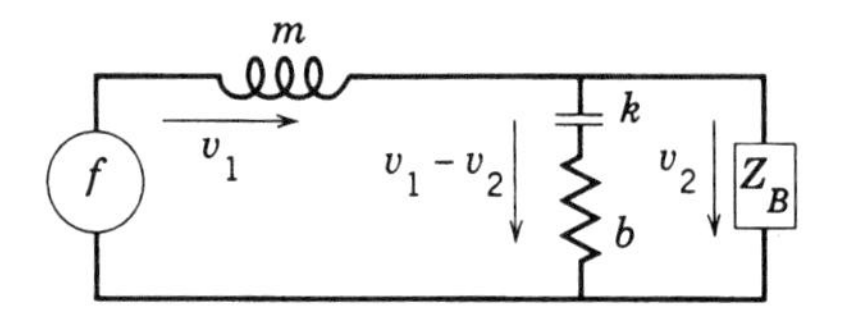

Fig. 7 Equivalent circuit example for a single-degree-of-freedom oscillator on a beam.

mechanical model results in a parallel arrangement in the equivalent circuit, and the parallel arrangement of spring and damper in the mechanical model results in a series arrangement in the equivalent circuit. In this equivalent circuit analogy, force is analogous to voltage and velocity is analogous to current. Under some circumstances it is more appropriate to formulate an equivalent circuit in which force is analogous to current and velocity is analogous to voltage. For that equivalent circuit analogy parallel or series arrangements of elements in the mechanical system will result in the same arrangement in the electrical system. (See Chapter 11.)

Fourier Transform Since the systems considered here are linear, one can apply the principle of superposition to use the frequency response derived above to compute the response of the system to forcing functions that are not harmonic. To do so one utilizes the Fourier transform and its inverse,[3] which are defined as

$$x(t) = \int_{-\infty}^{\infty} X(\omega)e^{j\omega t}\, d\omega, \tag{17}$$

$$X(\omega) = \frac{1}{2\pi}\int_{-\infty}^{\infty} x(t)e^{-j\omega t}\, dt, \tag{18}$$

where $x(t)$ is a real function and $X(\omega)$ is, in general, a complex function of ω. If one substitutes the Fourier transform representation of the external force $f(t) = \int_{-\infty}^{\infty} F(\omega)e^{j\omega t}\, d\omega$ and the Fourier transform of the velocity of the oscillator from its equilibrium position $du/dt = v(t) = \int_{-\infty}^{\infty} V(\omega)e^{j\omega t}\, d\omega$ into Eq. (1) and solves for $V(\omega)$ in terms of $F(\omega)$, one obtains

$$\begin{aligned} V(\omega) &= \left\{ j\omega m\left[1 - \left(\frac{\omega_0}{\omega}\right)^2 - 2j\zeta\left(\frac{\omega_0}{\omega}\right)\right]\right\}^{-1} F(\omega) \\ &= Y(\omega)F(\omega). \end{aligned} \tag{19}$$

An expression for the time response of the mass can then be obtained by taking the inverse Fourier transform of Eq. (19) as follows:

$$v(t) = \int_{-\infty}^{\infty} Y(\omega)F(\omega)e^{j\omega t}\, d\omega. \tag{20}$$

In Eq. (20) the external force has been broken up into a continuous distribution of infinitesimal forces of complex amplitude $F(\omega)\, d\omega$ at each frequency ω and the resulting velocity response computed by integrating the response to a force at a single frequency over all frequency so as to sum the contributions from each force. The function $F(\omega)$ can be obtained by utilizing the inverse Fourier transform of $f(t)$,

$$F(\omega) = \frac{1}{2\pi}\int_{-\infty}^{\infty} f(t)e^{-j\omega t}\, dt. \tag{21}$$

As an example consider a force of the form

$$f(t) = \begin{cases} F_0 e^{\beta t}, & t < 0, \\ F_0 e^{-\beta t}, & t \ge 0. \end{cases}$$

Substituting this function into Eq. (21) and carrying out the integration, one obtains

$$F(\omega) = \frac{F_0\beta}{\pi(\beta^2 + \omega^2)}.$$

If one substitutes this equation into Eq. (20) and carries out the indicated integration using contour integration methods,[3] one obtains

$$v(t) = \begin{cases} \dfrac{F_0\beta e^{\beta t}}{m(\beta^2 + \omega_0^2)}, & t < 0, \\ \dfrac{F_0\beta}{m(\beta^2 + \omega_0^2)}(-e^{-\beta t} + 2e^{-\zeta\omega_0 t}\cos\omega_0 t), & t > 0. \end{cases}$$

The result, which is valid for light damping ($\zeta^2 \ll 1$), shows that for $t < 0$ the velocity increases exponentially with time as the force increases. At $t = 0$ there is a discontinuity in the increase of the force, which excites the transient response of the oscillator (the second term in the equation for $t > 0$). As expected, however, the velocity, displacement, and acceleration are continuous at $t = 0$. Another approach to the estimation of the response of dynamic systems to nonharmonic forces is the use of the Laplace transform, a variation on the Fourier transform, which is especially useful for excitation that is zero up to $t = 0$. The interested reader is referred to Ref. 3 for further information.

Impulse Response (Green's Function) The Fourier transform is a frequency-domain approach. It is also possible to estimate dynamic system response to nonharmonic excitation in the time domain by using the impulse response function. For the simple oscillator of Fig. 1 the impulse response can be obtained from Eq. (3) for the transient response. It is only necessary to estimate the initial conditions. It is straightforward to show that if a unit impulse (delta function) is applied to the mass in Fig. 1 at $t = t_0$, at $t = t_0^+$ the initial displacement will be zero, and the initial velocity will be

$$\left(\frac{du_0}{dt}\right)_{t=t_0^+} = v_0 = \frac{1}{m}.$$

This equation does not mean that velocity has the units of $1/m$, where m is mass. Since we used a unit impulse, the 1 appearing in the equation has the units of momentum or mv.

By substituting this velocity into Eq. (3), one can obtain the displacement response of a lightly damped oscillator to a unit impulse as follows:

$$h(t - t_0) = \begin{cases} \dfrac{1}{m\omega_d} e^{-\zeta\omega_0(t-t_0)} \sin[\omega_d(t - t_0)], & t > t_0, \\ 0, & t < t_0 \end{cases} \tag{22}$$

Causality requires that physically realizable systems not respond before the impulse is applied. Consequently the impulse response function is zero for $t < t_0$. If one now takes a force $f(t_0)$, divides the force up into a number of infinitesimal impulses of magnitude $f(t_0)\,dt_0$, one can obtain the displacement of the oscillator at any time $t > t_0$ by integrating the product of the impulse amplitude and impulse response over all t_0 as follows:

$$u(t) = \int_{-\infty}^{\infty} h(t - t_0) f(t_0)\,dt_0 = \int_{-\infty}^{t} h(t - t_0) f(t_0)\,dt_0, \tag{23}$$

where the second form is equivalent to the first because $h(t - t_0) = 0$ for $t_0 > t$. Equation (23) is referred to as the convolution integral and has an equivalent form

$$u(t) = \int_{-\infty}^{\infty} h(t_0) f(t - t_0)\,dt_0 = \int_{0}^{\infty} h(t_0) f(t - t_0)\,dt_0.$$

If one takes the Fourier transform of both sides of Eq. (23), one obtains

$$U(\omega) = \frac{V(\omega)}{j\omega} = H(\omega)F(\omega), \tag{24}$$

where

$$H(\omega) = \frac{1}{2\pi}\int_{-\infty}^{\infty} h(t)e^{j\omega t}\,dt$$

is the Fourier transform of the impulse response. Here, $U(\omega)$, $V(\omega)$, and $F(\omega)$ are the Fourier transforms of the displacement, velocity, and force, respectively. By comparing Eq. (24) and Eq. (19), it is clear that

$$H(\omega) = \frac{Y(\omega)}{j\omega}$$

or that the Fourier transform of the impulse response is the frequency response.

3 MULTI-DEGREE-OF-FREEDOM SYSTEMS

3.1 Basic Equations

Figure 8 shows a simplified multi-degree-of-freedom lumped-parameter oscillator. One equation of the form

$$m_i \frac{d^2 u_i}{dt^2} - b_i \frac{du_{i+1}}{dt} + (b_i + b_{i-1}) \frac{du_i}{dt} - b_i \frac{du_{i-1}}{dt} - k_i u_{i+1} + (k_i + k_{i-1})u_i - k_i u_{i-1} = f_i \tag{25}$$

can be written for each mass in the system. The resulting N equations in the N unknown displacements u_i can be written in matrix form to yield

$$[M]\left\{\frac{d^2u}{dt^2}\right\} + [B]\left\{\frac{du}{dt}\right\} + [K]\{u\} = \{f\}, \tag{26}$$

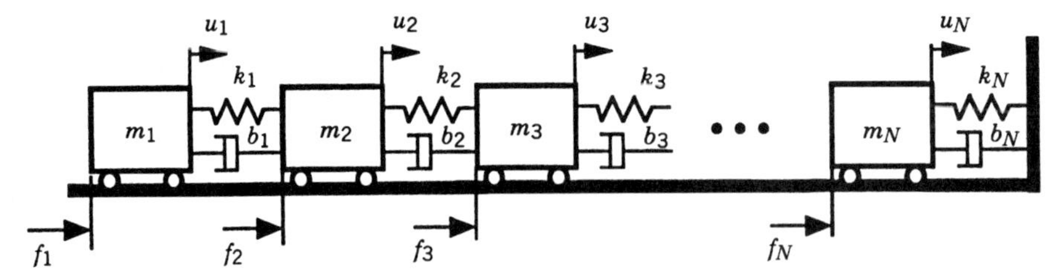

Fig. 8 Example of a multi-degree-of-freedom oscillator.

where the mass matrix is given by

$$[M] = \begin{bmatrix} m_1 & 0 & 0 & \cdots & 0 \\ 0 & m_2 & 0 & \cdots & 0 \\ \vdots & \vdots & \vdots & \ddots & \vdots \\ 0 & \cdots & \cdots & \cdots & m_N \end{bmatrix},$$

the damping matrix is given by

$$[B] = \begin{bmatrix} b_1 & -b_1 & 0 & 0 & \cdots \\ -b_1 & b_1 + b_2 & -b_2 & 0 & \cdots \\ 0 & -b_2 & b_2 + b_3 & -b_3 & \cdots \\ 0 & \ddots & & & \\ \vdots & & & & \end{bmatrix},$$

and the stiffness matrix is given by

$$[K] = \begin{bmatrix} k_1 & -k_1 & 0 & 0 & \cdots \\ -k_1 & k_1 + k_2 & -k_2 & 0 & \cdots \\ 0 & -k_2 & k_2 + k_3 & -k_3 & \cdots \\ 0 & \ddots & & & \\ \vdots & & & & \end{bmatrix}.$$

All matrices are square and symmetric. The notation $\{d^2u/dt^2\}$, $\{du/dt\}$, $\{u\}$, and $\{f\}$ means that these quantities are column vectors containing the acceleration, velocity, displacement, and force, respectively, associated with each mass. In general the three square matrices are fully populated, although because of the limited coupling between masses in Fig. 8, the mass matrix is diagonal and the damping and stiffness matrices are sparcely populated.

3.2 Modal Decomposition

Undamped System If the system is undamped and a harmonic force of the form $\{f\}e^{j\omega t}$ is applied to the system, the response can be readily calculated by assuming a response of the form $\{u\}e^{j\omega t}$ and simply inverting the resulting matrix to obtain

$$\{u\} = ([K] - \omega^2[M])^{-1}\{f\}, \tag{27}$$

where $[A]^{-1}$ means the inverse of the matrix $[A]$ defined by $[A][A]^{-1} = [I]$, where $[I]$ is the unit diagonal matrix.[3] The response of the system to nonharmonic forcing can then be obtained using the Fourier or Laplace transformation techniques described in the previous section. While this approach is certainly possible, it provides little insight into the dynamics of the system. A useful alternate approach is to decompose the system into normal modes.

If a solution of the form $\{u\}e^{j\omega t}$ is assumed for the undamped system, one can write

$$\{f\} = ([K] - \omega^2[M])\{u\}. \tag{28}$$

When the forcing is zero, in order for this equation to have a solution it is necessary that $\det([K] - \omega^2[M]) = 0$, where $\det([A])$ means the determinant of the matrix $[A]$. Setting the determinant to zero results in a polynomial equation in ω of the form

$$a_{2N}\omega^{2N} + a_{2N-2}\omega^{2N-2} + \cdots + a_0 = 0, \tag{29}$$

which has N real solutions of the form $\pm\omega_i$ where N is the number of degrees of freedom of the system. These natural frequencies or eigenvalues of the system are analogous to the natural frequency of the single-degree-of-freedom system in Eq. (1). Sometimes not all of the solutions to Eq. (29) are different. For this so-called degenerate case the reader is referred to any of the standard texts on matrix algebra.

If one substitutes the natural frequencies from Eq. (29) into the system equations, Eq. (28), with zero external force, and solves for the amplitudes $\{u\}$, one obtains one mode shape $\{\phi_i\}$ (or eigenvector) for each natural frequency ω_i. It can be easily shown[3,4] that if $[K]$ and $[M]$

are real and symmetric, these mode shapes are composed of all real elements and are orthogonal to one another with respect to the mass and stiffness matrices, that is,

$$\{\phi_i\}^{\mathrm{T}}[M]\{\phi_j\} = 0 \quad \text{for } i \neq j,$$
$$\{\phi_i\}^{\mathrm{T}}[K]\{\phi_j\} = 0 \quad \text{for } i \neq j, \tag{30}$$

where $\{\phi\}^{\mathrm{T}}$ means the transpose of the vector $\{\phi\}$. The mode shapes defined as described above are not unique but are defined only to within an arbitrary scale factor. One commonly used scaling factor is to set the maximum value in any mode shape to unity. With that scaling for the mode shapes the generalized mass and stiffness for mode i are defined as

$$\{\phi_i\}^{\mathrm{T}}[M]\{\phi_i\} = m_i,$$
$$\{\phi_i\}^{\mathrm{T}}[K]\{\phi_i\} = k_i \tag{31}$$

and are related to the natural frequency of mode i by

$$\omega_i = \left(\frac{k_i}{m_i}\right)^{1/2}. \tag{32}$$

If one now combines the mode shapes defined above into a matrix such that each modal column vector $\{\phi_i\}$ forms a column of the matrix, that is, $[\Phi] = [\{\phi_1\}, \{\phi_2\} \cdots \{\phi_N\}]$, it is straightforward to show that

$$[\Phi]^{\mathrm{T}}[M][\Phi] = [m_i],$$
$$[\Phi]^{\mathrm{T}}[K][\Phi] = [k_i], \tag{33}$$

where $[m_i]$ and $[k_i]$ are diagonal matrices. If one assumes a solution to Eq. (26) of the form

$$\{u\} = [\Phi]\{a(t)\}, \tag{34}$$

where the $a(t)$ are the modal amplitudes, substitutes this solution into the Eq. (26) with $[B] = 0$, and premultiplies the equation by $[\Phi]^T$, one obtains

$$[m_i]\left\{\frac{d^2a}{dt^2}\right\} + [k_i]\{a\} = [\Phi]^{\mathrm{T}}\{f\}. \tag{35}$$

Since the matrices $[m_i]$ and $[k_i]$ are diagonal, the above represents N uncoupled equations for N independent oscillators. Each of these equations can be solved independently of the other equations by any of the techniques described in the previous section. In the case where the external forces are all harmonic forces of the form $\{f\} = \{F\}e^{j\omega t}$ the solution $\{a\} = \{A\}e^{j\omega t}$ simplifies to

$$A_i = \frac{F_i}{m_i[\omega_i^2 - \omega^2]}, \tag{36}$$

where $F_i = \{\phi_i\}^{\mathrm{T}}\{F\}$ is the modal force for mode i and A_i is the ith element of $\{A\}$.

From this result and Eq. (34) the vector of displacement amplitudes of the individual masses $\{U\}$ becomes

$$\{U\} = [\Phi]\left[\frac{1}{m_i(\omega_i^2 - \omega^2)}\right][\Phi]^{\mathrm{T}}\{F\}, \tag{37}$$

where $[1/m_i(\omega_i^2 - \omega^2)]$ is a diagonal matrix.

The matrix $[\Phi][1/m_i(\omega_i^2 - \omega^2)][\Phi]^{\mathrm{T}}$ is the multimodal reactance matrix. It is analogous to Eq. (15), the reactance equation for a single-degree-of-freedom system. Now instead of a single frequency response function there are N^2 of such functions, although because of symmetry there are only $(N^2 - N)/2 + N$ different ones. Similar equations can be written for the admittance and accelerance matrices and the inverses of these quantities, for example, apparent stiffness, apparent mass, and impedance, can be obtained by simply taking the matrix inverse. For example, the admittance matrix $[Y]$ can be obtained from

$$\{V\} = j\omega\{U\} = [\Phi]\left[\frac{j\omega}{m_i(\omega_i^2 - \omega^2)}\right][\Phi]^{\mathrm{T}}\{F\} \tag{38}$$

and

$$[Y] = [\Phi]\left[\frac{j\omega}{m_i(\omega_i^2 - \omega^2)}\right][\Phi]^{\mathrm{T}}.$$

The impedance matrix $[Z]$ is given by

$$[Z] = [Y]^{-1}.$$

Damped System With the inclusion of damping modal decomposition is still possible but in the general case considerably more complicated. One simplified damping model commonly used in conjunction with viscous damping leads to modal decomposition that is closely related to the undamped case. This simplified model is called *proportional* damping. With proportional damping the damping matrix is given by

$$[B] = \alpha[M] + \beta[K], \tag{39}$$

where α and β are constants. In this instance the modal decomposition is simply related to the undamped system. However, now the eigenvalues are complex and occur in pairs given by

$$\omega_i' = \omega_i[j\zeta_i \pm (1 - \zeta_i^2)^{1/2}],$$

$$\zeta_i = \frac{\alpha}{2\omega_i} + \frac{\beta\omega_i}{2}, \tag{40}$$

where ω_i is the undamped natural frequency of mode i. The mode shapes, however, are real and are the same as the undamped system mode shapes

$$\{\phi_i^{\text{damped}}\} = \{\phi_i^{\text{umdamped}}\}.$$

In this case the damped system modal decomposition and the undamped system modal decomposition are simply related. Unfortunately, the requirement that the damping matrix be a linear combination of the mass and stiffness matrices is very restrictive, and more general models of damping are often necessary.

Proportional damping can also be applied to the case of hysteretic damping. However, the general case of hysteretic damping is nearly as simple. For harmonic excitation of a system with hysteretic damping, the vectors of force amplitudes $\{f\}$ and response amplitudes $\{u\}$ are related by

$$(-\omega^2[M] + [K] + j[H])\{u\} = \{f\}, \tag{41}$$

where the system response has been assumed to be of the form $\{u\}e^{j\omega t}$. Proceeding as for the undamped case, one can define a set of N eigenvalues $\pm\omega_i'$ (N is the number of degrees of freedom of the system) by solving

$$\det(-\omega^2[M] + [K] + j[H]) = 0. \tag{42}$$

An eigenvector ϕ_i' can be obtained for each eigenvalue by substituting each ω_i' into Eq. (41), solving for $\{u\}$ with the force vector $\{f\}$ set to zero and normalizing the result as for the undamped case. In this instance, however, because the system matrix is complex and not real, the resulting eigenvalues and eigenvectors will be complex. Similar orthogonality conditions apply as for the undamped case, that is,

$$[\Phi']^{\mathrm{T}}[M][\Phi'] = [m_i'],$$
$$[\Phi']^{\mathrm{T}}[K][\Phi'] = [k_i'], \tag{43}$$

where $[\Phi']$ is the matrix whose columns are formed by the eigenvectors (mode shapes) $\{\phi_i\}$ and $[m_i']$ and $[k_i']$ are diagonal matrices. In addition it can be shown that

$$(\omega_i')^2 = \frac{k_i'}{m_i'}$$

where k_i' and m_i' are the diagonal elements of the matrices $[k_i']$ and $[m_i']$, respectively. The response can be estimated using the same equations as for the undamped case [Eq. (37)] with the complex eigenvalues ω_i' and complex eigenvectors $\{\phi_i'\}$ substituted for their real counterparts in the undamped equations. It should be noted that the complex mode shapes or eigenvectors simply allow for a general variation in phase between elements of the eigenvectors as opposed to requiring that every element be either in phase or out of phase as is the case in the undamped system.

For the case of viscous damping that is not proportional a slightly different modal decomposition procedure can be applied to obtain a set of $2N$ eigenvalues consisting of N complex eigenvalues and their complex conjugates. There will also be $2N$ eigenvectors consisting of N complex eigenvectors and their N complex conjugates. However, for estimating the forced response, it is more convenient to reformulate the equations and put them in standard state-space form. If a new vector, $\{p\}$, $2N$ elements long consisting of the displacements of each mass and its velocity is defined as

$$\{p\} = \begin{bmatrix} \{u\} \\ \left\{\dfrac{du}{dt}\right\} \end{bmatrix}, \tag{44}$$

then the equations of motion can be rewritten in the form

$$[R]\left\{\frac{dp}{dt}\right\} + [S]\{p\} = \{q\}, \tag{45}$$

where

$$[R] = \begin{bmatrix} [B] & [M] \\ [M] & [0] \end{bmatrix}, \qquad [S] = \begin{bmatrix} [K] & [0] \\ [0] & -[M] \end{bmatrix},$$

$$\{q\} = \begin{bmatrix} \{f\} \\ \{0\} \end{bmatrix}.$$

These equations can be modally decomposed into $2N$ complex eigenvalues ω_i' (N complex-conjugate pairs) and $2N$ eigenvectors $\{\phi_i'\}$ (N complex-conjugate pairs) of length $2N$. The eigenvectors possess orthogonality properties of the form

$$[\Phi']^{\mathrm{T}}[R][\Phi'] = [r_i],$$
$$[\Phi']^{\mathrm{T}}[S][\Phi'] = [s_i], \tag{46}$$

where $[\Phi]$ is a $2N \times 2N$ matrix the columns of which are the eigenvectors $\{\phi_i'\}$ and $\omega_i' = s_i/r_i$. As was found for the undamped case, if a solution of the form $\{p\} = [\Phi']\{a\}$ is substituted into Eq. (45), one obtains the solution

$$\{p\} = [\Phi'][A][\Phi']^{\mathrm{T}}\{q\}, \tag{47}$$

where $[A]$ is a diagonal matrix with elements given by

$$A_{ii} = [r_i(j\omega + \omega_i')]^{-1}.$$

Although this form appears quite different form previous cases, the fact that the eigenvalues and eigenvectors come in complex conjugate pairs allows Eq. (47), when it is expanded, to take an appearance similar to the undamped case.

Example As an example we will reexamine the two-degree-of-freedom system illustrated in Fig. 9. This system is similar to the single-degree-of-freedom system already analyzed in Fig. 4. Here, we will allow the mass at the end of the cantilever beam to have significant rotary inertia. If the beam is assumed to deform quasi-statically, the matrix equation relating force and moment at the end of the beam to the resulting deflection and rotation is given by

$$\begin{bmatrix} F_b \\ M_b \end{bmatrix} = \begin{bmatrix} \dfrac{12EI}{L^3} & \dfrac{-6EI}{L^2} \\ \dfrac{-6EI}{L^2} & \dfrac{4EI}{L} \end{bmatrix} \begin{bmatrix} u \\ \theta \end{bmatrix}, \tag{48}$$

where EI is the bending stiffness of the beam, θ is the angle of rotation of the mass about its center of gravity, u is the lateral displacement of the end of the beam where it joins the mass, F_b is the force and M_b is the moment at the interface between the mass and the end of the beam, and the other variables are as defined previously. A similar set of equations can be written for the mass

$$\begin{bmatrix} F_b \\ M_b \end{bmatrix} = \begin{bmatrix} -m & -mr \\ -mr & I_m + mr^2 \end{bmatrix} \begin{bmatrix} \dfrac{d^2u}{dt^2} \\ \dfrac{d^2\theta}{dt^2} \end{bmatrix} + \begin{bmatrix} F_0 \\ -F_0 r \end{bmatrix}, \tag{49}$$

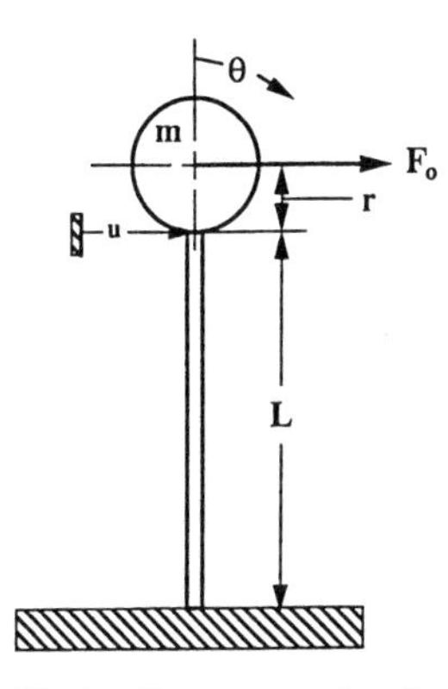

Fig. 9 Mass with significant rotary inertia on the end of a cantilever beam.

where F_0 is the force applied to the center of mass of the mass and I_m is the rotary inertia of the mass. Combining the two equations, we obtain the equation of motion of the system

$$\begin{bmatrix} m & mr \\ mr & I_m + mr^2 \end{bmatrix} \begin{bmatrix} \dfrac{d^2u}{dt^2} \\ \dfrac{d^2\theta}{dt^2} \end{bmatrix} + \begin{bmatrix} \dfrac{12EI}{L^3} & \dfrac{-6EI}{L^2} \\ \dfrac{-6EI}{L^2} & \dfrac{4EI}{L} \end{bmatrix} \times \begin{bmatrix} u \\ \theta \end{bmatrix} = \begin{bmatrix} F_0 \\ -F_0 r \end{bmatrix}. \tag{50}$$

As the equation clearly shows, the system has two degrees of freedom. To illustrate the solution of this system to harmonic excitation, consider solutions of the form $e^{j\omega t}$ for u and θ. The resulting equation becomes

$$\left\{ -\omega^2 \begin{bmatrix} 1 & r \\ r & \dfrac{I_m}{m} + r^2 \end{bmatrix} + \omega_0^2(1 + j\eta) \begin{bmatrix} 1 & -\dfrac{L}{2} \\ -\dfrac{L}{2} & \dfrac{L^2}{3} \end{bmatrix} \right\} \cdot \begin{bmatrix} u \\ \theta \end{bmatrix} = \begin{bmatrix} \dfrac{F_0}{m} \\ -\dfrac{F_0 r}{m} \end{bmatrix}, \tag{51}$$

where $\omega_0^2 = 12EI/mL^3$ and where hysteretic damping proportional to the stiffness terms in the equation has been assumed. This form of damping is commonly used if the modulus of the beam can be modeled as complex and of the form $E = E(1 + j\eta)$. If the physical properties of the system are given by

$$r = 0.001 \text{ m}, \qquad L = 0.1 \text{ m}, \qquad m = 0.1 \text{ kg},$$

$$EI = 0.0001 \text{ N} \cdot \text{m}^2, \qquad \frac{I_m}{m} = 0.0001 \text{ m}^2, \qquad \eta = 0.05.$$

Setting the determinant of the matrices in the brackets in Eq. (51) to zero and solving for the two natural frequencies result in

$$\begin{aligned} f_1 &= \frac{\omega_1}{2\pi} = 0.27 + 0.0067j \qquad \text{Hz}, \\ f_2 &= \frac{\omega_2}{2\pi} = 3.27 + 0.082j \qquad \text{Hz}. \end{aligned} \tag{52}$$

The natural frequencies are complex due to the damping, with the real parts only slightly different from the real natural frequencies that would result from the undamped system. Substituting these frequencies back into Eq. (51) and solving for θ in terms of u, one obtains the modal vectors (eigenvectors) associated with each natural frequency (eigenvalue):

$$\phi_1 = \begin{bmatrix} 0.0657 \\ 0.998 \end{bmatrix}, \qquad \phi_2 = \begin{bmatrix} -0.0025 \\ 1.000 \end{bmatrix}. \tag{53}$$

These modal vectors are real and the same as one would obtain from the undamped system because the damping matrix is proportional to the stiffness matrix. The system response to a harmonic force F_0 can be obtained by substituting the above results into Eqs. (33) and (36), which results in

$$\begin{bmatrix} \dfrac{u}{F_0} \\ \dfrac{\theta}{F_0} \end{bmatrix} = [\Phi] \begin{bmatrix} \dfrac{1}{m_1(\omega_1^2 - \omega^2)} & 0 \\ 0 & \dfrac{1}{m_2(\omega_2^2 - \omega^2)} \end{bmatrix} \cdot [\Phi]^{\mathrm{T}} \begin{bmatrix} 1 \\ -r \end{bmatrix},$$

where the modal masses ($m_1 = 4.55 \times 10^{-4}$, $m_2 = 1.02 \times 10^{-5}$) are computed using Eq. (33). The result of carrying out the indicated matrix multiplications is

$$\begin{aligned} \frac{u}{F_0} &= \frac{9.3}{\omega_1^2 - \omega^2} + \frac{0.86}{\omega_2^2 - \omega^2} \qquad \text{m/N}, \\ \frac{\theta}{F_0} &= \frac{142}{\omega_1^2 - \omega^2} - \frac{343}{\omega_2^2 - \omega^2} \qquad \text{rad/N}, \end{aligned} \tag{54}$$

where ω_1 and ω_2 are given by Eq. (52).

4 EXPERIMENTAL METHODS

4.1 Material Property Estimation Using Simple Oscillators

A common approach to the measurement of material properties is the use of *resonance testing*, so called because it typically involves the coupling of a known mass with the unknown material, and the response of the mass on the material near resonance is used to infer the stiffness and damping properties. A wide variety of arrangements of mass and material are commonly used for this type of testing. For example, in Fig. 10 two arrangements are shown, one appropriate for measuring the compressional properties of a pad of unknown material and one appropriate for measuring the shear properties. (See also Chapter 56.)

One also has a number of choices in how the sample is excited and in what quantities are measured. Figure 11 shows a schematic picture of a typical arrangement in which the pad is mounted on a rigid platform, the mass is mounted on top of the pad, a known force is applied to the mass, and the vertical motion of the mass is measured. The force is usually applied by an electromagnetic shaker as shown in the figure, and the force and acceleration are measured using an impedance head or force transducer and accelerometer. The influence of static load on the material properties is often of interest as well. The figure shows the commonly used approach for apply-

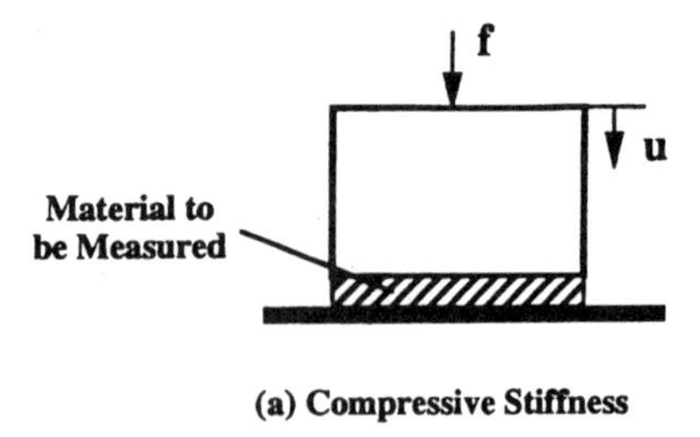

(a) Compressive Stiffness

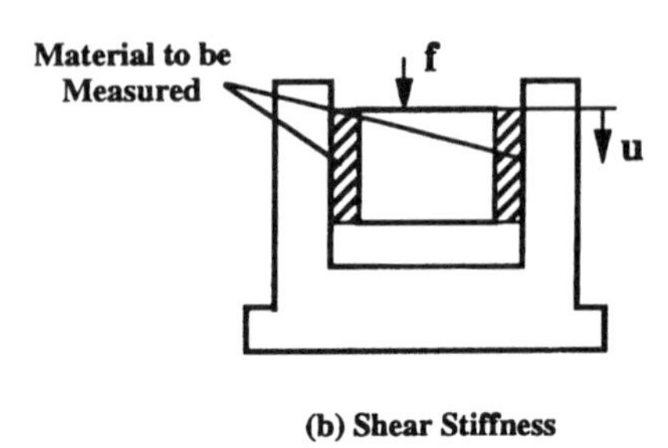

(b) Shear Stiffness

Fig. 10 Alternate arrangements for material property testing.

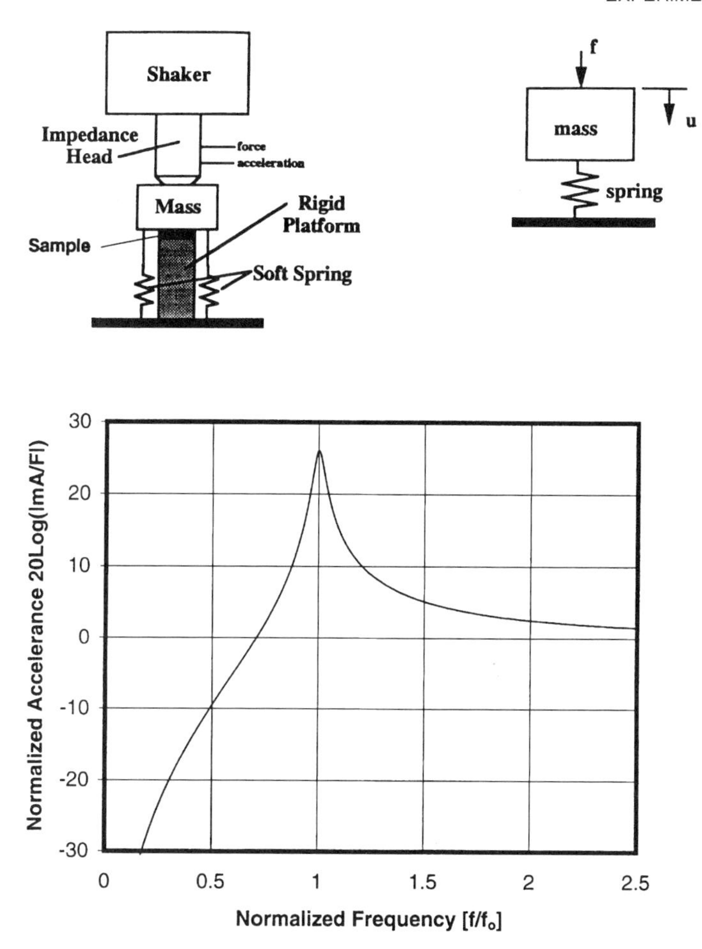

Fig. 11 Typical material property test arrangement including static loading of the sample.

ing a static load in which the load is applied through a spring with a stiffness much less than the pad stiffness. Properly designed such an arrangement will not affect the measured stiffness significantly. In practice several accelerometers are placed on the mass to ensure that the motion is purely vertical translation.

For the arrangement shown in Fig. 11, the ratio of acceleration to applied force (accelerance) is given by

$$\frac{A}{F} = \frac{\omega^2\{(k - \omega^2 m) - j\eta k\}}{\{(k - \omega^2 m)^2 + [\eta k]^2\}}, \tag{55}$$

where a hysteretic damping model for the pad has been assumed, that is, $k = k(1 + j\eta)$. A plot of the amplitude of this transfer function for $\eta = 0.05$ is given at the bottom of Fig. 11. Given measured data of the form of Fig. 11, it is the intent to estimate the stiffness k and the loss factor η. Theoretically, k and η can be obtained at any frequency by simply measuring the amplitude and phase of a/F and appropriately solving Eq. (55) for η and k. In practice, there are a number of approaches that make the process easier and less prone to error.

For estimating k the following approaches are typically applied:

- For $\omega \ll (k/m)^{1/2}$ the accelerance is dominated by the pad stiffness and

$$\left|\frac{A}{F}\right| \sim \frac{\omega^2}{k}. \tag{56}$$

From this equation k can be found at any frequency for which the above criterion holds.

- The $\mathrm{Re}\{A/F\} = 0$ when

$$\omega = \left(\frac{k}{m}\right)^{1/2}. \quad (57)$$

Consequently, if a multichannel fast Fourier transform (FFT) analyzer is available, one can excite the mass with band-limited white noise, use the analyzer to compute the $\mathrm{Re}\{A/F\}$ as a function of frequency, and search for the frequency at which it goes to zero. In addition when the $\mathrm{Re}\{A/F\} = 0$, the phase between A and F is 90°. Therefore this frequency can also be found by applying a harmonic force to the mass, varying the frequency and monitoring the phase between the force and acceleration until that phase becomes 90°. Once the frequency at which $\mathrm{Re}\{A/F\} = 0$ (phase of $A/F = 90°$), the stiffness at that frequency can be found from the above equation. To obtain k at other frequencies, one must use a different mass and retest, using lighter masses for higher frequencies and heavier masses for lower frequencies.

- The amplitude $|A/F|$ peaks at

$$\omega_p = \left\{\frac{k}{m[1+\eta^2]}\right\}^{1/2}. \quad (58)$$

Consequently, if the damping is very small, the dynamic stiffness at ω_p can be estimated approximately by

$$k \sim \omega_p^2 m. \quad (59)$$

As before the dynamic stiffness at other frequencies can be obtained by changing the mass and retesting the sample.

The following are a few examples of typical approaches for estimating the loss factor:

- At the frequency where $\mathrm{Re}\{A/F\} = 0$ and the phase between A and F is 90° [$\omega = (k/m)^{1/2}$], the amplitude of A/F becomes

$$\left|\frac{A}{F}\right| = \frac{\omega^2}{k\eta} \quad (60)$$

from which η can be readily calculated once k has been determined. To obtain estimates of η at other frequencies, the mass must be changed as before and the sample retested.

- The half-power bandwidth at resonance offers another technique for estimating the damping. If the frequency at which $|A/F|$ peaks is given by ω_p, then there is a frequency ω_h above ω_p and a frequency ω_l below ω_p at which $20\log(|A/F|)$ decreases from its maximum by 3 dB. If the damping is small, $\eta \ll 1$, then

$$\eta \approx \frac{\omega_h - \omega_l}{\omega_p}. \quad (61)$$

Once again to obtain η at other frequencies, the mass must be changed to change ω_p and the sample must be retested.

Two of these techniques have been used to estimate the mechanical properties of a number of cork and glass fiber pads for which measured accelerance data with a mass on the pads under two different static loads were available.[5] These data were obtained using an apparatus similar to that shown in Fig. 11. The dynamic stiffness was estimated using direct measurement well below the resonance frequency [Eq. (56)] and resonance testing [Eq. (59)] when η is small. As Table 2 shows, the two approaches give comparable results.

4.2 Experimental Modal Analysis

The measurement and data analysis techniques available for estimating the natural frequencies, damping, and mode shapes of multimodal systems are commonly referred to as experimental modal analysis.[4–6] While applicable to the experimental analysis of any multimodal dynamic system, these techniques have been most commonly applied to structures.

Experimental modal analysis has been the subject of considerable attention in recent years. This interest was initially spurred by the introduction of fast, increasingly capable, multichannel FFT analyzers, many of which came packaged with modal analysis software, and by the availability of associated instrumentation (impedance

TABLE 2 Dynamic Stiffness Test Results on 5-cm Cubes of Material

Test Sample	Resonance Testing (N/m)	Direct Stiffness (N/m)
Cork		
High density	9.71×10^5	9.79×10^5
Standard density	4.17×10^5	4.92×10^5
Low density	3.17×10^5	3.14×10^5
Glass fiber pad	1.30×10^5	1.23×10^5

Source: Ref. 5. Reprinted with permission of the author.

hammers, etc.). Both of these developments eased the tedious process of acquiring the necessary transfer function data. Further development and use of experimental modal analysis techniques were encouraged by the availability of increasingly less expensive personal computers and workstations that provided greater flexibility to the analyst in the processing of the transfer function data.

Experimental modal analysis consists of two major steps:

- The acquisition of measured transfer function data relating point force excitation at one or more points on a structure to response at a number of other points
- The processing of the transfer function data to estimate the natural frequencies, damping factors, and mode shapes

Experimental modal analysis may be carried out using either frequency-domain or time-domain techniques. The earliest procedures, utilizing frequency domain techniques, involved the use of arrays of shakers operating at a single frequency. The shakers were tuned to exactly balance the dissipative focus in the structure and thereby allow the estimation of the homogeneous, undamped equations of motion of the structure.[7] These approaches eventually evolved into the frequency response function (FRF) methods that are most commonly used today.[8] Time-domain techniques utilize the transient response characteristics of the structure to obtain the modal parameters. Typical examples of time-domain techniques would be the Ibrahim time-domain method[9] and the polyreference time-domain method.[10] All of these techniques are in current use and new approaches are constantly evolving.[11]

Data Acquisition The goal of the FRF method data acquisition program is to obtain good-quality frequency response functions relating the force applied to a number of points on the structure to the response (displacement, velocity, or acceleration) at a number of points on the structure, where at least some of the force and response points are the same. As with any carefully designed experimental program, experimental modal analysis must begin with the proper selection of transducers, signal conditioning electronics, and analysis hardware. Since these issues are not unique to modal analysis and are discussed adequately elsewhere in this handbook, they will be dealt with no further in this section.

A serious consideration is the mounting of the structure to be analyzed. Two types of mountings are often considered: free and fixed. In a free mounting the structure is suspended with very soft attachments. The attachments are generally designed to be soft enough, such that any resonances of the mass of the structure (in any of its rigid body degrees of freedom) on the suspension are at frequencies well below the frequency range of interest for the measurements. In the fixed mounting the structure is firmly attached to a *rigid* foundation, designed to be essentially motionless compared to the structure during the test. It is generally much easier to design a free suspension than to design one that is fixed. If the suspension is not properly designed, that is, free or fixed, then, the dynamics of the mounting system may contaminate the modal test results.

Once the structure has been properly mounted, a grid work of measurement points must be laid out on it. This is typically accomplished by carrying out a series of preliminary transfer function measurements at a number of points on the structure. These measurements allow the experimentalist to select the excitation levels to obtain good coherence in the measured transfer functions. Coherence is a data processing function available on most multichannel FFT analyzers that provides information on the quality of the data used to form a transfer function, for example, signal-to-noise ratio and linearity of the signals. The measurements also help to establish the required spacing between measurement points in the measurement grid. If a finite-element model of the structure is available, then, the initial definition of the measurement grid can be based on the predicted mode shapes.

Once the measurement grid has been laid out, the measurements may follow one of two possible forms. The experimenter may fix the location where the response is measured and apply the desired force to each of the grid points, or he may fix the point of application of the force and measure the response at each of the grid points. Usually more than one fixed response point or more than one fixed force point is used in order to minimize the likelihood that the fixed point will be a node of one of the modes of interest, that is, a point where the modal response is zero.

A variety of forms of excitation are used to excite the structure. If an electrodynamic shaker or similar device in conjunction with a force gauge is used, the excitation is usually swept sine, stepped sign, pseudo-random noise (digitally generated random noise), or random noise. For impulsive excitation a device usually referred to as an impedance hammer is used. The impedance hammer is a hammer with a force gauge fitted to the head. It is especially convenient for modal analysis measurements in which a few fixed response points are used with the excitation force applied to all of the grid points.

Data Analysis The analysis of the FRF data is usually but not always split into two parts: (1) the estimation of the natural frequencies, modal damping, and modal coefficients and (2) the estimation of the mode shape functions.

Software for performing both of these functions is generally available for PCs, workstations, and mainframe computers and in some cases comes bundled with multichannel FFT analyzers.

In principle, the first step requires only one FRF. In practice, however, because the point of force application or the point of response measurement may be at a node of one or more modes of the structure, data from most if not all of the FRFs are normally utilized.

In the simplest case of lightly damped, well-separated modes, the natural frequencies can be determined from a simple inspection of the FRF curve, and the modal damping coefficients can be estimated based on the half-power bandwidth of each resonance divided by the resonant frequency [see Eq. (61)]. The modal constants $(A_{kl})_q$ can be obtained by noting that

$$Y_{kl}(\omega_a) = \sum_{q=n1}^{N+ni} \frac{(A_{kl})_q}{\omega_q^2 - \omega_a^2}, \tag{62}$$

where $Y_{kl}(\omega_a)$ is a frequency response function at frequency ω_a relating the response measured at point k to the force applied at point l. The summation over q is over the modes of the structure. By forming an equation of the form of Eq. (62) for each frequency where measured data are available, one can form a matrix equation of the form

$$\begin{bmatrix} Y_{kl}(\omega_a) \\ Y_{kl}(\omega_b) \\ \vdots \end{bmatrix} = \begin{bmatrix} (\omega_1^2 - \omega_a^2)^{-1} & (\omega_2^2 - \omega_a^2)^{-1} & \cdots \\ (\omega_1^2 - \omega_b^2)^{-1} & (\omega_2^2 - \omega_b^2)^{-1} & \cdots \\ \vdots & \vdots & \cdots \end{bmatrix} \cdot \begin{bmatrix} (A_{kl})_1 \\ (A_{kl})_2 \\ \vdots \end{bmatrix}$$

or

$$\{Y_{kl}\} = [B]\{A_{kl}\}, \tag{63}$$

in which the matrix $[B]$ can be readily inverted to obtain the unknown modal constants in terms of the measured FRFs $\{Y_{kl}\}$ provided the number of data points is at least as great as the number of modal coefficients to be determined.

In practice, however, overlap between modes can distort the FRF sufficiently so as to obscure the response of individual modes. In such a case more sophisticated approaches must be employed. One of the original approaches to obtaining a simultaneous estimate of all modal parameters is based on a least-mean-square curve-fitting procedure.[12] While more sophisticated procedures are available today, this approach is still useful especially when computational resources are limited, and it is worthwhile spending a moment examining it. One begins with a set of FRFs, $Y_{kl}^{(m)}(\omega_i)$, measured at a number of frequencies ω_i. If the FRFs are receptances, and if we choose to use a structural damping model, the mathematical model to be fitted to the measured data is given by

$$Y_{kl}^{(e)}(\omega_i) = \sum_{q=n1}^{N+ni} \frac{(A_{kl})_q}{\omega_q^2 - \omega_i^2 + j\eta_q\omega_q^2} + \frac{1}{k_{kl}} - \frac{1}{\omega_i^2 m_{kl}}, \tag{64}$$

where the summation is the usual one over N modes, and the final two terms account for the influence of modes not included in the summation. The intent is to determine the modal parameters $\omega_{n1}, \omega_{n1+1}, \ldots, \eta_{n1}, \eta_{n1+1}, \ldots, (A_{kl})_{n1}, (A_{kl})_{n1+1}, \ldots, k_{kl}, m_{kl}$ so that the equation fits the measured data points with the minimum mean-square error. In mathematical terms, if the mean-square error is given by

$$e^2 = \sum_i |Y_{kl}^{(e)}(\omega_i) - Y_{kl}^{(m)}(\omega_i)|^2, \tag{65}$$

then it is desired to select the above-mentioned modal parameters so as to minimize e^2. To estimate the parameters, one differentiates Eq. (65) with respect to each modal parameter and sets the resulting equations equal to zero,

$$\frac{d(e^2)}{db} = 0, \tag{66}$$

where

$$b = \omega_{n1}, \omega_{n1+1}, \ldots, \eta_{n1}, \eta_{n1+1}, \ldots, (A_{kl})_{n1}, (A_{kl})_{n1+1}, \ldots, k_{kl}, m_{kl}.$$

Unfortunately Eq. (66) results in $4N + 2$ nonlinear simultaneous equations [the modal constants $(A_{kl})_q$ are complex and hence two parameters are required for each] in the unknown parameters. Such a set of equations cannot, in general, be solved directly. As a result iterative approaches (typically Newton–Raphson procedures) have been developed in which an initial estimate of the desired parameters must be provided. The initial estimate may be based on inspection of the real and imaginary parts of the FRF or more sophisticated measures may be employed.[12]

In the time domain a comparable procedure exists[4] that is called the complex exponential method. To apply it, one must have a time series representing the impulse response

$$h_{kl}(\Delta t), h_{kl}(2\,\Delta t), \ldots, h_{kl}(p\,\Delta t),$$

where, as before, l refers to the point of force application and k is the response point. It is desired to fit these data with a mathematical model of the form

$$h_{kl}(t) = \sum_{q=1}^{N} (A_{kl})_q e^{s_q t} + (A_{kl})_q^* e^{s_q^* t},$$

$$s_q = \omega_q[-\zeta_q + j(1 - \zeta_q^2)^{1/2}].$$

Using this equation at each time point, one obtains

$$\begin{bmatrix} h_{kl}(0) \\ h_{kl}(\Delta t) \\ h_{kl}(2\,\Delta t) \\ h_{kl}(3\,\Delta t) \\ \vdots \\ h_{kl}(p\,\Delta t) \end{bmatrix} = \begin{bmatrix} 1 & 1 & 1 & & 1 \\ V_1 & V_1^* & V_2 & \cdots & V_N^* \\ V_1^2 & (V_1^*)^2 & V_2^2 & \cdots & (V_N^*)^2 \\ \vdots & \vdots & \vdots & \vdots & \vdots \\ V_1^p & & & & (V_N^*)^p \end{bmatrix} \cdot \begin{bmatrix} (A_{kl})_1 \\ (A_{kl})_1^* \\ (A_{kl})_2 \\ \vdots \\ (A_{kl})_N^* \end{bmatrix}, \tag{67}$$

where $V_q = e^{s_q \Delta t}$.

One can solve Eq. (67) for V_1 to V_N using the Prony method[13] provided the number of data points exceeds $4N$. From the V's one obtains ω_i and ζ_i, and by substituting back into Eq. (67) and inverting the matrix of V's, one can obtain the modal constants $(A_{kl})_i$.

The parameter estimation procedures described so far operate on a single FRF or impulse response function (IRF). In practice, the procedures are applied to a number of different FRF/IRFs, and some average over the resulting modal parameters is used to obtain the desired estimates. Today, more sophisticated procedures are available that simultaneously fit a number of FRF/IRFs at once.[9,10] In fact the Ibrahim time-domain method[10] not only accepts all IRFs at once, it automatically estimates not only the modal frequencies and damping but the mode shape functions as well.

An example of applying a multi-degree-of-freedom curve-fitting process to the measurement of mobility on a substructure of a helicopter is shown in Fig. 12.[4] In the figure the modes were well separated enough and the damping light enough that the process for well-separated lightly damped modes could be applied quite successfully.

Once the natural frequencies, modal damping, and modal constants have been obtained by the curve-fitting process, the next step is the derivation of the mode shape functions. If the measured FRFs are receptances, then the relationship between modal constant and mode shape function is given by

$$(A_{kl})_i = \frac{\phi_i(x_k)\phi_i(x_l)}{m_i}. \tag{68}$$

For those measurements in which the force application point and the response measurement point are co-located, the relationship becomes

$$(A_{kk})_i = \frac{\phi_i(x_k)^2}{m_i}. \tag{69}$$

With Eqs. (68) and (69), one can solve for the mode shape functions $\phi_i(x)$.

Once the modal parameters have been estimated, they can be used in two general ways:

- To validate numerical models of the structure (e.g., finite element) through comparison of predicted and measured natural frequencies and mode shapes
- To generate a mathematical model of the structure

Once a mathematical model has been generated from a modal analysis test, there are a variety of uses to which it can be put. A few examples follow:

- To model a subassembly of a larger structure for integration into the total structural model
- To estimate the effects of structural modifications
- To estimate the excitation force applied to a structure when direct measurement of the force is difficult

In short, experimental modal analysis is a powerful tool that can be used alone or in conjunction with analytical models of structures to better understand the dynamics of complex structures and to aid in their design to ensure that they will function properly in demanding environments.

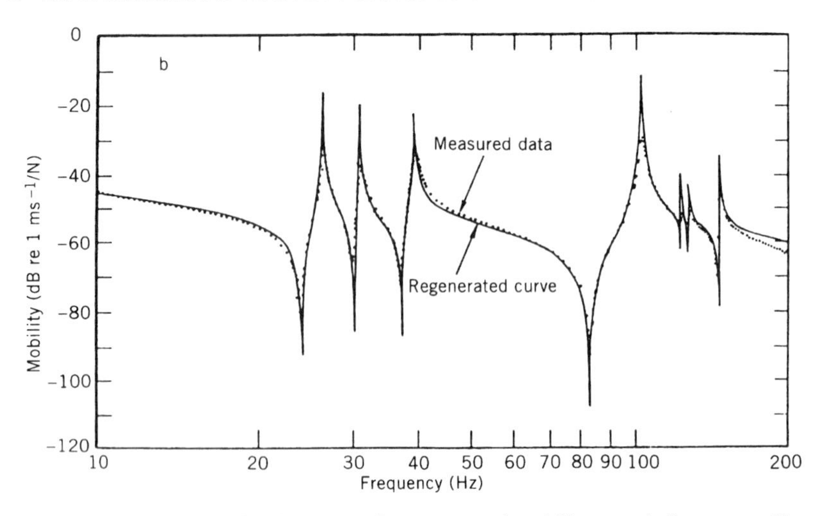

Fig. 12 Multi-degree-of-freedom curve fit to measured mobility on a helicopter auxiliary structure.[4] (Reprinted with permission of the author.)

REFERENCES

1. P. M. Morse, *Vibration and Sound*, McGraw-Hill, New York, 1948.
2. L. E. Kinsler and A. R. Frey, *Fundamentals of Acoustics*, Wiley, New York, 1962.
3. F. B. Hildebrand, *Advanced Calculus for Applications*, Prentice-Hall, Englewood Cliffs, NJ, 1976, Chap. 5.
4. D. J. Ewins, *Modal Testing Theory and Practice*, Wiley, New York, 1984.
5. I. L. Ver, "Measurement of Dynamic Stiffness and Loss Factor of Elastic Mounts as a Function of Frequency and Static Load," *Noise Control Eng.*, Vol. 3, No. 3, pp. 37–43, 1974.
6. R. J. Allemang, D. L. Brown, and R. W. Rost, "Experimental Modal Analysis and Aerodynamic Component Synthesis," Air Force Wright Aeronautical Laboratory Report No. 87-3069, 1987.
7. R. C. Lewis and D. L. Wristley, "A System for the Excitation of Pure Natural Modes of a Complex Structure," *J. Aeronaut. Sci.*, Vol. 17, No. 11, 1950, pp. 705–722.
8. D. L. Brown, R. D. Zimmerman, R. J. Allemang, and M. Mergeay, "Parameter Estimation Techniques for Modal Analysis," SAE Paper No. 790221, Society of Automotive Engineers Transactions, Vol. 88, 1979, pp. 828–846.
9. S. R. Ibrahim and E. C. Mikulcik, "A Method for the Direct Identification of Vibration Parameters for the Free Response," *Shock Vib. Bull*, Vol. 47, No. 4, 1977, pp. 183–198.
10. H. Vold and T. Rocklin, "The Numerical Implementation of a Multiple Input Modal Estimation Algorithm for Minicomputers," *Proceedings of the International Modal Analysis Conference (IMAC)*, Orlando, FL, 1982, pp. 542–548.
11. Excellent sources of information on all aspects of modal analysis are in the *Proceedings of the International Modal Analysis Conference (IMAC)* held annually since 1982 and sponsored by Union College, Schenectady, NY, 12308.
12. G. R. Gaukroger, C. W. Skingle, and K. H. Heron, "Numerical Analysis of Vector Response Loci," *J. Sound Vib.*, Vol. 29, No. 3, 1973, pp. 341–353.
13. R. Prony, "Essai experimental et analytique sur les lois de la delatabilite des fluides et sur celles de la force expansive del la vapeur de l'eau el de l'alkool differentes temperatures," *J. l'Ecole Polytech. (Paris)*, Vol. 1, No. 2, 1795, pp. 24–76.

49

VIBRATIONS OF ONE- AND TWO-DIMENSIONAL CONTINUOUS SYSTEMS

MANFRED A. HECKL

1 INTRODUCTION

Although all real structures are definitely three dimensional, the investigation of one- and two-dimensional systems always has been a field of active scientific interest.

The reasons for that are first that in the audio frequency range many structures such as beams, bars, columns, plates, membranes, shells, fuselages, ship hulls, pipes, and rings are "thin"; that is, at least in one dimension they are small compared to the relevant wavelength. Second, mathematically the treatment of structures in two dimensions is much simpler than in three dimensions.

In this text the important equations of motion are given. They follow from very basic principles of mechanics such as Newton's laws or Hamilton's principle and Hooke's law. The influence of damping usually is then "added on" in a somewhat heuristic way. From the equations of motion, wave speeds, different types of impedances, resonances, energy transmission quantities, impulse response functions, and so on, can be found. These quantities are very useful for noise control work, especially in vehicles and buildings.

Since all two-dimensional results are approximations, the limits of applicability are given whenever possible.

Throughout this chapter linearity is assumed; that is, material nonlinearities (e.g., in rubber) and geometric nonlinearities (e.g., when the amplitude of the motion is comparable with the thickness) are not taken into acount. (See Chapter 49.)

Note: References to chapters appearing only in the *Encyclopedia* are preceded by *Enc.*

2 IMPORTANT RELATIONS FOR PLANAR SYSTEMS

In the following the most important parameters describing mechanical vibrations will be listed. Since there are several idealizations and approximations, depending on the geometry of the system and the direction of motion, only two cases—longitudinal motion in a bar and bending of a plate—will be discussed at some length in the main text. For other cases the relevant equations are given in the appendix.

2.1 Equations of Motion

Different types of wave motion are sketched in Fig. 1. They are approximations that are based on the assumption that the structure is sufficiently thin. In homogeneous media the wave types are independent; therefore they can be treated independently.

Longitudinal Waves in a Bar Longitudinal waves in a one-dimensional structure are of some importance for the vibration transmission over large distances in buildings, ships, and so on. Usually it is difficult to excite a bar or a column to pure longitudinal waves because their input impedance is quite high. As a consequence, slight unsymmetries cause additional bending waves that have a much lower input impedance. For sound radiation pure longitudinal waves are of minor importance because only the very small lateral motion [see Eq. (3)] that is normal to the surface generates sound.

For bars, rods, or columns of cross section S, density ρ, and Young's modulus E the equation of motion for the (longitudinal) displacement ξ_l in the direction of wave propagation (see Fig. 1a) is

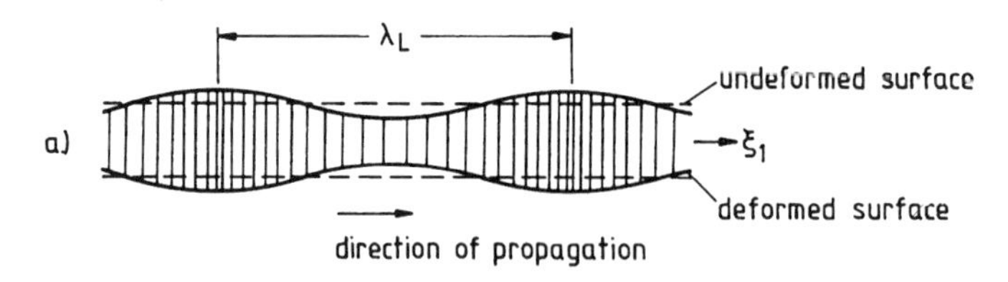

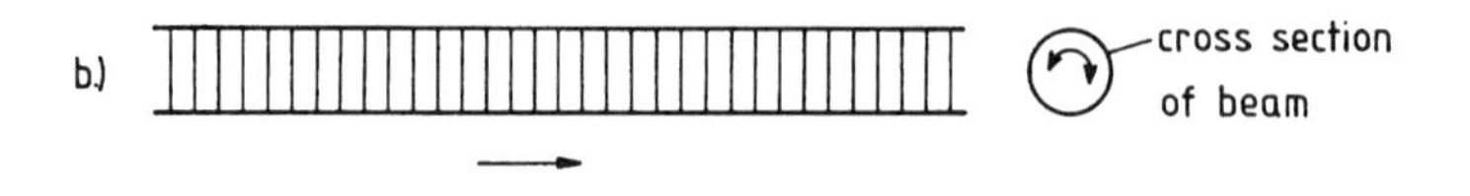

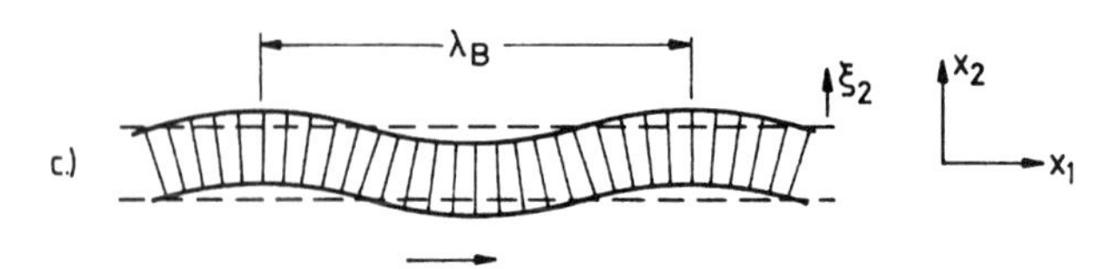

Fig. 1 Different wave types: (*a*) longitudinal with lateral contraction; (*b*) torsional; (*c*) bending.

$$ES\,\frac{\partial^2 \xi_1}{\partial x_1^2} - \rho S\,\frac{\partial^2 \xi_1}{\partial t^2} = 0\ [= -p_1(x_1)]. \qquad (1)$$

The term $p_1(x_1)$ represents forces per unit length that may be acting on the bar. (A rather unlikely case; therefore it is in brackets). As usual, t represents time.

Equivalent to Eq. (1) and sometimes more useful are the equations

$$F = -ES\,\frac{\partial \xi_1}{\partial x_1}, \qquad -\rho S\,\frac{\partial^2 \xi_1}{\partial t^2}\,[+p_1(x_1)] = \frac{\partial F}{\partial x_1}. \qquad (2)$$

Here F is the force acting at x_1 over the cross section S.

The motion ξ_2 in the x_2-direction, which is due to lateral contraction in the case of harmonic motion, is given at the surface by

$$|\xi_2| = \frac{|\xi_1|\mu\pi h}{\lambda_L}. \qquad (3)$$

(A similar relation holds for the x_3-direction.) Poisson's ratio μ relates Young's modulus E and shear modulus G via the equation

$$G = \frac{E}{2 + 2\mu}. \qquad (4)$$

Here, h is the thickness and λ_L the longitudinal wavelength [see Eq. (8a)]. Some values for E and G are given in Table 1 in Chapter 47. The above equations are valid as long as $\lambda_L > 10h$. In the two-dimensional case lateral excitation leads not only to longitudinal but also to shear motion. This case is treated in Section A.3.

Bending Waves in Plates* For most practical applications bending waves in plates or platelike structures are the most important type of motion of two-dimensional structures from the acoustic point of view. Because of their comparatively low input impedance, they can be excited quite easily. In addition, they are good sound radiators (if their wavelength is not too short, see Chapter 10) because their motion is normal to the plate surface. Bending waves are very important for sound isolation problems in connection with walls, ceilings, enclosures, and so on.

For homogeneous plates in bending motion the displacement ξ_2 in the direction normal to the surface is given by[1–4]

$$D\left[\frac{\partial^2}{\partial x_1^2} + \frac{\partial^2}{\partial x_3^2}\right]^2 \xi_2 + \rho h\,\frac{\partial^2 \xi_2}{\partial t^2}$$

$$= D\Delta\Delta\xi_1 + \rho h\,\frac{\partial^2 \xi_2}{\partial t^2} = p(x_1, x_3), \qquad (5)$$

*For bending waves in beams see Section A.2 in the Appendix.

where $D = Eh^3/[12(1 - \mu^2)]$ is the bending stiffness of the plate of thickness h, Δ is the Laplace operator, and $p(x_1, x_3)$ represents the external driving pressure (pressure difference between the two sides of the plate). The equation is valid in the range $\lambda_B > 6h$, where λ_B is the bending wavelength [see Eq. (9)]. For localized excitation one also has to demand that $p(x_1, x_3)$ does not change too rapidly in space because the effects of local elasticity are not included in Eq. (5). So-called point excitation should therefore be distributed over an area comparable to h^2. Otherwise, good correlation with measurements cannot be expected. The range of the applicability of bending theory can be somewhat extended by including the effects of shear stiffness and rotary inertia, as is done in Section A.4. In Section A.5 the related problem of orthotropic plates is also treated.

2.2 Dispersion Relations and Wave Speeds

From here on longitudinal waves in a bar (index L) and bending waves in a plate (index B) are treated simultaneously. They are considered to be the most important examples of one- and two-dimensional systems. For other examples some relevant equations are given in the Appendix. If there is no external excitation (i.e., the right-hand side of the equation of motion is zero) or if the excitation is outside the region of interest, there is still a nontrivial solution to the wave equations. It is called the free-wave solution. It can exist only when there is a certain relation (the dispersion relation) between frequency and wavenumber or wavelength or wave speed.

It is known that a linear superposition of exponential functions is the proper solution of linear differential equations in rectilinear coordinates. In one dimension such a function is

$$\mathrm{Re}\{\xi_1^v e^{j\omega t} c^{-jk_1 x_1}\}. \quad (6a)$$

In two dimensions one has

$$\mathrm{Re}\{\xi_2^v c^{j\omega t} e^{-jk_1 x_1} c^{-jk_3 x_3}\}. \quad (6b)$$

In the following the symbol Re $\{\cdot\}$ is not written explicitly and quite often the factor $\exp(j\omega t)$ is omitted. The terms ξ_1^v and ξ_2^v are complex amplitudes. To distinguish them from the time- and space-dependent quantities, they have the superscript v.

Equations (6a) and (6b) constitute a very convenient notation, especially when damping is involved. Care has to be taken, however, when energies, intensities, or similar quantities are calculated. In physical terms Eqs. (6a) and (6b) represent plane, harmonic waves with angular frequency ω and wavenumbers k_1, k_3. If Eqs. (6a) and (6b) are inserted into Eq. (1) or (5) with the right-hand side vanishing, the results are

$$Ek_{1f}^2 = \omega^2 \rho,$$

$$k_{1f}^2 = \frac{\omega^2 \rho}{E} = \frac{\omega^2}{c_1^2} \quad \text{(bar, longitudinal)}, \quad (7a)$$

$$D(k_{1f}^2 + k_{3f}^2)^2 = \omega^2 \rho h,$$

$$(k_{1f}^2 + k_{3f}^2)^2 = \frac{\omega^2}{c_L^2} \frac{12(1 - \mu^2)}{h^2} \quad \text{(plate, bending)}. \quad (7b)$$

These are the so-called dispersion relations, which give the free wavenumbers k_{1f}, k_{3f} for a given frequency. Free wavenumbers are characterized by the fact that they can exist even for vanishing excitation, or equivalently that the wave impedance Z_τ (see Section 2.5) becomes zero. Dispersion relations for other structures are given in the Appendix. In one dimension there is only one free wavenumber. It is related to the free wavelength λ_L and the phase speed c_{Ph} of propagation by

$$\lambda_L = 2\pi/k_{1f} = \frac{c_l}{f}, \qquad c_{\mathrm{Ph}} = \frac{\omega}{k_{1f}} = c_L. \quad (8a)$$

In two dimensions the situation is a little more complicated. For the plate the dispersion relations following from Eq. (7b) are

$$k_{1f}^2 + k_{3f}^2 = \frac{\pm\omega}{c_L h}\sqrt{12(1 - \mu^2)} = \pm k_B^2. \quad (8b)$$

When k_{1f}^2 and k_{3f}^2 are sufficiently small, one can write them as $k_{1f} = k_B \cos\vartheta$ and $k_{3f} = k_B \sin\vartheta$. Thus Eq. (8b) shows that, when the plus sign is taken, there are free waves with a wavenumber k_B that can travel in any direction ϑ. When either k_{1f}^2 or k_{3f}^2 is larger than k_B^2 or when the minus sign is taken, either k_{1f} or k_{3f} or both are imaginary. In this case there is a so-called near-field or evanescent wave that decays exponentially.

Near fields are only important in the immediate vicinity of the excitation region or near boundaries because they decay by approximately 55 dB within one wavelength.

Wavelength λ_B and wave speed c_B in the direction of propagation are given by

$$\lambda_B = \frac{2\pi}{k_B} = 2\pi\sqrt{\frac{c_1^2h^2}{[12\omega^2(1-\mu^2)]}}, \quad (9)$$

$$c_{\mathrm{Ph}} = c_B = \frac{\omega}{k_B} = \sqrt[4]{\frac{\omega^2h^2c_L^2}{12(1-\mu^2)}} = \sqrt[4]{\frac{\omega^2h^2E}{12\rho(1-\mu^2)}}. \quad (10)$$

Since the phase speed for bending waves depends on frequency, the group velocity c_{Bg}, which is the speed of propagation of the peak in a wave packet, is different. It is given by

$$c_g = \frac{d\omega}{dk_f} = \left(\frac{dk_f}{d\omega}\right)^{-1}. \quad (11)$$

For bending waves this yields

$$c_{Bg} = 2c_B.$$

2.3 Damping and Decay Rates

The simplest way to include material damping in the calculations is to introduce the loss factor η and to use complex moduli, complex wavenumbers, and complex amplitudes. In doing this, one is restricted to harmonic motion of frequency ω. If the complex Young's modulus

$$\underline{E} = E(1 + j\eta) \quad (12)$$

is introduced into Eqs. (1) and (5), the only change is that all quantities become complex. In particular, one finds

$$\underline{k}_L^2 = \frac{\omega}{c_L} = \frac{\omega^2\rho}{E(1+j\eta)},$$

$$\underline{k}_L = k_L\left(1 - \frac{j\eta}{2}\right) \quad \text{(longitudinal)} \quad (13a)$$

$$\underline{k}_B^2 = \sqrt{\frac{12\omega^2\rho(1-\mu^2)}{Eh^2}},$$

$$\underline{k}_B = k_B\left(1 - \frac{j\eta}{4}\right) \quad \text{(bending).} \quad (13b)$$

The approximations for k_L and k_B hold if the damping is small; that is, $\eta^2 \ll 1$. (Here, underlined quantities are complex; quantities that are not underlined are real). Combination of Eqs. (13a) and (13b) with Eqs. (1) and (6) leads to decaying waves with amplitude dependence: For longitudinal waves

$$|\xi_1| \sim e^{-\eta k_1 x_1/2}; \quad (14a)$$

for bending with $\vartheta = 0$, that is, for plane waves in the direction of propagation,

$$|\xi_2| \sim e^{-\eta k_n x_1/4}. \quad (14b)$$

On a decibel scale this corresponds to a level difference of ΔL for a separation Δx:

$$\Delta L = \begin{cases} 27.28\eta\,\dfrac{\Delta x}{\lambda_L} & \text{(longitudinal)} \quad (15a) \\[2ex] 13.64\eta\,\dfrac{\Delta x}{\lambda_B} & \text{(bending)} \quad (15b) \end{cases}$$

The formulas hold when there is propagation in just one direction without any reflections or other disturbances. In practice, this is achieved only on large, damped structures. When point loads on plates generate circular waves, there is an additional decay ΔL_g due to geometric spreading. When R_1 and R_2 are the distances to the center of a localized excitation, it is given by

$$\Delta L_g = 10\log_{10}\left(\frac{R_2}{R_1}\right) \quad \text{(dB).} \quad (15c)$$

The exponential decay of energy on a very large structure with losses is just one consequence of damping. Other effects are the additional phase shift between force (stress) and displacement, the reduction of the mean-square velocity of a finite structure (see statistical energy analysis in Chapters 60 and 62), the decrease of reverberation time and resonance amplitude, and so on.

2.4 Energy

The energy density W_e consists of kinetic energy $W_{e,\mathrm{kin}}$ and potential energy $W_{e,\mathrm{pot}}$. For the two cases considered here they are:

$$W_e = W_{e,\mathrm{kin}} + W_{e,\mathrm{pot}} = \frac{1}{2}\rho S\dot{\xi}_1^2 + \frac{1}{2}ES\left(\frac{\partial\xi_1}{\partial x_1}\right) \quad \text{(longitudinal),} \quad (16a)$$

$$W_e = W_{e,\mathrm{kin}} + W_{e,\mathrm{pot}}$$

$$= \frac{1}{2}\rho h \dot{\xi}_2^2 + \frac{1}{2} D \left\{ \left(\frac{\partial^2 \xi_2}{\partial x_1^2} + \frac{\partial^2 \xi_2}{\partial x_3^2} \right)^2 - 2(1-\mu) \left[\frac{\partial^2 \xi_2}{\partial x_1^2} \frac{\partial^2 \xi_2}{\partial x_3^2} - \left(\frac{\partial^2 \xi_2}{\partial x_1 \partial x_3} \right)^2 \right] \right\}$$

(bending). (16b)

Here $\dot{\xi}_1$ and $\dot{\xi}_2$ are the particle velocities, that is, the time derivatives of ξ_1 and ξ_2. Note that Eq. (16a) has the dimension energy per length (watts by seconds per metre), and Eq. (16b) has energy per area (watts by seconds per metre square).

If in Eqs. (16a) and (16b) the potential energy is not added but subtracted, the Lagrangian

$$L = W_{e,\mathrm{kin}} - W_{e,\mathrm{pot}} \tag{16c}$$

is obtained. This is a very useful quantity when it is applied in connection with Hamilton's principle. See Chapter 47.

Another closely related quantity, the intensity, which determines the energy flow in structures, is not treated here because it is the subject of Chapter 58.

2.5 Wave Impedance for Excitation by Plane Waves

Since one- and two-dimensional structures can be excited in many different ways, it seems appropriate to consider only a few idealized cases. One of them is the excitation of an infinite homogeneous structure by a plane wave with given frequency and wavenumber (or wave speed). The advantage of this approach is that it can be generalized, because all types of plane excitation can be expressed as a sum of plane waves; that is, the Fourier transform of the excitation can be taken. There remains, however, the restriction that the excited structure has to be very large and homogeneous. The general idea underlying the procedure is the following[2,3]:

(a) Because of linearity, all quantities are expressed as a sum of plane progressive waves, that is, as Fourier transforms.

(b) If the response of the structure to one plane wave is known (this is the purpose of the so-called wave impedance), the total response is simply an integration of many such responses.

If one applies Fourier transforms, the external pressure and the displacement can be written as

$$p_1(x_1) = \frac{1}{2\pi} \int p_1^v(k_1) e^{-jk_1x_1}\, dk_1,$$

$$\xi_1(x_1) = \frac{1}{2\pi} \int \xi_1^v(k_1) e^{-jk_1x_1}\, dk_1 \tag{17a}$$

for one dimension and

$$p(x_1, x_3) = \frac{1}{4\pi^2} \int p^v(k_1, k_3) e^{-jk_1x_1} e^{-jk_3x_3}\, dk_1\, dk_3,$$

$$\xi_2(x_1, x_3) = \frac{1}{4\pi^2} \int \xi_2^v(k_1, k_3) e^{-jk_1x_1} e^{-jk_3x_3}\, dk_1\, dk_3. \tag{17b}$$

for two dimensions. (All integrations are in the range $-\infty$ to $+\infty$.)

If these expressions are inserted in Eqs. (1) and (5), the result is

$$\xi_1^v(k_1) = \frac{p_1^v(k_1)}{ESk_1^2 - \omega^2 \rho S} = \frac{p_1^v(k_1)}{j\omega} \frac{1}{j\omega\rho S - jESk_1^2/\omega} = \frac{p_1^v(k_1)}{j\omega} \frac{1}{Z_{\tau L}} \tag{18a}$$

for longitudinal motion of a bar and

$$\xi_2^v(k_1, k_3) = \frac{p^v(k_1, k_3)}{D(k_1^2 + k_3^2)^2 - \omega^2 \rho h} = \frac{p^v(k_1, k_3)}{j\omega} \frac{1}{j\omega\rho h - j(k_1^2 + k_3^2)^2 D/\omega} = \frac{p^v(k_1, k_3)}{j\omega} \frac{1}{Z_{\tau B}} \tag{18b}$$

for bending motion of a plate. The quantities

$$Z_{\tau L} = j\left(\omega\rho S - \frac{ESk_1^2}{\omega}\right) \quad \text{(for bars)}, \tag{19a}$$

$$Z_{\tau B} = j\left[\omega\rho h - \frac{(k_1^2 + k_3^2)^2 D}{\omega}\right] \quad \text{(for plates)} \tag{19b}$$

are the wave impedances. Sometimes they are called transmission impedances. They relate the amplitude of an exciting plane pressure wave to the amplitude of the plane velocity wave that is generated this way in the structure.

The equations show that there are certain wavenumbers for which the wave impedance becomes zero; that is, a very small exciting force is sufficient to cause large vibrations. This phenomenon is called coincidence.[6] It occurs when the wavenumber of the excitation coincides with a free wavenumber of the structure or when the excitation moves with a speed that is very close to a free wave speed in the structure. Coincidence is of practical importance for the sound isolation of walls because sound waves that have a trace wave speed equal to the bending wave speed in a wall are transmitted almost unattenuated. Coincidence phenomena can also be important for the excitation of structures by turbulent boundary layers when the flow speed equals a free wave speed. Another example of coincidence can occur when the rolling speed (e.g., of a tire) is equal to a free wave speed in a structure that is in contact.

The existence of coincidence is shown here only for infinite structures. In finite structures it is also found but is less pronounced, especially when the dimensions of the structure are less than a wavelength.

2.6 Sound Transmission Loss

If the procedure of the last section is applied to the sound transmission problem, the driving pressure $p(x_1, x_3)$ consists of the incoming wave p_i plus the reflected wave p_r minus the transmitted wave p_t.[5,6] With the symbols used in Fig. 2 this gives the following expressions for the transmission coefficient τ and the sound reduction index R (also known as transmission loss TL):

$$\tau = \left|\frac{p_t}{p_i}\right|^2 = \left|1 + \frac{Z_\tau}{2\omega\rho_0}\sqrt{k_0^2 - k_1^2 - k_3^2}\right|^{-2},$$

$$R = 10\log_{10}\frac{1}{\tau} \qquad \text{(dB)}. \tag{20}$$

Here Z_τ is the wave impedance of the structure. Complete transmission with $\tau = 1$ and $R = 0$ occurs at coincidence; that is, when the free wavelength of the structure coincides with the trace wavelength $\lambda_0/\sin\vartheta$ of the incoming sound. See also Chapter 10.

In practical situations, when sound is coming from all directions to a wall, Eq. (20) has to be averaged over all angles of incidence. This has to be done with care, because τ changes extremely rapidly as the angle of incidence changes near coincidence. For a large, plane, homogeneous plate in bending the average result is

$$R \approx \begin{cases} 20\log_{10}\dfrac{\omega\rho h}{2\rho_0 c_0} - 5 \\ \qquad \text{if} \quad \omega\rho h \gg \rho_0 c_0, \quad \omega < \omega_g, \\ 20\log_{10}\dfrac{\omega\rho h}{2\rho_0 c_0} + 10\log_{10}\dfrac{\omega\eta}{\omega_g} - 3 \\ \qquad \text{if} \quad \omega\rho h \gg \rho_0 c_0, \quad \omega > 2\omega_g. \end{cases} \tag{21}$$

Here

$$\omega_g = c_0^2\sqrt{\frac{\rho h}{D}} \tag{22}$$

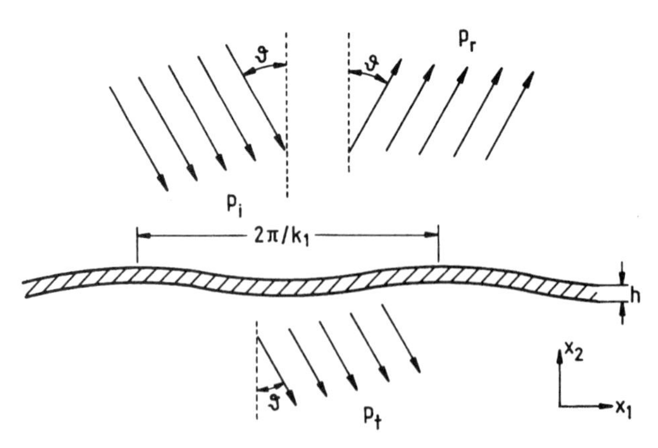

Fig. 2 Sound transmission through a plane structure of infinite extent: $k_0 = \omega/c_0$; $\cos\vartheta = \sqrt{k_0^2 - k_1^2 - k_3^2}/k_0$; ρ_0 and c_0 are density and speed of sound of the surrounding gas or fluid.

is the limiting frequency of coincidence where bending wave speed and speed of sound in the surrounding medium are equal.

The formulas agree fairly well with measured data, even for finite walls with dimensions of several wavelengths, provided the boundary losses are somehow included in the loss factor η. More on sound isolation is presented in Chapter 76.

Equation (20) does not include the effect of shear stiffness and rotary inertia. If this were done, it would result in slightly smaller values of R at high frequencies.

2.7 Point Impedances

In Section 2.5 the general method for calculating the motion of very large plane structures was described. Since it involves at least one integration, it is useful to give results for special cases. One such special case is point excitation with a localized force of frequency ω. Localized (i.e., "point") excitation in this context means that the excited area is small compared with the wavelength but not small compared with the thickness.

Results obtained in this way for excitation with a harmonic force of amplitude F_0 are

$$Z_{FL} = \frac{F_0}{v_0} = \epsilon\rho c_L S \quad \text{(bar in longitudinal motion)}, \tag{23a}$$

$$Z_{FB} = \frac{F_0}{v_0} = 8\sqrt{\rho h D} = \frac{8\omega\rho h}{k_B^2} \quad \text{(plate in bending)}. \tag{23b}$$

Here Z_F is the point impedance, which is defined as the (complex) ratio of the force amplitude to the velocity amplitude v_0 at the excitation point. $\epsilon = 1$ for a semi-infinite bar and $\epsilon = 2$ for an infinite bar. For a semi-infinite plate excited at the edge the factor 8 in Eq. (23b) has to be replaced by 2.35.

The point impedances for beams in bending and for several other structures are given in the Appendix.

The mechanical power that is transmitted by a force is [see also Eq. (14)]

$$P_F = \overline{\text{Re}\,\{F_0 e^{j\omega t}\}\,\text{Re}\{v_0 c^{j\omega t}\}} = \frac{1}{2}\,\text{Re}\{F_0 v_0^*\} = \frac{1}{2}\,|F_0^2|\text{Re}\left\{\frac{1}{Z_F}\right\}. \tag{24}$$

The asterisk indicates complex-conjugate values. The overbar indicates time averaging.

When a homogeneous plate is excited by two opposite forces that are a distance apart, there exists "moment excitation." In this case a moment impedance Z_M can be defined by

$$Z_M = \frac{M_0}{w_0} = \frac{2Z_F}{k_B^2}\left[1 + j\,\frac{4}{\pi}\,\ln\,\frac{1.1}{k_B a}\right],$$

$$P_M = \frac{1}{2}\,|M_0^2|\,\text{Re}\left\{\frac{1}{Z_M}\right\} \approx \frac{1}{8}\,|M_0^2|\,\text{Re}\left\{\frac{1}{Z_F}\right\}k_B^2. \tag{25}$$

Here M_0 is the exciting moment and $w_0 = j\omega\,\partial\xi_2/\partial x_1$, the amplitude of the angular velocity at the excitation point in the direction of the force separation.

The point impedances given here hold for infinite structures. But it can be shown that the quantity $\text{Re}\{1/Z_F\}$, which is very important for power transmission, is also applicable for finite structures, provided that frequency averages are taken that contain at least two resonance frequencies. For practical purposes it may be useful to note that impedances become bigger when the mass and/or the stiffness is increased.

For engineering applications the *power transfer quantity* $\text{Re}\{1/Z_F\}$ can be estimated in a rather crude way by

$$\text{Re}\left\{\frac{1}{Z_F}\right\} \approx \frac{1}{\omega\rho h S_q}. \tag{26}$$

Here ρ is the density of the material and S_q the size of the area that moves with the excitation. It is given by

$$S_g \approx \begin{cases} b\lambda_1/3 & \text{for one-dimensional structures,} \\ (\lambda_1/3)(\lambda_3/3) & \text{for two-dimensional structures.} \end{cases}$$

Here b is the width of the structure; λ_1, λ_3 are the wavelengths of the relevant structural waves in the two directions. It also can be shown that the real part of the inverse moment impedance (i.e., the moment mobility) is usually proportional to the real part of the inverse force impedance (i.e., the force mobility) divided by the square of the wavelength. Thus moment mobilities tend to become bigger with frequency. Consequently the relative importance of moment excitation as compared to force excitation increases with frequency.

2.8 Green's Functions

With the methods described in Section 2.5 it is also possible to find the motion of a structure when it is excited by a localized source. The function obtained this way is the Green's function. For a semi-infinite bar in longitudinal motion excited at one end by a harmonic force, the result is

$$v(x_1) = j\omega\xi_1(x_1) = F_0\rho c_L S\, e^{-jk_L x}. \tag{27a}$$

If the excitation is due to a sudden pulse of type $F_0(t) = I\delta(t)$, the response is

$$v(x_1, t) = \dot{\xi}_1(x_1, t) = I\rho c_L S\delta\left(t - \frac{x_1}{c_1}\right). \tag{27b}$$

This shows that the pulse I travels unchanged with speed c_L through the bar until it hits some sort of discontinuity or boundary where it gets fully or partially reflected.

The corresponding results for an infinite plate in bending are

$$v(r) = j\omega\xi_2(r) = \frac{F_0[H_0^{(2)}(k_B r) - H_0^{(2)}(-jk_B r)]}{Z_F}$$

$$\text{(harmonic excitation)}, \tag{27c}$$

$$v(r, t) = j\omega\xi_2(r, t) = \frac{I}{4\pi\sqrt{D\rho h}\, t} \sin\left(\frac{r^2}{4t}\sqrt{\frac{\rho h}{D}}\right)$$

$$\text{(pulse excitation)}. \tag{27d}$$

Here $\delta(\cdot)$ is the Dirac delta function, $H_0^{(2)}(\cdot)$ the Hankel function, I the transmitted momentum, and r the distance between excitation and observation.

Equation (27c) shows that for a given frequency bending waves on a very large plate decay with the square root of the distance, as is required by energy considerations. Because of the dependence on $k_B r = 2\pi r/\lambda_B$, the decay is more pronounced at high frequencies.

Equation (27d) shows that a short pulse gets completely distorted as it travels away from the source point. After some travel time it consists of small, rather rapid (high-frequency) fluctuations followed by larger and slower vibrations.

Note that in the harmonic case [i.e., Eqs. (27a) and (27c)] the time dependence $\exp(j\omega t)$ has to be taken into account and the real part has to be taken if the time history of the velocity is the desired quantity.

Due to the limits of applicability of the equations of motion, the results hold only below a certain frequency limit and not for very short times. For $k_B r > 2$, Eq. (27c) becomes

$$v(r) \approx \frac{F_0}{Z_F}\sqrt{\frac{2}{\pi k_B r}}\, e^{-jk_B r}\, e^{j\pi/4}. \tag{27e}$$

This resembles the well-known Green's function for sound waves in a two-dimensional free space.

2.9 Normal Modes and Resonance Frequencies

There are several possibilities to treat the influence of boundaries on the propagation of elastic waves. One method is to follow the paths of the waves including all reflections. This is equivalent to adding up the waves that are generated by the original source and all its images (see Chapter 74). This way the problem is reduced to investigating an infinite structure with many sources. Resonances occur when the waves originating from the images add up in phase. If the integral in Eq. (17b) is replaced by a Fourier series, that is, if there is a summation symbol instead of the integral, $(2\pi)^2/l_1 l_3$ instead of $dk_1\, dk_3$, and $2\pi n_1/l_1$, $2\pi n_3/l_3$ instead of k_1, $k_3 (n_1, n_3 = 1, 2, 3, \ldots)$, one is in essence dealing with a configuration that repeats itself with periodicity l_1 and l_3. Thus one has exactly the same situation that is obtained by image formation. The problem, however, is that one is restricted to rectangular structures and there is no way to include boundary conditions other than perfect reflection.

A second method is to express the wave field on a finite, homogeneous structure as a sum of possible propagating waves and near fields in the position and negative directions. For a beam in bending this would lead to

$$\xi(x) = A_+ e^{-jk_f x} + A_- e^{jk_f(x-1)} + B_+ e^{-k_N x} + B_- e^{k_N(x-1)}. \tag{28}$$

The two ends of the beam are at $x = 0$ and $x = 1$. Here, k_f is the free wavenumber for propagating waves and k_N is the near-field wavenumber. When the simple Bernoulli–Euler theory of bending is used (see Section A.2), the wavenumbers are given by $k_f = k_N = k_B$. When the more-accurate Timoshenko theory is used, one has to use the equations in Section A.4 of the Appendix and set $k_f = k_I$ and $jk_N = k_H$. The terms A_+, A_-, B_+, B_- are unknown coefficients that depend on the boundary conditions.

The corresponding method for plates consists in expressing the velocity field as a sum of propagating cylindrical waves with wavenumber k_F and cylindrical near fields with wavenumber jk_N. The amplitudes of these waves and near fields is then adjusted in such a way that a best fit to the boundary conditions is obtained.

A third method is the analog to the "solution" of the sound wave equation by the Kirchhoff–Helmholtz integral equation. In this method the wave field inside a boundary is expressed by a surface integral over the driving (outside) pressure $p(x_1, x_3)$ and a line integral over the field quantities along the boundary.[7]

The fourth and at present probably most often used method[8] to deal with the vibrations of one- and two-dimensional continuous systems is the finite element method (FEM) (see Chapters 12 and 48). This method, which can be seen as an application of Hamilton's principle, is described elsewhere.

For analytical calculations it is sometimes useful[9–11] to have simple expressions for the normal modes of a structure. For one-dimensional beams it is always possible to find the normal modes by using Eq. (28) and looking for solutions that fulfil the boundary conditions and that can exist without external excitation. For rods or bars exhibiting longitudinal or torsional motion, the results are the same as those for sound waves in tubes (see Chapter 6); the only difference is that the speed of sound has to be replaced by the longitudinal or torsional wave speed. For specific results on resonance frequencies and mode shapes the reader is referred to Refs. 1–3 and 9–11.

3 THIN, CURVED STRUCTURES

3.1 Cylinders

For the thin cylindrical shell shown in Fig. 3 there exist in the literature many slightly different equations of motion, depending on the required degree of accuraccy.[1–11] A simple and for most cases sufficiently accurate equation is

$$
\begin{aligned}
a_{11}u + a_{12}v + a_{13}w &= 0,\\
a_{12}u + a_{22}v + a_{23}w &= 0,\\
a_{13}u + a_{23}v + a_{33}w &= \frac{1-\mu^2}{Eh}\,p_r,
\end{aligned}
$$

$$
\begin{aligned}
a_{11} &= \frac{\partial^2}{\partial z^2} + \frac{1-\mu}{2a^2}\frac{\partial^2}{\partial \vartheta^2} - \frac{1}{c_L^2}\frac{\partial^2}{\partial t^2},\\
a_{12} &= \frac{1+\mu}{2a}\frac{\partial^2}{\partial z\,\partial \vartheta},\\
a_{13} &= \frac{\mu}{a}\frac{\partial}{\partial z},\\
a_{22} &= \frac{1-\mu}{2}(1+\beta^2)\frac{\partial^2}{\partial z^2} + \frac{1}{a^2}(1+\beta^2)\frac{\partial^2}{\partial \vartheta^2}\\
&\quad - \frac{1}{c_L^2}\frac{\partial^2}{\partial t^2},\\
a_{23} &= \frac{1}{a^2}\frac{\partial}{\partial \vartheta} - \beta^2\left(\frac{\partial^3}{\partial z^2\,\partial \vartheta} + \frac{1}{a^2}\frac{\partial^3}{\partial \vartheta^3}\right),\\
a_{33} &= \frac{1}{a^2} + \frac{h^2}{12a^2}\left(a^2\frac{\partial^4}{\partial z^4} + 2\frac{\partial^4}{\partial z^2\,\partial \vartheta^2} + \frac{1}{a^2}\frac{\partial^4}{\partial \vartheta^4}\right)\\
&\quad + \frac{1}{c_L^2}\frac{\partial^2}{\partial t^2},
\end{aligned}
\tag{29}
$$

where u, v, w are the displacements in the axial, tangential, and radial directions; p_r is the pressure that excites the cylinder; and c_L is given by $c_L^2 = E/\rho(1-\mu^2)$.

For many applications one can neglect terms with $\beta^2 = h^2/12a^2$. In this way the Donell–Mushtari shell equation is obtained. It is the simplest shell equation that, in addition to the membrane stresses, takes into account bending stresses in an approximate way.

The free wavenumbers (dispersion relation) of Eq. (29) can be obtained by inserting

$$
\begin{aligned}
u &= U\cos n\vartheta\; e^{-jk_z z}e^{j\omega t},\\
v &= V\sin n\vartheta\; e^{-jk_z z}e^{j\omega t},\\
w &= W\cos n\vartheta\; e^{-jk_z z}e^{j\omega t}
\end{aligned}
\tag{30}
$$

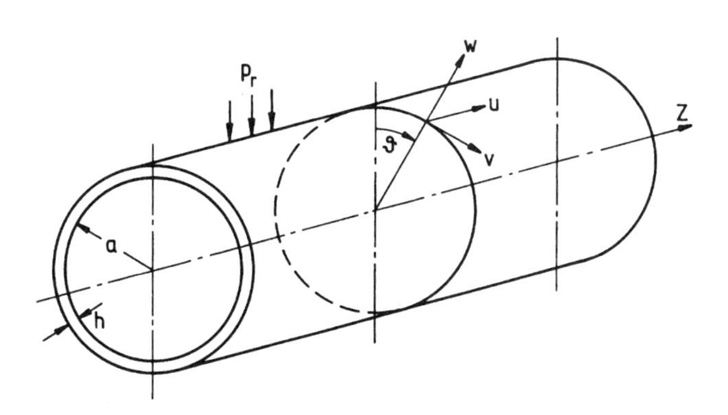

Fig. 3 Coordinate system for cylindrical shell.

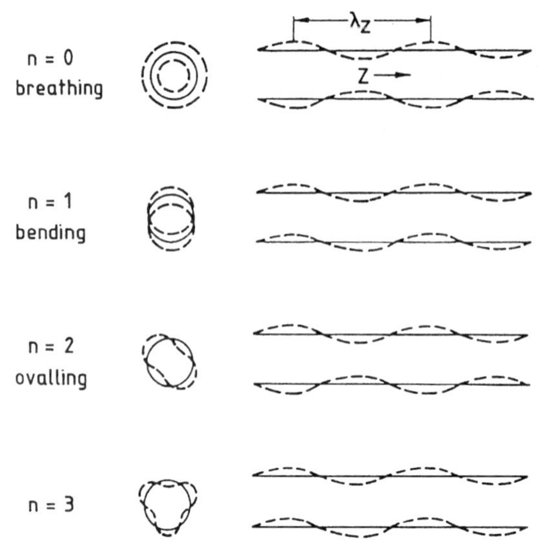

Fig. 4 Examples of mode shapes of a cylinder, $k_Z = 2\pi/\lambda_Z$.

into Eq. (29). In these equations $2n$ is the number of circumferential nodal lines and $2\pi/k_z$ the wavelength in the axial direction. See Fig. 4. One can view Eq. (30) also as two waves spinning in opposite directions around the cylinder. See also Section 4 in Chapter 10, where cylindrical shells are treated in connection with their sound radiation.

Combination of Eqs. (29) and (30) gives an expression of order $(k_z^2)^4$. Therefore there are four different waves or near fields. Closer examination shows that there are at most three different propagating waves for a particular value of n. For the case $a = 15h$ the wave speeds $c_z = \omega/k_z$ are plotted in Fig. 5. From this plot and several additional calculations the following can be seen:

(a) In the frequency range

$$\frac{\omega a}{c_L} < \frac{6\beta}{\sqrt{5}} \approx 0.775h/a$$

there are two waves with $n = 0$, one corresponding to a longitudinal motion with breathing, the other to a torsional motion. The only other possible (and very important) cylinder motion in this frequency range is bending ($n = 1$) as if the cylinder were a beam.

(b) In the range

$$\left(\frac{\omega a}{c_L}\right)^2 < \frac{1}{2}\,n^2(1-\mu)$$

there is one wave for each $n \le 2$; it starts with an infinite phase velocity (and a zero group velocity) at the ring frequency

$$\frac{\omega_n a}{c_L} = \frac{h}{\sqrt{12}a}\,\frac{n(n^2-1)}{\sqrt{n^2+1}}. \tag{31}$$

These waves are closely related to bending waves of a plate of thickness h made out of the cylinder material.

(c) Starting from $(\omega a/c_L)^2 = \frac{1}{2}n^2(1-\mu)$ there is a group of waves that correspond to shear waves in a plate.

(d) Starting from $(\omega a/c_L)^2 = n^2 + 1$ there is a group of waves corresponding to longitudinal waves in a plate.

If in Eq. (29) the substitution $x = a\vartheta$ is made, it is seen that the equations decouple for $a \to \infty$ giving the wave equations for in-plane waves (see Section A.3) and for bending waves [see Eq. (5)]. Thus in a good approximation one can say that for $\omega a/c_L > 2$ the curvature effects are of minor importance and a cylinder behaves almost like a flat plate of thickness h.

3.2 Rings

The equation of motion for rings is obtained by setting all terms with $\partial/\partial z$ in Eq. (29) to zero. This way equa-

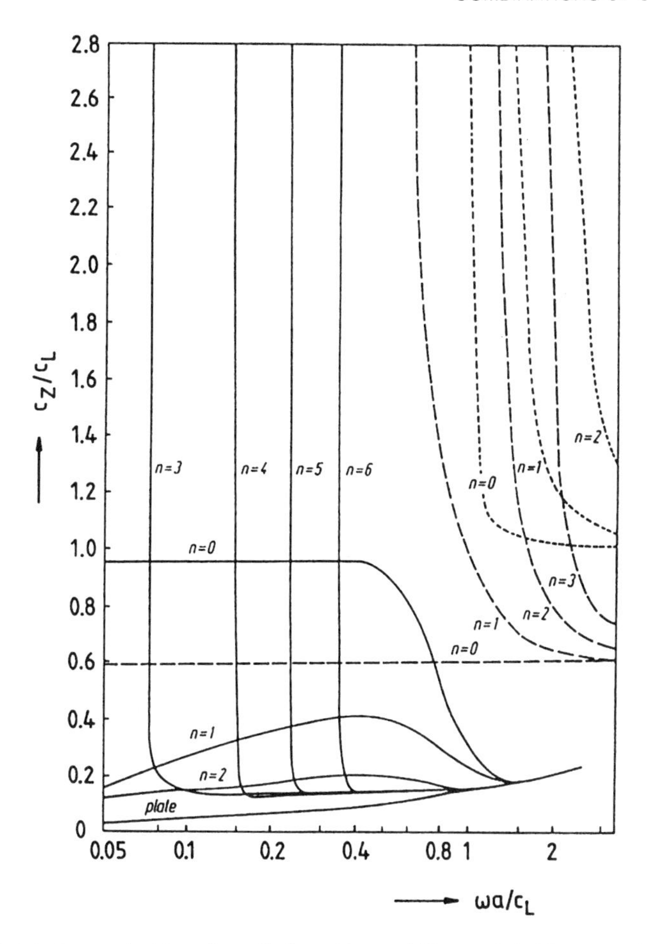

Fig. 5 Axial phase velocities of a cylinder with $a = 15h$; $\mu = 0.3$: The curve designated by "plate" corresponds to the phase velocity of flexural waves on a flat plate of thickness h. (-----) Related to shear motion. (-------) Related to longitudinal motion. (From Ref. 2.)

for the circumferential and radial motion (v, w) and for shear motion (u) in the axial direction are obtained. The ring resonance frequencies that can be found from these expressions are

$$\omega_n = \frac{hc_L}{\sqrt{12a^2}} \frac{n(n^2 - 1)}{\sqrt{n^2 + 1}} \tag{32}$$

and (of much less interest because twisting is not included)

$$\omega_n = \frac{nc_r}{a}$$

4 COMBINATIONS OF SUBSTRUCTURES

4.1 Beams

Beams that are connected with each other can be considered as one-dimensional models of intersecting walls. For their mathematical treatment one can use the following method[3]:

(a) For each substructure the motion is expressed in terms of the free propagating waves and near fields, similar to Eq. (28).

(b) The boundary conditions at the intersection points and at the ends are used to calculate the unknown amplitudes.

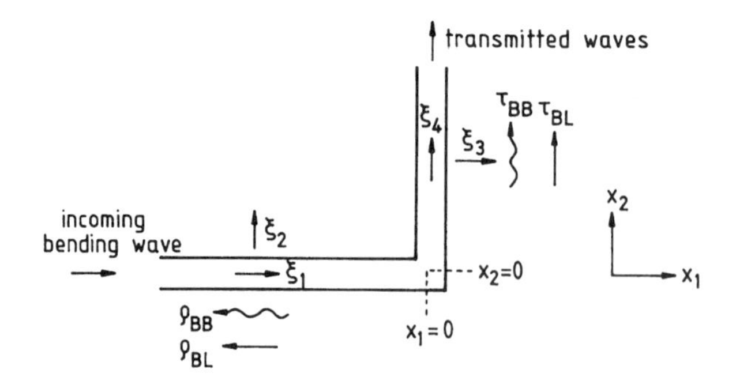

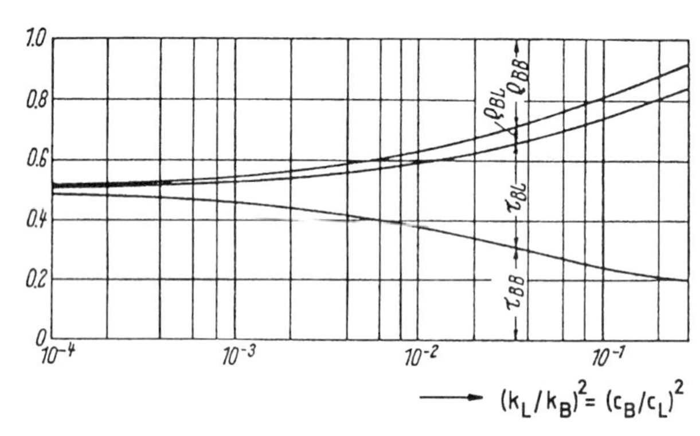

Fig. 6 Transmission of waves over a corner. The diagram was calculated for the case of identical beams.

(c) From these amplitudes the transmission coefficients can be found. For the example shown in Fig. 6 the following relations have to be used:

$$\xi_1(x_1) = r_{BL}e^{+jk_{L1}x_1},$$

$$\xi_2(x_1) = e^{-jk_{B1}x_1} + r_Be^{jk_{B1}x_1} + r_Ne^{k_{B1}x_1} \qquad \text{(horizontal)},$$

$$\xi_4(x_2) = t_{BL}e^{-jk_{L2}x_2},$$

$$\xi_3(x_2) = t_Be^{-jk_{B2}x_2} + L_Ne^{jk_{B2}x_2} \qquad \text{(vertical)}. \qquad (33)$$

The boundary conditions at $x_1 = 0$, $x_2 = 0$ are

$$\xi_1 = \xi_3,$$

$$\xi_2 = \xi_4 \qquad \text{(equality of displacements)},$$

$$\frac{\partial \xi_2}{\partial x_1} = \frac{\partial \xi_3}{\partial x_2} \qquad \text{(equality of angles)},$$

$$B_1 \frac{\partial^2 \xi_2}{\partial x_1^2} = B_2 \frac{\partial^2 \xi_3}{\partial x_2^2} \qquad \text{(equality of moments)},$$

$$-B_1 \frac{\partial^3 \xi_2}{\partial x_1^3} = -E_2S_2 \frac{\partial \xi_4}{\partial x_2},$$

$$-E_1S_1 \frac{\partial \xi_1}{\partial x_1} = -B_2 \frac{\partial^3 \xi_3}{\partial x_2^3} \qquad \text{(equality of forces)}.$$

These six conditions are sufficient to find t_B, $r_B, \ldots$. The curves in Fig. 6 were obtained this way. They show the reflection and transmission coefficients for the reflected bending energy, the transmitted bending energy, the reflected longitudinal energy, and the transmitted longitudinal energy. All are normalized by the incoming energy:

$$\rho_{BB} = |r_B^2|, \qquad \tau_{BB} = \frac{\rho_2S_2k_{B1}}{\rho_1S_1k_{B2}}\,|t_B^2|,$$

$$\rho_{BL} = \frac{k_{B1}}{2k_{L1}}\,|r_{BL}^2|, \qquad \tau_{BL} = \frac{\rho_2S_2k_{B1}}{2\rho_1S_1k_{L2}}\,|t_{BL}^2|. \qquad (34)$$

It is interesting to note that, at least on a decibel scale, τ_{BB} is not very small. Thus corners are not much of an obstacle (typically 3 dB) for bending waves. At low frequencies there is almost no transition of bending wave energy to longitudinal waves, but at higher frequencies this effect can be substantial. An alternative method of treating beam combinations is to apply Hamilton's principle, which in essence corresponds to the substructure technique (with each beam being an element) in the FEM.

4.2 Multilayer Plates

Multilayer plates (see, e.g., Fig. 7) are very common as plywood, sandwich plates, or plates with damping layers. If it is assumed that the core is reasonably stiff so that for the frequencies of interest the motion $\xi_2(x_1, x_3)$ in the normal direction is the same over the whole thickness, such plates can be treated by standard techniques. A convenient method to find the equation of motion for multilayer plates is again Hamilton's principle. A typical result of such a calculation has the following general structure:

$$\alpha_{11}\xi_1 + \alpha_{12}\xi_2 = 0, \qquad \alpha_{12}\xi_1 + \alpha_{22}\xi_2 = p_2. \tag{35}$$

Here the α_{ij} are rather lengthy linear differential operators that contain the different material properties and dimensions.

An interesting feature of Eq. (35) is the fact that in general the equations cannot be decoupled by introducing a frequency-independent neutral fiber for which $\xi_1 = 0$. Thus, in contrast to homogeneous beams, there is always a (small) coupling between bending motion and longitudinal motion.

By introducing a general plane-wave solution [see Eq. (6)], the dispersion relation, the free wavenumbers, and the wave speeds are obtained. By setting $\underline{G}_2 = G_2(1 + j\eta_{2G})$ and $\underline{E}_2 = E_2(1 + j\eta_{2E})$, where η_{2G} and η_{2E} are loss factors of the inner material, it is also possible to find the loss factor and decay rate of the multilayer beam. The results are practically the same as those given in *Enc.* Ch. 149, although there a quasi-static theory is used.

5 PERIODIC STRUCTURES

Periodic structures such as frame-stiffened ship hulls, airplane fuselages, certain types of ceiling constructions, and railway rails on ties are very common. For low frequencies, when the wavelength is much longer than the distance 1 between two discontinuities (see Fig. 8), periodic structures can be treated like homogeneous ones with modified mass (and rotary inertia). For discontinuities with a force impedance Z_{FD} on a bar, this can be done by replacing the mass per unit area ρS by the modified mass

$$\rho S \left(1 + \frac{Z_{FD}}{j\omega\rho S l}\right).$$

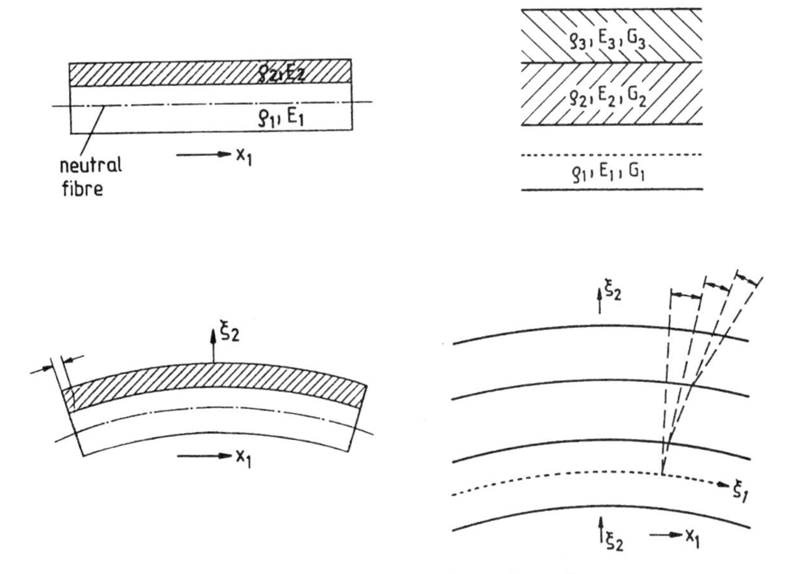

Fig. 7 Geometry of two- and three-layer plates.

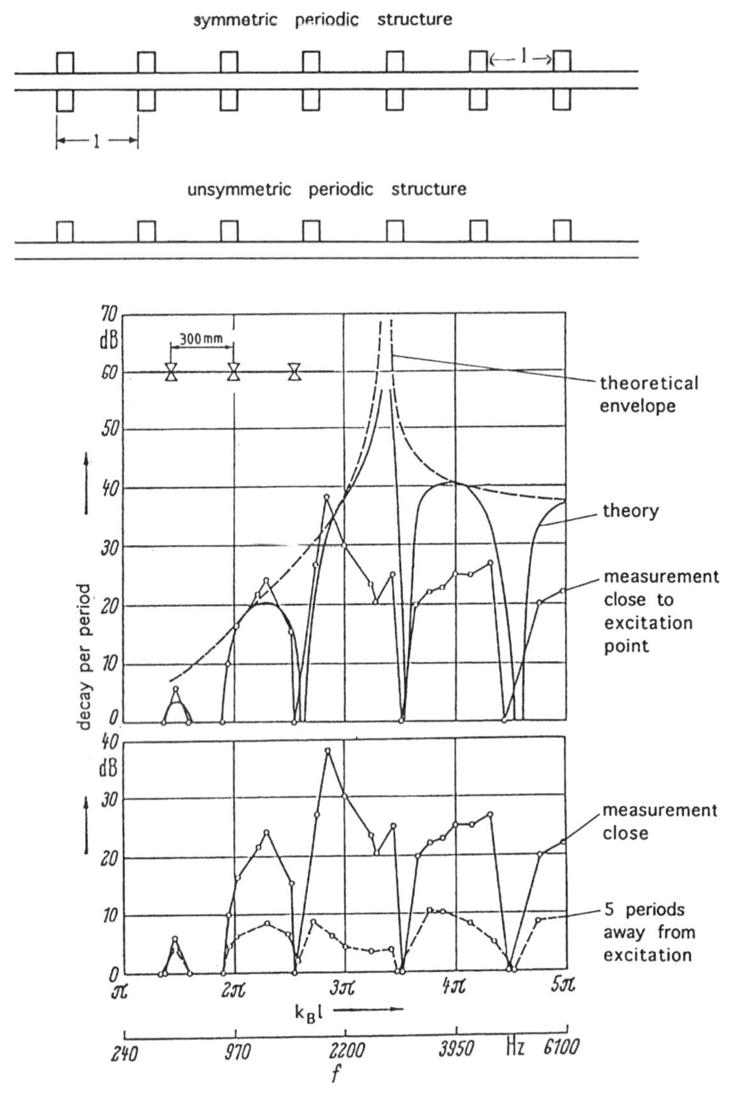

Fig. 8 Periodic system. The frequency curve holds for "blocking masses" of 600 g on a steel beam of cross section 10 × 30 mm.

For longitudinal waves on bars [see Eqs. (1) and (2)] this leads to the longitudinal, free wavenumber [instead of Eq. (7a)]

$$k_{1fD}^2 = \frac{\omega^2 \rho S}{ES}\left(1 + \frac{Z_{FD}}{j\omega\rho Sl}\right)$$

$$= k_{1f}^2\left(1 + \frac{Z_{FD}}{j\omega\rho Sl}\right). \qquad (36)$$

This method works for sinusoidal motion of frequency ω, provided the discontinuities are almost symmetric (otherwise new wave types are excited). The impedance may consist of a single mass m_D (i.e., $Z_{FD} = j\omega m_D$) or a combination of mass, spring, and damper elements. It also may be the impedance of a continuous structure.

For beams in bending Eq. (36) is a poor approximation in certain frequency bands. It is better to include also the effect of rotary inertia. This leads to the following equation for the low-frequency wavenumber k_{BD} of a beam with symmetric periodic discontinuities of force impedance Z_{FD} and moment impedance Z_{MD}:

$$k_{BD}^2 \approx k_B^2 \left[\alpha \pm \sqrt{1 + \frac{Z_{FD}}{j\omega\rho Sl} + \alpha^2} \right], \qquad (37)$$

with

$$\alpha = \frac{Z_{MD} k_B^2}{2j\omega\rho Sl}.$$

Here k_B is the free bending wavenumber of the beam without the discontinuities. It is given in the Section A.2.

It can be easily seen that impedances of inertia type (i.e., proportional to $+j\omega$) decrease the wavelength; impedances of spring type (i.e., proportional to $-j/\omega$) increase it or may even prevent waves from propagating. Real-valued impedances cause damping.

When the frequency is high enough to make $|k_{BD} l| > 2$ or $|k_{LD} l| > 2$, Eqs. (36) and (37) are no longer valid. In this case one has to apply the well-developed theory of periodic systems.[2,12,13] Some interesting results obtained this way and verified by experiments are as follows:

(a) When the wavelength is comparable with or smaller than the period spacing, there are stopping and passing bands, that is, frequency ranges in which the waves decay exponentially in space and alternate frequency ranges in which they propagate almost without attenuation. An example is shown in Fig. 8.

(b) The actual location of the stopping and passing bands is a rather complicated function of the force and moment impedances of the added frames, ties, and so on.

(c) When the periodic structure is not symmetric, wave conversion occurs; for example, at the discontinuities bending waves are converted to longitudinal waves and vice versa. A consequence of that is that within the stopping bands the vibration decay may be rather pronounced near the point of excitation and much smaller in some distance. The reason for this phenomenon is that after a certain distance the wave type with the smallest decay rate is responsible for the transport of energy.[2,14]

(d) The resonance frequencies of finite periodic structures always lie within the passing bands.

(c) For infinite periodic structures the input impedance for point or moment excitation is purely imaginary within the stopping bands.

APPENDIX: SOME EQUATIONS RELATED TO PLANE ONE- AND TWO-DIMENSIONAL STRUCTURES WITHOUT BOUNDARIES

A.1 Torsional Waves in Rods

1. *Equation of Motion*

$$GI \frac{\partial^2 w}{\partial x_1^2} - \rho\Theta \frac{\partial^2 w}{\partial t^2} = 0[= -M_d(x_1)],$$

where GI is torsional stiffness, Θ the moment of inertia (both per unit length), G the shear modulus, w the angle of torsion, and M_d the driving moment. For circular cross section $\Theta = \frac{1}{2}\pi(a_1^4 - a_2^4)$; $I = \frac{1}{2}\pi(a_1^4 - a_2^4)$. For rectangular cross section $\Theta = \frac{1}{12}(bh^3 + hb^3)$; $I = \alpha b^3 h$ (for $h = b$, $\alpha = 0.141$; for $h = 2b$, $\alpha = 0.229$; for $h = 10b$, $\alpha = 0.312$).

The following equations are equivalent:

$$M = -GI \frac{\partial w}{\partial x_1}, \qquad -\rho\Theta \frac{\partial^2 w}{\partial t^2} [+M_d(x_1)] = \frac{\partial m}{\partial x_1}.$$

2. *Dispersion Relation*

$$GIk_{1f}^2 = \rho\Theta\omega^2, \qquad c_T = (GI/\rho\Theta)^{1/2}.$$

3. *Energy Density*

$$W_{e,\text{kin}} = \frac{1}{2}\rho\Theta w^2, \qquad W_{e,\text{pot}} = \frac{1}{2} GI\left(\frac{\partial w}{\partial x_1}\right)^2.$$

4. *Moment Impedance Semi-infinite, Excited at One End*

$$Z_M = \frac{M_d}{w(0)} = \sqrt{\rho\Theta GI} = \rho\Theta c_T.$$

A.2 Bending Waves in Beams (Bernoulli–Euler)

1. *Equation of Motion (See Fig. 1c)*

$$B \frac{\partial^4 \xi_2}{\partial x_1^4} + \rho S \frac{\partial^2 \xi_2}{\partial t^2} = p_2,$$

where $B = ES\kappa^2$ is the bending stiffness, S the cross section, κ the radius of gyration [rectangular: $\kappa = h/\sqrt{12}$; circular with outer radius a_1 and inner radius a_2: $\kappa = 0.5(a_1^2 + a_2^2)^{1/2}$].

The following first-order equations are equivalent:

$$w = \frac{\partial \xi_2}{\partial x_1}, \qquad M = B\,\frac{\partial^2 \xi_2}{\partial x_1^2}, \qquad F_2 = \frac{\partial M}{\partial x_1},$$

$$\frac{\partial F_2}{\partial x} = -\rho S\,\frac{\partial^2 \xi_2}{\partial t^2},$$

where w is the bending angle, M the bending moment, F_2 the force in the x_2-direction.

2. *Dispersion Relation*

$$Bk_B^4 = \omega^2 \rho S, \qquad k_B = \omega / c_B = \left(\frac{\omega^2 \rho S}{B}\right)^{1/4}.$$

Group velocity $c_g = 2c_B$,

3. *Energy Density*

$$W_{e,\text{kin}} = \frac{1}{2}\,\rho S \dot{\xi}_2^2, \qquad W_{e,\text{pot}} = \frac{1}{2}\,B\left(\frac{\partial^2 \xi_2}{\partial x_1^2}\right)^2.$$

4. *Point Impedance*

$$Z_F = \begin{cases} \dfrac{2\rho S \omega (1+j)}{k_B} & \text{(infinite, excited in the middle)},\\[2ex] \dfrac{\rho S \omega (1+j)}{2k_B} & \text{(semifinite, excited at end)}. \end{cases}$$

A.3 In-plane Waves in Isotropic Plates

1. *Equations of Motion*

$$\left(\frac{2}{1-\mu}\,\frac{\partial^2}{\partial x_1^2} + \frac{\partial^2}{\partial x_3^2} - \frac{\rho}{G}\,\frac{\partial^2}{\partial t^2}\right)\xi_1 + \frac{1+\mu}{1-\mu}\,\frac{\partial^2 \xi_3}{\partial x_1\,\partial x_3}$$

$$= \frac{-1}{G}\left[\frac{\mu}{2(1-\mu)}\,\frac{\partial p_2}{\partial x_1} + \frac{p_1}{h}\right],$$

$$\frac{1+\mu}{1-\mu}\,\frac{\partial^2 \xi_1}{\partial x_1\,\partial x_3} + \left(\frac{\partial^2}{\partial x_1^2} + \frac{2}{1-\mu}\,\frac{\partial^2}{\partial x_3^2} - \frac{\rho}{G}\,\frac{\partial^2}{\partial t^2}\right)\xi_3$$

$$= \frac{-1}{G}\left[\frac{\mu}{2(1-\mu)}\,\frac{\partial p_2}{\partial x_3} + \frac{p_3}{h}\right],$$

where p_2 is the force per unit area acting in the x_2-direction (normal to plate), p_1 and p_3 the forces per unit area acting in the x_1- and x_3-directions (parallel to the plate), G the shear modulus, and μ Poisson's ratio.

2. *Dispersion Relations*

$$k_{1S}^2 = \frac{\omega^2 \rho}{G} - k_3^2 \quad \text{(shear waves)},$$

$$k_{1L}^2 = \frac{\omega^2 \rho}{2G}\,(1-\mu) - k_3^2 \quad \text{(longitudinal waves)}.$$

The plane-wave solution (for k_3 arbitrary) is given as

$$\xi_1(x_1, x_3) = [A_{S+}e^{-jk_{1S}x_1} + A_{S-}e^{jk_{1S}x_1} + A_{L+}e^{-jk_{1L}x_1} + A_{L-}e^{jk_{1L}x_1}]e^{-jk_3x_3}$$

$$\xi_3(x_1, x_3) = \left[-A_{S+}\,\frac{k_{1S}}{k_3}\,e^{-jk_{1S}x_1} + A_{S-}\,\frac{k_{1S}}{k_3}\,e^{jk_{1S}x_1} + A_{L+}\,\frac{k_3}{k_{1L}}\,e^{-jk_{1L}x_1} - A_{L-}\,\frac{k_3}{k_{1L}}\,e^{jk_{1L}x_1}\right] \cdot e^{-jk_3x_3}.$$

3. *Energy Density*

$$W_{e,\text{kin}} = \frac{1}{2}\,\rho h(\dot{\xi}_1^2 + \dot{\xi}_3^2),$$

$$W_{e,\text{pot}} = \frac{Eh}{2(1-\mu^2)}\left[\left(\frac{\partial \xi_1}{\partial x_1}\right)^2 + \left(\frac{\partial \xi_3}{\partial x_3}\right)^2 + 2\mu\,\frac{\partial \xi_1}{\partial x_3}\,\frac{\partial \xi_3}{\partial x_1} + \frac{1-\mu}{2}\left(\frac{\partial \xi_1}{\partial x_3} + \frac{\partial \xi_3}{\partial x_1}\right)^2\right].$$

4. *Point Impedance*[15]

$$\frac{1}{Z_F} = \frac{\omega(3-\mu)}{16hG} + j\,\frac{\omega}{8\pi hG}\left[(1-\mu)\ln\frac{2c_L}{\omega a} + 2\ln\frac{2c_S}{\omega a}\right],$$

$$c_S = \sqrt{\frac{G}{\rho}}, \qquad c_L = \sqrt{\frac{E}{[(1-\mu^2)\rho]}}.$$

A.4 Bending Waves in Isotropic Plates Including Shear Stiffness and Rotatory Inertia (Timoshenko–Mindlin Theory)[2,3]

1. *Equation of Motion*

$$D\Delta\Delta\xi_2 - \left(\frac{D\rho}{G} + \rho I\right)\Delta\frac{\partial^2\xi_2}{\partial t^2} + \rho h\frac{\partial^2\xi_2}{\partial t^2} + \rho I\frac{\rho}{G}\frac{\partial^4\xi_2}{\partial t^4} = p_2 - \frac{D}{Gh}\Delta p_2 + \frac{\rho I}{Gh}\frac{\partial^2 p_2}{\partial t^2},$$

where D is the plate bending stiffness, G the shear modulus, $I = h^3/12$, and p_2 the force per unit area in the x_2-direction (normal to plate).

An equivalent form is

$$Gh(\Delta\xi_2 + \Psi) = \rho h\frac{\partial^2\xi_2}{\partial t^2} - p_2,$$

$$D\Delta\Psi - Gh(\Delta\xi_2 + \Psi) = \rho I\frac{\partial^2\Psi}{\partial t^2},$$

$$\Psi = \frac{\partial w_1}{\partial x_1} + \frac{\partial w_3}{\partial x_3},$$

where w_1 and w_3 are bending angles.

2. *Dispersion Relation*

$$k_{\text{I}}^2 = (k_{1f}^2 + k_{3f}^2)_I = \beta_{SL}^2 + \left(\beta_{SL}^4 + \frac{\alpha\omega^2\rho h}{D}\right)^{1/2}$$

(propagating),

$$k_{II}^2 = (k_{1f}^2 + k_{3f}^2)_{II} = \beta_{SL}^2 - \left(\beta_{SL}^4 + \frac{\alpha\omega^2\rho h}{D}\right)^{1/2}$$

(near field),

with

$$\beta_{SL}^2 = \frac{\omega^2}{2}\left(\frac{1}{c_T^2} + \frac{1}{c_L^2}\right), \qquad \alpha = 1 - \frac{\omega^2 h^2}{12G}$$

and $c_T = (G/\rho)^{1/2}$ is the shear wave speed and $c_L = (D/\rho I)^{1/2} = (E/\rho(1-\mu^2))^{1/2}$ the longitudinal wave speed. For $\alpha > 0$ there are propagating waves and near fields; for $\alpha < 0$ there are only propagating waves.

3. *Energy Density*

$$W_{e,\text{kin}} = \frac{\rho h}{2}\left[\dot{\xi}_2^2 + \frac{h^2}{12}(\dot{w}_1^2 + \dot{w}_3^2)\right],$$

$$W_{e,\text{pot}} = \frac{E}{2(1-\mu^2)}\frac{h^3}{12}\left[\left(\frac{\partial w_1}{\partial x_1}\right)^2 + \left(\frac{\partial w_3}{\partial x_3}\right)^2 + 2\mu\left(\frac{\partial w_1}{\partial x_3}\frac{\partial w_3}{\partial x_1}\right)\right] + \frac{Gh}{2}\left[\left(w_1 + \frac{\partial\xi_2}{\partial x_1}\right)^2 + \left(w_3 + \frac{\partial\xi_2}{\partial x_3}\right)^2 + \frac{h^2}{12}\left(\frac{\partial w_1}{\partial x_3} + \frac{\partial w_3}{\partial x_1}\right)^2\right].$$

4. *Point Impedance*

$$Z_F = \frac{8D}{\omega}\frac{k_I^2 - \beta_{SL}^2}{A_R + jA_I},$$

$$A_R = \begin{cases} \alpha + 2\dfrac{D}{Gh}(k_I^2 - \beta_{SL}^2) & \text{for } \alpha > 0, \\ 2\dfrac{D}{Gh}(k_I^2 - \beta_{SL}^2) & \text{for } \alpha < 0, \end{cases}$$

$$A_1 = \frac{-2}{\pi}\left[\alpha\ln\left(\frac{\omega}{k_I^2}\sqrt{\frac{\rho h|\alpha|}{D}}\right) + \frac{D}{Gh}(k_I^2\ln k_I a - k_{II}^2\ln k_{II}a)\right],$$

where a is the radius of the excited area.

A.5 Bending Waves in Orthotropic Plates (Bernoulli–Euler)

1. *Equation of Motion*[16]

$$D_1\frac{\partial^4\xi_2}{\partial x_1^4} + 2(D_\mu + 2D_G)\frac{\partial^4\xi}{\partial x_1^2\partial x_3^2} + D_3\frac{\partial^4\xi_2}{\partial x_3^4} + \rho h\frac{\partial^2\xi_2}{\partial t^2} = p_2,$$

where D_1, D_3 are the bending stiffnesses in the x_1- and x_3-direction. For anisotropic materials $D_1 = E_1h^3/12(1-\mu^2)$, $D_3 = E_3h^3/12(1-\mu^2)$, $D_\mu = \frac{1}{12}E_\mu h^3$, and $D_G = \frac{1}{12}Gh^3$.

2. *Dispersion Relation*

$$D_1 k_{1f}^4 + 2(D_\mu + 2D_G)k_{1f}^2 k_{3f}^2 + D_3 k_{3f}^4 - \omega^2 \rho h = 0.$$

Here, the angle dependence is obtained by setting $k_{1f} = k_f \sin \vartheta$; $k_{3f} = k_f \cos \vartheta$.

3. *Point Impedance*

$$Z_F = \frac{4\pi}{K(\alpha)} (\rho^2 h^2 D_1 D_3)^{1/4},$$

$$\alpha = \left\{ \frac{1}{2} \left[1 - \frac{D_\mu + 2D_G}{(D_1 D_3)^{1/2}} \right] \right\}^{1/2},$$

where $K(\alpha)$ is the complete elliptic integral of the first kind. Useful approximations are $D_\mu + 2D_G \approx (D_1 D_3)1/2$ and $K(\alpha) = \frac{1}{2}\pi$.

REFERENCES

1. Lord Rayleigh, *The Theory of Sound*, Macmillan, London, 1877.
2. L. Cremer, M. Heckl, and E. E. Ungar, *Structure-Borne Sound*, Springer, Berlin, 1988.
3. M. C. Junger and D. Feit, *Sound Structures and Their Interaction*, 2nd ed., MIT Press, Cambridge, MA.
4. F. J. Fahy, *Sound and Structural Vibration*, Academic, London, 1989.
5. L. Cremer, "Theorie der Schalldämmung dünner Wände bei schrägem Einfall," *Akustische Zeitschrift*, Vol. VII, 1942, p. 81.
6. L. L. Vér, "Interaction of Sound Waves with Solid Structures," L. L. Beranek, I. L. Vér (Eds.), *Noise and Vibration Control Engineering*, Wiley, New York, 1992, Chapter 9.
7. J. Scheuren, "Application of the Method of Integral Equations to the Vibrations of Plates," *Recent Advances in Structural Dynamics*, Vol. I, ISVR, University of Southampton, Southampton, 1984, pp. 171–177.
8. O. L. Zienkiewicz, *The Finite Element Method*, McGraw-Hill, London, 1977.
9. R. D. Blevins, *Formulas for Natural Frequency and Mode Shape*, Van Nostrand, New York, 1979.
10. A. W. Leissa, "Vibration of Plates," NASA SP-160, Washington, DC, 1969.
11. A. W. Leissa, "Vibration of Shells," NASA SP-288, Washington, DC, 1973.
12. L. Brillouin, *Wave Propagation in Periodic Structures*, Dover, New York, 1953.
13. D. J. Mead, "Response of Periodic Structures to Noise Fields," in R. G. White (Ed.), *Noise and Vibration*, Ellis Horwood, Chichester, 1982, pp. 285–306.
14. V. D. Belov, S. A. Rybak, B. A. Tartakovskii, "Propagation of Vibrational Energy in Absorbing Structures," *Sov. Phys. Acoust.*, Vol. 23, 1977, pp. 115–118.
15. S. Ljunggren, "Generation of Waves in an Elastic Plate by a Torsional Moment and a Horizontal Force," *J. Sound Vib.*, Vol. 93, 1984, pp. 161–187.
16. M. Heckl, "Untersuchungen an orthotropen Platten," *Acustica*, Vol. 10, 1960, pp. 109–115.

50

NONLINEAR VIBRATION

LAWRENCE N. VIRGIN AND EARL H. DOWELL

1 INTRODUCTION

Problems in nonlinear mechanical vibration are considerably more difficult to analyze than their linear counterparts. The presence of nonlinear forces in many physical systems results in a spectacular array of possible behavior even for relatively low-order models. Mechanisms of instability and sudden changes in response are of fundamental importance in the behavior of nonlinear systems. The closed-form solutions familiar from linear vibration theory are of limited value and give little hint as to the complexity and sometimes unpredictability of typical nonlinear behavior.

Solutions to the governing nonlinear equations of motion are obtained either using approximate analytical methods or numerical simulation guided by dynamical systems theory. Due to the inherent complexity of nonlinear vibration, the use of the geometric, qualitative theory of ordinary differential equations play a central role in the classification and analysis of such systems.

This chapter is divided into two main sections. The first describes the effects of nonlinear stiffness and damping on free oscillations resulting in multiple equilibria, amplitude-dependent frequencies, basins of attraction, and limit cycle behavior. The second describes the effects of external periodic forcing resulting in nonlinear resonance, hysteresis, subharmonics, and chaos. Low-order, archetypal examples with relevance to mechanical and electrical oscillators are used to illustrate the phenomena.

Note: References to chapters appearing only in the *Encyclopedia* are preceded by *Enc.*

2 NONLINEAR FREE VIBRATION

2.1 Autonomous Dynamical Systems

Nonlinear free vibration problems are generally governed by sets of n first-order ordinary differential equations of the form[1]

$$\dot{\mathbf{x}} = \mathbf{f}(\mathbf{x}), \tag{1}$$

where the function $\mathbf{f}$ defines a vector field in the n-dimensional phase space, and an overdot represents a time derivative. In single-degree-of-freedom oscillators position and velocity are the two state variables. Often the stiffness terms are a function of displacement only and can be derived from a potential energy function. Many mechanical systems have the property that for small displacements this function will be linear in x, and furthermore the damping is often assumed to be governed by a simple, viscous energy dissipation giving linear terms in $\dot{x}$. In this case standard vibration theory (see Chapter 48) can be used to obtain exact closed-form solutions giving an exponentially decaying (sinusoidal or monotonic) response.[2,3] Linear algebra techniques and modal analysis can be used for multi-degree-of-freedom systems. For systems where the induced nonlinearity is small, linearization and certain approximate analytical techniques can be used, but if no restriction is placed on nonlinearity, then numerical methods must generally be used to solve Eq. (1). Specific examples will illustrate typical transient behavior.

2.2 Simple Pendulum

Small-Amplitude Linear Behavior As an introduction to typical nonlinear behavior, consider the simple

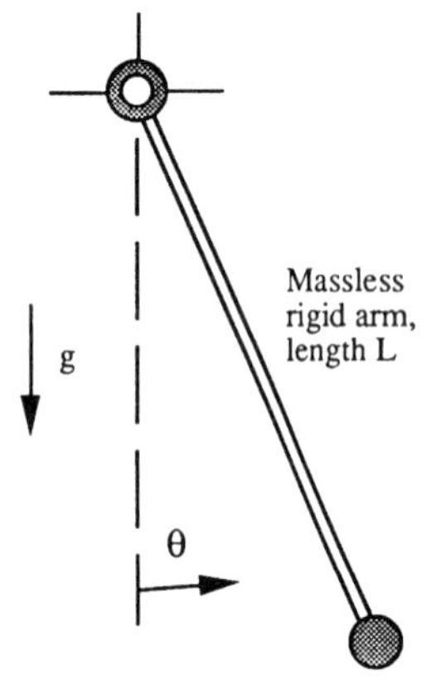

Fig. 1 Schematic diagram of the simple rigid-arm pendulum.

rigid-arm pendulum shown schematically in Fig. 1. It is a simple matter, using Newton's laws or Lagrange's equations,[4] to derive the governing equation of motion

$$\ddot{\theta} + b\dot{\theta} + \frac{g}{L}\sin\theta = 0, \tag{2}$$

where b is the (viscous) damping coefficient. Note that the linear, undamped natural frequency is a constant $\omega_0 = \sqrt{g/L}$. For small angles, $\sin\theta \cong \theta$, and given the two initial conditions θ_0 and $\dot{\theta}_0$, the pendulum will undergo periodic motion that decays with a constant (damped) period as $t \rightarrow \infty$, coming to rest at $\theta = 0$, the position of static stable equilibrium (Fig. 2a). This equilibrium position acts as a point attractor for all local transients.

It is very useful in nonlinear vibrations to look at phase trajectories in a plot of displacement against velocity. This is shown in Fig. 2b for the pendulum started from rest at $\theta_0 = \pi/8$. The near elliptical nature of these curves (indicating sinusoidal motion) is apparent as the motion evolves in a clockwise direction.

The motion of the pendulum can be thought to be occurring within the potential energy well shown in Fig. 3, that is, the integral of the restoring force. The assumption of small angles effectively means that the restoring force is assumed to be linear about the origin (Hooke's law), with a locally parabolic potential energy well. Using the analogy of a ball rolling on this surface, it is easy to visualize the motion shown in Fig. 2.

Large-Amplitude Nonlinear Behavior For large-angle swings of the pendulum the motion is no longer linear. The restoring force induces a softening spring effect. The natural frequency of the system will now be a function of amplitude as the unstable equilibrium at $\theta = \pi$ comes into effect. Linear theory can also be used about an unstable equilibrium, and local trajectories will tend to diverge away from a saddle point. Physically this is the case of the pendulum balanced upside down. The process of linearization involves the truncation of a power series about the equilibrium positions and characterizes the local motion. Often this information can then be pieced together to obtain a qualitative impression of the complete phase portrait.[5]

Consider the undamped ($b = 0$) pendulum started from rest at $\theta(0) = 0.99\pi$ and shown as a velocity time series in Fig. 4a based on direct numerical simulation. These oscillations are far from sinusoidal. The pendulum slows down as it approaches its inverted position, with a natural period of approximately 3.5 times the linear natural

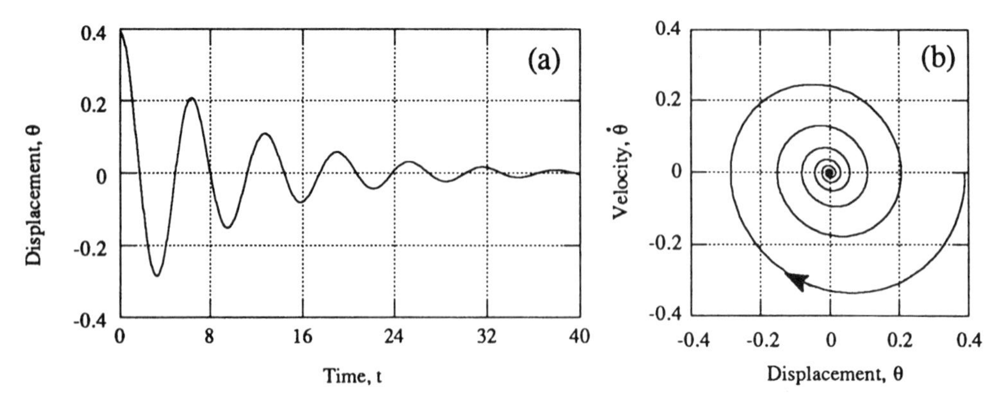

Fig. 2 Small-amplitude swings of the damped pendulum illustrating linear motion: (a) time series; (b) phase portrait ($\theta_0 = \pi/8$, $\omega_0 = 1$, $b = 0.1$).

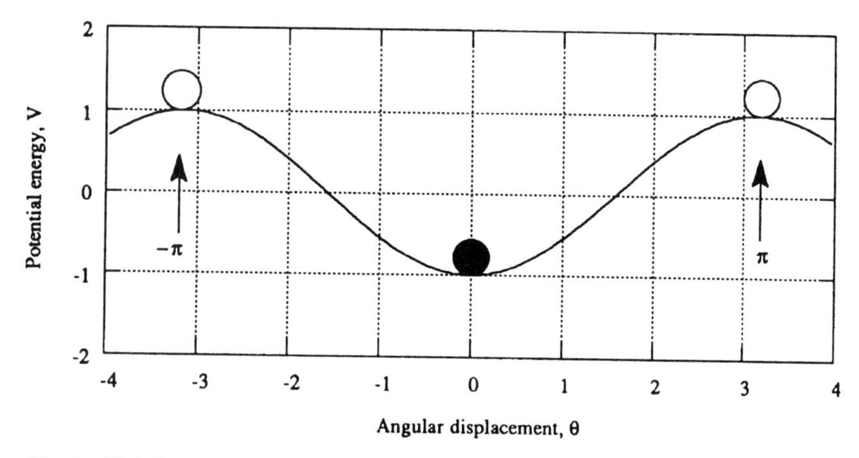

Fig. 3 Underlying potential energy for the pendulum showing the analogy of a rolling ball. The black ball represents stable equilibrium.

period of 2π. Exact analytical solutions are available here in the form of elliptic integrals,[6] and again this motion can be illustrated as a phase portrait (Fig. 4*b*). Since the phase space for this system is the surface of a cylinder, the saddle points corresponding to the hilltops in Fig. 3 represent the same inverted position. Here the phase trajectories (for this undamped system) can be considered as contours of constant total energy, that is,

$$E = \tfrac{1}{2}\dot{\theta}^2 - \omega_0^2 \cos\theta, \tag{3}$$

where the initial conditions $(\theta_0, \dot{\theta}_0)$ determine E. The phase trajectories described by Eq. (3) will be closed (i.e., periodic) for relatively small initial conditions, but open orbits corresponding to a spinning pendulum result if the starting conditions provide sufficient initial energy.

2.3 Duffing's Equation

Another archetypal example of an oscillator with a nonlinear restoring force is an autonomous form of Duffing's equation:

$$\ddot{x} + b\dot{x} + \alpha x + \beta x^3 = 0, \tag{4}$$

which can, for example, be used to study the large-amplitude motion of a pre- or postbuckled beam[4] or plate or the moderately large-amplitude motion of the pendulum. The free vibration of a beam loaded axially beyond

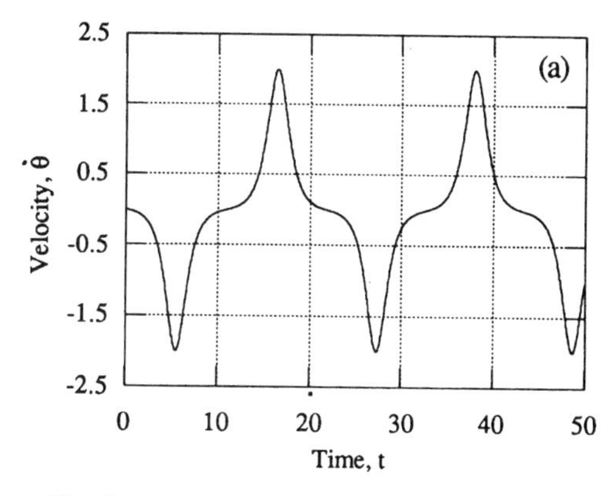

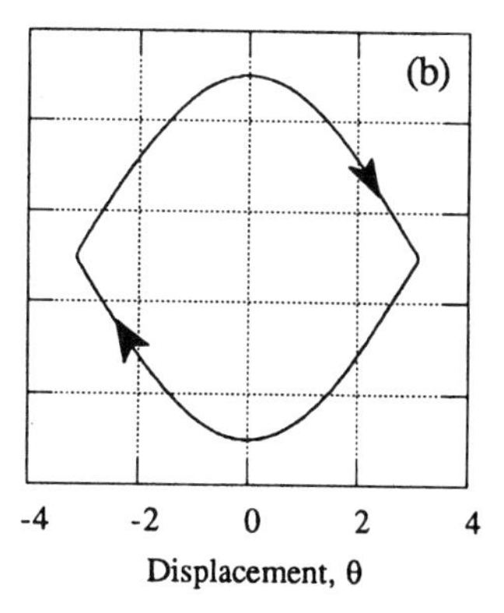

Fig. 4 Large undamped swings of the undamped pendulum: (*a*) velocity time series; (*b*) phase portrait ($\theta_0 = 0.99\pi$, $\omega_0 = 1$).

its elastic critical limit can be modeled by Eq. (4) with negative linear and positive cubic stiffness. In this case the origin (corresponding to the straight configuration) is unstable with two stable equilibrium positions at $x = \pm 2$, for example, when $\alpha = -1$ and $\beta = 4$. Now, not only does the natural period depend on initial conditions but so does the final resting position as the two stable equilibria compete to attract transients. Each equilibrium is surrounded by a domain of attraction. These interlocking domains, or catchment regions, are defined by the separatrix, which originates at the saddle point. It is apparent that some transients will traverse the (double) potential energy well a number of times before sufficient energy has been dissipated and the motion is contained within one well. Clearly it is difficult to predict which long-term resting position will result given a level of uncertainty in the initial conditions.[4,7]

2.4 Van der Pol's Oscillator

In the previous section the nonlinearity in the stiffness resulted in behavior dominated by point attractors (equilibria). Periodic attractors (limit cycles) are also possible in autonomous dynamical systems with nonlinear damping. A classical example relevant to electric circuit theory is Van der Pol's equation,[1,8]

$$\ddot{x} - h(1 - x^2)\dot{x} + x = 0, \tag{5}$$

where h is a constant. Here, the damping term is positive for $x^2 > 1$ and negative for $x^2 < 1$, for a positive h. When h is negative, local transient behavior simply decays to the origin rather like in Fig. 2. However, for positive h, a stable limit cycle behavior occurs where the origin is now unstable and solutions are attracted to a finite-amplitude steady-state oscillation. This is shown in Figure 5, for $h = 1$ and $x_0 = \dot{x}_0 = 0.1$, as a time series in 5a and a phase portrait in 5b. Initial conditions on the outside of this limit cycle would be similarly attracted. This phenomenon is also known as a self-excited or relaxation oscillation.[8,9] Related physical examples include mechanical chatter due to dry friction effects between a mass and a moving belt (stick-slip)[10] and certain flow-induced aeroelastic problems including galloping and flutter.[11]

2.5 Instability

Nonlinear free vibration problems are generally dominated by the influence of equilibria on transient behavior. There are several definitions of stability, but generally if a small perturbation causes the system to move away from equilibrium, then this is a locally unstable state. This is familiar from linear vibration theory where transient behavior is described by the characteristic eigenvalues of the system. For example, the linear oscillator in Fig. 2 has two complex conjugate characteristic eigenvalues with negative real parts. A linearization about the inverted position for the pendulum would lead to one positive real eigenvalue resulting in divergence. Similarly, behavior in the vicinity of equilibrium for the Van der Pol oscillator ($h > 0$) is characterized by complex conjugate eigenvalues but with positive real parts and hence a growth of unstable oscillations. The amplitude of the oscillation in this case is limited by nonlinear effects.

Such systems tend to be sensitive to certain control parameters, for example, h in Eq. (5). More generally a system of the form

$$\dot{\mathbf{x}} = \mathbf{f}(\mathbf{x}, \boldsymbol{\mu}) \tag{6}$$

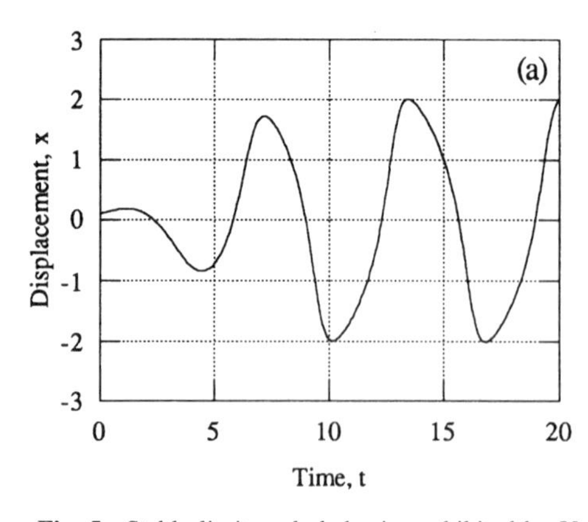

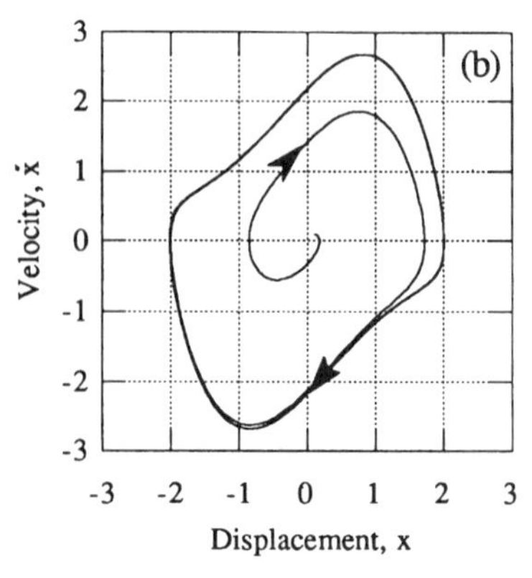

Fig. 5 Stable limit cycle behavior exhibited by Van der Pol's equation: (a) time series; (b) phase portrait ($x_0 = 0.1$, $\dot{x}_0 = 0.1$, $h = 1.0$).

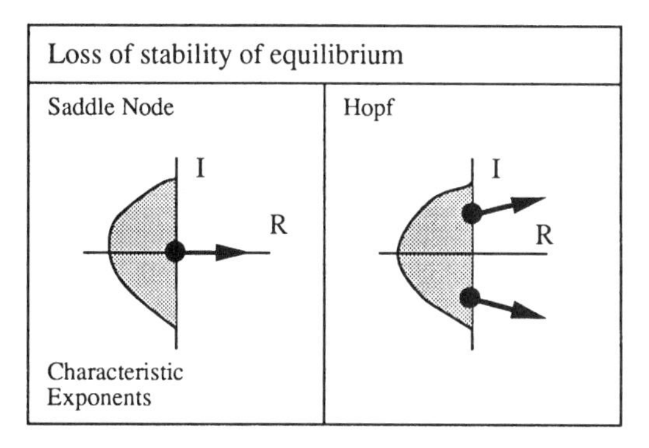

Fig. 6 Two generic instabilities of equilibria under the operation of a single control.

must be considered where the external parameters μ allow a very slow (quasi-static) evolution of the dynamics. In nonlinear dynamic systems such changes in these parameters may typically result in a qualitative change (bifurcation) in behavior, and for systems under the operation of one control parameter, these instabilities take place in one of two well-defined, generic ways. Both of these instability mechanisms have been encountered in the previous sections. The loss of stiffness is associated with the passage of an eigenvalue in to the positive half of the complex plane as shown in Fig. 6. This stability transition is known as the saddle-node but is also referred to as a fold bifurcation and is encountered, for example, in limit-point snap-through buckling in structural mechanics.[4] The loss of damping is associated with the passage of a pair of complex conjugate eigenvalues in to the positive half of the complex plane. This instability mechanism is known as a Hopf bifurcation.[12,13]

3 NONLINEAR FORCED VIBRATION

3.1 Nonautonomous Dynamical Systems

When a dynamic system is subjected to external excitation, a mathematical model of the form

$$\dot{\mathbf{x}} = \mathbf{f}(\mathbf{x}, t) \tag{7}$$

results. This system can be made equivalent to Eq. (1) by including the dummy equation $\dot{t} = 1$ to give an extra phase coordinate. In forced nonlinear vibration primary interest is focused on the case where the input (generally force) is assumed to be a periodic function. For a single-degree-of-freedom oscillator, the forcing phase effectively becomes the third state variable. If the system is linear, then the solution to Eq. (7) will consist of a superposition of a complementary function, which governs the transient free-decay response, and a particular integral, which governs the steady-state forced response (see Chapter 48). The steady-state oscillation generally responds with the same frequency as the forcing and acts as a periodic attractor to local transient behavior.

If Eq. (7) is nonlinear, then it is generally not possible to obtain an exact analytical solution. A number of choices are available: (i) linearize about the static equilibria and obtain solutions that are valid only in a local sense, (ii) use an approximate analytical method and assume relatively small deviation from linearity, for example, use a perturbation scheme,[14] (iii) simulate the equation of motion using numerical integration.[15] The first of these approaches leads to the familiar resonance results from standard vibration theory (see Chapter 48). A large body of research has been devoted to the second approach[1,14,16,17]. However, due to recent advances in, and availability of, digital computers and sophisticated graphics, the third approach has achieved widespread popularity in the study of nonlinear vibrations. Furthermore the numerical approach has been enhanced by the development of the qualitative insight of dynamical systems theory.[18]

3.2 Nonlinear Resonance

The steady-state response of nonlinear vibration problems modeled by Eq. (7) with external forcing of the form

$$G \sin(\omega t + \phi) \tag{8}$$

exhibit some interesting differences from their linear counterparts. Consider the harmonically forced Duffing equation, that is, Eq. (8) added to the right-hand side of Eq. (4). For sufficiently large G the cubic nonlinearity is induced and a typical resonance amplitude response is plotted as a function of forcing frequency in Fig. 7 for $G = 0.1$ and three different damping levels. These curves were obtained using the method of harmonic balance[1,7] and illustrate several nonlinear features. For relatively small amplitudes, that is, in a system with relatively small external forcing, or heavy damping, and generally away from resonance, the response is not significantly different from the linear case. However, for larger amplitudes the softening spring effect causes a bending over of the curves. This is related to the lengthening effect of amplitude on the natural period of the underlying autonomous system and causes the maximum amplitude to occur somewhat below the natural frequency, and for some frequencies multiple solutions are possible. For example, for the case $b = 0.18$, there are three steady-state solutions near $\omega = 0.8$. Two are stable and are separated by an unstable solution (not shown).

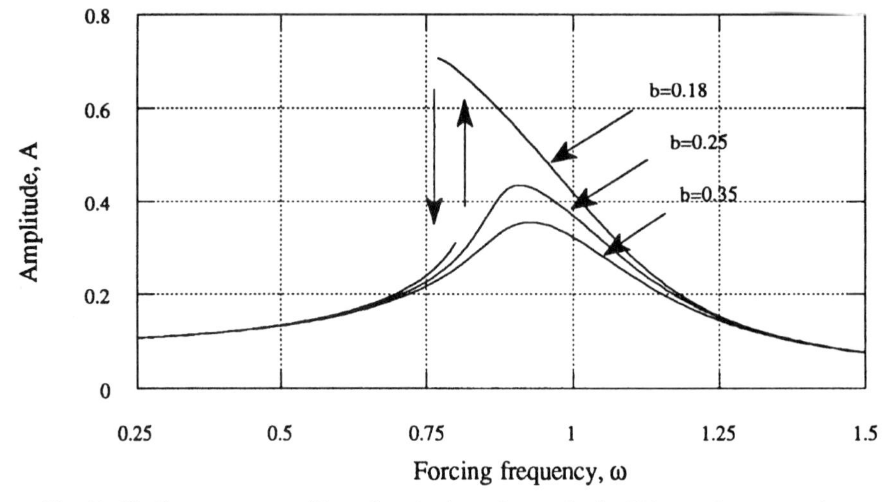

Fig. 7 Nonlinear resonance illustrating the jump in amplitude. This steady-state motion occurs about the offset static equilibrium position (b = 0.18, 0.25, 0.35, $\alpha = -1$, $\beta = 1$, $G = 0.1$).

The two stable steady-state cycles are periodic attractors and compete for the capture of transient trajectories, that is, in this region different initial conditions lead to different persistent behavior. The jump phenomenon is observed by gradually (quasi-statically) changing the forcing frequency ω while keeping all the other parameters fixed. Starting from small ω and slowly increasing, the response will suddenly exhibit a finite jump in amplitude and follow the upper path. Now starting with a large ω and gradually reducing, the response will again reach a critical state resulting in a sudden jump down to the small-amplitude solution. These jumps occur at the points of vertical tangency and bound a region of hysteresis.

A hardening spring exhibits similar features except the resonance curves bend to the right. A physical example of this type of behavior can be found in the large-amplitude lateral oscillations of a thin elastic beam or plate. In this case the induced in-plane (stretching) force produces a hardening (cubic) nonlinearity in the equations of motion, that is, α and β both positive, and for certain forcing frequencies the beam can be perturbed from one type of motion to another. The domains of attraction often consist of interlocking spirals, and hence it may be difficult to determine which solution will persist given initial conditions to a finite degree of accuracy in a similar manner to the unforced case.

3.3 Subharmonic Behavior

Periodic solutions to ordinary differential equations need not necessarily have the same period as the forcing term. It is possible for a system to exhibit subharmonic behavior, that is, the response has a period of n times the forcing period. Subharmonic motion can also be analyzed using approximate analytical techniques.[1,14,16]

Figure 8 illustrates a typical subharmonic of order 2 obtained by numerically integrating Duffing's equation:

$$\ddot{x} + b\dot{x} + \alpha x + \beta x^3 = G \sin \omega t \tag{9}$$

for $b = 0.3$, $\alpha = -1$, $\beta = 1$, $\omega = 1.2$, and $G = 0.65$. Subharmonic behavior is generally a nonlinear feature and is often observed when the forcing frequency is close to an integer multiple of the natural frequency. Also, subharmonics may result due to a bifurcation from a harmonic response, and thus may complicate the approximate scenario shown in Fig. 7. Superharmonics corresponding to a response that repeats itself a number of times within each forcing cycle are also possible especially at low excitation frequencies.

3.4 Poincaré Sampling

A useful qualitative tool in the analysis of periodically forced nonlinear systems is the Poincaré section.[5] This technique consists of stroboscopically sampling the trajectory every forcing cycle or at a defined surface of section in the phase space. The accumulation of points will then reflect the periodic nature of the response. A fundamental harmonic response appears as a single point simply repeating itself. The location of this point on the phase trajectory and hence in the $(x, \dot{x})$ projection depends on the initial conditions. For example, consider

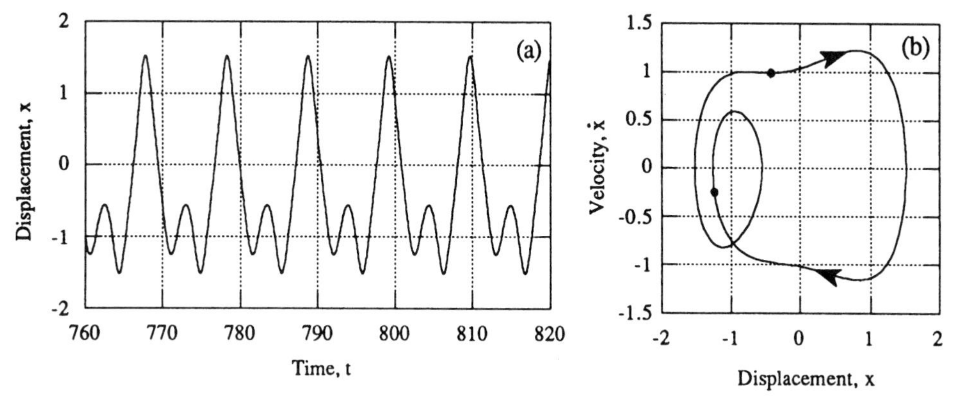

Fig. 8 Subharmonic oscillations in Duffing's equation: (*a*) time series; (*b*) phase projection ($b = 0.3$, $\alpha = -1$, $\beta = 1$, $\omega = 1.2$, $G = 0.65$).

the subharmonic response of Fig. 8. The forcing period is $T = 2\pi/\omega = 5.236$, and if the response is inspected whenever t is a multiple of T then two points are visited alternately by the trajectory, as shown by the dots in Fig. 8*b*. The Poincaré sampling describes a mapping and can also be used to study the behavior of transients, and hence stability. Further refinements to this technique have been developed including the reconstruction of attractors using time-delay sampling.[5] This is especially useful in experimental situations where there may be a lack of information about certain state variables or in autonomous systems where there is no obvious sampling period.

3.5 Quasi-periodicity

Another possible response in forced nonlinear vibration problems is the appearance of quasi-periodic behavior. Although the superposition of harmonics, including the beating effect, is commonly encountered in coupled linear oscillations, it is possible for nonlinear single-degree-of-freedom systems to exhibit a relatively complicated response where two or more incommensurate frequencies are present. For a two-frequency response local transients are attracted to the surface of a toroidal phase space, and hence Poincaré sampling leads to a continuous closed orbit in the projection because the motion never quite repeats itself. Although a quasi-periodic time series may look complicated, it is predictable. The fast Fourier transform (FFT) is a useful technique for identifying the frequency content of a signal and hence distinguishing quasi-periodicity from chaos.

3.6 Piecewise Linear Systems

There are many examples in mechanical engineering where the stiffness of the system changes abruptly, for example, where a material or component acts differently in tension and compression, or a ball bouncing on a surface. Although they are linear within certain regimes, the stiffness is dependent on position, and their behavior is often strongly nonlinear. If such systems are subjected to external excitation, then a complex variety of responses are possible.[19,20] Related problems include the backlash phenomenon in gear mechanisms where a region of free-play exists.[4] Nonlinearity also plays a role in some coulomb friction problems including stick-slip where the discontinuity is in the relative velocity between two dry surfaces.[1] Feedback control systems often include this type of nonlinearity.[21]

3.7 Chaotic Oscillations

The possibility of relatively simple nonlinear systems exhibiting extremely complicated (unpredictable) dynamics was known to Poincaré about 100 years ago and must have been observed in a number of early experiments on nonlinear systems. However, the relatively recent ability to simulate highly nonlinear dynamical systems numerically has led to a number of interesting new discoveries with implications for a wide variety of applications, especially chaos, that is, a fully deterministic system that exhibits randomlike behavior and an extreme sensitivity to initial conditions. This has profound implications for the concept of predictability and has stimulated intensive research.[22]

Figure 9 shows a chaotic time series obtained by the numerical integration of Eq. (9) for appropriately selected b, α, β, G, and ω using a fourth-order Runge–Kutta scheme.[14] Here, transient motion is allowed to decay leaving the randomlike wave form shown (Fig. 9*a*). This response, which traverses both static equilibria ($x_e = \pm 1$), may be considered to have an infinite period. An important feature of chaos, in marked

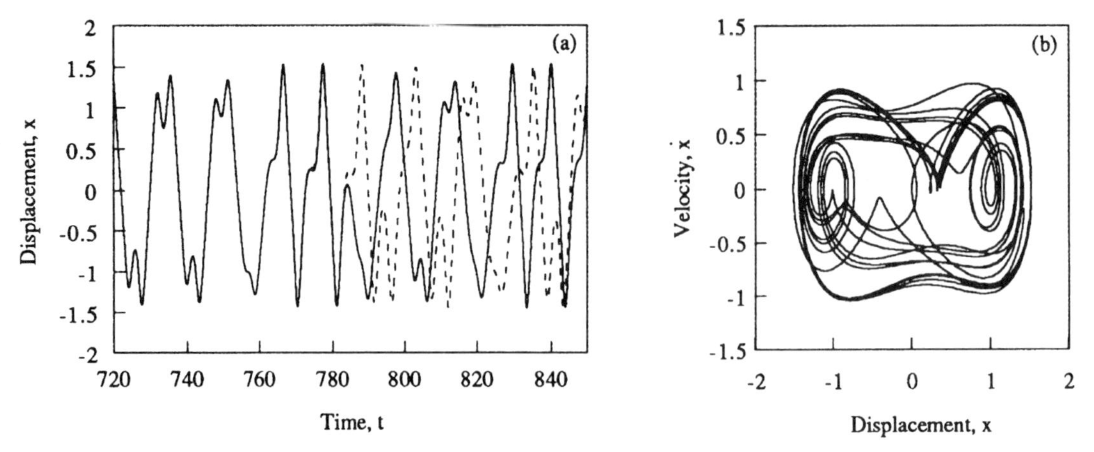

Fig. 9 Chaotic response of Duffing's equation including exponential divergence of initially close trajectories, parameters are the same as Fig. 8, $G = 0.5$: (a) time series, the initial separation in x is 0.001 at time $t = 750$; (b) phase projection.

contrast to linear vibration, is a sensitive dependence on initial conditions, and this figure also shows (dashed line) how initially adjacent chaotic trajectories diverge (exponentially on the average) with time. Given a very small difference in the initial conditions, that is, a perturbation of 0.001 in x at time $t = 750$, the mixing nature of the underlying (chaotic) attractor leads to a large difference in the response. Figure 9b shows the data for the original time series as a phase projection. Although the response remains bounded, the loss of predictability for increasing time is apparent.

However, unlike truly random systems, the randomness of a chaotic signal is not caused by external influences but is rather an inherent dynamic (deterministic) characteristic. Poincaré sampling can be used effectively to illustrate the underlying structure of a chaotic response. A number of experimental investigations of chaos have been made in a nonlinear vibrations context including the buckled beam,[4] although early work on nonlinear circuits made a significant contribution.[7] Consider the example of a physical analog of the twin-well potential system described by Eq. (9), that is, a small cart rolling on a curved surface.[23] A chaotic attractor based on experimental data is shown in Fig. 10 where

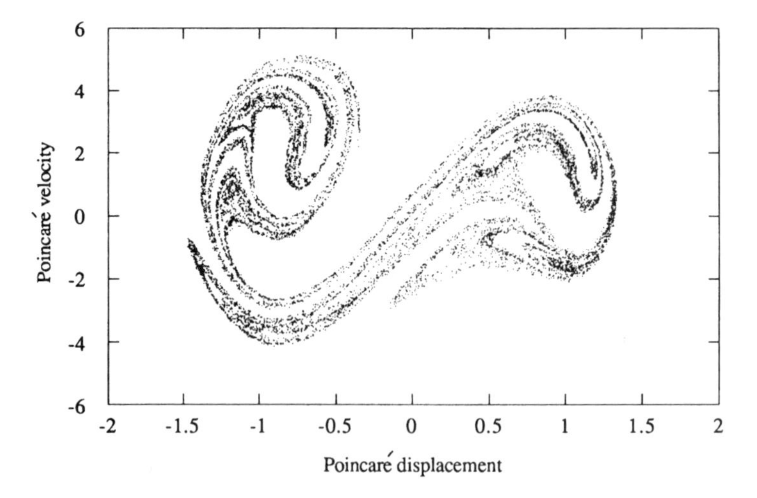

Fig. 10 Poincaré section of the chaotic attractor exhibited by an experimental (mechanical) analogue[23] of Eq. (9).

10,000 Poincaré points (i.e., forcing cycles) are plotted in the phase projection. A similar pattern is obtained by taking a Poincaré section on the data of Fig. 9. The evolution of the trajectory follows a stretching and folding pattern, exhibiting certain fractal characteristics[4], including a noninteger dimension, and can be shown to give a remarkably close correlation with numerical simulation.[23]

For certain parameter ranges catchment regions in forced oscillations also have fractal boundaries, that is, self-similarity at different length scales, although these can occur for periodic attractors. An example is given in Fig. 11 based on numerical integration of Eq. (9) with $b = 0.168$, $\alpha = -0.5$, $\beta = 0.5$, $G = 0.15$, and $\omega = 1$. Here a fine grid of initial conditions is used to determine the basins of attraction for two periodic oscillations, one located within each well. Investigation of all initial conditions is a daunting task, especially for a higher-order system, but certain efficient techniques have been developed.[24] A relatively rare analytical success used to predict the onset of fractal basin boundaries based on homoclinic tangencies is Melnikov theory.[18]

A further remarkable feature of many nonlinear systems is that they often follow a classic, universal route to chaos. Successive period-doubling bifurcations can occur as a system parameter is changed, and these occur at a constant geometric rate, often referred to as Feigenbaum's ratio. This property is exhibited by a large variety of systems including simple difference equations and maps.[5,13,18] Other identified routes to chaos include quasi-periodicity, intermittency and crises, and chaotic transients and homoclinic orbits.[13]

Two major prerequisites for chaos are (i) the system must be nonlinear and (ii) the system must have at least a three-dimensional phase space. A number of diagnostic numerical tools have been developed to characterize chaotic behavior, other than the Poincaré map. The randomlike nature of a chaotic signal is reflected in a broadband power spectrum with all frequencies participating.[25] The widespread availability of FFT software makes power spectral techniques especially attractive in an experimental situation. Also, the divergence of adjacent trajectories can be measured in terms of Lyapunov exponents.[26] The autocorrelation function also has been used to illustrate the increasing loss of correlation with time lag. Various measures of dimension have been developed as well to establish the relation between fractals, chaos, and dynamical behavior.

3.8 Instability

Analogous to the instability of (static) equilibria under the operation of one control parameter, periodic (dynamic) cycles also lose their stability in a small number of generic ways. Figure 12 summarizes the typical stability transitions encountered in nonlinear forced vibration problems. Here, Poincaré sampling is used to obtain information on local transient behavior in terms of char-

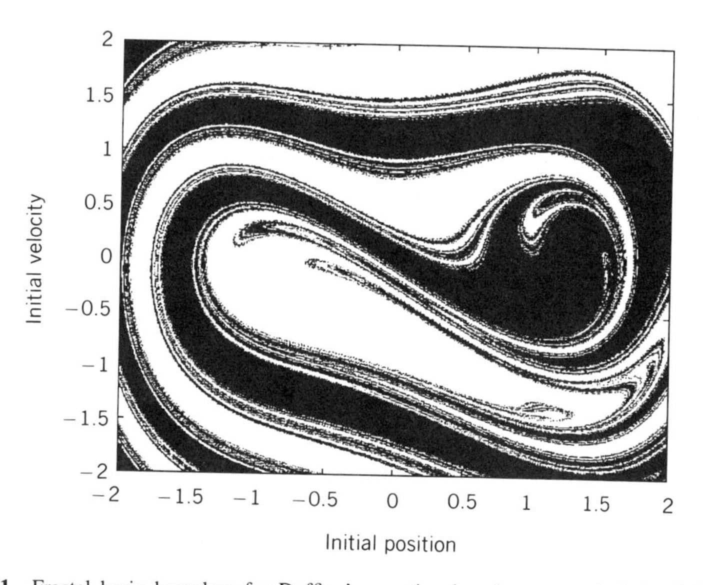

Fig. 11 Fractal basin boundary for Duffing's equation based on numerical simulation ($b = 0.168$, $\alpha = -0.5$, $\beta = 0.5$, $\omega = 1$, $G = 0.15$). The black (white) regions represent initial conditions that generate a transient leading to the periodic attractor in the right (left) hand well.

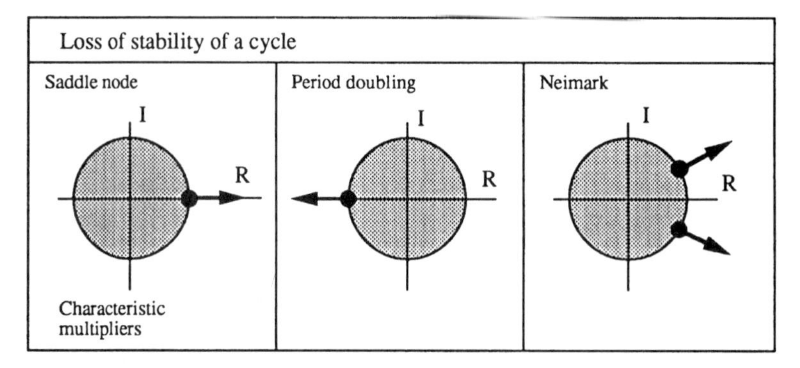

Fig. 12 Generic instabilities of cycles under the operation of a single control.

acteristic multipliers that describe the rate of decay of transients onto a periodic attractor, and hence, in the complex plane characterizes the stability properties of a cycle. This is familiar as the root locus in control theory based on the z-transform.[21] Slowly changing a system parameter may cause penetration of the unit circle and hence the local growth of perturbations. The cyclic fold is the underlying mechanism behind the resonant amplitude jump phenomenon and the flip bifurcation leads to subharmonic motion and may initiate the period-doubling sequence. The Neimark bifurcation is much less commonly encountered in mechanical vibration. Also, other types of instability are possible for nongeneric systems, for example, a perfectly symmetric, periodically forced pendulum may exhibit a symmetry-breaking pitchfork bifurcation.[18]

Again these instability phenomena can also be analyzed using approximate analytical techniques. Small perturbations about a steady-state solution are used to obtain a variational (Mathieu-type) equation, the stability of which can then be determined using Floquet theory or Routh–Hurwitz criteria[1,14,16].

4 SUMMARY

Although most of the foregoing examples are drawn from relatively simple single-degree-of-freedom systems, in principle, many of these techniques can be extended to systems of higher order. However, classification, analysis, and computation becomes increasingly difficult, especially for spatially extended systems modeled by partial differential equations. The study of nonlinear problems in vibrations has a solid foundation, especially based on classic analytical techniques,[7,16,17] but the scope of chaotic vibration has opened up a whole new range of techniques and applications, and even some potential uses for chaos based on control.[27] However, many problems are still ill-understood: This is even more true for nonlinear systems under random influences.[13,25]

REFERENCES

1. D. W. Jordan and P. Smith, *Nonlinear Ordinary Differential Equations*, Clarendon Press, Oxford, 1987.
2. R. E. D. Bishop, *Vibration*, Cambridge University Press, Cambridge, 1979.
3. W. T. Thomson, *Theory of Vibration with Applications*, Prentice-Hall, Englewood Cliffs, NJ, 1981.
4. F. C. Moon, *Chaotic and Fractal Dynamics*, Wiley, New York, 1993.
5. J. M. T. Thompson and H. B. Stewart, *Nonlinear Dynamics and Chaos*, Wiley, London 1986.
6. J. L. Singe and B. A. Griffith, *Principles of Mechanics*, McGraw-Hill, New York, 1942.
7. C. Hayashi, *Nonlinear Oscillations in Physical Systems*, McGraw-Hill, New York, 1964.
8. Van der Pol, B., "On Relaxation Oscillations," *Phil. Mag.*, Vol. 7, No. 2, 1926, pp. 978–992.
9. J. W. S. Rayleigh, *The Theory of Sound*, Dover, New York, 1896.
10. J. P. Den Hartog, *Mechanical Vibrations*, McGraw-Hill, New York, 1934.
11. E. H. Dowell, "Flutter of a Buckled Plate as an Example of Chaotic Motion of a Deterministic Autonomous System," *J. Sound Vib.*, Vol. 147, 1982, pp. 1–38.
12. A. B. Pippard, *Response and Stability*, Cambridge University Press, Cambridge, 1985.
13. A. J. Lichtenberg and M. A. Lieberman, *Regular and Chaotic Dynamics*, Springer-Verlag, New York, 1992.
14. A. H. Nayfeh and D. T. Mook, *Nonlinear Oscillations*, Wiley, New York, 1979.
15. C. W. Gear, *Numerical Initial-Value Problems in Ordinary Differential Equations*, Prentice-Hall, Englewood Cliffs, NJ, 1971.

16. N. Minorsky, *Nonlinear Oscillations*, Van Nostrand, Princeton, 1962.

17. J. J. Stoker, *Nonlinear Vibrations*, Interscience, New York, 1950.

18. J. Guckenheimer and P. J. Holmes, *Nonlinear Oscillations, Dynamical Systems and Bifurcations of Vector Fields*, Springer-Verlag, New York, 1983.

19. S. W. Shaw and P. J. Holmes, "A Periodically Forced Piece-Wise Linear Oscillator," *J. Sound Vib.*, Vol. 90, 1983, pp. 129–144.

20. P. V. Bayly and L. N. Virgin, "An Experimental Study of an Impacting Pendulum," *J. Sound Vib.*, Vol. 164, 1993, pp. 364–374.

21. A. I. Mees, *Dynamics of Feedback Systems*, Wiley, Chicester, 1981.

22. Y. Ueda, "Steady Motions Exhibited by Duffing's Equation: A Picture Book of Regular and Chaotic Motions," in *New Approaches to Nonlinear Problems in Dynamics*, P. J. Holmes (Ed.), SIAM, Philadelphia, 1980, pp. 311–322.

23. J. A. Gottwald, L. N. Virgin, and E. H. Dowell, "Experimental Mimicry of Duffing's Equation," *J. Sound Vib.*, Vol. 158, 1992, pp. 447–467.

24. C. S. Hsu, *Cell-to-Cell Mapping*, Springer-Verlag, New York, 1987.

25. D. E. Newland, *An Introduction to Random Vibrations and Spectral Analysis*, 2nd ed., Longman, London, 1984.

26. A. Wolf, J. B. Swift, H. L. Swinney, and J. Vastano, "Determining Lyapunov Exponents from a Time Series," *Phys. D*, Vol. 16, 1985, pp. 285–317.

27. E. Ott, C. Grebogi, and J. A. Yorke, "Controlling Chaos," *Phys. Rev. Lett.*, Vol. 64, 1990, pp. 1196–1199.

51

RANDOM VIBRATION

DAVID E. NEWLAND

1 INTRODUCTION

1.1 Random Vibration

Random vibration is an essentially unpredictable phenomenon. Three examples are the response of an offshore structure to excitation from the waves, the response of an aircraft wing to its turbulent boundary layer, and the response of a wheeled vehicle to an uneven road. The assumption that excitation from the waves or a turbulent boundary layer or an uneven road is random is a recognition of our lack of knowledge of the underlying characteristics of this excitation. For example, the surface topography of a given length of road may be measured to any desired accuracy. The excitation from that length of road is then deterministic (not random) because its surface may be precisely described. But it is not practicable to measure the topography of miles of roads precisely, and it is convenient to characterize road surfaces generally by statistical parameters that can be estimated from experimental measurements on particular representative lengths of road. In other cases, for example excitation by sea waves, we may not yet have the physical modeling skills to predict wave loading at any given time. However, measurements can be made that, when taken over a long enough period, allow wave loading to be represented statistically, and accurate predictions then become possible using probabilistic theory.

For constant-parameter, damped, linear systems there is a comprehensive body of theory for calculating how a dynamical system responds to random excitation. Much of this chapter describes the results of this linear theory, including the analysis of peaks and crossings and first-passage and fatigue problems. There are many important results of great practical value.

For time-varying and nonlinear systems, the theory is much less complete and remains a field of active research. Section 8 touches on this by describing three of the principal methods of theoretical analysis for such problems. Where theoretical predictions are impracticable, numerical experiments have become increasingly important. These use pseudorandom excitation generated by so-called Monte Carlo methods. The chapter ends with a brief introduction to this approach.

First it is necessary to describe the statistical structure of a continuously varying random signal.

1.2 Sample Averages

Experimental data on a quantity that is changing randomly, for example the pressure at a point in a turbulent boundary layer, may be represented as a time history in which the magnitude of the quantity concerned $x(t)$ is plotted as a function of time t. As time passes, the character of what is happening may be changing, as for instance when an aircraft descends from high altitude to land. In that case, the time history is from a *nonstationary* random process whose statistical characteristics are changing with time. For a *stationary* random process, the character of the time history does not change with time. Then average values may be obtained by integrating over time so that, for example, the *mean-square* value of $x(t)$ is given by

$$\langle x^2 \rangle = \frac{1}{T} \int_0^T x^2(t)\, dt, \qquad T \to \infty. \tag{1}$$

This is called a *sample average* because it is a result derived from a single time history, or *sample function*,

Note: References to chapters appearing only in the *Encyclopedia* are preceded by *Enc.*

of the random process. Although conceptually straightforward, the sample average formula (1) is only exact when the integration time T is infinite. This raises theoretical difficulties that have been overcome by introducing the idea of an ensemble of sample functions.

1.3 Ensemble Averages

An *ensemble* is a collection of infinitely many sample functions that together make up a *random process*. These separate sample functions are different time histories recorded for the same system: for example, recordings of the boundary layer pressure at the same point on identical aircraft all flying the same route simultaneously. An *ensemble average* is defined as the average value calculated for all the sample functions at a specified time t. The expectation operator E is often used to denote an ensemble average, and $E[x^2(t)]$ is then the ensemble-averaged mean-square value of x at time t.

If a random process is *stationary*, all its ensemble averages are independent of time. If, in addition, the process is *ergodic*, then each sample function has the same statistical characteristics as its ensemble, and then

$$E[x^2(t)] = E[x^2] = \langle x^2 \rangle = \frac{1}{T}\int_0^T x^2(t)\, dt, \qquad T \to \infty. \tag{2}$$

2 PROBABILITY DISTRIBUTIONS AND AVERAGES

2.1 Probability Density Functions

The distribution of values at a given time for all the sample functions or *realizations* of a random process is defined by its *first-order probability density function* $p(x)$. The probability that, for any arbitrarily chosen sample function, the value of $x(t)$ at the chosen time t lies in the band of values x to $x + dx$ is then given by $p(x)\, dx$. Since the probability that x will lie somewhere between $+\infty$ and $-\infty$ is unity (it is a certainty), it follows that

$$\int_{-\infty}^{\infty} p(x)\, dx = 1. \tag{3}$$

This concept can be extended to distributions of more than one random variable. For example, the joint distribution of the values of all the sample functions of a random process at two different times is defined by a *second-order probability density function*. The probability that, for any arbitrarily chosen sample function, the value of $x(t_1) = x_1$ at time t_1 lies in the band of values x_1 to $x_1 + dx_1$ and that, simultaneously, the value of $x(t_2) = x_2$ at time t_2 (for the same sample function) lies in the band x_2 to $x_2 + dx_2$ is given by $p(x_1, x_2)\, dx_1\, dx_2$. Since x_1 and x_2 must both have values between $+\infty$ and $-\infty$, it follows as before that

$$\int_{-\infty}^{\infty}\int_{-\infty}^{\infty} p(x_1, x_2)\, dx_1\, dx_2 = 1. \tag{4}$$

Ensemble averages can be calculated directly from the relevant probability density functions, so that, for example,

$$E[x^2] = \int_{-\infty}^{\infty} x^2 p(x)\, dx \tag{5}$$

and

$$E[x_1, x_2] = \int_{-\infty}^{\infty}\int_{-\infty}^{\infty} x_1 x_2 p(x_1, x_2)\, dx_1\, dx_2. \tag{6}$$

The second result gives the average value of the product of two values of the same sample function measured at times t_1 and t_2.

2.2 Gaussian Distributions

When a random process results from the combined action of infinitely many independent infinitesimal contributions (e.g., rain on the roof), then according to the *central limit theorem* the probability distributions for the process will be *Gaussian* (or *normal*). Then the probability density functions have the well-known bell shape. The first-order Gaussian probability density function (Fig. 1*a*) is defined by

$$p(x) = \frac{1}{\sqrt{2\pi}\sigma} \exp\left[-\frac{(x-m)^2}{2\sigma^2}\right] \tag{7}$$

and the second-order function (Fig. 1*b*) by

$$p(x_1, x_2) = \frac{1}{2\pi\sigma_1\sigma_2\sqrt{1-\rho_{12}^2}} \cdot \exp\left[-\frac{1}{2(1-\rho_{12}^2)}\left\{\frac{(x_1-m_1)^2}{\sigma_1^2} + \frac{(x_2-m_2)^2}{\sigma_2^2} - \frac{2\rho_{12}(x_1-m_1)(x_2-m_2)}{\sigma_1\sigma_2}\right\}\right]. \tag{8}$$

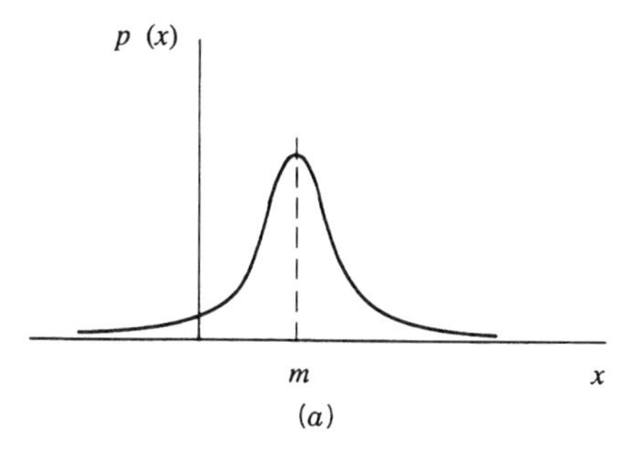

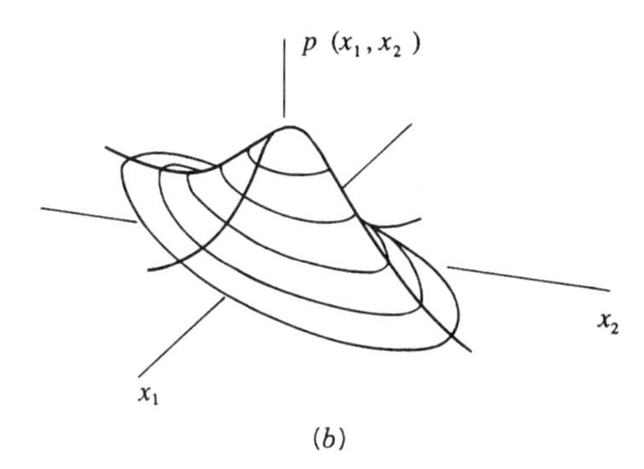

Fig. 1 (*a*) First-order Gaussian probability density curve. (*b*) Second-order Gaussian probability density surface.

In these formulas, the constants are defined as

$$m = E[x], \tag{9}$$

$$\sigma^2 = E[(x - m)^2] = E[x^2] - m^2, \tag{10}$$

and similarly for m_1 and σ_1 when x_1 replaces x and for m_2 and σ_2 when x_2 replaces x. In Eq. (8), the *correlation coefficient* ρ_{12} is defined as

$$\rho_{12} = \frac{E[(x_1 - m_1)(x_2 - m_2)]}{\sigma_1 \sigma_2}. \tag{11}$$

In Eq. (10), the right-hand side may be obtained by noting that

$$E[(x - m)^2] = E[x^2 - 2mx + m^2] = E[x^2] - 2mE[x] + m^2 \tag{12}$$

and using Eq. (9).

2.3 Rayleigh Distribution

Subject to various simplifying assumptions, it can be shown that the distribution of peak values of the sample functions of certain random processes (see Section 6) satisfy a Rayleigh distribution. This describes a random variable whose values lie in the range $0 \le x \le \infty$ (no negative values) and whose probability density function is defined by

$$p(x) = \frac{x}{x_0^2} \exp\left(-\frac{x^2}{2x_0^2}\right), \qquad 0 \le x \le \infty. \tag{13}$$

The shape of this function is shown in Fig. 2*a*. There is a low probability density for very small and very large values of x, signifying that these values are uncommon and that most of the values lie in the region of $x = x_0$.

2.4 Weibull Distribution

Experimental measurements is practical cases (e.g., the vibration of structures under wind loading and the wave-induced bending of ships) have shown that the Weibull distribution may be a better representation of the distribution of peak values of the sample functions of some random processes. The Weibull probability density function is defined by

$$p(x) = k(\ln 2) \frac{x^{k-1}}{X_0^k} \exp\left[-(\ln 2)\left(\frac{x}{X_0}\right)^k\right], \tag{14}$$

where the coefficient k is chosen so that the shape of the resulting probability density curve (Fig. 2*b*) matches the experimental data as closely as possible. In Fig. 2*b*, the curves are plotted against x/X_0 as abscissa, so that $p(x/X_0)$ is drawn instead of $p(x)$, defined by Eq. (14). In order to ensure that the area under the probability density curve remains unity, according to Eq. (3), it is necessary to put

$$p\left(\frac{x}{X_0}\right) = X_0 p(x). \tag{15}$$

When $k = 2$, the Weibull distribution becomes a Rayleigh distribution with the constants X_0 and x_0 related by

$$X_0^2 = 2(\ln 2)x_0^2. \tag{16}$$

As the value of the coefficient k is increased, it can be seen from Fig. 2*b* that most of the values of x lie close to X_0, which can be shown[1] to be the median value of x.

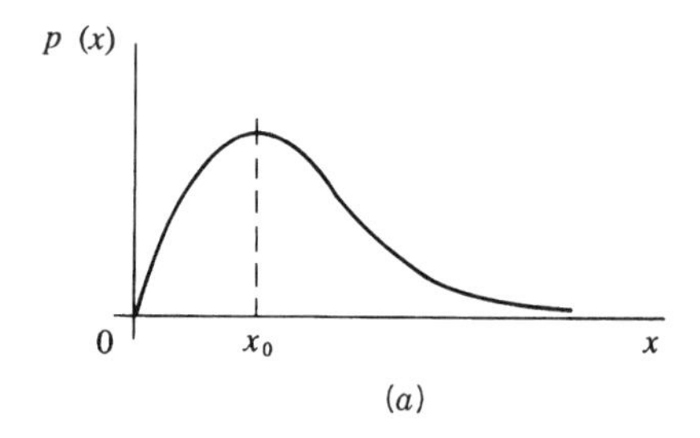

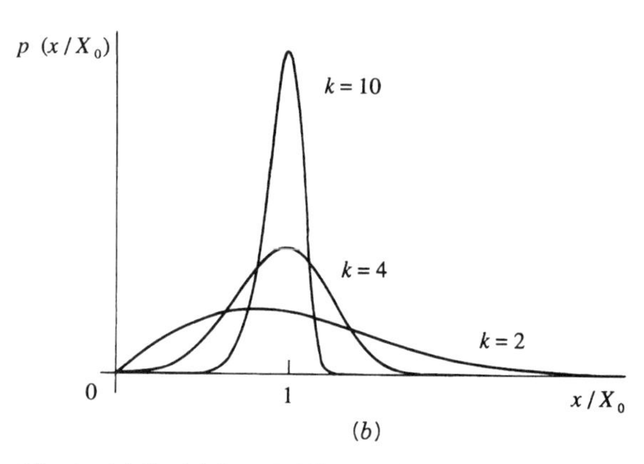

Fig. 2 (*a*) Rayleigh probability density curve. (*b*) Weibull probability density curves.

When $k = \infty$, all the values are X_0. This would be the case for a random process each of whose sample functions consists of a harmonic wave of constant amplitude X_0 but different phases randomly varying from sample function to sample function.

2.5 Distribution Functions

The *probability distribution function* $P(x)$ is defined as the integral of its corresponding density function $p(x)$ by the equation

$$P(x) = \int_{-\infty}^{x} p(\xi)\, d\xi, \tag{17}$$

where ξ is a dummy variable for x. Hence $P(x)$ gives the probability that a random variable lies in the range $-\infty$ to x. For a Gaussian distribution, $P(x)$ does not have a closed form. Values have to be obtained by computation (many vibration analysis software packages include suitable algorithms) or may be obtained from the extensive printed tables of results.[2] For the Weibull distribution

$$P(x) = 1 - \exp\left[-(\ln 2)\left(\frac{x}{X_0}\right)^k\right]. \tag{18}$$

This result may easily be checked against Eq. (14) by using Eq. (17), which, after differentiation, becomes

$$\frac{dP(x)}{dx} = p(x). \tag{19}$$

Gaussian and Weibull distribution functions are illustrated in Figs. 3*a* and 3*b*.

2.6 Conditional Probability

Calculations on the statistics of random vibration response use the notion of *conditional probability*. This describes the probability of one event occurring on the condition that a second event also occurs. For example, suppose that a sample function $x(t)$ from a random process is examined at times t_1 and t_2. We may want to know what the probability is that $x(t_1)$ lies in the band of values x_1 to $x_1 + dx_1$ on the condition that $x(t_2)$ lies in the band x_2 to $x_2 + dx_2$. Many of the sample functions will not satisfy either of these conditions. We are interested only in those samples that satisfy both the conditions.

From the ensemble of all sample functions, we separate out a subensemble of sample functions that satisfy one of the two conditions, for example $x_2 < x(t_2) < x_2 + dx_2$. Theoretically, since we have an infinite number of sample functions in the original ensemble, this subensemble will also have an infinite number of sample functions. The *conditional probability density* function, which is represented as $p(x(t_1)|x(t_2))$, then describes the distribution of the random variable $x(t_1)$ within the subensemble for which all the $x(t_2)$ lie in the band x_2 to $x_2 + dx_2$.

The probability that a sample chosen at random from the complete ensemble lies in the subensemble is $p(x_2)\, dx_2$, where $p(x_2)$ means the probability density function for the distribution of $x(t)$ at $t = t_2$. The probability that a sample from this subensemble lies in the band x_1 to $x_1 + dx_1$ is $p(x_1|x_2)\, dx_1$, where $p(x_1|x_2)$ means the probability density for the distribution of $x(t)$ at $t = t_1$ for samples in the subensemble. But the probability that a sample from the complete ensemble lies in both bands is $p(x_1, x_2)\, dx_1\, dx_2$, where $p(x_1, x_2)$ is the second-order probability density function for the joint distribution of $x(t)$ at t_1 and t_2. Hence it follows[1] that

$$p(x_1|x_2)p(x_2) = p(x_1, x_2). \tag{20}$$

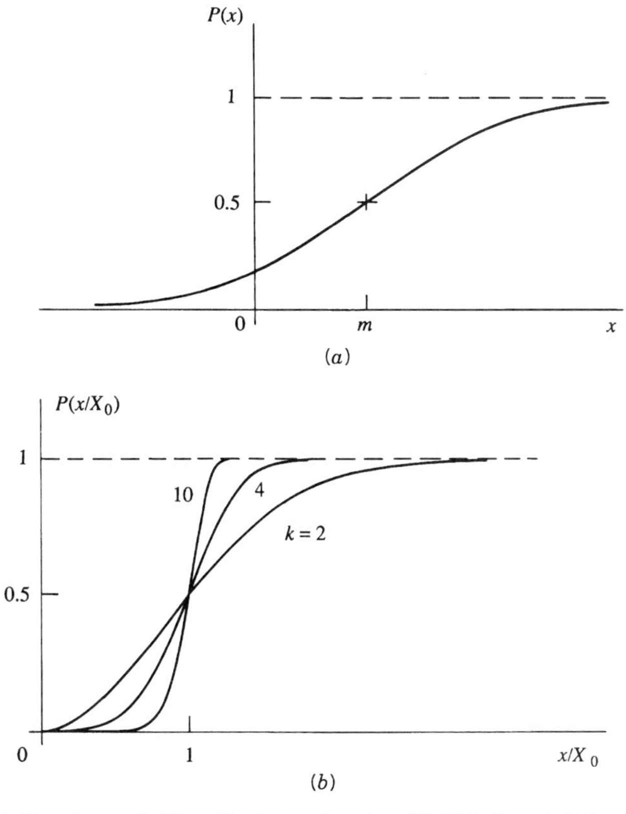

Fig. 3 (*a*) Gaussian probability distribution function. (*b*) Weibull probability distribution functions.

If the conditional probability density function is independent of its condition, so that

$$p(x_1|x_2) = p(x_1), \tag{21}$$

then it follows that

$$p(x_1, x_2) = p(x_1)p(x_2). \tag{22}$$

When a second-order probability density function can be factored into two first-order probability density functions, as in Eq. (22), the two random variables involved are said to be *statistically independent* of each other.

3 CORRELATION FUNCTIONS

3.1 Autocorrelation

The starting point for studying the characteristics of the sample functions that make up a random process is to examine how they change with time (or with whatever is the independent base parameter). This is done by using the *autocorrelation function*, which is the ensemble average of the product of values of each sample function at two different times, for example, $E[x(t_1)x(t_2)]$. This average is usually denoted by $R_{xx}(t_1, t_2)$, where the double subscript denotes the quantity being correlated and the arguments t_1, t_2 denote the times at which $x(t)$ has been sampled.

Using Eq. (6) gives

$$R_{xx}(t_1, t_2) = E[x(t_1)x(t_2)]$$

$$= \int_{-\infty}^{\infty}\int_{-\infty}^{\infty} x_1 x_2 p(x_1, x_2)\, dx_1\, dx_2, \tag{23}$$

where x_1 means $x(t_1)$ and x_2 means $x(t_2)$.

For a stationary process, the autocorrelation function must be independent of absolute time, so that if

$$\tau = t_2 - t_1, \tag{24}$$

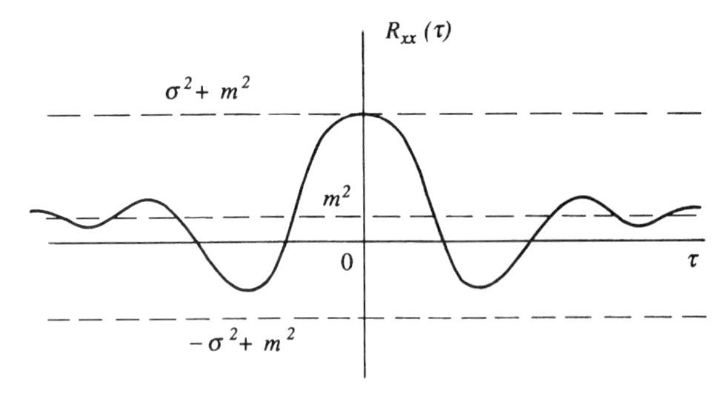

Fig. 4 Properties of the autocorrelation function $R_{xx}(\tau)$ for a stationary random process $x(t)$.

then $R_{xx}(t_1, t_2)$ becomes $R_{xx}(\tau)$ and is a function only of the separation between the two measuring points t_1 and t_2.

It can be shown[1] that, for a stationary process, $R_{xx}(\tau)$ has the important properties illustrated in Fig. 4. Its upper bound is its value at $\tau = 0$, which, from Eq. (23), is

$$E[x^2(t_1)] = E[x^2] = \sigma^2 + m^2 \tag{25}$$

from Eq. (10). Its lower bound is $-\sigma^2 + m^2$. When $|\tau| \to \infty$, $R_{xx}(\tau) \to m^2$. And since $R_{xx}(\tau)$ depends only on the magnitude of the time separation, it follows that $R_{xx}(\tau) = R_{xx}(-\tau)$, so that $R_{xx}(\tau)$ is an even function of τ.

3.2 Cross-correlation

When there are two corresponding random processes, for example, $x(t)$ and $y(t)$, in which each sample function of $x(t)$ is associated with a corresponding sample function of $y(t)$, then the *cross-correlation functions* for the two processes are defined as

$$R_{xy}(t_1, t_2) = E[x(t_1)y(t_2)] \tag{26}$$

and

$$R_{yx}(t_1, t_2) = E[y(t_1)x(t_2)] \tag{27}$$

If both processes $x(t)$ and $y(t)$ are stationary, then the cross-correlation functions depend only on τ [Eq. (24)], and

$$R_{xy}(\tau) = E[x(t)y(t+\tau)] = R_{yx}(-\tau)], \tag{28}$$

$$R_{yx}(\tau) = E[y(t)x(t+\tau)] = R_{xy}(-\tau)]. \tag{29}$$

It can be shown[1] that the upper and lower bounds of $R_{xy}(\tau)$ and $R_{yx}(\tau)$ are $\sigma_x\sigma_y + m_x m_y$ and $-\sigma_x\sigma_y + m_x m_y$, respectively, where m_x, m_y are the means and σ_x, σ_y the standard deviations of x and y according to Eqs. (9) and (10). The asymptotic value to which the cross-correlation functions converge as $|\tau| \to \infty$ is $m_x m_y$.

These properties are illustrated in Fig. 5.

The main importance of correlation functions is that they provide the theoretical basis for spectral analysis (Section 4).

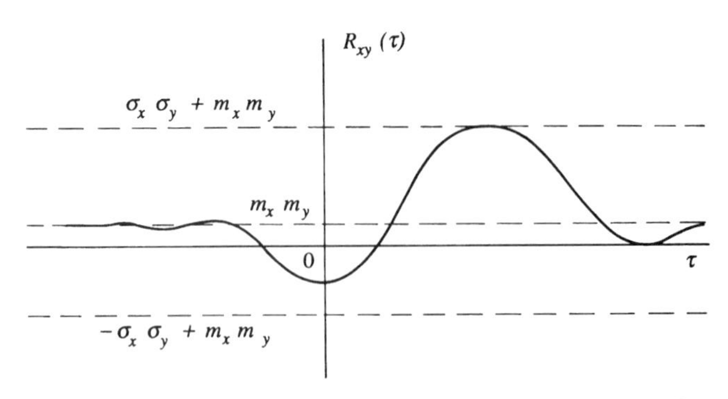

Fig. 5 Properties of the cross-correlation function $R_{xy}(\tau)$ for two stationary random processes $x(t)$ and $y(t)$.

4 SPECTRAL ANALYSIS

4.1 Spectral Density

The (auto) *spectral density* of a stationary random process $x(t)$ is defined as the Fourier transform of its autocorrelation function, so that

$$S_{xx}(\omega) = \frac{1}{2\pi} \int_{-\infty}^{\infty} R_{xx}(\tau) e^{-i\omega\tau}\, d\tau. \qquad (30)$$

The inverse Fourier transform allows the autocorrelation function to be recovered from the spectral density, according to

$$R_{xx}(\tau) = \int_{-\infty}^{\infty} S_{xx}(\omega) e^{i\omega\tau}\, d\omega. \qquad (31)$$

The spectral density $S_{xx}(\omega)$ is a function of angular frequency ω. It can be shown[1] that it is a real, even, and nonnegative function of ω and therefore has the form illustrated in Fig. 6. The term $S_{xx}(\omega)$ is often called the *mean-square spectral density* because, by putting $\tau = 0$ in Eq. (31),

$$E[x^2] = R_{xx}(\tau = 0) = \int_{-\infty}^{\infty} S_{xx}(\omega)\, d\omega, \qquad (32)$$

so that the shaded area in Fig. 6 is equal to the magnitude of the mean-square value of $x(t)$. The name *power spectral density* (or PSD) is also used and derives from electrical applications, where mean square was a measure of power but, strictly, this is a misnomer in the general case.

4.2 Cross-spectral Density

The cross-spectral densities $S_{xy}(\omega)$ and $S_{yx}(\omega)$ are defined as the Fourier transforms of their corresponding cross-correlation functions $R_{xy}(\tau)$ and $R_{yx}(\tau)$, so that

$$S_{xy}(\omega) = \frac{1}{2\pi} \int_{-\infty}^{\infty} R_{xy}(\tau) e^{-i\omega\tau}\, d\tau, \qquad (33)$$

$$S_{yx}(\omega) = \frac{1}{2\pi} \int_{-\infty}^{\infty} R_{yx}(\tau) e^{-i\omega\tau}\, d\tau. \qquad (34)$$

On account of Eqs. (28) and (29), it follows from these definitions that

$$S_{xy}(\omega) = S_{yx}^*(\omega) \quad \text{and} \quad S_{yx}(\omega) = S_{xy}^*(\omega), \qquad (35)$$

where the asterisk denotes complex conjugate. The cross-correlation functions may be regained from $S_{xy}(\omega)$ and $S_{yx}(\omega)$ by the inverse transform relations

$$R_{xy}(\tau) = \int_{-\infty}^{\infty} S_{xy}(\omega) e^{i\omega\tau}\, d\omega, \qquad (36)$$

$$R_{yx}(\tau) = \int_{-\infty}^{\infty} S_{yx}(\omega) e^{i\omega\tau}\, d\omega. \qquad (37)$$

Usually cross-spectral densities are complex (have real and imaginary parts) and they do not have the simple properties of (auto) spectral density functions.

5 RESPONSE OF LINEAR SYSTEMS

5.1 Frequency Response Functions

Consider a time-invariant linear dynamical system with a typical input $x_s(t)$, for example an applied load or voltage, and a typical output $y_r(t)$, for example a displacement or an electrical current. If only input $x_s(t)$ is excited with a unit harmonic excitation, then using phasor nota-

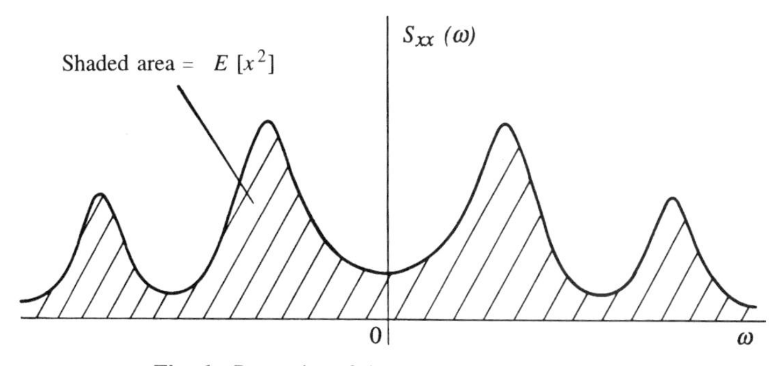

Fig. 6 Properties of the spectral density function.

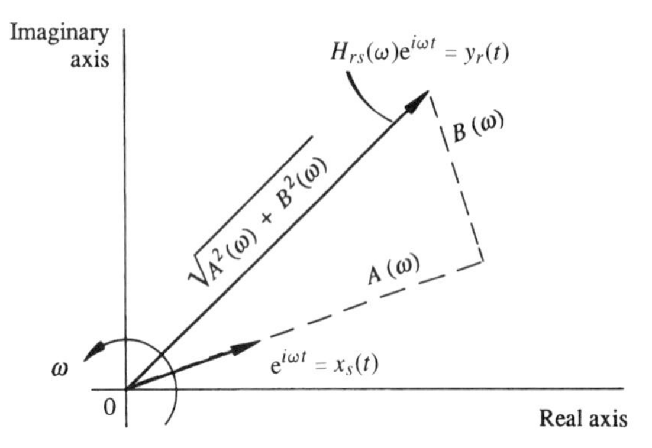

Fig. 7 Phasor representation of the frequency response function $H_{rs}(\omega)$ of a time-invariant, passive, linear system.

tion,

$$x_s(t) = e^{i\omega t}. \tag{38}$$

After steady conditions have been reached (the system is assumed to be stable so that transients always die away eventually), the response $y_r(t)$ can be represented by

$$y_r(t) = H_{rs}(\omega)e^{i\omega t}, \tag{39}$$

where $H_{rs}(\omega)$ is the *frequency response function* that describes the connection between input s and output r. It is a complex function of frequency that relates the amplitude and phase of output r to the assumed unit amplitude and arbitrary phase of input s. This relationship is illustrated in the phasor diagram of Fig. 7. The input $x_s(t)$ is represented as a phasor of unit length rotating counterclockwise at angular speed ω. Then, assuming that

$$H_{rs}(\omega) = A(\omega) + iB(\omega), \tag{40}$$

where $A(\omega)$ and $B(\omega)$ are real and positive, the output $y_r(t)$ is a rotating phasor of length $\sqrt{A^2(\omega) + B^2(\omega)}$ rotating in the same direction at speed ω and leading the excitation by the angle $\tan^{-1}[B(\omega)/A(\omega)]$.

The time dependence of $x_s(t)$ and $y_r(t)$ is given by the projection of the rotating phasors onto either the real (horizontal) axis or the imaginary (vertical) axis in Fig. 7. The selection of which one is arbitrary.

If there are n inputs and m outputs, then there are $n \times m$ frequency response functions that together make up the *frequency response matrix*[3] $\mathbf{H}(\omega)$.

5.2 Impulse Response Functions

When input $x_s(t)$ is given a unit impulse of excitation at $t = 0$ so that

$$x_s(t) = \delta(t), \tag{41}$$

with all the other inputs zero and with the system initially settled, the response $y_r(t)$ is represented by

$$y_r(t) = h_{rs}(t). \tag{42}$$

This real function of time $h_{rs}(t)$ is called the *impulse response function* of output r to input s. It is zero for $t < 0$, and since the linear system is assumed to be passive, it decays to zero as $t \to \infty$.

When there are n inputs and m outputs, there are $n \times m$ impulse response functions that together make up the *impulse response matrix*[3] $\mathbf{h}(t)$.

5.3 Fourier Transform Relationship

It can be shown[3] that each frequency response function is the Fourier transform of its corresponding impulse response function, so that

$$\mathbf{H}(\omega) = \int_{-\infty}^{\infty} \mathbf{h}(t)e^{-i\omega t}\, dt \tag{43}$$

and, conversely, that

$$\mathbf{h}(t) = \frac{1}{2\pi}\int_{-\infty}^{\infty} \mathbf{H}(\omega)e^{i\omega t}\, d\omega. \tag{44}$$

In this case, the factor $1/2\pi$ appears in the inverse transform equation rather than in the transform equation as, for example, in Eq. (30).

Both $\mathbf{H}(\omega)$ and $\mathbf{h}(t)$, and the relationship between them, are needed in establishing how linear systems respond to random excitation.

5.4 Response of a SISO System

The simplest case is of a *single-input, single-output* (SISO) time-invariant passive linear system. This is assumed to be excited by stationary random excitation $x(t)$ with mean level $E[x]$, autocorrelation $R_{xx}(\tau)$, and spectral density $S_{xx}(\omega)$.

It can be shown[1] that the corresponding statistics of the response are

$$E[y] = H(\omega = 0)E[x], \tag{45}$$

$$R_{yy}(\tau) = \int_{-\infty}^{\infty}\int_{-\infty}^{\infty} h(t_1)h(t_2)R_{xx}(\tau + t_1 - t_2) \cdot dt_1\, dt_2, \tag{46}$$

$$S_{yy}(\omega) = |H(\omega)|^2 S_{xx}(\omega), \tag{47}$$

where $H(\omega = 0)$ is the frequency response function evaluated at $\omega = 0$, $|H(\omega)|$ is the modulus of the frequency response function, and $h(t_1)$, $h(t_2)$ are the impulse response functions evaluated at two different times t_1 and t_2.

For comparison with the *multiple-input, multiple-output* (MIMO) case, it is convenient to write Eq. (45) as

$$m_y = H(\omega = 0)m_x. \tag{48}$$

5.5 Response of a MIMO System

When there are n inputs and m outputs, the corresponding results are[1]

$$m_{y_r} = \sum_{s=1}^{n} H_{rs}(\omega = 0)m_{x_s}, \tag{49}$$

$$R_{y_q y_r} = \sum_{s=1}^{n}\sum_{t=1}^{n}\int_{-\infty}^{\infty}\int_{-\infty}^{\infty} \cdot h_{qs}(t_1)h_{rt}(t_2)R_{x_s x_t}(\tau + t_1 - t_2)\,dt_1\,dt_2, \tag{50}$$

$$S_{y_q y_r}(\omega) = \sum_{s=1}^{n}\sum_{t=1}^{n} H_{qs}^*(\omega)H_{rt}(\omega)S_{x_s x_t}(\omega), \tag{51}$$

where now $R_{y_q y_r}$ is the cross-correlation function and $S_{y_q y_r}$ is the cross-spectral density for the two outputs $y_q(t)$ and $y_r(t)$.

These results can be expressed in matrix notation by defining the following:

$\mathbf{m}_x = n \times 1$	vector of the mean values of the n inputs
$\mathbf{m}_y = m \times 1$	vector of the mean values of the m outputs
$\mathbf{h}(t) = m \times n$	matrix of the $m \times n$ impulse response functions relating m outputs to the n inputs
$\mathbf{H}(\omega) = m \times n$	matrix of the $m \times n$ frequency response functions relating m outputs to the n inputs
$\mathbf{R}_{xx}(\tau) = n \times n$	matrix of the $n \times n$ auto- and cross-correlation functions for the inputs
$\mathbf{R}_{yy}(\tau) = m \times m$	matrix of the $m \times m$ auto- and cross-correlation functions for the outputs
$\mathbf{S}_{xx}(\omega) = n \times n$	matrix for the $n \times n$ auto- and cross-spectral density functions for the inputs
$\mathbf{S}_{yy}(\omega) = m \times m$	matrix of the $m \times m$ auto- and cross-spectral density functions for the outputs

The input–output results then become

$$\mathbf{m}_y = \mathbf{H}(\omega = 0)\mathbf{m}_X, \tag{52}$$

$$\mathbf{R}_{yy}(\tau) = \int_{-\infty}^{\infty}\int_{-\infty}^{\infty} \mathbf{h}(t_1)\mathbf{R}_{xx}(\tau + t_1 - t_2) \cdot \mathbf{h}^{\mathrm{T}}(t_2)\,dt_1\,dt_2, \tag{53}$$

$$\mathbf{S}_{yy}(\omega) = \mathbf{H}^*(\omega)\mathbf{S}_{xx}(\omega)\mathbf{H}^{\mathrm{T}}(\omega). \tag{54}$$

In addition to these formulas for the output statistics, it is also possible to calculate the correlation that there is between a typical output $y_r(t)$ and one of the inputs $x_s(t)$. This leads to the following results[1]:

$$\mathbf{R}_{yx}(\tau) = \int_{-\infty}^{\infty} \mathbf{h}(t)\mathbf{R}_{xx}(\tau + t)\,dt \tag{55}$$

$$\mathbf{S}_{yx}(\omega) = \mathbf{H}^*(\omega)\mathbf{S}_{xx}(\omega). \tag{56}$$

where $\mathbf{R}_{yx}(\tau)$ is the $m \times n$ matrix of the $m \times n$ cross-correlation functions between the m outputs $y_r(t)$, $r = 1, \ldots, m$, and the n inputs $x_s(t)$, $s = 1, \ldots, n$, and $\mathbf{S}_{yx}(\omega)$ is the $m \times n$ matrix of the corresponding $m \times n$ cross-spectral density functions between the outputs and inputs.

Because of the simplicity of Eqs. (47), (54), and (56), calculations in the frequency domain offer the main computational approach for studying many practical problems. A great deal of research has therefore been done on digital spectral processing,[1,4,5] which is now the major tool for practical random vibration calculations.

5.6 Coherence Calculations

If there are unknown sources of noise affecting a system's response, or if the system is nonlinear, the measured spectral densities of the output will differ from those calculated by using Eq. (54). Consider only one output, so that $m = 1$ in Eq. (54) and $S_{yy}(\omega)$ has only one term. If the measured output spectral density is $S_{yy}(\omega)$ and the calculated value is $\tilde{S}_{yy}(\omega)$, then their ratio

$$\frac{\tilde{S}_{yy}(\omega)}{S_{yy}(\omega)} = \eta_{yx}^2(\omega) \tag{57}$$

is called the *multiple coherence function* between the output process $y(t)$ and the input processes $x_s(t)$, $s = 1, \ldots, n$. It can be shown[6] that

$$0 \le \eta^2_{yx}(\omega) \le 1 \tag{58}$$

and that $\eta^2_{yx}(\omega)$ is only unity if the system is time invariant, linear, and noise free.

On substituting for the calculated spectral density from Eqs. (54) into (57),

$$\eta^2_{yx}(\omega) = \frac{\mathbf{H}^*(\omega)\mathbf{S}_{xx}(\omega)\mathbf{H}^{\mathrm{T}}(\omega)}{S_{yy}(\omega)}, \tag{59}$$

and substituting for the frequency response matrix $\mathbf{H}(\omega)$ from Eq. (56) give

$$\eta^2_{yx}(\omega) = \frac{\mathbf{S}_{yx}(\omega)\mathbf{S}^{-1}_{xx}(\omega)\mathbf{S}^{\mathrm{T}}_{xy}(\omega)}{S_{yy}(\omega)}. \tag{60}$$

If there is only one input, so that all the matrices have only one term, Eq. (60) becomes

$$\eta^2_{yx}(\omega) = \frac{S_{yx}(\omega)S_{xy}(\omega)}{S_{xx}(\omega)S_{yy}(\omega)}. \tag{61}$$

This gives the single or *ordinary coherence function* between output $y(t)$ and input $x(t)$ at frequency ω. If the system is linear and time invariant and there is only one input, then if the spectra are measured accurately, η^2 will be exactly unity.

Because of the need for accurate measurements, there are practical problems when computing coherence functions by digital analysis.[1,4]

5.7 Probability Distributions

When a time-invariant passive linear system is subjected to Gaussian excitation, it can be shown[7] that the system's response will also be Gaussian. For MIMO systems, the separate inputs have to be jointly Gaussian and the outputs are then also jointly Gaussian (as well as first-order distributions for the individual signals being Gaussian, higher order distributions between two or more signals are also Gaussian).

For non-Gaussian inputs, the output probability distributions can be expressed in terms of a series expansion involving the input probabilities and the system's response characteristics,[8] but the calculations are extremely complex. There are no general closed-form solutions.

The upshot is that, although the spectral density of the output can always be found, calculations on the amplitude distribution of the output may not be practicable unless the excitation is Gaussian. Fortunately, because of the central limit theorem, many naturally occurring processes are Gaussian to a good approximation.

The probability distributions for a Gaussian process are defined if the mean, mean-square, and appropriate correlation functions are known [see Eqs. (7) and (8)]. According to linear system theory, these statistics can always be calculated from the formulas in Section 5.5.

5.8 Continuous Systems

There are many applications where a continuous, stable, time-invariant, linear system, for example a mechanical structure, is subjected to spatially distributed time-stationary random excitation, for example a fluctuating pressure field. Figure 8 shows a notional continuous system subjected to distributed excitation (e.g., pressure) $p(\mathbf{s}, t)$ at position $\mathbf{s}$ and time t and that has a distributed response (e.g., displacement) $y(\mathbf{r}, t)$ at position $\mathbf{r}$ and time t. Corresponding to Eqs. (49)–(51), the response of this system is given by[1]

$$m_y(\mathbf{r}) = \int_R d\mathbf{s}\, H(\mathbf{r}, \mathbf{s}, \omega = 0) m_p(\mathbf{s}), \tag{62}$$

$$\begin{aligned} R_{yy}(\mathbf{r}_1, \mathbf{r}_2, \tau) = \int_R d\mathbf{s}_1 \int_R d\mathbf{s}_2 \int_{-\infty}^{\infty} dt_1 \int_{-\infty}^{\infty} \\ \cdot\, dt_2\, h(\mathbf{r}_1, \mathbf{s}_1, t_1) h(\mathbf{r}_2, \mathbf{s}_2, t_2) \\ \cdot\, R_{pp}(\mathbf{s}_1, \mathbf{s}_2, \tau + t_1 - t_2), \end{aligned} \tag{63}$$

and

$$\begin{aligned} S_{yy}(\mathbf{r}_1, \mathbf{r}_2, \omega) = \int_R d\mathbf{s}_1 \int_R d\mathbf{s}_2\, H^*(\mathbf{r}_1, \mathbf{s}_1, \omega) \\ \cdot\, H(\mathbf{r}_2, \mathbf{s}_2, \omega) S_{pp}(\mathbf{s}_1, \mathbf{s}_2, \omega). \end{aligned} \tag{64}$$

The finite summations in Eqs. (48)–(51) have been replaced by continuous integrals over the region involved, R. The definitions of terms correspond to those in Section 5.5, where now, $R_{yy}(\mathbf{r}_1, \mathbf{r}_2, \tau)$ denotes the cross-correlation between $y(\mathbf{r}_1, t)$ and $y(\mathbf{r}_2, t + \tau)$ and replaces $R_{y_q y_r}(\tau)$, and similarly for the other terms.

For the above results the excitation is assumed to be stationary in time. It may also be stationary in space (or *homogeneous* in space). In that case the correlation of $p(\mathbf{s}, t)$ at two different positions $\mathbf{s}_1$ and $\mathbf{s}_2$ depends only on the vector difference between these positions ($\mathbf{s}_2 - \mathbf{s}_1$). The cross-spectral density of the input can then be written as[1]

$$\int_\infty d\gamma\, S_{pp}(\gamma, \omega) \exp\, i\{\gamma \cdot (\mathbf{s}_2 - \mathbf{s}_1)\}, \tag{65}$$

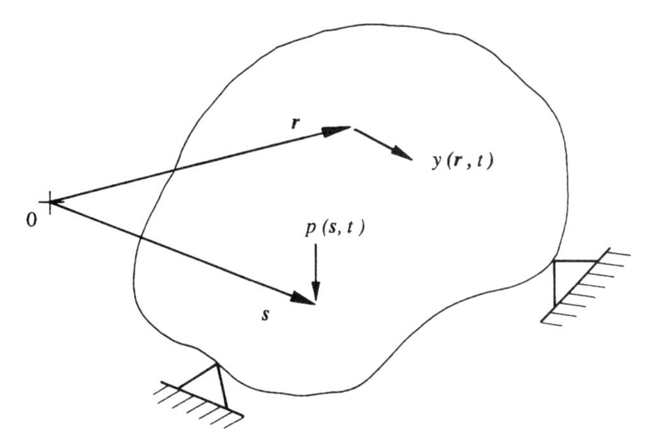

Fig. 8 Continuous dynamic system subjected to distributed excitation $p(s, t)$ in a specified reference direction and having distributed response $y(r, t)$, also in a specified reference direction.

where $S_{pp}(\gamma, \omega)$ is the multidimensional spectral density of the distributed excitation process $p(\mathbf{s}, t)$ in terms of the (vector) wavenumber γ and the angular frequency ω.

Substituting Eq. (65) into Eq. (64) leads to the result that

$$S_{yy}(\mathbf{r}_1, \mathbf{r}_2, \omega) = \int_\infty d\gamma\ S_{pp}(\gamma, \omega) G^*(\mathbf{r}_1, \gamma, \omega) G(\mathbf{r}_2, \gamma, \omega), \tag{66}$$

where

$$G(\mathbf{r}, \gamma, \omega) = \int_R H(\mathbf{r}, \mathbf{s}, \omega) \exp\ i(\gamma \cdot \mathbf{s})\ d\mathbf{s}. \tag{67}$$

This function $G(\mathbf{r}, \gamma, \omega)$ is called the *sensitivity function* for the system. It gives the sensitivity of the system measured at position $\mathbf{r}$ to distributed excitation that extends over the whole system and is harmonic in space and time at wave vector γ and time frequency ω.

5.9 Normal-Mode Analysis

Normal-mode analysis depends on the experimental result that, in many practical vibration problems, the response of interest is confined mainly to a limited number of natural modes (see Chapter 48). This provides an approximate method of calculating $H(\mathbf{r}, \mathbf{s}, \omega)$.

If $\mathbf{y}(\mathbf{r}, t)$ is a column vector of the geometric components of the response of a continuous system at position $\mathbf{r}$ and time t and if $\mathbf{U}_j(\mathbf{r})$ is a column vector of the corresponding components of the normal modes, then the normal-mode expansion of the response is given by

$$y(\mathbf{r}, t) = \sum_{j=1}^{\infty} q_j(t) \mathbf{U}_j(\mathbf{r}), \tag{68}$$

where $q_j(t)$, $j = 1, \ldots, \infty$, are the normal coordinates. The system is assumed to have no energy transfer across the boundaries and classical damping so that a full set of real, orthogonal, normal-mode functions exists. In that case the general vector frequency response function $\mathbf{H}(\mathbf{r}, \mathbf{s}, \omega)$ can be written in the expansion

$$\mathbf{H}(\mathbf{r}, \mathbf{s}, \omega) = \sum_{j=1}^{\infty} H_j(\omega) \mathbf{U}_j(\mathbf{r}) \mathbf{U}_j^{\mathrm{T}}(\mathbf{s}) \mathbf{e}_s, \tag{69}$$

where $H_j(\omega)$ is the modal frequency response function and $\mathbf{e}_s$ is a unit vector identifying the direction of the unit harmonic excitation applied at position $\mathbf{s}$.[3] The components of $\mathbf{H}(\mathbf{r}, \mathbf{s}, \omega)$ are the frequency responses in the three coordinate directions at $\mathbf{r}$ for unit harmonic excitation applied at $\mathbf{s}$ in direction $\mathbf{e}_s$.

For classical damping[3] the normal-mode functions are the same as those for the undamped problem and may, in principle, be obtained by solving for the undamped free vibration of the system. In practice, this may prove impossible, and the usual approach is then to establish an approximate model with a finite number of degrees of freedom, usually using the finite-element method, and to solve this new problem exactly.

6 NARROW-BAND PROCESSES

6.1 Properties of Narrow-Band Random Processes

When a strongly resonant system responds to broadband random excitation, its output spectral density falls mainly in a narrow band of frequencies close to the resonant frequency. Since this output is derived by filtering a broadband process, many nearly independent events contribute to it. Therefore, on account of the central limit theorem, the probability distribution of a narrow-band process approaches that of a Gaussian distribution even if the excitation is not Gaussian.

Assuming that the narrow-band process is Gaussian and stationary, then if its spectral density $S_{yy}(\omega)$ is specified, all its other statistics can be derived from $S_{yy}(\omega)$. The corresponding autocorrelation function $R_{yy}(\tau)$ can be calculated by using the inverse Fourier transform relation (31) and the variance σ_y^2 and mean m_y found from the properties of $R_{yy}(\tau)$ illustrated in Fig. 4. Also the statistics of the $\dot{y}(t)$ process can be derived by using the result that, for any stationary process,[1]

$$R_{\dot{y}\dot{y}}(\tau) = -\frac{d^2}{d\tau^2} R_{yy}(\tau). \tag{70}$$

Hence $\sigma_{\dot{y}}^2$ and $m_{\dot{y}}$ can be found.

For any stationary random process it can be shown[1] that $y(t)$ and $\dot{y}(t)$ are uncorrelated, so that, from Eq. (22),

$$p(y, \dot{y}) = p(y)p(\dot{y}). \tag{71}$$

The first-order Gaussian distributions are known when σ and m are specified, so that $p(y)$ and $p(\dot{y})$ are known and, from Eq. (71), so is the second-order Gaussian distribution $p(y, \dot{y})$.

6.2 Crossing Analysis

Figure 9 shows a sample function from a stationary, narrow-band random process. The ensemble average number of up-crossings (crossings with positive slope) of the level $y = a$ in time T, denoted by $N^+(a, T)$, will be proportional to T, so that

$$N^+(a, T) = \nu^+(a)T, \tag{72}$$

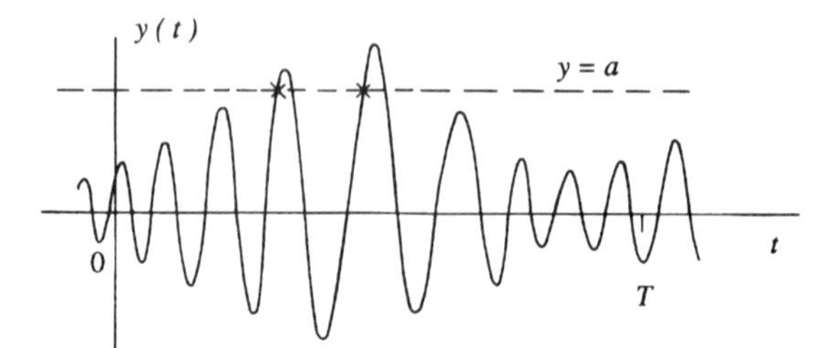

Fig. 9 Up-crossings of a sample function from a narrow-band process $y(t)$.

where $\nu^+(a)$ is the frequency of up-crossings of level $y = a$. It can be shown[1] that

$$\nu^+(a) = \int_0^\infty p(a, \dot{y})\dot{y}\, d\dot{y}, \tag{73}$$

where $p(a, \dot{y})$ is the joint probability density function for the distribution of y and $\dot{y}$ evaluated at $y = a$. By using Eq. (71) and for the Gaussian probability density function (7) with zero mean, this leads to the result that

$$\nu^+(a) = \frac{1}{2\pi} \frac{\sigma_{\dot{y}}}{\sigma_y} \exp\left(-\frac{a^2}{2\sigma_y^2}\right), \tag{74}$$

where σ_y^2 is the variance of $y(t)$ and $\sigma_{\dot{y}}^2$ is the variance of $\dot{y}(t)$.

For a process with zero mean, from Eqs. (10) and (32),

$$\sigma_y^2 = \int_{-\infty}^{\infty} S_{yy}(\omega)\, d\omega. \tag{75}$$

Also it can be shown from Eq. (70) that[1]

$$S_{\dot{y}\dot{y}}(\omega) = \omega^2 S_{yy}(\omega), \tag{76}$$

so that

$$\sigma_{\dot{y}\dot{y}}^2 = \int_{-\infty}^{\infty} \omega^2 S_{yy}(\omega)\, d\omega. \tag{77}$$

Using these results, the average frequency for the $y(t)$ process, which is obtained by putting $a = 0$ in Eq. (74), can be written as

$$\nu^+(0) = \frac{1}{2\pi}\left[\frac{\int_{-\infty}^{\infty} \omega^2 S_{yy}(\omega)\, d\omega}{\int_{-\infty}^{\infty} S_{yy}(\omega)\, d\omega}\right]^{1/2}. \tag{78}$$

This is the ensemble average frequency for the $y(t)$ process. It is only the same as the average frequency along the time axis if the process is ergodic.

6.3 Distribution of Peaks

For a narrow-band process that has one positive peak for every zero crossing, the proportion of cycles whose peaks exceed $y = a$ is $\nu^+(a)/\nu^+(0)$. This is the probability that any peak chosen at random exceeds a and leads to the result that[1,9]

$$p_p(a) = \frac{a}{\sigma_y^2} \exp\left(-\frac{a^2}{2\sigma_y^2}\right), \qquad a \geq 0, \tag{79}$$

which is the Rayleigh distribution shown in Fig. 2*a*.

6.4 Limit on Spectral Bandwidth

The result (79) depends on the assumption that there is only one positive peak for each zero crossing. It can be shown[10] that Eq. (73) is true for any random process $y(t)$ that can be differentiated at least once, and Eqs. (74) and (78) are true if the process is also Gaussian. If $y(t)$ can be differentiated at least twice, its frequency of maxima will be the same as the frequency of zero crossings of the derived process $\dot{y}(t)$. If the frequency of maxima is μ, then, by analogy with Eq. (78),

$$\mu = \frac{1}{2\pi}\left[\frac{\int_{-\infty}^{\infty} \omega^4 S_{yy}(\omega)\,d\omega}{\int_{-\infty}^{\infty} \omega^2 S_{yy}(\omega)\,d\omega}\right]^{1/2}. \tag{80}$$

Only if the spectral bandwidth is vanishingly small will μ calculated from Eq. (80) be the same as $\nu^+(0)$ calculated from Eq. (78). In other cases there are irregularities in the narrow-band response of Fig. 9 that give rise to additional local peaks.[1,9]

6.5 General Peak Distribution

The probability density function $p_p(a)$ for the distribution of peaks of a general stationary Gaussian process with zero mean, $y(t)$, provided this can be differentiated twice, is[10]

$$p_p(a) = \frac{(1-\xi)^{1/2}}{\sqrt{2\pi}\sigma_y} \exp\left\{-\frac{a^2}{2\sigma_y^2(1-\xi^2)}\right\} + \frac{\xi a}{2\sigma_y^2} \cdot \left[1 + \operatorname{erf}\left\{\frac{\xi a}{\sqrt{2}\sigma_y(1-\xi^2)^{1/2}}\right\} \cdot \exp\left(-\frac{a^2}{2\sigma_y^2}\right)\right], \tag{81}$$

where

$$\xi = \frac{\sigma_{\dot{y}}^2}{\sigma_y \sigma_{\ddot{y}}}, \tag{82}$$

which is the ratio of the expected rate of zero crossings to the expected rate of peaks, and erf denotes the error function defined as

$$\operatorname{erf} x = \frac{2}{\sqrt{\pi}} \int_0^x \exp(-w^2)\,dw. \tag{83}$$

In the limiting case of a narrow-band process with vanishingly small bandwidth, there is one peak for each zero crossing so that $\xi = 1$. In that case Eq. (81) reduces to the Rayleigh distribution (79). In contrast, a broadband random process has many peaks for each zero crossing, so that $\xi \rightarrow 0$. In this case $p_p(a)$ is just a normal distribution that is the same as the amplitude distribution.

6.6 Envelope Properties

The *envelope* $A(t)$ of a random process $y(t)$ may be defined in various different ways. For each sample function it consists of a smoothly varying pair of curves such that $|A(t)| \geq |y(t)|$ for all t and for which $|A(t)| = |y(t)|$ at, or nearly at, the peaks of $y(t)$. The various definitions differ in respect of their interpretation of the phrase "or nearly at." One common definition[9] is to say that the envelope is the pair of curves $A(t)$ and $-A(t)$ given by

$$A^2(t) = y^2(t) + \frac{\dot{y}^2(t)}{\nu^+(0)^2}, \tag{84}$$

where $\nu^+(0)$ is the average frequency of zero crossings of the $y(t)$ process.

When $y(t)$ is stationary and Gaussian, it can then be shown that this definition leads to the following envelope probability density function:

$$p(A) = \frac{A}{\sigma_y^2} \exp\left(-\frac{A^2}{2\sigma_y^2}\right), \qquad A \geq 0. \tag{85}$$

This is the same as the probability distribution for the peaks of a narrow-band, stationary, Gaussian process, [Eq. (79)]. However the two distributions differ if the process $y(t)$ is not both narrow band and Gaussian.

6.7 Clumping of Peaks

The response of a narrow-band random process is characterized by a slowly varying envelope, so that peaks

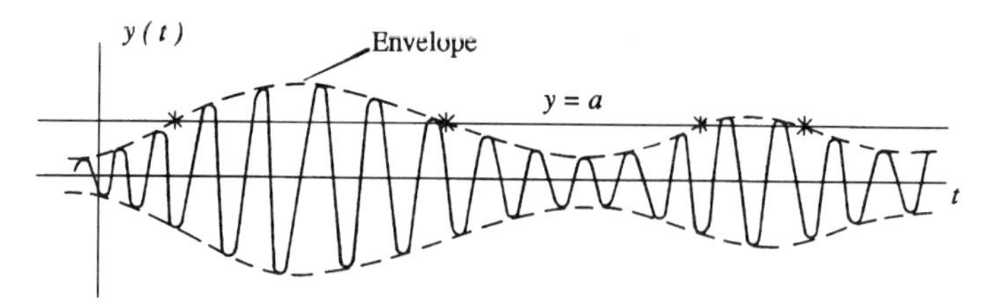

Fig. 10 Each clump of peaks greater than a is begun and ended by an envelope crossing of $y = a$.

occur in clumps (Fig. 10). Each clump of peaks greater than a is begun and ended by its envelope crossing $y = a$. If M denotes the number of up-crossings of $y = a$ of the sample function $y(t)$ for each up-crossing of the same level for its envelope $A(t)$, then

$$E[M] = \frac{\nu_y^+(a)}{\nu_A^+(a)}, \tag{86}$$

which can be shown[10] to be given by

$$E[M] = \frac{\sigma_y}{a}\left[2\pi\left(1 - \frac{\lambda^2}{\sigma_y^2\sigma_{\dot{y}}^2}\right)\right]^{-1/2}, \tag{87}$$

where

$$\lambda = 2\int_0^\infty \omega S(\omega)\, d\omega. \tag{88}$$

For a process whose spectral density is infinitesimally narrow, $\lambda^2 = \sigma_y^2\sigma_{\dot{y}}^2$, and then $E[M]$ becomes infinite as the clumps of peaks are infinitely long. In other cases, $E[M]$ depends on the level a by the factor σ_y/a and on the shape of the spectral density function through the term in square brackets in Eq. (87).

The result (87) does not take account of the fact that a clump cannot have less than one peak. It has been possible to identify and eliminate from the calculation those envelope crossings that are not followed by at least one up-crossing of $y(t)$. This leads to a revised expression for $E[M]$ that is asymptotic to Eq. (87) for narrow-band processes but diverges as the spectral band is widened.

6.8 Nonstationary Processes

The distribution of peaks of a nonstationary random process can also be calculated, and these results then depend on time. If the distribution of $y(t)$ remains Gaussian as it evolves, its peak distribution is close to a Weibull distribution (see Section 2.4) with the parameters X_0 and k [see Eq. (14)] depending on the characteristics of the nonstationary process and on its duration.[10]

The properties of the envelope of a nonstationary process can also be calculated and the level-crossing properties of this envelope may be derived from its probability description.[10]

7 FAILURE DUE TO RANDOM VIBRATION

7.1 Fractional Occupation Time

Failure under vibration may depend on the *fractional occupation time* or *time spent off-range*, which is defined as the proportion Y_a of total time that a random process $y(t)$ spends above the level $y = a$. If $y(t)$ is stationary, Gaussian, and with zero mean, the ensemble average of Y_a is[1]

$$E[Y_a] = \frac{1}{2}\left(1 - \text{erf}\left(\frac{a}{\sqrt{2}\sigma_y}\right)\right). \tag{89}$$

When the level $y = a$ is large, so that $a \geq \sigma_y$, this may be approximated by

$$E[Y_a] \approx \frac{1}{\sqrt{2\pi}}\frac{\sigma_y}{a}\exp\left(-\frac{a^2}{2\sigma_y^2}\right). \tag{90}$$

The variance of Y_a is also available.[10] In addition, both the mean value and variance of Y_a may be calculated when $y(t)$ is nonstationary provided that the probability density functions $p(y(t))$ for the mean value and $p(y(t_1), y(t_2))$ for the variance are available for the total interval of time to be considered.

7.2 First-Passage Time

The *first-passage time* for $y(t)$ is the time at which $y(t)$ first crosses a chosen level, $y = a$, after being started at $t = 0$. If the crossings of level a are a rare event, they may be assumed to be independent of each other and randomly distributed along the time axis. In that case, it can be shown that the probability of a first passage in time T is given by

$$\text{Probability of first passage in time } T = 1 - \exp[-\nu^+(a)T], \tag{91}$$

were, as in Section 6.2, $\nu^+(a)$ is the average frequency of up-crossings of the level a. The probability density function for first-passage time is obtained by differentiating this expression to obtain

$$p(T) = \nu^+(a)\exp[-\nu^+(a)T], \qquad T > 0, \tag{92}$$

which gives the following results for the mean and variance of the first-passage time:

$$E[T] = \frac{1}{\nu^+(a)} \quad \text{and} \quad \text{var}[T] = \frac{1}{[\nu^+(a)]^2}. \tag{93}$$

These results are only correct for crossings randomly distributed along the time axis. Because of clumping, the intervals between clumps will be longer than the average spacing between crossings and the probability of failure in time T will therefore be less than indicated by Eq. (91). The approximation will be particularly poor for a narrow-band process. For this it is more realistic to assume that failure may occur when the envelope $A(t)$ of $y(t)$ first crosses the level a. Assuming that the envelope crossings are independent and have average up-crossing rate $\nu_A^+(a)$, the statistics of the first-passage time are still given by Eq. (93) but with $\nu_A^+(a)$ replacing $\nu^+(a)$.

A general exact solution for the first-passage problem has not yet been found but there are a variety of approximate solutions of considerable complexity. Allowance can be made for $y(t)$ being nonstationary, and upper and lower bounds for the probability of first-passage failure have been calculated.

7.3 Maximum Value

The maximum value of a random process in a fixed interval of time is a random variable. If y_m is the maximum value of a stationary, narrow-band, Gaussian process $y(t)$ in time interval T, then y_m must be the highest peak in interval T. It can be shown that, assuming the peaks are independent, the probability that y_m is less than a is given by

$$P_{y_m}(a) = \exp\left[-\nu^+(0)T\exp\left(-\frac{a^2}{2\sigma_y^2}\right)\right], \tag{94}$$

where $\nu^+(0)$ is the average frequency of zero crossings of $y(t)$.

The mean and variance of the maximum value y_m can be calculated from $P_{y_m}(a)$ and are given by[10]

$$E[y_m] = \left(C + \frac{0.5772}{C}\right)\sigma_y, \tag{95}$$

$$\text{var}[y_m] = 1.638\,\frac{\sigma_y^2}{C^2}, \tag{96}$$

where

$$C = \{2\ln(\nu^+(0)T)\}^{1/2}. \tag{97}$$

7.4 Fatigue Failure

The calculation of fatigue damage accumulation is based on the notion that individual cycles of stress can be identified and that each stress cycle advances a fatigue crack. When the crack has reached a critical size, failure occurs. One cycle of stress of amplitude S is assumed to generate $1/N(S)$ of the damage needed to cause failure.

For a stationary, narrow-band, random process with average frequency $\nu^+(0)$, the number of cycles in time T will be $\nu^+(0)T$. If the probability density for the distribution of peaks is $p_p(S)$, then the average number of stress cycles in the range S to $S + dS$ will be $\nu^+(0)Tp_p(S)\,dS$.

The damage done by this number of stress cycles is

$$\nu^+(0)Tp_p(S)\,dS\,\frac{1}{N(S)},$$

and so the average damage done by all stress cycles together will be

$$E[D(T)] = \nu^+(0)T\int_0^\infty \frac{1}{N(S)}\,p_p(S)\,dS. \tag{98}$$

Failure occurs when the accumulated damage $D(T)$ is unity.

The variance of the accumulated fatigue damage, $\sigma_{D(T)}^2$, can also be calculated.[9]

When the stress cycles occur as a result of the response of a lightly damped oscillator to broadband Gaussian excitation, the damping ratio of the oscillator being ζ, it can be shown that

$$\frac{\sigma_{D(T)}}{E[D(T)]} \propto \frac{1}{[\zeta\nu^+(0)T]^{1/2}}. \tag{99}$$

For light damping, $\zeta \ll 1$, which is necessary for the narrow-band assumption to be correct, the number of stress cycles $\nu^+(0)T$ must be large if $\sigma_{D(T)}$ is to be small. In that case, an estimate of the average time to failure, $E[T_F]$, can be made by assuming that this is the value of T in Eq. (98) when $E[D(T)] = 1$.

The practical difficulty with this approach is that of modeling the fatigue crack propagation mechanism sufficiently accurately. The use of more complicated fracture models makes the statistical analysis a good deal more difficult.

8 RESPONSE OF NONLINEAR SYSTEMS

8.1 Introduction

The analysis of how dynamical systems respond to random excitation has been most fully developed for time-invariant, passive, linear systems. Their behavior can be represented by linear differential equations with constant coefficients, and their response is stable, so that a system with no excitation settles eventually to rest. The general excitation–response relationships for such systems are described by the results in Section 5.

In recent years there has been growing interest in the response of parametrically excited linear systems and of nonlinear systems subjected to random excitation. The response of these have not yet been reduced to generally applicable results, and problems of these types are usually solved either by exact methods specific to a particular problem (which, for random vibrations, are rare) or by approximate methods.

In principle, exact solutions for the response of any dynamical system (linear or nonlinear) subjected to broadband, Gaussian excitation (theoretically, white, Gaussian excitation) can be obtained from the theory of *continuous Markov processes*. This method, which is described in Section 8.2, determines the probability structure of a system's response as a function of time. Inherently it is extremely powerful. In practice, the diffusion equation describing how the probability distribution evolves is too difficult to be solved exactly and approximate methods have to be used to obtain a solution. This method forms the basis for one of the main approaches to parametric (time-varying) problems[11] as well as to nonlinear problems. For nonlinear problems there are many different methods, of which two of the more important are *perturbation methods* (Section 8.3) and *statistical linearization* (Section 8.4).

8.2 Markov Methods

For a one-dimensional random process $y(t)$, let $p(y, t|y_0, t_0)\, dy$ be the conditional probability that $y(t)$ has a value between y and $y + dy$ at time $t(t > t_0)$ if it had the value y_0 at time t_0. Then, for a continuous Markov process, the probability density function $p \equiv p(y, t|y_0, t_0)$ satisfies the *Fokker–Planck equation*[10]

$$\frac{\partial p}{\partial t} = -\frac{\partial}{\partial y}[m(y,t)p] + \frac{1}{2}\frac{\partial^2}{\partial y^2}[\sigma^2(y,t)p], \qquad (100)$$

where $m(y,t)$ is called the *drift coefficient* and $\sigma^2(y,t)$ is the *diffusion coefficient*. These are defined as follows. If Δy is the change in response $y(t)$ that occurs in time increment Δt, then

$$m(y,t) = \lim_{\Delta t \to 0} \frac{E[\Delta y]}{\Delta t} \qquad (101)$$

and

$$\sigma^2(y,t) = \lim_{\Delta t \to 0} \frac{E[\Delta y^2]}{\Delta t}. \qquad (102)$$

For the response of a multidimensional system that has a state vector $y(t)$ with n elements, a corresponding form of Eq. (100) exists with n first-order derivatives $\partial/\partial y_i$ and n^2 second-order derivatives $\partial^2/\partial y_i\,\partial y_j$.

The definition of a *continuous* Markov process[10] requires (i) that the future states of the process $y(t)$ depend only on the present state, not on how that state was arrived at, so that $p(y, t|y_0, t_0)$ is unique, and (ii) that higher order derivate moments like $\lim_{\Delta t \to 0}(E[\Delta y^n]/\Delta t)$ and $\lim_{\Delta t \to 0}(E[\Delta y_i\, \Delta y_j\, \Delta y_k]/\Delta t)$ all vanish when their order is greater than 2.

It can be shown that these conditions always apply when $y(t)$ is the response of a system to excitation that is white and Gaussian.

An alternative version of the Fokker–Planck equation (100) is called the *Kolmogorov equation*, and this expresses the same result but in a slightly different form.

In order to determine the drift and diffusion coefficients in the Fokker–Planck equation, it is necessary to refer to the equations for the dynamical system being studied. For example, consider the one-dimensional (linear) system

$$\dot{y} + cy = x(t), \qquad (103)$$

where $x(t)$ is stationary, white, Gaussian excitation. Since $y(t)$ is continuous but $x(t)$ is not, this may be written in the incremental form

$$\Delta(y) = -cy\,\Delta t + \int_t^{t+\Delta t} x(t)\, dt. \qquad (104)$$

Taking the ensemble average of this equation, subject to the condition that we are considering only samples for which $y(t)$ lies in the range y to $y + \Delta y$, gives

$$\lim_{\Delta t \to 0} \frac{E[\Delta y]}{\Delta t} = -cy \tag{105}$$

because $E[x(t)]$ is zero, and leads to the result that

$$\lim_{\Delta t \to 0} \frac{E[\Delta y^2]}{\Delta t} = 2\pi S_0, \tag{106}$$

where S_0 is the spectral density of $x(t)$.

Exact solutions to the Fokker–Planck equation have been obtained only for a few first-order systems. Even for stationary solutions, occurring as $t \to \infty$ when the time derivative $\partial p/\partial t$ has become zero, there are only solutions for certain specific classes of nonlinear equations.[10] As a result, various methods for obtaining approximate solutions have been developed.

For a lightly damped, one-degree-of-freedom oscillator, the method of *stochastic averaging* allows the two-dimensional Markov process describing the response $y(t)$ to be replaced approximately by a one-dimensional Markov process $a(t)$ describing the envelope of $y(t)$. This method has been used with considerable success for studying first-passage problems.[12] However, its application to systems with more than one degree of freedom has not been successful.

Other approximate methods for solving the *Fokker–Planck* equation include *Galerkin's method*, in which the probability density function is expanded in a set of trial functions that can be calculated approximately by analytical or numerical methods. Also there are *iterative* and *variational* methods that have worked well in particular cases.

8.3 Perturbation Methods

Classical *perturbation methods* have been used successfully to determine the approximate response statistics of weakly nonlinear systems. The basic idea is to expand the solution as a power series in terms of a small scaling parameter. For example, the solution of the system

$$\ddot{y} + \eta\dot{y} + \omega^2 y + \epsilon f(y, \dot{y}) = x(t), \tag{107}$$

where $|\epsilon| \ll 1$, is assumed to have the form

$$y(t) = y_0(t) + \epsilon y_1(t) + \epsilon^2 y_2(t) + \cdots. \tag{108}$$

After substituting Eq. (108) into Eq. (107) and collecting terms, the coefficients of like powers of ϵ are then set to zero. This leads to a hierarchy of linear second-order equations that can be solved sequentially by linear theory. The excitation for the second equation is derived from the response of the first equation, and so on. Using these results, it is then possible to calculate $R_{yy}(\tau)$, for example, since, from Eq. (108),

$$\begin{aligned} R_{yy}(\tau) &= E[y(t)y(t+\tau)] \\ &= R_{y_0y_0}(\tau) + \epsilon\{R_{y_0y_1}(\tau) + R_{y_0y_1}(-\tau)\} + O(\epsilon^2) \end{aligned} \tag{109}$$

and, by taking its Fourier transform,

$$S_{yy}(\omega) = S_{y_0y_0}(\omega) + 2\epsilon\ \mathrm{Re}\{S_{y_0y_1}(\omega)\} + O(\epsilon^2). \tag{110}$$

In practice, results have generally been obtained to first-order accuracy only, because of the complexity of the algebra for higher order calculations. The method can in principle be extended to multi-degree-of-freedom systems but subject to the same problem of algebraic complexity. Provided that $|\epsilon| \ll 1$, results have been found to agree with those from other methods, but a general proof of convergence is not currently available.

8.4 Statistical Linearization

This method involves replacing the governing set of nonlinear differential equations by a set of linear differential equations that are equivalent in some way.[13] In control theory, it is called the *describing function method* and, in random process theory, *equivalent linearization* or *stochastic linearization* as well as *statistical linearization*. The parameters of the equivalent linear system are obtained by minimizing the *equation difference*.

For example, suppose that the linear system that is equivalent to Eq. (107) is

$$\ddot{y} + \eta_e\dot{y} + \omega_e^2 y = x(t). \tag{111}$$

The equation difference is

$$e(y, \dot{y}) = \epsilon f(y, \dot{y}) + (\eta - \eta_e)\dot{y} + (\omega^2 - \omega_e^2), \tag{112}$$

where η_e and ω_e^2 are unknown parameters. They are chosen so as to minimize some measure of $e(y, \dot{y})$. The usual method is to minimize the mean square of $e(y, \dot{y})$ with respect to the parameters η_e and ω_e^2. This is done by setting

$$\frac{\partial E[e^2(y,\dot{y})]}{\partial \eta_e} = 0 \quad \text{and} \quad \frac{\partial E[e^2(y,\dot{y})]}{\partial \omega_e^2} = 0 \tag{113}$$

and leads to two equations for η_e and ω_e^2.

These equations include terms such as $E[y^m \dot{y}^n]$, which are not known unless the joint probability density function $p(y, \dot{y})$ is known, which it is not. This difficulty is overcome by assuming that the probability structure of $y(t)$ and $\dot{y}(t)$ is that of the equivalent system (111) when $x(t)$ is Gaussian. The argument is that, for a lightly damped system, the response $y(t)$ will be approximately Gaussian even if $x(t)$ is non-Gaussian. This enables $E[y^m \dot{y}^n]$ to be obtained in terms of the response statistics of (111), which are expressed in terms of η_e and ω_e^2. On substituting the results into Eq. (113), there are two nonlinear algebraic equations for η_e and ω_e^2. Solving these equations gives η_e and ω_e^2 and the equivalent linear system is determined.

The statistical linearization method can be generalized to multi-degree-of-freedom systems, including those with nonstationary excitation. In the latter case, or when considering the initial response to stationary excitation, the equivalent linear system has time-varying coefficients that are functions of the time-varying response statistics.

There has been considerable research on the statistical linearization method,[13] and it has been applied to many practical response problems, including problems in earthquake engineering with hysteretic damping that occurs due to yielding or slipping. The restoring force is then a function of the previous history of the response, not just its instantaneous state, and the response calculation is more complicated still. Statistical linearization has proved particularly successful in such cases.

8.5 Monte Carlo Simulation

An experimental approach for generating response statistics uses the digital simulation of random processes. Random excitation with the required properties is generated artificially, and the response it causes is found by numerically integrating the equations of motion. Provided that a sufficiently large number of experiments are conducted by generating new realizations of the excitation and integrating its response, so that an ensemble of sample functions is created, response statistics can be obtained by averaging across the ensemble. This permits the statistics of nonstationary responses to be estimated by averaging data from several hundred sample functions. For problems with a stationary response, the necessity for a large number of records can be avoided by averaging over time along a single sample provided that the response statistics required can be assumed to be ergodic.

There are various computational strategies for generating realizations of pseudorandom processes but the most common approach is to use the *inverse discrete Fourier transform*, which is computed by the fast Fourier transform (FFT) algorithm. In order to generate two correlated sample functions $x(t)$ and $y(t)$, with specified spectral densities $S_{xx}(\omega)$, $S_{xy}(\omega) = S_{yx}^*(\omega)$ and $S_{yy}(\omega)$, the procedure is as follows. Over a time interval T, each of $x(t)$ and $y(t)$ is assumed to be replaced by the series of equally spaced discrete values x_r, y_r, $r = 0, \ldots, N-1$. According to the *discrete Fourier transform*, there are corresponding (complex) Fourier coefficients X_k, Y_k, $k = 0, \ldots, N-1$. These are related to the spectral densities as follows[1]:

$$|X_k|^2 = \frac{2\pi}{T} S_{xx}\left(\omega = \frac{2\pi k}{T}\right),$$
$$|Y_k|^2 = \frac{2\pi}{T} S_{yy}\left(\omega = \frac{2\pi}{T}\right), \tag{114}$$

and

$$|X_k||Y_k| \exp\{i(\phi_k - \theta_k)\} = \frac{2\pi}{T} S_{xy}\left(\omega = \frac{2\pi k}{T}\right) = \frac{2\pi}{T} S_{yx}^*\left(\omega = \frac{2\pi k}{T}\right), \tag{115}$$

where ϕ_k, θ_k are the phase angles of the complex coefficients X_k, Y_k.

Therefore the phase differences $(\phi_k - \theta_k)$, $k = 0, \ldots, N-1$, are specified but not the absolute values of ϕ_k and θ_k. One or the other, say ϕ_k, is chosen at random using a random-number generator. To generate Gaussian functions, the ϕ_k must be randomly distributed between 0 and 2π. Then, since the sample values x_r, y_r are generated by the inverse discrete Fourier transform relations

$$x_r = \sum_{k=0}^{N-1} X_k \exp\left[i\left(\frac{2\pi k r}{N}\right)\right],$$
$$y_r = \sum_{k=0}^{N-1} X_k \exp\left[i\left(\frac{2\pi k r}{N}\right)\right], r = 0, \ldots, N-1, \tag{116}$$

each term x_r, y_r is the result of summing many statistically independent contributions and so, by the central limit theorem, tends to have Gaussian properties. If a non-Gaussian process has to be simulated, this will generally be done by first generating an artificial Gaussian process and then subjecting the Gaussian process to a nonlinear transformation to obtain the required non-Gaussian properties. Numerical integration of the equations of motion to obtain their response to artificial pseudorandom excitation is generally computed by using one of the Runge–Kutta algorithms.[3,5]

REFERENCES

1. D. E. Newland, *An Introduction to Random Vibrations, Spectral and Wavelet Analysis*, 3rd ed., Longman, Harlow, and Wiley, New York, 1993.
2. R. A. Fisher and F. Yates, *Statistical Tables for Biological, Agricultural and Medical Research*, 6th ed., Oliver and Boyd, Edinburgh, 1963.
3. D. E. Newland, *Mechanical Vibration Analysis and Computation*, Longman, Harlow, and Wiley, New York, 1989.
4. J. S. Bendat and A. G. Piersol, *Random Data Analysis and Measurement Procedures*, 2nd ed., Wiley, New York, 1986.
5. W. H. Press, B. P. Flannery, S. A. Teukolsky, and W. T. Vetterling, *Numerical Recipes*, Cambridge University Press, Cambridge, 1986.
6. G. M. Jenkins and D. G. Watts, *Spectral Analysis and Its Applications*, Holden-Day, San Francisco, 1968.
7. W. B. Davenport, Jr. and W. L. Root, *An Introduction to the Theory of Random Signals and Noise*, McGraw-Hill, New York, 1958.
8. M. Ohta, K. Hatakeyama, S. Hiromitsu, and S. Yamaguchi, "A Unified Study of the Output Probability Distribution of Arbitrary Linear Vibratory Systems with Arbitrary Random Excitation," *J. Sound Vib.*, Vol. 43, 1975, pp. 693–711.
9. S. H. Crandall and W. D. Mark, *Random Vibration in Mechanical Systems*, Academic Press, New York and London, 1963.
10. N. C. Nigam, *Introduction to Random Vibrations*, MIT Press, Cambridge, MA, 1983.
11. R. A. Ibrahim, *Parametric Random Vibration*, Research Studies Press, Letchworth, U.K., and Wiley, New York, 1985.
12. J. B. Roberts and P. D. Spanos, "Stochastic Averaging: An Approximate Method for Solving Random Vibration Problems," *Int. J. Non-Linear Mech.*, Vol. 21, 1986, pp. 111–134.
13. J. B. Roberts and P. D. Spanos, *Random Vibration and Statistical Linearization*, Wiley, Chichester, 1990.

52

SHOCK ANALYSIS AND DESIGN

THOMAS L. GEERS

1 INTRODUCTION

A shock excitation is *transient*, that is, it has a beginning and (usually) an end. Mechanical shock is conveyed to a mechanical system through forces and motions applied at one or more points. Most often, the system is in a quiescent state when the shock is applied; hence, the response of the system to the shock is also transient. Generally, the aspect of greatest interest in a shock problem is the possibility of system damage or failure.

Damage or failure in a mechanical system is caused by internal stresses and/or relative displacements that exceed the system's capacity to tolerate them. Thus, a shock *analysis* focuses on determining the peak responses that might produce damage and/or failure, and a shock *design* concentrates on controlling those peak responses. The central importance of response peaks that are often highly localized in both time and space constitutes a challenging aspect of shock problems. An equally challenging aspect is relating the response peaks to actual damage or failure.

The chapter is organized as follows. The next four sections focus on *shock analysis*, with Sections 2 and 3 discussing equations of motion for mathematical models and Sections 4 and 5 addressing the analytical and numerical solution of those equations. The following four sections focus on *shock design*, with Sections 6 and 7 describing design tools and Sections 8 and 9 discussing design practices. The chapter continues with two sections on the special topic of transient structure–medium interaction and concludes with a few remarks on the state of the technology.

Note: References to chapters appearing only in the *Encyclopedia* are preceded by *Enc.*

2 FORMULATION OF SHOCK PROBLEMS

In the vast majority of cases, shock problems are mathematically formulated in terms of time-dependent differential equations of motion. If the system model is *discrete* (characterized by lumped elements such as masses, springs, and dampers), these equations are ordinary differential equations in time. They are supplemented with a set of initial conditions that specify the state of the system just before the shock is applied. If the system model is *continuous* (characterized by spatially varying material properties such as mass density and Young's modulus), the equations of motion are partial differential equations in time and space. They are supplemented with initial conditions that specify the state of the system over all space just before the shock is applied and with boundary conditions that specify the state of the system over all time at given boundaries of the system.

A surprising number of shock problems may be satisfactorily formulated in terms of the single-degree-of-freedom, linear oscillator of Fig. 1. The equation of motion for this model when both force and displacement excitations are present is[1]

$$m\ddot{u}(t) + b\dot{u}(t) + ku(t) = f(t) + b\dot{w}(t) + kw(t), \qquad (1)$$

where m, b, and k are the oscillator mass, damping constant, and spring constant, respectively, $u(t)$ is the displacement of the oscillator mass from its static equilibrium position, $f(t)$ is the transient external force acting on the oscillator mass, and $w(t)$ is the transient external displacement acting at the base of the oscillator; an overdot denotes an ordinary differentiation in t. This ordinary differential equation must be supplemented with the initial conditions $u(0) = u_0$ and $\dot{u}(0) = \dot{u}_0$, where, without loss of generality, $t = 0$ is taken as the time just before

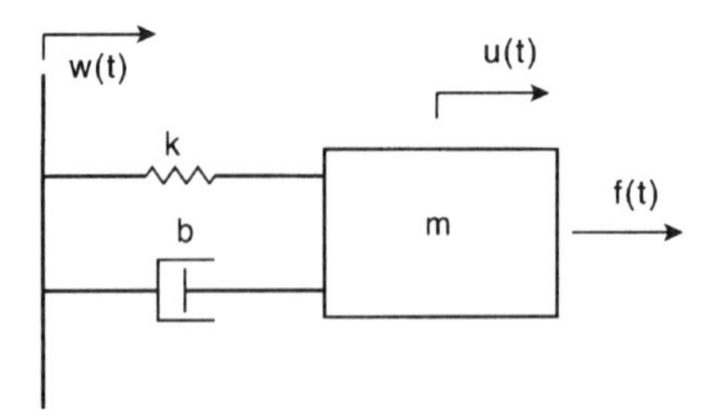

Fig. 1 Linear oscillator subjected to external force and base-displacement excitations.

the shock is applied; usually quiescent initial conditions apply, that is, u_0 and $\dot{u}_0$ are both zero. Many mechanical systems require multi-degree-of-freedom representations, which may be produced by direct modeling or by finite-difference or finite-element models of continuous systems. In such cases, Eq. (1) takes the form of a matrix equation, with m, b, and k becoming matrices and $u(t)$, $f(t)$, and $w(t)$ becoming computational column vectors.

At times, a continuous model is the most appropriate representation of a shock-excited system. A simple example of such a model is the undamped elastic bar of Fig. 2, the equation of motion for which is[2]

$$\rho(x)A(x)\ddot{u}(x,t) - [E(x)A(x)u'(x,t)]' = f(x,t), \quad (2)$$

where $\rho(x)$ and $E(x)$ are, respectively, the mass density and Young's modulus for the bar material, $A(x)$ is its cross-sectional area, $u(x,t)$ is the displacement of an infinitesimally thin disk of bar material from its static equilibrium position, and $f(x,t)$ is the transient external force distribution acting along the bar; a dot or prime denotes a partial differentiation in t or x, respectively. This partial differential equation must be supplemented with the initial conditions $u(x,0) = u_0(x)$ and $\dot{u}(x,0) = \dot{u}_0(x)$. It must also be supplemented with boundary conditions, which, for the configuration shown in Fig. 2, are $u(0,t) = w(t)$ and $u'(l,t) = 0$. There are even times when a mixed discrete/continuous model is the most appropriate representation: an example would be the elastic bar with a lumped mass on the end.

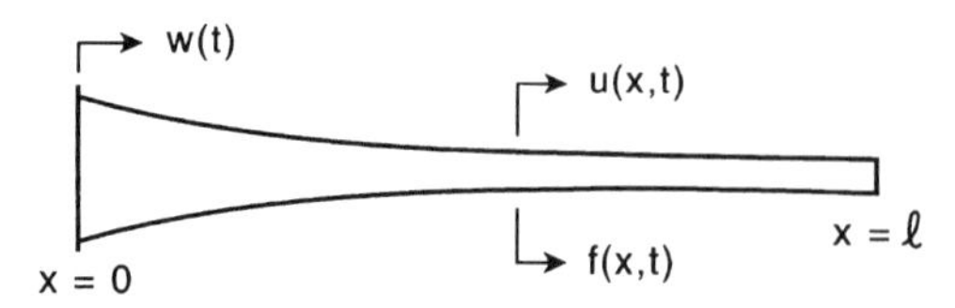

Fig. 2 Elastic bar subjected to external distributed-force and base-displacement excitations.

Equations (1) and (2) constitute equations of motion for *linear* models of mechanical systems. In many cases, nonlinear system behavior determines survivability. *Constitutive nonlinearity* typically appears in Eq. (1) when $k = k[r(t), \dot{r}(t)]$, where $r(t)$ is the relative displacement $u(t) - w(t)$; clearly, the spring "constant" is no longer constant. It typically appears in Eq. (2) when $E = E[x, \epsilon(x,t), \dot{\epsilon}(x,t)]$, where $\epsilon(x,t) = u'(x,t)$ is longitudinal strain. Constitutive nonlinearity is much more complicated in two and three dimensions. *Geometric nonlinearity* appears when motions exceed certain tolerances, such as the geometric limits of linear deformation and the "rattlespace" existing in the system. A rattlespace violation would occur in the system of Fig. 1 if the mass contacted another body. The appearance of nonlinear behavior in a previously linear system fundamentally alters system response and greatly complicates analysis and design (see Chapter 50). Fortunately, dynamic finite-element computer codes that treat both constitutive and geometric nonlinearity are widely available.

3 SHOCK EXCITATIONS

As indicated above, a shock excitation $e(t)$ can be quite general, constrained only by the fact that it must be transient, which can be stated as $e(t) = 0$ for $t < 0$. A variety of *idealized excitations* are useful in both theory and practice, such as rectangular, triangular, and exponential pulses, which are pictured below in Fig. 5, as well as the half-sine pulse, given by $A \sin(\pi t/T)$ for $0 \le t \le T$ and zero otherwise.

Two idealized excitations are especially useful. The first of these is the Heaviside step function[3] $H(t - \tau)$, where $\tau \ge 0$. The step function is equal to zero for $t < \tau$ and to unity for $t > \tau$; its value at $t = \tau$ is not precisely defined but may be conveniently taken as $\frac{1}{2}$. Note that any function, when multiplied by $H(t)$, becomes a transient function; for example, a pure sinusoid, when multiplied by $H(t - \tau)$, becomes a step sinusoid starting at $t = \tau$.

The second is the Dirac delta function[3] $\delta(t - \tau)$, which is the conceptual derivative of $H(t - \tau)$. The delta function is equal to zero for both $t < \tau$ and $t > \tau$ and is equal to infinity at $t = \tau$; furthermore, the area beneath it is unity. With these unconventional characteristics, the delta function is best viewed as the limit of a smoother function characterized by a parameter that can make the function simultaneously higher and narrower while maintaining unit area. An example is the rectangular pulse of width Δt and height $1/\Delta t$ with $\Delta t \to 0$; another is the isosceles triangular pulse with base Δt and height $2/\Delta t$ with $\Delta t \to 0$. Note that the physical dimension of $\delta(t - \tau)$ is reciprocal time. The most valuable characteristic of the delta function is its *filtering property*:

$$\int_0^\infty \delta(t-\tau)v(\tau)\,d\tau = v(t), \tag{3}$$

where $v(t)$ is any smooth function.

4 ANALYTICAL SOLUTION OF SHOCK PROBLEMS

When the system and excitation are sufficiently simple, shock response solutions may be obtained analytically. A useful mathematical tool in this regard is the *Laplace transform*,[3] defined by

$$V(s) = L[v(t)] \equiv \int_0^\infty v(t)e^{-st}\,dt, \tag{4}$$

in which the transform variable s is complex, that is, $s = s_R + is_I$, where s_R and s_I are real with physical dimension of reciprocal time. Laplace transformation is useful because it transforms differential expressions into algebraic expressions; for example, it is easily shown through integration by parts that $L[\dot{v}(t)] = sV(s) - v(0)$ and $L[\ddot{v}(t)] = s^2V(s) - sv(0) - \dot{v}(0)$. Thus Eq. (1) transforms from an ordinary differential equation to an algebraic equation and Eq. (2) from a partial differential equation to an ordinary differential equation; note that initial conditions are automatically incorporated. Solution is thereby simplified, provided that inverse Laplace transformation may be subsequently performed with reasonable effort.

As an example of solution by Laplace transformation, consider Eq. (1) for $w(t) = 0$ and quiescent initial conditions (see Fig. 1); transformation yields

$$U(s) = \frac{F(s)}{s^2m + sb + k}. \tag{5}$$

Thus, one may transform a known $f(t)$ to get $F(s)$, insert the result into Eq. (5), and inverse transform to find $u(t)$. Inverse transformation may be done analytically or through the use of widely available tables. For instance, consider the excitation $f(t) = \delta(t)$, for which, from Eqs. (4) and (3), $F(s) = 1$; in this case, $U(s)$ in Eq. (5) becomes the *transfer function* for the linear oscillator, and its inverse transform, the *impulse response*, is given for $b < 2\sqrt{mk}$ by

$$h(t) = (m\Omega_n)^{-1}e^{-\zeta\omega_n t}\sin\Omega_n t, \tag{6}$$

where the *undamped natural frequency* $\omega_n = \sqrt{k/m}$, the *damping ratio* $\zeta = b/2m\omega_n$, and the *damped natural frequency* $\Omega_n = \sqrt{1-\zeta^2}\,\omega_n$. For a more extensive discussion of this topic, see Chapter 48.

Another useful property of the Laplace transform is that *multiplication* in the s-domain corresponds to *convolution* in the t-domain (Borel's theorem). Thus, for any transient force $f(t)$, one can write from Eq. (5) the displacement solution in the form of the convolution integral

$$u(t) = \int_0^t h(t-\tau)f(\tau)\,d\tau. \tag{7}$$

This formula portrays the response $u(t)$ as a superposition of impulse responses, each delayed in time and weighted by the value of the force at the appropriate time. The convolution relation (7) is often expressed formally as $u(t) = h(t) * f(t)$.

As a second example, consider Eq. (2) with $\rho(x) = \rho_0$, $A(x) = A_0$, $E(x) = E_0$, and $f(x,t) = 0$, with initial conditions $u(x,0) = 0$, $\dot{u}(x,0) = 0$, and with boundary conditions $u(0,t) = w(t)$, $u'(l,t) = 0$ (see Fig. 2). Of course, $w(t) = 0$ for $t < 0$. Laplace transformation then yields, with $c = \sqrt{E_0/\rho_0}$ as the speed of sound in the bar,

$$c^2U''(x,s) - s^2U(x,s) = 0, \tag{8}$$

the solution to which is $U(x,s) = C_+(s)e^{sx/c} + C_-(s)e^{-sx/c}$. Here, $C_+(s)$ and $C_-(s)$ follow from the boundary conditions, which yield $C_+(s) + C_-(s) = W(s)$ and $C_+(s) = C_-(s)e^{-2sl/c}$. Yet another useful property of the Laplace transform is the *shifting property* $L[v(t-\tau)] = V(s)e^{-s\tau}$. Thus, the solution for $u(x,t)$ in this example may be obtained by inverse transformation as

$$u(x,t) = w\left(t - \frac{x}{c}\right) + w\left(t + \frac{x}{c} - \frac{2l}{c}\right) - u\left(x, t - \frac{2l}{c}\right). \tag{9}$$

This is a *traveling-wave* solution, with the first term on the right defining a wave propagating to the right and the second term a delayed wave propagating to the left; the last term subtracts from the two traveling waves the displacement field that existed at time $2l/c$ earlier.

It is interesting to compare Eq. (9) with the *standing-wave* solution for this problem.[4] To obtain this solution, it is convenient to work with the relative displacement $r(x,t) = u(x,t) - w(t)$, for which Eq. (8) yields

$$c^2R''(x,s) - s^2R(x,s) = s^2W(s). \tag{10}$$

One begins by setting $W(s) = 0$ in this equation, introducing the change of variable $s = i\omega$, and solving the resulting equation to obtain $R(x, i\omega) = S(i\omega)\sin(\omega x/c) + C(i\omega)\cos(\omega x/c)$. The terms $S(i\omega)$ and $C(i\omega)$ follow from the *homogeneous* boundary conditions $r(0, t) = 0$ and $r'(l, t) = 0$, which yield $C(i\omega) = 0$ and $\cos(\omega l/c) = 0$. The latter generates an infinite number of *natural frequencies* $\omega_n = n\pi c/2l$ for $n = 1, 3, 5, \ldots$. Thus, returning to the transform variable s, one can express the solution for $R(x, s)$ in terms of the *mode shapes* $\sin(\omega_n x/c) = \sin(n\pi x/2l)$ as the *modal sum*

$$R(x, s) = \sum_{n=1,3,5,\ldots}^{\infty} R_n(s) \sin\frac{n\pi x}{2l}. \qquad (11)$$

The unknown coefficients $R_n(s)$ are determined through exploitation of the *orthogonality property* of the mode shapes:

$$\int_0^l \sin\frac{m\pi x}{2l} \sin\frac{n\pi x}{2l}\, dx = \begin{cases} 0, & m \neq n, \\ \frac{1}{2}l, & m = n \end{cases}. \qquad (12)$$

One does this by introducing Eq. (11) into Eq. (10), multiplying both sides of the resulting equation by $\sin(m\pi x/2l)$, integrating over the length of the bar, and employing Eq. (12) to obtain

$$R_n(s) = -\frac{4}{n\pi}\frac{s^2 W(s)}{s^2 + \omega_n^2}. \qquad (13)$$

Thus, introduction of the Laplace transform of $w(t)$ into Eq. (13) and inverse transformation of the result yields $r_n(t)$. The solution for $u(x, t) = w(t) + r(x, t)$ then follows from Eq. (11) as

$$u(x, t) = w(t) + \sum_{n=1,3,5,\ldots}^{\infty} r_n(t) \sin\frac{n\pi x}{2l}. \qquad (14)$$

The solution may be neatly expressed in convolution form by writing $s^2/(s^2 + \omega_n^2) = 1 - \omega_n^2/(s^2 + \omega_n^2)$, so that inverse transformation of Eq. (13) gives $r_n(t) = -(4/n\pi)[w(t) - \omega_n \sin\omega_n t * w(t)]$; when introduced into Eq. (14), this yields

$$u(x, t) = 2\frac{c}{l} \sum_{n=1,3,5,\ldots}^{\infty} \left[\sin\left(\frac{n\pi c}{2l}\right) t * w(t)\right] \sin\frac{n\pi x}{2l}. \qquad (15)$$

For further discussion of solution by modal superposition see Section 4.3 of Chapter 47, Section 2.2 of Chapter 48, and Section 2.9 of Chapter 49.

Figure 3 contains snapshots of the velocity field $\dot{u}(x, t)$ produced by Eqs. (9) and (14) for a ramp displacement input $w(t) = v_0 t$, for which Eq. (13) yields $r_n(t) = -(4v_0/n\pi\omega_n)\sin\omega_n t$; the series in Eq. (14) is truncated after 10 terms ($n_{\max} = 19$). It is interesting that the standing-wave, or modal-summation, solution cannot reproduce the jump in the exact traveling-wave solution, generating instead spurious oscillations near the wave-

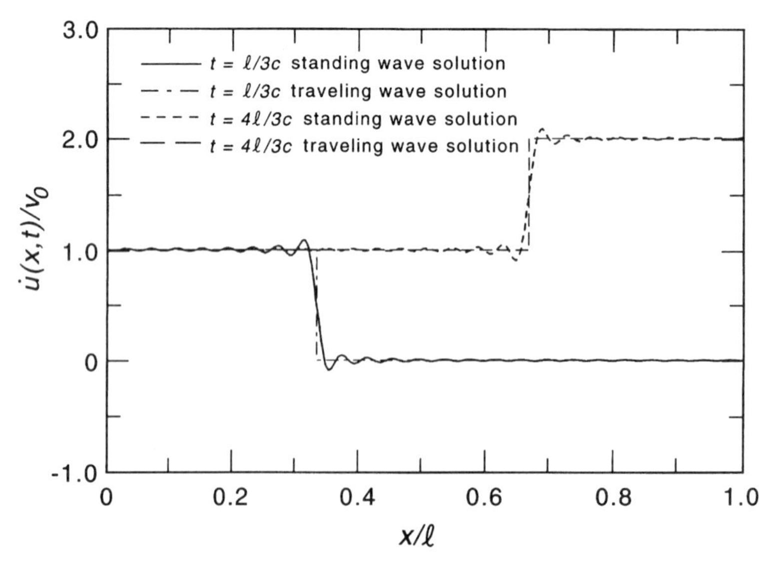

Fig. 3 Velocity-field snapshots in a uniform bar for a ramp-displacement base excitation.

front. This artifact, known as Gibbs's phenomenon,[5] cannot be eliminated by increasing the value of $n_{\max}$; doing so pulls the oscillations in closer to the jump, but does not reduce their magnitude. From Eq. (9), strain snapshots $\epsilon(x,t) = u'(x,t)$ and stress snapshots $\sigma(x,t) = E_0 u'(x,t)$ would exhibit the same characteristics.

5 NUMERICAL SOLUTION OF SHOCK PROBLEMS

Numerical methods offer a means for solving shock problems when systems and/or excitations are complicated. The approach is that so often followed to obtain analytical solutions, namely, reducing differential equations to algebraic equations.

For discrete-system models, the equations of motion are typically ordinary differential equations in time, as discussed in Section 2. For continuum models, the finite-element method[6] is widely used to reduce the partial differential equations of motion to ordinary differential equations. For example, the displacement field in the bar of Fig. 2 is approximated as $u(x,t) \approx N(x)u(t)$, where $N(x)$ is a row vector of spatially local shape functions and $u(t)$ is a column vector of nodal responses. Then a variational principle or weighted-residual method is applied to yield Eq. (1) with m as the mass matrix, b as the damping matrix, k as the stiffness matrix, $f(t)$ as the external force vector, and $w(t)$ as the external displacement vector. The systematic process of reducing spatially continuous models of dynamic systems to spatially discrete models is called *semidiscretization*. Because semidiscretization is akin to standing-wave solution, finite-element solutions for traveling waves with abrupt wavefronts exhibit Gibb's phenomenon.

For discrete and semidiscretized models, the ordinary differential equations in time (which may be thousands in number) are usually integrated by means of finite-difference schemes.[7] With the time continuum sampled evenly over one or more intervals at points $t_j = jh$, where h is a small time increment (not an impulse response!), response histories are obtained within each interval as $u(jh) = u_j$.

One of the most popular time-integration schemes is the *central-difference* scheme, in which displacement and velocity at discrete points in time are approximated as $u_j \approx u_{j-1} + h\dot{u}_{j-1/2}$, $\dot{u}_j \approx \frac{1}{2}(\dot{u}_{j+1/2} + \dot{u}_{j-1/2})$, and $\dot{u}_{j+1/2} \approx \dot{u}_{j-1/2} + h\ddot{u}_j$. These may be combined with Eq. (1) expressed at time t_j to obtain

$$u_{j+1} \approx D^{-1}[h^2 g_j + (2m - h^2 k)u_j - (m - \tfrac{1}{2}hb)u_{j-1}], \quad (16)$$

where $D = (m + \frac{1}{2}hb)$, $g_j = f_j + b\dot{w}_j + kw_j$, and D^{-1} denotes matrix inversion of D, which is implemented in large computations through matrix factorization. Discrete response histories are obtained by advancing j in unit steps, with starting points determined from the initial conditions as $u_0 = u_0$ and $u_{-1} = u_0 - h\dot{u}_0 + \frac{1}{2}h^2\ddot{u}_0$, where $\ddot{u}_0 = m^{-1}(g_0 - b\dot{u}_0 - ku_0)$. This scheme works quite well as long as the selected time increment is no larger than the critical time increment $h_c = 2/\omega_N$, where ω_N is the largest natural frequency in the discrete system; when $h > h_c$, the scheme is numerically unstable. Such conditional stability is the hallmark of an *explicit* numerical integration scheme. A group of equally popular explicit schemes is the family of Runge–Kutta algorithms.

In contrast to explicit schemes, some (but not all) *implicit* schemes are unconditionally stable with respect to time increment. Perhaps the most popular of these is the *trapezoidal rule*, which is one member of the family of Newmark algorithms. The trapezoidal rule approximations for displacement and velocity are $u_{j+1} \approx u_j + \frac{1}{2}h(\dot{u}_j + \dot{u}_{j+1})$ and $\dot{u}_{j+1} \approx \dot{u}_j + \frac{1}{2}h(\ddot{u}_j + \ddot{u}_{j+1})$, which may be combined with Eq. (1) expressed at t_{j+1} and t_j to obtain

$$u_{j+1} \approx E^{-1}[\tfrac{1}{4}h^2(g_{j+1} + 2g_j + g_{j-1}) + (m + \tfrac{1}{2}hb - \tfrac{1}{4}h^2 k) \cdot (u_j + u_{j-1}) + 2m(u_j - u_{j-1})] - u_j, \quad (17)$$

where $E = m + \frac{1}{2}hb + \frac{1}{4}h^2 k$. Discrete response histories are obtained by advancing j in unit steps, with starting points determined from the initial conditions, as discussed above.

Equations (16) and (17) indicate that the implicit trapezoidal rule is the better choice for integrating linear equations of motion with few changes in time increment, in that the costs of securing unconditional stability are a modest increase in algorithmic complexity and infrequent inversion (factorization) of a more complicated matrix. They indicate that the explicit central-difference scheme is the better choice for integrating linear equations with frequent time increment changes when $b = 0$ or when m and b are diagonal and for integrating equations with stiffness nonlinearity; in these situations, matrix inversion (factorization) is minimized by the central-difference method.

Some shock problems consist primarily of high-frequency response over a short period of time and tend to exhibit traveling-wave behavior. Others consist primarily of low-frequency response over a long period of time and tend to exhibit standing-wave behavior. Still other problems exhibit both types of behavior, for example, when a short, rapidly applied load produces long-lived free vibration. An explicit time-integration scheme is indicated for high-frequency problems because the time increment required for accurately tracking the response may be smaller than that required for numerical stability. An implicit scheme is indicated for low-frequency problems because the time increment required for accuracy may be far larger than that required for stability of explicit integration; problems of this type are called *stiff problems*.

Often, the excitation employed to perform a shock response calculation is a *standard pulse form*[8] that represents in a general sense the anticipated shock environment. An example is the half-sine pulse, which is characterized by a maximum value and a pulse duration. More complicated forms involve several parameters that control rise time, maximum value, attenuation rate, and oscillatory behavior. Parameter values are usually selected on the basis of experience with existing experimental data.

6 SHOCK SPECTRUM

Designers are often less interested in obtaining the solution to a particular shock problem than in characterizing a shock environment in such a way that system response is easily estimated. The most widely used design tool for accomplishing the latter is the *shock spectrum*,[9] which is a graph that characterizes a shock *excitation* in terms of the peak response to that excitation of an undamped linear oscillator. Although applicable to force excitations, the shock spectrum is usually employed to characterize motion excitations in terms of the oscillator's relative displacement $r(t) = u(t) - w(t)$.

To construct a shock spectrum for a motion excitation $w(t)$, one determines the maximum absolute value of $r(t)$ as a function of the natural frequency $\omega_n = \sqrt{k/m}$ from the following form of Eq. (1) with $b = f(t) = 0$:

$$\ddot{r}(t) + \omega_n^2 r(t) = -\ddot{w}(t). \tag{18}$$

Replacing the right side of this equation with $\delta(t)$, one finds an impulse response given by Eq. (6) with $m = 1$ and $\zeta = 0$; hence, from Eq. (7),

$$\begin{aligned} r(t) &= -\omega_n^{-1} \int_0^t \sin \omega_n(t-\tau)\, \ddot{w}(\tau)\, d\tau \\ &= \omega_n^{-1}[C(\omega_n, t)\cos \omega_n t - S(\omega_n, t)\sin \omega_n t], \end{aligned} \tag{19}$$

where

$$\begin{aligned} C(\omega_n, t) &= \int_0^t \ddot{w}(\tau)\sin \omega_n\tau\, d\tau, \\ S(\omega_n, t) &= \int_0^t \ddot{w}(\tau)\cos \omega_n\tau\, d\tau. \end{aligned} \tag{20}$$

The solution is conveniently plotted as a *pseudovelocity* shock spectrum $V_{Ps}(\omega_n) = \omega_n |r(t)|_{\max}$, that is,

$$V_{Ps}(\omega_n) = [C^2(\omega_n, t) + S^2(\omega_n, t)]_{\max}^{1/2}. \tag{21}$$

With $V_{Ps}(\omega_n)$ thus determined for a given motion excitation, one can immediately calculate, for a mass–spring oscillator subjected to that excitation, the maximum relative displacement across the spring as $|r(t)|_{\max} = \omega_n^{-1} V_{Ps}(\omega_n)$ and, from Eq. (18), the maximum acceleration of the oscillator's mass as $|\ddot{u}(t)|_{\max} = \omega_n V_{Ps}(\omega_n)$. The inclusion of oscillator damping spoils this symmetry, but because $\zeta \ll 1$ in most mechanical systems, the effects of damping are usually neglected.

The pseudovelocity shock spectrum for a rectangular acceleration pulse described by $\ddot{w}(t) = A[H(t) - H(t-T)]$ is readily determined as

$$V_{Ps}^{\text{rect}}(\omega_n) = AT \begin{cases} (\frac{1}{2}\omega_n T)^{-1} \sin \frac{1}{2}\omega_n T, & 0 \le \omega_n T \le \pi, \\ (\frac{1}{2}\omega_n T)^{-1}, & \omega_n T \ge \pi. \end{cases} \tag{22}$$

This function is plotted in Fig. 4. Also plotted is the magnitude of the pulse's Fourier transform,[10] given by

$$\mathcal{F}_{\text{Mag}}^{\text{rect}}(\omega) = AT(\tfrac{1}{2}\omega T)^{-1} |\sin \tfrac{1}{2}\omega T|, \tag{23}$$

where the circular frequency ω is the Fourier transform variable. Note that, even though the Fourier transform vanishes at $\omega T = 2\pi, 4\pi, \ldots$, oscillators with $\omega_n T = 2\pi, 4\pi, \ldots$ respond fully to the excitation. This demonstrates that a shock spectrum is different from the magnitude of a Fourier transform.

The format of Fig. 4 clearly lends itself to quick assessment of peak relative-displacement and absolute-acceleration values. The shock spectrum for the rectangular acceleration pulse approaches a line of constant velocity (AT) at low natural frequencies and a line of constant acceleration ($2A$) at high natural frequencies. In contrast, the shock spectrum for an excitation characterized by a finite maximum displacement approaches a constant-displacement line at low natural frequencies; if the excitation is also free of acceleration jumps and characterized by a peak acceleration, its shock spectrum approaches a constant-acceleration line at high natural frequencies. These more realistic characteristics form the basis of *shock design spectra*, an example of which is included in Fig. 4.

In many (if not most) situations, the system of interest is actually a subsystem attached to a larger system and $w(t)$ is the motion at the interface between the two. If the subsystem is small relative to the larger system, then a shock spectrum based on the motion of the inter-

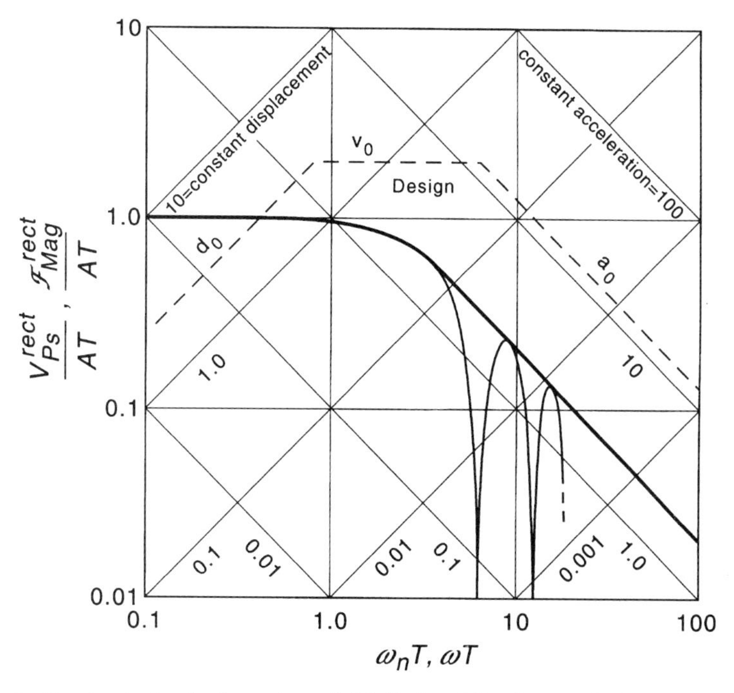

Fig. 4 Pseudovelocity shock spectrum (thick line) and Fourier magnitude spectrum (thin line) for a rectangular acceleration pulse; representative shock design spectrum (dashed line).

face in the absence of the subsystem is appropriate. If this is not the case, however, the subsystem can generate considerable feedback, thereby modifying the interface motion. This produces what is known as the *shock spectrum dip*,[11] which is produced by the subsystem's high resistance to base motion at excitation frequencies near ω_n. The result is that, in the vicinity of ω_n, $V_{Ps}(\omega_n)$ decreases with increasing oscillator mass in a complicated way that depends on the mechanical characteristics of the larger system. Nevertheless, it is often possible to generate shock design spectra that are representative of the general environment, either from experimental data or from numerical simulations.[12]

A limitation of the shock spectrum is the difficulty of extending it to multi-degree-of-freedom linear systems. A modal analysis of a base-excited system permits one to represent it as an array of parallel, uncoupled modal oscillators resting on the base[12]; the peak response of each modal oscillator then follows from the shock spectrum for the motion excitation. However, the maximum response at any physical point in the system is a superposition of modal response contributions at just the right time, not a sum of modal shock-spectrum contributions. There exists an upper bound for maximum response at a physical point as a sum of modal shock spectrum contributions, but it is generally far too conservative. A popular approximation is to determine from the shock spectrum the maximum contribution of each mode to the point response of interest and then to estimate the maximum response there as the vector sum (i.e., the square root of the sum of the squares) of the modal contributions.

For example, suppose one wished to estimate, by vector summation of modal strain contributions, the maximum strain at the root of the bar in Fig. 2 when it is excited by the base motion $w(t) = v_0 t$. Introduction of this motion's Laplace transform into Eq. (13), multiplication of both sides of the resulting equation by $(s^2+\omega_n^2)$, and inverse transformation yield

$$\ddot{r}_n(t) + \omega_n^2 r_n(t) = -\frac{4v_0}{n\pi}\,\delta(t). \tag{24}$$

But this is identical to Eq. (18) with $r(t) = r_n(t)$ and $\ddot{w}(t) = (4v_0/n\pi)\delta(t)$. Thus, Eqs. (20) and (21) yield $C(\omega_n, t) = 0$, $S(\omega_n, t) = 4v_0/n\pi$, $V_{Ps}(\omega_n) = 4v_0/n\pi$, and $|r_n(t)|_{\max} = \omega_n^{-1} V_{Ps}(\omega_n) = 8v_0 l/(n\pi)^2 c$.

Now from Eq. (14), the modal superposition expression for the strain field is

$$\epsilon(x, t) = u'(x, t) = \sum_{n=1,3,5,\ldots}^{\infty} \frac{n\pi}{2l}\, r_n(t) \cos\frac{n\pi x}{2l}. \tag{25}$$

Thus, the maximum modal strain at the base is $|\epsilon_n(0,t)|_{\max} = (n\pi/2l)|r_n(t)|_{\max} = 4v_0/n\pi c$, and the vector sum of the modal strain contributions is

$$\bar{\epsilon}(0) = \frac{4v_0}{\pi c}\left(\sum_{n=1,3,5,\ldots}^{\infty} n^{-2}\right)^{1/2} = \sqrt{2}\,\frac{v_0}{c}. \qquad (26)$$

This is 41% higher than the exact value of v_0/c, which may be obtained from Eq. (9).

Another limitation is the difficulty of extending the shock spectrum concept to nonlinear systems. For a single-degree-of-freedom model, the difficulty is not too great, once a means for characterizing the nonlinearity has been established. The system parameters required to describe different types and degrees of nonlinearity tend to proliferate, however, and the concept loses its simplicity of application. Further extension into nonlinear multi-degree-of-freedom systems appears impractical.

7 ISORESPONSE CURVE

Another shock design tool that should be mentioned is the isoresponse curve,[13] which is not restricted either to single-degree-of-freedom systems or to linear systems. To construct an isoresponse curve for *force excitation of a particular system*, one selects a system response quantity of interest and a specific type of loading. The peak force $F = \max|f(t)|$ and the force impulse $I = |\int_0^\infty f(t)\,dt|$ of the loading are allowed to vary. Computations are carried out, or experiments are performed, that enable one to determine several combinations of F and I that produce a certain peak level of the selected system response.

For example, consider the displacement response of an undamped linear oscillator (Fig. 1) subjected to rectangular-pulse forcing. The equation of motion is Eq. (1) with $b = w(t) = 0$. For an infinitely short rectangle with impulse $I_\delta, f(t) = I_\delta \delta(t)$. This produces $u(t) = I_\delta h(t)$, where $h(t)$ is given by Eq. (6) with $\zeta = 0$; peak displacement response is therefore $|u(t)|^\delta_{\max} = I_\delta/m\omega_n$. For an infinitely long rectangle with force value F_H, $f(t) = F_H H(t)$. This produces $u(t) = (F_H/m\omega_n^2)(1 - \cos\ \omega_n t)$; peak displacement response is therefore $|u(t)|^H_{\max} = 2F_H/m\omega_n^2$. Thus, to obtain a certain peak displacement of interest $|u|_{\max}$, the required delta-loading impulse is $I_\delta = m\omega_n|u|_{\max}$ and the required step-loading force is $F_H = \frac{1}{2}m\omega_n^2|u|_{\max}$.

Now the peak displacement response of the undamped linear oscillator to a rectangular loading of finite magnitude F and finite duration T is readily determined. Thus, for a particular value of $I = FT > I_\delta$, it is obviously possible to find the corresponding value of $F > F_H$ that produces $|u(t)|_{\max} = |u|_{\max}$. This exercise can be repeated for several different values of I to obtain an ensemble of isoresponse I, F pairs, of which $I = I_\delta$, $F = \infty$ and $I = \infty$, $F = F_H$ are two members. Finally, all the I-values and F-values are normalized to I_δ and F_H, respectively, and the locus of the normalized isoresponse pairs is plotted to obtain the isoresponse curve marked "rectangular load" in Fig. 5. As indicated, the other curves in this figure are isoresponse curves for triangular and exponential loadings. Because I_δ and F_H *are independent of the loading profile*, Fig. 5 shows that the rectangular (exponential) profile is the most brutal (gentle) of the three.

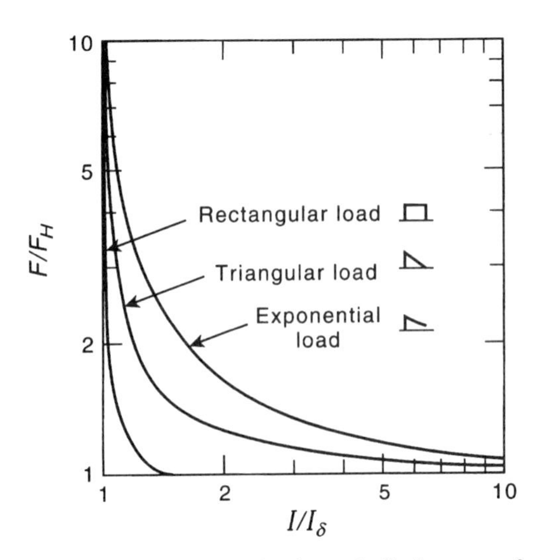

Fig. 5 Isoresponse curves for the peak displacement of an undamped linear oscillator excited by external force pulses. (From G. R. Abrahamson and H. E. Lindberg, *Nuclear Engineering and Design*, Vol. 37, 1976, pp. 35–46. Copyright ©1976 Elsevier Science S. A. Used with permission.)

Isoresponse curves have been generated for a variety of linear and nonlinear systems of widely varying complexity. Analytical, computational, and experimental techniques have all been used.

8 SHOCK SURVIVAL

The design of a mechanical system to survive a shock environment generally employs one or more of the following three strategies: *hardening the system*, *mitigating the shock*, and *increasing the rattlespace*. Hardening the system involves obvious measures such as reducing stress by enlarging cross-sectional areas, smoothing corners and utilizing tougher materials; brittle materials are adverse to shock hardness. One measure that is not so obvious consists of increasing material volume to accommodate greater strain energy. For example, a bolt in uniform tension can hold a maximum elastic strain energy given by $\frac{1}{2}E_0A_0l\epsilon_y^2$, where E_0 and A_0 are the uniform bar properties discussed in Section 4, l is the length of the bolt, and ϵ_y is the yield strain of the bolt material. Thus, a designer might alter the hold-down configuration of a system in order to incorporate longer bolts, thereby increasing their shock hardness.

The most common type of shock mitigation is cushioning. Suppose, for example, that a designer wishes to reduce the peak acceleration of the mass in Fig. 1 to a level below that characterizing a rectangular base acceleration pulse of height A and width T. For $b = 0$, Eq. (22) yields

$$|\ddot{u}(t)|_{\max} = 2A\begin{Bmatrix} \sin \frac{1}{2}\omega_n T, & 0 \le \omega_n T \le \pi \\ 1, & \omega_n T \ge \pi \end{Bmatrix},$$

$$|r(t)|_{\max} = \omega_n^{-2}|\ddot{u}(t)|_{\max}. \tag{27}$$

Clearly, $|\ddot{u}(t)|_{\max} < A$ only for $\omega_n < \pi/3T$, so the natural period of the oscillator, $T_n = 2\pi/\omega_n$, must be at least six times the pulse width T to achieve acceleration reduction. Over this range, $\sin \frac{1}{2}\omega_n T \approx \frac{1}{2}\omega_n T$, so $|\ddot{u}(t)|_{\max} \approx \omega_n AT$ and $|r(t)|_{\max} \approx \omega_n^{-1}AT$; thus, the price of reduced acceleration is increased rattlespace.

A useful measure of cushion effectiveness is[14]

$$e_c = \frac{V^2}{|\ddot{u}(t)|_{\max}|r(t)|_{\max}}, \tag{28}$$

where $V = |\int_0^\infty \ddot{w}(t)\,dt|$ is the acceleration impulse. In the preceding example, $e_c \approx 1$; however, if the damper in Fig. 1 is exploited ($b > 0$), performance can be markedly improved. This is shown in Fig. 6, where the damping ratio $\zeta = 0.4$ emerges as optimum, producing $e_c \approx 1.9$. Note that peak relative displacement steadily decreases with increasing ζ, but peak acceleration of the mass grows linearly with ζ for $\zeta \ge 0.5$. Actually, the curves in Fig. 6 are exact only in the limit $T \to 0$ but are quite accurate for $(\omega_n T)^2 \ll 1$. Cushion design is, of course, not limited to linear mass and damper assemblies; many cushions are nonlinear. However, a calculation of e_c is helpful in evaluating almost any cushion.

In some applications, damage or failure is threatened by contact between fragile systems, which is addressed by increasing rattlespace or increasing attachment stiffness; the latter, however, tends to increase peak acceleration. In other applications, the excitation is oscillatory, which could occur when the system of interest is attached to a large, flexible base structure. In this situation, care must be taken to avoid resonance excitation by keeping the fixed-base natural frequencies of the system away from the natural frequencies of the base structure.

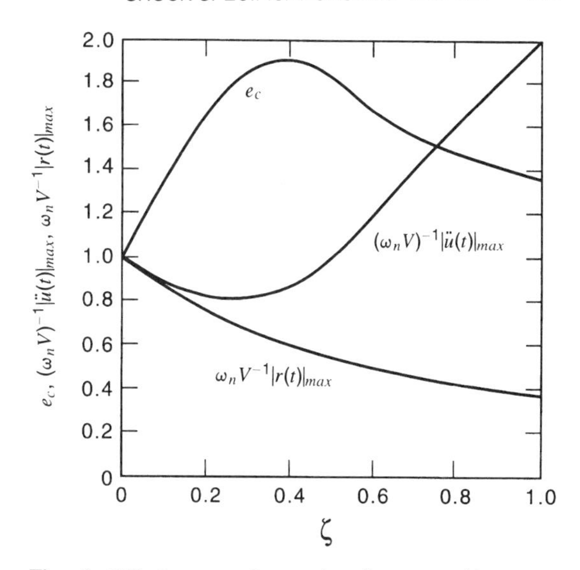

Fig. 6 Effectiveness of a spring–damper cushion as a function of damping ratio.

9 SHOCK SPECIFICATIONS AND TESTING

As might be expected, the most stringent shock specifications are typically associated with military requirements. Representative are the U.S. specifications contained in MIL-STD-810, which addresses such environments as gunfire, pyrotechnic shock and transportation. U.S. Navy specifications for underwater shock environments are contained in MIL-S-901D. International shock standards are contained in ISO 2017-1982 (isolators), ISO 8568-1989 (shock machines), and ISO 9688-1990 (analytical methods).

Shock tests range from a simple package drop test to a series of underwater shock tests conducted against an operating ship. A common shock qualification method involves subjecting a piece of equipment to one or more tests on a shock machine.[15] Such a machine generally consists of a mounting platform that is abruptly set into motion by a hammer, projectile, or quick release from a preloaded condition; or, the motion of a moving platform may suddenly be arrested, as in a drop test. A versatile shock machine may be calibrated to approximate pulses or shock spectra representative of actual environments.

The most common instruments for measuring shock response are accelerometers, velocity meters (or seismometers) and strain gages.[16] The most common accelerometer configuration is that of Fig. 1, in which the spring is a piezoelectric disk in parallel with another (precompressed) spring; the disk generates electric charge in response to the force across it. Because the damping is usually small and no external force is applied to the mass, the equation of motion is (18). For motions with frequency components substantially smaller than ω_n, the force across the piezoelectric disk is $f_r(t) = k_d r(t) \approx -\alpha m \ddot{w}(t)$, where k_d is the spring constant of the disk and α is the ratio of k_d to the sum of the two spring constants. From Eq. (18), this relation is accurate to within 4% for frequency components less than $0.2\omega_n$, where ω_n typically exceeds $2\pi(10^5\text{Hz})$.

Most velocity meters consist of a bar magnet suspended at the axis of a rigidized wire coil through which an electric current flows. The governing equation of motion is again (18), but with ω_n substantially smaller than the frequency components of interest. As the coil moves with acceleration $\ddot{w}(t)$, the lines of magnetic flux inside the coil are cut by the nearly stationary bar magnet, which produces a signal in the coil that is proportional to the relative velocity $\dot{r}(t) \approx -\dot{w}(t)$. Velocity meters are generally accurate over the frequency range 10–1000 Hz; their records can be seismically corrected by means of Eq. (18) to decrease the low-frequency limit.

A modern strain gage consists of a thin metal foil etched to form a long switchback wire that exhibits a change in resistance due to a change in gage length. Thus, it directly measures the average strain over its length by generating, in a dynamic environment, fluctuations in an electric current running through it. When securely bonded to a firm surface, a strain gage produces a highly accurate measurement over a frequency range that is limited by the associated electronic components. Strain gages may be used in the construction of other kinds of transducers for measuring displacement, acceleration, force, and pressure. See Chapter 56 for a more extensive discussion of instrumentation.

10 STRUCTURE–MEDIUM INTERACTION

Many important shock problems involve the interaction between a structure and a surrounding fluid or solid medium; typical examples pertain to airblast, underwater shock, and earthquake environments. The most challenging aspect of these problems is the unboundedness of the surrounding medium. There is some irony in this, because there is usually little interest in the medium per se; the only real interest is in the medium's effect on the structure.

A key strategy in treating structure–medium interaction is *separation* of the total dynamic field in the medium into an *incident-wave* field and a *scattered-wave* field. The incident-wave field is that which exists when the structure is removed and the resulting void is filled with medium. The scattered-wave field is the difference, at every point in the medium at any time, between the total field with the structure in place and the incident-wave field. Determination of the incident-wave field is a separate exercise; its definition is analogous to the specification of $f(x,t)$ in Eq. (2). Determination of the scattered-wave field is difficult in all but the simplest problems.

When the medium is air, the structure may be considered rigid, that is, the scattered-wave field is essentially unaffected by structural motion; exceptional cases involve very flexible structures, such as balloons. When the rigid-body approximation is applicable, the scattered-wave field may be regarded as having been reduced to a *reflected-wave* field. For incident-wave overpressures less than about 35,000 Pa (5 psi), as occur in sonic booms, air may be treated as an acoustic medium, and the fluid response problem reduces to that of rigid-body transient acoustic scattering. Higher overpressures introduce the complexities of nonlinear gas dynamics. In either case, structural response may be determined by applying the sum of the incident-wave and reflected-wave pressures at the surface of the structure.

When the medium is water or soil, however, complete dynamic interaction between the structure and medium occurs: motion of the medium forces the structure to respond, and motion of the structure alters the scattered-wave field in the medium. This is illustrated by the following example: A uniform semi-infinite bar is attached to the mass of the linear oscillator in Fig. 1; a transient incident wave propagates along the bar from right to left, exciting the attached oscillator.

The equations of motion for this example are (1) with $w(t) = 0$ and (2) with $\rho(x) = \rho_0$, $A(x) = A_0$, $E(x) = E_0$, and $f(x,t) = 0$. The boundary conditions for the bar are $u(0,t) = u(t)$ and $A_0\sigma(0,t) = f(t)$, where $\sigma(0,t)$ is the stress at the end of the bar. The initial conditions for the oscillator are $u(0) = \dot{u}(0) = 0$ and those for the bar are $u_S(x,0) = \dot{u}_S(x,0) = 0$; $u_S(x,t)$ is the scattered-wave displacement field. The shock-response solution is efficiently sought as follows.

First, one Laplace transforms the simplified Eq. (2) to obtain Eq. (8), the solution to which is $U(x,s) = U_I(x,s) + U_S(x,s)$, where the transformed incident-wave displacement $U_I(x,s) = C_+(s)e^{sx/c}$ and the transformed scattered-wave displacement $U_S(x,s) = C_-(s)e^{-sx/c}$. Also, because $\sigma(x,t) = E_0\epsilon(x,t) = \rho_0 c^2 u'(x,t)$ in the bar, $\Sigma_S(x,s) = \rho_0 c^2 U'_S(x,s)$ is identified as the Laplace transform of the scattered-wave stress field. Second, one uses these expressions for $\Sigma_S(x,s)$ and $U_S(x,s)$ to obtain $\Sigma_S(x,s) = -\rho_0 c s U_S(x,s)$, inverse transforms, and expresses the result at $x = 0$ to obtain the *temporal impedance relation*

$$\sigma_S(0,t) = -\rho_0 c \dot{u}_S(0,t). \tag{29}$$

Third, one introduces this relation and the separation statement $\sigma(x,t) = \sigma_I(x,t) + \sigma_S(x,t)$ into the second boundary condition to obtain $f(t) = A_0[\sigma_I(0,t) - \rho_0 c \dot{u}_S(0,t)]$. Also, the first boundary condition and the separation statement $u(x,t) = u_I(x,t) + u_S(x,t)$ are used to obtain $\dot{u}_S(0,t) = \dot{u}(t) - \dot{u}_I(0,t)$. Finally, using this last equation to eliminate $\dot{u}_S(0,t)$ in the $f(t)$ equation and introducing the result into Eq. (1) with $w(t) = 0$, one obtains the equation of motion for the "medium-encumbered" structure

$$m\ddot{u}(t) + (b + \rho_0 c A_0)\dot{u}(t) + ku(t) = A_0[\sigma_I(0,t) + \rho_0 c \dot{u}_I(0,t)]. \tag{30}$$

For an incident wave prescribed as $u_I(x,t) = \mu_I(t + x/c)$, $\sigma_I(x,t) = \rho_0 c^2 u_I'(x,t) = \rho_0 c \mu_I(t + x/c)$, Eq. (30) is readily solved by analytical or numerical methods, as discussed in Sections 4 and 5.

It is interesting to examine in Eq. (30) the nature of the encumberance imposed by the medium (the bar) on the structure (the oscillator). First, the medium provides acoustic damping through the term $\rho_0 c A_0$; if $\rho_0 c A_0$ is small (an "air bar"), the term $\rho_0 c A_0 \dot{u}(t)$ may be neglected, which produces rigid-body scattering as discussed above. The term $\rho_0 c A_0 \dot{u}_I(0,t)$ on the right is the reflected-wave force and cannot be neglected; in fact, the equations immediately following Eq. (30) produce, on the right side of Eq. (30), $\sigma_I(0,t) + \rho_0 c \dot{u}_I(0,t) = 2\sigma_I(0,t)$.

11 DOUBLY ASYMPTOTIC APPROXIMATIONS

Because determination of the transient scattered-wave field is so difficult in practical shock problems involving structure–medium interaction, a variety of *temporal impedance approximations* have been proposed. Among these are doubly asymptotic approximations (DAAs),[17] which approach exactness in both the early-time (high-frequency) and late-time (low-frequency) limits. A second-order DAA for a boundary-element model[18] of the wet surface of a structure submerged in an *acoustic fluid* is

$$\ddot{p}_S(t) + c\Gamma_1 \dot{p}_S(t) + c^2 \Gamma_2 \Gamma_1 p_S(t) = \rho c \dddot{u}_S(t) + \rho c^2 \Gamma_2 \ddot{u}_S(t), \tag{31}$$

where $p_S(t)$ and $u_S(t)$ are, respectively, computational column vectors of nodal scattered-wave pressures and fluid velocities normal to the surface; ρ and c are, respectively, the mass density of and speed of sound in the fluid; and Γ_1 and Γ_2 are constant matrices, each having the dimension of reciprocal length. Equation (31) constitutes a set of ordinary differential equations that are solved simultaneously with the semidiscretized structural equations of motion discussed in Section 5.

If, in Eq. (31), only the first term on the left and the first term on the right are retained, the second-order DAA reduces to $p_S(t) \approx \rho c \dot{u}_S(t)$, which corresponds to the temporal impedance relation obtained in the example of the previous section. This is the *plane-wave approximation*, which is accurate at early time during the structure–medium interaction. If only the last term on the left and the latter term on the right are retained, the second-order DAA reduces to $p_S(t) \approx \rho \Gamma_1^{-1} \ddot{u}_S(t)$. This is the *added-mass approximation*, which characterizes the incompressible flow of fluid at late time. The second-order DAA for an acoustic fluid has been found to yield satisfactory results in analytical studies and shock tests. Analogous DAAs for *elastic and poroelastic media* have also been formulated that reduce to plane-wave approximations at early time and quasi-static approximations at late time.

Temporal impedance approximations have been discussed above as being applied at the surface of the structure. An alternate approach is to semidiscretize a portion of the surrounding medium along with the structure and apply the temporal impedance approximation at the outer surface of that portion. This makes the structural response computations less subject to the inaccuracies of the approximation and permits the treatment of local nonlinearities in the medium and the structure–medium interaction. The price paid, however, is greater computational cost as well as the possible introduction of Gibbs oscillations, as discussed in Sections 4 and 5.

12 CONCLUSION

In recent decades, shock analysis and design have benefited greatly from the rapid development of computer hardware and software. It is now possible, through finite-element and boundary-element semidiscretization, to construct detailed mathematical models of highly complex problems and to generate copious response data from them. Corresponding capabilities exist in the acquisition and processing of experimental data.

To be useful, however, response data must be translated into damage estimates and failure predictions. For example, short-duration exceedance of yield strain within a small volume of reasonably ductile material hardly causes structural failure; but if the degree, duration, and extent of yield exceedance increase sufficiently, failure does occur. Hence, understanding the impact of temporally and spatially localized response peaks on system damage and failure remains the central challenge of shock response analysis and design.

REFERENCES

1. S. Rao, *Mechanical Vibrations*, 2nd ed., Addison Wesley, Reading, MA, 1986, pp. 128, 144.
2. S. Rao, *Mechanical Vibrations*, 2nd ed., Addison Wesley, Reading, MA, 1986, pp. 383, 384.
3. E. Kreyszig, *Advanced Engineering Mathematics*, 7th ed., Wiley, New York, 1993, Chapter 6.
4. S. Rao, *Mechanical Vibrations*, 2nd ed., Addison Wesley, Reading, MA, 1986, pp. 384–387.
5. E. Kreyszig, *Advanced Engineering Mathematics*, 7th ed., Wiley, New York, 1993, pp. 602–603.
6. O. Zienkiewicz and R. Taylor, *The Finite Element Method*, 4th ed., McGraw-Hill, New York, 1989.
7. T. Belytschko and T. Hughes (Eds.), *Computational Methods for Transient Analysis*, Elsevier Science, 1983, Chapters 1 and 2.
8. C. Harris and C. Crede (Eds.), *Shock and Vibration Handbook*, 2nd ed., McGraw-Hill, New York, 1961, Chapter 8.
9. C. Harris and C. Crede (Eds.), *Shock and Vibration Handbook*, 2nd ed., McGraw-Hill, New York, Chapter 23.
10. E. Kreyszig, *Advanced Engineering Mathematics*, 7th ed., Wiley, New York, 1993, Chapter 10.
11. G. O'Hara and P. Cunniff, "The Shock Spectrum Dip Effect," *J. Sound Vib.*, Vol. 103, 1985, pp. 311–321.
12. P. Cunniff and G. O'Hara, "A Procedure for Generating Shock Design Values," *J. Sound Vib.*, Vol. 134, 1989, pp. 155–164.
13. G. Abrahamson and H. Lindberg, "Peak Load-Impulse Characterization of Critical Pulse Loads in Structural Dynamics," *Nucl. Eng. Design*, Vol. 37, 1976, pp. 35–46.
14. C. Harris and C. Crede (Eds.), *Shock and Vibration Handbook*, 2nd ed., McGraw-Hill, New York, 1961, Chapter 31.
15. C. Harris and C. Crede (Eds.), *Shock and Vibration Handbook*, 2nd ed., McGraw-Hill, New York, 1961, Chapter 26.
16. E. Doebelin, *Measurement Systems*, 4th ed., McGraw-Hill, New York, 1990, Chapter 4.
17. T. Geers and P. Zhang, "Doubly Asymptotic Approximations for Submerged Structures with Internal Volumes," *J. Appl. Mech.*, Vol. 61, 1994, pp. 893–906.
18. J. Kane, *Boundary Element Analysis*, Prentice-Hall, Englewood Cliffs, NJ, 1994.

53

ACOUSTIC EMISSION

KANJI ONO

1 INTRODUCTION

Acoustic emission (AE) refers to the generation of transient elastic waves due to the rapid release of stored energy from a localized source or sources within a material or structure.[1,2] Such a transient wave is also called acoustic emission and originates at cracks, defective welds, yielding regions, and so forth in metals and at delamination, fiber, and matrix fracture in composite materials.

The generation of AE is a mechanical phenomenon. Acoustic emission from the stress relaxation at the crack tip due to crack propagation is the most representative, but it can originate from a number of mechanisms in various materials, including ceramics, concrete, plastics, and wood. Mechanical deformation and fracture are the primary sources of AE, while phase transformation, corrosion, friction, and magnetic processes among others also give rise to AE. The elastic waves then travel through the structure to a sensor. The manner of wave propagation greatly influences the nature of AE signals at the sensor; for example, when a section size is comparable to the wavelength, the wave travels as a guided wave (Chapter 49). In thick-walled structures, the surface wave is the dominant mode of propagation. Weak electrical signals produced by the sensor are amplified and characterized according to the waveform shape, intensity, frequency spectrum, and other features.[2]

From the detected AE signals, the locations and activity levels of the sources are determined in real time and utilized for the evaluation of the structure or component. Since AE signals can travel over a long distance, the entire structure can be examined in a single test sequence, for example, during preservice or periodic inspection. Acoustic emission can continuously monitor the integrity of a structure in service under favorable conditions. Materials research is another area where AE is useful because of its sensitivity to dynamic microscopic processes. Acoustic emission is a passive test method in that no direct excitation is applied, unlike the ultrasonic test method. However, AE testing requires the application of a certain stimulus to a test piece. In pressure vessels and piping, hydrostatic pressurization is usually employed to activate AE sources. Acoustic emission has become the inspection method of choice for glass-fiber-reinforced composite vessels and piping and more recently for metallic vessels, including industrial gas transport trailers and railroad tank cars.[3]

It is often difficult to establish a mechanical phenomenon at the source of a detected AE signal. This is due to the high sensitivity of AE techniques (since no other means exist to independently observe most dynamic source phenomena) and to the presence of resonance of the structures and sensors.

2 DETECTION

2.1 AE Signals and Source Functions

Two types of AE signals, burst-type (pulselike) and continuous-type (random noiselike) signals, are usually observed. Burst-type emissions arise from distinct events of elastic energy release, such as crack advances, inclusion fracture, fiber fracture, and delamination. A representative example is shown in Fig. 1*a*. It has a sharp rise, followed by an exponential decay. This ringing pattern arises from various resonances in a structure and

Note: References to chapters appearing only in the *Encyclopedia* are preceded by *Enc*.

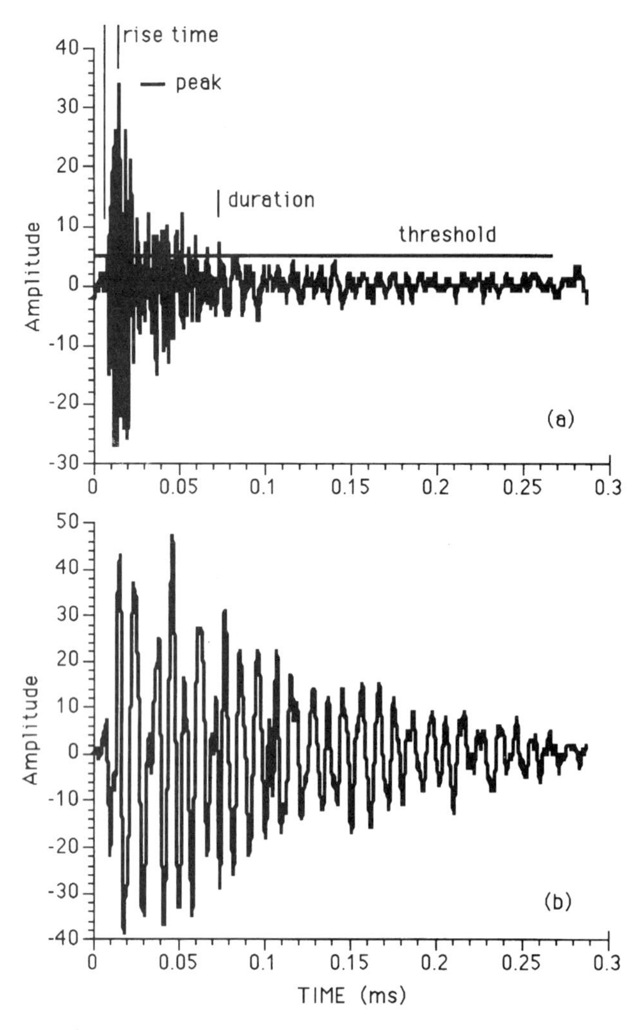

Fig. 1 (*a*) Waveform of a typical (extensional) AE signal from a unidirectional carbon fiber composite. (*b*) Another (flexural) waveform from the same sample. The rise time, duration, and peak amplitude are defined. Amplitude in millivolts after 46 dB amplification. Time axis is 512 μs long.

the resonance of the sensor. Continuous-type emissions are produced by many overlapping events and observed from plastic deformation of metals and from liquid or gas leaks. For typical structural monitoring, ultrasonic frequencies of 30 kHz to 2 MHz are detected. Airborne noise interferes with AE measurements at lower frequencies, while signal attenuation makes the higher frequency range difficult to use. In composite testing, the 30–150 kHz range is common. For applications involving concrete and rocks, a range of several kilohertz to 300 kHz is used, and for geotechnical monitoring the signal frequencies lie below several kilohertz.[1]

Acoustic emission signals contain various information on the AE sources and the propagation paths. The detected signal from a sensor is, in principle, given by convolving the source function (force–time history), the impulse response of the mechanical system to the source (Green's function), and the impulse response of the sensor. For simple geometry, source functions can be recovered by deconvolution when wide-band sensors are employed, and the first motions of signals are discernible. Unfortunately, this is impractical for most applications. Acoustic emission signals detected with a wide-band sensor can be classified according to the generating source types by means of pattern recognition analysis. This scheme primarily depends on the differences in the excitation of various modes of resonance and wave propagation and on their time history.[4]

2.2 Sensors

Most sensors used for AE detection employ a disk of piezoelectric ceramic or crystal and respond to velocity or acceleration normal to the face of a sensor. One type has a broadened frequency response using a backing material behind a transducer element. This construction is essentially identical to that of the ultrasonic sensors (*Enc.* Ch. 52). Another type enhances the sensitivity by using the thickness and/or radial resonance of a transducer element without using a backing material. These are resonant or narrow-band sensors and are employed most often in AE testing because of their high sensitivity. Typical resonant AE sensors have peak velocity sensitivities of a few kilovolts by seconds per metre; these are 20–40 dB more sensitive than those used for conventional ultrasonic testing. The frequency responses of AE sensors are usually calibrated (face-to-face) against a standard using compressional waves at normal incidence. For many applications, however, surface or plate waves are dominant, and their movements normal to the surface are detected, and the size of a sensor affects its responses. Proper sensor selection for frequency ranges or wave modes enhances the sensitivity of AE detection.

High-fidelity capture of signal waveforms is needed for specialized studies of AE sources. Surface displacements are measured using a capacitive sensor or a special sensor with a conical piezoelectric element with matched backing. Laser interferometers are still not practical due to large size, high cost, and low sensitivity. A sensor calibration system has been constructed using a capacitive sensor in combination with a large steel block and a mechanism to break a fine glass capillary [~0.2 mm outside diameter (OD)]. The breaking of the glass capillary produces a sudden release of force (~20 N within a few tenths of a microsecond) and provides a reproducible source of body waves and surface waves that match well with theoretical predictions. Peak surface displacements of several tenths of a nanometre are generated at 0.1 m from the capillary break. Figure 2 shows the surface wave calibration apparatus, in which a sensor under test is placed at a symmetric position to the capacitive sensor standard.[5] For a working calibration of AE measuring system, breaking of pencil lead (0.3–0.5 mm OD) has been used effectively. This generates ~1 N force drop over a 1-μs period. For concrete, rocks, and geotechnical applications, low-frequency sensors, such as geophones, hydrophones, and accelerometers, are also used. These can have flat frequency responses and some have good directionality.

Given a resonant AE sensor with 3.5 kV·s/m peak sensitivity at 200 kHz, a 1-μV signal corresponds to a 0.6-fm surface displacement over 2 μs. Since a preamplifier (150–300 kHz bandpass) has the input noise of over 0.6 μV rms, this signal cannot be clearly distinguished from noise but is approximately the detection limit. Because AE sensors used in practice have similar sensitivities, the present case may be considered as a ref-

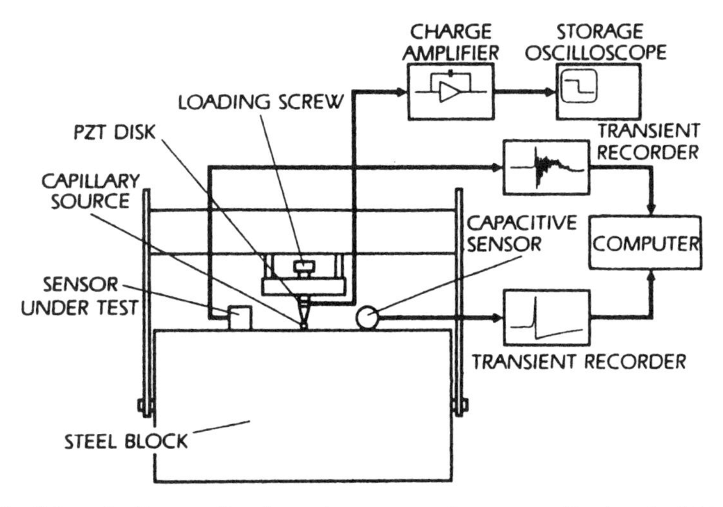

Fig. 2 Schematic diagram of surface pulse apparatus for sensor calibration. Steel block is 90 cm diameter and 43 cm high. Glass capillary source is broken by tightening loading screw and PZT disk measures the breaking force. (Reprinted, by permission, from Ref. 5, p. 88.)

erence in dealing with AE signal amplitude (1 μV at the sensor output is often set at 0 dB).

2.3 AE Signal Characteristics

Characteristics of AE signals in common use[2] are (a) the number (and rates) of burst-type emissions or AE event count (and AE event rates) (in multisensor AE detection systems, the term "hit" is used in place of "event" since a single AE event can reach more than one sensor and be counted each time), (b) the averaged signal intensity of continuous-type emissions [e.g., root-mean-square (rms) voltages of amplified signals], (c) the cumulative number or rates of oscillations that cross a prefixed (or noise-compensated) threshold value (called total AE counts or AE count rates), (d) the peak amplitude of burst emissions, (e) the rise time and duration of burst emissions, and (f) the signal strength of burst emission (area of the signal envelope).[3] See Fig. 1*a* for the definition of (d) and (e). In addition, the distributions of peak amplitude, rise time, and duration are also obtained, as well as arrival time differences of burst emissions at different sensors. These signal parameters can be acquired in real time with the advances of electronics. Since the duration of typical burst emissions is 0.1-1 ms, however, at most several thousand signals per second can be distinguished. For higher rates of emissions, average intensity measurements must be utilized. By recording AE signals digitally, the frequency spectrum (assuming the use of wideband sensors) and shape parameters of the signal waveforms can be determined. Peak intensity and position in the frequency domain, shifts in dominant frequency over time, and rates of rise and decay of the waveform are the features of importance. These features are also used in constructing intelligent classifiers of pattern recognition analysis.[4] Combinations of these and other test parameters, such as applied force or pressure, time under load, ground tilt, and displacement, are used to evaluate the nature of AE sources.

3 SOURCES OF EMISSIONS AND MATERIALS BEHAVIOR

3.1 Plastic Deformation of Metals

Plastic deformation of most structural alloys generates AE that reaches a maximum at or near the yield stress and diminishes with work hardening.[6] Figure 3 shows an example of this behavior of a low-alloy steel.[7] The load–time plot shows a typical discontinuous yielding,

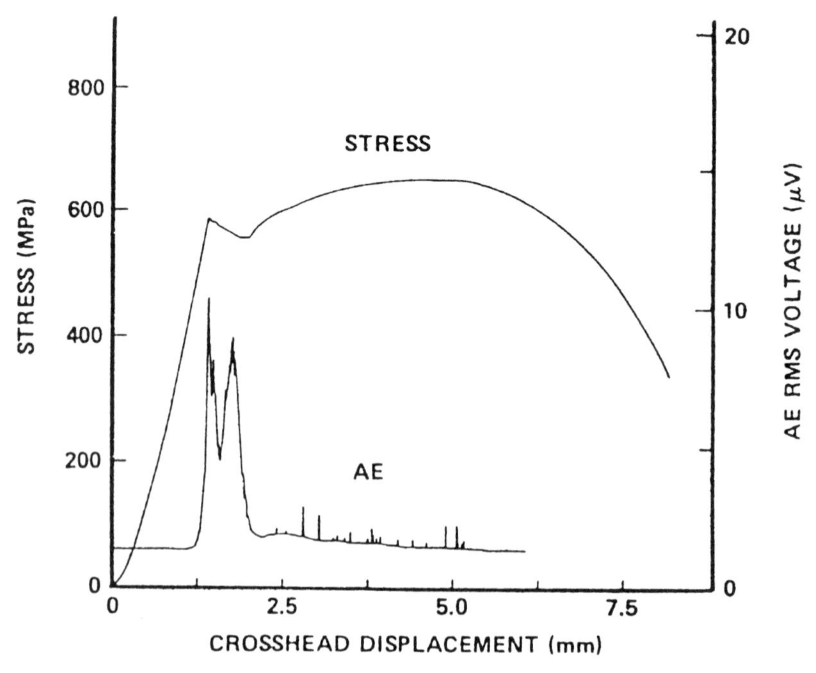

Fig. 3 Load and AE data plotted against the crosshead movement of A533B steel specimen (rolling direction); oil quenched and tempered at 650°C. (Reprinted, by permission, from Ref. 7, p. 21.)

during which AE activities indicated by the rms voltage of AE signals were several times higher than the background noise. This AE upon discontinuous yielding is due to dislocation motion, and the signals are of the continuous type. Purely elastic deformation produces no AE. Once the work hardening begins, the AE activities subside, and the rms voltage is only slightly above the background with occasional spikes due to burst emissions. These burst emissions originate from microcracks or from the fracture and decohesion of nonmetallic inclusions. In some metals, twinning produces burst-type acoustic emission. In metals and alloys showing continuous yielding, AE activities typically start at one fourth to one half of the yield stress and reach a peak at the yield, decreasing gradually with further deformation. At high temperatures where work hardening is low, the AE signal intensity remains strong even at high strains.

Acoustic emission phenomena are irreversible.[6] When a sample is deformed, unloaded, and reloaded, it emits no AE until the previously applied load is exceeded. This is known as the Kaiser effect. This behavior is illustrated in Fig. 4*a* and 4*b* where a sample was loaded, unloaded, and reloaded immediately.[7] The Kaiser effect is useful for the determination of a prior loading level, but is not permanent. Extended holding and/or heating before the second loading may reduce the load at which AE starts to be emitted again.

Microstructural variation affects the AE behavior.[6] When a metal or alloy is cold worked, the AE activity is suppressed; often it is eliminated completely. Martensitic and bainitic steels, especially untempered martensites, show very low AE activities as these have high dislocation densities. When the grain size is reduced, AE intensity often increases along with the strength. Other factors may, however, override the grain size effect. In most precipitation-hardened alloys, AE intensity decreases with aging. In Al alloys, the peak AE intensity at yield decreases by more than a factor of 3 from the solution-treated condition to the fully aged condition. When alloys are strengthened by the addition of dislocations or hard precipitates, the mean-free path of mobile dislocations is reduced. When an individual glide motion of the mobile dislocations is restricted, the AE intensity diminishes even though the number of such mobile dislocations increases.

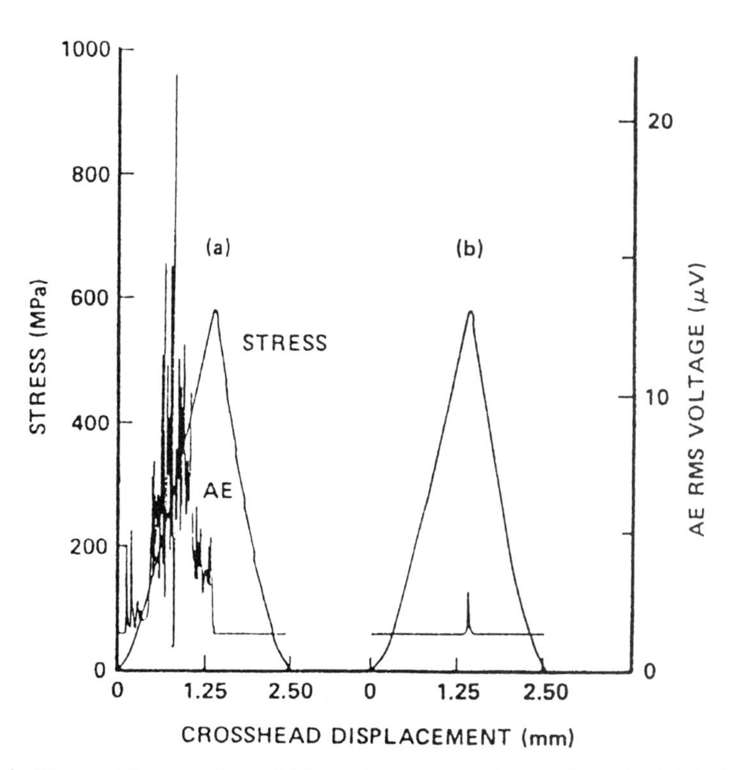

Fig. 4 Stress and the rms voltage of AE signals against time for a steel sample: (*a*) Elastically loaded to 560 MPa and unloaded, (*b*) immediately reloaded after unloading, showing no AE until prior load is reached. (Reprinted, by permission, from Ref. 7, p. 21).

Effects of solid solution impurities and alloying additions vary depending on the strength of solute–dislocation interaction. In alloy systems where the solute pinning of dislocations is weak, the solute additions suppress the AE intensity observed during yielding; cf. dilute Al-Mg, Cu-Al, and Al-Si systems. Quench aging effects occur in Al-Mg alloys, Ni-containing carbon, martensitic steels, and so forth. The AE levels are low in as-quenched states of all these alloys but increase substantially (by as much as 50 times) when solute atoms are allowed to migrate to dislocations, pinning them. When stressed, the pinned dislocations suddenly move and produce dislocation avalanches, generating strong AE activities upon yielding. The strongest AE activities are found in alloys that develop short-range ordering. These include Cu-Zn, Cu-Al, and Ni-Fe. The peak AE levels in these alloys are 10–100 times those of pure metals or usual solid-solution alloys. Dynamic strain aging also produces a series of strong AE peaks. It is found in a number of alloys over a certain range of temperature and strain rate. The AE peaks correspond to load drops. Small load drops are difficult to detect except by AE. Sudden releases of dislocations that had been immobilized are the basic origin of the load drops and bursts of AE activities.

Test temperature alters the AE behavior of materials undergoing plastic deformation. In pure metals and dilute alloys, the peak AE level at yield initially increased by 50–100% with test temperature, but decreased above T/T_m of 0.3–0.4, where T_m is the absolute melting temperature. In normalized steels, the AE level at yield increased fivefold over −150–150°C, while the yield strength decreased by 30%. Acoustic emission from austenitic stainless steels increased 10–20 times from room temperature to 1000 K, where the AE level reached a maximum. The large changes in stainless steels appear to reflect an increase in the stacking fault energy, thus altering the slip mode from planar to nonplanar with increasing test temperature.

Inclusions and second-phase particles are another important origin of AE during deformation. Inclusions are the significant AE sources during tensile loading of steels in the short-transverse direction and that of most high-strength Al alloys. In steels, the decohesion of MnS inclusions is the most important while the fracture of silicide inclusions causes the AE observed in Al alloys. In steels, inclusion-induced AE starts to appear during the elastic loading (see Fig. 4a) and continues into the work-hardening range. These are burst-type emissions and their number is proportional to the inclusion content. The inclusion-induced AE exhibits anisotropic behavior, that is, the highest number of emissions are found in the short-transverse direction and the least in the longitudinal direction.

3.2 Fracture and Crack Growth

During final fracture, most materials and structures produce large, audible sounds. In high-strength materials, strong elastic waves are generated during fracture. Acoustic emission can be detected long before final fracture and can be utilized in preventing catastrophic failures of engineering structures. This application has been and remains to be the primary impetus of the development of AE technology.

Brittle solids under tension or in bending, including ultra-high-strength steels and ceramics, often generate only a small number (<100) of AE signals just before final fracture. Subcritical crack growth is minimal, limiting the AE activities. This behavior is unfortunate because these materials are most likely to fail suddenly, and the need to prevent such failures is great. However, all the microfracture mechanisms that are operative in these materials, that is, cleavage, quasi-cleavage, and intergranular fracture, can be detected easily since a large fraction of the AE events have peak amplitudes above 1 mV or 60 dB in reference to 1 μV at the sensor. In some brittle materials, subcritical crack growth (or fracture process zone ahead of the crack tip) is found.[8] During the subcritical crack growth, AE event counts increase in proportion to mth power of stress intensity factor K_I, which is equal to $Y\sigma\sqrt{a}$, where Y is a geometrical factor, σ is stress, and a is the crack length;

$$N_c = A(K_I)^m, \tag{1}$$

where N_c is the cumulative number of AE events and A is a constant. The cumulative amplitude distribution of such AE signals can be approximated by a power-law distribution of the form

$$N_c = B(V_p)^{-n}, \tag{2}$$

where V_p is the peak amplitude and B and n are constants. In low toughness materials (e.g., alumina) showing cleavage and intergranular fracture, $m = 4$ and $n = {\sim}0.5$ (see Fig. 5).[8]

When the fracture toughness values increase to 10–100 MPa · m$^{1/2}$, as in high-strength Al and Ti alloys and in high-strength steels, AE signals are produced as the plastic zone ahead of the crack tip expands. Acoustic emission activity becomes significant with many high amplitude emissions (>60 dB) when the crack starts to grow. In this stage, various mechanisms of microfracture may be involved. That is, in addition to those mentioned above, low-energy tear, alternating shear, and microvoid are observed. Acoustic emission event counts increase with stress intensity factor according to Eq. (1) with

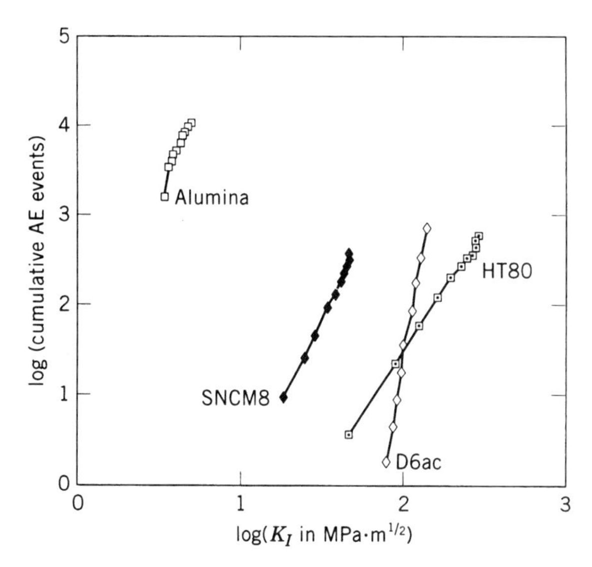

Fig. 5 Log–log plot of AE total events vs. stress intensity factor for alumina and steels. (Data from Refs. 8–11.)

$m = 2$–20 and the power-law amplitude distribution with $n = 0.7$–1.5.[9–11] The exponent m often increases as K_I approaches the fracture criticality. Transition in microfracture mechanisms can be reflected more clearly by plotting cumulative AE energy against J_I (cf. Fig. 6).

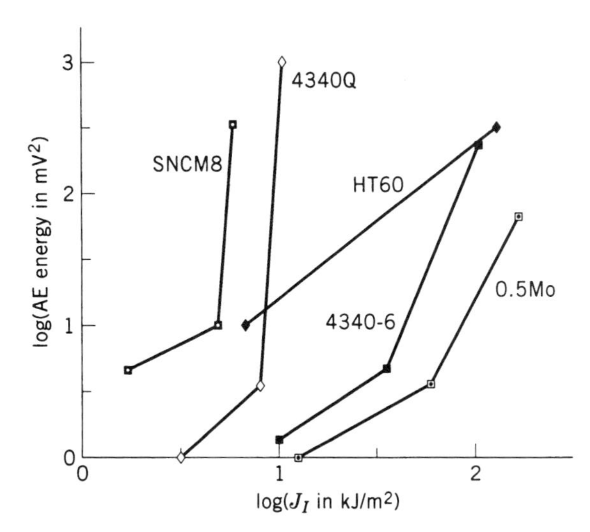

Fig. 6 Log–log plot of cumulative AE energy vs. stress intensity factor for steels. (Data from Refs. 9, 10.) SNCM8 (tempered at 200°C); 4340Q and 4340-6 (4340 steel, as-quenched and quenched and tempered at 600°C); HT60 and 0.5Mo are low-carbon structural steels.

Here, AE energy is defined as the square of peak amplitude, and J_I is given by $(1 - \nu)K_I^2/E$, where E and ν are Young's modulus and Poisson's ratio. The transition is sharp when cleavage microcracks initiate as in SNCM8 and as-quenched 4340 steels but is also visible when fibrous microfracture begins as in 0.5Mo and quenched and tempered (600°C) 4340 steels.

Higher fracture toughness materials fail typically under plane stress conditions by the microvoid coalescence and shear mechanisms with some tearing. In these ductile solids, the expansion of plastic zones ahead of the crack tip initially generates AE, with AE activity reaching a maximum at general yield just like the AE behavior of plastic deformation. This is the weakest among fracture-related AE with the peak amplitude of < 40 dB. Beyond the general yield, stable crack growth processes begin. It is often difficult to detect this by AE as the AE amplitude is still low (< 60 dB) reflecting high microscopic ductility of the materials. In ductile HT60 steel, the transition in AE energy was absent (cf. Fig. 6).

In moderate to high fracture toughness materials, nonmetallic inclusions exert substantial influence on the AE behavior.[6] MnS inclusions in steels and silicate inclusions in Al alloys have strong effects on AE. The decohesion of MnS inclusions emits moderately strong AE (peak amplitude of < 55 dB) from the early part of elastic loading and is the primary source of strong emissions in ductile steels. Plastic flow and fracture of inclusions (especially in Al alloys) also produce AE events. Still, even the maximum AE intensity due to inclusions is less than that from brittle fracture in low toughness materials. This inclusion effect is strongest in samples stressed in the short-transverse direction. The total counts at the maximum load for the short-transverse samples are 20–50 times that of longitudinal samples of the same steel.

Fatigue leads to eventual fracture in a number of engineering structures under repeated loading.[1] Acoustic emission accompanies fatigue crack initiation and subsequent growth, showing rapid rise just before final fracture. Often, the detection of AE precedes an optical observation of a fatigue crack. In cyclically loaded structures, similar trends have been found. Once a crack develops, different types of AE are emitted as a function of loading cycle phase.

(a) Near a peak tensile load, AE due to crack growth and inclusion fracture is observed.

(b) As the load is reduced to zero, crack closure noise is detected as AE; that is AE originates from crack face fretting and crushing of oxide particles.

(c) As the load increases, crack faces that had stuck together separate, producing AE.

Peak-load AE contributes less than 10% of the total AE activity in some cases, but more when high stress ranges are used. Initially, AE events are detected over a wide range of loading but tend to concentrate at the maximum load toward the end of fatigue loading. For the detection of fatigue cracking, frictional AE due to the fretting and crack face separation are important. In addition, particles of oxides and various corrosion products often form between the crack faces. Their subsequent fracture also contributes to the AE due to fatigue. Under high crack growth (plane stress) conditions, more AE is observed, but AE activity per unit crack area is reduced.

3.3 Fiber Fracture and Delamination of Composite Materials

Most composite materials in current use are reinforced by glass fibers, which take various forms, such as short random fibers, mat, woven rovings, and continuous fibers. Aerospace composites use higher modulus carbon and aramid fibers. The matrix materials include thermosetting plastics (polyester and epoxy) and thermoplastics (nylon and polysulfone). These resin–matrix composites are difficult to inspect with conventional nondestructive test methods and AE has been used widely.[12] Main sources of emissions are fiber fracture, delamination (matrix cracks between reinforcement layers), splitting (matrix cracks parallel to fibers), and transverse matrix cracks.

When unidirectionally reinforced composites are stressed, fiber fracture and matrix cracks contribute to AE.[3,4,12] Glass fibers generate high amplitude emissions (>70 dB in reference to 1 μV at the sensor) just prior to composite fracture (>90% of the fracture load). Carbon fibers start to fracture above 50% of the fracture load of the composites and produce low-amplitude (30–60 dB) emissions, reflecting their smaller diameters and lower fracture strain. The rate of emissions increases rapidly as the final fracture load is approached, similar to the behavior of brittle solids. Near the final fracture, high-amplitude, long-duration signals from splitting are also observed. Delamination occurs when reinforcement layers (called plies) have different fiber orientations or notches. Acoustic emission signals from delamination are the strongest (50–130 dB peak amplitude) and have the longest duration (0.1–10 ms).

Most composite structures are fabricated with multiple plies with various combinations of reinforcement orientations. When they are stressed, complex stress patterns develop between reinforcing fibers and plies and copious emissions are generated, mainly from the transverse matrix cracks and delaminations. Discrimination of individual AE signals relies on peak amplitude, rise time, and duration. Indication of damages in glass fiber com-

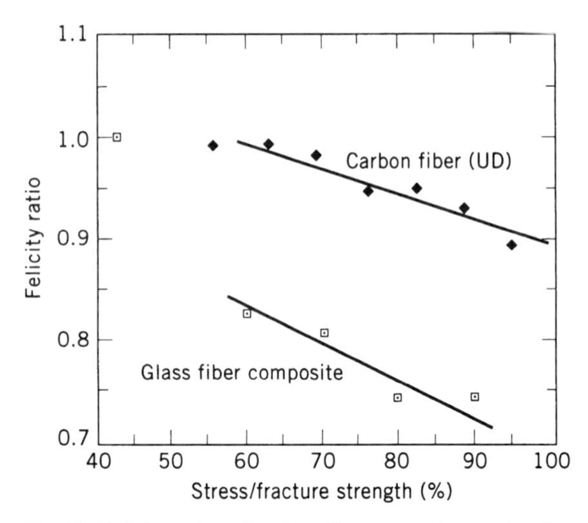

Fig. 7 Felicity ratios of a glass fiber composite panel and a unidirectional carbon fiber composite sample. (Glass fiber data from Ref. 13.)

posites include high-amplitude emissions (>70 dB), long-duration signals, and emissions during load hold periods. However, signal discrimination is often difficult because of high rates of emissions and strong signal attenuation.[3]

When a composite contains damages, the Kaiser effect is no longer observed. The Felicity ratio is defined as a ratio of the load at which AE is observed upon reloading to the maximum prior load. The Felicity ratio is 1.0 when the composite is sound and decreases to lower values (0.6–0.8) after severe loading. As shown in Fig. 7, the Felicity ratio drops to 0.8 by stressing above 70% of the fracture load in glass-fiber-reinforced composite.[13] On the other hand, a unidirectional carbon fiber composite retains high Felicity ratios (>0.9) up to 95% of the fracture load.

3.4 Other Emission Sources

In a corrosive environment, AE is observed from hydrogen evolution, pitting, and exfoliation, but anodic metal dissolution produces no AE. When materials are subjected to stress corrosion or hydrogen environments, they produce AE by a number of different mechanisms, including film breakage, hydrogen-induced cracking, inclusion and particle fracture at the crack tip, and plastic deformation.[14] The detection of hydrogen embrittlement is one of the early applications of AE techniques. During crack propagation due to hydrogen embrittlement, high-amplitude emissions are observed and can be distinguished easily from an active path corrosion process. The latter generates only a small number of low-

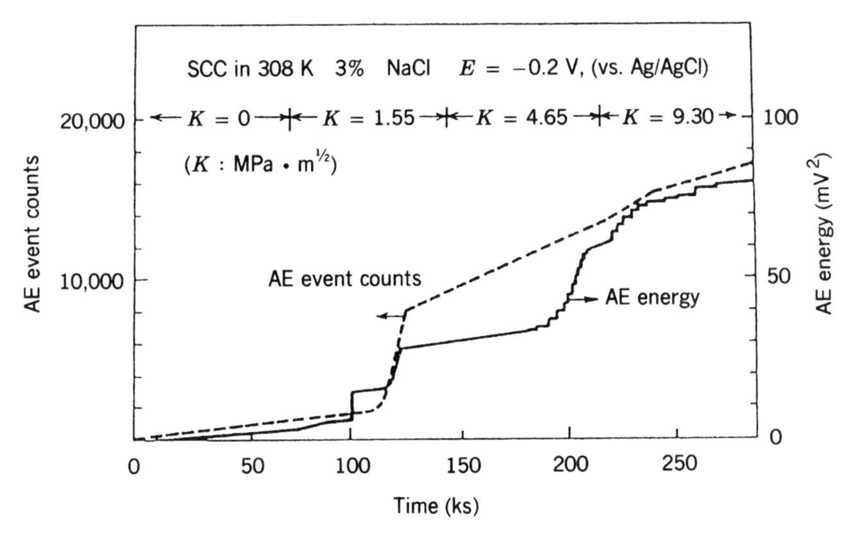

Fig. 8 Cumulative AE event counts and AE energy during stress corrosion cracking test under four different stress intensity factor levels; 304 stainless steel was exposed to a saline solution. (Reprinted, by permission, from Ref. 14, p. 75.)

amplitude emissions. The crack growth rate for stress corrosion cracking shows the three-stage behavior with respect to the applied K_I; that is, an initial increase above a threshold K_I, followed by a region with a nearly constant rate and the final rise near the critical K_I. The AE activities show a similar behavior. Cumulative AE event counts and AE energy are plotted against time in Fig. 8, where 304 stainless steel was exposed to a saline solution with varying levels of stress intensity factor K_I. When $K_I = 0$, AE is due to hydrogen evolution and has low amplitude only. For K_I above 1.5 MPa $\cdot$ $m^{1/2}$, microcracking due to stress corrosion is detected with high-amplitude AE signals.

Acoustic emission monitoring of welding processes can locate weld cracks and slag inclusions and find the depth of weld penetration and mistracking. Expulsion and lack of weld in spot-welding can also be detected. These in-process monitoring methods may avoid the electrical welding noise by using guard sensors and gating circuits synchronized to welding steps. Evaluation of weldments has been the primary goal of AE monitoring of welded structures, such as pressure vessels, tanks, and highway bridges. Defective welds can be located by AE and further evaluated by other nondestructive test methods.

Monitoring of various machining or metal removal processes detects tool wear, fool fracture, and cutting conditions. These are based on AE due to friction between tool and chips or a workpiece, abnormal signals from chipped tools, and plastic deformation and fracture of chips. This application is increasing its importance with automated manufacturing processes.

Sudden movements of magnetic domain walls in ferromagnetic materials produces AE. This is called magneto-AE, and its origin is related to magnetostriction. The intensity and waveform of magneto-AE depend on applied or residual stress, chemical composition, heat treatment, prior cold working, and test temperature. This is an acoustic equivalent of the Barkhausen effect and can be used to evaluate residual stress and microstructural changes of a ferromagnetic material.

The following are other important AE sources of AE: in metals, martensitic transformation, solidification (hot cracking), and oxidation (oxide cracking and spalling); gas and fluid leakage; corona discharges in power transformers; and friction and rubbing noise from rotating machinery.

4 SOURCE LOCATION

An important function of AE testing is to locate sources in a structure.[1] Acoustic emission testing uses the structure itself to discover discontinuities through the generation and propagation of AE signals, which can be continuous (e.g., fluid leaks and plastic deformation of metals) or burst emission (e.g., crack propagation and delamination of composites). Two sensors are used to locate the source in a linear structure such as a pipe and a tube (linear location). More sensors are needed when the

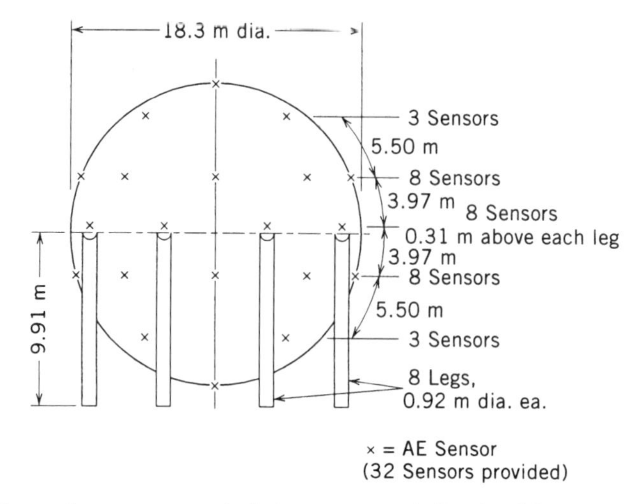

Fig. 9 Sensor placements on a spherical pressure vessel. (Reprinted from Ref. 15, p. 436.)

linear structure becomes longer and when three-dimensional structures are examined. Figure 9 shows typical sensor placements on a spherical pressure vessel.[15] In most applications, the structures can be regarded as a two-dimensional shell structure and the surfaces can be laid out on a plane. For example, a spherical tank can be represented by an icosahedron and by 20 triangles on a planar display.

Two general approaches are used in source location; that is, zone location and discrete source location. In the zone location method, a source is presumed to exist within the zone that belongs to the sensor receiving the strongest AE signal (typically in peak amplitude) or receiving the first hit signal. This method is suited for locating leaks (continuous emissions) and for sources in a highly attenuating medium (fiber composites and metals coated with viscous insulation). The method can be refined by determining which sensor receives the second strongest signal or the second hit signal. In this method, the zone around each sensor is subdivided, and the number of the subdivision is equal to that of surrounding sensors. When the amplitudes of the received signals are measured, further improvements in source location accuracy are possible through an interpolation technique by taking the attenuation into account.

In discrete source locations, the full spatial coordinates of an AE source are defined by measuring the differences in arrival times of AE signals at multiple sensors. The AE signals must be burst type and strong enough to reach three or more sensors. In attenuating media, the sensor spacing has to be sufficiently small. Generally, the wave velocity is assumed constant and triangulation techniques are used. A pair of sensors define an arrival time difference and a hyperbola passing through a source. The location of the source can be obtained as the intersection of two such hyperbolae. Hundreds of sources per second can be located by typical AE instruments. For this method to be practical, the attenuation through the structure cannot be high to avoid the use of excessive sensors and signal processing channels. Emission rates must be moderate so that the arrival time differences of a single event can be detected without interference of the next event.

In order to evaluate the severity of an AE source, the intensity and activity of the source and proximity to neighboring sources are determined. This requires the collection of additional data on the AE sources and typically posttest analysis. Such data include the presence of high-amplitude emissions, AE activities during load hold, those during stress increases, and Felicity ratios during load cycling. Statistical analysis (e.g., amplitude distribution) of AE signals received at individual sensors is also useful.

5 STRUCTURAL INTEGRITY MONITORING

5.1 Pressure Vessels, Tanks, and Piping: Fiber-Reinforced Composite

No satisfactory test for determining the structural adequacy of fiber-reinforced plastic equipment existed until AE testing of fiber-reinforced plastic vessels and piping has been developed and standardized.[3] Test vessels and piping are typically pressurized to 110% of the maximum allowable working pressure to locate substantial flaws. Using the zone location method, AE activities of flaws within each zone are detected and the zone represents the approximate position of these flaws. Sensors are positioned to detect structural flaws at critical sections of the test vessel (i.e., high-stress areas, geometrical discontinuities, nozzles, and manways). Pressurization in AE testing of a vessel proceeds in steps with pressure hold periods with depressure increments. A test is terminated whenever a rapid increase in AE activities indicates an impending failure. Detected flaws are graded using several criteria, including emissions during pressure hold periods, Felicity ratio, historic index (average signal strength of recent events divided by that of all events), severity (average signal strength of 10 largest events), and high-amplitude events. Emissions during hold indicate continuing permanent damage and a lack of structural integrity. For in-service vessels, Felicity ratio criterion (when it is less than 0.95) is an important measure of previous damage. Historic index is a sensitive measure to detect a sudden increase in signal strength. Large severity values result from delamination, adhesive bond failure, and crack growth. High-amplitude events indicate structural (glass-fiber) damages, especially in new vessels.

5.2 Pressure Vessels, Tanks, and Piping: Metals

Acoustic emission testing of metallic tanks and pressure vessels is conducted during pressurization. The tanks and vessels are pressurized following applicable code specifications. The maximum pressure is typically up to 110% of the operating pressure. Sources of emissions are usually crack growth, yielding, corrosion effects (cracking or spalling of corrosion product, local yielding of thinned section), stress corrosion cracking, and embrittlement. The most likely locations of flaws are at various weldments.[15] When metals are ductile, AE activities are low and AE test results should be evaluated carefully. Once the metals are embrittled by environment or at low temperatures, even early stages of stressing activate AE sources. In fact, acoustic emission is the best method for detecting hydrogen-induced cracking and stress corrosion cracking. Highway gas-trailer tube, chemical, and petrochemical vessels are monitored with AE during and between service without emptying the vessels, thus minimizing the cost of testing. According to the current industrial practices, the located flaws are usually confirmed by other nondestructive test methods. Increasingly, however, AE testing alone is used to evaluate tanks and pressure vessels. For certain classes of metallic vessels, the use of AE examination alone can be used to satisfy code-mandated inspection requirements in lieu of ultrasonics or radiography.

After AE testing indicates the locations of active AE sources, these are graded according to criteria similar to those for composite vessels. These provide real-time indications of defective areas of the pressure vessel being tested and prevent catastrophic failure of the vessel. The identified defective areas are then inspected using other test methods. Acoustic emission testing is applied on various types of vessels, including ammonia spheres and tanks, hydroformer reactors, and catalytic cracking units.

5.3 Aerospace

The first application of AE testing was to verify the structural integrity of filament-wound Polaris rocket motorcases in 1961.[16] Hydrostatic testing of rocket motorcases was instrumented using accelerometers, audio recording, and sound analysis equipment. Crack initiation and propagation were detected prior to catastrophic failure. Later, burst pressure was successfully predicted based on AE data during proof testing. The composite rocket motorcases are still successfully inspected using three (or four) pressurization cycles. The first cycle is to find leaks at a low pressure. The second goes to 100% (or 80%) of the mean expected operating pressure, followed by the third to 80% level. The AE behavior during the third cycle is used to examine the Kaiser effect and is the determining factor in the evaluation of the rocket motorcase integrity.

Acoustic emission testing is as yet unable to monitor the integrity of aircraft structures in flight because of substantial noise generated from the structures that are joined by numerous fasteners and from engines. For specific components, however, AE can detect the presence of cracks and incipient fracture. For surveillance of structural and fatigue damages of aircraft components during limit load and fatigue testing, AE has demonstrated its utility. All significant cracks were located in the main wing-spar of a Mirage jet-fighter during full-scale fatigue testing.[17] Some of the cracks emanated from bolt holes amid a wide range of spurious AE sources. Acoustic emission indications of damage severity vary depending on the particular component, but include load hold emissions, Felicity ratio, AE event rates, and high-amplitude emissions. In structural tests, wave propagation characteristics and the nature of crack-related (crack growth and crack face fretting) sources are also essential in evaluating AE observations.

5.4 Geological and Civil Engineering Structures

Acoustic emission from rocks in underground mines was first detected in the 1930s by Obert and Duvall and has been called microseismism, as it is a microscopic earthquake.[18] Acoustic emission monitoring has been studied to promote mine safety by detecting incipient rock bursts and gas outbursts and by estimating the areas of stress concentration ahead of mining regions. For this purpose many AE sensors are placed in and around mine tunnels (some from holes dug into the ground) and monitor low-frequency (below several kilohertz) AE signals. A mining operation itself can be a stimulus as underground rocks have numerous preexisting flaws and AE under geological pressure. Drilling exploratory holes ahead of a mining region can also activate the areas of high local stresses. Multichannel AE sensing has also been applied to the evaluation of hydrofracturing in geothermal wells. The stability of underground storage caverns is another area of AE applications.

By using the Kaiser effect, underground stresses may be estimated.[19] Compressive stresses are applied on a sample bored out from underground. When the previously existing stress is exceeded, AE activities increase and the preexisting stress can be estimated. Since a high compressive pressure always exists, the directionality of stresses must be taken into account. Loading methods and relaxation after coring may have effects on the results, however.

Acoustic emission techniques have been applied to monitoring the stability of highway slopes, tunnels, and landslide-prone areas. Acoustic emission measurements utilize waveguides that are driven into the ground. By mapping high AE activity areas, potential instability regions can be predicted. Concrete with steel reinforcement is another material used widely in civil engineering structures.[20] Under bending, compressive or tensile loading, AE activities increase as the fracture load is approached. Bending and shear modes of fracture can clearly be distinguished via AE. Because of the large size of concrete structures and high attenuation rates of AE signals in concrete, suitable AE techniques for structural integrity monitoring are yet to be perfected.

6 SUMMARY

Acoustic emission is the generation of transient elastic waves due to the rapid release of energy from a localized source or sources within a material or structure, originating from mechanical deformation and fracture as well as from phase transformation, corrosion, friction, and magnetic processes. From AE signals detected by sensors attached to the structure, the locations and activity levels of the sources are evaluated. Acoustic emission has become an indispensable and effective nondestructive inspection method and a valuable tool of materials research, delving into dynamic microscopic processes of materials. A survey of recent AE literature is available.[21]

REFERENCES

1. R. K. Miller and P. McIntire (Eds.), *Nondestructive Testing Handbook*, 2nd ed., Vol. 5, *Acoustic Emission Testing*, American Society for Nondestructive Testing, Columbus, OH, 1987.
2. A. G. Beattie, "Acoustic Emission, Principles and Instrumentation," *J. Acoust. Emission*, Vol. 2, 1983, pp. 95–128.
3. T. J. Fowler, "Chemical Industry Applications of Acoustic Emission," *Mat. Eval.*, Vol. 50, 1992, pp. 875–882.
4. K. Ono and Q. Huang, "Pattern Recognition Analysis of Acoustic Emission Signals," in T. Kishi et al. (Eds.), *Progress in Acoustic Emission VII*, Japan Soc. Nondestructive Inspection, Tokyo, 1994, pp. 69–78.
5. F. R. Breckenridge, "Acoustic Emission Transducer Calibration by Means of the Seismic Surface Pulse," *J. Acoust. Emission*, Vol. 1, No. 2, 1982, pp. 87–94.
6. C. R. Heiple and S. H. Carpenter, "Acoustic Emission Produced by Deformation of Metals and Alloys—A Review, Part I and II," *J. Acoust. Emission*, Vol. 6, 1987, pp. 177–204, 215–237.
7. I. Roman, H. B. Teoh, and K. Ono, "Thermal Restoration of Burst Emissions in A533B Steel," *J. Acoust. Emission*, Vol. 3, No. 1, 1984, pp. 19–26.
8. S. Wakayama, T. Kishi, and S. Kohara, "Microfracture Analysis in Al_2O_3 Evaluated by AE Source Characterization," in K. Yamaguchi et al. (Eds.), *Progress in Acoustic Emission III*, Japan Soc. NDI, Tokyo, 1986, pp. 653–660.
9. K. Kuribayashi, et al., "Fracture Toughness and AE of Structural Low Alloy Steel (SNCM8)," *Hihakai Kensa* (*J. Nondestructive Testing*), Vol. 30, 1981, pp. 842–846.
10. H. Takahashi, M. A. Khan, M. Kikuchi, and M. Suzuki, "Acoustic Emission Crack Monitoring in Fracture-Toughness Tests for AISI 4340 and SA533B Steels," *Exp. Mech.*, Vol. 21, 1981, pp. 89–99.
11. Y. Nakamura, C. L. Veach, and B. O. McCauley, "Amplitude Distribution of Acoustic Emission Signals," in *Acoustic Emission*, ASTM STP 505, Amer. Soc. Testing and Materials, Philadelphia, 1972, pp. 164–186.
12. J. Summerscales, "NDT of Advanced Composites—An Overview of the Possibilities," *Br. J. Non-Destructive Testing*, Vol. 32, No. 11, 1990, pp. 568–77.
13. T. J. Fowler and E. Gray, "Development of an Acoustic Emission Test for FRP Equipment," Preprint 3583, American Soc. of Civil Engineers, New York, 1979.
14. S. Yuyama, T. Kishi, and Y. Hisamatsu, "AE Analysis during Corrosion, Stress Corrosion Cracking and Corrosion Fatigue Processes," *J. Acoust. Emission*, Vol. 2, 1983, pp. 71–93.
15. T. J. Fowler, "Acoustic Emission Testing of Chemical Industry Vessels," in M. Onoe et al. (Eds.), *Progress in Acoustic Emission II*, Japan Soc. Nondestructive Inspection, Tokyo, 1984, pp. 421–449.
16. A. T. Green, C. S. Lockman, and H. K. Haines, "Acoustic Verification of Structural Integrity of Polaris Chambers," *Mod. Plastic*, Vol. 41, No. 11, 1964, pp. 137–139.

17. C. M. Scala, J. F. McCardle, and S. J. Bowles, "Acoustic Emission Monitoring of a Fatigue Test of an F/A-18 Bulkhead," *J. Acoust. Emission*, Vol. 10, Nos. 3/4, 1991/92, pp. 49–60.
18. H. R. Hardy, Jr. (Ed.), *Proc. the Fifth Conference on AE/Microseismic Activity in Geologic Structures,* Trans. Tech. Publ., Clausthal-Zellerfeld, Germany, 1991.
19. T. Kanagawa, M. Hayashi, and H. Nakasa, "Estimation of Spatial Geo-stress Components in Rock Samples Using the Kaiser Effect of Acoustic Emission," *Proc. Japan Soc. Civil Eng.*, No. 258, 1977, pp. 63–75.
20. M. Ohtsu, "Acoustic Emission Characteristics in Concrete and Diagnostic Applications, *J. Acoust. Emission*, Vol. 6, No. 2, 1987, pp. 99–108.
21. K. Ono, "Trends of Recent Acoustic Emission Literature," *J. Acoust. Emission*, Vol. 12, Nos. 3/4, 1994, pp. 177–198.

54

SEISMIC EXPLORATION: ACOUSTIC IMAGING FOR GEOLOGICAL INFORMATION

A. J. BERKHOUT

1 INTRODUCTION

Exploration seismology is based on the analysis of acoustic waves reflected from different rock layers in the earth's subsurface. (Here *acoustic* is used in the elastodynamic sense.) Impulsive acoustic energy, emitted into the subsurface, encounters discontinuities between the layers and is partially reflected back to the surface. The returning reflections are detected and their strengths and arrival times are recorded (*seismic acquisition*). After the recorded data are processed (*seismic processing*), geophysicists and geologists derive information from the result concerning the position, thickness, rock type, and porefill of each layer (*seismic interpretation*).

Most of the recent advances in seismic exploration techniques have been made possible by new digital technology. This holds true in all three seismic disciplines: acquisition, processing, and interpretation. The ability to record and store very large volumes of seismic data, the ability to reduce noise on the recordings and then improve the resolution properties, the ability to manipulate vast quantities of data with advanced imaging algorithms to arrive at complex structural interpretations, and finally the ability to apply multidisciplinary inversion techniques to arrive at the rock properties of potential reservoirs have all been made possible by improved digital technology. And as seismic exploration techniques continue to expand, new systems such as data robots, massive parallel processors, fast workstations with large internal memories, and advanced graphics will be taken for granted as everyday equipment in the exploration world.

Note: References to chapters appearing only in the *Encyclopedia* are preceded by *Enc.*

The chapter starts with a discussion on the close relationship between acoustic imaging and seismic exploration. This discussion is followed by an explanation of the principles of seismic exploration in terms of ray theory. Next a review is given of the seismic forward model in terms of wave-theory-based operators. This model is used to explain modern seismic processing in terms of *preprocessing* of the raw measurements, image *formation* (*seismic migration*) and image *characterization* (*lithologic inversion*).

2 ACOUSTIC IMAGING AND SEISMIC EXPLORATION

Acoustic reflection techniques aim at high-resolution images of objects under investigation. Figure 1 shows three important commercial applications, together with the corresponding media. Later in this chapter we will show that propagation properties can be quantified by propagation matrices $\boldsymbol{W}^{+}$, $\boldsymbol{W}^{-}$ and reflection properties can be quantified by reflection matrices $\boldsymbol{R}^{+}$, $\boldsymbol{R}^{-}$.

In *medical diagnostics* the objective is to collect information on the tissues inside the human body. In *seismic exploration* the geologic layers below the earth's surface are of interest. In *ultrasonic inspection* defects in construction materials need to be detected. Although the instrumentation in those applications is significantly different, they have the underlying acoustical principles in common.

In acoustic reflection techniques, the medium under investigation is illuminated with acoustic waves (*insonification*). The induced acoustic wave field (source wave field) propagates from a data acquisition surface into the medium, reflects at the inhomogeneities, propagates back

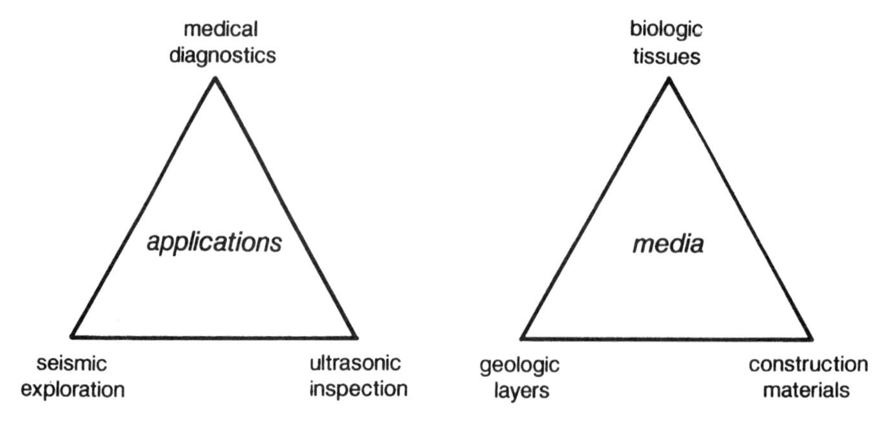

Fig. 1 Three important commercial applications of acoustic reflection techniques, together with the corresponding media.

to the surface (reflection response), and is measured by a range of acoustic detectors. Information about the structural and material properties of the interior is derived from the recorded detector signals. For many practical problems it is necessary to illuminate the medium from different locations at the data acquisition surface (different source positions), and the registered reflection responses of the different acoustical experiments may be optimally combined (processing). Later in this chapter we will see that reflection responses from different experiments can be conveniently represented by measurement matrix $\boldsymbol{P}$.

The properties of the detected waves are determined by two properties of the medium. The first are the *propagation* (or transmission) properties, which primarily depend on the *global* acoustic properties of the medium such as average velocity and average absorption; if applicable, the average values should be supplemented by gradients (trend model). The second are the *reflectivity* properties, which primarily depend on the *local* acoustic properties of the medium such as detailed variations in the modulus of elasticity (Fig. 2). The importance of the trend model concept can be readily appreciated by realizing that the relation between the reflection arrival times and the position of the corresponding reflection points is determined by the velocity trend model of the object. Note that in the inversion literature the trend model is generally referred to as the *background medium*.

In medical diagnostics the background is taken to be a homogeneous half-space; in ultrasonic inspection the background is represented by a *bounded* homogeneous medium; in seismic exploration the background is a half-space that consists of a number of macrolayers, where each macrolayer is characterized by its own trend parameters. In seismic exploration the background is generally referred to as the macromodel.

In principle, acoustic imaging aims at determination of both trend and detail. Particularly in seismic exploration, the estimation of the macromodel is of utmost importance before imaging for the detail can start. However, in medical as well as inspection applications an

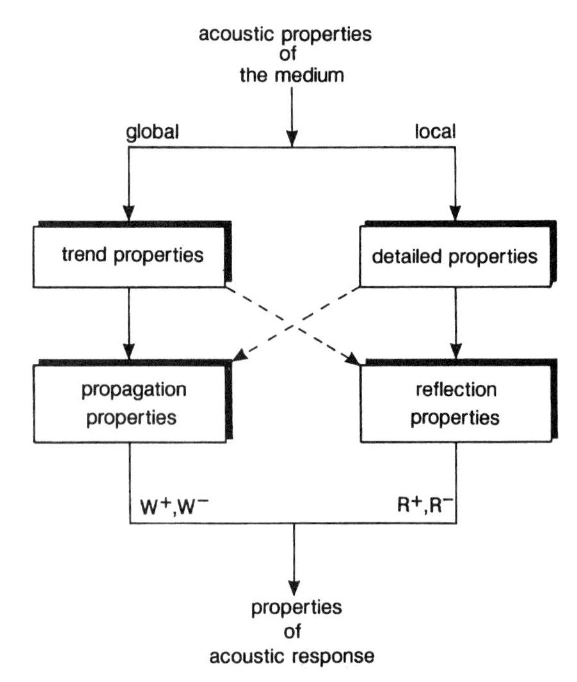

Fig. 2 Relatively slow trend properties of the medium determine the propagation (transmission) properties of the response, i.e., primarily the travel time information. The relatively fast detailed properties of the medium determine the reflection (backscatter) properties of the response, i.e., primarily the amplitude information.

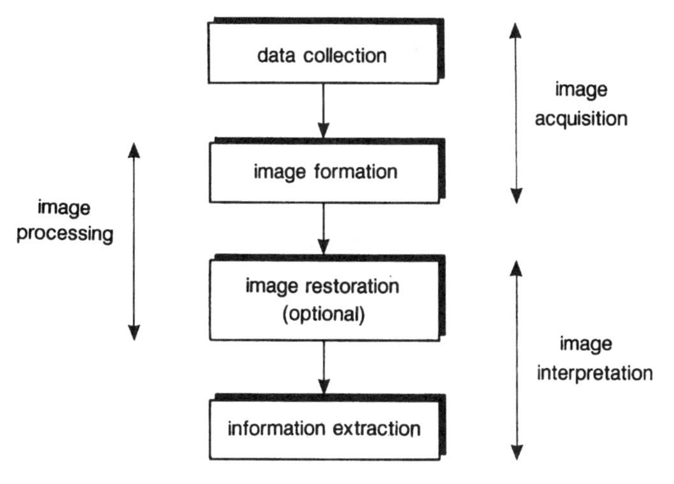

Fig. 3 Basic processes in reflection imaging.

estimate of the trend parameters of the medium is already available but local variations are unknown; the objective of acoustic imaging techniques in those applications is to obtain information on the local variations only.

Figure 3 shows the basic processes in acoustic reflection techniques. In its most general form reflection data are generated and digitally recorded (data collection) and optionally preprocessed. Next, the preprocessed data is used to form the actual image numerically (image formation). However, in *focused* experiments image formation is part of the data collection and images are directly recorded. Since data collection and image formation are never perfect in practice, the imperfections are optionally reduced by *image restoration*. The result is used as input to the final information extraction process (interpretation). Note that in the imaging literature numerical image formation is often referred to as *aperture synthesis*. Note also that in seismic exploration numerical image formation is always referred to as seismic migration.

In acoustical reflection imaging one has to deal with two fundamental problems:

1. Optimization of the signal-to-noise ratio
2. Optimization of the resolution

Noise is not only determined by statistical background signals but also by undesired wave types and multiple scattering.

The resolution along the time axis (temporal resolution) is determined by the range of frequencies that are present in each individual echo (temporal bandwidth). A large temporal resolution requires a large temporal bandwidth. The spatial resolution is determined by the size of the involved wavelengths together with the range of immergence and emergence angles that are present in the recorded echoes from each individual inhomogeneity (spatial bandwidth). Small wavelengths and a large range of angles are required for images with a high spatial resolution. Note that a large *temporal* resolution favors a large *spatial* resolution.[1]

Imperfect image formation, for example, due to the use of wrong macrovelocity information or limited focusing instrumentation, will decrease the resolution more than could be expected from the limited temporal and spatial bandwidth. It is the objective of the image restoration process to identify and repair the damage to the resolution (*deblurring*), taking the presence of noise into account. In those cases where *additional* information of the object can be specified, image restoration may yield results even beyond the resolution specified by the available temporal and spatial bandwidth.

In practical situations the signal-to-noise ratio decreases with increasing depth due to attenuation (penetration problem). Similarly, resolution decreases with increasing depth due to the relatively large attenuation for high frequencies (small wavelengths). In addition, the aforementioned range of immergence and emergence angles generally decrease with increasing depth, causing an extra problem for *lateral* resolution. Figure 4 provides a view of the applied frequencies in the different applications. Every frequency range is a compromise between resolution and penetration: By shifting toward the high frequencies, the resolution increases but the penetration decreases because of increased attenuation. Note particularly the large difference in utilized frequencies, as implemented in low-frequency seismic exploration and high-frequency ultrasonic inspection. Deep seismics is

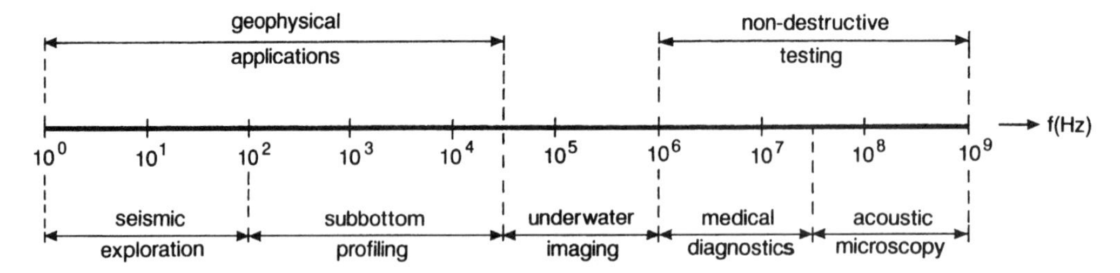

Fig. 4 Approximate frequency ranges as used in different acoustic imaging applications. Resolution increases with frequency; penetration decreases with frequency.

particularly used for hydrocarbon finding; the penetration is on the order of kilometres. High-frequency inspection occurs with the acoustic microscope; the penetration is on the order of millimetres.

In the following sections, we will concentrate on *seismic exploration*, that is, acoustic imaging of the geologic layers below the earth's surface. First, the principles will be reviewed in terms of ray theory (Section 3). Next, modern concepts are discussed in terms of wave theory (Section 4). Finally, a summary is given on seismic processing (Section 5).

3 PRINCIPLES OF SEISMIC EXPLORATION

Exploration seismology is based on analysis of acoustic waves reflected from different rock layers in the earth's subsurface. Impulsive acoustic energy, emitted into the subsurface, encounters discontinuities between the rock layers and is partially reflected back to the surface. The returning reflections are detected and their strengths and arrival times are recorded. After the recorded data are processed, geophysicists and geologists derive information from the result concerning the position, thickness, and rock property of each layer.

The seismic medium consists of layers of porous rock, such as sand, shale, and limestone. The pores of the rock generally contain water but may also contain gas and/or oil. In the oil and gas industry it is the objective of seismic exploration to identify structures that are favorable for containing a commercial amount of hydrocarbons.

Seismic reflection energy consists of compressional waves (P-waves) and shear waves (S-waves). Typical P-wave velocities are 500 m/s for the weathered surface layer at land (*land seismics*), 1500 m/s for the water surface layer at sea (*marine seismics*), 2500 m/s for consolidated sands and shales, and 4000 m/s for limestone layers. For most rock layers the ratio between P-wave and S-wave velocity (V_p–V_s ratio) varies between 1.5 and 2. This means that the Poisson ratio varies between 0.1 and 0.3, respectively. For shallow gas sands the V_p–V_s ratio may be as low as 1.4. For the weathered layer the V_p–V_s ratio may be as high as 2.5 (dry) to 5 (water filled). Of course, in the case of marine seismics $V_s = 0$ for the surface layer.

Due to the low-frequency content of seismic waves, the rock material and pore fluid move in phase as in a mixture. This means that the wave theory for *solids* may be used, the compressibility being given by the Gassmann equation.[2] The latter is consistent with the low-frequency approximation of the Biot theory.[3]

3.1 Seismic Acquisition

Figure 5 shows the conventional data acquisition geometry. The seismic source may consist of dynamite (land), a hydraulic vibrator (land and marine), or compressed air (marine). The subsurface response is measured by a number of detector stations. Each detector *station* consists of a number of electrically connected single detectors (*field pattern*). Each detector measures pressure (marine data) or the vertical component of the particle velocity (land data). The reflection sequence measured by one detector station is called a seismic trace. All traces related to one source position define one seismic *shot record*. Typically, there are 500 traces in one shot record. The data of one seismic line consist of a number of seismic shot records (at least several hundreds). Typical numerical quantities are as follows:

Distance between two adjacent shot positions	50 m
Distance between two adjacent detector–station positions	25 m
Length of a seismic trace	8 s
Time-sampling interval	4 ms

In a two-dimensional seismic survey the distance between adjacent seismic lines may be several kilometres. In a three-dimensional seismic survey adjacent seismic lines may be as close as 50 m. There is a general ten-

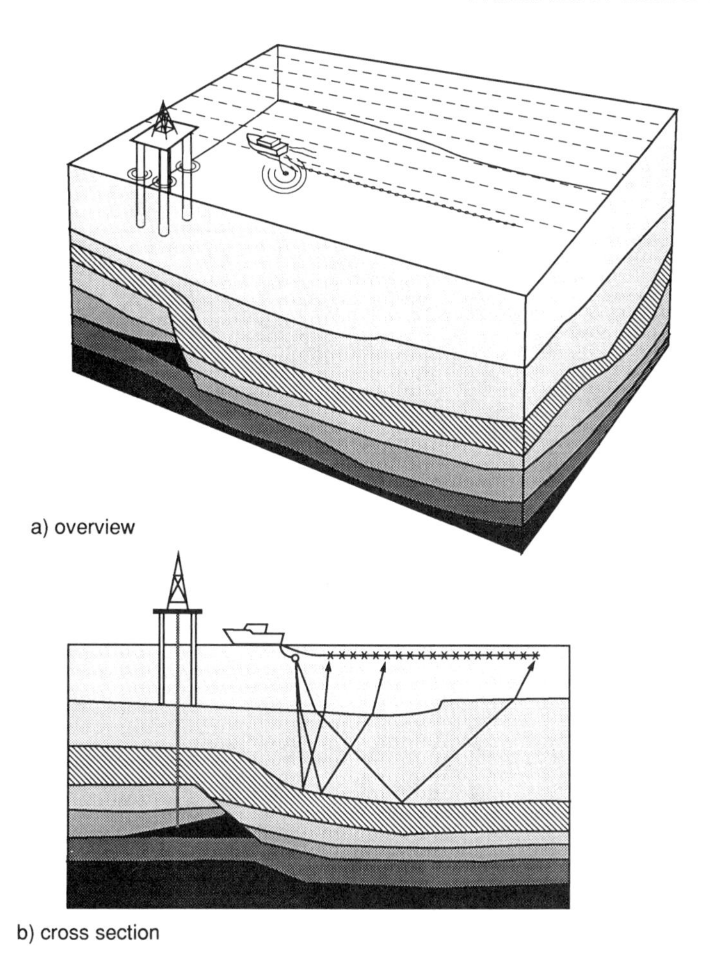

Fig. 5 Conventional seismic data acquisition geometry; the marine single-streamer situation.

dency in the seismic industry to decrease shot and detector spacings, to decrease the length of field patterns, and to increase the number of detector stations.

Note that for a shot record of 500 traces and a recording length of 8 s the number of data bytes equals $500 \times 2000 \times 4 = 4.10^6$ (4 megabytes per shot record). Hence, for a three-dimensional seismic survey of 250,000 shots, the total number of recorded bytes equals 10^{12} (1 terabyte per survey).

3.2 Seismic Processing

For the two-dimensional situation Fig. 6*a* illustrates with the aid of rays how a primary seismic wave field propagates through the subsurface (the multiple-scattering rays are not shown). The measurements at the surface define the shot record for the related source position (Fig. 6*b*). In the field the data are, after analog-to-digital conversion, stored on tape. Next, in the seismic processing centre data processing starts with several multioffset preprocessing steps: spatial resampling (*regularization*), removal of waves that travel *along* the surface (particularly for land data), attenuation of multiple scattering (particularly for marine data), correction for these source and receiver properties, and in case of land data, correction for the laterally irregular weathered surface layer.

Until now, the nucleus of any conventional seismic processing package consists of common midpoint (CMP) processing.[4] This means that *after* the above preprocessing steps the data are reordered such that all traces with the same midpoint between source and detector position are grouped in one gather, the so-called CMP gather. Figure 7*b* shows traces of one CMP gather. From the rays of the primary wave field (Fig. 7*a*) it can be seen that

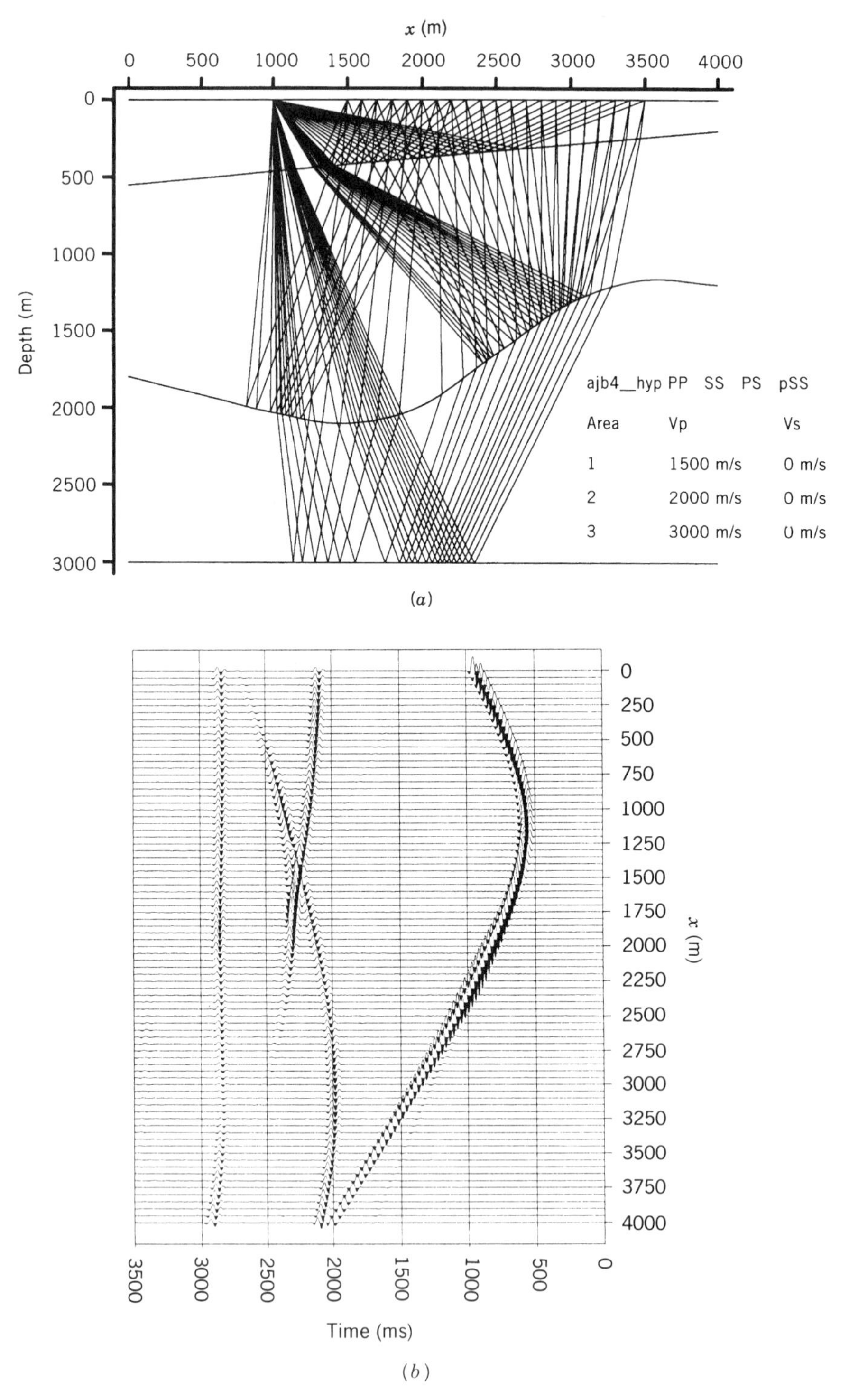

Fig. 6 Seismic ray paths and seismic measurements related to one seismic experiment (shot record). For simplicity only primary longitudinal waves are shown.

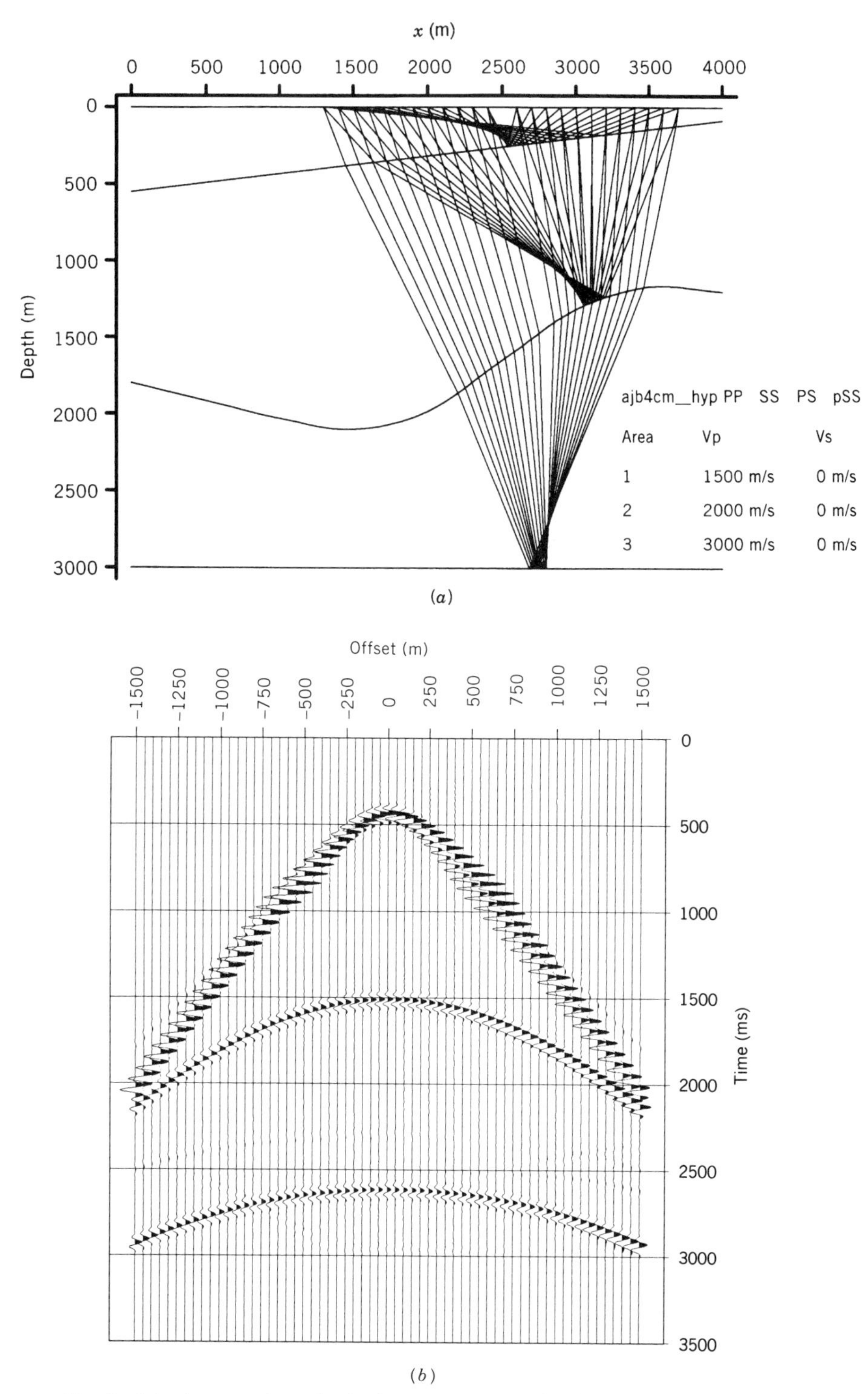

Fig. 7 Seismic ray paths and seismic measurements related to one CMP gather with its midpoint at $x = 2500$ m.

the reflection points for all traces in one CMP gather are grouped closely together. This property determines the usefulness of CMP processing. Even in fairly complicated geologic situations the arrival times of the reflections from one reflector in one CMP gather can be well approximated by a hyperbolic relationship,

$$t_{k,n}^2 = t_k^2 + \frac{(2n\,\Delta h)^2}{v_{st}^2(t_k)}, \qquad (1)$$

where t_k is the zero-offset ($n = 0$) reflection time of the kth reflector and $h_n = 2n\,\Delta h$ represents the distance (*offset*) between source and detector of the nth trace in the CMP gather under consideration; $v_{st}(t_k)$ is the so-called stacking velocity for the kth reflector.

Using a multitrace coherency criterion, v_{st} is estimated for a number of reflecting boundaries. From these values interval velocities for the layers between these boundaries can be computed. Figure 8 gives a contour plot of the coherency values related to the CMP data of Fig. 7*b*. Maxima are related to principal reflections. Together, they define v_{st} as a function of reflection time for the midpoint under consideration. For a comprehensive summary on the determination of stacking velocities the reader is referred to Ref. 5.

If one aims at high resolution in the $v_{st}(t)$ space, use of the generalized radon transform is very effective. For more information the reader is referred to Ref. 6.

Given a stacking velocity distribution for a given CMP gather, the arrival times of all traces within this CMP gather are corrected (Fig. 9),

$$t_{k,n} \rightarrow t_k,$$

such that the travel time effect of the offset (*normal move-out*) is eliminated (*NMO correction*). Finally, all NMO-corrected traces are added together to form one so-called stack trace. In this way a considerable improvement in signal-to-noise ratio occurs and, last but not least, a significant data reduction is achieved. The *CMP stacking* process is characteristic of conventional seismic processing.

A stack trace is considered to be a zero-offset trace and all stack traces of a seismic line define a stack or pseudo-zero-offset section. Figure 10 shows a stack section together with the related zero-offset rays.

The pseudo-zero-offset section must be transformed to a zero-offset reflectivity image (Fig. 11). This means that each reflection is "migrated" to its own reflection point as defined by the endpoint of its zero-offset ray path. In addition, the reflection amplitude is corrected according to the propagation losses between surface and reflection point. The imaging process of repositioning and amplitude correction is called seismic migration, in this particular situation *zero-offset seismic migration*. To generate the migration operator for each reflection point, the macrovelocity model must be determined first. This can be done with the aid of the stacking velocity information.[7]

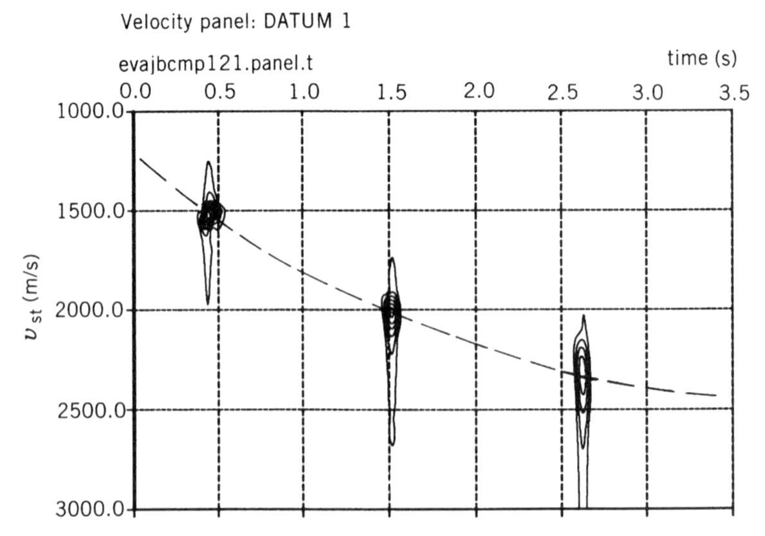

Fig. 8 Contour plot of the coherency values related to the CMP gather of Fig. 7 (velocity plot).

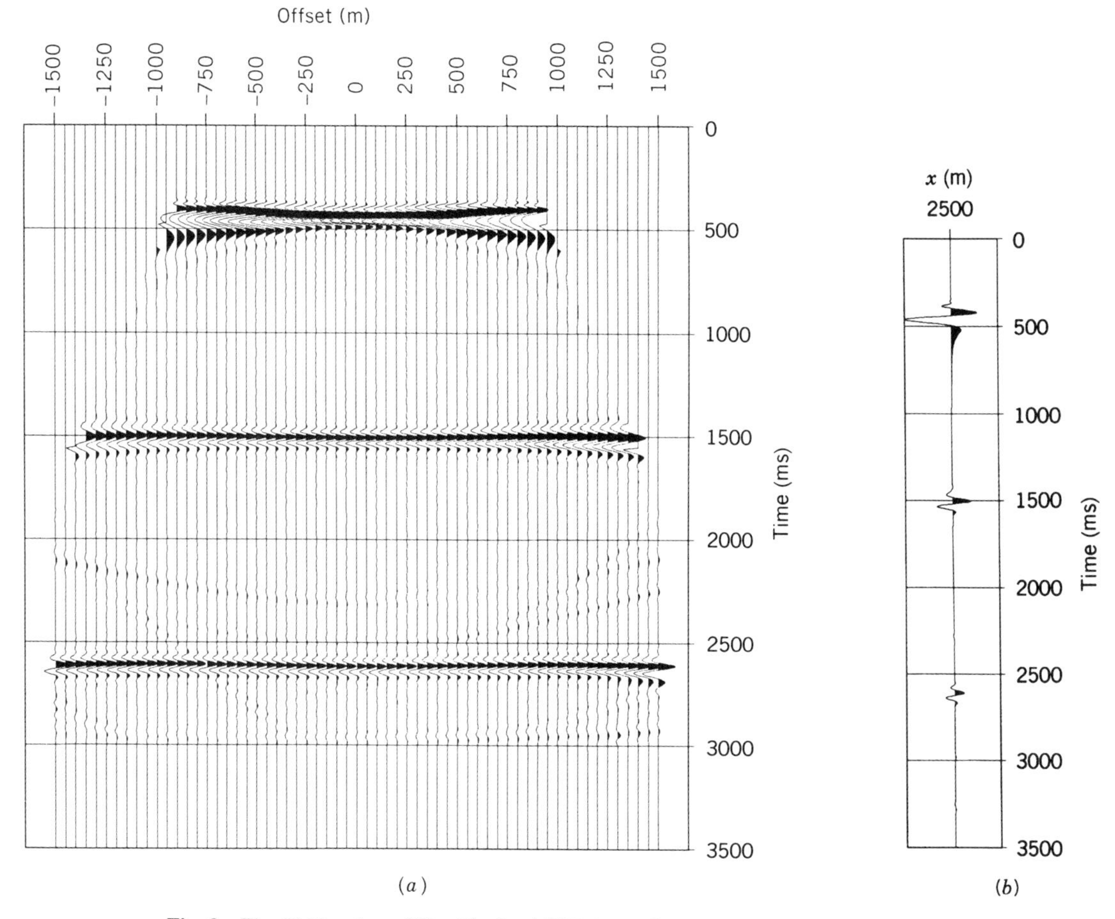

Fig. 9 The CMP gather of Fig. 7*b* after NMO (normal move-out) correction, together with its stacked version.

3.3 Seismic Interpretation

In the interpretation procedure the migrated reflections are tracked and integrated to a subsurface model. In this process the knowledge of *geological* processes (tectonics, sedimentation) is essential. In the early days tracking was pure hand work but nowadays the tracking process is largely automatic and carried out on interactive seismic workstations.

From the interpretation result potential hydrocarbon traps are indicated and well locations are proposed to verify and refine the seismic interpretation.

3.4 Summary

In the foregoing, the principles of seismic exploration have been explained in terms of ray theory. We have seen that, after data acquisition, the basic seismic processes multioffset preprocessing, CMP stacking, and zero-offset migration are applied (conventional method). In the preprocessing phase the attenuation of multiple scattering is an important step, particularly for marine data, where multiple scattering may be significant due to a large amount of trapped energy in the water layer. Actually, the CMP gather of Fig. 8 was already treated for multiple scattering. Figure 12*a* shows the same CMP gather, but now all multiples are included. Figure 12*b* illustrates the noticeable difference in stacking velocity between the primary reflection events and the multiples. This property can be used to remove the multiples.

CMP stacking can be formulated as a numerical focusing process in the midpoint-offset domain, the result being positioned at the zero-offset times of each involved midpoint. Zero-offset migration can be formulated as a numerical focusing process in the midpoint

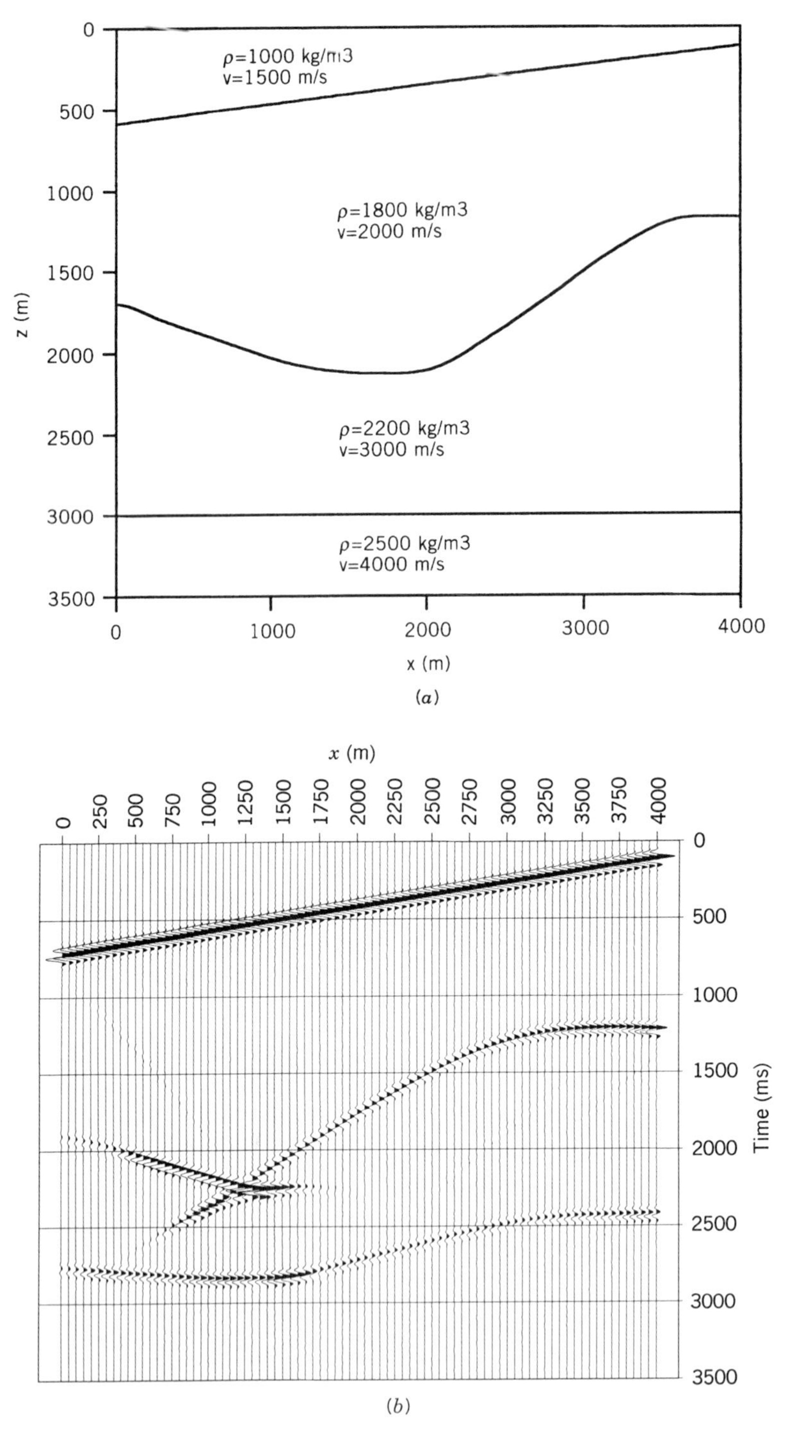

Fig. 10 Stack section (pseudo zero-offset section), consisting of all stacked CMP gathers: (*a*) subsurface model; (*b*) result of stacking. Note that CMP stacking defines a numerical focusing process.

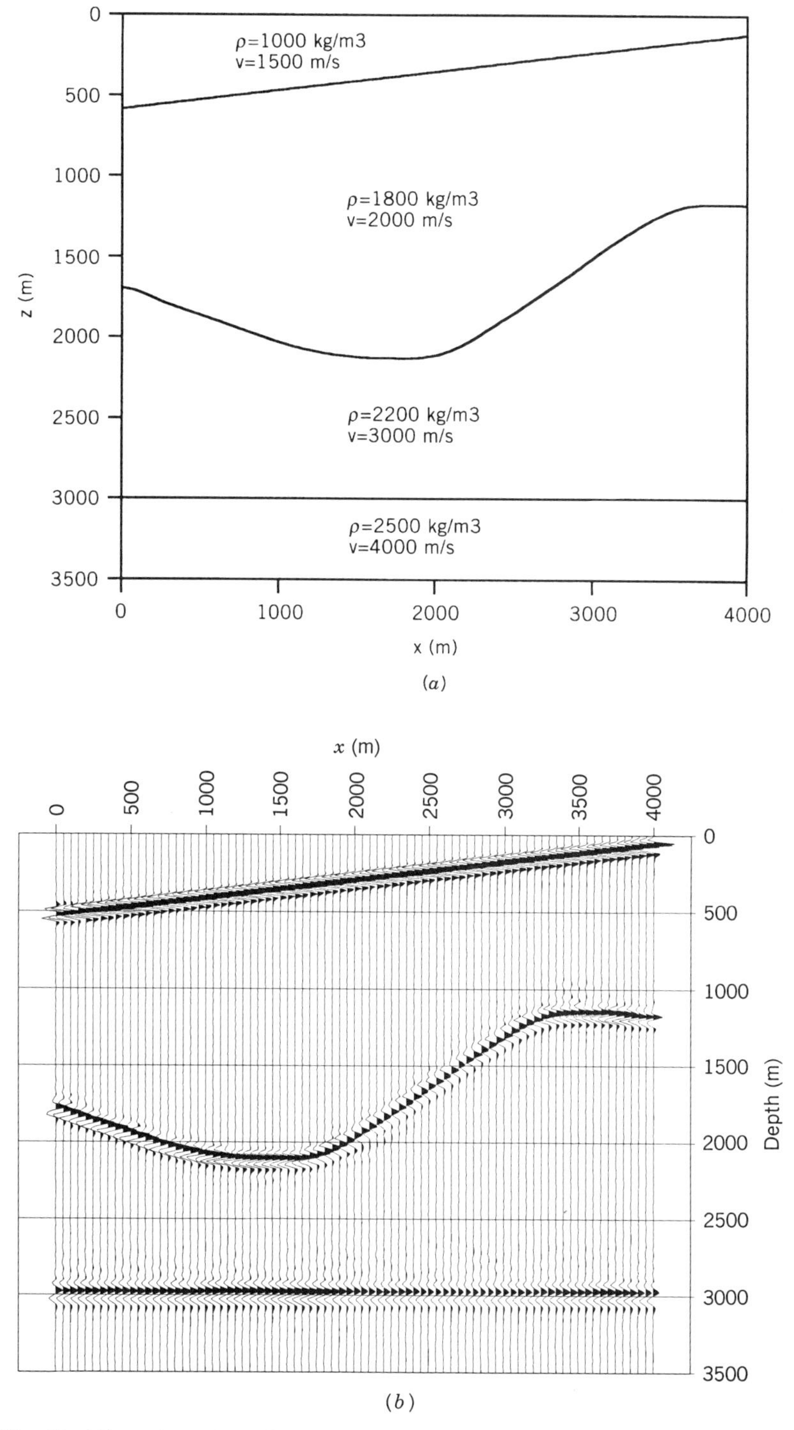

Fig. 11 Migrated version of Fig. 10, yielding the zero-offset reflectivity image: (*a*) subsurface model; (*b*) result of migration. Note that zero-offset migration defines a numerical focusing process.

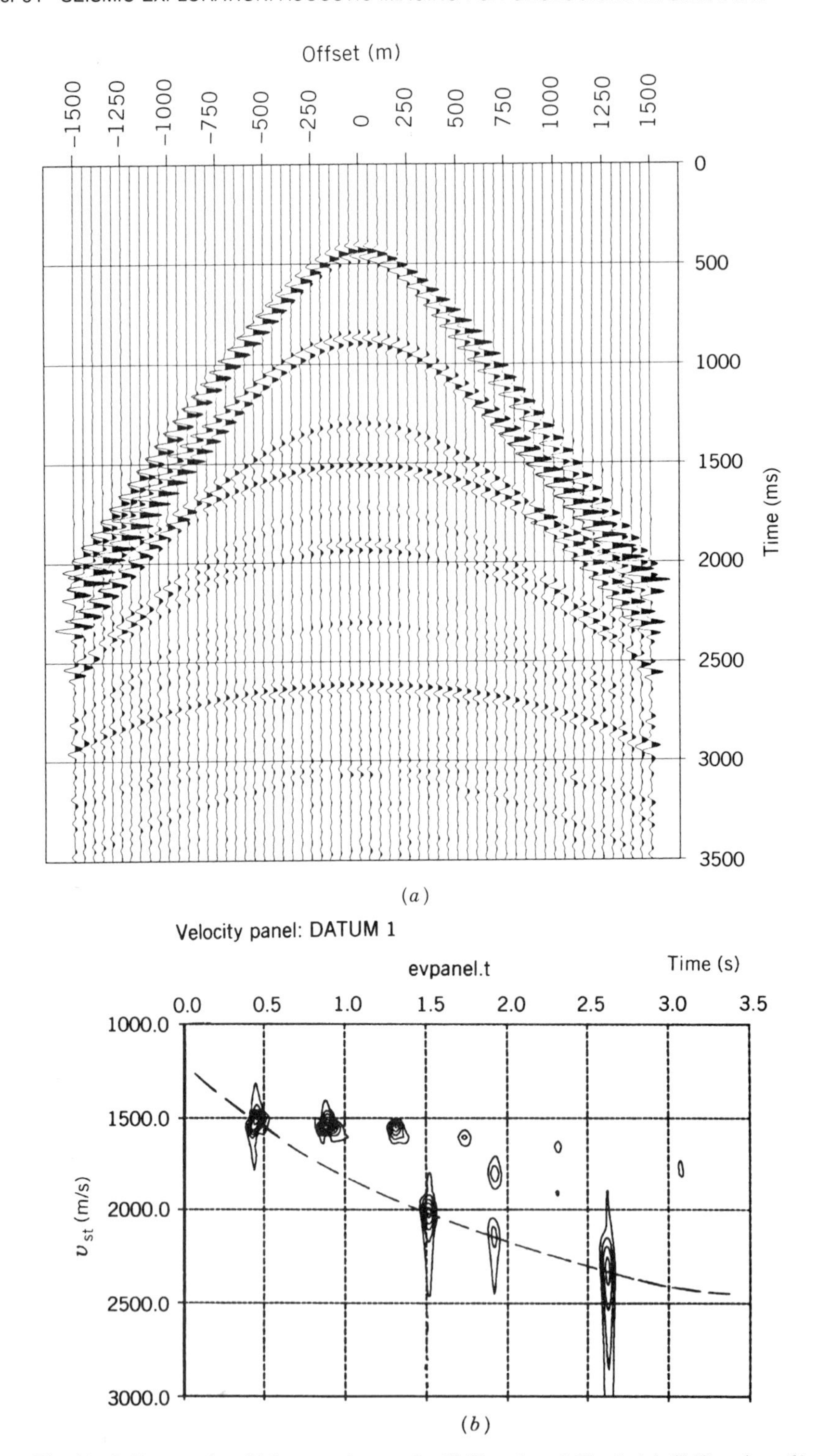

Fig. 12 Influence of multiple scattering on the CMP gather of Fig. 7: (*a*) CMP gather; (*b*) velocity analysis.

domain. Hence, conventional seismic processing consists of two consecutive focusing steps. The first focusing step translates multioffset measurements to zero-offset data (CMP stacking). The second focusing step translates zero-offset data to a subsurface image (zero-offset migration).

4 SEISMIC FORWARD MODEL

Today, the seismic exploration problem is approached from the *elastic wave theory* point of view: measurements are not considered as signals but as a superposition of waves. During preprocessing the different type of wave fields (surface wave field, multiple wave field, primary wave field) are separated, also referred to as wave field decomposition. Next, the pulse shape (*seismic wavelet*) of the primary wave field is shortened by model-based deconvolution and the output may be directly used in the image formation process (*multioffset migration*), yielding the structural information of the subsurface. In the target area the *angle-dependent* reflection property is estimated and used in the image characterization process (lithologic inversion), yielding the rock and pore information of the target.

In the following the seismic forward model is introduced in terms of a system of wave-theory-based operators.[8] We will see that the introduction of such a system makes the discussion on the seismic processing problem very straightforward.

4.1 Measurement Matrix

In practice acoustical measurements are always discrete in time and space. Consequently, imaging is always a discrete process and the theory should be discrete. Therefore, the forward model for seismic measurements is presented as a *discrete* model.

For linear wave theory in a time-invariant medium the problem may be described in the temporal frequency domain. Moreover, as our recording has a finite duration (T), we are dealing with a finite number of frequencies (N) per recorded signal (seismic trace), where

$$N = (f_{\max} - f_{\min})T,$$

$f_{\max} - f_{\min}$ being the temporal frequency range of interest. A typical number for N is 250. Taking into account the discrete property on the one hand and the allowed representation by independent frequency components on the other hand, vectors and matrices are preeminently suited for the mathematical description of seismic measurements. For instance, considering one seismic experiment (one source position), one element of the so-called measurement vector $\mathbf{P}(z_0)$ contains the complex number (defining amplitude and phase for the Fourier component under consideration) related to the recorded signal at one location of acquisition plane $z = z_0$ (one detector position). For a number of experiments the related measurement vectors can be grouped in a so-called measurement *matrix* (Fig. 13). Note that the main diagonal represents zero-offset data, the off-diagonals represent common offset data and the antidiagonals represent common midpoint data. In the next section the forward model will be derived to explain the data in the seismic measurement matrix. The symbol z will be used for depth.

4.2 Forward Model for Reflection Measurements

If the vector $\mathbf{S}^+(z_0)$ represents one Fourier component of the downward-traveling source wave field at the data acquisition surface $z = z_0$, then we may write

$$\mathbf{P}^+(z_m) = \boldsymbol{W}^+(z_m, z_0)\mathbf{S}^+(z_0), \tag{2}$$

where the vector $\mathbf{P}^+(z_m)$ represents the monochromatic downward-traveling source wave field at depth level z_m and $\boldsymbol{W}^+(z_m, z_0)$ defines the downward propagation operator from z_0 to z_m. Operator $\boldsymbol{W}^+$ is represented by a complex-valued matrix, where each column equals one Fourier component of the response at depth level z_m due to one dipole at the surface. Note that for one-dimensional media (no lateral variations) $\boldsymbol{W}^+$ becomes a convolution matrix.

At depth level z_m reflection occurs. For each Fourier component reflection at z_m may be described by a general linear operator $\boldsymbol{R}^+(z_m)$,

$$\mathbf{P}_m^-(z_m) = \boldsymbol{R}^+(z_m)\mathbf{P}^+(z_m), \tag{3}$$

where the vector $\mathbf{P}_m^-(z_m)$ represents the monochromatic upward-traveling reflected wave field at depth level z_m due to the inhomogeneities at depth level z_m only. Reflection operator $\boldsymbol{R}^+(z_m)$ is represented by a matrix, where each column describes the monochromatic angle-dependent reflection property of each grid point at z_m. If there exists no angle dependence, $\boldsymbol{R}^+(z_m)$ is a diagonal matrix with angle-independent reflection coefficients. Finally, the reflected wave field at z_m travels up to the surface,

$$\mathbf{P}_m^-(z_0) = \boldsymbol{W}^-(z_0, z_m)\mathbf{P}_m^-(z_m). \tag{4}$$

In Eq. (4) vector $\mathbf{P}_m^-(z_0)$ represents one Fourier component of the reflected wave field at data acquisition surface z_0 and $\boldsymbol{W}^-(z_0, z_m)$ defines the upward propagation operator from z_m to z_0. Each column of the matrix $\boldsymbol{W}^-$ contains

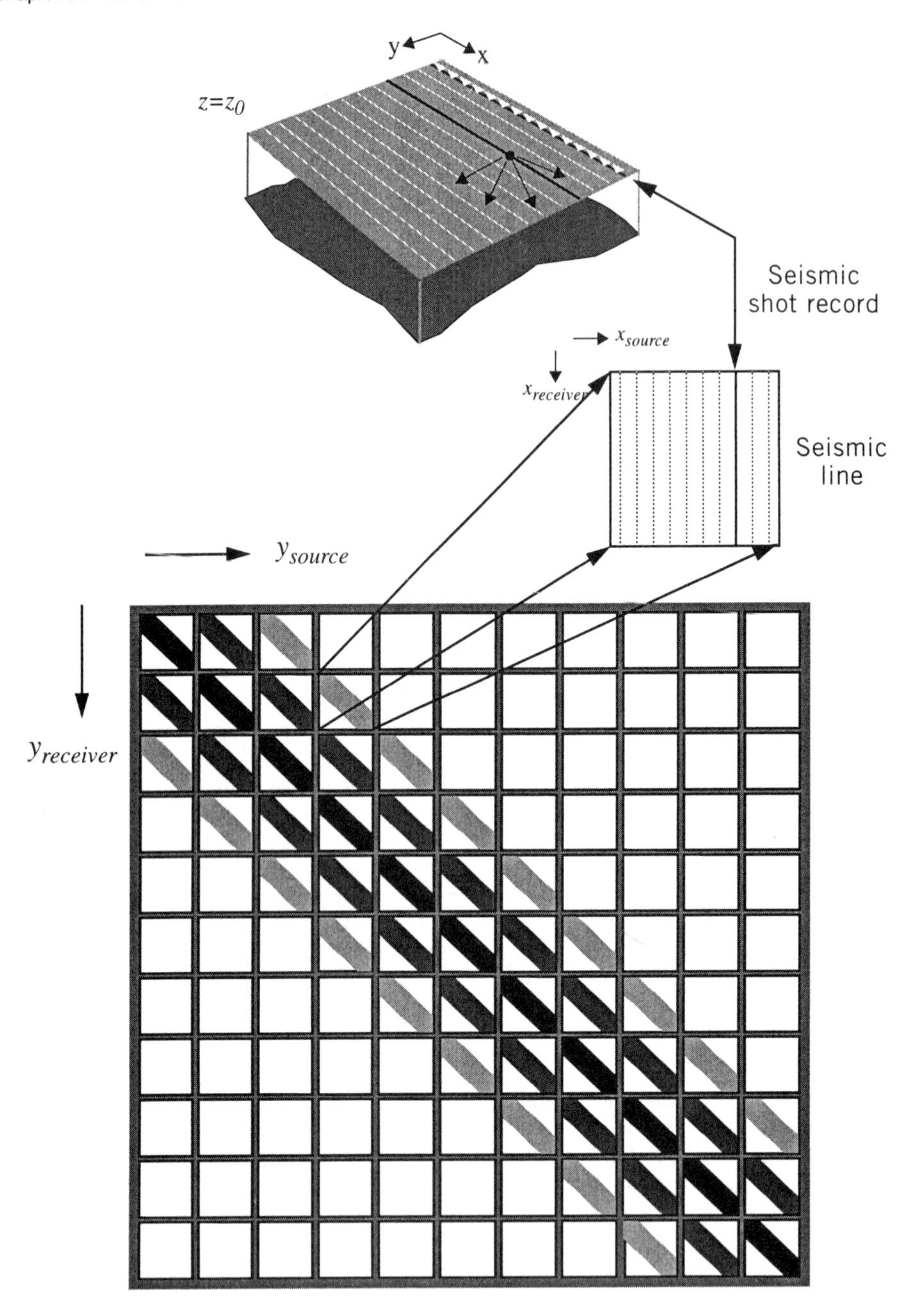

Multi line measurement matrix;
for $Y_{source} \neq Y_{receiver}$ the source has an offset with respect to the receiver line.

Fig. 13 Data matrix for three-dimensional seismic measurements. One column represents one multistreamer shot record. For an irregular dataset the columns are irregularly filled.

one Fourier component of the response at z_0 due to one dipole at depth level z_m.

Expressions (2), (3), and (4) may now be combined to one matrix equation for the reflection response:

$$\begin{aligned}\mathbf{P}^-(z_0) &= \sum_{m=1}^{\infty} \mathbf{P}_m^-(z_0)\\ &= \sum_{m=1}^{\infty} \boldsymbol{W}^-(z_0, z_m)\mathbf{P}_m^-(z_0)\\ &= \sum_{m=1}^{\infty} \boldsymbol{W}^-(z_0, z_m)\boldsymbol{R}^+(z_m)\mathbf{P}^+(z_m)\\ &= \sum_{m=1}^{\infty} [\boldsymbol{W}^-(z_0, z_m)\boldsymbol{R}^+(z_m)\\ &\quad \cdot \boldsymbol{W}^+(z_m, z_0)]\mathbf{S}^+(z_0),\end{aligned} \tag{5a}$$

or, for a continuous formulation in z,

$$\mathbf{P}^-(z_0) = \int_{z_0}^{\infty} [\boldsymbol{W}^-(z_0, z)\boldsymbol{R}^+(z)\boldsymbol{W}^+(z, z_0)dz]\mathbf{S}^+(z_0). \tag{5b}$$

Note that in expression (5) the surface reflectivity is not yet included $[\mathbf{R}^+(z_0) = 0]$. Note also that Eq. (5) contains the primary reflection response only; multiples and surface waves have been omitted.

If we define the half-space reflection operator at depth level z_m due to inhomogeneities at $z \geq z_m$ by matrix $\boldsymbol{X}^+(z_m, z_m)$, then it follows from Eq. (5a) that

$$\boldsymbol{X}^+(z_m, z_m) = \sum_{n=m}^{\infty} \boldsymbol{W}^-(z_m, z_n)\boldsymbol{R}^+(z_n)\boldsymbol{W}^+(z_n, z_m), \tag{6a}$$

where $m = 1, 2, \ldots$.

Matrix $\boldsymbol{X}^+(z_m, z_m)$ may be considered as a multidimensional *transfer function*; matrix element $\mathbf{X}_{ij}^+(z_m, z_m)$ represents one Fourier component of the reflection response at position i on surface $z = z_m$, due to a unit point source at position j on the same surface ($z = z_m$).

Finally, if we define

$$\boldsymbol{X}_0^+(z_m, z_m) = \sum_{n=m+1}^{\infty} \boldsymbol{W}^-(z_m, z_n)\boldsymbol{R}^+(z_n)\boldsymbol{W}^+(z_n, z_m), \tag{6b}$$

then it follows from Eq. (5) that

$$\mathbf{P}^-(z_0) = \boldsymbol{X}_0^+(z_0, z_0)\mathbf{S}^+(z_0). \tag{6c}$$

4.3 Including the Acquisition Surface

To expressions (5) the relations should be added between the *induced-source function* and the downgoing wave field on the one hand and the recorded detector signals and the upgoing wave field on the other hand:

$$\mathbf{S}^+(z_0) = \boldsymbol{D}^+(z_0)\mathbf{S}(z_0), \tag{7a}$$

$$\mathbf{P}(z_0) = \boldsymbol{D}^-(z_0)\mathbf{P}^-(z_0). \tag{7b}$$

The matrix operators $\boldsymbol{D}^+$ and $\boldsymbol{D}^-$ represent the local interaction with the surface, the directivity properties due to the array effects, and the type of sources and detectors, respectively.

In the seismic situation the data acquisition surface is the earth's surface. It is highly reflective and its significant influence on the seismic response can be easily included in the forward model by replacing in expression (5) downgoing source wave field $\mathbf{S}^+(z_0)$ by

$$\mathbf{P}^+(z_0) = \boldsymbol{R}^-(z_0)\mathbf{P}^-(z_0) + \mathbf{S}^+(z_0), \tag{7c}$$

where the vector $\mathbf{P}^+(z_0)$ represents the *total downgoing* wave field at the surface and matrix $\boldsymbol{R}^-(z_0)$ defines the reflectivity of the surface for *upgoing* waves (Fig. 14). Expression (7c) formulates a special (and for practical seismic situations the most important) form of multiple scattering.

Finally, if we include noise in the measurements, then Eq. (7b) needs to be generalized to

$$\mathbf{P}(z_0) = \boldsymbol{D}^-(z_0)\mathbf{P}^-(z_0) + \mathbf{N}(z_0). \tag{7d}$$

Here, noise is defined as all events that are not described by the model equations (e.g., waves that travel from source to detectors *along* the surface).

In Fig. 14 the foregoing theory is summarized in a diagram, showing the involved operators and data flow. Not

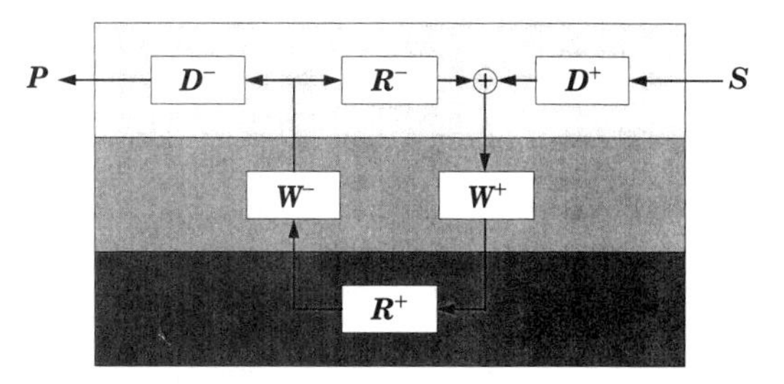

Fig. 14 Conceptual model of seismic data, consisting of surface operators ($\boldsymbol{D}^+$, $\boldsymbol{R}^-$, $\boldsymbol{D}^-$), propagation operators ($\boldsymbol{W}^+$, $\boldsymbol{W}^-$), and reflectivity operator $\boldsymbol{R}^+$.

only does the presented *conceptual* view on the problem yield a concise formulation but also the involved expressions provides theoretical detail. In addition, the involved basic algorithms are included if the operators are represented by matrices with predefined structure (Fig. 13). For instance, multiplication of a seismic dataset with the matrix operator $\boldsymbol{W}^+$ (or $\boldsymbol{W}^-$) defines a *forward* wave field extrapolation process, the (optimized) extrapolation operator for lateral position x_k being represented by the kth column of $\boldsymbol{W}^+$ (or $\boldsymbol{W}^-$). Similarly, multiplication of a seismic dataset with matrix operator $\boldsymbol{R}^+$ defines an angle-dependent reflection process, the (optimized) reflection operator for lateral position x_k being represented by the kth column of $\boldsymbol{R}^+$. In the opinion of the author, conceptual formulations are the way that experts of different disciplines should communicate with each other in interdisciplinary project teams. Moreover, as conceptual formulations are principally different from simplified ones, conceptual formulations appear be very fruitful for the development of new algorithms.[8] Examples of particular interest are surface-related multiple removal[9] and controlled illumination.[10]

4.4 Multiexperiment Data Sets

So far, we have formulated the equations for one experiment. However, our matrix formulation is very suited to represent multiexperiment data acquisition:

$$\boldsymbol{S}^+(z_0) = \boldsymbol{D}^+(z_0)\boldsymbol{S}(z_0), \tag{8a}$$

$$\boldsymbol{P}^+(z_0) = \boldsymbol{R}^-(z_0)\boldsymbol{P}^-(z_0) + \boldsymbol{S}^+(z_0), \tag{8b}$$

$$\boldsymbol{P}^-(z_0) = \sum_{m=1}^{\infty} [\boldsymbol{W}^-(z_0, z_m)\boldsymbol{R}^+(z_m) \cdot \boldsymbol{W}^+(z_m, z_0)]\boldsymbol{P}^+(z_0), \tag{8c}$$

$$\boldsymbol{P}(z_0) = \boldsymbol{D}^-(z_0)\boldsymbol{P}^-(z_0) + \boldsymbol{N}(z_0), \tag{8d}$$

where one column of the source matrix $\boldsymbol{S}(z_0)$ defines the induced source function of one monochromatic experiment and the related column of the measurement matrix $\boldsymbol{P}(z_0)$ defines the monochromatic versions of the measured signals (seismic traces) of that experiment.

If the source directivity and the surface-related multiple scattering have been removed, then we may write $\boldsymbol{P}^+(z_0) = S(\omega)\boldsymbol{I}$, and it can be easily verified from Eq. (8c) that the elements of $\boldsymbol{P}^-(z_0)$ can be written as

$$P_{ij}^-(z_0) = S(\omega) \sum_m \sum_k \sum_l \cdot W_{ik}^-(z_0, z_m)R_{kl}^+(z_m)W_{lj}^+(z_m, z_0), \tag{9a}$$

$P_{ij}^-(z_0)$ being the reflected wave field at surface position i due to a dipole at surface position j. Hence for zero-offset data ($i = j$), where each grid point may be characterized by a zero-offset reflection coefficient ($k = l$), we may write

$$P_{ii}^-(z_0) = S(\omega) \sum_{m=1}^{\infty} \sum_k W_{ik}(z_0, z_m)R_{kk}^+(z_m). \tag{9b}$$

In Eq. (9b) we have defined

$$W_{ik}(z_0, z_m) = W_{ik}^-(z_0, z_m)W_{ki}^+(z_m, z_0) \tag{9c}$$

and $R_{kk}^+(z_m)$ equals the zero-offset reflection coefficient at grid point x_k of depth level z_m.

4.5 Formulation for both P- and S-Waves

As mentioned before, seismic wave fields contain P as well as S reflections and, therefore, the seismic forward model should include both types of waves. This is particularly important for land data. We will see that the forward model (8) can be elegantly extended to address both P and S reflections:

(a) Each scalar element of $\boldsymbol{S}(z_0)$ should be replaced by a vector element representing the source stress components τ_{xz}, τ_{yz}, and τ_{zz}.

(b) Each scalar element of $\boldsymbol{P}(z_0)$ should be replaced by a vector element representing the detector particle velocity components v_x, v_y, v_z.

(c) Each scalar element of $\boldsymbol{S}^+(z_0)$ should be replaced by a vector element representing the P and S potentials ϕ^+, Ψ_x^+, Ψ_y^+ at z_0. This means that surface operator $\boldsymbol{D}^-(z_0)$ decomposes the downgoing source wave field in separate P and S contributions as well.

(d) Each scalar element of $\boldsymbol{P}^-(z_0)$ should be replaced by a vector element, representing the P and S potentials ϕ^-, Ψ_x^-, Ψ_y^- at z_0. This means that surface operator $\boldsymbol{D}^-(z_0)$ composes the upgoing P and S wave fields in particle velocity components as well.

Taking the above generalizations of Eqs. (8a) and (8d) into account, expressions (8b) and (8c) can be easily generalized as well. For simplicity the two-dimensional situation is formulated:

$$\underset{\substack{\downarrow\\ \boldsymbol{P}^+(z_0)}}{\begin{bmatrix}\phi^+\\ \Psi^+\end{bmatrix}} = \underset{\substack{\downarrow\\ \boldsymbol{R}^-(z_0)}}{\begin{bmatrix}\boldsymbol{R}_{pp}^- & \boldsymbol{R}_{ps}^-\\ \boldsymbol{R}_{sp}^- & \boldsymbol{R}_{ss}^-\end{bmatrix}} \underset{\substack{\downarrow\\ \boldsymbol{P}^-(z_0)}}{\begin{bmatrix}\phi^-\\ \Psi^-\end{bmatrix}} + \underset{\substack{\downarrow\\ \boldsymbol{S}^+(z_0)}}{\begin{bmatrix}S_\phi^+\\ S_\Psi^+\end{bmatrix}}, \tag{10a}$$

$$\underset{\boldsymbol{P}^-(z_0)}{\begin{bmatrix}\phi^-\\ \Psi^-\end{bmatrix}} = \sum_{m=1}^{\infty} \underset{\boldsymbol{W}^-(z_0, z_m)}{\begin{bmatrix}\boldsymbol{W}_{pp}^- & \boldsymbol{W}_{ps}^-\\ \boldsymbol{W}_{sp}^- & \boldsymbol{W}_{ss}^-\end{bmatrix}} \underset{\boldsymbol{R}^+(z_m)}{\begin{bmatrix}\boldsymbol{R}_{pp}^+ & \boldsymbol{R}_{ps}^+\\ \boldsymbol{R}_{sp}^+ & \boldsymbol{R}_{ss}^+\end{bmatrix}} \cdot \underset{\boldsymbol{W}^+(z_m, z_0)}{\begin{bmatrix}\boldsymbol{W}_{pp}^+ & \boldsymbol{W}_{ps}^+\\ \boldsymbol{W}_{sp}^+ & \boldsymbol{W}_{ss}^+\end{bmatrix}} \underset{\boldsymbol{P}^+(z_0)}{\begin{bmatrix}\phi^+\\ \Psi^+\end{bmatrix}}. \tag{10b}$$

Note that the matrices $\boldsymbol{R}_{sp}$ and $\boldsymbol{R}_{ps}$ quantify wave type conversion during *reflection*. In many cases it is acceptable to neglect wave type conversion during *propagation*, meaning that $\boldsymbol{W}_{sp}$ and $\boldsymbol{W}_{ps}$ may be neglected. Note also that in seismic marine exploration $\Psi = 0$ in the surface layer and receivers detect pressure just below z_0.

In conventional seismic land acquisition, stress sources generate τ_{zz} only (vertical vibrator) and detectors measure v_z only (single-component geophone). In the common situation of a low velocity surface layer ray paths are near to vertical at the detectors and, therefore, *vertical* component geophones measure mainly *compressional* reflection energy.

It can be easily verified that if $\boldsymbol{W}_{sp}$ and $\boldsymbol{W}_{ps}$ are neglected, Eq. (10b) can be written as four independent equations (two without conversion and two with conversion at reflection). This has the important practical advantage that the multicomponent *elastic* problem can be formulated in terms of four (two-dimensional) or nine (three-dimensional) independent single-component *acoustic* problems.

5 SEISMIC IMAGING

In the *forward* problem all details about the data acquisition procedure are known, the acoustic properties of the surface and medium (trend and detail) are available, and the multioffset measurements need to be computed (numerical simulation). In case we start with reflectivity, simulation means

$$\boldsymbol{R}^+(z) \rightarrow \boldsymbol{X}_0^+(z_0, z_0) \rightarrow \boldsymbol{P}(z_0).$$

In the *inverse* problem, all details about the data acquisition procedure should be known, the measurements are available, and the medium parameters need to be computed. If the spatial reflectivity distribution is aimed for, inversion means multioffset preprocessing and multioffset image formation:

$$\boldsymbol{P}(z_0) \rightarrow \boldsymbol{X}_0^+(z_0, z_0) \rightarrow \boldsymbol{R}^+(z).$$

Generally, in image formation the *diagonal* elements of $\boldsymbol{R}^+(z)$ are computed only, meaning that the angle dependence information of reflection is not aimed for (one reflection coefficient per medium grid point).

If all elements of the reflection matrix are computed, then for each grid point angle-dependent reflection information is available and image *formation* can be followed by image *characterization*:

$$\boldsymbol{R}^+(z) \rightarrow \text{lithology}.$$

Lithologic information includes types of rock (sand, shale, etc.), types of pore fill (water, oil, gas), porosity, and permeability.

Hence, full seismic processing consists of the three hierarchical inversion steps, i.e., multioffset preprocessing, multioffset image formation, and multioffset image characterization:

$$\boldsymbol{P}(z_0) \rightarrow \boldsymbol{X}_0^+(z_0, z_0) \rightarrow \boldsymbol{R}^+(z) \rightarrow \text{lithology}.$$

5.1 Preprocessing

From the forward model (8) it can be seen that if both the total measurement matrix and the total source matrix are available (multisource, multioffset data acquisition), then the transfer matrix $\boldsymbol{X}_0^+(z_0, z_0)$ can be estimated by removing the following surface-related influences from the measurement matrix $\boldsymbol{P}(z_0)$:

(a) Surface wave field $\boldsymbol{N}(z_0)$
(b) Transducer operators $\boldsymbol{D}^-(z_0)$, $\boldsymbol{D}^+(z_0)$
(c) Induced source function $S(\omega)$
(d) Multiple scattering operator $\boldsymbol{R}^-(z_0)$.

Removal of the surface wave field (particularly for land measurements) occurs with the least-squares linear radon transformation. For details the reader is referred to Ref. 11.

Removal of transducer operators $\boldsymbol{D}^+(z_0)$ and $\boldsymbol{D}^-(z_0)$ transforms the source functions into *downgoing* wave fields at z_0 and the measurements into *upgoing* reflected wave fields at z_0. In addition, directivity effects are compensated for. For details see Ref. 12.

Removal of $S(\omega)$ is referred to as seismic source deconvolution; it means pulse compression, improving the seismic resolution along the time axis. Seismic source deconvolution is limited by the signal-to-noise ratio and the accuracy of the source signal estimate (see Refs. 13 and 14).

Removal of $\boldsymbol{R}^-(z_0)$ means elimination of surface-related multiple scattering. From Eq. (8) it can be easily verified that the elimination process is defined by

$$\boldsymbol{P}_0(z_0) = \boldsymbol{P}(z_0)[\boldsymbol{I} + \boldsymbol{A}(z_0)\boldsymbol{P}(z_0)]^{-1}, \tag{11a}$$

$\boldsymbol{P}_0(z_0)$ representing the seismic data without surface-related multiples and

$$\boldsymbol{A}(z_0) = [\boldsymbol{S}^+(z_0)]^{-1}\boldsymbol{R}^-(z_0)[\boldsymbol{D}^-(z_0)]^{-1}. \tag{11b}$$

For details the reader is referred to Ref. 9

Figure 15 shows a result of preprocessing. The broadband version of one column of the matrix $\boldsymbol{P}(z_0)$ is shown before and after preprocessing (data from the North Sea). Note the significant amount of multiple scattering energy.

5.2 Image Formation

Removal of propagation operators $\boldsymbol{W}^+(z_m, z_0)$ and $\boldsymbol{W}^-(z_0, z_m)$ transforms $\boldsymbol{X}_0^+(z_0, z_0)$ to $\boldsymbol{X}^+(z_m, z_m)$; it means simulation of the data acquisition process at depth level z_m, $m = 1, 2, \ldots$. In the seismic literature this numerical process is referred to as *downward extrapolation*. From $\boldsymbol{X}^+(z_m, z_m)$ the reflection matrix $\boldsymbol{R}^+(z_m)$ can be easily extracted. If the diagonal elements of $\boldsymbol{R}^+(z_m)$ are computed only, then the twofold downward extrapolation process can be rewritten as double numerical focusing, simulating dynamic focusing in emission and dynamic focusing in detection for each grid point at depth level z_m. This is exactly the process that is carried out in multioffset migration. The multioffset migration process can be considered as the most advanced image formation process, addressing both the multisource and multioffset problems and avoiding the simplifying assumptions of CMP stacking in conventional seismic processing. For further reading see Refs. 8 and 15–17.

Figure 16 shows a result of the multioffset migration process using all columns of the preprocessed North Sea measurement matrix. Note the well-defined faults at a depth of 3 km.

5.3 Image Characterization

In the last inversion step of seismic processing angle-dependent reflection information, as given by $\boldsymbol{R}^+(z)$, is

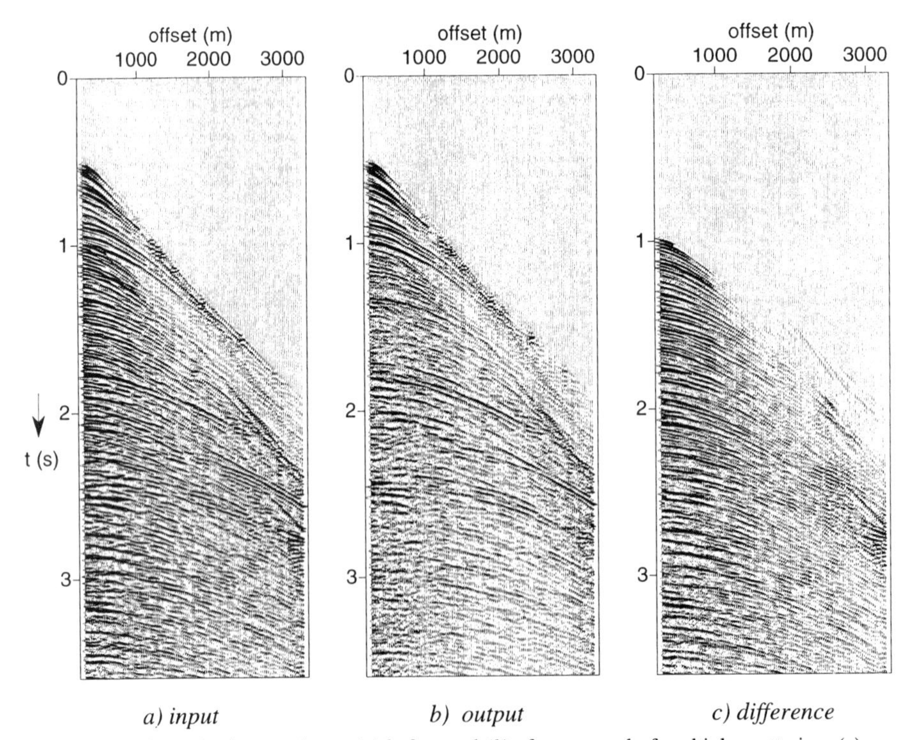

Fig. 15 One seismic experiment (*a*) before and (*b*) after removal of multiple scattering. (*c*) Note the significant amount of multiple scattering energy.

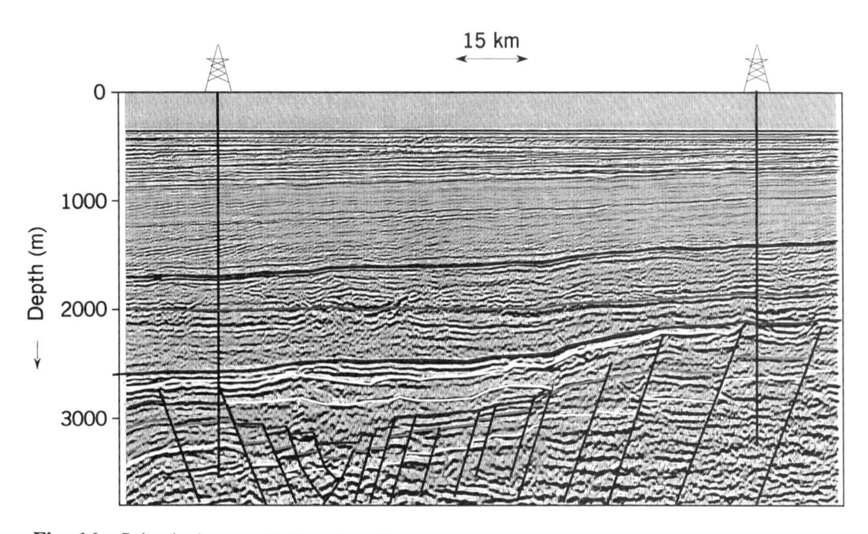

Fig. 16 Seismic image of the subsurface after preprocessing and multisource-multioffset image formation. Note that the latter may be formulated as a double focusing process: focusing in emission and focusing in detection.

inverted to a layered geological model, each layer being characterized by its rock and pore parameters. For this inversion step three forward models are used:

(a) Boundary conditions for angle-dependent reflection, e.g., as given by the Zoeppritz equations for solids[18]

(b) Gassmann equation for the P and S velocities of porous rocks[2]

(c) Density equation for porous rocks.

Using a constrained data-fitting technique,[19] the parameters of the forward models are adjusted such that the seismic reflectivity image(s) are explained as well as possible. Experience of seismic inversion for litho-parameters shows that the problem is seriously ill-posed. Success can only be realized if other, nonseismic, information is included as well (multidisciplinary approach).

Lithologic inversion needs, apart from seismic data, borehole information and a range of plausible geological models. The inversion process chooses the most likely one, together with limits of confidence.[20] Lithologic inversion is still largely in the research phase.

REFERENCES

1. A. J. Berkhout, *Seismic Resolution: Resolving Power of Acoustic Echo Techniques*, Geophysical Press, London, 1984.
2. F. Gassmann, "Elastic Waves through a Packing of Spheres," *Geophysics*, Vol. 16, 1951, pp. 673–685.
3. M. A. Biot, "Theory of Propagation of Elastic Waves in a Fluid Saturated Porous Solid," *J. Acoust. Soc. Am.*, Vol. 28, 1956, pp. 168–178.
4. O. Yilmaz, *Seismic Data Processing*, Society of Exploration Geophysicists, Tulsa, OK, 1987.
5. M. T. Taner and F. Koehler, "Velocity Spectra—Digital Computer Derivation and Application of Velocity Functions," *Geophysics*, Vol. 34, 1969, pp. 859–881.
6. D. Hampson, "Inverse Velocity Stacking for Multiple Elimination," *J. CSEG*, Vol. 22, No. 1, 1986, pp. 44–55.
7. P. Hubral, "Interval Velocities from Surface Measurements in the Three Dimensional Plane Layer Case," *Geophysics*, Vol. 41, 1976, pp. 223–242.
8. A. J. Berkhout, *Seismic Migration: Imaging of Acoustic Energy by Wave Field Extrapolation. A. Theoretical Aspect*, 3rd ed., Elsevier Science, Amsterdam, 1985.
9. D. J. Verschuur, "Surface-Related Multiple Elimination: An Inversion Approach," Ph.D. Thesis, 1991.
10. W. E. A. Rietveld, A. J. Berkhout, and C. P. A. Wapenaar, "Optimum Seismic Illumination of Hydrocarbon Reservoirs," *Geophysics*, Vol. 57, 1992, pp. 1334–1345.
11. D. J. Fyfe and P. G. Kelamis, "Removing Coherent Noise Using the Linear Radon Transform: Expanded Abstracts," 54th Annual Meeting of SEG, 1992, pp. 1465–1468.
12. C. P. A. Wapenaar and A. J. Berkhout, *Elastic Wave Field Extrapolation*, Elsevier Science, New York, 1989.
13. A. J. Berkhout, "Least-Squares Inverse Filtering and Wavelet Deconvolution," *Geophysics*, Vol. 42, No. 7, pp. 1369–1383.

14. A. Ziolkowski, *Seismic Deconvolution*, IHRDC, Boston, 1984.
15. J. F. Claerbout, *Fundamentals of Seismic Processing*, Blackwell Scientific.
16. R. H. Stolt and A. K. Benson, *Seismic Migration, Theory and Practice*, Geophysical Press, London, 1988.
17. A. J. Berkhout, "A Unified Approach to Acoustical Reflection Imaging, Part I: The Forward Model," *J. Acoust. Soc. Am.*, Vol. 93, No. 4, Pt. 1, 1993, pp. 2005–2017.
18. K. Aki and P. G. Richards, *Quantitative Seismology*, Freeman, San Francisco, 1980.
19. A. Tarantola, *Inverse Problem Theory, Method for Data Fitting and Model Parameter Estimation*, Elsevier Science, Amsterdam, 1987.
20. G. J. Lörtzer and A. J. Berkhout, "An Integrated Approach to Lithologic Inversion. Part I: Theory," *Geophysics*, Vol. 57, 1992, pp. 233–245.

55

VIBRATION ISOLATION AND DAMPING

ERIC E. UNGAR

1 INTRODUCTION

Isolation and structural damping constitute the two most widely applicable means for the control of vibration or structure-borne sound, particularly in the audio frequency range. This chapter delineates the most important considerations pertaining to passive vibration isolation and damping; active vibration control—achievement of the effects of isolation and damping by means of control systems and sources of energy that are external to the vibrating system—is covered in Chapter 59.

Vibration isolation in essence involves use of a resilient connection between a source of vibration and an item to be protected, so that this item vibrates less than it would if a rigid connection were used. In some typical situations the source consists of a vibrating machine or structure and the item to be protected is a supporting or connecting structure; in other typical situations the source consists of a vibrating structure and the item to be protected is a sensitive piece of equipment supported from that structure. Many salient features of vibration isolation can be analyzed in terms of a simple model consisting of a rigid mass that is connected to a support via a massless linear spring and that is constrained to translate along a single axis. The mass here may represent either the body of a vibration producing machine or a sensitive item to be protected from support vibrations, with the spring representing the isolator. More complex models are needed to address situations where the magnitude of the excitation depends on the motions, where an additional spring–mass system is inserted between the primary one and the support, and/or at comparatively high frequencies where the isolator mass plays a significant role or where the isolated items do not behave as rigid masses. Other complications arise because of non-uniaxial motions and nonlinearities.

Damping refers to the removal of mechanical energy from a vibration of concern. This removal may result from transfer of energy to structural components, fluids, or vibration modes that are not of concern or from conversion of mechanical energy into other forms. Damping affects the decay of free (unforced) vibrations and of those forced responses of systems that are controlled by energy balances. Various measures of damping are in use; most are based on observation of the effects that damping has on the motions of simple systems. Data on the damping inherent in materials cover a wide range, extending from comparatively low damping for high-strength structural materials to very high damping for some viscoelastic materials (typically, plastics or elastomers) with limited strength. Structural components that are acceptably strong and that also exhibit relatively high damping may be obtained by combining high-damping viscoelastic materials with structural materials in the form of added layers or in sandwich arrangements.

2 VIBRATION ISOLATION

2.1 Linear Model with Single Degree of Freedom

The basic idealized system of Fig. 1 represents the simplest model on the basis of which vibration isolation has been analyzed. The combination of the linear spring of stiffness (force/displacement) k and the dashpot with viscous damping coefficient (force/velocity) b represents an isolator (or an array of isolators); the isolator is taken to be massless and the mass m is assumed to be rigid

Note: References to chapters appearing only in the *Encyclopedia* are preceded by *Enc.*

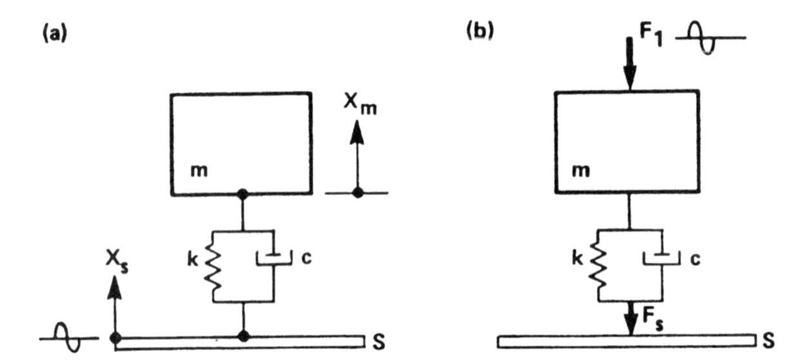

Fig. 1 Mass–spring–dashpot system: (*a*) excited by support motion; (*b*) excited by force acting on mass.

and to be capable of only vertical, purely translational motion. In the case of Fig. 1*a*, where the support is taken to vibrate vertically with a prescribed displacement amplitude X_s at a given frequency, the mass represents an item to be protected from support-induced vibrations. In the case of Fig. 1*b*, where a force of a prescribed amplitude F_1 at a given frequency is taken to act on the mass, the mass represents the frame of a machine from whose vibrations the support is to be protected. The equations of motion for this system are given in Chapter 48 and Ref. 2.

Transmissibility and Isolation Efficiency For the case of support-induced vibrations, the (dimensionless) motion transmissibility $T_v = X_m/X_s$ relates the displacement amplitude X_m of the oscillation of the mass (at the support motion frequency) to the displacement amplitude X_s of the support vibration. For the case where an oscillatory force acts on the mass, the (dimensionless) force transmissibility $T_f = F_s/F_1$ relates the amplitude F_s of the force that the isolator exerts on the support to the amplitude F_1 of the exciting force. If the support in Fig. 1*b* is immobile, then the two transmissibilities are given by the same expression,* namely

$$T_v = T_f = T = \left[\frac{1 + \eta^2}{(1 - r^2)^2 + \eta^2} \right]^{1/2}, \qquad (1)$$

where $r = f/f_n$ is the ratio of the excitation frequency f to the natural frequency f_n of the spring–mass system and η denotes the loss factor of the isolator (see below).

*Equality of the two transmissibilities, which refer to two different physical situations, is a consequence of the reciprocity principle and holds for any linear uniaxial system, no matter how complex,[1] and not just for the single-degree-of-freedom system of Fig. 1.

The natural frequency f_n (in hertz) obeys

$$f_n = \frac{1}{2\pi}\sqrt{\frac{k}{m}} = \frac{1}{2\pi}\sqrt{\frac{g}{X_{st}}} \approx \frac{3.13}{\sqrt{X_{st}\,(\text{in.})}}$$
$$= 15.7/\sqrt{X_{st}\,(\text{mm})} \qquad (2)$$

where g denotes the acceleration of gravity and X_{st} represents the static deflection induced in the isolator (spring) by the mass resting on it.

The (dimensionless) loss factor $\eta = D/2\pi W$ relates the energy D that the isolator dissipates per cycle in the steady state to the timewise maximum strain energy W stored in the isolator in a cycle. Although η may vary with amplitude and frequency in general, it is approximately constant for most metals and in certain limited frequency and temperature ranges also for plastics and elastomers. For viscous damping, in which motion is opposed by a force that is proportional to velocity, the loss factor obeys $\eta = 2rb/b_c$, where b represents the viscous damping coefficient (the constant relating the retarding force to velocity) and $b_c = 2\sqrt{km}$ denotes the so-called critical damping coefficient. See also Ref. 2 and Section 1 in Chapter 48.

Since transmissibility represents the fraction of the disturbing motion or force that is transmitted through the isolator, good isolation (i.e., large vibration reduction) corresponds to small values of transmissibility. A measure of isolation performance that takes on greater values for better isolation is the dimensionless *isolation efficiency* $I = 1 - T$, which represents the fraction of the disturbance that is *not* transmitted. The isolation efficiency generally is expressed in percent. For example, a transmissibility of 0.0073 corresponds to an isolation efficiency of 0.9927, or 99.27%.

From Eq. (1), it is evident that for small damping

($\eta^2 \ll 1$), which usually is of greatest practical interest,

$$T \approx 1 \qquad \text{for } r \ll 1, \tag{3a}$$

$$T \approx \frac{1}{\eta} \qquad \text{for } r \approx 1, \tag{3b}$$

$$T \approx \frac{1}{r^2} \text{ or } \frac{2(b/b_c)}{r} \qquad \text{for } r \gg 1, \tag{3c}$$

where the first expression of Eq. (3c) applies for the case of a constant, frequency-independent loss factor, and the second applies for viscous damping, in which the loss factor is proportional to frequency.

In order to obtain good isolation, one generally wants a system that operates in the range where $f \gg f_n$; one thus wants to make the natural frequency f_n as small as possible compared to the excitation frequency f. In view of the various complications that limit the applicability of Eq. (1), as discussed later, transmissibilities that are smaller than $\frac{1}{100}$ over extended frequency ranges can be obtained only rarely with practical conventional simple isolation systems. Reduced transmissibility in the resonance region ($f \approx f_n$) can be achieved only by increasing the damping, that is, the loss factor η.

Effect of Damping. Although Eqs. (1) and (3c) indicate that increased *viscous* damping may result in significant increases in transmissibility at high frequencies, this fact tends to be of little practical importance, not only because the damping of practical isolators generally is relatively small, but also because their damping behavior typically corresponds more nearly to hysteretic damping (as characterized by a frequency-independent loss factor) than to viscous damping.

Machines with Unbalance.[2] In a rotating or reciprocating machine, the dominant excitation force F_u generally is due to unbalance. This force occurs at the frequency f that corresponds to the machine's rotational speed and is given by $F_u = m_u e_u (2\pi f)^2$, where m_u denotes the unbalance mass and e_u its eccentricity (i.e., its distance from the axis of rotation). For such a frequency-dependent force, the force that acts on the structure that supports the machine is given by $F_s = F_u T_F$. At frequencies f that are much greater than f_n, this force approaches $F_s = k e_u m_u / m$. In contrast to the situation where the excitation force is frequency independent, the force transmitted to the support here does not decrease with increasing excitation frequency; the improvement of isolation with increasing frequency in essence is canceled by the increase in the excitation force.

Inertia Bases Vibration-producing machines are often mounted on massive supports, called "inertia bases," to increase the isolated mass. If the total stiffness of the isolators that support the machine is not changed as the isolated mass is increased, then the natural frequency f_n of the system is reduced and the force transmissibility at high frequencies is reduced accordingly. However, limitations on the loads that practical isolators can support often make it necessary to employ more and/or stiffer isolators if an inertia base is added, thus limiting the natural frequency changes and the transmissibility reductions achievable by the addition of inertia bases. However, the increased isolated mass still serves to reduce the vibrational excursions of the machine produced by the forces that act on it.

Limitations of Linear Spring–Mass Model with Prescribed Excitation The classical linear spring–mass model of Fig. 1 provides some insights but does not account for all practical physical phenomena. Real isolators are not massless, may exhibit internal resonances, and may be nonlinear. Real machine frames and supporting structures are not rigid; they may deflect, exhibit multimodal responses and resonances, and rock or otherwise move along paths that are more complex than simple straight lines. In classical analyses, the magnitude of the excitation is taken as independent of the resulting motion, whereas in real vibration sources the motions and forces they produce are interdependent. Some of the effects of these complications are discussed in the following sections of this chapter.

2.2 Isolation of Three-Dimensional Mass

In the simple model of Fig. 1 the mass is constrained to move vertically without rotation, the spring–mass system has a single natural frequency, and good isolation is obtained if this natural frequency is much lower than the excitation frequency. In contrast, a three-dimensional mass can translate and rotate with respect to three axes; such a spring-supported rigid mass has six natural frequencies, and good isolation is obtained only if all six of these fall considerably below the excitation frequency. Means for calculating all of the natural frequencies and the vibration responses of general rigid masses are available in Ref. 3; the design of isolation systems is addressed in Refs. 4–6.

2.3 General Linear Uniaxial Isolation[7]

Results obtained from the simple model may suffice for many cases where only relatively low frequencies are of interest, but for higher frequencies complications must be taken into account that result from the lack of rigidity of supports and machine frames, from mass and wave effects in isolators, and from loading effects

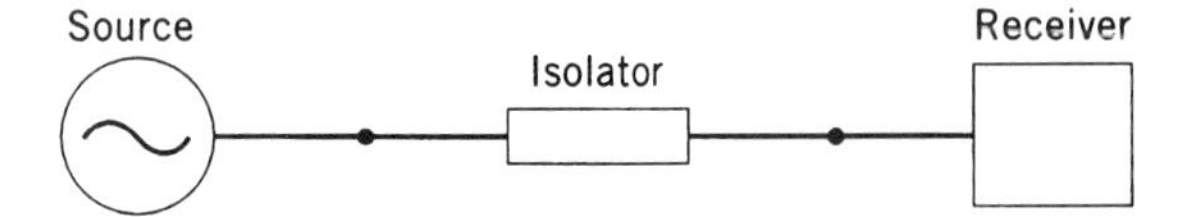

Fig. 2 Schematic representation of general uniaxial isolation arrangement.

on sources. The general three-dimensional situation is extremely complex; fortunately, many practical problems can be treated by focusing on one source-to-receiver connection point and one vibration direction at a time and by considering all components to be mathematically linear. The corresponding general uniaxial isolation problem may be visualized in terms of the diagram of Fig. 2, which shows a vibration source connected to a receiver via an isolator, whose purpose is to limit the magnitude of the vibrations that reach the receiver.

Source Characterization Loading of a source refers to reduction of the source's motion due to forces that oppose this motion. The loading susceptibility of a source may be evaluated by causing the source to act on structures with different dynamic characteristics (impedances) and measuring the velocities and forces produced by the source. For a mathematically linear source, the source's output velocity V_o is related to the force F_o it produces as

$$V_o = V_{\text{free}} - M_s F_o, \qquad M_s = \frac{V_{\text{free}}}{F_{\text{blocked}}}, \tag{4}$$

where all symbols represent complex frequency-dependent quantities. The term V_{free} denotes the source's velocity in the absence of any force generated by the source; F_{blocked} denotes the force produced by the source if its motion is fully blocked (so that the velocity vanishes). The *source mobility* M_s is zero for a *velocity source*, that is, one that produces the same velocity, regardless of its ouput force F_o, and is infinite for a *force source*, that is, one that produces the same force for all output velocities V_o.

General Linear Receiver A general linear receiver may be characterized by its frequency-dependent mobility M_R, which is defined as the ratio of the complex velocity (phasor) at the driving point to the complex force acting at that point. Here, $M_R = 1/Z_R$, where Z_R denotes the receiver's driving point impedance.

Isolation Effectiveness

Definition. Because transmissibility is defined in terms of prescribed excitations, it cannot take account of loading of the source. A measure of isolation performance that is useful in the presence of loading effects is the dimensionless *isolation effectiveness*, which, for a linear receiver, obeys

$$E = \frac{V_{Rr}}{V_{Ri}} = \frac{F_{Rr}}{F_{Ri}}. \tag{5}$$

Here V_R and F_R denote the amplitudes, respectively, of the velocity of the receiver and of the force acting on the receiver, with subscript i referring to the situation where the receiver is connected to the source via an isolator and subscript r referring to the condition where the isolator is replaced by a rigid and massless connection. For *velocity* sources, $E = 1/T$, where the transmissibility T is given by Eq. (1).

Massless Linear Isolator. An isolator may be considered as massless at frequencies that are considerably lower than the isolator's first internal resonance frequency; at such frequencies, the isolator transmits whatever force is applied to it. An isolator is linear if its deformation in the direction of the applied force is proportional to that force. A massless linear isolator may be characterized by its mobility M_I, defined as the ratio of the velocity difference across the isolator to the force applied to it. The effectiveness of such an isolator is given by

$$E = \left| 1 + \frac{M_I}{M_S + M_R} \right| \tag{6}$$

and thus depends on the mobilities of the source and receiver.

For a force source, for which M_S is infinite, $E = 1$. This result indicates that use of a massless isolator in the presence of a force source results in no less receiver motion than use of a rigid connection; in the presence of a more resilient connection the force source generates a greater velocity, keeping the transmitted force unchanged.

Generalized Linear Isolator. An isolator with finite mass cannot be characterized by a single mobility, because it does not simply transmit a force applied to one of its ends. The effectiveness of such an isolator is given by[7]

$$E = \left| \frac{\alpha}{M_S + M_R} \right| \left| 1 + \frac{M_S}{M_{Isb}} + \frac{M_R}{M_{Irb}} \left(1 + \frac{M_S}{M_{Isf}} \right) \right|, \tag{7}$$

where

$$\frac{1}{\alpha^2} = \frac{1}{M_{Irb}} \left(\frac{1}{M_{Isb}} - \frac{1}{M_{Isf}} \right). \tag{7a}$$

Here, M_{Isb} denotes the isolator's driving point mobility measured on its source side with its receiver side blocked (i.e., constrained not to move), M_{Irb} denotes the isolator's mobility measured on its receiver side with its source side blocked, and M_{Isf} denotes the isolator's mobility measured on its source side with its receiver side free to move.

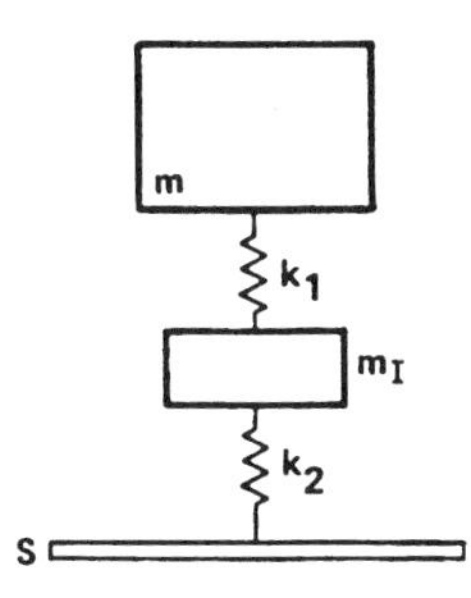

Fig. 3 Mass m and support S isolated from each other by two-stage system consisting of mass m_I between two ideal springs.

2.4 Two-Stage Isolation

Whereas conventional single-stage isolation employs a single resilient element between source and receiver, two-stage isolation employs two resilient elements with a mass between them, as shown in Fig. 3. The purpose of the included mass in essence is to provide an inertia force that opposes the imposed vibratory motion, producing better isolation at high frequencies. However, addition of the included mass results in an additional low-frequency resonance at which the isolation performance is degraded.

Mass Isolated by Two-Stage System with Massless Springs

Natural Frequencies. The dynamic system of Fig. 3 has two natural frequencies f_b, given by

$$f_b = f_0(U \pm \sqrt{U^2 - B^2})^{1/2}, \tag{8}$$

where

$$U = \frac{1}{2}\left(B^2 + 1 + \frac{k_2}{k_1}\right), \qquad B = \frac{f_I}{f_0}, \tag{8a}$$

$$2\pi f_I = \sqrt{\frac{k_1 + k_2}{m_I}}, \qquad 2\pi f_o = \left[m\left(\frac{1}{k_1} + \frac{1}{k_2}\right)\right]^{-1/2}, \tag{8b}$$

with k_1 and k_2 representing the stiffnesses of the two isolator springs, m_I the included mass, and m the mass to be isolated; B and U are auxiliary dimensionless quantities. The frequency f_o is the natural frequency that would be obtained if m_I were removed; f_I is the natural frequency that would be obtained if m were fixed. The system's natural frequency f_b obtained with the plus sign in Eq. (8) is greater than both f_o and f_I; the natural frequency f_b obtained with the minus sign in Eq. (8) is less than both f_o and f_I.

Transmissibility. For the system of Fig. 3, the force and motion transmissibilities are equal and obey

$$\frac{1}{T} = \frac{1}{B^2}\left(\frac{f}{f_o}\right)^4 - \left(1 + \frac{1 + k_2/k_1}{B^2}\right)\left(\frac{f}{f_o}\right)^2 + 1. \tag{9}$$

For excitation frequencies that are considerably greater than both f_o and f_I, the above reduces to $T \approx (f_o f_I / f^2)^2$. At high excitation frequencies, the transmissibility of a two-stage system varies inversely as the fourth power of the driving frequency, whereas that of a single-stage system varies as only the second power. The transmissibility becomes very large (infinite, for an undamped system) if the excitation frequency matches either of the two natural frequencies f_b. For most practical situations, where the damping is not very great, the effects of damping may be neglected, as they are in Eq. (9), if this frequency matching does not occur.

General Uniaxial System Isolated by Two-Stage System with Massless Isolators

Effectiveness. Figure 4 is a schematic sketch showing a general linear receiver with mobility M_R and a general linear source with mobility M_S connected by an isolation system consisting of two massless isolation elements with a mass between them. If M_{IS} and M_{IR} denote the mobilities of the isolation elements located on the source and the receiver sides of the included mass, respectively, and if M_m denotes the mobility of the included mass, the effectiveness of this two-stage isolation system may be written as

$$E = 1 + \left| \frac{M_{IS} + M_{IR}}{M_S + M_R} + \frac{(M_{IS} + M_S)(M_{IR} + M_R)}{M_m(M_S + M_R)} \right|. \tag{10}$$

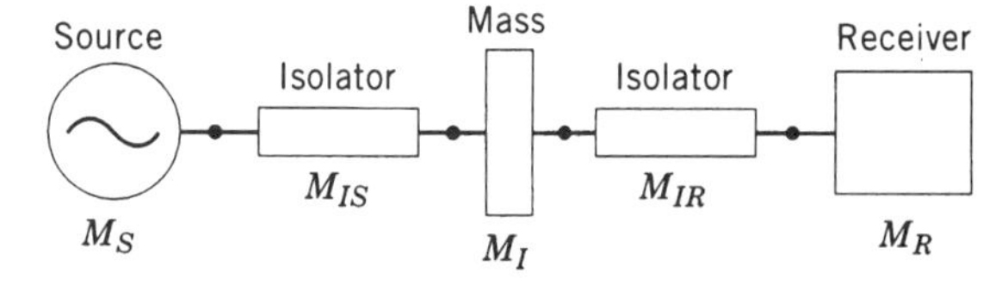

Fig. 4 Schematic representation of general uniaxial two-stage isolation arrangement.

The last fraction in this equation represents the effectiveness increase resulting from inclusion of the mass m_I.

Optimization. If p represents the fraction of the total mobility $M_I = M_{IS} + M_{IR}$ of the isolation elements that is present on the source side, so that the isolator on the source side has mobility $M_{IS} = pM_I$ and that on the receiver side has mobility $M_{IR} = (1 - p)M_I$, then the greatest effectiveness results for

$$p \approx \frac{1}{2}\left(1 + \frac{M_R - M_S}{M_I}\right). \tag{11}$$

In many practical cases the mobility of each of the isolation elements is much greater than the source and receiver mobility; then the greatest effectiveness is obtained for $p = \frac{1}{2}$, for which

$$E \approx \left|1 + \frac{M_I}{M_S + M_R}\left(1 + \frac{M_I}{4M_m}\right)\right|. \tag{12}$$

3 DAMPING OF STRUCTURES

3.1 Effects, Measures, and Measurement

Dynamic System Responses Affected by Damping

Damping refers to the removal of mechanical energy from a vibration of concern. Although damping has some effect on virtually all responses of structures and other dynamic systems, it has predominant effects on those system motions that are controlled by energy balances rather than by force balances. The system motions that are controlled by energy balances include those associated with excitation at resonance, broadband excitation where the excitation spectrum encompasses one or more system natural frequencies, trace matching (spatial coincidence of exciting pressures with structural surface velocities associated with propagating waves*), freely decaying vibrations, freely propagating waves, and self-excited vibrations (i.e., where the oscillatory exciting force is produced or controlled by the motion).

*This occurs when the *wave impedance* Z_t defined in Chapter 49 vanishes.

Measures of Damping

Viscous Damping. In the simple linear mass–spring–dashpot system of Fig. 1, which can also be taken to represent a mode of a vibrating structure, the dashpot represents an element that produces a retarding force that is proportional to the velocity of the mass. The ratio of this force to the velocity is called the *viscous damping coefficient b*. For a system with mass m and spring constant k, the *critical damping coefficient* b_c is given by $b_c = 2\sqrt{km}$; b_c is the smallest value of b for which the aforementioned simple linear system oscillates when it is deflected and released. (If such a system with $b > b_c$ is deflected and released, it drifts toward its equilibrium position without oscillating.)[2] The *fraction of critical damping*, or *damping ratio* ζ, is defined as $\zeta = b/b_c$.

Decaying Free Vibrations. An "underdamped" system (i.e., one with a damping ratio ζ of less than unity) that is disturbed from equilibrium and then permitted to vibrate freely will oscillate with decreasing amplitude, as sketched in Fig. 5. If X_i denotes a peak excursion from equilibrium and X_{i+N} denotes a peak excursion N cycles later, then the *logarithmic decrement* δ is defined as $\delta = (1/N)\ln(X_i/X_{i+N})$.

The rate at which the vibration (displacement, velocity, or acceleration) level decreases with time is called the *decay rate* Δ and usually is expressed in decibels per second. The *reverberation time* T_{60} is defined as the time it takes for the vibration level to decrease by 60 dB, corresponding to a decrease in amplitude by a factor of 1000.

Steady Forced Vibrations. If X denotes the amplitude of the simple system of Fig. 1 produced by a given steady sinusoidal force, then the corresponding amplification A

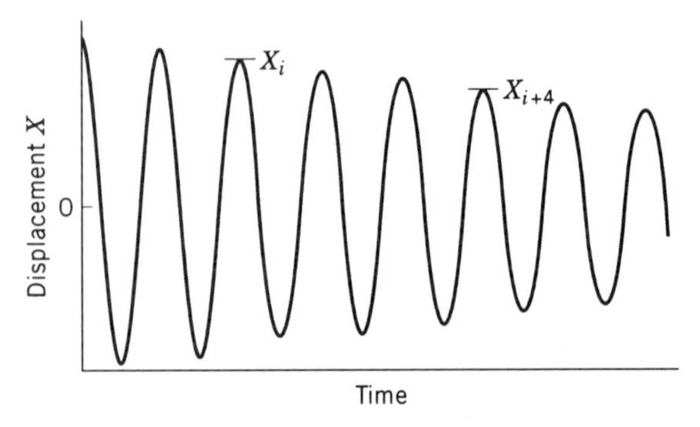

Fig. 5 Time variation of displacement of mass–spring–dashpot system in decaying free vibration. Excursions X_i and $X_i + 4$ illustrate determination of logarithmic decrement.

is defined as the dimensionless ratio $A = X/X_{st}$, where X_{st} denotes the amplitude that would be obtained if the same force were applied quasi-statically. The *amplification at resonance*, usually called "the Q" of the system and designated by Q, is the value of the amplification that results at resonance, that is, if the excitation frequency matches the system's natural frequency.

The bandwidth of a resonance is defined as the difference between the two excitation frequencies (one above and one below the natural frequency) at which the amplification is equal to $Q/\sqrt{2}$, or, equivalently, at which the amplification is 3 dB lower than Q. The *relative bandwidth* B_f is defined as the ratio of the bandwidth to the natural frequency.

The *phase angle* ϕ by which the displacement of the mass of Fig. 1 lags the excitation force may also be used to characterize the system's damping. For example, at radian frequencies ω much below the system's natural frequency, $\tan\phi = \omega b/k = k_i/k$. Here k_i is the imaginary part of a complex stiffness $k + ik_i$; this complex stiffness, which is equal to the ratio of the phasor of the applied force to that of the resulting displacement, characterizes both the energy storage and the energy dissipation of the system.

Energy Dissipation. The *damping capacity* Ψ of a vibrating system is defined as $\Psi = D/W$, where D represents the energy that is removed from the system per cycle and W denotes the vibrational energy stored in the system. The *loss factor* η, defined by $\eta = \Psi/2\pi$, represents the ratio of the energy removed per *radian* to the stored vibrational energy. In most practical situations the loss factor is less than 0.2 and one may take the stored energy to be equal to the total vibrational (kinetic plus potential energy).

Interrelations. The various measures of damping are related to each other by

$$\eta = \frac{\Psi}{2\pi} = \tan\phi = \frac{k_i}{k} \approx \frac{2.20}{f_n T_{60}} \approx \frac{\Delta}{27.3 f_n} \approx \frac{\delta}{\pi} \quad \text{(in general)}, \tag{13a}$$

$$\eta = 2\zeta = B_f = \frac{1}{Q} \quad \text{(at resonance)}, \tag{13b}$$

where f_n denotes the natural frequency in hertz and T_{60} is in seconds.

Measurement Approaches

Application of Simple Model. Most approaches to measuring the damping of a structure are based on considering a structural mode as a simple mass–spring–dashpot system and applying the foregoing definitions to corresponding parameters measured in freely decaying vibrations or steadily excited sinusoidal motions.

For viscous or approximately viscous damping, the magnitude of the damping present in a system may also be deduced from the phase angle between a sinusoidal force and the resulting motion. This generally is done most conveniently by the use of *Nyquist plots*—plots of the real versus imaginary parts of the system responses at a number of frequencies near resonance[8]—and by means of available software.

Decay Rate in Frequency Band. In applying any approach based on a single-degree-of-freedom model to a structure, care must be taken that the measurement reflects the response of only a single structural mode, with no significant contributions from adjacent modes. An approach that permits measurement of damping in the presence of several modes whose damping does not differ too greatly consists of measuring the rate of decay of a bandpass filtered signal obtained from a vibration sensor that is attached to a freely vibrating structure. (The filter's response must be fast enough so that it can follow the decaying signal. Filters with wider passbands must be used to observe the more rapid signal decays corresponding to higher damping.) For such a measurement it is useful to observe the logarithm of the rectified signal; for a simple viscously damped system, the upper bound envelope of this logarithm is a straight line from whose slope one may determine the damping. Deviations of this envelope from a straight line indicate the presence of several modes with different amounts of damping and/or the presence of other than viscous damping. Most modal analysis and spectrum analysis software includes means for assessing damping from vibration decay measurements based on essentially the foregoing concept.

Steady-State Energy Dissipation Rate. If one applies an oscillatory force at a given frequency f to a structure via an impedance head, one may use the product of the force and velocity signals derived from the impedance head to determine the time-average rate of energy input to the structure. In the steady state, this input rate is equal to the rate of energy loss, which is equal to fD, where D denotes the energy loss per cycle. One may calculate the structure's peak kinetic energy, which is equal to its total vibrational energy W, from measured velocities of the structure and its mass distribution. One may then compute the damping capacity or loss factor by applying their definitions.[9]

3.2 Damping Mechanisms and Magnitudes

Energy Dissipation *Material damping*, also called *mechanical hysteresis* or *internal friction*, refers to the conversion of mechanical energy into heat that occurs in materials as they are deformed. A variety of phenomena may be involved on the molecular, crystal lattice, and metal grain level. Figure 6 indicates the ranges of the loss factors reported for some common materials at small amplitudes and at various frequencies and temperatures.

A vibrating structure may also experience damping as the result of *friction* associated with relative motion between it and structures or fluids in contact with it. An electrically conductive structure moving in a magnetic field is subject to damping due to *eddy currents* generated in it. *Granular materials* in contact with a vibrating structure produce damping as the result of mechanical hysteresis, predominantly at grain asperities, and also as the result of impacts of grains against the structure and against each other. *Impact dampers*, consisting of elements that can rattle against a vibrating structure, essentially extract energy from the structure by converting the structure's vibrational energy to vibrational energy at higher frequencies.

Energy Transport Energy that is transported from a vibrating structure to adjacent structures or fluids (including energy transported in the form of radiated

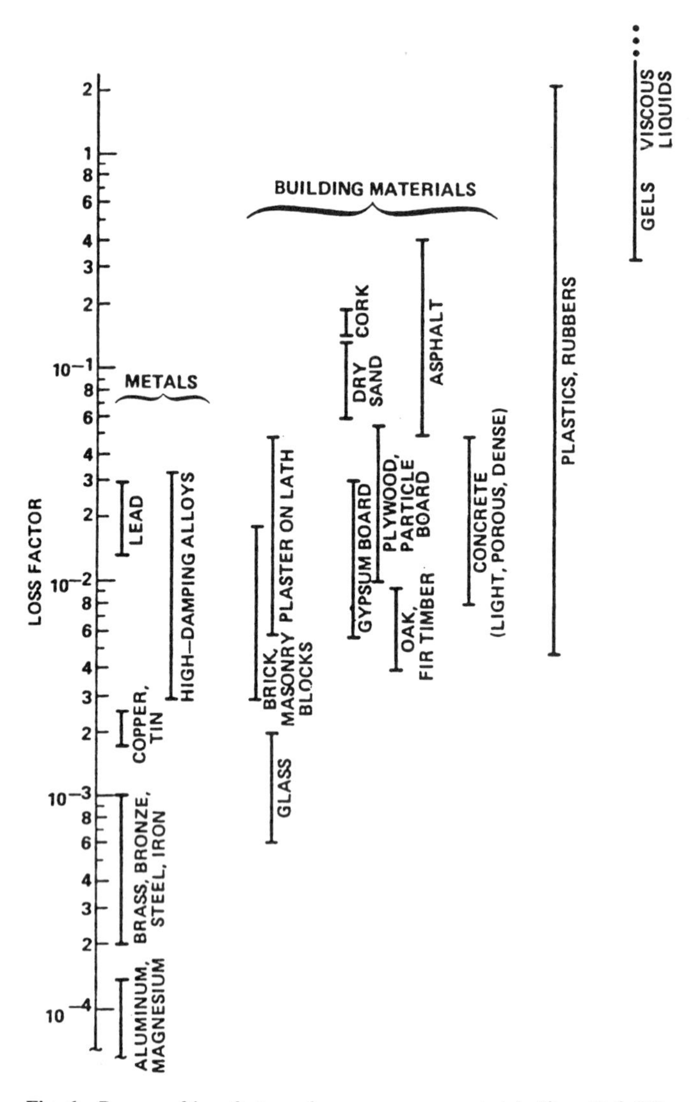

Fig. 6 Ranges of loss factors of some common materials (from Ref. 10).

sound) constitutes a loss of the structure's vibrational energy and thus contributes damping. For a structural component that is in intimate contact with others (e.g., for a panel that is part of an aircraft fuselage), energy transport to adjacent components can contribute considerable damping and generally makes it difficult to measure the component's energy dissipation itself. Energy transport via, for example, supports and cables, usually makes it very difficult in practice to measure the energy dissipation of lightly damped structures.

Boundaries and Reinforcements. For a structural component that behaves essentially as a uniform plate, the loss factor η_b due to energy transport across its boundaries and/or due to dissipation at one-dimensional discontinuities (such as stiffeners or seams) may be estimated from[10]

$$\eta_b = \frac{\lambda}{\pi^2 A} \sum \gamma_i L_i. \tag{14}$$

This relation applies at frequencies that are high enough so that the flexural wavelength λ on the plate is considerably shorter than a plate edge. Here, A denotes the plate area (one side) and γ_i represents the absorption coefficient of the ith boundary or discontinuity element; L_i represents the length of the ith element if that element is along a plate edge or twice that length if the element is in the interior of the plate area; the summation extends over all elements.

The absorption coefficients γ_i may be obtained by applying Eq. (14) to data obtained from loss factor measurements on plates with only one type of element attached. For riveted, bolted, or spot-welded seams or stiffeners, γ_i may be estimated by use of Fig. 7. For seams or stiffeners that are joined by means of continuous welding or by a rigid adhesive, the absorption coefficients generally are negligibly small.

Point-Attached Systems. If a system or structure that initially is free of vibration is attached to a vibrating struc-

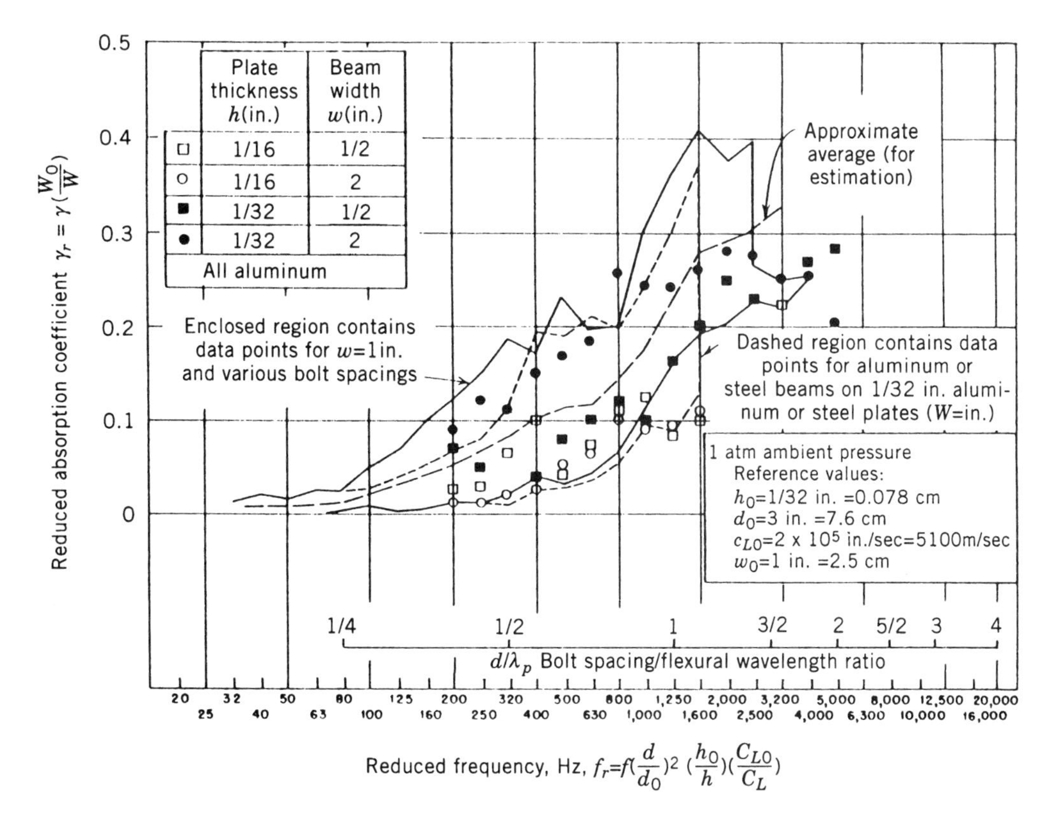

Fig. 7 Reduced data for estimation of absorption coefficients of stiffeners or seams in plates fastened by rivets, bolts, or spotwelds (from Ref. 10): f = frequency, d = fastener spacing, h = plate thickness, w = width of beam in contact with plate or of seam overlap, c_L = longitudinal wave speed in plate material. Subscript 0 refers to reference values.

ture at a point, then the energy D that is transported to the attached structure at frequency f is given by

$$D = \frac{V_s^2}{2f} \operatorname{Re}\{Z_A\} \left| 1 + \frac{Z_A}{Z_S} \right|^{-2}, \qquad (15)$$

where V_s denotes the amplitude of the vibrating structure's velocity at the attachment point before the second structure is attached, Z_A denotes the driving point impedance of the attached structure, and Z_S denotes the impedance of the vibrating structure at the attachment point (with both impedances measured in the direction of V_s).

A *waveguide absorber*[11] essentially consists of a structural element along which waves can travel and which includes means for dissipating the energy transported by these waves. For example, a long slender beam made of a highly damped plastic or coated with such a material may serve as a waveguide absorber at frequencies that are considerably above its fundamental resonance. The damping performance of such an absorber may be evaluated with the aid of the foregoing equation. To be effective, a waveguide absorber must be attached to a structure where it vibrates with considerable amplitude and it must not reduce the structural motion at the attachment point excessively.

Acoustic Radiation. Sound radiated from a structure removes energy from it and thus contributes damping. Acoustic radiation into air rarely results in significant damping, but radiation into liquids may produce considerable damping. The energy D removed from a vibrating structure per cycle at frequency f is given by $D = A\rho c v^2 \sigma / f$, where ρ denotes the density of the ambient fluid, c the speed of sound in it, A the structural surface area in contact with the fluid, v^2 the mean-square normal velocity of the surface area (averaged over both time and space), and σ the radiation efficiency as discussed in Refs. 6 and 12, for example. For a homogeneous plate of thickness h and material density ρ_p, the loss factor corresponding to radiation from one of its sides is given by $\eta = \sigma(\rho/\rho_p)(c/2\pi f h)$. For radiation from both sides, the loss factor is twice the foregoing value.

3.3 Viscoelastic Damping Treatments

Viscoelastic Materials and Material Combinations

Loss Factor of Combinations of Components. A *viscoelastic material* is one that has considerable energy dissipation capability in addition to strain energy storage capability. As evident from Fig. 6, high-strength structural materials generally have little inherent damping, whereas rubbery materials tend to have lesser strength and greater damping. Combinations of structural and viscoelastic materials can take advantage of the strength of one and the damping of the other.

The loss factor η of a structure that is made up of a number of components and that vibrates with a given deformation distribution obeys[10,13]

$$\eta = \sum \frac{\eta_{ij} W_{ij}}{W_T}, \qquad W_T = \sum W_{ij}, \qquad (16)$$

where W_{ij} denotes the strain energy stored in the ith deformation mode (e.g., extension, torsion, shear, flexure) of the jth component and η_{ij} denotes the loss factor associated with that deformation mode and component; the summations extend overall deformation modes and components. The contribution made to the loss factor η of the composite structure by an element deformation is significant only if its loss factor η_{ij} is considerable *and* if its energy storage W_{ij} is a significant fraction of the total energy storage W_T.

Behavior of Viscoelastic Materials. In analogy to the definition of complex stiffness, the complex modulus of elasticity of a material may be defined as $E_c = E + jE_I = E(1 + j\eta_e)$. This frequency-dependent complex modulus is equal to the ratio of the phasor of the applied stress to that of the resulting strain. The real part E is associated with energy storage, the imaginary part E_I with energy dissipation. The material's loss factor η_e associated with the modulus of elasticity is equal to E_I/E. Completely analogous definitions apply for the complex shear modulus.

The viscoelastic materials of greatest practical interest for damping applications are plastics and elastomers. For most practical damping materials of this type, the loss factors associated with all types of deformations are practically equal and the value of the shear modulus is very nearly one-third of the value of the corresponding modulus of elasticity. The various moduli typically vary little with strain amplitude, preload, and aging but vary markedly with frequency and temperature.[13] The moduli typically take on relatively high values at low temperatures and/or high frequencies but take on comparatively small values at high temperatures and/or low frequencies; the greatest loss factors occur in a transition region at intermediate frequencies and temperatures. The modulus values and the transition frequency and temperature ranges vary considerably from material to material. Data on commercially available materials generally may be obtained from their suppliers; a compilation of data for a number of materials appears in Ref. 13.

Beams and Plates with Viscoelastic Additions[10]

Viscoelastic layers typically may be applied to a structure with or without a covering "constraining" structure. In the absence of a constraining structure, the viscoelastic layer stores and dissipates energy primarily in extension/compression; in the presence of a constraining structure the viscoelastic layer acts primarily in shear. A viscoelastic layer system with a constraining structure often can produce greater damping than a viscoelastic layer of the same total weight without a constraining structure, at least in a limited frequency range, but the damping system must be designed for the frequency and structural wavelengths or deflection shapes of concern. On the other hand, a viscoelastic layer without a constraining structure usually can produce reasonable damping performance without taking wavelength into account in the design process.

Extensional Viscoelastic Additions. Figure 8 shows cross-sectional configurations of beams and plate strips where a layer or insert of viscoelastic material is attached to a uniform structural beam or plate so that flexure of the structure causes the viscoelastic component primarily to extend or contract axially. For flexure of such configurations, the loss factor increase η resulting from addition of the viscoelastic materials obeys

$$\eta = \beta\left[1 + \frac{q^2(1+\beta^2) + (r_1/H_{12})^2\alpha}{q[1+(r_2/H_{12})^2\alpha]}\right]$$

$$\approx \frac{\beta E_2}{E_1}\,\frac{A_2(r_2^2 + H_{12}^2)}{I_2}, \tag{17}$$

where $q = E_2A_2/E_1A_1$; $\alpha = (1+q)^2 + (\beta q)^2$; $r_1 = \sqrt{I_1/A_1}$, $r_2 = \sqrt{I_2/A_2}$; β represents the loss factor of the viscoelastic material; and H_{12} denotes the distance between the neutral fibers of the two components. Here, E denotes the (real) modulus of elasticity, A the cross-sectional area, and I the centroidal moment of inertia; subscript 1 refers to the structural component and 2 to the viscoelastic component. The simpler, approximate form of Eq. (17) applies to the often encountered case where $E_2A_2 \ll E_1A_1$.

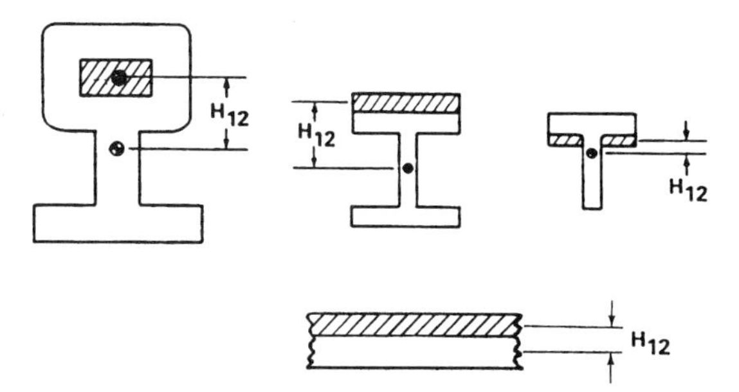

Fig. 8 Cross sections of some beam and plate configurations with attached viscoelastic layers (shown cross-hatched) that extend as the structures flex (adapted from Ref. 10).

Figure 9 shows how the loss factor of a plate with an added viscoelastic layer varies with the layer thickness and loss factor. In general, in order to obtain a relatively large damping increase by adding an extensional viscoelastic element, one needs to use a damping material with a comparatively large βE_2 product in a configuration that produces a significant separation H_{12} between the neutral fibers. In most practical cases the loss factor obtained by adding viscoelastic layers to both sides of a plate is approximately equal to the sum of the loss factors obtained with the layer added separately.

Viscoelastic Additions in Shear. Figure 10 shows generic cross sections of beams and plate strips that include viscoelastic components in configurations for which flexure of the composite causes the viscoelastic element (designated by subscript 2), which is rigidly adhered to two separate structural elements (subscripts 1 and 3), to deform primarily in shear.

The loss factor contribution η resulting from shear of the viscoelastic material may be calculated from

$$\eta = \frac{\beta XY}{1 + (2+Y)X + (1+\beta^2)(1+Y)X^2}, \tag{18}$$

where, for a beam,

$$X = \frac{SG_2b}{p^2H_2}, \qquad Y = \frac{H_{13}^2}{SN}, \qquad S = \frac{1}{E_1A_1} + \frac{1}{E_3A_3},$$
$$N = E_1I_1 + E_3I_3, \tag{19}$$

and for a plate,

$$X = \frac{SG_2}{p^2H_2}, \qquad Y = \frac{12H_{13}^2}{SN}, \qquad S = \frac{1}{E_1H_1} + \frac{1}{E_3H_3},$$
$$N = \frac{1}{12}(E_1H_1^3 + E_3H_3^3). \tag{20}$$

In all cases, β, G_2, and H_2 represent, respectively, the loss factor, shear modulus, and thickness of the viscoelastic material; b represents its average length as measured on a beam cross section; E_i, A_i, I_i, and H_i denote, respectively, the modulus of elasticity, cross-sectional area, centroidal moment of inertia of cross-sectional area, and thickness of component i; and H_{13} denotes the distance between the neutral surfaces of the two structural elements indicated by subscripts 1 and 3. Additionally, p represents the wavenumber of the flexural vibration of the composite beam or plate under consideration; it is

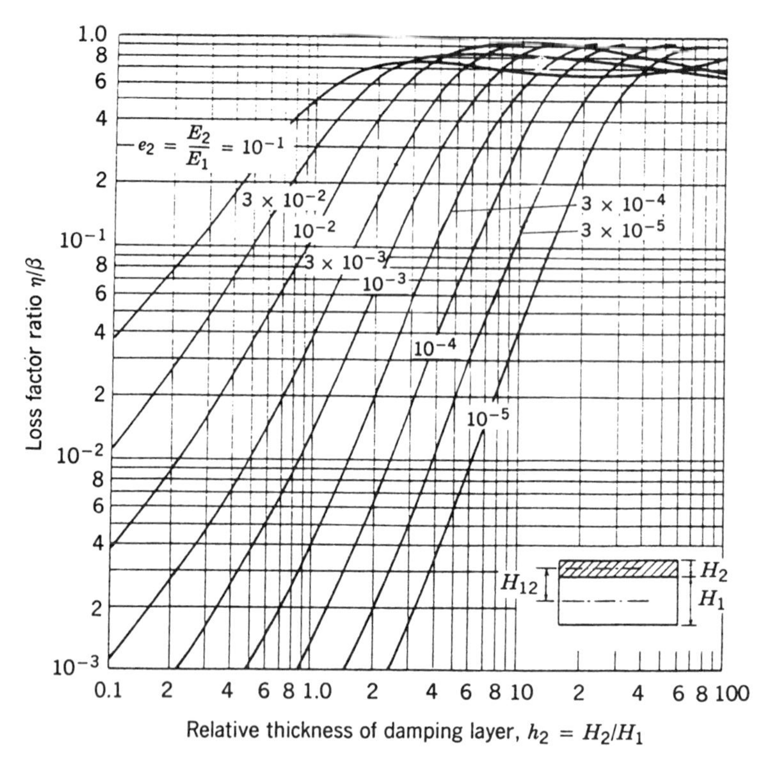

Fig. 9 Dependence of loss factor of plate with extensional viscoelastic layer on relative thickness and relative modulus of layer (from Ref. 10): η = loss factor of plate with attached layer, β = loss factor of viscoelastic material, E = modulus of elasticity, H = thickness. Subscript 1 refers to structural plate, 2 to viscoelastic layer.

related to the wavelength λ at frequency f by

$$\frac{1}{p^2} = \left(\frac{\lambda}{2\pi}\right)^2 = \sqrt{\frac{B/\mu}{2\pi f}}, \qquad (21)$$

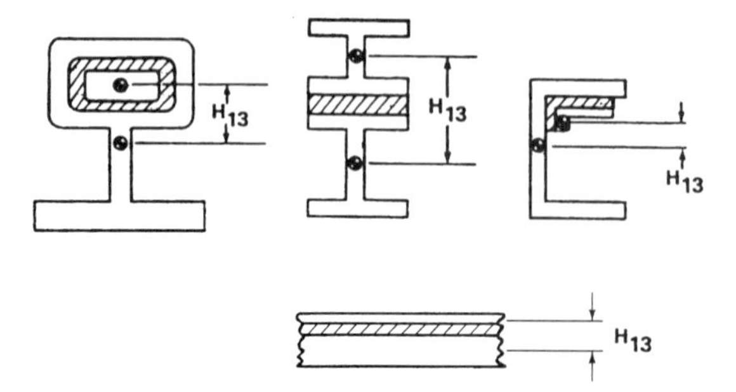

Fig. 10 Cross sections of sandwich plate and of some beam configurations that incorporate viscoelastic elements (cross-hatched) that shear as the structures flex (adapted from Ref. 10).

where B denotes the flexural rigidity of the composite beam (or the flexural rigidity per unit width of the composite plate) and μ represents the mass per unit length of the beam (or the mass per unit area of the plate). The flexural rigidity may be obtained from

$$B = N\left|1 + \frac{YX(1 + j\beta)}{1 + X(1 + j\beta)}\right|, \qquad (22)$$

where $j = \sqrt{-1}$ and the viscoelastic material property values to be used are those that correspond to the temperature and frequency of interest. Where p is not determined by boundary or other constraints, it may be found by solving Eqs. (19) and (20) simultaneously, either explicitly or by a numerical iterative approximation process.[10]

A composite with high damping at a given temperature and frequency may be obtained by use of a configuration with a large value of Y and a viscoelastic material with a high loss factor β, provided the thickness H_2 of that material is selected so that $X = X_{\text{opt}}$, where X is given by Eq. (19) or (20) and $X_{opt} = [(1 + \beta^2)(1 + Y)]^{1/2}$. Then the loss factor takes on its greatest value for the applicable Y and β values, namely, $\eta_{max} = \beta Y / [2 + Y + 2/X_{opt}]$.

REFERENCES

1. E. E. Ungar, "Equality of Force and Motion Transmissibilities," *J. Acoust. Soc. Am.*, Vol. 90, 1991, pp. 596–597.
2. W. T. Thomson, *Theory of Vibration with Applications*, 2nd ed., Prentice-Hall, Englewood Cliffs, NJ, 1981.
3. H. Himelblau, Jr., and S. Rubin, "Vibration of a Resiliently Supported Rigid Body," in C. M. Harris (Ed.), *Shock and Vibration Handbook*, McGraw-Hill, New York, 1988, Chapter 3.
4. J. N. MacDuff and J. R. Curreri, *Vibration Control*, McGraw-Hill, New York, 1958.
5. C. E. Crede and J. E. Ruzicka, "Theory of Vibration Isolation," in C. M. Harris (Ed.), *Shock and Vibration Handbook* 3rd ed., McGraw-Hill, New York, 1988, Chapter 30.
6. P. M. Norton, *Fundamentals of Noise and Vibration Analysis for Engineers*, Cambridge University Press, Cambridge, 1989.
7. E. E. Ungar and C. W. Dietrich, "High-Frequency Vibration Isolation," *J. Sound Vib.*, Vol. 4, 1966, pp. 224-241.
8. D. J. Ewins, *Modal Testing: Theory and Practice*, Research Studies Press, Letchworth, Hertfordshire, England, 1986.
9. B. L. Clarkson and R. J. Pope, "Experimental Determination of Modal Densities and Loss Factors of Flat Plates and Cylinders," *J. Sound Vib.*, Vol. 77, No. 4, 1981, pp. 535–549.
10. E. E. Ungar, "Structural Damping," I. L. Ver and L. L. Beranek (Eds.), *Noise and Vibration Control Engineering*, Wiley, New York, 1991, Chapter 12.
11. E. E. Ungar and L. G. Kurzweil, "Structural Damping Potential of Waveguide Absorbers," *Transactions, Inter-Noise 84*, Institute for Noise Control Engineering, Poughkeepsie, NY, 1984, pp. 571–574.
12. B. L. Clarkson and K. T. Brown, "Acoustic Radiation Damping," American Society of Mechanical Engineers, Publication 85-DET-24, 1985.
13. A. D. Nashif, D. I. G. Jones, and J. P. Henderson, *Vibration Damping*, Wiley, New York, 1985.

56

VIBRATION MEASUREMENTS AND INSTRUMENTATION

MICHAEL MÖSER

1 INTRODUCTION

Measurements on vibrating mechanical structures cover a broad band of applications ranging from simple laboratory setups to the vibrational characterization of whole bridges, buildings, trains, aircrafts, and so forth. This indicates that vibrational measurements are often complex enough to require a fundamental and detailed understanding of the instrumentation, the sensoring, and the apparatures involved. Normally, measurements are useful only if they are accompanied by some expectation regarding the physical behavior of the tested mechanical system (see all chapters in Part VI and Refs. 1–3).

This chapter will give an overview of the measurement quantities and the different types of sensors. It will address their fundamental principle, their specific use, their limits and inherent problems, and deal with some of the most important applications of vibrational measurements. As detailed overviews Refs. 4 and 5 are recommended for further reading.

2 MEASUREMENT QUANTITIES

The most commonly used measurement quantities are designated the descriptor q for the spatial vibration field at a certain point. These measurement quantities are the displacement $q = \xi(t)$, the velocity $q = v(t)$, the acceleration $q = a(t)$ with the relationships

$$v(t) = \frac{d\xi(t)}{dt} \quad \text{and} \quad a(t) = \frac{dv(t)}{dt}. \tag{1}$$

Note: References to chapters appearing only in the *Encyclopedia* are preceded by *Enc.*

The list continues with the angular velocity $q = w(t)$ on the structure, if relevant, the exciting force $q = F(t)$, and the sound pressure $q = p(t)$ in the surrounding medium. The appropriate levels (defined in Chapter 63) are usually based on the root-mean-square (rms) value

$$q_{\text{rms}} = \sqrt{\frac{1}{T}\int_0^T q^2(t)\,dt}. \tag{2}$$

Sometimes the peak amplitude is used instead of the rms value (e.g., evaluations with Fourier techniques, see Chapter 82). The two values differ by a factor of $\sqrt{2}$. Levels may be expressed by linear or A-weighted filtering methods (Chapter 105), the latter being more convenient for the measurement of emission levels. The usual time weighting (fast, slow, impulse, see Chapter 105) may also be employed.

Since the kinetic measurement quantities displacement, velocity, and acceleration are related by time differentiations and integrations [(Eq. 1)], their electric representation may be converted into one another by electrical networks or numerically after analog-to-digital conversion (Chapter 82). Integration networks of RC type are most commonly used. Their frequency response equals that of an ideal integrator, but only above some cutoff frequency (typically 10 Hz). Therefore a corresponding signal distortion of transients may cause difficulties. Numerical time integration and differentiation can be performed both in the time and frequency domain, the latter by (repeated) division and multiplication with $\pm j\omega$ (note that manufacturers might use different time conventions). In both domains, the mathematical operation causes significant frequency-dependent amplification of the signals noise component.

The above measurement quantities q are the only ones that can be measured directly by using a single (electromechanical) transducer. Indirect methods (based on theoretical assumptions and the use of several transducers) are necessary for intensity and power flow measurements (sound intensity, Chapter 106; structural intensity[6,7]).

3 MEASURING SENSORS

An instrument for the measurement of vibrations consists of one or more electromechanical transducers with a linear relationship between the mechanical input and the electrical output. The electrical path contains a circuit or system to process the transducer's output into a properly defined and correlated indication of the measurement quantity. The sensors should provide the lowest possible sensitivity to environmental effects such as high sound levels (for vibrations), transverse fields, temperature fluctuations, and electromagnetic or other radiation fields.

The dynamic range of the instrument must allow distortion-free measurements of the highest encountered vibration level. The lowest signal-to-noise ratio must exceed 10 dB for an accurate measurement result.

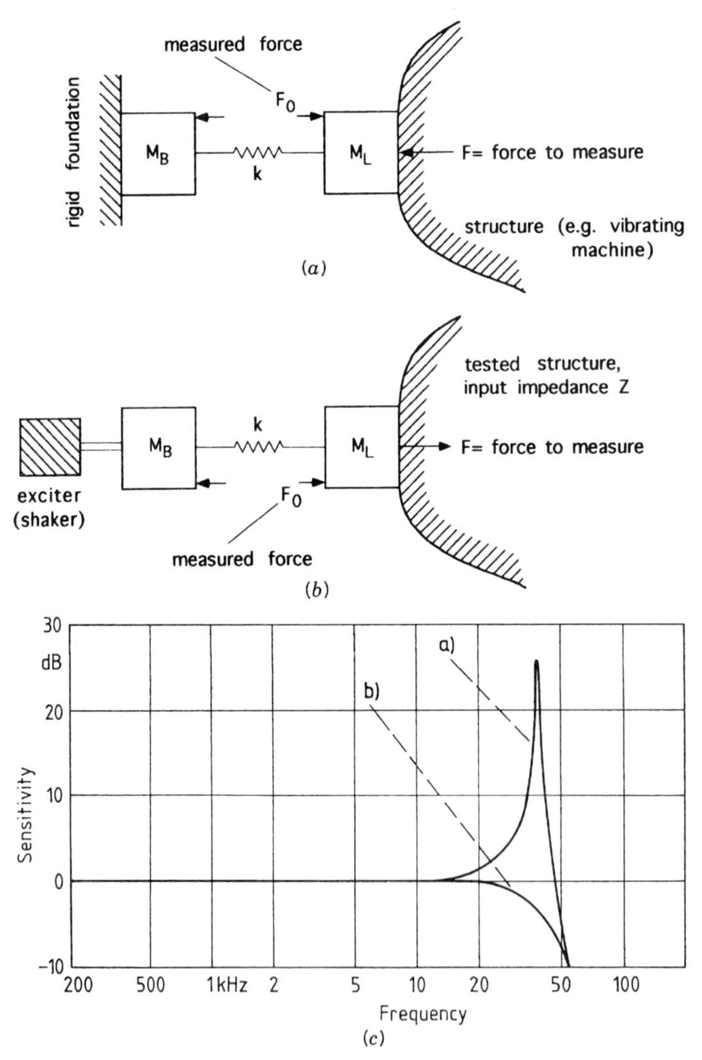

Fig. 1 Frequency responses of force transducers: (*a*) Measurement of forces produced by some vibrating structure and transmitted to a foundation, (*b*) measurement of forces exciting some structure, and (*c*) typical frequency responses in case (*a*) and (*b*).

3.1 Piezoelectric Transducers

The most commonly used vibration transducers are of the piezoelectric type. The deformation of a crystal element, consisting of materials such as quartz or artificially polarized ceramics, results in an electrical charge that is proportional to the acting force.[8,10]

Force transducers are this principle directly. Mechanically, they may be considered to be a spring (the piezoelectric element) with a mass on each end (base M_B and load M_L, which are parts of the housing). Their frequency response depends on the application. For the measurements of force F delivered through the sensor from a structure to a foundation (Fig. 1*a*), the sensor behaves like a simple resonator, giving a measured force $F_0 = F/(1 - \omega^2/\omega_0^2)$, shown in Fig. 1*c* (upper curve). In this case, the resonant frequency ω_0 depends on the piezoelectric spring with the spring constant k and M_L : $\omega_0^2 = k/M_L$.

For the measurement of forces that are delivered through the sensor to a structure (Fig. 1*b*), the top of the transducer is loaded with an impedance Z, which represents the structure including the sensor's mounting. The measured force F_0 is the difference between the force F acting on the structure and the inertia due to the load M_L :

$$F_0 = F - M_L a_L = F(1 + j\omega M_L/Z), \tag{3}$$

where a_L denotes the acceleration of M_L. Since the mechanical behavior of Z can be associated with a local spring constant k_L at higher frequencies, the measured force in this case is $F_0 = F(1 - \omega^2/\omega_0^2)$, where ω_0 denotes the resonant frequency of the local stiffness–load system alone: $\omega_0^2 = k_L/M_L$ (lower curve in Fig. 1*c*). Normally, the relevant mass M_L is minimized in the sensor's construction; foot and head should therefore not be exchanged in the mounting, even if the polarity change is irrelevant.

A piezoelectric accelerometer is made up of a mass (the so-called seismic mass) that is connected to the piezoelectric element (Fig. 2). The piezoelectric element in this case acts as a spring. The force transmitted by the spring due to Newton's law is proportional to the acceleration of the mass. The electrical charge is therefore given by

$$Q = Q_0 \frac{1}{d\omega_0^2} \frac{a}{1 - \omega^2/\omega_0^2}, \tag{4}$$

where a denotes the acceleration of the housing, Q_0 is a constant, $\omega_0 = 2\pi f_0 = \sqrt{k/M}$ is the resonant frequency of the mass–spring system, and k denotes the spring constant of the spring with the thickness d. A typical frequency-response curve is given in Fig. 3.

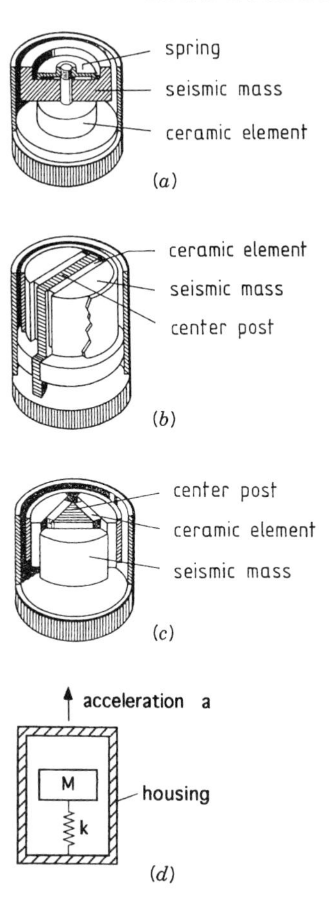

Fig. 2 Types of piezoelectric accelerometers: (*a*) Compression type; (*b*) planar–shear type; (*c*) triangular–shear type; (*c*) triangular–shear type; and (*d*) principle of mechanical behavior, M denotes the seismic mass and k the stiffness of the ceramic element with thickness d.

Below resonance, the foot of the sensor and the mass are strongly coupled. The charge then is directly proportional to the acceleration and to the seismic mass M. This is essentially the reason for the fact that the sensitivity of accelerometers increases linearly with their weight, giving them the disadvantage of lower resonant frequencies. The usable frequency region ends at the point where the frequency response rises to 10% above the flat-line level. Equation (4) gives $f = f_0/3.2$ for that limit.

Transducers with high sensitivity are heavy. They have lower resonant frequencies and a smaller usable frequency range as opposed to light-weight broadband accelerometers with lower sensitivity. Due to the different needs of the various measuring purposes, a range of accelerometers exists, making it possible to choose the correct accelerometer for a given purpose (see Chapter 56). Examples are given in Table 1.

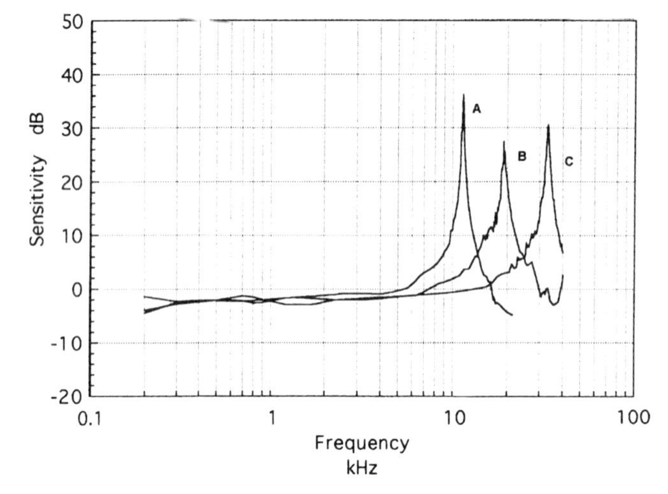

Fig. 3 Typical measured frequency responses of piezoelectric accelerometers, seismic masses: A, 43*g*; B, 30*g*; C, 11.3*g*.

Piezoelectric accelerometers may be designed as compression-type or as shear-type sensors (Fig. 2). The first collect the deformation-induced electrical charge in the material's polarization direction; the second collect the charge perpendicular to the polarization direction.[9] Extra charge due to temperature fluctuations is not picked up with the shear type. This is an important advantage of shear-type sensors, which reduces the temperature sensitivity by a factor on the order of 0.01. Sometimes, the sensitivity to transverse accelerations perpendicular to the sensors main axis causes problems. One example is a composed vibration field with different wave types to be observed. The transverse sensitivity of piezoelectric accelerometers is in the 5% range of the main axis sensitivity. The directivity of the transverse sensitivity depends strongly on the sensors construction (compression, shear, planar shear, and others).

The next element in the measurement system is a charge amplifier of low-input impedance chosen because of the small charge produced by the electromechanical transducer.

Care must be taken in the choice and mounting of the connecting cable to the sensor. The isolation material inside the coaxial cable may locally lose intimate contact to a conductor during cable bending. This, due to the piezoelectric properties of the isolator, may generate a triboelectric charge flow in the conductor across the electrical termination. Cables with noise reduction treatment (a coating technique) prevent triboelectric flow. Furthermore, relative movement between the electrical connec-

TABLE 1 Technical Data of Some Sensors

Type[a]	Sensitivity	Weight (g)	Frequency Limit (kHz)	Dynamic Range (*g*)
PE, S	1.1 pC/g	0.65	26	0.003–5000
PE, DS	3.1 pC/g	2.4	16.5	0.0008–5000
PE, DS	9.8 pC/g	11	12.6	0.0003–5000
PE, DS	31 pC/g	17	8.4	0.00015–2000
PE, DS	98 pC/g	54	4.8	0.00003–2000
PE, DS	310 pC/g	175	3.9	0.000015–500
ICP	10 mV/g	2	10	0.01–500
ICP	100 mV/g	55	3	0.001–50
ED	14 V/ms^{-1}	246	0.2	—
ED	30 V/ms^{-1}	450	2	—

[a]PE, piezoelectric; S, shear type; DS, delta-shear type; ICP, integrated circuit piezoelectric; ED, electrodynamical.

tion and the structure should be avoided by fixing the cable on the tested structure. The cables should leave the object at a nonvibrating point. They should not be longer than necessary because the noise produced by the charge amplifier increases with the cable length.

Integrated circuit piezoelectric (ICP) accelerometers are more comfortable. Their concept is based on the combination of the piezoelectric element and a miniature electrical impedance converter in a common housing with neglectable increase of weight. This combination offers a low-cost, robust measurement channel with fixed sensitivity (independent of cable length), high output voltage (up to 10 V), high signal-to-noise ratio (up to more than 90 dB), and "normal" coaxial cable connections.

3.2 Other Electromechanical Transducers

Corresponding with the various principles for electromechanical transducers used in the measurement of vibrational quantities,[8–10] for example, the variation of inductances or capacitances, there are two more commonly used types commercially available: the geophones, which use the electrodynamic principle of induction, and the strain gauges.

Electrodynamical Sensors Electrodynamic transducers[10] consist of a permanent magnet in the form of a hollow cylinder, in the center of which a coil with a large number of turns is suspended by means of light springs (Fig. 4). The output voltage $U \approx \text{const } x V_R$ due to electromagnetic induction is proportional to the relative velocity V_R between the coil and the magnet, the latter being fixed in the housing. Since the coil is dynamically decoupled from the magnet above the resonance of the coil-spring oscillatory system, the relative velocity $V_R \approx V$ equals that of the transducers foot V in the current frequency range. Below resonance, the relative velocity tends to zero. This is because of the strong mechanical coupling, which gives a fast decay of sensitivity with decreasing frequency. The resonant frequencies can be achieved at low values, typically between 2 and 20 Hz. To suppress the mechanical resonance, the sensors may be electrically damped by a shunt and by winding the coil on a metal form. This provides damping through induced eddy currents.

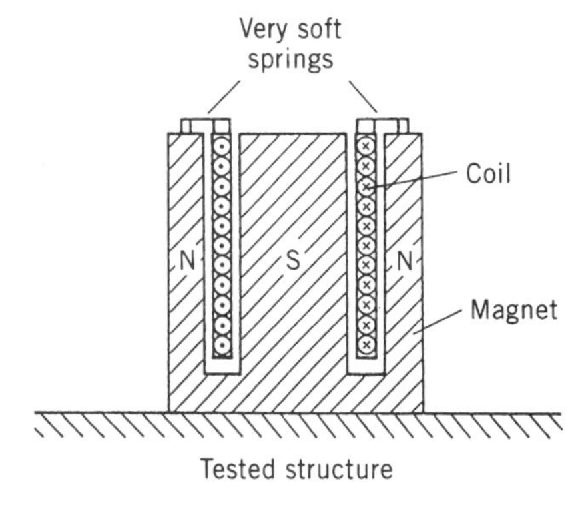

Fig. 4 Construction of electrodynamical sensors.

For the acoustical frequency range, electrodynamic transducers do not have a very constant sensitivity. This is due to their electrical network, which essentially forms a slow decaying low-pass filter. For this reason and due to their weight, electrodynamic transducers serve only for measurements on heavy structures such as buildings, bridges, and for ground vibration measurements. They therefore are frequently referred to as geophones or seismometers.

Strain Gauges This kind of sensor is purely passive. It consists of a metal foil or wire that is fixed on the surface of a test structure. The change of the sensors electrical resistance caused by the strain modulates the output voltage of a suitable electrical network, the gauge being part of it.[10,11] The transducer therefore is nonreversible.

To achieve sensitivities high enough, the gauges consist of many parallel turns (Fig. 5) of foil or wire having small cross sections. The foil or wire can be made of metal or a semiconductor. The axis of main sensitivity is parallel to the lines. Since the changes in resistance are very small, the evaluation of the resistance requires a Wheatstone bridge (Fig. 6), in which the nonactive control resistances serve to equalize the bridge.

The diagonal voltage is proportional to the strain on the gauge. This is represented by R_1 for the so-called quarter bridge with one active branch. The sensitivity can be increased with a half bridge and altering the equalization control resistances. Special use of a series circuit of two transverse gauges with the same cross sections and a length ratio equal to Poisson's ratio of the material under test allows the measurement of tension instead of strain (a result that can also be achieved using two equal gauges forming a specific angle). Measurement of torsional and radial strains are made with special gauge constructions. To compensate for temperature fluctuations, passive gauges may be incorporated in the control resistances. The frequency range of strain measurements is largely limited not by the gauge itself but by the electrical equipment and other considerations, including the vibrational wavelength with respect to the transducer dimensions. Commercially available instruments for use with strain gauges may use a DC or AC voltage to supply the Wheatstone bridge. In the latter case, in which the supply voltage is modulated by the strain, the measurement frequency range is limited to 0.1–0.2 times the car-

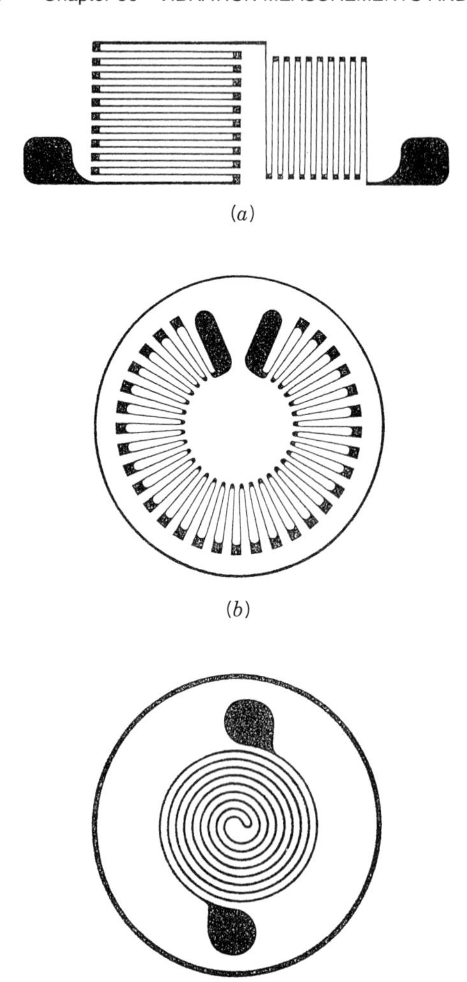

Fig. 5 Examples of strain gauges for the measurement of (*a*) tension, (*b*) radial, and (*c*) torsional strain.

rier frequency. Mainly due to their better stability, carrier systems are preferred. The handling of instruments and mounting of gauges is quite time consuming and requires very careful preparation. Thus, typical applications are long-term observations, for example, at critical points on dynamically loaded shafts and levers or on bridges and other buildings.

Strain gauges made of piezoelectric material are used to provide so-called piezoresistive accelerometers.[10] Their construction principle (similar to that of piezoelectric accelerometers) consists of a seismic mass connected to the housing by an arrangement with a finite spring constant (e.g., by a cantilever beam), carrying the strain gauge. Because of the piezoresistive accelerometer's low sensitivity, no small accelerations can be measured (lower range typically more than $1g$).

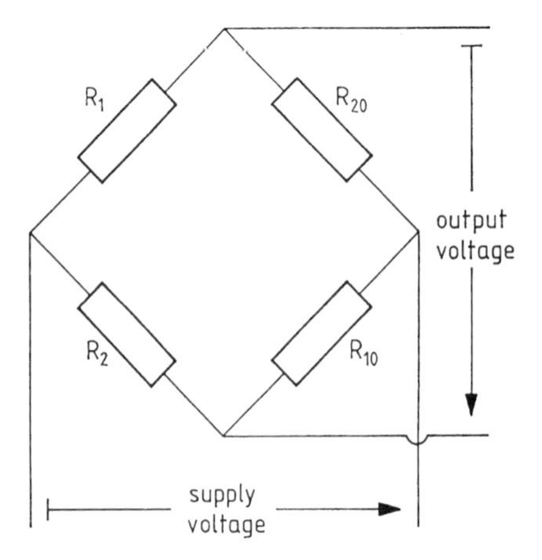

Fig. 6 Wheatstone bridge for strain measurements. R_1, R_{10}, R_2, and R_{20} may represent active or passive resistances (active resistances: quarter-bridge R_1, half-bridge R_1 and R_2, full-bridge: all).

3.3 Choice and Mounting

The correct choice of an accelerometer, or any other vibration sensor, is a question of matching the sensor to the tested structure and ensuring that it does not significantly alter it.

Generally, the mechanical input impedances (moment and force) of the structure at the measured point should be altered as little as possible by the connected sensor. For plates the highest measurement frequency depends on the mass M of the sensor and the impedance of the plate. The highest frequency is given by

$$f < \frac{0.36\rho c_L h^2}{M\sqrt{10^{\Delta L/10} - 1}}, \tag{5}$$

where ΔL denotes the relative error (dB); ρ, c_L, and h are the density, (longitudinal) wave speed, and thickness of the plate, respectively. If $\Delta L = 1$ dB, then Eq. (5) states that the sensor's mass M should be smaller than the mass of a plate disk with the radius of one-sixth of the bending wavelength. On a metal sheet of thickness of 0.5 mm (as is found in cars), a 2.4-g sensor can be used up to 750 Hz.

Clearly, for measurements on light-weight structures, such as metal sheets, small electromechanical sensors are needed. Heavy structures, like walls, allow the use of larger transducers with higher sensitivity. Often vibration levels of thin structures are quite high and broad

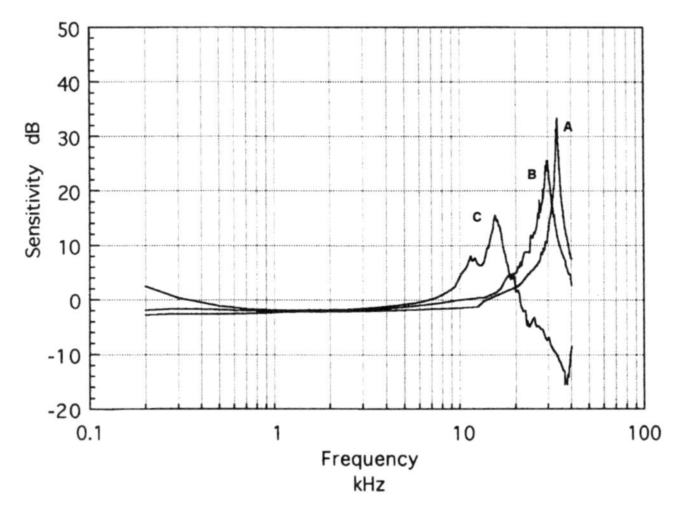

Fig. 7 Influence of sensor (43*g*) mounting, with (A) stud, (B) beeswax, and (C) handheld.

band, thus the need for transducers with high sensitivity occurs mostly for heavy structures, where normally the frequency range of interest is not too high. In most cases the broad range of available sensors (Table 1) allows a correct choice. For special purposes, such as for measurements on membranes or moving parts such as crankshafts, a sensoring is necessary where the structure is not touched (see Chapter 56). A topic is often discussed in the proper attachment of sensors to the structure. Clearly, a screwed connection is preferable. Figure 7 shows that small sensors can be attached using beeswax with almost the same result. If high frequencies are not of interest, double-sided self-adhesive tape or magnetic attachments will suffice. Heavier transducers can be mounted using good (two-component) adhesives. Hand-held monitoring often gives poor, irreproducible results.

3.4 Calibration Techniques

Vibration transducers may be calibrated using one of the following methods:

1. *Use of calibrators.* The sensitivity results from the output signal of the sensor, which is attached to an electromechanical vibration exciter with a previously known single-frequency vibration level. Exciters are available in the form of pocket-sized, battery-powered instruments that can be used where high precision is not required or as exciter tables.
2. *Comparison of the transducer's output* with that of a sensor with known sensitivity, measuring the same vibration. Care must be taken that vibrational differences between the two sensors are excluded. The influence of local mounting, such as stiffness variations, should be avoided. This can be achieved by arranging the transducers "back to back."
3. *Absolute methods* involve the measurement of the sensor's "true" vibration or transmission factor. For example, this could be done using a microscope and observing the amplitude of the (sine wave) displacement. Normally, a Michelson interferometer is used to determine the correct sensitivity. Since the instrumentation needed is so specialized, this kind of calibration is performed only for factory-certified sensitivities or for other high-precision requirements. Another absolute method is the reciprocity calibration[10] (Chapter 16). The standards[12–14] should be noted.

3.5 Optical Methods

Laser Vibrometers Laser vibrometers consist of three essential elements: (1) a radiation unit that produces one (or two) light beam(s) (the width of which is typically 5 μm) to illuminate the measuring point (or points); (2) an optical sensor, whose purpose is the monitoring of the reflected light; and (3) a signal processor, which converts the vibration information, hidden in the back scattered light, into an analog (or digital) output voltage. The basic principles for this conversion are:

1. Doppler effect, which changes the frequency of the reflected light due to the velocity of the observed scatterer

2. The evaluation of the interference stripes in the light pattern by counting the bright–dark transitions

These methods are often referred to as the corresponding codings of frequency modulation (FM) and amplitude modulation (AM), respectively. It should be noted that neither method uses the intensity of the back-scattered light as information. For FM, the output voltage is proportional to the observed velocity (two beams: difference of velocities), for AM to the displacement. For FM, the lowest measurable signal is a question of noise in the electrical path. The typical resolution is on the order of 1 μm/s. For AM, by simply counting the stripes in the pattern, the resolution is half a wavelength of the laser (approx. 350 nm for red light). Therefore a laser vibrometer may not have enough resolution to measure high-frequency vibrations. An enhancement (up to 30 dB in practice) is possible by using the continuous density information in the light pattern. Since the measurement characteristics merely depend on the internal signal processor, signal-to-noise ratios of 80 dB are often realized. For the same reason, the frequency response may be looked at as completely constant in the chosen frequency range. This is only due to the analog–digital converter.

Theoretically, the operation has no lower bound in frequency. In practice, measurements down to 0.1 Hz are possible. For this method calibration is not necessary.

Laser vibrometers have many convenient features and advantages. They provide very easy handling. Precise location of mirrors and other optical equipment is unnecessary (as is the case with holographic methods, see next section). With vibrometers the distances between the optical parts and the observed structure can vary from centimetres to metres. Using glass-fiber connections, the vibrometers allow applications in inaccessible environments and circumstances, including the observation of rotating and hot structure. Since the mechanical characteristics of the object under test are not influenced by the measurement transducer, very thin shells, membranes, needles, and so forth can be observed. Most optical sensors are of such high sensitivity that even coloring of the structure's surface is unnecessary. Note that the relative motion between the instrument and the illuminated point represents the measured quantity. Laser vibrometers therefore are not useful in situations with ambient motion (e.g., on aircrafts, trains, submarines).

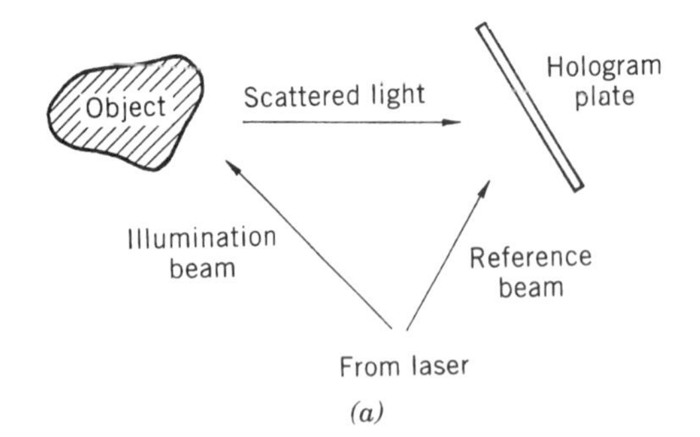

Fig. 8 (*a*) Principle of holography. (*b*) An interference pattern on a vibrating railway wheel model (time-average method used).

Holographic Methods In general, steady-state optical holography (see, e.g., Refs. 15 and 16) is a method that rearranges the scattered laser light field of an undeformed and unmoved body by the illumination of a prefabricated photographic hologram plate (Fig. 8*a*). Therefore, a three-dimensional picture of the body is produced. The hologram plate is exposed before with the originally scattered laser light from the body and a reference laser wave. The processed photographic plate contains the information about amplitude and phase of the originally scattered light, which is necessary to reproduce the spatial light distribution when illuminating the plate with the former reference wave.

Vibrational deformations of a body can be made visible if the hologram plate, for example, is exposed twice in two states of the body's displacements. Then the illumination of the plate simply delivers the sum of the current light distributions, providing an interference pattern with dark and light lines on the picture of the body's surface (see Fig. 8*b*), each line describing a contour of equal displacement. One dark–light–dark transition corresponds to a difference $\Delta\xi$ in the displacement that is proportional to the laser wavelength $\lambda_c(\Delta\xi = \lambda_c/2)$. For periodic, not necessarily single-frequency vibrations, the best resolution is obtained by exposing at the two inversion points. Normally, measurements are made at resonance.

Another method is to expose the hologram plate for a time longer than a cycle of vibration. The reconstructed picture is a result of the time mean of all wavefronts that were scattered from the object during the exposure time. The time of rest in a state of deformation is decisive. This time is longest in the inversion points. The hologram therefore is determined more or less by these two states, the resolution is $\Delta\xi = \lambda_c/4$. A related third method is the exposure of the nonvibrating object and illumination of the hologram plate by both the scattered light from the vibrating body and the reference laser wave. Since the eye (and a photographic picture) integrates only over the absolute intensity, again a pattern of dark and light lines is provided, whose resolution is $\Delta\xi = \lambda_c/2$. The very specialized equipment needed for optical holography makes it an instrument to be used only in special cases. Since the numerical evaluation of the results implies a quite complicated handling, this optical form of modal analysis is normally used to give an illustrated survey of the vibrational shapes.

4 SPECIAL VIBRATION MEASUREMENTS

Frequently, one is confronted with the simple problem of investigating the vibrational characteristics of a structure by measuring suitable quantities. Some examples follow:

- Vibration control of highly sensitive optical or medical equipment
- Surveillance of the vibration of structures in the vicinity of intense construction works (e.g., close to critical pressure vessels or pipelines)
- Investigation of vibration transmission paths
- Determination of acoustical radiation factor

In some cases the forces introduced into a system by an external excitation are of interest but are not directly measureable (e.g., elevators in buildings). This can be determined by an indirect method, using the principle of reciprocity.[1]

4.1 Measurement of Material Parameters

Important mechanical material parameters are those characterizing the specific stiffness (Young's modulus, shear modulus, bending stiffness, etc.) and a quantity describing the internal losses. These parameters are often determined experimentally using resonators, which contain a probe of the material (Table 2). From the resonant frequencies, the stiffness quantities can be computed.

With the same setups, the losses can be found by measuring the half-power bandwidth or, in the case of low loss factors and frequencies, the reverberation time. Other quantities describing losses can be computed by means of Table 3. If the frequency response curves are found with fast Fourier transform (FFT) methods, the half-power bandwidth should contain at least four to six frequency samples.

Only rather small loss factors can be found using resonance experiments, the limit depending on the resonator used. Other methods are available for higher damping. These include:

- Locally sampling the wave field on beams, similar to measurements of absorption coefficients in impedance tubes. This method can be used for weak damping as well.
- Measurement of local level reduction D per metre along beams ($D = 27.2\,\eta/\lambda_L$ for longitudinal waves with length $\lambda_L, D = 13.6\,\eta/\lambda_B$ for bending waves with length λ_B).
- Input impedance measurements (e.g., of the simple mass–spring system).

All these methods are based on a mathematical model for the specific structure with a number of unknown parameters, which are to be determined by the measurements. Generally, the list of mechanical structures, of which the resonators mentioned above are examples, may be enlarged with any structure for which a precise mathematical description is well known. The material parameters then result from one or more fits between measurements and the theoretical model, which, again, is the basic idea of modal analysis (see Chapters 48 and 56). Errors occur due to violations of the assumptions, inherent in the model, or from external influences not taken into account. For example, experiments with the simple resonator must be made under the condition that only the relevant "whole-body mode" is excited and analyzed. For experiments based on specific wave types, transformations into other wave types must be avoided. Losses result not only from internal damping but often from the effects of energy transport over the boundaries or supports. This influence must be reduced or avoided

TABLE 2 Estimation of Mechanical Stiffness from Resonators[a]

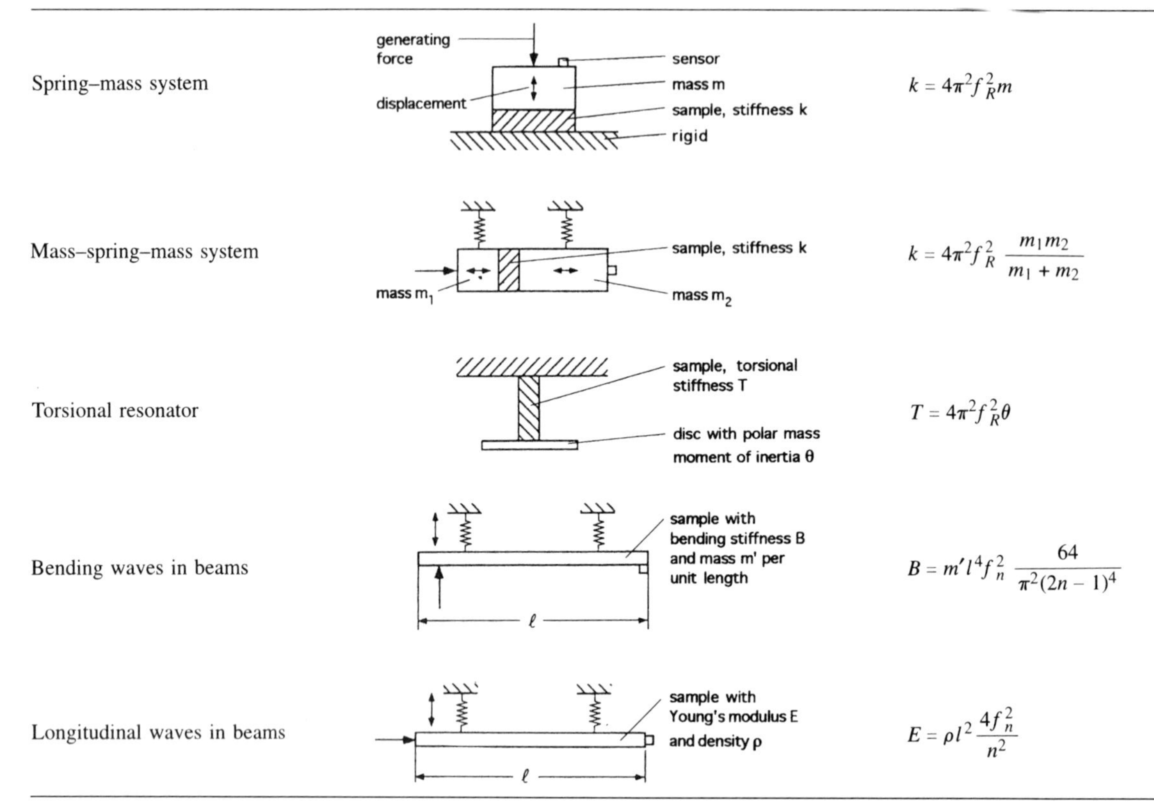

System	Formula
Spring–mass system	$k = 4\pi^2 f_R^2 m$
Mass–spring–mass system	$k = 4\pi^2 f_R^2 \dfrac{m_1 m_2}{m_1 + m_2}$
Torsional resonator	$T = 4\pi^2 f_R^2 \theta$
Bending waves in beams	$B = m' l^4 f_n^2 \dfrac{64}{\pi^2 (2n-1)^4}$
Longitudinal waves in beams	$E = \rho l^2 \dfrac{4 f_n^2}{n^2}$

[a] f_R, resonance frequency; f_n, resonance frequency with n nodes; S, cross-sectional area; d, thickness of sample; k, SE/d; E, Young's modulus; ρ, density; l, length.

TABLE 3 Descriptors for Losses

	Loss Factor η	Half-Power Bandwidth Δf	Degree of Damping ϑ	Reverberation Time T	Q Factor	Damping Coefficient c	Log. Decrement λ
$\eta =$	η	$\Delta f/f_0$	2ϑ	$\dfrac{2.2}{f_0 T}$	$\dfrac{1}{Q}$	$c/\sqrt{km}$	$\dfrac{2\lambda}{\sqrt{4\pi^2+\lambda^2}}$
Δf	$f_0\eta$	Δf	$2f_0\vartheta$	$\dfrac{2.2}{T}$	$\dfrac{f_0}{Q}$	$\dfrac{c}{2\pi m}$	$\dfrac{2f_0\lambda}{\sqrt{4\pi^2+\lambda^2}}$
$\vartheta =$	$\eta/2$	$\Delta f/2f_0$	ϑ	$\dfrac{1.1}{f_0 T}$	$\dfrac{1}{2Q}$	$\dfrac{c}{2\sqrt{km}}$	$\dfrac{\lambda}{\sqrt{4\pi^2+\lambda^2}}$
$T =$	$\dfrac{2.2}{f_0\eta}$	$\dfrac{2.2}{\Delta f}$	$\dfrac{1.1}{f_0\vartheta}$	T	$\dfrac{2.2Q}{f_0}$	$\dfrac{13.82m}{c}$	$\dfrac{1.1\sqrt{4\pi^2+\lambda^2}}{f_0\lambda}$
$Q =$	$1/\eta$	$f_0/\Delta f$	$\dfrac{1}{2\vartheta}$	$\dfrac{f_0 T}{2.2}$	Q	$\dfrac{\sqrt{km}}{c}$	$\dfrac{\sqrt{4\pi^2+\lambda^2}}{2\lambda^2}$
$c =$	$\eta\sqrt{km}$	$2\pi m\,\Delta f$	$2\vartheta\sqrt{km}$	$\dfrac{13.82m}{T}$	$\dfrac{\sqrt{km}}{Q}$	c	$2\lambda\sqrt{\dfrac{km}{4\pi^2+\lambda^2}}$
$\lambda =$	$\dfrac{\pi\eta}{\sqrt{1-(\eta/2)^2}}$	$\dfrac{\pi\Delta f}{\sqrt{f_0^2-(\Delta f/2)^2}}$	$\dfrac{2\pi\vartheta}{\sqrt{1-\vartheta^2}}$	$\dfrac{6.91}{\sqrt{f_0^2T^2-1.21}}$	$\dfrac{\pi}{Q\sqrt{1-(\frac{1}{2}Q)^2}}$	$\dfrac{2\pi c}{\sqrt{4km-c^2}}$	λ

f_0 = undamped natural frequency = $\sqrt{k/m}/2\pi$, k = spring constant, m = mass.

by the correct choice of supports to decouple the structure under test from the environment. Radiation losses also contribute to the vibration damping. Their loss factor is approximately

$$\eta = \frac{\rho c}{m'' \omega} \sigma \qquad (6)$$

for plates with mass m'' per unit area where σ describes the radiation efficiency, and ρc denotes the specific wave impedance in the surrounding medium. Finally, when interpreting the results, it must be remembered that these results are sometimes based on certain assumptions and are hence not easily transferable to more general cases (e.g., the loss factors of sandwich plates differ significantly for longitudinal and bending waves).

4.2 Modal Analysis

The object of modal analysis is the determination of eigenmode shapes for the object under investigation.[17] This implies that the description of the structure's vibration in terms of modes is meaningful, which is not the case for high damping. No energy should be lost to the surrounding structures and medium, and the internal damping should not be high. Free boundary conditions should be sought when appropriate. This is usually achieved by suspending the test object by very light springs, which results in low-frequency rigid-body modes (modes without elastic deformation of the structure itself). The application of modal analysis ranges from small structures, such as engine crankcases, engine blocks, and other technical components, to large constructions such as an entire car chassis, an aeroplane, or a railway carriage.

In principle, the structure under investigation is set in motion by one or more broadband force inputs, and the resulting motion is observed. One method is to excite the structure in a certain point (e.g., using a shaker) and to measure the structure's response at many points on the surface. Alternatively, the reciprocity principle allows the measurement of the response in one point only with excitation in many points one after another (usually using an impulse hammer). The input and output signals are converted into transfer functions with an FFT analyzer. The transfer functions are then fed into a computer with the modal analysis software. This software is used to calculate resonant frequencies, damping, and the mode shapes by "curve-fitting" methods, for which several different procedures are available (a more detailed description of the theory of modal analysis is given in Chapter 48). Besides the obvious applications of the modal analysis technique (examination of the sound radiation, vibrational discomfort in vehicles, aimed noise control measures, etc.), it also can be utilized to assist in modeling structural modifications. The orthogonality of the modal forms allows for the calculation of the so-called mass and stiffness matrices of the structure. These can be altered by the investigator in a number of different ways and the resulting changes can be seen in the newly calculated parameters and mode shapes.

Modal analysis has a number of inherent problems. They can be caused, for example, by the local properties of the test structure (e.g., local distribution of damping), which cannot be described by a single global value. It also is possible that the use of different curve-fitting methods leads to noticeable differences between modal forms, especially for strongly coupled modes. Therefore, in practice, an error observation analysis is often unavoidable (e.g., the construction of synthetic transfer functions through the predetermined modal parameters and the comparison with the measured curves, test of orthogonality, evaluation of modal parameters and shapes using different curve-fitting methods, etc.).

REFERENCES

1. L. Cremer and M. Heckl, *Structure-Borne Sound: Structural Vibrations and Sound Radiation at Audio Frequencies*, translated and revised by E. E. Ungar, Springer, Berlin, 1988.
2. F. J. Fahy, *Sound and Structural Vibration: Radiation, Transmission and Response*, Academic, London, 1985.
3. M. C. Junger and D. Feit, *Sound, Structures and Their Interaction*, MIT Press, Cambridge, MA, 1986.
4. M. P. Blake and W. S. Mitchell, *Vibration and Acoustic Measurement Handbook*, Spartan Books, New York, 1972.
5. C. M. Harris, *Handbook of Acoustical Measurements and Noise Control*, 3rd ed., McGraw-Hill, New York, 1991.
6. G. Pavic, "Structural Surface Intensity: An Alternative Approach in Vibration Analysis and Diagnosis," *J. Sound Vib.*, Vol. 115, 1987, pp. 405–422.
7. *Proc. 3rd Int. Congress on Intensity Techniques. Structural Intensity and Vibrational Energy Flow*, Centre Technique des Industrie Mécaniques, Senlis, 1990.
8. J. T. Broch, *Mechanical Vibration and Shock Measurement*, 2nd ed., Brüel & Kjaer, Naerum, Denmark, 1988.
9. M. Serridge and T. R. Licht, *Piezoelectric Accelerometers and Vibration Preamplifiers*, Brüel & Kjaer, Naerum, Denmark, 1987.
10. C. M. Harris, *Shock and Vibration Handbook*, 3rd ed., McGraw-Hill, New York, 1988.
11. J. Vaughan, *Strain Measurement*, Brüel & Kjaer, Naerum, Denmark, 1975.
12. American National Standards Institute, "Selection of Calibration and Tests for Electrical Transducers Used for Measuring Shock and Vibration," ANSI S2.11, New York, 1969; rev. 1978.
13. American National Standards Institute, "Methods for the

Calibration of Shock and Vibration Pickups," ANSI S2.2, New York, 1959, rev. 1981.

14. International Standardization Organization, "Methods of Calibration of Vibration and Shock Pickups," ISO 5347, 1985.
15. G. M. Brown, R. M. Grant, and G. W. Stroke, "Theory of Holographic Interferometry," *J. Acoust. Soc. Am.*, Vol. 45, 1969, pp. 1166–1179.
16. J. W. Goodman, *Introduction to Fourier Optics*, McGraw-Hill, New York, 1968.
17. D. J. Ewins, *Modal Testing Theory and Practice*, Wiley, New York, 1984.

57

MACHINERY CONDITION MONITORING

GOVINDAPPA KRISHNAPPA

1 INTRODUCTION

Mechanical condition monitoring is concerned with evaluation of the operating condition of a machinery system or its components and is accomplished through the analysis and interpretation of signals acquired from sensors and transducers. The main purpose of condition monitoring is to detect the presence of faults of damage in machinery during operation. It encompasses both diagnosis and prognosis in order to determine the remaining safe operating life of a machine before a breakdown or failure occurs. *On-condition* operation of machinery leads to efficient maintenance by preventing unexpected failures and improving safety.

Vibration analysis is a reliable technique that is used extensively for condition monitoring by machinery users and operators. Dynamic forces operating within the machinery generate vibratory forces that carry information on the condition of individual components. These vibrations are transmitted via complex paths to the casing, producing external vibration and noise radiation. Sources of vibration include impacts, forces in gears and cams due to machining variations and contact stiffness, electromagnetic forces, combustion pressures, and forces due to air and liquid flow. Detailed discussions on vibrating systems can be found in Chapters 47 and 48. In very simple machines, peak and root-mean-square (rms) values of vibration energy are trended to monitor system condition. Frequency spectra play a valuable role in fault detection, especially in rotating machines, and contain information on operating conditions that reveal malfunctions such as rotor imbalance, shaft misalignment, and bearing and gear damage. In complex machines containing multiple components, more elaborate signal processing and analysis techniques are required to extract information on the component of interest from the overall signal to enable effective detection of early damage.

Machinery diagnostics involve extracting the properties of vibration signals (temporal waveform or frequency spectra) of machinery components operating in good, as well as faulty, conditions. Under both conditions, the properties of signals generated by the individual components are recovered from the overall signature. In addition to energy levels, phase information can be used effectively to diagnose machinery condition in many instances. Statistical descriptors of the wave or spectral forms are useful to identify faults generated by impulsive events.

This chapter describes vibration analysis techniques for condition monitoring of mechanical components. Familiarity with general vibration theory and sensors and instrumentation for acquiring vibration data are required. Some knowledge of the basic tools of signal processing theory and statistical concepts would be helpful. Since this area is the subject of further research and development, only well-established techniques and methods are presented.

2 VIBRATION CHARACTERISTICS OF MECHANICAL FAULTS

Any machinery operation involves the generation of forces that produce vibrations. Even a machine in good running condition produces its own characteristic vibrations, caused by various dynamic forces associated with its operation. Sources of various dynamic forces that pro-

Note: References to chapters appearing only in the *Encyclopedia* are preceded by *Enc.*

duce vibrations include: imbalance, reciprocating action, misalignment, impact, sliding, electromagnetic forces, combustion, pressure, fluctuating forces due to air and liquid flows, variations in surface finish of machinery parts, and variations in local contact stiffness.

As machines wear, the play between the parts increases and the individual parts impact much harder on each other, producing stronger vibrations. Bearing wear displaces a shaft from its original axis of rotation to produce imbalance. Worn rotating and sliding surfaces show fatigue cracks. Vibration levels increase due to irregularities in the surface finish of the contacting machine part.

As a machine ages, vibration levels tend to gradually increase with only a slight change in the vibration pattern. Vibrations increase rapidly when machinery components approach the end of their useful life, producing drastic changes in vibration levels and patterns. Although individual faults produce their own characteristic vibration patterns, identification of a fault can be extremely difficult because of vibrations produced by other components in the machine and the presence of complex transmission paths between the component and the measurement point. Some faults on less complex machines can be identified by placing the transducer close to the mechanical component.

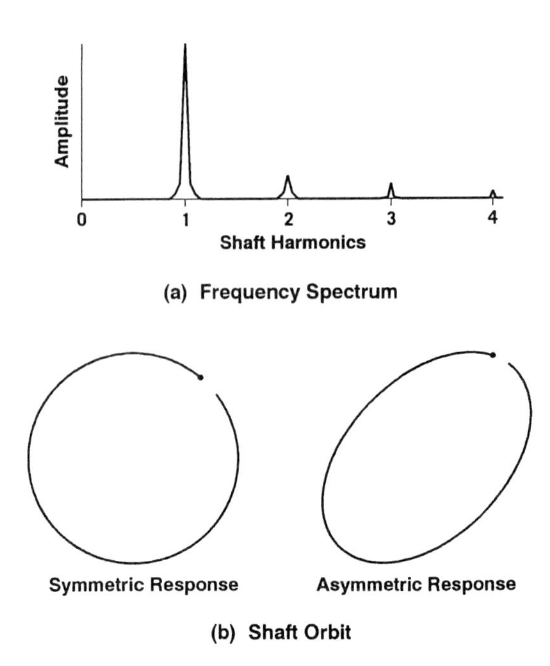

Fig. 1 Vibration characteristics of imbalance.

2.1 Imbalance

Imbalance is a frequent cause of failure in machines. The most common cause of imbalance is a physical difference between the mass center of gravity and the rotating center of the rotor system.[1] Magnetic forces and temperature gradients also cause shifts in the rotor system that produce imbalance. Imbalance can be characterized by a high-amplitude vibration signal at the shaft rotation frequency, frequently with an elliptical orbit (Fig. 1).[2] Excessive imbalance will also produce an abnormally high peak amplitude at the critical speed of the rotor system, decreasing in amplitude beyond the critical speed. Figure 2 shows that the vibration amplitude will remain higher than that of a balanced rotor at corresponding speeds.

2.2 Misalignment

Shaft misalignment is usually caused by the incorrectly adjusted bearings or couplings. Severe damage could result if care is not taken to correct this fault. Misalignment is expected in all but the most precisely manufactured equipment. Misaligned shaft bearings typically produce vibration characteristics similar to those shown in Fig. 3. Vibration amplitude levels at the first four harmonics of the shaft rotation frequency are high, with a peak at the second harmonic. Misalignment is usually associated with a peak vibration level at twice the shaft rotation frequency. The characteristics of shaft motion for minor, moderate, and severe misalignments are also shown in the figure. Poor alignment can cause bearings to be unloaded to the point where instability is induced. Other problems can include coupling lockup, resulting in thrust bearing failure.

2.3 Mechanical Looseness

Mechanical looseness, the improper fit between component parts, is generally characterized by a long series of high-amplitude harmonics of the shaft frequency (Fig.

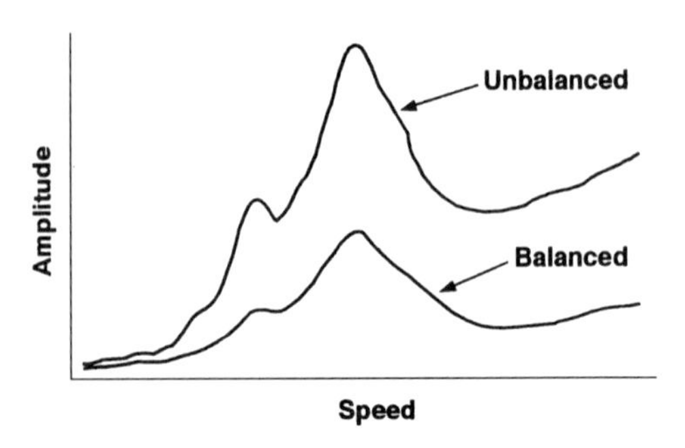

Fig. 2 Vibration amplitude response of balanced and unbalanced rotors.

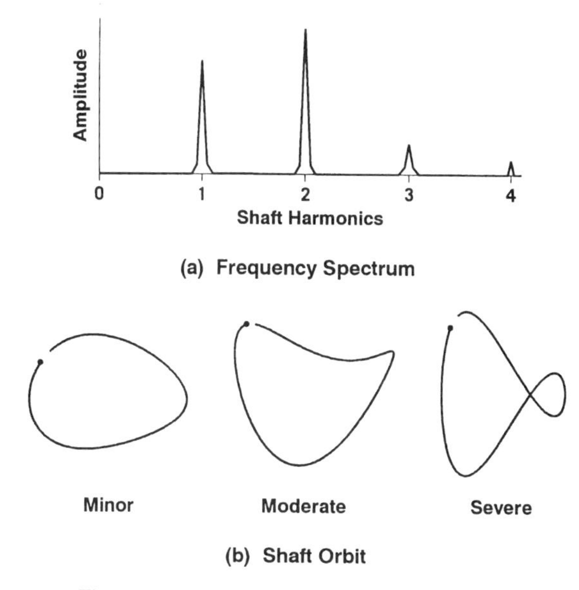

Fig. 3 Vibration characteristics of misalignment.

4). Although the exact method by which the harmonics are generated is not well understood, they probably originate from a nonlinear response of the loose part to dynamic inputs from the rotor. Looseness of bearings is also likely to cause vibrations at subharmonics of the shaft rotation frequency (i.e., $f/2, f/3, \ldots, f/n$).[2]

2.4 Rubs

A rub is generally a transitory phenomenon caused by several mechanisms, the most common of which are rotor and seal rubs. Blade rubs induced by insufficient clearance between the blade tips and the casing wall

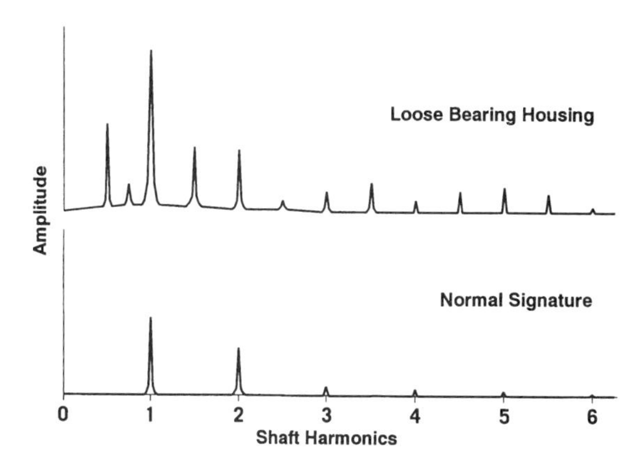

Fig. 4 Vibration characteristics of mechanical looseness.

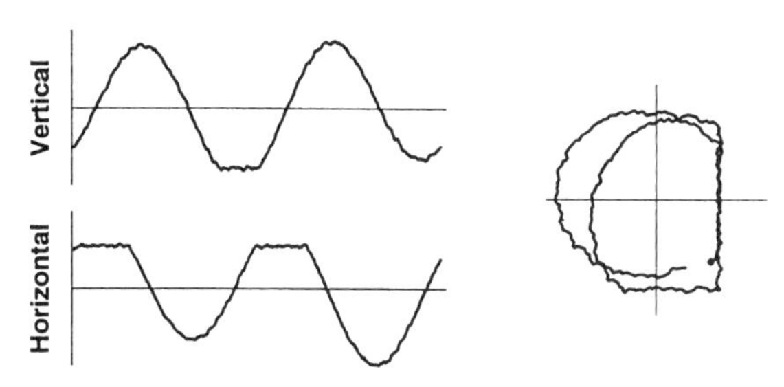

Fig. 5 Waveform and orbit of a rub.

are common in turbomachinery. Bearing wear, caused by problems such as lubrication starvation, imbalance, or vibrations, can also lead to rotor rubs. Rub changes the vibration characteristics of the rotor and its orbit. In the simple case shown in Fig. 5, a flattened waveform with subsynchronous frequency components and a flattened orbit diagram are indications of rub.[2] Because rub is a nonlinear phenomenon, it is not possible to characterize vibration signatures for all cases.

2.5 Whirling in Rotating Machinery

One frequent failure mode in rotating machinery is caused by whirling or whipping of the shaft due to self-excited vibrations. The energy for this excitation is provided by a uniform source of power associated with the machinery system itself. This source of power can give rise to oscillating forces from various mechanisms inherent in the system. The rotor system will vibrate at its own natural or critical frequency, essentially independent of any external stimulus. The direction of whirl may be in the same direction as the shaft rotation (forward whirl) or opposite to it (backward or reverse whirl). The direction of the whirl is dependent on the direction of the destabilizing force.[3]

In rotating machines, whirling is caused by the generation of a tangential force, normal to an arbitrary radial deflection of a rotating shaft, whose magnitude is proportional to the deflection. At the onset of whirl, such a force will overcome any stabilizing external damping forces that are generally present. Whirling motions are produced with an ever-increasing amplitude, limited only by nonlinearities that ultimately limit the maximum shaft deflection. All the self-excited systems involve friction or fluid energy mechanisms in the generation of destabilizing forces. A simple mathematical analysis shows that the trajectory of the center of gravity of a whirling rotor is an exponential spiral (Fig. 6). Some important examples of whirling and whirling instabilities are hysteretic whirl, fluids trapped in hollow rotors, dry friction whip, and fluid bearing whip.

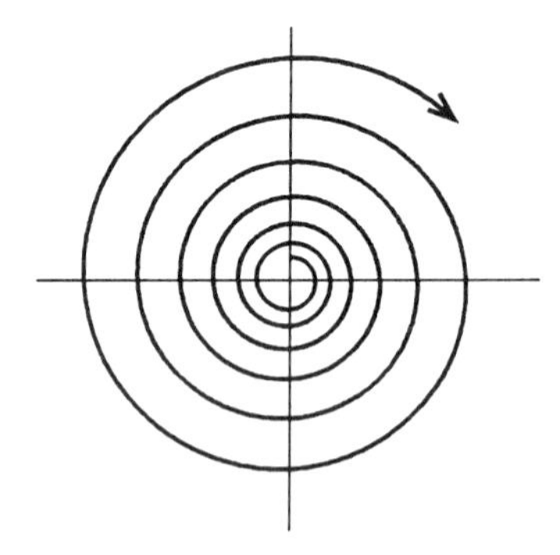

Fig. 6 Trajectory of the center of gravity of a rotor's mass due to whirling.

2.6 Hysteretic Whirl

Hysteretic whirl is caused by internal friction between the fibers within the rotating shaft or between fitted components, producing a phase shift between the neutral stress axis and the neutral strain axis. A displacement in angular orientation between the two axes gives rise to a resultant force, not parallel to the deflection of the shaft. The net effect is that the resultant force has a tangential component normal to the deflection, a prerequisite for whirling to occur. The tangential component is in the direction of rotation, inducing a forward whirl motion in the direction of rotation. This increases the centrifugal forces on the deflected rotor, thereby increasing the deflection. It has generally been recognized that hysteretic whirl can only occur at rotational speeds above the first shaft critical speed. Once whirl has started, the critical whirl speed that will be induced will have a frequency approximately half that of the onset speed.

2.7 Whirl due to Fluid Trapped in Rotor

This occurs in hollow rotors, mostly in high-speed rotating machinery, where liquids (such as oil in bearing sumps or steam condensates) may inadvertently be trapped in the internal cavity of the rotors. The spinning surface of the cavity drags fluid in the direction of rotation. The deflection of the shaft and the centrifugal forces on the trapped fluid produce a force having a tangential component, inducing a forward whirl. It has been shown that the onset speed for instability is always above the critical speed of the rotor and below twice critical speed.

2.8 Dry Friction Whip

Dry friction whip is experienced when the surface of the rotating shaft comes in contact with an unlubricated stationary surface. This can be caused by inadequate journal bearing lubrication, contact between rotors and labyrinth seals, or turbomachinery blade rubs. The contact between the surface of the rotating shaft and the stationary surface will induce a tangential force on the rotor, which then produces a whirling motion. This generates a larger centrifugal force on the rotor. The whirl produced is counter to the direction of shaft rotational (backward whirl). No first-order interdependence of whirl speed with rotational speed is established. It has been suggested that the whirl frequency is half the rotational speed of the rotor.

2.9 Fluid Bearing Whip

This effect appears in journal bearings, such as those found in high-speed turbomachinery. Radial deflection of a shaft rotating in a fluid-filled clearance entrains viscous fluid, which will circulate with an average velocity of about half the shaft speed. Pressure on the upstream side of the reduced clearance zone will be higher than that on the downstream side. The resultant bearing force will exert a tangential force component in the direction of rotation, inducing forward whirl. Any induced whirl produces increased centrifugal forces, which further reduce clearances and generates ever-increasing destabilizing tangential forces. This instability normally occurs at shaft speeds greater than twice the critical speed, with a whirling frequency at approximately half the shaft speed.

2.10 Rolling Element Bearings

Faults in a rolling element bearing start as a single defect caused by metal fatigue in either the raceways or the rolling elements. A normal bearing in good condition produces vibration signals that are essentially random, with only a few minor impulsive events present. The vibration signature of a damaged bearing is dominated by impulsive events related to ball or roller passing frequencies.[4,5] Figure 7 shows the general characteristics of a vibration signal at various stages of damage, in both the time and frequency domains. As damage spreads around the bearing, significant increases in the amplitude of both the broadband noise and the defect frequencies occur. Formulas for rolling element, inner and outer race ball passing, and train frequencies can be found in Ref. 6.

In a simple machine where the signal from the other components does not contaminate or mask the vibration signal, bearing condition can be monitored by measuring the crest factor. This is the ratio of the peak to rms values of the wide-band vibration signal and can be obtained from an accelerometer mounted directly on the bearing. Defects in the bearing components produce sharp peaks

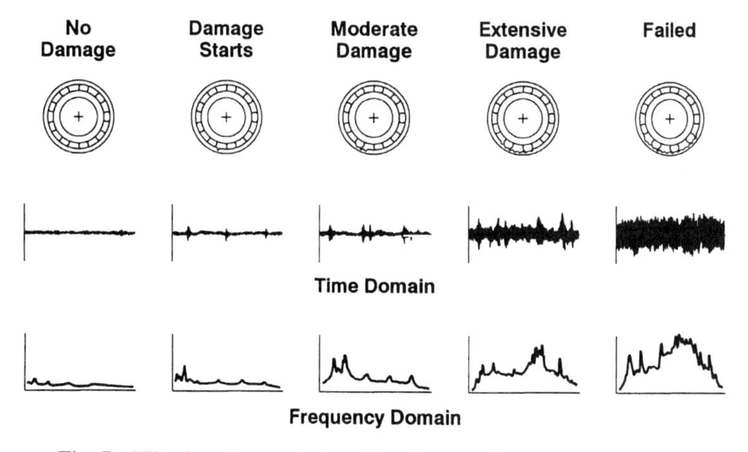

Fig. 7 Vibration characteristics of bearing at various stages of damage.

with higher crest factors. A crest factor of 4–6 is indicative of normal operation; a range of 6–12 marks the presence of damage.

The impulsive nature of the vibration signature of a defective bearing can be quantified using the normalized fourth moment of the time-domain signal. This statistical value is commonly called the kurtosis. A formula for calculating the kurtosis is given in the next section. A bearing in good operating condition will produce a kurtosis value of around 3 (indicative of random noise). Minor damage, such as a single fatigue spall, increases the impulsiveness of the signal and produces a kurtosis of about 6. As the damage spreads, the signal becomes less coherent and the kurtosis value falls. Finally, the damage becomes extensive and spreads around the periphery of the loaded region. At this stage, the vibration signature again becomes essentially random in nature because of the large number of impulsive events present, and the kurtosis values fall back to around 3. A kurtosis value of 3, for both a good and a damaged bearing, appears rather confusing at first. However, a significant increase in energy levels, as shown in Figure 7, is associated with advanced stages of damage.

Most bearing monitoring techniques rely heavily on the ability to place the transducer on the bearing housing, close to the loaded zone of the stationary outer race. The measured signal should be free from other sources of vibrations, otherwise the low-energy level of the bearing signal will be masked by background vibration signals. Using advanced signal processing techniques, the bearing signal can be recovered from the overall signal of the machinery system. One of the techniques used is called the high-frequency resonance technique and is described in the next section.

2.11 Gears

Vibration analysis is a powerful technique for the early detection of failure in gearboxes and transmissions.[7] A nominally perfect gear produces almost sinusoidal signals in the time domain, with prominent peaks at the tooth meshing frequency and lower-order harmonics in the frequency domain. Faults, such as a localized tooth defect or misalignment, alter the signal in the time domain and change the pattern of the frequency spectrum, yielding diagnostic information. Common faults that occur in gears can be identified from the time- and frequency-domain information. In a complex gearbox, the vibration signature obtained from an accelerometer placed on the outside surface contains contributions from all the components of the system. The spectrum of the overall signature cannot always give diagnostic information, as the signature of the gear of interest is often masked by the signatures of other components. Using synchronous time-domain averaging, explained in the next section, the time waveform of individual gears can be recovered from the overall signature.

Time-domain averaged waveforms and frequency spectra for some common gears faults are shown in Fig. 8.[8] A nominally perfect gear exhibits a sinusoidal waveform in the time domain, with a dominant peak in the spectrum at the meshing frequency. A misaligned gear produces a once-per-rev amplitude modulation in the time domain. A broken or fractured gear tooth fails to transmit its share of the load; the gears accelerate to engage the next pair of teeth and produce an impact upon engagement. This impulsive behavior produces sidebands about the mesh frequency and its harmonics in the frequency domain. Spalled teeth and soft

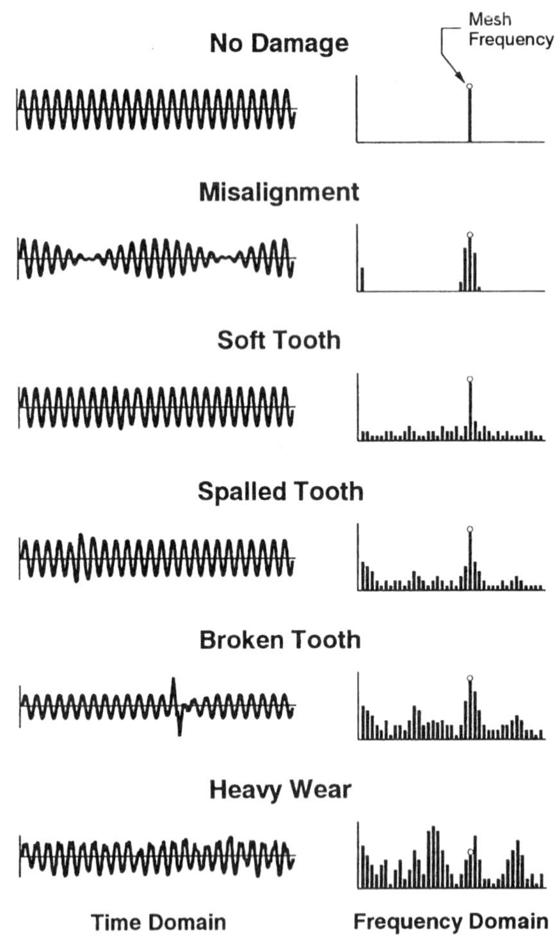

Fig. 8 Waveforms and spectra produced by common gear faults.

teeth behaved in a similar manner. A heavily worn gear, with wear all around or on a large number of teeth, produces a signature that is quite random in character.

Faults in gears can also be detected by the application of amplitude and phase modulation analysis techniques on the time-domain-averaged vibration signal. Signal modulations produced by a defective gear can be analyzed in the time domain itself, by narrow-band filtering about the mesh frequency and harmonics.[9–11] An example of this is shown in Fig. 9.

3 SIGNAL ANALYSIS AND PROCESSING TECHNIQUES

Condition monitoring, by definition, is the act of extracting information from a specific system, by means of

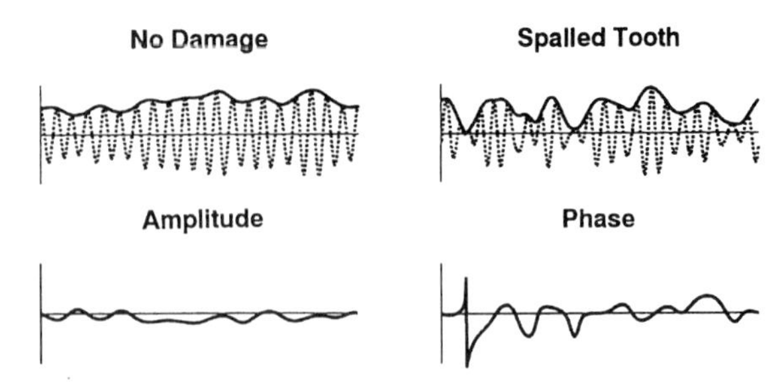

Fig. 9 Amplitude and phase time-domain signatures of undamaged and damaged gears.

appropriate measurements. Vibration monitoring consists of acquiring vibration data for use as information carriers on the condition of the machinery system/component. The raw vibration data must be processed and analyzed to determine whether or not a fault exists.[12–14] It can also be important to determine the nature of the fault; for example, to know whether the fault is local tooth damage or misalignment in gear condition monitoring. The processing techniques and analysis methods frequently depend on the machinery component and the nature of the fault to be detected.

3.1 Data Processing and Analysis

The basic operation of a data acquisition and processing system is described in this section. The analog signal from the sensor is routed through an antialiasing filter (low-pass filter), before performing analog-to-digital conversion. The main objective in signal processing is to enhance the desired signal and remove any undesirable features. This is done to improve the signal-to-noise ratio and facilitate the recovery of the signal generated by the machinery component of interest.

The next phase of signal processing and analysis is concerned with extracting features associated with a particular type of fault or component damage. In simple machines, these could be based on signal energy level measurements, such as rms, peak, crest factor, or spectral composition involving frequency-domain analysis. Complex machines with several components require more specific diagnostic techniques and tools to extract the features associated with particular faults. Analysis methods[15,16] use both time- and frequency-domain data, applying several advanced signal processing techniques. Pattern recognition techniques and statistical methods are also extremely useful in feature extraction. Detailed discussions on instrumentation and signal processing procedures can be found in Chapters 56 and 82.

Frequency-domain analysis methods dominate most mechanical signal analysis applications. The prevalence of spectral analysis has increased even more with the emergence of digital analyzers. The potential value of time-domain methods might have been overlooked. The recognition and analysis of patterns in the time domain is useful in identifying faults related to the kinematic motion of the machinery. In the following paragraphs we will discuss some of the time- and frequency-domain methods. Commercial software packages are available for most of the analysis techniques and some have been implemented in instrumentation hardware.

3.2 Orbit Displays

Many common problems in rotating machinery, such as imbalance and whirling, can be identified from the shape of a shaft orbit.[2] Orbit diagrams are generated using two noncontact vibration displacement transducers spaced 90° apart in the radial direction of the shaft. Various shaft orbit patterns are shown in Fig. 10. A circular orbit is obtained from two sinusoidal waveforms equal in amplitude. Two sinusoidal waveforms of different amplitudes produce an elliptical orbit, possibly indicating misaligned coupling restraints. If the two waveforms contain more than one discrete frequency, the resulting orbit pattern will increase in complexity.

3.3 Time-Domain Averaging

One of the advantages of time-domain averaging lies in the possibility of relating signal changes to specific kinematic events. Time-domain analysis can readily be performed on rotating machinery, as it is relatively easy to access reference time signals from sensors monitoring the main drive shaft. Synchronous time averaging with respect to the rotation of a component eliminates any unrelated external signals. Figure 11 shows how the synchronous time average is derived for one of the gears in a gearbox. The time record length of the averaged signal corresponds to one complete mesh cycle of the gear (the time taken for all of the teeth on a gear to pass through mesh).

3.4 Amplitude-Domain Analysis

In wear-induced failures, localized defects appear first. As these grow and spread, the damage becomes more distributed in nature. The vibration signature often is impulsive in nature at the onset of a localized defect, tending to change with time to a more continuous function. The sharp peaks at the onset of a defect affect the tails of the probability density function and the moments of the distribution. The crest factor and kurtosis function are used to describe the shape characteristics of the signal. Crest factor and kurtosis are defined by

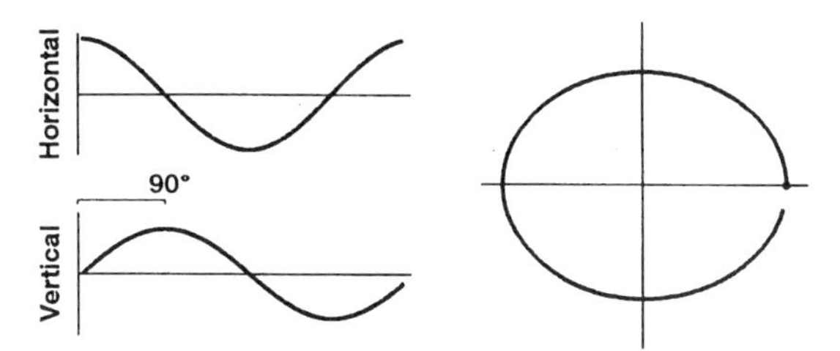

(a) Orbit produced by two sinusoids with a phase difference of 90°.

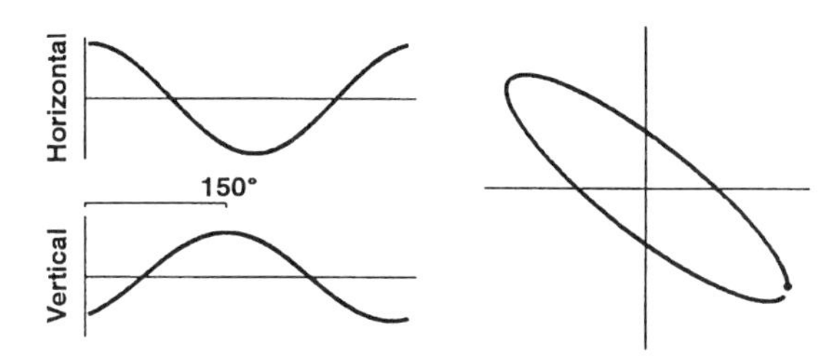

(b) Orbit produced by two sinusoids with a phase difference of 150°.

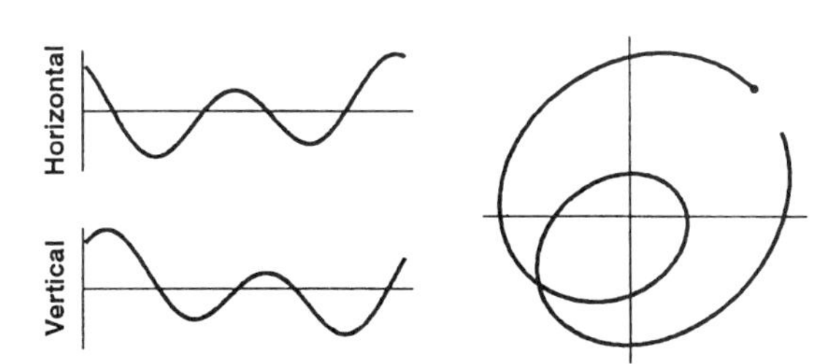

(c) Orbit produced by two complex periodic waveforms.

Fig. 10 Orbit displays.

$$\text{Crest factor} = \frac{\text{peak}}{\text{rms}}, \tag{1}$$

$$\text{Kurtosis} = \frac{1}{N} \frac{\sum_{n=1}^{N} (x_n - \overline{x})^4}{\left(\sum_{n=1}^{N} (x_n - \overline{x})^2\right)^2}. \tag{2}$$

The kurtosis is sensitive to the impulsiveness or "spikiness" of the data. The kurtosis of random or Gaussian noise is 3.0. The vibration signal of a rolling element bearing in good condition is expected to have this value. The kurtosis level rises sharply to around 6 at the onset

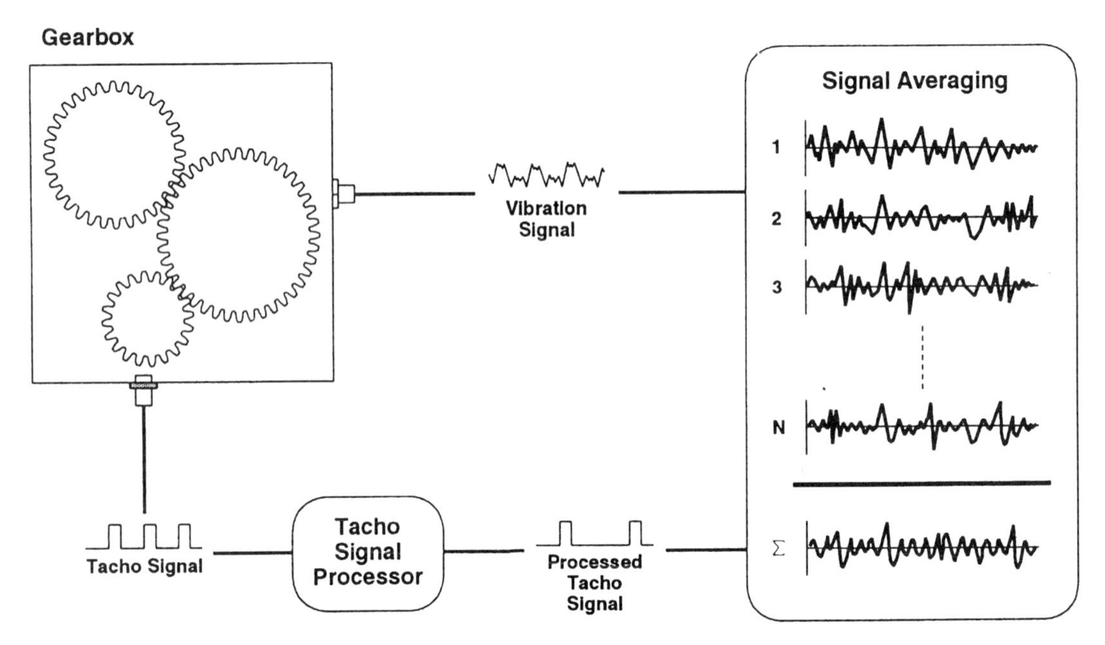

Fig. 11 Time-domain averaging for gear signal processing.

of discrete damage (i.e., a spall on the rolling elements or raceways). The kurtosis value falls as the damage spreads, indicating that the damage characteristics have become more random in nature.

3.5 Envelope Spectrum Analysis Technique

Most bearing monitoring techniques rely on the ability to place the vibration transducer on the bearing housing, close to the outer raceway load zone, to reduce contamination of the measured signal by extraneous sources of vibrations. However, it is not always possible to mount the transducer close enough to the bearing. The envelope spectrum analysis technique can be used to analyze signals from a transducer mounted at a more remote location, containing extraneous vibration signals from other machinery components. This technique makes use of high-frequency resonances that act as mechanical amplifiers and as carriers of the lower frequency fault signals. The initial process of locating a suitable resonant carrier frequency is the most difficult part of the envelope spectrum technique. Envelope analysis is not restricted to remote monitoring applications, as many bearing faults can be identified by modulations in the vibration signature.

The input signal from an accelerometer is routed through a bandpass filter centered on the resonant frequency. This produces a time-varying signal modulated by the transient generated by the defect. The filtered signal is rectified and then demodulated using an envelope detector. Alternatively, the amplitude demodulation techniques described in Section 3.6 can be employed to extract the envelope. The resulting signature is dominated by rolling element defect frequencies. The power spectrum of the envelope is then calculated. Any changes in defect-related frequency levels or patterns are assessed for fault evaluation.

3.6 Amplitude and Phase Modulation Techniques for Gear Analysis

A fault in any rotating machine is likely to introduce sidebands, due to amplitude and frequency modulations, around prominent frequency bands associated with the machine. In a simple pair of gears, it is relatively easy to identify sidebands around tooth meshing frequencies. In complex gear systems comprising large numbers of gear sets (e.g., helicopter transmission systems), it is more difficult to identify any specific gear because of the large number of meshing frequencies and sidebands present.

The vibrations produced by gears can be characterized as the normal gear meshing vibrations, plus additional effects producing sidebands around the mesh frequency harmonics. Defects can be identified by applying amplitude and phase modulation analysis techniques. These were originally proposed by McFadden and Smith[9] and further developed by Nicks and Krishnappa.[10,11] Under ideal conditions, the time-domain average of the vibration of a pair of gears is a periodic function that consists solely of tones at the fundamental and harmonics of the mesh frequency. A discrete gear fault, such as spall or a fatigue crack, modifies this equation by introducing periodic phase or amplitude modulations in the vibration signal. The equation for the modulated vibration signal can be written as

$$y(t) = \sum_{m=0}^{M} X_m[1 + a_m(t)] \cos[2\pi m T f t + \theta_m + b_m(t)], \quad (3)$$

where X_m is the vibration amplitude at the mesh frequency harmonic m $(0, 1, 2, \ldots, M)$, $a_m(t)$ is the amplitude modulation function, T is the number of teeth on the gear, f is the rotational frequency, θ_m is the phase lag, and $b_m(t)$ is the phase modulation function.

In the frequency domain, the Fourier transform of the above function comprises the fundamental and harmonics of the meshing frequency, surrounded by modulation sidebands. The function $y_m(t)$ can be approximated by bandpass filtering the time-domain averaged vibration signal. The amplitude envelope and phase modulation trace are extracted from the complex analytical function $z_m(t)$, given by

$$z_m(t) = y_m(t) + jH[y_m(t)], \quad (4)$$

where H is the Hilbert transform of $y(t)$. Examples of this are shown in Fig. 12.

3.7 Cepstrum Analysis

The application of cepstrum analysis to machinery diagnostics is mainly based on its ability to detect periodicity in the spectrum. In simple terms, the cepstrum can be considered to be the spectrum of a spectrum. It was originally defined to be the power spectrum of the logarithm of the power spectrum. The terminology used in cepstrum analysis is derived from that of frequency analysis: spectrum ⇔ cepstrum, frequency ⇔ quefrency, and harmonics ⇔ rahmonics.

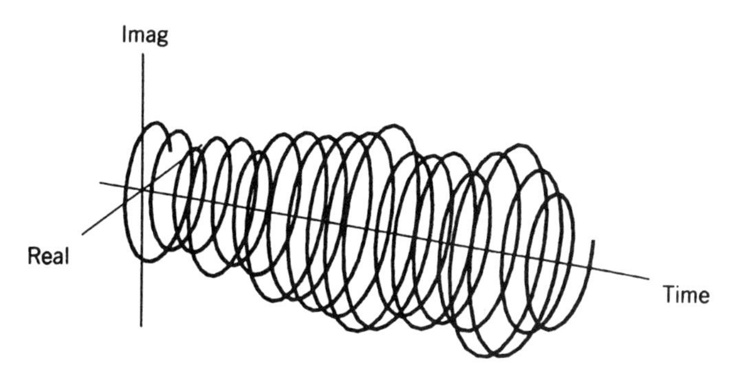

(*a*) Gear in good condition.

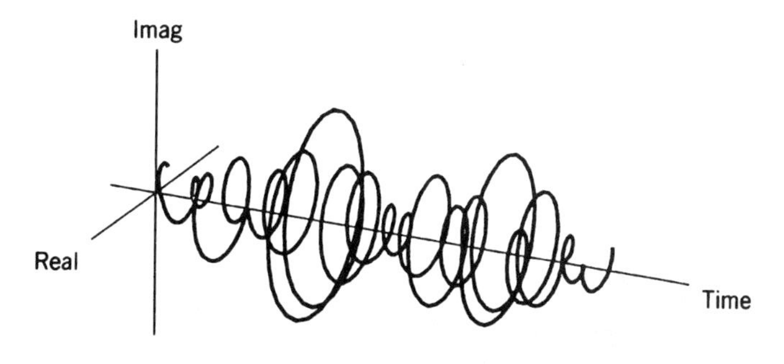

(*b*) Surface-damaged gear.

Fig. 12 Complex analytical vibration signatures of gears.

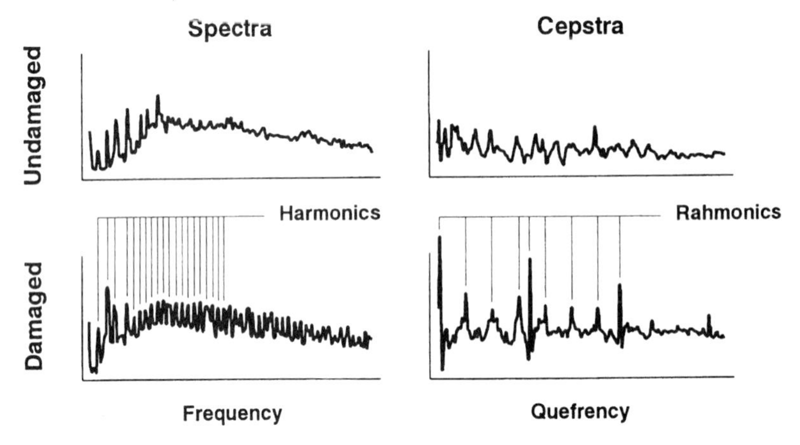

Fig. 13 Spectra and cepstra of undamaged and damaged gearboxes.

Cepstrum analysis can be used as a tool for the detection of families of harmonics with uniform spacing. The logarithmic amplitude scale emphasizes the harmonic structure of the spectrum and reduces the influence of the transmission path by which the signal travels from the source to the measurement point. Faults of various kinds in a gearbox are indicated by groups of harmonics and their sidebands. Although the frequency patterns can be seen by the eye, they may be quite complex and difficult to interpret. Cepstrum analysis is used to reduce these groups to a few rahmonics that can be much easier to monitor. A more detailed description of cepstrum analysis can be found in Ref. 17.

The cepstrum is now defined as the inverse transform of the logarithm of the power spectrum, or

$$C(\tau) = \mathcal{F}^{-1}(\log |X(f)|^2), \tag{5}$$

where $X(f)$ is the Fourier transform of a time series $x(t)$. One of the reasons for this new definition is to highlight the connection between the cepstrum and the autocorrelation function, defined as the inverse transform of the power spectrum:

$$R_x(\tau) = \mathcal{F}^{-1}(|X(f)|^2). \tag{6}$$

The independent variable τ of the cepstrum has the dimensions of time but is known as quefrency. A high quefrency indicates rapid fluctuations in the spectrum (small frequency spacings), and a low quefrency indicates slow changes with frequency (large spacings). The quefrency of the peak represents the periodic time of modulations in the time domain (i.e., sideband spacing in the frequency domain). The quefrency only identifies frequency-domain spacing—it tells nothing about absolute frequencies.

Spectra and cepstra for two gearboxes are shown in Fig. 13. The spectrum of the gearbox in good condition does not show any significant periodicity, and this is reflected in the cepstrum where few discrete rahmonics are present. In the damaged gearbox, however, a large number of discrete peaks are present in the spectrum. The cepstrum clearly shows that these are periodic and identifies the fundamental period.

REFERENCES

1. E. L. Thearle, "Dynamic Balancing of Rotating Machinery in the Field," *Trans. ASME*, Vol. 56, No. 10, 1934, pp. 745–753.
2. J. S. Mitchell, *An Introduction to Machinery Analysis and Monitoring*, Penwell, 1981, pp. 107–204.
3. C. M. Harris and C. E. Crede, *Shock and Vibration Handbook*, 2nd ed., McGraw-Hill, New York, 1976, pp. 5.1–5.24.
4. R. H. Bannister, "A Review of Rolling Element Bearing Monitoring Techniques, I. Mech. E. Condition Monitoring Machinery Plant," *Symposium of Condition Monitoring Machinery Plant*, 1985, pp. 11–24.
5. N. S. Swanson and S. C. Favalord, *Applications of Vibration Analysis to the Condition Monitoring of Rolling Element Bearings*, Aeronautical Research Laboratory, Aero Propulsion Report 163, 1984.
6. P. Eschman et al., *Ball and Roller Bearings*, 2nd ed., Wiley, New York, 1985, pp. 78–138.
7. J. D. Smith, *Gears and Their Vibration*, Marcel Dekker, New York, 1983.

8. R. M. Stewart, "Some Useful Data Analysis Techniques for Gearbox Diagnostics," *Proceedings of the Symposium "Applications of Signal Processing,"* Institute of Sound and Vibration Research, Southampton, UK.

9. P. D. McFadden and J. D. Smith, "A Signal Processing Technique for Detecting Local Defects in a Gear from the Signal Average of the Vibration," *Proceedings of the I. Mech. E.*, Part C, Vol. 199, 1985, pp. 287–292.

10. J. E. Nicks and G. Krishnappa, "A Preliminary Study into the Use of Phase Demodulation Techniques for the Analysis of Gear Vibration Data," National Research Council, Ottawa, TR-ENG-005, 1989, NRC Report No. 32083.

11. J. E. Nicks G. and Krishnappa, "Amplitude and Phase Modulation Analysis Methods for Early Detection of Gear Faults," *Proceedings of the International Congress on Recent Developments in Air- and Structure-Borne Sound and Vibration*, Auburn, Alabama, 1990, pp. 263–270.

12. R. H. Lyon, *Machinery Noise and Diagnostics*, Butterworth, Stoneham, MA, 1987, pp. 17–47.

13. S. Braun, *Mechanical Signature Analysis, Theory and Applications*, Academic, London, 1986.

14. C. Cempel and H. D. Haddad, *Vibroacoustic Condition Monitoring*, Ellis Horwood, 1991.

15. J. S. Bendat and A. G. Piersol, *Random Data Analysis and Measurement Procedures*, Wiley-Interscience, New York, 1971.

16. E. O. Brigham, *The Fast Fourier Transform and its Applications*, Prentice-Hall, Englewood Cliffs, NJ, 1988.

17. R. B. Randall and J. Hee, Cepstrum Analysis, *Brüel and KjærTech. Rev.*, No. 3, 1981, pp. 1–32.

58

STRUCTURE-BORNE ENERGY FLOW

G. PAVIC

1 INTRODUCTION

1.1 Structural Intensity and Energy Flow

In an elastic body, structural vibration induces the flow of mechanical energy through the body. The energy flow (EF) results from the interaction between vibration-generated stresses and movements throughout the body. The EF at any single point represents the instantaneous rate of energy transfer per unit area in a given direction and is called the structural intensity (SI). Defined in such a way, the intensity becomes a vectorial quantity associated with the observed point. Its physical unit is watts per metre squared. When integrating intensity across an area within the body, the total mechanical energy flow through this area is obtained. The unit of energy flow is the watt.

The intensity vector changes both its magnitude and direction in time. Variations of intensity at different points within the structure need not coincide mutually. In order to compare energy flows at various locations in a meaningful way, their time-averaged (net) value must be analyzed. Thus, the intensity concept is best associated with vibration fields that exhibit temporal stationarity.

Once the vibrational energy flow through a structure is known, it becomes possible to identify the positions of vibratory sources, propagation paths, and absorption areas. The nature of this information, which is not present in data on vibration levels, implies that energy flow techniques should be used primarily in the fields of source characterization, system identification, and vibration diagnostics.

The SI or EF of structures can be examined either by measurement or by computation, depending on the nature of the technical problem being considered. Measurements are used for diagnostic or source identification purposes, while computation is suitable for noise and vibration control during the design stage.

1.2 Energy Flow Concepts

The characteristics of energy transfer by structural vibration are such that three complementary concepts are viable in representing vibration-induced EF:

(a) The direct intensity concept utilizes SI (energy flow per unit area) as the final result in an analysis. This concept is completely analogous to sound intensity, as it represents the same physical quantity. The intensity concept is a tool for three-dimensional visualization of energy flow within an elastic body exhibiting structural vibration. It is more suitable for computational than measurement work, as any measurement of practical use made on an object of complex shape has to be restricted to the exterior surfaces only. The latter condition leads to a specific quantity, the surface structural intensity (SSI). In the majority of cases, surface intensity values can be used to qualify rather than to quantify the energy flow within the object under examination.

(b) Structures of uniform thickness (plates, shells) can be more appropriately analyzed if the intensity is integrated across the thickness. In this way a unit energy flow (UEF) is obtained, that is, the flow of energy per unit length normal to any line on the surface. If the energy exchange between the outer surfaces and the surrounding medium is small in comparison with the flow tangential to the surface, as it is for objects in contact with air, then the UEF vector lies in the tangential plane of the middle surface. For thin-walled structures, where

Note: References to chapters appearing only in the *Encyclopedia* are preceded by *Enc.*

the wavelengths are much larger than the thickness, the outer surface motions are related in a simple way to those in the interior (Kirchhoff's assumptions). This fact makes it possible to express UEF in terms of outer surface quantities, which in turn makes it possible to measure it directly. The UEF concept is best utilized for the needs of source localization and fault diagnostics.

(c) Structural parts that form waveguides or simple structural joints can most appropriately be analyzed by means of the concept of total energy flow (TEF). Some such parts, such as beams and elastic mountings, are of a simple enough shape to allow expression of the EF by simple formulas in terms of vibrational displacements. The TEF in waveguides of a more complex mechanical character, such as pipes filled with fluid, can then be evaluated by more complex procedures such as the wave decomposition technique.

2 PHYSICAL BACKGROUND OF STRUCTURAL ENERGY FLOW

Expressed in Cartesian coordinates, the three components of an SI vector read

$$\begin{aligned} I_x &= -\sigma_{xx}\dot{u}_x - \sigma_{xy}\dot{u}_y - \sigma_{xz}\dot{u}_z, \\ I_y &= -\sigma_{yx}\dot{u}_x - \sigma_{yy}\dot{u}_y - \sigma_{yz}\dot{u}_z, \\ I_z &= -\sigma_{zx}\dot{u}_x - \sigma_{zy}\dot{u}_y - \sigma_{zz}\dot{u}_z. \end{aligned} \tag{1}$$

Here σ are dynamic stresses and u the particle displacements at the observation point, their subscripts indicating component orientation. The dots above the symbols for displacements indicate time derivative; that is, intensity components are products of stresses and particle velocities.

Since the shear stress components obey symmetry, for example, $\sigma_{xy} = \sigma_{yx}$, the complete SI vector is built up of nine different quantities, six stresses and three velocities. Each of the SI components is a time-varying quantity. Therefore SI is a time-varying vector. Its magnitude and direction both change in time. It should be pointed out that the SI vector does not necessarily have to be normal to the wavefront surface.

The SI formulas (1) are similar to those for sound intensity components but exhibit two important differences. First, the SI formulas are more complex, since each component consists of three terms instead of a single one, as in the case of sound intensity. In gases and liquids the shear stresses due to sound motions are very small and can be disregarded when compared with normal stresses. Another difference is in the sign of each product, which merely results from the sign convention. Therefore, sound intensity is simply a special case of Eqs. (1) when shear effects vanish.

At a free surface, the intensity formulas (1) become simpler. If the coordinate z is taken as the normal to the surface, then all z-dependent terms vanish, reducing thus the intensity vector to I_x and I_y components each containing two terms only.

To determine the SI or EF values in a particular case, time-varying components of both the stresses and the velocities have to be known at the observed point. Stresses are not measurable in a direct way; in order to determine them, relationships are used that link them to (measurable) strains. For homogeneous and isotropic materials behaving linearly, such as metals, most plastics, rubber, and so on, the stress–strain relationships are fairly simple. These relationships contain material constants: Young's and shear moduli.[1]

Equations (1) refer to instantaneous values of intensity, the knowledge of which is not useful for most analytical purposes. Similarly, instantaneous values of sound pressure or vibration velocity are of no particular analytical use. It is therefore desirable to operate with averaged values of intensity and energy flow.

Time-averaged SI represents net energy flow per unit surface area at the observed point. It is termed "active" intensity, because it refers to the portion of total energy flow taken away from a given region without being returned. The remaining portion of the energy flow, which fluctuates to and fro but has zero mean value, is accordingly termed "reactive" energy.

3 FORMULAS FOR ENERGY FLOW

The basic SI formulas (1) are not sufficient for measurement of either unit or total energy flow. Using these formulas, the expressions for UEF or TEF can be evaluated for some simpler type of structures: rods, beams, plates, and shells. As a rule, a different governing formula will apply to each type of structure.

In this section formulas for EF are listed. The formulas are given in terms of strains and velocities, that is, the measurable quantities. Both the strains and the velocities can be detected by various techniques, but only at the outer surface of the object under investigation. In order to determine these quantities within the body, some knowledge is required of the internal velocity and strain distribution. For certain structures (rods, beams, plates, and shells) very simple relationships are sufficient to express internal quantities in terms of the external ones, as long as the wavelengths of the structural motion are much larger than the lateral dimensions of the object.

3.1 Longitudinally Vibrating Rod

In a longitudinally vibrating rod, only the axial component of the stress differs from zero; thus only the axial normal strain exists. The SI distribution in the cross section is uniform (Fig. 1*a*):

$$I_x = -E\varepsilon_{xx}\dot{u}_x. \tag{2}$$

The total EF is obtained by multiplying the intensity with the area of the cross section S:

$$P_x = I_x S. \tag{2a}$$

3.2 Flexurally Vibrating Beam

In this case the intensity varies over the cross section due to variations of both the stresses and the velocities caused by flexion (Fig. 1*b*). Two components of stress exist: the axial normal stress σ_{xx} and the lateral shear stress σ_{xz}. The axial stress varies linearly with distance from the neutral plane, while the variation of the shear stress depends on the shape of the cross section. For a rectangular cross section this variation is parabolic with the maximum value in the middle plane. The axial component of particle velocity, $\dot{u}_x$, exhibits the same form of variation as the axial stress, while the lateral component, $\dot{u}_z$, remains constant. Integration of axial intensity across the lateral section leads to expressions for the total EF. Since the axial and lateral displacements are coupled through a simple expression $u_x = -z\,\partial u_z/\partial x$, various formulations of the EF are possible:

$$P_x = JE\left(\frac{\partial^3 u_z}{\partial x^3}\,\dot{u}_z - \frac{\partial^2 u_z}{\partial x^2}\,\frac{\partial \dot{u}_z}{\partial x}\right) = \frac{JE}{\delta^2}\left(\delta\,\frac{\partial \varepsilon_{xx}}{\partial x}\,\dot{u}_z + \varepsilon_{xx}\dot{u}_x\right), \tag{3}$$

where J is the area moment of inertia about the bending axis and δ the distance from the neutral plane to the outer surface, which the quantities in Eq. (3) refer to. The first formulation[2] contains only lateral motions, displacements, and velocities, while the second one contains strains and velocities.[3]

3.3 Torsionally Vibrating Bar

Torsion in a bar produces shear stresses that rise linearly from the center of the bar. If the bar is of circular or annular cross section, these shear stress components are the only nonvanishing ones. In such a case the EF through the bar, obtained by integrating the SI through the cross section, reads (Fig. 2)

$$P_x = -T\,\frac{\partial \varphi}{\partial x}\,\dot{\varphi} = -\frac{T}{R_0}\,\varepsilon_{x\varphi}\dot{\varphi}. \tag{4}$$

Here φ is the angular displacement of the bar, $\varepsilon_{x\varphi}$ is the shear strain at the outer radius R_0 and T is the torsional stiffness. For a bar of an annular cross section, the torsional stiffness is equal to $T = \pi/2G(R_0^4 - R_1^4)$.

The torsional stiffness of a solid shaft is obtained by setting the inner radius R_1 to zero. If the bar is of rotationally nonsymmetrical cross section, Eq. (6) applies in the same way but in addition axial motions and stresses

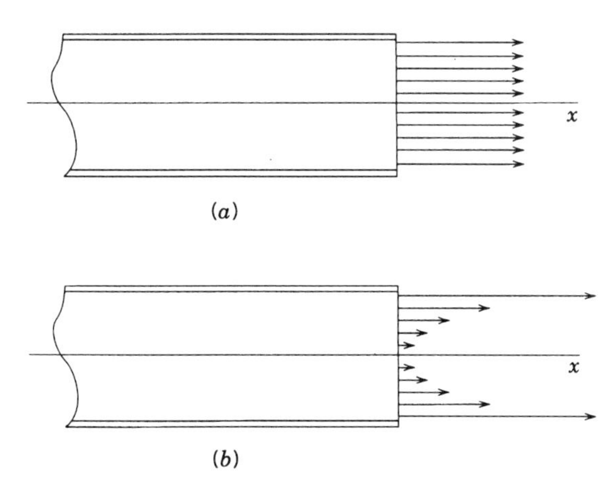

Fig. 1 Intensity distribution in a rod (beam): (*a*) longitudinal; (*b*) flexural vibration.

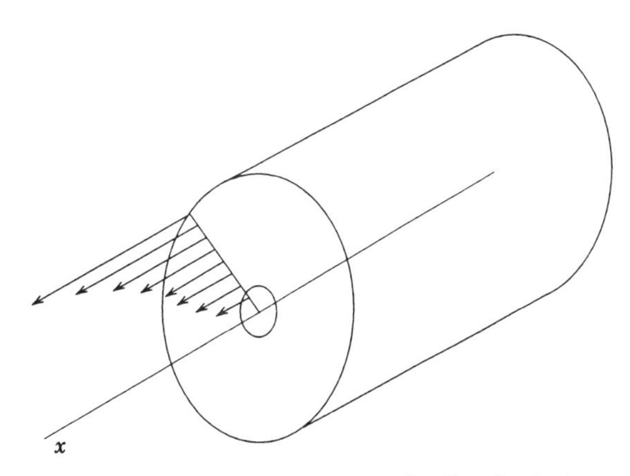

Fig. 2 Intensity distribution in a torsionally vibrating bar.

take place that can generate the flow of energy. The mechanism of such an additional EF very much depends on the shape of the cross section and should be analyzed separately for each individual case.

3.4 Longitudinally Vibrating Thin Plate

For a thin plate exhibiting in-plane motion only, the intensity is constant throughout the thickness. The SI component in the x-direction reads[3] (Fig. 3)

$$I_x = -E(1-\nu^2)[\varepsilon_{xx}\dot{u}_x + \tfrac{1}{2}(1-\nu)\varepsilon_{xy}\dot{u}_y],$$
$$P'_x = I_x h, \tag{5}$$

while for the y-direction an analogous expression applies, obtained by interchanging the x and y subscripts. The UEF P' is simply obtained by multiplying the intensity with the plate thickness h. The expression is similar to that for the intensity in the rod [Eq. (2a)], the difference being an additional term due to shear stresses carrying out work on motions perpendicular to the observed direction.

3.5 Flexurally Vibrating Thin Plate

As for the flexurally vibrating beam, the normal component of vibration in a flexurally vibrating plate is of a much higher level than the in-plane component. It is therefore appropriate to formulate the intensity expressions in terms of the normal component u_z. With respect to intensity distribution throughout the cross section, a situation applies analogous to that for the beam, with the exception of an additional mechanism of energy flow generation, due to twisting deformations of the plate. The UEF in the x-direction reads[2,4]

$$P'_x = \frac{Eh^3}{12(1-\nu^2)}\left[\frac{\partial(\Delta u_z)}{\partial x}\dot{u}_z - \left(\frac{\partial^2 u_z}{\partial x^2} + \nu\frac{\partial^2 u_z}{\partial y^2}\right) \cdot \frac{\partial \dot{u}_z}{\partial x} - (1-\nu)\frac{\partial^2 u_z}{\partial x \partial y}\frac{\partial \dot{u}_z}{\partial y}\right], \tag{6}$$

where Δ is the Laplacian. The UEF in the y-direction is obtained by x–y subscript interchange. Both components of the in-plane motion are related to normal motion, which makes possible an alternative expression for the UEF:

$$P'_x = -\frac{Eh}{3(1-\nu^2)} \cdot \left[\frac{h}{2}\frac{\partial(\varepsilon_{xx}+\varepsilon_{yy})}{\partial x}\dot{u}_z + (\varepsilon_{xx}+\nu\varepsilon_{yy})\dot{u}_x + \frac{1-\nu}{2}\varepsilon_{xy}\dot{u}_y\right]. \tag{6a}$$

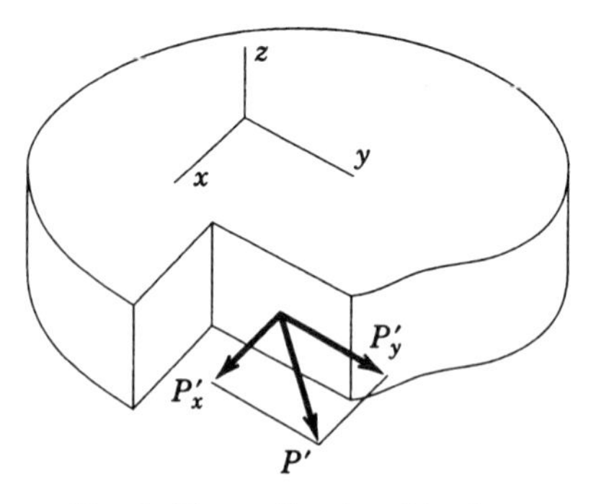

Fig. 3 Energy flow in a thin plate.

3.6 Composite Vibrations of Beams and Plates

In a beam (rod, bar) all three types of vibration discussed in Sections 3.1–3.3 can exist simultaneously. The same applies to longitudinal and flexural vibrations in a flat plate. In both cases, the different types of vibration are uncoupled, and consequently the intensities due to each of the different types of vibration should be summed algebraically. The problem here is that both longitudinal and flexural vibrations produce longitudinal motions. In order to apply intensity formulas correctly to a measurement, the origins of the motions must be identified. This can be done by connecting the measuring transducers in such a way as to suppress the effects of a particular type of motion. It should be borne in mind that not only the longitudinal vibration velocities but also the longitudinal strains are affected by flexural motion. The following formulas are used to extract the longitudinal velocity and strain components (denoted by subscript zero) from the total velocity and strain at the outer surface:

$$\dot{u}_{x0} = \dot{u}_x + \frac{h}{2}\frac{\partial \dot{u}_z}{\partial x}, \qquad \varepsilon_{xx0} = \varepsilon_{xx} + \frac{h}{2}\frac{\partial^2 \dot{u}_z}{\partial x^2},$$
$$\varepsilon_{xy0} = \varepsilon_{xy} + h\frac{\partial^2 \dot{u}_z}{\partial x \partial y}. \tag{7}$$

Again, components of velocity and strain in the y-direction are obtained from Eq. (7) by interchanging the x and y subscripts. It can be seen that measurements of EF due to longitudinal vibrations in the presence of flexural vibrations necessitate correction of the governing formulas (2) and (5) according to Eq. (7).

3.7 Simplified Governing Formulas for Flexurally Vibrating Beams and Plates

In regions far from vibration sources, such as discontinuities and boundaries, where the vibration field is essentially a superposition of plane waves, a simplified formula applies for net EF (TEF) due to flexural motion[5]:

$$P_x = C\omega \left\langle u_z \frac{\partial \dot{u}_z}{\partial x} \right\rangle = -\frac{C}{\omega} \left\langle \ddot{u}_z \frac{\partial \dot{u}_z}{\partial x} \right\rangle, \tag{8}$$

where

$$C = \begin{cases} h^2 \rho c/\sqrt{3} & \text{(plate)}, \\ 2\sqrt{JS}\rho c & \text{(beam)}. \end{cases}$$

The constants ρ and c represent mass density and longitudinal wave velocity, respectively; ω is the radian frequency.

3.8 Thin Circular Cylindrical Shells

Due to the curvature of the shell, the in-plane and the normal motions are coupled.[6] The expressions for the UEF along the wall of the shell are the same as for the flat plate, with the addition of a UEF term P'_c to take account of the curvature:[7]

$$P'_{\text{shell}} = (P'_{\text{longitudinal}} + P'_{\text{flexure}})_{\text{plate}} + P'_c. \tag{9}$$

The two terms in parentheses are given by Eqs. (5) and (6). The curvature term is fairly complex; however, for thin shells it reduces to a simple form.[8] If the shell is positioned axially in the x-direction (Fig. 4), with the local x–y–z coordinate system attached to the observation point, the axial and tangential components of the simplified curvature term read

$$P'_{cx} \approx -\frac{\nu}{a} u_z \dot{u}_x, \qquad P'_{cy} \approx -\frac{1}{a} u_z \dot{u}_y. \tag{9a}$$

Here, as in the case of composite plate vibrations, the in-plane (longitudinal) motions due to bending must be separated from the extensional motions before expressions (5) and (9a) can be applied. Corrections for the effect of bending [Eq. (7)], which affects longitudinal motion measured at the outer surface, apply for thin shells in the same way as for thin plates.

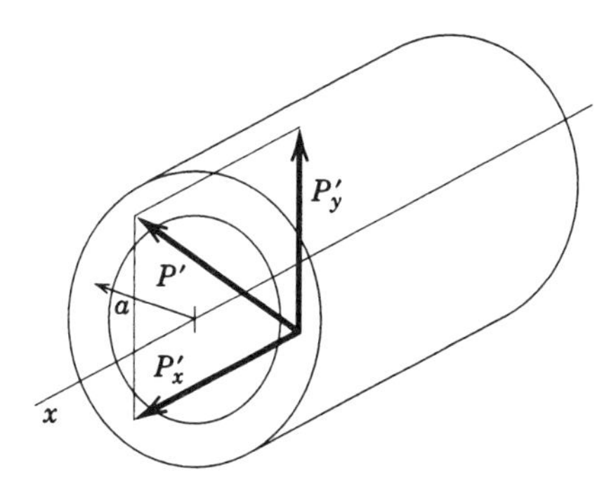

Fig. 4 Energy flow in a thin cylindrical shell.

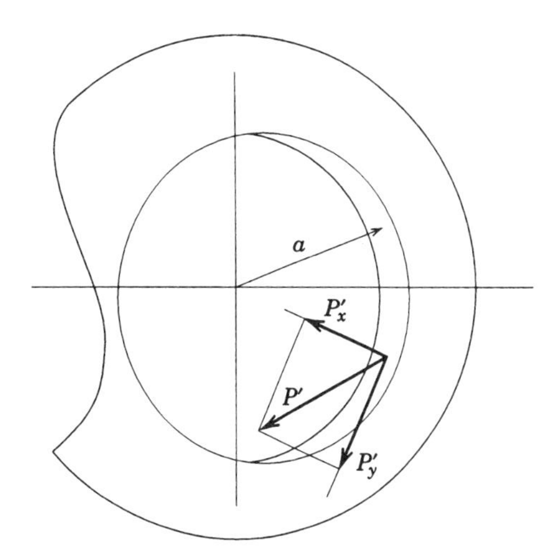

Fig. 5 Energy flow in a thin spherical shell.

3.9 Thin Spherical Shells

Figure 5 shows a section of a spherical shell with the local Cartesian coordinate system fixed to the observation point. The same UEF expression applies as for the cylindrical circular shell, Eq. (9). The only difference is in the curvature-dependent term, which is approximately equal to[9]

$$P'_x \approx -\frac{1+\nu}{a} u_z \dot{u}_x. \tag{10}$$

In this case there are no differences in the form of the x- and y-components of the curvature terms, as was the case for the cylindrical shell. This results from the properties of a spherical shell, which are centrally symmetrical about the normal to the surface.

3.10 Pipes

At not too high frequencies where higher order cylindrical waves cannot yet propagate, three types of pipe vibration occur: longitudinal, torsional, and flexural. The limiting frequency corresponding to such conditions is given by

$$f_{\text{lim}} \approx \frac{f_{\text{ring}}(6h/d)}{\sqrt{15+12\mu}}, \qquad h \ll d, \tag{11}$$

where $f_{\text{ring}} = c/\pi d$ is the pipe ring frequency, μ is the mass ratio of the contained fluid and the pipe wall, c is the velocity of longitudinal waves in the wall, and d is the mean diameter.

It turns out that the three types of motion contribute independently their respective energies to the total energy flow.[6] This enables straightforward measurements.[10,11] These three contributions should be determined separately using the formulas in Sections 3.1–3.3 and then added together.

3.11 Resilient Elements

A typical resilient element, used for vibro-isolation, consists of an elastic layer between two rigid surfaces. The forces acting at the two sides of the element stand in a firm relationship with the resulting velocities at the terminal points. This relationship is frequency dependent and is usually expressed in terms of the direct and transfer impedances of each of the endpoints.[12] Once the force–velocity relationship has been determined, either by computation or by measurement, knowledge of the end velocities of the mounted element is sufficient for evaluation of the TEF throughout the element. Due to inevitable internal losses in the element, the input EF must be larger than the output EF.

A simplified diagram of a resilient element is given in Fig. 6.

On the assumption of pure translational motion in the direction of its axis, the element is characterized by four impedances, Z_{11}, Z_{22}, Z_{12}, and Z_{21}, where

$$Z_{mn} = \left.\frac{F_m}{\dot{u}_n}\right|_{\dot{u}_m=0}, \qquad m, n = 1, 2.$$

The impedances Z are complex, frequency-dependent quantities. In sinusoidal vibration, the TEF at the endpoints 1 and 2 read

$$\begin{aligned} P_1 &= \tfrac{1}{2}\,\mathrm{Re}\{|\dot{u}_1|^2 Z_{11} + \dot{u}_1^* \dot{u}_2 Z_{12}\}, \\ P_2 &= \tfrac{1}{2}\,\mathrm{Re}\{|\dot{u}_2|^2 Z_{22} + \dot{u}_1 \dot{u}_2^* Z_{21}\}. \end{aligned} \qquad (12)$$

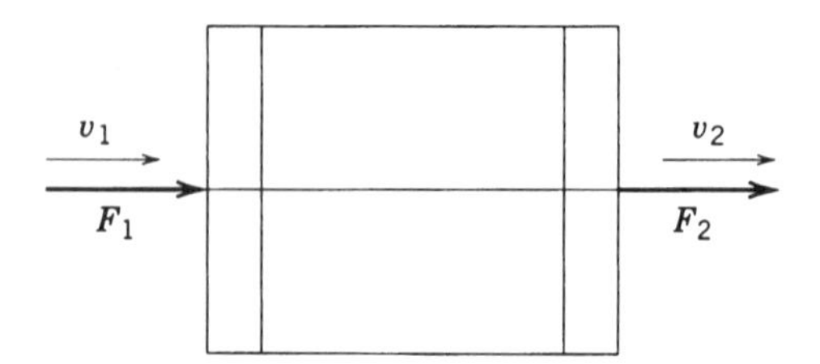

Fig. 6 Energy flow through a resilient element.

Here $|\cdot|$ and the asterisks denote absolute and complex-conjugate values, respectively, while Re denotes the real part of a complex quantity. If the motion of the endpoints is random stationary, the TEF spectra at the two points can be obtained simply by replacing the velocity products with the appropriate power and cross-spectral density functions and suppressing the common factor $\frac{1}{2}$.

At low frequencies, much below the first resonance of the element where it behaves essentially like a massless spring of stiffness K, the following formula applies for the TEF[13]:

$$P = K(u_1 - u_2)\dot{u}_1. \qquad (12a)$$

The average net value of the TEF in this case is

$$\langle P \rangle = K\langle u_1 \dot{u}_2 \rangle. \qquad (12b)$$

Equations (12a) and (12b) apply for an element with negligible damping. If damping is significant, a correction has to be added to the existing TEF expression (12a). For viscous damping, defined by a damping constant C, the correction reads

$$dP = C(\dot{u}_1 - \dot{u}_2)\dot{u}_2. \qquad (12c)$$

A resilient element is often subjected to motion of a more general nature than the simple translatory motion assumed so far. Both forces and moments are generated at the endpoints and both contribute to the TEF. Moreover, coupling of motions usually takes place, which means that excitation in one direction produces additional movements in other directions as well. In such circumstances, evaluation of energy flowing through the element becomes increasingly difficult. The basic guideline for evaluation of the TEF in such a case is the following:

- Determine the appropriate impedance matrix (i.e., relationships between all the relevant forces + moments and translational + angular velocities).
- For each particular case, reconstruct the resultant force + moment vectors at the endpoints from the known motions of these points.
- The TEF is the scalar product of the force vector and the translational velocity vector plus the product of the moment vector and the angular velocity vector. This applies to each of the end points separately.

If the resilient element is highly nonlinear in the operating range, or if the provisions for attachment cannot be considered as points (e.g., rubber blocks of extended length, sandwiched between steel plates), the preceding approach becomes invalid. Some resilient elements (e.g., ones made of elastomers) have mechanical properties that are highly dependent not only on frequency but also on static preloading, temperature, and dynamic strain. All of these parameters should be taken into consideration when determining the impedances of the element.

4 MEASUREMENT PROCEDURES

Formulas for SI or EF [Eqs. (2)–(6) and (8)–(10)] contain constants (such as $E, \nu, h, \ldots$) and time-varying quantities that have to be measured (velocities, displacements, strains). Time-varying quantities are mutually linked as products. In order to obtain instantaneous values of SI or EF, all the varying quantities must be measured simultaneously. Net values of SI or EF can be obtained by evaluating the average value of each constituent product separately (where more than one product appears in the governing equation) and adding partial results off-line. Spectral representation of SI or EF can be achieved in a similar way using spectral densities instead of product averages.

The main practical problem is measurement of quantities that are spatial derivatives of velocities, displacements, and strains.

In order to measure these quantities, finite-difference approximations of spatial derivatives can be employed. Finite-difference approximations cause errors that can be analytically evaluated and sometimes kept within acceptable limits by an appropriate choice of parameters influencing the measurement accuracy.

Another way of measuring intensity/energy flow consists of establishing a wave model of the structure analyzed and fitting it with measured data. This inverse technique is known as the wave decomposition approach. In such a way all of the variables entering the intensity formulas can be identified.

4.1 Transducers for Measurement of SI (EF)

The quantities that need to be measured for evaluation of SI or EF, velocities (displacements), or strains can be detected, in principle, in a variety of ways. In this respect, two basic approaches are possible: (a) use of transducers physically mounted on the object to be investigated and (b) detection of movement of the object by noncontact means.

Structural vibrations at a particular point in an object generally consist of both normal and in-plane (tangential) components, which can be translational and/or rotational. Measurement of SI or EF requires separate detection of some of these component motions, without any perceptible effect from other motions. This requirement poses the major problem in practical measurements. Another major problem results from the nonideal behavior of individual transducers such as phase shift between the physical quantity measured and the electrical signal that represents it. Both effects can severely degrade accuracy, since the measured quantities make up the product(s) where these effects matter more than in any straightforward measurement of the constituent quantities.

Seismic Accelerometer An accelerometer is easy to mount and to use. However, when using accelerometers for SI (FE) work, it is important to consider the following:

- Although the cross sensitivity of an accelerometer is usually low (up to a few percent), its effect can still be important in cases where one component of motion is measured in the presence of a strong component of cross motion (e.g., measurement of in-plane motion).
- Due to its finite size, the sensitivity axis of an accelerometer can never be positioned at the measurement surface but will always be somewhere above it. In this way, any in-plane rotation of the surface affects the output of the accelerometer. Compensation for this drawback can be effected by introducing a correction signal based on the measured rotation of the surface.
- An accelerometer dynamically loads the object to which it is attached. This effect is of importance for SI measurement on lightweight structures.

Strain Gauge Strain gauges measure normal strain elongation. Conventional (resistive) strain gauges exhibit few of the drawbacks associated with accelerometers: cross sensitivity is low and the sensitivity axis is virtually at the surface of the object, due to vanishingly small thickness of the gauge, while dynamic loading is practically zero. However, the conventional gauges exhibit low sensitivity where typical levels of strain induced by structural vibration are concerned, which means that typical signal-to-noise values could be low. Semiconductor gauges have a much higher sensitivity than the conventional type, typically two orders of magnitude higher. High dispersion of sensitivity makes the semiconductor gauges unacceptable for intensity work.

Shear strain at the free surface can be measured by two strain gauges positioned at 45° and 135° with respect to the direction to be considered. The shear strain is then equal to the difference between the two normal strains measured by the gauges. The difference signal can be obtained directly from the gauge conditioner by wiring the two gauges in adjacent branches of the gauge bridge.

Noncontact Transducers A range of various contactless transducers are available for the detection of structural motion such as inductive, capacitive, eddy current, light intensity, and light-modulated transducers. The majority of contactless transducers are of a highly nonlinear type, where linearity could be achieved by compensating electronic circuitry.

Optical transducers based on the interference of coherent light such as LDV (laser Doppler velocimeter) devices are well suited for SI work. The most promising are however optical techniques for whole-field vibration detection, such as holographic or speckle-pattern interferometric methods.[14] Applications of these are limited for the time being to plane surfaces only.

Another interesting approach for noncontact whole-field SI measurement consists in using acoustical holography.[15,16] This is again possible when the structure analyzed is of a simple regular shape such as a plate or cylinder.

4.2 Transducer Configurations

In the measurement of SI (EF), three situations arise where more than one transducer is needed for measurement of a single physical quantity:

- When separation of a particular type of motion from the total motion is required
- When detection of a spatial derivative is required
- When different wave components need to be extracted from the total wave motion

The first case arises with beams, plates, and shells where longitudinal and flexural motions take place simultaneously (in the case of shells, this is always the case) and where corrections of the type of Eq. (7) are not appropriate. The second case occurs when a direct measurement implementation of SI (EF) governing equations is performed. The third case applies to waveguides (beams, pipes etc.) where the wave decomposition approach is used for EF evaluation.

Extraction of a Single Type of Motion It has been shown in Section 3.6 how the longitudinal component of either motion or strain, denoted by 0, can be separated from the total motion measured at the surface.

An elegant way to circumvent the difficulty of measuring the correcting term consists in placing two identical transducers at opposite sides of the neutral plane and adding or subtracting the outputs from the transducers. In this way longitudinal and flexural strains can be separated (Fig. 7*a*). Figure 7*b* shows a similar scheme for separating pulsation-induced from flexural-induced motions of a vibrating pipe.

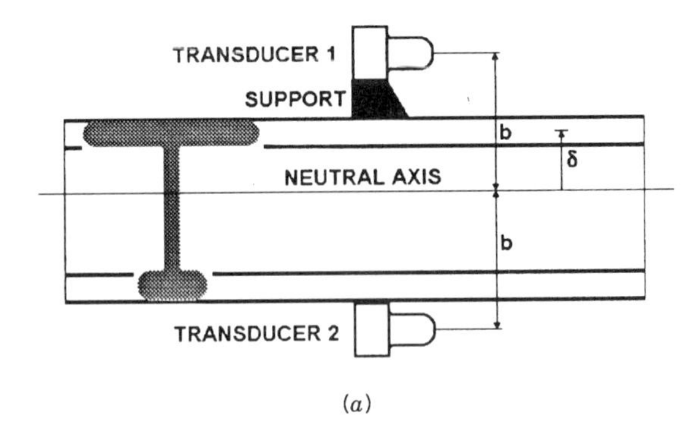

(*a*)

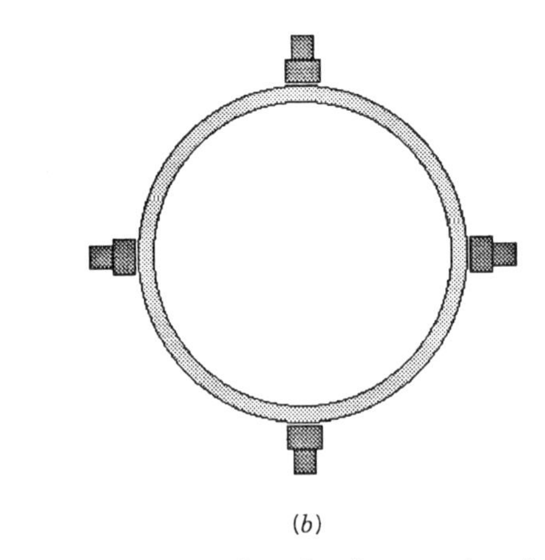

(*b*)

Fig. 7 Transducer configuration for extraction of a simple type of motion.

Detection of Spatial Derivatives Formulas for SI(EF) contain spatial derivatives that cannot be measured directly. The usual way of measuring spatial derivatives is by using finite-difference concepts.

The spatial derivative of a function q in a certain direction (e.g., the x-direction) can be approximated by the difference between the values of q at two adjacent points, 1 and 2, located on a straight line in the direction concerned:

$$\frac{\partial q}{\partial x} = \frac{q_2 - q_1}{d}, \qquad d \ll \text{wavelength}. \qquad (13)$$

The spacing d should be small in order to make the approximation (13) valid. The term "small," however, is not related to any absolute value, but rather to a value relative to the wavelength of the structural motion. The larger the wavelength (i.e., the lower the frequency), the better the approximation (13).

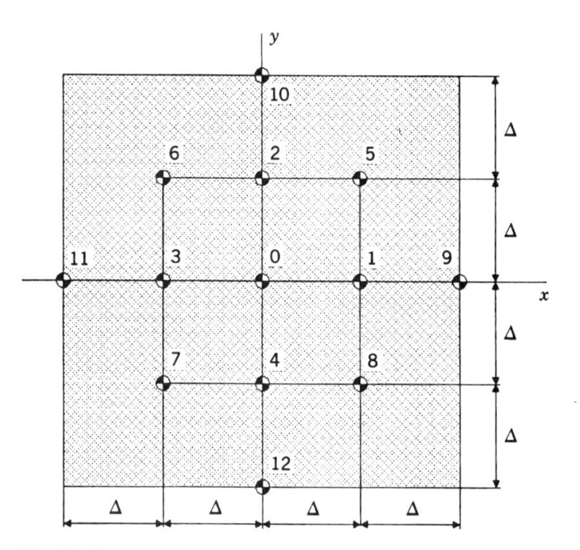

Fig. 8 Transducer configuration for measurement of spatial derivatives of vibration.

Using the principle applied in Eq. (13), higher order derivatives can be determined from suitable finite-difference approximations. Referring to the scheme shown in Fig. 8, the spatial derivatives appearing in the EF formulas can be determined from the finite-difference approximations listed below:

$$\frac{\partial q}{\partial x} \approx \frac{1}{2\Delta}(q_1 - q_3),$$

$$\frac{\partial q}{\partial y} \approx \frac{1}{2\Delta}(q_2 - q_4),$$

$$\frac{\partial^2 q}{\partial x^2} \approx \frac{1}{\Delta^2}(q_1 - 2q_0 + q_3),$$

$$\frac{\partial^2 q}{\partial y^2} \approx \frac{1}{\Delta^2}(q_2 - 2q_0 + q_4),$$

$$\frac{\partial^2 q}{\partial x\,\partial y} \approx \frac{1}{4\Delta^2}(q_5 - q_6 + q_7 - q_8), \tag{13a}$$

$$\frac{\partial(\Delta q)}{\partial x} \approx \frac{1}{2\Delta^3}(q_5 + q_8 + q_9 - 4q_1 - q_6 - q_7 - q_{11} + 4q_3),$$

$$\frac{\partial(\Delta q)}{\partial y} \approx \frac{1}{2\Delta^3}(q_5 + q_6 + q_{10} - 4q_2 - q_7 - q_8 - q_{12} + 4q_4).$$

Measurement of higher order derivatives, such as those appearing in the expressions for plate and shell flexural vibration, becomes difficult in practice due to the large number of transducers required. For example, to measure the Laplacian derivative, eight transducers are needed. In addition, measurement errors increase exponentially with the order of derivative. For these reasons, measurement of higher order spatial derivatives should be avoided. Simplified measurement techniques, such as the two-transducer method described in Section 3.7, may seem far more suitable for practical purposes. It should be pointed out, however, that simplifications assume far-field conditions, that is, locations far from sources or discontinuities. Very often the near-field regions will be of prime interest from the point of view of the applicability of results. Therefore simplified procedures should be used with care.

Wave Decomposition Technique This approach is suitable for waveguides (rods, beams, and pipes) where the vibration field can be described by a simple wave model in the frequency domain.[17]

Nondispersive vibration, longitudinal and torsional, will then be represented by two oppositely propagating waves:

$$\xi(x, \omega) = [A_+ \exp(-jkx + \varphi_+) + A_- \exp(jkx + \varphi_-)] \cdot \exp(j\omega t), \qquad j = \sqrt{-1}. \tag{14}$$

In such a representation the amplitudes A and phases φ of the two waves are unknown but the wavenumber k is supposed to be known: $k = \omega/c$, where c is the wave velocity. For longitudinal vibration $c = \sqrt{E/\rho}$, while for torsional vibration $c = \sqrt{G/\rho}$, where G is the shear modulus.

When flexural vibrations are concerned, the model (14) is valid only in the far field. If the far-field conditions are not met, it should be appended by two additional terms representing evanescent waves that decay from the two boundaries, $D_+ \exp(-kx + \phi_+)$ and $D_- \exp(kx + \phi_-)$.

Once the model of vibratory motion has been established, it can be substituted in the appropriate EF governing formula (Section 3). The EF of nondispersive vibration then becomes proportional to the difference of squares of two amplitues:

$$P_x = \rho c S \omega^2 (A_+^2 - A_-^2). \tag{15}$$

The EF of flexural waves will be contributed additionally by the evanescent terms:

$$P_{x,\text{flexural}} = JEk^3\omega[A_+^2 - A_-^2 - D_+D_- \sin(\phi_+ - \phi_-)], \tag{15a}$$

where $k = (\rho S/JE)^{1/4}\sqrt{\omega}$ is the flexural wavenumber. All that needs to be done by measurement is to determine the unknown amplitudes A_+ and A_- and, if needed, the evanescent components D_+, D_- and $\phi_+ - \phi_-$.

If two transducers are placed at two arbitrary points 1 and 2, the amplitude square difference will be proportional to the imaginary part of their product $u_1 u_2^*$, which corresponds to the cross spectral density between the two signals:

$$A_+^2 - A_-^2 = \frac{\text{Im}\{u_1 u_2^*\}}{\sin(\omega d/c)}, \qquad (16)$$

where the asterisk indicates the complex-conjugate form and d is the distance between the two points. It should be noted that Eqs. (15) and (16) represent an exact measure of EF in a nondispersive vibration field.

To measure flexural EF in the presence of a nonnegligible evanescent field, four instead of two transducers will be needed at different locations to determine all four wave amplitudes as well as phase difference $\phi_+ - \phi_-$. The evanescent vibration components will not be negligible in the frequency range where the wavelengths of vibration are larger than approximately one-half of the length of the measured structure.

4.3 Measurement Errors

Instrumentation Errors In making SI (EF) measurements, various types of errors may occur. In the preceding section, three different sources of inaccuracy were outlined: sensitivity to cross motions, off-set of the sensitivity axis due to finite transducer size, and dynamic loading due to finite transducer mass.

The measurement inaccuracies caused by transducer imperfections are not the only ones related to measurement equipment. An electrical signal from the transducer may be subjected to distortion and noise, which can greatly affect the resulting intensity readings.

The most significant source of signal distortion where SI (EF) is concerned is the phase shift between the physical quantity measured and the conditioned electrical signal that represents it. Since SI (EF) evaluation always required multiplication of signals, the phase shift will directly affect the result, altering the true physical phase angle between the original quantities. This effect becomes particularly significant in connection with finite-difference representation of some of the quantities.

The relative error in measuring the first spatial derivative due to the instrumentation phase shift $\Delta\varphi$ is

$$\varepsilon_{\Delta\varphi} = \frac{\Delta\varphi}{kd\cos\alpha}, \qquad (17)$$

where k is the wavenumber, d the distance between transducers, and α the angle between the direction of the derivative and the wave direction.

The error is linearly proportional to the instrumentation phase shift and inversely proportional to the product of the wavenumber and the transducer spacing. The error due to phase mismatch in measurement of higher order derivatives will be of a form similar to Eq. (19), providing $\Delta\varphi$ remains small:

$$\varepsilon_{\Delta\varphi} = \frac{\Delta\varphi}{(kd\cos\alpha)^n}, \qquad n = \text{order of derivative.} \qquad (17a)$$

To keep the phase mismatch error small, the instrumentation phase shift between channels must be extremely small where measurement of higher order derivatives is concerned. This is because the product kd, which appears in the denominator of Eq. (19a), has itself to be small in order to make the finite-difference approximation valid. For example, a two-transducer method (i.e., $n = 1$) was found to be appropriate when the phase shift was of the order of $0.1°$.[18]

Another important error that can affect measurement accuracy can be caused by the mass loading excited by the measuring transducer.[19] The transducer mass should be kept well under the value obtained by dividing the impedance of the structure by the radian frequency.

Finite-Difference Convolution Error By making approximations of spatial derivatives using finite differences, systematic errors are necessarily produced that depend on the spacing between the transducers and the nature of the vibratory field concerned. For example, the first spatial derivative of the x-direction, approximated by finite difference according to Eq. (18), will contain an error equal to

$$\varepsilon_{\partial u/\partial x} \approx \frac{d^2}{24}\frac{\partial^3 u/\partial x^3}{\partial u/\partial x}. \qquad (18)$$

In plane-wave motion, with the wave direction occurring at an angle α to the axis, the error reads approximately

$$\varepsilon_{\partial u/\partial x} \approx -\tfrac{1}{24}k^2 d^2\cos^2\alpha, \qquad \varepsilon \ll 1. \qquad (18a)$$

For flexural vibration, the wavenumber is proportional to the square root of the frequency. The error in measuring the first spatial derivative will therefore be proportional to the product of the frequency and the square of the transducer spacing, d. It depends on the angle of wave propagation α. The error is seen always to be negative; that is, the finite-difference technique underestimates the true value.

The finite-difference error for the second derivative is

approximately twice the error for the first derivative:

$$\varepsilon_{\partial_u^2/\partial_x^2} \approx 2\varepsilon_{\partial u/\partial x} \approx -\tfrac{1}{12}k^2 d^2 \cos^2 \alpha. \tag{18b}$$

The errors of higher order derivatives rise in proportion to the derivative order, in contrast to instrumentation errors, which exhibit an exponential rise.

The finite-difference error for any spatial derivative can be shown to be proportional to the square of the transducer spacing d. In addition, where plane waves are concerned, the error is proportional to the square of the wavenumber.

Wave Decomposition Coincidence Error The technique of wave decomposition is essentially a solution to an inverse problem. This solution can always be represented as a ratio of two frequency-dependent functions. The problem arises at frequencies where the denominator approaches zero, that is, where the solution is badly conditioned. At these particular coincidence frequencies, which correspond to a specific relationship between the geometry of the transducer array used and the wavelength, even very small measurement imperfections can cause significant measurement error.

For the simple case of nondispersive vibration motion [Eq. (14)], the frequency singularities occur for $\omega d/c = n\pi$, that is, each time half the wavelength becomes an integer fraction of the spacing between the two transducers. More complex transducer arrays will result in larger number of coincidence frequencies that need not be equidistant. The larger are the distance(s) between the transducers, the more numerous are the coincidence frequencies within a given frequency band.

Total Measurement Error For small error values due to instrumentation imperfections and measurement method imperfections, finite-difference convolution or wave decomposition coincidence, the total error will be the simple sum of the two errors. The total error is therefore subject to the same spacing/wavenumber dependence as errors in the measurement of individual quantities.

REFERENCES

1. S. P. Timoshenko and J. N. Goodier, *Theory of Elasticity*, Int. Student Ed., McGraw-Hill, New York, 1970.
2. D. Noiseux, "Measurement of Power Flow in Uniform Beams and Plates," *J. Acoust. Soc. Am.*, Vol. 47, 1970, pp. 238–247.
3. G. Pavic, "Determination of Sound Power in Structures: Principles and Problems of Realisation," *Proc. 1st Congress on Intensity Techniques*, Senlis (France), September–October 1981, pp. 209–215.
4. G. Pavic, "Measurement of Structure-Borne Wave Intensity, Part 1: Formulation of Methods," *J. Sound Vib.*, 1976, Vol. 49, pp. 221–230.
5. P. Rasmussen and G. Rasmussen, "Intensity Measurements in Structures," *Proc. 11th Int. Congress on Acoustics*, Paris, July 1983, Vol. 6, pp. 231–234.
6. C. R. Fuller and F. J. Fahy, "Characteristics of Wave Propagation and Energy Distribution in Cylindrical Elastic Shells Filled with Fluid," *J. Sound Vib.*, 1982, Vol. 81, pp. 501–518.
7. G. Pavic, "Vibrational Energy Flow in Elastic Circular Cylindrical Shells," *J. Sound Vib.*, 1990, Vol. 142, pp. 293–310.
8. G. Pavic, "Measurement of Structure-Borne Acoustical Intensity in Thin Shells," *Proc. of Inter-Noise 92*, Toronto, July 1992, Vol. 1, pp. 509–514.
9. I. Levanat, "Vibrational Energy Flow in Spherical Shell," *Proc. 3rd Congress on Intensity Techniques*, Senlis (France), August 1990, pp. 83–90.
10. J. W. Verheij, "Cross-Spectral Density Methods for Measuring Structure-Borne Power Flow on Beams and Pipes," *J. Sound Vib.*, 1980, Vol. 70, pp. 133–139.
11. C. A. F. de Jong and J. W. Verheij, "Measurement of Energy Flow Along Pipes," *Proc. Int. Congress on Recent Develop. in Air and Structure-Borne Sound and Vibration*, Auburn (USA), 1992, pp. 577–584.
12. R. J. Pinnington and R. G. White, "Power Flow through Machine Isolators to Resonant and Non-Resonant Beams," *J. Sound Vib.*, 1981, Vol. 72, pp. 179–197.
13. G. Pavic and G. Oreskovic, "Energy Flow through Elastic Mountings," *Proc. 9th Int. Congress on Acoustics*, Madrid, 4–9 June 1977, Vol. 1, p. 293.
14. J. C. Pascal, X. Carniel, V. Chalvidan, and P. Smigielski, "Energy Flow Measurements in High Standing Vibration Fields by Holographic Interferometry," *Proc. Inter-Noise 95*, Newport Beach, July 1995, Vol. 1, pp. 625–630..
15. E. G. Williams, H. D. Dardy, and R. G. Finck, "A Technique for Measurement of Structure-Borne Intensity in Plates," *J. Acoust. Soc. Am.*, 1985, Vol. 78, pp. 2061–2068.
16. J. C. Pascal, T. Loyau, and J. A. Mann III, "Structural Intensity from Spatial Fourier Transform and Bahim Acoustic Holography Method," *Proc. 3rd Int. Congress on Intensity Techniques*, Senlis (France), August 1990, pp. 197–206.
17. C. R. Halkyard and B. R. Mace, "A Wave Component Approach to Structural Intensity in Beams," *Proc. 4th Int. Congress on Intensity Techniques*, Senlis (France), August–September, 1993, pp. 183–190.
18. G. P. Carroll, "Phase Accuracy Considerations for Implementing Intensity Methods in Structures," *Proc. 3rd Congress on Intensity Techniques*, Senlis (France), August 1990, pp. 241–248.
19. R. J. Bernhard and J. D. Michol, "Probe Mass Effects on Power Transmission in Lightweight Beams and Plates," *Proc. 3rd Congress on Intensity Techniques*, Senlis (France), August 1990, pp. 307–314.

59

ACTIVE VIBRATION CONTROL

C. R. FULLER

1 INTRODUCTION

Passive control of vibrations in structures including isolation mounts has been discussed in Chapter 55 and Ref. 1. These methods work well at high frequencies or in a narrow frequency range but often have the disadvantage of added weight and poor low-frequency performance. Active vibration control has demonstrated the potential to solve many of these problems. Although the potential of active vibration control has been well known for many years, recent advances in fast digital signal processing (DSP) computer chips have made these systems realizable.

In active vibration control secondary vibration inputs are applied to the structure in order to modify its response in a desired manner. The major components of an active vibration control system are the plant, actuators, sensors, and a controller. The plant represents the physical system to be controlled. Error sensors are needed to measure the system response while control actuators provide the necessary inputs to the plant to modify its response. The controller implements the chosen control algorithm to ensure that the controlled (or closed-loop) system behaves as required. The closed-loop system consists of the open-loop (uncontrolled) system dynamics combined with the dynamics of the controller and thus behaves in a modified form. These components will be discussed separately below and then unified into a coupled system behavior by use of example applications. Basic theory for each of the components is also discussed in order to provide the framework for analyzing controlled system behavior.

The arrangement of an active control system is usually based upon the physics of the system to be controlled and thus, as illustrated below, is often application dependent. There are many texts on active control of vibration, mainly focused on the control theory. The text by Meirovitch[2] provides an excellent theoretical basis while the text of Fuller et al.[3] describes in detail both feedforward and feedback control theory, actuators and sensors, as well as many relevant applications. Chapter 67 also discusses the related area of active noise control. It should be noted that active control is not a panacea for every vibration problem, and its application should be chosen with care depending upon a number of important characteristics such as the temporal nature of the disturbance, the spatial nature of the noise field, and the dynamic characteristic of the disturbance.[4]

2 CONTROL ACTUATORS

Actuators are used to introduce control forces into the plant in order to modify its behavior, and their design/selection (in conjunction with the sensors) is often the most important step in an active noise control project. Actuators can take various forms that are dependent upon system requirements such as the required control authority (amount of control force, moment, strain, or displacement), power, frequency response, and physical constraints such as size, mounting requirements, and so on. Actuators are generally classified into two main categories: *fully active* actuators that apply a secondary vibrational response to the structure and *semiactive* actuators that are passive elements and can be used to adaptively adjust the mechanical properties of the system.

Note: References to chapters appearing only in the *Encyclopedia* are preceded by *Enc.*

2.1 Electrodynamic Actuators

Electrodynamic actuators or shakers consist of a moving wire coil mounted inside a permanent magnet. As they are readily available and their behavior is relatively well understood (see Ref. 5 and Chapters 48 and 56), electrodynamic shakers are presently the most common control input as a fully active device. They are usually installed by either attaching directly to the structure by a stinger or placing in series or parallel with passive mounts, as in the case of active isolation (see Section 5.1). While electrodynamic actuators have the advantage of relatively large control displacement, they are usually bulky and require some form of support structure. However, there are many applications at low frequencies in which their use is more than adequate. Electrodynamic actuators are usually modeled in terms of control inputs as point forces and they are thus "spectrally white" in terms of the spatial wavenumber response of the excited structure.

Typical structural components encountered in practice can often be simply represented as beams or plates. The transverse response w of a simply supported thin beam to an oscillating point force $Fe^{i\omega x}$, representative of a control shaker, can be described as[6]

$$w(x,t) = \frac{-2F}{M} \sum_{n=1}^{\infty} \frac{\sin(n\pi x_f/L)\sin(n\pi x/L)}{\omega^2 - \omega_n^2 + i2\zeta_n\omega_n\omega} e^{i\omega t}, \quad (1)$$

where x is the axial coordinate, x_f is the location of the force, L is the length of the beam, ω is the frequency of excitation, M is the total mass of the beam, and ζ_n is the modal damping ratio. In Eq. (1), ω_n is the natural frequency of the nth mode of the beam.[6] The transverse vibration displacement w of a simply supported thin panel excited by a point force $Fe^{i\omega t}$ is also described by[6]

$$w(x,y,t) = \frac{-4F}{M} \sum_{m=1}^{\infty}\sum_{n=1}^{\infty} \cdot \frac{\sin(m\pi x_f/L_x)\sin(n\pi y_f/L_y)\sin(m\pi x/L_x)\sin(n\pi y/L_y)e^{i\omega t}}{\omega^2 - \omega_{mn}^2 + i2\zeta_{mn}\omega_{mn}\omega}, \quad (2)$$

where the force is located at (x_f, y_f), ω_{mn} is the natural frequency of the (m,n)th mode, and M is the total mass of the plate. Response of other more complex systems to point forces can be found in well-known texts such as Ref. 6.

2.2 Piezoelectric Actuators

As discussed in Ref. 7, piezoelectric transducers consist of material that expands or contracts when an electric field is applied over it. By applying an oscillating voltage to the piezoelectric element, its strain can be made to oscillate at the same frequency as the input. Three major types of piezoelectric material are readily available: (1) a ceramic form such as PZT, which has relatively high control strain but is brittle; (2) a polyvinyl form such as PVDF, which is flexible but has less control strain for the same configuration; and (3) a piezoelectric rubber, which is useful for underwater applications. Piezoelectric material is configured in the two main forms of *stack* and *wafer*, as shown in Fig. 1. Piezoelectric stacks are configured so that when a voltage is applied across the electrodes, the stack usefully expands in its long or 3–3 axis.[7] Stack arrangements such as Fig. 1 are thus suitable for actuators in vibration isolation in the two configurations of parallel and series shown in Figs. 1*a* and 1*b*.

For a given applied voltage V the net static displacement δ of the piezoelectric ceramic actuator shown in the parallel configuration of Fig. 1*a* can be calculated from

$$\delta = \frac{d_{33}V + F/K_a}{1 + K/K_a}, \quad (3)$$

where d_{33} is the strain constant of the piezoelectric material in the 3–3 axis,[7] K is the external spring stiffness, K_a is the actuator stiffness ($K_a = E_aA_a/L_a$, where L_a, A_a, and E_a are the actuator length, cross sectional area, and Young's modulus, respectively); and F is the external load force. Note Eq. (3) can be used as the basis of an approximate dynamic analysis, as discussed in Ref. 3. The significant advantage of the piezoelectric stack is that it can provide high force; however, its displacement is limited (relative to electrodynamic actuators), which, as discussed later, implies that its use is primarily in series active isolation implementations.

The other common form of piezoelectric actuator is the wafer arrangement shown in Fig. 1*c*. In this form the transducer usefully strains in its thin transverse axis (3–1 or 3–2) when a voltage is applied across the electrodes. The wafer is usually bonded or embedded directly into the structure and actuates the structure by applying surface or interior strains. Figure 1*c* shows piezoelectric elements arranged to create pure bending. In this configuration two colocated elements are employed and wired 180° out of phase so that as one element expands, the other contracts. As discussed in Ref. 8, the surface-mounted colocated actuator for a perfect bonding layer (which is a good approximation if the glue layer is thin) effectively applies a line moment to the structure at the boundaries of the actuator. The magnitude of this linear

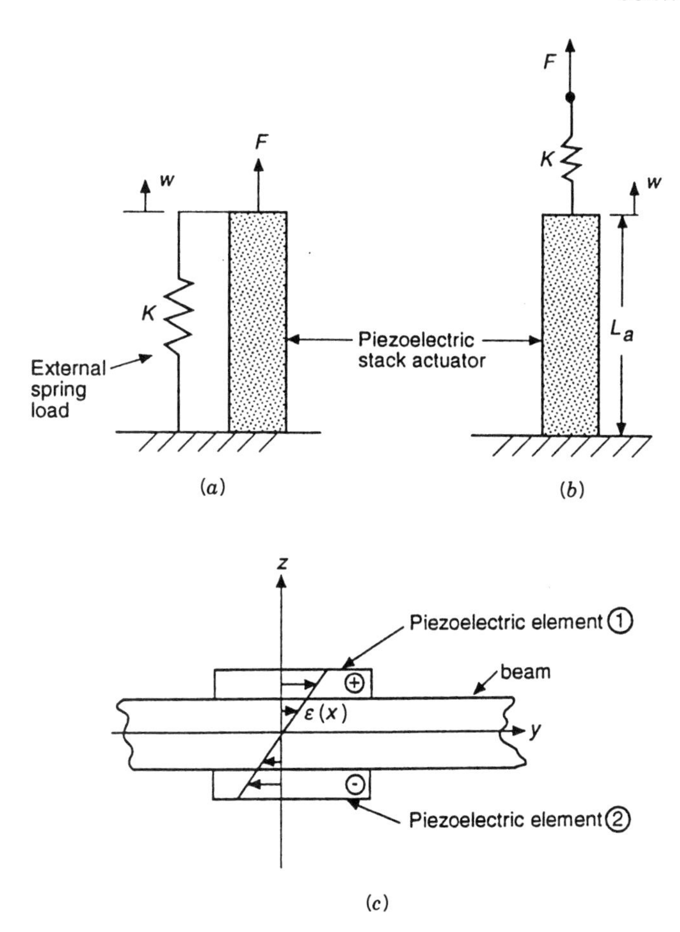

Fig. 1 Piezoelectric actuator configurations: (*a*) stack in parallel; (*b*) stack in series; (*c*) colocated wafer.

moment density for a one-dimensional structure such as a simply supported beam is given by[9]

$$m_x = \frac{-E_b 2Ph^2\epsilon_{pe}}{3(1-P)}, \qquad (4)$$

where the piezoelectric element unconstrained strain $\epsilon_{pe} = d_{31}V/t$. The constant P is defined as

$$P = \frac{-3th(2h+t)}{2(h^3+t^3)+3ht^2}\,\frac{E_{pe}}{E_b}. \qquad (5)$$

In Eqs. (4) and (5) E_b and E_{pe} are the Young's elastic moduli of the beam material and piezoelectric element, respectively, while h is the half thickness of the beam, t is the thickness of the piezoelectric element, V is the applied voltage, and d_{31} is the piezoelectric transverse strain constant. For a two-dimensional patch system in spherical bending of a plate the effective moments m_x, m_y of the actuator are given by[10]

$$m_x = m_y = \frac{-E_p(1+\nu_{pe})}{3(1-\nu_p)}\,\frac{2P'h^2\epsilon_{pe}}{(1+\nu_p-(1+\nu_{pe})P')}, \qquad (6)$$

where

$$P' = P\,\frac{1-\nu_p^2}{1-\nu_{pe}^2}. \qquad (7)$$

In Eqs. (6) and (7) ν_{pe} and ν_p are the Poisson's ratios of the piezoelectric and the plate material, respectively. More information of piezoelectric elements and their properties as well as definitions of important terms can be found in Ref. 7. Excitation of beam and plate structures with piezoelectric patches are discussed in Refs. 9 and 10. Piezoelectric actuators can be used either in a semiactive or a fully active form.

2.3 Advanced Actuators

The desire for an optimum actuator has stimulated development of new materials and configurations. The requirements, depending upon application, are typically high force and/or displacement, large bandwidth of operation, lower power consumption, and ease of use.

Shape memory alloy (SMA) shows much potential for high-force actuation at DC to very low frequencies. The mechanism by which SMA fibers or films exhibit their characteristic shape memory effect can be described very basically as follows: An object in the low-temperature martensitic condition, when plastically deformed and the external stresses removed, will regain its original (memory) shape when heated. The process is the result of a martensitic phase transformation taking place during heating. When SMA is heated through its transformation temperature, the elastic modulus of the material changes dramatically, and if the material is initially stretched, it will shrink back to its original size. If the SMA is constrained during this procedure, very high restoring forces are generated. More information on SMA can be found in Ref. 11.

The SMA systems are usually configured by embedding the material in a structue in a similar manner to layups of composite structures. It can be activated in a steady-state manner, applying steady-state in-plane forces that can be used to tune the resonant frequencies or change the mode shapes of the system (i.e., as semiactive actuator.[12] Over a very limited frequency range it can be used as an oscillating control input (i.e., a fully active actuator) by positioning the SMA off the central axis of the structure and driving with an oscillating voltage.[12] The advantage of the SMA is the extremely high force and deflection that it can provide at very low frequencies. Disadvantages are high power consumption, a need for heat dissipation, and a limited frequency response, which depends entirely upon the cooling rate. Use of SMA in the control of transient vibration of a cantilevered beam as well as sound radiation control is discussed in Ref. 12.

Magnetostrictive materials also show much possibility as advanced fully active actuators in that they fill the performance gap between low-force, high-displacement electromagnetic actuators and high-force, low-displacement piezoelectric devices. A typical actuator using the material Terfenol-D and approximately 11 cm length and 5 cm diameter can produce forces in excess of 450 N over a frequency range of 0–2.5 kHz.[13] However, the power requirements tend to be somewhat higher than electrodynamic and piezoelectric devices.

Electrorheological (ER) fluids are suspensions of highly polarized fine particles dispersed in an insulating oil.[14] The viscosity and the elasticity of the ER fluid can be changed several orders of magnitudes when an electric field is applied to the medium. The ER fluid can thus be embedded in structural systems and used as a "semiactive" actuator to tune the system's properties such as damping and stiffness by varying the voltage applied to the ER fluid.[15]

3 ERROR SENSORS

Error sensors are employed to measure the motion of the system to be controlled. This information either is used directly as the variable(s) to be minimized or is used to calculate a related state of the system to be controlled. Sensors come in three main forms: point sensors, arrays of point sensors, and distributed sensors. Choice of the particular configuration is dependent upon the system variable to be controlled, as discussed below.

3.1 Accelerometers, Force Transducers, and Impedance Heads

Accelerometers are commonly used as error sensors due to the ease of use and reliabilty of performance. Their implementation and behavior is discussed in more detail in Chapter 56. When attached to a vibrating structure, they can provide estimates of the time-varying acceleration (as well as velocity and displacement with signal processing) at the point of attachment. Due to their small size, at low frequencies accelerometers provide a point sampling of the structural motion and thus are "spectrally white" in wavenumber response. Thus an individual accelerometer will equally sense all structural wavenumber or modally weighted contributions. Use of an accelerometer output as an error signal will result in direct control of the structural motion at that point.

Force transducers can also be used as error sensors, particularly in applications such as active isolation, where they may be placed in series with passive isolation. Minimization of the force transducer output will result in zero dynamic force being applied to the receiving structure. The point input impedance of a structure such as at a disturbance location can be estimated using an impedance head that consists of a colocated accelerometer and force transducer. The output of the impedance head can then be used as a narrow-band error sensor variable in order to minimize the real part of the structural input impedance and thus disturbance input power flow into the structure.

Choice of one of the above transducers as an error sensor is dependent upon the form of application. This will be illustrated in the later example applications.

3.2 Arrays of Point Sensors

Usually it is more effective to use accelerometers as error sensors in an array configuration whose output signal is processed to provide some global or distributed state associated with the system.

The *kinetic energy* of a thin vibrating system can be minimized by using a distributed array of accelerometers positioned over the required controlled domain. An estimate proportional to out-of-plane kinetic energy of the structure is given by

$$E = \sum_{i=1}^{N} |\dot{w}_i|^2, \tag{8}$$

where N is the total number of accelerometers and $\dot{w}$ is the out-of-plane velocity obtained by integrating the acceleration signal. The variable E can be minimized, or if an N-channel controller is available, each accelerometer can be used as an individual control channel. The required number and spacing of accelerometers are determined by the Nyquist sampling criterion applied in the spatial domain: two accelerometers must be spaced within at least a half-wavelength of the highest mode required to be observed and controlled. Alternatively, the accelerometers can be appropriately distributed throughout the surface whose global motion is to be controlled. Use of a lower number of accelerometers (relative to the number of actuators) often leads to large minimization at their locations but increased levels of vibration at other locations. This effect is termed control spillover. In general, to ensure controllability, a system state or mode has to be individually observed.

Wavenumber components of structural motion can be obtained using arrays of equispaced accelerometers arranged along a coordinate axis. The spatial one-dimensional Fourier transform can be approximated using a discrete Fourier transform such as[3]

$$\hat{W}(k) = \sum_{n=1}^{K} w_m(x_n) e^{ikx_n} \Delta x, \tag{9}$$

where $w_m(x_n)$ is the measured out-of-plane complex displacement at position x_n. In Eq. (9) there are N measurement points equally spaced at a distance Δx. Note Eq. (9) can also be expanded into a time–space transform if necessary. The outpt of Eq. (9) can be used to provide an error signal. Minimization of spectral components $W(k)$ then will provide control of selected wavenumber values. For example, in a finite structure vibrating at a single frequency, the wavenumber components will be discrete and associated with particular modes or waves traveling in the positive or negative directions. In many situations it is appropriate to control these components directly.

Modal decomposition of the structural motion is also a useful processing technique. To achieve this, an array of accelerometers is used to measure out-of-plane motion at a number of points. Thus, knowing the structural mode shapes and using a pseudoinverse technique, the modal amplitudes can be estimated from[16]

$$[A] = ([L]^T [L]^{-1})[L]^T [w_{me}], \tag{10}$$

where $[A]$ is a vector of m unknown modal amplitudes and $[w_{me}]$ is a vector of n measured displacements at (x_n, y_n). In Eq. (10) $[L]$ is an $n \times m$ matrix associated with the mode shapes Ψ of the system, defined as

$$[L] = \begin{bmatrix} \Psi_1(x_1, y_1) & \Psi_2(x_1, y_1) & \cdots & \Psi_m(x_1, y_1) \\ \vdots & & \ddots & \\ \Psi_1(x_n, y_n) & & & \end{bmatrix}. \tag{11}$$

Equation (10) can also be readily modified to decompose narrow-band traveling-wave fields into individual wave components and can be implemented for broadband random disturbances, as outlined in the work of Gibbs et al.[17] Equation (10) can thus provide indirect error information of a global nature associated with each mode. Minimization of a modal estimate will ensure that a particular mode is globally controlled in a vibrating structure. This is of use when particular modes are more important than others in a control strategy.

Kalman filtering techniques are used extensively in linear quadratic (LQ) control problems,[3] where all the internal state variables of the system to be controlled must be known. When all the states are not measurable, they have to be estimated using a mathematical model of the dynamic system and measurements of the output variables. For Gaussian disturbances, the Kalman filter is the optimal estimator that minimizes the variance of the estimation error.[18]

Power flow can be estimated from an array of accelerometers mounted on the structure. For the simplest case of traveling waves on a thin beam the net narrow-band bending power flow can be calculated by using two measurement points and finite-difference theory, as described in Chapter 58 and Ref. 19:

$$P(\omega) = \frac{2EIk_f^2}{\Delta x \omega^3} \operatorname{Im}(S_{21}). \tag{12}$$

In Eq. (12) k_f is the flexural wavenumber at frequency ω,[6] Δx is the spacing of the two acceleration measurement points, and S_{21} is the cross spectrum between the output of the two accelerometers. Use of Eq. (12) as an error signal will thus result in minimization of net power flow in the structure. Extension of Eq. (12) to include near fields is also discussed in Ref. 19 (see also Chapter 58). As discussed above, it is also possible to decompose the motion of the structure into wave component amplitudes using a pseudoinverse technique. The power flow associated with each direction of propagation or wave type can be then estimated and used as an error signal. Using approaches such as these, it is possible to actively control termination impedances of beams.[20]

3.3 Distributed Piezoelectric Sensors

Piezoelectric material, when strained from its free state, also creates a charge.[7] Thus, when configured properly, piezoelectric material with attached electrodes to collect the charge can be used as an error sensor. The most common arrangement is to bond the piezoelectric material to one side of the surface of the structure whose motion is to be sensed. If the piezoelectric material is very thin, then it can be assumed that its presence has no effect on the motion of the structure. For a two-dimensional structure covered with a finite layer of piezoelectric material the charge output $q(t)$ of the sensor (in rectangular coordinates) can be calculated from[21]

$$q(t) = (h_p + h_s) \iint_S \Gamma(x, y) \cdot \left[\epsilon_{31} \frac{\partial^2 w}{\partial x^2} + \epsilon_{32} \frac{\partial^2 w}{\partial y^2} + 2\epsilon_{36} \frac{\partial^2 w}{\partial x \partial y} \right] dx\, dy, \tag{13}$$

where $w(x, y)$ is the plate out-of-plane response, $\Gamma(x, y)$ is the sensor shape function, h_p and h_s are the thickness of the plate and sensor, and $\epsilon_{31}, \epsilon_{32}$, and ϵ_{36} are the piezoelectric stress–charge coefficients where the superscripts 1, 2, 3 refer to the z, x, and y directions of a three-dimensional coordinate system.[7] In effect, the piezoelectric material integrates the surface strain of the structure over its area S (i.e., in the 1–2 plane) and thus provides continuous distributed sensing. The actual voltage output of the piezoelectric material is dependent upon the electrical circuitry used to measure the piezoelectric output. Shaped piezoelectric sensors are surface-mounted sensors that are of specified finite size and shape. Particular motions or modal components of the vibrating structure can be observed by cutting the piezoelectric material into a specialized shape. The charge output of an example rectangular piezoelectric strip mounted on a simply supported plate so that it extends completely over the plate in the x-direction but is narrow in the y-direction is given by[22]

$$q(t) = K \sum_{m=1}^{\infty} \sum_{n=1}^{\infty} \frac{L_x L_y}{nm\pi^2} [\cos(m\pi) - 1] \cdot \left[\cos\left(\frac{n\pi d}{L_y}\right) - \cos\left(\frac{n\pi c}{L_y}\right) \right], \tag{14}$$

where L_x, L_y is the dimension of the plate in the x, y directions, K is a constant that modifies the charge output due to the external electrical load impedance, m and n are modal orders, and d and c are the vertical locations of the sensor boundaries in the y-direction. Note Eq. (14) shows that for all even modes in the x-direction ($m = 2, 4, 6, \ldots$) the sensor output is zero. Thus the sensor acts as a spatial wavenumber or modal filter only observing odd modes in the x-direction. Positioning and sizing of the piezoelectric strip can be adjusted to observe required modal or wavenumber components.

In order to observe individual particular modes of motion, the piezoelectric strip can be cut into a shape corresponding to a system characteristic function. The output of a shaped sensor positioned on a finite simply supported beam is given by

$$q(t) = K \sum_{m=1}^{\infty} \int_0^L \Gamma(x) \Psi_m \, dx, \tag{15}$$

where the mode shape is $\Psi_m = \sin(m\pi/L)$. If the shape function $\Gamma(x)$ is chosen equal to a mode shape $\Psi_m(x)$, then, by orthogonality, Eq. (15) predicts that the sensor will only observe that mode. Figure 2 presents an example of a sensor shaped to only observe the $m = 2$ mode of the simply supported beam. Note that the sensor width is largest where the surface strain of the beam is highest. Thus, shaped sensors can provide a time-domain estimate of modal amplitudes of structural response, with minimal signal processing, in contrast to the previous method using arrays of accelerometers. However, once the sensor is shaped, its estimate is fixed, whereas the

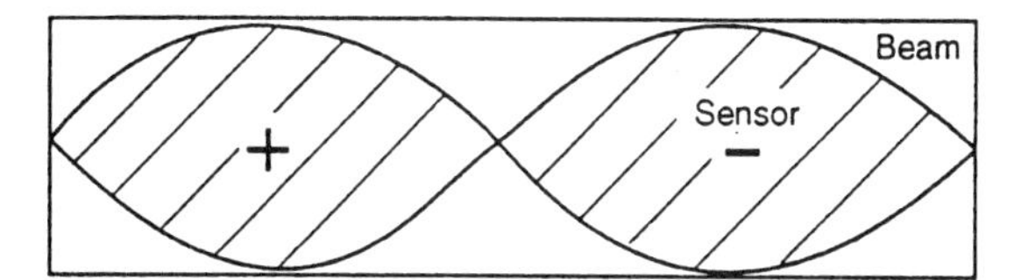

Fig. 2 Shaped distributed piezoelectric strain sensor on a simply supported beam.

output of the accelerometer array can be manipulated by signal processing to observe various modes or waves.

The most commonly used piezoelectric sensing material is PVDF. It can easily be cut or etched into particular shapes. It is readily attached to the surface with double-sided tape or other means. Care must be taken to attach leads to the bottom of the element, and when the shaped sensor has phase changes throughout its segment (as in Fig. 2), each segment output should be wired together with the correct relative polarity in order to give the required sensor output.

4 CONTROL APPROACHES

The function of the controller is to process the error information received from the error sensors in order to calculate control signals so as to cause the controlled or closed-loop system to behave in a required manner. Control approaches can be divided into two main categories: *feedforward* and *feedback* approaches, as discussed by Fuller et al.[3]

In general, feedforward control has found application when the designer has direct access to information about the disturbance signal to the system. On the other hand, feedback control has primarily been applied when the disturbance cannot be directly observed. In this case the control signals are obtained from the sensor(s) whose output is affected by both disturbance source and the control actuator(s). Chapter 67 also has relevant material on the active control paradigms discussed here.

4.1 Feedforward Control

Figure 3 presents a simple, illustrative arrangement of single-channel feedforward control; in this particular case the object is to minimize the vibration of a mechanical system at the error sensor, e. The most common form of electrical feedforward controller H is an adaptive finite impulse response (FIR) filter whose coefficients are updated by a control algorithm in order to drive the error signal to a minimum. A critically important aspect of the feedforward approach is obtaining a reference or training signal that provides a coherent estimate of the dis-

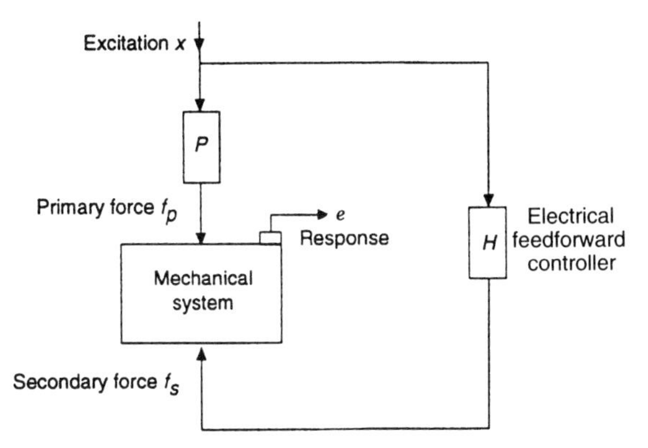

Fig. 3 Simple feedforward control arrangement.

turbance signal and is "fed forward" through the adaptive filter to provide the control signal. The most commonly used update equation for dynamic systems is the time-domain Filtered X version of the Widrow-Hoff least mean squared (LMS) algorithm written in single-channel form as[23]

$$w^i(n+1) = w^i(n) - 2\mu e(n) f(n), \tag{16}$$

where $w^i(n)$ is the ith coefficient of the adaptive filter at time n, μ is a convergence parameter, and $e(n)$ is the error signal at time n. The filtered reference signal $f(n)$ is given by

$$f(n) = \sum_{j=1}^{P} c^j x(n-j), \tag{17}$$

where $c^j, i = 1, 2, \ldots, P$, are the weights of a fixed FIR digital filter that models the plant dynamics from the actuator signal to the error signal and $x(n)$ is the reference signal taken directly from the disturbance. Equation (17) essentially accounts for the delay between the control system input and the error sensor output. The adaption equation (16) also assumes no feedback of control signal to the reference signal. The fixed FIR filter is usually constructed before switching on the controller or with a low-level probe signal, as described in Refs. 3 and 4. In essence the Filtered X LMS algorithm is a gradient-descent technique that finds the minimum of a quadratic cost function defined as the squared modulus of the error signal.

The choice of the filter format is dependent upon the noise disturbance. For a narrow-band signal it is necessary to use only two filter coefficients. Control of broadband signals requires a larger filter in terms of coefficients, and causality aspects associated with relative delay through the control path become more important.[3] The Filtered X algorithm can be readily extended to a multichannel configuration, as discussed in more detail in Ref. 24.

Note that the use of a reference signal estimate from the disturbance implies that the system will respond only to information coherent with the reference signal. The system will thus appear open loop to all other disturbances uncorrelated with the error signal. The maximum achievable attenuation of the power of the error signal for a single-input, single-output feedforward control arrangement can be calculated from[3,4]

$$\Delta \mathrm{dB} = -10\log(1 - \gamma_{de}^2), \tag{18}$$

where γ_{de}^2 is the coherence between the disturbance and the error signals.

Similar to feedback control, it can be demonstrated that closed loop systems under feedforward control have new eigenvalues and eigenfunctions[25] and that, for narrow-band disturbances, feedforward and feedback approaches are essentially equivalent.[3] Due to convergence time requirements (the adaptive filter has to search for the minimum of the error surface), the feedforward algorithm does not presently work well with transients applied to nonmodeled systems but is suitable for pure tones or steady-state broadband signals. As the control filter is adaptive and "learns" the system by minimizing a cost function, it is not necessary to have an accurate model of the open-loop system behavior from the control actuator to the error sensor. For a single-channel time-domain controller applied at a single frequency, it is only required to estimate the fixed transfer function between the control input and error sensor within an accuracy of ±90° to ensure controller convergence.

The behavior of feedforward controlled systems can also be analyzed using linear quadratic optimal control theory. For the system of Fig. 3 a quadratic cost function is constructed by squaring the modulus of the error signal (or summing the squared modulus of the error signals). The optimal control input is found by setting the gradient of the cost function to zero (i.e., at the minimum).

For a system with L error sensors the quadratic cost function is formed by summing the squared modulus of the error signals,

$$J = \sum_{l=1}^{L} |e_l|^2. \tag{19}$$

The optimal control inputs to minimize the cost function are found by setting the gradient of the cost function to zero, as described by Nelson and Elliott.[26] For a multiple-input, multiple-output arrangement the optimal vector q_{s0} of control inputs to minimize the cost function is specified by

$$q_{s0} = A^{-1} Z^H w_p, \tag{20}$$

where Z is a matrix of complex transfer impedances relating the control force inputs to the response at the error sensors, the matrix $A = Z^H Z$ and w_p is the noise or primary complex field at the error sensors. More details on how to use Eq. (20) as well as calculating the minimum of the cost function can be found in Ref. 26. The above analysis is for a single-frequency, steady-state excitation and represents the maximum achievable performance. For random signals, the actual control performance will depend upon whether the control filters are physically realizable and other related issues, as described in Ref. 4.

4.2 Feedback Control

Figure 4 presents a typical arrangement of single-channel feedback control of motion of a general mechanical system. In this arrangement the control signal is derived from a sensor on the structure and is then "fed back" through a controller (or compensator) to provide a control input. The optimal controller is derived in order to perform a specified control task. Hence, generally speaking, for feedback control the system changes its dynamic characteristics for *all* disturbances applied to it, in contrast to the feedforward approach discussed previously.

The open-loop equation of motion of a single-degree-of-freedom system excited by a disturbance force $f_p(t)$ can be written in the form[3]

$$m\ddot{w}(t) + c\dot{w}(t) + kw(t) = f_p(t), \tag{21}$$

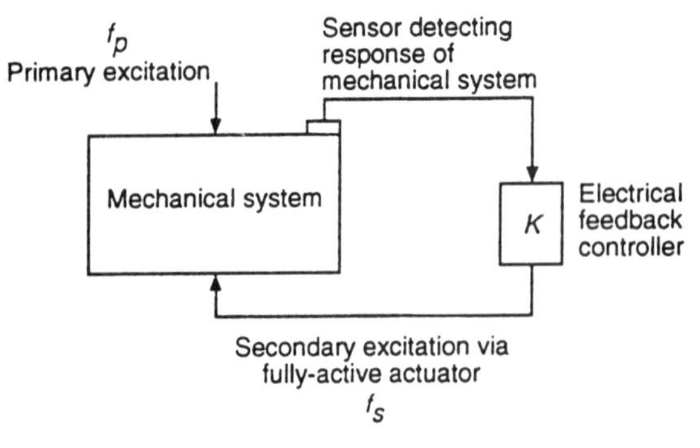

Fig. 4 Simple feedback control arrangement.

where m, c, and k are the system mass, damping, and stiffness, respectively. For the mechanical system in Fig. 4 this may correspond to the response of one mode. Extensions to distributed systems are described in Refs. 2 and 3. The open-loop transfer function (i.e., the ratio of the response to the disturbance) of this system can be evaluated using the Laplace transform[3] and is given by

$$G(s) = \frac{W(s)}{F_p(s)} = \frac{1}{ms^2 + cs + k}. \tag{22}$$

We now apply a secondary control force to the structure by feeding back a signal taken from the response of the system through a compensator with a transfer function K to drive the control force $f_s(t)$. The closed-loop transfer function of the system with stiffness control (for example) is now given by

$$\frac{W(s)}{F_p(s)} = \frac{G(s)}{1 + G(s)K(s)} = \frac{1}{ms^2 + cs + (K + k)s}, \tag{23}$$

where s is the Laplace variable and K is the compensator gain or transfer function. Note that the closed-loop system effectively has a new effective stiffness $(K + k)$ and thus new poles.

For transient control the objective is often to increase the damping of the system. In this case the closed-loop poles are moved further into the left s-plane by varying the compensator gain K. However, an important aspect of feedback control is stability, which can be determined by inspection of the position of the closed-loop poles as described in Ref. 27. Too much compensator gain K, for example, can lead to an unstable control system. Often, in practice, the stability is assessed from input–output measurements made on the system before control. The Nyquist stability criterion is then applied to check if the system is stable, as described in Ref. 26. The outcome is that the closed-loop system is stable only if the polar plot of the open-loop frequency response function does not enclose the point (−1, 0) in the complex Laplace plane s. Different forms of the compensator K to the simple example discussed above can be chosen, depending upon the required control performance, and these are discussed in Ref. 3.

If it is possible to achieve a stable feedback system, then the error signal power is approximately reduced by an amount proportional to the loop gain,[4] and the attenuation is given by

$$\Delta\text{dB} = -10\log\left|\frac{1}{1 + GK}\right|^2. \tag{24}$$

In Eq. (24), $G(s)$ is the transfer function through the plant and $K(s)$ is the transfer function through the compensator.

In general, in order to calculate the optimal controller K, it is required to have an accurate model of the system to be controlled. This can be done through calculations such as finite-element methods or more usually through estimates using system identification approaches as described in Ref. 3. Although multi-input, multi-output feedback control can have a larger implementation requirement than feedforward control, it ensures that the system is controllable for all forms of linear noise disturbances if a realizable controller can be defined.

Extension of feedback control to multi-input, multi-output arrangements can be achieved using state space methods as outlined in Refs. 27 and 3. Other important factors affecting the performance and stability of feedback controllers are reviewed in Ref. 4.

5 EXAMPLE APPLICATIONS OF ACTIVE VIBRATION CONTROL

The following examples illustrate some of the wide diversity of present applications of active vibration control. Other implementations of active vibration control are discussed in Ref. 3.

5.1 Active Isolation of Vibration

A classic approach to reducing unwanted machinery noise and vibration is to control these disturbances at transmission bottle necks (e.g., the machinery mount). Active vibration isolation has been studied by a number of authors, as summarized in Ref. 3, and a few representative examples will be presented here.

There are four possible configurations for active isolation: The two most common are *series*, in which the actuator opposes the load path, and *parallel*, where the active force is applied in parallel with the load path. As discussed by Scribner et al.,[28] if the actuator is placed in parallel, it must overcome the stiffness of the passive mount and the actuator deflection requirements are relatively larger. If the actuator is in series, then while the deflection requirements are lower than the parallel implementation, the actuator must carry the load and, as will be seen later, the force requirements at resonance are higher. Two other, less common implementations are when the active force is directly applied to the exciting system and when the active or secondary force is applied directly to the receiving structure at the mount location. This latter implementation is termed *opposed*. Both implementations have the disadvantage of requir-

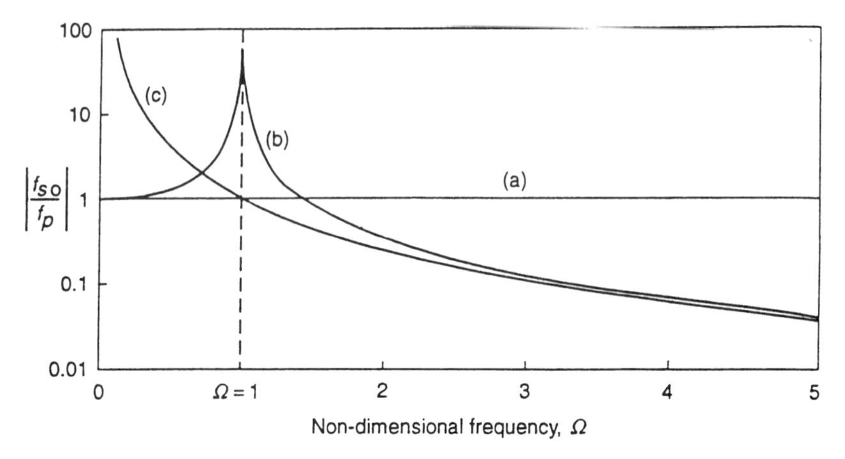

Fig. 5 Control force required for (*a*) direct, (*b*) opposed, and (*c*) parallel active vibration isolation systems.[29]

ing some form of inertial exciter (e.g., an electrodynamic shaker that is provided with a mass against which it can react).

Figure 5 presents the magnitude of the ratio of the secondary control force f_{s0} to the primary disturbance force f_p required for a "parallel" and "opposed" implementation calculated by Nelson et al.[29] for a lumped-parameter single-degree-of-freedom mass–spring system. The nondimensional frequency $\Omega = \omega/\omega_n$, where ω_n is the natural frequency of the mass–spring system on a rigid base. Also shown is the relatively trivial case of when the control force is directly applied to the exciting mass and thus exactly equals the disturbance magnitude in order to drive the motion to zero. The results of the opposed case are similar in trend to the series implementation, which is discussed in much more detail in Ref. 30. The results show that at resonance the force requirements of the opposed system are much larger while below resonance the converse is time. Well above resonance the force requirements of both arrangements are similar. For control systems designed to operate near the plant resonance the above characteristics usually imply that electrodynamic actuators (i.e., relatively large deflection, lower force) are used for the parallel implementation while piezoelectric ceramic actuators (i.e., relatively larger force, low deflection) are used for the series arrangement.

Scribner et al. has theoretically and experimentally studied the use of piezoelectric ceramic actuators in a "series" implementation of active isolation using a classical feedback approach[28] focused upon sinusoidal disturbances. Figure 6 presents typical values of open-loop and closed-loop behavior of the ratio of transmitted to applied force versus frequency. In this example the compensator natural frequency was set to 877 Hz and a large enhancement in transmitted force is observed near this design frequency. However, in the frequency range of $825 \leq f \leq 850$ Hz, high attenuation in the transmitted force is achieved. Scribner et al.[28] also demonstrated how the controller could be implemented in a self-tuning implementation so as to track a disturbance frequency, even through structural resonance.

Nelson et al.[3,29] have also analytically and experimentally investigated narrow-band isolation of the vibration of a rigid raft from a flexible receiving structure using multiple active isolators configured in the parallel arrangement. In this case the active inputs were provided by four electromagnetic coil actuators used in conjunction with a multichannel adaptive LMS feedforward control approach. The error signals were taken from either four accelerometers located adjacent to the mount points or eight accelerometers distributed over the surfce of the receiving structure. The results demonstrate that the use of eight accelerometers was more effective than four in attenuating the total energy of the receiving structure over much of the frequency range. This behavior illustrates the advantage in feedforward control of using more sensors than actuators; a "square" feedforward control system with equal number of actuators and sensors will always drive the response to zero at the error sensors. However, it is often more beneficial to use more error sensors than actuators when global reduction is required. The results presented in Ref. 3 also show that the passive isolation (i.e., no active control) produced an *increase* in the receiving structure vibration at some of its resonance frequencies. This characteristic illustrates the importance of the receiving structure dynamics on the isolation performance.

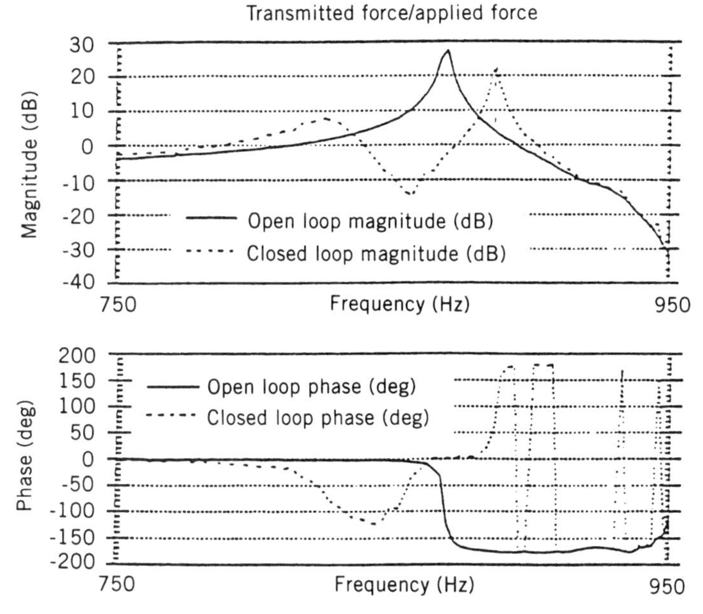

Fig. 6 Performance of a series active vibration isolation system.[28]

5.2 Active Control of Power Flow in Beams

Many structures are composed of basic elements that can be usefully approximated as beams and can transmit unwanted vibrations. The vibrational energy eventually spreads out through the ever increasing complexity of the structure making it more difficult to control. Thus direct reduction of vibrational power flow in the support beams is an area of great interest. In effect this problem is a subset of the active isolation problem discussed above.

Figure 7 shows an experimental arrangement where a semi-infinite beam is driven by a shaker at a single frequency positioned at one end (the disturbance). Piezoelectric ceramic actuators and sensors are positioned on the beam as shown and are wired so as to produce and sense pure bending, as discussed in Section 2.2. The out-of-plane velocity is measured with a laser vibrometer. The control objective is to minimize the power flow transmitted in the beam. Also shown in Fig. 7 is a schematic of a single-channel feedforward control implementation designed to minimize the square of the output of the piezoceramic error sensor which, for a semi-infinite thin beam system, is proportional to total power flow.[31]

The control algorithm used in the arrangement of Fig. 7 was the single-channel Filtered X paradigm of Eq. (16). The controller, indicated by the dashed boundary, was implemented on a TMS320C25 digital signal processing chip with a reference signal taken directly from the disturbance oscillator. Net power flow in the beam was also calculated using the finite-difference equation (12) with the laser vibrometer data. Figure 8 presents a typical result for the measured power flow before and after control is implemented.[31]

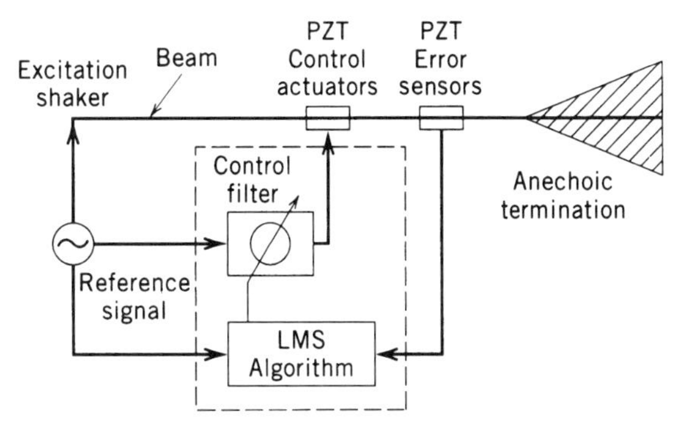

Fig. 7 Experimental arrangement and control block diagram for feedforward active reduction of narrow-band power flow in a beam with piezoelectric actuators and sensors.[31]

Figure 8 shows that when control is applied, the power flow is attenuated by approximately 35 dB at the error sensor. The results of Fig. 8 are net power flow and thus

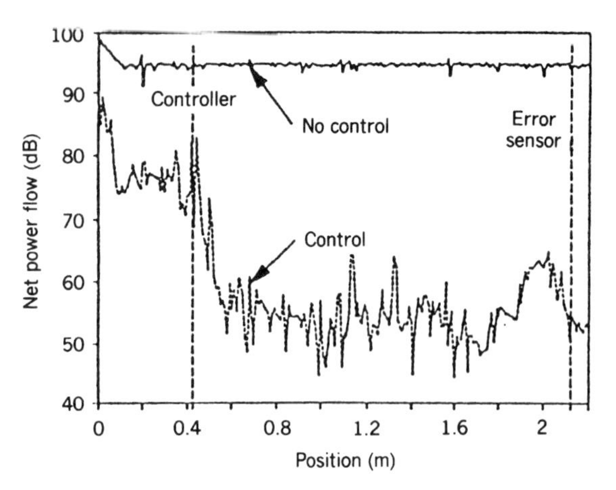

Fig. 8 Power flow distribution with and without control for the system of Fig. 7; f = 302 H.[31]

imply that power flow is highly attenuated at all points downstream of the error sensor. Two other observations of interest are found in the figure. First, the input power at the drive point is attenuated by around 20 dB. Thus the control is achieved mainly by the control actuator creating a reactive discontinuity at its location, reflecting waves back to the drive point and lowering its resistive input impedance. Second, the net power falls past the distributed piezoelectric control actuator, and this implies that the control actuator has also *absorbed* some vibrational power. Extension of the above approach to control of power flow in finite beams, simultaneous control of flexural and extensional waves in beams, and broadband disturbances are discussed in Refs. 31, 32, and 33, respectively. The method has also been successfully applied to attenuation of power flow in struts used to mount the engine and gearbox of a helicopter to the fuselage structure.[34]

5.3 Distributed Active Vibration Control of Plates

The control of vibration of distributed structural elements such as plates is also an important problem. Figure 9 is an experimental arrangement used by Rubenstein et al.[35] to implement linear quadratic Gaussian state space feedback control[3] of the vibration of a simply supported plate. The response of the system is sensed using an array of 12 accelerometers of the plate. The output of the accelerometers is passed through modal filters to obtain modal amplitudes and then through a Kalman filter to obtain the states of the system. The optimal controller $K(\hat{Q})$ can then be obtained using a Riccati equation solution of the state space feedback control arrangement.[3] In this experiment the steel panel was 0.6 × 0.5 m and 3 mm thick. The disturbance and control were implemented by electrodynamic shakers. Figure 10*a* shows a typical closed-loop response for mode 1 when the disturbance is narrow-band, continuous excitation, while Fig. 10*b* gives closed-loop response to a transient disturbance applied at near t = 0.65 s. In both cases high modal attenuation of the disturbance is achieved. Thus state space feedback methods show much potential for distributed active vibration control of a wide range of input disturbances. Note that state space feedback control implies complete control of specified states of the system as opposed to feedforward control, which usually controls a direct output of the plant at the error sensors (displacement, velocity, etc.). As discussed previously, the state space approach also guarantees control of the system to a wide range of disturbances without requiring access to a reference signal.

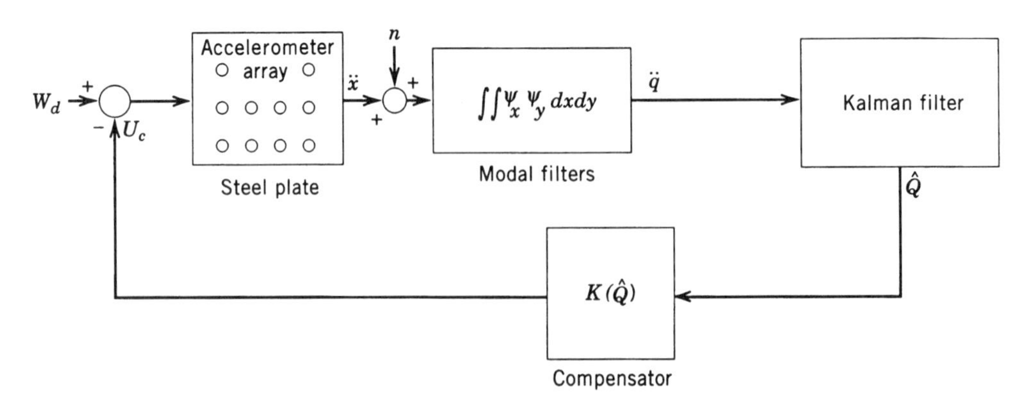

Fig. 9 Linear Quadratic Gaussian (LQG) feedback control system for active reduction of plate vibrations.[35]

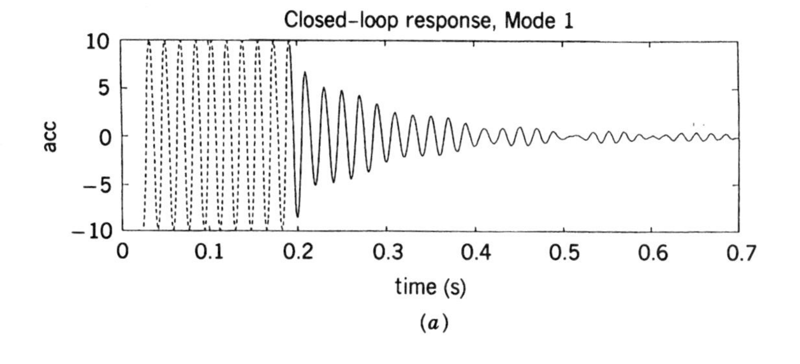

(a)

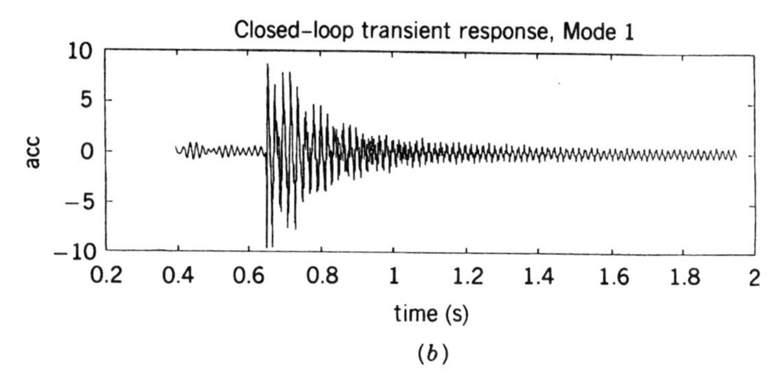

(b)

Fig. 10 Closed-loop response of the system of Fig. 10 for mode 1 with (a) narrow-band steady disturbance and (b) transient disturbance at $t = 0.65$ s.[35]

5.4 Active Structural Acoustic Control

Vibrating structures often radiate or transmit unwanted sound. An effective technique to control the structural sound radiation is to apply control inputs directly to the structure while minimizing the radiated sound or components of structural motion associated with the sound radiation.[3,36] The technique called active structural acoustic control (ASAC) thus falls in the class of general active vibration control techniques. Figure 11 shows an experimental arrangement in which the sound radiation from a simply supported plate is to be minimized.[22] The noise disturbance is provided by a point force shaker attached to the plate and driven by a pure tone. The control inputs are multiple colocated piezoelectric ceramic actuators bonded to the plate surface and wired to produce pure bending. The error sensors are either microphones in the radiated far field or shaped PVDF sensors attached to one side of the plate. The shaped PVDF sensors are rectangular strips positioned on the plate, as discussed in Section 3.3, so as to observe only the odd–odd modes (the efficient modal radiators at low frequencies). The control arrangement is the multichannel feedforward implementation of the Filtered X algorithm discussed previously. The reference signal is taken from the noise input signal generator and is filtered as outlined in Eq. (17) before it is used in the LMS update equation, as described in Section 4.1.

Figure 12 presents results at 349 Hz, which is near the (3, 1) resonant frequency of the plate. The dotted line is for a case where three piezoelectric ceramic actuators are used in conjunction with three error microphones in the radiated far field, and it is apparent that global attenuations of approximately 25 dB are achieved. The dashed line is a similar case except that now two piezoceramic actuators are used in conjunction with two PVDF strips arranged in the x- and y-direction on the plate, as outlined in Section 3.3. The results again demonstrate good global control of approximately 20 dB in the radiated sound levels. These results and others discussed in Ref. 3 demonstrate the potential of using ASAC to reduce radiated sound. For the ASAC technique there are two main mechanisms of control.[16] *Modal suppression* corresponds to a direct reduction of the dominantly radiating modes or a fall in the plate wavenumber spectrum. *Modal restructuring* corresponds to the residual plate response having a lower overall radiation efficiency while its total magnitude of response is reduced

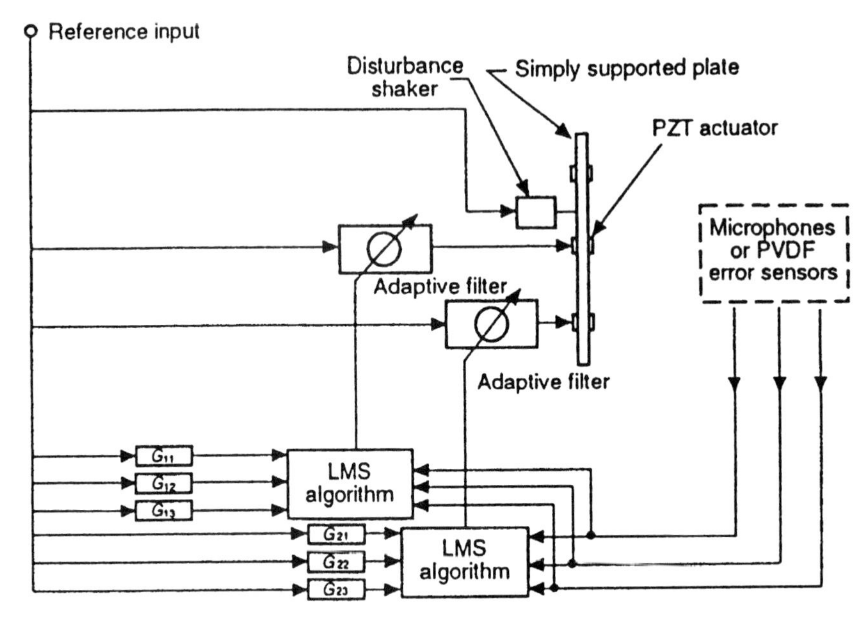

Fig. 11 Experimental arrangement and control block diagram for active reduction of sound radiation from a plate (ASAC).[22]

only slightly or may even increase. This second case corresponds to attenuation of only supersonic components of the plate wavenumber spectrum while subsonic, nonradiating components are left unaffected. Thus the ASAC technique takes maximum advantage of the natural structural–acoustic coupling behavior to reduce the dimensionality of the controller. The technique has been successfully applied to the control of interior noise in aircraft and the reduction of transformer radiated noise, as discussed in Refs. 3 and 4.

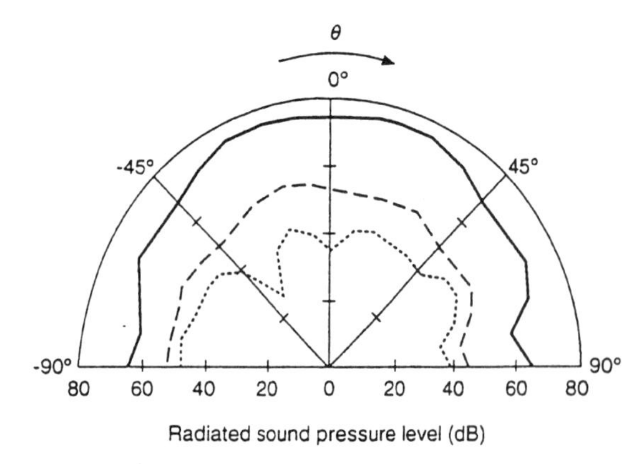

Fig. 12 Active structural acoustic control performance for the system of Fig. 11: (——) uncontrolled; (-------) controlled with PVDF structural error sensors; (.........) controlled with far-field error microphones; $f = 349$ Hz.[22]

From a system point of view, when multiple integrated actuators and sensors such as piezoelectric devices are used in conjunction with a non-model-based, adaptive controller, the application (such as ASAC discussed above) can be seen to be closely related to the field of *smart, adaptive*, or *intelligent* structures (see Ref. 37 for a review of this field). Developments in the field of smart structures are likely to be used in advanced active vibration control and ASAC because of the inherent compactness, light weight, and adaptability of this class of systems.

REFERENCES

1. L. L. Beranek, *Noise and Vibration Control*, INCE, Washington, DC, 1988.
2. L. Meirovitch, *Dynamics and Control of Structrues*, Wiley, New York, 1990.
3. C. R. Fuller, S. J. Elliott, and P. A. Nelson, *Active Control of Vibration*, Academic, London, 1996.
4. C. R. Fuller and A. H. Von Flotow, "Active Control of Sound and Vibration," *IEEE Control Systems Mag.*, December, 1995, pp. 9–19.
5. Ling Dynamic Systems, *User Manual*, Ling Dynamic Systems, Herts, England, 1973.

6. L. Cremer, M. Heckl, and E. E. Ungar, *Structural-Borne Sound*, Springer-Verlag, Berlin, 1988.
7. A. J. Moulson and J. M. Herbert, *Electroceramics: Materials, Properties, Applications*, Chapman and Hall, London, 1990.
8. E. F. Crawley and J. de Luis, "Use of Piezoelectric Actuators as Elements of Intelligent Structures," *AIAA J.*, Vol. 25, No. 10, 1989, pp. 1373–1385.
9. R. L. Clark, C. R. Fuller, and A. L. Wicks, "Characterization of Multiple Piezoceramic Actuators for Structural Excitation," *J. Acoust. Soc. Am.*, Vol. 90, No. 1, 1991, pp. 346–357.
10. E. K. Dimitriadis, C. R. Fuller, and C. A. Rogers, "Piezoelectric Actuators for Distributed Vibration Excitation of Thin Plates," *J. Vib. Acoust.*, Vol. 113, 1991, pp. 100–107.
11. C. M. Jackson, H. J. Wagner, and R. J. Wasilewski, "55-Nitinol-The Alloy with a Memory: Its Physical Metallurgy, Properties and Applications," NASA-SP-5110, Washington, DC, 1972.
12. C. A. Rogers, "Active Vibration and Structural Acoustic Control of Shape Memory Alloy Hybrid Composites: Experimental Results," *Proceedings of International Congress on Recent Developments in Air-And Structure-Borne Sound and Vibration*, Auburn University, AL, 1990, pp. 695–707.
13. C. G. Miller, "High Force, High Strain, Wide Bandwidth Linear Actuators Using the Magnetostrictive Material Terfenol-D," in C. A. Rogers and C. R. Fuller (Eds.), *Proceedings of Recent Advances in Active Control of Sound and Vibration*, (supplement) Technomic, Lancaster, PA, 1991.
14. J. E. Stangroom, "Electrorheological Fluids," *J. Phys. Tech.*, Vol. 14, 1983, pp. 290–296.
15. M. V. Gandhi and B. S. Thompson, "Dynamically-Tunable Smart Composite Featuring Electro-rheological Fluids," *Proceedings of the SPIE Conference on Fiber Optic Smart Structures and Skins II*, Boston, MA, 1989, pp. 294–304.
16. C. R. Fuller, C. H. Hansen, and S. D. Snyder, "Active Control of Sound Radiation from a Vibrating Rectangular Panel by Sound Sources and Vibration Inputs: An Experimental Comparison," *J. Sound Vib.*, Vol. 145, No. 2, 1991, pp. 195–215.
17. G. P. Gibbs, C. R. Fuller, and R. J. Silcox, "Active Control of Flexural and Extensional Power Flow in Beams using Real Time Wave Vector Sensors," *Proceedings of the 2nd Conference on Recent Advances in Active Control of Sound and Vibration*, Technomic, Lancaster, PA, 1993, pp. 909–925.
18. K. J. Astrom and B. Wittenmark, *Computer Controlled Systems*, Prentice-Hall, Englewood Cliffs, NJ, 1984.
19. J. M. Downing and K. P. Shepherd, "Power Flow in Beams Using a 5-Accelerometer Probe," *Proceedings of Noise-Con 88*, Purdue University, IN, 1988, pp. 335–340.
20. J. Scheuren, "Active Control of Bending Waves in Beams," *Proceedings of Inter-Noise 85*, Munich, 1985, pp. 591–595.
21. C. K. Lee, W. W. Chiang, and T. C. O'Sullivan, "Piezoelectric Modal Sensors and Actuators Achieving Critical Damping on a Cantilever Plate," AIAA Paper, No. 89-1390, 1989.
22. R. L. Clark and C. R. Fuller, "Modal Sensing of Efficient Radiators with PVDF Distributed Sensors in Active Structural Acoustic Approaches," *J. Acoust. Soc. Am.*, Vol. 91, No. 6, 1992, pp. 3321–3329.
23. B. Widrow and S. D. Stearns, *Adaptive Signal Processing*, Prentice-Hall, Englewood Cliffs, NJ, 1985.
24. S. J. Elliott, I. M. Stothers, and P. A. Nelson, "A Multiple Error LMS Algorithm and Its Application to the Active Control of Sound and Vibration," *IEEE Trans. Acoust., Speech and Signal Process.*, ASSP-35, 1987, pp. 1423–1434.
25. R. A. Burdisso and C. R. Fuller, "Theory of Feedforward Controlled Systems Eigenproperties," *J. Sound Vib.*, Vol. 153, No. 3, 1992, pp. 437–451.
26. P. A. Nelson and S. J. Elliott, *Active Control of Sound*, Academic, London, 1992.
27. W. L. Brogan, *Modern Control Theory*, Prentice-Hall, Englewood Cliffs, NJ, 1982.
28. K. B. Scribner, L. A. Sievers, and A. H. von Flotow, "Active Narrow Band Vibration Isolation of Machinery Noise From Resonant Substructures," *J. Sound Vib.*, Vol. 167, No. 1, 1993, pp. 17–40.
29. P. A. Nelson, M. D. Jenkins, and S. J. Elliott, "Active Isolation of Periodic Vibrations," *Proceedings of Noise-Con 87*, Penn State University, State College, PA, 1987, pp. 425–430.
30. A. H. von Flotow, "An Expository Overview of Active Control of Machinery Mounts," *Proceedings of the 27th Conference on Decision and Control*, Austin, TX, 1988.
31. G. P. Gibbs and C. R. Fuller, "Experiments on Active Control of Vibrational Power Flow Using Piezoceramic Actuators and Sensors," *AIAA J.*, Vol. 30, No. 2, 1992, pp. 457–463.
32. C. R. Fuller, G. P. Gibbs, and R. J. Silcox, "Simultaneous Active Control of Flexural and Extensional Waves in Beams," *J. Intell. Mat. Systems Struct.*, Vol. 1, No. 2, 1990, pp. 235–247.
33. S. J. Elliott and L. Billet, "Adaptive Control of Flexural Waves Propagating in a Beam," *J. Sound Vib.*, Vol. 163, 1993, pp. 295–310.
34. S. J. Elliott, T. J. Sutton, M. J. Brennan, and R. J. Pinnington, "Vibration Reduction by Active Wave Control in a Strut," *Proceedings of the IUTAM Symposium on the Active Control of Vibration*, Bath, England, 1994, pp. 1–8.
35. S. Rubenstein, W. R. Saunders, G. K. Ellis, H. H. Robertshaw, and W. T. Baumann, "Demonstration of a LQG Vibration Controller for a Simply Supported Plate," in C. A. Rogers and C. R. Fuller (Eds.), *Proceedings of Recent Advances in Active Control of Sound and Vibration*, Technomic, Lancaster, PA, 1991, pp. 618–630.
36. C. R. Fuller, "Analysis of Active Control of Sound Radiation from Elastic Plates by Force Inputs," *Proceedings of Inter-Noise 88*, Avignon, France, 1988, pp. 1061–1064.
37. B. K. Wada, J. L. Fanson, and E. F. Crawley, "Adaptive Structures," *Mech. Eng.*, November, 1990, pp. 41–46.

PART VII

STATISTICAL METHODS IN ACOUSTICS

60

INTRODUCTION

RICHARD H. LYON

1 INTRODUCTION

The following two chapters deal with some particular uses of statistics in acoustics. They have to do with the response of systems—sound fields and structures—to excitation. The main point to be made is that we can consider the excitation and/or response of systems to be random (or both).

Studies of sound fields in rooms have for a long time been considered statistical descriptions of the sound field, even when the excitation and response are deterministic. Modal descriptions used concepts such as modal density and resonance frequency spacing statistics. Wave descriptions, particularly, called ray acoustics at "high frequencies" have been statistical and have used descriptors such as mean free path or diffuseness. If the excitation is random, so much the better for the statistics.

The development of statistical ideas applied to structural response has proceeded in quite a different way. Vibration problems are often dominated by the very lowest modes of a structure. Consequently, engineers have thought of structures as deterministic, but the excitation and response of structures are often random.

2 RANDOM RESPONSE OF STRUCTURES AND SOUND FIELDS

The details of the random response of structures have been more important in structures than for sound fields. The room acoustician has usually been happy to know the frequency spectrum of the sound pressure and possibly its variance as the source and/or receiver move from one location to another. The structural engineer wants to know how likely a certain exceedance will occur. (Will one part of the structure bang into another?) Or, he or she may be calculating fatigue life and be interested in certain high-order moments of the response. We could describe the acoustician as interested in simple statistics on complicated systems, while the structural engineer is interested in complicated statistics on simple systems.

This dichotomy more or less describes the situation in the late 1950s and early 1960s. But that changed with increasing concern about the vibrational characteristics of large and complex structures, large launch rockets, and the broadband noise of engines that could not be analyzed by the few fundamental mode procedures. The sound on ships produced by their machinery, or by the turbulent flow over null structures could not be understood by conventional models.

It became apparent that new analytical techniques were needed to deal with large complex structures excited by broadband noise. In the early 1960s, new techniques using finite-element methods (FEMs) appeared to offer a solution, and large-scale computer programs such as NASTRAN were developed. As computers became more powerful over time, the FEM codes and the problems tackled by them increased in complexity. But it is safe to say that the goal of solving realistic large-scale structural response problems with these methods continues to be a tantalizing but unachieved goal.

3 STATISTICAL ENERGY ANALYSIS

At about the same time, a different approach was developed in which both sound fields and structures were statistically described. Although room acoustics had long

Note: References to chapters appearing only in the *Encyclopedia* are preceded by *Enc.*

used such descriptions, such an idea was not easily assimilated by structural engineers, who, after all, have detailed mechanical drawings for their structures. How could something built to such a prescription be considered statistical? This new approach was termed statistical energy analysis (SEA) and has since become a fairly standard, if somewhat misunderstood, tool in the analyst's bag of tricks.

But there is a fundamental conceptual issue in population statistics applied to system response. Statistical energy analysis and other related disciplines such as statistical room acoustics and statistical mechanics employ the ergodic hypothesis. Simply stated, this means that any one member of the population will obey the statistics of the population as a whole as realizations of its response are repeated. The practical consequence is that one can calculate population statistics using the SEA model and apply the results to a particular individual member of the population, to wit, the one sitting in the laboratory. This assumption, while reasonable, cannot be fully justified for constructed systems like structures and rooms.

4 NEW DEVELOPMENTS

Just as structural systems have had to employ a viewpoint traditionally associated with room acoustics, new developments in room acoustics have led to analysis techniques developed for structural dynamics. Signal processing applications in machine diagnostics has led to an interest in the phase properties of structural transfer function and to the group delay in particular. New room acoustics interests in environment simulation for sound reproduction and in reverberation suppression in group telephony require a much more detailed understanding of the statistical properties of sound fields. Coherence and group delay statistics of small- and intermediate-size rooms are matters of importance.

These new concerns for the modeling and analysis of the phase properties of transfer functions has led to a new methodology we might call statistical phase analysis (SPA). Some readers will be struck by the notion that SEA and SPA should be related by the fact that energy and phase are conjugate variables in classical mechanics (like displacement and momentum). Certainly, the analysis of group delay statistics has been shown to have some very close parallels to response amplitude calculations.

It will be clear to the reader that the methods and models to be applied to sound fields and structures have become closer. It would be a mistake, however, to conclude that the results from one area can be assumed to be directly applicable to the other, since there are some important differences in system parameters. The fact that structures tend to be one and two dimensional (beam- or platelike) while rooms are three dimensional is one difference. Another is the fact that sound waves in rooms are nondispersive while the bending waves are dispersive. One consequence of these differences is that a very important statistical determinant, the system modal overlap, in structures tends to be small while it tends to be large in rooms.

The discussions in the following two chapters will add flesh to the bones of this introductory discussion.

61

RESPONSE STATISTICS OF ROOMS

MIKIO TOHYAMA

1 INTRODUCTION

This chapter describes the statistical characteristics of sound fields in rooms. The reverberation behavior is described by introducing the concept of mean free path[1] into the sound reflection process, similar to the random motion theory of gas. The sound path from a source to a listener is, in principle, predictable in a deterministic way in a closed space. The time history of the sound reflections is expressed by the impulse response between a source and a receiving point. The impulse response, however, is very sensitive to the locations of both the source and receiver even in a simply formed rectangular room. The variabilities of the impulse responses are difficult to predict in any practical sense. Thus, in many cases, it may be more appropriate to assume that the samples of those impulse responses due to different locations make an ensemble of stochastic processes.

The reverberation decay properties in temporal statistics, the transfer functions in the frequency-domain statistics, and spatial distribution statistics are described. Modern reverberation formulas are expressed as well as the conventional ones. The transfer function statistics are extended into the phase analysis based on the distribution of poles and zeros in the frequency domain. The transfer function phase is important in sound field control such as active noise control. Spatial correlations of reverberation sounds are described as the spatial sampling statistics. These statistics are fundamental for acoustic measurement and design in a reverberant space, sound field control, and machine diagnostics using transfer functions.

Note: References to chapters appearing only in the *Encyclopedia* are preceded by *Enc.*

2 REVERBERATION SOUND IN ROOMS

Reverberation is a transient and nonstationary response of sound in rooms. The reverberation process is fairly slow; thus we can describe the process as a random process.

2.1 Multinomial Distribution of Reflections and Eyring's Formula

The reverberation behavior is described by introducing the concept of mean free path[1] into the sound reflection process, similar to the random motion theory of gases. Suppose that the boundaries of a room are composed of K different surfaces, α_k is the absorption coefficient of the kth surface, and n_k is the number of reflections that occur when a sound ray travels from the kth surface in the time interval t. The reverberation process is written as

$$E_{(n_1,n_2,\ldots,n_K)}(t) = E_0 \prod_{k=1}^{K} (1-\alpha_k)^{n_k} \qquad (\mathrm{J/m^3}), \qquad (1)$$

where E_0 is the energy density in the room at steady state ($t = 0$, i.e., the time when the sound source stops). The ensemble-averaged reverberation process is expressed by

$$E(t) = E_0 \sum_{n_1,n_2,\ldots} p(n_1,\, n_2, \ldots, n_k) \prod_{k=1}^{K} (1-\alpha_k)^{n_k} \qquad (\mathrm{J/m^3}), \qquad (2)$$

where $p(n_1, n_2, \ldots, n_k)$ is the joint probability distribution for the n_k. Following Kuttruff,[2] if we assume a multinomial distribution, we obtain Eyring's decay formula:

$$E(t) = E_0 \exp\left[\frac{cS}{4V} t \ln(1-\overline{\alpha})\right] \qquad (\text{J/m}^3), \quad (3)$$

where $\overline{\alpha}$ is the averaged absorption coefficient and the number of reflections N is given by

$$N = \frac{cSt}{4V}, \quad (4)$$

where $4V/S$ denotes the mean free path for the diffuse-field condition, V is the room volume (in cubic metres), S is the surface area (in square metres) of the room, and c is the sound speed (in metres per second). Thus the reverberation time is given by

$$T_R \approx 0.161 \frac{V}{-\ln(1-\overline{\alpha})S} \qquad (\text{s}). \quad (5)$$

2.2 Poisson Distribution and Sabine's Formula

If we assume independent Poisson distributions for the n_k with the average $\overline{n}_k$,[3]

$$\overline{n}_k = \frac{cS_k}{4V} t, \qquad p(n_1, n_2, \ldots, n_k) = \prod_{k=1}^{K} \left(\frac{\overline{n}_k^{n_k}}{n_k!}\right) e^{-\overline{n}_k}, \quad (6)$$

then the historically important Sabine formula is obtained from Eq. (2) as

$$E(t) = E_0 \exp\left(-\frac{c}{4V} t \sum_k \alpha_k S_k\right) \qquad (\text{J/m}^3). \quad (7)$$

We can also obtain the reverberation time T_R,

$$T_R = \frac{4V}{c\overline{\alpha}S} (\ln 10^6) \approx 0.161 \frac{V}{\overline{\alpha}S} \qquad (\text{s}). \quad (8)$$

2.3 Decay Formula for a Two-Dimensional Field

When the image sources are arranged in a two-dimensional plane including a sound source (Fig. 1), the reverberant energy decay in the two-dimensional field is expressed as a hyperbolic type of function.[4] The average

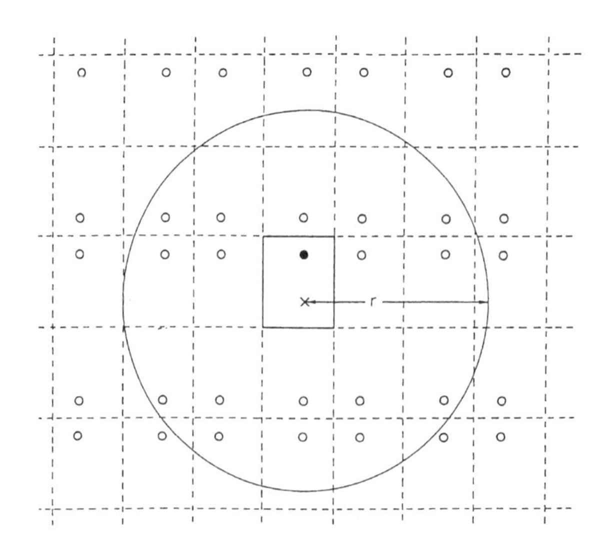

Fig. 1 Two-dimensional array of image sources. (From M. Tohyama, *The Nature and Technology of Acoustic Space*, Academic, London, 1995.)

number of the image sources from which the "reflected" impulse waves travel to a receiving point in a time interval dt after t seconds from the arrival of the direct sound is given by

$$n_{2d}(t) = \frac{2\pi c^2 t}{S_{2d}} dt, \quad (9)$$

where ct is the distance (in metres) between the receiving point and the location of the image source and S_{2d} denotes the area (in square metres) of the two-dimensional space. The average number of reflections N_{2d} is given by

$$N_{2d} = \frac{ct}{m_{2d}} = \frac{L_{2d}ct}{\pi S_{2d}}, \qquad m_{2d} = \frac{\pi S_{2d}}{L_{2d}} \qquad (\text{m}) \quad (10)$$

at t (seconds) after the direct sound arrives at the receiving point, where m_{2d} denotes the mean free path in a two-dimensional diffuse field[1] and L_{2d} (metres) shows the surrounding length of the two-dimensional space. The reverberant decay curve is expressed as

$$E_{2d}(t) = \int_t^{\infty} \frac{P}{2S_{2d}} \frac{dt}{t} \exp\left[\ln(1-\alpha) \frac{cL_{2d}t}{\pi S_{2d}}\right] \qquad (\text{W/m}^2), \quad (11)$$

where P denotes the power output of the point source. Equation (11) is asymptotically written as

$$E_{2d}(t) \rightarrow \begin{cases} \displaystyle\int_t^\infty \frac{dt}{t}, & (t \rightarrow 0), \\ \displaystyle\int_t^\infty \exp\left[\ln(1-\alpha)\frac{cL_{2d}t}{\pi S_{2d}}\right] dt, & (t \rightarrow \text{large}). \end{cases} \quad (12)$$

The reverberation time determined by the *later part of the energy decay* becomes

$$T_{R,2d} = \frac{0.128 S_{2d}}{-\ln(1-\alpha) L_{2d}} \quad \text{(s)}. \quad (13)$$

2.4 Decay Characteristics in a Nearly Two-Dimensional Field

Suppose that the sound field is assumed to be composed of both tangential waves and oblique waves that are close to tangential waves, called here nearly tangential waves. The reverberation process has a typical frequency characteristic in the midfrequency range, since the three-dimensional field is dominant at low frequencies while the two-dimensional field is significant at high frequencies.[5,6]

Average Number of Reflections and Wavenumber Constant The average number of reflections of an oblique wave in a rectangular room, ν, is given by[7]

$$\nu = \frac{c(L_y L_z|\alpha| + L_z L_x|\beta| + L_x L_y|\gamma|)}{V} = \nu_x + \nu_y + \nu_z \quad (\text{s}^{-1}), \quad (14)$$

where L_x, L_y, L_z are the lengths of the sides of the rectangular room (in metres); α, β, γ are direction cosines,

$$\alpha = \frac{k_x}{k_0}, \quad \beta = \frac{k_y}{k_0}, \quad \gamma = k_z/k_0; \quad (15)$$

and k_0 is the wavenumber constant of the oblique wave. When $\nu_z = r$, the z-component of the wavenumber constant k_z becomes k_{zr} by introducing

$$k_{zr} = \frac{k_0 r L_z}{c} = \frac{n_r \pi}{L_z} \quad (\text{m}^{-1}),$$

then the equation

$$n_r = 2\left(\frac{L_z}{c}\right)^2 f_0 r \quad (16)$$

can be obtained, where f_0 is the frequency of the oblique wave. The average number of reflections at the z-walls, ν_{zr}, where the z-components of the wavenumber range from 0 to k_{zr}, is given by

$$\nu_{zr} = \tfrac{1}{2} r \quad (\text{s}^{-1}). \quad (17)$$

Reverberation Time in a Nearly Two-Dimensional Diffuse Field Suppose that there are n_r groups of nearly xy-tangential waves whose z-components of the wavenumber range from 0 to k_{zr}. Introducing the mean free path m_{xy} in the two-dimensional diffuse field, the average number of reflections at the x- and y-walls in the nearly two-dimensional field, ν_{xyr}, becomes

$$\nu_{xyr} = \frac{n_r c}{m_{xy}} \quad (\text{s}^{-1}). \quad (18)$$

The ratio of the average number of reflections at the z-walls to that of the other side walls (x- and y-walls), μ, becomes

$$\mu = \frac{r/2}{n_r c/m_{xy}} = \frac{m_{xy} c}{L_z^2 4 f_0}. \quad (19)$$

Following Eq. (13), the reverberation time in a nearly two-dimensional diffuse field is given by

$$T_{\text{Al}-xy} = \frac{0.128 S_{2d}}{-\ln(1-\overline{\alpha}_{\text{Al}-xy}) L_{2d}} \quad \text{(s)}, \quad (20)$$

where the averaged absorption coefficient $\overline{\alpha}_{\text{Al}-xy}$ is given by

$$\overline{\alpha}_{\text{Al}-xy} = \overline{\alpha}_{xy}\left(1 - \frac{\nu_{zr}}{\nu_{xyr} + \nu_{zr}}\right) + \overline{\alpha}_z\left(\frac{\nu_{zr}}{\mu_{xyr} + \nu_{zr}}\right) \approx \overline{\alpha}_{xy}(1-\mu). \quad (21)$$

Here $\overline{\alpha}_z$ is the averaged absorption coefficient of the z-wall. Figure 2 illustrates an example[6] of the frequency charactristic that follows Eq. (20). The frequency characteristic of the reverberation time at the midfrequency range is not due to the absorption material properties but to the geometric conditions of the sound field.

2.5 Power Law Decay Formula for a Diffuse Field

A nonexponential decay formula is derived even in a three-dimensional diffuse field.[8] In wave theory, many

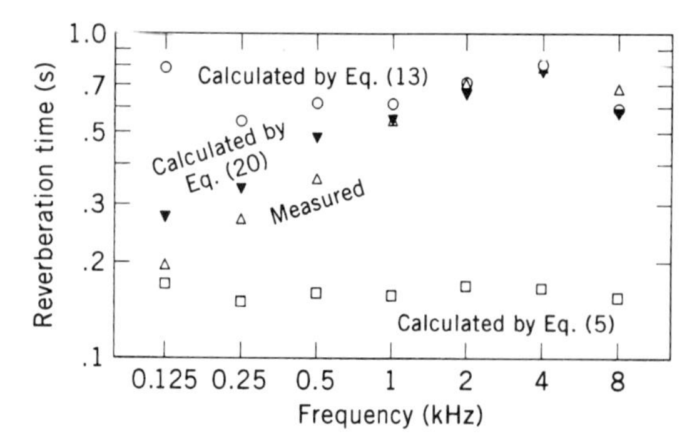

Fig. 2 Sample of frequency characteristics of reverberation time in a nearly two-dimensional diffuse field where the side walls perpendicular to the floor are hard and both floor and ceiling are soft. (From Ref. 6.)

wave modes are excited in a given frequency range. The energy decay is given by an average of exponential functions with different decay rates:

$$E(t) = E_0 \int_0^{\infty} P(\gamma) e^{-\gamma t}\, d\gamma \qquad (\mathrm{J/m^3}), \qquad (22)$$

where γ denotes a decay constant and $P(\gamma)$ is the distribution function for the decay rate.

Suppose that there is only one type of absorber on the surfaces of the room. The decay constant is proportional to the energy incident on the absorber. If we take an area S covered by an absorbing material, the incident energy flux can be expressed as

$$J_{\mathrm{in},S} = \int_S J_{\mathrm{in}}(r)\, dr \sim \sum_{n=1}^{N} p_n^2 \qquad (\mathrm{W}), \qquad (23)$$

where p_n^2 denotes the squared sound pressure data sampled at a point on the area S, N is the total number of "independent" samples (see Sections 4.2 and 4.3) in the data, $N \approx S/(\frac{1}{2}\lambda)^2$, and λ is the wavelength of the sound wave of interest.

As we see in Section 3.1, the square pressure data sampled in a diffuse field are distributed following an exponential distribution. Thus, the decay constant, written as a summation of N independent exponentially distributed variables, follows an Nth-order gamma distribution. Introducing this gamma distribution function into Eq. (23), we have a "power law" for the decay formula,

$$E(t) = E_0 \left(1 + \frac{\overline{\gamma} t}{N}\right)^{-N} \approx E_0 e^{-\overline{\gamma} t} \qquad (\mathrm{J/m^3}) \qquad (24)$$

as N becomes large, so that $\overline{\gamma} t/N \ll 1$, where $\overline{\gamma}$ denotes the average decay constant.[8]

3 TRANSFER FUNCTION IN A REVERBERATION FIELD

3.1 Transfer Function

The transfer function (TF) between a pair of source and receiving positions is obtained as the Fourier transform of the impulse response between the pair of positions. The TF in the space with a simple harmonic source at X_s and observer X_0 is given by a modal expansion form as[2]

$$H(\omega) = \mathrm{const} \sum_M \frac{\Psi_M(x_s)\Psi_M(x_0)}{\omega^2 - \Omega_M^2} \qquad (\mathrm{Pa \cdot s/m^3}), \qquad (25)$$

where Ψ_M are eigenfunctions (modal functions), $\omega = 2\pi f$, f is the frequency in hertz, and Ω_M are the eigenfrequencies.

3.2 Poles and Zeros of Transfer Functions

The TF can be rewritten as

$$H(\omega) = K \frac{(\omega - \omega_a)(\omega - \omega_b)\cdots}{(\omega - \omega_1)(\omega - \omega_2)\cdots} \qquad (\mathrm{Pa \cdot s/m^3}), \quad (26)$$

where ω_1, ω_2 are the poles—the singularities of the transfer function—ω_a, ω_b are the zeros of the function, and K is a constant. The poles are independent of the location of the source or observation point. The zeros are dependent on the locations in the space.[9] Since an $\exp(j\omega t)$ time dependence is assumed, all the poles must be in the upper half-plane because of the stability or causality of the system (Fig. 3).

3.3 Modal Density and Modal Overlap of the TF

The number of eigenfrequencies in a unit frequency interval is called the modal density. The modal density is given by[10]

$$n_p(\omega) \approx \frac{V\omega^2}{2\pi^2 c^3} \qquad (\mathrm{s/rad}), \qquad (27)$$

where V is the volume (in cubic metres). Many modes (modal functions) are excited simultaneously even by a

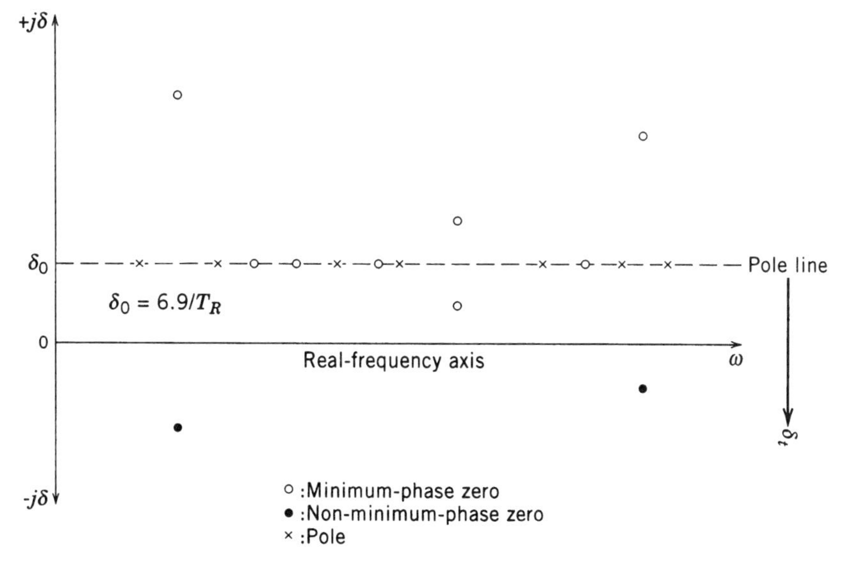

Fig. 3 Pole–zero pattern in the complex-frequency plane. (From Ref. 17.)

single frequency in a room. The number of such simultaneously excited modes is called the modal overlap,[10] which is defined as

$$M = Bn_p(\omega) = \pi\delta_0 n_p(\omega), \tag{28a}$$

where B is the modal bandwidth defined by

$$B = \frac{1}{H_{0,\max}^2}\int_0^\infty |H_0(\omega)|^2\, d\omega = \pi\delta_0 \qquad \text{(rad/s)}. \tag{28b}$$

H_0 is the frequency response of a single mode, and $H_{0,\max}$ is the maximum of the absolute value of the response. We assume here that all the modal functions have the same shape.

3.4 Distribution of Eigenfrequencies

The poles are determined by the eigenfrequencies and lie along the pole line.[11] The eigenfrequencies of a rectangular room of dimensions (L_x, L_y, L_z) with rigid boundaries are given by

$$\omega_{lmn} = \frac{c\pi}{L}\sqrt{(al)^2 + (bm)^2 + (cn)^2} \qquad \text{(rad/s)}, \tag{29}$$

where l, m, n are integer numbers; the lengths of the sides (in metres) are $L = aL_x = bL_y = cL_z$; and a, b, c are the ratios of the lengths. A Poisson model that results in an exponential distribution of the eigenfrequency spacings is not adequate to practical cases.[10] Groups of modes that have degenerate eigenfrequencies will suffer a splitting of the eigenfrequencies by the perturbations in practical systems.[12]

3.5 Distribution of Zeros

The poles are distributed along the pole line; however, the zeros are distributed two dimensionally throughout the complex-frequency plane.

Zeros in a Law Modal Overlap Condition The TF is approximately written as[13,14] (using a dimensionless form)

$$H(\omega) \approx \frac{A}{\omega - \omega_1} + \frac{B}{\omega - \omega_2} + R. \tag{30}$$

The primary part will dominate the TF in the modal expansion, with the rest of the terms producing a small remainder function R. A minimum-phase zero is produced between the two adjacent poles when they have residues of the same sign. Such a zero is located on the line connecting the poles (the pole line), as shown in Fig. 4. However, if the poles have residues of opposite signs, the occurrence of zeros will depend strongly on the remainder function R.[13,14]

Two cases of zeros are produced from residues of opposite signs and the remainder function, as shown in Fig. 5. One is a double zero between two adjacent poles.

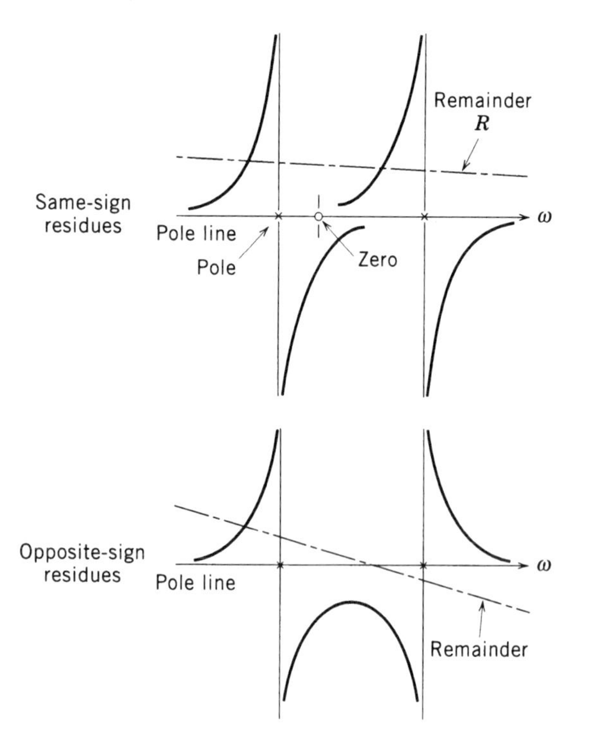

Fig. 4 Possibility of formation of a zero in the interval, depending on the relative signs of residues.

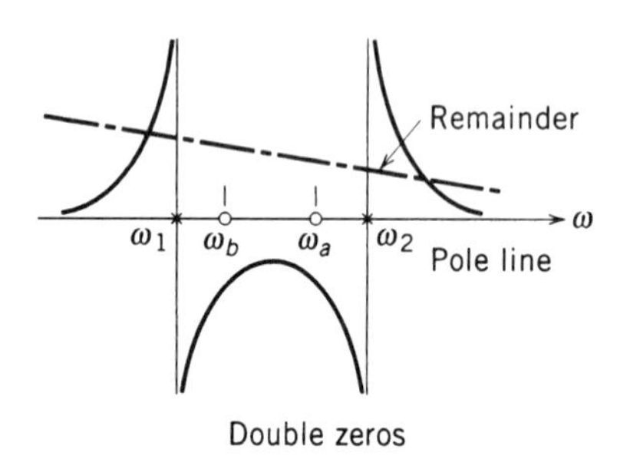

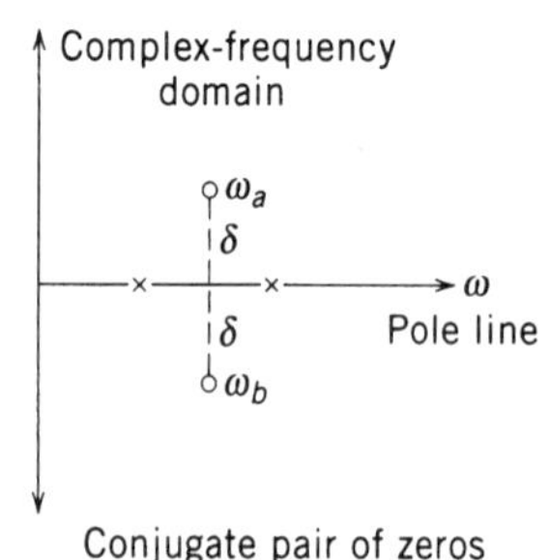

Fig. 5 Zeros from opposite-sign residues and remainder. (From Ref. 11.)

Both of these double zeros are minimum phase because they are located on the pole line. Another case produces a conjugate pair of zeros in the complex-frequency plane. Those zeros are located symmetrically with respect to each other at equal distances from the pole line.[11] One of these zeros can be non-minimum phase.

Zeros under a High Modal Overlap Condition

We take again a TF written as a modal expansion form by Eq. (25). When the source and observer are at the same location, all the residues are positive. Thus, the poles interlace with zeros, and the number of poles and zeros are equal. As source and receiver move apart, the zeros migrate (poles do not move). Some move above the line, an equal number move symmetrically below the line, and the remainder stay on the line.

If the modal functions are real, then, by symmetry,[11] the remaining zeros are evenly distributed above and below the pole line, which means that one-fourth of them are below the pole line. The possible number of zeros below the real frequency axis is $\frac{1}{4}N_p$, where $N_p = N_z^+ + N_z^-, N_z^+$, and N_z^- denote the number of non-minimum- and minimum-phase zeros, respectively. But this possible number of zeros is reduced as $\delta_0 = 6.9/T_R$ increases.

We will consider the TF for a reverberant field to be a random process in the complex-frequency domain.[15] Following the random process theory,[16] we postulate that the zeros are distributed around the pole line following a Cauchy distribution,[17]

$$p(\beta) = \frac{4/\pi}{1 + 4\beta^2}, \tag{31}$$

where $\beta = \delta_t/\overline{\delta\omega}$, $\overline{\delta\omega} = 1/n_p(\omega)$ being the average pole spacing (Fig. 6*a*).

The average number of zeros below the test frequency line in the lower half plane is illustrated in Fig. 6*b*[17] for four measured TFs. The result demonstrates that the distribution of zeros followed the solid line, which decreases following the Cauchy distribution by Eq. (31). For high modal overlap, the Cauchy distribution reduces to

$$p(\beta) = \frac{1}{\pi\beta^2}. \tag{32}$$

The number of non-minimum-phase zeros in a frequency band $\Delta\omega$ is

$$N_z^+ = \frac{n_p(\omega)\,\Delta\omega/4}{M} = \frac{\Delta\omega/4}{\pi\delta_t}. \tag{33}$$

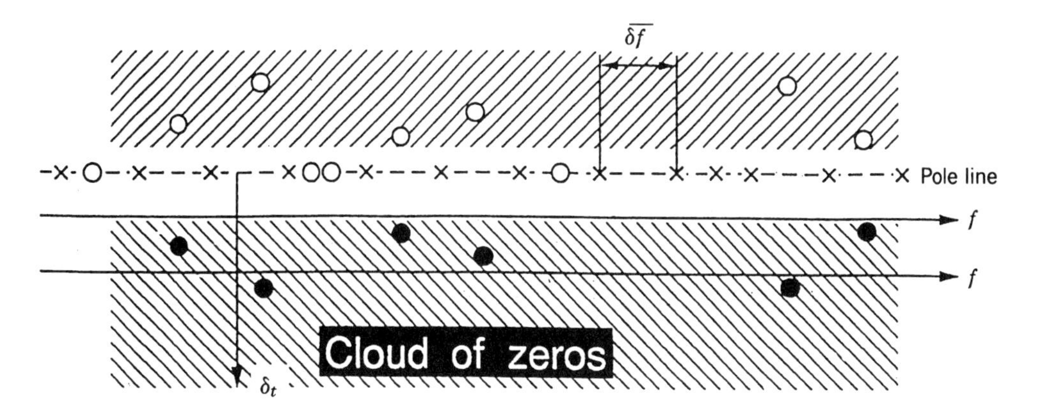

Probability density of cloud of zeros

$$p(\beta) = \frac{4/\pi}{1 + 4\beta^2}$$

$$\beta = \frac{\delta_t}{2\pi} / \overline{\delta f}$$

Cauchy distribution

(a)

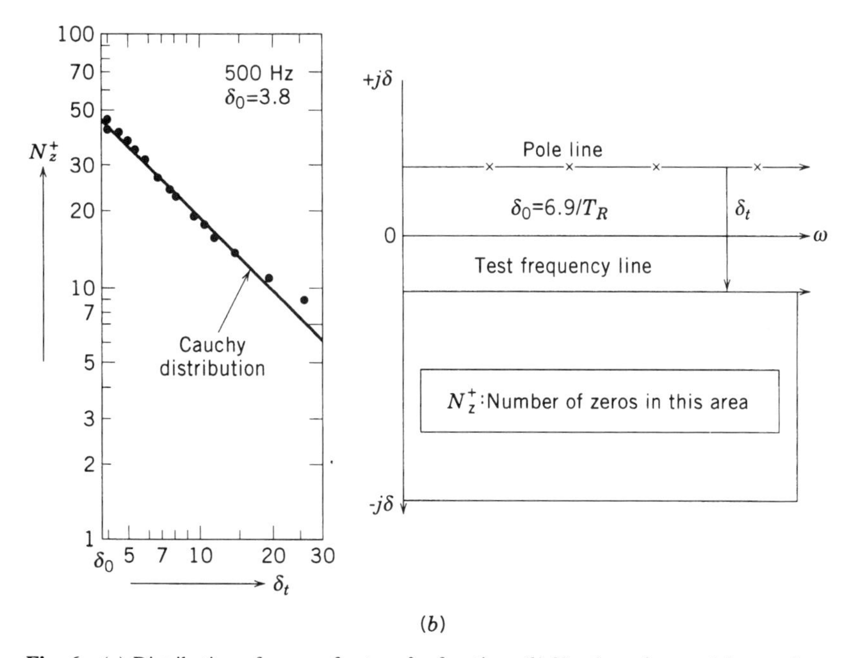

(b)

Fig. 6 (*a*) Distribution of zeros of a transfer function. (*b*) Number of non-minimum-phase zeros in 500-Hz 1/1 octave band. (Part *b* from Ref. 17.)

The densities of the non-minimum-phase zeros is independent of the frequency, while the number of poles increases as the frequency becomes high.

3.6 Magnitude and Phase of Reverberation Transfer Functions

Distribution Function of Magnitudes The real and imaginary parts of the TF can be assumed to be mutually independent Gaussian variables.[15] The magnitude of the TF is distributed following a Rayleigh distribution[18]:

$$f(|H| = x) = \frac{2x}{\overline{|H|^2}} \exp\left(-\frac{x^2}{\overline{|H|^2}}\right), \qquad (34)$$

where $\overline{|H|^2}$ denotes the average of $|H|^2$.

Propagation Phase The growth of phase between source and observation locations in a one-dimensional system is simply visualized.[13,14] As shown in Fig. 7, suppose that at the nth eigenfrequency the x_0 and x_s are separated by p nodes so that $\text{int}[(k/\pi)(x_0 - x_s)] = p$, where k is the wavenumber (in reciprocal metres). For higher eigenfrequencies, these locations stay in phase until p increases by unity. Then $\text{int}[(k/\pi)(x_0 - x_s)] = p + 1$ and the phase between observation and source locations has increased by π. The phase trend in a one-dimensional case where the number of nodes increases in an orderly fashion between source and receiver is[13,14]

$$\Phi = k(x_0 - x_s) \qquad \text{(rad).} \qquad (35)$$

The accumulated phase of a one-dimensional system is just the same as the phase delay due to the direct wave propagation from source to observer. We call this the *propagation phase*.[13,14]

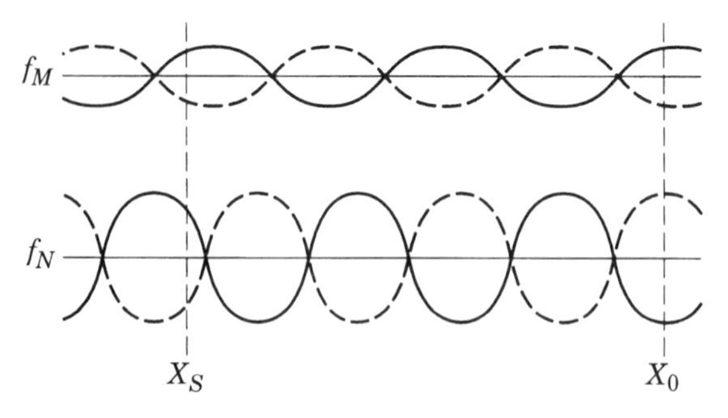

Fig. 7 Standing-wave pattern in a one-dimensional system showing phase advances as nodal pattern changes. (From Ref. 14.)

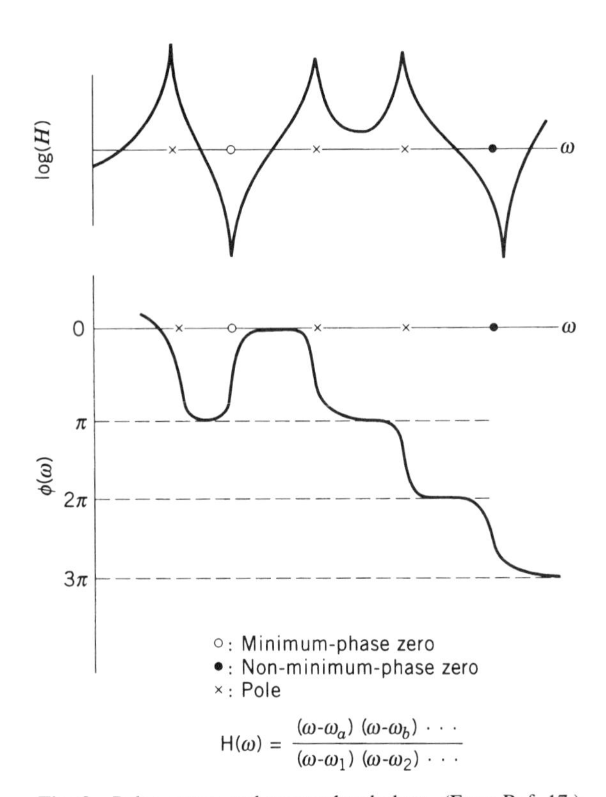

Fig. 8 Poles, zeros, and accumulated phase. (From Ref. 17.)

Reverberation Phase The two- and three-dimensional cases are not as orderly as the one-dimensional case, however. The accumulated phase can be determined by both the poles and zeros (Fig. 8). If there are N_p poles, and if there are N_z^- and N_z^+ zeros in the upper and lower half planes, respectively, then the accumulated phase is

$$\Phi = \pi(N_p + N_z^+ - N_z^-) = 2\pi N_z^+ \qquad \text{(rad),} \qquad (36)$$

where $N_p = N_z^+ + N_z^-$. We can expect a large number of zeros to lie in the lower half plane; thus the accumulated phase by Eq. (36), which we call the *reverberation phase*, is much greater than the propagation phase in a one-dimensional system. Figure 9 illustrates a sample of the reverberation phase of the TF. We can see that the accumulated phase is approximately estimated by Eqs. (36) and (33) at various conditions of damping (δt).

3.7 Modulation Transfer Function

The modulation transfer function (MTF) was introduced as a measure in room acoustics for assessing the effect of

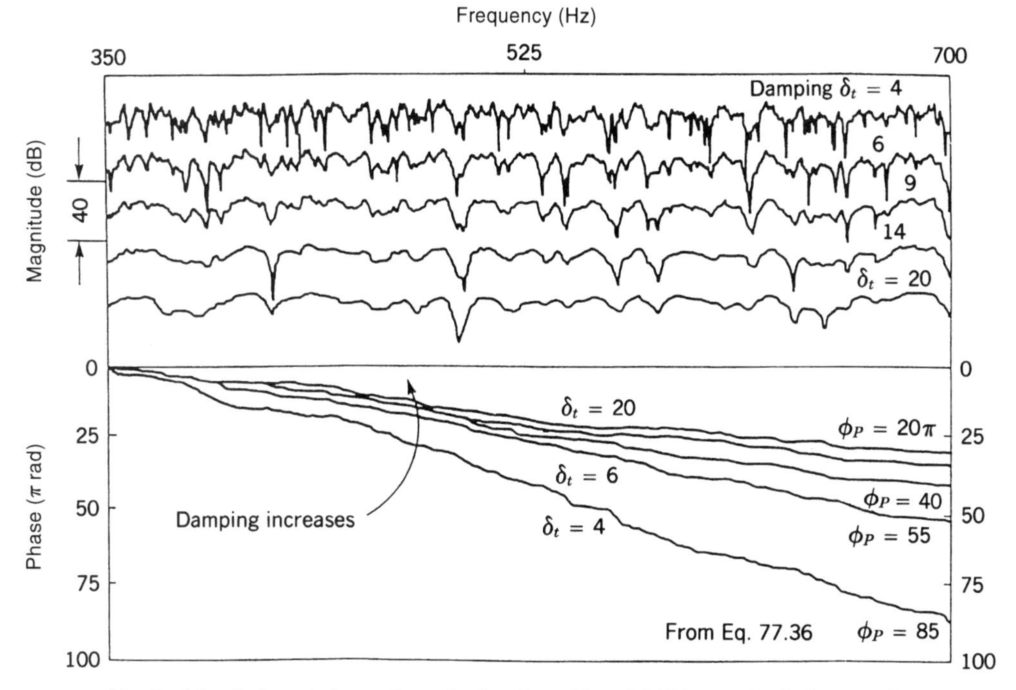

Fig. 9 Magnitude and phase of transfer functions. (From M. Tohyama, R. H. Lyon, and J. Koike, *J. Acoust. Soc. Am.*, Vol. 95, 1994, pp. 286–296.)

an enclosure on speech intelligibility.[19] A speech signal is nonstationary, but the temporal dynamic changes of speech signals are fairly slow; thus the temporal feature can be modeled by a modulated white noise. As shown in Fig. 10, for an input signal with a varying intensity of a white noise,

$$\bar{I}_i[1 + \cos 2\pi F t],$$

with F the modulation frequency; the output signal is given as

$$\bar{I}_0[1 - m \cos 2\pi F(t - \tau)],$$

with m the modulation index and τ the time lag due to the transmission. We define the function $m(F)$ as the MTF. The MTF is only of interest for the range of F relevant for speech. When the entire relevant range, from 0.4 Hz up to 20 Hz, is transmitted faithfully, intelligibility will be excellent.

When considering only the influence of reverberation, the MTF is the Fourier transform of the squared impulse response.[20] The input intensity is rewritten as

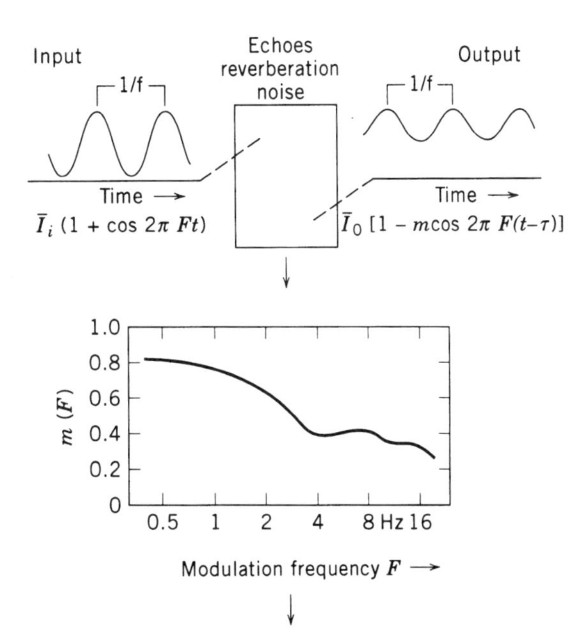

Fig. 10 Schematic illustration of the modulation transfer function. (From Ref. 19.)

$$I_i(t) = \bar{I}_i[1 + \mathrm{Re}(e^{j2\pi Ft})] \qquad (\mathrm{W/m^2}). \tag{37}$$

The output intensity is given by

$$I_0(t) = \int_0^\infty I_i(t - t')h^2(t')\,dt' = aI_i\left[1 + \frac{1}{a}\,\mathrm{Re}(be^{j2\pi Ft})\right] \qquad (\mathrm{W/m^2}), \tag{38}$$

where

$$a = \int_0^\infty h^2(t')\,dt', \qquad b = \int_0^\infty e^{-j2\pi Ft'}h^2(t')\,dt'. \tag{39}$$

The MTF is obtained as

$$m(F) = \frac{\left|\displaystyle\int_0^\infty e^{-j2\pi Ft'}h^2(t')\,dt'\right|}{\displaystyle\int_0^\infty h^2(t')\,dt'}. \tag{40}$$

The Fourier transform of the squared impulse response equals the complex autocorrelation function of the TF.[15]

4 SPATIAL DISTRIBUTION OF REVERBERATION SOUNDS

4.1 Potential Energy Distribution of Reverberation Sounds

There are no regular patterns of loops and nodes in a reverberant space because of the high modal overlap. Suppose the mean-squared pressure (proportional to the potential energy density) is measured at different points in the room. The sound pressure at a point is

$$p(t) = p_0 \sum_{i=1}^{n} \cos(\omega t + \alpha_i). \tag{41}$$

At this point, there is a set of n phase angles α_i distributed as $0, \ldots, 2\pi$ and a fixed value of mean-squared pressure at the point,

$$\overline{p_1^2} = \frac{1}{2}\,p_0^2\left[\left(\sum_{i=1}^{n}\cos\alpha_i\right)^2 + \left(\sum_{i=1}^{n}\sin\alpha_i\right)^2\right] = \frac{1}{2}\,p_0^2(r_1^2 + s_1^2) \qquad (\mathrm{Pa^2}). \tag{42}$$

Sampling at many well-spaced points gives the distribution of values of $\overline{p_k^2}$.

When α_i takes on any value between 0 and 2π, the quantities $\cos\alpha_i$ are distributed between the limits ± 1. The distribution of $\sum_{i=1}^{n}\cos\alpha_i = r_k$ is not normal, but for a large number of n, we can treat r_k and s_k as normally distributed variables. The mean-squared pressure is a random variable given by the squared sum of two normally distributed variables which have zero mean and variance $\frac{1}{2}n$ and are mutually independent.[18,21] Normalizing so that the expected value of $\overline{p_k^2}$ (ensemble average of $\overline{p_k^2} = \langle\overline{p_k^2}\rangle$) is unity, which corresponds to putting

$$\langle\overline{p_k^2}\rangle = \tfrac{1}{2}np_0^2 = 1 \qquad (\mathrm{Pa^2}), \tag{43}$$

the distribution of the sampled value of $\overline{p_i^2} = \overline{p^2}$ has the probability density function[18,21]

$$P(\overline{p^2} = x) = e^{-x}. \tag{44}$$

The probability density function of $\overline{p^2}$ follows an exponential curve independent of exciting frequency. Even in a large reverberation room, we can no longer expect a uniform distribution of squared sound pressure under the condition that a sound source radiates a pure sinusoidal wave. The exponential distribution belongs to a family of gamma distributions.

4.2 Spatial Correlations for Reverberation Sounds

Suppose that many observations are taken with separation r between two points of r_i and r_j, where $r = |r_i - r_j|$ in a reverberation field. Since the plane waves are uniformly distributed in all directions in the ensemble, the spatial cross correlation of the sound pressure can be written as an ensemble average,[18,22]

$$\langle\overline{p(r_i,t)p(r_j,t)}\rangle = \frac{1}{4}\int_0^\pi \cos\{kr\cos\theta\}\sin\theta\,d\theta = \frac{1}{2}\,\frac{\sin kr}{kr}, \tag{45a}$$

where the overbar denotes the time average. Consequently, the normalized spatial cross correlation is

$$R_0(r) = \frac{\sin kr}{kr}. \tag{45b}$$

4.3 Spatial Correlations for Squared Reverberant Sounds

To ensure independent sampling $\overline{p^2}$, it is necessary that spatial correlation between paired samples be negligibly small. Similar to the discussions above, taking the ensemble average of the paired mean-square pressure data, we have the normalized correlation coefficient for the mean-squared pressure data sampled at two locations whose separation is r[22]:

$$R_0(r) = \left(\frac{\sin kr}{kr} \right)^2. \tag{46}$$

Sound fields are described deterministically in principle based on the wave equation. But there are many parameters in sound fields, most of which are quite difficult to be precisely controlled and formulated, such as room temperature, boundary conditions, and sound source characteristics. The statistical approach is a practical way of predicting the sound field characteristics in rooms.

REFERENCES

1. C. W. Kosten, "The Mean Free Path in Room Acoustics," *Acustica*, Vol. 10, 1960, pp. 245–250.
2. H. Kuttruff, *Room Acoustics*, 3rd ed., Elsevier Science, New York 1991, pp. 114–120.
3. M. R. Schroeder, and D. Hackman, "Iterative Calculation of Reverberation Time," *Acustica*, Vol. 45, 1980, pp. 269–273.
4. M. Barron, "Growth and Decay of Sound Intensity in Rooms According to Some Formulae of Geometric Acoustics Theory," *J. Sound Vib.*, Vol. 27, 1973, pp. 183–196.
5. Y. Hirata, "Dependence of the Curvature of Sound Decay Curves and Absorption Distribution on Room Shapes," *J. Sound Vib.*, Vol. 84, 1982, pp. 509–517.
6. M. Tohyama and A. Suzuki, "Reverberation Time in an Almost-Two-Dimensional Diffuse Field," *J. Sound Vib.*, Vol. 111, 1986, pp. 391–398.
7. L. Batchelder, "Reciprocal of the Mean Free Path," *J. Acoust. Soc. Am.*, Vol. 36, 1964, pp. 551–555.
8. F. Kawakami and K. Yamaguchi, "A Systematic Study of Power-Law Decays in Reverberation Rooms," *J. Acoust. Soc. Am.*, Vol. 80, 1986, pp. 543–554.
9. R. H. Lyon, *Machinery Noise and Diagnostics*, Butterworth, Boston and London, 1987.
10. R. H. Lyon, "Statistical Analysis of Power Injection and Response in Structures and Rooms," *J. Acoust. Soc. Am.*, Vol. 45, 1969, pp. 545–565.
11. M. Tohyama and R. H. Lyon, "Zeros of a Transfer Function in a Multi-Degree-of-Freedom Vibrating System," *J. Acoust. Soc. Am.*, Vol. 86, 1989, pp. 1854–1863.
12. J. L. Davy, "The Distribution of Modal Frequencies in a Reverberation Room," *Proc. Inter-Noise '90*, 1990, pp. 159–164.
13. R. H. Lyon, "Progressive Phase Trends in Multi-Degree-of-Freedom Systems," *J. Acoust. Soc. Am.*, Vol. 73, 1983, pp. 1223–1228.
14. R. H. Lyon, "Range and Frequency Dependence of Transfer Function Phase," *J. Acoust. Soc. Am.*, Vol. 76, 1984, pp. 1435–1437.
15. M. R. Schroeder, "Frequency-Correlation Functions of Frequency Responses in Rooms," *J. Acoust. Soc. Am.*, Vol. 34, 1962, pp. 1819–1823.
16. A. Papoulis, *Probability, Random Variables, and Stochastic Processes*, 3rd ed., McGraw-Hill, New York, 1991.
17. M. Tohyama, R. H. Lyon, and T. Koike, "Reverberant Phase in a Room and Zeros in the Complex Frequency Plane," *J. Acoust. Soc. Am.*, Vol. 89, 1991, pp. 1701–1707.
18. K. J. Ebeling, "Statistical Properties of Random Wave Fields," in W. P. Mason, and R. N. Thurston (Eds.), *Physical Acoustics*, Academic, New York, 1984, pp. 223–310.
19. T. Houtgast, H. J. M. Steeneken, and R. Plomp, "Predicting Speech Intelligibility in Rooms from the Modulation Transfer Function. I. General Room Acoustics," *Acustica*, Vol. 46, 1980, pp. 60–72.
20. M. R. Schroeder, "Modulation Transfer Functions: Definition and Measurement," *Acustica*, Vol. 49, 1981, pp. 179–182.
21. R. V. Waterhouse, "Statistical Properties of Reverberant Sound Fields," *J. Acoust. Soc. Am.*, Vol. 43, 1968, pp. 1436–1444.
22. W. T. Chu, "Note on the Independent Sampling of Mean-Square Pressure in Reverberant Sound Fields," *J. Acoust. Soc. Am.*, Vol. 72, 1982, pp. 196–199.

62

STATISTICAL MODELING OF VIBRATING SYSTEMS

JEROME E. MANNING

1 INTRODUCTION

In statistical modeling, the properties of the vibrating system are assumed to be drawn from a random distribution. This allows great simplification in the analysis, whereby modal densities, average mobility functions, and statistical energy analysis (SEA) can be used to obtain response estimates and transfer functions. To demonstrate the benefits of a statistical approach, the point conductance is derived using a modal analysis, with the resonance frequencies and mode shapes assumed to be random variables. The average conductance is found to be a simple function of the modal density. A graphical representation of the modes in wavenumber space is used to calculate the modal density. The use of finite-element models to determine the modal density is also discussed.

Statistical energy analysis combines statistical modeling with an energy flow formulation. Using SEA, the energy flow between coupled subsystems is defined in terms of the average modal energies of the two subsystems and a coupling factor. Equations of motion are obtained by equating the time-average power input to each subsystem with the sum of the power dissipated and the power transmitted to other subsystems. Techniques to determine the coupling and dissipation factors needed for the SEA analysis are discussed.

2 STATISTICAL APPROACH TO DYNAMIC ANALYSIS

Earlier chapters have presented a variety of techniques to study the dynamic response of complex acoustic and structural systems. By and large, these techniques have used a deterministic approach. In the analytical techniques, it has been assumed that the system being studied can be defined in terms of an idealized mathematical model. Techniques based on the use of measured data have assumed that the underlying physical properties of the system are well defined and time invariant. Although the excitation of the system was considered to be a random process in several of the earlier chapters, the concept that the system itself has random properties was not pursued.

The most obvious sources of randomness are manufacturing tolerances and material property variations. Although variations in geometry may be small and have a negligible effect on the low-frequency dynamics of the system, their effect at higher frequencies cannot be neglected. Another source of randomness is uncertainty in the definition of the parameters needed to define a deterministic model. For example, the geometry and boundary conditions of a room will change with the arrangement of furnishings and partitions. Vehicles will also change due to different configurations of equipment and loadings. Finally, during the early phases of design, the details of the product or building being designed are not always well defined. This makes it necessary for the analyst to make an intelligent guess for the values of certain parameters. If these parameters do have a major effect on the response being predicted, the consequences of a poor guess are not serious. On the other hand, if the parameters do have a major effect, the consequences of a poor guess can be catastrophic. Although deterministic methods of analysis have the potential to give exact solutions, they often do not, because of errors and uncertainties in the definition of the required parameters.

In Chapter 48, the vibratory response of a dynamic system was represented in terms of the modes of vibration of the system. Both theoretical and experimental

Note: References to chapters appearing only in the *Encyclopedia* are preceded by *Enc.*

techniques to obtain the required mode shapes and resonance frequencies were described. A statistical approach will now be followed in which resonance frequencies and mode shapes are assumed to be random variables. This approach will result in great simplifications where average mobility functions and power inputs from excitation sources can be defined simply in terms of the modal density and structural mass or acoustic compliance of the system. Modal densities in turn can be expressed in terms of the dimensions of the system and a dispersion relation between wavenumber and frequency.

If the properties describing a dynamic system can be accurately defined, the modes of the system can be found mathematically in terms of the eigenvalues and eigenfunctions of a mathematical model of the system, or they can be found experimentally as the resonance frequencies and response amplitudes obtained during a modal test. The availability of finite-element models and computer software makes it possible today to determine the modes of very complex and large structures. Although it might be argued that the computational cost to determine a sufficient number of modes to analyze acoustic and structural systems at high frequencies is too high, that cost is dropping each year with the development of faster and less expensive computers. Thus, many vibration analysts believe that analysis based on modes determined from a finite-element model can provide them with the answers they need. Similarly, many engineers who are more inclined toward the use of measured data believe that analysis based on modes determined from a modal test will suffice.

The accuracy of a modal analysis depends on the accuracy with which the modal parameters can be determined and on the number of modes that are included in the analysis. Since the modes form a mathematically complete set of functions, a high degree of accuracy can be obtained by including a large number of modes in the analysis. In practice, the accuracy with which the model parameters can be determined decreases for the higher order modes. Thus, the accuracy of the modal analysis depends to a large extent on the number of modes needed to describe the vibration of the system. At low frequencies, for lightly damped systems, the response can be accurately described in terms of the response of a few modes. On the other hand, at high frequencies and for highly damped systems, a large number of modes are required, and the accuracy of the modal analysis decreases. In this case, a statistical description of the system is warranted. Using a statistical description avoids the need for an accurate determination of the resonance frequencies and mode shapes, thereby eliminating the main source of error in the modal analysis. Of course, in using a statistical description, the ability to determine the exact response at specific locations and frequencies is lost. Instead, a statistical description of the response is obtained.

2.1 Mobility Formulation

To illustrate the use of analysis and the application of a statistical approach, the response of a linear system to a point excitation with harmonic $e^{j\omega t}$ time dependence is considered. For a structural system, the ratio of the complex amplitude of the response velocity to the complex amplitude of the excitation force is defined as the point mobility for the system. A transfer mobility is used if the location of the response point is different than that of the excitation point. A drive point mobility is used if the locations of the response and drive points are the same. The drive point mobility at coordinate location x can be written as a summation of modal responses,

$$Y_{\text{pt}}(x,\omega) = \sum_{i=1}^{\infty} \frac{1}{M_i} \frac{j\omega\psi_i^2(x)}{(\omega_i^2 - \omega^2) + j\omega\,\omega_i\eta_i}, \tag{1}$$

where $Y_{\text{pt}}(x,\omega)$ is the drive point mobility for the structure, M_i is the modal mass, $\psi_i(x)$ is the mode shape for the ith mode, ω is the radian frequency of excitation, ω_i is the resonance frequency of the ith mode, η_i is the damping loss factor, $j = \sqrt{-1}$, and the summation is over all modes of the system. The mobility consists of a real part (the conductance) and an imaginary part (the susceptance). The conductance is often of greater interest, since the product of the conductance and the mean-square force is the input power to the system. The conductance and susceptance for a typical lightly damped system are shown in Fig. 1. For light damping, the conductance as a function of frequency shows a large peak at each resonance frequency. The amplitude of the peak is governed by the damping and by the value of the mode shape at the drive point.

Frequency Averages A frequency-averaged conductance is found by integrating the real part of the mobility over frequency and dividing by the bandwidth,

$$\langle G_{\text{pt}}(x,\omega)\rangle_{\Delta\omega} = \frac{1}{\Delta\omega}\int_{\Delta\omega} \text{Re}\{Y_{\text{pt}}(x,\omega)\}\Delta\omega, \tag{2}$$

where $\langle\cdot\rangle$ signifies an average, $\text{Re}\{\cdot\}$ signifies the real part of a complex number, and the integral is over the frequency band, $\Delta\omega$. For any particular frequency band, the average conductance is largely determined by the number of modes with resonance frequencies within the band. For light damping, the contribution to the integral

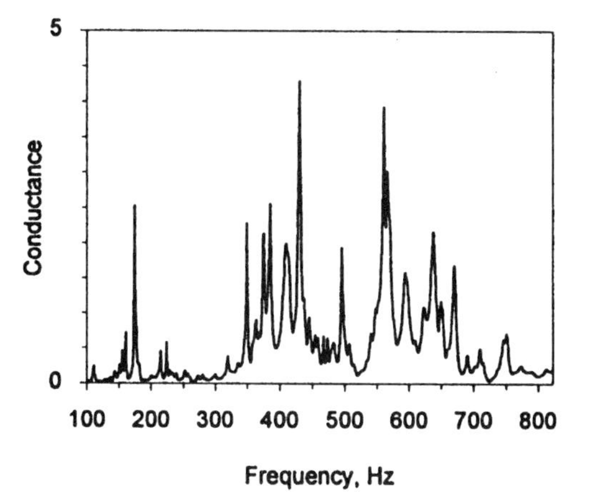

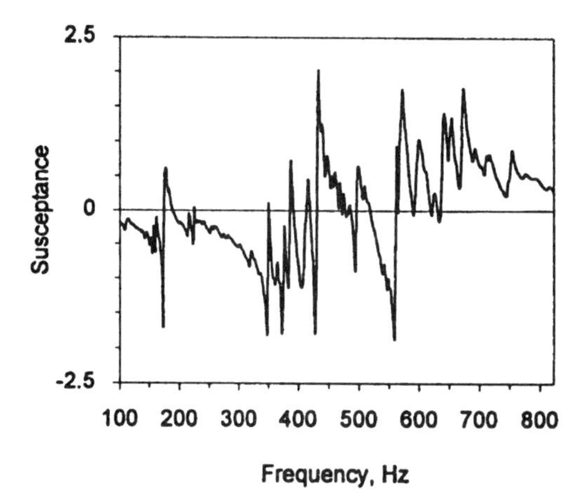

Fig. 1 Drive point conductance and susceptance for a typical structure.

in Eq. (2) from a single mode with a resonance frequency within the band is approximately $\psi_i^2\pi/2M_i$. If the resonance frequency of the mode is outside the band, $\Delta\omega$, the contribution to the integral is very small and can be ignored. Thus, the frequency-averaged conductance is given by

$$\langle G_{\text{pt}}(x,\omega)\rangle_{\Delta\omega} = \frac{1}{\Delta\omega}\,\frac{\pi}{2}\sum_{\substack{i\\ \text{modes}\\ \text{in}\,\Delta\omega}} \frac{\psi_i^2(x)}{M_i}, \tag{3}$$

where the summation is over all modes with resonance frequencies in the band. Note that the average conductance does not depend on the precise values for the resonance frequencies or the damping but only on the number of modes within the band and their mode shapes at the drive point.

Spatial Averages The statistical approach can be extended one step further by averaging the conductance over the spatial extent of the system. For a homogeneous system, the spatial-average value of the mode shape squared is equal to the modal mass divided by the physical mass of the system, M_i/M. Thus, the drive point conductance averaged both over a band of frequencies and over the spatial extent of the system is given by

$$\langle G_{\text{pt}}(x,\omega)\rangle_{\Delta\omega,x} = \frac{\pi}{2M}\,\frac{N_{\Delta\omega}}{\Delta\omega}, \tag{4}$$

where $N_{\Delta\omega}$ is the number of modes with resonance frequencies in the band $\Delta\omega$ (the "resonant mode count") and M is the physical mass of the structure. Equation (4) applies only to homogeneous structures. However, it can also be used for the general case if we replace M in Eq. (4) by the dynamic mass M_d, where

$$M_d(\omega) = \frac{1}{N_{\Delta\omega}}\sum_{\substack{i\\ \text{modes}\\ \text{in}\,\Delta\omega}} \frac{\langle\psi_i^2(x)\rangle_x}{M_i}. \tag{5}$$

The definition of a dynamic mass allows the expression given in Eq. (4) to be used for the average conductance of both homogeneous structures and nonhomogeneous structures, such as framed plates and structures loaded with components.

Acoustic Systems The formulation above is for a structural system in which the equations of motion are formulated in terms of a response variable such as displacement. For an acoustic system, the equations of motion are typically formulated in terms of pressure, a stress variable. In this case, a similar formulation can be carried out. However, the acoustic resistance (pressure divided by volume velocity) is obtained rather than the conductance, and the structural mass is replaced by the bulk compliance of the acoustic space,

$$\langle R_{\text{pt}}^A(x,\omega)\rangle_{\Delta\omega,x} = \frac{\pi}{2C_a}\,\frac{N_{\Delta\omega}}{\Delta\omega}, \tag{6}$$

where R_{pt}^A is the point acoustic resistance for the acoustic space and C_a is the bulk compliance of the space ($V/\rho c^2$ for a space with rigid walls).

Ensemble Averages An average mobility can also be obtained by averaging over an ensemble of structures with random parameters. This type of average has the advantage that it can be useful both for random excitation and for single-frequency excitation. In this case, an average is taken out over bands of frequency but over an ensemble of systems with randomly varying resonance frequencies and mode shapes. The average conductance for this ensemble of systems is given by

$$\langle G_{\mathrm{pt}}(x, \omega) \rangle_{\Delta\omega, x} = \frac{\pi}{2} \frac{n(\omega)}{M_d}, \tag{7}$$

where $n(\omega)$ is the modal density for the system (the ensemble-average number of modes per unit radian frequency).

The frequency-average and ensemble-average conductances are approximately equal, since the product of the modal density and the bandwidth gives an estimate for the ensemble-average resonant mode count. The equality of the frequency average and ensemble average is analogous to the ergodic hypothesis in random process theory.

The susceptance or imaginary part of the mobility can also be determined from a modal summation. However, since the imaginary part of each term in the summation exhibits both positive and negative values, the number of terms that must be included in the summation to obtain a good estimate of the susceptance is much greater than required to obtain a good estimate of the average conductance.

2.2 Modal Density

The previous section shows how the use of a statistical description of a dynamic system can lead to a great simplification in the determination of the average structural conductance or acoustic resistance. The exact resonance frequency and mode shape for each individual mode are no longer needed. Instead, the resonant mode count or modal density can be used. The distinction between these two variables is often quite subtle. The term *mode count* is used to describe the number of modes with resonance frequencies within a given frequency band. The modal density, on the other hand, is a mathematical quantity that gives a statistical estimate of the average number of resonant modes per unit frequency for an ensemble of systems. If we define an ensemble of systems for which the geometry and material parameters vary randomly within manufacturing tolerances or design limits, the mode count will vary from system to system within the ensemble. The average mode count over the ensemble can be estimated from the modal density,

$$N_{\Delta\omega} = \int_{\omega_1}^{\omega_2} n(\omega)\, d\omega, \tag{8}$$

where ω_1 and ω_2 are the lower and upper frequencies of the band, $\Delta\omega$. In many cases the modal density will be a fairly smooth function of frequency. The average mode count can then be obtained simply as the product of the modal density at the band center frequency and the frequency bandwidth.

Asymptotic Modal Densities In most cases, the mode count or modal density can be determined analytically using asymptotic modal densities that are valid at high frequencies, where many resonant modes exist even in narrow bands of frequency. The general use of these asymptotic modal densities to obtain the mode count has led to the idea that statistical modeling can only be used at high frequencies. This is indeed a misconception. As long as correct values are used for the modal density and the dynamic mass, the statistical modeling can be extended to low frequencies, where the number of modes is small. However, at these low frequencies the variance of the estimates may become quite large, so that the average value is not a good estimate for any individual structure.

One-Dimensional Systems The modes of a one-dimensional system have mode shapes that are functions of a single spatial coordinate. For a system with uniform properties, the mode shapes at interior positions away from any boundary are in the form of a sinusoid,

$$\psi_i(x) = A_i \sin(k_i x + \varphi_i), \tag{9}$$

where k_i is a wavenumber describing the rate of variation of the mode shape with the spatial coordinate x and φ_i is a spatial phase factor that depends on boundary conditions. The wavenumber for a mode is related to the number of half-wavelengths within the length of the system,

$$k_i = \frac{\pi}{L}(i + \delta), \tag{10}$$

where L is the length of the system, i is an integer, and δ is a small constant $(-\frac{1}{2} \leq \delta \leq \frac{1}{2})$ whose value depends on the boundary conditions at each end of the system.

The modes can be represented graphically in wavenumber space. For a one-dimensional system, wavenumber space is a single axis. Each mode can be represented by a point along the axis at the value k_i, as shown in Fig. 2. If the value of δ is constant from mode to mode, as

Fig. 2 Wavenumber space for a one-dimensional system.

would be the case for clamped or free boundary conditions, the spacing between modes in wavenumber space, δk, is simply the ratio π/L. If, on the other hand, the value of δ is random due, for example, to a random boundary condition, the average spacing between modes is also π/L. Although the exact position of a mode in wavenumber space depends on the value of δ, each mode will "occupy" a length of the wavenumber axis defined by the average spacing δk.

To determine the modal density, a relationship between wavenumber and frequency is needed. This relationship is the dispersion relation or characteristic equation for the system. For a simple one-dimensional acoustic duct the dispersion relation is $k = \omega/c$, where c is the speed of sound. For bending deformations of a one-dimensional beam (without shear deformations or rotational inertia) the dispersion relation is $k^4 = \omega^2 m/EI$, where m is the mass per unit length, E is Young's modulus for the material, and I is the bending moment of inertia. The dispersion relation allows mapping of the frequency range $\Delta\omega$ to the corresponding wavenumber range Δk. The average value of the mode count is then obtained by dividing Δk by δk. The modal density can now be written as the limit,

$$n(\omega) = \lim_{\Delta\omega \to 0} \frac{\Delta k}{\Delta\omega} \frac{1}{\delta k}. \tag{11}$$

The ratio $\Delta k/\Delta\omega$, in the limit as the frequency range approaches zero, becomes the derivative of the wavenumber with respect to frequency. This derivative is the inverse of $d\omega/dk$, which is the group speed for the system. Thus, the modal density for the general one-dimensional system becomes

$$n(\omega) = \frac{L}{\pi c_g}. \tag{12}$$

The modal density of the general one-dimensional system is simply related to the length of the system and the group speed. In this asymptotic result, which becomes exact as the number of modes increases, boundary conditions are not important. For the lower order modes, it is possible to correct the modal density for specific types of boundary conditions. However, the increase in accuracy may not be worth the effort.

The general result shows that the modal density depends on the group speed. Thus, the modal density of a one-dimensional acoustic lined duct should take into account the effect of the lining on the group speed. If the acoustic media is a fluid, the wall compliance can have a significant effect on the group speed and should be considered.

Two-Dimensional Systems The modes of a two-dimensional system have mode shapes that vary along two spatial coordinates. Like the one-dimensional system, the modes of the two-dimensional system can be represented graphically in wavenumber space. However, wavenumber space becomes a plane defined by two axes. Each mode can be represented by a point on the plane, so that the modes of the system can be represented by a lattice of points, as shown in Fig. 3. For a rectangular system, the lattice forms a rectangular grid of points. The average spacing between the points along the k_x axis is π/L_x, and the spacing between the points along the k_y axis is π/L_y, where L_x and L_y are the dimensions of the system. Thus, a mode occupies a small area in wavenumber space equal to π^2/A, where A is the area of the panel.

The dispersion relation for a two-dimensional system relates the frequency to the two wavenumber components k_x and k_y. For a simple rectangular acoustic layer, the dispersion relation is $k_x^2 + k_y^2 = \omega^2/c^2$, where c is the speed of sound. For bending deformations of a rectangular plate (without shear deformations or rotational inertia) the dispersion relation is $(k_x^2 + k_y^2)^2 = \omega^2 \mathrm{m}/EI'$, where m is the mass per unit area and I' is the bending moment of inertia of the plate. The dispersion relation allows lines of constant frequency to be drawn in wavenumber space for the upper and lower frequencies of the band, $\Delta\omega$. For the simple systems above, these lines are quarter-circles forming an annular region, as shown in Fig. 3. The average value of the mode count is obtained by dividing the area of this region by the area occupied by a single mode. The modal density can now be written as the limit.

$$n(\omega) = \lim_{\Delta\omega \to 0} \frac{\Delta k}{\Delta\omega} \frac{kA}{2\pi}, \tag{13}$$

where $k^2 = k_x^2 + k_y^2$. The modal density for the general

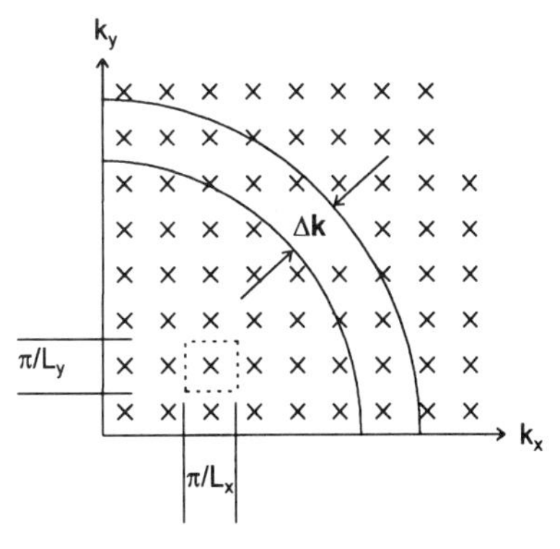

Fig. 3 Wavenumber space for a two-dimensional system.

two-dimensional system becomes

$$n(\omega) = \frac{kA}{2\pi c_g}. \tag{14}$$

Both the wavenumber and the group speed can be found from the dispersion relation.

The formulation of modal density using wavenumber space allows extension of the results to more complicated systems. For example, the modal density for bending modes of a fluid-loaded plate can be obtained by adjusting the group speed to account for the fluid loading. Similarly, the modal density for bending modes, including the effects of transverse shear deformations and rotary inertia, can be obtained by adjusting the group speed to account for these effects. The modal densities for cylindrical shells and orthotropic plates can be obtained using wavenumber space. For these cases, however, the lines of constant frequency will not be circular.

Three-Dimensional Systems The modes of a three-dimensional system have mode shapes that vary along three spatial coordinates. In this case, wavenumber space becomes a volume defined by three axes. Like the one-dimensional and two-dimensional systems, each mode can be represented by a point in wavenumber space, so that the modes of the system form a lattice of points in three dimensions. Each mode occupies a small volume in wavenumber space equal to π^3/V, where V is the volume of the system. The exact location of the mode within this volume will depend on the exact boundary conditions.

The dispersion relation for the three-dimensional system gives the frequency as a function of three wavenumbers. For a uniform acoustic space the dispersion relation is

$$\omega = c\sqrt{k_x^2 + k_y^2 + k_z^2}, \tag{15}$$

where c is the speed of sound. The dispersion relation allows two-dimensional surfaces of constant frequency to be drawn in wavenumber space. The average cumulative mode count can now be obtained by dividing the total volume under these surfaces by the average volume occupied by a single mode. The dispersion relation for the uniform acoustic space results in spherical surfaces with a volume of $\pi k^2/2$ (one-eighth sphere with radius k). Each mode occupies a volume equal to π^3/V, so that the cumulative mode count can be written as

$$N(\omega) = \frac{k^3 V}{6\pi^2}. \tag{16}$$

The modal density is found from the derivative of the cumulative mode count,

$$n(\omega) = \frac{k^2 V}{2\pi^2 c_g}. \tag{17}$$

The asymptotic modal density of the general three-dimensional space gives an accurate estimate of the mode count at high frequencies where many modes occur. It can also be used at lower frequencies as the ensemble average for systems with random boundary conditions. In room acoustics, the walls are commonly assumed to be rigid. In this case corrections to the asymptotic modal density can be introduced that improve the accuracy of the mode count at low frequencies. However, these corrections are only valid for a space in which the assumption of rigid walls is valid. Such an assumption would not be valid, for example, for a fluid-filled tank.

Finite-Element Modeling Finite-element models show great potential as a means to determine the correct mode count for complex structures and acoustic spaces at lower frequencies. Although these models may lose some accuracy in defining the resonance frequencies for the higher order modes, their use to determine the number of modes within defined frequency bands is justified. Two procedures can be used for determining a mode count from a finite-element model. In the first an eigenvalue analysis is used to obtain the resonance frequencies. The mode count is obtained by dividing the frequency range of interest into bands and counting the number of resonance frequencies in each band. In the second technique, the resonance frequencies are used to determine the frequency spacing between modes. The average frequency spacing is found by averaging over a set number of spacing intervals rather than over a set frequency band. Finally, the modal density is obtained from the inverse of the average spacing. Although the first technique is commonly used, the second technique is preferred, since it provides an estimate of the average modal density with a constant statistical accuracy.

3 STATISTICAL ENERGY ANALYSIS

Since its introduction in the early 1960s, statistical energy analysis, or SEA as it is commonly called, has gained acceptance as a method of analysis for structural–acoustic systems.[1,2] Statistical energy analysis draws on many of the fundamental concepts from statistical mechanics, room acoustics, wave propagation, and modal analysis.[3–8] At first, SEA appears to be a very simple method of analysis. However, because of the diversity of concepts used in formulating the basic SEA equations, the method quickly becomes very complex. For this reason, analysts have recommended caution in using SEA. However, when used properly, SEA is a powerful method of vibration and acoustic analysis.

In SEA, the system being analyzed is divided into a set of coupled subsystems. Each subsystem represents a group of modes with similar characteristics. The SEA subsystems can be considered to be "control volumes" for vibratory or acoustic energy flow. Under steady-state conditions, the time-average power input to a subsystem from external sources and from other connected subsystems must equal the sum of the power dissipated within the subsystem by damping and the power transmitted to the connected subsystems

Consider, for example, a piece of machinery located within an enclosure in a large equipment room, as shown in Fig. 4. The noise in the equipment room due to operation of the machine is of concern. A simple SEA model for this problem is shown in Fig. 5. In this model three subsystems are used: one for the acoustic modes of the interior space within the enclosure; one for bending modes of the enclosure walls; and one for the acoustic modes of the exterior space in the equipment room. The airborne and structure-borne noise from the machine are specified as power inputs to the model. The input power to the enclosure acoustic space, W_a^{in}, is taken to be the airborne noise radiated by the machine, which can be determined using acoustic intensity measurements. The input power to the enclosure wall, W_s^{in}, is determined from the vibration of the machine at its attachment to the enclosure base. The time-average power dissipated within each subsystem is indicated by the terms W_a^{diss}, W_s^{diss}, and W_r^{diss}.

Following the usual definition of the damping loss factor, the time-average power dissipated within the subsystem can be written in terms of the time-average energy of the system and the radian frequency of vibration,

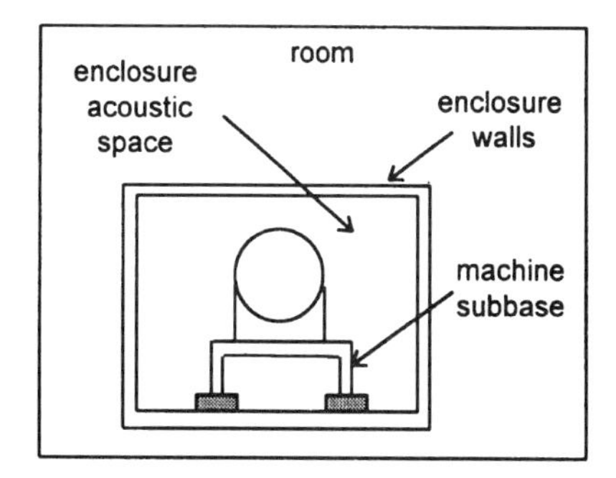

Fig. 4 Machinery noise problem.

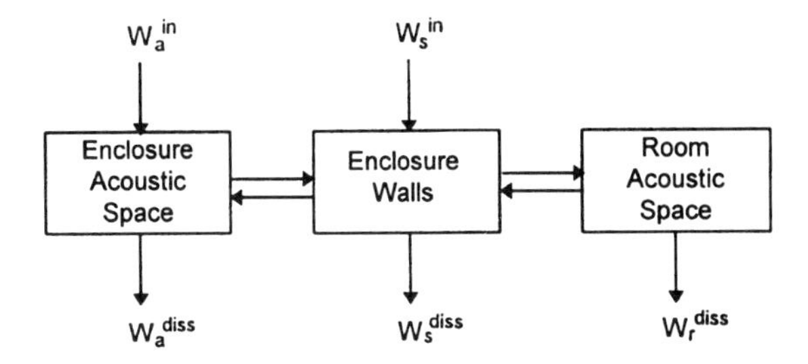

Fig. 5 Simple three-element SEA model of equipment enclosure.

$$W_s^{\text{diss}} = \omega \eta_{s;\text{diss}} E_s, \tag{18}$$

where ω is the radian frequency (typically, a one-third-octave band center frequency), $\eta_{s;\text{diss}}$ is the damping loss factor for subsystem s, and E_s is the time-average energy for subsystem s.

The energy transmitted between the connected subsystems can also be assumed to be proportional to the energy in each system. By analogy to the dissipated power, the factor of proportionality for transmitted power is called the coupling loss factor. However, since energy flow between the two systems can be in either direction, two coupling loss factors must be identified, so that the net energy flow between two connected subsystems is given by

$$W_{a;s}^{\text{trans}} = \omega \eta_{a;s} E_a - \omega \eta_{s;a} E_s, \tag{19}$$

where $\eta_{a;s}$ and $\eta_{s;a}$ are the coupling loss factors between subsystem a and s and between s and a. These two coupling loss factors are not equal. A power balance can now be performed on each subsystem to form a set of linear equations relating the energies of the subsystems to the power inputs:

$$\omega \begin{bmatrix} \eta_{a;d} + \eta_{a;s} + \eta_{a;r} & -\eta_{s;a} & -\eta_{r;a} \\ -\eta_{a;s} & \eta_{s;d} + \eta_{s;a} + \eta_{s;r} & -\eta_{r;s} \\ -\eta_{a;r} & -\eta_{s;r} & \eta_{r;d} + \eta_{r;a} + \eta_{r;s} \end{bmatrix} \cdot \begin{bmatrix} E_a \\ E_s \\ E_r \end{bmatrix} = \begin{bmatrix} W_a^{\text{in}} \\ W_s^{\text{in}} \\ W_r^{\text{in}} \end{bmatrix}. \tag{20}$$

Note that the subscript notation typically used in SEA is not conventional matrix notation. Also note that the loss factor matrix is not symmetric.

3.1 SEA Reciprocity

The coupling loss factors used in SEA are generally not reciprocal, that is, $\eta_{s;r} \neq \eta_{r;s}$. If it is assumed, however, that the energies of the modes in a given subsystem are equal, at least within the concept of an ensemble average, and that the responses of the different modes are uncorrelated, a reciprocity relationship can be developed. The assumptions for this relationship are more restrictive than required for a general statement of reciprocity, so that the term SEA reciprocity should be used.

Statistical energy analysis reciprocity requires that the coupling loss factors between two subsystems be related by the modal densities:

$$n(\omega)_s \eta_{s;r} = n(\omega)_r \eta_{r;s}. \tag{21}$$

Using this relationship, a new coupling factor, β, can be introduced that allows the energy balance equations to be written in a symmetric form,

$$\omega \begin{bmatrix} \beta_{a;d} + \beta_{a;s} + \beta_{a;r} & -\beta_{a;s} & -\beta_{a;r} \\ -\beta_{a;s} & \beta_{s;d} + \beta_{s;a} + \beta_{s;r} & -\beta_{s;r} \\ -\beta_{a;r} & -\beta_{s;r} & \beta_{r;d} + \beta_{r;a} + \beta_{r;s} \end{bmatrix} \cdot \begin{bmatrix} \dfrac{E_a}{n(\omega)_a} \\ \dfrac{E_s}{n(\omega)_s} \\ \dfrac{E_r}{n(\omega)_r} \end{bmatrix} = \begin{bmatrix} W_a^{\text{in}} \\ W_s^{\text{in}} \\ W_r^{\text{in}} \end{bmatrix}. \tag{22}$$

where

$$\beta_{s;r} = \omega \eta_{s;r} n(\omega)_s = \omega \eta_{r;s} n(\omega)_r = \beta_{r;s}. \tag{23}$$

The ratio of total energy to modal density has the units of power and can be called "modal power."

3.2 Coupling Loss Factor Measurement

The coupling loss factor, or coupling factors, cannot be measured directly. However, a power injection technique can be used to infer the coupling factor from measured values of power input and response energy. Using this technique, each subsystem is excited in turn with a unit power input, and the response energy of the subsystems is measured to form a matrix of measured energies. Each column in the matrix corresponds to the measured response energies when one subsystem is excited. For example, the second column contains the measured energies when the second subsystem is excited. The coupling loss factor matrix is determined by inverting the matrix of measured subsystem energies,

$$[\eta] = [E]^{-1}. \tag{24}$$

The off-diagonal terms are the negative values of the coupling loss factors, while the sum of terms for each row gives the damping loss factors. This measurement technique has been successfully used to "measure" in situ coupling and damping loss factors. However, errors in the energy measurement can result in large errors in the measured loss factors. Systems containing highly coupled subsystems will result in energy matrices that are poorly conditioned, since two or more columns will be nearly equal. Thus, the success of the measurement technique requires careful identification of the subsystems. The best results are obtained for light coupling, when the coupling loss factors are small compared to the damping loss factors, so that the loss factor matrix is diagonally dominant.

The measurement of subsystem energy is particularly difficult for subsystems with in-plane compression and shear modes. Because of the high stiffness of the in-plane modes, a small amount of motion results in a large amount of energy. The measurement of subsystem energy is also difficult for subsystems in which the mass is nonuniformly distributed. For these subsystems an effective or dynamic mass must be determined at each measurement point.

3.3 Coupling Loss Factor Theory

Coupling loss factors can be predicted analytically using wave and mode descriptions of the subsystem vibrations. Waves are used when the number of dimensions of the subsystem is greater than the number of dimensions of the connection: for example, a beam connected at a point, a plate connected at a point or along a line, and an acoustic space connected at a point, line, or area. Modes are used when the number of dimensions of the subsystem is equal to the number of dimensions of the connection: for example, a beam connected along a line and a plate connected over an area.

When a wave description can be used for all subsystems at the connection, the coupling loss factor between subsystems can be written in terms of a power transmission coefficient. For a point connection between beams, the coupling factor between subsystem s and subsystem r can be written as

$$\beta_{s;r} = \omega \eta_{s;r} n_s(\omega) = \frac{1}{2\pi} \tau_{s;r}, \tag{25}$$

where $\tau_{s;r}$ is the power transmission coefficient. The power transmission coefficient must take into account energy transmitted by all degrees of freedom at the connection: three translational degrees of freedom and three rotational degrees of freedom. For a point connection with a single degree of freedom (all other degrees of freedom are constrained), the transmission coefficient is given by

$$\tau_{s;r} = \frac{4R_s R_r}{|Z_j|^2}, \tag{26}$$

where R is the subsystem resistance (real part of the impedance) for the unconstrained degree of freedom and Z_j is the junction impedance—the sum of the impedances of all subsystems connected at the point. The coupling factor given by Eqs. (25) and (26) can also be used for two- and three-dimensional subsystems connected at a point with a single degree of freedom, as long as the correct impedances are used. For point connections with multiple degrees of freedom, an estimate of the coupling factor can be obtained by summing the power transmission coefficients for each degree of freedom.

The coupling factor between two-dimensional subsystems connected along a line of length L can also be written in terms of a power transmission coefficient. However, for this case an integration must be performed over all angles of incidence. The coupling factor is given in terms of the angle-averaged transmission coefficient as

$$\beta_{s;r} = \omega\eta_{s;r} n_s(\omega) = \frac{1}{2\pi}\frac{k_s L}{\pi}\overline{\tau_{s;r}} \tag{27}$$

where k_s is the wavenumber of the source subsystem and $\overline{\tau_{s;r}}$ is given by

$$\overline{\tau_{s;r}} = \frac{1}{2}\int_{-\pi/2}^{+\pi/2} \tau_{s;r}(\theta_s)\cos(\theta_s)\, d\theta_s \tag{28}$$

and θ_s is the angle of incidence for a wave in the source subsystem. The parameter $k_s L/\pi$ is the effective number of points for the line connection. For a line connection, the power transmission coefficent must take into account the energy transmitted by four degrees of freedom: three translational and one rotational. For a single degree of freedom, the transmission coefficient for an incident angle θ_s can be expressed in terms of the line impedances of the source and receiver subsystems as

$$\tau_{s;r}(\theta_s) = \frac{4R_s(k_r)R_r(k_t)}{|Z_j(k_t)|^2}, \tag{29}$$

where k_t is the trace wavenumber given by $k_s\cos(\theta_s)$ and $R(k_t)$ is the real part of the line impedance for the unconstrained degree of freedom.

The formulation above can also be used to predict the coupling loss factor between three-dimensional subsystems coupled along a line if the integration is performed over all solid angles of incidence. For this case, the angle-averaged transmission coefficient is written as

$$\overline{\tau_{s;r}} = \int_{-\pi/2}^{+\pi/2} \tau_{s;r}(\theta_s)\sin(\theta_s)\cos(\theta_s)\, d\theta_s. \tag{30}$$

For an area connection between three-dimensional subsystems, the coupling factor is given in terms of the angle-averaged transmission coefficient as

$$\beta_{s;r} = \omega\eta_{s;r} n_s(\omega) = \frac{1}{2\pi}\frac{k_s^2 S}{4\pi}\overline{\tau_{s;r}}, \tag{31}$$

where S is the area of the connection. The effective number of points for the area connection is given by the parameter $k_s^2 S/4\pi$.

When the number of dimensions of a subsystem is equal to the number of dimensions of the coupling, modes are used to calculate the coupling loss factor. For example, the coupling loss factor between a two-dimensional system such as a plate or shell and a three-dimensional system such as an acoustic space is obtained by calculating the radiation efficiency for each mode of the plate and averaging over all modes with resonance frequencies in the analysis bandwidth,

$$\eta_{s;r} = \frac{\rho_r c_r S}{\omega M_s}\frac{1}{N_s}\sum_i \sigma_i^{\mathrm{rad}}, \tag{32}$$

where $\rho_r c_r$ is the characteristic impedance of the acoustic space, M_s is the mass of the plate, N_s is the mode count for the plate, and σ_i^{rad} is the radiation efficiency for mode i of the plate. Approximations to the summation can be made by grouping the modes into "edge" and "corner" modes.[9]

The power transmission coefficients and radiation efficiencies can be calculated with great accuracy. However, the relationship between these parameters and the SEA coupling loss factors requires that some assumptions be made regarding the vibration fields in the connected subsystems. First, the vibrations of the two subsystems are assumed to be uncorrelated. Second, the vibrations of the two subsystems are assumed to be "diffuse"—waves are incident on a point within the subsystem from all angles with equal intensity. Although these assumptions are difficult to prove, even for idealized structures and acoustic spaces, they are generally valid for lightly coupled systems at high frequencies, where many modes participate in the vibration response. The validity of the assumptions for highly coupled subsystems is open to question. Fortunately, the errors incurred using the above equations for highly coupled subsystems are generally small. The assumptions are also open to question at low frequencies, where only a few modes participate in the response. At these frequencies, the equations above may predict coupling loss factors that are too large. However, it is difficult to quantify the error. In spite of the limited validity of the assumptions, the equations above provide useful estimates of the coupling loss factor, even for highly coupled subsystems at low frequencies.

3.4 Damping Loss Factor Theory

The damping loss factors can be predicted analytically for free-layer and constrained-layer treatments. The analysis approach is described in Chapter 55. The damping for an acoustic space is often specified by the average absorption coefficient for the space rather than a damping loss factor. The power dissipated within the acoustic space can be written in terms of the time-average energy and the absorption coefficient as

$$W_a^{\text{diss}} = \omega \frac{S_a}{4k_a V_a} \alpha_{a;\,\text{diss}} E_a = \omega \eta_{a;\,\text{diss}} E_a, \tag{33}$$

where S_a is the area of the absorbing surface, V_a is the volume of the acoustic space, and k_a is the acoustic wavenumber. It follows that the damping loss factor for an acoustic space can be obtained from the average absorption coefficient by the relation

$$\eta_{a;\,\text{diss}} = \frac{S_a}{4k_a V_a} \alpha_{a;\,\text{diss}}, \tag{34}$$

where the constant $4V/S$ is commonly referred to as the mean free path.

3.5 Energy and Response

The SEA power balance equations can be solved to obtain the modal energy or modal power of each subsystem. The final step in the analysis is to relate these variables to the subsystem response. For structural subsystems, the spatial-average mean-square velocity is calculated from the kinetic energy. For resonant vibrations, the time-average kinetic energy is equal to the time-average potential energy. Thus, the average mean-square velocity in a band of frequencies is given by

$$\langle v^2 \rangle_{x,t} = \frac{E}{M}, \tag{35}$$

where E is the total energy of all modes in the band and M is the mass of the subsystem. For acoustic subsystems, the spatial-average mean-square acoustic pressure is calculated from the potential energy,

$$\langle p^2 \rangle_{x,t} = \frac{E}{C_a}, \tag{36}$$

where C_a is the compliance of the subsystem, $V/\rho c^2$, for an acoustic space with rigid walls.

3.6 Variance

Statistical energy analysis provides a statistical description of the subsystem response. However, in many cases, SEA is used only to obtain an estimate of the mean. Although the mean provides the "best estimate" of the response, this estimate may differ significantly from the response measured for a single member of the ensemble of dynamic systems. The variance or standard deviation of the response provides a method to quantify the expected confidence intervals for the SEA prediction. When the variance is high, the confidence intervals will be large, and the mean does not provide an accurate estimate of the response.

Using SEA in design requires that a confidence interval be established for the response prediction, so that an upper bound or "worst-case" estimate can be compared with design requirements. If the mean response is used for design, half the products produced will fail to meet the design requirements. Use of the mean plus 2 times the standard deviation (square root of the variance) provides a reasonable upper bound for the response prediction.

Methods to predict the variance of the SEA prediction are not well established. Often an empirical estimate of the variance or confidence interval is used. In other cases, an estimate based on the modal overlap parameter and the frequency bandwidth of the analysis is used. The modal overlap parameter M_{overlap} is the ratio of the average damping bandwidth for an individual mode to the average spacing between resonance frequencies. This parameter can be written in terms of the damping loss factor and the modal density as

$$M_{\text{overlap}} = \tfrac{1}{2}\pi\omega\eta_d n(\omega),$$

where η_d is the effective total damping loss factor for the subsystem. Large values of the product of the modal overlap parameter and the analysis bandwidth result in low variance and a narrow confidence interval. In this case, the mean is a good estimate of the response. Small values of the product result in high variance and wide confidence intervals. In this case, the mean does not give a good estimate of the response, and the variance should be determined so that an upper bound to the prediction can be obtained.

Failure to include an estimate of the variance in the SEA analysis leads to some misunderstandings regarding the capabilities of SEA. First, SEA is not limited to high frequencies and high modal densities. However, at low frequencies and for low modal densities, the confidence interval for the SEA predictions will be large, so that an estimate of the variance must be made. Second, SEA is not limited to broadband noise analysis. However, for a single-frequency or narrow-band analysis, the confidence interval for the SEA predictions will be larger than for a one-third-octave or octave band analysis.

REFERENCES

1. J. E. Manning, "Statistical Energy Analysis—An Overview of Its Development and Engineering Applications," *Proc. 59th Shock and Vibration Symposium*, Albuquerque, New Mexico, 1988, pp. 25–38.
2. K. H. Hsu and D. J. Nefske (Eds.), *Statistical Energy Analysis*, Vol. 3, Am. Soc. of Mech. Eng., NCA, New York, 1987.
3. R. H. Lyon and R. G. DeJong, *Theory and Application of Statistical Energy Analysis*, 2nd ed., Butterworth-Heinemann, Newton, MA, 1995.
4. M. P. Norton, "Statistical Energy Analysis of Noise and Vibration," in *Fundamentals of Noise and Vibration Analysis for Engineers*, Cambridge University Press, Cambridge, U.K., 1989, Chap. 6.
5. F. J. Fahy, "Statistical Energy Analysis," in R. G. White and J. G. Walker (Eds.), *Noise and Vibration*, Ellis Horwood, Chichester, U.K., 1982, Chap. 7.
6. I. L. Ver, "Statistical Energy Analysis," in L. L. Beranek and I. L. Ver (Eds.), *Noise and Vibration Control Engineering Principles and Applications*, Wiley, New York, 1992, Sect. 9.8.
7. E. E. Ungar, "Statistical Energy Analysis," in L. Cremer, M. Heckl, and E. E. Ungar (Eds.), *Structure-Borne Sound*, 2nd ed., Springer-Verlag, Berlin, 1988, Sect. V.8.
8. W. G. Price and A. J. Keane (Eds.), "Statistical Energy Analysis," *Phil. Trans. Royal Soc. London*, Vol. 346, No. 1681, 1994, pp. 429–552.
9. G. Maidanik, "Response of Ribbed Panels to Reverberant Acoustic Fields," *J. Acoust. Soc. Am.*, Vol. 34, 1962, pp. 809–826.

PART VIII

NOISE: ITS EFFECTS AND CONTROL

63

NOISE CONTROL

MALCOLM J. CROCKER

1 INTRODUCTION

Noise is usually defined as unwanted sound. Its main effects on people include interference with speech and sleep. Noise causes annoyance and if sufficiently intense it can cause hearing loss (Chapter 92). In the United States it is estimated that 5.5 million workers in manufacturing and utilities are exposed to noise conditions that are hazardous to hearing.[1] According to a survey in Germany in 1994, 70% of the citizens are annoyed by environmental noise caused by road traffic and a further 40% by aircraft. In addition over 20% are annoyed by rail traffic and industrial noise.[2,3] Noise is normally considered in the *source–path–receiver* framework.[4] Part VIII describes some of the many machinery noise sources (Chapters 66 and 69) and methods of control, both passive (Chapter 66) and active (Chapter 67). The ways in which noise affects people and criteria for determining human response to noise are discussed in Chapter 64. Methods of measuring noise are discussed briefly in this part, both in *Enc.* Ch. 82 and Chapter 69. In most developed countries the major noise problems are industrial noise, aircraft and airport noise, surface transportation noise, and community noise. These topics have received special coverage in Chapters 69–71. In addition the control of noise and vibration sources in buildings are described in Chapter 68. In cases where noise is very intense (and cannot be reduced sufficiently by engineering controls) the use of hearing protection devices must be considered (see Chapter 65). The control of noise requires a good understanding of many concepts including sound sources, both steady state and transient (Chapters 8 and 9), and sound propagation inside buildings (Chapter 75) and through the atmosphere (Chapter 28). The reader concerned with noise will also find many other chapters in this book useful, including those concerned with the ear and hearing (Parts XI and XII) and those in Part XV, where acoustical measurements (*Enc.* Ch. 147), sound intensity measurements (*Enc.* Ch. 150) and analyzers (*Enc.* Ch. 151) are described. Since vibration often results in airborne noise, readers will also find the discussions in Part VI useful, including vibration measurements and instrumentation (Chapter 56) and vibration isolation and damping (Chapter 55). For readers who require further in-depth coverage of noise problems and methods for their solution, these are several books that give a comprehensive treatment.[4–8]

2 CRITERIA AND RATING MEASURES FOR NOISE

Noise disturbs people in a variety of ways. If it is sufficiently intense, it can cause hearing damage. Very intense noise can fracture the eardrum and cause immediate damage. If the noise is less intense but experienced over a considerable period of time, such as in some industrial situations, it will also cause damage.[1] This situation is described in Chapter 92. Noise has other effects too. It interferes with speech and communications in general, spoils sleep, and causes annoyance.[9] A large number of noise measures have been produced to evaluate noise. The effects of noise are to a certain extent dependent on level, duration, frequency content, variation in level with time, presence of pure tones in the noise, existence of background noise, and so on. Thus some of the measures make allowance for some of these factors. There are a bewildering number of measures that have been proposed to evaluate noise. Some are no longer in

Note: References to chapters appearing only in the *Encyclopedia* are preceded by *Enc.*

use. The search to find one single measure that could be used to evaluate a large number of noise situations has largely been abandoned. However the day–night equivalent sound level, usually A-weighted, has become very widely used to evaluate community noise particularly near airports. Chapter 64 describes a number of these noise measures that have been widely used. Included are some measures that have gone out of favor, but from which other measures have developed and for which a large amount of data exist in the literature.

3 HEARING PROTECTION DEVICES

As already discussed, hearing damage can be experienced in situations when the noise is intense. Hearing protectors offer a "first line of defense," but should really only be used as a last resort when engineering and other control measures fail. Hearing protectors come in three main types: earplugs, earmuffs, and helmets. Although they have been used for thousands of years (Odysseus used wax earplugs to prevent him from hearing the Sirens' song), they are now mainly used to reduce industrial noise damage or to cut out noise to aid in sleeping. Earmuffs usually consist of rigid earcups that contain acoustic absorbing material and are held in place against the head by a flexible band. The earmuffs avoid the unpleasant pressure on the ear canal made by earplugs, but can be uncomfortable to wear in hot weather. Helmets are sometimes used. Helmets not only protect the head against impacts, but also provide ear protection as well since they frequently contain internal earcups. All hearing protection devices should be fitted carefully to achieve full hearing protection. If leaks exist in earplugs, the degradation in performance can be considerable. Likewise leaks with earmuffs, caused for instance by hair or glasses that break the earmuff seal, also reduce performance. Chapter 65 describes the main types of hearing protectors and also their fitting and attenuation. This chapter also describes some of the many ways available to measure the attenuation provided by hearing protectors and provides information on the effects of hearing protectors on speech intelligibility, the ability to hear warning signals, and localization of sound.

4 NOISE EMISSION OF SOUND SOURCES

To determine the acceptability of the noise produced by a noise source, its noise emission must be determined. The noise emitted by moving sources such as aircraft or vehicles is usually described by the sound pressure level at a certain defined location. Stationary sources, however, particularly those which operate in a well-defined environment or are small enough to be moved into one, are usually described by their sound power output. *Enc.* Ch. 82 describes methods to determine the sound power of a source from sound pressure measurements. Several are available. If the source is small enough and portable, it can be placed in a reverberation room if one is available and then an accurate estimate of its sound power can be obtained. Alternatively, if the source cannot be moved and only its normal environment rather than a reverberant room is available, then its sound power can still be estimated from sound pressure measurements made on a hemispherical or rectangular enveloping measurement surface. Since the sound pressure measured on the enveloping surface will be increased by background noise and/or reverberation, a correction factor must be applied in such cases. The correction factor to account for reverberation can be determined by measuring the reverberation time. The use of a reference sound source, however, can enable the correction factor for either background noise or reverberation (or both) to be obtained by using a substitution method. If the background noise is too high, then the correction is inaccurate and can no longer be applied. In such cases it is advisable to obtain the sound power of the source using sound intensity measurements (see Chapter 106). Accurate sound power determinations can be made in cases where the background noise is very high and where use of the mean square sound pressure approach described in *Enc.* Ch. 82 becomes impossible. Of course the sound intensity measurement equipment may not be available, or it may be considered to be too expensive to use.

5 NOISE GENERATION IN MACHINERY AND ITS CONTROL

If the sound pressure level or the sound power level of a source has been determined and found to exceed certain criteria, then it may be decided to reduce the noise emission of the source by engineering methods. At the end of Chapter 66, methods to identify sources of noise in a complicated machine or system are described. These methods include selective operation, selective lead wrapping, frequency analysis, sound intensity measurements, multiple coherence measurements, and cepstrum among others. It is normally necessary to use more than one method to obtain more confidence in the results. After good information is obtained about the noise sources on a machine, then noise control methods can be used. Chapter 66 also describes the main passive noise control approaches that have been useful in reducing the noise of machinery. These include vibration isolation, use of sound-absorbing materials, machinery enclosures, and use of barriers and application of damping materials. Although there are many different types of machines in existence, a few of the main types of machines or machine elements and their noise-generating mechanisms are described in Chapter 66. Some of these machine noise discussions may be helpful even if the reader is faced with a noise problem with a different type of machine.

6 ACTIVE NOISE CONTROL

The main passive noise control approaches described in Chapter 66 (vibration isolation, absorption, enclosure, barriers, and damping), although well understood and usually implemented with only moderate cost, generally do not work very well at low frequency. This has led to the search for alternative methods that can be used in the low-frequency region. Active noise control seems to offer a possible solution. In recent years low-cost digital signal processors have become available. Together with the wave nature of sound they have made active noise control feasible. In this approach the sound field is sensed with a microphone, processed by a microprocessor, and fed back through a loudspeaker so as to destructively interfere with the sound field of the primary source. An error microphone situated away from the source whose signal is minimized as much as possible can also be used. There are two main approaches that are described in this chapter: feedback control and feedforward control. In general it is easier to control pure-tone sources than broadband noise sources. Active control technologies are only now beginning to emerge from the laboratory and become used in real engineering applications. The design of active control systems still remains a task for specialists, however. Active noise control has been successfully applied to several practical cases such as (1) the low-frequency *broadband* noise in ear defenders, air conditioning ducts, the exhaust pipes of vehicle engines, and the interior noise inside road vehicles, and (2) the low-frequency *pure-tone* noise in the cabins of propeller-driven aircraft, engine-generated pure-tone noise in vehicle interiors, and the pure-tone noise generated by electrical transformers. Active vibration control has also begun to be applied to the control of vibration in the form of active vibration isolation and distributed sensors and actuators in extended structures such as plates and cylinders. PVDF layers have frequently been used as sensors and PZT as actuators. Of course structural systems are frequently coupled to acoustic systems, and if the structural vibration can be controlled, so can the noise in the coupled acoustic system. Chapter 67 reviews both the theory for active noise control systems and practical applications.

7 NOISE CONTROL IN AIR CONDITIONING SYSTEMS

One of the main noise sources in buildings is the noise produced by the air conditioning system. The main source is of course the fan, but additional sound may also be generated by fittings and take-offs. Sound is naturally attenuated as it travels along ducts by transmission through duct walls (break-out), by energy division at take-offs, and by end reflection when the duct terminates at the room. If the sound transmitted to a room is too high, however, it can be reduced by lining the interior surfaces of the duct with sound-absorbing material or by using packaged attenuators or both. Care should be taken in air conditioning systems to minimize aerodynamic noise generation at elbows, turns, take-offs, silencers, and so on. The noise generated depends not only on the duct geometry, but also on the flow velocity. If the flow velocity is minimized, so also is the noise. Chapter 68 deals with the problems of air conditioning noise and, in particular, fan noise, duct attenuation, the effects of duct lining, turbulent flow generation, and sound generation and control in outlet devices. Chapter 79 also deals with air conditioning noise and, in particular, with the sound generated by the fan or blower.

8 SOUND POWER LEVELS OF MACHINERY

As discussed in Section 4, the sound power level is the preferred way to describe the noise emission of sources. Fixed machinery such as that used in industry is normally described by its sound power level. From this the sound pressure level in the interior of a building or at a certain specified distance from the source(s) in the community can be calculated. Chapter 69 presents procedures to calculate the sound power levels of equipment, including boilers, electric motors, gas turbines, reciprocating engines, steam turbines, transformers, compressors, pumps, fans, gears, generators, valves and piping, air conditioning equipment, cooling towers, chillers, construction equipment, and oil field equipment. The equations presented in Chapter 69 either are based on measured data or are semi-empirical. In construction projects, it is frequently necessary to make acoustical predictions. In that case either measurements should be made on the machinery to be used, or sound power information should be obtained from the manufacturers of the machinery. If neither of these options is available, then the procedures for estimating the sound power presented in this chapter can be used.

9 AIRCRAFT AND AIRPORT NOISE

Civilian air transport continues to expand worldwide. Although the latest passenger jets with their bypass turbofan engines are significantly quieter than the first generation of jet aircraft, which used pure turbojet engines, passenger aircraft jet noise remains a problem, particularly near airports. Although the noisiest passenger aircraft are being retired in some countries, airport noise is likely to remain a serious problem because in many countries the frequency of aircraft operations continues to increase and because of the public opposition to noise voiced by some citizens living near airports. This opposition has prevented extensions to some airports and the development of some other airports entirely because of environmental concerns. Chapter 64 describes some of the aircraft noise descriptors that have been used in the United States and overseas, such as CNR, NEF, and NNI. The NEF descriptor has now been replaced in the United States by the day–night equivalent sound level L_{dn}. However, a considerable variety of older varied descriptors remain in use in many other countries.[3] Chapter 70 discusses the way in which aircraft noise is monitored at airports in the United States and describes how to calculate airport noise and its impact on the population. Also discussed is noise abatement at the national level. Specific airport noise control measures that can be taken are described, including optimization of flight tracks, definition of preferential runways to minimize noise, setting of standards to restrict the noisiest aircraft, use of noise abatement procedures during take-off and landing to minimize noise. In addition, airport layout changes to reduce noise, soundproofing of schools, churches, and apartment buildings, and the erection of barriers to reduce the noise impact on the population are discussed.

10 SURFACE TRANSPORTATION NOISE

Surface transportation is really a bigger noise problem than aircraft noise. It affects more people than aircraft noise in most countries. This chapter describes noise sources on a diesel engine truck. Chapter 71 also discusses vehicle noise sources and describes in considerable detail noise regulations and limits for vehicle noise that apply to new highway vehicles sold in the United States, Canada, the European Union, Japan, and other countries. Chapter 71 also describes test procedures to measure vehicle noise. These test procedures normally require the vehicle to be operated in a full-acceleration condition, but in Japan fixed-speed passby tests at 60% rated engine speed and 65 km/hr and a stationary test to determine exhaust noise are also required. The chapter also includes discussion on locomotive noise regulations and on interior vehicle noise.

11 COMMUNITY NOISE

Noise remains a serious problem in communities. Chapter 64 discusses some of the descriptors that have been used. There has been a move to the use of the A-weighted L_{eq} in many countries for the assessment of road traffic noise, although L_{10} remains in use in Australia, Hong Kong, and the United Kingdom for target values and insulation regulations for new roads and planning values for new residential areas.[3] L_{dn} has become the main descriptor of community noise in the United States since its adoption by several federal agencies in the 1970s. Chapter 72 describes the current practice of evaluating community noise problems in the United States. Shaw has written a useful recent review of community noise.[10]

REFERENCES

1. A. H. Suter and D. L. Johnson, "Programs in Controlling Occupational Noise Exposure," *Noise Control Engineering Journal*, Vol. 44, 1996, pp. 121–126.
2. Institute fur proxisorientierte Sozial forschung, Meinungen zur Umweltpolitik 1994, Mannheim, Germany, 1994.
3. D. Gottlob, "Regulations for Community Noise," *Noise/News International*, Vol. 3, 1995, pp. 223–236.
4. C. M. Harris (Ed.), *Handbook of Acoustical Measurements and Noise Control*, 3rd ed., McGraw-Hill, New York, 1991.
5. L. L. Beranek and I. L. Ver (Eds.), *Noise and Vibration Control Engineering*, Wiley, New York, 1992.
6. L. H. Bell and D. A. Bell, *Industrial Noise Control*, 2nd ed., Marcel Dekker, New York, 1994.
7. C. H. Hansen and D. A. Bies, *Engineering Noise Control: Theory and Practice*, 2nd ed., E & FN Spon, 1996.
8. M. J. Crocker and A. J. Price, *Noise and Noise Control*, Vol. 1, CRC Press, Boca Raton, 1975, and M. J. Crocker and F. M. Kessler, *Noise and Noise Control*, Vol. II, CRC Press, Boca Raton, 1982.
9. H. E. von Gierke and K. McK. Eldred, "Effects of Noise on People," *Noise/News International*, Vol. 1, No. 2, 1993, pp. 67–89.
10. E. A. G. Shaw, "Noise Environments Outdoors and the Effects of Community Noise Exposure," *Noise Control Engineering Journal*, Vol. 44, No. 3, 1996, pp. 109–120.

64

RATING MEASURES, DESCRIPTORS, CRITERIA, AND PROCEDURES FOR DETERMINING HUMAN RESPONSE TO NOISE

MALCOLM J. CROCKER

1 INTRODUCTION

People are exposed to noise during daytime and nighttime hours. During the day the noise can interfere with various activities and cause annoyance, and at night it can affect sleep. Very intense noise can even lead to hearing damage (Chapter 92). In the daytime the activities most affected are communications that involve speech between individuals, speech in telephone communications, and speech and music on radio and television. If the noise is more intense, it is normally more annoying, although there are a number of other attitudinal and environmental factors too that affect annoyance. There are a large number of ways to measure and evaluate noise, each normally resulting in a different noise measure, descriptor, or scale. The different measures and descriptors mainly result from the different sources (aircraft, traffic, construction, industry, etc.) and the different researchers involved in producing them. From these measures and descriptors, criteria have been developed to decide on the acceptability of the noise levels for different activities. These criteria are useful in determining whether noise control efforts are warranted to improve speech communication, reduce annoyance, and lessen sleep interference. This chapter contains a review and discussion of some of the most important noise measures and descriptors. In the past 20–30 years these measures and descriptors have undergone some evolution and change as researchers have attempted to find descriptors that best relate to different human responses and are more easily measurable with improved instrumentation. For completeness this evolution is traced and some measures and descriptors are described that are no longer in use since knowledge of them is needed in the study of the results of various noise studies reported in the literature.

Note: References to chapters appearing only in the *Encyclopedia* are preceded by *Enc.*

2 LOUDNESS AND ANNOYANCE

As the level of the noise is increased, it is accompanied by an apparent increase in loudness. Loudness may be considered to be the subjective evaluation of the intensity of a noise when this evaluation is divorced from all the attitudinal, environmental, and emotional factors that may affect the listener's assessment of the annoying properties of the noise. Generally, if a noise is louder it is judged to be more annoying and vice versa, although there are exceptions. Table 1 shows some of the acoustical and nonacoustical factors that can contribute to the annoyance caused by noise. Some of the factors shown in Table 1 are also important in considerations of the effects of noise on speech communication (see Sections 6 and 7) and on sleep (see Section 15.1). The annoyance caused by noise is discussed further in Section 15.2.

3 LOUDNESS AND LOUDNESS LEVEL

As discussed in Chapter 91, which contains an in-depth discussion of the loudness of sound, the human ear does

TABLE 1 Some Acoustical and Nonacoustical Factors That Contribute to Annoyance Caused by Noise

Acoustical Factors	Sound pressure level
	Frequency spectrum
	Duration
	Pure-tone content
	Impulsive character
	Fluctuation in level
Nonacoustical Factors	Time of day
	Time of year
	Necessity for noise
	Community attitudes
	Past experience
	Economic dependence on source

not have a uniform sensitivity to sound as its frequency is varied. Figures 2 and 3 of Chapter 91 show equal-loudness-level contours. These contours connect together pure-tone sounds that appear equally loud to the average listener. Similar contours have been determined experimentally for bands of noise instead of pure tones. The unit of (linear) loudness is the *sone*. A sone is defined as the loudness of a pure tone with a sound pressure level of 40 dB at 1000 Hz. A sound that is twice as loud is said to have a loudness of 2 sones and so on. The *loudness level* of a 1000 Hz pure tone of 40 dB is defined as 40 *phons*. If the 1000 Hz pure tone is raised in level by 10 dB, it appears to be about twice as loud to the average listener. Thus a doubling of the loudness has been defined to be equivalent to an increase in loudness level of 10 phons. The relationship between loudness S and loudness level P (for both pure tones and bands of noise) is thus given by Eq. (1) and shown in Fig. 1:

$$S = 2^{(P-40)/10}. \qquad (1)$$

The preceding discussion has concerned the loudness of pure tones. Equal loudness contours for bands of noise have been determined experimentally, independently, by

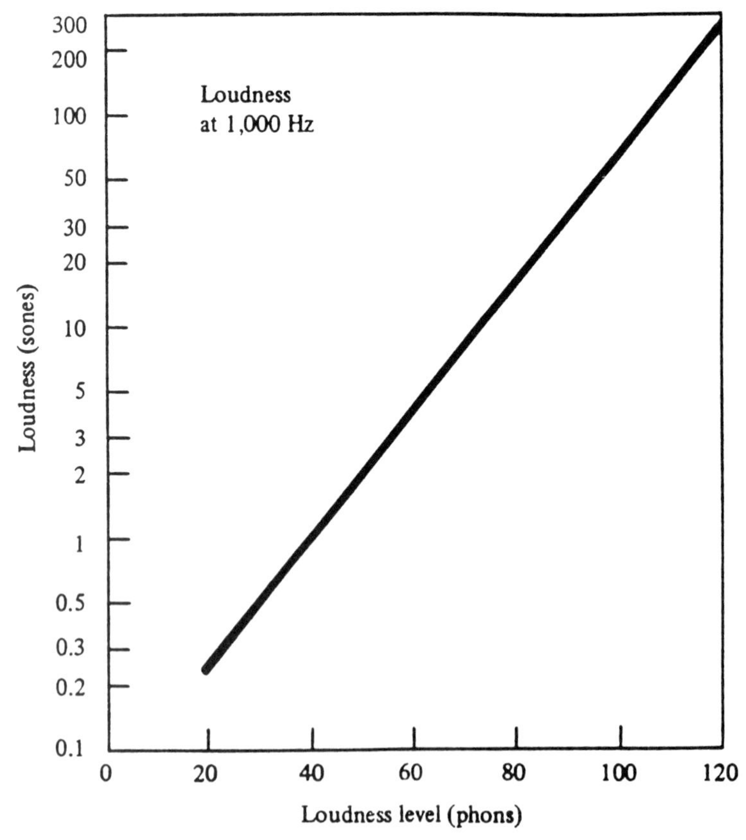

Fig. 1 The relationship between the loudness (in sones) and the loudness level (in phons) of a sound.

Stevens and Zwicker and standardized.[1–4] These can be used to evaluate the loudness (in sones) of noise sources. The procedure used in the Stevens mark VI method is to plot the noise spectrum in either octave or one-third-octave bands onto the loudness index contours. The loudness index (in sones) is determined for each octave (or one-third-octave) band and the total loudness S is then given by

$$S = S_{\max} + 0.3\left(\sum S - S_{\max}\right), \quad (2)$$

where $S_{\max}$ is the maximum loudness index and $\sum S$ is the sum of all the loudness indices. The 0.3 constant (used for octave bands) in Eq. (2) is replaced by 0.15 for one-third-octave bands. The Stevens method assumes that the sound field is diffuse and does not contain any prominent pure tones. The Zwicker method is based on the critical band concept and, although more complicated than Stevens', can be used either with diffuse or frontal sound fields. Complete details of the procedures are given in the ISO standard[4] and Kryter[5] has discussed the critical band concept in his book.

4 NOISINESS AND PERCEIVED NOISE LEVEL

4.1 Noisiness

Although the level of noise or its loudness is very important in determining the annoyance caused by noise, there are other acoustical and nonacoustical factors that are also important. In laboratory studies, people were asked to rate sounds of equal duration in terms of their noisiness, annoyance, or unacceptability.[6,7] Using octave bands of noise, Kryter and others have produced *equal noisiness index* contours. These equal noisiness contours are similar to those for equal loudness, except that at high frequency less sound energy is needed to produce equal noisiness and at low frequency more is needed. The unit of noisiness index is the *noy* N. Equal noisiness index curves are shown in Fig. 2. The procedure to determine the logarithmic measure, the *perceived noise level* (PNL) is quite complicated and has been standardized.[8] It has also been described in several books.[9–11] Briefly, it may be stated as follows. Tabulate the octave band (or one-third-octave band) sound pressure levels of the noise. Calculate the noisiness index in noise for each band from Fig. 2. Then calculate the total noisiness index N_t from

$$N_t = N_{\max} + 0.3\left(\sum N - N_{\max}\right) \quad (3)$$

where $N_{\max}$ is the maximum noisiness index and $\sum N$ is the sum of all the noisiness indices. If one-third-octave bands are used, the constant 0.3 for octave bands in Eq. (3) is replaced by 0.15.

The total perceived noisiness index N_t (summed over all frequency bands) is converted to the perceived noise level PNL or L_{PN} from

$$L_{\text{PN}} = 40 + (33.22) \log N_t. \quad (4)$$

The procedure is similar to that for calculating loudness level (phons) from loudness (sones). Some have questioned the usefulness of this procedure since listeners in laboratory experiments do not seem to be able to distinguish between (1) loudness or (2) noisiness and annoyance. However, this procedure has been widely used in assessing single-event aircraft noise. In the United States the Federal Aviation Administration (FAA) has adopted the effective perceived noise level for the certification of new aircraft.

4.2 Effective Perceived Noise Level

Noise which is of long duration is normally judged to be more annoying than noise of short duration. In addition, if the noise contains pure tones buried in a broadband noise spectrum, it is also judged to be noisier than without such tones. To account for these effects *the effective perceived noise level* (EPNL) or L_{EPN} has been defined as

$$L_{\text{EPN}} = L_{\text{PN}} + C + D \quad (5)$$

where C is the correction factor for pure tones (between 0 and 3 dB) and D is a correction for duration.[9,10] The procedure for calculating L_{EPN} is quite complicated, and its description is beyond the scope of this chapter. It is fully described in standards and in some other books.[8,10,12]

5 SOUND LEVELS

In Chapter 91, Fig. 3 shows that the ear is most sensitive to sounds in the midfrequency range around 4000 Hz. It has a particularly poor response to sound at low frequency. It became apparent to scientists in the 1930s that electrical filters could be designed and constructed with a frequency response approximately equal to the inverse of these equal loudness curves. Thus A-, B-, and C-weighting filters were constructed to approximate the inverse of the 40-, 70-, and 90-phon contours (i.e. for low-level, moderate and intense sounds), respectively. (See Fig. 3.) In principle, then, these filters, if placed between the microphone and the meter display of an instrument such as a sound level meter should give some indication of the loudness of a sound (but for pure tones only).

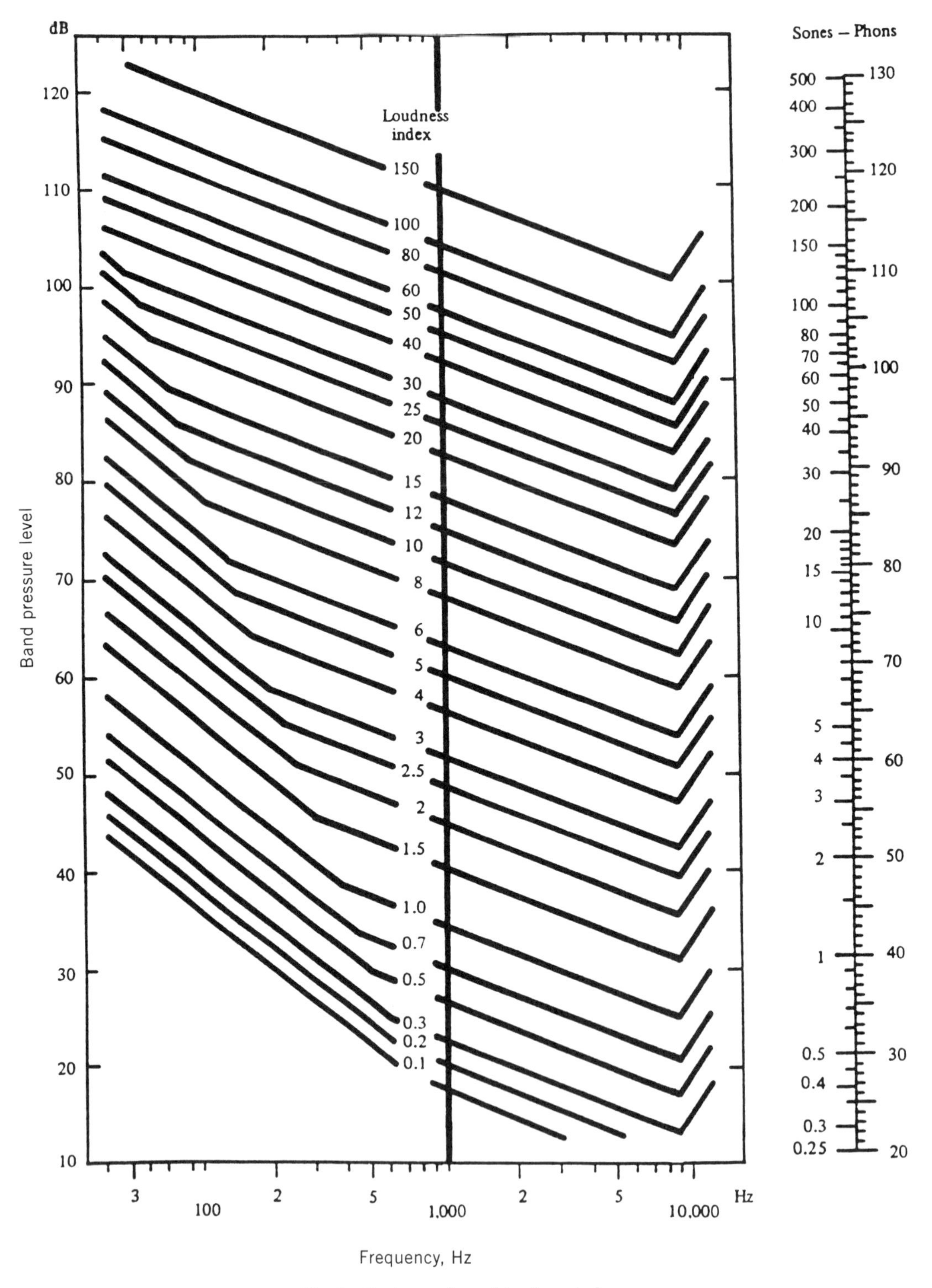

Fig. 2 Contours of equal loudness index.

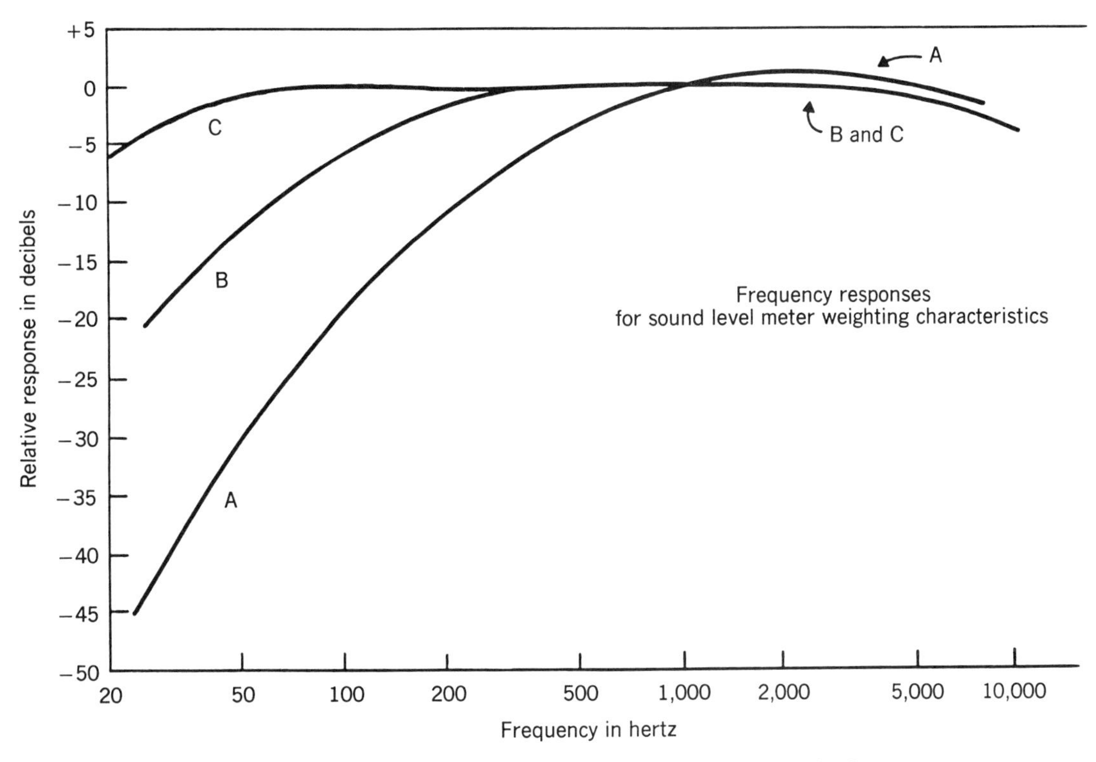

Fig. 3 A-, B-, and C-weighting filter characteristics used with sound level meters.

The levels measured with the use of these filters are commonly called the A-, B-, and C- weighted *sound levels*. The A-weighting filter has been much more widely used than either the B- or C-weighting filter and the *A-weighted sound pressure level* measured with it is known simply as the *sound level* (unless the use of some other filter is specified). Because it is simple, giving a single number, and it can be measured with a low-cost sound level meter, the A-weighted sound level has been used widely to give an estimate of the loudness of noise sources such as vehicles, even though these produce moderate to intense noise. Beranek has reviewed the use of the A-weighted sound level as an approximate measure of loudness level.[13] The A-weighted sound levels are often used to gain some approximate measure of the loudness of broadband sounds and even of the acceptability of the noise. Figure 4 shows that there is reasonable correlation between the subjective response of people to vehicle noise and A-weighted sound levels for vehicle noise. The A-weighted sound level forms the basis of many other descriptors described later in this chapter and of the *sound exposure level* (SEL), *single-event noise exposure level* (SNEL), and *noise reduction* (NR) discussed in Chapter 77. The A-weighted sound level descriptor is also used as a limit for new vehicles (Chapter 71) and noise levels in buildings (*Enc.* Ch. 97) in several countries.

6 ARTICULATION INDEX

The *articulation index* (AI) is a measure of the intelligibility of speech in a continuous noise. The AI was first proposed by French and Steinberg[14] and was extended later by Beranek.[13] Speech has a dynamic range of about 30 dB in each one-third-octave band from 200 to 6000 Hz, and the long-term rms overall sound pressure level at the speaker's lips is about 65 dB. In speech, vowels and consonants are joined together to produce not only words but sounds that have a distinctive personal nature as well. The vowels usually have greater energy than consonants and give the speech its distinctive characteristics. This is due to the fact that vowels have definite frequency spectra with superimposed short-duration peaks. The articulation index ranges from AI = 0 to 1.0 corresponding to 0% and 100% intelligibility, respectively. If the AI is less than about 0.3, speech communication is unsatisfactory (only about 30% of monosyllabic words understood);

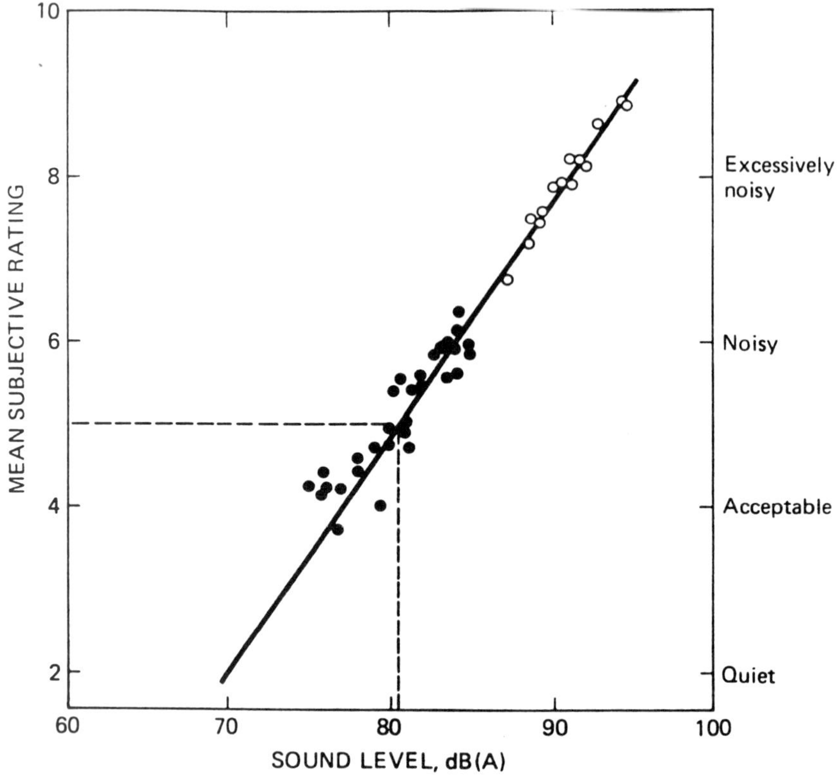

Fig. 4 Relation between subjective response and A-weighted sound level for diesel engine trucks undergoing an acceleration test; ● values measured in 1960, ○ values measured in 1968. (From C. H. G. Mills and D. W. Robinson, "The Subjective Rating of Motor Vehicle Noise," *The Engineer*, June 30, 1961; Ref. 39; and T. Priede, "Origins of Automotive Vehicle Noise," *J. Sound Vib.*, Vol. 15, No. 1, 1971, pp. 61–73.)

while if the AI is greater than about 0.6 or 0.7, speech communication is generally satisfactory (with more than 80% of monosyllabic words understood). Methods to calculate the AI are somewhat complicated and are given in an American National Standard[15] and explained in several books.[9,13] Since the calculation of AI is complicated, it will not be explained in detail here. The calculation of AI is described briefly in Chapter 77. The use of AI has somewhat declined in favor of the use of the *speech interference level.*

7 SPEECH INTERFERENCE LEVEL

The *speech interference level* (SIL) is a measure used to evaluate the effect of background noise on speech communication.[13] The speech interference level is the arithmetic average of the sound pressure levels of the interfering background noise in the four octave bands with center frequencies of 500, 1000, 2000, and 4000 Hz. (see ANSI S3.14-1977(R-1986)). If the SIL of the background noise is calculated, then this may be used in conjunction with Fig. 5 to predict the sort of speech required for satisfactory face to face communication with

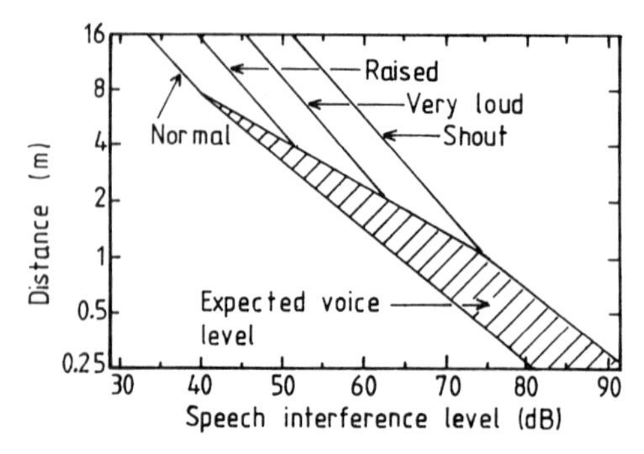

Fig. 5 Talker-to-listener distances for speech communication to be just reliable (reprinted from American Standard ANSI S3.14-1977).

male voices (i.e., for at least 95% sentence intelligibility). As an example, if the SIL is 40 dB and the speakers are males, they should be able to communicate with normal voices at 8 m. If the SIL increases to 50, raised voices must be used at 8 m. For females the SIL should be decreased by 5 dB (or the x-axis moved to the right by 5 dB.) The shaded area of Fig. 5 shows the range of speech levels that normally occur as people raise their voices to overcome the background noise. Speech interference level is also discussed in Chapter 77. Because it is simpler to measure than the SIL, the A-weighted sound level is sometimes used as a measure of speech interference, but with somewhat less confidence. Webster has produced a comprehensive diagram (see Fig. 6) which summarizes speech levels required for communication (at various distances) with 97% intelligibility of sentences for both outdoor and indoor situations.[16] Figure 6 is similar to Fig. 5, but contains some additional information concerning voice levels in different situations. With noise levels above 50 dB, people tend to raise their voice levels as shown by the "expected line" (at the left) for nonvital communication and the "communicating line" (at the right) of the diagonal shaded area for essential communication.

8 INDOOR NOISE CRITERIA

The speech interference level is mainly used to evaluate the effect of noise on speech in situations outdoors or indoors where the environment is not too reverberant. The A-weighted sound level can be used as a guide on the acceptability of noise in indoor situations, but it gives no indication which part of the frequency spectrum of the noise is of concern. A number of families of noise-weighting curves have been devised to evaluate the acceptability of noise in indoor situations. These include *noise criterion curves* (NC), *noise rating curves* (NR), *room criterion curves* (RC), and *balanced noise criterion curves* (NCB). The curves have resulted from the need to either specify acceptable noise levels in buildings or determine the acceptability of noise in existing build-

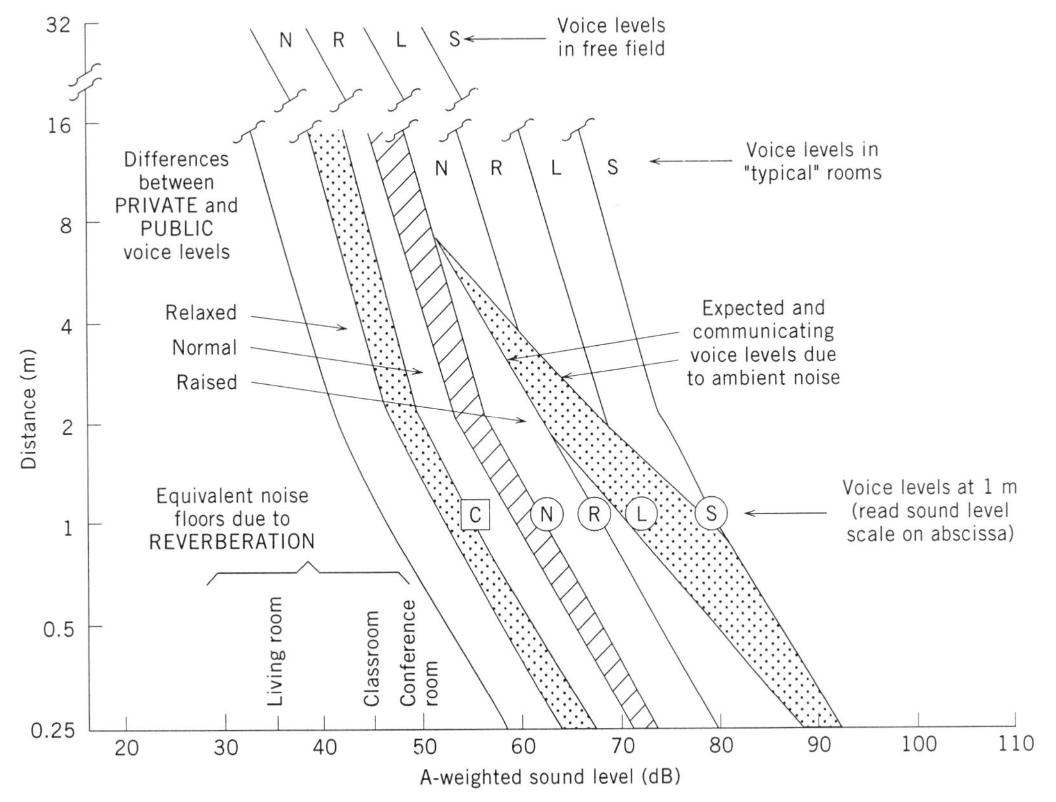

Fig. 6 Comprehensive diagram summarizing speech levels for communication (at various distances) for 97% intelligibility of sentences on first presentation to listeners for both outdoors (free field) and indoor situations.[16]

ing spaces. A major concern has been to determine the acceptability of air-conditioning noise. Beranek[17–20] has been a major contributor to the development of the NC and NCB noise criterion curves. Blazier[18,21] was mainly responsible for the development of the RC room criterion curves. NR curves were devised by Kosten and Van Os[22] and are similar to the NC curves. They have been standardized and adopted by the International Standards Organization. These noise-weighting curves are now reviewed briefly.

8.1 Noise Criterion Curves

The NC curves (Fig. 7) were developed from the results of a series of interviews with people in offices, public spaces, and industrial spaces.[17,18] These results showed that the main concern was the interference of noise with speech communication and listening to music, radio, and television. In order to determine the NC rating of the noise under consideration, the octave-band sound pressure levels of the noise are measured, and these are then plotted on the family of NC curves (Fig. 7). The noise spectrum must not exceed the particular NC curve specified in any octave band in order for it to be assigned that particular NC rating.[13] NC curves are also discussed in Chapter 77.

8.2 Noise Rating Curves

The *noise rating* NR curves are similar to the NC curves (see Fig. 7). They were originally produced to develop a procedure to determine whether noise from factories heard in adjacent apartments and houses is acceptable.[22] The noise spectrum is measured and plotted on the family of NR curves (Fig. 8) in just the same way as with the NC curves. One difference from the NC curves, however, is the use of corrections for time of day, intermittency, audible pure tones, fraction of time the noise is heard, and type of neighborhood. It has been found that in the range of NR or NC of 20 to 50 there is little difference between the results obtained from the two approaches.

8.3 Room Criterion Curves

NC curves are not defined in the low-frequency range (16- and 31.5-Hz octave bands) and are also generally regarded as allowing too much noise in the high-frequency region (at and above 2000 Hz). Blazier based his derivation of the *room criterion* RC curves on an extensive study conducted for ASHRAE by Goodfriend of generally acceptable background spectra in 68 unoccupied offices.[21] The A-weighted sound levels were mostly in the range of 40 to 50 dB. Blazier found that the curve

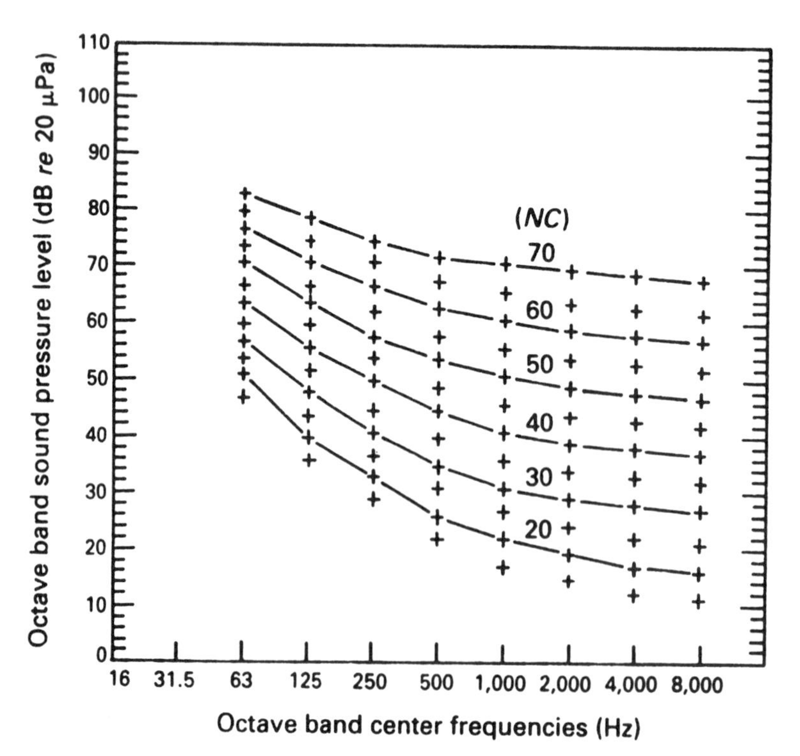

Fig. 7 Noise criterion (NC) curves.

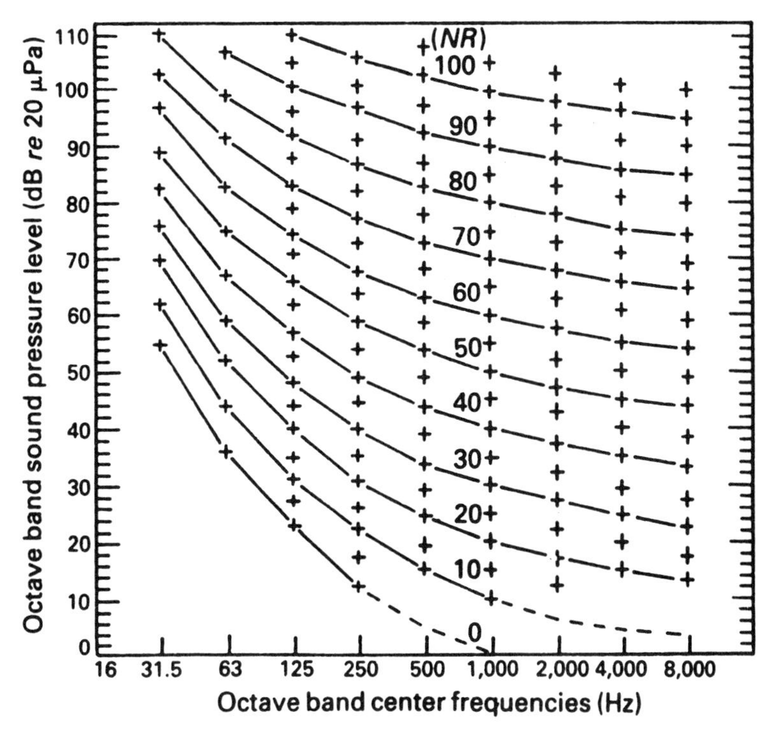

Fig. 8 Noise rating (NR) curves.

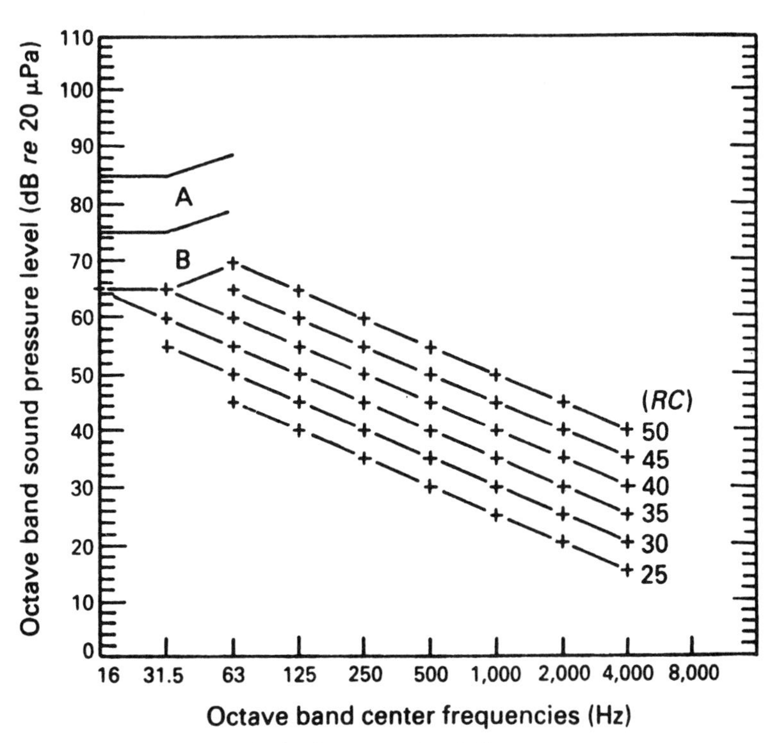

Fig. 9 Room criterion (RC) curves.

that he obtained from the measured data had a slope of about −5 dB/octave, and he thus drew a family of straight lines with this slope (see Fig. 9). He also found that intense low-frequency noise with a level of 75 dB or more in region A is likely to cause mechanical vibrations in lightweight structures (including rattles), while noise in region B has a low probability to cause such vibration. The value of the RC curve is the arithmetic average of the levels at 500, 1000, and 2000 Hz. Since these curves were obtained from measurements made with air-conditioning noise only, they are mostly useful in rating the noise of such systems. RC curves are also discussed in Chapter 77.

8.4 Balanced Noise Criterion Curves

In 1989, Beranek[19,20] modified the NC curves to include the 16- and 31.5-Hz octave bands and changed the slope of the curves so that it is now −3.33 dB/octave between 500 and 8000 Hz. He also incorporated the A and B regions as specified by Blazier in the RC curves. The rating number of a *balanced noise criterion* NCB curve is the arithmetic average of the octave band levels with midfrequencies of 500, 1000, 2000, and 4000 Hz. The result is a set of rating curves that are useful to rate air-conditioning noise in buildings (see Fig. 10). As an example of its use, a background noise spectrum from

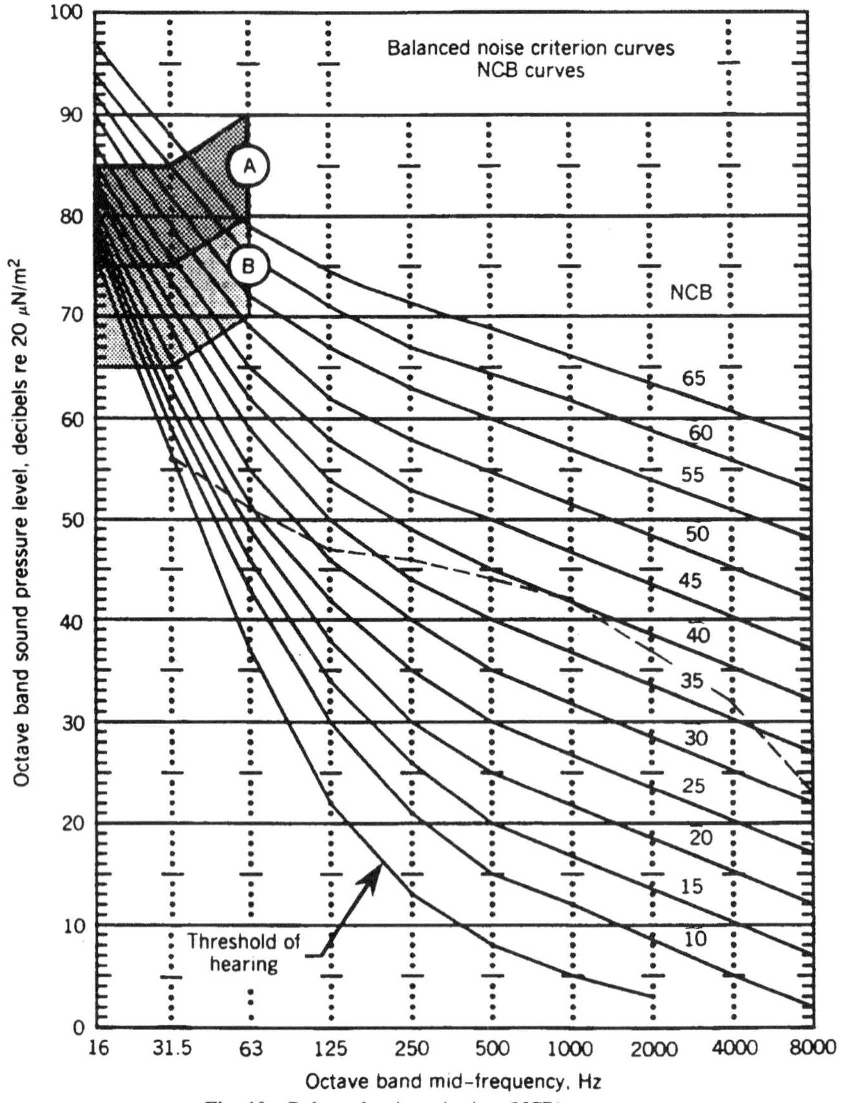

Fig. 10 Balanced noise criterion (NCB) curves.

air-conditioning is plotted in Fig. 10. It is seen that it is tangent to the NCB-40 curve and no octave-band level exceeds this curve. Thus this background noise spectrum can be assigned a rating of NCB 40. Such a noise spectrum might be just acceptable in a general office, and barely acceptable in a bedroom and living room in a house, but not acceptable at all in a church, concert hall, or theater. NCB curves are also described in Chapter 77.

9 EQUIVALENT SOUND LEVEL

For noise that fluctuates in level with time it is useful to define the *equivalent sound level* L_{eq} which is the A-weighted sound pressure level averaged over a suitable period T. This average A-weighted sound level is also sometimes known as the *average sound level* L_{AT}, so that $L_{eq} = L_{AT}$. The equivalent sound level is given by

$$L_{eq} = 10 \log \left(\frac{1}{T} \int_0^T p_A^2 dt \Big/ p_{ref}^2 \right) \quad \text{dB} \qquad (6)$$

where p_A is the instantaneous sound pressure measured using an A-weighted frequency filter and p_{ref} is the reference sound pressure 20 μPa. The averaging time T can be specified as desired to range from seconds to minutes or hours.

The average sound level (or the equivalent sound level L_{eq}) can be conveniently measured with an integrating sound level meter or some other similar device. Since it accounts both for magnitude and the duration, L_{eq} has become one of the most widely used measures for evaluating community (environmental) noise from road traffic, railways and industry.[23–25] L_{eq} has also been found to be well correlated with the psychological effects of noise.[26,27] For community noise, a long-period T is usually used (often 24 hours).

10 DAY–NIGHT EQUIVALENT SOUND LEVEL

In the United States during the 1970s, the Environmental Protection Agency developed a measure, from the equivalent sound level, known as the *day–night equivalent level* (DNL) or L_{dn} that accounts for the different response of people to noise during the night.[27]

$$L_{dn} = 10 \log \frac{15(10^{L_d/10}) + 9(10^{(L_n+10)/10})}{24}, \qquad (7)$$

where L_d is the 15-hour daytime A-weighted equivalent sound level (from 0700 to 2200 hrs) and L_n is the 9-hour nighttime equivalent sound level (from 2200 to 0700 hrs). The nighttime noise level is subjected to a 10 dB penalty because noise at night is deemed to be much more disturbing than noise during the day. This 10-dB nighttime penalty is analogous to the 10-dB nighttime penalty applied in both the composite noise rating CNR and the noise exposure forecast NEF as described in Section 12 which follows. The day–night equivalent level has become increasingly used in the United States and some other countries to evaluate community noise and in particular airport noise.[25,28] In 1980 the U.S. Federal Interagency Committee on Urban Noise (FICON) adopted L_{dn} as the appropriate descriptor of environmental noise in residential situations.[25,29] Its use is also discussed in Chapter 77, where it is termed day–night average sound level.

11 PERCENTILE SOUND LEVELS

The equivalent sound level discussed above accounts for the fluctuation in noise level of an unsteady noise by forming an average sound pressure level to find an equivalent steady A-weighted sound pressure level. There is, however, some evidence that unsteady noise (e.g., from noise sources such as passing road vehicles or aircraft movements) is more disturbing than steady noise. To try to better account for fluctuations in noise level and the intermittent character of some noises, *percentile sound levels* are used in some measures, in particular those for community and traffic noise.[23,30,31] The level L_n is defined to represent the sound level exceeded n% of the time, and thus L_{10}, for example, represents the sound level exceeded 10% of the time.

Figure 11 gives an example of L_{10}, L_{50}, and L_{90} levels and a cumulative distribution. It is seen in this schematic example figure that the level exceeded 10% of the time L_{10} is 85 dB(A), while L_{50} is sometimes termed the *median* noise level, since for half the time the fluctuating noise level is greater than L_{50} and for the other half it is less. L_{50} is used in Japan for road traffic noise. Levels such as L_1 or L_{10} are used to represent the more intense short-duration noise events. L_{10} is used in Australia and the United Kingdom (over an 18-hour 0600–2400 period) as a target value for new roads and for insulation regulations for new roads. Levels such as L_{90} or L_{99} are often used to represent the minimum noise level, the residual level from a graphic level recorder or the average minimum readings observed when reading a sound level meter. Figure 12 shows the outdoor A-weighted sound levels recorded in 1971 at 18 locations in the United States. Values of $L_{99}, L_{90}, L_{50}, L_{10}$, and L_1 are shown for the period 0700 to 1900 hours.[23,28] The small range in levels in recordings 1 and 4 (urban situations)

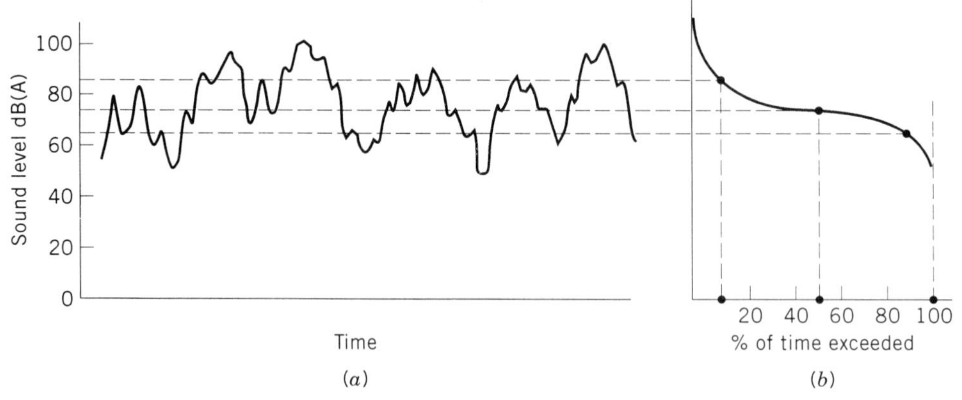

Fig. 11 (*a*) Percentile levels and (*b*) cumulative probability distribution function of percentile levels.

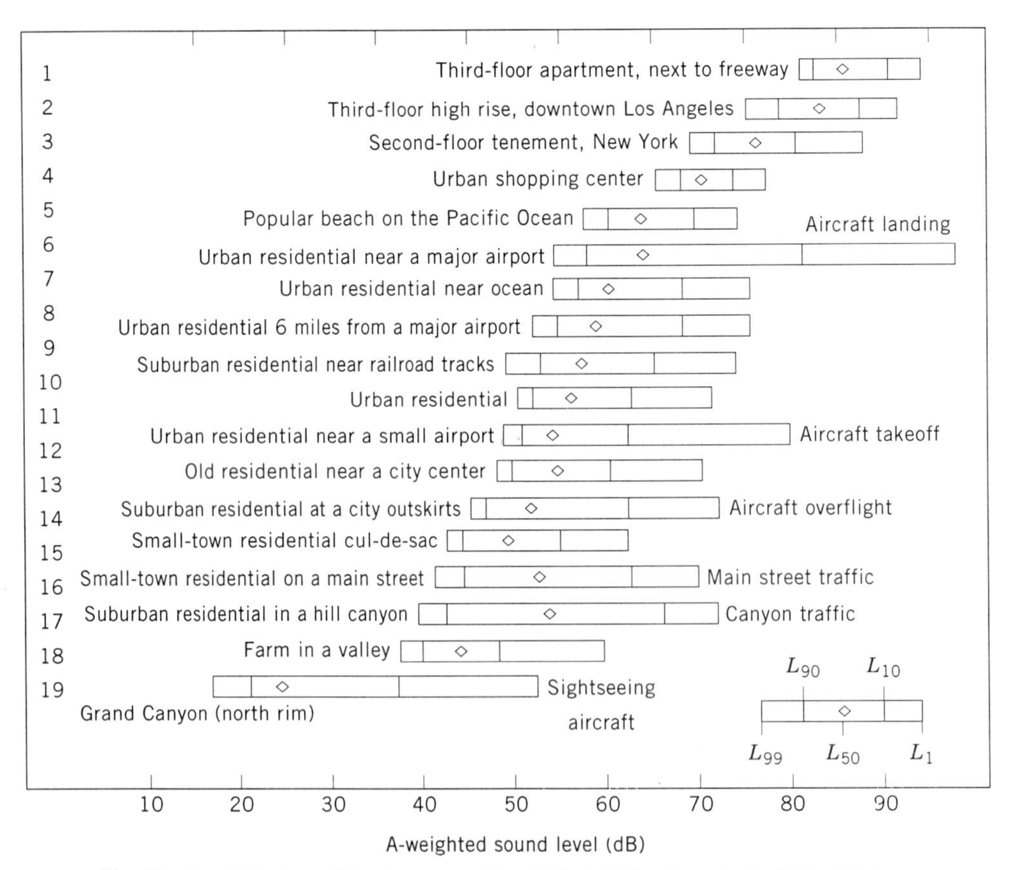

Fig. 12 A-weighted sound levels measured in 1971 at 18 locations in the United States. Values of the percentile levels $L_{99}, L_{90}, L_{50}, L_{10}$, and L_1, where L_n is the A-weighted sound level exceeded n percent of the time are shown for the period 0700–1900 hours. (From Refs. 23, 24, and 28.)

and the large range in levels in recordings 6, 11, 13, and 18 (situations involving aircraft overflights) are very evident. Obviously road traffic usually creates more steady noise, while aircraft movements lead to more extreme variations in noise levels. Percentile sound levels are also discussed in Chapter 77.

12 EVALUATION OF AIRCRAFT NOISE

The noise levels around airports are of serious concern in many countries. Several attempts have been made to produce measures to predict and assess the annoyance caused by aircraft noise in the community. A study of rating measures in 1994 showed 11 different measures in use in the 16 countries studied.[25] The following measures merit brief discussion.

12.1 Composite Noise Rating

The *composite noise rating* (CNR) has a long history dating back to the early 1950s.[32–34] Originally the basic measure it used was the level rank—a set of curves placed about 5 dB apart in the midfrequency range, rather similar to the NC and NR curves described earlier. The level rank was obtained by plotting the noise spectrum on the curves and finding the highest zone into which the spectrum protruded. The rank found initially plus the algebraic addition of corrections gave the CNR. The corrections[32] were for spectrum character, peak factor, repetitive character, level of background noise, time of day, adjustment to exposure, and public relations. The value of CNR obtained was associated with a range of community annoyance categories found from case histories—ranging from no annoyance, through mild annoyance, mild complaints, strong complaints, and threats of legal action, to vigorous community response.

In the late 1950s, the CNR was adapted to apply to the noise of military jet aircraft[35] and later of commercial aircraft.[36] Instead of assigning a level rank, the military aircraft noise was converted to an equivalent sound pressure level (SPL) in the 300–600-Hz range according to a set of curves similar to the level-rank curves.[31,35] The time-varying SPL was time-averaged and then modified by corrections similar to those mentioned above to give the aircraft CNR. The calculation was further modified later when applied to commercial aircraft by using the perceived noise level (PNL) instead of the level rank or the SPL just referred to. The final version of CNR is of the form

$$\mathrm{CNR} = \mathrm{PNL}_{\max} + N + K, \tag{8}$$

where $\mathrm{PNL}_{\max}$ is the average maximum perceived noise level for individual aircraft flyover events (landing or take-off) for a 24-hour period, N is a correction for the number of aircraft flyover events, and K is an arbitrary constant. The CNR has been found to be useful in predicting where the community reaction to aircraft noise will be on a scale ranging from *no reaction* to *vigorous community response*. The CNR for a certain ground point CNR_{ij} is given by

$$\mathrm{CNR}_{ij} = \mathrm{PNL}_{ij} + 10\log(N_{d,ij} + 16.7N_{n,ij}) - 12, \tag{9}$$

where $N_{d,ij}$ and $N_{n,ij}$ are the number of daytime and nighttime events for each aircraft class i and flight path j. The nighttime flights are penalized by 10 dB. The factor of 16.7 arises because there are fewer nighttime hours than daytime. Then the total CNR for the ground point is given by

$$\mathrm{CNR} = 10\log\sum_i\sum_j \operatorname{antilog}\frac{\mathrm{CNR}_{ij}}{10}. \tag{10}$$

The final version of CNR does not contain any corrections for background noise, previous experience, public relations, or other factors such as the presence of pure tones. Although CNR is no longer used, it is discussed here for completeness and because it has formed the basis for some of the other noise measures and descriptors, such as NEF, which follows.

12.2 Noise Exposure Forecast

The *noise exposure forecast* (NEF) is a similar measure to CNR but it uses the effective perceived noise level EPNL instead of PNL.[37,38] Thus NEF automatically takes account of the annoying effects of pure tones and the duration of the flight events. The NEF is of the form

$$\mathrm{NEF} = \mathrm{EPNL} + N + K, \tag{11}$$

where EPNL is the average effective perceived noise level for individual aircraft flyover events over a 24-hour period, N is a correction for the number of flyover events, and K is an arbitrary constant. The NEF at a certain ground point NEF_{ij} is calculated in a similar way to the CNL just described:

$$\mathrm{NEF}_{ij} = \mathrm{EPNL}_{ij} + 10\log(N_{d,ij} + 16.7N_{n,ij}) - 88, \tag{12}$$

where EPNL_{ij} is the effective perceived noise level for the aircraft class i and flight path j and $N_{d,ij}$ and $N_{n,ij}$ are as already defined for CNR. The constant 88 was chosen to make the NEF values distinct from CNR values.

Finally

$$\text{NEF} = 10 \log \sum_i \sum_j \text{antilog} \frac{\text{NEF}_{ij}}{10}. \quad (13)$$

The use of NEF has been superseded in the United States by the day–night level L_{dn}. However, in 1995 NEF remained in use in Canada and in a modified form in Australia.[25]

12.3 Noise and Number Index

The *noise and number index* (NNI) is a subjective measure of aircraft noise annoyance developed and used in the United Kingdom. The NNI was the outcome of a survey in 1961 of noise in the residential districts within 10 miles of London (Heathrow) Airport.[39] It is defined by

$$\text{NNI} = \langle \text{PNL} \rangle_N + 15 \log N - 80, \quad (14)$$

where $\langle \text{PNL} \rangle_N$ is the average peak noise level of all aircraft operating during a day and is given by

$$\langle \text{PNL} \rangle_N = 10 \log \left(\frac{1}{N} \sum_{n=1}^{N} \text{antilog} \frac{\text{PNL}}{10} \right). \quad (15)$$

Here PLN is the peak perceived noise level produced by an individual aircraft during the day and N is the number of aircraft operations over a 24-h period. Since no annoyance appeared to occur at levels below 80 PNdB, a constant of 80 is used, so that the value of NNI of 0 corresponds to no annoyance.

A second survey[40] in 1967 found that for the same noise exposure the reported annoyance was less than in 1961, though it was not clear whether this was because of noise-sensitive people leaving the area or noise-insensitive people arriving, or because people adapted or became resigned or apathetic to the noise, or because the background noise from traffic, construction, and industry had increased enough to mask more of the aircraft noise.

In 1988 NNI was superseded in the United Kingdom by a measure based on the A-weighted L_{eq}. The L_{eq} is averaged over the period 0700 to 2300 h since in the UK nighttime flights are restricted. However, in 1995 NNI remained in use in Ireland and Switzerland.[25]

12.4 Equivalent Sound Level L_{eq} and Day–Night Level L_{dn}

Some countries continue to use NEF or NNI or similar noise measures or descriptors related to these that include a weighting based on the number of aircraft movements.[25] However, because they are much simpler to measure and seem to give adequate correlation with subjective response, there has been a move in several countries towards the use of L_{eq} and L_{dn}.[25] In Germany, Luxembourg, and the United Kingdom L_{eq} has been adopted: (1) in Germany and Luxembourg with day (0600–2200) and night (2200–0600) periods, and (2) in the United Kingdom with an 18-h period only (0700–2300) (because nighttime flights are normally restricted).

In the United States since publication of the Environmental Protection Agency (EPA) "Levels Document"[27] and other similar publications, the use of CNR and NEF has been largely superseded by the day–night equivalent level DNL for the assessment of the potential impacts of noise and for planning recommendations and land-use management near civilian and military airports. In fact, in 1977 the Office of the Secretary of Defense adopted a new Part 256 of Title 32 of the Code of Federal Regulations, which contains policies for Air Installation Compatible Use Zones (AICUZ). Section 256.3 states that the day–night average sound level DNL will be used to assess the impact of noise with air installations, and Section 256.10 prohibits a further use of the composite noise rating CNR or the noise exposure forecast NEF.

13 EVALUATION OF TRAFFIC NOISE

13.1 The Traffic Noise Index

In an attempt to develop acceptability criteria for traffic noise from roads in residential areas, Griffiths and Langdon[41] produced a unit for rating traffic noise, the *traffic noise index* (TNI):

$$\text{TNI} = 4(L_{10} - L_{90}) + L_{90} - 30. \quad (16)$$

They measured A-weighted traffic noise at 14 sites in the London area and interviewed 1200 people at these sites in the process. Griffiths and Langdon excluded sites with noise sources other than traffic. They then used regression analysis to fit curves to the data. This indicated that L_{10} was better at predicting dissatisfaction than L_{50} or L_{90}, and that TNI was also superior to L_{10}, L_{50}, and L_{90}.

Use of the traffic noise index has not been widespread. The index attempts to make an allowance for the noise variability (with the first term in the above equation) since fluctuating noise is commonly assumed to be more annoying than steady noise. The second term, L_{90}, represents the background noise level, while the third term is simply a constant chosen to yield more convenient numbers.

Some doubt has been cast on the conclusions of Griffiths and Langdon, and it has been suggested that the very short sample times (100 seconds in each hour) used may have resulted in underestimates of L_{10} and overestimates of L_{90}.[42] TNI is not considered today to be significantly superior to either L_{10} or L_{eq} and has not been widely used.

Instead of TNI, the British government has adopted the A-weighted L_{10}, averaged over 18 hours from 0600 to 2400 hrs, as the noise index to be used to implement planning and remedial measures to reduce the impact on people of road traffic noise.[43–45] In addition the British government uses a 16-hour L_{eq} and an 8-hour L_{eq} for the case of land used for residential development.

13.2 Noise Pollution Level

In a later survey, Robinson[46] again concluded that, with fluctuating noise, L_{eq}, the equivalent continuous sound level on an energy basis, was an insufficient descriptor of the annoyance caused by fluctuating noise. He included another term in his *noise pollution level* NPL or L_{NP}, which he defined as

$$L_{NP} = L_{eq} + k\sigma, \tag{17}$$

where k is a constant and σ is the standard deviation of the sound level. He also offered an alternative expression:

$$L_{NP} = L_{eq} + a(L_{10} - L_{90}). \tag{18}$$

Robinson examined the available Griffiths and Langdon data[41] and concluded that $a = 1.0$ and $k = 2.56$ were good choices for the constants in these equations. He then examined the aircraft noise experiments of Pearsons[47] and found that L_{NP} predicted very well Pearsons's data points and the trade-off between duration and level for individual flyover events. A-weighted levels were used in L_{NP} with traffic noise, and perceived noise levels (PNdB) were used with aircraft noise. The noise pollution level is potentially attractive because, in principle, it allows annoyance from aircraft, traffic, and perhaps other sources such as industrial noise to be determined. However, the superiority of L_{NP} over all other forms of noise rating has not been proved in practice, and it has not been widely used.

13.3 Equivalent Sound Level

Figure 13 shows the annoyance results of Pearsons and co-workers, using six different noise ratings: L_1, L_{10}, L_{50}, L_{90}, L_{eq}, and L_{NP}.[48] As expected for all of the noise measures, annoyance increases with level. The shapes of the curves, however, do vary considerably when the annoyance is less than very annoying. In particular, the L_{10}

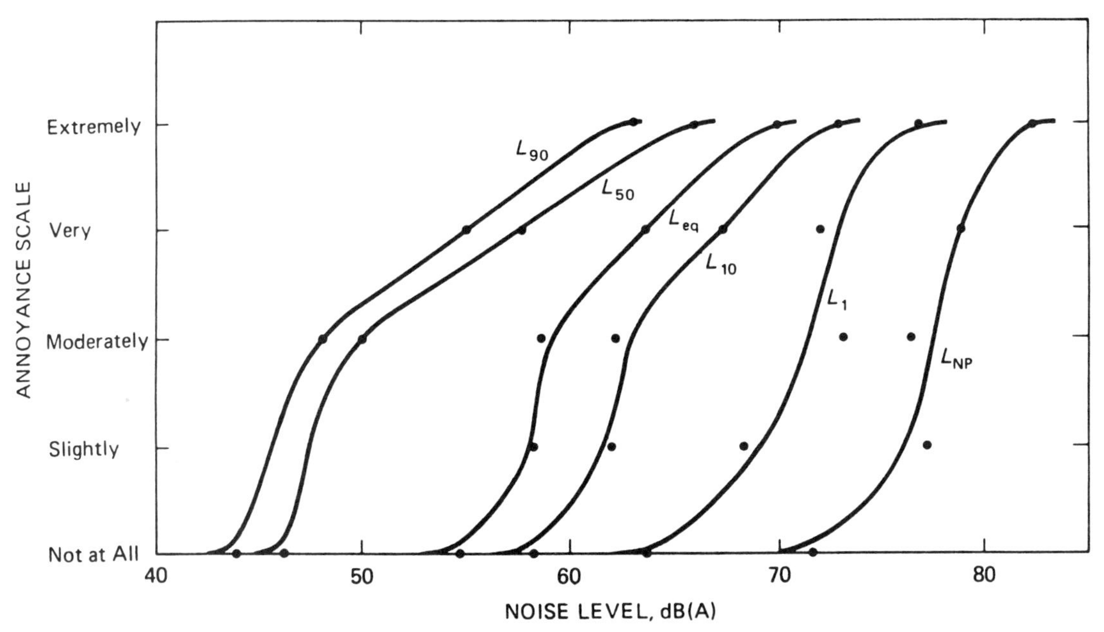

Fig. 13 Annoyance as a function of noise level. (From Ref. 48.)

and L_{eq} measures exhibit a very steep rise in annoyance from the categories of *slightly* to *moderately annoying* for no increase in noise level, presumably one of their drawbacks. However, except for the case of L_{NP} in the *extremely annoying* category, the standard deviation of L_{eq} for a specified response category was in all cases less than or equal to the standard deviation of the other noise measures. This is an advantage in the use of L_{eq} since there is more confidence in the annoyance scores predicted. There is one clear advantage of L_{eq} over L_{10}, however, in the case of noise containing short-duration, high-level single events. If the events do not occur for more than 10% of the time, then L_{10} will be insensitive to these high-level events and will tend to represent the "background" noise. In fact, for a noise measure L_n to be useful, the intruding noise events must be present for *more* than n% of the time. This suggests that L_{10} would be unsuitable as a measure of aircraft noise annoyance and that it might be a possible source of error in Griffiths and Langdon's results[41] for low traffic flows.

The equivalent sound level L_{eq} (often now denoted as L_{Aeq}) has become the measure most commonly used to assess and regulate road traffic (and railroad noise).[25] In the United States L_{dn} (a similar measure) is used for road traffic noise assessment.

14 EVALUATION OF COMMUNITY NOISE

In some community noise measures, corrections are applied to community noise levels to account for pure-tone components or impulsive character, seasonal corrections (summer or winter when windows are always closed), type of district (rural, normal suburban, urban residential, noisy urban, very noisy urban) and correction for previous exposure (such corrections are similar to those for NR already discussed).

Figure 14 gives three examples of the A-weighted *sound* levels measured in a community over a 24-hour period.[23] The triangles in the three figures are the maximum levels read from a graphic level recorder. The continuous lines are percentile levels measured for hour-long periods throughout the 24 hours. The highest percentile measure L_1 does not represent the maximum levels well, which are presumably mostly higher-level, short-duration sounds (occurring less than 1% of the time). It is observed in examples (*a*) and (*b*) in Fig. 14 that although the day–night levels L_{dn} are only 3 dB different (86 and 83) there is a much greater fluctuation in sound level with time in example (*b*). This example is for a location near a major airport and the use of either TNI [Eq. (16)] or NPL [Eqs. (17) or (8)] would suggest that the noise environment in location 6 would be much more annoying than in location 1 (near a freeway). Example (*c*) illustrates another quite different distribution of sound levels with time.

In the United States the A-weighted L_{eq} and L_{dn} are normally used to characterize community noise. However, HUD also recognizes the usefulness of L_{10} in some instances. In other countries the A-weighted L_{eq} is mainly used for community noise studies rather than L_{dn}. The international standards on the description and measurement of environmental noise[49] recommend the use of A-weighted L_{eq} and rating levels (which are A-weighted L_{eq} values to which tone and impulse adjust-

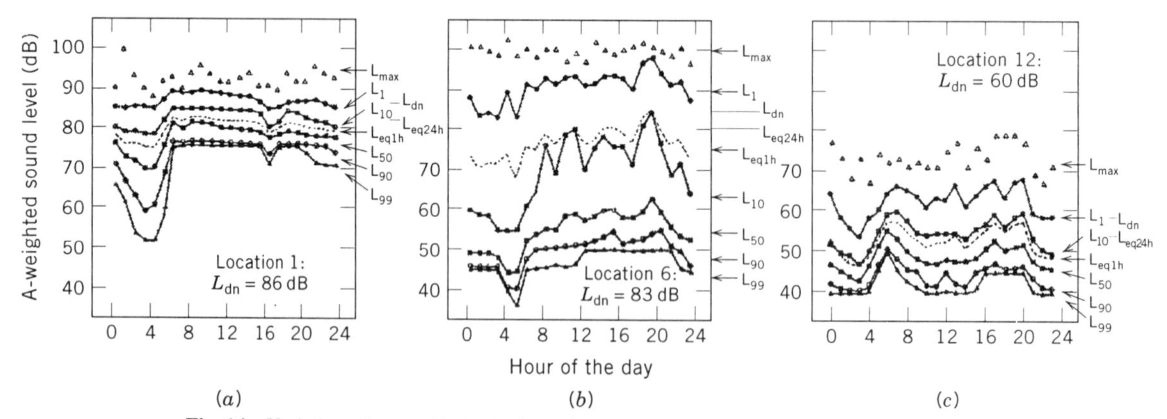

Fig. 14 Variation of percentile levels throughout 24 hour periods recorded at three residential locations (numbers 1, 6, and 12 in Fig. 12) in Los Angeles in 1971, (*a*) third-floor apartment near freeway, (*b*) urban residential location near major airport, and (*c*) old residential location near city center. The values of L_n represent the A-weighted sound levels exceeded n percent of the time during one-hour periods. Hourly maximum Levels L_{max}, 1-hour and 24-hour values of L_{eq}, and the day–night level L_{dn} are also shown. (From Refs. 23 and 28).

ments have been made). These standards also recommend that in some circumstances it may be useful to determine the distribution of A-weighted sound pressure levels by determining percentile levels such as L_{95}, L_{50}, and L_5.

In California some state legislation requires the use of the *community noise equivalent level* (CNEL) rather than the day–night level L_{dn}. The two descriptors are similar except that the CNEL has three periods instead of two. Besides the night penalty of 10 dB, an evening penalty of 5 dB is applied with CNEL. CNEL is defined[50] as

$$\text{CNEL} = 10\log(\textstyle\sum_{0800}^{1900} 10^{(\text{HNLD}(i))/10} + 3\times 10^{(\text{HNLE}+5)/10} + 9\times 10^{(\text{HNLN}+10)/10})/24 \qquad (19)$$

where HNLD(i), HNLE, and HNLN are the 12 one–hour average day, average evening, and average night hourly sound levels (HNL). The hourly noise level is given by

$$\text{HNL} = 10\log\frac{\int 10^{L/10}\,dt}{3600}, \qquad (20)$$

where L is the instantaneous A-weighted sound level. The integral is calculated and summed. HNL is usually computed electronically. As is discussed in Chapter 70 the CNEL has also been extensively used to evaluate airport noise in California.[51] In California it has also been used to assess environmental noise transmission into buildings (see Chapter 77).

15 HUMAN RESPONSE

15.1 Sleep Interference

Various investigations have shown that noise disturbs sleep.[52–56] It is well known that there are several stages of sleep and that people progress through these stages as they sleep.[52] Noise can change the progression through the stages and if sufficiently intense can awaken the sleeper. Most studies have been conducted in the laboratory under controlled conditions using brief bursts of noise similar to aircraft fly-overs or the passage of heavy vehicles. However, some have been conducted in the participants' bedrooms. Different measures have been used such as A-weighted *maximum sound level* $L_{A\max}$, A-weighted *sound exposure level* (ASEL), *effective perceived noise level* (EPNdB) and *day–night sound level* (DNL) and most studies have concentrated on the percentage of the subjects that are awakened. Recently Pearsons et al.[57] have reassessed data from 21 sleep disturbance studies (drawn from the reviews of Lucas[52] and Griefhan[53] and seven additional studies). From these data, Finegold et al.[58] have proposed sleep disturbance

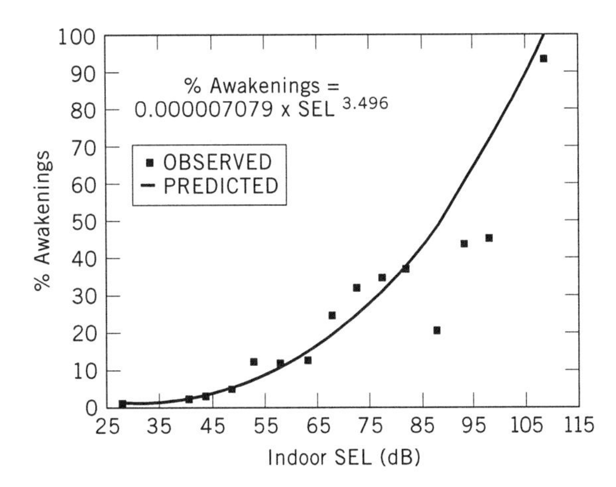

Fig. 15 Proposed sleep disturbance curve based on data of Pearsons et al.[57] (from Ref. 58 with permission). Curve represents percentage of subjects awakened against A-weighted sound exposure level ASEL.

criteria based on the indoor A-weighted *sound exposure level* (ASEL). In the reanalysis, because of the extremely variable and incomplete data bases, the data were averaged in 5-dB intervals to reduce variability. A regression fit to these data gave the following expression (which is also shown graphically in Fig. 15):

$$\%\text{Awakenings} = 7.1\times 10^{-6} L_{\text{AE}}^{3.5} \qquad (21)$$

where L_{AE} is the indoor A-weighted sound exposure level ASEL. Although the authors recognize that there are concerns with the existing data and there is a recognition that additional sleep disturbance data are needed, Finegold et al.[58] have proposed that Fig. 15 be used as a practical sleep disturbance curve until further data become available.

15.2 Annoyance

In 1978 Schultz published an analysis of 12 major social surveys of community annoyance caused by transportation noise.[59] This Schultz analysis which relates the percentage of the population that report they are "highly annoyed" by transportation noise to the day–night average sound level DNL, has become widely used all over the world as an important curve for describing the average community response to environmental noise. Since the Schultz curve was published, additional data have become available. Fidell et al.[60] used 453 exposure response data points compared to the 161 data points originally used by Schultz. This resulted in almost tripling the database for predicting noise annoyance from

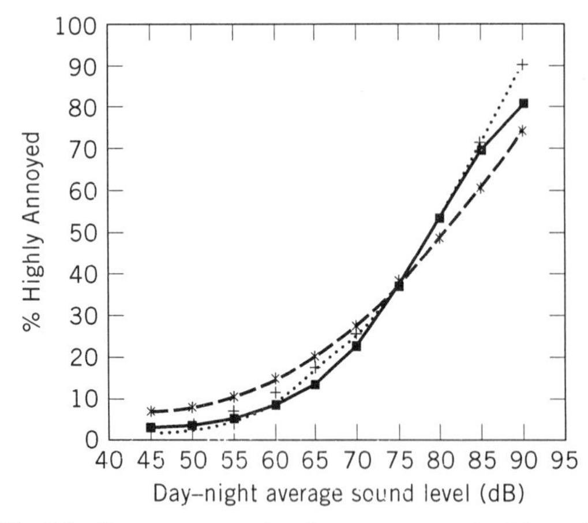

Fig. 16 Curves representing the percentage of subjects that are highly annoyed by noise against A-weighted day–night average sound level DNL: (—■—) New logistic USAF curve (400 data points), (· · · · + · · · ·) Schultz[59] third-order polynomial (161 data points), and (--*--) Fidell et al.[60] quadratic curve (453 data points). (From Ref. 58 with permission.)

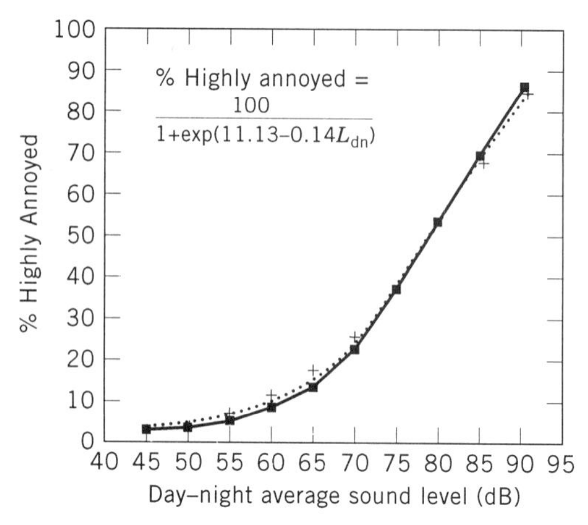

Fig. 17 Curves representing the percentage of subjects that are highly annoyed by noise against A-weighted day–night average sound level DNL. (—■—) Logistic fit to 400 community annoyance social survey data points and (· · · + · · ·) 1978 Schultz[59] curve. (From Ref. 58 with permission.)

transportation noise. A later study by the U.S. Air Force eliminated 53 of these data points because there was insufficient correlation between the DNL and the percentage of the population that was highly annoyed, %HA.[58] The results of these two studies and the original Schultz curve are given in Fig. 16. Finegold et al.[58] recommend a logistic fit,

$$\%\text{HA} = \frac{100}{1 + \exp(11.13 - 0.14L_{\text{dn}})}, \qquad (22)$$

rather than the quadratic fit used by Fidell et al.[60] or the third-order polynomial fit used by Schultz.[59] This results in a very close agreement between the curve obtained with the 400 data points and the original 1978 Schultz curve.[59] See Fig. 17. The differences between the curves in Figs. 16 and 17 are not very significant; however, there are several advantages to the use of the logistic fit given in Eq. (22), including (1) the same predictive utility in both the original Schultz curve and the Fidell et al. curve, (2) it allows prediction of annoyance to approach but does not reach 0% or 100%, (3) it approaches a 0% community annoyance prediction for a DNL of approximately 40 dB rather than the anomaly of an increase in annoyance for levels of less than 45 dB as predicted by the Fidell et al. curve, (4) use of a logistic function has had a history of success with U.S. federal environmental impact analyses, and (5) it is based on the most defensible social survey database.[58]

Most community noise impact studies since the late 1970s have been based on a combination of aircraft and surface transportation noise sources. However, there has been a continuing controversy over whether all types of transportation should be combined into one general curve for predicting community annoyance to transportation noise.[58,61–67] Some researchers have suggested that people find aircraft noise more annoying than traffic noise or railroad noise.[25,58,61,65]

The differences have been discussed in the literature[25,28,58] and can perhaps be explained by a variety of causes such as (1) methodological differences, (2) variability in the criterion for reporting high annoyance, (3) inaccuracy in some of the acoustical measurements, (4) community response biases, and (5) aircraft noise entering homes through parts of the building structure with less transmission loss (such as the roof rather than the walls with aircraft noise). Figure 18 shows logistic fits to 400 final data points from a total of 22 different community annoyance surveys. It can be seen that aircraft noise appears to produce somewhat more annoyance than railroad or traffic noise particularly for the higher DNL values. Miedema has made a recent reanalysis of data from selected social surveys which also shows that aircraft noise appears to cause more annoyance than other surface transportation noise sources.[61] However, the results from Miedema's study seem to suggest that for high values of DNL (over 60 to 70 dB), although aircraft noise is by far the most annoying source, railroad noise is also more annoying than traffic noise (in contrast to the results of Ref. 58).

If the five cases discussed above can be dismissed as responsible for the apparent greater annoyance of aircraft

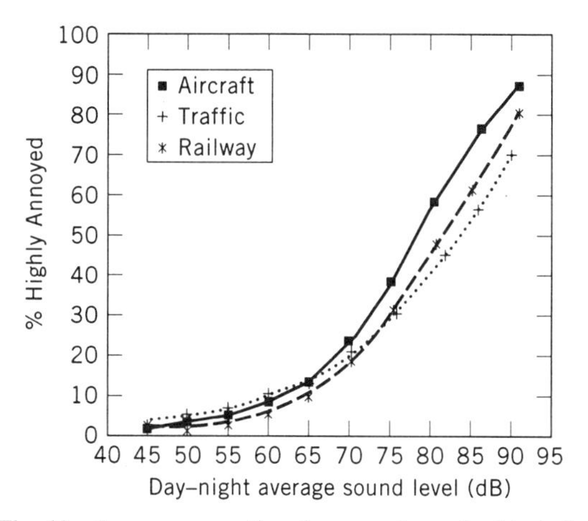

Fig. 18 Curves representing the percentage of subjects that are highly annoyed by noise against day–night average sound level DNL for different sources: —■— aircraft, · · · + · · · traffic, and --*-- railway. Curves based on data from Fidell et al.[60] (From Ref. 58 with permission.)

noise, then it may be that aircraft noise is more annoying than surface transportation noise for reasons such as the greater variation in level with time and the different frequency spectra from other types of transportation noise sources. As already discussed in the text accompanying Fig. 14 in Section 14, aircraft noise does generally exhibit a much greater variation in level from traffic noise and other sources of surface transportation. If such variation is indeed more annoying and is one of the main causes of the difference in annoyance caused by these different forms of transportation noise, this suggests that it may be advisable to re-examine such measures that account for variation in level such as the traffic noise index TNI or the noise pollution level NPL discussed in Sections 13.1 and 13.2.

16 NOISE CRITERIA AND NOISE REGULATIONS

Using some of the noise measures and descriptors discussed and surveys and human response studies, various criteria have been proposed so that noise environments can be determined that are acceptable for people, for speech communication, for different uses of buildings, for sleep and for different land uses. In some countries such criteria are used to write noise regulations for new machinery, vehicles, traffic noise, railroad noise, aircraft and airport noise, community noise, and land use and planning, etc. It is beyond the scope of this chapter to give a comprehensive summary of all these criteria and regulations. Instead just a few are described in this section. The interested reader is referred to the chapters fol-

TABLE 2 Recommended Values of NCB Curves for Different Uses of Spaces in Buildings[a]

Type of Space (and Acoustical Requirements)	NCB Curve	Approximate L_A, dB
Broadcast and recording studios	10	18
Concert halls, opera houses, and recital halls	10–15	18–23
Large auditoriums, large drama theaters, and large churches	<20	28
Broadcast, television, and recording studios	<25	33
Small auditoriums, small theaters, small churches, music rehearsal rooms, large meeting and conference rooms	<30	38
Bedrooms, sleeping quarters, hospitals, residences, apartments, hotels, motels, etc.	25–40	38–48
Private or semiprivate offices, small conference rooms, classrooms, and libraries	30–40	38–48
Living rooms and drawing rooms in dwellings	30–40	38–48
Large offices, reception areas, retail shops and stores, cafeterias and restaurants	35–45	43–53
Lobbies, laboratory work spaces, drafting and engineering rooms, and general secretarial areas	40–50	48–58
Light maintenance shops, industrial plant control rooms, office and computer equipment rooms, kitchens and laundries	45–55	53–63
Shops, garages, etc. (for just acceptable speech and telephone communication)	50–60	58–68
For work spaces where speech or telephone communication is not required, but where there must be *no risk* of hearing damage	55–70	63–78

Source: Based in part on Ref. 13.

[a]Also given are the approximate equivalent A-weighted sound levels L_A.

lowing in Part VIII and to Refs. 25, 28, and 62 for more complete summaries of criteria, regulations, and legislation. For instance, Chapter 71, provides information on limits for the noise of new vehicles in different countries (where acceleration noise tests are used). Such limits are based on results such as those presented in Fig. 4.

16.1 Noise Criteria

An example of noise criteria is given in Table 2, which is based on those suggested by Beranek[13] and gives recommended NCB curve values (and approximate A-weighted levels) for various indoor functional activity areas. The NCB curves are given in Fig. 10. For example, the air-conditioning unit chosen to supply air to bedrooms (used in residences, apartments, hotels, hospitals, etc.) should have a spectrum corresponding to no more than an NCB curve of 25–40 (or an A-weighted sound level of no more than about 38–48 dB).

Another example of noise criteria are the guidelines recommended by EPA,[27] WHO,[64] FICON,[29] and various European road traffic regulating bodies. See Table 3. As already mentioned L_{eq} is very widely used to evaluate road traffic, railroad, and even aircraft noise.[25] Interestingly, railroad noise has been found to be less annoying than traffic noise in several surveys.[25,68,69] This has resulted in noise limits (using L_{eq}) that are 5 dB lower for railroad noise than road traffic noise in Austria, Denmark, Germany, and Switzerland and 3 dB lower in The Netherlands.[25] Gottlob terms this difference the "railway bonus."[25] This seems to contradict the results shown in Fig. 18 and suggests the need for further research.

An example of national noise exposure criteria is the guidance given in the recent British government guidelines adopted in 1994 for land development given in Table 4. This table shows guidelines in A-weighted sound levels L_{eq} for four noise exposure categories.[70] The noise exposure categories can be interpreted as

TABLE 3 Guidelines from EPA,[27] WHO,[64] FICON,[29] and Various European Agencies for Acceptable Noise Levels

Authority	Specified Sound Levels	Criterion
EPA Levels Document[27]	$L_{dn} \leq 55$ dB (outdoors) $L_{dn} \leq 45$ dB (indoors)	Protection of public health and welfare with adequate margin of safety
WHO Document (1995)[64]	$L_{eq} \leq 50/55$ dB (outdoors; day) $L_{eq} \leq 45$ dB (outdoors; night) $L_{eq} \leq 30$ dB (bedroom) $L_{max} \leq 45$ dB (bedroom)	Recommended guideline values
U.S. Interagency Committee (FICON)[29]	$L_{dn} \leq 65$ dB $65 \leq L_{dn} \leq 70$ dB	Considered generally compatible with residential development Residential use discouraged
Various European road traffic regulations[25]	$L_{eq} \geq 65$ or 70 dB (day)	Remedial measures required

Source: Based on Ref. 28.

TABLE 4 Guidelines Used in the United Kingdom for A-Weighted Equivalent Sound Levels for Different Noise Exposure Categories

Noise Source		Noise Exposure Category			
		A	B	C	D
Road traffic	(07.00–23.00)	<55	55–63	63–72	>72
	(23.00–07.00)	<45	45–57	57–66	>66
Rail traffic	(07.00–23.00)	<55	55–66	66–74	>74
	(23.00–07.00)	<45	45–59	59–66	>66
Air traffic	(07.00–23.00)	<57	57–66	66–72	>72
	(23.00–07.00)	<48	48–57	57–66	>66
Mixed sources	(07.00–23.00)	<55	55–63	63–72	>72
	(23.00–07.00)	<45	45–57	57–66	>66

Source: Based on Ref. 70.

follows[71]: (a) Noise need not be considered as a determining factor in granting planning permission, although the noise level at the high end of the category should not be regarded as a desirable level; (b) noise should be taken into account when determining planning applications, and where appropriate, conditions should be imposed to ensure an adequate level of protection against noise; (c) planning permission should not normally be granted; where it is considered that permission should be given, for example because there are no alternative quieter sites available, conditions should be imposed to ensure a commensurate level of protection against noise; (d) planning permission should normally be refused.

REFERENCES

1. S. S. Stevens, "Procedure for Calculating Loudness: Mark VI," *J. Acoust. Soc. Am.*, Vol. 33, 1961, p. 1577.
2. E. Zwicker, "Ein Verfahren zur Berechnung der Lautstarke (A Means for Calculating Loudness)," *Acustica*, Vol. 10, 1960, p. 304.
3. E. Zwicker, "Subdivision of the Audible Frequency Range into Critical Bands (Frequensgruppen)," *J. Acoust. Soc. Am.*, Vol. 33, 1961, p. 248.
4. Anon., "Method for Calculating Loudness Level," ISO Recommendation R 532-1967, 1967.
5. K. D. Kryter, *The Effects of Noise on Man*, 2nd ed. Academic Press, Orlando, 1985.
6. K. D. Kryter, "Scaling Human Reactions to the Sound from Aircraft," *J. Acoust, Soc. Amer.*, Vol. 31, No. 11, 1959, pp. 1415–1429.
7. K. D. Kryter and K. S. Pearsons, "Some Effects of Spectral Content and Duration on Perceived Noise Level," *J. Acoust Soc. Amer.*, Vol. 35, No. 6, 1963, pp. 866–883.
8. ISO, "Procedure for Describing Aircraft Noise Heard on the Ground," ISO 3891-1978, Int. Std. Org., Geneva, Switzerland, 1978.
9. D. N. May, "Basic Subjective Responses to Noise," in D. N. May (Ed.), *Handbook of Noise Assessment*, Van Nostrand Reinhold, New York, 1978.
10. C. M. Harris (Ed.), *Handbook of Acoustical Measurements and Noise Control*, 3rd ed., McGraw-Hill, New York, 1991, Chapter 47.
11. J. S. Anderson and M. Bratos-Anderson, *Noise, Its Measurement, Analysis, Rating, and Control*, Avebury Technical, Aldershot, UK, 1993.
12. M. J. Crocker, "Noise of Air Transportation to Non-travellers," in D. N. May (Ed.), *Handbook of Noise Assessment*, Van Nostrand Reinhold, New York, 1978.
13. L. L. Beranek and I. L. Ver, *Noise and Vibration Control Engineering*, Wiley, New York, 1992.
14. N. R. French and J. C. Steinberg, "Factors Governing the Intelligibility of Speech Sounds," *J. Acoust. Soc. Am.*, Vol. 19, 1947, pp. 90–119.
15. ANSI, "Methods for the Calculation of the Articulation Index," ANSI S3.5-1969 (R1976), American National Standards Institute, New York, NY, 1976.
16. J. C. Webster, "Communicating in Noise in 1978–1983," in E. Rossi (Ed), *Proceedings of the 4th International Congress on Noise as a Public Health Problem*, Torino, Italy, Volume I, 1983, pp. 411–424.
17. L. L. Beranek, "Revised Criteria for Noise Control in Buildings," *Noise Control*, Vol. 3, 1957, pp. 19–27.
18. L. L. Beranek, W. E. Blazier, and J. J. Figiver, *J. Acoust. Soc. Amer.*, Vol. 50, 1971, pp. 1223–1228.
19. L. L. Beranek, "Balanced Noise Criterion (NCB) Curves," *J. Acoust. Soc. Amer.*, Vol. 86, pp. 650–664.
20. L. L. Beranek, "Applications of NCB Noise Criterion Curves," *Noise Control Engineering Journal*, Vol. 33, 1989, pp. 45–56.
21. W. E. Blazier, "Revised Noise Criteria for Application in the Acoustical Design and Rating of HVAC Systems," *Noise Control Engineering*, Vol. 16, 1981, pp. 64–73.
22. C. W. Kosten and G. J. van Os, "Community Reaction Criteria for External Noise, in the Control of Noise," NPL Symposium No. 12, HMSO, London, 1962, pp. 373–382.
23. EPA, "Community Noise," Report No. NTID 330.3, U.S. Environmental Protection Agency, Washington, DC, 1971.
24. K. M. Eldred, "Assessment of Community Noise," *Noise Control Engineering*, Vol. 3, No. 2, 1974, pp. 88–95.
25. D. Gottlob, "Regulations for Community Noise," *Noise/News International*, Vol. 3, No. 4, 1995, pp. 223–236.
26. EPA, "Public Health and Welfare Criteria for Noise," Report No. 550/9-73-002, U.S. Environmental Protection Agency, Washington, DC, 1973.
27. EPA, "Information on Levels of Environmental Noise Requisite to Protect Public Health and Welfare with an Adequate Margin of Safety," Report No. 550/9-74-004, U.S. Environmental Protection Agency, Washington, DC, 1974.
28. E. A. G. Shaw, "Noise Environments Outdoors and the Effects of Community Noise Exposure," *Noise Control Engineering Journal*, Vol. 44, N. 3, 1996, pp. 109–119.
29. "Guidelines for Considering Noise in Land Use Planning and Control," Federal Interagency Committee on Urban Noise, Document 1981-338-006/8071, U.S. Government Printing Office, Washington, DC, 1980.
30. W. E. Scholes and G. H. Vulkan, *Applied Acoustics*, Vol. 2, 1967, pp. 185–197.
31. T. J. Schultz, *Community Noise Ratings*, 2nd Ed., Applied Science, London, 1982.
32. W. A. Rosenblith, K. N. Stevens, and the staff of Bolt, Beranek, and Newman Inc., "Handbook of Acoustic Noise Control, Vol 2-Noise and Man," WADC TR-52-204, Wright Air Development Center, Wright Patterson Air Force Base, Ohio, 1953, pp. 181–200.
33. K. N. Stevens, W. A. Rosenblith, and R. H. Bolt, "A Community's Reaction to Noise: Can It Be Forecast?" *Noise Control*, Vol. 1, 1955, pp. 63–71.

34. W. J. Galloway and D. E. Bishop, "Noise Exposure Forecasts: Evolution, Evaluation, Extensions, and Land Use Interpretations," BBN Report No. 1862 for the FAA/DOT Office of Noise Abatement, Washington, DC, December 1969.
35. K. N. Stevens, A. C. Pientrasanta, and the staff of Bolt, Beranek, and Newman Inc., "Procedures for Estimating Noise Exposure and Resulting Community Reactions from Air Base Operations," WADC TN-57-10, Wright Air Development Center, Wright Patterson Air Force Base, Ohio, 1957.
36. Bolt, Beranek, and Newman Inc., "Land Use Planning Relating to Aircraft Noise," FAA Technical Report, October, 1964; also issued as Report No. AFM86-5, TM-5-365, NAVDOCKS P-38, Department of Defense, 1964.
37. D. E. Bishop and R. D. Horonjeff, "Procedures for Developing Noise Exposure Forecast Areas for Aircraft Flight Operations," FAA Report DS-67-10, Department of Transportation, Washington, DC, August 1967.
38. D. E. Bishop, "Community Noise Exposure Resulting from Aircraft Operations: Application Guide for Predictive Procedure," AMRL-TR-73-105, November 1974.
39. Committee on the Problem of Noise, "Noise–Final Report," HMSO, London, 1963.
40. "Second Survey of Aircraft Noise Annoyance around London Heathrow Airport," HMSO, London, 1971.
41. I. D. Griffiths and F. J. Langdon, "Subjective Response to Road Traffic Noise," *J. Sound Vib.*, Vol. 8, No. 1, 1968, pp. 16–33.
42. T. J. Schultz, "Some Sources of Error in Community Noise Measurement," *Sound and Vibration*, Vol. 6, No. 2, 1972, pp. 18–27.
43. Department of the Environment, "Motorway Noise and Dwellings," Digest 153, Building Research Establishment, Garston, Watford, England, May 1973.
44. Department of Transport, "Calculation of Road Traffic Noise," HMSO, London, 1988.
45. P. G. Abbott and P. M. Nelson, "The Revision of Calculation of Road Traffic Noise 1988," *Acoustics Bulletin* (Institute of Acoustics), Vol. 14, No. 1, 1989, pp. 4–9.
46. D. W. Robinson, "The Concept of Noise Pollution Level," NPL Aero Report AC 38, National Physical Laboratory, Aerodynamics Division, March 1969.
47. K. S. Pearsons, "The Effects of Duration and Background Level on Perceived Noisiness," Report FAA ADS-78, April 1966.
48. Bolt, Beranek, and Newman, Inc. "Establishment of Standards for Highway Noise Levels (Final Report)," Volume 5, Prepared for Transportation Research Board, National Co-operative Highway Research Program, National Academy of Sciences, NCHRP 3-7/3, November 1974.
49. ISO, "Acoustics—Description and Measurement of Environmental Noise—Part 1: Basic Quantities and Procedures, ISO 1996-11 : 1982; Part 2: Acquisition of Data Pertinent to Land Use, ISO 1996-22 : 1987; Part 3: Application to Noise Limits, ISO 1996-3 : 1987," Int. Std. Org., Geneva, Switzerland.
50. "Noise Standards for California Airports," California Administrative Code, Title 4, Subchapter 6.
51. J. Goldstein, "Descriptors of Auditory Magnitude and Methods of Rating Community Noise," in R. J. Peppin and C. W. Rodman (Eds.), *Community Noise*, ASTM, Philadelphia, 1979, pp. 38–72.
52. J. S. Lucas, "Noise and Sleep: A Literature Review and a Proposed Criterion for Assessing Effect," *J. Acoust. Soc. Am.*, Vol. 58, No. 6, 1975, pp. 1232–1242.
53. B. Griefahn, "Research on Noise-disturbed Sleep since 1973," *Proceedings of the Third International Congress on Noise as a Public Health Problem*, ASHA Report No. 10, 1980.
54. C. J. Jones and J. B. Ollerhead, "Aircraft Noise and Sleep Disturbance: A Field Study," in R. Lawrence (Ed.), *Proceedings of EURO-NOISE '92*, London, 1992, pp. 119–127.
55. J. B. Ollerhead et al., "Report of a Field Study of Aircraft Noise and Sleep Disturbance," Civil Aviation Authority, London, England, December 1992.
56. K. S. Pearsons, D. S. Barber, B. G. Tabachnik, and S. Fidell, "Predicting Noise-induced Sleep Disturbance," *J. Acoust. Soc. Am.*, Vol. 97, 1995, pp. 331–338.
57. K. S. Pearsons, D. S. Barber, and B. G. Tabachnik, "Analyses of the Predictability of Noise-induced Sleep Disturbance," Technical Report HSD-TR-89-029, Human Science Division (HSDY/YAH U.S. Air Force Systems Command), Brooks Air Force Base, TX, October 1989.
58. L. S. Finegold, C. S. Harris, and H. E. von Gierke, "Community Annoyance and Sleep Disturbance: Updated Criteria for Assessment of the Impacts of General Transportation Noise on People," *Noise Control Engineering Journal*, Vol. 42, No. 1, 1994, pp. 25–30.
59. T. J. Schultz, "Synthesis of Social Surveys on Noise Annoyance," *J. Acoust. Soc. Am.*, Vol. 64, 1978, pp. 377–405.
60. S. Fidell, D. S. Barber and T. J. Schultz, "Updating a Dosage Effect Relationship for the Prevalence of Annoyance Due to General Transportation Noise," *J. Acoust. Soc. Am.*, Vol. 89, 1991, pp. 221–233.
61. H. Miedema, "Response Functions For Environmental Noise in Residential Areas," Publication No. 92.021, Nederlands Instituut voor Praeventieve Gezondheidszorg TNO, Leiden, The Netherlands, June 1993.
62. H. E. Von Gierke and K. McK. Eldred, "Effects of Noise on People," *Noise/News International*, Vol. 1, No. 2, 1993, pp. 67–89.
63. W. Passchier-Vermeer, "Noise and Health," Publication No. A93/02, Health Council of Netherlands, The Hague, 1993.
64. B. Berglund and T. Lindvall (Eds.), "Community Noise," Document prepared for the World Health Organization, Center for Sensory Research, Stockholm, Sweden, 1995.

65. K. D. Kryter, "Community Annoyance from Aircraft and Ground Vehicle Noise," *J. Acoust. Soc. Am.*, Vol. 72, No. 4, 1982, pp. 1222–1242.

66. T. J. Schultz, "Comments on K. D. Kryter's paper, 'Community Annoyance from Aircraft and Ground Vehicle Noise,'" *J. Acoust. Soc. Am.*, Vol. 72, No. 4, 1982, pp. 1243–1252.

67. K. D. Kryter, "Rebuttal by Karl D. Kryter to Comments by T. J. Schultz," *Journal of the Acoustical Society of America*, Vol. 72, No. 4, 1982, pp. 1253–1257.

68. U. Möhler, "Community Response to Railway Noise: A Review of Social Surveys," *J. Sound Vib.*, Vol. 120, 1988, pp. 321–332.

69. J. Lang, "Schallimmission von Schienenverkerhstrecken. Forschungsarbeiten aus dem Verkehrswesen," Band 23, Vienna, Austria, 1989.

70. British Government Planning Policy Guidance PPG, "Planning and Noise," London, England, September 1994.

71. Private Communication with Rupert Thornley-Taylor, April 29, 1996.

65

HEARING PROTECTION DEVICES

ELLIOTT H. BERGER AND JOHN G. CASALI

1 INTRODUCTION

A hearing protection device (HPD) is a personal safety product that is worn to reduce the harmful auditory and/or annoying subjective effects of sound. Hearing protectors are often a method of last resort, when other means such as engineering controls or removal of the person from the noisy environment are not practical or economical.

To a large extent, research and development in hearing protection began during World War II in response to the tremendous hearing loss caused by military weapons. Military and industrial hearing conservation programs and the use of hearing protection followed in the early 1950s, with use proliferating in the early 1970s. Interest was spurred again in the 1980s due to federal regulations mandating use of hearing protection in occupational settings.[1]

This chapter, which begins by describing HPDs and how to dispense and use them, continues with the more technical issues of hearing protector physics, measurement, and performance assessment. It also includes guidance on how to estimate protection and the effects of HPDs on auditory perception and concludes with a discussion of special types of HPDs and relevant standards and regulations.

2 TYPES OF HEARING PROTECTION DEVICES

Hearing protection devices may be broadly categorized into earplugs, which are placed into or at the entrance of the earcanal to form a seal and block sound, earmuffs, which fit over and around the ears (circumaural) to provide an acoustic seal against the head, and helmets, which normally encase the entire head. Although in certain cases acoustical concerns may dictate the selection of a particular type of HPD, normally ergonomic considerations, personal preference, and/or compatibility with other safety gear and job requirements are the deciding factors.

2.1 Earplugs

Earplugs are made from materials such as vinyl, silicone, elastomer formulations, slow-recovery closed-cell foam, spun fiberglass, and cotton–wax combinations. They are available in a wide variety of types and styles, some of which are shown in Fig. 1.

Premolded earplugs are formed from flexible materials into conical, bulbous, or other shapes, often including flanges or sealing rings. They usually are available in a range of sizes to fit most ears and are pushed into place in the earcanal, whereupon a seal is made against the canal walls. By contrast, *formable earplugs*, which are made from materials such as slow-recovery foam, fiberglass, and silicone putty, either are formed prior to insertion into the earcanal (e.g., foam plugs are rolled into thin cylinders and reexpand in the canal after insertion) or are pressed into the canal and forced to (de)form to seal the canal at its entrance. *Custom-molded earplugs* are manufactured from an individual impression of the earcanal, which must precisely match the shape of the canal to create a seal and block the noise.

A final type of earplug is the *semi-insert* (also called semiaural, canal caps, or concha seated), which consists of soft pods that are held in place against and/or slightly inside the rim of the earcanal by a lightweight band. They are easily hung around the neck when not in use.

Note: References to chapters appearing only in the *Encyclopedia* are preceded by *Enc.*

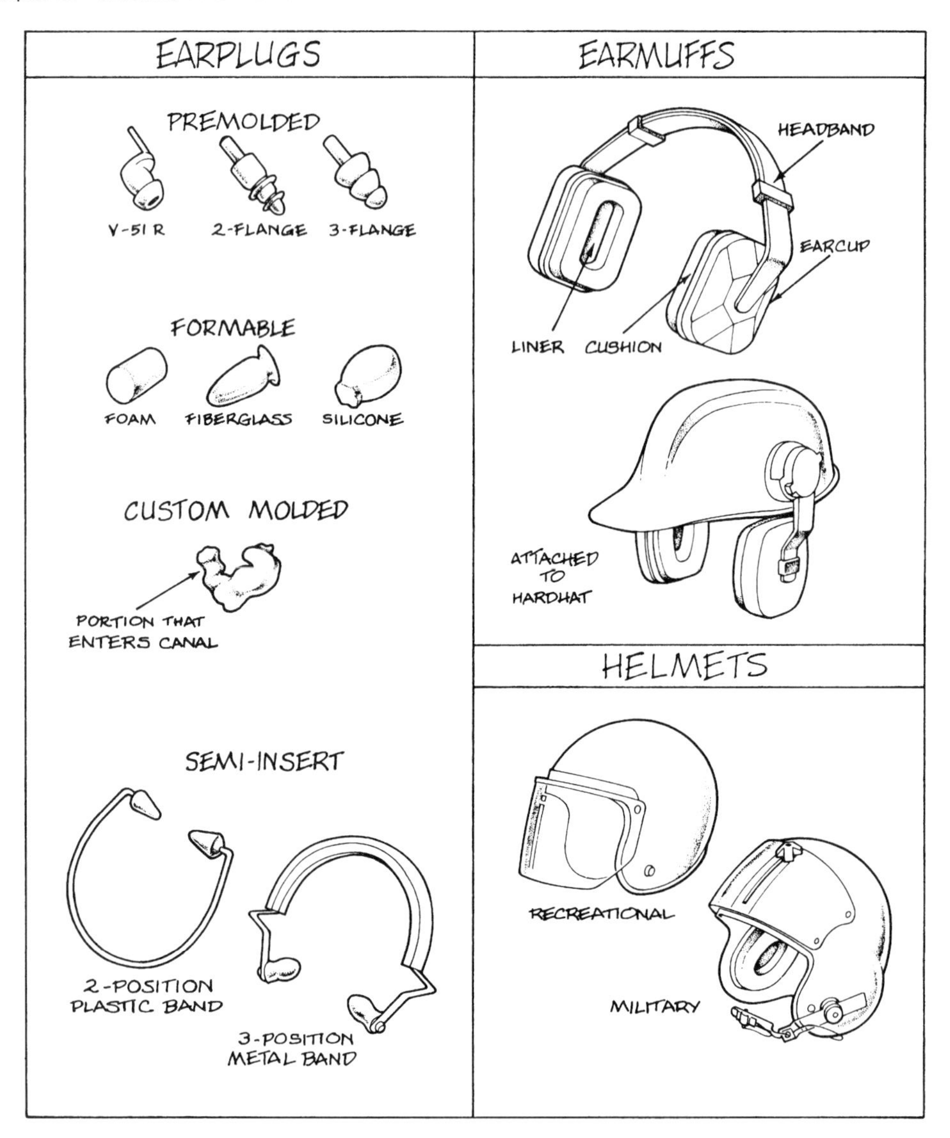

Fig. 1 Personal hearing protection devices. From C. W. Nixon and E. H. Berger, "Hearing Protection Devices," in C. M. Harris (Ed.), *Handbook of Acoustical Measurements and Noise Control*, McGraw-Hill, New York, 1991, pp. 21.1–21.24.

2.2 Earmuffs

Earmuffs normally consist of rigid molded plastic earcups that completely enclose and seal around the outer ear (also called the pinna) using foam- or fluid-filled cushions. The earcups are held in place by an adjustable headband or by short spring-loaded arms attached to a hard hat or other headgear. Headbands, which may be of plastic or metal construction, may function in only a single position or be "universal," suitable for use over the head, behind the neck, or under the chin. The earcups are lined with acoustical material, typically foam, to absorb high-frequency (>2 kHz) energy within the cup.

2.3 Helmets

Helmets enclose a substantial portion of the head and are usually designed primarily for impact protection. When they contain circumaural earcups or a dense liner to seal around the ears, they can also provide beneficial amounts of hearing protection.

3 DISPENSING, FITTING, USE, AND CARE

3.1 Initial Dispensing or Purchase

Initial dispensing of HPDs within an occupational hearing conservation program, individually or in small groups, can have a substantial positive effect on the efficacy of the overall hearing conservation effort.[2]

With the aid of a focused light source (such as an otoscope, earlight, or penlight) a visual inspection of the external ear and head should be made to identify medical or anatomical conditions that might interfere with or be aggravated by the hearing protector (e.g., inflammation, tenderness, excessive or impacted cerumen). When such conditions are present, medical consultation and/or corrective treatment should be obtained.[3]

When consumers purchase hearing protectors for occasional nonoccupational use, it is usually neither feasible nor necessary to have an ear examination prior to purchase or use. However, if ear pain or other aural symptoms develop or discomfort persists, use should be discontinued until audiological or medical advice is obtained. Other brands or types of HPDs may need to be evaluated until the one best suited to the individual is found.

3.2 Fitting Tips

Earplugs generally require more skill and attention than earmuffs during the initial issue, but a common mistake is to presume that earmuffs are foolproof and to dispense them indiscriminately, without providing assistance.

When possible, wearers should be exposed to noise during the fitting procedure so they can hear the noise diminish to the lowest perceived level while adjusting the HPD. A portable cassette player can be used to present recordings of broadband noise or representative industrial sounds.

For all types of earplugs, insertion is easier when the pinna is pulled *outward and upward* with the hand opposite the ear being fitted. Earplugs should be inserted into the right ear using the right hand and into the left ear using the left hand. This allows the hand inserting the earplug to have the best line of approach to the earcanal.

A properly inserted premolded earplug will generally create a blocked-up feeling due to the requisite airtight seal. When the seal is present, suction should be felt if an attempt is made to withdraw the plug from the canal.

Foam earplugs are prepared for insertion by rolling them into a *very thin* crease-free cylinder. This is accomplished by squeezing lightly as one begins rolling and then applying progressively greater pressure as the plug becomes more tightly compressed. Unlike other types of earplugs, foam earplugs should not be readjusted while in the ear. If the initial fit is unacceptable, they should be removed, rerolled, and reinserted.

3.3 Occlusion Effect and Its Use as a Test for Fit

The occlusion effect is the increase in efficiency with which bone-conducted sound is transmitted to the ear at frequencies below 2000 Hz when the earcanal is occluded and sealed. This effect causes wearers of HPDs to perceive a change in the loudness and quality of their own voices and other body-generated sounds/vibrations such as those caused by chewing, breathing, and walking.

The effect is easily demonstrated by plugging one's ears with the fingers while reading aloud. The voice will appear to have added fullness, hollowness, or resonant bassiness. The occlusion effect is often cited as an objectionable characteristic of wearing hearing protection. Its magnitude varies with the way in which the ear is occluded, and may be reduced by adjusting earplugs toward a deeper insertion or selecting earmuffs with larger volume earcups. The maximum effect occurs for semi-inserts. Listening for the presence (*not* the magnitude) of the occlusion effect is a useful technique for fitting nearly all types of HPDs, since it indicates the presence of an acoustical seal and proper fit.[4]

3.4 Comfort

Comfort is a critical feature of a hearing protection device, usually equivalent in importance to the attenuation that the device can provide. Although comfort assessments are, of necessity, subjective in nature, research in recent years has been directed at quantifying the important parameters.[5]

An HPD that is uncomfortable will not be worn consistently and correctly or, worse, will not be worn at all. Comfort can be improved by properly matching the device to the wearer and by providing instruction in its proper use. Common sources of discomfort include improperly sized or inserted earplugs, overly tight earmuff headbands, worn-out devices with hardened/cracked cushions or stiffened/ragged flanges, and the use of earmuffs in hot environments.

3.5 Hygiene, Maintenance, and Safety

Hearing protection devices should be cleaned regularly according to manufacturers' instructions. Normally, warm water and mild soap are satisfactory cleansing

agents. Earplugs should be washed and dried thoroughly before reuse or storage. Earmuff cushions should be periodically wiped or washed clean. When they cannot be adequately cleaned or no longer retain their original appearance or resiliency, earmuff cushions and earplugs should be replaced.

Although the likelihood of hearing protection devices increasing the incidence of outer ear infections is minimal, such ear infections may occur.[3] Earplugs or earmuffs that may be implicated are commonly found to be contaminated with caustic or irritating substances or embedded with abrasive materials.

The headbands on earmuffs and semi-insert earplugs may lose their force with time or be purposely sprung to reduce force and increase comfort. This will degrade attenuation. Regular inspection of HPDs and proper instructions to the users are necessary to overcome this problem.

Commercially available HPDs are safe to wear as long as the manufacturers' recommendations are followed; however, some precautions are advisable. For example, earplugs that create an airtight seal, such as premolded inserts, should be removed with a slow twisting motion to gradually break the seal as they are withdrawn, to assure that removal does not cause pain or harm to the ear. When selecting earmuffs, smaller, lighter weight (lower profile), dielectric versions will normally be less likely to interfere with workers' activities or create hazards.

4 SOUND PATHWAYS TO THE OCCLUDED EAR

In the *un*occluded ear, the dominant sound path for external sounds is along the earcanal to the eardrum. However, in the occluded ear four distinct and important paths by which sound can penetrate the HPD can be identified, as described below.[4]

4.1 Air Leaks

For maximum protection, HPDs must make virtually an airtight seal with the walls of the earcanal or the circumaural regions surrounding the pinna. Air leakage paths can reduce attenuation by 5–15 dB over a broad range of frequencies depending upon the size of the leak. Typically the loss is most noticeable at low frequencies.

4.2 Hearing Protector Vibration

Due to the flexibility of the earcanal flesh, earplugs can vibrate in a pistonlike manner, thus limiting their low-frequency attenuation. Earmuffs, too, vibrate as a mass spring system, the stiffness of the spring depending upon the dynamic characteristics of the earmuff cushion and the circumaural flesh as well as the volume of the air entrapped inside the earcup. These actions limit low-frequency attenuation. Representative maximum values at 125 Hz for earmuffs, premolded earplugs, and foam earplugs are 25, 30, and 40 dB, respectively.

4.3 Transmission through Materials

In much the same way as sound will propagate through any barrier, so too will it pass through an HPD, albeit with diminished intensity. The transmission depends upon the mass, stiffness, and internal damping of the HPD materials. The limitations caused by this transmission path are most pronounced above 1 kHz.

4.4 Bone and Tissue Conduction

Even if an HPD were completely effective in blocking the preceding pathways, the sound would still reach the inner ear by bone and tissue conduction, thus imposing a limit on the maximum real-ear attenuation that a device can provide. However, since that limit is 40–50 dB below the sound level reaching the ear through the open earcanal, it does not become important unless attenuation of the HPD itself approaches such values (see Fig. 2). This is not often the case unless dual HPDs are worn (see Section 6.3).

5 MEASURING ATTENUATION

More than a dozen different methods of measuring HPD attenuation have been described in the literature, but only a few have been found practical and reliable and are commonly utilized.[6]

Historically, standards committees have been primarily concerned with the precision of HPD attenuation measurements, that is, the repeatability of the data, both within and between laboratories. Although this is certainly an important issue, experience has shown that undue emphasis on precision, with a consequent lack of regard for accuracy or "realism," may lead to data of questionable utility; by definition, "useful" data predict with reasonable accuracy the protection that can typically, or even optimally, be expected to be achieved in practice.[7]

5.1 Laboratory Procedures

Hearing protector research is usually conducted under laboratory conditions. This allows the experimenter to exert the most control over the relevant parameters of

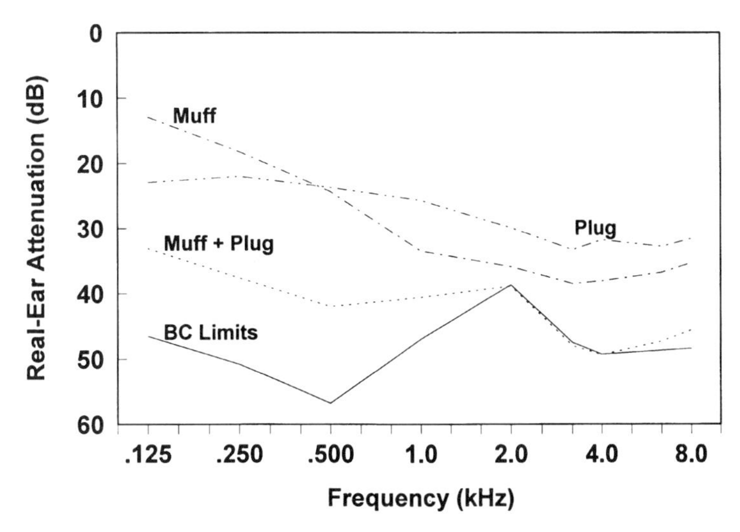

Fig. 2 Bone conduction (BC) limits to hearing protector attenuation and an example of attenuation provided by an earplug, an earmuff, and the two devices worn together.

which he or she is aware and provides the best assurance of generating repeatable test data. However, even the most accurate of laboratory methods will only provide data representative of field (real-world) protection if (a) the HPD is fit and worn by the subjects in the same manner as by wearers in practice and (b) the test population is representative of the actual users.

Real-Ear Attenuation at Threshold The most common and one of the most accurate HPD attenuation tests is the measurement of real-ear attenuation at threshold (REAT).[8,9] Virtually all available manufacturers' reported data have been derived via this method, and it is the procedure required by the Environmental Protection Agency (EPA) to obtain data for computation of the Noise Reduction Rating (NRR).

The REAT is a measurement of the difference between the minimum level of sound that a subject can hear without wearing an HPD (open threshold) and the level needed when the HPD is worn (occluded threshold). The difference is a measure of the real-ear attenuation provided by the device. This method accurately represents the performance of all conventional HPDs (those that provide constant attenuation regardless of sound level, i.e., level-*in*dependent HPDs). For those HPDs whose attenuation increases with level (i.e., amplitude-sensitive HPDs), REAT usually provides a lower bound estimate.

Microphone in Real Ear The microphone-in-real-ear (MIRE) method utilizes physical (acoustical) measurements to determine the difference, termed insertion loss, between the sound pressure levels in the human ear with and without the HPD in place. The levels are measured via a miniature or probe microphone at the entrance to or sometimes in the earcanal itself.[10] Alternatively, a noise reduction measurement utilizes two microphones to simultaneously measure the levels inside and outside the HPD. The subject who wears the HPD is required to sit still and behave as an inanimate acoustic test fixture.

This method can be used to measure attenuation for all types of circumaural HPDs including those fitted with electronics; difficulties can arise with earplugs because of the problems of inserting both a microphone and an earplug simultaneously into the earcanal. The MIRE method is much less time consuming than REAT testing as well as an excellent alternative, especially when measurements are required at high sound levels such as for investigation of the response of HPDs to weapons fire and explosions.

Acoustical Test Fixture The acoustical test fixture (ATF) method is directly analogous to the MIRE method, except that an inanimate fixture is utilized.[10] The ATF incorporates a microphone at the approximate eardrum location and, depending upon its sophistication, may match the impedance of the eardrum (via a suitable acoustic coupler) and also include simulations of the circumaural and earcanal tissues.

The ATF measurements provide a quick assessment for product development, quality control, and acceptance testing of earmuffs and helmets. In general, ATF measurements of earplugs are still ill defined, and for all devices the data are generally not suitable for direct prediction of real-ear attenuation values.

5.2 Field Procedures

Generally, the most accurate laboratory procedures, modified to reduce cost and improve portability and ruggedness, are those utilized under field conditions. The purpose is to derive more accurate estimates of the attenuation provided by HPDs in use. The procedures that have been most successfully utilized are sound-field REAT in a manner similar to that described in the existing standards,[8,9] REAT conducted using large circumaural cups for the stimulus presentation instead of loudspeakers in the test room, and the MIRE approach with microphones simultaneously mounted on the inside and outside of earmuffs.[6]

When the REAT procedures are implemented, workers are usually selected without warning from their place of work and accompanied to a test site, with care taken to assure they do not readjust their HPDs prior to testing. For the MIRE procedure, the instrumented earmuffs are utilized to simultaneously measure the protected and unprotected noise doses received by the exposed workers.

6 ATTENUATION CHARACTERISTICS OF HEARING PROTECTION DEVICES

Especially for insert-type devices, differences among wearer or test subject fitting procedures lead to a wide variation in reported attenuation,[11] even in the case of purportedly "optimum-fit" data measured in the laboratory.[4] Comparison of published NRRs and rank ordering of HPDs is strongly influenced by the particular data available for comparison purposes. Sources of interlaboratory variability are (1) differences in interpretation and implementation of the standardized measurement methodology, (2) uncertainty of obtaining the proper fit to avoid acoustic leaks, (3) differences in subject selection and training, and (4) differences in data reduction techniques. The repeatability of attenuation measurements in the same laboratory is also subject to variation, but to a lesser extent than between laboratories.

Differences in the labeled NRRs of competing products of less than 3 dB have *no practical importance*, as a consequence of the inter- and intralaboratory variability discussed above, and even 4–5-dB differences are of questionable significance unless data from closely controlled tests are being compared.

6.1 Optimum Laboratory Performance

The values reported in this section are representative of those obtained in laboratory-based REAT measurements, wherein the HPDs are fitted either by or under the close supervision of the experimenter.

Earplugs Attenuation values are similar for various types of premolded earplugs, clustering closely around 25 dB at frequencies up to 1000 Hz, and increasing to about 40 dB at the higher frequencies. Formable cotton–wax or silicone earplugs provide a wider range of attenuation across brands or types, with values as low as 10 dB for devices with the least attenuation and ranging from 30 to 45 dB above 2000 Hz.

Formable slow-recovery foam earplugs provide among the highest overall protection of any single device. Attenuation can range from 30 to 45 dB at and above 2000 Hz and, depending upon the depth of insertion, from about 20 to 40 dB below that frequency.

Attenuation of custom-molded earplugs varies widely due to impression-taking and manufacturing procedures as well as differences in materials. On average, the attenuation of these devices tends to be less than found for premolded earplugs.

Generally, earplugs of the premolded and foam varieties provide better attenuation than earmuffs below 500 Hz and equivalent or greater protection above 2000 Hz. At the intermediate frequencies of 500 and 1000 Hz, earmuffs usually provide the highest attenuation.

Semi-insert devices display a wide range of performance at all frequencies. The best devices provide an average attenuation of about 20 dB at frequencies below and 35 dB at frequencies above 2000 Hz.

Earmuffs The attenuation of earmuffs is substantially controlled by earcup volume and mass, the area of the cup opening, headband force, and the size, shape, and material of the cushion. Attenuation usually increases above 9 dB/octave from 125 to 1000 Hz and, for most earmuffs, at 2000 Hz, approaches the limit imposed by bone conduction of approximately 40 dB. Above that frequency, attenuation averages around 35 dB.

When eyeglasses are worn with earmuffs, the temples should fit close to the sides of the head and be as thin as practical to reduce their effect on the ability of the cushions to seal around the ears. The loss in attenuation that eyeglasses create, with cushions in good condition, is normally 3 to 7 dB, but the effect varies widely among earmuffs and eyeglasses.

Helmets Unless helmets seal well around the head and/or contain internal circumaural cups, the attenuation they provide is very low, especially at the frequencies below 2000 Hz. For example, typical motorcycle helmets provide less than 10 dB of protection below 1000 Hz and only from 10 to 30 dB at the higher frequencies. Some military helmets (e.g., flight helmets), which possess a more effective acoustical design, yield attenuation that is comparable to conventional earmuffs, but still about 5–10 dB less at the frequencies below 2000 Hz. From 4 to 8 kHz, military helmets can surpass earmuff attenuation by about 5 dB if they fully enclose the head, since this decreases bone conduction transmission at high frequencies.

6.2 Laboratory versus Real-World Attenuation

Real-world HPD attenuation is typically much less than found in the laboratory wherein measurements usually reflect the maximum performance that can be expected with a device that is worn under ideal circumstances. Such conditions rarely if ever are duplicated in the workplace. This is partly because of the emphasis in most laboratory testing on obtaining maximum attenuation, disregard of comfort during fitting of protectors (primarily earplugs), and the short time periods that the test subjects wear the devices (5–10 min). But the disparity is also due to the fact that workplace performance usually falls short of what could reasonably be expected to be achieved, because of various deficiencies in hearing conservation programs.

Numerous studies have been conducted in the field to assess real-world performance. The data in the next section are drawn from 22 such studies spanning seven countries, greater than 90 different industrial sites, and approximately 2900 worker/subjects.[12] The results provide an indication of the performance of HPDs, circa 1980–1990, that can be anticipated in the better industrial and military hearing conservation programs.

Comparative Octave Band Data As shown in Fig. 3, attenuation measured in the workplace with actual workers as subjects shows lower mean attenuation values (most notably for earplugs) and higher standard deviations (by a factor of 2–3) than do manufacturers' reported laboratory data. Earmuff performance in the workplace is better relative to earplugs than would be predicted from laboratory data,[13] but regardless of the HPD that is selected, individual noise exposure can be substantially underestimated when it is computed from laboratory attenuation values.

Comparative NRRs Whether attenuation is estimated using laboratory octave band data or via a single-number rating such as the NRR that is computed therefrom (see Section 7.2), the discrepancies with respect to field data are the same.[12] The field NRRs attained by 84% of wearers are less than or equal to 7 dB for all types of earplugs, except the foam variety, which yields about 13 dB, a value roughly comparable to that of most earmuffs.

Attenuation versus Use Time The effective protection of a device (based upon a user's total accumulated noise dose) is substantially reduced when it is removed by an employee for even short periods of time during the workday, as illustrated in Fig. 4. An HPD with a nominal NRR of 25 dB that is removed by the user

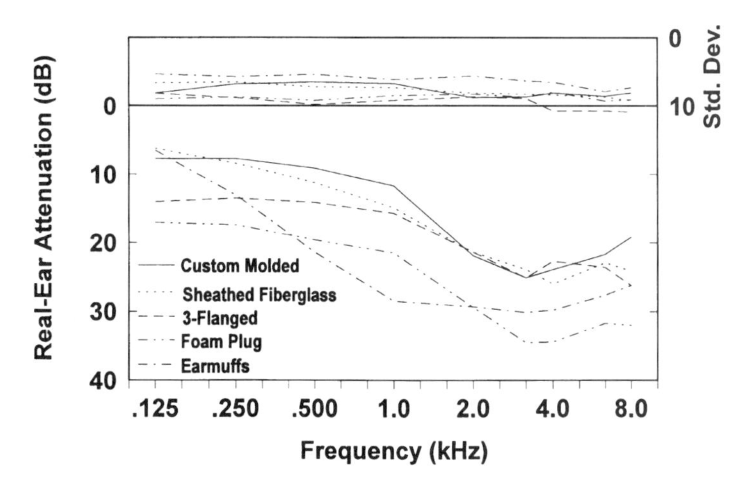

Fig. 3 Summary of real-world data for hearing protectors separated into four earplug and one earmuff categories (from Ref. 12.)

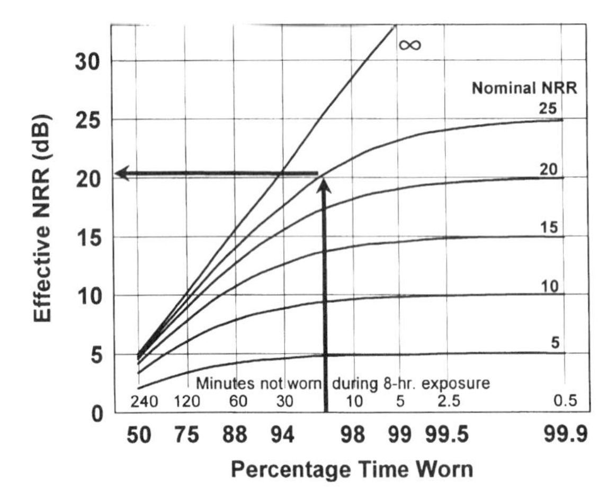

Fig. 4 Corrections to nominal Noise Reduction Ratings (NRRs) as a function of the time during a work day that a hearing protection device is not worn based on a 5-dB trading relationship.

for 15 min during an 8-h workday provides an effective NRR of only 20 dB, for a loss of 5 dB. Greater reductions in effectiveness are observed for higher NRRs.

6.3 Dual Protection

Dual protection, such as earplugs worn in combination with earmuffs, helmets, or communications headsets, typically provides greater attenuation than either device alone.[14] However, the attenuation of the combination is *not* equal to the sum of the separate attenuations, as illustrated in Fig. 2. At individual frequencies the incremental gain in performance varies from approximately 0 to 15 dB over the better single device, but at 2000 Hz the gain is limited to only a few decibels. Attenuation changes very little when different earmuffs are used with the same earplug, but for a given earmuff the choice of earplug is critical for attenuation at frequencies below 2000 Hz. At and above 2000 Hz, essentially all dual-protection combinations provide attenuation approximately equal to the limitations imposed by the bone conduction pathways, approximately 40–50 dB, depending upon frequency.

6.4 Infrasound/Ultrasound

At infrasonic frequencies (i.e., below about 20 Hz), well-fitted earplugs provide attenuation that is about equal to that in the 125-Hz one-third-octave band. However, earmuffs provide very little protection from infrasound and may even cause it to be amplified.

At ultrasonic frequencies (i.e., above about 20,000 Hz), conventional earplugs and earmuffs generally provide adequate protection, with attenuation exceeding 30 dB for frequencies from about 10,000 to 30,000 Hz.

6.5 Weapons Noise

Higher attenuation earmuffs and well-seated earplugs usually provide adequate attenuation of small-arms fire (impulses composed mainly of middle- and high-frequency energy from pistols and standard rifles), reducing the peak sound pressure levels to less than 140 dB. Earmuff attenuation decreases as the peak energy in the impulse moves to lower frequencies, as is the case for larger weapons. The attenuation of peak sound levels provided by a typical earmuff is about 30 dB for pistol fire, about 18 dB for rifle fire, and as little as 5 dB for cannons. Foam earplugs provide attenuation similar to that of earmuffs against pistol fire but are substantially more effective than earmuffs against cannons and bazookas, providing protection greater than 15 dB.[15]

The use of dual HPDs can afford extra protection, especially for low-frequency impulses that have high peak values. Attenuation values of 20–25 dB can be achieved.

7 USING ATTENUATION DATA TO ESTIMATE PROTECTION

The accuracy with which the laboratory or real-world data can predict the protection for an individual, or group of users, is critically dependent upon the correspondence between the test and actual usage conditions. The accuracy can be very poor, as is clearly shown when optimum laboratory data are used to estimate protection in a typical hearing conservation program (see Section 6.2). Undue emphasis on protection estimates is unwarranted.

7.1 Octave Band Method

Potentially the most accurate computational procedure for applying test data to estimate protected exposures is the octave band (OB) method, as illustrated in Table 1. At each frequency, the HPD's mean attenuation less a standard deviation (SD) correction is subtracted from the measured A-weighted OB sound pressure levels. Two SDs are normally used in the United States; one SD is used in Europe and Australia. The protected levels are then logarithmically summed to determine the A-weighted sound level under the HPD. This computation requires that the user have OB noise data available and that the computations be made individually for each HPD–noise spectrum combination.

A critical conceptual error that is often made is to pre-

TABLE 1 Octave Band Calculation of HPD Noise Reduction for a Known Noise Environment[a]

Steps in Procedure	125 Hz	250 Hz	500 Hz	1000 Hz	2000 Hz	4000 Hz	8000 Hz	dB (A)[b]
1. Measured sound pressure levels	85.0	87.0	90.0	90.0	85.0	82.0	80.0	
2. A-weighting correction	−16.1	−8.6	−3.2	0.0	+1.2	+1.0	+1.1	
3. A-weighted sound levels (step 1 + step 2)	68.9	78.4	86.8	90.0	86.2	83.0	78.9	93.5
4. Typical premolded earplug attenuation	27.4	26.6	27.5	27.0	32.0	46.0[c]	44.2[d]	
5. Standard deviation × 2	7.8	8.4	9.4	6.8	8.8	7.3[c]	12.8[d]	
6. Estimated protected A-weighted sound levels (step 3 − step 4 + step 5)	49.3	60.2	68.7	69.8	63.8	44.3	47.5	73.0

[a]The estimated protection for 98% of the users in the noise environment assuming they wear the device in the same manner as did the test subjects and assuming they are accurately represented by the test subjects is 93.5 − 73.0 = 20.5 dB (A).
[b]Logarithmic sum of seven octave band levels in the row.
[c]Arithmetic average of 3150- and 4000-Hz data.
[d]Arithmetic average of 6300- and 8000-Hz data.

sume that the SD correction adjusts laboratory data to estimate real-world values. The actual purpose of the SD is to adjust the mean test data to reflect the attenuation to be expected in 84% (for a one-SD correction) or 98% (for a two-SD correction) of the test subjects.

7.2 Noise Reduction Rating

A single-number descriptor is convenient and often sufficiently accurate to estimate protected exposures. One such descriptor, precalculated by manufacturers in the United States and provided on their packaging, is the Noise Reduction Rating (NRR).[16] The NRR is an attenuation index that represents the overall average A-weighted noise reduction, in decibels, that an HPD will provide in an environment with a known C-weighted sound level. A similar descriptor has been standardized as the single-number rating (SNR) by the International Organization for Standardization (ISO).[17]

The NRR is calculated in a manner analogous to the OB approach, except that a pink noise spectrum (equal energy in each OB) is used instead of the actual spectrum. The computation includes a two-SD adjustment for percentage of population protected, as with the OB method, and an additional spectral safety factor of 3 dB.[4] The NRR computed for the HPD example in Table 1 is 20.7 dB.

The NRR is used to estimate wearer noise exposures by subtracting it from the C-weighted sound levels:

$$\text{Estimated exposure[dB(A)]} = \text{workplace noise level[dB(C)]} - \text{NRR}. \quad (1)$$

The practice of subtracting the NRR from the C-weighted sound level may seem illogical; however, it is justified on theoretical and empirical grounds. Considerable accuracy is lost when the NRR is subtracted from A-weighted sound levels, in which case an additional safety factor of 7 dB must be included, as shown in Eq. (2). An alternative to the 7-dB correction has been provided by Berger[4]:

$$\text{Estimated exposure[dB(A)]} = \text{workplace noise level[dB(A)]} - (\text{NRR} - 7\text{dB}). \quad (2)$$

7.3 Comparison of Octave Band and NRR Estimates

Although errors may arise in using the NRR to estimate results computed using the OB method, the NRR is usually of sufficient accuracy considering the *in*accuracies in the basic OB data from which either set of computations are made. For the example shown in Table 1, with exterior C- and A-weighted sound levels of 95.2 and 93.5 dB, respectively, the OB-computed A-weighted protected exposure level is 73.0 dB, as compared to the following NRR-computed values:

$$95.2 \text{ dB(C)} - 20.7 = 74.5 \text{ dB(A)}, \quad (3)$$

or

$$93.5 \text{ dB(A)} - (20.7 - 7) = 93.5 - 13.8 = 79.8 \text{ dB}. \quad (4)$$

Note the substantially larger discrepancy between the OB value and the estimate from Eq. (4) than that from Eq. (3). This is due to the conservative 7-dB adjustment required when only A-weighted sound levels are available.

The primary reason for examining the OB data is to better match the attenuation curve of the device to the noise spectrum. For example, both laboratory and workplace data show that earmuffs are a poor choice for noises with significant low-frequency energy (125–250 Hz), in which case a foam or premolded earplug is preferred, but are best when strong midfrequency energy is present (primarily around 1000 Hz).

7.4 Derating Laboratory Data

The NRR, or the laboratory data from which it is computed, must be reduced (derated) to provide attenuation values more realistic for the workplace. A derating of the NRR by 10 dB is the minimum average correction necessary to achieve this goal;[4] however, even larger deratings or a percentage (such as 50%) instead of constant-decibel approaches may be justified and may be more appropriate. Some authors have suggested that a percentage derating should be less for the easier to fit HPDs such as earmuffs than for the harder to fit devices such as earplugs.[13]

Whatever derating is selected, it must be applied equally to the OB calculations of attenuation; that is, if a percentage derating is used for NRRs, that same percentage should also be taken from the OB attenuation values before they are entered in row 4 of Table 1.

7.5 OSHA's Procedure for Derating Laboratory Data

The Occupational Safety and Health Administration (OSHA) specifies use of manufacturers' labeled data derived from laboratory measurements to assess adequacy of HPDs for the noise exposures in which they are worn.[1] Subsequent to publication of the Hearing Conservation Amendment, OSHA recognized that laboratory data required derating and recommended reducing published NRRs by 50% (e.g., an NRR of 24 dB would be reduced to 12 dB), but only for the purpose of evaluating the *relative efficacy* of HPDs and engineering noise controls.[18] The derated NRRs are not applicable for determining compliance with the hearing protection requirements of the Hearing Conservation Amendment.

8 EFFECTS OF HPDs ON AUDITORY PERCEPTION AND INTELLIGIBILITY

8.1 Speech Communications

Conventional hearing protectors cannot differentiate and selectively pass speech versus noise at a given frequency; thus the devices do not improve the speech-to-noise ratio, which is the most important factor in achieving reliable intelligibility. However, hearing protectors sometimes do afford intelligibility improvements in intense noise by lowering the total energy of both speech and noise incident on the ear. Overload distortion in the cochlea is thereby reduced, establishing more favorable conditions for discrimination.

Prediction of the effects of protectors on speech intelligibility in noise is a complex issue that depends on a host of factors, including the listener's hearing abilities, absolute speech and noise levels and speech-to-noise ratio, whether or not the talker is occluded, HPD attenuation, reverberation time of the environment, facial cues, and content/complexity of the message.

A distillation of the evidence concerning speech communication for *normal hearers* with HPDs suggests that conventional passive HPDs have little or no degrading effect on intelligibility in noise above about 80 dB(A) but cause considerable misunderstanding at lower levels, for which the use of protection is not typically needed anyway. At ambient noise levels greater than about 85 dB(A), several studies have reported slight intelligibility improvements with specific HPDs,[19] while others attempting to simulate on-the-job conditions have reported small intelligibility decrements, especially when the talker's ears are occluded.[20]

When wearing HPDs in noise, one tends to lower the voice level because the bone-conducted voice feedback inside the head is amplified by the presence of the protector, especially at low frequencies. In comparison, attenuation of the ambient noise occurs primarily through interruption of air conduction by the HPD. Thus, one's own voice is perceived as louder in relation to the noise than is actually true, causing a compensatory lowering of about 2–4 dB unless a conscious effort is made to "speak up."[4]

For individuals with high-frequency hearing losses, the effects of HPDs on communications are not clear-cut, although these persons are certainly at a disadvantage due to the fact that their already elevated thresholds for mid- to high-frequency speech sounds are further raised by the protector. As the amount of hearing loss increases, the chances of degrading communications with HPDs also increases. Though there is not consensus among studies, it does appear that hearing-impaired individuals often experience no improvement or even reduced communication abilities with HPDs in noise environments from 80 to 95 dB(A).

8.2 Auditory Warning Signals and Other Sounds

As with speech communication effects, the same HPD influences on signal-to-noise ratio and reduction of aural distortion apply to the detection and recognition of nonverbal signals such as warnings, target annunciators, and machinery sounds. Due to the high-frequency bias in attenuation of conventional HPDs, coupled with the typically elevated high-frequency thresholds of the neurally hearing-impaired, signals above about 2000 Hz are the most likely to be missed. However, warning signal parameters, such as frequency, intensity, and temporal profile, may be designed to help alleviate detection problems and increase perceived urgency. For instance, low-frequency (i.e., below 500 Hz) warning signals more readily diffract around barriers and are less attenuated by HPDs, thereby aiding detection.

Although aural cues may sound spectrally different under an HPD, the bulk of empirical studies indicate that signal detection will not be compromised by HPDs for normal-hearing individuals.[21] While the evidence is less extensive for hearing-impaired listeners, they can be expected to experience some detection and recognition difficulty, depending on the particular signal, ambient noise, and hearing protector worn. When an occluded/unoccluded detection disadvantage does occur, it is more common with earmuffs than earplugs, perhaps due to the muffs' stronger tendency to attenuate high-frequency signals and to pass more low-frequency noise, which effects an upward spread of masking.

8.3 Localization

Perceptual judgments of sound direction and distance may be influenced by HPDs, at least partially because some of the high-frequency binaural cues (above about 4000 Hz) that depend on the pinnae are lost or altered. Earmuffs and helmets, which completely obscure the pinnae, interfere most with localization in the vertical plane and also tend to cause contralateral (left-right) and ipsilateral (front-back) judgment errors. Earplugs generally induce fewer localization problems than muffs.

8.4 General Remarks

In most hazardous noise exposure situations, even though the quality of the auditory experience may change under HPDs, detection and intelligibility performance is not hampered for individuals of normal hearing. While hearing-impaired individuals may exhibit a reduced communication ability under HPDs in noise, use is not contraindicated; instead, it is essential that the remaining auditory sensitivity be preserved by proper protector use. All individuals may benefit from visually or tactually presented cues to augment auditory information, or in certain situations from the use of HPDs incorporating special communications features as described next.

9 HPDs WITH SPECIAL FEATURES

Special hearing protector designs have been developed to improve speech communication and auditory perception capabilities of the noise-exposed wearer. These devices may be categorized into passive (i.e., without electronics), active (i.e., incorporating electronics), and communications headsets.

9.1 Passive HPDs

Frequency-Sensitive HPDs Efforts to improve communications under earplugs have involved the use of apertures or channels through an earplug body, sometimes opening into an air-filled cavity encapsulated by the earplug walls. For example, a small (~0.5-mm) longitudinal channel through a custom-molded earplug causes an air leak that creates a low-pass filter characteristic. Attenuation is negligible below about 1000 Hz, increasing up to about 30 dB at 8000 Hz. Because most of the speech frequencies critical to intelligibility lie in the 1000–4000-Hz range, the potential communication benefits of the low-pass feature may be relatively small, especially in low-frequency noise, which causes upward masking into the critical speech band. Furthermore, the attenuation may be insufficient for many industrial noise environments.

Amplitude-Sensitive HPDs Because a conventional HPD provides a constant amount of attenuation that is independent of incident sound level, hearing ability is compromised during the quiet periods of intermittent sound exposures. Amplitude-sensitive HPDs address this problem by providing reduced attenuation at low sound levels, with increasing protection at high levels.

Passive HPDs utilize a nonlinear component, such as a sharp-edged orifice opening into a duct, to effect the change in attenuation. This technique, used with success in both earmuffs and earplugs, takes advantage of the fact that low-intensity sound waves predominantly exhibit laminar airflow and pass relatively unimpeded through the aperture, whereas high-intensity waves involve turbulent flow and are attenuated due to increasing acoustic resistance.[22]

A critical performance parameter is the transition sound level, normally 110–120 dB, above which insertion loss increases at a rate of about 1 dB for each 2–4-dB increase in sound level. Because this transition level is so high, passive amplitude-sensitive HPDs are best suited for outdoor impulsive blasts and gunfire exposures. At lower but still hazardous sound levels, most passive devices exhibit vented earplug behavior, affording very weak protection at frequencies less than 1000 Hz. One exception is an orifice-type earmuff that provides approximately 25 dB attenuation from 400 to 8000 Hz.[22]

Uniform Attenuation HPDs Because the attenuation of conventional HPDs increases with frequency, not only are sounds reduced in level, but the wearer's perception of spectral balance is distorted as well. This problem contributes to protector nonuse where, for instance, a machine operator's auditory feedback from a cutting tool is muffled or a musician's perception of timbre is compromised. To counter such problems, flat or uniform attenuation HPDs are designed to impose attenuation that is nearly constant from about 100 to 8000 Hz. One successful technique employs a Helmholtz resonator and a sound channel through a custom-molded earplug to provide a flat attenuation profile of about 15 dB.[23]

Properly fit uniform attenuation HPDs provide adequate protection and better hearing perception in low to moderate noise levels of about 90 dB(A) or less. Workers with high-frequency hearing losses and professional musicians may find them particularly beneficial.

9.2 Active HPDs

Amplitude-Sensitive Sound Transmission HPDs

Active sound transmission hearing protectors consist of modified conventional earmuffs or earplugs that incorporate microphone and limiting amplifier systems to transmit external sounds (often filtered to include the critical speech frequencies) to earphones mounted inside the earcups. Typically, the amplifier maintains a predetermined (in some cases user-adjustable) earphone level, often at about 85 dB(A), until the ambient noise is so intense that direct transmission through the earcup becomes the controlling factor. In comparison to both conventional and passive amplitude-sensitive earmuffs, sound transmission earmuffs are more expensive (upward of $100) but offer a viable alternative for use in intermittent noises, especially those with impulse-type (e.g., gunfire) or short-duration on-segments. However, in continuous, high-level noise, the distortion products of some systems may cause annoyance and compromise hearing ability.

Auditory perception and noise level under sound transmission earmuffs depends on numerous electronic design parameters. The microphones may be diotic, wherein a single microphone in one earcup feeds both earphones, or dichotic, in which each earcup has an independent microphone. The latter approach provides better localization performance.

Active Noise Reduction HPDs Recently integrated with conventional passive earmuffs, active noise reduction (ANR) electronics rely on the principle of destructive interference to cancel noise. A microphone senses the sound inside the earcup, which is then fed back through a phase compensation filter to a processing circuit and amplifier (Fig. 5). The resultant "antinoise" signal is presented through an earphone at equal amplitude to but 180° out of phase with the original noise, causing energy cancellation.

The ANR is most effective against repetitive or continuous noises that are relatively invariant in spectrum or level, which allow the system to stabilize and fine-tune the phase and amplitude parameters needed for cancellation. The ANR is limited to the reduction of low-frequency noises below about 1000 Hz, with maximum attenuation of 20–25 dB occurring below 300 Hz. Like conventional earmuffs, circumaural ANR devices are also susceptible to leakage under the earcup cushions such as caused by eyeglass temples, which can result in a reduction in active attenuation.[24]

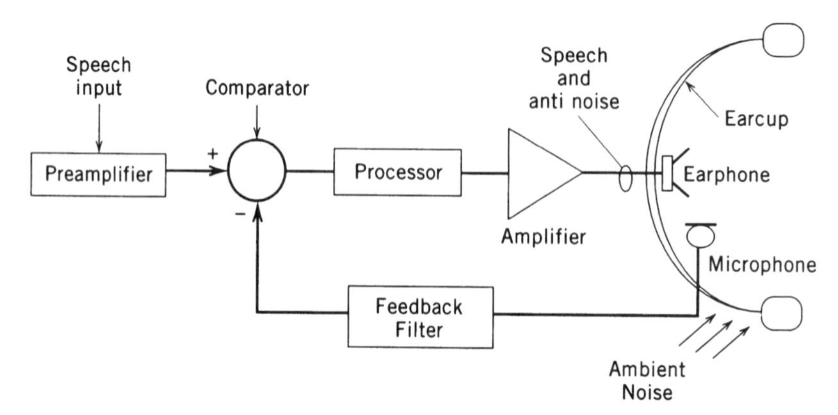

Fig. 5 Block diagram of an active noise reduction (ANR) communications headset.

The low-frequency effectiveness of ANR is fortuitous in that it compensates the typically high-frequency biased protection of a passive earmuff. If the noise environment consists of *only* low-frequency energy, open-back ANR headphones may offer communications and perhaps comfort advantages over circumaural ANR earmuffs. Consisting of lightweight supra-aural earphones that rest on the pinnae, these devices have the disadvantage of offering no passive protection if the ANR circuit malfunctions.

Attenuation tests for ANR hearing protectors are as yet not standardized. A combination procedure may be required, that is, REAT evaluations[8] to obtain the muff's passive attenuation and MIRE tests on human subjects[10] to obtain the active attenuation. The benefits of ANR devices come at a relatively high price of $150–$1000 per unit.

9.3 HPDs with Communications Features

For provision of communication or music signals at the ear, small loudspeakers have been integrated into HPDs. Headsets (including ANR examples) consist of earphones housed in earmuff earcups, which support a directional microphone (often noise canceling and/or voice activated) in front of the mouth. Small receivers can also be remotely located on a hard hat or behind the pinna and coupled to the ear via tubing through earplugs. An alternative is an earplug-like unit, called an ear microphone, which consists of a receiver button and a microphone that picks up the wearer's voice as a result of sound radiation from the bone-conduction-excited earcanal walls. Each of these approaches is available as one- or two-way systems using wireless (radio frequency or infrared) or wired technology.

It is important that communications earphones and receiver buttons be output limited so that amplified signal levels do not pose a hearing hazard. Furthermore, care must be exercised when selecting such devices for hearing protection purposes, since although some circumaural devices provide passive attenuation equivalent to a standard earmuff, certain ear microphones have been shown to provide less protection than comparable conventional earplugs.[25]

In cases where the attenuation afforded by a circumaural headset is inadequate, so that communications signals are noise masked, improvements can be realized by wearing an earplug under the headset. While the earplug will reduce the communications signal as well as the noise, the signal-to-noise ratio in the earcanal can be improved if the earphone provides sufficient distortion-free gain to compensate for the insertion loss of the earplug. Other enhancements may be provided in the system electronics, such as peak clipping and signal conditioning, to enhance the acoustic power of consonants, which are critical to word discrimination.

As diagrammed in Fig. 5, ANR has been integrated into communications headsets. The sound (speech and noise) signal from the earcup microphone is fed back to a comparator through a filter that reverses the phase. The comparator's difference output is the undesirable noise component that is out of phase with the noise inside the earcup. Since the low-frequency components of the (desired) speech signal are also canceled by the ANR, a speech preamplifier is used for compensation. For certain noise situations, ANR headsets result in an improvement in speech intelligibility over the use of passive devices alone.[24]

10 HPD STANDARDS, REGULATIONS, AND GOVERNMENT LISTINGS

10.1 United States

National standards in the United States specify experimental procedures for measuring HPD attenuation, either via REAT[8] or a supplemental physical method using an acoustical test fixture.[10] The U.S. Department of Commerce has established a National Voluntary Laboratory Accreditation Program (NVLAP) that can accredit laboratories for conformance with the aforementioned REAT standard.

10.2 International Organization for Standardization

The ISO provides methods corresponding to the two ANSI documents cited above: a REAT procedure[9] and an acoustical test fixture method[26] as well as a standard for using REAT data to compute effective noise levels when HPDs are worn.[17] Numerous additional ISO and European standards on methods of approval for earmuffs, earplugs, and other HPDs, and guidelines for selection and care of HPDs are expected to be promulgated in the late 1990s.

10.3 Environmental Protection Agency

The EPA promulgated a regulation in 1979[16] that specifies that any device or material capable of being worn on the head or in the earcanal that is sold wholly or in part on the basis of its ability to reduce sound entering the ear is to be evaluated according to the REAT portion of ANSI S3.19[27] (a currently out-of-date standard) and to be labeled with an NRR. All commercially available HPDs sold in the United States must be labeled in accordance with this regulation.

10.4 Occupational Safety and Health Administration

OSHA requires that HPDs reduce the effective sound level at the ear to a time-weighted average level of less than or equal to 90 dB(A), except in the case of employees demonstrating a standard threshold shift, in which case the level must be less than or equal to 85 dB(A).[1] The methods of computing noise reduction conform to those described in Sections 7.1, 7.2, and 7.5.

10.5 Other Agencies

The National Institute for Occupational Safety and Health publishes compendia, which are periodically updated, of hearing protector data and specifications that are obtained from the manufacturers of the devices.[28] The Mining Safety and Health Administration also publishes a compilation, albeit more abbreviated.[29]

REFERENCES

1. OSHA, "Occupational Noise Exposure; Hearing Conservation Amendment," Occupational Safety and Health Administration, *Fed. Regist.*, Vol. 48, No. 46, 1983, pp. 9738–9785.
2. L. H. Royster and J. D. Royster, "Hearing Protection Devices," in A. S. Feldman and C. T. Grimes (Eds.), *Hearing Conservation in Industry*, Williams and Wilkins, Baltimore, 1985, pp. 103–150.
3. E. H. Berger, "EARLog #17—Ear Infection and the Use of Hearing Protection," *J. Occup. Med.*, Vol. 27, No. 9, 1985, pp. 620–623.
4. E. H. Berger, "Hearing Protectors—Specifications, Fitting, Use, and Performance," in D. M. Lipscomb (Ed.), *Hearing Conservation in Industry, Schools, and the Military*, College-Hill Press, Boston, 1988, pp. 145–191.
5. J. G. Casali, "Comfort: The 'Other' Criterion for Hearing Protector Design and Selection," in *Proceedings of Hearing Conservation Conf.*, Off. Eng. Serv., Univ. Kentucky, Lexington, KY, 1992, pp. 47–53.
6. E. H. Berger, "Review and Tutorial—Methods of Measuring the Attenuation of Hearing Protection Devices," *J. Acoust. Soc. Am.*, Vol. 79, No. 6, 1986, pp. 1655–1687.
7. E. H. Berger, "Development of a Laboratory Procedure for Estimation of the Field Performance of Hearing Protectors," in *Proceedings of Hearing Conservation Conf.*, Off. Eng. Serv., Univ. Kentucky, Lexington, KY, 1992, pp. 41–45.
8. ANSI, "Method for the Measurement of the Real-Ear Attenuation of Hearing Protectors," Am. Natl. Stds. Inst., S12.6-1984, New York, 1984.
9. ISO, "Acoustics—Hearing Protectors—Part 1: Subjective Method for the Measurement of Sound Attenuation," ISO 4869-1 : 1990, Geneva, Switzerland, 1990.
10. ANSI, "Microphone-in-Real-Ear and Acoustic Test Fixture Methods for the Measurement of Insertion Loss of Circumaural Hearing Protection Devices," S12.42-1995, Am. Natl. Stds. Inst., New York, 1995.
11. J. G. Casali and S. T. Lam, "Effects of User Instructions on Earmuff/Earcap Sound Attenuation," *J. Sound Vib.*, Vol. 20, No. 5, 1986, pp. 22–28.
12. E. H. Berger, J. R. Franks, and F. Lindgren, "International Review of Field Studies of Hearing Protector Attenuation," in A. Axelsson, H. Borchgrevink, R. P. Hamernik, L. Hellstrom, D. Henderson, and R. J. Salvi (Eds.), *Scientific Basis for Noise-Induced Hearing Loss*, Thieme Medical, New York, 1996.
13. M. Y. Park and J. G. Casali, "A Controlled Investigation of In-Field Attenuation Performance of Selected Insert, Earmuff, and Canal Cap Hearing Protectors," *Hum. Factors*, Vol. 33, No. 6, 1991, pp. 693–714.
14. E. H. Berger, "Laboratory Attenuation of Earmuffs and Earplugs Both Singly and in Combination," *Am. Ind. Hyg. Assoc. J.*, Vol. 44, No. 5, 1983, pp. 321–329.
15. J. Starck and J. Pekkarinen, "Objective Measurements of Hearing Protector Attenuation for Weapon's Noise Exposure, in *Proceedings of Hearing Conservation Conf. 1992*, Off. Eng. Serv., Univ. Kentucky, Lexington, KY, 1992, pp. 89–92.
16. EPA, "Noise Labeling Requirements for Hearing Protectors," Environmental Protection Agency, *Fed. Regist.*, Vol. 44, No. 190, 40CFR Part 211, 1979, pp. 56130–56147.
17. ISO, "Acoustics—Hearing Protectors—Part 2: Estimation of Effective A-Weighted Sound Pressure Levels When Hearing Protectors Are Worn," ISO 4869-2 : 1994, Geneva, Switzerland, 1994.
18. OSHA, "OSHA Instruction CPL 2-2.20A, change 2, March 1," in *Industrial Hygiene Technical Manual*, U.S. Government Printing Office, Washington, D.C., 1987, pp. VI-13–VI-20.
19. J. G. Casali and M. J. Horylev, "Speech Discrimination in Noise: The Influence of Hearing Protection," in *Proceedings of the Human Factors Society—31st Annual Meeting*, Human Factors Soc., New York, 1987, pp. 1246–1250.
20. H. Hormann, G. Lazarus-Mainka, M. Schubeius, and H. Lazarus, "The Effect of Noise and the Wearing of Ear Protectors on Verbal Communication," *Noise Control Eng. J.*, Vol. 23, No. 2, 1984, pp. 69–77.
21. P. Wilkins and A. M. Martin, "Hearing Protection and Warning Sounds in Industry—A Review," *Appl. Acoust.*, Vol. 21, No. 4, 1987, pp. 267–293.
22. C. H. Allen and E. H. Berger, "Development of a Unique Passive Hearing Protector with Level-Dependent and Flat Attenuation Characteristics," *Noise Control Eng. J.*, Vol. 34, No. 3, 1990, pp. 97–105.
23. M. Killion, E. DeVilbiss, and J. Stewart, "An Earplug with Uniform 15-dB Attenuation," *Hearing J.*, Vol. 41, No. 5, 1988, pp. 14–17.

24. C. W. Nixon, R. L. McKinley, and J. W. Steuver, "Performance of Active Noise Reduction Headsets," in A. L. Dancer, D. Henderson, R. J. Salvi, and R. P. Hamernik (Eds.), *Noise-Induced Hearing Loss*, Mosby-Year Book, St. Louis, 1992, pp. 389–400.

25. D. Mauney and J. G. Casali, "Preliminary Noise Attenuation Assessment of AICOMM's AIMic: Current Fitting Device Performance and Improved Ear Couplers," Virginia Tech. Ind. Eng. and Operations Res., Rept. 9001, Blacksburg, VA, 1990.

26. ISO, "Acoustics—Hearing Protectors—Part 3: Simplified Method for the Measurement of Insertion Loss of Ear-Muff Type Protectors for Quality Inspection Purposes," ISO/TR 4869-3 : 1989, Geneva, Switzerland, 1989.

27. ANSI, "Method for the Measurement of Real-Ear Protection of Hearing Protectors and Physical Attenuation of Earmuffs," S3.19-1974 (ASA STD 1-1975), Am. Natl. Stds. Inst., New York, 1974.

28. J. R. Franks, C. L. Themann, and C. Sherris, "The NIOSH Compendium of Hearing Protection Devices," U.S. Dept. of HHS, Pub. No. 95-105, Cincinnati, OH, 1994.

29. MSHA, "Hearing Protector Factor List," Mining Safety and Health Administration, Pittsburgh, PA., 1996.

66

GENERATION OF NOISE IN MACHINERY, ITS CONTROL, AND THE IDENTIFICATION OF NOISE SOURCES

MALCOLM J. CROCKER

1 INTRODUCTION

Machines are used for a variety of purposes. The noise of appliances is often simply just annoying. The noise of machines in industry, however, is frequently intense enough to cause hearing damage. Although each machinery noise problem is somewhat different, a systematic approach and use of several well-known methods often produce sufficient noise reduction and acceptable acoustical conditions.[1] This chapter begins with a discussion of the *source–path–receiver* model, continues with a description of the most useful passive noise control approaches, and concludes with a discussion of the sources of noise in several important classes of machines and methods of identifying noise sources.

2 SYSTEMATIC APPROACH TO NOISE PROBLEMS

Noise control should always be attempted at the design stage wherever possible because there are more low-cost options and possibilities rather than to make individual machines or installations quieter.[2,3] After machines are built or installations completed, noise control approaches can still be achieved through various modifications and add-on treatments, but these are frequently more difficult and expensive to implement.

Noise problems can be described using the simple *source–path–receiver* model shown in Fig. 1.[4] The *sources* are of two main types: (1) airborne sound sources caused by gas fluctuations (as in the fluctuating release of gas from an engine exhaust) or (2) structure-borne machinery vibration sources that in turn create sound fields (for example, engine surface vibrations). Moreover, these sound pressure and vibration sources are of two types: (1) steady state and (2) impulsive. Both steady-state and impulsive vibrations (caused by impacting parts) are commonly encountered in machines. The *paths* may also be airborne or structure-borne in nature.

Source modifications are the best practice, but are sometimes difficult to accomplish. Often changes in the *path* or at the *receiver* may be the only real options available. The model shown in Fig. 1 is very simple. In reality there will be many sources and paths. The dominant source should be treated first, then the secondary one, and so on. The same procedure can also be applied to the paths. Finally, when all other possibilities are exhausted, the receiver can be treated. If, as in most noise problems, the receiver is the human ear, earplugs or earmuffs or even complete personnel enclosures can be used.

Measurements, calculations, and experience all play a part in determining the dominant noise sources and paths. The dominant sources (and paths) can sometimes be determined from careful experiments. In some cases, parts of a machine can be turned off or disconnected to help identify sources. In other cases, parts of a machine can be enclosed, and then sequential exposure

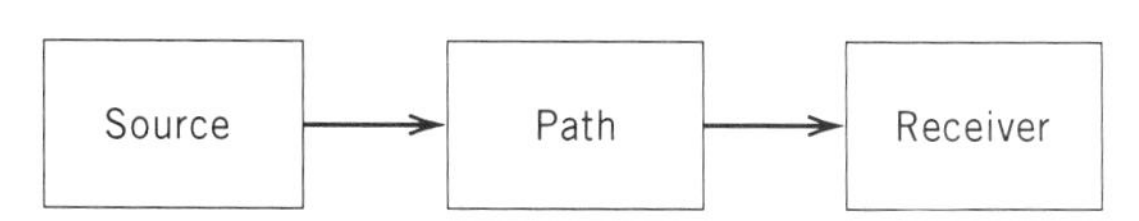

Fig. 1 Source–path–receiver model for noise problems.

Note: References to chapters appearing only in the *Encyclopedia* are preceded by *Enc.*

of machine parts can be used to identify major sources. Frequency analysis of machines can also be used as a guide to the causes of noise, as with the case of the firing frequency in engines, the pumping frequency of pumps and compressors, and the blade-passing frequency of fans. More sophisticated methods are also available involving the use of coherence, cepstrum, and intensity methods. Methods for determining the sources of noise in machinery are discussed in Section 6 of this chapter.

3 NOISE REDUCTION TECHNIQUES

A study of the literature reveals many successful well-documented methods used to reduce the noise of machines. These can be classified using the source–path–receiver model. Some of the most useful approaches can generally be used only at the source or in the path. Others, such as *enclosure*, can be adapted for use at any location. For instance, a small enclosure can be built inside a machine around a gear or bearing, or a larger enclosure or room can be built around a complete machine. Finally, an enclosure or personnel booth can be built for the use of a machine operator. Table 1 summarizes a large number of approaches that have been found useful in practice.

4 MAIN PASSIVE NOISE CONTROL APPROACHES

In Section 4 the main passive noise control approaches are briefly summarized. These include the use of (1) vibration isolators, (2) acoustical absorbing material, (3) enclosures, (4) barriers, and (5) vibration damping material.

4.1 Use of Vibration Isolators

The design of vibration isolators has been discussed frequently in the literature. Vibration isolators are used in two main situations: (1) where a machine source is producing vibration that it is desired to prevent flowing to supporting structures and (2) where a delicate piece of equipment (such as an electronics package or precision grinder) must be protected from vibration in the structure. It is the first case that will receive attention here. Primary emphasis is placed on reducing the force trans-

TABLE 1 Various Passive Noise Control Approaches That May Be Considered for Source, Path, or Receiver

Source	Choose quietest machine source available Reduce force amplitudes Apply forces more slowly Use softer materials for impacting surfaces Balance moving parts Use better lubrication Improve bearing alignment Use dynamic absorbers Change natural frequencies of machine elements Increase damping of machine elements Isolate machine panels from forces Put holes in radiating surfaces Stagger time of machine operations in a plant
Path	Install vibration isolators Use barriers Install enclosures Use absorbing materials Install reactive or dissipative mufflers Use vibration breaks in ductwork Mismatch impedances of materials Use lined ducts and plenum chambers Use flexible ductwork Use damping materials
Receiver	Provide earplugs or earmuffs for personnel Construct personnel enclosures Rotate personnel to reduce exposure time Locate personnel remotely from sources

mitted from the machine source to the supporting structure, but a secondary consideration is to reduce the vibration of the machine source itself.

It is often found that machines are attached to metal decks, grills, and sometimes lightweight wood or concrete floors. The machine on its own is usually incapable of radiating much noise (particularly at low frequency). The supporting decks, grills, and floors, however, tend to act like sounding boards, just as in musical instruments, and amplify the machine noise. Properly designed vibration isolators can overcome this noise problem.

Vibration isolators are of three main types: (1) spring, (2) elastomeric, and (3) pneumatic. Spring isolators are durable but have little damping. Elastomeric isolators are less durable and are subject to degradation due to corrosive environments. They have higher damping and are less expensive. Pneumatic isolators are used where very low frequency excitation is present.

Often the exciting forces are caused by rotational out-of-balance forces in machines or machine elements or are caused by magnetic or friction and other forces. Usually, these forces are simple harmonic in character. Such forces occur in electric motors, internal combustion engines, bearings, gears, and fans. Sometimes, however, the exciting forces may be impulsive in nature (e.g., in the case of punch presses, stamping operations, guillotines, tumblers, and any machines where impacts occur). The design of vibration isolators for a machine under the excitation of a simple harmonic force is considered here.

Theory of Vibration Isolation A machine may be considered, for simplicity, to be represented by a rigid mass m. If the machine is attached directly to a large rigid massive floor, as shown in Fig. 2, then all the periodic force $F_m(t)$ applied to the mass is directly transmitted to the floor. We will assume that the force on the mass

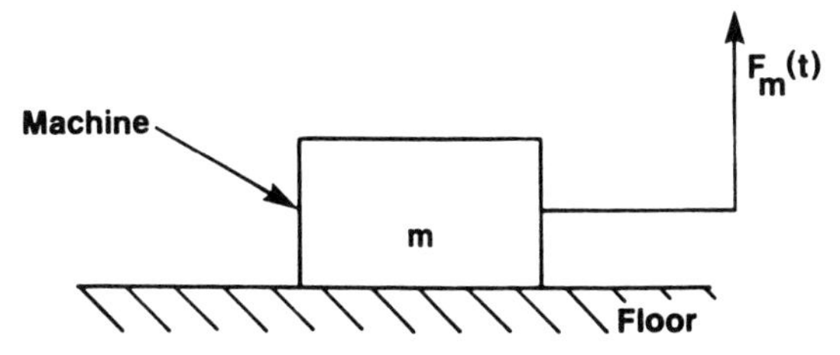

Fig. 2 Rigid machine of mass m attached to a rigid massive floor.

is vertical and $F_m(t) = F_m \sin 2\pi f t$, where F_m is the amplitude of this force, and f is its frequency (Hz).

If a vibration isolator is now placed between the machine and the floor, we can model this system with the well-known single-degree-of-freedom system shown in Fig. 3.

Suppose, for the moment, that the periodic force is stopped; then both the applied force $F_m(t)$ and the transmitted force $F_t(t)$ will be zero. If the mass m is displaced from its equilibrium position and released, then it will vibrate with a *natural frequency* of vibration f_n given by

$$f_n = \left(\frac{1}{2\pi}\right)\sqrt{\frac{K}{m}} \quad \text{Hz,} \tag{1}$$

where, in SI units, K is the stiffness (N/m) and m is the mass (kg). Alternatively in English units, K is the stiffness [lb (force)/in.] and m is the mass [lb (mass)]. (Note: lb (force) = g lb (mass), where g = acceleration of gravity = 386.4 in./s^2.)

If the exciting force $F_m(t)$ is now resumed, then a force $F_t(t)$ will be transmitted to the rigid floor. This force $F_t(t)$ will be out of phase with $F_m(t)$, but it is sim-

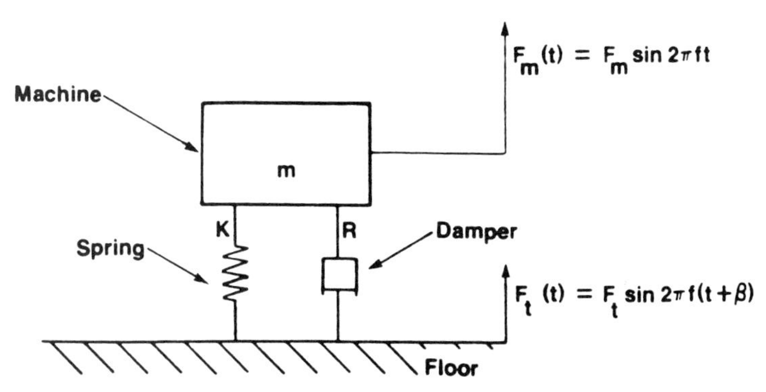

Fig. 3 Rigid machine of mass m separated from rigid massive floor by vibration isolator of stiffness K and damping R.

ple to show that the ratio of the amplitudes of the forces is given by

$$T_F = \frac{F_t}{F_m} = \sqrt{\left\{\frac{1 + 4(f/f_n)^2(R/R_c)^2}{[1 - (f/f_n)^2]^2 + 4(f/f_n)^2(R/R_c)^2}\right\}}, \tag{2}$$

where f is the frequency of the exciting force (Hz), f_n is the natural frequency of vibration of the mass m on the spring (Hz), R is the coefficient of damping, and R_c is the coefficient of critical damping, [$= 2\sqrt{mK}$]. The ratio F_t/F_m is known as the force transmissibility T_F.

The vibration amplitude A of the machine mass m is given by

$$\frac{A}{F_m/K} = \frac{1}{\{[1 - (f/f_n)^2]^2 + 4(f/f_n)^2(R/R_c)^2\}^{1/2}}. \tag{3}$$

The ratio $A/(F_m/K)$ is known as the *dynamic magnification factor* (DMF). This is because F_m/K represents the static displacement of the mass if a static force of value F_m is applied, while A represents the dynamic displacement amplitude that occurs due to the periodic force of amplitude F_m. Note that the ratio R/R_c is known as the *damping ratio* δ, which can vary from 0 to ∞. If $\delta = 1.0$, the damping is called *critical damping*. With vibration isolators, δ may be in the range from about 0.01 to 0.2.

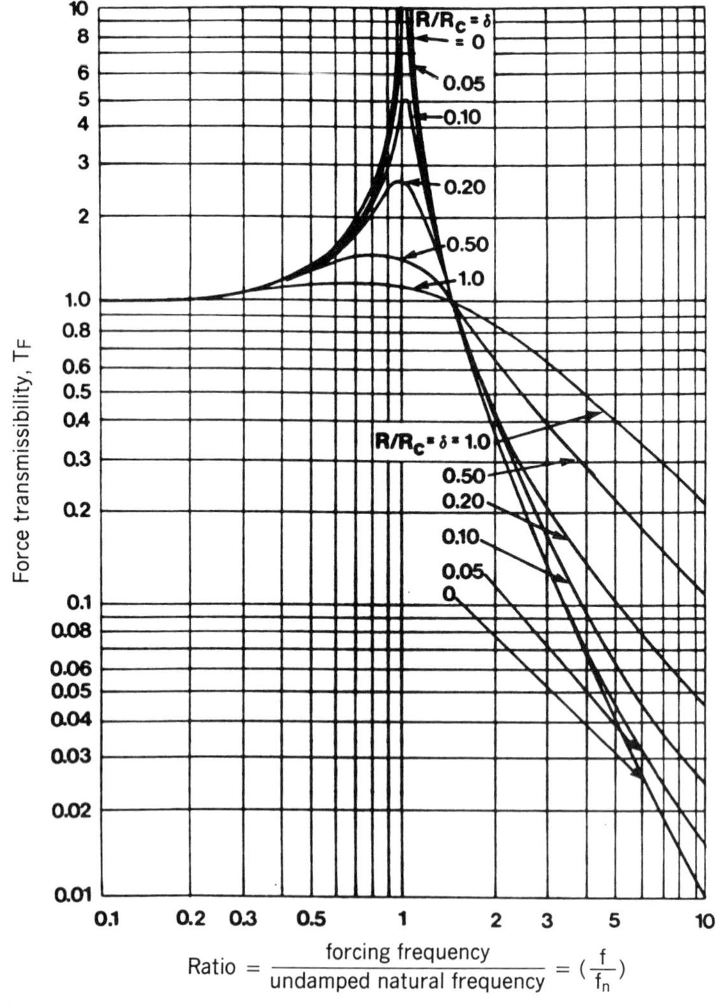

Fig. 4 Force transmissibility T_F for the rigid machine–isolator–rigid floor system.

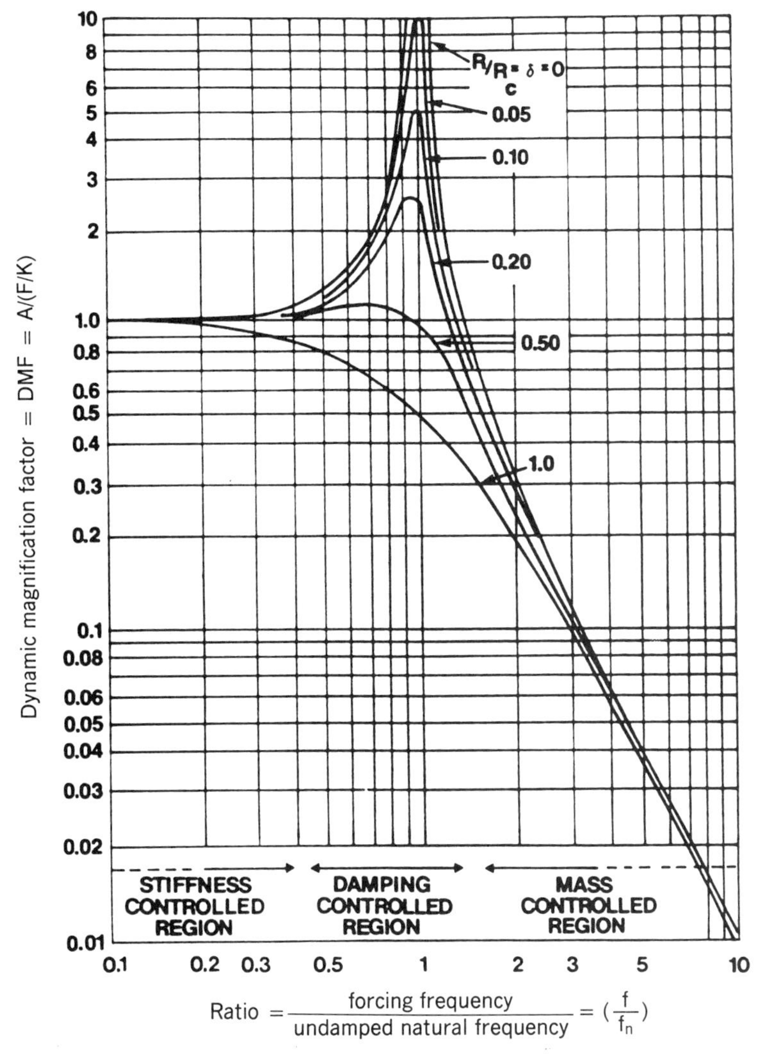

Fig. 5 Dynamic magnification factor DMF for the rigid machine–isolator–floor system.

Equations (2) and (3) are plotted in Figs. 4 and 5, respectively.

If the machine is run at the natural frequency f_n, we see from Fig. 4 that f/f_n is 1.0 and the force amplitude transmitted to the floor is very large, particularly if the damping in the isolator support is small. If the machine is operated much above the natural frequency, however, then the force amplitude transmitted to the floor will be very small.

Example 1. Suppose we wish to isolate the 120-Hz vibration of an electric motor. If we choose isolators so that the system has a natural frequency of 12 Hz, then the ratio $f/f_n = 10$. If the damping in the isolator system is $R/R_c = 0.1$, the force transmissibility will be only about 0.025, or 2.5%.

We define the efficiency η of the isolator as

$$\eta = (1 - T_F) \times 100\%. \tag{4}$$

Thus in Example 1 the isolator efficiency is 97.5%.

In order to reduce the force transmissibility still further, we could use softer isolators and choose a still lower resonance frequency. There is some danger in doing this, however, because the static deflection of the machine will naturally increase if we use softer isolators. Since a large static deflection may be undesirable (it may interfere with the operation of the machine), this restricts the softness of the isolator and thus how low we can make the natural frequency f_n. The static deflection d produced in the isolator by the gravity force on the mass m is given by $d = mg/K$. We have already seen that the natural frequency is related to K and m by $f_n = (1/2\pi)\sqrt{K/m}$. Hence we can relate d and f_n:

$$d = \frac{g}{4\pi^2 f_n^2}, \tag{5}$$

where the static deflection d is given in centimetres (or inches) depending on whether the acceleration of gravity g is 981 cm/s^2 or 386.4 in./s^2.

The relationship given in Eq. (5) has been plotted in Fig. 6. The greatest static deflection d that can be allowed from operational considerations should be chosen. This will then allow a determination of the lowest allowable f_n. It should be noted that excessive static deflection may interfere with the operation of the machine by causing alignment problems. Also, using isolators with a small vertical stiffness usually means that they will have a small horizontal stiffness, and there will be stability problems. Both these considerations limit the lowest allowable f_n.

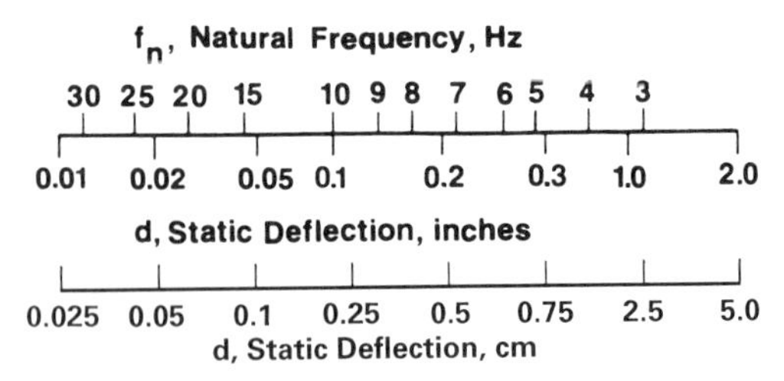

Fig. 6 Relationship between natural frequency f_n of machine–isolator–floor system and static deflection d of machine.

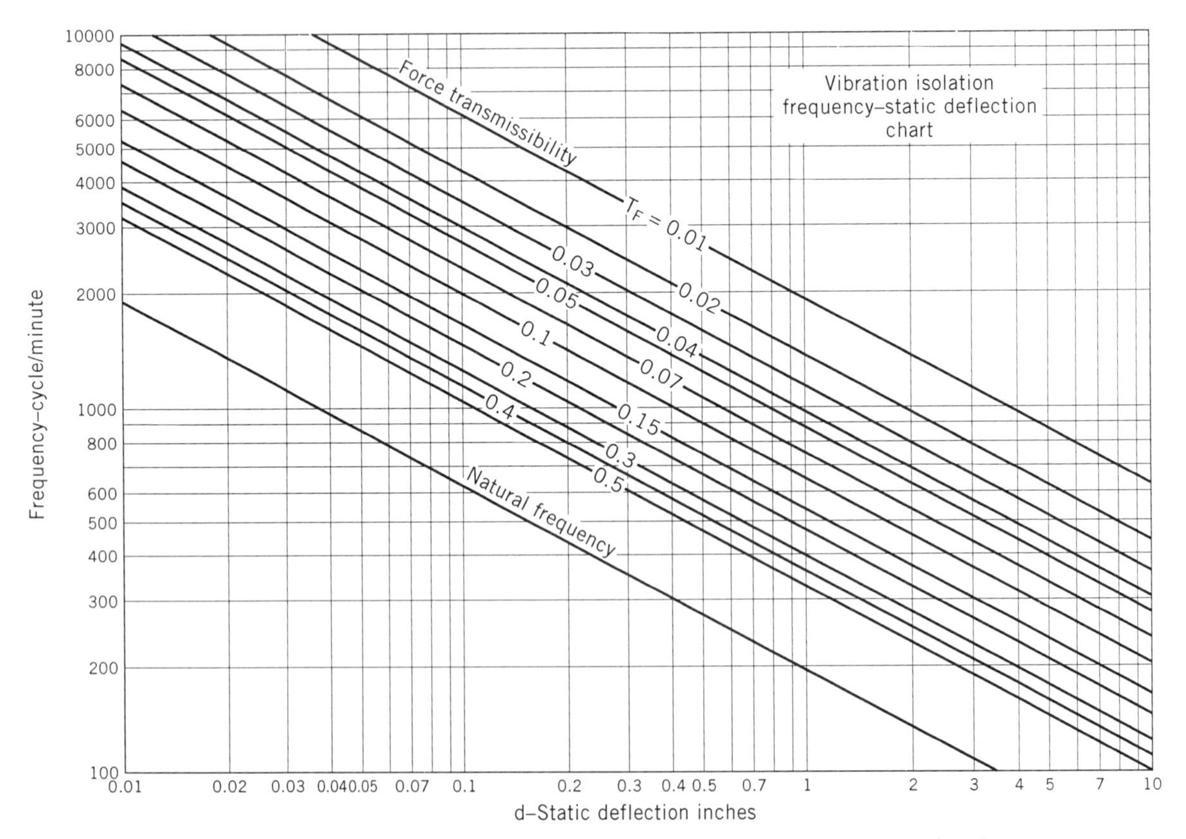

Fig. 7 Relationship between forcing frequency f_n, force transmissibility T_F, and static deflection d of machine–isolator–floor system, without damping ($R = 0$).

Simple Design Procedure and Worked Example

With springs, the damping is small (usually $\delta < 0.1$), and we may use Eqs. (2) and (3) and Figs. 4 and 5, assuming that $\delta = 0$. With $\delta = 0$ (or, equivalently, $R = 0$), Eqs. (2) and (5) give

$$T_F = \frac{F_t}{F_m} = \frac{1}{1 - 4\pi^2 f^2 d/g}. \tag{6}$$

This result is plotted in Fig. 7. The suggested procedure is as follows.

1. Establish the total weight of the machine and the lowest forcing frequency experienced.
2. From Fig. 7 select the force transmissibility allowable (this determines the static deflection, given the lowest forcing frequency).
3. From the spring constants given by the manufacturer, the machine weight, and the static deflection, choose the appropriate vibration isolator.

Example 2. An electric motor weighing 200 lb and a reciprocating compressor weighing 1000 lb are mounted on a common support. The motor runs at 2400 r/min and by a belt drives the compressor at 3000 r/min. The vertical force fed to the support is thought to be excessive. Choose six spring mounts to provide a force transmissibility not exceeding 5%.

1. The total machine weight is 1200 lb, and the lowest forcing frequency is 2400 r/min (or 40 Hz).
2. From Fig. 7, the static deflection required is 0.13 in.
3. Thus, since there are six spring isolators, each must support a weight of 200 lb. This requires a spring constant for each isolator of 200/0.13 = 1550 lb/in.

As a check on the calculation, we see from Fig. 6 that a static deflection of 0.13 in. requires a natural frequency of about 9 Hz. Thus, the ratio of forcing frequency to natural frequency $f/f_n = 40/9 \cong 4.5$. From Fig. 4, with zero damping, $T_F = 0.05$, which agrees with the design requirement.

Machine Vibration There are several other factors that should be considered in isolator design. First, we notice from Figs. 4 and 5 that when a machine is started or stopped, it will run through the resonance condition and then, momentarily, $f/f_n = 1$. When passing through the resonance condition, large vibration amplitudes can exist on the machine, and the force transmitted will be large, particularly if the damping is small. In order to provide small force transmissibility at the operating speed, small damping is required; however, to prevent excessive machine vibration and force transmission during stopping and starting, large damping is needed. These two requirements are conflicting. Fortunately, some forces, such as out-of-balance forces, are much reduced during starting and stopping. However, it is normal to provide a reasonable amount of damping ($\delta = 0.1$ to 0.2) in spring systems to reduce these starting and stopping problems. The severity of the machine vibration problem can be gauged from Fig. 8.

Use of Inertia Blocks Inertia blocks are normally made from concrete poured onto a steel frame. If the mass supported by the vibration isolators is increased by mounting a machine on an inertia block, the static deflection will be increased. If the isolator stiffness is correspondedly increased to keep the static deflection the same, then there is no change in the resonance frequency or the force transmissibility T_F. The use of an inertia block does, however, result in a reduced vibration amplitude of the machine mass. It also has additional advantages including (1) *improving stability* by providing vibration isolator support points that are farther apart, (2) *lowering the center of gravity* of the system, thus improving stability and reducing the effect of coupled modes and rocking natural frequencies, (3)

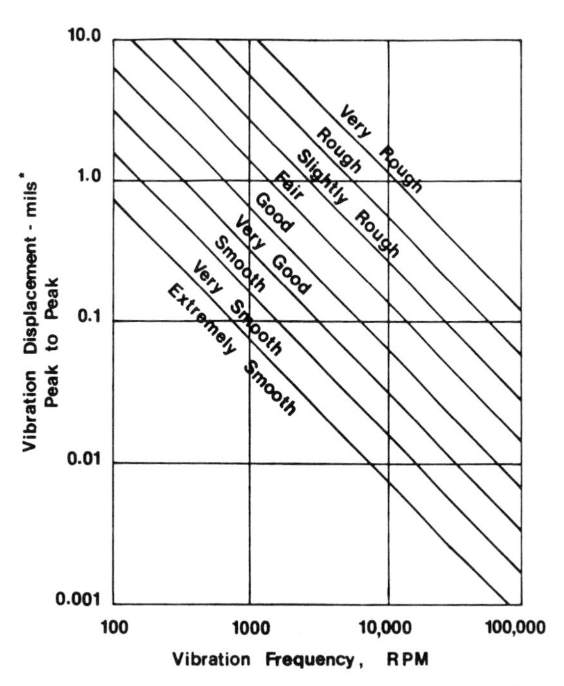

Fig. 8 Machine vibration severity chart showing peak-to-peak vibration (2A) mils (1 mil = 0.001 in. = 0.0254 mm).

producing more even weight distribution for machines often enabling the use of symmetrical vibration isolation mounts, (4) *functioning as a local acoustic barrier* to shield the floor of an equipment room from the noise radiated from the bottom of the machine and reducing its transmission to rooms below, and (5) *reducing the effect on the machine of external forces* such as transient loads or torques caused by operation of motors or fans or rapid changes in machine load or speed.

Other Considerations The performance of vibration isolators in the high-frequency range ($f/f_n \gg 1$) is often disappointing. There are several possible reasons, but often they all result in an increase in force transmissibility T_F for $f/f_n \gg 1$. Usually, the reasons are because the simple single-degree-of-freedom model neglects various other effects mentioned now.[4]

(a) **Support Flexibility.** If the assumption that the support is rigid is poor, then the support will also deflect (e.g., a flexible floor). If the isolator stiffness is similar in magnitude to that of the supporting floor, then additional resonances in the floor–machine system will occur for $f/f_n > 1$. It is normal practice to choose isolators (assuming a rigid support or floor) to have a natural frequency well below the fundamental natural frequency of the floor itself. (If possible there should not be any machine-exciting frequencies in the frequency range 0.8 to 1.3 times the floor fundamental natural frequency.) The fundamental natural frequency of a wood floor is usually in the 20–30-Hz frequency range, while that of a concrete floor is in the 30–100-Hz range.

(b) **Machine Resonances.** Internal resonances in the machine structure will also increase the force transmissibility T_F in a similar manner to support flexibility. Increasing the stiffness and/or the damping of machine members can help to reduce this effect.

(c) **Standing-Wave Effects.** Standing-wave effects in the vibration isolators can significantly decrease their performance and increase force transmissibility for $f/f_n \gg 1$, particularly when the mass of the isolators becomes appreciable compared to the machine mass. This effect can be reduced by increasing the damping in the isolators. Springs particularly suffer from this problem because of their low inherent damping. Soft materials (such as felt or rubber) placed between the spring and the support can alleviate the problem. Elastomeric isolators are less prone to this problem because of their higher damping.

(d) **Shock Isolation.** If the forces in the machine are impulsive in character (such as caused by repeated impacts), then the discussion for isolation of the single-degree-of-freedom model presented so far can still be used. It is normal practice to choose an isolator that provides a natural period T $(= 1/f_n)$ much greater than the shock pulse duration but less than the period of repetition of the force.

Example 3. A nail-making machine cuts nails 5 times each second ($T = 0.2$ s). The shock pulse duration is approximately 0.015 s. What is a suitable choice of vibration isolator for the machine? Elastometric isolators that provide a natural period of vibration of 0.1 s (natural frequency of 10 Hz) would be a good choice. They have high damping.

4.2 Use of Sound-Absorbing Materials

Sound-absorbing materials have been found to be very useful in the control of noise. They are used in a variety of locations: close to sources of noise (e.g., close to sources in electric motors), in various paths, (e.g., above barriers) and sometimes close to a receiver (e.g. inside earmuffs). When a machine is operated inside a building, the machine operator usually receives most sound through a direct path, while people situated at greater distances receive sound mostly through reflections (see Fig. 9).

The relative contributions of the sound reaching people at a distance through direct and reflected paths are determined by how well the sound is reflected and absorbed by the walls in the building. The *absorption coefficient* of a material $\alpha(f)$, which is a function of frequency, has already been defined in Chapter 1:

$$\alpha(f) = \frac{\text{sound intensity absorbed}}{\text{sound intensity incident}}. \tag{7}$$

Thus $\alpha(f)$ is the fraction of incident sound intensity that is absorbed, and it can vary between 0 and 1. Materials that have a high value of α are usually fibrous or porous. Fibrous materials include those made from natural or artificial fibers including glass fibers. Porous materials made from open-celled polyurethane foams are widely used.

It is believed that there are two main mechanisms by which the sound is absorbed in materials: (1) viscous dissipation of energy by the sound waves as they propagate in the narrow channels of the material and (2) energy losses caused by friction as the fibers of the material rub together under the influence of the sound waves. These mechanisms are illustrated in Fig. 10. In

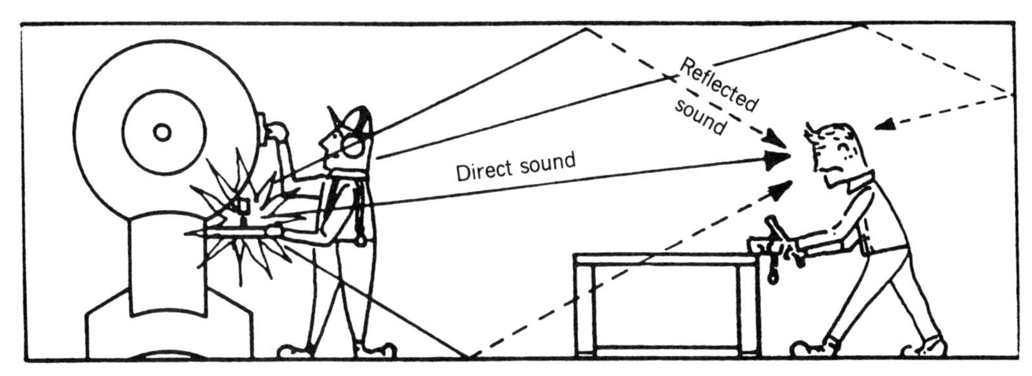

Fig. 9 Paths of direct and reflected sound emitted by a machine in a building.

both mechanisms sound energy is converted into heat. The sound absorption coefficient of most acoustic materials increases with frequency (see Fig. 11), and coefficients of some common sound-absorbing materials and construction materials are shown in Tables 2 and 3. The noise reduction coefficient NRC of a sound-absorbing material is defined as the average of the absorption coefficients at 250, 500, 1000, and 2000 Hz (rounded off to the nearest multiple of 0.05).

In the case of machinery used in reverberant spaces such as factory buildings, the reduction in sound pressure level L_p in the reverberant field caused by the addition of sound-absorbing material placed on the walls or under the roof (see Fig. 12) can be estimated for a source of sound power level L_w from the so-called room equation:

$$L_p = L_w + 10 \log \left(\frac{D}{4\pi r^2} + \frac{4}{R} \right), \tag{8}$$

where the *room constant* $R = S\overline{\alpha}/(1 - \overline{\alpha})$, $\overline{\alpha}$ is the surface average absorption coefficient of the walls, D is the source directivity (see Chapter 1), and r is the distance in metres from the source. The surface average absorption coefficient $\overline{\alpha}$ may be estimated from

$$\overline{\alpha} = \frac{S_1\alpha_1 + S_2\alpha_2 + S_2\alpha_2 + \cdots}{S_1 + S_2 + S_3 + \cdots}, \tag{9}$$

where S_1, S_2, $S_3, \ldots$ are the surface areas of material with absorption coefficients α_1, α_2, $\alpha_3, \ldots$. For the sus-

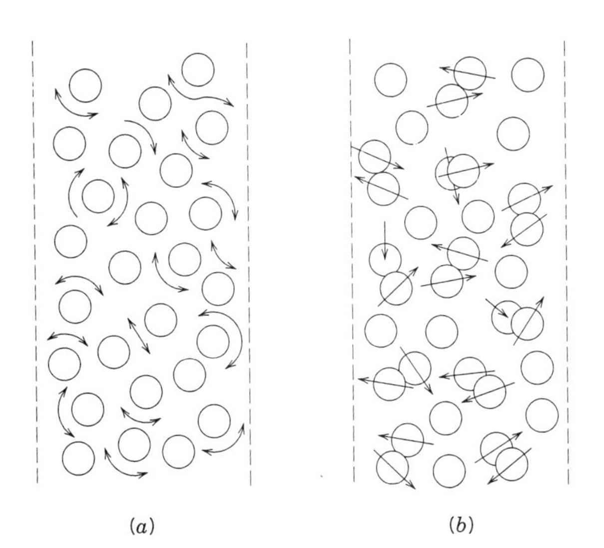

Fig. 10 The two main mechanisms believed to exist in sound-absorbing materials: (*a*) viscous losses in air channels and (*b*) mechanical friction caused by fibers rubbing together.

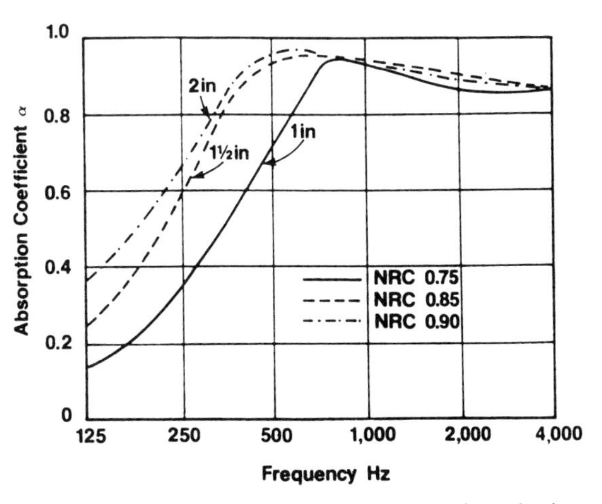

Fig. 11 Sound absorption coefficient α and noise reduction coefficient NRC for typical fiberglass formboard.

TABLE 2 Sound Absorption Coefficients $\alpha(f)$ of Common Acoustic Materials

Materials	Frequency (Hz) 125	250	500	1000	2000	4000
Fibrous glass (typically 4 lb/cu ft) hard backing						
1 in. thick	0.07	0.23	0.48	0.83	0.88	0.80
2 in. thick	0.20	0.55	0.89	0.97	0.83	0.79
4 in. thick	0.39	0.91	0.99	0.97	0.94	0.89
Polyurethane foam (open cell)						
$\frac{1}{4}$ in. thick	0.05	0.07	0.10	0.20	0.45	0.81
$\frac{1}{2}$ in. thick	0.05	0.12	0.25	0.57	0.89	0.98
1 in. thick	0.14	0.30	0.63	0.91	0.98	0.91
2 in. thick	0.35	0.51	0.82	0.98	0.97	0.95
Hairfelt						
$\frac{1}{2}$ in. thick	0.05	0.07	0.29	0.63	0.83	0.87
1 in. thick	0.06	0.31	0.80	0.88	0.87	0.87

TABLE 3 Sound Absorption Coefficient $\alpha(f)$ of Common Construction Materials

Material	Frequency f (Hz) 125	250	500	1000	2000	4000
Brick						
Unglazed	0.03	0.03	0.03	0.04	0.04	0.05
Painted	0.01	0.01	0.02	0.02	0.02	0.02
Concrete block, painted	0.10	0.05	0.06	0.07	0.09	0.03
Concrete	0.01	0.01	0.015	0.02	0.02	0.02
Wood	0.15	0.11	0.10	0.07	0.06	0.07
Glass	0.35	0.25	0.18	0.12	0.08	0.04
Gypsum board	0.29	0.10	0.05	0.04	0.07	0.09
Plywood	0.28	0.22	0.17	0.09	0.10	0.11
Soundblox concrete block						
Type A (slotted), 6 in.	0.62	0.84	0.36	0.43	0.27	0.50
Type B, 6 in.	0.31	0.97	0.56	0.47	0.51	0.53
Carpet	0.02	0.06	0.14	0.37	0.60	0.66

pended absorbing panels shown in Fig. 12, both sides of the panel are normally included in the surface area calculation.

If the sound absorption is increased, then from Eq. (8) the change in sound pressure level ΔL in the reverberant space (beyond the critical distance r_c) (see Chapter 1) is

$$\Delta L = L_{p1} - L_{p2} = 10 \log \frac{R_2}{R_1}. \qquad (10)$$

If $\overline{\alpha} \ll 1$, then the reduction in sound pressure level (sometimes called the *noise reduction*) is given by

$$\Delta L \approx 10 \log \frac{S_2 \overline{\alpha}_2}{S_1 \overline{\alpha}_1}, \qquad (11)$$

where S_2 is the total surface area of the room walls, floor, and ceiling and any suspended sound-absorbing material, $\overline{\alpha}_2$ is the average sound absorption coefficient of these surfaces after the addition of sound-absorbing material, and S_1 and $\overline{\alpha}_1$ are the area and the average sound absorption coefficient before the addition of the material.

Example 4. A machine source operates in a building of dimensions 30 m × 30 m with a height of 10 m. Suppose the average absorption coefficient is $\overline{\alpha} = 0.02$ at 1000 Hz. What would be the noise reduction in the reverberant field if 100 sound-absorbing panels with dimensions 1 m × 2 m each with an absorption coefficient of $\alpha = 0.8$ at 1000 Hz were suspended from the ceiling (assume both sides absorb sound)? The room surface area = 2(900) + 4(300) = 3000 m^2, therefore $R_1 = (3000 \times 0.02)/0.98 = 60/0.98 = 61.2$ sabins (m^2). The new aver-

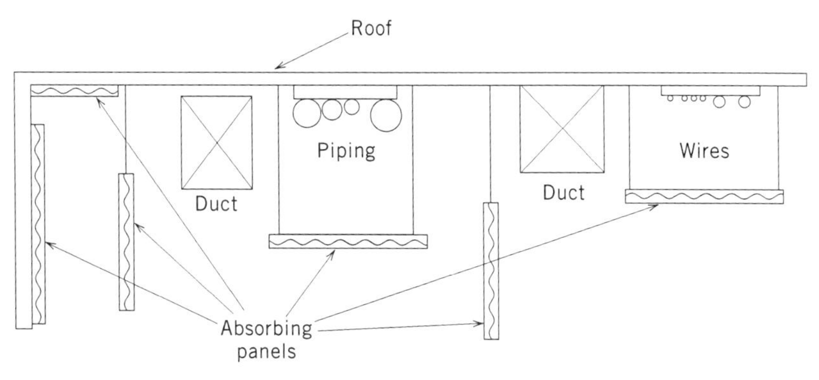

Fig. 12 Sound absorbing material placed on the walls and under the roof and suspended as panels in a factory building.

age absorption coefficient $\overline{\alpha}_2 = [3000 \times 0.02 + 200 \times 2 \times 0.8]/3400 = 60 + 320 = 380/3400 = 0.11$. The new room constant is $(3400 \times 0.11)/0.89 = 420$ sabins (m^2). Thus from Eq. (10) the predicted noise reduction $\Delta L = 10 \log (420/61.2) = 8.3$ dB. This calculation may be repeated at each frequency for which absorption coefficient data are available. It is normal to assume that about 10 dB is the practical limit for the noise reduction that can be achieved by adding sound absorbing material in industrial situations.

4.3 Acoustic Enclosures

Acoustic enclosures are used wherever containment or encapsulation of the source or receiver are a good, cost-effective, feasible solution. They can be classified in four main types: (1) large loose-fitting or room-size enclosures in which complete machines or personnel are contained, (2) small enclosures used to enclose small machines or parts of large machines, (3) close-fitting enclosures that follow the contours of a machine or a part, and (4) wrapping or lagging materials often used to wrap pipes, ducts, or other systems.

The performance of such enclosures can be defined in three main ways[4]: (1) *noise reduction* (NR), the difference in sound pressure levels between the inside and outside of the enclosure,* (2) *transmission loss* (TL, or equivalently the *sound reduction index*), the difference between the incident and transmitted sound intensity levels for the enclosure wall, and (3) *insertion loss* (IL), the difference in sound pressure levels at the receiver point *without* and *with* the enclosure wall in place. Enclosures can either be complete or partial (in which some walls are removed for convenience or accessibility). Penetrations are also often necessary to provide access or cooling.

**Note*: The *noise reduction* (NR) is not the same as the *noise reduction* discussed in Section 4.2.

The transmission coefficient τ of a wall may be defined as

$$\tau = \frac{\text{sound intensity transmitted by wall}}{\text{sound intensity incident on wall}}, \tag{12}$$

and this is related to the transmission loss TL (or sound reduction index) by

$$\text{TL} = 10 \log \frac{1}{\tau}. \tag{13}$$

If the sound fields can be considered to be reverberant on the two sides of a complete enclosure, then

$$\text{NR} = L_{p1} - L_{p2} = \text{TL} + 10 \log \frac{A_2}{S_e}, \tag{14}$$

where L_{p1} and L_{p2} are the sound pressure levels on the transmission and receiving sides of the enclosure, $A_2 = S_2\overline{\alpha}_2$ is the absorption area in square metres or feet in the receiving space material where $\overline{\alpha}_2$ is the average absorption coefficient of the absorption material in the receiving space averaged over the area S_2, and S_e is the enclosure surface area in square metres or feet.

Equation (14) can be used to determine the transmission loss TL of a partition of surface area S_e placed between two isolated reverberation rooms in which a noise source creates L_{p1} in the source room and this results in L_{p2} in the receiving room. Also Eq. (14) can be used to design a personnel enclosure. If the enclosure is located in a factory building in which the reverberant level is L_{p1}, then the enclosure wall TL and interior absorption area A_2 can be chosen to achieve an acceptable value for the interior sound pressure level L_{p2}. In the case that the surface area S_2 of the interior absorbing material $S_2 = S_e$, the enclosure surface area, then Eq.

(14) simplifies to:

$$\mathrm{NR} = \mathrm{TL} + 10 \log \overline{\alpha}_2 \tag{15}$$

We see that generally the noise reduction NR achieved is less than the transmission loss TL. We note that when $\overline{\alpha}_2$, the average absorption coefficient of the absorbing material in the receiving space, approaches 1, then NR $\rightarrow$ TL (the expected result), although when $\overline{\alpha}_2 \rightarrow 0$, then the theory fails.

Example 5. If the reverberant level in a factory space is 90 dB in the 1000-Hz one-third octave band, what values of TL and $\overline{\alpha}$ should be chosen to insure that the interior level inside a personnel enclosure is below 60 dB? Assuming that $S_2 = S_e$, then if TL is chosen as 40 dB and $\overline{\alpha} = 0.1$, NR = 40 + 10 log 0.1 = 40 − 10 = 30 dB, and $L_{p2} = 60$ dB. If $\overline{\alpha}$ is increased to 0.2, then NR = 40 + 10 log 0.2 = 40 − 7 = 33 and $L_{p2} = 57$ dB, meeting the requirement. Note that in general, since TL varies with frequency (see Eq. 18) then this calculation would have to be repeated for each one-third octave band center frequency of interest. At low frequency, since large values of TL and $\overline{\alpha}$ are difficult to achieve, it may not be easy to obtain large values of NR.

When an enclosure is used to contain a source, it works by reflecting the sound field back towards the source causing an increase in the sound pressure level inside the enclosure. From energy considerations the insertion loss would be zero if there were no acoustical absorption inside. The buildup of sound energy inside the enclosure, however, can be reduced significantly by the placement of sound-absorbing material inside the enclosure. It is also useful to place sound-absorbing materials inside personnel protective booths for similar reasons. In general, it is difficult to predict the insertion loss of an enclosure. For one installed around a source with a considerable amount of absorbing material used inside to prevent *any* interior reverberant sound energy buildup, then from energy considerations:

$$\mathrm{IL} \approx \mathrm{TL}$$

if the receiving space is quite absorbent and if IL and TL are wide-frequency-band averages (e.g., at least one octave). If insufficient absorbing material is placed inside the enclosure, the sound pressure level will build up inside the enclosure by multiple reflections, and the enclosure effectiveness will be degraded. From energy considerations it is obvious that if there is no sound absorption inside ($\overline{\alpha} = 0$), the enclosure will be useless, and its insertion loss IL will be zero.

An estimate of the insertion loss of intermediate cases $0 < \overline{\alpha} < 1$ can be obtained by assuming that the sound field inside the enclosure is reverberant and that the interior surface of the enclosure is lined with absorbing material of average absorption coefficient $\overline{\alpha}$. With the assumptions that (1) the average absorption coefficient in the room containing the noise source is not greater than 0.3 (which is true for most reverberant factory or office spaces), (2) the noise source does not provide direct mechanical excitation to the enclosure, and (3) the noise source occupies less than about 0.3 to 0.4 of the enclosure volume, then it may be shown that the insertion loss of a loose-fitting enclosure made to contain a noise source situated in a reverberant room is[5]

$$\mathrm{IL} = L_{p1} - L_{p2} = \mathrm{TL} + 10 \log \frac{A_e}{S_e} \tag{16}$$

where L_{p1} is the reverberant level in the room containing the noise source (without the enclosure), L_{p2} is the reverberant level at the same point (with the enclosure), A_e is the absorption area inside the enclosure $S_i\overline{\alpha}_i$, and $\overline{\alpha}_i$ = surface average absorption coefficient of this material. In the case that the surface area of the enclosure $S_e = S_i$, the area of the interior absorbing material, then Eq. (16) simplifies to

$$\mathrm{IL} = \mathrm{TL} + 10 \log \overline{\alpha}_i. \tag{17}$$

We see that in general the insertion loss of an enclosure containing a source is less than the transmission loss TL. We note that when $\overline{\alpha}_i$ the average absorption coefficient of the internal absorbing material approaches 1, the IL $\rightarrow$ TL (the expected result), although when $\overline{\alpha} \rightarrow 0$, then this theory fails.

Example 6. If the sound pressure level L_{p1} in a reverberant factory space caused by a machine source in the 1000 Hz one-third octave band is 85 dB, what values of TL and $\overline{\alpha}$ should be chosen to ensure that the reverberant level is reduced to be below 60 dB? Assuming that $S_i = S_e$, then if TL is chosen to be 30 dB and $\overline{\alpha} = 0.1$, IL = 30 +10 log 0.1 = 30 − 10 = 20 dB and $L_{p2} = 65$ dB. If $\overline{\alpha}$ is increased to 0.2, then IL = 30 +10 log 0.2 = 30 − 7 = 23 dB and $L_{p2} = 62$ dB. If $\overline{\alpha}$ is increased to 0.4, then IL = 30 +10 log 0.4 = 30 − 4 = 26 dB and $L_{p2} = 59$ dB, thus meeting the requirement. As in Example 5, we note that since TL varies with frequency f (see Eq. 18) this calculation should be repeated at each frequency of interest. At low frequency, since large values of TL and $\overline{\alpha}$ are hard to obtain, it may be difficult to obtain large values of IL.

Thus the transmission loss TL can be used as a rough guide to the IL of a sealed enclosure only if allowance is made for the sound absorption inside the enclosure. The transmission loss of an enclosure is normally dominated by the mass/unit area m of the enclosure walls (except in the coincidence-frequency region). This is because

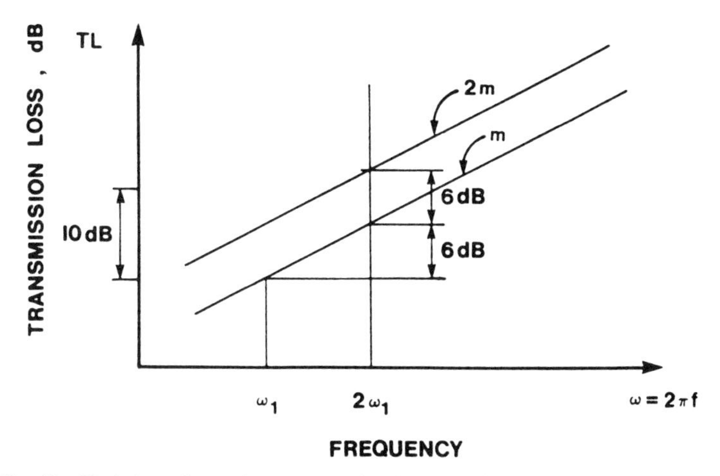

Fig. 13 Variation of mass law transmission loss TL of a single panel (see Eq. 18).

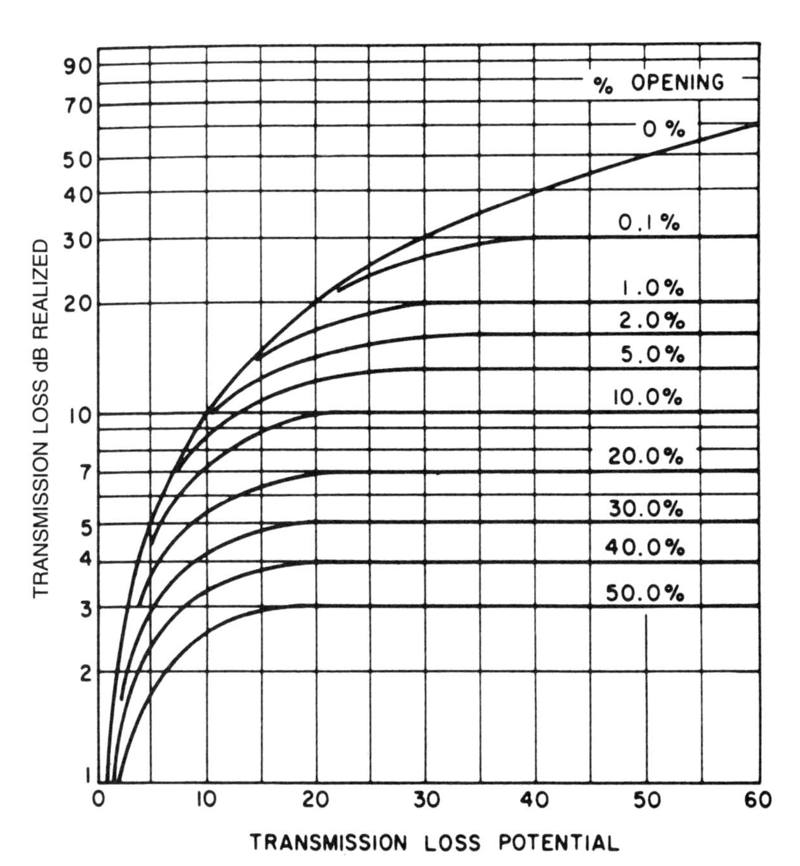

Fig. 14 Transmission loss of enclosure walls with holes as a function of transmission loss TL of enclosure wall without holes and percentage open area of holes.

when the stiffness and damping of the enclosure walls are unimportant, the response is dominated by inertia $m(2\pi f)$ where f is the frequency in hertz. The transmission loss of an enclosure wall for sound arriving from all angles is approximately

$$\mathrm{TL} = 20\log(mf) - C \quad \mathrm{dB} \tag{18}$$

where m is the surface density (mass/unit area) of the enclosure walls and $C = 47$ if the units of m are kg/m^2 and $C = 34$ if the units are lb/ft^2. Figure 13 shows that the transmission loss of a wall (given by Eq. 18) theoretically increases by 6 dB for doubling of frequency or for doubling of the mass/unit area of the wall. If some areas of the enclosure walls have a significantly poorer TL than the rest, then this results in a degradation in the overall transmission loss of the enclosure. Where the enclosure surface is made of different materials (e.g., wood walls and window glass), then the average transmission loss $\mathrm{TL_{ave}}$ of the composite wall is given by

$$\mathrm{TL_{ave}} = 10\log\frac{1}{\bar{\tau}}, \tag{19}$$

where the average transmission coefficient $\bar{\tau}$ is

$$\bar{\tau} = \frac{S_1\tau_1 + S_2\tau_2 + \cdots + S_n\tau_n}{S_1 + S_2 + S_3 + \cdots + S_n}, \tag{20}$$

where τ_i = transmission coefficient of the ith wall element and S_i = surface area of ith wall element (m^2). If holes or leaks occur in the enclosure walls and if the TL of the holes is assumed to be 0 dB, then Fig. 14 shows the degradation in the average TL of the enclosure walls. If the penetrations in the enclosure walls are lined with absorbing materials as shown in Fig. 15, then the degradation in the enclosure TL is much less significant.

Enclosures are available from manufacturers in a wide variety of ready-made modular panels (see Fig. 16).

Close-fitting Enclosures If the noise source occupies no more than about one-third of the volume of a sealed enclosure, then the simple theory described by Eqs. (16) and (17) can be used. However, in many cases when machines are enclosed it is necessary to locate the enclosure walls close to the machine surfaces, so that the

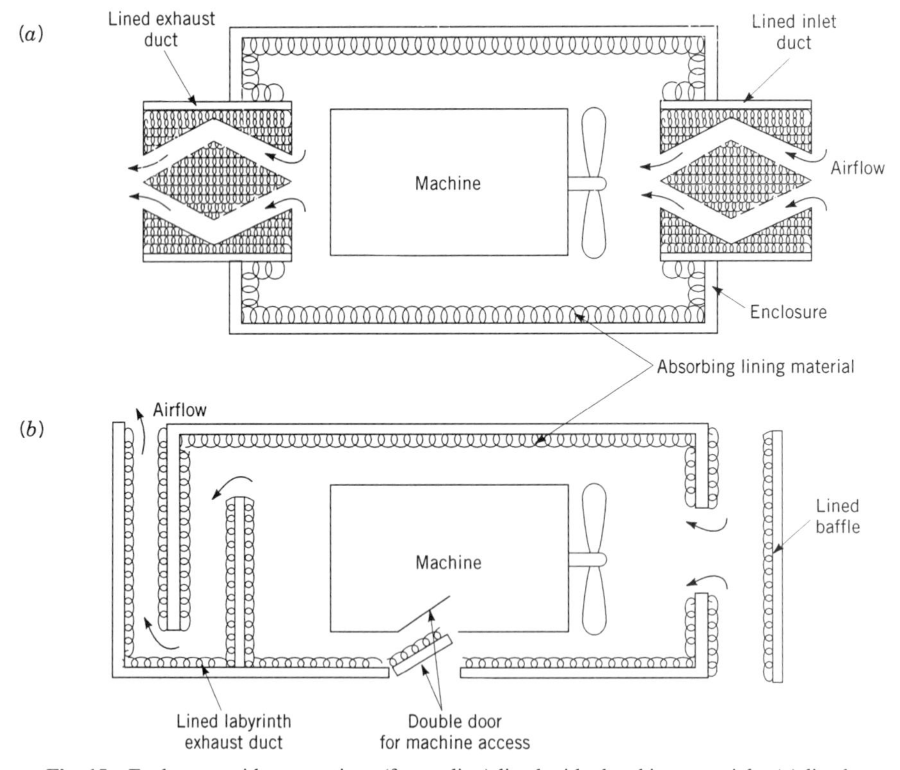

Fig. 15 Enclosures with penetrations (for cooling) lined with absorbing materials: (*a*) lined ducts, (*b*) lined baffles with double door access provided to interior of machine.

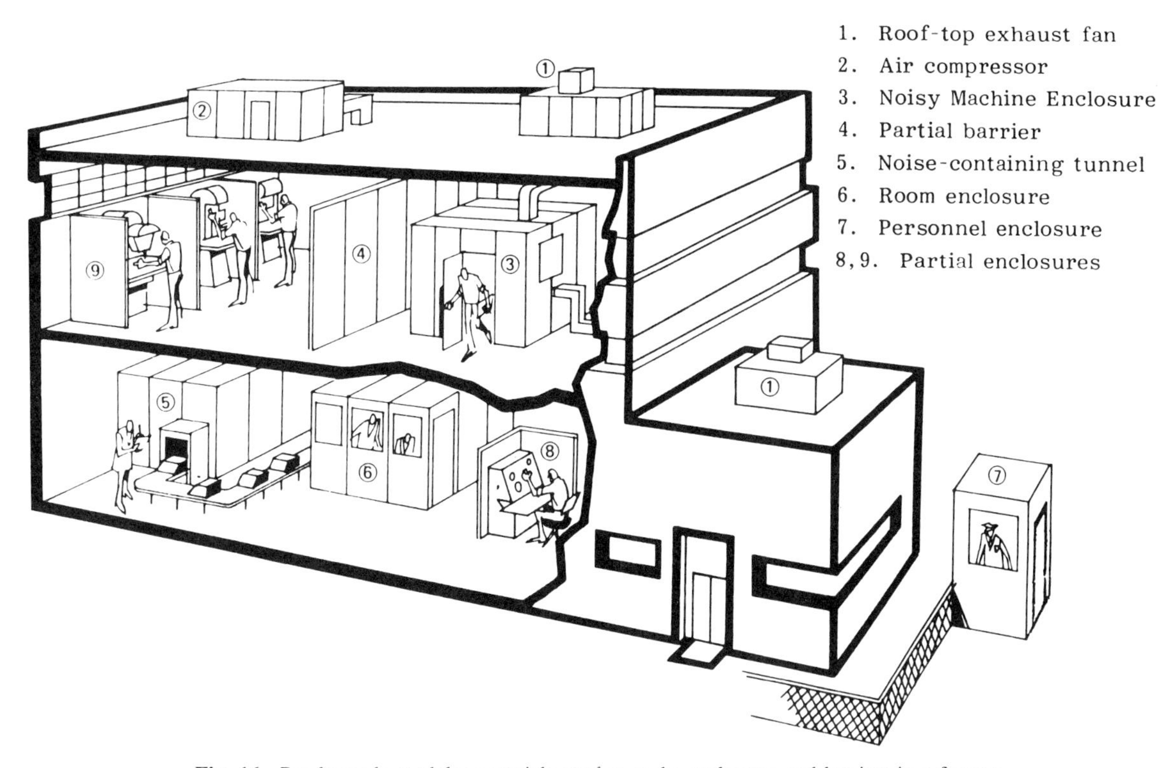

Fig. 16 Ready made modular materials used to make enclosures and barriers in a factory building. (Courtesy of Lord Corporation, Erie, PA.)

resulting air gap is small. Such enclosures are termed *close-fitting enclosures*. In such cases the sound field inside the enclosure is not reverberant or diffuse and the theory discussed so far can be used for a first approximation of the insertion loss of an enclosure. There are several effects that occur with close-fitting enclosures.

First, if the source has a low internal impedance, then in principle the close-fitting enclosure can "load" it so that it produces less sound power. However, in most machinery noise problems, the internal impedance of the source is high enough to make this effect negligible. Second, and more importantly, are the reductions in the insertion loss IL that occur at certain frequencies (when the enclosure becomes "transparent"). See the frequencies f_0 and f_{sw} in Fig. 17. When an enclosure is *close-fitting*, then to a first approximation the sound waves approach the enclosure walls at normal incidence instead of random incidence. When the air gap is small, then a resonant condition at frequency f_0 occurs where the enclosure wall mass is opposed by the wall and air gap stiffness. In addition standing wave resonances can occur in the air gap at frequencies f_{sw}. These resonances can be reduced by the placement of sound absorbing material in the air gap.[6,7] Jackson has produced simple theoretical models for close-fitting enclosures which assume a uniform air gap.[6,7] However, in practice the air gap varies with real enclosures and these simple theoretical mod-

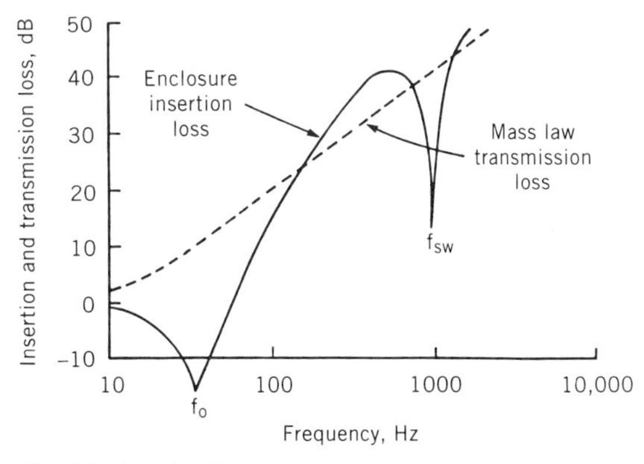

Fig. 17 Insertion loss and transmission loss of an enclosure, (mass/unit area = 16 kg/m^2, air gap = 16 cm) (from Ref. 6).

els and some later ones can only be used to give some guidance of the insertion loss to be expected in practice. Finite element and boundary element approaches can be used to make insertion loss predictions for close-fitting enclosures with complicated geometries.

4.4 Barriers

An obstacle placed between a noise source and a receiver is termed a *barrier* or *screen*. As explained in Chapter 1 when a sound wave approaches the barrier, some of the sound wave is reflected and some is transmitted past (see Fig. 18). At high frequency barriers are quite effective, and a strong acoustic "shadow" is cast. At low frequency (when the wavelength can equal or exceed the barrier height), the barrier is less effective, and some sound is diffracted into the shadow zone. Indoors, barriers are usually partial walls. Outdoors, the use of walls, earth berms, and even buildings can protect residential areas from traffic and industrial noise sources. Empirical charts are available to predict the attenuation of a barrier.[4] Figure 19 shows the *insertion loss* or reduction in sound pressure level expected after installation of a semi-infinite barrier in free space between a source and receiver. If barriers are used inside buildings, their performance is often disappointing because sound can propagate into the shadow zone by multiple reflections. To produce acceptable attenuation it is important to suppress these reflections by the use of sound absorbing material, particularly on ceilings just above the barrier. In Fig. 19, N is the dimensionless Fresnel number related to the shortest distance over the barrier $d_1 + d_2$ and the straight line distance d between the source S and receiver R:

$$N = 2(d_1 + d_2 - d)/\lambda, \qquad (21)$$

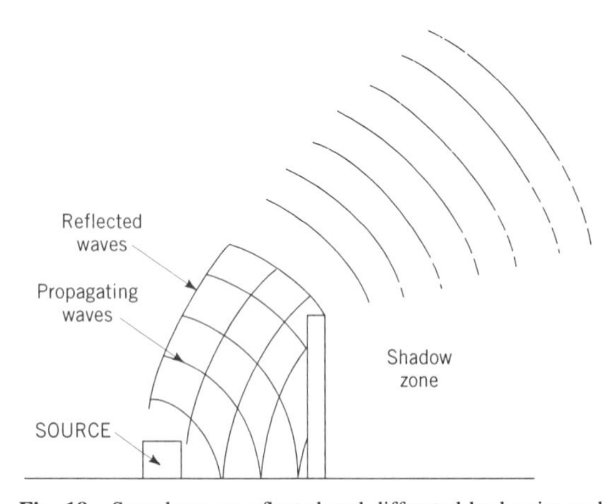

Fig. 18 Sound waves reflected and diffracted by barrier and acoustic shadow zone.

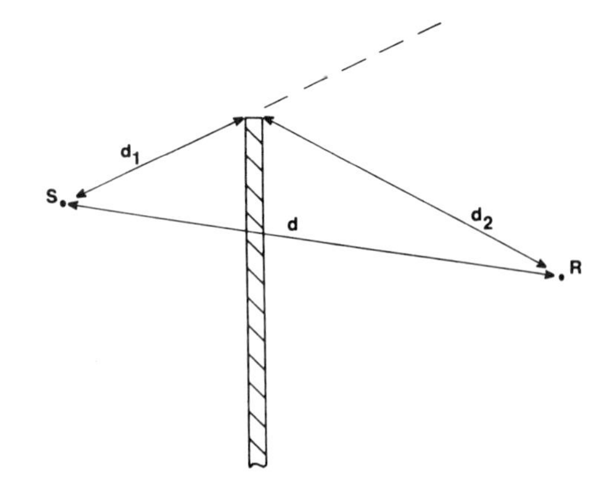

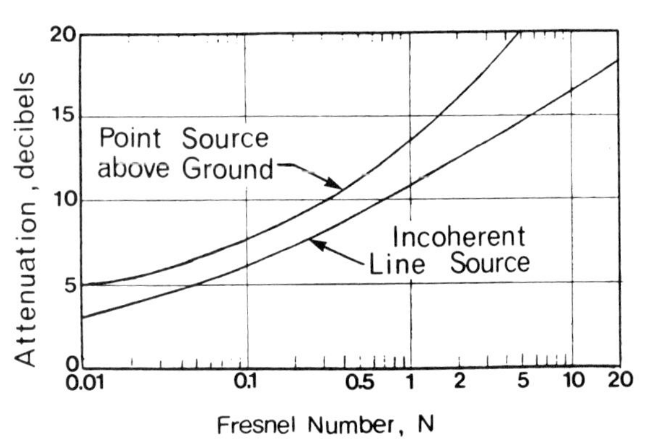

Fig. 19 Attenuation of a barrier as a function of Fresnel number N for point and incoherent line source.

where λ is the wavelength and d_1 and d_2 are shown in Fig. 19.

4.5 Damping Materials

Load-bearing and non–load-bearing structures of a machine (panels) are excited into motion by mechanical machine forces resulting in radiated noise. Also, the sound field inside an enclosure excites its walls into vibration. When resonant motion dominates the vibration, the use of damping materials can result in significant noise reduction. In the case of machinery enclosures, the motion of the enclosure walls is normally mass controlled (except in the coincidence-frequency region), and the use of damping materials is often disappointing. Damping materials are often effectively employed in machinery in which there are impacting parts since these impacts excite resonant motion.

The mechanisms involved in damping all involve the conversion of mechanical energy into heat and include friction (rubbing) of parts, air pumping at joints, sound

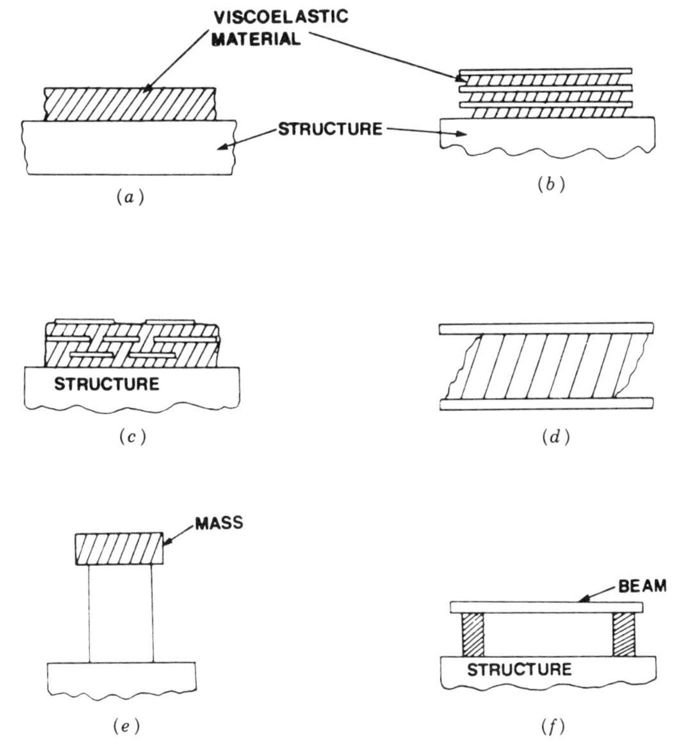

Fig. 20 Different ways of using vibration damping materials: (*a*) Free layer, (*b*) multiple constrained layer, (*c*) multi-layer spaced treatment, (*d*) sandwich panel, (*e*) tuned damper, (*f*) resonant beam damper.

radiation, viscous effects, eddy currents, and magnetic and mechanical hysteresis. Rubbery, plastic, and tarry materials usually possess high damping. During compression, expansion or shear, these materials store elastic energy. When the motion is reversed, some viscous flowing occurs, and the energy stored is converted into heat. The damping properties of such materials are temperature dependent. Damping materials can be applied to structures in a variety of ways. Figure 20 shows some common ways of applying damping materials and systems to structures.

5 MACHINERY NOISE SOURCES

Many different machines are used in appliances, in transportation vehicles, and in industry. But some components such as bearings, gears, and fans are used in many more complex machines. It is impossible to discuss every type of machine here. The discussion following will of necessity be brief and concern some of the most common machine components. Several references discuss machinery noise mechanisms and sources in more detail than is possible here.[8–14]

5.1 Bearings

There are two main types of bearings: (1) *rolling contact* and (2) *sliding contact*.[15,16] Rolling contact bearings are more commonly used, but sliding contact bearings are usually quieter than rolling contact bearings, if properly manufactured and maintained. Proper lubrication is essential for both rolling and sliding contact bearings.

Rolling contact bearings consist of the rolling elements contained between the *inner* and *outer raceways*. The rolling elements are normally kept from touching each other by a *cage*. The rolling elements may be spherical, cylindrical, tapered, or barrel-shaped.[15] Figure 21 shows a bearing with spherical rolling elements. The noise made by a rolling contact bearing is normally caused by vibration from two main sources: (1) rotation of bearing elements and (2) resonances in the elements, raceways, or cage. Shahan and Kamperman have identified seven discrete frequencies (and their harmonics) that

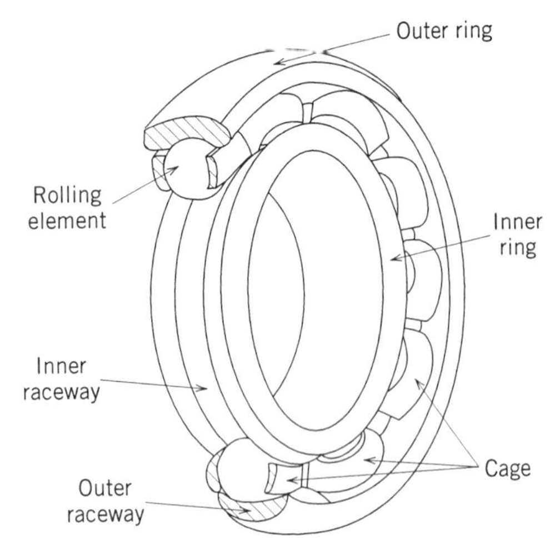

Fig. 21 Bearing with spherical rolling elements.

are related to bearing geometry and rotational speed.[13] The first is the shaft rotational frequency f_s, where

$$f_s = N/60, \tag{22}$$

where N is shaft rotational speed rpm.

The other six frequencies are related to the shaft frequency f_s by factors that depend on the roller diameter, the pitch diameter of the bearing, the contact angle between the roller element and the raceway, and the number of rolling elements. Manufacturing imperfections cause bearing noise, which can be increased further by wear. Even a perfect bearing will make a noise when loaded.[9] If a rolling bearing is manufactured to a higher grade of precision (smaller tolerance), then it normally becomes quieter (and more expensive). (Classes of tolerance are specified in ISO Standard 492-1986. Methods to test bearings for noise and vibration are given in ANSI/AFBMA Standard 13-1970. The (American) Military Specification MIL-B-17913D defines permissible vibration limits for bearings.)

Sliding contact bearings are of three main types: (1) journal, (2) thrust, and (3) guide. Journal bearings are cylindrical in shape and allow rotation (see Fig. 22). Thrust bearings are used to prevent motion along a shaft axis, while guide bearings are normally used for motion of a part in one direction without rotation (for example, a piston sliding in a cylinder of an internal combustion engine). When shaft rotation occurs with a journal bearing, the shaft rides on a film of lubricant. Under some conditions, however, this film can break down causing metal-to-metal contact and noise and vibration. A well-known instability condition known as *oil whirl* can occur

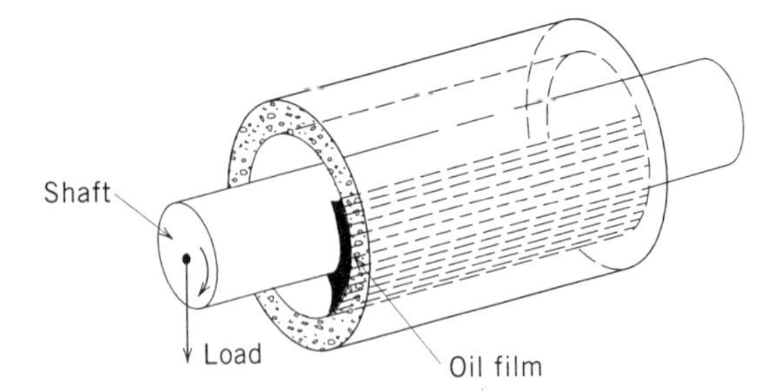

Fig. 22 Sliding contact bearing.

causing noise at a frequency of about half of the shaft rotational speed. This is because the average speed of the lubricant film is about half of the shaft speed. Oil-whirl noise can be magnified if a shaft resonance frequency occurs near to half the shaft rotational speed. To minimize sliding-bearing noise, proper attention should be paid to lubricant viscosity, pressure, alignment, and structural stiffness. References 9, 13, 14, and 16 contain more detailed discussion on sliding-bearing noise.

5.2 Gears

Most modern gear teeth have an involute profile, although some have circular-arc profiles.[8,9] Figure 23*a* shows some of the terms used with involute gears and Fig. 23*b* shows two parallel-axis spur-type gears meshing together.

There are several different types of gears in common use, as shown in Figure 24. They are of two main classes, either having: (1) parallel axes (spur, helical) or (2) nonparallel axes (straight bevel, spiral bevel, hypoid). These spur gears and straight-bevel gears are usually the noisiest, while the helical spiral-bevel gears are usually the quietest of their respective classes. This is because the load between gear wheel teeth is transferred more gradually with helical and spiral-bevel gears rather than more abruptly with spur and straight-bevel gears. Gear noise can arise from a variety of causes. As the gear teeth mesh, a pulsating force occurs at the gear-tooth meshing frequency f_m and its harmonics. Harmonics are present because the pulsating force on the gear teeth is not purely sinusoidal. The strength of the harmonics depends on the force pulse shape and also on other impulsive forces caused by tooth deformation, machinery errors, bearing misalignment, pinion wheel deformation, and so on. At high speeds, air or lubricating fluids can be expelled from between the meshing teeth at almost sonic speeds and can even become the dominant source of noise. At low speed and load the sound pressure level from a gear increases by about 3 dB for a doubling of load, while at higher speeds and loads the

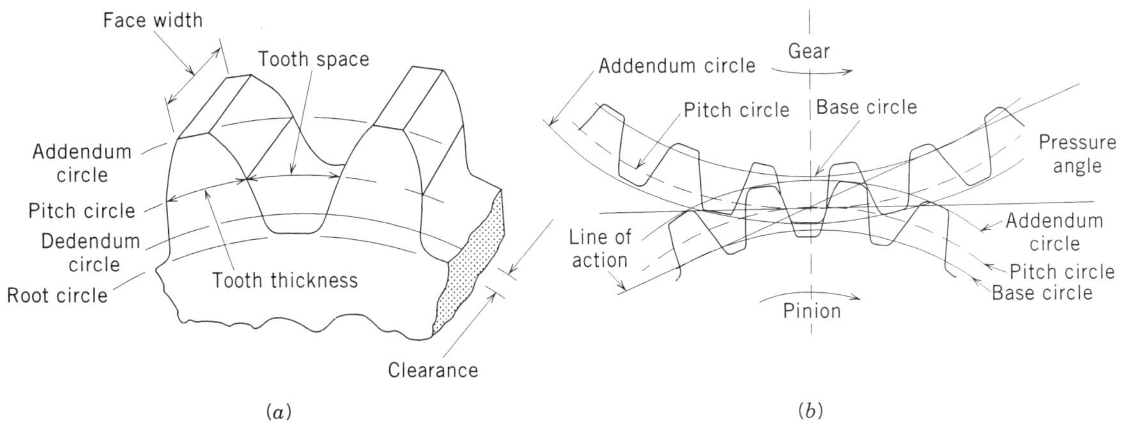

Fig. 23 Terms used with gears: (*a*) involute gear, (*b*) meshing of two parallel-axis spur gears. (Based in part on figures in Refs. 2, 8, and 13.)

sound pressure level increases by about 6 dB for a doubling of speed or load.[18] The noise of a gear set is quite dependent on the quality of manufacture and the tolerances achieved. Thus gear sets can be chosen with very low noise although the cost is high. In many cases where low noise levels are required, it is more cost effective to choose a gear set of moderate cost and to vibration isolate the pinion and gear bearings, apply damping to the gear housing, and, if necessary, completely enclose it. In gears used to transmit only small loads (for exam-

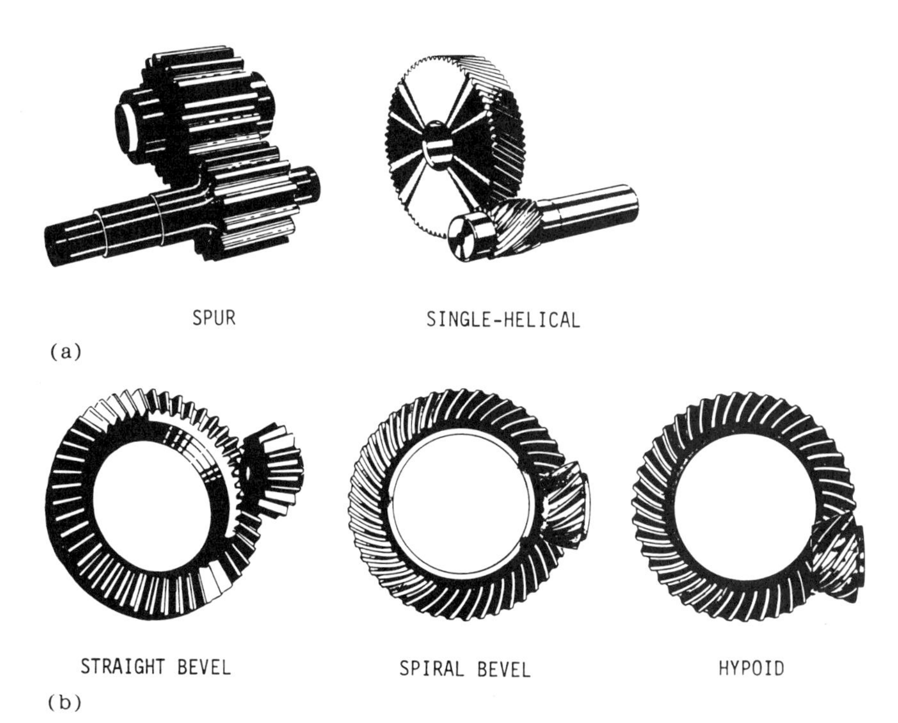

Fig. 24 The main types of gear in use: (*a*) parallel axis (spur, helical); (*b*) nonparallel axis (straight bevel, spiral bevel, and hypoid). (Reprinted with permission from R. G. Schegel, R. J. King, and H. R. Mull, "How to Reduce Gear Noise," *Machine Design*, pp. 134–142, 2/27/64, a Penton Publication.)

ple, those in electric clocks), very low noise levels can be achieved by the use of soft gears (which reduce the gear tooth force pulses), and other means.[19] Manufacturing deficiencies can result in variations in pitch and profile from tooth to tooth and eccentricity of the gear wheel, which causes increased noise and vibration.[2,8,13] There would, however, be some noise generated even if the gears were without any imperfections. The frequency of this noise (and vibration) would only occur at the gear meshing frequency f_m and its harmonics:

$$f_m = \frac{N_p n_p}{60} \qquad \text{Hz},$$

where N_p is the number of pinion teeth and n_p is the pinion speed in rpm.

5.3 Fans and Blowers

Fans and blowers are used in appliances, in buildings, in air distribution systems for heating and cooling, and in industry for a variety of purposes.[4] There are two main types of fan design: axial and centrifugal (see Fig. 25). The first three centrifugal fan types: airfoil, backward-curved, and radial are mostly used in industrial applications. The airfoil fan is the most efficient, but it is only suitable for clean-air industrial applications because dust and other particles can adhere to the fan blades and cause malfunctioning. The forward-curved fan is least efficient and usually made of lightweight low-cost materials. The radial fan is the noisiest and least efficient, but is useful for dirty or corrosive flows. Axial fans have the disadvantage that the discharged air is rotating, unless downstream stators are installed; they are mainly used in low- or medium-pressure air conditioning systems. The vaneaxial fan is the most efficient axial type and can operate at high pressure. The propeller type is the least expensive of the axial type. It has the lowest efficiency and is normally limited to low-pressure, high-volume flow applications.

The primary purpose of a fan is to move a required volume flow rate of air against a given back pressure with maximum efficiency and low first cost. There may be additional requirements such as high resistance to abrasion, ability to transport dusty air, and ease of manufacture or maintenance and repair.

Fan noise has pure-tone and broadband frequency components. Although the mathematical theory of fan noise is well developed, it is beyond the scope of this chapter, and physical explanations are presented instead. Each time a blade passes a point in space, an impulsive force fluctuation is experienced. If a fan has n blades and the rotational speed is N rpm, then the number of impulses experienced per second f_B is

$$f_B = \frac{nN}{60} \qquad \text{Hz}. \tag{23}$$

The frequency f_B is known as the *blade-passing frequency* or often the *blade frequency* for short. Since the time history of the impulsive force at the point will not be completely sinusoidal, harmonics will appear. The strength of the harmonics will be expected to be affected by upstream or downstream solid-body flow obstructions. Fan sound power level data are normally provided by manufacturers in eight one-third-octave bands from 63 to 8000 Hz. If noise is of major concern and the fan has been properly installed, it will probably be necessary to install intake and discharge sound attenuators (with flexible vibration breaks at attachment points to duct systems, if present).

It is also possible to provide further noise attenuation by the use of ducting, elbows, and plenum chambers lined with sound-absorbing material. Care must be taken that significant noise is not generated in this ductwork.

5.4 Electrical Equipment

Examples of electrical equipment that cause noise and vibration include motors, generators and alternators,[12,16] transformers, relays, solenoids, and circuit breakers. Electric motors are used widely in appliances, vehicles, and industry in a variety of types and sizes. They may be commutated, synchronous, or induction types. Electrical energy is converted into mechanical energy, and in the process some heat is produced. Fans are often provided to remove the heat and are the main sources of noise in electric motors. Because of the requirement for most motors that they should operate in either direction, they are usually provided with axial fans, the noisiest type. The sources of noise and vibration in electrical equipment are mostly aerodynamic, mechanical, and electromagnetic in nature. They are summarized in Table 4.

5.5 Control Valves

In industrial plants the aerodynamic noise generated by control valves, regulators and orifices is a major noise source in piping systems. In addition hydrodynamic noise (which is mostly due to cavitation) is of some importance. However in most situations, the aerodynamic noise is dominant.[20] The main noise-generating mechanisms include turbulent mixing, turbulence interaction with boundary, shock waves, interaction of turbulence with shocks, cavity resonances, flow separation, vortex shedding and mechanical vibration.[20] The main noise-generating processes can be divided into two

	Fan Type	Description	Design details
CENTRIFUGAL TYPE FANS	Airfoil	Uses 10 to 16 airfoil shape blades; has highest efficiency of all centrifugal types; used where horsepower savings will be important; can be used on low, medium, and high-pressure systems	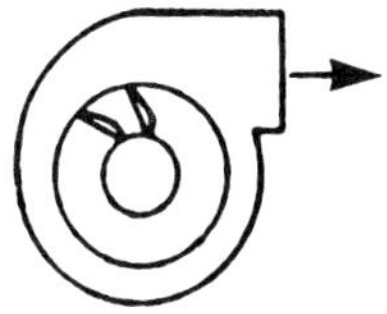
	Backward inclined backward curved	Uses 10 to 16 blades; used for similar applications as airfoil fan; gas flow should be clean but need not be as clean as gas flow with airfoil fans	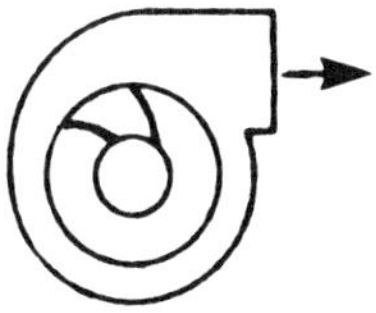
	Industrial (radial)	Uses 6 to 10 blades of either radial (R) or modified radial (M) type; has lowest efficiency of centrifugal types; used mainly in industrial applications where gas is hot and dirty	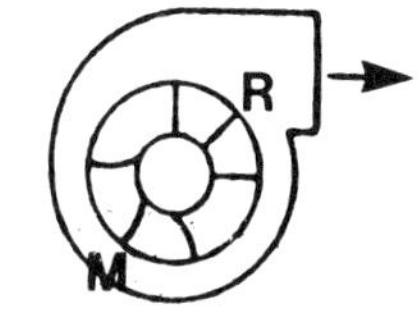 R, M
	Forward curved	Uses 24 to 64 blades; construction is usually low-cost and lightweight; efficiency is lower than airfoil or backward curved fans; usually operates at lowest speed and is smallest of centrifugal types; used in low-pressure heating, ventilating, and air-conditioning (HVAC) systems	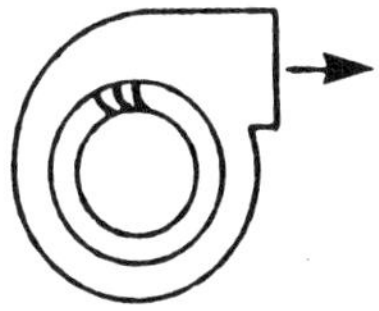
AXIAL TYPE FANS	Vaneaxial	Uses 3 to 16 blades; high efficiency airfoil blades may be fixed or adjustable; used in low-, medium-, and high-pressure systems in HVAC and other industrial applications	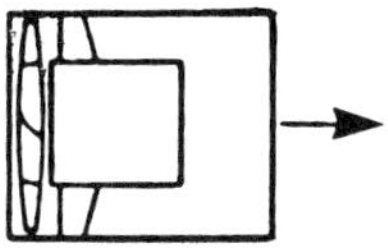
	Tubeaxial	Uses 4 to 8 blades; usually more efficient than propeller type below and can operate at a higher pressure; does not use guide vanes as does vaneaxial type; Used in low- and medium-pressure systems in HVAC and other applications	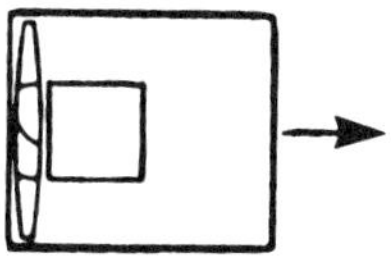
	Propeller type	Uses 2 to 8 blades usually in a circular ring or orifice plate; efficiency low, this type usually limited to low-pressure, high-volume flow applications such as exhaust or attic fans	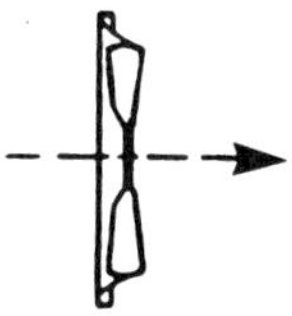
	Tubular centrifugal	The fan wheel is usually similar in design to that of the airfoil or backward curved type used in centrifugal fans; since the air is discharged radially and must turn 90° in the guide vane section, its efficiency is lower than similar centrifugal fans; normally used in low-pressure HVAC return air systems	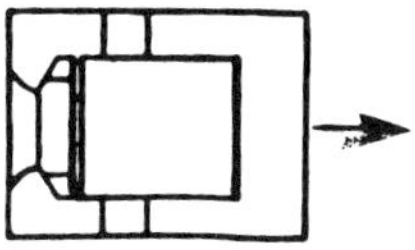

Fig. 25 The main types of fans with descriptions of their use and design.

TABLE 4 The Main Sources of Noise in Electric Motors

Mechanical	Excessive bearing clearance
	Nonround bearings
	Rotor unbalance
	Rotor eccentricity
	Crooked shaft
	Brush and brush holder vibration
	Misalignment
	Loose laminations
Electromagnetic	Magnetostriction
	Torque pulsations
	Air gap eccentricity
	Air gap permeance variation
	Dissymmetry
Aerodynamic	Fan blade-passing frequency
	Turbulence
	Noise due to airflow path restrictions

regimes: *subsonic*, consisting of turbulence-boundary interaction noise, and *supersonic*, consisting of broadband shock noise. Although the flow is confined by the piping, the turbulent mixing noise is similar to the noise of a free jet discussed in Chapters 24 and 25 and is essentially quadrupole in nature. The shocks are caused by incorrectly expanded flow after the valve which is operating above the critical pressure ratio. The shock noise has two main parts: screech and broadband noise. The screech is discrete in nature and is caused by a feedback mechanism and is not often encountered with valves and regulators. The broadband shock noise is common, however, and has been shown to be mostly independent of flow velocity and to be a function of pressure ratio. Valve noise can be reduced by a number of approaches including design of valves with multiple streams, arranging for the pressure drop to occur through several stages, and using absorptive silencers and thicker pipe walls and pipe lagging. The multiple stream valve designs and absorptive silencers can become plugged by solid particles and moisture can be a problem in the silencer material so that effectiveness is lost. Several authors have given good reviews of valve and piping system noise and the reader is referred to these for more detailed discussion.[20–22]

5.6 Internal Combustion Engines

The internal combustion engine is a major source of noise in transportation and industrial use. The intake and exhaust noise can be effectively silenced. However, the noise radiated by engine surfaces is more difficult to control. In gasoline engines a fuel–air mixture is compressed to about one-eighth to one-tenth of its original volume and ignited by a spark plug. In diesel engines air is compressed to about one-sixteenth to one-twentieth of its original volume and liquid fuel is injected in the form of a spray, then spontaneous ignition and combustion occurs. Because the rate of pressure rise is initially more abrupt with a diesel engine than with a gasoline engine, diesel engines tend to be noisier than gasoline engines. The noise of diesel engines has consequently received more attention from both manufacturers and researchers. The noise of engines can be divided into two main parts: combustion noise and mechanical noise. The combustion noise is caused mostly by the rapid pressure noise caused by ignition, and the mechanical noise is caused by a number of mechanisms with perhaps piston slap being one of the most important, particularly in diesel engines. The noise radiated from the engine structure has been found to be almost independent of load, although dependent on cylinder volume and even more dependent on engine speed. Measurements of engine noise over a wide range of cylinder capacities have suggested that engine noise increases by about 17.5 dB(A) for a tenfold increase in cylinder capacity.[10] Engine noise has been found to increase at an even greater rate with speed than with capacity (at least at twice the rate) with about 35 dB(A) for a tenfold increase in speed. Engine noise can be reduced by attention to details of construction. In particular stiffer engine structures have been shown to reduce radiated noise. Partial add-on shields and complete enclosures have been demonstrated to reduce diesel engine noise on the order of 3 to 10 dB(A). Priede has given a good review of internal combustion engine noise with an emphasis on diesel engine noise (see Chapter 19 in Ref. 10).

6 NOISE SOURCE IDENTIFICATION

6.1 Introduction

In most machinery noise control problems, a knowledge of the dominant noise sources in order of importance is very desirable so that suitable modifications can be made. In a complicated machine, such information is often difficult to obtain and many noise reduction attempts are made based on inadequate data so that frequently expensive or inefficient noise reduction methods are employed. Machine noise can also be used to diagnose increased wear. The methods used to identify noise sources will depend on the particular problem and the time and resources (personnel, instrumentation, and money) and expertise available and on the accuracy required. In most noise source identification problems, it is usually the best practice to use more than one method to ensure greater confidence in the identification procedure. In Section 6 of this chapter both well-tried and more novel methods are reviewed.

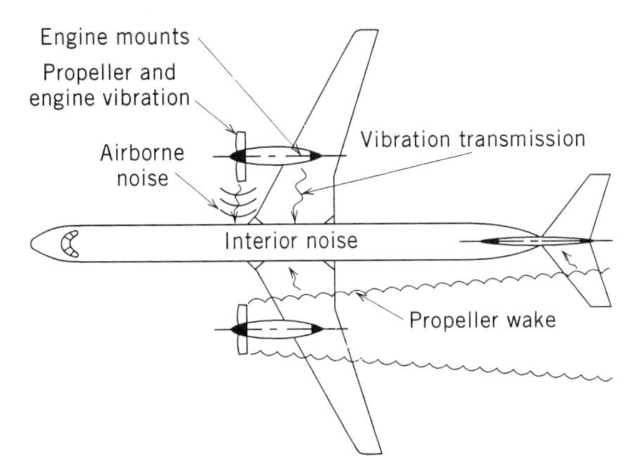

Fig. 26 Sources and paths of airborne and structure-borne noise resulting in interior noise in an airplane cabin. (Based in part on Fig. 37 in Ref. 33.)

6.2 Source–Path–Receiver

As discussed in Section 2 of this chapter, noise energy must flow from a *source* through one or more *paths* to a *receiver* (usually the human ear). Figure 1 shows the simplest model of a *source–path–receiver* system. Noise sources may be *mechanical* in nature (impacts, out-of-balance forces in machines, vibration of structural parts) or *aerodynamic* in nature (pulsating flow, flow–structure interaction, jet noise, turbulence). Noise will flow through a variety of airborne and structure-borne paths to the receiver. Figure 26 shows an example of a propeller-driven airplane. This airplane situation can be idealized as the much more complicated source–path–receiver system shown in Fig. 27. In some cases it is not easy to separate neatly the sources from the paths, and sources and paths must be considered in conjunction. In other cases the distinction between the source(s) and path(s) is not completely clear. However, the source–path–receiver model is a useful concept that is widely used. In Section 6 of this chapter we will mainly concern ourselves with noise sources. However, we shall also consider the fact that *cutting* or *blocking* one or more noise *paths* often gives invaluable information about the noise sources.

6.3 Clues to Noise Sources

There are various characteristics of a sound field that give indications of the type of noise sources present and their spatial distribution. These include the frequency distribution of the sound pressure level, the directional properties of the sound field, and the variation of the sound pressure level with distance from the sources. A knowledge of the variation in the sound pressure level with time can also be instructive. For more sophisticated measurements it is useful to have some theoretical understanding of the propagation of sound from the idealized models—monopole, dipole, quadrupole, line source, piston—and the generation of sound caused by mechanical systems—mechanical impacts and vibrating beam and plates. See Chapters 1, 8, and 10. Such theoreti-

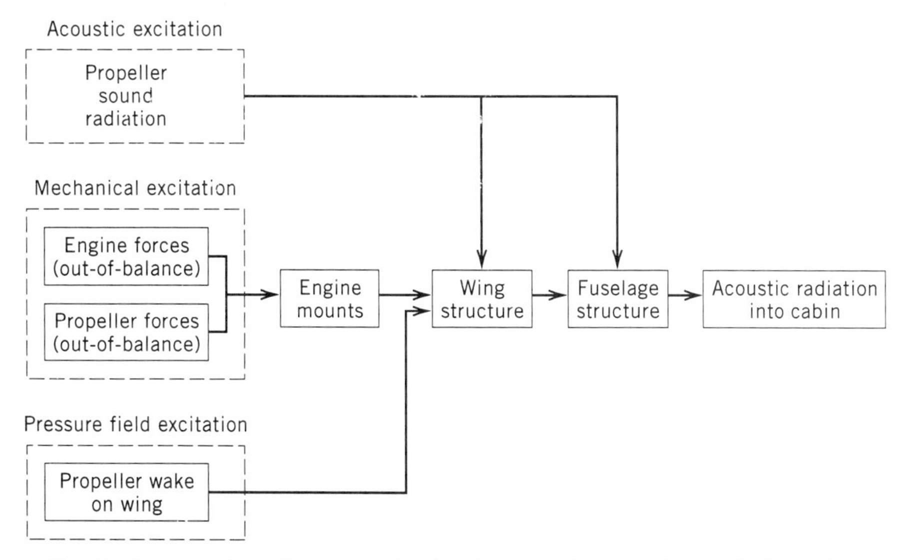

Fig. 27 Source–path–receiver system showing airborne and structure-born paths for a twin-engine propeller-driven airplane.

cal understanding not only guides us in deciding which measurements to make, but also aids in interpreting the results. In fact as the measurements made become more sophisticated, increasing care must be taken in their interpretation. There is a danger with sophisticated measurements of collecting large amounts of data and then either being unable to interpret them or reaching incorrect conclusions. With the advent of minicomputers and other sophisticated signal processing equipment, a knowledge of random process and signal processing theory is very important. See Chapters 81 and 82.

7 CLASSICAL METHODS OF IDENTIFYING NOISE SOURCES

Some of the elementary or classical methods of identifying noise sources have been reviewed in detail in the literature[23] and do not require a profound theoretical understanding of sound propagation. They will only be briefly reviewed here.

Subjective assessment of noise is very useful because the human ear and brain can distinguish between different sounds more precisely than the most sophisticated measurement system. With practice, an operator can tell by its sound if a machine is malfunctioning; a noise consultant can accurately estimate the blade passage frequency of a fan. In identifying sources, we should always listen to a machine first. In order to localize the sources and cut out extraneous noise, stethoscopes can be used, or a microphone–amplifier–earphone system. Such a system also sometimes allows one to position the microphone near a source in an inaccessible or dangerous place where the human ear cannot reach.

Selective operation of a complicated machine is often very useful. As long as the operation of a machine is not severely changed by disconnecting some of the parts sequentially, such a procedure can indicate the probable contribution of these parts to the total machine noise when all parts are operating simultaneously. For instance, engine noise can be measured with and without the cooling fan connected, and this can then give an estimate of the fan noise. Note, however, that if the fan noise is lower than the engine noise in level, the accuracy of the estimate will be poor. The fan noise estimate can be checked by driving the fan separately by a "quieted" electric motor or other device. We should be aware that in some cases disconnecting some parts may alter the operation of the other parts of the machine and give misleading results. An example is an engine driving a hydraulic transmission under load. Disconnecting the transmission will give a measure of the engine-structure radiated noise, provided the inlet and exhaust noise are piped out of the measuring room and the cooling fan is disconnected. But note that the engine is now unloaded, and its noise may be different from the loaded case. Also an unfueled engine may be driven by a "quieted" electric motor so that the so-called engine "mechanical" noise is measured and the "combustion" noise is excluded. However, in this procedure the mechanical forces in the engine will be different from the normal fueled running situation, and we cannot say that this procedure will give the "true" mechanical noise.

Selective wrapping or enclosure of different parts of a machine is a useful procedure often used. If the machine is completely enclosed with a tight-fitting sealed enclosure and the parts of the enclosure are sequentially removed exposing different machine surfaces, then, in principle, the noise from these different surfaces can be measured. This method has the advantage that it is not normally necessary to stop any part of the machine, and thus the machine operation is unchanged. However, some minor changes may occur if any damping of the machine surfaces occurs or because sound will now diffract differently around the enclosure.

Measurements of the sound power radiated from different surfaces of a diesel engine using this selective wrapping approach are discussed in Refs. 24 and 25. An array of 30 microphone positions on a spherical surface surrounding the engine was used for both the lead-wrapping and bare-engine sound measurements.

Overall and one-third octave sound pressure level data were gathered. These data were then converted to sound intensity by assuming (1) spherical propagation, (2) no reflections, and (3) that the intensity is equal to the mean square pressure divided by the air impedance ρc. Then the sound intensity data were integrated numerically over the surface of the spherical measuring surface enclosing the engine to obtain the sound power.

The lead was wrapped on the engine so that later a particular surface, for example the oil pan, could be exposed. The one-third octave band results for the fully wrapped engine compared with the results of the bare engine at the 1500 engine speed, 542-N-m torque operating condition are shown in Fig. 28*a*. This comparison shows that the lead wrapping is ineffective at frequencies at and below the 250 Hz one-third octave band.

Eight individual parts of the engine were chosen for noise source identification and ranking purposes using the measurement technique with the spherical microphone array and the engine lead-wrapped. The measurements were made by selectively exposing each of the eight parts, one at a time, while the other seven parts were encased in the 0.8-mm foam-backed lead. Sound power level measurements were then made for each of the eight parts for three separate steady-state operating conditions. One-third-octave band comparisons with the fully wrapped engine are shown in Figs. 28*b* and 28*c* for

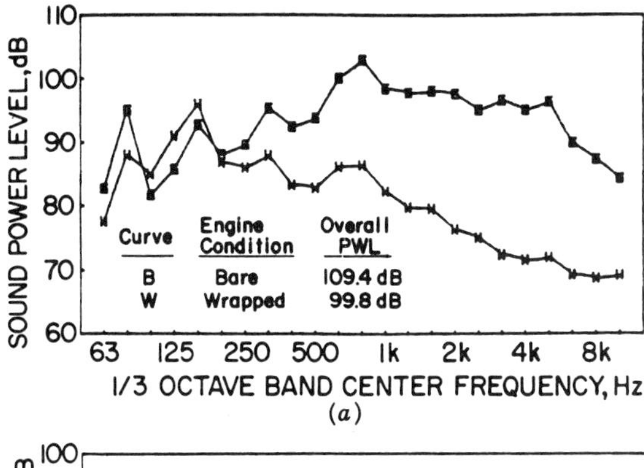

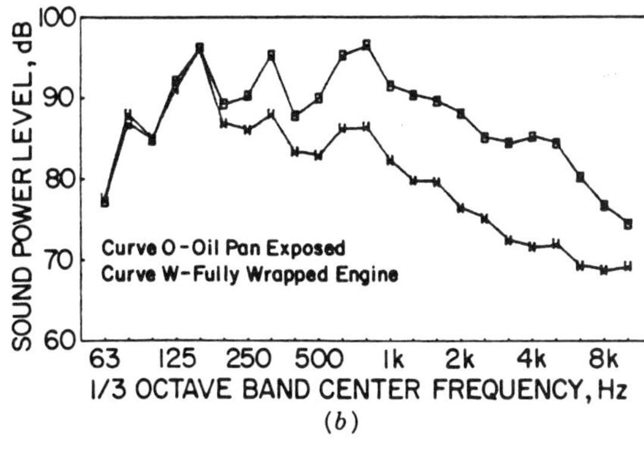

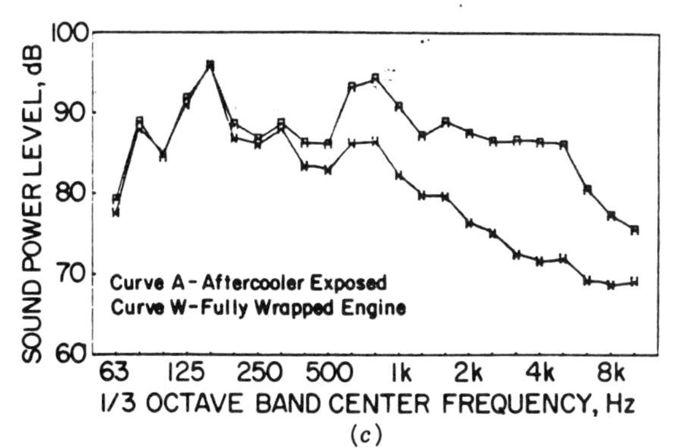

Fig. 28 Sound power level of engine and parts at 1500 rpm and 542-N-m torque: (*a*) bare and fully wrapped engine, (b) oil pan and fully wrapped engine, (c) aftercooler and fully-wrapped engine. (From Ref. 24.)

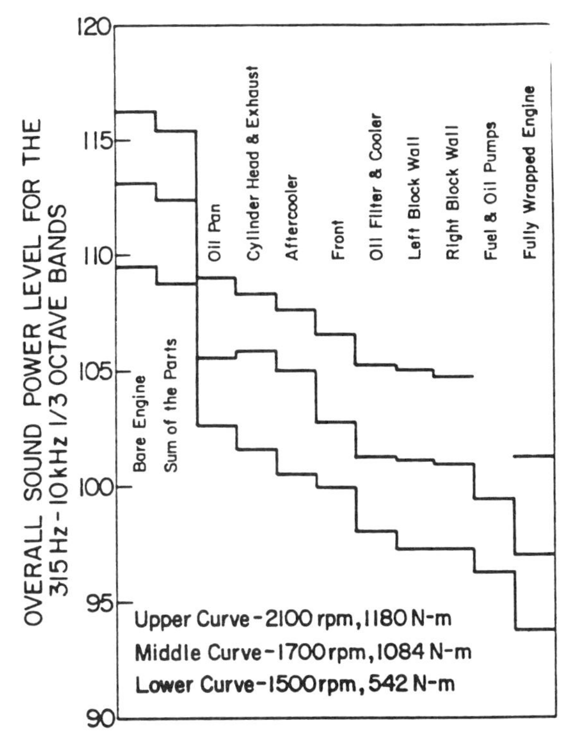

Fig. 29 Noise source ranking in terms of sound power levels from lead-wrapping far-field spherical array method for the 315 Hz to 10 kHz one-third-octave bands. (From Ref. 24.)

two of the eight parts at the 1500 rpm operating condition. These plots show that the sound power measurements using lead wrapping are not always accurate for the weaker parts until the frequency exceeds 1000 Hz. A plot of the noise source ranking using overall sound power levels for the eight parts under investigation for the 315-Hz to 10-kHz one-third-octave bands is given in Fig. 29.

Similar selective wrapping techniques have been applied to a complete truck (Fig. 30). Exhaust noise was suppressed with an oversize muffler, the engine was wrapped, and even the wheel bearings were covered! The results of this investigation[26] are shown in Figures 31*a* and 31*b*. Although the selective wrapping technique is very instructive, it is time consuming and this has led to the search for other techniques that do not require complete enclosure, as described later.

8 FREQUENCY ANALYSIS

If the noise (or vibration) of either the complete machine or of some part (obtained by selective operation or

Fig. 30 Selective wrapping and oversize muffler used on a diesel engine truck.

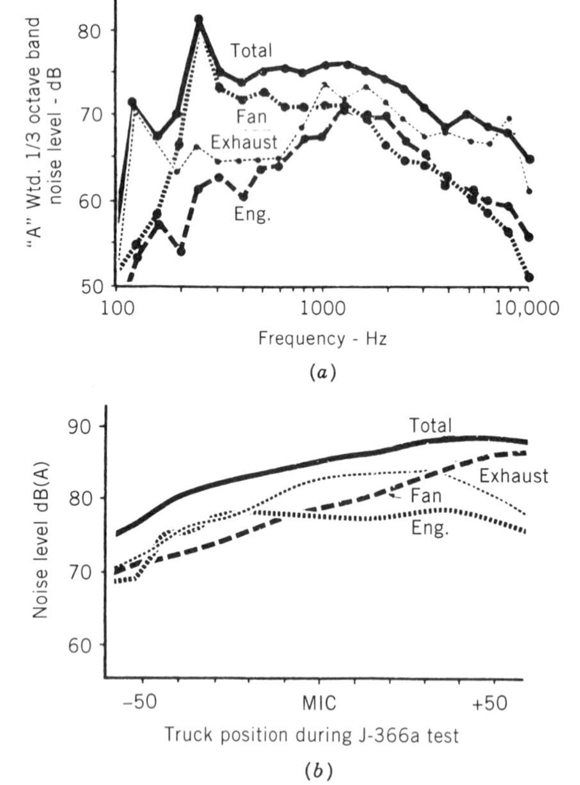

Fig. 31 A-weighted sound levels obtained from selective wrapping and selective operation techniques applied to truck in Fig. 30: (*a*) one-third-octave analysis, and (*b*) A-weighted overall sound level contributions of truck sources as a function of truck drive by position (in feet). (From Ref. 26.)

sequential wrapping) is measured, then various analyses may be made of the data. The most common analysis performed is *frequency analysis*. The reason probably is because one of the most important properties of the ear is performing frequency analysis of sounds. Analog analyzers (filters) have been available for many years. More recently real-time analyzers and fast Fourier transform (FFT) analog/digital minicomputer analyzers have become available.

The analog and real-time analyzers may be either of the parallel or the swept filter type, and these filters may be either of the constant percentage or the constant bandwidth type. With analog filters, we should note that the standard error in the measurements is inversely proportional to the square root of the bandwidth and the averaging time. Thus care should be taken to use large averaging times with narrow-band measurements. With real-time and FFT analyzers different problems arise including aliasing and leakage errors.

The real-time and FFT analyzers speed up the frequency analysis of the data, but because of the analysis errors that are possible (particularly with the FFT) great care should be taken. With FFT analyzers antialiasing filters can be used to prevent frequency folding, and if reciprocating machine noise is analyzed, the analysis period must comprise the repetition period (or multiples) or incorrect frequencies are diagnosed.

Assuming the frequency analysis has been made correctly then it can be an important tool in identifying noise sources. With reciprocating and rotating machinery, pure tones are found that depend on the speed of rotation N in rpm, and geometrical properties of the systems. As discussed in Section 5 of this chapter, several frequencies can be identified with rolling element bearings[13,15,27] including the fundamental frequency $N/60$ Hz, the inner and outer race flaw defect frequencies and the ball flaw defect, and the ball resonance and the race resonance frequencies. With gears, noise occurs at the tooth contact frequency $tN/60$ Hz and integer multiples (where t is the number of gear teeth). Often the overtone frequencies are significant. In fan noise, the fundamental blade-passing frequency $nN/60$ (where n is the number of fan blades) and integer multiples are important noise sources. Further machine noise examples could be quoted. Since these frequencies just mentioned are mainly proportional to machine rotational speed, a change of speed or load will usually immediately indicate whether a frequency component is related to a rotational source or not. A good example here is diesel engine noise. Figure 32*a* shows the time history and power spectrum of the cylinder pressure of a six-cylinder direct injection engine,[28] measured by Seybert. Ripples in the time history are thought to be caused by a resonance effect in the gas above the piston producing a broad peak in the frequency spectrum in the region of 3–5000 Hz. See Fig. 32*b*. That this resonance frequency *is* related to a gas resonance and not to the rotational speed can clearly be demonstrated since if the speed is changed the frequency is almost unchanged (for constant load), although if the load is increased (at constant speed) the frequency increases. On further investigation the resonance frequency is found to be proportional to the square root of the absolute temperature of the gas (the gas temperature increases with load) and thus as the load increases, so does the resonance frequency.

Frequency analysis can sometimes be used with advantage to reveal information about noise or vibration sources and/or paths by carefully changing the source or the path in some controlled way. The following discussion gives some examples for an air conditioner and an airplane.

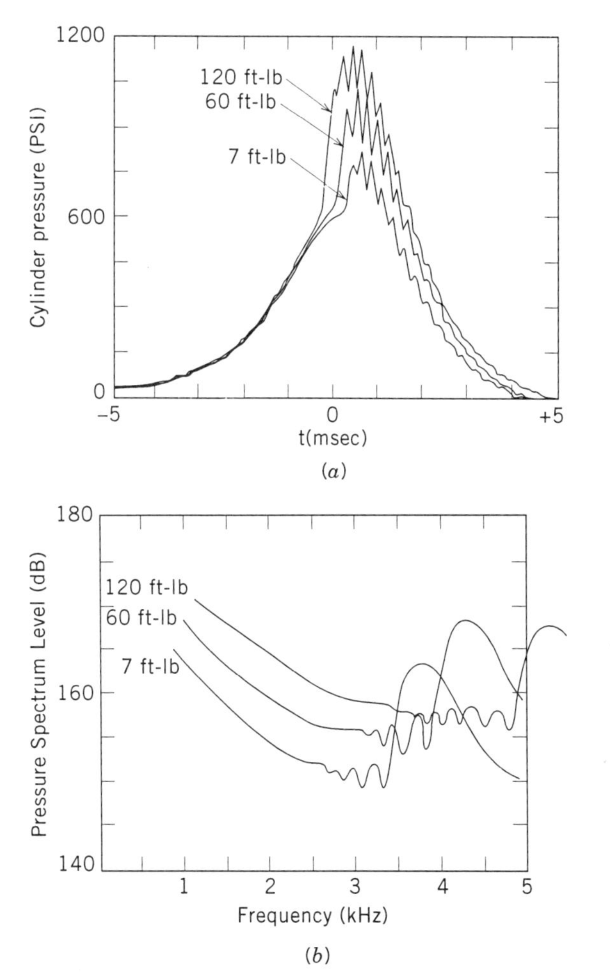

Fig. 32 (*a*) Diesel engine cylinder combustion pressures versus crank angle for three engine load conditions at 2400 rpm; (*b*) combustion pressure spectra at 2400 rpm. (From Ref. 28.)

Change of Excitation Frequency This can give guidance on sources of noise and resonances in a system, as already described with a diesel engine, when the load (and speed) are changed. In some cases (e.g., electrical machines) changing speed is difficult. In the case of the noise from an air conditioner, mechanical resonances in the system can be important.[29] These could be examined by exciting the air conditioner structure at different frequencies with a vibrator. However, this means changing the mechanical system in an unknown way. The approach finally adopted in Ref. 29 was to drive the air conditioner electric motors with an alternating current that could be varied from 50 to 70 Hz from a variable speed DC motor–AC generator set. Since the noise problem was caused by pure-tone magnetic force excitation at twice line (mains) frequency, this was monitored with a microphone at the same time as the motor acceleration. Figure 33 shows the results. This experiment showed that the motor mounts were too stiff. Reducing the stiffness eliminated the pure-tone noise problem. Hague[30] has shown how this technique can be improved further in air-handling systems by using an electric motor in which rotational speed, line frequency, and torque magnitude (both steady and fluctuating) can be varied independently.

In the case just discussed with the air conditioner, it was shown that the transmission path was through the motor mount (which was too stiff). In cases where airborne and structure-borne noise are both known to be problems and the source is known (an engine), the dominant path can sometimes be determined quantitatively by cutting one or more of the paths or by modifying the paths in a known way. Examples include engine noise

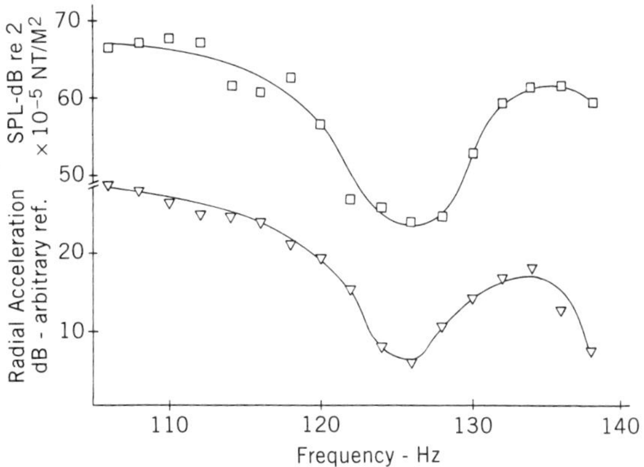

Fig. 33 Sound pressure level and motor acceleration of condenser fan. (From Ref. 29.)

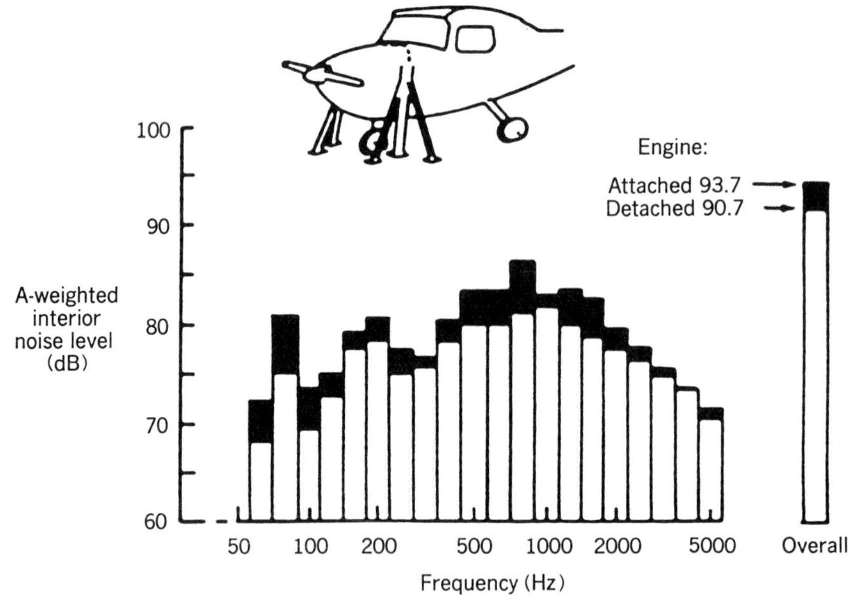

Fig. 34 Determination of structure-borne engine noise by engine detachment (from Ref. 31 with permission).

problems and airborne and structure-borne transmission in cars, farm tractors, and airplanes.

Path Blocking One example is shown here of the evaluation of the structure-borne path of engine noise transmission by detachment of the engine from the fuselage in ground tests.[31] Figure 34 shows the A-weighted sound pressure levels of the cabin noise with and without the engine connected to the fuselage. When the engine was detached, it was moved forward about 5 cm to prevent any mechanical connections. The overall reduction of 3 dB suggests that the contributions of airborne and structure-borne noise to the cabin noise are about equal for this situation. There is a variation in the structural noise contribution with frequency, however, and at some frequencies the structure-borne contribution is seen to exceed the airborne contribution. McGary has presented a method of evaluating airborne and structure-borne paths in airplane structures.[32]

Airborne paths can be reduced or "blocked" by the

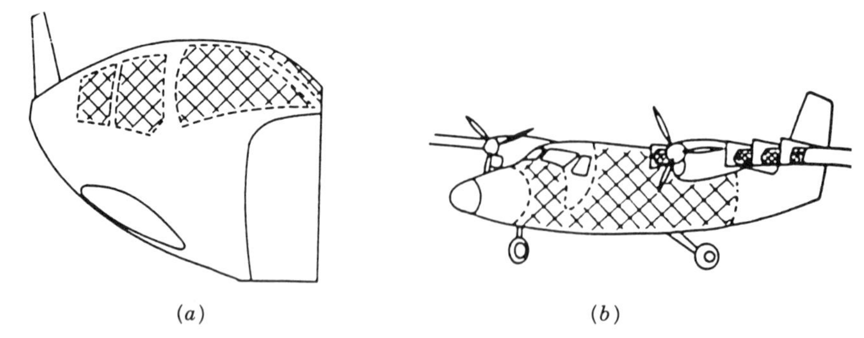

Fig. 35 Surfaces of airplane walls covered to aid in identifying importance of airborne and structure-borne paths. [From Ref. 33, based on figures from V. L. Metcalf and W. H. Mayes, "Structure borne Contribution to Interior Noise of Propellor Aircraft," *SAE Trans.*, Sect. 3, Vol. 92, 1983, pp. 3.69–3.74 (also SAE Paper 830735) and S. K. Jha and J. J. Catherines, "Interior Noise Studies for General Aviation Types of Aircraft, Part II: Laboratory Studies," *J. Sound Vib.*, Vol. 58, No. 3, 1978, pp. 391–406, with permission.]

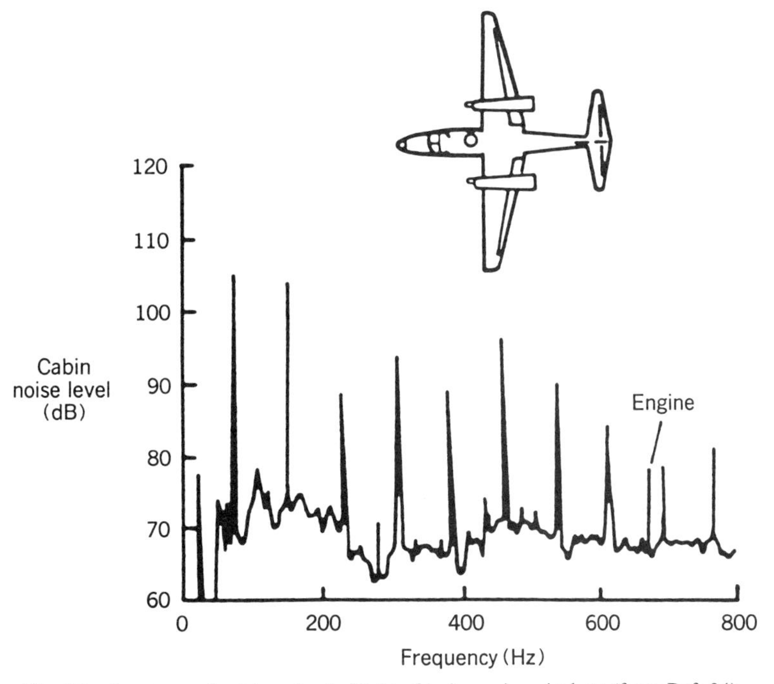

Fig. 36 Spectrum of cabin noise in flight of twin engine airplane (from Ref. 34).

addition of lead sheets or mass-loaded vinyl sheets of 5 to 10 kg/m^2. Figure 35, for example, shows areas of the cabin walls of airplanes that were covered to reduce the airborne transmission.[33]

The spectrum of the noise in the cabin of a light twin-engine airplane is shown in Figure 36. Discrete frequency tones associated with the blade passage frequency and harmonics are evident with the first two being dominant. The broadband noise at about 70 dB is associated with boundary layer noise. The tone at about 670 Hz can be related to the engine turbine speed and suggests the presence of engine-generated structure-borne noise.[34]

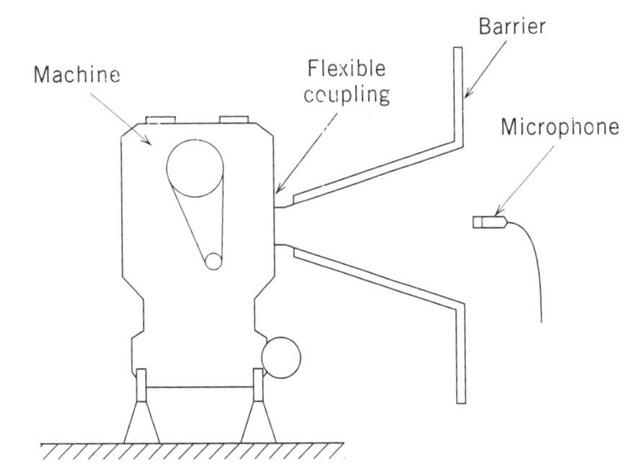

Fig. 37 Acoustical duct used to identify sources.

9 OTHER CONVENTIONAL METHODS OF NOISE SOURCE IDENTIFICATION

Acoustic ducts have been successfully used by Thien[35] as an alternative to the selective wrapping approach (see Fig. 37). One end of the duct is carefully attached to the part of the machine under examination by a "sound-proof" elastic connection. Thien claims accurate repeatable results from this method. However, one should note various difficulties. Sealing the duct to the machine may be difficult. The surroundings should be made as anechoic as possible to prevent contamination from any reverberant field. Theoretically the duct may alter the acoustic radiation from the machine part examined except for the frequency region much above coincidence. Fortunately this frequency region is the one normally encountered with heavy machinery, and since in this region radiation is mainly perpendicular to the surface, it may be little affected in practice.

Acoustic ducts or guides have also been used to investigate paths of noise. Figure 38 shows an example where

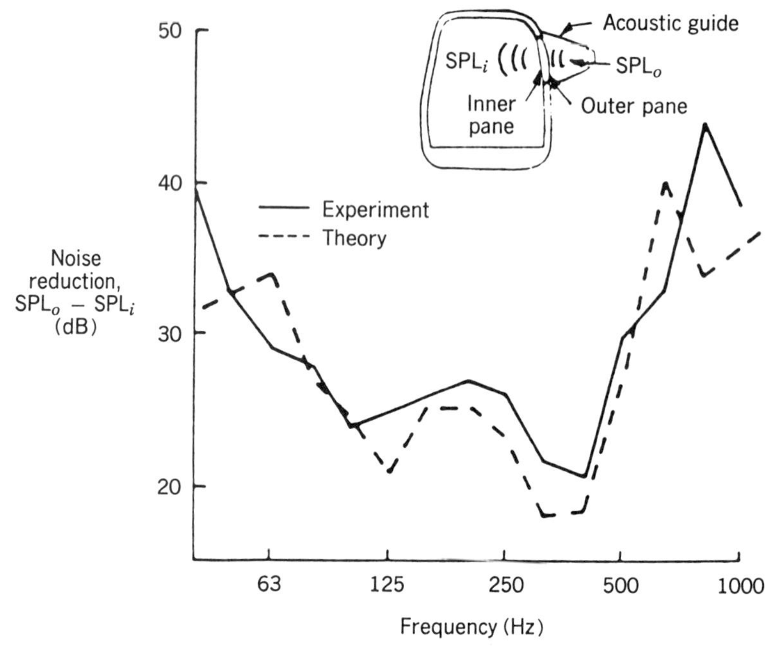

Fig. 38 Investigation of airborne noise path through cabin wall of light airplane using acoustic guide or duct. (From Ref. 36.)

airborne noise was directed onto the cabin wall of a light airplane. The noise reduction through the wall of the airplane was measured and compared with theory. The agreement between the experimental results and the modal theory was excellent at frequencies below about 250 Hz.[36]

Near-field measurements are very often used in an attempt to identify major noise sources on machines. This approach must be used with extreme care. If a microphone is placed in the acoustic near field where kr is small (k is the wavenumber = 2π frequency $f \div$ wavespeed c, and r is the source to microphone distance), then the sound intensity is not proportional to sound pressure squared, and this approach gives misleading results. This is particularly true at low frequency. The acoustic near field is reactive: The sound pressure is almost completely out-of-phase with the acoustic particle velocity.

Thus, this near-field approach is unsuitable for use on relatively "small" machines such as engines. However, for "large" machines such as vehicles with several major sources including an engine cooling fan, exhaust, or inlet, the near-field approach has proved useful.[37] In this case where the machine is large, of characteristic dimension l (length or width), it is possible to position the microphone so that it is in both the near geometric field, where r/l is small, and the far acoustic field, where kr is reasonably large (except at low frequency). In this case the microphone can be placed relatively close to each major noise source. Now the sound intensity is proportional to the sound pressure, and the well-known inverse square law applies. Besides the acoustic near-field effect already mentioned there are two other potential problems with this approach: (i) source directivity and the need to use more than one microphone to describe a large noise source and (ii) contamination of microphone signals placed near individual sources by sound from other stronger sources. Unfortunately, *contamination* cannot be reduced by placing the microphones closer to the sources because then the acoustic near field is increased. However, *contamination* from other strong sources can be allowed for by empirical correction using the inverse square law and the distance to the contaminating source.

Despite the various potential problems with the near-field method, it has been used quite successfully to identify the major noise sources on a large diesel engine truck.[37] By placing microphones near each major noise source (Fig. 39) and then extrapolating to a position 50 ft. (14.5 m) from the truck (Fig. 40) good agreement could be obtained with separate *selective operation* and *selective wrapping* drive-by measurements (Fig. 41). Averaging over five positions near each source and correcting for contaminating sources gave quite close agreement (Fig. 42).

Surface velocity measurements have been used by

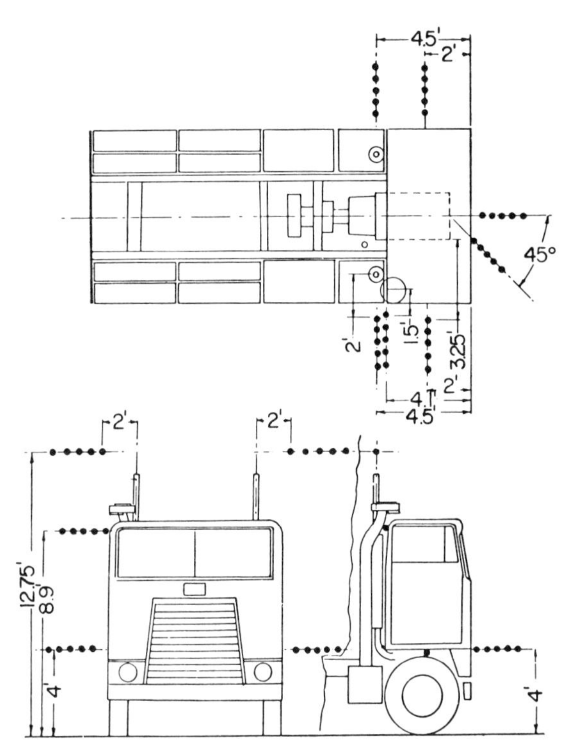

Fig. 39 Microphone positions used in near-field measurements on a truck. (From Ref. 37.)

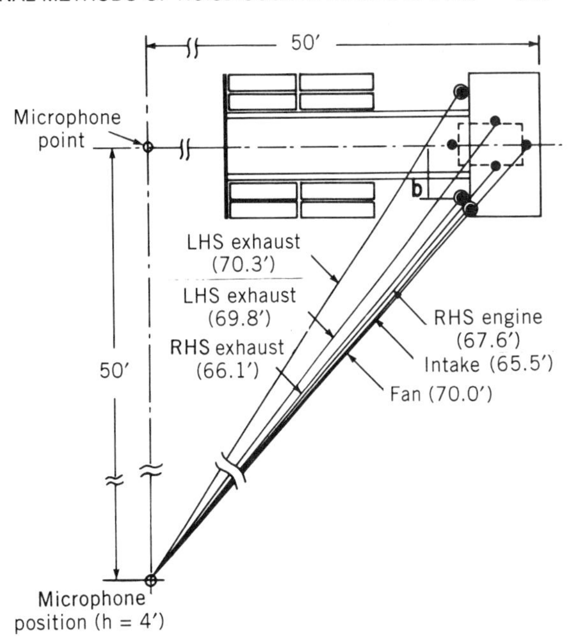

Fig. 40 The position (in feet) of truck (in Fig. 39) in drive-by test when truck engine reached its governed speed of 2100 rpm in sixth gear. Extrapolations to 50 ft measurement location shown. (Note: 1′ = 1 ft, LHS = left-hand side, RHS = right-hand side). (From Ref. 37.)

several investigators to try to determine dominant sources of noise on engines and other machines. The sound power W_{rad} radiated by a vibrating surface of area S is given by

$$W_{rad} = \rho c S \langle v^2 \rangle \sigma_{rad}, \tag{24}$$

where ρc is the air characteristic impedance, $\langle v^2 \rangle$ the space-average of the mean-square normal surface velocity, and σ_{rad} the radiation efficiency. Chan and Anderton[38] have used this approach. By measuring the sound power and the space-averaged mean-square velocity on several diesel engines they calculated the radiation efficiency σ_{rad}. They concluded that above 400 Hz for most diesel engines σ_{rad} was approximately 1. There was a scatter of ±6 dB in their results, although this was less for individual engines. Since σ_{rad} is difficult to calculate theoretically for structures of complicated geometry such as a diesel engine, the assumption that σ_{rad} = 1 can thus give an approximate idea of the sound power radiated by each component.

Mapping of the sound level around a machine has been shown[39] to be a useful simple procedure to identify

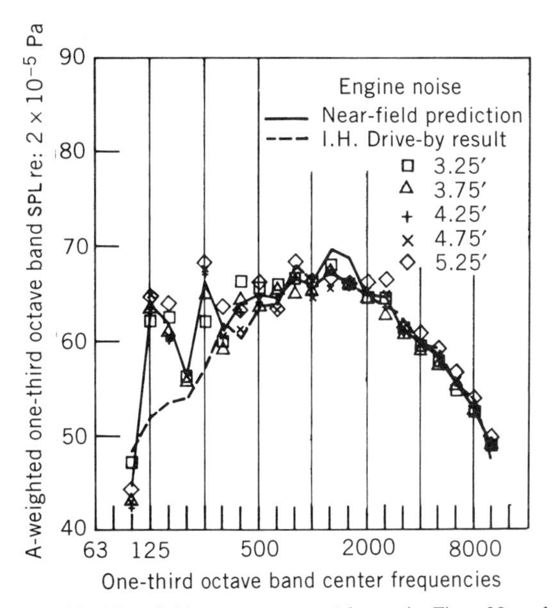

Fig. 41 Near-field measurements (shown in Figs. 39 and 40) extrapolated to 50 ft for comparisons with drive-by noise source identification procedure based on SAE J366a test. (From Ref. 37.)

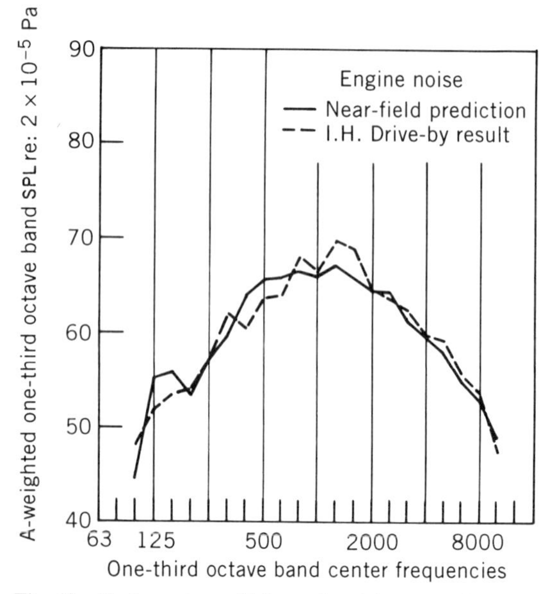

Fig. 42 Engine noise at 50 ft. predicted from near-field measurements (see Fig. 41) after corrections are made to eliminate the contribution from the exhaust system noise, compared with the drive-by result. (From Ref. 37.)

major noise sources. For sources radiating hemispherically, it is easy to show that the floor area within a certain level contour is proportional to the sound power output of the machine. This method is graphic and probably more useful to identify the noisiest machines in a workshop than the noisiest parts on the same machine because of the contamination problems between different sources in the latter case. Reverberation in a workshop will affect and distort the lower sound level contours.

10 USE OF CORRELATION AND COHERENCE AND TECHNIQUES TO IDENTIFY NOISE SOURCES AND PATHS

Correlation and coherence techniques in noise control date back over 40 years to work of Goff[40] who first used the correlation technique to identify noise sources. Although correlation has often been used in other applications, it has not been widely used in noise source identification. One exception is the work of Kumar and Srivastava, who reported some success with this technique in identifying noise sources on diesel engines.[41] Since the ear acts as a frequency analyzer, the corresponding approach in the frequency domain (coherence) instead of the time domain is usually preferable. Crocker and Hamilton have reviewed the use of the coherence technique in modeling diesel engine noise.[42] Such a coherence model can also be used, in principle, to give some information about noise sources on an engine. Figure 43 shows an idealized model of a multiple-input, single-output system. It can be shown[42] that the auto spectrum of the output noise S_{yy} is given by

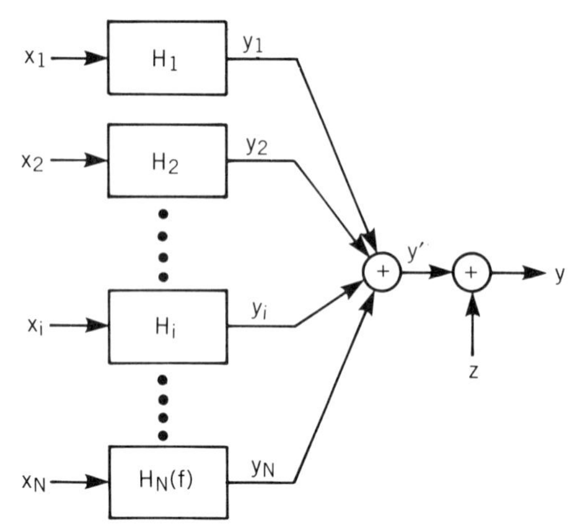

Fig. 43 Multiple-input, single-output system with uncorrelated noise $z(t)$ in output.

$$S_{yy} = \sum_{i=l}^{N} \sum_{i=l}^{N} S_{ij} H_i H_j + S_{zz} \tag{25}$$

where N is the number of inputs, the S_{ij} are cross-spectral densities between inputs, H_i and H_j are frequency responses, and S_{zz} is the autospectral density of any uncorrelated noise z present at the output. Note that the S and H terms are frequency dependent.

If there are N inputs, then there will be N equations:

$$S_{iy} = \sum_{i=l}^{N} H_j S_{ij} \qquad i = 1, 2, 3, \ldots, N \tag{26}$$

The frequency responses H_1, H_2, $H_3, \ldots, H_N$ can be found by solving the set of N equations (26).

Several researchers have used the coherence approach to model diesel engine noise.[42–45] This approach appears to give useful information on a naturally aspirated diesel engine, which is combustion-noise dominated and where the N inputs are $x_1, \ldots, x_N$, the cylinder pressures measured by pressure transducers in each cylinder. In this case H_i, the frequency response (transfer function) between the ith cylinder and the far-field microphone can be calculated. Provided proper frequency averaging is performed, the difficulty of high coherence between the cylinder pressures can be overcome because the phasing between the cylinder pressures is exactly known.[46] In this case the quantity $S_{ii}|H_i|^2$ may be regarded as the far-field output noise contributed by the ith cylinder. This may be useful noise source information for the engine designer.

Hayes et al.[47] in further research have extended this coherence approach to try to separate combustion noise from piston-impact noise in a running diesel engine.

Wang and Crocker showed that the multiple coherence approach could be used successfully to separate the noise from sources in an idealized experiment such as one involving three loudspeakers, even if the source signals were quite coherent.[48,49] However, when a similar procedure was used on the more complicated case of a truck which was modeled as a six-input system [fan, engine, exhaust(s), inlet, transmission] the method gave disappointing results and the simpler near-field method appeared better.[37,48] The partial coherence approach was also used in the idealized experiment and the truck experiments. It is believed that contamination between input signals and other computational difficulties may have been responsible for the failure of the coherence method in identifying truck-noise sources.

11 USE OF SOUND INTENSITY FOR NOISE SOURCE AND PATH IDENTIFICATION

Most studies[25,40] have shown that sound intensity measurements are quicker and more accurate than the selective wrapping approach in identifying and measuring the strength of noise sources.

The sound intensity I is the net rate of flow of acoustic energy per unit area. The intensity I_r in the r direction is

$$I_r = \langle p u_r \rangle, \qquad (27)$$

where p is the instantaneous sound pressure, u_r is the acoustic particle velocity in the r direction, and the angle brackets denote a time average. The sound power W radiated by a source can be obtained by integrating the component of the intensity I_n normal to any surface enclosing the source:

$$W = \int I_n \, dS. \qquad (28)$$

The sound pressure p and particle velocity are normally determined with two closely spaced microphones (1 and 2) and the surface S is a surface enclosing the source.

Chung first used this technique with the surface of a diesel engine subdivided into $N = 98$ radiating areas.[50] The power was determined from each area in about 2 min each. The engine was source-identified in less than a day, which involved much less time than the selective wrapping technique. Since the intensity measurements were made only 20 mm above the engine surfaces, an anechoic or semi-anechoic room was unnecessary. Many other researchers have used this sound intensity approach to identify noise sources since then, as is discussed in more detail in Chapter 106.

REFERENCES

1. R. H. Bolt and K. U. Ingard, "System Considerations in Noise Control Problems," in C. M. Harris (Ed.), *Handbook of Noise Control*, McGraw-Hill, New York, 1957, Chapter 22.
2. R. H. Lyon, *Machinery Noise and Diagnostics*, Butterworths, Boston, 1987.
3. D. A. Bies and C. H. Hansen, *Engineering Noise Control*, 2nd ed., Chapman & Hall, London, 1996.
4. M. J. Crocker and F. M. Kessler, *Noise and Noise Control*, Volume II, CRC Press, Boca Raton, 1982, Chapter 1.
5. I. Sharland, *Woods Practical Guide to Noise Control*, Woods of Colchester Limited, printed by Waterlow and Sons Ltd, London, 1972.
6. R. S. Jackson, "The Performance of Acoustic Hoods at Low Frequencies," *Acustica*, Vol. 12, 1962, p. 139.
7. R. S. Jackson, "Some Aspects of the Performance of Acoustic Hoods," *J. Sound Vib.*, Vol. 3, No. 1, 1966, p. 82.
8. L. H. Bell and D. H. Bell, *Industrial Noise Control*, 2nd ed., Marcel Dekker, New York, 1994.
9. C. M. Harris (Ed.), *Handbook of Acoustical Measurements and Noise Control*, 3rd ed., McGraw-Hill, New York, 1991.
10. L. L. Beranek and I. L. Ver (Eds.), *Noise and Vibration Control Engineering*, Wiley, New York, 1992.
11. S. Skaistis, *Noise Control of Hydraulic Machinery*, Marcel Dekker, New York, 1988.
12. P. L. Timar (Ed.), *Noise and Vibration of Electrical Machines*, Elsevier, Amsterdam, 1989.
13. J. E. Shahan and G. Kamperman, "Machine Element Noise," in *Handbook of Industrial Noise*, Industrial Press, 1976, Chapter 8.
14. M. P. Norton, *Fundamentals of Noise and Vibration Analysis for Engineers*, Cambridge University Press, Cambridge, 1989.
15. P. Eschmann, L. Hasbargen, and K. Weigand, *Ball and Roller Bearings: Theory, Design, and Application*, Wiley, Chichester, 1985.
16. W. B. Rowe, *Hydrostatic and Hybrid Bearings Designs*, Butterworths, London, 1983.

17. D. D. Fuller, *Theory and Practice of Lubrication for Engineers*, 2nd ed. Wiley, Chichester, 1984.
18. L. D. Mitchell, "Gear Noise: The Purchaser's and the Manufacturer's Views. Noise and Vibration Control Engineering," *Proceedings of the Purdue Noise Control Conference*, M. J. Crocker (Ed.), Purdue University, July 1971, pp. 95–106.
19. W. M. Viebrock and M. J. Crocker, "Noise Reduction of a Consumer Electric Clock," *Sound and Vibration*, Vol. 7, No. 3, 1973, pp. 22–26.
20. K. M. Ng, "Control Valve Noise," *ISA Trans.*, Vol. 33, 1994, pp. 275–286.
21. G. Reethof, "Turbulence-Generated Noise in Pipe Flow," *Ann. Rev. Fluid Mech.*, 1978, pp. 333–367.
22. W. F. Blake, *Mechanics of Fluid-Induced Sound and Vibration*, Academic Press, New York, 1986.
23. C. E. Ebbing and T. H. Hodgson, "Diagnostic Tests for Locating Noise Sources," *Noise Control Eng.*, Vol. 3, 1974, pp. 30–46.
24. M. C. McGary and M. J. Crocker, "Surface Intensity Measurements on a Diesel Engine," *Noise Control Eng.*, Vol. 16, No. 1, 1981, pp. 26–36.
25. T. E. Reinhart and M. J. Crocker, "Source Identification on a Diesel Engine Using Acoustic Intensity Measurements," *Noise Control Eng.*, Vol. 18, No. 3, 1982, pp. 84–92.
26. R. L. Staadt, "Truck Noise Control," in *Reduction of Machinery Noise*, (rev. ed.), Short Course Procs., Purdue University, December 1975, pp. 158–190.
27. R. A. Collacott, "The Identification of the Source of Machine Noises Contained within a Multiple-Source Environment," *Appl. Acoustics*, Vol. 9, No. 3, 1976, pp. 225–238.
28. A. F. Seybert, "Estimation of Frequency Response in Acoustical Systems with Particular Application to Diesel Engine Noise," Ph.D. Thesis, Purdue University, Herrick Laboratories Report HL 76-3, December 1975.
29. A. F. Seybert, M. J. Crocker, et al., "Reducing the Noise of a Residential Air Conditioner," *Noise Control Eng.*, Vol. 1, No. 2, 1973, pp. 79–85.
30. J. M. Hague, "Dynamic Vibration Exciter of Torsional Axial, and Radial Modes from Modified D.C. Motor," Paper No. 962 presented at ASHRAE, Semi-Annual Meeting, Chicago, Illinois, February 13–17, 1977.
31. J. F. Unruh et al, "Engine Induced Structural-Borne Noise in a General Aviation Aircraft," NASA CR-159099, 1979 (see also SAE Paper 7900626).
32. M. C. McGary and W. H. Mayes, "A New Measurement Method for Separating Airborne and Structureborne Aircraft Interior Noise," *Noise Control Eng.*, Vol. 20, No. 1, 1983, pp. 21–30.
33. J. S. Mixon and J. F. Wilby, "Interior Noise," in H. H. Hubbard (Ed.), *Aeroacoustics of Flight Vehicles: Theory and Practice*, Volume 2: Noise Control, NASA Reference Publication 1258, August 1991, Chapter 16.
34. J. F. Wilby et al., "In-Flight Acoustic Measurements on a Light Twin-Engined Turbo prop Airplane," NASA CR-178004, 1984.
35. G. E. Thien, "The Use of Specially Designed Covers and Shields to Reduce Diesel Engine Noise," SAE Paper 730244, 1973.
36. R. Vailaitis et al., "Transmission Through Stiffened Panels," *J. Sound Vibr.*, Vol. 70, No. 3, 1980, 413–426.
37. M. J. Crocker and J. W. Sullivan, "Measurement of Truck and Vehicle Noise," 1978, SAE Paper 780387, see also SAE Transactions.
38. C. M. P. Chan and D. Anderton, "Correlation Between Engine Block Surface Vibration and Radiated Noise of In-Line Diesel Engines," *Noise Control Eng.*, Vol. 2, No. 1, p. 16, 1974.
39. P. Francois, "Isolation et Revetments," *Les Carte de Niveaux Sonores*, Jan.–Feb. 1966, pp. 5–17.
40. K. W. Goff, "The Application of Correlation Techniques to Source Acoustical Measurements," *J. Acoust. Soc. Am.*, Vol. 27, No. 2, 1955, pp. 336–346.
41. S. Kumar and N. S. Srivastava, "Investigation of Noise Due to Structural Vibrations Using a Cross-Correlation Technique," *J. Acoust. Soc. Am.*, Vol. 57, No. 4, 1975, pp. 769–772.
42. M. J. Crocker and J. F. Hamilton, "Modeling of Diesel Engine Noise Using Coherence," SAE Paper 790362.
43. J. Y. Chung, M. J. Crocker, and J. F. Hamilton, "Measurement of Frequency Responses and the Multiple Coherence Function of the Noise Generation System of a Diesel Engine," *J. Acoust. Soc. Am.*, Vol. 58, No. 3, 1975, pp. 635–642.
44. A. F. Seybert and M. J. Crocker, "The Use of Coherence Techniques to Predict the Effect of Engine Operating Parameters On Noise," *Trans. ASME, J. Eng. Ind.*, Vol. 97B, 1976, p. 13.
45. A. F. Seybert and M. J. Crocker, "Recent Applications of Coherence Function Techniques in Diagnosis and Prediction of Noise," *Proc. INTER-NOISE 76*, 1976, pp. 7–12.
46. A. F. Seybert and M. J. Crocker, "The Effect of Input Cross-Spectra on the Estimation of Frequency Response Functions in Certain Multiple-Input Systems," *Arch. Acoustics*, Vol. 3, No. 3, 1978, pp. 3–23.
47. P. A. Hayes, A. F. Seybert, and J. F. Hamilton, "A Coherence Model for Piston-Impact Generated Noise," SAE Paper 790274.
48. M. E. Wang, "The Application of Coherence Function Techniques for Noise Source Identification," Ph.D. Thesis, Purdue University, 1978.
49. M. E. Wang and M. J. Crocker, "Recent Applications of Coherence Techniques for Noise Source Identification, *Proc. INTER-NOISE 78*, 1978, pp. 375–382.
50. J. Y. Chung et al., "Application of Acoustic Intensity Measurement to Engine Noise Evaluation," SAE Paper 790502, 1979.

67

ACTIVE NOISE CONTROL

P. A. NELSON AND S. J. ELLIOTT

1 INTRODUCTION

Active techniques for controlling noise have developed rapidly during the 1980s and 1990s, largely because of the developments during this period in modern electronics. Relatively inexpensive digital signal processors have become widely available that enable analog audio frequency signals to be converted into digital form, processed via a digital filter, and then converted back into an analog signal with very little time delay. This electronic capability, together with the wave nature of sound, has made the implementation of active noise control a feasible practical proposition.

Relatively simple single-channel systems are now commercially available and have been used successfully, for example, for the control of low-frequency sound in air conditioning systems. In this case, the sound in the air ducts propagates only as plane waves (below a certain frequency) and destructive interference caused by the field produced by a control loudspeaker leads to successful suppression of the sound field in the duct downstream of the loudspeaker. The possibilities for extending these techniques to other types of acoustic field have stimulated considerable research in recent years, much of it summarized in Ref. 1. It is the physical nature of sound fields that most often dictates the limits to the performance of active techniques, and it is with this in mind that we review below the possibilities for and recent practical applications of active noise control. No attempt has been made here to deal with the suppression of noise through the active control of vibration. This is dealt with comprehensively in Ref. 2 and is also summarized in Chapter 59.

Note: References to chapters appearing only in the *Encyclopedia* are preceded by *Enc.*

At the time of writing, active control has been successfully applied to the suppression of low-frequency duct-borne broadband sound, low-frequency periodic sound generated by electrical power transformers, aircraft propellers and automobile and turbofan engines, and low-frequency broadband road-generated noise inside road vehicles. However, the design of active control systems remains a specialist activity and is only just reaching the stage of commercialization in a few, mostly high technology, applications. Perhaps the application that has achieved the most widespread commercial production is the "noise-canceling headset," which uses a simple feedback loop to suppress low-frequency sound in the region of the listener's ears. This and other possibilities will be described further below.

2 PRINCIPLES OF ACTIVE CONTROL

A simple single-channel active noise control system is illustrated in Fig. 1. Sound is detected by a microphone and processed via a digital filter (implemented on a special-purpose microprocessor) prior to being input to a loudspeaker that radiates the sound that is intended to destructively interfere with the "primary" unwanted sound. The success of the system is monitored by an "error microphone" placed downstream of the loudspeaker. The characteristics of the digital filter are designed (and often continuously adapted) in order to minimize the time-averaged signal at the error microphone.

In the following description of active techniques we will work from first principles and concentrate initially on the limitations imposed by the physical nature of sound fields. It is of course the principle of superposition that enables active control to be achieved. Thus, any

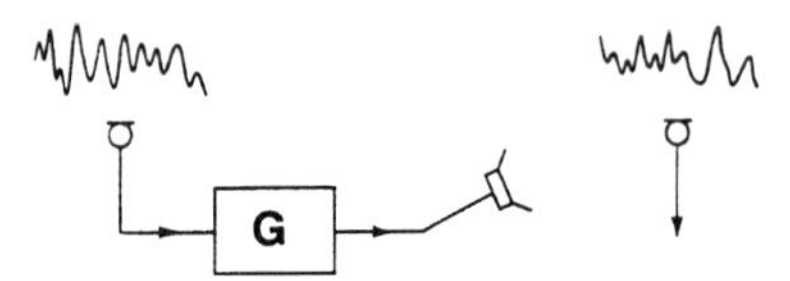

Fig. 1 Single-channel active noise control system showing a detection sensor whose output is passed through a signal processor (G) before being input to a loudspeaker. The success of the system is monitored by an error sensor.

number of sound fields that individually satisfy the governing (linear) wave equation can simply be superposed (in space and time) in order to produce a more desirable sound field. In evaluating the extent to which this can be achieved in practice, it is often useful to work at a single frequency and deal with the spatially dependent complex pressure $p(\mathbf{x})$, which satisfies the homogeneous Helmholtz equation

$$(\nabla^2 + k^2)p(\mathbf{x}) = 0, \tag{1}$$

where the complex pressure $p(\mathbf{x})$ is the sum of the complex pressures $p_p(\mathbf{x})$ and $p_s(\mathbf{x})$ in the primary and secondary sound fields. The wavenumber k is given by ω/c_0, where ω is the angular frequency of the fluctuation. While it is often convenient to evaluate the potential for active control by working at a single frequency, it should always be recognized that no constraint is necessarily placed on the causal relationship between the secondary and primary fields.[1] Nevertheless, calculations at a single frequency always reveal the best that can be achieved by active techniques.

3 ACTIVE CONTROL OF FREE-FIELD SOUND

3.1 Control of Sound with Continuous Source Layers

Figure 2 shows a primary-source distribution in an unbounded medium specified by the complex volume velocity per unit volume $q_{p,\text{vol}}(\mathbf{y})$, where the vector $\mathbf{y}$ is used as a position vector in the source coordinate system. The primary complex pressure field $p_p(\mathbf{x})$ satisfies the Kirchhoff–Helmholtz integral equation

$$\int_S [g(\mathbf{x}|\mathbf{y})\, \nabla_y p_p(\mathbf{y}) - p_p(\mathbf{y})\, \nabla_y g(\mathbf{x}|\mathbf{y})] \cdot \mathbf{n}\, dS = \begin{cases} p_p(\mathbf{x}), & \mathbf{x} \text{ within V}, \\ 0, & \mathbf{x} \text{ outside V}, \end{cases} \tag{2}$$

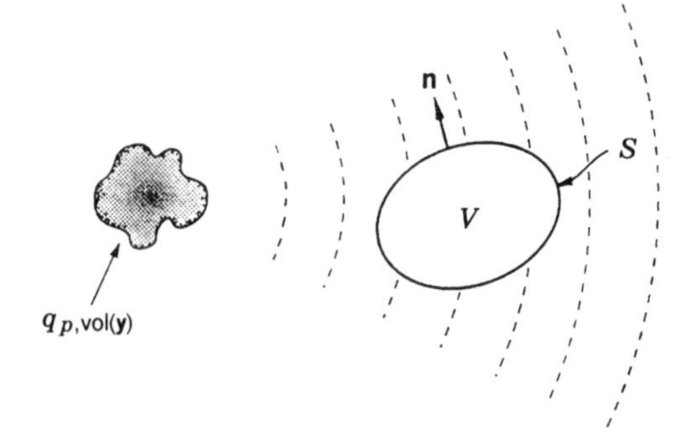

Fig. 2 Continuous monopole–dipole source layer placed on the surface S can be used to produce perfect cancellation of the sound field inside V while leaving the field outside V completely unchanged. Reprinted with kind permission of Academic Press.

where V is a restricted volume of the medium bounded by the surface S and $g(\mathbf{x}|\mathbf{y})$ is the free-space Green function $e^{-jk|\mathbf{x}-\mathbf{y}|}/4\pi|\mathbf{x}-\mathbf{y}|$. The symbol ∇_y represents the del operator in the $\mathbf{y}$-coordinate system. The terms in the surface integral on the left side of this equation can be interpreted as a surface distribution of monopole and dipole sources.[1] Thus, if one places on the surface S a secondary distribution of monopole and dipole sources whose strengths are respectively determined by $\nabla_y p_p(\mathbf{y})$ and $p_p(\mathbf{y})$, then one can generate a field equal to $-p_p(\mathbf{x})$ inside the volume V while leaving the field outside V unchanged. The possibilities for active control that this solution represents were first pointed out by Jessel[3] and Malyhuzinets.[4] While offering a very general solution to the problem of canceling a sound field over a spatial volume, the need to appropriately activate a continuous monopole/dipole source layer renders this technique extremely difficult to implement in practice.

3.2 Control of a Compact Primary-Source Distribution with a Series of Point Monopoles

Another possibility for producing complete cancellation of the far-field sound of a primary-source distribution is that suggested by Kempton.[5] For a primary-source distribution that is restricted to a spatial region of dimensions small compared to the acoustic wavelength, the radiated field can be represented by that of a series of point multipoles located at a given position in or close to the source region. It can be shown[6] that the monopole, dipole, quadrupole, and so on, moments associated with the multipole series are determined exactly by the properties of the primary-source distribution. It follows, therefore, that with the multipole moments appropriately determined, a series of multipole sources of opposite

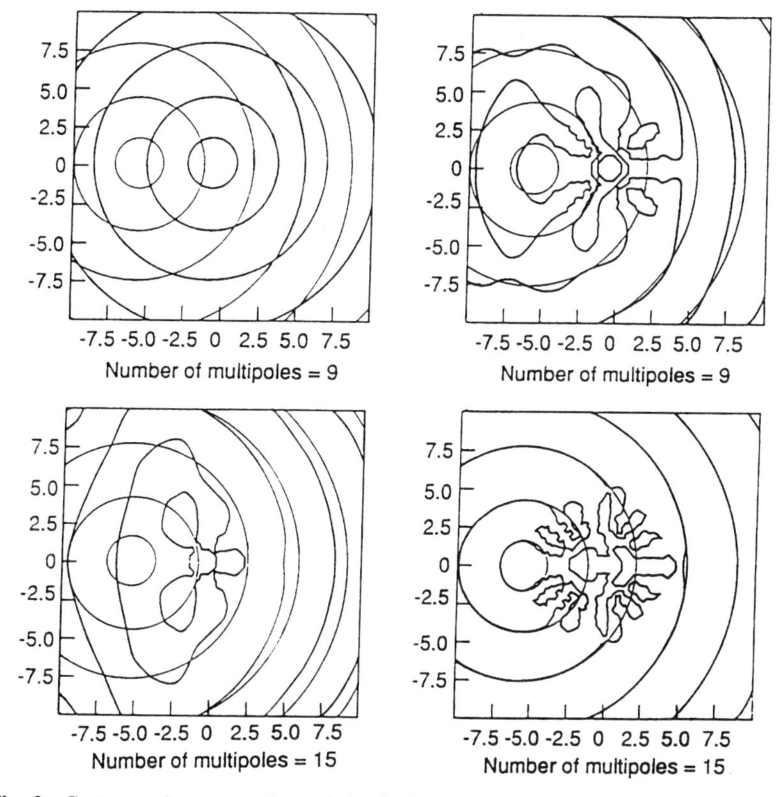

Fig. 3 Contours of constant phase of the fields due to a point monopole $(-\varepsilon, 0, 0)$ and a series of multipoles at the origin. The above results were computed by Kempton[5] for a value of $k\varepsilon = 5$. Reprinted with kind permission of Academic Press.

strength when placed at a position close to the primary source will produce cancellation of the far field. The results of some of Kempton's computer simulations[5] are shown in Fig. 3. Again, the difficulty in realizing a practical approximation to a multipole series of such complexity precludes the practical application of the technique.

3.3 Active Control with Discrete Monopole Secondary Sources

A more realistic approach to the active control of free-field sound is to accept that the secondary sources that one can use in practice consist of discrete monopole (loudspeaker) elements. The strengths of these secondary sources are adjusted in order to minimize a quadratic cost function such as the sum of the squared pressures at a discrete number of sensor locations. This approach has been applied by Kido and Onoda,[7] Piraux and Nayroles,[8] and Nelson et al.[9–11] Figure 4 shows L sensors at which the complex pressure produced by the primary source can be specified by the complex vector $\mathbf{p}_p$. The M secondary sources have complex strengths (volume velocities) specified by the complex vector $\mathbf{q}_s$. The total com-

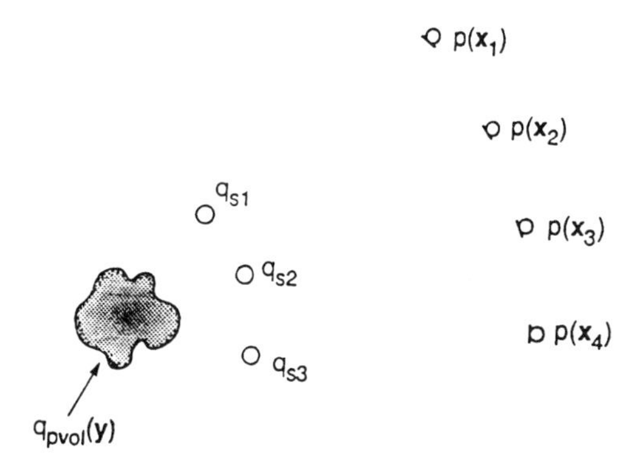

Fig. 4 Active control of sound with discrete secondary sources. The strengths of the secondary sources q_{sm} are adjusted to minimize the sum of the squared pressures $p(\mathbf{x}_l)$.

plex pressure at the sensors is thus given by the vector

$$\mathbf{p} = \mathbf{p}_p + \mathbf{Z}\mathbf{q}_s, \tag{3}$$

where $\mathbf{Z}$ is the $L \times M$ complex transfer impedance matrix that determines the pressure produced at the sensors by the secondary sources. For $M = L$, $\mathbf{p}$ can be made equal to zero by choosing $\mathbf{q}_s = -\mathbf{Z}^{-1}\mathbf{p}_p$ for a nonsingular $\mathbf{Z}$. Another approach, when $L > M$ is to determine $\mathbf{q}_s$, in order to minimize the sum of squared pressures,

$$J_p = \sum_{l=1}^{L} |p(\mathbf{x}_l)|^2 = \mathbf{p}^{\mathrm{H}}\mathbf{p}, \tag{4}$$

where H denotes the Hermitian transpose (complex-conjugate transpose) operation. Substitution of Eq. (3) into Eq. (4) yields

$$J_p = \mathbf{q}_s^{\mathrm{H}}\mathbf{A}\mathbf{q}_s + \mathbf{q}_s^{\mathrm{H}}\mathbf{b} + \mathbf{b}^{\mathrm{H}}\mathbf{q}_s + c, \tag{5}$$

which is a Hermitian quadratic function of the vector $\mathbf{q}_s$ with the matrix $\mathbf{A} = \mathbf{Z}^{\mathrm{H}}\mathbf{Z}$, the complex vector $\mathbf{b} = \mathbf{Z}^{\mathrm{H}}\mathbf{p}_p$, and the scalar $c = \mathbf{p}_p^{\mathrm{H}}\mathbf{p}_p$. If the matrix $\mathbf{A}$ is positive definite, it can be shown[9] that this function has a unique minimum J_{p0} corresponding to an optimal vector $\mathbf{q}_{s0}$. These are given by

$$J_{p0} = c - \mathbf{b}^{\mathrm{H}}\mathbf{A}^{-1}\mathbf{b}, \tag{6}$$

$$\mathbf{q}_{s0} = -\mathbf{A}^{-1}\mathbf{b}. \tag{7}$$

These equations provide an unequivocal definition of the best that can be done to reduce the sum of the mean-squared pressures in the sound field. One important result that follows from an analysis of this type is the minimum total power output to the sound field when the field of a point monopole primary source is controlled with a point monopole secondary source, that is, the minimum of the sum of the squared pressures produced at an infinite number of sensors on a surface in the far field of the source pair. It can be shown[9] that no appreciable reductions in power output can be achieved if the sources are separated by a distance that is greater than one-half wavelength. This result serves to emphasize that active techniques are most often successful at low frequencies.

4 ACTIVE CONTROL OF SOUND IN DUCTS

4.1 Active Control of Sound Propagation in an Infinite Rigid Waveguide

The complex pressure in the infinite rigid-walled waveguide illustrated in Fig. 5 can be expressed in terms of the eigenfunctions (mode shape functions) $\psi_n(x_1, x_2)$ of the duct cross section. Thus, for $x_3 \gg y_3$ we may write[12]

$$p(\mathbf{x}) = \sum_{n=0}^{N} \frac{\omega\rho_0\psi_n(x_1, x_2)}{2\kappa_n L_1 L_2} e^{-j\kappa_n y_3} \cdot \int_V q_{\mathrm{vol}}(\mathbf{y}) e^{j\kappa_n y_3}\psi_n(y_1, y_2)\, dV, \tag{8}$$

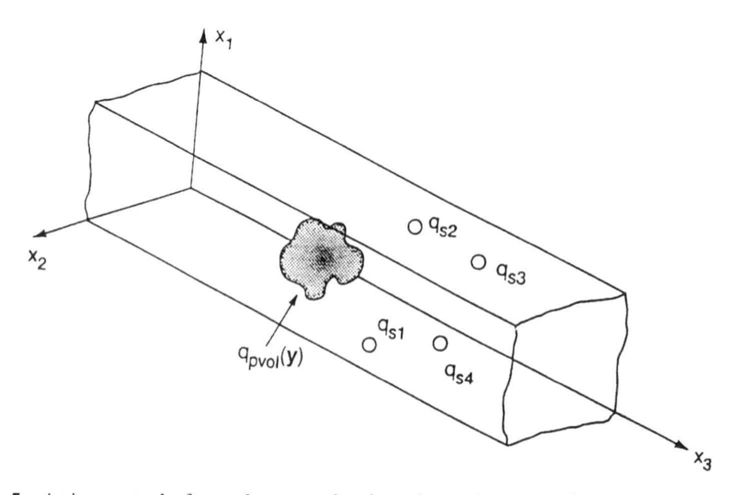

Fig. 5 Active control of sound propagating in a duct using a number of point secondary sources.

where ρ_0 is the density, the wavenumber

$$\kappa_n = \sqrt{k^2 - \left(\frac{n_1\pi}{L_1}\right)^2 - \left(\frac{n_2\pi}{L_2}\right)^2}$$

(the integers n_1 and n_2 characterizing a given mode), and the summation is taken over only the N propagating modes for which κ_n is real. If, as illustrated in Fig. 5, the source distribution is given by some primary-source distribution $q_{p,\text{vol}}(\mathbf{y})$ plus a number M of discrete secondary sources, the source strength distribution may be written as

$$q_{\text{vol}}(\mathbf{y}) = q_{p,\text{vol}}(\mathbf{y}) + \sum_{m=1}^{M} q_{sm}\delta(\mathbf{y} - \mathbf{y}_m), \tag{9}$$

where $\delta(\cdot)$ represents the three-dimensional Dirac delta function. Substitution into Eq. (8) shows that

$$p(\mathbf{x}) = \sum_{n=1}^{N} a_n\psi_n(x_1, x_2) = \mathbf{a}^{\mathrm{T}}\boldsymbol{\psi}, \tag{10}$$

where $\boldsymbol{\psi}$ is a vector of mode shape functions and $\mathbf{a}$ is a vector of complex-mode amplitudes given by

$$\mathbf{a} = \mathbf{a}_p + \mathbf{B}\mathbf{q}_s. \tag{11}$$

The vector $\mathbf{a}_p$ defines the complex-mode amplitudes produced by the primary source only, while the $N \times M$ matrix $\mathbf{B}$ determines the mode amplitudes produced by the secondary sources whose complex strengths are characterized by the vector $\mathbf{q}_s$. In the case $N = M$, with the secondary sources located to ensure that $\mathbf{B}$ is nonsingular, it follows that choosing $\mathbf{q}_s = -\mathbf{B}^{-1}\mathbf{a}_p$ ensures that $\mathbf{a} = 0$ and thus $p(\mathbf{x}) = 0$. Thus, if at a given frequency the number of secondary sources is chosen to equal the number of propagating modes, the sound field well downstream of the secondary sources can in principle be made equal to zero. Successful controller implementations based on this strategy have only recently been reported.[13,14]

4.2 Active Control of Plane-Wave Sound

In ducts excited at frequencies less than the lowest cut-on frequency $c_0\pi/L_1$ (assuming $L_1 > L_2$), only the plane-wave mode will propagate. The active control of such sound fields has been the subject of extensive research (see, e.g., Refs. 15–18). It follows from the above analysis that a single secondary source, when appropriately driven, will be sufficient to ensure zero pressure well downstream of the secondary source. It can be shown[1] that such a strategy ensures that downstream propagating plane waves are reflected back upstream; the secondary source provides a pressure-release boundary condition. A standing-wave field is then produced between the primary and secondary sources. A means of avoiding this was suggested by Swinbanks,[18] who showed that two secondary sources spaced apart along the duct axis would be used to provide perfect absorption of incident plane-wave sound over a limited bandwidth. It can also be shown in the low-frequency limit[1] that the secondary-source strengths required to accomplish this give a monopole–dipole source combination that follows from the general theory presented in Section 3.1.

5 ACTIVE CONTROL OF ENCLOSED SOUND FIELDS

5.1 Control at Low Modal Densities

For acoustic wavelengths that are appreciable compared to the dimensions of the enclosure, it is appropriate to use a modal model of the sound field. The complex pressure can be written as[1,19]

$$p(\mathbf{x}) = \sum_{n=0}^{\infty} \frac{\omega\rho_0 c_0^2 \psi_n(\mathbf{x})}{V[\omega c_0 D_{nn} + j(\omega^2 - \omega_n^2)]} \int_V q_{\text{vol}}(\mathbf{y})\psi_n(\mathbf{y})\, dV, \tag{12}$$

where $\psi_n(\mathbf{x})$ is the mode shape function corresponding to the nth mode, V is the enclosure volume, D_{nn} is a damping term determined by the admittance of the enclosure walls,[1,19] and ω_n is the natural frequency of the nth mode. It is now assumed, as in the case of duct-borne sound described in Section 4.1, that there is some primary source distribution in the enclosure and that the complex pressure can similarly be expressed as the *finite* summation

$$p(\mathbf{x}) = \sum_{n=0}^{N} a_n\psi_n(\mathbf{x}) = \mathbf{a}^{\mathrm{T}}\boldsymbol{\psi}. \tag{13}$$

If it is also assumed that M secondary sources are introduced to control the field, the vector $\mathbf{a}$ can again be expressed in the form

$$\mathbf{a} = \mathbf{a}_p + \mathbf{B}\mathbf{q}_s. \tag{14}$$

At first sight it appears that this problem can also be

treated by ensuring that $M = N$ and choosing $\mathbf{q}_s = -\mathbf{B}^{-1}\mathbf{a}_p$ to give $\mathbf{a} = 0$. However, in order to give an adequate representation of the sound field, the number N of modes required in the summation will necessarily be very much greater than the number M of secondary sources. The approach taken[10] is therefore to find the secondary-source strengths that minimize the total acoustic potential energy in the enclosure. This is given by

$$E_p = \frac{1}{4\rho_0 c_0^2} \int_V |p(\mathbf{x})|^2 \, dV = \frac{V}{4\rho_0 c_0^2} \mathbf{a}^{\mathrm{H}}\mathbf{a}, \quad (15)$$

where the second of these expressions results from the orthogonality property of the acoustic modes. Substitution of Eq. (14) then shows that

$$E_p = \frac{V}{4\rho_0 c_0^2} [\mathbf{q}_s^{\mathrm{H}}\mathbf{B}^{\mathrm{H}}\mathbf{B}\mathbf{q}_s + \mathbf{q}_s^{\mathrm{H}}\mathbf{B}^{\mathrm{H}}\mathbf{a}_p + \mathbf{a}_p^{\mathrm{H}}\mathbf{B}\mathbf{q}_s + \mathbf{a}_p^{\mathrm{H}}\mathbf{a}_p], \quad (16)$$

which is a quadratic function of $\mathbf{q}_s$ of the type defined by Eq. (5) and has a unique minimum value associated with an optimal vector of secondary-source strengths. Figure 6 shows the results of some computer simulations undertaken by Bullmore et al.[20] Global reductions in energy are produced by a single secondary source placed close to the primary source. Three secondary sources placed appropriately are also shown to produce global reduc-

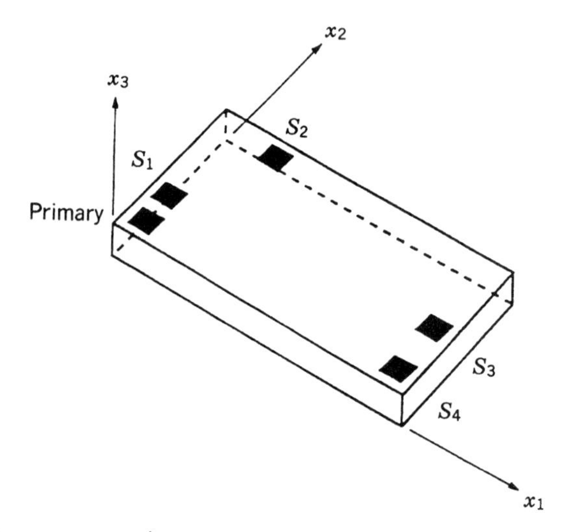

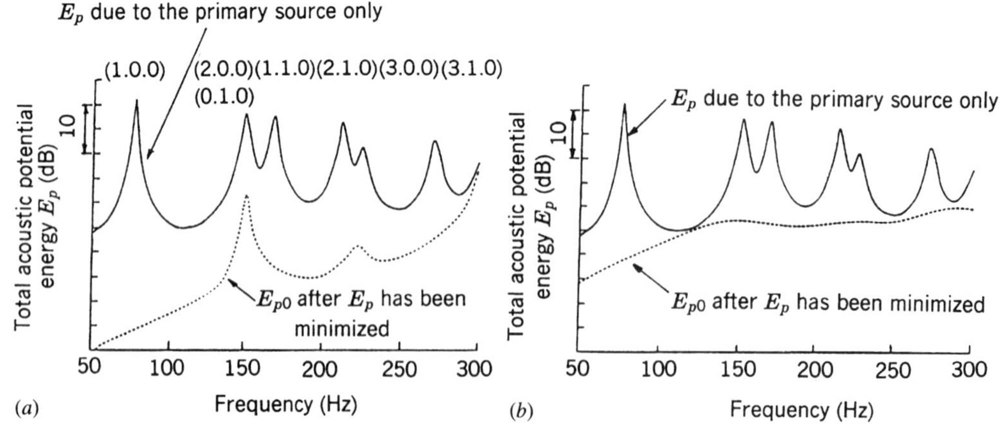

Fig. 6 Shallow rectangular enclosure in which the total acoustic potential energy is minimized by (*a*) source S_1 and (*b*) sources S_2, S_3, and S_4 together. (Results after Ref. 20.) Reprinted with kind permission of Academic Press.

tions in the energy of the enclosure at frequencies close to the natural frequencies of lightly damped modes.

5.2 Control at High Modal Densities

For frequencies of excitation that correspond to acoustic wavelengths that are much smaller than the enclosure dimensions, global control of an enclosed sound field is far more difficult to accomplish. In fact, for frequencies above the Schroeder frequency[21] it can be shown[10] that global reductions in energy in the enclosure can only be produced if secondary sources are separated from the primary source by a distance that is much less than one-half wavelength from the primary source. In practice, in the absence of a highly localized primary source, it may only be possible to produce localized "zones of quiet" around sensors at which the total pressure is minimized by the action of secondary sources. This strategy has been investigated by Elliott et al.,[22] who used computer simulations together with an analysis based on diffuse-field theory to investigate the size of the quiet zone produced around a sensor at which the pressure was driven to zero. The *average* size of the zone in which the sound pressure was reduced to at least 10 dB below that due to the primary source was found to be described by a sphere having a diameter of only one-tenth of an acoustic wavelength.

6 SINGLE-CHANNEL FEEDBACK CONTROL SYSTEMS

One of the earliest attempts to automatically control a sound field was that described by Olson and May,[23] who used a simple high-gain feedback loop to drive the pressure to a low value in the region of a sensor placed close to a loudspeaker. More recently, this approach has been found useful in controlling the sound field in ear defenders[24,25] and in controlling plane-wave sound propagation in ducts.[26,27] Its use in the former case is illustrated in Fig. 7. The electroacoustic frequency response function relating the voltage input to the loudspeaker to the voltage output to the microphone is given by $C(j\omega)$, and the output of the microphone is fed back to the input of the loudspeaker via a filter having the frequency response function $G(j\omega)$. Using the superposition principle applied to the electrical variables involved, the Fourier transform of the microphone output signal can be written as

$$E(\omega) = D(\omega) + G(j\omega)C(j\omega)E(\omega), \tag{17}$$

where $D(\omega)$ is the Fourier transform of the microphone signal produced by the primary field alone. Thus

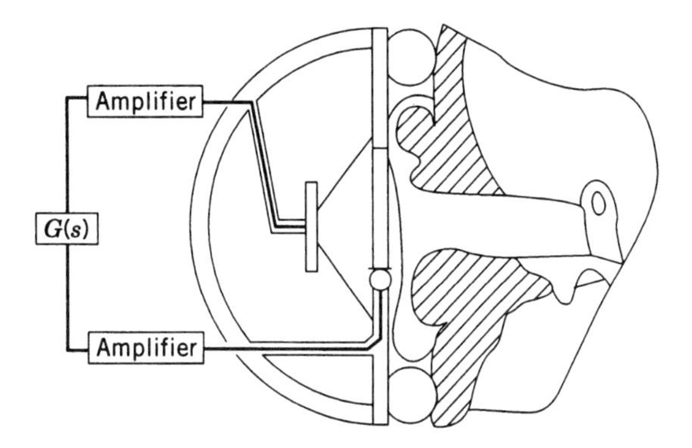

Fig. 7 Feedback control of the sound field at the entrance to the ear canal (after Ref. 25). Reprinted with kind permission of Academic Press.

$$E(\omega) = D(\omega)\left[\frac{1}{1 - G(j\omega)C(j\omega)}\right], \tag{18}$$

and the spectral densities of $E(\omega)$ and $D(\omega)$ are related by[1]

$$S_{ee}(\omega) = S_{dd}(\omega)\left[\frac{1}{|1 - G(j\omega)C(j\omega)|^2}\right]. \tag{19}$$

This expression shows that $|1 - G(j\omega)C(j\omega)|^2$ must be made as large as possible if the spectral density of the microphone signal is to be reduced. Thus, in general terms, given a value of $C(j\omega)$, the frequency response function $G(j\omega)$ must be manipulated in order to ensure as high as possible a value of $|G(j\omega)C(j\omega)|$ over as broad a frequency range as possible while still maintaining a stable system. The system stability is governed by the Nyquist stability criterion, which, roughly speaking,[1] requires that the loop gain be less than unity when the loop phase shift is an integer multiple of 2π. Considerable attention must therefore be given to the design of the phase response of the filter $G(j\omega)$, and it is worth noting[1] that many of the successful implementations to date[24,25] have used a minimum-phase filter $G(j\omega)$ with the same number of poles and zeros. Figure 8 shows the results of the system devised by Carme,[25] who used a second-order compensation filter $G(j\omega)$ with considerable success.

7 MULTICHANNEL FEEDFORWARD CONTROL SYSTEMS

7.1 Equivalent Block Diagram

The wave nature of sound leads more naturally to a feedforward approach to control system design, and it is this

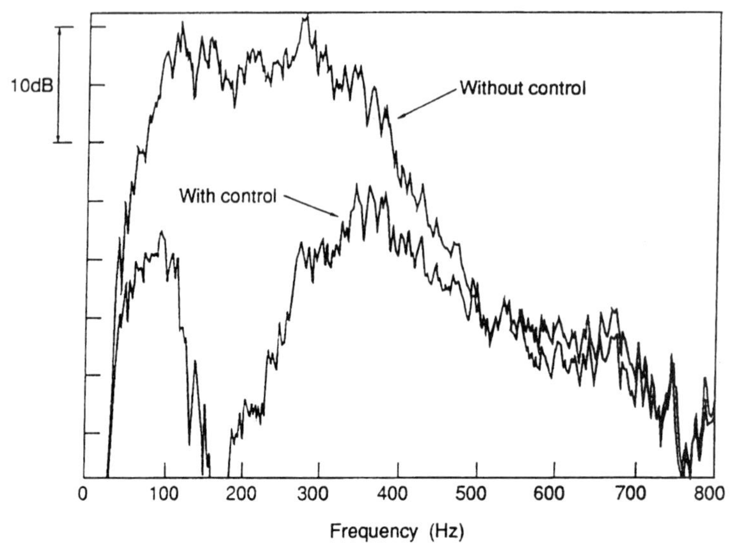

Fig. 8 Reduction in the power spectral density of broadband noise at the entrance to the ear canal produced by the closed-loop system devised by Carme.[25] Reprinted with kind permission of Academic Press.

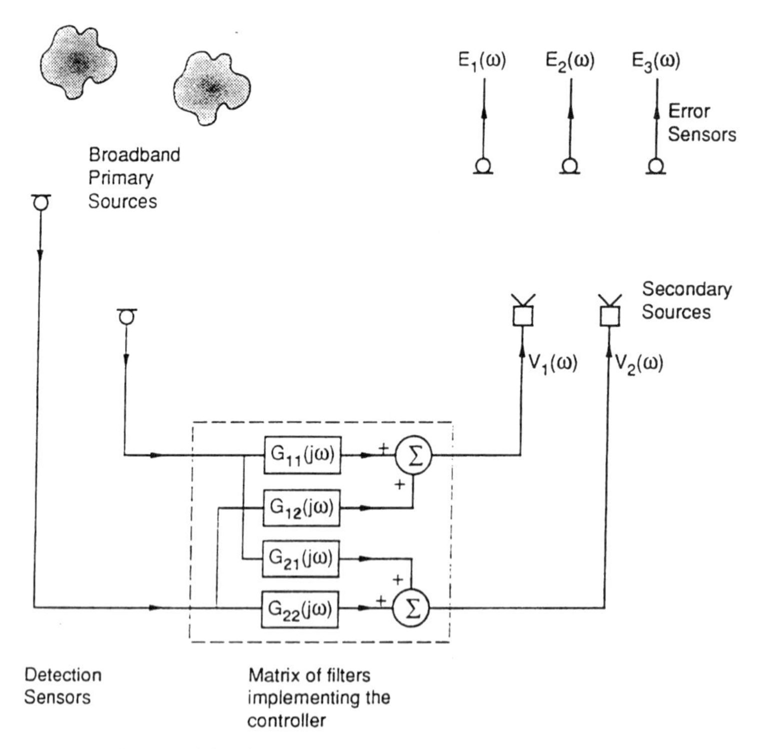

Fig. 9 The multichannel feedforward control problem. The sound radiated by a number of primary sources is detected by K detection sensors, processed via a matrix of control filters and output via M secondary sources. The success of control is monitored by L error sensors. Note that the output of the secondary sources may also be fed back to the detection sensors. Reprinted with kind permission of Academic Press.

approach that has recently received more attention than feedback control. The general structure of a multichannel feedforward controller is sketched in Fig. 9 and is shown in block diagram form in Fig. 10.

In the block diagram it has been assumed that we can apply the principle of superposition to the electrical variables involved. The vector **e** represents the vector of Fourier transforms of the signals detected by L error sensors. These signals are the sum of the signals **z** produced by the secondary sources and the signals **d** produced by the primary sources. The latter are also assumed to contain measurement noise. The control filters are specified by the matrix **G** of frequency response functions. The M controller output signals, specified by the vector **v** of Fourier transforms, reach the error sensors after passing via the electroacoustic paths specified by the matrix **C** of frequency response functions. They also reach the K detection sensors after passing via the "feedback" electroacoustic paths specified by the matrix **F** of frequency response functions. In the absence of the secondary sources, the detection sensors produce a vector of output signals **u** that contain contributions from the primary sources and measurement noise.

In determining the matrix **G**, it is useful to first determine the matrix **H** of frequency response functions relating **v** to **u**, as depicted in Fig. 10. Once the optimal value

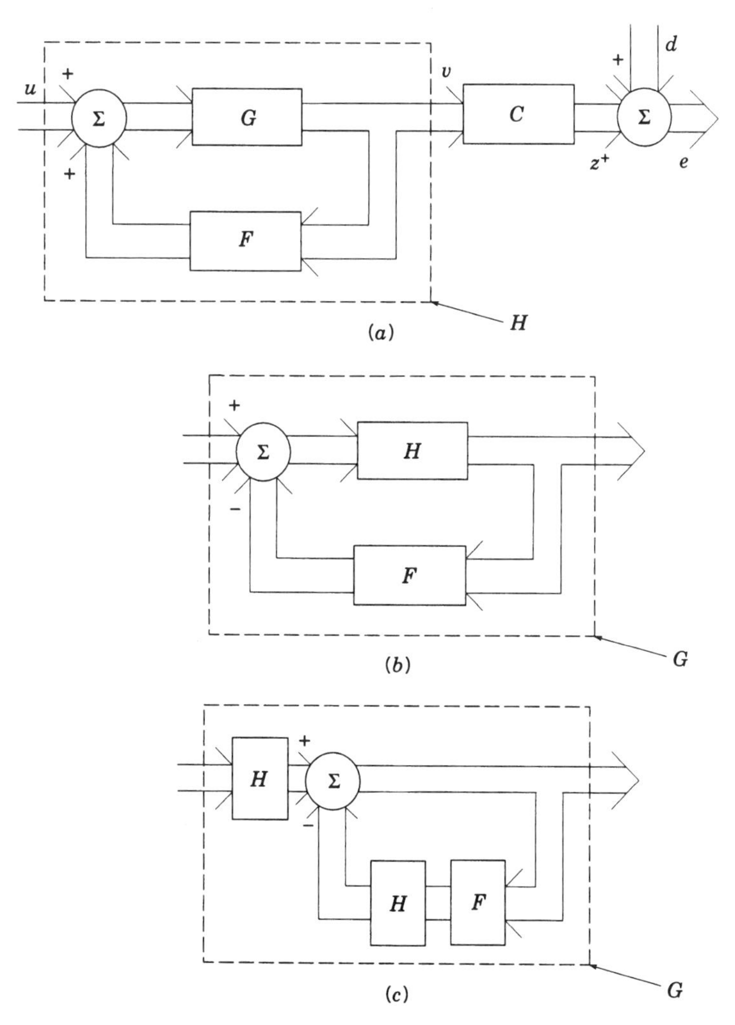

Fig. 10 Block diagram of the multichannel feedforward control problem. The elements of the matrix **H** are first determined. The matrix of control filters **G** can be implemented either using the "feedback cancellation" architecture shown in (*b*) or the "recursive" structure shown in (*c*). Reprinted with kind permission of Academic Press.

of **H** is determined, assuming **F** is known (it can easily be measured), the optimal value of **G** can also in principle be found. It follows from Fig. 10 that

$$\mathbf{v} = \mathbf{G}[\mathbf{u} + \mathbf{F}\mathbf{v}] \tag{20}$$

and therefore that

$$\mathbf{v} = [\mathbf{I} - \mathbf{G}\mathbf{F}]^{-1}\mathbf{G}\mathbf{u}, \tag{21}$$

where **I** is the identity matrix. Thus the matrix **H** can be identified as

$$\mathbf{H} = [\mathbf{I} - \mathbf{G}\mathbf{F}]^{-1}\mathbf{G}. \tag{22}$$

This expression can be rearranged[28] to give two alternative solutions for the matrix **G**. Thus

$$\mathbf{G} = [\mathbf{I} + \mathbf{H}\mathbf{F}]^{-1}\mathbf{H} = \mathbf{H}[\mathbf{I} + \mathbf{F}\mathbf{H}]^{-1}. \tag{23}$$

These expressions can be represented by the filter structures depicted in Fig. 10. The first shows the "feedback cancellation" architecture suggested by the first expression for the matrix **G**, while the second illustrates that the same can be achieved using a "recursive"-type arrangement of filter matrices.

The equivalent block diagram consisting of the cascade of matrices **H** and **C** can also be redrawn by effectively reversing the order of operation of the elements of the matrices. Since the matrices themselves do not commute, we have to first consider the signal produced by the secondary source at the lth sensor. Thus, the Fourier transform of the signal is given by

$$Z_l(\omega) = \sum_{k=1}^{K} \sum_{m=1}^{M} H_{mk}(j\omega) R_{lmk}(\omega), \tag{24}$$

where $R_{lmk}(\omega)$ is the Fourier transform of the *filtered reference signal*, given by

$$R_{lmk}(\omega) = C_{lm}(j\omega) U_k(\omega), \tag{25}$$

which is the signal produced by passing the kth detected signal through the lmth element of the matrix **C**.

7.2 Frequency-Domain Analysis

With the filtered reference signals appropriately defined, it is now possible to undertake an analysis in the frequency domain. This enables the deduction of the best results that can be achieved using a controller that is not constrained to be causal. However, it will be assumed that the signals involved are random with stationary statistical properties, the equivalent results for single frequency sound being readily obtained as a special case. The Fourier transform $Z_l(\omega)$ can be written in the form

$$Z_l(\omega) = \mathbf{h}^{\mathrm{T}}\mathbf{r}_l = \mathbf{r}_l^{\mathrm{T}}\mathbf{h}, \tag{26}$$

where the vectors $\mathbf{r}_l$ and $\mathbf{h}$ are defined by

$$\mathbf{r}_l^{\mathrm{T}} = [R_{l11}(\omega)\cdots R_{lM1}(\omega)|R_{l12}(\omega)\cdots R_{lM2}(\omega)| \cdots |R_{l1K}(\omega)\cdots R_{lMK}(\omega)], \tag{27}$$

$$\mathbf{h}^{\mathrm{T}} = [H_{11}(j\omega)\cdots H_{M1}(j\omega)|H_{12}(j\omega)\cdots H_{M2}(j\omega)| \cdots |H_{1K}(j\omega)\cdots H_{MK}(j\omega)]. \tag{28}$$

The vector **h** thus contains all the elements of the matrix **H** and the vector $\mathbf{r}_l$ consists of the filtered reference signals defined in Eq. (25). The Lth-order vector of Fourier transforms $Z_l(\omega)$ can now be written as

$$\mathbf{z} = \mathbf{R}\mathbf{h}, \tag{29}$$

where **R** is the matrix given by

$$\mathbf{R} = [\mathbf{r}_1^{\mathrm{T}}\mathbf{r}_2^{\mathrm{T}}\cdots\mathbf{r}_L^{\mathrm{T}}]^{\mathrm{T}}. \tag{30}$$

The vector of error signal Fourier transforms $E_l(\omega)$ can now be written as

$$\mathbf{e} = \mathbf{d} + \mathbf{R}\mathbf{h}. \tag{31}$$

A cost function is defined that is given by the sum of the error signal power spectra. Thus

$$J(\omega) = \sum_{l=1}^{L} E[E_l^*(\omega)E_l(\omega)], \tag{32}$$

where the expectation operator $E[\cdot]$ formally implies the ensemble average of the Fourier transforms of signals of infinite duration, although in practice suitable spectral estimates will be made. It follows that

$$J(\omega) = E[\mathbf{e}^{\mathrm{H}}\mathbf{e}], \tag{33}$$

and substitution of Eq. (31) into this expression yields a quadratic function of the type defined by Eq. (5) with the parameters

$$\mathbf{A} = E[\mathbf{R}^{\mathrm{H}}\mathbf{R}], \qquad \mathbf{b} = E[\mathbf{R}^{\mathrm{H}}\mathbf{d}], \qquad c = E[\mathbf{d}^{\mathrm{H}}\mathbf{d}]. \quad (34)$$

Assuming positive definiteness of $\mathbf{A}$, the optimal vector of filter frequency response functions that minimizes the sum of the error sensor power spectra is thus given by $\mathbf{A}^{-1}\mathbf{b}$, and the corresponding minimum value is given by $c - \mathbf{b}^{\mathrm{H}}\mathbf{A}^{-1}\mathbf{b}$.

The case of multiple detected signals and a single error sensor is of particular interest. In this case it follows that the ratio of the minimum value $J_0(\omega)$ of the error sensor power spectrum to its value $J_d(\omega)$ in the absence of control is given by[28]

$$\frac{J_0(\omega)}{J_d(\omega)} = 1 - \eta_{nd}^2(\omega), \quad (35)$$

where $\eta_{nd}^2(\omega)$ is the multiple coherence function[29] between the detection sensor inputs and the error sensor output. In the case of a single detection sensor, this result reduces to the ordinary coherence function. The significance of these coherence functions in the context of active control was pointed out by Ross,[30] and they provide a convenient means for assessing the potential for active control in a practical application. Measurement of these functions can be used to define an upper bound on the performance of an active controller without the necessity for implementing the control system.

7.3 Time-Domain Analysis

The best that can be achieved with a controller that is constrained to be causal can be calculated by conducting the analysis in the discrete-time domain. This has obvious practical application when the controller is implemented as a matrix of finite impulse response (FIR) digital filters, but in undertaking such an analysis the filters are also constrained to have a finite as well as a causal impulse response. Analyses of active control problems can be undertaken in order to determine the structure of optimal causal controllers in continuous time, but this requires the solution of a Wiener Hopf integral equation.[31] Working in discrete time leads naturally to the numerical solution of this equation. The nth sample of the signal due to the secondary sources at the lth error sensor can be written as the convolution

$$\begin{aligned} z_l(n) = {} & \mathbf{h}^{\mathrm{T}}(0)\mathbf{r}_2(n) + \mathbf{h}^{\mathrm{T}}(1)\mathbf{r}_l(n-1) + \cdots \\ & + \mathbf{h}^{\mathrm{T}}(I-1)\mathbf{r}_l(n-I+1), \end{aligned} \quad (36)$$

where it is assumed that each filter in the matrix $\mathbf{H}$ has I coefficients and a composite tap weight vector and a sampled reference signal vector are defined respectively by

$$\begin{aligned} \mathbf{h}^{\mathrm{T}}(i) = {} & [h_{11}(i)\cdots h_{M1}(i)|h_{12}(i)\cdots h_{M2}(i)| \\ & \cdots|h_{1K}(i)\cdots h_{MK}(i)], \end{aligned} \quad (37)$$

$$\begin{aligned} \mathbf{r}_l^{\mathrm{T}}(n) = {} & [r_{l11}(n)\cdots r_{lM1}(n)|r_{l12}(n)\cdots r_{lM2}(n)| \\ & \cdots|r_{l1K}(n)\cdots r_{lMK}(n)]. \end{aligned} \quad (38)$$

A further composite tap weight vector $\mathbf{w}$ can now be defined that consists of all the I tap weights of all the $L \times M$ filters such that

$$\mathbf{w}^{\mathrm{T}} = [\mathbf{h}^{\mathrm{T}}(0)\mathbf{h}^{\mathrm{T}}(1)\cdots\mathbf{h}^{\mathrm{T}}(I-1)]. \quad (39)$$

The Lth-order vector of sampled error signals is now written as

$$\mathbf{e}(n) = \mathbf{d}(n) + \mathbf{R}(n)\mathbf{w}, \quad (40)$$

where the matrix $\mathbf{R}(n)$ is defined by

$$\mathbf{R}(n) = \begin{bmatrix} \mathbf{r}_1^{\mathrm{T}}(n) & \mathbf{r}_1^{\mathrm{T}}(n-1) & \cdots & \mathbf{r}_1^{\mathrm{T}}(n-I+1) \\ \mathbf{r}_2^{\mathrm{T}}(n) & \mathbf{r}_2^{\mathrm{T}}(n-1) & \cdots & \mathbf{r}_2^{\mathrm{T}}(n-I+1) \\ \vdots & \vdots & \ddots & \vdots \\ \mathbf{r}_L^{\mathrm{T}}(n) & \mathbf{r}_L^{\mathrm{T}}(n-1) & \cdots & \mathbf{r}_L^{\mathrm{T}}(n-I+1) \end{bmatrix}. \quad (41)$$

The relevant cost function is now the sum of the time-averaged error signals given by

$$J = E[\mathbf{e}^{\mathrm{T}}(n)\mathbf{e}(n)]. \quad (42)$$

Substitution of Eq. (40) then shows that the cost function can be written in the quadratic form

$$\begin{aligned} J = {} & \mathbf{w}^{\mathrm{T}}E[\mathbf{R}^{\mathrm{T}}(n)\mathbf{R}(n)]\mathbf{w} + 2\mathbf{w}^{\mathrm{T}}E[\mathbf{R}^{\mathrm{T}}(n)\mathbf{d}(n)] \\ & + E[\mathbf{d}^{\mathrm{T}}(n)\mathbf{d}(n)]. \end{aligned} \quad (43)$$

Again, this function has a unique minimum value associated with an optimal value of the tap weight vector $\mathbf{w}$. In a single-input, single-output controller, the techniques for searching this quadratic performance surface are well established.[32] The application of the method of steepest descent, together with the use of an instantaneous estimate of the gradient of the surface, leads directly to the filtered-x LMS algorithm.[32] With the quadratic function given by Eq. (43), identical arguments can be applied. Thus the gradient of the cost function given by Eq. (43) can be written as

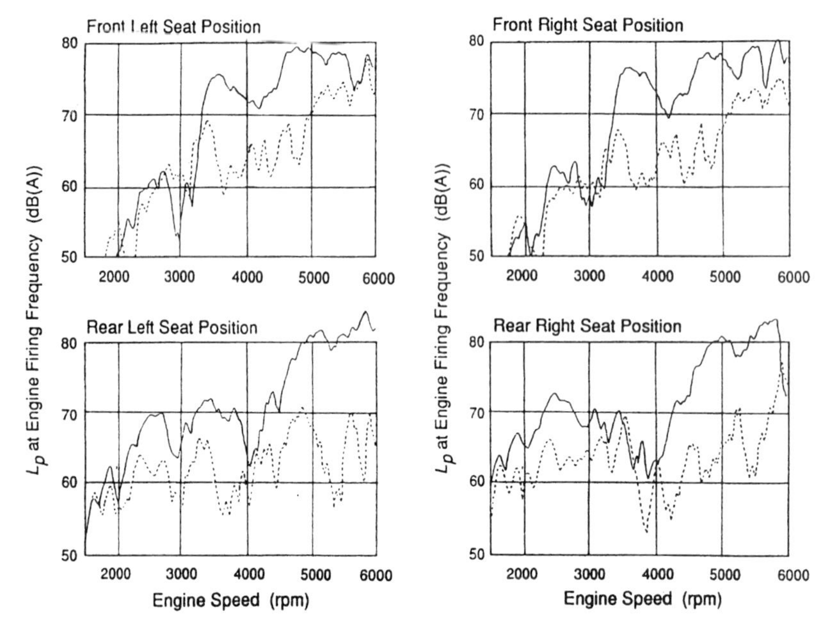

Fig. 11 A-weighted sound pressure level at the engine firing frequency measured at head height in the four seat positions of a 1.1-L small saloon car as it was accelerated hard in second gear with (- - - -) and without (—) a four-loudspeaker, eight-microphone active noise control system in operation. Reprinted with kind permission of Academic Press.

$$\frac{\partial J}{\partial \mathbf{w}} = 2E[\mathbf{R}^{\mathrm{T}}(n)\mathbf{R}(n)\mathbf{w} + \mathbf{R}^{\mathrm{T}}(n)\mathbf{d}(n)], \qquad (44)$$

which, on use of Eq. (40), reduces to

$$\frac{\partial J}{\partial \mathbf{w}} = 2E[\mathbf{R}^{\mathrm{T}}(n)\mathbf{e}(n)]. \qquad (45)$$

The method of steepest descent suggests that the vector $\mathbf{w}$ should be updated by an amount proportional to the negative of the gradient $\partial J/\partial \mathbf{w}$. The use of an instantaneous estimate of this gradient then leads to the tap weight update equation

$$\mathbf{w}(n+1) = \mathbf{w}(n) - \alpha \mathbf{R}^{\mathrm{T}}(n)\mathbf{e}(n), \qquad (46)$$

where α is a convergence coefficient. This has become known as the multiple-error LMS algorithm[33] and has proved most successful when used as the basis for the adaptation of digital filters used in multichannel control systems for the suppression of periodic sound fields in aircraft[34] and automobiles.[35] An illustration of the performance of a typical multichannel active controller used for the suppression of engine-induced noise in automobiles is shown in Fig. 11.

REFERENCES

1. P. A. Nelson and S. J. Elliott, *Active Control of Sound*, Academic, London, 1992.
2. C. R. Fuller, S. J. Elliott, and P. A. Nelson, *Active Control of Vibration*, Academic, London, 1996.
3. M. J. M. Jessel, "Sur les absorbeurs actifs," *Proceedings 6th ICA*, Tokyo, 1968, Paper F-5-6, p. 82.
4. G. D. Malyuzhinets, "An Inverse Problem in the Theory of Diffraction," *Sov. Phys. Doklady*, Vol. 14, 1969, pp. 118–119.
5. A. J. Kempton, "The Ambiguity of Acoustic Sources—A Possibility for Active Control?" *J. Sound Vib.*, Vol. 48, 1976, pp. 475–483.
6. P. E. Doak, "Fundamentals of Aerodynamic Sound Theory and Flow Duct Acoustics," *J. Sound Vib.*, Vol. 28, 1973, pp. 527–561.
7. K. Kido and S. Onoda, "Automatic Control of Acoustic Noise Emitted from Power Transformer by Synthesising Directivity," *Science Reports of the Research Institutes, Tohoku University (RITU), Sendai, Japan*, Series B: Tech-

nology. Part I: Reports of the Institute of Electrical Communication (RIEC) 2397-110, 1972.

8. J. Piraux and B. Nayroles, "A Theoretical Model for Active Noise Attenuation in Three-Dimensional Space," *Proceedings of Inter-Noise '80*, Miami, 1980, pp. 703–706.
9. P. A. Nelson, A. R. D. Curtis, S. J. Elliott, and A. J. Bullmore, "The Minimum Power Output of Free Field Point Sources and the Active Control of Sound," *J. Sound Vib.*, Vol. 116, No. 3, 1987, pp. 397–414.
10. P. A. Nelson, A. R. D. Curtis, S. J. Elliott, and A. J. Bullmore, "The Active Minimization of Harmonic Enclosed Sound Fields, Part I: Theory," *J. Sound Vib.*, Vol. 117, No. 1, 1987, pp. 1–13.
11. P. A. Nelson, A. R. D. Curtis, and S. J. Elliott, "On the Active Absorption of Sound," *Proceedings of Inter-Noise '86*, 1986, pp. 601–606.
12. P. M. Morse and K. U. Ingard, *Theoretical Acoustics*, McGraw-Hill, New York, 1968.
13. R. J. Silcox, "Active Control of Multi-modal Random Sound in Ducts," *Proceedings Institute of Acoustics Spring Meeting*, Southampton, England, 1990.
14. L. J. Eriksson, M. C. Allie, R. H. Hoops, and J. V. Warner, "Higher Order Mode Cancellation in Ducts Using Active Noise Control," *Proceedings of Inter-Noise '89*, Newport Beach, CA, 1989.
15. L. J. Eriksson, M. C. Allie, and R. A. Greiner, "The Selection and Application of an IIR Adaptive Filter for Use in Active Sound Attenuation," *IEEE Trans. Acoust., Speech Signal Process.*, Vol. ASSP-35, No. 4, 1987, pp. 433–437.
16. A. Roure, "Self Adaptive Broad Band Active Sound Control System," *J. Sound Vib.*, Vol. 101, No. 3, 1985, pp. 429–441.
17. C. F. Ross, "An Algorithm for Designing a Broad Band Active Sound Control System," *J. Sound Vib.*, Vol. 80, No. 3, 1982, pp. 373–380.
18. M. A. Swinbanks, "The Active Control of Sound Propagation in Long Ducts," *J. Sound Vib.*, Vol. 27, No. 3, 1973, pp. 411–436.
19. A. D. Pierce, *Acoustics: An Introduction to its Physical Principles and Applications*, McGraw-Hill, New York, 1981.
20. A. J. Bullmore, P. A. Nelson, A. R. D. Curtis, and S. J. Elliott, "The Active Minimization of Harmonic Enclosed Sound Fields, Part II: A Computer Simulation," *J. Sound Vib.*, Vol. 117, No. 1, 1987, pp. 15–33.
21. M. R. Schroeder and K. H. Kuttruff, "On Frequency Response Curves in Rooms. Comparison of Experimental Theoretical and Monte Carlo Results for the Average Frequency Spacing between Maxima," *J. Acoust. Soc. Am.*, Vol. 34, No. 1, 1962, pp. 76–80.
22. S. J. Elliott, P. Joseph, A. J. Bullmore, and P. A. Nelson, "Active Cancellation of a Point in a Pure Tone Diffuse Sound Field," *J. Sound Vib.*, Vol. 120, No. 1, 1988, pp. 183–189.
23. H. F. Olson and E. G. May, "Electronic Sound Absorber," *J. Acoust. Soc. Am.*, Vol. 25, 1953, pp. 1130–1136.
24. P. D. Wheeler, "Voice Communications in the Cockpit Noise Environment—The Role of Active Noise Reduction," Ph.D. Thesis, University of Southampton, England, 1986.
25. C. Carme, "Absorption Acoustique Active Dans Les Cavites," Ph.D. Thesis, l'Universite d'Aix-Marseille II, 1987.
26. W. K. W. Hong, Kh. Eghtesadi, and H. G. Leventhall, "The Tight-coupled Monopole (TCM) and Tight-coupled Tandem (TCT) Attenuators: Theoretical Aspects and Experimental Attenuation in an Air Duct," *J. Acoust. Soc. Am.*, Vol. 81, 1987, pp. 376–388.
27. M. C. J. Trinder and P. A. Nelson, "Active Noise Control in Finite Length Ducts," *J. Sound Vib.*, Vol. 89, No. 1, 1983, pp. 95–105.
28. P. A. Nelson, T. J. Sutton, and S. J. Elliott, "Performance Limits for the Active Control of Random Sound Fields from Multiple Primary Sources," *Proceedings of IOA Spring Conference*, Southampton, England, 1990.
29. D. E. Newland, *An Introduction to Random Vibrations and Spectral Analysis*, 2nd ed., Longman Scientific and Technical, New York, 1984.
30. C. F. Ross, "Active Control of Sound," Ph.D. Thesis, University of Cambridge, England, 1980.
31. P. A. Nelson, J. K. Hammond, P. Joseph, and S. J. Elliott, "Active Control of Stationary Random Sound Fields," *J. Acoust. Soc. Am.*, Vol. 87, No. 3, 1990, pp. 963–975.
32. B. Widrow and S. D. Stearns, *Adaptive Signal Processing*, Prentice-Hall, Englewood Cliffs, NJ, 1985.
33. S. J. Elliott, I. M. Stothers, and P. A. Nelson, "A Multiple Error LMS Algorithm and Its Application to the Active Control of Sound and Vibration," *IEEE Trans. Acoust., Speech Signal Process.*, ASSP-35, No. 10, 1987, pp. 1423–1434.
34. S. J. Elliott, P. A. Nelson, I. M. Stothers, and C. C. Boucher, "In-flight Experiments on the Active Control of Propeller-induced Cabin Noise," *J. Sound Vib.*, Vol. 140, 1990, pp. 219–238.
35. S. J. Elliott, I. M. Stothers, P. A. Nelson, A. M. McDonald, D. C. Quinn, and T. Saunders, "The Active Control of Engine Noise inside Cars," *Proceedings of Internoise 88*, Avignon, France, 1988, pp. 987–990.

68

NOISE SOURCES AND PROPAGATION IN DUCTED AIR DISTRIBUTION SYSTEMS

HOWARD F. KINGSBURY

1 INTRODUCTION

This chapter deals with the generation, propagation, and control of noise in a ducted air distribution system. First, fan noise control is considered, with reference to a simple system, with complications added later. The discussion refers to octave band noise generation and control. Included is discussion of "natural attenuation," that is, the noise reduction that occurs due to the transmission of sound through the duct walls, the division of noise due to duct branches, and the effect of duct fittings. Also treated are the effects of porous duct linings and packaged attenuation units (silencers). Consideration is also given to the conversion of sound power levels to sound pressure levels in the room served by the fan system. The methods and effects of varying the air volume delivered by the fan is discussed. Second, turbulent flow noise generation and control in the ducts are covered. The generation and control of "break-out noise," that is, noise that is radiated through the duct wall, is discussed, as are the differences between round and rectangular duct in the control of this noise. The effects of volume control dampers located near outlets are included. Special consideration is given to roof-top units, where turbulent flow is a common cause of noise generation. Third, noise generation and control in the outlet devices are considered. Both constant-air-volume and variable-air-volume (VAV) types of terminals are included, including so-called fan-powered terminals. Information on the sound generated by fans is in Chapter 79.

Note: References to chapters appearing only in the *Encyclopedia* are preceded by *Enc.*

2 FAN NOISE CONTROL IN DUCTED SYSTEMS

Understanding of noise control in a ducted air distribution system can best be developed by beginning with a simple system and then adding the complications. Figure 1 shows a ducted system reduced to its barest essentials. It is a constant-volume system, with a single fan running at constant speed, enclosed in a plenum, both supplying and returning the air from the room being served. Eliminated from this system are such accessories as heating or cooling coils, filters, and so on. Usually, the most critical part of the system is the room closest to the fan serving that system unless other rooms downstream require lower sound levels.

The recommended noise control procedure is as follows, working in octave bands:

1. Determine the sound power output of the fan(s) that is emitted into the duct system.
2. Determine the so-called natural attenuation of the duct system. While doing so, determine if the sound generated by fittings and take-offs may add to that being transmitted down the duct.
3. Convert from the sound power level at the outlet to the resulting sound pressure level in the space being served (so-called room factor or space effect) and compare it with the chosen sound level criterion for that space.
4. Add any required sound attenuation measures to the system needed to meet the criterion in 3.
5. Determine if sound transmission through the duct wall (break-out noise) may be a problem and, if so, apply proper sound control procedures.

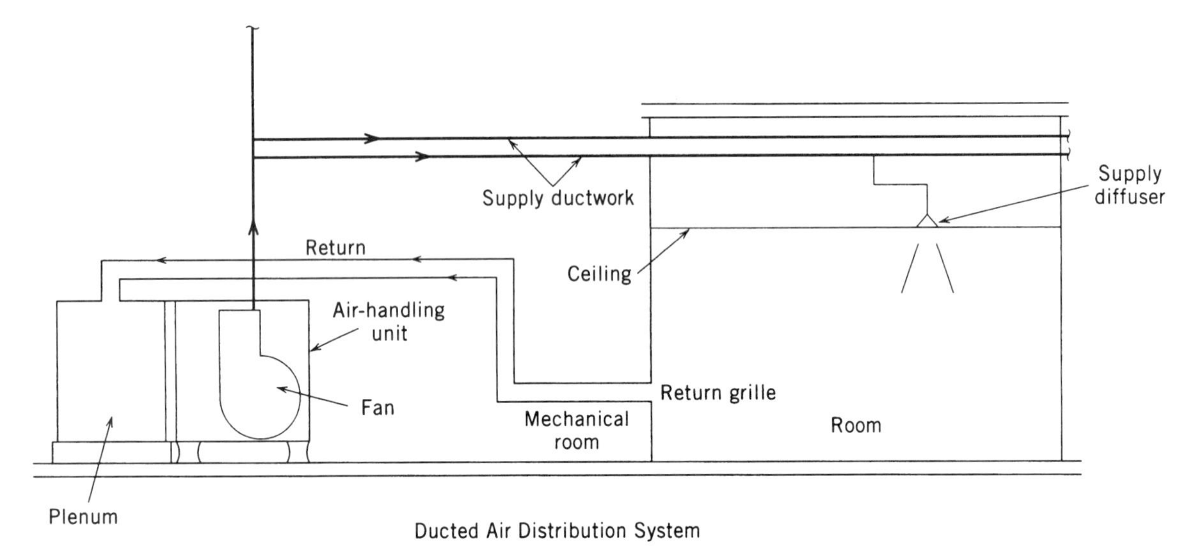

Fig. 1 Ducted air distribution system.

6. Repeat the entire procedure for the return side of the system.

In addition to these considerations of fan noise, note that the final sound level in the room will be a combination of the noise from the fan and noise from the chosen outlet device(s), which will be discussed later.

2.1 Natural Attenuation

The first aspect of noise control in a duct distribution system is the so-called natural attenuation, which has four components. First, some sound is radiated through the duct wall, causing the duct wall to vibrate; this is sensed as a reduction of energy inside the duct (see Table 1).[1] (Round and flat oval ducts show less of this effect than rectangular ducts due to their shape. For round ducts, use an attenuation of 0.1 dB/m up to 1000 Hz, rising to 0.4 dB/m at 4000 Hz.) A second mechanism is the division of energy between the main duct and the take-offs. In the absence of specific information to the contrary, it is assumed the energy division is proportional to the ratio of the branch duct area to that of the main duct (see Table 2). A third common loss occurs at the end of the duct, where it enters the room. This is called end-reflection loss (some of the sound energy is reflected back into the duct system) and is most pronounced when there is a sudden transition from the duct to the room. Diffusers reduce this effect; it is suggested that it be reduced by 6 and 4 dB in the 63- and 125-Hz octave bands for duct sizes 300 mm and under when diffusers are on the end of the duct (see Table 3).[1] Finally, a fourth component is the conversion from sound power in the duct system to the sound pressure in the room (the room factor noted above). For most practical rooms, the classical equation (Chapter 74), which assumes both a direct and reverberant field, does not apply. A room having an acoustical

TABLE 1 Attenuation of Unlined Sheet Metal Ducts, (dB/unit length) as Function of P/A by Octave Band Center Frequency

P/A (mm/mm^2)	Attenuation (dB/m)		
	63 Hz	125 Hz	≥250 Hz
Over 0.012	0.05	1.0	0.33
0.012–0.005	1.0	0.33	0.33
Under 0.005	0.33	0.33	0.33

Note: If duct is externally insulated, double above listed attenuations; P = duct perimeter; A = cross-sectional area.
Source: Ref. 1, p. 42.11.

TABLE 2 Division of Sound Power at Branch Take-offs

Ratio	dB Reduction
1.0	0
0.5	3
0.25	6
0.10	10

Note: Ratio = branch duct/continuing duct area.

TABLE 3 Duct End Reflection Loss by Octave Band Center Frequency

Mean Duct Width (mm)	Reflection Loss (dB)				
	63 Hz	125 Hz	250 Hz	500 Hz	1000 Hz
150	18	12	8	4	1
250	14	9	5	1	0
500	9	5	1	0	0
1000	4	1	0	0	0

Note: If duct terminates in a diffuser, deduct 6 dB (63 Hz) and 4 dB (125 Hz) for duct sizes under 300 mm, 3 dB (63 Hz) for duct sizes over 300 mm. Do *not* apply to linear diffusers or diffusers tapped directly into primary duct work.

Source: Ref. 1, p. 42.16.

ceiling and in which the ceiling height is significantly less than either of the horizontal dimensions, the Schultz equation (Chapter 74) should be used for this conversion.

2.2 Lined Duct Attenuation

In many systems additional attenuation is required. This may be obtained by lining the interior surfaces of the duct with sound-absorbing material or by use of packaged attenuators or both. Duct lining serves two purposes: It is a thermal insulant and a sound absorber. While 13 mm thickness may be sufficient for thermal reasons, for acoustical purposes the thickness should be a minimum of 25 mm, and 50 mm is often desirable. This is due to the minimal low-frequency attenuation provided by thin linings, and it is in the low-frequency region that most duct system noise control problems occur.

The acoustical characteristics of glass fiber duct lining in rectangular ducts have been rather extensively investigated; the same is not true for round ducts and in the absence of further experimental evidence, Table 4[1] should be used for round-duct attenuation.

The attenuation of fiber glass lined rectangular duct sections may be calculated using Eqs. (1) and (2).[1]

For frequencies below 800 Hz

$$\text{Total attenuation} = \frac{t^{1.08} h^{0.356} (P/A)/L f^{(1.17+Kd)}}{K_1 d^{2.3}}, \qquad (1)$$

where $K_1 = 5.46 \times 10^{-3}$, $K = 0.0119$, h is the smaller inside dimension of the duct in millimetres, f is frequency in hertz, P is the inside duct perimeter in millimetres, A is the inside area of the duct in square millimetres, and L is the duct length in metres.

Note that since the attenuation is strongly frequency dependent, it is suggested that low-frequency octave band attenuation be evaluated for the one-third-octave band below the octave band center frequency of interest, that is, 100 Hz for the 125-Hz octave band. Also, add the attenuation of unlined duct (Table 1) to the attenuation provided by this calculation for the 63- and 125-Hz octave bands.

For frequencies above 800 Hz

$$\text{Total attenuation} = K_4 \frac{(P/A) L f^{(K_5 - 1.611 \text{ of } P/A)}}{w^{2.5} h^{2.7}}, \qquad (2)$$

where w is the larger inside dimension of duct in millimetres, $K_4 = 3.32 \times 10^{18}$, and $k_5 = 3.79$.

Note that the above equation may be used for straight duct lengths of 3 m or less; for straight lengths (i.e., no elbows) longer than 3 m, limit the total attenuation to 40 dB. The limit of 40 dB for any straight duct length applies to both equations, since structural flanking through the duct wall imposes the upper limit. Further, do not use either equation when the duct is constructed of glass fiber with a lightweight, low-transmission-loss cov-

TABLE 4 Insertion Loss for Lined Circular Ducts with 25- or 50-mm-Thick Lining by Octave Band Center Frequency

	Insertion Loss, dB/m											
	63 Hz		125 Hz		250 Hz		500 Hz		1000 Hz		2000 Hz	
Diameter (mm)	25 mm	50 mm	25 mm	50 mm	25 mm	50 mm	25 mm	50 mm	25 mm	50 mm	25 m	50 mm
150	1.3	1.8	1.9	2.6	3.1	4.5	5.0	7.4	7.2	7.2	7.6	7.6
300	0.76	1.4	1.5	2.2	2.7	4.1	4.8	7.2	7.2	7.2	6.3	6.3
600	0.23	0.83	0.83	1.5	1.9	3.3	4.2	6.6	5.6	5.6	4.1	4.1
1200	0	0.36	0	0.50	0.59	2.1	2.1	4.5	.86	.86	1.1	1.1

Source: Ref. 1, p. 42.14.

TABLE 5 Insertion Loss for Lined, Square Elbows with Turning Vanes by Octave Band Center Frequency

	Insertion Loss (dB)					
Duct Width (mm)	63 Hz	125 Hz	250 Hz	500 Hz	1000 Hz	2000 Hz
125–250	0	0	1	2	3	4
260–510	0	1	2	3	4	6
520–1020	1	2	3	4	5	6
1030–2030	2	3	3	5	6	8

Note: Duct width is dimension in plane of the turn.
Source: Ref. 1, p. 42.15.

ering, such as a metal foil. Sound transmission through the covering limits the total attenuation, since the sound reenters the duct downstream.

2.3 Lined Rectangular Elbows

Elbows with turning vanes also provide worthwhile attenuation. Round elbows, lined or unlined, provide relatively little attenuation, in general 5 dB or less. Table 5 provides typical values for rectangular elbows.[1]

2.4 Passive Attenuators or Silencers

These items are available in a variety of sizes and lengths. Commonly available lengths of rectangular silencers are 0.9, 1.5, 2.1, and 3.1 m. The length of most circular cross-sectional silencers is a function of their diameter. In the selection of silencers, three interrelated factors must be considered: (1) insertion loss, (2) pressure drop, and (3) self-generated noise. The primary controlling element is the airflow velocity through the face area of the silencer. Increasing velocity increases both the pressure drop and the self-noise; it reduces the insertion loss slightly. Manufacturers' literature should be consulted for details on these items. A recognized test method for determining these properties is ASTM E-477, which measures the insertion loss by the duct-to-reverberant-room method with and without airflow.

2.5 Active Sound Attenuation

Recent developments in computer technology have led to the development of phase-canceling systems having application to low-frequency noise in ducted systems. Both random noise and tones are accounted for, with an upper frequency limit of about 400 Hz and insertion losses typically of 20–30 dB.

2.6 Attenuation of Noise in Lined Plena

The noise reduction offered by lined fan plena, as shown in Fig. 2, should be taken into account. The following approximate expression may be used to calculate the effects of such a lined plenum[2]:

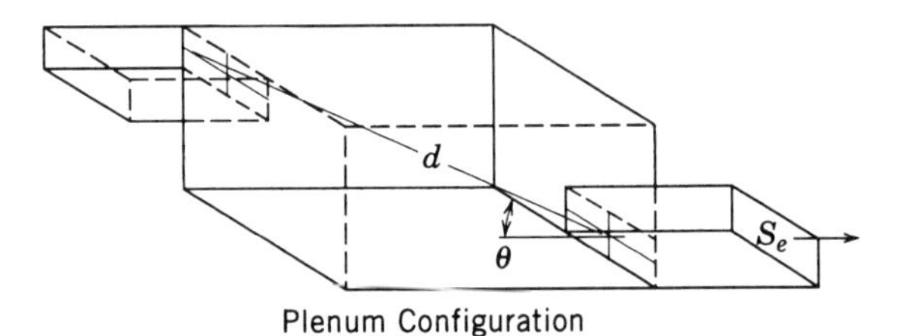

Fig. 2 Diagram of sound-absorbing plenum.

$$\text{Attenuation} = 10 \log \left[S_e \left(\frac{\cos \Theta}{2\Pi d^2} + \frac{1-\alpha}{\alpha S_w} \right) \right]^{-1}, \quad (3)$$

where α is the absorption coefficient of lining at octave band center frequencies, S_e the plenum exit area in square metres, d the slant distance between entrance and exit in metres, S_w the plenum wall area in square metres, and Θ the angle of incidence at exit, that is, the angle that d makes with the normal to the exit. The results of this calculation are accurate to within a few decibels for frequencies where the wavelength is much less than the plenum dimensions. At low frequencies (63 and 125 Hz) it is conservative by 5–10 dB due to sound reflections at the entrance and exit of the plenum.

3 AIR TERMINAL DEVICES

In most ducted systems, a device is placed at the end of the duct to deliver the air to the room. These are most commonly diffusers, grilles, or registers and may contain volume control devices. Since they radiate sound directly into the room, they must be selected to meet the room noise level criterion. For the most part, their noise emissions occur in the octave bands above 250 Hz. Manufacturers' literature should be consulted for appropriate

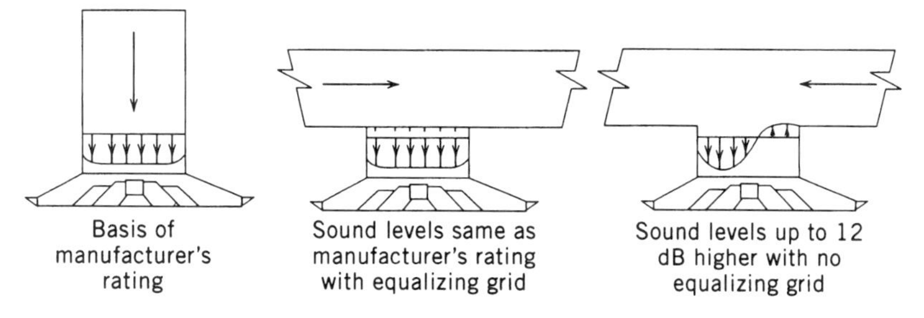

Fig. 3 Noise effect of poor diffuser flow condition. (From Ref. 1, p. 42.19.)

sound ratings. Care should be taken that the conditions for which the data are presented match the application at hand and any adjustments made as necessary. The noise radiated by a terminal device is dependent not only on the volume and velocity of the air flow through the element but also on the upstream flow conditions. Manufacturer's data are taken under essentially ideal flow conditions; poor flow conditions may add 10–15 dB to the catalog data. A variety of upstream airflow conditions and associated additional sound levels are shown in Fig. 3.[1] Frequently, flexible duct is used to connect the terminal device to the branch duct. The length of this flexible duct should be kept as short as possible to avoid additional flow noise, with 1 m being a suggested maximum. Misalignment of the flexible duct can cause increased sound levels, as shown in Fig. 4.[1]

Volume control dampers may be another source of increased noise levels, typically if located close to the outlet. The noise they create is a function of the additional pressure drop they introduce into the system; the additional noise is 20 log of the ratio of the additional pressure drop to that of the diffuser itself. In acoustically critical spaces, volume control dampers should be located well back in the duct system (5–10 duct diameters from the outlet), with the lined duct between the damper and the outlet.

3.1 Variable-Air-Volume Systems

For energy conservation reasons, many current air conditioning systems modulate the airflow to the conditioned space to match the required heating or cooling load, as it varies throughout the day and year. These VAV systems have two distinguishing characteristics from the acoustical point of view: (1) a terminal device to modulate the airflow and (2) some method of varying the flow volume from the fan.

3.2 Variable-Air-Volume Terminals

In its simplest form, the terminal device consists of an automatically operating damper in a sheet metal box. It is commonly connected directly to the supply duct and with one or more outlets feeding diffusers. Noise is created by the damper as a function of the pressure drop

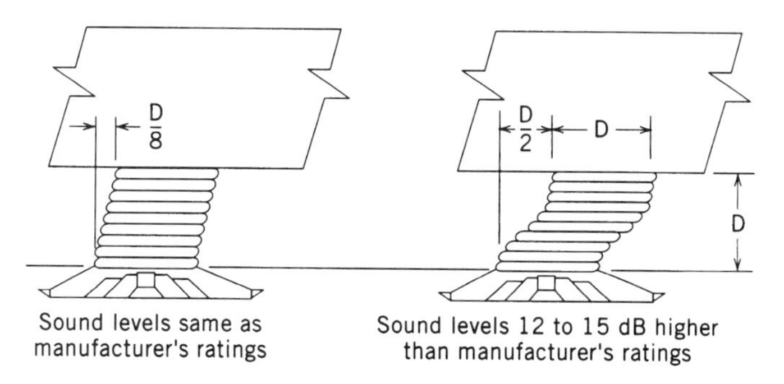

Fig. 4 Effect of proper and improper alignment of flexible duct connector. (From Ref. 1, p. 42.23.)

across the damper. This noise can be radiated by two paths: (1) directly through the wall of the unit and (2) downstream through the connected duct(s). This noise is frequently controlled by a section of lined duct between the box and the outlet(s). A more difficult control case is presented when the outlet and VAV terminal are combined. In some situations, the casing radiated sound must be controlled. In most cases, such terminal units are located above acoustical panel ceilings, which have limited sound transmission loss properties and may contain acoustically transparent areas such as return air grilles and light fixtures. In general, VAV boxes that handle more than about 0.8 m^3/s should not be located over acoustical environments such as offices; they should be located over more noncritical spaces such as corridors, storage areas, or toilet rooms. In some cases, where such acoustically less sensitive spaces are not available and large box sizes must be used, it may be necessary to increase the transmission loss characteristics of the ceiling by incorporating a layer of 12.5-mm gypsum board between the unit and the acoustical ceiling or encasing the box with a 25-mm layer of glass fiber and 12.5-mm gypsum board.

3.3 Fan-Powered VAV Terminals

A special case is a VAV terminal that incorporates a fan to mix the air from the plenum above the ceiling with the primary air. These are called fan-powered terminals (FPTs) and are of two kinds: one in which the fan cycles on and off, as required, and a second in which the fan runs full time. Since the fan in the first goes on and off, the occupants are more aware of the control unit sound. The noise emitted by these units consists of both the noise radiated by the casing and the fan and that through the room air outlet. By reason of the fan contribution to the total direct radiated noise, units handling more than about 0.8 m^3/s commonly produce noise levels in the space below corresponding to a noise criteria (NC) rating of 40 or above and hence should not be located over spaces requiring a lower sound level. Some possible noise control measures for these units include, in addition to moving them to more remote locations, keeping them at least 2 m from any open return grilles in the ceiling, installing sound attenuators if offered as an option by the manufacturer, or field fabricating lined elbows for installation on the induced air opening of the FPT. Consult manufacturers' literature for appropriate information on the noise levels produced by the above units and how to determine the noise levels in the occupied space served by them. Data on these units are taken in accordance with the Air-Conditioning & Refrigerating Institute (ARI) standard 880-89 (estimation of the sound level in the occupied space by ARI 885).

3.4 Fan Volume Control

As the airflow volume through the terminals varies to meet the required heating or cooling load, the volume of air produced by the fan must also vary. There are three basic methods of doing so: (1) use of inlet vanes on the fan, (2) motor speed controls, and (3) varying the pitch of the fan blades:

Inlet vanes are positioned at the inlet to the fan scroll and close down to reduce airflow. They add resistance to the system and, depending on the fan type, may cause significant additional low-frequency fan sound, which must be dealt with further downstream. Since resistance to flow is added, there is also energy usage inefficiency when the vanes are in their partially open position.

Variable-speed controls reduce the speed of the fan and thus the volume flow; they also reduce the fan noise generation. Variable-speed controls have a significant energy savings over inlet vanes; this should be taken into account in the selection process.

Variable-pitch blades are available for some axial flow fans; this also results in quieter operation at reduced flow.

4 SOUND TRANSMISSION THROUGH DUCT WALLS

Sound radiated through duct walls may be a problem if not sufficiently attenuated before the duct passes over or through occupied space. This is most frequently a problem close to the fan room, where the noise in the duct may be composed of both that from the fan and that aerodynamically generated due to poor flow conditions. Typically, this noise is concentrated in the octave bands below 250 Hz and may extend down through the 31.5-Hz band. It is commonly called *breakout noise*, and its level is dependent on the noise level within the duct, the duct cross section, and the length of the duct in or over the occupied space. The transmission loss characteristics of rectangular, flat oval, and round ducts have been determined; representative values for round and rectangular ducts are shown in Table 6.[1] The transmission loss of rectangular ducts does not vary significantly as a function of duct size or construction; the transmission loss of round ducts does vary as a function of the duct diameter and values for several typical diameters are included. Note that at low frequencies the transmission loss of round ducts is markedly higher than for rectangular ones. Using round ducts is an excellent method of controlling low-frequency breakout noise. Since the noise is

TABLE 6 Transmission $Loss_{out}$ for Rectangular and Round Duct by Octave Band Center Frequency (Hz)

Rectangular Sizes (300–1200 mm)	63Hz 20	125 Hz 23	250 Hz 26	500 Hz 29	1000 Hz 32	2000 Hz 38
Round Duct Dia. (mm)						
200	> 48	> 65	> 75	> 72	56	56
350	> 48	> 53	55	33	34	35
650	> 48	> 53	36	32	32	28
800	> 48	42	38	25	26	24

Source: Ref. 1, p. 42.23.

better contained, additional duct system attenuation may be needed downstream for appropriate control of room sound levels. The transmission loss of flat oval duct is intermediate between rectangular and round duct, but due to the relatively large flat surfaces, it tends to follow the same characteristics as that of rectangular duct, that is, lower values at low frequencies.

5 TURBULENT AIRFLOW NOISE CONTROL

Although fans are usually the major source of noise in a duct system, aerodynamic noise may be generated at elbows, turns, take-offs, silencers, and other duct elements. The generated noise levels depend on the element geometry as well as the airflow velocity. Since the intensity of flow generated sound is proportional to between the fifth and sixth power of the local velocity in the duct, the best methods of control are to (1) keep the duct velocities as low as possible and (2) avoid closely spaced elements (turns, take-offs) or abrupt changes in duct area that can contribute to turbulent flow. In general, if the flow velocities are below about 10 m/s and elements can be separated by about three to five duct diameters, most aerodynamically induced noise control problems are minimized.

6 ROOFTOP UNITS

Rooftop units (packaged, factory assembled units containing fans and associated heating and cooling equip-

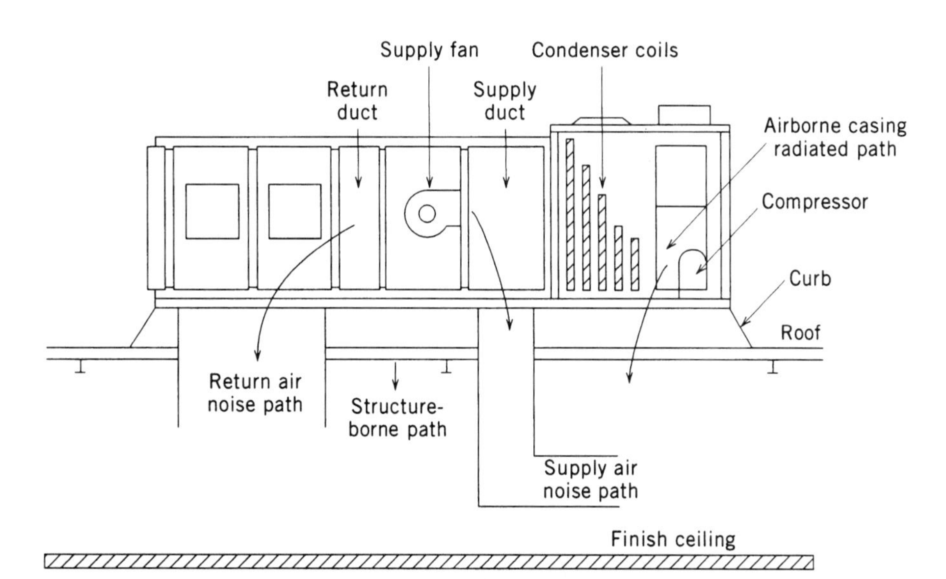

Fig. 5 Noise paths in typical rooftop installations. (From Ref. 1, p. 42.27.)

TABLE 7 Duct Breakout Insertion Loss: Potential Low-Frequency Improvement over Bare Duct Elbow

Discharge Duct Configuration of Horizontal Supply Duct	Duct Breakout Transmission Loss at Low Frequencies (dB)			Side View	End View
	63 Hz	125 Hz	250 Hz		
Rectangular duct: no turning vanes (reference).	0.0	0.0	0.0	AIR FLOW	
Rectangular duct: two dimensional turning vanes.	0.0	0.5	1.0	TURNING VANES	
Rectangular duct: wrapped with foam and lead.	4.0	3.0	4.5	SEE END VIEW	FOAM INSUL. W/ 2 LAYERS LEAD
Rectangular duct: wrapped with glass fiber, one layer of gypsum board.	4.0	7.0	5.5	SEE END VIEW	GLASS FIBER PRESSED FLAT AGAINST DUCT.
Rectangular duct: wrapped with glass fiber, two layers of gypsum board.	7.5	8.5	9.0	SEE END VIEW	GYPSUM BOARD SCREWED TIGHT
Rectangular to multiple drop: round mitered with turning vanes, three parallel round supply ducts.	17.5	11.5	13.0	0.6 mm	1.9 mm
Rectangular to multiple drop: round mitered elbow with turning vanes, three parallel round lined double wall supply ducts	18.0	13.0	16.0	0.6 mm	1.9 mm

Source: Ref. 1, p. 42.28.

ment) pose important noise control problems. They are frequently mounted on curbs supported by lightweight roof structure directly above occupied space. Because of their proximity to the building interior, they should be located over acoustically noncritical spaces. The primary noise paths that need consideration, as shown in Fig. 5, are[1] (1) the casing-radiated noise, (2) the supply and return air ducts, and (3) structure-borne noise.

6.1 Casing-Radiated Noise

This noise is best controlled by making certain the roof structure inside the curb has adequate mass to block the sound being radiated by the fan casing and compressor, if included. This construction may consist of concrete or multiple layers of gypsum board. Sealing around the perimeter of the curb and around the supply and return duct penetrations is important. The supply and return ducts are commonly directly above the ceiling of the occupied space, and often there is insufficient space for easy application of adequate noise control measures. On the supply side, low-frequency breakout noise is a common problem. Control methods consist of proper configuration of the fan outlet and ducts. A common problem is the duct elbow between the fan outlet and the horizontal duct above the ceiling. Round elbows and round duct are a very important way of controlling this noise. See Table 7.[11]

6.2 Structure-Borne Noise

This noise is controlled by proper vibration isolation. Internal isolation of fans in units without compressor/condenser units will usually be sufficient if the isolators have adequate deflection; steel springs of 50–75 mm deflection are usually required. Additionally, several manufacturers can supply curbs with high deflection springs for such isolation.

An alternate noise control technique is to raise the unit above the roof on a structure (commonly called dunnage). In this configuration, the ducts can be run above the roof before entering the building, creating the opportunity to include noise control measures before the ducts enter the building; casing-radiated noise through the bottom of the unit is also less difficult. Vibration transfer to a lightweight roof is avoided, since the dunnage is usually supported from heavier roof-supporting structure.

REFERENCES

1. *ASHRAE Handbook, HVAC Applications*, ASHRAE, Atlanta, GA, 1991.
2. R. J. Wells, "Acoustical Plenum Chambers," *Noise Control*, July, 1958.

69

SOUND POWER LEVEL PREDICTIONS FOR INDUSTRIAL MACHINERY

ROBERT D. BRUCE AND CHARLES T. MORITZ

1 INTRODUCTION

Procedures for calculating the sound power level or sound pressure level of industrial machinery, such as boilers, fans, motors, pumps, and turbines, are presented in this chapter. These procedures can be used for modeling the sound levels in a space or for developing purchase specifications for new equipment. With any project, acoustical data for individual equipment, specifically sound pressure or sound intensity measurements and sound power level calculations in accordance with recognized standards, should be obtained. Many manufacturers provide the estimated sound power level or the measured sound pressure level at 1 m from their equipment as well as offer special low-noise options. If information from the manufacturer is unavailable, efforts should be made to measure a similar unit in operation at an installation. If this is not possible or if an estimate for a preliminary study is required, then the material in this chapter can be used. The equations presented are either entirely based on measured data or are semiempirical and tend to be conservative, usually predicting higher sound levels than are measured in the field. For some equipment, a large database has not yet been established so estimates of the sound power level for similar equipment are presented.

2 POWER SOURCES

2.1 Boilers[1]

Main Steam Boilers The sound power level for main steam boilers (between 125 and 800 MWe) can be calculated using the equation

$$L_w = 84 + 15 \log \text{ MWe}, \tag{1}$$

where L_w is the overall sound power level and MWe is the electrical generating rating of the unit.

The octave band sound power levels can be obtained by subtracting the values shown in Table 1.

Auxiliary Boilers The noise produced by auxiliary boilers is often due primarily to the blower and the burner and not the walls of the boiler. An estimate of the sound power level for auxiliary boilers between 50 and 2000 boiler horsepower (bhp) can be calculated using the equation

$$L_w = 95 + 4 \log \text{ bhp}, \tag{2}$$

where 1 bhp = 15 kg steam/h. The octave band sound power levels can be obtained by subtracting the values shown in Table 1.

2.2 Electric Motors

Motors under 750 kW[2] For totally enclosed fan-cooled (TEFC) motors, the sound power level can be cal-

*The information from Ref. 1 is reprinted by permission of the Edison Electric Institute (EEI) from EEI's *Electric Power Plant Environmental Noise Guide*, 2nd ed., Copyright 1984, all rights reserved. Neither EEI nor any member of EEI (a) makes any warranty, express or implied, with respect to the use of this information or that such use may not infringe privately owned rights or (b) assumes any liability with respect to the use of or for damages resulting from the use of this information.

Note: References to chapters appearing only in the *Encyclopedia* are preceded by *Enc.*

TABLE 1 Octave Band and A-Weighted Sound Power Level Adjustments

Source	31.5	63	125	250	500	1000	2000	4000	8000	*A*
Main steam boiler	4	5	10	16	17	19	21	21	21	12
Auxiliary boiler	6	6	7	9	12	15	18	21	24	9
TEFC motors under 750 kW	14	14	11	9	6	6	7	12	20	1
Drip-proof motors under 750 kW	9	9	7	7	6	9	12	18	27	4
Gas turbine casing	10	7	5	4	4	4	4	4	4	2
Gas turbine exhaust	12	8	6	6	7	9	11	15	21	4
Gas turbine intake	19	18	17	17	14	8	3	3	6	0
Reciprocating engines (<600 rpm)	12	12	6	5	7	9	12	18	28	4
Reciprocating engines (600–1500 rpm)	14	9	7	8	7	7	9	13	19	3
Reciprocating engines (600–1500 rpm with blower)	22	16	18	14	3	4	10	15	26	1
Reciprocating engines (>1500 rpm)	22	14	7	7	8	6	7	13	20	2
Reciprocating engine air inlet (turbocharged)	4	11	13	13	12	9	8	9	17	3
Reciprocating engine exhaust	5	9	3	7	15	19	25	35	43	12
Steam turbines	11	7	6	9	10	10	12	13	17	5
Steam turbine generator units	9	3	5	10	14	18	21	29	35	12
Transformers	−3	3	5	0	0	−6	−11	−16	−23	0
Centrifugal compressor casing	10	10	11	13	13	11	7	8	12	2
Centrifugal compressor air inlet	18	16	14	10	8	6	5	10	16	0
Rotary and reciprocating air compressors	11	15	10	11	13	10	5	8	15	2
Feed pumps (1000–9000 kW)	11	5	7	8	9	10	11	12	16	4
Feed pumps (9500–18,000 kW)	19	13	15	11	5	5	7	19	23	1
Centrifugal fan	11	9	7	8	9	9	13	17	24	[a]
Centrifugal fan casing	3	6	7	11	16	18	22	26	33	[a]
Gas recirculation fan casing	10	7	4	7	18	20	25	27	31	12
Axial-flow fan	11	10	9	8	8	8	10	14	15	3
Generators	11	8	7	7	7	9	11	14	19	4
Motor-driven pumps	13	12	11	9	9	6	9	13	19	2
Mechanical-draft cooling towers (full speed)	9	6	6	9	12	16	19	22	30	10
Mechanical-draft cooling towers (half speed)	9	6	6	10	10	11	11	14	20	5
Chiller with reciprocating compressor	—	19	11	7	1	4	9	14	—	0
Chiller with rotary-screw compressor	25	19	15	3	6	10	15	20	22	5
Centrifugal chillers, internal geared	—	8	5	6	7	8	5	8	—	0
Centrifugal chillers, direct drive	—	8	6	7	3	4	7	12	—	0
Centrifugal chillers, >1000 tons	—	11	11	8	8	4	6	13	—	0
Diesel powered, mobile equipment	—	11	6	3	8	10	13	19	25	5

[a]Dependent on blade passage frequency.

culated using the following equations:

$$L_w = 17 + 17\log \text{kW} + 15\log \text{rpm} + 10\log A \quad \text{(under 40 kW)}, \tag{3}$$

$$L_w = 28 + 10\log \text{kW} + 15\log \text{rpm} + 10\log A \quad \text{(over 40 kW)}, \tag{4}$$

where kW is the nameplate motor rating, rpm is the speed at which the motor is operating, and A is the conformal surface area (in square metres) at 1 m from the motor. For TEFC motors between 300 and 750 kW, use the value 300 kW in Eq. (4).

The octave band sound power levels can be obtained by subtracting the values shown in Table 1.

For drip-proof motors, the sound power level can be calculated using the following equations:

$$L_w = 12 + 17\log \text{kW} + 15\log \text{rpm} + 10\log A \quad \text{(under 40 kW)}, \tag{5}$$

$$L_w = 23 + 10\log \text{kW} + 15\log \text{rpm} + 10\log A \quad \text{(over 40 kW)}. \tag{6}$$

For drip-proof motors between 300 and 750 kW, use the value 300 kW in Eq. (6).

The octave band sound power levels can be obtained by subtracting the values shown in Table 1.

TABLE 2 Octave Band and A-Weighted Sound Power Levels

Source	31.5	63	125	250	500	1000	2000	4000	8000	*A*
Electric motor 750–4000 kW (1800–3600 rpm)	94	96	98	98	98	98	98	95	88	104
Electric motor 750–4000 kW (1200 rpm)	88	90	92	93	93	93	98	88	81	101
Electric motor 750–4000 kW (900 rpm)	88	90	92	93	93	96	96	88	81	101
Electric motor 750–4000 kW (<720 rpm)	88	90	92	93	93	98	92	83	75	100
Electric motor 750–4000 kW (250 and 400 rpm vertical)	86	87	88	88	88	98	88	78	68	99
1200 kW Triplex mud pump	102	104	108	105	104	99	93	86	83	105
Mud reconditioning degasser	78	83	82	87	94	94	95	88	81	99
Drawworks	96	104	109	112	112	108	105	102	96	113
Top drive	98	97	100	110	105	100	100	96	82	108
Natural draft cooling tower, rim noise	—	—	105	104	106	108	110	112	110	117
Natural draft cooling tower, discharge noise	—	—	100	99	101	103	105	107	105	112

Motors between 750 and 4000 kW[1] The sound power level for large drip-proof electric motors (between 750 and 4000 kW) can be estimated by using Table 2.

2.3 Gas Turbines[2]

Manufacturers of gas turbines will likely have sound power level data for the exhaust and inlet. Obtaining accurate casing data can be difficult due to contributions from the inlet, exhaust, or other equipment. The overall sound power level for gas turbines can be estimated using the following expression:

$$L_w = 127 + 15 \log \text{MW} \quad \text{(intake)}, \tag{7}$$

$$L_w = 133 + 10 \log \text{MW} \quad \text{(exhaust)}, \tag{8}$$

$$L_w = 120 + 5 \log \text{MW} \quad \text{(casing)}, \tag{9}$$

where MW is the power rating of the turbine. The octave band sound power level for each of these sources can be obtained by subtracting the values shown in Table 1.

2.4 Reciprocating Engines[2]

The major noise sources of natural-gas and diesel reciprocating engines are the engine casing, exhaust noise, and for turbocharged engines, the air inlet.

Engine Casing Noise The sound power level for engine casing noise can be calculated using the equation

$$L_w = 94 + 10 \log \text{kW} + A + B + C + D, \tag{10}$$

where L_w is the overall sound power level of the casing, kW is the full-load rating of the engine, and A, B, C, and D are correction factors obtained from Table 3. The octave band sound power levels can be obtained by subtracting the values shown in Table 1.

Air Inlet For those engines with a turbocharger, the air inlet noise can be calculated using the equation

$$L_w = 95 + 5 \log \text{kW} - \frac{l_{\text{inlet}}}{1.8}, \tag{11}$$

where L_w is the overall sound power level of the air inlet, kW is the full-load rating of the engine, and l_{inlet} is the length of the inlet ducting in metres. For engines under about 340 kW, the calculation for the casing noise also takes into account the air inlet noise for both turbocharged and naturally aspirated engines.

The octave band sound power levels can be obtained by subtracting the values shown in Table 1.

Engine Exhaust The sound power level for the unmuffled exhaust can be calculated using the equation.

$$L_w = 120 + 10 \log \text{kW} - T - \frac{l_{\text{exhaust}}}{1.2}, \tag{12}$$

where L_w is the overall sound power level of the exhaust, kW is the full-load rating of the engine, T is a correction term for engines with turbochargers ($T = 6$ if the engine has a turbocharger, $T = 0$ if there is no turbocharger), and l_{exhaust} is the length of the exhaust in metres. The octave band sound power levels can be obtained by subtracting the values shown in Table 1.

2.5 Steam Turbines[1]

Manufacturers of steam turbines will likely have sound pressure level measurements, but they may not have been

TABLE 3 Sound Power Level Correction Terms for Casing Noise of Reciprocating Engines

Correction Term	Correction Type	Qualifying Condition	dB
A	Speed	Less than 600 rpm	−5
		600–1500 rpm	−2
		Over 1500 rpm	0
B	Fuel	Diesel only	0
		Diesel and/or natural gas	0
		Natural gas only (with small amount of "pilot oil")	−3
C	Cylinder arrangement	In-line	0
		V-type	−1
		Radial	−1
D	Air intake	Unducted air inlet to unmuffled Roots blower	+3
		Ducted air from outside the engine compartment enclosure	0
		Ducted air to muffled Roots blower	0
		All other types of inlets (with or without a turbocharger)	0

measured in accordance with a recognized standard. The overall sound power level for steam turbines can be estimated using the expression

$$L_w = 93 + 4 \log \text{kW}. \quad (13)$$

The octave band sound power levels can be obtained by subtracting the values shown in Table 1.

2.6 Steam Turbine–Generator Units[1]

The sound power level for main steam turbine–generator units (between 200 and 1100 MWe) can be calculated using the equation

$$L_w = 113 + 4 \log \text{MWe}, \quad (14)$$

The octave band sound power levels can be obtained by subtracting the values shown in Table 1. Equation (14) takes into account sound contributions from the high- and low-pressure turbines, generators, and shaft-driven exciters. Noisy couplings and steam control valves can cause higher sound power levels than those predicted with this equation.

2.7 Transformers[1]

Noise from the body of the transformer is made up of tones at the even harmonics of the line frequency (120, 240, 360, 480, ... Hz) with the 120-Hz tone being the dominant sound at normal receiving distances. When additional cooling of the transformer is needed, noise from the cooling fans can become the dominant noise source. Lower fan speeds and optimal blade shapes have helped make newer fans much quieter, which is particularly helpful in noise-sensitive applications.

The National Electrical Manufacturers Association (NEMA) sound pressure level rating is the A-weighted sound pressure level one foot from the transformer and can be estimated from the following formulas:

$$\text{NEMA sound rating} = 55 + 12 \log \text{MV(A)} \quad \text{for a standard transformer,} \quad (15)$$

$$\text{NEMA sound rating} = 45 + 12 \log \text{MV(A)} \quad \text{for a quieted transformer.} \quad (16)$$

These equations are valid for transformers between 20 and 450 MV(A). The A-weighted sound power level can be calculated using the formula

$$L_w = \text{NEMA sound rating} + 10 \log S, \quad (17)$$

where L_w is the A-weighted sound power level and S is the surface area of the four side walls in square metres.

The term $10 \log S$ may be estimated from the MV(A) rating by using the formula

$$10 \log S = 14 + 2.5 \log \text{MV(A)}. \quad (18)$$

Octave band sound power and pressure levels can be

obtained by *adding* the appropriate octave band correction factors shown in Table 1.

3 DRIVEN EQUIPMENT

3.1 Air Compressors

Centrifugal Compressors[3] The sound power level radiated from the casing or discharge piping for large centrifugal compressors can be calculated using the equation

$$L_w = 3 + 20\log \text{kW} + 50\log U - 17\log(mf), \qquad (19)$$

where U is impeller tip speed in metres per second ($30 < U < 230$), m is the surface weight of the casing or pipe wall in kilograms per square metre, and f is the octave band center frequency.

The frequency of the maximum sound power is given as

$$f_p = 4.1U. \qquad (20)$$

For the octave band containing f_p, the sound power level is the value from Eq. (19) minus 4.5 dB. The adjacent octave bands above and below f_p should roll off at a rate of 3 dB/octave.

Reciprocating Compressors[3] The octave band sound power levels radiated from the casing or discharge piping for large reciprocating compressors can be calculated using the equation

$$L_w = 154.5 + 10\log \text{kW} - 17\log(mf), \qquad (21)$$

where kW is the power of the driver motor, m is the surface weight of the casing or pipe wall in kilograms per square metre, and f is the octave band center frequency.

This equation assumes that the sound power is radiated from 15 m of discharge piping. To determine the octave band levels, first calculate the fundamental frequency [$f_p = B$ (rpm)/60, where B is the number of cylinders]. For the octave band containing f_p, the sound power level is the value from the equation minus 4.5 dB. The adjacent octave bands above and below f_p should roll off at a rate of 3 dB/octave.

If only the rated power of the compressor is known, the following may be used to calculate the sound power level of centrifugal, rotary, and reciprocating compressors.

Centrifugal Air Compressors[1] The sound power level for the casing noise of centrifugal compressors between 1100 and 3700 kW can be calculated using the equation

$$L_w = 79 + 10\log \text{kW}, \qquad (22)$$

where L_w is the overall sound power level of the casing and kW is the rated power of the compressor. The octave band sound power levels can be obtained by subtracting the values shown in Table 1.

The sound power level for the unmuffled air inlet of centrifugal compressors can be calculated using the equation

$$L_w = 80 + 10\log \text{kW}. \qquad (23)$$

The octave band sound power levels can be obtained by subtracting the values shown in Table 1.

Rotary and Reciprocating Air Compressors[1] The sound power level for rotary and reciprocating air compressors, including partially muffled air inlets, can be calculated using the equation

$$L_w = 90 + 10\log \text{kW} \qquad (24)$$

where kW is the rated power of the compressor. The octave band sound power levels can be obtained by subtracting the values shown in Table 1.

3.2 Boiler and Reactor Feed Pumps[1]

The sound power level for large boiler and reactor feed pumps driven by either motors or steam turbines can be estimated by using Table 4.

The larger pumps, which are generally driven by steam turbines, have high tonal noise components that increase the noise levels by several decibels in the

TABLE 4 Sound Power Levels of Boiler and Reactor Feed Pumps

Pump Power Rating (kW)	Sound Power Level	
	Linear	A-Weighted
1,000	108	104
2,000	110	106
4,000	112	108
6,000	113	109
9,000	115	111
9,500	113	112
12,000	115	114
15,000	119	118
18,000	123	122

250–2000-Hz octave bands. Thus, the sound power ratings in Table 4 rise sharply for pumps with high power ratings.

The octave band sound power levels can be obtained for two ranges of power ratings by subtracting the values shown in Table 1.

3.3 Fans[1]

Centrifugal-Type Fans with Airfoil or Backward-Curved Blades The sound power level of the inlet noise of forced-draft fans with inlet control vanes or the discharge noise of induced-draft fans can be calculated using the equation

$$L_w = 10 + 10 \log Q + 20 \log \delta P, \qquad (25)$$

where Q is the air flow in cubic metres per minute and δP is the pressure drop across the fan in newtons per square metre. The octave band sound power levels can be obtained by subtracting the values shown in Table 1.

Five decibels should be added in the octave band containing the blade passage frequency for forced-draft fans used in high-pressure applications. For induced-draft fans, add 10 dB in the bands containing the blade passage frequency and its second harmonic. This correction term is called the blade frequency increment. The blade passage frequency is calculated as follows:

$$\text{Blade passage frequency} = (\text{fan rpm} \times \text{number of blades})/60. \qquad (26)$$

Centrifugal-Type Fan Casing If there is ducting on the inlet to the fan, the overall sound power level of the uninsulated fan casing may be calculated using the equation

$$L_w = 1 + 10 \log Q + 20 \log \delta P. \qquad (27)$$

The octave band sound power levels can be obtained by subtracting the values shown in Table 1.

Add 5 dB in the octave band containing the blade passage frequency for high-pressure applications.

The noise generated at the uninsulated fan breaching (discharge ductwork) is approximately 6 dB lower than than that generated by the uninsulated fan housing. The frequency spectrum is similar to that for the uninsulated fan housing.

Gas Recirculation Fan Casing The overall casing sound power level for centrifugal fans used for gas recirculation service (1300–4100 kW) can be calculated using the equation

$$L_w = 13 + 10 \log Q + 20 \log \delta P. \qquad (28)$$

The octave band sound power levels can be obtained by subtracting the values shown in Table 1.

Axial-Flow Fans The overall sound power level of the inlet noise of forced-draft fans or the discharge noise of induced-draft fans can be calculated using the equation

$$L_w = 24 + 10 \log Q + 20 \log \delta P. \qquad (29)$$

The octave band sound power levels can be obtained by subtracting the values shown in Table 1, except for the octave band containing the blade passage frequency, where 6 dB should be added rather than subtracting the value in the table. For vane axial fans, inlet guide vanes cause much higher blade passage noise than outlet guide vanes.

3.4 Gears[2]

The following equation provides an estimate of gearbox noise levels based on studies of gearboxes rated from 200 to 17,500 kW:

$$L_w = 79 + 3 \log \text{rpm} + 4 \log \text{kW} + 10 \log A, \qquad (30)$$

where L_w is the octave band sound power level, rpm is the speed of the slower gear shaft, kW is the power transmitted through the gearbox, and A is the conformal surface area (in square metres) at 1 m from the gearbox.

Since the octave bands containing the gear-meshing and ringing frequencies are different according to the particular gearbox being used, the level predicted by the equation is an estimate for all bands equal to or greater than 125 Hz. For the 31-Hz band subtract 6 dB, and for the 63 Hz band, subtract 3 dB from the level obtained from Eq. (30).

3.5 Generators[3]

The sound power level for generators (not including the driver noise) can be calculated from the equation

$$L_w = 84 + 10 \log \text{MW} + 6.6 \log \text{rpm}, \qquad (31)$$

where L_w is the overall sound power level, MW is the power rating of the generator, and rpm is the speed of the generator. The octave band sound power levels can be obtained by subtracting the values shown in Table 1.

3.6 Pumps[2]

The sound power level for motor-driven pumps under 2000 kW can be calculated using Table 5. In this table, kW is the nameplate motor rating and A is the conformal surface area (in square metres) at 1 m from the pump. The octave band sound power levels can be obtained by subtracting the values shown in Table 1. These equations are based on sound pressure level measurements at 1 m from the pump and therefore include some sound contributions from the driver and suction and discharge piping.

For high-pressure applications, the level in the band that contains the blade passage frequency and its second harmonic may be 5–10 dB higher than the calculated levels.

4 VALVE AND PIPING NOISE

The sound level produced by gas control valves is dependent on whether the flow through the valve is subcritical or supercritical (choked-valve condition). For subcritical flow, the valve noise is due to turbulent mixing and turbulence–boundary interaction. For supercritical flow, the valve noise is due primarily to broadband shock noise. Many valve manufacturers, such as Fisher or Masoneilan, have semiempirical models to predict the sound pressure level at a position 1 m downstream of the valve and 1 m out from the piping. These models are expected to predict the A-weighted sound level to within 5 dB of the actual level. For situations when the manufacturer cannot provide data or as an alternative, Ng[4] provides a control valve calculation procedure. All of these methods assume some number of straight piping sections downstream of the valve. The presence of a tee, mitered bend, or an elbow near the valve discharge can significantly increase the sound level from quiet trim valves.

Turbulence generated in piping can also be a significant source of noise. Concowe[5,6] and Norton[7] have published procedures by which flow-induced sound power levels can be predicted. These procedures require a fairly detailed knowledge of the flow conditions and piping and are beyond the scope of this chapter but are expected to predict the A-weighted sound level to within 3 dB.

5 INDUSTRY-SPECIFIC EQUIPMENT

5.1 Air Conditioning for Buildings

Cooling Towers[1]

Mechanical-Draft Cooling Towers The sound power level of mechanical-draft propeller-type cooling towers can be calculated using the equations

$$L_w = 96 + 10 \log \text{kW} \quad \text{(for full-speed operation)}, \tag{32}$$

$$L_w = 88 + 10 \log \text{kW} \quad \text{(for half-speed operation)}, \tag{33}$$

where L_w is the overall sound power level and kW is the full-speed power rating of the fan. The octave band sound power levels can be obtained by subtracting the values shown in Table 1.

When calculating sound pressure levels at distances greater than 10–20 m from the tower, *add* the values shown in Table 6 to account for directional effects.

Natural-Draft Cooling Towers The sound power level of the rim, located at the base, and the discharge, located at the top, of large natural-draft cooling towers can be estimated by using the values shown in Table 2. Since the low-frequency noise levels of the water flow are not significant, sound power levels are not listed for the 31.5- and 63-Hz octave bands.

In the horizontal direction and 30° above and below the horizontal direction, the noise from the rim may be assumed to radiate uniformly. Most of the noise from the discharge is radiated upward. When calculating sound pressure levels from the discharge, *add* the values shown in Table 6 to account for directional effects.

TABLE 5 Overall Sound Power Level of Motor-Driven Pumps

Operating Speed (rpm)	For Motor Ratings under 75 kW	For Motor Ratings above 75 kW
450–900	68 + 10 log kW + 10 log A	81 + 3 log kW + 10 log A
1000–1500	70 + 10 log kW + 10 log A	83 + 3 log kW + 10 log A
1600–1800	75 + 10 log kW + 10 log A	88 + 3 log kW + 10 log A
3000–3600	72 + 10 log kW + 10 log A	85 + 3 log kW + 10 log A

TABLE 6 Octave Band and A-Weighted Sound Power Level Adjustments for Directional Effects of Cooling Towers

Source	31.5	63	125	250	500	1000	2000	4000	8000	A
Mechanical draft, air inlet side	0	0	0	1	2	2	2	3	3	2
Mechanical draft, enclosed side	−3	−3	−3	−3	−3	−3	−4	−5	−6	−4
Mechanical draft, top	3	3	3	3	3	4	4	3	3	3
Natural draft, top	8	8	8	8	9	9	10	10	10	10
Natural draft, horizontal	−2	−3	−4	−6	−8	−10	−12	−14	−16	11
Natural draft, 30° below horizontal	−3	−4	−5	−7	−9	−12	−14	−16	−18	13

Packaged Chillers with Compressors

Packaged Chillers with Reciprocating Compressors[8] The A-weighted sound power level can be calculated from the equation below as a function of tons of refrigeration (1 ton of cooling capacity is equal to 3.52 kW of heat removal). The standard error of this equation is estimated to be 5 dB. The noise levels are due primarily to the compressor component and not the drive motor:

$$L_w = 71 + 9\log(\text{tons of cooling capacity}) + 10\log A, \quad (34)$$

where A is the conformal surface area (in square metres) at 1 m from the unit. The octave band sound power levels are obtained by subtracting the values shown in Table 1.

Packaged Chillers with Rotary-Screw Compressors[2] The overall sound power level of rotary-screw compressors between 100 and 300 tons of cooling capacity and operating at approximately 3600 rpm is given by the equation

$$L_w = 95 + 10\log A. \quad (35)$$

The octave band sound power levels can be obtained by subtracting the values shown in Table 1.

Packaged Chillers with Centrifugal Compressors[8] The A-weighted sound power level can be calculated from the equation below. The standard error of this equation is estimated to be 4 dB. For larger units with built-up assemblies, use the conformal area of the compressor section only:

$$L_w = 60 + 11\log(\text{tons of cooling capacity}) + 10\log A. \quad (36)$$

The octave band sound power levels can be obtained by subtracting the values shown in Table 1.

5.2 Construction Equipment[1]: Diesel-Powered, Mobile Equipment

The sound power level for diesel powered, mobile equipment such as crawler tractors, dozers, tractor shovels, front-end loaders, backhoes, graders, mobile cranes, and trucks can be calculated using the equation

$$L_w = 99 + 10\log \text{kW}, \quad (37)$$

where L_w is the overall sound power level and kW is the power rating of the engine. This equation is for turbocharged or naturally aspirated engines with conventional exhaust mufflers. In the typical application, the sound level will average about 4 dB lower than the calculated level since the engine is frequently not operated in the maximum-power condition. The octave band sound power levels can be obtained by subtracting the values shown in Table 1.

6 OIL FIELD EQUIPMENT

Increased governmental noise regulations and concerns about drilling in environmentally sensitive locations have created a need for drilling rig sound level analyses. Example sound power level data for a standard degasser, drawworks, mud pump, and top drive are presented in Table 2. Some manufacturers offer special noise control options that can provide a reduction of 5–10 dB(A) from the values provided in the table. The sound power for the degasser is typical for a unit rated at 227 m^3/h and includes the contribution from the drive motor. The sound power levels for the mud pump and drawworks do not include the drive motors.

APPENDIX: CALCULATION OF CONFORMAL SURFACE AREA

A conformal surface is a hypothetical surface located a distance d from the nearest point on the envelope of the reference box. It is different from a rectangular surface because a conformal surface has rounded corners. For small surfaces, the difference in area between a rectangular surface and a conformal surface is typically small ($10 \log A < 1$dB); however, for large surfaces the difference in areas can be significant ($10 \log A > 2$ dB). The area of a conformal surface can be calculated from the equation

$$S = 4(ab + bc + ca)\,\frac{a + b + c}{a + b + c + 2d},$$

where $a = 0.5L + d$, $b = 0.5W + d$, and $c = H + d$, with d the measurement distance, L the length of the reference box, W the width of the reference box, and H the height of the reference box.

REFERENCES

1. L. N. Miller, E. W. Wood, R. M. Hoover, A. R. Thompson, and S. L. Patterson, *Electric Power Plant Environmental Noise Guide*, rev. ed., Edison Electric Institute, Washington, DC, 1984, Chapter 4.
2. *Noise and Vibration Control for Mechanical Equipment*, Manual TM5-805-4/AFM 88-37/NAVFAC DM-3.10, manual prepared by Bolt, Beranek and Newman for the Joint Department of the Army, Air Force, and Navy, Washington, DC, 1980, Chapter 7.
3. D. A. Bies and C. H. Hansen, *Engineering Noise Control*, Unwin Hyman, London, 1988, Chapter 11.
4. K. W. Ng, "Control Valve Aerodynamic Noise Generation and Prediction," *Proceedings of NOISEXPO*, *80, Acoustical Publications, Inc., Bay Village, OH, 1980, pp. 49–54.
5. J. N. Pinder, "The Study of Noise From Steel Pipelines," CONCAWE Report No. 84/55, 1984, Brussels, Belgium.
6. P. H. M. Corbet and P. J. van de Loo, "Experimental Verification of the ISVR Pipe Noise Model," CONCAWE Report No. 84/64, 1984, Brussels, Belgium.
7. M. P. Norton and A. Pruiti, "Universal Prediction Schemes for Estimating Flow-Induced Industrial Pipeline Noise and Vibration," *Appl. Acoust.*, Vol. 33, 1991, pp. 313–336.
8. ASHRAE, "Sound and Vibration Control," in *HVAC Applications*, American Society of Heating, Refrigerating and Air-Conditioning Engineers, Inc., Atlanta, 1991, Chap. 42.

70

AIRPORT NOISE

KENNETH MCK. ELDRED

1 INTRODUCTION

The noise associated with civil aircraft operations in the vicinity of airports has been the major environmental noise problem in the United States and in many other countries since the late 1950s, when turbojet aircraft entered service. The public opposition to increases in aircraft noise is the major factor that has prevented most development of new airports. This problem is worldwide, and the environmental restrictions on airport development are found in Japan, Germany, England, and many other countries.

The civil aviation jet age in the United States opened with a roar in 1958, with the beginning of transcontinental and transoceanic flight operations by Boeing 707 and Douglas DC-8 turbojet aircraft. Their takeoff noise, expressed in maximum A-weighted sound level, was about 20 dB greater than that of the propeller aircraft that they replaced. The Port Authority of New York promulgated its takeoff noise regulation with limits of 112 Pndb[1] at locations at 2.7–4.2 statute miles from the start of takeoff roll with 24-h noise monitoring at several locations to determine compliance. London Heathrow, with its jet operations, also including the Comet and Caravelle, imposed night and noise level restrictions. Soon other major airports developed rules to attempt to control the noise.[2]

The noise of the early jets was controlled by a variety of multiple and corrugated nozzle schemes that shifted much of the sound energy to higher frequencies. However, these devices had significant performance penalties of about 1% thrust loss per decibel reduction. Within 2 years, engine manufacturers introduced low bypass turbofan engines by adding a bypass co-annular flow fan stage to the turbojets used in the first generation of civil engines. This development significantly lowered the full power jet noise and added takeoff thrust, which improved performance. Unfortunately, the fans sounded like sirens, resulting in trading the improvement in jet noise during takeoff for a large increase in approach noise. Finally, by the late 1960s, the world's leading engine manufacturers had developed the familiar high-bypass-ratio engines, which were much quieter than the earlier engines, particularly on takeoff, and provided much higher propulsion efficiency. Also, concerted government and industry research led to the development of quieter fans and turbines, providing a significant reduction in the pure tones that were characteristic of the earlier turbofan engines.

This chapter addresses the assessment and control of civil airport noise in the United States. Then, it reviews the principal national and airport-specific actions undertaken to reduce the noise and concludes with a summary of the changes in the estimated aircraft noise population impact in the United States from 1960 to 2000. Although the principal examples and discussion of this topic are oriented to commercial civil aviation in the United States, the general problems and approaches to solutions apply at civil, military, and general aviation airports throughout the world.

2 ASSESSMENT OF AIRPORT NOISE

The assessment of aircraft noise in the vicinity of an airport involves measurement and/or calculation of the noise and estimating its impact, primarily on people who reside in areas that are subject to the noise. In these assessments it is usually necessary to describe the noise

Note: References to chapters appearing only in the *Encyclopedia* are preceded by *Enc.*

of a single aircraft flight (single event sound) and to quantify the cumulative noise of many flights throughout a 24-h day. The reference pressure is 20 μPa. The descriptors for the sound of a single event include the following:[3,4]

Sound Level. Usually level of the ratio of the mean-square A-weighted sound pressure obtained with the slow time constant to a squared reference sound pressure in decibels.

Maximum Sound Level (MXSL). Greatest slow (1-s) sound level within a stated time interval or during a specific single event in decibels.

Perceived Noise Level (PNL). A frequency-weighted sound pressure level that combines the sound pressure levels in 24 one-third-octave bands from 50 Hz to 10 kHz in decibels.[5]

Sound Exposure Level (SEL). Usually the level of the ratio of a given time integral of the squared instantaneous A-weighted sound pressure over a stated time interval or event to the product of the squared reference sound pressure of 20 μPa and reference duration of 1 s in decibels.

Effective Perceived Noise Level (EPNL). Level of the time integral of antilogarithm of one-tenth of the tone corrected perceived noise level over the duration of an aircraft flyover and with a reference duration of 10 s in decibels.[5]

The descriptors for the cumulative noise of many events include the following:

Equivalent Sound Level (LEQ). Usually the level of the ratio of the time average integral of the squared instantaneous A-weighted sound pressure during a stated time interval (often 1 h or 1 day) to the product of the squared reference pressure and the duration of the time interval in seconds in decibels.

Day/Night Sound Level (DNL). Usually the level of the ratio of the time average integral of the squared instantaneous A-weighted sound pressure during a 24-h period after multiplying by 10 the values from 10 p.m. to 7 a.m. to the product of the squared reference pressure and 86,400 s in decibels.

Community Noise Equivalent Level (CNEL). Usually the level of the ratio of the time average integral of the squared instantaneous A-weighted sound pressure during a 24-h period after multiplying by 3 the values between 7 p.m. and 10 p.m. and multiplying by 10 the values between 10 p.m. and 7 a.m. to the product of the squared reference pressure times 86,400 s in decibels. *Note:* The CNEL is used in California instead of the DNL.[6]

The primary descriptor used in the United States for airport noise is the annual average value of the DNL. For airport projects that are funded partially or wholly by the federal government, the environmental assessment procedures of the Federal Aviation Administration (FAA) require use of the DNL, supplemented by other descriptors, where necessary. The magnitude of impact is generally measured by counting the number of people estimated to reside in areas where the DNL from airport operations exceeds 65 dB (for some projects 60 dB) and extending to the highest value of DNL, usually in intervals of 5 dB.

2.1 Monitoring Airport Noise

Airport noise may be monitored by several methods, including continuous or intermittent sound measurements, with or without an observer present, combined sound measurement with radar flight track data, or estimates based on calculations using aircraft operations data in a computer model.

Permanently installed and/or portable field microphone systems are used to monitor aircraft noise at many airports. The first permanent system was installed in 1959 by the Port Authority of New York and New Jersey.[7] But it was not until the early 1970's, when the State of California promulgated an Airport Noise Standard,[6] that permanent noise-monitoring systems began to be installed at many other airports in the United States. The California Standard addressed both permanent and portable noise measurements. For the latter, it specified that measurements be made for 4 weeks during the year, 1 week during each season. The California standard became a de facto specification for airport noise-monitoring systems.

Typical airport noise-monitoring systems have from 5 to 25 remote sound monitors that provide continuous 24-h coverage. The remote monitors are usually connected to the central system by standard telephone lines and transmit data either continuously or when dialed by the central system for a data download. Some systems also have a direct telephone line that may be used to listen to the noise. Most systems can provide noise samples at intervals of 1 s or less. They also can provide the maximum sound levels and sound exposure levels for sound events that meet preset criteria (e.g., sound level thresholds, duration, rise and fall time). These sound event criteria are usually set to discriminate against nonaircraft sound events so that the reported measured sound events have a high probability to be aircraft related. Concurrently, or at a later time, the true sources for the sound events can be determined by the system's operator or automatically by a computer system by matching other data on the time of occurrence, runways in use, flight track data, if available, and identity of aircraft arrivals and departures. Most systems provide the hourly equivalent sound level and the day–night sound level, sometimes subdivided into aircraft and nonaircraft noise components. Often, other quantities are available such as the statistical sound levels exceeded for stated percentage of time, the maximum sound level, and the number of sound events in hourly and 24-h periods.

The locations selected for remote monitors depend on the purpose of the measurements. For airports in California, many of the monitors are located in the vicinity of the noise impact boundary, which approximates the CNEL of 65 dB noise contour in areas where the land use is residential or in other noise-sensitive uses. Often sound monitors are located either directly under flight paths or to each side of the expected ground track to bracket the flight paths. Sound monitors used to check aircraft adherence to noise abatement procedures often are located in the vicinity of the nearest residences affected by noise from operations on the flight path to be monitored. They are also used at greater distances from the airport to monitor flight path procedures that incorporate turns to avoid flying near to specified residential areas or to determine the sounds that cause complaints about aircraft noise in distant communities. In some cases monitors are located under the flight path at specific distances from the start of roll for departures or landing threshold for arrivals to compare measured values with noise limits in an airport's noise regulations.

The data gathered from the sound monitors may be used for many purposes, including monitoring the cumulative noise and its change with time, the maximum noise at specific locations with respect to noise regulations, and the noise of specific flights by aircraft type, airline, and so on, to compare the relative effectiveness of flight procedures for noise abatement. At many airports selected summary data are published on an annual basis to keep the public informed of the quantitative progress of the airport's noise control program.* The availability of factual data is often of significant importance in providing a basis for substantive discussion of noise issues.

2.2 Monitoring Fight Paths

At the majority of airports in the United States used for commercial turbojet aircraft, the FAA uses an Aircraft Radar Tracking System (ARTS) for real-time monitoring of aircraft position so that air traffic controllers can see the relative positions of all aircraft in their assigned control sectors. The system (currently ARTS-3 at most airports) determines the X, Y-position of the aircraft by radar echoes at intervals of about 4 s. It determines the height and identity of each aircraft by the radio return data from a transponder on the aircraft that is activated by the radar signal. Each aircraft is assigned a transponder code for a specific flight plan, so that the controller radar display can give the height as well as the identity of the aircraft (airline flight number and type) and other data from the flight plan associated with the transponder code.

These radar data are retained for a limited time by the FAA on digital disk or tape and are accessed by some airports[8] to obtain flight path data and aircraft identity after a 15-day waiting period. In many other countries the airport radar is available in real time to a system that correlates flight track data with sound event data measured at remote monitors. A few airports obtain aircraft path data from other passive systems that use the radar and its triggered transponder to independently calculate path data. These passive systems operate in real time but cannot directly obtain an aircraft's identity, only its transponder code.

Radar-derived flight path data may be used to determine the sources of measured sound events, providing a greater certainty of the aircraft flight-related events. Further, they may be used to evaluate the effectiveness of flight track rules and control procedures and for the definition of representative noise paths for computer model analysis.

2.3 Calculating Airport Noise and Population Impact

The calculation of airport noise[10] and potential population impact must account for the major factors that control airport noise impact given in Table 1. Contours of cumulative noise in DNL, or other metrics, for an average day during a specified time period may be computed by the use of the FAA integrated noise model (INM)[10] or the Air Force noise map model.[11] See example in Fig. 1.[12] These models can also be utilized to generate a sound exposure level (SEL) contour for a single flight of a specific type of aircraft on a specific or generalized flight track.

These computer programs contain an aircraft noise data base specifying for each aircraft type the relationship between sound exposure level, slant distance from receiver to the aircraft, and engine thrust setting. In addition, the INM contains a set of flight profiles for each aircraft type that give the height, engine thrust, and speed for takeoffs at various aircraft weights as a function of distance from the start of takeoff roll and a standard pro-

*The California regulation requires quarterly submittals to the state from airports that are required to monitor noise. These records are one of the largest public sources of data regarding the effect of weather and operational changes on the propagation and generation of aircraft noise.

TABLE 1 Principal Factors That Control Airport Noise Impact

Aircraft Noise Characteristic
Noise vs. distance by aircraft type
Number of operations by aircraft type
Ratio of the number of daytime to nighttime operations by aircraft type
Airport Noise Impact Potential
Flight procedures (throttle and flap management) used for landing and takeoff by aircraft type
Stage lengths (takeoff and landing weights by aircraft type)
Spatial configuration and relative utilization of the flight tracks by aircraft type
Spatial distribution of population

file for landing. The input data for each specific case includes the runway and flight track layout; the number of takeoff and landing operations in day and night, distance to departure destination, and runway/track utilization, all by aircraft type; and supplemental data for special situations.

The noise contours can be superimposed over maps containing land use or population data, enabling determination of the land areas with specified uses or the residential population within an interval between any two noise contours. An alternative method for estimating the population residing at various noise levels is to calcu-

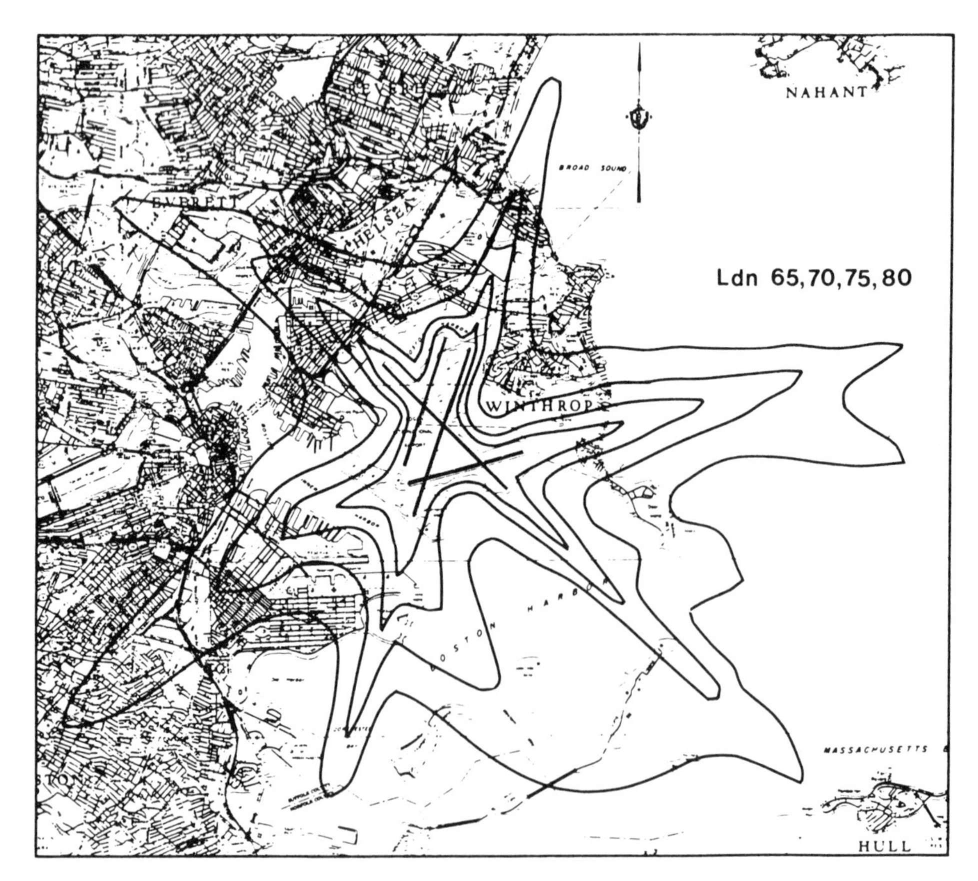

Fig. 1 Example of DNL noise contours at Boston Logan International Airport for 1980 operations.[12]

late the noise level at the centroids of population census blocks or block groups. The results can be summarized in appropriate intervals for display and may be used to ascertain the estimated change in noise at each population centroid for various study alternatives.

The calculations may be used to supplement the measured noise exposure data and to provide insight on the causes of the noise at various locations. The measured data may be used to validate the calculations that usually are within ±2 dB for cumulative noise level if the measurement time period is sufficiently long for the types of variation encountered at the airport; if the operation is accurately defined as to aircraft type, stage length, time of day, and runway use; and if the flight tracks are modeled in sufficient detail. Once a model of the noise at an airport is verified and accepted by the local authority, it becomes the basis for studying the potential effects of proposed future noise abatement actions.

3 NOISE ABATEMENT AT THE NATIONAL LEVEL

The U.S. national effort to reduce the noise around airports is led by the FAA. It has a comprehensive set of programs including regulations pertaining to noise limits for aircraft types, time schedules for phasing older, noisy aircraft out of the fleet, detailed environmental assessment procedures, and the funding and management of airport specific projects for the control of noise. These programs are addressed in the following section.

3.1 Aircraft-Type Certification for Noise and Phaseout of Noisy Aircraft

In 1958 The Port Authority of New York promulgated a noise rule[1] that contained a perceived noise level limit of 112 dB, applicable to the noise received at specific locations from a departing aircraft. The four current controlling locations at John F. Kennedy International Airport are at an average distance of 5900 m (4770–6690-m range) from the beginning end of the runway along the extension of its centerline. This noise level limit was difficult to meet on hot summer days by the early intercontinental models of the Boeing 707 and Douglas DC-8 aircraft powered by turbojet engines, often requiring a reduction of payload. This rule provided the de facto design requirement for departure noise until 1969, when the FAA promulgated its aircraft-type certification requirements for noise in FAR Part 36.[13]

In FAR Part 36 the FAA specified three locations for the measurement of noise: takeoff, sideline, and approach. The original locations in FAR Part 36 have evolved with international consensus as follows. The takeoff noise measurement position is at 6500 meters from the start of takeoff roll along the extended centerline, the sideline noise measurement position is at the point of maximum takeoff noise along a line parallel to the runway centerline and displaced by 450 m, and the approach noise measurement position is at 2000 m from the runway threshold along the extended centerline.

The Part 36 noise limits are defined in terms of the EPNL descriptor in units of decibels. The noise is measured for each aircraft type under tightly controlled test conditions that are defined in the regulation or is calculated from previous measurements of an earlier model of the same basic aircraft. The original FAA limits[13] for subsonic fixed-wing aircraft are now referred to as stage 2 noise limits, and aircraft certified prior to FAR Part 36 are referred to as stage 1 aircraft (no noise limits). The currently applicable FAA limits, referred to as stage 3 noise limits, were promulgated in 1975.[14] Figures 2–4 illustrate the stage 2 and 3 noise limits for the approach, sideline, and takeoff measurement locations using examples of some of the noisiest models certified. Note that stage 1 aircraft over 75,000 lb maximum takeoff weight have not been allowed to operate at U.S. civil airports since 1985.[15]

The FAR Part 36 regulations have made major improvements in the noise radiated from civil aircraft. Figure 2 shows the stage 2 and 3 noise limits for arrival noise at a location under a 3° sloped flight path and located 2000 m from the landing threshold. It also shows 79 certification data points[15] for selected heavier models of 31 basic aircraft types, representative of each of the three noise stages. It shows that stage 2 put a ceiling on arrival EPNL with reductions up to 13 dB. The stage 3 limit was 4 dB lower than stage 2 and the mean of the stage 3 data is about 4 dB lower than the limit.

Figure 3 contains a similar presentation of the maximum EPNL during takeoff obtained along the sideline located about 450 m from the runway centerline. This maximum level is usually found after the aircraft has climbed to several hundred feet above the ground and is operating with *full takeoff power* and flaps. The stage 3 "mean modern > 75K" line for the aircraft with high-bypass turbofan engines is about 5 dB below the stage 3 limit and 13 dB below the stage 2 limit. One aircraft, the 355,000-lb stage 3 DC8-73, is 16 dB below the stage 1 DC8-63 after reengining with high-bypass-ratio turbofan engines.

However, there are six stage 3 aircraft weighing between 150,000 and 200,000 lb that are just at the noise limit and two aircraft weighing between 700,000 and 750,000 lb that are about 1 dB above the noise limit.*

*FAR Part 36 allows for tracking levels up to 2 dB at one location and 1 dB at a second location for 3 dB under limit at the third location.

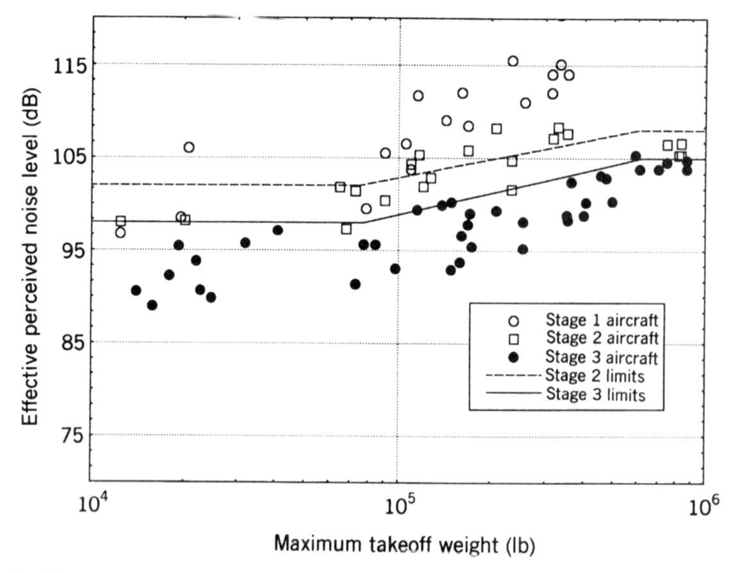

Fig. 2 FAR Part 36 stage 2 and 3 approach noise limits with data for some of the noisier models of various aircraft types.[15]

Four of the six aircraft are Boeing 727-100 and 727-200 with Valsan or Fedex hushkits. The other two aircraft in this group are McDonald Douglas MD-80's. The two heavier aircraft are the Boeing 747-100 and 747-SP recertified to stage 3. Obviously, these hushkitted or recertified aircraft do not represent the same degree of noise reduction technology as do the "mean modern > 75K," which are 5 dB below the limit. Conversely, the

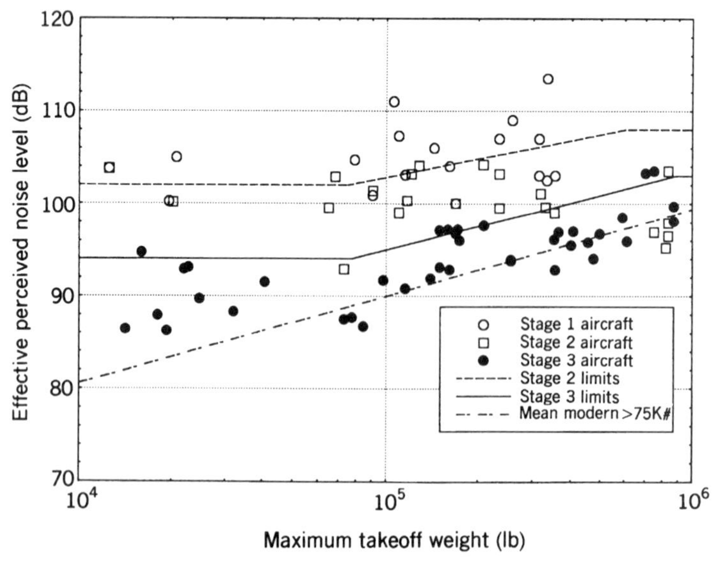

Fig. 3 FAR Part 36 stage 2 and 3 sideline noise limits with data for some of the noisier models of various aircraft types.[15]

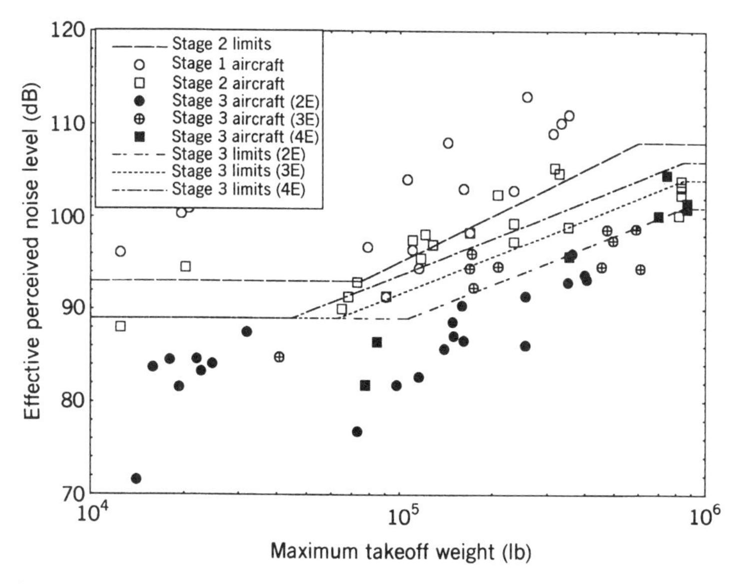

Fig. 4 FAR Part 36 stage 2 and 3 takeoff noise limits with data for some of the noisier models of various aircraft types.[15]

modern engine data demonstrate the technology feasibility of promulgating a rule that lowers the EPNL limits for future aircraft.[16]

Figure 4 is a similar presentation of the maximum EPNL at the takeoff location. The figure is more complex because different stage 3 limits apply to aircraft with two, three, or four engines. These differing limits are the result of the higher climb rates of aircraft with fewer engines, resulting from their higher thrust-to-weight ratio, which in turn is caused by the FAA safety requirements for performance with one engine out. In general, at this location the stage 3 aircraft with modern high-bypass-ratio turbofan engines are still operating at full takeoff thrust and are significantly better than the requirements, whereas the 727 aircraft with low-bypass-ratio JT8D engines and hushkits operate with engines at cutback thrust to just meet the requirements.

Because the stage 3 aircraft with modern turbofan engines have such significant reductions of noise over stage 2 aircraft, many airports have promulgated regulations that restrict operations by stage 2 aircraft during certain hours. The concerns over possible impacts such regulations might have on interstate commerce and the public concern to phase out the noisy stage 2 aircraft led to the passage of the Airport Noise and Capacity Act of 1990.[17] As a result, the FAA has recently amended FAR Part 91[18] to regulate an orderly phaseout of stage 2 aircraft with a maximum certified weight greater than 75,000 lb from the civil fleet by the year 2000. The FAA estimates that this phaseout of stage 2 aircraft will reduce the population living in neighborhoods where the DNL exceeds 65 dB by 85% from the 1990 estimated values.

3.2 Environmental Assessment and Mitigation Policies and Procedures

In response to the requirements of the National Environmental Policy Act of 1969,[19] the FAA established in its Order 1050.1D,[20] a set of policies and procedures for the preparation of environmental impact statements (EISs) and findings of no significant impact (FONSIs) and for preparing and processing environmental assessments (EAs) of FAA actions. For noise, these policies and procedures respond to other legislative requirements, including the Noise Control Act of 1972[21] and the Aviation Safety and Noise Abatement Act of 1979.[22] Additional guidance for preparation and processing the required environmental analyses and statements is contained in the FAA's Airport Environmental Handbook, Order 5050.4A.[23]

In Order 1050.1D the FAA specifies the use of DNL for analysis of the cumulative noise at an airport and the maximum sound level or the sound exposure level for analysis of single-event noise. Further, it prescribes the use of its INM (Noise Map or other FAA approved model) for construction of DNL contours, contains a land use noise compatibility table principally derived from the Federal Inter-Agency Guidelines,[24] and provides additional guidance to the preparer. For evaluation of noise impacts, Order 1050.1D defines the FAA's threshold of "significance" for initial analysis of impacts at specific noise-sensitive areas to be a DNL increase of 1.5 dB when the DNL is greater than 65 dB and a DNL increase of 3 dB when the DNL is 60–65 dB.[25] Proposed actions that result in increases that equal or exceed this threshold require additional analysis as part of the EIS process. However, this threshold does not apply to categorically excluded actions such as changes to air traffic control procedures at altitudes exceeding 3000 ft above ground level.[19]

As a result of these orders, EAs are routinely made on most U.S. airport projects that involve possible noise effects and that are partially or wholly funded by the FAA. Many of these assessments, particularly those related to airport expansion projects, lead to the preparation of EISs. Additionally, many states have similar environmental analysis and reporting requirements.

The FAA has initiated a comprehensive program of airport planning for noise compatibility. This program is codified under FAR Part 150[26] and additional information is provided in FAA Advisory Circular AC 150/5020-1.[27] The performance of this program through 1990 was assessed by the FAA in a report to Congress.[28] The program's objective is to reduce existing noncompatible land uses and to prevent introduction of additional noncompatible uses around an airport. Compatibility is assumed to be achieved when the noise level at a location does not exceed an acceptable value for the land use in that area.

To achieve the program objectives, Part 150 provides the following[26]:

- Establishes standard noise methodologies and units
- Establishes the INM as the standard computer model
- Identifies land uses that normally are compatible or incompatible with various levels of airport noise
- Provides for the voluntary development of noise exposure/Maps (NEMs) and noise compatibility programs (NCPs) by airport operations in consultation with all affected parties
- Provides for review of NEMs to ensure compliance with Part 150 regulations
- Provides for review and approval or disapproval of Part 150 NCPs submitted to the FAA by airport operators

In its approval of NCPs, the FAA has exhibited considerable concern and resistance to proposed airport regulations that restrict airport access. The restrictions of concern include aircraft noise limits at various times of day, limits on operation of aircraft certified under FAR Part 36 stage 2 noise limits, and nighttime curfews. In response to the Airport Noise and Capacity Act of 1990,[17] the FAA issued a new Part 161 entitled Notice and Approval of Airport Noise and Access Restrictions.[29] This regulation contains detailed requirements on the airport operator for providing notice of intent to regulate and stringent requirements for noise and economic impact analyses to be submitted to the FAA for approval.

4 AIRPORT-SPECIFIC NOISE CONTROL MEASURES

Almost all major civil airports in the United States have active noise abatement programs. Most of these utilize the Part 150 process that involves all directly interested parties in the development of consensus solutions. Table 2 summarizes the principal actions that can be taken to abate noise and its perceived problems.

4.1 Optimize Flight Tracks

In developing noise reduction alternatives for a specific airport, one of the first factors to examine is the location of its flight tracks with respect to land use. It is desirable that these tracks be located to make maximum use of noise compatible land such as water, industrial, commercial, and agricultural. Where opportunities exist for changing flight tracks to reduce their noise impact potential, they should be carefully studied with respect to noise benefits, local air space management, fuel, time, cost, and other factors. The results can be dramatic, particularly when large bodies of water are next to the airport.

The Boston Logan runway 22R departure EIS[30] contains one of the more detailed studies of flight track alternatives. Six alternative tracks were studied, comparing potential population impacts for days in which all departures were on runway 22 (220° magnetic). After attaining sufficient altitude, the aircraft turned to one of six headings. These headings varied over 120°, ranging from 220° flying over densely populated urban areas to 100° flying out to sea through the harbor entrance. The populations estimated to be living in areas where the DNL on a maximum day exceeded 65 dB ranged from 91,000 for the base case to 3000 for the harbor entrance alternative that was finally selected.

Although noise analyses are often made with nominal tracks, the tracks of actual flights vary from the nominal track. The amount of this variation for straight tracks may be of the order of ±5°, but the variation is usually much greater for tracks with significant turns. Several airports monitor the actual tracks from the FAA radar data, including the Port Authority of New York and New

TABLE 2 Possible Airport Actions to Abate Noise

Flight tracks	Get aircraft away from people.
Preferential runways	Increase use of runways with least impact.
Restrict noisy aircraft	Minimize operations in day or night.
Noise abatement flight procedures	Require use of noise abatement throttle and flap management procedures for takeoff and/or approach.
Airport layout	Extend or build new runways and taxiways to make best use of compatible land and water.
Shielding barriers	Shield people from noise of ground operations.
Soundproof	Soundproof schools, homes, and churches.
Land use control	Assure compatible land use through acquisition of property or other rights.
Monitor and model	Monitor airport noise and flight tracks to provide data to the public and for evaluating proposed alternatives.
Communication	Listen to complaints and suggestions; develop and institutionalize continuing effective dialogue and information transfer among all concerned parties.

Jersey, which has a system to provide continuous routine track monitoring at its three airports.[8] In most of these situations, defined flight tracks are a principal noise abatement measure.

A variant of the natural flight track dispersion is the use of fanning, a deliberate controlled program of dispersing the flight tracks. There are a variety of techniques employed. For example, in the mid-1970s at Detroit sequential departures on one runway were asked to execute a 10° right turn, 10° left turn, and "no" turn; and presently for runway 31 at New York's LaGuardia Airport each of three tracks at 10° intervals is used for approximately 1 h during non–peak traffic periods prior to switching to the next track. The practice of fanning appears to be accepted as desirable when the noise associated with each track is below the threshold value of significant annoyance for the exposed population. However, if the resulting noise is greater than this threshold, fanning tends to lead to greater numbers of people impacted.

4.2 Define Preferential Runways

Once the flight tracks have been optimized at a specific airport, opportunities for reduction in the noise-impacted population may exist by optimizing the relative utilization of these tracks. The potentials for utilization of the tracks associated with each runway depend upon the wind and weather conditions, the capacity of each runway (or runway combinations) to meet demand, and runway length. In general, preferential runway assignments can only be made in weather conditions that meet the visibility, wind velocity and direction, and runway conditions established by FAA Order 8400.9.[31] The proportion of time that preferential utilization is available is critically dependent on the limit values for tail and cross winds. Small propeller aircraft are typically certificated to lower cross-wind velocities than are jet aircraft and may restrict preferential utilization below that which would result from the FAA order.[31]

The results of implementing preferential runways can also be dramatic in some situations. For example, the nighttime preferential approach to Los Angeles International Airport is from the west over the ocean, rather than from the east over densely populated urban areas. With this procedure, both departures and approaches overfly the ocean. It essentially eliminates nighttime aircraft approach noise in those urban areas for 90–95% of the year when wind direction and speed permit the procedure. Similarly, the nighttime use of runway 15 for departures and 33 for arrivals at Logan International Airport keeps the aircraft over the harbor, rather than over populations in urban areas, when the traffic is only a few aircraft per hour and the wind direction and speed permit the procedure.

The techniques for the implementation of preferential runway systems vary from describing a simple list of runways in order of preference, as above, to sophisticated systems that attempt to achieve multiple goals. If the variable level of demand throughout the day requires more complex runway configurations to provide adequate capacity, then the runway configurations, rather than single runways, must be listed in order of pure preference. If consensus on the order of runways for pure preference cannot be obtained, then it may be possible to obtain agreement among the concerned parties on annual and short-term goals for runway utilization. Implementation of a runway utilization goal system can be based on the expected weather distribution over a long-term data base period, such as 10 years. Preferential use rules can then be formulated that enable long-term average achievement of these runway utilization goals accounting for the expected variation of demand and the relative capacities of the runway configuration. Such a system may also have algorithms to compensate for variations of the weather in any one year from the long-term average weather to enable a closer approximation of the annual goal in each year. The initial Preferential Runway Advisory System for Logan[32] operated on a microcomputer (with manual backup) to attain consensus annual goals and short-term goals to minimize persistent use of a runway combination in a 3-day period and dwell during a 24-h day.

4.3 Restrict Noisy Aircraft

Any restriction on the use of an airport by noisy aircraft must be nondiscriminatory and it must not constitute an interference with interstate commerce. One approach is to pick a maximum A-weighted sound level applicable to overflights at some reasonable location. For example, Santa Monica, California, has a 95-dB maximum limit and has investigated establishing a set of lower limits by aircraft class. A second approach is to require use of the quietest technology. For example, Boston Logan and San Francisco International airports and many others require that night operations be conducted by aircraft that meet FAR Part 36 stage 3 noise limits. A third approach is to combine a single sound level limit with a stage 3 requirement, so that the maximum noise (i.e., aircraft size) is controlled and the noise control technology requirement is consistent for all smaller aircraft. A fourth approach, used by several airports, is to give noise budgets to each airport user that allow the user to trade off between its number of operations and the noise associated with its various aircraft types. These budgets can be designed to be increasingly restrictive in future years.

A basic difference between the two pure approaches—maximum sound level limit and requirement to be certified to a specific FAR Part 36 stage—is that the former only addresses the noisiest aircraft, presumably the largest, most powerful, and/or oldest types. All other aircraft are permitted to make the same amount of noise. On the other hand, a regulation requiring certification to a specific stage should mean that all aircraft that meet the stage noise limits must have incorporated the same level of noise control technology regardless of size and have met the same criterion for safety, technological practicability, economic reasonableness, and appropriateness to type. Thus, the latter approach tends to minimize the total noise exposure, whereas the single limit generally cannot minimize total exposure. An FAA advisory circular[33] gives estimated sound levels at the FAR Part 36 approach and takeoff certification measurement locations that can be of assistance in choosing a specific level appropriate for the airport and time of day. Noise data for both certified and uncertified aircraft are provided by the FAA.[15,34] The data in any of these FAA advisory circulars can be used to determine which aircraft can meet the noise limits and/or stage requirements.

The nighttime noise penalty of 10 dB in the DNL descriptor means that the noise from 1 nighttime operation is counted the same as is the noise from 10 daytime operations of similar character. Thus, for airports that have nighttime operations (10 p.m. to 7 a.m.), their reduction often appears to have considerable potential to reduce noise impact. Often, considerable progress may be achieved through working with the airport's user-operators to eliminate nighttime flights that have only marginal benefit, rescheduling other operations and providing quieter aircraft for nighttime. This latter result also can be achieved for all nighttime operations by choice of a suitable nighttime noise limit. Generally, a noise limit is preferable to a total curfew because a curfew may impede airport service sufficiently to interfere with interstate commerce and may prove very difficult to remove at a later date when aircraft are sufficiently quiet and more service is desired.

4.4 Use Noise Abatement Flight Procedures

Potential noise reductions can be obtained from the use of improved throttle and aircraft configuration techniques for both approach[35,36] and takeoff.[37] For example, lower flap settings reduce aircraft drag, enabling lower engine power settings and, hence, lower noise levels. If these procedures could be properly tailored to specific airport requirements, the gains could even be greater. However, there is considerable controversy over the notion of changing cockpit management procedures from airport to airport, particularly with respect to pilot workload and safety implications, and there are questions of responsibility for adopting specific procedures for specific aircraft types and runways at individual airports.

4.5 Improve Airport Layout

Changes to airport layouts are usually initiated to improve capacity and/or safety. They are rarely made to improve the noise environment because any proposed change to the airport is usually viewed with suspicion by the noise-impacted public.

However, changes in airport layout by extending existing runways, reorienting runways to new directions, or building new parallel runways can in some situations provide significant noise benefits. For example, extensions may be used to have aircraft at greater heights over populated areas. Reoriented runways may be designed to make better use of existing compatible land (or water) for new flight tracks. New parallel runways may be used to move sideline noise further from existing communities and to make better use of existing compatible land. These and other concepts should be reviewed carefully in designing solutions to airport noise problems for their potential applicability.

4.6 Soundproofing and Noise Barriers

The use of physical noise reduction techniques to increase the attenuation in the path between source and receiver is a standard engineering approach. Its implementation in airport noise problems is usually in the sound insulation of schools and residential dwelling units or the construction of sound barrier walls.

In the early 1990s, major sound insulation programs were under way at many airports, including Atlanta, Boston, Denver, Los Angeles, New York, Providence, San Francisco, and Seattle. Most of the programs involve both schools and homes. Usually the goal in schools is to keep the aircraft noise hourly equivalent sound level under 45 dB inside classrooms and the maximum sound level under 60 dB. The goal in homes is usually to attain a stated A-weighted noise reduction for aircraft noise. Typically, the stated amount of noise reduction is between 25 and 35 dB, varying with airport program and sometimes at various locations within a program. For the most part, the sound insulation is accomplished by replacing the windows with double glazing designed to provide good sound attenuation, improving ventilation as required, improving some doors, and sealing cracks and gaps with suitable caulking or gaskets. In many cases, sound insulation of homes is accompanied by an avigation easement that prevents the homeowner from suing the airport unless the noise increases significantly in the future.

Sound barriers around airports are usually earth berms with shrubbery and trees, free-standing walls, or buildings. For example, Minneapolis Airport has an extensive landscaped earth berm to reduce the noise from taxiing and idling aircraft at the near neighbor homes. It also improves the view toward the airport, translating an airport industrial complex scene into tranquil shrubs and trees. Walls are employed, particularly where distance between sound and receiver is small and land is intensively used. New York LaGuardia, Los Angeles International, and Boston Logan airports have sound barrier walls. Airport walls are usually much higher than highway barriers because of the height of the engines on the aircraft. At Logan, the walls are over 30 feet in height and a portion of the barrier consists of a multistory office building, sited for noise control purposes. Typical barriers are designed to attain a 10-dB or more reduction of the noise from ground operations at locations near to the barrier.

4.7 Change Land Use

Land use may be controlled through many means, including zoning restrictions, conversion to noise-compatible activities, purchase with compatible redevelopment, and purchase of avigation easements.[27]

New airports are typically built in areas where land is relatively cheap and for which the pre-airport land uses were low density and low value such as agricultural, surface mining, forestry, or otherwise not used. Once the airport is constructed, it creates a demand for commercial, industrial, and residential use. Most of the initial industrial use is often on the airport property and the commercial use is usually on and immediately adjacent to the airport property. Consequently, the highest use for the majority of the remaining surrounding land is residential development. Any attempt to restrain landowners from residential development is usually seen as depriving them of their rights. Therefore, zoning has proved to be ineffective in preventing residential development near airports.

Overlay zoning that imposes sound insulation requirements for new construction may be effective. However, purchase and redevelopment have proven most effective. In many cases, the land is incorporated into airport expansion or used for the development of industrial parks for commercial use. In others it may be used for recreational purposes such as public golf courses.

4.8 Monitor and Model

The implementation of a noise-monitoring program utilizing permanent and/or portable sound monitors with identification of aircraft and documentation of cumulative noise from aircraft and nonaircraft sources is essential to provide data for public discussion and validation of noise modeling. Where possible, this noise measurement program should be coordinated with a flight track monitoring program that provides aircraft identification and tracks to determine the causes of the various measured sounds. When a combined program is in place, designed with public participation and providing significant regular data output, it improves the possibility of improving public confidence in both the process and the results. These data then can be used as a basis for discussing the existing noise situation at an airport and examining alternatives that might provide future mitigation of noise.

4.9 Communication

All of the steps and progress toward the reduction of the physical measurable quantities of noise will not reduce the public's perception of the noise at an airport unless

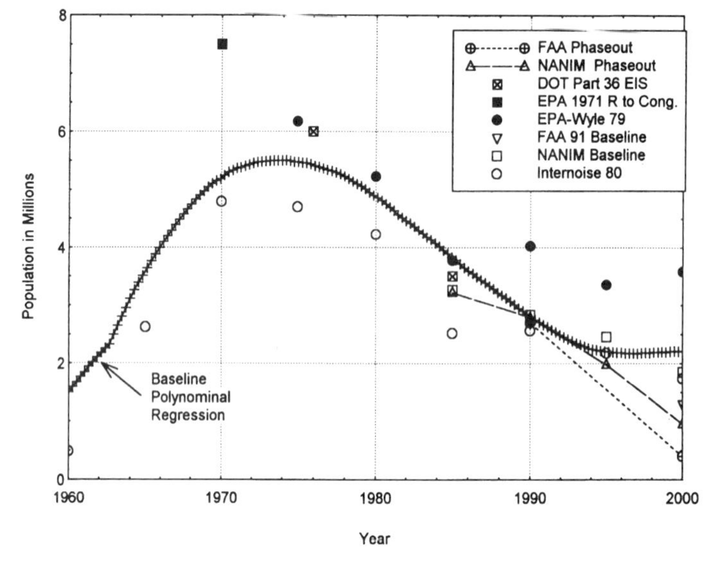

Fig. 5 Various estimates of the U.S. national population residing in neighborhoods where the L_{dn} exceeds 65 dB, illustrating the potential effect of the FAR Part 36 stage 2 phaseout in 2000 based on FAA[18] and NANIM[44] calculations and the estimates for the period 1960–2000 without the stage 2 phaseout from various sources (Refs. 42, 38, 43, 18, 44, and 40, respectively).

communication channels are established between the airport authorities, the local FAA Air Traffic Control personnel, the airport users, and the airport's neighbors. One of the most important outcomes of a successful Part 150 study is the development of a capability for the concerned parties to talk and to begin to understand each other. Open dialogue, consensus decision making, and teamwork in seeking creative solutions can build the beginnings of trust and respect among the parties and provide a basis for institutionalizing the process so that the parties continue to work together and develop a practical and rational decision-making capability.

5 CHANGES IN POPULATION IMPACT 1960–2000

The number of people residing in neighborhoods where the DNL exceeded 65 dB has been used as a measure of the total population impact of civil aircraft noise. This number rose from a small value in 1958 to a maximum in the early 1970s, then decreased as a result of the introduction of quieter aircraft and the implementation of airport-specific noise control measures. The maximum population impact is estimated to have been between 5 and 7.5 million people, occurring in the early 1970s.[38–42]

Figure 5 illustrates various estimates of population impact above a DNL of 65 dB.[38,40–43] The two phaseout curves are estimates of the effect of phasing out stage 2 aircraft in the year 2000, one made by the FAA in 1991[18] and the other by the National Airport Noise Impact Model[44] in 1987.[18,44] The baseline polynomial regression curve represents the points that do not assume a stage 2 phaseout in 2000. It shows a maximum of about 5.5 million people in 1973, a reduction to 2.2 million in 1995, with a slight increase thereafter. Both the FAA phaseout and National Airport Noise Impact Model (NANIM) phaseout curves coincide with the baseline polynomial regression in 1990 but decrease to 400,000–900,000 people in the year 2000. The FAA estimate[18] of 400,000 is probably the most accurate since it is based on the most recent data, particularly the results of the Part 150 studies at the nation's airports.

These estimates indicate that the extent of most intense noise impacts is in the process of being controlled. However, they do not address the number of people who have significant remaining noise impact at lower values of DNL, such as 60 dB, nor do they address the inevitable increase after the year 2000 after the noisiest aircraft are retired. Thus, unless the noise levels of new aircraft are reduced below the stage 3 noise limits, the growth of the fleet and its number of daily operations will ensure that the population suffering noise impact will increase.

REFERENCES

1. PNYA, "Air Terminals—Rules and Regulations II," Port of New York Authority adopted 1951 with Amendments, 1958.
2. "Noise Final Report," Wilson Report Cmnd.2056, London, 1966.
3. ANSI S12.9-1988, "American National Standard Quantities and Procedures for Description and Measurement of Environmental Sound, Part 1," Acoustical Society of America, New York, 1988.
4. ANSI S1.1-1994, "American National Standard Acoustical Terminology," Acoustical Society of America, New York, 1994.
5. "Environmental Protection—Annex 16 to the Convention on International Civil Aviation, Vol. 1, Aircraft Noise," International Civil Aviation Organization, as amended, November 1985.
6. California, "Subchapter 6. Noise Standards," Register 79, No. 21-5-26-79, Public Utilities Code (Regulation of Airports), California Department of Transportation, Division of Aeronautics, with Amendments to March 1990.
7. S. Goldstein and A. H. Odell, "Comments on the Problem of Jet Aircraft Noise," Port of New York Authority, 1965.
8. J. P. Muldoon, "The Development of a New Aircraft Noise Abatement Monitoring System Using Air Traffic Control Radar Data," *Internoise 84 Proceedings*, Vol. 1, December 1984, pp. 689–692.
9. SAE AIR1845, "Procedure for Calculation of Airplane Noise in the Vicinity of Airports," Society of Automotive Engineers, A-21, March 1986.
10. FAA, "INM Integrated Noise Model Version 3 User's Guide," Federal Aviation Administration, FAA EE-81-17, October 1982.
11. C. L. Moulton, "Air Force Procedure for Predicting Aircraft Noise around Air Bases: Noise Exposure Model (Noise Map) User's Manual," U.S. Air Force Systems Comand, AAMRL-TR-90-011, February 1990.
12. Massport, "The Results of Four Years of Noise Abatement at Boston's Logan International Airport," Massport Noise Abatement Office, 1981.
13. Federal Aviation Regulations Part 36, "Noise Standards: Aircraft Type and Airworthiness Certification," Federal Aviation Administration, November 1969.
14. Federal Aviation Regulations Part 36, "Noise Standards: Aircraft Type and Airworthiness Certification," Federal Aviation Administration, Amendment 7, February 1977.
15. FAA, Advisory Circular AC36-1F, "Noise Levels for U.S. Certificated and Foreign Aircraft," Federal Aviation Administration, June 1992.
16. K. M. Eldred, "Airport Noise: Solving a World Class Problem," *Proceedings, Internoise 92*, Vol. 1, July 1992, pp. 3–12.
17. U.S. Congress, "Airport Noise and Capacity Act of 1990," PL101-508, November 1990.
18. FAA, "Transition to an All Stage 3 Fleet Operating in the 48 Contiguous United States and the District of Columbia," Federal Aviation Administration, 14 CFR Part 91 Amendment, September 1991.
19. U.S. Congress, "National Environmental Protection Act of 1969," PL91-190, 42 U.S.C. 24321, 1969.
20. FAA Order 1050.1D, "Policies and Procedures for Considering Environmental Impacts," Federal Aviation Administration, December 1983.
21. U.S. Congress, "Noise Control Act of 1972," PL92-574, 42 U.S.C. 4901, with amendments, 1977.
22. U.S. Congress, "Aviation Safety and Noise Abatement Act of 1979," PL101-508, 1990.
23. FAA Order 5050.4A, "Airport Environmental Handbook," Federal Aviation Administration, October 1985.
24. FICUN, "Guidelines for Considering Noise in Land Use Planning and Control," Federal Interagency Committee on Urban Noise," June 1980.
25. FICON, "Federal Agency Review of Selected Airport Noise Analysis Issues," Federal Interagency Committee on Noise, Environmental Protection Agency, Washington, D.C., 1992.
26. FAA, "Airport Noise Compatibility Planning," 14 CFR Part 150, as amended, March 1988.
27. FAA Advisory Circular 150/5020-1, "Noise Control and Compatibility Planning for Airports," Federal Aviation Administration, August 1983.
28. FAA, "Report to Congress Part 150 Airport Noise Compatibility Planning," Federal Aviation Administration, November 1989.
29. FAA, "Notice and Approval of Airport Noise and Access Restrictions," FAR Part 161, Federal Aviation Administration, September 1991.
30. FAA, ANE-500-79-3, "Final EIS Departure Procedures, Runway 22 Right, Logan International Airport," March 1980.
31. FAA Order 8400.9, "National Safety and Operational Criteria for Runway Use Programs," November 1981.
32. K. M. Eldred, "Massport Phase II Preferential Runway Study for Logan International Airport," Ken Eldred Engineering Report 82-80, Concord, MA, November 1980.
33. FAA Advisory Circular 36-3F, "Estimated Airplane Noise Levels in A-Weighted Decibels," Federal Aviation Administration, August 1990.
34. FAA Advisory Circular 36-2C, "Measured or Estimated (Uncertificated) Airplane Noise Levels," Federal Aviation Administration, February 1986.
35. ICAO Circular 157-AN/101, "Assessment of Technological Progress Made in Reduction of Noise from Subsonic and Supersonic Jet Aeroplanes," Montreal, 1981.
36. "The Noise Benefits Associated with the Use of Continuous Descent Approach and Low Power/Low Drag Approach Procedures at Heathrow Airport, CAA Paper 78006.
37. FAA Advisory Circular AC91-53.A, "Noise Abate-

ment Departure Profile," Federal Aviation Administration, 1994.

38. EPA, "Report to the President and Congress on Noise," Senate Document No. 92-63, February 1972.
39. DOT-NASA, "Civil Aviation Research and Development Policy Study," Joint DOT-NASA Report DOT TST 10-5, NASA SP266, March 1971.
40. K. M. Eldred, "Model for Airport Noise Exposure on a National Basis—1960 to 2000," *Proceedings Internoise 80*, December 1980, pp. 803–808.
41. J. E. Wesler, "Airport Noise Abatement—How Effective Can It Be?" *J. Sound Vib.*, February 1975, pp. 16–21,
42. DOT, FAA, "FAR Part 36 Compliance Regulation, Final EIS," Federal Aviation Administration, November 1976.
43. C. Bartel and L. C. Sutherland, "Noise Exposure of Civil Air Carrier Airplanes Through the Year 2000," EPA 550/9-79-313-1, February 1979.
44. K. M. Eldred, "The National Airport Noise Impact Model and Its Application to Regulatory Alternatives," FAA EE88-3, July 1987.

71

SURFACE TRANSPORTATION NOISE

ROBERT HICKLING

1 INTRODUCTION

Surface transportation noise is caused by highway vehicles such as trucks, buses, passenger cars, and motorcycles, together with railroad vehicles, both above and below ground. Also included is the noise of off-highway vehicles, used for construction, agri- and horticulture, and recreation. The noise of boats, ships, and hover craft can also be regarded as surface transportation noise. Although the noise of wheeled vehicles in cities has been of concern for centuries,[1] it is only in recent decades that quantitative noise measurements have been developed. Major government reports in the 1960s and 1970s[2,3] indicated the need for noise standards. The principal burden of meeting the vehicle noise standards that eventually developed fell on vehicle manufacturers. Highway vehicle manufacturers with worldwide markets are required to meet a diversity of exterior or community noise standards. Interior noise standards, developed as a result of studies of hearing loss due to noise exposure in the workplace (see Chapters 69 and 92), are by their nature less diverse. Since noise in the workplace is covered in Chapter 69, the emphasis here is on exterior noise.

2 NOISE SOURCES IN SURFACE TRANSPORTATION

The same noise sources occur in different forms of surface transportation. These are shown, for example, in Fig. 1 for a heavy-duty truck and a motorcycle. Primarily there is the noise associated with the power source, which usually is an air-breathing diesel or gasoline engine. This consists of noise from the air intake, exhaust, cooling system, engine structural vibrations, pumps for water, fuel and lubricating oil, and electrical and other ancillary systems. Diesel (compression ignition) engines are noisier than gasoline (spark ignition) engines because compression ignition involves a greater impulsive force. Diesel engines are built more strongly than gasoline engines to withstand this force. Diesels, especially direct-injection diesels, are more efficient and are used for heavy-duty trucks and diesel–electric locomotives, where they power the electric motors that drive the wheels on the rails. Gasoline engines have a wider speed range than diesels, requiring fewer gears to operate them. Diesels in passenger cars generally have indirect injection into a precombustion chamber, which reduces noise and provides a speed range comparable to gasoline engines. Indirect-injection diesels are more efficient than gasoline engines but less efficient than direct-injection diesels. Gasoline engines are preferred in passenger cars because they are less noisy and lack the characteristic diesel odor.

Noise from the drive train comes from the transmission, drive axle, differential gears, and wheel bearings. There is also noise associated with surface propulsion, that is, the interaction of tires with the road, and of wheels with rails, track systems with the ground, and marine propellers and hulls in water. Noise originates also from electrical generators and alternators and from air conditioning, power steering, and braking systems. Brake squeal continues to be a chronic problem.[4,5] In addition, there is noise due to airflow over the vehicle. This occurs principally in interior noise; however, it can be a significant part of exterior noise, especially at higher speeds.

To develop methods of controlling the noise of surface transportation, it is necessary to understand how

Note: References to chapters appearing only in the *Encyclopedia* are preceded by *Enc.*

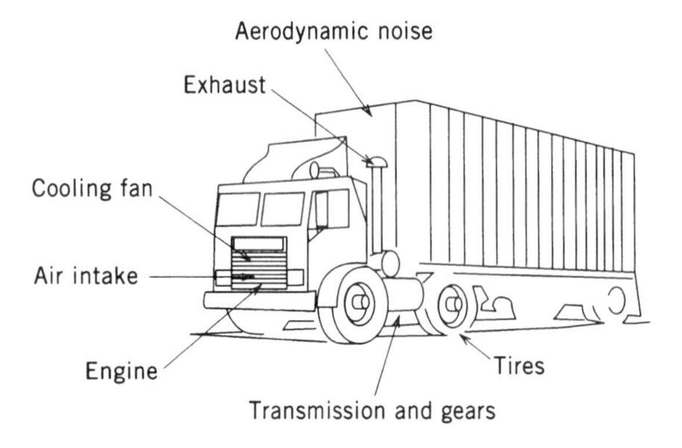

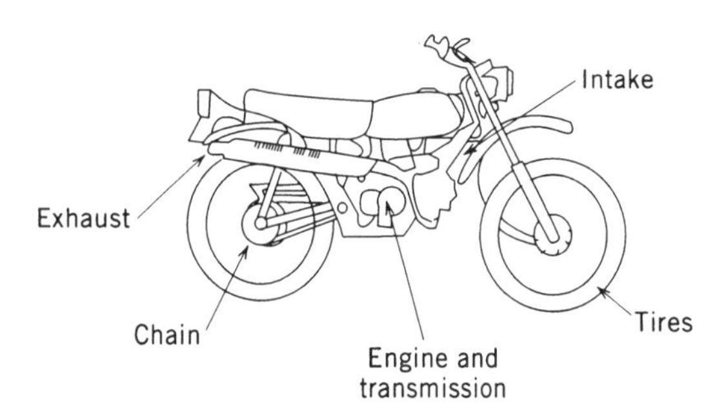

Fig. 1 Noise sources in a heavy-duty truck and a motorcycle. (After Ref. 1.)

the different vehicle components generate noise and how they are designed. The total sound power emanating from a vehicle is miniscule compared to the power involved in the operation of the vehicle. For example, the sound power of a passenger car at constant speed under road load is several milliwatts, whereas the power involved in operating the vehicle is of the order of kilowatts. Noise reduction is therefore not always of prime importance in the design of surface transportation, which is unfortunate because the understanding required to reduce noise can result in other improvements. For example, the design of a quieter passenger car cooling fan resulted in improved fan efficiency[6] with no significant increase in cost. Control of engine and machinery noise is discussed in a number of texts.[7–9] The development of sound intensity techniques within the past two decades (see Chapter 106) is important for vehicle noise diagnostics and standard tests. Also active control (see Chapter 59) provides a new approach to reducing noise and vibration, for example, reducing vehicle interior noise.[10]

With new lower limits in vehicle passby noise standards, tire noise has become an increasingly important contributor to the overall noise of a vehicle, comparable to intake and exhaust noise. The noise due to tire–road interaction has received much attention.[11,12] Regulation of tire noise separate from the rest of the vehicle has been proposed.[13] The relation between tire noise and the exterior aerodynamic noise of a vehicle has been discussed.[14] Aerodynamic noise is roughly proportional to the sixth power of air speed. Hence, as vehicle speed increases, aerodynamic noise eventually predominates over tire noise. Figure 2 shows how the aerodynamic noise of a vehicle can be separated from tire noise. In the figure, the tire noise of a coasting stake-bed truck is kept constant by keeping the ground speed at 80 km/h, while the air speed is varied by coasting the vehicle with and against a prevailing wind. Sound intensity measurement close to a tire on a roadway has provided useful information on how tire noise is generated.[15] This type of measurement could also be used as a standard tire noise test to replace coast-by tests.[15,16] Standardization of test track surfaces is being proposed.[17] Tire noise has been compared under various conditions of cruise and acceleration.[18]

Wheel–rail interaction is an important contributor to railroad noise,[19] caused mainly by the weight of rolling stock and metal-to-metal contact. Roughness of the wheel and rail play[5] an important role. It is believed that most of the noise is radiated by the rails.[20] A significant part of the noise can be caused by vibrations transmitted to rolling-stock structures because the suspension system generally does not provide good isolation from wheel–rail interaction. Structures connected to the rails also generate noise. Elevated railroads are notoriously noisy, as are trains passing over bridges. In addition there is rumble in buildings near a railroad caused by the transmission of low-frequency vibrations through the ground.[21]

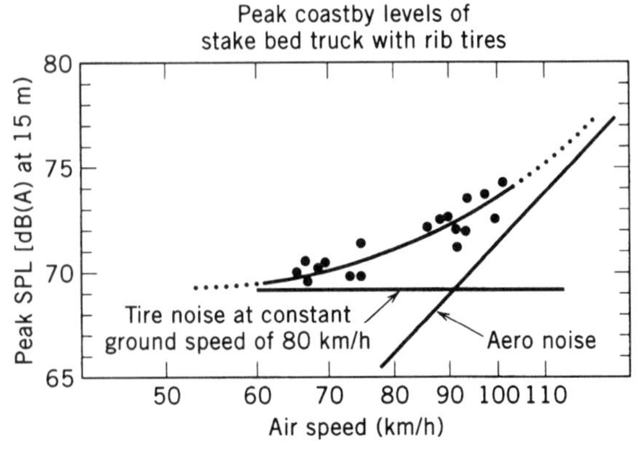

Fig. 2 Separation of aerodynamic noise from tire noise of a coasting vehicle using the prevailing-wind method. (From Ref. 2.)

3 TEST PROCEDURES AND LIMITS FOR EXTERIOR NOISE

Noise regulations are outlined for three main economic regions, the United States and Canada, the European Union, and Japan. Vehicle noise regulations in other countries are also briefly discussed. Vehicle noise regulations overseas are much more stringent than in the United States.

3.1 Noise Regulations in the United States and Canada

Table 1 lists the noise limits that apply to new highway vehicles sold in the United States and Canada. The U.S. government regulates the noise of medium- and heavy-duty trucks, including motor homes, and also motorcycles and mopeds. Passenger cars, buses, and other light vehicles are regulated by certain states and local jurisdictions whose combined effect determines the noise level of new light vehicles sold in the United States. Light vehicles are generally defined as having a gross vehicle weight rating (GVWR) less than 4.5 tonnes. The GVWR is the loaded weight of a vehicle, as specified by the manufacturer. The passby test procedure for light vehicles in the United States and Canada is the SAE J986.[22] For heavy vehicles over 4.5 t, the test is the SAE J366, equivalent to Section 40, Part 205, Subpart B, of the Code of Federal Regulations. For motorcycles and mopeds, the test is the SAE F76a, equivalent to Section 40 CFR, Part 205, Subpart D, of the Code of Federal Regulations; Subpart E presents noise-labeling requirements for motorcycle replacement exhaust silencers. All these tests are maximum-acceleration tests past a microphone 15 m from the center of the lane of travel of the vehicle and 1.2 m above the ground.

The passby site for the SAE J986 test is shown in Fig. 3. At the start point, maximum acceleration is applied, and the approach to the start point is chosen so that the vehicle can reach maximum rated engine speed in the end zone. Maximum noise during the passby is recorded. Originally, the test was run in first (lowest) gear; however, test constraints now force many vehicles to be tested in second gear. Ambient noise has to be at least 10 dB(A) below the passby reading, and wind speeds have to be less than 19 km/h. The test has to be run on a dry roadway, without dirt or powdery snow. Both sides of the vehicle are tested. Four "good" runs are required, that is, readings within 2 dB(A) of each other for each side of the vehicle, which are averaged arithmetically. The average of the two highest readings for the noisiest side of the vehicle is the test result.

The regulations for railroad noise are contained in Section 40, Part 201, and Section 49, Part 210, of the Code of Federal Regulations. Test measurements are made within a clear level area 30 m from the track center line. Locomotives are tested either stationary using remote load cells or in motion. Noise limits for locomotives manufactured in 1980 and thereafter are as follows:

Stationary, idle throttle setting	70 dB(A)
Stationary, all other throttle settings	87 dB(A)
Moving	90 dB(A)

For rail cars, the limits are as follows:

Moving at speeds 83 km/h or less	88 dB(A)
Moving at speeds greater than 83 km/h	93 dB(A)

Weather and ambient noise restrictions apply as stated above for the J986 test.

3.2 Noise Regulations in the European Union

Table 2 lists the exterior limits that apply to new highway vehicles sold in the European Union (EU), or, as it was previously known, the European Economic Community (EEC). These limits are mandatory and should not be confused with the regulations of the United Nations Economic Commission for Europe (UN/ECE, or ECE), which are voluntary on the part of participating countries. It is planned to harmonize EU and ECE noise regulations in the future. The EU vehicle noise test procedure is provided in Council Directives 92/97/EEC and 81/334/EEC, as presented in the Official Journal of the European Communities. For automatic-drive and high-power vehicles, this was amended to 84/372/EEC. The test is based on ISO 362, which is equivalent to SAEJ1470.[22] A 1-dB(A) tolerance is permitted in the test. This test and its derivatives are widely used throughout the world. A good discussion of the test is presented in Ref. 23. The test site configuration is shown in Fig. 4. It is used for all classes of highway vehicles, including motorcycles. It is important to note the difference in the size of the test site, compared to the larger J986 test site in Fig. 3. In particular, the measurement distance is 7.5 m compared to 15 m in the J986 test. The J986 test is designed to measure the maximum noise output of a vehicle, whereas the EU test is designed to measure the noise of a typical urban acceleration. As the limits in Table 2 are reduced, however, the two measurement objectives approach each other. The fact that the measurement distance is half that of the J986 test is considered to imply that, for the same vehicle, the EU test gives results that are about 6 dB(A) higher than the results of the J986 test. However, the difference between the tests is more complicated than this, and in practice the correlation is sometimes quite poor.

TABLE 1 Summary of Highway Vehicle Noise Limits in the United States and Canada

Jurisdiction	PC	MPV	LT	HT	BUS	SBUS	MH	MC	MOP	GVRW (T)	Static dB(A)	Passby dB(A)
									x			70
U.S. federal								x				80
				x			x			> 4.5		80
Canada[a]								x				80
	x	x	x							<= 2.7		80
		x	x	x	x	x	x			<= 4.5		83
				x	x	x	x			> 4.5		83
California	x	x	x							<= 4.5	95	80
				x	x	x	x			> 4.5		80
Colorado	x	x	x							< 2.7		84
		x	x	x	x	x	x			>= 2.7		86
Florida[b]	x	x	x							<= 4.5	95	80
		x								<= 4.5	95	83
				x	x	x	x			> 4.5		83
Maryland[b]	x	x	x							<= 4.5		80
				x	x	x	x			> 4.5		83
Nebraska[c]				x						>= 4.5		80
					x	x				> 4.5		83
Nevada	x	x	x							< 2.7		84
		x	x	x	x	x	x			>= 2.7		86
Oregon[b]	x	x	x							<= 4.5	93/95[d]	80
				x	x	x	x			> 4.5		80
Washington	x	x	x							<= 4.5		80
				x	x	x	x			> 4.5		80
Cook County, IL[b]	x	x	x							<= 4.5		80
				x	x	x	x			> 4.5		83
Boston, MA	x	x	x							<= 4.5		80
				x	x	x	x			> 4.5		80
Chicago, IL	x	x	x							<= 4.5		80
(w/in Cook County)				x	x	x	x			> 4.5		80
Des Plaines, IL	x	x	x							<= 4.5		80
(w/in Cook County)[b]				x	x	x	x			> 4.5		84
Grand Rapids, MI	x	x	x							<= 4.5		80
				x	x	x	x			> 4.5		84
Madison, WI	x	x	x							<= 2.7		84
		x	x	x	x	x	x			> 2.7		86
Prairie Village, KS	x	x	x							< 3.6		75
		x	x	x	x	x	x			>= 3.6		75
Urbana, IL	x	x	x							< 3.6		80
		x	x	x	x	x	x			>= 3.6		74
Washington, DC	x	x	x							<= 4.5		80
				x	x	x	x			> 4.5		80

Abbreviations: PC, passenger car; MPV, multipurpose vehicle; LT, light truck; HT, heavy truck; MOP, moped; BUS, transit/intercity bus; SBUS, school bus; MH, motor home; MC, motorcycle.

[a]Canadian noise regulations allow a 2-dB (A) tolerance on noise limits.

[b]Certification required

[c]Applies to diesel engine vehicles only.

[d]2500 r/min; 93 dB(A) and front-engine vehicles; 95 dB(A), rear and midengine vehicles.

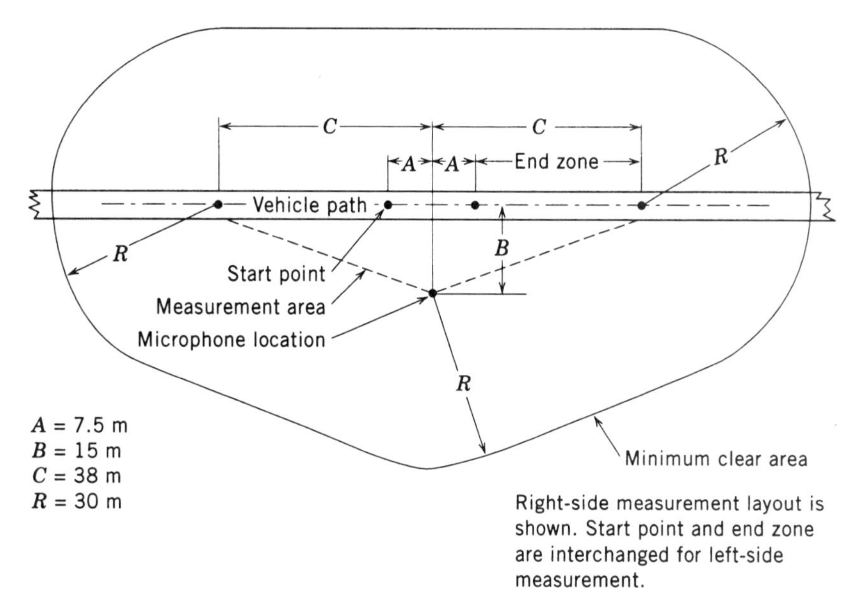

Fig. 3 Plan view of test site for the SAE J986 test. (From SAE Standard J986. Reproduced with permission from the Society of Automotive Engineers, Warrendale, PA.)

At present the EU has no exterior noise standards for railroad equipment because of intermingling with equipment from outside the EU and differences in gauge size. Railroads use a number of noise standards in an ad hoc manner, such as the ORE E82/RP.4 developed by the International Union of Railways and the ISO 3095: 1975.

An important development in EU noise standards is the use of sound power as a noise measure, instead of sound pressure at a single point in space, for construction site and earth-moving equipment. Sound power measures the total noise emanating in all directions from a source. Details are given in directives 95/27/EU, 89/514/EEC, 86/662/EEC, 84/533-536/EEC, and 79/113/EEC. From 79/113/EEC, the prescribed distribution of measurement positions on a hypothetical measurement hemisphere enclosing a stationary piece of earth-moving machinery is shown in Fig. 5. Sound pressure is measured at the measurement points, which is converted to sound intensity, using the far-field approximation and integrated to obtain sound power. The sound power determination would be more accurate[24–26] if sound intensity measurements were used (see Chapter 106) instead of sound pressure. In its present form, the test uses only half the measurement points shown in Fig. 5. A similar sound power test was developed for lawn mowers in 84/538/EEC. New noise limits in the EU for earth-moving machines of

TABLE 2 European Union Highway Vehicle Noise Limits

Vehicle Description	Unloaded Weight (t)	Power (kW)	Limit [dB(A)]
Passenger car	—	—	74
Mini bus >9 seats	<2	—	76
	>2 < 3.5	—	77
Bus >9 seats	>3.5	<150	78
Light truck/van	<2	—	76
	>2 < 3.5	—	77
Medium truck/van	>3.5	<75	78
	>3.5	>75 < 150	78
Heavy trucks	>12	>150	80
Motorcycles			
≤80 cc			75
>80 ≤ 175 cc			77
>175 cc			80

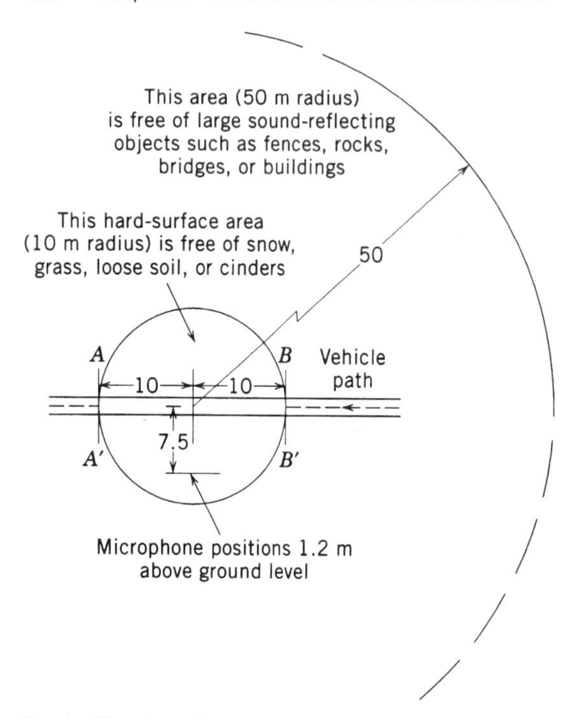

Fig. 4 Plan view of test site for the ISO 362 test. All dimensions in metres. (From SAE Standard J1470. Reproduced with permission from the Society of Automotive Engineers, Warrendale, PA.)

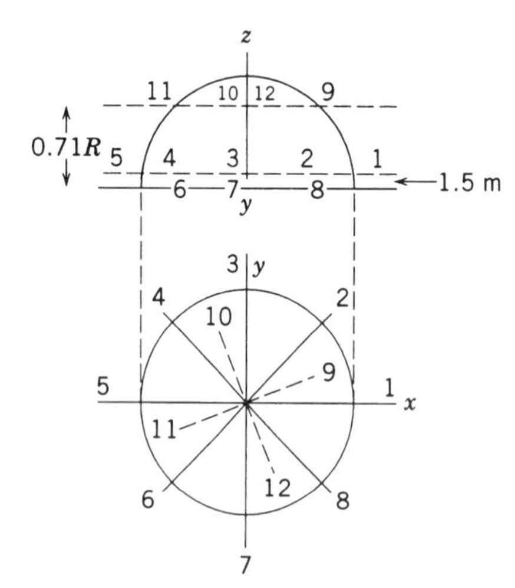

Fig. 5 Distribution of measurement points on hypothetical measurement hemisphere for determining the sound power of earth-moving equipment, as prescribed in 79/13/EEC.

net installed power less than 500 kW came into effect at the end of 1996. Machines with power greater than 500 kW operate in quarries and mines and are considered to have a negligible effect on community noise. Between 1996 and 2001, permissible sound power levels L_{WA}, in A-weighted decibles relative to 1 pW, are given by the following formulas:

1. Tracked vehicles (except excavators):

$$L_{WA} = 87 + 11 \log P \quad \text{for } L_{WA} \geq 107.$$

2. Wheeled dozers, loaders, excavator-loaders:

$$L_{WA} = 85 + 11 \log P \quad \text{for } L_{WA} \geq 104.$$

3. Excavators:

$$L_{WA} = 83 + 11 \log P \quad \text{for } L_{WA} \geq 96.$$

In the preceding P is the net installed power of the vehicle in kilowatts. Below the lower limit on the right of the formulas the machine automatically passes the test. After 2001 all the numerical A-weighted decibel values in the above formulas, including lower limits, are reduced by 3 dB(A). The coefficient 11 is unchanged.

3.3 Vehicle Noise Regulations in Japan

In Japan, exterior noise tests for road vehicles were used as far back as 1971.[27] The present tests are very thorough and consist of three parts[28]:

1. A fixed-speed passby test at 7 m, 60% rated engine speed or 65 km/h
2. An acceleration passby test at 7.5 m
3. A stationary test behind the exhaust outlet at 20 m

In all the tests the microphone is 1.2 m above ground level. The acceleration passby test is essentially the same as the EEC test; however, there is no 1-dB(A) tolerance and the test applies only to the left side of the vehicle. Noise limits for the acceleration test are given in Table 3. The limits for passenger cars, medium- and heavy-duty trucks, and buses will be reduced by 2 dB(A) in 1998. The stationary test limit is 103 dB(C). Japan also has a close-in stationary exhaust noise test based on ISO 5130. Reductions in these limits are planned for 1998 and 2002.

TABLE 3 Noise Limits for New Highway Vehicles in Japan

Vehicle Description	GVWR (t)	Power (kW)	Limit [dB(A)]
Passenger car	—	—	78
Light truck	<3.5	—	81
Medium truck, bus	>3.5	<150	83
Heavy bus	>3.5	>150	83
Heavy truck	>3.5	>150	83
Moped < 50 cc			72
Motorcycle			
>50 cc <125 cc			72
>125 cc <250 cc			75
>250 cc			78

TABLE 4 Sample of Passenger Car Noise Limits in Different Countries

Country	Limit[a] [dB(A)]
Austria	74
Switzerland	74
Finland	74
Norway	74
Sweden	74
Australia	77
Czechoslovakia	74
Hungary	80
Israel	80
Korea	75
Gulf States	82

[a]No 1-dB(A) tolerance.

3.4 Vehicle Noise Regulations in Other Countries

Many other countries have vehicle noise regulations, but it is not possible to give these here in detail. The information can be obtained from appropriate government agencies. A sampling of passenger car exterior noise limits is presented in Table 4. These are based on the EEC/ECE test procedures at 7.5 m. As the allowable limit drops, tire noise makes a proportionately increased contribution to the passby noise level.

4 VEHICLE INTERIOR NOISE

As previously stated, regulation of vehicle interior noise is based on regulation of noise in the workplace (see Chapter 69) and is applied principally to noise inside truck and locomotive cabs. In the United States, locomotive cab noise is regulated by the Federal Railroad Administration, as specified in 49 CFR 229.121, with allowed levels of noise exposure as given in Table 5. Tests are conducted 150 mm from a crew member's ear using the procedures described in Chapter 69. Truck cab noise is regulated by the Federal Highway Administration, as specified in 49 CFR 393.94. The test is conducted with the engine operated in neutral at maximum rated speed. The sound level limit is 90 dB(A), measured 150 mm to the right of the driver's ear. By current standards (see Chapter 92), these levels are too high for adequate protection of hearing. In passenger vehicles, both on highways and on railroads, control of noise is governed principally by customer complaint and commercial competition. Recently concern has developed over exterior noise from auto stereo systems.

The sources in a vehicle that excite interior noise are the same as those discussed in Section 2. Generally a distinction is made between airborne or structure-borne excitation. The fact that interior noise occurs within a closed structural shell makes a significant difference. In particular, interior noise is aggravated when an external excitation couples with a cavity resonance of the structure. This occurs principally at low frequencies. Airflow over the vehicle is a significant contributor to interior

TABLE 5 Duration Limits for Locomotive Crew

Duration (h)	Sound Level [dB(A)]
12	87
8	90
6	92
4	95
2	100
1.5	102
0.5	110
0.25 or less	115

noise, especially in light vehicles. Wind tunnel tests to reduce the aerodynamic drag of light-vehicle prototypes are usually combined with tests to measure aerodynamic interior noise.

5 CONCLUDING COMMENTS

Surface transportation is the major contributor to community noise. In coming to grips with the problem, a diversity of standard test procedures has developed throughout the world. Vehicle manufacturers would prefer to test indoors to avoid delays and additional cost due to bad weather. Large expensive anechoic facilities have been built to simulate the passby test indoors. The test vehicle is operated on a dynamometer and the passby acceleration is simulated by recording sequentially along a row of microphones. This simulation has been accepted by regulatory agencies as a substitute for the outdoor passby test. Passby tests determine a somewhat ill-defined quantity that depends on the acoustic radiation pattern of the vehicle and on the type of test. Obviously it would be better to standardize on a common, well-defined quantity, such as sound power. Indoor sound power tests of passenger cars and a truck have been performed in the United States.[24–26] Such tests provide data with much less variability than the outdoor passby test. Also they provide diagnostic information that can be used to identify and correct noise problems. In addition, the sound power of the tires on the dynamometer can be subtracted from the total sound power of the vehicle. Indoor sound power tests are performed routinely in the United States on earth-moving equipment, to meet European noise standards.[29]

REFERENCES

1. M. J. Crocker, "Noise of Surface Transportation to Nontravelers," in D. N. May (Ed.), *Handbook of Noise Assessment*, Van Nostrand Reinhold, New York, 1978.
2. Wilson Committee, *Noise—Final Report*, H. M. Stationery Office, London, 1963.
3. Administrator of the Environmental Protection Agency, *Report to the President and Congress on Noise*, U.S. Government Printing Office, Washington, D.C., 1972.
4. R. A. C. Fosbery and Z. Holubecki, "Disc-Brake Squeal: Its Mechanisms and Suppression," Research Rept. No 1961/2, Motors Industries Research Association (MIRA), Nuneaton, Warks, UK, 1961.
5. G. D. Liles, "Analysis of Disc-Brake Squeal Using Finite-Element Methods," SAE Paper 891150, Society of Automotive Engineers, Warrendale, PA, 1989.
6. R. E. Longhouse, "Control of Tip-Vortex Noise of Axial-Flow Fans Using Rotating Shrouds," *J. Sound Vib.*, Vol. 58, No. 1, 1978, pp. 201–214.
7. R. Hickling and M. M. Kamal (Eds.), *Engine Noise: Excitation, Vibration and Radiation*, Plenum, New York–London, 1982.
8. D. Baxa (Ed.), *Noise Control in Internal Combustion Engines*, Wiley, New York, 1982.
9. R. H. Lyon, *Machinery Noise and Diagnostics*, Butterworths, Boston, 1987.
10. P. Rogers and C. R. Fuller (Eds.), *Recent Advances in Active Control of Sound and Vibration*, Technomic, Lancaster, PA, 1991.
11. R. Hillquist (Ed.), "Highway Tire Noise," *Proceedings of SAE Symposium*, San Francisco, CA, Society of Automotive Engineers, Warrendale, PA, 1976.
12. K. Samberg, "Tire/Road Noise," *Proceeding of 2nd International Tire/Road Noise Conference*, Gothenburg, Sweden, Vols. I and II, STU/Information Nos. 794, 795/1990, Swedish National Board for Technical Development, Stockholm, 1990.
13. W. H. Close, "Should There Be Truck Tire Noise Regulations?" *Sound and Vibration*, Feb. 1975.
14. R. Hickling and L. J. Oswald, "A Proposed Revision of the SAE J57a Tire Noise Testing Procedure," General Motors Research Laboratories, Warren, MI, Research Publication No. GMR 2816, 1978.
15. P. Donavan and L. J. Oswald, "Identification and Quantification of Truck Tire-Noise Sources under Road-Operating Conditions," *Proceedings INTER-NOISE 80*, Miami, FL, 1980.
16. J. S. Bolton and H. H. Hall, "Correlation of Tire Intensity Levels and Passby Sound Pressure Levels," SAE Paper No 951355, Society of Automotive Engineers, Warrendale, PA, 1995.
17. U. Sandberg, "Standardization of a Test Track Surface for Use during Vehicle Noise Testing," SAE Paper No. 911048, Society of Automotive Engineers, Warrendale, PA, 1991.
18. P. R. Donavan, "Tire-Pavement Interaction Noise Measurement Under Vehicle Operating Conditions of Cruise and Acceleration," SAE Paper 931276, Society of Automotive Engineers, Warrendale, PA, 1993.
19. C. Stanworth, "Sources of Railway Noise," in P. M. Nelson (Ed.), *Transportation Noise Reference Book*, Butterworths, Boston, 1987, Chap. 14.
20. P. J. Remington, "Wheel/Rail Noise: The State of the Art," *NOISE CON 77*, Hampton, VA, 1977, pp. 257–284.
21. P. J. Remington, L. G. Kurzweil, and D. A. Towers, "Low-Frequency Noise and Vibration from Trains," in P. M. Nelson (Ed.), *Transportation Noise Reference Book*, Butterworths, Boston, 1987, Chap. 16.
22. *Handbook of the Society of Automotive Engineers*, Warrendale, PA, 1991.
23. D. Morrison, "Road Vehicle Noise-Emission Legislation,"

in P. M. Nelson (Ed.), *Transportation Noise Reference Book*, Butterworths, Boston, 1987, Chap. 9.

24. R. Hickling, "Narrow-band Indoor Measurement of the Sound Power of a Complex Mechanical Noise Source," *J. Acoust. Soc. Am.*, Vol. 87, 1990, pp. 1182–1191.

25. J. Pope, R. Hickling, D. A. Feldmaier, and D. A. Blaser, "The Use of Sound-Intensity Scans for Sound-Power Measurement and for Noise-Source Identification in Surface Transportation Vehicles," SAE Paper No. 810401, Society of Automotive Engineers, Warrendale, PA, 1981.

26. R. Hickling, P. Lee, and W. Wei, "Investigation of Integration Accuracy of Sound-Power Measurement Using an Automated Sound-Intensity System," *Appl. Acoust.*, in press.

27. Automobile Type Approval System in Japan, Sept. Traffic Safety and Nuisance Research Institute, Ministry of Transport, Japan, TRIAS 20, 1971, 1977, p. 579.

28. Japan Ministry of Transport, Automobile Type Approval Test Division, Partial Revision of Noise Test Procedure for Road Vehicles, TRIAS 20, 1980.

29. L. Tweed, "Implementation of a Second-Generation Sound-Power Test for Production Testing of Earthmoving Equipment," *Proceedings of SAE Noise and Vibration Conference*, Society of Automotive Engineers, SAE Publication P222, 1989, pp. 185–192.

72

COMMUNITY RESPONSE TO ENVIRONMENTAL NOISE

SANFORD FIDELL AND KARL S. PEARSONS

1 INTRODUCTION

Transportation (air, road, and rail) noise is the dominant and most common source of noise exposure* in residential neighborhoods, and annoyance is its most consequential effect. Since noise is generally defined as "unwanted sound," it follows almost as a matter of definition that annoyance is among the most immediate and obvious effects of noise exposure on people: who, after all, enjoys "unwanted" exposure to any agent? No other effect of noise on communities (e.g., speech, sleep, and task interference) is as widespread, long-lasting, and well documented as annoyance. Residential exposure to community noise sources poses no meaningful risk of hearing damage nor any credible risk of extraauditory health effects.

*The term *exposure* is commonly used in two ways. One use of the term implies the time integral of intensity, while the other use implies the average sound intensity over a specific time period. Intensity is the rate of flow of sound energy per unit area per second. At distances from sound sources that are of interest in environmental analyses, sound intensity is directly proportional to the square of sound pressure. Thus, sound exposure is usually represented as the time integral of squared sound pressure. This process is often referred to informally as *energy summation*. Magnitudes are reported in logarithmic terms. For example, sound exposure level is 10 times the logarithm to the base 10 of the ratio of sound exposure to a reference exposure of 400 $\mu Pa^2 \cdot s$. In this logarithmic form, squared sound pressure is called sound level and expressed in units of decibels. Sound level in decibel notation is often expressed as an average (equivalent) sound level over a specified time interval (usually 1 h or 24 h). Single events are often described by their sound exposure level (SEL) with a reference time interval of one second.

Note: References to chapters appearing only in the *Encyclopedia* are preceded by *Enc*.

Annoyance is not an immediate sensation such as loudness, governed solely by characteristics of acoustic signals, nor is it an overt behavior such as a complaint. Annoyance is instead an attitude with both acoustic and nonacoustic determinants. If noise is defined less elliptically as "acoustic energy that someone else considers too inconvenient or too expensive to control," it is apparent that an attitude engendered by noise exposure must have both acoustic and nonacoustic determinants. A brief review of community noise research provides background useful for understanding modern approaches to predicting the prevalence of annoyance in communities.

2 BACKGROUND

It has long been understood that the acoustic determinants of annoyance can include all of the factors that affect the audibility of sounds by people. The first community noise survey[1] provided an early demonstration of this principle by establishing the need for frequency weighting networks to account for human response to differing distributions of acoustic energy in the frequency domain. This pioneering work showed that measurement of noise in a manner relevant to its effect on people cannot be arbitrarily based on the bandwidth, frequency response, and dynamic range of available instrumentation. The noise worth measuring is the noise that people can hear. Creation and standardization of the A, B, and C weighting networks in the 1930s by inversion of the Fletcher–Munson contours[2] was the first embodiment of this principle.

Interest in systematic means for predicting community response to aircraft noise can be traced at least to the early 1950s. Rosenblith and Stevens[3] developed a Community Noise Rating (CNR) based on two tacit assumptions: (1) that the prevalence of annoyance grows monotonically with the loudness of noise exposure and (2) that both acoustic and nonacoustic factors affect annoyance. CNR attempted to account for community reaction via "level ranks," an estimate of the effective loudness of noise exposure. These level ranks were modified by more or less arbitrary adjustments of 5 dB for nonacoustic factors such as novelty and economic dependence on noise sources. A nonquantitative role was even provided for ambient noise levels, on the assumption that masking of low-level sounds could reduce their audibility and hence annoyance. The pragmatic approach of this study set the style for investigations in this field for many years.

Concern about noise pollution in residential areas intensified greatly when jet transport aircraft began commercial operation in the late 1950s. Questions were raised in airport neighborhoods about the compatibility of airports and communities, about limits on the growth of the air transport industry, about measures that could be taken to mitigate airport noise impacts, and about tolerable levels of noise exposure. Within a few years of the start of jet air transport service, an initial round of large-scale social surveys had been conducted in the vicinity of major international airports. (See Schultz[4] for a comprehensive discussion of early survey results and noise measurement methods.)

Little standardization of community response survey methods, questionnaire items, noise measurements, or analysis techniques was practiced throughout the 1960s and 1970s. Nonetheless, interpretations made in the early 1960s of the findings of the first generation of social surveys alarmed many. Aircraft noise exposure was viewed by some as seriously impairing important aspects of residential living. At the same time, many laboratory studies were undertaken of the annoyance, speech, and sleep interfering properties of noise. There was essentially no agreement on which acoustic and which social variables were worth measuring, little understanding of their relationships to one another, and little communication among researchers in different countries.

The set of acoustic properties suggested to predict annoyance is a tribute to the collective imagination of several generations of researchers. In addition to many different frequency weighting schemes, the physical properties thought to be useful in predicting annoyance have included tonality, impulsiveness, rise time, onset time, periodicity, time of day, and temporal variability. Pearsons and Bennett[5] and Schultz (see Ref. 4) have cataloged dozens of physical measures of sound that have been seriously considered as predictors of the annoyance of noise exposure.

Data dredging (misuse of regression techniques to identify and interpret random correlations among variables in large data sets) was a popular approach to inferring acoustic and other factors affecting annoyance judgments through the 1970s. Multiple regression equations so developed tended to work reasonably well for individual data sets, but poorly when applied to other data sets, so that one researcher's main effect is merely another researcher's error variance. The predictable outcome of this approach to estimating community response to noise exposure was an embarrassment of riches—not one preferred metric of noise exposure but rather an alphabet soup of measures. Botsford's[6] summary of annoyance prediction scales (which he called the "weighting" game) identifies conflicting interests in prolonging the search for the holy grail of the acoustic noise measurement procedure: the one that perfectly predicts the annoyance of a wide range of noise exposure on the sole basis of its acoustic content.

Two observations may be made about predicting the annoyance of noise exposure on the basis of measurements of properties of sound alone. The first observation is that most noise metrics correlate highly with each other, if not with self-reports of annoyance.[7] This high correlation among the various noise metrics is the basis for a common view that the choice of metric for measuring community noise is relatively unimportant. Adoption of any reasonable noise metric for regulatory purposes would have been more productive than decades of research searching for an optimal measure. For this reason, agreement by U.S. federal agencies on a set of equivalent energy noise metrics (SEL, L_{eq}, L_{dn}) in the early 1970s was an important accomplishment. This agreement was paced by the requirements of the National Environmental Policy Act in 1969 and the Noise Control Act of 1972.

The second observation is that a search for a single, purely physical metric that can serve as a universal predictor of the annoyance of noise exposure is unlikely to succeed. There are as many ways to measure noise as there are purposes for the measurement. In the absence of a theory of annoyance, the search for an optimal acoustic measure of annoyance places the cart before the horse. The fundamental problem is not, after all, measurement but rather deciding what is worth measuring. After decades of study of community reaction to noise exposure, it is now clear that it is unrealistic to expect purely physical measures to provide a complete account of the prevalence of noise-induced annoyance in communities.

Figure 1 illustrates the futility of the search for the perfect acoustic predictor of community response to noise exposure. The range and variance in this data set are so great that the shape of a dosage–response relationship intended to summarize these findings is not greatly constrained by proximity to the data points.

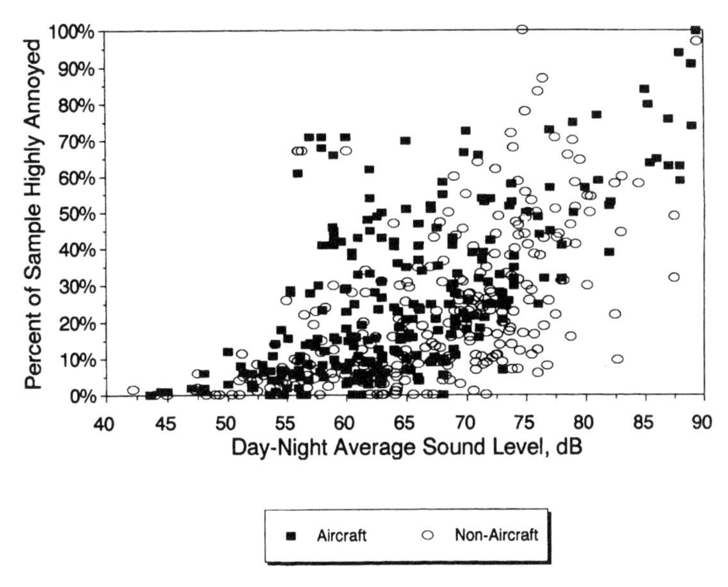

Fig. 1 Results of 545 determinations of the prevalence of noise-induced annoyance in communities exposed to transportation noise.

3 METHODS OF ASSESSING ANNOYANCE

3.1 Direct Measurement of the Prevalence of Annoyance

The least ambiguous way to estimate the prevalence of noise-induced annoyance in a community is to measure it directly by means of a social survey. Survey techniques for such purposes are well developed; Fields[8] has documented hundreds of surveys of varying sophistication on community noise effects. Considerable care must be taken in survey design, sampling, questionnaire construction, administration of interviews, and in data analysis to avoid potential errors and biases in reaching inferences on the basis of survey findings. Fidell and Green[9] provide a brief overview of the major procedural concerns in the conduct of noise effects surveys.

Less direct methods of estimating the prevalence of annoyance due to noise exposure in a community (described in the next sections) are available if the conduct of a well-designed social survey is unaffordable or otherwise impractical.

3.2 Prediction of Annoyance via an Empirical Dosage–Response Relationship

Schultz's[10] production of a well-documented, widely acknowledged quantitative dosage–response relationship via meta-analysis was a major step toward development of a standard method for prediction of transportation noise effects. Schultz fitted a third-order polynomial to a data set relating day–night average sound level (DNL) to the prevalence of a consequential degree of annoyance in communities. A simple quadratic fitting function[11] provides an updated but still purely empirical basis for estimating the prevalence of annoyance in communities:

$$\% \text{ Highly annoyed} = 0.036L_{\text{dn}}^2 - 3.27L_{\text{dn}} + 79.14. \quad (1)$$

This relationship is simply an atheoretical fitting function that produces meaningless predictions when evaluated outside the range of approximately 45 dB $< L_{\text{dn}} < 85$ dB. The Federal Interagency Committee on Noise prefers a logistic fit to a subset of the data reported by Fidell, Barber, and Schultz (see Ref. 11) that omits observations from several studies in which relatively low levels of noise exposure were associated with relatively high proportions of reported annoyance:

$$\% \text{ Highly annoyed} = 100/[1 + \exp(11.13 - 0.141L_{\text{dn}})]. \quad (2)$$

This function predicts somewhat lower levels of annoyance at lower noise exposure levels than the quadratic fit: For example, a prevalence rate of only 12.3% is predicted at $L_{\text{dn}} = 65$ dB, in contrast with 18.8% for the quadratic fit to the full data set.

A CHABA Working Group on Assessment of Community Response to High-Energy Impulsive Sounds (Fidell[12]) has identified two dosage–response relationships for predicting the annoyance of impulsive sounds with CSEL values in excess of 85 dB (75 dB at night). One of these predicts the proportion of a community highly annoyed by high-energy impulsive noise exposure as

$$P = \exp(-10^{0.045(D* - L_{Cdn})}) \tag{3}$$

where D^* is the constant 61.1 chosen to minimize the variance of the set of field observations of the annoyance of high-energy impulsive noise exposure.

An alternative method recognized in the CHABA report for predicting the prevalence of a consequential degree of annoyance due to impulsive noise exposure requires a level-dependent summation of CSEL values of individual, C-weighted impulsive noise events in annoyance units that are numerically equal to units of sound exposure for nonimpulsive sounds, measured in A-weighted units. Readers are referred to Schomer[13] for further details of this method.

Limitations of the "Equal Energy Hypothesis" as a Predictor of Annoyance Day–night level was adopted after much discussion in the 1970s by many U.S. federal agencies as a convenient general descriptor of long-term environmental noise exposure, particularly in residential settings. Although DNL was intended originally as a convenient, single-number index of complex community noise environments, its use as a predictor (and sometimes even a surrogate) of annoyance was not long delayed. This latter use of DNL has probably become its major use; after all, if noise exposure did not bother people, there would be little purpose in community noise measurement for its own sake, and little reason for measuring noise at all.

The DNL index is a time-weighted average (in effect, the average acoustic energy per second with an arbitrary nighttime weighting) variant of a family of so-called integrated energy metrics,* all of which are equally sensitive to the duration and magnitude of individual noise events and directly proportional on an energy (10 log N) basis to number of events. Reliance on any integrated energy metric to predict or explain individual and community response to aircraft noise is based on the "equal energy" hypothesis.

The equal energy hypothesis expresses the notion that the number, level, and duration of noise events are fully interchangeable determinants of annoyance as long as their product (energy summation) remains constant. Thus, whether acknowledged or not, quantification of noise exposure in units of DNL for purposes of predicting annoyance reflects a tacit theory: People are indifferent between the annoyance of small numbers of very high level noise events of short duration and the annoyance of large numbers of compensatingly lower level noise and/or longer duration noise events. The hypothesis is the underpinning of a convenient method for measuring noise exposure for purposes of predicting annoyance.

The equal energy hypothesis has provided an adequate account for data on the prevalence of annoyance due to more or less continuous (e.g., urban) noise exposure over a range of roughly 20 dB, from about L_{dn} = 55 dB to L_{dn} = 75 dB. The plausibility of the hypothesis is clearly limited in extreme cases, however. For example, no community is likely to tolerate even infrequent operation of a noise source powerful enough to damage hearing. Likewise, it is difficult to assert without substantial empirical evidence that the prevalence of long-term annoyance in a community can be predicted by the simple expedient of annualizing the exposure created by rare or infrequent noise events. Doubts about the limits of the equal energy hypothesis are often expressed as challenges to the universality of DNL as a predictor of annoyance.

Although DNL predicts the prevalence of annoyance over a 20-dB range of exposure levels with useful precision (accounting for about half of the variance in the data set), there is no evidence that the relationship between DNL and the prevalence of annoyance in a community is a causal one. Since most other reasonable metrics of noise exposure are highly correlated with DNL in airport neighborhoods, and since no other metric of noise exposure alone accounts for appreciably more variance, there are ample grounds for accepting the equal energy hypothesis in urban neighborhoods.

The utility of DNL as a predictor of the prevalence of annoyance also depends on a number of additional tacit assumptions that are most relevant to urban areas about the nature and circumstances of noise exposure. For example, reliance on equivalent energy descriptors to predict the annoyance associated with sporadic (i.e., infrequent, unexpected, and intermittent) operation of noise sources strains the hypothesis considerably. Furthermore, when noise exposure is composed of small numbers of noise intrusions of very different levels, the maximum level of a single event can control the daily

*Other members of this family of noise measures include sound exposure level (SEL), equivalent continuous level L_{eq} and variants of DNL such as community noise equivalent level (CNEL) and C-weighted DNL. Further detail about these and other metrics in common use for characterizing environmental noise may be found in the various parts of the American National Standard "Quantities and Procedures for Description and Measurement of Environmental Sound" (ANSI S12-9).

integrated exposure. Thus, when dealing with truly sporadic noise exposures, it becomes essentially irrelevant whether people's reactions are based on a perfect integration of exposure, sensitivity to the maximum level, or anything in between.

Common sense suggests that the integration of individual noise exposures on which long-term annoyance judgments are based cannot be in perfect, loss-free proportion to the exposure created by multiple noise intrusions. If people based their annoyance judgments on a perfect accumulation of the noise exposure created by multiple events over months and years, then everyone would eventually find continued noise intrusions intolerable, and would remain fully annoyed indefinitely, even following a lengthy hiatus in exposure. Since long-term annoyance asymptotes even in heavily noise-exposed areas, the temporal integration of exposure on which annoyance judgments are based must be "leaky" in some sense. In other words, an individual's annoyance cannot ratchet indefinitely, but must at any point in time represent a balance between a rate of accumulation and a rate of dissipation.

Other Limitations of Empirical Prediction Methods None of the empirical dosage–response relationships is grounded in detailed understanding of the mechanisms by which noise exposure generates annoyance. None is fully applicable to all situations in which residential or other populations are exposed to noise. None accounts for the greater part of the variance in the social survey data.

Furthermore, an empirical dosage–response relationship does not in itself provide answers to the following sorts of questions:

- How can the prevalence of annoyance in two communities exposed to the same level of noise differ greatly?
- How accurately can the prevalence of noise-induced annoyance be predicted from knowledge of noise exposure alone?
- Decibel for decibel, is noise exposure produced by different sources equally annoying?
- What are the relative contributions of acoustic and other factors to the prevalence of annoyance in a community?
- How is the immediate annoyance created by individual events linked to the long-term annoyance of cumulative noise exposure?
- What are the time constants of arousal and decay of community annoyance?

Incomplete understanding of the origins of community annoyance can have costly consequences, including adoption of potentially ineffective means for mitigating effects of noise exposure on communities. Fidell and Silvati,[14] for example, have shown that there is no clear evidence for a reduction in the prevalence of long-term aircraft noise annoyance among homeowners whose dwellings had been treated as part of an airport noise mitigation program to increase their acoustic transmission loss. In fact, one paradoxical consequence of acoustically treating a portion of the heavily noise-exposed housing stock in the immediate vicinity of an airport could be exposure of greater numbers of untreated homes to yet higher noise exposure. This situation could arise if the numbers of flight operations increase following relief from the threat of aircraft noise litigation following relatively inexpensive acquisition of aviation easements (via home insulation projects) in neighborhoods near airports.

3.3 Theory-Based Prediction of Prevalence of Annoyance

An alternate approach to predicting the prevalence of annoyance in communities[15,16] treats annoyance not as an inexplicable concomitant of noise exposure but rather as a rational process, capable of description if not understanding in probabilistic terms. This approach permits inference of the shape of a dosage–response relationship from first principles, rather than in a purely ad hoc manner.

Fidell et al.[15] view noise exposure as a form of treatment administered to a community. The response of the community to the treatment is determined by applying a criterion for reporting annoyance to the average dose. The effective dose produced by exposure is assumed, as in the CNR system, to grow at a rate similar to the growth of loudness with level. Variability in the prevalence of annoyance in communities that is not accounted for by this compressive transform on DNL is attributed to response bias. Among the many nonacoustic factors that may affect the prevalence of noise-induced annoyance in communities are (a) various attitudes toward noise sources and their operators (approval, fear, distrust, etc.), (b) socioeconomic levels of individuals, and (c) economic dependence on operation of noise sources.

This approach follows from recognition of two separable components in self-reports of annoyance: a component directly linked to noise exposure and an entirely independent component associated with individual willingness to describe oneself as annoyed in some degree. The latter component is referred to as *response bias*. The two components are confounded in a verbal report of the form "I'm very annoyed by that aircraft flyover," since the self-report alone provides no way to distinguish the contributions of the acoustically related factors from the contributions of response bias to the expressed degree of annoyance. For lack of adequate means of distinguishing the two components, most interpretations of self-reports of annoyance tacitly assume that the contributions of response bias to annoyance are negligible.

Given that acoustic and nonacoustic determinants are confounded in individual self-reports, it follows that they are also confounded in the proportion of respondents in a neighborhood who describe themselves as highly annoyed by noise exposure. Two communities in which 20% of the residents describe themselves as highly annoyed might have quite different noise exposures. Such differences are attributed to difference in response biases among the residents. Civic action groups, political or media attention, and other factors can make the residents of the quieter community more likely to describe themselves as highly annoyed even by relatively low noise exposures.

In communities in which the prevalence of annoyance is affected primarily by noise exposure, reductions in noise exposure can be expected to lead to reductions in prevalence of annoyance. In communities in which the prevalence of annoyance is controlled by nonacoustic factors, there may be little or no reduction in annoyance associated with reductions in noise exposure.

More specifically, reactions of individuals in the community to noise exposure are assumed to be exponentially distributed with a *mean population value*, m. The value of m is assumed to be related to the day–night average sound level by

$$10 \log m = 0.3 L_{\text{dn}}. \tag{4}$$

Thus, noise exposure creates a distribution of reactions within a community with a mean value that increases with the level of noise exposure. Individuals describe themselves as highly annoyed when their reactions to noise exposure exceed a fixed value of a criterion value (A) for reporting annoyance. The net effect of the nonacoustic factors on the decision-making process may be regarded as a form of response bias. The proportion of the population describing itself as highly annoyed is predicted as follows. Suppose

$$P = e^{-(A/m)}, \tag{5}$$

where P is the probability of reporting high annoyance, m is defined as in Eq. (2), and A is the criterion value for reporting annoyance. The value of A may vary from neighborhood to neighborhood for any of a number of nonacoustic reasons. For example, this criterion value may differ because the residents of one neighborhood value the commerce or convenience associated with operation of a noise source more highly than residents of another neighborhood, or because greater media or political attention has been focused on environmental problems in one neighborhood than in another, or because nonenvironmental problems are more pressing to residents of one neighborhood than of another, and so forth.

The criterion value for reporting high annoyance, D, may also be expressed in exposure-like units. If D is defined as

$$D = \frac{1}{0.3}\, 10 \log A, \tag{6}$$

then Eq. (4) may be rewritten as

$$10 \log (-\ln P) = 0.3(D - L_{\text{dn}}) \tag{7}$$

or as

$$P = \exp\,[-10^{0.03(D - L_{\text{dr}})}], \tag{8}$$

where 10 log D, or D^*, is thus a decibel-like quantity that represents the average value of the criterion for reporting annoyance in units directly comparable with L_{dn} values.

The average value of the criterion for reporting high annoyance can also depend on the source of the noise intrusion. For example, people seem to adopt a more stringent criterion for reporting annoyance due to noise from diffusely owned noise sources, such as surface street traffic, and a less stringent criterion for reporting annoyance due to aircraft operations (see Ref. 16). Thus, L_{dn} values created by aircraft and surface transportation noise can differ by 5 dB without affecting the percentage of residents who describe themselves as highly annoyed.

Equation (8) shows that if $L_{\text{dn}} = 60$ dB and the criterion for annoyance $D^* = 70$ for aircraft, then the 13.6% of the population exposed to aircraft noise will report high annoyed. If $D^* = 75$ for surface transportation noise, only 6% of the residents with equivalent traffic noise exposure will describe themselves as highly annoyed.

4 SPECIAL SITUATIONS

4.1 Dynamics of Community Response to Noise Exposure

No widely applicable methods are available for predicting the prevalence of noise-induced annoyance in communities in other than steady-state conditions, since little is known about the rates at which communities accommodate temporal variability in noise exposure. Although many changes in community noise exposure are gradual, step changes can occur, especially in noise produced by aircraft operations. While short-term opinions may track large exposure transients with little latency, it is clear that people distinguish between short- and long-term consequences of noise exposure. Fidell and co-workers[17] have shown that the prevalence of long-term annoyance only sluggishly reflects changes in noise exposure. At least tens of days must elapse before short-term opinions about noise exposure are transformed into long-term attitudes.

There are limits to the usefulness of viewing people as sound level meters with enormous integration times, however, particularly when exposure is intermittent and dominated by occasional high-level events. Fidell, Sneddon, and Green[18] suggest an alternative approach to modeling the annoyance of this sort of intermittent noise exposure, based on an integration of reactions to noise exposure rather than integration of noise exposure per se.

4.2 Annoyance of Low-Level Noise Intrusions

The influence of ambient noise levels on annoyance are generally neglected in assessments of the effects of high-level noise intrusions on communities. In airport neighborhoods, for example, levels of individual aircraft overflights often exceed ambient levels by 30 dB or more. In other settings, such as intrusions of en route aircraft noise in quiet suburban or rural settings, ambient noise levels can have a considerable influence on the audibility and noticeability of noise intrusions.

The audibility of a steady-state, broadband acoustic signal is determined by its bandwidth-adjusted signal-to-noise ratio, as described in many studies.[19–24] Since audibility is a matter of signal-to-noise ratio rather than signal level alone, a signal of low audibility is not necessarily a signal of low absolute level. Furthermore, the same signal can be differentially audible in different ambient noise environments.

The common unit of measurement of audibility is the scalar (dimensionless) quantity known as d'. [The audibility of an acoustic signal is given in terms of its bandwidth-adjusted signal to noise ratio: $d' = \eta S/N (BW)^{1/2}$. Such calculations can be conducted in one-third octave bands for most community noise applications.] An acoustic signal characterized by a d' value of 0 is physically impossible to detect, since no information is available with which to discriminate it from the background noise environment in which it occurs. Under controlled laboratory conditions, in which well-trained observers pay close attention to a structured acoustic detection task, a signal with a d' value of 2.3 will be reliably reported half of the time. A "reliable" report of the presence of a signal is one with a low false alarm rate: in the present context, a false alarm rate of 1% or less. [A "false alarm" is a report of the presence of a signal that has not in fact occurred. False alarms and correct detections co-vary in a specifiable manner for any observer (cf. Ref. 23).] Under the same highly controlled listening conditions, the presence of signals characterized by d' value of 4 is reported consistently with a negligible false alarm rate.

Under less constrained listening conditions, acoustic signals must attain slightly higher signal-to-noise ratios before observers reliably report their presence. A d' value of about 5 suffices for field observers actively listening for sound sources, while an additional 10 dB of signal-to-noise ratio ($d' = 50$) is a reasonable estimate of the audibility required to bring a signal to the attention of someone who is not actively listening for it.[25]

5 REGULATORY AND POLICY INTERPRETATIONS OF NOISE EXPOSURE EFFECTS

The pressure for standard prediction techniques and nationwide guidelines for interpreting noise exposure effects in the United States became intense following passage of the National Environmental Policy Act of 1969 and the Noise Control Act of 1972. More recent concerns in prediction of the annoyance of en route aircraft noise have arisen due to increased low-altitude, high-speed military flight activity in rural areas: U.S. Public Law 100-91 mandates assessment of impacts in nonresidential, short-term, outdoor recreational settings; development of advanced military and civil hypersonic aircraft; and unexpectedly intense adverse reactions to commercial air traffic noise in low ambient noise residential communities.

A number of organizations have developed methods and policies for community noise measurement, recommendations for land-use planning, general guidelines for noise mitigation, and various interpretations of noise effects on communities. These generic policies, recommendations, and guidelines have been developed for nationwide application and are neither intended nor guaranteed to be applied to any specific community. Guidance of this sort is fully useful only in settings unconstrained by existing land uses, such as a site selection study for a new airport.

In practice, when applied to an existing community with established land uses, the recommended planning and noise mitigation measures serve largely to render a community compatible with an airport (rather than vice versa). For example, rezoning residential land to commercial or industrial uses might in fact reduce the size of a heavily noise-impacted population. However, such rezoning might also affect a community's character, composition, quality of life, and tax base in ways that some would consider intolerable.

The best known and most well established of noise effects guidelines in the United States are those of the Federal Inter-Agency Committee on Noise (FICON).[26,27] FICON is composed of several federal agencies with interests in environmental noise exposure, including some with aviation-related interests (i.e., FAA and DoD). FICON's guidelines and recommendations for "land-use compatibility" are expressed as ranges of values of DNL. Identification of DNL values associated with "acceptable" and "unacceptable" noise exposure levels are based not on acoustic considerations per se, but rather on the expected tolerance of community effects of noise exposure on communities and on political and economic concerns. In the absence of an explicit and systematic supporting rationale for the selection of particular values of DNL as "compatible" or "incompatible" with particular land uses, this is an oddly circular approach: Both the cause (noise exposure) and the effect ("incompatibility") are defined in the same units.

FICON considers noise exposure levels lower than L_{dn} = 65 dB to be "compatible with most residential land uses." However, FICON also states that "for populated areas, there may be appreciable numbers of persons highly annoyed by exposure below L_{dn} = 65 dB." FICON then draws the conclusion that "thus, evaluation of the noise impact in such areas, in terms of highly annoyed, may be appropriate." In other words, if appreciable numbers of people are exposed to levels below L_{dn} = 65 dB, it may be "appropriate" to characterize and evaluate the effects of noise exposure upon them not in acoustic terms but rather in terms of annoyance.

One implication of FICON's statement that noise exposure levels lower than L_{dn} = 65 dB are "compatible with most residential land uses" is that it is reasonable for roughly 15% of a residential population to be highly annoyed by noise exposure [cf. Eqs. (1) and (2)]. Although it may be acceptable from some perspectives for 15% of a community to be highly annoyed by aircraft noise exposure, it may not be acceptable from other perspectives.* The U.S. Environmental Protection Agency[28] (p. 25) notes that "decisions about how much noise is too much noise for whom, for how long, and under what conditions demand consideration of economic, political, and technological matters." EPA goes on to note "the need to reconcile local economic and political realities with scientific information. People who formulate local noise abatement programs cannot escape the responsibility of making such economic and political compromises for their constituencies."

*FICON explicitly acknowledges that its recommendations are merely advisory and that its policy perspectives may differ from those of local communities. In fact, FICON does not consider its guidelines as thresholds of noise impacts and encourages local government agencies to consider economic, technical, and political ramifications of noise impacts. At least one community's land-use planning criteria that are more stringent than those recommended by FICON has survived challenge in a U.S. federal appeals court.

Acknowledgments

Portions of the text of this chapter paraphrase material from Fidell[29,30] Fidell and Green,[9] Fidell et al.[11,15] and Green and Fidell.[16]

REFERENCES

1. H. Fletcher, A. H. Beyer, and A. B. Duel, "Noise Measurement in City Noise," Report of the Noise Abatement Commission, Department of Health, City of New York, 1930.
2. H. Fletcher and W. A. Munson, "Loudness, Its Definition, Measurement and Calculation," *J. Acoust. Soc. Am.*, Vol. 5, 1933, pp. 82–108.
3. W. A. Rosenblith, K. N. Stevens, and the staff of Bolt Beranek and Newman, *Handbook of Acoustic Noise Control, Vol. 2, Noise and Man*, WADC TR-52-204, Wright Air Development Center, Wright-Patterson Air Force Base, Ohio, 1953.
4. T. J. Schultz, *Community Noise Rating*, Elsevier, New York, 1982.
5. K. S. Pearsons and R. Bennett, *Handbook of Noise Ratings*, NASA CR-2376, Washington, D.C., 1974.
6. J. H. Botsford, "Using Sound Levels to Gauge Human Response to Noise," *J. Sound Vib.*, Vol. 3, No. 10, 1969, pp. 16–28.
7. S. Fidell, "Community Response to Noise," in D. M. Jones and A. J. Chapman (Eds.), *Noise and Society*, Wiley, Chichester, 1984, Chap. 10.
8. J. M. Fields, *An Updated Catalog of 318 Social Surveys of Residents' Reactions to Environmental Noise (1943–1989)*, NASA TM-187553, 1991.
9. S. Fidell and D. M. Green, "Noise-Induced Annoyance of Individuals and Communities," in C. Harris (Ed.), *Handbook of Acoustical Measurements and Noise Control*, 3rd ed., McGraw-Hill, New York, 1991, Chap. 23.
10. T. J. Schultz, "Synthesis of Social Surveys on Noise Annoyance," *J. Acoust. Soc. Am.*, Vol. 64, No. 2, 1978, pp. 377–405.
11. S. Fidell, D. Barber, and T. Schultz, "Updating a Dosage-Effect Relationship for the Prevalence of Annoyance Due to General Transportation Noise," *J. Acoust. Soc. Am.*, Vol. 89, No. 1, 1991, pp. 221–233.
12. S. Fidell (Ed.), "Community Response to High-Energy Impulsive Sounds: An Assessment of the Field Since

1981," Report of CHABA Working Group 102, National Academy Press, Washington, DC, 1996.

13. P. Schomer, "New Descriptor for High-Energy Impulsive Sounds," *Noise Control Eng. J.*, Vol. 2, No. 5, 1994, pp. 179–191.
14. S. Fidell and L. Silvati, "An Assessment of the Effect of Residential Acoustic Insulation on Prevalence of Annoyance in an Airport Community," *J. Acoust. Soc. Am.*, Vol. 89, No. 1, 1991.
15. S. Fidell, T. Schultz, and D. M. Green, "A Theoretical Interpretation of the Prevalence Rate of Noise-Induced Annoyance in Residential Populations," *J. Acoust. Soc. Am.*, Vol. 84, No. 6, 1988.
16. D. M. Green and S. Fidell, "Variability in the Criterion for Reporting Annoyance in Community Noise Surveys," *J. Acoust. Soc. Am.*, Vol. 89, No. 1, 1991, pp. 234–243.
17. S. Fidell, R. Horonjeff, J. Mills, E. Baldwin, S. Teffeteller, and K. Pearsons, "Aircraft Noise Annoyance at Three Joint Air Carrier and General Aviation Airports," *J. Acoust. Soc. Am.*, Vol. 77, No. 3, 1985, pp. 1054–1068.
18. S. Fidell, M. Sneddon, and D. M. Green, "Relationship Between Short and Long Term Annoyance of Noise Exposure," USAF NSBIT TOR No. 22, July 1990.
19. S. Fidell, "Effectiveness of Audible Warning Signals for Emergency Vehicles," *Human Factors*, Vol. 20, No. 1, 1978, pp. 19–26.
20. S. Fidell, K. Pearsons, and R. Bennett, "Prediction of Aural Detectability of Noise Signals," *Human Factors*, Vol. 16, No. 4, 1974, pp. 373–383.
21. S. Fidell, S. Teffeteller, R. Horonjeff, and D. M. Green, "Predicting Annoyance from Detectability of Low Level Sounds," *J. Acoust. Soc. Am.*, Vol. 66, No. 5, 1979, pp. 1427–1434.
22. S. Fidell and S. Teffeteller, "Scaling the Annoyance of Intrusive Sounds," *J. Sound Vib.*, Vol. 78, No. 2, 1981, pp. 291–298.
23. D. M. Green and J. A. Swets, *Signal Detection Theory and Psychophysics*, Wiley, New York, 1966.
24. L. Hutchings and S. Fidell, "Documentation for Version 7 of Tank-Automotive Command's Acoustic Detection Range Prediction Model (ADRPM-7)," Warren, MI, 1989.
25. S. Fidell, K. Pearsons, and M. Sneddon, "Evaluation of the Effectiveness of SFAR 50-2 in Restoring Natural Quiet to Grand Canyon National Park," BBN Report 7197, February 1992.
26. American National Standard S12.40-1990, "Sound Level Descriptors for Determination of Compatible Land Use," Standards Secretariat, Acoustical Society of America, New York, August 1990.
27. Federal Interagency Committee on Noise (FICON), "Final Report: Airport Noise Assessment Methodologies and Metrics," Washington, D.C., 1992.
28. United States Environmental Protection Agency, "Protective Noise Levels," Condensed Version of EPA Levels Document, EPA 550/9-79-100, November 1978.
29. S. Fidell, "Community Response to Noise," in C. Harris (Ed.), *Handbook of Noise Control*, 2nd ed., McGraw-Hill, New York, 1976, Chap. 36.
30. S. Fidell, "An Historical Perspective on Predicting the Annoyance of Noise Exposure," presented at NOISE-CON 90, Austin, Texas, October 15–16, 1990.

PART IX

ARCHITECTURAL ACOUSTICS

73

INTRODUCTION

A. HAROLD MARSHALL AND WILLIAM J. CAVANAUGH

1 INTRODUCTION

Architectural acoustics encompasses the acoustical dimensions of the human-built environment. Related expressions that may be used to define this field or parts of it are building acoustics and noise control, architectural engineering (acoustics), architectural science (acoustics), room acoustics, and acoustical design. In this chapter an attempt is made to provide a general framework within which aspects of architectural acoustics can be seen to fit. The term *architectural*, in particular, merits some consideration. The chapter starts with a summary of related book sections underlining the cross-disciplinary scope of architectural acoustics. *Foreground* and *background* in the context of architectural design for both noise control and communication lead to a discussion of the relationship between architectural acoustics and human hearing processes and its implications for the preferred properties of rooms for both speech and music. Acoustical art forms as a special case of human communication lead to the expectation that the signal will be enhanced by the room in which it takes place. The idea of "acoustical scale" as a useful way of distinguishing appropriate design concepts is introduced, followed by sections on active architectural acoustics—the interaction of electronics and rooms—and predictive design techniques. Finally there is a brief history of the research in this century upon which the foregoing ideas are based and of the open questions that remain.[1–4]

Because the field includes human subjective response to sounds, a book cannot be usefully prescriptive on all points (humans are varied in their preferences), but in spite of the nonlinearity of human responses to different acoustic stimuli at different times, it is possible to see all auditory events as lying on continua subject to the signal processing provided by the binaural auditory tract and the higher functions of the brain. These continua are signal/noise, clarity/unclarity, certainty/uncertainty of localization, and communication/privacy.

2 RELATED PARTS

Other parts of this work are also relevant to problems in architectural acoustics, including the following:

- Part VI (Mechanical Vibration and Shock), particularly Chapter 55 (Vibration Isolation and Damping) and Chapter 59 (Active Vibration Control).
- Part VII (Statistical Methods in Acoustics), notably Chapter 61 (Response Statistics of Rooms).
- Part VIII (Noise Control), especially Chapter 64 (Rating Measures, Descriptors, Criteria, and Procedures for Determining Human Response to Noise).
- Part XII (Psychological Acoustics), particularly Chapters 89 (Auditory Masking), Chapter 90 (Frequency Analysis and Pitch Perception), *Enc.* Ch. 117 (Functions of the Binaural System), and Chapter 91 (Loudness).
- Part XIII (Speech Communication). In particular, Chapter 94 (Introduction) is noteworthy since so many problems in architectural acoustics involve human voice communication as well as speech privacy.
- Part XIV (Music and Musical Acoustics). Chapter 99 (Introduction) is a worthwhile guide to the

Note: References to chapters appearing only in the *Encyclopedia* are preceded by *Enc.*

remaining chapters, 100–103, *Enc.* Ch. 132, 135, 137, 138, 139, which cover practically all musical instruments including the human singing voice as well as electronic music and computer music. Nearly every auditorium or church application in architectural acoustics involves music sources that must be considered in order to design optimum environments for their performance and enjoyment.

- Part XV (Acoustical Measurements and Instrumentation). Since all architectural acoustics applications are based on criteria, guidelines, or standards for the measurement of sound sources and sound transmission paths in buildings, Chapter 104 (Introduction) is particularly relevant.
- Part XVI (Transducers). Chapter 111 (Loudspeaker Design) and *Enc.* Ch. 163 (Public Address and Sound Reinforcement Systems) deal with many of the applications basic to building spaces.

3 ARCHITECTURAL ACOUSTICS: FOREGROUND OR BACKGROUND

The extent to which the architecture of a building is determined by acoustical considerations depends on the program for the building. Buildings affect the acoustic signal that arrives at the ears, both spectrally and temporally and both within spaces and between them, but in most cases the form of the building is determined by factors other than sound. Acoustics and noise control are then background issues to be dealt with in the choice of functional criteria and the engineering of the fabric of the building and its plant to achieve them. Chapters 74–79, *Enc.* Ch. 95, 97 deal with such topics.

It is only when these considerations become the determinants of the built form that the term "architectural" takes its design-centered meaning. Such considerations may take precedence either for noise control, as in the case of a self-protecting, naturally ventilated building designed to protect its opening windows by turning away from freeway noise, or, more often, for the acoustical design of spaces in which communication is important. For such buildings and rooms the acoustical considerations are foreground, in that if they are not correctly included in the design thinking from the start, there may not be a solution available that is not in conflict with the architectural intentions for the space. The chances of achieving excellence in the quality of either the building or the communication within it are not great in this case. "Correct" inclusion implies that the design is based on the realities of the physics of sound and the properties of human hearing and includes quantitative prediction of the sought-for outcome at the design stage of the project.

4 ARCHITECTURAL ACOUSTICS AND COMMUNICATION

The idea of communication underlies architectural acoustics and gives rise to several of its units of measurement. The signal-to-noise ratio, for example, is the fundamental measure for all auditory tasks. A derivative measure, the articulation index (AI), takes the specifics of speech intelligibility, masking, and vocal power to produce a measure as useful for rating privacy as it is for ease of communication with speech. Many of the other measures for the noise environment are single sided in that they rate only difficulty of communication against noise levels but in each case speech communication is the yardstick. Examples of such measures in common use are the noise criteria family of curves (NC, PNC, RC, and NCB, see Chapter 64) and the noise-rating numbers (NR). Inspection of the shape of these curves and the rating basis for partition performance (sound transmission class) reveals the common factor between them in the intelligibility of transmitted speech.

In a fundamental way that is not recognized in these units, concepts that are useful take account of the fact of binaural hearing. It is only in extreme situations for speech communication such as a poor signal-to-noise ratio or of several competing sources that the benefits of a binaural system become apparent. Then humans are able to achieve a high degree of subjective acoustic imaging that enables localization of the source, signal enhancement, and its converse, discrimination against the competing sources. Further, sound from the source (a person or a loudspeaker), reflected to arrive within about 50 ms of the direct sound, is usefully added by human hearing to it, effectively increasing its level and clarity—though the extent to which this is a binaural process clearly depends on the arrival direction of the reflected energy. Several useful measures for the ratio of early to late energy are good predictors of the adequacy of speech communication, and these are all single-channel measures (D50, C80, STI). Placing reflecting surfaces to take advantage of reflections is as fundamental to the design of excellent rooms for speech communication as the more conventional control of echo and reverberation. Strong, late reflections and excessive reverberation are as destructive of clarity in speech communication as the early reflections are useful. The integration function for reflections of speech and the threshold for the occurrence of echo as a function of the level and delay of reflections have been known since the 1960s.

The binaural properties of hearing and its integration function are even more significant in the design of rooms for the performance of the other principal human communication, music. There, however, the variability of the signal and the variety of listener preference require several measures on the sound fields, and there is not yet complete unanimity as to what these should be. The problem arises in that musical style ranges from the cerebral to the emotional and is appreciated differently according to the aptitudes and sensibilities of the audience. Clarity, and with it the precise imaging of the source, which is a universally desired property of a speech communication channel, is undesired on the emotional axis of music. Thus some weakening of localization is preferred, especially with music of the Romantic genre, which currently dominates the Western concert programs. That weakening is achieved by decreasing the early–late ratio and increasing the relative strength of lateral reflections. The former increases reverberance (nondirectional), while the latter increases apparent source width and the sense of envelopment in the sound (weaker localization).

The common ground among the competing music room measures is that strength and reverberance are essential, and these are measured monaurally. Various measures for the early–late ratio are current, also measured with a single channel, among them the early–late energy ratio C and the center time T_c. The lateral reflected sound is measured either by some version of the interaural correlation coefficient (IACC) or directly by the lateral-to-frontal energy ratio in the arriving sound. It now appears that source broadening depends on the early lateral sound while the sense of envelopment is highly correlated with the late (after 80 ms) lateral strength. Other contenders as contributors of excellence are a spectral measure and the delay of the first strong reflections related to the autocorrelation function of the music. The latter means that the rate of change of the music determines the preference for the delay of the reflections after the direct sound.

5 ACOUSTICAL ART FORMS

It is plain, then, that enclosures such as rooms affect the signal transmitted acoustically through them in ways that may be regarded as either positive or negative. Architectural acoustics seeks to address these issues so that the outcomes are positive in terms of the functions for which the room is intended. The most demanding level of acoustical communication occurs in the art forms—theater, poetry, music—and rooms for these make correspondingly high demands for acoustical excellence. The acoustical attributes of enclosure have come in this century to be recognized as potential enhancements of the signal, particularly for music and less obviously also for speech communication. That this is a recent phenomenon is illustrated by the preference studies of the past two decades and the fact that in 1941 F. R. Watson could write that the objective of acoustical design was to make room conditions as similar to outdoors as possible.

6 ACOUSTICAL SCALE

The audible effects of rooms are markedly different in small spaces from those observed in large spaces. The concept of "acoustical scale" is helpful in accounting for these differences. Acoustical scale relates size (of rooms or building elements) to wavelength and so to frequency. It follows that a given room may behave as a small room up to some frequency and above it as a large room. Small-room response to sound is dominated by modal behavior. In large rooms, modes are so numerous that reasonably reliable statements about average energy density, decay rates, and reverberation times and the like can be made. The useful notion of sound "rays" is also available to track early reflected sound from surfaces that are large relative to the wavelength. The transition frequency between the two types of behavior, called the Schroeder frequency, fs, after its proposer M. R. Schroeder, is given by $fs = 2000/TV$, where T is the reverberation time in seconds and V is the room volume in cubic metres. Below this frequency it makes little sense to calculate the reverberation time.

7 ACTIVE ARCHITECTURAL ACOUSTICS

For many years electronic devices have been available to assist the natural acoustical properties of enclosure. These consisted initially of reinforcement and amplification systems with, from the mid-1950s, a variety of artificial reverberation devices. Synthetic "reflections" could be achieved with tape-delay devices but with all the disadvantages inherent in a mechanical tape transport system. The advent of digital signal processing has overcome these problems, and it seems certain that more versatile and sophisticated systems will continue to develop. In parallel with these developments active control of noise using cancellation techniques has become a reality in fields as diverse as automotive design and dentistry, with a considerable application in architectural acoustics. Again one sees a continuum, in this case device based, that extends from the control of noise to the enhancement of communication.

8 PREDICTIVE TECHNIQUES DURING DESIGN

While some quantities may be calculated easily during design,[5] others such as energy fractions in rooms dominated by the early reflection sequence are less accessible. The advent of low-cost digital computers has given rise to numerous programs for these predictions from digitized representations of the space, allocating absorption and diffusion coefficients to the various surfaces and either tracing rays from a source or projecting image sources in the surfaces. Depending on the complexity of the room, measures derived from the impulse response may be obtained by this means during the design stage. The more complex the room, especially if it includes curved surfaces that are difficult to digitize, the greater the uncertainty in the results. A third, hybrid method makes use of the advantages of digital data acquisition in a scale physical model of the room, compensating for the absorption of the real air conditions numerically during the signal processing. The same process is then available at full size during commissioning for measurements on the finished room. Scales commonly chosen for the model range from 1: 50 up to 1: 10, the only advantage of the latter being the availability of scaled music reproduction on a multispeed tape recorder.

Recent developments in auralization may obviate even that advantage.

9 HISTORY OF RESEARCH IN ARCHITECTURAL ACOUSTICS

Development of knowledge in architectural acoustics falls into three phases. Starting with the pioneering work of Wallace Clement Sabine and leading in the late years of the nineteenth century to the well-known relationship $T = kVA$, there was a period of consolidation during which all acoustical design was based solely on the control of reverberation time. Sabine's discovery was so far-reaching and such an advance on the previous state of knowledge that it appeared that all architectural acoustical issues were resolved in it. The subtleties of the perception of reflections as components of reverberation processes did not appear as important factors for nearly 50 years. Indeed, the preoccupation of the Western world with two world wars and the depression of the 1930s provided limited scope for major communications spaces during this period. The second phase, starting after World War II, addressed the significance of the early reflections, to account among other matters for the observed fact that measured reverberation time did not always represent the subjective impression adequately. Further, the notion that the gratifying listening experience in concert halls depends on the temporal details of the early reflections was significantly extended by the discovery that lateral reflections made an important contribution to the sense of spaciousness found in the most preferred halls. That is, both the spatial and temporal properties of the early reflections must be controlled in the design in addition to the reverberation time. The shape of the room, ignored in the Sabine relationship, must be taken into account in acoustical design. The integration time and absolute threshold of perceptibility for the early reflections and the threshold for echo for speech and other signals became the focus of research. Even more significant was the application of factor analysis to rigorously designed preference experiments, the demonstration that preference is multidimensional, and the isolation of four principal subjective factors determining it. Objective correlates for these factors were proposed: reverberation time (RT) and early decay time (EDT) for reverberance, the total level referred to the level of the direct sound at 10 m from the source (G) for "strength," the 80-ms energy fraction (C80) and related measures such as Cremer's center time T_c for clarity, and either the 80-ms lateral fraction (LF) or the IACC for spatial impression. Space here precludes naming the key authors during this period, and the reader is referred to Refs. 6–9 for a comprehensive bibliography.

In this century the third phase of architectural acoustics research has centered on refinements of listener preference criteria.

Open areas of research are changing preference frame and changing art forms. The third phase of research is on-going and will remain so especially for the art forms, as the media themselves change the preference frame within which judgments of excellence are made. One has only to see the way views have changed in 50 years since Floyd Watson's dictum in 1941 to guess that listening habits follow developments in equipment. Perhaps of even greater significance will be the selection of the classical concert repertoire, which has been dominated by the Romantic composers of recent centuries—only a blip in the history of musical communication. Finally the art forms themselves change, again following availability of technical development. The last two decades of the twentieth century have seen the widespread availability of digital computers that lead to entirely new possibilities in what constitutes music, including interaction of the visual and audible. This trend is accelerating without any slackening in the popularity of other forms of auditory entertainment.

In this introductory chapter a rationale for the concerns of architectural acoustics has been sought in the way human binaural hearing deals with sounds. The fact that, on the basis of the sheer weight of practice time, noise control issues[10] in this field may appear more important than communication issues has tended to obscure this rationale. Noise that intrudes, annoys, interferes, and to which humans find difficulty in adaptation is, according to the picture presented here, precisely what is inherent in music. The acoustical properties of enclosure that accentuate the noise enhance the art.

REFERENCES

1. F. V. Hunt, *Origins in Acoustics* (with foreword by R. E. Apfel); American Institute of Physics, Woodbury, NY, 1992, originally published 1978, Yale University Press.
2. J. W. Kopec, *The Sabines of Riverbank*. Peninsular, Woodbury, NY, 1997.
3. T. D. Northwood, Ed., *Architectural Acoustics*, Vol. 10, Benchmark Papers in Acoustics Series, Dowden, Hutchinson and Ross, Stroudsburg, PA, 1977.
4. W. C. Sabine, *Collected Papers on Acoustics*, Peninsular, Woodbury, NY, 1994, originally published 1921 Harvard University Press and 1964 Dover.
5. V. O. Knudsen and C. M. Harris, *Acoustical Designing in Architecture*, American Institute of Physics, Woodbury, NY, 1980, originally published 1950, Wiley, New York.
6. M. Barron, *Auditorium Acoustics and Architectural Design*, E & FN Spon Publisher, an imprint of Chapman & Hall, London and New York, 1993.
7. L. L. Beranek, *Concert and Opera Halls ... How They Sound*, American Institute of Physics, Woodbury, NY, 1996; originally published, *Music, Acoustics and Architecture*, 1962, Wiley, New York.
8. L. Cremer and H. A. Muller, *Principles and Applications of Room Acoustics*, Vols. I and II, Applied Science Publishers, London and New York, 1978 (translated by T. J. Schultz).
9. M. D. Egan, *Architectural Acoustics*, McGraw-Hill, New York, 1988.
10. C. M. Harris, Ed. *Noise Control in Buildings*, McGraw-Hill, New York, 1994.

74

SOUND IN ENCLOSURES

K. HEINRICH KUTTRUFF

1 INTRODUCTION

This chapter deals with the physics of sound propagation in enclosures, that is, in spaces filled with a homogeneous fluid—for instance, with air—that are completely surrounded by a passive boundary.

We restrict the discussion to enclosures with dimensions larger than all sound wavelengths of interest. Although most results may be applied to small spaces such as living rooms or the interior of passenger cars, our main interest is the acoustical behavior of large rooms such as performance halls or noise propagation in all kinds of working environments. In such cases, the sound field is made up of many components whether we try to describe it in terms of normal modes or of sound rays or particles.

This chapter presents three different approaches to describe sound fields. The first one involves the acoustic wave equation and solutions of it. This procedure yields the physically correct theory of sound in enclosures. Although it is not particularly well suited for solving practical problems, it is the basis of a more than superficial understanding of sound fields in enclosures. A more illustrative description is obtained by restricting the discussion to the propagation and distribution of sound energy. This is effected by introducing sound rays and image sources that yield a simple method of computing the distribution of sound energy.

For many practical purposes it is reasonable to compute the sound field in an enclosure only to that degree of accuracy needed to evaluate certain characteristic parameters of it, such as the stationary energy density and the reverberation time. Closed expressions are obtained by considering energetic averages rather than the energies of individual sound rays or sound particles.

2 WAVE ACOUSTICS

2.1 Normal Modes and Eigenfrequencies

The basis of a physically exact description of sound propagation in an enclosure is the acoustic wave equation. If dissipation in the medium is neglected and if harmonic wave motions with angular frequency ω are considered, the wave equation transforms into the Helmholtz equation

$$\nabla^2 p + k^2 p = 0. \tag{1}$$

In this differential equation p is the complex sound pressure amplitude and $k = \omega/c$ denotes the (angular) wavenumber; c is the sound velocity.

The effect of the boundary on the sound field can be accounted for by imposing certain boundary conditions on the possible solutions of Eq. (1). If the boundary is locally reacting to the sound field, these conditions can be expressed in terms of its complex wall impedance Z, which is usually a function of the sound frequency ω and the location:

$$\omega\rho_0 p + iZ\frac{\partial p}{\partial n} = 0 \quad \text{along the boundary.} \tag{2}$$

Here, $\partial p/\partial n$ is the component of the sound pressure gra-

Note: References to chapters appearing only in the *Encyclopedia* are preceded by *Enc.*

dient in the direction of the outward-pointing wall normal and ρ_0 is the density of the medium.

The Helmholtz equation (1) supplemented by the boundary conditions [e.g., as expressed by Eq. (2)] can only be solved for discrete values k_n of the wavenumber, the so-called eigenvalues. (The subscript stands for three integers, e.g., n_x, n_y, n_z.). They are related to *allowed frequencies*, or *eigenfrequencies*, $f_n = ck_n/2\pi$, or when expressed as angular frequencies, $\omega_n = ck_n$. Each eigenvalue is associated with an *eigenfunction*, or *characteristic function*, $p_n(\mathbf{r})$, the vector $\mathbf{r}$ symbolizing the space coordinates. (Strictly speaking, more than one eigenfunction may belong to one eigenvalue. However, this case, known as *degeneracy*, is of no importance in practical situations.) Another common expression for an eigenfunction is *normal mode*. It can be thought of as a three-dimensional standing wave the nodes of which are situated on nodal surfaces. Of course, the distinctness of these standing waves depends on the reflective properties of the boundary, in an anechoic room, for instance, there will be mainly free-wave propagation superimposed by faint standing waves.

If the wall impedance is complex, which is usually the case, the eigenvalues are also complex: $k_n = (\omega_n + i\delta_n)/c$. Accordingly, the time dependence of the sound pressure is $\exp(i\omega_n t - \delta_n t)$, indicating that the imaginary parts can be interpreted as damping constants. In this case the standing waves are not persistent but will die out unless a sound source will continuously support the energy absorbed by the boundary. This is the basis of sound decay or reverberation, which will be dealt with later in more detail.

If the boundary of the enclosure is not locally reacting, that is, if waves propagating in or behind the walls can interact with the sound field inside the considered enclosure, the boundary conditions cannot be formulated by such a general expression as Eq. (2). In this case, which has been considered by Pan and Bies,[1] the normal modes of the enclosure may be coupled to modes within or outside the boundary.

Once the normal modes of an enclosure are known, its acoustical response to any excitation by a sound source can be formally expressed by them. However, closed expressions for the eigenfunctions and eigenfrequencies can be worked out only for enclosures with simple geometry and boundary conditions (see below). In most practical cases where the shape of the room and the acoustical properties of its walls are complex, it may turn out quite difficult to determine even a moderate number of normal modes. In general, one has to resort to numerical methods such as the finite-element method, and even then only small enclosures and/or the low-frequency range where the density of eigenfrequencies is relatively small can be treated in this way. For large rooms a geometric analysis based on sound rays may be more useful (see Section 3).

2.2 Example: Normal Modes of a Rectangular Enclosure

Let us consider a rectangular room bounded by three pairs of parallel planes that are perpendicular to each other and to the axis of a Cartesian coordinate system (see Fig. 1) the origin of which coincides with one of the room corners. The side lengths of this box are L_x, L_y, and L_z.

Then Eq. (1) reads

$$\frac{\partial^2 p}{\partial x^2} + \frac{\partial^2 p}{\partial y^2} + \frac{\partial^2 p}{\partial z^2} + k^2 p = 0. \tag{3}$$

Furthermore, we assume that the walls are rigid, that is,

$$\frac{\partial p}{\partial x} = 0 \quad \text{for } x = 0,\ x = L_x. \tag{4}$$

Two similar boundary conditions hold for the y- and z-directions.

The solutions of Eq. (3) satisfying these conditions are given by

$$p_{n_x n_y n_z}(x, y, z) = A \cos\left(\frac{n_x \pi x}{L_x}\right) \cos\left(\frac{n_y \pi y}{L_y}\right) \cdot \cos\left(\frac{n_z \pi z}{L_z}\right) \tag{5}$$

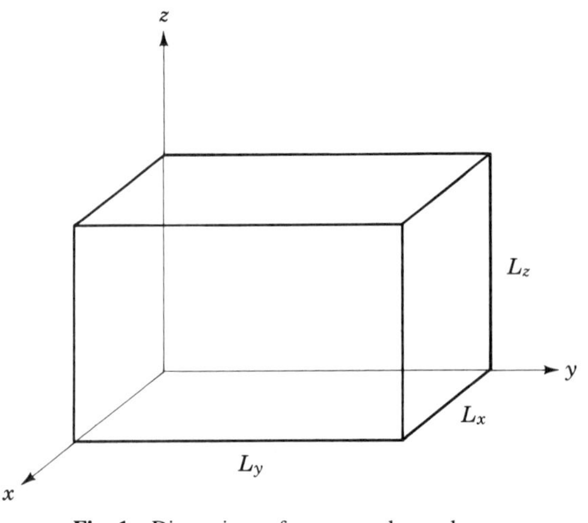

Fig. 1 Dimensions of a rectangular enclosure.

with an arbitrary constant A; the eigenfrequencies are

$$f_{n_x n_y n_z} = \frac{c}{2}\left[\left(\frac{n_x}{L_x}\right)^2 + \left(\frac{n_y}{L_y}\right)^2 + \left(\frac{n_z}{L_z}\right)^2\right]^{1/2} \tag{6}$$

with n_x, n_y, and n_z denoting integers including zero. The normal mode is an *axial* one if two of these indices vanish; it is formed by successive sound reflections between one pair of parallel walls. If two pairs of walls are involved, we speak of a *tangential* mode, for which one of the three indices is zero. All remaining modes are *oblique modes*.

Figure 2 shows the curves of constant sound pressure at a given moment in the plane $z = 0$ for $n_x = 3$, $n_y = 3$, and arbitrary n_z. Within adjacent rectangles the pressure has opposite signs at the moment considered; half a period later all signs will have changed. Along the dotted lines the sound pressure is zero for any time. These lines are the intersection of the plane $z = 0$ with two systems of equidistant nodal planes; a third system is parallel to $z = 0$. The numbers of nodal planes perpendicular to the x-, y-, and z-axis are given by the indices n_x, n_y, and n_z, respectively.

The number of eigenfrequencies within the frequency range from 0 to an upper limit f can be shown to be given approximately by

$$N_f \approx \frac{4\pi}{3} V\left(\frac{f}{c}\right)^3 + \frac{\pi}{4} S\left(\frac{f}{c}\right)^2 + \frac{L}{8}\frac{f}{c}. \tag{7}$$

In this expression V is the volume of the room and S is the area of its boundary; $L = 4(L_x + L_y + L_z)$ is the sum of the lengths of all edges of the room. It is noteworthy that in the limit of very high frequencies the first term of this formula is valid for any enclosure, that is, not only for rectangular enclosures with rigid walls. The same holds for the average spacing of adjacent eigenfrequencies, which is obtained by differentiating the first term of Eq. (7) with respect to the angular frequency:

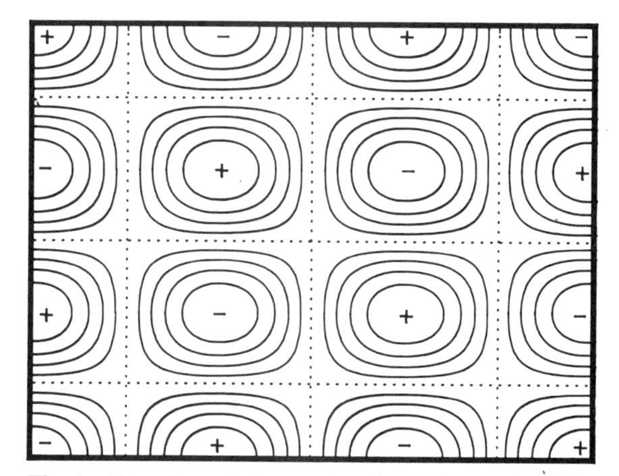

Fig. 2 Normal mode in a rectangular enclosure (see Fig. 1): curves of equal sound pressure amplitude in a plane perpendicular to the z-axis ($n_x = 3, n_y = 3$). (From H. Kuttruff, *Room Acoustics*, Chapman & Hall, London, 1991.)

$$\langle \delta f \rangle \approx \frac{c^3}{4\pi V f^2}. \tag{8}$$

Equation (7) shows that in larger rooms the number of eigenfrequencies needed to cover the audio frequency range may be quite large. This is another reason why the eigenfunction concept is of little use for solving problems in practical room acoustics.

2.3 Steady-State Sound Field

If all eigenvalues $k_n = (\omega_n + i\delta_n)/c$ and the associated eigenfunctions $p_n(\mathbf{r})$ are known for a given enclosure, the complex sound pressure amplitude in any inner point $\mathbf{r}$ can be expressed in terms of them. Let us suppose an enclosure the normal modes of which have been determined in some way and that is excited by a point source at a location $\mathbf{r}'$ operating at an angular frequency ω. Then, under the assumption $\omega_n \ll i\delta_n$, the complex sound pressure amplitude in a point $\mathbf{r}$ is obtained as

$$p_n(\mathbf{r}) = C\sum_{n=0}^{\infty} \frac{p_n(\mathbf{r})p(\mathbf{r}')}{K_n(\omega^2 - \omega_n^2 - 2i\delta_n\omega_n)}, \tag{9}$$

where the constant C contains the strength of the sound source; K_n is a normalization constant: For rectangular rooms it is 1/2 for axial modes, 1/4 for tangential modes, and 1/8 for oblique modes. Each term of the sum in Eq. (9) reaches its maximum when the driving frequency ω is close to the corresponding angular eigenfrequency ω_n, that is, if it represents a resonance of the enclosure.

Concerning the frequency dependence of the pressure amplitude, two limiting cases have to be distinguished: Suppose the average spacing of eigenfrequencies after Eq. (8) is substantially smaller than the average resonance half-width $\langle \Delta f \rangle = \langle \delta \rangle / 2\pi$, with $\langle \delta \rangle$ denoting the average of the damping constants δ_n. Then the transfer function of the enclosure consists of an irregular succession of well-separated resonance curves. In the opposite case, which is typical for large rooms, several or even many eigenfrequencies are situated within the average half-width. Then the corresponding resonance curves show considerable overlap. Hence several or many terms

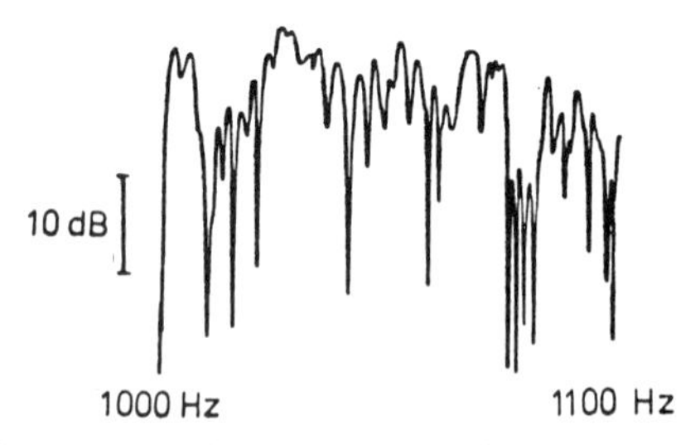

Fig. 3 Measured frequency curve of an enclosure in the range of 1000–1100 Hz. (From H. Kuttruff, *Room Acoustics*, Chapman & Hall, London, 1991.)

of the sum in Eq. (9) contribute to the total sound pressure even if the room is excited with a sine signal. Since these contributions are superimposed with virtually random phases, the sound pressure can be considered as randomly distributed. The condition for this case is[2]

$$f > \frac{5000}{(V\langle\delta\rangle)^{1/2}} \approx 2000\sqrt{\frac{T}{L}}, \qquad (10)$$

where T is the so-called reverberation time of the enclosure, which will be defined in the next section. As an example, Fig. 3 represents the steady-state sound pressure level measured in a particular room as a function of the driving frequency; the recording covers only a relatively narrow frequency interval. The irregular shape of this curve is typical for all such *frequency curves* of an enclosure, no matter where the sound source or the receiving point is located, and their general appearance is the same even for virtually all enclosures.[3] This means that no information on acoustical properties or peculiarities of a performance hall can be derived from the statistical quantities of such recordings.

The following facts hold under the assumption that the distance between the sound source and the observation point is much larger than the so-called reverberation distance [see Eq. (33)]. Then the magnitudes $|p|$ of the sound pressures encountered in any large enclosure follow a Rayleigh distribution: Let q be the magnitude $|p|$ divided by its average; then the probability that this quantity lies within the interval from q to $q+dq$ is given by

$$P(q)\,dq = \frac{\pi}{2}\exp\left(-\frac{\pi}{2}q^2\right) q\,dq \qquad (11)$$

(see Fig. 4). From this distribution density it can be concluded that the sound pressure level fluctuates within a range of about 10 dB on the average. Furthermore, the mean spacing between the maxima of a frequency curve is approximately[2]

$$\langle\delta f_{\max}\rangle \approx \frac{\langle\delta_n\rangle}{\sqrt{3}} \approx \frac{4}{T}. \qquad (12)$$

As before, T is the reverberation time. When the driving frequency is continuously altered, not only the amplitude of the sound pressure in the receiving point $\mathbf{r}$ but also its phase varies in an irregular way. The average phase variation per hertz is[4]

$$\left\langle \frac{d\Phi}{df} \right\rangle \approx \frac{\pi}{\langle\delta\rangle} \approx \frac{T}{2.2}. \qquad (13)$$

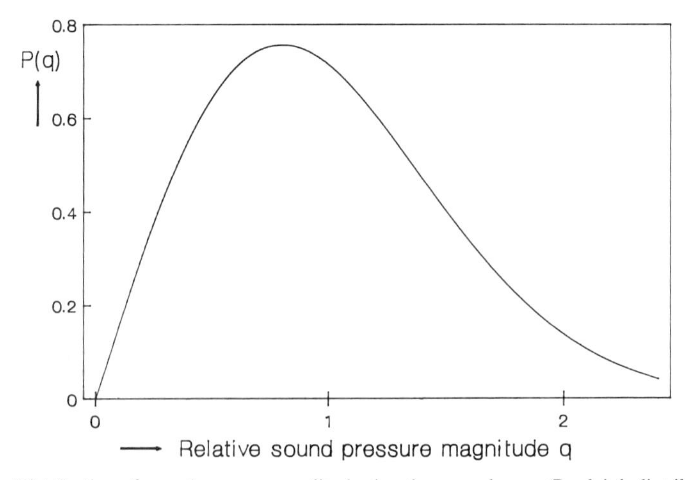

Fig. 4 Distribution of sound pressure amplitudes in a large enclosure (Rayleigh distribution).

If not the driving frequency but the location **r** of the receiving point is continuously altered while the frequency is kept constant, similar fluctuations of the sound pressure amplitudes occur. Since the samples of the sound pressure are taken from the same ensemble as before when the frequency was varied, the distribution of the sound pressure magnitudes is again given by Eq. (11).

2.4 Transient Response: Reverberation

As noted in Section 2.1, any normal mode of an enclosure with complex wall impedance will decay unless a sound source compensates for the wall losses. Therefore, if the room is excited by some sound source that is abruptly stopped at time $t = 0$, all excited normal modes will die out with their individual damping constants δ_n. Accordingly, if we assume $\delta_n \ll \omega_n$, the sound pressure in any point can be expressed by

$$h(t) = \sum_{n=0}^{\infty} a_n e^{-\delta_n t} e^{i\omega_n t} \quad \text{for } t \geq 0. \tag{14}$$

The complex coefficients a_n contain the eigenfunctions p_n and hence the spatial dependence of the sound pressure. Furthermore, they depend on the location of the sound source and the signal is emitted in $t < 0$. Equation (14) describes what is called *reverberation* in room acoustics.

Very often the damping constants δ_n do not differ too much from each other and may be therefore replaced without much error by their mean value $\langle\delta\rangle$. Then the energy density in a certain point of the sound field will decay at a uniform rate:

$$u(t) = u_0 e^{-2\langle\delta\rangle t} \quad \text{for } t \geq 0. \tag{15}$$

In room acoustics, the decay rate is usually characterized not by the mean damping constant but by the *reverberation time*, or the *decay time*, T, defined as the time interval (see Fig. 5) in which the sound energy or energy density drops to one millionth of its initial value. This is tantamount to the requirement that the sound pressure level falls down by 60 dB. From Eq. (15) it follows that

$$T = \frac{3 \ln 10}{\langle\delta\rangle} = \frac{6.91}{\langle\delta\rangle}. \tag{16}$$

The reverberation time is one of the most important quantities in room acoustics. It can be measured with good accuracy, and the available formulas predict it with reasonable accuracy. More will be said about this matter in Section 4.3.

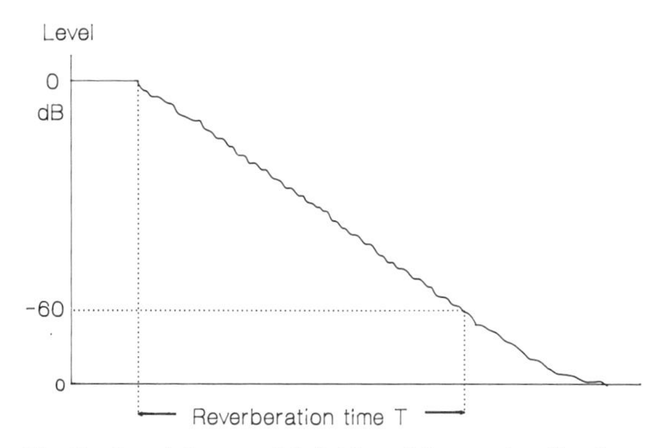

Fig. 5 Sound decay and definition of the reverberation time.

3 GEOMETRIC ACOUSTICS

Although the formal solution of the wave equation as outlined in the preceding section yields many interesting and important general results, its practical application turns out to be too complicated, except for small enclosures and low frequencies.

We arrive at a much simpler, although less rigorous, description if we restrict the discussion to mere energy propagation. This means that all effects involving phase differences such as interference or refraction are neglected. This is admissible if the sound signals of interest are not sinusoids or other signals with small-frequency bandwidth but are composed of many spectral components covering a wide frequency range. Then it can be assumed that constructive or destructive phase effects cancel each other when two or more sound field components are superimposed at a point, and the total energy in the considered point is simply obtained by adding their energies. Components with this property are often referred to as *mutually incoherent*.

Now for the discussion of propagation of sound energy, the wave concept would lead to unnecessary complications. Instead, it is more convenient and illustrative to employ the concept of sound rays or bundles of sound rays. A sound ray can be thought of as a small sector that has been cut out of a spherical wave and subtends a vanishingly small solid angle. These sound rays, which are straight lines in a homogeneous medium, are the paths along which the sound energy travels with constant velocity, and from their definition it follows that this energy does not depend on the distance it has traveled provided the medium is free of dissipation. The intensity within a sound ray, however, decreases proportionally to l/r^2, where r denotes the distance from the sound source. In this picture, uniform radiation of sound from a point source is represented as a diverging bundle of rays originating from it.

Furthermore, we assume that all sound-reflecting objects, in particular all walls of a room, are large compared to the acoustic wavelength. Then the reflection of sound obeys the same laws as those known from optics; that is, if a reflecting surface is smooth, the usual reflection law can be applied.

3.1 Sound Reflection: Image Sources

When a sound ray falls upon a plane and smooth wall of infinite extension, it will be at least partially reflected from it. The law of specular reflection states that the angle under which the reflected ray leaves the wall equals the angle of incidence where both angles are measured against the wall normal (see Fig. 6*a*) and that the incident ray, the reflected ray, and the wall normal are lying in a plane. This law can also be applied to walls of finite size if its dimensions are much larger than wavelengths of all spectral components involved. Consequently, any diffraction effects from, for example, wall edges are neglected in this "high-frequency" limit of sound propagation.

If the incident ray is emitted by a point source, the reflected ray can be thought of as originating from a virtual sound source that is the mirror image of the original source with respect to the reflecting wall (see Fig. 6*b*). This secondary source, which is assumed to emit the same signal as the original one, will be referred to as the *image source*. It is particularly useful for constructing the reflection of a ray bundle from a plane wall portion or for connecting a given sound source and receiving point by a path that includes wall reflections. Its full potential is developed, however, in the discussion of sound propagation in enclosures.

The fact that only a fraction of the incident sound energy will be reflected from the wall can be accounted for by its absorption coefficient α, that is, the fraction of the incident sound energy absorbed by the wall. Accordingly, the reflection process reduces the energy of the sound ray by the factor $1 - \alpha$. This is tantamount to operating the image source at a power reduced by this factor. In this way also frequency-dependent absorption coefficients can be taken into regard; the dependence of the absorption coefficient on the direction of sound incidence could be accounted for by modifying the directivity of the image source. If the sound source itself has a certain directivity, the same but symmetrically inverted directivity pattern must be assigned to the image source.

The law of specular reflection is valid for walls that are plane and smooth. It can also be applied to curved walls, provided their radius of curvature is very large compared to the wavelength. However, the concept of image sources fails in this case. Instead, the reflection of each ray has to be treated separately by constructing the reflected ray path with respect to the tangential plane in the point where the incident ray hits the wall.

If there are two or more plane reflecting walls, an image source can be attributed to each of them. Multiple reflections can be accounted for by higher order image sources that are constructed by mirroring the previously obtained images. This is illustrated by Fig. 7, which depicts an edge formed by two adjacent plane walls. In addition to the two first-order image sources A_1' and A_2', there are two images of second order, A_1'' and A_2''. Viewed from the receiving point, an image source may turn out to be "invisible," as is the case for A_2''. This happens whenever a ray intersects the plane of a wall outside its physical boundary.

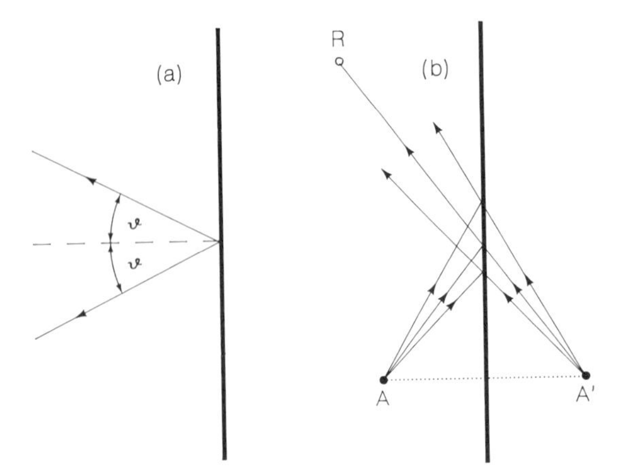

Fig. 6 Specular sound reflection from a plane wall for (*a*) one ray and (*b*) several rays: *A*, original sound source; *A′*, image source.

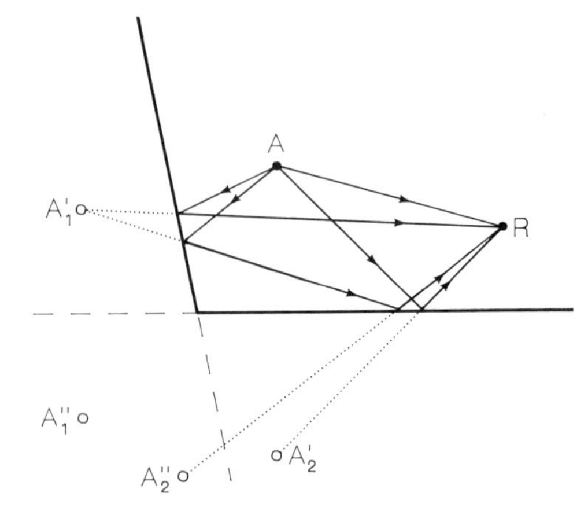

Fig. 7 Sound reflections from an edge formed by two adjacent walls: *A*, original sound source; *A′*, first-order image sources; *A″*, second-order image sources.

For a space that is completely enclosed by plane walls, this process is repeated over and over, leading to images of increasing order. If the enclosure is made up of N plane walls, the total number of image sources of up to order m is

$$n_m = N \frac{(N-1)^m - 1}{N-2}. \tag{17}$$

These images form a three-dimensional pattern that depends on the geometry of the enclosure. However, most of these image sources are invisible from the given observation point. Therefore, it has to be checked which of them will really contribute to the total energy in a given point and which do not.[5,6] As before, the absorption coefficients of the walls are taken into account by attributing properly reduced powers to the image sources. Then, for discussing the energy distribution within an enclosure with plane and smooth walls, the boundary itself is no longer needed, and the energy density (or the sound pressure) in a certain point is obtained just by adding the contributions of all visible image sources provided they are not too faint. The effect of air attenuation is easily included if desired.

In this way, not only the steady-state energy density but also transient processes can be calculated. Suppose the original sound source would produce the energy density $u(t)$ in some point of an unbounded space. Under the assumption of frequency-independent absorption, each image source produces a weaker replica of this energy signal that will be received with some delay τ_n, depending on its distance. Thus the received energy signal is

$$u'(t) = \sum_n a_n u(t - \tau_n). \tag{18}$$

If the signal emitted by the sound source is a very short impulse represented by a Dirac function $\delta(t)$, the result of this superposition is the energetic *impulse response* of the enclosure with respect to a particular source–receiver arrangement. A schematic example is shown in Fig. 8. Each reflection is represented by a vertical line the length of which indicates its relative energy, whereas the abscissa is its delay time τ_n with respect to the direct sound, which is represented by the first line. Measured impulse responses differ from this simple scheme since the acoustical properties of real walls usually depend on the sound frequency.

It should be mentioned that the image source model is not limited to the treatment of energy propagation but can be used as well to obtain the sound pressure in a receiving point for any source signal, at least in the limit of high frequencies.

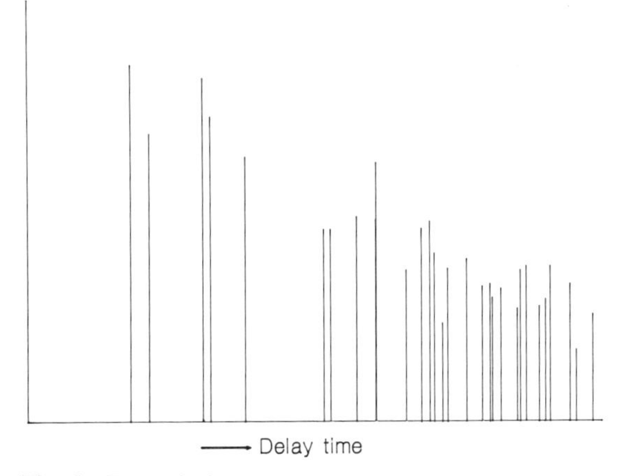

Fig. 8 Energetic impulse response of an enclosure (schematically).

3.2 Two Examples: Flat Room and Rectangular Room

For certain enclosures with simple geometry, all image sources are visible and are arranged in a regular pattern. Accordingly, the total sound energy density in a given point can be represented by a closed formula.

As a first example, we consider the space bounded by two parallel infinite planes with distance a, as depicted in Fig. 9. This enclosure can be considered as the model of many factory spaces or open-plan bureaus the height of which is small compared to their lateral dimensions. Since most points are far from the side walls, the contributions of the latter can be neglected. The image sources

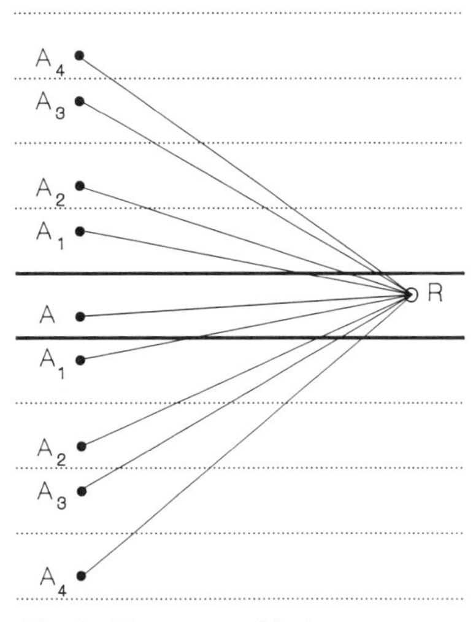

Fig. 9 Flat room and its image sources.

are regularly arranged along a vertical line; if the primary source lies exactly in the middle between the floor and the ceiling, they are equidistant. They can be used to calculate the sound pressure level in flat rooms (see *Enc.* Ch. 95).

If perpendicular side walls are added that are also perpendicular to each other, we arrive at the rectangular room already considered in Section 2.2. Figure 10 represents one plane of the pattern of image rooms and image sources. Of course, this lattice has to be completed in the third dimension; that is, there are infinitely many patterns of this kind one upon the other. Each image of the room contains exactly one image source, and each of them is visible from all positions within the enclosure. A general property of the energy impulse response can be easily derived from this picture: All contributions arriving in the time interval between t and $t + dt$ are due to image sources located in a spherical shell with radius $R = ct$. Since the volume of a thin shell increases with the square of its radius, the temporal density of reflected components in a diagram similar to Fig. 8 grows proportionally with the square of the time delay t. This increase is balanced by the inverse-square law of spherical spreading, and the influence of wall absorption (and attenuation in the medium) leads to an exponential decrease of the energy received per second.

3.3 Enclosures with Diffusely Reflecting Walls

Very often some or all boundaries of an enclosure have a structured surface with irregularities the dimensions of which are not very small compared with the acoustic wavelength. Then at least a fraction of the incident sound energy will not be reflected specularly but will be scattered into many different directions. This process is called *diffuse reflection*. In the limiting case all the energy is diffusely reflected and no specular component will occur at all. In this case it is often convenient to apply Lambert's law, according to which the scattered intensity is proportional to the cosine of the scattering angle, and accordingly the polar diagram of the scattered intensity (see Fig. 11) is a circle (or rather a sphere). This is expressed by the following formula, in which B is the *irradiation strength* of a wall element dS, that is, the energy arriving at the wall per second and unit area. Then the scattered intensity at a point P a distance r from the wall element is

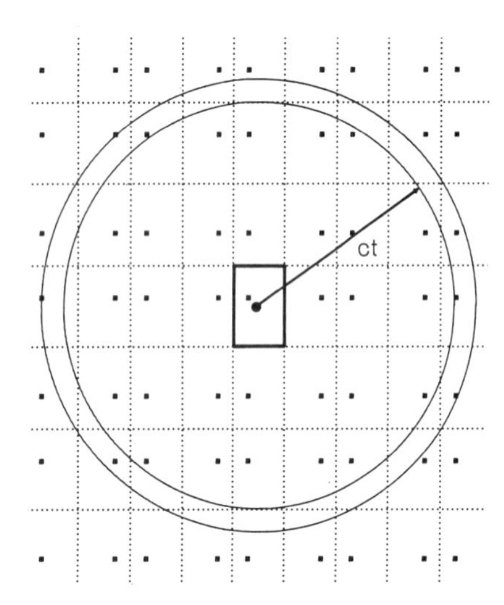

Fig. 10 Rectangular room and its image sources (one plane).

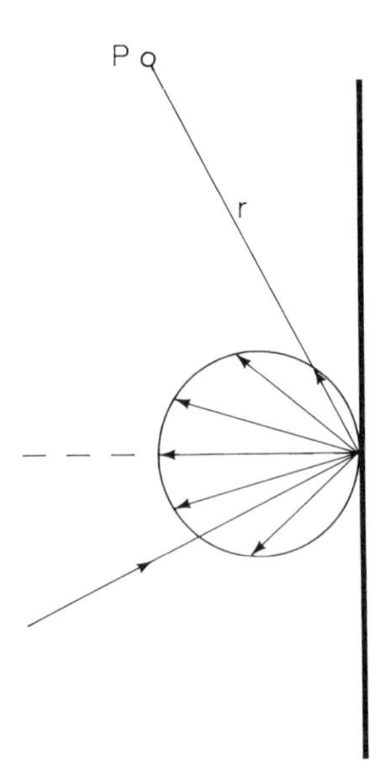

Fig. 11 Diffuse reflection according to Lambert's law.

$$dI(r, \vartheta) = B(1 - \alpha)\, dS\, \frac{\cos\vartheta}{\pi r^2}. \tag{19}$$

As before, α denotes the absorption coefficient while ϑ is the scattering angle.

Neither the concept of sound rays nor that of image sources can be applied to diffusely reflecting boundaries because each sound ray will be split up into infinitely many secondary rays when it hits the boundary. The scattering leads to strong mixing of the sound energy, and as a result, the distribution of sound energy will be much more uniform than with specularly reflecting walls.

Let us now suppose that Lambert's law according to Eq. (19) can be applied to all wall reflections. Then the following integral equation for the *irradiation strength* B

can be set up[7,8]:

$$B(\mathbf{r}, t) = \iint_S K(\mathbf{r}, \mathbf{r}')\rho(\mathbf{r}')B\left(\mathbf{r}', t - \frac{R}{c}\right) dS' + B_0(\mathbf{r}, t). \quad (20)$$

In this equation $\mathbf{r}$ and $\mathbf{r}'$ denote the coordinates of the boundary elements dS and dS', respectively (see Fig. 12), and R is their distance. The term $B_0(\mathbf{r})$ is the "direct" irradiation strength produced by a sound source, and $\rho = 1 - \alpha$ denotes the *reflection coefficient* of the wall at $\mathbf{r}$. The kernel $K(\mathbf{r}, \mathbf{r}')$ of the integral equation contains Lambert's law and also the inverse-square law of sound propagation. It is given by

$$K(\mathbf{r}, \mathbf{r}') = \frac{\cos\vartheta \cos\vartheta'}{\pi R^2}, \quad (21)$$

where ϑ and ϑ' are the angles between the boundary in dS and dS' and the straight line connecting them. The double integral in Eq. (20) is extended over the whole boundary S.

Equation (20) means simply that the sound energy arriving at a certain wall element dS at time t is the sum of the contributions that have been reemitted previously (at time $t - R/c$) by all other wall elements dS'. The effects of sound attenuation in the medium can be included by an additional factor $\exp(-mR)$ on the right side of Eq. (21), where m denotes the intensity-related attenuation constant.

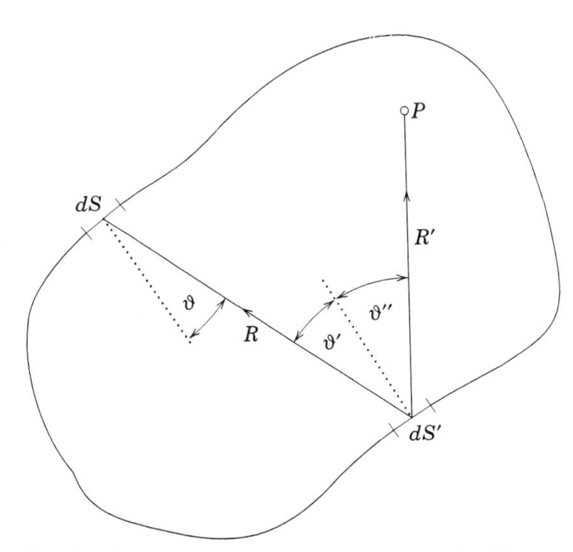

Fig. 12 Sound propagation in an enclosure with diffusely reflecting boundary: dS, receiving wall element; dS', emitting wall element; P, interior point.

If the sound power produced by the sound source is constant in time, neither B_0 nor B will depend on time; hence the steady-state version of Eq. (20) reads

$$B(\mathbf{r}) = \iint_S K(\mathbf{r}, \mathbf{r}')\rho(\mathbf{r}')B(\mathbf{r}')\, dS' + B_0(\mathbf{r}). \quad (22)$$

On the other hand, for $B_0 = 0$, Eq. (20) describes the sound decay in the enclosure. If the irradiation strength B on the boundary is known, the energy density in any point $\mathbf{r}''$ within the enclosure can be calculated[9]:

$$u(\mathbf{r}'', t) = \frac{1}{\pi c} \iint_S \rho(\mathbf{r}')B\left(\mathbf{r}', t - \frac{R'}{c}\right) \frac{\cos\vartheta''}{R'^2}\, dS' + u_0(\mathbf{r}'', t), \quad (23)$$

where ϑ'' denotes the angle between the wall normal and the line with length $R' = |\mathbf{r}'' - \mathbf{r}'|$ that connects the wall element dS' with the inner point. The function u_0 is the energy density due to the direct sound. For steady-state conditions, the time dependence of B is just neglected. A similar expression holds for the sound intensity.

Generally, Eq. (20) or (22) must be numerically solved, which is not very difficult since standard procedures are available. The steady-state case has been discussed by Carroll and Miles,[10] who presented a general, yet formal solution to Eq. (22). Numerical calculations for the rectangular room have been carried out by Miles,[11] who treated the sound distribution under steady-state conditions as well as in decaying sound fields. Closed solutions can be worked out only for simple room shapes such as spherical enclosures.[8] Another example for which a closed solution can be obtained, at least for stationary sound propagation, is for the infinite flat room[12] already considered in Section 3.2. But now both walls or one of them is assumed as diffusely reflecting, which yields a more realistic model of a real hall than that with smooth walls since in factories or bureaus there are machines, piles of materials, furniture, and so on, which cause noticeable scattering of sound energy. For this flat room, the kernel $K(\mathbf{r}, \mathbf{r}')$ in Eq. (22) is just $a^2/\pi R^4$. More about the solution may be found in *Enc.* Ch. 95.

4 STATISTICAL ACOUSTICS

In this section closed formulas for the energy density in an enclosure both at steady-state conditions and during sound decay are presented. Such expressions are of considerable importance in practical room acoustics, for example, to predict noise levels in working environments or to assess the suitability of a room for certain types of performances.

As in the preceding section, only energy propagation will be discussed here; that is, no phase relations between sound waves will be taken into account. Therefore the sound field components we consider are sound rays; alternatively, even the somewhat hypothetical notion of "sound particles" will be used. (These should not be identified with "phonons"; their physical meaning is rather that of short sound impulses or bursts with wide bandwidth traveling along a sound ray.) However, in contrast to the discussions of the preceding section, energy densities are not obtained by adding the energies of individual sound rays or sound particles but by considering their "average fate."

Most of the laws to be discussed in this section are valid under the assumption of a homogeneous or diffuse sound field, a condition that is only incompletely fulfilled in practical situations. Therefore, the formulas and laws derived under this assumption cannot be expected to be quite accurate. In fact, deviations from complete diffusivity influence the reliability of the expressions differently, depending on whether we regard steady-state parameters or reverberation time; in the latter case the requirements concerning the diffusion or diffusivity of the sound field are less stringent.

Because of the central importance of the diffuse sound field, in the following some basic properties of diffuse sound fields will be presented.

4.1 Basic Properties of Diffuse Sound Field

We speak of a perfectly diffuse sound field if the sound energy arriving at any point inside the enclosure is uniformly distributed over all possible directions of incidence. In this case the intensity vector in the field is zero everywhere. It is clear that this is an ideal condition that cannot be exactly fulfilled because in a real enclosure there are losses in the medium and at the boundary therefore at least some energy transport from the sound source into the medium and toward the walls must occur.

It is easy to show that, under the assumption of a diffuse field, the energy density is locally constant. Furthermore, each point of the boundary is hit by the same amount of energy per second and unit area, and if we call the latter quantity irradiation strength and denote it by B as in Section 3.3, it can be shown that

$$B = \frac{c}{4}\, u. \tag{24}$$

From this relation the average number n of boundary reflections that a sound ray undergoes per second can be obtained:

$$\overline{n} = \frac{cS}{4V}, \tag{25}$$

where V is, as before, the volume of the enclosure and S is the area of its boundary. Then, the average length of a sound ray between two reflections from the boundary, or the mean free path length of a sound particle, is

$$\bar{l} = \frac{4V}{S}. \tag{26}$$

At each boundary point the sound energy arrives uniformly from all directions of the half space. Therefore the effective absorption coefficient α_d is obtained by suitably averaging the angle-dependent absorption coefficient $\alpha(\vartheta)$:

$$\alpha_d = \int_0^{\pi/2} \alpha(\vartheta)\, \sin(2\vartheta)\, d\vartheta. \tag{27}$$

This relation is known as *Paris's formula*. For locally reacting boundaries, that is, if the wall impedance Z is independent of the direction of sound incidence, the result of this averaging can be expressed in a closed formula. Its content is shown in Fig. 13, which represents the contours of constant absorption coefficient α_d in the plane formed by the magnitude and the phase angle of the specific wall impedance ϑ. It is noteworthy that α_d will never reach unity but has an absolute maximum 0.951 for the real impedance $\zeta = 1.567$.

Finally a remark may be in place on the conditions under which a diffuse sound field can be expected. The degree of sound field diffusion depends not only on the shape of the enclosure but also on the surface structure of its boundary and the magnitude and the spatial distribution of the wall absorption. In an enclosure with the shape of an irregular polyhedron the sound rays will be mixed up to some extent and thus establish a sound field with at least some diffusion. On the contrary, we can hardly expect a diffuse sound field within a sphere with a specularly reflecting boundary. The situation will be changed, however, if the sphere has a diffusely reflecting boundary, which will equalize the energy distribution. Generally, diffuse wall reflections increase the diffusion of the sound field in the enclosure. (A similar effect will have sound-scattering objects within the enclosure.) But even then, the absorption distribution along the walls influ-

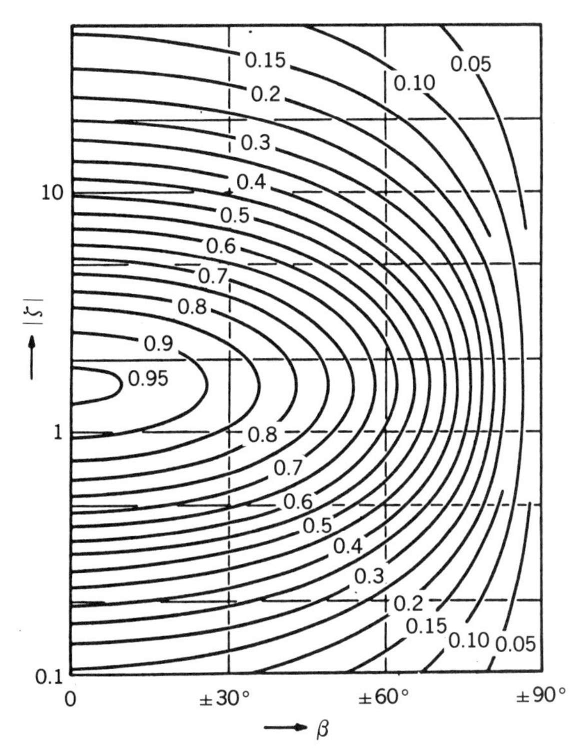

Fig. 13 Curves of equal absorption coefficient of a locally reacting wall for random sound incidence. The abscissa and the ordinate are the phase angle and the magnitude, respectively, of the specific wall impedance $\zeta = Z/\rho_0 c$.

ences the degree of diffusion. Therefore, the condition of diffuse reflections must be distinguished from that of a diffuse sound field.

4.2 Stationary Energy Density

We imagine an enclosure in which a sound source with arbitrary directivity emits the sound power P. It is assumed that the produced sound field is a diffuse one. Therefore, all absorption coefficients are those for random incidence, averaged according to Eq. (27). (For the sake of simplicity, the index d is omitted in the following.) Now suppose that the boundary can be subdivided into portions with areas S_i and with absorption coefficients α_i. Each of these portions absorbs the energy $\alpha_i S_i B = \alpha_i S_i cu/4$ per second. Under these assumptions, the temporal change of the energy density within the enclosure is

$$\frac{du}{dt} = \frac{1}{V}\left[P(t) - \frac{c}{4} Au\right]. \tag{28}$$

In this expression the quantity

$$A = \sum_i \alpha_i S_i \tag{29}$$

has been introduced, the so-called equivalent absorption area or total absorption of the enclosure. It has the dimension of square metres and can be imagined as the area of a totally absorbing boundary portion, for instance, an open window, in which the total absorption of the enclosure is concentrated. Absorption coefficients of typical wall materials may be found in Chapter 75.

Since we discuss the steady-state energy density u_s in this section, we assume constant power output P; accordingly the time derivative of u will vanish. Then Eq. (28) reduces to

$$u_s = \frac{4P}{cA}. \tag{30}$$

This relation is the basis of an important method to determine experimentally the total power output P of a sound source by measuring the energy density. For this purpose it may be convenient to convert the above relation into the logarithmic scale and thus to connect the sound pressure level SPL with the power level PL $= 10\log_{10}(P/10^{-12}\text{ W})$:

$$\text{SPL} = \text{PL} - 10\log_{10}\left(\frac{A}{1\text{m}^2}\right) + 6\text{ dB}. \tag{31}$$

The equivalent absorption area A of the enclosure used for this procedure is obtained by measuring its reverberation time (see next section).

Equations (30) and (31) are valid as long as the distance r of the observation point from the sound source is relatively large. In the vicinity of the sound source, however, the contribution of the direct component to the energy density is predominant. For a nondirectional sound source this is

$$u_d = \frac{P}{4\pi c r^2}. \tag{32}$$

In a certain distance, the so-called reverberation distance r_r, both components, namely the direct and the reverberant parts of the energy density, are equal (see Fig. 14):

$$r_r = \frac{1}{4}\sqrt{\frac{A}{\pi}} \approx 0.057\sqrt{\frac{V}{T}}. \tag{33}$$

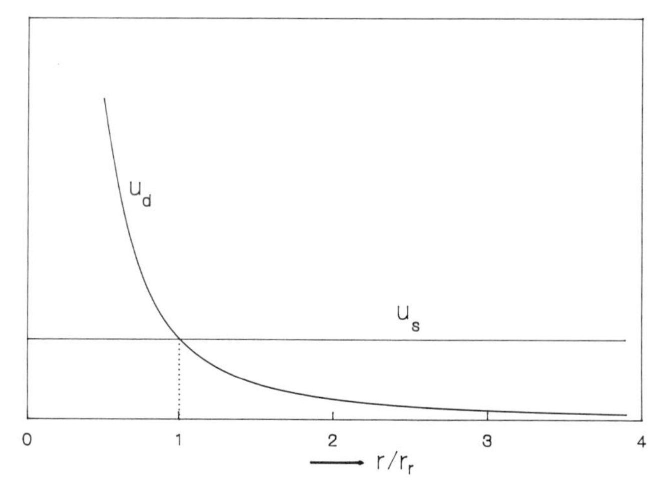

Fig. 14 Direct energy density u_d and reverberant energy density u_s in an enclosure as a function of the ratio r/r_r.

Hence, a somewhat more general expression for the total energy density including the direct component is

$$u = \frac{P}{4\pi c}\left(\frac{1}{r^2} + \frac{1}{r_r^2}\right). \tag{34}$$

If the sound source has a certain directivity, the first term in the parentheses has to be multiplied with its *directivity factor* Γ, which compares the intensity radiated into the direction of observation with the average intensity, both for the same distance. Hence, the corresponding expression for the sound pressure level reads

$$\text{SPL} = \text{PL} - 20\log_{10}\left(\frac{r}{1\text{m}}\right) + 10\log_{10}\left(\Gamma + \frac{r^2}{r_r^2}\right) - 11 \text{ dB}. \tag{35}$$

The above relations should be applied with some caution. For sine signals or other signals with small-frequency bandwidth they yield an average, at best, since we know from the discussions in Section 2.3 that the sound pressure amplitudes as well as the energy densities in an enclosure are far from being constant but are distributed over a wide range. But even for wide-band excitation of the enclosure, the energy density is usually not constant throughout the enclosure. One particular deviation known as the *Waterhouse effect* is due to the fact that any reflecting wall enforces certain phase relations between incident and reflected waves no matter from which directions the waves arrive at the wall.[13] As a consequence, there are fluctuations of the energy density in the vicinity of a wall that depend on its acoustical properties and on the frequency spectrum of the sound signal. In any case, the energy density in front of a rigid wall is twice its average value far away from it. Since the range of these fluctuations is about one-quarter of a wavelength, they can be neglected in the high-frequency limit.

A relatively high degree of sound field diffusion can be achieved in reverberation chambers with highly reflective walls and with an irregular shape that supports the mixing and blending of sound. For such a room Eq. (31) may be applied with some precautions. In other cases, it is preferable to compute the steady-state energy density by applying the image source model described in Section 3 if the walls are specularly reflecting or by numerically solving Eq. (22) and inserting the result into Eq. (23), provided the boundary reflection is at least partially diffuse.

4.3 Sound Decay

Now we consider decaying sound fields. For our discussion we imagine that a sound source exciting the enclosure until time $t = 0$ is abruptly stopped. (As an alternative, such a decaying sound field can be produced by a short impulse emitted at $t = 0$.) If the sound field can be expected to be diffuse, Eq. (28) with $P = 0$ can be applied, which produces the energy density

$$u(t) = u_0 \exp\left(-\frac{cA}{4V}t\right) \quad \text{for } t \geq 0, \tag{36}$$

where u_0 is the initial energy density at $t = 0$. From this equation, the reverberation time T of the enclosure, that is, the time in which the energy density has dropped by a factor of 10^6 (see Section 2.4), is obtained:

$$T = 0.163 \frac{V}{A}. \tag{37}$$

The numerical constant in this formula is $24 \ln 10/c$ and has the dimension seconds per metre. Accordingly, all lengths have to be expressed in metres.

Equation (37) is the famous Sabine reverberation formula and is probably the most important relation in room acoustics. Despite its simplicity, its accuracy is sufficient for many practical applications provided the correct value for the equivalent absorption A is inserted. However, for enclosures with relatively high absorption, it fails: For an absorption coefficient 1 on all walls, it predicts a finite reverberation time, although it is clear that in this case there will be no reverberation at all.

Therefore more accurate reverberation formulas have been derived, taking into account that sound decay is not a continuous process as assumed in Eq. (28) but involves stepwise reduction of a sound particle's energy whenever it is reflected from a wall.[14] Let t be the time that has elapsed since the beginning of the decay process. Then a typical sound particle will have undergone $\overline{n}t$ reflections from the boundary, with $\overline{n}$ denoting the average number of wall collisions per second, according to Eq. (25), and its energy will have diminished by the factor $(1-\alpha)^{\overline{n}t}$. Because of the assumed diffusivity of the sound field, this particle can be regarded representative for all sound particles. Therefore the same reduction factor holds for the whole energy within the enclosure:

$$u(t) = u_0(1-\alpha)^{\overline{n}t} \quad \text{for } t \geq 0. \tag{38}$$

For nonuniform distribution of the absorption along the boundary, the absorption coefficient α is replaced with its arithmetic average:

$$\overline{\alpha} = \frac{A}{S} = \frac{1}{S} \sum_i \alpha_i S_i. \tag{39}$$

The reverberation time obtained from this relation is

$$T = 0.163 \frac{V}{-S \ln(1-\overline{\alpha})}. \tag{40}$$

This formula is usually known as Eyring's formula, although it was described earlier by Fokker (1924) and by Waetzmann and Schuster (1929). It is obvious that it yields the reverberation time $T = 0$ for a totally absorbing boundary. For $\overline{\alpha} \ll 1$, it is identical to Sabine's formula [Eq. (37)].

Sound attenuation in the free medium can be taken into account by an additional factor $\exp(-mct)$ in Eq. (36). This modifies the above formula for the reverberation time in the following way:

$$T = 0.163 \frac{V}{4mV - S \ln(1-\overline{\alpha})}. \tag{41}$$

In Sabine's formula, Eq. (37), the attenuation by the medium can also be included by adding the term $4mV$ in the denominator.

Several authors[15,16] proposed not to average the absorption coefficients arithmetically, as in Eq. (39), but to average the logarithm of the *reflection coefficients* $1-\alpha_i$. This leads to

$$T = 0.163 \frac{V}{-\sum_i S_i \ln(1-\alpha_i)}. \tag{42}$$

This is the reverberation formula of Millington and Sette. It has the strange consequence that the reverberation time will be zero if any part of the boundary has the absorption coefficient 1. A more rigorous derivation shows, however, that the averaging procedure prescribed by Eq. (39) of Chapter 61 is the correct one.

It should be recalled that the derivation of all the formulas presented in this section was based upon the assumption of a diffuse sound field that ensures that each element of the enclosure is hit by the same energy amount per second. On the other hand, it was pointed out in Section 4.1 that sound fields in real rooms will fulfill this condition at best approximately because of the inevitable wall and medium losses. Even stronger violations are brought about by particular room shapes, such as those of very long or very flat rooms, or by a highly nonuniform distribution of the absorption along the boundary. In an occupied hall, for instance, the absorption is concentrated onto the area where the audience is seated. In such cases we must expect that the application of one of the reverberation formulas above would lead to unpredictable errors.

If the assumption of a diffuse sound field is dropped, the energy decay could be computed by evaluating Eq. (20), which reduces to a simple recursion if it is expressed in discrete form. In any case, the final sound decay follows an exponential law with the same decay rate at each location, as has been shown by Miles.[11] Several effective iteration schemes have been developed on the basis of Eq. (20)[9,17] to determine this decay rate and hence the reverberation time of the enclosure. Strictly speaking, however, Eq. (20) holds only for enclosures with diffusely reflecting boundaries. Nevertheless, the mentioned iteration schemes can be applied as well to practical situations where wall reflections are only partially diffuse. In this case the condition of diffuse reflections is reached in several steps, so to speak, since the conversion of specularly reflected into diffusely reflected sound is irreversible. Moreover, it has been shown by Hogson[18] that in nearly all halls there is at least some diffuse wall reflection.

To summarize the discussions of this section, it can be stated that the reverberation time of an enclosure can be predicted or calculated quite well. In many cases it is sufficient to apply one of the simple formulas for the decay time. But even if this is not the case, more reliable methods can be applied with still reasonable computing effort. Concerning the steady-state sound level or energy density, the situation is less satisfactory, since the reliability of simple formulas such as Eq. (34) depends more critically on the diffusion of the sound field than the relations for the decay time. Therefore, for computation of the steady-state level, often more complicated schemes are needed, such as the image source model.

REFERENCES

1. J. Pan and D. A. Bies, "The Effect of Fluid-Structural Coupling on Sound Waves in an Enclosure," *J. Acoust. Soc. Am.*, Vol. 87, 1990, pp. 691–707 (theoretical part), pp. 708–717 (experimental part).
2. M. R. Schroeder and H. Kuttruff, "On Frequency Response Curves in Rooms. Comparison of Experimental, Theoretical and Monte Carlo Results for the Average Spacing between Maxima," *J. Acoust. Soc. Am.*, Vol. 34, 1962, pp. 76–80.
3. H. Kuttruff and R. Thiele, "Über die Frequenzabhängigkeit des Schalldrucks in Räumen," *Acustica*, Vol. 4 (Beihefte), 1954, pp. 614–617.
4. M. Schröder, "Die statistischen Parameter der Frequenzkurven von großen Räumen," *Acustica*, Vol. 4 (Beihefte), 1954, pp. 594–600.
5. J. Borish, "Extension of the Image Source Model to Arbitrary Polyhedra," *J. Acoust. Soc. Am.*, Vol. 75, 1978, pp. 1827–1836.
6. H. Lehnert, *Binaurale Raumsimulation: Ein Computermodell zur Erzeugung virtueller auditiver Umgebungen*, Shaker, Aachen, 1992.
7. H. Kuttruff, "Simulierte Nachhallkurven in Rechteckräumen," *Acustica*, Vol. 25, 1971, pp. 333–342; "Nachhall und effektive Absorption in Räumen mit diffuser Wandreflexion," *Acustica*, Vol. 35, 1976, pp. 141–153.
8. W. B. Joyce, "Effect of Surface Roughness on the Reverberation Time of a Uniformly Absorbing Spherical Enclosure," *J. Acoust. Soc. Am.*, Vol. 64, 1978, pp. 1429–1436.
9. H. Kuttruff, "A Simple Iteration Scheme for the Computation of Decay Constants in Enclosures with Diffusely Reflecting Boundaries," *J. Acoust. Soc. Am.*, Vol. 98, 1995, pp. 288–293.
10. M. M. Carroll and R. N. Miles, "Steady-State Sound in an Enclosure with Diffusely Reflecting Boundary," *J. Acoust. Soc. Am.*, Vol. 64, 1978, pp. 1424–1428.
11. R. N. Miles, "Sound Field in a Rectangular Enclosure with Diffusely Reflecting Boundaries," *J. Sound Vib.*, Vol. 92, 1984, pp. 203–213.
12. H. Kuttruff, "Stationäre Schallausbreitung in Flachräumen," *Acustica*, Vol. 57, 1985, pp. 62–70.
13. R. V. Waterhouse, "Interference Patterns in Reverberant Sound Fields," *J. Acoust. Soc. Am.*, Vol. 27, 1955, pp. 247–258.
14. C. F. Eyring, "Reverberation Time in 'Dead' Rooms," *J. Acoust. Soc. Am.*, Vol. 1, 1930, pp. 217–241; "Methods of Calculating the Average Coefficient of Sound Absorption," *J. Acoust. Soc. Am.*, Vol. 4, 1993, pp. 178–192.
15. G. Millington, "A Modified Formula for Reverberation," *J. Acoust. Soc. Am.*, Vol. 4, 1932, pp. 69–82.
16. W. J. Sette, "A New Reverberation Time Formula," *J. Acoust. Soc. Am.*, Vol. 4, 1933, pp. 193–210.
17. E. N. Gilbert, "An Iterative Calculation of Reverberation Time," *J. Acoust. Soc. Am.*, Vol. 69, 1981, pp. 178–184.
18. M. R. Hodgson, "Evidence of Diffuse Surface Reflection in Rooms," *J. Acoust. Soc. Am.*, Vol. 89, 1991, pp. 765–771.

75

SOUND ABSORPTION IN ENCLOSURES

DAVID A. BIES AND COLIN H. HANSEN

1 INTRODUCTION

Sound absorption in enclosures occurs when sound waves strike objects in the enclosure and the enclosure boundaries as well as during propagation through the acoustic medium (usually air) that fills the enclosing space.

Sound absorption in enclosures plays an important part in the determination of sound pressure levels resulting from the operation of sound sources of known sound power output as well as in determining the amount of reverberation, or "liveliness," of the enclosure, which is quantified in terms of its reverberation time. Thus this chapter will be concerned with the measurement and calculation of sound absorption in enclosures as well as the estimation of the effect of sound absorption on sound levels and reverberation times. Relationships between sound absorption, acoustic impedance, and flow resistance for acoustically absorptive materials will also be discussed.

The physical mechanism associated with sound absorption at the enclosure boundaries may take one of two forms, dependent upon the vibroacoustic properties of the room boundaries, the frequency range of interest, and the size of the enclosure. One of the mechanisms is referred to as *local reaction*, meaning that the response of the boundary to an incident sound wave (and thus the absorption properties) is entirely dependent upon the local properties of the boundary at the point of incidence. In this case the boundary absorption may be described in terms of a sound absorption coefficient that is the ratio of energy absorbed to the energy incident.

Note: References to chapters appearing only in the *Encyclopedia* are preceded by *Enc.*

The second mechanism of absorption that can occur when a sound wave is incident on a boundary is one of modal coupling, which means that the sound field in the enclosure excites the boundaries into vibration, which in turn results in removal of energy from the sound field and effective absorption by the boundaries. Whether the absorption mechanism is of the first or second kind will determine the type of analysis needed to evaluate its effect on the enclosed sound field. Conditions that need to be satisfied for each of the absorption mechanisms to occur will be discussed in the following paragraphs.

2 MODAL PROPERTIES OF ENCLOSURES

Sound absorption in enclosures is dependent upon the modal response of the enclosed space, and thus what is meant by a *mode* is important to the understanding of sound absorption. Modal response in enclosures is discussed in Chapter 74. As shown in the latter chapter, the case of a rectangular enclosure allows an exact solution of the wave equation in terms of the enclosure dimensions. It should be noted that the conclusions drawn from consideration of a rectangular enclosure are, however, by no means restricted to rectangular enclosures but apply to enclosures of any shape. The analysis leads to a simple expression for calculating the frequencies of the normal modes of a rectangular enclosure [see Chapter 74, Eq. (6)] as well as the total number of modes within a frequency band [see Chapter 74, Eq. (7)]. It may be shown that the number of modes N resonant up to and including frequency f in a room of any shape is a function only of the room volume V, the room surface area S, the room perimeter P (when properly defined), and speed of sound c. Thus the modal density, defined as the number of modes per unit frequency, $\partial N/\partial f$, in a room of any shape is given by[1]

$$n_A = \frac{4\pi V}{c^3} f^2 + \frac{\pi S}{2c^2} f + \frac{P}{8c}. \qquad (1)$$

The absorption mechanism involving enclosure–boundary modal coupling is important when the enclosure modal density is low and wall damping is also small. This mechanism thus characterizes vehicles of various kinds (e.g., aircraft and automobiles); it is not amenable to generalization, has received only limited investigation, and is beyond the scope of this chapter, except for the discussion of panel absorbers at the end.

If the enclosure boundaries are well damped or if the modal density is sufficiently high that the sound field may be regarded as diffuse, then the absorption mechanism will be dominated by the local reaction effect described earlier. The modal density may be regarded as sufficiently high when the quadratic term in frequency of Eq. (1), which describes the modal density of oblique modes, is larger than the linear and constant terms, indicating that oblique modes dominate the enclosure response.

A more precise definition of the conditions under which the sound field may be considered diffuse is related to damping of the modes in the room as well as to the modal density. The modal damping can be quantified in terms of the modal bandwidth Δf, as shown in Fig. 1. As can be seen from the figure, the modal bandwidth for a particular mode is the width in hertz of the response curve of the mode measured 3 dB down from the peak response. The modal bandwidth (averaged over all modes in a frequency band B) is defined in terms of the room reverberation time T_{60} for the frequency band B as follows[2]:

$$\Delta f = \frac{2.2}{T_{60}}, \qquad (2)$$

where the reverberation time is defined as the time taken for the sound field to decay by 60 dB after the sound source is shut off.

Referring to Fig. 1, the modal overlap (or the extent to which the modes result in a uniform pressure distribution as a function of frequency) is defined as

$$M = \frac{\Delta f_1 + \Delta f_2 + \Delta f_3}{B}, \qquad (3)$$

or in more general terms as[3]

$$M = \Delta f n_A. \qquad (4)$$

For the purposes of the present discussion, a value of modal overlap greater than 3 will mean that the sound field is sufficiently diffuse that the local reaction mechanism of sound absorption will dominate. The frequency range for which this occurs will be referred to as the high-frequency range. It is of interest to note that most auditoria and industrial spaces are sufficiently large and sufficiently well damped that the entire audio range generally of interest (63 Hz to 8000 Hz) may be considered as the high-frequency range.

The discussion to follow will be restricted to enclosures for which no dimension exceeds the others by more than about a factor of 3. Such enclosures will be described here as Sabine enclosures; enclosures that do not satisfy this restriction are discussed in *Enc.* Ch. 95. It is important to note that the Sabine room can be characterized by a reverberant sound pressure level L_{pR} and a reverberation time T_{60} and as such provides a means for the definition, measurement, and use of the Sabine absorption coefficient of materials.

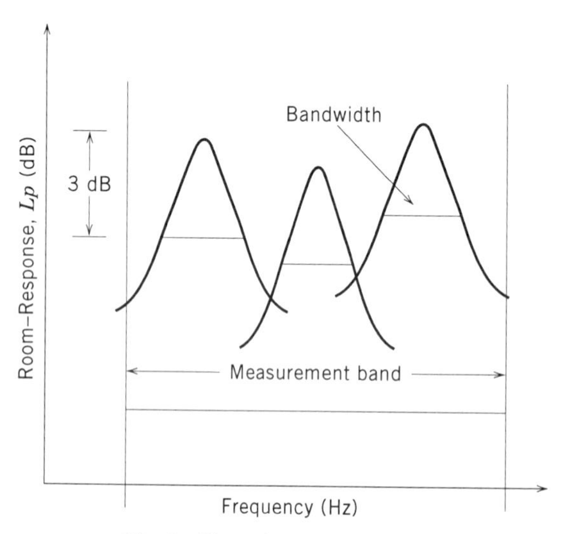

Fig. 1 Illustration of modal overlap.

3 RELATIONSHIP BETWEEN SOUND ABSORPTION AND ENCLOSURE SOUND PRESSURE LEVEL

In the high-frequency range, the sound field in a Sabine room may be though of as consisting of two fields superimposed upon each other. These fields are, respectively, the direct field of the source and the reverberant field of the room. The latter field is the result of multireflections at the boundaries of the room and is in an average sense uniform throughout. The direct field of a source is greatest near the source and dominates in some area about the source, but as it spreads out into the room, it necessarily decays until it in turn is dominated by the reverberant field in the room.

The simplification will be made that a source may be represented as a point source multiplied by a directivity factor D_i and a sound power weighting factor d_i. When account is taken of n sources, an expression for the sound pressure level L_p at any point in a Sabine room in terms of the total sound power level L_w of all sources is as follows[3]:

$$L_p = L_w + 10 \log_{10} \left[\sum_{i=1}^{n} \frac{d_i D_i}{4\pi r_i^2} + \frac{4}{R} \right]. \tag{5}$$

In Eq. (5), the weighting factor d_i and the directivity factor D_i for the ith source have been introduced. The weighting factor is calculated as $d_i = W_i/W$, where W_i is the sound power of the ith source and W is the arithmetic sum of all sources, both in consistent units. The directivity factor D_i may be taken as 1 for a source raised up off the floor and 2 for a source on the floor. Alternatively, if the directional properties of the source are known, then this information may be used instead. The quantities r_i are the linear distances from the assumed acoustical centers of the several sources to the point of observation.

The room constant R introduced in Eq. (5) is calculated in terms of the mean Sabine absorption coefficient $\overline{\alpha}$ and the total area S of all absorbing surfaces S_i as follows.[3] The determination of $\overline{\alpha}$ is discussed in Section 5:

$$R = \frac{S\overline{\alpha}}{1 - \overline{\alpha}}, \tag{6}$$

where values of Sabine absorption coefficients $\overline{\alpha}_i$ for areas S_i are known, perhaps by reference to a table of measured values, and the mean Sabine absorption coefficient $\overline{\alpha}$ is calculated as the area-weighted average as

$$\overline{\alpha} = \sum_{i=1}^{n} \frac{S_i}{S} \overline{\alpha}_i. \tag{7}$$

The quantity $S_i\alpha_i$ has the units of square feet or square metres, but in either case it has the name *sabin*. Here it will be called a metric sabin when it has the units of square metres.

The direct field has been described as the field close to a source where it dominates the reverberant field. Close to the ith source, the distance r_i will usually be small compared to the distances r_j of all other sources. Consequently, the ith term will usually be larger than all other terms in the sum [see Eq. (7)] and also larger than the reverberant field term. Close to the ith source, r_i satisfies the requirement

$$\frac{1}{4}\sqrt{\frac{Rd_iD_i}{\pi}} \gg r_i. \tag{8}$$

In the range satisfied by Eq. (8), the sound field is controlled by the ith source, and consequently any absorptive treatment that decreases the reverberant field will have an insignificant effect upon the observed sound level in this range. For example, if an employee works close to a machine in its direct field, the placement of absorptive treatment upon the walls or ceiling of the room will have no significant effect upon the noise to which the employee is exposed. Alternatively, if the employee's machine is close to a reflecting wall, treatment of the wall will reduce the reflected sound from the wall, but this could at most decrease the employee's exposure on a sound energy basis by half, resulting in a 3-dB sound reduction. This reduction would barely be noticeable.

The reverberant field exists well removed from all sources. For example, when the point of observation is closest to the ith source and in the reverberant field, the criterion that is met is as follows:

$$r_i \gg \frac{1}{4}\sqrt{\frac{Rd_iD_i}{\pi}}. \tag{9}$$

In the reverberant field that satisfies Eq. (9), the sound pressure level is controlled by the room constant R, which in turn is controlled by the mean Sabine absorption coefficient. The placement of sound absorbents on the surfaces of the room will reduce the sound in the reverberant field to the extent that the mean Sabine absorption coefficient is reduced. Reference to Eq. (7) shows that the mean Sabine absorption coefficient is an area-weighted quantity. Consequently, best results are obtained when least absorptive surfaces are treated with absorbent material.

The reduction in sound pressure level L_p in the reverberant field due to treatment with sound-absorbing material may be calculated using the following equation[1]:

$$\Delta L_p = 10 \log_{10} \frac{R_f}{R_i}. \tag{10}$$

Here R_i is the initial value and R_f is the final value of the room constant after the addition of absorbent.

4 EFFECT OF SOUND ABSORPTION ON REVERBERATION TIME

When sound is introduced into a room, the reverberant field level will increase until the sound energy introduction is just equal to the sound energy absorption. If the sound source is abruptly shut off, the reverberant field will decay at a rate determined by the rate of sound energy absorption. The time required for the reverberant field to decay by 60 dB is called the reverberation time and is the single most important parameter describing the acoustical properties of a room.

These observations may be summarized as follows, where it will be assumed that sound is absorbed at the boundaries of the room. The power W_a absorbed by the room boundaries is equal to the product of the boundary area S, the mean Sabine absorption coefficient $\overline{\alpha}$, and the amplitude of the sound intensity component normally incident at the boundary surface. Thus,[3]

$$W_a = IS\overline{\alpha} = \frac{\psi S c\overline{\alpha}}{4}. \tag{11}$$

Note that although the intensity averaged over all directions (the net intensity) is zero[1], the sound intensity component in a particular direction in a diffuse field is nonzero and is as given above.

The introduced sound power W less the sound power absorbed is equal to the rate of change of stored sound energy in the room. Thus,

$$W - \frac{\psi S c\overline{\alpha}}{4} = V\frac{\partial \psi}{\partial t}. \tag{12}$$

In Eq. (12), $\psi(t)$ is the sound energy density at time t, S is the surface area of the walls of the room, c is the speed of sound, V is the volume of the room, and $\overline{\alpha}$ (the fraction of incident energy absorbed) is the mean sound absorption coefficient for the walls of the room. The sound energy density is related to the mean-square sound pressure $\langle p^2 \rangle$, gas density ρ, and speed of sound c and is a function of time t as follows:

$$\psi(t) = \frac{\langle p^2(t)\rangle}{\rho c^2}. \tag{13}$$

At time zero, when equilibrium exists between power input and power absorbed, the source is abruptly shut off, so that $W = 0$, and the sound field decays. The mean-square sound pressure at time t at any point in the room is then[1]

$$\langle p^2(t)\rangle = \langle p^2(0)\rangle e^{-Sc\overline{\alpha}t/4V}. \tag{14}$$

The time required for the sound field to decay by 60 dB, called the reverberation time T_{60}, may be determined by solving Eq. (14). The resulting equation, originally derived by Sabine,[4] gives the relation between the reverberation time and the mean sound absorption coefficient. The latter quantity is given the name *Sabine absorption coefficient*[4]:

$$T_{60} = \frac{55.25V}{Sc\overline{\alpha}}. \tag{15}$$

To gain some insight into the assumptions implicit in the derivation of Eq. (15), it is useful to repeat the analysis using a modal description of the reverberant field within an enclosure.[5] In fact, if any enclosure is driven at a frequency slightly off resonance and the source is abruptly shut off, the frequency of the decaying field will be observed to shift to the resonance frequency of the driven mode. The modal analysis requires the introduction of the modal mean free path L_i and the modal mean sound reflection coefficient β_i. The modal mean free path L_i is the mean distance between reflections of a sound wave traveling around a closed modal circuit. The modal mean reflection coefficient is the product of the reflection coefficients encountered by a wave traveling around a modal circuit.

It may be shown that the mean of the modal mean free paths is given by the equation[3]

$$\frac{1}{N}\sum_{i=1}^{N} L_i = \frac{4V}{S}. \tag{16}$$

The assumption will be made that on average, at zero time, the energy distribution among the modes of the room will be about equal; thus,

$$\langle p_i(0)^2\rangle = \frac{\langle p(0)^2\rangle}{N}. \tag{17}$$

The following equation may be shown to describe a modally decaying sound field:

$$\langle p(t)^2\rangle = \langle p(0)^2\rangle \frac{1}{N}\sum_{i=1}^{N} \exp\left(\frac{ct}{L_i}\log_e \beta_i\right). \tag{18}$$

Some simplifications will now be made. In Eq. (18), all L_i will be replaced with the mean free path $4V/S$ [see Eq. (16)] and all modal reflection coefficients β_i will be replaced with $1 - \alpha_{st}$, where α_{st} is the statistical absorption coefficient. Equation (18) may be rewritten as

$$\langle p(t)^2 \rangle = \langle p(0)^2 \rangle \exp\left(\frac{Sct}{4V} \log_e(1 - \alpha_{st}) \right). \tag{19}$$

Equation (19) may be solved for the reverberation time and the equation of Norris Eyring is obtained as

$$T_{60} = \frac{55.25V}{-Sc \log_e(1 - \alpha_{st})}. \tag{20}$$

The statistical absorption coefficient α_{st} for the entire enclosure is an area-weighted quantity. Where α_{st_i} is the statistical coefficient of subarea S_i, and the total room area S is the sum of all n subareas, the statistical absorption coefficient for the entire room is determined as

$$\alpha_{st} = \sum_{i=1}^{n} \frac{S_i \alpha_{st_i}}{S}. \tag{21}$$

When $\alpha_{st} < 0.4$, an error of less than 0.5 dB is made in the determination of the sound pressure level in Eq. (19) if the following approximation is made:

$$\alpha_{st} = -\log_e(1 - \alpha_{st}). \tag{22}$$

Use of Eq. (22) allows Eq. (20) to be rewritten in the form of Eq. (15), suggesting that the Sabine absorption coefficient is the same as the statistical absorption coefficient. In fact, when edge diffraction is taken into account, the suggestion appears to be true.[5]

Alternatively, if the mean free path replacement is made in Eq. (18), but all of the modal mean reflection coefficients β_i are replaced with the mean statistical reflection coefficient β_{st}, then the equation of Millington[7] and Sette[8] is obtained:

$$T_{60} = \frac{55.25V}{-Sc \log_e \beta_{st}}. \tag{23}$$

The statistical reflection coefficient is defined as

$$\beta_{st} = \left(\prod_{i=1}^{n} \beta_i \right)^{S_i/S}. \tag{24}$$

Experience has shown that Eq. (23) provides more accurate results than Eq. (15) or (20) for predicting reverberation times for large auditoria. Note that the statistical sound absorption coefficient α_{st} must be calculated from measurements of the material flow resistance or normal impedance, whereas the Sabine absorption coefficient $\overline{\alpha}$ is a measured quantity (see Table 1).

The preceding considerations of the reverberation time are based upon the assumption that the reverberant field is diffuse. However, enclosures are often encountered in which the field is obviously not diffuse. In this case a proposal due to Fitzroy has had some success.[9] Fitzroy proposes a modification to the Norris Eyring equation (20) as

$$T_{60} = \frac{55.25V}{-Sc} \sum_{i=1}^{3} \frac{S_i}{S \log_e(1 - \alpha_i)}. \tag{25}$$

In Eq. (25), S_i and α_i ($i = 1-3$) are respectively the total side-wall area S_{sw} and mean side-wall absorption coefficient α_{sw}, the total end-wall area S_{ew} and mean end-wall absorption coefficient α_{ew}, and the total floor and ceiling area S_{fc} and mean floor and ceiling absorption coefficient α_{fc}. The total room area $S = S_{sw} + S_{ew} + S_{fc}$. The equation should be used with caution as it has only been shown to provide reliable results for the cases considered by Fitzroy and there are other cases for which it is incorrect.

Optimum values of the reverberation times for various uses of an enclosed space may be calculated as[10]

$$T_{60} = K(0.0118V^{1/3} + 0.1070). \tag{26}$$

The constant K of Eq. (26) takes the following values according to the proposed use: For speech $K = 4$, for orchestra $K = 5$, and for choirs and rock bands $K = 6$. At frequencies in the 250-Hz octave band and lower frequencies an increase over the value calculated by Eq. (26), ranging from 40% at 250 Hz to 100% at 63 Hz, is suggested.

5 MEASUREMENT OF THE SABINE ABSORPTION COEFFICIENT

Equation (15) forms the basis for the definition of the Sabine absorption coefficient $\overline{\alpha}$ and its measurement. It is important to note that it is the result of several approximations, and consequently it should come as no surprise that values of $\overline{\alpha}$ are sometimes determined that are greater than unity, suggesting that the material tested is capable of absorbing more sound energy than is incident

TABLE 1 Octave Band Average Sabine Sound Absorption Coefficients for Some Commonly Used Materials

Material	63 Hz	125 Hz	250 Hz	500 Hz	1000 Hz	2000 Hz	4000 Hz
100% occupied audience (orchestra and chorus areas); upholstered seats	0.34	0.52	0.68	0.85	0.97	0.93	0.85
Unoccupied; average well-upholstered seating areas	0.28	0.44	0.60	0.77	0.89	0.82	0.70
Unoccupied; leather-covered upholstered seating areas		0.40	0.50	0.58	0.61	0.58	0.50
Wooden pews; 100% occupied		0.57	0.61	0.75	0.86	0.91	0.86
Wooden chairs; 100% occupied		0.60	0.74	0.88	0.96	0.93	0.85
Wooden chairs; 75% occupied		0.46	0.56	0.65	0.75	0.72	0.65
Fiberglass or rockwool blanket 60 kg/m^3							
25 mm thick (typical)		0.18	0.24	0.68	0.85	1.00	1.00
50 mm thick (typical) (consult manufacturer)		0.25	0.83	1.00	1.00	1.00	1.00
Floors							
Wood platform with large space beneath		0.40	0.30	0.20	0.17	0.15	0.10
Concrete or terrazzo		0.01	0.01	0.01	0.02	0.02	0.02
Linoleum, asphalt, rubber, or cork tile on concrete		0.02	0.03	0.03	0.03	0.03	0.02
Varnished wood joist floor		0.15	0.12	0.10	0.07	0.06	0.07
Carpet, heavy, on concrete		0.02	0.06	0.14	0.37	0.60	0.65
Carpet, heavy, on 40-oz (1.35 kg/m^2) hair felt or foam rubber		0.08	0.24	0.57	0.69	0.71	0.73
Carpet, 5 mm thick, on hard floor		0.02	0.03	0.05	0.10	0.30	0.50
Cork floor tiles ($\frac{3}{4}$ in. thick); glued down		0.08	0.02	0.08	0.19	0.21	0.22
Glazed tile/marble		0.01	0.01	0.01	0.01	0.02	0.02

upon it. When values of $\overline{\alpha}$ greater than unity are encountered, it is customary to arbitrarily set the measured quantity to unity.

Measured values of Sabine absorption coefficients for some commonly used acoustic materials are given in Table 1. More extensive tables are available in the literature[11–13].

In presenting absorption coefficient information, the term *noise reduction coefficient* (NRC) is sometimes used.[11] The latter quantity is defined as the arithmetic average of absorption coefficients determined at 250, 500, 1000, and 2000 Hz. In such presentation it is customary to round off the average to the nearest multiple of 0.05.

The Sabine absorption coefficient of a material is measured using a reverberation chamber, in which reverberation decay times are determined, generally in one-third-octave bands, with and without the material under test. The differences in measured decay times with the absorbent material in place, T'_{60}, and absent, T_{60}, allow determination of the absorption due to the presence of the test material using the following equation [derived from Eq. (15)]:

$$\overline{\alpha} = \frac{55.3V}{Sc}\left(\frac{1}{T'_{60}} - \frac{S' - S}{S'T_{60}}\right). \tag{27}$$

In Eq. (27), S' is the total area of all room surfaces, including the sample when in place, and S is the area of material (usually between 10 and 12 m^2, with a width-to-length ratio of between 0.7 and 1.0) exposed to the sound field. As described in the relevant standards,[14–17] the edges of the sample must be isolated from the sound field using appropriate rigid edging. An absorption coefficient is calculated using Eq. (27) for each one-third-octave band.

It is generally recognized that the Sabine equation

TABLE 1 ***(Continued)***

Material	63 Hz	125 Hz	250 Hz	500 Hz	1000 Hz	2000 Hz	4000 Hz
Walls							
Plush curtain, deeply folded		0.15	0.45	0.90	0.92	0.95	0.95
Polyurethane foam, 27 kg/m^3, 15 mm thick		0.08	0.22	0.55	0.70	0.85	0.75
Acoustic plaster, 10 mm thick, sprayed on solid wall		0.08	0.15	0.30	0.50	0.60	0.70
Hard surfaces (brick walls, plaster, hard floats, etc.)		0.02	0.02	0.03	0.03	0.04	0.05
Slightly vibrating walls (suspended ceilings, etc.)		0.10	0.07	0.05	0.04	0.04	0.05
Strongly vibrating surface (wooden paneling over airspace, etc.)		0.40	0.20	0.12	0.07	0.05	0.05
Plaster, gypsum or lime, smooth finish							
On brick		0.013	0.015	0.02	0.03	0.04	0.05
On concrete block		0.012	0.09	0.07	0.05	0.05	0.04
On lath		0.014	0.10	0.06	0.04	0.04	0.03
Curtains							
Light velour, 10 oz/yd^2, hung	0.03	0.04	0.11	0.17	0.24	0.35	
Medium velour, 14 oz/yd^2 draped to half area	0.07	0.31	0.49	0.75	0.70	0.60	
Heavy velour, 18 oz/yd^2 draped to half area	0.14	0.35	0.55	0.72	0.70	0.65	
Glass							
Glass, heavy plate	0.18	0.06	0.04	0.03	0.02	0.02	
Ordinary window	0.35	0.25	0.018	0.12	0.07	0.04	
Stage openings	0.30	0.40	0.50	0.60	0.60	0.50	
Water (surface of pool)	0.01	0.01	0.01	0.015	0.02	0.03	

(15) is a gross simplification of a complex problem. However, it has recently been proposed[5] that, in fact, the Sabine equation is exact and that, when edge diffraction is taken into account in the computation of Eq. (21), the Sabine absorption coefficient of Eq. (15) is equal to the statistical absorption coefficient of Eq. (21). The effect of edge diffraction is to increase the effective area of the material.[1] Unfortunately, edge diffraction is ignored both in the standard measurement of Sabine absorption coefficients and in their application. Currently available procedures for the computation of the Sabine absorption coefficient are not adequate for general use.[18]

One problem resulting from neglecting edge diffraction in the standard test is that absorption coefficients greater than 1 are often obtained, whereas in principle only values less than 1 are permitted by the basic assumption that the absorption coefficient is a measure of the incident energy that is absorbed. A second problem is that results for the same test material can be highly variable. The basic problem remains that no theoretical solution has been accepted relating statistical absorption to reverberation decay in a Sabine room, although one has recently been proposed.[5] However, conditions necessary for agreement among results for the determination of the Sabine absorption coefficient of the same test material tested in several rooms have been determined.[19] Reverberation chambers ranging in size from 106 to 607 m^3 have been shown to give essentially the same absorption coefficients for a 10.5-m^2 test sample of Silan when measured using the reverberation decay method, and the results are interpreted using Sabine's equation provided that the reverberation chamber contains auxiliary diffusing elements of total area (both sides) greater than 1.4 times the chamber floor area. The problem of variability of test results can be overcome with adequate auxiliary diffusion.

In the same series of investigations test samples ranging in size from 22.5 to 10.5 m^2 in the large 607-m^3 room and from 10.5 to 5.0 m^2 in the small 106-m^3 room all gave similar results. Absorption coefficients greater than 1 in the frequency range of the 500-Hz one-third-octave band and higher frequency bands were reported, although the predicted statistical absorption coefficient has a maximum value of 0.954. Only a small tendency for the smallest test sample to be associated with the greatest sound absorption coefficient is to be observed in the range of 500 Hz in the largest room test data and in data of an intermediate-size 208-m^3 room. The tendency is not observed in the data of the smallest room. It is to be noted that an edge effect would predict that greater absorption coefficients would be observed in smaller test samples tested in the smallest room, contrary to what has been reported.

In rooms with inadequate auxiliary diffusion a decrease in effective absorption is observed at high frequencies and a tendency for smaller test samples to give larger absorption coefficients than larger samples is quite evident. However, the latter effects tend to disappear or to be greatly reduced when there is adequate auxiliary diffusion in the room. Alternatively, diffraction at the test sample edges, which is ignored in the standard test and has been assumed to account for the greater effectiveness of small test samples compared to large samples, seems to be overridden when diffusion is adequate. Adequate auxiliary diffusion seems to suppress any edge effects attributed to diffraction.

If the reverberation room is considered to be a model for all Sabine-type rooms, then the problems of data interpretation, which have been described, are equally true for application of measured absorption coefficient data to the design of Sabine-type rooms. Until reliable analysis is provided, Sabine absorption coefficient data must be used with care.

6 EFFECT OF AN AUDIENCE

In occupied auditoria the audience will represent the major contribution to sound absorption. In considering the effect of an audience upon the sabins of absorption they contribute, two possibilities have been investigated. One method is for some effective area of total absorption to be attributed to a person, for example, some number of sabins (square metres) per person and the area per person multiplied by the expected total number of people, to obtain an estimate of the expected absorption. Alternatively, an effective absorption coefficient may be attributed to the area occupied by the expected audience. The latter procedure has been shown to give more accurate results than the former, which can be in serious error. Beranek has investigated this problem and finds an effective absorption coefficient of about 0.92 for the 500- and 1000-Hz octave bands, giving an effective absorption in sabins of between 2.3 and 5.7 m^2 per occupied seat.[11]

7 AIR ABSORPTION

In addition to energy loss on reflection at the boundaries of an enclosure, some energy is dissipated while the sound wave propagates through the air. Such propagational loss is small and is generally only important in very large rooms and at frequencies above about 500 Hz. In considering the absorption of sound as it travels through air, it will be assumed that the fraction of sound energy lost during propagation is linearly related to the mean free path between reflections. As may readily be shown,[3] the total mean Sabine absorption coefficient becomes

$$\overline{\alpha} = \overline{\alpha}_w + 9.21 \times 10^{-4} m V/S. \tag{28}$$

Values of the absorption of sound m expressed in decibles per thousand metres as a function of frequency and relative humidity, given in Table 2, were calculated using Sutherland's method.[20] In Eq. (28) the Sabine absorption coefficient for the boundary of a room has been identified with a subscript w. It is to be noted that what is implied by the mean Sabine absorption coefficient $\overline{\alpha}$ in Eq. (15) is the quantity given by Eq. (28).

8 STATISTICAL SOUND ABSORPTION COEFFICIENT

In a diffuse sound field, sound energy will be incident upon a boundary from all directions in the adjacent half space. As the sound absorbed at the boundary can be expected to be dependent upon the angle of incidence, it is convenient to define a statistical sound absorption coefficient α_{st}, which is the average value of sound absorption coefficient, assuming all angles of incidence with equal intensity to be equally probable. The statistical sound absorption coefficient, together with Eq. (20), is recommended for use in the design of large auditoria in place of the Sabine absorption coefficient and Eq. (15).

In principle, the Sabine absorption coefficient should be the same as the statistical sound absorption coefficient, but this is not the case. Though similar in concept, the former is defined by Eq. (15) and the associated measurement of Eq. (27), which is not necessarily the definition of the latter. In fact, no way of making

TABLE 2 Attenuation due to Atmospheric Absorption[a]

Relative Humidity (%)	Temperature (°C)	m (dB/1000 m)							
		63 Hz	125 Hz	250 Hz	500 Hz	1000 Hz	2000 Hz	4000 Hz	8000 Hz
25	15	0.2	0.6	1.3	2.4	5.9	19.3	66.9	198.0
	20	0.2	0.6	1.5	2.6	5.4	15.5	53.7	180.5
	25	0.2	0.6	1.6	3.1	5.6	13.5	43.6	153.4
	30	0.1	0.5	1.7	3.7	6.5	13.0	37.0	128.2
50	15	0.1	0.4	1.2	2.4	4.3	10.3	33.2	118.4
	20	0.1	0.4	1.2	2.8	5.0	10.0	28.1	97.4
	25	0.1	0.3	1.2	3.2	6.2	10.8	25.6	82.2
	30	0.1	0.3	1.1	3.4	7.4	12.8	25.4	72.4
75	15	0.1	0.3	1.0	2.4	4.5	8.7	23.7	81.6
	20	0.1	0.3	0.9	2.7	5.5	9.6	22.0	69.1
	25	0.1	0.2	0.9	2.8	6.5	11.5	22.4	61.5
	30	0.1	0.2	0.8	2.7	7.4	14.2	24.0	68.4

[a]Calculated using Sutherland's method (Ref. 20).

a direct measurement of the statistical sound absorption coefficient of a material has been demonstrated, though predicted dependence upon angle of incidence has been experimentally verified. On a theoretical basis, it may be shown that the statistical sound absorption coefficient α_{st} has a maximum value of 0.954.

The statistical sound absorption coefficient of porous materials may be calculated based upon either a measurement of the normal impedance using an impedance tube or a measurement of the material flow resistance.[3,21] Both methods will be reviewed and the results of some calculations will be presented in the following sections.

8.1 Impedance Tube Method

Any heavy-walled tube may be used to construct an impedance tube. A source of sound is placed at one end and the sample to be investigated at the other. Investigation is limited to the frequency range in which plane-wave propagation in the tube may be assured. For a circular cross-sectional tube of diameter d, this frequency range is between zero and an upper frequency defined by $f = 0.586c/d$. The normal impedance of the sample may be determined from investigation of the sound field in the tube. Test methods are described in various standards.[22,23]

When the material under test is in place and the sound field in the tube is explored, a series of maxima and minima will be observed. The maxima L_{max} will be effectively constant in level, but the minima will be observed to increase in level as the mesurement point is moved away from the sample material. Straight-line extrapolation of the minima back to the surface of the sample is recommended to determine an estimate of the minimum level L_{min}.[1] The standing-wave ratio L_0 is given as

$$L_0 = L_{max} - L_{min}. \tag{29}$$

The location D of the first minimum relative to the surface of the sample is determined. The minima are located one half-wavelength ($\lambda/2$) apart, from which the ratio D/λ may be formed. The statistical sound absorption coefficient may then be calculated as

$$\alpha_{st} = \frac{8\cos\beta}{\xi} \times \left[1 - \frac{\cos\beta}{\xi}\log_e(1 + 2\xi\cos\beta + \xi^2) + \frac{\cos 2\beta}{\xi\sin\beta}\tan^{-1}\frac{\xi\sin\beta}{1 + \xi\cos\beta}\right], \tag{30}$$

where

$$K_0 = 10^{L_0/20}, \tag{31a}$$

$$M = 0.5(K_0 + K_0^{-1}), \tag{31b}$$

$$N = 0.5(K_0 - K_0^{-1}), \tag{31c}$$

$$\theta = 360\left(\frac{2D}{\lambda} - 0.5\right)\text{ deg}, \tag{32a}$$

$$R = (M - N\cos\theta)^{-1}, \tag{32b}$$

$$X = RN\sin\theta, \tag{32c}$$

$$\xi = (R^2 + X^2)^{1/2}, \tag{33a}$$

$$\beta = \tan^{-1}\frac{X}{R}. \tag{33b}$$

8.2 Flow Resistance Measurement Method

If a constant differential pressure is imposed across a layer of bulk porous material of open cell structure, a steady flow of gas will be induced through the material. Provided that the flow velocity is small, the differential pressure P and the induced normal velocity U of gas at the surface of the material (volume velocity per unit surface area) are linearly related for a wide range of materials. The ratio of differential pressure in pascals to normal velocity in metres per second is known as the flow resistance R_f (MKS rayls) of the material.

If the material is generally of uniform composition, then the flow resistance is proportional to the material thickness L, and the flow resistivity may be defined as $R_1 = R_f/L$. The measurement of flow resistance is described in a standard[21] and may be calculated for a sample of area A from the measured quantities as

$$R_f = \frac{PA}{U}. \tag{34}$$

As has been shown, the quantities R and X of Eqs. (32b) and (32c) may be calculated in terms of the dimensionless parameter $\rho f/R_1$, where ρ is the density of the gas (air) and the other quantities have been defined. By this means, Fig. 2 has been prepared. It is found that the calculated statistical sound absorption coefficient is close to the optimal value of 0.95 when the thickness of material is greater than 0.1 wavelengths and the material flow resistance is between $2\rho c$ and $5\rho c$.

The effective thickness of the absorptive material may be increased by providing a backing cavity equal in depth to the thickness of the porous material.[24] In this case, the composite of material and cavity behaves similar to a layer of material twice as thick. No advantage is gained

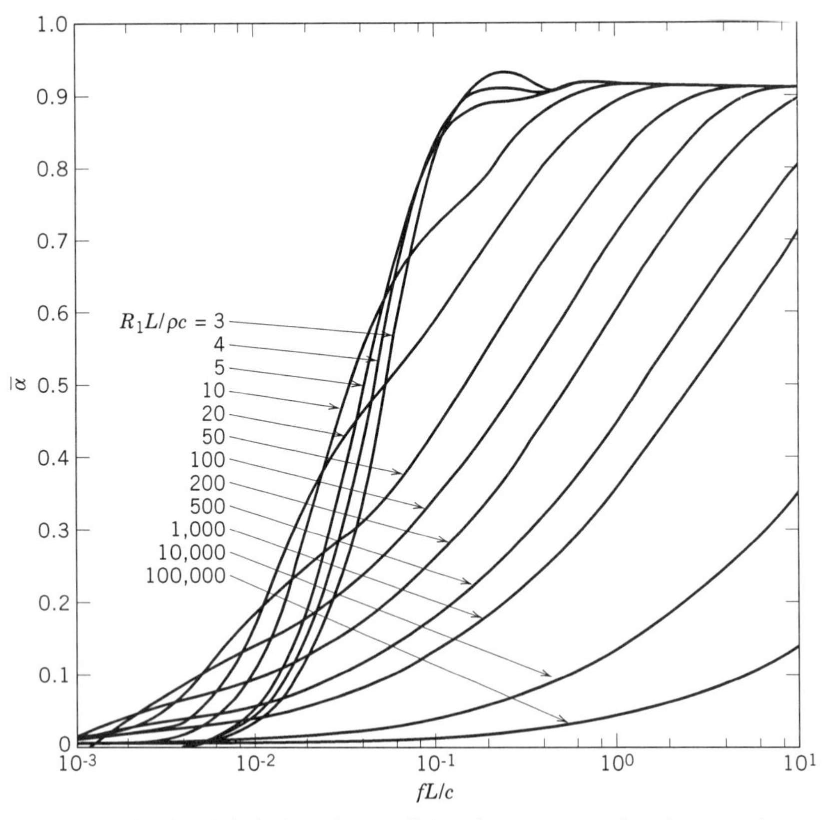

Fig. 2 Calculated statistical absorption coefficient for a porous surface in a reverberant field as a function of a frequency parameter for various indicated values of normalized flow resistance: R_1L, flow resistance of porous material (MKS rayls); ρ, density of air (kg/m^3), c, speed of sound in air (m/s); f, frequency (Hz) of incident sound; L, porous material thickness (m).

with a deeper backing cavity, and care must be taken in its construction to provide septa to prevent propagation in the cavity parallel to the material surface, which will degrade the performance of the treatment.

9 EFFECT OF PROTECTIVE FACINGS ON ABSORBING MATERIAL

When mechanical protection is needed for the porous liner, it may be covered using perforated wood, plastic, or metal panels. If the open area provided by the perforations is greater than about 25%, the expected absorption is entirely controlled by the properties of the porous liner, and the panel has no effect.

Alternatively, if the facing panel has an open area of less than 25%, then the frequency at which absorption is maximum may be calculated by treating the facing as an array of Helmholtz resonators. This procedure leads to the approximate equation

$$f_{\max} = \frac{c}{2\pi}\left[\frac{P/100}{L[t + 0.85d(1 - 0.22d/q)]}\right]^{1/2}, \quad (35)$$

where P is the percentage open area of the panel, L is the depth of the backing cavity including the porous material layer, t is the panel thickness, d is the diameter of the perforations, and q is the spacing between holes. If the porous material fills the entire cavity so that the thickness of the porous material is also L, then the speed of sound c should be replaced with $0.85c$ to account for isothermal rather than adiabatic propagation of sound in the porous material at low frequencies. The condition $fL/c < 0.1$ must be satisfied for this equation to give results with less than 15% error. Measured data for a panel of 10% open area are presented in Fig. 3.

If a perforated facing is used over a porous liner wrapped in an impervious membrane, care must be taken to ensure that the facing and liner are not in contact; otherwise, the sound-absorbing properties of the combination will be severely degraded. Noncontact is usually achieved in practice by placing a 12-mm-square wire mesh spacer between the liner and perforated sheet.

10 MODALLY REACTIVE BOUNDARIES

Modal coupling has received attention in the literature concerned with sound level prediction in vehicles but generally not in rooms. However, one case of some importance for the consideration of sound absorption in rooms has been considered and may be reported here.

10.1 Panel Absorption: Analytical Approach

When low-frequency absorption is required, panel absorbers are often the best option. To achieve absorp-

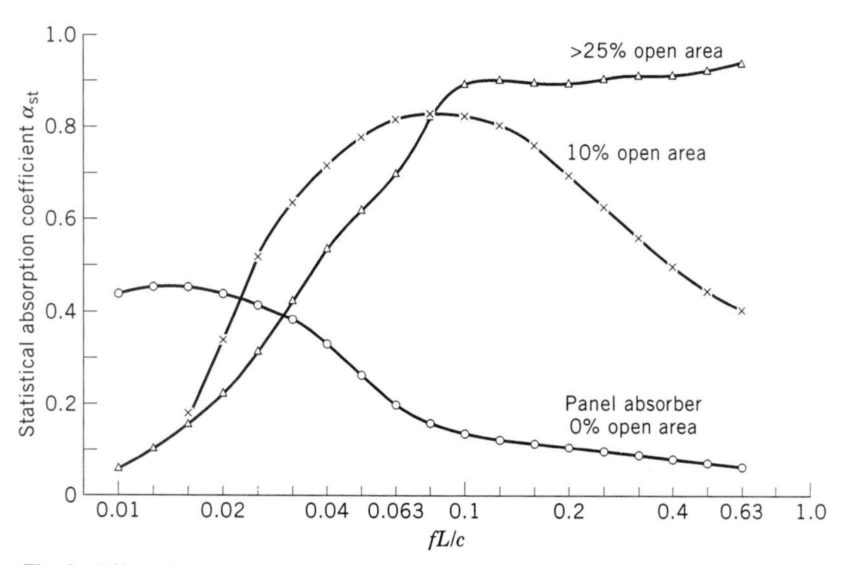

Fig. 3 Effect of perforations on the sound absorption of a plywood panel, 3 mm thick and 2.5 kg/m^2 surface weight, backed by a cavity 50 mm thick and filled with a flexible porous liner of flow resistance $5\rho c$.

tion, panels are mounted spaced out from the wall providing a shallow backing cavity and resonant response in the frequency range of interest. Sometimes the cavity is filled with a porous material to improve the absorption.

The Sabine absorption coefficient of n resonant panels of total surface area S and individual surface area A_p may be calculated by using the following formulas.[25] It is to be noted that in this case the Sabine absorption coefficient is explicitly a function of the properties of the room as well as the properties of the panel, and consequently fairly good results can be expected in using the formula. Unfortunately, the price paid for good results will be quite a few measurements to determine the properties of both room and panels:

$$\overline{\alpha} = \frac{4V\pi f}{Sc}\left\{\left|\left(\eta_A + \frac{n_p}{n_A}\eta_{pA} + \eta_{pT}\right) - \left[\left(\eta_A + \frac{n_p}{n_A}\eta_{pA} - \eta_{pT}\right)^2 + 4\frac{n_p}{n_A}\eta_{pA}^2\right]^{1/2}\right| - 2\eta_A\right\}, \tag{36}$$

where the quantities in the above equation are as follows.

The room loss factor with the panel absent is

$$\eta_A = \frac{2.20}{fT_{60A}}. \tag{37}$$

The panel loss factor as measured in an anechoic room is

$$\eta_{pT} = \frac{2.20}{fT_{60p}}. \tag{38}$$

The panel coupling loss factor is

$$\eta_{pA} = \frac{\rho c \sigma_{\text{rad}}}{2\pi m_p}. \tag{39}$$

The panel modal density is

$$n_p = \frac{\sqrt{3}A_p}{c_L h}. \tag{40}$$

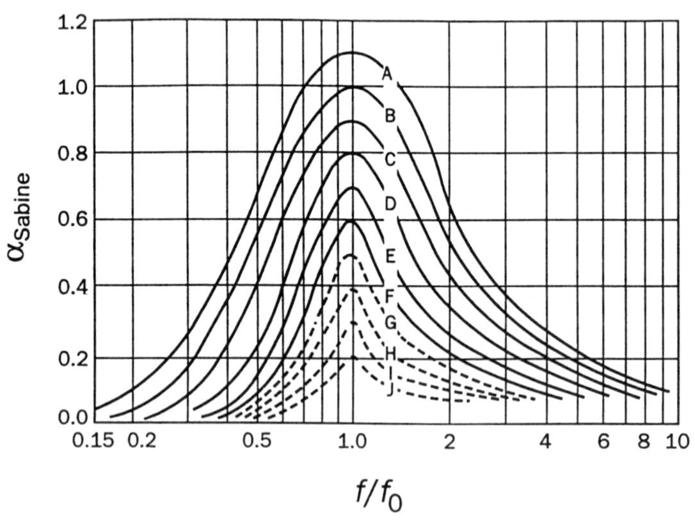

Fig. 4 Sabine absorption coefficients for resonant plywood panels. The panel configurations corresponding to the curves labeled A–J may be identified using Fig. 5. Dashed curves (G–J) represent configurations with no absorptive material in the cavity behind the panel. Configurations A–F require a sound-absorbing blanket between the panel and backing wall. The blanket must not contact the panel and should be between 10 and 50 mm thick and consist of glass or mineral fiber with a flow resistance between 1000 and 2000 MKS rayls. Panel supports should be at least 0.4 m apart. Courtesy of Hardwood Plywood & Veneer Association.

The surface area of all panels is

$$S = \sum_{i=1}^{n} A_{pi}. \tag{41}$$

The modal density n_A for any room may be calculated using Eq. (1). The quantities T_{60A} and T_{60p} are, respectively, the reverberation times of the room (without panels) and the panels (in a semianechoic space), c_L is the longitudinal wave speed in the panel, f is the center frequency of the frequency band of interest, h is the panel thickness, ρ is the density of air, and m_p is the mass per unit surface area of the panel.

The panel radiation ratio σ_{rad} at frequency f used in the preceding equations may be calculated by using the equations

$$\sigma_{\text{rad}} = 0.367\gamma \frac{c_L}{c} \frac{Ph}{A_p} \sin^{-1}\sqrt{\frac{f}{f_c}}, \qquad f < f_c, \tag{42}$$

$$\sigma_{\text{rad}} \simeq \sqrt{\frac{Pf_c}{2c}}, \qquad f \simeq f_c, \tag{43}$$

$$\sigma_{\text{rad}} = \sqrt{1 - \frac{f_c}{f}}, \qquad f > f_c, \tag{44}$$

$$f_c = 0.551 \frac{c^2}{c_L h}. \tag{45}$$

In the preceding equations the quantities are the panel perimeter P and area A_p. The panel is assumed to be isotropic of uniform thickness h and longitudinal wave speed c_L. For steel and aluminum c_L takes the value of about 5200 m/s, while for wood the value lies between 3800 and 4500 m/s. For simply supported panels γ takes the value of 1 while for clamped edge panels γ takes the value of 2. All other conditions will lie between these extremes.

10.2 Panel Absorption: Empirical Approach

A less accurate but much simpler empirical prediction scheme for determining the Sabine absorption coefficient is presented in Figs. 4 and 5. First the Sabine absorption required is selected from curves *A–J* in Fig. 4. The solid curves refer to panels for which the backing cavity contains a porous material 25 mm thick, with a flow resis-

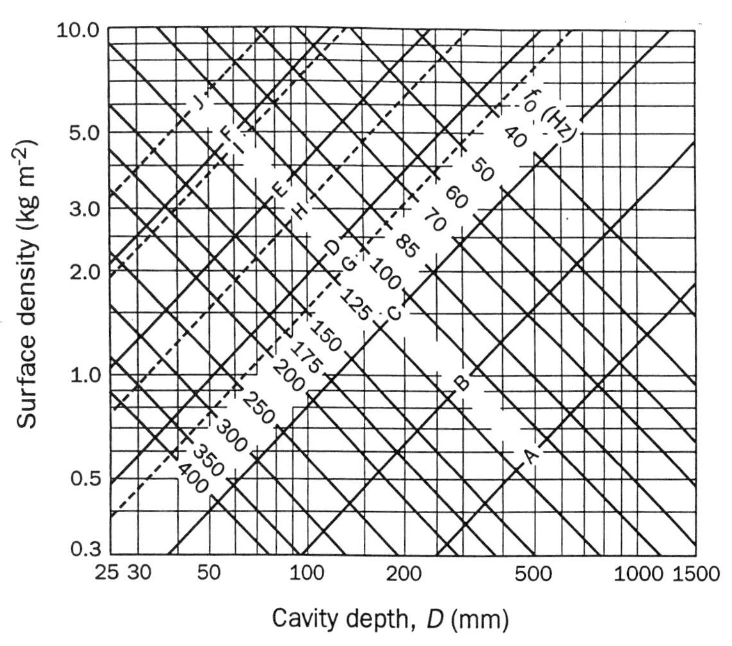

Fig. 5 Design curves for resonant plywood panels, to be used in conjunction with Fig. 4. The quantity f_0 is the frequency at which maximum sound absorption is required. Courtesy of Hardwood Plywood & Veneer Association.

tance of between $2\rho c$ and $5\rho c$, while the dashed curves refer to panels for which there is no porous material in the backing cavity. The frequency at which maximum absorption is required, f_0, is then used to enter Fig. 5, from which the required depth of backing cavity and surface density of the panel may be determined.

Acknowledgments

The authors thank Fergus Fricke for helpful discussions and contructive criticisms.

REFERENCES

1. P. M. Morse and R. H. Bolt, "Sound Waves in Rooms," *Rev. Modern Phys.*, Vol. 16, 1944, pp. 65–150.
2. L. Cremer, M. Heckl, and E. E. Ungar, *Structure-Borne Sound*, Springer-Verlag, New York, 1973.
3. D. A. Bies and C. H. Hansen, *Engineering Noise Control: Theory and Practice*, 2nd. ed., E. and F. N. Spon, London, 1996.
4. W. H. Sabine, *Collected Papers on Acoustics*, Dover, New York, 1964.
5. D. A. Bies, "Notes on Sabine Rooms," *Acoust. Austral.*, Vol. 23, 1995, pp. 97–103.
6. C. F. Eyring, "Reverberation Time in 'Dead Rooms,'" *J. Acoust. Soc. Am.*, Vol. 1, 1930, pp. 217–41.
7. G. Millington, "A Modified Formula for Reverberation," *J. Acoust. Soc. Am.*, Vol. 4, 1932, pp. 69–82.
8. W. J. Sette, "A New Reverberation Time Formula," *J. Acoust. Soc. Am.*, Vol. 4, 1933, pp. 193–210.
9. D. Fitzroy, "Reverberation Formula Which Seems to be More Accurate with Nonuniform Distribution of Absorption," *J. Acoust. Soc. Am.*, Vol. 31, 1959, pp. 893–97.
10. R. W. B. Stephens and A. E. Bate, *Wave Motion and Sound*, Edward Arnold, London, 1950.
11. M. J. Crocker and A. J. Price, *Noise and Noise Control*, CRC Press, Cleveland, OH, 119 Volume I, 1975, and Volume II, 1985.
12. H. Kuttruff, *Room Acoustics*, 2nd ed., Applied Science, London, 1979.
13. L. J. Doelle, *Environmental Acoustics*, McGraw-Hill, New York, 1972.
14. International Standardization Organization, "Measurement of Sound Absorption in a Reverberation Room," ISO R354-1963, 1963.
15. Australian Standard, "Measurement of Sound Absorption in a Reverberation Room," AS 1045-1988, 1988.
16. American Society for Testing and Materials, "Test Method for Sound Absorptoin and Sound Absorption Coefficients by the Reverberation Room Method," ASTM C423-81a, 1981.
17. American Society for Testing and Materials, "Practice for Mounting Test Specimens during Sound Absorption Tests," ASTM E795-81, 1981.
18. T. D. Northwood, "Absorption of Diffuse Sound by a Strip or Rectangular Patch of Absorptive Material," *J. Acoust. Soc. Am.*, Vol. 38, 1963, pp. 1173–1177.
19. W. A. Davern and P. Dubout, "First Report on Australian Comparison Measurements of Sound Absorption Coefficients," CSIRO—Division of Building Research, Highett, Victoria, Australia, 1980.
20. J. Piercy, T. Embleton, and L. Sutherland, "Review of Noise Propagation in the Atmosphere," *J. Acoust. Soc. Am.*, Vol. 61, 1977, p. 1403.
21. American Society for Testing and Materials, "Test Method for Airflow Resistance of Acoustical Materials," ASTM C522-80, 1980.
22. American Society for Testing and Materials, "Test Method for Impedance and Absorption of Acoustic Materials by the Impedance Tube Method," ASTM C384-77, 1977.
23. Australian Standard, "Method for Measurement of Normal Incidence Sound Absorption Coefficient and Specific Normal Acoustic Impedance of Acoustic Materials by the Tube Method," AS 1935–1976, 1976.
24. D. A. Bies and C. H. Hansen, "Flow Resistance Information for Acoustical Design," *Appl. Acoust.*, Vol. 13, 1980, pp. 357–391.
25. J. Pan and D. A. Bies, "The Effect of Fluid-Structural Coupling on Acoustical Decays in a Reverberation Room in the High-Frequency Range," *J. Acoust. Soc. Am.*, Vol. 87, 1990, pp. 718–27.

76

SOUND INSULATION: AIRBORNE AND IMPACT

A. C. C. WARNOCK AND WOLFGANG FASOLD

1 INTRODUCTION

Problems of sound propagation in buildings can be considered as the classic questions of architectural acoustics. Although early publications date back to the beginning of the twentieth century,[1] theoretical solutions sufficient to describe the sound insulation of building elements and of buildings did not appear until between 1930 and 1950.[2–6]

In buildings, it is customary, to differentiate between airborne and structure-borne sound excitation. Airborne sound originates from a source radiating sound waves into the air that then impinge on the surfaces of building elements. Examples are human speech or music from a loudspeaker. Problems of airborne sound insulation occur between adjacent rooms, between neighboring buildings, and with respect to outdoor sound sources entering the building.

Structure-borne sound is generated by direct mechanical excitation of building elements, for example, by machines and by people walking. Footstep noise is an example of impact sound, a subset of structure-borne sound. This chapter only deals with impact sound, not with structure-borne sound in general. (See Chapters 49 and 79.) Problems of impact sound insulation primarily occur with floors and stairs in apartment houses, schools, hospitals, and hotels. While perhaps not so frequent, problems can also arise with slamming doors in cupboards and in hallways.

Note: References to chapters appearing only in the *Encyclopedia* are preceded by *Enc.*

2 AIRBORNE SOUND INSULATION

2.1 Sound Reduction Index or Sound Transmission Loss

Sound transmission through a wall separating two adjacent rooms is shown in principle in Fig. 1.[7] The sound power W_1 that impinges on the surface of the wall in room 1 (the source room) partly reflects back into the room, partly propagates to other connecting structures (W_3 and W_4), partly dissipates as heat within the material of the wall, and partly transmits into room 2 (the receiving room) with sound power W_2.

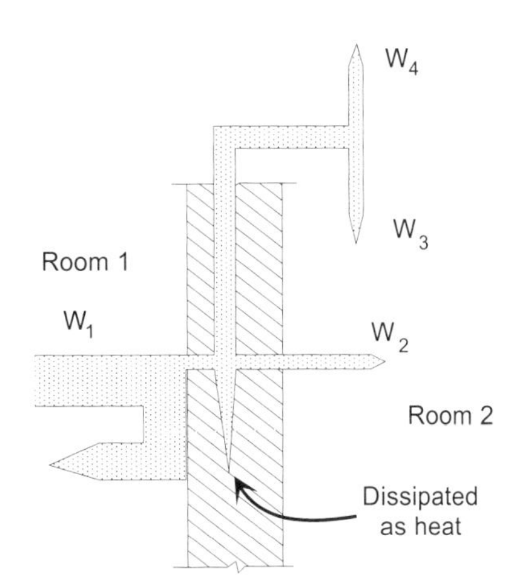

Fig. 1 Principle of sound transmission through a wall: W_3 and W_4 represent flanking sound transmitted to other parts of the structure; W_3 is eventually radiated into room 2, W_4 is not.

The sound transmission coefficient τ is defined as

$$\tau = \frac{W_2}{W_1}. \quad (1)$$

Typically, sound transmission coefficients are very small numbers, and it is more convenient to use the sound reduction index R, which is defined as

$$R = 10 \log \frac{1}{\tau} \quad \text{dB}. \quad (2)$$

The sound reduction index is also known as sound transmission loss (TL), and both terms are used in this chapter. Sound transmission varies with the frequency of the incident sound waves, and measurements are usually made in one-third-octave bands.

2.2 Airborne Sound Reduction Measurement

Methods for measuring sound insulation are specified in national and international standards. The most common method of determining airborne sound insulation is the two-room method.[8–10] These measurements are performed in laboratory test facilities in which the only path for the transmission of sound is through the specimen; transmission along other paths is suppressed.[11] In some cases, however, testing is carried out with well-defined flanking conditions[12] that are appropriate for the building types used in a region. *Flanking transmission* is transmission along some path other than that directly through the main partition. It is discussed further in Section 2.15. In Fig. 1 W_3 is an example of flanking transmission.

The test specimen is mounted between a source room and a receiving room, both of which are hard surfaced and reverberant. In the source room an approximately diffuse sound field is generated using loudspeakers, usually fed with white noise. The average sound pressure level is measured in each room in one-third-octave bands. The sound pressure level in the receiving room is determined by the sound reduction index (transmission loss) R of the test specimen, the area of the test specimen, and the equivalent sound absorption area of the receiving room. The relationship between the measured quantities is

$$R = L_1 - L_2 + 10 \log \frac{S}{A} \quad \text{dB}, \quad (3)$$

where L_1 is the sound pressure level in the source room and L_2 the sound pressure level in the receiving room (both in decibels) and S is the area of the test specimen and A the equivalent sound absorption area (also called sound absorption) of the receiving room (in square metres).

Here, A can be determined from

$$A = \frac{0.921 V d}{c} \quad \text{m}^2, \quad (4)$$

or from

$$A = 0.16 \frac{V}{T} \quad \text{m}^2, \quad (5)$$

where V is the volume of the receiving room (in cubic metres), d is the rate of decay of the sound pressure level in the room (in decibels per second), $T = 60/d$ is the reverberation time of the receiving room (in seconds), and c is the speed of sound in air (in metres per second). The velocity of sound in air is temperature dependent and is given by $c = 20.047\sqrt{273.15 + t}$ metres per second, where t is temperature in degrees Celsius.

Laboratory methods are used to measure the sound insulation of building elements for comparison with building code requirements and to allow the comparison of different systems. Similar methods are also available to check the quality of sound insulation achieved in buildings.[13,14] In the International Standardization Organization (ISO) building acoustics standards a prime after a symbol, for example R'_w, means that flanking sound transmission is either suspected or known to be present.

2.3 Sound Intensity Measurements

Since 1981,[15] sound intensity measurements have been used to determine sound insulation. The use has been standardized[16] but not widely. Many researchers[17,18] have found differences at low frequencies between results from the standard two-room test procedures and those from intensity measurements. The test procedure consists of measuring the average sound pressure in a room on one side of a test specimen, $\langle p^2 \rangle$, and then measuring the average sound intensity over the surface of the specimen on the other side, $\langle I \rangle$. The method offers the possibility of obtaining the sound reduction index without the influence of flanking transmission in laboratories or in buildings, but great care is needed to obtain accurate results.

In a diffuse sound field, the intensity incident on a specimen is $\langle p^2 \rangle / 4\rho c$. Thus the transmission coefficient is given by

$$\frac{1}{\tau} = \frac{\langle p^2 \rangle}{4\rho_0 c \langle I \rangle}, \quad (6)$$

or in decibels,

$$R = L_p - L_I - 6 \text{ dB}, \tag{7}$$

where L_p is the sound pressure level re 2×10^{-5} Pa and L_I the sound intensity level re 10^{-12} W/m^2 (both in decibels).

2.4 Short Test Methods

Short test methods for airborne and impact sound insulation of building elements have lost their importance since advanced instrumentation is available to simplify measuring and evaluating procedures. Methods that have been used in practice are based on the fact that an approximate single number rating can be obtained by measuring a frequency-weighted sound pressure level in the receiving room. The weighting curve depends on the spectrum of the sound source. Problems arise because formulas defining the sound reduction index and normalized impact sound pressure level (see Sections 2.1 and 3.1) require knowledge of the reverberation time in the receiving room. Therefore, such short test methods may be efficient only in cases in which an estimate of reverberation time is permissible.

2.5 Facade Sound Reduction

To measure the airborne sound insulation of facade elements and facades,[19] special methods are in use. For measurement with loudspeaker noise, the loudspeaker is placed outside the building at a distance of about 10 m radiating at an angle of about 45° to the test specimen. Another measuring method especially suitable for buildings near roads with high traffic density uses the traffic noise that impinges on the surface of the test specimen. If the test specimen can be fully opened (e.g., window or door) and is mounted in a heavy wall, the airborne sound insulation can be determined by measuring the average sound pressure level in the receiving room with the specimen in the open and in the closed position.

2.6 Single-Number Ratings

Examples of sound transmission measurements for some common building materials are shown in Fig. 2. Such plots contain more information than is convenient for use in building codes. Methods have been standardized whereby the frequency-dependent values of airborne sound insulation can be converted into a single-number rating by means of a reference curve.[20–22] These rating systems can be applied to many different types of sound reduction measurements, not just transmission loss or sound reduction index.

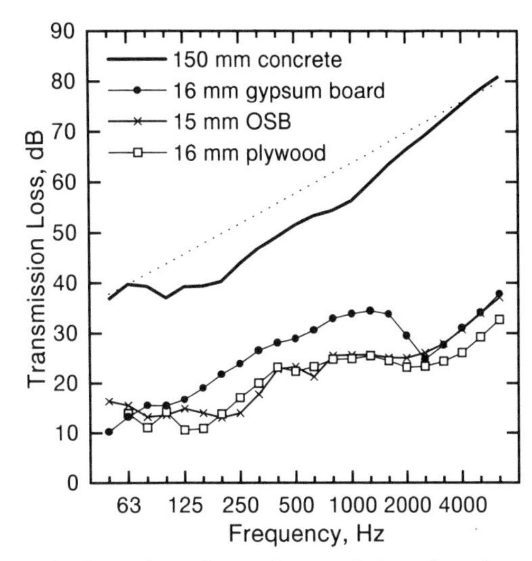

Fig. 2 Examples of sound transmission through some common building materials: 150-mm concrete (345 kg/m^2), 16-mm plywood (7.5 kg/m^2), 16-mm gypsum board (11.3 kg/m^2), 15-mm oriented strand board (OSB, 8.8 kg/m^2). Dashed line represents mass law predictions for 345 kg/m^2.

The ISO and American Society for Testing and Materials (ASTM) reference curves are shown in Fig. 3 fitted to measured curves. To evaluate the results of a measurement, the ISO reference curve (100–3150 Hz) is shifted

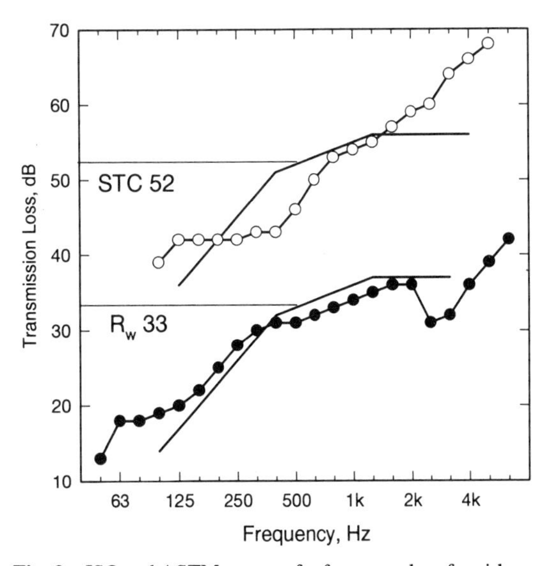

Fig. 3 ISO and ASTM curves of reference values for airborne sound compared with measurement results: (●) two layers of 13-mm gypsum board (20 kg/m^2); (○) 100-mm concrete (230 kg/m^2).

in steps of 1 dB toward the measured curve until the mean unfavorable deviation is as large as possible but not more than 2 dB. An unfavorable deviation at a particular frequency occurs when the measured result is less than the reference value. The mean unfavorable deviation is calculated by dividing the sum of the unfavorable deviations by 16, the total number of measurement frequencies. The value, in decibels, of the reference curve at 500 Hz after shifting it according to this procedure is denoted the *weighted sound reduction index* R_w. (In ISO terminology, a subscript w after a symbol means that the value presented is a weighted single-number rating of a one-third-octave band spectrum. The weighted sound reduction index R_w was formerly known as the airborne sound insulation index I_a.)

A similar procedure is used to calculate the *sound transmission class* (STC). Note that the frequency range of the STC reference curve is from 125 to 4000 Hz. During the evaluation procedure attention has to be paid to the maximum unfavorable deviation. If it exceeds 8 dB at any frequency, the ISO standard requires that it be recorded and reported. The ASTM standard does not allow any unfavorable deviation to exceed 8 dB. Once this value is attained, the reference curve can no longer be adjusted. Despite the difference in procedures, the numerical values of the STC and R_w are usually very close, although when the 8-dB rule controls the STC rating, differences of several decibels may occur. Where only one rating is given in this chapter, assume the value holds for both ratings.

Although the shapes of the reference curves were derived from a mean spectrum of noise inside dwellings, they have been also used for outdoor noise, especially for traffic noise; the standardized evaluation procedure is applied to exterior walls, windows, and doors as well as to interior partitions. This use provoked criticism and led to proposals for specific reference curves with a changed slope and frequency range that take into account the higher importance of low frequencies in the traffic noise spectrum.[23,24]

2.7 Single-Leaf Building Elements

The sound reduction index of a homogeneous, limp, nonporous plate for plane-wave incidence is given by the *mass law*[2,25] as

$$R = 10 \log \left[1 + \left(\frac{\pi f M \cos \theta}{\rho_0 c} \right)^2 \right] \quad \text{dB}, \qquad (8)$$

where f is frequency (in hertz); M is mass per unit area of the plate (in kilograms per square metre); θ is the angle of incidence; ρ_0 is the density of air, typically 1.2 kg/m^3; and c is the velocity of sound in air (in metres per second). For $t = 20°\text{C}$, $c = 343$ m/s and $\rho_0 c = 412\ \text{N} \cdot \text{s/m}^3$. The mass law predicts an increase in sound reduction index of about 6 dB for each doubling of mass per unit area or frequency.

Bending waves in panels with finite stiffness give rise to the *coincidence effect*.[5] The velocity of bending waves in panels increases with increasing frequency according to

$$c_B = \left(\frac{B\omega^2}{m} \right)^{1/4}, \qquad (9)$$

where B is the bending stiffness per unit width, $\omega = 2\pi f$ is the angular frequency (in reciprocal seconds), and m is the panel surface density (in kilograms per square metre). At every frequency above a certain *critical frequency* f_c there is an angle of incidence for which the wavelength of the bending wave becomes equal to the wavelength of sound in air projected on to the plate; the waves coincide. Because of the coincidence effect, a region of relatively low sound insulation occurs in some frequency range in the curve of the sound reduction index. For example, the curve for gypsum board in Fig. 2 shows a coincidence dip with a minimum at 2500 Hz. The critical frequency for a thin panel can be calculated from

$$f_c = \frac{c^2}{2\pi} \sqrt{\frac{\rho h}{B}} \quad \text{Hz}, \qquad (10)$$

where c is the velocity of sound in air (in metres per second), ρ is the density of material (in kilograms per cubic metre), h is the thickness of the panel (in metres), and B is the bending stiffness per unit width (in kilograms by square metres per square second). The bending stiffness is related to Young's modulus by

$$B = \frac{Eh^3}{12} (1 - \nu^2) \quad \text{kg} \cdot \text{m}^2/\text{s}^2, \qquad (11)$$

where E is Young's modulus (in newtons per square metre) and ν is Poisson's ratio. Thus for a given material the product $f_c h$ is a constant that depends only on the physical properties of the material. Table 1 gives values of this product for common building materials. The resonance dip due to the coincidence effect usually begins about an octave below the critical frequency. The depth of the resonance dip depends on the damping of the panel. Below the frequency range of coincidence the sound reduction index is determined by the mass law.

TABLE 1 Surface Mass for 1-mm Thicknesses and Constant $f_c h$ for Some Common Building Materials

Material	Surface Mass ρh(kg/m^2)	$f_c h$ (Hz · mm)
Aluminum	2.7	12,900
Concrete, dense poured	2.3	18,700
Hollow concrete block	1.1	20,900
Fir timber	0.55	8,900
Glass	2.5	15,200
Lead	11.0	55,000
Plexiglass or Lucite	1.15	30,800
Steel	7.7	12,700
Gypsum board	0.7	39,000
Oriented strand board[a]	0.6	15,000
Plywood[a]	0.5	16,000

[a]These materials are orthotropic and show very broad, shallow coincidence dips. The values given estimate the frequency at the middle of this dip.

Above the region of coincidence, the sound reduction index depends on frequency and is given by

$$R = R_{\perp} + 10 \log \frac{2\eta f}{\pi f_c} \quad \text{dB}, \qquad (12)$$

where η is the loss factor and $R_{\perp}$ is the sound reduction index for normally incident sound ($\theta = 0$), which is given by

$$R_{\perp} = 20 \log \frac{\pi f M}{\rho_0 c} \quad \text{dB}. \qquad (13)$$

Thus, for a constant loss factor, the sound reduction index increases by 9 dB/octave above the critical frequency. The data shown in Fig. 1 for a 150-mm concrete slab increase about 8 dB/octave. In reality, damping in materials is frequency dependent.

The coincidence effect can greatly influence the sound insulation, and where possible, it seems best to position the critical frequency below or above the important frequency range where single-number ratings are calculated. To place f_c at a very low frequency requires a building element with high thickness and stiffness. Heavy walls and floors of brick or concrete usually fulfill this requirement. In any case, it is not usually practical to change the thickness or material properties of such walls. The data for 150-mm concrete in Fig. 2 show that most of the frequency range is above coincidence. The dotted line in the figure is the mass law prediction for a limp layer of the same weight as the concrete.

To place f_c at a high frequency requires a thin and heavy plate of low stiffness. Because of the shape of the reference contours for R_w and STC, unless the coincidence dips at high frequencies are very deep, they do not often reduce the single-number rating.

In some cases, because of the coincidence effect, using a thicker layer of material may not improve R_w or STC. Although mass per unit area is increased, the coincidence dip may be shifted into the frequency range most important for determining the single-number rating. This can be avoided by using two or more layers with the same total thickness. When the layers are attached normally using screws or dabs of glue, the bending stiffness of the double layer is not significantly altered from that of a single layer. Thus the benefits of additional mass per unit area are retained without incurring the loss due to coincidence. The data for the double layer of gypsum board in Fig. 5 give one example of this. The ratings for laminated glass in Fig. 17 give others.

For a random distribution of angles of incidence (diffuse field) the sound reduction index R_{diff} is given by

$$R_{\text{diff}} = R_{\perp} - 10 \log 0.23 R_{\perp} \quad \text{dB}. \qquad (14)$$

In practice, sound fields in rooms do not generate an exactly random distribution of angles of incidence on the surface of a building element. There is a lack of waves at grazing incidence, and real panels are not infinite in size. A commonly used approximation is the *field incidence mass law*

$$R = R_{\perp} - 5 = 20 \log Mf - 47 \quad \text{dB}. \qquad (15)$$

2.8 Orthotropic and Corrugated Materials

Many common building materials are not isotropic, and the bending stiffness varies for different directions along the panels. Plywood and oriented strand board are two such materials that are quite strongly orthotropic. Such materials show a greatly broadened coincidence dip (see Fig. 1). Methods for calculating the sound insulation of such materials are given in Refs. 26 and 27.

Corrugating a sheet of material or adding ribs to it increases the bending stiffness in one direction and creates a new coincidence frequency. This can result in significant reductions in sound insulation, as shown in Fig. 4. Methods for estimating sound transmission through corrugated sheets are given in Refs. 26 and 28.

2.9 Single-Leaf Poured Concrete

To avoid a reduction in sound insulation due to transmission through the pores of materials, porous building elements such as foamed lightweight concrete should be plastered on at least one side. Inhomogeneous elements

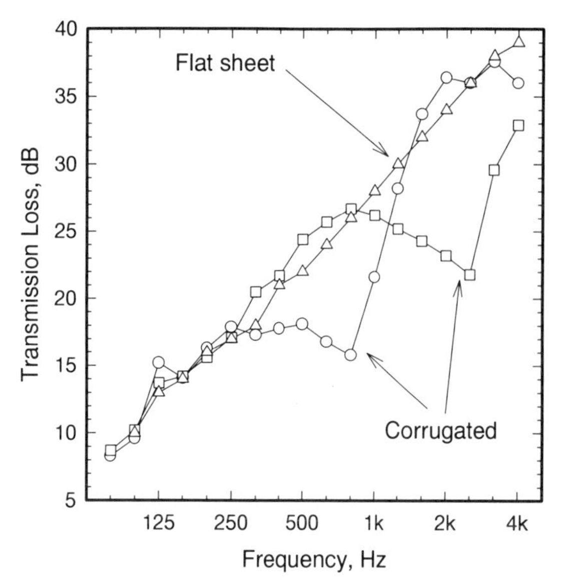

Fig. 4 Sound transmission loss for 0.8-mm steel sheets (6 kg/m²), flat and with two types of corrugations.

such as floors and exterior masonry with large cavities may provide lower sound insulation because of resonance effects.[29] Small cavities, however, such as in hollow bricks, mostly do not influence sound insulation.

Table 2 gives representative STC values for some common concrete walls or floors measured in laboratories. It will be seen that simple concrete partitions can provide STC ratings from about 45 to 58. To achieve an STC much greater than 58, the weight required is likely to be prohibitive in modern buildings, because the surface mass must be doubled for an increase of 6 dB. The usual solution where high STC ratings are needed is to use multilayer partitions—a central massive wall with one or more leafs attached to each side. This is discussed in Section 2.12.

TABLE 2 Sound Transmission Class for Some Common Concrete Constructions

Concrete Construction	Surface Mass (kg/m²)	STC
Solid slab		
50 mm	115	43
70 mm	161	47
100 mm	230	52
150 mm	345	55
200 mm	460	58
Hollow-core slab		
150 mm	220	48
200 mm	278	50
250 mm	312	50
Precast double T		
355 mm deep with 50 mm topping	366	54

2.10 Single-Leaf Masonry Walls

In the thicknesses commonly used in buildings, hollow-core concrete block provides a few decibels less sound insulation than a solid concrete wall or brick wall of the same weight per unit area, provided all cracks, openings, and voids are properly sealed.

Solid and hollow-core concrete blocks are manufactured in a variety of thicknesses, core sizes, aggregates, and densities. Representative values of the STC for block walls are given in Table 3. These values are only valid if the wall surfaces are properly sealed and the mortar joints are sound.

The sound insulation provided by a hollow-core block depends not only on the density of the block material but also on its porosity. The more porous the block, the more sound will leak through the block structure and the greater will be the improvement if the surfaces of the block are sealed (e.g., by concrete paint, epoxy paint, or a skim coat of plaster). It is enough to apply the sealant on one face of the block only. In fact, when gypsum board is

TABLE 3 Sound Transmission Class Ratings for Standard Hollow Normal and Lightweight Block Walls Sealed on at Least One Side

Nominal Thickness (mm)	Lightweight			Normal Weight		
	kg/block	kg/m²	STC	kg/block	kg/m²	STC
100	8	105	43	10	130	44
150	10	130	44	15	190	46
200	14	180	46	18	225	48
250	17	215	47	21	260	49
300	20	250	49	25	310	51

to be added on resilient supports to finish the wall, this can be an advantage (see Section 2.12). Improvements of 5–10 STC points are not uncommon after sealing. A layer of gypsum board attached directly with screws or dabs of glue to the block surface is *not* an effective seal because it is still able to vibrate as a separate layer. Even when gypsum board on studs or resilient supports is to be used to finish a block or concrete wall, the wall should be sealed. Covering a flaw with gypsum board does not eliminate the detrimental effects of the flaw.

The porosity of acoustic materials is characterized by the airflow resistivity. This quantity is calculated from the volume velocity of air flowing through a specimen and the air pressure drop across it.[30,31] Blocks with airflow resistivities greater than about 2×10^5 Pa · s/m^2 show no improvement after sealing. Normal-weight hollow blocks and solid blocks are not likely to need sealing. Lightweight blocks usually do. In the absence of information on block airflow resistivity, to ensure the maximum sound reduction from a block or masonry wall, it is safer to seal the wall properly on one side. In any case, all the mortar joints must be properly finished and free from leakage.

Table 4 gives approximate sound insulation values for single-leafed building elements. These values apply when flanking transmission exists that is typical of that for construction having a mean mass per unit area of 300 kg/m^2.

2.11 Double-Leaf Building Elements

When high sound insulation is required without using heavy single-leaf constructions, the best approach is to construct a wall consisting of two solid leaves with an airspace or resilient layer between them. When properly designed, such a construction provides sound insulation greater than that for a single layer of the same mass per unit area and thickness. The total sound insulation of a double-leafed building element is, however, lower than the sum of the sound insulation of each leaf alone. This is because of the coupling through the air in the cavity and through any physical ties between the layers.

Figure 5 shows the sound reduction index through 16-mm gypsum board for four cases: (1) installed as a single layer, (2) installed as two layers screwed together, (3) installed as two layers with a cavity between them, and (4) installed as two layers with a cavity filled with glass fiber batts. This figure shows typical behavior for double-layer walls where the studs are resilient or there are independent studs for each layer of gypsum board. At the frequency marked f_r there is a dip in the transmission loss curve that reduces the sound insulation below

TABLE 4 Sound Insulation of One-Leafed Building Elements with Mass per Unit Area *M* and Thickness *d* with Flanking Typical of Construction with Mass per Unit Area of 300 kg/m^2

	R' (dB) at f (Hz)								
Material	125	250	500	1000	2000	4000	M (kg/m^2)	d (mm)	R_w^1
Hollow brick with	36	42	48	55	56	59	350	270	52
15 mm plaster on both sides	34	40	40	46	51	56	210	145	46
Brick with 15 mm plaster	40	46	51	54	59	62	480	270	55
on both sides	36	42	42	48	53	58	260	145	48
	31	39	39	37	49	53	170	105	42
	31	37	37	34	47	52	150	95	37
Dense concrete,	39	43	50	55	62	66	430	190	54
without plaster	38	42	47	54	61	64	350	150	52
	34	38	48	53	61	63	300	120	50
	30	33	37	44	51	59	170	70	42
	31	29	27	36	43	48	95	40	37
Lightweight concrete	30	37	42	50	56	60	190	250	46
with 10 mm plaster	31	32	37	45	50	56	130	220	42
on both sides	31	32	34	43	50	55	100	170	40
	28	32	30	36	46	55	85	110	36
	25	29	29	28	38	44	60	70	32
Gypsum block, without	39	40	41	50	55	58	315	260	47
plaster	31	33	34	42	50	60	105	100	41
	32	30	26	34	43	49	85	70	33

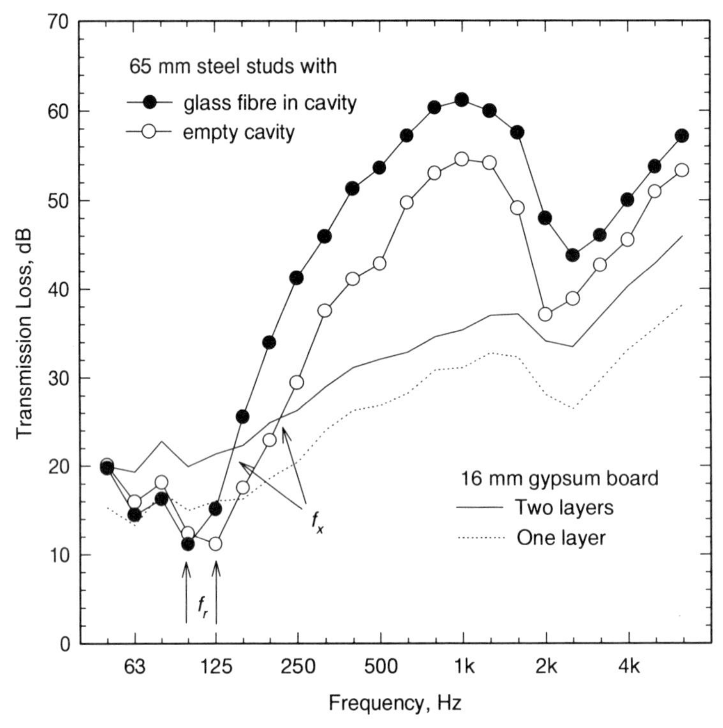

Fig. 5 Sound reduction index of double- and single-leafed gypsum board walls. The 25-gauge (0.5-mm) steel studs in the double wall are 38 × 65 mm deep. The sound-absorbing material is glass fiber with a thickness of 65 mm and a flow resistivity of 3600 N · s/m^2.

that for the two layers in contact and even below that for the single layer. The transmission loss of the double-leafed walls becomes equal to that for two leaves in contact below the resonant frequency at the lowest frequency measured. Above the resonance, the sound insulation rises steeply, crosses the curve for the two layers in contact, and increases for several bands before following a more gradual increase and then turning down into the coincidence dip.

For an air-filled cavity, the mass–air–mass resonance frequency f_r can be calculated from

$$f_r = \frac{1}{2\pi}\sqrt{\frac{\rho_0 c^2}{d M_{\text{eff}}}}, \qquad (16)$$

where d is the distance between the inner surfaces (in metres) and M_{eff} is the effective mass, defined as

$$\frac{1}{M_{\text{eff}}} = \frac{1}{M_1} + \frac{1}{M_2}, \qquad (17)$$

M_1, M_2 being the masses per unit area of the two leaves (in kilograms per square metre).

For a given total mass, this frequency has its minimum value when both masses are equal. Adding sound-absorbing material to the cavity lowers the frequency at which the mass–air–mass resonance occurs. In the cavity, the propagation constant will be quite different from that in air; so the effective value of $\rho_0 c^2$ changes. The shift in the mass–air–mass resonance depends on the flow resistivity and thickness of the sound-absorbing material. As an approximate guide, when the wall cavity contains sound-absorbing material, multiply the frequency calculated from Eq. (16) by a factor of 0.7. Losses in the sound-absorbing material will influence the depth of the resonance.

Prediction of sound transmission through double-layer walls has been studied by several authors.[32–34] Two frequencies are important in making predictions: the mass–air–mass resonance and the first cross-cavity resonance f_1, which is given by

$$f_1 = \frac{c}{2\pi d} \quad \text{Hz}. \qquad (18)$$

Higher cross-cavity resonances are integer multiples of this frequency. Another advantage of having sound-absorbing material in the cavity is that it damps resonances at f_1 and higher cavity modes.

When the two panels are completely isolated from each other, the sound reduction index can be predicted approximately by

$$R = \begin{cases} 20\log[(m_1 + m_2)f] + 47, & f < 2f_r/3, \\ R_1 + R_2 + 20\log fd - 29, & f_r < f < f_1, \\ R_1 + R_2 + 6, & f > f_1, \end{cases} \quad (19)$$

where R_1, R_2 are the sound reduction indexes for each layer of the double wall measured or calculated separately. Thus, high sound insulation can be attained by selecting a high mass per unit area of the leaves and a large cavity depth to ensure a low mass–air–mass resonance frequency, for example, below 100 Hz.

Procedures have been given for calculating sound reduction indices when the panels are connected along a line or at points,[32] but no reduction for the mass–air–mass resonance is considered. Modified procedures to deal with resilient line connections (light gauge metal studs) and the mass–air–mass resonance have also been suggested.[34] The formulas in Eq. (19) combine the ideas from Refs. 32 and 34 and give an approximate way of dealing with the low-frequency resonances that invariably limit the sound insulation of lightweight double-panel walls.

In practice, it is very difficult to construct a double-panel wall where the two panels are not connected in some way, and the sound reductions predicted by Eq. (19) are seldom achieved. Except in exceptional cases, there is often a solid connection around the edges. The more rigid the connections between the panels, the lower the sound reduction. Wood studs joining the two panels provide line connections between them. Such solid connections should be avoided where possible. Resilient metal channels on one or both sides of solid wood or stiff metal studs reduce the transmission of vibration from one panel to the other and give better sound insulation but still not as good as completely isolated studs. If connections must be solid, then it is better to connect only at points rather than along a line. In all cases, connections to panels should be minimized; the closer the spacing of studs or resilient metal channels, the lower the sound reduction.[35]

Typical methods for providing mechanical isolation between double-wall layers are illustrated in Fig. 6. Non-load-bearing steel studs [typically made from 25-gauge (0.5-mm) sheet steel] are usually flexible enough to reduce substantially the vibration transfer between the layers on each side.

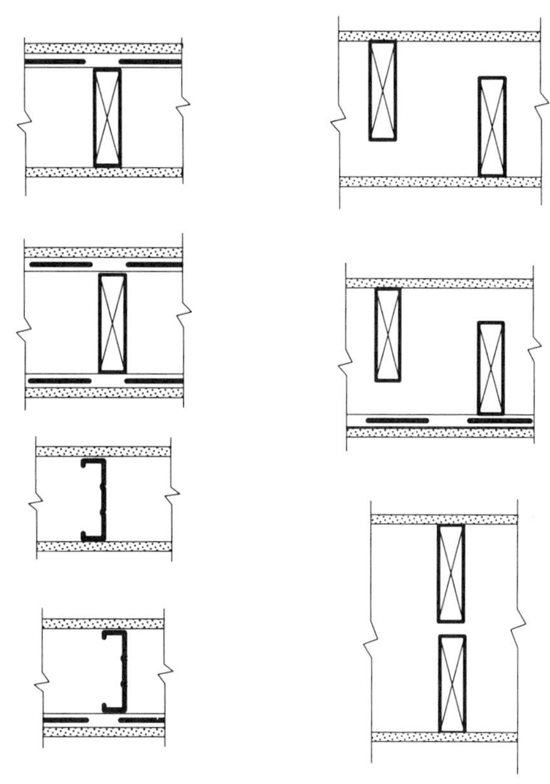

Fig. 6 Methods for providing mechanical isolation between partition layers: staggered wood studs with and without resilient metal channels and sharing a common sole and header plate, separate rows of wood or steel studs with independent sole and header plates, a single row of wood or load-bearing studs with resilient metal channels, and a single row of light-gauge metal studs.

The more sound-absorbing material in the cavity, the higher the sound insulation, but the greatest increases occur for the first 25% or so of the cavity depth.[35,36,37] Incremental increases after the cavity is about 75% full are usually small. Figure 7 shows an example of measurement results showing this effect. The absorbent material should have[38] an airflow resistivity of more than about 5 kPa · s/m^2, a value satisfied by most common fibrous materials. Higher values of resistivity do not significantly increase the STC or R_w, although there are significant decreases in sound transmission at higher frequencies.

If the cavity is tightly filled with a fairly dense absorptive material, there is some risk that this will cause connections of high stiffness between the leaves and lower the sound insulation; sound will propagate through the body of the sound-absorbing material. Overfilling the cavity should be avoided.

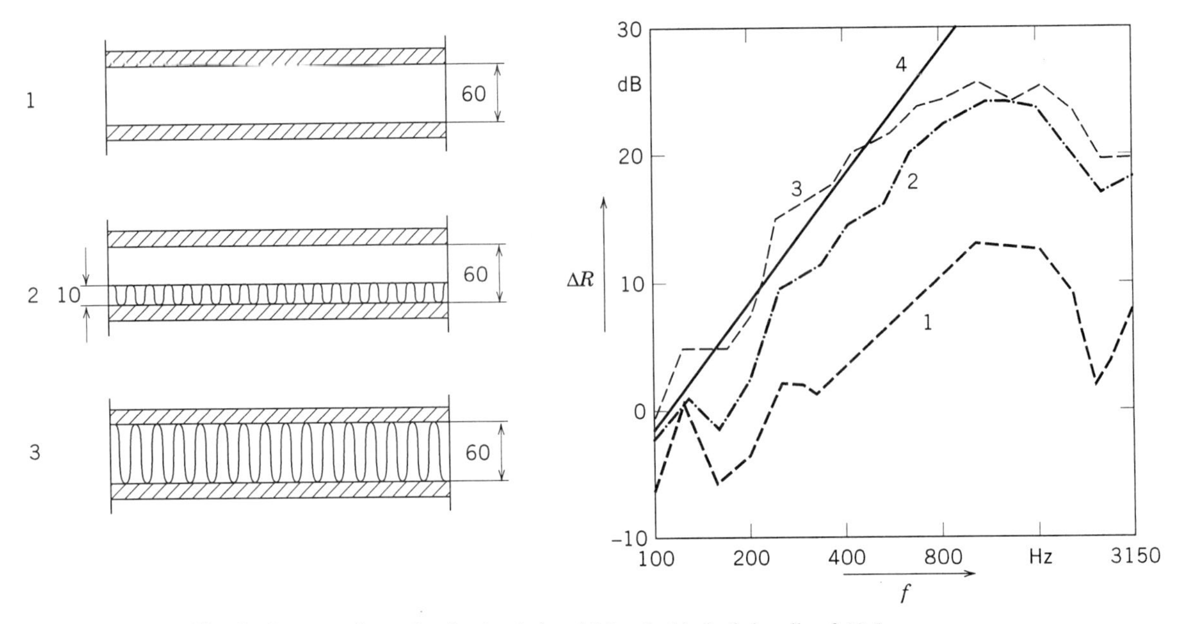

Fig. 7 Increase of sound reduction index ΔR by double-leafed walls of 12.5-mm gypsum board (cavity width 60 mm) relative to a single-leaf wall of the same mass per unit area: (1) without absorbent lining; (2) 10 mm mineral wool; (3) cavity filled with mineral wool; (4) calculated values.

The same acoustical and construction principles have to be taken into account when using a suspended ceiling to improve the sound insulation of a floor. The suspended ceiling will increase both airborne and impact sound insulation (see Sections 3.3 and 3.6). Materials like wood-wool slabs, plastering on expanded metal. gypsum board, and chipboard are in use for suspended ceilings. Best results are obtained with resilient suspensions or independent ceiling joists.

Double-Leaf Gypsum Board Walls Table 5 gives measured STC values for some walls constructed from wood or steel studs and 13- or 16-mm gypsum board.[39] The surface mass of the fire-rated gypsum board was respectively about 10 and 11 kg/m^2. Other lighter types of board were used in the study and gave correspondingly lower results. In each case, the cavity was about three-fourths to completely full of glass fiber batts. The type of sound-absorbing material has only a small effect on the single-number rating.[40] Other constructions with similar structural connections and cavity depth should give similar values of sound insulation.

In determining the sound reduction of cavity walls, the two major factors are the total mass of the layers and the depth of the cavity. For typical gypsum board cavity walls it has been found that doubling the total mass of the cavity depth increases the STC or R_w by about 10 points.[40]

Adding a single, solid, nonporous layer to one set of studs in the middle of a double stud wall with sound-absorbing material in both stud cavities reduces the sound insulation by about three to four points. A second layer added internally to the other set of studs would make the sound insulation even worse. In practical situations, for a given total weight, single-cavity walls perform better overall than walls with multiple cavities.

A casual examination of the cross section for a staggered wood stud wall suggests that there will be no solid connection between the layers of gypsum wallboard on each side. This conclusion overlooks the connection through the sole and header plates. Sound transmission around the periphery of such walls reduces the sound reduction for the wall. Table 5 shows that adding resilient metal channels to one side of the studs reduces this peripheral transmission and increases STC ratings. While the channels increase the depth of the cavity, this is not enough to account for the increase in STC over the case with no channels. The improvement comes from reducing the transmission through the plates around the edges of the wall.

For stud walls, even for a double stud wall, the sound reduction is influenced by the spacing of the studs and

TABLE 5 Approximate STC Ratings for Walls with 13- and 16-mm Gypsum Board on Both Faces

	Gypsum Board 16 mm			Gypsum Board 13 mm		
	1|1	1|2	2|2	1|1	1|2	2|2
38 × 89 mm wood studs, 400 mm oc	34 (37)	36 (40)		34 (39)	37 (41)	38 (43)
38 × 89 mm wood studs, 400 mm oc, resilient metal channels on one side, 600 mm oc	46 (47)	53 (52)	59 (57)	46 (48)	51 (52)	57 (57)
Staggered 38 × 89 mm wood studs	47 (48)	52 (52)	56 (56)	47 (49)	50 (52)	55 (56)
Staggered 38 × 89 mm wood studs, resilient metal channels on one side, 600 mm oc	51 (52)	56 (57)	62 (64)	49 (52)	54 (57)	60 (62)
Double row of 38 × 89 mm wood studs, 400 mm oc	58 (58)	62 (64)	67 (69)	58 (58)	62 (63)	66 (68)
Double row of 38 × 89 mm wood studs, 600 mm oc	59 (60)	64 (65)	69 (69)	57 (58)		
65-mm steel studs, 25 ga, 400 mm oc	39 (42)	45 (47)	52 (52)	36 (41)	42 (46)	48 (50)
65-mm steel studs, 25 ga, 600 mm oc	44 (44)	51 (49)	55 (53)	45 (44)	51 (49)	55 (54)
90-mm steel studs, 25 ga, 400 mm oc	49 (47)	52 (50)	56 (55)	46 (45)	51 (50)	55 (53)
90-mm steel studs, 25 ga, 600 mm oc	49 (48)	54 (53)	57 (55)	48 (47)	52 (52)	55 (54)
150-mm steel studs, 25 ga, 600 mm oc	51 (51)			52 (52)		
90-mm steel studs, ≤20 ga, 400 mm oc	49 (49)			48 (49)	53 (53)	59 (59)

Note: Numbers in parentheses are R_w ratings. The notation 1|1, 1|2, 2|2 indicates the number of layers of gypsum board on each face, oc = on center, spacing between studs; ga = gauge.

resilient channels that support the gypsum board. Within practical limits, it is best if these are as far apart as possible. Thus a 600-mm spacing gives better results than 400-mm spacing. For double wood studs the increase is about two points.

When resilient metal channels are used to support gypsum board on wood studs and the construction is unbalanced (two layers of gypsum wallboard on one side and one on the other), it does not matter which side the channels are on; the STC rating is the same. However, the assembly with two layers of gypsum wallboard attached to the resilient metal channels gives a better fire resistance rating.

In concrete buildings, there is invariably a certain amount of flanking sound transmission that reduces the overall sound insulation. This reduction can be calculated as discussed in Section 2.15. Figure 8 gives some typical values of sound insulation for double-leafed lightweight walls in situ. When joist construction is used exclusively in a building, it is very much more difficult to calculate or predict flanking, and the data in Fig. 8 are not applicable to such buildings.

Double-Leaf Masonry or Concrete Walls Two-leaf masonry or concrete walls can, in principle, provide very high sound insulation because they comprise two very heavy layers separated by an airspace. The maximum insulation attained is usually limited, however, by the practical difficulties of constructing two leafs that are structurally isolated. Depending on the height of a block wall, metal ties are required for structural reasons or to meet applicable building codes. Solid metal ties transmit sound energy from one leaf to the other, but special ties, designed to reduce sound transmission, can minimize the detrimental effects of mechanical ties.

Perhaps more important is transmission of energy along the floor and ceiling, along walls abutting the periphery of the cavity wall, and through other parts

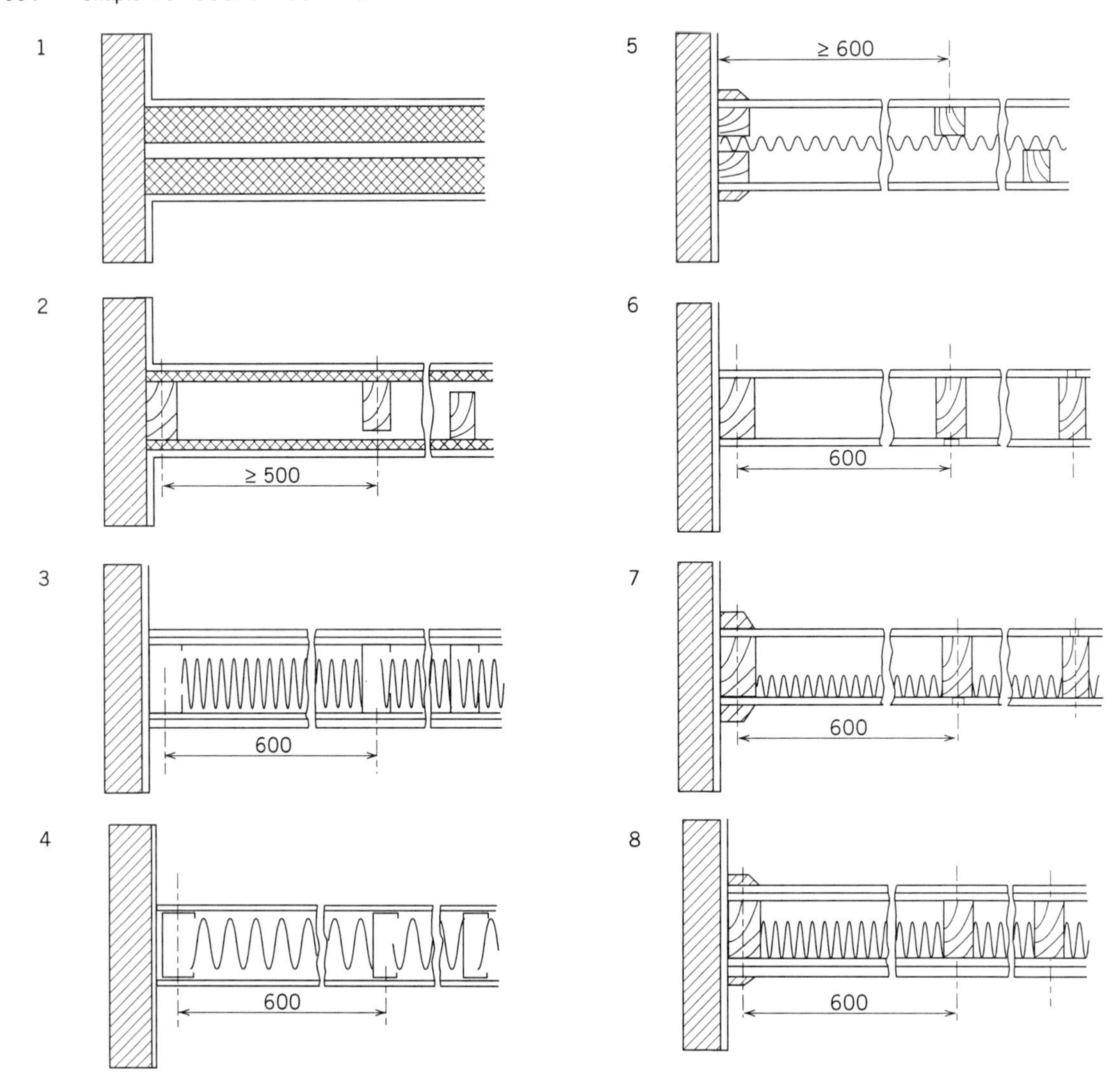

Fig. 8 Sound insulation for double-leafed lightweight walls in concrete buildings. (1) Wood–wool (50 mm), plastered; cavity (30 mm); $R'_w = 52$ dB. (2) Wood–wool (25 mm), plastered double-studs; cavity (100 mm); $R'_w = 52$ dB. (3) Plasterboard (2 × 12.5 mm); cavity (100 mm); absorptive quilt; sheetmetal studs; $R'_w = 52$ dB. (4) Like 3, but plasterboard (1 × 12.5 mm); $R'_w = 47$ dB. (5) Like 2, but plasterboard (12.5 mm) and added absorptive quilt (40 mm); $R'_w = 47$ dB. (6) Plasterboard (12.5 mm); wooden studs; cavity (100 mm); $R'_w = 37$ dB. (7) Like 6, but added absorptive quilt (40 mm); $R'_w = 42$ dB. (8) Like 7, but plasterboard (2 × 12.5 mm); $R'_w = 48$ dB.

of the structure. These *flanking paths* bypass the cavity wall and reduce its apparent sound insulation. This kind of flanking is illustrated in Fig. 9 and discussed further in Section 2.15. Physical breaks in the floor, ceiling, and abutting walls are needed to reduce transmission along such paths. However, even if gaps are included in the design, during construction the airspace between leafs may become bridged by mortar droppings or rubble that will reduce the sound insulation. Such defects are concealed and impossible to correct after the wall is complete. In practical installations careful design of the whole system, good supervision, and care during con-

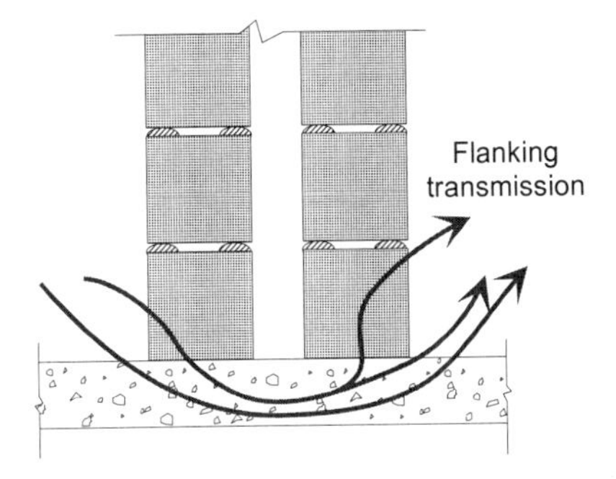

Fig. 9 Flanking paths under a cavity block (or concrete) wall resting on a continuous slab of concrete. Any solid connections between the block walls will also reduce the sound insulation.

struction are needed for two-leaf block walls to achieve their potential.

Figure 10 shows laboratory sound transmission loss results for two-leaf walls formed from two isolated 90-mm block layers with a cavity of 125 mm between them. The figure also shows results for a double-layer wall formed from 90- and 190-mm blocks with a cavity of 165 mm between them. The cavity between the blocks had a 65-mm-thick layer of glass fiber, 50-mm-thick layer of polystyrene foam, and only air in the cavity. The benefits of the glass fiber batts are clear. Table 6 gives sound insulation values for double-leafed heavy walls assuming typical conditions of flanking transmission. Although sound-absorbing material is very effective inside lightweight constructions or in heavy cavity walls in the laboratory, because of flanking, it has almost no effect on heavy concrete or masonry walls in field situations.

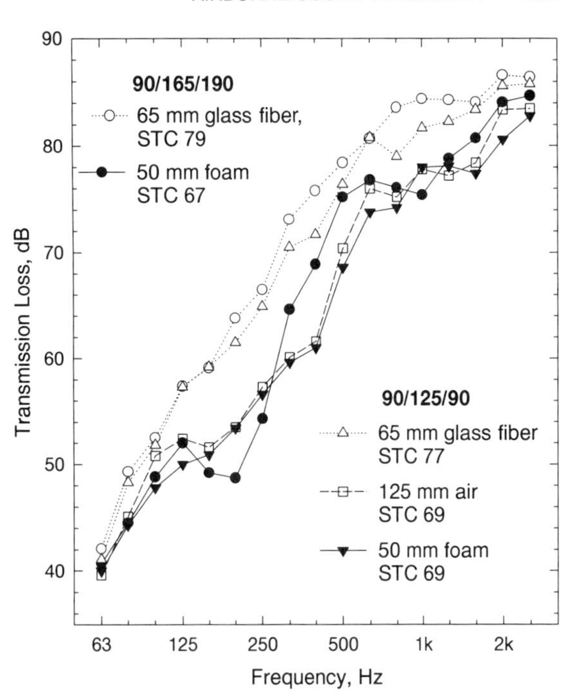

Fig. 10 Double-layer concrete block walls measured in a laboratory. One wall was supported on the edge of one of the reverberation rooms and the other on an independently supported frame. One wall was formed from two isolated 90-mm block layers with a cavity of 125 mm between them. The figure also shows results for a double-layer wall formed from 90- and 190-mm blocks with a cavity of 165 mm between them. The cavity between the blocks had a 65-mm-thick layer of glass fiber, 50-mm-thick layer of polystyrene foam, and only air in the cavity.

TABLE 6 Apparent Sound Reduction Index of Double-leafed Building Elements

Construction	1st layer (mm)	Air gap (mm)	2nd layer (mm)	Sound Reduction Index by Frequency 125 Hz	250 Hz	500 Hz	1000 Hz	2000 Hz	4000 Hz	M (kg/m^2)	Total width (mm)	R'_w (dB)
Dense concrete, without	40	25	70	33	38	43	50	57	55	275	135	49
absorbent material in	70	10	70	43	44	50	54	55	60	340	150	53
the cavity	40	50	70	35	42	45	53	58	60	275	160	50
	70	50	70	44	42	48	54	59	58	340	190	52
	40	100	70	44	42	47	55	58	62	275	210	52
	70	100	70	43	41	48	54	59	65	340	240	53
Lightweight concrete,	70	110	120	42	44	46	48	53	60	175	300	51
mineral wool in the	70	160	70	38	41	42	44	52	60	135	300	47
cavity (50 mm)	70	50	70	37	43	41	44	55	63	135	190	47
	115	80	115	45	42	46	59	56	64	190	310	52
Gypsum, without absorbent	70	60	70	32	40	39	45	53	64	170	200	46
material in the cavity	80	60	80	38	42	41	46	54	62	210	220	47
Gypsum, mineral wool in	60	30	60	39	40	40	48	55	64	100	150	48
the cavity (30 mm)	70	60	70	35	40	41	46	56	63	160	200	47
	80	30	80	36	41	39	43	52	67	170	190	46

2.12 Block or Concrete Walls with Added Gypsum Board

The highest sound insulation that can be attained in practice from a single-leaf wall is limited by the mass. To get STC or R'_w ratings much greater than about 58 requires surface weights of 500 kg/m^2 or more. This sound insulation may not be enough or the weight may be unacceptable. In practice, the sound insulation of a heavy wall is often improved by adding layers of gypsum board, or something similar, supported independently or on resilient mounts to form a cavity construction.

Gypsum board is usually added on both sides of the wall, so the construction becomes a triple wall. The important parameters—mass–air–mass resonance, separate studs or resilient furring to avoid direct transmission between the gypsum board and the blocks, and the use of sound-absorbing material in the cavity—are doubly important. Reductions in sound insulation due to the mass–air–mass resonance are doubled but, conversely, improvements are doubled too.

Experiments with 190-mm-thick concrete blocks[41,42] showed the importance of avoiding small cavity depths that result in the mass–air–mass resonance being within or near the range used in calculating the STC or R_w (Fig. 11). Adding sound-absorbing material to the cavity clearly lowered the mass–air–mass resonance frequency (Fig. 12). Adding gypsum board on both sides of the blocks caused the mass–air–mass resonance to be deeper than it was when the gypsum board was applied on just one side (Fig. 13). When the cavity depth was only 13 mm, the STC was reduced by one point, but the detrimental effects on sound insulation below 160 Hz were much more serious.

Some concrete blocks are sufficiently porous that the cavity formed behind the gypsum board has an effec-

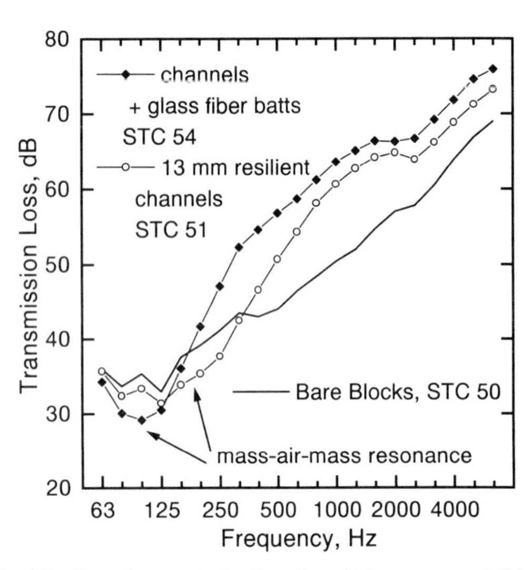

Fig. 12 Sound transmission loss for a 190-mm concrete block wall with 16-mm gypsum board attached on 13-mm resilient metal channels on one side of the wall with and without sound-absorbing material in the cavity.

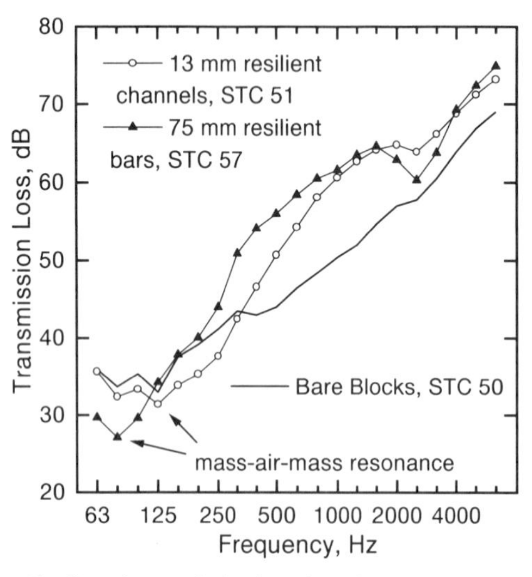

Fig. 11 Sound transmission loss through a 190-mm concrete block wall with 16-mm gypsum board attached on 13-mm resilient metal channels and 75-mm-deep resilient metal Z-bars.

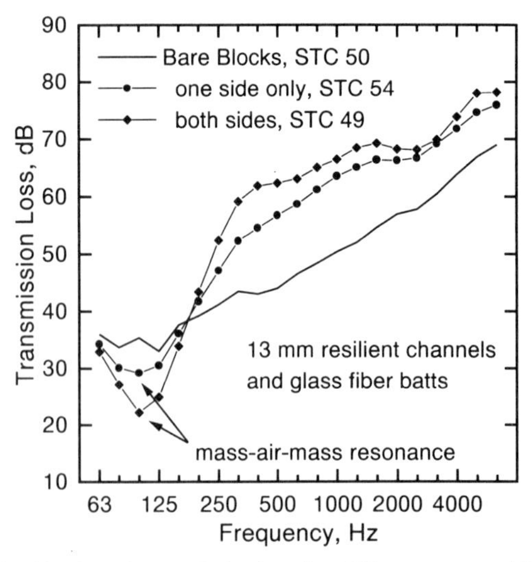

Fig. 13 Sound transmission loss for a 190-mm concrete block wall with 16-mm gypsum board attached on 13-mm resilient metal channels to one side and to both sides of the wall with sound-absorbing material in the cavity.

tive depth that is greater than the distance from the surface of the blocks to the rear face of the gypsum board; the air in the cavity penetrates the block.[43] This lowers the mass–air–mass resonance and increases sound insulation, but the blocks do not provide the full sound insulation expected from their mass. When the blocks are extremely porous, the sound insulation obtained is determined almost completely by the outer layers of gypsum board. As pointed out above (Section 2.10), the reduction in sound insulation can be prevented by sealing one face of the blocks with plaster or latex paint. On the sealed side of the blocks, of course, there is no longer an increase in effective cavity depth because of porosity, but on the unsealed side there can be considerable benefit (see Fig. 14).

The measured improvement in STC rating that resulted from the attachment of 16-mm gypsum board in a variety of ways to a concrete block wall is given in Table 7. The STC ratings for complete wall systems can be calculated by adding the improvements given in Table 7 to the STC rating of the masonry walls from Table 3 or the concrete walls from Table 2. Note again that when the cavity behind the gypsum board is too small, the STC is actually lowered relative to the bare-block case. Note also the very large increases in STC when the cavity is large and filled with sound-absorbing material (see Fig. 14).

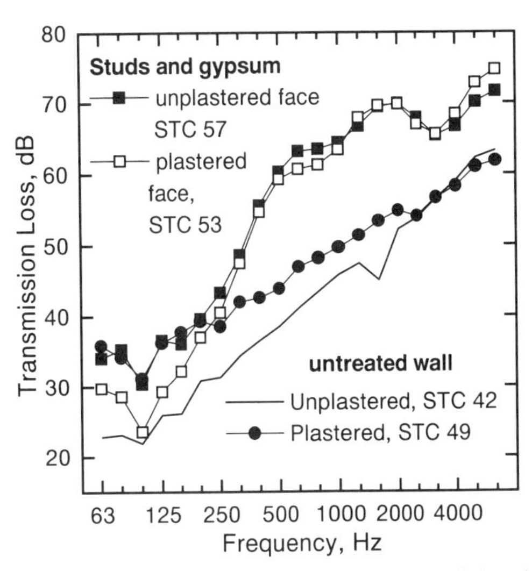

Fig. 14 Effect of plastering one face of a 190-mm lightweight block and of adding 16-mm gypsum board and 40-mm steel studs on each face. Note the reduced effect of the mass–air–mass resonance when gypsum board is applied on the porous face.

2.13 Double Walls Containing Resonators

Some research has been done to find a means of counteracting the mass–air–mass resonance to improve the low-frequency sound insulation of double walls by including Helmholtz resonators within the double-wall cavity. Data presented for double-panel constructions where there were Helmholtz resonators around the edge of the cavity[44,45] showed that a correctly chosen Helmholtz frequency improves sound insulation around the mass–air–mass resonance frequency.

Certain types of concrete blocks have slots in one face opening into the internal block cavities and are designed to act as Helmholtz resonators. The blocks absorb sound quite effectively at the resonance. These resonator blocks offer a way of counteracting the deleterious effects of the mass–air–mass resonance in block walls. By constructing the core wall so the slots are exposed equally on both faces of the wall, each layer of gypsum board will have a resonant structure behind it. In Fig. 15 the sound transmission losses for normal-weight and resonant blocks

TABLE 7 Increase in STC Ratings with Single Layer of 16-mm Gypsum Board Added to One or Both Sides of Concrete Block or Poured Concrete Wall with and without Fiberglass Batts Filling Cavity between Gypsum Board and Concrete

	Without Fiberglass in Cavity		With Fiberglass in Cavity	
Gypsum Board Attachment	One Side	Both Sides	One Side	Both Sides
Directly on concrete block	+0	−1		−1
On 13-mm resilient steel channels	+2	−1	+4	+14
On 50-mm resilient steel furring	+2	+2	+9	+22
On 65-mm steel studs	+8	+7	+10	
On 75-mm resilient steel furring	+7		+11	

Note: The same values may be used for 13-mm gypsum board without serious error.

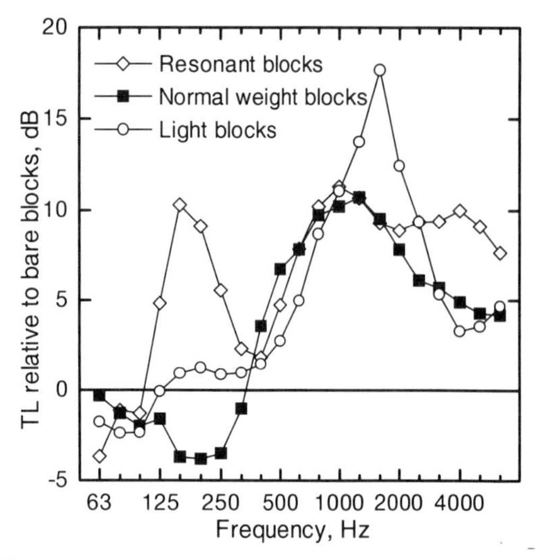

Fig. 15 Transmission loss relative to untreated concrete blocks for three kinds of blocks with gypsum board and 13-mm resilient metal channels on one face only.

with gypsum board supported on resilient metal channels are compared.[46] To eliminate the differences between the basic blocks, the transmission losses for the composite walls are shown as differences from those for the basic blocks in each case. The differences in the curves around 200 Hz are quite striking. For the normal-weight blocks, the mass–air–mass resonance causes a reduction in sound transmission loss. The resonant blocks show a strong absorptive resonance at 160 Hz, the Helmholtz resonance for these blocks, that counteracts the mass–air–mass resonance and leads to increased STC and R_w. The cost of a construction using such blocks might be justifiable in some cases.

2.14 Doors and Windows

Since doors and windows must be operable and therefore light, the sound insulation available from mass alone is limited. High insulation can be obtained, however, by the use of double-leafed elements, that is, double doors or double windows. As with any double-leafed system, the

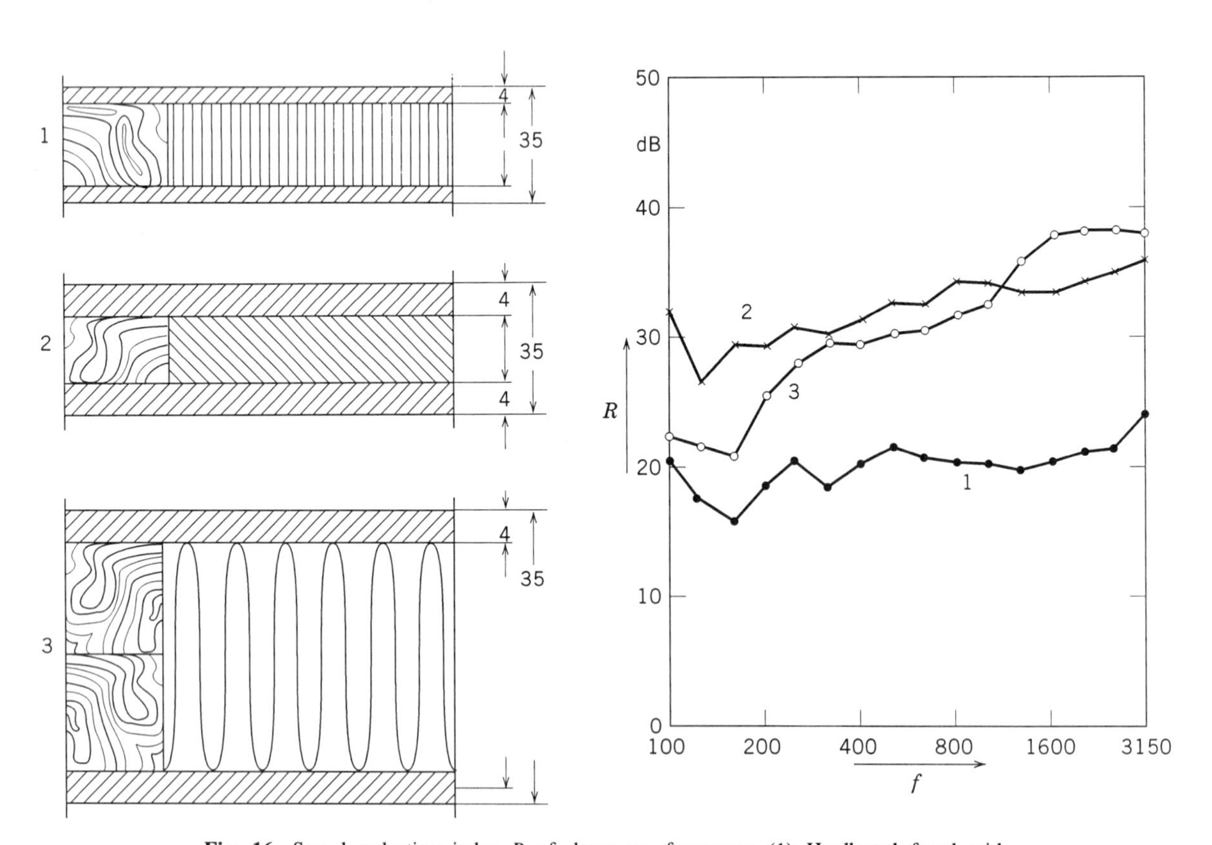

Fig. 16 Sound reduction index R of doors vs. frequency. (1) Hardboard faced with honeycomb skeleton: $M = 11\ \text{kg/m}^2$; $R_w = 12$ dB. (2) Gypsum board between wood chipboard: $M = 42\ \text{kg/m}^2$; $R_w = 35$ dB. (3) Double-leafed door with wood chipboard; mineral wool lining: $M = 23\ \text{kg/m}^2$; $R_w = 32$ dB.

width of the airspace should be as large as possible. In the case of doors the distance should be increased to form a lobby or corridor whenever possible. An absorbent lining will improve sound insulation.

In Fig. 16 the sound reduction index of three typical door leaves sealed at the edges is shown. For lightweight doors made, for example, of hardboard-faced flush skeleton ($M \approx 10$ kg/m^2), a weighted sound insulation index of 20–25 dB cannot be exceeded. An increase in the door weight ($M \approx 45$ kg/m^2), attained by using cored panels or two or more solid layers such as gypsum board or wood chipboard, improves sound insulation to about 35–40 dB. Higher sound insulation can be gained with well-designed double-leafed panels (see Section 2.11). The use of double doors may result in an improvement of about 10 dB, provided there is a sufficient distance (> 100 mm) between them.[47]

Proper seals and weather stripping are essential. A decrease of sound insulation of about 10–20 dB can be expected if the doors or windows have poorly fitting seals. Figure 17 shows some examples of single- or double-sealed door openings. Of course, the higher sound insulation of the panel, the better the sealing must be. If a threshold at the foot of the door must be avoided, sealing can be particularly difficult. Figure 18 gives some effective solutions to meet sound insulation requirements.

The sound insulation of single-glazed windows is influenced by coincidence with the critical frequency f_c, which may be calculated from the information given in Table 1. The sound insulation does not increase with increasing thickness as rapidly as might be expected from mass law, but some improvement can be obtained by using laminated glass (security glass). This consists of thin glass panes with an intervening plastic layer that minimizes the increase of stiffness due to increasing thickness. Figure 19 gives the weighted sound reduction index obtainable by single-glazed windows both with normal and laminated panes. The values in the figure may be calculated using

$$R_w = 11.7 \log d + 23.1 \quad \text{for solid panes} \tag{20}$$

or

$$R_w = 17.9 \log d + 19.2, \qquad d \geq 4 \quad \text{for laminated panes,} \tag{21}$$

where d is the total thickness of the pane in millimetres.

To obtain an improvement in sound insulation by using double-glazed windows requires a cavity depth of about 20 mm or more. An airspace smaller than 100 mm is inadvisable if a weighted sound reduction index of more than about 40 dB is to be obtained. In Fig. 20 the approximate weighted sound reduction index of double-glazed window systems is shown for different total thicknesses of both glass panes. Higher values can be obtained by using panes of different thicknesses to reduce the influence of coincidence. A ratio of about 1 : 2 in thickness is recommended. As much absorption as possible can be provided in the cavity by lining the perimeter (reveals, head, and sill if possible) with an efficient sound-absorbing material. This gives most benefit at high frequencies. Further improvements can be made using resonant cavities around the perimeter.[44,45] The data in the figure may be calculated using $R_w = 25(1 + \log d)$.

Fig. 17 Examples of single- and double-sealed door openings.

With double windows it is even more important than with single windows to ensure proper sealing. Just as with door openings, resilient materials of high durability must be used. To have the seals fit well, the frame

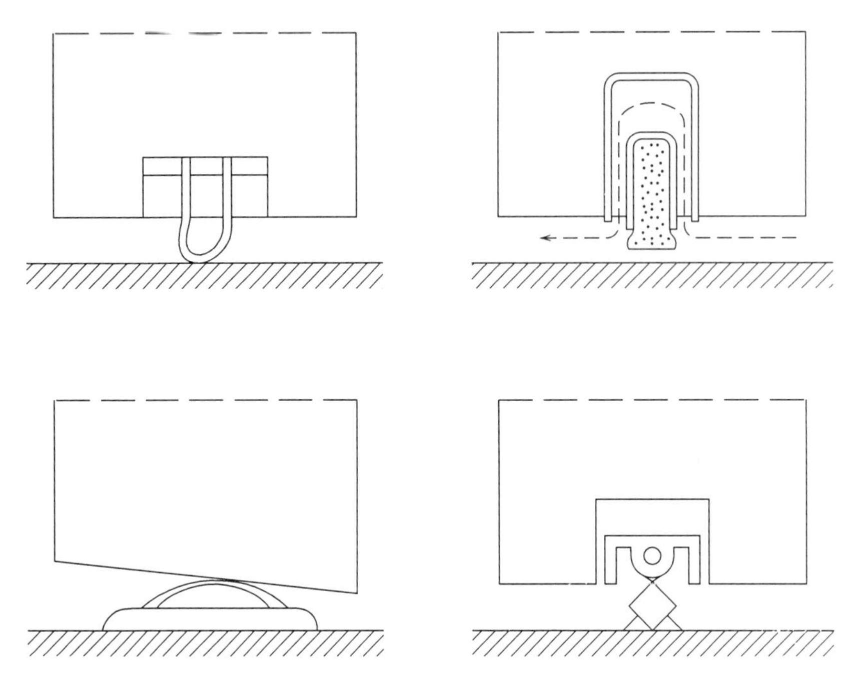

Fig. 18 Sealing of door openings on the ground without threshold.

must be sufficiently stiff. Otherwise, there is no significant dependence of sound insulation on the frame material whether of timber, metal, or plastic. Poor sealing increases sound transmission, especially in the high-frequency range. The maximum energy of traffic noise occurs at middle and low frequencies; therefore, for traffic noise, cavity depth and thickness of panes is more important than airtight joints.

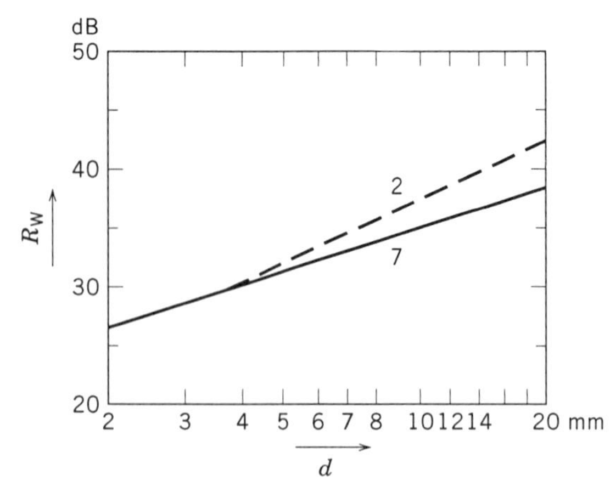

Fig. 19 Weighted sound reduction index R_w of single glazings vs. thickness d: (1) normal pane; (2) laminated pane.

Thermal glazing, consisting of two panes joined at the perimeter and placed into a single frame usually has a small cavity. Therefore, it would require very thick panes to obtain high sound insulation. A more practical solu-

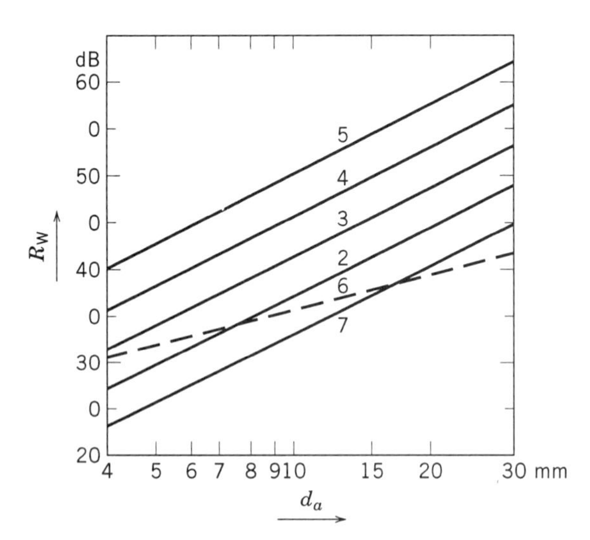

Fig. 20 Weighted sound reduction index R_w of double glazings vs. total thickness d_a of both panes. Distance: (1) 10 mm; (2) 20 mm; (3) 40 mm; (4) 80 mm; (5) 160 mm; (6) single glazing.

TABLE 8 Representative Values of Sound Insulation of Window Constructions

	Type of Window Construction					
	Single Frame with Single or Double Panes		Connected Double Frame with Single Pane and Double Pane or Two Double Panes		Large-Cavity Double Frame with Single Pane and Double Pane or Two Double Panes	
R'_w (dB)	Panes	Seals	Panes	Seals	Panes	Seals
<21	3	None	2 × 3	None	2 × 3	None
21–25	6 or 3/8/3	None	2 × 3	None	2 × 3	None
26–30	9 or 3/12/6	One	3/30/3	One	3/50/3	None
31–35	5/16/5 or 4/16/6	One	4/40/4 or 3/40/4	One	3/50/5 or 3/75/3	One
36–40	8/16/6 or 8/24/4	Two	8/50/4 or 8/50/6/12/4	Two	4/100/4 or 6/100/4/12/4	Two
41–45	No		10/60/8 or 8/60/8/12/4	Two	6/100/6 or 8/100/6/12/4	Two

tion is to add an additional pane (storm windows) to form a large cavity with the thermal glazing. An improvement of about 5 dB is possible by filling the cavity with a suitable heavy gas, such as sulfur hexafluoride (SF_6).

Table 8 gives some representative information about window constructions that meet different requirements on sound insulation. More information can be found in Refs. 48–50.

2.15 Flanking Transmission

Except in special laboratories for the measurement of sound transmission through building elements[11] there is practically always a certain amount of transmission of sound energy along paths other than the direct path shown in Fig. 1. These paths are called *flanking paths.*

Typically in buildings, in addition to the sound power W_2 transmitted through the separating wall, W_3, the sound power transmitted into the receiving room by flanking elements or by leaks, is significant. In this case, sound insulation is expressed by the *apparent sound reduction index*

$$R' = 10 \log \frac{W_1}{W_2 + W_3} \quad \text{dB}. \tag{22}$$

The expression *apparent transmission loss* is also in use and is equivalent to the apparent sound reduction index.

Assuming that diffuse sound fields exist in the two adjacent rooms, the apparent sound reduction index may be evaluated from

$$R' = L_1 - L_2 + 10 \log \frac{S}{A} \quad \text{dB}, \tag{23}$$

which is the same as Eq. (3) except that, because L_2 includes sound energy from all surfaces, we are dealing with the apparent sound reduction index. (In the case of staggered rooms, S is that part of the area of the partition common to both rooms.)

In typically furnished dwellings, the equivalent sound absorption area of sitting rooms and bedrooms is about 10–20 m^2; often the area of partition walls and floors has about the same magnitude. Thus the 10 log S/A term is very small, and there is little difference between apparent sound reduction index and level difference, $L_1 - L_2$.

The direct and various possible flanking transmission paths between two adjacent rooms are shown in Fig. 21. The sound power transmitted into the receiving room can be assumed to consist of the sum of the following components:

W_{Dd}, which has entered the partition directly and is radiated from it directly (= W_2 in Fig. 1);

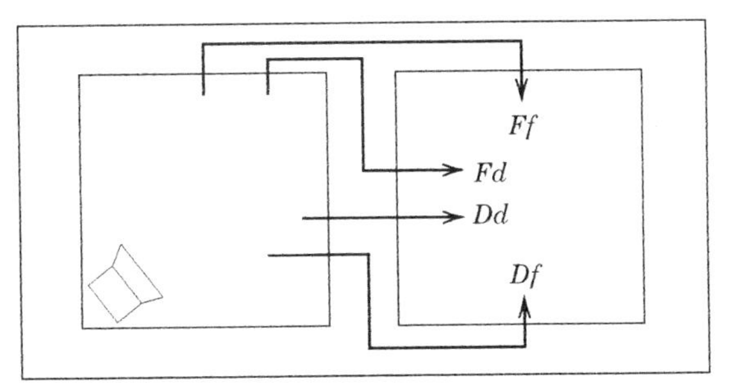

Fig. 21 Transmission paths between two adjacent rooms: Dd is the main path; Ff, Df, and Fd are flanking paths.

W_{Df}, which has entered the partition directly but is radiated from flanking constructions;

W_{Fd}, which has entered flanking constructions and is radiated from the partition; and

W_{Ff}, which has entered flanking constructions and is radiated from flanking constructions.

The total power transmitted along flanking paths is thus

$$W_3 = W_{Df} + W_{Fd} + W_{Ff}. \tag{24}$$

For each flanking path a sound reduction index (R_{Df}, R_{Fd}, R_{Ff}) can be defined, for example,

$$R_{Ff} = 10 \log \frac{W_1}{W_{Ff}} \quad \text{dB}. \tag{25}$$

The index R_{Ff} (also denoted R_L) is of special interest because it depends only on flanking constructions. Improvements in the sound insulation of the primary separating element have little or no influence on R_{Ff}. It can be thought of as an upper limit to the sound insulation in any construction type.

The apparent sound reduction index R' between two adjacent rooms can be calculated from

$$R' = -10 \log \left(10^{-R/10} + \sum_i 10^{-R_i/10} \right) \quad \text{dB}, \tag{26}$$

where the index i stands for Df, Fd, and Ff as appropriate.

At each junction in the building there is a certain amount of attenuation that depends on the type of junction, mass ratios, and damping in the various components. Some methods for calculating the effect of flanking have been proposed,[51,52] but the subject is still developing. Probably the most accurate results are obtained using Statistical Energy Analysis techniques. In a concrete apartment building about 50% of the incident sound power is transmitted to the adjacent rooms along flanking paths. Great attention has to be paid to flanking paths if high sound insulation is to be obtained in a building.

In the case of a lightweight wall perpendicular to a heavy floor, the attenuation of vibration in the floor due to the wall is very low. The importance of this is clearly evident in Fig. 22 for the example of a double-leafed lightweight wall of high sound insulation (more than $R_w = 55$ in the laboratory) resting on a concrete slab. Depending on the mean mass per unit area of the slab, the apparent sound reduction index of the wall increases from 42 up to 54 dB. In the vertical direction the trans-

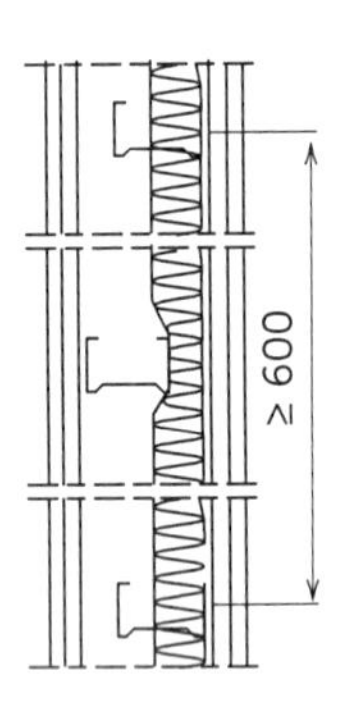

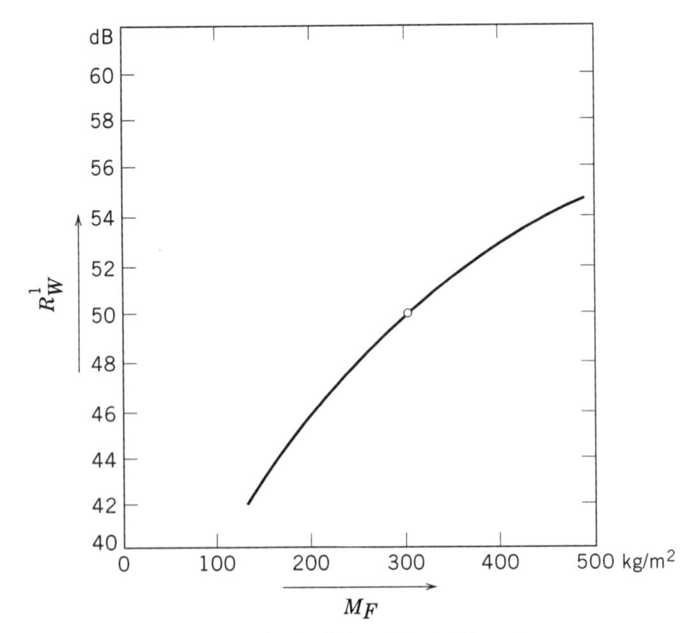

Fig. 22 Weighted apparent sound reduction index R'_w of a lightweight double wall vs. mean mass per unit area M_F of the flanking constructions.

mission is controlled by the floor and the lightweight walls have little effect. Lightweight wall systems are often used in apartment buildings. Thus structure-borne sound propagation in the floor slab and side walls is of great importance in such a building.

2.16 Multielement Partitions, Apertures, and Leaks

The total sound energy passing through each component of a multielement partition depends on the product of the area and sound transmission coefficient of each component. The equation governing the process is

$$A_t\tau_t = \sum_{i=1}^{N} A_i\tau_i, \tag{27}$$

where A_i and τ_i represent the area and transmission coefficient of the ith component of the partition. The quantities A_t and τ_t represent the total area and effective composite transmission coefficient of the multielement partition. The transmission coefficients for the individual components may be calculated from their sound reduction indices by rearranging Eq. (2) as

$$\tau = 10^{-R/10}. \tag{28}$$

The sound insulation of a multielement building partition is usually controlled by the component with the lowest sound insulation. In case of an opening, the sound reduction index is zero (the transmission coefficient is 1), and only the ratio of the area of the opening to the area of the rest of the partition determines the sound insulation. An opening of 10%, for example, an open door or open window in a wall, limits the total sound reduction index to 10 dB.

The approximation $\tau = 1$ is not correct, however, for small apertures and narrow slits with dimensions small compared with the wavelength. The sound transmission coefficient of those openings can be much higher than 1. Due to resonant effects, this is especially evident for depths of apertures and slits near the half of the wavelength. Apparent transmission coefficients $\tau > 100$ have been measured.[53,54]

Figure 23 shows an example of measurement results[55] of sound insulation for a plate with slits of various breadths. The decrease of sound insulation in the high-frequency range is typical. Sound transmission through an aperture depends on its position in the building element. The sound pressure at an edge of wall is higher than that in the middle, and the sound transmission through an aperture there is correspondingly higher.

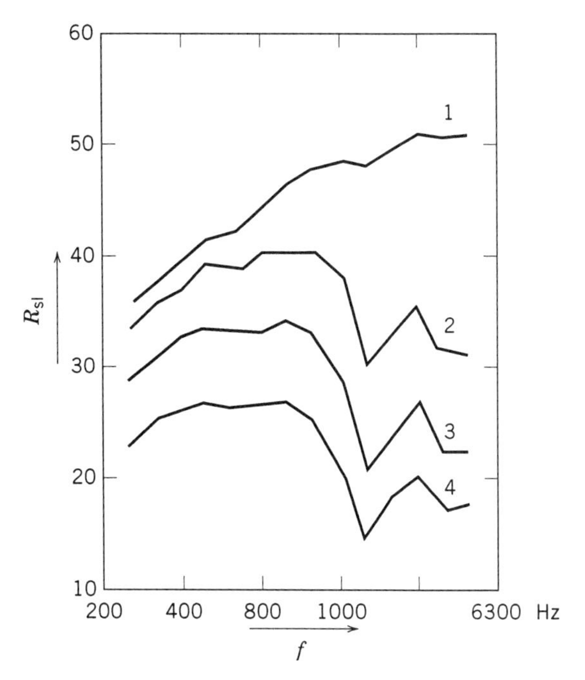

Fig. 23 Sound reduction index R_{sl} of slits (length 1 m) of various breadths relative to 1 m^2 area in a plate of 75 mm thickness: (1) without a slit; (2) breadth 2 mm; (3) breadth 3 mm; (4) breadth 7 mm.

3 IMPACT SOUND INSULATION

3.1 Definitions

Impact sound was defined above as a special kind of structure-borne sound. Perhaps the most common source is footsteps, but there are many others, such as furniture being moved and cleaning and other equipment operating directly on the surface of floors, stairs, and landings. Impact sound is radiated into rooms below, but besides this generally important case, horizontal transmission between apartments or from staircases or even transmission to the room above can cause annoyance.

The definition of impact sound pressure level is based on excitation using a standard tapping machine.[56–59] This machine, shown in Fig. 24, consists of five hammers placed along a line with a distance between adjacent hammers of 100 mm. The mass of each hammer is 500 g, its diameter is 30 mm, and it falls freely from a height of 40 mm. The time between successive impacts is 100 ms. For determination of impact sound insulation, the tapping machine is placed on the floor under test in different positions. In the receiving room the average sound pressure level L_i and the reverberation time are measured in

Fig. 24 Standard tapping machine.

one-third-octave bands, and the normalized impact sound pressure is calculated.

The *normalized impact sound pressure level* L_n is the level corresponding to a reference value of equivalent sound absorption area in the receiving room. It is, by definition,

$$L_n = L_i + 10 \log \frac{A}{A_0} \quad \text{dB,} \qquad (29)$$

where A is the equivalent sound absorption area in the receiving room and A_0 is the reference equivalent sound absorption area (10 m^2). Where it is uncertain that results are obtained without flanking transmission, the normalized impact sound pressure level is denoted by L'_n.

The impact sound insulation of concrete floors is generally unsatisfactory unless improved by a soft floor covering or a floating floor. Such measures cause a reduction of impact sound pressure level denoted by

$$\Delta L = L_{n0} - L_n \quad \text{dB,} \qquad (30)$$

where L_{n0} is the normalized impact sound pressure level in the absence of the floor covering and L_n is the normalized impact sound pressure level when the floor covering is in place. If a floating floor (see Section 3.5) is covered by a flexible covering, both ΔL values can be added in each one-third-octave band. That is not possible when two soft coverings are put one upon the other.

3.2 Single-Number Ratings

The impact sound pressure levels defined above depend on frequency, and standards require that they be measured in the frequency range 100–3150 Hz. To obtain a single-number rating of impact sound, a reference curve is used[60,61] in a procedure analogous to that for airborne sound insulation. The reference curve is shown in Fig. 25 compared with examples of measured curves for a wood joist floor with and without a concrete topping. In contrast to airborne sound, an unfavorable deviation at a particular frequency occurs when the measured value is higher than the reference value. The weighted values are impact insulation class, denoted by IIC,[61] and the weighted index $L_{n,w}$.[60] In this case, there is no difference between the ISO and the ASTM reference curves or fitting procedures. The ASTM procedure includes one more step, however. When the fitting is complete, the IIC is calculated by subtracting the value of the reference contour at 500 Hz from 110 (STC = 110 $-L_{n,w}$). Thus the IIC rating, similar to the STC and R_w, *increases* as the impact sound insulation improves. The $L_{n,w}$ rating *decreases* as the impact sound insulation improves.

In Fig. 25, the reference contour is fitted to the floor with the concrete topping. The single-number rating is controlled by the high-frequency levels. Despite the increased weight, the impact sound insulation is reduced. There is substantial improvement at low frequencies, however, and the addition of a soft covering on the concrete or a resilient support between it and the plywood would greatly improve the rating.

Since its introduction, there has been considerable debate about the usefulness of ISO tapping machine data

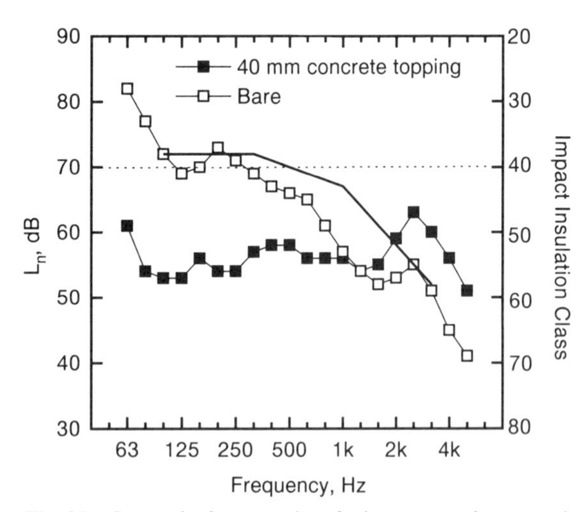

Fig. 25 Curve of reference values for impact sound compared with measurement results. The floor consists of 16-mm plywood resting on 240-mm deep wood joists (400 mm on center). The space between the joists contains a layer of glass fiber 90 mm thick. The ceiling is 16-mm gypsum board suspended on resilient metal channels (□). (■) For the same floor with a layer of concrete 40 mm thick placed on top. The reference contour is fitted to this curve.

obtained on different types of floors.[62–68] The debate continues.

In many European countries, most apartment buildings are constructed from concrete. There are several standard test procedures that make it more convenient to evaluate concrete floors and toppings for such construction. The procedures are not always in use in other countries but are of some interest.

The reduction of impact sound pressure level by a floor covering is, to a large extent, dependent on the floor on which it has been measured. Therefore, to get a single-number rating for the reduction of impact sound pressure due to a floor covering, the measured values of ΔL have to be related to a reference floor.[69] The reference floor is defined by the curve $L_{n,r,0}$, which can be found in Ref. 60. Procedures for making the measurements and calculations are dealt with in the appropriate standards.[56,60]

The resulting curve $L_{n,r} = L_{n,r,0} - \Delta L$ has to be evaluated following the standard procedure[60] to get the weighted normalized impact sound pressure level $L_{n,r,w}$. Thus the weighted impact sound improvement index ΔL_w of the floor covering can be calculated by

$$\Delta L_w = L_{n,r,0,w} - L_{n,r,w} = 78 - L_{n,r,w}, \quad (31)$$

where $L_{n,r,0,w} = 78$ dB is the weighted normalized impact sound pressure level of the reference floor.

Another procedure is in use to evaluate the impact sound pressure level of bare concrete floors. These floors are invariably inadequate and any impact sound insulation is due to the covering. Therefore a more realistic rating of the bare concrete floor is obtained by taking into account the influence of a reference floor covering. The weighted sound improvement index of the reference floor[60] covering is $\Delta L_{r,w} = 19$ dB.

To evaluate a bare floor, the first step is to calculate the normalized impact sound pressure level of the floor with the reference covering from

$$L_{n,1} = L_{n,0} - \Delta L_r. \quad (32)$$

From these values of $L_{n,1}$ a weighted value $L_{n,1,w}$ is calculated. The equivalent weighted normalized impact sound pressure level of the bare concrete floor can be found from

$$L_{n,w,\mathrm{eq}} = L_{n,1,w} + \Delta L_{r,w} = L_{n,1,w} + 19 \text{ dB}. \quad (33)$$

The definition of single-number ratings for floor coverings, ΔL_w, and bare concrete floors, $L_{n,w,\mathrm{eq}}$, allows designers to calculate the weighted normalized impact pressure level of a completed floor, $L_{n,w}$, using the equation

$$L_{n,w} = L_{n,w,\mathrm{eq}} - \Delta L_w. \quad (34)$$

Thus requirements for impact sound insulation can be specified for floor coverings to match the bare concrete floor to be used.

The above deals with the most important situation of disturbance by impact sound where the receiving room is directly below the excited floor. For concrete constructions, when the receiving room is immediately adjacent on the same level, the impact sound insulation will be about 5 dB greater, about 10 dB greater if it is the second room on the same level, and about 10–20 dB greater, depending on flanking constructions, if the receiving room is above the source room. Such generalizations do not apply to lightweight joist construction using double-layer assemblies.

3.3 Concrete Floors

For a homogeneous reinforced concrete floor, results of calculations[17,70] and measurements[22] show a small increase of normalized impact sound pressure level with frequency (1.5 dB/octave). Doubling the thickness of such a floor causes an improvement of impact sound insulation of about 10 dB; nevertheless, in practice, concrete floors with no coverings or hard finishes adhered directly to the floor are never satisfactory.

This is also true for concrete floors with large or small cavities. Somewhat lower values of $L_{n,w,\mathrm{eq}}$ of floors with small cavities (such as hollow-core concrete slabs), represented in Fig. 26, curve 2, are based on the increased thicknesses of these floors for the same mass per unit area as homogeneous floors. Higher values of $L_{n,w,\mathrm{eq}}$ in curve 3 of Fig. 26 occur because of resonant effects in large hollows such as those in coffered floors with directly fastened ceilings and without absorbent material.

Suspended ceilings generally reduce impact and airborne sound transmission to a certain extent. About 5 dB reduction of $L_{n,w,\mathrm{eq}}$ may be obtained, depending on the elasticity of the hangers. But in buildings, due to the flanking transmission in the walls, the improvement gained is generally not sufficient to warrant the cost.

3.4 Soft Floor Coverings

An obvious solution to the problem of impact sound is to reduce the excitation of the structural floor by covering it with a resilient layer. Of course, such layers do not significantly improve airborne sound insulation as well; any improvements are usually only at high frequencies.

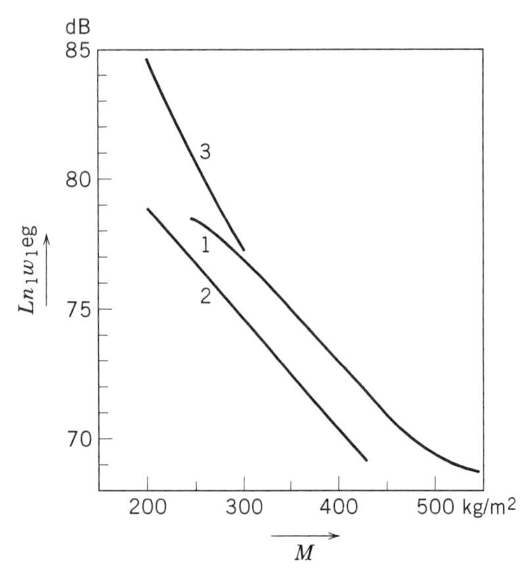

Fig. 26 Equivalent weighted normalized impact sound pressure level $L_{n,w,\text{eq}}$ of concrete floors vs. mass per unit area M: (1) homogeneous material; (2) small cavities; (3) large hollows.

Soft floor coverings are effective in reducing the impact sound pressure level due to the standard tapping machine beginning at a resonant frequency f_0 given by

$$f_0 = \frac{1}{2\pi}\sqrt{\frac{A_h E}{mh}}$$

$$= 5.98 \times 10^{-3}\sqrt{\frac{E}{h}} \quad \text{Hz}, \tag{35}$$

where A_h is the striking area of the tapping machine hammer (in square metres), E is the dynamic Young's modulus of the covering (in newtons per metre), m is the mass of the hammer (in kilograms), and h is the thickness of the covering (in metres).

To estimate the improvement due to a soft floor covering, the acceleration of a hammer of the tapping machine striking upon the covering has to be measured.[71] The weighted impact sound improvement index is

$$\Delta L_w = 71.5 - 17.3 \log a \quad \text{dB}, \tag{36}$$

where a is the maximum acceleration of the hammer (in metres per square second).

The test specimen should have dimensions of about 200 × 200 mm and must be placed on a solid, heavy, horizontal surface.

Flexible layers of a uniform material such as plastic or rubber provide only a low weighted impact sound improvement index of about 5–10 dB. A higher impact sound insulation (ΔL_w about 15–20 dB) is obtained by use of soft floors composed of a hard upper layer (rubber, plastic, or linoleum) and a resilient lower layer (soft felt board, fiber board, or foamed plastics). Usually those layers are glued to the floor. Thus the influence of the adhesive on impact noise has to be considered. Variation of ΔL_w based on the type of adhesive may be around ±3 dB. The impact sound improvement index for a carpet depends on thickness (about 20 dB for 4–5 mm and about 25 dB for 6–8 mm). The combination of a soft underpad and a carpet can give an increase of around 50 dB or more.

3.5 Floating Floors

When a soft floor covering may not be used, the most practical means of obtaining high impact sound insulation is to use a floating-floor construction. This type of floor has the additional advantage that it will also improve airborne sound insulation.

As shown in Fig. 27, a floating floor consists of a load-distributing slab resting on the structural floor but separated from it by a resilient layer (e.g., mineral wool, glass fiber, foamed plastics). Instead of a continuous resilient layer, a number of resilient pads may be used. The cavity between the floating slab and the structural floor is still best filled with soft sound-absorbing material. Most floating floors are made of concrete (about 30–60 mm thick), but they may also consist of gypsum, asphalt, or wooden floorboards. In the latter case the boards are nailed to battens that rest on resilient material. Special attention has to be paid to avoid solid connections (sound

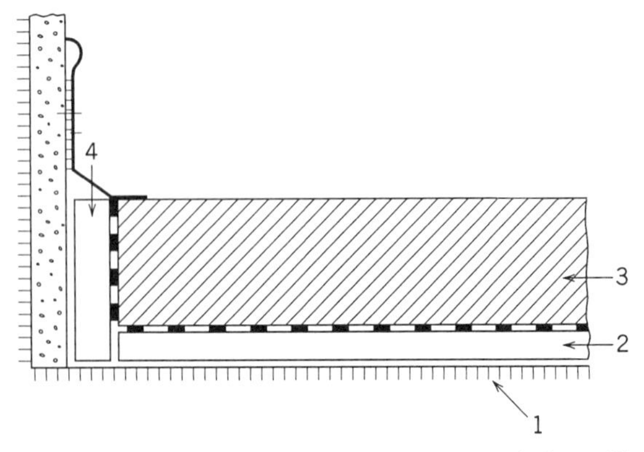

Fig. 27 Floating floor essentials: (1) structural floor; (2) resilient layer; (3) load-distributing slab; (4) insulation strip around edges.

bridges) between the floating slab and the structural floor. Therefore the elastic layer should be covered by plastic, roofing felt, building paper, or a similar material before pouring concrete. Resilient strips placed around all slab edges are necessary to avoid reduction of the impact sound insulation caused by sound transmission from the load-distributing slab to the walls.

As with other mass–spring–mass systems, the floating floor has a fundamental resonance frequency, f_{sl}, given by

$$f_{sl} = \frac{1}{2\pi}\sqrt{\frac{ns}{M_{sl}}}, \tag{37}$$

where s is the dynamic stiffness of the elastic mounts (in newtons per metre), n is the number of mounts per unit area (for a continuous layer of material n is 1 and s is the dynamic stiffness per unit area of the material), and M_{sl} is the mass per unit area of the floating floor (in kilograms). As with other resonant systems, there is usually a decrease in sound insulation around the resonance frequency.

Where the floating slab is thick, rigid, and lightly damped, the tapping machine will generate an approximately reverberant bending wave field in the slab. In this case the improvement in impact insulation is

$$\Delta L_n \approx 10 \log \frac{2.3 M_{sl}\omega^3 \eta c_L h}{ns^2} \quad \text{dB}, \tag{38}$$

where η is the loss factor in the slab, c_L is the longitudinal wave speed in the slab (in metres per second), and h is the thickness of the slab (in metres). This equation predicts an increase in ΔL of 30 dB/decade in frequency if the loss factor is independent of frequency. The dynamic stiffness per unit area of a resilient layer consists of the dynamic stiffness per unit area of the material and of that of the enclosed air, which includes the effects of airflow resistivity.[30,31,72] When separate resilient mounts are used, the stiffness of the entrapped air must also be considered.[73]

The improvements predicted in Eq. (38) are not always achieved in practice. Improvements in L_n for two types of floating slabs on two types of floors are shown in Fig. 28. When the slabs rest on the 150-mm-thick concrete, the negative ΔL_n around the resonance frequencies is evident. There are no negative values of ΔL_n when the slabs rest on the wood joist floor and, overall, the improvement from the floating slabs is not as great. The floating slab, F3, which is made of wood, has a more resilient surface than the concrete slab, so the improvements at high frequency are greater.

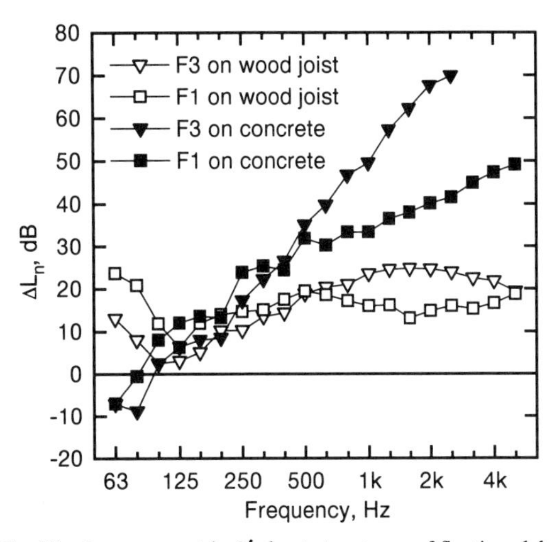

Fig. 28 Improvement in L'_n due to two types of floating slabs, F1 and F3, on 150-mm concrete slab and the wood joist floor described in Fig. 25. F1 is a 40-mm thick concrete slab floating on a glass fiber (25 mm thick, 50 kg/m^3). F3 is a 19-mm thick wood slab attached to 38-mm-thick strapping resting on glass fiber (25 mm thick, 114 kg/m^3).

The dependence of the weighted impact sound improvement index ΔL_w on the dynamic stiffness per unit area of the supporting resilient material is given in Fig. 29. It shows, for example, that a concrete slab with a mass of 50 kg/m^2 (about 25 mm thick) resting on a resilient layer with a dynamic stiffness per unit area of

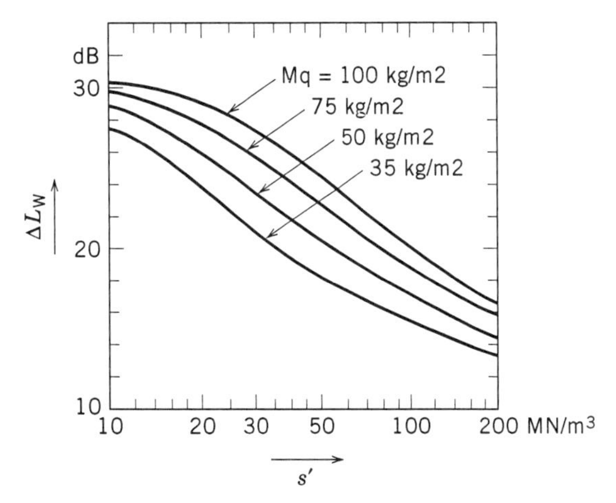

Fig. 29 Weighted impact sound improvement index ΔL_w of floating floors vs. dynamic stiffness per unit area s' of the resilient layer for different values of mass per unit area M_1 of the load-distributing slab. The structural slab is assumed to be heavy and rigid.

30 MN/m · m^2 provides an improvement in the weighted impact sound index of 23 dB. From Fig. 26 we see that a homogeneous concrete floor 150 mm thick (330 kg/m^2) has $L_{n,w,\mathrm{eq}} = 76$ dB. Thus $L_{n,w} = 53$ dB (IIC = 57). A dynamic stiffness of 30 MN/m · m^2 can be obtained by using a resilient layer of mineral wool or glass fiber about 10 mm thick (density 100–150 kg/m^3).

The dynamic stiffness per unit area of elastic layers to be used under floating floors can be measured by a resonance method.[37] In this method, the fundamental resonant frequency of the vertical vibration of a spring and mass system is measured. The spring is the test specimen of the resilient material under test (200 × 200 mm) and the mass is a load plate placed on it. The load plate is excited by sinusoidal signals. Once the dynamic stiffness per unit area of the elastic layer has been determined, the weighted impact sound improvement index of the floating floor can be found in Fig. 29.

3.6 Joist and Truss Floors

The same factors that control sound transmission through double-layer walls are as important in double-layer floors. To avoid having the floor and ceiling layers rigidly connected usually requires separate joists or a resilient suspension system for the ceiling.

Double-layer joist or truss floors are generally much lighter than solid concrete floors, and it is very difficult to reduce the transmission of impact sound at the low frequencies below 100 Hz. The IIC or $L_{n,w}$ rating might meet all code requirements but building occupants may still complain about thumping sounds from people walking.

As much weight as can be tolerated and some kind of floating floor offer the best available solution. Floating floors take three main forms:

1. A solid slab of concrete (typically 40–50 mm thick), wood, or some other material resting on a resilient sheet.
2. Wooden flooring nailed to battens to form a raft. The raft rests on a sheet of resilient material laid on a subfloor with the battens directly above and parallel to the floor joists. Instead of a sheet of resilient material, pads or strips might be placed under the battens and sound-absorbing material placed in the cavity between the battens.
3. Where allowed by building codes, a wood raft may also rest on resilient layers draped over the joists with no intervening subfloor.

In all cases the raft must not be nailed down to the joists at any point or contact the side walls anywhere. It should be isolated from the surrounding walls to avoid flanking transmission, for example by leaving a gap round the edges that can be filled with resilient material and covered by the skirting (baseboard).

A heavy pugging (a loose filling usually poured between the joists) is advantageous for adding mass to a floor. In older wooden joist constructions, puggings are placed on separate pugging boards inserted between the joists. Such puggings may consist of granular or fibrous materials such as sand or ashes. In reconstructed wooden joist floors often these separate pugging boards have been removed, in which case the floor should be reconstructed to have a resiliently supported ceiling with sound-absorbing material placed in the cavity.

The impact sound insulation of wooden joist floors cannot be calculated by dividing into the acoustical effects of floor coverings and structural floors, as is the case with concrete. The weight and structure of the upper layer can vary quite widely, and it is best to find measured data for specific types of floors. Figure 30 shows airborne and impact sound insulation for some types of wooden joist floors that might be used in Europe. Table 9 shows data for construction more typical of North America.

Table 9 makes it plain that it is relatively simple to get high airborne sound insulation ratings with joist construction. Similar results would be expected for wood I-beams or trusses. The joists in the table were 240 mm deep. Deeper joists, trusses, or I-beams would give better results. High impact sound ratings are not so easy to attain. The impact sound ratings are very much controlled by the resilience of the surface layer or by the presence of a soft layer under a floating slab. An impact insulation class greater than 55 with a hard-surfaced floor results in an STC rating over 60. The last floor in the table serves as a warning. It shows the price to be paid if the floor layer is rigidly connected to the ceiling and the danger of allowing small air cavities in the ceiling. Resilient metal channels should never be used between layers of gypsum board in a floor or a wall.

3.7 Standard Tapping Machine Compared to a Walker

The standard tapping machine test procedure does not deal with impact sound generated at frequencies below 100 Hz. With lightweight joist floors, low-frequency impact sound can be very annoying. Figure 31 shows tapping machine and walker sound levels generated under a 150-mm concrete slab with and without a carpet. For the tapping machine, for all but the very low frequencies, the carpet has a very significant effect. The walker was wearing shoes with rubber-tipped heels and so generated less high-frequency sound than the tapping

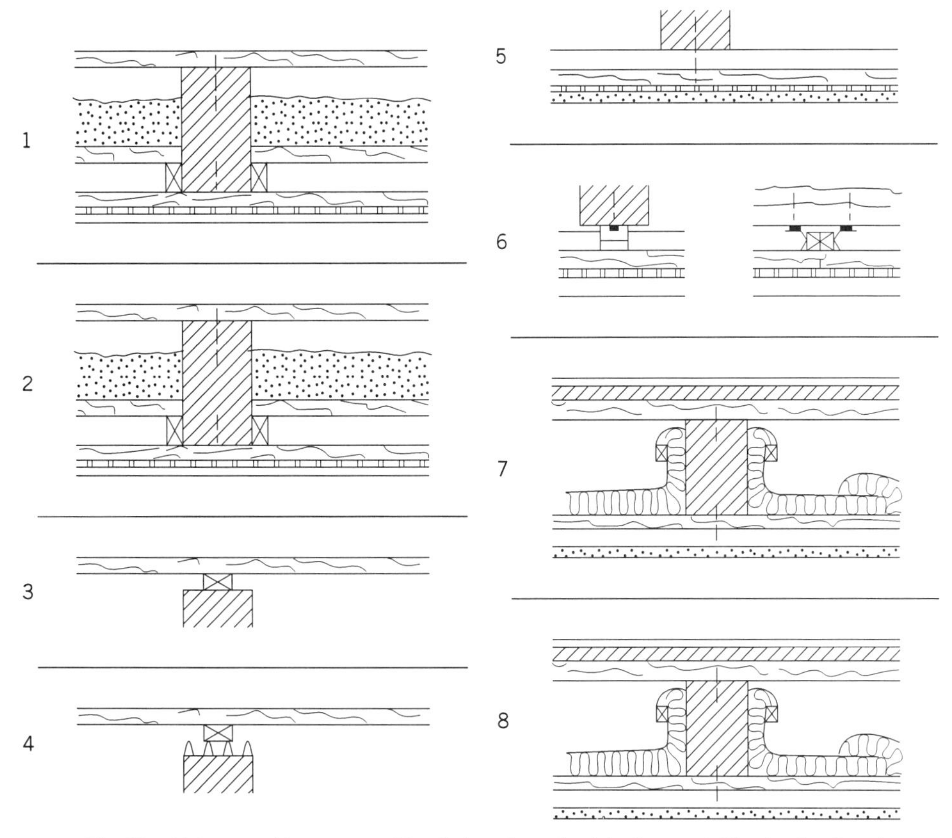

Fig. 30 Airborne and impact sound insulation of wooden joist floors. (1) Wooden floorboards (25 mm) directly nailed to the joists (200 mm); pugging of slag on separate pugging boards; plastering on expanded metal below ceiling boards, directly fastened to the joists: $R'_w = 45$ dB; $L'_{n,w} = 68$ dB. (2) Like 1, but mineral wool (60 mm) instead of pugging; gypsum board (12.5 mm) instead of plastering: $R'_w = 49$ dB; $L'_{n,w} = 65$ dB. (3) Like 1, but floorboards nailed to battens; $R'_w = 49$ dB; $L'_{n,w} = 63$ dB. (4) Like 3, but raft resting on layers of mineral wool (10 mm): $R'_w = 51$ dB; $L'_{n,w} = 58$ dB. (5) Like 1, but ceiling fixed to battens perpendicular to the joists: $R'_w = 49$ dB; $L'_{n,w} = 63$ dB. (6) Like 3, but ceiling fixed with resilient metal hangers: $R'_w = 51$ dB; $L'_{n,w} = 58$ dB. (7) Wooden floorboards (25 mm) floating on mineral wool (20 mm); wooden floorboards (25 mm) directly nailed to the joists (200 mm); absorbent material (50 mm); ceiling of gypsum board (2 × 12.5 mm), fixed with crossing battens: $R'_w = 53$ dB; $L'_{n,w} = 53$ dB. (8) Like 7, but floating concrete slab (40 mm) instead of wooden floorboards: $R'_w = 55$ dB; $L'_{n,w} = 51$ dB.

machine on the bare floor, but the addition of the carpet still makes a significant improvement for frequencies over 100 Hz. At the very low frequencies, the carpet provides no improvement, but the sound levels there are relatively low anyway.

In contrast, the same kind of data are shown in Fig. 32 for a wood joist floor. Again the carpet makes little difference at low frequencies, but for this floor the low-frequency levels are much higher and more likely to be annoying. The data for the walker also show indications of floor creaks that are not present in the more rigid concrete floor or with the tapping machine.

3.8 Staircases

In apartments, there is usually no need for special measures to control impact sound insulation in stairways or landings if staircases will not be directly adjacent to sitting rooms or bedrooms. There will be at least two junctions between the stairs and sensitive areas, so impact

TABLE 9 Impact and Airborne Sound Ratings for Selected Floors

Rating Code	Floor Surface	Ceiling Surface	IIC	STC	R_w
A	One layer of 15-mm oriented strand board (OSB)	One layer of 16-mm gypsum board	46	51	51
B	Two layers of 15 mm OSB	As A	47	55	56
C	One layer of 15 mm OSB	Two layers of 16-mm gypsum board	49	55	55
D	Two layers of 15 mm OSB	As C	53	60	60
E	One layer of 16 mm plywood	As C	41	52	53
F	As E + 13 mm hardwood on 3-mm shredded rubber	As C	50	57	61
H	As E + 9 mm wood parquet, 40-mm normal concrete and 1-mm-thick asphalt paper	As C	55	69	69
I	As E + 9 mm wood parquet, 40-mm cellular concrete and 1-mm-thick asphalt paper	As C	53	67	67
J	One layer of 16-mm pylwood or OSB	Two layers of 16-mm gypsum board with 13-mm resilient metal channels between	31	38	38

Note: Constructed using 38 × 240 mm wood joists, 400 mm on center (oc). All gypsum board layers are supported on resilient metal channels spaced 600 mm oc.

sound insulation in the horizontal direction should be sufficient (see Section 2.15) even for stairs with a concrete surface.

Otherwise, gaps between the stairway and the landing and staircase wall are a suitable means of improving impact sound insulation. Gaps may be open or filled with any resilient material and, naturally, can be used only where structural considerations for the stairway or landing do not necessitate support on the partition wall. Gaps should be covered to prevent accumulation of debris that might bridge the gap.

In some cases, the only option available is an elastic support of the stairways or landings. Rubber pads (thickness not less than 20 mm) may be a suitable material for

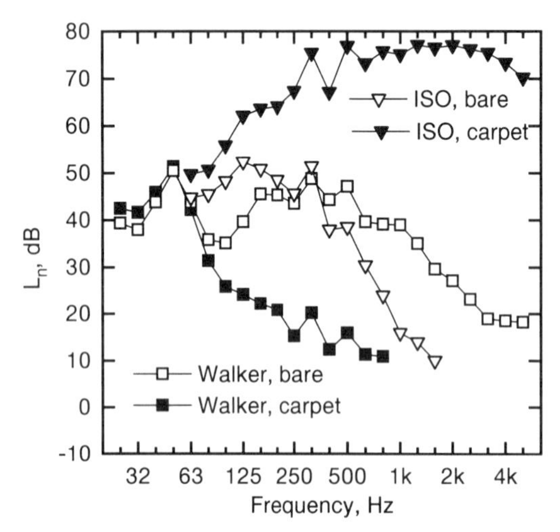

Fig. 31 Impact sound transmission through a 150-mm-thick concrete floor with and without a carpet and underpad. The levels are generated by a person and ISO standard tapping machine.

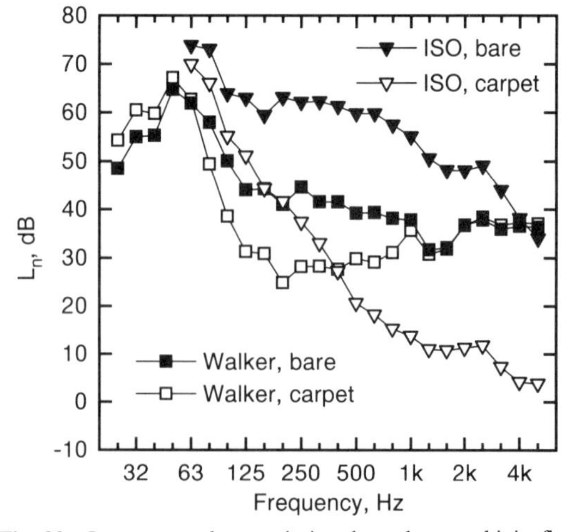

Fig. 32 Impact sound transmission through a wood joist floor with and without a carpet and underpad. The floor is similar to that in Fig. 25 but with two layers of 16-mm gypsum board and 270 mm of glass fiber batts in the cavity. The levels are generated by a person and the standard tapping machine.

this purpose. A soft or floating floor can also be used for stairs and landings. Because of the attenuation across junctions in the horizontal direction, any requirements on weighted impact sound improvement index may be diminished by about 5 dB. Nevertheless, construction of an effective floating floor needs the greatest care for it to work properly. In practice it is best to combine a floating floor on the landings with an elastic support of stairways upon the landings and gaps between stairways and walls of the staircase.

4 ACOUSTICAL REQUIREMENTS IN BUILDING CODES

National building codes and standards usually contain requirements for sound insulation aimed at meeting the demands of different functions of rooms and buildings. Internationally, the minimum requirements may be of similar magnitude, but in detail there are important differences. In apartment houses, especially, a compromise is necessary between the demand for a sufficiently quiet atmosphere on the one hand and economic practicalities on the other, the latter causing national differences. Therefore this section only deals with the kind of requirements and gives some examples.

Of major importance are the requirements on interior building elements that separate two apartments, but there are also requirements between particular rooms in public and office buildings. For this, the apparent weighted sound reduction index (R'_w), or STC, and the weighted normalized impact sound pressure level ($L'_{n,w}$), or IIC, are in use (see Sections 2.6 and 3.2).

For example, in Germany the requirements are $R'_w \geq$ 53 dB for partition walls and $R'_w \geq 54$ dB and $L'_{n,w} \leq 53$ dB for floors between dwellings.[74] These values, though corresponding with a good and fairly costly sound insulation, do not ensure that loud voices or music, household machinery, or children jumping in the adjacent flat cannot be heard. Therefore two levels giving a higher level of privacy are recommended[75]: $R'_w \geq 56$ dB and $R'_w \geq 59$ dB for partition walls and $R'_w \geq 57$ dB, $L'_{n,w} \leq 46$ dB and $R'_w \geq 60$ dB, $L'_{n,w} \leq 39$ dB for floors.

To protect against noise from staircases in apartment houses, requirements on impact sound insulation of stairs and landings ($L'_{n,w} \leq 58$ dB) and on airborne sound insulation of doors ($R'_w \geq 37$ dB, without corridors) have to be met.[59]

Higher demands have to be met if dwellings are adjacent to noisy rooms (restaurants, shops, workshops, heating stations, etc.). The required sound insulation depends on the average equivalent sound pressure level in the source room. For example, for $L_{eq} \leq 80$ dB(A), $R'_w \geq$ 57 dB is reqired; for $L_{eq} \leq 85$ dB(A), $R'_w \geq 62$ dB is required; and for $L_{eq} \leq 95$ dB(A), $R'_w \geq 72$ dB is required.[74]

Requirements for exterior walls in apartment houses and public buildings, especially for windows, depend on outside noise level and spectrum and on the permissible inside noise level [e.g., $L_{eq} \leq 30$ dB(A) in bedrooms and in living rooms]. The problem then is to select an appropriate value of R'_w for the facade when given an equivalent sound pressure level $L_{eq,o}$ caused by outside noise (about 1–2 m in front of the building) and a legally permissible equivalent sound pressure level $L_{eq,i}$ in the receiving room (both A weighted). The equation used is

$$R'_w = L_{eq,o} - L_{eq,i} + 10 \log \frac{S}{A} + C \quad \text{dB}, \qquad (39)$$

where S is the area of the facade, A is the equivalent sound absorption area of the receiving room (approximation for living rooms $A \approx 0.8 \times$ floor area), and C is a coefficient taking into account the spectrum of the outside noise (in decibels). The following values of C can be used in practice[75]:

Railway noise	0–3 dB
Road traffic noise	3–6 dB
Aircraft noise	6–8 dB

For typical housing, outside equivalent sound pressure levels $L_{eq,o}$ of 61 to 65 dB(A) from road traffic noise require an apparent weighted sound reduction index of $R'_w = 40$ dB for the facade. This value increases or decreases in steps of 5 dB corresponding to changes of 5 dB(A) of the outside level.

For most housing construction the sound insulation of the facade generally depends entirely on the windows. Sound transmission through the wall has to be considered only in few cases, for example, when an exterior wall is a lightweight sandwich component or the like, which will be found mainly in office buildings. The required sound insulation of windows can be calculated, taking into account the area ratio of the window to the whole facade (see Section 2.16).

Demands on the airborne sound insulation in industrial buildings are also influenced by kind of noise source and by permissible sound pressure level. Usually requirements are given in A-weighted sound pressure level. For transmission to interior building elements the necessary apparent weighted sound reduction index can be found using Eq. (23) (see Section 2.15).

If the concern is sound transmission to a receiving point outside an industrial hall, the following equation can be used to get the necessary sound reduction index of the facade:

$$R' = L_i - L_p + 10\log \frac{S}{r^2} \quad 14\ \text{dB}, \qquad (40)$$

where L_i is the sound pressure level within a frequency band (octave or one-third octave) inside the building, L_p is the permissible sound pressure level in a frequency band at the receiving point, S is the area of the radiating facade, and r is the distance of the receiving point to the facade ($r \geq \sqrt{S}$). For facades consisting of components of different sound insulation, the information in Section 2.16 has to be used.

4.1 Sources of Information on Sound Transmission

Many manufacturers provide collections of sound transmission data using their products. As well, other agencies have printed collections of information.[76–80] When comparing nominally identical structures, in such publications it is important to be sure that all details are identical. Small changes in mass, screw, or stud spacing or some other detail can have a large effect on the sound transmission. The density of gypsum board, for example, is not a constant.

REFERENCES

1. Lord Rayleigh, *The Theory of Sound*, Dover, New York, 1945.
2. L. Cremer, "Theorie der Schalldammung dunner Wande bei schrägem Einfall," *Akustische Zeitschrift*, Vol. 3, 1942, pp. 81–93.
3. A. London, "Transmission of Reverberant Sound through Double Walls," *J. Acoust. Soc. Am.*, Vol. 44, 1950, pp. 270–278.
4. H. Cremer and L. Cremer, "Theorie der Entstehung des Klopfschalls," *Frequenz*, Vol. 2, 1948, p. 61.
5. A. London, "Transmission of Reverberant Sound through Single Walls," *J. Res. Natl. Bureau Stand.*, RP1998, Vol. 42, June 1949, p. 605.
6. A. London, "Transmission of Reverberant Sound through Double Walls," *J. Res. Natl. Bureau Stand.*, RP2058, Vol. 44, January 1950, p. 605.
7. W. Fasold, E. Sonntag, and H. Winkler, *Bauphysikalische Entwurfslehre, Bau- und Raumakustik*, Verlag für Bauwesen, Berlin, 1987.
8. International Standardization Organization, "Acoustics, Measurement of Sound Insulation in Buildings and of Building Elements, Laboratory Measurements of Airborne Sound Insulation of Building Elements," ISO 140-3, 1978.
9. International Standardization Organization, "Acoustics, Measurement of Sound Insulation in Buildings and of Building Elements, Measurement of Sound Insulation of Small Building Elements," ISO 140-10, 1989.
10. American Society for Testing and Materials, "Standard Test Method for Laboratory Measurement of Airborne Sound Transmission Loss of Building Partitions," ASTM E90, 1990.
11. International Standardization Organization, "Acoustics, Measurement of Sound Insulation in Buildings and of Building Elements, Requirements for Laboratories," ISO 140-1, 1990.
12. "Bauakustische Prüfungen, Luft- und Trittschalldamm-Messungen an Bauteilen," Deutsches Institut für Normung, DIN 52210, T.2, 1984.
13. International Standardization Organization, "Acoustics, Measurement of Sound Insulation in Buildings and of Building Elements, Field Measurements of Airborne Sound Insulation between Rooms," ISO 140-4, 1978.
14. American Society for Testing and Materials, "Standard Test Method for Measurement of Airborne Sound Insulation in Buildings," ASTM E336, 1990.
15. M. J. Crocker, P. K. Raju, and B. Forssen, "Measurement of Transmission Loss of Panels by Direct Determination of Acoustic Intensity," *Noise Control Eng.*, Vol. 17, 1981, p. 6.
16. "Building Elements: Sound Insulation, Intensity Scanning under Laboratory Conditions," Nordtest method NT ACOU 084.
17. R. E. Halliwell and A. C. C., Warnock, "Sound Transmission Loss, Comparison of Conventional Techniques with Sound Intensity Techniques," *J. Acoust. Soc. Am.*, Vol. 77, 1985, pp. 2094–2103.
18. J. C. S. Lai, M. A. Burgess, P. P. Narang, and K. Miki, "Transmission Loss Measurements: Comparisons between Sound Intensity and Conventional Methods," *Proceedings of InterNoise 91*, 1991, p. 1029.
19. International Standardization Organization, "Acoustics, Measurement of Sound Insulation in Buildings and of Building Elements, Field Measurement of Airborne Sound Insulation of Facade Elements and Facades," ISO 140-5, 1978.
20. International Standardization Organization "Acoustics, Rating of Sound Insulation in Buildings and of Building Elements, Airborne Sound Insulation in Buildings and of Interior Building Elements," ISO 717-1, 1982.
21. International Standardization Organization "Acoustics, Rating of Sound Insulation in Buildings and of Building Elements, Airborne Sound Insulation of Facade Elements and Facades," ISO 717-3, 1982.
22. American Society for Testing and Materials, "Classification for Rating Sound Insulation," ASTM E413, 1994.
23. W. Fasold, Zur Bewertung der Schalldämmung von Fenstern, 5th International Congress of Acoustics, Lüttich, 1965.
24. American Society for Testing and Materials, "Classification for Determination of Outdoor–Indoor Transmission Class," ASTM E1332, 1994.
25. L. L. Beranek and I. L. Vér, *Noise and Vibration Control Engineering*, Wiley, New York, 1992.

26. D. A. Bies and C. H. Hansen, *Engineering Noise Control*, Unwin Hyman, London, 1988.
27. M. Heckl, "Investigations of Orthotropic Plates," *Acustica*, Vol. 10, 1960, p. 109 (in German).
28. L. L. Beranek and I. L. Vér, *Noise and Vibration Control Engineering*, Wiley, New York, 1992.
29. K. Gösele, "Verringerung der Luftschalldammung von Wänden durch Dickenresonanzen," *Bauphysik*, Vol. 12, 1990, pp. 187–191.
30. International Standardization Organization, "Acoustics, Material for Acoustical Applications, Determination of Airflow Resistance," ISO 9053, 1990.
31. American Society for Testing and Materials, "Standard Test Method for Airflow Resistance of Acoustical Materials. American ASTM Method C522, 1993.
32. B. H. Sharp, "A Study of Techniques to Increase the Sound Insulation of Building Elements," NTIS PB 222 829/4, U.S. Department of Housing and Urban Development, Washington, DC; "Prediction Methods for the Sound Transmission of Building Elements," *Noise Control Eng.*, Vol. 11, 1978, p. 53.
33. K. Gösele, "Prediction of the Sound Transmission Loss of Double Partitions (without Structureborne Connections)," *Acustica*, Vol. 45, 1980, p. 218 (in German).
34. G. Qiangguo and W. Jiquing, "Effect of Resilient Connection on Sound Transmission Loss of Metal Stud Double Panel Partitions," *Chinese J. Acoust.*, Vol. 2, No. 2, 1983, p. 113.
35. J. D. Quirt and A. C. C. Warnock, "Influence of Sound-Absorbing Material, Stud Type and Spacing, and Screw Spacing on Sound Transmission through a Double-Panel Wall System", *Proceedings Inter-Noise 93*, Vol. 2, 1993, pp. 971–974.
36. W. Loney, "Effect of Cavity Absorption on the Sound Transmission Loss of Steel Stud Gypsum Wallboard Partitions," *J. Acoust. Soc. Am.*, Vol. 49, 1971, p. 385.
37. P. P. Narang, "Effect of Fiberglass Density and Flow Resistance on Sound Transmission Loss of Cavity Plasterboard Walls," *Noise Control Eng. J.*, Vol. 40, No. 3, 1993, pp. 215–220.
38. K. Gösele and U. Gösele, "Einflußder Hohlraumdämpf ung auf die Steifigkeit von Luftschichten bei Doppelwänden," *Acustica*, Vol. 38, 1977, pp. 159–166.
39. "Summary Report for the Consortium on Gypsum Board Walls: Sound Transmission Loss Results," Internal Report IR693, Institute for Research in Construction, NRCC, Canada.
40. A. C. C. Warnock and J. D. Quirt, "Sound Transmission through Gypsum Board Walls," *Proceedings Inter-Noise 95*, 1995, pp. 727–730.
41. A. C. C. Warnock, "Sound Transmission through Concrete Blocks with Attached Drywall," *J. Acoust. Soc. Am.*, Vol. 90, 1991, pp. 1454–1463.
42. A. C. C. Warnock, "Sound Transmission Loss Measurements through 190 mm and 140 mm Blocks with Added Gypsum Board and through Cavity Block Walls," NRCC Internal Report IR586, 1990.
43. A. C. C. Warnock, "Sound Transmission through Two Kinds of Porous Concrete Blocks with Attached Drywall," *J. Acoust. Soc. Am.*, Vol. 92, 1992, pp. 1452–1460.
44. J. Enger and T. E. Vigran, "Transmission Loss of Double Partitions Containing Resonant Absorbers," *Proc. Inst. Acoust.*, Vol. 7, Part 2, 1985, p. 125.
45. J. M. Mason and F. J. Fahy, "The Use of Acoustically Tuned Resonators to Improve the Sound Transmission Loss of Double-Panel Partitions," *J. Sound Vib.*, Vol. 124, No. 2, 1988, p. 367.
46. A. C. C. Warnock, "Sound Transmission through Slotted Concrete Blocks with Attached Gypsum Board," *J. Acoust. Soc. Am.*, Vol. 94, No. 5, 1993, pp. 2713–2720.
47. J. D. Quirt, "Sound Transmission through Double Doors," *Can. Acoust.*, Vol. 14, No. 4, 1986, p. 3.
48. J. D. Quirt, "Measurements of the Sound Transmission Loss of Windows," Building Research Note No. 172, Institute for Research in Construction, NRCC, 1981.
49. J. D. Quirt, "Sound Transmission through Windows. I. Single and Double Glazing," *J. Acoust. Soc. Am.*, Vol. 72, No. 3, 1982, pp. 834–844.
50. J. D. Quirt, "Sound Transmission through Windows. II. Double and Triple Glazing," *J. Acoust. Soc. Am.*, Vol. 74, No. 2, 1983, pp. 534–542.
51. E. Gerretsen, "Calculation of Sound Transmission between Dwellings by Partitions and Flanking Structures," *Appl. Acoust.*, Vol. 12, 1979, p. 413.
52. E. Gerretsen, "Calculation of Airborne and Impact Sound Insulation between Dwellings," *Appl. Acoust.*, Vol. 19, 1986, p. 245.
53. M. C. Gomperts and T. Kihlmann, "The Sound Transmission Loss of Circular and Slit-Shaped Apertures in Walls," *Acustica*, Vol. 18, 1967, pp. 144–150.
54. G. P. S. Wilson and W. W. Soroka, "Approximation to the Diffraction of Sound by a Circular Aperture in a Rigid Wall of Finite Thickness," *J. Acoust. Soc. Am.*, Vol. 37, 1965, pp. 286–297.
55. K. Gösele and W. Schüle, Schall, Wärme, Feuchtigkeit 4. Auflage, Bauverlag Wiesbaden, Berlin 1977.
56. International Standardization Organization, "Acoustics, Measurement of Sound Insulation in Buildings and of Building Elements, Laboratory Measurements of Impact Sound Insulation of Floors," ISO 140-6, 1978.
57. International Standardization Organization, "Acoustics, Measurement of Sound Insulation in Buildings and of Building Elements, Field Measurement of Impact Sound Insulation of Floors," ISO 140-7, 1978.
58. American Society for Testing and Materials, "Standard Method of Laboratory Measurement of Impact Sound Transmission through Floor-Ceiling Assemblies using the Tapping Machine, ASTM E492, 1996.
59. American Society for Testing and Materials, "Standard

Test Method for Field Measurement of Tapping Machine Impact Sound Transmission through Floor-Ceiling Assemblies and Associated Support Structures," ASTM E1007.

60. International Standardization Organization, "Acoustics, Rating of Sound Insulation in Buildings and of Building Elements, Impact Sound Insulation," ISO 717-2, 1982.
61. American Society for Testing and Materials, "Classification for Determination of Impact Insulation Class (IIC)," ASTM E989, 1994.
62. W. Fasold, "Untersuchungen über den Verlauf der Sollkurve für den Trittschallschutz im Wohnungsbau," *Acustica*, Vol. 15, 1965, pp. 271–284.
63. T. Mariner and W. W. Hehmann, "Impact Noise Rating of Various Floors," *J. Acoust. Soc. Am.*, Vol. 41, 1967, pp. 206–214.
64. R. Josse, "How to Assess the Sound-Reducing Properties of Floors to Impact Noise (Footsteps)," *Appl. Acoust.*, Vol. 5, 1972, pp. 15–20.
65. E. Gerretsen, "A New System for Rating Impact Sound Insulation," *Appl. Acoust.*, Vol. 9, 1976, pp. 247–263.
66. K. Bodlund, "Alternative Reference Curves for Evaluation of the Impact Sound Insulation between Dwellings," *J. Sound Vib.*, 1985, pp. 381–402.
67. A. C. C. Warnock, "Floor Impact Noise and Foot Simulators," *Proceedings Inter-Noise 83*, 1983, p. 1127.
68. A. C. C. Warnock, "Low Frequency Impact Sound Transmission through Floor Systems," *Proceedings Inter-Noise 92*, 1992, p. 743.
69. International Standardization Organization, "Acoustics, Measurement of Sound Insulation in Buildings and of Building Elements, Laboratory Measurement of the Reduction of Transmitted Impact Noise by Floor Covering on a Standard Floor," ISO 140-8, 1978.
70. L. Cremer, M. Heckl, and E. E. Ungar, *Structure-Borne Sound*, Berlin, Heidelberg, New York, Springer-Verlag, 1988.
71. E. Sonntag, "Kurzprüfverfahren für das Verbesserungsmaß des Trittschallschutzes bei Weichbelägen, Bauforschung," *Baupraxis*, Vol. 7, 1977, pp. 81–89.
72. International Standardization Organization, "Acoustics, Determination of Dynamic Stiffness, Materials Used under Floating Floors in Dwellings," ISO 9052-1, 1989.
73. E. E. Ungar, "Design of Floated Slabs to Avoid Stiffness of Entrapped Air," *Noise Con. Eng.*, p. 12, July–August, 1975.
74. DIN 4109, Schallschutz im Hochbau, Anforderungen und Nachweise, 1989.
75. VDI 4100, VDI-Richtlinien, Schallschutz von Wohnungen, Kriterien für Planung und Beurteilung, E 1989.
76. R. A. Hedeen, "NIOSH Compendium on Materials for Noise Control, DHEW Publication No. 80-116, U.S. Department of Health, Education and Welfare, Cincinnati, OH.
77. R. B. Dupree, "Catalog of STC and IIC Ratings for Wall and Floor/Ceiling Assemblies," Office of Noise Control, California Department of Health Services, Berkeley, CA.
78. A. C. C. Warnock and D. W. Monk, "Sound Transmission Loss of Masonry Walls: Tests on 90, 140, 190, 240 and 290 mm Concrete Block Walls with Various Surface Finishes," Building Research Note 217, NRCC.
79. T. D. Northwood and D. W. Monk, "Sound Transmission Loss of Masonry Walls: Twelve-Inch Lightweight Concrete Blocks—Comparison of Latex and Plaster Sealers," Building Research Note 93, NRCC.
80. T. D. Northwood and D. W. Monk, "Sound Transmission Loss of Masonry Walls: Twelve-Inch Lightweight Concrete Blocks with Various Surface Finishes," Building Research Note 90, NRCC.

77

RATINGS AND DESCRIPTORS FOR THE BUILDING ACOUSTICAL ENVIRONMENT

GREGORY C. TOCCI

1 INTRODUCTION

Generally, ratings for acoustical environments either set limits on background sound to achieve an acceptable venue or set requirements for audio signal levels to achieve acceptable functionality. In addition, architectural acoustical criteria set limits for sound transmission through building partition constructions, for example, outdoor sound transmitted indoors and indoor sound transmitted between rooms.

This chapter chiefly describes descriptor and rating methods used in establishing acoustical criteria in buildings. Acoustical descriptors characterize in a single number two or more of the three fundamental characteristics of sound: (1) spectral content, (2) level, and (3) time-varying character. Many of the descriptors briefly discussed in this chapter are discussed in more detail elsewhere in this book.

The chapter begins with a description of common acoustical descriptors used to quantify sound in the environment. These include sound pressure level, speech interference level, equivalent sound level, and so forth. Many of these are discussed in more detail elsewhere, in particular in Chapter 64. Section 2 of this chapter discusses ratings used to quantify and characterized sound inside and outside of buildings. These include day–night average sound level, noise criteria curves, and so forth. Section 3 discusses descriptors and ratings for the acoustical performance of materials and systems used in the construction of buildings. These generally break down into ratings quantifying the ability of building materials to absorb sound and the ability of building materials and systems to resist the transmission of sound through them. Section 4 discusses descriptors used to assess the quality of sound in building spaces. The discussions draw heavily on standards and methods of the American National Standards Institute (ANSI) and American Society for Testing and Materials (ASTM). Finally, Section 5 discusses additional ratings and descriptors used in International Standards Organization (ISO) standards and standards of selected European and Asian countries.

Note: References to chapters appearing only in the *Encyclopedia* are preceded by *Enc*.

2 GENERAL ACOUSTICAL DESCRIPTORS

2.1 A-Weighted Sound Pressure Level

The human ear is sensitive to sound extending from approximately 20 to 20,000 Hz but is most sensitive to sound in the 500–4000 Hz frequency range. Above and below this most sensitive frequency range the ear is progressively less sensitive to sound. Since sound measurement instrumentation is typically more uniformly sensitive to sound over a considerably wider frequency range than human hearing, sound level meters use electronic filtering to reproduce the varying sensitivity of the ear to sound at different frequencies. Using this filtering, sound level meters indicate a sound level that is more representative of what the ear hears. This filtering is called A weighting, and the sound pressure levels measured using A-weighted filtering are often called *sound levels* and typically denoted as dB(A) (abbreviation, AL; letter symbol, L_A).

The *A-weighted sound pressure level* is often used to evaluate general human exposure to sound. A-weighted sound pressure levels are most widely used as they are a convenient single-number quantifier of sound over the audible frequency range (as discussed in Chapter 64). Originally, A weighting was intended to represent the varying sensitivity of the ear to sound at sound pressure levels ranging between 40 and 60 dB. Subsequently, B and C weightings were developed to represent the varying sensitivity of the ear to sound over higher sound pressure level ranges. Both ratings allow more low-frequency sound energy with C weighting being nearly flat across the audible frequency range. However, B and C weightings are no longer widely used, with the exception that C weighting is used by some agencies and researchers in assessing transient sound. Also, in the absence of spectral information (octave band sound levels), the difference between A-weighted and C-weighted sound levels can indicate the relative content of low-frequency energy in the sound spectrum.

2.2 Speech Interference Levels

The *speech interference level* (SIL) is the arithmetic average of the sound pressure level in octave bands centered at 500, 1000, 2000, and 4000 Hz.[1] SIL is also discussed in Section 7 of Chapter 64. Because the definitions and octave bands used in the past have changed, ANSI S3.14 recommends that the abbreviation SIL be followed by an indication of the octave bands used for computation, for example, SIL (0.5, 1, 2, 4).

Figure 1 shows the relationship between speech interference level and vocal effort required at various distances to achieve just-reliable speech communication. The crosshatch range indicates *expectant* voice level, that is, the natural inclination of speakers to raise their voices in the presence of high background sound levels. The rate at which the voice is raised ranges between 3 and 6 dB for every 10-dB increase in backaground noise level above 50 dB(A).

The SIL criteria presented in Fig. 1 are applicable to both indoor and outdoor environments and simply relate to speech communication, without any accounting for other circumstances or activities in a space. This figure also includes instructions for its use for determining minimum vocal effort and/or maximum distance to achieve "just reliable" speech articulation (articulation index of 0.45) in the presence of broadband (*flat spectrum*) background noise. The method is not valid for (1) situations where the background sound spectrum departs significantly from being "flat," (2) situations where the background or the signal sound levels vary significantly with time, (3) communications in highly reverberant environments, and (4) situations when speech is distorted.

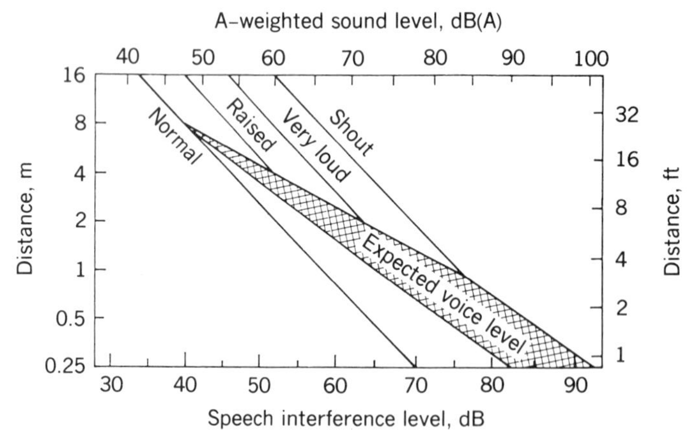

Fig. 1 Talker-to-listener distances for just-reliable communication. The curves show maximum permissible talker-to-listener distances for just-reliable speech communication. The parameter on each curve indicates the relative voice level. Since a talker will raise his or her voice in noise, typically at the rate of 3–6 dB for every 10-dB increase in noise level above 50 dB(A), the expected voice level will increase with increasing noise level. The cross-hatched area shows the range of permissible talker-to-listener distances under these conditions. The lower bound of the cross-hatched area is for voice level being raised at the rate of 3 dB per 10-dB increase in noise level; the upper bound is for a rate of increase of 6 dB per 10-dB increase in noise level. (1) Measure the A-weighted sound level of the background noise using the *slow* response of the sound level meter. (2) Locate this sound level on the upper abscissa. (3) A perpendicular dropped from this point will intersect several curves, indicating the maximum distance between the talker and listener for just-reliable speech communication for various levels of vocal effort.

2.3 Loudness Level

The *loudness level* (LL) is a single-number rating expressed in units of *phons*. It is described more fully in Chapters 64 and 91. The loudness level in phons is the level of a 1000-Hz reference tone with a loudness that appears to be equal to that of the spectrum being rated.[2] The loudness level of an octave band spectrum is obtained by a procedure described in ANSI S3.4-1980(R 1986).[3] The procedure involves overlaying an octave band spectrum of sound to be rated on contours of approximate equal loudness and interpolating *loudness index* values for each octave band level of the spectrum. These loudness index values are then used with specified formulas to arrive at a single-number loudness level rating in phons. Phons are a logarithmic quantity. The corresponding linear quantity is *sones* where 1 sone is the loudness of a sound whose loudness level is 40 phons. A doubling of amplitude in sones corresponds to a factor of 10 increase in loudness level expressed in phons.

2.4 Equivalent Sound Level

As discussed in Chapters 64, 70, and 72, the *equivalent sound level* (QL; symbol, L_{eq}) is the energy average sound level occurring at a particular location over a given time interval. The equivalent sound level is also discussed in Section 9 of Chapter 64. Often, the ambient sound level is monitored for 1-h intervals for environmental sound assessments. When appropriate, hourly equivalent sound levels may be combined in order to determine the equivalent sound level for a 24-h period, or equivalent sound levels for shorter time intervals can be combined to determine the hourly equivalent sound level. Most often, the equivalent sound level is an A-weighted sound pressure level, and the L_{eq} symbol implies A weighting. Some agency criteria incorporate the C-weighted equivalent sound level, symbolized as L_{Ceq}.

2.5 Sound Exposure Level

The *sound exposure level* (SEL; symbols, L_{ET} or L_{AE}) quantifies the total A-weighted acoustic energy integrated over a time interval for a given acoustical event. To simplify the units to decibels, the SEL of an event is described as the hypothetical equivalent sound pressure level enduring for a period of one second that would have the same amount of acoustic energy as the specific transient event for which the SEL was measured. The SEL is used to quantify the sound energy associated with a transient event such as an aircraft over flight. Its practical use requires setting a measurement start and a finish time, but if the maximum sound level occurring during an event is high enough in amplitude compared to levels before and after the event, defining precise start and finish times has little influence on the SEL value reached.

2.6 Single-Event Noise Exposure Level

The *single-event noise exposure level* (SENEL) is similar to the SEL except that it incorporates the integration of sound energy from and to defined points in time before and after a transient event. These points in time are defined as the moments during which the event sound pressure level rises above and falls below a specific sound level threshold. For environmental sound analysis, a commonly used threshold is 65 dB(A). This threshold is not set by any standard, but by measurement personnel seeking a clear point at which the sound pressure level associated with a transient event is sufficiently above the ambient so that lesser transients in the ambient do not trigger a SENEL event. The concept of SENEL has its roots in instrumentation developed to measure transient events from sources such as aircraft and trains.

2.7 Percentile Sound Levels

Percentile sound levels ($L_{XX\%}$) are A-weighted sound pressure levels exceeded an indicated (by the *XX*% value) percent of a time interval. Percentile sound levels are also discussed in Section 11 of Chapter 64. Strictly speaking, the descriptor should also indicate the interval length, for example, L_{90}(1 h). The L_{90}(1 h) is the A-weighted sound pressure level exceeded a total of 54 min out of a continuous 60-min interval. Accordingly, the L_{10}(20 min), is the A-weighted sound pressure level exceeded 2 min out of a continuous 20-min interval.

The L_{10} is representative of intrusive sounds, sounds of short duration, but high level. The L_{90}, on the other hand, is representative of nearly the lowest levels of sound occurring during quiet interludes and is often referred to as the background or residual sound level.

The Federal Highway Administration (FHWA) allows the alternative use of L_{10} or L_{eq} for assessing traffic sound levels and uses these descriptors to set guideline limits for traffic noise exposure at wayside areas according to various categories of land use.[4] The FHWA regulation for assessing the impact of highway noise allows the use of either L_{10} or L_{eq}, but not both.

Some states and municipalities set environmental noise limits based on percentile sound level limits. For example, the Commonwealth of Massachusetts Department of Environmental Protection requires that sound generated by a commercial facility may not cause an ambient sound level that exceeds the preexisting background sound level (L_{90}) without the facility operating by more than 10 dB(A).[5] The City of Cambridge, Massachusetts, requires that construction sound levels not exceed an L_{10}(20 min) of 75 dB(A) at nearest residential property.[6]

3 RATINGS USED TO ASSESS THE ACCEPTABILITY OF AMBIENT SOUND

3.1 Day–Night Average Sound Level

The *day–night average sound level* (DNL; symbol, L_{dn}) is a 24-h average A-weighted sound level where a 10-dB "penalty" is applied to sound occurring between the hours of 10 : 00 p.m. and 7 : 00 a.m. The 10-dB penalty accounts for the heightened sensitivity of a community to noise occurring at night. This descriptor is also discussed in Section 10 of Chapter 64 of named day–night equivalent level. Day–night average sound level has become the primary descriptor for general environmental sound and is often used to assess sound from transportation systems. Among agencies using the day–night average sound level in their criteria and regulations are the U.S. Environmental Protection Agency (EPA),[7] the Federal Aviation Administration (FAA),[8] and the U.S. Department of Housing and Urban Development (HUD).[9]

The U.S. EPA has taken the lead among all federal agencies in unifying usage of environmental sound level descriptors. The EPA has fostered the development of the day–night average sound level but has not enacted regulations controlling general environmental noise. Instead, it has issued guidelines that identify yearly L_{dn} sound levels "sufficient to protect public health and welfare from the effects of environmental noise." Table 1 presents EPA's suggested levels to protect public health and welfare. The EPA specifically cautions that these tabulated levels are not to be used as regulations by other agencies without addressing economic and other considerations associated with sound level restrictions. Of these levels, the most widely cited are a day–night average sound level of 55 dB for outdoor residential areas and a day–night average sound level of 45 dB for indoor residential spaces. Again, these are only to be used as levels with a margin of safety incorporated and not as EPA's recommendations for agency limits.

3.2 Community Noise Equivalent Level

The *community noise equivalent level* (CNEL) is similar to the DNL (L_{dn}) except that, in addition to the 10-dB(A) penalty applied between the hours of 10 : 00 p.m. to 7 : 00 a.m., there is a 5-dB(A) penalty applied to sound between the hours of 7 : 00 p.m. and 10 : 00 p.m. CNEL is defined in Chapter 64. Use of this descriptor seems to be declining nationally, but is most commonly used in California standards, where it is allowed as an alternative environmental sound level descriptor to the day–night average sound level for assessing environmental sound transmission into building spaces.[10,11] It is also used to assess community noise in particular near airports (Chapter 70).

3.3 Noise Criterion Curves

Speech interference level and loudness level are not widely used to evaluate ambient sound in engineering acoustics. More widely used descriptors are noise criterion (NC) curves. Noise criterion curves are a simpler alternative to the evaluation of perceived loudness than the loudness level computation procedure. NC curves are also discussed in Section 8 of Chapter 64. Figure 2 contains the set of NC curves.[12] The NC curves are smoothed versions of the loudness level index curves. The NC value shown for each curve is the

TABLE 1 Yearly L_{dn} Values That Protect Public Health and Welfare with a Margin of Safety

Effect	Level	Area
Hearing	$L_{eq(24)} \leq 70$ dB	All areas (at the ear)
Outdoor activity interference and annoyance	$L_{dn} \leq 55$ dB	Outdoors in residential areas and farms and other outdoor areas where people spend widely varying amounts of time and other places in which quiet is a basis for use.
	$L_{eq(24)} \leq 55$ dB	Outdoor areas where people spend limited amounts of time, such as school yards, playgrounds, etc.
Indoor activity interference and annoyance	$L_{dn} \leq 45$ dB	Indoor residential areas
	$L_{eq(24)} \leq 45$ dB	Other indoor areas with human activities such as schools, etc.

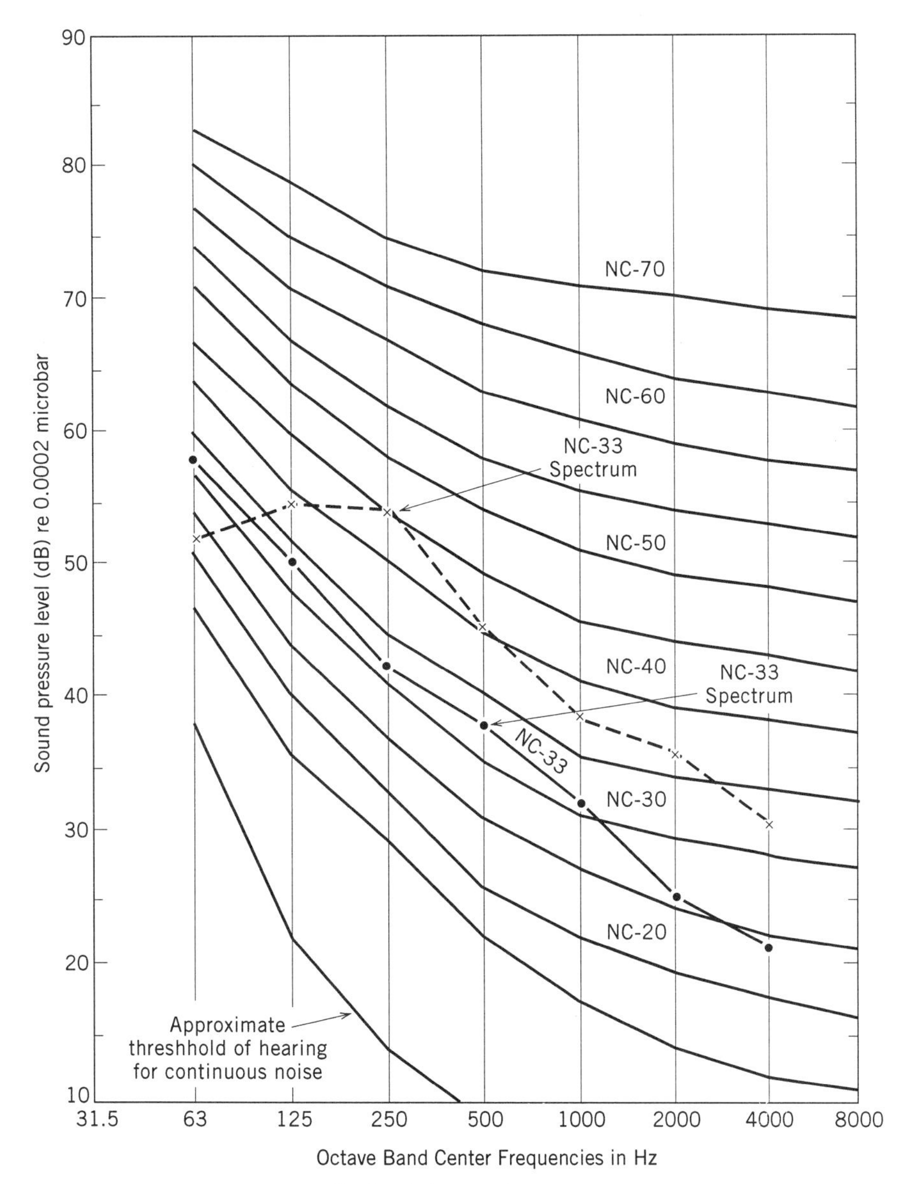

Fig. 2 Noise criterion curves.

SIL (600–1200, 1200–2400, and 2400–4800 Hz), that is, in the "old" octave bands, for each curve. Noise criterion values are commonly determined for an octave band spectrum of sound by superposing the NC curves over the spectrum. The highest NC curve reached by the measured spectrum (tangent to) is the NC rating of the sound pressure level spectrum. Figure 2 shows two typical spectra and the resulting NC values.

Note that the NC curves are defined in terms of the "old" frequency bands 20–75, 75–50, 50–300 cps, and so on. Schultz[13] simply overlaid the "new" preferred octave bands on the NC curves previously defined in old octave bands, added an NC-15 curve, and corrected the approximate threshold of hearing curve originally shown with the NC curves when they were first published. This "updating" of NC curve assessment has not changed the method for evaluating the NC value of a sound pressure spectrum plotted in preferred octave bands.

Often, the NC value of a spectrum is considered within a limited frequency range, such as 500–4000 Hz in the case of concerns for speech intelligibility or 63–125 Hz in the case of low-frequency "rumble"-type sound levels, typically associated with heating, ventilation, and air conditioning (HVAC) systems.

It should also be noted that NC curves are generally used in evaluating continuous sound inside building spaces produced by mechanical systems or environmental sound transmitted into building spaces. Guidelines for assessing the acceptability of sound in buildings generally presume that the background sound does not have tonal or temporal characteristics that lend a distinctive feature to sound, such as transformer hum or fan blade passage tones. Using these criteria to evaluate tonal noise may not be appropriate as it may underestimate the interference of sound on occupant use of spaces.

In the United States background sound in building spaces produced by mechanical systems is not limited by any specific regulation or agency. Instead, the building design profession has, through various organizations, established design criteria for noise in architectural spaces. The most commonly used criteria are the design guidelines for HVAC system noise in unoccupied spaces established by the American Society of Heating, Refrigerating, and Air-Conditioning Engineers (ASHRAE). For most of the many years that ASHRAE has published recommended design criteria for spaces, it has expressed them as acceptable ranges of NC curves.

In recent years, ASHRAE has introduced a new set of curves called room criterion (RC) curves, which are discussed below. On their first introduction, ASHRAE publications suggested that RC curves could be used in lieu of NC curves. As the years progressed, the use of RC curves was cited as preferred to the use of NC curves. Finally, the *1995 ASHRAE Applications Handbook* lists recommended ranges of background sound in building spaces expressed only as ranges of RC(N) curves (Ref. 14, p. 43.5). No listing of criteria using NC curves is given in the *1995 Handbook* except that the guidelines state that "if the quality of the sound in the space is of secondary concern, the criteria may be specified in terms of NC levels" (footnote b in Table 2, p. 43.5). Although RC curves represent a better quality of background sound, NC curves are still in wider use at this time as they represent an acceptable compromise between economy and quality of background sound. Table 2 presents the ASHRAE recommended design guidelines for HVAC system noise in unoccupied spaces. These are expressed in RC(N) levels. For use with NC curves, substituting NC levels equal to the RC levels shown is acceptable, subject to the cautionary footnotes in Table 2.

3.4 Room Criterion Curves

ASHRAE has adopted the use of room criterion (RC) curves for evaluating the background sound in building spaces produced by mechanical systems. RC curves are also described in Section 8 of Chapter 64. Unlike NC curves, RC curves are straight lines, as shown in Fig. 3. These curves are used by superposing them on a measured octave band spectrum. The RC value of the spectrum is determined through the following procedure:

1. Determine the arithmetic average, to the nearest whole number, of the sound pressure levels in the 500-, 1000-, and 2000-Hz octave frequency bands. This is the RC level associated with the room sound level spectrum.
2. Draw a line having a −5 dB/octave slope through the RC level at 1000 Hz determined from step 1.
3. Classify the subjective quality or character of the room sound level spectrum as follows:
   a. *Neutral.* A spectrum classified as neutral is free of tonal quality and would be perceived as unobtrusive or bland. A neutral sound spectrum falls exactly along or close to a single RC contour. If the octave band data do not exceed the RC curve determined in step 2 by more than 5 dB at and below 500 Hz and do not exceed the RC curve by more than 3 dB at and above 1000 Hz, the spectrum is considered neutral, and the designator (N) is placed after the RC level.
   b. *Rumble.* A sound spectrum that is perceived to have a "rumbly" quality has an excess of low-frequency sound energy. A rumbly spectrum is characterized as one with octave band sound levels that exceed the RC curve determined in step 2 by more than 5 dB at and below 500 Hz. For such spectra the designator (R) is placed after the RC level.
   c. *Hiss.* A sound spectrum that is perceived to have a "hissy" quality has an excess of high-frequency energy. A hissy spectrum is characterized as one with octave band sound levels that exceed the RC curve determined in step 2 by more than 3 dB above 500 dB. For such spectra the designator (H) is placed after the RC level.
   d. *Acoustically Induced Perceptible Vibration.* The cross-hatched region of the RC curves in Fig. 3 indicates sound pressure levels in the 16- to 63-Hz octave bands at which perceptible vibration in building walls and ceilings can occur. These sound levels often produce rattles in cabinets, doors, pictures. lighting fixtures, and so forth.

TABLE 2 Design Guidelines for HVAC System Noise in Unoccupied Spaces

Space	RC(N) Level[a,b]
Private residences, apartments, condominiums	25–35
Hotels/Motels	
Individual rooms or suites	25–35
Meeting/banquet rooms	25–35
Halls, corridors, lobbies	35–45
Service/support areas	35–45
Office buildings	
Executive and private offices	25–35
Conference rooms	25–35
Teleconference rooms	25 (max)
Open plan offices	30–40
Circulation and public lobbies	40–45
Hospitals and clinics	
Private rooms	25–35
Wards	30–40
Operating rooms	25–35
Corridors	30–40
Public areas	30–40
Performing arts spaces	
Drama theaters	25 (max)
Concert and recital halls	—[c]
Music teaching studios	25 (max)
Music practice rooms	35 (max)
Laboratories (with fume hoods)	
Testing/research, minimal speech communication	45–55
Research, extensive telephone use, speech communication	40–50
Group teaching	35–45
Churches, mosques, synagogues	25–35
With critical music programs	—[c]
Schools	
Classrooms up to 750 ft^2	40 (max)
Classrooms over 750 ft^2	35 (max)
Lecture rooms for more than 50 (unamplified speech)	35 (max)
Libraries	30–40
Courtrooms	
Unamplified speech	25–35
Amplified speech	30–40
Indoor stadiums and gymnasiums	
School and college gymnasiums and natatoriums	40–50[d]
Large seating capacity spaces (with amplified speech)	45–55[d]

[a]The values and ranges are based on judgment and experience, not on quantitative evaluations of human reactions. They represent general limits of acceptability for typical building occupancies. Higher or lower values may be appropriate and should be based on a careful analysis of economics, space usage, and user needs. They are not intended to serve by themselves as a basis for a contractual requirement.
[b]When the quality of sound in the space is important, specify criteria in terms of RC(N). If the quality of the sound in the space is of secondary concern, the criteria may be specified in terms of NC levels.
[c]An experienced acoustical consultant should be retained for guidance on acoustically critical spaces (below RC 30) and for all performing arts spaces.
[d]Spectrum levels and sound quality are of lesser importance in these spaces than overall sound levels.
(Reprinted with permission from the *1995 ASHRAE Handbook—HVAC Applications*.)

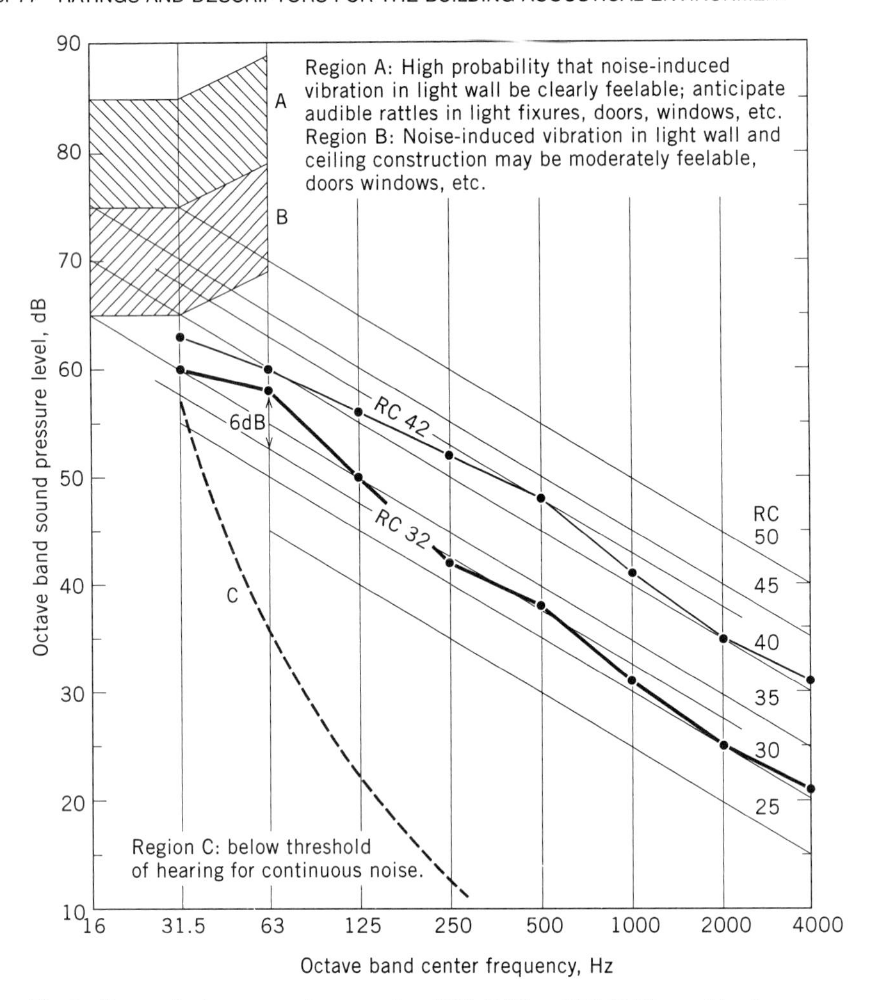

Fig. 3 Room criterion curves plus examples of RC 42(N) and RC 32(R) sound level spectra. (Reprinted with permission from the *1995 ASHRAE Handbook—HVAC Applications*.

For spectra with levels that fall into this range, the designator (RV) is placed after the RC level.

Figure 3 also provides examples of two types of spectra. In one case, the spectrum follows along the RC 42 curve (i.e., the average of the sound pressures in the 500-, 1000-, and 2000-Hz octave bands is 42 dB). In this spectrum, all of the octave band levels are within the prescribed limits, and is therefore designated an RC 42(N) spectrum. The spectrum labeled RC 32(R) is generally 10 dB lower than the RC 42(N) spectrum, except in the low frequencies where the 63-Hz octave band exceeds the RC 32 contour by 6 dB, thus constituting a low-frequency rumble component.

3.5 Balanced Noise Criterion Curves

Beranek[15,16] has drawn upon low-frequency hearing considerations to extend NC curves down to low frequencies. He has also sloped downward the high-frequency ends of the NC curves to reduce the subjectively hissy nature of sound spectra that conform to NC spectra shapes, thereby creating a "balanced" spectrum subjectively perceived as more uniform and devoid of significant tonal content. A set of balanced noise criterion curves (NCBs) are presented in Fig. 4. NCB curves are also discussed in Section 8 of Chapter 64. As with RC curves, NCB curves are accompanied by a procedure for assessing the perceived balance of a sound spectrum, that is, whether or not a spectrum will be perceived as rumbly

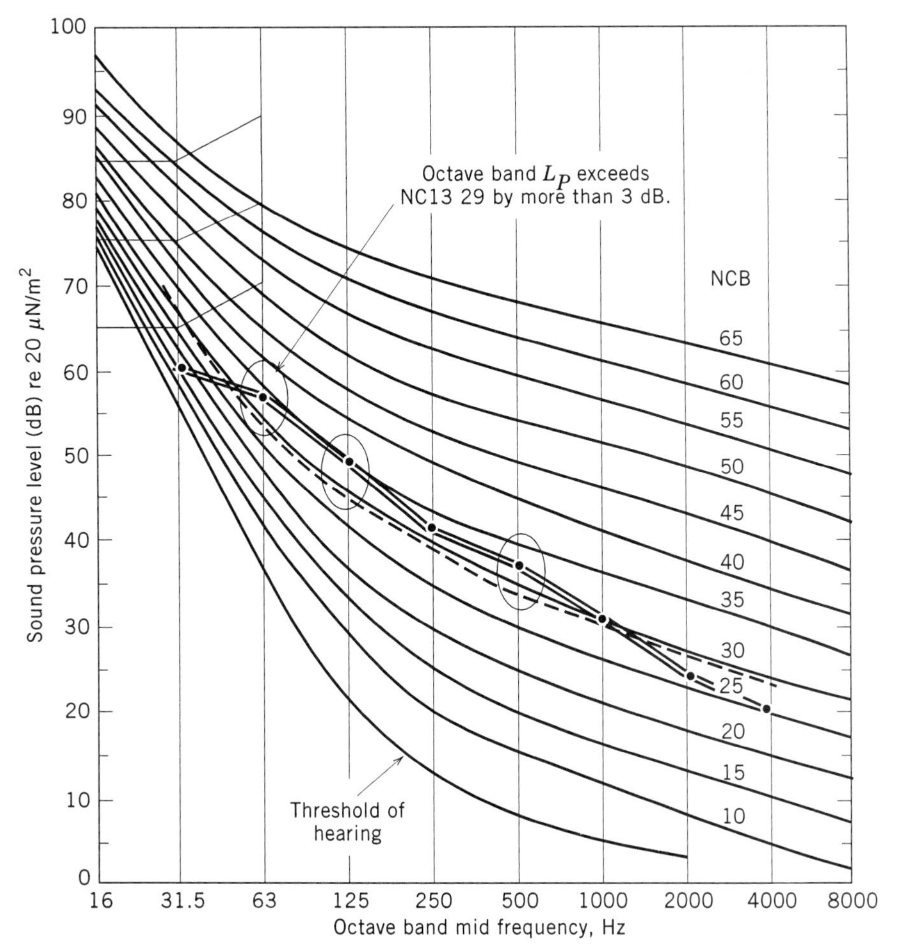

Fig. 4 Balanced noise criterion curves for occupied rooms. Octave-band sound pressure levels of the magnitudes indicated in regions *A* and *B* may induce audible rattles or feelable vibrations in lightweight partitions and ceiling constructions (e.g., thin plaster or gypsum board on metal framing) as follows (taken from Ref. 12): region *A*, clearly noticeable vibrations; region B, moderately noticeable vibrations.

or hissy, and so forth. This procedure is much like that used with the RC curves, but designators (N), (R), (H), and (RV) are not used as part of the descriptor as is the case with RC values. Rules for assessing rumble and hissy qualities of sound spectra using NCB curves are as follows:

1. Determine the L_{SIL} for the spectrum being evaluated. The L_{SIL} is the arithmetic average of sound levels in the 500-, 1000-, 2000-, and 4000-Hz octave bands rounded to the nearest decibel, for example, XX. This spectrum would then be denoted an NCB-XX spectrum.
2. The spectrum will be perceived as rumbly if any octave band sound level at or below 1000 Hz is above the NCB-YY curve. YY is equal to the XX value in step 1 plus 3 dB.
3. The spectrum will be perceived as hissy if any octave band sound level at frequencies above 500 Hz exceeds the NCB-ZZ curve. To determine the value ZZ, first determine the arithmetic average of sound pressure levels in the three octave bands 125 through 500 Hz. Then determine which NCB curve has this sound pressure level at 250 Hz. This is the NCB-ZZ curve.

Figure 4 also illustrates the evaluation of a spectrum using NCB curves. The example spectrum is equal to the lower of the two spectra shown in Fig. 3 used to illustrate the use of RC curves. Note that using the NCB curves, this spectrum exhibits rumbly characteristics in the 63-, 125-, and 500-Hz octave bands where octave band sound levels exceed the NCB 29 curve by more than 3 dB.

4 DESCRIPTORS AND RATINGS FOR MATERIALS AND BUILDING CONSTRUCTIONS THAT AFFECT SOUND INSIDE BUILDING SPACES

4.1 Sound Absorption Coefficient

The ability of a material to absorb sound is characterized by the sound absorption coefficient (α). This is the ratio of the amount of sound energy absorbed to the amount incident on the surface of a sound-absorbing material. Theoretically, this can range between 0 and 1.0. The higher the sound absorption coefficient the better the ability of the material to absorb sound. Normally, materials considered to be sound absorbing have sound absorption coefficients greater than or equal to 0.5 (i.e., 50%).

Because sound-absorbing materials have differing sound absorption performances at different frequencies, sound absorption coefficients are generally determined via measurements and reported in octave band frequencies—generally 125, 250, 500, 1000, 2000, and 4000 Hz. Sound absorption coefficient is measured in a reverberation chamber[17] using any one of a variety of specified mounting methods.[18] Mounting methods, especially clearance between the material and hard backing surface (reverberation room walls, floor, or ceiling) can affect the measured sound absorption coefficient significantly.

4.2 Noise Reduction Coefficient

To simplify comparison between sound-absorbing materials, a single-value descriptor called the noise reduction coefficient (NRC)[17] is used. It is the arithmetic average of the sound absorption coefficients at 250, 500, 1000, and 2000 Hz. Generally, for sound-absorbing materials, the NRC falls between 0.55 and 0.95. Again, the higher the NRC, the higher the sound absorption performance of a material. Table 3 also presents noise reduction coefficients for the common building materials listed.

Often, the sound absorption of building components, such as upholstered chairs in an auditorium or room sound absorption devices sold as units, are quantified in sabins, either metric (square metres) or English (square feet). This is the case for chairs, pews, and seated audience in Table 3. One sabin is an amount of room absorption equivalent to one square foot (or one square metre in metric units) of material having a sound absorption coefficient of 1.0.

4.3 Noise Reduction

Rather than specifying a specific sound level limit inside building spaces, standards and regulations set by government agencies and organizations often control interior sound levels by specifying minimum required noise reductions (NRs) of exterior facades or interior building partitions. Noise reduction is defined as the difference between the sound pressure level produced by a sound source on one side (the source side) and the sound pressure level on the other side (the receiver side) of a construction being evaluated. Noise reduction is most often measured in octave bands, but sometimes it is expressed as an A-weighted sound pressure level difference. Note that the A-weighted noise reduction depends on the source spectrum. For example, the A-weighted noise reduction for a building exterior wall exposed to aircraft sound may be different from the A-weighted noise reduction for traffic sound. This is because aircraft and traffic sound have their respective energies concentrated in different frequency ranges, and the sound isolation performance for a wall in those two ranges may differ. There are no widely available standards for measuring noise reduction; however, ASTM has formalized the procedure for measuring the noise reduction between interior spaces in buildings in ASTME 597-81 (R1987).[19]

Noise reduction is not only related to the ability of a wall construction to resist the transmission of sound, but it is also related to the size of the transmitting wall and the reverberant character of the receiver room. Hence, a wall construction used in different circumstances may result in different noise reductions. Most importantly, noise reduction is significantly affected by the quality of construction, specifically with respect to sound leakage paths associated with gaps and penetrations in the construction, including the perimeter of doors and windows, electrical outlets, fixtures, ductwork, and so forth.

4.4 Sound Transmission Loss

Sound transmission loss (TL) is 10 times the logarithm of the ratio between the sound energy incident on a material or sound isolation construction, divided by the sound energy transmitted through this material. Unlike noise reduction, sound transmission loss is measured in a laboratory according to standards established in ASTM E 90.[20] The standard requires that sound transmission loss be measured in one-third octave frequency bands in a source room–receiver room test suite that complies with certain standards of sound isolation and room dimensions. The result of the test standard is a one-third octave

TABLE 3 Coefficients of General Building Materials and Furnishings

Material	Sound Absorption Coefficient (α)						
	125 Hz	250 Hz	500 Hz	1000 Hz	2000 Hz	4000 Hz	NRC
Brick, unglazed	0.03	0.03	0.03	0.04	0.05	0.07	0.05
Carpet, heavy, on concrete	0.02	0.06	0.14	0.37	0.60	0.65	0.45
Same, on 40-oz hairfelt or foam rubber	0.08	0.24	0.57	0.69	0.71	0.73	0.55
Concrete block, painted	0.10	0.05	0.06	0.07	0.09	0.08	0.05
Fabrics							
Light velour, 10 oz/yd^2, hung straight, in contact with wall	0.03	0.04	0.11	0.17	0.24	0.35	0.15
Medium velour, 14 oz/yd^2, draped to half area	0.07	0.31	0.49	0.75	0.70	0.60	0.55
Heavy velour, 18 oz/yd^2, draped to half area	0.14	0.35	0.55	0.72	0.70	0.65	0.60
Floors							
Concrete or terrazzo	0.01	0.01	0.015	0.02	0.02	0.02	0
Linoleum, asphalt, rubber or cork tile on concrete	0.02	0.03	0.03	0.03	0.03	0.02	0.05
Wood	0.15	0.11	0.10	0.07	0.06	0.07	0.1
Glass, large panes of heavy plate	0.18	0.05	0.04	0.03	0.02	0.02	0.05
Gypsum board, $\frac{1}{2}$ in. nailed to 2 × 4's 16 in. on center	0.29	0.10	0.05	0.04	0.07	0.09	0.05
Marble or glazed tile	0.01	0.01	0.01	0.01	0.02	0.02	0
Openings							
Stage, depending on furnishings				0.25–0.75			
Deep balcony, upholstered seats				0.50–1.00			
Grilles, ventilating				0.15–0.50			
Plaster, gypsum or lime, smooth finish on tile or brick	0.13	0.15	0.02	0.03	0.04	0.05	0.05
Plaster, gypsum or lime, rough finish on lath	0.02	0.03	0.04	0.05	0.04	0.03	0.05
Same, with smooth finish	0.02	0.02	0.03	0.04	0.04	0.03	0.05
Plywood paneling, $\frac{7}{8}$ in. thick	0.28	0.22	0.17	0.09	0.10	0.11	0.05
Water surface, as in a swimming pool	0.008	0.008	0.013	0.015	0.020	0.025	0
Air, sabins per 1000 ft^3	—	—	—	—	2.3	7.2	—

Type of Seats/Audience	Absorption[a]					
	125 Hz	250 Hz	500 Hz	1000 Hz	2000 Hz	4000 Hz
Audience, seated in upholstered seats, ft^{-2} of floor area	0.60	0.74	0.88	0.96	0.93	0.85
Unoccupied cloth-covered upholstered seats, ft^{-2} of floor area	0.49	0.66	0.80	0.88	0.82	0.70
Unoccupied leather-covered upholstered seats, ft^{-2} of floor area	0.44	0.54	0.60	0.62	0.58	0.50
Wooden pews, occupied, ft^{-2} of floor area	0.57	0.61	0.75	0.86	0.91	0.86
Chairs, metal or wood seats, each, unoccupied	0.15	0.19	0.22	0.39	0.38	0.30

[a]Values given are in sabins per square foot of seating area or per unit.

band sound transmission loss spectrum extending from 100 (80 optionally) to 5000 Hz.

Unlike noise reduction, the TL is influenced only by mass, stiffness, and damping; and how a construction is configured. Noise reduction is influenced equally by these but also by the surface area size of the partition and the acoustical conditions of the source and receiver rooms. Specifying the performance of a wall construction is awkward because NR is most often influenced by reverberant room characteristics and/or sound flanking conditions that are not part of the wall. Specifying the performance of a wall construction in terms of TL is more logical in that TL is only influenced by characteristics directly related to properties of the material or wall.

4.5 Sound Transmission Class Rating

Since sound TL is frequency dependent, it is generally reported in the third octave frequency bands between 125 and 4000 Hz. As a convenience, a single-number rating method has been developed that allows a single value to be ascribed to a *transmission loss spectrum* that is the set of 16 one-third octave band TL values. This rating is referred to as the sound transmission class (STC), which has been defined in the ASTM Standard E413.[21] This standard defines a procedure for determining the STC value for a TL spectrum that involves fitting a specified contour to one-third octave band TL data.

The STC value was originally developed to approximate the performance of a material in reducing the transmission of intelligible speech sounds. The STC characterization of TL is useful for a quick comparison of materials, but it is not necessarily useful for assessing the performance of a material with respect to other nonspeech sounds such as music and transportation systems sources such as aircraft, highway traffic, train passbys, and so forth. For this purpose, it is necessary to consider octave or third octave band TL spectra with specific attention, particularly, to the low-frequency range, typically at frequencies below 500 Hz. The STC contour is shown in Fig. 5. Determining the STC for a partition construction and/or material involves fitting this contour to the 16 one-third octave band TL values. This procedure involves raising or lowering the contour until the following two rules are met:

1. The contour may not be raised above the point at which the TL in any one-third octave band falls more than 8 dB below the contour.
2. The contour may not be raised above the point at which the total number of deficiencies is greater than 32.

A deficiency occurs when the TL data in any one one-third octave band falls below the contour. For example, three deficiencies would be when the TL data in any one one-third octave fall 3 dB below the contour, or when the data in three one-third octave bands falls 1 dB below the contour. Figure 5 further illustrates the meaning of deficiency.

The STC rating from the contour fitting procedure is the TL value of the contour at 500 Hz. This is illustrated in Figs. 5 and 6.

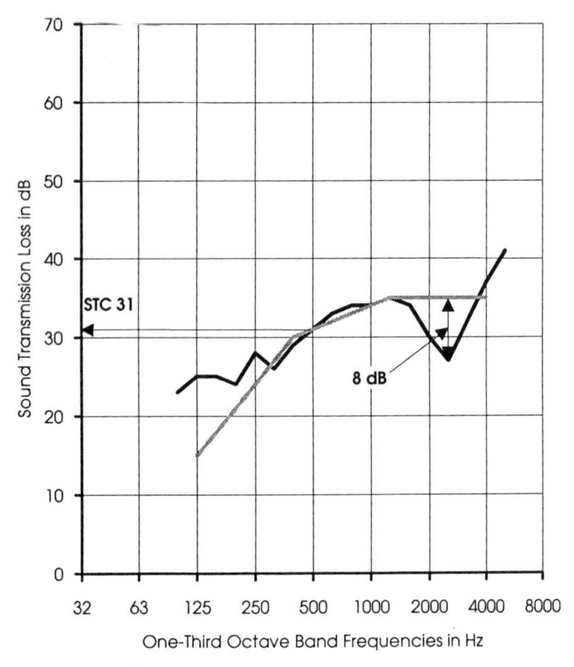

Fig. 5 A $\frac{1}{4}$-in. glass sound transmission loss and STC contour.

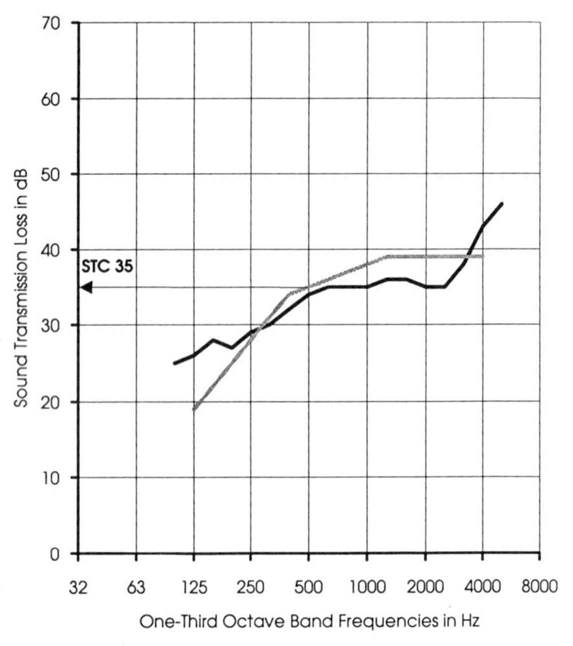

Fig. 6 A $\frac{1}{4}$-in. laminated glass sound transmission loss and STC contour.

In Fig. 5, $\frac{1}{4}$-in. monolithic glass is shown to have an STC rating of 31. In this example, the STC contour placement is constrained by the maximum allowed 8 dB deficiency at 2500 Hz (rule 1).

In Fig. 6, which shows TL data for $\frac{1}{4}$-in. laminated glass, the dip in the TL data, characteristic of $\frac{1}{4}$-in. monolithic glass, is removed by the damping interlayer. In the case of $\frac{1}{4}$-in. laminated glass, the STC contour placement is constrained by the maximum 32-dB deficiency requirement (rule 2).

4.6 Field Sound Transmission Class Rating

The field sound transmission class (FSTC) rating is used for in situ wall and floor/ceiling sound isolation performance assessment.[22] The standard requires the measurement of sound transmission loss and includes a required procedure to show that the FSTC rating, as it has been determined by the test procedure, was not influenced by flanking of sound around the partition intended to be tested. Sound transmission class and FSTC ratings are intended by standard to be equivalent; however, practical experience indicates that FSTC ratings tend to be up to five rating points less than laboratory-measured STC ratings.

4.7 Noise Isolation Class

The noise isolation class (NIC) rating is similar to the STC and FSTC. However, the standard STC rating contour is applied to the one-third octave band noise reduction data measured in a field situation, rather than the transmission losses measured in the field.[21] No correction to the measured noise reduction data is made to account for partition size, receiving room absorption, or sound flanking. Like the STC and FSTC ratings, the field measured NIC rating of a noise reduction spectrum is equal to the value of the contour crossing at 500 Hz. In the absences of sound flanking, the NIC is generally within five points of the laboratory STC rating for typical building partition constructions. The NIC rating is used to assess the sound isolation performance of in situ partition constructions, especially complicated ones that involve multiple sound transmission paths that are not suited for laboratory testing. There are no widely used standards using the NIC rating; however, the NIC rating is often used in lieu of STC and FSTC ratings.

4.8 Normalized Noise Isolation Class

The normalized noise isolation class (NNIC) is the same as the NIC rating except the receiving room absorption is normalized to correspond to a 0.5-s reverberation time.[21]

4.9 Outdoor–Indoor Transmission Class

The STC rating was developed to assess the sound isolation performance of partitions and materials with respect to speech and household sounds, that is, sounds produced by home appliances and entertainment equipment.[23] "Household sound" tends to be rich in mid- and high-frequency sound energy. However, environmental sound, typically associated with transportation systems and equipment, is rich in low- and mid-frequency sound energy. The outdoor–indoor transmission class (OITC) rating is a single-number rating that can be used for comparing the sound isolation performance of building facades and facade elements. The rating has been devised to quantify the ability of these to reduce the perceived loudness of ground and air transportation noise transmitted into buildings. This standard is contained in ASTM E 1332-90 Standard Classification for Determination of Outdoor–Indoor Transmission Class.

The standard establishes a single-number rating, the OITC, by defining a standard spectrum of ground and air transportation noise. This spectrum is used with sound transmission loss data measured in a laboratory using the ASTM E 90 method and a mathematical relationship given in the standard to calculate the OITC rating.

The OITC rating is similar to the STC rating in that it uses ASTM E 90 TL data and uses this data to derive a single-number rating that increases with increasing sound isolation ability. It differs in that the OITC does not involve a contour fitting process but instead uses a standard spectrum and a mathematical relationship. The mathematical relationship is as follows:

$$\mathrm{OITC} = 100.14 - 10\log\sum_{f} 10^{(L_f - TL_f + A_f)/10} \quad \mathrm{dB}.$$

Table 4 presents a worksheet summarizing the method for calculating the OITC rating for $\frac{1}{4}$-in. glass. As indicated in Table 2, the OITC rating for $\frac{1}{4}$-in. glass is 29. Note that the TL in the 80-Hz band in the worksheet of Table 4 is not shown since it was not measured. The actual TL value in the 80-Hz band, were it included, would not affect the OITC rating indicated in this table.

4.10 Impact Isolation

In residential buildings, quantifying the impact isolation of floor/ceiling constructions with respect to footfall is defined in ASTM E 492-90[25] and ASTM E 1007-90[26] for laboratory and field situations, respectively. Both standards involve placing a mechanical impacting device called a tapping machine on the floor of the construction to be tested and measuring resulting sound levels below.

TABLE 4 Worksheet for Calculating the OITC Rating of $\frac{1}{4}$-in. Glass

Band Center Frequency, Hz (1)	Reference Sound Spectrum, dB (2)	A-Weighting Correction, dB (3)	Column 2 plus Column 3, dB (4)	Specimen TL, dB (5)	Column 4 minus Column 5, dB (6)	$10^{(\text{Column } 6/10)}$
80	103	−22.5	80.5			
100	102	−19.1	82.9	23	59.9	977,237
125	101	−16.1	84.9	25	59.9	977,237
160	98	−13.4	84.6	25	59.6	912,011
200	97	−10.9	86.1	24	62.1	1,621,810
250	95	−8.5	86.4	28	58.4	691,831
315	94	−6.6	87.4	26	61.4	1,380,384
400	93	−4.8	88.2	29	59.2	831,764
500	93	−3.2	89.8	31	58.8	758,578
630	91	−1.9	89.1	33	56.1	407,380
800	90	−0.8	89.2	34	55.2	331,131
1000	89	0.0	89.0	34	55.0	316,228
1250	89	0.6	89.6	35	54.6	288,403
1600	88	1.0	89.0	34	55.0	316,228
2000	88	1.2	89.2	30	59.2	831,764
2500	87	1.3	88.3	27	61.3	1,348,963
3150	85	1.2	86.2	32	54.2	263,027
4000	84	1.0	85.0	37	48.0	63,096
						12,317,071
						10 log(Sum) = 70.91
						OITC = 100.14 − 10 log(Sum) = 29

From Ref. 24.

The design and use of the tapping machine are described in ISO 140/6.[27] The result of the test is a one-third octave band impact sound pressure level in the room beneath the floor/ceiling construction being tested.

4.11 Impact Insulation Class

Similar to the STC, the impact insulation class (IIC) rating is a single-number rating that facilitates the comparison of the impact sound isolation performance of floor/ceiling constructions. The IIC rating is determined by fitting a standard contour to the one-third octave band sound pressure level data measured in accordance with ASTM E 492.

4.12 Field Impact Insulation Class

The field impact insulation class (FIIC) rating is the same as the IIC rating except it is used to rate the impact sound insulation performance of in situ floor/ceiling constructions and is used in conjunction with the ASTM E 1007 test method. The standard includes procedures for assessing flanking conditions.

5 DESCRIPTORS ASSESSING THE QUALITY OF SOUND INSIDE BUILDING SPACES

5.1 Reverberation Time

The reverberation time (T_{60}) is the time it takes for sound to decay by 60 dB(A). Reverberation time is controlled in interior building spaces to enhance speech intelligibility or enhance music quality. Generally, speech intelligibility benefits from a short reverberation time (typically 1 s) while orchestral concert and organ music benefit from long reverberation times (2–5 s). Figure 7 presents criteria for reverberation time in spaces according to use for music or speech and according to room volume.

There are four generally accepted methods for calculating the reverberation time. The first and probably most widely used is the *Sabine equation*. It is more appropriately used in moderately to highly reverberant spaces. The Sabine equation is as follows:

$$T_{60} = \frac{0.049V}{A} \quad \text{(English units).}$$

The second method for determining room reverbera-

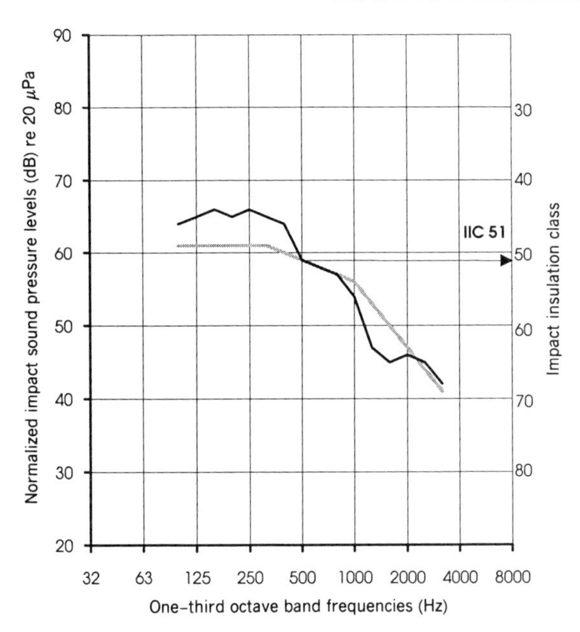

Fig. 7 Typical impact sound level spectrum and IIC contour.

tion is the *Norris Eyring equation*. It is most appropriately used in highly absorptive rooms. The Norris Eyring equation is as follows:

$$T_{60} = \frac{0.049V}{(-S)\ln(1-\overline{\alpha})} \quad \text{(English units)}.$$

The third method for determining room reverberation is the *Fitzroy equation*. It is used primarily for rooms where the absorption of parallel surfaces is approximately the same. The Fitzroy equation is as follows:

$$T_{60} = \frac{X}{S}\left(\frac{0.049V}{S\overline{\alpha}_x}\right) + \frac{Y}{S}\left(\frac{0.049V}{S\overline{\alpha}_y}\right) + \frac{Z}{S}\left(\frac{0.049V}{S\overline{\alpha}_z}\right) \quad \text{(English units)}.$$

In the above equations, V is room volume, S is room surface area, A is room absorption, $\overline{\alpha}$ is average sound absorption coefficient (the x, y, and z subscripts are the average sound absorption coefficients of walls normal to the x-, y-, and z-axes), and X, Y, and Z are the surface areas of walls normal to the x-, y-, and z-axes. If units of volume, surface area, and absorption are metric, then the 0.049 constant in each equation should be 0.161.[28]

5.2 Room Constant

Room constant (R) is defined as the ratio of room sound absorption to the quantity 1 minus the average room absorption coefficient. Using the above symbol definitions, the relationship is expressed as

$$R = \frac{A}{1-\overline{\alpha}}.$$

There are no criteria that make use of room constant; however, it can be useful in assessing the *critical distance*, defined as the distance from a source at which the reverberant field sound pressure level equals the directly radiated sound pressure level.

5.3 Articulation Index

The articulation index (AI) is a weighted proportion of a speech signal usable to convey information. The procedure[29] for estimating articulation index involves accounting for signal strength, background or masking noise, time-varying character of background noise, speech signal clipping, and reverberation. The estimation procedure is carried-out in one-third or full octave frequency bands. The calculated articulation index is used to estimate percent of syllables, words, or sentences understood correctly. This analysis can be used to assess both speech intelligibility or its inverse, speech privacy. AI is also described in Section 6 of Chapter 64.

5.4 Speech Transmission Index

The speech transmission index (STI) is similar to the articulation index. However, the STI uses a modulation transfer function[30,31] in its evaluation and incorporates a procedure for expressing the loss in articulation produced by room reverberation as an equivalent background noise contribution, which can be added to the actual measured background noise level.[32]

5.5 Rapid Speech Transmission Index

The rapid speech transmission index (RASTI) is similar to the STI, uses fewer modulation frequencies, and incorporates only speech and background sound levels in only two octave bands—500 and 2000 Hz. The RASTI value is considered sufficiently accurate only if no nonlinearities in speech transmission exist and if the background sound level is devoid of strong tonal components.

6 DESCRIPTORS AND RATINGS USED IN ISO AND IN SELECTED EUROPEAN AND ASIAN STANDARDS AND REGULATIONS

The definitions of acoustical terms discussed below are based on ISO 140 and ISO 717 standards. Most European and Asian standards and regulations relating to acoustics are based on ISO terminology and methods defined below. Further information regarding the use of these criteria and descriptors is contained in *Enc.* Ch. 97.

6.1 Average Sound Pressure Level in a Room

Ten times the common logarithm of the ratio of the space and time average of the sound pressure squared, to the square of the reference sound pressure.[33] The spatial average is over the entire room with the exception of those areas of the room significantly influenced by direct radiation of a sound source or the near field of the boundaries (wall, etc.). This quantity is denoted by L in decibels:

$$L = 10 \log \frac{p_1^2 + p_2^2 + \cdots + p_n^2}{n p_0^2} \quad \text{dB},$$

where $p_1, p_2, \ldots, p_n$ are the root-mean-square (rms) sound pressures at n different positions in the room and $p_0 = 20\ \mu\text{Pa}$ is the reference sound pressure.

6.2 Sound Reduction Index and Transmission Loss

Ten times the common logarithm of the ratio of the sound power W_1 incident on a test specimen to the sound power W_2 transmitted through the specimen.[33] This quantity is denoted by R:

$$R = 10 \log \frac{W_1}{W_2} \quad \text{dB}.$$

The sound reduction index depends on the angle of incidence. If the sound fields are diffuse and if the sound is transmitted only through the specimen, the sound reduction index for diffuse incidence may be evaluated from

$$R = L_1 - L_2 + 10 \log \frac{S}{A} \quad \text{dB},$$

where L_1 is the average sound pressure level in the source room, L_2 is the average sound pressure level in the receiving room, S is the area of the test specimen, which is normally equal to the free test opening, and A is the equivalent absorption area in the receiving room. (*Note:* If the sound fields are not completely diffuse, this equation is an approximation.)

6.3 Apparent Sound Reduction Index and Apparent Transmission Loss

Ten times the common logarithm of the ratio of the sound power W_1 incident on a partition under test to the total sound power W_3 transmitted into the receiving room.[33] This quantity is denoted by R':

$$R' = 10 \log \frac{W_1}{W_3} \quad \text{dB}.$$

In general, the sound power transmitted into the receiving room consists of the sum of the following components:

- W_{Dd}, which has entered the partition directly and is radiated from it directly;
- W_{Df}, which has entered the partition directly but is radiated from flanking constructions;
- W_{Fd}, which has entered flanking construction and is radiated from the partition directly;
- W_{Ff}, which has entered flanking constructions and is radiated from flanking constructions; and
- W_{leak}, which has been transmitted (as airborne sound) through leaks, ventilation ducts, and so on.

Also in this case, under the assumption of diffuse sound fields in the two rooms, the apparent sound reduction index may be evaluated from

$$R' = L_1 - L_2 + 10 \log \frac{S}{A} \quad \text{dB}.$$

The 10 log(S/A) correction term eliminates the receiving room reverberation effect and the effect of partition size, thus leading to the apparent sound reduction index that is not influenced by conditions surrounding the partition's use.

6.4 Traffic Noise Reduction Index

For exterior building facades and facade components, in situ noise reductions can be determined using traffic noise as the source.[34] The traffic sound reduction index R_{tr} is obtained from the equivalent sound pressure levels measured as a function of frequency on both sides of the test specimen (exterior facade or facade component). The traffic noise reduction index is denoted by R_{tr}:

$$R_{tr} = L_{eq,1} - L_{eq,2} + 10 \log \frac{S}{A} \quad \text{dB},$$

where $L_{eq,1}$ is the equivalent sound pressure level 2 m in front of the test specimen, including the reflection effect of the test specimen, $L_{eq,2}$ is the equivalent sound pressure level in the receiving room averaged over the room, S is the area of the test specimen, and A is the equivalent absorption area in the receiving room in metric sabins. The above equation is applicable when the line of traffic is sufficiently long and straight to ensure a uniform distribution of incident sound. When the angle of elevation (observed from the point of least distance between the test specimen and the line of traffic) is more than about 20°, somewhat different results may occur because of oblique angles of incidence. (In this context, an angle of incidence of 0° is perpendicular to the facade. When the angle of elevation exceeds 50°, this equation should not be used.

6.5 Level Difference

The difference in the space and time average sound pressure levels produced in two rooms by one or more sound sources in one of them.[35] This quantity is denoted by D:

$$D = L_1 - L_2$$

where L_1 is the average sound pressure in the source room and L_2 is the average sound pressure level in the receiving room.

6.6 Standardized Level Difference

The adjusted level difference.[35] The adjustment is 10 times the logarithm of the ratio of the receiving room reverberation time and a reference reverberation time. This quantity is denoted by D_{nT}:

$$D_{nT} = D + 10 \log \frac{T}{T_0} \quad \text{dB},$$

where D is the level difference, T is the reverberation time in seconds in the receiving room, and T_0 is the reference reverberation time in seconds, which for dwellings is 0.5 s.

Normalizing (standardizing) of the level differences to a revelation time of 0.5 s takes into account the field experience that furnished rooms of dwellings have an average reverberation time of 0.5 s. It has also been observed that this reverberation time is nearly independent of room volume and frequency. This adjustment will result in the noise reduction, D_{nT}, for a pair of rooms of dissimilar volumes joined by a common wall to be different depending on whether the source and receiver rooms are the larger and smaller, respectively, or *vice versa*. The D_{nT} for the larger source room and small receiving room will be smaller than the D_{nT} for the smaller source room and larger receiving room.

Normalizing the level difference to a 0.5-s reverberation time in the receiving room is equivalent to normalizing the level difference to that with a receiving room having an equivalent absorption area of

$$A_0 = 0.32V,$$

where A_0 is the equivalent absorption area in square metres and V is the volume of the receiving room in cubic metres.

6.7 Standard Traffic Level Difference

Similarly, the in situ standardized traffic noise level difference $D_{nT,tr}$ is defined as follows[36]:

$$D_{nT,tr} = L_{eq,1} - L_{eq,2} + 10 \log \frac{T}{T_0} \quad \text{dB},$$

where T is the measured reverberation time in seconds in the receiving room and T_0 is the reference reverberation time, 0.5 s for dwellings.

6.8 Impact Sound Pressure Level

The average sound pressure level in a specific frequency band in the receiving room when the floor under test is excited by the standardized impact sound source.[37] This quantity is denoted by L_i.

6.9 Normalized Impact Sound Pressure Level

The impact sound pressure level L_i adjusted by a correction term in decibels equal to 10 times the common logarithm of the ratio between the measured equivalent absorption area A of the receiving room and the reference absorption area A_0.[37] This quantity is denoted by L_0:

$$L_0 = L_1 + 10 \log \frac{A}{A_0} \quad \text{dB},$$

where $A_0 = 10 \text{ m}^2$.

In all cases, where it is uncertain whether results are obtained without flanking transmission, the normalized impact sound pressure level should be denoted by L_0'.

6.10 Standardized Impact Sound Pressure Level

The impact sound pressure level L_i adjusted by a correction term in decibels equal to 10 times the common logarithm of the ratio between the measured reverberation time T of the receiving room and the reference reverberation time T_0.[38] This quantity is denoted by L'_{nT}:

$$L'_{nT} = L_i - 10 \log \frac{T}{T_0}.$$

For dwellings T_0 is given by

$$T_0 = 0.5 \text{ s}.$$

Normalizing the impact sound pressure level to a reverberation time of 0.5 s accounts for the field experience that reverberation times in dwellings is nearly independent of the volume and frequency, and is equal to 0.5 s.

Normalizing of the impact sound pressure level to a reverberation time of 0.5 s is equivalent to normalizing the impact sound pressure level to an equivalent absorption area of

$$A_0 = 0.32V,$$

where A_0 is the equivalent area in square metres and V is the volume of the receiving room in cubic metres.

6.11 Reduction of Impact Sound Pressure Level (Improvement of Impact Sound Insulation)

The difference between the average sound pressure levels in the receiving room before and after installation of, for example, a floor covering; see ISO 140/VIII.[37] This quantity is denoted by ΔL.

6.12 Measurement and Evaluation of the Equivalent Absorption Area

The correction term for the above equation, containing the equivalent absorption area, may preferably be evaluated from the reverberation time measured according to ISO/R 354 and evaluated using Sabine's formula.[37]

$$A = \frac{0.163V}{T},$$

where A is the equivalent absorption area in square metres, V is the receiving room volume in cubic metres, and T is the reverberation time in seconds.

An alternative method of determining the equivalent absorption area is to measure the average sound pressure level produced by a known, sufficiently stable sound source.

6.13 Average Sound Pressure Level in a Room

Ten times the common logarithm of the ratio of the space and time average of the sound pressure squared to the square of the reference sound pressure.[35] The spatial average is over the entire room with the exception of those areas significantly influenced by direct radiation of a sound source or the near field of the boundaries (wall, etc.). This quantity is denoted by L:

$$L = 10 \log \frac{p_1^2 + p_2^2 + \cdots + p_n^2}{np_0^2} \quad \text{dB},$$

where $p_1, p_2, \ldots, p_n$ are the rms sound pressures at n different positions in the room and $p_0 = 20\ \mu\text{Pa}$ is the reference sound pressure.

REFERENCES

1. American National Standards Institute, "Rating Noise with Respect to Speech Interference," S3.14-1977 (R1986).
2. C. M. Harris (Ed.), *Handbook of Acoustical Measurements and Noise Control*, 3rd ed., McGraw-Hill, New York, 1991, p. 17.9.
3. American National Standards Institute, "Procedure for the Computation of Loudness of Noise," S3.4-1980 (R1986).
4. "Procedures for Abatement of Highway Traffic Noise and Construction Noise," in Federal-Aid Program Manual, U.S. Department of Transportation, Federal Highway Administration, Washington, DC, August 9, 1982, Volume 7, Chapter 7, Section 3.
5. Massachusetts Department of Environmental Protection, Bureau of Waste Prevention–Air Quality, "Supplemental Form for Survey of Noise Potential," BWP AQ SFP 03.
6. City of Cambridge, Massachusetts, Noise Ordinance, Ordinance No. 877, 1977.
7. "Protective Noise Levels—Condensed Version of EPA Levels Documents," U.S. Environmental Protection Agency, Publication No. EPA 550/9-79-100, November 1978.
8. "Noise Control and Compatibility Planning for Airports," U.S. Department of Transportation, Federal Aviation Administration, Advisory Circular AC 150/5020-1, August 3, 1983.

9. 24 CFR Part 51, U.S. Department of Housing and Urban Development, Environmental Criteria and Standards, 44 FR 40860, July 12, 1979; amended by 49 FR 880, January 6, 1984.
10. California Noise Insulation Standards, State Building Code (Part 2, Title 24, CCR), UBC Appendix Chapter 35, "Sound Transmission Control," December 1988.
11. Appendix Chapter 35, "Sound Transmission Control," Uniform Building Code, International Conference of Buildings Officials, 1988.
12. L. L. Beranek (Ed.), *Noise Reduction*, McGraw-Hill, New York, 1960, pp. 518–520.
13. T. J. Schultz, "Noise-Criterion Curves for Use with the USASI Preferred Frequencies," *J. Acoust. Soc.*, Vol. 43, No. 3, 1968, pp. 637–638.
14. *1995 ASHRAE Applications Handbook*, American Society of Heating, Refrigerating, and Air-Conditioning Engineers, New York.
15. L. L. Beranek, "Application of NCB Noise Criterion Curves," *Noise Control Eng. J.*, Vol. 33, No. 2, 1989.
16. L. L. Beranek, "Balanced Noise-Criterion (NCB) Curves," *J. Acoust. Soc. Am.*, Vol. 86, No. 2, 1989.
17. American Society for Testing and Materials, *Section 4—1991 Annual Book of ASTM Standards*, "Standard Test Method Sound Absorption and Sound Absorption Coefficients by the Reverberation Room Method," C 423-90a.
18. American Society for Testing and Materials, *Section 4—1991 Annual Book of ASTM Standards*, "Standard Practices for Mounting Test Specimens during Sound Absorption Tests," E 795-91.
19. American Society for Testing and Materials, *Section 4—1991 Annual Book of ASTM Standards*, "Standard Practice for Determining a Single-Number Rating of Airborne Sound Isolation for use in Multi-unit Building Specifications," E 90-90.
20. American Society for Testing and Materials, *Section 4—1991 Annual Book of ASTM Standards*, "Standard Test Method for Laboratory Measurement of Airborne Sound Transmission Loss of Building Partitions," E 90-90, p. 678.
21. American Society for Testing and Materials, *Section 4—1994 Annual Book of ASTM Standards*, ASTM E 413-87 (R1994), p. 715.
22. American Society for Testing and Materials, *Section 4—1991 Annual Book of ASTM Standards*, "Standard Test Method for Measurement of Airborne Sound Insulation in Buildings, Annex A1. Measurement of Field Sound Transmission Loss," ASTM E 226-90.
23. G. C. Tocci, "A Comparison of STC and EWR for Rating Glazing Noise Reduction," *Sound Vib.*, October 1987, p. 32.
24. *Monsanto Acoustical Glazing Design Guide*, 3rd ed., 1995.
25. American Society for Testing and Materials, *Section 4—1991 Annual Book of ASTM Standards*, "Standard Test Method for Laboratory Measurement of Impact Sound Transmission through Floor-Ceiling Assemblies Using the Tapping Machine," E 492-90, p. 721.
26. American Society for Testing and Materials, *Section 4—1991 Annual Book of ASTM Standards*, "Standard Test Method for Field Measurement of Tapping Machine Impact Sound Transmission through Floor-Ceiling Assemblies and Associated Support Structures," E492-90, p. 794.
27. International Standards Organization, "Acoustics—Measurement of Sound Insulation in Buildings and of Building Elements: Part 6—Laboratory Measurements of Impact Sound Insulation of Floors," Standard 140/6.
28. K. A. Hoover, *An Appreciation of Acoustics*, Berklee School of Music, Boston, MA.
29. American National Standards Institute, "Methods for the Calculation of the Articulation Index," ANSI S3.5-1969 (R1986).
30. T. Houtgast and H. J. M. Steeneken, "A Review of the MTF Concept in Room Acoustics and its Use for Estimating Speech Intelligibility in Auditor," *J. Acoust. Soc. Am.*, Vol. 77, No. 3, March 1985.
31. D. Davis and C. Davis, *Sound System Engineering*, 2nd ed., Howard W. Sams & Co., Indianapolis, IN, 1987, p. 235.
32. C. M. Harris (Ed.), *Handbook of Acoustical Measurements and Noise Control*, 3rd ed., McGraw-Hill, New York, 1991, p. 16.16.
33. International Standardization Organization, "Acoustics—Measurement of Sound Insulation in Buildings and of Building Elements—Part III: Laboratory Measurements of Airborne Sound Insulation of Building Elements, ISO 140/III.
34. International Standardization Organization, "Acoustics—Measurement of Sound Insulation in Buildings and of Building Elements—Part V: Field Measurements of Airborne Sound Insulation of Facade Elements and Facades," ISO 140/V.
35. International Standardization Organization, "Acoustics—Measurement of Sound Insulation in Buildings and of Building Elements—Part IV: Field Measurements of Airborne Sound Insulation between Rooms," ISO 140/IV, 1978.
36. International Standardization Organization, "Acoustics—Measurement of Sound Insulation in Buildings and of Building Elements—Part V: Field Measurements of Airborne Sound Insulation of Facade Elements and Facades," ISO 140/V.
37. International Standardization Organization, "Acoustics—Measurement of Sound Insulation in Buildings and of Building Elements—Part VI: Laboratory Measurements of Impact Sound Insulation of Floors," ISO 140/VI.
38. International Standardization Organization, "Acoustics—Measurement of Sound Insulation in Buildings and of Building Elements—Part VII: Field Measurements of Impact Sound Insulation of Floors," ISO 140/VII.

78

ACOUSTICAL GUIDELINES FOR BUILDING DESIGN

EWART A. WETHERILL

1 INTRODUCTION

This chapter summarizes the basic requirements of building acoustics and enumerates suitable construction and interior finish materials for the situations most likely to be encountered. It provides a working guide to enable a designer or builder with little or no background in acoustics—but with the patience to follow a sequence of well-defined steps—to distinguish between problems of *noise intrusion* and *room acoustics*, to evaluate the potential for an acoustical problem, to consider alternative solutions, and to recognize when more rigorous evaluation is needed.

Successful acoustical design of buildings requires familiarity with appropriate construction methods, and also with building codes and other local restrictions, to ensure full compatibility before construction begins. When a conflict in priorities becomes apparent during construction, the builder will seldom place hard-to-prove acoustical needs ahead of better understood and more tangible construction needs.

Most acoustical questions can be resolved by following the steps illustrated in Fig. 1 for which the designer selects appropriate criteria for the particular building or space to determine the required type of construction. This review procedure also defines plan relationships that should be avoided and provides a checklist for coordinating the work of other disciplines and examining the appropriateness of possible design revisions. Detailed references are found in other chapters as noted herein.

Note: References to chapters appearing only in the *Encyclopedia* are preceded by *Enc.*

2 ACOUSTICAL DESIGN OF BUILDINGS

Architectural acoustics encompasses the process through which acoustical conflicts can be *avoided*, unwanted sounds *controlled*, and wanted sounds *enhanced* by the building design. Any building may contain many spaces, each with its own requirements for acoustical quality and for isolation from noise and vibration.

The acoustical design for a particular building, or for a space within the building, must be compatible with other construction requirements and must be incorporated in the architectural drawings and specifications that form the basis for the awarding of a construction contract. This process requires close coordination and detailed cross checking because virtually any architectural design decision can have acoustical implications. The design of a building is generally a series of compromises in which benefits for one discipline must be weighed against disadvantages for another. The designer should know to what extent acoustical goals may be compromised without significantly affecting the usefulness of the building. Basic design assumptions should be firmly established before consideration of detailed requirements. For example, how important is total control of exterior noise or speech privacy between spaces? Or, should the interior be designed to *enhance* or *suppress* projection of sound? Clarification of such design issues at the outset will simplify the detailed selection process.

Section 6 summarizes general concerns and acoustical design requirements for current building types in several categories. If the proposed occupancy encompasses two or more categories, each one should be considered and the more stringent acoustical requirements should govern the design.

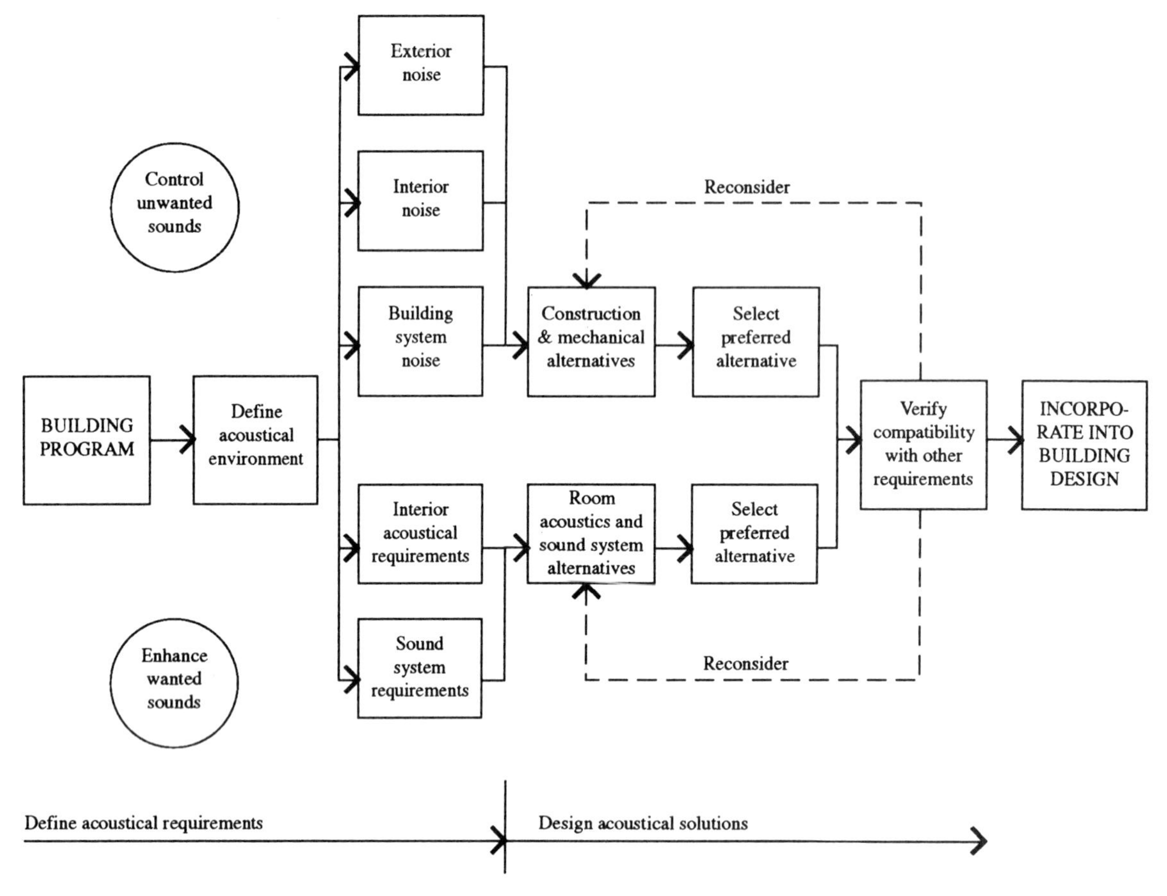

Fig. 1 Diagram illustrating steps to be used to solve acoustical questions.

3 CRITERIA

3.1 Selection of Acoustical Criteria

The acoustical criteria for a facility—or for individual spaces within it—can be stated *objectively* in terms of physical properties required to meet a specified need. However, they may be conditioned by a wide range of *subjective* requirements such as a listener's hearing acuity or individual sensitivity to intruding noise, as well as by cost and other nonacoustical requirements. Basic acoustical concerns are background noise level, reverberation, and noise levels generated within the listening space (see Chapter 77).

The desired background noise level for each space is typically specified by reference to noise criteria or other rating systems (see Chapter 64). Selection of desired background level is determined by the following:

(a) Required degree of communication between users
(b) Speech privacy required from adjacent spaces or nearby occupants of the same space (e.g., open-plan offices)

Actual background level is governed by the following:

(a) Control of intruding noises (from outdoors or from adjacent spaces) by appropriate construction (see Chapter 76)
(b) Control of noise from building systems (including air conditioning, plumbing, elevators, etc.) (see Chapter 79)
(c) Use of sound masking system to control minimum sound level (see *Enc.* Ch. 163)

Reverberation time (see Chapters 75 and 77) is an adequate single descriptor for the acoustical quality of most spaces. It is governed by the following:

(a) Volume and shape of space

(b) Amount and location of sound absorption in space

A quick estimate of the reverberation time inherent in a space of a given size will enable the designer to approximate the amount of sound absorption that is needed. Section 5.1 shows how to calculate approximate reverberation time, while Table 4 defines recommended mid-frequency (i.e., 500–1000 Hz) reverberation times for most common uses. Detailed calculations of reverberation time may be required in the final design and will be required during preliminary design for spaces where acoustical quality is of particular importance.

For acoustically sensitive uses such as music performance and for other special requirements, reverberation time alone may not be an adequate criterion; thus a more detailed analysis is needed. Spaces in which *projection* of sound is required, such as auditoria, must be treated differently from spaces in which *nonprojection* of sound is required (e.g., library reading room) (see *Enc.* Ch. 95 and 163).

3.2 Anticipated Noise Levels within the Space

In addition to the need for freedom from intruding noises, the likelihood that loud sounds could be generated within the space should be considered. For example, in a band rehearsal room special sound-absorbing treatment will be required to control the sound of the band (see Chapter 75).

3.3 Selection of Acoustical Requirements

Sections 4–6 enable the designer to select the acoustical conditions appropriate for a specific use, to assess the intruding noise or vibration that must be controlled, and to define the interior acoustical requirements of the space as summarized in Fig. 1. Using the work sheet in Section 7 and illustrative examples in Section 8, the designer selects the type of construction and interior treatment required to meet the design goals. To accommodate nonacoustical criteria, the designer applies a judgment factor, based on familiarity with the requirements for the particular building or space, to choose between possible alternatives.

Recommended values for background noise and reverberation are presented, in addition to subjective criteria, which are summarized for typical occupancies in Section 6. Each step in the evaluation process requires the designer to select the most appropriate condition from a range of five alternatives on the basis of familiarity with the project requirements and site conditions. These alternatives will be readily understood by most designers or can be approximated by reference to existing buildings. Where extreme conditions are encountered, or if considerable doubt exists as to the appropriate selection, the designer is advised to seek the assistance of an experienced consultant.

While this summary is based on generally accepted practice in acoustical design and on current North American building codes, the designer should also check on local requirements and on special conditions such as defined by the Americans with Disabilities Act (ADA), Occupational Safety and Health Act (OSHA), and so forth (see Chapters 92 and 93).

4 SELECTION OF CONDITIONS FOR SPACE BEING ANALYZED

This section presents criteria for noise and vibration (Table 1) and sources of intruding noise and vibration (Table 2).

5 CONSTRUCTION ALTERNATIVES FOR SPACE BEING ANALYZED

This section presents criteria for the control of intruding noise and vibration (Table 3).

5.1 Estimation of Reverberation Time

Approximate reverberation time (RT) in seconds can be derived using the Sabine equation (for detailed analysis refer to Chapters 61 and 75):

$$\mathrm{RT} = \frac{KV}{A_1 + A_2 + A_3 + \cdots},$$

where K is 0.16 (SI units) or 0.05 (English units), V is the volume in cubic metres or cubic feet, and A is the absorption coefficient α of each material multiplied by its surface area. For a preliminary estimate, absorption coefficients can be simplified:

For hard materials (e.g., concrete, wood) assume $\alpha = 0.1$.

For soft materials (thick carpet, "acoustical" materials, etc.) $\alpha = 0.8$.

As an example consider an auditorium with a volume of 807.5 m^3. Also assume the following:

Area 1			
Ceiling, walls, exposed floor	417.8 m^2	$\alpha_1 = 0.1$	$A_1 = 41.78$
Area 2			
Empty seating (not upholst.)	133.8 m^2	$\alpha_2 = 0.1$	$A_2 = 13.38$
Area 3			
Sound-absorbing material	25.6 m^2	$\alpha_3 = 0.8$	$A_3 = 20.48$
		Total empty	75.64

TABLE 1 Criteria for Noise and Vibration Based on a Range of Five Alternatives

1	2	3	4	5
		a. Sensitivity to noise: desired background noise levels (Chapter 77)		
Not critical, e.g., shopping mall (NC 45–50)	Restaurant, open office (NC 35–45)	Office, quiet recreation (NC 25–35)	Sleep, worship, performance (NC 20–30)	Special uses: recording, test laboratory (<NC 15–20)
		b. Sensitivity to vibration (Chapters 68 and 76)		
Not critical	Possible distraction	Sensitive to traffic vibration, footfalls	Laboratories/ optical processes	Extremely vibration sensitive
		c. Speech privacy requirements (Chapter 77)		
No concern	Conversational	Residential	Confidential	Security
		d. Anticipated sound level within space (Chapters 69 and 77)		
Conversation, general activity	Teaching, laboratory, computer use	Quiet recreation, music performance and playback	Indoor sports, mechanical equipment, industrial	Rock music, very high sound levels
		e. Anticipated vibration within space (Chapters 68 and 76)		
Footfalls	Plumbing, home appliances	Mechanical equipment and piping	Aerobics, dancing	Industrial processes, heavy machinery

TABLE 2 Sources of Intruding Noise and Vibration Based on a Range of Five Alternatives

1	2	3	4	5
		a. Exterior noise and vibration sources (Chapters 70 and 71)		
Rural site, extremely quiet	Suburban, residential	Suburban, near major street, outdoor equipment	Urban site near heavy-traffic streets	Urban, truck traffic, industry, rail, airport
		b. Interior sound sources: adjacent space (Chapters 69 and 77)		
Conversation, general activity	Teaching, laboratory, computer	Quiet recreation, music performance and playback	Indoor sports, mechanical equipment, industrial	Rock music, very high sound levels
		c. Interior vibration sources (Chapters 68 and 76)		
Footfalls	Plumbing, home appliances	Mechanical equipment and piping	Aerobics, dancing	Industrial processes, heavy machinery
		d. Noise and vibration from HVAC[a] systems (Chapter 79)		
Remote central system, low velocity air	Central equipment room, medium- to high-velocity air system	Equipment suspended above ceiling: fan-coil, air terminal boxes	Adjacent mechanical room, roof-mounted equipment above	Special-use systems, high noise and vibration
		e. Other building equipment (Chapters 68 and 76)		
Plumbing, domestic water systems	Pneumatic tube systems, automatic door openers	Elevators (especially hydraulic), escalators	Adjacent mechanical equipment room	Special equipment, e.g., fountain pump, emergency generator

[a]Heating, ventilation, and air conditioning.

TABLE 3 Control of Intruding Noise and Vibration Based on a Range of Five Alternatives

1	2	3	4	5
		a. Exterior walls		
No acoustical concerns, open windows	Lightweight construction, windows closed	Stucco or heavier, laminated or double glazing	Masonry walls, reduced window area	Double-wall construction, no windows
		b. Interior walls		
No acoustical concerns, studs and gypsum board	Lighweight construction, no openings	Meet code requirements for multiunit dwellings	Composite wall, preferably masonry	Double-wall construction, (consider relocation)
		c. Roof/ceiling		
No acoustical concerns, ceiling optional	Lightweight construction, acoustic tile or gypsum board suspended ceiling	Lightweight concrete deck with suspended gypsum board ceiling	Concrete deck with resiliently-suspended ceiling, minimum 36-in. cavity	Heavy concrete deck with fully isolated heavy ceiling
		d. Floor/ceiling		
No acoustical concerns, ceiling optional	Lightweight construction, acoustic tile or gypsum board suspended ceiling	Meet code requirements for multiunit dwellings	Concrete deck with resiliently-suspended heavy ceiling	Double (floated) floor with resiliently-suspended heavy ceiling (consider relocation)
		e. Control of vibration/impacts		
No acoustical concerns	Lightweight construction, with suspended ceilings	Meet code requirements for multiunit dwellings	Vibration sources resiliently isolated	Complete isolation from structure with vibration source
		f. Control of HVAC noise/vibration (Chapter 79)		
Quiet fan selection, resilient mounting	Duct liner, low air velocity, resiliently isolated equipment	Equipment enclosed, resilient isolators, main ducts remote	Special construction and resilient isolators, ducts lagged or enclosed	Equipment completely remote from occupied spaces
		g. Control of noise from building systems (Chapters 68 and 69)		
Resilient sleeving at all piping connections	Remote equipment, enclosed and resiliently isolated	Elevators, trash chutes remote, equipment/noise and vibration isolated	Heavy construction, equipment fully isolated	Equipment completely remotely from occupied spaces

Note: See Chapter 76.

Add a full audience:

133.8 m^2	$\alpha = 0.7(0.8 - 0.1)$	Add	93.66
		Total full	169.30

Then

$$\text{RT empty} = \frac{0.16 \times 807.5}{75.64} = 1.71 \text{ s.}$$

$$\text{RT full } \frac{0.16 \times 807.5}{169.30} = 0.76 \text{ s.}$$

Table 4 presents criteria for the control of sound within a space.

TABLE 4 Control of Sound within the Space Based on a Range of Five Alternatives

1	2	3	4	5
		a. Spaces where nonprojection of sound required (Chapter 75)		
Dwelling; RT 0.5–0.7 s	Work, study; RT 0.8–1.0 s	Dining, assembly, multiuse spaces; RT 0.8–1.0 s	Industrial processes, high noise levels; RT 0.8–1.0 s	Special low noise, recording, anechoic; RT 0.4–0.6 s
		b. Spaces where projection of sound required (Enc. Ch. 95)		
Drama, high speech clarity; RT 0.4–0.8 s	Conference, lecture, classroom; RT 0.6–0.8 s	Multipurpose audience, musicals; RT 1.2–1.6 s	Opera, chamber orchestra, ballet; RT 1.4–1.7 s	Orchestra, organ, chorus; RT 1.7–2.2 s

5.2 Shaping of Space for Optimum Sound Distribution

This important aspect of auditorium design requires much greater detail than can be provided in this book. The reader is referred to the references listed at the end of Chapter 73 for further information.

6 BUILDING CATEGORIES

6.1 Dwelling

The most important acoustical requirement in dwellings is control of intruding noise, particularly at night, when the ambient sound level is lowest. Private homes typically rely upon open windows for ventilation and so are sensitive to outdoor noises. Reverberation is generally controlled by household furnishings, and other internal acoustical conditions are seldom of concern. (See Table 5.)

For new dwellings in noisy urban settings or close to heavy traffic, airports, and so forth, some jurisdictions require special construction for noise control, which also entails mechanical ventilation so that occupants can leave windows closed to limit noise intrusion. These are typical requirements for apartment/condominium construction, in addition to control of airborne and structure-borne noise from adjacent dwelling units and from building systems (e.g., plumbing, heating, trash chutes). However, noise control standards in current building codes do not ensure satisfactory noise and impact isolation between dwelling units. Recent studies suggest that noise and impact isolation required to avoid disturbance of neighbors cannot be satisfied by light frame construction.

Hotel noise control requirements are frequently exacerbated by proximity to airports or major highways. Internally, adequacy of noise isolation between guest rooms, building systems noise control, and isolation of guest rooms from meeting and convention facilities must be considered.

Hospital wards require stringent noise control measures because bed-ridden patients cannot escape from intruding noises. Special attention should be given to

TABLE 5 Acoustical Requirements for Dwelling Spaces

	Requirements for Noise Control[a]					Requirements for Interior Design[b]		
						Reverberation Time		
Required Use of Space	Desired Background Level	Sensitivity to Vibration	Speech Privacy Required	Maximum Sound Level in Space	Maximum Vibration in Space	Nonprojection of Sound	Projection of Sound Required	Is Sound Amplification Required?
Private Residence	4	1	3	1	2	1		
Apartment/Condo	4	1	3	1	2	1		
Hotel guest rooms	4	1	3	1	3	1		
Hospital Ward	4	2	3	1	3	1		

[a] See Table 1.
[b] See Table 4.

planning and design for control of noise and vibration from hospital activities, building systems, shared bathrooms, and so forth. Despite the need for washable wall and floor surfaces, a substantial amount of sound-absorbing treatment is needed in corridors, nurses' stations, and public spaces to allow speech privacy and to minimize disturbance of patients.

6.2 Work/Study

Acoustical requirements for work or study facilities depend on the tasks to be accommodated (Table 6). For open-plan offices, the background level should be low enough for telephone use yet high enough for some degree of speech privacy between workstations. Since the background level from variable air volume (VAV) ventilation systems can change significantly as air quantity changes, it may be necessary to add an electronic background sound system to maintain a constant sound level to enhance speech privacy. Floor and ceiling finishes should be sound absorbing to minimize sound projection, while acoustical separation between workstations should be provided by sound-absorbing partial-height partitions. Enclosed offices may require a higher degree of privacy.

Large open-plan spaces such as library reading rooms and museum galleries require nonprojection of sound, that is, localization of activity sounds and control of reverberation. In general, a sound-absorbing ceiling is a basic requirement, with a carpeted floor or wall-mounted sound-absorbing treatment and with a constant background sound level for effective masking of intruding noises. Quiet areas for serious study and meditation should be as remote as possible from circulation spaces, reference desks, and other noisy activity areas.

Mechanical equipment for air conditioning and ventilation should be well removed from occupied spaces wherever possible. However, for cost-saving reasons, roof-mounted air conditioning units and noisy VAV boxes may be planned directly above acoustically sensitive spaces. In such situations, noise and vibration could be virtually impossible to control. Main ducts close to large fans should be kept away from occupied spaces for control of "breakout" noise through duct walls. As a general rule, roof-mounted units and VAV boxes should be located above storage or washrooms, with acoustical lining in ducts serving occupied spaces.

6.3 Meeting/Hospitality

Meeting spaces may vary from fixed-seating, special-purpose facilities such as lecture and demonstration rooms to flat-floor, divisible rooms that are easily adapted to many types of events (Table 7). For either extreme, the size may vary from a classroom for 25–30 people to a convention facility seating over a thousand people, with amplified sound and projection systems. In general, sound amplification is desirable for all except the very smallest spaces, while standards such as ADA make hearing-impaired systems a special requirement for many uses. Where video demonstrations and sound playback are required, projection and lighting systems should be integrated with sound amplification. Microphones may be required in the seating area for audience

TABLE 6 Acoustical Requirements for Work/Study Spaces

Use of Space	Requirements for Noise Control[a]					Requirements for Interior Design[b]		
						Reverberation Time		
	Desired Background	Vibration Sensitivity	Speech Privacy	Maximum Sound	Maximum Vibration	Nonprojection	Projection	Sound Amplification
Office								
Open plan	2	1	1	2	1	2	—	—[c]
Closed plan	3	1	2	1	1	2		
In noisy building	3	1	1	2	3	2	—	—[c]
High security	3	1	5	1	1	2		
Library								
Reading	3	2	2	1	1	2		
Reference	3	1	2	1	1	2		
Classroom	3	1	2	1	1	—	2	
Museum	3	1	1	1	1	2		
Meditation	2	2	4	1	1	1		

[a]See Table 1.
[b]See Table 4.
[c]Special evaluation needed.

TABLE 7 Acoustical Requirements for Meeting/Hospitality Spaces

	Requirements for Noise Control[a]					Requirements for Interior Design[b]		
						Reverberation Time		
Use of Space	Desired Background	Vibration Sensitivity	Speech Privacy	Maximum Sound	Maximum Vibration	Nonprojection	Projection	Sound Amplification
Lecture/demonstration rooms	4	1	3	3	1	—	2	Y
Conference	4	1	4	3	1	—	2	Y
Dining	2	1	2	1	1	3	—	—
Assembly	3	2	2	3	1	3	—	Y
Divisible space	3	2	3	3	1	3	—	Y

[a] See Table 1.
[b] See Table 4.

response. Fixed lecture/demonstration spaces could also enhance unamplified speech or music if sound reflecting wall and ceiling surfaces are suitably shaped for optimum natural projection of reflected sounds. Reverberation times should be low for speech and amplified sound but could be lengthened for music performance by use of retractable curtains or other removable sound-absorbing materials.

By contrast, multiuse spaces such as divisible convention rooms should be designed for nonprojection of unamplified sound, should have substantial areas of fixed sound absorption on ceiling and upper walls for control of crowd noise, and should rely upon sound amplification for most events. It is practically impossible to design for natural sound reinforcement from walls and ceiling if the room size and sound source location may vary and if large movable wall surfaces must be flat and acoustically untreated. The sound system layout should be coordinated with the divisible sections of the facility so that each subspace can function either independently or as part of the larger space.

6.4 Performance

The most difficult of acoustical challenges, performing arts facilities range from low-cost flexible spaces such as community centers or drama/music classrooms to highly specialized auditoria seating 3000 or more (Table 8). These may be single-purpose spaces or may have complex stages that can be changed mechanically to accommodate from orchestra and chorus to opera/ballet or drama. Each type of event has well-defined requirements for sight-lines and lighting. Reverberation requirements may vary from under 1.0 s for drama to over 2.0 s for choral/orchestral and organ works. Very few existing major facilities provide truly optimum conditions for the full range of performing arts, mostly because of the high cost of large-scale adjustability. Detailed information on

TABLE 8 Acoustical Requirements for Performance Spaces

	Requirements for Noise Control[a]					Requirements for Interior Design[b]		
						Reverberation Time[c]		
Use of Space	Desired Background	Vibration Sensitivity	Speech Privacy	Maximum Sound	Maximum Vibration	Nonprojection	Projection	Sound Amplification
Community center	4	2	2	5	4	3	2	Y
Theatre/opera	5	3	4	5	4		1/4	—[c]
Multipurpose auditorium	4	2	4	5	4		3	Y
Orchestral	5	3	4	5	2		5	—[c]
Rehearsal	4	2	4	5	4		1/2	
Outdoor performance	4	1	1	5	4			Y

[a] See Table 1.
[b] See Table 4.
[c] Special evaluation needed.

design of performing arts facilities can be found in the references in Chapter 74.

The primary acoustical requirement for a performance space is a low background noise level, free from distracting noise intrusions, to allow the audience to hear stage whispers and pianissimos. This usually determines the construction of the building enclosure and may also lead to rejection of a building site that is too noisy. The ventilation system must be conservatively designed, with remote mechanical equipment spaces and very low, constant air velocity so that diffusers are completely silent. These requirements must be preserved even when cost cutting is necessary.

After performance and audience areas are established (including seat spacing, aisle widths, and other code-mandated requirements), plan and section configurations should be reviewed to ensure the best compromise between sight lines, lighting, and acoustics. The total volume of the space (e.g., ceiling height) must allow for the required maximum reverberation time, avoiding deep underbalcony seating areas that do not share the acoustical characteristics of the main volume. In general, the best contribution to cost saving is a compact, space-efficient seating plan because this minimizes the total sound absorption of the audience and thus the room volume needed for the specified reverberation time.

The interior shaping of wall and ceiling surfaces (including balcony overhangs) establishes the acoustical character of the space: clarity and loudness of unamplified sounds, balance between sound sources (e.g., sections of orchestra or opera singer relative to pit orchestra), envelopment of sound, and so forth. Detailed analysis of the many variables for a large performance space is complicated, but the basic requirements are straightforward. All potentially useful sound reflecting surfaces should be as hard as possible, located and oriented to distribute a balanced sequence of reflected sounds to each listener. A very short time interval (ideally less than 0.03 s) should occur between the arrival of the "direct" (i.e., unreflected) sound and the first reflections at the listener's ears. Surfaces that would focus sound reflections or return long-delayed echoes (over 0.05 s of delay) to any listener should be reoriented to provide useful reflections. In extreme cases, surfaces that create echoes could be controlled by covering with sound-absorbing material. However, since this added absorption robs the space of reverberation and could necessitate a compensating increase in volume, interior shaping for sound distribution should be the primary goal.

Well-understood techniques allow a performing arts hall to be adapted acoustically to accommodate from a large-scale orchestra plus organ and chorus to musicals and opera within a limited time. Movable sound reflectors, including a demountable orchestra enclosure of variable size, modify the strength and balance of sound reflections; the orchestra pit lift can either serve as a thrust stage, create an orchestra pit, or add prime audience seating; electrically operated sound-absorbing curtains change reverberation characteristics to suit a particular work or to deaden the response of the empty hall during rehearsals. If done effectively, this added flexibility may be the difference between a fixed concert hall with limited use and a multiuse space that will be used many more times per year.

Finally, a high-quality sound amplification system designed to complement the natural acoustics of the hall will provide clarity and intelligibility for a soloist or speaker. It can also accommodate a Broadway musical, including an infrared broadcast system for listeners with impaired hearing, playback of recorded speech or music, and sound effects from any location.

6.5 Worship

Traditional forms of worship have comprised one or more of the following: quiet prayer and meditation, sermon and discussion, choral and instrumental (frequently organ) music. Worship spaces range from small chapels to large cathedrals, many of which have borrowed the complex forms of the Romanesque or Gothic church. The use of sophisticated sound amplification systems for music as well as speech, and the trend to very large congregations, has led to spaces seating 7000–10,000 people in a roughly semicircular "arena" configuration. Extreme examples of this form are too large for serious consideration of natural acoustics and should be considered a special case of "arena" design.

For more conventional sizes and programs, the acoustical requirements for a worship space are similar to those for a performing arts facility (Table 9): well-defined locations of celebrants and congregation, wide range of reverberation requirements, the need for speech clarity, with the added requirement of congregational singing. However, lay readers are seldom trained speakers, and liturgical constraints frequently preclude the use of movable elements or acoustically appropriate wall and ceiling shaping.

Most worship needs can be satisfied fairly well by designing the space to meet music requirements (i.e., long reverberation time) and designing the sound amplification system to ensure intelligibility of speech. If the congregation can afford to install a pipe organ, particular emphasis should be placed on a long reverberation time, enhancement of sound projection from choir and organ, and acoustical support of congregational singing. The location of the organ and shaping of walls and ceiling are particularly important for projection of organ sound and for maximizing the acoustical response of the sanc-

TABLE 9 Acoustical Requirements for Worship Spaces

	Requirements for Noise Control[a]					Requirements for Interior Design[b]		
						Reverberation Time		
Use of Space	Desired Background	Vibration Sensitivity	Speech Privacy	Maximum Sound	Maximum Vibration	Nonprojection	Projection	Sound Amplification
Speech	4	2	3	3	1	—	2	Y
Music	4	2	3	3	1	—	5	Y
Choir rehearsal	4	1	2	3	1	—	1/2	—

[a]See Table 1.
[b]See Table 4.

tuary. These conditions entail a large volume, the use of heavy construction materials to minimize absorption of bass sounds, and the avoidance or restricted use of carpet. If music quality is considered less important, adequate loudness and good sound coverage for speech and amplified music can be attained in a light frame construction of modest volume with carpeting and upholstered pews.

Since most worship facilities tend to rely upon natural lighting and ventilation, the choice of building site is of particular importance in ensuring a quiet background for worship. If a completely quiet site is not feasible, the worship space should preferably be shielded from traffic by other buildings. If not, it should be mechanically ventilated, with as few openings as possible facing toward exterior noise sources.

6.6 Industrial/Transport

Acoustical requirements for industrial and transport spaces are presented in Table 10.

Industrial Industrial machinery may create high noise levels but often cannot be enclosed because of the need for accessibility. Sound-absorbing materials on walls and ceiling provide little or no benefit for workers close to the machinery, so special worker enclosures may be needed for hearing protection and voice communication. Possible alternatives include modifications to machinery, rotation of personnel to limit noise exposure, and use of special hearing protectors.

In general, the noisiest items of equipment should be separated and enclosed to limit worker exposure, with extensive use of sound-absorbing materials on ceiling and walls.

Transport Either air or ground transport systems may create serious noise problems for neighboring areas. Initial planning studies should include evaluation of compatibility and may entail preparation of an Environmental Impact Report (EIR) (see Chapter 72). An extreme example is a medical emergency helicopter, which typically

TABLE 10 Acoustical Requirements for Industrial/Transport Spaces

	Requirements for Noise Control[a]					Requirements for Interior Design[b]		
						Reverberation Time		
Use of Space	Desired Background	Vibration Sensitivity	Speech Privacy	Maximum Sound	Maximum Vibration	Nonprojection	Projection	Sound Amplification
Factory								
Light industrial	1	1	1	4	3	4	—	Y
Heavy industrial	1	1	1	5	5	4	—	Y
Transport								
Terminal	1	1	1	5	5	3	—	—
Waiting	2	1	1	2	1	3	—	Y
Ticketing	2	1	1	2	1	3	—	Y
Hospital Heli pad	1	1	1	5	5			

[a]See Table 1.
[b]See Table 4.

creates a severe noise impact for both hospital patients and neighboring residential areas. Truck loading areas may be less noisy but could be in use much longer, and at any hour unless time limits are imposed.

Control of noise within terminals, for passengers and employees of transport systems, typically requires acoustically treated walls and ceilings to make waiting areas and workspaces suitable for telephone use and listening to amplified announcements.

6.7 Arena/Stadium

Enclosed arenas and stadiums are too large for unamplified speech or music such as orchestral or choral concerts (Table 11). A basic requirement is sound-absorbing wall and ceiling material for control of reverberation, crowd noise and echoes (long-delayed sound reflections). Design of ventilation systems should be reviewed carefully to ensure that location of fans, layout of diffusers, and supply air velocities are compatible with the desired background noise level.

Amplification of sound requires highly specialized systems that can accommodate a wide range of events and audience sizes. Outdoor stadiums and tent enclosures require added consideration of noise intrusion—if close to an airport, for example—and of the potential for disturbing residential areas up to several miles from the stadium.

TABLE 11 Acoustical Requirements for Arenas and Stadiums

	Requirements for Noise Control[a]					Requirements for Interior Design[b]		
						Reverberation Time		
Use of Space	Desired Background	Vibration Sensitivity	Speech Privacy	Maximum Sound	Maximum Vibration	Nonprojection	Projection	Sound Amplification
Arena	2	1	1	5	4		—[c]	Y
Enclosed stadium	1	1	1	5	4		—[c]	Y
Outdoor stadium	1	1	1	5	4		—[c]	Y

[a]See Table 1.
[b]See Table 4.
[c]Special evaluation needed.

TABLE 12 Acoustical Requirements for Critical Uses

	Requirements for Noise Control[a]					Requirements for Interior Design[b]		
						Reverberation Time[c]		
Use of Space	Desired Background	Vibration Sensitivity	Speech Privacy	Maximum Sound	Maximum Vibration	Nonprojection	Projection	Sound Amplification
Dance rehearsal	3	1	2	5	4	—	1/2	Y
Music/dance teaching	2	2	3	5	4	—	1/2	Y
Indoor athletics/ aerobics/ swimming pool	1	1	1	4	4	4	—	Y
Building in high-noise environment	2	1	2	4	5	4	—	—[c]
Acoustic/vibration-sensitive space	5	4	4	3	1	5	—	—
Acoustic labs/ Special standards	5	5	4	4	1	5	—	—
Recording/ broadcast	5	3	5	5	1	5	—	—[c]

[a]See Table 1.
[b]See Table 4.
[c]Special evaluation needed.

6.8 Acoustical Requirements for Critical Uses

This category encompasses facilities that must be fully isolated from any intruding noise or vibration as well as facilities with very high noise or vibration levels (Table 12). They typically require highly specialized construction, complete structural separation from other facilities, and possibly restricted access. Particular care is needed to ensure that criteria for acceptable use are clearly defined, and that the best plan locations are selected.

Examples of acoustically sensitive spaces include broadcast/recording studios and anechoic chambers. Vibration-sensitive uses include optical systems with high magnification (e.g., electron microscopes) and manufacturing facilities for computer chips or other high-precision systems.

Facilities with very high noise levels must comply with standards such as established in the United States by OSHA. Examples range from band rehearsal rooms to engine test facilities and shooting ranges. High vibration levels should be anticipated from aerobics and dance rehearsal, indoor running tracks, and heavy machinery such as printing presses, air conditioning equipment, and so forth.

7 ACOUSTICAL ANALYSIS OF SPACE

In this section a worksheet for acoustical analysis of spaces is provided.

Project ______________________ Title/No. of Space ______________________

Control of Noise and Vibration

1. Summary of noise control requirements (from space summaries in Section 6)

Desired background	Vibration sensitivity	Speech privacy	Code requirements	Maximum sound level in space	Maximum vibration in space

2. Exterior noise control Noise sources ______________________

Intruding noise (Table 2, a)	Construction alternatives (Table 3)				Judgment factor	Final selection	
	Walls		Roof				
	a	b	a	b		Walls	Roof

3. Interior noise control Noise sources ______________________

Intruding noise (Table 2, b)	Construction alternatives (Table 3)				Judgment factor	Final selection	
	Walls		Floor/ceiling				
	a	b	a	b		Walls	Floor/ceiling

4. Interior vibration control Vibration sources ______________________

Intruding vibration (Table 2, c)	Construction alternatives (Table 3)		Judgment factor	Final selection
	a	b		

5. HVAC system Noise/vibration sources ______________________

Type of system (Table 2, d)		Construction alternatives (Table 3)		Judgment factor	Final selection
		a	b		
Own					
Other					

6. Other building systems Noise sources ______________________

Type of System (Table 2, e)	Construction alternatives (Table 3)		Judgment factor	Final selection
	a	b		

Control of Sound within the Space

1. Spaces for nonprojection of sound

Required reverberation time	Acoustical treatment			Sound Amplification/announce system
	Ceiling	Walls	Floor	

2. Spaces for projection of sound

Required reverberation time	Acoustical treatment					Sound amplification system
	Stage	Ceiling	Walls	Floor	Fixed seating	

3. Detailed requirements for auditorium

Seating Capacity		Stage facilities				Adjustable volume	Adjustable absorption
Main floor	Balcony	Pit	Lift	Fly loft	Orchestra shell		

4. Requirements for sound amplification system

Central loudspeaker cluster	Distributed system	Pew-back system	Reverberation enhancement	Sound effects	Audience response

5. Room shaping assumptions for auditorium (see references at end of Chapter 74) (use extra sheets as required)

8 WORK SHEET EXAMPLES

In the following examples of worksheets crit means acoustically critical, special evaluation needed; R means sound reflecting, Y is yes, and N is no:

CONTROL OF NOISE AND VIBRATION: APARTMENT/CONDO—URBAN

line 1	—	4	1	3	UBC	3	2
line 2	—	4	2/3	2/3	cost	2	2
line 3	—	3	3/4	3/4	cost	3	3
line 4	—	2	3	4	cost	3	
line 5							
	own	—	—	—	—	—	
	other	1	1	—	—	1	
line 6	—	3	3	—	—	3	

Note—Control of sound within each space not applicable for apartment/condominium.

CONTROL OF NOISE AND VIBRATION: MULTIPURPOSE AUDITORIUM—URBAN

line 1	—	4	3	4	UBC	5	4
line 2	—	5	5/–	5/–	crit	5	5
line 3	—	5	5/4	5/4	crit	5	5
line 4	—	4	4	5	crit	5	
line 5							
	own	2	4	5	crit	4	
	other	2	4	5	crit	4	
line 6		3	3	5	crit	5	

CONTROL OF SOUND WITHIN THE SPACE

line 1	—	Not applicable						
line 2	—	3	R	R	R	R	Y	Y
line 3	—	1500/750	Y	Y	Y	Y	crit	crit
line 4	—	Y	N	N	N	Y	Y	

79

NOISE CONTROL FOR MECHANICAL AND VENTILATION SYSTEMS

R. M. HOOVER AND R. H. KEITH

1 INTRODUCTION

This chapter deals with the level and character of sound for the principal types of electrical and mechanical equipment found in the mechanical equipment rooms of typical buildings. Sound pressure (L_p) and sound power (L_w) data for mechanical equipment are presented. Where possible, the data have been correlated with some of the more obvious equipment operational parameters, such as type, speed, power rating, and flow conditions. The levels quoted are suggested for design uses; these levels represent approximately the 80–90 percentile values and assume no extraordinary noise control. Specific information on equipment description and operation is not presented, however, this information is readily available.[1,2]

2 SOUND LEVEL FOR TYPICAL EQUIPMENT

These data are presented in the form of octave band sound power levels (at 10–12 W) or normalized octave band pressure levels. The normalized octave band sound pressure data are representative of a distance at 0.9 m and a room constant of 74 m^2. Sound pressure level data measured at other distances and room constants may be converted to these normalized conditions by using the methodologies found in Chapters 61 and 73. A-weighted sound levels are also given. A-weighted sound levels are useful for comparing the noise output of competitive equipment. For a complete analysis octave band levels should be used.

2.1 Chillers

Chillers are used to provide cooling water for air conditioning systems. The three principal types are (1) packaged chillers with reciprocating compressors, (2) packaged chillers with rotary-screw compressors, and (3) packaged chillers with centrifugal compressors. In addition absorption machines, which are heat-operated refrigeration machines, are sometimes used to provide cooling water. Normalized sound pressure level data for each of these machines are given in Table 1 (Ref. 3, pp. 7-3–7-6). Specific considerations for each classification are as follows:

(a) *Packaged Chillers with Reciprocating Compressors.* The data for these chillers have been reduced to the normalized distance from the acoustical center of the assembly. When cooling requirements exceed 700 kW, centrifugal compressors become more economical, so there are few reciprocating units rated above 700 kW. Although major interest is concentrated here on the compressor component of a refrigeration machine, an electric motor is usually the drive unit for the compressor. The noise levels attributed here to the compressor will encompass the drive motor most of the time.

(b) *Packaged Chillers with Rotary-Screw Compressors.* The octave band sound pressure levels represent near-maximum noise levels for the size range of 350–1000 kW cooling capacity, operating at or near 3600 rpm.

Note: References to chapters appearing only in the *Encyclopedia* are preceded by *Enc.*

TABLE 1 Normalized Sound Pressure Levels, at 0.9 m, for Chillers

Octave Frequency Band (Hz)	Reciprocating Compressors		Rotary-Screw Compressors, 350–1000 kW Cooling Capacity	Centrifugal Compressors		Absorption Machines, All Capacities
	35–175 kW Cooling Capacity	175–700 kW Cooling Capacity		Under 1750 kW Cooling Capacity	1750 kW More Cooling Capacity	
31	79	81	70	92	92	80
63	83	86	76	93	93	82
125	84	87	80	94	94	82
250	85	90	92	95	95	82
500	86	91	89	91	93	82
1000	84	90	85	91	98	81
2000	82	87	80	91	98	78
4000	78	83	75	87	93	75
8000	72	78	73	80	87	70
A-weighted	89	94	90	97	103	86

(c) *Packaged Chillers with Centrifugal Compressors.* Noise levels for packaged chillers with centrifugal compressors are representative for both hermetic and nonhermetic units. These compressors range in size from 351 to 14,067 kW. The noise levels may be influenced by the motors, gears, or turbine drives, but the data provided emphasize the compressor noise. The low-frequency noise levels reflect the increase found for off-peak loads for most centrifugal machines. These data may be used for packaged chillers, including their drive units. For built-up assemblies, these data should be used for the centrifugal compressor only.

(d) *Absorption Machines.* The sound levels of absorption machines are generally masked by other noises in a mechanical equipment room. The machine usually includes one or two small pumps; steam flow noise or steam valve noise may also be present.

(e) *Built-up Refrigeration Machines.* The noise of packaged chillers, as presented in the preceding paragraphs, includes the noise of both the compressor and the drive unit. If a refrigeration system is built up of separate pieces, then the noise level estimate should include the noise of each component making up the assembly. Compressor noise levels should be taken from the packaged chiller data. Sound level data for the drive units (motors, gears, steam turbines) should be taken from the appropriate tables in this chapter or obtained from the manufacturers.

2.2 Boilers

The wide variety of blower assemblies, air and fuel inlet arrangements, burners, and combustion chambers provides such variability in the noise data that it is impossible to correlate noise with heating capacity. Noise levels at the normalized 0.9-m distance may be as high for the smallest as for the largest units. The estimated noise levels given in Table 2 (Ref. 3, p. 7-6) are believed applicable for all boilers in the size range of 50–2000 boiler horsepower (1 bhp = 33,500 Btu/h or 9818 W). These noise levels apply to the front of the boiler, so when other distances are of concern, the distance should always be

TABLE 2 Normalized Sound Pressure Levels for Boilers

Octave Frequency Band (Hz)	Sound Pressure Level (dB) 50–2000 bhp
31	90
63	90
125	90
250	87
500	84
1000	82
2000	80
4000	76
8000	70
A-weighted, dB(A)	88

taken from the front surface of the boiler. Noise levels are lower off the side and rear of the typical boiler.

2.3 Cooling Towers

The generalizations drawn here may not apply exactly to all cooling towers and condensers, but the data are useful for laying out cooling towers and their possible noise control treatments. For aid in identification, four general types of cooling towers are sketched in Fig. 1: centrifugal-fan blow-through type, axial-flow blow-through type (with the fan or fans located on a side wall, induced-draft propeller type, and "underflow" forced-draft propeller type (with the fan located under the assembly).

(a) *Sound Power Level of Propeller-Type Cooling Towers.* The approximate overall and A-weighted sound power levels of propeller-type cooling towers are given by (Ref. 2, p. 7-9):

$$L_w = 96 + 10\log(\text{fan kW}) \quad \text{for overall } L_w, \tag{1}$$

$$L_{wa} = 87 + 10\log(\text{fan kW}) \quad \text{for A-weighted } L_w, \tag{2}$$

where fan kW is the nameplate power rating of the motor that drives the fan. Octave band sound power levels can be obtained by subtracting the values of Table 3 from the overall L_w obtained by Eq. (1).

(b) *Sound Power Level of Centrifugal-Fan Cooling Towers.* The approximate overall and A-weighted sound power levels of centrifugal-fan cooling towers are given by (Ref. 3, p. 7-10):

$$L_w = 86 + 10\log(\text{fan kW}) \quad \text{for overall } L_w, \tag{3}$$

$$L_{wa} = 79 + 10\log(\text{fan kW}) \quad \text{for A-weighted sound power level.} \tag{4}$$

When more than one fan or cooling tower is used, the fan power rating should be the total motor drive rating of all fans or towers. Octave band sound power levels can be obtained by subtracting the values of Table 3 from the overall L_w obtained with Eq. (3).

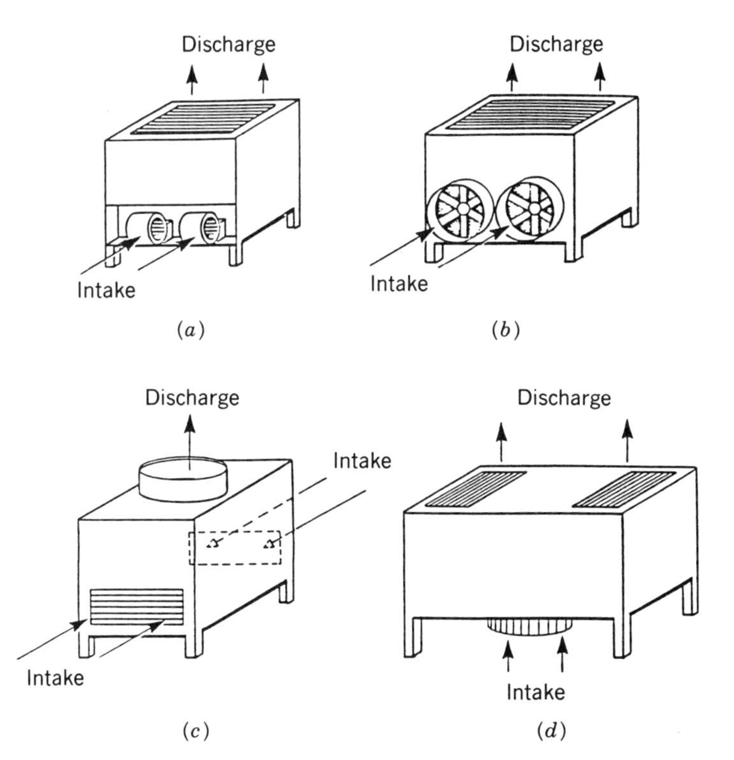

Fig. 1 Principal types of cooling towers: (*a*) centrifugal-fan blow-throw type; (*b*) axial-flow blow-through type (*c*) induced-draft propeller type; (*d*) forced-draft propeller-type "underflow."

TABLE 3 Frequency Adjustments for Cooling Towers to Obtain Octave Band PWLs

Octave Frequency Band (Hz)	Centrifugal Fan	Propeller Fan
31	6	8
63	6	5
125	8	5
250	10	8
500	11	11
1000	13	15
2000	12	18
4000	18	21
8000	25	29
A-weighted, dB(A)	7	9

(c) *Directionality of Cooling Towers.* Cooling towers usually radiate different amounts of sound in different directions, and the directional corrections of Table 4 should be made to the octave band sound power levels. These corrections apply to the five principal directions from a cooling tower, that is, in a direction perpendicular to each of the four sides and to the top of the tower. If it is necessary to estimate the sound pressure level at some direction other than the principal directions, it is common practice to interpolate between the values given for the principal directions.

(d) *Close Proximity Sound Levels of Cooling Towers.* Sound power level data usually will not give accurate calculated sound pressure levels at very close distances to large-size sources, such as cooling towers. The data of Table 5 may be used where it is required to estimate close-in sound pressure levels at nearby walls, floors, or in closely confined spaces (at 0.9–2 m distances from inlet and discharge openings).

(e) *Noise Reduction with Reduced Fan Speed.* When it is practical to do so, the cooling tower fan can be reduced to half speed in order to reduce noise. Half speed produces approximately two-thirds cooling capacity and approximately 8- to 10-dB noise reduction in the octave bands that contain most of the fan-induced noise. For half-speed operation, the octave band L_p or L_w of full-speed cooling tower noise may be reduced by the following amounts, where fB is the blade passage frequency and is calculated from the relation fB = number of fan blades × shaft rpm/60:

TABLE 4 Correction to Average PWLs for Directional Effects of Cooling Towers

Cooling Tower Type	31 Hz	63 Hz	125 Hz	250 Hz	500 Hz	1000 Hz	2000 Hz	4000 Hz	8000 Hz
					Centrifugal-Fan Blow-Through Type				
Front (fan inlet)	+3	+3	+2	+3	+4	+3	+2	+2	+2
Side (enclosed)	0	0	0	−2	−3	−4	−5	−5	−5
Rear (enclosed)	0	0	−1	−2	−3	−4	−5	−6	−6
Top (discharge)	−3	−3	−2	0	+1	+2	+3	+4	+5
					Axial-Flow Blow-Through Type				
Front (fan inlet)	+2	+2	+4	+6	+6	+5	+5	+5	+5
Side (enclosed)	+1	+1	+1	−2	−5	−5	−5	−5	−4
Rear (enclosed)	−3	−3	−4	−7	−7	−7	−8	−11	−8
Top (discharge)	−5	−5	−5	−5	−2	0	0	+2	+1
					Induced-Draft Propeller-Type				
Front (air inlet)	0	0	0	+1	+2	+2	+2	+3	+3
Side (enclosed)	−3	−3	−3	−3	−3	−3	−4	−5	−6
Top (discharge)	+3	+3	+3	+3	+3	+4	+4	+3	+3
					"Underflow" Forced-Draft Propeller Type				
Any side	−1	−1	−1	−2	−2	−3	−3	−4	−4
Top	+2	+2	+2	+3	+3	+4	+4	+5	+5

TABLE 5 Approximate Close-in SPLs near the Intake and Discharge Openings of Various Cooling Towers (0.9–2 m distance)

Octave Band	31 Hz	63 Hz	125 Hz	250 Hz	500 Hz	1000 Hz	2000 Hz	4000 Hz	8000 Hz
Centrifugal-Fan Blow-Through Type									
Intake	85	85	85	83	81	79	76	73	68
Discharge	80	80	80	79	78	77	76	75	74
Axial-Flow Blow-Through Type (including "underflow" type)									
Intake	97	100	98	95	91	86	81	76	71
Discharge	88	88	88	86	84	82	80	78	76
Propeller-Fan Induced-Draft Type									
Intake	94	96	94	92	88	83	78	72	76
Discharge	99	103	99	95	90	85	81	76	70

Octave Band That contains	Noise Reduction Due to Half Speed
$\frac{1}{8}$ fB	3 dB
$\frac{1}{4}$ fB	6 dB
$\frac{1}{2}$ fB	9 dB
1 fB	9 dB
2 fB	9 dB
4 fB	6 dB
8 fB	3 dB

If the blade passage frequency is not known, it may be assumed to fall in the 63-Hz band for propeller-type cooling towers and in the 250-Hz band for centrifugal cooling towers. Waterfall noise usually dominates in the upper octave bands, and it would not change significantly with reduced fan speed.

(f) *Close Confined Locations.* Most of the preceding discussion assumes that cooling towers will be used in outdoor locations. If they are located inside enclosed mechanical equipment rooms or within courts formed by several solid walls, the sound patterns will be distorted. In such instances, the L_w of the tower (or appropriate portions of the total L_w) can be placed in that setting, and the enclosed or partially enclosed space can be likened to a room having certain estimated amounts of reflecting and absorbing surfaces. In the absence of a detailed analysis of cooling tower noise levels inside enclosed spaces, it is suggested that the close-in noise levels of Table 5 be used as approximations.

2.4 Pumps

The overall and A-weighted normalized sound pressure levels for pumps are given in Table 6 (Ref. 3, p. 7-16) and are representative of most pumps encountered in commercial situations. The pump power rating is taken as the nameplate power of the drive motor. For pump ratings under 75 kW, the radiated noise increases with 10 log of the pump rating, but above 75 kW the noise changes more slowly with increasing power (i.e., 3 log of the pump rating). Octave band sound pressure lev-

TABLE 6 Overall and A-Weighted Sound Pressure Levels for Pumps

Speed Range (rpm)	Drive Motor Nameplate Power: Under 75 kW	Drive Motor Nameplate Power: Above 75 kW
Overall Sound Pressure Level (dB)		
3000–3600	72 + 10 log kW	85 + 3 log kW
1600–1800	75 + 10 log kW	88 + 3 log kW
1000–1500	70 + 10 log kW	83 + 3 log kW
450–900	68 + 10 log kW	81 + 3 log kW
A-Weighted Sound Level [dB(A)]		
3000–3600	70 + 10 log kW	82 + 3 log kW
1600–1800	73 + 10 log kW	86 + 3 log kW
1000–1500	68 + 10 log kW	81 + 3 log kW
450–900	66 + 10 log kW	79 + 3 log kW

TABLE 7 Octave Band Frequency Adjustments for Pumps

Octave Frequency Band (Hz)	Value Substracted from Overall SPL (dB)
31	13
63	12
125	11
250	9
500	9
1000	6
2000	9
4000	13
8000	19
A-weighted, dB(A)	2

els are obtained by subtracting the values of Table 7 from the overall sound pressure level of Table 6. Pumps intended for high-pressure operation have smaller clearances between the blade tips and the cutoff edge and, as a result, may have higher noise peaks in the octave bands containing the impeller blade passage frequency and its first harmonic. These would usually fall in the 1000- and 2000-Hz octave bands.

2.5 Fans

Manufacturers often furnish in-duct sound power level data of their fans on request. However, in early designs the specific model or manufacturer is not known. In these cases the octave band sound power of a fan can be estimated by[4]

$$L_w = \text{Kw} + 10 \log Q + 20 \log P + \text{BFI} + C, \tag{5}$$

where L_w is the total in-duct sound power level of the fan (inlet and outlet), Kw is the specific sound power level for the particular fan design, Q is the volume flow rate in cubic metres per second, and P is the static pressure produced by the fan (in kilopascals). Values of Kw for the octave bands and for various basic fan blade designs are given in Part A of Table 8. The blade passage frequency of the fan is obtained from

$$\frac{\text{Fan rpm} \times \text{number of blades}}{60}, \tag{6}$$

and the *blade frequency increment* (BFI) (in decibels) is added to the octave band sound power level in the octave in which the blade passage frequency occurs. It is best to obtain the number of blades and the fan rotational speed from the manufacturer to calculate the blade passage frequency. In the event this information is not available, Part B of Table 8 provides the usual blade passage frequency. The estimates given by this method assume ideal inlet and outlet flow conditions and operation of the fan at its peak design condition. The noise is quite critical to these conditions and increases significantly for deviations from ideal. Part C of Table 8 provides a correction factor for off-peak fan operation. The operating static efficiency of a fan may be calculated by

$$\text{Static efficiency} = \frac{Q \times P}{1002 \times \text{BkW}} \tag{7}$$

where Q is air volume in liters per second, P is total static pressure in kilopascals, and BkW is the brake power in kilowatts. The ratio of calculated static efficiency to peak efficiency determines the correction factor C. For cursory design the maximum peak efficiency can be taken as 65% for centrifugal fans with forward curved blades, 80% for centrifugal fans with backward inclined blades, and 70% for vaneaxial fans.

The fan housing and its nearby connected ductwork radiate fan noise into the fan room. Because of the number of variables involved, there exists no simple analysis procedure for estimating the L_w of the noise radiated by the housing and ductwork. However, Table 9 offers a method to estimate this type of noise. These are simply deductions, in decibels, from the induct fan noise.

2.6 Air Compressors

Two types of air compressors are frequently found in buildings: One is a relatively small compressor, usually under 7 kW, used to provide a high-pressure air supply for operating the controls of the ventilation system, and the other is a medium-size compressor, possibly up to 75 kW, used to provide "shop air" to maintenance shops, machine shops, and laboratory spaces, or to provide ventilation system control pressure for large buildings. Normalized sound pressure levels for air compressors are given in Table 10 (Ref. 3, p. 7-19).

2.7 Emergency Generators

Emergency generators are typically driven by reciprocating engines or small gas turbines. As a general rule the sound level of the generator is not significant with respect to the driver:

TABLE 8 Specific Sound Power Level Kw, Blade Frequency Increment, and Off-Peak Correction for Use in Eq. (5)

Part A. Specific Sound Power Levels for Inlet or Outlets and Blade Frequency Increments (BFI) for Fans[a]

Fan and Blade Type	Size and Operation	Octave Band Center Frequency (Hz)								
		63	125	250	500	1000	2000	4000	8000	BFI
Centrifugal										
Airfoil, Backward	Over 750 mm dia.	85	85	84	79	75	68	64	62	3
Curved and Inclined	Under 750 mm dia.	90	90	88	84	79	73	69	65	3
Centrifugal										
Forward Curved	All sizes	98	98	88	81	81	76	71	66	2
Centrifugal										
Radial over 1,000 mm	Low pressure, SP 1–2.5 kPa	101	92	88	84	82	77	74	71	7
	Mid pressure, SP 2.5–5 kPa	103	99	90	87	83	78	74	71	8
	High pressure, SP 5–15 kPa	106	103	98	93	91	89	86	83	8
Vaneaxial	Hub ratio 0.3–0.4	94	88	88	93	92	90	83	79	6
	Hub ratio 0.4–0.6	94	88	81	88	86	81	75	73	6
	Hub ratio 0.6–0.8	98	97	96	96	94	92	88	85	6
Tubeaxial	Over 1000 mm dia.	96	91	92	94	92	91	84	82	7
	Under 1000 mm dia.	93	92	94	98	97	96	88	85	7
Propeller	All sizes	93	96	103	131	100	97	91	87	5

Part B. Octave Band in Which Blade Frequency Increment (BFI) Occurs[b]

Fan Type	Octave Band in Which BFI Occurs (Hz)
Centrifugal	
Airfoil, backward curved and backward inclined	250
Forward curved	500
Radial blade	125
Vaneaxial	125
Tubeaxial	63
Propeller	63

Part C. Correction Factor C for Off-peak Operation

Static Efficiency % of Peak	Correction Factor (dB)
90–100	0
85–89	3
75–84	6
65–74	9
55–64	12
50–54	15

[a] Kw for total subtract 3 dB for inlet or outlet level.
[b] Use for estimating purposes only. Where actual fan is known, use manufactures' data.

(a) *Reciprocating Engines.* Reciprocating engines have three sound sources of interest; the engine casing, the air inlet into the engine, and the exhaust from the engine. The L_w of the noise radiated by the casing of a natural-gas or diesel reciprocating engine is given by (Ref. 3, p. 7-20)

$$L_w = 94 + 10\log(\text{rated kW}) + A + B + C + D, \quad (8)$$

where L_w is the overall sound power level (in decibels);

TABLE 9 Octave Band Adjustments for Estimating the PWL of Noise Radiated by a Fan Housing and Its Nearby Connected Duct Work

Octave Frequency Band (Hz)	Value Subtracted from In-Duct Fan Noise (dB)
31	—
63	0
125	0
250	5
500	10
1000	15
2000	20
4000	22
8000	25

TABLE 10 Normalized Sound Pressure Levels for Air Compressors

Octave Frequency Band (Hz)	Air Compressor Power Range		
	0.7–1.5 kW	2–7 kW	8–55 kW
31	82	87	92
63	81	84	87
125	81	84	87
250	80	83	86
500	83	86	89
1000	86	89	92
2000	86	89	92
4000	84	87	90
8000	81	84	87
A-weighted, dB(A)	91	94	97

"rated kW" is the engine manufacturer's continuous full-load rating for the engine, and *A*, *B*, *C*, and *D* are correction terms given in Table 11. Octave band sound power levels can be obtained by subtracting the Table 12 values from the overall L_w given by Eq. (8). The octave band corrections are different for the different engine speed groups. For small engines, under about 335 kW, the air intake noise is usually radiated close to the engine casing, so it is not easy or necessary to separate these two sources; and the engine casing noise may be considered as including air intake noise (from both naturally aspirated and turbocharged engines).

Most large engines have turbocharges at their inlet to provide pressurized air into the engine for increased performance. Turbine configuration and noise output can vary appreciably, but an approximation of the L_w of the

TABLE 11 Correction Terms Applied to Eq. (8) for Estimating the Overall PWL of the Casing Noise of a Reciprocating Engine

Speed correction term *A*
- Under 600 rpm, −5 dB
- 600–1500 rpm, −2 dB
- Above 1500 rpm, 0 dB

Fuel correction term *B*
- Diesel fuel only, 0 dB
- Diesel and/or natural gas, 0 dB
- Natural gas only (may have small amount of "pilot oil"), −3 dB

Cylinder arrangement term *C*
- In-line, 0 dB
- V-type, −1 dB
- Radial, −1 dB

Air intake correction term *D*
- Unducted air inlet to unmuffled roots blower, +3 dB
- Ducted air from outside the room or into muffled roots blower, 0 dB
- All other inlets to engine (with or without turbochargers), 0 dB

TABLE 12 Frequency Adjustments for Casing Noise of Reciprocating Engines

Octave Frequency Band (Hz)	Value Subtracted from Overall PWL (dB)			
	Engine Speed under 600 rpm	Engine Speed 600–1500 rpm: Without Roots Blower	Engine Speed 600–1500 rpm: With Roots Blower	Engine Speed over 1500 rpm
31	12	14	22	22
63	12	9	16	14
125	6	7	18	7
250	5	8	14	7
500	7	7	3	8
1000	9	7	4	6
2000	12	9	10	7
4000	18	13	15	13
8000	28	19	26	20
A-weighted, dB(A)	4	3	1	2

turbocharged inlet noise is given by

$$L_w = 95 + 5 \log(\text{rated kW}) - \frac{L}{1.8}, \tag{9}$$

where L_w and rated kW are as previously defined and *L* is the length, in metres, of a ducted inlet to the tur-

TABLE 13 Frequency Adjustments for Turbocharger Air Inlet Noise and Engine Exhaust

Octave Frequency Band (Hz)	Turbocharger Air Intake	Unmuffled Exhaust
31	4	5
63	11	9
125	13	3
250	13	7
500	12	15
1000	9	19
2000	8	25
4000	9	35
8000	17	43
A-weighted, dB(A)	3	12

TABLE 14 Frequency Adjustments (in dB) for Gas Turbine Engine Noise Sources

Octave Frequency Band	Value Subtracted from Overall PWL (dB)		
(Hz)	Casing	Inlet	Exhaust
31	10	19	12
63	7	18	8
125	5	17	6
250	4	17	6
500	4	14	7
1000	4	8	9
2000	4	3	11
4000	4	3	15
8000	4	6	21
A-weighted, dB(A)	2	0	4

bocharger. For many large engines, the air inlet may be ducted to the engine from a fresh air supply or a location outside the room or building. The octave band values given in Table 13, for air inlet, are subtracted from the overall L_W of Eq. (9) to obtain the octave band sound power levels of turbocharged inlet noise.

The sound power level of the noise radiated from the unmuffled exhaust of an engine is given by

$$L_W = 120 + 10\log(\text{rated kW}) - T - \frac{L}{1.2}, \qquad (10)$$

where T is the turbocharger correction term ($T = 0$ dB for an engine without a turbocharger and $T = 6$ dB for an engine with a turbocharger) and L is the length, in metres, of the exhaust pipe. The octave band sound power levels of unmuffled exhaust noise are obtained by subtracting the values of Table 13, for exhaust, from the overall L_W derived from Eq. (10). If the engine is equipped with an exhaust muffler, the final noise radiated from the end of the tailpipe is the L_W of the unmuffled exhaust minus the insertion loss, in octave bands, of the muffler.

(b) *Small Gas Turbine.* As with reciprocating engines, the three principal sound sources of turbine engines are the engine casing, the air inlet, and the exhaust. Most gas turbine manufacturers will provide sound power estimates of these sources. However, when these are unavailable, the overall sound power levels of these three sources, with no noise reduction treatments, are as given in the following equations (Ref. 3, p. 7-24):

$$L_W = 120 + 5\log(\text{rated MW}) \quad \text{for engine casing noise}, \qquad (11)$$

$$L_W = 127 + 15\log(\text{rated MW}) \quad \text{for air inlet noise}, \qquad (12)$$

$$L_W = 133 + 10\log(\text{rated MW}) \quad \text{for exhaust noise}, \qquad (13)$$

where rated MW is the maximum continuous full-load rating of the engine in megawatts.

For casing and inlet noise, particularly strong high-frequency sounds may occur at several of the upper octave bands. However, which bands contain the tones will depend on the specific design of the turbine and, as such, will differ between models and manufacturers. Therefore, the octave band adjustments of Table 14 allow for possible peaks in several different bands. Because of this randomness of peak frequencies, the A-weighted levels may also vary from the values quoted.

The turbine manufacturer often provides the turbine casing with a protective thermal wrapping or an enclosing cabinet, either of which can give some noise reduction. Table 15 suggests the approximate noise reduction for casing noise that can be assigned to different types of engine enclosures. Refer to the notes of Table 15 for a broad description of the enclosures. The values of Table 15 may be subtracted from the octave band sound power levels of casing noise to obtain the adjusted sound power levels of the covered or enclosed casing. An enclosure specifically designed to control casing noise may provide larger noise reduction values than those in the table. However, it should be noted that the performance of enclosures that are supported on the same structure as the gas turbine will be limited by structure-borne sound. For this reason care should be used in applying laboratory data of enclosure performance to the estimation of sound reduction of gas turbine enclosures.

The directivity of the exhaust stack should be considered in the assessment of noise control. Table 16

TABLE 15 Approximate Noise Reduction of Gas Turbine Engine Casing Enclosures

Octave Frequency Band (Hz)	Noise Reduction (dB)				
	Type 1[a]	Type 2[b]	Type 3[c]	Type 4[d]	Type 5[e]
31	2	4	1	3	6
63	2	5	1	4	7
125	2	5	1	4	8
250	3	6	2	5	9
500	3	6	2	6	10
1000	3	7	2	7	11
2000	4	8	2	8	12
4000	5	9	3	8	13
8000	6	10	3	8	14

[a]Glass fiber or mineral wool thermal insulation with lightweight foil cover over the insulation.
[b]Glass fiber or mineral wool thermal insulation with minimum 20 gauge aluminum or 24 gauge steel or 12-mm-thick plaster cover over the insulation.
[c]Enclosing metal cabinet for the entire packaged assembly, with *open* ventilation holes and with *no* acoustical absorption lining inside the cabinet.
[d]Enclosing metal cabinet for the entire packaged assembly, with *open* ventilation holes and *with* acoustical absorption lining inside the cabinet.
[e]Enclosing metal cabinet for the entire packaged assembly, with all ventilation holes into the cabinet muffled and with acoustical absorption lining inside the cabinet.

TABLE 16 Approximate Directivity Effect of a Large Exhaust Stack Compared to a Nondirectional Source of the Same Power

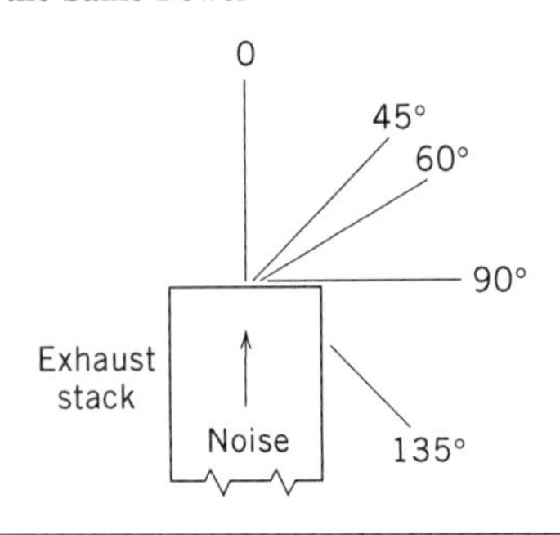

Octave Frequency Band (Hz)	Relative Sound Level for Indicated Angle from Axis				
	0°	45°	60°	90°[a]	135° and Larger[a]
31	8	5	2	−2	−3
63	8	5	2	−3	−4
125	8	5	2	−4	−6
250	8	6	2	−6	−8
500	9	6	2	−8	−10
1000	9	6	2	−10	−13
2000	10	7	0	−12	−16
4000	10	7	−1	−14	−18
8000	10	7	−2	−16	−20

[a]For air intake openings subtract 3 dB from the values in the 90° and 135° columns, i.e., −2 − 3 = −5 dB for 31 cps at 90°.

gives the approximate directivity effect of a large exhaust opening. This can be used for either a horizontal or vertical stack exhausting hot gases. These values also apply for a large-area intake opening into a gas turbine for the 0°–60° range; for the 90°–135° range, subtract an additional 3 dB from the already negative-valued quantities. For small openings in a wall, such as for ducted connections to a fan intake or discharge, use approximately one-half the directivity effect of Table 16 (as applied to intake openings) for the 0°–90° region. For angles beyond 90° estimate the effect of the wall as a barrier.

2.8 Electric Motors

Electric motors used in commercial buildings are usually either a total enclosed fan-cooled (TEFC) motor or an open drop-proof motor (DRPR):

(a) *TEFC Motors.* The normalized unweighted sound pressure level for TEFC motors follow approximately the following relationships (Ref. 3, p. 7-28):

$$L_p = 17 + 17 \log \text{kW} + 15 \log \text{rpm} \quad \text{for power ratings under 37 kW,} \tag{14}$$

$$L_p = 28 + 10 \log \text{kW} + 15 \log \text{rpm} \quad \text{for power ratings above 37 kW,} \tag{15}$$

where kW is the nameplate motor range and rpm is the motor shaft speed. For motors above 300 kW, the calculated noise value for a 300-kW motor should be used. Equation (15) is not applicable to large commercial motors in the power range of 750–4000 kW. The octave band corrections for TEFC motors are given in Table 17. Some TEFC motors produce strong tonal sounds in the 500-, 1000-, or 2000-Hz octave bands because of the cooling fan blade frequency. The octave band frequency adjustments of Table 17 allow for a moderate amount of these tones, but a small percentage of motors may still exceed these calculated levels by as much as 5–8 dB. When specified, motors that are quieter than these calculated values by 5–10 dB can be purchased.

(b) *DRPR Motors.* The normalized unweighted sound pressure level for DRPR motors follow approximately the relationships (Ref. 3, p. 7-29):

TABLE 17 Frequency Adjustments for TEFC and DRPR Electric Motors

Octave Frequency Band (Hz)	TEFC	DRPR
31	14	9
63	14	9
125	11	7
250	9	7
500	6	6
1000	6	9
2000	7	12
4000	12	18
8000	20	27
A-weighted, dB(A)	1	4

$$L_p = 12 + 17 \log \text{ kW} + 15 \log \text{ rpm} \quad \text{for power ratings under 37 kW,} \tag{16}$$

$$L_p = 23 + 10 \log \text{ kW} + 15 \log \text{ rpm} \quad \text{for power ratings above 37 kW.} \tag{17}$$

For motors above 300 kW, the calculated noise value for a 300-kW motor should be used. The octave band corrections for DRPR motors are given in Table 17.

2.9 Steam Turbines

Steam turbines are sometimes used as primary or backup drivers for chillers, pumps, and air compressors. The noise levels of steam turbines are found generally to increase with increasing power rating, but it has not been possible to attribute any specific noise characteristics with speed or turbine blade passage frequency. Suggested normalized sound pressure levels for steam turbines, with a power range of 370–11,000 kW, are given in Table 18 (Ref. 3, p. 7-31).

2.10 Gears

Speed reduction or increasing gear units are used to transfer power between the primary driver and driven equipment, when there is a speed difference between the two. It is generally true that the noise of gears increase with increasing speed and power, but it is not possible to predict in which frequency band the gear tooth contact rate or the "ringing frequencies" will occur for any unknown gear. The possibility that these frequency components may occur in any of the upper octave bands is covered by Eq. (18), which gives the normalized octave band sound pressure level estimate for all bands at and about 125 Hz (Ref. 3, p. 7-31):

$$L_p = 79 + 3 \log(\text{rpm}) + 4 \log(\text{kW}), \tag{18}$$

where rpm is the speed of the slower gear shaft and kW is the horsepower rating of the gear or the power transmitted through the gear. For the 63-Hz band, 3 dB is deducted; and for the 31-Hz band, 6 dB is deducted from the Eq. (18) value.

2.11 Generators

The noise of generators, in general, can be quite variable, depending on speed, the presence or absence of air cooling vanes, clearances of various rotor parts, and so on,

TABLE 18 Normalized Sound Pressure Levels for Steam Turbines

Octave Frequency Band (Hz)	Steam Turbine Power Range		
	370–1100 kW	1100–3700 kW	3700–11,000 kW
31	86	88	90
63	91	93	95
125	91	93	95
250	88	90	92
500	85	87	89
1000	85	88	91
2000	87	91	95
4000	84	88	92
8000	76	81	86
A-weighted, dB(A)	92	95	99

TABLE 19 Approximate Overall PWL of Generators, Excluding Noise of Driver Unit

	Electrical Power Rating of Generator							
Generator Speed (rpm)	0.2 MW (dB)	0.5 MW (dB)	1 MW (dB)	2 MW (dB)	5 MW (dB)	10 MW (dB	20 MW (dB)	50 MW (dB)
600	95	99	102	105	109	112	115	119
1200	97	101	104	107	111	114	117	121
1800	98	102	105	108	112	115	118	122
2400	99	103	106	109	113	116	119	123
3600	100	104	107	110	114	117	120	124
4800	101	105	108	111	115	118	121	125

but, most of all, on the driver mechanism. Table 19 gives a reasonable approximation of the overall sound power level. It is to be noted that the L_w of the generator is usually less than that of the drive gear and less than that of the untreated engine casing. Octave band corrections to the overall L_w are given in Table 20.

2.12 Transformers

Transformer manufacturers commonly provide an average A-weighted sound level for their products. Typically this is an average of the sound levels, on a reference sound producing surface space at a distance of 0.3 m from the outline of the transformer. On the basis of field studies of many transformer installations, the sound power level in octave bands has been related to the average A-weighted sound level and the area of the four side walls of the unit. This relationship is expressed by

$$L_w = \text{average } L_{pa} + 10 \log A + C + 10, \qquad (19)$$

TABLE 20 Frequency Adjustments for Generators, Without Drive Unit

Octave Frequency Band (Hz)	Value Subtracted from Overall SPL (dB)
31	11
63	8
125	7
250	7
500	7
1000	9
2000	11
4000	14
8000	19
A-weighted, dB(A)	4

where A is the total surface of the four side walls of the transformer in cubic metres and C is an octave band correction that has different values for different uses, as shown in Table 21 (Ref. 3, p. 7-34). If the exact dimensions of the transformer are not known, an approximation will suffice. If in doubt, the area should be estimated on the high side. An error of 25% in area will produce a change of 1 dB in the sound power level. The most nearly applicable C value from Table 21 should be used. The C_1 value assumes normal radiation of sound. The C_2

TABLE 21 Octave Band Corrections Used in Eq. (20) for Obtaining PWL of Transformers in Different Installation Conditions

Octave Frequency Band	Octave Band Corrections (dB)		
(Hz)	C_1 [a]	C_2 [b]	C_3 [c]
31	−1	−1	−1
63	+5	+8	+8
125	+7	+13	+13
250	+2	+8	+12
500	+2	+8	+12
1000	−4	−1	+6
2000	−9	−9	+1
4000	−14	−14	−4
8000	−21	−21	−11

[a] Use C_1 for outdoor location or for indoor location in a large mechanical equipment room (over about 5000 ft^3) containing many other pieces of mechanical equipment that serve as obstacles to diffuse sound and break up standing waves.
[b] Use C_2 for indoor locations in transformer vaults or small rooms (under about 5000 ft^3) with parallel walls and relatively few other large-size obstacles that can diffuse sound and break up standing waves.
[c] Use C_3 for any location where a serious noise problem would result if the transformer should become noisy above its NEMA rating, following its installation and initial period of use.

value should be used in regular-shaped confined spaces where standing waves will likely occur, which typically may produce 6 dB higher sound levels at the transformer harmonic frequencies of 120, 240, 360, 480, and 600 Hz (for 60 Hz line frequency; or other sound frequencies for other line frequencies). The C_3 value is an approximation of the noise of a transformer that has grown noisier (by about 10 dB) during its lifetime. This happens occasionally when the laminations or tie bolts become loose, and the transformer begins to buzz or rattle. In a highly critical location, it would be wise to use this value. Field measurements have shown that transformers may actually have A-weighted sound levels that range from a few decibels (2 or 3 dB) above to as much as 5 or 6 dB below the quoted A-weighted sound level. Quieted transformers that contain various forms of noise control treatments can be purchased at as much as 15–20 dB below normal A-weighted ratings.

3 PROPAGATION OF AIRBORNE NOISE FROM MECHANICAL EQUIPMENT ROOMS

The analysis of the impact of mechanical equipment, on surrounding spaces is relatively straightforward. Once the sound pressure level within the mechanical equipment room has been established, the degree of transmission to adjacent spaces can be determined with knowledge of the transmission loss properties of the walls, floor, and ceiling of the mechanical room (see Chapter 76) and the acoustical properties of the adjacent rooms (see *Enc.* Ch. 95).

(a) *Transmission Loss of Mechanical Room Partitions.* Transmission loss data for different partitions can be found in many publications. As a general rule this information is derived from laboratory measurements (e.g., ASTM E 90). However, due to measurement limitations, transmission loss data below 100 Hz is rarely reported. Large mechanical equipment will often produce significant acoustic energy below 100 Hz. Therefore there is a need to obtain transmission loss performance below 100 Hz, or alternatively, estimate the low-frequency performance (see Chapter 76). Quite often a single number rating, such as the Sound Transmission Loss (STC) (ASTM E 413-87) is provided. However, most of these single-number classifications are heavily weighted toward the 500–2000 Hz frequency range. While this range is suitable for evaluation of isolation for speech, some music, and most transportation noise sources, it is not suitable for the evaluation of mechanical equipment noise sources. There is growing usage of an alternative rating call the MTC (for mechanical or music transmission loss) [Roller 85]. The determination of the MTC is similar to the STC in that it uses the same reference curve and measured one-third-octave band transmission loss data. The determination of the MTC rating differs from the determination of the STC rating in that:

1. No deficiencies are allowed in the 125- and 160-Hz one-third-octave bands.
2. And if there are any surpluses above the STC contour in the 125- and 160-Hz one-third-octave bands the rating is increased by one-third of the sum of the surpluses.

Studies have indicated that, when the A-weighted sound level within the mechanical equipment room is less than the sum of the MTC rating of the partition and the RC rating of the background sound within the adjacent room, the intrusive noise should be acceptable. MTC ratings are useful as a cursory evaluation technique. Final selection of partition types should be based on a more complete analysis (e.g., octave or one-third-octave band analysis).

(b) *Openings in Walls.* An opening, such as a door, window, or louvered vent, in an exterior wall of a noisy room will allow noise to escape from that room and perhaps be disturbing to neighbors. The sound power of the sound that passes through the opening can be estimated from the equation.

$$L_w = L_p + 10 \log A, \qquad (20)$$

where L_p is the sound level in the room at the location of the opening and A is the area, in cubic metres, of the opening. For normal openings (windows or vents) without ducted connections to the noise source, it may be assumed that the sound radiates freely in all directions in front of the opening. For ducted connections from a sound source to an opening in the wall, the sound is somewhat "beamed" out of the opening and may be assumed to have a directivity effect of above one-half the amount given in Table 16 for air intake openings of large stacks.

4 VIBRATION ISOLATION OF MECHANICAL EQUIPMENT

If mechanical equipment is not provided with proper vibration isolation, acoustic energy will be transmitted into the supporting structure resulting in unwanted vibration and structure-borne sound. The isolator types and isolation guidelines presented in this chapter are based on experience with successful installation of mechanical equipment in commercial buildings.

4.1 Isolator Types and Transmissibility

A transmissibility curve is often used to indicate the general behavior of a vibration-isolated system. Transmissibility is roughly defined as the ratio of the force transmitted through the isolated system to the supporting structure to the driving force exerted by the piece of vibrating equipment (see Chapter 55). Strict interpretation of transmissibility data and isolation efficiencies, however, must be adjusted for real-life situations. Usually, floors that have large column spacing will have larger deflections than floors of shorter column spacing.

The *static deflection* of a mount is simply the difference between the free-standing height of the uncompressed, unloaded isolator and the height of the compressed isolator under its static load. This difference is easily measured in the field or estimated from the manufacturer's catalog data. For this reason it is often most beneficial to specify the static deflection for the vibration isolation system, taking into account the type of isolator, the linearity of the isolator system and the rigidity of the structure of the equipment base and supporting structure.

Table 22 provides a suggested schedule for achieving various degrees of vibration isolation in normal construction. This table is based on the transmissibility curve (see Fig. 2 of Chapter 55), but suggests operating ranges of the ratio of driving frequency to natural frequency. The terms "low," "fair," and "high" are merely word descriptors, but they are more meaningful than such terms as 95 or 98% isolation efficiency, which are clearly erroneous if they do not take into account the mass and stiffness of the floor slab. Vibration control recommendations given in this chapter are based on the application of this table. Table 23 lists the principal types of vibration isolators used in the isolation of building mechanical equipment and their general range of applications:

TABLE 22 Suggested Schedule for Estimating Relative Vibration Isolation Effectiveness of a Mounting System

Ratio of Driving Frequency of Source to Natural Frequency of Mount	Degree of Vibration Isolation
Below 1.4	Amplification
1.4–3	Negligible
3–6	Low
6–10	Fair
Above 10	High

(a) *Steel Spring Isolators.* Steel springs are used to support heavy equipment and to provide isolation for the typical low-frequency range of about 3–60 Hz (180- to 3600-rpm shaft speed). Steel springs have natural frequencies that fall in the range of about 1 Hz (for approximately 254 mm static deflection) to about 6 Hz (for approximately 6 mm static deflection). Springs can transmit high-frequency vibrational energy. For this reason springs are often used in series with a neoprene, compressed glass fiber, or cork-pad-type isolator, when

TABLE 23 General Types and Applications of Vibration Isolators

			Vibration Isolation Applications—Nonspecific			
			Noncritical Locations		Critical Locations	
Isolator Type	Typical Range of Static Deflection (mm)	Corresponding Approximate Range of Natural Frequency (Hz)	Rotary Equipment	Reciprocating Equipment	Rotary Equipment	Reciprocating Equipment
Steel spring	6–12	1–6	Yes	Yes	Yes	Special
Neoprene-in-shear, double deflection	6–12	4–6	Yes	Yes	Yes	No
Neoprene-in-shear, single deflection	3–6	6–10	Yes	Yes	Yes	No
Compressed block of glass fiber, 2-in. thick	0.5–4	8–20	Yes	No	Yes	No
Neoprene pad, ribbed or waffle pattern, 1–4 layers	0.15–6	6–20	Yes	No	Yes	No
Felt or cork pads or strips	0.07–2.5	1–30	See text for applications and limitations			
Air Spring	—	1–10	See text for applications and limitations			

used to support equipment directly over critical locations in a building. Unhoused "stable" steel springs are preferred over housed unstable or stable springs. Stable steel springs have a diameter that is about 0.8–1.2 times their compressed height. They have a horizontal stiffness that is approximately equal to their vertical stiffness; therefore, they do not have a tendency to tilt sideways when a vertical load is applied.

(b) *Neoprene-in-Shear Isolators.* Neoprene is a long-lasting material that, when properly shaped, can provide good vibration isolation for the conditions shown in Table 23. The mount usually has an interior hollow space that is conically shaped. The total effect of the shaping is that for almost any direction of applied load, there is a shearing action on the cross section of neoprene. A solid block of neoprene in compression is not as effective as an isolator. Manufacturers' catalogs will show the upper limit of load-handling capability of neoprene-in-shear mounts. Two neoprene-in-shear mounts are sometimes constructed in series in the same supporting bracket to provide additional static deflection. This is the double deflection mount referred to in Table 23.

(c) *Compressed Glass Fiber.* Blocks of compressed glass fiber serve as vibration isolators when properly loaded. The manufacturers have several different densities available for a range of loading conditions. Typically, a block is about 50 mm thick and has an area of about 60–130 cm^2, but other dimensions are available. These blocks attenuate high-frequency structure-borne noise, and they are often used alone, at various spacings, to support floating concrete floor slabs. The manufacturer's data should be used to determine the density and area of a block required to achieve the desired static deflection. Unless otherwise indicated, a static deflection of about 5–10% of the uncompressed height is normal. With longtime use, the material may compress an additional 5–10% of its height. This gradual change in height must be kept in mind during the designing of floating floors to meet floor lines of structural slabs.

(d) *Ribbed Neoprene Pads.* Neoprene pads with ribbed or waffle pattern surfaces are effective as high-frequency isolators. In stacks of 2–4 thickness, with each layer separated by a metal plate, they are also used for vibration isolation of low-power rotary equipment. The pads are usually about 6–10 mm thick, and they compress by about 20% of their height when loaded at about 140–350 kPa. Higher durometer pads may be loaded up to about 700 kPa. The pads are effective as isolators because the ribs provide some shearing action, and the spaces between the ribs allow lateral expansion as an axial load is applied. The manufacturer's literature should be used for proper selection of the material (load–deflection curves, durometer, surface area, height, etc.).

(e) *Felt Pads.* Felt strips or pads are effective for reducing structure-borne sound transmission in the mounting of piping and vibrating conduit. One or more layers of 3- or 6-mm-thick strips should be wrapped around the pipe under the pipe clamps that attach the piping to building structures. Felt pads will compress under long duration and high load application and should not be used alone to vibration isolate heavy equipment.

(f) *Air Springs.* Air springs are the only practical vibration isolator for very low frequencies, down to about 1 Hz or even lower for special applications. An air mount consists of pressurized air enclosed in a resilient reinforced neoprene chamber. Since the air chamber is subject to slow leakage, a system of air mounts usually includes a pressure sensing system and external air supply. A group of air mounts can be arranged to maintain very precise leveling of a base by automatic adjustment of the pressure in the individual mounts. Specific design and operation data should be obtained from the manufacturer.

4.2 Mounting Assembly Types

In this section, five basic mounting systems are described for the vibration isolation of equipment. These mounting systems are applied to specific types of equipment in Section 4.3.

(a) *Type I Mounting Assembly.* The specified equipment should be mounted rigidly on a concrete inertia block. The length and the width of the inertia block should be at least 30% greater than the length and width of the supported equipment. Mounting brackets for stable steel springs should be located off the sides of the inertia block at or near the height of the vertical center of gravity of the combined completely assembled equipment and concrete block. The clearance between the floor and the concrete inertia block shall be at least 100 mm, and provision should be allowed to check this clearance at all points under the block. The inertia block adds stability to the system and reduces motion of the system in the vicinity of the driving frequency.

(b) *Type II Mounting Assembly.* This mount is the same as the type I mount in all respects except that the mounting brackets and the top of the steel springs should be located as high as practical on the concrete inertia block but not necessarily as high as the vertical center-of-gravity position of the assembly. The steel springs can be recessed into pockets in the concrete block, but clearances around the springs should be large enough to assure no contact between any spring and any part of the mounted assembly. Provision must be made to allow positive visual inspection of the spring clearance. When this type of mounting is used for a pump, the concrete inertia block can be given a T-shape plan, and the pipes to and from the pump can be supported rigidly with the pump onto the wings of the T. In this way, the pipe elbows will not be placed under undue stress.

(c) *Type III Mounting Assembly.* The equipment or the assembly of equipment should be mounted on a steel frame that is stiff enough to allow the entire assembly to be supported on flexible point supports without fear of distortion of the frame or misalignment of the equipment. The frame should then be mounted on resilient mounts, steel springs, neoprene-in-shear mounts, or isolation pads, as the static deflection would require. If the equipment frame itself already has adequate stiffness, no additional framing is required, and the isolation mounts may be applied directly to the base of the equipment.

(d) *Type IV Mounting Assembly.* The equipment should be mounted on an array of "pad mounts." The pads may be of compressed glass fiber or of multiple layers of ribbed neoprene or waffle pattern neoprene of sufficient height and of proper stiffness to support the load while meeting the static deflection recommended in the applicable accompanying tables.

(e) *Type V Mounting Assembly (for Propeller-Type Cooling Towers).* Large, low-speed propeller-type cooling towers located on roof decks of large buildings may produce serious vibration in their buildings if adequate vibration isolation is not provided. In extreme cases, the vibration may be evident two or three floors below the cooling towers. It is recommended that the entire cooling tower assembly be isolated on stable steel springs. The springs should be in series with at least two layers of ribbed or waffle pattern neoprene if there is an acoustically critical area immediately below the cooling tower (or within about 10 m horizontally on the floor immediately under the tower). It is necessary to provide

TABLE 24 Recommended Vibration Isolation Mounting Details for Centrifugal and Axial-Flow Fans

Equipment Conditions			Mounting Recommendations				
					Min. Defl. w/Col. Spacing		
Equipment Location	Power Range (kW)	Speed Range (rpm)	Mounting Type	Inertia Base Wt. Ratio	9 m	12 m	15 m
On grade slab		Under 600					
	Under 2	600–1200				No isolation required	
		Over 1200					
		Under 600				25 mm	
	2–19	600–1200	III	—		12 mm	
		Over 1200				12 mm	
		Under 600				40 mm	
	20–150	600–1200	III	—		25 mm	
		Over 1200				12 mm	
On upper floor above noncritical area		Under 600	III	—	25 mm	40 mm	50 mm
	Under 2	600–1200	III	—	12 mm	20 mm	25 mm
		Over 1200	III	—	12 mm	12 mm	20 mm
		Under 600	II	2	25 mm	40 mm	50 mm
	3–19	600–1200	III	—	40 mm	50 mm	75 mm
		Over 1200	III	—	25 mm	40 mm	50 mm
		Under 600	II	2	50 mm	75 mm	100 mm
	70–150	600–1200	II	2	40 mm	50 mm	75 mm
		Over 1200	II	2	25 mm	40 mm	50 mm
On upper floor above critical area		Under 600	II	2	40 mm	50 mm	75 mm
	Under 2	600–1200	III	—	40 mm	50 mm	75 mm
		Over 1200	III	—	25 mm	40 mm	50 mm
		Under 600	II	3	50 mm	75 mm	100 mm
	3–19	600–1200	II	2	40 mm	50 mm	75 mm
		Over 1200	II	2	25 mm	40 mm	50 mm
		Under 600	II	3	75 mm	100 mm	125 mm
	20–150	600–1200	II	2	50 mm	65 mm	75 mm
		Over 1200	II	2	25 mm	40 mm	50 mm

limit stops on these springs to limit movement of the tower when it is emptied and to provide limited movement under wind load.

4.3 Tables of Recommended Vibration Isolation Details

Tables 24–34 (Ref. 3, pp. 9-10–9-21) provide suggested vibration isolation for the most commonly found mechanical equipment in commercial buildings. A common format is used for all the tables that summarize the recommended vibration isolation details for the various types of equipment. Additional comments are as follows:

(a) *Centrifugal and Axial-Flow Fans.* Ducts should contain flexible connections at both the inlet and discharge of the fans, and all connections to the fan assembly should be clearly flexible. The entire assembly should bounce with little restraint when one jumps up and down on the unit. Large ducts

TABLE 25 Recommended Vibration Isolation Mounting Details for Reciprocating Compressor Refrigeration Equipment Assembly (Including Motor, Gear, or Steam Turbine Drive Unit)

Equipment Conditions			Mounting Recommendations				
	Power Range	Speed Range	Mounting	Inertia Base Wt.	Min. Defl. w/Col. Spacing		
Equipment Location	(kW)	(rpm)	Type	Ratio	9 m	12 m	15 m
On grade slab		600–900	III			50 mm	
	35–175	901–1200	III			40 mm	
		1201–2400	III			25 mm	
		600–900	II	2–3		50 mm	
	175–600	901–1200	III			50 mm	
		1201–2400	III			40 mm	
On upper floor above noncritical area		600–900	II	2–3	50 mm	75 mm	100 mm
		901–1200	II	2–3	40 mm	50 mm	70 mm
		1201–2400	II	2–3	40 mm	40 mm	50 mm
		Under 600	II	3–4	75 mm	100 mm	125 mm
		600–1200	II	3–4	50 mm	75 mm	100 mm
		Over 1200	II	2–3	50 mm	50 mm	75 mm
On upper floor above critical area		600–900	II	3–4	75 mm	100 mm	125 mm
		901–1200	II	3–4	50 mm	75 mm	100 mm
		1201–2400	II	2–3	50 mm	50 mm	75 mm
		600–900	I	4–6	75 mm	100 mm	125 mm
		901–1201	II	3–5	50 mm	75 mm	100 mm
		1201–2400	II	3–4	50 mm	50 mm	75 mm

TABLE 26 Recommended Vibration Isolation Mounting Details for Rotary Screw Compressor Refrigeration Equipment Assembly (Including Motor Drive Unit)

Equipment Conditions			Mounting Recommendations				
	Power Range	Speed Range	Mounting	Inertia Base Wt.	Min. Defl. w/Col. Spacing		
Equipment Location	(kW)	(rpm)	Type	Ratio	9 m	12 m	15 m
On grade slab	350–1700	2400–4800	III			25 mm	
On upper floor above noncritical area	350–1700	2400–4800	III		25 mm	40 mm	50 mm
On upper floor above critical area	250–1700	2400–4800	II	2–3	25 mm	40 mm	50 mm

TABLE 27 Recommended Vibration Isolation Mounting Details for Centrifugal Compressor Refrigeration Equipment Assembly[a]

Equipment Conditions			Mounting Recommendations				
	Power Range	Speed Range	Mounting	Inertia Base Wt.	Min. Defl. w/Col. Spacing		
Equipment Location	(kW)	(rpm)	Type	Ratio	9 m	12 m	15 m
On grade slab	350–1750	Over 3000	III			20 mm	
	1750–1400	Over 3000	III			20 mm	
On upper floor above noncritical area	350–1750	Over 3000	III		25 mm	40 mm	50 mm
	1750–1400	Over 3000	III		40 mm	50 mm	75 mm
On upper floor above critical area	350–1750	Over 3000	II	2–3	40 mm	50 mm	75 mm
	1750–1400	Over 3000	II	3–5	40 mm	50 mm	75 mm

[a]Including condenser and chiller tanks and motor, gear, or steam turbine drive unit.

TABLE 28 Recommended Vibration Isolation Mounting Details for Absorption-Type Refrigeration Equipment Assembly

Equipment Conditions			Mounting Recommendations				
	Cooling	Speed Range	Mounting	Inertia Base Wt.	Min. Defl. w/Col. Spacing		
Equipment Location	Capacity	(rpm)	Type	Ratio	9 m	12 m	15 m
On grade slab	All sizes		IV			6 mm	
On upper floor above noncritical area	All sizes		III		12 mm	20 mm	25 mm
On upper floor above critical area	All sizes		III		25 mm	40 mm	50 mm

TABLE 29 Recommended Vibration Isolation Mounting Details for Boilers

Equipment Conditions			Mounting Recommendations				
	Heating Capacity	Speed Range	Mounting	Inertia Base Wt.	Min. Defl. w/Col. Spacing		
Equipment Location	(bhp)	(rpm)	Type	Ratio	9 m	12 m	15 m
On grade slab	Under 200		—			Not required	
	200–1000					Not required	
	Over 1000					Not required	
On upper floor above noncritical area	Under 200		III		3 mm	6 mm	12 mm
	200–1000		III		6 mm	12 mm	25 mm
	Over 1000		III		6 mm	12 mm	25 mm
On upper floor above critical area	Under 200		III		12 mm	25 mm	40 mm
	200–1000		III		25 mm	40 mm	50 mm
	Over 1000		III		25 mm	40 mm	50 mm

(cross-section area over 1.5 m^2) that are located within about 10 m of the inlet or discharge of a large fan (over 15 kW) should be supported from the floor or ceiling with resilient mounts having a static deflection of at least 6 mm.

(b) *Reciprocating-Compressor Refrigeration Equipment.* These recommendations apply also to the drive unit used with the reciprocating compressor. Pipe connections from this assembly to other equipment should contain flexible connections.

(c) *Centrifugal-Compressor Refrigeration Equipment.* The recommended vibration isolation details for this equipment include the drive unit and the condenser and evaporator tanks.

TABLE 30 Recommended Vibration Isolation Mounting Details for Propeller-Type Cooling Towers[a]

Equipment Conditions			Mounting Recommendations		
Equipment Location	Power Range (kW)	Speed Range (rpm)	Mounting Type	Inertia Base Wt. Ratio	Min. Defl. w/Col. Spacing — All Spacings
On grade slab			Vibration Isolation Usually not Required		
On upper floor above noncritical area	Under 20	150–300		125 mm	
		301–600	v Install on	75 mm	Springs may be located
		Over 600		75 mm	
	20–100	150–300		150 mm	
		301–600	v Dunnage attached to	100 mm	Under drive assembly
		Over 600		75 mm	
	Over 100	150–300		150 mm	
		301–600	v Building columns only	125 mm	Or under tower base
		Over 600		100 mm	
On upper floor above critical area	Same as for location above noncritical area, except install ribbed or waffle pattern neoprene between tower and building				

[a]Where several towers are placed at the same general location, the power range for total power of all towers should be used.

TABLE 31 Recommended Vibration Isolation Mounting Details for Centrifugal-Type Cooling Towers[a]

Equipment Conditions			Mounting Recommendations				
					Min. Defl. w/Col. Spacing		
Equipment Location	Power Range (kW)	Speed Range (rpm)	Mounting Type	Inertia Base Wt. Ratio	9 m	12 m	15 m
On grade slab			Vibration isolation usually not required				
On upper floor above noncritical area	Under 20	450–900			25 mm	40 mm	50 mm
		901–1800	III		20 mm	50 mm	40 mm
		Over 1800			20 mm	50 mm	40 mm
	20–100	450–900			40 mm	50 mm	75 mm
		901–1800	III		25 mm	40 mm	50 mm
		Over 1800			20 mm	25 mm	40 mm
	Over 100	450–900			50 mm	75 mm	100 mm
		901–1800	III		40 mm	50 mm	75 mm
		Over 1800			25 mm	40 mm	50 mm
On upper floor above critical area	Under 20	450–900			40 mm	100 mm	75 mm
		901–1800	III		25 mm	40 mm	50 mm
		Over 1800			20 mm	25 mm	40 mm
	20–100	450–900			50 mm	75 mm	100 mm
		901–1800	III		40 mm	50 mm	75 mm
		Over 1800			25 mm	40 mm	50 mm
	Over 100	450–900			75 mm	100 mm	150 mm
		901–1800	III		40 mm	50 mm	75 mm
		Over 1800			25 mm	40 mm	50 mm

[a]Power is total of all fans at the same general location.

(d) *Boilers.* The recommended vibration isolation for boilers apply for boilers with integrally attached blowers. Table 24 should be followed for the support of blowers that are not directly mounted on the boiler. A flexible connection or a thermal expansion joint should be installed in the exhaust breaching between the boiler and the exhaust stack.

TABLE 32 Recommended Vibration Isolation Mounting Details for Motor–Pump Assemblies

Equipment Conditions			Mounting Recommendations				
					Min. Defl. w/Col. Spacing		
Equipment Location	Power Range (kW)	Speed Range (rpm)	Mounting Type	Inertia Base Wt. Ratio	9 m	12 m	15 m
On grade slab		450–900					
	Under 15	901–1800	Vibration isolation usually not required for acoustic purposes				
		1801–3600					
		450–900	II	2–3		40 mm	
	15–75	901–1800	II	1.5–2.5		25 mm	
		1801–3600	II	1.5–2.5		20 mm	
		450–900	II	2–3		50 mm	
	Over 75	901–1800	II	2–3		40 mm	
		1801–3600	II	1.5–2.5		25 mm	
On upper floor above noncritical area		450–900	II	2–3	40 mm	40 mm	75 mm
	Under 20	901–1800	II	1.5–2.5	25 mm	40 mm	50 mm
		1801–3600	II	1.5–2.5	20 mm	25 mm	40 mm
		450–900	II	2–3	40 mm	50 mm	75 mm
	20–100	901–1800	II	2–3	25 mm	40 mm	50 mm
		1801–3600	II	1.5–2.5	25 mm	40 mm	50 mm
		450–900	II	3–4	100 mm	75 mm	100 mm
	Over 100	901–1800	II	2–3	40 mm	50 mm	75 mm
		1801–3600	II	2–3	25 mm	40 mm	50 mm
On upper floor above critical area		450–900	II	3–4	40 mm	100 mm	75 mm
	Under 20	901–1800	II	2–3	25 mm	75 mm	50 mm
		1801–3600	II	2–3	20 mm	50 mm	40 mm
		450–900	II	3–4	50 mm	100 mm	100 mm
	20–100	901–1800	II	2–3	40 mm	75 mm	75 mm
		1801–3600	II	2–3	20 mm	50 mm	50 mm
		450–900	II	3–4	75 mm	100 mm	125 mm
	Over 100	901–1800	II	2–3	50 mm	75 mm	100 mm
		1801–3600	II	2–3	40 mm	50 mm	75 mm

(e) *Motor-Pump Assemblies.* Electrical connections to the motors should be made with long "floppy" lengths of flexible armored cable, and piping should be resiliently supported. For most situations, a good isolation mounting of the piping will overcome the need for flexible connections in the pipe. An important function of the concrete inertia block (type II mounting) is its stabilizing effect against undue bouncing of the pump assembly at the instant of starting. This gives better long-time protection to the associated piping. These same recommendations may be applied to other motor-driven rotary devices such as centrifugal-type air compressors and motor-generator sets in the power range up to a few hundred kilowatts.

(f) *Steam Turbines.* Table 33 provides a set of general isolation recommendations for steam-turbine-driven rotary equipment, such as gears, generators, or centrifugal-type gas compressors. The material given in Table 27 applies when a steam turbine is used to drive centrifugal-compressor refrigeration equipment. The recommendations given in Table 25 apply when a steam turbine is used to drive reciprocating-compressor refrigeration equipment or reciprocating-type gas compressors.

(g) *Gears.* When a gear is involved in a drive system, vibration isolation should be provided in accordance with recommendations given for either the main power drive unit or the driven unit, whichever imposes the more stringent isolation conditions.

(h) *Transformers.* Power leads to and from the transformers should be as flexible as possible. In outdoor locations, earth-borne vibration to nearby neighbors is usually not a problem, so no vibration isolation is suggested. If vibration should become a problem, the transformer could be installed on neoprene or compressed glass fiber pads having 6 mm static deflection.

TABLE 33 Recommended Vibration Isolation Mounting Details for Steam-Turbine-Driven Rotary Equipment, Such as Gear, Generator, or Gas Compressor[a]

Equipment Conditions			Mounting Recommendations				
	Power Range	Speed Range	Mounting	Inertia Base Wt.	Min. Defl. w/Col. Spacing		
Equipment Location	(kW)	(rpm)	Type	Ratio	9 m	12 m	15 m
On grade slab	370–110	Over 3000	III			12 mm	
	1100–3700	Over 3000	III			20 mm	
	3700–11,000	Over 3000	III			25 mm	
On upper floor above noncritical area	370–1100	Over 3000	III		25 mm	40 mm	50 mm
	1100–3700	Over 3000	III		40 mm	50 mm	75 mm
	3700–11,000	Over 3000	III		50 mm	75 mm	100 mm
On upper floor above critical area	350–1100	Over 3000	II	2–3	25 mm	40 mm	50 mm
	1100–3700	Over 3000	II	2–3	40 mm	50 mm	75 mm
	3700–1100	Over 3000	II	2–3	50 mm	75 mm	100 mm

[a]Use Table 25 for reciprocating compressor driven by steam turbine; use Table 26 for centrifugal compressor driven by steam turbine.

TABLE 34 Recommended Vibration Isolation Mounting Details for Transformers

Equipment Conditions			Mounting Recommendations				
	Power Range	Speed Range	Mounting	Inertia Base Wt.	Min. Defl. w/Col. Spacing		
Equipment Location	(kW)	(rpm)	Type	Ratio	9 m	12 m	15 m
On grade slab	Under 10	N/A	IV	N/A		3 mm	
	10–100	N/A	IV	N/A		3 mm	
	Over 100	N/A	IV	N/A		6 mm	
On upper floor above noncritical area	Under 10	N/A	IV	N/A	3 mm	6 mm	6 mm
	10–100	N/A	III	N/A	6 mm	12 mm	12 mm
	Over 100	N/A	III	N/A	6 mm	12 mm	25 mm
On upper floor above critical area	Under 10	N/A	III	N/A	6 mm	12 mm	20 mm
	10–100	N/A	III	N/A	6 mm	20 mm	25 mm
	Over 100	N/A	III	N/A	6 mm	25 mm	40 mm

[a]N/A = Not available.

(i) *Air Compressors.* Recommended mounting details for centrifugal-type air compressors of less than 75 kW are the same as those given for motor-pump units in Table 32. The same recommendations would apply for small (under 8 kW) reciprocating-type air compressors. For reciprocating-type air compressors (with more than two cylinders) in the 8–49 kW range, the recommendations given in Table 25 apply for the particular conditions. For 8–80 kW, one or two cylinder, reciprocating-type air compressors, the recommendations of Table 35 apply. This equipment is a potentially serious source of low-frequency vibration in a building if it is not isolated. In fact, the compressor should not be located in certain parts of the building, even if it is vibration isolated. When these compressors are used, all piping should contain flexible connections, and the electrical connections should be made with flexible armored cable.

4.4 Vibration Isolation: Miscellaneous

It is good practice to isolate all piping in the mechanical equipment room that is connected to vibrating equipment with resilient ceiling hangers or from floor-mounted resilient supports. As a general rule, the first three pipe supports nearest the vibrating equipment should have a static deflection of at least one-half the static deflection of the mounting system used with that equipment. Beyond the third pipe support, the static deflection can be

TABLE 35 Recommended Vibration Isolation Mounting Details for One- or Two-Cylinder Reciprocating-Type Air Compressors

Equipment Conditions			Mounting Recommendations				
	Power Range	Speed Range	Mounting	Inertia Base Wt.	Min. Defl. w/Col. Spacing		
Equipment Location	(kW)	(rpm)	Type	Ratio	10 m	12 m	15 m
On grade slab		300–600	I	4–8		100 mm	
	Under 15	601–1200	I	2–4		50 mm	
		1201–2400	I	1–2		25 mm	
		300–600	I	6–10		125 mm	
	15–75	601–1200	I	3–6		75 mm	
		1201–2400	I	2–3		40 mm	
On upper floor above		300–600	Not recommended (NO[a])				
noncritical area	Under 15	601–1200	I	3–6	100 mm	NO[a]	NO[a]
		1201–2400	I	2–3	50 mm	4 mm	NO[a]
		300–600	Not recommended				
	15–75	601–1200	Not recommended				
		1201–2400	I	3–6	75 mm	150 mm	NO[a]
On upper floor above		300–600	II	3–4	40 mm	100 mm	75 mm
critical area	Under 15	601–1200	II	2–3	25 mm	75 mm	50 mm
		1201–2400	II	2–3	20 mm	50 mm	40 mm
	15–75	300–2400			Not Recommended		

reduced to 6 or 12 mm for the remainder of the pipe run in the mechanical equipment room. When a pipe passes through the mechanical equipment room wall, a minimum 25 mm clearance should be provided between the pipe and the hole in the wall. The pipe should be supported on either side of the hole, so that the pipe does not rest on the wall. The clearance space should then be stuffed with fibrous filler material and sealed with a nonhardening caulking compound at both wall surfaces. Vertical pipe chases through a building should not be located beside acoustically critical areas. If they are located beside critical areas, pipes should be resiliently mounted from the walls of the pipe chase for a distance of at least 3 m beyond each such area, using both low-frequency and high-frequency isolation materials. Pipes to and from the cooling tower should be resiliently supported for their full length between the cooling tower and the associated mechanical equipment room. Steam pipes should be resiliently supported for their entire length of run inside the building. Resilient mounts should have a static deflection of at least 12 mm.

Whenever a steel spring isolator is used, it should be in series with a neoprene isolator. For ceiling hangers, a neoprene washer or grommet should always be included; and if the pipe hangers are near very critical areas, the hanger should be a combination hanger that contains both a steel spring and a neoprene-in-shear mount.

During inspection, the hanger rods should be checked to ensure they are not touching the sides of the isolator housing and thereby shorting-out the spring.

To be at all effective, a flexible pipe connection should have a length that is approximately 6–10 times its diameter. Tie rods should not be used to bolt the two end flanges of a flexible connection together. Flexible connections are either of the bellows type or are made up of wire-reinforced neoprene piping, sometimes fitted with an exterior braided jacket to confine the neoprene.

When a mechanical equipment room is located directly over or near a critical area, it is usually desirable to isolate most of the "nonvibrating equipment" with a simple mount made up of one or two pads of neoprene, or a 25- or 50-mm layer of compressed glass fiber. Heat exchangers, hot water heaters, water storage tanks, large ducts, and some large pipe stands may not themselves be noise sources, yet their pipes or their connections to vibrating sources transmit small amounts of vibrational energy that they then may transmit into the floor. A simple minimum isolation pad will usually prevent this noise transfer.

REFERENCES

1. *Heating Ventilating and Air-Conditioning Systems and Equipment*, American Society of Heating, Refrigerating and Air-Conditioning Engineers, Atlanta, GA. 1992. Chap-

ters 18 ("Fans"), 28 ("Boilers"), 35 ("Compressors"), 37 ("Cooling Towers"), 39 ("Centrifugal Pumps"), and 41 ("Engines and Turbine Drives").

2. R. Jorgensen (Ed.), *Fan Engineering*, 8th ed., Buffalo Forge Company, Buffalo, NY, 1983, Part II, Chapters 9–19.
3. "Noise and Vibration Control for Mechanical Equipment," Army Manual TM 5-805-4, Headquarters Department of Army, The Air Force and Navy, Washington, DC, 1983.
4. B. J. Graham and R. M. Hoover, "Fan Noise," in C. M. Harris (Ed.), *The Handbook of Acoustical Measurements and Noise Control*, 3rd ed., McGraw-Hill, New York, 1991, pp. 41.4–41.20.

PART X

ACOUSTICAL SIGNAL PROCESSING

80

INTRODUCTION

A. F. Seybert

1 INTRODUCTION

In the present context, acoustical signal processing is the analysis of sound or vibration data for the purpose of extracting information important for understanding the underlying physical mechanisms of noise or vibration. While the data are usually signals obtained from experiments via transducers, the data may also be obtained from computer simulations.

In this chapter and the two that follow, the basic concepts of acoustical signal processing are discussed. In Chapter 81, the statistical theory of signals for stationary and nonstationary signals is presented. Of primary importance here is the concept of the frequency spectrum, probably the single most important quantity that can be obtained from an acoustical signal. In Chapter 81 the frequency spectrum for periodic, random, and transient signals is discussed. Also presented in Chapter 81 are the time-domain (autocorrelation) function, the probability density function, the cross-correlation function, the coherence function, and the frequency response function.

In Chapter 82, considerations of the practical implementation of the acoustical signal processing are discussed. The need for this discussion is apparent when one considers that in practice one does not analyze continuous signals of infinite duration. Modern acoustical signal processing utilizes a finite amount of discrete (sampled) acoustical data from which all required information must be obtained. As one is now working with limited data, the information obtained becomes fuzzy, that is, one must tolerate a certain amount of error or uncertainty. Chapter 82 deals with the implications of practical realities such as signal truncation, sampling, windowing of the signal, random and bias errors, and confidence limits.

An important example of acoustical signal processing is presented in *Enc.* Ch. 102. In that chapter, the fundamentals of acoustical holography are discussed. Acoustical holography is an intensive use of acoustical signal processing techniques. Other applications of acoustical signal processing may be found in Chapter 106 (Sound Intensity).

2 AN EXAMPLE OF ACOUSTICAL SIGNAL PROCESSING

The most common type of signal in acoustics is that obtained from a measurement of sound pressure using, for example, a microphone or a hydrophone. The sound pressure fluctuation is converted to a voltage signal by the transducer and is amplified. At this point the signal may be displayed as a function of time or analyzed in any number of ways to reveal its frequency content, its amplitude distribution, and so on, depending on what information is desired.

In most applications, one is interested in the frequency content of the signal, or the *frequency spectrum*. There are two primary reasons for obtaining frequency information about a signal. First, the response of the ear and the sensation of sound in humans is strongly dependent on frequency. Second, the physical processes of sound emission, propagation, diffraction, and transmission are all frequency dependent.

In studies of sound emission, for example, it is usually very difficult to identify and quantify the contribution of individual noise sources or noise mechanisms (such as gear noise) by examining the time history of the

Note: References to chapters appearing only in the *Encyclopedia* are preceded by *Enc.*

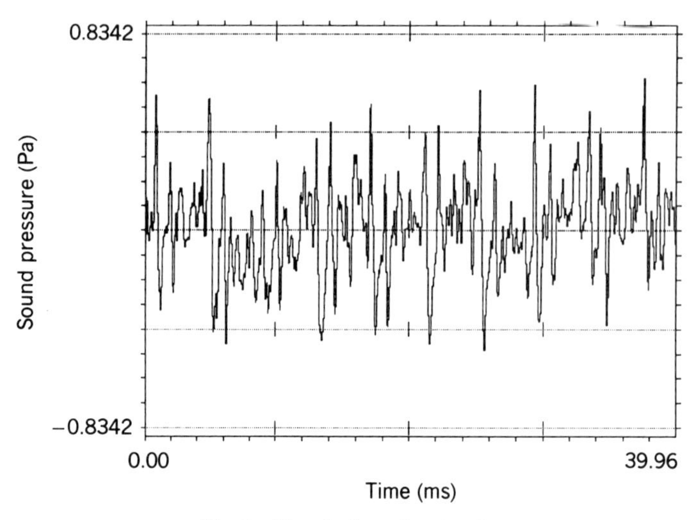

Fig. 1 Electric fan noise signal.

microphone signal. However, by analyzing the frequency spectrum of the microphone signal, a noise control engineer can identify offending components of the noise and correlate this information to underlying mechanisms of noise emission in the system under study. For example, it is well-known that most fans will produce a spectrum of noise that includes a component located at a frequency equal to the number of blades multiplied by the rotational speed of the fan in revolutions per second. This information may be used to diagnose a noise problem and assist in its solution.

Acoustic signals may be classified as either *deterministic* or *random*. Deterministic signals are those that originate from processes that either repeat (e.g., rotating machinery) or are transient (e.g., impacts or explosions). Random signals are those that originate from processes so complex that they have no apparent deterministic nature. The sound produced by a turbulent boundary layer or by high-speed airflows are typical examples of random signals.

All practical signals in acoustics are combinations of the deterministic and random signals in varying degrees. A good example is the noise produced by an electric-motor-driven fan such as found in electric appliances and

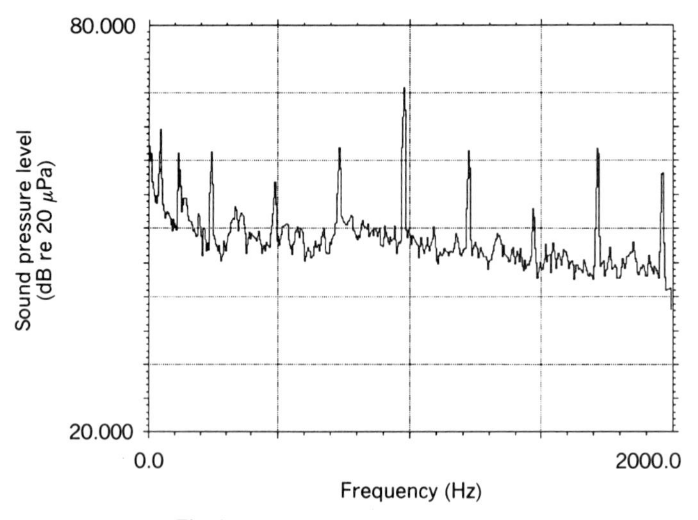

Fig. 2 Spectrum of electric fan noise.

computers. The signal from such a noise source is shown in Fig. 1. The frequency spectrum of the electric fan noise is shown in Fig. 2. Note that the spectrum consists of several distinct peaks (the deterministic part of the signal) superimposed on a broadband background (random) component. The peaks are produced by the repetitive processes of the fan, including the shaft rotational frequency (first peak, 49 Hz), the twice-line-frequency electric motor hum (second peak, 120 Hz), the blade passage frequency–shaft rotational frequency times number of fan blades (third peak at 245 Hz), and harmonics of the blade passage frequency (peaks at 490, 735, 980 Hz, etc.). The broadband background portion of the spectrum is due to the noise induced by the random turbulent flow of air through the fan and fan housing.

Signal processing is an important tool in acoustics and noise control. However, like all tools, signal processing has limitations and must be used correctly. The previous example shows that one must have an understanding of the physical mechanisms in order to accurately interpret signal processing data. Second, signal processing techniques can only analyze and display physical effects, for example, those that can be measured with a transducer, while the subjective effects of sound must be established by a skilled professional. Figure 2, for example, shows that most of the sound energy is contained in the broadband random portion of the spectrum, and this is what one would strive to reduce if the objective is to reduce the overall sound emission of the fan. However, it is the deterministic components of the signal, which contain little energy, that one must reduce to make the fan less objectionable from the standpoint of noise.

81

STATISTICAL THEORY OF ACOUSTIC SIGNALS

ALLAN G. PIERSOL

1 INTRODUCTION

This chapter details the basic statistical functions used to describe deterministic, transient, and random acoustic signals and summarizes the applications of these functions to general acoustical problems. The first section following this introduction addresses magnitude domain descriptions (average values and probability functions); the second section covers time-domain descriptions (correlation functions); and the third section details frequency-domain descriptions (spectral functions). The fourth section discusses joint-function descriptions (cross-correlation and cross-spectral functions), while the final section summarizes the statistical sampling errors associated with the estimation of the various functions from time history measurements of acoustic signals. All functions are defined in analog terms, but the equations can be readily converted to discrete terms by simply replacing the continuous signal, $x(t)$; $0 \leq t \leq T$, by the discrete-valued signal, $x(n\,\Delta t)$; $n = 1, 2, \ldots, N$, where Δt is the time interval between the discrete values and $T = N\,\Delta t$. The actual computations of the various functions are addressed in Chapter 82 and the cited references. Analysis hardware and software are covered in Part XV.

2 MAGNITUDE-DOMAIN PARAMETERS AND FUNCTIONS

A *stationary signal* is either a deterministic or random signal whose average properties do not vary with translations in time, for example, all periodic signals are stationary. Magnitude-domain descriptions of acoustic signals are generally limited to stationary acoustic signals, although the various descriptions can sometimes be applied to a nonstationary acoustic signal by assuming the signal is piecewise stationary and computing the desired parameters and functions over each of a series of contiguous, short time segments.[1] The following definitions apply to a stationary acoustic signal $x(t)$ defined over the time interval $0 \leq t \leq T$.

2.1 Mean and rms Values

In general, various different average values (moments) of a stationary signal $x(t)$ might be computed, although the *mean value* μ_x and the *standard deviation* σ_x (or the *rms value* ψ_x) are usually the only average values of interest. These average values are defined by

$$\mu_x = \lim_{T \to \infty} \frac{1}{T} \int_0^T x(t)\,dt \tag{1}$$

$$\sigma_x = \lim_{T \to \infty} \left\{ \frac{1}{T} \int_0^T [x(t) - \mu_x]^2\,dt \right\}^{1/2} \tag{2}$$

$$\psi_x = \lim_{T \to \infty} \left\{ \frac{1}{T} \int_0^T x^2(t)\,dt \right\}^{1/2} \tag{3}$$

where T is the linear averaging time. It can be shown[1] that the three quantities defined in Eqs. (1)–(3) are interrelated by

$$\psi_x^2 = \mu_x^2 + \sigma_x^2. \tag{4}$$

Hence, a knowledge of any two quantities determines the

Note: References to chapters appearing only in the *Encyclopedia* are preceded by *Enc.*

third. The mean value μ_x defines the central tendency of the signal, while the standard deviation σ_x defines the dispersion, each with the same units as the signal (usually pressure in pascals for acoustic signals). The rms value is a function of both central tendency and dispersion. For most acoustic measurements, the transducer does not sense the static or DC component of the signal. For this case, the mean value of the signal is zero ($\mu_x = 0$) and $\psi_x = \sigma_x$, as will be assumed henceforth.

The limiting operations on T in Eqs. (1)–(3) cannot be accomplished in practice. For the case of stationary deterministic (periodic) signals, there will be little or no error in the computed average values as long as either $T = iT_p$; $i = 1, 2, 3, \ldots$, or $T \gg T_p$, where T_p is the period of the signal. For stationary random signals, however, there will be a statistical sampling (random) error in the computed average values that diminishes with an increasing value of T, as well as with an increasing value of the signal bandwidth B^2. The statistical sampling errors for mean and rms value estimates are summarized in Section 7.

2.2 Probability Distribution Functions

For periodic signals, which can be described by an equation, the exact value of the signal at any future time is determined simply by substituting the time of interest into the equation. For stationary random signals, however, no equation exists, so the value of the signal at a future time can be predicted only in probability terms. The basic function used to describe the potential value of a random signal at a future time is called the *probability distribution function* $P(x)$, which defines the probability that the signal $x(t)$ will be equal to or less than a specific value x at any arbitrary time t, that is,

$$P(x) = \text{Prob}[x(t) \le x]. \tag{5}$$

From a computational viewpoint, the probability distribution function is given by

$$\lim_{T \to \infty} \frac{T_{x(t) \le x}}{T}, \tag{6}$$

where $T_{x(t) \le x}$ is the time spent by the signal during the duration T with a magnitude of $x(t) \le x$. The plot of a typical probability distribution function for a stationary random signal with a mean value of zero and a standard deviation of unity is illustrated in Fig. 1*a*. The units of the plot are probability versus instantaneous magnitude (usually pressure in pascals for acoustic signals). The interpretation of probability distribution functions is straightforward, namely, $P(x)$ defines the probability that

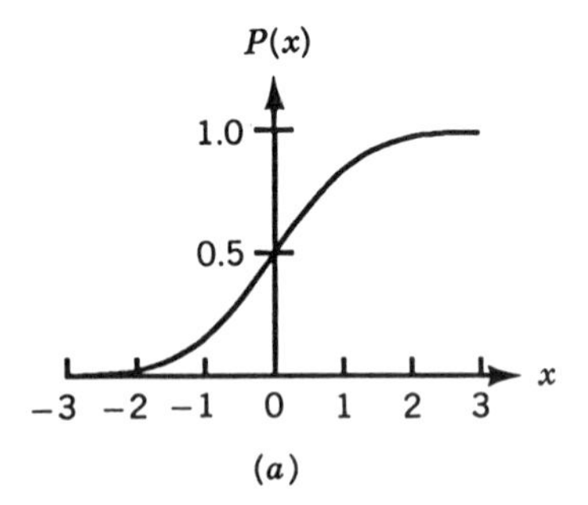

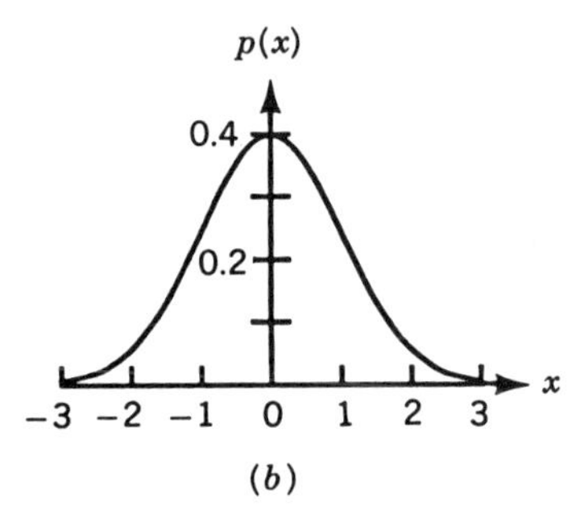

Fig. 1 Typical probability distribution and density functions for random signal ($\mu_x = 0, \sigma_x = 1$): (*a*) distribution function; (*b*) density function.

the magnitude of the signal at a future time (assuming the signal remains stationary) will have a value equal to or less than x.

2.3 Probability Density Functions

Referring to Fig. 1*a*, the value of a probability distribution function always starts at zero for magnitudes approaching $-\infty$ and ends at unity for magnitudes approaching $+\infty$; it is the manner in which the function varies from zero to unity that differentiates random signals. Hence, it is common to look at the slope of the probability distribution function, which gives the *probability density function* $p(x)$, defined as[2]

$$p(x) = \frac{dP(x)}{dx}. \tag{7}$$

From a computational viewpoint, the probability density function is given by

$$p(x) = \lim_{T \to \infty} \lim_{\Delta x \to 0} \frac{T_{x(t) \in \Delta x}}{T\,\Delta x}, \tag{8}$$

where $T_{x(t)\in\Delta x}$ is the time spent by the signal during the duration T in a narrow magnitude window Δx centered at the magnitude value x, and it is assumed that T approaches infinity faster than Δx approaches zero. The plot of the probability density function associated with the probability distribution function in Fig. 1*a* is shown in Fig. 1*b*. The units of the plot are probability density (probability per unit magnitude) versus instantaneous magnitude (usually pressure in pascals for acoustic signals). From the viewpoint of interpretations, the integral of the probability density function (the area under the function) between any two magnitude values, x_1 and x_2, defines the probability that the magnitude of the signal at a future time (assuming the signal remains stationary) will have a value between x_1 and x_2; that is,

$$\int_{x_1}^{x_2} p(x)\,dx = \text{Prob}[x_1 \le x(t) \le x_2]. \qquad (9)$$

It follows from Eq. (5) that

$$\int_{-\infty}^{x} p(x')\,dx' = P(x) \quad \text{and} \quad \int_{-\infty}^{\infty} p(x)\,dx = 1. \qquad (10)$$

Since the limiting operations in Eq. (8) cannot be accomplished in practice, the computation of probability density functions will involve both a statistical sampling error due to the finite measurement duration T, and a magnitude resolution bias error due to the finite magnitude interval Δx^2. These statistical sampling and bias errors are summarized in Section 7.

2.4 Normal (Gaussian) Probability Density Function

Due to the practical implications of the Central Limit Theorem in statistics,[3] there is a strong tendency for a wide class of stationary random signals, including random acoustic signals, to have a probability density function that is closely approximated by the *normal*, or *Gaussian, probability density function*, given by[4]

$$p(x) = \frac{1}{\sqrt{2\pi}\sigma_x} \exp\left[\frac{-(x-\mu_x)^2}{2\sigma_x^2}\right], \qquad (11)$$

where μ_x and σ_x are the mean value and standard deviation, respectively, of the signal, as defined in Eqs. (1) and (2). The probability distribution and density functions shown in Fig. 1 are of the normal form. There sometimes are significant deviations from normality in random acoustic signals, particularly when the random signals are mixed with periodic components[2] or when nonlinearities are involved in the mechanisms producing the signals,[5] which may have a direct bearing on their interpretations and applications.

3 TIME-DOMAIN FUNCTIONS

For a deterministic signal, the linear relationship between the signal values at two different times, t and $t+\tau$, can be determined directly from the equation for the signal. For stationary random signals, however, this linear relationship is defined by the *autocorrelation function* $R_{xx}(\tau)$, which is given by[2]

$$R_{xx}(\tau) = \lim_{T\to\infty} \frac{1}{T-\tau} \int_0^{T-\tau} x(t)x(t+\tau)\,dt. \qquad (12)$$

The plot of a typical autocorrelation function for a wide bandwidth random signal is shown in Fig. 2. The units of the plot are squared magnitude (usually squared pressure for acoustic signals, i.e., squared pascals) versus time displacement in seconds. For stationary random signals, the inability to accomplish the limiting operation in Eq. (12) will produce a statistical sampling error in the computed autocorrelation values,[2] as summarized in Section 7.

It is seen from Eq. (12) that autocorrelation functions are always even, that is, $R_{xx}(-\tau) = R_{xx}(\tau)$. Also, it is clear from Eq. (3) that $R_{xx}(0) = \psi_x^2$. For the usual case of acoustic signals where $\mu_x = 0$, $R_{xx}(0) = \sigma_x^2$. For sta-

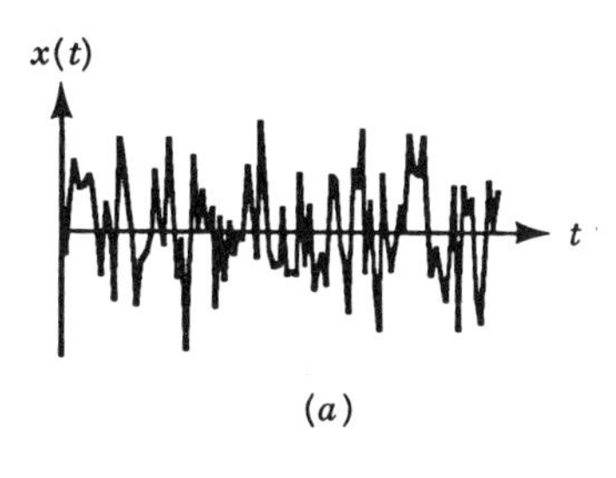

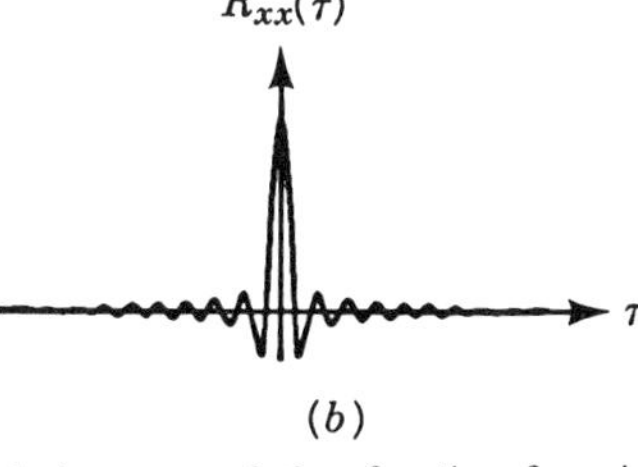

Fig. 2 Typical autocorrelation function for wide-bandwidth random signal: (*a*) signal; (*b*) autocorrelation function.

tionary random signals, the magnitude of the autocorrelation function will be less than $R_{xx}(0)$ as τ increases in either the positive or negative direction. The envelope of the autocorrelation function can be directly interpreted as a measure of how well future values of the signal can be predicted based on a knowledge of past values.[2,6] However, in most cases, the information contained in the autocorrelation function is more easily interpreted in terms of the Fourier transform of the autocorrelation function, called the power spectral density function discussed in Section 4.2.

4 FREQUENCY-DOMAIN FUNCTIONS

The most common and useful presentation for individual acoustic signals is some form of frequency decomposition. As noted in other chapters of this book, a spectral description in terms of rms values (usually expressed in decibels referenced to 20 μPa in air and 1 μPa in water) in one-third-octave frequency bands is a universal standard for the presentation of general acoustical data.[7] For detailed studies, however, a more comprehensive spectral description of acoustic signals is often required.

4.1 Spectral Description of Periodic Signals

Given a periodic signal, $x(t)$, $0 \leq t \leq T$, where $T = iT_P$, $i = 1, 2, 3, \ldots$, the frequency decomposition of the signal is given by the *linear (line) spectrum* $|P_x(f)|$, which is defined for nonnegative frequencies (a one-sided spectrum) by

$$|P_x(f)| = \begin{cases} \dfrac{2}{T}\,|X(f,T)|, & f > 0, \\ \dfrac{1}{T}\,|X(f,T)|, & f = 0, \\ 0, & f < 0, \end{cases} \qquad (13)$$

where $|X(f,T)|$ is the magnitude of the finite Fourier transform of $x(t)$, $0 \leq t \leq T$, defined by

$$\begin{aligned} X(f,T) &= \int_0^T x(t) e^{-j2\pi ft}\,dt \\ &= \int_0^T x(t)\cos(2\pi ft)\,dt \\ &\quad - j\int_0^T x(t)\sin(2\pi ft)\,dt, \\ &\quad -\infty \leq f \leq \infty. \end{aligned} \qquad (14a)$$

For a discrete-valued signal, $x(n\,\Delta t)$, $n = 0, 1, \ldots, N-1$ (it is convenient to index the signal values for Fourier transforms from $n = 0$ rather than $n = 1$),

$$X(k\,\Delta f) = \Delta t \sum_{n=0}^{N-1} x(n\,\Delta t)\exp[-j2\pi kn/N], \quad k = 0, 1, \ldots, N-1. \qquad (14b)$$

where $\Delta f = 1/(N\,\Delta t)$. The discrete-valued formulation in Eq. (14b) is usually referred to as a *discrete Fourier transform* (DFT). The DFT of a discrete-valued signal can be computed very rapidly using algorithms commonly referred to as *fast Fourier transform* (FFT) procedures,[8] as discussed in Chapter 82.

In Eq. (14), note that $X(f,T)$ or $X(n\,\Delta f)$ is defined for both positive and negative frequencies (in the discrete-valued formulation, the spectral components at $k > \frac{1}{2}N$ correspond to negative frequency components). This is why a factor of 2 appears in the linear spectrum defined for positive frequencies only in Eq. (13). A typical plot of the linear spectrum for a periodic signal is presented in Fig. 3. The units of the plot are magnitude (usually pressure in pascals for acoustic signals) versus frequency in hertz. Assuming $T = iT_P$, $i = 1, 2, 3, \ldots$, the spectrum $|P_x(f)|$ in Eq. (13) represents the magnitude of the Fourier series coefficients of $x(t)$. In most cases, the phase of the Fourier coefficients is omitted, although phase information can be retained from Eq. (14)

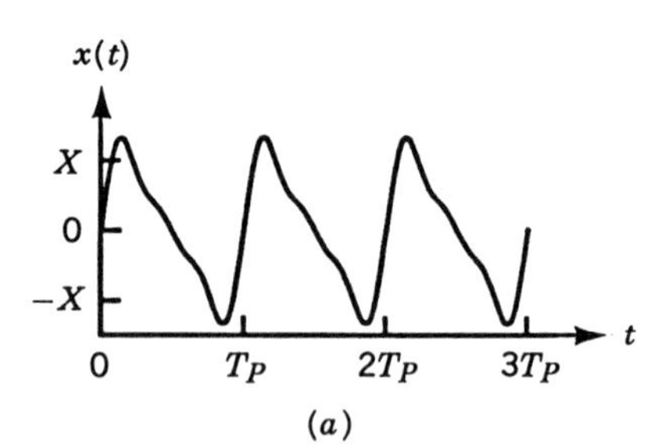

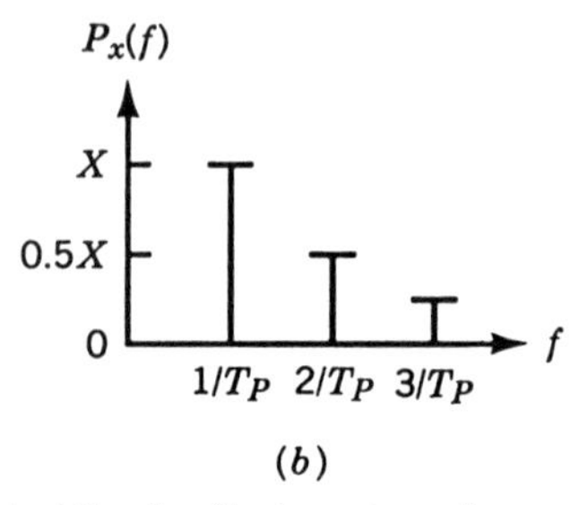

Fig. 3 Typical Fourier (line) spectrum for a periodic signal: (*a*) signal; (*b*) spectrum.

if desired (the phase for the Fourier coefficients plotted in Fig. 3 is zero).

4.2 Spectral Description of Stationary Random Signals

Given a stationary random acoustic signal $x(t)$, $0 \leq t \leq T$, the frequency decomposition of the signal is given by the *power spectral density function* $G_{xx}(f)$ (also called the *power spectrum, autospectral density function*, or *autospectrum*), which is defined for nonnegative frequencies (a one-sided spectrum) by

$$G_{xx}(f) = \begin{cases} \lim\limits_{T \to \infty} \dfrac{2}{T} E|X(f,T)|^2, & f > 0, \\ \lim\limits_{T \to \infty} \dfrac{1}{T} E|X(f,T)|^2, & f = 0, \\ 0, & f < 0, \end{cases} \tag{15}$$

where $X(f,T)$ is the finite Fourier transform defined in Eq. (14), and $E[\cdot]$ denotes the expected value, that is, an average over an infinite ensemble of squared Fourier transform computations from statistically independent blocks of data.

It can be shown[2] that the power spectral density function is equal to the Fourier transform of the autocorrelation function defined in Eq. (12); that is,

$$\begin{aligned} G_{xx}(f) &= 2\int_{-\infty}^{\infty} R_{xx}(\tau)e^{-j2\pi f\tau}\, d\tau \\ &= 2\int_{-\infty}^{\infty} R_{xx}(\tau)\cos(2\pi f\tau)\, d\tau, \\ &\qquad f \geq 0, \end{aligned} \tag{16}$$

where the factor of 2 appears because $G_{xx}(f)$ is a one-sided spectrum, and the second equality occurs because $R_{xx}(\tau)$ is an even function of τ. A typical plot of the power spectrum for a stationary random signal with a wide bandwidth is shown in Fig. 4. The units of the plot are squared magnitude per unit frequency (usually squared pressure per hertz for acoustic signals, e.g., squared pascals per hertz) versus frequency in hertz. It should be mentioned that some academic references define the power spectrum in terms of angular frequency, where the units are squared magnitude per radian per second versus radian per second.

The interpretation of the power spectrum is facilitated by two important properties. First, the integral of the power spectrum (the area under the function) between any two frequencies, f_1 and f_2, defines the mean-square value (the square of the rms value) of the signal between f_1 and f_2; that is,

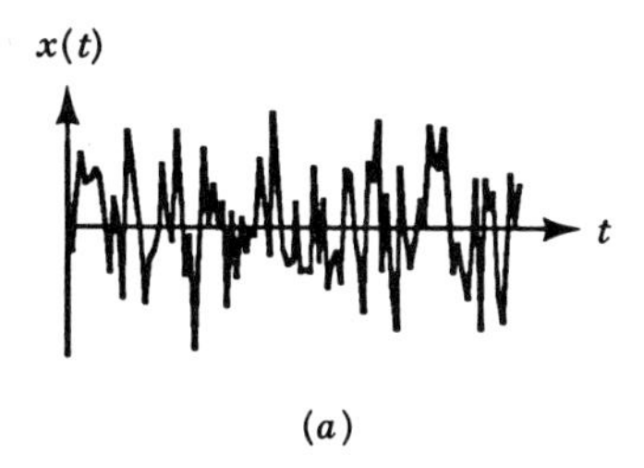

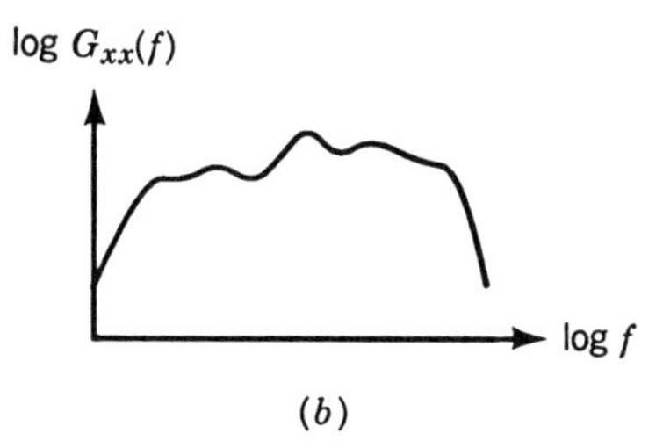

Fig. 4 Typical power spectral density function for wide-bandwidth random signal: (*a*) signal; (*b*) spectrum.

$$\int_{f_1}^{f_2} G_{xx}(f)\, df = \psi_x^2[f_1 \leq f \leq f_2], \tag{17}$$

where $\psi_x^2 = \sigma_x^2$ if either $\mu_x = 0$ or $f_1 > 0$. It follows that

$$\int_0^{\infty} G_{xx}(f)\, df = \psi_x^2 = R_{xx}(0). \tag{18}$$

Second, given two or more statistically independent signals with power spectra $G_{ii}(f)$, $i = 1, 2, \ldots, M$, the power spectrum for the sum of the signals is the sum of their individual power spectra; that is,

$$G_{xx}(f) = \sum_{i=0}^{M} G_{ii}(f). \tag{19}$$

Referring back to Eq. (15), neither the expected value operation nor the limiting operation on T can be accomplished in practice. However, the expected value can be approximated by averaging the individual power spectra computed from each of a finite number of disjoint (statistically independent) blocks of data, often created by subdividing one long measurement into n_d contiguous segments, each of finite length T (see Chapter 82). The number of segments n_d for the averaging operation determines the statistical sampling error of the resulting power spectrum estimate,[2] while the inability to accomplish the limiting operation on the duration T produces a frequency resolution bias error[2,9] in the estimate. These statistical sampling and bias errors are summarized in Section 7.

4.3 Spectral Description of Nonstationary Signals

A *nonstationary signal* is a signal that is ongoing but whose average properties vary with translations in time. A spectral representation for a nonstationary acoustic signal is sometimes generated by simply computing the Fourier transform in Eq. (14) over short, contiguous time segments, each of duration T, and plotting the results with time as the ordinate axis, frequency as the abscissa axis, and magnitude as the darkness of the plotted data point. Such plots are sometimes called *spectrograms*. A time-varying spectral representation for a nonstationary signal can also be produced by repeatedly computing a short time-averaged rms value for the output of each of a set of contiguous narrow bandpass filters.[1,6] However, a more rigorous description of the spectral content of nonstationary acoustic signals is given by the *instantaneous power spectral density function* $G_{xx}(f,t)$ (also called the *instantaneous power spectrum*), which is defined for nonnegative frequencies (a one-sided spectrum) by[2]

$$G_{xx}(f,t) = \begin{cases} 2\int_{-\infty}^{\infty} R_{xx}(\tau,t)e^{-j2\pi f\tau}\,d\tau, & f > 0, \\ \int_{-\infty}^{\infty} R_{xx}(\tau,t)e^{-j2\pi f\tau}\,d\tau, & f = 0 \\ 0, & f < 0, \end{cases} \qquad (20)$$

where $R_{xx}(\tau,t)$ is the *instantaneous autocorrelation function* given by

$$R_{xx}(\tau,t) = E\left[x\left(t - \frac{\tau}{2}\right)x\left(t + \frac{\tau}{2}\right)\right], \qquad (21)$$

where, again, $E[\cdot]$ denotes the expected value, that is, an average over an infinite ensemble of product computations from statistically independent blocks of data. For deterministic data, there are no statistical sampling errors in the product computation and, hence, the expected value operation is not needed. The instantaneous power spectrum in this case is often called the *Wigner–Ville distribution*. For random signals, some form of averaging over small time and/or frequency increments is often employed to suppress statistical sampling errors in instantaneous power spectra estimates.[1,6]

The interpretation of the instantaneous power spectrum at any time t is the same as discussed for the time-averaged power spectrum in Section 4.2, except the instantaneous power spectrum can have negative values at certain combinations of time and frequency. In all cases, however, the integral of the instantaneous power spectrum over frequency at any time yields the mean-square value of the signal at that time, that is,

$$\int_0^{\infty} G_{xx}(f,t)\,df = \psi_x^2(t). \qquad (22)$$

Furthermore, the integral of the instantaneous power spectrum over time yields

$$\int_0^{\infty} G_{xx}(f,t)\,dt = E_{xx}(f), \qquad (23)$$

where $E_{xx}(f)$ is the energy spectral density function defined in Section 4.4.

4.4 Spectral Description of Transient Signals

There are two related functions used to describe transient acoustic signals in the frequency domain; (a) the Fourier spectrum, which is appropriate for deterministic transients and (b) the energy spectral density function, which is appropriate for random transients.

Fourier Spectrum For the case of a deterministic transient signal $x(t)$, $0 \leq t \leq T$, a spectral description of the transient is provided by its *Fourier spectrum* $F_x(f)$, which is defined for nonnegative frequencies (a one-sided spectrum) by

$$F_x(f) = \begin{cases} 2X(f,T), & f > 0, \\ X(f,T), & f = 0, \\ 0, & f < 0, \end{cases} \qquad (24)$$

where $X(f,T)$ is the finite Fourier transform defined in Eq. (14). Assuming the block duration T equals or exceeds the duration of all significant values of the transient, the finite Fourier transform will essentially yield sample values of the exact Fourier transform of the transient $x(t)$. This follows because the values of the transient outside the time interval of the computation must be zero and, hence, all values of the transient are known from

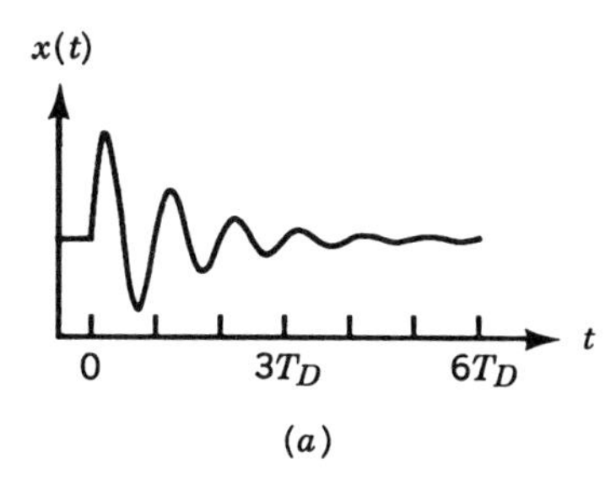

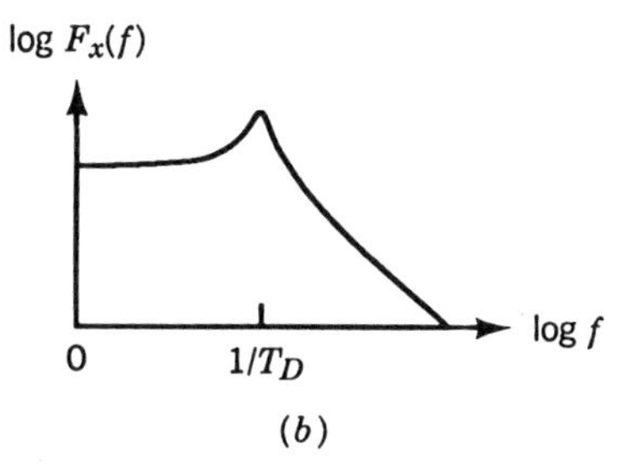

Fig. 5 Typical Fourier spectrum for transient signal: (*a*) signal; (*b*) spectrum.

minus to plus infinity. For many applications, the phase portion of the Fourier transform is ignored and only the Fourier magnitude spectrum, $|F_x(f)|$, is presented.

A typical plot of the Fourier spectrum for a deterministic transient is shown in Fig. 5. The units of the plot are the product of magnitude and time (usually pascals by seconds for acoustic signals) versus frequency in hertz. The interpretation of the Fourier spectrum for a transient signal is similar to that of a linear spectrum for the periodic signal discussed in Section 4.1, except the Fourier spectrum is theoretically continuous over all frequencies, rather than a line spectrum at discrete frequencies.

Energy Spectral Density Function For the case of a transient signal $x(t)$, $0 \le t \le T$, that is random in character, it is common to describe the frequency content of the signal in terms of an *energy spectral density function* $E_{xx}(f)$ (also called the *energy spectrum*), which is defined for nonnegative frequencies (a one-sided spectrum) by

$$E_{xx}(f) = \begin{cases} 2E[|X(f,T)|^2], & f > 0, \\ E[|X(f,T)|^2], & f = 0, \\ 0, & f < 0, \end{cases} \qquad (25)$$

where $X(f,T)$ is the finite Fourier transform defined in Eq. (14), and $E[\cdot]$ denotes the expected-value operation, as discussed for the power spectrum in Eq. (15). In this case, however, the expected value is approximated by averaging over the energy spectra computed from repeated transient measurements. As for the Fourier spectrum in Eq. (24), if the duration T in Eq. (25) equals or exceeds the duration of all significant values of the transient, the finite Fourier transform will essentially yield sample values of the exact Fourier transform of the transient $x(t)$. Plots of energy spectra often look much like the plots of power spectra, except the units of the energy spectrum plot are squared magnitude per unit frequency multiplied by time (usually squared pressure per hertz multiplied by time for acoustic signals, e.g., pascals squared by seconds per hertz) versus frequency in hertz. The interpretations of the energy spectrum and the statistical sampling error in its estimation are similar to those for the power spectrum, except the integral of an energy spectrum over frequency yields a mean-square value multiplied by time (e.g., pascals squared by seconds), rather than just a mean-square value.

5 JOINT FUNCTIONS

Many applications of acoustic data are served by the functions for individual signals discussed in Sections 2–4. For more advanced applications involving random acoustic signals, however, joint functions of two or more signals may be required. There are some applications where it is convenient to present the desired information in the time domain using cross-correlation and related functions. However, it is generally more common to work directly in the frequency domain using cross-spectral density and related functions.

5.1 Time-Domain Functions

There are three related time-domain descriptions of stationary random signals that are of common interest: (a) cross-correlation functions, (b) correlation coefficient functions, and (c) unit impulse response functions.

Cross-Correlation Functions Given two stationary random signals $x(t)$ and $y(t)$, $0 \le t \le T$, any linear relationship between these two signals will be extracted by the *cross-correlation function* $R_{xy}(\tau)$, defined by

$$R_{xy}(\tau) = \lim_{T \to \infty} \frac{1}{T - \tau} \int_0^{T-\tau} x(t)y(t+\tau)\, dt. \qquad (26)$$

A typical plot of a cross-correlation function between two correlated random signals with wide bandwidths is shown in Fig. 6. In this plot, the signals could be inter-

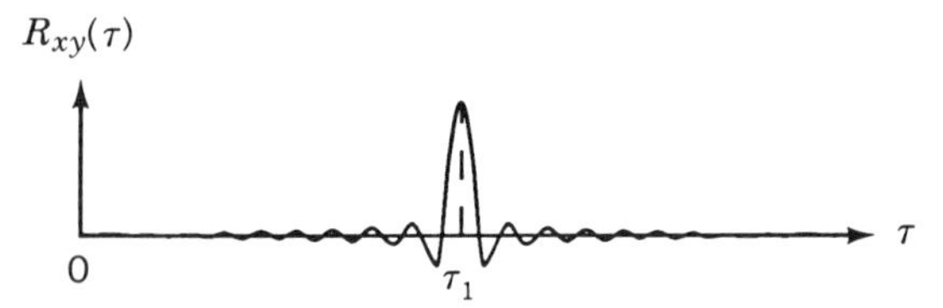

Fig. 6 Typical cross-correlation function between wide-bandwidth random signals representing measurements at two points along a propagating acoustic wave (propagation time $\tau = \tau_1$).

preted as two measurements along the path of a propagating acoustic wave, where the propagation time is τ_1 seconds. The units of the plot are the product of the magnitudes of $x(t)$ and $y(t)$ (usually pascals squared for acoustic signals) versus time displacement in seconds.

It can be shown[2] that the cross-correlation function is the inverse Fourier transform of the cross-spectral density function defined later in Eq. (28) and, hence, contains no information that is not available from a cross-spectrum. However, the time-domain character of the cross-correlation function may sometimes present information in a more desirable format for engineering evaluations. This is particularly true of applications involving time delay estimates,[6] as illustrated in Fig. 6. For stationary random signals, the inability to accomplish the limiting operation in Eq. (26) will produce a statistical sampling error in the computed cross-correlation values,[2] as summarized in Section 7.

Correlation Coefficient Functions For some applications, it is more convenient to estimate a normalized cross-correlation function, called the *correlation coefficient function* $\rho_{xy}(\tau)$, given by

$$\rho_{xy}(\tau) = \frac{R_{xy}(\tau)}{\sigma_x \sigma_y}, \tag{27}$$

where $x(t)$ and/or $y(t)$ has a zero mean value, and σ_x and σ_y are the standard deviations of $x(t)$ and $y(t)$, respectively, as defined in Eq. (2). The quantity $\rho_{xy}^2(\tau)$ is bounded by zero and unity and defines the fraction of the variance (the square of the standard deviation) of $y(t)$ that is linearly related to $x(t)$; that is, $\rho_{xy}^2(\tau) = 0$ means there is no linear relationship and $\rho_{xy}^2(\tau) = 1$ means there is a perfect linear relationship between $x(t)$ and $y(t)$ at the time displacement τ.

Unit Impulse Response Functions The *unit impulse response function* $h_{xy}(\tau)$ of a physical system is defined as the response of that system to a delta function input, that is, letting $y(t)$ be the response of a system to an input $x(t)$, $h_{xy}(\tau) = y(t)$ when $x(t) = \delta(t)$. The unit impulse response function is the inverse Fourier transform of the frequency response function defined in Section 5.2, and is more easily interpreted in that context.

5.2 Frequency-Domain Functions

There are three related frequency-domain descriptions of stationary random signals that are of common interest: (a) cross-spectral density functions, (b) coherence functions, and (c) frequency response functions.

Cross-Spectral Density Functions Given two signals $x(t)$ and $y(t)$, $0 \le t \le T$, any linear relationship between these two signals at various different frequencies will be extracted by the *cross-spectral density function* $G_{xy}(f)$ (also called the *cross-spectrum*), which is defined for nonnegative frequencies (a one-sided spectrum) by

$$G_{xy}(f) = \begin{cases} \lim_{T \to \infty} \dfrac{2}{T} E[X^*(f,T)Y(f,T)], & f > 0, \\ \lim_{T \to \infty} \dfrac{1}{T} E[X^*(f,T)Y(f,T)], & f > 0, \\ 0, & f < 0, \end{cases} \tag{28}$$

where $X^*(f,T)$ is the complex conjugate of the finite Fourier transform of $x(t)$, and $Y(f,T)$ is the finite Fourier transform of $y(t)$, as defined in Eq. (14). The expected-value notation $E[\cdot]$ has the same significance as discussed in Section 4.2. It should be mentioned that the cross-spectral density function is defined in some references as the average of $X(f,T)Y^*(f,T)$, which gives the complex conjugate of the definition in Eq. (28). Also, it can be shown[2] that the cross-spectral density function is equal to the Fourier transform of the cross-correlation function defined in Eq. (26); that is,

$$G_{xy}(f) = 2\int_{-\infty}^{\infty} R_{xy}(\tau) e^{-j2\pi f\tau}\, d\tau, \qquad f \ge 0, \tag{29a}$$

where the factor of 2 appears because $G_{xy}(f)$ is a one-sided spectrum.

The cross-spectral density function is generally a complex number that can be written as

$$G_{xy}(f) = C_{xy}(f) - jQ_{xy}(f), \tag{29b}$$

where the real part, denoted by $C_{xy}(f)$, is called the *coincident spectral density function* or *cospectrum*, and the imaginary part, denoted by $Q_{xy}(f)$, is called the *quadrature spectral density function* or *quad-spectrum*. The cross-spectrum may also be expressed in terms of magnitude and phase by

$$G_{xy}(f) = |G_{xy}(f)|e^{-j\theta_{xy}(f)}, \tag{29c}$$

where

$$|G_{xy}(f)| = [C_{xy}^2(f) + Q_{xy}^2(f)]^{1/2},$$

$$\theta_{xy}(f) = \tan^{-1}\left[\frac{Q_{xy}(f)}{C_{xy}(f)}\right]. \tag{29d}$$

A typical plot of the cross-spectral density function between two correlated random signals with wide bandwidths, having the same origin as discussed in Fig. 6, is presented in Fig. 7. The units of the plot are the product of the magnitudes of $x(t)$ and $y(t)$ per unit frequency (usually pascals squared per hertz for acoustic signals) versus frequency in hertz. The cross-spectrum may be interpreted as a measure of the linear dependence between two signals as a function of frequency, although the coherence function discussed next is more useful for this application. The statistical sampling and bias errors in cross-spectra estimates are summarized in Section 7.

Coherence Functions It is often convenient to normalize the cross-spectral density magnitude to obtain a quantity called the *coherence function* $\gamma_{xy}^2(f)$ (sometimes called coherency squared), given by

$$\gamma_{xy}^2(f) = \frac{|G_{xy}(f)|^2}{G_{xx}(f)G_{yy}(f)}, \tag{30}$$

where $G_{xy}(f)$ is the cross-spectrum defined in Eq. (28) and $G_{xx}(f)$ and $G_{yy}(f)$ are the power spectra defined in Eq. (15). The coherence function is bounded by $0 \leq \gamma_{xy}^2(f) \leq 1$ at all frequencies, and essentially identifies the fractional portion of linear dependence (or correlation) between two signals $x(t)$ and $y(t)$ as a function of frequency. Specifically, $\gamma_{xy}^2(f) = 0$ means there is no linear relationship, and $\gamma_{xy}^2(f) = 1$ means there is a perfect linear relationship between $x(t)$ and $y(t)$ at frequency f. For values between zero and unity, the coherence can be interpreted as the fractional portion of $G_{yy}(f)$ that can be determined from a knowledge of $G_{xx}(f)$. A plot of $\gamma_{xy}^2(f)$ is a dimensionless number versus frequency. The statistical sampling and bias errors in coherence function estimates are summarized in Section 7.

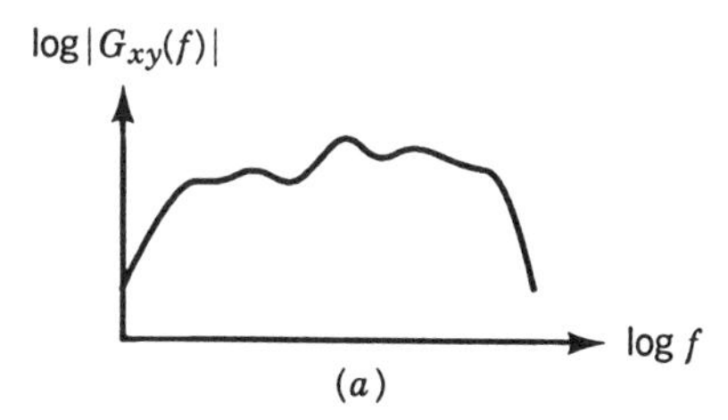

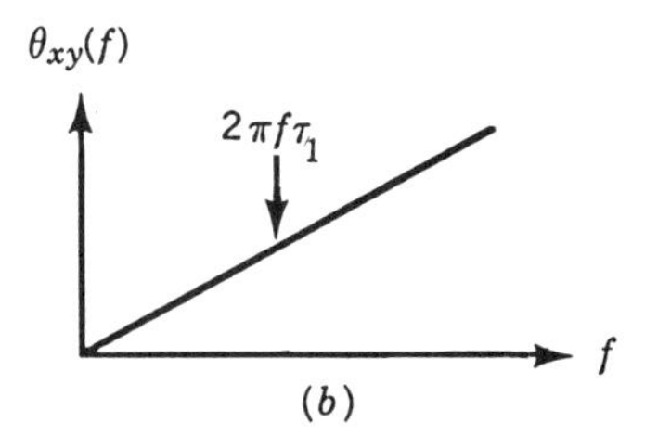

Fig. 7 Typical cross-spectral density function between wide-bandwidth random signals representing measurements at two points along a propagating acoustic wave (propagation time $\tau = \tau_1$): (*a*) magnitude; (*b*) phase.

Frequency Response Functions For two signals $x(t)$ and $y(t)$ representing the input and response, respectively, of a physical system, the ultimate goal of a cross-spectral density function computation is often the determination of a *frequency response function* $H_{xy}(f)$ for the system, which is given by[2]

$$H_{xy}(f) = \frac{G_{xy}(f)}{G_{xx}(f)}, \tag{31a}$$

where $G_{xy}(f)$ is the cross-spectrum defined in Eq. (28), and $G_{xx}(f)$ is the power spectrum defined in Eq. (15). Noting that $G_{xy}(f) = C_{xy}(f) - jQ_{xy}(f)$, a complex quantity, the frequency response function can be written in terms of a magnitude and phase, as follows:

$$H_{xy}(f) = |H_{xy}(f)|e^{-j\phi_{xy}(f)}, \tag{31b}$$

where

$$|H_{xy}(f)| = \frac{[C_{xy}^2(f) + Q_{xy}^2(f)]^{1/2}}{G_{xx}(f)},$$

$$\phi_{xy}(f) = \tan^{-1}\left[\frac{Q_{xy}(f)}{C_{xy}(f)}\right]. \tag{31c}$$

The frequency response function of a physical system is the Fourier transform of the unit impulse response of the system defined in Section 5.1; that is,

$$H_{xy}(f) = \int_0^\infty h_{xy}(\tau)e^{-j2\pi f\tau}\, d\tau. \tag{31d}$$

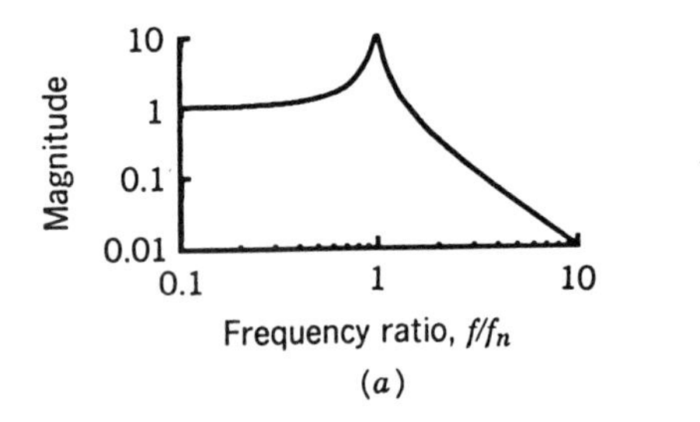

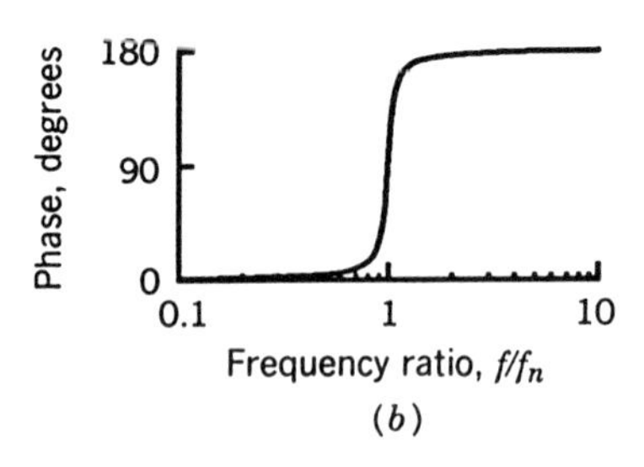

Fig. 8 Typical frequency response function for single-degree-of-freedom system ($\zeta = 0.05$): (*a*) magnitude; (*b*) phase.

A typical plot of the frequency response function for a single degree-of-freedom system with 5% damping (see Chapter 48) is shown in Fig. 8. The units of the plot are the ratio of the magnitudes of $y(t)$ over $x(t)$ (e.g., metres per newton) versus frequency in hertz. Frequency response functions are often called transfer functions, even though the transfer function is a more detailed description of the physical system given by the Laplace

TABLE 1 Summary of Estimation Errors for Common Signal Analysis Estimates

Parameter or Function, ϕ	Estimation Formula, $\hat{\phi}$	Normalized Random Error, $\epsilon_r(\hat{\phi})$	Normalized Bias Error, $\epsilon_b(\hat{\phi})$
μ_x	$\frac{1}{T}\int_0^T x(t)\,dt$	$\frac{1}{\sqrt{2BT}}\left(\frac{\sigma_x}{\mu_x}\right)$	0
$\psi_x(\mu_x = 0)$	$\left\{\frac{1}{T}\int_0^T x^2(t)\,dt\right\}^{1/2}$	$\frac{1}{2\sqrt{BT}}$	0
$p(x)$	$\frac{T_{x(t)\in\Delta x}}{T\Delta x}$	$\frac{c}{\sqrt{2BT\Delta x}}$	$\frac{\Delta x^2}{24}\left(\frac{d^2p(x)/dx^2}{p(x)}\right)$
$R_{xx}(\tau)$	$\frac{1}{T-\tau}\int_0^{T-\tau} x(t)x(t+\tau)\,d\tau$	$\frac{1}{\sqrt{2BT}}\,[1+\rho_{xx}^{-2}(\tau)]^{1/2}$	0
$G_{xx}(f)$	$\frac{2}{n_dT}\sum_{i=1}^{n_d} \lvert X_i(f,T)\rvert^2$	$\frac{1}{\sqrt{n_d}}$	$\frac{\Delta f^2}{24}\left(\frac{d^2G_{xx}(f)/df^2}{G_{xx}(f)}\right)$
$R_{xy}(\tau)$	$\frac{1}{T-\tau}\int_0^{T-\tau} x(t)y(t+\tau)\,d\tau$	$\frac{1}{\sqrt{2BT}}\,[1+\rho_{xy}^{-2}(\tau)]^{1/2}$	0
$\lvert G_{xy}(f)\rvert$	$\left\lvert\frac{2}{n_dT}\sum_{i=1}^{n_d} X_i^*(f,T)Y_i(f,T)\right\rvert$	$\frac{1}{\lvert\gamma_{xy}(f)\rvert\sqrt{n_d}}$	$\frac{\Delta f^2}{24}\left(\frac{d^2\lvert G_{xy}(f)\rvert/df^2}{G_{xy}(f)}\right)$
$\gamma_{xy}^2(f)$	$\frac{\lvert\hat{G}_{xy}(f)\rvert^2}{\hat{G}_{xx}(f)\hat{G}_{yy}(f)}$	$\frac{\sqrt{2}(1-\gamma_{xy}^2(f))}{\lvert\gamma_{xy}(f)\rvert\sqrt{n_d}}$	$\frac{(1-\gamma_{xy}^2(f))^2}{\gamma_{xy}^2(f)n_d}$
$\lvert H_{xy}(f)\rvert$	$\frac{\lvert\hat{G}_{xy}(f)\rvert}{\hat{G}_{xx}(f)}$	$\frac{(1-\gamma_{xy}^2(f))^{1/2}}{\lvert\gamma_{xy}(f)\rvert\sqrt{2n_d}}$	—[a]

Symbols:
B = frequency bandwidth of signal assuming a uniform spectrum within B, Hz
T = total duration of signal or duration of data blocks for spectral analysis,
Δf = frequency resolution bandwidth for spectral analysis ($1/T$),
n_d = number of disjoint (statistically independent) data blocks for spectral analysis
Δx = magnitude resolution window width for probability density analysis, in units of x
c = coefficient with value $0.3 < c < 1$, depending on sampling rate
[a] See Refs 2 and 9.

transform of the unit impulse response function.[2] Also, the phase factor is often defined as the negative of the value given by Eq. (31c). Frequency response functions can be interpreted as the gain and phase of a filter that produces an output $y(t)$ due to an input $x(t)$. The statistical sampling error in a frequency response function estimate is summarized in Section 7.

6 MISCELLANEOUS FUNCTIONS

Other functions that may sometimes be required to properly describe acoustic signals for specialized applications include the following:

(a) Correlated or coherent output power functions for source identification studies[2,6]
(b) Higher order moments, such as the third (skewness) moment and the fourth (kurtosis) moment, for nonlinear signal studies[5]
(c) Higher order spectral density functions (e.g., bispectra) and coherence functions for nonlinear signal studies[5]
(d) Conditioned spectral density functions, including multiple and partial coherence functions, for correlated source studies[2,6]
(e) Hilbert transforms for the generation of envelope functions and causality studies[2]
(f) Wavelets for the analysis and simulation of transients[10]
(g) Parametric spectral analysis procedures for short duration measurements[11]

7 ESTIMATION ERRORS

The definitions for the various descriptive functions of random signals in Sections 2–5 involve limiting operations that cannot be accomplished in practice. It follows that the actual computation of each of these functions will provide only an estimate of the function, which will include a statistical sampling (random) error, and in some cases a bias error as well. It is convenient to quantify these estimation errors in normalized terms; that is,

$$\text{Random error} = \epsilon_r(\hat{\phi}) = \frac{\sigma_{\hat{\sigma}}}{\phi},$$

$$\text{Bias error} = \epsilon_b = \frac{b_{\hat{\phi}}}{\phi}, \tag{32}$$

where ϕ is a function of interest, $\hat{\phi}$ is an estimate of function ϕ, $\sigma_{\hat{\phi}}$ is the standard deviation of estimate $\hat{\phi}$, and $b_{\hat{\phi}}$ is the bias of the estimate $\hat{\phi}$.

The interpretation of the normalized errors is as follows. If $\epsilon_r(\hat{\phi}) = 0.1$, the standard deviation of the estimate $\hat{\phi}$ is 10% of the function ϕ being estimated. In terms of a confidence interval, it can be said that there is about 68% confidence that the estimate $\hat{\phi}$ is within 10% of the true value ϕ, assuming the estimate is unbiased. If $\epsilon_b(\hat{\phi})$ equals +0.1 or −0.1 (bias errors carry a sign), the estimate $\hat{\phi}$ is, on the average, 10% higher or 10% lower than the function ϕ being estimated (assuming $\phi \neq 0$).

Approximate expressions for the normalized estimation errors associated with the magnitudes of the functions defined in Sections 2–5 are summarized in Table 1. All formulas in Table 1 are presented in analog terms, but may be converted to discrete terms by substituting $x(n\,\Delta t)$ for $x(t)$ and $N\,\Delta t$ for T. The formulas are generally acceptable approximations for $\epsilon < 0.2$. For those cases where $\phi = 0$, the expressions for the normalized errors in Table 1 should be multiplied by ϕ to obtain error expressions in engineering units. See Ref. 2 for more general formulas and confidence intervals.

REFERENCES

1. H. Himelblau, *Handbook for Data Acquisition and Analysis*, IES-RP-DTE012.1, Institute of Environmental Sciences, Mount Prospect, IL, 1994.
2. J. S. Bendat and A. G. Piersol, *RANDOM DATA: Analysis and Measurement Procedures*, 2nd ed., Wiley, New York, 1986.
3. M. Loeve, *Probability Theory*, 4th ed., Springer-Verlag, New York, 1977.
4. J. K. Patel and C. B. Read, *Handbook of the Normal Distribution*, Vol. 40: *Statistics: Textbooks and Monographs*, Marcel Dekker, New York, 1982.
5. J. S. Bendat, *Nonlinear System Analysis from Random Data*, Wiley, New York, 1990.
6. J. S. Bendat and A. G. Piersol, *Engineering Applications of Correlation and Spectral Analysis*, 2nd ed., Wiley, New York, 1993.
7. Anonymous, "American National Standard, Specification for Octave-Band and Fractional-Octave-Band Analog and Digital Filters," ANSI Std S1.11-1986.
8. J. W. Cooly and J. W. Tukey, "An Algorithm for the Machine Calculation of Complex Fourier Series," *Math. Comput.*, Vol. 19, 1965, pp. 297–301.
9. H. Schmidt, "Resolution Bias Errors in Spectral Density, Frequency Response and Coherence Function Estimates," *J. Sound Vib.*, Vol. 101, 1985, pp. 347–427.
10. D. E. Newland, *Random Vibrations, Spectral and Wavelet Analysis*, 3rd ed., Wiley, New York, 1993.
11. S. M. Kay, *Modern Spectral Estimation: Theory and Application*, Prentice-Hall, Englewood Cliffs, NJ, 1988.

82

PRACTICAL CONSIDERATIONS IN SIGNAL PROCESSING

JOHN C. BURGESS

1 INTRODUCTION

Practical signal processing means planning to acquire valid data and selecting methods for data reduction to achieve useful results. Practical methods usually are based on an FFT (fast Fourier transform), an efficient implementation of the DFT (discrete Fourier transform). (As used in this chapter, DFT implies computation with an FFT.) The objective of this chapter is to provide insight into methods based on the DFT and their practical use.

This chapter includes discussion of the four principal factors that must be specified:

1. The maximum frequency $f_{\max}$ for information in a signal sets the sampling rate $f_s > 2f_{\max}$ (Section 2.5).
2. The minimum acceptable data record length T_r sets the minimum acceptable DFT block size $N = T_r f_s$. Also, T_r sets the nominal spectral resolution $\Delta f = 1/T_r = f_s/N$ (Sections 2.3 and 2.7).
3. A data window may be required to control leakage (Sections 2.6 and 4.2).
4. For random signals, block averaging normally is required to achieve desired statistical stability (Section 6.4). Lag windows may also be used (Section 6.6).

Additionally, where an excitation signal can be specified, chirps and choips can be designed with the twin advantages of low peak factor and energy concentration in a desired power band.[1]

Note: References to chapters appearing only in the *Encyclopedia* are preceded by *Enc.*

There are three general methods for practical signal processing: covariance, parametric, and direct. The *covariance* method[2] has primary value for power spectrum estimation where very limited amounts of data are available (e.g., sun-spots). *Parametric* methods[3] have primary value where the amount of data is somewhat limited (e.g., quasi-stationary signals such as speech) and where models are desired.

Direct methods are based on the DFT. They can be used with limited or unlimited amounts of data and can reduce significantly the computation time required by covariance and parametric methods. However, a DFT *always* treats a single data record as if the underlying analog signal were periodic (deterministic). This means that care is required to recover correct values for periodic signal parameters (Section 4.3) and that the treatment of random signals must account for the deterministic nature of single DFTs (Section 6).

2 TRANSFORMS AND SPECTRA

Transform means any of several equations that can provide a spectrum given a signal. *Inverse transform* means the corresponding equations that can provide a signal given its spectrum. *Transform pair* means the two equations that define a transform and its inverse.

Different nomenclature has developed within three technologies—electrical engineering, mechanical engineering, and statistics—to describe equivalent signal processing artifacts (Table 1). *System*, *filter*, and *process* are synonyms. Commonly used filter synonyms are "transversal" for finite impulse response (FIR) and "recursive" for infinite impulse response (IIR).

Relationships among the four transform types used most often are summarized in Table 2. Fourier trans-

TABLE 1 Comparative Nomenclature for System Types[a]

Electrical Engineering	Mechanical Engineering	Statistics
Filter	System	Process
IIR filter (all pole)	System	AR process
FIR filter (all zero)	None	MA process
IIR filter (pole zero	System	ARMA process

[a]*Abbreviations:* IIR, infinite impulse response; FIR, finite impulse response; AR, autoregressive; MA, moving average; ARMA, autoregressive moving average.

Source: Adapted with permission from Ref. 4.

TABLE 2 Relationships among Transform Types

	Time Continuous: $x(t)$, Frequency Aperiodic	Time Discrete: $x(n)$, Frequency Periodic
Frequency continuous: $\mathbf{X}(f)$, time aperiodic	Laplace transform, Fourier transform	z-Transform
Frequency discrete: $\mathbf{X}(k)$, time periodic	Fourier series	DFT, FFT

Source: Adapted with permission from Ref. 4.

forms, Fourier series, and z-transforms are widely used for theoretical representations. Only finite Fourier transforms, Fourier series, and the DFT involve data records. Only the DFT is normally used to compute spectra.

2.1 Discrete Fourier Transform

The DFT equations normally used for signal processing,

$$x(n) = \frac{1}{N} \sum_{k=0}^{N-1} \mathbf{X}(k) e^{j2\pi kn/N}, \qquad 0 \le n \le N-1, \quad \text{(1a)}$$

$$\mathbf{X}(k) = \sum_{n=0}^{N-1} x(n) e^{-j2\pi kn/N}, \qquad 0 \le k \le N-1, \quad \text{(1b)}$$

are based on a z-transform derivation,[5,6] where N is both the record length and the DFT size. Another transform pair often encountered is based on a Fourier series derivation.[7]

An FFT is an efficient computer algorithm used to compute a DFT.[5,6] There are many such algorithms,[8] referred to collectively as "the FFT." The computation time required for an N-point FFT is approximately $N \log_2 N$, while it is approximately N^2 for a direct DFT computation. The time savings are substantial with the relatively long data records typical of acoustical applications.[9] *DFT* represents Eq. (1b), and *IDFT* represents the inverse transform Eq. (1a).

2.2 Linear Spectrum

Equations (1) can be used to define a *discrete linear spectrum*,

$$\mathbf{X}(k) = X(k) e^{j\phi(k)}, \qquad (2)$$

where $X(k) = |\mathbf{X}(k)|$ represents a *magnitude spectrum*, $\phi(k)$ represents a *phase spectrum*, and k is an integer frequency index (Section 2.3). Spectra represented by Eq. (2) are called "linear," where necessary to distinguish them from "energy" or "power" spectra. Discrete linear and power spectra are also called "discrete Fourier" and "Fourier line" spectra.

Figures 1*b* and 2 present a data record. Figures 3 and 4 show two views of the linear spectrum of the data record that appears in Figs. 1*b* and 2 (signal "seen" by a DFT). Comparison of Figs. 4 and 5 shows that the signal analyzed usually is not the true signal. (Figure 5 shows the linear spectrum of the periodic signal with period T that appears in Fig. 1*a* ("true" signal).) A major analysis objective is to estimate true signal values from the DFT values (Section 4.3). The dimensions of spectra are shown in Table 3.

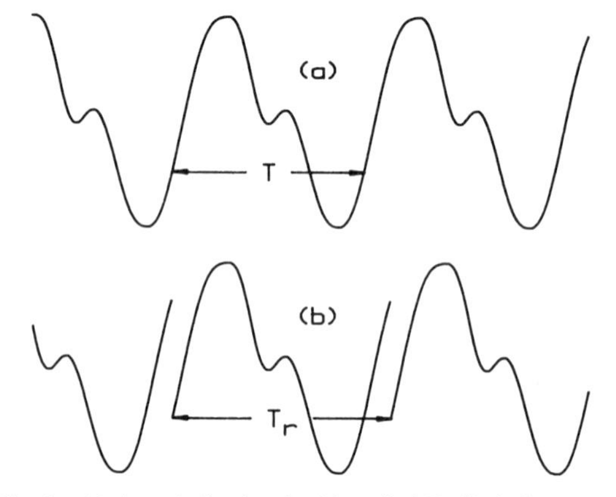

Fig. 1 (*a*) A periodic signal with period T. (*b*) A data record of length T_r and the periodic signal "seen" by repeating it.

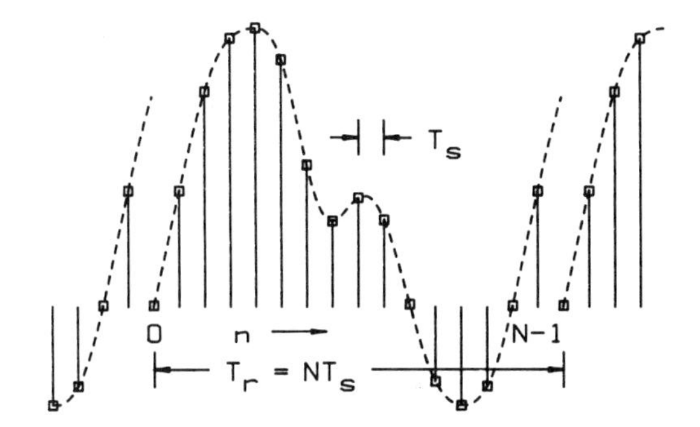

Fig. 2 Sampled periodic data record from Fig. 1*b*, $N = 16$.

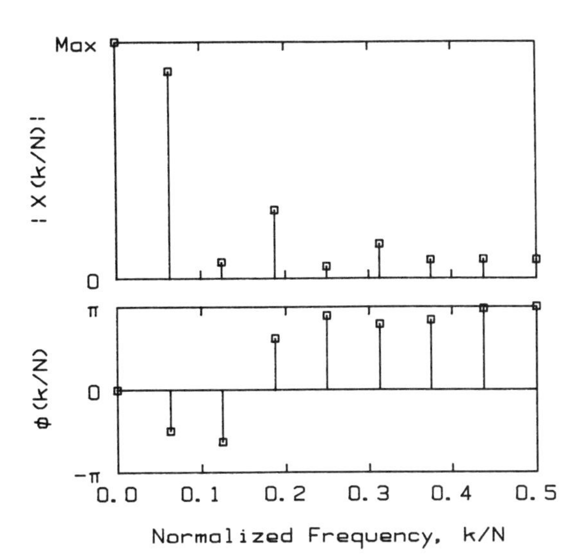

Fig. 4 Positive half of the two-sided spectrum from Fig. 3 with normalized frequency.

All spectra discussed here are "two sided," meaning that they range over both negative and positive frequencies. This has particular value when analyzing the effects of data windows. The magnitude spectrum of a physical signal is an even function of frequency, and its phase spectrum is an odd function of frequency (Fig. 3). Since the negative-frequency part of such two-sided spectra can always be obtained from the positive part, it is common practice to show only the positive half, usually with normalized frequency [Eq. (9)] as the abscissa (Fig. 4). One-sided and two-sided magnitude spectra differ by a factor of 2; their phase spectra are the same.

For practical applications, power spectra are often given as "one sided," meaning that all the power is shown in the positive-frequency range.

2.3 Signal and Spectrum Definitions

Discrete time and frequency are defined by

$$t(n) = n\,\Delta t = nT_s, \tag{3a}$$

$$f(k) = k\,\Delta f, \tag{3b}$$

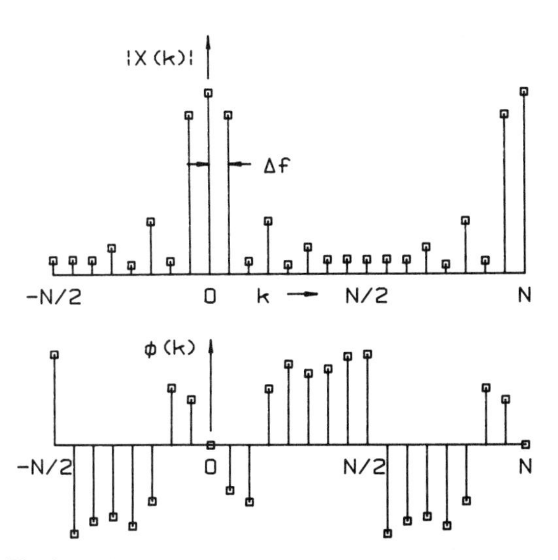

Fig. 3 Periodic discrete spectrum of the sampled data record in Fig. 2 calculated with a DFT.

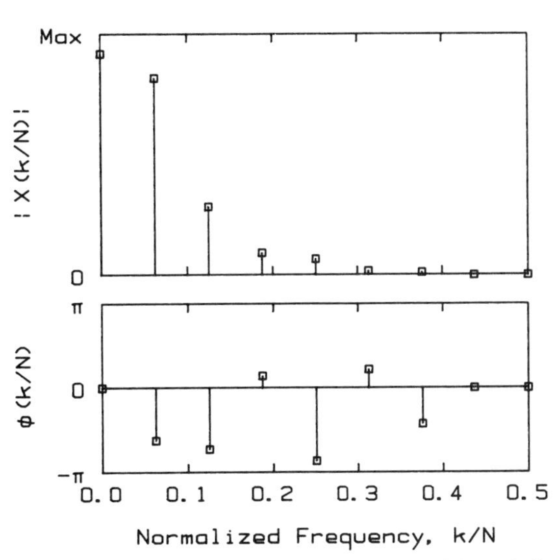

Fig. 5 Positive half of the two-sided spectrum of the periodic signal shown in Fig. 1*a* plotted with the same magnitude scale as Fig. 4.

TABLE 3 Dimensions of Linear, Energy, and Power Spectrum Magnitudes with Signal Dimensions of Volts

	Linear		Energy and Power	
Transform	Symbol	Dimensions	Symbol	Dimensions
Fourier series	$X(k)$	V	$X^2(k)$	V^2
DFT	$X(k)$	V	$X^2(k)$	V^2
z-Transform	$X(f)$	V	$X^2(f)$	V^2
Fourier transform	$X(f)$	V/Hz	$S(f)$	V^2/Hz

Source: Adapted with permission from Ref. 4.

where n and k are integers, called time and frequency *indices* (Figs. 2 and 3). The time between discrete data values, $T_s = \Delta t$, is the *sampling interval* (also called *sampling period* and *Nyquist interval*), and Δf is the spectral *frequency interval.*

The *sampling frequency* f_s (also called *sampling rate*) is the time rate at which discrete data values are taken, usually from an analog signal. The *Nyquist frequency* is defined here as

$$f_{Ny} = \tfrac{1}{2} f_s. \tag{4}$$

Nyquist rate was defined originally and is still used to mean *sampling rate*[10] (i.e., Nyquist *rate* means f_s and Nyquist *frequency* means $\frac{1}{2} f_s$. Only "Nyquist frequency" is used here. *Nyquist range* means the frequency range $-f_{Ny} \leq f \leq f_{Ny}$, or, alternately, the positive half range $0 \leq f \leq f_{Ny}$. Figures 1*b* and 2 show a signal as "seen" by a DFT. An analog data record $x(t)$ of duration T_r is taken from an original signal. The digital data record $x(n)$ of length N is related to T_r by Eq. (3a) and

$$T_r = NT_s, \tag{5}$$

where the sampling interval is

$$T_s = \frac{1}{f_s} = \frac{1}{2f_{Ny}}. \tag{6}$$

The analog data record extends from $t = 0$ to $t = T_r - \epsilon$, where ϵ is infinitesimally small, while unique discrete data values extend only from $t(0)$ to $t(N-1)$ (Fig. 2).

If T_r is measured in seconds,

$$\Delta f = f(1) = \frac{1}{T_r} = \frac{f_s}{N} \quad \text{Hz}. \tag{7}$$

The frequency corresponding to frequency index k is

$$f(k) = k\,\Delta f = \frac{kf_s}{N} \quad \text{Hz}. \tag{8}$$

The sampling frequency is often used to define a dimensionless *normalized frequency* (Figs. 4 and 5)

$$f_{\text{norm}} = \frac{f(k)}{f_s} = \frac{k}{N}. \tag{9}$$

The normalized frequency range $0 \leq f_{\text{norm}} \leq 0.5$ corresponds to the Nyquist range.

2.4 Periodicity and Circular Functions

Equations (1) make both $x(n)$ and $\mathbf{X}(k)$ periodic with period N (Figs. 2 and 3). Functions having this characteristic are called *circular.*

A computational advantage of this periodicity is that, after a data record has been acquired, sums taken over any sequential N values of $x(n)$ or $\mathbf{X}(k)$ are identical. For example,

$$\sum_{0}^{N-1} (\cdot) = \sum_{-N/2+1}^{N/2} (\cdot). \tag{10}$$

2.5 Aliasing

Shannon's information theorem[11] states that all information (signal components) in a continuous signal at frequencies $f \leq f_{Ny}$ will be preserved when the signal is sampled at the constant rate $f_s \geq 2f_{Ny}$. *Aliasing* refers to the incorrect identification of components in a continuous signal at frequencies higher than f_{Ny} as low-frequency components in the sampled signal (Fig. 6). Components at frequencies higher than f_{Ny} are "aliased" to

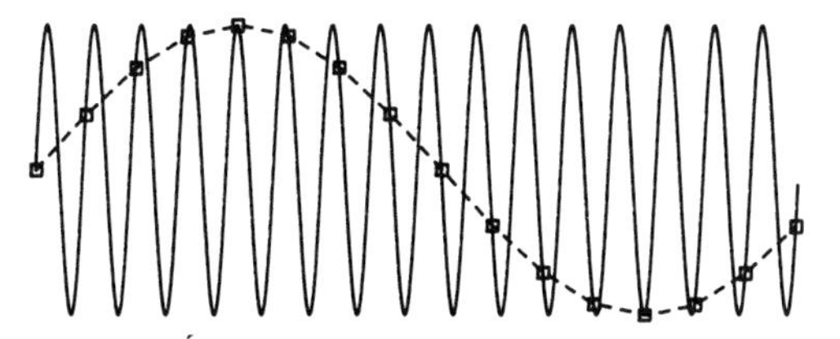

Fig. 6 An aliasing example.

lower frequencies as if the spectrum of the continuous signal were folded about the Nyquist frequency (f_{Ny} is sometimes called the "folding frequency"). For example, an 8-kHz tone sampled at a 10-kHz rate will appear as a 2-kHz signal:

> *There is only one way to prevent aliasing. Before sampling, be sure there are no components in the analog signal at frequencies higher than the Nyquist frequency.*

There are two practical ways to prevent aliasing:

- Use a sampling rate $f_s > 2f_{\max}$.
- Use an *anti-aliasing filter* (AAF) (Section 3.1).

2.6 Leakage and Data Windows

Leakage refers to spectral spreading in a transform that results from applying a transform to a finite-length data record. It does not result from sampling, and it does not depend on signal type.

Leakage is most easily observed and described with a sinusoidal signal. An infinitely long sampled signal with frequency f_m,

$$x(n) = \frac{A \sin 2\pi n f_m}{f_s}, \qquad -\infty \le n \le \infty, \tag{11}$$

has a line spectrum given by its z-transform (Fig. 7). If the signal is truncated so that only a finite length N is nonzero, each of the ideal spectral values shown in Fig. 7 is spread, or "leaked," over the entire (continuous) spectrum (Fig. 8).

The two parts of the leaked spectrum can be called[12] a *direct contribution*, centered on f_m, and an *image contribution*, centered on $-f_m$. Figure 8 appears to show the magnitudes of the spectral contributions. Actually, each contribution is twisted helically about the frequency axis, and the complex spectrum (Fig. 9) is the frequency-dependent vector sum of the two contributions. The lack of symmetry about the frequencies $\pm f_m$ in Fig. 9 is a result of leakage from direct and image contributions into each other.

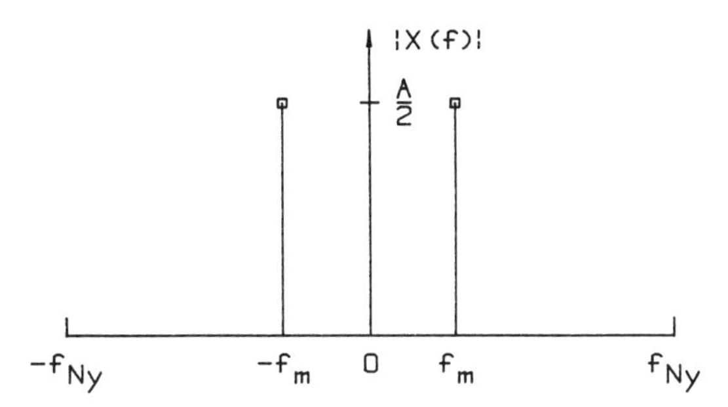

Fig. 7 Ideal spectrum of an infinite-length sinusoidal signal (z-transform); also the DFT spectrum of a sinusoidal signal that is periodic in the data record length.

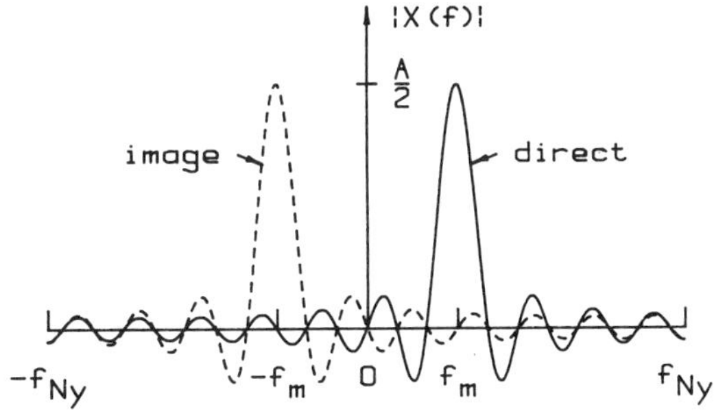

Fig. 8 Direct and image contributions to the magnitude spectrum of a finite-length sinusoidal signal (z-transform). Smearing of the spectrum is called "leakage."

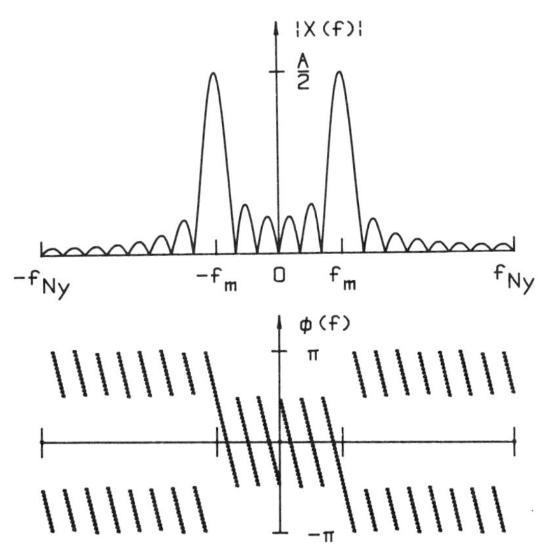

Fig. 9 Magnitude and phase spectra of a finite-length sinusoidal signal (z-transform).

Applying a z-transform to a signal truncated to finite length N is mathematically equivalent to multiplying the infinite-length signal by the weighting function

$$b(n) = \begin{cases} 1.0, & 0 \leq n \leq N-1, \\ 0, & \text{elsewhere.} \end{cases} \tag{12}$$

This function is called a *data window*. The specific data window shape given by Eq. (12) is usually called a *rectangular window* (also "gate" or "boxcar" function). As used in Eq. (12), "rectangular" has literal meaning only when used with a finite Fourier transform or a z-transform. When this window shape is used with a DFT, *open window* is more descriptive, since Eq. (12) periodically repeated looks like a constant over all time.[12] "Open" implies that everything is seen through the window that can be seen.

The spectrum of a data window is called a *spectral window*. Spectral windows typically have a *main lobe* and *side lobes* on each side of the main lobe. The main lobe of the open spectral window has a width $2/T_r$, while the side lobes have half that width (Fig. 10). Some authors use "function" or "weighting" to mean "data window" and "window" to mean "spectral window."

When a data window is used, the signal analyzed by a DFT is the product

$$\tilde{x}(n) = b(n)x(n). \tag{13a}$$

The spectrum of $\tilde{x}(n)$ is the convolution

$$\tilde{\mathbf{X}}(k) = \mathbf{B}(k) * \mathbf{X}(k). \tag{13b}$$

Thus the leaked spectrum of a truncated sinusoidal signal (Fig. 9) is equivalent to an ideal two-sided spectrum (Fig. 7) convolved with the spectrum of the open data window (Fig. 10).

Leakage means that the appearance of a DFT spectrum can be misleading. If a sinusoid is periodic in the data record length, the DFT spectrum looks like the true

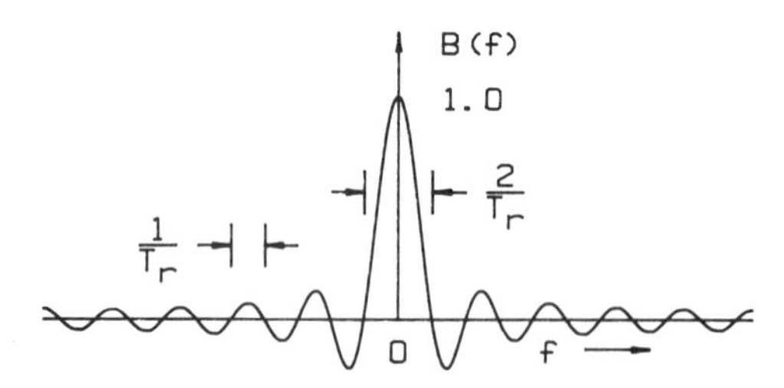

Fig. 10 Spectral window for a rectangular data window (z-transform).

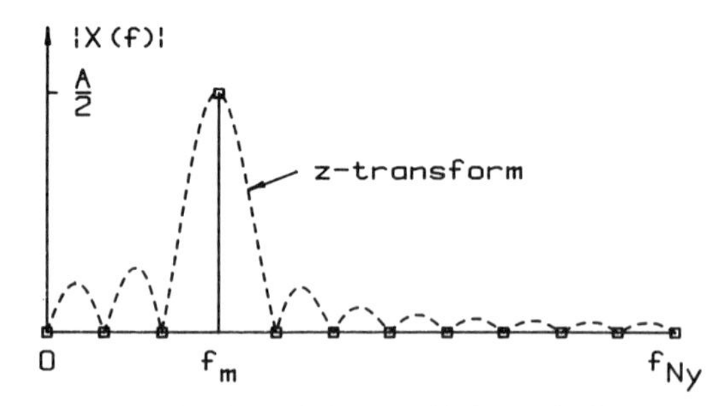

Fig. 11 Positive half of the two-sided magnitude spectrum for a data record that has exactly 3.0 periods taken from a sinusoidal signal. A z-transform (- - -) is superposed on the DFT values.

spectrum (compare Figs. 7 and 11). But leakage is not avoided; it is hidden and will appear if the spectrum is interpolated.

The usual case is that sinusoids are not periodic in the data record length. The worst case occurs when the data record has an extra half period of a sinusoidal signal[12] (Fig. 12). This effect is independent of N. While the main-lobe maximum is now hidden, since it corresponds to none of the DFT magnitudes, it can be computed (Section 4.3) or estimated by interpolation (Section 2.7).

Figure 12 also illustrates the possibility of a subtle error. When a signal has periodic components, connecting DFT magnitude and phase values with straight-line segments can imply an incorrect spectrum.

The misleading concept of "bins" is sometimes used to describe DFT components. The concept implies that a DFT acts like an array of narrow-band filters. While this has merit if the signal is band-limited white noise, is periodic in the data record length, or is a pulse captured entirely within the data record, it is not correct in the presence of leakage, which results in spectral spreading into all the bins. In particular, fractional octave band analysis can be corrupted by leakage.

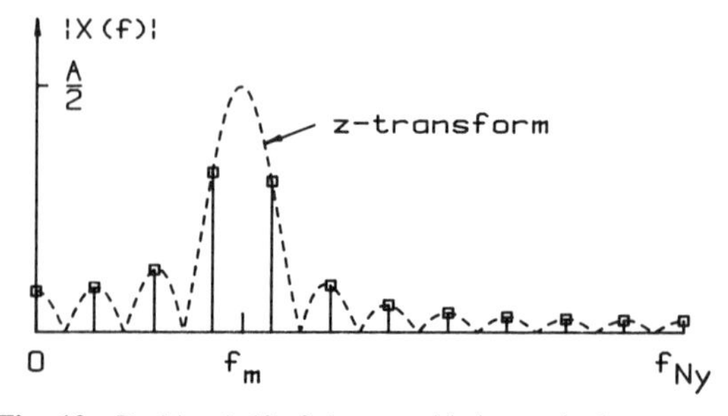

Fig. 12 Positive half of the two-sided magnitude spectrum for a data record that has 3.5 periods taken from a sinusoidal signal. A z-transform (- - -) is superposed on the DFT values.

2.7 Interpolation

Interpolation means finding values of a function between known discrete values, and it implies that a continuous function exists from which the discrete values were taken. There are two general procedures: formal interpolation and padding with zeros. Interpolation works with the data record, not the original signal. Aliasing and leakage are preserved.

Signal Interpolation A data record of size L can be interpolated to size $N = mL$, m any integer, by rewriting Eq. (1a) with Eq. (10) to get

$$x(n) = \frac{1}{N}\left\{X(0) + 2\sum_{k=1}^{L/2-1} X(k)\cos\left(\frac{2\pi kn}{N} + \phi(k)\right)\right.$$
$$\left. + X\left(\frac{L}{2}\right)\cos\left[\frac{\pi Ln}{N} + \phi\left(\frac{L}{2}\right)\right]\right\},$$
$$0 \le n \le N. \tag{14}$$

In the last term, the argument $\pi Ln/N$ must be $\pm\pi$, and the phase angle $\phi(\frac{1}{2}L)$ is required, since it may be $\pm\pi$ (e.g., Fig. 4). Signal interpolation is also called *signal reconstruction*.[13] Only if $X(\frac{1}{2}L) = 0$ can interpolation in the time domain also be accomplished by padding with zeros in the frequency domain (evenly spaced about $\frac{1}{2}L \rightarrow \frac{1}{2}N$).

Spectrum Interpolation Calculating a DFT by using Eq. (1b) is equivalent to sampling a continuous z-transform at regular frequency intervals.[5] This means that a data record has a continuous spectrum and that details of this spectrum can be revealed by interpolating the discrete DFT spectrum. Formal interpolation requires summing a sequence of basis functions [see Ref. 6, Eq. (2.142)]. A simpler way is to pad the data record with zeros (append zero values to the end of the data record).

3 DIGITAL MEASUREMENT SYSTEM

Figure 13 represents the major components in a typical digital data acquisition and analysis system.

- A time-varying physical phenomenon of any type, $p(t)$, is sensed by a transducer assumed to have an electrical output linearly proportional to $p(t)$. With transducers having a high electrical impedance, such as condenser microphones and piezoelectric accelerometers, considerable care is required in setting up the required instrumentation to avoid problems resulting from line capacitance, stray electromagnetic noise, and ground loops.[14,15]
- Signal conditioning implies electronic devices to assure that an adequate voltage $x(t)$ is created proportional to $p(t)$ (e.g., analog-to-digital converters normally require input voltages in the range 10 mV to 10 V).
- Normally, $x(t)$ is further conditioned by an anti-aliasing filter (AAF, see Section 3.1).
- The analog signal $x(t)$ is then converted to a digital signal $x(n)$ by an analog-to-digital converter (A/D, see Section 3.2).
- The data can then be stored for later processing or used on-line (e.g., for adaptive sound control).

Every step of the data acquisition process from the physical signal $p(t)$ to the digital signal $x(n)$ is subject to error. The signal sensed by the transducer is not identical to the signal that would exist in the absence of the transducer (e.g., refraction effects at a microphone, mass loading of a structure by an accelerometer). Transducers, signal conditioning, AAFs, and A/Ds can introduce magnitude and phase distortion. Sample-and-hold (S/H) modules in the A/Ds are subject to amplitude and timing errors (e.g., jitter: small variations in the sampling interval that depend on signal amplitude[10]). Quantization and granularity errors result from the A/D process. Harmonic distortion can be introduced by any analog process, including the analog front end of an A/D converter.

Taken as a whole, these measurement errors limit the effective dynamic range it is possible to achieve and therefore limit the nominal resolution it is reasonable to specify for an A/D converter. A 16-bit normal resolution is near the optimum for most practical purposes.[10] The effective resolution can be improved by oversampling (Section 3.3).

3.1 Anti-Aliasing Filters

There is some art in selecting an AAF. Its magnitude, phase, and transient response characteristics must be considered relative to the signal being acquired.

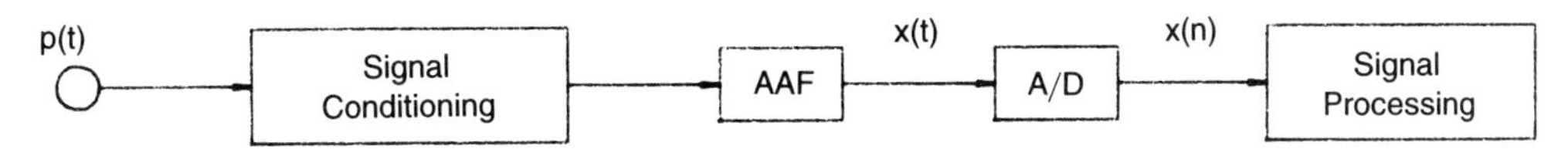

Fig. 13 Data flow in a typical digital data acquisition system.

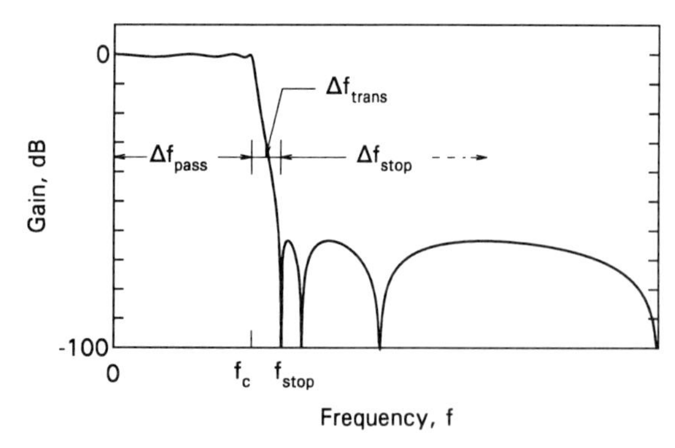

Fig. 14 Magnitude response (gain) of a typical anti-aliasing filter, showing a passband, transition band, and stopband.[13]

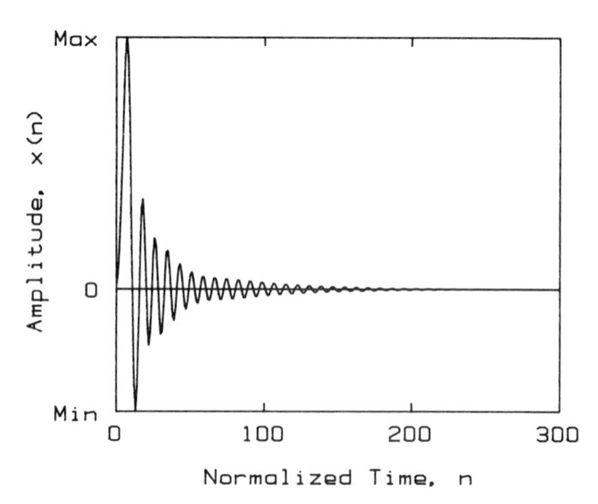

Fig. 16 Impulse response of the anti-aliasing filter of Figs. 14 and 15.

Magnitude Response An AAF's magnitude response is characterized by a passband Δf_{pass}, a transition region (also called "guard band") Δf_{trans}, and a stopband Δf_{stop} (Fig. 14). Both the passband and the stopband normally have a "ripple" (Figs. 14 and 15), meaning that the gain varies between upper and lower limits within the band. The upper frequency limit of the passband, f_c (cuttoff frequency), and the lower frequency limit of the stopband, f_{stop}, are defined as shown in Figs. 14 and 15. An optimum AAF should fulfill the conditions $f_{\max} \leq f_c$ and $f_s \geq 2f_{\text{stop}}$, where $f_{\max}$ is the highest frequency at which information is desired in the signal.

For use with stationary signals, flat response in the passband (usually low ripple) and rapid decay in the transition region are desirable criteria. These criteria are adequate if the application is not sensitive to phase distortion. An AAF's transient response (Fig. 16) must die out before a stationary input–output relationship can exist.

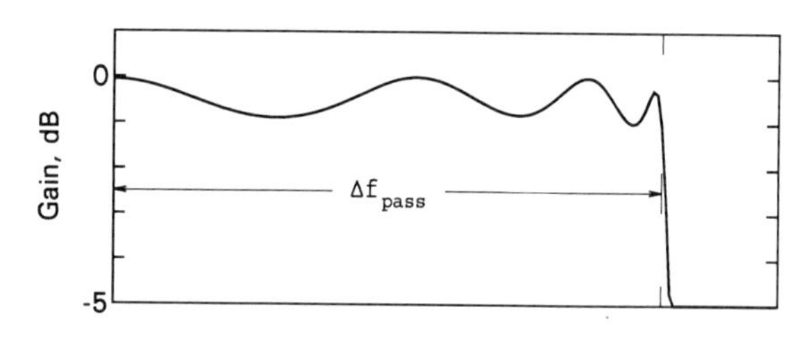

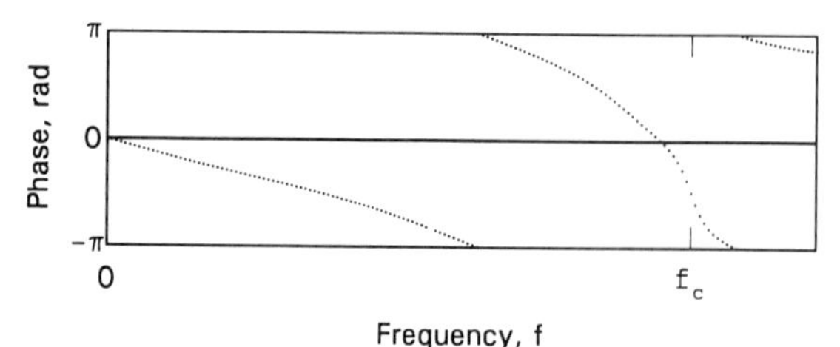

Fig. 15 Magnitude and phase response in the passband of the anti-aliasing filter of Fig. 14.

Phase Response Generally, AAFs have poor phase response, typically being nonlinear over a large part of the passband (Fig. 15). The resulting phase distortion can affect the shape of a signal. When preservation of signal shape is important, deconvolution (Section 5.3), oversampling (Section 3.3), or an all-pass filter[16] can be used to decrease phase distortion. If the signal is minimum phase, a correct signal phase can be re-created by using a Hilbert transform.[17]

Transient Response If the signal to be acquired is a transient (pulse), the AAF's output will be the convolution of the signal and the impulse response of the AAF (Fig. 16). There are three ways to decrease the effects of AAF transient response on a signal: (1) use a Bessel filter,[18] (2) deconvolve the AAF impulse response from the AAF output (Section 5.3), and (3) oversample (Section 3.3).

3.2 Analog-to-Digital Converter

An A/D converter is designed to convert an analog signal $x(t)$ into a digital signal $x(n)$ at a constant sampling rate f_s. Each discrete signal value $x(n)$ is stored in computer memory as the bits of a digital word. The signal $x(t)$ is acquired by a sample-and-hold (S/H) module, which operates under computer control to hold snapshot values of $x(t)$ long enough for the A/D converter to set all the bits it is designed to resolve and to transfer them to a digital word.

For data acquisition in acoustics, A/D converters usually have a hardware resolution of 12, 14, or 16 bits, called *nominal resolution* here. When 1 bit is used as a sign bit, *actual resolution* is reduced by 1 bit. The *dynamic range* of the digitizing process is set by quantizing noise. If the quantizing noise is uniformly distributed over the Nyquist range (band-limited white noise), the maximum dynamic range can be expressed in terms of a signal-to-noise ratio[19]:

$$(\mathrm{S/N})_{\max} \simeq 6.02(M + 0.5L) + 1.76\ \mathrm{dB}, \qquad (15)$$

where M is the actual resolution of an A/D converter in bits and L is the number (not necessarily integer) of octaves of oversampling.[19] For example, if the actual resolution is $16 - 1 = 15$ bits and there is no oversampling, $(\mathrm{S/N})_{\max} = 92$ dB. Dynamic range can be adversely affected by signal level and form (Section 3.3). It may be improved by oversampling or, if noisy, by averaging (Section 6.7).

3.3 Oversampling and Resolution Enhancement

Oversampling means using a sampling rate intentionally greater than the minimum required to prevent aliasing. As a general rule, it is advisable to operate an A/D converter at the highest possible sampling rate, so that L in Eq. (15) is large. Oversampling provides a number of practical benefits.[10,18,19] The three results most important to users are (1) AAF requirements can be relaxed, (2) dynamic range can be increased, and (3) phase response can be improved. The greater the oversampling, the less will be the need for expensive AAFs. A simple *RC* filter or even no filter can often suffice.

Oversampling increases the maximum dynamic range (i.e., effective resolution) of an A/D converter. Equation (15) implies an increase of 3 dB in $(\mathrm{S/N})_{\max}$ for each doubling of f_s beyond $(f_s)_{\min}$. The actual increase is signal dependent. The ideal signal is broadband with amplitudes that use all the bits of the converter. Doubling the sampling rate then halves the power of the quantization noise in the original Nyquist range. The ideal result is degraded if the signal is low frequency or activates only the lowest bits of the converter. Adding a dither (small-amplitude, high-frequency signal) to $x(t)$ can improve the resulting spectrum for such nonideal signals.[19] The effective S/N is decreased by 6 dB for each bit not used.

4 SPECTRAL ANALYSIS OF SIGNALS

The three classes of signal usually encountered are pulse, periodic, and random. There is no aliasing if $f_s \geq 2f_{\max}$. The two major analysis problems encountered concern leakage and noise. Spectral analysis of pulse and periodic signals is discussed in this section. Spectral analysis with noise and spectral analysis of random signals are discussed in Section 6.

4.1 Pulse Signals

A pulse is defined as a signal with a finite time duration T_p. There is no leakage if $T_r \geq T_p$. Except for the effects of noise, a DFT then gives a complete and accurate spectrum of the pulse. The DFT can be interpolated as desired to disclose the fine structure of the spectrum (Section 2.7). Useful broadband pulses for system identification include chirps[1] and a finite-length pulse taken from a white band-limited random process.

4.2 Stationary Signals, Windows, and Leakage Control

Almost always, leakage (Section 2.6) will corrupt the DFT spectrum of a stationary signal. *Leakage control* is the process by which spectral side-lobe amplitudes resulting from finite-length data records are decreased by using a shaped data window. Some authors use "tapering" to describe this process. While there seems to be a semi-infinite number of data windows,[20] very few are needed to control leakage. These few are expressed by a Fourier series,

$$b(n) = \sum_{i=0}^{M-1} B_i \cos \frac{2\pi i n}{N}, \qquad 0 \leq n \leq N - 1, \qquad (16)$$

where i is an integer, M is the number of (one-sided) Fourier coefficients, and N is even. Only four coefficients are needed to reduce side-lobe amplitudes by 98 dB.[21,22]

The open window is a one-coefficient window ($M = 1$) defined by $B_0 = 1$. Two well-known two-coefficient windows ($M = 2$) are the Hann (often called Hanning) and the Hamming windows (Table 4). Hamming window coefficients are often shortened to $B_0 = 0.54$, $B_1 = -0.46$.

Figures 17*a–d* show normalized spectra for several windows. They illustrate the three most important aspects of window spectra (Table 4): main-lobe width, side-lobe depression, and rate of side-lobe decay. Main-lobe width generally increases as side lobes are depressed and as side-lobe decay rate is increased. The practical advantage of selecting a data window with side lobes depressed into a signal's noise level is that the significant effects of leakage are then concentrated in a frequency band equal to the main-lobe width.

For many applications in acoustics, it is desirable to use data windows having the narrowest possible

TABLE 4 One-Sided Fourier Series Coefficients for Several Data Windows[a]

	SAR (dB)	S (k)	SLDR (dB/octate)	B_0	B_1	B_2	B_3
Open	−13.46	1.626	6	1.0			
Hann	−31.47	3.743	18	0.5	−0.5		
Hamming	−43.19	3.846	6	0.53836	−0.46164	0.0	
Opt59	−59.72	5.053	6	0.461445	−0.491728	0.046827	0.0
Opt71	−71.48	5.907	6	0.424379	−0.497340	0.078281	0.0
Opt98	−98.17	7.929	6	0.363579	−0.489179	0.136602	−0.010640

[a]Coefficients have signs that place the data window maximum at $n = \frac{1}{2}N$ and are normalized so that $\sum(-1)^i B_i = 1$. The selectivity amplitude ratio (SAR) is the ratio of the highest side-lobe to main-lobe amplitudes. Selectivity (S) is the width of the main lobe at the SAR. SLDR is the side-lobe decay rate. Values of B_i given as 0.0 are zero in fact ("most efficient" windows); those not given are unknown but taken to be zero.

Source: Adapted with permission from Ref. 21.

main lobe for a specified side-lobe depression. While Dolph–Chebyshev windows have this ideal characteristic, they have the disadvantage that N different coefficients must be computed for each value of side-lobe depression and for each data record size N.

The narrow main lobes of Dolph–Chebyshev windows can be closely approximated by optimized windows described by Eq. (16) with $M \leq 4$ coefficients (Figs. 17c, d). Fourier series coefficients B_i are given in Table 4 for several "most efficient" optimum windows.[21]

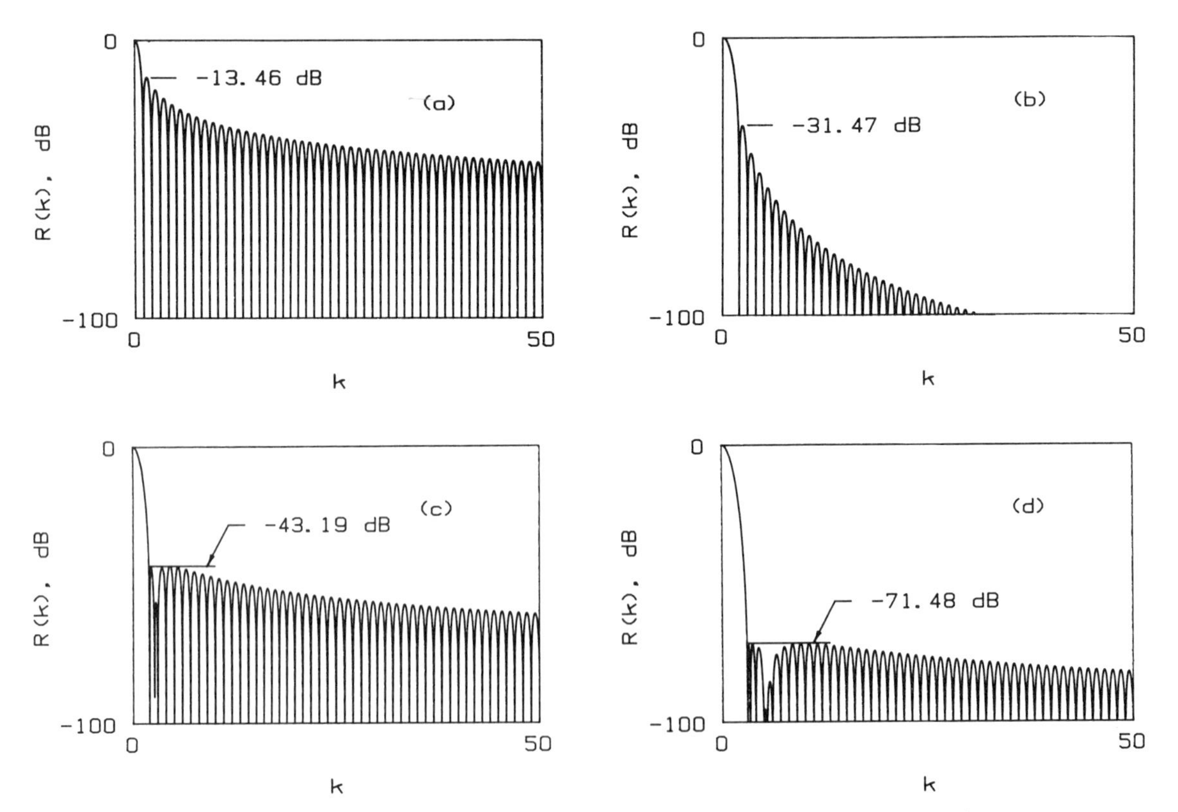

Fig. 17 Normalized spectral windows for (a) open (rectangular) data window, (b) Hann data window, (c) Hamming data window, and (d) Opt71 data window (Table 4).

4.3 Accurate Estimation of Sinusoidal Signal Parameters

It is often desirable to estimate accurate values for the amplitude, frequency, and phase of one or more sinusoids in a signal using a DFT. This requires only two complex DFT values, called *principal values*, for each sinusoid, one on each side of the frequency at which the main-lobe maximum is expected. Computation details are available.[12,23]

5 SYSTEM IDENTIFICATION

System means an analog linear, time-invariant (LTI) or discrete linear shift-invariant (LSI) operation that transforms an input signal into an output signal (Fig. 18). "System" does not refer to hardware, but to the operations performed by hardware or software.

The practical problem is to model a real LSI system with a computer-simulated LSI model using experimentally measured data. This usually means representing a distributed-parameter physical system by a lumped-parameter model.

5.1 System Identification

The operation of a discrete system (model) can be represented by the backward-difference equation

$$\sum_{i=0}^{N-1} A(i)y(n-i) = \sum_{i=0}^{M-1} B(i)x(n-i), \qquad (17)$$

where $x(\cdot)$ and $y(\cdot)$ are input and output sequences indexed backward from present values $x(n)$ and $y(n)$ and

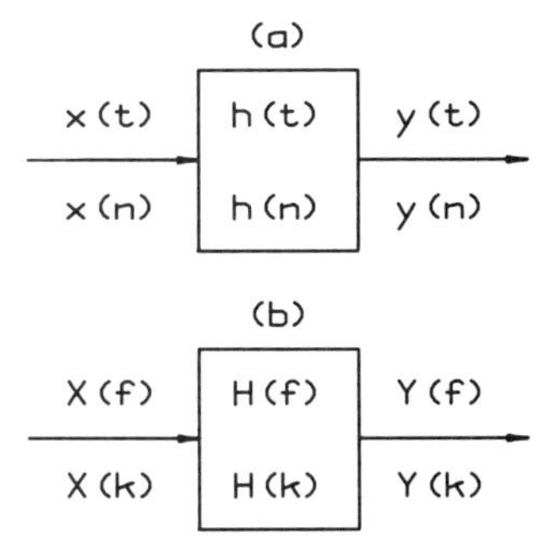

Fig. 18 Block diagrams of input/output relationships for analog (LTI) and digital (LSI) systems in the (*a*) time and (*b*) frequency domains.

M and N represent the number of coefficients (sequence lengths). The corresponding system function is[5] (z-transform)

$$H(z) = \frac{\sum_{i=0}^{M-1} B(i)z^{-i}}{\sum_{i=0}^{N-1} A(i)z^{-i}}, \qquad (18)$$

where the system gain is $G = B(0)/A(0)$. The $M-1$ roots of the numerator polynomial are called the *zeros* of the system, and the $N-1$ roots of the denominator polynomial are called the *poles* of the system. Complex-conjugate poles define resonances, and complex-conjugate zeros define antiresonances.

System identification often means estimating numerical values for the system parameters $B(i)$ and/or $A(i)$ for use in an FIR or IIR model.[3,24,25] A system can have many parametric models, and the number of parameters and their values can be wildly different (e.g., an FIR model of a one-coefficient IIR system may need 20–100 coefficients). A system can be identified also by its impulse response, frequency response, or pole–zero values. The four ways are equivalent.

A system is *causal* if it cannot produce an output prior to receiving an input (it cannot predict). All real physical systems are causal. A causal discrete system is *stable* if all its poles are inside the unit circle. A causal, stable discrete system is *minimum phase* if all its zeros are inside the unit circle.

The (continuous) frequency response is given by evaluating Eq. (18) on the unit circle, $z = e^{j\theta}$, $\theta = 2\pi k/N$, $k/N = f_{\text{norm}}$, where N can be selected to provide any Δf. If N is selected for an FFT, fast deconvolution (Section 5.3) and interpolation can be used.

5.2 Convolution

Convolution is the process by which an LTI or LSI system transforms an input $x(\cdot)$ into an output $y(\cdot)$ (Fig. 18). For an LSI system initially at rest or after transient response has disappeared, convolution is defined by the sequence

$$y(n) = \sum_{i=0}^{n} x(i)h(n-i), \qquad n \geq 0, \qquad (19)$$

or, equivalently, by the *response function*

$$\mathbf{Y}(k) = \mathbf{H}(k)\mathbf{X}(k). \qquad (20)$$

When a DFT is used, Eq. (20) results in a circular response function $\mathbf{Y}^c(k)$. A circular convolution sequence is given by

$$y^c(n) = \text{IDFT}\{\mathbf{Y}^c(k)\}. \tag{21}$$

The process is called *fast convolution* when an FFT is used.

When signals are applied to physical systems, the process is called *linear convolution*. While circular convolution does not usually give the same results as linear convolution, a linear convolution can often be recovered (Section 5.4).

5.3 Deconvolution

Deconvolution means undoing the effects of convolution. It can be used to estimate $x(n)$ or $h(n)$. Both are implemented in the frequency domain by rearranging Eq. (20). Deconvolution can be implemented also by cepstral analysis (Ref. 5, Chap. 12). The effects of noise can be reduced (Section 6.7).

System Input Signal Estimation A system's input can be estimated when its output and impulse responses are known. From Eq. (20),

$$\mathbf{X}(k) = \frac{\mathbf{Y}(k)}{\mathbf{H}(k)}, \tag{21a}$$

$$x(n) = \text{IDFT}\{\mathbf{X}(k)\}. \tag{21b}$$

Deconvolution, in seeking an input given an output, amounts to prediction, a noncausal process, and it is therefore impossible to implement with analog systems.

System Impulse Response Estimation A system's frequency response and impulse response can be expressed by

$$\mathbf{H}(k) = \frac{\mathbf{Y}(k)}{\mathbf{X}(k)}, \tag{22a}$$

$$h(n) = \text{IDFT}\{\mathbf{H}(k)\}. \tag{22b}$$

Equations (22) are most useful with broadband deterministic signals and low measurement system noise levels. Chirps and choips[1] are particularly useful, although care must be taken to avoid an excitation data record (not necessarily the true signal) with mean value zero or near to it. A DC bias can always be added.

Phase computed with a DFT may not be unique, since DFT phase has only the range $-\pi \le \phi \le \pi$. For example, a true phase 2.5π will be computed as 0.5π. Recovering the true phase is called *phase unwrapping*.[5,26]

Deconvolution involves inverse filtering.[17] If a filter has any finite zeros, they become poles of its inverse, $H_{\text{if}}(k) = 1/H(k)$. The practical result is that a data record length N satisfactory for $H(k)$ may not be long enough to adequately represent $H_{\text{if}}(k)$. Errors can result if the inverse filter has small spectral values, especially if they are affected by noise.

5.4 Linear Convolution and Correlation

Special care is required to realize a linear convolution or correlation when using the intrinsically circular DFT operation. The general procedure is to assure that at least one of the two sequences has an adequate number of zero values.

If the sequences are pulses of lengths L and P, data record lengths $N \ge L + P - 1 + D$ (Fig. 19) will result in a linear realization, where D represents output delay relative to input.

With continuing waveforms, truncation and zero padding are required.

A typical objective is to convolve a long sequence with a short sequence (Fig. 20). For example, $h(n)$ may represent an impulse response of length P. If a data record $x(n)$ can be taken with length $N > P$, a linear convolution $y(n)$ of length $N - P$ can be realized by setting to zero the last P values of $x(n)$ to get $x'(n)$, then convolving $x'(n)$ with $h(n)$. When the original $x(n)$ cannot be entirely contained within a single record length,

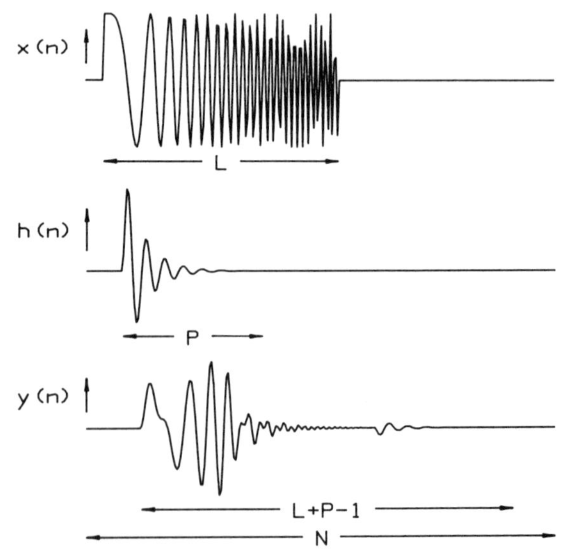

Fig. 19 Length relationships among a pulse excitation $x(n)$ (chirp), a system impulse response $h(n)$, and a system response $y(n)$ to allow linear convolution/deconvolution to be recovered from circular convolution/deconvolution.

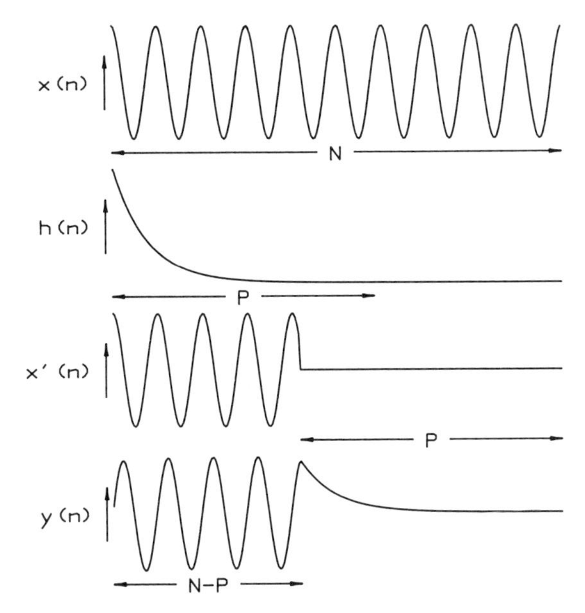

Fig. 20 Length relationships between a data record from an extended excitation $x(n)$ and a system impulse response $h(n)$ to allow recovery of a linear convolution $y(n)$ from circular convolution.

an extended $y(n)$ can be found using overlap-save or overlap-add methods.[5,6]

Another typical objective is to find the linear autocorrelation sequence $r_{xx}(\tau)$ of a signal $x(n)$ (Fig. 21). The procedure is to create a signal $y(n)$ by setting to zero the last P values of $x(n)$, calculate the energy spectrum

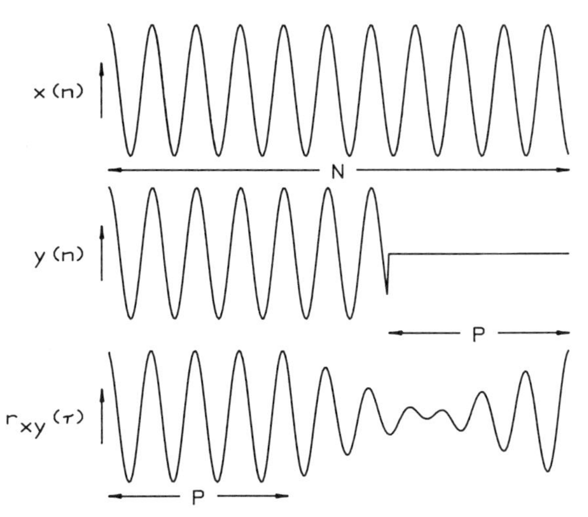

Fig. 21 Length relationships for a data record to allow recovery of a linear correlation from a circular correlation.

$\mathbf{S}^c_{xy}(k) = \mathbf{X}^{c*}(k)\mathbf{Y}^c(k)$ (or power spectrum, Section 6.3), and take the IDFT to get $r^c_{xy}(\tau)$. The first P values of $r^c_{xy}(\tau)$ are identical to those of the linear $r_{xx}(\tau)$. Since an autocorrelation sequence is even $[r_{xx}(-\tau) = r_{xx}(\tau)]$, the result can be used to get a linear autocorrelation for $|\tau| \leq P$. A linear cross-correlation sequence can be calculated in a similar way, with care taken to calculate separately for $\tau < 0$ and $\tau > 0$, since cross-correlation sequences usually are not even. When $x(n)$ is a random signal, $P \geq \tau_{cl}$ and averaging are necessary (Section 6).

6 NOISE AND RANDOM SIGNALS

Analysis objectives for random-signal components include reducing the effects of unwanted noise, finding the frequency ranges where signal power is significant, and identifying systems. The practical problem is to achieve analysis objectives using a DFT, a method that treats each data record as if the underlying process were periodic.

Random-signal analysis requires working with bias and random errors. Emphasis in this section is placed on limiting the effects of random error. The procedures discussed are based on first- and second-order statistics (mean, variance, correlation), stationary random processes, and single-input, single-output (SISO) systems. Higher order statistics,[27] nonstationary signals,[28–30] and multiple input–output systems[9] are discussed elsewhere. Much information is available concerning random-signal analysis.[2,3,9,13,31,32]

6.1 Ideal Random Signal

An *ideal* random signal $x(n)$ is defined here as stationary, with a band-limited white spectrum, a normal amplitude distribution, and an infinite duration. A major effect of this assumption is to provide a base measure for the amount of random data to acquire.

Random signal means that $x(n)$ is a single realization, $-\infty \leq n \leq \infty$, from a *random process* (also called *stochastic process*).

White means that $x(n)$ has a constant-magnitude power spectrum. It means also that $x(n)$ is an *independent* random variable, that no value of the sequence is influenced by any other. *Band limited* means that $x(n)$ has power only in the Nyquist range. When practical random signals do not originate in a white process, they can often be regarded as the result of passing a white random signal through a linear filter. *Practical* also means finite-length data records and sample spectra. Sample spectra taken from a white process will not be white, and spurious spectral peaks and valleys can be nasty sources of interpretational error (Ref. 2, Chap. 7).

Stationary means that the statistical moments of the process do not change with time. For practical purposes, a stationary random process may be assumed to be *ergodic*, meaning that its ensemble moments and time moments have the same descriptive properties. *Weakly stationary* means that only the first and second moments (mean, variance) are stationary. *Strongly stationary* means that all moments are stationary. Practical random signals are not stationary, but are often nearly so over finite time intervals.

Normal means that a normal (Gaussian) probability density function (pdf) describes the amplitude distribution of $x(n)$. The mean, variance, standard deviation, and pdf of $x(n)$ are represented by μ, σ^2, σ, and $p(x)$ (Fig. 22*a*). Conventional practice is to replace $x(n)$, by a *standardized normal random variable*

$$z - x - \frac{\mu}{\sigma}, \tag{23}$$

where z has a zero mean, unit variance, and pdf $p(z)$. Tables of values for $p(z)$ are widely available (e.g., see Ref. 33). While second-order analysis is adequate for normal processes, most real-world signals are not normal. Much additional information can be extracted by using higher order moments.[27]

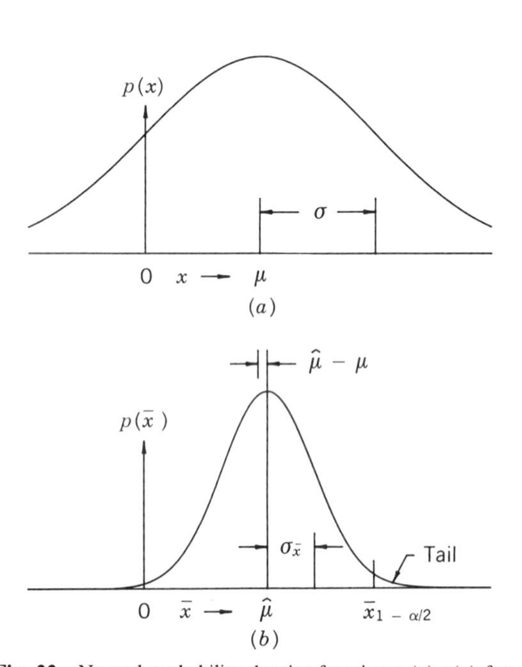

Fig. 22 Normal probability density functions: (*a*) $p(x)$ for a random variable x; (*b*) $p(\bar{x})$ for the sampling distribution of the mean.

6.2 Degrees of Freedom

Averaging is the key to achieving (statistical) *stability*, meaning the relative believability (error or precision) of an estimate. Whether random signals are averaged in the time or frequency domains, it is essential to acquire data having a sufficient number of (statistical) *degrees of freedom* ν, meaning the number of statistically independent random variables.

When enough data can be acquired to assure $\nu \gtrsim 30$, *large-sampling theory*[31] can be used in planning for data acquisition. Otherwise, *small- (exact) sampling theory* may be necessary. The emphasis in this section is on large-sampling theory.

6.3 Correlation and the Power Spectrum

Random signals can be represented by their correlation sequences or their power spectra. Normally, a power autocorrelation sequence dies out with increasing lag time τ [τ is defined in Eqs. (24)]. A useful parameter is the *correlation length* τ_{cl}, defined as the value of τ for which $r(\tau)$ decays to ϵ, an arbitrarily small quantity (e.g., system noise level).

Linear *power autocorrelation* and *power cross-correlation* sequences can be defined by

$$r_{xx}(\tau) = \frac{1}{L - \tau_{cl} + 1} \sum_{l=0}^{L-\tau_{cl}} x(l)x(l+\tau),$$

$$0 \le \tau \le \tau_{cl} - 1, \tag{24a}$$

$$r_{xy}(\tau) = \frac{1}{L - \tau_{cl} + 1} \sum_{l=0}^{L-\tau_{cl}} x(l)y(l+\tau),$$

$$0 \le \tau \le \tau_{cl} - 1, \tag{24b}$$

where the amount of data required is at least $L \ge \tau_{cl}$ and $L \gg \tau_{cl}$ represents averaging necessary to improve stability. Equations (24) provide equal stability over the entire range of τ.

Circular *discrete power spectra*

$$\mathbf{S}^c_{xx}(k) = \frac{1}{N} \mathbf{X}^*(k)\mathbf{X}(k) = \frac{1}{N} |\mathbf{X}(k)|^2,$$

$$-\frac{N}{2} + 1 \le k \le \frac{N}{2}, \tag{25a}$$

$$\mathbf{S}^c_{xy}(k) = \frac{1}{N} \mathbf{X}^*(k)\mathbf{Y}(k),$$

$$-\frac{N}{2} + 1 \le k \le \frac{N}{2} \tag{25b}$$

are calculated using DFTs, where the asterisk means complex conjugate. These spectra, computed from single data records $x(\cdot)$ and $y(\cdot)$, are called *periodograms*. In effect, a periodogram is a *sample spectrum*. When the Fourier components $\mathbf{S}^c(k)$ are independent random variables (ideal random process), each has only $\nu = 2$ degrees of freedom for $1 \le |k| \le \frac{1}{2}N - 1$ and $\nu = 1$ degree of freedom for $|k| = 0, \frac{1}{2}N$. The stability of a periodogram is notoriously poor, its standard deviation being equal to its mean, regardless of size N [Eq. (32)].

Average circular *power density spectra* can be estimated by dividing the discrete spectral values by the frequency interval, $\mathbf{S}^c(k)/\Delta f$.

Circular *power correlation* sequences

$$r_{xx}^c(\tau) = \mathrm{IDFT}\{\mathbf{S}_{xx}^c(k)\}, \tag{26a}$$

$$r_{xy}^c(\tau) = \mathrm{IDFT}\{\mathbf{S}_{xy}^c(k)\} \tag{26b}$$

can be calculated from Eqs. (25). Linear sequences can be recovered (Section 5.4).

6.4 Block Averaging: Sample Mean, Variance, and Spectrum

When a DFT is used, analysis objectives for stability (i.e., degrees of freedom, confidence[14]) can be accomplished by *block averaging*, meaning to average across M data records or spectra, each of size N.

Time-Domain Averaging Averaging in the time domain is done over the M block values $x_m(n)$. The sequences of N unbiased and consistent estimators are

$$\overline{x}(n) = \frac{1}{M} \sum_{m=0}^{M-1} x_m(n), \qquad 0 \le n \le N-1, \tag{27a}$$

$$s^2(n) = \frac{1}{M-1} \sum_{m=0}^{M-1} [x_m(n) - \overline{x}(n)]^2, \qquad 0 \le n \le N-1, \tag{27b}$$

where the N *sample means* $\overline{x}(n)$ are estimates of μ and the N *sample variances* $s^2(n)$ are estimates of $\sigma^2(n)$.

Data values in the sequence $x_m(n)$ are usually at least N data values apart in the sequence $x(n)$, and the sequence $x_m(n)$ will be uncorrelated if the correlation length $\tau_{cl} < N$. If the effects of correlation are not important, the sequences $\overline{x}(n)$ and $s^2(n)$ can be further averaged over N to obtain sample mean values $\overline{x}$ and s^2 over $M \times N$ data values.

Power-Domain Averaging Averaging in the power domain means averaging sample spectra. This results in a sequence of N averaged estimates of spectral values,

$$\overline{\mathbf{S}}(k) = \frac{1}{M} \sum_{m=0}^{M-1} \mathbf{S}_m(k), \qquad 0 \le k \le N-1, \tag{28}$$

where $\mathbf{S}_m(k)$ is the kth spectral component from the mth periodogram and the $\overline{\mathbf{S}}(k)$ are sample spectral values (sample variances) [the true spectrum would be represented by $\mathbf{S}(k)$]. Since each spectral component $\mathbf{S}_m(k)$ is influenced by all values $x(n)$ in a data record, the sequence $\mathbf{S}_m(k)$ will be uncorrelated only if the M data records are separated by at least τ_{cl} data values. Even if $x(n)$ is taken from a white process, the practical restriction to a finite amount of data means that the sequence $\mathbf{S}_m(k)$ will have a random correlation.

When the variance of a random variable ϕ [e.g., $\overline{\mathbf{S}}(k)$] can be calculated, a useful relationship is the *normalized variance* (also called *dimensionless variability*[32] and *normalized mean-square random error*[9]), defined by

$$\mathrm{Var}_{\mathrm{norm}}\{\phi\} = \frac{\mathrm{Var}\{\phi\}}{[\mathrm{Ave}\{\phi\}]^2}, \tag{29}$$

where Var means "variance" and Ave means "mean value."

6.5 Some Sampling Statistics

Sampling Distribution of the Mean Sample means $\overline{x}(n)$ [Eq. (27a)] have a pdf $p(\overline{x})$ called the *sampling distribution of the mean* (Fig. 22*b*). The central limit theorem assures that $p(\overline{x})$ is very nearly normal for $\nu \gtrsim 30$ whether the pdf $p(x)$ is normal or not. When $p(x)$ is also normal, $p(\overline{x})$ can be nearly normal for $\nu < 30$, possibly[9] for $\nu \gtrsim 4$. The mean, variance, and standard deviation of $p(\overline{x})$ are represented here by $\hat{\mu}$, $\sigma_{\overline{x}}^2$, and $\sigma_{\overline{x}}$.

When both $p(\overline{x})$ and $p(x)$ are normal, the relationship between the variances of $\overline{x}$ and x is

$$\frac{\sigma_{\overline{x}}^2}{\sigma^2} = \frac{1}{\nu}. \tag{30}$$

Sampling Distribution of the Variance and the Spectrum Sample variances s^2 and $\overline{\mathbf{S}}(k)$ [Eqs. (27b) and (28)] have a pdf $p(s^2)$ called the *sampling distribution of the variance* (Fig. 23). This pdf has a chi-squared distribution with ν degrees of freedom, χ_ν^2. The mean and variance of a χ_ν^2 distribution are

$$\mu_\chi = \nu, \tag{31a}$$

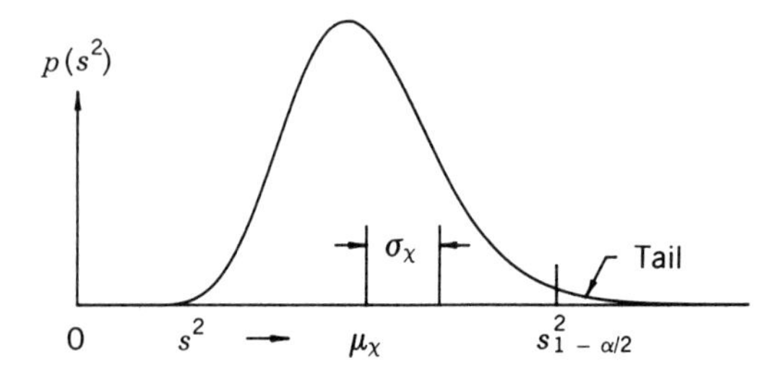

Fig. 23 Sampling distribution of the variance; chi-squared distribution for $\nu = 30$ degrees of freedom.

$$\sigma_\chi^2 = 2\nu. \tag{31b}$$

The ratio of the chi-square standard deviation to the mean,

$$\frac{\sigma_\chi}{\mu_\chi} = \left[\frac{2}{\nu}\right]^{1/2}, \tag{32}$$

is often used as a measure of stability.

When $p(x)$ is normal, the relationship between the sample and true variances is

$$\frac{s^2}{\sigma_\chi^2} = \frac{\chi_\nu^2}{\nu}. \tag{33}$$

The pdf $p(s^2)$ is very nearly normal[33] for $\nu \gtrsim 30$ (Fig. 23), and

$$\chi_\nu^2 \simeq \tfrac{1}{2}[z + (2\nu - 1)^{1/2}]^2 \tag{34}$$

can be used to calculate χ_ν^2 when a value for z (standardized normal random variable) can be specified. For example, if $z = 1.645$ (corresponds to 90% confidence[14]) and $\nu = 30$, $\chi_{30}^2 \simeq 43.5$ (true value is 43.8).

6.6 Averaging over Frequency: Windows

Spectral resolution for a random process is influenced by a data window (Sections 2.6 and 4.2), and both resolution and stability are influenced by a lag window. *Lag windows* are time-domain weighting functions similar to data windows except that they are applied to correlation sequences rather than to data records.

With a random process, using any formed window is equivalent to performing a moving average over frequency. A time-domain window (data or lag) creates a *spectral window* $W(k)$ that is convolved with a sample spectrum $\overline{\mathbf{S}}(k)$. The process is often called *smoothing*. While the effect is similar to that of a constant-bandwidth analog swept-frequency filter, the results are not identical. When a DFT is used, the result is a circular moving average, and values near zero and Nyquist frequencies will be "aliased."

Because a spectral window acts like a swept-frequency filter, it is common practice to define *resolution bandwidth* W_e as the effective bandwidth of the "filter"[9] (also called *statistical bandwidth*).

Resolution bandwidth is determined by the data window Fourier coefficients B_i and/or the lag window coefficients W_i. It can be defined in terms of normalized variance [Eq. (29)] as (see Ref. 32, Sections 8 and 9)

$$W_e = \frac{1}{\mathrm{Var}_{\mathrm{norm}}} = \frac{[\mathrm{Ave}\{\overline{\mathbf{S}}(k)\}]^2}{\mathrm{Var}\{\overline{\mathbf{S}}(k)\}}. \tag{35a}$$

If the power spectrum has no sharp magnitude changes, resolution bandwidth can be approximated by (see Ref. 2, Section 6.4, and Ref. 32, Section 9)

$$W_e = \frac{[\sum B_i^2]^2}{\sum B_i^4} \tag{35b}$$

for a data window and by

$$W_e = \frac{[\sum W_i]^2}{\sum W_i^2} \tag{35c}$$

for a full-length lag window. As used here, bandwidth is defined in units of k, the frequency index. When no formed windows are used (the open-window case), $W_e = 1$ is the base resolution bandwidth set by the data record length. Multiplying W_e by $\Delta f = 1/T_r$ [Eq. (7)] gives the bandwidth in hertz.

Resolution bandwidths created by data windows are narrow, while those created by lag windows can be quite broad. Resolution bandwidths greater than the widths of peaks and valleys in a true spectrum will bias resonances negatively and antiresonances positively.

When enough data cannot be acquired to achieve a desired stability by block averaging, smoothing with a lag window is the only way to improve stability. The effect of using a window with random data is quite distinct from its effect with deterministic data (Sections 2.6 and 4.2). Data windows control leakage with both random and deterministic data. While lag windows normally do not control leakage, a unique procedure may do so.[34]

Note that the values of W_i for data and lag windows are not the same.

Smoothing with a Data Window The (power) spectral window coefficients corresponding to a data window are

$$W_i = B_i^2. \tag{36}$$

For example, the Fourier coefficients for the two-sided Hann data window $B_i = \pm\frac{1}{4}, \frac{1}{2}, \pm\frac{1}{4}$ correspond to spectral window coefficients $W_i = \frac{1}{16}, \frac{1}{4}, \frac{1}{16}$.

Unless the coefficients B_i are normalized so that $\sum W_i = 1$, the power spectrum will be biased. This normalization is different from that used for leakage control of deterministic signals (Table 4). For example, if a Hann data window is used, bias can be removed by multiplying the power spectrum by the scaling factor $1/\sum B_i^2 = \frac{8}{3}$. Using Eq. (35b), the resolution bandwidth for the Hann data window is $W_e = 2$.

The different normalizations required for leakage control and smoothing result in different calibration factors for periodic and random components in a signal[12].

A peculiar characteristic of data windows is that they increase stability by averaging over frequency, but they lose the same amount of stability by discarding useful data.[35] (The "lost" stability can be recovered by increasing the number of data records and overlapping them.[34]) Thus the degrees of freedom remain unchanged when a data window is used,

$$\nu = \begin{cases} 2 \cdot M, & 1 \le k \le (N/2) - 1, \\ M, & k = 0,\ N/2, \end{cases} \tag{37}$$

where M is the number of periodograms averaged.

Smoothing with a Lag Window The procedure is to calculate correlation sequences [Eqs. (24) or (26)], multiply by a lag window, and apply a DFT to get a smoothed spectrum.

A lag window can be represented by Eq. (16), with b, B, n, and N replaced by w, W, τ, and L, where the signs of W_i must make $w(\tau)$ an even function of τ. For a full-length lag window, $L = N$. Normally,[2,32] $L \ll N$, with $w(\tau) = 0$, $|\tau| > \frac{1}{2}L$.

The resolution bandwidth for a full-length lag window is given by Eq. (35c). For example, the Fourier coefficients for the two-sided Hann lag window of length $L = N$ are $W_i = \frac{1}{4}, \frac{1}{2}, \frac{1}{4}$, and its resolution bandwidth is $\frac{8}{3}$. The signs (all positive) place the window maximum at $\tau = 0$. Bias is avoided by using the normalization $\sum W_i = 1$.

Increasing the resolution bandwidth corresponds to increasing the number of degrees of freedom to

$$\nu = \begin{cases} 2 \cdot M \cdot W_e, & 1 \le k \le (N/2) - 1, \\ M \cdot W_e, & k = 0,\ N/2. \end{cases} \tag{38a}$$

When lag window length $L \ll N$, a conservative approximation for the degrees of freedom is given by[32]

$$\nu \simeq \begin{cases} \dfrac{M \cdot N}{L}, & 1 \le k \le (N/2) - 1, \\ \dfrac{M \cdot N}{2L}, & k = 0,\ N/2. \end{cases} \tag{38b}$$

6.7 Noise Added to a Deterministic Signal

If a deterministic signal $x(n)$ can be repeated exactly (e.g., pulse, choip[1]), its average over M data records will not change. When the signal is corrupted by random noise $n(n)$, the result

$$y(n) = x(n) + n(n) \tag{39}$$

exhibits random amplitude variations. The effects of $n(n)$ can be reduced by averaging in either the time or frequency domain.

Time-Domain Averaging A block average over M data records in the time domain is

$$\bar{y}(n) = x(n) + \frac{1}{M}\sum_{m=0}^{M-1} n(n), \qquad 0 \le n \le N - 1. \tag{40}$$

If $n(n)$ has a nonzero mean, $\bar{y}(n)$ will be displaced from $x(n)$ approximately by this value. Otherwise, $\bar{y}(n) \to x(n)$. An advantage of this method is that results do not depend on the spectrum of the noise. Also, $\bar{n}(n)$ can be further averaged over N.

Averaging improves the *signal-to noise-ratio*, S/N, defined as the ratio of powers in signal and noise. Equation (30) can be used to estimate an enhanced S/N,

$$\text{S/N} = 10 \log_{10} M + (\text{S/N})_{\text{orig}} \quad \text{dB}. \tag{41}$$

Frequency-Domain Averaging Block averaging in the frequency domain results in

$$\overline{\mathbf{Y}}(k) = \mathbf{X}(k) + \frac{1}{M}\sum_{m=0}^{M-1} \mathbf{N}(k), \qquad 0 \le k \le N - 1, \tag{42}$$

where $\mathbf{X}(k)$ and $\mathbf{N}(k)$ represent circular or linear spectra of deterministic and random components in $y(n)$. A practical difficulty is that the noise may not be white, and $\overline{\mathbf{Y}}(k)$ will have a frequency-dependent bias.

The spectral estimate $\overline{\mathbf{Y}}(k)$ can be improved if $\overline{\mathbf{N}}(k)$ can be separately estimated. This requires careful use of an A/D converter to assure that a noise estimate is not adversely affected by quantization and granular noise (Section 3) or by different electrical impedances in the measurement system with and without the signal source impedance. When two signals are required (e.g., deconvolution, frequency response), the two noise signals may be affected differently.

Averaging can also be done in the power domain, but phase information for $\overline{\mathbf{X}}(k)$ is lost. If $x(n)$ is minimum phase, the phase spectrum can be estimated with a discrete Hilbert transform.[17]

6.8 System Identification

Frequency response can be more reliably estimated by exciting a system with a deterministic signal than with a random signal (Section 5). If random-signal excitation is necessary, well-established procedures are available.[9,13,32]

The system to be analyzed can be represented by Fig. 24, where $n(n)$ represents noise added to the unknown true system $H(z)$ anywhere between input $x(n)$ and measured output $y(n)$, $\tilde{H}(z)$ represents an FIR model, and $\epsilon(n) = y(n) - \tilde{y}(n)$ is the residual. Minimizing $\epsilon(n)$ in the least mean-squares sense is a classic optimization problem.[24] The "optimum" frequency response is estimated by

$$\hat{H}(k) = \frac{\overline{S}_{xy}(k)}{\overline{S}_{xx}(k)}, \qquad (43)$$

where the $\overline{\mathbf{S}}(\cdot)$ represent block average sample spectra [Eq. (28)]. Care must be taken to account for the effects of noise and possible small values in the inverse filter, $1/\overline{\mathbf{S}}_{xx}(k)$ (Section 5.3).

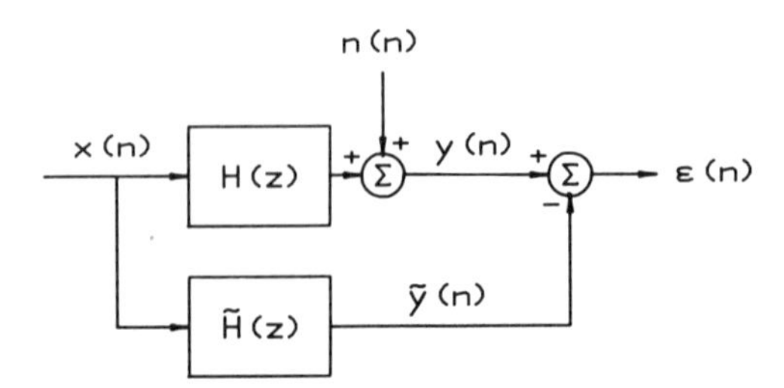

Fig. 24 Block diagram representing system identification with additive noise $n(n)$.

Equation (43) results from the optimization process whether or not there is additive noise $n(n)$ [additive noise is hidden in $\overline{\mathbf{S}}_{xy}(k)$]. The practical problem is how much of $\hat{H}(k)$ to believe. In the *coherency-squared spectrum*[2]

$$\gamma_{xy}^2(k) = \frac{|\overline{\mathbf{S}}_{xy}(k)|^2}{\overline{\mathbf{S}}_{xx}(k)\overline{\mathbf{S}}_{yy}(k)}, \qquad 0 \le \gamma_{xy}^2 \le 1.0, \qquad (44)$$

1.0 means that there is no additive noise; $\gamma_{xy}^2(k) = 0$ means that $\hat{H}(k)$ is all noise. The term $\gamma_{xy}^2(k)$ treats nonlinear response as if it were noise, since it is interpreted as a degradation of perfect linear response. When $\gamma_{xy}^2(k)$ shows degradation at resonant peaks, nonlinear response may be the cause. For a single data record, $\gamma_{xy}^2(k) = 1$ for all k, since a single data record is treated as deterministic by the DFT. A coherency-squared spectrum requires block-averaged spectra.

REFERENCES

1. J. C. Burgess, "Chirp Design for Acoustical System Identification," *J. Acoust. Soc. Am.*, Vol. 91, 1992, pp. 1525–1530.
2. G. M. Jenkins and D. G. Watts, *Spectral Analysis and Its Applications*, Holden-Day, San Francisco, 1969.
3. D. G. Childers (Ed.), *Modern Spectral Analysis*, Institute of Electrical and Electronics Engineers, New York, 1978.
4. J. C. Burgess, "Digital Spectral Analysis with Application to Identification of Mechanical Systems and Planning for Measurement," unpublished class notes, University of Hawaii, 1987.
5. A. V. Oppenheim and R. W. Schafer, *Discrete-Time Signal Processing*, Prentice-Hall, Englewood Cliffs, NJ, 1989.
6. L. R. Rabiner and B. Gold, *Theory and Application of Digital Signal Processing*, Prentice-Hall, Englewood Cliffs, NJ, 1975.
7. R. N. Bracewell, *The Fast Fourier Transform and Its Applications*, 2nd rev. ed., McGraw-Hill, New York, 1986.
8. Digital Signal Processing Committee (Ed.), *Programs for Digital Signal Processing*, Institute of Electrical and Electronics Engineers, New York, 1979.
9. J. S. Bendat and A. G. Piersol, *Random Data, Analysis and Measurement Procedures*, Wiley, New York, 1986.
10. B. A. Blesser, "Digitization of Audio: A Comprehensive Examination of Theory, Implementation, and Current Practice," *J. Audio Eng. Soc.*, Vol. 26, 1978, pp. 739–771.
11. C. E. Shannon, "Communication in the Presence of Noise," *Proc. IRE*, Vol. 37, 1949, pp. 10–31.
12. J. C. Burgess, "Digital Spectrum Analysis of Periodic Signals," *J. Acoust. Soc. Am.*, Vol. 58, 1975, pp. 556–567.
13. J. G. Proakis and D. G. Manolakis, *Introduction to Digital Signal Processing*, Macmillan, New York, 1988.

14. A. P. G. Peterson, *Handbook of Noise Measurement*, 9th ed., GenRad, Concord, MA, 1980.
15. E. B. Magrab and D. S. Blomquist, *The Measurement of Time-Varying Phenomena*, Wiley, New York, 1971.
16. J. Meyer, "Time Correction of Anti-Aliasing Filters in Digital Audio Systems," *J. Audio Eng. Soc.* (Engineering Reports), 1984, pp. 132–137.
17. R. Kuc, *Introduction to Digital Signal Processing*, McGraw-Hill, New York, 1988.
18. T. F. Darling and M. O. J. Hawksford, "Oversampled Analog-to-Digital Conversion for Digital Audio Systems," *J. Audio Eng. Soc.*, Vol. 38, 1990, pp. 924–943.
19. M. W. Hauser, "Principles of Oversampling A/D Conversion," *J. Audio Eng. Soc.*, Vol. 39, 1991, pp. 3–26.
20. F. J. Harris, "On the Use of Windows for Harmonic Analysis with the Discrete Fourier Transform," *Proc. IEEE*, Vol. 66, 1978, pp. 51–83.
21. J. C. Burgess, "Optimum Approximations to Dolph-Chebyshev Data Windows," *IEEE Trans. Signal Proc.*, Vol. 40, 1992, pp. 2592–2594.
22. A. H. Nuttall, "Some Windows with Very Good Sidelobe Behavior," *IEEE Trans. Acoust. Speech Signal Proc.*, Vol. ASSP-29, 1981, pp. 84–91.
23. J. C. Burgess, "Accurate Analysis of Periodic Signals Using a DFT," in preparation.
24. J. Makhoul, "Linear Prediction: A Tutorial Review," *Proc. IEEE*, Vol. 63, 1975, pp. 561–584; reprinted with corrections in D. G. Childers (Ed.), *Modern Spectral Analysis*, Institute of Electrical and Electronics Engineers, New York, 1978.
25. L. J. Eriksson, M. C. Allie, C. D. Bremigan, and J. A. Gilbert, "Active Noise Control on Systems with Time-Varying Sources and Parameters," *Sound Vib.*, July 1989, pp. 16–21.
26. S. Braun and J. K. Hammond, "Additional Techniques," in S. Braun (Ed.), *Mechanical Signature Analysis*, Academic, Orlando, FL, 1986, Chapter 6.
27. C. L. Nikias and J. M. Mendel (Eds.), "Special Section on Higher Order Spectral Analysis," *IEEE Trans. Acoust. Speech Signal Proc.*, Vol. ASSP-38, 1990, pp. 1236–1317.
28. O. Rioul and M. Vetterli, "Wavelets and Signal Processing," *IEEE Signal Proc. Mag.*, Vol. 8, 1991, pp. 14–38.
29. F. Hlawatsch and G. F. Boudreaux-Bartels, "Linear and Quadratic Time-Frequency Signal Representatives," *IEEE Signal Proc. Mag.*, Vol. 9, 1992, pp. 21–67.
30. W. A. Gardner, "Exploitation of Spectral Redundancy in Cyclostationary Signals," *IEEE Signal Proc. Mag.*, Vol. 8, 1991, pp. 14–36.
31. M. R. Spiegel, *Theory and Problems of Statistics*, Schaum's Outline Series, McGraw-Hill, New York, 1961.
32. R. B. Blackman and J. W. Tukey, *The Measurement of Power Spectra*, Dover, New York, 1958.
33. R. C. Weast and S. M. Selby (Eds.), *Handbook of Tables for Mathematics*, 4th rev. ed., CRC Press, Cleveland, OH, 1975.
34. A. H. Nuttall and G. C. Carter, "Spectral Estimation Using Combined Time and Lag Weighting," *Proc. IEEE*, Vol. 70, 1982, pp. 1115–1125.
35. E. A. Sloane, "Comparison of Linearly and Quadratically Modified Spectral Estimates of Gaussian Signals," *IEEE Trans. Audio Electroacoust.*, Vol. 17, 1969, pp. 133–137.

PART XI

PHYSIOLOGICAL ACOUSTICS

83

INTRODUCTION

JOSEPH E. HIND

1 INTRODUCTION

Physiological acoustics (*PA*) is a subdivision of acoustics involved in the study of structure and function of the auditory system in animals ranging from insects to humans. Although originally defined more narrowly, PA now includes the entire auditory system, that is, all of the acoustical and neural events that occur when a sound stimulus evokes the spatial and temporal patterns of neuronal activity in the brain that underlie sensation and perception in hearing. From a disciplinary standpoint, PA embraces many of the facets of neuroscience, including the subdivision designated neurobiology, which emphasizes basic biological processes, especially cellular and molecular mechanisms. Contemporary research in PA typically involves interdisciplinary teams whose members come from such diverse disciplines as acoustics, anatomy, audiology, biochemistry, biophysics, computer science, electrical engineering, neurophysiology, molecular biology, otology, psychology and pharmacology.

2 ORIENTATION OF CHAPTERS IN PART XI

This and the 4 following chapters in Part XI review knowledge and working hypotheses in PA from several perspectives. "Working hypotheses" are included to emphasize the relatively short half-life of beliefs and ideas in PA in comparison with, for example, the laws of physical acoustics. Several chapters restrict consideration to a specific locus in the auditory pathway (e.g., external ear, middle ear, cochlea) while others are oriented according to discipline (anatomy, biochemistry, pharmacology). *Enc.* Ch. 104 and 113 have been written in a style and with an objective not typically found in a book. Whereas each of the other chapters is primarily intended to act as a source of facts for a relatively restricted segment of PA, *Enc.* Ch. 104 and 113 draw upon data from a broad range of observations and, with the aid of some plausible speculation, formulate a mutually consistent set of working hypotheses that explain specific complex functions of auditory systems. *Enc.* Ch. 113 also illustrates the great advantages that can accrue from choosing, as an experimental subject, an animal that manifests such highly specialized behavior and sensory systems as the echo-locating bats.

3 ORGANIZATION OF THIS CHAPTER

This chapter is primarily intended to introduce the field of PA to readers who are not specialists in this area. To provide an indication of the scope of research in PA, the chapter begins by outlining some of the broad categories of *methods* in current use. This is followed by a brief overview of the ways in which the peripheral auditory system responds to sound, tracing the stream of stimulus information from the outer ear, through the middle ear, to the cochlea where the information is coded in the discharges of auditory nerve fibers. *Cochlear function* is one of the most active foci of ongoing research in PA; this topic is addressed in much greater detail in Chapters 86 and 87, but a short summary is included here, both because of its fundamental importance to understanding of the auditory system and as an example of a highly dynamic field in which changes in basic concepts over the past two decades have been nothing less than revolutionary. It is suggested that Chapter 86 be

Note: References to chapters appearing only in the *Encyclopedia* are preceded by *Enc.*

read before consulting any other material on the peripheral auditory system. All readers are encouraged to utilize the detailed information and extensive bibliographic references in each of the remaining 4 chapters in Part XI.

The present chapter next points out some relationships between physiological acoustics and *psychological acoustics*, including the use of *standardized terminology*, which succinctly and unambiguously describes the manner in which auditory stimuli are applied to one or both ears. For purposes of orientation, a short commentary on each of the 4 chapters is then presented. This chapter concludes with a few broadly based references that review and summarize some of the major areas of PA research.

4 METHODS

From the foregoing description of the breadth of PA, it follows that the methods used in present-day research include, in addition to acoustical measurements and theory, virtually the full complement of techniques employed in neuroscience at all levels: molecular, cellular, and systems. A brief listing of some of the more prominent methods follows.

4.1 Acoustics

Acoustical methodology informs and provides crucial support to all areas of research and practical applications in PA. From steadily increasing sophistication in the generation and monitoring of auditory stimuli to characterization of the biological components of the auditory system in terms of systems analysis, very few research strategies in contemporary PA could be pursued in the absence of acoustical instrumentation and theory. As a minor example, a growing number of investigators are studying the merit of measuring the energy absorbed by the ear as an index of auditory stimulus strength rather than specifying only sound pressure as has commonly been done (see Chapters 84 and 85). On a related topic, it may come as a surprise to learn that for high frequencies there is still uncertainty as to the exact modes of vibration of the eardrum and the attached malleus. The malleus is the first in the chain of three ossicles in the mammalian middle ear that conduct the vibratory energy of a stimulus from the eardrum to the fluid of the inner ear (cochlea). Even more challenging problems in vibration measurement arise within the cochlea, ranging from studies of multicellular structures such as the basilar membrane to the stereocilia of an individual hair cell. The most commonly used methods for the measurement of vibration in the ear involve some form of laser interferometry although the Mössbauer technique remains useful in some situations.

A final example of PA research in which acoustical methodology is playing a vital role is the development of virtual acoustic space paradigms. In this technique, earphones are used to generate stimuli at the two eardrums that reproduce the sounds recorded previously at these sites for a free-field sound source at a specified direction with respect to the listener. In areas of applied research the utility of virtual space methods is being investigated in such diverse fields as the entertainment industry and in military efforts to facilitate interaction between an airplane pilot and the surrounding environment. In basic research, virtual space stimuli are proving valuable in studies of directional hearing in human listeners and in electrophysiological studies of the spatial receptive fields of single neurons in experimental animals. Such earphone-generated stimuli are not only much more convenient to use than their free-field counterparts but the method also enables independent control of each of the attributes of the stimuli (e.g., interaural time and intensity differences) in a manner that would not be possible under actual free-field conditions.

4.2 Electrophysiology

Recordings of electrical potentials in both experimental animals and human subjects have occupied a prominent position in PA since the origin of the field. Since a detailed description of electrophysiological studies of the central auditory nervous system is presented in *Enc.* Ch. 111, only a brief listing of types of electrodes and a few examples of the kinds of potentials they record will be given here. *Macroelectrodes*, which are large in size in comparison with a single neuron or receptor cell, record the spatially summed potentials from many cells. In or near the cochlea such electrodes pick up two classes of potentials, one known as the *cochlear microphonic (CM)*, which represents the spatially summed sensory receptor potentials of the activated hair cells and which also roughly reproduces the stimulus waveform. The other potential, originally designated N_1 or the "first neural" potential, represents the spatially summed, temporally synchronized impulses in auditory nerve fibers in response to a transient signal such as a click (see Chapter 87). In general, recordings with macroelectrodes reveal synchronized neural discharges generally referred to as *evoked potentials*. Evoked potentials at the level of the brainstem are utilized in the clinic to evaluate the function of the middle ear and cochlea in patients. In experimental animals such recordings can be used to "map" auditory areas at many levels of the nervous system and to study the organization of the responsive neural fields, including what is referred to as *tonotopicity*, which is the orderly representation of stimulus frequency along one or more axes within the field.

Detailed studies of *stimulus coding* in the discharges of a single neuron became possible with the introduction of *extracellular microelectrodes*, which have tips small enough to pick up the external field potentials from only one cell body or nerve fiber. The propagated trains of all-or-none discharges in a single auditory nerve fiber in response to varying stimuli reveal how the attributes of a stimulus (e.g., frequency, level, duration) are represented in the activity of the fiber. The *temporal patterns* of discharges in the entire array of fibers activated by a stimulus, together with the *spatial pattern* of cochlear locations (or *places*) from which the activated fibers project, represents all of the information about that stimulus that is available to the central auditory nervous system (see *Enc.* Ch. 109). Similar studies of single neuron discharges at higher levels of the auditory system reveal how more complex stimulus attributes are encoded, such as the binaural interaction observed with differences in intensity and arrival time of sounds at the two ears; such interaction is believed to be related to mechanisms involved in sound localization (see *Enc.* Ch. 110, 111, and 113).

While studies of stimulus coding are typically carried out on single neurons, the investigator generally attempts to record from a large enough sample of cells to estimate the behavior of the whole population of similar neurons; such studies thus represent a kind of *systems analysis* with the objective of learning how the entire system of related neurons behaves. In contrast, studies of individual neurons using microelectrodes small enough to penetrate the cell membrane without causing significant alteration in cellular function yield so-called *intracellular recordings*. Such recordings can reveal many details of cellular function, including *excitatory and inhibitory postsynaptic potentials*, which, together with the intrinsic properties of the cell membrane, determine how the neuron *processes* incoming information before releasing *neurotransmitter* in synaptic transmission to the next cell or cells in the neural circuit. Continuing advances in molecular biology and biophysics are yielding fundamental new insights into membrane physiology through development of the technique of *patch-clamping* in which an electrode is sealed to an area of cell membrane so tiny that movements of ions into and out of the cell in a single channel can be recorded while the membrane potential is held at any desired value. Among many other possibilities, this technique holds promise for increased understanding of the gating mechanisms that control the flow of ions in hair cell mechano-transduction.

In summary, electrophysiological studies constitute one of the most important sources of information on the function of the auditory system, whether that function be the mechanisms by which a single hair cell in the cochlea responds to movement of its stereocilia or the behavior of the vast population of neurons that underlie sound localization.

4.3 Imaging

From the very beginning of research in PA, images of structure have played a vital role, starting with the first sketches of the gross anatomy of the ear, continuing with light microscopy and its depiction of cellular groupings and their interconnections, followed by transmission electron microscopy (TEM) and the ultrastructure it reveals and culminating in scanning electron microscopy (SEM), which yields such striking three-dimensional views of hair cells with their orderly patterns of stereocilia. The principles of systems analysis clearly require identification of the components of the system and their interconnections while cellular and molecular studies of single cells of all varieties are critically dependent upon knowledge of their ultrastructure. It is for these reasons that neurophysiologists and neuroanatomists have long recognized the rewards to be gained by closely correlated studies of structure and function. Today that correlation is often implemented by the injection of dye via an intracellular recording microelectrode into a single neuron whose function has been studied earlier. Subsequent to appropriate histological staining of the tissue, all of the processes of the cell, including cell body, dendritic tree, and axon, will be clearly marked and may be traced for entry into a computer, together with a record of the cell's response properties. When a collection of similar cells has accumulated, it is frequently possible to draw conclusions that relate structure and function. This approach has been applied to several of the auditory nuclei, especially the cochlear nucleus (CN), which is the first way-station along the auditory projection pathway (see *Enc.* Ch. 110 and 111). Several discrete categories of temporal discharge patterns have been identified in CN neurons, and these patterns have been found to correlate well with cell types defined on the basis of morphology.

Another kind of imaging tool that is experiencing steady growth in usage in PA research is nuclear magnetic resonance imaging (NMRI), which produces images of brain sections that show localized areas of change in brain activity in response to sensory stimuli or tasks such as mental arithmetic. Since the technique is noninvasive and easily tolerated by human subjects, NMRI has the potential to provide much more straightforward data on the localization of various classes of brain function than the methods currently in use.

4.4 Additional Methods

It is obvious that the methods outlined above represent but a fraction of the techniques currently in use in auditory neuroscience. One other area of increasing importance is described in *Enc.* Ch. 112 on the biochemistry and pharmacology of the auditory system. Much progress has been made in identifying neurotransmitters that are the chemical *messengers* through which neurons communicate at synapses. Yet another area of great current interest is that of developmental studies of the nervous system, including the question of regeneration of damaged or destroyed sensory receptor cells as well as the role of genetic linkages in normal and abnormal development. Lastly, two methods that increase access to individual cells under direct visualization, improve mechanical stability for recording purposes, and control the biochemical and pharmacological milieu are tissue culture and slice preparations of brain tissue, both of which are enjoying increasing usage.

5 THE "ACTIVE" COCHLEA

The remarkable capabilities of the auditory system continue to pose intriguing challenges to our understanding. For example, at the threshold of hearing, human listeners can detect sounds that cause displacements in the cochlear sensory organ that are subatomic in size, of the same order, or less than the displacements calculated to arise from thermal noise agitation. Full details of the transduction process, in which mechanical displacement of the cilia of receptor cells triggers impulses in afferent nerve fibers, remain an elusive goal, central to the understanding of auditory function. Another capability that has long puzzled theorists is the capacity of listeners to discriminate changes in the frequency of a pure tone of less than 0.5% for which any known design of electrical filter of comparable bandwidth would have a response time-constant many times greater than the time required for the auditory system to discriminate differences in the arrival time of a sound at the two ears which may be as little as 10 μs.

Stimulated by such puzzles and fueled by rapid advances in neurobiology over the past two decades, the pace of research on cochlear function has steadily quickened. In the overall scheme of sensory systems, hearing is considered to be one of the *special senses* or *modalities*, so designated because it involves highly specialized mechano-receptors known as *hair cells*, which are activated by bending of their *hair bundles* or *ciliary tufts*. Hair cells are found in a wide variety of sensory organs ranging from the *lateral line* organ in fish to the *inner ear* of mammals. The latter contains six different sensory organs. Five of the six encode head-tilt and linear and angular acceleration, information essential to the maintenance of balance. The sixth is the so-called *organ of Corti* in the fluid-filled, snail-shaped *cochlea* that contains two kinds of hair cells, the *outer hair cells* (*OHC*s) and the *inner hair cells* (*IHC*s). Omitting many important details, the OHCs receive information from the central nervous system (CNS) via *efferent* nerve fibers while the IHCs are innervated by the vast majority of *afferent* fibers that transmit information about an auditory stimulus to the CNS (see Chapter 86 for details of cochlear anatomy).

Throughout the cochlear spiral, the organ of Corti is supported by the *basilar membrane* and covered, in part, by the overlying *tectorial membrane*. The modern era of understanding of mechanical events in the cochlea originated over half a century ago with the Nobel prize-winning work of Georg von Bekesy[1]; he showed that vibration of the *stapes* in the *oval window* generates sound pressure in the cochlear fluid, causing the basilar membrane to undergo a *traveling wave* of deformation that begins at the *base* and proceeds toward the *apex*. The mechanical tuning of the basilar membrane varies systematically from high frequencies at the base to low frequencies at the apex, primarily due to a decrease in stiffness from base to apex, resulting in a frequency-to-place mapping along the cochlear spiral. The cochlea thus functions as a kind of hydromechanical frequency analyzer in which the mechanical input to the hair cells varies according to their locus or place. Afferent nerve fibers innervating the basal IHCs have tuning curves with high *best* or *characteristic frequencies* (CFs) while those closer to the apex exhibit correspondingly lower CFs.

Until about two decades ago it was almost universally believed that cochlear hair cells function as *passive* transducers that undergo depolarization upon mechanical deformation of their stereocilia. This causes release of excitatory neurotransmitter at synapses with afferent auditory nerve fiber endings, triggering propagated all-or-none impulses in the fibers. In the early 1970s evidence began to accumulate suggesting that the stereocilia of some types of hair cells might exhibit motility so that the cell could function as both a sensor and a motor element. The initial evidence was anatomical, showing that some stereocilia contain actin, a component of muscle fibers associated with their contractile mechanisms. Later, visual observation of isolated hair cells that had been removed from living animals confirmed that the cells did indeed exhibit motile behavior when subjected to electrical stimulation. Continuing investigation and theorizing gave rise to the concept of the "*active*" cochlea[2] with increased sensitivity and greater sharpness of tuning in comparison with the passive models previously accepted (see Chapters 86 and 87). Although differing widely in detail, many hypotheses have been advanced which assume that the OHCs in the active cochlea change their mechanical properties under control of the CNS via the efferent projections, thereby functioning as a feedback system.

Major advances in our understanding of the details of transduction have resulted from remarkably delicate experiments on hair cells in lower forms. Changes in the intracellular potential of hair cells from the bullfrog ampulla have been directly measured as their hair bundles were displaced.[3] Another series of fundamental studies on hair cells in the turtle cochlea has shown that the mechanical frequency tuning exhibited by the cell is determined by resonance of the equivalent electrical circuit of the cell membrane impedance rather than from mechanical resonance of the surrounding structures as is the case in mammalian cochleas.[4]

While hair cell physiology is yielding radically new concepts about function, investigators are developing increasingly sensitive instruments to measure basilar membrane vibration, including refinements in the Mössbauer technique and laser interferometry (see Chapter 87). Whereas Bekesy, working in cadaver cochleas at the very high stimulus levels required by his optical measurement technique, had reported linear basilar membrane vibration and broad tuning, recent studies demonstrated that basilar membrane vibration near the best frequency at the point of observation is *nonlinear* and that mechanical tuning could be as *sharp* as the tuning curves of single auditory nerve fibers, provided the cochlea was in good physiological condition and only moderate stimulus levels were used. As the state of the cochlea deteriorates or when higher stimulus levels are presented, the tuning curves broaden and the system becomes linear, similar to Bekesy's observations. Thus most students of cochlear function now postulate that some kind of active feedback mechanism mediated by the OHCs is superimposed on the passive cochlear behavior described by Bekesy. The result of this active mechanism is an increase in sensitivity of as much as 40 dB, together with substantial sharpening of the tuning curves of auditory nerve fibers.

The contractile behavior of hair cells in active cochleas has also provided one basis for explaining the origin of so-called otoacoustic emissions, which are sounds *generated in the cochlea* and transmitted *backward* via the ossicles and tympanic membrane *to the outer ear canal*, presumably because excessive positive feedback results in oscillation (see Chapter 87).

6 RELATION TO OTHER FIELDS IN ACOUSTICS

If attention shifts from physiological activity to *sensation* and *perception*, a disciplinary boundary is crossed into the realm of psychological acoustics (see Part XII), another subdivision of acoustics closely related to PA. The border between physiological and psychological acoustics has long been fertile ground for interdisciplinary interaction; the findings on one side of the boundary often stimulate and guide research and theoretical interpretation on the other. Although there are important exceptions, up to the present time studies in physiological acoustics have been carried out primarily on subhuman species while psychological experiments have typically utilized human subjects. The ongoing rapid development of noninvasive imaging techniques [such as NMRI and positron emission tomography (PET) scans] holds great promise that, in the future, many experimental observations will be easily performed on humans, thus eliminating the need to assume that the results in an experimental animal apply directly to humans. Physiological acoustics also has ties to other areas of acoustics, including speech processing and perception, musical acoustics, and bioacoustics.

7 STANDARDIZED TERMINOLOGY FOR STIMULUS AND RESPONSE PARAMETERS

Several conventions in terminology have been adopted in both physiological and psychological acoustics in order to minimize ambiguity by distinguishing between stimulus and response properties. *Psychological attributes* of sound include qualities such as *pitch*, *loudness*, and *timbre*, which are used only to describe *sensation* and/or *perception*, that is, "states" or "processes" internal to a listener. *Stimulus attributes* include such physically measurable quantities as *frequency*, *phase*, *sound pressure*, and *spectrum level*. Other stimulus parameters of importance in both physiological and psychological acoustics are specifications concerning the precise manner in which stimuli are applied to the two ears. If a stimulus is presented to only one ear via an earphone or through blockage of the other ear, the stimulus is referred to as *monaural* or *monotic*. *Binaural* means that both ears are stimulated (as distinguished from monaural), but the signals to the two ears may be the same or different. The term *diotic* is used when an identical stimulus is presented to both ears (usually via earphones), while *dichotic* describes the situation in which the stimulus to each ear is different and individually controlled (again via earphones).

Not infrequently, authors in both the physiological and psychoacoustic literature are guilty of imprecise use of the term *intensity* to describe the magnitude of an auditory stimulus. With a few important exceptions (see Chapters 84 and 85), acoustic intensity is not often actually measured because of the complexity of the measurement technique. Instead, most investigators use signal-generating equipment that controls the electrical input to a transducer, the acoustic output of which is calibrated in terms of *sound pressure* (linear scale) or *sound pressure level* (log scale). Although calibrated microphones are now widely used in PA, an occasional publication will be encountered in which stimulus strength is expressed in *decibels of attenuation* below the maximum available output level, which is often not specified. Another way to describe the relative magnitudes of stimuli is to indicate the amount (usually in decibels) by which the level of a specified stimulus exceeds that of a reference stimulus; when the reference stimulus is the threshold of a behavioral response, the magnitude of the specified stimulus is expressed as *dB SL* (*sensation level*).

8 COMMENTARY ON CHAPTERS IN PART XI

1. *Chapter 84: Acoustical Characteristics of the Outer Ear.* The anatomy of the outer ear is described in detail, together with a rigorous analysis of the acoustical properties resulting from its structure. The focus is primarily on the human ear although information on other species is included. The transformation of sound pressure from the free field in the absence of a listener to the pressure measured at the eardrum with the listener in the field is described and interpreted in terms of acoustical theory; this transform varies with frequency and source direction and thus provides cues that could be of use in sound localization. The overall efficiency of the human outer ear as a collector of sound energy is estimated in terms of the diffuse field absorption cross section for a model ear, the values of which are close to the theoretical limit for frequencies above about 2.5 kHz. The properties of some artificial ears, earphone couplers and ear simulators are also discussed. The chapter should be of interest to readers seeking a general understanding of the role of the external ear in hearing as well as to specialists in need of detailed, quantitative information on this structure.

2. *Chapter 85: Acoustic Properties of the Middle Ear.* The chapter begins with a thorough description of the structures in the mammalian middle ear, followed by an equally thorough discussion of their functions. Starting with the vibratory representation of an auditory stimulus in the motion of the eardrum and the attached malleus, the discussion follows the conduction of stimulus energy to the movement of the stapes footplate in the oval window of the cochlea, which creates sound pressure in the perilymphatic fluid. The transfer function of the middle ear is modeled as an acoustic two-port network involving the acoustic impedance of the eardrum and that of the stapes footplate, both quantities having been measured in humans and in several experimental animals. The discussion considers alterations in middle ear transmission due to ear pathology, to changes in static pressure in the middle ear cavity, and to contractions of the middle ear muscles. Under normal circumstances and in the absence of muscle contractions, the middle ear operates as a linear system up to very high sound pressures. The ratio of sound pressure measured in the cochlear fluid to that measured in the air in the outer ear canal near the eardrum was found to be between 20 and 30 dB for frequencies from 1 to 8 kHz for cat, guinea pig, and chinchilla; these values are consistent with the ratio of the area of the eardrum to that of the stapes footplate in each of these species, showing that most of the pressure gain produced in the middle ear is due to this area ratio.

3. *Chapter 86: Anatomy of the Cochlea and Auditory Nerve.* This chapter presents a detailed description of the morphology of the cochlea and its innervation. Additionally, interwoven into the stream of structural information is an equally detailed description of function at a qualitative level. The result is a concise, coherent blend of cochlear anatomy and physiology that should be of interest to both novices and experts. It is recommended that readers lacking a background in PA read Chapter 86 before consulting other chapters in Part XI.

4. *Chapter 87: Cochlear Mechanics and Biophysics.* This chapter presents a comprehensive review of cochlear function, from the passive, linear mechanics described by Bekesy for cadaver ears and intense stimuli to the current concept of the active cochlea with its physiologically vulnerable feedback system involving outer hair cell motility. The coverage is very broad, including information on hair cell transduction, cochlear potentials, and otoacoustic emissions in addition to basilar membrane vibration. The effects of different kinds of stimuli are also considered, together with the generation of distortion products. Sufficient information on cochlear anatomy is provided that the chapter can be consulted on a stand-alone basis for readers having a background in this topic.

REFERENCES

1. G. von Bekesy, *Experiments in Hearing*, McGraw-Hill, New York, 1960.
2. P. Dallos, "The Active Cochlea," *J. Neurosci.*, Vol. 12, 1992, pp. 4575–4585.
3. A. J. Hudspeth, "How the Ear's Works Work," *Nature*, Vol. 341, 1989, pp. 397–404.
4. R. Fettiplace, "Electrical Tuning of Hair Cells in the Inner Ear," *Trends Neurosci.*, Vol. 10, 1987, pp. 421–425.

84

ACOUSTICAL CHARACTERISTICS OF THE OUTER EAR

EDGAR A. G. SHAW

1 INTRODUCTION

The outer ears and the head are components of a complex acoustical antenna system that couples the eardrums to the external sound field. This system is best characterized in terms of the free-field response measured at the eardrum and the interaural differences in response, as functions of source direction, that provide the physical basis for sound localization. The outer ear protects the eardrum from mechanical damage but also enhances the coupling between the eardrum and the sound field and contributes substantially to the directionality of the system, especially at high frequencies where the normal modes of the concha come into play. The acoustical functions of the various components of the outer ear have been elucidated through the development of physical models with characteristics designed to match those of the average human ear. When sounds are presented to the ear through earphones, an artificial system is created with complex acoustical characteristics that point to the need for real ear calibrations based on sound pressure measurements in the ear canal.

2 ANATOMICAL FEATURES

A descriptive diagram of the human outer ear is presented in Fig. 1 that also shows the principal positions at which sound pressure measurements have been made. In real ears, the canal is an irregular tubular structure with a sinuous central axis and a cross-sectional area that is sharply tapered in the region approaching and adjoining the eardrum, which, in fact, defines the shape of the canal at the inner end.[1] For this reason, and since there is no clear boundary between the concha and the canal, it is difficult to assign a precise meaning to canal length. However, at frequencies below 8 kHz, the primary acoustical characteristics of the average canal are well represented by a simple cylindrical cavity 7.5 mm in diameter and 22.5 mm in length (volume 1 cm^3) terminated by a pistonlike eardrum that is perpendicular to the axis.[2]

The concha, with a volume of approximately 4 cm^3, is a broad shallow cavity that is tightly coupled to the canal. This cavity is partially divided by the crus helias and the upper part, the cymba, is connected to the fossa. These features clearly have specific acoustical attributes[3] whereas the structures extending from the concha, such as the helix, the antihelix, and the lobule, seem to function collectively as a simple flange.[4] In this respect, the human ear is markedly different from many other mammalian ears in which the pinna extension is a significant conical structure under muscular control. Other differences are indicated in Table 1, which brings together some average dimensions for human, cat, and guinea pig ears. As can be seen, the concha volume is relatively large in the human ear. The outer ear is very shallow in infant ears but attains adult characteristics within 2 years.[5]

The pinna is a flexible cartilaginous structure that can generally be treated as a hard boundary. However, when the human pinna is brought into contact with a supra-aural earphone, the finite compliance of the concha wall is evident in an apparent increase in enclosed volume at frequencies below 300 Hz.[2] Similarly, the sound attenuation attainable with earplugs may be reduced when the plug is confined to the outer part of the canal where the skin is supported by cartilage rather than bone.

Note: References to chapters appearing only in the *Encyclopedia* are preceded by *Enc.*

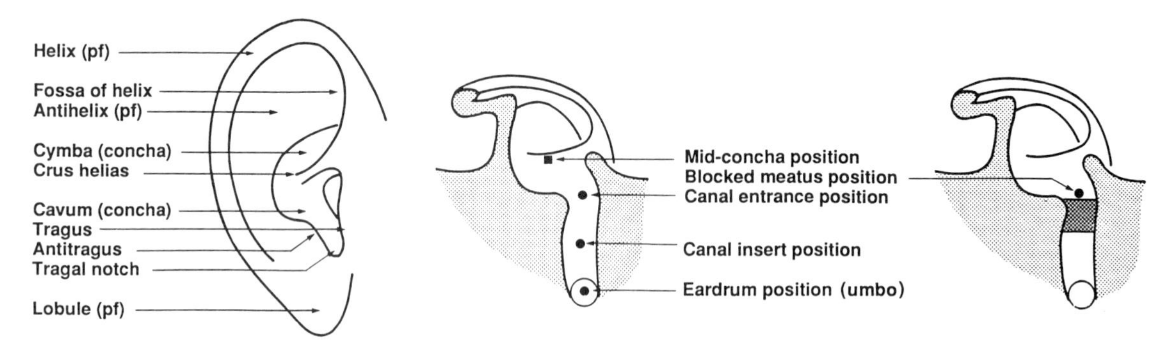

Fig. 1 Full view of human outer ear and horizontal cross sections showing five measurement positions.

3 TRANSFORMATION OF SOUND PRESSURE FROM FREE FIELD TO EARDRUM

The outer ear, head, neck, and torso are components of a complex acoustical antenna system that couples the middle and inner ear to the external sound field. This system can best be characterized in terms of the free-field response measured at the eardrum.

3.1 Transformation in the Horizontal Plane

Figure 2 shows average values of sound pressure level at the human eardrum, with respect to the free-field level at the center-head position, for plane waves approaching the head in the horizontal plane. This self-consistent family of response curves is based on a synthesis of experimental data from many laboratories[6,7] and is supported by other data.[8,9] Common to all the curves is the peak at 2.6 kHz, the primary resonance frequency of the average outer ear. Here the pressure gain attains its largest values. On the high-frequency side of the peak, resonance in the concha sustains the response for almost one octave. The steady decline between 5 and 10 kHz is peculiar to the horizontal plane and is associated with the mode structure of the concha.[3]

In the anterior sector, the directionality displayed in Fig. 2 is in broad agreement with calculations based on elementary diffraction theory where the outer ear is treated as a simple pressure detector at the surface of a hard spherical head.[2,6] At other angles of incidence, the agreement is at best approximate, especially at high frequencies. The difference in response between 45° and 135° azimuth, amounting to 11 dB at 4.5 kHz, is particularly significant. This lack of front–back symmetry, which provides a physical basis for localization in the lateral sector, can be attributed to diffraction by the pinna extension.[4]

The pure-tone transformation curves for individual subjects can be expected to be distributed about the average values with standard deviations of 1 dB or less below 500 Hz rising to 5 dB or more above 5 kHz. Particularly large variations are to be expected at high frequencies and in the shadow zone. Substantial smoothing occurs where pure tones are replaced by third-octave bands of noise.

3.2 Directionality in the Human Ear

Families of transformation curves comparable to those shown in Fig. 2 are not available for angles of incidence above and below the horizontal plane. It is, however, clear that variations in high-frequency response with

TABLE 1 Average Outer Ear Dimensions

	Canal Length (cm)	Canal Volume (cm^3)	Concha Depth (cm)	Concha Volume (cm^3)	Pinna Length (cm)	Interaural Distance (cm)
Man	2.25	1.0	1.0	3.8	6.7	17.5
Cat	1.5	0.23	1.0	0.3	4	10
Guinea pig	0.9	0.045	0.28	0.055	2.5	4

Source: After Ref. 2.

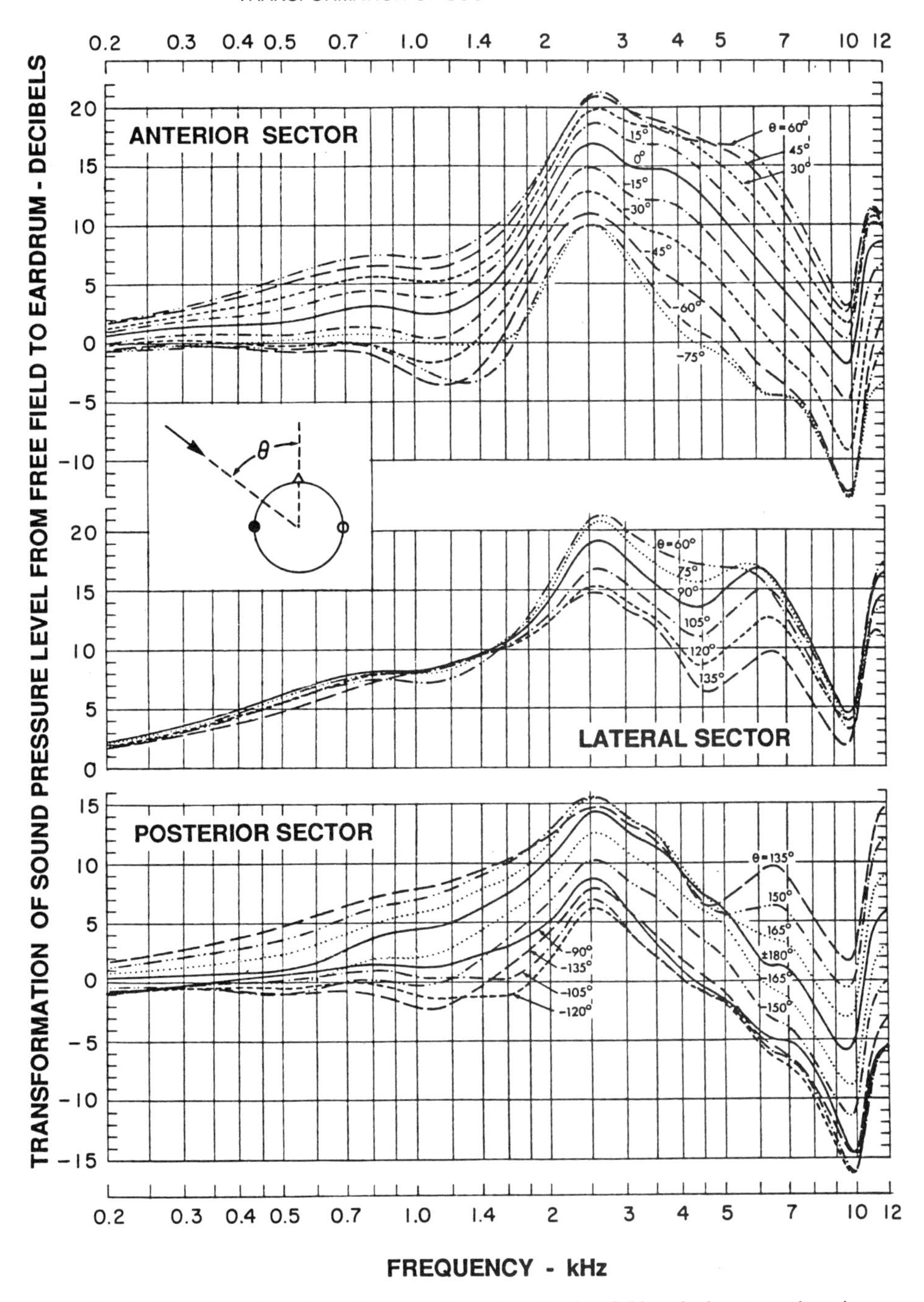

Fig. 2 Transformation of sound pressure level from the free field to the human eardrum in the horizontal plane as a function of frequency at 15° intervals of azimuth θ. Curves show average values based on data from 12 studies. (After Ref. 6).

TABLE 2 Directions of High Sensitivity in Human Ear: Approximate Angular Positions of Zone Centers and Angular Magnitudes of Zones

Frequency (kHz)	4	6	8[a]	8[b]	10	12[a]	12[b]	14
Azimuth, deg	40	60	80	30	70	10	90	15
Elevation, deg	10	20	30	−40	−20	−10	0	10
Magnitude, deg	±40 azimuth, ±60 elevation	±40	±30	±20	±25	±20	±15	±20

[a]Primary zone.
[b]Secondary zone.
Source: Based on Ref. 10.

source elevation[3,8] provide substantial spectral cues that are available for localization, especially in the median plane.

Some major high-frequency characteristics have recently been brought into focus in terms of contours of equal sensitivity presented on a spherical surface.[10] Similar patterns of directionality were found with all six subjects studied though the frequency associated with any particular pattern varied appreciably from subject to subject. Table 2 shows the approximate average directions and sizes of the primary and secondary zones of high sensitivity found in this study, where the levels at the zone boundaries are 5 dB below the highest sound pressure levels measured at each frequency. At 4 kHz, high sensitivity is maintained from 0° to 80° azimuth and from −40° to +80° in elevation. At 6 kHz, there is a decrease in zone size and movement of the zone center to greater angles of azimuth and elevation. At 8 kHz, the pattern becomes bipolar with zones of high sensitivity above and below the horizontal plane. The pattern is also bipolar at 12 kHz with the zones of high sensitivity now widely separated in azimuth and centered near the horizontal plane. These features provide substantial spectral cues that are available for sound localization.

3.3 Sound Pressure Transformation in Various Mammalian Ears

In many respects, the transformation characteristics measured in cats, guinea pigs, chinchillas, and other mammals are similar to those found in humans taking into account the differences in pinna dimensions.[2] At high frequencies, however, the patterns of directionality associated with the conical pinna extension are notably different from those of the human ear. For example, in cats with pinnas in the "alert" position, the direction of maximum sensitivity generally lies close to the frontal axis (e.g., azimuth ~ 30°, elevation ~ 20°), varying only modestly with frequency, and the sensitive zone systematically decreases in size between 4 and 20 kHz consistent with reception through a simple aperture of appropriate diameter (6 cm).[11] Nevertheless, refined measurements of sound pressure transformation from the free field to the cat eardrum display fine structure in the response curves not unlike that observed with human subjects.[12]

3.4 Interaural Time Differences

In calculations of interaural phase differences (IPDs) and interaural time differences (ITDs) based on diffraction theory, it is customary to treat the human head as a rigid sphere and the ears as simple pressure detectors at opposite ends of a diameter representing the interaural axis.[13] This treatment leads to the following asymptotic values of ITD at low and high frequencies:

$$\tau_{\rm lf} = \frac{a}{c}\,(3 \sin\,\theta), \tag{1}$$

$$\tau_{\rm hf} = \frac{a}{c}\,(\sin\,\theta + \theta), \tag{2}$$

where a is the radius of the sphere (~87.5 mm), c is the velocity of sound in air, and θ is the azimuth of the incident waves. The high-frequency asymptote agrees with the frequency-independent value of ITD stemming from geometrical acoustics where sound travels in a straight line to the left ear but bends to follow the spherical head as it approaches the right ear. In the transition region (0.5–2 kHz), it is necessary to distinguish between phase delay and group delay since the system is dispersive. (See also *Enc.* Ch. 117.)

Measurements of IPD with low-frequency tones and measurements of ITD with broadband high-frequency signals, such as clicks, have provided strong support for Eqs. (1) and (2).[2] Nevertheless, the actual high-frequency ITD for individual subjects must surely contain substantial frequency-dependent and direction-dependent variations in response corresponding to the sharp peaks and valleys seen in the amplitude domain.[14,15] Recent measurements of interaural envelope delays with signals of moderate bandwidth (0.22 octave) and center frequencies between 4 and 12 kHz provide mean values in broad agreement with Eq. (2) with individual values clustering about the mean with standard deviations of the order of 30%.[15]

4 ACOUSTICAL PARAMETERS, WAVE CHARACTERISTICS, AND PHYSICAL MODELS OF THE HUMAN OUTER EAR

Sound pressure measurements in real ears (Refs. 3, 6, 8, 10, 16, 17), experiments with replicas and physical models,[2–4] impedance measurements,[2,3] mathematical modeling,[1,18] and instrument technology and the quest for artificial ears[2,3,9] have all played a part in the formation of a comprehensive acoustical description of the human outer ear.

4.1 Eardrum Impedance

The impedance presented to the ear canal by the eardrum strongly affects the free-field response of the ear in the vicinity of resonance and is crucial when the ear is driven by an earphone inserted into the canal.

The parameter values presented in Table 3 are updated estimates of average values based on data from more than 20 studies.[2,3,16,19] Below 2 kHz, the original data are the mean and median values of R_d and X_d obtained in direct measurements of eardrum impedance. From 3 to 10 kHz, the primary data are the mean and median values of the standing-wave ratio (SWR) obtained from measurements of sound pressure distribution in the ear canal. The values of SWR presented in Table 3 indicate the general trend of these data. The corresponding values of R_d and X_d are calculated values based on the simplifying assumptions that the human ear canal can be treated as a uniform cylinder with a diameter of 7.5 mm (characteristic acoustic impedance $R_0 = 94$ g/cm^4 · s), and that R_d is essentially constant between 3 and 10 kHz. In the transition region (2–3 kHz), the values of R_d and X_d are consistent with the free-field response of the human ear at resonance.

At frequencies greater than 1 kHz, the eardrum impedance curves for individual subjects can be expected to display appreciable complexity reflecting the acoustical characteristics of the middle ear cavities and the dynamics of the eardrum, ossicular chain, and cochlea.

4.2 Normal Modes

The normal modes of the human outer ear provide the key to its high-frequency characteristics especially its directionality. It is, in fact, sufficient to study the modes of the concha and pinna extension in isolation since the directionality measured at the center of a rigid airtight plug closing the entrance to the ear canal (Fig. 1, blocked-meatus position) is essentially the same as that measured at the eardrum.[17]

Figure 3 shows the average characteristics of five modes based on data from 10 subjects measured under such conditions.[3] Mode 1, at 4.3 kHz, is a simple depth resonance with uniform pressure across the base of the concha.[4] All other modes have pressure distributions that are essentially transverse with nodal surfaces separating zones that are in phase opposition. At grazing incidence, mode 2, at 7.1 kHz, and mode 3, at 9.6 kHz, are best excited at approximately 75° elevation. In contrast, mode 4, at 12.1 kHz, and mode 5, at 14.4 kHz, are strongly excited from the front. In effect, the ear behaves as a vertical dipole under the control of modes 2 and 3 and as a horizontal dipole under the control of modes 4 and 5. These characteristics are consistent with the free-field directionality patterns measured in human subjects[10] (see Table 2).

4.3 Physical Models

The data presented in Fig. 3 pave the way for the construction of physical models with relatively simple geometry that have approximately the same mode frequencies, pressure distributions, directionality, and excitation as the average human ear.[3]

The essential characteristics of the depth resonance of the concha (mode 1 in Fig. 3) can be matched with a cylindrical cavity 22 mm in diameter and 10 mm in depth inclined at 30° to a rigid plane representing the

TABLE 3 Estimated Average Values of Ear Canal Standing-Wave Ratio (SWR), Eardrum Resistance (R_d), and Eardrum Reactance (X_d) for Normal Human Ears at Various Frequencies

Parameter	0.1 kHz	0.2 kHz	0.3 kHz	0.5 kHz	0.7 kHz	1 kHz	2 kHz	3 kHz	5 kHz	7 kHz	10 kHz
SWR, dB								13	14.5	19	21
R_d, g/cm^4 · s	490	430	390	350	330	320	390	420	400	400	400
X_d, g/cm^4 · s	−2800	−1400	−950	−550	−360	−170	−80	−30	200	410	510

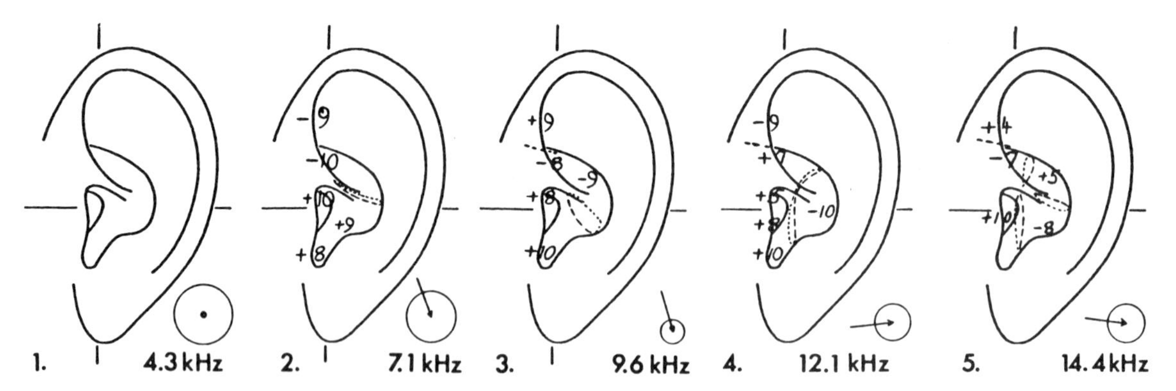

Fig. 3 Average transverse pressure distributions and resonance frequencies of five normal modes in human ears with ear canal closed at entrance showing nodal surfaces (broken lines) and relative pressure (numerals). Circles indicate relative degrees of excitation and arrows show directions of maximum response at grazing incidence. (After Ref. 3.)

surface of the head surrounding the ear (the *circumaural plane*). The cavity opening is partially surrounded by a simple flange, representing the pinna extension, which improves the coupling between the cavity and the sound field while suppressing reception from the rear.[4] To proceed beyond 5 kHz, greater geometrical complexity is required.[3] As a step in this direction, it is convenient to replace the cylindrical cavity with one that is rectangular in form. The first horizontal resonance (mode 4) is then readily tuned by adjusting the breadth of the cavity. To tune mode 2, with its vertical pressure distribution and remarkably low resonance frequency, it is essential to introduce a partial barrier representing the crus helias. This substantially increases the effective length of the cavity without increasing its volume. The presence of mode 3 is dependent on the introduction of a small auxiliary channel, connected to the main cavity, to represent the fossa. The final shape of the concha–fossa cavity is shown inset in panel (*a*) of Fig. 4.

As can be seen, the five resonance frequencies in the model, in the absence of the canal [Panel (*b*)], are almost identical with the average values for human subjects shown in Fig. 3. Successful representations of the human concha and fossa based on other geometrical forms, such as the semicircle, can also be built.[3]

In the complete physical model featured in panel (*a*), the eardrum is represented by a condenser microphone and a two-branch acoustical network (NRC Network D) with characteristics that are compatible with the data given in Table 3. The canal is represented by a cylindrical cavity 7.5 mm in diameter with an adjusted length ($\simeq$25 mm) chosen to place the principal resonance frequency at 2.6 kHz in agreement with the average value for the human ear.[3] Evidently, the enfolding geometry at the junction between the "canal" and the "cavum" produces an effective canal length ($\lambda/4 \simeq 32$ mm) that is considerably greater than the length of the canal proper. The addition of the canal increases the number of resonances below 16 kHz from five to eight. However, at most frequencies within this range, the presence of the canal has little effect on the sound pressure distribution in the concha except in the vicinity of the ear canal entrance.

4.4 Response of Outer Ear Measured in Isolation

The families of response curves presented in Fig. 4 display the high-frequency characteristics of a particular physical model designated "IRE" (inclined rectangular cavity: Stage E) in terms of the response measured at the eardrum position [panel (*a*)] and at the center of a rigid airtight plug closing the canal entrance [panel (*b*)]. Such measurements are made with a special sound source designed to irradiate the human ear with clean progressive waves at grazing incidence and in virtual isolation from the head and torso.[3] All angles of incidence lie in the circumaural plane with 0° and 90° representing waves approaching the ear from the front and from above, respectively. As can be seen, the directionality measured at the "blocked meatus" is almost identical with the directionality at the eardrum.

The curves shown in panel (*b*), though taken with a physical model, could readily be included in a collection of response curves obtained with real human ears under comparable conditions.[3,17] In particular, in panel (*b*) as in the human data, with sound waves approaching from high angles of incidence (60°–90°), the response remains

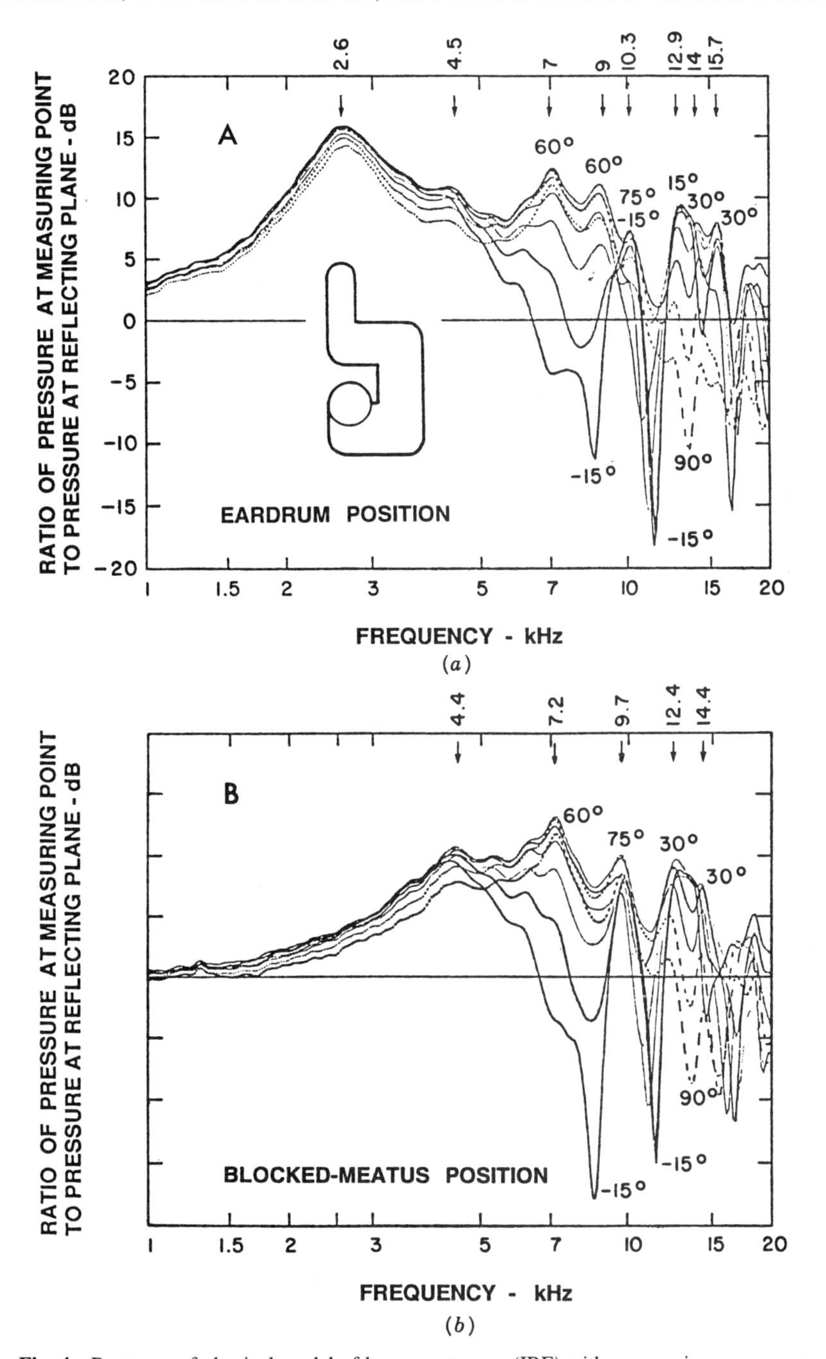

Fig. 4 Response of physical model of human outer ear (IRE) with progressive wave source at eight angles of incidence in circumaural plane: (*a*) complete ear; (*b*) with ear canal closed at entrance (blocked meatus). Arrows indicate frequencies of normal modes. Shape of concha–fossa representation is shown in panel (*a*). (After Ref. 3.)

strong between 6 and 9 kHz but is weak between 11 and 15 kHz, whereas, with frontal incidence (-15° to $+15^\circ$), the response is weak between 6 and 9 kHz but strong between 11 and 16 kHz. While these and other major acoustical features appear to be characteristic of human ears in general, the pattern of response for each individual ear is rich in detail and distinctive in character.[17]

4.5 Sound Pressure Measurements in the Ear Canal

The presence of longitudinal standing waves in the ear canal makes it necessary to measure sound pressure within 2 mm of the umbo if the input to the human eardrum is to be determined with an absolute accuracy of 1 dB at 10 kHz. It is also necessary to use a fine probe tube. Useful measurements can, however, be made at appreciable distances from the eardrum, as indicated in Fig. 1, provided that the transfer function between the point of measurement and the eardrum position can be measured or estimated.[2,6] Where a relative measure of sound pressure is sufficient, the quarter-wavelength minimum in the standing-wave pattern can be used as a position marker to ensure accurate placement and replacement of the probe microphone orifice in the canal. As noted earlier, measurements at the convenient and highly reproducible blocked-meatus position also have substantial value as indicators of directionality (see also Ref. 20).

At frequencies greater than 5 kHz, the wave system in the canal may be contaminated with evanescent transverse modes generated in the concha.[2,18] These modes are unlikely to be present at significant amplitudes, except in the vicinity of the entrance to the ear canal and at minima in the longitudinal standing-wave pattern, since they have high attenuation constants [e.g., 3 dB/mm for the (1, 1) mode at 20 kHz assuming a canal diameter of 7.5 mm].

The human ear canal can be treated as an acoustical horn in which the cross-sectional area is measured perpendicular to a curvilinear central axis that follows the shape of the canal. The length of this axis is appreciably greater than the superficial length of the canal.[1] This technique has also been used to interpret sound pressure measurements in the cat ear canal at frequencies as great as 30 kHz.[21]

5 DIFFUSE-FIELD RESPONSE AND OVERALL PERFORMANCE

By invoking the acoustical reciprocity principle, it can be shown that the diffuse-field response of the outer ear, p_d^2/p_f^2, is essentially determined by two impedances, Z_d and Z_a, as follows:

$$\frac{p_d^2}{p_f^2} = \left(\frac{c}{\pi\rho\nu^2}\right)\frac{\eta R_a|Z_d|^2}{|Z_a + Z_d|^2}. \tag{3}$$

In this expression, p_f^2 is the mean-square pressure in the diffuse sound field, p_d^2 is the mean-square pressure generated at the eardrum, Z_d is the load impedance presented to the outer ear by the eardrum, Z_a is the impedance presented to the eardrum by the outer ear (in essence, the radiation impedance), R_a is the real part of Z_a, and η is the acoustical radiation efficiency of the outer ear, while ν, ρ, and c are the frequency, density of air, and velocity of sound, respectively.[22]

Curve *A* in Fig. 5 is an estimated median diffuse-field response curve for the human ear based on synthesized response data for the mode-matched model ear featured in Fig. 4 and for replicas of several human ears, as well as measured values of Z_a and Z_d for the model ear.[3] This estimate is in broad agreement with other data.[23] The frontal incidence response (curve *B*) is well below the diffuse-field response between 5 and 10 kHz reflecting the relative insensitivity of the human ear in the horizontal plane at these frequencies (see Fig. 4). The intersubject differences in a diffuse sound field are generally much smaller than those found at any particular angle of incidence in the free field.

The overall performance of the outer ear as a "collector" of sound energy can also be quantified in terms of Z_d, Z_a, and η.[22,24,25] Hence, it can be shown that the maximum amount of sound power that can be received from a diffuse sound field is

$$W_d = \frac{\lambda^2}{4\pi}\frac{p_f^2}{\rho c}, \tag{4}$$

where λ is the wavelength, which is also the power that would flow into a perfectly absorbing sphere of cross-sectional area $\lambda^2/4\pi$ (radius $\lambda/2\pi$). This limit is attained where Z_a is the conjugate of Z_d and where there is no loss of energy during transmission to the eardrum (η = 1). More generally, the performance of any outer ear can be expressed in terms of its diffuse field absorption cross-section $A \le \lambda^2/4\pi$.

Figure 6 shows the value of *A* as a function of frequency for the mode-matched model ear, neglecting radiation losses that are probably of the order of 1 dB.[3,24] Also shown is the value of *A* when the outer ear is eliminated and the "eardrum" (i.e., Z_d) is placed at the surface of a sphere representing the head. As can be seen,

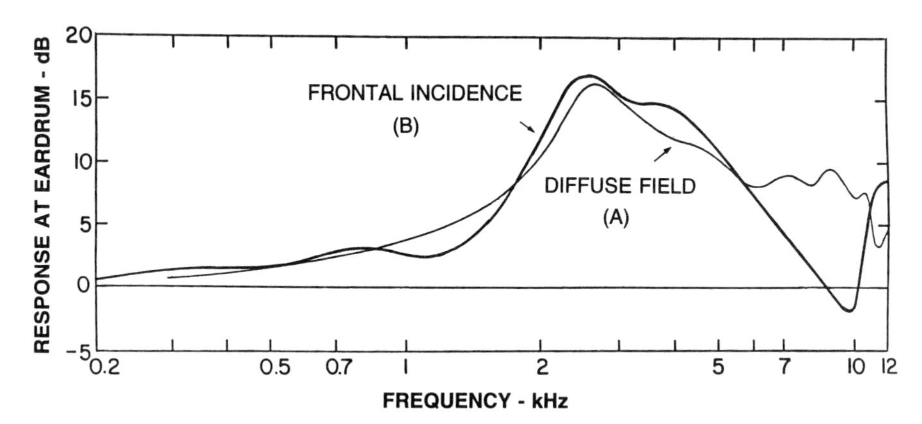

Fig. 5 Curve *A*: Estimated median transformation of sound pressure level from diffuse field to human eardrum. Curve *B*: Average transformation with frontal incidence: 0° azimuth in Fig. 2. (After Ref. 3.)

the absorption cross section of the human ear is not far below the theoretical limit at the principal resonance frequency (2.6 kHz) and at higher frequencies. Comprehensive measurements of Z_a made on outer ears excised from cats indicate that, in this species also, the value of A approaches the theoretical limit at the principal resonance frequency.[25]

6 EARS COUPLED TO EARPHONES

When sounds are presented to the outer ear through earphones, the free-field characteristics of the ear are replaced by the characteristics of an artificial system that are dependent on complex acoustical interactions between the individual earphone and the individual ear.

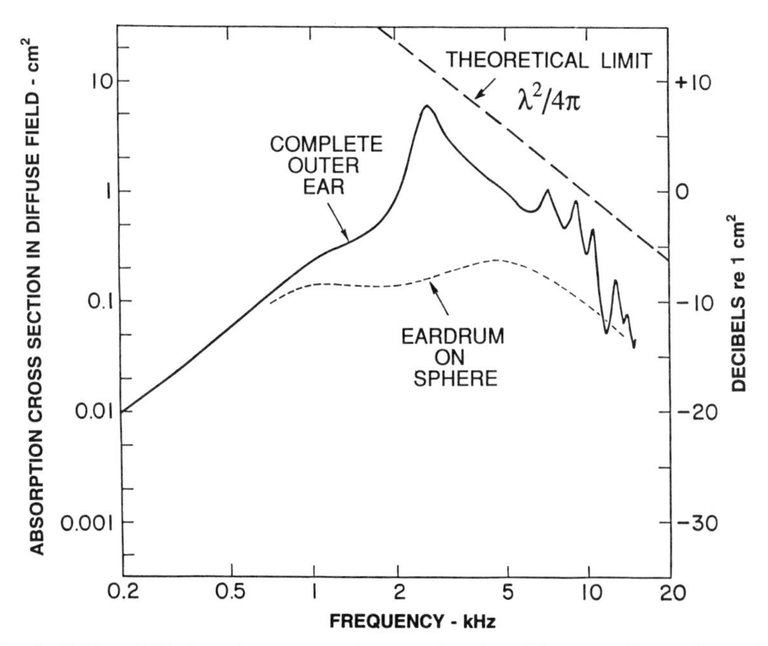

Fig. 6 Diffuse-field absorption cross section as a function of frequency for mode-matched physical model of human outer ear. (After Ref. 24.)

These interactions may include wave effects that are under poor control, especially at high frequencies.

Artificial ears, earphone couplers, and ear simulators reproduce selected acoustical characteristics of the human ear, with particular reference to the calibration of sound sources such as audiometric earphones and hearing aid receivers under standard conditions.[2,3,26,27] The simplest couplers are cylindrical cavities engaging with an earphone at one end and closed by a microphone at the other. In the NBS9A coupler, a single cavity represents the effective volume presented to a supra-aural earphone by the average human ear and the microphone provides an approximate measure of the sound pressure generated in the concha. In the more advanced IEC Artificial Ear, the main cavity, which includes the microphone, represents the volume of the concha when compressed by a supra-aural earphone, while auxiliary cavities represent the effective volume of the ear canal and eardrum, and the compliance of the pinna flange.[26] In occluded ear simulators, a cylindrical cavity, representing the inner portion of the ear canal, is terminated by a microphone surrounded by highly developed acoustical networks representing the impedance of the eardrum.[27] These simulators are well suited to the calibration of insert earphones and can be embodied in realistic "manikins" that include replicas or representations of the human concha, pinna flange, head, and torso.[9]

The limitations inherent in coupler methods of calibration are avoided when earphones are calibrated on real ears. The sound pressures generated by plane waves approaching the head from one or many directions in a free-field are first measured with a probe microphone, at a selected position in the ear canal, to create a set of *head-related transfer functions* (HRTFs).[14,20] These functions are then compared with the sound pressure generated at the same canal position by sounds presented through the earphone. Ideally, separate sets of HRTF and earphone response functions are created for each listener. Recently, it has been shown that some earphones can be successfully calibrated with measurements made at the blocked-meatus position shown in Fig. 1.[28]

REFERENCES

1. M. R. Stinson and B. W. Lawton, "Specification of the Geometry of the Human Ear Canal for the Prediction of Sound-Pressure Level Distribution," *J. Acoust. Soc. Am.*, Vol. 85, 1989, pp. 2492–2503.
2. E. A. G. Shaw, "The External Ear," in W. D. Keidel and W. D. Neff (Eds.), *Handbook of Sensory Physiology*, Vol. V/1, Springer-Verlag, Berlin, 1974, pp. 455–490.
3. E. A. G. Shaw, "The Acoustics of the External Ear," in G. A. Studebaker and I. Hochberg (Eds.), *Acoustical Factors Affecting Hearing Aid Performance*, University Park Press, Baltimore, 1980, pp. 109–124.
4. R. Teranishi and E. A. G. Shaw, "External-Ear Acoustic Models with Simple Geometry," *J. Acoust. Soc. Am.*, Vol. 44, 1968, pp. 257–263.
5. B. Kruger and R. J. Ruben, "The Acoustic Properties of the Infant Ear," *Acta Otolaryngol. (Stockh.)* Vol. 103, 1987, pp. 578–585.
6. E. A. G. Shaw, "Transformation of Sound Pressure Level from the Free Field to the Eardrum in the Horizontal Plane," *J. Acoust. Soc. Am.*, Vol. 56, 1974, pp. 1848–1861.
7. E. A. G. Shaw and M. M. Vaillancourt, "Transformation of Sound-Pressure Level from the Free Field to the Eardrum Presented in Numerical Form," *J. Acoust. Soc. Am.*, Vol. 78, 1985, pp. 1120–1123.
8. S. Mehrgardt and V. Mellert, "Transformation Characteristics of the External Human Ear," *J. Acoust. Soc. Am.*, Vol. 61, 1977, pp. 1567–1576.
9. M. D. Burkhard and R. M. Sachs, "Anthropometric Manikin for Acoustic Research," *J. Acoust. Soc. Am.*, Vol. 58, 1975, pp. 214–222.
10. J. C. Middlebrooks, J. C. Makous and D. M. Green, "Directional Sensitivity of Sound-Pressure Levels in the Human Ear Canal," *J. Acoust. Soc. Am.*, Vol. 86, 1989, pp. 89–108.
11. M. B. Calford and J. D. Pettigrew, "Frequency Dependence of Directional Amplification at the Cat's Pinna," *Hearing Res.*, Vol. 14, 1984, pp. 13–19.
12. A. D. Musicant, J. C. K. Chan, and J. E. Hind, "Direction-dependent Spectral Properties of Cat External Ear: New Data and Cross-Species Comparisons," *J. Acoust. Soc. Am.*, Vol. 87, 1990, pp. 757–781.
13. G. F. Kuhn, "Model for the Interaural Time Differences in the Azimuthal Plane," *J. Acoust. Soc. Am.*, Vol. 62, 1977, pp. 157–167.
14. J. Blauert, *Spatial Hearing*, MIT Press, Cambridge, 1983.
15. J. C. Middlebrooks and D. M. Green, "Directional Dependence of Interaural Envelope Delays," *J. Acoust. Soc. Am.*, Vol. 87, 1990, pp. 2149–2162.
16. M. R. Stinson, "Revision of Estimates of Acoustic Energy Reflectance at the Human Eardrum," *J. Acoust. Soc. Am.*, Vol. 88, 1990, pp. 1773–1778.
17. E. A. G. Shaw, "External Ear Response and Sound Localization," in R. W. Gatehouse (Ed.), *Localization of Sound: Theory and Applications*, Amphora Press, Groton, CT, 1982, pp. 30–31.
18. R. D. Rabbitt and M. T. Friedrich, "Ear Canal Cross-Sectional Pressure Distributions: Mathematical Analysis and Computation," *J. Acoust. Soc. Am.*, Vol. 89, 1991, pp. 2379–2390.
19. M. R. Stinson, E. A. G. Shaw, and B. W. Lawton, "Estimation of Acoustical Energy Reflectance at the Eardrum from Measurements of Pressure Distribution in the Human Ear Canal," *J. Acoust. Soc. Am.*, Vol. 72, 1982, pp. 766–773.
20. H. Møller, M. F. Sorenson, D. Hammershoi, and C. B.

Jensen, "Head-Related Transfer Functions of Human Subjects," *J. Audio Eng. Soc.*, Vol. 43, 1995, pp. 300–321.

21. M. R. Stinson and S. M. Khanna, "Sound Propagation in the Ear Canal and Coupling to the Eardrum, with Measurements on Model Systems," *J. Acoust. Soc. Am.*, Vol. 85, 1989, pp. 2481–2491.
22. E. A. G. Shaw, "Diffuse Field Response, Receiver Impedance, and the Acoustical Reciprocity Principle," *J. Acoust. Soc. Am.*, Vol. 84, 1988, pp. 2284–2287.
23. G. F. Kuhn, "The Pressure Transformation from a Diffuse Sound Field to the External Ear and to the Body and Head Surface," *J. Acoust. Soc. Am.*, Vol. 65, 1979, pp. 991–1000.
24. E. A. G. Shaw, "1979 Rayleigh Medal Lecture: The Elusive Connection," in R. W. Gatehouse (Ed.), *Localization of Sound: Theory and Applications*, Amphora Press, Groton, CT, 1982, pp. 13–29.
25. J. J. Roskowski, L. H. Carney, and W. T. Peake, "The Radiation Impedance of the External Ear of Cat: Measurements and Applications," *J. Acoust. Soc. Am.*, Vol. 84, 1988, pp. 1695–1708.
26. *American National Standard Method for Coupler Calibration of Earphones*, S3.7-1995, Acoustical Society of America, New York, 1995.
27. *American National Standard for an Occluded Ear Simulator*, S3-25-1989, Acoustical Society of America, New York, 1995.
28. H. Møller, D. Hammershoi, C. B. Jensen, and M. F. Sørensen, "Transfer Characteristics of Headphones Measured on Human Ears," *J. Audio Eng. Soc.*, Vol. 43, 1995, pp. 203–216.

85

ACOUSTIC PROPERTIES OF THE MIDDLE EAR

William T. Peake and John J. Rosowski

1 INTRODUCTION

The middle ear couples acoustic signals from the external ear canal to the inner ear. Measurements of its acoustic properties include acoustic input impedance and transfer ratios. As a signal transmission system the middle ear is essentially linear; the pressure transfer ratio is bandpass with a gain of 20–30 dB in the midband. The gain can be reduced by contraction of either of two small muscles or by static-pressure difference between the middle-ear air space and the ear canal. Acoustic measures are useful for diagnosis of some hearing disorders.

2 DESCRIPTION OF THE MIDDLE EAR

2.1 Structure

The middle ear consists of the tympanic membrane, the ossicular chain with its attached ligaments and muscles, and the air-filled cavity that contains the ossicles (see Fig. 1*a*). The tympanic membrane is the boundary between the external ear and the middle ear; the oval and the round windows of the cochlea are boundaries between the middle ear and the inner ear. The ossicles link motion of the tympanic membrane to the inner ear through motion of the stapes in the oval window. This description is appropriate for all terrestrial vertebrate species except those with no tympanic membrane (e.g., salamanders[1] and snakes[2]).

In mammals the ossicular chain consists of a linkage of three bones, the malleus (hammer), the incus (anvil), and the stapes (stirrup). Although the configuration of these bones varies among species,[1,3,4] some features occur regularly. An elongated process of the malleus (the manubrium, or handle of the hammer) is embedded between layers of the tympanic membrane. The malleus and the incus are connected to the bony walls of the middle ear by ligaments and to each other through a joint. The incus connects through another joint to the stapes, the inner face of which (the stapes footplate) is attached by an annular ligament in the oval window of the cochlea. Two muscles attach to the ossicles, the tensor-tympani muscle to the malleus and the stapedius to the stapes. The configuration of the middle-ear cavity varies among mammalian species[5,6]; in the human ear the main (tympanic) cavity is connected through a narrow region (the aditus) to a second cavity (the antrum), which is coupled to numerous smaller cavities that make up the "aerated" mastoid region of the temporal bone.[7] The eustachian tube, which connects the tympanic cavity to the nasopharynx, is usually closed, but it can be opened by contraction of attached muscles.[8,9]

2.2 Function

The function of the middle ear is to transmit acoustic signals from the ear canal to the cochlea. This signal coupling takes place primarily through the action of the tympanic membrane (TM) and the ossicular chain. For the human ear, except for the upper part of the audio-frequency range (i.e., above 5 kHz), the transverse dimensions of the ear canal are small relative to sound wavelengths and the sound pressure is approximately constant across the TM.[10,11] In this case the acoustic interactions at the middle-ear input (i.e., at the tympanic membrane) can be described in terms of two variables, the sound pressure P_T at the TM and the TM volume velocity U_T. The motion of the TM and the attached malleus is coupled through the three ossicles. The incudomallealar and

Note: References to chapters appearing only in the *Encyclopedia* are preceded by *Enc.*

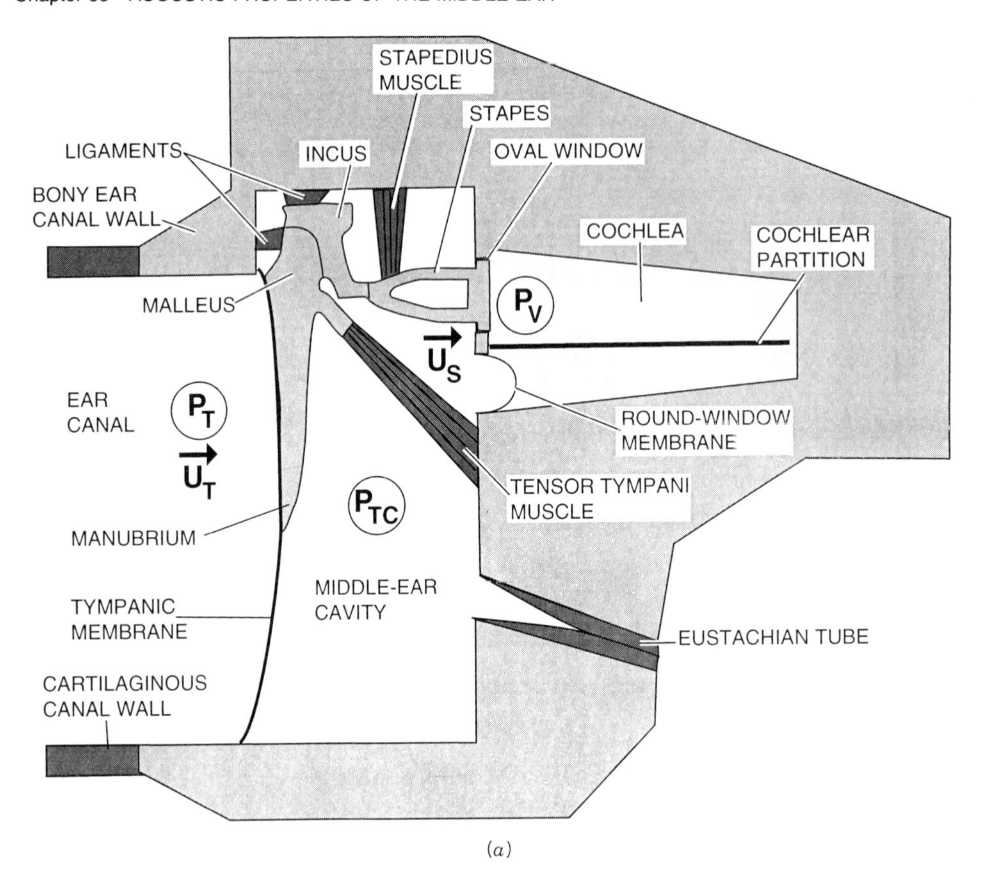

(*a*)

Fig. 1 (*a*) Schematic representation of the structures of the mammalian middle ear with acoustic variables indicated that will be used in the text. The five labeled acoustic variables in the figure represent complex amplitudes of sinusoidal functions of time t with frequency f. For instance, the pressure in the ear canal $p_T(t) = |P_T| \cos(2\pi f t + \angle P_T)$. Other variables are U_T = the volume velocity at the tympanic membrane, P_{TC} = the sound pressure in the tympanic cavity, U_S = the volume velocity of the stapes footplate, and P_V = the sound pressure in the inner-ear vestibule just inside the stapes footplate. Components of the middle-ear cavity configuration are not represented in this picture. (*b*) An analog-circuit representation of the signal-transmission function of the middle ear. The "impedance analogy" is used: that is, volume velocity is analogous to current and sound pressure is analogous to voltage. The middle ear is represented by a two-port network with input variables P_T, U_T and output variables P_V, U_S. External and inner ears are represented by Thévenin equivalents. The equivalent external source $P_{EX}(\phi, \theta)$ is a function of the sound propagation direction (ϕ, θ) relative to the head; Z_E is the output impedance of the external ear; Z_C is the input impedance of the cochlea. A Thévenin equivalent source for otoacoustic emissions is indicated as P_{OAE}; if the switch with this source is thrown from the position drawn here, this source is removed, which produces a good approximation of most middle-ear behavior. Influences of other variables on the middle ear are indicated by arrows below the two-port.

incudostapedial joints are thought to introduce little relative motion in response to sound,[12–16] but they flex appreciably during muscle contraction[17] and with large static pressures across the tympanic membrane.[18] The approximately pistonlike motion of the stapes into the fluid of the cochlea, which can be described by a volume velocity U_S, generates a sound pressure in the vestibule P_V.

In the classical view the lever mechanism produced by the TM with rigid-body rotational motion of the malleus and incus increases the sound pressure from external ear

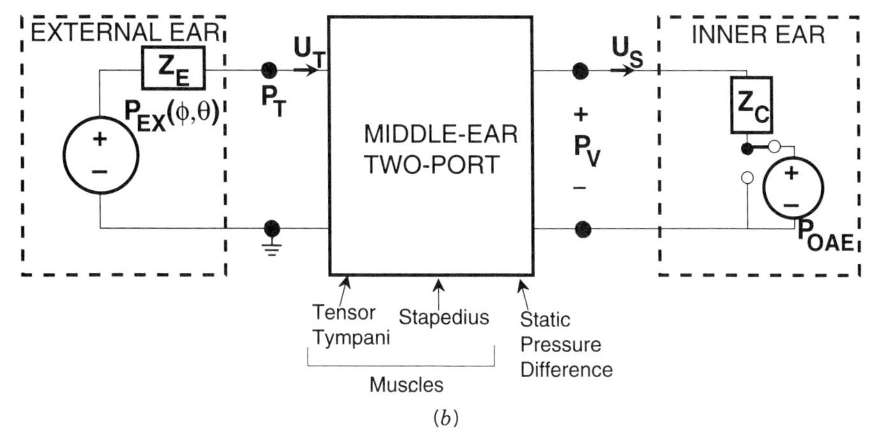

(*b*)

Fig. 1 (*Continued*)

to inner ear.[19,20,21] Recent measurements of displacements along the manubrium of the malleus in response to sound indicate that the motion is more complicated than pure rotation, at least for some conditions, and the manubrium does not move as a rigid body at higher frequencies.[22,23] Thus, it seems clear that mechanisms other than a rigid lever are involved.[24]

The acoustic transmission of the middle ear can be represented by an acoustic two-port (Fig. 1*b*), in which the input terminal pair is connected to the output of the external ear and the output terminal pair is connected to the cochlea (Fig. 1*b*). Measurements (and theory[25]) indicate that the acoustic input impedance of the inner ear Z_C is approximately resistive at least in the mid-frequency range.[16,26,27] In normal operation transmission of acoustic signals through the middle ear is altered (usually reduced) by contractions of the middle-ear muscles[28] and by static pressure difference across the tympanic membrane,[29] which occurs either because of changes in ambient pressure (e.g., as a result of a change in altitude) or because of exchange of gas between the middle-ear cavity and its soft-tissue lining.[30] These static pressure differences are normally removed by momentary openings of the eustachian tube, which equalize the middle ear and ambient pressures. Pathological conditions in the middle ear also produce reductions in transmission through the ossicular chain. For instance, with middle-ear inflammation (otitis media) fluid builds up in the middle-ear cavity that can alter the mobility of the TM and ossicles[31,32]; in otosclerosis abnormal growth of bone around the footplate of the stapes can stiffen the suspension of the stapes so that less signal is coupled to the cochlea.[33] Reduced hearing sensitivity that results from interference with the mechanical operation of the middle ear is called *conductive hearing loss*. Such hearing deficits may be alleviated either through external amplification (hearing aids) or through medical and surgical procedures.[34–36]

In recent years measurements have shown that a variety of acoustic signals can be recorded in the external ear that seem to originate in the cochlea (P_{OAE} in Fig. 1*b*). These signals, which originate in the inner ear, referred to collectively as *otoacoustic emissions* (OAE),[37–40] are coupled to the external ear by the middle ear. As OAEs are relatively small and apparently have little effect on inward transmission through the middle ear, they will not be considered further. That is, we will assume that the switch in Fig. 1*b* is thrown so as too disconnect P_{OAE} from the circuit and the ear canal.

3 TERMINAL PERFORMANCE OF THE MIDDLE EAR

3.1 Linearity

In response to sound stimuli (and in the absence of reflex contractions of the middle-ear muscles) the middle ear operates approximately as a linear acoustic system up to very high sound pressure levels (SPL). In ears of anesthetized cats response amplitudes are proportional to stimulus amplitude for tonal stimuli with P_T up to 140 dB SPL.[13,26,41,42]

One kind of nonlinear behavior of the middle ear is commonly experienced when static pressure builds up in the middle ear. The reduction in acoustic sensitivity that accompanies static pressure differences of the order of 10^{-2} atmospheres (e.g., during aircraft ascent or descent) demonstrates that these large pressures (relative to normal *sound* pressures) can produce substantial variations from linear behavior. Another kind of nonlinearity has been demonstrated in which the input impedance varies with the level of the input sound pressure.[43] These rather small deviations from linearity, which occur in narrow frequency bands at low stimulus levels, are apparently generated by cochlear processes and are called *stimulus-frequency otoacoustic emissions.*

3.2 Acoustic Input Impedance

Measurements of acoustic impedance (or admittance) at the TM have been used to assess the acoustic behavior of the middle ear for scientific,[44] medical,[33] and engineering[45] purposes. Measurements in live humans are complicated by uncertainty about the representation of the coupling space between the measurement site and the tympanic membrane, which has a large effect at the higher frequencies.[46] In human cadaver ears measurements can be made near to the TM, the coupling space can be assessed more accurately, and results have been extended to somewhat higher frequencies (5 kHz[47]). In ears of anesthetized animals measurements have been reported to above 10 kHz.[44,48,49] Measurements in human ears also extend to higher frequencies, but the location of the measurements relative to the tympanic membrane is not known accurately.[50–52]

Impedance measurements in human ears (Fig. 2*a*) have some consistent features. At low frequencies (i.e., below 0.3 kHz) the angle is near −0.25 periods and the magnitude is inversely proportional to frequency; this is the behavior of an acoustic compliance (or mechanical spring). At higher frequencies (1–3 kHz) the impedance angle is near zero and the magnitude does not show large changes with frequency, which is resistance-like behavior. Interear variations in magnitude of somewhat less than a factor of 10 are consistent with measurements over larger population of ears at low frequencies.[53–55] These features are also generally seen in measurements from other mammalian species (Fig. 2*b*).

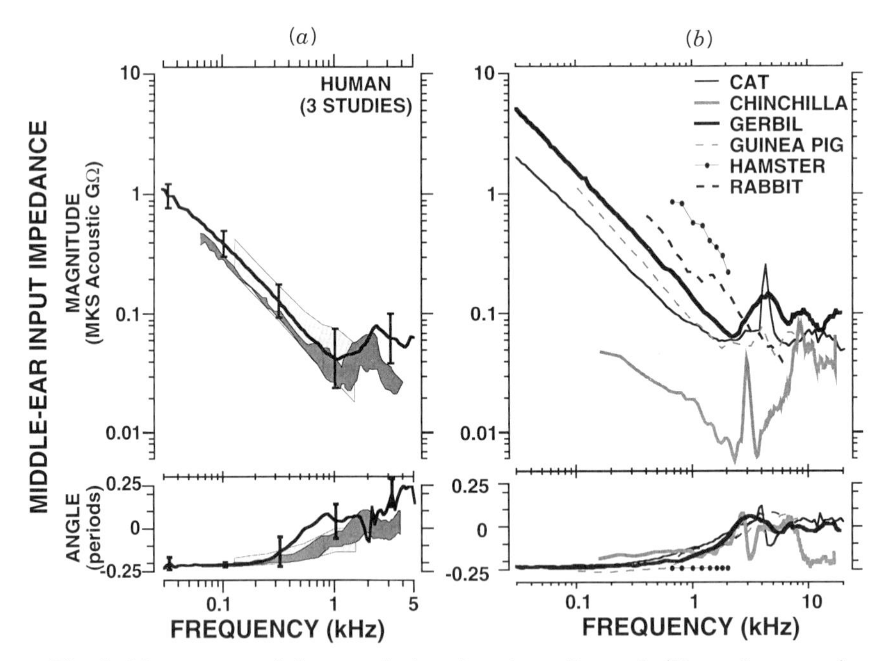

Fig. 2 Measurements of the acoustic input-impedance $Z_T = P_T/U_T$ at the tympanic membrane of (*a*) human ears and of (*b*) some other mammals. Magnitudes and angles of the measured impedances are plotted vs. frequency. The vertical scale for the magnitude plots is given in MKS acoustic $G\Omega = 10^9$Pa-s/m^3. For the angle plots the indicated scale is in periods where 1 period = 2π radians = 360°. (*a*) A summary of human ear measurements from three studies. The lighter shaded area represents the 10th through the 90th percentile ranges of 33 subjects with normal hearing.[33] The darker shaded area is the range for 4 young males with normal hearing.[46] The solid line (with error bars) represents the median and standard deviation from measurements in 15 cadaver ears.[47] (*b*) Measurements of acoustic input impedance of the middle ear in six mammalian species. For each species average magnitudes and phase angles are plotted.[44]

The acoustic impedance of ears with pathological conditions that cause conductive hearing losses can deviate substantially from normal.[33,35] However, in spite of numerous investigations of the clinical application of such results,[56–58] measurements of impedance magnitude are not now widely used in diagnosis of middle ear pathologies, partly because their ability to define the nature and location of the abnormality is too limited and partly because surgeons find other diagnostic methods preferable.

3.3 Transfer Functions

Two kinds of middle-ear transfer ratio are shown in Fig. 3. In Fig. 3*a* the ratio of output volume velocity U_S to input pressure P_T is shown. At low frequencies this transfer admittance's magnitude increases proportionally with frequency and its angle leads the applied pressure by about 0.25 periods. This result indicates that the middle-ear structures are displaced in phase with each other and in phase with the input pressure, and suggests that motion is primarily determined by elastic forces. At the highest frequencies the angle of the transfer function decreases, generally becoming more negative than −0.25 periods; the total change in angle with frequency of more than 0.5 periods indicates that an analytical representation of the middle-ear transmission function must be of higher order than second; that is, if a wide frequency range is considered, the middle ear is more complex than a simple harmonic oscillator.

The input–output pressure ratio (Fig. 3*b*) shows that in the midfrequency range the middle ear produces a pressure gain of 20–30 dB. Presumably the mechanism responsible is primarily the ratio of TM area to stapes-

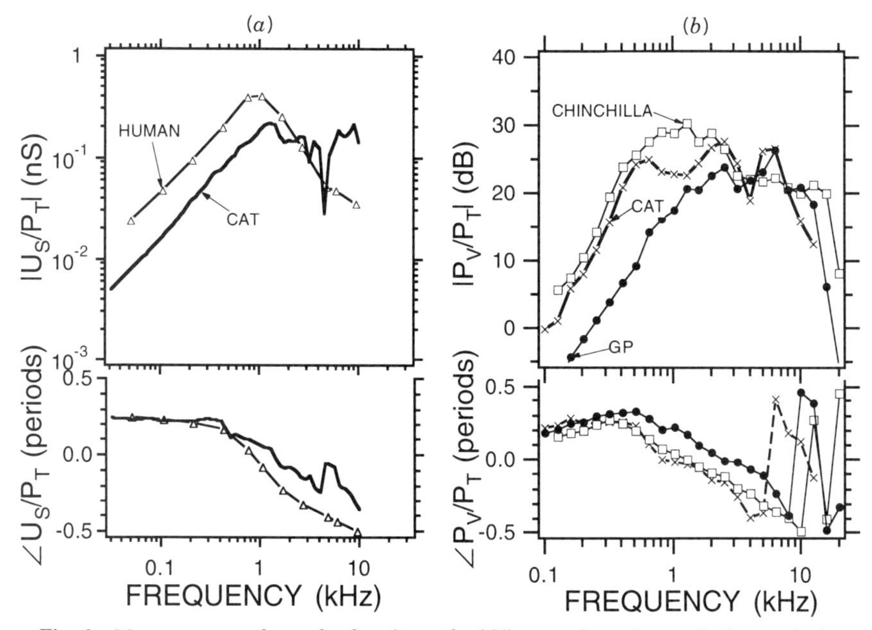

Fig. 3 Measurements of transfer functions of middle ears. In each case both magnitudes (upper) and angles (lower) are plotted versus frequency. (*a*) The acoustic transfer admittance U_S/P_T from human and cat ears. The vertical scale is acoustic nanoSiemens (i.e., 1 nS = $10^{-9} m^3/Pa\text{-}s$). The cat measurements were made on 25 anesthetized animals with observation of the stapes motion through a microscope[13]; corrections were made for the effects of opening the middle-ear cavities.[59] The human results[81] were obtained on 20 cadaver ears through acoustic measurement of volume velocity of the round window, which is assumed to equal the volume velocity of the stapes.[82] (*b*) Averaged measurements of the pressure transfer ratio P_V/P_T for cat (N = 6), guinea pig (GP) (N = 5), and chinchilla (N = 4). In each case the sound pressure in the vestibule of the inner ear was determined by a pressure transducer that was introduced through a hole drilled into the bone of the inner ear.[15] Middle-ear cavities are "intact." In the angle plot (lower, right) each of the plotted values jumps at a high frequency by about 1 period because the plotted angles are restricted to the ±0.5 period range. (From Ref. 15. Used with permission.)

footplate area, which is in the range 26–30 dB for many species (see Ref. 4, Fig. 6.15). The frequency dependence of the pressure gain demonstrates that the TM-ossicular chain does not act as an ideal transformer.[4,24]

From measurements of input impedance and transfer ratios it is possible to determine a power transmission efficiency for the middle ear. Available measurements on cat ears indicate that over the audio-frequency range the efficiency is a maximum of about 50%; that is, at least half the power absorbed into the middle ear is not delivered to the inner ear.[59]

3.4 Alterations in Middle-Ear Transmission

Static Air Pressure Differences between static pressures in the ear canal and in the middle-ear cavity can change the acoustic input impedance, forward sound transmission, and outward transmission of cochlear acoustic emissions. Pressure differences of 10 cm of H_2O (0.01 atm = 1 kPa) produce detectable changes in humans ears[60–63] and in other mammals (e.g., Fig. 4*a* and *b*). Positive pressures produce reductions that increase with the size of the pressure difference and are generally larger for the lower frequencies. Negative pressures have a more complex pattern with some pressures producing an increase in admittance for higher frequencies. The lack of equality of the input admittance change (Fig. 4*a*) and the transmission change (Fig. 4*b*) indicates that the stiffening caused by static pressures changes the coupling of TM velocity to stapes velocity.

The effect of static pressure is the basis of a clinical diagnostic procedure. Measurements of the input admittance magnitude (usually at a low frequency, such as 226 Hz) as a function of the static pressure (e.g., ±0.2 kPa) applied in the ear canal are called tympanograms.[57,58] In normal ears tympanograms have a sharp maximum (in admittance magnitude) when the ear canal static pressure equals that in the middle ear and the magnitude decreases for positive or negative pressures differences. Deviations from this pattern are evidence for conductive abnormalities in the middle ear. For instance, flat tympanograms (i.e., an input admittance magnitude that is relatively insensitive to static-pressure changes) indicate the presence of fluid in the middle ear.[53]

Middle-Ear Muscle Contractions The middle ear muscles contract under a variety of circumstances including high intensity sound stimulation (the "acoustic reflex"), before and during vocalization, after tactile stimulation to the skin of the head, and during general body motion.[64] The stapedius muscle has been more completely characterized than the tensor tympani, but the effects of their contractions appear to be similar, at least in rabbit, cat, and guinea pig.[29,65,66]

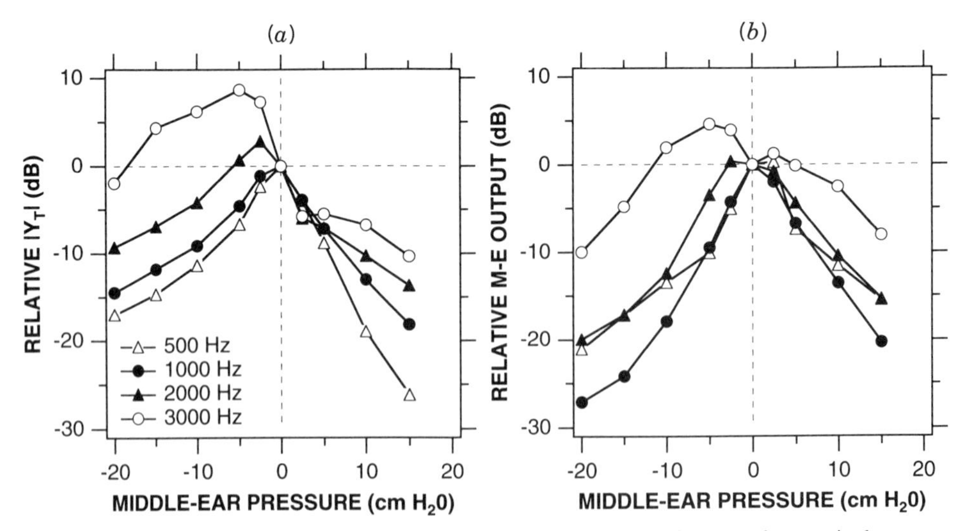

Fig. 4 Effects of static pressure difference across the tympanic membrane on the magnitudes of both the acoustic input admittance $Y_T = U_T/P_T = (Z_T)^{-1}$ and the transmission of a middle ear of cat. The abscissa in both plots is the static pressure in the middle-ear airspace P_{ME} relative to ambient. Measurements are shown for four frequencies. In (*a*) the ordinate is $20 \log_{10} |Y_T(P_{ME})|/|Y_T(0)|$. The transmission changes in (*b*) were measured in the cochlear electric potential recorded in response to tonal stimuli. (Data from Ref. 12; similar, but more extensive transmission change results from cat ears are reported in Ref. 19.)

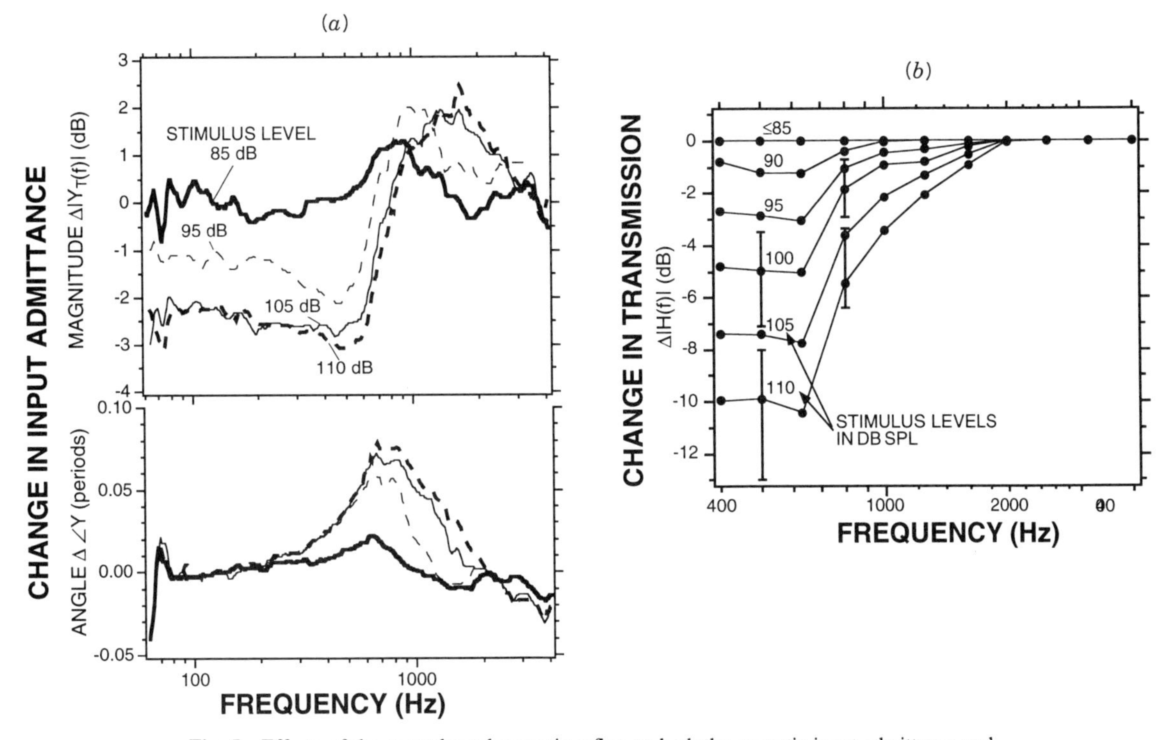

Fig. 5 Effects of the contralateral acoustic reflex on both the acoustic input admittance and the transmission of the human middle ear. The acoustic reflex was produced by relatively intense bursts of 2–4 kHz bandpass noise delivered (through an earphone) to the contralateral ear. The duration of each noise burst was 0.65 s with the impedance measurements made during the last 0.25 s, when the reflex contraction appeared to be approximately constant. (*a*) Reflex-induced changes in the magnitude and angle of the acoustic input admittance vs. frequency with the level of the reflex-producing noise as a parameter. The $\Delta|Y_T|$ ordinate is plotted as $20\log_{10}|Y_T(during\ reflex)|/|Y_T(without\ reflex)|$. From one subject (of 4) whose reflex changes were nearest to the mean. (*b*) The average change in the magnitude of the middle-ear transmission is shown for the four subjects. These measurements (made at 11 frequencies) are based on a psychophysical method in which the rapid change with input level of the phase of a perceptual combination tone, $2F_1–F_2$, is used.[77,83] (From Ref. 77, pp. 196, 209. Used with permission.)

In human ears changes in the middle-ear acoustic behavior following brief high-intensity sound stimulation are primarily the result of reflex contraction of the stapedius.[28] These changes can be studied by applying the reflex-evoking sound to one ear while measuring changes in the opposite ear. Results of such measurements are shown in Fig. 5. In Fig. 5*a* the changes in magnitude and angle of the input admittance are plotted. For frequencies below 0.5 kHz the admittance magnitude decreases by a factor that is approximately independent of frequency and the angle is roughly unchanged. These results indicate that the compliance of the ear is decreased by the reflex. A change is seen for levels of the evoking stimulus greater than 85 dB SPL, and the effect ceases growing with level between 105 and 110 dB. At higher frequencies (1–3 kHz) the admittance magnitude is increased during the reflex.

The corresponding changes in transmission (Fig. 5*b*) are reductions in magnitude at low frequencies again with approximately constant (with frequency) changes for frequencies below 0.63 kHz. The transmission changes appear first for stimulus levels above 85 dB SPL as do the admittance changes, but the transmission continues to decrease with increasing stimulus level up to the highest level (110 dB).

The action of the stapedius contraction is to displace the head of the stapes in a direction normal to its motion in response to sound. Measurements in cats show that transmission decreases of 10 dB (as in Fig. 5*b*) result from a displacement of 40 μm. With muscle contraction the stapes head slides parallel to the incudo-stapedial joint surface and does not produce a comparable motion of the incus.[17]

Various consequences of the acoustic reflex have been shown: (1) The absence of the reflex, in patients with Bell's palsy, can have detrimental effects on the discrimination of speech sounds presented at high levels.[67] (2) The reflex can reduce the loss of hearing sensitivity that occurs from prolonged high-intensity sound stimulation.[68] (3) Activation of the stapedius muscle can have a large effect (e.g., 40 dB) in reducing masking by low-frequency sound of high-frequency stimuli.[69,70] (4) In clinical practice detection of reflex-evoked changes in acoustic impedance provides objective evidence of normal operation of the lower part of the central auditory system.[71]

4 MODELS OF THE MIDDLE EAR

Knowledge of the acousto-mechanical relationships involved in the middle ear's operation have been represented in network models, usually presented as electric-circuit analogs.[66,72–76] The effects of the muscles may be represented by changes of impedances when the muscle contracts.[66]

Circuit models of this kind can be an efficient summary of our knowledge of the middle ear. For instance, the model of Kringlebotn[76] can be used to represent the input impedance and middle-ear transfer function of normal and pathological human ears,[74] and transmission measurements. This kind of model can also explain the observation (Fig. 5) that the effect of the stapedius reflex on the input admittance saturates with increasing stimulus, even as the transmission continues to decrease.[77] However, we must remember that we have built assumptions into the models. For instance, most models assume that only one variable is required to represent the motion of the malleus. If, as has been reported,[22] the components of mallear motion or the location of the rotational axis changes (e.g., with frequency or with contraction of the muscle), or the manubrium bends,[23,78] the model must be modified to include these features. Another kind of behavior such models do *not* represent is the effects of coupling to the cochlea through mechanisms other than the motion of the ossicular chain.[79,80]

REFERENCES

1. O. W. Henson, "Comparative Anatomy of the Middle Ear," in W. D. Keidel and W. D. Neff (Eds.), *Handbook of Sensory Physiology, Vol. V/1, Auditory System. Anatomy. Physiology (Ear)*, Springer-Verlag, New York, 1974, pp. 39–110.
2. E. G. Wever, *The Reptile Ear*, Princeton University Press, Princeton, NJ, 1979.
3. G. Fleischer, "Evolutionary Principles of the Mammalian Middle Ear," in *Advances in Anatomy Embryology and Cell Biology*, Hill-Fay Associates, Winnetka, IL, 1978, pp. 3–69.
4. J. J. Rosowski, "Outer and Middle Ears," in R. R. Fay and A. N. Popper (Eds.), *Comparative Hearing: Mammals*, Springer-Verlag, New York, 1994, pp. 172–247.
5. I. Kirikae, *The Structure and Function of the Middle Ear*, University of Tokyo Press, Tokyo, 1960.
6. M. J. Novacek, "Aspects of the Problem of Variation, Origin and Evolution of the Eutherian Auditory Bulla," *Mammal Rev.*, Vol. 7, 1977, pp. 131–150.
7. B. J. Anson and J. A. Donaldson, *The Surgical Anatomy of the Temporal Bone and Ear*, Saunders, Philadelphia, 1967.
8. J. Sadé (Ed.), *The Eustachian Tube*, Kugler, Amsterdam/Berkeley, 1978.
9. I. Honjo, *Eustachian Tube and Middle Ear Diseases*, Springer-Verlag, Tokyo, 1988.
10. M. R. Stinson, "The Spatial Distribution of Sound Pressure within Scaled Replicas of the Human Ear Canal," *J. Acoust. Soc. Am.*, Vol. 78, 1981, pp. 1596–1602.
11. M. R. Stinson and S. M. Khanna, "Sound Propagation in the Ear Canal and Coupling to the Eardrum, with Measurements on Model Systems," *J. Acoust. Soc. Am.*, Vol. 85, 1989, 2481–2491.
12. A. R. Møller, "Transfer Function of the Middle Ear," *J. Acoust. Soc. Am.*, Vol. 35, 1963, pp. 1526–1534.
13. J. J. Guinan, Jr. and W. T. Peake, "Middle-Ear Transmission in the Anesthetized Cat," *J. Acoust. Soc. Am.*, Vol. 41, 1967, pp. 1237–1261.
14. W. R. Rhode, "Some Observations on Cochlear Mechanics," *J. Acoust. Soc. Am.*, Vol. 64, 1978, pp. 158–176.
15. L. Décory, "Origine des Différences Interspecifiques de Susceptibilité an Bruit," Thése de Doctorat de l'Université Bordeaux II, France, 1989.
16. M. A. Ruggero, N. C. Rich, L. Robles, and B. G. Shivapuja, "Middle-Ear Response in the Chinchilla and Its Relationship to Mechanics at the Base of the Cochlea," *J. Acoust. Soc. Am.*, Vol. 87, 1990, pp. 1612–1629.
17. X. D. Pang and W. T. Peake, "How Do Contractions of the Stapedius Muscle Alter the Acoustic Properties of the Ear?" in J. B. Allen, J. L. Hall, A. Hubbard, S. T. Neely, and A. Tubis (Ed.), *Peripheral Auditory Mechanisms*, Springer-Verlag, New York, 1986.
18. K. B. Hüttenbrink, "The Mechanics of the Middle-Ear at Static Air Pressure," *Acta Otolaryngol.*, Suppl. 45, 1988, pp. 1–35.
19. E. G. Wever and M. Lawrence, *Physiological Acoustics*, Princeton University Press, Princeton, NJ, 1954.

20. P. Dallos, *The Auditory Periphery: Biophysics and Physiology*, Academic, New York, 1973.
21. J. J. Zwislocki, "The Role of the External and Middle Ear in Sound Transmission," in D. B. Tower and E. E. Eagles (Eds.), *The Nervous System, Vol. 3, Human Communication and Its Disorders*, Raven, New York, 1975, pp. 45–56.
22. W. F. Decraemer, S. M. Khanna, and W. R. J. Funnell, "Malleus Vibration Mode Changes with Frequency," *Hear. Res.*, Vol. 54, 1991, pp. 305–318.
23. W. F. Decraemer, S. M. Khanna, and W. R. J. Funnell, "A Method for Determining Three-Dimensional Vibration in the Ear," *Hear. Res.*, Vol. 77, 1994, pp. 19–37.
24. R. L. Goode, M. Killion, K. Nakamura, and S. Nishihara, "New Knowledge about the Function of the Human Ear: Development of an Improved Model," *Am. J. Otol.*, Vol. 15, 1994, pp. 145–154.
25. J. J. Zwislocki, "Analysis of Some Auditory Characteristics," in R. D. Luce, R. R. Bush, and E. Galanter (Eds.), *Handbook of Mathematical Psychology*, Wiley, New York, 1965.
26. A. Dancer and R. Franke, "Intracochlear Sound Pressure Measurements in Guinea Pigs," *Hear. Res.*, Vol. 2, 1980, pp. 191–205.
27. T. J. Lynch, III, V. Nedzelnitsky, and W. T. Peake, "Input Impedance of the Cochlea in Cat," *J. Acoust. Soc. Am.*, Vol. 72, 1982, pp. 108–130.
28. E. Borg, "A Quantitative Study of the Effect of the Acoustic Stapedius Reflex on Sound Transmission through the Middle Ear of Man," *Acta Otolaryngol.*, Vol. 66, 1968, pp. 461–472.
29. A. R. Møller, "An Experimental Study of the Acoustic Impedance of the Middle Ear and Its Transmission Properties," *Acta Otolaryngol.*, Vol. 60, 1965, pp. 129–149.
30. W. J. Doyle and J. T. Serok, "Middle Ear Gas Exchange in Rhesus Monkeys," *Ann. Otol. Rhinol. Laryngol.*, Vol. 103, 1994, pp. 636–645.
31. R. J. Nozza, C. D. Bluestone, D. Kardatzke, and R. Bachman, "Identification of Middle-Ear Effusion by Aural Acoustic Admittance and Otoscopy," *Ear Hearing*, Vol. 15, 1994, pp. 310–323.
32. M. Von Unge, W. F. Decraemer, J. J. Dirckx, and D. Bägger-Sjobäck, "Shape and Displacement Patterns of the Gerbil Tympanic Membrane in Experimental Otitis Media with Effusion," *Hear. Res.*, Vol. 82, 1995, pp. 184–196.
33. J. J. Zwislocki and A. Feldman, "Acoustic Impedance of Pathological Ears," ASHA Monographs Number 15, American Speech and Hearing Association, Washington, D.C., 1970.
34. J. B. Booth (Ed.), *Otology, Vol. 3*, in *Scott-Brown's Otolaryngology*, 5th ed., Butterworths, London, 1987.
35. J. B. Nadol, Jr., and H. F. Schucknecht (Eds.), *Surgery of the Ear and Temporal Bone*, Raven, New York, 1993.
36. R. T. Sataloff and J. Sataloff, *Hearing Loss*, 3rd ed., M. Dekker, New York, 1993.
37. D. T. Kemp, "Stimulated Acoustic Emissions from within the Human Auditory System," *J. Acoust. Soc. Am.*, Vol. 64, 1978, pp. 1386–1391.
38. P. Zurek, "Acoustic Emissions from the Ear: A Summary of Results from Humans and Animals," *J. Acoust. Soc. Am.*, Vol. 78, 1985, pp. 340–344.
39. B. L. Lonsbury-Martin, F. P. Harris, B. B. Stagner, B. A. Hawkins, and G. K. Martin, "Distortion-Product Emissions in Humans I. Basic Properties in Normally-Hearing Subjects," *Annals ORL*, Vol. 99, 1990, pp. 3–14.
40. B. P. Kimberley, I. Hernadi, A. M. Lee, and D. K. Brown, "Predicting Pure Tone Thresholds in Normal and Hearing Impaired Ears with Distortion Product Emissions and Age," *Ear Hearing*, Vol. 15, 1994, pp. 199–209.
41. V. Nedzelnitsky, "Sound Pressure in the Basal Turn of the Cat Cochlea," *J. Acoust. Soc. Am.*, Vol. 68, 1980, pp. 1676–1689.
42. T. J. F. Buunen and M. S. M. G. Vlaming, "Laser-Doppler Velocity Meter Applied to Tympanic Membrane Vibrations in Cat," *J. Acoust. Soc. Am.*, Vol. 69, 1980, pp. 744–750.
43. E. Zwicker and E. Schloth, "Interrelation of Different Oto-Acoustic Emissions," *J. Acoust. Soc. Am.*, Vol. 75, 1984, pp. 1148–1154.
44. M. E. Ravicz, J. J. Rosowski, and H. F. Voigt, "Sound-Power Collection by the Auditory Periphery of the Mongolian Gerbil *Meriones unguiculatus*: I. Middle-Ear Input Impedance," *J. Acoust. Soc. Am.*, Vol. 92, 1992, pp. 157–177.
45. J. J. Zwislocki, "An Ear-like Coupler for Earphone Calibration," Special Report LSC-S-9 of Laboratory of Sensory Communication, 1971.
46. W. M. Rabinowitz, "Measurement of the Acoustic Input Immitance of the Human Ear," *J. Acoust. Soc. Am.*, Vol. 70, 1980, pp. 1025–1035.
47. J. J. Rosowski, P. J. Davis, S. N. Merchant, K. M. Donahue, and M. D. Coltrera, "Cadaver Middle Ears as Models for Living Ears: Comparisons of Middle-Ear Input Impedance," *Ann. Otol. Rhinol. Laryngol.*, Vol. 99, 1991, pp. 403–412.
48. J. B. Allen, "Measurements of Eardrum Acoustic Impedance," in J. B. Allen, J. L. Hall, A. Hubbard, S. T. Neely, and A. Tubis (Eds.), *Peripheral Auditory Mechanisms*, Springer-Verlag, New York, 1986, pp. 44–51.
49. T. J. Lynch, III, W. T. Peake, and J. J. Rosowski, "Measurements of the Acoustic Input Impedance of Cat Ears: 10 Hz to 20 kHz," *J. Acoust. Soc. Am.*, Vol. 96, 1994, pp. 2184–2209.
50. H. Hudde, "Measurement of the Eardrum Impedance of Human Ears," *J. Acoust. Soc. Am.*, Vol. 73, 1983, pp. 242–247.
51. D. H. Keefe, J. C. Bulen, K. H. Arehart, and E. M. Burns, "Ear-Canal Impedance and Reflection Coefficient

in Human Infants and Adults," *J. Acoust. Soc. Am.*, Vol. 94, 1993, pp. 2617–2638.

52. S. E. Voss and J. B. Allen, "Measurements of Acoustic Impedance and Reflectance in the Human Ear Canal," *J. Acoust. Soc. Am.*, Vol. 95, 1994, pp. 372–384.
53. J. Jerger, S. Jerger, and L. Mauldin, "Studies in Impedance Audiometry. I. Normal and Sensorineural Ears," *Arch. Otolaryngol.*, Vol. 96, 1972, pp. 513–523.
54. A. S. Feldman and L. A. Wilber, *Acoustic Impedance & Admittance—The Measurement of Middle Ear Function*, Williams & Wilkens, Baltimore, 1976.
55. R. H. Margolis and J. E. Shanks, "Tympanometry," in J. Katz (Ed.), *Handbook of Clinical Audiology*, 3rd ed., Williams and Wilkens, Baltimore, 1985. Chap. 23.
56. Working Group on Aural Acoustic-immitance Measurements, Committee on Audiologic Evaluation, "Tympanometry," *J. Speech Hear. Disord.*, Vol. 53, 1988, pp. 354–377.
57. R. H. Margolis and J. E. Shanks, "Tympanometry: Basic Principles and Clinical Applications," in William F. Rintelmann (Ed.), *Hearing Assessment*, 2nd ed., Pro-Ed, Austin, TX, 1991, pp. 179–245.
58. J. W. Hall, III and D. Chandler, "Tympanometry in Clinical Audiology," in J. Katz (Ed.), *Handbook of Clinical Audiology*, Williams and Wilkens, Baltimore, 1994, Chap. 20.
59. J. J. Rosowski, L. H. Carney, T. J. Lynch, III, and W. T. Peake, "The Effectiveness of the External and Middle Ears in Coupling Acoustic Power into the Cochlea," in J. B. Allen, J. L. Hall, A. Hubbard, S. T. Neely, and A. Tubis (Eds.), *Peripheral Auditory Mechanisms*, Springer-Verlag, New York, 1986, pp. 3–12.
60. H. Rasmussen, "Studies on the Effect on the Air Conduction and Bone Conduction from Changes in Meatal Pressure in Normal Subjects and Otosclerotic Patients," *Acta Otolaryngolog.* Suppl. 74, 1948, pp. 54–64,
61. E. H. Huizing, "Bone Conduction—The Influence of the Middle Ear," *Acta Otolaryngol.* Suppl. 155, 1960, pp. 1–99.
62. H. Wada and T. Kobayashi, "Dynamical Behavior of Middle Ear: Theoretical Study Corresponding to Measurement Results Obtained by a Newly Developed Measuring Apparatus," *J. Acoust. Soc. Am.*, Vol. 87, 1990, pp. 237–245.
63. S. L. Naeve, R. H. Margolis, S. C. Levine, and E. M. Fournier, "Effect of Ear-Canal Air Pressure on Evoked Oto-Acoustic Emissions," *J. Acoust. Soc. Am.*, Vol. 91, 1992, pp. 2091–2095.
64. S. Silman (Ed.), *The Acoustic Reflex: Basic Principles and Clinical Applications*, Academic Press, New York, 1984.
65. E. Teig, "Differential Effect of Graded Contraction of Middle Ear Muscles on the Sound Transmission of the Ear," *Acta Physiol. Scand.*, Vol. 88, 1973, pp. 382–391.
66. A. L. Nuttall, "Tympanic Muscle Effects on Middle-Ear Transfer Characteristic," *J. Acoust. Soc. Am.*, Vol. 56, 1974, pp. 1239–1247.
67. E. Borg and J-E. Zakrisson, "Stapedius Reflex and Speech Features," *J. Acoust. Soc. Am.*, Vol. 54, 1973, pp. 525–527.
68. F. B. Simmons, "Middle Ear Muscle Protection from the Acoustic Trauma of Loud Continuous Sound," *Ann. Otol. Rhinol. Laryngol.*, Vol. 69, 1960, pp. 1063–1072.
69. E. Borg and J-E. Zakrisson, "Stapedius Reflex and Monaural Masking," *Acta Otolaryngolog.*, Vol. 78, 1974, pp. 155–161.
70. X. D. Pang, "Effects of Stapedius-Muscle Contractions on Masking of Tone Responses in the Auditory Nerve," RLE Tech. Rep. No. 544, Research Laboratory of Electronics, M.I.T. Cambridge, MA, 1989.
71. R. H. Wilson and R. H. Margolis, "Acoustic-Reflex Measurement," in William F. Rintelmann (Ed.), *Hearing Assessment*, 2nd ed., Pro-Ed Inc., Austin, TX, 1991, pp. 247–319.
72. Y. Onchi, "Mechanism of the Middle Ear," *J. Acoust. Soc. Am.*, Vol. 33, 1961, pp. 794–805.
73. A. R. Møller, "Network Model of the Middle Ear," *J. Acoust. Soc. Am.*, Vol. 33, 1961, pp. 168–176.
74. J. Zwislocki, "Analysis of the Middle-Ear Function. Part I: Input Impedance," *J. Acoust. Soc. Am.*, Vol. 34, 1962, pp. 1514–1523.
75. J. W. Matthews, "Modeling Reverse Middle Ear Transmission of Acoustic Distortion Signals," in E. de Boer and M. A. Viergever (Eds.), *Mechanics of Hearing*, Martinus Nijhoff, The Hague, 1983, pp. 11–17.
76. M. Kringlebotn, "Network Model for the Human Middle Ear," *Scand. Audiol.*, Vol. 17, 1988, pp. 75–85.
77. W. M. Rabinowitz, "Acoustic-Reflex Effects on the Input Admittance and Transfer Characteristics of the Human Middle Ear," Ph.D. Thesis, Massachusetts Institute of Technology, Cambridge, MA, 1977.
78. W. R. J. Funnell, S. M. Khanna, and W. F. Decraemer, "On the Rigidity of the Manubrium in a Finite Element Model of the Cat Eardrum," *J. Acoust. Soc. Am.*, Vol. 91, 1992, pp. 2081–2090.
79. W. T. Peake, J. J. Rosowski, and T. J. Lynch, III, "Middle-Ear Transmission: Acoustic versus Ossicular Coupling in Cat and Human," *Hear. Res.*, Vol. 57, 1992, pp. 245–268.
80. C. A. Shera and G. Zweig, "Middle-Ear Phenomenology: The View from the Three Windows," *J. Acoust. Soc. Am.*, Vol. 92, 1992, pp. 1356–1369.
81. M. Kringlebotn and T. Gundersen, "Frequency Characteristics of the Middle Ear," *J. Acoust. Soc. Am.*, Vol. 77, 1985, pp. 159–164.
82. M. Kringlebotn, "The Equality of Volume Displacements in the Inner Ear Windows," *J. Acoust. Soc. Am.*, Vol. 98, 1995, pp. 192–196.
83. J. L. Goldstein, "Auditory Nonlinearity," *J. Acoust. Soc. Am.*, Vol. 41, 1967, pp. 676–689.

86

ANATOMY OF THE COCHLEA AND AUDITORY NERVE

NORMA B. SLEPECKY

1 INTRODUCTION

The cochlea contains the sensory cells, supporting cells, and nerve fibers of the inner ear that are responsible for detection of sound. When sound pressure waves vibrate the tympanic membrane, the movement is conducted through the middle-ear space by the middle-ear bones (the malleus, incus, and stapes). The stapes passes these vibrations into the fluid-filled cochlea. The vibrations are distributed in a frequency-specific manner along the length of the sensory organ to stimulate individual hair cells and nerve fibers in the sensory organ of Corti.

It was thought initially that the organ of Corti was purely a passive detector of mechanical vibration and that normal frequency selectivity depended largely on the structural gradients of the component parts. However, it is now known that the hair cells resond to stimulation, and this response actively increases the threshold sensitivity and frequency selectivity of this exquisitely designed mechanoreceptive organ.

In order to understand the crucial steps in the process by which mechanical vibrations are transduced into electrical responses and neural impulses, it is necessary to understand the anatomy and cell biology of the sensory organ involved. The goal of this chapter is to introduce the cochlea—the sensory and supporting cells, the nerve fibers, and the cells of the stria vascularis and spiral ligament that are responsible for the production of endolymph and the maintenance of the endocochlear potential.

2 THE COCHLEA

The cochlea is actually a cavity in the temporal bone having the shape of a coiled tube. The outside of the spiral cochlea is defined by the otic capsule (Fig. 1). Within the cochlea, the membranous labyrinth contains the cells, structures, and fluids necessary for the detection of sound. The membranous labyrinth is subdivided into three anatomically defined compartments or scalae: scala vestibuli, scala media, and scala tympani (Fig. 2*a*). These three compartments separate two fluids; perilymph, which is high in sodium and typical of the type of fluids found outside of cells elsewhere in the body, and endolymph, which is high in potassium similar in composition to fluids found inside of cells. It is commonly stated that perilymph is contained in both scala vestibuli and scala tympani while endolymph is found only within the region of scala media (Fig. 2*a*). However, the actual boundaries of the fluid-filled compartments are not the same as the anatomically defined compartments described above. The endolymph containing space is in fact limited by the tight junctions between the adjacent cells of Reissner's membrane, the top surfaces of the sensory and supporting cells of the organ of Corti (forming the reticular lamina), and the marginal cells of the stria vascularis where the blood vessels run along the lateral wall (Fig. 2*b*). Thus, the sensory and supporting cells of the receptor organ of Corti have endolymph on one surface and a perilymphlike fluid on the other. The nerve terminals and fibers are bathed in a perilymphlike fluid (sometimes referred to as Cortilymph) that is high in sodium.

3 THE ORGAN OF CORTI

In the organ of Corti, the sensory and supporting cells rest on the basilar membrane. The sensory organ contains (depending on the species studied) approximately 20,000 sensory hair cells, which are in contact with various types of supporting cells. The sensory cells are described as *inner* hair cells, which form one row along the length of

Note: References to chapters appearing only in the *Encyclopedia* are preceded by *Enc.*

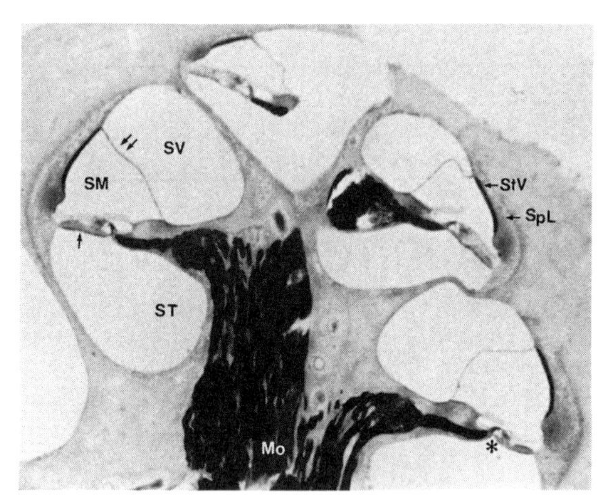

Fig. 1 Midmodiolar section of the rabbit cochlea, showing the membranous labyrinth encased in the bone of the otic capsule. Reissner's membrane (double arrow) and the basilar membrane (single arrow) form the boundaries of scala media (SM) and separate it from scala vestibuli (SV) and scala tympani (ST). The stria vascularis (StV) and the spiral ligament (SpL) lie along the lateral wall. The organ of Corti (*) is located on the basilar membrane and the nerve fibers supplying it run in the modiolus (Mo).

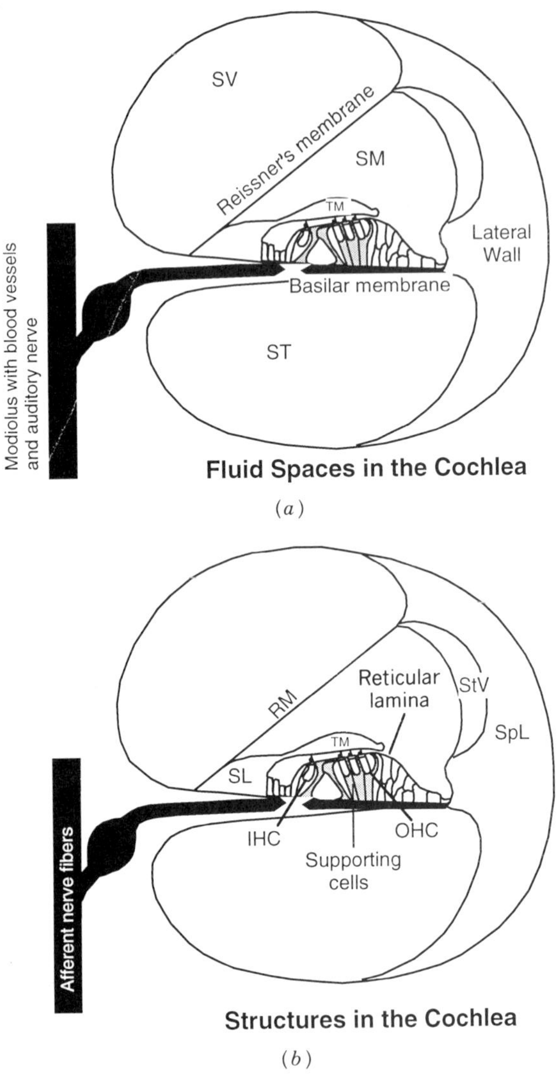

Fig. 2 (*a*) Boundaries of the endolymph containing compartment called scala media (SM) are Reissner's membrane, the basilar membrane, and the lateral wall. Both scala vestibuli (SV) and scala tympani (ST) contain perilymph. (*b*) Reticular lamina of the organ of Corti is made up of the apical surfaces of the inner (IHC) and outer (OHC) hair cells and the supporting cells. The tectorial membrane (TM) projects out from the spiral limbus (SL) and overlies the organ of Corti. The stria vascularis (StV) and the spiral ligament (SpL) line the lateral wall. Afferent nerve fibers from IHC and OHC run in the modiolus.

the sensory epithelium closer to the central axis of the spiral and the modiolus in which the auditory nerve fibers are found, and *outer* hair cells, which are found in three rows and which are separated from the inner hair cells by the inner and outer pillar supporting cells (Fig. 2*b*). During stimulation by sound, movement of the basilar membrane is transmitted through the supporting cells so that the top surfaces of the sensory and supporting cells vibrate at the same frequency as the stimulus. Shearing of the stereocilia, the hairs on the sensory hair cells, against the gelatinous mass of the overlying tectorial membrane results in movement of the stereocilia and is the event by which mechanical movement is transduced into an electrical response in the sensory cells.

The organ of Corti appears to be loosely structured because of the tremendous amount of fluid-filled spaces surrounding the sensory and supporting cells. However, cell structure and organization appear to function optimally in the transmission of movement of the basilar membrane to the apical surface of the sensory hair cells in response to both low- and high-frequency sound. The sensory cells are designed for maximal sensitivity to small motions, and the supporting cells are structured for rigid support. The cellular and structural arrangement of cells within the organ of Corti have been well studied in many different species of animals, and for the most part they appear similar.[1–3] In most species, the size of the organ of Corti changes continuously along its length from apex to base. In the apex of the cochlea, cells are usually larger, stereocilia are longer and less stiff, the basilar membrane is wider and less stiff, and the tectorial membrane has a greater mass than at the base. The absolute size varies from species to species and is probably related to the tuning capacities and sensitivities of the receptor organs. In some species there are regions with exaggerated structural features that appear only in localized regions and are correlated with specialized functions.

3.1 Sensory Hair Cells

The sensory hair cells are specialized to detect mechanical stimulation and to transmit information to nerve fibers that connect with the central nervous system. On their apical surface, the sensory hair cells have stereocilia,[4] which appear as stiff extensions present in rows arranged in a U or W shape. The stereocilia are graduated in length (Figs. 3 and 4), with the shortest in a row on the side of the hair cell nearer the central region of the cochlear spiral and the tallest in a row on the side nearer the lateral wall. All the stereocilia on one hair cell are linked together.[5] These links include side-to-side links that connect stereocilia within one row and tip links that connect the top of one stereocilium with its taller neighbor in the adjacent row (Fig. 5). Since actin filaments are present in the stereocilia and crosslinked by actin-binding proteins that form the filaments into rigid bundles, it is thought that the stereocilia are stiff and pivot at their base where they are inserted into the cuticular plate. Deflection of the stereocilia in one direction opens ion-selective channels at the tips of the stereocilia; potassium ions flow into the cell, the cell becomes depolarized, and neurotransmitter is released from the cell at an increased rate.

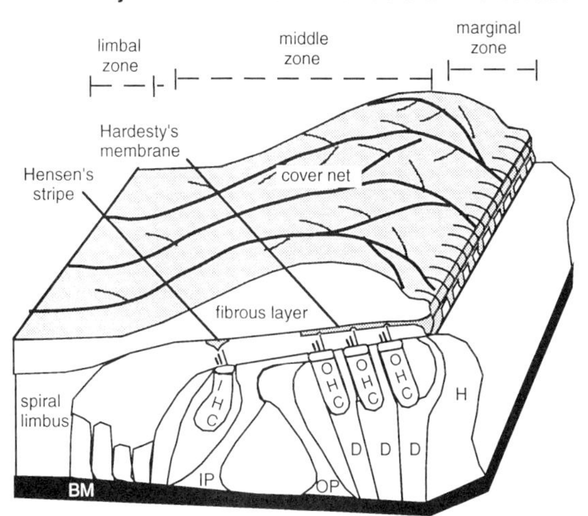

Fig. 3 Organ of Corti rests on the basilar membrane (BM). In the organ of Corti there are both IHC and OHC sensory hair cells. The inner (IP) and outer (OP) pillar cells form the tunnel of Corti. The Deiters (D) cells support the OHCs. The stereocilia from the IHC and OHC project into the tectorial membrane, which can be subdivided into several zones and layers. The limbal zone covers the spiral limbus. The middle zone overlies the sensory cells of the organ of Corti. The marginal zone is at the lateral edge of the TM. The cover net forms the superficial layer of the TM and becomes contiguous with Hardesty's membrane in the region of the marginal zone. The body of the TM is comprised mainly of the fibrous layer. Hensen's stripe and Hardesty's membrane are structures of the TM that form continuous regions (from the base of the cochlea to the apex) that are most closely associated with the tips of the stereocilia of the sensory hair cells.

Based on careful analysis of sensory hair cells, supporting cells and nerve fibers, it is now clear that inner hair cells (IHCs) differ significantly from outer hair cells (OHCs). Inner hair cells are goblet shaped with a centrally placed nucleus. Their cytoplasm contains a network of filaments and tubules that act to support the cell and maintain the cell shape. Each cell contains all of the machinery for protein synthesis. The cells have stereocilia at their top surface, which are anchored into a filament network at one end within the cell and that project into the endolymphatic fluid at their other end. The sides of the cells are in contact with supporting cells, which span the distance between the basilar membrane and the reticular lamina. The bottom ends of the cells make direct contact with the nerve fibers projecting to the central nervous system. Because the IHC afferent nerve fibers comprise 95% of the fibers within the auditory nerve, it is thought that IHCs are the primary receptor cells in the cochlea. Efferent nerve fibers carrying information from the brain to the organ of Corti make direct contact with the afferent nerve fibers in this region, but not with the IHCs.

Outer hair cells are long and cylindrical in shape, with a more basally placed nucleus (Fig. 5). They form three rows along the length of the organ of Corti, they are slightly smaller in diameter than IHCs, and they number almost four times as many. At their top surface, the OHCs contact supporting cells and at their bottom surfaces they are surrounded by supporting cells. Unlike IHCs, they contact only fluid along their lateral surfaces. Stereocilia project out from the top surface of the OHCs, and the tips of the stereocilia from the tallest row are embedded in the tectorial membrane. Above the nucleus, there is some evidence for the machinery required for protein synthesis. Along the sides of the cylinder, there is a prominent layer of membranous sacs lined with mitochondria that provide energy to the cell. At the bottom of the OHCs, the cells make direct contact with both affer-

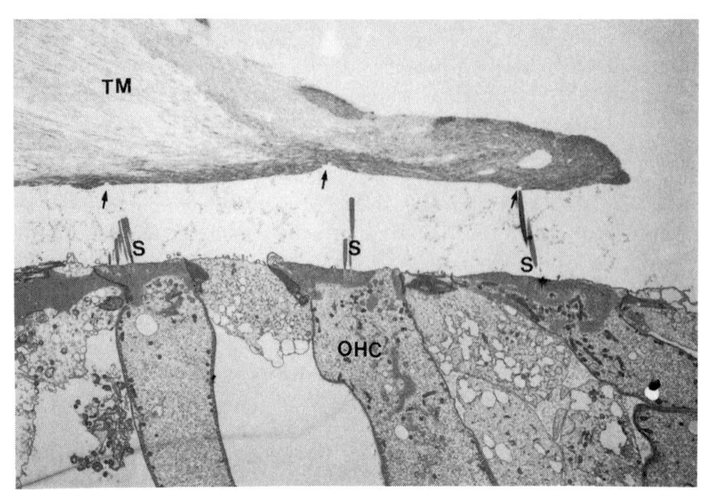

Fig. 4 In a higher magnification micrograph, it can be seen that there are indentations in the bottom surface (arrows) of the tectorial membrane (TM) where the tips of the stereocilia (S) of the outer hair cells (OHC) are embedded.

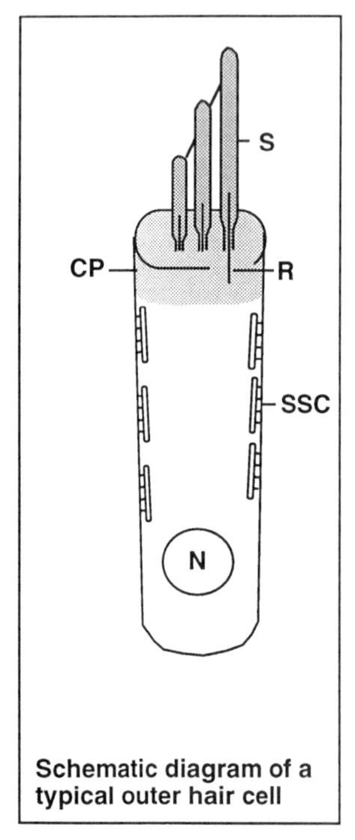

Fig. 5 Schematic diagram of a typical outer hair cell. Along the apical surface, stereocilia (S) extend out from the cuticular plate (CP). At its base, each stereocilium has a rootlet (R) that projects into the cuticular plate, and a tip link that connects it to its taller neighbor. Along the lateral wall there are membranous sacs, or subsurface cisternae (SSC). The nucleus (N) is located near the base of the cell.

ent nerve fibers going to the brain and efferent nerve fibers carrying information from the brain back out to the organ of Corti.

During stimulation by sound, OHCs (but not IHCs) shorten and elongate at both a slow and a fast rate. It has been suggested that these motile responses result in active processes that have been described in the cochlea, including the establishment of sharp frequency tuning and the generation of otoacoustic emissions. Two distinct structural differences between OHCs and IHCs could be responsible for their functional differences in the cochlea. First, efferent nerve fibers directly contact OHCs while they synapse only on afferent nerve fibers below the IHCs. Second, the OHCs appear to have a specialized cell membrane that is responsible for the fast OHC shortening and elongation.

3.2 Supporting Cells

Supporting cells in the organ of Corti have not been described in as great detail as the sensory cells; however, they have critical roles in providing structural and metabolic support. Based on their anatomy as seen at the light microscope level, they can be divided into two groups—cells with filaments and tubules in bundles forming a rigid scaffolding (to provide physical support) and cells with less organized skeletal components yet containing organelles necessary for protein synthesis and ion transport (to maintain a normal balance of ions in the cells and fluids).

The inner tunnel pillar cells, outer tunnel pillar cells, and Deiters cells are structurally similar and contain large numbers of filaments and tubules in bundles.[6] These cells are found in the region of the organ of Corti where it rests on the basilar membrane and is not directly supported by any bone. This region is critical for stimulation of the hair cells, since vibration of the basilar membrane is transmitted through the supporting cells to the stereocilia, and movement of the stereocilia results in transduction of mechanical stimulus into an electrical response in the sensory hair cells. Thus, it is not surprising that these supporting cells are specialized and contain microfilaments and microtubules to reinforce them and allow them to better withstand mechanical stress. Unlike the sensory cells, these supporting cells also contain an additional type of filaments, intermediate in size between the microtubles and microfilaments, which play a role in the maintenance of stable cell shape.[7]

Less is known about the supporting cells, which lack organized filament and tubule bundles. Hensen cells are directly adjacent to the third row of Deiters cells. In the apical turns of the cochlea, they reach from the reticular lamina to the basilar membrane. In the basal turns of the cochlea they rest on Boettcher cells. While they contain no filament or tubule bundles, they may be distinguished from the other cells in this region by their prominent oil droplets (the function of which is not yet known). Other cells that are part of the supportive system include Claudius cells, which extend from the organ of Corti to the lateral wall of the cochlea. External sulcus cells, or root cells, are located at the junction of the basilar membrane with the cells of the lateral wall. All of these supporting cells appear similar to each other morphologically,[8] and it is believed that they participate in the formation of the fluid surrounding the hair cells rather than providing structural support.

4 ASSOCIATED STRUCTURES

4.1 Tectorial Membrane

Although the exact mechanism that initiates sensory cell excitation in the organ of Corti is unknown, it is generally accepted that moving the stereocilia on the apical surface of the sensory hair cells by a shearing movement of the tectorial membrane (TM) causes sensory cell excitation. The structure and composition of the TM is of particular interest because it is in a unique position to affect the transduction process. The TM is anchored at the spiral limbus (Fig. 6), and overlies the stereocilia of the sensory hair cells. Contrary to the early anatomical findings, which suggested that the TM was an amorphous mass with no internal organization, it is now known that the TM is composed of at least two distinct types of fibers.[9] Type A fibers are straight, unbranched, and 10 nm in diameter, running radially in parallel bundles within the fibrous layer of the TM. Type B fibers

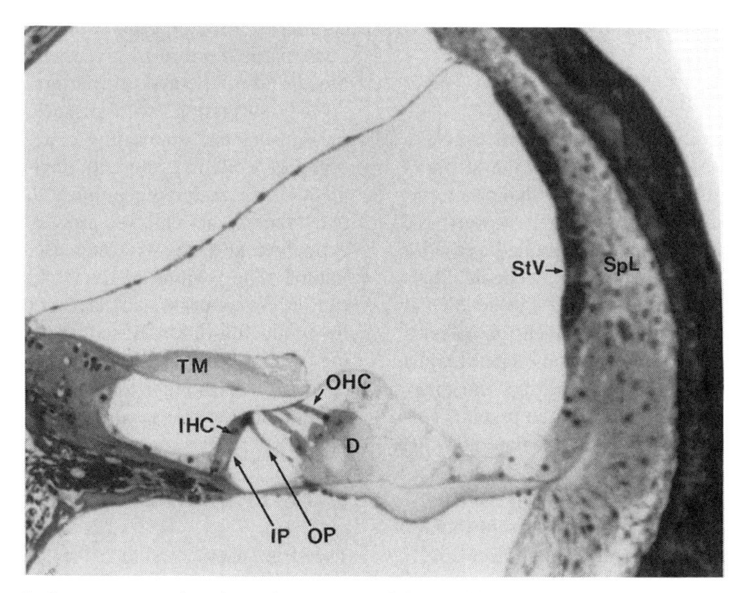

Fig. 6 In a cross section through one turn of the cochlea, the tectorial membrane (TM), inner hair cell (IHC), outer hair cells (OHCs), and supporting cells (IP, OP, and D) of the organ of Corti can be seen even at the light microscopic level. Along the lateral wall, the stria vascularis (StV) appears more darkly stained than does the spiral ligament (SpL).

are tightly coiled 4-nm filaments that are wavy, highly branched, and interconnected.[10]

The distribution of the type A and type B fibers and the different packing arrangement of type B fibers divide the TM into three layers (running parallel to the reticular lamina) and three zones[11] (Fig. 3). The regions over the sensory hair cells are specialized for contact with the stereocilia. A structure called Hardesty's membrane forms a continuous layer over the region of the OHCs, and there are indentations where the tips of the longest stereocilia insert into it. Hensen's stripe is the thickened region that overlies the IHCs, but consistent demonstration of direct contact of the tips of the IHC stereocilia with the TM is still lacking.

It is now clear that the TM contains at least three types of the protein collagen.[12–14] The type A fibers are composed of type II and type IX collagens in bundles that are embedded in a network of type B fibers that contain type V collagen. Thus, the TM appears to be composed of materials (collagen fibers and other proteins associated with carbohydrates that surround the collagen fibers) that are similar in composition to those found in cartilage. Type II collagen fibrils are thought to contribute to the structural stabilization and tensile strength of the TM and provide support as the TM is subjected to stress during vibration of the organ of Corti. The other components can influence the amount of water within the fiber matrix of the TM and influence the amount of crosslinking of the component materials. Thus, the components of the TM will determine not only the structure but also the biomechanical properties of the TM.

4.2 Basilar Membrane

The organ of Corti rests on a structure called the basilar membrane (BM), which contributes to the stiffness and mass of the cochlear partition. In normal adult animals, the stiffness increases progressively toward the base while the mass decreases, and this gradient provides the physical basis for the largely passive mechanical tuning seen in the cochlea. In the apex of the cochlea it is wide (spanning a large distance between the osseous spiral lamina and the spiral ligament at the lateral wall) and thin while at the base of the cochlea it becomes narrow (spanning a short distance between the osseous spiral lamina and the spiral ligament at the lateral wall) and thick. It is composed mainly of bundles and networks of fibers that are produced by cells in the area.[15] In this region there are several proteins present including type II collagen, type IX collagen[16] and fibronectin.[17] The region called the pars tecta (arcuate zone) extends from the spiral limbus to the region under the outer pillar cells. The region called the pars pectinata (pectinate zone) extends from the region under the outer pillar cells to the spiral ligament along the lateral wall. The side of the BM facing scala tympani is covered with tympanic border cells. Because there are large spaces between adjacent cells in this region, it is thought that the BM is in direct contact with perilymph and that perilymph permeates through the BM to contribute to the fluid surrounding the sides and bottom surfaces of the sensory and supporting cells.

5 BLOOD SUPPLY

The blood supply to the cochlea is compartmentalized into at least three large and spatially separate capillary beds. Extensive studies[18] have determined that there are similarities in the blood supply to the cochleas in different mammalian species. Large arteries and veins spiral in the modiolus. Small arteries (arterioles) branch off to form capillary beds in the region of the cochlea between the modiolus and the organ of Corti. Some arterioles radiate out through the bony partition of each turn, over the scala vestibuli to form distinct capillary beds in the spiral ligament and stria vascularis. The capillaries are drained by small veins (venules) that empty into one or two larger veins running spirally around the modiolus. The capillaries in the stria vascularis form the most dense network in the lateral wall. Cells making up the walls of the capillaries are tightly coupled so that substances in the blood cannot pass out into the cochlear tissue between adjacent cells. It has been noted that the capillaries form a barrier between the blood and the strial cells that is more restrictive than the barrier between the blood and the neurons of the brain.

It is surprising, with respect to the metabolism of the sensory and supporting cells, that the area near the organ of Corti is essentially devoid of capillaries. Thus, metabolites and oxygen must diffuse in the fluids to reach the sensory cells—a process that might not be rapid enough to keep up with increased demand during noise trauma. The regions of the cochlea that do have dense capillary networks also have cells that are specialized for producing fluids. Based on the cell types present and the proximity to capillaries and fluid-filled spaces of the cochlea, it is thought that these regions are involved in maintaining and/or producing perilymph.

6 LATERAL WALL

The stria vascularis and the spiral ligament are the cells along the lateral wall of the cochlea that are thought to play a critical role in the production of cochlear fluids (maintaining the high concentration of potassium in endolymph) and maintaining the endolymphatic potential in scala media. Both of these functional roles are required for optimal transduction and our perception of sound.

6.1 Stria Vascularis

The stria vascularis is the region along the lateral wall of the cochlea where there is a high density of capillaries and morphological specializations characteristic of fluid transporting cells. Three cell types are present in the stria vascularis (Fig. 7): marginal cells, intermediate cells, and basal cells. The marginal cells and the basal cells are tightly coupled along their lateral surfaces so that they form a separate fluid-filled compartment. Marginal cells line the endolymphatic surface of the scala media. These cells are specialized for high rates of metabolism and transport of fluids and ions.[19,20] The intermediate cells form a discontinuous layer beneath the marginal cells. They do not contact the endolymph, and their intermediate distribution between the marginal and basal cells allows them to bridge the other two cell types and play a role in generating the positive component of the endocochlear potential. These cells do not show evidence of high rates of metabolism but are capable of synthesizing melanin pigment granules.[21] In mutants where intermediate cells are lacking, the endocochlear potential is close to zero suggesting that intermediate cells are necessary either for the development or activity of marginal cells or for providing potassium for movement into endolymph.[22] The basal cells are flat, overlapping cells that form a continuous cell layer separating the marginal cells, intermediate cells, and capillaries of the stria vascularis from the spiral ligament and perilymph.

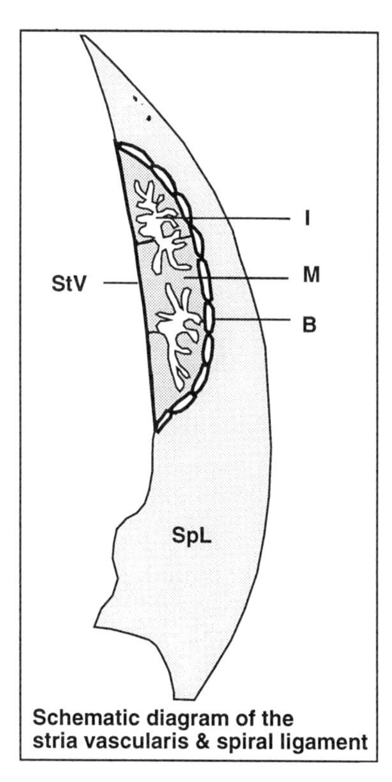

Fig. 7 Schematic diagram of the stria vascularis and spiral ligament along the lateral wall of the cochlea. The stria vascularis (StV) is composed of marginal cells (M), intermediate cells (I), and basal cells (B) as well the major capillary bed. The spiral ligament (SpL) contains much extracellular fibrous material and specialized ion transporting fibroblasts.

6.2 Spiral Ligament and Spiral Prominence

The spiral ligament is a region of the cochlea adjacent to the otic capsule (Figs. 1 and 7), where there are only a small number of cells embedded in a network of fibers. It provides support for the capillaries in the stria vascularis and may absorb stresses that are generated by vibration of the BM, with which it is structurally linked. It can be divided into five regions based on the type and shape of cells present[23]—some of which are morphologically and biochemically specialized for transporting fluids and ions.[19,20] The spiral prominence is a region that contains a large number of capillaries and it bulges into scala media near the point at which the BM inserts.

7 INNERVATION

Early studies in the cochlea showed that there was more than one type of nerve fiber running between the organ of Corti and the central nervous system. One set of nerves contacting the sensory cells of the cochlea link them directly to the central nervous system. Neurons projecting from the sensory hair cells in each cochlea go only to the cochlear nuclear complex on the same (ipsilateral) side of the brain. It is here that nerve fibers branch, and there is clear evidence for the establishment of parallel processing of the primary sensory information carried in the auditory nerve. Another set of nerve fibers project from the central nervous system out to the organ of Corti, where they are in a position to influence the responses of the auditory nerve fibers and the OHCs directly. Anatomical studies were instrumental in identifying the fibers as afferent or efferent and in determining the number of afferent and efferent nerve endings contacting each type of hair cell in each turn of the cochlea. Since then, many reports on the innervation pattern, structure of the contact site between the hair cell and the nerve fiber, and the chemical that signals the transmission of information have appeared.[24]

7.1 Afferents

The point of contact where information is passed from one cell to another is called the synapse. The synapse between the hair cell and the afferent nerve fiber is characterized by a localized thickening of the hair cell membrane and the presence in the hair cell of small vesicles thought to contain the chemical that will be released from the cell to signal the adjacent nerve fiber. The synapse between an efferent nerve fiber and a hair cell is characterized by accumulations in the nerve fiber terminal of small vesicles. In regions of the cochlea away from the hair cells, the two populations of nerve fibers are difficult to distinguish because the bundles contain both types of nerve fibers. Thus, analysis has been accomplished by techniques that selectively identify fiber types (observing degeneration of the nerve fibers after cutting them in regions where the populations are physically separated, following the pathway of individual fibers in adjacent sections over long distances, injecting tracer substances at one place and seeing where they end up, use of specific markers for the different chemical transmitters used by the different types of nerve fibers).

Based on these results, it has been confirmed that the afferent fibers are subdivided into at least two systems. Early studies[25,26] showed that 90–95% of the fibers in the auditory nerve are type I fibers. Type I fibers originate from type I ganglion cells, are thick, surrounded by myelin sheaths, and contact only IHCs. They are unbranched so that each fiber terminates only once and with only one IHC. Each IHC contacts approximately 20 different afferent nerves and these afferent nerves innervate all the major regions of the cochlear nucleus.

Type II fibers originate from type II ganglion cells (5% of the population) and are thin and lack myelin sheaths. These fibers are highly branched and contact multiple OHCs. The type II ganglion cells are smaller than the type I cells, and processes from these cells have been traced to the small cap regions of the cochlear nucleus.

7.2 Efferents

Efferent fibers coming from the central nervous system[27] enter the cochlea near the basal and middle turns of the cochlea. They spiral in both apical and basal directions as the intraganglionic spiral bundle, running through the spiral ganglion cells in Rosenthal's canal. At a given point, depending on the frequency specificity of the individual cell bodies of origin, the efferent fibers turn out toward the organ of Corti in a radial direction, and run with the afferent fibers in the osseous spiral lamina. They pass through to the organ of Corti with the afferent nerve fibers, and like them are unmyelinated within the organ of Corti. Although the divisions of the efferent innervation to the cochlea are at present based on which type of hair cell the fibers contact (IHC or OHC), it must be noted that subdivisions are possible based on the presence of different chemical substances that are used as neurotransmitters.[24]

The fibers to the IHC region are thin, unmyelinated, and originate from cell bodies associated with the lateral superior olivary complex on either the same side or opposite side of the brainstem. The efferent fibers that remain in the IHC region spiral in bundles beneath the IHCs or spiral in the tunnel of Corti, where they branch and contact the afferent fibers present beneath the IHCs (not usually making contact directly with IHCs). In cat, the lateral IHC efferents are most often found synapsing with low spontaneous rate, high-threshold afferent fibers.[28]

The efferent nerve fibers contacting the OHCs are from large diameter, myelinated efferent fibers originating from cell bodies close to the medial superior olivary nuclei. Projections may be either from the same side or opposite side of the brain stem, although a small component of the OHC efferent fibers arise from neurons projecting to both cochleas. The nerve fibers enter the organ of Corti and run, without branching, near the IHCs. There they can spiral for a short distance or pass through the tunnel of Corti at a midlevel to reach the OHC region. The fibers branch almost immediately in all directions and run in bundles that spiral under the OHCs prior to contacting the OHCs directly. There are differences between the innervation of the OHCs in the different rows, with more efferent contacts on the first row of OHCs at each place along the length of organ of Corti. In most species, there is a base-to-apex gradation of efferent innervation with a larger number of efferent terminals ending on each OHC at the base of the cochlea.

8 SUMMARY

The observations of many anatomists, cell biologists, and biochemists have provided us with clues as to the roles played by the different cell types in the cochlea. It is clear that the sensory hair cells are not only passive mechanoreceptors but provide active responses to stimulation that in turn affects cochlear function. Supporting cells provide both structural and metabolic support, and cells within the stria vascularis and spiral ligament interact in complex ways to transport ions and water to maintain cochlear fluid balance and endocochlear potentials. However, cells do not act alone and hearing depends critically on interactions between them. Many of the most interesting problems in these areas of cochlear research are now focused on the nature of these interactions and their role in establishing and maintaining basic peripheral mechanisms in hearing.

REFERENCES

1. S. M. Echteler, R. R. Fay, and A. N. Popper, "Structure of the Mammalian Cochlea," in R. R. Fay and A. N. Popper (Eds.), *Comparative Hearing: Mammals*, Springer-Verlag, New York, 1994, pp. 134–171.
2. D. J. Lim, "Functional Structure of the Organ of Corti: A Review," *Hear. Res.*, Vol. 22, 1986, pp. 117–146.
3. N. B. Slepecky, "Structure of the Mammalian Cochlea," in P. Dallos, A. N. Popper, and R. R. Fay (Eds.), *The Cochlea: Springer Handbook of Auditory Research*, Springer-Verlag, New York, 1996.
4. D. W. Nielsen and N. B. Slepecky, "Stereocilia," in R. A. Altschuler, R. P. Bobbin, and D. W. Hoffman (Eds.), *Neurobiology of Hearing: The Cochlea*, Raven Press, New York, 1986, pp. 23–46.
5. J. O. Pickles, S. D. Comis, and M. P. Obsorne, "Cross-links between Stereocilia in the Guinea Pig Cochlea," *Hear. Res.*, Vol. 18, 1984, pp. 177–188.
6. C. Angelborg and H. Engstrom, "Supporting Elements in the Organ of Corti," *Acta Otolaryngol.*, Vol. Suppl. 301, 1972, pp. 49–60.
7. Y. Raphael, G. H. Marshak, A. Barash, and B. Geiger, "Modulation of Intermediate-Filament Expression in Developing Cochlear Epithelium," *Differentiation*, Vol. 35, 1987, pp. 151–162.
8. R. S. Kimura, "The Ultrastructure of the Organ of Corti," *Int. Rev. Cytol.*, Vol. 42, 1975, pp. 173–222.
9. A. Kronester-Frei, "Ultrastructure of the Different Zones of the Tectorial Membrane," *Cell Tiss. Res.*, Vol. 193, 1978, pp. 11–23.
10. J. A. Hasko and G. P. Richardson, "The Ultrastructural Organization and Properties of the Mouse Tectorial Membrane Matrix," *Hear. Res.*, Vol. 35, 1988, pp. 21–38.
11. D. J. Lim, "Fine Morphology of the Tectorial Membrane," *Arch. Otolaryngol.*, Vol. 96, 1972, pp. 199–215.
12. N. B. Slepecky, J. E. Savage, L. K. Cefaratti, and T. J. Yoo, "Electron Microscopic Localization of Type II, IX and V Collagen in the Organ of Corti," *Cell Tiss. Res.*, Vol. 267, 1992, pp. 413–418.
13. G. P. Richardson, I. J. Russell, V. C. Duance, and A. J. Bailey, "Polypeptide Composition of the Mammalian Tectorial Membrane," *Hear. Res.*, Vol. 25, 1987, pp. 45–60.
14. P. Munyer and B. Schulte, "Immunohistochemical Identification of Proteoglycans in Gelatinous Membranes of Cat and Gerbil Inner Ear," *Hear. Res.*, Vol. 52, 1991, pp. 369–378.
15. S. Iurato, *Submicroscopic Structure of the Inner Ear*, Pergamon Press, Oxford, 1967, 1.
16. I. Thalmann, "Collagen of Accessory Structures of Organ of Corti," *Connective Tiss. Res.*, Vol. 29, 1993, pp. 199–201.
17. E. M. Keithley, A. F. Ryan, and N. K. Woolf, "Fibronectin-like Immunoreactivity of the Basilar Membrane of Young and Aged Rats," *J. Comp. Neurol.*, Vol. 327, 1993, pp. 612–617.
18. A. Axelsson and A. Ryan, "Comparative Study of the Vascular Anatomy in the Mammalian Cochlea," in A. F. Jahn and J. R. Santos-Sacchi (Eds.), *Physiology of the Ear*, Raven Press, New York, 1988, pp. 295–316.
19. T. Kikuchi, R. S. Kimura, D. L. Paul, and J. C. Adams, "Gap Junction Systems in the Rat Cochlea: Immunohistochemical and Ultrastructural Analysis," *Anat. Embryol.*, Vol. 191, 1995, pp. 101–118.
20. B. A. Schulte and J. C. Adams, "Distribution of Immunoreactive Na+, K+ -ATPase in Gerbil Cochlea," *J. Histochem. Cytochem.*, Vol. 37, 1989, pp. 127–134.
21. D. A. Hilding and R. D. Ginzburg, "Pigmentation of the Stria Vascularis," *Arch. Otolaryngol.*, Vol. 84, 1977, pp. 24–37.
22. B. A. Schulte and K. P. Steel, "Expression of α and β Subunit Isoforms of Na, K-ATPase in the Mouse Inner Ear and Changes with Mutations at the Wv or Sld Loci," *Hear. Res.*, Vol. 78, 1994, pp. 259–260.
23. M. M. Henson and O. W. Henson, "Tension Fibroblasts and the Connective Tissue Matrix of the Spiral Ligament," *Hear. Res.*, Vol. 35, 1988, pp. 237–258.
24. M. Eybalin, "Neurotransmitters and Neuromodulators of the Mammalian Cochlea," *Physiol. Rev.*, Vol. 73, 1993, pp. 309–373.
25. H. Spoendlin, "The Innervation of the Cochlear Receptor," in A. Moller (Ed.), *Basic Mechanisms in Hearing*, Academic Press, New York, 1983, pp. 182–230.
26. A. Schwartz, "Auditory Nerve and Spiral Ganglion Cells; Morphology and Organization," in R. A. Altschuler, R. P. Bobbin, and D. W. Hoffman (Eds.), *Neurobiology of Hearing: The Cochlea*, Raven Press, New York, 1986, pp.271–282.
27. W. B. Warr, "Organization of Olivocochlear Efferent Systems in Mammals," in D. B. Webster, A. Popper, and R. R. Fay (Eds.), *Mammalian Auditory Pathways: Neuroanatomy*, Springer-Verlag, New York, 1992, pp. 410–448.
28. M. C. Liberman, L. W. Dodds, and S. Pierce, "Afferent and Efferent Innervations of the Cat Cochlea: Quantitative Analysis with Light and Electron Microscopy," *J. Comp. Neurol.*, Vol. 301, 1990, pp. 443–460.

87

COCHLEAR MECHANICS AND BIOPHYSICS

MARIO A. RUGGERO AND JOSEPH SANTOS-SACCHI

1 INTRODUCTION

The present chapter focuses on the function of the mammalian hearing organ, the cochlea of the inner ear (whose structure is described in Chapter 86). All groups of vertebrate animals possess hearing organs, but their morphology and physiology differ greatly across the evolutionary tree (*Enc.* Ch. 147; for more complete review see Ref. 1). "Cochlear mechanics and biophysics" refers to the processes whereby the inner ear encodes sound into electrical signals. Figure 1 presents a block diagram of the peripheral auditory system of mammals, with arrows indicating the direction of signal transmission. Sound arriving at the external ear (Chapter 84) sets in motion tiny bones (ossicles) in the middle ear (Chapter 85); these, in turn, create pressure fluctuations in the cochlear fluids, causing a displacement wave to propagate along the basilar membrane. The displacement wave stimulates cellular transducers, the inner and outer hair cells (IHCs and OHCs), to generate, much like tiny microphones, electrical receptor potentials that mimic the acoustic stimulus. Finally, these receptor potentials produce chemically mediated excitation in the peripheral terminals of cochlear afferent neurons, generating trains of electrical pulses (action potentials) that travel, via the auditory nerve (*Enc.* Ch. 109), to the cochlear nucleus, the first station of the auditory central nervous system (*Enc.* Ch. 110 and 111).

At several stages of the auditory periphery signal transmission proceeds not only inward but also outward. In Fig. 1, outward-pointing thin arrows indicate that the ear broadcasts sound into the environment. The sound emanates from the cochlea, where the outer hair cells behave as motors that enhance basilar membrane motion.

2 BASILAR MEMBRANE MOTION

2.1 Traveling Wave and Spatial Frequency Analysis

When the middle ear ossicles vibrate, they generate pressure waves that propagate very rapidly (in a few microseconds) throughout the cochlear fluids but also cause a secondary and much slower (transverse) displacement wave that travels on the basilar membrane, from the base of the cochlea toward its apex, in a few milliseconds[3] (Fig. 2*a*). Although the usual physiological route for airborne sound is through vibration of the oval window by the stapes, identical base-to-apex propagation patterns of the displacement wave are generated when sound stimulates the cochlea via an artificial apical window or via bone conduction. The speed of travel of the displacement wave is not uniform as a function of distance: At the basal end propagation is fast and, therefore, for sinusoidal stimulation, the wavelength is long; as the wave proceeds toward the apex, it slows and the wavelength shortens (Fig. 2*a*). The finite speed of travel implies an accumulating delay as a function of distance from the stapes; for sinusoidal stimulation this translates into accumulating phase lag (Fig. 2*b*). Regardless of stimulus frequency, the shape of the vibration envelope is similar in that the basal portion (corresponding to the region of long wavelengths) has a shallow spatial gradient, there is a maximum at a particular longitudinal location, and the maximum is followed on the apical side by a relatively steep slope (Fig. 2*a*). The displacement envelopes are confined to progressively more basal locations as fre-

Note: References to chapters appearing only in the *Encyclopedia* are preceded by *Enc.*

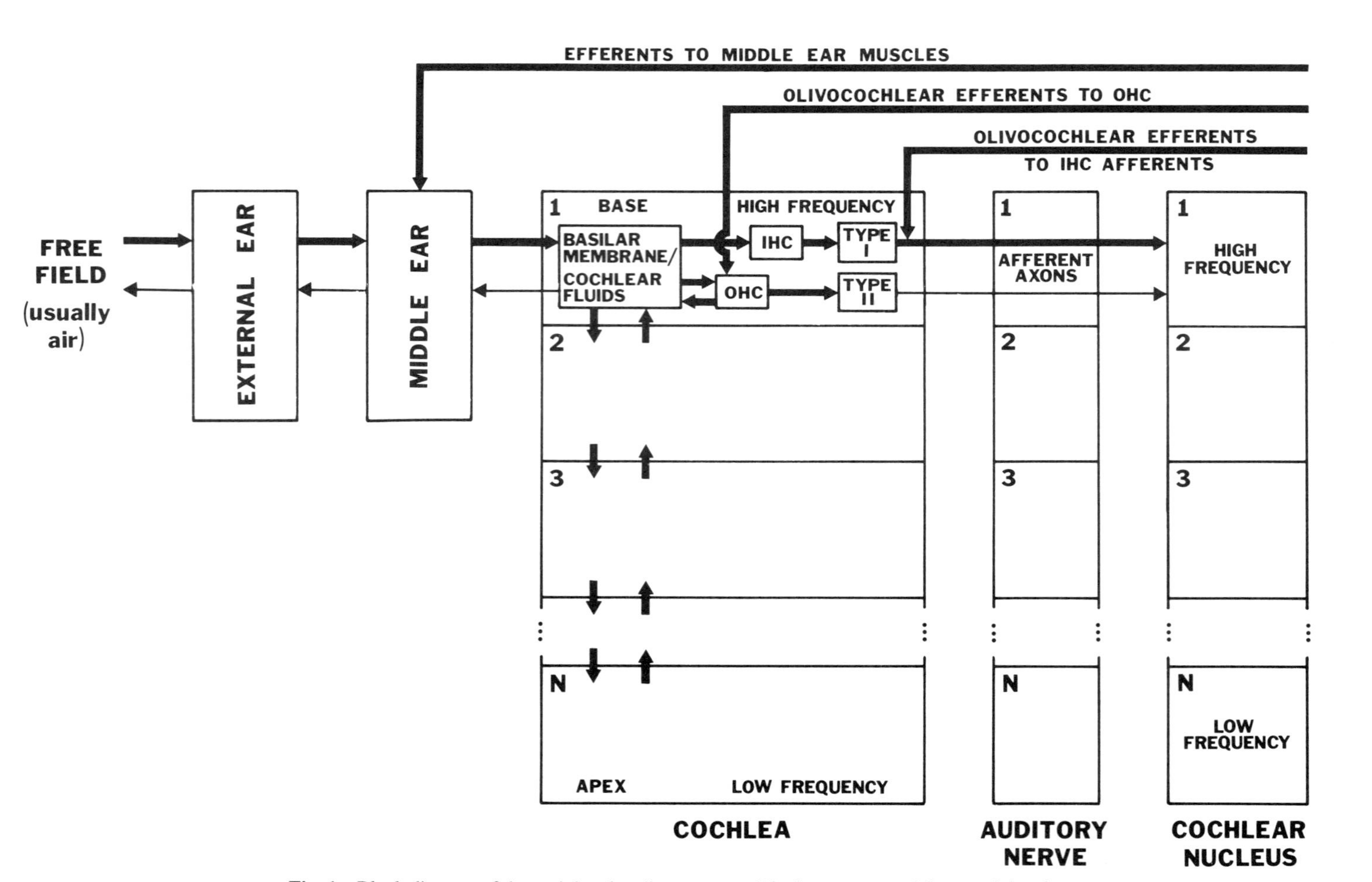

Fig. 1 Block diagram of the peripheral auditory system. Blocks are arranged from peripheral (i.e., closer to the environment, left) to central (i.e., closer to the brain, right). Arrows indicate the direction of signal transmission. Inner hair cells and outer hair cells are represented by the blocks labeled IHC and OHC, respectively. The two types of cochlear afferent neurons are indicated by the blocks labeled Type I and Type II. While the external ear and the middle ear are each represented as a single, undivided box, the cochlea, auditory nerve, and cochlear nucleus are shown subdivided into segments, indicating the systematic mapping of frequency into space. For simplicity, the interconnections within and between the cochlea, the auditory nerve, and the cochlear nucleus are only shown in the segments labeled 1. (Modified with permission from Ref. 2.)

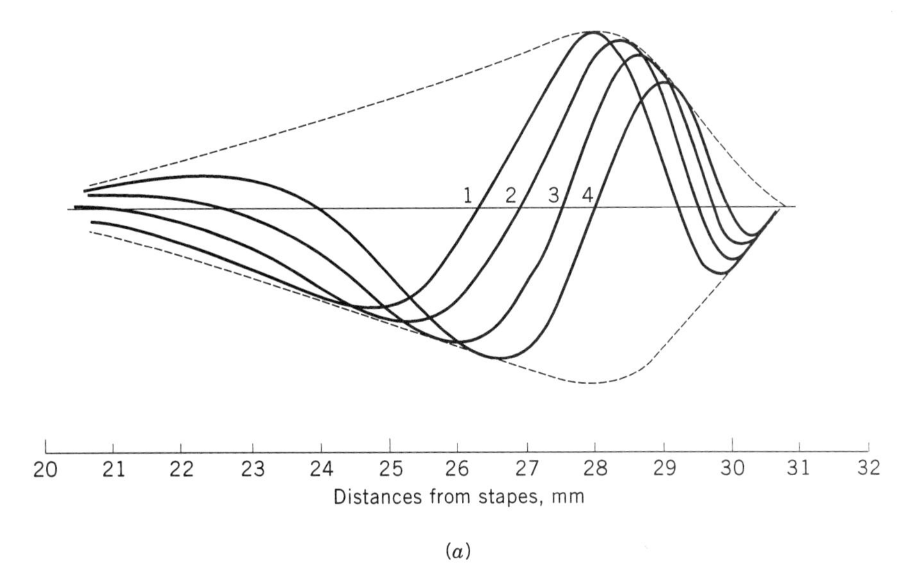

(*a*)

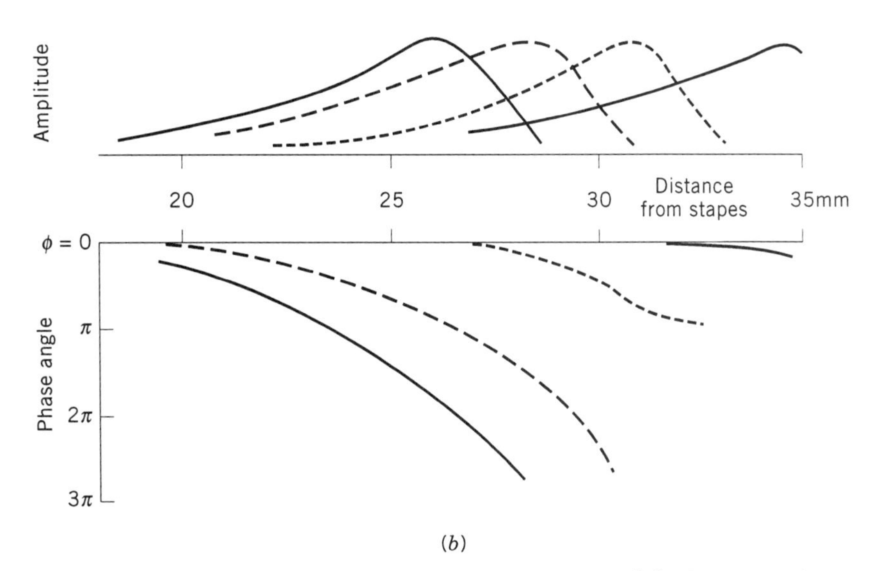

(*b*)

Fig. 2 The basilar membrane traveling wave, as measured by Békésy in human cadavers. (*a*) "Snapshots" at four times (1 = earliest) of the basilar membrane displacement patterns elicited by a 200-Hz tone; the scale of the ordinate is greatly exaggerated relative to that of the abscissa. The dashed lines indicate the peak displacements of the basilar membrane toward scala vestibuli and scala tympani at each cochlear location. (*b*) Peak-to-peak displacements as a function of distance for responses to tones with four different frequencies (upper section), and the corresponding phase patterns (lower section). (Reproduced by permission, with modifications, from Ref. 3.)

quency increases (Fig. 2*b*). Thus, a frequency map is laid out along the cochlea, in which each longitudinal location of the basilar membrane vibrates maximally at its *characteristic frequency* (CF).

The basilar membrane is stiffest at the cochlear base, where it is narrow and thick, and becomes progressively more elastic toward its (wider and thinner) apical end.[3] The variation of basilar membrane elasticity as a function of position (amounting to a ratio of 1 to 100 between the extremes of the human cochlea) is thought to be the fundamental physical characteristic that generates frequency tuning and spatial frequency mapping in the cochlea. Accordingly, mathematical and hardware models of the cochlea that consist of little more than two fluid-filled chambers separated by a membrane with smoothly varying elasticity are able to mimic, at least qualitatively, the main features of basilar membrane motion observed in the ears of dead animals and human cadavers (reviewed in Ref. 4). Such models appropriately exhibit displacement waves that travel from base to apex and whose envelopes are tonotopically organized, with high-frequency responses localized to the stiffer end. However, these simple models are incapable of reproducing realistic basilar membrane responses in healthy cochleae (reviewed in Ref. 5).

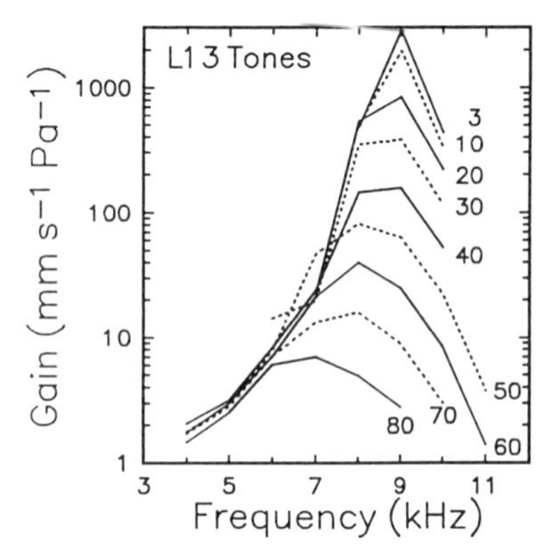

Fig. 3 Frequency response of the chinchilla basilar membrane at a site 3.5 mm from the cochlear base. Gains (i.e., velocity magnitudes divided by pressure magnitudes) are expressed in millimeters per second per Pascal (mm/s/Pa). The parameter is stimulus intensity, expressed in decibels re 20 μPa. (Modified with permission from Ref. 10.)

2.2 Nonlinearity and Lability of Basilar Membrane Motion

The basilar membrane responds to stimulation by single tones with a transverse AC sinusoidal vibration. At the basal region of normal cochleae, basilar membrane responses to near-CF tones grow with stimulus intensity at rates lower than linear, while response growth is linear at other frequencies[6–8] (reviewed in Ref. 9). Because of the frequency-selective nature of the compressive nonlinearity, basilar membrane responses are more sharply tuned for low-level than for high-level stimuli. Figure 3 shows response magnitude curves, as a function of stimulus frequency and intensity, for a healthy chinchilla basilar membrane site located 3.5 mm from the basal end. To facilitate comparison of tuning and sensitivity as a function of stimulus level, the curves have been normalized to stimulus magnitude; that is, they are presented as gain curves, with the parameter being stimulus level. Responses at frequencies well below CF are linear and remain constant as stimulus level is varied. At low stimulus levels CF responses are sharply tuned and sensitive, but at intense stimulus levels they become insensitive and frequency tuning deteriorates. The normal frequency selectivity of the responses of inner and outer hair cells (Section 3.3) and auditory-nerve fibers (*Enc.* Ch. 109) derives, largely or entirely, from mechanical (i.e., basilar membrane) frequency tuning. This is illustrated in Fig. 4, which compares the frequency tuning of the basilar membrane at a site 3 mm from the basal end of the guinea pig cochlea with that of a cochlear afferent with appropriate CF (about 18 kHz). Although it is not yet certain whether neural thresholds more directly reflect velocity or displacement of the basilar membrane, it is unquestionable that, at least in basal regions of the cochlea, the frequency selectivity displayed by cochlear afferents near CF can be fully accounted for by the frequency tuning of basilar membrane responses. Comparison between neural and mechanical tuning curves for the guinea pig 18-kHz cochlear site suggests that at neural CF threshold the basilar membrane peak displacement is about 0.35 nm, which corresponds to a velocity of 40 μm/s; for the chinchilla 8- to 10-kHz basilar membrane site, the values are 2 nm or 100 μm/s.

In the same manner that the sharp tuning of cochlear afferent responses reflects the frequency selectivity of the basilar membrane, mechanical counterparts of CF-specific nonlinearities in auditory-nerve responses to two-tone stimuli also exist at the level of the basilar membrane (reviewed in Ref. 10). Rate suppression in the auditory nerve (see *Enc.* Ch. 109), characterized by the ability of two-tone stimuli to elicit a response rate lower than that evoked by a single tone with frequency near CF, has a clear mechanical analog: The basilar membrane response to a CF tone can be reduced by introduction of a second tone, which by itself elicits smaller responses.[11,12] All the main features of two-tone suppression in the basilar membrane faithfully mirror those of rate suppression in auditory-nerve fibers.

Two-tone distortion products have been unambigu-

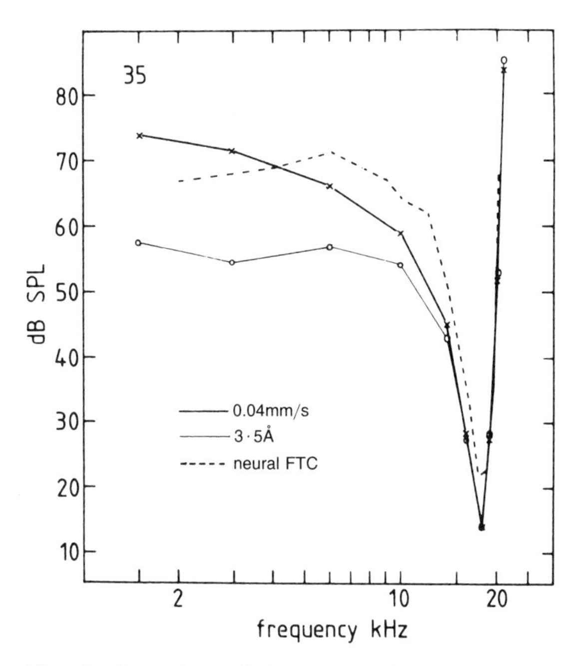

Fig. 4 Comparison of frequency tuning in the basilar membrane and in the auditory nerve. The basilar membrane data, presented as both isovelocity (0.04 mm/s, crosses) and isodisplacement (0.35 nm, open symbols) curves, were recorded from guinea pigs at a site about 3 mm from the basal end of the cochlea. The neural curve (dashed line) is representative of recordings from cochlear afferents with CF similar to that of the basilar membrane recording site. (Reproduced with permission from Ref. 7.)

ously demonstrated in the basilar membrane of several mammalian species.[13–15] Upon stimulation with two primary tones, f_1 and f_2, with frequency near the CF of the basilar membrane site, a variety of combination tones are elicited, with frequencies equal to $2f_1–f_2$, $3f_1–2f_2$, $2f_2–f_1$, and so forth, and magnitudes that vary inversely with separation from CF. At low stimulus levels, distortion product intensity may be no more than 20 dB below primary-tone intensity. If the primary tones are chosen with frequencies such that $2f_1–f_2$ corresponds to CF, it is possible to elicit a distortion product with magnitude far surpassing the response to the primaries. This is strong evidence that the distortion products originate near the sites of the two primary tones and subsequently propagate to the basilar membrane sites with CF corresponding to the distortion product frequency.

Following cochlear damage or death of the animal, basilar membrane responses to tones far from CF remain unchanged while responses to near-CF tones are drastically affected: They become linear and less sensitive, their magnitude being reduced by as much as 40–60 dB[7,8] (reviewed in Ref. 16). Cochlear damage also abolishes two-tone suppression.[12] The fact that the loss of mechanical sensitivity and nonlinearities occurs exclusively at frequencies near CF provides a unifying explanation for the CF specificity of neural two-tone suppression and distortion (*Enc.* Ch. 109) and of the effects of hair cell loss[17] and olivocochlear efferent activation[18] on auditory-nerve responses: All of these effects must reflect alterations of the basilar membrane mechanical response. However, since the basilar membrane is largely acellular and thus probably relatively insensitive to metabolic insults, the lability of its responses suggests that its mechanical properties are influenced by cellular processes.

The sensitivity and frequency tuning of basilar membrane responses to sound can be reversibly reduced by systemic injection of furosemide (Fig. 5), an ototoxic diuretic whose effects on cochlear function result principally from abolishment of the endocochlear potential and a consequent reduction of transduction currents in inner and outer hair cells.[19] The mechanical effect of furosemide strongly implies that the CF-specific sensitivity and nonlinearity of basilar membrane responses depend on hair cell receptor currents or potentials. Further, since the effects of DC currents applied intracellularly to inner hair cells[20] are not CF specific (in contrast with currents applied extracellularly between scala media and scala tympani), the effects of furosemide almost certainly must be mediated by outer hair cells.

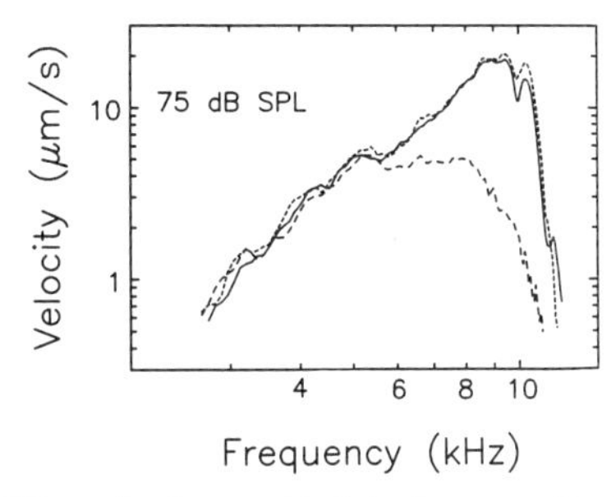

Fig. 5 Effect of a furosemide injection upon the frequency spectrum of basilar membrane responses to clicks. The frequency spectra were obtained by Fourier transformation of velocity responses to 75-dB SPL (sound pressure level re 20 μPa) clicks. The abscissa indicates frequency; the ordinate indicates spectral velocity in units of micrometre per second (μm/s). Three curves are displayed, representing responses immediately preceding (continuous line) and following (long-dash line) a furosemide injection, and when full response recovery had occurred (short dash line). (Reproduced with permission from Ref. 19.)

3 HAIR CELLS

3.1 Apical Membrane: Stereocilia and Transduction

The inner and outer hair cells of the organ of Corti are responsible for the translation of basilar membrane motion into electrical responses in mammalian ears. Hair cells are characterized by apical membrane specializations—rigid microvilli called stereocilia—which underlie their transducer functions (reviewed in Ref. 21). The stereocilia are arranged into a few rows that are graded in height and are aligned into a W or U pattern (see Chapter 86). When the stereocilia are in their normally erect position, a small standing ionic current flows from the scala media into the hair cells through a small percentage of open transduction channels. This current is modulated by deflection of the stereociliar bundle along an axis of maximum sensitivity: Movement toward the tallest row causes an increase in apical membrane conductance, while movement in the opposite direction decreases the conductance.[22,23] Movements of the bundle orthogonal to this axis produce no response, whereas intermediate angles of deflection produce responses whose magnitude depends on the size of the vectoral component along the most sensitive axis. The resultant modulation of the transduction current elicits receptor potentials measurable across the basolateral membrane of the cells; depolarization and hyperpolarization correspond, respectively, to transducer conductance increase and decrease. The changes in membrane conductance are believed to arise from the gating of mechanically sensitive ion channels located near the tips of the stereocilia.[24,25] The gating mechanism may involve direct interactions between interstereociliar microfilaments (so-called tip links[26]) and the transduction channel, which is nonspecifically selective for cations. These tip links, elastic filaments ("springs") that link the top of each stereocilium with a dense membranous plaque on the upper side of the adjacent taller stereocilium, occur in line with the axis of maximal bundle sensitivity. They are believed to provide the tension required to open transduction channels during bundle deflection, one end of the filament being anchored and the other pulling on the channel gate as adjacent stereocilia shear past one another. Since endolymph, the fluid surrounding the stereocilia, has a high K^+ concentration (150 mM), this ion is probably the major charge carrier of the transduction current.

Regardless of sensory organ or animal species, hair cell transduction is nonlinear.[21] Opposite but equal displacements of the hair bundle from the resting position generate unequal conductance changes, resulting in asymmetrical voltage responses. The response to bundle movement toward the tallest stereociliar row, that is, depolarization, is greater (Fig. 6). This asymmetry is highly significant because the depolarizing receptor potentials induce a Ca^{2+} influx, which in turn leads to neurotransmitter release and neural excitation (*Enc.* Ch. 109).

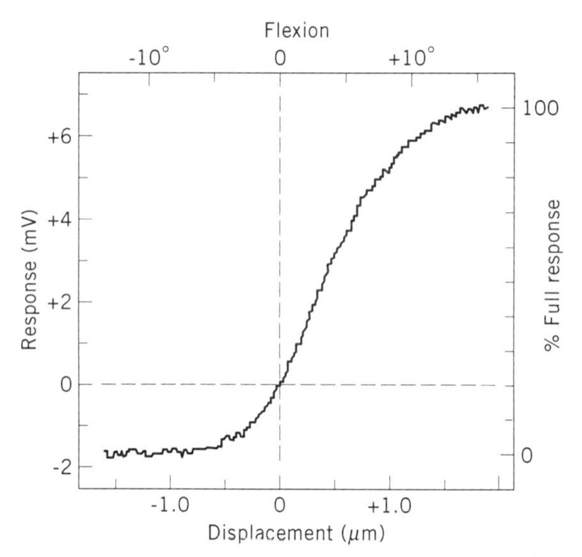

Fig. 6 Sigmoidal stereociliar displacement-response function of hair cell. Due to the offset of the function along the displacement axis, symmetrical sinusoidal displacement of stereocilia produces AC and DC receptor potential components. (Reproduced with permission from Ref. 27.)

3.2 Basolateral Membrane

Hair cell receptor potentials evoked by stereociliar transduction are heavily influenced by the properties of the basolateral membrane, namely, its passive resistance and capacitance and the activity of ligand- and voltage-gated ion channels. Whereas it is possible that the only ion channel in the apical membrane is the one involved in transduction, several types of channels exist in the basolateral membrane. Their conductances have been most completely described in hair cells of nonmammalian species (e.g., those from the frog sacculus and the turtle and chick basilar papillae; reviewed in Ref. 28).

Common among all hair cell types are large outward K^+ currents that activate at membrane potentials more positive than −60 to −50 mV. In vivo, the normal resting potential of the inner hair cell is near −40 mV and the outer hair cell near −70 mV. In inner hair cells isolated from the guinea pig, two types of potassium conductances have been described.[29] The larger of the two has very rapid onset kinetics, approaching a 150-μs onset time constant at large depolarizations. While depolarization induces an inward Ca^{2+} current in inner hair cells, the K^+ currents do not require extracellular Ca^{2+}.

Outer hair cells also display depolarization-induced outward K^+ currents, but their kinetics are much slower than the dominant IHC current: Peak currents occur from 20 to 40 ms after a voltage step.[30,31] There exist at least two K^+ channels.[30] The larger conductance channel (230 pS) is activated at voltages more positive than −35 mV. The smaller (45 pS) is probably active at near-resting membrane potentials (−90 to −50 mV).[32] A non-inactivating inward Ca^{2+} current is evoked by depolarizations greater than −40 mV.[31,33,34] Although both the K^+ channels appear to be Ca-dependent,[30] the bulk of depolarization-induced outward K^+ currents is unaffected by removal of extracellular Ca^{2+}.[33] In addition to voltage-activated ion channels, OHCs possess at least two ligand-activated currents: an ACh-induced slowly activating K^+ current that is dependent on extracellular Ca^{2+}[35] and an ATP-induced inward cationic current, carried most efficiently by Ca^{2+}.[36,37] The ACh-induced K^+ current probably contributes to the cochlear effects of the medial efferent system, whose terminals synapse at the basal membrane of OHCs. The role of the ATP-induced inward cationic current is obscure.

3.3 In Vivo Responses to Sound

The translation of basilar membrane motion into stereociliar deflection underlies sound reception in the mammal. The traditional view of stereocilia excitation is that bending and shearing occur as a result of transverse displacements of the basilar membrane. Since the tips of OHC stereocilia insert into an acellular superstructure, the tectorial membrane, maximum deflection of the basilar membrane toward scala vestibuli should and does coincide with maximal displacement of the stereociliary bundle toward the tallest stereocilia and with maximal OHC depolarization, at least at low stimulus frequencies. However, at high frequencies, differing basilar membrane and outer hair cell responses are measurable and may be attributable to complex micromechanical interactions between stereocilia and the tectorial membrane.[38] Dynamic effects on OHC membrane RC (resistance–capacitance) filter characteristics due to voltage-dependent conductances and capacitance (see below) may also be influential.

The mode of mechanical stimulation of inner hair cells, whose stereocilia may not contact the tectorial membrane, differs from that of the outer hair cell. Inner hair cell stereocilia are thought to be deflected by fluid motion, rather than by shearing against the tectorial membrane.[39] This idea is supported by low-frequency recordings from inner hair cells in both apical and basal cochlear regions.[40,41] However, recording from auditory-nerve afferents, which maintains cochlear integrity, indicates that basal inner hair cells may actually be maximally stimulated when the basilar membrane moves toward scala tympani.[42]

As a consequence of the asymmetrical transducer function of hair cells, sound-induced sinusoidal stimulation of the stereociliary bundle produces a depolarizing DC response in addition to an AC response. This can be seen in the responses of an inner hair cell recorded in vivo from the basal (high-frequency) region of the cochlea (Fig. 7). With increasing stimulus frequency, AC

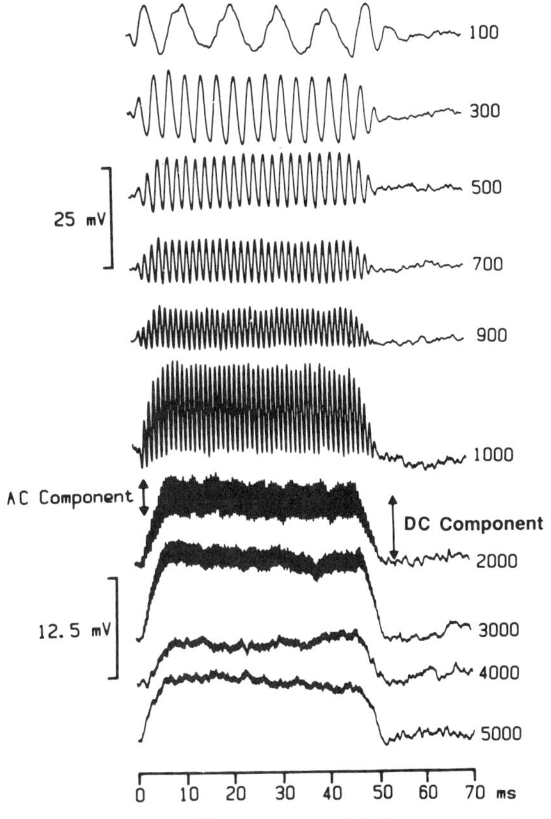

Fig. 7 Inner hair cell receptor potentials in response to tone pips. The hair cell was located at the base of a guinea pig cochlea and had a CF close to that of the tuning curves in Fig. 4. The frequency (in hertz) of the tone-pip stimulus is indicated at the right of each trace. The lower ordinate scale applies to the bottom five traces. Note that the AC component is insignificant in responses to the higher-frequency tone pips. (Reproduced with permission from Ref. 43.)

voltage responses are markedly attenuated by the electrical properties of the basolateral membrane: Its conductances ($1/R$), in combination with its capacitance (C, $1 \mu F/\text{cm}^2$), act as a low-pass frequency filter with a cutoff equal to $\frac{1}{2}\pi$RC (Fig. 7). The frequency extent of phase locking in auditory-nerve fibers (*Enc.* Ch. 109) is probably primarily determined by the roll-off of IHC AC responses,[43] although it may also be influenced by the kinetics that control Ca entry and its intracellular concentration.[44] If it were not for the reduced membrane time constant resulting from the activation of a significant proportion of basolateral conductances at the IHC's resting potential,[29] phase locking would probably roll off at even lower frequencies.

Despite the frequency-dependent roll-off of AC responses, the asymmetrical transducer function of IHCs ensures that high-frequency acoustic information is signaled as DC responses. The polarity of the DC component of IHCs is always positive (excitatory) regardless of frequency of stimulation or location of the cell along the cochlear duct.[45,46] In contrast, the sign of the DC receptor potential of OHCs depends on frequency relative to CF and on stimulus magnitude. The DC response of apical OHCs is hyperpolarizing at frequencies below CF at low stimulus intensities but reverses polarity as intensity increases.[46] The DC responses of basal OHCs to low-frequency stimulation also show a level-dependent sign reversal, but the reversal occurs at much higher stimulus levels than in apical cells.[47] It is puzzling that DC responses of basal OHCs to CF tones can only be measured at very intense sound levels.

In vivo responses of mammalian hair cells show no evidence of adaptation, with receptor potentials remaining uniform in magnitude throughout the duration of acoustic stimulation.[45] This contrasts with transduction in nonmammalian hair cells, where adaptation is ubiquitous.[48–50] However, in vitro, receptor potentials from outer hair cells do exhibit adaptation during static displacements of their stereocilia.[51,52]

At any particular cochlear location, the frequency tuning of inner and outer hair cells is similar[45,46,53] and closely resembles the tuning displayed in the responses of auditory nerve fibers and the basilar membrane. Furthermore, most, probably all, CF-specific response properties of auditory-nerve fibers have clear counterparts in the voltage responses of inner hair cells (*Enc.* Ch. 109).

A variety of experimental manipulations, including electrical stimulation of the olivocochlear efferent system[54] and injection of electrical currents into the cochlea,[20] alters the properties of inner hair cell responses to sound. The effects of these manipulations indicate that outer hair cells are able to influence inner hair cells, probably via an effect on basilar membrane vibration (see Section 2.2).

4 COCHLEA AS A SOUND GENERATOR

4.1 Otoacoustic Emissions

The discovery by Kemp[55] of otoacoustic emissions—that is, sounds produced by the cochlea—was one of the central developments that revolutionized our understanding of inner-ear physiology during the last decade. Kemp showed that when the normal human ear is exposed to clicks or short tone pips, the ear responds, after a fairly long delay (about 10–15 ms), by broadcasting low-frequency (0.5–4 kHz) sounds (Fig. 8). Click-evoked otoacoustic emissions, whose frequency spectra contain several peaks, are unique to each ear and remain remarkably stable over time.

In addition to the click-evoked emissions first measured by Kemp, four other types of otoacoustic emissions can be recognized according to their mode of stimulation (reviewed in Ref. 57). Stimulus-frequency otoacoustic emissions, distortion-product otoacoustic emissions, and electrically evoked otoacoustic emissions are elicited, respectively, by single continuous tones, pairs of tones, and electrical AC currents. Spontaneous otoacoustic emissions[58] are narrow-band sounds that emanate continuously from the ears of a substantial proportion (about 30%) of humans and other primates, and also from the ears of some other animals. While it is doubtful that spontaneous otoacoustic emissions play any role in normal hearing, their existence is highly significant: It demonstrates that, in the absence of external energy inputs (i.e., acoustic stimuli), the cochlea can transduce biological (biochemical) energy into mechanical vibrations. Electrically evoked otoacoustic emissions are produced by injecting AC electrical currents into scala media.[59] These otoacoustic emissions, acoustic analogs of the electrical stimulus, can interact with simultaneously presented external acoustic stimulation. For example, a low-frequency electric current f_{el} concurrent with an acoustic tone f_{ac} can produce otoacoustic emissions at frequencies $(f_{ac} + f_{el})$ and $(f_{ac} - f_{el})$ as well as at f_{el}. Such interactions constitute evidence not only for the existence in the cochlea of a process of reverse transduction, which converts energy from an electrical to a mechanical form, but also suggest that reverse transduction occurs on a cycle-by-cycle basis. In other words, the motile elements in the cochlea appear to act as AC, rather than DC, motors.

A clue to the cellular origin of cochlear reverse transduction is that subjects affected by sensorineural deafness (usually caused by hair cell loss) in a given frequency range lack otoacoustic emissions (OAEs) over that range,[60] suggesting that hair cells are necessary for the generation of otoacoustic emissions. Other clues come from experimental manipulation of forward trans-

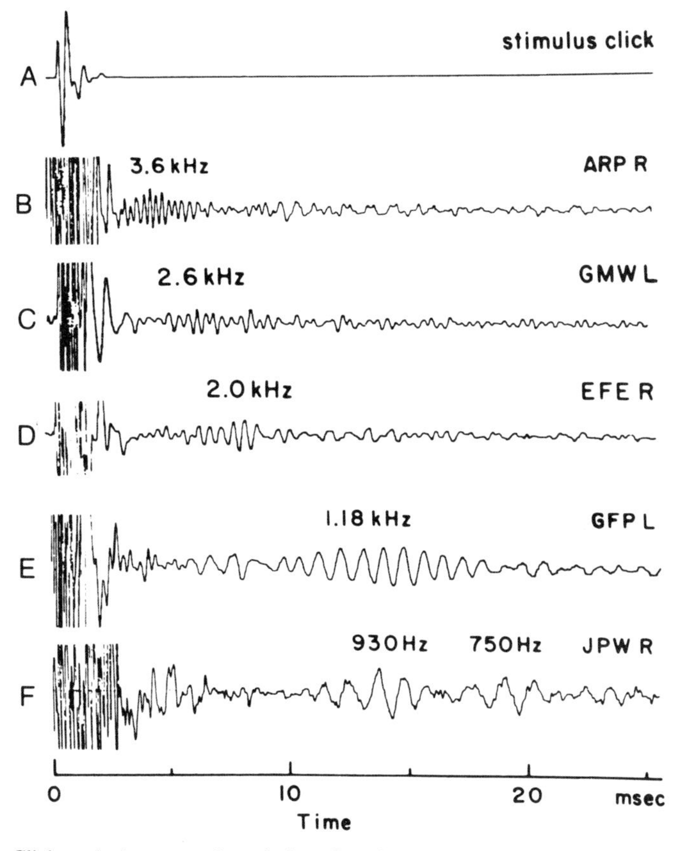

Fig. 8 Click-evoked otoacoustic emissions from human ears. The upper trace indicates the time waveform of the stimulus. The lower five traces show averages of the acoustic waveforms recorded by microphones inserted in the external ear canals of the subjects (identified by name initials and L or R for left and right ear, respectively). Note that each ear produces different emissions and that lower-frequency emissions have longer latencies. (Reproduced with permission from Ref. 56.)

duction in hair cells. Most types of OAEs can be suppressed by tones in a frequency-selective manner[58] and can be altered by changing the endolymphatic potential[61] or by systemic injection of ototoxic drugs.[62] Most significantly, activation of the medial olivocochlear efferent system can also alter the level of distortion-product otoacoustic emissions.[61] Since the medial olivocochlear efferents terminate mostly upon outer hair cells (Chapter 86), it is concluded that the synaptic activity of the efferent terminals can modulate the organ of Corti vibrations, which give rise to otoacoustic emissions. A corollary is that outer hair cells must give rise to otoacoustic emissions and, therefore, must be capable of generating mechanical forces and/or displacements. However, not all otoacoustic emissions are produced by outer hair cells: Otoacoustic emissions can be recorded from the ears of many nonmammalian vertebrates (including frogs, reptiles, and birds) that lack outer hair cells.[63–65]

4.2 Outer Hair Cell Motility

In 1983, Brownell discovered that isolated OHCs are capable of electrically evoked length changes. This form of electrical-to-mechanical transduction may underlie the exquisite sensitivity and frequency selectivity of basilar membrane responses to sound. Brownell et al.[66] determined that transcellular or intracellular current stimulation causes a change in the shape of the OHC; depolarizing currents shorten the cell and hyperpolarizing currents elongate the cell. The mechanical response is not based on a musclelike mechanism, since it does not directly require metabolic substrates (e.g.,

ATP) or calcium.[30,31,67,68] Nor are cytoskeletal elements involved.[68]

Although the underlying mechanism of OHC electrically evoked motility is unknown, it is clear that its effective stimulus is the transmembrane potential, rather than current.[31,69,70,71] The voltage-dependent nature of OHC motility and the existence of a nonlinear charge movement (up to 2.5 pC per OHC or equivalently a voltage-dependent capacitance of 17 pF) indicate that the mechanical response is dependent on discrete sensor-motor elements within the plasma membrane.[70] The number of voltage sensors within the plasma membrane is estimated to be about 4000–7500/μm^2, similar to the number of intramembranous 10-nm particles observed with electron microscopy.[72–74] Indeed, Dallos et al.[75] have shown that the total length change is the sum of many independent elements along the basolateral surface of the cell. The voltage sensors likewise have been shown to be distributed within the central extent of the cell's length.[76] Furthermore, both sensor and motor reside solely within the plasma membrane.[77,78]

The voltage-to-movement function is sigmoidal[79,80] and begins to saturate in the depolarizing direction at voltages well above physiologically meaningful values.[31] However, saturation in the hyperpolarizing direction occurs near normal in vivo resting potentials (Fig. 9). Consequently, whereas mechanical responses as large as 30 nm/mV have been observed,[31] responses occurring at physiological potentials are much smaller.[79] The significance of this asymmetry is similar to that of the hair cell transducer asymmetry—sinusoidal voltage stimulation at the resting potential produces AC and DC mechanical response components. The DC component is in the contraction-depolarizing direction. Unlike the DC component of the receptor potential, which is unaffected by the RC time constant of the basolateral membrane of the cell, the mechanical DC component is immensely, though indirectly, vulnerable to the effects of the cell's low-pass characteristics.

The physiological consequence of the voltage-dependent nature of OHC motility is substantial. Since receptor potentials drive the mechanical response in vivo, the magnitude and phase of the mechanical response must be governed by the nonlinear RC characteristic of the cell. In vitro, the measured frequency response of motility is limited by the speed of the voltage delivery system.[71]

Fairly flat responses have been measured out to about 20 kHz.[81] Even this value probably underestimates the true mechanical capabilities of the OHC. Nevertheless, estimates of OHC motility derived from measures of OHC receptor potentials in the high-frequency region of the guinea pig cochlea[47] indicate that the AC component would be substantially smaller than basilar membrane motion.[79] The corresponding DC mechanical component

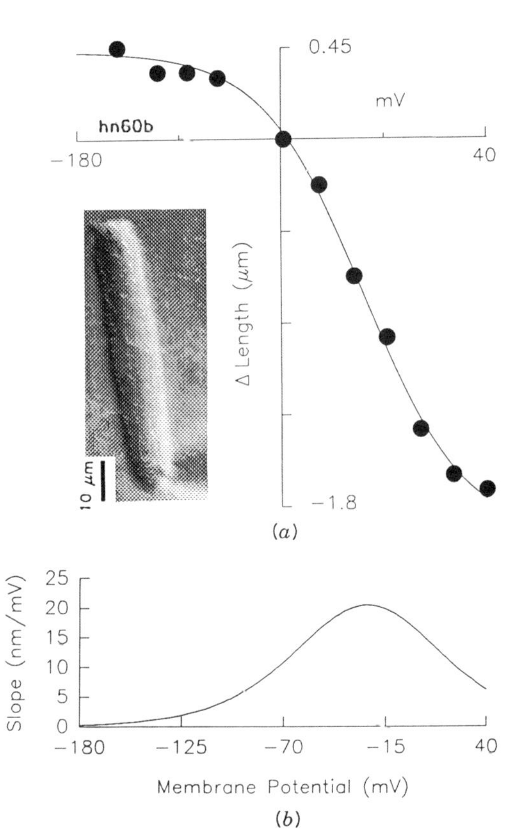

Fig. 9 Mechanical response of mammalian outer hair cell under voltage clamp. (*a*) The OHC changes its length when the cell is held at different membrane potentials. (*b*) The slope of the sigmoidal input–output function defines the cell's responsiveness to membrane potential change. (Reproduced with permission from Ref. 71.)

would be insignificant.[71] Clearly, it is difficult to reconcile these observations with current concepts of the role that OHCs play in the intact cochlea.

Nonetheless, the voltage-dependent mechanical response may be dynamically modulated by a variety of physiologically important factors that can modify the mechanical gain of the cell. One of the most effective means of influencing the motility function is by exerting tension on the plasma membrane.[82] Reducing membrane tension increases the gain at the normal resting potential.[83] Since intracellular turgor pressure effectively alters membrane tension, the control of cell turgor may be physiologically significant. Ligand gated ionic channels that can alter the resting membrane potential may also influence the mechanical gain of the cell.[35–37] Finally, it cannot be ruled out that the characteristics of voltage-dependent OHC motility are substantially different in the intact cochlea so as to account for its presumed amplifier activity.

Acknowledgments

M. A. R. was supported by NIH Grants DC-00110 and DC-00419. J. S.-S. was supported by NIH Grants DC-00273 and DC-02003. Many thanks to Nola Rich, John Guinan, and Thomas Weiss for their comments on previous versions of the chapter and to Alberto Recio and Andrei Temchin for help in the preparation of figures.

REFERENCES

1. E. R. Lewis, E. L. Leverenz, and W. S. Bialek, *The Vertebrate Inner Ear*, CRC Press, Boca Raton, FL, 1985.
2. M. A. Ruggero and M. N. Semple, "Acoustics, Physiological," in G. L. Trigg, (Ed.) *Encyclopedia of Applied Physics*, VCH Publishers, Weinheim, Germany, 1991, pp. 213–259.
3. G. von Békésy, *Experiments in Hearing*, McGraw-Hill, New York, 1960.
4. C. D. Geisler, "Mathematical Models of the Mechanics of the Inner Ear," in W. D. Keidel and W. D. Neff (Eds.), *Handbook of Sensory Physiology. Vol. V: Auditory System. Part 3: Clinical and Special Topics*, Springer Verlag, Berlin, 1976, pp. 391–415.
5. E. De Boer, "Auditory Physics. Physical Principles in Hearing Theory. III," *Phys. Rep.*, Vol. 203, 1991, pp. 125–231.
6. W. S. Rhode, "Observations of the Vibration of the Basilar Membrane in Squirrel Monkeys Using the Mössbauer Technique," *J. Acoust. Soc. Am.*, Vol. 49, 1971, pp. 1218–1231.
7. P. M. Sellick, R. Patuzzi, and B. M. Johnstone, "Measurement of Basilar Membrane Motion in the Guinea Pig Using the Mössbauer Technique," *J. Acoust. Soc. Am.*, Vol. 72, 1982, pp. 131–141.
8. L. Robles, M. A. Ruggero, and N. C. Rich, "Basilar Membrane Mechanics at the Base of the Chinchilla Cochlea. I. Input-Output Functions, Tuning Curves, and Response Phases," *J. Acoust. Soc. Am.*, Vol. 80, 1986, pp. 1364–1374.
9. M. A. Ruggero, "Responses to Sound of the Basilar Membrane of the Mammalian Cochlea," *Curr. Opin. Neurobiol.*, Vol. 2, 1992, pp. 449–456.
10. M. A. Ruggero, L. Robles, N. C. Rich, and A. Recio, "Basilar Membrane Responses to Two-Tone and Broadband Stimuli," *Phil. Trans. Roy. Soc. Lond. B*, Vol. 336, 1992, pp. 307–315.
11. R. Patuzzi, P. M. Sellick, and B. M. Johnstone, "The Modulation of the Sensitivity of the Mammalian Cochlea by Low Frequency Tones. III. Basilar Membrane Motion," *Hear. Res.*, Vol. 13, 1984, pp. 19–27.
12. M. A. Ruggero, L. Robles, and N. C. Rich, "Two-Tone Suppression in the Basilar Membrane of the Cochlea: Mechanical Basis of Auditory-Nerve Rate Suppression," *J. Neurophysiol.*, Vol. 68, 1992, pp. 1087–1099.
13. A. L. Nuttall, D. F. Dolan, and G. Avinash, "Measurements of Basilar Membrane Tuning and Distortion with Laser Velocimetry," in P. Dallos, C. D. Geisler, J. W. Matthews, M. A. Ruggero, and C. R. Steele (Eds.), *The Mechanics and Biophysics of Hearing*, Springer Verlag, Berlin, 1990, pp. 288–295.
14. L. Robles, M. A. Ruggero, and N. C. Rich, "Two-Tone Distortion Products in the Basilar Membrane of the Chinchilla Cochlea," in P. Dallos, C. D. Geisler, J. W. Matthews, M. A. Ruggero, and C. R. Steele (Eds.), *The Mechanics and Biophysics of Hearing*, Springer Verlag, Berlin, 1990, pp. 304–311.
15. L. Robles, M. A. Ruggero, and N. C. Rich, "Two-Tone Distortion in the Basilar Membrane of the Cochlea," *Nature*, Vol. 349, 1991, pp. 413–414.
16. M. A. Ruggero, Robles, N. C., L. Robles, and A. Recio, "The Effects of Acoustic Trauma, Other Cochlear Injury and Death on Basilar-Membrane Responses to Sound," in A. Axelsson, H. Borchgrevink, P. A. Hellström, D. Henderson, R. P. Hamernik, and R. Salvi (Eds.), *Scientific Basis of Noise-Induced Hearing Loss*, Thieme Medical Publishers, New York, 1996, pp. 23–35.
17. P. Dallos and D. Harris, "Properties of Auditory Nerve Responses in Absence of Outer Hair Cells," *J. Neurophysiol.*, Vol. 41, 1978, pp. 365–382.
18. M. L. Wiederhold, "Physiology of the Olivocochlear System," in R. A. Altschuler, D. W. Hoffman, and R. P. Bobbin (Eds.), *Neurobiology of Hearing: the Cochlea*, Raven Press, New York, 1986, pp. 349–370.
19. M. A. Ruggero and N. C. Rich, "Furosemide Alters Organ of Corti Mechanics: Evidence for Feedback of Outer Hair Cells upon the Basilar Membrane," *J. Neurosci.*, Vol. 11, 1991, pp. 1057–1067.
20. A. L. Nuttall, "Influence of Direct Current on DC Receptor Potentials from Cochlear Inner Hair Cells in the Guinea Pig," *J. Acoust. Soc. Am.*, Vol. 77, 1985, pp. 165–175.
21. J. Howard, W. M. Roberts, and A. J. Hudspeth, "Mechanoelectrical Transduction by Hair Cells," *Ann. Rev. Biophys. Biophys. Chem.*, Vol. 17, 1988, pp. 99–124.
22. D. P. Corey and A. J. Hudspeth, "Ionic Basis of the Receptor Potential in a Vertebrate Hair Cell," *Nature*, Vol. 281, 1979, pp. 675–677.
23. I. J. Russell, G. P. Richardson, and A. R. Cody, "Mechanosensitivity of Mammalian Auditory Hair Cells In Vitro," *Nature*, Vol. 321, 1986, pp. 517–519.
24. A. J. Hudspeth, "How the Ear's Works Work," *Nature*, Vol. 341, 1989, pp. 397–404.
25. J. O. Pickles and D. P. Corey, "Mechanoelectrical Transduction by Hair Cells," *T.I.N.S.*, Vol. 15, 1992, pp. 254–259.
26. J. O. Pickles, S. D. Comis, and M. P. Osborne, "Cross-

Links between Stereocilia in the Guinea Pig Organ of Corti and their Possible Relation to Sensory Transduction," *Hear. Res.*, Vol. 15, 1984, pp. 103–112.

27. A. J. Hudspeth and D. P. Corey, "Sensitivity, Polarity and Conductance Change in the Response of Vertebrate Hair Cells to Controlled Mechanical Stimuli," *Proc. Natl. Acad. Sci. USA*, Vol. 74, 1977, pp. 2407–2411.
28. P. A. Fuchs, "Ionic Currents in Cochlear Hair Cells," *Prog. Neurobiol.*, Vol. 39, 1992, pp. 493–505.
29. C. J. Kros and A. C. Crawford, "Potassium Currents in Inner Hair Cells Isolated from the Guinea Pig," *J. Physiol. (Lond.)*, Vol. 421, 1990, pp. 263–291.
30. J. F. Ashmore and R. W. Meech, "Ionic Basis of the Resting Potential in Outer Hair Cells Isolated from the Guinea Pig Cochlea," *Nature*, Vol. 322, 1986, pp. 368–371.
31. J. Santos-Sacchi and J. P. Dilger, "Whole Cell Currents and Mechanical Responses of Isolated Outer Hair Cells," *Hear. Res.*, Vol. 35, 1988, pp. 143–150.
32. G. D. Housley and J. F. Ashmore, "Ionic Currents of Outer Hair Cells Isolated from the Guinea-Pig Cochlea," *J. Physiol. (Lond.)*, Vol. 448, 1992, pp. 73–98.
33. J. Santos-Sacchi, "Calcium Currents, Potassium Currents and the Resting Potential in Isolated Outer Hair Cells," Association for Research in Otolaryngology, Midwinter Meeting Abstracts, St. Petersburg, FL, February, Vol. 12, 1989, pp. 81–82.
34. T. Nakagawa, S. Kakehata, N. Akaike, S. Komune, T. Takasaka, and T. Uemura, "Calcium Channel in Isolated Outer Hair Cells of Guinea Pig Cochlea," *Neurosci. Lett.*, Vol. 125, 1991, pp. 81–84.
35. G. D. Housley and J. F. Ashmore, "Direct Measurement of the Action of ACh on Isolated OHCs of the Guinea Pig Cochlea," *Proc. Royal Soc. London*, Vol. 244, 1991, pp. 161–167.
36. T. Nakagawa, N. Akaike, T. Kimitsuki, S. Komune, and T. Arima, "ATP-Induced Current in Isolated Outer Hair Cells of Guinea Pig Cochlea," *J. Neurophysiol.*, Vol. 63, 1990, pp. 1068–1074.
37. J. F. Ashmore and H. Ohmori, "Control of Intracellular Calcium by ATP in Isolated OHCs of the Guinea Pig Cochlea," *J. Physiol. (Lond.)*, Vol. 428, 1990, pp. 109–131.
38. J. J. Zwislocki, "Analysis of Cochlear Mechanics," *Hear. Res.*, Vol. 22, 1986, pp. 155–169.
39. P. Dallos, M. C. Billone, J. D. Durrant, C.-Y. Wang, and S. Raynor, "Cochlear Inner and Outer Hair Cells: Functional Differences," *Science*, Vol. 177, 1972, pp. 356–358.
40. P. Dallos and J. Santos-Sacchi, "AC Receptor Potentials from Hair Cells in the Low Frequency Region of the Guinea Pig Cochlea," in W. R. Webster and L. M. Aitkin, (Eds.), *Mechanisms of Hearing*, Monash University Press, Clayton, Australia, 1983, pp. 11–16.
41. A. L. Nuttall, M. C. Brown, R. I. Masta, and M. Lawrence, "Inner Hair Cell Responses to the Velocity of Basilar Membrane Motion in the Guinea Pig," *Brain Res.*, Vol. 211, 1981, pp. 171–174.
42. M. A. Ruggero, "What Is the Mechanical Stimulus for the Inner Hair Cell? Clues from Auditory-Nerve and Basilar-Membrane Responses to Low-Frequency Sounds," in Å. Flock, D. Ottoson and M. Ulfendahl (Eds.), *Active Hearing*, Elsevier (Pergamon), Oxford, UK, 1995, pp. 321–336.
43. A. R. Palmer and I. J. Russell, "Phase-Locking in the Cochlear Nerve of the Guinea-Pig and Its Relation to the Receptor Potential of Inner Hair Cells," *Hear. Res.*, Vol. 24, 1986, pp. 1–16.
44. R. C. Kidd and T. F. Weiss, "Mechanisms That Degrade Timing Information in the Cochlea," *Hear. Res.*, Vol. 49, 1990, pp. 181–208.
45. I. J. Russell and P. M. Sellick, "Intracellular Studies of Hair Cells in the Mammalian Cochlea," *J. Physiol. (Lond.)*, Vol. 284, 1978, pp. 261–290.
46. P. Dallos, J. Santos-Sacchi, and Å. Flock, "Intracellular Recordings from Outer Hair Cells," *Science*, Vol. 218, 1982, pp. 582–584.
47. I. J. Russell, A. R. Cody, and G. P. Richardson, "The Responses of Inner and Outer Hair Cells in the Basal Turn of the Guinea-Pig Cochlea and in the Mouse Cochlea Grown in vitro," *Hear. Res.*, Vol. 22, 1986, pp. 199–216.
48. R. A. Eatock, D. P. Corey, and A. J. Hudspeth, "Adaptation of Mechanoelectrical Transduction in Hair Cells of the Bullfrog's Sacculus," *J. Neurosci.*, Vol. 7, 1987, pp. 2821–2836.
49. A. C. Crawford, M. G. Evans, and R. Fettiplace, "Activation and Adaptation of Transducer Currents in Turtle Hair Cells," *J. Physiol. (Lond.)*, Vol. 419, 1989, pp. 405–434.
50. N. Hacohen, J. A. Assad, W. J. Smith, and D. P. Corey, "Regulation of Tension on Hair-Cell Transduction Channels: Displacement and Calcium Dependence," *J. Neurosci.*, Vol. 9, 1989, pp. 3988–3997.
51. I. J. Russell, G. P. Richardson, and M. Kössl, "The Responses of Cochlear Hair Cells to Tonic Displacements of the Sensory Hair Bundle," *Hear. Res.*, Vol. 43, 1989, pp. 55–70.
52. J. F. Ashmore, "The Cellular Machinery of the Cochlea," *Exp. Physiol.*, Vol. 79, 1994, pp. 113–134.
53. A. R. Cody and I. J. Russell, "The Responses of Hair Cells in the Basal Turn of the Guinea Pig Cochlea to Tones," *J. Physiol. (Lond.)*, Vol. 383, 1987, pp. 551–569.
54. M. C. Brown and A. L. Nuttall, "Efferent Control of Cochlear Inner Hair Cell Responses in the Guinea Pig," *J. Physiol. (Lond.)*, Vol. 354, 1984, pp. 625–646.
55. D. T. Kemp, "Stimulated Acoustic Emissions from within the Auditory System," *J. Acoust. Soc. Am.*, Vol. 64, 1978, pp. 1386–1391.
56. J. P. Wilson, "Evidence for a Cochlear Origin for Acoustic Re-emissions, Threshold Fine Structure and Tonal Tinnitus," *Hear. Res.*, Vol. 2, 1980, pp. 233–252.
57. R. Probst, B. L. Lonsbury-Martin, and G. K. Martin, "A Review of Otoacoustic Emissions," *J. Acoust. Soc. Am.*, Vol. 89, 1991, pp. 2027–2067.
58. D. T. Kemp, "Evidence of Mechanical Nonlinearity and

Frequency-Selective Wave Amplification in the Cochlea," *Arch. Otorhinolaryngol.*, Vol. 224, 1979, pp. 37–45.

59. A. E. Hubbard and D. C. Mountain, "Alternating Current Delivered into the Scala Media Alters Sound Pressure at the Eardrum," *Science*, Vol. 222, 1983, pp. 510–512.
60. P. M. Zurek, W. W. Clark, and D. O. Kim, "The Behavior of Distortion Products in the Ear Canals of Chinchillas with Normal or Damaged Ears," *J. Acoust. Soc. Am.*, Vol. 72, 1982, pp. 774–780.
61. D. C. Mountain, "Changes in Endolymphatic Potential and Crossed Olivocochlear Bundle Stimulation Alter Cochlear Mechanics," *Science*, Vol. 210, 1980, pp. 71–72.
62. S. D. Anderson and D. T. Kemp, "The Evoked Cochlear Mechanical Response in Laboratory Primates," *Arch. Otorhinolaryngol.*, Vol. 224, 1979, pp. 47–54.
63. G. Manley, M. Schulze, and H. Oeckinghaus, "Otoacoustic Emissions in a Song Bird," *Hear. Res.*, Vol. 26, 1987, pp. 257–266.
64. A. R. Palmer and J. P. Wilson, "Spontaneous and Evoked Otoacoustic Emissions in the Frog Rana Esculenta," *J. Physiol. (Lond.)*, Vol. 324, 1981, p. 64P.
65. C. Köppl and G. A. Manley, "Spontaneous Otoacoustic Emissions in the Bobtail Lizard. 1. General Characteristics," *Hear. Res.*, Vol. 71, 1993, pp. 157–169.
66. W. E. Brownell, C. R. Bader, D. Bertrand, and Y. De Ribaupierre, "Evoked Mechanical Response of Isolated Cochlear Outer Hair Cells," *Science*, Vol. 227, 1985, pp. 194–196.
67. B. Kachar, W. E. Brownell, R. Altschuler, and J. Fex, "Electrokinetic Shape Changes of Cochlear Outer Hair Cells," *Nature*, Vol. 322, 1986, pp. 365–368.
68. M. C. Holley and J. F. Ashmore, "On the Mechanism of a High Frequency Force Generator in Outer Hair Cells Isolated from the Guinea Pig Cochlea," *Proc. R. Soc. Lond. B*, Vol. 232, 1988, 413–429.
69. J. F. Ashmore, "Transducer Motor Coupling in Cochlear Outer Hair Cells," in J. P. Wilson and D. J. Kemp (Eds.), *Cochlear Mechanisms: Structure, Function and Models*, Plenum Press, New York, 1989, pp. 107–113.
70. J. Santos-Sacchi, "Reversible Inhibition of Voltage Dependent Outer Hair Cell Motility and Capacitance," *J. Neurosci.*, Vol. 11, 1991, pp. 3096–3110.
71. J. Santos-Sacchi, "On the Frequency Limit and Phase of Outer Hair Cell Motility: Effects of the Membrane Filter," *J. Neurosci.*, Vol. 12, 1992, 1906–1916.
72. R. L. Gulley and T. S. Reese, "Regional Specialization of the Hair Cell Plasmalemma in the Organ of Corti," *Anat. Rec.*, Vol. 189, 1977, pp. 109–124.
73. A. Forge, "Structural Features of the Lateral Walls in Mammalian Cochlear Outer Hair Cells," *Cell Tiss. Res.*, Vol. 265, 1991, pp. 473–483.
74. M. C. Holley, F. Kalinec, and B. Kachar, "Structure of the Cortical Cytoskeleton in Mammalian Outer Hair Cells," *J. Cell Sci.*, Vol. 102, 1992, pp. 569–580.
75. P. Dallos, B. N. Evans, and R. Hallworth, "On the Nature of the Motor Element in Cochlear Outer Hair Cells," *Nature*, Vol. 350, 1991, pp. 155–157.
76. G.-J. Huang and J. Santos-Sacchi, "Mapping the Distribution of the Outer Hair Cell Motility Voltage Sensor by Electrical Amputation," *Biophys. J.*, Vol. 65, 1993, pp. 2228–2236.
77. F. Kalinec, M. C. Holley, K. H. Iwasa, D. J. Lim, and B. Kachar, "A Membrane-Based Force Generation Mechanism in Auditory Sensory Cells," *Proc. Natl. Acad. Sci. USA*, Vol. 89, 1992, pp. 8671–8675.
78. G.-J. Huang and J. Santos-Sacchi, "Motility Voltage Sensor of the Outer Hair Cell Resides within the Lateral Plasma Membrane," *Proc. Natl. Acad. Sci. USA*, Vol. 91, 1994, pp. 12268–12272.
79. J. Santos-Sacchi, "Asymmetry in Voltage Dependent Movements of Isolated Outer Hair Cells from the Organ of Corti," *J. Neurosci.*, Vol. 9, 1989, pp. 2954–2962.
80. B. N. Evans, P. Dallos, and R. Hallworth, "Asymmetries in Motile Responses of Outer Hair Cells in Simulated In Vivo Conditions," in J. P. Wilson and D. J. Kemp (Eds.), *Cochlear Mechanisms: Structure, Function and Models*, Plenum Press, New York, 1989, pp. 205–206.
81. P. Dallos and B. N. Evans, "High-Frequency Motility of Outer Hair Cells and the Cochlear Amplifier," *Science*, Vol. 267, 1995, pp. 2006–2009.
82. K. H. Iwasa, "Effect of Stress on the Membrane Capacitance of the Auditory Outer Hair Cell," *Biophys. J.*, Vol. 65, 1993, pp. 492–498.
83. S. Kakehata and J. Santos-Sacchi, "Membrane Tension Directly Shifts Voltage Dependence of Outer Hair Cell Motility and Associated Gating Charge," *Biophys. J.*, Vol. 68, 1995, pp. 2190–2197.

PART XII

PSYCHOLOGICAL ACOUSTICS

88

INTRODUCTION

DAVID M. GREEN AND DENNIS MCFADDEN

1 INTRODUCTION

Traditionally, the study of the auditory system has been pursued by scientists of two general types using different experimental subjects and procedures. On the one hand are auditory physiologists, who typically study nonhumans. They use all the modern tools of scientific technology: electrical recordings of various sorts, neural staining, labeling, biochemical tracers, electron microscopy, and so on. Their goal is to better describe the construction, organization, and function of individual elements and subsystems at the various stages of processing in the auditory nervous system.

The other group consists of auditory psychophysicists, or psychoacousticians. They generally study humans, using experimental paradigms and acoustic stimuli that are chosen for their ability to reveal how the human listener senses and perceives auditory stimuli. The methods are behavioral, and the products of this research include numerical scales of the major auditory attributes and measures of the limits of auditory performance as well as the effects of various stimulus parameters on those limits. The aim of psychological acoustics is to provide a functional description of the hearing process. The approach is to treat the auditory system as an information-processing system and to describe the salient features of that system. The chapters in this section describe the current status of knowledge in several representative topic areas of psychological acoustics.

One of the purposes of introductory chapters of this kind is to inform curious readers from other disciplines of what might be of interest to them in the chapters that follow. Our presumption is that many readers of the following section will be engineers and other physical scientists who are generally knowledgeable about acoustics but somewhat unsure about the scientific questions or issues that are addressed by psychological acousticians. Accordingly, this introduction begins with some background information and then discusses some of the specific topics taken up in the following chapters.

2 HUMAN AUDITORY SYSTEM

The human auditory system is impressive for a number of reasons, but two outstanding features are its absolute sensitivity and its enormous dynamic range. Normal listeners can detect sounds that produce an average displacement of the eardrum smaller than the diameter of a hydrogen atom. They can also hear sounds six orders of magnitude larger than this displacement without the familiar, protracted dark and light adaptation that characterizes the visual system. Listeners are also reasonably sensitive to relative change over this entire range; they can discriminate changes in sound amplitude of less than 10%. The auditory system consists of a mechanically sophisticated "front end," the middle ear and cochlea, which transforms variations in air pressure into minute displacements of the basilar membrane. There is increasing evidence that the exquisite sensitivity of the system arises from certain nonlinearities that are present in the basilar membrane. These displacements produce graded potentials in the transducer elements, the hair cells, which in turn lead to neural impulses in the individual neurons constituting the VIIIth cranial nerve. These neurons all terminate (synapse) in the first way-station in the brainstem, the cochlear nucleus. The basilar mem-

Note: References to chapters appearing only in the *Encyclopedia* are preceded by *Enc.*

brane is only about 3.5 cm in length and contains about 12,000–16,000 of the sensory receptors called hair cells. The hair cells are arranged in rows, four cells abreast, each inner hair cell flanked by three outer hair cells. It is believed that the approximately 4000 inner hair cells are responsible for most of the information processing that takes place in the cochlea, because about 90% of the afferent neurons make contact with these inner hair cells.

A great deal is also known about the coding of the neural impulses traveling in the VIIIth nerve and about the complexity of neural interaction in the cochlear nucleus. Our understanding of how information is coded grows increasingly more sketchy as we move up the brainstem to the central regions of the auditory system. For example, much less is known about the processing in the auditory cortex than in the visual cortex.

3 PSYCHOACOUSTICAL STUDIES

Psychoacoustical studies try to discover the functional significance of this system by presenting controlled stimuli and measuring the reactions of human listeners. Quantifying these reactions is a matter of some subtlety. One major effort is to use broad categories of responses to develop numerical scales of the major attributes. For example, Chapter 91 describes how scales of loudness are constructed from numerical estimates given by the listeners. Another source of data is a kind of balance judgment—for example, adjusting the frequency of one sound until its pitch is equal to the pitch of some other sound, as described in Chapter 90.

A final source of psychophysical information is the smallest amount of change in some stimulus variable that can be reliably detected or discriminated. For example, how much does one need to change the frequency of a sinusoid in order for listeners to discriminate a difference?

Historically, measures of detection and discrimination were plagued by the problem of uncontrolled response biases. Some subjects—by virtue of their personality characteristics and/or past experience in decision-making situations—would be strongly inclined toward giving one of the two possible responses (signal present or absent, stimuli same or different, etc.) when uncertain about which stimulus condition occurred, as is often the case in experiments designed to measure the limits of sensitivity. The consequence was that classical measures of sensitivity were confounded by response bias. A major advance occurred when it was demonstrated that certain forms of forced-choice experiment had the potential to yield measures of sensitivity that are independent of the subject's response bias or criterion for response. The widespread use of forced-choice procedures in the past 30 years has led to far greater replicability of results in detection and discrimination tasks than was previously true. Modern forced-choice procedures and measures of sensitivity are discussed by Green and Swets,[1] and the results from discrimination experiments comprise much of the material discussed in Chapter 89, *Enc.* Ch. 117 and 121. Discriminating the presence or absence of a sinusoidal signal in quiet (the absolute threshold of hearing) is the entire topic of *Enc.* Ch. 123.

These different methods contribute to our understanding of how auditory signals are processed and perceived. Like most biological measurements, there is considerable variability in the measurements. A typical discrimination value, which provides the most precise measurements in the area, often has a standard deviation that is as much as 25% of the mean, even if the measurements are made on the same listener at different times. This variability is even larger when measurements are taken from different individuals. Some investigators believe that understanding these differences between individual listeners will become a major topic of future psychoacoustical research. Certainly there is much to be learned in this area.

Once psychoacoustical data have been collected and analyzed, models or theories are formulated in an attempt to explain the experimental results. These theories are not basically different from scientific theories in other disciplines, with the possible exception that the inferences are often deep—meaning that "black-box" theorizing is common. Psychoacoustical investigators follow with great interest the research of their physiological colleagues, because it often provides insight into the mechanisms that process the stimulus information and convert it to measured responses. Progress is made when a physiological finding helps us understand some psychophysical result, and often there is a gratifying correspondence between the physiological and psychological data.

The past century has witnessed enormous advancements in our understanding of the auditory system. In large part, these advancements stem from our utilization of modern electronic technology. It should be realized that control of the stimulus was virtually impossible in the early part of this century. Controlling the pitch of a sinusoidal sound was comparatively easy, because, in principle, different tuning forks could be constructed to produce vibrations of any desired frequency. But controlling the amplitude of a sinusoidal sound, which is a critical physical parameter in most auditory tasks, was, for the most part, either difficult or impossible. Controlled changes in relative amplitude are possible only in an anechoic room where the inverse-square law operates. Thus, the distance between the source and listener could be used as a means of systematically altering the relative level of two sounds. The measurement of the absolute level of a sound wave was first possible in 1882 with Rayleigh's disc. The refinements of inexpensive earphones and electronics made it possible to achieve convenient control of the two most salient properties of an auditory stimulus—frequency and level—which correspond roughly to pitch and loudness.

4 INTENSITY AND FREQUENCY DISCRIMINATION

The modern era of psychoacoustics was heralded by quantitative studies of two important auditory capacities: intensity discrimination[2] and frequency discrimination.[3] At about the time of these studies, Georg von Bekesy, a Hungarian telephone engineer, began a series of investigations about how the basilar membrane vibrates, for which he was awarded a Nobel Prize in 1961. (For an excellent review of that work and a translation of some of his earlier papers, see the work of Bekesy.[4]) Perhaps Bekesy's most interesting discovery was that the basilar membrane supports an unusual form of traveling waves (their amplitude increases and speed decreases with distance from the stapes). From a functional point of view, Bekesy's most important observation was that the place of maximum vibration along the basilar membrane systematically changes with stimulus frequency, as would occur in a resonant system. This confirmed the earlier conjecture of Ohm and Helmholtz that there is a strong tonotopic organization in the cochlea. High-frequency tones stimulate the part of the basilar membrane near the entrance to the canal, whereas low-frequency tones produce their maximum vibrations at the more apical parts.

Another telephone engineer, Harvey Fletcher, working at the Bell Telephone Laboratories in the United States, developed the concept of a critical band, namely, that the first stage in auditory processing can be likened to a set of filters or frequency channels, each tuned to slightly different center frequencies and having a bandwidth of less than a third of an octave.[5,6] Measurements of the detectability of a sinusoidal signal in noises of differing bandwidths suggested that the critical quantity was the ratio of signal power to noise power present in this filter or frequency band. These experiments provided the basic data for this concept, and its applicability to a variety of other psychoacoustical phenomena was soon appreciated. Eberhard Zwicker, in Germany, demonstrated the effects and importance of such filtering in such diverse areas as masking, the ability to discriminate amplitude and frequency modulation, and loudness judgments. Feldtkeller and Zwicker[7] provided a good summary of that research. More recent measurements have led to more precise estimates of the filter shape[8] and the equivalent rectangular bandwidths (ERBs) of these peripheral filters.[9] The importance of this first-stage filter concept is evident in the topics discussed in Chapters 89–91.

5 PITCH

While analysis in the frequency domain dominated the early research in hearing, Jan Schouten in Holland began the first of a series of experiments that challenged the ideas of Ohm and Helmholtz as to how the pitch of a complex sound is determined.[10] The primary puzzle was the pitch of the missing fundamental. A periodic sound rich in harmonics is perceived as having a pitch equal to the fundamental, whether or not energy is present at the fundamental frequency. Ohm and Helmholtz argued that the fundamental energy did not need to be present in the stimulus, because the nonlinearity of the transmission process would produce energy at the difference in frequency between the overtones, which is equal to the fundamental frequency. Thus, they could assume that pitch arose from a resonance process causing stimulation of the cochlea at the location tuned to the fundamental frequency of a complex sound. Schouten showed that these ideas were wrong. For example, masking noise can be present at the fundamental frequency (making it inaudible), yet the pitch created by the audible overtones is unchanged. Also, when all components of an harmonic sequence are increased in frequency by the same amount (so the difference frequency remains the same), the pitch of the complex increases slightly. These observations destroyed the rigid connection between place and pitch that was the crux of the Ohm–Helmholtz resonance theory and have led to various theories emphasizing the importance of the periodicity information contained in both the displacement pattern of the basilar membrane and the firing pattern of VIIIth nerve fibers. A review of this fascinating area along with more recent research is the topic of Chapter 87.

A cliché among students of the brain is that the nervous system is primarily concerned with stimuli that change—that is, with stimuli that contain spatial or temporal transients. A corollary of this view is that the brain rapidly adapts to steady-state stimuli. Common examples include the acute awareness of a tight piece of clothing when it is first put on and the obliviousness to it shortly thereafter. The visual system is known for the emphasis it places on edges, boundaries, and onsets. What about the auditory system? While many aspects of loudness, pitch, masking, and so on, involve relatively steady-state stimuli and constant perceptions, there are also situations in which these perceptions can change over time. For example, the detectability of a brief tonal signal can be quite different, depending upon whether it occurs at the beginning, middle, or end of a noise burst. Some of these changes are thought to be related to the dynamic changes in neural firing rate that are observed physiologically when sounds are turned on and off. These adaptation-like phenomena are discussed in *Enc.* Ch. 122.

6 LOCALIZATION OF SOUND

A topic of long-standing interest to many acousticians is how sound is localized in space, an ability that is remarkably acute. Over the years, psychoacoustical research has documented this skill and has examined the nature and range of the acoustical cues underlying it. Nearly 100 years ago, Lord Rayleigh[11] advanced what is known as the duplex theory, namely, that different cues are used in different frequency regions. One set of cues depends on differences in the level of the sound at the two ears (because of the sound shadow thrown by the head); the second cue set depends on differences in time of arrival of the sound at the two ears (because of the slight difference in path length to the further ear). Our ability to use these two cues continues to receive attention. The limitation that Rayleigh presumed to exist on the usefulness of the time cue at high frequencies is now known not to exist when the waveform is complex, as it invariably is in the real world. Our physiological colleagues have established that the cues of interaural time and intensity are processed by different populations of cells at various levels in the auditory system, and they have demonstrated evolutionary trends that link head size, cue effectiveness, and the upper frequency limit of hearing. Other discoveries have revealed that the mechanisms underlying sound localization can also aid in the extraction of certain signals from background noise. In recent times, there has been considerable interest in simulating acoustical environments with headphones or small numbers of loudspeakers (a psychoacoustical virtual reality). Success at this task requires knowledge of the acoustical cues that provide information about the elevations of real-world sound sources as well as their azimuths. These topics, along with the basic facts and theories of sound localization, are discussed in *Enc.* Ch. 117.

7 PERCEPTION OF COMPLEX SPECTRA

Recent psychoacoustical experiments have begun to explore stimuli more complicated than single sinusoids or noise. Data obtained with multitonal complexes, amplitude-modulated noise bands, and other complex spectra have produced results that are inconsistent with the idea that detection is determined solely by the signal-to-noise ratio in a single critical band. In one experiment, Hall, Haggard, and Fernandes[12] showed that Fletcher's original critical-band experiment produced very different results when the entire noise band was amplitude modulated together. As the bandwidth of the noise increased, the threshold actually improved by 10 dB or more. It is now appreciated that the temporal modulation of a spectrally remote noise band can provide a strong cue to the detection of a signal in another noise band if the modulation is correlated in the two bands. In other work, Green[13] has shown that some signals are detected by comparing the relative amplitude of components widely spaced in the spectrum, not just by monitoring the activity in a single frequency channel centered at the signal frequency. From their increasing experience with more complex waveforms, psychoacousticians now realize that the abilities and performance of individual listeners can differ considerably, and the bases for these individual differences are receiving increasing attention. *Enc.* Ch. 121 explores these and other related topics that have expanded our awareness of how more complex waveforms are perceived.

8 IMPAIRED HEARING

Finally, the topic of impaired rather than normal hearing is considered. One important definition of impaired hearing is still based largely on the audibility of single sinusoidal signals presented in quiet—the audiogram. Degradation in speech intelligibility is clearly another important aspect of impaired hearing. Unfortunately, the available data indicate that speech intelligibility correlates with audiometric thresholds only modestly. How the absolute threshold is defined and measured is the topic of *Enc.* Ch. 123. Chapter 93 discusses the different types of hearing impairment and the tests used to determine these categories. It has been clear for some time that exposure to environmenal noise can have profound short- and long-term effects on hearing. While such effects have been recognized for at least a century, quantitative data relating noise exposure and hearing level have only begun to be acquired. *Enc.* Ch. 121 relates what is known and not known in this important area.

The aim of this chapter was to provide the curious outsider with an overview of psychological acoustics and of the material contained in the following chapters. The newcomer to psychoacoustics is encouraged to also study the chapters written by our colleagues in physiological acoustics, for, in our view, the information contained there is crucial to our common, ultimate goal of understanding the behavior of the auditory system. This has always been a joint effort, and it is becoming increasingly so. Pleasant reading.

Lastly, on a more personal note, we wish to thank the authors who wrote chapters for this section. Most were punctual, and all were cooperative and pleasant to deal with. We appreciated that.

REFERENCES

1. D. M. Green and J. A. Swets, *Signal Detection Theory and Psychophysics*, Wiley, New York, 1966; reprinted 1988 Penisular Publishing, Los Altos, CA.
2. R. R. Riesz, "Differential Sensitivity of the Ear for Pure Tones," *Phys. Rev.*, Vol. 31, 1928, pp. 867–875.
3. E. G. Shower and R. Biddulph, "Differential Pitch Sensitivity of the Ear," *J. Acoust. Soc. Am.*, Vol. 3, 1931, pp. 275–287.
4. G. von Bekesy, *Experiments in Hearing*, McGraw-Hill Series in Psychology, translated and edited by E. G. Wever, McGraw-Hill, New York, 1960.
5. H. Fletcher, "Auditory Patterns," *Rev. Modern Phys.*, Vol. 12, 1940, pp. 47–65.
6. H. Fletcher, *Speech and Hearing in Communication*, Van Nostrand, New York, 1953.
7. R. Feldtkeller and E. Zwicker, *Das Ohr als Nachrichtenempfanger*, S. Hirzel Verlag, Stuttgart, 1956.
8. R. D. Patterson, "Auditory Filter Shapes Derived with Noise Stimuli," *J. Acoust. Soc. Am.*, Vol. 59, 1976, pp. 640–646.
9. B. C. J. Moore and B. R. Glasberg, "Suggested Formulae for Calculating Auditory-Filter Bandwidths and Excitation Patterns," *J. Acoust. Soc. Am.*, Vol. 74, 1983, pp. 750–753.
10. J. F. Schouten, "The Residue Revisited," in R. Plomp and G. F. Smoorenburg (Ed.), *Frequency Analysis and Periodicity Detection in Hearing*, A. W. Sijthoff, Leiden, 1970. An excellent reference to Schouten's early experiments is J. F. Schouten, "Five Articles on the Perception of Sound (1938–1940)," Instituut voor Perceptie Onderzoek, Eindhoven, The Netherlands.
11. Lord Rayleigh, "On Our Perception of Sound Direction," *Philos. Mag.*, Vol. 13, 1907, pp. 214–232.
12. J. W. Hall, III, M. P. Haggard, and M. A. Fernandes, "Detection in Noise by Spectro-temporal Pattern Analysis," *J. Acoust. Soc. Am.*, Vol. 76, 1984, pp. 50–56.
13. D. M. Green, *Profile Analysis: Auditory Intensity Discrimination*, Oxford University Press, New York, 1988.

89

AUDITORY MASKING

SØREN BUUS

1 INTRODUCTION

The purpose of this chapter is to review masking as it applies to hearing. In hearing, masking generally is defined as the interference with the perception of one sound (the signal) by another sound (the masker). The interference may decrease the loudness of the signal, may make a given change in the signal less discriminable, or may make the signal inaudible. The first two cases are often called partial masking. Some cases of partial masking are considered in Chapters 64 and 91, *Enc.* Ch. 122.

This chapter considers only how a masker changes the audibility of the signal. The change of audibility is usually measured as the *amount of masking*, which is the increase of threshold, in decibels, caused by the presence of the masker. Thus, the amount of masking is defined as the sound pressure level (SPL) at which the signal is just audible in the presence of the masker minus the SPL at which it is just audible in the quiet. However, many aspects of masking are best summarized by the SPL or signal-to-noise ratio at which the signal is just audible. (Here and in the following, "just audible" means that listeners' performance meets some fixed criterion in a detection experiment. The criterion may be defined implicitly or explicitly by the experimental procedure. Whatever criterion is used, however, it should be the same with and without the masker. See Chapter 64 for further discussion of signal detection theory, detectability, and audibility.)

Because quiet thresholds set a lower limit on thresholds obtained under masking and serve as a reference for calculating the amount of masking, they are discussed briefly in the beginning of this chapter (see also *Enc.* Ch. 123). Then follows a discussion of *simultaneous masking*, which is the increase in threshold caused by a masker that is present throughout the duration of the signal. Next, the chapter discusses *nonsimultaneous masking*, which is the increase in threshold caused by a masker presented before the signal (*forward masking*) or after it (*backward masking*). The final section of this chapter considers special aspects of masking, which cannot be explained easily by conventional models of masking. These phenomena include *interaural masking* (also called *central masking*), *comodulation masking release* (CMR), and "*overshoot*." Other phenomena that are not explained by conventional models of masking are *profile analysis* (see *Enc.* Ch. 121) and the *binaural masking level difference* (see *Enc.* Ch. 117).

2 ABSOLUTE THRESHOLDS

Absolute thresholds depend on frequency (see *Enc.* Ch. 123). As discussed below, they also depend on the bandwidth and duration of the signal.

2.1 Effect of Bandwidth

The threshold for a multitone complex with closely spaced components is determined by the total intensity as long as all the components are within some limiting bandwidth.[1] Adding tones outside that bandwidth does not reduce the threshold (in terms of level per tone). As shown in Fig. 1, the limit is about 160 Hz below 1100 Hz and about 320 Hz below 2000 Hz. These bandwidths are closely related to the auditory filter bandwidths, which are discussed in detail later. A recent study replicated these findings.[2]

Note: References to chapters appearing only in the *Encyclopedia* are preceded by *Enc.*

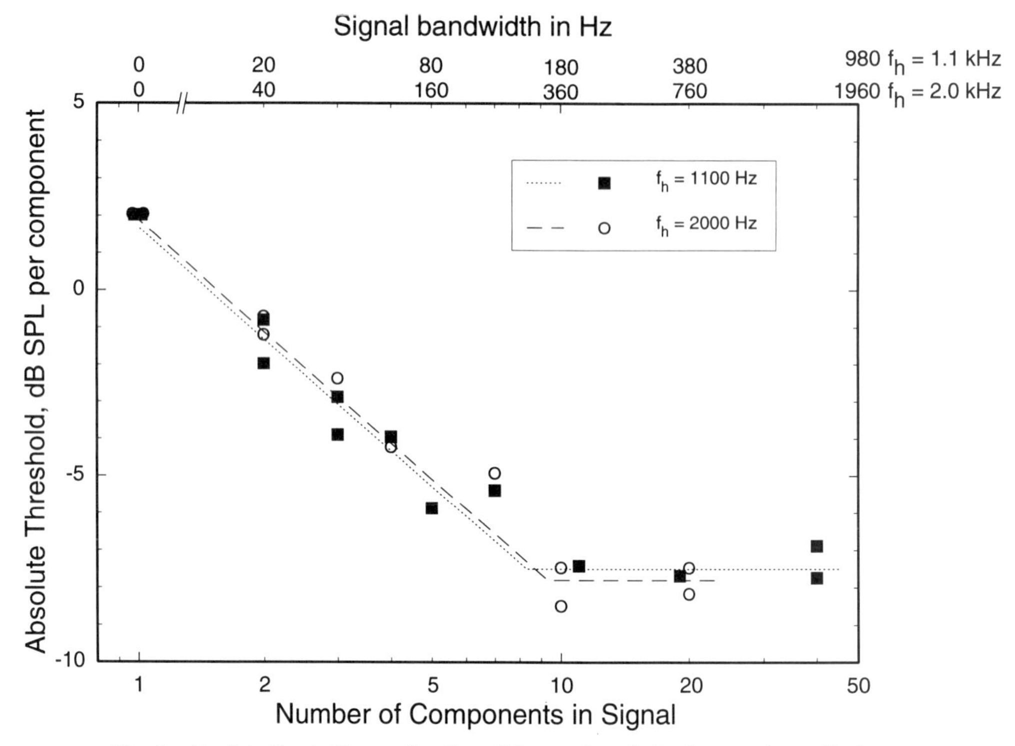

Fig. 1 Absolute thresholds as a function of the number of closely spaced, equally intense tones contained in the signal.[1] Data are shown for tones added every 20 Hz below an 1100-Hz upper frequency (■, dotted line) and for tones added every 40 Hz below a 2000-Hz upper frequency (○, dashed line). The corresponding bandwidths are indicated on the top abscissa. The lines are Gässler's summary of the data. The sloping line decreases 3 dB per doubling of the number of components, corresponding to perfect energy summation of signal components within the critical bandwidth. The horizontal lines indicate the thresholds that would be obtained if components more than 160 Hz below 1100 Hz and 320 Hz below 2000 Hz did not aid detection in the quiet. (Adapted with permission from Ref. 1.)

2.2 Effect of Duration

The threshold values discussed in *Enc.* Ch. 123 apply to signals with durations of 500 ms or more. Generally, thresholds increase for shorter durations.[3] For bursts of white noise with durations between 0.1 ms and 1 s, the intensity at threshold is well described by a power function of duration with an exponent of about −0.75.[4] This indicates that thresholds decrease about 7.5 dB for each 10-fold increase in duration. Tones with durations between the reciprocal of the auditory filter bandwidth and 500 ms yield similar results.[3] However, only part of the rise and fall may contribute to detection. Temporal integration data for multiple tone bursts follow a power function of duration more closely when only part of the rise and fall is included in the calculation of stimulus duration.[5]

Although power functions provide a good description of the effect of duration, they fail at very short and at long durations. Most studies show a faster rate of threshold decrease for durations less than about five periods of the tone frequency and a slower rate for durations longer than 200 ms.[3,6] For short durations, spectral splatter and auditory filtering can account for the faster rate.[6] The changes at extreme durations are evident in Fig. 2, which shows a summary of normal listeners' data for tones at or near 0.25, 1, 4, and 14 kHz obtained in several studies.

Whether the effect of duration changes with frequency is not clear. Figure 2 indicates that the frequency dependence of temporal integration, if any, is small. The best-fitting time constant for an exponential integrator model decreases with increasing signal frequency in some studies,[6] whereas little or no effect of frequency is evident in other studies.[3] The shallow slope at 14 kHz

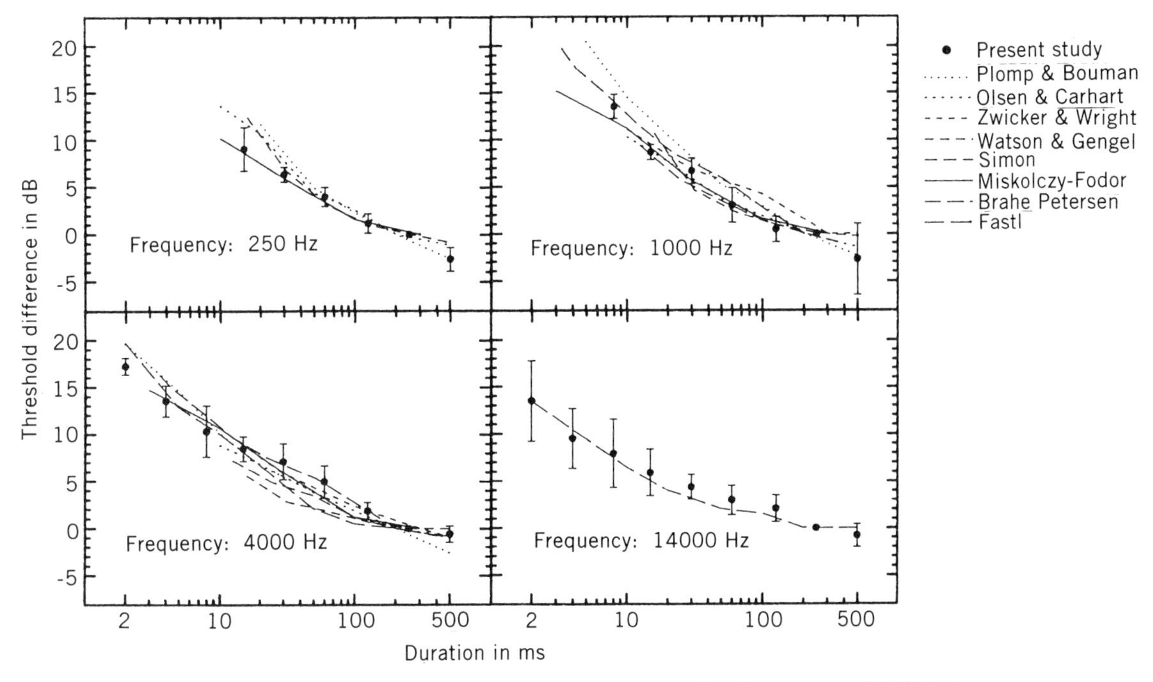

Fig. 2 Change in absolute thresholds as a function of duration for tones near 0.25 kHz (top left), 1 kHz (top right), 4 kHz (bottom left), and 14 kHz (bottom right) from Florentine et al.[3] The filled points show data for five normal listeners tested by Florentine et al.; the bars show plus and minus one standard deviation across listeners. The various lines show data obtained in other studies, as indicated on the right. (Reprinted with permission from Ref. 3.)

in Fig. 2 may be caused by elevated thresholds at this frequency in two of the five normal listeners tested.[3]

Data for detection of multiple brief tone bursts generally indicate that thresholds decrease as the number of bursts increases and increase as the presentation rate decreases, whether the overall signal duration or overall energy is constant.[7] However, one study found virtually no increase in thresholds as the interburst interval between four 8-ms bursts increased from 4 to 67 ms.[5] The reason for this discrepancy is unclear. Thresholds for two broadband clicks appear to increase from about −4 dB (relative to the threshold for a single click) for an interclick interval of 1 ms to an asymptote of approximately −1.5 dB for interclick intervals between 10 and 200 ms.[8]

The decrease in thresholds with increasing duration is often described as the result of a neural integration of stimulus power.[6] Plomp and Bouman's exponential integrator model predicts most of the data for multiple tone bursts and provides a good description of the effect of duration. In particular, it predicts that long-duration tones should show a gradual approach to the asymptotic threshold for continuous tones, which agrees with most data.[6]

An alternative account of the shallower slope at long durations comes from signal detection theory, which predicts that multiple independent observations of a long-duration tone should cause thresholds to decrease by 5 dB per decade of duration[9]—up to some limit set by decay of memory. The latter approach has the advantage that it easily accounts for findings that detection of a tone in noise improves up to a duration of 3 s.[9] The exponential integrator model cannot account for this finding unless the time constant is made considerably longer than the 100–300 ms normally used to fit the data.

However, models that integrate the stimulus intensity, such as the exponential integrator model, are inconsistent with the relatively small contribution to detection by the rise and fall. Moreover, this model cannot predict impaired listeners' data for either single or multiple tone bursts, although predictions obtained with a 10–30-ms time constant rarely differ more than 1 or 2 dB from the data. In part, this failure may reflect that impaired listeners can improve their performance by multiple observations on a stimulus longer than the integration time, but the threshold decrease is diminished due to these listeners' abnormally steep psychometric functions.[7]

Integration of stimulus intensity over long durations is also inconsistent with findings that the effect of interclick interval is asymptotic at around 10 ms.[8] This finding indicates that intensity is integrated only over a period less than 10 ms and that the decrease in threshold for longer durations reflects the effect of multiple observations. Such a model is consistent with the two-click data and with the finding that the threshold for two 10-ms tone bursts presented in temporal gaps of a broadband noise is largely unaffected by 6-dB changes in the level of the noise during 50 ms in the middle of a 100-ms interval that separates the gaps.[8] It also can predict the effect of duration for a single tone burst, but only if it is assumed that the weight on the signal increases with duration.

A somewhat different model entails short-term integration of a compressive power function transformation of the stimulus intensity.[4] The short integration time is compatible with the good temporal resolution shown in a variety of experiments (see Refs. 10 and 11 for reviews) and obviates the need to postulate different integration mechanisms for temporal integration and temporal resolution experiments. This model provides a good description of the effects of stimulus duration and interburst interval, but the parameters necessary to predict the latter effect differ from those used to predict the former. The model also predicts accurately the small effect of interburst interval obtained for impaired listeners[7] when its parameters are altered to reflect these listeners' abnormally fast growth of loudness and diminished temporal integration for single bursts. However, this model is inconsistent with the relatively small contribution to detection by the rise and fall.

None of the models specify the locus of temporal integration, which is most likely to be within or beyond the inferior colliculus because temporal integration is similar for acoustical stimulation and for electrical stimulation of the inferior colliculus.[12] Surprisingly, however, noise-induced hearing loss, which is thought to affect primarily the transduction from acoustical stimulus to neural impulses within the cochlea, changes temporal integration for both electrical and acoustical stimulation.[12] How lesions to the peripheral auditory system can result in changes to the apparently central process of integration is a mystery.

3 SIMULTANEOUS MASKING

A masker that is present throughout the duration of the signal produces simultaneous masking if it is sufficiently intense and its spectral energy is not too far removed from the signal. On-frequency masking occurs when the masker contains significant power in the frequency region of the signal.

3.1 On-Frequency-Noise Maskers

Effect of Masker Level Experiments on masking by white noise provide important information about the effects of masker level and signal frequency. The noise power density of white noise is independent of frequency. Thus, it has significant power in any frequency region—at least if the overall level is sufficiently high—and yields a simple relation between masker level and masked threshold. Comprehensive measurements show that the masked threshold varies with frequency and increases 10 dB for every 10-dB increase in masker level.[13] This linear relation is also obtained for narrow-band maskers whose passbands encompass the signal frequency.[14–16]

Effect of Signal Frequency (Critical Ratios and Critical Bands) The masked thresholds of tones in white noise vary with frequency because on-frequency masking is determined primarily by the noise power within a narrow band of frequencies surrounding the signal and the width of that masking band varies with frequency, as discussed below. The idea that masked threshold is determined by the power at the outputs of the auditory filters is often referred to as the power spectrum model of masking. An elaborate version of such a model is discussed in Section 3.4.

Fletcher suggested that the effective signal-to-noise ratio within the masking band is equal to 0 dB at masked threshold.[14] According to this suggestion, the ratio of the spectral power density of the noise to the tone's intensity at masked threshold provides a direct measure of the range of frequencies that contribute to masking. Fletcher referred to this frequency range as a critical band.[14] However, other authors have reserved the term *critical band* for more direct estimates obtained in a wide variety of psychoacoustical experiments that vary the bandwidth of the stimulus (for review, see Ref. 17) and use the term *critical ratio* for the bandwidths measured by the ratio of the spectral power density of a white masking noise to the tone's intensity at masked threshold.

The critical band provides a basis for an auditory frequency scale—the Bark scale (named in honor of the German physicist Heinrich Barkhausen, who was a pioneer in the study of loudness). One bark is equal to the width of one critical band. Analytical expressions for the critical bandwidth as a function of physical frequency, $W_z(f)$, auditory frequency in barks as a function of physical frequency, $z(f)$, and physical frequency as a function of auditory frequency, $f(z)$, facilitate model calculations and planning of experiments. One such set of internally consistent expressions is given below. The critical bandwidth is

$$W_z(f) = \frac{(f + a)^b}{c}, \qquad (1)$$

where $a = 1750$ Hz, $b = 1.81$, and $c = 9800\ \mathrm{Hz}^{0.81}$. (The unit for c is $\mathrm{Hz}^{0.81}$ to make the unit for $W_z(f)$ hertz.) The expression for $z(f)$ is obtained by integrating the reciprocal of Eq. (1):

$$z(f) = \frac{c}{1-b}\,\frac{1}{(f + a)^{b-1}}\ \text{barks} + k, \qquad (2)$$

where the integration constant k is set 28.32 barks, such that z is zero when $f \approx 20$ Hz. Solving $z(f)$ for f yields

$$f(z) = \left(\frac{c \cdot 1\ \text{bark}}{(1-b)\cdot(z-k)}\right)^{1/(b-1)} + a. \qquad (3)$$

Equation (1) yields values for $W_z(f)$ that are within ±10% of Zwicker's[18] table and Eq. (2) yields values for $z(f)$ that are within ±0.2 barks. Other expressions show similar deviations but are not internally consistent because they were derived by fitting each function separately.[19]

Figure 3 shows how the critical bandwidth and the critical ratio vary with frequency. The critical bandwidths are from Eq. (1) and the table.[18] The data for the critical ratios (converted to bandwidth under the assumption of a 0-dB signal-to-noise ratio at threshold) are for a signal duration of 100 ms and $d' = 1$.[20] The solid line is predicted by the model in Section 3.4. The critical ratios shown here are roughly equal to half the critical bandwidth, which is about 25% larger (1 dB in terms of thresholds) than those reported by Hawkins and Stevens.[13] The model indicates that this difference may reflect the different signal durations and threshold criteria used in the two studies. The difference between critical bandwidths and critical ratios usually is thought to reflect that the signal-to-noise ratio at threshold typically is −3 to −4 dB, rather than 0 dB.[17] Finally, the dashed line shows the equivalent rectangular bandwidth (ERB), which is calculated as $0.11(f + 165\ \text{Hz})$, where f is the frequency in hertz (cf. Ref. 21). The ERB describes modern measurements of the auditory filter bandwidths in notched-noise masking experiments (discussed in *Enc.* Ch. 38). The function for the ERB shows less curvature than the critical-band and critical-ratio functions. In the midfrequency range, the ERB is about 20% less than the

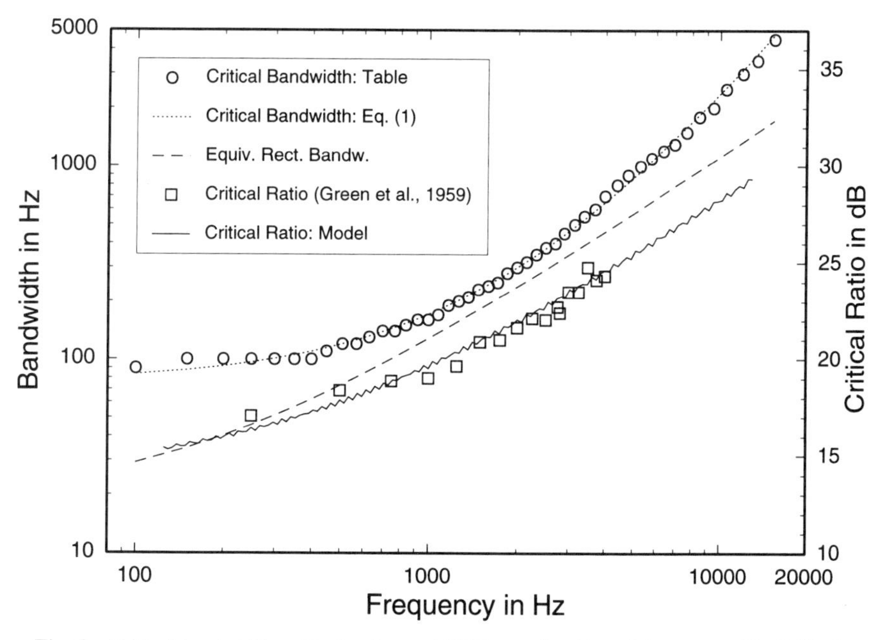

Fig. 3 Critical bandwidth, critical ratio, and ERB as a function of frequency. (○) Critical bandwidths from Zwicker[18] are compared to those calculated from Eq. (1) (dotted line) and to ERBs calculated according to a cochlear map function (dashed line).[21] (□) Critical ratios from Green et al.[20] are compared to predictions from the multiband masking model discussed in Section 3.4 (solid line).

critical bandwidth; the difference increases at higher and lower frequencies.

Effect of Masker Bandwidth Fletcher envisioned the masking band as a bandpass filter centered on the signal frequency and stated that, as an approximation, it might be considered rectangular in shape.[14] If the signal-to-noise ratio at masked threshold is constant, this approximation predicts that thresholds in a variable-bandwidth masker with constant spectrum level should increase 3 dB for each doubling of bandwidth up to the bandwidth of the rectangular filter; further increases of masker bandwidth should have no effect. Several studies have obtained approximately such results.[14,22–24] However, the effect of masker bandwidth and the estimate of filter bandwidth vary considerably with the assumed shape of the filter. Moreover, signal detection theory indicates that the signal-to-noise ratio at masked threshold ought to increase as the masker bandwidth decreases because the relative variability of the noise energy in an observation increases.[22,25] The exact prediction depends on the relative magnitudes of variability in the auditory system and variability in the masker. Modern experiments show that thresholds for tones presented in narrow-band maskers with constant spectrum level decrease only about 1.5–2 dB per halving of masker bandwidth.[23,24]

Effect of Signal Bandwidth Like quiet thresholds, masked thresholds for complex signals also depend on their bandwidth. The masked threshold for a complex sound with closely spaced components is determined by the overall intensity as long as the bandwidth is less than that of the auditory filter.[1,2,26] For wider signal bandwidths, some studies indicate that detection is determined by the auditory filter with the best signal-to-noise ratio,[1,2] but most studies show that components in separate auditory filters are detected more readily than each component alone.[20,26–29] When the duration exceeds about 50 ms, the detectability of tone complexes with equally detectable components spaced more than a critical band apart increases as the square root of the number of components N. This $\sqrt{N}$ rule has been shown to hold for 2-tone complexes,[27] 12- and 16-tone complexes,[20] and 18-tone complexes.[28] Similar results have been obtained with noise bands as signals.[30]

The $\sqrt{N}$ (or square-root bandwidth) rule is easily understood by considering that the auditory system comprises many channels tuned to different frequencies. For tone complexes, it follows from a model using an optimal combination of observations in independent, frequency-selective channels.[27] For noise signals, the square-root bandwidth rule follows from signal detection theory because the relative stimulus variability decreases with increasing noise bandwidth. However, level discrimination of noise bands is not affected appreciably by stimulus variability unless it is quite large as when the bandwidth duration product (WT) is less than about 30.[31] Thus, it seems that the bandwidth effects obtained for noise signals are more likely to result from variations in internal noise and reflect the same multiband decision rules used to account for the $\sqrt{N}$ behavior obtained with complex-tone signals.

Serious departures from the $\sqrt{N}$ rule have been reported for signals with short durations, however. Complex signals composed of tone pulses with Gaussian envelopes and durations that vary with frequency to confine the energy of each component within a one-third octave yield almost complete integration of signal energy across three octaves.[29] The "efficient across-spectral integration" obtained for short-duration signals is difficult to explain by the multiband model.

3.2 Spread of Masking

If the auditory filters were rectangular, a band-limited masker should only mask signals whose frequencies are within its passband, but it is well known that intense low-frequency tones may mask high-frequency tones. This shows that the filters are not rectangular. The gradual attenuation on the skirts of the auditory filters is clearly shown by measurements of thresholds for tones at various frequencies in the presence of a fixed-frequency pure-tone masker.[32] As shown in Fig. 4, the results illustrate the fundamental properties of auditory filtering but are complicated by the many auditory percepts that arise from interactions of the masker and signal tones. The masked thresholds indicated by the solid line show relatively little masking at frequencies below the 1200-Hz masker but considerable masking above it. The notches around 2400 and 3600 Hz indicate that the ear creates harmonics by nonlinear distortion. When only a single sinusoid is presented, the harmonic distortion generally is not detectable, because it is masked by the stronger fundamental. However, adding a signal with frequency near a harmonic frequency produces beating, which reveals the existence of the aural harmonics. This beating causes the masked thresholds to show notches near the masker frequency and its second and third harmonics. Other aural distortion products—cubic-difference tones—enhance the detectability of signals between the masker frequency and its second harmonic.[33] The distortion products also affect perception well above the masked thresholds, and at higher signal levels, the percept of the masker plus signal becomes one of many components.

The asymmetry of masking indicated in Fig. 4 depends on the masker level. More masking is obtained

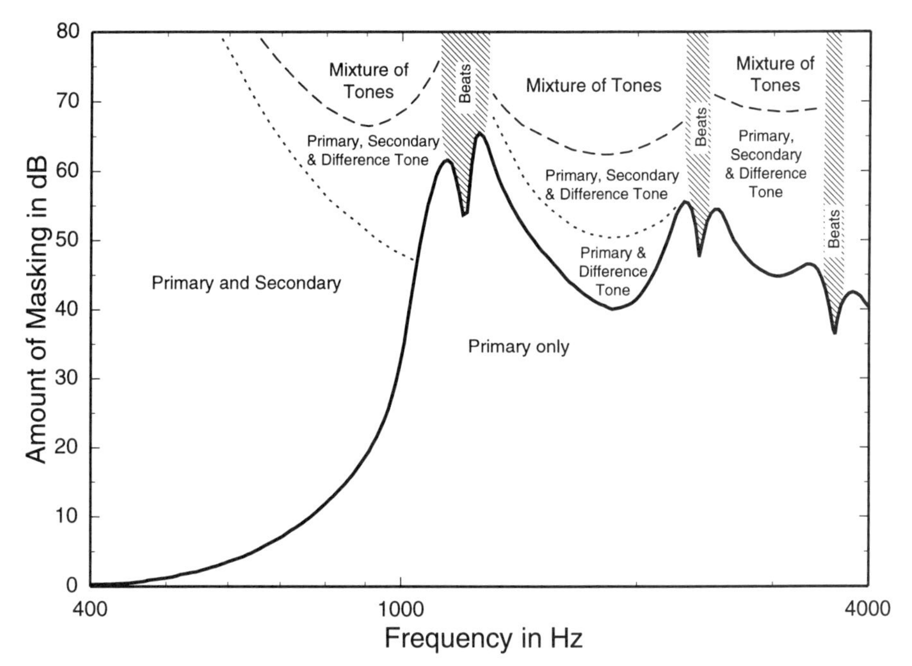

Fig. 4 Masked thresholds for tones at various frequencies in the presence of a 1200-Hz, 80-dB SPL tonal masker.[32] The solid line shows the masked thresholds as a function of frequency. Dashed and dotted lines show the approximate limits among the various percepts that result when the fixed masker is presented together with a signal of the frequency and level indicated. The hatched areas show regions of beating between the signal and the masker or masker harmonics created by aural distortion. (Reproduced with permission from Ref. 32.)

above than below a high-level masker, but the opposite holds for maskers below 40 dB SPL.[34] This reflects that masking by a pure-tone masker grows more slowly than the masker level for signal frequencies at or below the masker but faster than the masker level for signal frequencies above it.[15,16,32] For signal frequencies well above a pure-tone masker, masking may grow as much as 3 dB for each 1-dB increase in masker level. This faster-than-linear growth of masking probably reflects fundamental properties of basilar membrane mechanics. The amplitude of the basilar membrane vibration grows faster with amplitude of a tone when the frequency is below the characteristic frequency than when the frequency is near it (see Chapter 87).

Noise maskers with bandwidths less than a critical band generally yield masked thresholds quite similar to those obtained with a pure-tone masker with a level equal to the overall level of the noise,[15,16] but some differences are apparent. First, the random frequency and amplitude fluctuations of the noise masker render beats imperceptible. Second, the influence of aural distortion products is apparent for frequencies about one-half octave above the center frequency of a narrow-band masker, but it is considerably less than that obtained with pure-tone maskers. Third, aural distortion produces bands of noise below the masker by intermodulation distortion among the spectral components of the noise masker.[33] These bands elevate masked thresholds for signal frequencies somewhat below the masker's passband and probably are responsible for the "remote masking" that is observed far below the passband of intense octave-band maskers.[35] Finally, growth of masking for signal frequencies above that of the masker tends to be slower for noise maskers than for pure-tone maskers, and at high masker levels a pure-tone masker yields more masking of such signals than does a narrow-band noise.[15,36] The amplitude fluctuations of a narrow-band masker yield brief periods of low masker intensity, during which the listener may detect the tonal signal.[36] Certainly, thresholds for brief tones are lower when presented during periods of low masker intensity than during periods of high masker intensity of narrow-band maskers.[16]

In the experiments discussed so far, the level of the masker was kept constant and the signal varied to find the masked threshold. Other experiments fix the signal and vary the masker to find the level that just masks the signal for various masker frequencies.[37] When the signal is a pure tone at a low sensation level, the resulting plots of masker level as a function of masker frequency yield psychoacoustical tuning curves, which bear a striking resemblance to physiological tuning curves measured on the basilar membrane and in single auditory nerve fibers (see Chapter 87, *Enc.* Ch. 109 for review of physiological tuning curves).[37] The low-level signal causes excitation only in auditory channels tuned to frequencies close to the signal and, insofar as masking is excitatory (i.e., is caused by activity due to the masker swamping the activity due to the signal), the psychoacoustical tuning curve might be considered an isoexcitation curve for the channels excited by the signal.

This simple view of psychoacoustical tuning curves is not tenable, however. Listeners can improve their detection performance by listening to frequencies other than that of the signal, for example, by detecting spectral splatter at the onset or offset of a tonal signal.[38] Even if the signal excites only a narrow range of auditory channels, listeners can improve detection of the signal by listening to channels tuned slightly above the signal frequency when the masker is below the signal and tuned slightly below it when the masker is above the signal, that is, "off-frequency listening."[39] Owing to the rounded shape of the auditory filter (see Chapter 90), the signal-to-masker ratio generally is best within a filter centered slightly away from the signal frequency. Furthermore, simultaneous masking may not be exclusively excitatory. Auditory nerve measurements indicate that it is excitatory for maskers above the signal frequency, but at least partly suppressive for intense maskers below the signal frequency.[40] Thus, it is likely that simultaneously masked tuning curves do not reflect an isoactivity contour for the masker's excitation in the channels tuned to the signal frequency. Rather, they may reflect an isosuppression contour for maskers below the signal frequency and an isoexcitation contour for maskers above it.

Although tone-on-tone masking and psychoacoustical tuning curves reveal basic properties of the auditory filters, they do not provide precise estimates of their shapes. A much more useful psychoacoustical characterization of auditory filtering is afforded by notched-noise maskers.[41] Briefly, the auditory filter shape is derived from masked thresholds for tones at frequencies between the proximal cutoff frequencies of two noise bands under the assumption that the thresholds represent a constant signal-to-noise ratio within the filter providing the best signal-to-noise ratio. This procedure takes the effect of off-frequency listening into account if detection is based on the linearly filtered signal energy, but not if detection is based on auditory distortion products arising from nonlinear interactions between the masker and the signal.[42] However, such nonlinear effects appear to be small in the conditions generally used to measure auditory filter shapes.[43] The properties of the auditory filters are discussed extensively in Chapter 90. As discussed in Section 3.4, rather precise estimates of masked thresholds for a variety of maskers and signals can be calculated on the basis of the outputs of such filters.

3.3 Additivity of Simultaneous Masking

The previous sections considered primarily maskers encompassing a single contiguous spectral region. What if two or more such maskers are combined? Except for effects of off-frequency listening, the power spectrum model of masking predicts that the total masking by the composite masker should be equal to the sum of the masking powers of the individual maskers. For example, consider two maskers that provide the same amount of masking. One might expect the amount of masking to increase 3 dB (i.e., the masking power to be doubled) when both maskers are presented together, but data show 10 or even 20 dB.[16,44,45] Whereas this *nonlinear additivity* or excess masking points to some limitations in the power spectrum model, a closer examination of the conditions used to produce it shows that it results from three factors: (a) off-frequency listening, (b) combination tone detection, and (c) changes in the masker envelope when two maskers are combined.[46]

The power spectrum model readily accommodates restrictions in off-frequency listening and can also account for the effects of distortion products if they are included as part of the effective stimulus. However, excess masking also occurs when off-frequency listening and auditory distortion products are ruled out, if at least one of the maskers fluctuates relatively slowly,[46] or if one masker is a low-frequency tone.[16] In these cases, it is likely that detectability of the signal in a single masker varies substantially over the course of a long-duration signal and that masked thresholds are based on brief periods of high detectabilty.[36,46] This effect cannot be explained by the power spectrum model, unless the weight assigned to the signal channel varies over time and depends on the short-term masker intensity.

3.4 Calculation of Simultaneous Masking

Although the traditional power spectrum model of masking fails in cases when the masker amplitude fluctuates slowly, it provides rather precise predictions of masked thresholds in many cases. Therefore, it is worthwhile to consider a general version that can account for many of the data discussed above. The model is based on an energy detector model with internal noise.[25,30] It assumes that the stimulus is filtered by auditory filters with the characteristics discussed in Chapter 90 and that detection is based on the outputs of these filters, which are integrated over time and corrupted by internal noise. When the signal duration exceeds the integration time, multiple observations distributed over time are used to optimize performance. Likewise, multiple observations distributed across frequency-selective auditory channels are included to account for off-frequency listening and detection of complex signals.

To avoid having to select channels to suit a particular masker–signal combination, it is convenient to use channels that are spaced 0.5 ERBs apart. This is sufficient to keep the predicted thresholds for a fixed set of channels from deviating more than about 0.3 dB across different possible sets. Characterizing each channel by its center frequency $f_c(i)$, it can be shown that the overall sensitivity d'_{total} of such a multiband energy detector model is given as

$$d'_{\text{total}} = \left(\sum_{i=1}^{80} d'^2(f_c[i]) \right)^{1/2}, \tag{4}$$

where the center frequencies are

$$f_c(i) = (165\,\text{Hz}) \times (10^{0.024i} - 0.85) \tag{5}$$

and $d'(f_c)$ is the sensitivity to the signal within the auditory channel tuned to f_c, which can be calculated as

$$d'(f_c) = \begin{cases} \dfrac{T_s}{\tau}\,\dfrac{I_s(f_c)}{I_m(f_c)} \cdot \left(\dfrac{W(f_c)\tau}{\sigma_I^2 W(f_c)\tau + 2[1 + T_s I_s(f_c)/\tau I_m(f_c)]} \right)^{1/2} & \text{if } T_s \le \tau, \\[2ex] \sqrt{\dfrac{T_s}{\tau}\,\dfrac{I_s(f_c)}{I_m(f_c)}} \cdot \left(\dfrac{W(f_c)\tau}{\sigma_I^2 W(f_c)\tau + 2[1 + I_s(f_c)/I_m(f_c)]} \right)^{1/2} & \text{if } t_s > \tau, \end{cases} \tag{6}$$

where T_s is the signal duration, τ is the 100-ms auditory integration time for the rectangular integration window in the model, $I_s(f_c)$ is the signal intensity and $I_m(f_c)$ the masker intensity at the output of the auditory filter centered on f_c, $W(f_c)$ is the ERB at f_c (see Chapter 90), and σ_I is a constant that determines the amount of internal noise. If σ_I is set to 0.65, this model yields predictions in excellent agreement with a wide variety of data. (The factor of 2 in front of the second term of the denominators accounts for the correlation between the overlapping channels.)

The model agrees with a large body of masking data, but this does not indicate that detection of a signal in noise necessarily is based on the stimulus energy. The critical ratio is nearly unaffected by random variations in the masker level between presentations, and the thresholds obtained with a 32-dB range are considerably lower than those predicted by an ideal energy detector.[47] Similar results have been obtained with narrow-band maskers.[48] These findings indicate that an across-interval comparison of energy in the signal band is an unlikely cue for detection of a tone in noise, contrary to the basic tenet of the power spectrum model of masking. Thus, these findings present a fundamental problem for the model and provide a strong basis for rejecting it. Nevertheless, the power spectrum model described by Eqs. (4)–(6) appears very useful as an engineering tool, and provides remarkably accurate predictions of listeners' performance in a wide variety of masking situations.

4 NONSIMULTANEOUS MASKING

The previous section discussed simultaneous masking as obtained when the masker is essentially continuous. However, a masking noise of limited duration also masks brief signals presented before the onset of the masker (backward masking) and, especially, following the offset of the masker (forward masking). Nonsimultaneous masking depends strongly on the temporal relations between the masker and the signal, as well as on the parameters that govern simultaneous masking.

4.1 Forward Masking

Temporal Parameters Forward masking decays as the delay between the masker and the signal offsets is increased, and little masking occurs beyond 200 ms.[49–51] The masker–signal delay is specified between offsets, because many investigators argue that this is the relevant variable.[49,52] The amount of masking is nearly independent of signal duration when the offset-to-offset delay Δt is fixed.[49] Thus, temporal effects in forward masking are most easily described by the amount of masking as a function of Δt.

The rate of decay in forward masking increases with the amount of masking produced for short delays. In other words, masked thresholds decrease faster with increasing masker–signal delay as the masker level and

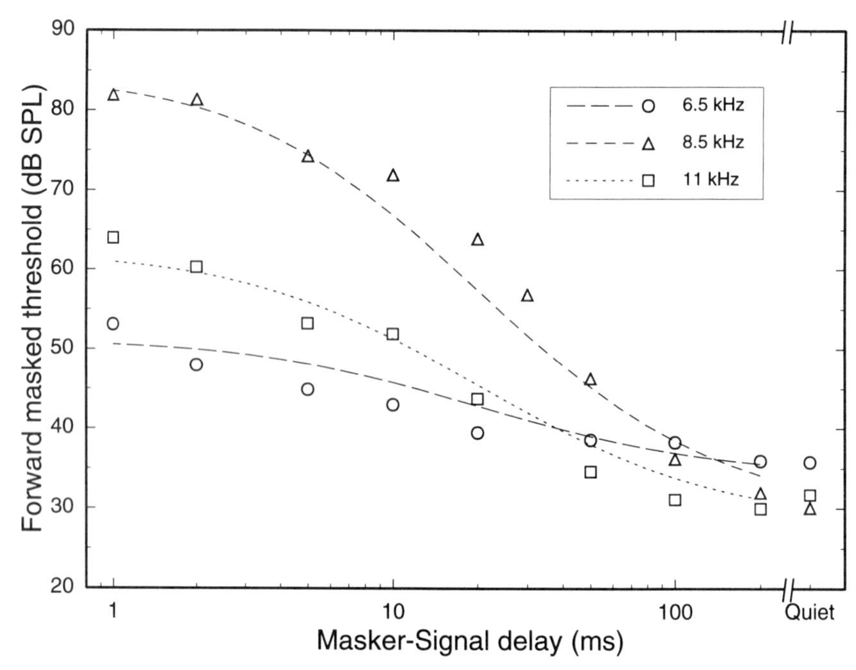

Fig. 5 Forward masked thresholds for tones at (○) 6.5, (△) 8.5, and (□) 11 kHz as a function of the delay between the offset of the 500-ms narrow-band masker at 8.5 kHz and the offset of the 1-ms signal.[50] Quiet thresholds are shown by the points to the far right. The dashed lines indicate masked thresholds calculated according to Eq. (7). (Adapted with permission from Ref. 50.)

spectral proximity of the masker and signal increase. This is illustrated in Fig. 5, which shows that forward masking by a long-duration masker lasts approximately 200 ms regardless of the initial amount of masking.[50] Generally, the decay of forward masking is well approximated by a straight line when the masked threshold in decibels is plotted as a function of delay on a logarithmic scale,[51,53] except for very brief delays, where masking shows almost no decay.[16,49,50] The rate of decay increases as the masker duration decreases below about 300 ms,[49,54,55] and forward masking ceases for signal delays exceeding about 50 ms when the masker duration is 5 ms.[55]

Because masking decays faster for high than for low masker levels, the effect of masker level depends on the masker–signal delay. For pure-tone maskers and signals with the same frequency, the interaction of masker level and temporal parameters in a forward-masking experiment can be expressed as a product of two independent factors.[51] The amount of masking increases proportionally with the masker level, and the proportionality constant is determined only by the temporal parameters. For example, the amount of masking at 1000 Hz increased about 0.6 dB for each 1-dB increase in masker level when Δt was 25 ms and about 0.25 dB when it was 60 ms.[51] These growth-of-masking slopes also depend on masker duration.[54]

The proportionality between masker level and amount of masking and the effects of temporal parameters may be summarized by a descriptive formula. Such a formula must reflect that forward masking degenerates into simultaneous masking as Δt goes to zero and into no masking as Δt becomes large; that is, the rate-of-growth multiplier describing the temporal effects in forward masking should decrease from unity (independent of masker duration T_m) for $\Delta t = 0$ to zero for Δt greater than about 200 ms. It should also reflect that the amount of masking obtained for a given Δt does not increase with T_m above 200–400 ms[49,50,54] and account for the effect of auditory filtering on forward masking.[11,56] One expression for the amount of masking M that fulfills these requirements is

$$M = \frac{dT_m^a}{dT_m^a + b(d + T_m)^a \dfrac{\Delta t^2}{\Delta t + c/\mathrm{ERB}(f_t)}} L_m^*, \qquad (7)$$

where a, b, c, and d are constants (d = 390 ms, a = 0.431, b = 0.495, and c = 0.693), ERB(f_t) is the auditory filter bandwidth at the test frequency f_t and L_m^* is the amount of simultaneous masking produced by the masker, which may be calculated from Eqs. (4)–(6). (The values of a, b, c, and d were obtained from a set 179 data points compiled from five studies.[16,49–51,54])

As shown by the lines in Figs. 5 and 6, this formula provides a very reasonable account of a variety of data on forward masking. The calculated amounts of masking rarely differ more than 3 or 4 dB from those measured. In particular, Eq. (7) describes quite well the interaction of masker duration and masker–signal delay shown in Fig. 6.[55] Both the data and Eq. (7) indicate that forward masking should be less than 3 dB when Δt exceeds 200 ms and the masker duration is 300–500 ms, and when Δt exceeds 50 ms and the masker duration is 5 ms.

Equation (7) also yields growth-of-masking slopes close to those observed in several studies.[50,51,54,57] Thus, Eq. (7) appears useful to estimate the amount of masking obtained in a variety of experiments, although its theoretical significance is unclear.

Frequency Selectivity Measured in Simultaneous and Forward Masking Frequency selectivity as measured by tuning curves, masking patterns, and auditory filter shapes generally is more acute in forward than in simultaneous masking. For example, forward-masked tuning curves have Q_{10} values (the signal frequency divided by the bandwidth of the tuning curve measured 10 dB above its tip) about twice those obtained in simultaneous masking and two to three times steeper slopes.[50,58] Likewise, forward-masked auditory filter characteristics have slopes almost twice those obtained under simultaneous masking, but the ERBs are only about 20% smaller.[41] In contrast, masking patterns are broader in forward than in simultaneous masking owing to a flattening of the peak that results because forward masking decays faster when the amount of simultaneous masking is large than when it is small.[50]

In part, the difference between forward and simultaneous masking may reflect the operation of suppression as has been observed in physiological measurements. In the auditory nerve, the response to a probe tone can be reduced or eliminated by introducing a second tone (the suppressor) at a frequency above or below that of the probe (see Chapter 87). Thus, a simultaneous masker may suppress the response to the signal. A forward masker cannot suppress the response to the signal because suppression appears to reflect instantaneous nonlinear interaction at the level of the basilar

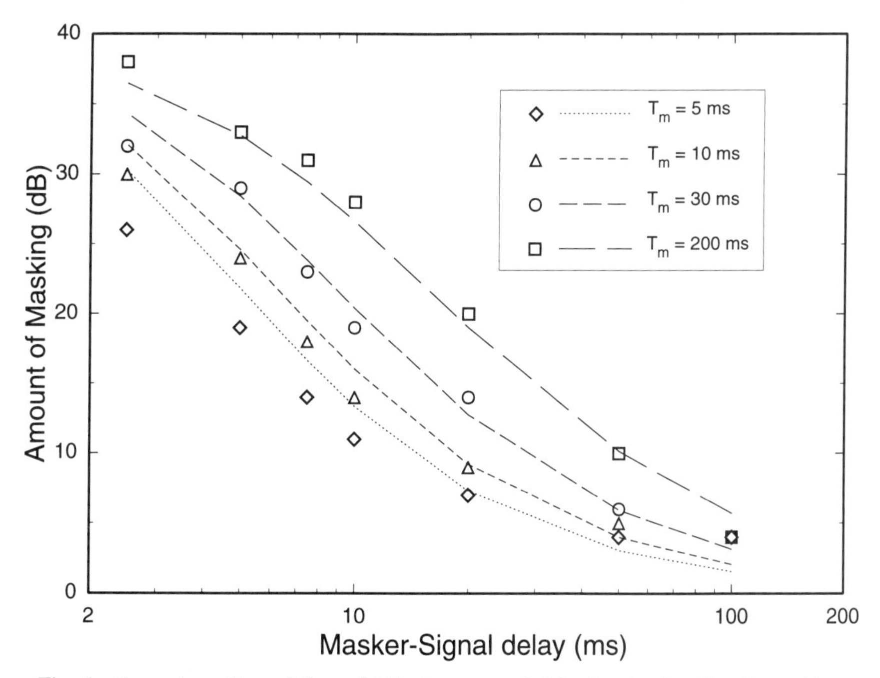

Fig. 6 Forward masking of 5-ms, 2-kHz tones preceded by bursts of uniformly masking noise are plotted as a function of the delay between masker and signal offsets. The parameter is masker duration as indicated. The symbols are data from Zwicker.[55] The lines indicate masked thresholds calculated according to Eq. (7). (Adapted with permission from Ref. 55.)

membrane (see Chapter 87). Accordingly, the forward-masked tuning curve may correspond to an isoexcitation pattern more closely than the simultaneously masked tuning curve.

The listener's task in forward masking is quite complex, however, and a large part of the differences between forward and simultaneous masking is likely to reflect off-frequency listening and cueing.[59] In particular, listeners may have great difficulty in detecting a brief signal following an on-frequency forward masker. Excessive masking can occur if the delay between the masker and signal is too short to make the silence between them audible or if the delay and signal duration combine to make the signal confusible with natural fluctuations in a narrow-band masker.[60] Cueing and off-frequency listening are highly likely to affect forward-masked psychoacoustical tuning curves but are unlikely to affect the auditory filter measurements in forward masking, which employ maskers that have components both above and below the signal frequency and bandwidths much wider than that of the auditory filter centered on the signal. Consequently, the difference between forward and simultaneous masked auditory filter shapes probably is primarily due to suppression. This conclusion is consistent with the finding that the difference between forward and simultaneous masking is pronounced on the skirts of the auditory filters but modest in the central part of the filter characteristic, which provides the major contribution to the ERB.

Composite Maskers: Multiple Spectral Components Why is suppression not observed in simultaneous masking? One explanation is that, in simultaneous masking, the suppressor reduces activity due to the signal in the same proportion as that due to the masker.[61] Thus, suppression does not alter the effective signal-to-masker ratio in simultaneous masking and the masked threshold is not altered. In forward masking, the suppressor is presented together with the masker only, and thus, it does not suppress the signal. Consequently, suppression of the activity due to the masker improves the detectability of the signal. For example, forward masking by an on-frequency masker is reduced if a relatively intense suppressor is added at frequencies around one octave below or about 15% above the on-frequency masker.[62] The area of frequencies and SPLs for which suppression is obtained is qualitatively similar to the suppression areas observed in auditory nerve recordings. Suppression also occurs for maskers with broad spectra. Thresholds for a 2-kHz signal can be 20 dB lower with a 4000-Hz-wide masker than with a 500-Hz-wide masker,[63] but a large part of the decrease is likely to be caused by cueing effects as discussed above. An extensive review of the effects of cueing on estimates of suppression leads to the conclusion that the true contribution of suppression toward forward-masked thresholds rarely exceeds 5 dB.[59]

4.2 Backward Masking

A masker elevates thresholds not only for signals presented simultaneously with or after it but also for signals presented before the masker. The time interval for such backward masking rarely exceeds 25 ms[49] but can reach 100 ms.[64] Backward masking is frequency selective[50,63,64] and shows suppression,[63] which may reflect temporal smearing of the masker and signal by the auditory filters.[56] However, it may also reflect confusion between the masker and signal, because highly practiced listeners often show little or no backward masking.[65]

4.3 Nonadditivity of Forward, Simultaneous, and Backward Masking

Unusual masking effects can be obtained when masker components are separated in time. Two equally effective maskers presented without temporal overlap can produce 10 dB more masking than either masker alone.[66] Essentially similar results are obtained for two forward maskers, two backward maskers, and combinations of forward and backward maskers. Excess masking also has been shown for combinations of forward and simultaneous maskers (for a review, see Ref. 67). The excess masking may result because temporal integration in the auditory system acts on a compressive transformation of the stimulus power.[66] Compression by a modified power law can account for the nonaddivity of masking obtained with combinations of maskers that do not overlap temporally and/or spectrally,[67] but the exponent of the compressive power function is a free parameter that varies among listeners and experiments and has no obvious relation to other auditory phenomena.

5 SPECIAL EFFECTS IN MASKING

5.1 Interaural (Central) Masking

The masking effects discussed previously all were obtained with the masker and the signal presented to the same ear. Masking also can be produced when the masker is in one ear and the signal in the other. This interaural or central masking typically is measured using special earphones with more than 80 dB interaural attention, which ensures that the masking is not caused by the sound crossing from the masking ear to the signal ear (for review and references, see Ref. 52). Interaural masking grows very slowly (<0.2 dB/dB) with increasing masker level and is highly dependent on the delay between the masker and signal onsets. It is large when they are simultaneous but decreases with a time constant of 50 ms as the signal onset is delayed. When the masker is continuous, interaural masking is nearly absent for a pulsed signal but large for a continuous signal. Frequency selectivity is quite sharp. With 12–15 dB on-frequency interaural masking, the main peak of the masking pattern is about one critical band wide and masking decreases to less than 5 dB for signal frequencies more than about 250 Hz away from the masker. Unlike ipsilateral masking, interaural masking patterns are quite symmetrical up to a masker level of 60 dB SL (sensation level) but some asymmetry is apparent for 70-dB-SL maskers.

5.2 Comodulation Masking Release

The model of simultaneous masking presumes that detection of the signal depends only on the signal-to-noise ratio in channels responding to the signal. Thus, masking should depend primarily on the noise power in channels tuned to frequencies at or near a pure-tone signal. Whereas this simple view of masking is accurate in most instances, it fails under special circumstances. For example, the amount of masking may decrease from 80 to 40 dB when a continuous one-octave masker at 90 dB SPL is pulsed (32 ms on, 32 ms off)[68] but the masker energy decreases by only 3 dB. Because the masker excitation in the signal channel is reduced dramatically during the off periods, this release of masking may be thought to reflect temporal resolution.[68] According to this view, however, the same release should be obtained when a narrow-band masker is pulsed or modulated, which is not true.[69,70] This indicates that the release from masking depends, at least in part, on modulation at frequencies remote from the signal.

The release of masking is called comodulation masking release (CMR) because it occurs primarily when the modulation at remote frequencies is similar to that around the signal frequency, that is, when the on frequency and remote masker bands are comodulated.[23] As shown in Fig. 7, the effect of masker bandwidth changes dramatically when a Gaussian noise masker is multiplied by a 50-Hz low-pass noise, which produces envelopes that are approximately the same within each 100-Hz-wide band of noise. The data for random noise follows the pattern established by Fletcher.[14] As the bandwidth is increased, masked thresholds increase until some critical bandwidth is reached and then remain constant. The modulated noise yields results similar to those for the random noise up to a bandwidth of about 100 Hz. As the modulated noise is widened further, however, thresholds

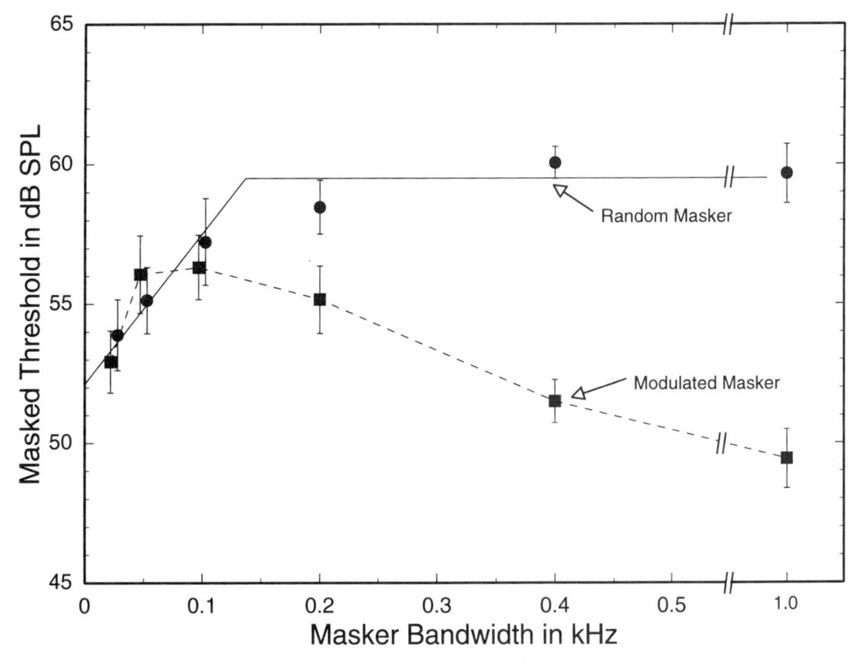

Fig. 7 Thresholds for 350-ms tones at 1 kHz masked by bands of random (●) or narrow-band noise-modulated (■ connected by dashed lines) noise plotted as a function of bandwidth of the masker.[23] The maskers were centered at 1 kHz. The bars show plus and minus one standard deviation across listeners. (Reproduced with permission from Ref. 23.)

start to decrease and at the 1-kHz bandwidth the common modulation results in a 10-dB reduction of masking.

The existence of CMR demonstrates that masking not only depends on the signal-to-noise ratio in channels excited by the signal but also may depend on channels excited only by the masker. When the masker consists of two or more bands with similar envelopes, masking is less than that obtained for a single narrow band centered on the signal. (The masker bands that are added away from the signal frequency are usually called cue bands.) The advantage afforded by CMR is most evident for detection. Level discrimination thresholds for partially masked tones of equal SPLs are approximately the same whether the masker is random or comodulated, except when the tones are at or below detection threshold in the random noise.[71] Likewise, filtered speech signals consistently yield a CMR for detection, but not for recognition.[72]

More than two dozen papers describe the effects of the many parameters in CMR experiments. Hall and Grose[70] provide an excellent review. Briefly, CMR generally decreases as the frequency separation between the on-frequency masker band and the cue band(s) increases.[73] In part, this effect may reflect that within-channel cues exist when a single cue band is separated from the on-frequency masker by less than one or two ERBs.[24] Several studies show that CMR increases as the bandwidth (or modulation frequency) of the masker band decreases.[74,75] In part, this apparent increase in CMR may be attributed to within-channel processes, because the contribution to CMR by the cue bands is almost independent of modulation frequency up to 100 Hz.[69]

If the cue bandwidth measured in ERBs or critical bands is held constant, CMR is approximately independent of frequency.[74] The level of the cue band relative to the masker band is of little importance to the magnitude of CMR. Comodulation masking release is present when the masker and cue bands differ in level by 30 dB[76] or even 60 dB,[77] but CMR increases with the masker level.[68,77] The range of masker–cue level differences for which CMR is obtained appears to be limited primarily by masking. As long as the cue does not make the masker inaudible or vice versa, a CMR is obtained. The wider the frequency separation between the masker and cue, the larger the range of level differences yielding CMR and the largest ranges are obtained when the cue band(s) are presented contralaterally to the masker and signal.[77]

Exact synchrony between masker and cue envelopes appears very important for CMR. If the envelope of the cue is phase shifted or delayed relative to the masker envelope, CMR is absent or reduced substantially.[78] Auditory grouping (i.e., the perception of disparate spectral components as a complex sound emanating from a single source) is important for CMR. Adding two codeviant bands, whose identical envelopes differ from that of six cue bands and the masker band, reduces the CMR, but it increases again as the number of codeviant bands increases.[79] Apparently, lack of perceptual segregation of a small number of codeviant bands causes them to be grouped with the cue envelopes, which effectively reduces the correlation between the cue envelope and the masker envelope.

One possible explanation for CMR is that detection may be mediated by a simultaneous comparison of envelopes in different critical bands.[23] This view is supported by the finding that CMR generally does not depend on uncertainty about the cue correlation, except if both the cue bands are uncorrelated in the non–signal interval and comodulated in the signal interval.[80] The envelope comparison may be described by an equalization–cancellation (EC) model, which extracts envelopes in different critical bands, adjusts their root-mean-square (rms) amplitudes to be equal (equalization), and subtracts envelopes in channels remote from the signal from the envelope in the signal channel (cancellation).[36] In the comodulated condition, the envelopes are similar when the signal is absent but dissimilar when it is present. Thus, the presence of the signal is indicated by a large remainder after cancellation.

Detection of decorrelation between the envelopes in the signal and the cue bands may also explain CMR.[73,81] The predictions by this model and the EC model may be identical.[82] When the signal is absent, comodulation causes the correlation to be high; when the signal is present, it alters the envelope in the signal band, which in turn decreases the correlation. The decorrelation caused by a signal close to threshold in a typical CMR experiment is sometimes comparable to the just-detectable decrease in correlation of two bands of noise,[81] but the effects of masker bandwidth are opposite in CMR and decorrelation detection.[83]

Comodulation masking release may also be mediated by detection of a difference in modulation depth between the masker-plus-signal and the cue bands.[76] The tonal signal reduces the modulation depth in the signal band, because it tends to "fill in" the periods of low masker amplitude. Indeed, a CMR can occur when the signal changes the depth but not the pattern of modulation.[84] In addition, CMRs obtained in sinusoidally modulated maskers are also consistent with discrimination thresholds for changes in modulation depth.[85]

A rather different explanation for CMR is that the comodulated cue indicates the optimal times to listen and allows listeners to "listen in the valleys."[36] If periods with low masking power ("valleys") are weighted more heavily than periods with high masking power, quite efficient detection may result. Like the other models, listening in the valleys requires extraction of the envelopes in different frequency bands, but unlike them, it does not depend on direct across-frequency comparisons of envelopes.

Many experiments have attempted to distinguish among the various explanations for CMR. Experiments with signals consisting of multiple tones show that a CMR still is obtained if signal components are present in all noise bands, but the presence of a cue band without signal components improves the CMR.[86,87] Simulations with specific implementations of the EC model, the correlation model, and the listening-in-the-valleys model indicate that such results are most consistent with the EC model.[86] However, different implementations of the general concepts embodied in the various models may yield different results.

Indeed, different conclusions result from other experiments. For example, a CMR is also obtained when the signal is identical to the on-frequency masker.[70,88] This finding is consistent with a comparison of envelope amplitudes between the masker–signal band and the cue band. Decorrelation detection and the EC model would not be effective because the signal does not alter the correlation between the signal and cue bands. Likewise, listening in the valleys should be no better than applying time-invariant weights. On the other hand, CMR is also obtained when the level of the cue band is varied randomly between presentations.[88] In this case, direct comparisons of envelope amplitudes are ineffective and CMR may be mediated by decorrelation detection or listening in the valleys.

The most surprising evidence, however, comes from experiments comparing CMR for signals presented at the peaks of the masker envelope and signals presented in the valleys.[70,87] No CMR is obtained for signals at the peaks, whereas a copious CMR is obtained for signals in the valleys. Because either signal causes a substantial change to the envelope, the EC model and the decorrelation model predict that a CMR should be obtained in both cases. Thus, these experiments clearly favor listening in the valleys as an explanation for CMR.

In summary, a CMR can be obtained even if one or the other of the proposed mechanisms are rendered ineffective by some experimental manipulation. The auditory system clearly can process simultaneous envelopes in different frequency bands in several different ways, and probably, detection is based on whichever process yields the best performance.[70]

5.3 "Overshoot" and the Effect of Off-Frequency "Fringes" on On-Frequency Masking

The section on simultaneous masking did not consider possible effects of the temporal relation between a pulsed masker and the signal. Indeed, the power spectrum model of masking does not easily predict such an effect. However, thresholds for brief tones presented in a wide-band masker can be elevated more than 10 dB if the signal onset occurs within a few milliseconds after the onset of a pulsed masker.[89] This overshoot of masking gradually diminishes as the delay between the masker and signal onsets increases up to about 300 ms.[89,90]

The overshoot produced by a broadband noise is small or absent at low levels, increases to reach about 10–15 dB at spectrum levels around 20 or 30 dB SPL, and either stays constant[91,92] or decreases somewhat at higher levels.[93–95] The signal frequency may explain part of the difference between studies. One study found that the amount of overshoot decreased at high levels for a 4-kHz signal, but not for a 1-kHz signal.[94] Likewise, another study found that the overshoot obtained with a wide-band masker decreased at high levels for a 6.5-kHz signal but not for a 2.5-kHz signal.[92] Individual differences may also play a role because the amount of overshoot obtained for a given masker–signal combination varies considerably among listeners.[93]

Overshoot is also obtained when the frequency of a tonal masker differs from the signal frequency.[96] In contrast, no overshoot is obtained when the masker is centered on the signal frequency and has a bandwidth similar to that of the signal, whether the bandwidth of the signal is narrow[89,96] or wide.[91]

As shown in Fig. 8, the overshoot decreases systematically as the bandwidth of a noise masker decreases.[89,90] The overshoot is absent for masker bandwidths of one and two critical bands but increases at wider bandwidths to reach about 12 dB for a white-noise masker (24 barks wide). Generally, overshoot is obtained only when the masker contains energy outside the frequency range encompassed by the auditory filters responding to the signal.[89,90,96–98] In particular, overshoot seems to depend on the presence of masker energy in auditory filters tuned to frequencies above that of the signal.[99] One exception is that overshoot also is obtained in narrow-band maskers at high frequencies and medium levels,[92] which indicates that the overshoot for such stimuli may, at least in part, be due to mechanisms different from those operating at other frequencies and/or levels.

The importance of off-frequency masker energy for the overshoot is very clearly demonstrated by studies that divided a broadband masker into several spectral regions in which the onset could occur at different times. Such measurements show that a broadband masker does not produce overshoot when the masker onset occurs close to the signal onset only in a narrow band around the signal frequency,[98] except at high frequencies and medium levels.[92] On the other hand, if the masker onset is close to the signal onset only in frequency regions outside a

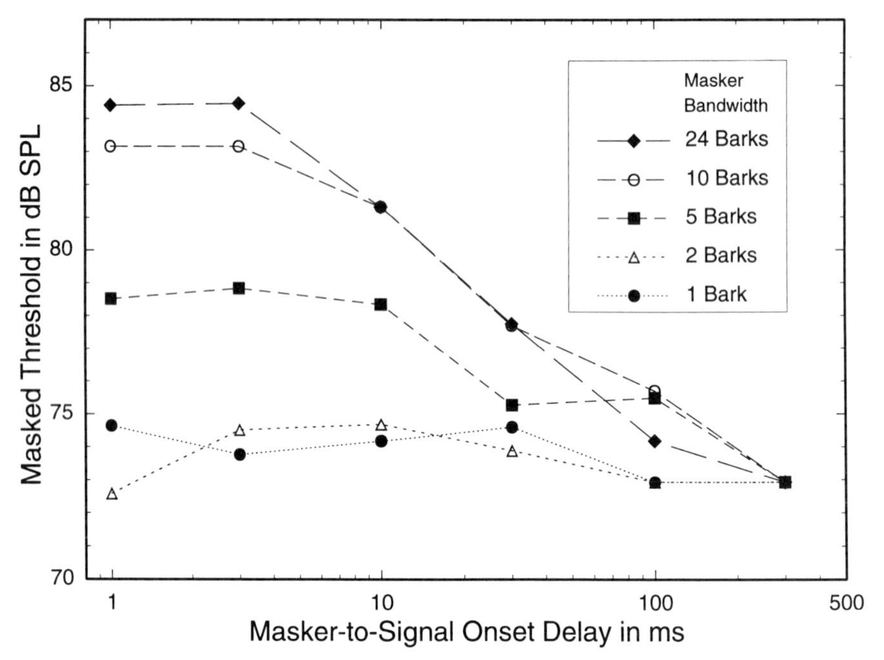

Fig. 8 Masked thresholds for a 2-ms tone at 4.8 kHz are plotted as a function of time between the onset of the 600-ms masker and the onset of the signal.[89] The parameter is masker bandwidth as indicated. (Reproduced with permission from Ref. 89.)

narrow band around the signal frequency, the overshoot is large.[92,98] Measurements with various cutoff frequencies between the on- and off-frequency bands indicate that the frequency regions important for producing overshoot with a 2.5-kHz signal are primarily between 1 and 2 kHz on the low side and between 3 and 4 kHz on the high side.[98] This corresponds to three to five ERB wide bands with proximal cutoff frequencies about one to two ERBs away from the signal frequency.

Probably several mechanisms are responsible for the overshoot. Several authors have suggested that overshoot is related to auditory nerve adaptation.[100] Although auditory nerve adaptation may contribute to the overshoot, it does not account for it completely because auditory nerve adaptation appears to be governed primarily by frequencies near the characteristic frequency,[101] whereas the overshoot is governed primarily by remote frequencies, at least at frequencies below 4 or 5 kHz.

Central mechanisms probably also are involved. Because one of the most important functions of the auditory system is to detect events in the environment as revealed by changes to the auditory input, it seems likely that any change in sound level and quality would attract attention. In terms of the power spectrum model, a sudden onset of the masker over a broad range of frequencies may temporarily force a high weighting to be applied to the channels remote from the signal. In essence, the sudden change in the masker channels distracts listeners from the task of monitoring the signal channel. As the onset in remote channels is removed in time from the signal onset, however, the change in the remote channels has been processed and attention can return to the signal channel. This hypothesis of time-varying channel weights is similar to the listening-in-the-valleys hypothesis for CMR and is consistent with findings that gating a 5-bark-wide masker causes a marked increase in the attention bandwidth as measured by the detectability of off-frequency probes.[102]

Auditory grouping may also contribute to the overshoot.[97] When the onsets of the masker and the signal occur together, their envelopes are quite similar and thus may be heard as a signal auditory object. This hypothesis is consistent with findings of slightly elevated thresholds for brief signals presented near the masker offset.[96] However, it may not be consistent with the overshoot's dependence on the off-frequency bands, because a signal should be about equally difficult to hear whether it is perceived as integral to a wide-band or a narrow-band auditory object.

Acknowledgments

Thanks are due to Mary Florentine, David M. Green, Armin Kohlrausch, Dennis McFadden, Fan-Geng Zeng, and two anonymous reviewers for helpful comments on an earlier draft. The preparation of this chapter was supported by NIDCD grant R01 DC00187.

REFERENCES

1. G. Gässler, "Über die Hörschwelle für Schallereignisse mit Verschieden Breitem Frequenzspektrum," *Acustica*, Vol. 4, 1954, pp. 408–414.
2. M. B. Higgins and C. W. Turner, "Summation Bandwidths at Threshold in Normal and Hearing-Impaired Listeners," *J. Acoust. Soc. Am.*, Vol. 88, 1990, pp. 2625–2630.
3. M. Florentine, H. Fastl, and S. Buus, "Temporal Integration in Normal Hearing, Cochlear Impairment, and Impairment Simulated by Masking," *J. Acoust. Soc. Am.*, Vol. 84, 1988, pp. 195–203.
4. M. J. Penner, "A Power-Law Transformation Resulting in a Class of Short-Term Integrators That Produce Time-Intensity Trades for Noise Bursts," *J. Acoust. Soc. Am.*, Vol. 63, 1978, pp. 195–201.
5. G. M. Gerken, V. K. H. Bhat, and M. Hutchison-Clutter, "Auditory Temporal Integration and the Power Function Model," *J. Acoust. Soc. Am.*, Vol. 88, 1990, pp. 767–778.
6. R. Plomp and M. A. Bouman, "Relation between Hearing Threshold and Duration for Tone Pulses," *J. Acoust. Soc. Am.*, Vol. 31, 1959, pp. 749–758.
7. R. P. Carlyon, S. Buus, and M. Florentine, "Temporal Integration of Trains of Tone Pulses by Normal and by Cochlearly Impaired Listeners," *J. Acoust. Soc. Am.*, Vol. 87, 1990, pp. 260–268.
8. N. F. Viemeister and G. H. Wakefield, "Temporal Integration and Multiple Looks," *J. Acoust. Soc. Am.*, Vol. 90, 1991, pp. 858–865.
9. D. M. Green, T. B. Birdsall, and W. P. Tanner, "Signal Detection as a Function of Signal Intensity and Duration," *J. Acoust. Soc. Am.*, Vol. 29, 1957, pp. 523–531.
10. D. M. Green, "Temporal Factors in Psychoacoustics," in A. Michelsen (Ed.), *Time Resolution in Auditory Systems*, Springer, Berlin, 1985, pp. 122–140.
11. S. Buus and M. Florentine, "Gap Detection in Normal and Impaired Listeners: The Effect of Level and Frequency," in A. Michelsen (Ed.), *Time Resolution in Auditory Systems*, Springer, Berlin, 1985, pp. 159–179.
12. G. M. Gerken, J. M. Solecki, and F. A. Boettcher, "Temporal Integration of Electrical Stimulation of Auditory Nuclei in Normal-Hearing and Hearing-Impaired Cat," *Hear. Res.*, Vol. 53, 1991, pp. 101–112.
13. J. E. Hawkins and S. S. Stevens, "The Masking of Pure Tones and of Speech by White Noise," *J. Acoust. Soc. Am.*, Vol. 22, 1950, pp. 6–13.
14. H. Fletcher, "Auditory Patterns," *Rev. Mod. Phys.*, Vol. 12, 1940, pp. 47–65.
15. J. P. Egan and H. W. Hake, "On the Masking Pattern of a Simple Auditory Stimulus," *J. Acoust. Soc. Am.*, Vol. 22, 1950, pp. 622–630.
16. E. Zwicker and H. Fastl, *Psychoacoustics: Facts and Models*, Springer, New York, 1990.
17. B. Scharf, "Critical Bands," in J. V. Tobias (Ed.), *Foundations of Modern Auditory Theory*, Vol. 1, Academic, New York, 1970, pp. 157–202.
18. E. Zwicker, "Subdivision of the Audible Frequency Range into Critical Bands (Frequenzgruppen)," *J. Acoust. Soc. Am.*, Vol. 33, 1961, p. 248.
19. E. Zwicker and E. Terhardt, "Analytical Expressions for Critical-Band Rate and Critical Bandwidth as a Function of Frequency," *J. Acoust. Soc. Am.*, Vol. 68, 1980, pp. 1523–1525.
20. D. M. Green, M. J. McKey, and J. C. R. Licklider, "Detection of a Pulsed Sinusoid as a Function of Frequency," *J. Acoust. Soc. Am.*, Vol. 31, 1959, pp. 1446–1452.
21. D. D. Greenwood, "Critical Bandwidth and Consonance in Relation to Cochlear Frequency-Position Coordinates," *Hear. Res.*, Vol. 54, 1991, pp. 164–208.
22. E. De Boer, "Note on the Critical Bandwidth," *J. Acoust. Soc. Am.*, Vol. 34, 1962, pp. 985–986; Erratum, p. 1278.
23. J. W. Hall, M. P. Haggard, and M. A. Fernandes, "Detection in Noise by Spectrotemporal Analysis," *J. Acoust. Soc. Am.*, Vol. 76, 1984, pp. 50–56.
24. G. P. Schooneveldt and B. C. J. Moore, "Comodulation Masking Release (CMR) as a Function of Masker Bandwidth, Modulator Bandwidth, and Signal Duration," *J. Acoust. Soc. Am.*, Vol. 85, 1989, pp. 273–281.
25. D. M. Green and J. A. Swets, *Signal Detection Theory and Psychophysics*, Academic, New York, 1966.
26. A. Langhans and A. Kohlrausch, "Spectral Integration of Broadband Signals in Diotic and Dichotic Masking Experiments," *J. Acoust. Soc. Am.*, Vol. 91, 1992, pp. 317–326.
27. D. M. Green, "Detection of Multiple Component Signals in Noise," *J. Acoust. Soc. Am.*, Vol. 30, 1958, pp. 904–911.
28. S. Buus, E. Schorer, M. Florentine, and E. Zwicker, "Decision Rules in Detection of Simple and Complex Tones," *J. Acoust. Soc. Am.*, Vol. 80, 1986, pp. 1646–1657.
29. W. A. C. Van den Brink and T. Houtgast, "Spectro-temporal Integration in Signal Detection," *J. Acoust. Soc. Am.*, Vol. 88, 1990, pp. 1703–1711.
30. C. E. Bos and E. De Boer, "Masking and Discrimination," *J. Acoust. Soc. Am.*, Vol. 39, 1966, pp. 708–715.
31. S. Buus, "Level Discrimination of Frozen and Random Noise," *J. Acoust. Soc. Am.*, Vol. 87, 1990, pp. 2643–2654.
32. R. L. Wegel and C. E. Lane, "The Auditory Masking of One Pure Tone by Another and Its Probable Relation to the Dynamics of the Inner Ear," *Phys. Rev.*, Vol. 23, 1924, pp. 266–285.
33. D. D. Greenwood, "Aural Combination Tones and Auditory Masking," *J. Acoust. Soc. Am.*, Vol. 50, 1971, pp. 502–543.

34. E. Zwicker and A. Jaroszewski, "Inverse Frequency Dependence of Simultaneous Tone-on-Tone Masking Patterns at Low Levels," *J. Acoust. Soc. Am.*, Vol. 71, 1982, pp. 1508–1512.
35. R. C. Bilger and I. J. Hirsh, "Masking of Tones by Bands of Noise," *J. Acoust. Soc. Am.*, Vol. 28, 1956, pp. 623–630.
36. S. Buus, "Release from Masking Caused by Envelope Fluctuations," *J. Acoust. Soc. Am.*, Vol. 78, 1985, pp. 1958–1965.
37. E. Zwicker, "On a Psychoacoustical Equivalent of Tuning Curves," in E. Zwicker and E. Terhardt (Eds.), *Facts and Models in Hearing*, Springer, Berlin, 1974.
38. B. Leshowitz and F. L. Wightman, "On-Frequency Masking with Continuous Sinusoids," *J. Acoust. Soc. Am.*, Vol. 49, 1971, pp. 1180–1190.
39. D. Johnson-Davies and R. D. Patterson, "Psychophysical Tuning Curves: Restricting the Listening Band to the Signal Region," *J. Acoust. Soc. Am.*, Vol.65, 1979, pp. 765–770.
40. B. Delgutte, "Physiological Mechanisms of Psychophysical Masking: Observations from Auditory-Nerve Fibers," *J. Acoust. Soc. Am.*, Vol. 87, 1990, pp. 791–809.
41. B. C. J. Moore and B. R. Glasberg, "Auditory Filter Shapes Derived in Simultaneous and Forward Masking," *J. Acoust. Soc. Am.*, Vol. 69, 1981, pp. 1003–1014.
42. D. D. Greenwood, "Critical Band and Consonance: Their Operational Definitions in Relation to Cochlear Nonlinearity and Combination Tones," *Hear. Res.*, Vol. 54, 1991, pp. 209–246.
43. R. A. Lutfi and R. D. Patterson, "Combination Bands and the Measurement of the Auditory Filter," *J. Acoust. Soc. Am.*, Vol. 71, 1982, pp. 421–423.
44. D. M. Green, "Additivity of Masking," *J. Acoust. Soc. Am.*, Vol. 41, 1967, pp. 1517–1525.
45. R. A. Lutfi, "Additivity of Simultaneous Masking," *J. Acoust. Soc. Am.*, Vol. 73, 1983, pp. 262–267.
46. B. C. J. Moore, "Additivity of Simultaneous Masking, Revisited," *J. Acoust. Soc. Am.*, Vol. 78, 1985, pp. 488–494.
47. G. Kidd, Jr., C. R. Mason, M. A. Brantley, and G. A. Owen, "Roving-Level Tone-in-Noise Detection," *J. Acoust. Soc. Am.*, Vol. 86, 1989, pp. 1310–1317.
48. V. M. Richards, "The Detectability of a Tone Added to Narrow Bands of Equal-Energy Noise," *J. Acoust. Soc. Am.*, Vol. 91, 1992, pp. 3424–3435.
49. H. Fastl, "Temporal Masking Effects: I. Broad Band Noise Masker," *Acustica*, Vol. 35, 1976, pp. 287–302.
50. H. Fastl, "Temporal Masking Effects: II. Critical Band Noise Masker," *Acustica*, Vol. 36, 1977, pp. 317–331.
51. W. Jesteadt, S. P. Bacon, and J. R. Lehman, "Forward Masking as a Function of Frequency, Masker Level, and Signal Delay," *J. Acoust. Soc. Am.*, Vol. 71, 1982, pp. 950–962.
52. J. J. Zwislocki, "Masking: Experimental and Theoretical Aspects of Simultaneous, Forward, Backward, and Central Masking," in E. C. Carterette and M. P. Friedman (Eds.), *Handbook of Perception, Vol. 4: Hearing*, Academic, New York, 1978, pp. 283–336.
53. R. Plomp, "Rate of Decay of Auditory Sensation," *J. Acoust. Soc. Am.*, Vol. 36, 1964, pp. 277–282.
54. G. Kidd, Jr., C. R. Mason, and L. L. Feth, "Temporal Integration and Forward Masking in Listeners Having Sensorineural Hearing Loss," *J. Acoust. Soc. Am.*, Vol. 75, 1984, pp. 937–944.
55. E. Zwicker, "Dependence of Post-masking on Masker Duration and Its Relation to Temporal Effects in Loudness," *J. Acoust. Soc. Am.*, Vol. 75, 1984, pp. 219–223.
56. H. Duifhuis, "Consequences of Peripheral Frequency Selectivity for Nonsimultaneous Masking," *J. Acoust. Soc. Am.*, Vol. 54, 1973, pp. 1471–1488.
57. B. C. J. Moore, P. W. F. Poon, S. P. Bacon, and B. R. Glasberg, "The Temporal Course of Masking and the Auditory Filter Shape," *J. Acoust. Soc. Am.*, Vol. 81, 1987, pp. 1873–1880.
58. B. C. J. Moore, "Psychophysical Tuning Curves Measured in Simultaneous and Forward Masking," *J. Acoust. Soc. Am.*, Vol. 63, 1978, pp. 524–532.
59. B. C. J. Moore and B. J. O'Loughlin, "The Use of Nonsimultaneous Masking to Measure Frequency Selectivity and Suppression," in B. C. J. Moore (Ed.), *Frequency Selectivity in Hearing*, Academic, New York, 1986, pp. 179–250.
60. D. L. Neff, "Confusion Effects with Sinusoidal and Narrow-Band Noise Forward Maskers," *J. Acoust. Soc. Am.*, Vol. 79, 1986, pp. 1519–1529.
61. T. Houtgast, "Auditory-Filter Characteristics Derived from Direct-Masking Data and Pulsation-Threshold Data with a Rippled-Noise Masker, *J. Acoust. Soc. Am.*, Vol. 62, 1977, pp. 409–415.
62. R. V. Shannon, "Two-Tone Unmasking and Suppression in a Forward-Masking Situation," *J. Acoust. Soc. Am.*, Vol. 59, 1976, pp. 1460–1470.
63. D. L. Weber, "Suppression and Critical Bands in Band-Limiting Experiments," *J. Acoust. Soc. Am.*, Vol. 64, 1978, pp. 141–150.
64. L. L. Elliott, "Development of Auditory Narrow-Band Frequency Contours," *J. Acoust. Soc. Am.*, Vol. 42, 1967, pp. 143–153.
65. B. C. J. Moore, *An Introduction to the Psychology of Hearing*, 3rd ed., Academic, London, 1989.
66. M. J. Penner and R. M. Shriffrin, "Nonlinearities in the Coding of Intensity within the Context of a Temporal Summation Model," *J. Acoust. Soc. Am.*, Vol. 67, 1980, pp. 617–627.
67. L. E. Humes and W. Jesteadt, "Models of the Additivity of Masking," *J. Acoust. Soc. Am.*, Vol. 85, 1989, pp. 1285–1294.
68. E. Zwicker, "A Device for Measuring Temporal Resolution of the Ear," *Audiol. Acoust.*, Vol. 19, 1980, pp. 94–108.
69. R. P. Carlyon, S. Buus, and M. Florentine, "Comodula-

tion Masking Release for Three Types of Modulator as a Function of Modulation Rate," *Hear. Res.*, Vol. 42, 1989, pp. 37–46.

70. J. W. Hall and J. H. Grose, "Relative Contributions of Envelope Maxima and Minima to Comodulation Masking Release," *Quart. J. Exp. Psych.*, Vol. 43A, 1991, pp. 349–372.
71. J. W. Hall and J. H. Grose, "Amplitude Discrimination in Masking Release Paradigms," *J. Acoust. Soc. Am.*, Vol. 98, 1995, pp. 847–852.
72. J. H. Grose and J. W. Hall, "Comodulation Masking Release for Speech Stimuli," *J. Acoust. Soc. Am.*, Vol. 91, 1992, pp. 1042–1050.
73. M. Cohen and E. D. Schubert, "Influence of Place Synchrony on Detection of a Sinusoid," *J. Acoust. Soc. Am.*, Vol. 81, 1987, pp. 452–458.
74. M. P. Haggard, J. W. Hall, and J. H. Grose, "Comodulation Masking Release as a Function of Bandwidth and Test Frequency," *J. Acoust. Soc. Am.*, Vol. 88, 1990, pp. 113–118.
75. D. A. Eddins and B. A. Wright, "Comodulation Masking Release for Single and Multiple Rates of Envelope Fluctuation," *J. Acoust. Soc. Am.*, Vol. 96, 1994, pp. 3432–3442.
76. J. W. Hall, "The Effect of Across-Frequency Differences in Masking Level on Spectrotemporal Pattern Analysis," *J. Acoust. Soc. Am.*, Vol. 79, 1986, pp. 781–787.
77. B. C. J. Moore and M. J. Shailer, "Comodulation Masking Release as a Function of Level," *J. Acoust. Soc. Am.*, Vol. 90, 1991, pp. 829–835.
78. B. C. J. Moore and G. P. Schooneveldt, "Comodulation Masking Release as a Function of Bandwidth and Time Delay between On-Frequency and Flanking-Band Maskers," *J. Acoust. Soc. Am.*, Vol. 88, 1990, pp. 725–731.
79. J. W. Hall and J. H. Grose, "Comodulation Masking Release and Auditory Grouping," *J. Acoust. Soc. Am.*, Vol. 88, 1990, pp. 119–125.
80. B. A. Wright and D. McFadden, "Uncertainty about the Correlation among Temporal Envelopes in Two Comodulation Tasks," *J. Acoust. Soc. Am.*, Vol. 88, 1990, pp. 1339–1350.
81. V. M. Richards, "Monaural Envelope Correlation Perceptions," *J. Acoust. Soc. Am.*, Vol. 82, 1987, pp. 1621–1630.
82. D. M. Green, "On the Similarity of Two Theories of Comodulation Masking Release," *J. Acoust. Soc. Am.*, Vol. 91, 1992, p. 1769.
83. B. C. J. Moore and D. S. Emmerich, "Monoaural Envelope Correlation Perception, Revisited: Effects of Bandwidth, Frequency Separation, Duration, and Relative Level of Noise Bands," *J. Acoust. Soc. Am.*, Vol. 87, 1990, pp. 2628–2633.
84. D. A. Fantini and B. C. J. Moore, "Comodulation Masking Release (CMR) and Profile Analysis: The Effect of Varying Modulation Depth," in Y. Cazals, L. Demany, and K. Horner (Eds.), *Auditory Physiology and Perception*, Pergamon, New York, 1992, pp. 479–488.
85. D. A. Fantini, "The Processing of Envelope Information in Comodulation Masking Release (CMR) and Envelope Discrimination," *J. Acoust. Soc. Am.*, Vol. 90, 1991, pp. 1876–1888.
86. J. W. Hall, J. H. Grose, and M. P. Haggard, "Comodulation Masking Release for Multicomponent Signals," *J. Acoust. Soc. Am.*, Vol. 83, 1988, pp. 677–686.
87. W. A. C. Van Den Brink, T. Houtgast, and G. F. Smoorenburg, "Signal Detection in Temporally Modulated and Spectrally Shaped Maskers," *J. Acoust. Soc. Am.*, Vol. 91, 1992, pp. 267–278.
88. J. W. Hall and J. H. Grose, "Comodulation Masking Release: Evidence for Multiple Cues," *J. Acoust. Soc. Am.*, Vol. 84, 1988, pp. 1669–1675.
89. E. Zwicker, "Temporal Effects in Simultaneous Masking and Loudness," *J. Acoust. Soc. Am.*, Vol. 38, 1965, pp. 132–141.
90. S. P. Bacon and M. A. Smith, "Spectral, Intensive, and Temporal Factors Influencing Overshoot," *Quart. J. Exp. Psych.*, Vol. 43A, 1991, pp. 373–400.
91. E. Zwicker, "Temporal Effects in Simultaneous Masking by White-Noise Bursts," *J. Acoust. Soc. Am.*, Vol. 37, 1965, pp. 653–663.
92. R. P. Carlyon and L. J. White, "Effect of Signal Frequency and Masker Level on the Frequency Regions Responsible for the Overshoot Effect," *J. Acoust. Soc. Am.*, Vol. 91, 1992, pp. 1034–1041.
93. S. P. Bacon, "Effect of Masker Level on Overshoot," *J. Acoust. Soc. Am.*, Vol. 88, 1990, pp. 698–702.
94. S. P. Bacon and G. A. Takahashi, "Overshoot in Normal-Hearing and Hearing-Impaired Subjects," *J. Acoust. Soc. Am.*, Vol. 91, 1992, pp. 2865–2871.
95. R. Von Klitzing and A. Kohlrausch, "Effect of Masker Level on Overshoot in Running- and Frozen-Noise Maskers," *J. Acoust. Soc. Am.*, Vol. 95, 1994, pp. 2192–2201.
96. S. P. Bacon and N. F. Viemeister, "The Temporal Course of Simultaneous Tone-on-Tone Masking," *J. Acoust. Soc. Am.*, Vol. 78, 1985, pp. 1231–1235.
97. R. P. Carlyon, "Changes in the Masked Thresholds of Brief Tones Produced by Prior Bursts of Noise," *Hear. Res.*, Vol. 41, 1989, pp. 223–236.
98. D. McFadden, "Spectral Differences in the Ability of Temporal Gaps to Reset the Mechanisms Underlying Overshoot," *J. Acoust. Soc. Am.*, Vol. 85, 1989, pp. 254–261.
99. S. Schmidt and E. Zwicker, "The Effect of Masker Spectral Asymmetry on Overshoot in Simultaneous Masking," *J. Acoust. Soc. Am.*, Vol. 89, 1991, pp. 1324–1330.
100. R. L. Smith and J. J. Zwislocki, "Short-Term Adaptation in Incremental Response of Single Auditory-Nerve Fibers," *Biol. Cybern.*, Vol. 17, 1975, pp. 169–182.
101. D. M. Harris and P. Dallos, "Forward Masking of Auditory Nerve Fiber Responses," *J. Neurophysiol.*, Vol. 42, 1979, pp. 1083–1107.
102. H. Dai and S. Buus, "Effect of Gating the Masker on Frequency-Selective Listening," *J. Acoust. Soc. Am.*, Vol. 89, 1991, pp. 1816–1818.

90

FREQUENCY ANALYSIS AND PITCH PERCEPTION

BRIAN C. J. MOORE

1 INTRODUCTION

Unlike the sinusoidal tones often used in the assessment of hearing, sounds encountered in everyday life are generally complex, containing many sinusoidal frequency components. One of the most important characteristics of the auditory system is that it acts as a limited-resolution frequency analyzer; complex sounds are broken down into their sinusoidal frequency components. The initial basis of this frequency analysis almost certainly depends upon the tuning observed in the cochlea (see Chapter 87). Indeed, it is possible that the tuning observed in the cochlea is sufficient to account for the frequency-analyzing capacity of the entire auditory system.[1] It is largely as a consequence of this frequency analysis that it is possible to hear one sound in the presence of another sound with a different frequency. This ability is known as frequency selectivity or frequency resolution.

This chapter starts by reviewing ways in which the frequency selectivity of the ear can be measured and quantified using masking experiments. It goes on to describe some other aspects of auditory perception that depend strongly on frequency analysis, specifically the ability to hear individual partials in complex sounds and the dependence of loudness on the bandwidth of sounds.

The second half of the chapter gives an account of the perception of pitch for pure tones and for complex tones. It argues that pitch perception depends partly on the frequency analysis that occurs in the peripheral auditory system. However, processes based on time pattern analysis also play an important role.

Note: References to chapters appearing only in the *Encyclopedia* are preceded by *Enc.*

2 CHARACTERIZATION OF FREQUENCY-ANALYZING CAPACITY OF THE EAR

2.1 Masking and the Auditory Filter

Masking can be used to measure the limits of frequency analysis; when two sounds with different frequencies are completely resolved by the ear, then neither will be masked by the other. One conception of auditory masking, which has had both theoretical and practical success, assumes that the auditory system includes a bank of overlapping bandpass filters[2] (see Chapter 89). It is assumed that the observer will "listen" to the filter whose output has the highest signal-to-masker ratio. The signal will be detected if that ratio exceeds a certain value. In experiments where the bandwidth of the signal or masker is varied, the results often change at a certain bandwidth, called the *critical bandwidth*[3,4] (CB). This bandwidth corresponds approximately to the bandwidth of the auditory filter.

The relative response of the auditory filter as a function of frequency is often called the filter "shape." Patterson et al.,[5] based on the results of experiments measuring the detection of a sinusoidal signal presented in a noise masker with a spectral notch, suggested that the shape of each side of the auditory filter could be described approximately by the expression

$$W(g) = (1 + pg)\exp(-pg), \qquad (1)$$

where g is the normalized frequency deviation from the center of the filter; $g = |f - f_c|/f_c$. The parameter p, calculated from the detection data, determines the sharpness of the auditory filter and its value may differ for the upper and lower halves. The equivalent rectangular bandwidth (ERB) of the auditory filter is equal to $2f_c/p_l + 2f_c/p_u$, where p_l and p_u are the values of p for the lower and upper halves of the filter. At moderate sound levels p_l and p_u are approximately equal, but at high sound levels p_l tends to decrease; that is, the low-frequency side of the filter becomes less steep.[6]

Auditory filters have ERBs that increase with increasing center frequency. At moderate sound levels, the ERB is well approximated by

$$\text{ERB} = 24.7(4.37F_c + 1), \qquad (2)$$

where F_c is frequency in kilohertz.[7] This expression is shown in Fig. 1, together with estimates of the ERB from several different experiments. (For quick mental calculations the reader might find the following approximate expression useful: ERB $= \frac{1}{9}f_c + 25$, where f_c is in hertz.) At a proportion of center frequency, the bandwidth tends to be narrowest at middle to high frequencies (1–5 kHz).

A scale that compensates for differences in frequency resolution across the audible range can be useful for gaining insight into the internal respresentation of the spectra of sounds. Such a scale is obtained by expressing a given frequency in terms of the numbers of CBs or ERBs needed to cover the audible range up to that frequency. One scale of this type, based on the traditional CB, is the Bark scale,[8] where the number of barks up to a given frequency is indicated by the symbol z. A good approximation to this scale is[9]

$$z = \frac{26.8}{1 + 1.96/F_c} - 0.53. \qquad (3)$$

A scale based on the ERB of the auditory filter and derived from Eq. (2), is[7]

$$\text{Number of ERBs, } E = 21.4 \log_{10}(4.37F_c + 1). \qquad (4)$$

2.2 The Excitation Pattern

The shape of any given auditory filter characterizes frequency selectivity at a particular center frequency, or,

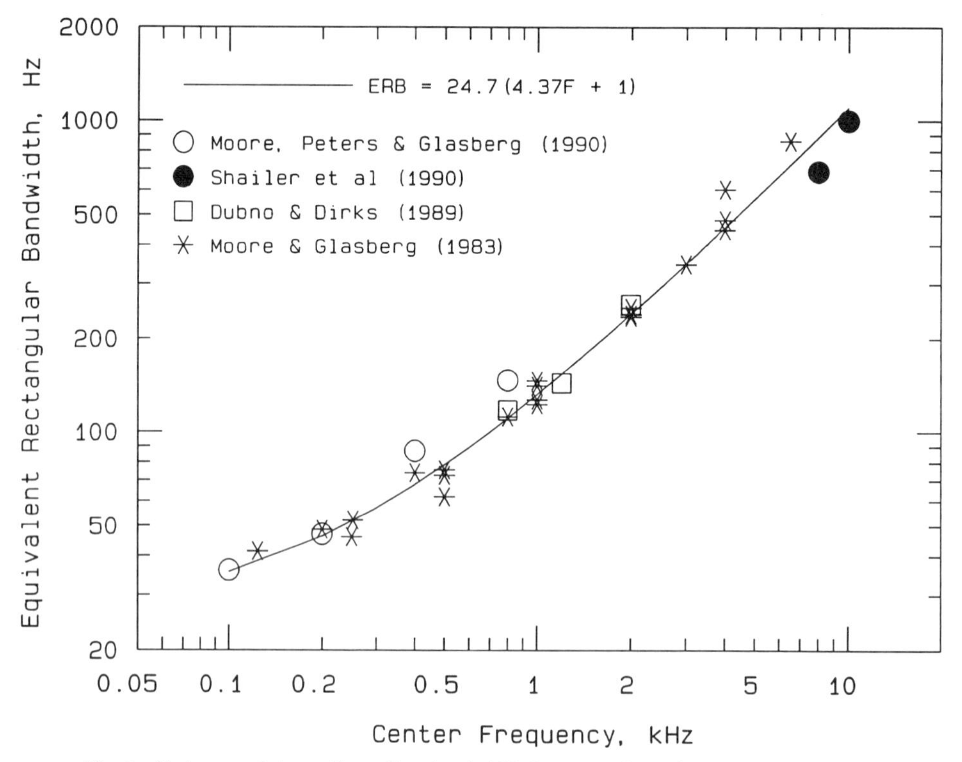

Fig. 1 Estimates of the auditory filter bandwidth from a variety of experiments plotted as a function of center frequency. The solid line represents the equation suggested by Glasberg and Moore.[7]

correspondingly, at a particular place on the basilar membrane. An alternative approach is to characterize the distribution of excitation across frequencies (or places along the basilar membrane) for a given sound: This distribution has been referred to as the excitation pattern of the sound.[10] In terms of the filter-bank analogy, the excitation pattern can be considered as the output of the auditory filters as a function of filter center frequency. Using this definition, it is possible to calculate the excitation pattern of any sound assuming that the auditory filters have the form given in Eq. (1) and have a bandwidth that varies with center frequency according to Eq. (2).[6,7,11] Figure 2 gives an example for a vowel sound. The frequency-related axis is scaled in this figure according to the number of ERBs using Eq. (4). Vowel sounds contain spectral prominences at particular frequencies. These are called formants, and they correspond to resonances in the vocal tract. However, the prominent peaks in the excitation pattern at low center frequencies correspond to individual harmonics, rather than to the first (lowest) formant. Note also that the second, third, and fourth formants, which are clearly seen in the spectrum of the sound, are not clearly separated in the excitation pattern.

2.3 Audibility of Partials in Complex Sounds

When listening to two sinusoids presented simultaneously, either a single sound may be heard, corresponding to the mixture of the two, or two tones may be heard, each with a pitch corresponding to one of the sinusoids. The minimum frequency separation at which the two sinusoids can be heard separately varies depending on their mean frequency. At 500 Hz it is about 35 Hz, while at 5000 Hz it is about 700 Hz.[12]

When listening to a complex tone containing many components, produced, for example, by a musical instrument, a single pitch is usually heard (see below). However, it is possible to "hear out" individual partials provided attention is directed in an appropriate way, for example, by presenting a comparison sinusoidal tone whose frequency coincides with that of one of the partials.

For multitone complexes, it is generally slightly more difficult to hear the individual components than when only two are present. Particularly for frequencies below 1000 Hz, the frequency separation between adjacent components required to hear out the components is greater than that required for a two-tone complex.[12–14] Figure 3 summarizes results from Plomp[12] and Plomp and Mimpen.[15] It shows the frequency separation between adjacent partials required for a given partial to be heard out from a complex tone with many equal-amplitude components when the components are harmonically related (circles) or nonharmonically related (asterisks). For comparison, the short-dashed line shows the traditional critical-band function and the solid line shows the ERB function described by Eq. (2). Moore and Ohgushi[14] showed that the components in a multicomponent complex had to be separated by about 1.25 ERBs in order for them to be heard out with 75% accuracy. The long-dashed line shows 1.25 times the ERB function; this fits the data rather well.

The harmonics of a periodic complex tone are equally spaced on a linear frequency scale. Thus, for a harmonic complex, the lower harmonics are more easily heard out, or resolved, than the higher harmonics (compare Fig. 2). For a complex tone with equal-amplitude components, only about the first five to eight harmonics are resolvable. This plays an important role in modern theories of pitch perception for complex tones (see below).

2.4 Critical Bandwidth for Loudness

If the total intensity of a complex sound is fixed, its loudness depends on the frequency range over which the sound extends. Consider, as an example, a noise whose total intensity is held constant while the bandwidth is varied. When the bandwidth of the noise is less than a certain value, the loudness is roughly independent of bandwidth. However, as the bandwidth is increased beyond a certain point, the loudness starts to increase. The bandwidth at which loudness starts to increase is known as the critical bandwidth for loudness summation.[16] Its value is slightly greater than the ERB of the auditory filter. The effect is described in more detail in Chapter 91, where its underlying basis is also described.

3 INTRODUCTION TO THE PERCEPTION OF PITCH

Pitch may be defined as "that attribute of auditory sensation in terms of which sounds may be ordered on a musical scale."[17] In other words, variations in pitch give rise to a sense of melody. Pitch is related to the repetition rate of the waveform of a sound; for a pure tone this corresponds to the frequency, and for a periodic complex tone to the fundamental frequency. There are, however, exceptions to this simple rule. Since pitch is a subjective attribute; it cannot be measured directly. Assigning a pitch value to a sound is generally understood to mean specifying the frequency of a sinusoid having the same subjective pitch as the sound. Sometimes a periodic complex sound, such as a pulse train, is used as a matching stimulus.

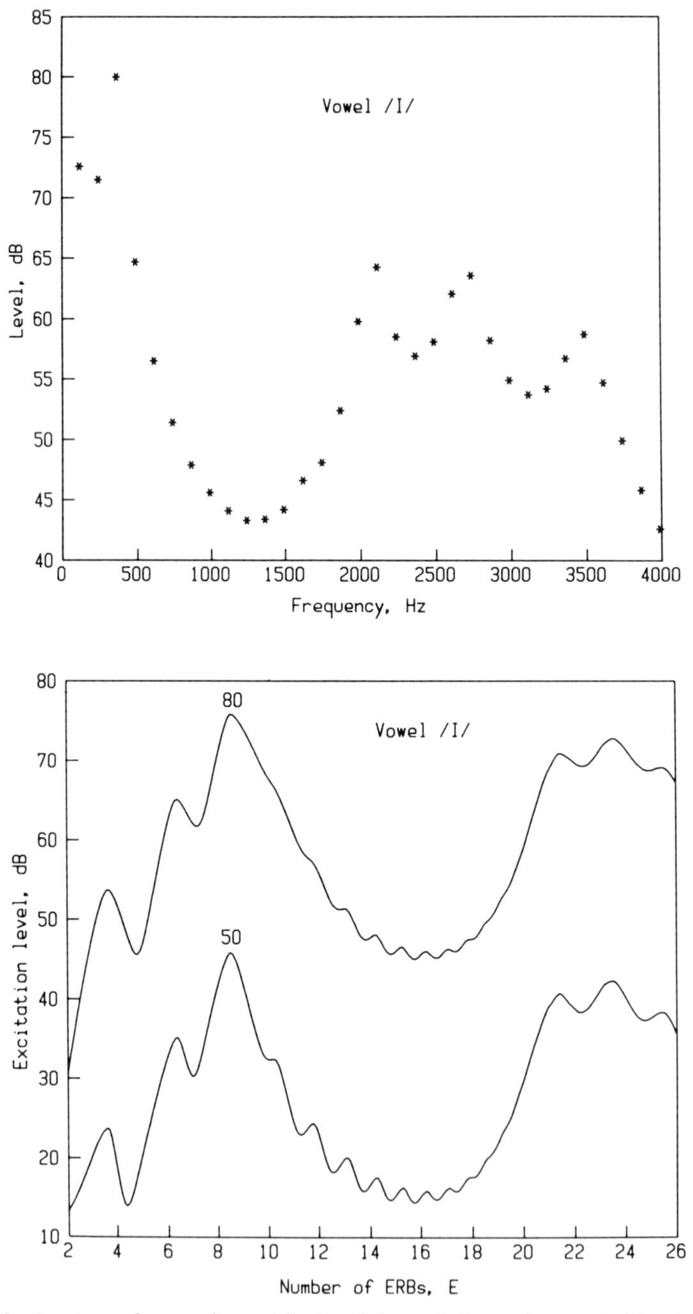

Fig. 2 Spectrum of a vowel sound (top) and the excitation pattern evoked by that sound (bottom) for two different overall sound levels. Note the linear frequency scale in the upper panel and the ERB scale in the lower panel.

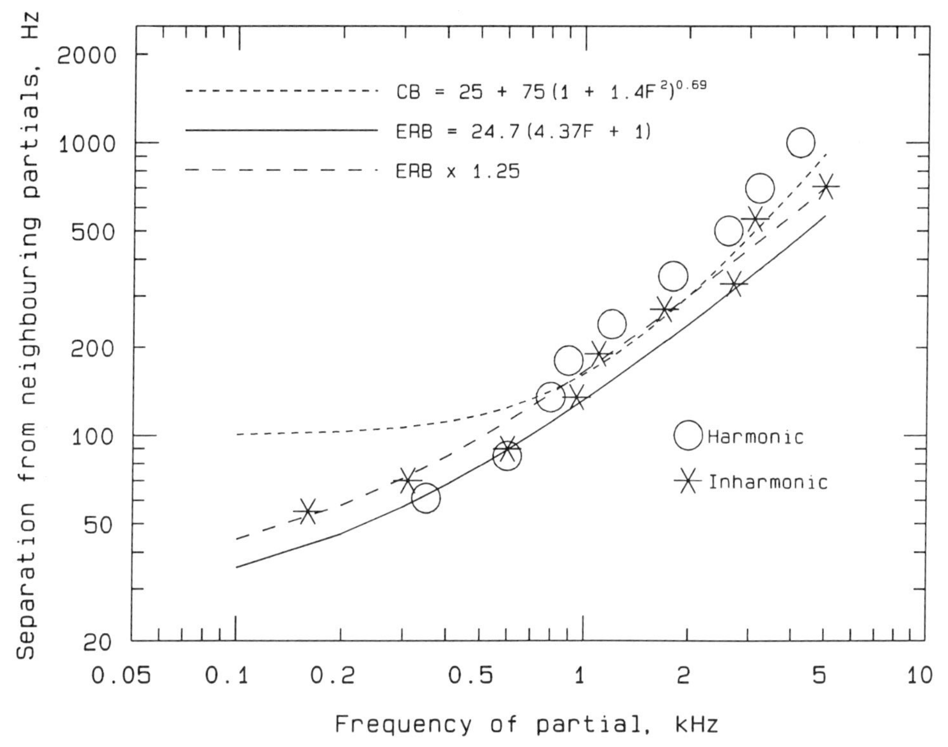

Fig. 3 Frequency separation between adjacent partials required for a given partial to be "heard out" from a complex tone with multiple harmonic (circles) or inharmonic (asterisks) equal-amplitude components, plotted as a function of the frequency of the partial. The short-dashed line shows the traditional CB function. The solid line shows the ERB of the auditory filter as a function of frequency [Eq. (2)]. The long-dashed line shows 1.25 times the ERB.

3.1 Traditional Theories of Pitch Perception

For many years there have been two competing theories of pitch perception. One, the "place" theory of hearing, has two distinct postulates. The first is that the stimulus undergoes a limited-resolution spectral analysis in the inner ear, so that different frequencies (or frequency components in a complex stimulus) excite different places along the basilar membrane. The second is that the pitch of a stimulus is related to the excitation pattern produced by that stimulus; for a pure tone the pitch is generally assumed to correspond to the position of maximum excitation. The first of these two postulates is now well established and has been confirmed in a number of independent ways, including direct observation of the movement of the basilar membrane. The second is still a matter of dispute.

An alternative to the place theory, called the "temporal" theory, suggests that the pitch of a stimulus is related to the time pattern of the neural impulses evoked by that stimulus. Nerve firings tend to occur at a particular phase of the stimulating waveform, and thus the intervals between successive neural impulses approximate integer multiples of the period of the stimulating waveform (this is called phase locking; see *Enc.* Ch. 109). The temporal theory cannot work at very high fundamental frequencies, since phase locking does not occur for frequencies above about 5 kHz. However, the tones produced by musical instruments, the human voice, and most everyday sound sources have fundamental frequencies below this range.

4 PERCEPTION OF THE PITCH OF PURE TONES

This section will discuss the perception and discrimination of the frequency of pure tones. It is important to distinguish between frequency selectivity and frequency discrimination. The former refers to the ability to resolve the frequency components of a complex sound, as discussed earlier. The latter refers to the ability to detect changes in frequency over time. The smallest detectable change is called the difference limen (DL). For place theories, frequency selectivity and frequency discrimination are closely connected; frequency discrimination depends upon the filtering that takes place in the cochlea. For temporal theories, frequency selectivity and frequency discrimination are not necessarily closely connected.

4.1 Frequency Discrimination of Pure Tones

There have been two common ways of measuring frequency discrimination. One measure involves the discrimination of two successive steady tones with slightly different frequencies. This measure will be called the DLF (difference limen for frequency). A second measure uses tones that are frequency modulated at a low rate (typically 2–4 Hz). The amount of modulation required for detection of the modulation is determined. This measure will be called the FMDL (frequency modulation detection limen).

Early studies of frequency discrimination mainly measured FMDLs.[18,19] Recent studies have concentrated more on the measurement of DLFs. A summary of the results of some of these studies is given in Fig. 4, taken from Wier et al.[20] Expressed in hertz the DLF is smallest at low frequencies and increases monotonically with increasing frequency. Expressed as a proportion of center frequency, the DLF tends to be smallest for middle frequencies and larger for very high and very low frequencies. Wier et al. found that the data describing the DLF as a function of frequency fell on a straight line when plotted as log(DLF) versus the square root of frequency; the axes are scaled in this way in Fig. 4. Nelson et al.[21] also found DLFs to be linearly related to the square root of frequency, and they found further that DLFs decrease as the sensation level (SL) of the signal is increased. They occur at different points along the frequency scale. Their results suggested that DLFs could be described by

$$\log(\mathrm{DLF}) = a\sqrt{f} + k + \frac{m}{\mathrm{SL}}, \qquad (5)$$

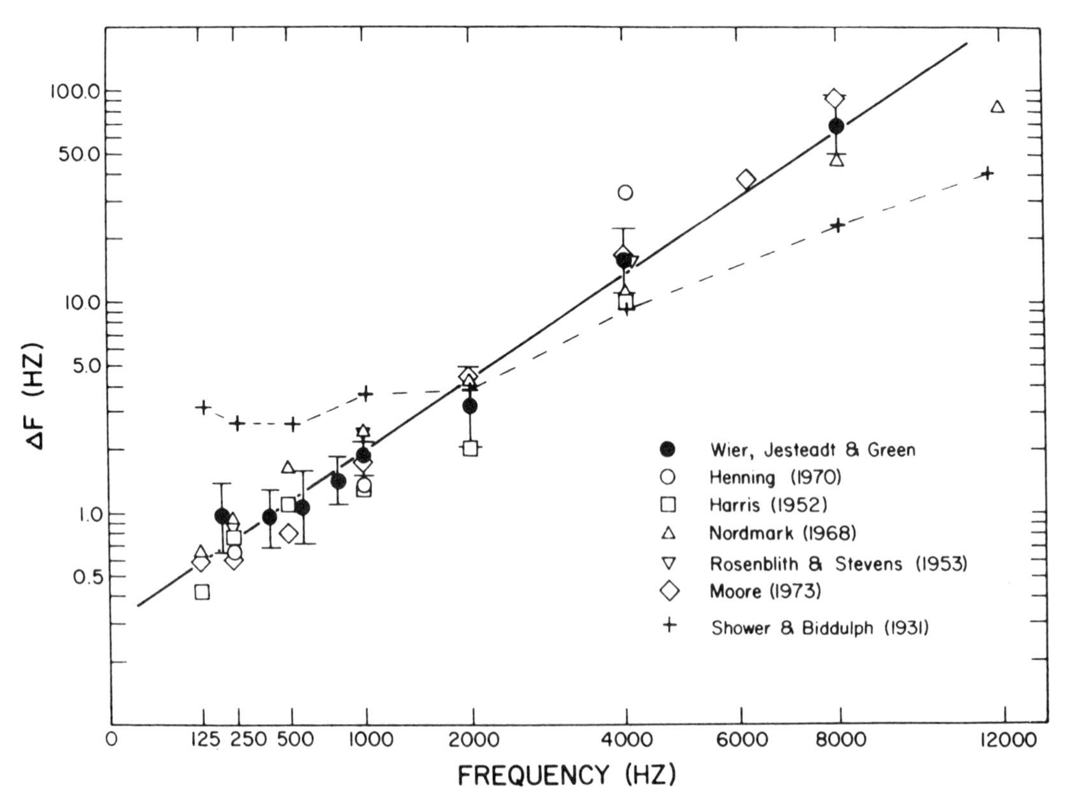

Fig. 4 Summary of the results of several studies measuring frequency DLs. The DLs, ΔF, are plotted in hertz as a function of frequency. All studies measured DLFs, except that of Shower and Biddulph,[18] which measured FMDLs. (From Ref. 20, by permission of the authors and the *Journal of the Acoustical Society of America*.)

where f is the signal frequency in hertz, SL is the sensation level of the signal (in decibels), and a, k, and m are constants. The values of the constants vary somewhat depending on factors such as stimulus duration and the degree of training of the subjects, but typical values are $a = 0.023, k = -0.25$, and $m = 4.3$. The FMDLs, shown as the dashed line in Fig. 4, tend to vary less with frequency than do DLFs.

Place models of frequency discrimination[10,22,23] predict that frequency discrimination should be related to frequency selectivity; the sharper the tuning of peripheral filtering mechanisms, the smaller should be the frequency DL, and the frequency DL should vary with frequency in the same way as the CB. In fact, this prediction fails for DLFs[24,25] but works reasonably well for FMDLs,[10,26] especially when the modulation rate is above about 5 Hz.[27] At low frequencies, DLFs are usually smaller than predicted by place models, but at high frequencies, above 4–5 kHz, they are not. This is consistent with the idea that DLFs are determined mainly by temporal information for frequencies up to about 4–5 kHz, and mainly by place information above that.

4.2 Perception of Musical Intervals

Two tones that are separated in frequency by an interval of one octave (i.e., one has twice the frequency of the other) sound similar. They are judged to have the same name on the musical scale (for example, C or D). This has led several theorists to suggest that there are at least two dimensions to musical pitch. One aspect is related monotonically to frequency (for a pure tone) and is known as *tone height*. The other is related to pitch class (i.e., name note) and is called *tone chroma*.[28]

If subjects are presented with a pure tone of a given frequency f_1 and are asked to adjust the frequency f_2 of a second tone so that it appears to be an octave higher in pitch, they will generally adjust f_2 to be roughly twice f_1. However, when f_1 lies above 2.5 kHz, so that f_2 would lie above 5 kHz, octave matches become very erratic.[29] It appears that the musical interval of an octave is only clearly perceived when both tones are below 5 kHz.

Other aspects of the perception of pitch also change above 5 kHz. A sequence of pure tones above 5 kHz does not produce a clear sense of melody. This has been confirmed by experiments involving musical transposition. For example, Attneave and Olson[30] asked subjects to reproduce sequences of tones (e.g., the NBC chimes) that showed an abrupt breakpoint at about 5 kHz above which transposition behavior was erratic. Also, subjects with absolute pitch (the ability to assign name notes without reference to other notes) are very poor at naming notes above 4–5 kHz.[31]

These results are consistent with the idea that the pitch of pure tones is determined by different mechanisms above and below 5 kHz, specifically, by temporal mechanisms at low frequencies and place mechanisms at high frequencies. It appears that the perceptual dimension of tone height persists over the whole audible frequency range, but tone chroma only occurs in the frequency range below 5 kHz.

4.3 Variation of Pitch with Level

The pitch of a pure tone is primarily determined by its frequency. However, sound level also plays a small role. On average, the pitch of tones below about 2000 Hz decreases with increasing level, while the pitch of tones above about 4000 Hz increases with increasing sound level. The early data of Stevens[32] showed rather large effects of sound level on pitch, but more recent data generally show much smaller effects.[33] For tones between 1 and 2 kHz, changes in pitch with level are generally less than 1%. For tones of lower and higher frequencies, the changes can be larger (up to 5%). There are also considerable individual differences both in the size of the pitch shifts with level and in the direction of the shifts. Shifts in pitch with level tend to be very small for complex (musical) tones.

4.4 Diplacusis

When a sinusoid of a given frequency is presented alternately to the two ears, using earphones, it may be perceived as having slightly different pitches in the two ears.[34] The effect can be quantified by asking the listener to adjust the frequency of the tone in one ear so that its pitch matches that of the tone in the other ear. Generally, the shifts measured in this way are less than 1%. This effect has been considered a problem for the temporal theory of pitch. However, adherents of temporal theory argue that the time pattern of neural spikes is analyzed and converted into another form of neural code at a relatively early stage in the auditory system, probably before the point of binaural interaction. If so, then small pitch differences between the two ears could arise at this stage of recoding.

5 PITCH PERCEPTION OF COMPLEX TONES

5.1 Phenomenon of the Missing Fundamental

The classical place theory has difficulty in accounting for the perception of complex tones. For such tones the pitch does not, in general, correspond to the position of maximum excitation on the basilar membrane. A striking illustration of this is provided by the "phenomenon of the missing fundamental." Consider, as an example, a sound consisting of short impulses (clicks) occurring 200 times per second. This sound has a low pitch, which is very close to the pitch of a 200-Hz pure tone, and a sharp timbre. It contains harmonics with frequencies 200, 400, 600, 800, ... Hz. However, it is possible to filter the sound so as to remove the 200-Hz component, and it is found that the pitch does not alter; the only result is a slight change in the timbre of the note. Indeed, all except a small group of midfrequency harmonics can be eliminated, and the low pitch still remains, although the timbre becomes markedly different.

Schouten[35] called the low pitch associated with a group of high harmonics the "residue." Several other names have been used to describe the pitch of a complex tone (whether or not the fundamental component is present), including *periodicity pitch* and *virtual pitch*. This chapter will use the term "virtual pitch." Schouten pointed out that the virtual pitch is distinguishable, subjectively, from a fundamental component that is physically presented or from a fundamental that may be generated (at high sound pressure levels) by nonlinear distortion in the ear. Thus it seems that the perception of a virtual pitch does not require activity at the point on the basilar membrane that would respond maximally to a pure tone of similar pitch. This is confirmed by a demonstration of Licklider[36] that the virtual pitch persists even when low-frequency noise is present that would mask any component at the fundamental frequency. Even when the fundamental component of a complex tone is present, the pitch of the tone is usually determined by harmonics other than the fundamental. Thus the perception of a virtual pitch should not be regarded as unusual. Rather, virtual pitches are normally heard when listening to complex tones.

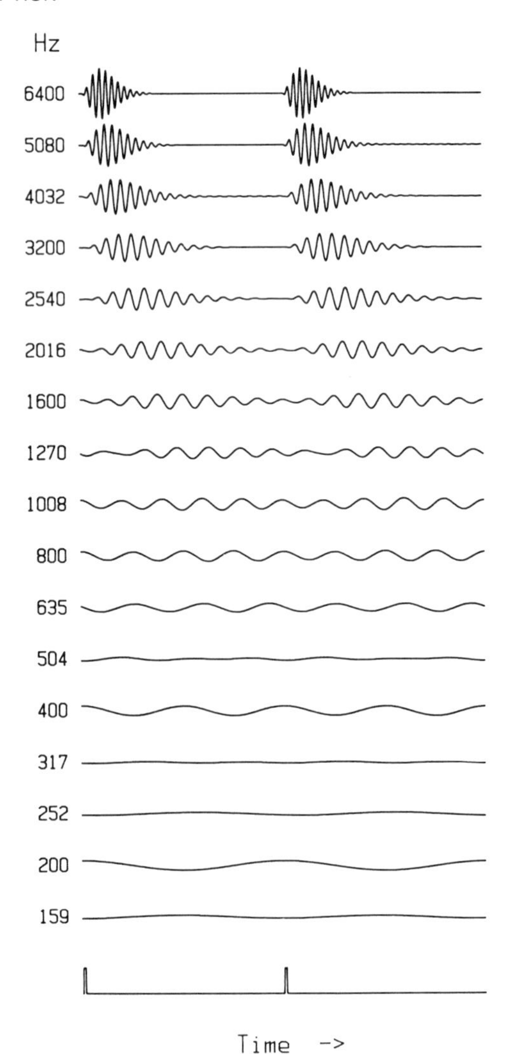

Fig. 5 Simulation of the responses on the basilar membrane to periodic impulses of rate 200 pulses per second. Each number on the left represents the frequency that would maximally excite a given point on the basilar membrane. The waveform that would be observed at that point is plotted opposite that number. (From Ref. 4.)

5.2 Discrimination of the Pitch of Complex Tones

When the waveform repetition rate of a complex tone changes, all of the components change in frequency by the same ratio, and a change in virtual pitch is heard. The ability to detect such changes in pitch is better than the ability to detect changes in a sinusoid at the fundamental frequency,[37] and it can be better than the ability to detect changes in the frequency of any of the sinusoidal components in the complex tone.[38] This indicates that information from the different harmonics is combined or integrated in the determination of virtual pitch. This can lead to very fine discrimination; changes in repetition rate of about 0.2% can often be detected for fundamental frequencies in the range 100–400 Hz.

5.3 Analysis of a Complex Tone in the Peripheral Auditory System

A simulation of the analysis of a complex tone in the peripheral auditory system is illustrated in Fig. 5. In this example, the complex tone is a periodic pulse train whose spectrum contains many equal-amplitude harmonics. The lower harmonics are partly resolved on the basilar membrane and give rise to distinct peaks in the pattern of activity along the basilar membrane. At a place tuned to the frequency of a low harmonic, the waveform on the basilar membrane is approximately a sine wave at the harmonic frequency. In contrast, the higher harmonics are not resolved and do not give rise to distinct peaks on the basilar membrane. The waveform at places on the membrane responding to higher harmonics is complex and has a repetition rate equal to the fundamental frequency of the sound.

5.4 Theories of Pitch Perception for Complex Tones

Several theories have been proposed to account for virtual pitch. Theories prior to 1980 may be divided into two broad classes. The first, spectral theories, proposed that the perception of the pitch of a complex tone involves two stages. The first stage is an analysis that determines the frequencies of some of the individual sinusoidal components of the complex tone; this depends partly on the resolution of lower harmonics on the basilar membrane. The second stage is a pattern recognizer that determines the pitch of the complex from the frequencies of the resolved components.[39,40] In essence, the pattern recognizer tries to find the harmonic series giving the best match to the resolved frequency components; the fundamental frequency of this harmonic series determines the perceived pitch. For these theories, the lower resolvable harmonics should determine the pitch that is heard.

The alternative, temporal, theories assumed that pitch is based on the time pattern of the waveform at a point on the basilar membrane responding to the higher harmonics. Pitch is assumed to be related to the time interval between corresponding points in the fine structure of the waveform close to adjacent envelope maxima.[41] Nerve firings tend to occur at these points (i.e., phase locking occurs), so this time interval will be present in the time pattern of neural impulses. For such theories, the upper unresolved harmonics should contribute substantially to the pitch that is heard.

Some recent theories (spectrotemporal theories) assume that both frequency analysis (spectral resolution) and time pattern analysis are involved in pitch perception.[4,42–44]

5.5 Physical Variables Influencing Virtual Pitch

Pitch of Inharmonic Complex Tones Schouten et al.[41] investigated the pitch of AM sine waves. If a carrier of frequency f_c is amplitude modulated by a modulator with frequency g, then the modulated wave contains components with frequencies $f_c - g, f_c$, and $f_c + g$. For example, a 2000-Hz carrier modulated 200 times per second contains components at 1800, 2000, and 2200 Hz and has a pitch similar to that of a 200-Hz sine wave.

Consider the effect of shifting the carrier frequency to, say, 2040 Hz. The complex now contains components at 1840, 2040, and 2240 Hz, which do not form a simple harmonic series. The perceived pitch, in this case, corresponds roughly to that of a 204-Hz sinusoid. The shift in pitch demonstrates that the pitch of a complex tone is not determined by the envelope repetition rate, which is 200 Hz, or by the spacing between adjacent components, which is also 200 Hz. In addition, there is an ambiguity of pitch; pitches around 185 Hz and 227 Hz may also be heard.

The perceived pitch can be explained both by spectral theories and by temporal theories. According to the spectral theories, a harmonic complex tone with a fundamental of 204 Hz would provide a good "match"; this would have components at 1836, 2040, and 2244 Hz. Errors in estimating the fundamental occur mainly through errors in estimating the appropriate harmonic number. In the above example the presented components were assumed by the pattern recognizer to be the 9th, 10th, and 11th harmonics of a 204-Hz fundamental. However, a reasonable fit could also be found by assuming the components to be the 8th, 9th, and 10th harmonics of a 226.7-Hz fundamental or the 10th, 11th, and 12th harmonics of a 185.5-Hz fundamental. Thus the theories predict the ambiguities of pitch that are actually observed for this stimulus.

The pitch shift is explained by temporal theories in the following way. For the complex tone with a shifted carrier frequency, the time intervals between corresponding peaks in the fine structure of the wave are slightly less than 5 ms; roughly 10 periods of the carrier occur for each period of the envelope, but the exact time for 10 periods of a 2040-Hz carrier is 4.9 ms, which corresponds to a pitch of 204 Hz. This is illustrated in Fig. 6. Nerve spikes can occur at any of the prominent peaks in the waveform, labeled 1, 2, 3 and $1', 2', 3'$. Thus the time intervals between successive nerve spikes fall into several groups, corresponding to the intervals 1–1′, 1–2′, 1–3′, 2–1′, ... This can account for the fact that the pitches of these stimuli are ambiguous. The intervals 1–1′, 2–2′, and 3–3′ correspond to the most prominent pitch of about 204 Hz, while other intervals, such as 1–2′ and 2–1′, correspond to the other pitches that may be perceived.

Existence Region of Virtual Pitch According to the spectral theories, a virtual pitch should only be heard when at least one of the sinusoidal components of the stimulus can be heard out. Thus a virtual pitch should not be heard for stimuli containing only very high unresolvable harmonics. Ritsma[45] investigated the audibility of virtual ptiches for AM sinusoids containing three components as a function of the modulation rate and the harmonic number (carrier frequency divided by modulation rate). Later,[46] he extended the results to complexes with a greater number of components. He found that the tonal character of the virtual pitch existed only within a limited frequency region, referred to as the "existence region."

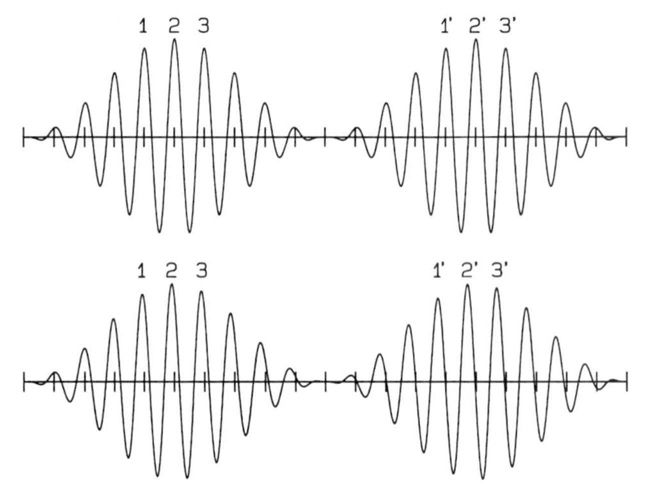

Fig. 6 Waveforms of two amplitude-modulated sinusoids. In the upper waveform, the carrier frequency (2000 Hz) is an exact multiple of the modulation frequency (200 Hz). In the lower waveform, the carrier frequency is shifted slightly upward, to 2040 Hz.

When the harmonic number was too high (above about 20), only a kind of high buzz of undefinable pitch was heard. However, virtual pitches could clearly be heard for harmonic numbers between 10 and 20, provided the frequencies of the harmonics were below about 5 kHz. These harmonics would not be resolvable.

Ritsma's results may have been influenced by the presence of combination tones in the frequency region below the lowest components in the complex tones, as has since been pointed out by Ritsma[47] himself. (See Chapter 87 for more information on combination tones.) However, more recent experiments, which have made use of background noise to mask combination tones, also indicate that virtual pitch can be perceived when no resolvable harmonics are present. Moore and Rosen[48] and Houtsma and Smurzynski[49] have demonstrated that stimuli containing only high, unresolvable harmonics can evoke a sense of musical interval and hence of pitch. The spectral theories cannot account for this.

Principle of Dominance Ritsma[50] carried out an experiment to determine which components in a complex sound are most important in determining its pitch. He presented complex tones in which the frequencies of a small group of harmonics were multiples of a fundamental that was slightly higher or lower than the fundamental of the remainder. The subject's pitch judgments were used to determine whether the pitch of the complex as a whole was affected by the shift in the group of harmonics. Ritsma found that "for fundamental frequencies in the range 100 Hz to 400 Hz, and for sensation levels up to at least 50 dB above threshold of the entire signal, the frequency band consisting of the third, fourth and fifth harmonics tends to dominate the pitch sensation as long as its amplitude exceeds a minimum absolute level of about 10 dB above threshold."

This finding has been broadly confirmed in other ways,[51] although the data of Moore et al.[38,52] show that there are large individual differences in which harmonics are dominant, and for some subjects the first two harmonics play an important role. Other data also show that the dominant region is not fixed in terms of harmonic number but also depends somewhat on absolute frequency.[51,53] For high fundamental frequencies (above about 1000 Hz), the fundamental is usually the dominant component, while for very low fundamental frequencies (around 50 Hz), harmonics above the fifth may be dominant.[54]

On the whole, these results are better explained by the spectral theories. The third, fourth, and fifth harmonics are relatively well resolved on the basilar membrane and are usually separately perceptible. However, for very low fundamental frequencies (around 50 Hz), harmonics in the dominant region (above the fifth harmonic) would not be resolvable.

Pitch of Dichotic Two-Tone Complexes Houtsma and Goldstein[55] investigated the perception of stimuli that comprised just two harmonics with successive harmonic numbers, for example, the fourth and fifth harmonics of a (missing) fundamental. They found that musically trained subjects were able to identify musical intervals between such stimuli when they were presented dichotically (one harmonic to each ear), presumably indicating that virtual pitches were heard. These results cannot be explained by temporal theories, which require an interaction of components on the basilar membrane. They imply that "the pitch of these complex tones is mediated by a central processor operating on neural signals derived from those effective stimulus harmonics which are tonotopically resolved."[55] However, it is not yet entirely clear that the findings apply to the pitches of complex tones in general. The pitch of a two-tone complex is not very clear; indeed, many observers do not hear a single low pitch but rather perceive the component tones individually.[56]

Pitch of Stimuli without Spectral Cues A number of workers have investigated pitches evoked by stimuli that contain no spectral peaks and therefore would not produce a well-defined maximum on the basilar membrane. Such pitches cannot arise on the basis of place information and so provide evidence for the operation of temporal mechanisms.

Miller and Taylor[57] used random white noise that was turned on and off abruptly and periodically at various rates. Such interrupted noise has a flat long-term magnitude spectrum. They reported that the noise had a pitchlike quality for interruption rates between 100 and 250 Hz. Burns and Viemeister[58,59] showed that noise that was sinusoidally amplitude modulated had a pitch corresponding to the modulation rate. As is the case for interrupted noise, the long-term magnitude spectrum of white noise remains flat after modulation. They found that the pitch could be heard for modulation rates up to 800–1000 Hz and that subjects could recognize musical intervals and melodies played by varying the modulation rate.

It should be noted that, although the long-term magnitude spectrum of interrupted or amplitude-modulated noise is flat, some spectral cues exist in the short-term magnitude spectrum.[60] Thus, the use of place cues to extract the pitch of these stimuli cannot be completely ruled out. This is not the case, however, for experiments showing pitch sensations as a result of the binaural interaction of noise stimuli. Huggins and Cramer[61] fed white noise from the same noise generator to both ears via headphones. The noise to one ear went through an all-pass filter, which passed all audible frequency components without change of amplitude but produced a phase change in a small frequency region. A faint pitch was heard corresponding to the frequency region of the phase transition, although the stimulus presented to each ear alone was white noise and produced no pitch sensation. The effect was only heard when the phase change was in the frequency range below about 1 kHz. A phase shift of a narrow band of the noise results in that band being heard in a different position in space from the rest of the noise. This spatial separation presumably produces the pitchlike character associated with that narrow band. Pitches produced by binaural interaction have also been reported by Fourcin[62] and Bilsen and Goldstein.[63]

Effect of Relative Phases of Components on Pitch Changes in the relative phases of the components making up a complex tone change the temporal structure of the tone but have little effect on the auditory representation of the lower, resolved harmonics. Thus, spectral theories predict that relative phase should not influence pitch, while temporal theories predict that relative phase should affect pitch, pitch being more distinct for phase relations producing "peaky" waveforms after auditory filtering (waveforms with high crest factors). For tones containing many harmonics, phase has only small effects on the value of the pitch heard,[64] but it does affect the clarity of pitch.[65] For tones containing only a few high harmonics, phase can affect both the pitch value and the clarity of pitch.[66] Phase can affect the discrimination of changes in virtual pitch, and the phase effects are larger in hearing-impaired subjects than in normal subjects, presumably because auditory filters are broader in impaired subjects and produce more interaction of the components in the peripheral auditory system.[54] Overall, these results suggest that the temporal structure of complex sounds does influence pitch perception.

6 SPECTROTEMPORAL MODELS OF PITCH PERCEPTION OF COMPLEX TONES

The evidence reviewed above indicates that neither spectral nor temporal theories can account for all aspects of the pitch perception of complex tones. Spectral theories cannot account for the following facts: A virtual pitch can be perceived when only high, unresolvable harmonics are present; stimuli without spectral peaks can give rise to a pitch sensation; and the relative phases of the components can affect pitch perception. Temporal theories cannot account for these facts: The lower, resolvable harmonics are dominant in determining pitch; and a virtual pitch may be heard when there is no possibility of the components interacting in the peripheral auditory system.

These findings have led several workers to propose theories in which both spectral and temporal mechanisms play a role; the initial place/spectral analysis in the cochlea is followed by an analysis of the time pattern of the waveform at each place or of the neural spikes evoked at each place.[4,42–44] The model proposed by Moore is based on the analysis of interspike intervals at each characteristic frequency (CF) and includes a device that compares the time intervals present at different CFs and searches for common time intervals. The device may also integrate information over time. In general the time interval that is found most often corresponds to the period of the fundamental component, and the perceived pitch corresponds to the reciprocal of this interval. Spectrotemporal theories can account for most existing data on the pitch perception of complex tones and so are more satisfactory than either spectral or temporal theories.

Acknowledgments

I thank David Green, Adrian Houtsma, Dennis McFadden and Dixon Ward for helpful comments on an earlier version of this chapter.

REFERENCES

1. J. O. Pickles, "The Neurophysiological Basis of Frequency Selectivity," in B. C. J. Moore (Ed.), *Frequency Selectivity in Hearing*, Academic, London, 1986, pp. 51–121.
2. H. Fletcher, "Auditory Patterns," *Rev. Mod. Phys.*, Vol. 12, 1940, pp. 47–65.
3. B. Scharf, "Critical Bands," in J. V. Tobias (Ed.), *Foundations of Modern Auditory Theory*, Academic, New York, 1970, pp. 157–202.
4. B. C. J. Moore, *An Introduction to the Psychology of Hearing*, 3rd ed., Academic, London, 1989.
5. R. D. Patterson, I. Nimmo-Smith, D. L. Weber, and R. Milroy, "The Deterioration of Hearing with Age: Frequency Selectivity, the Critical Ratio, and Audiogram, and Speech Threshold," *J. Acoust. Soc. Am.*, Vol. 72, 1982, pp. 1788–1803.
6. B. C. J. Moore and B. R. Glasberg, "Formulae Describing Frequency Selectivity as a Function of Frequency and Level and Their Use in Calculating Excitation Patterns," *Hear. Res.*, Vol. 28, 1987, pp. 209–225.
7. B. R. Glasberg and B. C. J. Moore, "Derivation of Auditory Filter Shapes from Notched-Noise Data," *Hearing Res.*, Vol. 47, 1990, pp. 103–138.
8. E. Zwicker and E. Terhardt, "Analytical Expressions for Critical Band Rate and Critical Bandwidth as a Function of Frequency," *J. Acoust. Soc. Am.*, Vol. 68, 1980, pp. 1523–1525.
9. H. Traunmuller, "Analytical Expressions for the Tonotopic Sensory Scale," *J. Acoust. Soc. Am.*, Vol. 88, 1990, pp. 97–100.
10. E. Zwicker, "Masking and Psychological Excitation as Consequences of the Ear's Frequency Analysis," in R. Plomp and G. F. Smoorenburg (Eds.), *Frequency Analysis and Periodicity Detection in Hearing*, Sijthoff, Leiden 1970, pp. 376–394.
11. B. C. J. Moore and B. R. Glasberg, "Suggested Formulae for Calculating Auditory-Filter Bandwidths and Excitation Patterns," *J. Acoust. Soc. Am.*, Vol. 74, 1983, pp. 750–753.
12. R. Plomp, "The Ear as a Frequency Analyzer," *J. Acoust. Soc. Am.*, Vol. 36, 1964, pp. 1628–1636.
13. B. C. J. Moore, "Some Experiments Relating to the Perception of Complex Tones," *Q. J. Exp. Psych.*, Vol. 25, 1973, pp. 451–475.
14. B. C. J. Moore and K. Ohgushi, "Audibility of Partials in Inharmonic Complex Tones," *J. Acoust. Soc. Am.*, Vol. 93, 1993, pp. 452–461.
15. R. Plomp and A. M. Mimpen, "The Ear as a Frequency Analyzer II," *J. Acoust. Soc. Am.*, Vol. 43, 1968, pp. 764–767.
16. E. Zwicker, G. Flottorp, and S. S. Stevens, "Critical Bandwidth in Loudness Summation," *J. Acoust. Soc. Am.*, Vol. 29, 1957, pp. 548–557.
17. American Standards Association, "Acoustical Terminology," SI, 1-1960, American Standards Association, New York, 1960.
18. E. G. Shower and R. Biddulph, "Differential Pitch Sensitivity of the Ear," *J. Acoust. Soc. Am.*, Vol. 2, 1931, pp. 275–287.
19. E. Zwicker, "Die elementaren Grundlagen zur Bestimmung der Informationskapazität des Gehörs," *Acustica*, Vol. 6, 1956, pp. 356–381.
20. C. C. Wier, W. Jesteadt, and D. M. Green, "Frequency Discrimination as a Function of Frequency and Sensation Level," *J. Acoust. Soc. Am.*, Vol. 61, 1977, pp. 178–184.
21. D. A. Nelson, M. E. Stanton, and R. L. Freyman, "A General Equation Describing Frequency Discrimination as a Function of Frequency and Sensation Level," *J. Acoust. Soc. Am.*, Vol. 73, 1983, pp. 2117–2123.
22. G. B. Henning, "A Model for Auditory Discrimination and Detection," *J. Acoust. Soc. Am.*, Vol. 42, 1967, pp. 1325–1334.
23. W. M. Siebert, "Frequency Discrimination in the Auditory System: Place or Periodicity Mechanisms," *Proc. IEEE*, Vol. 58, 1970, pp. 723–730.
24. B. C. J. Moore, "Relation between the Critical Bandwidth and the Frequency-Difference Limen," *J. Acoust. Soc. Am.*, Vol. 55, 1974, p. 359.
25. B. C. J. Moore and B. R. Glasberg, "The Role of Frequency Selectivity in the Perception of Loudness, Pitch and Time," in B. C. J. Moore (Ed.), *Frequency Selectivity in Hearing*, Academic, London 1986, pp. 251–308.
26. B. C. J. Moore and B. R. Glasberg, "Mechanisms Underlying the Frequency Discrimination of Pulsed Tones and the Detection of Frequency Modulation," *J. Acoust. Soc. Am.*, Vol. 86, 1989, pp. 1722–1732.
27. A. Sek and B. C. J. Moore, "Frequency Discrimination as a Function of Frequency, Measured in Several Ways," *J. Acoust. Soc. Am.*, Vol. 97, 1995, pp. 2479–2486.
28. A. Bachem, "Tone Height and Tone Chroma as Two Different Pitch Qualities," *Acta Psychol.*, Vol. 7, 1950, pp. 80–88.

29. W. D. Ward, "Subjective Musical Pitch," *J. Acoust. Soc. Am.*, Vol. 26, 1954, pp. 369–380.
30. F. Attneave and R. K. Olson, "Pitch as a Medium: A New Approach to Psychophysical Scaling," *Am. J. Psychol.*, Vol. 84, 1971, pp. 147–166.
31. K. Ohgushi and T. Hatoh, "Perception of the Musical Pitch of High Frequency Tones," in Y. Cazals, L. Demany, and K. Horner (Eds.), *Ninth International Symposium on Hearing: Auditory Physiology and Perception*, Pergamon, Oxford, 1991, pp. 207–212.
32. S. S. Stevens, "The Relation of Pitch to Intensity," *J. Acoust. Soc. Am.*, Vol. 6, 1935, pp. 150–154.
33. J. Verschuure and A. A. Van Meeteren, "The Effect of Intensity on Pitch," *Acustica*, Vol. 32, 1975, pp. 33–44.
34. G. Van den Brink, "Two Experiments on Pitch Perception: Diplacusis of Harmonic AM Signals and Pitch of Inharmonic AM Signals," *J. Acoust. Soc. Am.*, Vol. 48, 1970, pp. 1355–1365.
35. J. F. Schouten, "The Residue Revisited," in R. Plomp and G. F. Smoorenburg (Eds.), *Frequency Analysis and Periodicity Detection in Hearing*, Sijthoff, Lieden, The Netherlands 1970, pp. 41–54.
36. J. C. R. Licklider, "Auditory Frequency Analysis," in C. Cherry (Ed.), *Information Theory*, Academic, New York, 1956, pp. 253–268.
37. J. L. Flanagan and M. G. Saslow, "Pitch Discrimination for Synthetic Vowels," *J. Acoust. Soc. Am.*, Vol. 30, 1958, pp. 435–442.
38. B. C. J. Moore, B. R. Glasberg, and M. J. Shailer, "Frequency and Intensity Difference Limens for Harmonics within Complex Tones," *J. Acoust. Soc. Am.*, Vol. 75, 1984, pp. 550–561.
39. J. L. Goldstein, "An Optimum Processor Theory for the Central Formation of the Pitch of Complex Tones," *J. Acoust. Soc. Am.*, Vol. 54, 1973, pp. 1496–1516.
40. E. Terhardt, "Pitch, Consonance, and Harmony," *J. Acoust. Soc. Am.*, Vol. 55, 1974, pp. 1061–1069.
41. J. F. Schouten, R. J. Ritsma, and B. L. Cardozo, "Pitch of the Residue," *J. Acoust. Soc. Am.*, Vol. 34, 1962, pp. 1418–1424.
42. P. Srulovicz and J. L. Goldstein, "A Central Spectrum Model: A Synthesis of Auditory-Nerve Timing and Place Cues in Monaural Communication of Frequency Spectrum," *J. Acoust. Soc. Am.*, Vol. 73, 1983, pp. 1266–1276.
43. R. D. Patterson, "A Pulse Ribbon Model of Monaural Phase Perception," *J. Acoust. Soc. Am.*, Vol. 82, 1987, pp. 1560–1586.
44. R. Meddis and M. Hewitt, "Virtual Pitch and Phase Sensitivity Studied Using a Computer Model of the Auditory Periphery: Pitch Identification," *J. Acoust. Soc. Am.*, Vol. 89, 1991, pp. 2866–2882.
45. R. J. Ritsma, "Existence Region of the Tonal Residue. I," *J. Acoust. Soc. Am.*, Vol. 34, 1962, pp. 1224–1229.
46. R. J. Ritsma, "Existence Region of the Tonal Residue. II," *J. Acoust. Soc. Am.*, Vol. 35, 1963, pp. 1241–1245.
47. R. J. Ritsma, "Periodicity Detection," in R. Plomp and G. F. Smoorenburg (Eds.), *Frequency Analysis and Periodicity Detection in Hearing*, Sijthoff, Leiden, 1970, pp. 250–263.
48. B. C. J. Moore and S. M. Rosen, "Tune Recognition with Reduced Pitch and Interval Information," *Q. J. Exper. Psychol.*, Vol. 31, 1979, pp. 229–240.
49. A. J. M. Houtsma and J. Smurzynski, "Pitch Identification and Discrimination for Complex Tones with Many Harmonics," *J. Acoust. Soc. Am.*, Vol. 87, 1990, pp. 304–310.
50. R. J. Ritsma, "Frequencies Dominant in the Perception of the Pitch of Complex Sounds," *J. Acoust. Soc. Am.*, Vol. 42, 1967, pp. 191–198.
51. R. Plomp, "Pitch of Complex Tones," *J. Acoust. Soc. Am.*, Vol. 41, 1967, pp. 1526–1533.
52. B. C. J. Moore, B. R. Glasberg, and R. W. Peters, "Relative Dominance of Individual Partials in Determining the Pitch of Complex Tones," *J. Acoust. Soc. Am.*, Vol. 77, 1985, pp. 1853–1860.
53. R. D. Patterson and F. L. Wightman, "Residue Pitch as a Function of Component Spacing," *J. Acoust. Soc. Am.*, Vol. 59, 1976, pp. 1450–1459.
54. B. C. J. Moore and R. W. Peters, "Pitch Discrimination and Phase Sensitivity in Young and Elderly Subjects and Its Relationship to Frequency Selectivity," *J. Acoust. Soc. Am.*, Vol. 91, 1992, pp. 2881–2893.
55. A. J. M. Houtsma and J. L. Goldstein, "The Central Origin of the Pitch of Pure Tones: Evidence from Musical Interval Recognition," *J. Acoust. Soc. Am.*, Vol. 51, 1972, pp. 520–529.
56. G. F. Smoorenburg, "Pitch Perception of Two-Frequency Stimuli," *J. Acoust. Soc. Am.*, Vol. 48, 1970, pp. 924–941.
57. G. A. Miller and W. Taylor, "The Perception of Repeated Bursts of Noise," *J. Acoust. Soc. Am.*, Vol. 20, 1948, pp. 171–182.
58. E. M. Burns and N. F. Viemeister, "Nonspectral Pitch," *J. Acoust. Soc. Am.*, Vol. 60, 1976, pp. 863–869.
59. E. M. Burns and N. F. Viemeister, "Played Again SAM: Further Observations on the Pitch of Amplitude-Modulated Noise," *J. Acoust. Soc. Am.*, Vol. 70, 1981, pp. 1655–1660.
60. J. R. Pierce, R. Lipes, and C. Cheetham, "Uncertainty Concerning the Direct Use of Time Information in Hearing: Place Clues in White Spectra Stimuli," *J. Acoust. Soc. Am.*, Vol. 61, 1977, pp. 1609–1621.
61. W. H. Huggins and E. M. Cramer, "Creation of Pitch through Binaural Interaction," *J. Acoust. Soc. Am.*, Vol. 30, 1958, pp. 413–417.
62. A. J. Fourcin, "Central Pitch and Auditory Lateralization," in R. Plomp and G. F. Smoorenburg (Eds.), *Frequency Analysis and Periodicity Detection in Hearing*, Sijthoff, Leiden, 1970, pp. 319–328.
63. F. A. Bilsen and J. L. Goldstein, "Pitch of Dichotically Delayed Noise and Its Possible Spectral Basis," *J. Acoust. Soc. Am.*, Vol. 55, 1974, pp. 292–296.

64. R. D. Patterson, "The Effects of Relative Phase and the Number of Components on Residue Pitch," *J. Acoust. Soc. Am.*, Vol. 53, 1973, pp. 1565–1572.
65. C. Lundeen and A. M. Small, "The Influence of Temporal Cues on the Strength of Periodicity Pitches," *J. Acoust. Soc. Am.*, Vol. 75, 1984, pp. 1578–1587.
66. B. C. J. Moore, "Effects of Relative Phase of the Components on the Pitch of Three-Component Complex Tones," in E. F. Evans and J. P. Wilson (Eds.), *Psychophysics and Physiology of Hearing*, Academic, London, 1977, pp. 349–358.

91

LOUDNESS

BERTRAM SCHARF

1 INTRODUCTION

Loudness is the subjective magnitude or intensity of sound. (The word *intensity* is used here in its generic sense and is akin to the word *volume* as used by many engineers. Psychoacoustics reserves the word volume to designate the perceived size of a sound.) Every sound can be characterized as located on a scale of loudness going from nearly inaudible to intolerably loud. In English, *loudness* and its associated adjective *loud* are common, everyday words that apply directly to auditory experience. This readily available vocabulary has facilitated the psychoacoustical measurement of loudness. Other languages have coined special words for loudness, for example, *sonie* in French, *Lautheit* in German.

Historically, the measurement of loudness has involved three distinct tasks for listeners. One task has been to scale the loudness of a series of sounds, usually as a function of sound pressure level (SPL); another task has been to match for loudness sounds that differ along a single dimension, such as spectral content or duration; a third task is to discriminate between two sounds that differ only with respect to level. This chapter distills a large number of experiments devoted to one or another of these three types of measurement. The first two tasks are treated together, because *loudness functions* obtained by scaling different sounds (such as a tone and white noise) can be related to one another by a matching operation and matching, in turn, gives rise to *equal-loudness contours*. The third task is treated separately; it provides measures of *loudness discrimination*, the sensitivity of the auditory system to changes or differences in level.

Note: References to chapters appearing only in the *Encyclopedia* are preceded by *Enc.*

Loudness functions and contours are related to stimulus variables (sound pressure level, frequency, bandwidth, duration, temporal properties, background) and to listener variables (listening mode, whether monaural or binaural, prior sound exposure, auditory pathology). Discrimination is also related to many of these same variables. After the presentation of the basic data, psychophysical and physiological models of loudness and of loudness discrimination are discussed briefly.

2 LOUDNESS FUNCTIONS AND EQUAL-LOUDNESS CONTOURS

This section begins with the loudness function for a 1-kHz tone. This function, which relates loudness to sound pressure, is the standard used for the measurement of the loudness of all other sounds. Later sections show how the loudness of pure tones at other frequencies depends on SPL, how the loudness of complex sounds varies with bandwidth, how loudness depends on duration and other temporal properties, and how it is affected by the presence of other sounds.

2.1 Sound Pressure (Loudness Function for a 1-kHz Tone)

Figure 1 gives the loudness of a 1-kHz tone as a function of SPL. The straight line, including the dashed section, is from the international standard[1] and has a slope of 0.6, which means that for every 10-dB increase in level, loudness doubles. The straight line on these double-logarithmic coordinates describes a power function

$$L = kP^{0.6}, \quad (1)$$

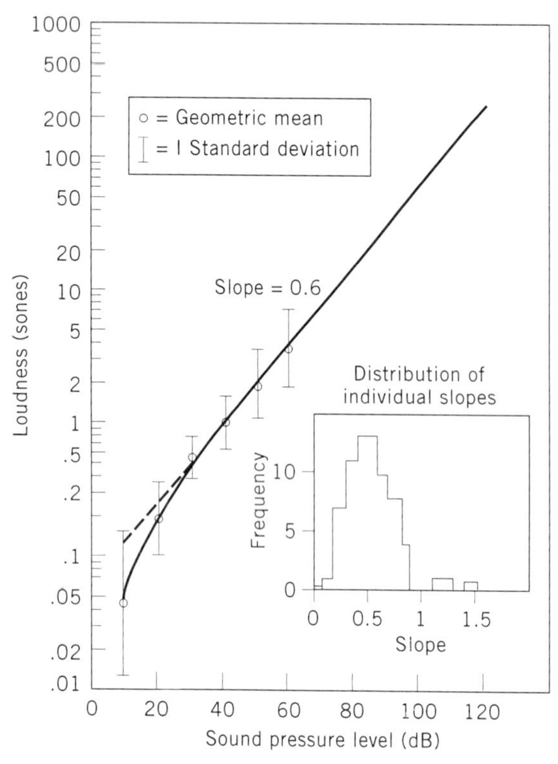

Fig. 1 Loudness function for a binaural, 1000-Hz tone. Loudness in sones is plotted as a function of the SPL in decibels measured at the center of the locus of the listener's head with the listener absent. The straight portion of the solid line is from the international standard for a 1-kHz tone[1] and has a slope of 0.6. The curved portion was taken from Ref. 26 and is based on a number of studies near threshold. The loudness of a 1000-Hz tone increases with an exponent of 0.6 according to the power law $L = 0.01(P - P_0)^{0.6}$, where L is loudness in sones, P is sound pressure in micropascals, and P_0 has the value of 45 μPa, which is the effective threshold at which loudness begins to increase with sound pressure; the subtractive constant is included to reflect the curvature of the loudness function at low levels. The slope of the straight line drawn in log–log coordinates is the exponent of the power function. An exponent of 0.6 means that loudness doubles whenever SPL increases by 10 dB. The data in the figure are the geometric means of magnitude estimates by 70 naive listeners who were run in groups of 4–11 in an anechoic room with the loudspeaker at a distance of about 3 m[7]. Curves fitted to individual data by the method of least squares yielded the 70 slopes plotted in the inset. Their mean is 0.6.

where L is loudness in sones, $k = 0.01$, and P is sound pressure in micropascals. The unit for loudness, the *sone*, is defined such that the loudness of a binaural 1-kHz tone at 40 dB SPL is equal to 1 sone. The standard loudness function was based on many investigations, most requiring listeners to evaluate loudness directly, by assigning a number or a ratio to the loudness of a tone. However, none of those investigations had made detailed measurements near threshold. Careful examination[2] of the function below about 40 dB sensation level (SL: the level above an individual listener's threshold or the SPL of the presented sound minus the SPL for that sound at the listener's threshold) showed that loudness changes more rapidly with level below about 30 dB SPL than above, as indicated by the solid line that curves down below 30 dB in Fig. 1. One way to modify the power law so as to represent this curvature is to deduct the effective threshold P_0 from P in Eq. (1). The modified power function is then

$$L = k(P - P_0)^{0.6}. \tag{2}$$

The effective threshold is that level at which the sound begins to have a sensory effect in the experimental context. A value for P_0 of 45 μPa is a reasonable value, the equivalent of a few decibels above the expected detection threshold for a binaural 1-kHz tone. However, other modifications to Eq. (1) can also yield a reasonable function.[3] One of these modifications, proposed by Zwislocki[4] (also see a closely related proposal by Humes and Jesteadt[5]), has the advantage that it fits loudness data when thresholds have been elevated by hearing impairment[6] or masking. Whatever the best fit, the important point is that the loudness function is steeper below approximately 30 dB than above.

Figure 1 also presents a set of data from 70 listeners who judged a 1-kHz tone presented from a loudspeaker in an anechoic room.[7] The circles are the geometric means for the 70 naive listeners who were run in groups of 4 to 11. The inset shows the distribution of the 70 slopes based on each listener's individual curve that had been determined by the method of least squares. The mean of these individual slopes is 0.6. The large variability in the individual slopes almost surely reflects largely judgmental differences among the listeners rather than sensory differences. If the individual slopes represented the true course of loudness, then the listener with the steepest slope of 1.5 would hear people's voices at loudnesses many thousand times greater than would the listener with the shallowest slope of 0.1. (To arrive at this relation, we need only assume that the two listeners have similar thresholds and similar loudness sensations near threshold.)

The function in Fig. 1 is defined for a tone presented from a loudspeaker in a free field, and the data shown were obtained under those listening conditions. The same function (except for a small threshold correction) applies to a tone presented binaurally through a pair of earphones. To apply to monaural presentation through a sin-

TABLE 1 Sone Values for a Binaural 1-kHz Tone and for a Binaural White Noise Listed as a Function of Sound Pressure Level

SPL (dB)	Tone	Noise	SPL (dB)	Tone	Noise	SPL (dB)	Tone	Noise
10	0.052	—	50	2.00	3.85	90	32.0	46.0
12	0.072	—	52	2.30	4.45	92	36.8	50.5
14	0.095	—	54	2.64	5.20	94	42.2	57.5
15	0.110	—	55	2.83	5.60	95	45.3	61.0
16	0.125	—	56	3.03	6.00	96	48.5	65.0
18	0.155	—	58	3.48	7.00	98	55.7	72.0
20	0.190	—	60	4.00	7.85	100	64.0	80
22	0.230	—	62	4.59	8.9	102	73.5	91
24	0.280	—	64	5.28	10.2	104	84.4	102
25	0.305	—	65	5.66	10.9	105	90.5	108
26	0.330	—	66	6.06	11.5	106	97.0	114
28	0.395	.450	68	6.96	13.0	108	111	128
30	0.460	.580	70	8.00	14.7	110	128	
32	0.550	.720	72	9.19	16.4	112	147	
34	0.640	.900	74	10.6	18.5	114	169	
35	0.700	1.00	75	11.3	19.5	115	181	
36	0.750	1.10	76	12.1	20.6	116	194	
38	0.860	1.36	78	13.9	23.2	118	223	
40	1.00	1.65	80	16.0	26.0	120	256	
42	1.15	2.00	82	18.4	29.0			
44	1.32	2.40	84	21.1	32.5			
45	1.41	2.60	85	22.6	34.8			
46	1.52	2.80	86	24.3	36.5			
48	1.74	3.28	88	27.9	41.0			

Note: A value of 1 sone is assigned to the loudness of a binaural 1-kHz tone at 40 dB SPL. Above 1 sone, the values listed for the tone are from the international standard (Ref. 1). Below 1 sone, the values are based on the curve in Fig. 1, which summarizes a number of studies of loudness at low intensities. The sone values for a white noise are based on measurements of the loudness of white noise. (Adapted with permission from Ref. 26.)

gle phone, the function would have to be moved to the right to take into account the higher monaural threshold and the reduced monaural loudness, as described below.

Table 1 presents the values of the solid-line function of Fig. 1. For a 1000-Hz tone, the sone value is given at even values of SPL from 10 to 120 dB and also at all levels ending with the digit 5. (Table 1 also gives the sone values for white noise.)

2.2 Frequency

The loudness function at 1 kHz is not valid at all frequencies. As can be seen from the curve relating threshold in the quiet to frequency (*Enc.* Ch. 123), our ears are most sensitive to the middle frequencies, including 1 kHz. When the level is increased above threshold, frequencies do not suddenly become equally effective and produce the same loudness. However, the differences among the lower frequencies do become smaller as level increases. These differences are quantified by having listeners match the loudness of tones at other frequencies to that of a tone at 1 kHz or by having them estimate loudness at other frequencies as a function of level, just as has been done for a 1-kHz tone. Much effort has been directed toward measuring the equal-loudness contours both under earphones and in the free field, so these are presented first. Relatively few data are available for directly measured loudness functions at frequencies other than 1 kHz.

Although measured over half a century ago, the contours published by Fletcher and Munson[8] are still the standard for earphone listening. Figure 2 plots SPL as a function of frequency with loudness level as the parameter. *Loudness level* is the SPL at which a 1-kHz tone is judged as loud as another sound. The unit of loudness level is a decibel, but to distinguish it from ordinary decibels of SPL, it is called the *phon.* Fletcher and Munson's subjects matched in loudness 1-s tones at each of many other frequencies to a tone at 1 kHz. Thus, each contour represents a given loudness level, ranging from 13 to 113 phons. The bottom-most contour is the threshold curve that was at 3 phons for these subjects, whose average threshold at 1 kHz was 3 dB SPL. Other than on the threshold curve, all points on a contour represent equally

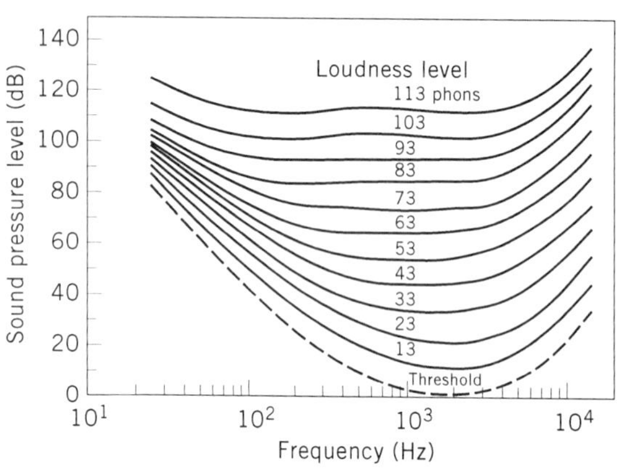

Fig. 2 Equal-loudness contours for pure tones presented through a pair of earphones. The ordinate gives the SPL at which a tone at the frequency given on the abscissa sounded equal in loudness to a 1-kHz reference tone. The level of the 1-kHz tone is the loudness level shown as the parameter on the curves. A method of constant stimuli was used in which the variable frequency was presented for 1 s followed 0.5 s later by the 1-kHz tone for 1 s. Rise–fall times were 100 ms. After two such presentations the subject responded whether the reference tone was louder or softer than the variable-frequency tone. Eleven experienced subjects each made 27 judgments at each loudness level for a total of 297 judgments, the median of which served in the construction of the curves in the figure. These curves were drawn through the median points and smoothed. The standard error of the combined results was between 1 and 2 dB, with the variability somewhat larger when the variable frequency was far from 1 kHz than when it was close. A given contour gives the combinations of frequency and SPL that yield the loudness level indicated on that contour. The bottom curve is based on threshold measurements obtained by a method of limits. Threshold at 1 kHz is assumed to be 3 dB. (Adapted from Ref. 18.)

loud tones. Thus on the 13-phon contour, a 100-Hz tone at 40 dB, a 1-kHz tone at 13 dB, and a 10-kHz tone at 33 dB are all equally loud. Most notable is the compression of the contours at the low frequencies, meaning that loudness grows most rapidly with sound pressure there; at higher frequencies the contours remain essentially parallel and loudness grows less rapidly. It is also noteworthy that the shape of the 13-phon contour is very similar to that of the threshold curve, although the two curves were obtained by very different procedures, one involving the detection of a tone, the other the equating in loudness of two tones alternating in time.

Equal-loudness contours have also been measured in the free field, with the tones presented from a loudspeaker. Figure 3 presents the contours measured by Robinson and Dadson[9] and adopted as an international standard.[10] The ordinate gives the SPL measured in the free field with the listener absent. The presence of the listener alters the sound field so that the sound pressure in the ear canal is greater around 4 kHz and smaller around 8 kHz than in earphone listening. Accordingly, the dip around 4 kHz and the peak near 8 kHz reflect primarily diffraction by the head and pinnae. Presumably, were the sound pressure at the eardrum measured in both earphone and free-field listening, the equal-loudness contours in Figs. 2 and 3 would be the same (except for an effect of the ear cushion, which may give rise to noise that reduces loudness of weak, low frequencies). (Some recent investigations have suggested that the contours in Fig. 3 should be higher at frequencies below 1 kHz; this would mean that to match the loudness of a 1-kHz tone, tones at the lower frequencies would have to be at higher levels than indicated in the figure.[11])

Having found a nice correspondence between threshold measurements and loudness matching, we now examine the relation between the outcome of loudness matching and direct estimations of loudness, which serve as the basis for the standard loudness function at 1 kHz.

From the equal-loudness contours for earphone listening (Fig. 2), we can calculate the loudness functions at frequencies other than 1 kHz. For earphone listening, we should expect loudness functions for frequencies above 1 kHz to be parallel to the 1-kHz function. Owing to threshold differences, the functions would be displaced to the left (lower thresholds) for frequencies between 1 and 4 kHz and to the right (higher thresholds) for frequencies above 4 kHz. On the other hand, functions for frequencies below 1 kHz will not be parallel because the contours are not parallel but are bunched together. For any given frequency, the loudness function can be calculated by taking the SPLs at which it is equal in loudness to a 1-kHz tone and the corresponding loudness in sones from Fig. 1 or Table 1. Some representative functions are shown in Fig. 4. These derived functions are in nice agreement with those measured by magnitude estimation at 100, 250, and 1000 Hz.[12] Schneider et al.[13] also reported reasonable agreement between directly measured loudness functions and published equal-loudness contours.

2.3 Bandwidth

The effect of signal bandwidth on loudness is revealed at its extreme by the difference between the loudness functions for a 1-kHz tone and for white noise. These differences can be gleaned from Table 1, which gives sone values for a 1-kHz tone and for white noise. At every level beginning at 28 dB SPL, a white noise is louder than an equally intense 1-kHz tone. At the midlevels, between about 50 and 60 dB, the noise is nearly twice as loud as a tone at the same level. In other terms, where the

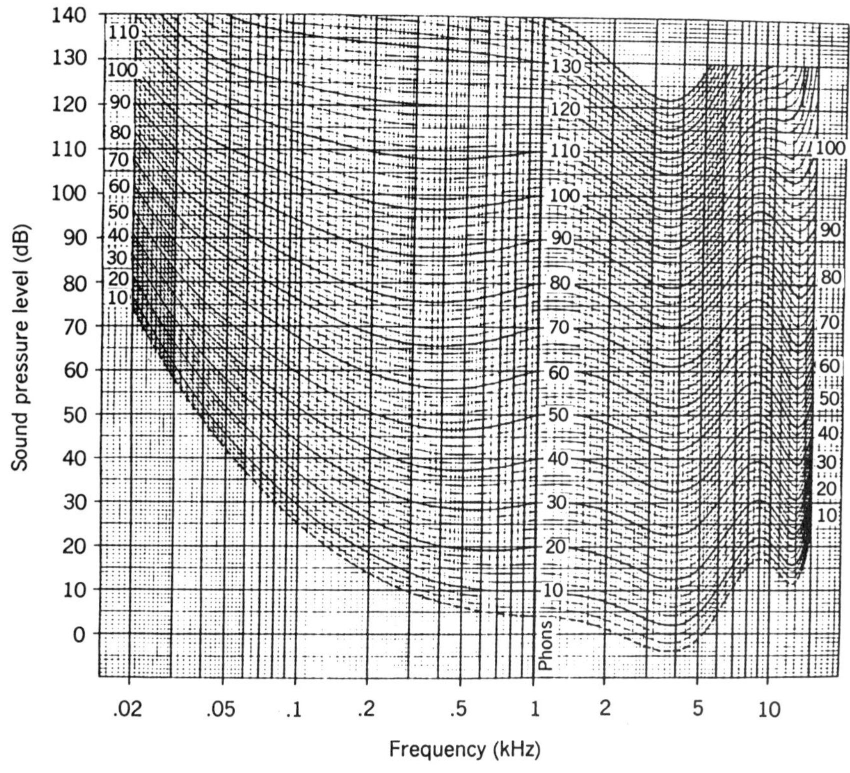

Fig. 3 Equal-loudness contours for pure tones presented in a free field. The ordinate, abscissa, and parameter are the same as in Fig. 2. Data were collected by the method of constant stimuli in a fashion similar to that described for Fig. 2. The results from a group of about 30 men and women aged 18–25 years were used to construct these contours, which constitute the international standard. Variability for the entire original group depended upon the difference in frequency between the variable tone and the reference 1-kHz tone and also on level. For moderate-frequency separations the standard deviation was about 5 phons, increasing to 10 phons for wide separations and also at high levels. The free-field contours differ from the earphone contours primarily at the high frequencies where the head, torso, and pinnae render the sound pressure generated at the ear canal entrance different from that measured in the sound field with the listener absent. Differences at low frequency are related more to leakage under the ear cushion and to greater influence of internal noise. (From Ref. 18, based on Ref. 9.)

difference is maximum, the tone must be 10 dB more intense than the noise to equal it in loudness; a 55-dB noise is just as loud as a 65-dB tone. Given that the difference between tone and noise depends on level, the white noise cannot follow the same simple power function above 30 dB that the tone does. Were the loudness function for white noise plotted on the double-logarithmic coordinates of Fig. 1, it would be bowed, with the loudness of noise growing more rapidly than that of the 1-kHz tone up to about 50 dB and more slowly above about 70 dB.

White noise contains all the audible frequencies and so is, for the human ear, the widest possible band of audible noise. When the bandwidth of the noise is narrowed, loudness begins to diminish even though the overall level of the noise is held constant. (However, given the nonlinear relationship between the loudness of white noise and the loudness of a 1-kHz tone, the change in loudness with bandwidth depends on overall level.) Figure 5 gives curves based on a matching procedure in which a 1-kHz tone was equated in loudness to a band of noise whose width is given on the abscissa. The level at which the tone was judged equal in loudness to the tone is the loudness level of the noise, shown on the

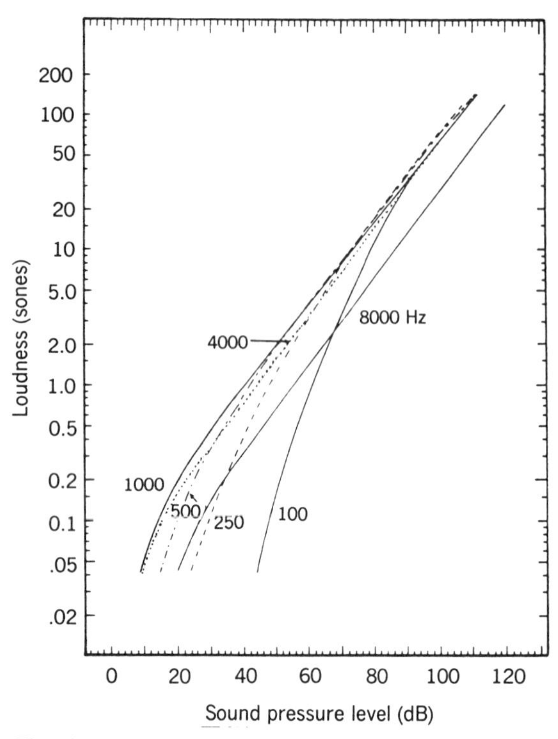

Fig. 4 Loudness of tones at different frequencies as a function of SPL. Functions are derived from the equal-loudness contours of Fig. 2 and the 1-kHz loudness functions of Fig. 1. (From Ref. 26.)

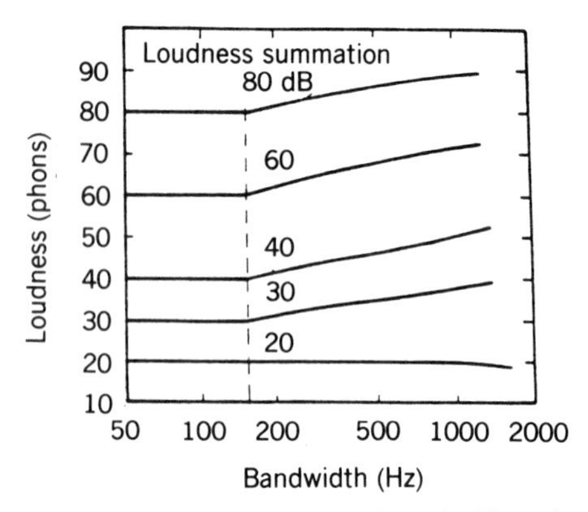

Fig. 5 Loudness level of a band of filtered white noise as a function of its width. Parameter on the curves is the overall SPL of the noise bands. The dashed line indicates the location of the critical band for these noises centered on 1000 Hz. (From Ref. 18.)

ordinate. Hence, the curves in Fig. 5 are not equal-loudness contours; rather, they show directly how loudness level changes as a function of bandwidth. The increase in loudness with bandwidth is called *loudness summation*. Loudness summation is greatest where the difference between the loudnesses of white noise and of a 1-kHz tone is greatest, around 60 dB SPL. At higher and lower levels, loudness increases less rapidly with bandwidth. Indeed, at 20 dB, loudness diminishes somewhat at very large bandwidths. The mechanisms underlying this complex set of relations are discussed in the section on models of loudness.

The increase in loudness with bandwidth does not begin until the *critical band* is exceeded. The critical band, which plays an essential role in masking (Chapter 89) and frequency analysis (Chapter 90), depends on the center frequency of the sound; it is 160 Hz at 1 kHz. At subcritical bandwidths, loudness does not change with width. However, at very narrow bandwidths, narrower than 40 or 50 Hz, fluctuations become so strong as to make loudness judgments difficult and highly variable. Indeed, at high frequencies and moderate levels, rapid amplitude fluctuations seem to enhance the loudness of very narrow bands of noise.[14]

The rules of loudness summation that apply to noises with continuous, flat spectra apply also to multitone complexes, that is, line spectra, sounds composed of two or more pure tones. Loudness depends primarily on the difference between the frequencies of the lowest and highest components, which expresses, in effect, the bandwidth of the multitone complex. Loudness appears to be independent of the number of components,[15,16] so that a two-tone or four-tone complex is as loud as an equally intense continuous-spectrum noise with the same bandwidth and overall level. Given a multitone complex, loudness is greatest when the components are evenly spaced across critical bands[16] and when they are set equally loud.[17]

In general, the loudness of a complex sound increases with bandwidth once the critical band has been exceeded. The increase in loudness is greatest at moderate SPLs. This level dependence means that the loudness of a broadband sound cannot be a power function of sound pressure, as shown above.

2.4 Duration

For the most part, loudness is independent of sound duration. However, the exceptions are important. At durations shorter than a few hundred milliseconds, loudness increases with sound duration. This increase is called *temporal summation* of loudness. At durations longer than a few seconds, loudness diminishes with duration up to 2 or 3 min but *only for weak* tones within about 30 dB of threshold or for tones at very high frequencies. This decrease is called *loudness adaptation*.

Temporal Summation A large number of studies have been devoted to measuring the dependence of loudness on durations ranging from a few milliseconds to a second or more. Results have been inconsistent, probably owing to the difficulty of judging loudness independently of perceived duration. Scharf and Houtsma[18] recently reviewed those studies. They suggested that the best summary of the data is the exponential

$$I(t) = \frac{I_\infty}{1 - e^{t/80}}, \tag{3}$$

where $I(t)$ is the sound intensity required to maintain constant loudness, I_∞ is the asymptotic intensity at long durations, and t is stimulus duration in milliseconds. The time constant of 80 ms is based on an extensive round-robin study[19] completed in the 1970s in which a number of different laboratories repeated the same measurements; the value of 80 ms is reasonably close to most of the time constants that can be estimated from other, less extensive investigations. That study focused on the loudness of 1-kHz tone bursts and found that up to 100 ms loudness depends on total energy. Although the round-robin studies used only 1-kHz signals, other investigations do not reveal any consistent divergences at other frequencies or for noise.

According to Eq. (3), loudness is kept constant by reducing intensity in nearly direct proportion as duration lengthens up to near 80 ms. This trading relation suggests that at very brief durations loudness depends on sound energy and not on intensity (or sound pressure). It also means that loudness level comes to within 1 phon of its asymptotic value when duration reaches about 125 ms and to within half a phon near 180 ms.

Loudness Adaptation Many earlier reports in the literature purported to show that loudness adaptation is a general and strong phenomenon. Almost all of them were based on interaural loudness matching, which gave rise to interaural interaction and to induced adaptation. Studies that avoid such interactions show that after approximately 200 ms loudness generally does not change with sound duration. Only when a steady sound is within about 30 dB of threshold does loudness adapt. The closer to threshold, the greater the decline in loudness. At 5 dB SL, a tone becomes inaudible for many listeners. Tones above about 10 kHz show loudness adaptation already at much higher levels.[20] Noises and intermittent tones show considerably less loudness adaptation than do steady tones.[21]

Although most sounds above 30 dB SL do not decline in loudness over time, they can be induced to do so by incrementing them intermittently (e.g., every 2 s for 1 s or every 20 s for 10 s) or by presenting the steady sound monaurally and introducing an intermittent sound in the contralateral ear.[22] (See also *Enc.* Ch. 44 on adaptation in the auditory system.)

2.5 Temporal Properties Other Than Duration

Rise–Fall Time A rapid onset and offset of a sound produces transients across a wide range of frequencies and thus enhances the loudness of the sound at most listening levels, especially if the sound is a brief, narrow-band stimulus such as a tone. The greater loudness results from the spread of energy beyond a single critical band. Once the rise–fall time exceeds a few milliseconds, it has a negligible effect on loudness.[23,24]

Temporal Pattern Kuwano et al.[24] reported small effects of envelope pattern on the loudness of impulsive sounds. Changes in the modulation rate or modulation frequency of a 1-kHz sinusoid resulted in loudness changes equivalent to less than 1 dB.

Intermittency Repeating a sound increases its loudness, provided, according to earlier studies from Germany, that the repetition rate is not less than 2 Hz.[25] At slower rates individual pulses are too far apart to permit temporal summation of loudness. Such an interpretation is in line with data on the loudness of double pulses for which time constants are of the order of 200 ms (see Ref. 26, pp. 210–211, for a brief review). However, a round-robin study conducted in 11 laboratories in Japan with 111 subjects suggests that loudness increases with the number of repetitions—all presented over nearly the same duration—whether the temporal separation between individual pulses is nearly 1 s or only 30 ms.[27] In other words, temporal summation seemed to be as good with a spacing of 1 s as with a spacing of 30 ms. The discrepancy with the earlier data may reflect cultural differences in the criteria used in judging loudness.

A somewhat different discrepancy relates to the temporal integration of the loudness of pulse trains that has been shown to continue to increase over durations up to and beyond 2 s.[28] Such integration times are much longer than those measured for continuous sounds. Berglund and Nordin[28] suggested that the off times in an intermittent sound may permit the longer integration times.

Time-Varying Sounds The assessment of sounds that vary in level and spectrum over long time periods requires continuous judgments. Namba[29] presented a model for such conditions and examples of various outcomes. It appears that the instantaneous loudness thus measured depends largely on the same stimulus and observer characteristics as a short-duration sound.[30] However, decreasing the level of a tone slowly over time provokes a more rapid decline in loudness than the physical decrease alone can account for, a kind of *decruitment*.[31,32] In this case, as in adaptation, the loudness depends not only on the immediate stimulus characteristics but also on the preceding history of the stimulus.

2.6 Background Sounds (Partial Masking)

When listeners are exposed to more than one sound at a time, they can judge the loudness of one sound to the exclusion of all other sounds or they can judge the overall loudness of all the sounds together. However, natural sounds coming from different directions with different time signatures as well as spectra are not readily judged together. Laboratory studies have concentrated on sounds presented via earphones with onset and offset disparities that permit the listener to judge one of two sounds despite strong interaction in the periphery of the auditory system. A typical set of results is shown in Fig. 6 for a 1-kHz tone presented and adjusted in level to match in loudness the same tone presented against a white noise.[33] The level of the tone in noise is given on the abscissa, and the level of the noise is the parameter on the curves. The noise was turned on for 2.8 s and the partially masked tone came on 0.4 s later and went off 1.9 s later, 0.5 s before the noise went off. During a 2.8-s interval with no noise, the tone was presented for 1.8 s and its level adjusted by the listener. The dashed line traces the levels at which both tones would be at the same level. All data points fall below the dashed line because for these levels of tone and noise the loudness of the tone was always reduced by the noise. This reduction in loudness caused by another sound is called *partial masking*. Hence, the level of the tone in noise is always greater than that of the equally loud tone in the quiet. However, the partial masking diminishes as the level of the tone in noise increases; the loudness level of the masked tone increases more rapidly than the dashed line, and the more intense the masker the more rapidly the loudness recovers. But even at the highest levels of the masked tone, some residual masking remains.

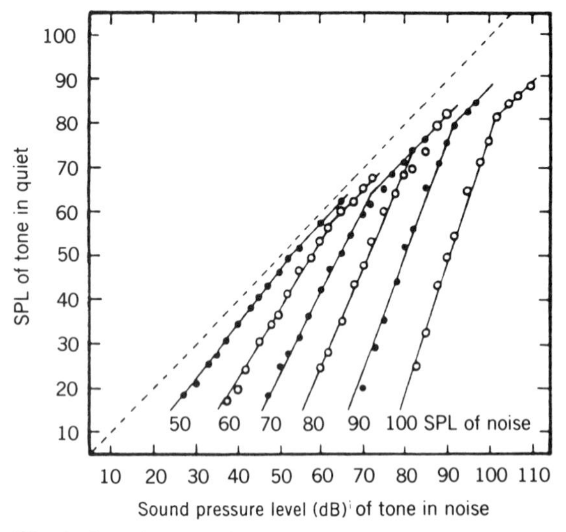

Fig. 6 Sound pressure level at which a 1-kHz tone is quiet is judged equal in loudness to a tone partially masked by white noise. Parameter on the curves is the level of the masking noise in decibels. Symbols are filled and unfilled for clarity. (From Ref. 26, based on Ref. 33.)

The rapid growth of loudness seen in Fig. 6 is called *loudness recruitment*; it is a characteristic of partial masking and also of sensorineural hearing impairment (see below). Loudness grows even more rapidly in the presence of narrow-band noise centered on the tone.[34,35] Whereas in wide-band noise, loudness grows rapidly up to 20–30 dB above masked threshold, in narrow-band noise it grows rapidly up to 10 or 15 dB above threshold, by which level the loudness almost always reaches its normal, unmasked loudness.

Partial masking, like complete masking, is greatest when the masker occupies the same frequency region as the signal.[36] Masking decreases when the masker and signal are in different frequency regions. For signals near masked threshold, loudness is reduced more by a masker at a lower frequency than the signal, just as in complete masking. For more intense signals, loudness is reduced more by a masker at a higher frequency.

2.7 Listener Variables

First data on binaural versus monaural loudness will be presented. Then, the effect on loudness of prior sound exposure and auditory pathology will be outlined. The available data on such factors as attention and individual differences do not warrant a treatment in this compendium.

Binaural versus Monaural Loudness Binaural and monaural loudness vary in the same way as a function of stimulus variables such as SPL, frequency, and bandwidth. The major difference is that a sound to both ears is judged to be nearly twice as loud as the same sound to one ear. Many studies based on equal-loudness matching and direct scaling support this conclusion. Nonetheless, the precise loudness ratio has not been agreed upon.[37] Agreement is better on the binaural advantage at threshold, which is 3 dB (*Enc.* Ch. 123). Equation (2) above for the binaural loudness function can be modified as follows to serve for the monaural function:

$$L = (0.01B^{-1}(P - 1.4P_0)^{0.6} \tag{4}$$

where, as before, L is in sones, P is in micropascals, P_0 is the effective threshold with a value 45 μPa, and B is the binaural-to-monaural loudness ratio, which may have a value between 1.7 and 2.0.

In terms of decibels, above about 30 dB SL where the loudness function becomes straight, a monaural sound must be approximately 10 dB more intense than a binaural sound to be judged equally loud. Below 30 dB, as the loudness function steepens, the difference in decibels decreases to 3 dB at threshold, but the binaural-to-monaural loudness ratio remains fairly constant.

Prior Sound Exposure (Auditory Fatigue) Exposure to an intense sound for a short period or to a weaker sound for a longer period may result in a temporary reduction in the loudness of sounds presented after cessation of the fatiguing sound just as such exposure results in higher thresholds, the temporary threshold shift (TTS). The loudness of a sound a few decibels above the elevated threshold is less than that of the same sound prior to the exposure to the fatiguing sound. As the level of the post exposure test sound is increased, loudness grows rapidly, showing loudness recruitment like that noted in the presence of a partially masking sound. For example, as TTS increases from 10 to 30 dB, the slope of the post exposure loudness function goes from 0.62 to 1.06.[38] After exposure to an intense sound, TTS is greatest about half an octave above the frequency of the fatiguing sound, whereas temporary loudness shift (TLS) is greatest at the fatiguing frequency. After exposure to a moderately intense sound of 65 dB SPL, both TTS and TLS are greatest at the fatiguing frequency, and TLS is greater (measured in decibels) than TTS.[39]

Auditory Pathology Patients with cochlear pathology such as noise-induced hearing loss or Ménière's disease usually show loudness recruitment, a rapid increase in loudness as a function of SPL. This pathological recruitment looks much like that measured against noise in normally hearing listeners. Such patients with a hearing loss in only one ear would yield functions like those shown in Fig. 6, but instead of matching a tone in noise to a tone in quiet, the functions would be based on matching a tone in the healthy ear to a tone in the impaired ear. The parameter on the curves would be hearing loss instead of noise level, and the data would be from a group of impaired listeners. Just as the loudness function in noise steepens as the noise is made more intense (and hence the masked threshold higher), so does it steepen as the hearing loss becomes greater. Generally, loudness grows so rapidly above the elevated threshold that loudness in the impaired ear is about equal to that in a normal ear at about 30 dB SL and above.[6]

Loudness summation over bandwidth is also affected by cochlear pathology so that loudness increases less than in normal hearing when stimulus bandwidth widens.[40,41]

3 LOUDNESS DISCRIMINATION

Loudness discrimination (also referred to as level or intensity discrimination) is the ability to detect a difference in the loudness of two sounds. The many published reports on level discrimination permit us to specify the physical variables that control discrimination. However, subjects vary greatly in their ability to discriminate level, and the causes of this variability are unclear. This section summarizes the data on the role of level, frequency, spectral bandwidth, duration, and sound context in discrimination. First, the major procedural issues are addressed.

Two general stimulus paradigms have been employed in experiments on discrimination. In the typical *burst-comparison* paradigm, a sound is presented successively first at one level and then at another, and the listener's task is essentially to report which sound is louder. In the *modulation* paradigm, a single sound is increased briefly in level; the listener reports whether the increment was heard. Any of the several psychophysical procedures described in Chapter 89 can be applied under these stimulus paradigms. The comparison paradigm is a better measure of discrimination because it forces the listener to base his or her decision almost entirely on the difference, whereas modulation always includes a single onset and offset that may provide the basis for the listener's decision. (Both bursts in the comparison paradigm have essentially the same onsets and offsets.) In many ways, modulation is more akin to a masking paradigm since it requires the detection of one sound—here, the increment—in the presence of another—here, the base or pedestal. However, the modulation paradigm or increment detection is generally included in discussions of discrimination rather than of masking, and so a brief section is devoted to it after discussion of the comparison paradigm.

3.1 Burst Comparison

Level and Frequency Level discrimination for pure tones improves with sensation level. At 1 kHz, the level difference ΔL required to discriminate between two levels decreases from approximately 2 dB near threshold to approximately 0.5 dB at 80 dB SL.[42] This relationship holds at frequencies up to about 6 kHz; it can be summarized as follows:

$$\Delta L = 1.644\text{dB} - 0.0141 \times \text{SL}, \qquad (5)$$

which accounted for 50% of the variance in one extensive set of data.[43] Because ΔL decreases with level instead of remaining constant as called for by Weber's law, this improvement in discrimination is called the near miss to Weber's law. The decrease in the size of ΔL with level also goes counter to the hypothesis that discrimination is a direct reflection of the slope of the loudness function, being better whenever the slope is steeper. In fact, discrimination is poorest near threshold, where the slope is steepest. Furthermore, a direct test of the hypothesis showed that level discrimination was the same for a tone heard in narrow-band noise as for an equally loud tone in broadband noise despite a large difference in the slopes of the partially masked loudness functions.[44]

At frequencies above 6 kHz or so, discrimination depends on level in more complex ways.[45,46] Between 8 and 12 kHz discrimination worsens up to moderate levels and then first begins to improve with further increases in level. At 14 kHz, discrimination changes little with level. Figure 7 plots ΔL as a function of tone level for eight studies over a range of frequencies from 0.25 to 14 kHz. The solid line from Florentine et al.[46] tends to be higher than results of many other studies, probably because their six listeners were not as well trained and motivated.

As implied above, level discrimination is independent of frequency only up to 6 kHz or so; it worsens (ΔL increases) with frequency starting at about 8 kHz, especially at levels greater than 50 dB SPL.[46] As discussed below, this deterioration can be ascribed to the restricted spread of excitation at high frequencies.

Duration Level discrimination improves as stimulus duration increases up to anywhere from 12 ms to 2 s, depending on the investigation and the sound frequency.[47] Reasons for the wide divergence among studies may be related to listener variables.

Bandwidth Level discrimination improves as a band of noise is widened. Unlike pure-tone discrimination, for a broadband stimulus such as white noise[48] or a widely spaced 18-tone complex,[49] discrimination improves with level only up to about 20 or 30 dB SPL. At higher levels ΔL is constant, meaning that Weber's law applies. Thus, near threshold, discrimination for broadband noise is no better than for a pure tone, at moderate levels discrimination is better, and above about 75 dB SPL it is worse.

Listener Variables Listeners differ markedly in their ability to discriminate between successive bursts. Florentine et al.[46] reported ratios of 2 : 1 and even 3 : 1 in the ΔL of one listener compared to another for the identical tone. Although Florentine et al.[46] were able to reduce the size of their listeners' ΔL's by giving them extensive practice and monetary reward, they did not report a concomitant reduction in intersubject variability.

Cochlear impairment has been thought to lead to improved level discrimination. However, the improvement is seen only when impaired listeners are compared to normally hearing listeners at the same SLs where the SPLs are higher for the impaired listeners. Compared at the same SPLs (lower SLs for the impaired), impaired listeners are no better and are often poorer than normally hearing listeners; just how much poorer depends on the shape of the impaired listener's threshold curve.[47]

3.2 Increment Detection

In the older literature, reference was often made to the measurements of Riesz,[50] who used sinusoidal amplitude modulation of continuous tones presented at many levels and frequencies. However, such measurements are contaminated by too many variables, such as adaptation, to give a clear picture of the ability to discriminate loudness differences.

Nonetheless, measurements of the minimum detectable increment in an ongoing sound are of importance in understanding the neural code for loudness. Measurements are available for tones at various frequencies and levels and for bands of noise. An important variable has been the time difference between the onset of the pedestal and the onset of the increment as well as such factors as the duration and rise–fall time of the increment. The interactions among these variables are complex, but a few general statements can be made.

Harris,[51] in an extensive series of measurements of increment detection, found that ΔL varied little with signal frequency. The increments lasted 150 ms and had a rise–fall time of 10 ms and occurred once a second. The effect of level was like that for burst comparison; ΔL decreased rapidly with level up to about 30 dB and more slowly thereafter. This presentation mode yields results more like that for burst comparison than for sinusoidal modulation. Harris also showed that the ΔL decreased as the duration of the increment increased from 20 to between 150 and 300 ms; the decrease was striking at sensation levels below 40 dB. At 40 dB and above, the ΔL decreased much less rapidly with duration. Duration would be expected to play less of a role in increment detection whenever onset is most important. Thus at low levels onset would seem to be less important than the excitation pattern evoked by the whole increment.

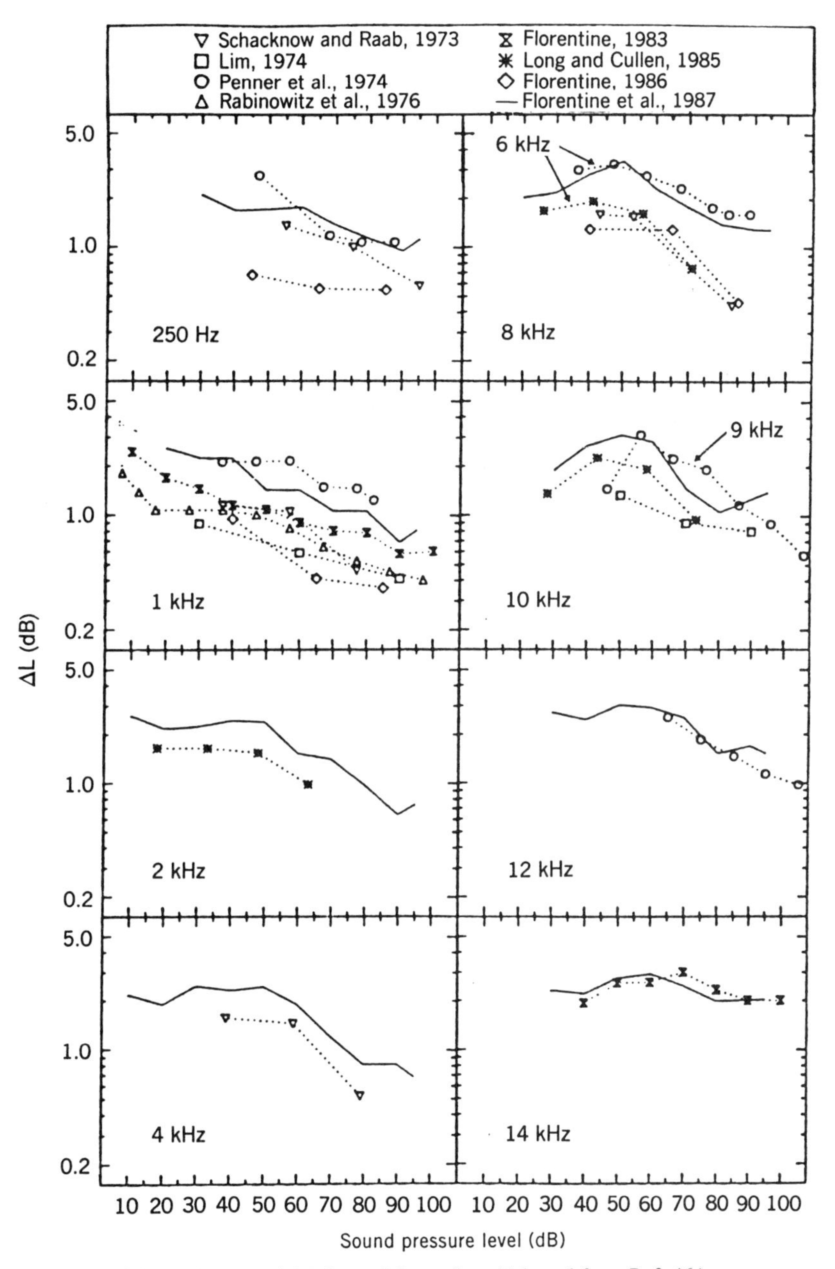

Fig. 7 Measured ΔL from eight studies. (Adapted from Ref. 46.)

The importance of the onset depends on frequency as well as level. Increment detection improves with increases in the delay between the onset of the base or pedestal sound and the onset of the increment;[52] the improvement is greater at 4 kHz than at 0.5 kHz and is greater at higher levels. Because detection continued to improve as the increment onset was delayed many seconds, some kind of long-term neural adaptation is likely to be involved; however, loudness adaptation is not directly implicated.

4 MODELS OF LOUDNESS AND LOUDNESS DISCRIMINATION

Although the neural code for loudness remains unclear, it is generally thought to be related to the quantity of neural activity in the auditory system. The more neurons that respond to a sound and the faster they respond, presumably the greater the loudness. Logical as this analysis may seem, the physiological data do not provide clear support and some even speak against it. The toughest problem has been to identify a physiological response capable of encoding the immense range of sound pressures to which organisms are sensitive, both with respect to loudness magnitude and discrimination; loudness increases from near inaudibility at threshold to intolerable levels over a range of sound pressures of more than a million to one. Yet, level discrimination remains extremely fine throughout. The demonstration by Liberman[53] that 10–15% of primary auditory nerve fibers have very high absolute thresholds, as much as 70 dB higher than the most sensitive fibers, provided another physiological basis for encoding intensity. These findings have also provided a basis for physiological models of intensity discrimination.[54] (Recent and comprehensive reviews of these matters are available.[55,56]) Nonetheless, because these physiological data have not yet led to quantitative models, the following discussion focuses on models that emphasize the quantity of assumed excitation in the auditory system.

Perhaps the most successful and best known of these psychophysical models are Zwicker's[35,57,58] for loudness and its extension by Florentine and Buus[49] to discrimination. The models are based on Fletcher and Munson's[59] proposal that masking patterns (such as revealed by measuring the threshold for a pure tone as a function of frequency in the presence of a narrow-band masker) reflect the level and spread of excitation in the auditory system. The excitation pattern thus derived determines loudness and provides the basis for loudness discrimination.

Hellman and Zwicker[60] provided an interesting, recent example of the model's application to loudness. They compared the measured and calculated loudness of sounds composed of broadband noise and a tonal component. Increasing the overall level of the complex sound by increasing the noise led to greater loudness, but increasing the overall level by increasing the level of the tone (for high tone-to-noise ratios of 20–30 dB re third-octave band of noise) and simultaneously decreasing the level of the noise diminished overall loudness. Figure 8 shows how the loudness of the complex increased with the overall level of tone plus noise. Also shown is the loudness calculated from Zwicker's model. (A convenient computer program for calculating loudness by this model is available.[61]) The points on the graph represent

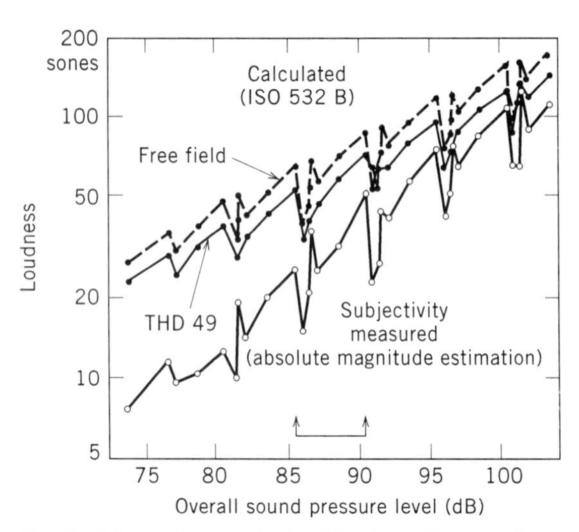

Fig. 8 Measured and calculated loudness functions for a tone–noise complex composed of 1-kHz tone and broadband noise. The dashed curve shows the calculated loudness function for listening in a free field, the dotted curve shows the calculated loudness function for listening with TDH-49 earphones, and the solid curve shows the measured loudness function. (From Ref. 60.)

different tone-to-noise ratios, ranging from +5 to +30 dB (as well as the overall levels shown on the abscissa). Calculated and measured values are placed so as to reveal their close agreement with respect to the nonmonotonic relation between loudness and overall SPL; they differ mainly in the rate at which loudness increases with overall level. In the model, loudness is calculated by integrating over all the sound frequencies in the excitation pattern. When overall level is increased by taking a great deal of energy from the broadband noise and concentrating it at a single frequency—as happens at high tone-to-noise ratios—the height of the loudness pattern at all frequencies except those near the tonal component is reduced. Owing to the slow rate at which loudness changes with amplitude, the increase in height around the tonal component is too small to compensate for the lowered height at all the other frequencies.

The excitation pattern model applied by Florentine and Buus[49] to discrimination data helped explain many of the effects of level and frequency on discrimination. For example, the near miss to Weber's law arises from the spread of excitation as a pure tone is increased in level, providing the system with more and more channels for the comparison of one sound to another. A white noise does not show the near miss because it already stimulates much of the system (except close to threshold where, indeed, ΔL does decrease with increasing level).

Moreover, Weber's law can be recovered, as Viemeister[62] has shown, by partially masking a tone with a high-pass masker that eliminates the high-frequency channels and so prevents discrimination from improving with level, in perfect accord with the excitation pattern model. Very high frequencies also obey Weber's law, presumably because few channels remain into which their excitation can extend with increasing level. Despite the success of the excitation pattern models in handling many psychophysical data for both loudness and loudness discrimination, some data require other assumptions about the physiology underlying loudness[63] and about the possible role of more complex variables such as stimulus context.[64]

Acknowledgments

I thank Rhona Hellman and Georges Canévet for helpful comments on this manuscript. Wise advice from the editors, David M. Green and Dennis McFadden, was most useful. Support for the completion of this chapter was provided in part by a grant, RO1NS07270, from the National Institutes of Health.

REFERENCES

1. International Standardization Organization (ISO), "Method for Calculating Loudness Level," R532, ISO, New York, 1966.
2. B. Scharf and J. C. Stevens, "The Form of the Loudness Function near Threshold," *Proceedings of the Third International Congress on Acoustics*, Vol. 1, Elsevier, Amsterdam, 1961, pp. 80–82.
3. L. E. Marks and J. C. Stevens, "The Forms of the Psychophysical Function near Threshold," *Perception and Psychophysics*, Vol. 4, 1968, pp. 315–318.
4. J. J. Zwislocki, "Analysis of Some Auditory Characteristics," in R. D. Luce, R. R. Bush, and E. Galanter (Eds.), *Handbook of Mathematical Psychology*, Vol. III, Wiley, New York, 1965.
5. L. E. Humes and W. Jesteadt, "Models of the Effects of Threshold on Loudness Growth and Summation," *J. Acoust. Soc. Am.*, Vol. 90, 1991, pp. 1933–1943.
6. R. P. Hellman and C. H. Meiselman, "Loudness Relations for Individuals and Groups in Normal and Impaired Hearing," *J. Acoust. Soc. Am.*, Vol. 88, 1990, pp. 2596–2606.
7. G. Canévet, R. Hellman, and B. Scharf, "Group Estimation of Loudness in Sound Fields," *Acustica*, Vol. 60, 1986, pp. 277–282.
8. H. F. Fletcher and W. A. Munson, "Loudness, Its Definition, Measurement and Calculation," *J. Acoust. Soc. Am.*, Vol. 5, 1933, pp. 82–108.
9. D. W. Robinson and R. S. Dadson, "A Re-determination of the Equal-Loudness Relations for Pure Tones," *Br. J. Appl. Phys.*, Vol. 7, 1956, pp. 166–181.
10. International Standardization Organization (ISO), "Normal Equal-Loudness Contours for Pure Tones and Normal Threshold of Hearing under Free Field Listening Conditions," R226, ISO, New York, 1962.
11. K. Betke and R. Weber, "Re-examination of Equal Loudness Contours and Minimum Audible Field," *Proceedings of the Thirteenth International Congress on Acoustics*, Vol. I, Dragan Srnic, Belgrade, 1989, pp. 483–486.
12. R. P. Hellman and J. J. Zwislocki, "Loudness Determination at Low Sound Frequencies," *J. Acoust. Soc. Am.*, Vol. 43, 1968, pp. 60–64.
13. B. Schneider, A. A. Wright, W. A. Edelheit, P. Hock, and C. Humphrey, "Equal Loudness Contours Derived from Sensory Magnitude Judgments," *J. Acoust. Soc. Am.*, Vol. 51, 1972, pp. 1951–1959.
14. E. Zwicker, "Procedure for Calculating Loudness of Temporally Variable Sounds," *J. Acoust. Soc. Am.*, Vol. 62, 1977, pp. 675–682.
15. B. Scharf, "Loudness of Complex Sounds as a Function of the Number of Components," *J. Acoust. Soc. Am.*, Vol. 31, 1959, pp. 783–785
16. E. Zwicker, G. Flottorp, and S. S. Stevens, "Critical Bandwidth in Loudness Summation," *J. Acoust. Soc. Am.*, Vol. 39, 1957, pp. 548–557.
17. B. Scharf, "Loudness Summation and Spectrum Shape," *J. Acoust. Soc. Am.*, Vol. 34, 1962, pp. 228–233.
18. B. Scharf and A. Houtsma, "Audition II," in K. Boff, L. Kaufman, and J. Thomas (Eds.), *Handbook of Perception and Human Performance*, Vol. 1, Wiley, New York, 1986.
19. O. J. Pedersen, P. E. Lyregaard, and T. E. Poulsen, "The Round Robin Test on Evaluation of Loudness Level of Impulsive Noise," Report No. 22, Acoustics Laboratory, Technical University of Denmark, Copenhagen, 1977.
20. A. Miskiewicz, B. Scharf, C. Meiselman, and R. Hellman, "Loudness Adaptation at High Frequencies," *J. Acoust. Soc. Am.*, Vol. 94, 1993, pp. 1281–1286.
21. B. Scharf, "Loudness Adaptation," in J. V. Tobias and E. D. Schuber (Eds.), *Hearing Research and Theory*, Vol. 2, Academic, New York, 1983.
22. G. Canévet, B. Scharf, and M.-C. Botte, "Simple and Induced Loudness Adaptation," *Audiology*, Vol. 24, 1985, pp. 430–436.
23. K. Gjaevenes and E. R. Rimstad, "The Influence of the Rise Time on Loudness," *J. Acoust. Soc. Am.*, Vol. 51, 1972, pp. 1233–1239.
24. S. Kuwano, S. Namba, H. Miura, and H. Tachibana, "Evaluation of the Loudness of Impulsive Sounds Using Sound Exposure Level Based on the Results of a Round Robin Test in Japan," *J. Acoust. Soc. Jpn.*, (E), Vol. 8, 1987, pp. 241–248.
25. E. Port, "Ueber die Lautstärke einzelner kurzer Schallimpulse," *Acustica*, Vol. 13, 1963, pp. 212–223.
26. Scharf, B. "Loudness," in E. C. Carterette and M. P. Fried-

man (Eds.), *Handbook of Perception*, Vol. 4, *Hearing*. Academic, New York, 1978, pp. 187–242.

27. T. Sone, et al., "Loudness and Noisiness of a Repeated Impact Sound: Results of Round Robin Tests in Japan (II)," *J. Acoust Soc. Jpn., (E)*, Vol. 8, 1987, pp. 249–261.
28. B. Berglund and S. Nordin, "Loudness and Brainstem Auditory Evoked Response for Intermittent Sound," *Proceedings of the Thirteenth International Congress on Acoustics*, Vol. I, Dragan Srnic, Belgrade, 1989, pp. 277–380.
29. S. Namba, "Loudness Varying with Time," *Proceedings of the Thirteenth International Congress on Acoustics*, Vol. I, Dragan Srnic, Belgrade, 1989, pp. 361–364.
30. S. Kuwano and H. Fastl, "Loudness Evaluation of Various Kinds of Non-Steady State Sound Using the Method of Continuous Judgment by Category," *Proceedings of the Thirteenth International Congress on Acoustics*, Vol. I, Dragan Srnic, Belgrade, 1989, pp. 365–368.
31. G. Canévet and B. Scharf, "The Loudness of Sounds That Increase or Decrease Continuously in Level," *J. Acoust Soc. Am.*, Vol. 88, 1990, pp. 2136–2142.
32. R. S. Schlauch, "A Cognitive Influence on the Loudness of Tones That Change Continuously in Level," *J. Acoust Soc. Am.*, Vol. 92, 1992, pp. 758–765.
33. S. S. Stevens and M. Guirao, "Loudness Functions under Inhibition," *Percept. Psychophys.*, Vol. 2, 1967, pp. 459–465.
34. R. P. Hellman, "Effect of Noise Bandwidth on the Loudness of a 1000 Hz Tone," *J. Acoust Soc. Am.*, Vol. 48, 1970, pp. 500–504.
35. E. Zwicker, "Ueber die Lautheit von ungedrosselten und gedrosselten Schallen" *Acustica*, Vol. 13 (Beiheft 1), 1963, pp. 194-211.
36. B. Scharf, "Partial Masking," *Acustica*, Vol. 14, 1964, pp. 16–23.
37. L. E. Marks, "Sensory and Cognitive Factors in Judgments of Loudness," *J. Exper. Psychol.: Human Percept. Perform.*, Vol. 5, 1979, pp. 426–443.
38. W. Riach, D. N. Elliott, and J. C. Reed, "Growth of Loudness and Its Relationship to Intensity Discrimination under Various Levels of Auditory Fatigue," *J. Acoust Soc. Am.*, Vol. 34, 1962, pp. 1764–1767.
39. M.-C. Botte and S. Monikheim, "New Data on Short-Term Effects of Tone Exposure," *J. Acoust Soc. Am.*, Vol. 95, 1994, pp. 2598–2605.
40. M. Florentine, S. Buus, B. Scharf, and E. Zwicker, "Frequency Selectivity in Normally-Hearing and Hearing-Impaired Observers," *J. Speech Hear. Res.*, Vol. 23, 1980, pp. 646–669.
41. B. Scharf and R. P. Hellman, "A Model of Loudness Summation Applied to Impaired Ears," *J. Acoust Soc. Am.*, Vol. 40, 1966, pp. 71–78.
42. W. M. Rabinowitz, J. S. Lim, L. D. Braida, and N. I. Durlach, "Intensity Perception. VI. Summary of Recent Data on Deviations from Weber's Law for 1000-Hz Tone Pulses," *J. Acoust Soc. Am.*, Vol. 59, 1976, pp. 1506–1509.
43. W. Jesteadt, C. C. Wier, and D. M. Green, "Intensity Discrimination as a Function of Frequency and Sensation Level," *J. Acoust Soc. Am.*, Vol. 61, 1977, pp. 169–177.
44. R. Hellman, B. Scharf, M. Teghtsoonian, and R. Teghtsooinian, "On the Relation between the Growth of Loudness and the Discrimination of Intensity for Pure Tones," *J. Acoust Soc. Am.*, Vol. 82, 1987, pp. 448–453.
45. R. P. Carlyon and B. C. J. Moore, "Intensity Discrimination: A Severe Departure from Weber's law," *J. Acoust Soc. Am.*, Vol. 76, 1984, pp. 1369–1376.
46. M. Florentine, S. Buus, and C. R. Mason, "Level Discrimination as a Function of Level for Tones from 0.25 to 16 kHz," *J. Acoust Soc. Am.*, Vol. 81, 1987, pp. 1528–1541.
47. B. Scharf and S. Buus, "Audition I," in K. Boff, L. Kaufman, and J. Thomas (Eds.), *Handbook of Perception and Human Performance*, Vol. 1, Wiley, New York, 1986.
48. A. J. Houtsma, N. I. Durlach, and L. D. Braida, "Intensity Perception XI. Experimental Results on the Relation of Intensity Resolution to Loudness Matching," *J. Acoust Soc. Am.*, Vol. 68, 1980, pp. 807–813.
49. M. Florentine and S. Buus, "An Excitation-Pattern Model for Intensity Discrimination," *J. Acoust Soc. Am.*, Vol. 70, 1981, pp. 1646–1654.
50. R. R. Riesz, "Differential Intensity Sensitivity of the Ear," *Phys. Rev.*, Vol. 31, 1928, pp. 867–875.
51. J. D. Harris, "Loudness Discrimination," *J. Speech Hear. Disord.*, Vol. 11, 1963, pp. 1–63.
52. B. Scharf, G. Canévet, and L. M. Ward, "On the Relation between Intensity Discrimination and Adaptation," in Y. Cazals, L. Demany, and K. Horner (Eds.), *Auditory Physiology and Perception*, Pergamon, Oxford, 1992, pp. 289–295.
53. M. C. Liberman, "Auditory-Nerve Response from Cats Raised in a Low-Noise Chamber," *J. Acoust Soc. Am.*, Vol. 63, 1978, pp. 442–455.
54. R. L. Winslow and M. B. Sachs, "Effect of Electrical Stimulation of the Crossed Olivocochlear Bundle on Auditory Nerve Response to Tones in Noise," *J. Neurophysiol.*, Vol. 75, 1987, pp. 1002–1021.
55. N. F. Viemeister, "Psychophysical Aspects of Auditory Intensity Coding," in G. M. Edelman, W. E. Gall, and W. M. Cowan (Eds.), *Auditory Function: Neurobiological Bases of Hearing*, Wiley, New York, 1988, pp. 213–241.
56. R. L. Smith, "Encoding of Sound Intensity by Auditory Neurons," in G. M. Edelman, W. E. Gall, and W. M. Cowan (Eds.), *Auditory Function: Neurobiological Bases of Hearing*, Wiley, New York, 1988, pp. 243–274.
57. E. Zwicker and B. Scharf, "A Model of Loudness Summation," *Psychol. Rev.*, Vol. 72, 1965, pp. 3–26.
58. B. C. J. Moore and B. R. Glasberg, "The Role of Frequency Selectivity in the Perception of Loudness, Pitch and Time," in B. C. J. Moore (Ed.), *Frequency Selectivity in Hearing*, Academic, London, 1986, pp. 251–308.

59. H. F. Fletcher and W. A. Munson, "Relation between Loudness and Masking," *J. Acoust Soc. Am.*, Vol. 9, 1937, pp. 1–10.

60. R. Hellman and E. Zwicker, "Why Can a Decrease in dB(A) Produce an Increase in Loudness?" *J. Acoust Soc. Am.*, Vol. 82, 1987, pp. 1700–1705.

61. E. Zwicker, H. Fastl, and C. Dallmayr, "Basic Program for Calculating the Loudness of Sounds from Their 1/3-oct-Band Spectra According to ISO 532B," *Acustica*, Vol. 55, 1984, pp. 63–67.

62. N. F. Viemeister, "Intensity Discrimination of Pulsed Sinusoids: The Effects of Filtered Noise," *J. Acoust Soc. Am.*, Vol. 51, 1972, pp. 1265–1269.

63. F-G. Zeng and C. W. Turner, "Binaural Loudness Matches in Unilaterally-Impaired Listeners," *Quart. J. Exper. Psychol.*, Vol. 43A, 1991, pp. 565–583.

64. L. E. Marks, "Recalibrating the Auditory System: The Perception of Loudness," *Percept. Psychophys.*, Vol. 20, 1994, pp. 382–396.

92

EFFECTS OF HIGH-INTENSITY SOUND

W. DIXON WARD

1 INTRODUCTION

Sounds of high intensity make speech communication difficult, if not impossible, and may render warning signals inaudible. They elicit changes in the autonomic system, such as dilation of the pupil of the eye, increased heart rate, and vasoconstriction of the extremities and, if their onset is sudden, evoke the startle response. They may degrade performance of certain tasks, interfere with sleep, and induce irritation ranging from annoyance to rage. At very high levels, sound becomes painful and may even directly affect the function of the visual and vestibular systems as well as of internal organs.

The main aftereffect of high-intensity sounds, however, is on the auditory system, manifested principally in the form of a temporary or permanent loss of sensitivity and acuity, although tinnitus and suppression of otoacoustic emissions may also occur. Permanent hearing loss can be caused by short exposures to extremely high sound levels or by repeated exposures to more moderate levels; the extent of the damage produced in a particular ear is determined not only by the duration and level of the exposure but also by its spectrum and temporal pattern as well as by individual characteristics of the auditory system concerned.

Exposures that result in permanent hearing loss can be regarded as stressors that may produce other aftereffects, particularly on the circulatory system. However, there is little evidence that other stressors such as vibration, heat, or exercise act synergistically with high-intensity sound to produce hearing loss.

Until a few years ago, an "intense" sound was commonly understood to mean one that overloaded the hearing mechanism, that is, an acoustic stimulus that elicited a nonlinear response from the auditory system. However, with the discovery of the essential nonlinearity of the cochlea whereby energy released at the basilar membrane enhances the vibratory motion of the cochlear partition at the lowest values of input energy, it has become clear that the auditory system as a whole is never really linear, so some other definition of *intense* must be found. A suitable alternative is suggested by a different characteristic of the auditory system, namely, a sound capable of arousing the acoustic reflex. Whether the stiffening of the ossicular chain produced by this reflex, which reduces the intensity of the sound reaching the inner ear, is regarded as providing protection against one's own voice or against other sounds, the auditory system apparently judges some sounds to be "too loud" and takes steps to reduce them.

2 CONCOMITANT EFFECTS

2.1 Middle-Ear Muscles

If the foregoing definition of intense sound is adopted, then one of its primary effects is, circularly, contraction of the stapedius and tensor tympani muscles. Because the strength of their contraction depends on the loudness of the sound rather than on its intensity, all of the physical characteristics of sound that determine loudness are relevant: not only overall intensity but also frequency, duration, and spectral width. However, for sounds of more than half a second in duration and in the audible range, strength of contraction increases with intensity, beginning at a threshold value of 85–90 dB sound pressure level (SPL) for single pure tones, 70–75 dB SPL

Note: References to chapters appearing only in the *Encyclopedia* are preceded by *Enc.*

for broadband noise, and reaching a maximum at levels about 30 dB higher in either case.[1]

Although the latency of the first muscular response to an arousal sound is only 10–15 ms, full contractile strength is not reached until as much as 150 ms after onset.[2] In the case of a sustained arousal sound, the muscles do not maintain full strength indefinitely, but relax (adapt) more or less rapidly depending on the frequency and temporal structure of the arouser. The lower the frequency of the sound, or the greater its fluctuation in time, the more slowly will the muscles adapt. As the increased muscular tension produces primarily an increase in the stiffness of the middle-ear mechanism, low frequencies (below 1000 Hz) are attenuated more than high frequencies by the reflex.

Other mechanisms do exist that may modify the sound reaching the inner ear. As sound level is increased, peak clipping eventually will occur when the motion of the eardrum or the stapes reaches a limit imposed by the annulus or annular ligament, respectively.[3] This apparently happens only for levels above 120–130 dB, at least in the anesthetized cat.[4] In addition, Bekesy's observations on human cadaver ears[5] revealed that at a similar level the axis of rotation of the stapes may suddenly shift by 90°, thereby reducing the sound transmitted to the cochlea. In both of these cases, however, the middle-ear muscles were inoperative, so just how all of these three influences interact in the normal ear is not known. One can expect, however, that some sort of discontinuity of function may occur at around 120 dB SPL.

2.2 Masking

High-intensity sounds may render other sounds inaudible. In the case of direct masking—that is, when the masking sound (the masker) is in the same frequency region as the sound that is masked (the signal)—the intensity of the masker is irrelevant, in the sense that the masked threshold of the signal will be linearly related to the level of the masker, whether the latter is weak or intense: An increase of X decibels in masker level will produce an X-decibel increase in the threshold of the signal. If the signal is speech, then its intelligibility will remain essentially constant if it were at the same time increased by that same X decibels.

The foregoing generalization is not true when the signal and masker frequencies f_s and f_m are different. In either upward spread of masking ($f_m < f_s$) or remote masking ($f_m > f_s$), the rate of growth with masker level increases with level, as the distortion products—aural harmonics, combination tones, and perhaps envelope detection—generated in the cochlea by the masker also grow.[6] So by the time the masking level reaches the intense range, considerable upward spread of masking

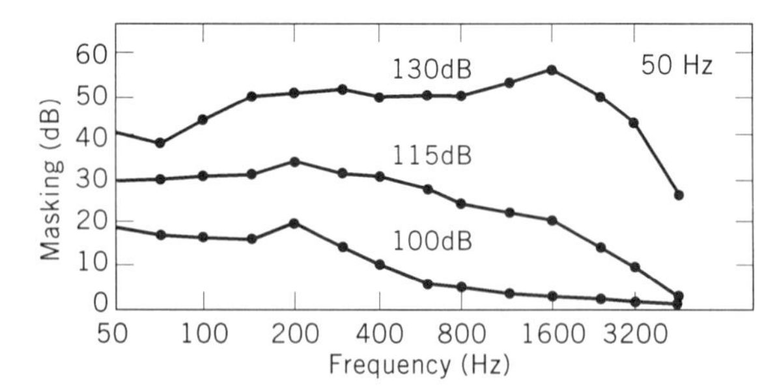

Fig. 1 Upward spread of masking produced by a 50-Hz pure tone at high intensities (100, 115, and 130 dB SPL). Note that the growth of masking is highly nonlinear, increasing at 1.6 kHz by more than 30 dB as the masking level is raised from 115 to 130 dB SPL. (From Ref. 7.)

already exists for any type of masker—pure tone, tonal complex, or narrow band of noise. Fig. 1 shows the masking pattern developed by a 50-Hz pure tone at 100, 115, and 130 dB SPL[7]; at the highest level, the pattern of excitation on the basilar membrane apparently covers the entire auditory range.

Remote masking by sounds above 100 dB has not been extensively studied beyond the original observations of Bilger and Hirsh.[6] Figure 2 shows their results for a masker that was a half-octave band of noise and illustrates the characteristics of spread of masking mentioned above.

The spread of cochlear activity at high intensities probably accounts for the fact that the masking of speech

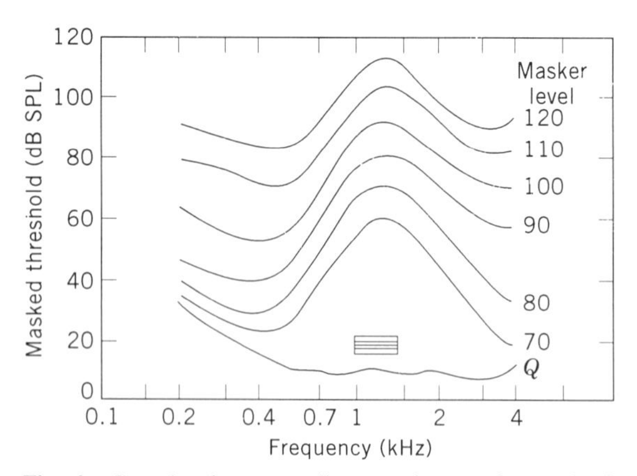

Fig. 2 Growth of remote, direct, and upward spread of masking with level of masker. The ordinate indicates the absolute thresholds at various frequencies (abscissa) as a function of masker level in decibels of SPL (parameter). Thresholds in quiet are represented by Q. The dark bar indicates the frequency range of the half-octave masker (1000–1420 Hz). (After Ref. 6.)

produced by loud noises depends very little on the precise spectral characteristics of the sound unless it is a pure tone. Without amplification, speech cannot be understood in any typical 90-dB(A) noise, even if the talker is shouting, unless the listener's ear is only a foot or so from the talker.[8] Even with amplification to keep the overall speech-to-noise ratio constant, speech intelligibility declines once the level exceeds 100 dB(A), dropping rapidly above 120 dB(A).[9]

2.3 Temporal Changes in Masking

A difference exists between types of masking in terms of the stability of masking by a sustained high-intensity sound. Direct masking shows essentially no change in masked threshold of a tone with time; remote masking, both ipsilateral and contralateral, displays gradual improvement (less masking) during a period of several minutes; however, speech intelligibility appears to deteriorate over at least the first 15 min of exposure to a broadband noise at 125–130 dB SPL.[9] The reason for these differences is not completely clear. Logically, direct masking would be expected to be invariant with time, since any time-linked process that would affect the noise should also affect the signal equally. The gradual loss of intelligibility of speech may be associated with the development of pronounced auditory fatigue. However, the decrease of remote masking with time has no simple explanation. A gradual relaxation of the middle-ear-muscle reflex, allowing an increasing amount of energy at the low frequency of the test signal to enter the cochlea, would produce an effect in the indicated direction; however, a change in remote masking is observed in auditory systems in which the middle-ear muscles are either excised[10] or paralyzed.[11] Furthermore, this change in remote masking occurs with sounds that are not intense enough to arouse the middle-ear muscles. Thus remote masking may be primarily caused by the cochlear excitation associated with distortion products developed in the ear, and its change with time should reflect some change in these distortion products, the nature of which is still obscure. However, the efferent system that may produce changes in the mechanical characteristics of the organ of Corti with sustained stimulation[12] is probably involved.

2.4 Physiological Reactions

Sudden intense sounds will evoke the startle response: an immediate contraction of the orbital eye muscles and the flexor muscles of the legs, arms, and back, manifested as an eyeblink and a crouching movement.[13] This startle response may be followed by an orienting reflex in which the head and eyes turn toward the sound source. The magnitude of the startle response is related directly to the loudness of the sound and inversely to its rise time but is also affected by its unexpectedness and by the level of the background ambient noise.[14,15] Davis et al.[14] found that for 2-s bursts of a 1-kHz tone the muscular response increased dramatically between levels of 90 and 120 dB. Repetition of the sound will usually result in a reduction of the response, although such habituation may not become complete. Even when no overt movement can be seen, transient muscular activity detected by electrophysiological techniques may interfere with an ongoing motor task involving those muscles. This lack of complete habituation, especially to gunfire involving peak levels of 140 dB or over, is probably responsible for the successful promotion of the use of hearing protectors at shooting ranges. Shooters accept earplugs and muffs not so much for protection against hearing loss as for an improvement in accuracy that occurs when the reports from other shooters' guns are attenuated enough that they no longer produce an involuntary flinch.

If the intense sound continues beyond a few minutes, other physiological changes will be manifested. Some of these are clearly linked to the startle response in the sense that the change was initiated by the abrupt onset of the sound but could be measured only several seconds later; these include a short-lived increase in blood pressure and heart rate and a decrease in rate of breathing.[14] Other changes, however, are more persistent. Of these, the two most easily measured and hence most extensively studied are a vasoconstriction of the extremities, indicated by measurements of finger volume, and a dilation of the pupil of the eye.[16] Others include decreased skin resistance and various biochemical and endocrinological reactions. Most of these are among those included in the "stress reaction," so their occurrence indicates that high-intensity sound can be considered to be a stressor. However, the magnitude of the observed effect (changes in characteristics of the circulatory systems, including blood components such as leukocytes, eosinophils, and basophils; composition of urinary excretions of endocrinic substances such as epinephrine, norepinephrine, and other corticoids) associated with sound exposures that do not cause hearing loss lie within the range of normal variability of these indicators in everyday life.[17] So it is not surprising that, despite innumerable studies of these stress-induced reactions, it has still not been shown convincingly that sound alone can produce any permanent aftereffects on the autonomic or circulatory systems.

2.5 Tactile Effects

At high levels, sounds will produce tactile sensations in the middle ear, the nature of which is relatively independent of the spectral characteristics (there are no tactile receptors in the inner ear). Early systematic studies of this phenomenon indicated that "discomfort" usually began at about 120 dB SPL, "tickle" at 130 dB SPL, and "pain" at 140 dB SPL.[18] However, individual differences in discomfort are large, with some persons reporting sounds to be "uncomfortably loud" at 80 or 90 dB SPL, although the average normal listener needs about 105 dB.[19] Totally deaf individuals without eardrums reported no pain even when exposed to jet noise at 170 dB SPL,[20] so the eardrum may be the primary source of pain.

Explosions may rupture the eardrum and/or disrupt the ossicular chain. Such events do not occur under controlled conditions in humans, so the sound levels involved have not been measured; however, the evidence indicates that peak levels under 170 dB SPL will not cause this form of damage.

2.6 Vestibular Effects

Because the vestibular sense organ (the semicircular canals) is connected to the cochlea, pressure transmitted to the cochlea will also reach the semicircular canals. High intensities of sound might therefore be expected to affect the sense of balance. However, nystagmus, vertigo, and loss of balance occur only in levels of 140 dB SPL or greater[21] as long as the stimulation is bilaterally symmetrical. If only one ear is exposed, on the other hand, balance can be affected by sound levels of 100 dB or lower.[22]

2.7 Annoyance

Any sound made by others may generate annoyance in a particular individual. The magnitude of that annoyance does depend to some extent on the level of the sound: On average, the higher the level, the greater the interference with speech communication, mental concentration, task performance, recreational activities, and sleep and hence the greater the annoyance. However, there are so many other factors that determine the annoyingness of a sound (its spectrum, duration, and temporal pattern; the time of day; perceived avoidability, expectedness, meaningfulness, inter alia) that the role of level per se remains expressible only in qualitative terms. Indeed, widespread agreement does not exist on how to quantify the degree of annoyance, so that it is expressed in terms of the fraction of a population that is "highly annoyed".[23] It seems safe to say that any sound regarded as intense according to the present definition (90 dB SPL or higher for pure tones, 75 dB or above for noises) will become "highly annoying" to the average person if it is sufficiently prolonged.[24]

2.8 Miscellaneous Concomitant Effects

When one is exposed to the noise from an immobilized jet aircraft (generating 140 dB SPL near the exhaust from the engine or 160 dB SPL if the afterburner is operating), even with hearing protectors, numerous effects associated with the direct action of high-intensity low-frequency components of the noise on the body become apparent. Vision may be blurred because of vibration of the eyeballs, the teeth may ache, the sinuses "buzz," and the chest cavity and viscera are clearly being shaken at different rates. However, these intersensory effects occur only in sound levels that are so high that a few minutes of exposure would produce irreversible damage to the auditory system, so they have not been studied systematically.

3 AFTEREFFECTS OF INTENSE SOUND

The concomitant effects of high-intensity sounds are largely independent of their duration. This is not true of aftereffects, however. Duration of exposure is always an essential parameter, especially in the case of production of hearing loss. Although there is a persistant popular belief that there is an intensity of sound above which permanent damage to the auditory system will result, regardless of the duration of the sound, no empirical support for this belief exists: Any level that is "critical" in this sense is critical only for a particular duration.[25] On the other hand, duration and level are not the only parameters that must be specified in establishing a relation between a given exposure to high-intensity sound and its aftereffects. The spectral distribution and temporal pattern of the sound are also important. Thus the adoption by the International Standardization Organization (ISO) of the definition of "exposure" (E) as the integral of the square of the A-weighted sound pressure over the time concerned,

$$E = \int_{t_1}^{t_2} p_A^2(t)\, dt,$$

represents a drastic oversimplification of a complex issue.[26]

The problem here is that this definition embodies three assumptions: two that are demonstrably false—that spectra are equalized in effectiveness by the use of A weighting and that temporal pattern is irrelevant—and one assumption that is true only for a restricted set of conditions—that the total energy (the product of intensity and time, It, since I is proportional to p^2) involved in an exposure is the sole determining factor in predicting its effect. For this reason, the ISO definition of exposure will be rejected here. The term *exposure* will retain its original meaning: the entire history of pressure during a specified time period measured at the position of the listener's ear but with the listener absent.

3.1 Temporary Threshold Shift

The most well-known aftereffect of exposure to high-intensity sound is the change in auditory sensitivity. If the auditory threshold is measured before and after an exposure, the difference in hearing threshold levels (HTLs) is by definition the threshold shift (TS). If the shift later completely disappears, then it was a temporary threshold shift (TTS); if not, the final shift is termed a permanent threshold shift (PTS), and the initial shift was a compound threshold shift (CTS), that is, a combination of TTS and PTS.

It would be desirable to categorize threshold shifts in terms of the underlying physiological changes. For example, all of the following could conceivably be involved in TTS: (1) residual middle-ear-muscle activity; (2) displacement of the tectorial membrane relative to the basilar membrane; (3) changes in the chemical environment of the hair cells; (4) swelling of the hair cells, making stimulation more difficult in a mechanical sense; (5) an increase in internal noise, as for example due to increased blood flow ("pounding in the ears") or an audible tinnitus; (6) changes in or results of efferent activity at the basilar membrane; and of course (7) ordinary poststimulatory decrease in nerve excitability that could occur in the eighth nerve, cochlear nucleus, lateral lemniscus, inferior colliculus, medial geniculate, or acoustic cortex. However, until more knowledge about the relative contribution of these factors is available, classification of TTSs may be based on the duration of their combined effect: less than 1 s, 1 s to 2 min, 2 min to 16 h, and greater than 16 h.[27]

A TTS that lasts less than 1 s has been called *residual masking* or *forward masking*, although there is no reason to think that masking is actually involved (see Chapter 93). The magnitude of this ultra-short-term TTS, measured immediately after exposure to a short one, is proportional to the SPL of the exposure but relatively independent of its duration. The recovery process is exponential, in the sense that the TTS in decibels decreases linearly with the logarithm of recovery time.[28]

A TTS lasting a minute or so will follow a short exposure (a few minutes) at levels up to 80 dB SPL but is nearly independent of that level; that is, the magnitude of the shift is the same following exposure at 30 dB as at 80 dB. This variety of TTS, *low-level TTS*, is critically dependent on the procedure for determining threshold; it disappears rapidly if an interrupted test tone is used.

Intense sounds, if sustained long enough, will produce "ordinary" TTS, also called *physiological fatigue*: a shift that requires up to 16 h to disappear, with a course of recovery that, as in the case of forward masking, is linear in the logarithm of time (recovers exponentially).

The first minute or so following intense sound generally involves a "rushing noise" tinnitus that operates in the opposite direction of an objective noise. When the noise seems loudest, the threshold is lowest, and when the noise disappears (leaving a striking silence), the threshold is elevated. Recent work on this paradoxical noise indicates that it does not originate only in the cochlea but may involve higher centers of the auditory chain.[29]

Finally, TTSs that persist for more than 16 h, sometimes termed *pathological fatigue*, are produced by more severe exposures. This TTS recovers linearly in time (i.e, a certain number of decibels per day) rather than in log time, suggesting that the TTS in this case has a different physiological correlate. As much as 3 weeks can be required for this "delayed recovery" to be complete.[30]

In all cases, individual differences in the magnitude of the TTS caused by a particular exposure are huge. The problem of individual variability and the prediction of susceptibility will be discussed later. However, because of the possibility that TTS produced by a moderate exposure to a given sound spectrum might be correlated with the PTS resulting from a subsequent more severe exposure (longer, more intense, or given repeatedly), extensive study of ordinary TTS has provided knowledge about the exposure parameters involved in its generation.

TTS from Continuous Exposures Ordinary TTS is customarily measured 2 min after termination of exposure, that is, after the shorter lived recovery processes, which may produce a diphasic recovery curve, have run their course. For a continuous steady sound, this TTS_2 grows linearly with the logarithm of the exposure duration until it reaches an asymptote [asymptotic threshold shift (ATS)] after about 8 h of exposure. Also, TTS_2 grows linearly with sound level once the level has exceeded a base value below which only short-duration effects are produced, no matter how long the exposure. This base SPL is termed *effective quiet* (EQ), and the difference between EQ and a given SPL may be called the *effective level* (EL). EQ is about 85 dB SPL for pink noise and 70–80 dB for sounds an octave or narrower in bandwidth[31]; the exact value mirrors the transfer function of the outer and middle ears, being lowest for sounds in the 2–4-kHz region, where the resonance of the outer ear canal provides a 10–15-dB amplification of the measured field SPLs. For the typical industrial noise, EQ will probably have a value of 80 dB SPL; the ATS_2 produced by 8 h exposure at 100 dB SPL (EL = 100 − 80 = 20 dB) will be twice that produced by one at 90 dB (EL = 10 dB) and half of that produced by 120 dB SPL (EL = 40 dB). Since ATS grows at the rate of 1.7 dB per decibel of EL[32], the typical industrial noise will produce an ATS of 17 dB at 90 dB and 34 dB at 100 dB.

TTS from Varying Levels Prediction of the TTS produced by time-varying and intermittent noise exposures is somewhat complicated, because of partial recovery of hearing during the quieter periods. However, for relatively rapid interruptions or fluctuations (periods at a given level of 5 min or less), the TTS is proportional to the average effective level during the exposure, that is,

$$TTS_2 \propto \frac{1}{T} \int_0^t EL(t)\, dt.$$

Thus for the typical industrial noise, the same TTS_2 will be produced by 8 h exposure to a 90-dB steady noise, a 100-dB noise that is on half the time and off half the time, or a noise that changes at regular intervals from 80 to 90 to 100 dB and back to 80, or a noise that fluctuates irregularly but for which the average number of decibels by which 80 dB is exceeded is 10 dB.[33]

Temporary threshold shifts produced by patterns of longer exposures and rest periods can be calculated by means of empirical equations describing alternate growth and recovery of TTS_2. In this case also, however, the TTS is considerably less than the TTS that would have been produced by a steady noise over the same time but whose level was such as to keep the total energy constant [$L_{eq}(t)$]. For example, in the case of the noise that is on half the time at 100 dB and off the other half, L_{eq} is 97 dB, yet it produces only the same TTS as a 90-dB noise. Thus TTS is not a consistent function of the total energy of an exposure; how the energy is distributed in time makes a considerable difference in the magnitude of the TTS produced.

TTS from Impulse Noise The same is true for TTS from impulse noise, such as gunfire. Although 30 impulses of simulated gunfire at 150 dB SPL peak level may produce a TTS_2 of 20 dB in a particular ear, 300 impulses of the same pulse shape but at 140 dB SPL (i.e., the same total energy) will usually produce no TTS_2 in this individual.[34] Here it appears that there is a critical level (which, however, is a function of pulse duration) below which no effect is produced and above which the TTS increases with level. Unlike the growth of TTS_2 with time for steady or interrupted noises, the TTS_2 from impulses appears to be proportional to the number of impulses, as if each impulse produced the same amount of TTS in decibels.[35] For this reason, no attempts have been able to determine the point at which the TTS from repeated impulse noise reaches an asymptotic value.

Delayed Recovery The foregoing discussion assumes that TTS_2 does not exceed 40 dB; if this value is reached, or if even somewhat lower values of TTS_2 are the result of prolonged intermittent exposure to intensities of 110 dB or higher, recovery may shift to the "delayed" mode, in which little recovery occurs during the first hours after exposure and more than 16 h is required for full recovery.[36] It has long been suspected that delayed recovery indicates damage that is more likely to become irreversible if the next exposure occurs before full recovery,[37] although no substantial supporting evidence has been advanced.

3.2 Permanent Threshold Shift

The most undesirable aftereffect of exposure to high-intensity sound is PTS. Sound-induced PTS is commonly divided into two categories depending on whether the loss was produced by a single short exposure at a very high level (acoustic trauma) or by repeated longer exposures at more moderate levels. It is clear from animal studies that in acoustic trauma the inner ear has been subjected to such stress that its elastic limit, so to speak, has been exceeded. Various structures of the organ of Corti, including the hair cells, may become partly or wholly detached, and one or more of the several membranes in the cochlea may be ruptured, allowing intermixture of fluids of different composition, thereby poisoning the hair cells that survived the mechanical stress. The end result is a pronounced loss of sensitivity at the frequencies correlated with the locus of this destruction.

The physiological correlates of gradually induced PTS are not as clear. A fairly large PTS may exist despite a near-normal appearance of the cochlear structures, although the locus of what damage does exist will generally correspond to the frequencies showing the greatest PTS. Clearly, subtle defects in the morphological, chemical, and neurological characteristics of the cochlea are involved; these have been summarized recently by Saunders et al.[38]

Acoustic Trauma Little is known about acoustic trauma in humans, although it is not at all rare. Victims of acoustic trauma seldom have had a recent audiogram that would enable the amount of PTS to be determined with certainty. Even then, the exposure level and duration can usually only be grossly estimated. Finally, differences among people in susceptibility to damage are so great that single cases show that acoustic trauma is possible from a given exposure but not that it is inevitable or even likely.

The difference between likelihood of safety and possibility of damage is illustrated in Fig. 3. The circled symbols indicate the most severe single exposures that have been administered under controlled conditions to a group of at least five normal-hearing young men but that produced no PTS, while the symbols in squares show the estimated level and duration of exposures that apparently produced a PTS of 15 dB or more in at least one single individual. The probably safe exposures were all to broadband noise. The exposures included in the figure are from the classic studies of Davis et al.[39] on the effect of noise: point *D*, 32 min at 130 dB; point *E*, 1 min at 135 dB[40]; and point *SN*, 0.4 s at 153 dB, an exposure designed to be equivalent to the sound produced by the opening of an air bag.[41]

By contrast, a PTS of 15 min or more followed 1 min exposure to a 2-kHz tone at 130 dB (point *M*) or 8 min exposure to a 4-kHz tone at 120 dB SPL (point *S*) in one of several men so exposed in the Davis et al. study.[39] Davis himself received a 30-dB increase in his preexisting high-frequency hearing loss after exposure for 20 min to a 500-Hz tone at 140 dB SPL (point *H*). Point *O* indicates the effect of the ring of a cordless telephone in which the same transducer was used for the ringer and for voice, an exposure that has produced a measurable PTS in some small but unknown fraction of those individuals exposed,[42] and point *L* is an exposure of "a few seconds" to a tone of around 138 dB SPL that was being used routinely in an attempt to elicit the acoustic reflex, one that produced additional damage in two individuals who already had considerable loss.[43]

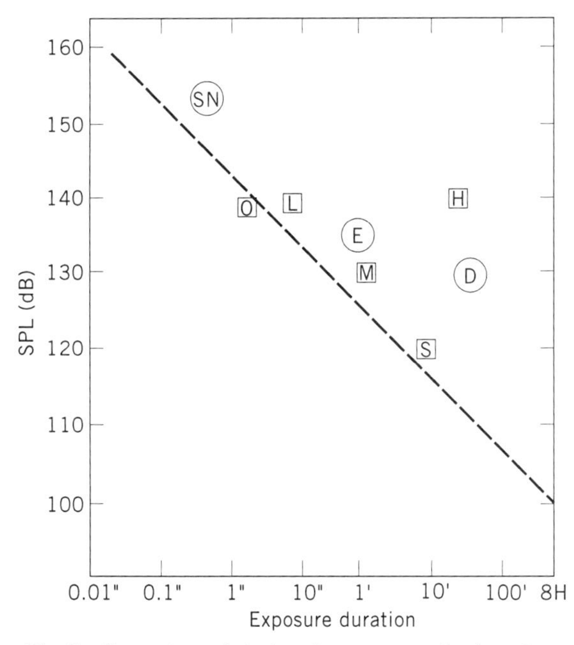

Fig. 3 Comparison of single noise exposures that have been shown to be without hazard to the average young healthy ear (circled symbols) and those that have apparently produced some permanent hearing loss in at least one person (symbols in squares). The dashed line, representing 8 h of exposure at 100 dB SPL or its energy equivalent, divides single exposures that are usually innocuous from those that are capable of causing permanent damage in the most highly susceptible individuals. See text for identification of data sources.

In Fig. 3, the spectrum of the various sounds is not considered. However, the difference between probably safe and possibly damaging points would be reduced if the ordinate were changed to "octave-band SPL," as the circled points would be reduced by a few decibels. In this case, the data are not in conflict with the generalization that acoustic trauma in humans is highly unlikely unless the single uninterrupted exposure exceeds 145 dB for 1 s or its energy equivalent, which is defined by the dashed line in Fig. 3.

Thus for single continuous steady exposures, the energy theory seems tenable: Exposures of equal energy give rise to approximately equal effects. Unfortunately, this equivalence breaks down for multiple exposures or a fluctuating sound level.

Gradual Growth of PTS Even more common and less understood than acoustic trauma is the gradual growth of hearing loss that accompanies repeated moderate exposures to sound, exposures that apparently only produce TTS. Cross-sectional studies of workers exposed daily to steady high-level industrial noise show that their average hearing is less sensitive than that of individuals who have quiet workplaces but whose age and exposure to noises outside the work situation are the same as for the noise-exposed workers. This difference in HTLs, which is by definition the INIPTS (industrial-noise-induced permanent threshold shift), grows rapidly with exposure time but reaches an asymptote in a few years. The terminal PTS increases linearly as level exceeds some base value (a level analogous to the EQ in the development of TTS) and is greater for the 3–6-kHz region than for lower frequencies.

Figure 4, from a synthesis by Passchier-Vermeer[44] of the literature involving continuous steady noises, shows the inferred INIPTS at the various audiometric frequencies as a function of workplace A-weighted sound level after 10 years of exposure. As a first approximation, it appears that (1) 80 dB(A) is the level that can be regarded as being innocuous; (2) 85 dB(A) will result

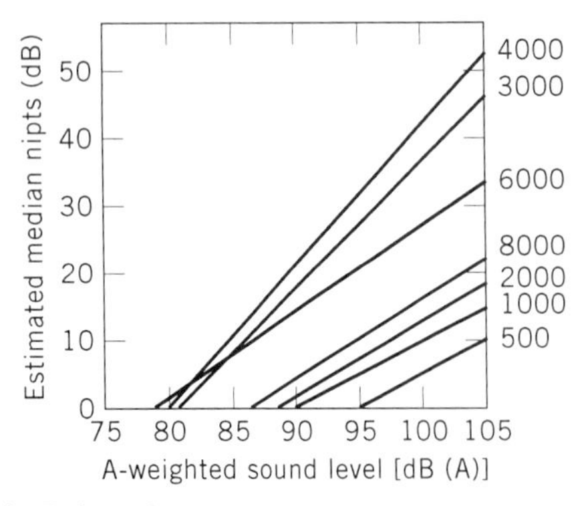

Fig. 4 Estimated average noise-induced permanent threshold shifts (ordinate) at various frequencies (parameter) after 10 years of exposure (8 h/day, 5 days/week) to steady A-weighted sound levels of 80–105 dB (abscissa). (After Ref. 44.)

in a loss of about 10 dB at the most noise-sensitive frequencies of 3, 4, and 6 kHz (10 dB is the smallest loss in the individual case that could be regarded as meaningful); (3) 90 dB(A) will generate a loss of 15–20 dB at these frequencies, although the traditional speech frequencies (0.5, 1, and 2 kHz) are still largely unaffected; only above 90 dB(A) does the noise adversely affect these lower frequencies; and (4) noises of 100 dB(A), which are still common, will produce severe losses at high frequencies and moderate losses at the low frequencies.

Thus Fig. 4 implies that 8 h exposure to a 90-dB(A) noise (plus, of course, whatever noise exposure is experienced outside the work situation) is capable of producing PTS when repeated daily for many weeks, even though the TTS at the end of a single workday will ordinarily recover completely. The value of TTS_2 of 17 dB that was estimated earlier to be the result of 8 h exposure to 90 dB(A) is just about equal to the 20 dB of PTS indicated in Fig. 4. This fact has lent support to the hypothesis that a daily exposure that produces a given TTS_2 in normal ears will eventually produce that same amount of PTS.[45] However, whether this agreement represents mere coincidence or proves a causal relation is still unclear.

There are two main hypotheses about how PTS grows with time in the individual. The more well-known theory posits that each day's exposure causes a change in the cochlea, but a change so small that it is undetectable by the rather gross and inherently variable audiometric procedures. This could be characterized as the *cumulative microtrauma* theory.[46] A 20-dB INIPTS displayed by a worker after 2 years of working in a 90-dB(A) noise (500 working days) could be regarded as the cumulative effect of 500 equal exposures, each of which produced a loss of gradually decreasing magnitude (e.g., the first exposure producing a loss of 0.30 dB, the second a loss of 0.28 dB, and so on, until perhaps the 500th exposure added only 0.001 dB).

An equally plausible scenario, however, rests on the fact that those 500 daily exposures were not identical. The exposure will vary from day to day not only during the 8-h workday but also during the intervening 16 h. It is possible, then, that damage will increase only when the total daily or weekly exposure is for some reason more severe than any preceding one. This might be termed the *progressive macrotrauma* theory.

Which of these two theories is more valid can be determined only by longitudinal studies in which the growth of PTS or some other indicator of permanent damage can be followed in individual experimental animals whose daily exposures can be held constant. At the moment, no clear decision has been reached; a study of guinea pigs indicated a gradual progressive growth of damage,[47] while chinchillas showed as much PTS after 1 week of exposure as after an additional 8 weeks.[48]

A conclusion regarding INIPTS that has been studied extensively enough that it can be regarded as a true principle is that INIPTS does not increase if noise exposure ceases. That is, INIPTS will not increase if a worker is removed from the noise.

Interrupted and Intermittent Exposures The damage produced by a schedule in which daily exposures of up to 8 h are separated (interrupted) by recovery periods of 16 h or more is less than that produced by an uninterrupted exposure. A further reduction of PTS in the chinchilla is effected if the daily exposure itself is intermittent rather than continuous,[49] just as in the case of TTS, although the reduction is not as great. The reduction in damage is greater, the shorter the noise bursts or the longer the intervening periods of relative quiet, although the former appears to be the more important factor.

3.3 Other Measures of Auditory Damage

As indicated earlier, physiological damage to the cochlea by high-intensity sound is not necessarily reflected in a measurable PTS. Evidence from both animal and human studies[50] implies that several hundred of the hair cells that have been presumed to be important in the process of hearing may be destroyed before a change in threshold is measurable. This clearly casts doubt on the adequacy of threshold sensitivity as the only valid index of damage. To date, however, efforts to find an indicator that is more sensitive to physiological damage have been unsuccessful, although many have been studied, including changes in the loudness, pitch, or maskability of suprathreshold tones, in difference limens for frequency and intensity, in temporal integration, and in strength of cochlear emissions. Although these indices are often abnormal when a threshold shift, either temporary or permanent, can be measured, they appear to return to normal if the threshold does.

Tinnitus A frequent aftereffect of intense stimulation is a "ringing in the ears." A tinnitus of this sort follows almost any impulse noise such as gunfire whose peak levels are above 150 dB; its duration ranges from a few seconds to several hours, depending on the severity of the exposure. Permanent tinnitus often but not universally accompanies acoustic trauma, and it may occur in an ear whose thresholds appear essentially normal.

3.4 Other Physiological Aftereffects

Cardiovascular System As mentioned earlier, there is only equivocal evidence that the changes in the cardiovascular system associated with high sound levels eventually lead to a significant enduring effect. While some epidemiological studies indicate a significant correlation between history of noise exposure or degree of hearing loss and hypertension or blood cholesterol content, such a relation has never been demonstrated when high-intensity sound is the only stressor involved. For example, although the incidence of cardiovascular disease was found to be higher than normal in steel mill workers,[51] their working environment included high temperature and noxious fumes as well as high sound levels, so the contribution to the coronary problems by the noise alone is indeterminate.

Miscellaneous Similarly, there is no convincing evidence that high-level sound alone can directly or even indirectly cause gastric ulcers, narrowing of the visual field, insomnia, abnormal somnolence, sexual impotence, birth defects, insanity, and death, inter alia, despite claims of antinoise alarmists whose allegations are amplified by the popular press. Of all of mankind's afflictions, only cancer and Alzheimer's disease seem never to have been attributed to noise exposure.

3.5 Susceptibility and Vulnerability

A given exposure, however it is measured, does not produce the same hearing loss in every ear. A particular ear may be more or less susceptible to damage—either to PTS or to TTS, which is assumed to be a precursor of PTS—because of relatively fixed structural and biochemical characteristics of the hearing mechanism and also more or less vulnerable to damage at a particular time because of temporary changes in one or more of those characteristics.

It is obvious that any factor that determines how much sound at a particular frequency reaches the inner ear will influence susceptibility, such as the dimensions of the pinna and the length of the auditory canal, the area and density of the eardrum, the mass of the ossicular chain, and the strength of the middle-ear muscles. Indeed, a conductive hearing loss renders the ear less susceptible to noise-induced hearing loss.[52] In the cochlea itself, one would expect characteristics such as the stiffness of the cochlear partition, the thickness and elasticity of the various membranes, and the adequacy of the blood supply, if these features could be measured, to be important in determining susceptibility to damage, either directly or, through their influence on recovery processes, indirectly. For example, there is some evidence that blue-eyed persons display higher thresholds than brown-eyed persons, which is explained if the melanin in the eye is highly correlated with the melanin in the cochlea and if the latter makes the cochlea more rigid.

Vulnerability depends on more labile characteristics of the auditory system. For example, an ear is less vulnerable to damage from sound when ear wax is blocking the ear canal or when a common cold has caused accumulation of fluid in the middle ear.[53] Vulnerability is higher in persons taking certain drugs that are themselves potentially ototoxic, such as kanamycin, neomycin, and streptomycin.

Although neither a particular exposure to noise nor a specific small dose of a drug may produce a measurable loss of hearing, their joint action may do so. Such synergistic action with noise has been demonstrated for only a few other substances or conditions, although dozens have been studied. Vulnerability to hearing damage seems to be affected by heat, cisplatin, aspirin, simultaneous whole-body vibration, and inhalation of carbon monoxide, although only the last named has been demonstrated to produce a true synergistic action.[54] Although attempts have been made to link susceptibility and vulnerability to general health, smoking habits, use of social drugs or stimulants, artificial food additives, industrial chemicals of various kinds, exercise, poor posture, and even mental attitude toward the source of sound, no reliable relation has yet emerged from these studies.[55]

A search for medications or procedures that will reduce TTS and PTS has been similarly unproductive. Although claims have been made that substances such as vitamin A, vitamin C, nicotinic acid, procaine, nylidrin, adenosine triphosphate, brain cortex gangliosides, gingko-biloba extract, amino-oxyacetic acid, and dextran have reduced PTS from acoustic trauma, subsequent studies failed to confirm this amelioration. The most promising substance at this time is carbogen (95% oxygen and 5% carbon dioxide), whose inhalation has been shown to reduce both TTS and PTS.[56,57]

Attempts to predict susceptibility by means of a single test in order to identify persons who will be at greatest risk have met with only limited success. It has always seemed reasonable to expect that the best predictor of PTS caused by extended exposure to a particular noise should be the TTS produced by a short exposure to that same noise, because both seem to depend on the same factors. However, studies in which both TTS and PTS were measured have generally shown correlations that, although statistically significant, are so low that they are of only minor practical importance, TTS accounting for only 15% of the variance in PTS among individuals.[58]

Although women on average have better hearing than men, even when working in the same noise, it is not clear whether this difference is due to some sex-linked structural difference or merely reflects differences in exposure to high-intensity sounds outside the workplace (socioacusis) and to other deleterious influences such as blows to the head and diseases (nosoacusis). In any event, the difference is not great enough to justify recommending that only women be employed in noisy occupations.

Age, experience, and hearing status all appear to be irrelevant as predictors of susceptibility. Despite many studies of these characteristics, it has not yet been demonstrated that the older person is more or less susceptible than the younger person, that those who work in noise gradually become more resistant to damage, or that workers with damaged hearing become more susceptible to further damage because of the injury.

REFERENCES

1. T. L. Wiley and R. S. Karlovich, "Acoustic-Reflex Response to Sustained Signals," *J. Sp. Hear. Res.*, Vol. 18, 1975, pp. 148–157.
2. W. D. Ward, "Studies on the Aural Reflex. III. Reflex Latency as Inferred from Reduction of Temporary Threshold Shift from Impulses," *J. Acoust. Soc. Am.*, Vol. 34, 1962, pp. 1132–1137.
3. G. R. Price, "Upper Limit to Stapes Displacement: Implications for Hearing Loss," *J. Acoust. Soc. Am.*, Vol. 56, 1974, pp. 195–197.
4. J. J. Guinan, Jr. and W. T. Peake, "Middle-Ear Characteristics of Anesthetized Cats," *J. Acoust. Soc. Am.*, Vol. 41, 1967, pp. 1237–1261.
5. G. von Bekesy, "Zur Physik des Mittelohres und über das Hören bei fehlerhaften Trommelfell," *Akust. Z.*, Vol. 1, 1936, pp. 13–23.
6. R. C. Bilger and I. J. Hirsh, "Masking of Tones by Bands of Noise," *J. Acoust. Soc. Am.*, Vol. 28, 1956, pp. 623–630.
7. A. Finck, "Low-Frequency Pure Tone Masking," *J. Acoust. Soc. Am.*, Vol. 33, 1961, pp. 1140–1141.
8. J. C. Webster, "Noise and Communication," in D. M. Jones and A. J. Chapman (Eds.), *Noise and Society*, Wiley, London, 1984, pp. 185–220.
9. I. Pollack and J. M. Pickett, "Masking of Speech by Noise at High Sound Levels," *J. Acoust. Soc. Am.*, Vol. 30, 1958, pp. 127–130.
10. R. C. Bilger, "Remote Masking in the Absence of Intra-aural Muscles," *J. Acoust. Soc. Am.*, Vol. 39, 1966, pp. 103–108.
11. W. D. Ward, "Further Observations on Contralateral Remote Masking and Related Phenomena," *J. Acoust. Soc. Am.*, Vol. 42, 1967, pp. 593–600.
12. M. C. Liberman, "Structural Basis of Noise-Induced Threshold Shift," in *Proceedings of the Fifth International Congress on Noise as a Public Health Problem*, Vol. 4, Stockholm, Sweden, 1990, pp. 17–30.
13. C. Landis and W. A. Hunt, *The Startle Pattern*, Farrar and Rinehart, New York, 1939.
14. R. C. Davis, A. M. Buchwald, and R. W. Frankman, "Autonomic and Muscular Stimuli and Their Relation to Simple Stimuli," *Psych. Monogr.*, Vol. 69, No. 405, 1955.
15. R. Rylander and A. Dancer, "Startle Reactions to Simulated Sonic Booms: Influence of Habituation, Boom Level and Background Noise," *J. Sound Vib.*, Vol. 61, 1978, pp. 235–243.
16. G. Jansen, "Physiological Effects of Noise," in C. M. Harris (Ed.), *Handbook of Acoustical Measurements and Noise Control*, 3rd ed., McGraw-Hill, New York, 1991, pp. 25.1–25.19.
17. M. Loeb, *Noise and Human Efficiency*, Wiley, New York, 1986, pp. 140–161.
18. S. R. Silverman, "Tolerance for Pure Tones and Speech in Normal and Defective Hearing," *Ann. Otol. Rhinol. Laryngol.*, Vol. 56, 1947, pp. 658–677.
19. R. A. Bentler and C. V. Pavlovic, "Comparison of Discomfort Levels Obtained with Pure Tones and Multitone Complexes," *J. Acoust. Soc. Am.*, Vol. 86, 1989, pp. 126–132.
20. H. W. Ades, A. Graybiel, S. Morrill, G. Tolhurst, and J. Niven, "Non-auditory Effects of High Intensity Sound Stimulation on Deaf Human Subjects," *J. Aviat. Med.*, Vol. 29, 1958, pp. 454–467.
21. E. D. D. Dickson and D. L. Chadwick, "Observations on Disturbances of Equilibrium and Other Symptoms Induced by Jet Engine Noise," *J. Laryngol. Otol.*, Vol. 65, 1951, pp. 154–165.
22. C. S. Harris, "Effects of Increasing Intensity Levels of Intermittent and Continuous 1000 Hz Tones on Asymmetrical Stimulation," *Perc. Motor Skills*, Vol. 35, 1972, pp. 395–405.
23. P. N. Borsky, "Review of Community Response to Noise," in *Proceedings of the Third International Congress on Noise as a Public Health Problem*, Freiburg, West Germany, ASHA Reports No. 10, 1980, pp. 453–474.
24. T. Schultz, "Synthesis of Social Surveys on Noise Annoyance," *J. Acoust. Soc. Am.*, Vol. 64, 1978, pp. 377–405.
25. W. D. Ward and D. Henderson, "The Dependence of Critical Level on Time," in *Proceedings of the Twelth International Congress on Acoustics*, Vol. 1, Toronto, Canada, July 24–31, 1986, N-4.3.

26. International Standardization (ISO), "Acoustics—Determination of Occupational Noise Exposure and Estimation of Noise-Induced Hearing Impairment," ISO 1990-01-15, ISO, Geneva, Switzerland, 1990.
27. W. D. Ward, "Adaptation and Fatigue," in J. Jerger (Ed.), *Modern Developments in Audiology*, Academic, New York, 1973, pp. 301–344.
28. O. Bentzen, "Investigations on Short Tones," Ph.D. Dissertation, Aarhus University, Denmark, 1953.
29. S. Kemp and R. N. George, "Masking of Tinnitus Induced by Sound," *J. Speech Hear. Res.*, Vol. 35, 1992, pp. 1169–1179.
30. W. D. Ward, "Recovery from High Values of Temporary Threshold Shift," *J. Acoust. Soc. Am.*, Vol. 32, 1960, pp. 497–500.
31. W. D. Ward, E. M. Cushing, and E. M. Burns, "Effective Quiet and Moderate TTS: Implications for Noise Exposure Standards," *J. Acoust. Soc. Am.*, Vol. 59, 1976, pp. 160–165.
32. J. H. Mills, R. M. Gilbert, and W. Y. Adkins, "Temporary Threshold Shifts in Humans Exposed to Octave Bands of Noise for 16-24 Hours," *J. Acoust. Soc. Am.*, Vol. 65, 1979, pp. 1238–1248.
33. W. D. Ward, A. Glorig, and D. L. Sklar, "Temporary Threshold Shift Produced by Intermittent Exposure to Noise," *J. Acoust. Soc. Am.*, Vol. 31, 1959, pp. 791–794.
34. H. McRobert and W. D. Ward, "Damage-Risk Criteria: The Trading Relation between Intensity and the Number of Nonreverberant Impulses," *J. Acoust. Soc. Am.*, Vol. 53, 1973, pp. 1297–1300.
35. W. D. Ward, W. Selters, and A. Glorig, "Exploratory Studies on Temporary Threshold Shift from Impulses," *J. Acoust. Soc. Am.*, Vol. 33, 1961, pp. 781–793.
36. W. D. Ward, "Temporary Threshold Shift and Damage-Risk Criteria for Intermittent Noise Exposures," *J. Acoust. Soc. Am.*, Vol. 48, 1970, pp. 561–574.
37. A. Peyser, "Gesundheitswesen und Krankenfürsorge. Theoretische und experimentelle Grundlagen des persönlichen Schallschutzes," *Deut. Med. Wochenschr.*, Vol. 56, 1930, pp. 150–151.
38. J. C. Saunders, Y. E. Cohen, and Y. M. Szymko, "The Structural and Functional Consequences of Acoustic Injury in the Cochlea and Peripheral Auditory System: A Five Year Update," *J. Acoust. Soc. Am.*, Vol. 90, 1991, pp. 136–146.
39. H. Davis, C. T. Morgan, J. E. Hawkins, Jr., R. Galambos, and F. W. Smith, "Temporary Deafness Following Exposure to Loud Tones and Noise," *Acta Otolaryngol.*, Suppl. 88, 1950.
40. K. M. Eldred, W. J. Gannon, and H. Von Gierke, "Criteria for Short Time Exposure of Personnel to High Intensity Jet Aircraft Noise," WADC Technical Note 55-355, Wright Air Development Center, U.S. Air Force, Wright-Patterson AFB, OH, September 1955.
41. H. C. Sommer and C. W. Nixon, "Primary Components of Simulated Air Bag Noise and Their Relative Effects on Human Hearing," AMRL-TR-73-52, Aerospace Medical Research Laboratory, Wright-Patterson AFB, OH, November 1973.
42. D. J. Orchik, D. R. Schraier, J. J. Shea, Jr., J. R. Emmett, W. H. Moreta, and J. J. Shea III., "Sensorineural Hearing Loss in Cordless Telephone Injury," *Otolaryngol. Head Neck Surg.*, Vol. 96, 1987, pp. 30–33.
43. T. Lenarz and J. Gülzow, "Akustisches Innenohrtrauma bei Impedanzmessung. Akutes Schalltrauma?" *Laryngol. Rhinol.*, Vol. 62, 1983, pp. 58–61.
44. W. Passchier-Vermeer, "Hearing Loss Due to Exposure to Steady-State Broadband Noise," *J. Acoust. Soc. Am.*, Vol. 56, 1974, pp. 1585–1593.
45. A. Glorig, W. D. Ward, and J. Nixon, "Damage Risk Criteria and Noise-Induced Hearing Loss," *AMA Arch. Otolaryngol.*, Vol. 74, 1961, pp. 413–423.
46. D. W. Granendeel and R. Plomp, "Micro-noise Trauma?" *Arch. Otolaryngol.*, Vol. 71, 1960, pp. 656–663.
47. J. Herhold, "Über tierexperimentelle Untersuchungen zur Beurteilung der gehörschädigenden Wirkung des Lärms," Ph.D. Dissertation, Technische University, Dresden, 1979.
48. Z. Zhou and W. D. Ward, "Is the Damaged Ear More Susceptible?" *J. Acoust. Soc. Am.*, Vol. 87, Suppl. 1, 1990, p. S102.
49. W. D. Ward, "The Role of Intermittance in PTS," *J. Acoust. Soc. Am.*, Vol. 90, 1991, pp. 164–169.
50. G. Bredberg, "Cellular Pattern and Nerve Supply of the Human Organ of Corti," *Acta Otolaryngol.*, Suppl. 236, 1968.
51. G. Jansen, "Vegetative Lärmwirkungen bei Industriearbeitern," *Lärmbekampfung*, Vol. 6, 1962, pp. 126–128.
52. R. Nilsson and E. Borg, "Noise-Induced Hearing Loss in Shipyard Workers with Unilateral Conductive Hearing Loss," *Scand. Audiol.*, Vol. 12, 1983, pp. 135–140.
53. D. Y. Chung, "The Effect of Middle Ear Disorders on Noise-Induced Hearing Loss," *J. Am. Audit. Soc.*, Vol. 4, 1978, pp. 77–80.
54. L. D. Fechter, J. S. Young, and L. Carlisle, "Potentiation of Noise Induced Threshold Shifts and Hair Cell Loss by Carbon Monoxide," *Hearing Res.*, Vol. 34, 1988, pp. 39–48.
55. W. D. Ward, "Endogenous Factors Related to Susceptibility to Damage from Noise," in *Occupational Medicine: State of the Art Reviews: Occupational Hearing Loss*, Hanley and Belfus, Philadelphia, 1995.
56. S. S. Joglekar, D. M. Lipscomb, and G. E. Shambaugh, Jr., "Effects of Oxygen Inhalation on Noise-Induced Threshold Shifts in Humans and Chinchillas," *Arch. Otolaryngol.*, Vol. 103, 1977, pp. 574–578.
57. M. Hatch, M. Tsai, M. J. Larouere, A. L. Nuttall, and J. M. Miller, "The Effects of Carbogen, Carbon Dioxide, and Oxygen on Noise-Induced Hearing Loss," *Hear. Res.*, Vol. 56, pp. 265–272.
58. W. Burns and D. W. Robinson, *Hearing and Noise in Industry*, Her Majesty's Stationery Office, London, 1970, pp. 183–210.

93

CLINICAL AUDIOLOGY: AN OVERVIEW

LARRY E. HUMES

1 INTRODUCTION

Approximately 9% of the U.S. population is hearing impaired.[1] The prevalence of hearing impairment in this country, moreover, varies considerably with age. Among those 65 years of age and older, for example, approximately 30–40% have a significant hearing impairment.[1] Furthermore, the 65-and-over age group represents one of the most rapidly growing segments of the U.S. population. By the year 2000, it is estimated that one of every five individuals living in this country will be 65 years of age or older.[2] Given the rapid growth of the elderly in this country and the high prevalence of hearing impairment among this segment of the population, it is clear that the problems of the hearing impaired will demand more attention from those working in acoustics in the years ahead.

The problems of the hearing impaired have been the focus of a professional discipline within the broad realm of acoustics for the past 50 years. This field, known as audiology, was born during the 1940s as the result of national efforts to deal with the large number of hearing-impaired American soldiers returning home from World War II. The field of audiology has undergone rapid growth and diversification in the intervening decades, but the primary focus of this field remains the rehabilitation of those with hearing impairment. Today, there are approximately 11,000 audiologists in this country working in settings that include public schools, hospitals, speech and hearing clinics, industry, and independent private practices.

Although clinical audiology is an increasingly diverse profession, there is a common core of information and skills shared by all audiologists. It is this common core that is reviewed in the remaining pages of this chapter. The remainder of this chapter has been divided into two sections: (1) diagnostic audiology and (2) rehabilitative audiology.

Note: References to chapters appearing only in the *Encyclopedia* are preceded by *Enc.*

2 DIAGNOSTIC AUDIOLOGY

One of the first steps in the rehabilitation of the hearing impaired is the establishment of an accurate diagnosis. The diagnosis is typically established with the assistance of a physician. The audiologist conducts a wide array of auditory tests to assist the physician in determining the extent of hearing difficulty and its underlying cause or etiology. A critical piece of information used in establishing the etiology of hearing impairment is determination of the location of the pathology within the auditory system. This is generally referred to as establishing the *site of lesion*. Much information about the extent of the impairment and its cause can be obtained from the tests included in the basic hearing evaluation, which is comprised of three types of tests: (1) pure-tone audiometry, (2) immittance measurements, and (3) speech audiometry.

2.1 Pure-Tone Audiometry

In pure-tone audiometry, sinusoidal signals presented to the listener's ears are adjusted in sound pressure level (SPL) to determine the level that can just be detected 50% of the time. These behavioral measurements are made for each ear at octave intervals from 250 through 8000 Hz using standardized transducers to deliver the stimuli to each ear. The most common types of transducer used clinically today are supra-aural earphones, insert earphones, and bone vibrators. When the sinusoidal stimuli are delivered with either of the first two electroacoustic transducers an acoustic signal is delivered to the outer ear of the listener and the term *air conduction* testing is used. For such signals to be perceived normally, the entire peripheral auditory system, from the outer ear through the auditory nerve, must be functioning normally.

The bone vibrator, on the other hand, is an electromechanical transducer that is usually coupled with a headband to the bony skull at the forehead or to the mastoid process of the temporal bone (the bony protuberance just behind the outer ear). At either location, the bone vibrator stimulates the bony skull surrounding the inner ear and (in a normal ear) gives rise to an auditory sensation identical to that of air-conducted stimuli. By stimulating the inner ear directly, the bone conduction measurements have bypassed the more peripheral outer and middle ears. Differences in the sound levels needed to detect a sinusoid via air and bone conduction can offer some insight into the location of pathology underlying a particular hearing loss.

The just-detectable sound levels, or hearing thresholds, measured at each frequency by air and bone conduction are plotted on a standardized graph referred to as an audiogram. The top panel of Fig. 1 displays an audiogram from both ears of a young adult with normal hearing, whereas the bottom panel displays the audiogram from both ears of an "average" 65-year-old male. Notice that the normal-hearing young adult has an audiogram that is near 0 dB across the entire range of frequencies. The decibel scale used on an audiogram is referred to as decibels hearing level (HL). The just-detectable SPLs at each frequency represent the reference for this scale. (That is, dB HL = dB SPL_m − dB SPL_{ref}, or 20 log P_m − 20 log P_{ref}, where the subscripts m and ref refer to the patient's measured values and standardized reference values, respectively.) Table 1 displays the reference threshold SPLs at each frequency that correspond to 0 dB HL on the audiogram. It is important to emphasize that the values in this table represent transducer-specific average threshold SPLs derived from several studies and hundreds of "otologically normal" adults (free of ear disease). Thus, this normal-hearing young adult with thresholds of 0 dB HL at all frequencies requires 45 dB SPL at 250 Hz and 7 dB SPL at 1000 Hz to detect these sinusoidal signals.

Notice that increasing hearing loss in decibels hearing level is plotted on the audiogram by moving down on the graph. Thus, contrary to what one might expect, more intense sounds are plotted *lower* on the graph than less intense sounds. (If considered in terms of hearing *loss*, however, the y-axis may not appear to be so counterintuitive.) The audiogram in the lower panel of Fig. 1, then, indicates that this 65-year-old man has more hearing loss in the high frequencies than in the low frequencies.

TABLE 1 Standard Reference Equivalent Threshold SPLs for a TDH-39 Earphone Measured on the NBS-9A Coupler

Frequency (Hz)	dB SPL
125	45.0
250	25.5
500	11.5
1000	7.0
2000	9.0
4000	9.5
8000	13.0

Source: From Ref. 3.

The audiograms in Fig. 1, moreover, serve to illustrate several other features of audiogram notation. First, a standardized set of symbols is used to graph the results. Air conduction (AC) thresholds from the right ear (RE), for example, are represented by open circles, whereas "X" symbols are used for the corresponding results from the left ear (LE). The left- and right-pointing arrowheads shown in Fig. 1, moreover, represent the results from bone conduction (BC) testing obtained from the right and left ears, respectively. (For reasons not necessary to review here, an audiologist would actually seldom obtain BC thresholds from each ear in the cases illustrated in this figure. Results from the nontest ear can often be inferred from those obtained from the test ear via BC.)

The audiogram obtained from pure-tone audiometry provides several pieces of information at a glance. First, the extent and severity of the hearing impairment is depicted. That is, the amount of hearing loss and the frequencies affected are both readily apparent from the audiogram. Second, comparisons across ears can be made readily. In both audiograms shown in Fig. 1, the hearing thresholds were essentially equivalent in both ears. This is known as a *bilaterally symmetrical* hearing loss. (Measurement error is 5 dB in clinical pure-tone audiometry so that between-ear differences must exceed 5 dB to be considered real differences.) Third, any difference in hearing threshold measured by AC and by BC will be readily apparent. This difference, referred to as the *air–bone gap*, is of diagnostic significance. Bone conduction thresholds are never significantly worse than AC thresholds, but they can be significantly better. Better hearing by BC implies a lower (or better) hearing threshold for the inner ear compared to that of the entire auditory periphery (outer, middle, and inner ears). If a higher hearing threshold is measured for air-conducted sound than for bone-conducted sound, then the extra hearing

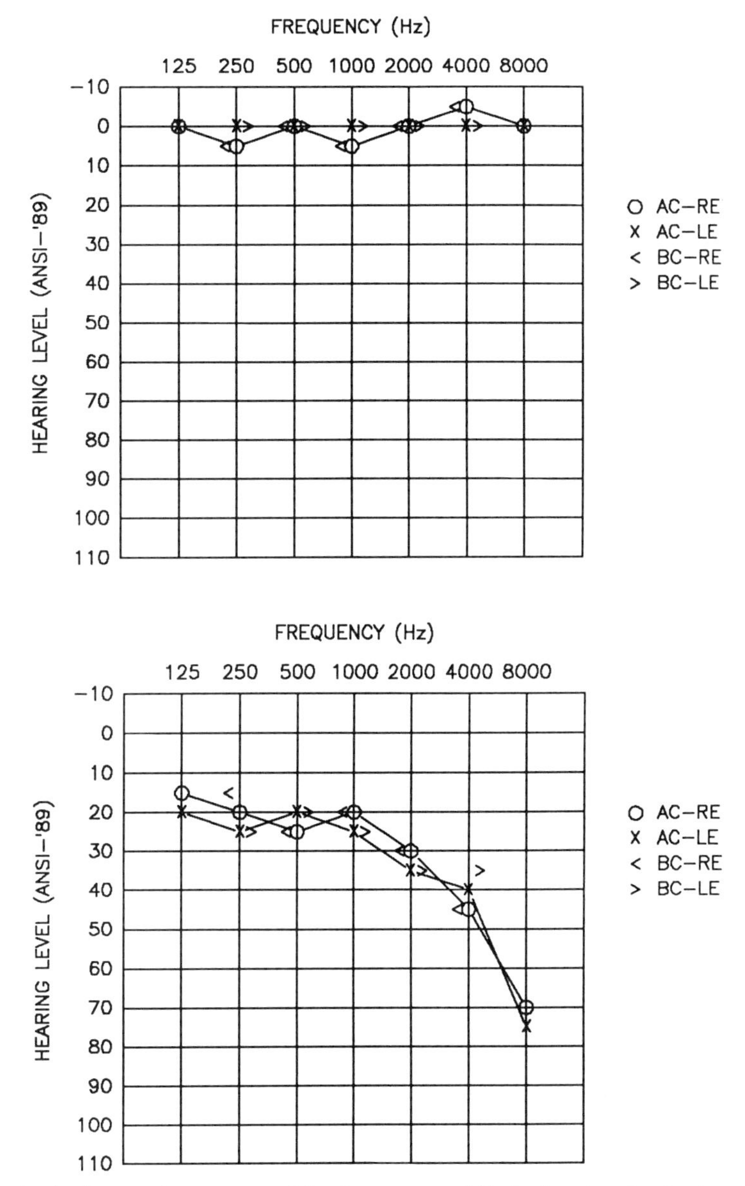

Fig. 1 Audiogram from a normal-hearing young adult (top) and a hearing-impaired 65-year-old male with the high-frequency sensorineural hearing loss (bottom) (AC, air conduction; BC, bone conduction; RE, right ear; LE, left ear).

loss associated with the air-conducted sound must be due to a disorder of the outer or middle ear. Air–bone gaps of 10 dB or more are considered significant clinically and indicate the presence of pathology in the outer or middle ear. Because the pathology usually attenuates the sound being "conducted" through the outer and middle ears to the inner ear, this type of hearing loss is referred to as *conductive* hearing loss. Examples of common causes of conductive hearing loss include excessive cerumen ("wax") in the outer ear, fluid in the middle ear associated with "ear infections" or "earaches" (otitis media), and ossification of the bones in the middle ear (otosclerosis). The hearing loss measured for AC signals in conductive pathology is usually flat across frequency or slightly worse in the low frequencies than in the high frequencies.

When hearing loss is present but the thresholds measured by air and bone conduction are equivalent (as in the lower panel of Fig. 1), the hearing loss is referred to as *sensorineural* hearing loss. Since the hearing threshold of the inner ear alone (bone conduction) is the same as that of the entire auditory periphery (air conduction), the presence of a sensorineural hearing loss implies a disorder of the sensory or neural portions of the inner ear.

The results of pure-tone audiometry could also reveal the presence of a hearing loss that was both conductive and sensorineural. This is referred to as a *mixed* hearing loss. If the 65-year-old male whose audiogram is shown in the bottom panel of Fig. 1 was retested after wax had accumulated in the outer ear, then the AC thresholds in the low and midfrequencies would be higher while the BC thresholds would remain the same. The appearance of this air–bone gap would indicate the presence of a conductive hearing loss superimposed on the existing sensorineural hearing loss.

In the top panel of Fig. 1, the normal-hearing young adult had hearing thresholds near 0 dB HL at all frequencies. There is actually a range of normal hearing from −10 to 25 dB HL. Although there is some debate about the upper limit of "normal" hearing, an individual with AC hearing thresholds less than or equal to 25 dB HL at all frequencies would be considered normal. Table 2 describes a classification system commonly used in audiology to categorize different degrees of hearing loss. According to this scheme, for example, the audiogram from the elderly individual in the bottom panel of Fig. 1 would represent a bilaterally symmetrical, moderately severe, high-frequency sensorineural hearing loss.

TABLE 2 Classification of Severity of Hearing Impairment Based on Air Conduction Pure-Tone Thresholds

Thresholds (dB HL)	Classification
−10–25	Normal
26–40	Mild
41–55	Moderate
56–70	Moderately severe
71–90	Severe
> 90	Profound

2.2 Immittance Measurements

Immittance measurements represent another major component of the basic hearing evaluation. Generally, these measurements are used to help establish the site of lesion within the auditory system and are not used to determine the extent or severity of hearing loss.

"Immittance" is actually a nonsensical term without a physical foundation that was adopted in an effort to create a uniform terminology for these measurements. The clinical instruments used to measure immittance actually measure some aspect of either the impedance of the middle ear or its reciprocal, admittance.

Immittance measurements are made with the assistance of a special transducer assembly that fits into the ear canal. Figure 2 illustrates a schematic cross-sectional diagram of the assembly sealed in the ear canal. Essentially, there are three tubes that terminate in the assembly. One tube is connected to a receiver used to deliver a sinusoidal signal into the sealed ear canal. A second tube is connected to a microphone used to measure the level

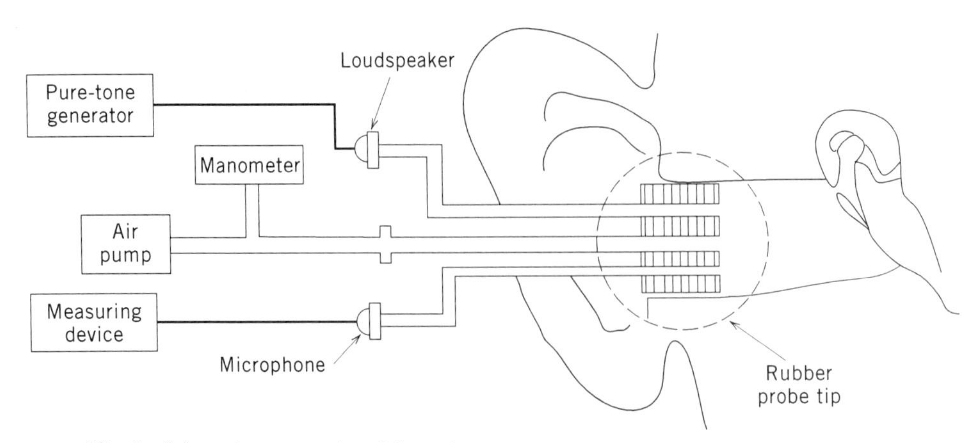

Fig. 2 Schematic cross section of the probe assembly of an immittance measurement system inserted in the ear canal.

of the sound in the sealed ear canal. Finally, a third tube is connected to a manometer and is used to vary the air pressure inside the sealed ear canal in a controlled fashion.

Several measurements are possible with this assembly, but the general principle behind them all is that as the impedance of the middle ear increases, less of the sound introduced into the ear canal by the receiver will be transmitted through the middle ear and more will be reflected back into the sealed ear canal cavity. The increased sound level in the sealed ear canal will be picked up by the microphone and displayed in one of several ways (needle deflection on a meter, pen displacement on a graphic level recorder, digital display, etc.).

One of the tests comprising the immittance test battery is typanometry. In tympanometry, the manometer is used to systematically vary the air pressure in the sealed ear canal from approximately −400 to +200 daPa. At the extreme negative and positive air pressures, the stiffness of a normal middle ear is increased and more sound is reflected back into the sealed cavity. This results in an increased sound level in the ear canal measured by the microphone. As the air pressure in the ear canal approaches that on the other side of the eardrum in the air-filled middle ear, the middle ear becomes more compliant and less sound is reflected back into the ear canal. This results in a decreased output from the microphone. When the output of the microphone is recorded graphically, the result is referred to as a tympanogram.

Figure 3 displays three tympanograms. The solid line represents a normal tympanogram. There are several features of the tympanogram used to define its normality, including the peak amplitude, the ear canal air pressure at which the peak occurs, and the general shape of the tympanogram. One common tympanogram abnormality is illustrated by the dashed tympanogram. Note that this tympanogram is nearly identical to that of the normal middle ear in amplitude and shape but is shifted horizontally so that the peak occurs at an air pressure of −200 daPa. This is referred to as negative middle-ear pressure and, if chronic, can be a precursor to the development of otitis media (ear infections or earaches). A short-lived, acute change in middle-ear pressure is frequently experienced by rapid changes in elevation, as in air travel. Finally, the dotted tympanogram in Fig. 3 illustrates a flat tympanogram associated with a fluid-filled middle ear, as in otitis media. The fluid has increased the stiffness of the middle ear and results in greater sound being reflected back into the sealed ear canal at all air pressures. It should be noted that because the vast majority of middle-ear disorders affect the stiffness of the middle ear, the sinusoid introduced into the sealed ear canal through the receiver tube is usually low in frequency

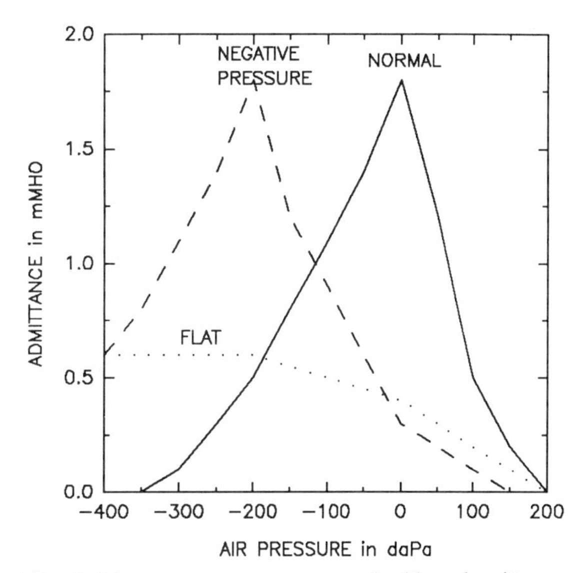

Fig. 3 Three tympanograms measured with an immittance measurement system. Tympanograms illustrated are normal (solid line), negative pressure (dashed line), and flat (dotted line).

(220 and 660 Hz are the most commonly used frequencies).

Tympanograms are used to assess the integrity of the middle ear. Another component of the immittance test battery, however, is sensitive to the presence of pathology in the middle ear, inner ear, auditory nerve, and brainstem. This test, referred to as acoustic-reflex measurement, involves the reflexive contraction of the middle-ear muscles in response to high-intensity sound. The two aspects of acoustic-reflex measurements that are of primary clinical importance are reflex threshold and reflex decay. In the measurement of reflex threshold, the intensity of a stimulating sound is adjusted to determine the level that will just elicit a muscle contraction. The transducer assembly used with tympanometry is also used to monitor the middle ear for the occurrence of a muscle contraction. When the muscles contract, the stiffness of the middle ear is increased slightly, which increases the ear canal SPL measured by the microphone in the assembly. The acoustic stimulus delivered to the ear to elicit the reflex either can be presented to the same ear with the transducer assembly through a fourth tube or it can be delivered to the opposite ear with an earphone. These two different stimulus presentation modes are referred to as ipsilateral and contralateral presentations, respectively. Contralateral presentation is possible because the middle-ear reflex is a consensual phenomenon. That is, for a normal auditory system, stimulation of one ear with high-level sound results in middle-ear muscle contractions on both sides.

Complete details regarding the diagnostic significance of acoustic reflex thresholds are beyond the scope of this overview. Essentially, reflex thresholds are measured for pure tones at octave intervals from 500–4000 Hz and are judged as being within the normal limits (<105 dB HL), elevated (>100 dB HL), or absent. Elevated or absent acoustic reflexes can be observed in individuals with normal pure-tone hearing (likely indicating an auditory nerve or brainstem lesion) or in individuals with conductive or sensorineural hearing loss. Thus, elevated or absent acoustic reflex thresholds, in and of themselves, do not pinpoint the site of lesion. When interpreted in conjunction with the tympanograms and the pure-tone audiogram, however, acoustic reflex thresholds can offer considerable insight into the location of the pathology within the auditory system.

If an acoustic reflex threshold can be obtained for a 500- or 1000-Hz stimulus, then measures of reflex decay are also usually obtained at one or both of these frequencies. Using a suprathreshold sound level, the magnitude of immittance change associated with the reflex is monitored for a 10-s presentation of the reflex-activating sound. The magnitude of the decay or adaptation of the reflex is then quantified and classified as abnormal (too rapid decay) or normal. Abnormal reflex decay is usually an indication of a problem in the auditory nerve or brainstem, such as an acoustic neuroma (tumor on the auditory nerve).

2.3 Speech Audiometry

The third primary component of the basic hearing evaluation is speech audiometry. There are several variations of speech audiometry, but the two measurements common to this area of testing in the United States are speech recognition threshold (SRT) and word recognition score (WRS). Briefly, measurement of the SRT involves the presentation of standardized lists of two-syllable words ("spondees") at various intensities to determine the sound level corresponding to 50% correct recognition of these words. Word recognition performance, on the other hand, typically makes use of standardized lists of one-syllable consonant–vowel–consonant words (CVC monosyllables) presented at a fixed level. The level used is typically 40 dB above the SRT. Whereas the SRT determines the softest level at which speech can be recognized, the WRS estimates how well speech can be understood at a comfortable level.

Word recognition performance is often also assessed in a background of competition. The competition is frequently white noise, speech-shaped noise, or a multitalker babble. The speech-to-competition ratio may vary from 0 to +10 dB, depending on the application and the particular set of speech materials. In addition, speech materials may be presented in an open (fill-in-the-blank) or closed (multiple-choice) response format. Finally, speech materials other than monosyllables may be used. Standardized lists are available for nonsense syllables as well as for meaningful sentences. Although standardized test materials exist, the presentation level of the materials, as well as the type and level of background competition, have not been standardized.

2.4 Basic Hearing Evaluation: A Synopsis

It is always desirable to obtain as much information from a given hearing-impaired individual as possible. In the basic hearing evaluation, results are needed from pure-tone audiometry, immittance measurements, and speech audiometry. This information, along with a detailed case history, can go a long way toward establishing an accurate diagnosis. It is not always possible, however, to obtain measures from all of the tests in this basic hearing evaluation battery. For example, it is seldom possible to obtain an accurate and complete pure-tone audiogram from very young children (less than 2 years of age), although alternative procedures are available to the audiologist to permit the assessment of hearing in children as young as 3–6 months of age. Nonetheless, for the vast majority of cases seen by the audiologist, results from all three components of the basic hearing evaluation are readily obtained.

The results of individual components of the basic hearing evaluation are frequently mutually reinforcing and point to the same site of lesion. Consider the case illustrated in Fig. 4. The top panel of this figure shows the AC and BC thresholds from the right ear of an individual with chronic otitis media. This disorder is very common in children and is often characterized by the accumulation of fluid in the middle ear. The left ear (not shown here for clarity) is assumed in this example to be normal in all respects. Note that inner ear (BC) hearing in the right ear is normal whereas AC thresholds indicate the presence of a mild, flat hearing loss. These large air–borne gaps are consistent with the presence of conductive pathology in this ear.

The immittance measurements summarized in the lower panel of this figure are also consistent with the presence of conductive pathology in the right ear. Note that the tympanogram from this ear is flat and that contralaterally activated acoustic reflexes are absent bilaterally. When the high-intensity reflex-activating sound is presented to the normal left ear, the extreme stiffness of the fluid-filled right ear prevents the detection of the small change in stiffness associated with reflex

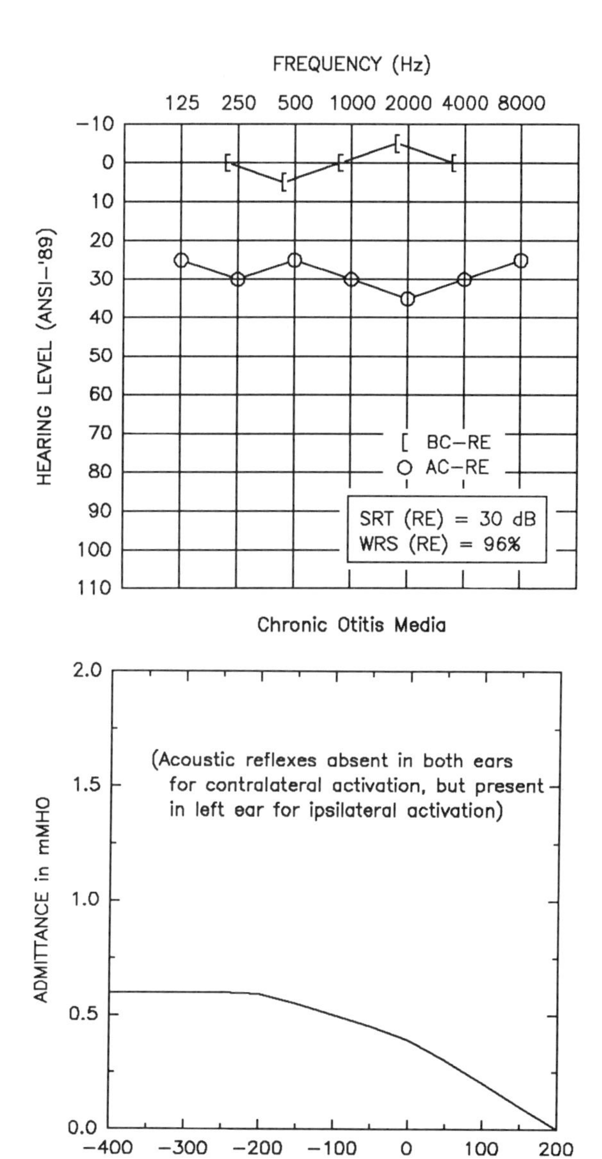

Fig. 4 Results of the basic hearing evaluation for a child with chronic otitis media in the right ear and normal hearing in the left (not shown). Results of pure-tone and speech audiometry are shown in the top panel and immittance measurements are shown in the bottom panel.

contraction. When the activating stimulus is presented to the right ear, however, the attenuation of the activating sound by the conductive hearing loss in the right ear results in a stimulus level that is below the reflex threshold and an acoustic reflex does not occur, even though the probe assembly is in the normal left ear. Ipsilateral presentation of the activating sound to the left ear does reveal the presence of a normal acoustic-reflex threshold, as expected for this healthy ear.

Finally, the results from speech audiometry are also consistent with the presence of conductive pathology in the right ear. The SRT, measured with headphones, agrees with the average hearing loss measured for pure tones delivered the same way (AC thresholds). In addition, as is true with most cases of conductive pathology, excellent WRSs are observed once the level of the speech is increased to compensate for the attenuation of the hearing loss. This is often *not* the case for many individuals with a sensorineural hearing loss.

The results of the basic hearing evaluation in this example all point to the same conclusion: the presence of conductive middle-ear pathology in the right ear. In many cases, however, the results from the basic hearing evaluation itself do not pinpoint the site of lesion. Advanced special tests, usually consisting of various measures of speech audiometry or measures of the electrical activity of the auditory portions of the brainstem and cortex evoked by acoustic stimuli, are frequently needed to determine the site of lesion.[4,5] Description of these advanced tests and their interpretation, however, is beyond the scope of this brief overview.

3 REHABILITATIVE AUDIOLOGY

As mentioned, the first step in the rehabilitation process is the determination of the extent and severity of hearing impairment and the site of lesion. In the vast majority of conductive hearing impairments, the underlying pathology can often be eliminated and hearing restored through medical treatment consisting of prescribed medication, surgery, or both. If the site of lesion is retrocochlear (pathology of the auditory nerve or auditory pathways in the brainstem or cortex), medical intervention (frequently surgery) is also necessitated. For those individuals with chronic untreatable middle-ear pathology or sensorineural hearing loss associated with a cochlear site of lesion, on the other hand, aural rehabilitation is the primary course of action.

A comprehensive review of aural rehabilitation is also beyond the scope of this overview of clinical audiology. Much of the early portion of the aural rehabilitation process, however, is centered on the selection and evaluation of a prosthetic device. The types of devices available, their function, and their advantages and limitations are described in the pages to follow. The types of devices reviewed, in the order of their presentation, are (1) conventional electroacoustic hearing aids, (2) assistive listening devices, (3) vibrotactile devices, and (4) cochlear implants. It should be emphasized, however, that the aural rehabilitation process *begins*, not *ends*, with the selection and evaluation of a prosthetic device. Following device selection, the hearing-impaired individual needs to participate in a variety of rehabilitation activities, including auditory training, training in effective communication strategies, and training in lipreading. Finally, the focus in this section is on the hearing-impaired adult. The rehabilitation process is considerably different for children, especially if the hearing loss is severe and occurs prior to the normal development of speech and language.

3.1 Conventional Electroacoustic Hearing Aids

The conventional electroacoustic hearing aid is essentially a miniature, head-worn, public address system, complete with microphone, amplifier, and loudspeaker (receiver). One of the primary differences between the typical hearing aid and a public address system is that the latter system places the microphone close to the desired sound source. This is not the case with the conventional hearing aid and, as will be demonstrated, remains one of the primary limitations of this system.

There have been significant advances in the conventional hearing aid over the past decade. The size of these instruments has been reduced from devices worn in the shirt pocket or on top of the ear to units that can be worn entirely inside the ear canal of the listener. At the same time, fidelity and electroacoustic flexibility have been enhanced. Many hearing aids available today offer directional microphones and various signal-processing options designed to reduce background noise. In addition, several contemporary conventional hearing aids are programmable, storing four to seven sets of electroacoustic characteristics in memory. Remote controls with infrared coupling to the hearing aid represent another recent advance that is of considerable benefit to elderly wearers lacking the manual dexterity to adjust the tiny controls on a head-worn hearing aid.

Most conventional hearing aids are selected for a particular hearing-impaired person on the basis of the extent and severity of the hearing loss. The primary problem experienced by most individuals with sensorineural hearing loss associated with a cochlear site of lesion is that speech can be heard, but not understood. The audiograms in Fig. 5 help to explain the reasons for this common complaint and how a well-fitted hearing aid can compensate for it. The "X" symbols on this audiogram represent the hearing thresholds from the left ear of a normal-hearing young adult. Superimposed on this audiogram is the approximate range of speech intensities at each frequency for speech at a conversational level (65 dB SPL at 1 m). This range, which encompasses the 10th to 90th

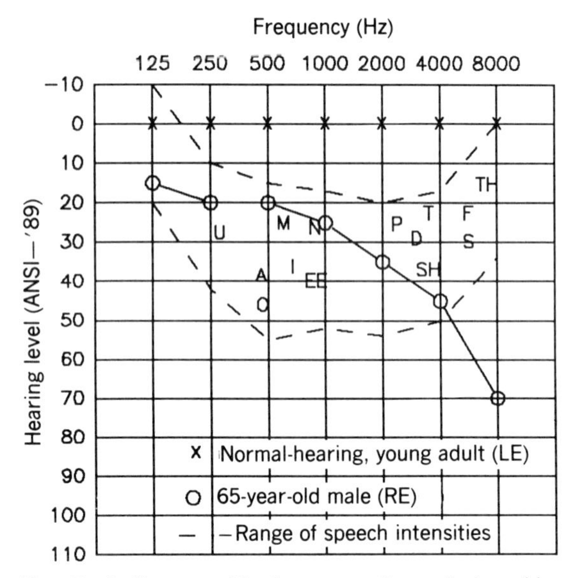

Fig. 5 Audiogram with the range of speech intensities (dashed lines) and various speech sounds superimposed on it. The crosses (×) represent hearing thresholds for a normal-hearing young adult and the circles represent hearing thresholds of a typical 65-year-old male.

percentiles of the amplitude distributions at each frequency, is represented by the dashed lines. Also shown are the approximate locations of various speech sounds plotted on these same coordinates. Note that all of the speech sounds are well above threshold for this hypothetical normal-hearing young adult.

Figure 5 also illustrates the audiogram from the right ear of an average 65-year-old male (circles). Note that the presence of a high-frequency sensorineural hearing loss renders many of the high-frequency, low-intensity consonants inaudible for this individual. Words such as *wife*, *white*, *wize*, *wipe*, and *wide* are essentially all heard by this listener as *why*. Thus, the listener *hears* speech when each word is spoken but cannot understand what has been said.

Consider now the effect of a well-fitted hearing aid on the audibility of speech. By selectively amplifying speech in the frequency region with hearing loss, the high-frequency consonants that were previously inaudible can now be heard comfortably by the listener. In the example shown in Fig. 5, for instance, little or no gain would be required below 1000 Hz, with increasing amounts of gain needed as frequency increased above 1000 Hz. Under these circumstances, a well-fitted conventional hearing aid typically results in large improvements in speech understanding.

There are three primary limitations, however, of a well-fitted conventional hearing aid that restrict the improvements that can be achieved. First, communication seldom takes place in quiet. Frequently there is noise or other conversation in the background. The conventional hearing aid amplifies the speech and the noise equally and does nothing to improve the speech-to-noise ratio. In fact, because of bandwidth limitations, distortion, and internal noise, the conventional hearing aid may often effectively *decrease* the speech-to-noise ratio. The incorporation of directional microphones and signal-processing circuits in conventional hearing aids in recent years represent two efforts to improve the speech-to-noise ratio.

The second limitation of conventional hearing aids has to do with the acoustic gain that can be achieved by the instrument without developing feedback. Acoustic feedback is a serious limitation for many hearing aids because of the close proximity of the microphone to the receiver, especially for the tiny instruments that fit entirely within the ear canal. Several feedback reduction circuits are currently being investigated as a means of alleviating this problem.

The third primary limitation has to do with the severity of hearing loss and the amount of amplification provided by the hearing aid. Although listeners with sensorineural hearing impairment cannot hear low-level sounds as well as normal-hearing listeners, high-intensity sounds are usually equally loud for both types of listener. Consequently, the amount of amplification required to make all of the low-intensity speech sounds audible cannot always be achieved without overamplifying other sounds of moderate or high intensity. The output of a hearing aid must be limited by peak clipping or amplitude compression to avoid presentation of sound levels that are uncomfortably loud or even potentially harmful to the hearing aid wearer. Amplitude compression continues to be investigated as an alternative to linear amplification with peak clipping. Compression schemes that have been investigated include variations in the number of channels, attack–release times, compression thresholds, compression ratios, and whether the system makes use of input- or output-dependent compression.[6,7]

The electroacoustic characteristics of a conventional hearing aid, such as its frequency–gain characteristics and its saturation SPL, are typically prescribed for a particular hearing-impaired listener using one of several available, theoretically based, prescriptive methods. The prescriptions are generated for a particular hearing-impaired person from knowledge of the individual's threshold and uncomfortable loudness level at several frequencies. The accuracy of the fit of the custom-ordered hearing aid to the prescription is then confirmed in the listener's ear canal using a tiny probe-tube microphone. Gain and maximum output characteristics are measured at several frequencies in a matter of a few minutes, and the measured characteristics are then compared to those prescribed for the individual and fine tuned to obtain a reasonable match.

3.2 Assistive Listening Devices

As mentioned, one of the primary limitations of the conventional hearing aid is the poor speech-to-noise ratio present at the microphone. To compensate for the inaudibility of some speech sounds, even with amplification, hearing-impaired listeners frequently require a better than normal speech-to-noise ratio. One simple way in which this can be accomplished is to move the microphone closer to the desired sound source, and many devices have been designed in recent years to do just that.[8] These devices, referred to as assistive listening devices, represent alternatives or supplements to conventional hearing aids. They must be considered as alternatives to hearing aids for those impaired listeners with mild hearing loss requiring only limited assistance of a specific nature or as supplements to hearing aids for severely hearing-impaired listeners requiring additional assistance beyond that available from their hearing aid.

As noted, one class of assistive listening device is designed to improve the speech-to-noise ratio by moving the microphone closer to the desired sound source. The desired sound source might be a talker, television, stereo, and so on. A detached microphone is placed in close proximity to the sound source, and its electrical output is sent to a body-worn amplifier and receiver. The output of the microphone can be sent to the amplifier via hardwire, FM transmission, or infrared transmission. The amplifier output can also be delivered to the ears of the listener in a variety of ways, including lightweight stereo headsets, stethoscope-type earphones, insert earphones, or hearing aids.

Other classes of assistive listening devices include special telephone listening aids that can be used with or without a hearing aid to improve telephone communication. Special visual or vibrotactile alerting devices are also available for profoundly impaired listeners. These devices are designed to flash lights that can be seen or produce vibration that can be felt by the hearing-impaired individual when specific acoustic events take place, such as the ring of a doorbell, an alarm clock, or a telephone. This latter type of device is clearly a supplement to the conventional hearing aid for those profoundly impaired individuals not receiving enough benefit from conventional amplification.

3.3 Vibrotactile Devices

For many individuals with profound sensorineural hearing loss in both ears, alternatives to conventional amplification have been made available that may generally be referred to as sensory substitution devices. That is, rather than delivering high-intensity sound to a damaged ear through a hearing aid or an assistive listening device, the sound is converted to an alternate form of energy and presented to the individual. For vibrotactile devices, the sound is converted to an electrical signal and delivered to the skin as a pattern of mechanical vibration by an array of vibrating mechanical contacts. Frequency can be encoded by both the rate of contact vibration and the location of the vibrating contact within the array. Amplitude may also be encoded by these devices in various ways, including the amplitude and rate of vibration. For most vibrotactile devices, the array of vibrating contacts is either placed on the forearm or incorporated into a belt worn around the midsection. An excellent overview of the relative advantages and disadvantages of conventional amplification, cochlear implants, and vibrotactile devices was provided recently by a working group of the Committee on Hearing, Bioacoustics and Biomechanics (CHABA) of the National Research Council.[9]

3.4 Cochlear Implants

Cochlear implants transduce the acoustical signal to an electrical one, process it, and deliver the electrical signal directly to the nerve fibers in the inner ear. Although a variety of implant types exist and each differs in some way from the others, the components of a "generic" cochlear implant are as follows. A small head-worn microphone transduces the acoustic stimulus to an electrical signal. This electrical signal is then sent to a speech-processing unit. This unit typically extracts certain relevant aspects of the speech signal, such as the fundamental frequency and the formant frequencies. The processing unit then sends an electrical code of this information to an external receiver worn behind the ear. The external receiver transmits this coded information across the skin to a surgically implanted internal receiver. The electrical output of the internal receiver is then distributed to the auditory nerve fibers in the inner ear via one or more electrodes that have been surgically inserted into the inner ear.

A wide variety of implant types have been investigated over the years with most fitting into a classification scheme based on the number of channels and number of electrodes. The number of channels refers to the number of parallel channels in the processing of the encoded information and the number of electrodes refers to the number of independent electrical connections carrying that information to the nerve fibers of the inner ear. In general, the research conducted on cochlear implants over the past decade has indicated that the multiple-channel, multiple-electrode devices are the most beneficial. Although there are some individuals that have demonstrated truly remarkable performance with the implants, such as understanding conversational speech on the telephone, the vast majority of implantees appear to primarily receive timing information that is a valuable supplement to lipreading (visual information). Although the implant often does not appear to present users with true auditory sensation, it does make them aware of sound. Sound awareness and supplementation of visual information represent the two most common improvements experienced by implant users. Continued research in this area, however, is likely to bring additional improvements in performance.

At present, the primary candidates for a cochlear implant are those individuals who have acquired a profound bilateral hearing loss after they have developed speech and language and who receive only limited benefit from other prosthetic devices, such as the conventional hearing aid. This represents a rather small segment of the hearing-impaired population.

3.5 Rehabilitation for Persons with Tinnitus

Concluding this section on rehabilitative audiology, it should be noted that the aural rehabilitation process is not always centered on the loss of hearing. One of the most troubling concomitants of sensorineural hearing loss is the presence of tinnitus. Tinnitus is a "ringing," "whistling," or "buzzing" sensation perceived by the listener in the absence of sound. There appear to be many causes of tinnitus, and a wide variety of treatments have been suggested.[10] Head-worn noise generators that physically resemble hearing aids, for example, are often recommended for tinnitus sufferers. These noise generators essentially bring relief to the tinnitus sufferer by masking the tinnitus with the external noise. Frequently, residual masking effects are observed in which the tinnitus is reduced or eliminated for prolonged periods of time after removal of the tinnitus masker. As mentioned, there are a wide variety of causes and treatments for tinnitus, and the audiologist will frequently play a role in the rehabilitation of the tinnitus sufferer.

4 SUMMARY

This chapter has provided a very brief overview of the field of clinical audiology. As mentioned, audiologists are the primary professionals involved in the rehabilitation of the hearing impaired. Initial steps in the rehabilitation process involve the determination of the extent and severity of hearing loss and the site of lesion. Once this information is obtained, the next step typically involves the selection and evaluation of an appropriate prosthetic device for the hearing-impaired individual. Devices available include conventional hearing aids, assistive listening devices, vibrotactile aids, and cochlear implants. Conventional hearing aids and assistive listening devices are beneficial to the vast majority of hearing-impaired individuals, whereas vibrotactile devices and cochlear implants are primarily reserved for those profoundly impaired individuals unable to benefit from conventional amplification. Following the selection and evaluation of the appropriate device, additional aural rehabilitation, such as training in lipreading, auditory training, and training in effective communication strategies, is often provided by the audiologist.

This brief overview provides little more than a thumbnail sketch of clinical audiology. Additional details regarding the many facets of clinical audiology can be found elsewhere.[11,12]

REFERENCES

1. A. J. Moss and V. L. Parsons, "Current Estimates from the National Health Institute Survey, United States, 1985." *Vital and Health Statistics*, Series 10, No. 160, DHHS pub. no. (PHS) 86-1588, Public Health Services, Washington, DC, 1986.
2. U.S. Department of Health and Human Services, "Aging America: Trends and Projections, 1987–1988 Edition," U.S. Government Printing Office, Washington, DC, 1988.
3. *Specification for Audiometers*, ANSI S3.6-1989. Acoustical Society of America, New York, 1989.
4. J. W. Hall, *Handbook of Auditory Evoked Responses*, Allyn & Bacon, Boston, 1992.
5. J. T. Jacobson and J. L. Northern, *Diagnostic Audiology*, Pro-Ed, Austin, TX, 1991.
6. L. Braida, N. Durlach, R. Lippmann, B. Hicks, W. Rabinowitz, and C. Reed, "Hearing Aids—a Review of Past Research on Linear Amplification, Amplitude Compression and Frequency Lowering," Asha Monograph No. 19, American Speech-Language-Hearing Association, Rockville, MD.
7. G. Walker and H. Dillon, "Compression in Hearing Aids: An Analysis, a Review and Some Recommendations," National Acoustic Laboratories Rept. No. 90, Australian Government Printing Service, Canberra, 1982.
8. C. L. Compton, *Assistive Devices: Doorways to Independence*, Assistive Devices Center, Gallaudet University, Washington, DC, 1989.
9. Working Group 95, Committee on Hearing, Bioacoustics and Biomechanics (CHABA), National Research Council, "Personal Speech Perception Aids for the Hearing Impaired: Current Status and Needed Research," *J. Acoust. Soc. Am.*, Vol. 90, 1991, pp. 637–685.
10. D. McFadden, *Tinnitus: Facts, Theories and Treatments*, National Academy, Washington, DC, 1982.
11. F. H. Bess and L. E. Humes, *Audiology, the Fundamentals*, Williams & Wilkins, Baltimore, MD, 1990.
12. J. Katz (Ed.), *Handbook of Clinical Audiology*, 4th ed., Williams & Wilkins, Baltimore, MD, 1994.

PART XIII

SPEECH COMMUNICATION

94

INTRODUCTION

J. L. FLANAGAN

1 HUMAN COMMUNICATION BY SPEECH

Since the beginning of time humans have had the need to exchange information. To meet this need, it was necessary to evolve a signaling means having sufficient transmission capacity to rapidly convey human thoughts and concepts. Acoustic signaling proved better matched to human information capacities than other modalities, such as visual hand signals or "body English"—and, speech and language developed. (The British phonetician Paget suggests that ancient man likely evolved speech when he found it inconvenient to "talk with his hands full.")

Humans produce sound using physiological apparatus designed to serve more life-sustaining functions—breathing and eating. But, the respiratory system in combination with vocal cords and vocal and nasal tracts can generate a rich diversity of sounds having distinctive energy patterns in the frequency spectrum audible to the human ear. These patterns constitute the signals of speech.[1,2] The agreed-upon conventions by which these sounds are issued in sequence, and the meanings attached to the sound sequences, constitute spoken language. The steps in the process of speech generation and perception are diagrammed according to qualitative functions in Fig. 1. Associated with these functions are counterparts that are representative of machine processes in synthesizing speech and in recognizing speech. Typical information rates necessary to these processes are also indicated, suggesting that more efficient (compact) coding of information exists at the higher sensory levels. In production of natural speech, the vocal system assumes a prescribed sequence of shapes. A unique sound spectrum is associated with each shape. Illustrative envelopes of sound spectra corresponding to several vocal-tract shapes for vowel sounds are shown in Fig. 2. Long-time average spectrum levels for conversational speech are shown in comparison to auditory thresholds in Fig. 3.

But acoustic waves do not propagate well over distances. Typically the energy spreads spherically and diminishes as the square of the distance between speaker and listener, eventually becoming too weak for detection by the human ear. For ages humans sought effective means for communicating at distances (including such efforts as Caesar's voice relays using towers spaced within hailing distance stretching from Gaul to Rome). Invention of the telephone in 1876 brought an electrical solution to enhance acoustical communication at distances. Since that time, scientific incentives have focused on still greater ranges (provided by undersea cable and satellite radio), greater efficiency in representing the perceptually important information of speech, and (as digital technology, microelectronics and computing have evolved) on voice communication with machines.[3–5]

The interest in efficient transmission and storage of speech information is traditionally served by research in *speech coding*. The interest in human/machine communication by voice is traditionally served by research in *speech synthesis* and automatic *speech recognition*.

2 SPEECH CODING

Invention of the telephone was based on the notion of producing an electric signal that is a facsimile of the acoustic time waveform. But it was recognized very early that the human ear does not require preservation of the time waveform for high-quality perception. Rather, owing to the frequency-analyzing properties of the auditory system, some phase detail need not be retained, and

Note: References to chapters appearing only in the *Encyclopedia* are preceded by *Enc.*

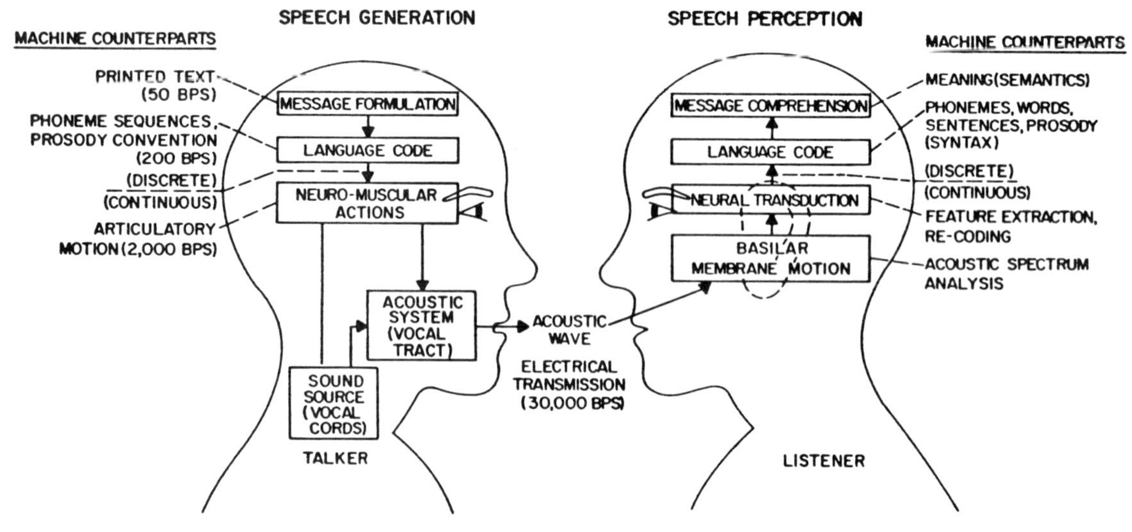

Fig. 1 Steps in the process of speech generation and perception.

intelligibility is preserved in the short-time amplitude spectrum of the signal. Further, the properties of auditory masking are such that an intense spectral component may render imperceptible a weaker component (especially if it is close and slightly higher in frequency), making it totally unnecessary to devote transmission capacity to those components that cannot be heard. As the capabilities and economies of digital processing have advanced, such effects can be computed moment by moment. And, high-quality signal representation (for transmission and

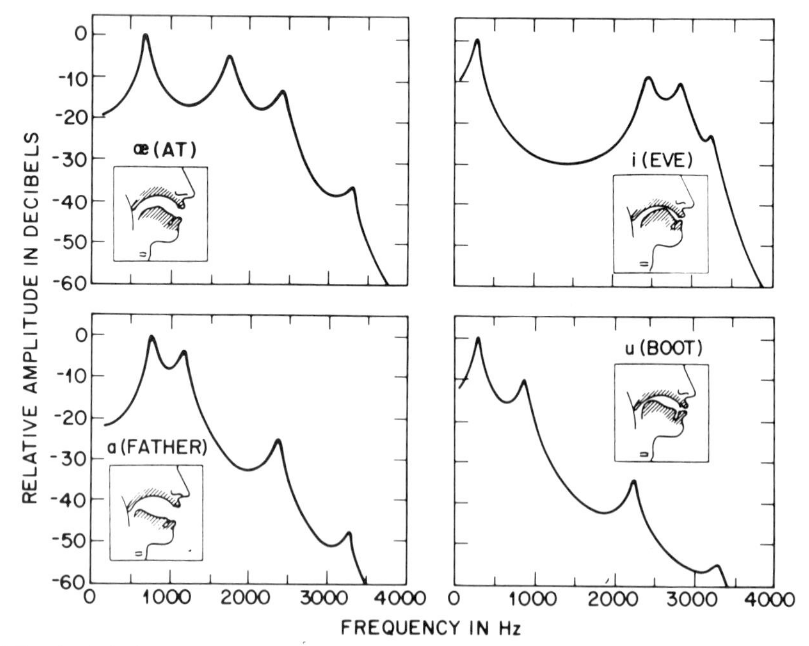

Fig. 2 Articulatory shapes and the corresponding sound spectra for four English vowel sounds: æ (at), i (eve), a (father), and u (boot). The maxima, or peaks, in the spectra reflect the distinctive acoustic resonances (or formants) corresponding to the vocal-tract shapes.

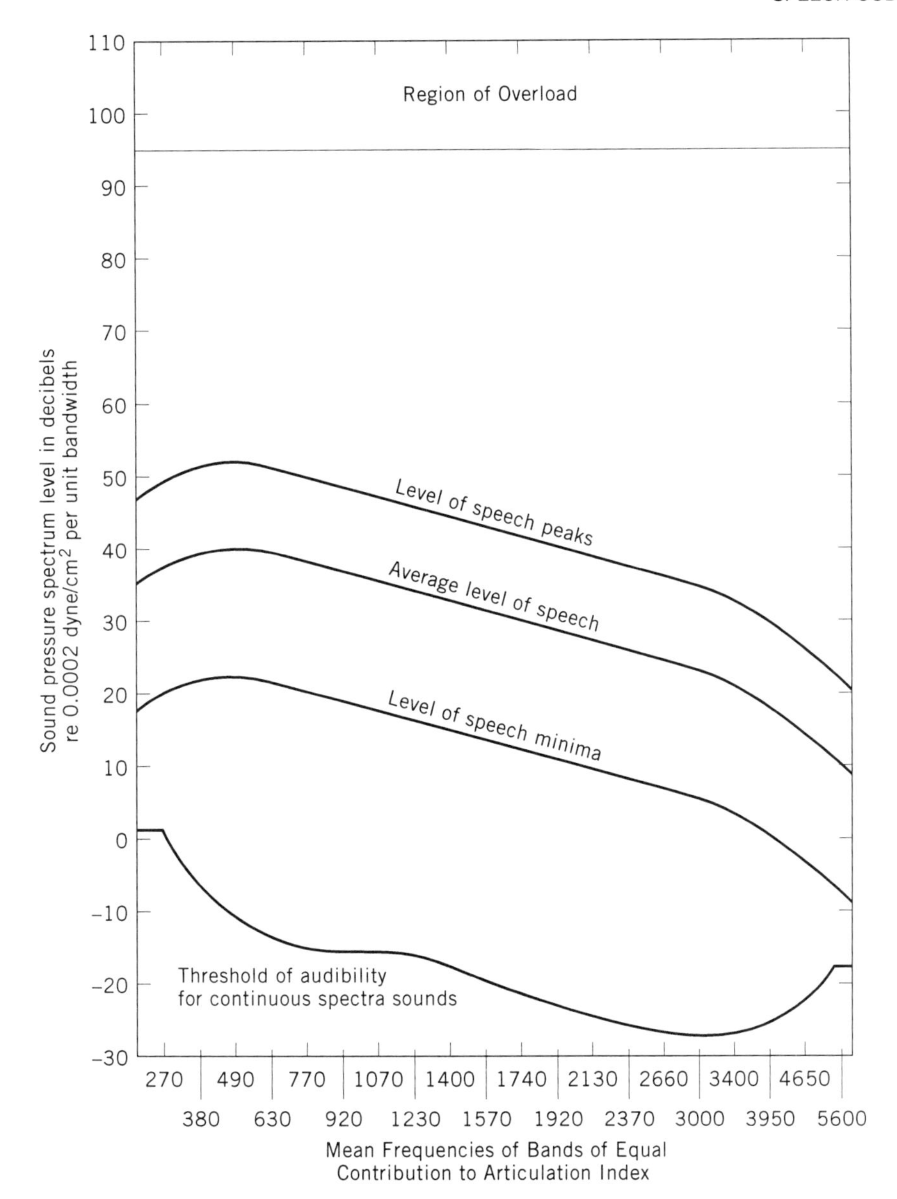

Fig. 3 Average spectrum levels for typical conversational speech. The thresholds for auditory detectability and overload are also shown. The frequency scale reflects the mean frequencies of the 20 bands of equal contribution to speech intelligibility.

storage) continues to incorporate more constraints that characterize auditory perception.

In the digital domain the quest for efficiency takes the form of obtaining the highest quality signal at the lowest digital bit rate.[6,7] Efficiency can be sought in two ways. One way is coding with a fidelity criterion appropriate to the receiver (the human ear), as mentioned above. Another is coding in terms of parameters that describe a specific class of signals, such as speech. In this case only the specific class is represented with accuracy. The mechanism of human speech production offers salient parameters in the form of vocal-tract resonances (formants) and sound generation by vocal-cord vibration (voicing at a fundamental "pitch" frequency)

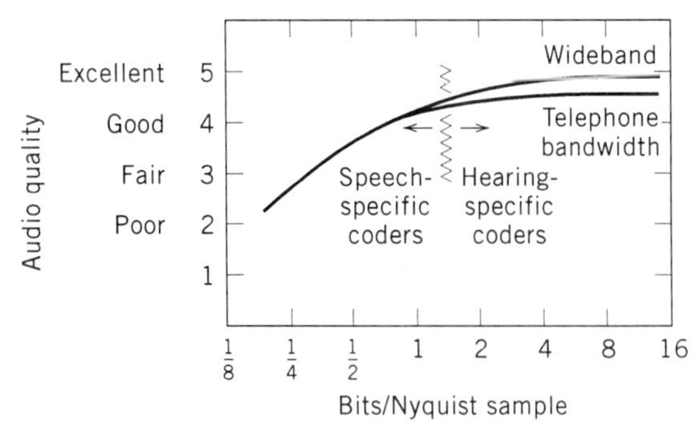

Fig. 4 Audio signal quality as a function of digital representation. Auditory criteria are incorporated in the higher bit rate coders, and speech and hearing criteria are incorporated in the speech-specific lower bit rate coders.

or flow turbulence at a constriction (voiceless sounds). These and related parameters have been central to the design of low bit-rate coders (or vocoders) for speech. There are experimental reasons to believe that high-quality coding of speech can be obtained at data speeds as low as 2000 bits/s. But complete solution of this problem remains clusive. Figure 4 reflects current understanding for "hearing-specific" coders, and additionally, for those that are "speech specific." Wideband high-fidelity sound (including music) can be accurately coded at between 1 and 2 bits/Nyquist sample. (Nyquist sampling is at twice the frequency of the highest spectral component.) Good-quality telephone bandwidth speech can be obtained at 8–16 kbits/s, corresponding to sampling at 8000 s^{-1} and quantization at 1–2 bits/sample. At lower rates some loss in quality occurs.

3 SPEECH SYNTHESIS AND RECOGNITION

Around 1970 accumulated advances in sampled data theory, computing, and microelectronics heightened the interest in enabling sophisticated machines to interact conversationally with humans.[4,5,8,9] This requires giving the machine a "mouth" to speak information to a human user (speech synthesis) and an "ear" to listen and (in some sense) understand human-spoken commands (speeh recognition). (See Fig. 5.)

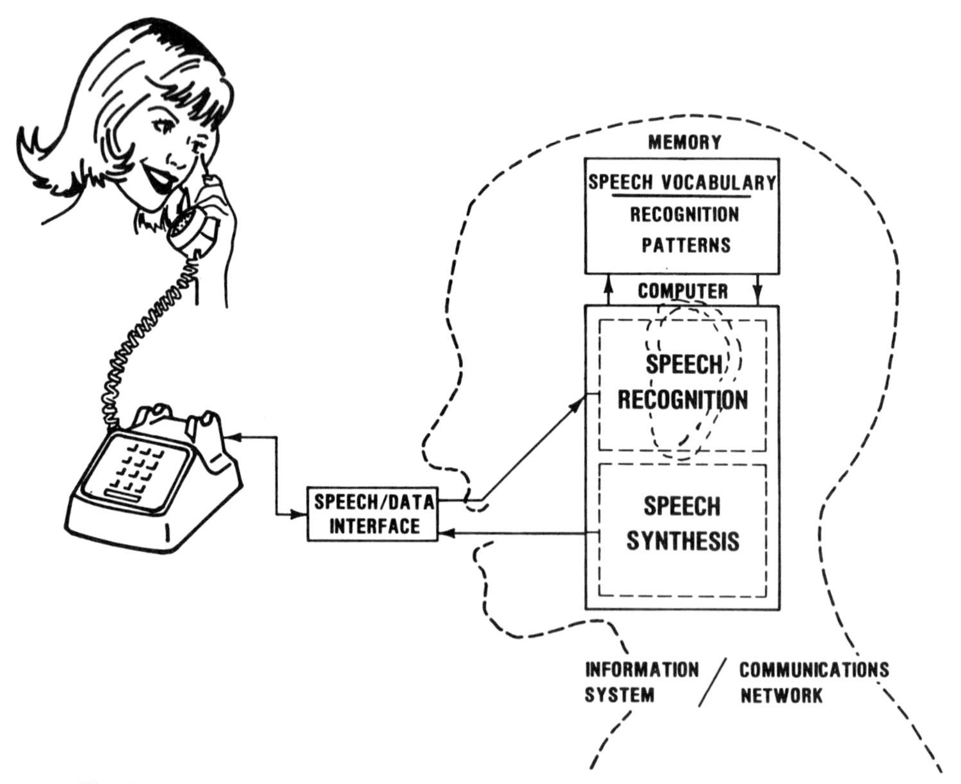

Fig. 5 Interactive voice communication with machines requires providing the machine the ability to speak (speech synthesis), and the ability to hear and, in a sense, understand (speech recognition).

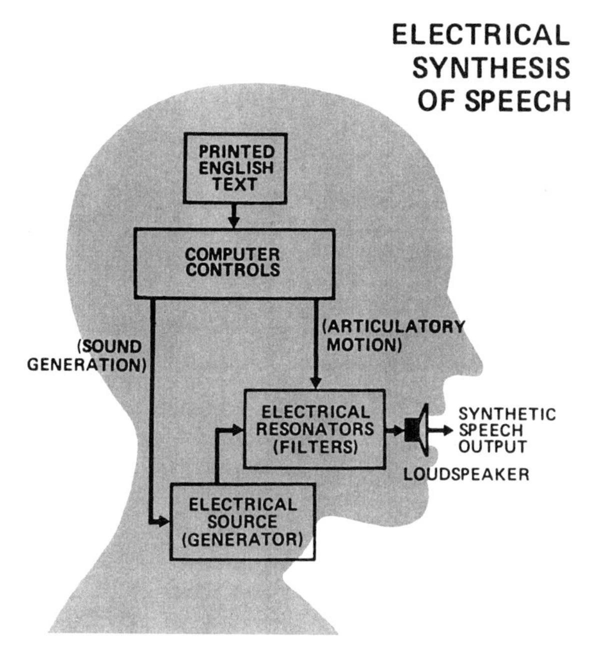

Fig. 6 Ingredients of speech generation for human and machine. Compare this diagram to the speech generation part of Fig. 1.

3.1 Speech Synthesis

Effective speech synthesis is based on fundamental understanding of the physics of speech production, and on the linguistic constraints that characterize a given language. The constituent parts of speech synthesis include the mechanisms of human sound generation, acoustic wave propagation in the vocal tract, dynamics and physiology of articulatory movement, and the language conventions that govern allowed sound sequences. These mechanisms must be simulated by the machine in producing synthetic speech, as illustrated in Fig. 6.

Ideally one would like to be able to synthesize the spoken equivalent of unrestricted printed text. This process is typically referred to as text-to-speech synthesis. With sufficient dictionary storage and computational power this is becoming possible (see Fig. 7). Intelligibility for unrestricted text synthesis is generally good, but signal quality is machinelike. The research frontiers lie in making the synthesis more natural sounding, duplicating varieties of voice characteristics (including personalizing the synthesis to individual voices) and synthesizing in a variety of different languages.

3.2 Speech Recognition

Ideally one would like to speak in a natural conversational voice to a machine and have it understand and respond as humans do. Current realities fall short of this capability.

The technology that is established permits high-reliability recognition of individual words and phrases chosen from vocabularies limited in size. A well-proven technique uses stored templates for recognition (Fig. 8). A vocabulary of templates, whose parameters describe the short-time amplitude spectra of the allowed commands, is stored digitally in the machine. An unknown input command is measured, and its spectrum is compared to all vocabulary entries. The comparison is made using an automatic time-alignment procedure, or *dynamic time warp*. The closest fitting match is asso-

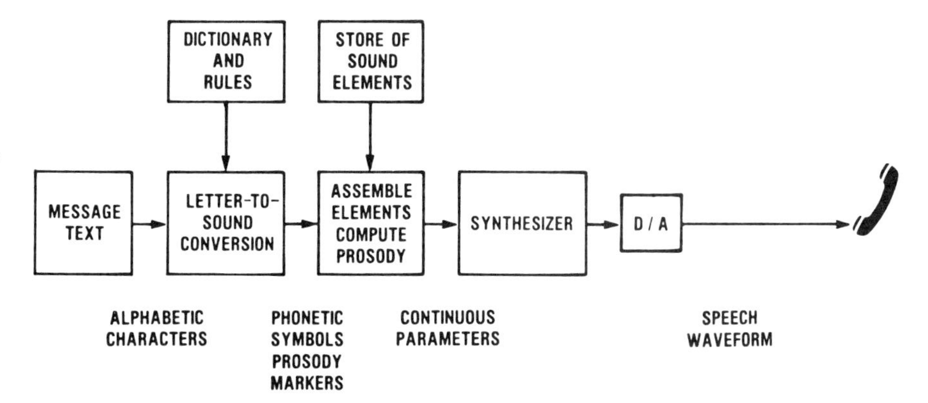

Fig. 7 Machine conversion of unrestricted printed text into natural-sounding speech requires transformation of the grapheme (alphabetic character) string into a phonetic equivalent, an understanding of the rules of language for the correct pronunciation of sounds and for voice inflection (pitch, intensity, and duration of successive sounds), and a means for controlling an electronic artificial vocal tract to yield an approximation to human speech.

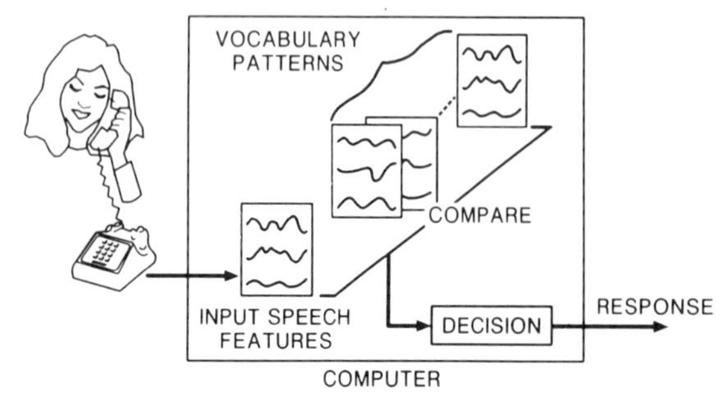

Fig. 8 Automatic speech recognition requires comparison of acoustic features measured for an unknown input with stored templates or statistical models for vocabulary items acceptable by the machine. The stored parameters describe short-time spectral characteristics of the vocabulary entries.

ciated with the input. The recognition choice typically is confirmed to the user by the synthetic voice of the machine. If no match is sufficiently close, the machine may decline to make a recognition, so announce, and possibly ask for a repetition of the input command.

Expanded sophistication and concomitant improvements in reliability and performance are being achieved by statistical models of sound sequences that go beyond the stored template approach. These statistical models provide probabilistic estimates of word patterns. A currently favored technique is based on the hidden Markov model (HMM) in which word and sound sequences are modeled in terms of states, probabilities of states, and probabilities of transitions among states. The models are deduced from observed sequences of spectra for speech.

The dimensions of the recognition task are at least three: (1) the number of commands that may be in the vocabulary, (2) the nature of the spoken input—isolated words or continuous speech, and (3) whether the stored templates or HMM models have to be specific to a designated user or can anyone talk to the machine—that is, speaker dependent (SD) or speaker independent (SI).

Recognition performance depends both on the size and the nature of the vocabulary. Performance generally diminishes with increasing vocabulary size, and recognition is more difficult if the items are brief and acoustically similar words (e.g., monosyllables). For small vocabularies, say several tens of entries, performance typically can be in the high 90% level. The spoken digits are both few and acoustically distinctive, and SI recognition accuracy exceeding 98% is typical. But some vocabularies may be similar acoustically—say, the spoken monosyllabic alphabetic symbols *a*, *b*, *c*, *d*, *e*, and so on, which might be used to spell names to an automated telephone directory-assistance system. The acoustical similarity makes recognition more difficult, and in this case additional leverage must be sought from syntactic constraints (such as the allowed names in a telephone directory).

The conventions of language constrain the allowed words in a sequence. By representing these statistical properties in the recognition algorithm, task accuracies can be maintained at usefully high levels (typically in the upper 90% levels), even though absolute recognition of individual words in a sequence is less reliable. A number of recognition products are now being deployed for communications applications such as voice repertory dialing (to establish a telephone connection simply by speaking the name of the called person), spoken number (digit) dialing, call routing, order entry, and database access and control.

Connected-word recognition requires about an order of magnitude more computation than does isolated-word recognition. State-of-the-art technology now permits SI recognition of connected 7-digit numbers with a string accuracy of about 98%. This is an SI application in which parallel computation and economical microprocessors are playing a key role. The nature of recognition from a vocabulary of templates or statistical (HMM) models is a natural fit to a parallel computing architecture. The individual processors can be assigned subsets of the total vocabulary and can make comparisons simultaneously to an unknown input utterance. The results of the comparisons made in parallel by the individual processors are therefore available nearly simultaneously for a final recognition decision.

The ultimate goal is to enable the machine to recognize fluent, unrestricted speech on any topic by any talker. This capability is not yet in sight. But continuous speech recognition for large vocabularies (on the order of several thousand words) on restricted or specified topic areas is in active research and significant advances can be anticipated.

Recognition using whole-word templates or models, which serve small vocabularies of isolated words so well, is not appropriate for the more ambitious capability of unrestricted fluent input because there are too many words. Rather, an approach based on subword phonetic units is more appropriate. There are a manageable number of these, and computationally tractable algorithms can make estimates of complete whole words from sequences of these units. A stored dictionary (lexicon) is helpful in making probabilistic estimates of the words. The estimates of whole-word candidates then must be followed by estimates of word sequences. Because the input is fluent whole sentences, a grammar and language model can be applied to the probabilistically hypothesized word sequences. This process identifies the most likely word candidates that compose a sentence acceptable in the grammar. Again the grammatical constraints, which strongly restrict word order, aid in obtaining the correct word string.

3.3 Talker Verification

Corollary to automatic speech recognition is differentiation of different talkers—that is, recognition of the talker rather than recognition of the utterance. Absolute identification of a talker from a large population is relatively difficult, and the identification error rate tends to increase with population size. But authentication or verification of a claimed identity is a more tractable problem, and one of much more interest in the business sector (for credit charging by telephone, for funds transfer, or for access to privileged information).

In this case, the unknown talker is asked to make an identity claim (say, provide a checking account number, which can either be keyed in by touch-tone dialing or spoken and SI recognized). The machine accesses from storage a spectral pattern corresponding to the claimed identity. This pattern has been created previously by analyzing training utterances provided by the authorized user in an earlier enrollment session. The pattern may also correspond to a confidential code phase typically of sentence length. Then the machine (by synthetic voice) requests the user to speak the assigned code phrase. It measures the spectral pattern of this response and compares it to the pattern corresponding to the claimed identity. If the match is close enough, the identity claim is accepted and the requested transaction is carried out (see Fig. 9). The gravity of the transaction can control the precision of match required by the machine (say, read-out of a checking account balance as compared to a payment of thousands of dollars). If the transaction is of great importance, the decision algorithm will keep the miss rate (acceptance of impostor) at a very small value at the expense of a somewhat higher false-alarm rate (rejection of authorized users).

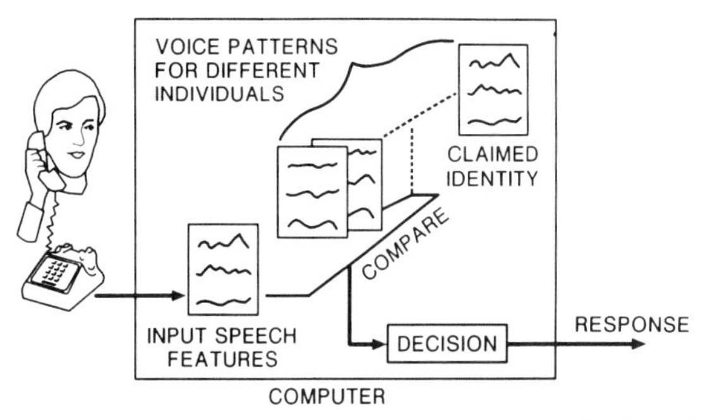

Fig. 9 Talker verification systems perform authentication of a claimed identity through measurements on the voice signal and comparison to a stored pattern known to characterize the claimed individual. A preassigned confidential code phase adds to the security. The machine measures the voice input, makes the comparison, and decides whether to accept or reject the identity claim.

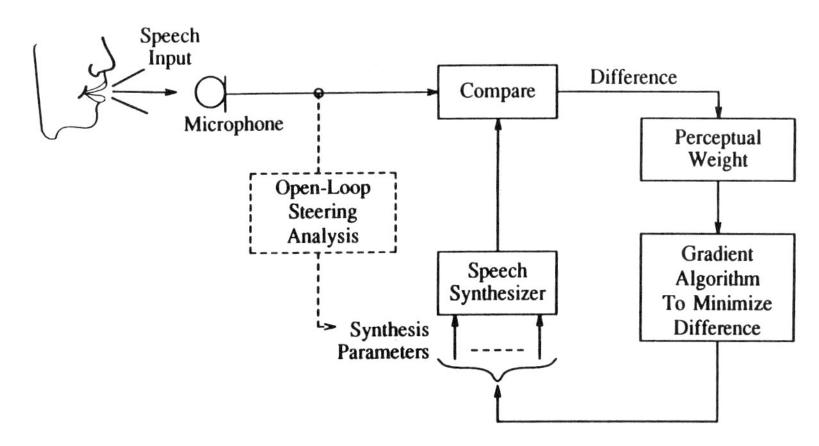

Fig. 10 The voice mimic approach to speech processing aims to coalesce the problems of speech coding, synthesis, and recognition and achieve a contemporaneous solution through determination of compact signal-defining parameters.

4 RESEARCH DIRECTIONS

Although speech coding, synthesis, and recognition have been discussed as separate topics, they partake of the same underlying knowledge of human communication. As this knowledge expands, there is the possibility that the problems may coalesce and be solved contemporaneously. One ambitious long-term project, a computer "voice mimic," aims in this direction (Fig. 10).

The parameters of a synthesizer are adjusted moment by moment to duplicate a natural input signal. The synthesizer models the physics of sound generation by the vocal cords, sound propagation and radiation by the vocal tract, and the dynamics of articulatory motion. Gradient-climbing algorithms, with perceptual weighting factors, attempt to minimize the difference between the speech generated by the synthetic mimic and the input natural speech signal. The adaptive parameter adjustment for the mimic can be attempted in several domains: text symbols, phonetic symbols, articulatory parameters, or short-time spectrum coefficients. The first two are discrete and the latter two are continuous. The first domain—using conventional text to make the adjustment—represents the ultimate solution—literally providing a voice typewriter! Simultaneously implied is the solution of speech coding with the greatest possible efficiency (namely, representation of the signal by the discrete printed equivalent of the information), as well as the achievement of speech synthesis from printed text with natural quality.

In laboratory experiments, the mimic algorithm can in fact follow arbitrary natural speech input. Presently, adaptation of continuous articulatory parameters provides the most favorable result, but experiments are extending the process to the discrete text domain. A sobering aspect is that the mimic algorithm requires about 1000 times real time to compute on a mainframe parallel processor. While this amount of computation is presently impractical for real-time implementation, the advances being made in parallel computation render the problem less formidable. And, the fundamental understanding that is being accumulated may one day lead to a truly coalesced solution for speech coding, synthesis, and recognition.

REFERENCES

1. G. C. M. Fant, *Acoustic Theory of Speech Production*, Mouton, Gravenhage, 1960.
2. J. L. Flanagan, *Speech Analysis, Synthesis and Perception*, 2nd ed., Springer Verlag, New York, 1972.
3. L. R. Rabiner and R. W. Schafer, *Digital Processing of Speech Signals*, Prentice-Hall, Englewood Cliffs, NJ, 1978.
4. S. Furui, *Digital Speech Processing, Synthesis, and Recognition*, Marcel Dekker, New York, 1989.
5. D. O'Shaughnessy, *Speech Communication: Human and Machine*, Addison-Wesley, New York, 1987.
6. N. S. Jayant and P. Noll, *Digital Coding of Waveforms*, Prentice Hall, Englewood Cliffs, NJ, 1984.
7. E. A. Lee and D. G. Messerschmitt, *Digital Communication*, Kluwer, Boston, 1988.
8. T. Parsons, *Voice and Speech Processing*, McGraw-Hill, New York, 1987.
9. M. Fagen and S. Millman, Eds., *A History of Science and Engineering in the Bell System*, 6 vols., AT&T Technologies, Indianapolis, IN, 1975–1985.

95

MODELS OF SPEECH PRODUCTION

KENNETH N. STEVENS

1 INTRODUCTION

This chapter reviews the mechanisms of generation of speech sounds as they occur in syllables, words, and sentences. The acoustic processes in speech production include the generation of sound sources by modulation of airflow in the vocal tract, either through vibration of the vocal folds or through turbulence in the flow. These sources are filtered by the vocal tract to produce a variety of vowel and consonant sounds. The filtering is characterized by resonances whose frequencies are determined by the shape of the vocal tract. Models that predict the sound from the properties of the sources and the transfer functions of the vocal-tract filters are described. When speech sounds occur in sequence to form words and sentences, the articulatory movements to produce the sounds are often influenced by the context, and hence the spectral and temporal attributes of the sounds may be modified relative to their properties in isolated syllables.

2 BASIC MECHANISMS OF SPEECH SOUND PRODUCTION

Speech sounds are produced by causing modulation of the airflow through constrictions in the airways between the larynx and the lips. This modulation of the flow gives rise to the generation of sound. One type of modulation arises from vibration of the vocal folds, which causes quasiperiodic changes in the space between the vocal folds (the glottis), and hence modulation of the volume flow through the glottis. Another type of modulation is a consequence of turbulence in the flow and hence the generation of turbulence noise. Transient sound sources can also be produced by raising or lowering the pressure behind a closure in the airway and then rapidly opening this constriction, causing an abrupt change in the pressure behind the constriction.

The sound sources cause acoustic excitation of the airways, and the filtering of the sources by the vocal tract gives rise to spectral prominences in the sound radiated from the lips or nose. These prominences are a consequence of the natural modes of the vocal tract. Depending on the locations and spectra of the sources, these modes can be excited in various amounts by the different types of sources.

The acoustic process involved in production of speech sounds can be modeled[1] as in Fig. 1. One or more sources, with spectrum $S(f)$, forms the excitation for an acoustic system with a transfer function $T(f) = U_o(f)/S(f)$, where $U_o(f)$ is the spectrum of the acoustic volume velocity at the mouth or nose, and a radiation characteristic $R(f) = p_r(f)/U_o(f)$, where $p_r(f)$ is the spectrum of the sound pressure at a distance r from the lips. Thus we have

$$p_r(f) = S(f) \cdot T(f) \cdot R(f). \qquad (1)$$

When an utterance is produced, the speaker manipulates the articulatory structures that control the shape of the airways at the larynx and above the larynx, and thereby controls the sources and the transfer functions that filter these sources. Throughout an utterance, the shape of the vocal tract alternates between a configuration that is relatively open and free of obstructions and configurations that produce narrow constrictions in the airway. Vowel sounds are generated during the open phase of this cycle, and consonants are generated during the more constricted phase.

Note: References to chapters appearing only in the *Encyclopedia* are preceded by *Enc.*

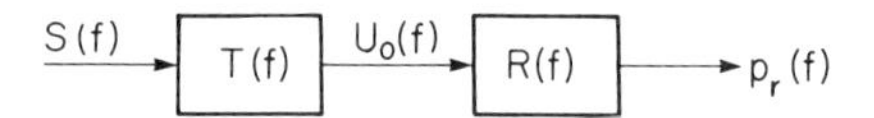

Fig. 1 Representation of speech sound production as a source $S(f)$ filtered by a transfer function $T(f)$ and a radiation characteristic $R(f)$ to give the spectrum of the radiated sound pressure $p_r(f)$.

This chapter begins with a brief review of some aspects of the anatomy and physiology of speech production. This review is followed by a discussion of models of sound source generation for speech, radiation of sound from the mouth, and vocal-tract transfer functions for vowels and consonants. Examples of spectra for several classes of vowel and consonant sounds in simple syllables are given. Finally there is a brief discussion of attributes of speech sounds when they occur in the context of words and sentences.

3 ANATOMY AND PHYSIOLOGY OF SPEECH PRODUCTION

3.1 Airways below the Larynx

A pressure is created in the lungs that causes the air to move through the various airways below and above the glottis. For almost all speech sounds, it is the outward flow of air from the lungs that is modulated to form the sound sources used to produce speech sounds. During an utterance consisting of a phrase or sentence, the pressure in the lungs is relatively constant, and the flow of air is controlled by manipulating one or more constrictions in the airway at the larynx or above the larynx. Maintenance of the constant lung pressure is achieved by contraction of muscles controlling the chest wall, the disphragm, and the abdomen, assisted by the pressure due to restoring forces that are a consequence of the elasticity of the pleural space when the lungs are inflated. Typical subglottal pressures during conversational speech are in the range 5–10 cm H_2O, but much higher pressures are used for producing loud speech or shouting. Although the total lung volume for an adult is 4000 cm^3 or more, the volume from one inspiration to the next during speech production is usually in the range 500–1000 cm^3. For most speech sounds, the fixed airways below the larynx do not have a significant influence on the radiated sound. However, resonances of this subglottal region can be observed in the speech of some individuals, particularly for sounds that are produced with a relatively open glottis. Typical frequencies for the first few subglottal resonances when they are observed in speech are roughly 700, 1500, and 2200 Hz for adult speakers.[2–4]

3.2 Laryngeal Anatomy and Physiology

The principal structures that constrict the airways in the laryngeal region are the vocal folds. The vocal folds are attached to the thyroid cartilage at their anterior ends and to paired arytenoid cartilages at their posterior ends. The arytenoid cartilages rest on the cricoid cartilage, a ring-like structure that surrounds the airway from the lungs. The length of the vocal folds is about 1.0 cm for a female adult to 1.6 cm for a male adult and is as short as 1 mm for a newborn child. The vocal folds consist of compliant tissue with a complex layered structure. For an adult, the thickness of the folds is in the range 2–3 mm, but the thickness decreases when the folds are stretched. When the vocal folds are appropriately positioned and the pressure is raised in the airways below the glottis, the folds are set into vibration and the airflow through the glottis is modulated periodically. The spectrum of this modulated flow is rich in harmonics. This periodic flow forms a monopole acoustic source that provides excitation for the airways above the larynx.

The vocal folds can be manipulated in two principal ways through contraction of sets of muscles that attach to the various cartilages of the larynx. One type of adjustment changes the stiffness of the vocal folds and modifies the frequency of vibration. The other manipulation changes the separation between the vocal folds, which can range from an adducted configuration, with the vocal folds pressed together, to a spread or abducted configuration. Vocal-fold vibration occurs only over particular ranges of stiffness and vocal-fold separation. Within the range of conditions for which vocal-fold vibration occurs, the waveform of the modulated flow can have a variety of shapes with different spectral characteristics, ranging from creaky voice (narrow glottal pulses, rich in harmonics) through modal voice to breathy voice (wide glottal pulses, weak high-frequency harmonics, accompanied by sound due to turbulent airflow near the glottis). The frequency of vibration of the vocal folds during normal speech production is usually in the range 170–340 Hz for adult females, 80–160 Hz for adult males, and 250–500 Hz for younger children. The frequencies can extend well beyond these ranges for the singing voice.

3.3 Articulatory Structures Controlling the Vocal-Tract Shape

The airways above the glottis can be shaped by several structures, as illustrated in Fig. 2. This shaping of the airways has two functions in the production of speech. One function is to form a narrow constriction at some point along the airway in order to produce a consonant sound. The formation or the release of the narrow constriction produces a particular type of discontinuity or

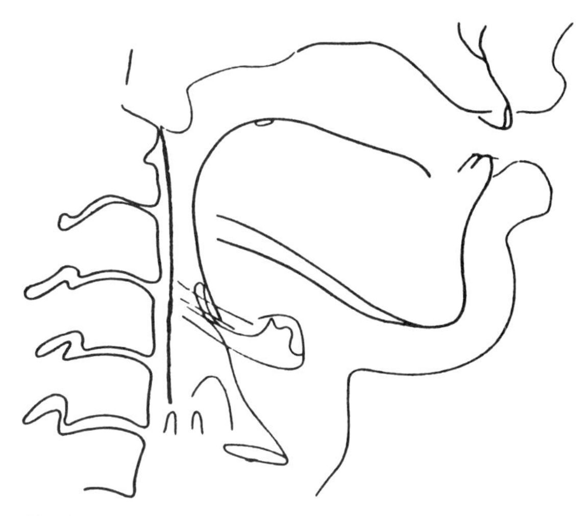

Fig. 2 Vocal tract and the articulatory structures surrounding the airways above the larynx. (From Ref. 5 by permission.)

abrupt modulation in the characteristics of the sound. A second function is to shape the airways to have particular natural frequencies so that the sound source or sources at or above the glottis are filtered by these resonances.

There are five principal anatomical structures that can be manipulated to change the vocal-tract shape. A set of constrictor muscles surrounds the pharyngeal part of the vocal tract, and contraction of these muscles can form a constriction in the airway in this region. The tongue body can be displaced vertically and horizontally through contraction of several groups of muscles. The tongue blade forms an extension of the tongue body and can be shaped or displaced upward to form a constriction in the region extending from the teeth to the hard palate. The lips can be shaped and displaced to form a narrow constriction or can be protruded to increase the effective length of the airway from the glottis to the lip opening. Finally, the soft palate can be raised or lowered to vary the cross-sectional area of the passage from the upper pharynx into the nasal cavity or to close this passage completely.

The total length of the vocal tract from the glottis to the lips for an adult is usually in the range 14–18 cm, with females being in the lower end of this range and males in the upper end. The pharyngeal, or vertical, portion of the vocal tract is roughly equal in length to the oral, or horizontal, portion. The volume of the airway is generally in the range 40–90 cm^3 with an average cross-sectional area of about 3.5 cm^2. As the vocal tract is shaped to produce different vowels and consonants, the cross-sectional area of the airway at different points along its length can vary from zero (i.e., a complete closure) to 10 cm^2 or more, with cross dimensions up to about 3 cm. These dimensions are such that sound propagation in the vocal tract is roughly one-dimensional in the frequency range up to about 5 kHz. That is, in this frequency range the acoustic behavior of the vocal tract can be approximated by considering it as a tube with variable cross-sectional area in which one-dimensional waves propagate.

The vocal-tract length for children is, of course, shorter than it is for adults, and in the first months of life the pharyngeal portion of the tract is considerably shorter than the oral portion. For a newborn infant, the overall vocal-tract length is about 8 cm.

4 AIRFLOWS, PRESSURES, AND SOUND SOURCES IN THE VOCAL TRACT

The flow of air in the vocal tract when there is a fixed subglottal pressure is normally controlled by adjusting the configuration of the larynx and/or of particular supralaryngeal structures to form one or more constrictions in the airway. The pressure drop ΔP across a constriction and the airflow U through the constriction are related approximately[6] by the equation

$$\Delta P = k\,\frac{\rho U^2}{2A^2}, \tag{2}$$

where A is the cross-sectional area of the constriction, ρ is the density of air, and k is a constant that is usually close to unity, but can vary by 10–20% depending on the shape of the construction.

When the vocal folds are placed close together and a subglottal pressure is applied, the vocal folds will vibrate. The main features of the mechanism of vibration can be understood by modeling each vocal fold as two coupled masses, one representing the lower portion of the fold and the other representing the upper portion.[7] The masses are coupled to each other and to the surrounding structures by elements that have stiffness, as shown in Fig. 3*a*. The figure also shows how the lower and upper masses move when a subglottal pressure is applied. The subglottal pressure exerts an outward force on the lower masses, and these masses accelerate in response to this force. When the displacement of the lower mass reaches a particular value (point A in Fig. 3*b*), the mechanical coupling to the upper mass causes the upper masses to separate (point B). As the upper masses move apart, air flows through the glottis, and the Bernoulli pressure drop results in a decrease in pressure within the lower portion of the glottis. The outward force on the lower masses decreases, and these masses

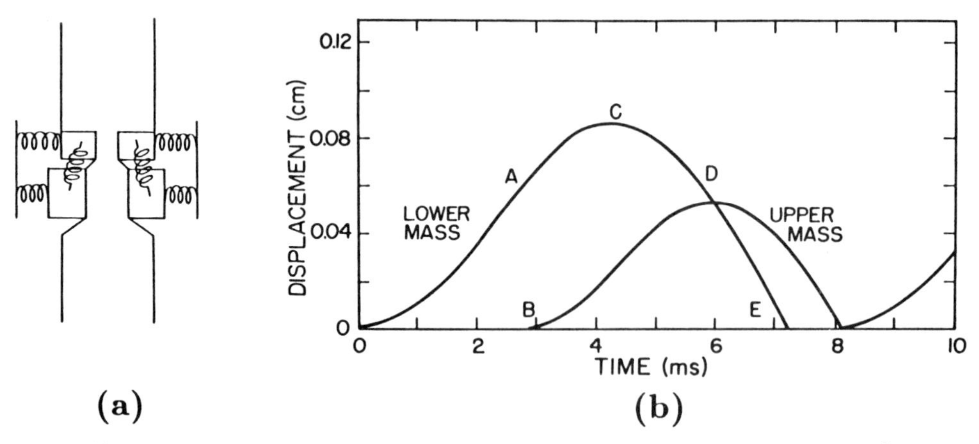

Fig. 3 (*a*) Representation of the vocal folds as two pairs of coupled masses with springs.[7] (*b*) Lateral displacement of lower and upper masses in (*a*) when a pressure is applied below the glottis.

return to a more closed position (trajectory *ACDE*) as a consequence of passive forces from the stiffness elements. Through mechanical coupling, the upper masses are drawn inward as the lower masses come together and abruptly cut off the flow of air (at point *E* in the figure). The process then repeats itself as the subglottal pressure once again exerts an outward force on the lower masses.

An approximation to the derivative of the waveform of the glottal airflow for a male voice is shown at the bottom of Fig. 4. [The derivative is shown rather than the airflow waveform, and thus includes the effect of the radiation characteristic $R(f)$ in Fig. 1, which is equivalent to differentiation, as will be shown later.] The differentiated waveform of each pulse has a negative peak, with an abrupt return to zero following this peak. The spectrum of this periodic flow is displayed above the waveform. The spectrum has a series of harmonics at multiples of the fundamental frequency. At high frequencies (above about 1 kHz), these amplitudes decrease at about 6 dB/octave or sometimes at a somewhat greater rate. A speaker controls the frequency of vocal-fold vibration by contracting muscles that cause a change in the tension on the vocal folds, and some variation in the waveform is achieved by adjusting the resting configuration of the folds. Changes in the subglottal pressure cause an increase or decrease in the amplitude of the glottal pulses. The waveform of the glottal airflow can be approximated as a volume–velocity source, since the impedance of the glottal opening is usually large compared to the impedance looking up from the glottis into the vocal tract.[1,8]

When the velocity of the airflow through a constriction in the vocal tract or at the larynx is sufficiently high, turbulence occurs in the airstream and can result in the generation of noise. Usually, this turbulence is a consequence of the rapid airflow impinging on a surface or an obstacle (such as the lower teeth). The noise in this case can be considered to be generated by fluctuating forces exerted by the obstacle on the airstream. This type of noise source can be modeled as a dipole or sound pressure source located in the airway in the region where the turbulence occurs. The spectrum of this source depends on the flow and on the configuration of the constriction and the obstacle, but experimental data with mechanical models[9] show a spectrum of the form given in Fig. 5, for flows in the range normally observed in speech. The amplitude of this source increases roughly as the third power of the flow velocity. Under some circumstances, random fluctuations occur in the flow through a constriction in the vocal tract. This type of turbulence creates a monopole source at the constriction.

Still another type of sound source that is used in speech is a transient source that occurs when pressure is built up behind a closure in the vocal tract and this closure is suddenly released. A transient source can also be produced by creating a partial vacuum in an enclosed space that is formed by the tongue and lips, and then releasing this vacuum. This type of source is used to generate clicks in some languages.

5 RADIATION OF SOUND FROM THE HEAD; RADIATION IMPEDANCE

For most speech sounds, acoustic energy is radiated from the mouth opening. If $p_r(f)$ is the spectrum of the sound

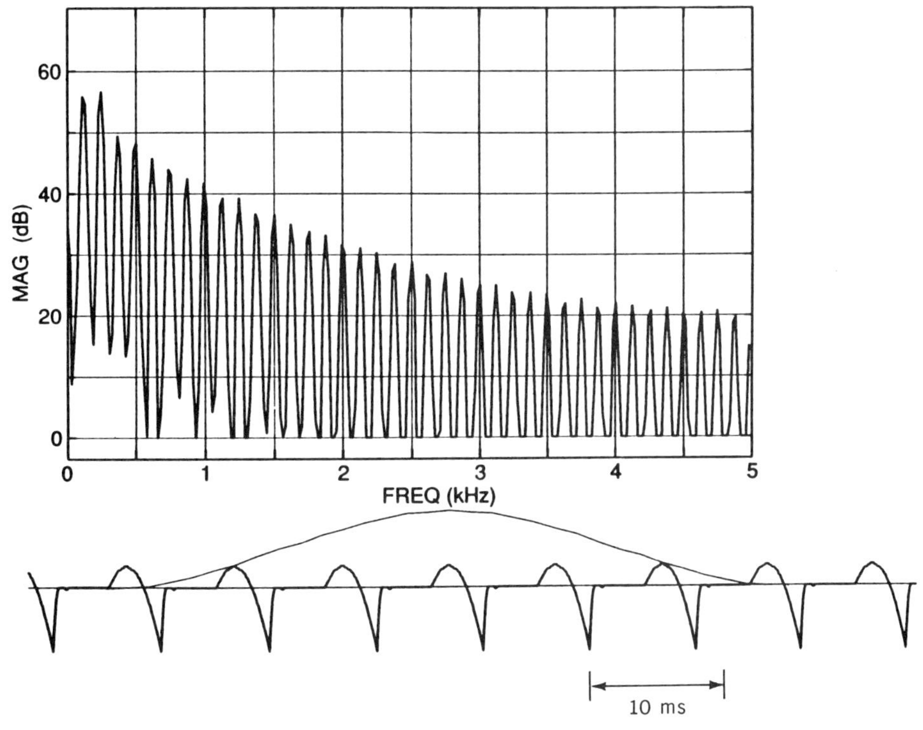

Fig. 4 Waveform at the bottom is a model of the derivative of the glottal airflow when the vocal folds are vibrating (based on a model of Klatt and Klatt[4]). The spectrum of this waveform is shown at the top.

pressure at a distance r from the mouth opening, and $U_o(f)$ is the acoustic velocity at the lips, then a radiation characteristic can be defined as

$$R(f) = \frac{p_r(f)}{U_o(f)}. \tag{3}$$

This function is approximately equal to the radiation characteristic of a piston in a sphere,[1,10] with the sphere representing the head and the piston the mouth opening. It is of the form

$$R(f) = K(f)\,\frac{j\rho f}{2r}\,e^{-jkr}, \tag{4}$$

where $k = 2\pi f/c, c$ is the velocity of sound, and $K(f)$ is a factor that depends on the head size and the direction relative to the midline of the piston.

At low frequencies, for which the wavelength is large compared with the head size, $K(f) \cong 1$, and the radiation characteristic reduces to that of a simple monopole source. The sound pressure amplitude is proportional to frequency, and it decreases at 6 dB per doubling of distance. In the time domain, the factor f in Eq. (4) is equivalent to differentiation of the waveform. At higher frequencies, the magnitude of the factor $K(f)$ in front of the mouth opening on the midline rises from 0 dB at low frequencies to about 5 dB in the frequency range 2–5 kHz. The factor $K(f)$ is a function of angle from the midline. For example, in the frequency range 1–5 kHz, there is a decrease in $K(f)$ of 1–5 dB at an angle of 90° in the

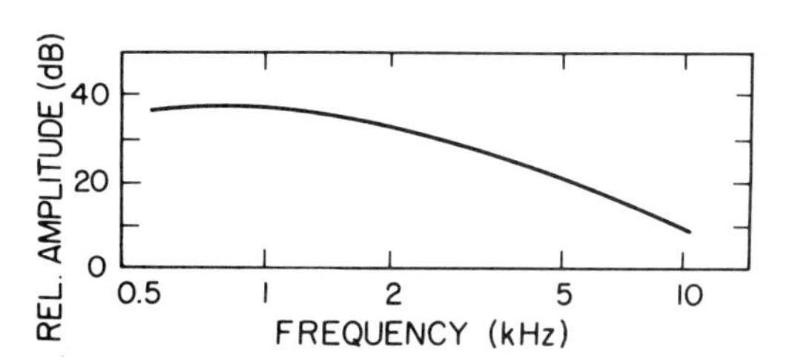

Fig. 5 Spectrum of sound pressure source due to turbulence in the vocal tract (based on measurements with mechanical models[9]).

horizontal plane, relative to its value in the midline.[8,10] Below 1 kHz, the variation with angle is less than 1 dB.

The radiation impedance at the mouth opening forms the acoustic termination of the vocal tract at its upper end. This radiation impedance can be approximated as the radiation impedance of a piston in a sphere with appropriate dimensions[1,10] and can be written as

$$Z_r = \frac{\rho c}{A_m}\left(\frac{\pi f^2}{c^2} A_m\right) K_s(f) + j2\pi f \frac{\rho(0.8a)}{A_m}, \qquad (5)$$

where A_m is the area of the mouth opening, a is the radius ($\pi a^2 = A_m$), and $K_s(f)$ is a dimensionless frequency-dependent factor that accounts for the baffling effect of the head. The factor K_s is unity at low frequencies, up to a few hundred hertz, and then rises to a maximum value of about 1.7 at 2000 Hz. Above 2000 Hz, the behavior of K_s with frequency depends on the size of the mouth opening; but for mouth areas of principal interest for most speech sounds, it is a reasonable approximation to take K_s as 1.5 in the frequency range 2–6 kHz. The reactive term in Z_r is written in a form to show that the radiation reactance is that of an acoustic tube of area A_m and length $0.8a$. The resistive part of Z_r in Eq. (5) increases as f^2, but at all frequencies of interest in speech is much smaller than $\rho c/A_m$, the characteristic impedance of a tube of area A_m.

6 VOWEL PRODUCTION

Vowel sounds are normally produced with a source at the glottis and with the airways above the glottis configured such that any narrowing in the airways is not sufficient to cause a buildup of pressure behind the constriction. For purposes of modeling the acoustic behavior of the vocal tract for vowels, the configuration is specified by the cross-sectional area of the tract as a function of distance from the glottis, called the area function.

The transfer function of the vocal tract for vowels is the ratio of the complex amplitudes of the volume velocity U_o at the lips to the volume velocity U_g at the glottis, that is, $T(f) = U_o(f)/U_g(f)$. This transfer function has a relatively simple form for the special case in which the area function is uniform, the acoustic losses are neglected, and the radiation impedance at the mouth opening is assumed to be small. For a tube of length l, this transfer function is

$$T(f) = \frac{1}{\cos(2\pi f l/c)}. \qquad (6)$$

The transfer function has only poles, and these are at frequencies $F_1 = c/4l$, $F_2 = 3c/4l$, $F_3 = 5c/4l, \ldots$. These are called formant frequencies, and they are the natural frequencies of the vocal tract when it is closed at the glottis end. The velocity of sound, c, at body temperature is 35,400 cm/s.

If the effects of losses and of the finite radiation impedance are included, the poles of the transfer function are perturbed slightly from the values given above. The reactive part of the radiation impedance in Eq. (5) causes a small downward shift in the natural frequencies, and the resistive part introduces a finite bandwidth to the poles. In addition, there are acoustic losses due to the finite impedance of the walls of the vocal tract, the effects of viscosity and heat conduction, and the impedance of the glottal opening. Experimental data, as well as theoretical analysis of the acoustic losses in the vocal tract,[8,11] show that the bandwidths of the prominences in $|T(f)|$ are usually in the range of 60–100 Hz for F_1, 80–150 Hz for F_2, and progressively greater for higher formants (cf. Chapter 96).

A plot of the transfer function $|T(f)|$ for a uniform vocal tract of length 17 cm (corresponding to a typical adult male vocal tract) with a mouth opening of 3.0 cm^2 is shown in Fig. 6*a*. The peak-to-valley ratio for $|T(f)|$ is about 18 dB at low frequencies and somewhat less than this at higher frequencies. For this vocal-tract length, the formant frequencies are at 498, 1494, 2490, ... Hz. The effect of the radiation reactance has been included in these estimates.

If this transfer function is combined with the source spectrum in Fig. 4 [which includes the effect of the radiation characteristic in Eq. (4)], the vowel spectrum in Fig. 6*b* is obtained. The amplitudes of the spectral prominences corresponding to the formants decrease with increasing frequency. The lowest spectral prominence, corresponding to the first formant, has the highest amplitude. The overall sound pressure level of the vowel is determined primarily by the spectrum amplitude of this prominence.

When the shape of the vocal tract is different from that of a uniform tube, the formant frequencies are shifted from the uniformly spaced values shown in Fig. 6. The frequencies of the formants for a nonuniform vocal tract can be calculated from the one-dimensional wave equation for a nonuniform tube. A rough indication of the way in which the formant frequencies shift when the shape of the vocal tract is modified relative to a uniform shape can be obtained using a perturbation method.[12,13] In this method, the distribution of potential and kinetic energy in the standing wave for each of the modes of the tube is calculated. The shape of the tube is now perturbed so that a small increase or decrease in the cross-sectional area is made in a local region of the tube. Perturbation

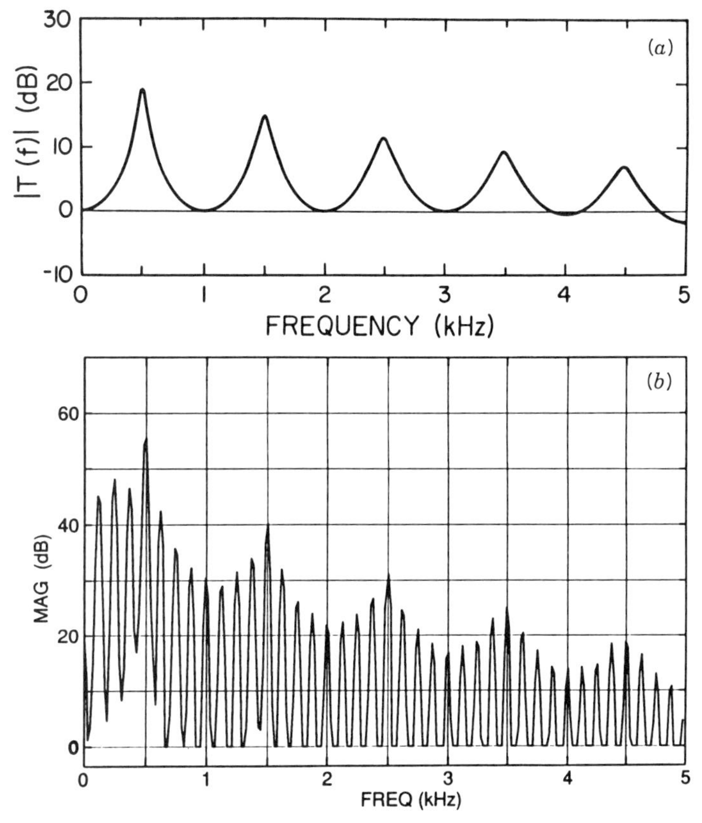

Fig. 6 (*a*) Transfer function of vocal tract with a configuration having a uniform cross-sectional area. (*b*) Calculated spectrum of a vowel produced with this uniform vocal tract.

theory shows that if a narrowing of the tube is made in a region that is near a maximum in the kinetic energy distribution for a particular mode, then the frequency of that mode will decrease. A similar reduction in area near a maximum in the potential energy distribution causes an increase in the frequency of the mode. The shifts in the frequencies are in the opposite direction if the perturbation in area is an increase rather than a decrease.

Figure 7 shows outlines of the vocal-tract shape in the midline for three different vowels: /i/ as in b*ea*t, /ɑ/ as in f*a*ther, and /u/ as in b*oo*t. Spectra for these three vowels are also sketched in the figure. The spectra show prominences at the frequencies of the formants. The three vocal-tract shapes illustrate three different ways in which the formant frequencies for a uniform tube can be perturbed: a lowered F_1 and a raised F_2 for /i/, a raised F_1 and a lowered F_2 for /ɑ/, and both F_1 and F_2 lowered for /u/. The tongue-body positions giving rise to these formant patterns are: a raised and fronted tongue body for /i/, a lowered and backed tongue body for /ɑ/, and a raised and backed tongue body together with protrusion and narrowing of the lips for /u/. There are differences in the relative amplitudes of the spectral prominences for the different vowels as a consequence of interactions between the contributions of individual formants in the all-pole transfer function.

Languages differ in the inventory of vowels that are used to distinguish between words, although most languages include the three vowels in Fig. 7 in their inventories. A listing of the vowels in English, together with their formant frequencies for adult males, adult females, and children, is given[14] in Table 1. In addition to these vowels, English also contains diphthongs or diphthongized vowels, as in the words *bite* or *bait*.

In some languages, a distinction is made between nasal and nonnasal vowels. To produce a nasal vowel, an opening is created between the oral cavity and the nasal cavity by lowering the soft palate. The resulting config-

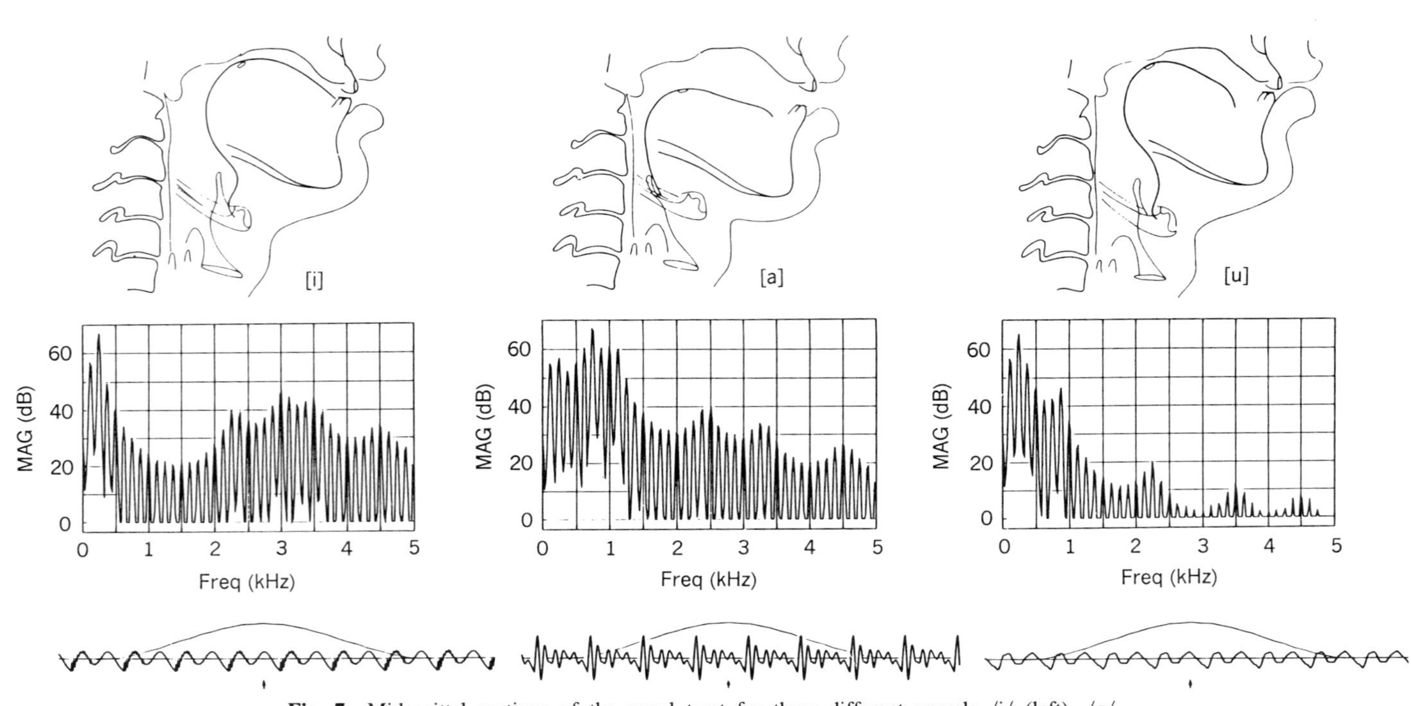

Fig. 7 Midsagittal sections of the vocal tract for three different vowels /i/ (left), /ɑ/ (middle), and /u/ (right) are shown at the top.[5] Below each configuration is the calculated spectrum for the vowel, together with the waveform.

TABLE 1 Frequencies of First Three Formants (Hz) for Vowels in English for Adult Males (M), Adult Females (F), and Children (Ch)

	heed	*hid*	*head*	*had*	*hawed*	*hod*	*hud*	*hood*	*who'd*
				Format F_1					
M	270	390	530	660	730	570	640	440	300
F	310	430	610	860	850	590	760	470	370
Ch	370	530	690	1010	1030	680	850	560	430
				Formant F_2					
M	2290	1990	1840	1720	1090	840	1190	1020	870
F	2790	2480	2330	2050	1220	920	1400	1160	950
Ch	3200	2730	2610	2320	1370	1060	1590	1410	1170
				Formant F_3					
M	3010	2550	2480	2410	2440	2410	2390	2240	2240
F	3310	3070	2990	2850	2810	2710	2780	2680	2670
Ch	3730	3600	3570	3320	3170	3180	3360	3310	3260

Note: The vowels are identified by the words at the top of the columns. From Ref. 14.

uration can be modeled by the interconnection of tubes shown in Fig. 8*a*. The acoustic coupling to the nasal cavity causes a modification of the transfer function from the glottal source to the output, which is now the sum of the volume velocities U_n and U_o at the nose and mouth. The principal modification is the introduction of additional poles and zeros in the transfer function, and there are also shifts in the frequencies of the formants relative to those for the nonnasal configuration. The lowest additional pole and zero has the most consistent and perceptually the most salient effect on the spectrum. The natural frequencies of the coupled system are the frequencies for which the sum of the susceptances looking in the three directions at the coupling point is zero. If most of the energy is radiated from the mouth, the zeros in the transfer function are at the frequencies for which the impedance looking into the nasal cavity is zero.

Figure 8*b* shows a typical spectrum that would be obtained if the vowel /i/ as in b*ee*t were nasalized. This spectrum has an extra pole at 810 Hz and a zero at 1260 Hz, and the first formant frequency is shifted upward relative to its value for the nonnasal configuration (Fig. 7*a*). For greater or smaller areas of the passage to the nasal cavity, there is an increased or decreased distance between the additional pole and zero, resulting in a greater or smaller perturbation of the spectrum.

7 PRODUCTION OF CONSONANTS

Consonants are produced with a relatively narrow constriction in some region of the airway above the larynx, whereas for vowels the airway is more open. Consonants can be classified along several dimensions depending on the articulator that is responsible for making the constriction in the vocal tract, the degree of constriction, the state of the glottis and the vocal folds when the constriction is formed, and whether or not pressure is built up behind the constriction. Thus the consonant /b/ is produced by forming a complete closure at the lips, by maintaining vibration of the vocal folds, and with a pressure buildup behind the constriction. For the consonant /n/, on the other hand, the closure is formed with the tongue blade, the vocal folds continue to vibrate, and no pressure is built up behind the constriction.

If the consonantal constriction is sufficiently narrow, there is a pressure drop across the constriction when a subglottal pressure is applied, assuming that there is no bypass path for the flow, such as through the nasal passages. Within this class of consonants, called obstruents, stop consonants are produced when a complete closure is formed, and airflow through the constriction occurs only after the closure is released. At this time of release, a brief burst of turbulence noise is formed by the rapid flow of air through the constriction. Fricative consonants, on the other hand, are produced with a narrow constriction, and turbulence noise is generated in the airflow throughout the time interval in which the constriction is in place.

As noted above, the turbulence noise source can be modeled as a sound pressure source in the vocal tract. An example of an obstruent consonant is the fricative sound /s/, for which the vocal-tract shape in the midline is shown in Fig. 9*a*. For this consonant, the airflow through

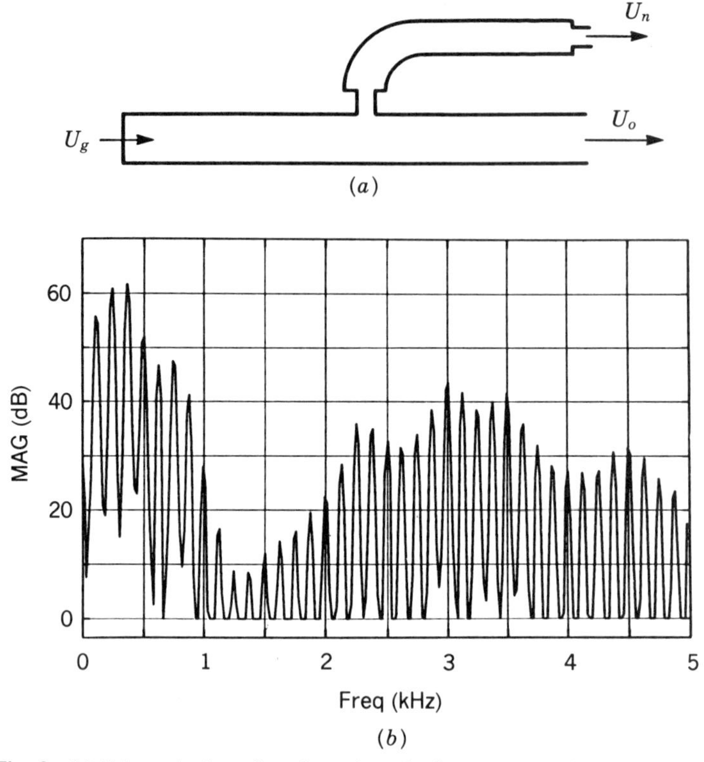

Fig. 8 (*a*) Schematization of configuration of tubes corresponding to a nasal vowel. (*b*) Calculated spectrum for a nasalized vowel /ĩ/, showing an additional pole at about 800 Hz and a zero at about 1300 Hz.

the constriction is directed against the lower incisors, and the principal source of turbulence noise is located in the vicinity of these teeth. The generation of sound for this consonant can be modeled by a constricted tube with an obstacle, shown in Fig. 9*b*. The noise source is represented as a pressure source p_s. Typical dimensions for the /s/ configuration are length of front cavity $l_f = 2$ cm (including correction for the radiation reactance), distance from constriction to obstacle 1 cm, and cross-sectional area of front cavity 1 cm^2. The magnitude of the transfer function U_o/p_s exhibits a peak corresponding to the front-cavity resonance, which is at about 4400 Hz in this case, and a zero in the transfer function that is close to zero frequency.

We now take the source spectrum in Fig. 5 and add (in decibels) the transfer function and the radiation characteristic [Eq. (4)] to obtain the spectrum in Fig. 9*c*. This spectrum derived from the model is similar to measured spectra of the frication noise for /s/ for adult male speakers. A similar spectrum is obtained for the brief noise burst that is generated at the release of the stop consonants /d/ and /t/, which are produced with a constriction at a location similar to that for /s/.

When an obstruent consonant is produced with a constriction at a different place in the vocal tract, the length of the front cavity of the model in Fig. 9*b* is different, and the source strength and its location in relation to the constriction may change. Thus the spectrum of the radiated sound will have peaks at different frequencies and with various degrees of prominence depending on the place of the constriction. This spectrum, then, provides some information concerning the location of the constriction in the vocal tract.

After the release of a consonant, as the articulators move toward configurations appropriate for the following vowel, the frequencies of the formants undergo changes. These transitions of the formants provide additional information concerning the place in the vocal tract where the consonantal constriction is located.[15] Obstruent consonants can also be distinguished from each other based on whether or not the vocal folds are caused to vibrate during the time interval when the

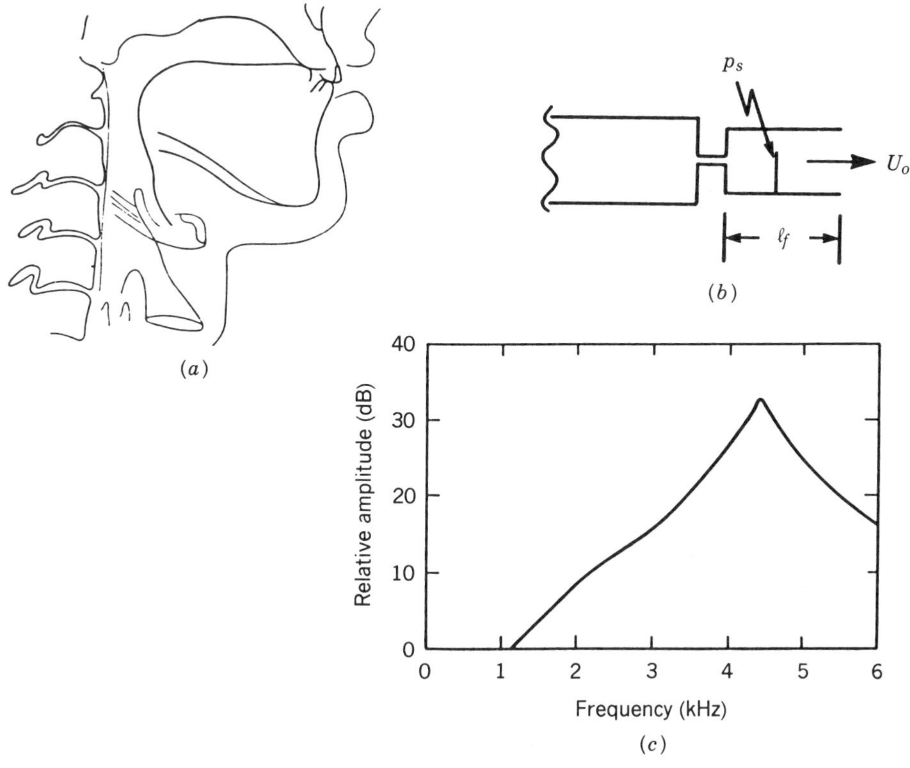

Fig. 9 (*a*) Midsagittal section of the vocal tract for the fricative consonant /s/. (*b*) Model of anterior part of the vocal tract for /s/, showing the location of the source p_s near the lower incisors. (*c*) Calculated spectrum of radiated sound pressure for /s/, based on the model in (*b*).

constriction is in place. Examples of spectrograms of syllables containing obstruent consonants /s/, /d/, and /g/ are given in Fig. 10. The transitions of the formants, particularly F_2 and F_3, are evident in the sound following the consonant release, as are the different spectra of the noise for the consonants /d/ and /g/, which are produced with constrictions at different places in the vocal tract.

In contrast to obstruent consonants, sonorant consonants are produced with essentially no pressure buildup in the vocal tract behind the constriction, and consequently little or no turbulence noise is generated. The principal source of acoustic energy is at the glottis for these consonants. For several of the consonants in this class, the vocal tract is shaped in such a way that the acoustic path from the glottis to the output (either the nose or the mouth) contains a side branch or is split into more than one path. For example, the nasal consonant /m/ is produced with an opening to the nasal cavity, as in the model of a nasal vowel in Fig. 8*a*, but with the oral cavity closed at the lips. In this configuration, the side branch to the main acoustic path is the mouth cavity anterior to the place where there is acoustic coupling to the nasal cavity. For a configuration of this kind, the transfer function from source to output contains zeros as well as poles. When the consonant is released into a vowel, there are rapid shifts in some of the poles and zeros in the transfer function—particularly the zeros—as the output shifts from the nose to the mouth, and these movements cause rapid changes in the spectrum of the sound. The abrupt change in the spectrum shape at the consonant release can be seen in the spectrogram of the syllable /mɑ/, at the right of Fig. 10.

8 SPEECH SOUNDS IN WORDS AND SENTENCES

When speech sounds are produced in utterances larger than consonant–vowel and vowel–consonant syllables, the articulatory movements that are implemented to generate sequences of sounds are often modified relative to the movements that occur for simple syllables, partic-

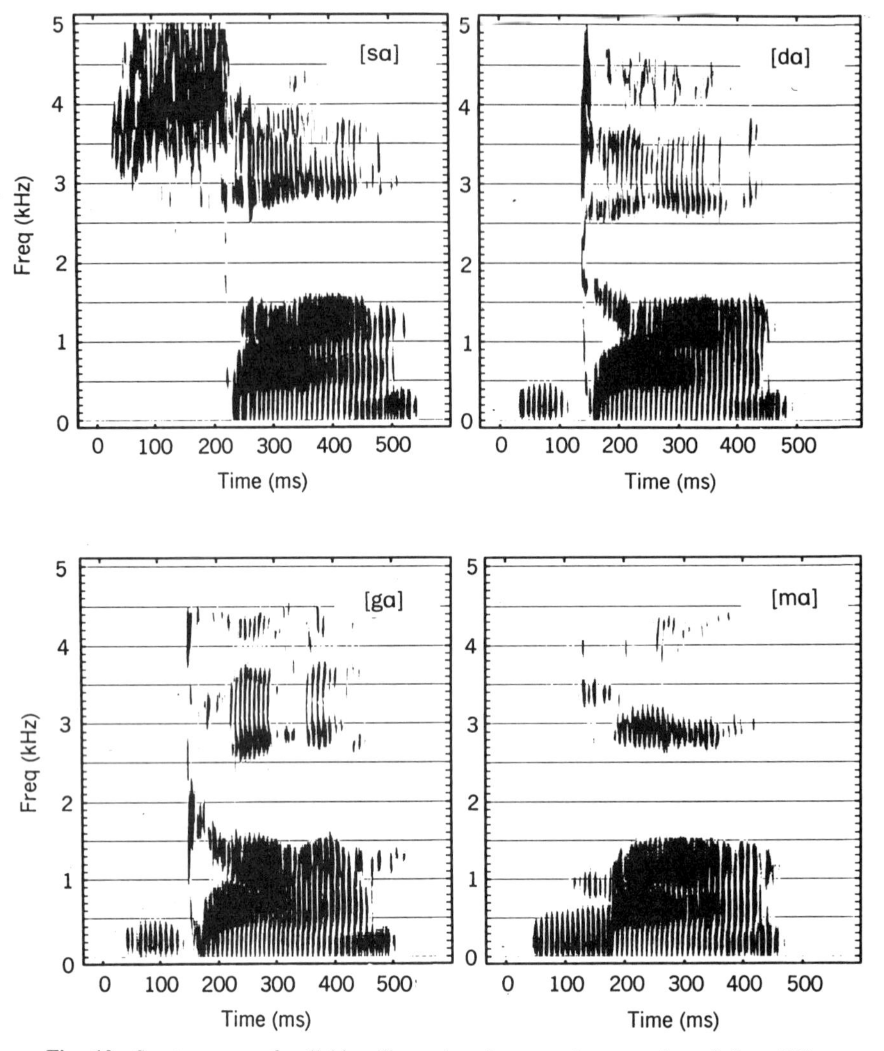

Fig. 10 Spectrograms of syllables illustrating the acoustic properties of four different consonants.

ularly when the speaker is talking in a casual fashion. Speech production models that incorporate and account for these sentence-level phenomena are not well developed, but a few examples for English can be given. A spectrogram of an utterance illustrating these effects is displayed in Fig. 11. The figure also shows the contour of frequency of vocal-fold vibration versus time.

There are several places in this utterance where sequences of two or three consonants occur, and some modification of these consonants takes place. These are indicated by the italicized consonants in the sentence: The ga*s sh*ortage was *felt* i*n th*at cou*ntr*y. For both consonants in the sequence *s–sh* (300–400 ms in Fig. 11), the tongue blade is the principal articulator forming the constriction, but the blade is normally shaped and positioned differently for the two consonants. For this sequence, speakers change the first consonant to take on the characteristics of the second. In the cluster *z–f* (in *was felt*, 750–880 ms), the first fricative is normally produced with vocal-fold vibration. When it is followed by a voiceless consonant, however, it can become devoiced, that is, both consonants are produced with no vocal-fold vibration. In the case of *l–t* in *felt* (950–1050 ms), there is no evidence that the tongue blade makes contact with the palate (as might be expected for /l/) before closure is made for the stop consonant /t/. Thus the sequence

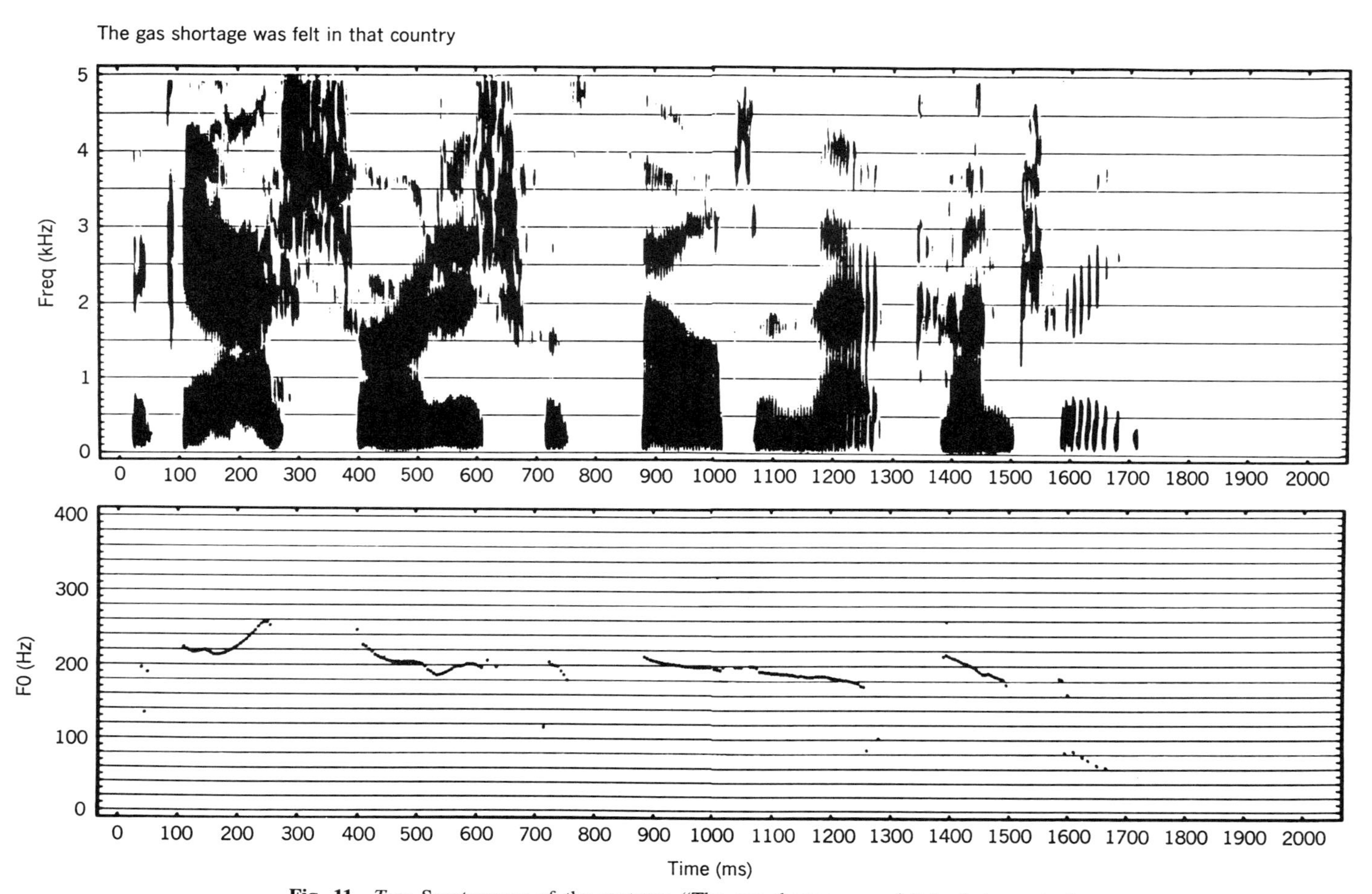

Fig. 11 *Top:* Spectrogram of the sentence "The gas shortage was felt in that country." *Bottom:* Plot of fundamental frequency vs. time for the sentence.

/ɛl/ in this word is produced as a sequence of vowel-like sounds, with no consonantal closure for /l/. The *th* sound in the words *in that* (1150 ms) is not produced as a fricative consonant, and the *n–th* becomes a long nasal consonant. Finally, for the consonant cluster *tr* in *country* (1520–1580 ms), the /t/ is produced with the tongue blade retracted, under the influence of the following /r/. For all of these consonant sequences, some attribute of a consonant is modified by an adjacent sound. These modifications are typical of everyday speech.

The spectrogram and fundamental frequency contour in Fig. 11 also show patterns of speech production that extend over time intervals beyond the levels of the speech sound and the syllable. Thus, for example, there are substantial differences in the durations of the vowels, ranging from about 20 ms for the first vowel in the sentence to 140 ms for the vowel in *gas*. Although some of these differences are inherent to particular vowels, the position of the vowel in a word or phrase can also influence the duration. For example, vowels in words with several syllables tend to be shorter than vowels in one-syllable words, and vowels in words near the end of an utterance are longer than vowels in the middle of the utterance.

The fundamental frequency contour of an utterance, such as the one shown in Fig. 11, is implemented by a speaker to mark the grouping of words within an utterance and to indicate words that are to receive special emphasis. In the fundamental frequency (F_0) contour in Fig. 11, for example, there is a peak in F_0 in *gas*, followed by a fall in the next few syllables. Another small peak occurs in the word *felt*, and there is a larger peak in the first syllable of *country*, again with a following fall in F_0. With a contour of this type, the speaker is grouping the first three words into one unit and is placing a kind of focus or emphasis on the last word.

REFERENCES

1. G. Fant, *Acoustic Theory of Speech Production*, Mouton, The Hague, 1960.
2. G. Fant, K. Ishizaka, J. Lindqvist, and J. Sundberg, *Subglottal Formants*, Speech Transmission Lab, Quarterly Progress and Status Report 1, Royal Institute of Technology, Stockholm, 1972, pp. 1–12.
3. K. Ishizaka, M. Matsudaira, and T. Kaneko, "Input Acoustic Impedance Measurements of the Subglottal System," *J. Acoust. Soc. Am.*, Vol. 60, 1976, pp. 190–197.
4. D. H. Klatt and L. C. Klatt, "Analysis, Synthesis, and Perception of Voice Quality Variations Among Female and Male Talkers," *J. Acoust. Soc. Am.*, Vol. 87, 1990, pp. 820–857.
5. J. S. Perkell, *Physiology of Speech Production: Results and Implications of a Quantitative Cineradiographic Study*, MIT Press, Cambridge, MA, 1969.
6. K. N. Stevens, "Airflow and Turbulence Noise for Fricative and Stop Consonants," *J. Acoust. Soc. Am.*, Vol. 50, 1971, p. 1180–1192.
7. K. Ishizaka and M. Matsudaira, "What Makes the Vocal Cords Vibrate?" in Y. Kohasi (Ed.), *The Sixth International Congress on Acoustics*, Vol. II, Elsevier, New York, 1968, pp. B9–B12.
8. J. L. Flanagan *Speech Analysis, Synthesis, and Perception*, Springer-Verlag, New York, 1972.
9. C. Shadle, "The Acoustics of Fricative Consonants," RLE Technical Report 506, Massachusetts Institute of Technology, Cambridge, MA, 1985.
10. P. M. Morse, *Vibration and Sound*, Acoustical Society of America, New York, 1976.
11. G. Fant, *Vocal Tract Wall Effects, Losses, and Resonance Bandwidths*, Speech Transmission Lab. Quarterly Progress and Status Report 2–3, Royal Institute of Technology, Stockholm, 1972, pp. 28–52.
12. M. R. Schroeder, "Determination of the Geometry of the Human Vocal Tract by Acoustic Measurements," *J. Acoust. Soc. Am.*, Vol. 41, 1967, pp. 1002–1010.
13. G. Fant, "The Relations between Area Functions and the Acoustic Signal," *Phonetica*, Vol. 37, 1980, pp. 55–86.
14. G. E. Peterson and H. L. Barney "Control Methods Used in a Study of the Vowels," *J. Acoust. Soc. Am.*, Vol. 24, 1952, pp. 175–184.
15. P. C. Delattre, A. M. Liberman, and F. S. Cooper, "Acoustic Loci and Transitional Cues for Consonants," *J. Acoust. Soc. Am.*, Vol. 27, 1955, pp. 769–773.

96

ACOUSTICAL ANALYSIS OF SPEECH

GUNNAR FANT

1 INTRODUCTION

The object of acoustical analysis of speech is the sound field emitted by a human speaker, usually picked up from a pressure-sensitive microphone. The purpose of the analysis is to perform a signal processing appropriate for sampling, storage, and visualization of relevant data. The most common procedure is spectrographic analysis, that is, to display the speech wave as a pattern in a time–frequency–intensity domain. Supplementary analyses of the temporal variation of the voice fundamental frequency F_0, formant frequencies, intensity, and voice source parameters may be synchronized with a spectrographic display to aid the extraction of information-bearing elements needed for general phonetic descriptive purposes as well as for applications in speech synthesis and recognition.

Acoustical analysis of speech also involves the acoustical aspects of the speech production mechanisms, see Chapter 95. The time–frequency–intensity patterns of speech waves are intimately related to the sound-generating and sound-shaping mechanisms. With this wider view of the acoustics of speech, we are in a better position to relate speech wave patterns to the framework of phonetics and linguistics, in other words, to derive acoustical correlates of vowels and consonants and of prosodic categories such as stress and intonation patterns, thus paving the way for a better understanding of the speech code. A more complete treatment of this subject is found in Refs. 1–3.

Note: References to chapters appearing only in the *Encyclopedia* are preceded by *Enc.*

2 SPECTRUM ANALYSIS

Basically two types of spectral representations are used in speech analysis, amplitude versus frequency and frequency–intensity versus time. The amplitude versus frequency analysis usually pertains to a frame of short duration, of the order of 3–30 ms, and is associated with a specific location in time. This is referred to as a spectral section. If the frame duration is extended to cover a larger portion of speech of the order of a sentence or more, the result is a long-time average spectrum, which mainly reveals speaker-specific characteristics. The short-time spectral section can provide quantitative data on specific parts of a speech sound. In order to visualize an utterance in all its details, we need a running short-time analysis with updating of spectrum calculations at very short intervals of the order of 1 ms. This is the basis of the time–frequency–intensity spectrogram commonly used in acoustic-phonetic studies of speech.

In a spectrogram, see Fig. 1, spectral intensity is represented by the density or by the color of the marking within areas of minimal time–frequency resolution. These are conceptually defined by an extension T in time and B in frequency. By the law of reciprocal spread, the product BT is a constant of the order of 1. Thus a fine frequency resolution, that is, a small B, implies a large T and thus a low temporal resolution. This is needed for portraying individual harmonics, which require that B is smaller than F_0, the fundamental frequency of voiced sounds, where F_0, defined as the inverse of the time interval between consecutive air pulses emitted through the vibrating vocal cords, typically varies in the range of 60–200 Hz in a male voice and somewhat less than an octave higher for an average female voice. If, on the other hand, the analysis bandwidth B is larger than

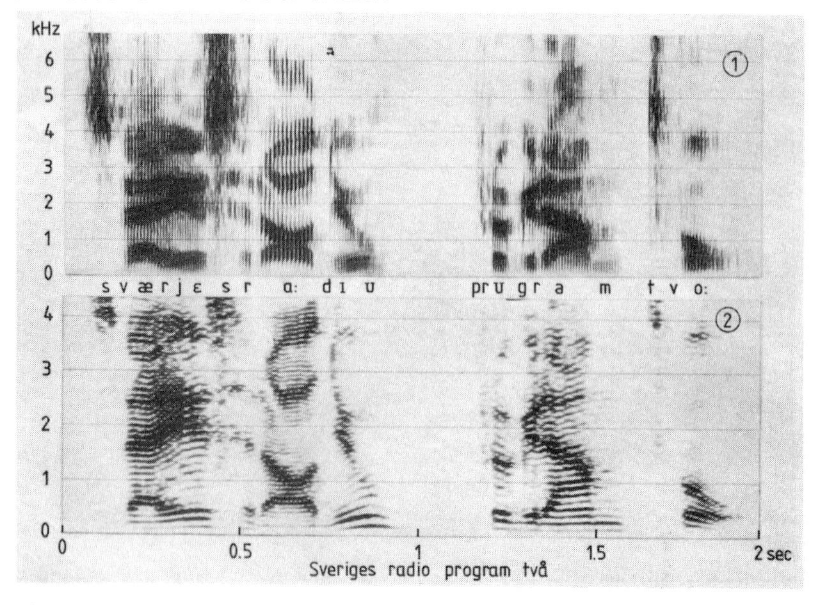

Fig. 1 Frequency–intensity–time spectrograms. Broadband analysis above and narrow-band analysis below.

F_0, the harmonics are no longer resolved, and the spectral pattern is dominated by individual formants, that is, peaks induced by vocal-tract resonances. The increased temporal resolution associated with the larger B makes it possible to follow rapid spectral variations such as the onset and decay of spectral intensity within a voice fundamental period. Thus, the narrow-band analysis, $B < F_0$, suits a frequency-domain harmonic representation, while the broadband analysis, $B > F_0$, brings out a time-domain view of voiced sounds as a sum and sequence of damped sinusoids representing the response of vocal-tract resonant modes to each successive excitation impulse from the glottal source. One such damped oscillation is thus the time-domain equivalent of a formant. (See also Chapter 95.)

The implication is that the bandwidth should be chosen to satisfy one of these two conditions. In practice, narrow-band analysis is mainly used for amplitude–frequency sectioning of vowels and other voiced sounds, while broadband analysis is commonly used in time–frequency–intensity spectrograms and for the sectioning of unvoiced fricatives and stops, as illustrated in connection with Fig. 2. An intermediate choice of B close to F_0 creates an ambiguity of what is a formant and what is a harmonic. The analysis bandwidth as well as the frequency spacing of data points may also be adapted to an auditory scale, such as the technical mel scale,[4]

$$1 \text{ mel} = 1000 \frac{\ln(1 + f/1000)}{\ln 2} \tag{1}$$

or an approximation to the Bark scale[5]

$$1 \text{ bark} = 7 \ln[x + (x + 1)^{0.5}], \tag{2}$$

where $x = f/650$.

3 TECHNICAL METHODS OF SPECTRUM ANALYSIS

Early attempts to perform spectrum analysis of speech employed a Fourier series analysis of speech waveforms recorded on an oscillograph. The breakthrough came in the middle of the 1940s with the Visible Speech sound spectrograph developed at the Bell Telephone Laboratories.[6] This became the prototype of the Kay Electric Sonagraph, a spectrum analyzer that produced printouts by passing a spectrally modulated current through an electrosensitive paper. Today, spectrographic analysis is generally performed digitally with a general-purpose computer and a laser printout. A high resolution sets demands both on the printer and on the rate of overlapping frames in the spectrum analysis and

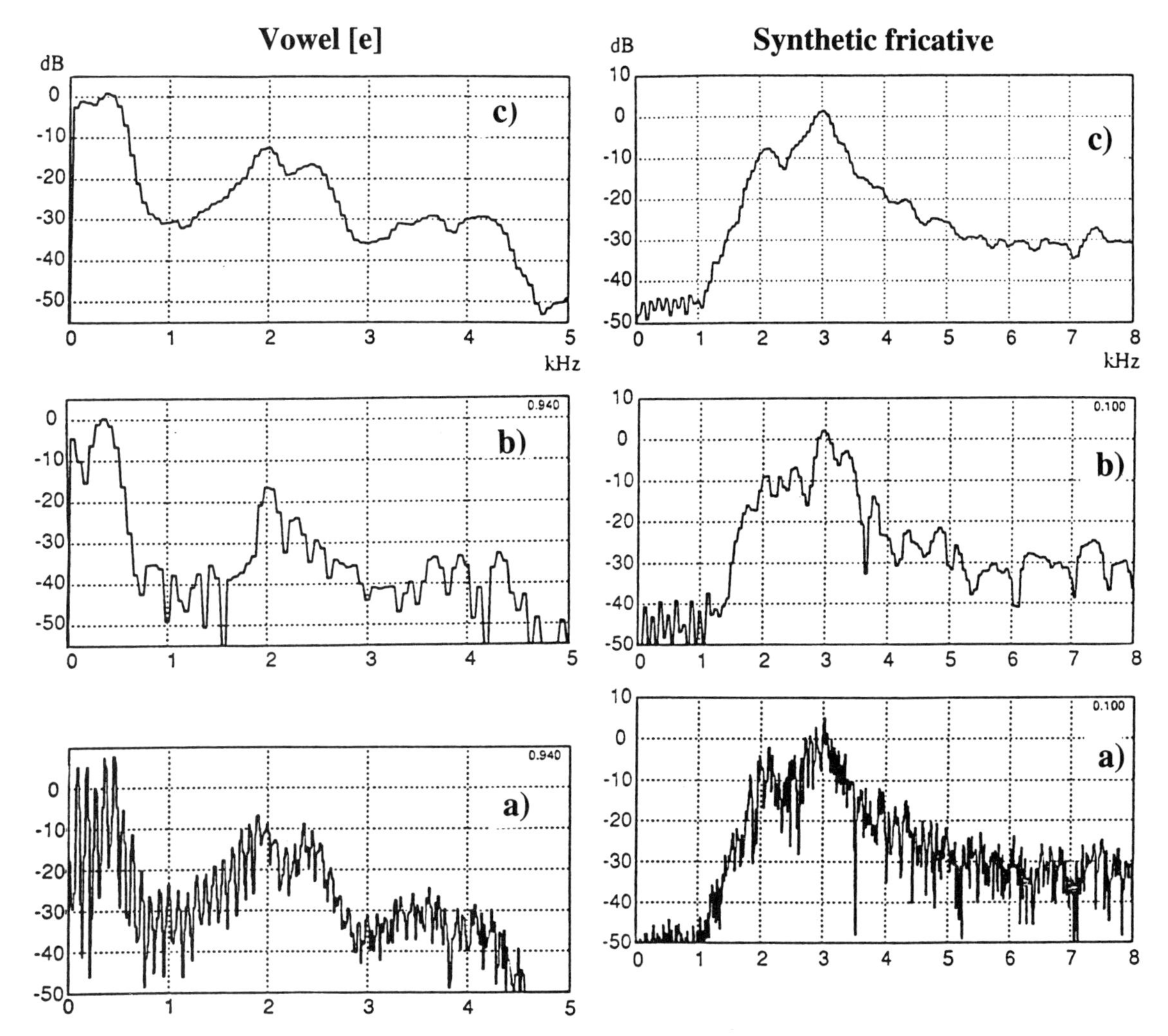

Fig. 2 Spectral sections of a vowel [e] (left) and of a synthetic fricative [ʃ] (right). Sampling frequency 16 kHz: (*a*) simple FFT with $N = 640$ sample points, $B = 25$ Hz, $df = 25$ Hz frequency spacing, $T = 40$ ms; (*b*) $N = 64$, $B = 250$ Hz, 576 zero samples added, $df = 25$ Hz, $T = 4$ ms; (*c*) same as (*b*) but average is taken over 10 successive 4-ms spectrum samples, $T = 40$ ms.

does not generally exceed the definition obtained in a special-purpose analog spectrograph.

Spectrum analysis may be performed either as a filtering process or by a digital Fourier analysis of the waveform based on the discrete fourier transform (DFT),[7–9] which usually is implemented by the fast Fourier transform (FFT).[10] The most common filtering method is by means of a filter bank. Another method is by means of a single bandpass filter, narrow-band or broadband, the effective center of which is varied continuously as in a wave analyzer. This method was used in the Sonagraph.

The conceptual analogy[4] between the output of a bandpass filter and a component of a Fourier series is basically related to the temporal weighting function of the envelope of the impulse response of the bandpass filter. A mathematical identity exists,[4] when we consider the speech signal to be convolved with the weighting function before being submitted to a formal Fourier analysis. The effective duration of the impulse response, which is of the order of $1/B$, serves as a functional equivalent to the frame duration T of a DFT defined by

$$V_n = \sum_{i=0}^{N-1} v_i \exp\left(\frac{-j2\pi ni}{N}\right), \qquad (3)$$

where N is the number of samples, v_i is the sample number i, and V_n the spectral component number n. For an approximate rectangular window function, the frame duration is $T = N/F_s$, where F_s is the sampling frequency. A typical combination for narrow-band analysis would be $N = 512$ waveform samples of $F_s = 16$ kHz, which provides $N/2 = 256$ spectrum samples up to the Nyquist frequency 8 kHz. These are spaced at $df = B = 1/T = 31.25$ Hz intervals, each representing a bandwidth of $B = 1/T = 31.25$ Hz. For broadband applications, say for an equivalent bandwidth of $B = 1/T = 250$ Hz, there is a need to calculate spectrum samples at smaller intervals than B. An improved definition of the spectrum envelope is attained by adding a frame of zero amplitude samples prior to the computation. Thus, with $N = 64$ waveform samples at $F_s = 16$ kHz appropriate for $B = 250$ Hz, we need to add $(8 - 1)64 = 448$ zero amplitude samples in order to achieve a spacing of spectral data points of 31.25 Hz. Similarly, for a temporal definition higher than that implied by successive frames at T-second intervals, one must repeat the calculation at shorter intervals than T. For a broadband spectrographic display, this is achieved by updating spectral calculations at intervals of the order of 1–2 ms or as densely as motivated by the definition of the computer screen or the recording medium.

An additional averaging of spectral data is needed for removing statistical uncertainties in amplitude versus frequency sections of unvoiced sounds. The randomly occurring striations, which are typical of time–frequency–intensity spectrograms of fricatives, are the natural consequences of the filtering of random noise. These striations occur at random time intervals that average $1/B$. For producing clean amplitude–frequency sections with an acceptable random error, a spectral section needs to be averaged over a time interval T_a substantially greater than the frame duration $T = 1/B$.

From statistical considerations it follows that the uncertainty in the estimate of the amplitude A_e of a spectral component is proportional to $(T/T_a)^{0.5}$. On a decibel scale and assuming T_a greater than T, the following approximate expression for the random ripple σ_e, superimposed on a spectral component A_e, has been empirically validated[1]:

$$20 \log_{10} \frac{\sigma_e}{A_e} = 4(BT_a)^{-0.5} \quad \text{(dB)}. \qquad (4)$$

A convenient method of averaging, especially suited for filter bank spectrum analysis, is by means of a low-pass filtering of the rectified signal in each channel in which case the substitution $T_a = 1/2W$, where W is the low-pass cutoff frequency, may be used.

The practical consequences of the choice of analysis parameters are illustrated in Fig. 2 pertaining to a human vowel [e] and a synthetic unvoiced fricative [ʃ]. In both cases the sampling frequency was 16 kHz. A direct FFT spectrum calculation with $N = 640$ waveform samples, covering an interval of $T = 40$ ms, produces 320 spectrum samples up to 8 kHz with a bandwidth $B = 25$ Hz and a spacing of $dF = 25$ Hz. This direct approach is adequate for portraying the harmonic structure of the vowel spectrum. We may identify the voice fundamental frequency $F_0 = 90$ Hz and formant peaks at $F_1 = 400$ Hz, $F_2 = 1900$ Hz, and $F_3 = 2400$ Hz. The weakly apparent peak at 3250 Hz is associated with F_4 and the peaks at 3700 Hz and 4200 Hz with F_5 and F_6, respectively.

The same direct approach applied to the fricative produces a random fine structure that in accordance with Eq. (4) is of the order of ±4 dB. Although the two formants at 2000 and 3000 Hz appear as expected, there remains an uncertainty about the possible occurrence of other spectral peaks. A more appropriate processing of a 40-ms sample of the fricative involves the following steps. First an analysis bandwidth of $B = 250$ Hz, requiring $N = 64$ waveform samples, is selected. In order to retain a spectral definition of $df = 25$ Hz, a number of $9 \times 64 = 572$ empty samples are added before the FFT is executed. However, the random component in the spectrum remains the same, as illustrated by the middle graph of Fig. 2. It is effectively reduced by taking the average of 10 successive FFT spectra covering an effective frame width of $T_e = 40$ ms, as illustrated by the top graph of Fig. 2. In agreement with Eq. (4), the ripple is now reduced to the order of ±1 dB.

The same operation applied to the vowel also enhances the outline of the spectrum envelope by effectively removing the harmonic fine structure. In these examples, the analysis embraced a 40-ms speech sample. For analysis of stop bursts, a 20-ms sample is recommended. It should also be kept in mind that the effective duration of a Hanning or Hamming window is about 40% less than the nominal value of a rectangular window.[8] In other words, the particular weighting function emphasizes the middle part of the window.

An alternative and more common method of deriving a smoothed spectrum envelope, as in Fig. 2*a*, is by means of a cepstrum deconvolution,[7–9] the first step of which involves calculating the log spectrum of a sufficiently long frame of the speech wave, usually of the order of 40 ms. This is submitted to an inverse Fourier transform that is windowed to retain a region up to a few milliseconds after which follows a new Fourier transform,[7,8] which is the cepstrum envelope. One may also apply a linear prediction (LPC) analysis to model the spectrum envelope as an all-pole system,[7,8] see the section on speech parameter extraction.

Another point often overlooked in digital spectrum analysis of speech is the difference between a conventional DFT or FFT and a proper Fourier series harmonic analysis. The relation is

$$A_n = \frac{2V_n}{T_0}, \tag{5}$$

where V_n is the Fourier integral component [Eq. (3)], T_0 is the voice fundamental period $T_0 = 1/F_0$, and A_n is the amplitude of the corresponding harmonic. The factor 2 derives from the transform from a double-sided to a single-sided spectrum. Scale factors are often lost in digital processing of speech. In this case, if lost, the F_0 scale factor would cause errors in the tracking of the amplitude of the voice fundamental or its harmonics within an utterance.

4 SPEECH PARAMETER EXTRACTION

One object of speech analysis is to extract essential parameters of the acoustical structure, which may be regarded as a process of data reduction and an enhancement of information-bearing elements. In order to attain an effective and reasonable complete specification, one has to rely on synthesis models. An important subsystem is that of voiced sounds with negligible coupling to the nasal and subglottal systems. In a source-filter decomposition,[11] the filter function is ideally described by the so-called F-pattern, F1, F2, F3, F4, ... These vocal-tract resonance modes are quantified by their formant frequencies $F_1, F_2, F_3, F_4, \ldots$, the corresponding formant bandwidths $B_1, B_2, B_3, B_4, \ldots$, formant amplitude levels $L_1, L_2, L_3, L_4, \ldots$. The labels F1, F2, F3, F4, ... may also be used to refer to formant frequencies. An alternative notation of F_0 is $F0$. For male speech the normal range of variation is F_1 = 180–800 Hz, F_2 = 600–2500 Hz, F_3 = 1200–3500 Hz, and F_4 = 2300–4000 Hz. The average distance between formants is 1000 Hz. Females have on the average 20% higher formant frequencies than males, but the relation between male and female formant frequencies is nonuniform and deviates from a simple scale factor.[2] Formant bandwidths may vary considerably. Typical values[12] are $B_n = 50(1 + f/1000)$ Hz. Formant amplitudes vary systematically with the overall pattern of formant frequencies and the spectral properties of the voice source. The basic parameter of the voice source is the fundamental frequency F_0, which determines the intonation.

Some of these parameters may be inferred from visual inspection and processing of a spectrogram or an oscillogram. Thus, in a narrow-band spectrogram, F_0 may be traced from one of the harmonics. Crude information on F_0 is also available in a broadband spectrogram from the distance $T_0 = 1/F_0$ between successive voice pulse striations. The time-domain extraction of F_0 is performed with greater accuracy in an oscillogram with expanded time scale. With suitable preprocessing to isolate a single formant, the oscillographic trace may also be used to determine formant frequencies and bandwidths, as illustrated in Fig. 3. The bandwidth is here determined from the exponential decay, $\exp(-\pi B_n t)$ of the formant oscillation envelope within the closed glottis interval of the voice cycle.

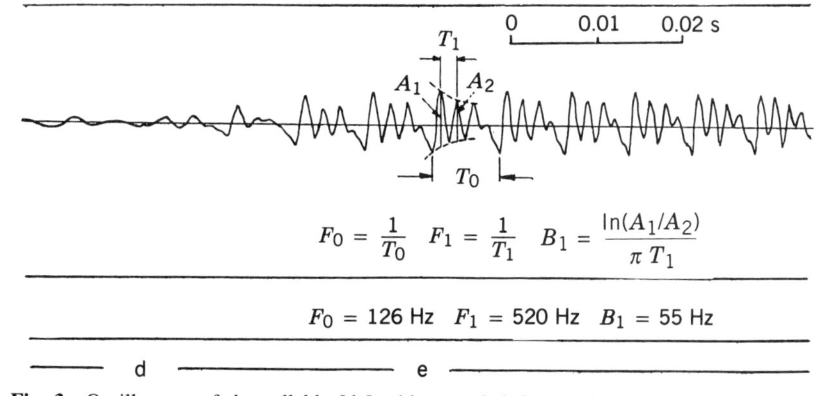

Fig. 3 Oscillogram of the syllable [de] with extended time scale and removal of energy above F_1, illustrating the time-domain extraction of voice fundamental frequency, F_0, first formant frequency F_1, and first formant bandwidth B_1. From Ref. 4.

Fully automatic methods of continuous tracking of speech parameters, such as F_0 and formant frequencies, still need to be established. The problem lies in the reliability of performance.[9] Any existing system of F_0 extraction is prone to fail at some instances depending on individual voice types and registers. Typical errors are jumps to the second harmonic and the indeterminacy of F_0 at aperiodicities such as in voice creak. A system may perform quite well on one type of voice but not on other voices.

An even more difficult task is the automatic extraction of formant frequencies and bandwidths. Linear prediction coding, which models voiced sounds as the response of an all-pole function to a simple sequence of excitation impulses,[13] has become a standard routine for formant tracking, see Chapter 97. However, the human voice source departs from the all-pole model in several ways and may show local spectral peaks and dips. Also, the vocal-tract transfer function generally contains both poles and zeros due to nasal and subglottal coupling, which is specially apparent in continuous speech. Even with more realistic and complex models of speech production, an automatic extraction of formant frequencies frequently breaks down. This is specially the case for high fundamental frequencies where the tracking system tends to pick harmonics rather than formants. For simple applications to vowel analysis, LPC techniques may give accurate estimates of formant frequencies, providing the results are carefully checked against spectrograms. Typical errors[14] are temporal jitter from one frame to the next, tendencies at high F_0 to synchronize on harmonics, and occasional jumps to a higher or a lower formant. Linear prediction coding determination of bandwidths is less reliable.

For many laboratory applications, the computer-derived spectrogram is combined with a synchronous oscillogram and an extracted F_0 curve. In addition, one may display the time course of overall intensity with or

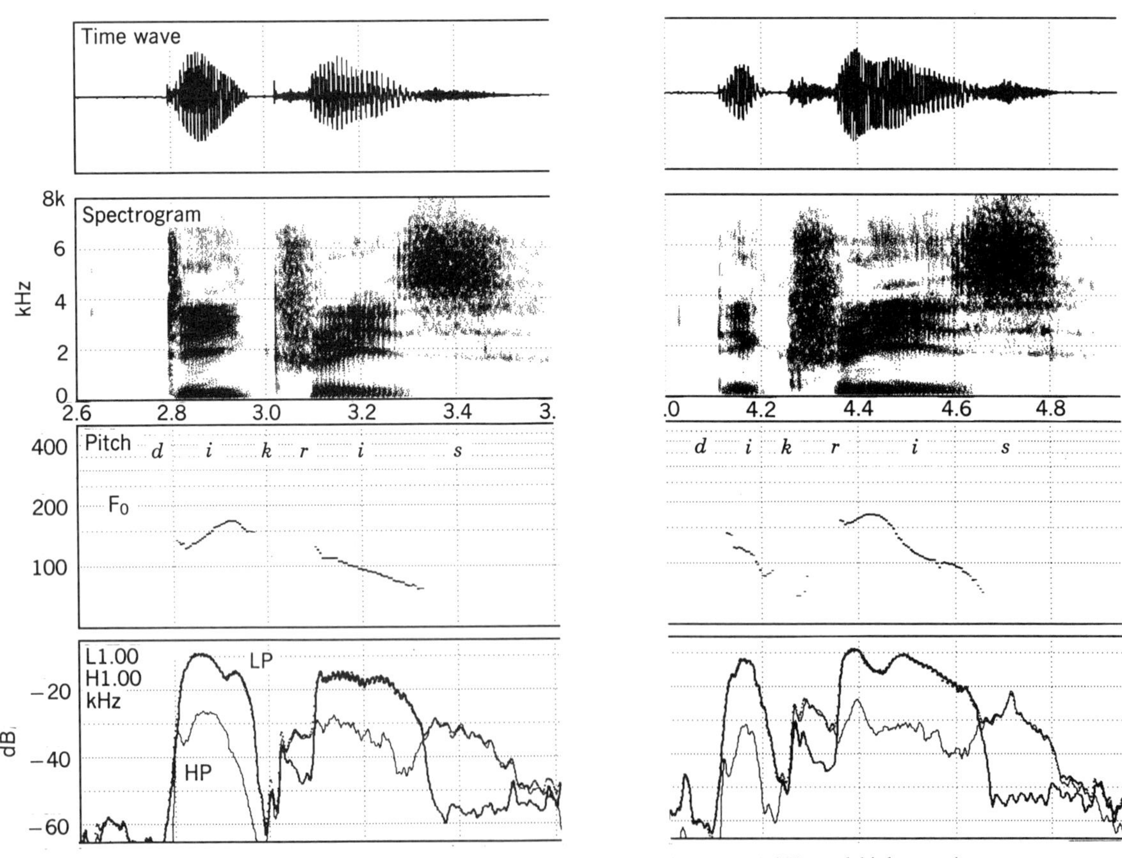

Fig. 4 Oscillogram, spectrogram, F_0 contour, intensity (low-pass 1 kHz and high-pass 1 kHz) illustrating contrasting stress patterns of the word *decrease* pronunced as noun (left) and as verb (right).

without special preemphasis or other filtering. This is exemplified in Fig. 4, which illustrates two words with contrasting stress patterns: the noun *decrease* with the stress on the first syllable and the verb *decrease* with stress on the second syllable. The two intensity curves at the bottom pertain to low-pass filtering and high-pass filtering, both with cutoff frequencies of 1000 Hz. The relative dominance of the first syllable of the noun and of the second syllable of the verb is reflected in a larger duration and in higher F_0 and intensity. The contrast in intensity comparing the first and the second syllable is specially apparent in the high-pass intensity contour, which reflects a relative greater dominance of higher partials of the voice source at a higher stress level.

The temporal and spectral shape of any speech sound is a function of both source and transfer function characteristics, which is illustrated in Fig. 5. The source of voiced sounds may be regenerated by submitting the speech wave to an inverse filtering, which in effect cancels the poles and zeros of the transfer function.[12,15] Since the transfer from volume velocity at the lips to the sound pressure in front of the speaker involves a differentiation, the net result of an inverse filtering without integration is to regenerate the time derivative of glottal flow, which appears in the lower left part of the figure. The negative-going spikes of this pulse train are the derivative of glottal flow at the instant of glottal closure, which is a measure of excitation strength.

A continuous inverse filtering is illustrated in Fig. 6, which pertains to the [jö] part of the word *adjö*. In the lower graph, a sample of the [ö] has been submitted to a routine analysis of voice source parameters, which shows an overall fair match with our reference LF model[16] except for time-domain and frequency-domain irregularities, for example, the doubled peaked positive part of the glottal flow derivative and the associated spectral minimum just above 1 kHz. Such perturbations are a natural consequence of superpositions and aerodynamic nonlinearities adding to speech naturalness.

A further complication in voice source analysis is the covariation of source and filter characteristics associated with glottal adduction-abduction gestures[17–19] and

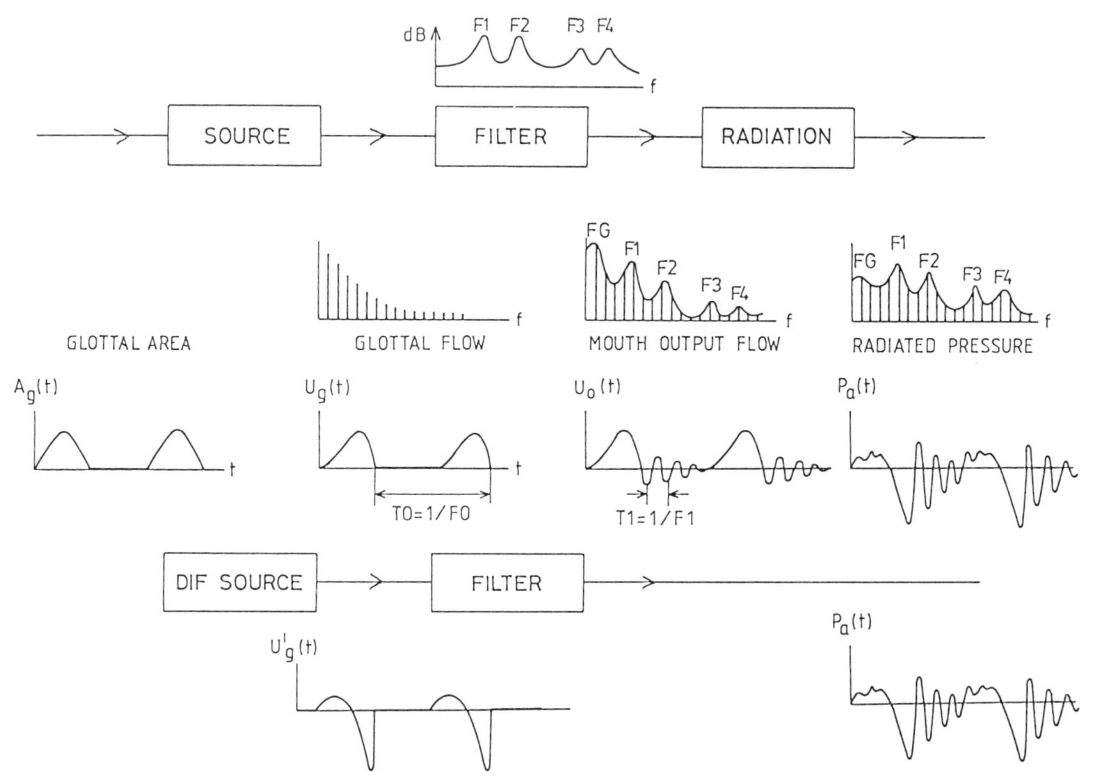

Fig. 5 Frequency- and time-domain view of source-filter decomposition of voiced sounds. The negative-going peak of glottal flow derivate source at the bottom is a scale factor for formant amplitudes.

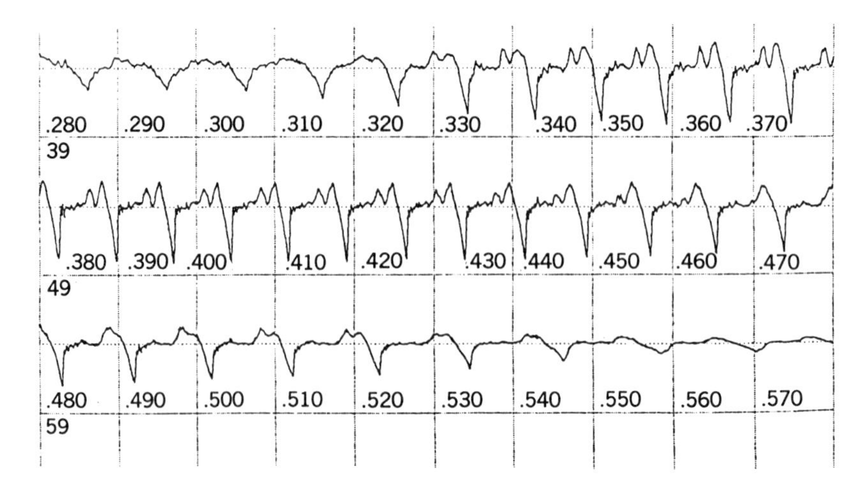

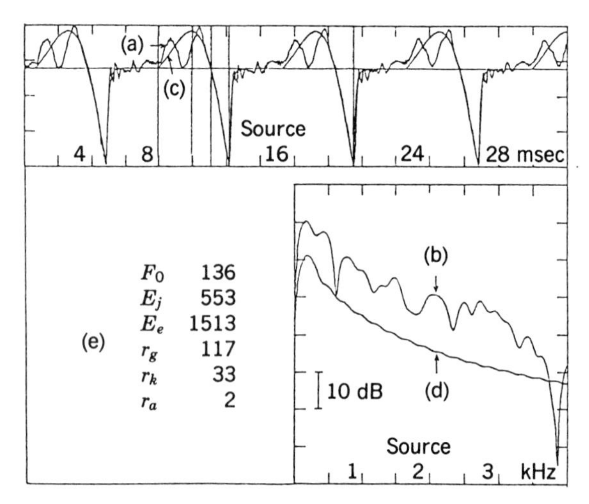

Fig. 6 Continuous inverse filtering of the [jö] part of the word *adjö*, spoken by a male subject. The [j] covers the first half of the top line. A routine LF source parameter analysis of a sample of the [ö] is illustrated in the lower graph, where the smooth curves pertain to the LF model. After Ref. 19.

the sensitivity of the source parameters to vocal-tract constrictions. The inhibitory influence of a supraglottal narrowing on the voice excitation amplitude explains the relative low amplitude and smoothed out waveform of the [j]-source function in Fig. 6. Similar interaction effects are also encountered in semiconstricted vocalic intervals.[15] The source is, thus, not invariant with articulation. The inherent intensity of a vowel is not only dependent on its formant pattern but also on its source. An emphatic stress on a syllable containing a vowel produced with a relatively constricted vocal tract, for example, [i:], [y:], [ʉ:], and [u:] in Swedish, may cause an additional decrease of the constriction area and, accordingly, a local decrease of source strength. The relation between stress and intensity thus becomes rather complex.

5 READING SPECTROGRAMS: THE SPEECH CODE

A spectrogram, examples of which were given in Fig. 1 and Fig. 4, provides the potential of a more complex insight into a specific utterance than what can be achieved from any direct set of observations of the production process. With appropriate knowledge, the spectrogram may thus serve as a window for inferring articulatory activity and organizational principles extending the view all the way up to the brain. It is our basic means of developing speech synthesis strategies and, more generally, of developing knowledge of the speech code,[3] which is the relation between messages and sound waves. A speech message is not only confined to a sequence of words, syllables, and phonemes, usually referred to as the segmental structure. A most important part is the prosodic structure, essentially signaled by patterns of F_0, relative duration, and intensity. In general, the prosodic pattern may be regarded as superimposed on the segmental structure. Besides grammatical distinctions, as already exemplified in Fig. 4, the prosodic pattern carries information about relative emphasis and semantic grouping and about speaker-specific denotations, attitudes, and speaking style.[2] A language representative prosody is essential for comprehension and for securing naturalness in speech synthesis. Pathological speech also displays recurrent patterns with their specific codes.

Human speech, as viewed from a spectrogram, is a mixture of continuous and discrete elements. The continuous elements reflect a system of overlaid simultaneous movements of speech articulators with often independent starting and ending points.[20] Coarticulation and reduction account for contextual adjustments of speech patterns and undershoot of targets.[3] The continuous elements are interrupted by discrete breaks in gross pattern type, for example, the switching between voiced and unvoiced patterns and between sound and silence. Both the discrete and the continuous elements are used in segmenting utterances into smaller units, *phones*, corresponding to phonemes or parts of phonemes. Thus, a stop consonant may be assigned an interval from the offset of sound at the initiation of an occlusion up to the onset of voicing after the release. This total interval may further be subdivided into separate parts, such as occlusion, release transient, frication, and a possible aspiration. In other instances, for example, when two voiced sounds occur in succession (e.g., two vowels or a vowel and a voiced consonant), the segmentation has to rely on continuous formant pattern variations to locate a boundary between two presumed target points. This is a problem in establishing routines for segmentation and labeling of speech.[21]

The relation between acoustic segments and phonemes is indeed complex. Thus, one phoneme on the message level may influence several successive acoustic segments. Conversely, one acoustic segment is usually influenced by a domain of a message that includes more than one phoneme as well as by specific prosodic attributes, such as stress and phrase junctures. However, an oversimplified but fairly potent rule of speech perception is that a speech sound may be identified by a specific segment type and by the pattern of formant transitions in and out of the segment. For a fuller account see Ref. 1.

Our knowledge of speech as a code, relating message contents and speaker characteristics to specific sound patterns, is still in a developing phase. Spectrogram-reading exercises serve a most important purpose of confronting models with reality, a continuous process of learning, revision, and updating of knowledge.

The ultimate level of performance of speech synthesis and recognition will depend on how deep we can penetrate the speech code including basic mechanisms, human behavior, language structure, and specific variations.[3] Synthesis architecture[22,23] is fairly well established with respect to basic acoustic theory[4,11,24] and principles of digital implementation.[8,25] An important trend is to employ articulatory encoding of speech processes[3,26] instead of formant coding or as an organizational principle.[27] However, to improve the quality of speech synthesis we need a deeper insight into the speech code. Ideally the computer should have some understanding of what it is going to say.

REFERENCES

1. G. Fant, "Analysis and Synthesis of Speech Processes," in B. Malmberg (Ed.), *Manual of Phonetics*, North Holland, Amsterdam, 1968, pp. 173–277.
2. G. Fant, A. Kruckenberg, and L. Nord, "Prosodic and Segmental Speaker Variations," *Speech Communication*, Vol. 10, 1991, pp. 521–531.
3. G. Fant, "What Can Basic Research Contribute to Speech Synthesis?" *J. Phonet.*, Vol. 19, 1991, pp. 75–90.
4. G. Fant, "Acoustic Analysis and Synthesis of Speech with Applications to Swedish," *Ericsson Technics*, No. 1, 1959, pp. 3–108.
5. G. Fant, "Feature Analysis of Swedishs Vowels—A Revisit," *STL-QPSR*, Vols. 2–3, 1983, pp. 1–19.
6. R. K. Potter, A. G. Kopp, and H. C. Green, *Visible Speech*, D. van Nostrand, New York, 1947.
7. L. R. Rabiner and R. W. Schaefer, *Digital Processing of Speech Signals*, Prentice-Hall, Englewood Cliffs, NJ, 1978.
8. S. Furui, *Digital Speech Processing, Synthesis, and Recognition*, Marcel Dekker, New York, 1989.
9. W. Hess, *Pitch Determination of Speech Signals. Algorithms and Devices*, Springer Verlag, Berlin, 1983.
10. J. D. Markel, "FFT Pruning," *IEEE Trans. Audio Electroacoust.*, Vol. AU-19, 1971, pp. 305–311.
11. G. Fant, *Acoustic Theory of Speech Production*, Mouton, 's-Gravenhage, Holland, 1960.
12. G. Fant, "The Acoustics of Speech," *Proc. 3rd Int. Congress of Acoustics*, Stuttgart, 1961, pp. 188–201.

13. J. D. Markel and A. H. Gray, *Linear Prediction of Speech*, Springer Verlag, Berlin, 1975.
14. H. Wakita, "Linear Prediction of Speech and Its Application to Speech Processing," in J.-P. Haton (Ed.), *Automatic Speech Analysis and Recognition*, Reidel, Boston, 1982, pp. 1–20.
15. G. Fant, "Some Problems in Voice Source Analysis," *Speech Communication*, Vol. 13, 1993, pp. 7–22.
16. G. Fant, J. Liljencrants, and Q. Lin, "A Four-Parameter Model of Glottal Flow," *STL-QPSR*, Vol. 4, 1985, pp. 1–13.
17. D. H. Klatt and L. C. Klatt, "Analysis, Synthesis and Perception of Voice Quality Variations Among Female and Male Talkers," *J. Acoust. Soc. Am.*, Vol. 87, No. 2, 1990, pp. 820–857.
18. G. Fant and Q. Lin, "Frequency Domain Interpretation and Derivation of Glottal Flow Parameters," *STL-QPSR*, Vols. 2–3, 1988, pp. 1–21.
19. C. Gobl, "Voice Source Dynamics in Connected Speech," *STL-QPSR*, Vol. 1, 1988, pp. 123–159.
20. G. Fant, "Descriptive Analysis of the Acoustic Aspects of Speech," *Logos*, Vol. 5, 1962, pp. 3–17.
21. R. Carlson and B. Granström, "A Search for Durational Rules in a Real-Speech Data Base," *Phonetica*, Vol. 43, 1986, pp. 140–154.
22. R. Carlson, B. Granström, and S. Hunnicutt, "Multilingual Text-to-Speech Development and Applications," in A. W. Ainsworth (Ed.), *Advances in Speech, Hearing and Language Processing*, JAI Press, London, 1991.
23. D. H. Klatt, "Software for a Cascade/Parallel Formant Synthesizer," *J. Acoust. Soc. Am.*, Vol. 67, 1980, pp. 971–995.
24. J. L. Flanagan, *Speech Analysis Synthesis and Perception*, Springer Verlag, New York, 1972.
25. B. Gold and L. R. Rabiner, "Analysis of Digital and Analog Formant Synthesizers," *IEEE Trans. Audio Electroacoust.*, Vol. AU-16, No. 1, 1968, pp. 81–94.
26. Q. Lin, "Speech Production Theory and Articulatory Speech Synthesis," Dr. Sc. Thesis, Dept. of Speech Communication and Music Acoustics, KTH, Stockholm, 1990.
27. K. N. Stevens and C. A. Bickley. "Constraints Among Parameters Simplify the Control of Klatt Formant Synthesizer," *J. Phonet.*, Vol. 19, No. 1, pp. 161–174.

97

TECHNIQUES OF SPEECH CODING

BISHNU S. ATAL

1 INTRODUCTION

Techniques for speech coding are designed to convert the speech waveform into digital codes with minimal loss of information. Speech is a signal full of redundant information, and speech coding techniques exploit the redundancies in speech to minimize the information rate needed to reproduce the signal at a specified level of accuracy.

Speech coding techniques can be broadly divided into two classes: waveform coding that aims at reproducing the speech waveform as faithfully as possible and vocoders that preserve only the spectral properties of speech in the encoded signal. The waveform coders are able to produce high-quality speech at sufficiently high bit rates. Vocoders produce intelligible speech at much lower bit rates, but the level of speech quality—in terms of its naturalness and uniformity for different speakers—is also lower. The applications of vocoders so far have been limited to communication over special low-bit-rate digital communication channels. The main focus of this chapter is the waveform coders that are suitable for speech communication over digital channels with bit rates ranging from 4 to 64 kb/s.

This chapter first reviews basic principles of waveform coding followed by a brief discussion of various techniques that have been found useful for reducing the bit rate in waveform coders. The chapter concludes with a discussion of subjective quality of digitally coded speech at various bit rates.

Note: References to chapters appearing only in the *Encyclopedia* are preceded by *Enc.*

2 BASIC CONCEPTS IN SPEECH CODING

The information rate of a digital channel is expressed in bits/s. The conversion of an analog signal into digital form consists of two steps: sampling and quantization. Sampling is a process of converting a continuous function of time to a discrete sequence representing the function at regularly spaced time intervals, whereas quantization converts continuous amplitude into discrete values.

2.1 Sampling

Sampling of a continuous time signal $s(t)$ at times nT, where T is the sampling interval, results in a sequence of discrete time samples denoted by $s(nT)$. The sampling frequency or sampling rate f_s is given by the reciprocal of the sampling interval T. One of the most important results in sampling is the sampling theorem,[1] which states that an analog signal $s(t)$ band limited to the highest frequency of B hertz can be exactly reconstructed from its samples $s(nT)$, if the sampling frequency is higher than $2B$. The perfect reproduction of a signal from its sampled values is not possible, if the signal waveform is sampled at a frequency lower than $2B$. The sampling results in periodic repetition of the signal spectrum at intervals of f_s hertz as shown in the middle panel of Fig. 1. Therefore, if f_s is less than $2B$, the spectrum will fold over in the baseband of the signal. The spectral distortion introduced by undersampling is illustrated in the bottom panel of Fig. 1. For speech transmitted through the traditional telephone channels, $B = 4$ kHz, $f_s = 8$ kHz, and $T = 125\ \mu$s.

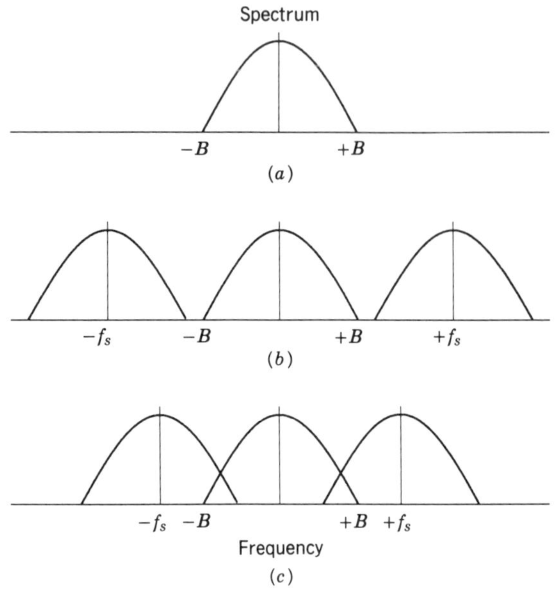

Fig. 1 Spectral distortion produced by undersampling: (*a*) assumed signal spectrum; (*b*) spectrum of sampled signal for $f_s > 2B$; (*c*) spectrum of sampled signal for $f_s < 2B$.

2.2 Quantization

Quantization is the procedure for representing the amplitude of an analog waveform by certain preassigned values. The process of quantization invariably introduces errors, or quantization noise, in the signal. The ratio of the signal power to the noise power is usually expressed as the signal-to-noise ratio (SNR) and serves as an important measure of the performance of quantizers.

Scalar and Vector Quantization Independent quantization of each sample of a continuous signal is called scalar quantization. Joint quantization of a block of samples of a signal is called vector quantization, and it results in lower error than scalar quantization with the same number of values.

Uniform Quantization In uniform quantization, the quantization levels are spaced at equal intervals. For a quantizer with the step size Δ, the quantization noise is distributed uniformly in the range from $-\Delta/2$ to $+\Delta/2$ with its power equal to $\Delta^2/12$. To quantize a range of signal values from $-x_{\max}$ to $+x_{\max}$ with an n-bit quantizer, the step size becomes

$$\Delta = \frac{2x_{\max}}{2^n}. \tag{1}$$

The mean-square quantization error is

$$\sigma_q^2 = \tfrac{1}{3}x_{\max}^2 2^{-2n}. \tag{2}$$

The ratio of signal power to noise power usually expressed in decibels is given by

$$10\log_{10}\text{SNR} = 4.77 + 6.02n + 20\log_{10}\frac{\sigma_x}{x_{\max}}, \tag{3}$$

where σ_x is the root-mean-square (rms) value of the signal. Equation (3) states that the SNR increases approximately 6 dB for each added bit in the quantizer.

Nonuniform Quantization In uniform quantization, the noise power stays constant independent of the signal level. The power in the speech signal varies over a considerable range of values, and therefore speech segments with low power levels are quantized with a poor SNR, as illustrated in Fig. 2. This problem is avoided if the quantization intervals are not set to be uniform but are spaced nonuniformly to increase with the signal value. The nonuniform quantization is equivalent to compression (companding) of the analog signal followed by uniform quantization and subsequent expansion of the output.

3 LOW COMPLEXITY WAVEFORM CODERS

3.1 Companded Pulse Code Modulation

One important compression law[2,3] is the μ-law defined as:

$$F(x) = \text{sgn}(x)x_{\max}\frac{\log(1+\mu|x|/x_{\max})}{\log(1+\mu)}, \tag{4}$$

where x is the input signal amplitude, $x_{\max}$ is the maximum signal amplitude, and μ is a parameter that controls the degree of compression. The values of μ in the range 100–200 are often used. The companded pulse code modulation (PCM) is usually referred to as μ-law PCM or log-PCM. Figure 3 shows an example of the power levels for the speech signal and the quantization noise using the μ-law compression. For μ-law PCM, the noise power is decreased considerably during low-level speech segments.

3.2 Adaptive Differential Pulse Code Modulation

Adjacent samples in speech waveforms are highly correlated. Differential PCM takes advantage of such cor-

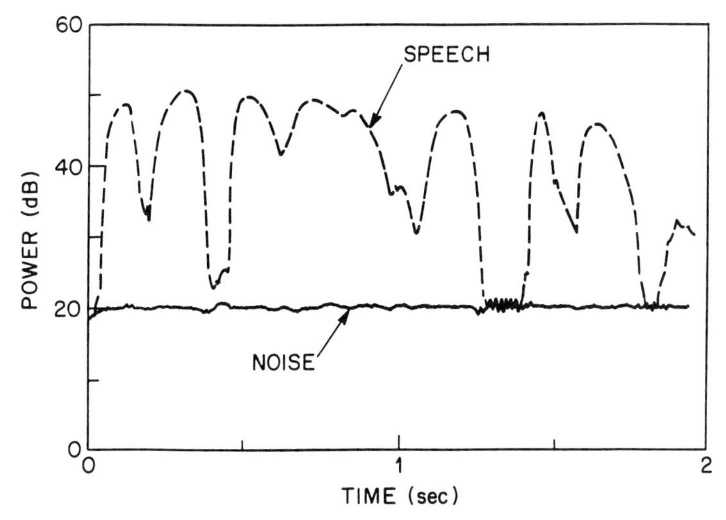

Fig. 2 Power levels of speech and quantizing noise with uniform quantization (linear PCM).

relations by quantizing the difference between the signal and its predicted value based on the previous quantized samples. The dynamic range of the difference signal is much smaller than the input speech signal, and it therefore needs a smaller number of bits for the same quantization error. In adaptive differential PCM (ADPCM) coders,[3] the difference signal is quantized with adaptive control of the quantization step size. The comparison of the SNR of ADPCM with that of μ-law PCM at different bit rates is shown in Fig. 4. The ADPCM shows a gain in SNR of 8 dB, which includes a 4-dB gain due to the differential encoding and a 4-dB gain due to the quantizer adaptation.

4 PREDICTIVE CODING

Linear predictive coding systems exploit sample-to-sample correlations in speech to predict the current speech sample from its past quantized values and to reduce the bit rate of the coded speech signal.[4] The difference between speech sample and its predicted value is gener-

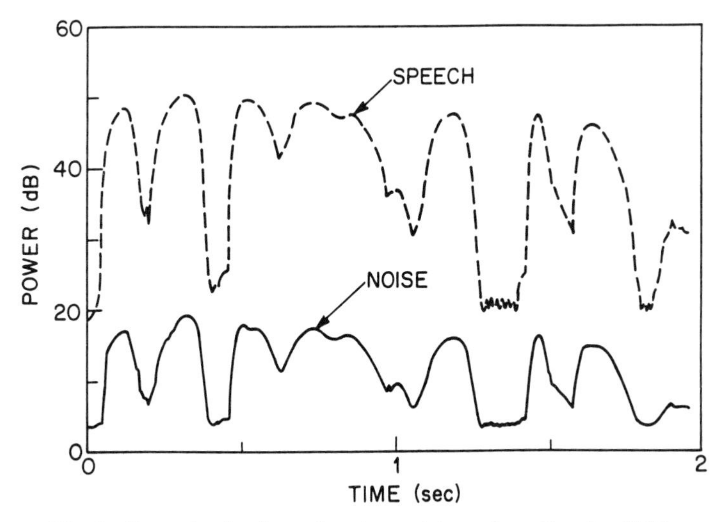

Fig. 3 Power levels of speech and quantizing noise with μ-law PCM.

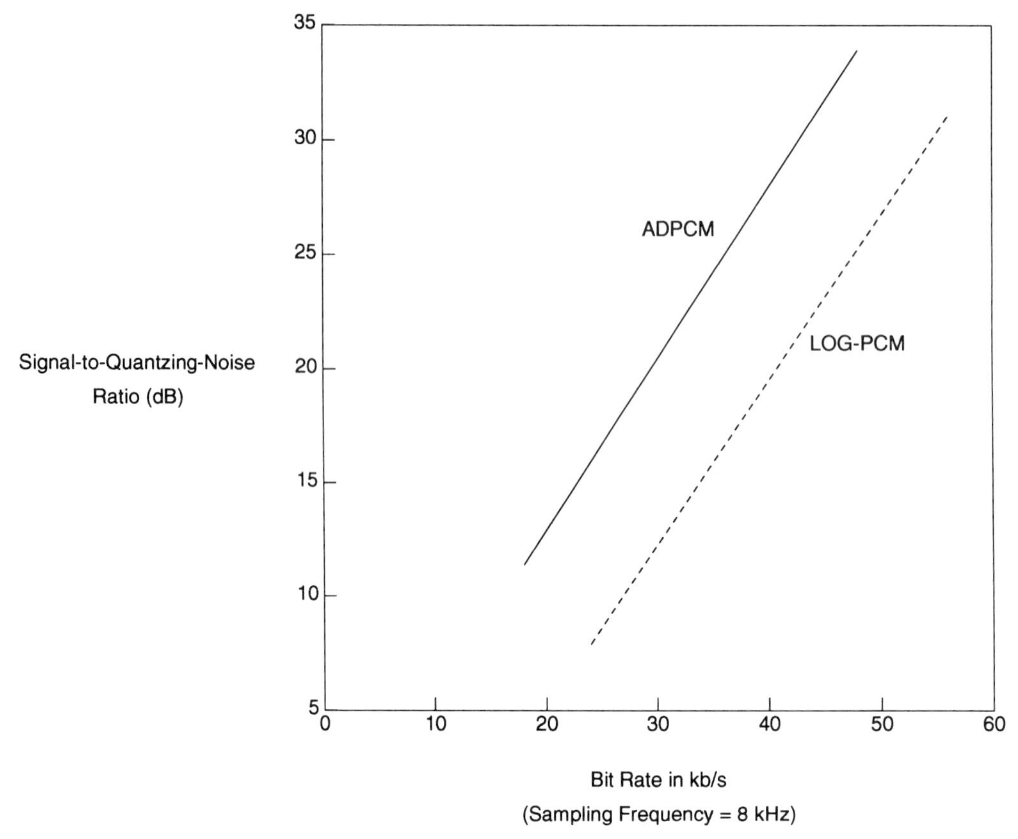

Fig. 4 Comparison of ADPCM with log PCM for band-limited (200–3200 Hz) speech input sampled at 8 kHz.

ally much smaller than the amplitude of the speech signal, and therefore it can be encoded with fewer bits.

4.1 Predictive Coding System

A block diagram illustrating the principle of predictive coding is shown in Fig. 5. The speech signal $s(t)$ is sampled, and the predictor forms an estimate $\hat{s}_n$ of the signal's present value based on the past samples of the quantized speech signal at the transmitter. The predicted value $\hat{s}_n$ is subtracted from the signal value s_n to form the difference δ_n, which is quantized, encoded, and transmitted to the receiver. The transmitted signal is decoded at the transmitter and used to predict the next speech sample using a predictor that is identical to one employed at the transmitter. At the receiver, the transmitted signal is decoded and added to the predicted value of the signal to generate samples of the decoded speech signal that are low-pass filtered to produce the analog speech signal.

In a predictive coding system, the error $e_n = s'_n - s_n$ between the original and decoded speech samples is identical to the error $\delta'_n - \delta_n$ introduced by the quantizer, encoder, and the decoder. Therefore, the SNR in the decoded speech signal *exceeds* the SNR of the decoded difference signal by a factor equal to the ratio of the mean-square value of the input speech signal to the mean-square value of the difference signal. By using predictive coding, one can expect improvement of about 20 dB in SNR over a PCM system using identical quantizing levels.

4.2 Linear Prediction of Speech Signals

Linear prediction is a well-known method of removing the redundancy in a signal. For speech, the predictor P includes two separate predictors, one based on the short-time spectral envelope and the other based on the quasi-periodicity of voiced speech. The predictor based on the

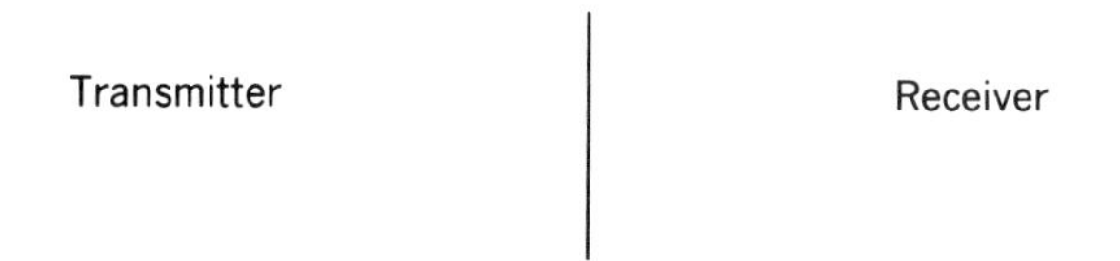

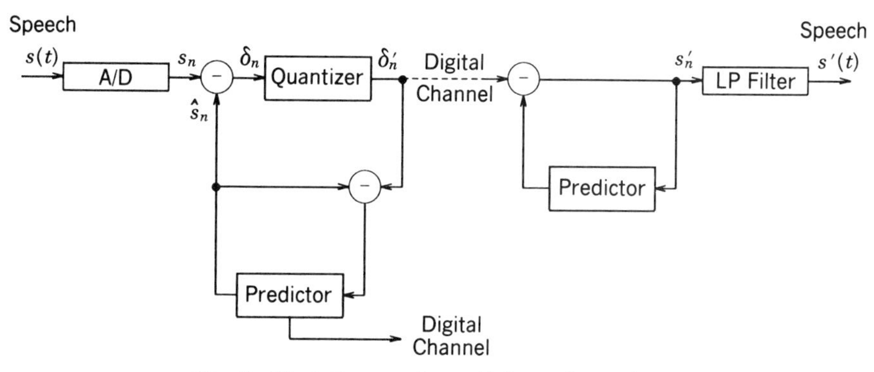

Fig. 5 Block diagram of a predictive coding system.

short-time spectral envelope is represented in z-transform notation as

$$P_1(z) = \sum_{k=1}^{p} a_k z^{-k}, \qquad (5)$$

where a_k are called predictor coefficients. The value of p, the number of past samples, is usually 10 for speech sampled at 8 kHz. The predictor based on the quasi-periodicity is represented as

$$P_2(z) = \beta z^{-M} \qquad (6)$$

where M represents the period and β represents the degree of periodicity.

The predictor coefficients, a_k, are determined by solving a set of linear equations[5] obtained by minimizing the mean-square prediction error. For the first predictor, the linear equations are

$$\sum_{k=1}^{p} c_{|i-k|} a_k = c_i, \qquad i = 1, 2, \ldots, p, \qquad (7)$$

where c_i are the short-time autocorrelations of the speech signal.

4.3 Spectrum of Quantization Noise

Traditionally, waveform coders attempt to minimize the rms error between the original and coded speech waveforms. It is now recognized that subjective perception of signal distortion is not based on the rms error alone but also on the short-time spectra of quantizing noise and speech. A simple method of providing flexibility in controlling the spectrum of the quantizing noise is to use a conventional predictive coding system with a prefilter and postfilter[5] as illustrated in Fig. 6. Ideally, the filter $1 - R$ could be adjusted to minimize the perceived distortion. But, a practical choice is given by

$$1 - R(z) = \frac{1 - \sum_{k=1}^{p} a_k z^{-k}}{1 - \sum_{k=1}^{p} a_k \alpha^k z^{-k}}, \qquad (8)$$

where α is a parameter controlling the shape of the noise spectrum. Typically, $\alpha = 0.80$ at a sampling frequency of 8 kHz. A large part of perceived noise in a coder comes from frequency regions where the signal level is low. By adjusting the shape of the noise spectrum, it is possible to shift the noise from the regions where the signal level is low to the regions where the signal levels are high (formant regions). The theory of auditory masking suggests that the quantizing noise in the formant regions would be partially (or totally) masked by the speech signal. An example of the quantizing noise spectrum together with the corresponding speech spectrum is shown in Fig. 7.

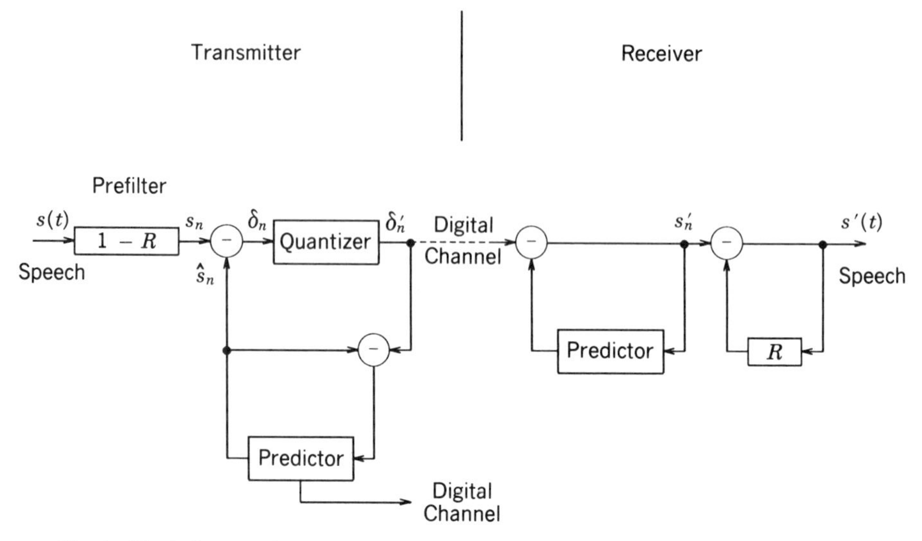

Fig. 6 Block diagram of a generalized predictive coder using pre- and postfiltering to achieve desired spectrum of the quantizing noise by selection of the filter $1 - R$.

Although with *fine* quantization, the generalized predictive coder shown in Fig. 6 can produce any desired shape of the quantizing noise spectrum by proper selection of the filter R, this is not the case with *coarse* quantization. This is a major shortcoming of predictive coders with instantaneous (sample-by-sample) quantizers. The problem is avoided by joint quantization of a block of samples (vector quantization).

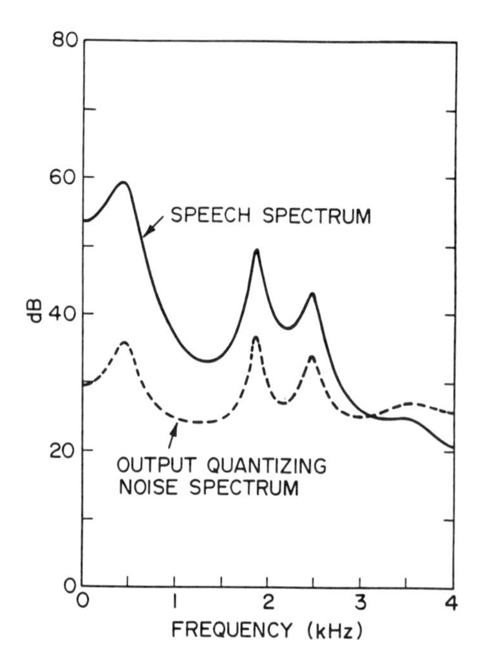

Fig. 7 An example showing the spectrum of quantizing noise (dotted curve) shaped to reduce the perceived distortion and the corresponding speech spectrum (solid curve).

5 CODE-EXCITED LINEAR PREDICTION

The joint quantization of a block of samples is accomplished by replacing the instantaneous quantizer in the predictive coder of Fig. 6 by a codebook and by selecting one code word from the codebook that minimizes the error averaged over the entire block.[6]

The block diagram illustrating the procedure for selecting the best code word is shown in Fig. 8. Each code word is scaled in amplitude and filtered through the predictive filter. The predictor P again includes two separate predictors as represented in Eqs. (5) and (6). The resulting speech signal is compared with the original signal to form an error signal that is next filtered through a perceptual error weighting filter, squared and averaged to provide an estimate of the perceptual error. The code word that produces the minimum perceptual error is selected as the quantizer output.

The purpose of the weighting filter is to provide a subjectively meaningful error measure, by attenuating the frequencies where the error is less important and by amplifying those frequencies where the error is more important. The transfer function of the perceptual

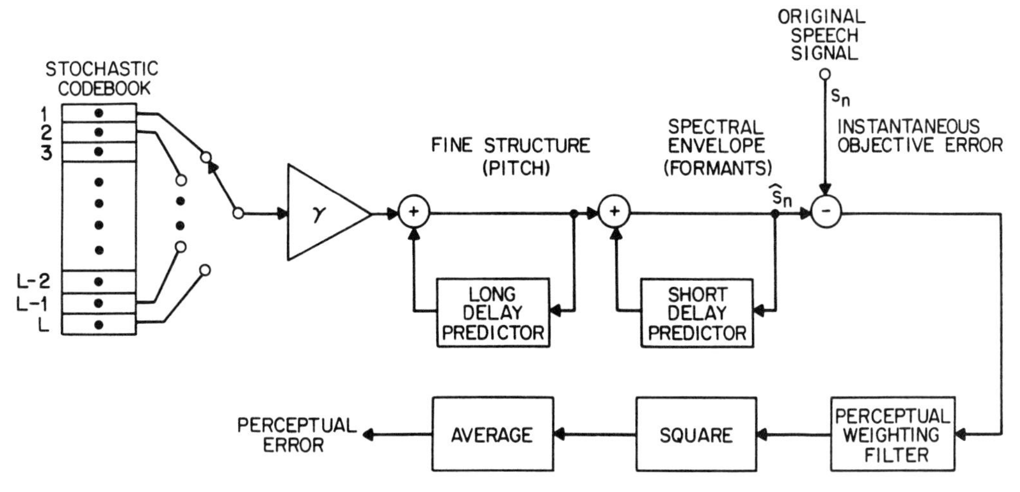

Fig. 8 Block diagram of a code-excited linear predictive coder.

weighting filter is the same as the transfer function of the prefilter used in the predictive coder of Fig. 6.

Efficient search procedures have been developed to minimize the time needed to search the codebook.[7] A recent federal standard for voice coding at 4.8 kb/s uses a fixed codebook containing sparse, overlapping, ternary-valued pseudo-random code words and an adaptive codebook in parallel.[8] The adaptive codebook performs a function similar to that of the predictor based on quasi-periodicity represented in Eq. (6). The predictive filter then includes only the predictor based on the short-time spectral envelope as represented in Eq. (5). The North American standard for digital cellular transmission uses a combination of an adaptive codebook and two fixed codebooks containing overlapping binary code words.[6] With several codebooks, the best code word from each codebook is selected sequentially to minimize the search time.

Code-excited linear predictive (CELP) coding is one of the most important techniques at present for coding high-quality speech at bit rates between 4 and 16 kb/s.[6]

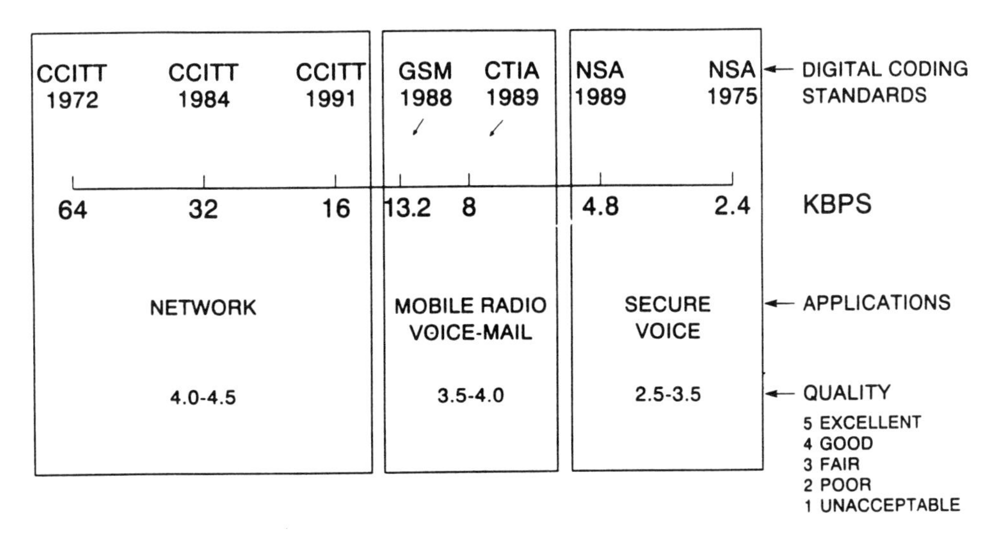

Fig. 9 Subjective quality at different bit rates for various coding standards and their application area (from Ref. 9).

6 SUBJECTIVE QUALITY OF CODED SPEECH

Speech coding invariably introduces distortion in the speech signal. It is important to determine the magnitude of this distortion by subjective assessment of speech quality. Although objective measures of speech quality can provide some measure of the distortion, the final assessment can only be done by proper subjective evaluation of speech quality.

6.1 Subjective Evaluation Methods

A number of methods for evaluating speech quality in subjective tests have been proposed.[6] The most common method for evaluating waveform coders is the absolute category judgment method that results in a mean opinion score (MOS) for the particular condition of the test coder. In this test, often called the MOS test, listeners are asked to choose a category—such as unsatisfactory, poor, fair, good, excellent—to describe their impression of the quality of the speech test material. Each category used in the test is given a number ranging from 1 for unsatisfactory to 5 for excellent, and the numbers are averaged to provide the mean opinion score.

6.2 Speech Quality at Different Bit Rates

Figure 9 shows the speech quality that is currently achieved at different bit rates. The figure provides a summary of the current capabilities of speech coders. A number of speech coding standards have been established to support digital transmission of speech. The highest speech quality, good to excellent (MOS > 4), is achievable only at bit rates above 16 kb/s and is suitable for telephone network applications. The speech quality at bit rates around 8 kb/s with MOS of about 4 is adequate for cellular and voice-mail applications. Speech coders operating at bit rates of 4.8 kb/s and below are able to deliver speech whose quality is only fair. These coders are suitable for special applications, such as for sending encrypted speech from one user to another over the public telephone network.

REFERENCES

1. L. R. Rabiner and R. W. Schafer, *Digital Processing of Speech Signals*, Prentice-Hall, Englewood Cliffs, NJ, 1978, pp. 24–26.
2. J. Bellamy, *Digital Telephony*, Wiley, New York, 1982, p. 99.
3. N. S. Jayant and P. Noll, *Digital Coding of Waveforms*, Prentice-Hall, Englewood Cliffs, NJ, 1984, pp. 307–311.
4. B. S. Atal and M. R. Schroeder, "Adaptive Predictive Coding of Speech Signals," *Bell System Tech. J.*, Vol. 49, No. 8, 1970, pp. 1973–1986.
5. B. S. Atal, "Predictive Coding of Speech at Low Bit Rates," *IEEE Trans. Commun.*, Vol. COM-30, No. 4, 1982, pp. 600–614.
6. J. R. Deller, J. G. Proakis, and J. H. L. Hansen, *Discrete-Time Processing of Speech Signals*, Macmillan, New York, 1993.
7. I. M. Trancoso and B. S. Atal, "Efficient Search Procedures for Selecting the Optimum Innovation in Stochastic Coders," *IEEE Trans. Acoust., Speech, Signal Proc.*, Vol. 38, No. 3, 1990, pp. 385–396.
8. J. P. Campbell, Jr., T. E. Tremain, and V. C. Welch, "The Federal Standard 1016 4800 bps CELP Voice Coder," *Digital Signal Proc.*, Vol. 1, 1991, pp. 145–155.
9. N. S. Jayant, "Signal Compression: Technology Targets and Research Directions," *IEEE J. Select. Areas Commun.*, Vol. 10, No. 5, 1992, pp. 796–818.

98

MACHINE RECOGNITION OF SPEECH

LAWRENCE R. RABINER

1 INTRODUCTION

Algorithms for machine recognition of speech can be characterized broadly as either pattern recognition approaches or acoustic-phonetic approaches. Pattern recognition approaches rely primarily on well-established methods of feature extraction, pattern training, and pattern comparison, and although they do not explicitly exploit some important properties of the speech signal (e.g., pitch, voiced–unvoiced nature of the signal, etc.), they provide high recognition performance for a wide range of interesting and useful applications. Acoustic-phonetic approaches try to uncover the fundamental linguistic units that comprise spoken language (e.g., phonemes) on the basis of observed acoustic properties of the speech. Because knowledge of the acoustic-phonetic properties of speech sounds is as yet inadequate, recognition performance of such systems falls considerably below that of comparable pattern recognition approaches. Hence this chapter will concentrate on explaining current techniques for speech recognition based on the pattern recognition framework.

2 RECOGNITION TECHNOLOGY HIERARCHY

The broad goal of speech recognition is to provide enhanced access to machines via voice commands. For most practical applications there are alternatives to voice for machine access, for example, touch screens, keyboards, track-ball, mouse, and so forth. Voice access and control is a natural means of communications with a machine and therefore will be considered a suitable means of machine access whenever voice recognition technology can provide a meaningful and workable user interface. A key aspect of the user interface is the means by which a user speaks to (or commands) the machine. The technology hierarchy for accessing a machine by voice has three distinct levels, namely:

- Isolated word (phrase) recognition whereby the machine learns individual words (or phrases) and responds appropriately when the words (or phrases) are spoken. This mode of access to a machine is often called *command and control* because usually each voice command consists of a single word (or phrase) to which the machine responds immediately. Applications of this technology include menu-based access, call routing, repertory dialing, and so forth.
- Connected word recognition whereby the machine learns a limited vocabulary of words and responds to a fluently spoken string of words from the learned vocabulary. The most obvious vocabulary for this technology is that of the spoken digits (zero to nine plus oh) with potential applications in the area of order entry, credit card validation, digit dialing, and so forth.
- Continuous speech recognition whereby the machine learns a vocabulary of fundamental speech units (subword) from which any spoken word can be created by an appropriate concatenation of subword units as specified within a word lexicon (dictionary). The area of continuous speech recognition is often subdivided into transcriptionlike systems (where every spoken word is recognized directly) and spoken language understanding systems (where the recognized speech is converted to a natural lan-

Note: References to chapters appearing only in the *Encyclopedia* are preceded by *Enc.*

guage query that need not agree word-for-word with the recognized speech). Transcriptionlike systems are used for voice dictation and similar applications; spoken language understanding systems are used for database management and access and language translation.

3 SOURCES OF VARIABILITY OF SPEECH

Speech recognition by machine is inherently difficult because of the variability in the signal being presented to the machine to be recognized. Sources of variability include the following:

- Within-speaker variability in maintaining consistent pronunciation and use of words/phrases
- Across-speaker variability due to regional accents, foreign languages, and so forth
- Transducer variability while speaking over different microphones/telephone handsets
- Variability introduced by the transmission system
- Variability in the speaking environment including extraneous conversations and acoustic background events (e.g., road noise, fans, door slams, etc.)

4 ISSUES IN IMPLEMENTATION OF SPEECH RECOGNITION SYSTEMS

There are three key issues involved in the design and implementation of a speech recognition system:

1. Speech detection, namely the problem of separating the speech to be recognized from the acoustic background (which may consist of the speech of other talkers). The solution to this problem is to build a dynamic (adaptive) model of the background and to continuously classify (via the recognition algorithm) the incoming signal as either background signal or the speech to be recognized. The methods used to perform this task are essentially those used in connected-word recognition.
2. Recognition technology used to classify the signal. Over time four technologies have evolved including template-based systems, statistical model-based systems, acoustic-phonetic model-based systems, and neural network approaches. The template and statistical model approaches have had the most success in practical systems and will be described below.
3. Human factors, namely a user-friendly interface that allows the speaker to make human errors (e.g., stammering, starting over, speaking too fast or too slow) without the system degrading to the point where all dialogue between the machine and the user is lost. The ways in which such a user-friendly interface are achieved are beyond the scope of this chapter.

5 STATISTICAL PATTERN RECOGNITION MODEL

Figure 1 shows the basic statistical pattern recognition model used for speech recognition. There are four key steps in the model, namely feature extraction (called parametric representation), pattern training, pattern classification, and the decision algorithm.

There are two types of reference patterns that can be

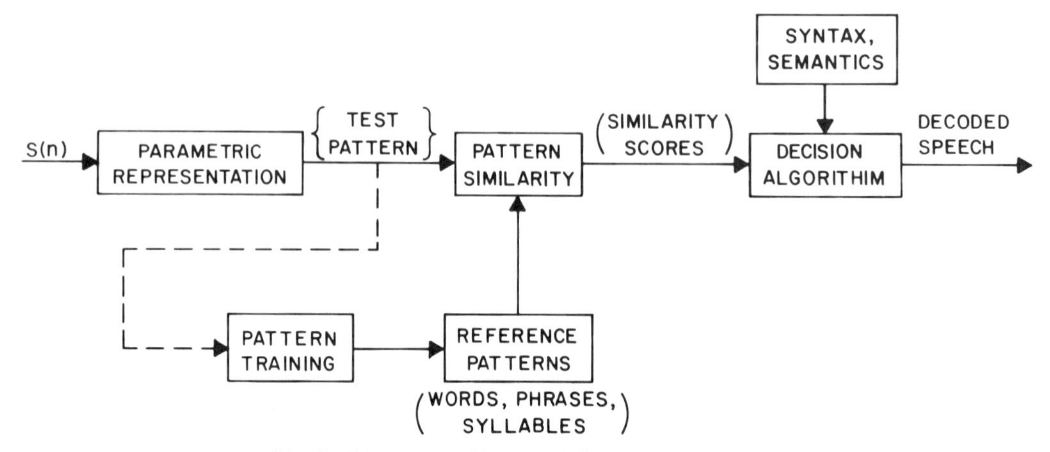

Fig. 1 Pattern recognition model for speech recognition.

used with the model of Fig. 1. The first type, called nonparametric reference patterns, are patterns created from one or more real-world tokens of the actual pattern. The second type, called statistical reference models, are created as a statistical characterization (via a fixed type of model) of the behavior of a collection of real-world tokens. Ordinary template approaches are examples of the first type of reference patterns; hidden Markov models are examples of the second type of reference patterns.

The model of Fig. 1 has been used (either explicitly or implicitly) for almost all commercial and industrial speech recognition systems for the following reasons:

1. It is invariant to different speech vocabularies, users, feature sets, pattern similarity algorithms, and decision rules.
2. It is easy to implement in either software or hardware.
3. It works well in practice.

We will concentrate on this model throughout this chapter. We now discuss the elements of the pattern recognition model and show how it has been used for isolated word, connected-word, and for continuous-speech recognition. Because of the tutorial nature of this chapter we will minimize the use of mathematics in describing the various aspects of the signal processing.

5.1 Feature Extraction

The purpose of the speech analysis system is to represent the spectral properties of the speech signal, over time, in an efficient and compact manner. The most widely used spectral representation is the linear predictive coding (LPC) analysis method in which a frame of the speech signal (a block of N consecutive samples of speech) is analyzed to provide a vector of spectral features. A block diagram of the LPC analysis model is shown in Fig. 2. The steps involved in obtaining a vector of LPC coefficients, for a given frame of N speech samples for a telephone bandwidth (200–3400 Hz) signal, are as follows:

1. Preemphasis by a first-order digital network in order to spectrally flatten the speech signal
2. Frame windowing, that is, multiplying the N speech samples within the frame by an N-point Hamming window so as to minimize the endpoint effects of chopping an N sample section out of the speech signal (window sizes on the order of 20–45 ms are typical for speech)
3. Autocorrelation analysis in which the windowed set of speech samples is autocorrelated to give a set of $(p + 1)$ coefficients, where p is the order of the LPC analysis (typically 8–12)
4. Linear predictive coding analysis in which the vector of LPC coefficients is computed from the autocorrelation vector using a Levinson or a Durbin recursive method.[1]

New speech frames are created by shifting the analysis window by M samples (typically $M < N$ with 10–15 ms shifts being most often used) and the above steps are repeated on the new frame until the entire speech signal has been analyzed.

The LPC feature model has been a popular speech representation because of its ease of implementation, and because the technique provides a robust, reliable, and accurate method for characterizing the spectral properties of the speech signal.

5.2 Pattern Training

Pattern training is the method by which representative sound patterns are converted into reference patterns for

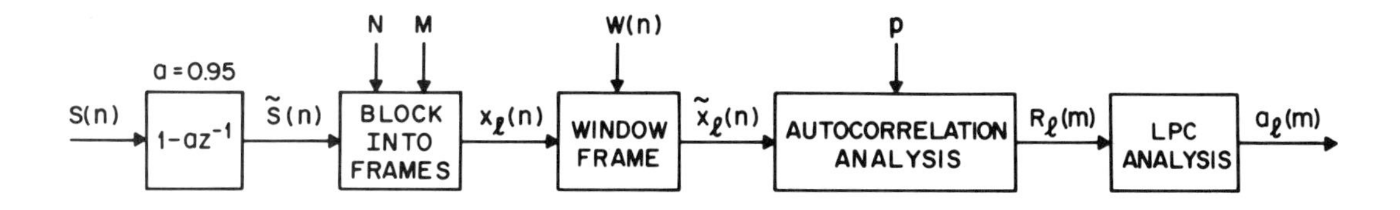

$$\tilde{s}(n) = s(n) - a s(n-1)$$

$$x_\ell(n) = \tilde{s}(M\ell + n), \quad \ell = 0, 1, 2, \cdots, L-1$$
$$n = 0, 1, 2, \cdots, N-1$$

Fig. 2 LPC analysis model.

use by the patten similarity algorithm. There are several ways in which pattern training can be performed, including:

1. Casual training in which each individual training pattern is used directly to create either a nonparametric reference pattern (often called a template) or a very crude statistical model (because of the paucity of data). Casual training is the simplest, most direct method of creating reference patterns.
2. Robust training in which several (i.e., two or more) versions of each vocabulary entry are used to create a single reference template or statistical model. Robust training gives increased statistical confidence to the reference patterns since multiple patterns are used in the training.
3. Clustering training in which a large number of versions of each vocabulary entry are used to create one or more templates or statistical models. A statistical clustering analysis is used to determine which members of the multiple training patterns are similar and hence are used to create one or more coherent reference patterns. Clustering training is generally used for creating speaker-independent reference patterns, in which case the multiple training patterns of each vocabulary entry are derived from a large number of different talkers.

The final result of the pattern training algorithm is the sete of reference patterns used in the recognition phase of the model of Fig. 1.

5.3 Pattern Similarity Algorithm

A key step in the recognition algorithm of Fig. 1 is the determination of similarity between the measured (unknown) test pattern and each of the stored reference patterns. Because speaking rates vary greatly from repetition to repetition, pattern similarity determination involves both time alignment (registration) of patterns and once properly aligned, distance computation along the alignment path.

Figure 3 illustrates the problem involved in time aligning a test pattern, $T(n), 1 \leq n \leq NT$ [where each $T(n)$ is a spectral vector], and a reference pattern $R(m), 1 \leq m \leq NR$. The goal is to find an alignment function, $m = w(n)$, that maps R onto the corresponding parts of T. The criterion for correspondence is that some measure of distance between the patterns be minimized by the mapping w. Defining a local distance measure, $d(n, m)$, as the spectral distance between vectors $T(n)$ and $R(m)$, then the task of the pattern similarity algorithm is to determine the optimum mapping, w, to minimize the total distance

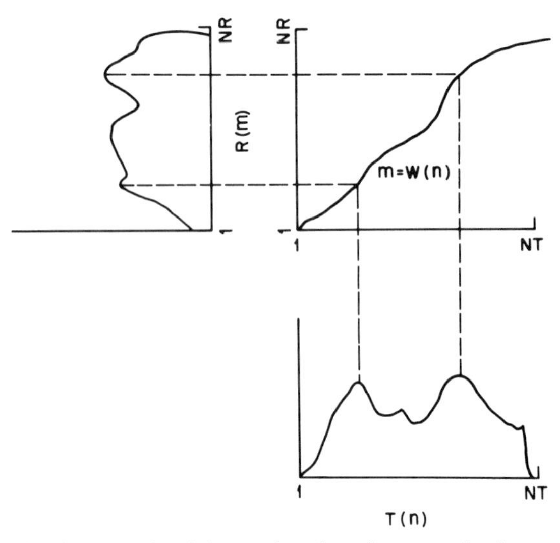

Fig. 3 Example of time registration of a test and reference pattern.

$$D^* = \min_{w(n)} \sum_{i=1}^{NT} d[i, w(i)]. \qquad (1)$$

The solution to Eq. (1) can be obtained in an efficient manner using the techniques of dynamic programming. In particular a class of procedures called dynamic time warping (DTW) techniques has evolved for solving Eq. (1) efficiently.[2] The above discussion has shown how to time align a pair of templates. In the case of aligning statistical models, an analogous procedure, based on the Viterbi algorithm, can be used.[3]

5.4 Decision Algorithm

The last step in the statistical pattern recognition model of Fig. 1 is the decision that utilizes both the set of pattern similarity scores (distances) and possibly other system knowledge, such as syntax and/or semantics, to decode the speech into the best possible transcription. The decision algorithm can (and generally does) incorporate some form of nearest-neighbor rule to process the distance scores to increase confidence in the results provided by the pattern similarity procedure. The system syntax helps to choose among the candidates with the lowest distance score by eliminating candiates that do not satisfy the syntactic constraints of the task or by deweighting extremely unlikely candidates. The decision algorithm can also have the capability of providing multiple decodings of the spoken string. This feature is especially useful in cases in which multiple candidates have indistinguishably different distance scores.

6 TEMPLATES VERSUS STATISTICAL MODELS

The basic speech recognition pattern can be represented as either a template or a statistical model. The template is created by averaging spectral vectors of the different training tokens, from which the pattern is created, along the time alignment path provided by the DTW procedure. Hence the template provides very fine temporal resolution (typically 10–15 ms) but only provides first-order statistics (mean values) of the spectral parameters of the reference.

An alternative representation is a statistical model in which the speech pattern is represented as a Markov model of the type shown in Fig. 4. The model has N (5 in the figure) states in each of which the speech signal is assumed to be a stationary process characterized by a mixture Gaussian density of spectral features (called the observation density) and an energy and a state duration probability. The states (which proceed in the example of Fig. 4 from left to right) represent the changing temporal nature of the speech signal; hence indirectly they represent the speech sounds within the pattern. The model of Fig. 4 is called a hidden Markov model (HMM) because the association of speech spectral vectors (observations) with model states is not observable directly but can only be estimated from the statistics of the model.

The HMM usually provides a coarser temporal resolution than the template (i.e., there are much fewer states in the model than frames within a template), but models the spectral vectors of the speech using both first- and second-order statistics (i.e., mean and covariance estimates), and therefore is generally a better representation of the variability of speech patterns. A complete description of the hidden Markov model is found in Ref. 4.

7 RESULTS ON ISOLATED WORD RECOGNITION

Using the pattern recognition model of Fig. 1, with an eighth-order LPC parametric representation, and using either the nonparametric template approach or the HMM method for reference patterns, a wide variety of tests of the recognizer have been performed on telephone speech with isolated word inputs in both speaker-dependent (SD) and speaker-independent (SI) modes. Vocabulary sizes have ranged from as small as 10 words (i.e., the digits zero–nine) to as many as 1109 words. Table 1 gives a summary of recognizer performance under the conditions discussed above. It can be seen that the resulting error rates are not strictly a function of vocabulary size but also are dependent on vocabulary complexity. Thus a simple vocabulary of 200 polysyllabic Japanese city names had a 2.7% error rate (in an SD mode), whereas a complex vocabulary of 39 alphadigit terms (in both SD and SI modes) had error rates of on the order of 4.5–7.0%.

8 CONNECTED-WORD RECOGNITION MODEL

The basic approach to connected-word recognition from discrete reference patterns is shown in Fig. 5. Assume

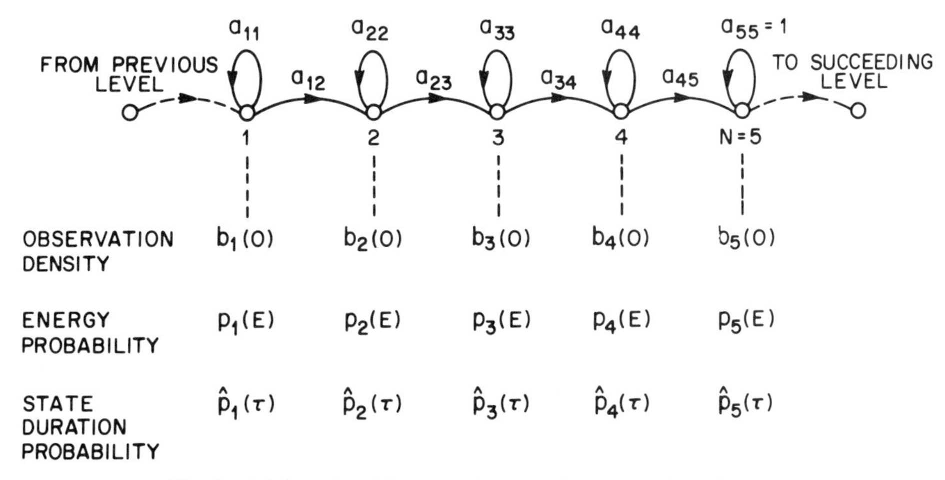

Fig. 4 A left-to-right hidden Markov model representation of speech.

TABLE 1 Performance of Template-Based and HMM-Based Isolated Word Systems

Number of Words in Vocabulary	Mode	Error Rate (%)
10 digits	SI	0.1
37 dialer words	SD	0
39 alphadigits	SD	4.5
	SI	7.0
54 computer terms	SI	3.5
129 airline words	SD	1.0
	SI	2.9
200 Japanese cities	SD	2.7
1109 basic English	SD	4.3

we are given a test pattern **T** that represents an unknown spoken word string, and we are given a set of V reference patterns, $\{R_1, R_2, \ldots, R_V\}$ each representing some word of the vocabulary. The connected-word recognition problem consists of finding the "super" refernce pattern $\mathbf{R}^s$ of the form

$$\mathbf{R}^s = R_{q(1)} \oplus R_{q(2)} \cdots R_{q(L)},$$

which is the concatenation of L reference patterns $R_{q(1)}, R_{q(2)}, \ldots, R_{q(L)}$ that best matches the test string **T** in the sense that the overall distance between **T** and $\mathbf{R}^s$ is minimum over all possible choices of $L, q(1), q(2), \ldots, q(L)$, where the distance is an appropriately chosen distance measure.

There are several problems associated with solving the above connected-word recognition problem. First we do not know L, the number of words in the string. Hence our proposed solution must provide the best matches for all reasonable values of L, for example, $L = 1, 2, \ldots, L_{\max}$. Second we do not know nor can we reliably find word boundaries, even when we have postulated L, the number of words in the string. The implication is that the word recognition algorithm must work without direct knowledge of word boundaries; in fact the estimated word boundaries can be shown to be a by-product of the matching procedure. The third problem with a whole-word matching procedure is that the word matches are generally much poorer at the boundaries than at frames within the word. In general this is a weakness of word-matching schemes that can be somewhat alleviated by the matching procedures that can apply lesser weight to the match at pattern boundaries than at frames within the word. A fourth problem is that word durations in long strings are often grossly different (shorter) than the durations of the corresponding reference patterns, which are determined from strings of all lengths. Finally the last problem associated with matching word strings is that the combinatorics of matching strings exhaustively (i.e., by trying all combinations of reference patterns in a sequential matter) is prohibitive.

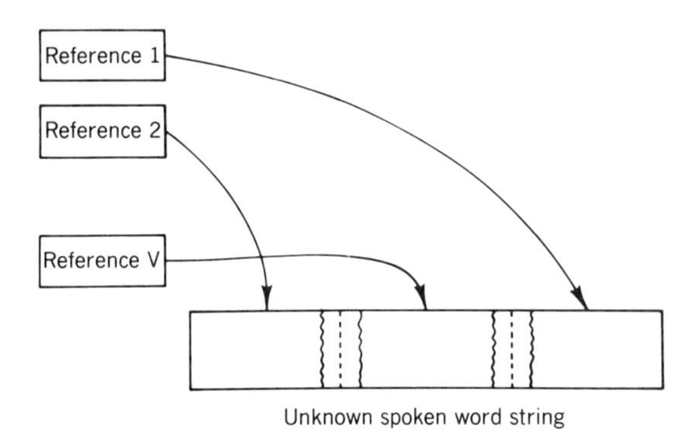

Fig. 5 Illustration of connected-word recognition from word patterns.

A number of different ways of solving the connected-word recognition problem has been proposed that avoids the plague of combinatorics mentioned above. Among these algorithms are the two-level DP approach of Sakoe,[5] the level-building approach of Myers and Rabiner,[6] the parallel single-stage approach of Bridle et al.,[7] and the nonuniform sampling approach of Gauvain and Mariani.[8] Although each of these approaches differs greatly in implementation, all of them are similar in that the basic procedure for finding $\mathbf{R}^s$ is to solve a time-alignment problem between **T** and $\mathbf{R}^s$ using DTW methods.

8.1 Performance of Connected-Word Recognizers

Typical performance results for connected-word recognizers, based on a level-building implementation, are shown in Table 2. Included in the table are both word and string error rates. To the extent that word errors are independent in each spoken string, the average string error rate is essentially the average word error rate multiplied by the average number of words in a string. When multiple word errors occur within a string, the average string error rate is smaller than what would be anticipated from the average word error rate. For a digits vocabulary, string error rates less than 1% have been obtained. For name retrieval, by spelling, from a 17,000-name directory, string error rates of from 4 to 10% have been obtained. Finally, using a moderate size vocabulary of 127 words, the error rate of sentences for obtaining information about airline schedules is between 1 and 10%. Here the average sentence length was close to 10 words. Many of the errors occurred in sentences with long strings of digits.

TABLE 2 Performance of Connected-Word Recognizers on Specific Recognition Tasks

Vocabulary	Mode	Word Error Rates	Task	String (Task) Error Rates
Digits (10 words)	Speaker dependent or speaker independent	0.2% SI, 0.1% SD	1–7 digit strings	0.8% SI[a]
			1–7 digit strings	0.4% SD[a]
Letters of the alphabet (26 words)	Speaker dependent or speaker independent	≈10%	Directory listing retrieval (17,000-name directory)	4% SD
				10% SI
Airline terms (129 words)	Speaker dependent or speaker independent	0.1% SD, 3% SI	Airline information and reservations	1% SD
				10% SI

[a]Known string length.

9 CONTINUOUS, LARGE VOCABULARY, SPEECH RECOGNITION

The area of continuous, large vocabulary, speech recognition refers to systems with at least 1000 words in the vocabulary, a syntax approaching that of natural English (i.e., an average branching factor on the order of 100), and possibly a semantic model based on a given, well-defined, task. For such a problem, there are three distinct subproblems that must be solved, namely choice of a basic recognition unit (and a modeling technique to go with it), a method of mapping recognized units into words (or, more precisely, a method of scoring words from the recognition scores of individual word units), and a way of representing the formal syntax of the recognition task (or, more precisely, a way of integrating the syntax directly into the recognition algorithm).

For each of the three parts of the continuous speech recognition problem, there are several alternative approaches. For the basic recognition unit, one could consider whole words, half syllables such as dyads, demisyllables, or diphones, or sound units as small as phonemes or phones. Whole-word units, which are attractive because of our knowledge of how to handle them in connected environments, are totally impractical to train since each word could appear in a broad variety of contexts. Therefore the amount of training required to capture all the types of word environments is unrealistic. For the subword units, the required training is extensive, but could be carried out using a variety of well-known, existing training procedures. A full system would require between 1000 and 2000 half-syllable speech units. For the phonelike units, only about 30–100 units would have to be trained.

The problem of representing vocabulary words, in terms of the chosen speech unit, has several possible solutions. One could create a network of linked-word unit models for each vocabulary word. The network could be either a deterministic (fixed) or a stochastic structure. An alternative is to do lexical access from a dictionary in which all word pronunciation variants (and possibly part of speech information) are stored, along with a mapping from pronunciation units to speech representation units.

Finally the problem of representing the task syntax, and integrating it into the recognizer, has several solutions. The task syntax, or grammar, can be represented as a deterministic state diagram, as a stochastic model (e.g., a model of word trigram statistics), or as a formal grammar. There are advantages and disadvantages to each of these approaches.

9.1 Performance of Large Vocabulary Systems

To illustrate the state of the art in large vocabulary speech recognition, consider the results shown in Table 3. The table shows results for two tasks, namely an office dictation system[9] and a naval resource management system.[10–14] The office dictation system uses phoneme-like units in a statistical model to represent words, where

TABLE 3 Performance of Large Vocabulary Speech Recognizers on Specific Recognition Tasks

Task	Syntax	Mode	Vocabulary	Word Error Rate
Office dictation (IBM)	Word trigram (perplexity = 100)	SD, isolated word input	5000 words	2%, prerecorded speech 3.1%, read speech 5.7, spontaneous speech
Naval resource management (DARPA)	Finite-state grammar (perplexity = 60)	SI, fluent speech input	991 words	4.5%

each phonemelike unit is a statistical model based on vector-quantized spectral outputs of a speech spectrum analysis. A third statistical model is used to represent syntax; thus the recognition task is essentially a Bayesian optimization over a triply embedded sequence of statistical models. The computational requirements are very large, but a system has been implemented using isolated word inputs for the task of automatic transcription of office dictation. For a vocabulary of 5000 words, in a speaker-trained mode, with 20 min of training for each talker, the average *word* error rates for 5 talkers are 2% for prerecorded speech, 3.1% for read speech, and 5.7% for spontaneously spoken speech.[9] The naval resource management system also uses a set of about 2000 phonemelike units (PLUs) to represent words where each PLU is a statistical model of a phoneme in specified contexts. The task syntax (that of a ship's database) is specified in the form of a finite-state network from a word pair grammar with average word branching factor (perplexity) of 60. When tested on a vocabulary of 991 words, in a speaker-independent mode, using continuous, fluently spoken, sentences, a word error rate of about 4.5% was obtained.[15]

10 SUMMARY

In this chapter we have reviewed and discussed the general pattern recognition framework for machine recognition of speech. We have discussed some of the signal processing and satistical pattern recognition aspects of the model and shown how they contribute to the recognition. A more complete discussion of all aspects of speech recognition can be found in the recent textbook by Rabiner and Juang.[16]

The challenges in speech recognition are many. As illustrated above, the performance of current systems is barely acceptable for large vocabulary systems, even with isolated word inputs, speaker training, and favorable talking environment. Almost every aspect of continuous speech recognition, from training to systems implementation, represents a challenge in performance, reliability, and robustness.

REFERENCES

1. J. D. Markel and A. H. Gray, *Linear Prediction of Speech*, Springer Verlag, New York, 1976.
2. F. Itakura, "Minimum Prediction Residual Principle Applied to Speech Recognition," *IEEE Trans. Acoust., Speech, Signal Proc.*, Vol. ASSP-23, Feb. 1976, pp. 67–72.
3. F. Jelinek, "Speech Recognition by Statistical Methods," *Proc. IEEE*, Vol. 65, April 1976, pp. 532–556.
4. L. R. Rabiner, "A Tutorial on Hidden Markov Models and Selected Applications in Speech Recognition," *Proc. IEEE*, Vol. 77, No. 2, Feb. 1989, pp. 257–286.
5. H. Sakoe, "Two Level DP Matching—A Dynamic Programming Based Pattern Matching Algorithm for Connected Word Recognition," *IEEE Trans. Acoust., Speech, Signal Proc.*, Vol. ASSP-27, Dec. 1979, pp. 588–595.
6. C. S. Myers and L. R. Rabiner, "Connected Digit Recognition Using a Level Building DTW Algorithm," *IEEE Trans. Acoust., Speech, Signal Proc.*, Vol. ASSP-29, No. 3, June 1981, pp. 351–363.
7. J. S. Bridle, M. D. Brown, and R. M. Chamberlain, "An Algorithm for Connected Word Recognition," in J. P. Haton (Ed.), *Automatic Speech Analysis and Recognition*, 1982, pp. 191–204.
8. J. L. Gauvain and J. Mariani, "A Method for Connected Word Recognition and Word Spotting on a Microprocessor," *Proc. 1982 ICASSP*, May 1982, pp. 891–894.
9. F. Jellinek, "The Development of an Experimental Discrete Dictation Recognizer," *Proc. IEEE*, Vol. 73, No. 11, Nov. 1985, pp. 1616–1624.
10. K. F. Lee, H. W. Hon, and D. R. Reddy, "An Overview of the SPHINX Speech Recognition System," *IEEE Trans. Acoust., Speech, Signal Proc.*, Vol. 38, 1990, pp. 600–610.
11. Y. L. Chow, M. O. Dunham, O. A. Kimball, M. A. Krasner, G. F. Kubala, S. Makhoul, S. Roucos, and R. M. Schwartz, "BBYLOS: The BBN Continuous Speech Recognition System," *Proc. IEEE Int. Conf. Acoust., Speech, Signal Proc.*, Apr. 1987, pp. 89–92.
12. D. B. Paul, "The Lincoln Robust Continuous Speech Recognizer," *Proc. ICASSP 89*, Glasgow, Scotland, May 1989, pp. 449–452.
13. M. Weintraub, et al., "Linguistic Constraints in Hidden Markov Model Based Speech Recognition," *Proc. ICASSP 89*, Glasgow, Scotland, May 1989, pp. 699–702.
14. V. Zue, J. Glass, M. Phillips, and S. Seneff, "The MIT Summit Speech Recognition System: A Progress Report," *Proc. Speech and Natural Language Workshop*, Feb. 1989, pp. 179–189.
15. C. H. Lee, L. R. Rabiner, R. Piereaccini, and J. G. Wilpon, "Acoustic Modeling for Large Vocabulary Speech Recognition," *Comput. Speech Language*, Vol. 4, 1990, pp. 127–165.
16. L. R. Rabiner and B. H. Juang, *Fundamentals of Speech Recognition*, Prentice-Hall, Englewood Cliffs, NJ, 1993.

PART XIV

MUSIC AND MUSICAL ACOUSTICS

99

INTRODUCTION

THOMAS D. ROSSING

1 INTRODUCTION

Musical acoustics is a broad interdisciplinary field that deals with the *production* of musical sound, the *transmission* of musical sound to the listener, and the *perception* of musical sound. Researchers in musical acoustics come from the disciplines of physics, music, psychology, physiology, electrical or mechanical engineering, and architecture. Conferences devoted to musical acoustics will usually attract performing musicians, musical instrument builders, and architects, as well as acoustics researchers. Many scientists and engineers are enjoying a "second career" in this challenging and fascinating field, which truly forms a bridge between science and art.

Chapters 100–103, *Enc.* Ch. 132, 135, 137–139 focus on the production of musical sound by various musical instruments, including the digital computer and the human voice. The transmission of musical sound to the listener in concert halls is discussed in Chapter 78, while transmission to the listener electronically is treated in Chapters 112–114, *Enc.* Ch. 161, 163, 166. Materials on the perception of musical sound will be found in Chapters 90 and 91, *Enc.* Ch. 117, 121.

2 PRODUCTION OF MUSICAL SOUND

The most common way of classifying musical instruments is according to the nature of the primary vibrator. Thus, we speak of string instruments, wind instruments, and percussion instruments (or the equivalent classifications by musicologists von Hornbostel and Sachs: chordophones, aerophones, idiophones, and membranophones[1,2]). To these, we add the electronic synthesizer, the digital computer, and the human voice.

Another interesting way of classifying instruments is according to the nature of the feedback. In the case of most percussion, plucked string, and struck string instruments, the player delivers energy to the primary vibrator (string, membrane, bar, or plate) and thereafter has little control over the way it vibrates. In the case of wind and bowed string instruments, however, the continuing flow of energy is controlled by feedback from the vibrating system. In the case of brass and reed woodwinds, pressure feedback opens or closes the input valve. In the case of flutes or flue organ pipes, the input valve is flow controlled. In the case of bowed string instruments, pulses on the string control the stick-slip action of the bow on the string.

Considerable attention has been given to the materials used in musical instruments. In some cases, traditional materials are becoming scarce (e.g., Brazilian rosewood for marimba bars) or expensive to process (e.g., calfskin for drumheads), and synthetic materials are replacing them. Given enough research, synthetic materials that duplicate any given property of natural materials can usually be developed. The development process may be long and costly, however, and attempts to reduce this may lead to inferior materials.

3 TRANSMISSION OF MUSICAL SOUND TO THE LISTENER

The topic of transmission of musical sound to the listener generally suggests the subject of concert hall acoustics. When I tell a new acquaintance that I work in musical acoustics, I am often asked the question "When will *they* learn to build good concert halls?" To the layperson,

Note: References to chapters appearing only in the *Encyclopedia* are preceded by *Enc.*

acoustics is the science of designing concert halls, and indeed it is an important part of the field.

Fortunately for listener and performer alike, great strides have been made in concert hall design, thanks in part to availability of digital computers and signal processing equipment. No longer are architects tied to "safe" designs that replicate proven halls (such as the long rectangular or "shoebox" design that gave relatively few listeners a really good view of the performers).

Transmission of musical sound to the listener these days often means via recorded medium, as more people presently listen to recorded music than attend live concerts. Fortunately, recording and reproduction of musical sound can now be done with a degree of fidelity hardly dreamed possible a few years ago. Compact disc digital audio has been an enormous success, and some observers are predicting similar success for digital audio tape recording.

4 PERCEPTION OF MUSICAL SOUND

The bottom line in the field of musical acoustics is how the listener perceives the transmitted sound. This depends, to some extent, on how the individual's ear–brain computer is programmed, both by heredity and by experience. The study of sound perception is the central theme of an interdisciplinary field called *psychoacoustics*.

Table 1 partly describes how perceived qualities such as loudness, pitch, and timbre depend on physical parameters such as sound pressure, frequency, and spectrum. Note the interdependence is indicated as strong (+++) to weak (+) but never zero.

TABLE 1 Dependence of Subjective Qualities on Physical Parameters

Physical Parameter	Subjective Quality			
	Loudness	Pitch	Timbre	Duration
Pressure	+++	+	+	+
Frequency	+	+++	++	+
Spectrum	+	+	+++	+
Duration	+	+	+	+++
Envelope	+	+	++	+

Note: The physical duration of a sound and its perceived (subjective duration), though closely related, are not the same. "Envelope" includes the attack, the decay, and variations in amplitude throughout the duration.

5 MUSICAL ACOUSTICS RESEARCH

Although there are few major centers for research in musical acoustics, a lot of enthusiastic researchers work alone, or with a few students, at universities, colleges, or in their own homes. Research in this area has been greatly facilitated by the availability of personal computers, fast Fourier transform (FFT) analyzers, and other digital instrumentation.

Modes of vibration in complex structures are now being studied theoretically by use of finite-element methods and experimentally by modal analysis and holographic interferometry, methods that were unknown a few years ago. Measurements in the time domain complement measurements in the frequency domain.

Electronic music and computer music studios are now becoming available in most music departments. Although primarily intended for teaching and for composing, they are being used for research in musical acoustics as well. Digital music synthesizers and keyboard instruments are affordable by individuals for their home studios and laboratories. Many composers of electronic music are using physical models of instruments as a basis for their algorithms for synthesizing musical sound.

A historical view of research on bowed string instruments has been written by Hutchins,[3] and a summary of musical acoustics research during the decade 1983–93 was prepared by Rossing.[4] Both of these reviews include extensive lists of references. A review of research leading to the development of an octet of violin-family instruments also has been prepared by Hutchins.[5]

REFERENCES

1. C. Sachs, *The History of Musical Instruments*, Norton, New York, 1940.
2. S. Macuse, *Survey of Musical Instruments*, Harper and Row, New York, 1975.
3. C. M. Hutchins, "A History of Violin Research," *J. Acoust. Soc. Am.*, Vol. 73, 1983, pp. 1421–1440.
4. T. D. Rossing, "Musical Acoustics Research Since SMAC83," *Proc. Stockholm Music Acoustics Conf.*, Swedish Academy of Music, Stockholm, 1995, pp. 10–19.
5. C. M. Hutchins, "A 30-Year Experiment in the Acoustical and Musical Development of Violin-Family Instruments," *J. Acoust. Soc. Am.*, Vol. 92, 1992, pp. 639–650.

100

STRINGED INSTRUMENTS: BOWED

J. WOODHOUSE

1 INTRODUCTION

Musical instruments based on the bowed string are an ancient invention and can be found today in all areas of the world. Although the detailed designs show a wide variety, the essential chain of events by which they function is always the same. The source of vibrational energy consists of one or more stretched strings, which are set into self-sustained oscillation by frictional interaction with a bow of some sort, which is drawn across the string. Stick-slip behavior is encouraged by coating the rubbing surface of the bow with a coniferous resin. The pitch of the note is controlled by the player using a finger to vary the vibrating length of the string. Since a vibrating string is a very poor sound radiator, the string is attached to some kind of mechanical amplifier or impedance-matching device, which allows a significant fraction of the energy to be converted to sound radiation. The most familiar form of this device is a wooden box, as in the orchestral stringed instruments, but stretched membranes, horn systems, and even plastic bottles have also been used to provide this function.

2 TERMINOLOGY

This discussion will concentrate on the modern orchestral stringed instruments, but many of the general ideas can be applied to the wide range of other bowed instruments. For brevity these instruments will all be called "violins," but the discussion may be taken to refer also to the viola, cello, and bass, which differ from the violin in pitch range and hence size, but only in rather minor details of construction otherwise.

First, the terminology of the main parts of a modern violin is introduced in Fig. 1. The only parts not obvious from an external examination of a violin are (i) the bass bar, a reinforcing strut running some three-fourths of the length of the top plate and passing beneath the foot of the bridge on the bass-string side; (ii) the soundpost, a wooden rod whose ends are carefully shaped to fit between the top and back plates, roughly beneath the other bridge foot, where it is held in by friction alone; and (iii) the corner blocks, end blocks, and lining strips, which serve to provide sufficient gluing area to hold the woodwork together.

Different timber species are used to make the various parts. The top plate is generally made from Norway spruce, *Picea abies*, grown at altitude to produce close and even annual rings. Some other softwood species are also used, but not many share the desirable properties of Norway spruce.[1] The back, ribs, and neck are usually made from European maple, *Acer platanoides*, the preferred wood having markings called "figure," arising from waviness in the grain lines as the tree grows. Blocks and linings are preferably made from willow, but various softwoods are also commonly used. The fingerboard is usually of ebony, while pegs, endpin, and tailpiece may be of ebony, rosewood, or boxwood. Finally, the bridge is made from maple, usually of a slow-grown and hard variety.

Although all the main component parts of the violin box are curved in one way or another, only in the case of the ribs is this curvature invariably achieved by bending the wood. The top and back plates are carved from the solid in almost all hand-made instruments, although factory instruments may use plates with arching imparted by pressing. For an account highly respected

Note: References to chapters appearing only in the *Encyclopedia* are preceded by *Enc.*

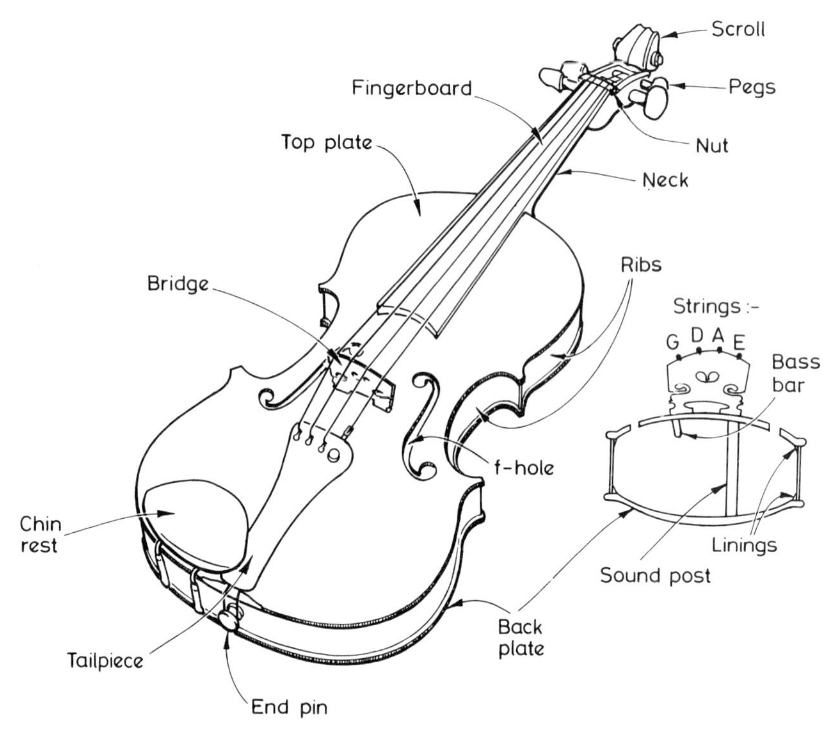

Fig. 1 Main component parts of a violin.

by many luthiers of some of the details and subtleties of the process of violin making, see Sacconi.[2] For a general account of the historical, constructional, and musical aspects of the violin, see Gill.[3]

3 OVERVIEW OF VIOLIN SCIENCE

To classify the scientific problems relevant to understanding the violin, it is first necessary to describe in more detail the chain of events when a violin is played. The player controls the vibration of the string, using the bow and the fingers of the left hand. Most of the energy of this vibration is in the form of transverse oscillations polarized in the bowing plane, although there is some transverse motion in the orthogonal plane, and also significant torsional motion and some excitation of longitudinal motion. The transverse string motion applies an oscillating force to the top of the bridge, at the string notch. This force is transmitted by the bridge to the violin top, being modified on the way by the dynamical behavior of the bridge itself. To a useful first approximation, the soundpost gives a strong constraint to the treble-side bridge foot, so that the bridge motion (at low frequencies, at least) consists largely of rocking about the treble foot, imparting normal motion to the top plate through the bass foot. This excites bending vibrations of the top plate, and hence of the rest of the box. The motion of these relatively large areas of wood then radiates sound waves, with a frequency-dependent directivity pattern. These interact with the acoustics of the room and finally reach the ears of a listener or the player's own ears. In the latter case, they are then processed by the player's brain to allow feedback to the fingers as an essential part of fine control over the intonation, articulation, and other musical aspects of the performance.

The large impedance jumps at the string–body interface and the body–air interface allow the problem to be studied in three largely independent stages. The major reference on the subject, Cremer,[4] is organized according to precisely these three stages. The string motion involves a nonlinear self-sustained oscillator, which exhibits complex behavior of a kind that has become qualitatively familiar through recent work on nonlinear systems and chaos. The body vibration is at sufficiently small amplitude that linear vibration theory seems quite adequate, at least for a good first approximation to the behavior. The problems here arise from the complex geometry of the instrument, coupled with the very strong anisotropy of elastic and damping behavior exhibited by wood. The sound radiation problem is again linear to a good approximation, but the complex shape and wide frequency range of interest make the details very hard to calculate. The interaction with room acoustics brings in statistical rather than deterministic concepts, as is usual in room acoustics because of the high modal overlap shown by even a small room at audio frequencies.

In addition to these three problem areas, there are other questions of physics or mechanical engineering that must be considered. The design and function of the bow raises questions of its own.[5] The violin itself seems to have evolved toward maximizing sound production while keeping size and weight minimal. Those ends are achieved by having relatively heavy stringing on a relatively light body. Such tendencies are limited not by acoustical considerations but by structural strength. The combined string tension of a typical set of modern violin strings is around 250 N. The angle of the strings over the bridge is such as to produce a downbearing force on the top plate of roughly one third of this amount. This is supported on a wooden structure with typical thicknesses 2–3 mm (top plate), 2.5–5 mm (back plate), and 1 mm (ribs). In the light of these figures, many of the design features of the violin should be seen in terms of structural engineering. For example, arched rather than flat plates, the bass bar, the soundpost, and the shape of the f-holes to minimize stress concentrations at the ends (which are vulnerable to crack initiation) all contribute to structural integrity.

To be logical, though, before any useful physics can be done, it is necessary to know what the right questions are. The criteria of quality judgment by violinists and musical listeners are subjective and are not readily correlated with quantities amenable to physical measurement. Thus there are important scientific questions in the areas of experimental psychology and psychoacoustics that must be addressed in parallel with physical studies. These are the hardest and the least studied questions, and in the absence of very much good data, much of the work on the physics of violins is based on anecdotal evidence from makers and players on what seems to work well and what does not. Such evidence carries weight and should not be ignored, but properly conducted experiments might well give some rather different answers. The most glaring example of this difficulty is the mystique surrounding old rather than new instruments, and in particular instruments made in Italy by a group of famous makers between the late-sixteenth and the eighteenth centuries. It is extremely hard to obtain objective data on the relative merits of these famous old instruments and the very best of modern instruments—the author's personal suspicion is that there is little to choose between them.

4 MOTION OF A BOWED STRING

4.1 Helmholtz Motion

The very unexpected form of the motion of a string during a normal steadily bowed note was first described by Helmholtz. The idealized form of this *Helmholtz motion* is illustrated in Fig. 2*a*. A rather sharp "corner" shuttles back and forth around the visible envelope of the string vibration. As the corner passes the bow, it triggers transitions between sticking friction and sliding friction and vice versa. The result is a velocity waveform at the bow like that shown in Fig. 2*b*, with one episode of slipping and one of sticking in every cycle. The role of the bow is to supply sufficient energy during the sticking phase to compensate for any energy losses to sound radiation and internal dissipation in the string, bow, or instrument body. The geometry of the Helmholtz motion dictates that the length of slipping time as a fraction of the cycle period is equal to the fractional distance of the bowed point along the length of the string. This is conventionally designated β: If the string length is L, the bowed point is a distance βL from the bridge. In normal violin playing β takes quite small values, in the range 0.2–0.02, more extreme values being used for special effects.

When the string is executing a Helmholtz motion, the force waveform applied to the bridge is a sawtooth, as shown in Fig. 2*c*. An interesting consequence is that the shape of the driving waveform for the instrument body is, to a first approximation, independent of the bowing parameters. Its amplitude is controlled by bow speed and β alone. The actual force waveform in practice approximates such a sawtooth quite closely. It differs from it in detail in two main ways. First, the Helmholtz corner is never ideally sharp but is rounded by bending stiffness of the string, the finite width of the bow, and the dispersive and dissipative properties of wave transmission along the string and reflection from the ends.[4,6,7] Second, the bow position does have a small influence arising from partial reflections of waveform perturbations at the bow during the sticking phase.[4] These effects are both visible on the measured bridge force waveform in Fig. 2*d*.

4.2 Bow Force Limits

Other motions of the string are also possible, which are usually not sought by players but that account for the range of undesirable noises a beginner makes on a violin. These were analyzed first by Raman, who called them "higher types" (Helmholtz motion being the "first type") and produced an extensive classification. The main interest in these other bowed string motions is to shed light on how the Helmholtz motion can break down when the player fails to control the bow correctly. The Helmholtz

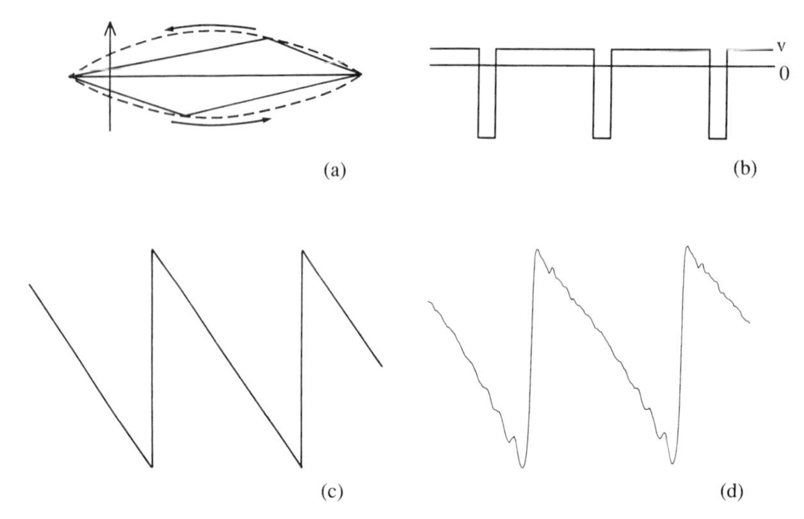

Fig. 2 Helmholtz motion of a bowed string: (*a*) snapshots of the string motion (solid) within the envelope of the vibration (dashed), in response to the movement of the bow (arrow); (*b*) velocity waveform at the bowed point on the string (in ideal form); (*c*) waveform of transverse force exerted on the violin bridge (in ideal form); (*d*) typical measured bridge-force waveform, to compare with (*c*).

motion is only possible in a certain region of the player's parameter space, and the boundaries of that region are determined by breakdown of the Helmholtz motion in one way or another.

For steady bowing, the player must control three things: the bow speed v, the position β, and the normal force between the bow and the string F. For given values of v and β, the bow force F must lie between certain limits. If F is too low, the string fails to continue sticking to the bow through the nominal sticking period. The motion turns into one with two or more slips per cycle, described by players as *surface sound*. If F is too large, on the other, the string does not release cleanly from the bow when the Helmholtz corner arrives. The accurate timekeeping provided by the corner is thus lost, and the motion degenerates into a raucous noise. Those two simple insights can be analyzed to give formulas for minimum and maximum bow force[4,7] as follows:

$$F_{\min} = \frac{vZ^2}{2\beta^2 R(\mu_s - \mu_d)}, \qquad F_{\max} = \frac{2vZ}{\beta(\mu_s - \mu_d)},$$

where μ_s and μ_d are, respectively, the static and dynamic coefficients of friction, Z is the characteristic impedance of the string, and R is an equivalent dashpot rate representing energy loss into the violin body (so Z/R is the string-to-body impedance ratio for a purely resistive model of the body).

This issue of bowing tolerance is very important to the player and merits closer study. The formulas given above are based on the simplest analysis and give the main dependence of the force limits on the playing parameters and the instrument string and body parameters. Both formulas have their limitations. Minimum bow force is indeed limited in practice by precisely the mechanism described above, but the theory models the energy loss from the string into the body very crudely, by a simple dashpot. A real instrument will have complicated behavior not captured by this approximation, so no single value for the dashpot rate tells the whole story.[7] For maximum bow force, the formula is based on a quite accurate model for the onset of raucous motion, but the snag is that skilled players never reach this regime. Maximum bow force in practice is limited by transitions to other undesirable vibration regimes, in which the note either plays unacceptably flat or is accompanied by an unacceptable level of "noise" due to the occurrence of small, irregular slips within the width of the ribbon of bow hair contacting the string.[4,6]

Musicians are not only concerned with steady bowing. A wide variety of bowing gestures are used to create different musical effects,[8] and the important question for the player is whether a given gesture gives rise to a Helmholtz motion, and if so how long the initial transient is. Understanding of such questions is being advanced by efficient algorithms for computer simulation of a family of reasonably realistic models of the physical processes involved.[6] These algorithms are being used in systematic studies of bowed-string behavior[7] and are also beginning to be explored with a view to simulation for the purposes of electronic music performance.

4.3 Wolf Note

A particular phenomenon associated with bowing tolerance is worthy of special mention. If one tries to play a note whose fundamental frequency matches a strong resonance of the instrument body, the resulting significant motion of the string termination at the bridge may interfere with the maintenance of a steady Helmholtz motion. Instead, a warbling or stuttering noise is produced, known as a *wolf note*. The effect is most obvious when trying to play near minimum bow force, and the warbling consists of an alternation between periods of Helmholtz motion and periods in which there are two or more slips per cycle, characteristic of playing below minimum bow force.[9]

The wolf is progressively more troublesome in the viola and the cello than it is in the violin. This fact was explained by Schelleng,[1] who pointed out that if the violin were scaled geometrically by the appropriate factors to correspond to the string tunings of the viola and cello, the result would be instruments much larger than the conventional viola and cello. They are presumably made smaller than this "ideal" size for the sake of easier handling by the player. In order to keep body resonances low enough to reinforce the lower notes of these instruments, the top and back plates are made relatively thin. The strings, on the other hand, are rather short, and thus need to be of relatively heavy gauge. The result is that the impedance mismatch between string and body is less in the viola and cello than in the violin. This exacerbates the wolf problem, which is caused by significant coupling between string and body resonances.

The effects of a troublesome wolf can often be ameliorated by fitting a *wolf eliminator*, a dynamic absorber of one kind or another that must be carefully tuned to the wolf frequency. It may take the form of a cantilever device glued to the underside of the top plate, or more simply a weight attached to one of the short lengths of string between the bridge and tailpiece so as to tune it to the relevant note.

5 INSTRUMENT BODY

The violin body is a multimodal resonant structure whose vibrational characteristics govern the sound quality and playing properties of the instrument. These characteristics are determined by the choice and seasoning of the wood, the many geometric parameters of the construction, and the details of the "setup," that is, the fitting and positioning of soundpost and bridge and the choice of strings. The modal deformations involve bending and stretching of the wooden shell of the body, standing pressure waves in the internal cavity, and also motion of the peripheral parts of the instrument, such as neck, fingerboard, and tailpiece. Modal densities can be estimated in the usual way from the area and material properties of the plates and the volume of the cavity.

Most published information on body vibration is in the form of frequency response curves of one kind or another. The commonest types are either transfer functions from force applied at the bridge to some measure of radiated sound or input admittance measurements at some point on the structure, such as the bridge. These latter determine the coupling of body motion to string motion, while the former relate to the perceived sound when a violin is played. A typical bridge input admittance curve for a violin is shown in Fig. 3. It is apparent that the system has low modal overlap up to about 1 kHz and significant overlap at higher frequencies.

The low-frequency resonances of a typical violin have been investigated by a variety of means. Mode shapes in the wooden structure have been visualized via holographic interferometry (see Ref. 10 and references therein) and experimental modal analysis.[11] Internal cavity modes have been mapped (see Ref. 12 and references therein). There have also been some finite-element computational studies of the vibration of the whole body or of components of it.[13] These latter have shown encouraging agreement with experiment, but the problems of a finite-element calculation encompassing the full details of the violin are formidable. All these approaches reveal a sequence of mode shapes that broadly follows expectation. At low frequencies the whole body deforms in relatively simple shapes. Going up the mode series, the mode shapes become more complicated as the wavelengths of deformation in the various component parts become progressively shorter. Precise details of mode shapes and frequencies are, inevitably, sensitive to structural parameters.

The variations of mode shapes, frequencies, and damping factors due to differences in wood, constructional details, and choice of varnish somehow control the variation of musical qualities among different instruments. Damping is perhaps the least obvious of these factors but is as important as the other two. It arises from a combination of internal dissipation associated with mechanical vibration and radiation losses. The former are generally larger.[1]

To a first approximation, many of the vibration modes can be labeled as primarily "wood" resonances or primarily "air" resonances, but there is almost always significant coupling that blurs this distinction. The most

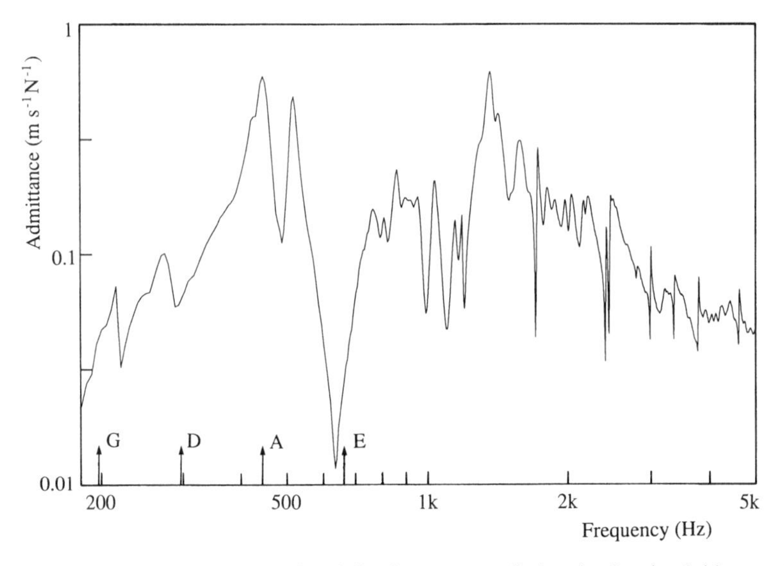

Fig. 3 Typical input admittance of a violin: force was applied at the G-string bridge notch in the plane of bowing, and the velocity response at the same point in the same direction was measured. The fundamental frequencies of the open strings of the violin are marked.

thoroughly studied case is that of the lowest useful resonance (typically around 270 Hz in a violin), which relies on the contained air volume and the relatively small f-holes, and has much in common with the behavior of a Helmholtz resonator. However, the nonrigidity of the top and back plates, and to a lesser extent the ribs, has a very significant influence on the frequency of this mode and on its ability to couple well to string motion: Cremer[4] gives a detailed discussion.

The instrument maker manipulates the body modes by choices of outline shape, arching profile, and wood thickness distribution. Most makers do not actively think in terms of frequencies and so on while building, relying more on established practices tempered by that elusive factor, intuition. However, there is a tradition of listening to the tap tones of the separate top and back plates in their unattached condition and using frequency and damping information about one or more modes of these free plates to guide the fine adjustment of thickness. Efforts have been made to systematize such procedures using Chladni patterns to examine frequencies and shapes of free plate modes during construction. In particular, the work of Hutchins[14] has led to some guidelines for makers that have produced promising results.

One important prerequisite for any modeling of the vibrational behavior of the instrument body is knowledge of the appropriate elastic and damping constants for the materials used. Wood is an extremely anisotropic material, and being a natural material any given species exhibits significant variation from tree to tree, and even among specimens from a single tree. Variations between species are even larger. The most thorough characterization of wood properties to be found in the published literature treats it as an orthotropic solid, having thus nine elastic constants (with nine associated damping factors). Only a very small number of measurements have been published of all nine elastic constants, and for internal damping no more than four have been determined for any single specimen.[15] A typical set of the nine elastic constants for instrument-quality spruce is listed in Table 1. The very high degree of anisotropy should be noted: the ratio of Young's modulus "along the grain" (axially in the tree) and "across the grain" (radially in the tree) may be as high as 30. The damping factors associated with bending in these two directions are also quite different, so that typical "long-grain Q factors" are around 120, while

TABLE 1 Typical Elastic Constants for Norway Spruce (*Picea abies*)

E_L = 11 GPa	E_R = 0.71 GPa	E_T = 0.43 GPa
G_{LR} = 0.50 GPa	G_{LT} = 0.61 GPa	G_{RT} = 0.023 GPa
ν_{LR} = 0.38	ν_{LT} = 0.51	ν_{RT} = 0.51

Note: The three directions in the tree are labeled L (longitudinal, parallel to the tree trunk), R (radial), and T (tangential, parallel to the annual rings). Young's moduli are denoted E, shear moduli G, and Poisson's ratios ν. Variation of the order of 20% will occur among specimens of the same species.

"cross-grain Q factors" are around 30. There have been efforts to design composite materials to replace spruce for musical instrument soundboards.[16] These must take into account the elastic anisotropy, the very large ratio of long-grain stiffness to area density that is characteristic of spruce soundboards, and the damping values. Musical instruments embodying such materials are now commercially available.

Before leaving the topic of body vibration, mention should be made of the role of the bridge. The bridge is not just a passive element, transmitting force by simple leverage. The curious shape of the modern bridge has dynamical significance. A violin or viola bridge has at least two significant in-plane resonances within the audible range (see Cremer[4]). These fall at around 3 and 6 kHz for the violin bridge. The cello bridge, with its longer legs, has somewhat different behavior, involving an additional in-plane resonance. Since all string forces have to pass through the bridge before they can excite the body, the filtering effect of these resonances influences the sound of the instrument profoundly. The violin maker knows this, at least intuitively, and is able to modify the sound of an instrument by choosing the right bridge and making small adjustments to its detailed shape.

6 SOUND RADIATION

The external air around the instrument body will provide both reactive loading, which will modify the modal structure somewhat, and sound radiation, which is of course the object of the body. The precise details will be influenced by shadowing effects from the player. This whole problem is discussed at some length by Cremer.[4] The lowest note of the violin, 196 Hz, corresponds to a wavelength in air of around 1.6 m. Thus at the lowest frequencies the body is a compact radiator, and simple radiation patterns would be expected. Provided there is some net volume fluctuation of the body and its internal air, monopole radiation with an omnidirectional pattern will result. (At very low frequencies the effective incompressibility of air means that there is no net monopole moment, and dipole radiation results.[17]) Many of the low vibration modes involve body motion with very little volume fluctuation. The relatively few modes that do involve such motion and are therefore efficient radiators are presumably the ones that it is most important for the violin maker to control.

An interesting role in low-frequency radiation efficiency is played by the soundpost. From the outside the violin body appears laterally symmetrical. This would lead one to expect that a lateral force applied to the bridge by transverse string motion would drive predominantly antisymmetric motion of the top plate, with no associated volume change. The asymmetric internal structure of soundpost and, to a lesser extent, bass-bar, prevents this from happening. The effect on low-frequency radiation efficiency is very clearly audible, if the soundpost is removed from a violin during playing of a low note.

Higher up the frequency range increasingly complex radiation patterns are found, since the body is no longer compact and its complicated mode shapes produce interference patterns in the radiated field. Of course, even for the lower notes as played with a bow there is significant energy in higher harmonics, so some radiational complexity will be present. (The Helmholtz motion drives the bridge with a force waveform that is approximately a sawtooth, so the Fourier spectrum of driving has only a $1/\omega$ fall-off with frequency, at least initially.) Almost certainly, these complex radiation patterns contribute significantly to the perceived sound quality of a violin. The best multichannel recording and reproduction systems cannot yet fool a blindfolded listener into thinking that a live instrument is being played. The most obvious explanation is that the directional characteristics of loudspeakers differ in an essential way from those of a live instrument. Possible mechanisms for this involve movements of the player during performance and rapid movement of radiation pattern lobes in space caused by frequency modulation from vibrato playing. In a reasonably reverberant room, suitable for musical performance, information about these movements will reach the listener via reflections from room surfaces.

7 PSYCHOPHYSICAL ASPECTS

The violin cannot be understood by studying physics alone. Some physical quantities such as radiation efficiency are undoubtedly relevant to the effectiveness, in musical terms, of an instrument. However, attempts to correlate perceived qualities with particular physically measurable quantities have in general met with only limited success. (Notable in this regard, though, is the work of Dünnwald.[18]) It has often been supposed that the "secret of Stradivari" could be exposed by measuring appropriate properties (such as response curves) of a few agreed good instruments and some poor ones and looking for the common factor among the good ones. The result of such tests has usually been that whatever parameter was chosen for analysis, the range of variation among the good instruments turned out to overlap strongly with the range of the poor ones, or indeed to coincide completely with it. This should not be too surprising. In all areas of sensory perception, human ability to perform certain tasks is remarkable. This has become increasingly apparent in recent years through the efforts to reproduce by computer human abilities at, for example, visual analysis. Our hearing system has evolved to discriminate subtle differences in certain types of sound, and musical instruments have been developed to exploit that ability to the full. It follows that the small differences between instruments that matter so much to a virtuoso violinist may well be near the limits of human discrimination and well beyond present abilities of signal processing algorithms. The problem lies not in matters of simple processing power but in identifying the psychoacoustically relevant "features" to try to detect in the sounds.

As an example, the sound from a normal violin, even during nominally steady bowing, is never precisely periodic. There are fluctuations from cycle to cycle, both systematic (as in vibrato) and apparently random (perhaps coming from irregularities in the distribution of rosin on the bow hairs). Between these extremes, there is also some correlated structure in the cycle-by-cycle variation, arising from human reflexes or bowed-string physics. Some efforts to analyze such phenomena have been made, for example by Schumacher.[19]

Nor is sound processing the end of the story. Violins are not chosen by listeners but by players. The player is inside the feedback loop by which the sound is produced and so has access to much more information than the listener. In particular, a sufficiently skilled violinist may be able to make a good sound on almost any violin. Although the listener may not hear much difference, the player is perfectly well aware that one instrument responds much more easily than another. The task of the violin maker is not just to reproduce the sound of the Stradivarius but the feel of it. The task of the scientist in determining the right questions to ask of physical models is to identify physical correlates of "feel" as well as sound. Recent work is beginning to tackle this difficult problem.[7]

REFERENCES

1. J. C. Schelleng, "The Violin as a Circuit," *J. Acoust. Soc. Am.*, Vol. 35, 1963, pp. 326–338.
2. S. F. Sacconi, *The 'Secrets' of Stradivari, Libreria de Convegno*, Cremona, Italy, 1979.
3. D. Gill (Ed.), *The Book of the Violin*, Phaidon Press, Oxford, 1984.
4. L. Cremer, *The Physics of the Violin*, (trans. J. S. Allen), MIT Press, Cambridge, MA, 1984.
5. R. T. Schumacher, "Some Aspects of the Bow," *Catgut Acoust. Soc.*, Newslett., Vol. 24, 1975, pp. 5–8.
6. M. E. McIntyre, R. T. Schumacher, and J. Woodhouse, "On the Oscillations of Musical Instruments," *J. Acoust. Soc. Am.*, Vol. 74, 1983, pp. 1325–1345.
7. J. Woodhouse, "On the Playability of Violins, Parts I and II," *Acustica*, Vol. 78, 1993, pp. 125–136, 137–153.
8. A. Askenfelt, "Measurement of Bow Motion and Bow Force in Violin Playing," *J. Acoust. Soc. Am.*, Vol. 80, 1986, pp. 1007–1015.
9. M. E. McIntyre and J. Woodhouse, "The Acoustics of Stringed Musical Instruments," *Interdisc. Sci. Rev.*, Vol. 3, 1978, pp. 157–173.
10. E. V. Jansson, N.-E. Molin, and H. O. Saldner, "On Eigenmodes of the Violin–Electronic Holography and Admittance Measurements," *J. Acoust. Soc. Am.*, Vol. 95, 1994, pp. 1100–1105.
11. K. D. Marshall, "Modal Analysis of a Violin," *J. Acoust. Soc. Am.*, Vol. 77, 1985, pp. 695–709.
12. C. M. Hutchins, "A Study of the Cavity Resonances of a Violin and Their Effects on Its Tone and Playing Qualities," *J. Acoust. Soc. Am.*, Vol. 87, 1990, pp. 392–397.
13. O. Rodgers, "Influence of Local Thickness Changes on Violin Top Plate Frequencies," *J. Catgut Acoust. Soc.*, Series II, Vol. 1, No. 6, 1990, pp. 6–10.
14. C. M. Hutchins, "The Acoustics of Violin Plates," *Sci. Am.*, Vol. 245, No. 4, 1981, pp. 170–186.
15. M. E. McIntyre and J. Woodhouse, "On Measuring the Elastic and Damping Constants of Orthotropic Plates and Shells," *Acta Metallurgica*, Vol. 36, 1988, pp. 1397–1416.
16. D. W. Haines, C. M. Hutchins, M. A. Hutchins, and D. A. Thompson, "A Violin and a Guitar with Graphite-Epoxy Composite Soundboards," *Catgut Acoust. Soc. Newslett.*, Vol. 24, 1975, pp. 25–28.
17. G. Weinreich, "Sound Hole Sum Rule and the Dipole Moment of the Violin," *J. Acoust. Soc. Am.*, Vol. 77, 1985, pp. 710–718.
18. H. Dünnwald, "Deduction of Objective Quality Parameters on Old and New Violins," *J. Catgut Acoust. Soc.*, Series II, Vol. 1, No. 7, 1991, pp. 1–5 (German version in *Acustica*, Vol. 71, 1990, pp. 269–276).
19. R. T. Schumacher, "Analysis of Aperiodicities in Nearly Periodic Waveforms," *J. Acoust. Soc. Am.*, Vol. 91, 1992, pp. 438–451.

101

WOODWIND INSTRUMENTS

NEVILLE H. FLETCHER

1 INTRODUCTION

The woodwind family comprises mouth-blown instruments excited by a single reed in a nearly cylindrical tube (clarinet), a single reed in a conical tube (saxophone), a double reed in a conical tube (oboe and bassoon), or an air jet in a nearly cylindrical tube (flute). Renaissance instruments such as recorders, shawms, and krummhorns and folk instruments such as panpipes and bagpipes are also woodwinds. In all these instruments, except for the panpipes and the drones of the bagpipes, the resonator tube is pierced by tone holes that are closed by the finger tips or by finger-actuated keys to change the pitch of the note. Most instruments are made of wood, with metal keys, but saxophones are always, and flutes now usually, made totally of metal. Surveys of the history[1] and acoustics[2,3] of all these instruments, and more detailed treatments of individual types,[4–7] are available in the literature. Important instruments of the family are shown in Fig. 1.

To understand the acoustics of woodwind instruments, it is convenient to consider first the passive linear behavior of the air column and then the operation of the active nonlinear reed or air jet generator that maintains the resonator in oscillation to produce sound. Reed generators are pressure-controlled valves that present a negative acoustic conductance under appropriate conditions. They therefore drive the air column from a point of maximum acoustic impedance, and the reed end of the tube resonator is effectively stopped. Air jets, on the other hand, are flow-controlled valves that can also present a negative acoustic conductance. They drive the air column at a point of minimum acoustic impedance so that the blowing end of the tube is effectively open. For both flutes and reed instruments, the generator produces a complete and precisely harmonic spectrum based on the fundamental frequency determined by the generator and resonator in combination. Resonances of the air column then serve to shape this spectrum.

2 AIR COLUMN RESONATOR

The fundamental frequency of the note sounded by a woodwind instrument is determined by the mode frequencies of the resonator, usually that of the lowest mode, and these frequencies are varied by opening finger holes to reduce the effective length of the air column. It is therefore important that the mode frequencies of the resonator maintain the same relationship to each other as the resonator is shortened, to give tonal coherence over the playing range. This is possible, in the case of a reed-driven instrument with an impedance maximum at the reed, if the instrument horn is either cylindrical or else conical with the reed at the apex.[8]

In the cylindrical case, adopted for the clarinet family, the modes with impedance maxima at the reed form an approximately odd-harmonic series, with the tube being one-quarter of a wavelength long at the fundamental resonance. This shows up in the radiated spectrum, illustrated in Fig. 2*a*, the very weak second harmonic being the principal timbre characteristic. Because the resonator is not ideally cylindrical, however, its higher resonances are appreciably inharmonic and higher even harmonics of the reed frequency appear. Above the radiation cutoff frequency for the tone holes and open end, whatever the bore shape, there is no reflected wave in the tube and all harmonics are radiated equally. This cutoff is typically between 1 and 2 kHz, depending on the type of instrument.[9]

Note: References to chapters appearing only in the *Encyclopedia* are preceded by *Enc.*

Fig. 1 Flute, oboe (without reed), clarinet, and (to a different scale and in rear view) bassoon. Note the simple integrated keywork of the flute, with one key per finger (except in the foot joint), the small holes and many extra keys in oboe and clarinet, and the large number of keys operated by the two thumbs in the bassoon.

The oboe, bassoon, and saxophone have conical bores for which the tube is half a wavelength long at the fundamental resonance. A conical-bore reed instrument must therefore be nearly twice as long as an instrument with a cylindrical bore to produce the same note—the bassoon is folded to accommodate its 2.6-m length. Because of partial truncation of the cone at the reed end, the acoustic length of the air column is rather greater than the geometric length. There is a further length correction from the acoustic admittance of the reed itself, and this applies also in the case of the clarinet. All harmonics are reinforced in the sound of conical-bore instruments, as shown in Fig. 2*b*.

The input impedance curve of a woodwind instrument with several open finger holes is very complex,[10] and a successful design for a reed instrument appears to depend on achieving a high input impedance at the frequency of at least one of the low harmonics of the note fundamental. This is accomplished by adjustment of tone hole sizes and small variations in bore profile.

In the case of the flute family, the jet generator must be at an impedance minimum for the tube, and the far end is usually open. Bore shapes satisfying the mode-similarity condition can be cylindrical or can have the shape of either flaring or tapering incomplete cones. In each case there is a complete set of harmonic modes, and the fundamental tube length is half a wavelength. There is generally a considerable end correction to be added at the embouchure hole because of its small diameter relative to that of the bore. In a side-blown flute there is also a cavity between the embouchure hole and the stopped end, the size of which is chosen so as to optimize the tuning of the pipe resonances. Before the mid-nineteenth century, flutes had a tapering conical bore and a cylindrical head joint. This geometry is retained in most orchestral piccolos, while the modern flute has a cylin-

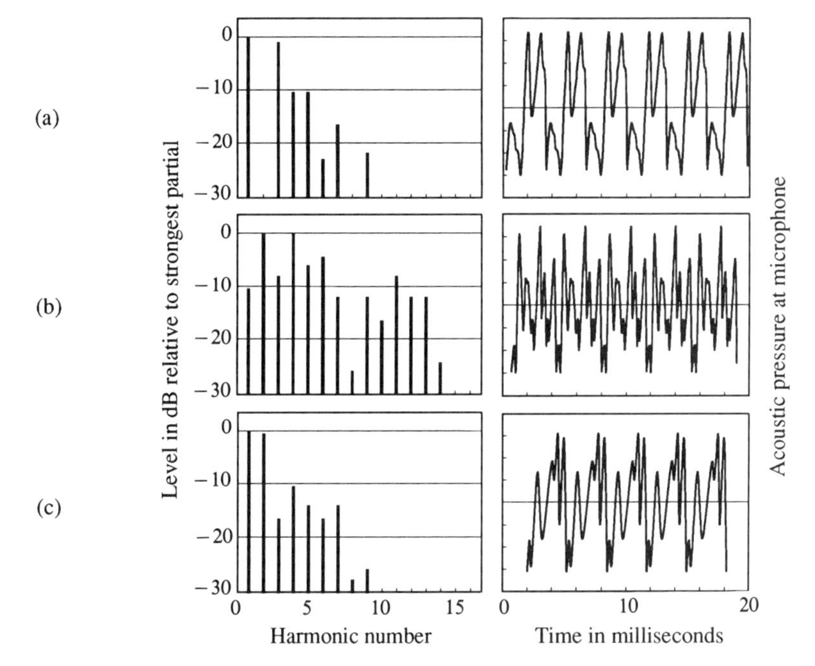

Fig. 2 Waveform and radiated spectrum of the note D_4 played with full tone on (*a*) a clarinet, (*b*) an oboe, and (*c*) a flute. Higher notes have less harmonic development for all instruments, and high notes on the clarinet do not have the characteristic missing second harmonic. Waveforms and spectra depend considerably upon the player and the microphone arrangement.

drical bore and a tapering head joint. In each case the arrangement is designed so as to flatten the lowest resonance frequency and so improve response.[11] The spectrum of a typical flute sound is shown in Fig. 2*c*. Upper harmonics are less prominent than in the oboe because of the more gentle nonlinearity, but are still quite well developed, particularly in low notes.

Geometrical details for typical models of common woodwind instruments[2] are given in Table 1. Note the increase in cone semiangle from the bassoon and oboe to the saxophone. This correlates with increasing relative power in the fundamental and low harmonics of the sound because of reduced wall damping and more efficient radiation. The clarinet, oboe, and saxophone all have flaring end sections or bells.

Most woodwind instruments are made from selected hardwood, though saxophones and most flutes are of metal. Acoustic evidence and listening trials[12] suggest that there is no direct influence of the construction material on radiated sound, provided the internal wall is smooth. The amplitude of wall vibrations is in any case very small, and much more affected by wall thickness than by changes in material. More subtle effects such as the influence of material on roughness of bore, sharpness of tone-hole edges, and fit of key pads may, however, be important. There are also questions of ease of machining, surface finish, durability, dimensional stability, and appearance, all of which define particular woods and metals as being specially suitable for instrument making. Plastics can also meet these requirements and are widely used in less expensive instruments.

3 KEY MECHANISMS

The fingering problem for woodwind instruments arises because the number of fingers on a pair of human hands is less than the number of notes in an octave, when semitones are considered. In a simple tone hole arrangement for a flute, oboe, or bassoon, there are six finger holes and one thumb hole to produce the seven notes of a diatonic (white-note) scale by simply opening them in sequence, with a further octave being available by "overblowing" to the next resonator mode, and a few notes above this by using special fingerings. Each semitone step involves reducing the acoustic length of the pipe by about 6%, and each full tone step by about 12%. Finger hole spacing is usually reduced for convenience by using small holes

TABLE 1 Typical Physical Dimensions of Woodwinds

	Musical Range	Tube Length (mm)	Top Diameter (mm)	Bell Flare Diameter (mm)	Cone Semiangle (deg)	Truncation Ratio
Piccolo in C (conical)	D_5–A_7	262[a]	11	8 → 9	−0.6	—
Flute in C	C_4–C_7	604[a]	17[b]	19[c]	0	—
Oboe in C	B_3–G_6	644[d]	3	17 → 37	0.7	0.13
Bassoon in C	B_1–C_5	2560[d]	4	40[e]	0.4	0.09
Clarinet in B♭	D_3–G_6	664[d]	14	16 → 60	0	—
Alto saxophone in E♭	D_3–A_6	1062[d]	12	70 → 123	1.6	0.12

[a]To center of embouchure hole.
[b]Head taper to embouchure hole.
[c]Cylindrical tube.
[d]To tip of reed.
[e]Uniform cone.

placed rather closer to the generator end than this simple proportion would suggest. This device is taken to an extreme in the bassoon, which has tone holes drilled at an angle through thick wood to bring them within reach of the fingers. The diameter and placing of the tone holes can be slightly varied to achieve optimal tuning in the two octaves.[9] In this simple fingering system, semitones are produced by closing one or more holes below the first open hole to increase the acoustic load and lower the pitch.

Musical developments in the eighteenth and nineteenth centuries required greater facility and tonal evenness for chromatic semitones, and extra keys, normally standing closed, were provided for this purpose between the original tone holes. Further keys added below the normal bottom note also served to extend the range downward. This device was essential in the clarinet, which overblows to the twelfth rather than the octave and must therefore have a basic fingering range of a twelfth, leaving four diatonic and three chromatic notes to be filled in between the two six-finger registers. Clarinets, oboes, and bassoons still retain small open finger holes, with many supplementary keys, as shown in Fig. 1, and incorporate some coupling mechanisms borrowed from the flute, as discussed below. These reed instruments are generally made of selected hardwoods, tradition dictating different wood for different instruments, though plastic materials are sometimes used in cheaper instruments. The saxophone, invented by Adolphe Sax in the mid-nineteenth century, has a clarinetlike single reed, a wide conical bore, and very large tone holes with a key mechanism related to that of the flute, though the fingering is somewhat different. Saxophones are almost always made of metal.

The flute, which in baroque times had a tapered conical body with a cylindrical head, was redesigned in the mid-nineteenth century by Theobald Boehm. His initial design retained this bore shape, but his revised model of 1847 used a cylindrical body, with a head joint tapering slightly toward the mouth hole, and with large tone holes placed in acoustically appropriate positions.[4,9,13] This arrangement required large padded keys to cover the tone holes and a mechanism of axles and clutches connecting all keys to give the player control. Boehm designed an elegant and enduring mechanism for this purpose. Many of his later flutes were made of silver rather than of wood. These changes increased greatly the acoustic power and tonal uniformity of the flute and are universally used today. Modern flutes are usually made of silver (typically 925 fine), gold (typically 14 carat) or even platinum, the alloy composition being chosen to give good strength, appearance, and manufacturing properties. Any acoustic difference between flutes made of these different materials is much less significant than the differences introduced by variations in head taper design, acoustic absorption of the key pads, and adjustments to the embouchure hole introduced during hand finishing. Student instruments are usually made of a cheaper cupronickel alloy with silver plating.

As well as changing the frequency of the resonator modes, the tone holes also have an influence on the tone quality of the instrument. While some of the sound is radiated from the first open tone hole, much of it, particularly the harmonics of higher frequency, propagates along the remainder of the bore and is radiated from open tone holes and from the open end, giving a rather complex directional characteristic.[14] The lattice of open tone holes along the lower part of the bore acts as a filter and modifies the upper part of the radiated spectrum.[9] The size of the tone holes is, of course, partly determined by the diameter and cone angle of the instrument tube, but small variations are possible in the design of instruments of one type. Small tone holes reduce the radiation of higher frequency components and make the instrument less bright than do large tone holes.[8]

The musical ranges of the more common members of the woodwind family are shown in Table 1. In addition there is a cylindrical version of the piccolo, and alto flute in G, and a bass flute in C, as well as a less common small flute in $E^\flat$. The oboe family is augmented by the oboe d'amore and cor anglais, both of which have a globular rather than a flaring bell, and there is a contra-bassoon an octave below the normal bassoon. The clarinet family has a high-pitched member in $E^\flat$ and an additional orchestral version in A, as well as lower pitched alto clarinet in F and bass clarinet in $B^\flat$. The saxophone family contains as many as six members, with a high soprano in $B^\flat$ and several larger relatives.

4 REED GENERATORS

The reed in a woodwind instrument is a pressure-controlled valve that is closely coupled to the air column resonator, which determines its vibration frequency. Reeds are made from thinned and carefully shaped cane. A single reed, bound against an aperture in a mouthpiece, is shown in Fig. 3*a* and double reeds in Figs. 3*b*, *c*. All are of the type that is blown closed by steady pressure in the mouth and have natural resonance frequencies well above the playing frequency. If the blowing pressure is p_0, the pressure in the mouthpiece of the instrument p, and the opening distance of the reed x, then the volume flow U into the instrument is determined by Bernoulli's law and has the approximate form

$$U \approx (2/\rho)^{1/2} W x (p_0 - p)^{1/2}$$
$$\approx (2/\rho)^{1/2} W [x_0 - C(p_0 - p)](p_0 - p)^{1/2}, \qquad (1)$$

where ρ is the density of air, x_0 is the static opening of the reed, W is its width, and C is proportional to its elastic compliance. The form of this relation is shown in Fig. 3*d*, the reed being closed and the flow zero to the right of B. Details are slightly different for a real instrument because of mouthpiece curvature in single reeds and reed arch in double reeds, so that the closure is not abrupt but more as indicated by the broken curve. Slight modifications can be made to Eq. (1) to take account of more complex reed geometry.[15]

In a linear small-signal approximation, the acoustic admittance presented to the resonator is $Y = -\partial U/\partial p$, and can be found as a real conductance by differentiating Eq. (1), provided the frequency is much less than the resonance frequency of the reed. This conductance is

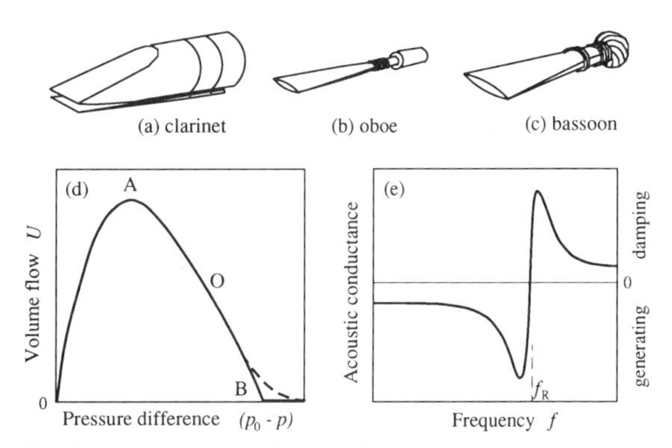

Fig. 3 (*a*)–(*c*) Reeds of the clarinet, oboe, and bassoon, to a similar scale. (*d*) Static flow curve for a reed generator; the operating point is set near O. (*e*) Acoustic conductance as a function of frequency for blowing pressure near the point O; f_R is the resonance frequency of the reed.

negative in the AB region of the curve, above a blowing pressure p_0^* that is sufficient to reduce the reed opening so that $x < 2x_0/3$. The blowing pressure p_0 is therefore set by the player to an operating point O close to that shown on the curve, where $x \approx x_0/3$. When the reed is treated in more detail as a mechanical vibrator,[16] its acoustic conductance for $p_0 > p_0^*$ has the form shown in Fig. 3*e*. This is negative at all frequencies below the reed resonance, so that the reed can act as an acoustic generator at these frequencies when coupled to a resonator of sufficiently high Q, the oscillation frequency being determined largely by the resonator. There is a negative conductance valley just below the reed resonance frequency f_R, but this does not dominate the behavior as it does in organ reed pipes, largely because the woodwind reed is damped by the player's lips, giving a broad low peak, while an organ pipe reed is clamped by a metal wire and has a high Q.

Note that the flow curve of Fig. 3*d* is nonlinear, so that exact harmonics of the basic reed motion appear in the flow into the resonator.[17] The reed can respond to several mode frequencies of the resonator, and behavior is complicated by the conductance peak just below the reed resonance. As with most highly nonlinear systems, the oscillation usually locks into a single periodic regime with high harmonic development. This is facilitated if the resonator has several resonances in approximately harmonic relationship (i.e., with frequencies in the ratios of small integers[10]). If there is no pipe resonance close to the frequency of a particular harmonic of the reed fundamental, as in the case of the second harmonic for a cylindrical clarinet pipe, then the reed flow at that frequency does not produce an appreciable acoustic pressure, and the radiated harmonic partial is weak. Figures

2*a*, *b* show waveforms and spectra of typical midrange notes on a clarinet and an oboe, illustrating the high harmonic content.

5 AIR JET GENERATORS

An air jet emerging from the windway of a recorder, as in Fig. 4*a*, or from a flute player's lips, as in Fig. 4*b*, is unstable against the influence of transverse acoustic flow, which produces a sinuous disturbance that propagates along the jet at about half the jet speed and grows exponentially in amplitude. If the jet strikes against the edge of a resonator a small distance away from the flue, then flow of the jet into the resonator excites oscillations of the air column that act back upon the motion of the jet. The mechanism by which the jet drives the pipe combines the effect of injecting a volume flow at a position just inside the end correction, where there is a small acoustic pressure, and injecting momentum, and thus pressure, at a position of large acoustic flow.[18]

The phase of the coupling determines whether or not oscillation can be maintained, and this depends on the length of the jet from lip aperture to resonator edge, on the jet speed or blowing pressure, and on the mode frequency of the resonator.[19] The acoustic conductance presented by the jet to the resonator has the form shown in Fig. 4*d* as a function of the parameter

$$\delta = 2lf[\rho/(2p_0)]^{1/2}, \tag{2}$$

where p_0 is the blowing pressure, l the jet length, and f the sounding frequency. The parameter δ is essentially equal to the propagation phase shift, measured in wavelengths, for transverse waves on the jet. Only the principal negative minimum, labeled O, near $\delta = \frac{1}{2}$ is used in normal playing. The player adjusts blowing pressure and lip position to place this minimum near one of the mode frequencies of the resonator, which then sounds.

The velocity profile across a planar jet has approximately the bell-shaped form $v_0 \operatorname{sech}^2(y/w)$, where y is the coordinate across the jet, w is the jet half-thickness, and v_0 is the centerplane velocity. The flow into the resonator mouth is then

$$U \approx \tfrac{1}{2}U_0[1 + \tanh(y_0 + y)], \tag{3}$$

where y is now the time-varying displacement of the center plane of the jet, as deflected by the acoustic flow, y_0 is the static offset of this plane relative to the plane of the sharp edge of the resonator, both measured in units of w,

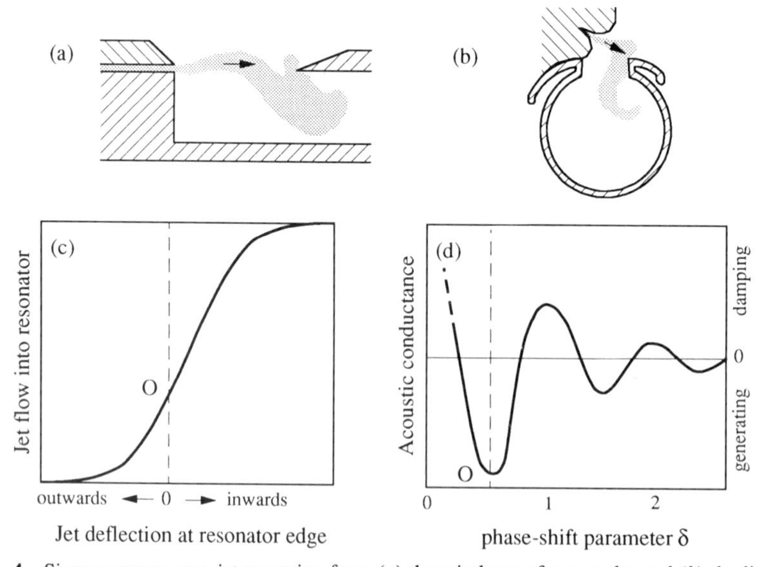

Fig. 4 Sinuous waves on a jet emerging from (*a*) the windway of a recorder and (*b*) the lips of a flute player. (*c*) Flow characteristic for an air jet generator with static jet position *O*. (*d*) Acoustic conductance of an air jet generator; the jet speed operating point *O* is set near the principal minimum.

and U_0 is the steady volume flow of the jet. This nonlinearity, shown in Fig. 4*a*, is softer than that of a reed generator, so that the relative strength of high harmonics is less.

6 TIME-DOMAIN ANALYSIS

While physical understanding of woodwind instruments is most easily expressed in terms of frequency components, numerical simulation and calculation is best carried out in the time domain. For such calculations, the reed or air jet generator is described by differential equations, and the air column resonator in terms of a reflection function that is related to the Fourier transform of its acoustic input impedance. Such numerical calculations give quite good agreement with experiment for both transient and steady states for both the clarinet[20] and the flute[21] but are limited at present by the reliability with which the generator, and particularly the airflow, can be understood and modeled.

7 PERFORMANCE PARAMETERS

In reed instruments, the major parameters available for adjustment are the blowing pressure p_0 and the static reed opening x_0, which is set by lip pressure. Little variation in p_0 is possible if the operating point O in Fig. 3*d* is to remain near the center of section *AB*. From Eq. (1), the magnitude of the negative conductance at O is proportional to $x_0^{1/2}$, the flow range *AB* to $x_0^{3/2}$, and the pressure range *AB* to x_0. Changes in lip pressure to alter the static opening x_0 can therefore be used to change the acoustic excitation magnitude and instrument loudness. Increase in the negative conductance for loud playing ensures that more of the nonlinear curve is sampled, and so increases relative harmonic content. This increase in harmonic content is more important than absolute power level in determining musical impact in loud playing.

For the clarinet and saxophone, the normal blowing pressure is 3–4 kPa for soft playing and 5–7 kPa for loud, with little systematic change with pitch through the compass. In double-reed instruments, the blowing pressure increases steadily with pitch and with loudness, ranging from 4 to 12 kPa for the oboe and from 1.5 to 6 kPa for the bassoon. Average acoustic output power varies with the pitch of the note played, but is typically a few milliwatts for all woodwind instruments. The dynamic range[22] of the oboe and the bassoon for a skilled player is typically 20–30 dB, varying somewhat with the pitch of the note played. The clarinet has a dynamic range of up to 40 dB over most of its playing compass. Conversion efficiency from pneumatic to acoustic power is typically 0.1–1% for reed instruments, being higher at higher power levels.

In flute instruments[19,23] the blowing pressure is determined by jet phase matching requirements, as discussed in relation to Eq. (2), and increases approximately linearly with the frequency of the note played, from about 0.3 kPa for C_4 to 2 kPa for C_7, an accompanying reduction being made in jet length for high notes by pushing the lips forward. The loudness is controlled by the total jet flow, which is varied primarily by changing the aperture between the player's lips. Average acoustic power output is again typically a few milliwatts, but the dynamic range is only about 20 dB for low notes and less than 10 dB for very high notes. Acoustic conversion efficiency is typically about 1%, depending on pitch and loudness level. Again the harmonic content increases at loud playing levels. More subtle changes in timbre can be effected by changing the direction of the air jet and hence the operating point O in Fig. 3*c*. If O is set to the inflection point of the curve, then even harmonics are suppressed and odd harmonics reinforced, while for other jet directions particular even harmonics can be emphasized.[24]

In reed instruments the player's mouth cavity is closely coupled to the reed, so that its acoustic properties can have a significant influence on sound production.[25] Analysis shows that the vocal-tract impedance is effectively in series with the impedance of the instrument as presented to the reed but, because the peak impedance of the vocal-tract resonances is low compared with that of the instrument tube resonances, the vocal tract has only a relatively small effect on the frequency and amplitude of the sound produced. It is necessary, however, to adjust mouth configuration to make the low notes on the bassoon sound at all, and mouth configuration has a significant influence on tone quality in all reed instruments. In the case of flutes the mouth cavity is only weakly coupled to the instrument air column, and vocal-tract effects are much smaller, except insofar as they affect the flow, shape, and turbulence of the air jet.

REFERENCES

1. A. Baines, *Woodwind Instruments and Their History*, Faber and Faber, London, 1967.
2. C. J. Nederveen, *Acoustical Aspects of Woodwind Instruments*, Frits Knuf, Amsterdam, 1969.
3. N. H. Fletcher and T. D. Rossing, *The Physics of Musical Instruments*, Springer Verlag, New York, 1991.
4. P. Bate, *The Flute*, Norton, New York, 1979.
5. P. Bate, *The Oboe*, Norton, New York, 1975.
6. L. G. Langwill, *The Bassoon and Contrabassoon*, Norton, New York, 1966.

7. F. G. Rendall, *The Clarinet*, Williams and Norgate, London, 1954.
8. A. H. Benade, "On Woodwind Instrument Bores," *J. Acoust. Soc. Am.*, Vol. 31, 1960, pp. 137–146.
9. A. H. Benade, "On the Mathematical Theory of Woodwind Finger Holes," *J. Acoust. Soc. Am.*, Vol. 32, 1960, pp. 1591–1608.
10. J. Backus, "Input Impedance Curves for the Reed Woodwind Instruments," *J. Acoust. Soc. Am.*, Vol. 56, 1974, pp. 1266–1279.
11. A. H. Benade and J. W. French, "Analysis of the Flute Head Joint," *J. Acoust. Soc. Am.*, Vol. 37, 1965, pp. 679–691.
12. J. Backus, "Effect of Wall Material on the Steady-State Tone Quality of Woodwind Instruments," *J. Acoust. Soc. Am.*, Vol. 36, 1964, pp. 1881–1887.
13. J. W. Coltman, "Acoustical Analysis of the Boehm Flute," *J. Acoust. Soc. Am.*, Vol. 65, 1979, pp. 499–506.
14. J. Meyer, *Acoustics and the Performance of Music*, Verlag Das Musikinstrument, Frankfurt am Main, 1978.
15. J. Backus, "Small-Vibration Theory of the Clarinet," *J. Acoust. Soc. Am.*, Vol. 35, 1963, pp. 305–313.
16. N. H. Fletcher, "Excitation Mechanisms in Woodwind and Brass Instruments," *Acustica*, Vol. 43, 1979, pp. 63–72.
17. N. H. Fletcher, "Nonlinear Theory of Musical Wind Instruments," *Appl. Acoust.*, Vol. 30, 1990, pp. 85–115.
18. N. H. Fletcher, "Jet Drive Mechanism in Organ Pipes," *J. Acoust. Soc. Am.*, Vol. 60, 1976, pp. 481–483.
19. J. W. Coltman, "Sounding Mechanism of the Flute and Organ Pipe," *J. Acoust. Soc. Am.*, Vol. 44, 1968, pp. 983–992.
20. R. T. Schumacher, "Ab Initio Calculations of the Oscillations of a Clarinet," *Acustica*, Vol. 48, 1981, pp. 71–85.
21. J. W. Coltman, "Time-Domain Simulation of the Flute," *J. Acoust. Soc. Am.*, Vol. 92, 1992, pp. 69–73.
22. J. Meyer, "Zur Dynamik und Schalleistung von Orchesterinstrumenten," *Acustica*, Vol. 71, 1990, pp. 277–286.
23. N. H. Fletcher, "Acoustical Correlates of Flute Performance Technique," *J. Acoust. Soc. Am.*, Vol. 57, 1975, pp. 233–237.
24. N. H. Fletcher and L. M. Douglas, "Harmonic Generation in Organ Pipes, Recorders and Flutes," *J. Acoust. Soc. Am.*, Vol. 68, 1980, pp. 767–771.
25. P. G. Clinch, G. J. Troup, and L. Harris, "The Importance of Vocal Tract Resonance in Clarinet and Saxophone Performance—A Preliminary Account," *Acustica*, Vol. 50, 1982, pp. 280–284.

102

BRASS INSTRUMENTS

J. M. BOWSHER

1 INTRODUCTION

Brass instruments are not necessarily made of brass. By universally accepted convention, a brass instrument is one sounded by the vibration of the player's lips when they are lightly pressed against a suitably shaped mouthpiece and the player blows between them. Most brass instruments are, however, made of the type of brass known as "yellow brass," a 70 : 30 alloy of copper and zinc; but "gold brass," an alloy of 85% copper, 15% zinc (sometimes 90 : 10) is very popular among symphonic players and on the European continent. "Nickel silver" (usually 63% copper, 27% zinc, and 10% nickel) instruments, especially French horns and trombone slide sections, are quite often seen. Pure copper and silver are occasionally used for the manufacture of bugles and ceremonial trumpets, but are too soft to make practically robust instruments.

Brass instruments may have side holes, like the woodwind instruments discussed in Chapter 101. This variety of brass instrument was common a few centuries ago and is played today by those interested in historically correct performances. The forms most likely to be seen are the cornett and its bass cousin, the serpent. These are conventionally made of leather-covered wood and have comparatively small side holes that can be covered by the player's fingers. The ophicleide and keyed bugle are of brass and have larger side holes, like those of the saxophone, and require mechanically operated pads; they were widely used until the mid-nineteenth century. Instruments with side holes have approximately conical bores; the other instruments are cylindrical over most of their length. Modern instruments use either slides or valves to add lengths of tubing to provide notes not readily obtainable on a simple tube. Sometimes a distinction between "cylindrical" (e.g., trumpet and trombone) and "conical" (e.g., cornet, French horn, and tuba) modern instruments is made; but, since the slide or valve pitch changing mechanism imposes the condition that a considerable length of the tubing must have constant diameter, the distinction is largely artificial and really refers to the rate of flare of the bell section.

This chapter is primarily concerned with measurements on brass instruments and the way they produce their sounds. Other aspects are discussed only briefly. To avoid complication, the trombone will mostly be used as an exemplar.

2 SUMMARY OF BASIC PARAMETERS AND OUTPUT PROPERTIES

Brass instruments are nowadays found in four general categories: French horn, trumpet, trombone, and tuba. Instruments are made in many different sizes and bores to suit players' requirements; for example, French horns may be single, double, or triple; trumpets are made in at least eight different pitches; smaller bored versions of the tuba appear as the baritone and euphonium. The output of any one brass instrument depends greatly on the player and on the measuring environment. For these reasons only the most general data can be quoted here; information on directivities, spectra, and power output may be found in Meyer[1] and Dickreiter,[2] but it must be remembered that these are average, overall data not applicable to any specific circumstance. Representative physical dimensions and output characteristics are given

Note: References to chapters appearing only in the *Encyclopedia* are preceded by *Enc.*

in Table 1; note that some players can exceed the compass quoted by an octave or more in both directions.

Graphs of output spectra found in the literature always require careful examination, especially if measurements are made in echoic conditions or in an anechoic room to which the player has not been accustomed. With mostly "cylindrical" instruments (Section 1), the spectrum depends strongly on the dynamic level, having some 5 or 6 significant components for quiet playing and in excess of 40 for loud playing of low notes. For instruments of more extended flare, the variation with level is less, being from 4 or 5 to about 20 or 25 components. Spectra depend greatly on the direction of measurement; off axis the higher components rapidly diminish. Broadly speaking, the variation of directivity with frequency follows that expected for a radiator of bell size, though there are some minor perturbations due to phase changes across the bell opening. It should be noted that the French horn and tuba do not normally "point at" the listener, so the interpretation of their directivities requires some care.

3 TERMINOLOGY

Much confusion is created in reports of earlier work on brass instruments by the use of poorly defined terminology and by inadequately precise specifications of measuring conditions. Fourier showed that any periodic oscillation may be described in terms of the amplitudes and phases of a number of *harmonics*, integral frequency multiples of the fundamental frequency of the oscillation. In this chapter the word *harmonic* will have that meaning only. The different oscillation frequencies allowed by the boundary conditions will be referred to as *modal frequencies* and each pattern of vibration as a *modal pattern*. The frequencies produced on a brass instrument by a player will be called the *playing frequencies*. It is important to realize that the playing frequencies are not the same as the modal frequencies nor the harmonic frequencies, even though they are often called harmonics by players and by writers on brass instrument acoustics.

4 BASIC ACOUSTICS

The representation of a wind instrument in terms of a single parameter, implying that, if the correct coordinate system could be chosen, the propagation is one-dimensional, is sufficiently good for a brief discussion.[3] In these terms, the acoustic field in the tube in any plane normal to the propagation is completely defined by one measurement of the pressure p and one of the volume velocity U, and it is possible to represent the whole instrument as a two-port network in which any convenient planes may be used to define the input port, at which p_i and U_i are measured, and the output port, at which p_o and U_o are measured; both velocities being considered positive when directed into the instrument. Unfortunately, many writers do not define the input and output planes with sufficient care, and comparison of results is often made needlessly difficult. Numerical data regarding brass instruments should always be closely examined for exact details of the input and output conditions applicable to the measurement before being used for the development of theories or other instruments. The measuring point for the output port is usually taken to be on axis, one bell radius away from the physical end of the instrument; this assumes that the wave front is hemispherical just outside the bell. The assumption is reasonably good for the lower and middle frequencies but is in error for those frequencies where the wavelength is appreciably less than the bell diameter. Inside the instrument the wave front has a nonuniform curvature that depends on the precise shape of the bell and on the frequency. Many authors use the plane of the mouthpiece rim to locate the input port; this plane has two significant disadvantages: (i) the mouthpiece itself is

TABLE 1 Representative Parameters of Brass Instruments

Instrument Category	Bore[a] (mm)	Mouthpiece Cup Diameter (mm)	Bell Diameter (mm)	Playing Compass	Output Power[b] (W)	Dynamic Range[c] Low	
						Low	High
French horn	12	17.5	295	$B_1 \to F_5$	0.2	52 → 73	88 → 118
Trumpet	11.5	16.8	125	$E_3 \to D_6$	2.5	75 → 105	110 → 125
Trombone	13	25.5	210	$E_2 \to D_5$	5.0	65 → 100	92 → 117
Tuba	18.5	32.5	400	$C_1 \to G_4$	1.0	78 → 112	98 → 118

[a]Measured through valve or slide section.

[b]Maximum power outputs are quoted (usually for fairly high notes in an instrument's compass).

[c]The dynamic range is quoted in terms of the minimum and maximum sound pressure level (SPL) playable measured 1 m from the bell; "low" applies to the lowest note readily playable, "high" to the highest.

included in the measurement, and so a different mouthpiece requires new measurements; (ii) the plane is not well-defined since the player's lips protrude into the mouthpiece by amounts depending on the size of the mouthpiece and the note being played. Thus measurements cannot exactly be referred to actual playing conditions. An attempt to use a well-defined and reproducible plane half way along the throat of the mouthpiece, and a measuring system[4] that was inherently insensitive to source conditions upstream of the measuring plane has not been widely accepted. However, recent developments in impulse techniques (see Section 5) have largely rendered the problem obsolescent.

4.1 Frequency-Domain Parameters

Once the input and output planes are defined, the following parameters may be defined: the input acoustic impedance, $\mathbf{Z}_i(f) = \mathbf{p}_i(f)/\mathbf{U}_i(f) \equiv R_i(f) + iX_i(f)$, the load impedance, $\mathbf{Z}_L(f) = -\mathbf{p}_o(f)/\mathbf{U}_o(f)$, and the output transfer function $\mathbf{OTF}(f) = \mathbf{p}_o(f)/\mathbf{p}_i(f)$. Bold characters are used to indicate complex quantities. The modal frequencies of an instrument are defined by the negative-going zeroes of X_i. By definition, implied but frequently not explicitly stated, impedance in the frequency domain must be measured using an infinite source impedance generator except for the special case when both p_i and U_i are measured; many reports show equipment with a non-infinite source impedance, their results are subject to an error that often cannot be corrected because insufficient information is supplied. Early investigations reported measurements of what was called the *resonance frequencies* of instruments; by examining the experimental equipment used, it is clear that these measurements were of an approximate $\mathbf{OTF}(f)$. "Approximate" because the characteristics of the source were not known (or are not reported), so the quantity p_i is not defined; and, in addition, the measurements of p_o were not carried out in an anechoic room and are severely distorted by standing waves. The input impedance and output transfer level of a symphonic tenor trombone of 13.9 mm bore are shown in Figs. 1*a*, *b*. Instruments of more extended bell flare such as the tuba exhibit fewer peaks and increasing departures from harmonicity.

The zero-frequency input resistance of an instrument is neither zero nor constant; due to turbulence it depends quite strongly on the magnitude of flow velocity in the mouthpiece; however, the resistance is much lower than the magnitude of a maximum of the alternating frequency impedance. Even for very loud playing, when dc flow velocities of around 20 or 30 m/s in the throats of a trombone or trumpet mouthpiece may occur, the zero-frequency resistance does not exceed 0.3 or 0.6 MΩ, about 1 or 2% of the peaks of the input impedance. Musical needs suggest that measurements of modal frequencies should be accurate to within one or two cents (hundredths of a semitone); this accuracy requires that the temperature of the air column be known to within about 0.05°C. Few reports address the need for such temperature control; it is salutary to note that the temperature under playing conditions varies between about 35°C at the mouthpiece and 22°C at the bell of the instrument.[5] The full consequences of this variation remain to be explored.

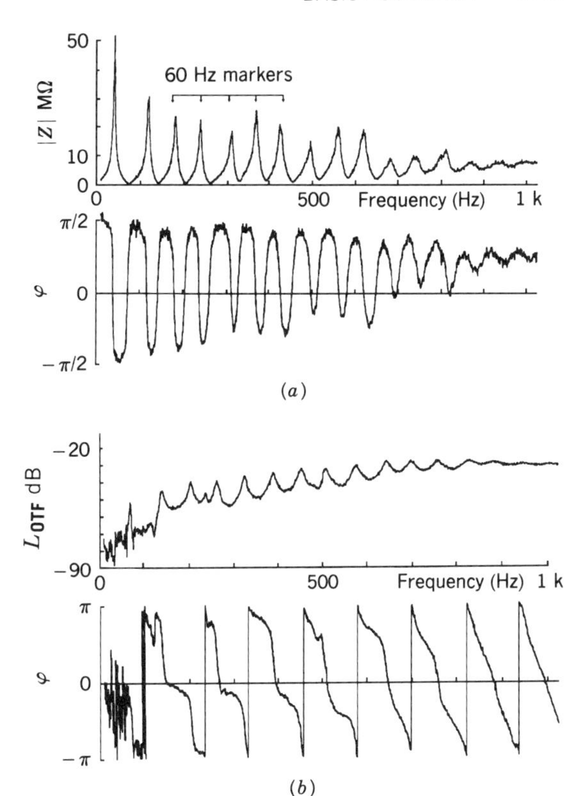

Fig. 1 (*a*) Input impedance (measured according to Ref. 3) of a symphonic tenor trombone of 13.9 mm bore; the units of impedance are megohms, where the ohm, officially the pascal second per cubic meter, equals 1 kg/m^4·s. (*b*) Output transfer response, level = $20 \log_{10} |\mathbf{OTF}|$, and phase, $\varphi = \arg \mathbf{OTF}$, for the same instrument as in Fig. 1*a*; the data are contaminated by external noise at low frequencies. Note the very low levels for low frequencies and that the level is below −30 dB even for the highest frequencies.

4.2 Simple Description of Flared Horn Behavior

As shown in Fig. 1*a*, the modal frequencies, apart from the first, are nearly evenly spaced by about 60 Hz. They therefore resemble the members of an harmonic series with a fictitious "fundamental" of 60 Hz. Simple theory for a largely cylindrical tube, based on the incontrovertible boundary conditions that one end is open and the other closed by the lips of the player, would lead one to expect a sequence of frequencies approximating to the odd members of the harmonic series. The reason for this apparent disagreement is the nonuniform bore of a practical brass instrument: the instrument flares at the bell end and there are small perturbations of nominal bore throughout the instrument. The result is that the modal patterns are changed by the trapping of energy within the bell flare and the variations of phase velocity introduced by the bore perturbations (Fig. 2). As the modal frequency rises, the modal pattern extends further and further into the bell until, eventually, there is no energy trapping and waves propagate freely through the instrument. The departures of phase velocity from the ambient value may be calculated, at least to first order, by using a perturbation theory; a convenient account of early work on perturbation theories, together with a general review of the subject and an excellent list of references, is available in Smith and Mercer.[6] An instrument must have a precisely determined bore in order to possess the desirable property of having a series of modes that is quite closely harmonic. Note that an exactly harmonic series is not desirable for several reasons: (i) a note for which the modes were exactly harmonic could not be controlled easily during changes of level[7]; (ii) the most easily produced different playing frequencies would not accord well with conventional tuning (Smith and Mercer[6]); (iii) it is not possible to achieve an exact series for all valve or slide positions. A designer aims to produce an instrument that has the highest possible number of reasonably controllable playing frequencies.

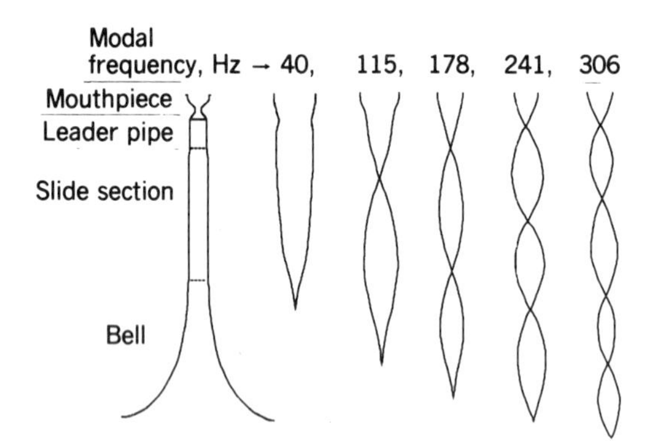

Fig. 2 Diagrammatic representation of modal patterns of pressure in a trombonelike instrument. Locations of salient constructional features are shown on the left and approximate modal frequencies above each diagram. The variations of phase velocity are shown by the differing loop lengths along the instrument, and the apparent position of the open end is shown by the terminating null whose position varies markedly with frequency.

4.3 Effect of the Mouthpiece

The mouthpiece has two main purposes: to provide a comfortable support for the player's lips and to increase the input impedance of the instrument in the playing range. The mouthpiece consists of a cup, a throat, and a backbore that combine to form a Helmholtz resonator in series with the input impedance. Since the coupling is close, the relative alignment of the Helmholtz resonance and the modal frequencies can produce amounts of "frequency pulling" which depend markedly on such parameters as the air temperature and the protrusion of the player's lips into the mouthpiece in addition to the basic physical parameters of the mouthpiece and instrument. Approximately, though, it is sufficient to state that the mouthpiece increases the magnitudes of the input impedance from the second or third to the fifth or seventh modes (Fig. 1*a*) and thus improves the matching of the lips to the instrument in this range. Players do not regard a mouthpiece as part of an instrument, and many will use the same one throughout many changes of instrument—this tendency inhibits the design of mouthpiece and instrument as a combination.

4.4 Deficiencies of Simple Models

By a very unfortunate coincidence, the playing frequencies of a well-designed instrument may be calculated with fair accuracy using a model wherein the instrument is approximated by a cylindrical open–open pipe of length slightly greater than the physical length of the instrument. An open–open cylindrical pipe gives, of course, a harmonic set of modal frequencies, so many early workers considered that the behavior of the instrument was well-represented by this wholly incorrect model and regarded the slight extra length as an end correction. The modal patterns of a system of greater physical reality: a closed–open pipe with a distribution of phase velocity along its length and frequency-dependent energy trapping in the bell flare are totally unlike those of a cylindrical open–open pipe. The incorrect model undoubtedly held back understanding of the physical processes within a brass instrument for many years. A similar problem will be shown later to have occurred in theories of note production, wherein an inaccurate, but at first sight plausible, model diverted attention from the underlying physical processes.

5 WALL EFFECTS

Folklore among players and manufacturers of brass instruments has it that the material from which the tubing is made is of very great importance; some even hold that the geographical place where the metal is rolled is vital. Many investigators have studied this belief, and the evidence is clear that within reasonable limits—the material used must be workable to the precision imposed by the bore requirements and sufficiently robust to stand up to use—the actual wall material has no inherent acoustical importance. The quality of internal finish affects the real part of the wave propagation constant, but there is no one answer as to what value is ideal. Some players prefer instruments having lower energy loss to the walls and consequently choose instruments with a very smooth internal finish; others prefer instruments with slightly higher loss and deliberately coat the inside of an instrument that is too "free." There are few reports of measurements of the internal loss function and even fewer of the efficiency of brass instruments in converting the player's input power to sound power. Conversion efficiency is higher for the higher notes of an instrument, but is certainly less than about 1%, even for the tuba, which appears to be the most efficient brass instrument. The measurement is awkward since conventional techniques cannot be applied to a system having a poorly known load termination; Watkinson, Shepherd, and Bowsher[8] developed a method that is at least as sensitive to changes in loss function as players. Figure 3 shows the real part of the propagation constant for a clean 12.7-mm bore trombone slide.

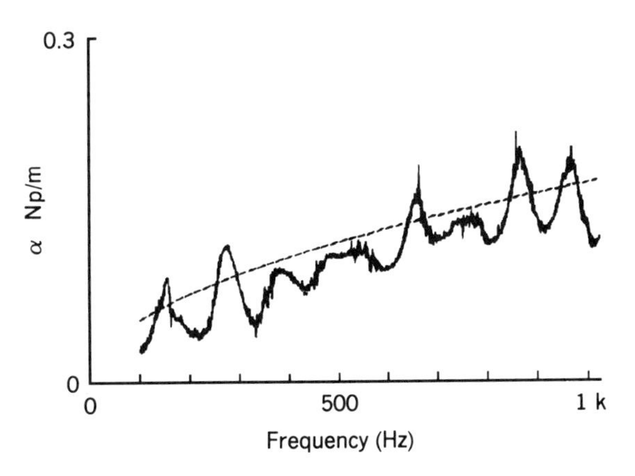

Fig. 3 The real part, α, of the propagation constant for a clean 12.7-mm bore trombone slide. The dashed line shows the best fitting curve having the equation $\alpha = G\sqrt{f}/a$; G is a constant and a the tube radius (= 6.35 mm). See Chapter 7.

Other parameters of the wall material are, however, of some importance. Coupling between an air column and the surrounding walls is extremely low for a perfectly axisymmetric structure, so departures from perfect symmetry, such as seams, supporting stays, or bends, are critical. Although detailed analyses have not been made, there is sufficient evidence[9] to indicate that the coupling is nonlinear and depends on the thickness of the wall material. Thinner materials encourage a nonlinear interaction regime between the air column and the walls; therefore thinner walled instruments exhibit a greater tendency for the tone quality to change at higher playing levels. The modal patterns of vibrating bells have been calculated using finite elements.[10] The resonances are sharply tuned and the response to the acoustic excitation is small except at the resonances. Radiation from the vibrating walls is also small except, possibly, at these resonances and is, moreover, likely to be heard only by the player due to the dimensions of the radiating structure. Thus it is probable that a psychological interaction with the player affects the sound quality of the instrument more than a direct physical cause.

6 IMPULSE MEASUREMENTS AND TIME-DOMAIN PARAMETERS

Impulse measurements on an instrument are directly useful as reflections are generated from acoustic impedance mismatches along the instrument and so indicate not only discontinuities in the bore such as valves, junctions, and slides, but also transitions from cylindrical to tapering sections, and so forth. Structural faults in an instrument may very easily be detected and located, and accurate comparisons between instruments can check consistency in manufacturing output. The impulse response of a system is defined as the response of the system to a delta function of the measured quantity applied at the input. For brass instruments the measuring point can be at the input (source tube or mouthpiece) giving the input impulse response, or at the output (outside the bell) giving the output impulse response. The input source conditions must be specified; practically these are either the nonreflecting source (by use of an appropriate source tube), where waves traveling out from the instrument do not return, or the totally reflecting source where the waves do return. Watson and Bowsher[11] present a self-consistent notation that avoids many of the confusions exhibited in the literature. The Fourier transform of a pressure impulse response is a pressure transfer function: transmitted or reflected pressure divided by the incident pressure.

Fourier transforming the input impulse response for

a reflecting source at the input gives the input transfer function, not the input impedance as is often stated. Time-domain input impedance can, however, be expressed in terms of the input impulse response.[12] When transformed to the frequency domain, this gives $\mathbf{Z}_i(f) = Z[\{1 + \mathbf{ITF}(f)\}/\{1 - \mathbf{ITF}(f)\}]$ for a nonreflecting source; and $\mathbf{Z}_i(f) = Z[1 + \mathbf{ITF}'(f))$ for a totally reflecting source at the input. $\mathbf{ITF}(f)$ and $\mathbf{ITF}'(f)$ are the input (pressure) transfer function measured with a nonreflecting source and with a totally reflecting source. Here, Z is the characteristic impedance of the instrument bore or source tube. Source conditions for impulse measurement may be matched by providing a source tube sufficiently long to ensure that no signals arrive in the measurement window apart from those from the instrument. Input impedance measurements in the frequency domain must be made with a reflecting source at the input; so the second equation is used to calculate this form of the impedance from the impulse response.

Most of the transient measurements of brass instruments reported in the literature are of the input pressure response to a pulse at the mouthpiece; few attempts seem to have been made to deconvolve the signal and remove the spectral characteristics of the pulse. If deconvolution is performed, though, an impulse response is obtained that may be used in suitable algorithms to convert the data to frequency domain; or, by solving the inverse problem, to reconstruct the bore profile that created the measured impulse response. Bore reconstruction is immensely useful to manufacturers, particularly when other ways of measuring the bore are impossible. For example, the serpent, being made of wood, is hard to x-ray accurately since the rays are scattered by amounts difficult to quantify, and the shape precludes the use of traditional measuring disks. Figures 4*a–c* show the input impulse response and the input impedance and internal bore profile calculated from it of a serpent made in 1988.[13]

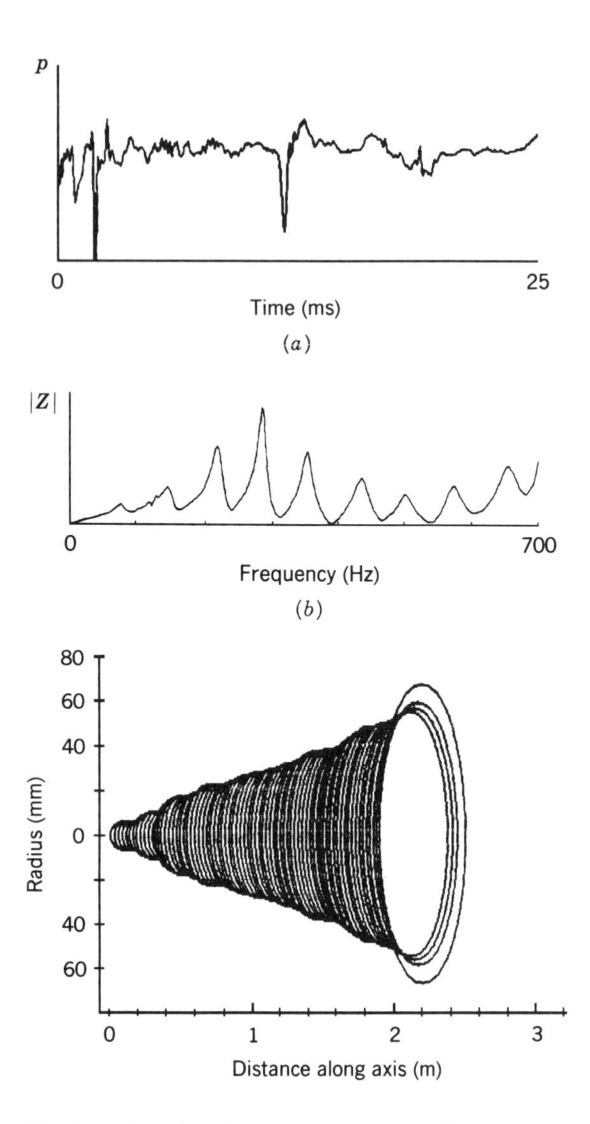

Fig. 4 (*a*) Input impulse response (measured with a nonreflecting source) of a serpent made in 1988. (*b*) Input impedance magnitude calculated from the data in (*a*); compare the different envelope for a predominantly cylindrical instrument in Fig. 1*a*. Note the poor positioning of some of the input impedance peaks—this explains the subjectively found relatively poor definition of certain playing frequencies. (*c*) Internal bore profile calculated from the data in (*a*) and artificially straightened by the presentation. Note the "wavy" profile at certain points.

7 REGENERATION MECHANISMS AND SOURCE DYNAMICS

The basic equation governing acoustic flow U through an orifice under the influence of a pressure difference $p_0 - p$ is given in Chapter 101, Eq. (1); for brass instruments, where the "reed" opens outward toward the instrument, the sign of the term representing the reed contribution must be reversed. The effect of this is that the reed resistance becomes positive (dissipative) for all frequencies except a narrow band just above the reed resonance.[14,15] Therefore, the player has complete control of which mode is excited by controlling the mechanical resonance of his lips. The important differences between brass and reed woodwind instruments are that: (i) for the former, the resonance frequency of the reed is slightly below that of the fundamental oscillation frequency, rather than much higher; (ii) the blowing pressure is not much greater than the peaks of the alternating pressure in the mouthpiece and is often less. Point (ii) was not known until recently when measurements were made of pressures within the mouthpiece and the player's mouth.

Figure 5 shows typical waveforms for a tenor trombone. The alternating mouth pressure is due to the pressure drop along the vocal tract and the mouth cavities as the alternating component of the flow passes through them; the effect may be quantified by considering the series Thévenin equivalent impedance of the source. There is information on the likely magnitude of this equivalent source impedance from work on speech; since there is much variability, an order of magnitude figure of 2 MΩ with zero phase shift at around 200 Hz may be considered typical. The mouthpiece waveform has a very characteristic nonsinusoidal shape—a result of the large amplitude of vibration of the lips which are far apart for a significant fraction of each period. When the lips come together, the flow into the mouthpiece is restricted, the turbulence level changes considerably, and a downward "spike" of pressure, exaggerated by the reactive components of the impedances in the system, is produced. This waveform is the essential basis of the harmonically rich radiated spectrum of brass instruments; nonsinusoidal motion of the lips or nonlinear effects in the air column are of lesser importance, even though values of L_p are around 155 dB in the mouthpiece and 170 dB in its throat. At higher frequencies, the lips vibrate less and the mouthpiece pressure waveform becomes more sinusoidal. The terms "higher" and "lower" frequency are not absolute but relative to the possible playing frequencies of the instrument under consideration.

The reed dynamics are under direct muscular control, and the influence of the player is therefore much stronger than in most other musical instruments. This influence is reflected in every parameter affecting the regeneration and capable of measurement; for instance, the very basic parameter of blowing pressure is greatly affected by the player blowing the test notes. For the woodwind instruments in Chapter 101, 3 kPa is quoted as typical for the clarinet; for brass instruments, some players use different pressures from others for similar notes on similar instruments. In general, blowing pressure increases as pitch and volume increase, but for any given note a range of over 2 : 1 may be found. Pressures outside the values 0.2–20 kPa are unusual. Records of lip motion suggest that the two lips behave differently during oscillation, but there are no data on the variability of lip motions and their relationship to the alternating pressures and flows in the system. The lip motion has significant components both parallel and normal to the plane of the mouthpiece rim, unlike the cane reed motion of woodwind instruments; additionally, it is possible for the motion on the mouth side to be different from that on the mouthpiece side since the vibrating parts are controlled by the player. Many of these effects may partially be combined into *equivalent reed impedances* for the mouth and mouthpiece sides as illustrated in the "circuit diagram" in Fig. 6; typical impedance magnitudes for low-to-medium register notes on a trombone are shown in the figure caption. Data on other instruments are lacking but may be expected to have the same relative magnitudes. The relatively small effect of the instrument on volume flows is clearly demonstrated by these data.

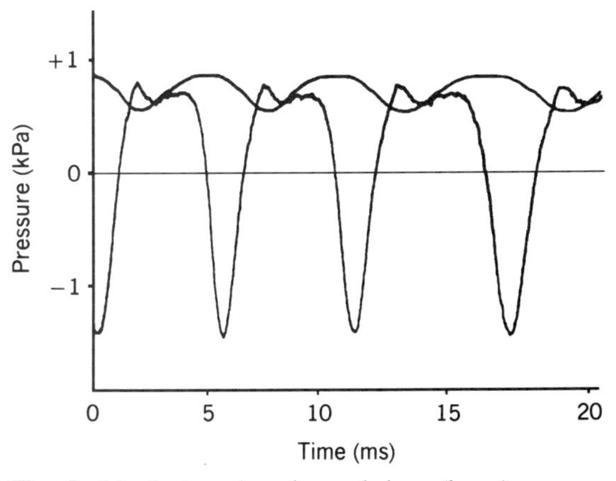

Fig. 5 Mouth (upper) and mouthpiece (lower) pressure waveforms for the note F_3 on a tenor trombone at moderate volume; the zero of the pressure scale is the ambient pressure. Features of the upper waveform are very variable and depend greatly on the jaw and tongue configuration. The lower waveform has a nonsinusoidal shape as a result of the large amplitude of vibration of the lips.

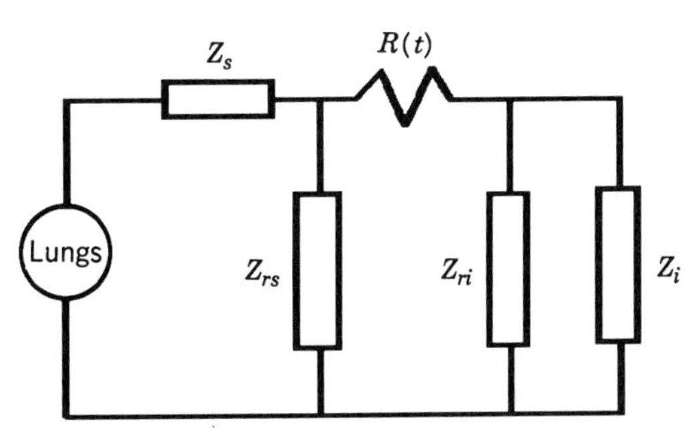

Fig. 6 Diagram showing the arrangement of the impedances associated with the regeneration process involving the player's lips and the instrument. Typical values for a trombone are instrument input impedance Z_i = 10 MΩ; Thévenin equivalent source impedance Z_s = 2 MΩ; equivalent reed impedance (mouthpiece size) Z_{ri} = 3 MΩ; equivalent reed impedance (mouth side) Z_{rs} = 0.5 MΩ; dynamic lip impedance $R(t)$ = 1 MΩ.

Since the basic waveform of the regeneration is nonsinusoidal, impedance components at harmonics of the oscillation frequency play a significant role in the interaction. The quantification of their effect is the basis of the "sumfunction" proposed by Wogram,[16] the maxima of which predict playing frequencies from measurements of the input impedance. The sumfunction may be written:

$$\mathrm{SF}(f) = \frac{1}{n_{\max}} \sum_{n=1}^{n_{\max}} R_i(nf) \times n^{-k}, \qquad (1)$$

where n is an integer, $n_{\max}$ is the maximum value of n for which $n_{\max} f <$ (the maximum frequency measured), and k is a weighting function allowing approximately for the differing degrees of nonlinearity at different playing levels. Reliable values of k cannot be given since their determination requires accurate values of the playing frequencies—inherently impossible to obtain given the variability between players and between occasions with the same player. It is best in practice to use different values of k to assess qualitatively how the playing frequencies drift with dynamic level. The sumfunction is a useful predictor of playing frequencies for well-designed instruments but cannot reliably be used when the modal frequencies are far from harmonic.

Most models of source behavior have assumed that (a) the steady component of the mouth pressure was much higher than the peaks of the alternating pressure in the mouthpiece; (b) the alternating pressure in the mouth was small; and (c) the airflow was basically a large dc flow toward the instrument with a small alternating component superimposed upon it. All these assumptions are incorrect for brass instruments. The flow, for example, reverses every cycle. The models have therefore been at variance with the physics of the interaction and their poor predictive performance is thus explained. A complete regeneration model has yet to be developed, but a beginning is reported in Elliott and Bowsher,[17] and a continuation in Fletcher.[14]

8 PHYSIOLOGICAL AND PSYCHOLOGICAL FACTORS

Since the player has a very great influence on the behavior of a brass instrument, physiological and psychological factors are actually more important than acoustical when attempting to assess differences between instruments of the same general configuration. Frequently, reports concern only listening tests on an extremely artificial set of notes and fail completely to take into account nonacoustic factors, even though social surveys of players' opinions clearly show that these are of prime importance. Mechanical aspects are supreme; if the valves do not respond quickly enough or the slide is "sticky," the instrument will not be used, irrespective of its acoustic performance. Many players are strongly influenced by fashion or the opinions of colleagues and can only perform well upon instruments of "approved" make and model. Subjective assessment tests are often carelessly designed and their results entirely nullified by nonacoustic factors. It may be stated with some certainty that an inadequately designed experiment aimed at finding differences between instruments will only confirm preexisting subjects' opinions.

The search for acoustical features related to instrument quality is paralleled by work on the computer recognition of speech; constraints derived from knowledge of acoustical structure, word structure, syntax topic, and meaning in a knowledge-based, adaptive computer program are now applied to recognition. The percepts of timbre or of instrument quality are similarly extremely complex and also involve some sort of pattern-matching algorithm together with an *expectation* in the player's or listener's consciousness of the qualities desired. This implies an expert degree of familiarity with the task, precisely the circumstances where conventional experimental psychology is at its weakest. The playing process involves an extremely complicated network of feedback paths: physical, neurophysiological and psychological, linear and nonlinear.[18] Unravelling is at an early stage, but it is already known that some players can adjust their sound more than others in response to changes of environment; however, no independent tests have been developed to test for these classes of players. The writer believes that studying the ease of obtaining the desired effect, that is, the ergonomics of the instrument, is the best approach to this problem, but little has been done.

REFERENCES

1. J. Meyer (trans. J. M. Bowsher and S. Westphal), *Acoustics and the Performance of Music*, Verlag Das Musikinstrument, Frankfurt am Main, 1978, pp. 37–49 and 78–80.
2. M. Dickreiter, *Mikrofon-Aufnahmetechnik*, S. Hirzel Verlag, Stuttgart, 1984, pp. 36 sqq.
3. S. J. Elliott and J. M. Bowsher, "Input and Transfer Response of Brass Wind Instruments," *J. Acoust. Soc. Am.*, Vol. 72, No. 6, 1982, pp. 1747–1760.
4. R. L. Pratt, S. J. Elliott, and J. M. Bowsher, "The Measurement of the Acoustic Impedance of Brass Musical Instruments," *Acustica*, Vol. 38, 1977, pp. 236–246.
5. J.-C. Goodwin, "Relations between the Geometry and Acoustics of Brass Instruments," Ph.D. Thesis, University of Surrey, Guildford, Surrey, U.K., 1981.
6. R. A. Smith and D. M. A. Mercer, "Recent Work on Musi-

cal Acoustics," *Rep. Progress Phys.*, Vol. 42, No. 7, 1979, pp. 1085–1129.

7. A. H. Benade, *Fundamentals of Musical Acoustics*, Oxford University Press, London, 1976, p. 402.
8. P. S. Watkinson, R. Shepherd, and J. M. Bowsher, "Acoustic Energy Losses in Brass Instruments," *Acustica*, Vol. 51, No. 4, 1982, pp. 213–221.
9. R. A. Smith, "Recent Developments in Trumpet Design," *Inter. Trumpet Guild J.*, Vol. 3, 1978, pp. 27–29.
10. P. S. Watkinson and J. M. Bowsher, "Vibration Characteristics of Brass Instrument Bells," *J. Sound Vibration*, Vol. 85, No. 1, 1982, pp. 1–17.
11. A. P. Watson and J. M. Bowsher, "Impulse Measurements on Brass Musical Instruments," *Acustica*, Vol. 66, 1988, pp. 170–174. See also N. Amir, U. Shimony, and G. Rosenhouse, "A discrete model of tubular systems with varying cross-section—the direct and inverse problems," *Acustica*, Vol. 81, 1995, Part 1, pp. 450–462, Part 2, pp. 463–474.
12. M. M. Sondhi and J. R. Resnick, "The Inverse Problem for the Vocal Tract: Numerical Methods, Acoustical Experiments and Speech Synthesis," *J. Acoust. Soc. Am.*, Vol. 73, 1983, pp. 985–1002.
13. J. M. Bowsher, A. P. Watson, and P. A. Drinker, "Impulse Measurements on Wind Instruments," in G. Widholm and M. Nagy (Eds.), *Das Instrumentalspiel*, Doblinger, Wien München, 1989, pp. 55–76.
14. N. H. Fletcher, "Nonlinear Theory of Musical Wind Instrument," *Appl. Acoust.*, Vol. 30, 1990, pp. 85–115.
15. N. H. Fletcher and T. D. Rossing, *The Physics of Musical Instruments*, Springer-Verlag, New York, 1991.
16. K. Wogram, "Ein Betrag zur Ermittlung der Stimmung von Blechblasinstrumenten," Dissertation, University of Braunschweig, Braunschweig, Germany, 1972.
17. S. J. Elliott and J. M. Bowsher, "Regeneration in Brass Wind Instruments," *J. Sound Vibration*, Vol. 83, No. 2, 1982, pp. 181–217.
18. J. M. Bowsher, "A Control Structure Approach to Playing a Wind Instrument," in G. Widholm and M. Nagy (Eds.), *Das Instrumentalspiel*, Doblinger, Wien München, 1989, pp. 43–54.

103

PIANOS AND OTHER STRINGED KEYBOARD INSTRUMENTS

GABRIEL WEINREICH

1 INTRODUCTION

Stringed keyboard instruments are conveniently classified by the method used to excite the string. In the piano it is struck by a hammer that immediately rebounds; in the harpsichord it is plucked by a plectrum; and in the clavichord it is struck by a metal *tangent*, rigidly attached to the key, which then remains in contact with it, thus serving not only as the source of vibratory energy but also as the termination of the string's vibrating length. In all three cases, the vibration of the string is transmitted to a soundboard, which serves as the main source of acoustic radiation.

There is a sharp gradation of mechanical complexity from the clavichord, whose mechanism is essentially trivial, to the harpsichord and finally to the piano. In the clavichord, the tangent is, in effect, part of the key. The harpsichord needs plucking plectra which can return past the string without plucking it again; it must also be equipped with *dampers* that terminate a note when the key is released. Finally, the piano action, which we shall discuss in some detail below, is by far the most complex of all.

Clavichords and harpsichords were well known and well developed by the beginning of the sixteenth century; by contrast, the piano is considered to have had its origin early in the eighteenth century, when the Florentine harpsichord maker Bartolommeo Cristofori described a new invention of his as *gravicembalo col piano e forte*, that is, as a harpsichord capable of playing both softly and loudly. The oldest existing piano dates from 1720.

Note: References to chapters appearing only in the *Encyclopedia* are preceded by *Enc.*

2 REVIEW OF STRING DYNAMICS

If one assumes that the string has a uniform linear density, and that the only restoring force on it is due to its tension (which acquires a transverse component when the slope of the string is different from zero), the equation of motion becomes

$$\frac{\partial^2 y}{\partial t^2} = \frac{T}{\rho} \frac{\partial^2 y}{\partial x^2} \tag{1}$$

in which T is the tension and ρ the linear density. Such a formulation is equivalent to attributing to the string a potential energy density proportional to the square of its slope. As is well known, solutions of Eq. (1) can be of arbitrary shape, moving with speed

$$c \equiv (T/\rho)^{1/2} \tag{2}$$

along the string in one direction or in the other without changing their form; or, of course, superpositions of such solutions, since the wave equation is linear.

Additional forces, which make the string deviate from the above assumptions, can be classified into two obvious categories: dissipative and nondissipative. The first can, in turn, be separated into internal forces due to the flexing of the string material (*internal friction*) and external forces due to interaction with the surrounding air; these involve both the ordinary viscous interactions and the direct radiation of sound by the string's motion. In practice, neither type of dissipative force seems to be of great importance (at least for unwound piano strings), the rate of energy loss being generally dominated by nonrigidity of the end supports, rather than dissipation in the strings themselves.

By contrast, the additional nondissipative forces can be quite important. They arise from a term in the potential energy density proportional to the square of the *second* space derivative of the displacement rather than the first, that is, the string's *curvature*; and their physical origin is in the *stiffness* of the string. It is obvious on simple dimensional grounds that, for a string of given properties held under a given tension, the relative importance of the two terms in the potential energy depends on the wavelength of the disturbance: The shorter the wavelength, the larger will be the contribution of the stiffness relative to that of the tension.

Since the equation that determines the eigenvalues and eigenfunctions of the vibrating string is now of fourth, rather than second, order, we must also expand our range of boundary conditions. In the theory of vibrating bars, of which piano strings are a slight generalization (in that tension is applied), one deals generally with three kinds of boundary conditions: (1) free, in which both the transverse force and the moment of the force vanish at the end; (2) hinged, in which the displacement and the moment of the force vanish, but not the force; and (3) clamped, in which the displacement and the slope vanish, the force and moment taking on appropriate nonzero values. The fact that none of these describe the situation in a piano very well is simply an indication that energy can flow out of the ends of a piano string. Nonetheless, an order-of-magnitude estimate of the relative importance of the stiffness (compared to the tension) can be made: If it were the only correction, stiffness would introduce an increase in the nth eigenfrequency proportional to n^2 (provided this correction is small).

The effect on the normal frequencies due to the finite admittances of the end supports is also easily calculated (assuming that it is small). If, for example, one end of the string is fixed (zero admittance of support), the driving point admittance of the string at the other end can be shown to be

$$Y = \frac{i}{Z_0} \tan \frac{\omega L}{c}, \qquad (3)$$

where L is the length of the string, Z_0 its characteristic impedance, c the speed of transverse waves along it, and ω the radian frequency. By equating this to zero, we find

$$\omega = \frac{n\pi c}{L}, \qquad n = 1, 2, \ldots \qquad (4)$$

for the normal frequencies of the string when both ends are fixed. If, on the other hand, the support presents a finite admittance Y_B, we find the normal frequencies by equating Y in Eq. (3) to $-Y_B$ and solving for ω. We find in this way that a springy admittance of the support will lower the eigenfrequencies, and a massy admittance will raise them. A more general, complex admittance (such as is always encountered in real life) will perturb the ω's by a complex amount, of which the real part signifies a shift in frequency and the imaginary part a damping of the mode.

In a piano the characteristic impedance of a string near the middle of the keyboard is of the order of magnitude of a few kilograms per second, whereas a typical soundboard impedance is of the order of 1000 kg/s. In a harpsichord, both numbers are very considerably smaller.

3 STRUCTURE OF A HARPSICHORD

The soundboard of a harpsichord is typically about 3 mm thick, reinforced with "ribs" that run perpendicular to the grain, and it is mounted on a "box" so that its bottom surface is not exposed. Very often the soundboard contains a "rose," or decorated hole; but (unlike, e.g., in a guitar) this opening has no great acoustical importance because the cavity formed between the soundboard and the bottom of the box is open to the atmosphere along a considerably larger area at the keyboard end of the assembly. The vibrating lengths of the strings are terminated by two raised bridges of which only the one glued to the soundboard is properly called the bridge, the other being referred to as the nut. The strings themselves, including their "dead" parts, extend from tuning pins at the keyboard end (which are themselves set into a block of wood called the wrest plank) to hitch pins set into the solid rim of the instrument box. If there is a (shorter) choir of 4-ft strings (i.e., a set of strings sounding an octave above the nominal pitch of the key), they normally have a separate bridge glued to the soundboard, and their hitch pins are often set directly into the soundboard itself (with a reinforcing rib just underneath them).

The playing machinery of a harpsichord consists of the keys themselves, which are pivoted near their center, and a series of jacks, or playing assemblies, one for each string. A jack is typically a rectangular parallelepiped, perhaps 9–15 cm high, 1 cm deep, and 0.4 cm wide (as seen from the keyboard side), which moves up and down in locating slots above each key. In one possible design, it does not quite rest on the key but hangs from the damper, which is a vertical piece of felt that extends like a flag sideways from the jack and rests on the string, kept there by the weight of the jack (Fig. 1). Unlike the damper of a piano, that of a harpsichord is quite thin in the direction of the string length, and thus unable to damp any modes that have a node at that location; but

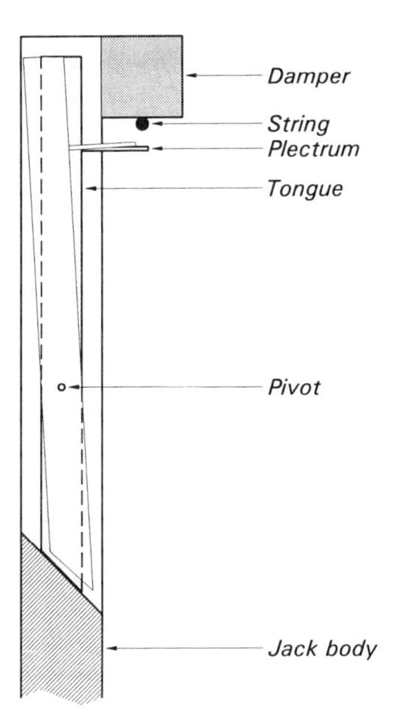

Fig. 1 Cross-sectional view of the upper section of one design of a harpsichord jack, as seen in a direction parallel to the string, in its rest position. When the key is depressed, the whole assembly moves upward, causing the damper to rise and the plectrum to pluck the string. As the jack falls back down, the string pushes the plectrum out of the way by tipping the tongue, as indicated by the lighter lines. Not shown is the thin wire spring that finally returns the tongue to its normal position.

since it is located in the same place where the string is plucked, such modes are not excited in the first place, and so do not need damping.

Each jack carries a plectrum that extends sideways from it underneath the string but is mounted to a spring-loaded tongue so that on the way down, once the string has been plucked, the plectrum simply moves out of the way, allowing the jack to fall back to its rest position. There is also a padded wooden rail above the jacks that prevents them from jumping out of their slots when the key is attacked with high velocity. Since the distance of plucking is determined by the geometry, the loudness of the sound is (except for small corrections) independent of the velocity with which the key is depressed.

Harpsichords with more than one choir of strings are equipped with more than one jack for each key. In such a case, each jack is guided by two slots, of which the upper one is in a rail that can be displaced to the right or to the left by an external stop lever. In this way, each set of jacks can be made to pluck, or to miss, the corresponding strings.

4 STRUCTURE OF A PIANO: STRINGS AND FRAME

In the modern grand piano, the tension of the strings is supported by a cast-iron frame. At the far end (as seen from the keyboard), the strings are looped around steel hitch pins set into the frame; at the near end, they are wound around steel tuning pins that are held in position by being driven into a laminated hardwood pin block (or wrest plank) that is attached to the frame by a number of very large wood screws. The tuning pins have squared ends to which a special wrench, traditionally called a tuning hammer, is applied to adjust the tuning. The "speaking" (that is, vibrating) length of the string is delimited at the keyboard end either by agraffes (special screws set into the frame with transverse holes through which the strings pass) or, in the treble region, by going underneath a transverse frame member called *capo d'astro*, which is shaped like a (not terribly sharp) knife edge. At the other end, the speaking lengths of the strings are limited by going over the wooden bridge glued to the soundboard. Each string takes a "slalom" path around two steel pins set into the bridge, effectively defining the speaking length.

Throughout most of the compass of a typical piano (which is somewhat more than 7 octaves, ranging from A_0 to C_8), there are three steel strings for each note, varying in diameter from about 0.8 mm at the upper end to about 1.2 mm at the "break," where triple stringing stops. The remaining notes (perhaps 20 on a large grand, more on a smaller piano) are furnished with strings that are wound with copper over a steel core, in order to increase their mass without undue increase in stiffness; at the same time, the number of strings per note changes from three to two, then to one in the deep bass.

5 STRUCTURE OF A PIANO: ACTION

A diagram of a typical modern grand piano action is shown in Fig. 2. Each hammer head is glued to a wooden shank that is, in turn, pivoted on a horizontal pin. A leather-covered "knuckle" is attached to the shank relatively near the pivot so as to provide a considerable velocity ratio from key motion to hammer motion. The rather complicated structure of the mechanism that intervenes between the two is best explained by its function, as follows:

1. *Escapement.* When the hammer hits the string, it must be free from any coupling to the lever system, which originally gave it its velocity; this is accomplished by the escapement mechanism. At the beginning of the

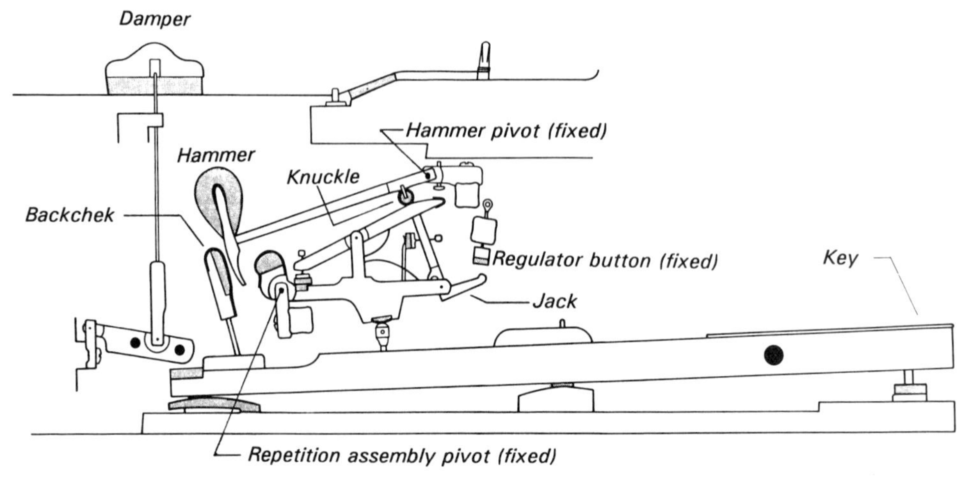

Fig. 2 Schematic of a modern grand piano action. When depressed, the key lifts the pivoted repetition assembly. The L-shaped jack then pushes up on the hammer knuckle, but is itself deflected when its toe encounters the fixed regulator button. [Reprinted by permission from A. Askenfelt (Ed.), *The Acoustics of the Piano*, Royal Swedish Academy of Music, Stockholm, 1990, p. 40.]

key stroke, the hammer knuckle is pushed upward by an L-shaped lever called a jack, which is itself pivoted. As the hammer approaches the string, the short arm of the jack encounters the regulating button, which causes it to swing out of the way, leaving the hammer "flying" freely toward the string, ready to bounce back without impediment.

2. *Backcheck.* Multiple bouncing of the hammer is prevented by the tail of the hammer encountering the leather-covered surface of the backcheck, which is oriented so that the impact is close to grazing. The friction between the two then quickly brings the hammer to rest. Since the backcheck is attached to the key, it moves out of the way, and releases the hammer, as soon as the key is released.

3. *Damper.* A soft felt damper, which normally rests on the string, is lifted when the key is depressed roughly half of its travel. It is lowered again, terminating the vibration of the string, when the key is released. The higher strings (above approximately F_6) are not equipped with dampers, since their vibration naturally dies down rapidly, so that notes cannot, in any case, be sustained for any length of time.

4. *Repetition.* When the hammer bounces back from the string, its knuckle (which was resting on the jack on its way up) descends on a spring-loaded lever (sometimes called the wippen), thus cocking a spring; when the backcheck releases the hammer, this cocked spring lifts the knuckle sufficiently for the jack to be able to slide under it again, allowing the note to be repeated.

As one might expect, there is considerable variation in nomenclature of piano action parts, even among English-speaking technicians.

6 INTERACTION BETWEEN STRING AND HAMMER

The mass of the hammers in a modern grand piano typically varies from about 10 g at the bass end to about 4 g at the treble end; appreciably larger variation is precluded by the very different forces that would then be required to operate the various keys. (In fact, the static forces are equalized by lead weights embedded in the keys, but the variation in inertia remains). On the other hand, the range of vibrating string mass is considerably larger, so that the ratio of hammer mass to string mass varies from a minimum of around 0.04 in the extreme bass to a maximum of close to 10 in the treble. (The string mass, as used here, is the total vibrating mass if two or three strings correspond to the same hammer.) In the middle of the keyboard, the ratio of hammer mass to string mass is near unity, indicating that the fundamental frequency of the hammer moving in contact with the string is comparable to the fundamental frequency of the string alone. Thus the time that the thrown hammer spends in contact with the string is by no means negligible compared to the period of the string's free vibration.

Since an infinitely long, perfectly flexible string presents an impedance to lateral impact that is purely resistive (equal, in fact, to twice the string's characteristic impedance), an impinging point mass would respond as to a viscous force and tend simply to come to rest in contact with the string. To the degree that the system is well represented by such a model, nothing more can happen until the first reflections arrive from the ends of the string; and since the hammer location is typically at approximately one-eighth of the string length, the contact time cannot be shorter than one-eighth of the string's period, even if the hammer mass (measured, of course, in units of the string mass) approaches zero. In fact, the description of the hammer's encounter with the string is considerably more complex, because (1) a more massive hammer will not have come completely to rest before the first reflections arrive; (2) the impedance presented by the string is, due to its stiffness, not completely resistive; and (3) the hammer is itself elastic. As a result, a useful description of the behavior of piano hammers is only practical in terms of computer simulations.[1]

In this connection, one must also note the great importance of the nonlinear elasticity of the hammer, which arises not only from the inherent properties of the layers of felt of which the hammer is made but also from the fact that the cross-sectional contact area increases radically with increasing impact force (at least when the hammer is in good condition, without deep grooves, which develop with wear and which make the contact area between hammer and string more constant)[2]. It has now been firmly established that nonlinearity of the hammer constitutes a central element of what we recognize as piano sound. Specifically, it is clear that a soft hammer will act as a low-pass filter in its interaction with the string, decreasing the amplitudes of higher partials. The nature of the observed nonlinearity is such as to make the hammer effectively "harder" at higher playing levels, as a result of which the spectrum shifts so as to be "brighter." A *fortissimo* piano passage that is recorded and played back *pianissimo* will, in other words, absolutely not match the same passage recorded *pianissimo* in the first place and played back at a similar level. (It is precisely these subtle properties of the hammer that a skilled technician can adjust in the process called *voicing*, which involves mechanical and/or chemical treatment of the felt after the hammers have been installed.)

7 PROMPT SOUND AND AFTERSOUND

If a piano key is struck and held (so that the damper does not return to the string), the level of sound does not decay at a steady logarithmic rate, as one would normally expect of a linear system; instead, it begins to decay at one rate, then "breaks" to another, slower one. A number of causes contribute to this phenomenon whereby the sound divides into a *prompt sound* and an *aftersound*. Before we examine the details, it is important to mention that (as is the case with the nonlinearity of the hammer) the resulting double decay characteristic is perceived as an important component of piano tone. Specifically, it appears that a listener tends to diagnose a tone as "loud" if it *starts out* to be loud, and to diagnose it as "well sustained" if *some part of it* persists for a long time, even if that part is relatively weak. This is important in all musical applications involving free vibrations, since the designer of an instrument always needs to steer between the Scylla of strong string-soundboard coupling, which tends to enhance the playing volume but shortens the free vibration time, and the Charybdis of weak coupling, which does the opposite. A piano obviates this dilemma by having a number of normal modes that have (almost) the same frequency but different damping rates. In this way the sound can begin with a strongly coupled mode (loud but short-lived) and then go over to a weakly coupled one (long-lived but quiet).

This multiplicity, and near degeneracy, of string vibration modes arises from two essential causes: (1) the two possible polarizations that a string vibration can have and (2) the presence of a number of strings for each note, whose motions are dynamically coupled by being mounted to the bridge at almost the same point.[3] The first factor leads to a splitting of decay times because, ordinarily, the admittance of the bridge to vertical string motion is larger than to horizontal motion, so that even a single string exhibits a division into prompt sound and aftersound. It should be noted that if the hammer struck the string in an exactly vertical direction, and if the bridge exhibited the kind of symmetry that would guarantee that the eigenmodes were exactly horizontal and vertical, there would be no aftersound from a single string because the horizontal mode would not be excited at all. In practice, however, neither of these hypotheses is exactly satisfied.

The second factor, having to do with the presence of more than one string, is more important but also more complicated because it depends critically on the exact way the individual strings are tuned. The simplest case is one in which there are two identical strings tuned in an identical way. The coupling through the bridge will then give rise to two (vertical) normal modes, in one of which the two strings move symmetrically, in the other antisymmetrically. It is clear in such a case that the amplitude of combined oscillatory force exerted by the strings on the bridge is much larger in the symmetric than in the antisymmetric case; and if the energy losses of the strings are due primarily to finite bridge admittance (as is ordinarily true in the piano), the damping of the symmetric mode will also be considerably larger. Since the excitation of the two strings by the hammer impact is near symmetric, the amplitude of the symmetric mode will be dominant in the beginning of the tone; but as time goes on the (originally weaker) antisymmetric mode will take over because of its slower decay rate.

This phenomenon is illustrated by the contrast between Figs. 3*a*, *b*. The first shows the vertical component of vibration amplitude of a string (on a logarithmic scale) as a function of time, when the other two strings of the unison triplet are damped; in the second, the same quantity is measured while one other unison string is free to vibrate. The dramatic decrease in damping rate is caused by the excitation of the antisymmetric mode.

When the two strings are not tuned identically, there will still be two coupled modes with different decay rates, but they will no longer be exactly symmetric or exactly antisymmetric. It has been suggested that this fact allows a skilled tuner to adjust the *level* of aftersound, making it appropriately even from note to note, instead of depending on accidental factors such as irregularities of the hammer.

The division of a piano tone into prompt sound and aftersound has sometimes been explained by stating that initially the vibrations of the two (or three) strings are in phase, but as time goes on they get out of phase due to slight differences in their tuning. This is not a correct way to describe the phenomenon, since it implies that some time later the vibrations will again get into phase and the decay rate will return to its "prompt" value. In fact, this approach ignores the dynamical coupling that the bridge introduces, as a result of which the vibration frequencies are not the same as when each string vibrates in isolation. In particular, for the case of identical tuning, the two modes will be split in complex frequency by an amount proportional to the admittance of the bridge, but in such a way that a pure imaginary—that is, purely reactive—bridge causes a splitting in the frequency but not in the damping, whereas a purely real—that is, purely dissipative—bridge causes a splitting in decay rate but not in frequency. The latter case is especially useful in making a conceptual separation between the beats resulting from mistuning and the phenomenon we are discussing here.

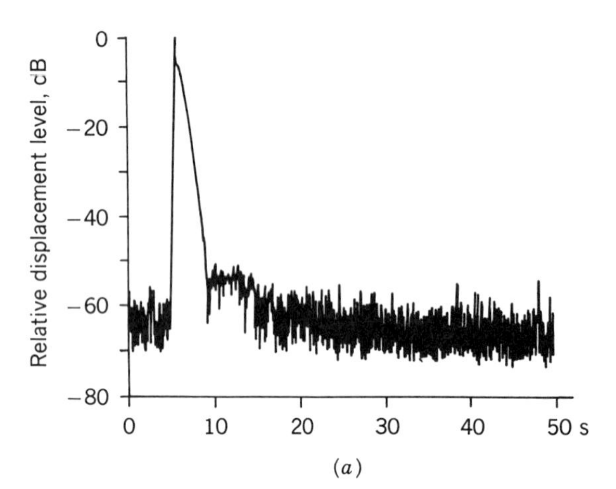

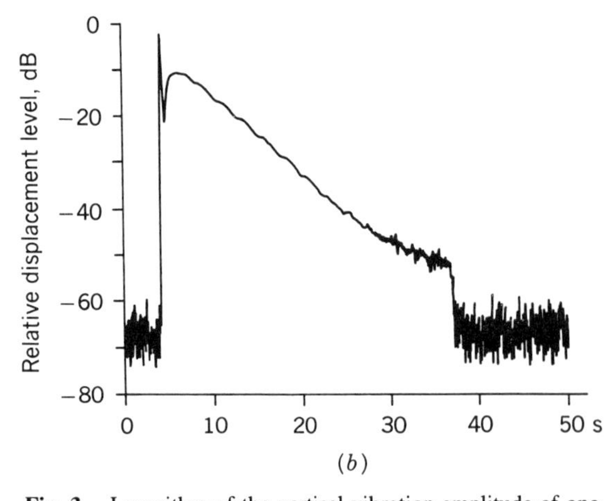

Fig. 3. Logarithm of the vertical vibration amplitude of one string of a unison triplet as a function of time, after all three have been struck by the hammer. (*a*) Remaining two strings of the triplet damped. (*b*) One other string of the triplet free to vibrate. (Reprinted by permission from G. Weinreich, *Coupled Piano Strings*, *J. Acoust. Soc. Am.*, Vol. 62, 1977, p. 1482.)

8 STRUCTURE OF A PIANO: SOUNDBOARD

The purpose of the soundboard is, of course, to provide a large enough vibrating area for appropriately efficient radiation of the string vibrations into the surrounding air. Piano soundboards are generally made of spruce, up to around 10 mm thick, assembled by gluing strips of wood together on edge and machining to the desired thickness profile. The resulting sheet is then reinforced and stiffened by ribs of a similar material glued on perpendicularly, or almost perpendicularly, to the grain of the sheet. At least approximately, the resulting soundboard has the same flexural stiffness along and across the grain, giving the board as a whole the higher mode density characteristic of a two-dimensional, rather than a one-dimensional, structure.

Flexural waves propagating along an isotropic sheet have a dispersion law of the form

$$\omega = ak^2, \tag{5}$$

where k is the magnitude of the propagation vector, ω the radian frequency, and a a constant depending on the material and its thickness. In order for such a wave, propagating in an infinitely large sheet, to radiate a sound wave into the surrounding fluid (i.e., into the air), both the frequency and the component of k parallel to the sheet must match. This leads to the following equation for the angle θ at which the sound wave is radiated:

$$\cos\theta = (c^2/a\omega)^{1/2}. \tag{6}$$

Here c is the speed of sound in air. The frequency $\omega_c = c^2/a$ is often called the *coincidence frequency*; at frequencies lower than that, $\cos\theta$ comes out greater than 1, signifying that no radiation is possible. Of course, this statement applies strictly only for an infinitely large sheet; nonetheless, it seems reasonable to require that a piano soundboard have a coincidence frequency not too much higher than the lowest frequencies that one is interested in radiating with any efficiency.

Another criterion concerns the frequency of the fundamental resonance of the soundboard, which serves to enhance its motion in the bass and increase the efficiency of radiation. Good piano soundboards have a fundamental well below 100 Hz. At frequencies lower than the fundamental resonance, the response of the soundboard is weak; above it, one hopes to encounter a large density of normal modes such as to give the board a good response without excessive variation with frequency. (It is this density of modes that is enhanced by having the elastic properties of the board near isotropic.)

Since the frequency of the fundamental resonance is determined by the geometry of the board together with the dispersion law of Eq. (5), it is not surprising that the constant a can be eliminated between the expression for the coincidence frequency and that for the fundamental mode, with the approximate result:

$$f_0 f_c = f_a^2, \tag{7}$$

where f_c and f_0 are, respectively, the coincidence frequency and the frequency of the fundamental mode, and f_a is the frequency of a sound wave whose wavelength is equal to twice a typical soundboard diameter. It follows that if, in designing a piano, we wish to keep both f_0 and f_c well below 100 Hz, we must also keep f_a well below 100 Hz, implying something like 1.8 m or more for the size of the soundboard. Such an estimate is, indeed, consistent with the size of good concert grands.

As we mentioned earlier, the mechanical impedance of a piano soundboard, measured at points on the bridge, is typically in the vicinity of 1000 kg/s. It shows the expected behavior—that is, a distinct drop in magnitude—at the frequency of its fundamental mode; above that, both its magnitude and its phase exhibit the irregular behavior characteristic of a mechanical system with a large number of normal modes. We see from this that the effect of the finite bridge admittance on the free vibration frequencies of a string will be rather irregular, and typically of the order of some parts in 1000 (since the characteristic impedance of a string is typically of the order of magnitude of a few kilograms per second). What matters here is not, of course, the shift of absolute frequency (which the tuner in any case adjusts empirically), but the contribution of this effect to inharmonicity and decay rate of the string.

REFERENCES

1. D. E. Hall and A. Askenfelt, "Piano String Excitation V: Spectra for Real Hammers and Strings," *J. Acoust. Soc. Am.*, Vol. 83, 1988, pp. 1627–1638. (This reference also lists earlier articles in the same series.)
2. H. Suzuki, "Modeling of a Piano Hammer," *J. Acoust. Soc. Am.*, Vol. 78, 1985, p. S33; X. Boutillon, "Model for Piano Hammers: Experimental Determination and Digital Simulation," *J. Acoust. Soc. Am.*, Vol. 83, 1988, pp. 746–754.
3. G. Weinreich, "Coupled Piano Strings," *J. Acoust. Soc. Am.*, Vol. 62, 1977, pp. 1474–1484.

PART XV

ACOUSTICAL MEASUREMENTS AND INSTRUMENTATION

104

INTRODUCTION

P. V. BRÜEL, J. POPE, AND H. K. ZAVERI

1 INTRODUCTION

Acoustical measurements are made for a wide variety of applications using many kinds of instruments. While any measurements in acoustics could be called acoustical measurements, for present purposes acoustical measurements shall be defined as physical measurements characterizing the acoustical phenomenon. This scope includes measurements of sound pressure, particle velocity, density fluctuation, and so on, and the related intensity and energy density. Amplitude, duration, frequency, phase, modulation, harmonic relationships, and temporal and spatial correlation, among other things, and how these vary with time or position may be of interest.

This chapter introduces acoustical measurements and instrumentation and the surprising complexity of measurement, analysis, and interpretation that arises from seemingly simple acoustical questions and issues. The remaining chapters of this part survey specific subjects of importance in acoustical measurements. Sound level meters are the subject of Chapter 105. Sound intensity and its measurement are covered in Chapter 106. The various kinds of analyzers and selection of an appropriate one for a specific application are the subject of Chapter 107. Critical to all these instruments is a suitable transducer. Chapter 108 discusses calibration of microphones.

2 ACOUSTICAL MEASUREMENTS[1–5]

The earliest acoustical measurements probably would be better characterized as observations: observations of sound in the context of music and architecture. Later, true measurements were made of parameters such as the speed of sound and frequency. Not until the nineteenth century, however, were quantitative measurements of sound pressure and particle velocity made. Such measurements became more practical with the development of electroacoustic knowledge accompanying telephony and developed further with the availability of modern electronic instrumentation.

The fundamental challenge in most acoustical measurements is that the acoustical phenomenon of interest consists of small and rapid fluctuations. Human hearing, for example, is more or less sensitive to pressure fluctuations of 20 Hz (i.e., 20 oscillations per second) to as much as 15,000 Hz (i.e., 15,000 oscillations per second) and more. The minimum sensible fluctuation is as small as 20 μPa, about 0.00000002% of atmospheric ambient static pressure. Although the fluctuations are small, they cover a wide range of smallness. The energy content varies by a factor on the order of 10^{12} between the lower (minimum sensible) and upper (pain) limits of human hearing.

Until recently, the majority of acoustical measurements were sound pressure measurements of one kind or another. Most were made using a sound level meter or spectrum analyzer. Further analysis, if necessary, would be carried out by hand or using a general-purpose computer. Modern developments in computer and microcomputer technology are moving additional, and more complex, analysis capability into instruments. In some cases, with the addition of a suitable transducer, the computer becomes the instrument. Digital instrumentation also has permitted improved intensity measurements, and instruments for the measurement of sound intensity are becoming common.

Fundamental to the development of modern measurement methods and instrumentation are standards developed by national and international organizations such as the American National Standards Institute

Note: References to chapters appearing only in the *Encyclopedia* are preceded by *Enc.*

(ANSI),[6] the American Society for Testing and Materials (ASTM),[7] the International Standardization Organization (ISO),[8] and the International Electrotechnical Commission (IEC).[9]

2.1 Classification of Acoustical Measurements

Three schemes for classifying acoustical measurements can be considered common: (1) by physical quantity measured, (2) by parameter measured, and (3) by application. Physical quantities include sound pressure, particle velocity, density, and so on. Parameters include amplitude, duration, frequency, and phase, determined for the raw signal, frequency-weighted signals, or in specified frequency bands. Applications include such things as measurement of noise levels, characterization of sound sources, and determination of acoustical properties.

The usefulness of such classification is to organize and categorize relationships among measurements. A particular instrument, for example, often will be suitable for a number of measurements within the classification.

2.2 Classification of Applications

Acoustical measurements are made for a multiplicity of applications. One scheme that is particularly useful for classifying applications is based on the categories: sound source characterization, transmission path analysis, evaluation of acoustical environments, and effects of sound. These categories comprise the range of physical considerations. Biological considerations may be included under categories of physiological or psychological effects, as appropriate. Basic measurements under these categories may be combined, and otherwise analyzed, to yield further results. Acoustical properties of materials, for example, often are determined from transmission path measurements in a controlled acoustical environment. Some such analysis is performed in multichannel analyzers; more complex or advanced analysis is performed by general-purpose digital computers.

The goal in source characterization is to determine quantities that are independent of the particular acoustical environment or installation or that permit prediction of characteristics in other environments or installations. Either sound pressure or sound intensity measurements may be used to determine source sound power. Given a suitable free-field environment, source directivity may be determined.

Transmission path analysis may include sound pressure or intensity measurements, comprising individual measurements or cross-channel correlation/time-delay determinations. Transmission path analysis may be performed to determine what modifications are necessary or desirable or can be used to characterize the acoustical properties of materials along the transmission path.

Measurements to evaluate the acoustical environment comprise a range of issues that can be characterized as suitability for intended use: for work, recreation, sleep, speech communication, and musical performance, among others. The issue may be comfort, health, or safety, the latter comprising both direct hazard, such as damage to hearing, and indirect hazard, such as inability to recognize audible warning information.

These topics are the subjects of the chapters in the present section as well as many chapters in other sections. Measurement, analysis, and interpretation are surprisingly complex for even seemingly simple questions and issues. To illustrate this complexity, the balance of this chapter will address some issues and approaches for basic measurements of sound pressure as they relate to noise.

3 SOUND PRESSURE AND SOUND PRESSURE LEVEL MEASUREMENTS

3.1 Approach and Rationale

Hearing describes the perception or sensing of sound by the ear(s). The ear itself is sensitive, more or less, to fluctuations of pressure, that is, *sound pressure*. For this reason the measurement and analysis of sound pressure are fundamental to understanding and describing sound that is heard. *Sound pressure level*, a logarithmic measure of certain specified characteristics of the amplitude of sound pressure, often is determined.

Hearing is necessary for many desirable purposes: for communication, the enjoyment of music, and to locate sound sources, among others. It is also the means by which we receive noise, that is, unwanted sound. The reception and analysis of sound make up a complicated process that still is not completely understood, and the ear itself is a complex instrument capable of excellent discrimination over a wide range of frequency, amplitude, and duration. Establishment of objective parameters for measuring sounds, and how we perceive them, involves not only considerations of physics but also considerations of psychoacoustics. A loud sound can be quite acceptable in the form of music or speech, whereas a weaker sound, for example that of a mosquito in the bedroom, can be disturbing and very annoying. The very same sound that is pleasant music to one listener can be annoying noise to another. The relationship between sound pressure level and the perception of sound is not simple.

Although the problem is difficult, for most applications it is important that the numerical values of the parameters measured be in reasonable agreement with the subjective impression of sound phenomena. Furthermore, the instrumentation developed for measurement of these parameters must be accurate to provide consistent values for comparison of results from different workers and measurement sites.

To facilitate objective measurements, it is necessary to limit the fields of application and to measure well-defined physical parameters. Nevertheless, problems quickly arise when sounds of differing frequency content (e.g., from an air moving fan and an industrial lathe) or rapidly fluctuating level (e.g., a hammer blow or gunshots) have to be measured and be represented by a single number. Before an instrument is designed or selected for these measurements, the purpose of the measurements ought to be addressed:

- Should the instrument measure the *amplitude* of the sound?
- Should it give a measure of the *annoyance* of the sound?
- Should it indicate how dangerous the sound is for causing *hearing damage*?

Note that the amplitude question implies a physical measurement only. The annoyance and damage questions require psychological and physiological correlations in addition to the physical measurement of sound pressure. For measurements requiring such correlations, the quality will depend on the accuracy of the correlation in addition to all of the concerns associated with physical measurements.

On the basis of extensive studies carried out on the subjective impression of various sounds on the human ear, several guidelines can be established. Because of the discrepancies among the results from the various researchers, however, one can in general conclude that the accuracy of the results is not very high. The wide variation in results of individual researchers does not help the situation either. In order to lay down specifications for instrumentation, therefore, it has been necessary to compromise. Some mean values that are within the distribution of results, and also are technically feasible, are stipulated.

Since measuring instruments often are used to establish whether a sound level is above or below some preset limit, the accuracy of the instrumentation must be very high even if the premise on which the instrument is developed may be questionable. In the case of aircraft noise certification, for example, 0.1 dB can decide whether an aircraft is permitted to operate from a large international airport: 0.1 dB can in this case account for millions of dollars of revenue lost or gained. The need for international standardization is especially important for sound-measuring equipment since it is evident that many parameters admit to a variety of arbitrary selections. Such standardization is achieved through organizations such as the ISO and IEC.

3.2 Building Blocks of a Sound Level Measuring System[2,10]

A wide variety of different systems, some consisting of a number of interconnected instruments, are available for the measurement of sound (pressure) level. Although very different in detail, every system consists essentially of a transducer (microphone) and associated signal conditioning, an analysis section, and a read-out unit, as shown schematically in Fig. 1. The analysis section includes optional frequency weighting and a detector.

***Transducer*[11,12]** A microphone is the usual transducer when the acoustic medium is air, converting the sound pressure into an electrical signal. Any transducer chosen for a particular measurement will have to fulfill two different groups of conditions, however. It must, of course, meet the technical requirements necessary for accurate and repeatable measurements, for example, frequency response, dynamic range, directivity, and stability. In addition, it must operate satisfactorily over a range of environmental conditions such as humidity, temperature, air pollution, and wind.

The *condenser microphone* usually is best able to meet all these conditions and therefore has become the most widely used type. It operates on the well-known principle that the capacitance of two electrically charged plates varies with their separation distance. The charge may be generated either by an external polarizing voltage, or in the case of the *prepolarized condenser microphone* (or *eletret microphone*) by the properties of the material itself. One of the plates is an extremely light diaphragm that moves in response to acoustic pressure variations. The resulting change in capacitance produces a voltage that is then measured. Figure 2 shows a view of a typical condenser microphone with the major components labeled. Because of practical design considerations, the size of the microphone usually is increased in order to achieve higher sensitivity. Unfortunately, this conflicts with the requirements for wide frequency range and omnidirectivity, both of which are better if the diaphragm is small compared to the wavelength of sound being measured.

The frequency response of a microphone suitable for accurate acoustic noise measurements should be flat over

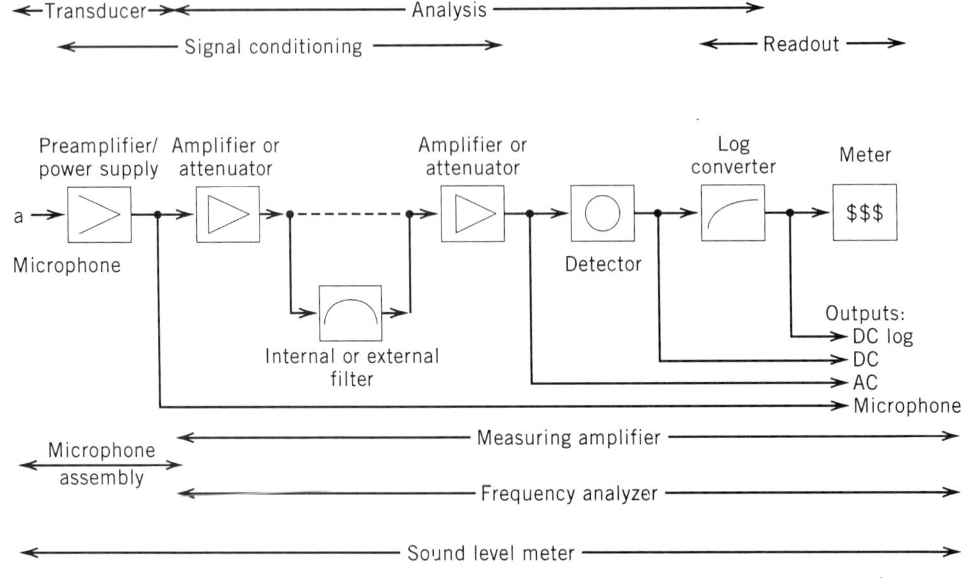

Fig. 1 Block diagram of a basic sound pressure measuring system: transducer, signal conditioning, analysis, and read-out. Sound level meter includes frequency-weighting networks. Measuring amplifier may include frequency-weighting networks and connect to external filter(s). Frequency analyzer includes bandpass filter(s) for frequency analysis. Detectors include peak and root-mean-square (rms) types. The rms detectors may be exponential or linear time-weighting type with selectable averaging time. Frequency-weighting networks are specialized filters.

the frequency range of interest. Usually this means the audible range of the human ear, that is, from about 20 Hz to 15 kHz, but for specific applications it could be extended above or below this range.

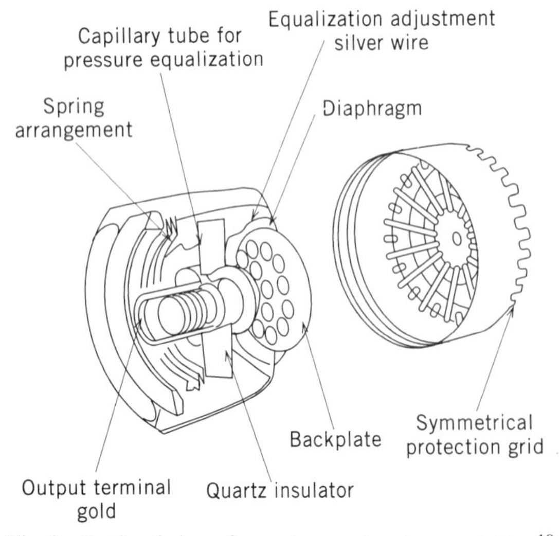

Fig. 2 Sectional view of a condenser microphone cartridge.[10]

Frequency Weighting The simplest noise measurement is the overall sound pressure level. Because this disregards both frequency content and variations with time, it is independent of the two factors that are known to affect subjective reaction to sound as much as the level itself. It is used, typically, mostly when data are being recorded for later laboratory analysis, providing a frequency response limited only by the instrumentation. By weighting the signal spectrum in a way that corresponds to the frequency response of the human ear, however, it is possible to describe a measured sound pressure level by a single value that is more representative of its subjective effects.

Standardized Frequency Weighting.[13,14] The ear is not equally sensitive at all frequencies, and furthermore, the frequency dependency also is a function of the sound level. As early as 1928 Fletcher and Munson[15] established *equal-loudness curves* for pure-tone sounds; later researchers found considerable dependence of the result on the details on the method of determination. Figure 3 shows the now-standardized curves from ISO 226.[16]

Equal-loudness curves provide the rationale for the standardized frequency weightings. Due to the inherent uncertainty, it was decided that just three weightings ought to be sufficient: A, B, and C, corresponding

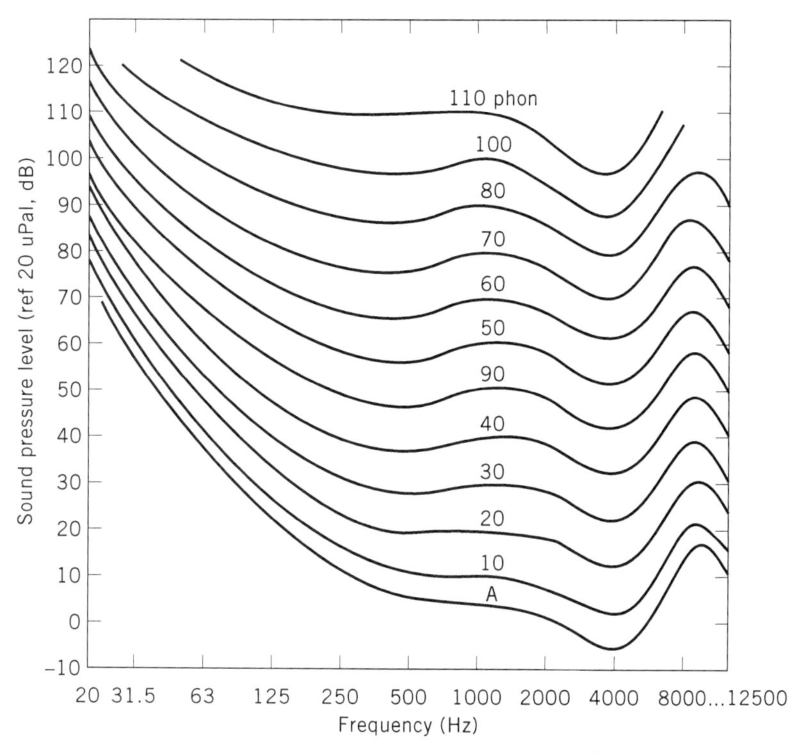

Fig. 3 Standardized equal-loudness curves.[10]

approximately to the inverse of the 40-phon, 70-phon, and 100-phon equal-loudness curves, respectively. The equal-loudness curves appear somewhat irregular, but again recalling the inherent uncertainty, Arnold Petersen in 1948 suggested that for standardization very simple curves be used. The C-curve, to be used for sounds above 85 phons, was to have flat frequency response between 20 Hz and 20 kHz. The B-curve, to be used for sounds in the range 55–85 phons, could be achieved by adding low-frequency attenuation with a filter made up of a single resistor and capacitor. The A-curve, to be used only below 55 phons, could be achieved by adding another *RC* filter to the B-curve.

The B-curve now is little used, but with minor modifications and clarifications the A- and C-weighting curves are the same in use today. Figure 4 shows the standardized frequency-weighting curves. When compared with the equal-loudness curves the discrepancies of the standardized weightings are obvious. More disconcerting, however, is the observation that modern practice is to use the A-curve for almost all sound level measurements, regardless of loudness.

Sound pressure levels that are frequency weighted are termed *sound levels*. Whenever sound level measurements are quoted, both the frequency-weighting and reference pressure should be stated so that no confusion can arise when making later comparisons. An example of a suitable statement is the following: "The A-weighted sound level was 76 dB re 20 μPa."

Detector The frequency-weighting network is followed by a detector that extracts the information of interest from the signal. For most measurements the

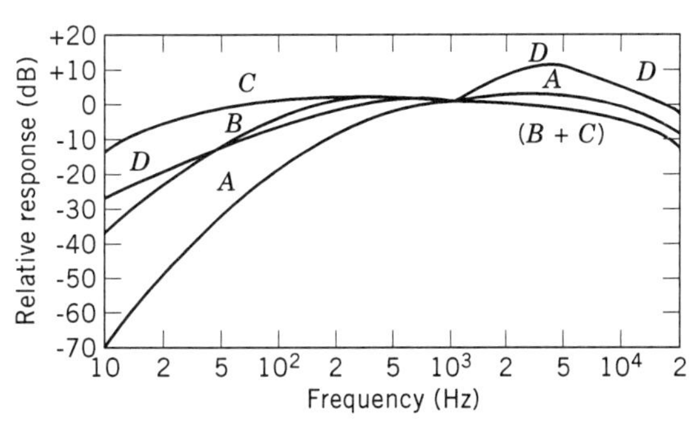

Fig. 4 Standardized frequency-weighting curves.[10]

detector is a root-mean-square (rms) detector that integrates all frequency components on a mean-square (often called "energy" for convenience of description) basis. For most measurements a *true rms detector* is required, although some low-cost equipment may employ a *rectified average detector* (often called simply an *average detector*). Unlike the true rms detector, however, the rectified average detector is sensitive to the phase relationships between any frequency components present in the signal. A *peak detector* is necessary if the highest instantaneous value of a signal is to be measured.

Time Weighting. The time weighting determines the effective limits of the signal on which the detector operates. Type and time usually must be specified. For rms detectors, exponential (*RC*) and linear averaging types are common. The time weighting of exponential averaging is specified by the *time constant,* while that of linear averaging is specified by *averaging time.*

Standard sound level meters provide exponential averaging rms detection with two time constants: 125 ms and 1 s, called "F" and "S" for fast and slow, respectively. Impulse sound level meters have an additional "I" time constant, discussed below. Integrating-averaging sound level meters provide linear averaging rms detection with specified *integration (averaging) times.*

It is instructive to note that the F and S time constants were selected for practical reasons, and not for any particular relevance to the hearing mechanism of the human ear. The 125 ms is simply the time constant of the fastest moving coil meter that was available in 1950, and 1 s was chosen exclusively to allow more time for accurate meter reading.

The I time constant is 35 ms for rising levels and 3 s for decreasing levels. The 35 ms time constant was selected as a compromise based on an elementary understanding of the hearing mechanism. The 3 s time constant was introduced for strictly practical reasons to allow reading the meter when a single impulse is measured.

4 SUMMARY

Continuing advancement in instrumentation has led to more sophisticated measurements utilizing ever-increasing quantities of data. There is every indication that this trend will continue. The challenges will remain making measurements of sufficient accuracy and extracting the relevant data.

REFERENCES

1. F. V. Hunt, *Origins in Acoustics*, Acoustical Society of America, Woodbury, NY, 1992.
2. L. L. Beranek, *Acoustical Measurements*, Acoustical Society of America, Woodbury, NY, 1988.
3. F. V. Hunt, *Electroacoustics*, Acoustical Society of America, Woodbury, NY, 1982.
4. J. W. S. Rayleigh, *The Theory of Sound*, Dover, New York, 1945.
5. A. Wood, *Acoustics*, Dover, New York, 1966.
6. American National Standards Institute, 1430 Broadway, New York, NY 10018.
7. American Society for Testing and Materials, 1916 Race Street, Philadelphia, PA 19103.
8. International Organization for Standardization, 1 rue de Varembé, Geneva, Switzerland.
9. International Electrotechnical Commission, 3 rue de Varembé, Geneva, Switzerland.
10. J. R. Hassall and K. Zaveri, *Acoustic Noise Measurements*, Brüel & Kjær, Nærum, Denmark, 1979.
11. G. S. K. Wong and T. F. W. Embleton (Eds.), *AIP Handbook of Condenser Microphones*, American Institute of Physics, Woodbury, NY, 1995.
12. Anonymous, *Microphone Handbook*, Brüel & Kjær, Nærum, Denmark, 1995.
13. ANSI S1.42, "American National Standard Design Response of Weighting Networks for Acoustical Measurements," American National Standards Institute, New York, 1982.
14. IEC 179, "Sound Level Meters," International Electrotechnical Commission, Geneva, Switzerland, 1973, IEC 651-1979 and IEC 1672-1997.
15. H. Fletcher and W. A. Munson, "Loudness, Its Definition, Measurement and Calculation," *J. Acoust. Soc. Am.*, Vol. 5, 1933, pp. 82–108.
16. ISO R 226, "Normal Equal Loudness Contours for Pure Tones and Normal Threshold of Hearing under Free Field Listening Conditions," International Standardization Organization, Geneva, Switzerland, 1987.

105

SOUND LEVEL METERS

ROBERT W. KRUG

1 INTRODUCTION

Sound level meters (SLMs) are designed to measure sound over a range of frequencies and levels comparable to the range of the human ear. The human ear may respond to frequencies from 20 to 20,000 Hz and pressure changes in excess of $1–1 \times 10^7$. The SLMs measure and display changes in acoustic pressures in a systematic and reproducible manner. Pressures are compressed logarithmically so that $1–1 \times 10^7$ is expressed as 0–140 dB. The display, whether a ballistic meter movement or a digital display that updates only a few times a second, does not display acoustic pressure changes instantaneously. Instead, it averages the changes in one of several methods to produce a readable number. The two principal methods are exponential-averaging and integrating-averaging. Optional features that may be included in sound level meters include peak level, maximum level, minimum level, noise dose, sound exposure, events, and exceedance levels. Meters may log data for statistical distributions and time histories. Filter sets for measuring sound over a restricted frequency range may be included.

1.1 Principles of Operation

Exponential-averaging meters (see Fig. 1) measure sound pressure from a microphone. The amplifier and weighting circuit limit the frequencies to a prescribed range. The signal is then squared, so positive and negative pressure changes are converted to the square of the input signal.

Note: References to chapters appearing only in the *Encyclopedia* are preceded by *Enc.*

The time constant is a single-pole low-pass filter with an exponential time prescribed time constant. The meter may be graduated directly in decibels, which compresses the low end of the scale and expands the high end, or the logarithm of the signal is taken to produce a linear display in decibels.

Integrating-averaging SLMs (see Fig. 2) detect, frequency weight, and square the sound pressure level similar to exponential-averaging SLMs. The squared signal is integrated. The logarithm of the integrated signal has the logarithm of time subtracted from it. The level is then displayed on a meter movement or digital display.

1.2 Standards

Standards are required to measure sound in a systematic and reproducible manner. Any number of frequency weightings, time constants, and integration methods could be used. Standards prescribe specific methods of measuring sound.

Principal standards applicable to sound level meters are as follows:

ANSI S1.4-1983[1]	Specification for sound level meters
ANSI S1.25-1978, 1991[2,3]	Specification for personal noise dosimeters
IEC 651-1979[4]	Sound level meters
IEC 804-1984[5]	Integrating-averaging sound level meters
IEC 1252-1992[6]	Specifications for personal sound exposure meters

All of the above standards are being revised and new standards will be issued.

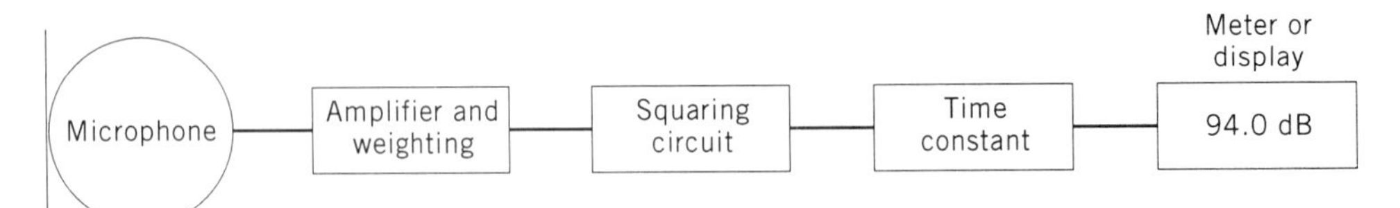

Fig. 1 Exponential-averaging SLM.

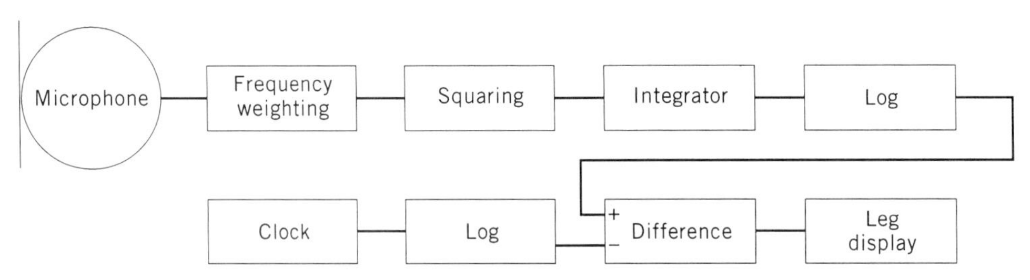

Fig. 2 Intergating-averaging SLM.

The American National Standards Institute ANSI S1.4-1978 and the International Electrotechnical Commission IEC 651-1979 both specify exponential-averaging SLMs. They are similar except for the directional specifications of the microphone (see Section 2.2). Both specify three accuracy types of meters: type 0, type 1, and type 2. Type 0 instruments are the most accurate and are intended as laboratory standards. Type 1 instruments are intended for laboratory or field use where the acoustic environment can be closely specified and/or controlled. Type 2 SLMs are suitable for general field applications. Type 3 SLMs are specified in IEC 651-1979. They are of low accuracy and are intended for sound surveys.

The IEC 804-1984 specifies type 0, type 1, type 2, and type 3 integrating-averaging SLMs. The personal noise dosimeter and the personal sound exposure meter specify instruments intended to be worn on a person to measure noise exposure. They specify only limited-range, single-frequency-weighting, type 2 instruments.

2 MICROPHONES

Microphones are transducers that convert changes in acoustic pressure to an electrical signal. As is typical of many transducers, the microphone is likely to be the most critical, fragile, error prone, and expensive part of the instrument. Its response is likely to change with frequency, direction, temperature, absolute pressure, time, and several other factors. As such, it is important to appreciate the limitations and work with them to achieve accurate and reliable response (see Chapters 108 and 112).

2.1 Types

Sound level meters use several different microphone types. Three types are most commonly used: piezoelectric (ceramic) and two types of air condensers (polarized and electret).

Ceramic Ceramic microphones operate on the principle that piezoelectric materials will generate a voltage when subject to force changes. Ceramics are relatively low cost, have low internal electrical noise, and compared to other microphones, are quite rugged. They are quite sensitive to vibrations. The frequency response of ceramic microphones is not as flat as other types, and as a result, they are often only used with type 2 SLMs.

Condenser The capacitance will change if the separation between two plates is changed. Condenser or capacitor microphones are constructed with one plate of the capacitor made with a very light material that can move when subjected to changes in air pressure. A high voltage is connected with a high impedance to the other plate of the capacitor. Since charge on a capacitor is equal to the capacitance times the voltage, if the capacitance changes and the charge remains constant, the voltage also will change.

Condenser microphones tend to be accurate and stable and have a wide, well-defined frequency bandwidth. They are fragile and must be protected from hostile environments and high humidity.

Polarized air condenser microphones require an external voltage often in the range of 200 V. Sensitivity is a function of this voltage so it must be well regulated.

Permanent charged or electret microphones have permanent charge built into a plastic membrane. Response is similar to the polarized microphones.

2.2 Microphone Response

Microphone response changes with frequency. Microphones roll off below a couple of hertz to zero sensitivity at 0 Hz. At high frequencies the microphone diaphragm is an appreciable part of a wavelength. As a result, the microphone rolls off at high frequencies and changes with the direction in which the microphone is pointed relative to the sound source.

Microphones can be designed to be close to flat with frequency for sounds arriving from one direction but will not be flat for sound arriving from another. It should not be surprising that microphones with smaller dimensions have better high-frequency response but also less sensitivity.

Free Field Free-field microphones are intended to measure sound in an open space free from reflections. The microphone should be pointed directly at the noise source at a 0° incidence. At 0° incidence the frequency response is close to flat over the widest frequency range. High-frequency sound arriving from other angles may be somewhat attenuated (see Fig. 3). The IEC specifies SLMs with a free-field microphone.

Random Incidence Random-microphone microphones are intended to measure sound in a diffuse field where the sound is arriving from all directions such as inside a noisy plant or in an area with many reflections. Random-incidence microphones have the flattest response if pointed at about a 70° angle to the noise source. High-frequency noise arriving at angles less than 70° will cause the SLM to read somewhat high, while angles greater than 70° will generally read somewhat low (see Fig. 3). The ANSI specifies that SLMs shall use random-incidence microphones.

Pressure Pressure microphones are intended to measure sound in a closed cavity. Its response is often similar to the response of a random-incidence microphone.

The SLMs are sometimes designed with special circuitry to correct a free-field microphone for random incidence.

3 FREQUENCY WEIGHTING

3.1 Standard Weightings

SLM standards IEC 651-1979 and ANSI S1.4-1983 specify standard frequency weighting networks. A number of describers have been assigned to frequency-weighting networks. A weighting, B weighting and C weighting are specified in IEC 651-1979 and ANSI S1.4-1983. D weighting is used to measure aircraft engines. A number of others have been proposed and used (see Fig. 4).

A weighting is intended to simulate a nominal human ear at 40 phons. It is also considered by many regulations in many countries to be the best weighting for predict-

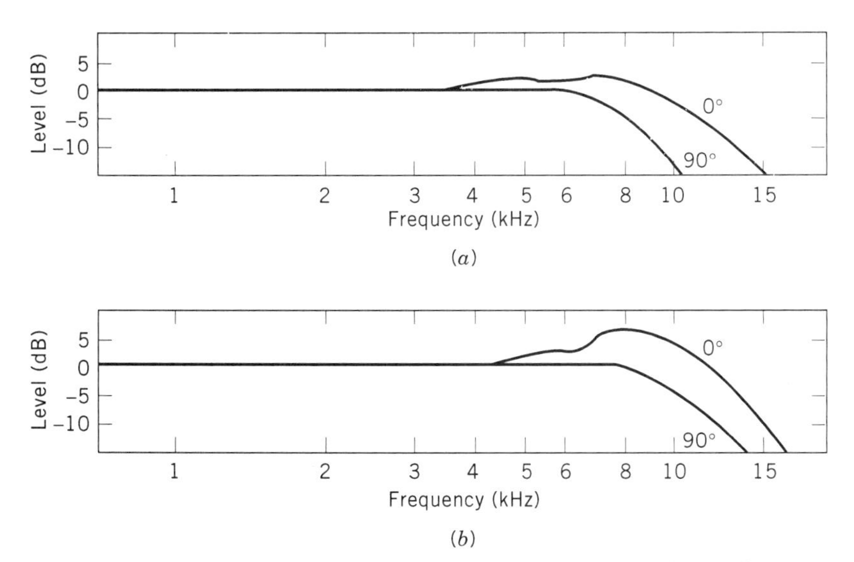

Fig. 3 (*a*) Free-field and (*b*) random-incidence microphones at 0° and 90°.

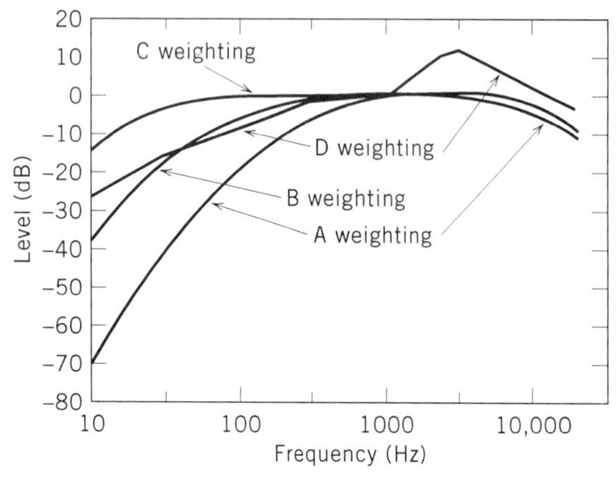

Fig. 4 Frequency-weighting curves.

ing hearing loss due to noise exposure. The SLMs, personal sound exposure meters, and noise dosimeters use A weighting to determine the effects of noise on humans. See Eq. (2).

B weighting is intended to simulate a nominal human ear at 70 phons. It is not widely used and may be dropped from future standards. See Eq. (3).

C weighting is intended to simulate the nominal human ear at 100 phons. It is flat over most of the audible frequencies and is down 3 dB at 31.6 Hz and 8000 Hz. Since it is flat over the audible range, it is often used to measure acoustic emission of machinery. It is also used to specify hearing protectors and to measure peak sound pressure level. See Eq. (1).

D weighting was developed to measure noise from jet aircraft that have a perceived noise level that is higher than the level measured with A weighting. It is not widely used and may be dropped from future standards. See Eq. (3).

Flat or linear response is sometimes included in SLMs. It is not a standard weighting network. Its frequency response is normally flat between two frequencies. The frequency response is often similar to C weighting, with the response rolling off at the low end at a couple of hertz and at the high end at several tens of kilohertz.

3.2 Equations

The equations for frequency responses for A weighting, B weighting, and C weighting are

$$W_C = 20\log\left(\frac{A_1(fF_4)^2}{(f^2+F_1^2)(f^2+F_4^2)}\right), \tag{1}$$

$$W_A = W_C + 20\log\left(\frac{A_2 f^2}{\sqrt{(f^2+F_2^2)(f^2+F_3^2)}}\right), \tag{2}$$

$$W_B = W_C + 20\log\left(\frac{A_3 f}{\sqrt{(f^2+F_5^2)}}\right), \tag{3}$$

where f is frequency in hertz,

$A_1 = 1.007152$, $A_2 = 1.249936$, $A_3 = 1.012482$,
$F_1 = 20.598997$ Hz, $F_2 = 107.65265$ Hz,
$F_3 = 737.86223$ Hz, $F_4 = 12194.217$ Hz,
$F_5 = 158.48932$ Hz,

and $A_1, \ldots, A_3$ were chosen such that $W_x = 0$ dB at 1000 Hz.

4 SQUARING AND AVERAGING

The instantaneous level at the input of the squaring circuit is converted to a level proportional to the square of the level. It is interesting to note that if the input pressure varies over 60 dB or 1000 : 1, the output must vary over a range of 1–1 × 10^6. Considering 60 dB is the minimum range for type 1 integrating SLMs, it is not surprising that designers use a number of clever designs in this critical circuit.

The parameter p_a at the output of the squaring circuit can be found as

$$L_{\text{inst}} = 10\log(p_a^2/p_0^2) \tag{4}$$

where L_{inst} is the instantaneous level in decibels and p_a^2/p_0^2 is the ratio of the squared instantaneous A-weighted sound pressure to the squared reference sound pressure. The reference sound pressure is 20 μPa.

4.1 Exponential Averaging

A low-pass filter is placed after the squaring circuit to smooth out the instantaneous fluctuation and make it possible to read the level on a meter or digital display. Fast and slow responses are specified by several SLM standards and impulse response is specified by IEC 651-1979. (See Fig. 5.)

Time Constants *Slow* (or S) is specified as a 1-s time constant. Slow is used for measuring sound where an estimate of the average sound level is needed and the fluctuations are too fast to follow with a fast time con-

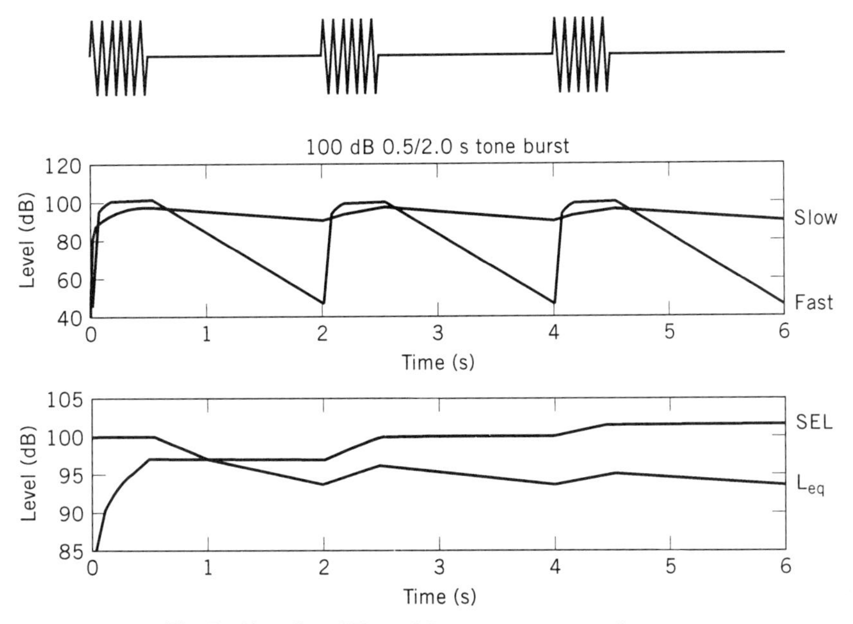

Fig. 5 Slow, fast, SEL, and L_{eq} responses to tone burst.

stant. Slow is specified by the Occupational Safety and Health Administration (OSHA) and the Department of Defense (DOD) in the United States for measuring noise dose.

Fast (or *F*) is specified as a 0.125-s time constant. Fast follows fluctuations in sound levels better than slow but may fluctuate too much to be read. Fast is often used to measure transient noise such as vehicle pass-by noise.

Impulse (or *I*) is specified in IEC 651-1979 as a 0.035-s time constant followed by a peak detector with a 1.5-s decay time such that the indicator will rise very rapidly to increasing levels but decay slowly when the level decreases. It is used primarily in Germany and a few other countries to measure highly impulsive noise.

4.2 Peak, Maximum, or Minimum Hold

The SLMs often include additional functions to measure the highest instantaneous level as well as the highest and lowest level after the time constant circuit. These functions can normally be reset to allow detection of the next level.

Peak is the highest instantaneous level. Peak is detected before the time constant circuit. It may have a different frequency weighting than the weighting used to calculate sound pressure level and other levels. Peak signals respond to pulses as short as 50 μs. For a sine wave, the peak level is 3 dB higher than the r.m.s. level. The tone burst shown in Fig. 5 would have a peak reading of 103 dB.

Maximum and minimum levels are the highest and lowest levels detected after the time constant circuit. The tone burst shown in Fig. 5 has, after a few cycles, a fast maximum of 100 dB and a slow maximum of 96.6 dB. The minimum level for fast is 65.3 dB and for slow is 90.1 dB.

Response to Tone Burst On a logarithmic scale, the time constant will cause the display to change faster for an increase in level than for a decrease in level [see Eqs. (5) and (6)].

The maximum rates of decay for a big decrease in level are 4.34 dB/s for slow, 34.74 dB/s for fast, and 2.90 dB/s for impulse. The maximum response to a single tone burst relative to the continuous level of the same tone is

$$L_{on} = L_{tb} - 10\log(1 - e^{-T_1/\tau}). \tag{5}$$

The response when a tone is turned off is

$$L_{off} = L_{on} - 10\log(e^{-T_2/\tau}),$$

$$L_{off} = L_{on} - 4.3429448\,\frac{T_2}{\tau}. \tag{6}$$

The integrated sound level of an integral number of tone bursts is

$$L_{eq} = L_{tb} - 10\log\left(\frac{T_1}{T}\right), \qquad (7)$$

where L_{tb} is the level of tone burst, T_1 the duration of tone burst, T_2 the time from the end of a continuous tone, T the time from the start of one tone burst to the start of the next tone burst (=1 per repetition frequency), and τ an exponential time constant.

4.3 Integrating Averaging

Equivalent Sound Pressure Level Integrating SLMs, combine all the sound pressures between the start and end of an integration period. [See Eq. (8).] The equivalent sound pressure level (L_{eq}) is the logarithmic average of the squared instantaneous pressures. While the exponential-averaging SLM display depends on what the sound level was immediately preceding the time it was read, integrating-averaging SLMs weighs equally all the sound from the start to the end of the integration time. As can be seen in Fig. 5, exponential-averaged SLMs keep changing when measuring discontinuous sound. Equivalent sound pressure level meters quickly settle to an average level. As a rule of thumb, the L_{eq} level should be within 0.1 dB of the final value after as many minutes as the time between impulses in seconds.

In theory, integrating SLMs do not have a time constant. In practice, some do exponentially average the squared pressure before doing the integration. If the integration time period is long compared to the time constant, the exponential circuit makes little difference. If a loud noise occurred shortly before the start time or shortly before the end time, part of that noise may be included or excluded:

$$L_{eq} = L_{A_{eq},T} = 10\log\frac{1}{T}\int_{T_1}^{T_2}\frac{p_a^2(t)}{p_0^2}\,dt, \qquad (8)$$

$$\mathrm{SEL} = L_{EA,T} = 10\log\int_{T_1}^{T_2}\frac{p_a^2(t)}{p_0^2}\,dt, \qquad (9)$$

where p_a is the instantaneous A-weighted sound pressure (for weighting other than A weighting the subscript is changed) and T is the time between the start and end times T_1 and T_2 (for the sound exposure level SEL the time is always in seconds).

Sound Exposure Level The SEL is similar to L_{eq} in that pressure is integrated over the measurement period [See Eq. (9).] The SEL measures the total energy and normalizes it to 1 s. The SEL is equal to L_{eq} after 1 s. If the L_{eq} is steady with time, the SEL will be 3 dB greater than L_{eq} after 2 s and 6 dB greater after 4 s, and increase by 3 dB whenever time doubles. The SEL is used to measure the total energy in an event independent of the time duration of the event.

Short Equivalent Sound Pressure Level Short L_{eq} is an L_{eq} value computed at very short intervals, perhaps every $\frac{1}{8}$ or $\frac{1}{16}$ s, and stored in memory for later analysis. Data in memory can then be used to calculate a number of sound describers such as L_{eq}, SEL, exposure, exceedance levels, sound pressure level, and maximum and minimum levels.

Equivalent Sound Pressure Level per Day and Time-Weighted Average The term $L_{eq,D}$ is the equivalent sound level per day, and TWA is the time-weighted average. Both $L_{eq,D}$ and TWA are similar to SEL except the integration is averaged over 8 h instead of 1 s. The level will be less than the L_{eq} level for time periods less than 8 h and greater than the L_{eq} level after 8 h. The $L_{eq,D}$ and TWA are used to measure worker noise exposure during a work day. It is assumed that workers who work less than 8 h can be exposed to higher noise level during the time they are working without increasing their risk of noise-induced hearing loss. The parameter $L_{eq,D}$ is used in the European community while TWA is used in the United States. If the level is constant, $L_{eq,D}$ will increase by 3 dB if the time is doubled. Also, $L_{eq,D}$ is 44.59 dB less than SEL. The TWA may use different doubling or exchange rates. For OSHA compliance, it will increase by 5 dB if the time is doubled.

5 OPERATING RANGE

Ideally, an SLM should accurately measure all sounds from the noise floor to overload regardless of the temporal nature of the sound. While SLMs are available that accurately measure all sounds over a wide range, many SLMs that meet standards have very limited range, particularly when measuring transients.

In the SLM standards IEC 651-1979 and ANSI S1.4-1983, the requirements for steady and transient response are very limited. Accurate steady-level response is only required over a 10-dB range, and transient response is only required for one tone burst. As many instruments made since these standards were issued barely meet the minimum requirements of the standards, much of the data taken with these instruments may be of little value.

The standards IEC 651-1979 and ANSI S1.4-1983 state that the accuracy of instruments shall be within

±0.4, ±0.7, and ±1.0 dB for type 0, 1, and 2 instruments, respectively. This accuracy is required only for steady-tone sounds arriving from the reference direction under reference conditions of pressure, temperature, and humidity. Also, it is required only on one primary indicator range over a 10-dB span.

The SLM standards requirements for tone burst require only that type 1 and type 2 instruments be within tolerance at one point. On fast response only 0.2-s tone bursts and on slow response only 0.5-s tone bursts are required to be accurately measured. Unfortunately, there are many instruments on the market that only meet the standard for this one tone burst. If a noise burst is shorter than specified by the standard, the instruments read low or ignore the burst completely.

The standard IEC 804-1984 expands the linearity range to 70, 60, and 50 dB for type 0, 1, and 2 instruments, respectively. It also requires tone bursts, as short as 0.001 s to be measured within ±0.5, ±0.5, and ±1.0 dB for type 0, 1, and 2 instruments, respectively. The proposed drafts for personal sound exposure meters and noise dosimeters follow the IEC 804-1984 type 2 specifications. While this range may not be accurate for very short duration sound burst or for wide level changes, it is sufficient for most noise measurements.

5.1 Linearity Range

The linearity range is specified for continuous sound pressure levels. Ideally the instrument should meet the linearity requirements on all ranges, although additional tolerance is allowed on ranges other than the primary indicator range. Instruments are required to be linear over a range of frequencies from 31.5 Hz to 8 kHz.

5.2 Dynamic Range

Dynamic range, often called pulse range, determines the meter's performance when the sound level is not continuous. Ideally it should be the same as the linearity range. Integrating SLMs are required to have 70, 60, or 50 dB pulse range for type 0, 1, and 2 m, respectively. Exponential-averaging meters may be very limited in dynamic range capabilities.

5.3 Overload Indicators

Integrating-averaging SLMs are required and exponential-averaging SLMs are requested to have overload detectors to indicate when an overload has occurred. The overload can occur at several places in the meter, and the overload detector may be required to measure several points. It is not unusual for the overload indicator to indicate an overload while the average sound level is below the upper limit of the range.

5.4 Noise Floor

The lowest level an instrument can read is determined by the noise floor. Noise may be generated in the microphone, its preamplifier, or the electronics of the meter. The lowest level an SLM can measure with 1-dB accuracy is 6 dB above the noise floor. For 0.1-dB accuracy the level must be 16 dB above the noise floor. If the noise floor is well behaved, it is possible to measure lower levels and calculate the actual noise level by logarithmically subtracting the noise from the signal. This technique is most useful when measuring sound that can be turned on and off, such as calibrating pure-tone audiometers [see Eq. (10)]:

$$L = 10 \log(10^{(L+N)/10} - 10^{N/10}), \qquad (10)$$

where N is the noise floor, $L + N$ the measured level including the noise, and L the level without the noise.

5.5 Threshold Circuits

It is sometimes desirable not to include sounds below a threshold level in the noise calculations. An example is excluding background noise when measuring the noise of an aircraft fly-over. The OSHA regulations allow exclusion of noise below certain threshold levels.

Thresholds of noise dosimeters are placed after the time constant circuit and are well defined. Other threshold levels are not well defined, and the user must consult the meter-operating manual to determine exactly how the threshold works.

6 DISPLAYS AND OUTPUTS

6.1 Analog and Digital

Conventional SLMs use an analog ballistic meter movement to measure the sound level. Historically, the conversion to decibels was accomplished by scale graduations. This resulted in a scale with the numbers very close together on the left or lower end and separated on the high or upper end. Such scales were only useful over about a 20-dB range. When meter manufacturers included the logarithmic conversion in the meter electronics, analog meter scales of 30 or more decibels are possible with equal space per decibel. Although digital displays are now widespread, many users still prefer the analog display for ease of reading and detecting changes.

Digital displays offer wide range, and the numbers are easy to read down to a fraction of a decibel. The display is difficult to read and interpret when used with exponential-averaging meters. If the sound level is not constant, the display will be different each time it updates. It is also necessary to consult the owner's manual to determine if the displayed number is the maximum level, the average level, or the level at the end of the update. Since a display that updates once a second can change 30 or more decibels during that second, it is important to know what is displayed.

6.2 AC and DC Outputs

Auxiliary outputs are sometimes provided to connect the SLM to other equipment. Consult the operating manual before connecting to make sure the two pieces of equipment are compatible.

An AC output connector can be used to connect a SLM to a recorder or headset. The AC connector is normally connected after the frequency-weighting filter. For the best recording, the flat or linear frequency response should be used. If this weighting is not available, use C weighting. Since the SLM may have a dynamic span of 100 dB or more, it may exceed the capabilities of all but the highest quality recorders.

A DC output may be provided. It may have an exponential-averaged time constant or be another time averaging. Traditionally, the DC output was to connect to a chart recorder. Logging SLMs are rapidly making chart recorders obsolete. For special applications, the DC output may be used to sense a level and operate an external switch or control. The DC output is normally in volts per decibel.

6.3 Printer and Computer Outputs

Digital electronics and microcomputer circuits now allow vast quantities of information to be stored in the SLM. Second by second, short L_{eq}'s and a complete summary of extensive sound calculations may be stored in the meters and transferred to a printer or computer. No SLM standards exist for transferring data from the meter. Some manufacturers use proprietary hardware and software methods. Others transfer data as ASCII text files over a RS-232 port for serial communications and Centronics parallel for parallel printer connections. Several manufacturers have extensive software libraries for post-processing.

7 CALIBRATION

It is recommended that SLMs be calibrated before every use and checked after every use with an acoustic calibrator. Manufacturers' recommendations for daily calibration should be followed. Normally, calibration consists of removing the windscreen, carefully sliding the calibrator over the microphone, and reading the level. Care must be taken to ensure there is a good seal between the microphone and the calibrator or errors will result. If calibration has changed by a few tenths of a decibel, it can normally be adjusted to the correct level.

Acoustic calibrators check only the calibration at one or a few frequencies and levels. It is possible for a calibration not to change at one frequency and be out of tolerance at another. If a microphone develops a pin hole air leak, the level may not change dramatically at 1000 Hz but may change considerably at other frequencies. Therefore, if daily calibrations require more than a few tenths of a decibel of corrections, it may be a sign of a much larger error at other frequencies.

Good procedure recommends SLMs be checked yearly to ensure the instrument remains within the tolerance required by standards.

8 OTHER TYPES OF SOUND LEVEL METERS

8.1 Dosimeters and Personal Sound Exposure Meters

A special type of meter for measuring noise is the dosimeter or personal sound exposure meter. A block diagram of a noise dosimeter is shown in Fig. 6. An exposure meter would have a similar block diagram except the time constant, threshold, and exponent circuits would be deleted.

The instruments measure noise according to

$$E = \int_0^T p_a^2 \, dt, \tag{11}$$

$$\text{Dose} = \frac{100}{T_C} \int_0^T 2^{(L - L_C)/\text{ER}} \, dt, \tag{12}$$

where E is exposure, p_a the A-weighted sound pressure, T_C the criterion time (normally 8 h), L_C the criterion level, and ER the exchange rate or doubling. In SI units, exposure is measured in pascal-squared seconds. For convenience, exposure in personal sound exposure meters is expressed in pascal-squared hours. Exposure to 85 dB for 8 h is approximately equal to one pascal-squared hour.

Dose is another method of measuring noise exposure, with the results expressed as a percentage of a criterion exposure. The Occupational Safety and Health Administration (OSHA), in the United States, defines 100% dose as the equivalent of a level of 90 dB for 8 h with a 5-dB exchange rate after the time constant. Other regulations may use different criterion levels and exchange rates. Since dose combines different levels with 5-dB (OSHA)

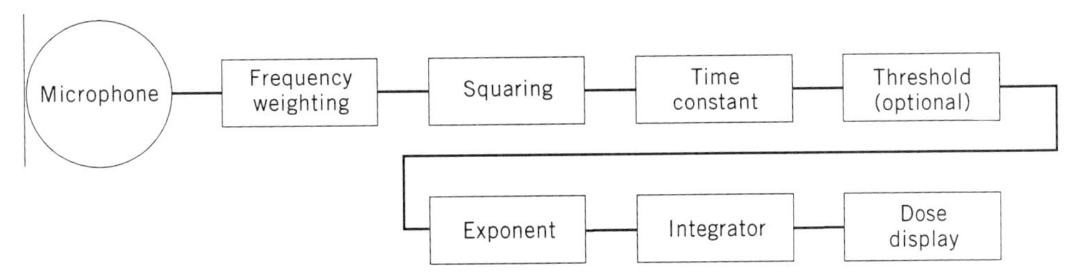

Fig. 6 Noise dosimeter.

or 4-dB (DOD) exchange rates, after the time constant circuit, the time constant will change the calculated dose. A slow time constant will produce a higher dose than a fast or no time constant if the sound level is not constant.

While Eqs. (11) and (12) appear quite different, they are similar. If the exchange rate is 3 dB, the criterion level is 90 dB, and criterion time is 8 h, then 100% dose = 3.2 $Pa^2 \cdot h$.

9 OPTIONS

9.1 Filter Sets

Filters are used to divide the audio spectrum into smaller frequency bands for analysis. The SLMs measure sound over the full audio spectrum but provide little information about what is happening at discrete frequencies. Filters may be added to some SLMs to measure sound in certain frequency bands and reject sound in others.

The most common filters sets used with SLMs are octave and one-third-octave band filter sets. The term octave is used because there are eight notes in a musical scale. One filter in a set has a center frequency of 1000 Hz. Other filters in an octave filter set are either 1000 Hz times a power of 2 or more commonly $1000 \times 10^{0.3N}$, where N is a positive or negative integer. Other filters in a one-third-octave filter set are either 1000 Hz $\times$ $2N/3$ or more commonly $10^{0.1N}$. One-third-octave band filters using base-10 separation have center frequencies of 10, 100, 1000, and 10,000 Hz, whereas base-2 filters are slightly different.

An ideal octave band filter with a center frequency of 1000 Hz would have no attenuation between 707 and 1414 Hz and infinite attenuation at all other frequencies. The next octave band would have a center frequency of 2000 Hz and pass frequencies between 1414 and 2828.

An ideal 1000-Hz, one-third-octave band filter would pass all frequencies between 891 and 1122 Hz and reject all others. The one-third-octave band above and below would have center frequencies of 794 and 1260 and pass frequencies of 707–891 and 1122–1414 Hz, respectively. Three one-third-octave band filters have the same bandwidth as a single-octave band filter.

Real filters do not have zero attenuation in the bandpass and infinite attenuation elsewhere. There are IEC and ANSI filter standards that prescribe shapes of filters and tolerances.

When a filter set is added to a sound level, the response may or may not be modified by the frequency weighting of the meter. The flat response on linear weighting is normally used unless there is a specific need to modify the response with A weighting or another weighting network.

9.2 Additional Options

A number of additional devices may be connected to SLMs to expand their performance. Some of these are as follows:

- Accelerometers
- Acceleration, velocity, displacement, and jerk modules
- Reverberation time module
- Special frequency-weighting networks
- Frequency counters
- Tachometers
- Power supplies
- Modems

10 MULTIFUNCTION SOUND LEVEL METERS

Historically, SLMs were analog devices that measured a single sound level and indicated that level on a meter. Microcomputers and digital electronics have expanded the capabilities to allow multiple functions to be calculated, stored, and displayed in parallel. A multifunction SLM as shown in Fig. 7 may have many of the individual blocks moved into the microcomputer.

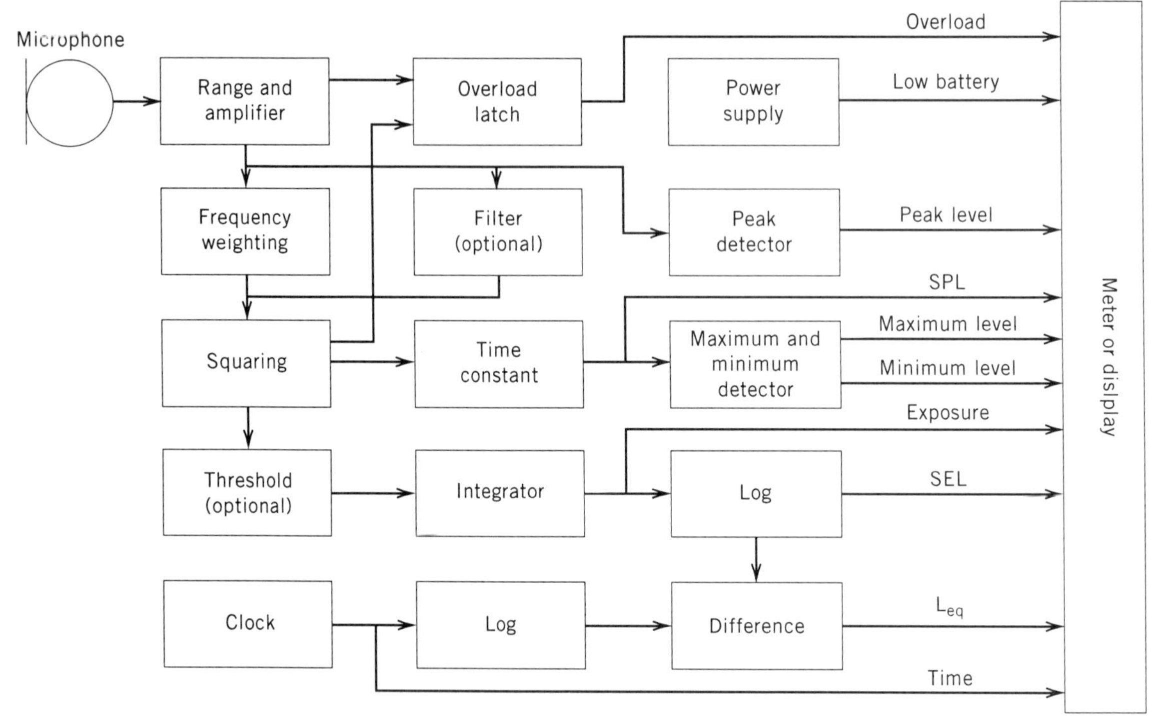

Fig. 7 Multifunction SLM.

10.1 Logging Sound Level Meters

Sound level meters may store levels in increments as short as a fraction of a second for periods in excess of 24 h. A time history graph may then be printed, plotted, or transferred to a computer for a detailed record. Computer software programs are available to reprocess and analyze the record.

10.2 Events

Sound level meters may store a series of events with multiple functions calculated for each event. An event may start and stop when the user presses a button, automatically at discrete times, or when the sound pressure level exceeds a threshold level. Each event may be printed or transferred to a computer for analysis.

10.3 Exceedance Levels and Statistical Distributions

Exceedance levels are the level exceeded a given percent of the time. The L_1 level indicates the level exceeded 1% of the time while L_{50} is the level exceeded 50% of the time. If exceedance levels are measured manually, the user must write down the sound level at discrete intervals until a large number of readings have been taken, arrange the levels in order, and select the individual levels. Exceedance level programs automate this process.

REFERENCES

1. ANSI S1.4-1983, "Specification for Sound Level Meters," Acoustical Society of America, New York, 1983.
2. ANSI S1.25-1978, "Specification for Personal Noise Dosimeters," Acoustical Society of America, New York, 1978.
3. ANSI S1.25-1991, "Specification for Personal Noise Dosimeters," Acoustical Society of America, New York, 1991.
4. IEC 651-1979, "Sound Level Meters," International Electotechnical Commission, Geneva, Switzerland, 1979.
5. IEC 804-1984, "Integrating-averaging Sound Level Meters," International Electrotechnical Commission, Geneva, Switzerland, 1984.
6. IEC 1252-1992, Electrocoustics, "Specifications for Personal Sound Exposure Meters," International Electrotechnical Commission, Geneva, Switzerland, 1992.

106

SOUND INTENSITY

MALCOLM J. CROCKER AND FINN JACOBSEN

1 INTRODUCTION

Sound intensity is a measure of the magnitude and direction of the flow of sound energy. Although acousticians have attempted to measure sound intensity since the 1930s, the first reliable measurement of sound intensity did not occur until the late 1970s, when the convergence of theoretical and experimental advances, including the derivation of the cross-spectral formulation for sound intensity and developments in digital signal processing, propelled sound intensity measurements from the laboratory into practical use. Most modern measurements of sound intensity are made using the simultaneous measurement of the sound pressure with two closely spaced microphones.

The sound intensity **I** is a vector quantity and is defined as the time average of the net flow of sound energy through a unit area in a direction perpendicular to the area. The dimensions of the intensity are sound energy per unit time per unit area (watts per square metre). For sound energy to be conserved, the sound power generated by a source must be equal to the normal component of the sound intensity integrated over any surface that completely encloses the source. This holds even in the presence of other sources outside the surface. A central point in noise control engineering is to determine the sound power radiated by sources. The value and relevance of determining the sound power radiated by a source is due to this quantity being largely independent of the surroundings of the source in the audible frequency range. Sound intensity measurements make it possible to determine the sound power of sources without the use of costly special facilities such as anechoic and reverberation rooms.

Sound intensity measurements seem to be most useful for the determination of the sound power of large machinery in situ, for noise source identification, and for measurement of the sound transmission loss of partitions. Care must be taken to sample the sound intensity in the sound field appropriately in order to reduce errors. Calibration should be undertaken as recommended by the manufacturers and standards bodies.

2 THEORETICAL BACKGROUND

Sound fields are usually described in terms of the sound pressure, which is the quantity we hear. However, sound fields are also energy fields, in which kinetic and potential acoustic energies are generated, transmitted, and dissipated. At any point in a sound field the instantaneous intensity vector $\mathbf{I}(t) = p(t)\mathbf{u}(t)$ expresses the magnitude and direction of the instantaneous flow of sound energy, as briefly shown in the following.

By combining the fundamental equations that govern a sound field, the equation of mass continuity, the relation between sound pressure and density change, and Euler's equation of motion (Newton's second law), one can derive the equation

$$\nabla \cdot \mathbf{I}(t) + \frac{\partial w(t)}{\partial t} = 0, \tag{1}$$

where $\nabla \cdot \mathbf{I}(t)$ is the divergence of the instantaneous sound intensity (that is, the instantaneous net outflow of sound energy per unit volume) and $w(t)$ is the total instantaneous energy density.[1,2] This is the equation of

Note: References to chapters appearing only in the *Encyclopedia* are preceded by *Enc.*

conservation of sound energy, which expresses the simple fact that the rate of increase (or decrease) of the sound energy density at a given position in a sound field (represented by the second term) is equal to the rate of the flow of converging (or diverging) sound energy (represented by the first term). The global version of this equation is obtained using Gauss's theorem:

$$\int_V \nabla \cdot \mathbf{I}(t)\, dV = \int_S \mathbf{I}(t) \cdot d\mathbf{S} = -\frac{\partial}{\partial t}\left(\int_V w(t)\, dV\right), \qquad (2)$$

where S is the area of a surface around the source and V is the volume contained by the surface. The three terms above represent, respectively, the local net outflow of sound energy integrated over the volume, the total net outflow of sound energy through the surface, and the rate of change of the total sound energy within the surface. It can be seen that the rate of change of the sound energy within a closed surface is identical with the surface integral of the normal component of $\mathbf{I}(t)$.

In practice the time-averaged intensity

$$\mathbf{I} = \langle p(t)\mathbf{u}(t)\rangle_t \qquad (3)$$

is more important than the instantaneous intensity. Examination of Eq. (2) leads to the conclusion that the time average of the instantaneous net flow of sound energy out of a given closed surface is zero unless there is generation (or dissipation) of sound power within the surface; in this case the time average of the net flow of sound energy out of a given surface enclosing a sound source is equal to the net sound power of the source. In other words,

$$\int_S \mathbf{I} \cdot d\mathbf{S} = 0, \qquad (4)$$

unless there is a steady source (or a sink) within the surface, irrespective of the presence of steady sources outside the surface, and

$$\int_S \mathbf{I} \cdot d\mathbf{S} = P_a \qquad (5)$$

if the surface encloses a steady source that radiates the sound power P_a, irrespective of the presence of other steady sources outside the surface.

If the sound field is simple harmonic with angular frequency $\omega = 2\pi f$, then the sound intensity in the r-direction is of the form

$$I_r = \langle p\cos(\omega t)u_r\cos(\omega t + \varphi)\rangle_t = \tfrac{1}{2} p u_r \cos\varphi, \qquad (6)$$

where φ is the phase angle between the sound pressure $p(t)$ and the particle velocity in the r-direction $u_r(t)$. (For simplicity we consider only the component in the r-direction here.) It is common practice to rewrite Eq. (6) as

$$I_r = \tfrac{1}{2}\mathrm{Re}\{p u_r^*\}, \qquad (7)$$

where both the sound pressure p and the particle velocity u_r here are complex exponential quantities, and u_r^* denotes the complex conjugate of u_r. The time averaging gives the factor of $\frac{1}{2}$. We note that the use of complex notation is mathematically very convenient and that Eq. (7) gives the same result as Eq. (6).

In a plane progressive wave traveling in the r-direction, the sound pressure p and the particle velocity u_r are in phase ($\varphi = 0$) and related by the characteristic impedance of the medium, ρc, where ρ is the density and c is the speed of sound:

$$p(t) = \rho c u_r(t). \qquad (8)$$

Thus for a plane wave the sound intensity is

$$I_r = \langle p(t)u_r(t)\rangle_t = \langle p^2(t)/\rho c\rangle_t = p_{\mathrm{rms}}^2/\rho c. \qquad (9)$$

In this case the sound intensity is simply related to the mean square sound pressure p_{rms}^2, which can be measured with a single microphone. In most practical cases, however, the sound intensity is not simply related to the sound pressure. Examination of Eq. (3) shows that both the sound pressure and the particle velocity must be estimated simultaneously and that their product must be time averaged. This requires the use of a more complicated device than a single microphone.

3 CHARACTERISTICS OF SOUND FIELDS

The diversity of sound fields encountered in practice is, of course, enormous. However, the sound field near to a sound source has certain well-known characteristics, sound fields generated by many independent sources have other characteristics, a reverberant sound field has different properties, and so on.

We have seen that the sound pressure and the particle velocity are in phase in a plane propagating wave. This is also the case in a free field, sufficiently far from the source that generates the field. Conversely, one of the characteristics of the sound field near a source is that the sound pressure and the particle velocity are partly out of phase (in quadrature). To describe such phenomena one may introduce the concept of *active* and *reactive* sound fields.

In a simple harmonic sound field the particle velocity may, without loss of generality, be divided into two components: one component in phase with the sound pressure and the other component out of phase with the sound pressure.

The *instantaneous active intensity* is the product of the sound pressure and the in-phase component of the particle velocity. This quantity fluctuates at twice the frequency of the sound wave and has a nonzero time average. The time-averaged quantity is the component that is generally referred to simply as *sound intensity*. The (active) sound intensity is associated with net flow of sound energy and has a direction normal to the wavefronts.[3]

The *instantaneous reactive intensity* is the product of the sound pressure and the out-of-phase component of the particle velocity. This quantity fluctuates at twice the frequency of the sound wave and has a time average equal to zero at any point in a sound field.

Very near a sound source, the reactive field is usually stronger than the active field. However, in the absence of reflections, the reactive field diminishes rapidly with increasing distance from sources. Therefore, at a moderate distance from sources, the sound field is dominated by the active field. The extent of the reactive field depends on the frequency and on the dimensions and the radiation characteristics of the sound source; however, in practice, the reactive field may be assumed to be negligible at a distance greater than, say, 0.6 m from the source, provided the reflected sound field is small.

Consider plane sound waves traveling along a hard-walled tube, as illustrated in Fig. 1*a*; only a single frequency is present. The tube is terminated at the right end with a perfect absorber; therefore there is no reflection of sound at the termination of the tube. Under these conditions the pressure and the particle velocity are in phase at every position in the tube, and the spatial distribution of the pressure is in phase with the spatial distribution of the particle velocity, as shown in the figure at two instants of time. The instantaneous intensity is always positive in the direction towards the termination.

In Fig. 1*b* the tube is terminated with material that is partly absorptive. There will be partial reflection at the termination in this example, so that a weaker wave returns from right to left. The two opposite traveling waves add together, giving the pressure distribution shown at two different instants of time. In this case the spatial distribution of the particle velocity is somewhat out of phase with the spatial distribution of the pressure. The two waves interact to give an active intensity that is less than shown in Fig. 1*a*. There is also a reactive component that flows back and forth from right to left; the time average of this component is zero at any point in the tube.

In Fig. 1*c* the tube is terminated with an infinitely hard material. Therefore the waves are perfectly reflected at the termination, and the reflected waves traveling back to the left have the same amplitude as the incident waves traveling to the right. The incident and reflected waves combine to give a standing wave pattern. In this sound field the pressure and the particle velocity are in quadrature at all positions; therefore the time average of the intensity is zero at any point (i.e., the sound field is completely reactive). The spatial distribution of the pressure is 90° out of phase with the spatial distribution of the particle velocity. The magnitude of the reactive intensity varies with the position: at some locations it is maximum, and at other locations it is zero.

In the general case where the sound field cannot be assumed to be simple harmonic, one cannot make an instantaneous separation of the particle velocity into components in phase and in quadrature with the pressure; under some conditions the particle velocity is not fully correlated with the sound pressure.[4] However, the concept of active and reactive sound fields is still useful. Figure 2 shows the result of a measurement at a position about 30 cm (one wavelength) from a small monopole source, a loudspeaker driven with a band of one-third-octave noise. The sound pressure and the particle velocity (multiplied by ρc) are almost identical; therefore the instantaneous intensity is always positive: this is an *active* sound field. In Fig. 3 the result of a similar measurement very near the loudspeaker is shown. In this case the sound pressure and the particle velocity are almost in quadrature (90° out of phase), and as a result the instantaneous intensity fluctuates about zero, that is, sound energy flows back and forth. This is an example of a strongly *reactive* sound field. Finally Fig. 4 shows the result of a measurement in a reverberant room several metres from the loudspeaker. Here the sound pressure and the particle velocity appear to be uncorrelated signals; this is neither an active nor a reactive sound field but rather a *diffuse* sound field.

4 MEASUREMENT OF SOUND INTENSITY

Many researchers have attempted to obtain quantitative measurements of sound intensity by different means in the past 60 years. One method that seems to have met with some success is sometimes known as the surface

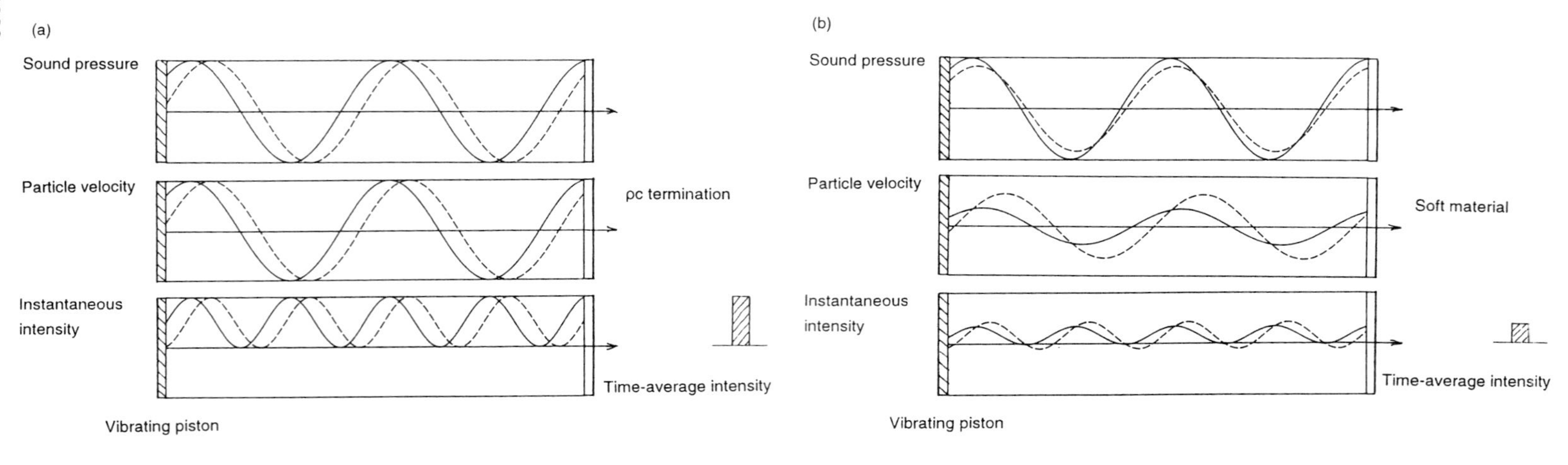

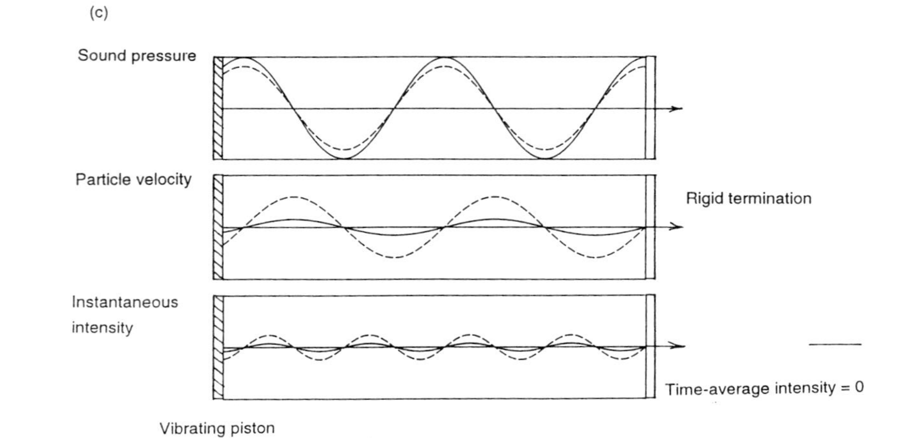

Fig. 1 Spatial distributions of instantaneous sound pressure, instantaneous particle velocity and instantaneous sound intensity for pure-tone one-dimensional sound field in a tube at two different instants of time. (*a*) Case with no reflection at the right end of the tube, (*b*) case with partial reflection at the right end of the tube, (*c*) case with perfect reflection at the right end of the tube.

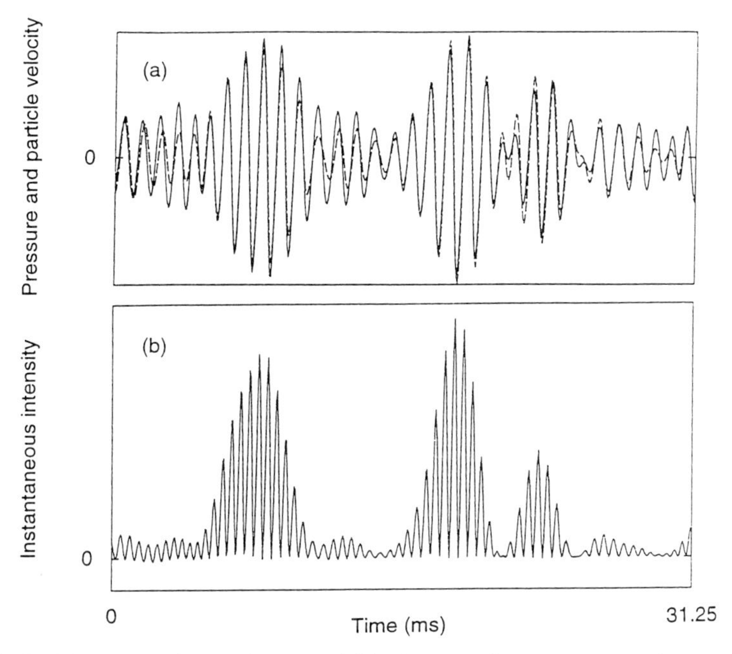

Fig. 2 Measurement in an active sound field: (*a*) ———, instantaneous sound pressure; - - -, instantaneous particle velocity and (*b*) instantaneous sound intensity. (After Jacobsen.[4])

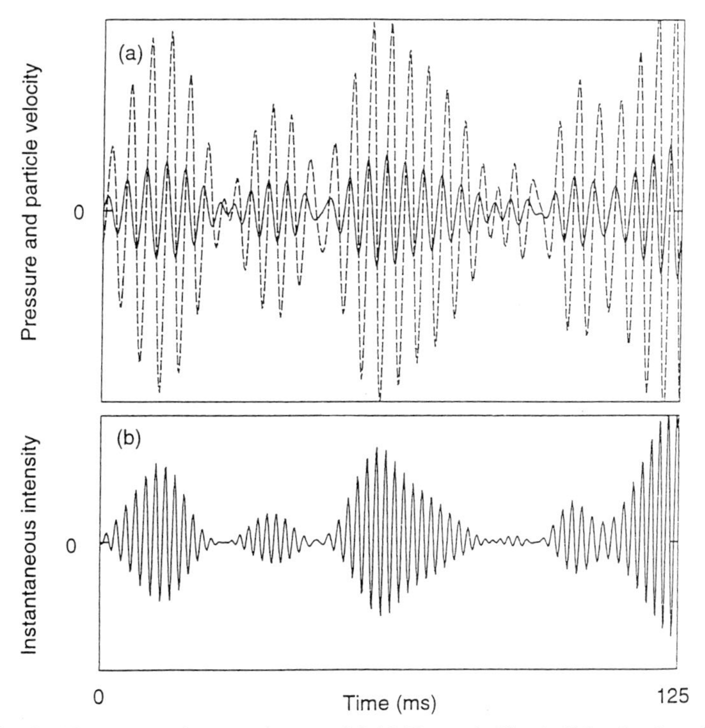

Fig. 3 Measurement in a reactive sound field. Key as in Fig. 2. (After Jacobsen.[4])

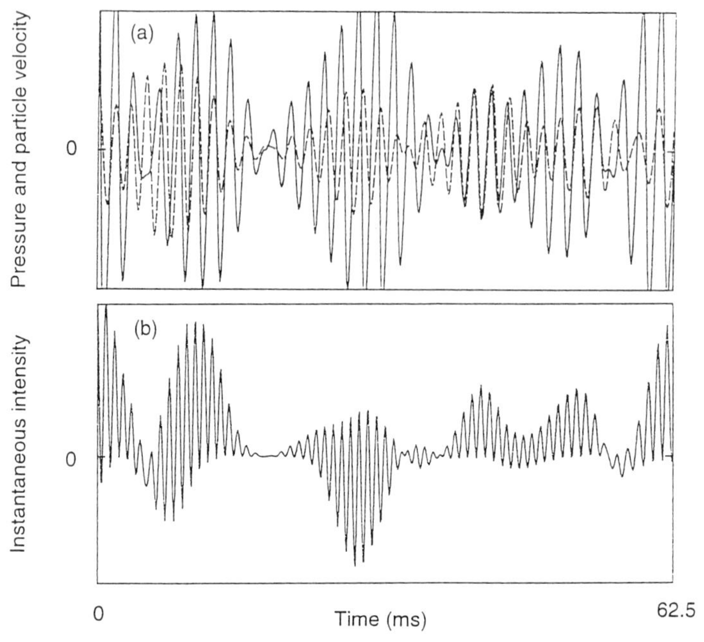

Fig. 4 Measurement in a diffuse sound field. Key as in Fig. 2. (After Jacobsen.[4])

intensity approach. The sound pressure is measured close to some point on a vibrating surface, and the velocity of the surface at the same point is sensed either with an accelerometer or with some noncontacting device such as a capacitance gauge or an optical system such as a laser.[5,6] Some success has been achieved using a microphone/accelerometer approach on the surface of a diesel engine.[7] However, sound fields near complex sources are often very complicated; therefore the technique using a microphone and an accelerometer is time-consuming since a fairly high density of the measurement points is necessary. Besides, the calibration of surface intensity devices remains a difficult problem.[8]

The most successful measurement principle employs two closely spaced pressure microphones.[9] The particle velocity is obtained through Euler's relation,

$$\nabla p(t) + \rho \frac{\partial \mathbf{u}(t)}{\partial t} = 0, \tag{10}$$

as

$$\hat{u}_r(t) = -\frac{1}{\rho} \int_{-\infty}^{t} \frac{p_2(\tau) - p_1(\tau)}{\Delta r}\, d\tau, \tag{11}$$

where p_1 and p_2 are the sound pressure signals from the two microphones, Δr is the microphone separation distance, and τ is a dummy time variable. The caret indicates the finite difference estimate obtained from the "two-microphone approach." The sound pressure at the center of the probe is estimated as

$$\hat{p}(t) = \frac{p_1(t) + p_2(t)}{2}, \tag{12}$$

and the time-averaged intensity component in the axial direction is, from Eqs. (3), (11), and (12),

$$\begin{aligned} \hat{I}_r &= \langle \hat{p}(t)\hat{u}_r(t)\rangle_t \\ &= \frac{1}{2\rho\Delta r} \left\langle (p_1(t) + p_2(t)) \int_{-\infty}^{t} (p_1(\tau) - p_2(\tau))\, d\tau \right\rangle_t . \end{aligned} \tag{13}$$

All intensity measurement systems in commercial production today are based on the two-microphone method. Some commercial intensity analyzers use Eq. (13) to measure the intensity in one-third-octave frequency bands. Another type calculates the intensity from the imaginary part of the cross spectrum G_{12} between the two microphone signals:

$$\hat{I}_r(\omega) = -\frac{1}{\omega\rho\Delta r} \operatorname{Im}\{G_{12}(\omega)\}. \tag{14}$$

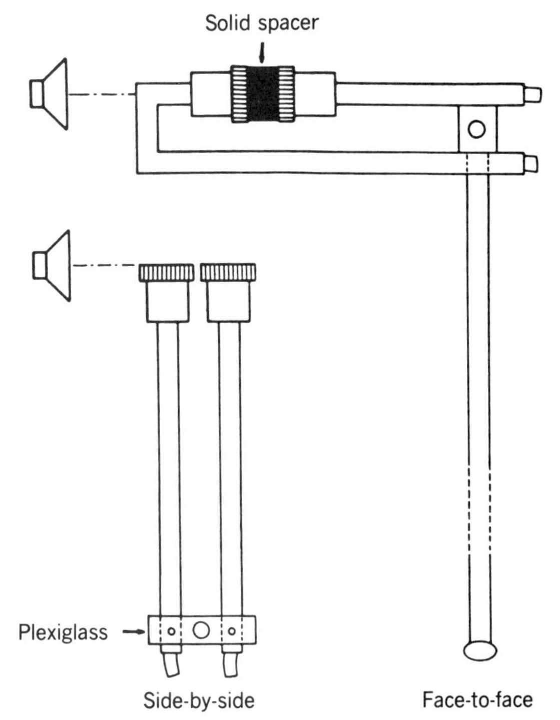

Fig. 5 Microphone arrangements used to measure sound intensity.

The time domain formulation and the frequency domain formulations, Eqs. (13) and (14), are equivalent. Equation (14), which makes it possible to determine sound intensity with a dual channel FFT analyzer, appears to have been derived independently by Fahy and Chung in the late 1970s.[10,11]

Figure 5 shows two of the most common microphone arrangements, "side-by-side" and "face-to-face." The side-by-side arrangement has the advantage that the diaphragms of the microphones can be placed very near a radiating surface, but the disadvantage that the microphones shield each other. At high frequencies the face-to-face configuration with a solid spacer between the microphones is superior.[12]

5 ERRORS AND LIMITATIONS IN MEASUREMENT OF SOUND INTENSITY

There are many sources of error in the measurement of sound intensity, and a considerable part of the sound intensity literature has been concerned with identifying and studying such errors, some of which are fundamental and others which are associated with technical deficiencies.[13,14] One complication is that the accuracy depends very much on the sound field under study; under certain conditions even minute imperfections in the measuring equipment will have a significant influence. Another complication is that small local errors are sometimes amplified into large global errors when the intensity is integrated over a closed surface.[15]

The following is an overview of some of the sources of error in the measurement of sound intensity. A more detailed discussion is given in Ref. 16. Those who make sound intensity measurements should know about the limitations imposed by

- the finite difference error,[12,17] and
- instrumentation phase mismatch.[11,14,18]

Other possible errors that are usually less serious are

- random errors associated with a given finite averaging time,[13,19] and
- bias errors caused by turbulent airflow.[20]

5.1 Errors due to the Finite Difference Approximation

One of the obvious limitations of the measurement principle based on two pressure microphones is the frequency range. The finite difference estimate given by Eq. (13) is accurate only if the distance between the microphones is much less than the wavelength, as suggested in Fig. 6, and this clearly implies an upper frequency limit that is inversely proportional to the distance between the microphones. The finite difference error, that is, the ratio of the measured intensity $\hat{I}_r$ to the true intensity I_r, can be shown to be

$$\hat{I}_r/I_r = \frac{\sin k\Delta r}{k\Delta r} \qquad (15)$$

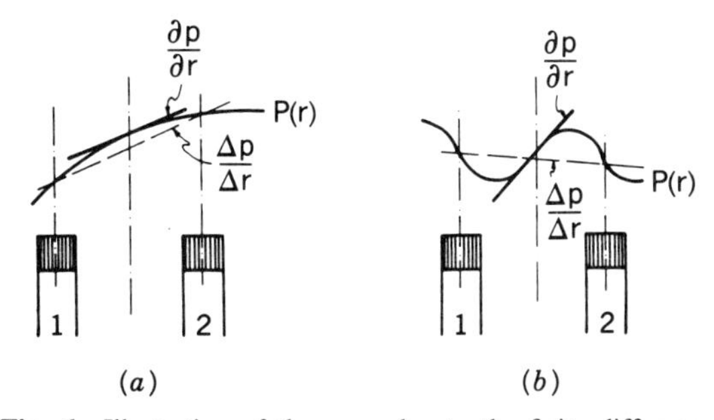

Fig. 6 Illustration of the error due to the finite difference approximation: (*a*) good approximation at a low frequency and (*b*) very poor approximation at a high frequency.

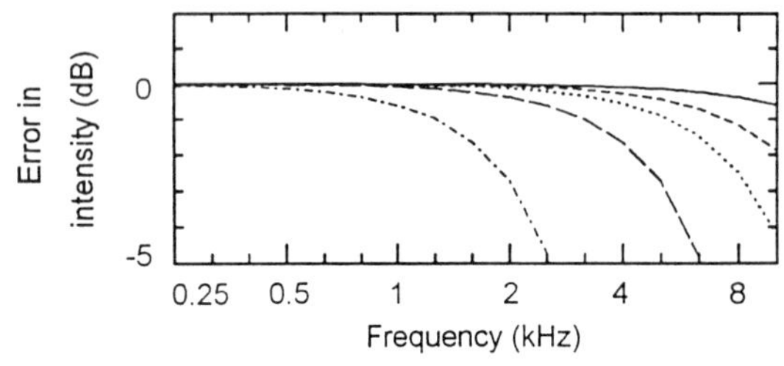

Fig. 7 Finite difference error of an ideal two-microphone sound intensity probe in a plane wave of axial incidence for different values of the separation distance: ——, 5 mm; ---, 8.5 mm; ···, 12 mm; — — —, 20 mm; — · — · —, 50 mm.

for an ideal intensity probe in a plane wave of axial incidence. This relation is shown in Fig. 7 for different values of the microphone separation distance. Although the finite difference error in principle depends on the sound field,[17] the upper frequency limit of intensity probes has generally been considered to be the frequency at which this error is acceptably small.[2] Note, however, that the interference of the microphones on the sound field has been ignored. A recent numerical and experimental study of such interference effects has shown that the upper frequency limit of an intensity probe with the microphones in the usual face-to-face configuration is an octave above the limit determined by the finite difference error if the length of the spacer between the microphones equals the diameter, because the resonance of the cavities in front of the microphones gives rise to a pressure increase that to some extent compensates for the finite difference error. Figure 8, which corresponds to Fig. 7, shows the error calculated for a probe with two 12-mm-long half-inch

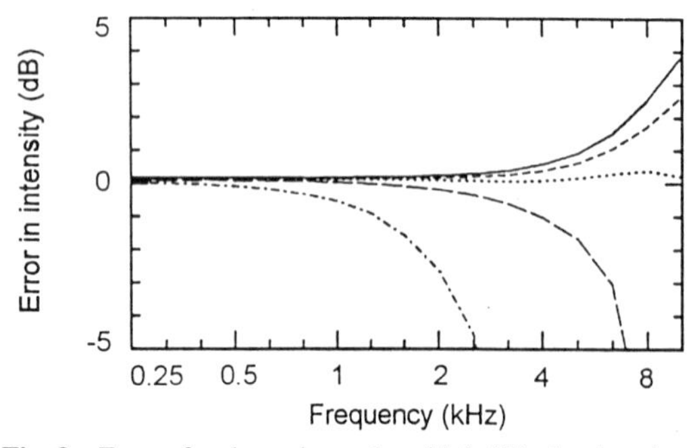

Fig. 8 Error of an intensity probe with half-inch microphones in the face-to-face arrangement in a plane wave of axial incidence for different spacer length: ——, 5 mm; ---, 8.5 mm; ···, 12 mm; — — —, 20 mm; — · — · —, 50 mm. (After Jacobsen et al.[12])

microphones. It is apparent that the optimum length of the spacer is about 12 mm and that a probe with this geometry performs very well up to 10 kHz.[12]

5.2 Instrumentation Phase Mismatch

Phase mismatch between the two measurement channels is the most serious source of error in the measurement of sound intensity, even with the best equipment that is available today. It can be shown that the estimated intensity, subject to a phase error φ_e, to a very good approximation can be written as

$$\hat{I}_r = I_r - \frac{\varphi_e}{k\Delta r}\frac{p_{\text{rms}}^2}{\rho c}; \tag{16}$$

that is, the phase error causes a bias error in the measured intensity that is proportional to the phase error and to the mean square pressure.[18] Ideally the phase error should be zero, of course. In practice one must, even with state-of-the-art equipment, allow for phase errors ranging from about 0.05° at 100 Hz to 2° at 10 kHz. Both the IEC standard and the North American ANSI standard on instruments for the measurement of sound intensity specify performance evaluation tests that ensure that the phase error is within certain limits.

Equation (16) is often written as

$$\hat{I}_r = I_r + I_0(p_{\text{rms}}^2/p_0^2), \tag{17}$$

where the residual intensity I_0 has been introduced. This quantity is the "false" intensity indicated by the instrument when the two microphones are exposed to the same pressure p_0, for instance in a small cavity. Phase mismatch is usually described in terms of the so-called pressure-residual intensity index,

$$\delta_{pI0} = 10\log\left(\frac{p_0^2/\rho c}{|I_0|}\right), \tag{18}$$

which is just a convenient way of measuring, and describing, the phase error φ_e. With a microphone separation distance of 12 mm the typical phase error mentioned above corresponds to a pressure-residual intensity index of 18 dB in most of the frequency range.

Practically all engineering applications that utilize sound intensity measurements involve integrating the normal component of the intensity over a surface. Integrating both sides of Eq. (17) over a measurement surface S gives the expression

$$\hat{P}_a = P_a + \frac{I_0 \rho c}{p_0^2} \int_S (p_{\rm rms}^2/\rho c)\, dS$$

$$= P_a \left(1 + \frac{I_0 \rho c}{p_0^2} \frac{\int_S (p_{\rm rms}^2/\rho c)}{\int_S \mathbf{I} \cdot d\mathbf{S}} \right), \qquad (19)$$

which shows that phase mismatch is of no consequence provided that

$$\Delta_{pI} \ll \delta_{pI0}, \qquad (20)$$

where

$$\Delta_{pI} = 10 \log \left(\int (p_{\rm rms}^2/\rho c)\, dS \Big/ \int \mathbf{I} \cdot d\mathbf{S} \right) \qquad (21)$$

is the global pressure intensity index of the measurement. It can be shown[21] that the resulting error is less than 1 dB if

$$\Delta_{pI} < \delta_{pI0} - 7 \text{ dB}. \qquad (22)$$

It is obvious that the presence of noise sources outside the measurement surface increases the mean square pressure on the surface, and thus the influence of a given phase error; therefore phase mismatch limits the range of measurement. Most modern sound intensity analyzers can determine the global pressure-intensity index concurrently with the actual measurement. Figure 9 shows examples of the index measured under various conditions.[22] In practice one should examine whether the inequality (22) is or is not satisfied if there is significant noise from extraneous sources. If the inequality is not satisfied it can be recommended to use a measurement surface somewhat closer to the source than is advisable in more favorable circumstances. It may also be necessary to modify the measurement conditions—to shield the measurement surface from strong extraneous sources, for example, or to increase the sound absorption present in the room.

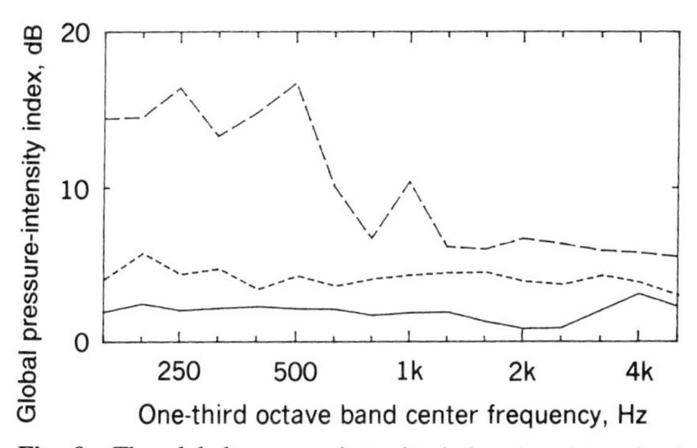

Fig. 9 The global pressure-intensity index Δ_{pI} determined under three different conditions: ———, measurement using a "reasonable" surface; ---, measurement using an eccentric surface; — — —, measurement with strong background noise. (After Jacobsen.[22])

6 APPLICATIONS

Some of the most common practical applications of sound intensity measurements are now discussed briefly.

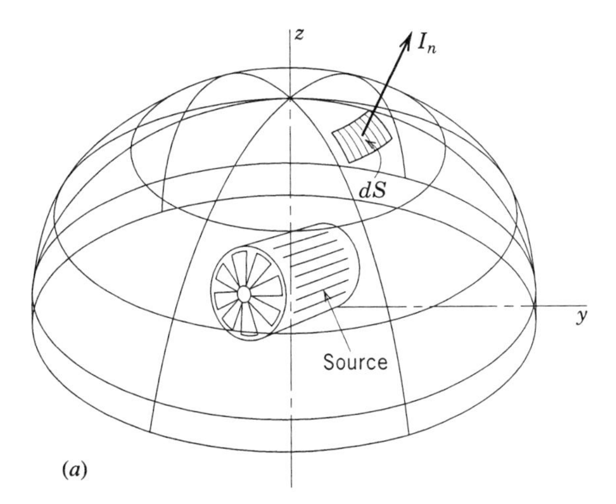

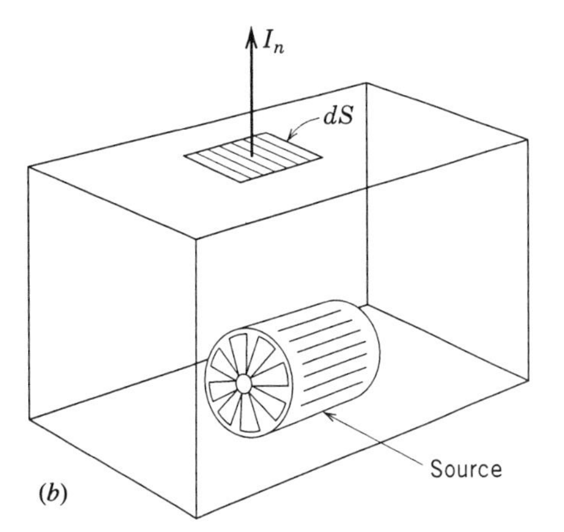

Fig. 10 Sound intensity measured on a segment of (*a*) a hemispherical measurement surface and (*b*) a rectangular measurement surface.

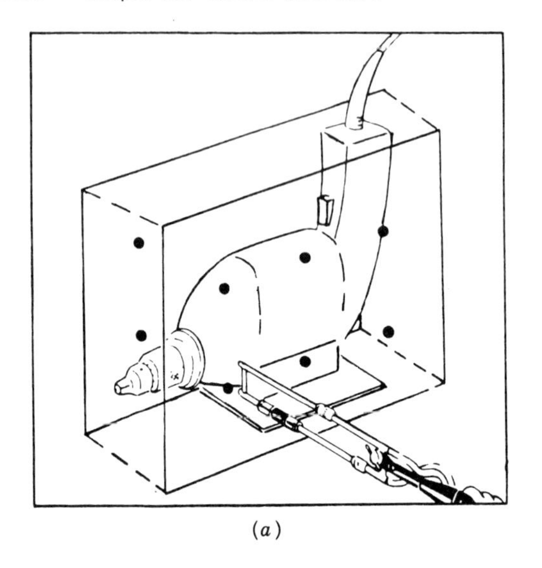

(a)

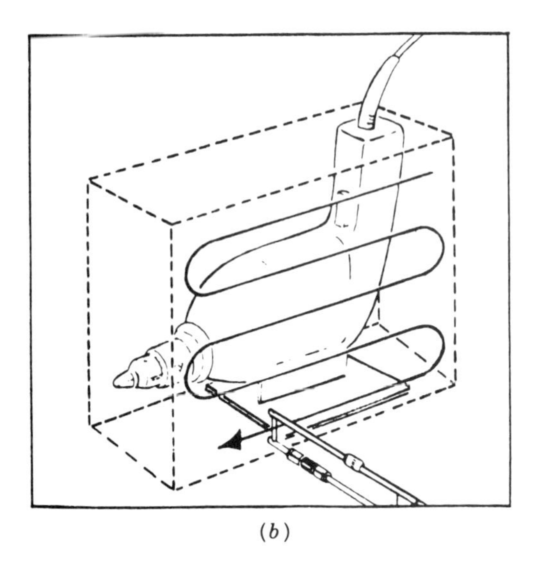

(b)

Fig. 11 Typical box surface used in sound power determination with the intensity method: (a) measurements at discrete points and (b) measurement path used in scanning measurement. (After Rasmussen.[26])

6.1 Sound Power Determination

One important application of sound intensity measurements is the determination of the sound power of operating machinery in situ. Sound power determination using intensity measurements is based on Eq. (5), which shows that the sound power of a source is given by the integral of the normal component of the intensity over a surface that encloses the source. Theoretical considerations seem to indicate the existence of an optimum measurement surface that minimizes measurement errors. In practice one uses a surface of a simple shape at some distance, say, 25–50 cm, from the source, as illustrated in Fig. 10. If there is a strong reverberant field or significant ambient noise from other sources, the measurement surface should be chosen to be somewhat closer to the source under study.

The surface integral can be approximated either by

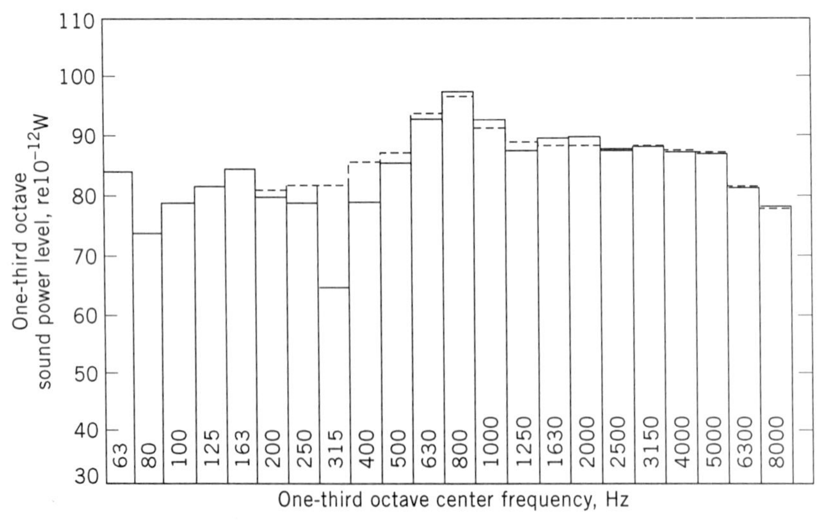

Fig. 12 The sound power of the oil pan of a diesel engine: ———, sound intensity method; ---, lead-wrapping results. (After Reinhart and Crocker.[25])

sampling at discrete points or by scanning manually or with a robot over the surface as shown in Fig. 11. With the scanning approach, the intensity probe is moved continuously over the measurement surface in such a way that the axis of the probe is always perpendicular to the measurement surface. The scanning procedure, which was introduced in the late 1970s on a purely empirical basis, was regarded by some with much skepticism for more than a decade,[23] but is now generally considered to be more accurate and far more convenient than the procedure based on fixed points.[24] A moderate scanning rate, say 0.5 ms^{-1}, and a "reasonable" scan line density should be used, say, 5 cm between adjacent lines if the surface is very close to the source, 20 cm if it is farther away.

6.2 Noise Source Identification

This is perhaps the most important application. Every noise reduction project starts with the identification and

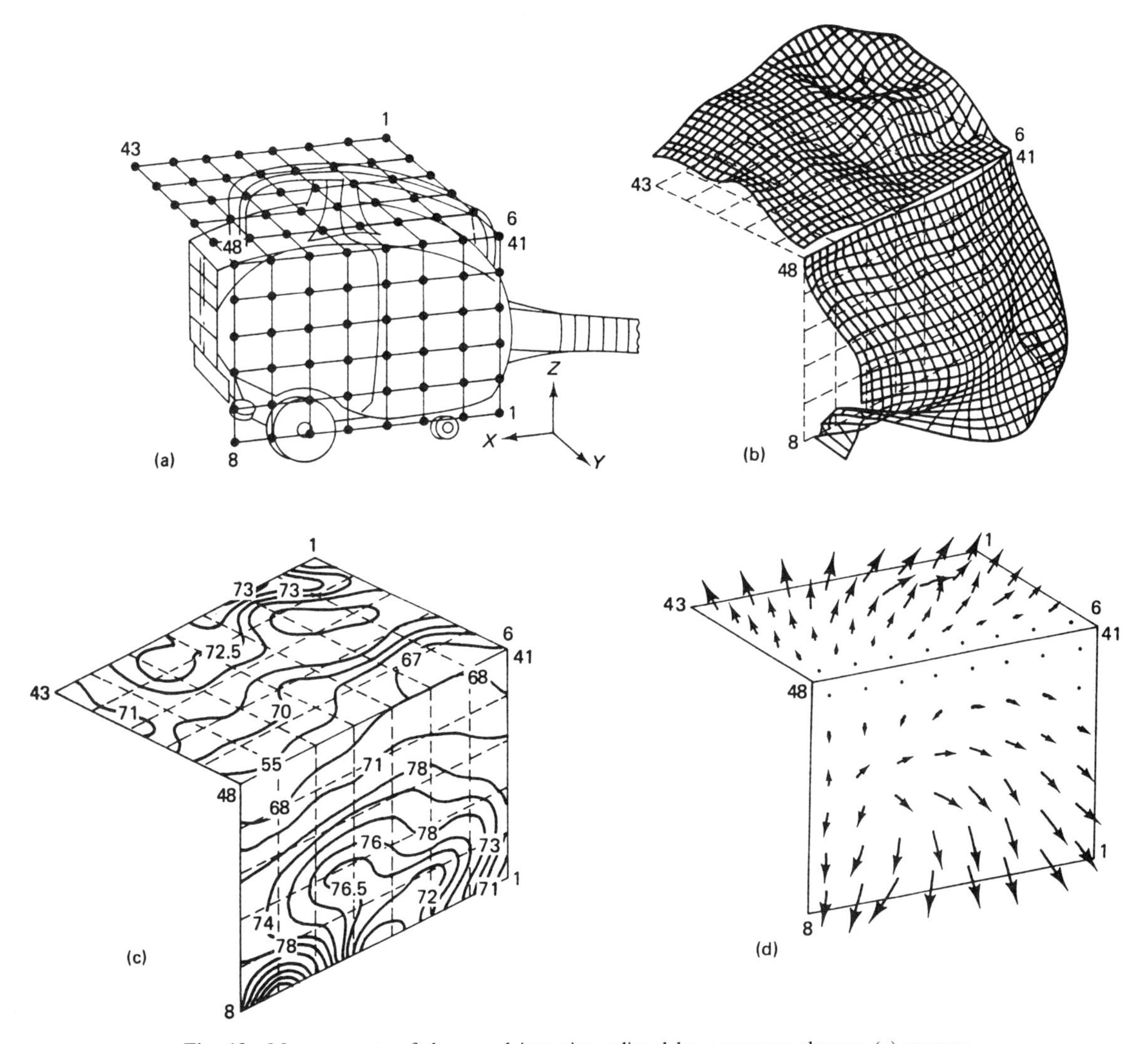

Fig. 13 Measurements of the sound intensity radiated by a vacuum cleaner: (*a*) vacuum cleaner and measurement points (two measurement surfaces each with 48 measurement points), (*b*) mesh diagram showing magnitude of normal intensity on the two surfaces, (*c*) contour plot showing contours of equal normal intensity on the two measurement surfaces, and (*d*) vector intensity flow diagram. (*Source:* RION Technical Notes.)

ranking of noise sources and transmission paths. In one study of diesel engine noise the whole engine was wrapped in sheet lead. The various parts of the engine were then exposed, one at a time, and the sound power was estimated with single microphones at discrete positions on the assumption that $I_r \simeq p_{\text{rms}}^2/\rho c$. Sound intensity measurements make it possible to determine the partial sound power contribution of the various components directly. Figure 12 shows the sound power of the oil pan of the diesel engine determined with the sound intensity scanning approach compared to the result of the old lead wrapping approach.[25]

Plots of the sound intensity measured on a measurement surface near to a machine can be used in locating noise sources. Figure 13*a* shows measurement surfaces near to a vacuum cleaner. Figures 13*b* and *c* show a mesh diagram and a contour plot of the normal intensity, and Fig. 13*d* gives a vector flow diagram of the intensity.

Magnitude plots using arrows are useful in visualizing sound fields. Figure 14 shows the sound intensity measured in the vicinity of a violoncello.[27]

6.3 Transmission Loss of Structures

The conventional measure of the sound insulation of panels and partitions is the transmission loss (also called

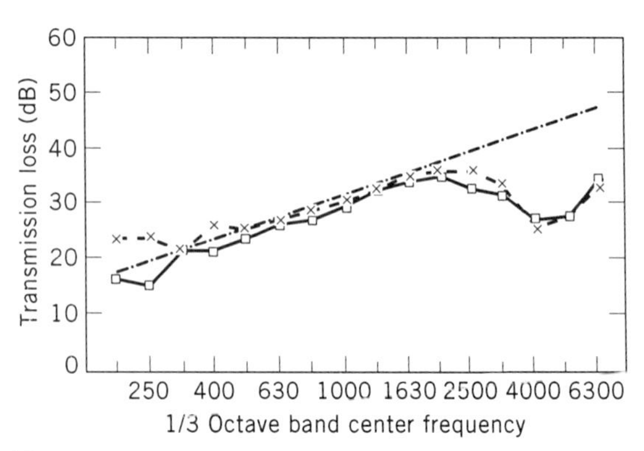

Fig. 15 Transmission loss of a 3.2 mm thick aluminum panel: □–□–□, Sound intensity method; × -- ×, conventional method; — · — · —, mass law. (After Crocker et al.[28])

sound reduction index), which is the ratio of incident and transmitted sound power in logarithmic form. The traditional method of measuring this quantity requires a transmission suite consisting of two vibration-isolated reverberation rooms; the incident sound power is deduced from an estimate of the spatial average of the mean square sound pressure in the source room on the assumption that the sound field is diffuse, and the transmit-

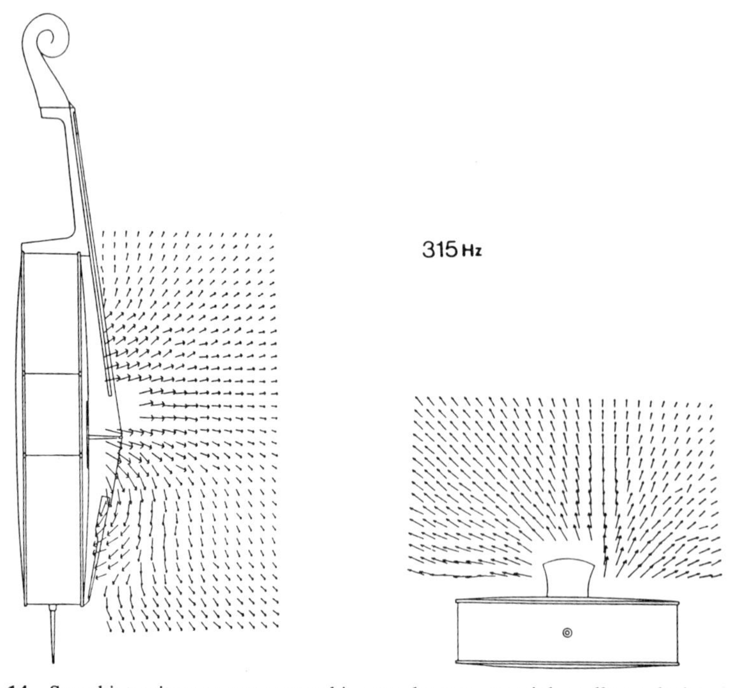

Fig. 14 Sound intensity vectors measured in two planes near a violoncello producing a note having a fundamental frequency of 316 Hz. (*a*) in a plane on the axis and (*b*) in a plane intersecting the bridge. (After Tachibana[27] with permission.)

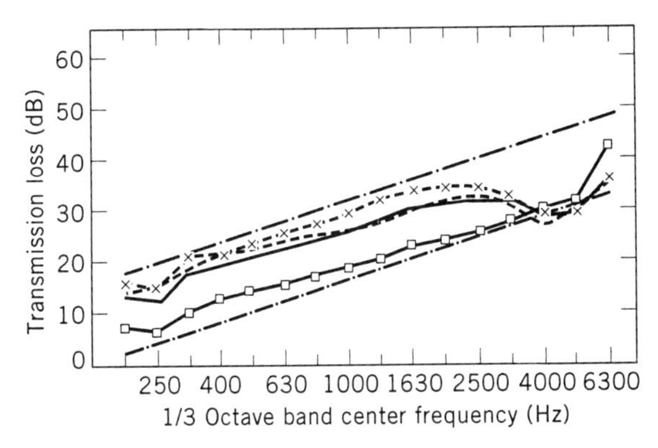

Fig. 16 Measured and calculated transmission loss of a composite aluminum-plexiglass panel. Measured values: × -- × -- ×, aluminum; □–□–□, plexiglass; - - -, total transmission loss. Calculated values; — · — · —, mass law, aluminum; – · – · – · –, mass law, plexiglass; ——— total transmission loss. (After Crocker et al.[28])

ted sound power is determined from a similar measurement in the receiving room where, in addition, the reverberation time must be determined. Alternatively, one can measure the transmitted sound power directly using a sound intensity probe. In this case it is not necessary that the sound field in the receiving room is diffuse, which means that only one reverberation room is necessary.[28] However, the main advantage over the conventional approach is that it is possible to evaluate the transmission loss of individual parts of the partition. The intensity approach can also be used on structures in situ.[29]

Figure 15 shows the transmission loss of an aluminum panel determined with the intensity approach and with the conventional two-room method, and Fig. 16 shows measured and calculated transmission loss values of a composite panel, an aluminum aircraft panel with a plexiglass window.[28]

7 STANDARDS FOR SOUND INTENSITY MEASUREMENTS

Several international and national standards for the measurement of sound intensity have been completed.

ISO (International Organization for Standardization) 9614-1 Acoustics—Determination of Sound Power Levels of Noise Sources Using Sound Intensity—Part 1: Measurement at Discrete Points, 1993.

ISO (International Organization for Standardization) 9614-2 Acoustics—Determination of Sound Power Levels of Noise Sources Using Sound Intensity—Part 2: Measurement by Scanning, 1996.

IEC (International Electrotechnical Commission) 1043 Electroacoustics—Instruments for the Measurement of Sound Intensity, 1993.

ANSI (American National Standards Institute) S12.12-1992 Engineering Method for the Determination of Sound Power Levels of Noise Sources Using Sound Intensity.

ANSI (American National Standards Institute) S1-12-1994 Instruments for the Measurement of Sound Intensity.

There are also several standards under development that give the intensity method as an alternative such as

DIS (Draft international standard) 140 Acoustics—Measurement of Sound Insulation in Buildings and of Building Elements—Part 5: Field Measurements of Airborne Sound Insulation of Facade Elements and Facades.

REFERENCES

1. A. D. Pierce, *Acoustics: An Introduction to Its Physical Principles and Applications.* 2nd ed., Acoustical Society of America, New York, 1989.
2. F. J. Fahy, *Sound Intensity*, 2nd ed., E&FN Spon, London, 1995.
3. J. A. Mann III, J. Tichy, and A. J. Romano, "Instantaneous and Time-Averaged Energy Transfer in Acoustic Fields," *J. Acoust. Soc. Am.*, Vol. 82, 1987, pp. 17–30.
4. F. Jacobsen, "A Note on Instantaneous and Time-Averaged Active and Reactive Sound Intensity," *J. Sound Vibr.*, Vol. 147, 1991, pp. 489–496.
5. J. A. Macadam, "The Measurement of Sound Radiation from Room Surfaces in Lightweight Buildings," *Appl. Acoust.*, Vol. 9, 1976, pp. 103–118.
6. T. H. Hodgson, "Investigation of the Surface Acoustical Intensity Method for Determining the Noise Sound Power of a Large Machine In-Situ," *J. Acoust. Soc. Am.*, Vol. 61, 1977, pp. 487–491.
7. M. C. McGary and M. J. Crocker, "Surface Intensity Measurements on a Diesel Engine," *Noise Control Eng. J.*, Vol. 16, 1981, pp. 27–36.
8. M. C. McGary and M. J. Crocker, "Phase Shift Errors in the Theory and Practice of Surface Intensity Measurements," *J. Sound Vibr.*, Vol. 82, 1982, pp. 275–288.
9. W. P. Waser and M. J. Crocker, "Introduction to the Two-Microphone Cross-Spectral Method of Determining Sound Intensity," *Noise Control Eng. J.*, Vol. 22, 1984, pp. 76–85.
10. F. J. Fahy, "Measurement of Acoustic Intensity Using the

Cross-Spectral Density of Two Microphone Signals," *J. Acoust. Soc. Am.*, Vol. 62, 1977, pp. 1057–1059.

11. J. Y. Chung, "Cross-Spectral Method of Measuring Acoustic Intensity without Error Caused by Instrument Phase Mismatch," *J. Acoust. Soc. Am.*, Vol. 64, 1978, pp. 1613–1616.
12. F. Jacobsen, V. Cutanda, and P. M. Juhl, "Numerical and Experimental Investigation of the Performance of Sound Intensity Probes at High Frequencies," *J. Acoust. Soc. Am.* 103(3), 1998.
13. F. Jacobsen, "Random Errors in Sound Intensity Estimation," *J. Sound Vibr.*, Vol. 128, 1989, pp. 247–257.
14. M. Ren and F. Jacobsen, "Phase Mismatch Errors and Related Indicators in Sound Intensity Measurement," *J. Sound Vibr.*, Vol. 149, 1991, pp. 341–347.
15. J. Pope, "Qualifying Intensity Measurements for Sound Power Determination," *Proc. Inter-Noise 89*, 1989, pp. 1041–1046.
16. F. Jacobsen, "An Overview of the Sources of Error in Sound Power Determination Using the Intensity Technique," *Applied Acoustics 50*, 1997, pp. 155–156.
17. U. S. Shirahatti and M. J. Crocker, "Two-Microphone Finite Difference Approximation Errors in the Interference Fields of Point Dipole Sources," *J. Acoust. Soc. Am.*, Vol. 92, 1992, pp. 258–267.
18. F. Jacobsen, "A Note on the Accuracy of Phase Compensated Intensity Measurements," *J. Sound Vibr.*, Vol. 174, 1994, pp. 140–144.
19. F. Jacobsen, "Sound Intensity Measurement at Low Levels," *J. Sound Vibr.*, Vol. 166, 1993, pp. 195–207.
20. F. Jacobsen, "Intensity Measurements in the Presence of Moderate Airflow," *Proc. Inter-Noise 94*, 1994, pp. 1737–1742.
21. S. Gade, "Validity of Intensity Measurements in Partially Diffuse Sound Field," *Brüel & Kjær Tech. Rev.*, Vol. 4, 1985, pp. 3–31.
22. F. Jacobsen, "Sound Field Indicators: Useful Tools," *Noise Control Eng. J.*, Vol. 35, 1990, pp. 37–46.
23. M. J. Crocker, "Sound Power Determination from Sound Intensity—To Scan or Not to Scan," *Noise Control Eng. J.*, Vol. 27, 1986, p. 67.
24. U. S. Shirahatti and M. J. Crocker, "Studies of the Sound Power Estimation of a Noise Source Using the Two-Microphone Sound Intensity Technique," *Acustica*, Vol. 80, 1994, pp. 378–387.
25. T. E. Reinhart and M. J. Crocker, "Source Identification on a Diesel Engine Using Acoustic Intensity Measurements," *Noise Control Eng. J.*, Vol. 18, 1982, pp. 84–92.
26. G. Rasmussen, "Intensity—Its Measurements and Uses," *Sound Vibr.*, Vol. 23, 1989, pp. 12–21.
27. H. Tachibana, "Visualization of Sound Fields by the Sound Intensity Technique" (in Japanese), *Proc. Second Symp. Acoustic Intensity*, Tokyo, 1987, pp. 117–127.
28. M. J. Crocker, P. K. Raju, and B. Forssen, "Measurement of Transmission Loss of Panels by the Direct Determination of Transmitted Acoustic Intensity," *Noise Control Eng. J.*, Vol. 17, 1981, pp. 6–11.
29. Y. S. Wang and M. J. Crocker, "Direct Measurement of Transmission Loss of Aircraft Structures Using the Acoustic Intensity Approach," *Noise Control Eng. J.*, Vol. 19, 1982, pp. 80–85.

107

ANALYZERS

J. Pope

1 INTRODUCTION

An analyzer is a device that extracts specified information from data signals. The goal, usually, is to present relevant data in a simpler form. Depending on the information to be extracted, various types of analyzers are used in acoustical applications.

A basic acoustic analyzer provides level detection and frequency analysis. The level of interest typically is root mean square or peak, determined over a specified time. Frequency analysis may be performed with a simple filter. Advanced frequency analyzers provide for analysis in multiple frequency bands simultaneously, often using digital techniques.

Specialized analyzers are available for acoustical applications that include statistical level analysis, loudness and other subjective ratings, order analysis, reverberation time determination, drive-by or fly-over measurements, and time delay spectroscopy, among other things. In general, these analyzers build upon and extend the techniques of basic analyzers.

No analyzer performs exactly the analysis theoretically desired. National and international standards have been developed to set minimum requirements that allow comparison of data analyzed using different equipment.

2 FUNDAMENTAL CONCEPTS

2.1 Analysis and Analyzers

Acoustic signals arise from many sources. An analyzer performs a specified analysis on a signal, that is, it extracts specified information, of interest. Often the signal to be analyzed is the output of a transducer, but sometimes an analyzer is deemed to include the transducer. Analyzers may consist of level detectors, frequency analyzers, and more.

The sound level meter (SLM) is the most common acoustic analyzer, although not always identified as such. An SLM allows for detection of the signal root mean square (rms), after broadband filtering (an A or C-weighting network).

A frequency analyzer is used to determine the level of a signal as a function of frequency. A simple frequency analyzer is obtained by addition of a filter set to the basic SLM. More advanced frequency analyzers provide for analysis in multiple frequency bands simultaneously, and often use digital analysis techniques. Frequency analyzers are not limited to filter sets, however. The fast Fourier transform (FFT) algorithm and developments in electronics have made possible the class of frequency analyzers popularly called FFT analyzers. Beyond frequency analysis, more specialized analyzers extend and sometimes combine the techniques of basic filter-set and FFT analyzers.

An analyzer may be a dedicated analog or digital device or a general-purpose computer equipped with appropriate means of signal input and suitable application software. The term analyzer implies simply a device configured to perform analysis. Depending on the analyzer, the user may have more or less ability to control the parameters of the analysis performed. For example, in principle, the parameters of spectrum analysis using an FFT can be selected to optimize the analysis. When using a specific FFT analyzer, however, the user may be limited to a few discrete frequency ranges, four or so data windows, and the particular antialiasing filter supplied. On the other hand, it usually is unreasonable to expect that a single setting of analysis parameters will

Note: References to chapters appearing only in the *Encyclopedia* are preceded by *Enc.*

be suitable for all signals. It is the responsibility of the analyst (i.e., the analyzer user) to ensure that the analysis performed is suitable.

2.2 Detectors[1]

Acoustic signals fluctuate in time. The detector extracts the specified amplitude characteristic of the fluctuation: often the rms, but possibly peak value or some other parameter.

Root-Mean-Square Detectors The rms is the effective average value of a signal. For a signal with zero mean, the usual case for acoustical signals, the rms is equal to the standard deviation (Chapter 81).

The rms is used in preference to other measures of "average" because of its unique properties. Simply stated, if the signal is a voltage, the rms is that voltage that would on a steady (DC) basis dissipate the same power in a resistor. Furthermore, for a signal with multiple fluctuating components, the rms value is independent of the phasing of the components.

Lower quality detectors sometimes employ peak or mean-rectified ("average") detectors calibrated for rms by assuming a sinusoidal signal. For most acoustical measurements, however, it is essential that a true rms detector be used when the rms is desired. For signals that are not sinusoids the relation between rms and other measures differs from that for sinusoids. Given a signal with multiple sinusoidal frequency components, for example, the composite signal is not a sinusoid, and while the peak and mean-rectified values depend on the phase relationship of the various components the rms value does not.

It should be noted that some so-called rms detectors compute the time average of a squared signal, and would more properly be called mean-square (ms) detectors. When the detector output is expressed as a level in decibels, however, the distinction is moot.

The process of rms detection implies a time-averaging operation. The two types of time averaging in common use are *linear* and *exponential*. Linear averaging is used when rms detection is to be restricted to a well-defined time interval. Exponential averaging permits continuous tracking of a signal with amplitude that varies with time. Fig. 1 illustrates rms detection schematically.

Linear Averaging The linear time average weights all data in the average equally. For a signal $x(t)$:

$$x_{\rm rms}(t;T) = \sqrt{\frac{1}{T}\int_{t-T}^{t} x^2 / \, dt}, \qquad (1)$$

where $x_{\rm rms}(t;T)$ is the rms computed at time t over previous interval (and averaging time) T, and $/$ is the dummy

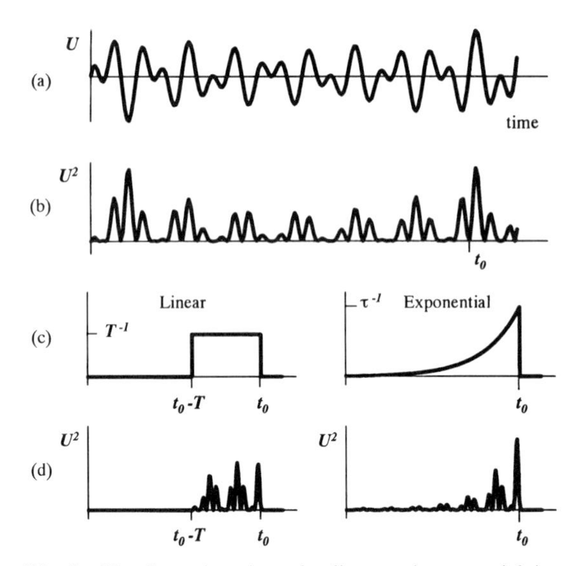

Fig. 1 Signal rms detection using linear and exponential time averaging: (*a*) signal; (*b*) squared signal; (*c*) time-weighting function, T and τ are averaging time and time constant, respectively (see text); and (*d*) squared signal multiplied by time-weighting function at time t_0, signal rms at time t_0 is the square root of the area under the curve. (Note: Time axis scale of *c* and *d* is compressed relative to *a* and *b*.)

variable of integration. While $x_{\rm rms}(t;T)$ is defined as a continuous function of time, due to hardware limitations most analyzers compute $x_{\rm rms}(t;T)$ only at a discrete set of times with spacing not less than T.

Exponential Averaging The exponential time average gives recent data more weight than older data. For a signal $x(t)$:

$$x_{\rm rms}(t;\tau) = \sqrt{\frac{1}{\tau}\int_{-\infty}^{t} x^2(t) e^{-(t-/)/\tau} \, dt}, \qquad (2)$$

where $x_{\rm rms}(t;\tau)$ is the rms computed at time t; τ is the time constant of the exponential average, and $/$ is the dummy variable of integration. Exponential averaging is implemented using the well-known *RC* (resistor–capacitor) circuit, or its equivalent. Sound level meter "F" and "S" time-averaging characteristics are based on exponential time averaging with τ equal to 0.125 s and 1.0 s, respectively. (The equivalent-average sound level $L_{\rm eq}$ measured by an integrating-averaging sound level meter is based on linear time averaging.)

Relation between Linear and Exponential Averaging For a random signal, rms computations using linear averaging with averaging time T and exponential averaging with time constant τ will yield the same statistical accuracy when $T = 2\tau$. For deterministic signals, it should be noted that rms computations using linear and exponential averaging can have substantially different ripple suppression characteristics.

Peak Detectors Peak describes the maximum value of a signal. For acoustic signals, peak is understood to mean the maximum absolute value of the signal unless a polarity is stated explicitly.

Peak detection most commonly is applied to wide-band signals with minimal filtering. When a signal is filtered, the Gibbs' phenomenon can increase the peak over that of the wide-band signal. Furthermore, waveform distortion introduced by a filter not having a linear phase characteristic can influence the detected peak and cause difficulty when attempting to compare results obtained using filters with nominally similar magnitude-response characteristics but that may have different phase-response characteristics.

Most peak detectors provide for peak hold. This permits determination of the maximum peak detected since the hold was last reset.

2.3 Filters[1]

A filter is a frequency-selective device. It often is desirable to filter a signal prior to rms detection.

Given an input signal $x(t)$, the output $y(t)$ of a filter is given by the convolution of the filter impulse response $h(t)$ with the input:

$$y(t) = \int_{-\infty}^{+\infty} h(t - \ell)x(\ell)\, d\ell, \tag{3}$$

where ℓ is the dummy variable of integration.

The filter is completely described in the time domain by its impulse response. Alternatively, and with equal validity, the filter can be described in the frequency domain by its frequency response. The output spectrum $Y(f)$ is given by the product of the frequency response $H(f)$ and the input spectrum $X(f)$:

$$Y(f) = H(f)X(f). \tag{4}$$

Here $H(f)$, $Y(f)$, and $H(f)$ are complex and given by the Fourier transforms of $x(t)$, $y(t)$, and $h(t)$, respectively. The impulse response and frequency response are a Fourier transform pair. While the action of a filter often is characterized in the frequency domain, most filters operate in the time domain.

Basic Filter Types Filters for use in acoustics usually fall into one of four general categories: high pass, low pass, bandpass, and band reject (or notch) (Fig. 2). Combinations of any or all of these basic filter types are used for signal conditions. (Bandpass and notch filters are themselves combinations of a high pass and a low pass stage.) Bandpass filters are the basic building blocks of a frequency analyzer.

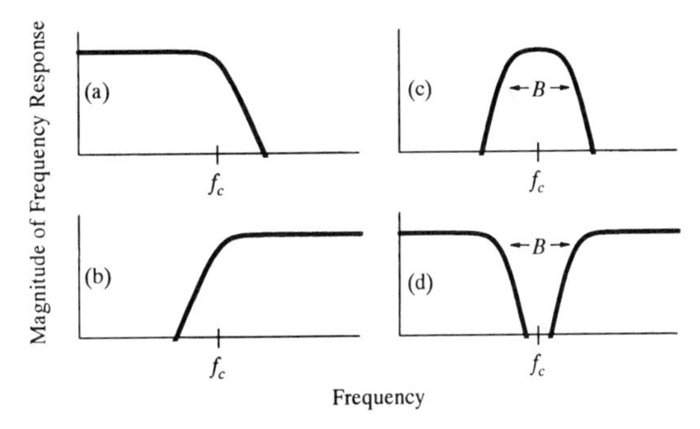

Fig. 2 Basic filter types: (*a*) low pass, (*b*) high pass, (*c*) bandpass, and (*d*) notch (band reject). Frequency f_c is cutoff frequency (*a*, *b*) or center frequency (*c*, *d*), B is bandwidth of passband (*c*) or notch band (*d*).

Bandpass Filter Specification Bandpass filter descriptive terminology is shown in Fig. 3. The *bandwidth* of a bandpass filter is the effective span of frequencies transmitted by the filter, that is, within the passband. Whereas an ideal bandpass filter has a well-defined passband, a practical filter has a transition band that leads to differing definitions of the bandwidth. In acoustics, usually the *noise bandwidth* is specified. The noise bandwidth is the bandwidth of an ideal filter passing the same mean-square signal when the input is white noise. Alternatively, the bandwidth is the interval between two fre-

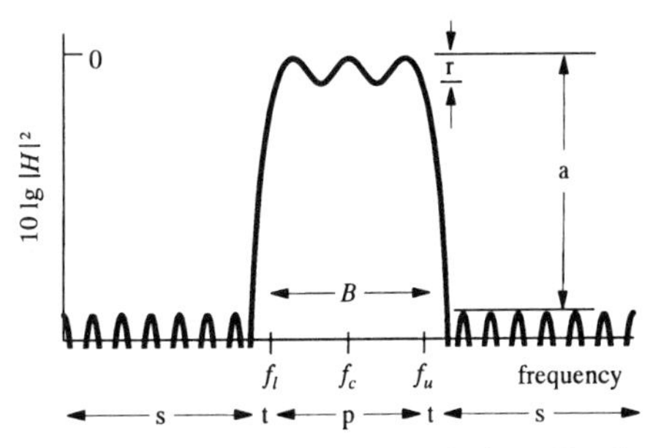

Fig. 3 Bandpass filter descriptive terminology: Bandpass filter relative response in decibels; p is passband; s is stop band; t is transition band; r is ripple; a is minimum attenuation of stop band; H is frequency response function; B is bandwidth, f_c is center frequency, f_l and f_u are lower and upper passband-edge frequencies (see text).

quency points defined to be the limits of the passband. The half-power (3-dB attenuation) points are used in this way to define the *3-dB bandwidth*, but other points are used also. For high-quality filters used in acoustics the noise bandwidth and 3-dB bandwidth, while different, usually are nearly equal.

Bandpass filters used for frequency analysis must cover the frequency range of interest with sufficient resolution. Depending on the analysis, a single filter can be tuned across the frequency range or a set of filters covering the frequency range may be used. Furthermore, the filter bandwidth may remain fixed or it may vary with the center frequency of the filter. The filters used for frequency analysis in acoustics usually are the constant bandwidth or the constant percentage bandwidth type.

Constant Bandwidth Filters These filters have a defined bandwidth that remains fixed as the center frequency of the filter varies. The center frequency is specified as the arithmetic mean of the upper and lower passband-edge frequencies. The frequency responses of a set of these filters have a uniform appearance on a linear frequency scale.

Constant Percentage Bandwidth Filters These filters have a bandwidth that is a defined fraction, or percentage, of the center frequency. The center frequency is specified as the geometric mean of the upper and lower passband-edge frequencies. The frequency responses of a set of these filters have a uniform appearance on a logarithmic frequency scale.

Octave and fractional-octave filters are of this type. The bandwidth of an octave filter is approximately 71% of the center frequency, a one-third octave filter approximately 23%.

2.4 Bandwidth–Time Product[1,2]

For rms detectors, statistical accuracy can be described in terms of the bandwidth–averaging time (BT) product.

For a random signal the normalized random error ϵ_r, that is, the ratio of the standard deviation of the estimated rms value to the true rms value is $\epsilon_r = (2\sqrt{\mathrm{BT}})^{-1}$. For accurate analysis the BT product must be sufficiently large such that the ϵ_r is acceptable. A consequence of this relation is that in order to preserve accuracy as frequency resolution is increased, the averaging time must be increased in proportion to the bandwidth reduction.

For deterministic signals other relations apply. The averaging time must be sufficiently long for the filter to respond and to suppress detector ripple. The former requires that T be at least the effective length of the filter impulse response. The latter depends on the number and spacing of tones within the filter bandwidth. In practice the required BT usually is much less than that required for random signals. (See also Chapter 81.)

2.5 Analog Analysis[1]

An *analog signal* is one that is continuous in time and that, at least in principle, takes on a continuous range of values. *Analog analysis* is continuous analysis of an analog signal. Typically a transducer is used to convert the acoustic signal of interest to an electrical analog (voltage, current, charge, and so on) that is then manipulated using suitable electronic components.

2.6 Digital Analysis[1–8]

A *digital signal* is a sequence of numbers obtained by quantizing an analog signal both in time and amplitude, and *digital analysis* is analysis of such a quantized signal. Digital analysis is useful to the extent that the digital signal being analyzed is representative of a physical analog signal of interest, that is, provided that the sampling and quantization processes do not introduce unacceptable distortions.

Amplitude quantization places an upper limit on the dynamic range of a digital analysis. The digital word size (number of bits) of the analog-to-digital converter (ADC) usually is considered the primary factor that determines quantization error, but the signal conditioning prior to the ADC and the accuracy of quantization influence linearity and distortion and also must be taken into account.

In a high-quality analyzer, the inherent error of quantization can be made very small through careful design and the use of high-quality components. To make full use of the available dynamic range, however, the user must be careful to set the analyzer signal conditioning so as to make optimum use of the digital conversion equipment.

Sampling (quantization) in time determines the useful frequency range of a digital analysis. In most applications sampling in time is at a uniform rate called the *sampling frequency*. If a continuous-time signal is band-limited to a frequency f_b (i.e., contains only frequencies in a band f_b wide, either inherently or due to signal conditioning filters), and if the sampling frequency f_s is such that $f_s \geq 2f_b$, then the continuous-time signal can be reproduced exactly from the time-sampled signal (except for error due to amplitude quantization). The minimum sampling rate that satisfies the inequality ($f_2 = 2f_b$) is called the *Nyquist rate*. To ensure that the sampled signal is properly band-limited, antialiasing filters are used. (See also Chapter 82.)

Nyquist Frequency The frequency $f_s/2$, one-half the sampling frequency, is called the *Nyquist frequency* and defines the maximum frequency content (bandwidth) that can be present in the analog signal and recovered after sampling. As implemented in digital analyzers, the Nyquist frequency is one-half the Nyquist rate, and to avoid ambiguity some authors deprecate the term Nyquist frequency. In the literature of commercially available acoustic analyzers, however, the term Nyquist frequency appears often while the term Nyquist rate is rare.

Aliasing and Antialiasing Filters If an analog signal contains frequency content above the Nyquist frequency, those frequencies will be aliased in the digital signal, that is, it will appear to have a frequency different from that of the actual signal. Due to aliasing the spectrum of a digital signal thus may differ from the spectrum of the analog signal from which it is derived.

The Nyquist frequency is often called the *folding frequency* because the spectrum of the digital signal appears to be made by folding the spectrum of the analog signal back onto itself at this frequency and its multiples. Analog signals with frequency $nf_s \pm f$, where n is an integer, all appear to have frequency f in the digital spectrum.

To ensure that aliasing does not occur, an antialiasing filter (AAF) is used to band-limit the signal. Typically AAF's are low-pass filters with a high rate of attenuation above the nominal cutoff frequency. Because practical filters have a finite attenuation rate, the cutoff must be set somewhat below the Nyquist frequency. Usually the cutoff is optimized so that any aliased frequencies fold back onto the cutoff frequency attenuated by an amount equal to the dynamic range. This approach works well for most signals but must be properly coordinated with other signal conditioning if the analog signal contains appreciable out-of-band frequency content.

Use of an AAF is essential for accurate spectrum analysis but because of phase and amplitude effects may be undesirable for analysis of a digital signal in the time domain. Signal rms is affected by an AAF, but only to the extent that the AAF excludes significant frequency content. Even when the rms is nearly unaffected, however, filters can alter the statistical distribution of the signal and have a profound effect on the apparent peak value. If both time and frequency analysis are to be performed on the same digital signal, it is important to use an AAF optimized for both.[9,10] Alternatively, the signal should be digitized twice: with the AAF for frequency analysis and without the AAF for time analysis. Modern digital analyzers often employ oversampling followed by digital low-pass filtering and decimation.[10] The effect of any digital filter must be taken into account along with that of an analog AAF, of course.

2.7 Analyzer Terminology[1–8,11,12]

Signal Terminology Analysis of acoustic signals draws on contributions from a wide range of disciplines. In many cases individual disciplines have developed terminology that facilitates communication within that field but that can be confusing, and even misleading, to other disciplines. The user of an analyzer must be careful to investigate thoroughly the definitions and conventions used with that analyzer.

By convention in signal processing, a signal is assumed to have physical units (generic unit: U) equivalent to voltage (unit: V). For purposes of analysis, it is not uncommon to treat all signals as voltages, with conversion to physical units at the last stage of analysis if at all. The time integral of a squared signal (or product of two signals) is called energy (unit: U^2s) regardless of missing scale factors. Similarly, the time average of a squared signal (or product of two signals) is called power (unit: U^2). These conventions provide a convenient shorthand for describing the signal processing but may conflict with clarity when applied to acoustic signals. The mean-square sound pressure is clearly different from sound power, yet both might be called "power" using this terminology.

Input Signal Conditioning Various transducers are used to convert acoustic signals of interest to electrical fluctuations. While transducers that produce digital signals are available for some applications, most acoustic transducers produce analog signals that must be conditioned in some way prior to analysis.

Amplifiers and attenuators are used to match the signal amplitude to that accepted by the analyzer input. In addition, a signal conditioning amplifier can be used to modify the output impedance of the transducer or convert one kind of electrical fluctuation to a more suitable one (current or charge to voltage, for example). Depending on the nature of the transducer and signal, AC or DC coupling can be used.

Filtering is performed to remove specifically undesirable frequency components, and in order to enhance frequency components of interest. Prefiltering to eliminate out-of-band signal, for example, can avoid an unnecessary limitation of the useful dynamic range of the analog-to-digital converter. In-band filtering is used to adjust signals with amplitudes that vary with frequency so as to fit the signal to the analyzer dynamic range, a process called *prewhitening*.

Dynamic Range The *dynamic range* of an analyzer is the ratio, usually expressed in decibels, between the largest and smallest signals that can be analyzed. When comparing the specifications of various analyzers, it is important to note that some interpretation of this definition may be needed. The smallest signal that can be analyzed usually is assumed to be equal to the noise floor. However, some analyzers introduce tonal distortion and aliasing components that are properly considered noise but that is not apparent unless a signal is present at the input. Such noise may have a spectrum level several orders of magnitude larger than the noise present without the signal. Also, some specifications include a presumption of range adjustment that may be automatic or not.

Other descriptors are used to indicate the ability of an analyzer to measure accurately within its dynamic range: *absolute accuracy* is usually stated as a percentage of the full-scale maximum, and *linearity* describes sensitivity to range changes.

Calibration and Spectrum Scaling Calibration expresses the relation between the output of the analyzer and the physical parameters of the signal being analyzed. Many analyzers permit calibration in physical units (e.g., pascals, decibels relative to 1 pW, and so on) or at least "engineering units" (units or U), although some are restricted to electrical voltage only. In the case of the latter, the user must supply the transducer voltage-to-unit calibration factor.

Basic calibration is accomplished by accounting for the previously determined calibration factors of each component of the measurement system, or by performing an "end-to-end" calibration using a known signal at the input. Many users do both and compare the results in order to verify proper operation of the equipment and avoid blunders in the calibration process. In performing a calibration, it is vital to be aware of the uncertainties involved so that calibration accuracy can be known. Much of what a user calls "calibration" might better be called a calibration check, calibration itself being performed in a metrology laboratory.

In performing frequency analysis a common source of calibration misunderstanding is spectrum scaling. Most spectrum analyzers produce the time-averaged value of a squared filter output, or the equivalent of this. With this scaling the spectrum usually is called a *power spectrum*. For many applications this scaling may be inappropriate (see Chapters 81 and 82), and other scalings are derived from this. Table 1 lists typical scalings.

TABLE 1 Spectrum Scaling Units

Parameter	Abbreviation	Unit[a]	Derivation[b]
Spectrum[c,d]	PWR	U^2	—
Spectrum[c,e]	RMS	U	$(PWR)^{0.5}$
Energy spectrum	EGY	U^2s	$(PWR) \cdot T$
Power spectral density[e,f]	PSD	U^2/Hz	$(PWR)/B$
Energy spectral density[f]	ESD	U^2s/Hz	$(PWR) \cdot T/B$

[a] U is an engineering unit equivalent to voltage, the unit of time is assumed to be the second, and the unit of frequency is assumed to be the hertz.
[b] T is the averaging time and B is the noise bandwidth.
[c] The bandwidth should be specified as a prefix, e.g., one-third octave band spectrum or 30 Hz bandwidth spectrum.
[d] Also called "power spectrum" because of its units and to distinguish this scaling from other spectrum scalings.
[e] Sometimes called "power spectrum with rms scaling" or simply "rms spectrum."
[f] By analogy to the RMS spectrum, the square root of this quantity is sometimes used instead. Sometimes shortened to "power spectrum" in context where this scaling will not be confused with U^2 scaling.

3 FILTER ANALYZERS

Numerous techniques and systems, both analog and digital, have been developed to perform frequency analysis using filters. Choice of an appropriate method usually depends on both the technical requirements and practical considerations. While in a particular application there may be an advantage to one or another approach, the technical requirements often permit a range of choice, and the advantage is found in the specific implementation of the approach rather than the method itself.

3.1 Standard Filter Sets[13–16]

To ensure comparability of data, national and international standards identify preferred frequencies for octave and fractional octave filter sets as well as performance requirements for the filter(s). Table 2 gives the nominal values for the filter frequencies usually used in acoustics.

3.2 Analog Filter Analyzers[1,17]

Three techniques are in common use for frequency analysis using analog filters: swept filter, stepped filter, and parallel filter. The designations here refer to the external behavior apparent to the user and not necessarily to the internal operating principle. A swept filter, for example, may use a variable-frequency filter to achieve the sweep, or a fixed-frequency filter with variable-frequency heterodyning.

Swept Filter A swept filter approach is used when continuous frequency resolution is required, perhaps because of the presence of pure tones to be identified. To produce useful results, it is essential that the signal being analyzed not change during the filter sweep time.

The analyzer consists of a tunable filter, a detector, and a recording device, typically a level recorder or pen plotter. The sweep (filter tuning) mechanism may be either mechanical or electrical. Sweep rate, filter bandwidth, and rms averaging time must be selected with reference to each other in order to achieve the desired anal-

TABLE 2 Octave and One-Third Octave Bands

Nominal Center Frequency[a] (Hz)	Approximate Octave Passband (Hz)	Approximate One-Third Octave Passband (Hz)
25	—	22.4–28
31	22.4–45	28–35.5
40	—	35.5–45
50	—	45–56
63	45–90	56–71
80	—	71–90
100	—	90–112
125	90–180	112–140
160	—	140–180
200	—	180–224
250	180–355	224–280
315	—	280–355
400	—	355–450
500	355–710	450–560
630	—	560–710
800	—	710–900
1,000	710–1,400	900–1,120
1,250	—	1,120–1,400
1,600	—	1,400–1,800
2,000	1,400–2,800	1,800–2,240
2,500	—	2,240–2,800
3,150	—	2,800–3,550
4,000	2,800–5,600	3,550–4,500
5,000	—	4,500–5,600
6,300	—	5,600–7,100
8,000	5,600–11,200	7,100–9,000
10,000	—	9,000–11,200
12,500	—	11,200–14,000
16,000	11,200–22,400	14,000–18,000
20,000	—	18,000–22,400

[a]Exact center frequency is calculated from $f_c = 10^{n/10}$ for base-10 series filters, and from $f_c = 2^{n/3}$ for the base-2 series filters, where n is an integer.

ysis accuracy. It should be noted in particular that, due to filter and detector response times, in the measured spectrum the maximum response to a tone does not occur when the center frequency of the filter is tuned to the tone frequency, but is shifted from the tone frequency by an amount that varies with the sweep rate.

Stepped Filter A stepped filter analyzer can be thought of as a special case of the swept filter analyzer. In essence the frequency spectrum is sampled at discrete frequencies, usually at a frequency spacing equal to the bandwidth of the filter.

This approach commonly is used with a sound level meter to provide a compact, field-portable frequency analyzer at reasonable cost.

Parallel Filters In a parallel filter analyzer a bank of fixed filters and detectors operates in parallel (simultaneously) replacing the variable-frequency filter of stepped filter analysis. Parallel filters offer the opportunity for frequency analysis of nonstationary signals, as well as more rapid analysis of stationary signals. The disadvantage is the increased expense of multiple filters and detectors, as well as the extensive calibration efforts required. Due to these economic considerations, most parallel filter analyzer implementations today are digital. The analog approach, however, allows independent adjustment of each frequency band for optimal dynamic range when analyzing a signal with a wide range of spectral content.

3.3 Digital Filter Analyzers[1,18,19]

Digital filter frequency analyzers for acoustical applications usually are parallel filter implementations. While there is no technical reason that swept or stepped filters cannot be implemented digitally, typically there is no practical reason to do so. For special-purpose analyses, however, sweeping or tracking filters may be useful. Two approaches are available: multiplexing and multiple filters.

Multiplexed Filters Multiplexing takes advantage of the relation between the effective frequency of a digital bandpass filter and the sampling frequency. For practical reasons, multiplexed filter analyzers are all of the octave or fractional-octave type. Coefficients are defined for each filter within the highest octave band of analysis. After the signal has been filtered in this octave, it is low pass filtered, resampled at half the original sampling frequency, then sent back into the digital filter. This process is continued, with each factor-of-2 reduction in sampling frequency reducing the effective filter frequency by a factor of 2 (i.e., an octave). This method is used because it makes efficient use of hardware when the available computational resources are sufficiently fast; it requires approximately twice the computation speed that would be required to filter only in the highest octave band.

Multiple Filters With the availability of inexpensive "digital filter on a chip" electronics, digital analyzers using multiple independent filters for parallel analysis have become more widely available. The approach is straightforward, and the advantage is the ability to tune each filter independently for frequency, bandwidth, and so forth.

Detectors When digital filters are used, typically digital detectors are used as well. As compared to analog techniques, digital detectors allow for more flexible implementation of rms detectors.

4 FFT SPECTRUM ANALYZERS

Fast Fourier transform analyzers are based on the discrete Fourier transform (DFT). The FFT is an efficient implementation of the DFT, several variations of which are available. An FFT analyzer comprises a number of practical considerations in addition to the mathematics of the DFT. (See also Chapter 82.)

4.1 Discrete Fourier Transform and Spectrum Estimation[1–3,17]

Except for a signal periodic in the period of the DFT, the DFT gives a poor estimate of the spectrum. The estimate is improved by windowing and averaging. The usual approach is to divide the signal into a number of blocks, apply a window (see below), and compute a DFT for each. The DFT for each block is conjugate multiplied with itself, then averaged over the ensemble of similar products from the other blocks. Scaling appropriate for the analysis is then applied.

Spectrum Scaling Spectrum scaling is used to make the result more or less independent of the analysis resolution. PWR or RMS scaling is appropriate for tonal and narrowband signals. PSD scaling is used for broadband random signals. ESD scaling is used for short transients.

Averaging Averaging has little effect on the accuracy of discrete tone amplitude estimates, but for random signals it significantly improves the spectrum estimate. The improvement depends on the effective number of statistically independent (disjoint) averages n_d, with $n_d \leq M$, where M is the actual number of averages. Details and formulas are given in Chapter 80. Approximate normalized random error is given in Table 3 for selected values of n_d. As a practical matter, in those applications where sufficient signal is available, statistical accuracy and signal stationarity often are evaluated experimentally, that is, by repeating the analysis several times on different parts of the signal.

TABLE 3 Spectrum Level Estimation Accuracy for Random Signals[a]

n_d	68% Confidence[b] (dB)	99% Confidence[c] (dB)
25	+0.8	+2.0
	−1.0	−4.0
50	+0.6	+1.5
	−0.7	−2.4
100	+0.4	+1.1
	−0.5	−1.5
200	±0.3	+0.8
		−1.0
500	±0.2	+0.5
		−0.6
1,000	±0.1	±0.4
10,000	±0.04	±0.1

[a] Approximate normalized random ms error ϵ_r tabulated in decibels for selected number of disjoint averages n_d: $\epsilon_r \approx 1/\sqrt{n_d}$. Decibel data is rounded to the nearest 0.1 dB or 1 significant figure.
[b] 68% confidence: ϵ_r.
[c] 99% confidence: $3\epsilon_r$.

4.2 Sampling and Resolution Parameter Relations[1–3,17]

The sampling frequency f_s, number of time samples per DFT N, and the record length T are linked by $T = N/f_s$. Thus fixing any two parameters determines the third.

The frequency spacing of DFT lines (line or bin spacing) Δf is related to the record length by $\Delta f = 1/T$. While Δf is sometimes called the resolution of the DFT, it is important to note that the bandwidth associated with each line depends on the time window used and need not be equal to the line spacing.

The Nyquist frequency is equal to $N\,\Delta f/2 = N/2T = f_s/2$. As discussed previously, the useful upper frequency limit $f_{\max}$ is somewhat less than the Nyquist frequency.

4.3 Zoom and Range Translation[20]

Analysis following the above relations in the frequency range $0 - f_{\max}$ is called *baseband analysis*. FFT analyzers use various techniques to increase resolution in a limited band of frequencies, that is, $f_1 - f_2$, with $f_1 > 0$ and $f_2 \leq f_{\max}$. Such an analysis usually is called *zoom*.

Zoom by range translation is the most common method. The signal first is heterodyned (multiplied) with a complex exponential that shifts the center frequency of the desired zoom band to DC. The heterodyned signal is then low pass filtered (digitally) to band-limit it, and is resampled to give a lower effective sampling frequency. Thus with the same N, a longer T and smaller Δf are achieved. The factor of the improvement in resolution is called that *zoom factor*, and for simplicity in resampling usually is a power of 2. With this method very large zoom factors can be achieved, but when spectrum averaging is performed most implementations do not permit repeated analyses of the same signal in different zoom bands. An alternative method uses a clever recombination of multiple short DFTs on a long time signal. This method permits repeated analyses in different zoom bands but usually is implemented with only modest zoom factors and without spectrum averaging.

4.4 Windows[1,21,22]

Windows are used to improve the quality of spectrum estimates from the DFT. The window may be applied as a weighting of the digital time signal block, or alternatively as a convolution with the DFT of the block.

When used for power spectrum estimation, the effect of a window can be understood as if each DFT line represents the detected output of a filter. The frequency response of the filter, determined by the Fourier transform of the time window, can, and usually does, extend many Δf's away from the center frequency. (This analogy oversimplifies by neglecting circularity and image frequency interactions. See Chapter 82.)

A wide range of windows is available, with most optimized for a specific signal processing application. Unfortunately, there is little agreement on terminology by the various manufacturers of commercially available analyzers. *Rectangle* (also *no window* or *open window*) and *Hann* (or *Hanning*) are nearly universal, but the details of other windows should be checked carefully in the analyzer specifications.

The rectangle window usually is the default window, used if none other is specified. It is appropriate for analysis of transients shorter than the window length and for analysis of signals known to be exactly periodic in the window. The Hann window is a general-purpose window that improves on most characteristics of the rectangle window for analysis of continuous signals. Other windows usually are optimized for a given parameter (main lobe width, side lobe magnitude, etc.). These often are supplied by the analyzer manufacturer or may be generated using appropriate convolution coefficients. (See Chapter 82.)

A cautionary note on the naming of windows is in order: Different users may apply the same name to the different windows. *Flat-top window*, for example, may apply to a window flat in the time domain or flat in the frequency domain. Check definitions carefully in order to avoid error.

4.5 Real-Time Analysis[1]

Analysis is said to be in *real time* when the analysis is completed during the time in which new data is collected. The term real-time analysis is applied to at least three situations using an FFT analyzer:

1. Real-time display, an analysis that shows a spectrum result within a reasonable period after time data have been sampled. The time data analyzed may or may not contain gaps.
2. Gapless analysis, an analysis where the time to compute and update a spectrum average is less than the time to collect a record for analysis. The time data is analyzed without gaps, but may be weighted unequally.
3. Gapless analysis with uniform weighting, a gapless analysis where unequal weighting due to application of a window is compensated, usually by overlap processing.

Real-Time Bandwidth *Real-time bandwidth* is a term used to describe the processing speed of an FFT analyzer. The usual definition is $B_{RT} = n_{max}/T_{calc}$, where B_{RT} is the real-time bandwidth, n_{max} is the number of useful frequency lines ($n_{max} \cdot \Delta f = f_{max}$), and T_{calc} is the time to compute and update one spectrum average. T_{calc} and the real-time bandwidth are a function of the internal resources of the analyzer, and often depend on such things as the number of channels being analyzed, the time window, and the display update rate.

When the analyzer range is set to a value less than the real-time bandwidth, gapless analysis is possible. When overlap processing is used, the *effective real-time bandwidth* is lower because the same data must be analyzed more than once. With overlap factor β, the effective real-time bandwidth B_{RTE} is $B_{RTE} = (1 - \beta)B_{RT}$. With appropriate selection of time window and the corresponding overlap, gapless analysis with uniform weighting is possible when the analyzer range is set to a value less than the effective real-time bandwidth.

4.6 Averaging, Triggers, Delays[1]

Triggers and delays are used to select specific signal segments for analysis. Triggers can be based on a signal itself or on an external event. The most flexible trigger implementations provide for variable hysteresis and other kinds of conditional triggering. Delay can be positive or negative time, negative delays being limited by the memory capability of the analyzer.

Spectrum averaging is used to reduce random error. (See previous discussion and Chapters 81 and 82.)

For deterministic signals, *synchronous-triggered time-domain averaging* can be used to improve signal-to-noise ratio prior to, or as a substitute for, spectrum analysis. This technique also is called *periodic signal averaging* or *evoked potential analysis*.

5 MULTICHANNEL ANALYZERS

Multichannel analyzers permit parallel analysis of multiple simultaneous signals, and/or determination of cross-channel properties. Multichannel analysis can be performed with filter or FFT analyzers. For parallel analysis, the principal advantages of a multichannel analyzer are convenience and the economy of hardware multiplexing when performance requirements permit. For cross-channel analysis, the well-defined synchronization between channels is essential.

The principal cross-channel spectrum functions are the *cross-power spectrum* (cross spectrum), *frequency response function*, and *coherence function*. (See Chapters 81 and 82.) Measurement and computation of cross spectrum and coherence are straightforward implementations of their definitions. A number of techniques are available for estimation of the frequency response function; selection is based on expected bias in the instrumentation setup.

Analyzer techniques for determination of cross-channel functions are extensions of the techniques used for signal-channel analysis.[1,23]

6 ADDITIONAL ANALYSES

Many variations on "basic" spectrum analysis can be useful for analysis of specific signals. In this section several of the more common techniques are introduced.

6.1 Order Analysis

Order (tracking) analysis is a technique used to study the fundamental and harmonics of what might be described as a periodic signal with time-varying period. To minimize confusion with the terminology of stationary signals, the term *order* is used in place of *harmonic*, and the *fundamental* becomes the *first order*. A typical application might be a signal from an engine, or other rotating machine, during a run-up/down. Key to the technique is the availability of a *tracking signal* from which the fundamental period can be determined.

Order tracking analysis is best described for a specific application. Consider the engine run-up suggested above. The signal of interest may be from a microphone near the engine. The tracking signal might be derived from the same signal if the signal-to-noise ratio is sufficient but more typically is obtained from a separate tachometer. The frequency of the tachometer signal, while varying in time, is constant if the independent variable is considered to be rotation angle. The essence of order tracking is to remove the time-varying frequency dependence by a change of variable. One approach is to use the tracking signal to control the analyzer sampling process so as to acquire a fixed number of samples per engine rotation. Many analyzers will accept an external sampling signal that replaces the internal sampling clock for just this purpose. The microphone signal thus is sampled uniformly in rotation angle, and nonuniformly in time. When the spectrum of this signal is computed, each signal component with a frequency proportional to engine speed remains fixed at a unique DFT line regardless of how much the actual frequency may vary. This approach is complicated by the need to provide for antialiasing protection with the now varying Nyquist frequency. A newer technique avoids this complication by using digital analysis after sampling both the tracking signal and the signal of interest in the usual way. Digital antialiasing and resampling of the signal of interest at the periodic rate derived from the sampled tracking signal produce the desired effect.

6.2 Intensity

Intensity analysis is available in both filter and FFT multichannel analyzers. In essence, an intensity spectrum is a special implementation of two-channel analysis. Specific measurement techniques are discussed in Chapter 106. The key analyzer requirement is good, or at least well known, interchannel phase and amplitude response matching.

6.3 Correlation

Auto and cross-correlation functions usually are available on FFT analyzers as the inverse DFT of the autopower and cross-power spectrum functions, respectively. Care should be taken to use time-signal zero padding and bow-tie correction, as appropriate, to mitigate circularity effects. Also the effect of the antialiasing filters on the time signal may need to be taken into account. (See Chapter 82.)

6.4 Cepstrum

The cepstrum (sometimes called the power cepstrum) often is available as the inverse DFT of a logarithmic spectrum. The complex cepstrum is obtained from the inverse DFT of a logarithmic complex spectrum (magnitude and phase). (See Chapter 82.)

7 TYPICAL MEASUREMENTS

7.1 Frequency Analysis of Signal

The steps for a typical frequency analysis of a stationary signal are outlined in Checklist 1. Note that often it is necessary to carry out a preliminary analysis in order to determine appropriate parameters for the final analysis. Proper frequency analysis of a nonstationary signal generally is more complicated. If the signal varies

CHECKLIST 1 Frequency Analysis of Stationary Signal

- Identify and select the bandwidth for the analysis, considering both the signal to be analyzed and the purpose of the analysis. Some preliminary investigation of the signal usually will be helpful.
- Identify and select a suitable frequency analyzer, considering the technical (signal to be analyzed, bandwidth and accuracy needed, etc.) and practical (analyzer availability, cost, portability, etc.) requirements.
- a. If using a filter and analyzer, measure in the required frequency bands. Select an averaging time to give a BT product yielding sufficient measurement accuracy.
  b. If using an FFT analyzer, measure the spectrum using a data window and number of spectrum averages selected to give sufficient measurement accuracy.
- Record the spectrum result with scaling appropriate for the signal: RMS, PWR, PSD, EGY, or ESD.

only slowly in time or can otherwise be treated as nearly stationary over the time of analysis; however, often the techniques of stationary signal analysis can be used with success. Judicious choice of the averaging time and use of appropriate triggers to capture the portion of the signal that is of interest are required.

Table 4 summarizes the suitability of various analyzer equipment for spectrum estimation of different signals. Choice of an analysis system often depends as much on practical considerations as on the technical suitability. Such practical considerations may include availability, cost, time and personnel requirements, size and weight (portability), and environmental robustness.

7.2 Frequency Response of System

While the frequency response of a system is defined as the ratio of system output to input, rarely is the ratio measured directly today. Most frequency response measurements use a ratio of cross- and auto-spectra obtained using a two- (or more) channel FFT analyzer, though in some applications a filter analyzer may be preferable. The preferred method of calculation of the frequency response function (FRF) from the measured data depends on the expected conditions.

Checklist 2 outlines the steps for a typical FRF deter-

CHECKLIST 2 System Frequency Response Using an FFT Analyzer

- Identify a suitable excitation signal and point of input to the system under test. Provide a transducer to measure the input into the system.
- Identify a suitable point at which to measure the response of the system under test. Provide a transducer to measure the system output.
- Using a two- (or more) channel analyzer, measure the auto- and cross-power spectra of the input and output signals. Use data windows appropriate for the signals. Estimate the FRF from the measured spectra (see text).
- Calculate and examine the coherence function spectrum to gain insight into the quality of the data signals and the FRF estimate.

TABLE 4 Spectrum Analyzer Selection Guide[a]

	Spectrum Analyzer Type[b]				
Signal Type	Swept Filter	Stepped Filter	Parallel Filter	FFT (real-time[c])	FFT (not real-time[c])
Broadband stationary	√	√	√	√	√
Broadband nonstationary			√	√	
Tonal stationary	√	√	√	√	√
Tonal nonstationary			√	√	
Random transient[d]			√	√	√
Periodic transient[d]		√	√	√	√

[a] A checkmark indicates that the spectrum analyzer system type is more or less suitable for analysis of the indicated signal type; the absence of notation indicates "not recommended." Technical and practical considerations may make some analyzers more suitable than others for the intended analysis. No attempt is made to evaluate practical considerations here. (See text.)
[b] Type of analyzer based on appropriate use of typical configurations commercially available. All analyzers may not be suitable.
[c] Real-time definition 3 (see text).
[d] Assumes use of suitable trigger or other method to capture transient. For FFT not real-time analysis assumes DFT record length exceeds duration of transient.

mination. It is important that the transducers used to measure the input and output not affect the system under test. If they do, their effect must be compensated.

Typical excitation signals include broadband random, impulse, pseudorandom, multitone, and swept sine. For many systems the exact signal used to excite the system is not of much importance; indeed, that is the point of measuring an FRF. Typically the excitation is chosen on the basis of measurement time required and convenience. For some systems, however, particularly those that may exhibit nonlinearity or considerable noise, selection of an appropriate excitation signal is critical to accurate determination of the FRF.

CHECKLIST 4 Synchronous Time Averaging Using a Digital Analyzer

- Identify a suitable synchronization (sync) signal and provide a suitable transducer for its measurement.
- If the signal to be averaged is periodic with the sync signal but the period varies in time, use order tracking to remove the time variations.
- Select a suitable analyzer record length and frequency range. If frequency analysis is not to be performed, it may be desirable to turn off any antialiasing filters.
- Perform triggered averaging with the number of averages chosen to provide the required rejection of nonsynchronous signal but also considering the stability of the signal of interest.

7.3 Correlation

While the cross-correlation function for two signals usually is defined through a convolution in the time domain, most measurements today are made with an FFT analyzer. In most cases, convolution via Fourier transform, multiplication, and inverse Fourier transform is significantly faster than the direct approach. Use of the DFT in this process, however, requires special care to avoid circular effects. Also, due to the finite analyzer record lengths and the significant delays often encountered in acoustical applications, it usually will be advantageous to introduce artificial interchannel delays into the analysis process.

Checklist 3 outlines the steps for cross correlating two time signals using an FFT analyzer. The procedure here is general but is not necessarily optimal for all signals. (See Chapter 82).

CHECKLIST 3 Cross Correlation Using an FFT Analyzer

- Identify signals to be correlated and perform preliminary investigation to determine parameters for the steps below.
- Select a suitable two- (or more) channel FFT analyzer and set appropriate frequency range and record length parameters.
- a. If the duration of the signals of interest is no more than one-half the record length, proceed to next step.
  b. If the duration of the signals of interest is at least as long as the record length, use zero padding in the second half of the record.
  c. If neither of the above apply, return to previous step and select a longer (preferred) or shorter record length.
- Set an interchannel delay equal to the delay time of the correlation peak of interest.
- Spectrum average to determine the cross spectrum, then inverse Fourier transform to obtain the cross correlation.
- Compensate the displayed result for zero padding (bow tie correction) and interchannel delay, if either has been used.

Autocorrelation can be considered to be a special case of cross correlation, the two signals simply being identical. The procedure for autocorrelation measurement is similar but with no need for introduction of an interchannel delay.

7.4 Synchronous Time Averaging for Signal Enhancement

Synchronous time averaging is used to enhance the waveform of synchronous signals relative to nonsynchronous ones. Most synchronous time averaging is performed with digital analyzers, but analog techniques can also be used.

Checklist 4 outlines the steps for synchronous time averaging using a digital analyzer. Key to the technique is establishment of a suitable synchronization (sync) trigger. If the signal to be averaged is periodic with the sync signal but the period varies in time, for example, sound produced by rotating equipment in nonsteady rotation, those portions of the signal away from the sync pulse will tend to be lost in averaging because of "smearing." Signal loss can be minimized by using order analysis techniques in addition to synchronous time averaging.

8 IMPLEMENTATION CONSIDERATIONS

The physical implementation of analyzers ranges from those that are hand held to those that are room size. Some are dedicated to a specific application; others are based on general-purpose computers. Depending on the application and the skill and knowledge of the user, the various implementations are found to offer unique advantages.

Dedicated analyzers are those that have traditionally

been thought of as "analyzers." Their principal strength is also their weakness: They do best what they are designed to do. Many users find them easy to use, the manufacturer providing setups and push buttons for standard functions. An analyzer can be designed for portability, cost, application, accuracy, speed, and use in harsh environments, virtually whatever the specifications call for.

Many dedicated analyzers are, of course, a kind of computer. It is useful, however, to discuss analyzers built around a general-purpose computer. This might be a portable personal computer or a laboratory-based network computer. The principal advantage here can also be seen as a weakness: complete flexibility to do virtually anything. Because anything can be done, it sometimes is difficult to know where to begin. Such a system usually consists of software with add-in cards or other signal processing subsystem. Typically it is the user's responsibility to ensure that everything works together. As compared to a dedicated system, hardware may be less expensive due to economies of scale, but it is also bulkier and perhaps not as rugged. For many users, a real advantage is the ability to integrate the analysis and analysis results with other requirements, for example, further analysis and report generation.

REFERENCES

1. R. B. Randall, *Frequency Analysis*, 3rd ed., Brüel and Kjær, Nærum, Denmark, 1987. (Much of this material is summarized in Chapter 13, C. M. Harris (Ed.), *Shock and Vibration Handbook*, 3rd ed., McGraw-Hill, New York, 1988.)
2. J. S. Bendat and A. G. Piersol, *Random Data: Analyses and Measurement Procedures*, 2nd ed., Wiley Interscience, New York, 1986.
3. L. R. Rabiner and B. Gold, *Theory and Applications of Digital Signal Processing*, PrenticeHall, Engelwood Cliffs, NJ, 1975.
4. A. V. Oppenheim and R. W. Schafer, *Digital Signal Processing*, Prentice-Hall, Engelwood Cliffs, NJ, 1975.
5. K. G. Beauchamp, *Signal Processing Using Analog and Digital Techniques*, Allen and Unwin, 1973.
6. J. S. Bendat and A. G. Piersol, *Engineering Applications of Correlation and Spectral Analysis*, 2nd ed., Wiley, New York, 1993.
7. S. L. Marple, *Digital Spectral Analysis with Applications*, Prentice-Hall, Engelwood Cliffs, NJ, 1987.
8. L. R. Rabiner, J. W. Cooley, H. W. Helms, L. B. Jackson, J. K. Kaiser, C. M. Rader, R. W. Schafer, K. Steiglitz, and C. J. Weinstein, "Terminology in Digital Signal Processing," *IEEE Trans. Audio Electroacoust.*, Vol. AU-20 (5), 1972, pp. 322–337.
9. J. Meyer, "Time Correction of Anti-Aliasing Filters Used in Digital Audio Systems," *J. Audio Engr. Soc.*, Vol. 32 (3), 1984, pp. 132–137.
10. M. W. Hauser, "Principles of Oversampled A/D Conversion," *J. Audio Engr. Soc.*, Vol. 39(1/2), 1991, pp. 3–26.
11. L. L. Beranek (Ed.), *Noise and Vibration Control*, rev. ed., Institute of Noise Control Engineering, Washington, DC, 1988.
12. L. L. Beranek and I. L. Ver (Eds.), *Noise and Vibration Control Engineering*, Wiley, New York, 1992.
13. ISO 226, "International Standard for Acoustics—Preferred Frequencies for Measurements," International Organization for Standardization, Geneva, 1975.
14. IEC 1260, "International Standard for Electroacoustics—Octave-Band and Fractional-Octave-Band Filters," International Electrotechnical Commission, Geneva, 1995.
15. ANSI S1.11, "American National Standard Specification for Octave-Band and Fractional-Octave-Band Analog and Digital Filters," American National Standards Institute, New York, 1986.
16. ANSI S1.6, "American National Standard Preferred Frequencies, Levels, and Band Numbers for Acoustical Measurements," American National Standards Institute, New York, 1984.
17. D. W. Steele, "Data Analysis," in L. L. Beranek (Ed.), *Noise and Vibration Control*, rev. ed., Institute of Noise Control Engineering, Washington, DC, 1988, Chapter 5.
18. O. Roth, "Digital Filters in Acoustic Analysis Systems," *Brüel & Kjær Tech. Rev.*, Vol. 1, 1977.
19. J. W. Waite, "A Multirate Bank of Digital Bandpass Filters for Acoustic Applications," *Hewlett-Packard J.*, Vol. 44, No. 2, 1993, pp. 56–64.
20. N. Thrane, "Zoom-FFT," *Brüel & Kjær Tech. Rev.*, Vol. 2, 1980.
21. F. J. Harris, "On the Use of Windows for Harmonic Analysis with the Discrete Fourier Transform," *Proc. IEEE*, Vol. 66 (1), 1978, pp. 51–83, and corrections in A. H. Nuttall, "Some Windows with Very Good Sidelobe Behavior," *IEEE Trans. Acoust., Speech, Signal Process.*, Vol. ASSP-29 (1), 1981, pp. 84–91.
22. J. C. Burgess, "On Digital Spectrum Analysis of Periodic Signals," *J. Acoust. Soc. Am.*, Vol. 58, 1975, pp. 556–566.
23. H. Herlufsen, "Dual Channel FFT Analysis," *Brüel & Kjær Tech. Rev.*, Vols. 1 and 2, 1984.

108

CALIBRATION OF PRESSURE AND GRADIENT MICROPHONES

VICTOR NEDZELNITSKY

1 INTRODUCTION

Calibration of a microphone establishes the quantitative relation between the signal (usually, the voltage) at the electrical terminals of the microphone and the acoustical signal at its diaphragm or other specified reference position. This chapter deals with microphones used in gases, principally in air, for which this signal may be sound pressure, its first or higher order gradient, or a combination of these quantities. Calibration is necessary for accurately measuring potentially hazardous noise, as well as desired acoustical signals. National and international legal, regulatory, and quality control measurements for health, safety, and commerce depend on various acoustical instruments and systems calibrated or characterized by calibrated microphones.

To accommodate the wide variety of microphone types, characteristics, and applications, various calibration procedures greatly differ in complexity, uncertainty, and the labor and equipment costs required to realize given frequency ranges and accuracies. This chapter briefly describes only some calibration methods of fundamental importance or currently in widespread use and some factors important in determining calibration uncertainties. Reciprocity and reciprocity-based primary laboratory methods of the highest accuracy are included, as well as less complicated, less costly, and, usually, less accurate methods based on comparisons, calibrators of the closed-coupler type, and electrostatic actuators. Selected references are provided.

Note: References to chapters appearing only in the *Encyclopedia* are preceded by *Enc.*

2 CATEGORIES OF MICROPHONES

To some degree, all microphones exhibit compromises in their essential mechanical and electroacoustical performance characteristics. These characteristics include size, frequency range, uniformity of sensitivity with frequency (usually for a specified type of sound field), dynamic range, directionality of response, stability, ruggedness, the degree to which performance characteristics are influenced by changes in ambient environmental conditions (e.g., static pressure, temperature, relative humidity), and sensitivity to extraneous influences such as electromagnetic fields, ionizing radiation, and vibration. Some of these compromises are related to cost, and most are closely related to the applications intended for a given microphone, as well as its transduction mechanism and design.

Microphones are frequently categorized by their intended application and transduction mechanism, as in Chapters 110 and 112. For calibration purposes, however, it is essential to consider the acoustical signal for which the microphone is intended, as well as certain general characteristics and applications closely related to calibration.

Four relatively broad categories are considered here: laboratory standard, pressure, pressure-gradient, and combination microphones. The first two categories are sometimes grouped together and termed pressure (or pressure-sensing) microphones, but here we consider laboratory standard microphones, which are specifically designed for the most accurate calibration purposes, to be a separate category. The categories pressure-gradient microphones and combination microphones are sometimes grouped together and termed pressure-gradient microphones, or, simply, gradient microphones. Each category has its distinctive characteristics and applications, although in some cases a given microphone may correctly be classified in more than one category. Some categories can also be divided into subcategories: for example, pressure microphones include working standard and other measuring microphones, as well as pressure-sensing communications microphones. Differences in microphone design and performance characteristics often, but not always, indicate the use of calibration methods particularly appropriate to given categories.

2.1 Laboratory Standard Microphones

Critical mechanical and electroacoustical characteristics have been standardized[1,2] for these microphones, which are intended for calibration by primary methods to the highest attainable accuracies. Laboratory standard microphones that are handled carefully are very stable, so that their sensitivity changes very little with passage of time. Consequently, after calibration, they can be used in comparison methods to calibrate other microphones and instruments. At present, every available laboratory standard microphone is a condenser microphone, to which a polarizing voltage, usually 200 V DC, is applied from an external source. These microphones respond to sound pressure; that is, the microphone output voltage is intended to be proportional to the sound pressure at the microphone diaphragm in a given type of acoustic field.

2.2 Pressure Microphones

Pressure microphones also respond to sound pressure but are not necessarily as stable in their sensitivity, with regard to time and changes in environmental conditions, as laboratory standard microphones. Working standard microphones are externally polarized condenser microphones that approach the stability of laboratory standard microphones and, after calibration, are used to calibrate other microphones and instruments, often on a frequent and regular basis. Working standard microphones are a subject of an international standard[2a] and of an American standard being developed by the Standards Committee S1, Acoustics, Working Group 1, Standard Microphones and Their Calibration, accredited by the American National Standards Institute (ANSI) and administered by the Acoustical Society of America (ASA). Measuring microphones are used in custom-designed systems, sound level meters, and personal sound exposure meters (noise dosimeters) to determine sound pressure or sound exposure in laboratory or field applications. These microphones include externally polarized condenser microphones, as well as electret, piezoelectric, and electrodynamic microphones. Many working standard and measuring condenser microphones are sufficiently similar to laboratory standard microphones to warrant calibration by reciprocity techniques. However, most are calibrated by secondary methods.

Some pressure microphones typically are not used to measure sound pressure but to transduce it to an electrical signal for communication or recording purposes, as in telephones, many hearing aids, and some radio broadcasting and audio recording applications. When necessary for design prototype testing, production quality control, comparative performance evaluation, and so forth, most of these microphones are calibrated by secondary techniques.

2.3 Gradient and Combination Microphones

For a gradient (also termed pressure-gradient) microphone, the electrical response corresponds to a spatial gradient of sound pressure. The response of a first-order gradient microphone corresponds to the difference in sound pressure between two points in space. Over much, if not all, of the intended range of operating frequencies, these points are usually separated by a sufficiently small distance compared to the wavelength that the pressure gradient corresponds to the acoustic particle velocity. Such microphones are often called velocity microphones. Higher-order gradient microphones are those in which the electrical response corresponds to a second, or higher, order spatial gradient of the sound pressure. Gradient microphones of the first and higher orders are directional, discriminate against random-incidence (diffuse) components of the sound field, and are useful in many applications for which such discrimination is often desirable,[3–5] as in sound recording, broadcasting, and communications systems.

In a combination microphone, desired directionality and frequency response characteristics related to those of both pressure and gradient microphones are achieved by a variety of methods.[3–5]

3 SELECTION OF CALIBRATION METHODS AND SCHEDULES

3.1 Selection of Calibration Methods

Practical microphones are seldom sufficiently small that the effects of diffraction can be considered negligible throughout the entire frequency range of operation. Consequently, a microphone is usually characterized by different frequency-dependent sensitivities when it is used in different sound fields, such as a spatially uniform sound pressure, a plane sound wave at a specified angle of incidence in the free field, or a diffuse field. Different calibration procedures have evolved so that an appropriate calibration is available for each field type. These issues, and many others, must be considered in selecting a calibration method.

Sufficiently detailed, rigorous general rules for selecting appropriate calibration methods for specific tasks in every nation cannot be given in this brief chapter. Despite international standardization, national standards and regulations are not always identical in different nations. Detailed procedures for many of the most accurate and expensive, as well as frequently performed and inexpensive, calibrations vary among laboratories and manufacturers, and are not fully documented in accessible form. The user of an instrument must consider the microphone category, its intended application, the range of operating frequencies, the dynamic range, and the frequency-dependent and level-dependent uncertainties that need to be achieved, so that the calibration complexity and cost are commensurate with the acceptable degree of uncertainty. A laboratory standard microphone or working standard microphone that is used by a calibration laboratory to calibrate other standard microphones, acoustical calibrators (including pistonphones), and instruments over a broad frequency range may need the best available reciprocity or reciprocity-based calibration, but if lesser accuracy is acceptable, a secondary method may be appropriate. Examples for which secondary methods are commonly employed include microphones used in sound level meters, integrating-averaging sound level meters, and personal sound exposure meters for which exhaustive data establishing the design integrity of the instrument models are already available. Applicable secondary methods include calibration by comparison with a calibrated working standard microphone, the use of a properly calibrated multifrequency acoustical calibrator, or the use of a single-frequency calibrator and an electrostatic actuator method. The manufacturer's recommendations and advice from national and private calibration laboratories, as well as regulatory and legal authorities, should be carefully considered, along with the experience of the instrument users.

There is one cardinal, succinct rule: The conditions of calibration must correspond, as closely as is necessary and practical, to the pertinent critical conditions of intended use of the microphone, so that the calibration method is appropriate to the applicable kind of measurement and sound field.

Most microphones contain a cavity behind the diaphragm. To prevent fluctuations in ambient barometric pressure from mechanically biasing the diaphragm, and thereby affecting the microphone sensitivity, this cavity is almost always equalized to this pressure by a high-acoustic-impedance vent that has only a small effect on the acoustical performance of the microphone. However, for purposes requiring high accuracy at low frequencies, this effect is usually not negligible. For such purposes, if the vent is exposed to the sound field, as in most free-field and diffuse-field measurements, the vent must also be exposed during calibration. If the vent is not so exposed, for example, during measurements of sound pressure in nearly all standardized couplers, the vent should not be exposed during calibration.

For some calibrations, especially those using laboratory standard microphones, the desired sensitivity involves the open-circuit output voltage of the microphone itself (sometimes called the microphone cartridge). In other calibrations, the sensitivity may involve the output voltage of the preamplifier, or the combined preamplifier and amplifier, to which the microphone is connected and used in a system. The Thevenin equivalent electrical network impedances (also called source impedances) of most condenser, electret, and piezoelectric microphones are not negligibly small compared with the input impedances of the preamplifiers with which they are used. Consequently, a system usually should be calibrated as it is used or reliable corrections should be applied for the loading and other effects (including gains) of the preamplifier and amplifier, which cause the voltages at the preamplifier and amplifier outputs to differ from the open-circuit output voltage of the microphone.

3.2 Selection of Calibration Schedules

Selection of calibration schedules depends on the microphone category, its application and treatment, the necessary accuracy of measurement, the cost or penalty of inaccurate measurement, and the stability of the microphone as demonstrated by a history of its calibrations, preferably performed at regular intervals. These issues are familiar to calibration laboratories and have been described in some detail.[6] For example, a laboratory standard microphone that is handled carefully, used only occasionally as one of several available reference standards, and has an excellent record of stability may be calibrated only every 12–18 months or more, with its calibration interval overlapping those of the other such microphones, so that at least one has been calibrated every 6–12 months. However, a sound level meter or personal sound exposure meter used daily in a hostile industrial environment such as underground mining may require weekly calibration with a multifrequency, multilevel acoustical calibrator and field checks at a single frequency and a single level at the beginning and end of each day. The recommendations of the microphone or instrument manufacturer, the national and private calibration laboratories, and regulatory and legal authorities in a nation, as well as practical experience acquired by the users of the instruments in specific circumstances, should be carefully considered.

4 PRIMARY CALIBRATION OF LABORATORY STANDARD MICROPHONES BY ELECTROACOUSTICAL RECIPROCITY TECHNIQUES

Nearly all of the major national standards laboratories of the world use these techniques. Except for the relatively undeveloped and less accurate determination of diffuse-field sensitivity by the reciprocity method discussed in Section 4.4, these are the most highly developed and cost-effective methods available for achieving the highest attainable absolute accuracies of laboratory standard microphone calibration over a wide range of operating frequencies. Such calibrated microphones are needed to perform the most accurate sound pressure measurements and to calibrate other microphones and acoustical instruments, so that nearly all critical measurements of sound pressure are traceable to these methods.

Electroacoustical reciprocity techniques permit calibration based on fundamental principles and constants, and typically involve measurements of AC voltages, transfer impedances, and the quantitative determination of the acoustical coupling between transmitting (electrically driven to serve as a sound source) and receiving microphone pairs.

4.1 Insert-Voltage Technique and Ground Shield Dimensions

The insert-voltage technique[7,8] is used to determine the open-circuit sensitivity of the microphone when it is connected to a practical preamplifier. Usually, the open-circuit sensitivity of a laboratory standard microphone is reported, so that highly accurate measurements can be made with this microphone at different laboratories, essentially independently of interlaboratory differences between preamplifiers. However, critical microphone-to-preamplifier mounting and ground-shield dimensions have been standardized[7,8] for these microphones to avoid the influence of significantly differing stray capacitances associated with differences in these dimensions.

4.2 Determination of Pressure Sensitivity by the Reciprocity Technique

In the method using two microphones and an auxiliary transducer the sensitivities of two microphones are determined.[7,8,9] One of these microphones must be reversible, that is, must be used as a transmitter, as well as a receiver. The auxiliary transducer, which may be a microphone more sensitive than the other two, is used only as a transmitter to determine the ratio of their sensitivities.

Another method determines the sensitivities of three microphones. Each is used as a transmitter as well as a receiver. From a sequence of pairwise measurements, the sensitivities of the microphones are determined.[8]

Uncertainties in Pressure Sensitivities Determined by the Reciprocity Technique Generalizations concerning these uncertainties are difficult because they are not only frequency-dependent but also critically dependent on particular choices made in different laboratories with regard to rather complicated details of method and apparatus. Furthermore, different laboratories have used different methods of combining individual systematic and random uncertainty components to estimate overall calibration uncertainty. Consequently, comparisons of calibration results obtained at different laboratories are invaluable for establishing the approximate uncertainties in primary calibrations. The 1986–1987 comparison[10] of pressure calibrations of IEC Type LS1P (ANSI Type L) laboratory standard microphones[1,2] among 17 laboratories in 17 different IEC member nations indicated that, at frequencies from 63 Hz to 10 kHz, agreements in calibration results of approximately 0.1 dB (often better at frequencies from a few hundred hertz to a few kilohertz, and sometimes a bit worse at the highest frequencies) were usually obtained. Particularly noteworthy was the agreement obtained from 63 Hz to 10 kHz (23 frequencies at one-third-octave intervals) between calibrations at the National Physical Laboratory (UK) (abbreviated NPL), which used an air-filled "plane-wave" coupler, and the National Bureau of Standards (U.S.A.) (renamed the National Institute of Standards and Technology in 1988, abbreviated NIST), which used a larger coupler filled with air at low frequencies and with hydrogen at higher frequencies. Other aspects of the apparatus and procedures, and consequently tradeoffs among individual uncertainty components during calibration, were also significantly different in the two laboratories. For both microphones in this comparison,[11] the absolute values of the differences between pressure response levels determined at NPL and at NIST were 0.02 dB or less at frequencies from 200 Hz to 4 kHz inclusive. At the remaining frequencies from 63 to 200 Hz, and from 4 to 10 kHz, these absolute values were no greater than 0.05 dB. This is perhaps the closest agreement in such calibrations ever achieved by laboratories using independent and significantly dissimilar apparatus and procedures, and is probably somewhat fortuitous, especially at high frequencies. However, these results demonstrate the kind of agreement that can be achieved under nearly ideal circumstances.

4.3 Determination of Free-Field Sensitivity by the Reciprocity Technique

Analogous to the corresponding method for pressure calibration using two microphones and an auxiliary transducer, the sensitivities of two microphones are determined, and an auxiliary transducer, which may be a microphone more sensitive than the other two, is used only as a transmitter.[7,9,12] Instead of being sealed into an acoustic coupler, however, each microphone pair, or each microphone and the auxiliary transducer, are placed in an anechoic chamber. A method using three microphones in a sequence of pairwise measurements, analogous to the corresponding procedure for pressure calibration, has also been standardized for free-field calibration,[12] and this IEC standard additionally considers IEC Type LS2P microphones.

Uncertainties in Primary Free-Field Calibration

Uncertainties in primary free-field calibration are very dependent on the type of microphone being calibrated, as well as on specific details of calibration apparatus and procedures. The quality of the anechoic chamber, signal-to-noise ratios, and electrical crosstalk are all particularly important. The best anechoic chambers introduce uncertainty components of approximately 0.1 dB, and the worst may introduce components of more than 1 dB. Probably the most useful laboratory microphones routinely given free-field reciprocity calibrations are 12.7 mm (0.5 in.) nominal diameter, primarily because the useful amplitude-frequency responses of these microphones are more nearly constant to higher frequencies than are the amplitude-frequency responses of the larger Type LS1P microphones. Among all interested IEC member nations, extensive interlaboratory comparisons involving the free-field response levels of 12.7-mm nominal diameter microphones such as the IEC Types LS2F, LS2aP, and LS2bP have not yet been conducted. For some essentially similar condenser microphones of this size, overall uncertainties of approximately 0.1–0.2 dB have been obtained at frequencies from about 1.25 to more than 20 kHz at NIST with microphone separation distances of about 0.2 m.[11,13,14]

4.4 Determination of Diffuse-Field Sensitivity

Reciprocity Method The reciprocity method in a diffuse field has been devised and performed at the PTB (Braunschweig, Germany) by Diestel.[15,16] The procedure is basically similar to the method of pressure or free-field calibration using two microphones, here denoted a and b, and an auxiliary transducer. However, the microphones and transducer are placed in a reverberation room during measurements, and the transmitter is excited by sequentially presented bands (typically one-third-octave) of random noise rather than sine-wave signals. Furthermore, to achieve an adequate signal-to-noise ratio, Diestel used a small electrostatic loudspeaker instead of a reversible microphone for frequency bands below 2 kHz. Diestel's English-language paper[16] expresses the diffuse-field sensitivities M_{dfa} and M_{dfb} of a and b (notation of this chapter) in terms of the diffuse-field reciprocity parameter, J_{df}, and the electrical voltages and currents measured during calibration. (Note that a typographical error occurs in Diestel's Eqs. (21) and (22): The entire right-hand side of each equation should be raised to the exponent $\frac{1}{2}$. The corresponding equations in the German-language paper[15] avoid this error.) Diestel expressed[15,16] $J_{df} = 2h_0/\rho_0 f$ by considering h_0 to be the "diffuse-field distance,"[16] that is, the distance from an idealized point source at which the energy density of directly radiated sound equals the average energy density in the reverberation room, and by denoting the frequency and ambient air density as f and ρ_0, respectively. Diestel also used the room volume V, the Sabine reverberation time T, and the speed of sound c, to determine $J_{df} = (2.1/\rho_0 f)(V/cT)^{1/2}$, another useful expression.

While the significance of Diestel's accomplishment has been recognized, this method has not been widely used or standardized, for at least two reasons. First, the reverberation room must provide a very good diffuse field and must be extremely well characterized. Even with a good room at PTB, Diestel apparently needed to average results from measurements using three different positions of the microphones/loudspeakers to estimate the microphone sensitivity level with a standard error less than 1 dB at frequency bands from 0.5 to 16 kHz. Second, because even the best reverberation rooms are imperfect, it is difficult to include the effects of sound absorption by the air in the room on calibration results, especially at high frequencies, where such absorption can be sufficiently large and dependent on environmental conditions (especially humidity) to compromise primary calibration accuracy. Consequently, there is ambiguity in the determination of J_{df}, and, therefore, in the determination of M_{dfa} and M_{dfb}.

Method from Free-Field Measurements of Directivity and Reference Sensitivity Of two important primary methods of diffuse-field calibration, this is the more widely used method, has long been available,[17] and has appeared in both an IEC standard[18] on diffuse-field calibration of sound level meters and in established

ANSI standards.[7,19] This IEC standard formally distinguishes between the random-incidence and diffuse-field sensitivities of a microphone, using the term "diffuse-field" differently than the ANSI standard[7] for microphone calibration. However, the IEC standard considers the random-incidence and diffuse-field sensitivities to be equivalent, so that they can be used as synonyms. Brinkmann and Goydtke[20] discuss this issue in detail and describe experimental procedures and apparatus for determining random-incidence and diffuse-field sensitivities in an anechoic chamber and reverberation room, respectively. From results of both kinds of microphone calibrations, they conclude that, when carefully performed, the two methods can be considered equivalent to an accuracy sufficient for most practical sound measurements.

For free-field measurements in an anechoic chamber, a sound source is excited by a sine-wave signal or band of noise. For a sufficient number of angles of incidence sampling the solid angle 4π (or a smaller angle, given a symmetry assumption about the microphone and sound field), the directivity of the microphone is measured. From these and the appropriate reference sensitivity measurements at the specified reference direction of incidence, the diffuse-field sensitivity is calculated, for example, as a correction to the reference free-field sensitivity. This reference sensitivity may be obtained by the reciprocity technique described in Section 4.3. The directivity measurements require a very good and well-characterized anechoic chamber, which a primary standards laboratory must already have for use in the primary and secondary free-field calibration of microphones. If all measurements of directivity patterns in a given calibration are conducted at the same, or nearly the same, ambient environmental conditions, the effects of atmospheric attenuation of sound need not be calculated. If necessary, these effects are much easier to calculate and to include in this method than in reverberation room measurements.

Uncertainties in Determination of Diffuse-Field Sensitivity Level Uncertainties in determination of diffuse-field sensitivity level are particularly dependent on the characteristics of the reverberation room or anechoic chamber in which the measurements are performed. Comparison of the results from different methods is invaluable. Brinkmann's and Goydke's results showed that random-incidence and diffuse-field sensitivity levels of the microphone carefully determined in an anechoic chamber and reverberation room, respectively, were equal with an uncertainty of about 0.1 dB for frequencies as high as about 12.5 kHz. Their methods and measurements in a reverberation room have based on free-field measurements of directivity evidently improved upon the accuracy of the diffuse-field reciprocity method devised by Diestel.

For what is now termed[1,2] an IEC Type LS1Po or ANSI Type L laboratory standard microphone, Diestel[16] compared the sensitivity levels measured by the diffuse-field reciprocity method using one-third-octave bands of noise and the sensitivity levels calculated from the directivity patterns and the free-field response at the midfrequency of each band. Throughout the frequency range 0.5–16 kHz, the agreement was within about 1 dB. At frequencies below 0.5 kHz, the sound wavelength is sufficiently large relative to the dimensions of this microphone-type that diffraction effects are practically negligible, and the pressure, free-field, and diffuse-field sensitivities may be considered essentially equal, provided that the effect of the ambient pressure equalization vent is also negligible.

5 TESTS OF MICROPHONE LINEARITY WITHIN OPERATING DYNAMIC RANGE

Laboratory standard and other condenser microphones are similar to single-sided electrostatic loudspeakers,[21] which do not behave as linear transducers at sound pressures so large that the diaphragm displacement is not small relative to the spacing between the diaphragm and the back electrode (backplate). Consequently, an upper limit to the microphone operating dynamic range is typically specified in terms of the maximum sound pressure level (SPL) that the microphone can measure for a given total harmonic distortion at its electrical output terminals.[1,2] For all microphones, the interaction of the microphone and preamplifier should be considered when measuring or interpreting the dynamic range. Important distinctions may exist between the upper limit of the operating dynamic range and the maximum SPL and static overpressure that the microphone system can withstand (but not necessarily measure) without damage. The lower limit of dynamic range typically is attributable to noise mechanisms, including thermal noise, in the microphone and preamplifier, as well as the type of signal processing used to measure a band-limited or periodic signal in the presence of noise.

The most accurate and convenient measurements of microphone linearity typically involve exciting a sound source such as a condenser microphone or electrodynamic transducer with a precise electrical signal that can be accurately attenuated in known increments. This source is acoustically coupled (in a coupler, or in a free-field, etc.) to the receiving microphone or microphone system, and the received output voltage level is compared with the excitation signal level for each value of attenuation. If the receiver has already been calibrated by a primary or secondary method at one or more given levels, the differential level linearity of the source–acoustical-coupling–receiver combination can be checked.[11]

6 SELECTED SECONDARY CALIBRATION METHODS

6.1 Direct-Comparison (Substitution) Method

A sound source is placed in an acoustical coupler (for pressure calibration) or in an anechoic chamber (for free-field calibration), and the transfer function relating the source excitation signal and the output voltage of a reference microphone (calibrated by a primary method) is determined.[7] The test microphone of unknown sensitivity is substituted for the reference microphone, and the corresponding transfer function for the source and test microphone is determined. From the ratio of these transfer functions, the known sensitivity of the reference microphone, and consideration of certain potentially significant differences between the test and reference microphones, the sensitivity of the test microphone is determined.

6.2 Reciprocity-Based (Reference Sensitivity and Impedance) Comparison Method

A reference microphone or other sound source for which both the receiving sensitivity and the modulus of the driving-point electrical impedance have been determined by the reciprocity method is used as a sound source, and the test microphone is used as a receiver, in an acoustical coupler or an anechoic chamber. The reference source and the test microphone occupy the same positions in the same coupler or anechoic chamber as have been used for the corresponding sources and microphones in the calibration of this reference. From the measured transfer function relating the output voltage of the test microphone to the voltage driving the reference source, the known sensitivity and electrical impedance of the source, and the known properties of the coupler or anechoic chamber (including the acoustic transfer impedance between the source and receiver), the sensitivity of the test microphone is determined.[22]

This method is particularly applicable to the pressure calibration of microphones in acoustical couplers, especially if hydrogen is used for measurements at high frequencies, because only a single assembly and sealing of the microphones into the coupler is needed, practically halving the labor required for a given calibration by the substitution method. Signal-to-noise ratio limitations in free-field comparison calibrations, due to the low output levels available from microphones used as sound sources, typically require the use of electrodynamic sound sources. These are often more directional, and in particular less stable, than microphone sources. Consequently, free-field calibrations are usually performed by the direct comparison (substitution) method, which relies on the short-term (not the long-term) stability of the electrodynamic source.

6.3 Uncertainties in Determination of Pressure Sensitivity by the Reciprocity-Based (Reference Sensitivity and Impedance) Comparison Method

Because the same couplers are used at the same frequencies in the comparison calibration and in the calibration of the reference microphone by reciprocity, some systematic uncertainty components of the comparison calibration that are associated with the acoustic transfer impedance between the transmitting and receiving microphones are partially canceled and are only as large as they would be in a reciprocity calibration. If the reference microphones are sufficiently stable throughout the time intervals between their calibrations by the reciprocity method, and if certain other conditions are met, the overall uncertainty of calibration approaches that of a reciprocity calibration.[11]

6.4 Uncertainties in Free-Field Calibration by the Direct-Comparison (Substitution) Method

Among the most critical factors in the direct-comparison method are the quality of the anechoic chamber and the limitations imposed by the characteristics of the sound source, including its directivity, output level, linearity, and short-term stability. The anechoic chamber and apparatus used for comparison calibrations are often distinct from, and less well characterized than, those used for free-field primary calibration by the reciprocity method. In this case, differences between the sensitivity levels of the test microphone determined in the comparison calibration and of the reference microphone determined in the primary calibration can be compared with the corresponding differences in sensitivity levels determined for both microphones by the primary calibration using the relatively well-characterized, dedicated anechoic chamber and reciprocity calibration apparatus. At each frequency, the difference of these differences provides a check of the degree to which comparison calibration results approach those of the primary calibration by the reciprocity method. Such checks have been performed with 12.7-mm nominal diameter condenser measuring microphones at NIST,[11] using in the comparison calibration a variety of source excitation signal types and signal processing methods. These signals and methods differed in the degrees to which they were influenced by reflections from the chamber walls, transducer/measurement system nonlinearities, signal-to-noise ratios, and so forth. For the case of a band-limited impulse excitation signal, with signal processing [fast Fourier transform (FFT) analysis] configured to eliminate the most significant of the relatively slight reflections from the interior surfaces of the chamber, the absolute values of these differences were about 0.2 dB or less at frequencies from 2 to 20 kHz and about 0.35 dB or less at frequencies from 20 to 40 kHz.

6.5 Calibrators of the Closed-Coupler Type

Closed-coupler type devices, which include pistonphones, as well as other calibrators incorporating electrodynamic, piezoelectric, or combination transducers, produce a sound pressure (usually a sine-wave signal), most commonly at a frequency or frequencies from 250 Hz to 1 kHz.[23,24] The microphone is inserted into these calibrators so that its diaphragm forms part of the walls of a coupler that is enclosed, except for a capillary tube or vent intended to equalize the coupler interior to ambient barometric pressure. Often, the equalization vent of the microphone is not exposed to the sound field. Most frequently, such calibrators are manufactured to serve as convenient, portable devices generating conveniently high SPL (most commonly from 94 to 124 dB, relative to 20 μPa) for secondary and tertiary laboratory and user field checks of the pressure sensitivity of a microphone system, sound level meter, personal sound exposure meter, or other instrument. Typically, calibrated laboratory standard microphones are used to measure the SPL produced by these calibrators, which are then capable of achieving uncertainties of about 0.15–0.3 dB at their primary calibration frequencies and levels in checking the pressure sensitivities of the instruments under test.

6.6 Electrostatic Actuator Method

An electrostatic actuator (typically a slotted or perforated electrically conductive plate), to which are applied an AC signal voltage and a much larger polarizing DC voltage, is placed in close (e.g., 0.5 mm) proximity to the diaphragm of the microphone to be calibrated. Usually, this diaphragm is at ground potential, and the resulting electrostatic force can be calculated and used to determine the approximate pressure sensitivity of the microphone.[25] Because such absolute determination requires precise measurement of the small separation distance between the actuator and microphone diaphragm, this method is usually performed to determine only the relative frequency response of the test microphone. This response is then combined with a single-frequency pressure calibration, for example, by means of a pistonphone or acoustic calibrator, to determine the approximate pressure sensitivity. This method is widely used by instrument manufacturers[26] and others because it can be applied over a broad frequency range to the rapid, relatively low-cost calibration of individual microphones, as in production line testing. However, the degree to which the actuator-determined response approximates a pressure calibration depends on the relation between the mechanical impedance of the microphone diaphragm and the effective mechanical radiation impedance loading the diaphragm in the presence of the actuator.[27] Consequently, the actuator-determined sensitivity of the microphone differs from its pressure sensitivity. For ANSI Type L laboratory standard microphones, this difference may be as large in absolute value as about 1.5 dB within the frequency range 5–20 kHz;[28,29] for condenser measuring microphones 12.7 mm in nominal diameter, this value may be as large as about 1.3 dB at frequencies from 10 to 20 kHz, and depends on both the actuator and microphone type.[11,22] Therefore, there are no current major international or ANSI standards for the primary or secondary calibration of laboratory standard microphones by the actuator method, and the use of a given electrostatic actuator with a given microphone type to approximate a pressure calibration should be validated by reciprocity or reciprocity-based calibrations performed in couplers. The large AC signal electrostatic fields produced by the actuator do not usually cause a serious problem of crosstalk to the high-input-impedance electrical input terminal of a condenser microphone preamplifier for microphones with electrically grounded metal diaphragms and microphone/preamplifier housings, because this terminal is effectively enclosed in a Faraday cage. However, there can be problems in the case of electret and semiconductor condenser microphones with diaphragms that are not good electrical conductors. The electrostatic actuator calibration of such microphones must always be validated by other calibration techniques that are far less susceptible to such crosstalk.

7 SPECIAL CONSIDERATIONS REGARDING GRADIENT AND COMBINATION MICROPHONES

7.1 Types of Sources, Microphones, and Separation Distance

For these microphones, the results obtained in a given application depend on the nature of the source (simple monopole, dipole, quadrupole, etc.), the frequency of the sound radiated by the source, the kind of microphone (first-order, second-order, or higher-order gradient, combination, etc.), and the distance between the microphone and the source. The source itself must be well characterized and its radiation characteristics must be known. Even in the case of a simple (monopole) source, the responses of gradient microphones, as well as the gradient components of the responses of combination microphones, show pronounced increases (relative to their far-field responses to the same source) at frequencies and distances for which the distance becomes significantly smaller than the sound wavelength.[3–5] This relative rise in low-frequency response is sometimes termed the proximity effect and is even more pronounced for high-order gradient microphones than for those of first order. Indeed, this effect is inherent in the design of certain noise-canceling microphones intended for "close-talking" applications.[5]

7.2 Free-Field Calibration by the Direct-Comparison (Substitution) Method

A common method of calibrating gradient and combination microphones is by comparison with a calibrated reference microphone, using a simple (monopole) source of sound at a given distance in the free field. For these conditions, the sound pressure gradients of first and higher orders can be determined from the measured sound pressure and distance from the source by using harmonic excitation of the source and invoking: (1) the solution of the wave equation in spherical coordinates and (2) the definitions of the sound pressure gradients of the first and higher orders. In such calibrations, the source behavior must be characterized, for example, by measuring its directionality and the sound pressure as a function of distance from the source, and by examining the degree to which measured behavior in the region about the microphone position used for calibration approaches the theoretical behavior of such a source.

7.3 Influence of Calibration Environment

Particular attention in the case of gradient and combination microphones must be given to the degree to which the source characteristics and acoustic field approximate the intended calibration situation. For example, if the sound field radiated by a source that is considered a monopole includes dipole, quadrupole, or even more complicated components, the interpretation of calibration results can be substantially in error, especially at frequencies for which the distance between the source and microphone is smaller than the sound wavelength. In this case, for free-field calibrations, anechoic chamber imperfections (such as reflections from the interior surfaces of the chamber) can cause particularly significant errors. These errors can occur both in the calibration at a specific angle of incidence and in the measurement of directionality, especially if the source and receiver are placed too close to these surfaces, so that the microphone effectively responds to the near-field components of image sources as well as the intended source. At such frequencies and distances for diffuse-field calibrations in a reverberation room, room imperfections can introduce very substantial errors, not only in diffuse-field calibration but also in the measurement of other properties such as the noise rejection provided by close-talking microphones.

To examine the calibration uncertainties associated with imperfections in an anechoic chamber or reverberation room, it may be helpful to perform calibrations at two or more different positions for several different source-receiver separations and to compare differences in the calibration results for different positions with the same separation distance, as well as differences in the results for different separation distances. Comparison of calibration results with results obtained in another chamber of known high quality can also be useful. In frequency ranges for which the microphone response does not vary too much with frequency, differences between calibration results using sinusoidal or other periodic signals that would be expected strongly to excite standing waves, and results using frequency bands of random noise that are much less likely to do so, may be valuable. For anechoic chambers, using transient signals and gating techniques to eliminate major sound reflections from interior surfaces of the chamber can also be helpful.

8 PHASE RESPONSE OF MICROPHONES

Numerous needs exist in experimental science, research and development, and engineering practice for determining the absolute and relative (e.g., to another microphone) sinusoidal-steady-state frequency response phase angles as well as the response moduli (magnitudes) of microphones.[30,31] The pressure and free-field sensitivities obtained during the primary calibration of laboratory standard microphones by reciprocity methods[7,8] can be expressed in terms of complex amplitudes. From these expressions and physical considerations resolving mathematically derived sign ambiguity,[14] the microphone phase response can be obtained.[32] However, there must be further development of these primary standard methods, ideally involving interlaboratory comparisons of calibration results, before these methods can be used readily to determine absolute phase response with well-documented frequency-dependent uncertainties.

Relative pressure and free-field phase response calibrations of microphones by secondary methods are particularly important for the measurement of sound intensity using one or more pairs of pressure microphones (often termed the "p-p intensity probe method").

9 RELATIVE RESPONSES OF PRESSURE MICROPHONES USED IN P-P INTENSITY PROBES

The p-p method for measuring sound intensity is critically dependent on microphone system response angles at low frequencies, at which for a given microphone spacing the phase difference due to the acoustic pressure gradient (and, consequently, the acoustic particle velocity) between the microphones is small. This dependence is especially critical for cases in which one must measure time-varying or nonstationary signals, for which switching techniques[33] intended to lessen this dependence are difficult or impossible to implement. Issues regarding the calibration of p-p acoustic intensity probes are discussed in Chapter 106, and are considered in an IEC standard[34] and an ANSI standard[35] that is nearing completion. The measurement of relative phase response remains an active area of ongoing research, but it appears at present that the most accurate measurements of relative amplitude and phase response can be obtained in couplers[36] and standing-wave tubes at low frequencies, and in the free field at high frequencies.

REFERENCES

1. IEC, *Measurement Microphones—Part 1: Specifications for Laboratory Standard Microphones*, IEC 1094-1, International Electrotechnical Commission, Geneva, Switzerland, 1992.
2. ANSI, *Specifications for Laboratory Standard Microphones*, ANSI S1.12-1967(R1986), American National Standards Institute, New York, 1967.

2a. IEC, *Measurement Microphones, Part 4: Specifications for Working Standard Microphones*, IEC 1094-4, International Electrotechnical Commission, Geneva, Switzerland, 1995.

3. H. F. Olson, *Acoustical Engineering*, Van Nostrand, Princeton, 1957, pp. 275–339.
4. L. L. Beranek, *Acoustics*, 2nd ed., American Institute of Physics for the Acoustical Society of America, New York, 1986, pp. 144–182.
5. H. F. Olson, *Modern Sound Reproduction*, Van Nostrand Reinhold, New York, 1972, pp. 67–104.
6. G. S. K. Wong, "Overview on Acoustical Calibration and Standards," *Canadian Acoust./Acoust. Canadienne*, Vol. 17, No. 3, 1989, pp. 15–22.
7. ANSI, *Method for the Calibration of Microphones*, ANSI S1.10-1966(R1986), American National Standards Institute, New York, 1966.
8. IEC, *Measurement Microphones, Part 2: Primary Method for Pressure Calibration of Laboratory Standard Microphones by the Reciprocity Technique*, IEC 1094-2, International Electrotechnical Commission, Geneva, Switzerland, 1992.
9. L. L. Beranek, *Acoustical Measurements*, rev. ed., American Institute of Physics for the Acoustical Society of America, New York, 1988, pp. 113–148.
10. G. R. Torr and D. R. Jarvis, "A Comparison of National Standards of Sound Pressure," *Metrologia*, Vol. 26, 1989, pp. 253–256.
11. V. Nedzelnitsky, "Laboratory Microphone Calibration Methods at the National Institute of Standards and Technology, U.S.A.," in G. S. K. Wong and T. F. W. Embleton (Eds.), *AIP Handbook of Condenser Microphones: Theory, Calibration, and Measurements*, American Institute of Physics Press, New York, 1995, pp. 145–161.
12. IEC, *Measurement Microphones, Part 3: Primary Method for Free-field Calibration of Laboratory Standard Microphones*, IEC 1094-3, International Electrotechnical Commission, Geneva, Switzerland, 1995.
13. E. D. Burnett and V. Nedzelnitsky, "Free-Field Reciprocity Calibration of Microphones," *J. Res. Natl. Bur. Stand. (U.S.)*, Vol. 92, No. 2, 1987, pp. 129–151.
14. V. Nedzelnitsky, "Primary Method for Calibrating Free-Field Response," in G. S. K. Wong and T. F. W. Embleton (Eds.), *AIP Handbook of Condenser Microphones: Theory, Calibration, and Measurements*, American Institute of Physics Press, New York, 1995, pp. 103–119.
15. H. G. Diestel, "Absolut-Bestimmung des Übertragungsfaktors Von Mikrophonen im Diffusen Schallfeld," *Acustica*, Vol. 10, 1960, pp. 277–280.
16. H. G. Diestel, "Reciprocity Calibration of Microphones in a Diffuse Sound Field," *J. Acoust. Soc. Am.*, Vol. 33, No. 4, 1961, pp. 514–518.
17. L. L. Beranek, *Acoustical Measurements*, rev. ed., American Institute of Physics for the Acoustical Society of America, New York, 1988, pp. 635–639.
18. IEC, *Electroacoustics—Random-Incidence and Diffuse-Field Calibration of Sound Level Meters*, IEC 1183, International Electrotechnical Commission, Geneva, Switzerland, 1994.
19. ANSI, *Procedures for Calibration of Underwater Electroacoustic Transducers*, ANSI S1.20-1988, American National Standards Institute, New York, 1988.
20. K. Brinkmann and H. Goydtke, "Random-Incidence and Diffuse-Field Calibration," in G. S. K. Wong and T. F. W. Embleton (Eds.), *AIP Handbook of Condenser Microphones: Theory, Calibration, and Measurements*, American Institute of Physics Press, New York, 1995, pp. 120–135.
21. F. V. Hunt, *Electroacoustics: The Analysis of Transduction and Its Historical Background*, American Institute of

Physics for the Acoustical Society of America, New York, 1982, pp. 168–212.

22. V. Nedzelnitsky, E. D. Burnett, and W. B. Penzes, "Calibration of Laboratory Condenser Microphones," in *Proceedings of the Tenth Transducer Workshop*, Transducer Committee, Telemetry Group, Range Commanders Council, Colorado Springs, 1979, pp. 27–48.
23. ANSI, *Specifications for Acoustical Calibrators*, ANSI S1.40-1984(R1990), American National Standards Institute, New York, 1984.
24. IEC, *Sound Calibrators*, IEC 942, International Electrotechnical Commission, Geneva, Switzerland, 1988.
25. S. Ballantine, "Technique of Microphone Calibration," *J. Acoust. Soc. Am.*, Vol. 3, 1932, pp. 319–360.
26. E. Frederiksen, "Electrostatic Actuator," in G. S. K. Wong and T. F. W. Embleton (Eds.), *AIP Handbook of Condenser Microphones: Theory, Calibration, and Measurements*, American Institute of Physics Press, New York, 1995, pp. 231–246.
27. G. B. Madella, "Substitution Method for Calibrating a Microphone," *J. Acoust. Soc. Am.*, Vol. 20, 1948, pp. 550–551.
28. W. Koidan, "Calibration of Standard Microphones: Coupler vs. Electrostatic Actuator," *J. Acoust. Soc. Am.*, Vol. 44, 1968, pp. 1451–1453.
29. G. Rasmussen, "The Free Field and Pressure Calibration of Condenser Microphones Using Electrostatic Actuator," *Proc. 6th International Congress on Acoustics D*, published for the International Council of Scientific Unions with the financial assistance of UNESCO, Tokyo, 1968, pp. 25–28.
30. V. Nedzelnitsky, "Development of Standards for Measuring the Phase Response of Microphones," in G. C. Maling (Ed.), *Proceedings of the Inter-Noise Conference 84*, Vol. 2, Noise Control Foundation, Poughkeepsie, NY, 1984, pp. 1323–1328.
31. D. Preis, "Phase Distortion and Phase Equalization in Audio Signal Processing—A Tutorial Review," *J. Audio Eng. Soc.*, Vol. 30, No. 11, 1982, pp. 774–794.
32. F. M. Wiener, "Phase Distortion in Electroacoustic Systems," *J. Acoust. Soc. Am.*, Vol. 13, 1941, pp. 115–123.
33. J. Y. Chung, "Cross-Spectral Method of Measuring Acoustic Intensity without Error Caused by Instrument Phase Mismatch," *J. Acoust. Soc. Am.*, Vol. 64, No. 6, 1978, pp. 1613–1616.
34. IEC, *Electroacoustics—Instruments for the Measurement of Sound Intensity—Measurement with Pairs of Pressure Sensing Microphones*, IEC 1043, International Electrotechnical Commission, Geneva, Switzerland, 1993.
35. ANSI, *Instruments for the Measurement of Sound Intensity*, ANSI S1.9-1996, American National Standards Institute, New York, 1996.
36. G. S. K. Wong, "Phase Match of Microphones," in G. S. K. Wong and T. F. W. Embleton (Eds.), *AIP Handbook of Condenser Microphones: Theory, Calibration, and Measurements*, American Institute of Physics Press, New York, 1995, pp. 247–254.

PART XVI

TRANSDUCERS

109

INTRODUCTION

HARRY B. MILLER

Transduction may be defined as the conversion of mechanical energy to electrical energy and of electrical energy to mechanical energy. A transducer might well have been called a "transformer," but that word was already appropriated. An example of the first type of transducer is the microphone or the phonograph pickup or the accelerometer. An example of the second type of transducer is the loudspeaker or the phonograph record-cutter or the electromechanical shaker.

The basic equations in electric circuit (or mechanical circuit) theory all are derived from Kirchhoff's two laws. Since a very sophisticated discipline eventually evolved in the electrical engineering community, namely linear circuit theory, it was considered expedient to translate the mechanical circuit equations into analogous electrical language, using "equivalent circuit" analogies.

In the impedance analogy for the mechanical circuit: force corresponds to voltage, f to v; linear velocity corresponds to current, u to i; and mechanical impedance corresponds to electrical impedance, z_m to z_e. In the impedance analogy for the acoustical circuit: pressure corresponds to voltage, p to v; volume velocity corresponds to current, U to i; and acoustical impedance corresponds to electrical impedance, z_{ac} to z_e. Now z_m is f/u (or N·s /m) and z_{ac} is p/U (or N·s/m^5).

So the ratio of $z_m : z_{ac}$ is m^4 or area^2. Thus if a mechanical piston were to drive a fluid load, say air or water, the two different equivalent circuits would have to be joined by a mechano-acoustical "transformer." At the junction a simple transformer can be inserted having the turns ratio $A:1$.

Then $z_m = z_{ac} \cdot A^2$ in N·s/m. Here A is the area of the piston or of the loudspeaker diaphragm.

Note: References to chapters appearing only in the *Encyclopedia* are preceded by *Enc.*

All this is discussed in Refs. 1–5. References 3–5 also discuss the mobility analogy, an alternate method for treating mechanical circuits where: velocity corresponds to voltage, u to v and force corresponds to current, f to i.

Once the mechanical circuit has been translated into the equivalent electrical circuit, one can see instantly how increasing a mass here will lower a resonant frequency, or adding a spring there will provide an additional resonance. A great many very interesting examples can be found in Ref. 1. And indeed many of the papers in this part are concerned with design problems involving equivalent circuits. This introductory chapter does not repeat the basic points given in the chapters that follow. Instead it will present comments on some of these chapters.

Thus the major portion of Chapter 110 is concerned with reciprocal devices. A smaller portion treats nonreciprocal devices. Nonreciprocal microphones are basically modulation devices. Hence a carrier wave is required. The carrier can be either DC or high-frequency AC. When an electric carrier (e.g., charge or voltage) is removed, the transducer, if it can function at all, will function only as a loudspeaker (or "projector"), not as a microphone or "hydrophone." These points are discussed at more length in Ref. 6. Note that when the transducer is a magneto-mechanical device, the carrier would be magnetic flux rather than electric charge.

In the discussion of electrostrictive devices the authors point out that force is proportional to charge squared, but that when a polarizing charge or DC carrier Q_0 is added, the force becomes primarily proportional to charge. So the polarizing charge is a linearizing device. Note that a similar thing happens with the ordinary electromagneto-mechanical telephone earphone, or "receiver." Here a permanent magnet's flux density B is the linearizing device, analogous to Q_0 in the electrical case. The action of the polarizing flux density B is discussed in Ref. 7.

In the linear piezoelectric discussion, the authors connect the mechanical domain with the electrical domain by using the e constant for their piezoelectric coefficient. The more usual tabulated piezoelectric coefficients are the d constant and the g constant. All these are discussed in Ref. 8. A useful memory aid device for the four piezoelectric constants can be constructed (Fig. 1).

In Fig. 1 S is strain in metres/metre; E is electric field in volts/metre; D is electric flux density in Coloumbs/metre squared; T is stress in newtons/metre squared. Note that when d = mech/el = S/E, the inverse relation, el/mech, is not $1/d$ and therefore E/S but is another *form* of d, namely D/T. Likewise g is S/D and E/T. If we wish to use D/S or T/E, we must go to another family, the e family. Similarly, if we want E/S or T/D, we are in the h-family. However, the most readily available data sheets (from manufacturers) supply only the g constant and d constant data. The other two constants then can be derived, using in addition Young's modulus Y or its inverse s.

A similar memory aid device, or "matrix," can be constructed for the piezomagnetic coefficients. The E is replaced by H, the magnetic intensity or magnetic field strength. And D is replaced by B, the magnetic flux density. Again, note that the inverse of $f = Bli$ is not $i = f/Bl$ but rather $v = Blu$, Faraday's great discovery.

Chapter 111 gives a thorough discussion of the factors that control the design of a loudspeaker. To obtain the maximum value from this chapter, it would be desirable for the reader to have at hand a pad and pencil in order to follow along with the author's exposition. Thus, after a systematic background has been presented of the basic requirements for a loudspeaker system, the author is ready to give some comparative evaluations. For example, his comparison of bass reflex systems with acoustic suspension systems (two types) is broad and thorough. Then the PAE (power available efficiency) and Fig. 15 are discussed at some length. Five distinct frequency regions (or bands) are treated in Fig. 15. Years ago, the response used to be described by three regions and treated a little differently. This treatment was easy to remember. Thus, after the superposing of Figs. 1 and 12 onto Fig. 15: Region 1, roughly corresponding to region A, covers all frequencies *below* f_{res} of the voice coil if they are simultaneously below $ka = 2$. This is because we are in the stiffness-controlled region of the voice coil or mechanical-resonance circuit, where the velocity has a slope of 6 dB/octave. And we have an additional 6 dB/octave due to the rising of the radiation resistance R_{rad} in this region. The two working together provide roughly a 12-dB/octave slope.

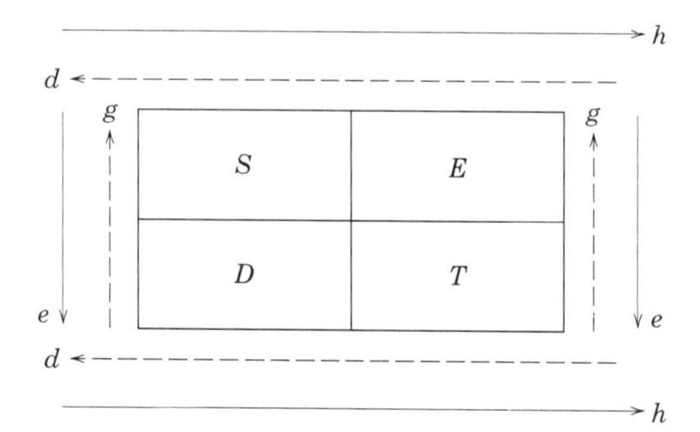

Fig. 1 Memory aid for the four piezoelectric constants g, d, e, h.

Region 2, corresponding to region C, covers frequencies *above* f_{res} of the voice coil if they are simultaneously below $ka = 2$. This is because we are in the mass-controlled region of the voice coil, where the velocity has a slope of -6 dB/octave. The still-rising radiation resistance of R_{rad} contributes a 6-dB/octave slope, which just nullifies the voice coil slope. So the result is a flat response. Recall that 6 dB means 4 : 1 for power or resistance.

Region 3, roughly corresponding to region D, refers to a coincidence of the mass-controlled region of the voice coil (-6 dB/octave) with the ka of Fig. 1 being greater than 2 (0 dB/octave for R_{rad}). This gives the -6 dB slope of region D when generator output is constant voltage.

Further refining of this analysis explains region B (resistance-controlled) and region E (generator output) dropping -6 dB/octave because of voice coil inductance, giving a total of -12 dB/octave.

The discussion on horns in *Enc.* Ch. 161 is written by Vincent Salmon, who is the inventor of a third family of horns, the catenoidal horn. The other two best-known families are the conical horn and the exponential horn. The three families are discussed in detail by Morse and by Salmon (see Salmon's Refs. 12 and 6 in *Enc.* Ch. 161).

In referring to Fig. 3 of *Enc.* Ch. 161, the horn's normalized input reactance vs. frequency, Salmon mentions in two places that the reactance above the cutoff frequency is masslike, but that the slope is primarily negative. It should be noted that when the slope of any masslike reactance is negative, this reactance always acts like a negative *stiffness*. Such an element is well behaved and can be used to reduce the value of an adjacent positive stiffness. The three elements are illustrated in Fig. 2. Note that *stiffness* s is merely the reciprocal of *compliance* c.

Another example of a negative stiffness reducing the value of a positive stiffness would be the stiff steel dia-

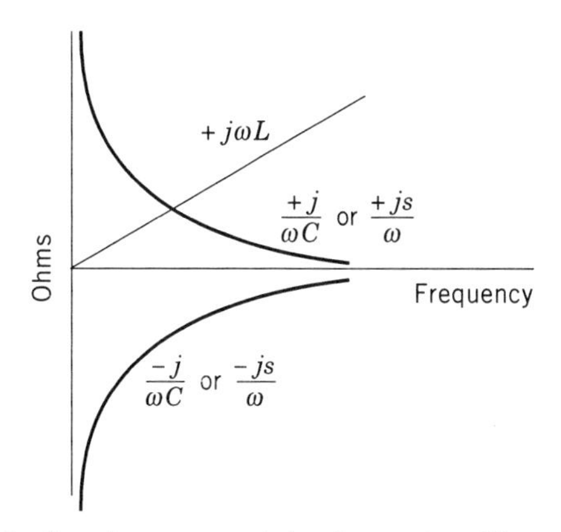

Fig. 2 How the reactance $+js/\omega$ of a negative stiffness varies with frequency.

phragm of an ordinary telephone receiver being pulled in by the magnetic field of the pole pieces until the diaphragm almost "oil-cans," or bottoms-out, against the pole pieces. In this case the magnetic field strength H can be modeled as a negative stiffness. This negative stiffness impedance is in series with the positive stiffness of the steel diaphragm. Hence it can be seen that the net stiffness could easily become zero. And indeed this is the situation just before "oil-canning" is allowed to occur.

A few comments on Salmon's references are in order. Hanna and Slepian[9] saw the behavior of nonresonant horns quite differently from Webster (see Ref. 3 of *Enc.* Ch. 161). Among other things Ref. 9 makes use of equivalent circuits and discusses the reciprocal use of horns as receivers of sound, in addition to comparing exponential horns and conical horns mathematically.

Musical wind instruments make use of resonant horns, which have quite different characteristics from nonresonant horns. This subject is discussed in Refs. 10 and 11.

For the reader interested in directionality, the best references outside of Olson,[2] probably reside in the antenna literature. An excellent paper by R. S. Elliott appears in Ref. 12. Another excellent treatment is Ref. 13. The material in Chapter 4, of Ref. 13 is especially useful.

Chapter 112 is concerned with types of microphones. This chapter discusses, among other things, condenser, carbon, and ribbon microphones, as well as calibration techniques.

The carbon microphone has been manufactured in the millions around the world, although the carbon particles "cake up" in high-humidity weather and then produce much distortion. The reason for its popularity is that it is a combination amplifier (simple and cheap) and microphone. The variable resistor created by the impinging sound wave can be modeled by a fixed resistor R_0 in series with an AC voltage generator e_{ac} and a DC bias voltage E_0. Then $e_{ac} = \alpha E_0 \sin \omega t$. So the sensitivity e_{ac}/p is proportional to E_0, which is usually about 6 V, safe and easy to implement. Note that this cheap amplifier was available long before vacuum tubes and transistors were invented. The action can also be thought of as a modulation process, with the DC bias voltage acting as the carrier. And, in fact, if the DC voltage were replaced with a high-frequency AC voltage, the microphone could be made to work equally well. This discussion could also be applied to the condenser microphone,[6] another modulation device, where the DC carrier is often 200 V. And here again, the DC carrier could be replaced with a high-frequency AC carrier, producing either an AM or an FM wave. Note that in any modulation situation, the audio wave is being amplified in proportion to the carrier voltage E_0.

The ribbon microphone or velocity or pressure-gradient microphone provides a strong bass boost to a close-talking speaker's voice. This is very noticeable when a radio crooner changes over from a pressure microphone to a pressure-gradient microphone. A good explanation is the following. The sensitivity of a pressure-gradient microphone can be rewritten from e/p to e/u, where u is the particle velocity of the air molecules. This is useful because the velocity microphone follows the particle velocity of the air more closely than it follows the pressure. The sound radiating from a projector, for example, the human mouth, passes through the near field to the far field. Using the far-field sound pressure p as a fixed reference, we can compare the near-field particle velocity ratio u spherical wave/p to the far-field particle velocity ratio u planar wave/p. Then for a given reference pressure p the ratio u sphere/u plane is

$$\sqrt{1 + \frac{1}{k^2 r^2}}$$

Here $k = \omega/c$ and r is the distance from the singer's mouth to the microphone.

As the frequency moves down from 1000 to 100 Hz at a distance of 6 in., the velocity magnification, and hence the bass boost, becomes approximately 11 dB.

Turning now to calibration, in the discussion *absolute direct calibration* the authors refer to the pistonphone, a reciprocating piston for changing the volume of a calibrated cavity in order to generate a standard sound pressure level. A pistonphone driven by a variable-speed motor can produce an absolute frequency response down to a fraction of 1 Hz. However, an interesting thing happens at such low frequencies. Recall that at 1000 Hz the relationship between pressure and volume is $\Delta p = (\gamma p_0/V_0)\,\Delta V$. This is a simple "spring equation" where $\gamma p_0/V_0$ is the cavity stiffness and $\gamma = C_p/C_v$, the ratio of the two specific heats of the working gas. In the case of air, $\gamma = 1.4$. This equation holds for any adiabatic compression/expansion. But, at frequencies lower than 1 Hz, the slow compression/expansion is no longer adiabatic but isothermal instead. Therefore, the factor 1.4 reduces to 1, which is a 3-dB decrease in the generated value of Δp. And so the flat frequency response gradually sinks to a new plateau 3 dB lower than that at 1000 Hz, even though the microphone's true sensitivity is 3 dB higher (i.e., flat all the way down to DC).

Magnetic recording and reproducing systems are discussed in Chapter 113 by M. Camras, who invented much of the art. Mostly, it is concerned with analog recording, which has more difficult problems than the fast-growing digital recording art.

It is possible to record on magnetic tape using longitudinal recording alone or perpendicular recording alone or a combination of the two (isotropic recording). Camras deals with longitudinal recording, where the particles (e.g. ferrite) on the tape have been magnetized so as to line up parallel to the direction of motion of the tape.

One of the great unresolved questions in the magnetic recording art was, for a long time, the detailed mechanism of high-frequency AC bias. This question was discussed perhaps best by Camras himself[17] and by Eldridge.[18] Briefly, the unbiased transfer function between recording field H and remanent flux density B_r is shown by the solid curve in Fig. 3. This is highly nonlinear in the region near the origin and would produce intolerable distortion in a simple speech or music recording made with neither DC bias nor AC bias. Note that this remanence curve should not be confused with the more familiar initial magnetization curve that proceeds to generate a family of hysteresis loops.

When an AC bias field of perhaps 150,000 Hz is superposed on a signal field of 15,000 Hz, the composite field is as shown in Fig. 4. Note that the appearance is quite different from that of an amplitude modulation curve, which is a multiplication. The bias field is an addition. The effect of the AC bias field is to produce the transfer function shown by the dotted curve in Fig. 3. It used to be said that the mechanism of the AC bias field was to forcibly straighten out the solid curve of Fig. 3. This gives a misleading picture. The best explanation of what is probably going on is given by Camras[17] and by Eldridge.[18] Eldridge shows that the dotted curve is approximately the same as the S-shaped cumulative distribution curve appearing in probability theory, which in turn is the integral of a normal distribution function. That curve describes the random distribution of the "interaction field" set up by thousands of ferrite particles. When the signal field is added to this interaction field, the calculated result gives the dotted curve of Fig. 3. It should be noted that these interaction fields have actually been measured and are found to agree with Eldridge's calculations. Moreover, his linearized remanence curve of Fig. 3 agrees well with experimental results.

The subject of digital audio is taken up in Chapter 114.

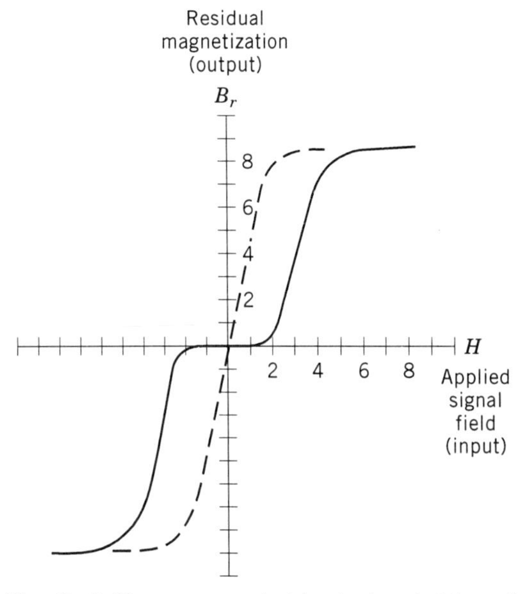

Fig. 3 B–H curve or output–input characteristics of a magnetic recording on tape, without bias (solid curve) and with AC bias (dashed curve).

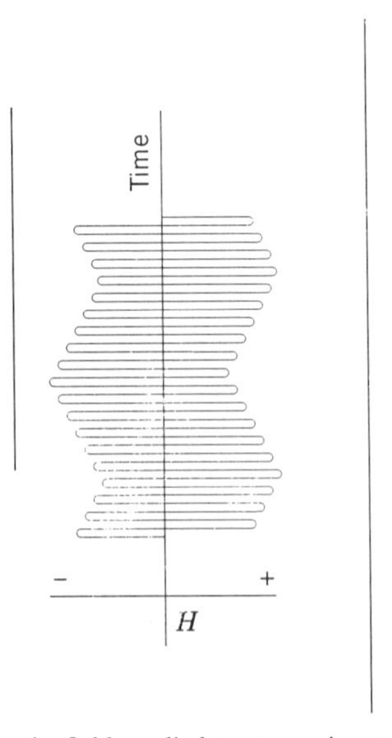

Fig. 4 Magnetic field applied to a moving tape, showing high-frequency bias field *added to* low-frequency signal field.

Although the compact disc (CD) is probably the most visible result of the digital revolution, the authors arrive at it methodically and carefully, first discussing sampling of analog signals in the time domain and frequency domain. This leads to discussions on aliasing, then to antialiasing filters. Then quantization noise is discussed, accompanied by good references. Digital-to-analog and analog-to-digital converters are explained and are followed by analyses of granulation noise, dither, and oversampling. This last leads to dicussions on noise reduction and delta-sigma converters. At this point the authors are ready to discuss commercial digital hardware including CDs, digital audio tapes (DATs), and other desirables. Then a deeper look into the working of digital devices such as finite impulse response (FIR) filters, reverberators, and sound synthesizers. Finally, mention is made of noise reduction and restoration of old recordings, the elimination of wow and flutter, and the ease of creating one's own digital audio workstation. The reference section is broad and deep.

REFERENCES

1. H. F. Olson, *Dynamical Analogies*, Van Nostrand, Princeton, NJ, 1943.
2. H. F. Olson, *Acoustical Engineering*, Van Nostrand, Princeton, NJ, 1957.
3. L. L. Beranek, *Acoustics*, McGraw-Hill, New York, 1954.
4. O. B. Wilson, *An Introduction to the Theory and Design of Sonar Transducers*, Superintendent of Documents, U.S. Government Printing Office, Washington, DC, 1985.
5. R. S. Woollett, *Sonar Transducer Fundamentals*, Naval Underwater Systems Center, Newport, RI, 1990.
6. H. B. Miller, *Acoustical Measurements*, Hutchinson Ross (now Van Nostrand, New York), 1982; see Chapters 12 and 14 and editor's notes.
7. W. L. Everitt, *Communication Engineering*, McGraw-Hill, New York, 1932, 1937, 1956; see chapter entitled "Electromechanical Coupling."
8. W. P. Mason, *Physical Acoustics*, Vol. 1, Part A, Academic, New York, 1964, see pp. 182–193.
9. C. R. Hanna and J. Slepian, "The Function and Design of Horns for Loudspeakers," *A.I.E.E.*, Vol. 43, 1924, pp. 393–411. Also MORSE.

10a. G. W. Stewart and R. B. Lindsay, *Acoustics*, Van Nostrand, New York, 1930.

10b. P. M. Morse, *Vibration and Sound*, 2nd ed., McGraw-Hill, New York, 1948.

11. A. H. Benade, *Fundamentals of Musical Acoustics*, Oxford University Press, New York, 1976.
12. R. C. Hansen, *Microwave Scanning Antennas*, Academic, New York, 1966.
13. B. D. Steinberg, *Principles of Aperture and Array System Design*, Wiley, New York, 1976.
14. J. Eargle, *Handbook of Recording Engineering*, Van Nostrand Reinhold, New York, 1986.
15. G. H. P. Giddings, "Audio System Design and Installation," Howard Sams, Indianapolis, IN, 1990.
16. A. H. Burdick, "The Clean Audio Installation Guide," Benchmark Media Systems, Syracuse, NY, 1985.
17. M. Camras, "Current Problems in Magnetic Recording," *Proc. IRE*, Vol. 50, 1962, pp. 751–761.
18. D. F. Eldridge, "The Mechanism of AC-Biased Magnetic Recording," *IRE Trans Audio*, AU-9, No. 5, 1961, pp. 155– 158.

110

TRANSDUCER PRINCIPLES

ELMER L. HIXSON AND ILENE J. BUSCH-VISHNIAC

1 INTRODUCTION

Transducers are devices that transform energy from one form to another. We will present devices that transform mechanical or acoustical energy to electrical energy and the reverse. Some are bidirectional or reciprocal and some are nonreciprocal.

Transducers have application as microphones (see Chapter 112) for receiving sound waves and as loudspeakers (see Chapter 111) for generating sound waves in gasses. In liquid media they are used for sending and receiving sonar signals (see *Enc.* Ch. 140) and sound waves to probe the human body (see *Enc.* Ch. 144). In solids, transducers are used to generate and sense stress waves in the earth (see Chapter 54) and for vibration testing of structures (see Chapter 56). In fact, the application of almost every facet of acoustics presented in this book requires a transducer.

Here we will develop the relationships between acoustical, mechanical, and electrical variables for both reciprocal and nonreciprocal devices. This will involve the physical properties of certain materials that relate these variables to each other. Mechanical stresses and strains can perturb the electrical, magnetic, or resistive properties of these materials. Some of these relationships are fundamentally linear and some square law. It will be shown that polarization can be used to linearize the later devices and make them effectively reciprocal.

After presenting the fundamental transducer relationships and modeling methods to predict their performance, various materials that can be used to make reciprocal transducers will be discussed.

Section 2 is limited to the discussion of reciprocal devices, and Section 3 will present nonreciprocal devices. These have electrical properties that are resistive, capacitive, or inductive. Finally, Section 4 will discuss devices and methods of transduction that are currently considered new technologies.

2 RECIPROCAL DEVICES

There is a large class of materials and devices whose electrical and mechanical properties are coupled. A number of these are inherently linear and reciprocal. Others are inherently square law and not necessarily reciprocal. However, by electrical polarization, they can be made to act in a linear and reciprocal fashion.

These materials and devices can be divided into dielectric types, in which electric fields are perturbed, and magnetic types, in which magnetic fields are perturbed. The linear electromechanical relationships may result from forces on bound charges or current flow in magnetic fields. Square-law characteristics result from material properties that change with density. Square-law characteristics also result from a change in electric or magnetic stored energy resulting from a mechanical deformation. Electric field devices will be discussed first.

2.1 Electric Field Types

Square-Law Characteristics

Capacitive: Varying Plate Separation. Transducers that utilize the relations between stored electrical and mechanical energy generally take the form of parallel-plate capacitors, as shown in Fig. 1. When a charge Q is applied, the stored electrical energy W is given as

$$W = \frac{1}{2}\frac{Q^2}{C} = \frac{Q^2 d}{2\epsilon A}, \qquad (1)$$

Note: References to chapters appearing only in the *Encyclopedia* are preceded by *Enc.*

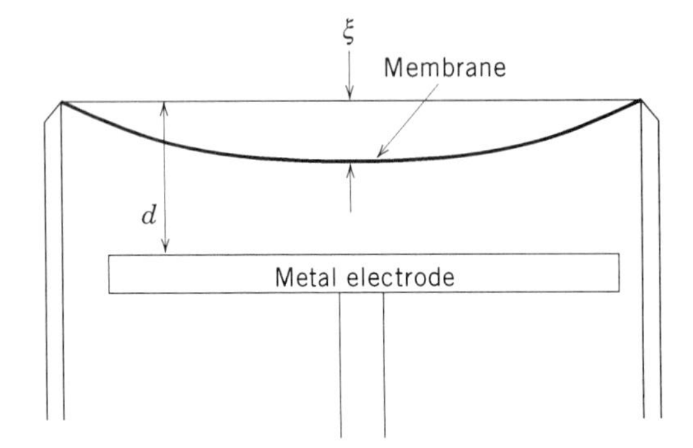

Fig. 1 Parallel-plate capacitive transducer.

where C is the capacitance, $\epsilon A/d$, and ϵ is the permittivity of the dielectric between the plates of area A and spacing d.

When d is decreased by a displacement ξ with constant Q, the stored energy becomes

$$W = \frac{Q^2}{2\epsilon A}(d - \xi). \tag{2}$$

By differentiating this energy equation with respect to displacement ξ and charge Q, the force and potential relations can be obtained.

The force on the plates is then

$$f = -\left.\frac{\partial W}{\partial \xi}\right|_Q = \frac{Q^2}{2\epsilon A}, \tag{3}$$

which is quadratic in Q.

With a polarizing charge Q_0 added to a small charge q, a static force $Q_0^2/2\epsilon A$ results that deflects the membrane and a nearly linear force–charge relation results (ignoring a small distortion term). (see Ref. 1, Chapter 3):

$$f = \frac{Q_0}{\epsilon A} q. \tag{4}$$

A nearly linear potential displacement relation also results if the term that is quadratic in q is ignored:

$$v = \frac{Q_0}{\epsilon A} \xi. \tag{5}$$

(See Ref. 2, Chapter 6.)

Using a frequency-domain representation, we can write two-port equations in terms of sinusoidal potentials V, current I, force F, and velocity U:

$$V = Z_e^u I + \alpha U, \qquad F = \alpha I + Z_m^i U. \tag{6}$$

Here $\alpha = Q_0/j\omega\epsilon A$, Z_e^u is the electrical impedance with the mechanical side clamped, and Z_m^i is the mechanical impedance with the electrical side open circuited.

Using the mobility analog with force analogous to current and velocity analogous to potential, the circuit representation of Fig. 2a can be drawn. Here Z' is the inverse of Z_m^i and is the mechanical mobility with the electrical side open circuited. Transforming to the electrical side, the simple form of Fig. 2b results. In the example of Fig. 2c a mechanical force F_s is applied. Then the potential V across the electrical load Z_L can be found by simple circuit theory. Of course, the mechanical output due to an electrical input can readily be calculated.

For circuit representations using the impedance analog, see Ref. 2, p. 108.

For the pressure sensor of Fig. 1, $F = PA$ and $U = \overline{U}/A$, where P is acoustic pressure, $\overline{U}$ is acoustic volume velocity, and A is the membrane area. The two-port equations become

$$V = Z_e^u I + \frac{\alpha}{A}\overline{U}, \qquad P = \frac{\alpha}{A} I + Z_a^i \overline{U}. \tag{7}$$

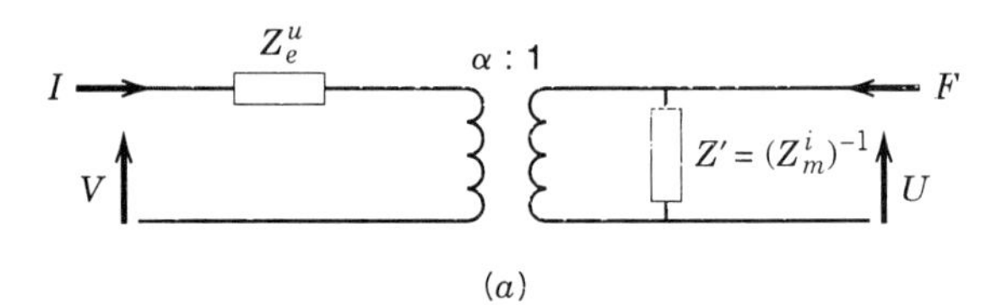

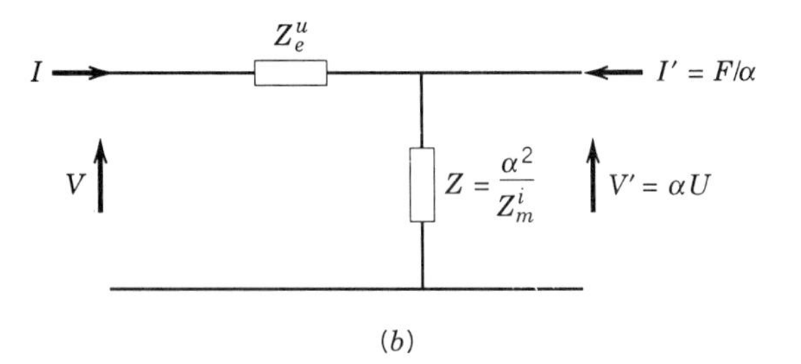

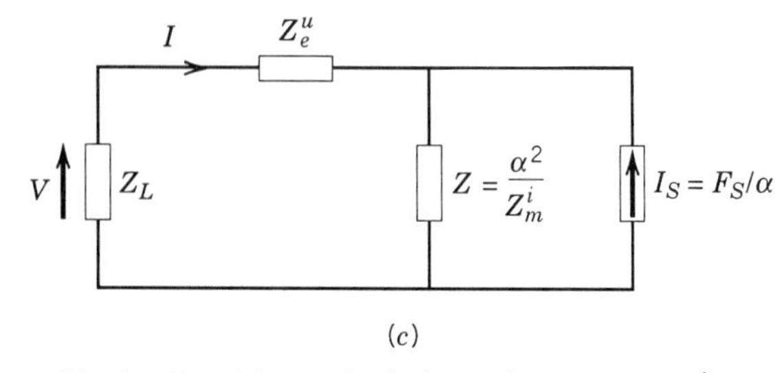

Fig. 2 Capacitive mechanical transducer representation.

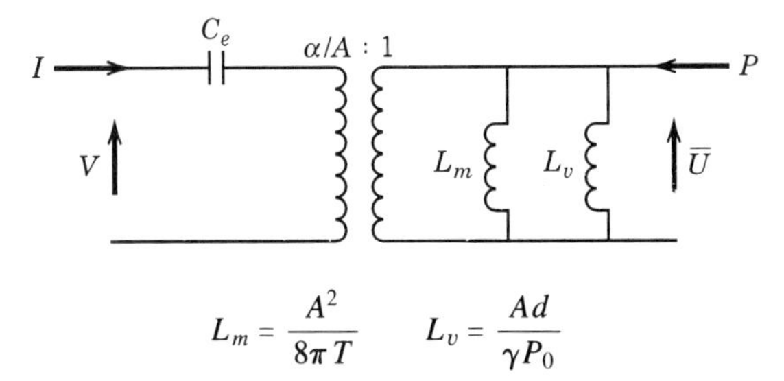

Fig. 3 Capacitive acoustic transducer representation.

Here $Z_e^u = 1/j\omega C_e$ and $Z_a^i = 1/j\omega C_m + 1/j\omega C_v, C_e$ is electrical capacitance, C_m is the compliance of the membrane under tension T, C_v is the compliance of the air with ratio of specific heats γ and static pressure P_0, and α is as in Eq. (6).

If P is chosen to be analogous to I and $\overline{U}$ analogous to V, a mobility analog, the circuit representation of Fig. 3 results, where Z_a^i has been replaced by the symbols for the compliance of the membrane, L_m, and the compliance of the air. With suitable electrical sources and/or impedances attached and suitable acoustical sources and/or impedances, using the simplification of Fig. 2, the response as a microphone or loudspeaker can be predicted.

An important class of capacitive transducers is of the type discussed above, but they use a membrane made of a dielectric material that will retain a charge after polarization. These are called electrets and require no external polarization. One side is metalized to provide the electrical connection. The operation and representation is then identical to the externally polarized transducers discussed above.[3]

Electrostrictive. Electrostrictive materials are those whose dielectric properties change with density changes caused by a deformation. When electric energy is stored in the material, a change in mechanical energy caused by deformation induces a change in stored electrical energy. (See Ref. 1, Chapter 2.)

When a cube of such material with side a is placed in an electrical field with flux density D, the electrical energy stored is

$$W = \frac{1}{2}\frac{a^3D^2}{\epsilon}. \tag{8}$$

If the electrical permittivity ϵ is a function of density ρ, the change in stored energy with change of density becomes

$$dW = -\frac{1}{2}a^3\frac{D^2}{\epsilon^2}\frac{\partial\epsilon}{\partial\rho}\,d\rho. \tag{9}$$

The force acting on a surface becomes

$$f = \frac{1}{2}\frac{\partial\epsilon}{\partial\rho}\frac{\rho}{a^2\epsilon^2}Q^2, \tag{10}$$

where Q is the total electrical charge. Thus force is proportional to charge squared.

If a polarizing charge Q_0 that is large compared to q is applied, a nearly linear force–charge relation results:

$$f = \frac{\partial\epsilon}{\partial\rho}\frac{\rho}{a^2\epsilon^2}Q_0q. \tag{11}$$

An approximately linear potential displacement relation can be obtained as well:

$$v = \frac{\partial\epsilon}{\partial\rho}\frac{\rho}{a^2\epsilon^2}Q_0\xi. \tag{12}$$

Again, using a frequency-domain representation, two-port equations can be written:

$$V = Z_e^uI + \alpha U, \qquad F = \alpha I + Z_m^iU, \tag{13}$$

where

$$\alpha = \frac{\partial\epsilon}{\partial\rho}\frac{\rho}{a^2\epsilon^2}\frac{Q_0}{j\omega}$$

and Z_e^u is electrical impedance, mechanically clamped, and Z_m^i is mechanical impedance, electrically open circuited.

The circuit representation is identical to that of Fig. 2.

Linear Relationships: Piezoelectric Crystals

Crystalline materials that lack symmetry exhibit the piezoelectric effect. That is, the application of a strain produces a charge and the application of a potential produces a stress. The lack of symmetry results in an anisotropic material in which the electrical and elastic properties depend on direction with respect to the crystal lattice or the direction of polarization for polycrystaline materials. Relationships between them can be represented by a set of state equations. For example, when strain S and electric field strength E are independent variables, the stress T and charge density D are given by

$$D = \epsilon^SE + eS, \qquad T = -e_tE + c^eS. \tag{14}$$

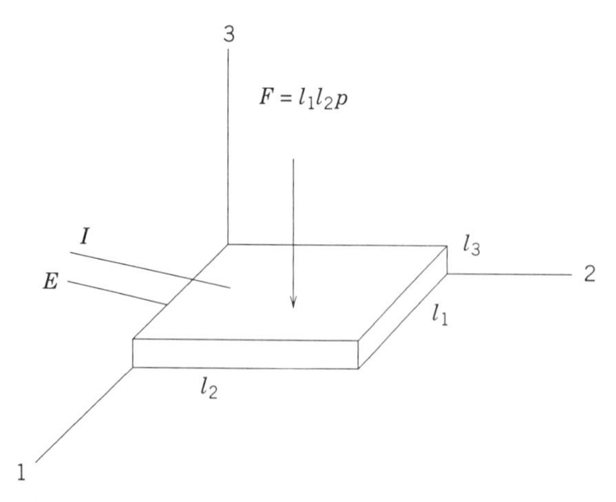

Fig. 4 Flat-plate piezoelectric sensor. Force and motion and potential and current in the 3-direction.

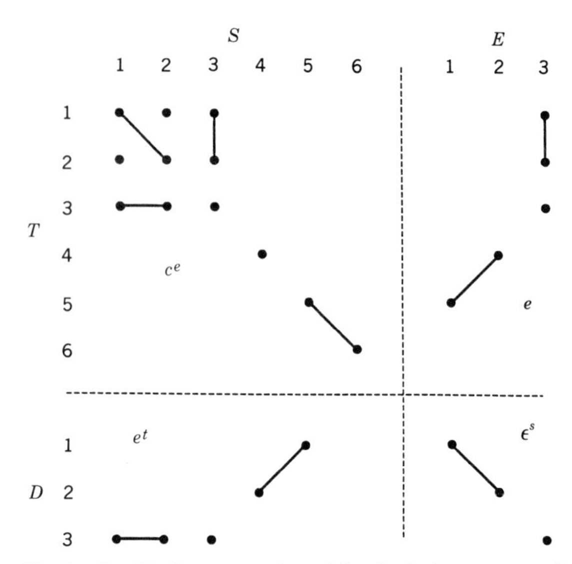

Fig. 5 Graphical representation of the physical parameters of barium titanate.

The dielectric constant ϵ^s, measured under fixed strain, can be represented by a 3×3 matrix of values. Values may differ depending on the three orthogonal directions 1, 2, 3, as in Fig. 4. Cross coupling between the directions may occur. The stiffness term c^e measured with fixed potential is a 6×6 matrix because of the three translational and three rotational directions. The coupling term e or its transpose e_t is a 3×6 or 6×3 matrix that represents coupling between the three electric field direction and six mechanical directions.

Because of symmetry and many zero values in these matrices, a representation shown in Fig. 5 is useful in applying these materials. The dots indicate which coefficients are nonzero and the connection lines indicate equal values. Tables of material properties will give values for the coefficients.[4]

When a thin plate of piezoelectric material as shown in Fig. 4 is subjected to forces and motion in only one direction and the electrical potential applied in the same direction by a thin conducting surface, only the properties in that direction are significant. Then the equations of state (14) become

$$D_3 = \epsilon_{33}^s E_3 + e_{33} S_3, \qquad T_3 = -e_{33} E_3 + c_{33}^e S_3, \qquad (15)$$

where polarization is in the 3-direction. A frequency-domain representation yields a set of two-port equations of terminal properties:

$$I = Y_e V + \alpha U, \qquad F = -\alpha V + Z_m U, \qquad (16)$$

where V and I are potential and current and F and U are force and velocity:

$$Y_e = j\omega C_e, \qquad C_e = \epsilon_{33}^s \left(\frac{l_1 l_2}{l_3} \right),$$

$$Z_m = j\omega M/3 + \frac{1}{j\omega C_m},$$

$$M = \rho l_1 l_2 l_3, \qquad C_m = l_3/(c_{33}^e l_1 l_2), \qquad \alpha = \frac{e_{33} l_1 l_2}{l_3}.$$

Using the impedance analog $V \sim F, I \sim U$, the transducer can be represented by the circuit of Fig. 6 at frequencies low enough to neglect wave motion.

When both sides of the transducer plates are available for external connection, the device becomes a three-port. When wave effects cannot be neglected, the circuit model of Fig. 7 may be used. See Ref. 5 for complete coverage of circuit representations.

As seen in the above representation, the coupling between mechanical and electrical quantities involves the properties of the piezoelectric materials and is a linear

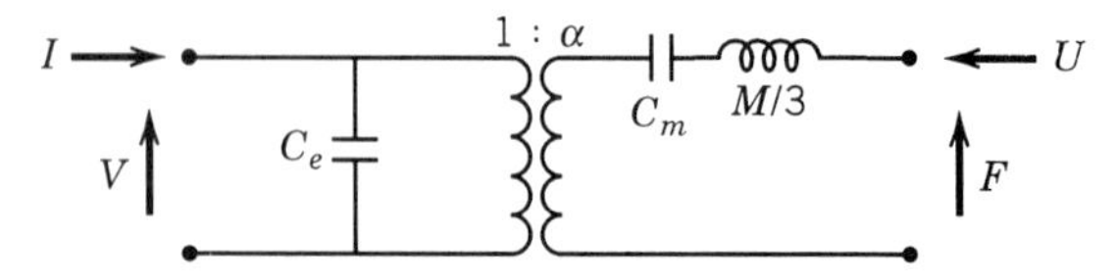

Fig. 6 Equivalent circuit of a transducer based on the two-port equations.

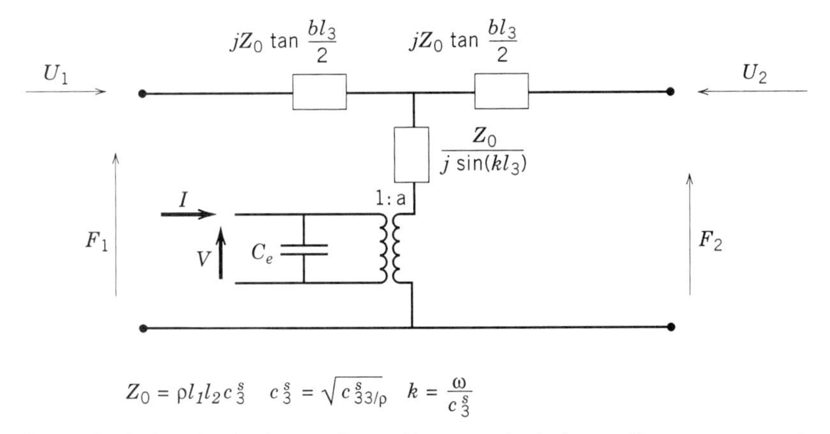

Fig. 7 Equivalent circuit of a transducer with two mechanical ports. Stress wave properties are included.

relationship. The application of an electrical source produces a source of mechanical energy and vice versa.

2.2 Magnetic Field Types

Square-Law Characteristics

Moving Armature or Electromagnetic. When magnetic flux passes through an area as shown in Fig. 8, the stored energy is

$$W = \frac{1}{2} \frac{\Phi^2}{\mu A} d, \tag{17}$$

where Φ is the total flux and μ is the permeability of the material in the volume Ad. When d is allowed to decrease by a small amount ξ with Φ constant, Eq. (17) becomes

$$W = \frac{1}{2} \frac{\Phi^2}{\mu A} (d - \xi). \tag{18}$$

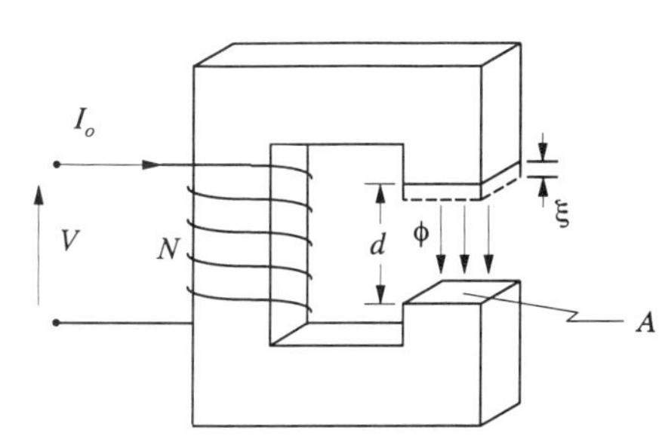

Fig. 8 Magnetic field stored energy sensor.

By differentiating energy with respect to displacement and flux, force and current relationships can be obtained.

Then the force required to move the distance ξ becomes

$$f = -\left.\frac{\partial W}{\partial \xi}\right|_{\Phi} = \frac{1}{2} \frac{\Phi^2}{\mu A}, \tag{19}$$

so f is quadratic in Φ. For a polarizing flux Φ_0 and a small dynamic flux ϕ, a nearly linear relationship results:

$$f = \frac{\Phi_0 \phi}{\mu A}. \tag{20}$$

When Φ_0 is produced by a current I_0 in a coil of N turns in the magnetic circuit, the force relation becomes

$$f = \frac{N}{d} I_0 \phi. \tag{21}$$

When ϕ is produced by a potential v applied to the N-turn coil, the linear force term becomes

$$f = \frac{I_0}{d} \int v \, dt. \tag{22}$$

When d is displaced by ξ, a current I flows in the N-turn coil:

$$I = \frac{\partial W}{\partial \Phi_0} = \frac{\Phi_0}{\mu A} (d - \xi). \tag{23}$$

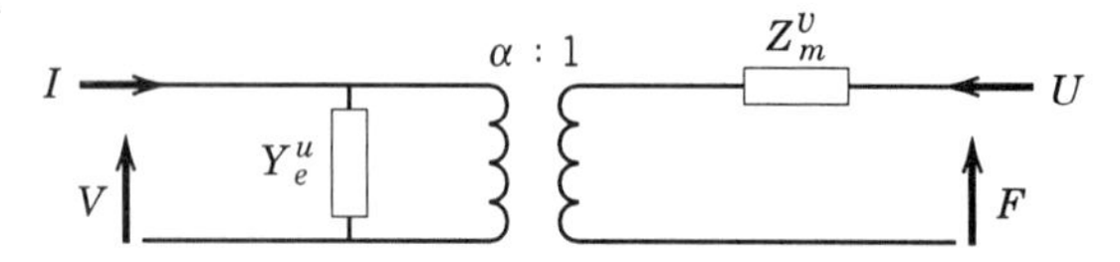

Fig. 9 Circuit representation of a magnetic field mechanical transducer.

The dynamic term then becomes

$$i = -\frac{\Phi_0}{\mu A}\,\xi = -\frac{NI_0}{d}\,\xi. \tag{24}$$

With a polarizing current I_0 the linear two-port representation results:

$$I = Y_e^u V - \alpha U, \qquad F = \alpha V + Z_m^V U, \tag{25}$$

where Y_e^u is the clamped electrical circuit admittance, Z_m^v is the shorted mechanical system impedance, and the electromechanical coupling term is $\alpha = NI_0/j\omega d$. When F is analogous to V and U analogous to I, the circuit representation of Fig. 9 can be drawn. Then the response to electrical or mechanical sources can be predicted.

When the magnetic circuit with an air gap is coupled to a diaphragm as in Fig. 10, this transducer principle can be used as a microphone or loudspeaker.

Magnetostrictive. In certain materials the magnetic permeability changes with a change in density caused by a mechanical deformation. When energy is stored in the material in the form of magnetic flux, a change in permeability due to a change in mechanical energy causes a change in the stored magnetic energy resulting in a change in the magnetic field. Likewise, a change in magnetic flux can cause a change in the mechanical fields.

If a cube of magnetostrictive material with side a is placed in a magnetic field with flux density B, the stored energy is

$$W = \frac{1}{2}\,a^3\,\frac{B^2}{\mu}. \tag{26}$$

Since μ is a function of density, the change in stored energy with ρ becomes

$$dW = -\frac{1}{2}\,a^3\,\frac{B^2}{\mu^2}\,\frac{\partial\mu}{\partial\rho}\,d\rho. \tag{27}$$

Then the surface stress becomes

$$T = \frac{1}{2}\,\frac{\partial\mu}{\partial\rho}\,\frac{\rho}{\mu^2}\,B^2.$$

When a rod of cross-sectional area A as in Fig. 11 is used, the force produced on the end is

$$F = \frac{1}{2}\,\frac{\rho}{\mu^2 A}\,\frac{\partial\mu}{\partial\rho}\,\Phi^2, \tag{28}$$

which is quadratic in Φ. A linear force term becomes

$$f = \frac{\rho}{\mu^2 A}\,\frac{\partial\mu}{\partial\rho}\,\frac{\Phi_0}{N}\int v\,dt, \tag{29}$$

where a small change in flux is produced by a potential v applied to a coil of N turns around the magnetostrictive rod. The polarizing flux Φ_0 can be produced by a current I_0 in the coil.

With the polarizing flux, a small displacement of the end of the rod, ξ, produces a current in the coil:

$$i = -\frac{\rho}{\mu^2 A}\,\frac{\partial\mu}{\partial\rho}\,\frac{\Phi_0}{N}\,\xi. \tag{30}$$

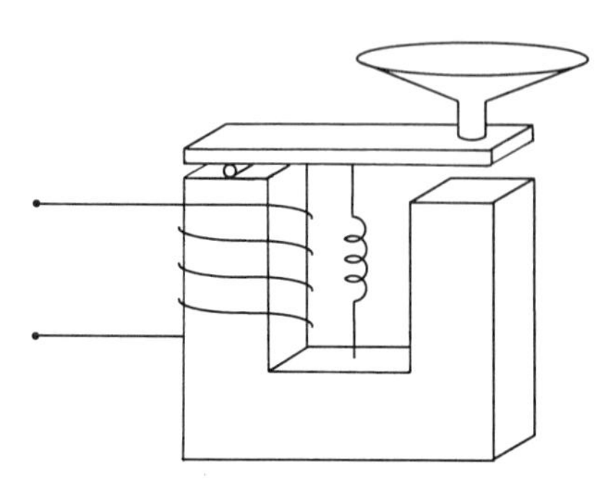

Fig. 10 Magnetic field acoustic sensor.

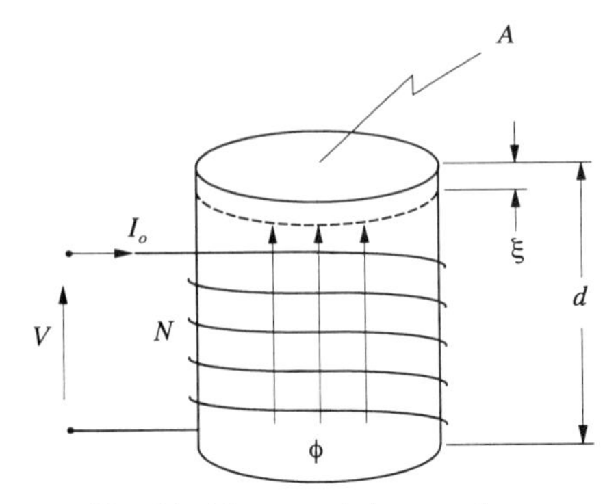

Fig. 11 Magnetostrictive transducer.

The two-port equations then become

$$I = Y_e^u V - \alpha U, \qquad F = \alpha V + Z_m^V U. \tag{31}$$

Here Y_e^u is the clamped electrical admittance and Z_m^v is the shorted mechanical impedance of the rod. The electromechanical coupling term is

$$\alpha = \frac{\rho}{\mu^2 A} \frac{\partial \mu}{\partial \rho} \frac{\Phi_0}{j\omega N}$$

Using the same analogies as in Eqs. (25), a circuit identical to Fig. 9 results.

In the polarized state it is advantageous to describe the transducer properties in terms of equations of state as in Eqs. (14):

$$B = \mu^s H + eS, \qquad T = -e_t H + c^H S, \tag{32}$$

where μ^s is the clamped permeability and c^H is the stiffness with constant H field. When the magnetic fields can be related to the potential and current in a coil around the magnetostrictive material, two-port equations like Eqs. (31) can be derived.

Linear Relationships: Moving Coil or Electrodynamic When a conductor of length l is placed in a magnetic field of flux density B, as in Fig. 12, the following linear relationship results. If a current I is passed through the conductor a clamped force acts on the conductor:

$$F_c = (I \times B)l. \tag{33}$$

If the conductor is passed through the flux with a velocity U, an open-circuit potential is developed across the conductor,

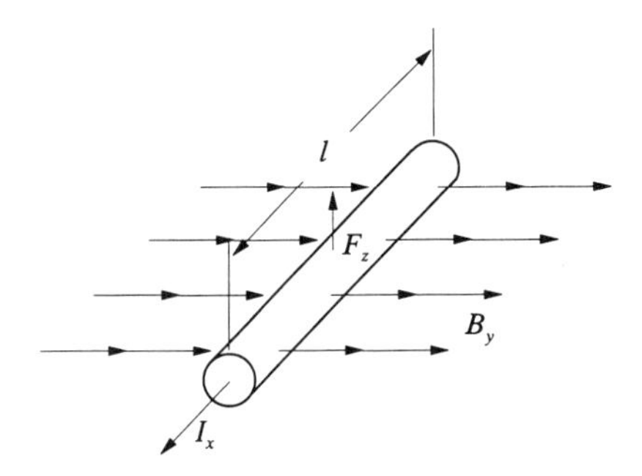

Fig. 12 Force–current relations for a linear electromagnetic transducer.

$$V_{oc} = (U \times B)l. \tag{34}$$

When the conductor and the flux are perpendicular, as in Fig. 12, the cross product is maximized. Then in a transducer sense, the clamped force becomes

$$\bar{a}_z F_c = \bar{a}_x I \times \bar{a}_y Bl \tag{35}$$

and the open circuit potential becomes

$$\bar{a}_x V_{oc} = -\bar{a}_z U \times \bar{a}_y Bl. \tag{36}$$

Here $\bar{a}_x, \bar{a}_y$, and $\bar{a}_z$ are unit vectors in the x, y, and z directions, respectively. The transducer two-port equations become

$$V = Z_e^u I - BlU, \qquad F = BlI + Z_m^i U, \tag{37}$$

where Z_e^u represents the electrical properties of the conductor with $U = 0$ and Z_m^i represents the mechanical properties of the conductor with $I = 0$. Using the same analogies as in Eqs. (6), a circuit identical to Fig. 2 results.

In many applications many turns of wire are wound on a circular form suspended in a circular air gap of a magnetic circuit that produces a radial flux. In this way a large Bl product is produced. Because of the resistance of the wire and the magnetic circuit, Z_e^u takes the form $Z_e^u = R_e + j\omega L$, where R_e is the electrical resistance and L the inductance. The coil has mass M and is usually suspended by a spring; then with damping possible, Z_m^i becomes $Z_m = R_m + j\omega M + 1/j\omega C_m$, where R_m is mechanical resistance and C_m is the compliance of the spring.

When the coil of wire suspended in the magnetic field is mechanically connected to a diaphragm, an acoustic transducer results. Then the force is the pressure times the diaphragm area and the acoustic volume velocity is velocity times the diaphragm area.

3 TRANSDUCER MATERIALS

Many solid materials exhibit electrical–mechanical coupling, some by natural linear behavior and others by change in stored energy that is linearized by applying a polarizing field. A large class of these are dielectric materials that exhibit capacitive effects at their terminals. Others are magnetic materials that have inductive characteristics. In general, both kinds of materials store energy with small loss of energy by dissipation.

3.1 Dielectrics

Many natural or artificially grown crystals have linear piezoelectric properties, and certain polycrystalline ceramics can be polarized to exhibit linear piezoelectric effects. Composites of polarized ceramics and polymers can be used to enhance transducer properties. Certain polymers that retain an imbedded charge and others that have a linear piezoelectric property are usable as sensors. Some of these will be discussed below.

Crystalline Materials Natural quartz crystals were used by the Curies[6] in their discovery of the piezoelectric effect. Artificially grown quartz crystals continue to have major application in high-frequency projectors and receivers, accelerometers, and oscillators for time keeping.

Other natural crystals such as Rochelle salts, tourmaline, and lithium sulfate have effective piezoelectric properties. Artificially grown crystals such as ammonium dihydrogen phosphate (ADP), potassium dihydrogen phosphate (KDP), zinc sulfate, and gallium arsenide (GaAs) have found considerable application.

The electromechanical coupling in crystals relates to the crystal lattice. The matrix of coefficients for Rochelle salt in Fig. 13 indicates that the coupling is only in the shear direction e_{14}, e_{25}, and e_{36}. The applications are thus limited to those couplings.

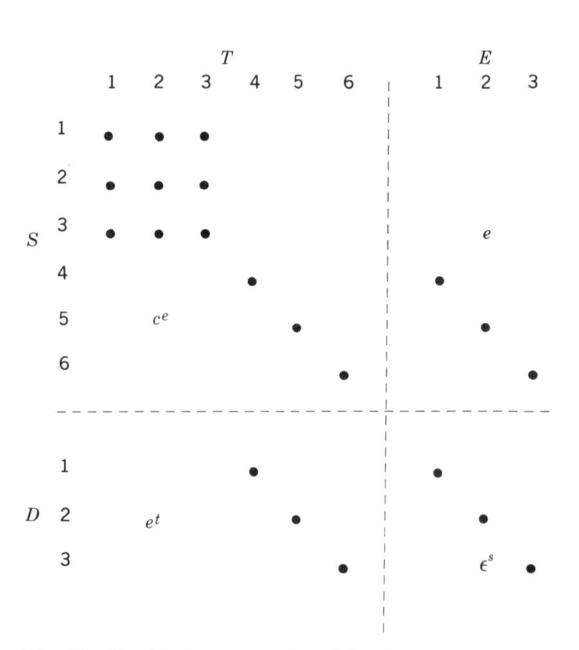

Fig. 13 Graphical representation of the physical properties of Rochelle salts.

The electrical, mechanical, and coupling terms are effected by thermal effects. In addition to change in dimension with temperature, the properties above change with temperature. A major consideration is the Curie temperature above which the thermal lattice vibrations destroy the piezoelectric properties.

Another consideration in choosing a material is the fraction of dielectric energy that can be extracted as elastic energy. This can be expressed as a quantity k:

$$k = \frac{W_m}{W_d W_e}, \qquad (38)$$

where W_m is the mutual energy, W_d is the dielectric energy, and W_e is the elastic energy. In terms of a shear coupling this becomes

$$k_{14} = \frac{e_{14}}{\sqrt{\epsilon_{11}^s c_{44}^e}}, \qquad (39)$$

which has terms that can be found in tables of material properties.

Finally dielectric materials may dissipate electrical and mechanical energy as heat. This is usually represented as a complex dielectric constant and a finite Q of the resonant mechanical structure. Electrically

$$\frac{\epsilon_{\text{im}}}{\epsilon_{\text{real}}} = \tan \delta, \qquad (40)$$

which can be found in tables. Here δ is the loss factor of the material.

Polarized Ceramics In searching for high-permeability materials to reduce the size of capacitors, it was discovered that ceramics of certain rare earths are electrostrictive. If was further discovered that applying a large DC potential when the temperature is raised above the Curie point leaves a remnant polarization. This leads to nearly linear operation, giving a polycrystalline piezoelectric material in which the direction of polarization is defined as the 3-axis.

Barium titanate with some controlled impurities is a material with density and stiffness very much like steel and a relative dielectric constant over 1000. However, the Curie temperature is 115°C. Lead zircornate titanate (PZT) can have a Curie point over 350°C and a relative dielectric constant over 3000. Various trace additives to PZT can control the stiffness, dielectric constant, and coupling and loss factors. The properties of many commercially available polarized ceramics are given by Kino.[4]

A unique advantage of polarized ceramics is the fact that the material can be formed in arbitrary shapes and polarized in any direction. Thus plates, cylinders, and spheres can be formed to produce specialized sound sources. Single-crystal materials do not give this freedom.

Composites In the polarized ceramics discussed above the stiffness and coupling to shear and cross directions is controlled by the ceramic properties. Newnham[7] and others have found that composites of polymers and active ceramic pieces can reduce stiffness and cross coupling while retaining the piezoelectric sensitivity. Of course, the electrical properties are also effected.

Choices of polymer material and piezoelectric additives provide a wide range of properties. Composites provide the possibility of designing a transducer material with properties optimum for particular applications.

Thin Films As discussed in Section 2.1, thin-film polymers that can retain a charge are known as electrets. These materials make effective membranes in capacitive transducers such as microphones.

Polyvinylidene fluoride (PVDF) polymer in thin-film form has linear piezoelectric properties. Because of small thickness, the material is effective to high frequencies in the thickness mode. The stiffness in the length mode is very much smaller than in the thickness mode. This makes it useful in membrane applications. The high flexibility is useful in some applications.

3.2 Magnetic Materials

Iron, nickel, and cobalt are the only metals found to be magnetostrictive. Alloys of nickel have been most often used in transducers. Ceramics of certain rare earths containing iron, nickel, and cobalt have large coupling coefficients. A material known as Terfenol-D ($Tb_{0.3}Dy_{0.7}Fe_{1.9}$) developed by Clark[8] can produce peak strains about three times that of polarized ceramics and an order of magnitude higher than nickel.

In general, magnetostrictive materials are physically more rugged than piezoelectric materials. However, electrical losses are high. Polarizing magnetic fields are produced by currents in coils with appreciable I^2R losses. Since the materials are conductors, eddy current losses are present. Thus thin laminations are usually used.

For more information and transducer material properties see Refs. 5 and 9.

4 NONRECIPROCAL DEVICES

The devices to be considered here have a means for a mechanical variable to produce a change in an electrical quantity. However, an applied electrical quantity will not produce a mechanical output. The electrical properties affected can be resistance, capacitance, or inductance.

4.1 Resistance

There are materials whose resistance changes appreciably with length. Thus they respond to strain and are called strain gages. Other materials exhibit a piezoresistive effect when strain deforms the crystal lattice.

When such a strain gage is attached to a diaphragm, as in Fig. 14, an applied pressure produces a strain that is read out as a potential. In the figure a constant current source is used. The strain gage relates strain to change in resistance:

$$\frac{\Delta R}{R} = KS, \tag{41}$$

where K is the gage factor and R is the nominal resistance. Then

$$E = I_0(R + \Delta R) \quad \text{or} \quad E = I_0R + I_0RKS. \tag{42}$$

The second term gives a linear measure of strain. The I_0R term may be removed electrically. Electrical bridge circuits may also be used to read ΔR without the I_0R term.

When a mechanical or acoustical variable can produce a strain, the strain gage can be a useful electrical read-out. Although strain gages are usually considered for static or low-frequency measurements, they are useful at frequencies up to those when the strain gage length becomes appreciable comparable to the wavelength of stress waves in the material to which it is attached. Microelectronic fabrication techniques have made it possible to manufacture extremely small strain

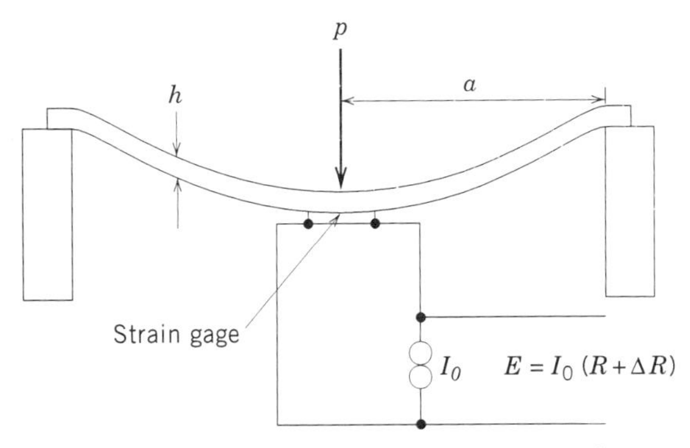

Fig. 14 Pressure sensor using a strain gage to produce an electrical output.

gages, increasing their frequency range over larger models.

4.2 Capacitance

Without the polarizing charge in the transducer of Fig. 1 the displacement of the membrane simply changes the capacitance of a parallel-plate capacitor. For a displacement ξ,

$$C = \frac{\epsilon_0 A}{d + \xi}, \tag{43}$$

where ϵ_0 is the permittivity of the medium. Then, for $\xi \ll d$,

$$\Delta C = -C_0 \frac{\xi}{d}. \tag{44}$$

It remains then to measure ΔC. This can be done by dynamically measuring the change of the AC impedance of the capacitor.

When the variable capacitor is part of a simple series *RC* circuit as in Fig. 15, for $\omega_0 R C_0 \gg 1$, the electrical output becomes

$$e_0 = -\frac{E_0 \cos \omega_0 t}{\omega_0 R C_0}\left(1 + \frac{\xi}{d}\right), \tag{45}$$

which is a sinusoid amplitude modulated by the ξ/d term. Then ξ can be retrieved by an AM detector.

When the variable capacitor is made part of a resonant circuit as in Fig. 16*a*, it can be used to determine the frequency of an electronic oscillator. Then, for $\xi \ll d$, the oscillator angular frequency becomes

$$\omega = \omega_0 + \frac{\omega_0}{2d}\,\xi, \tag{46}$$

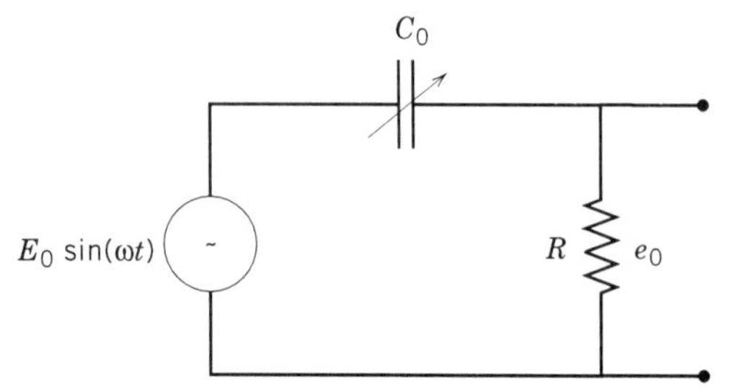

Fig. 15 Circuit for converting a capacitance charge to an electrical output.

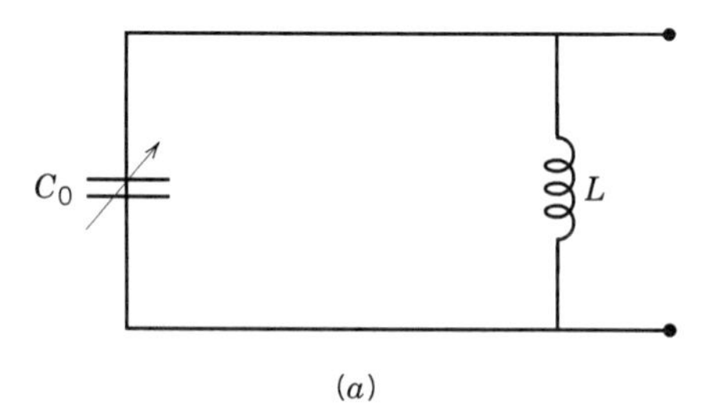

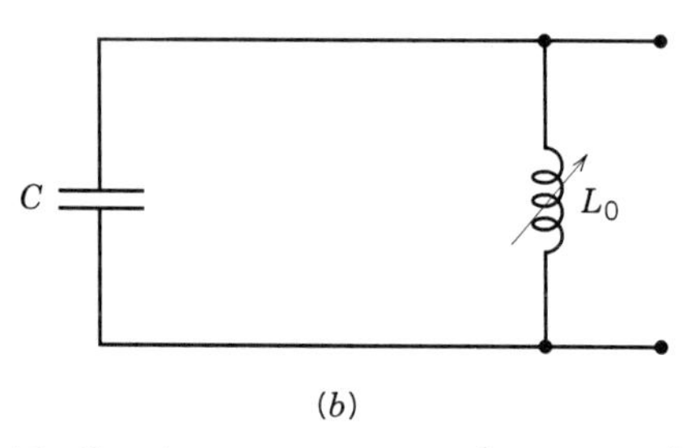

Fig. 16 Capacitance sensor as part of a resonant circuit.

where $\omega_0^2 = 1/L_0 C_0$. Thus ξ becomes a frequency modulation term that can be demodulated by conventional means. These two methods can be considered AC polarization rather than DC polarization.

4.3 Inductance

When there is an air gap in a magnetic circuit such as in Figs. 8 and 10, a change in d causes a change in the reluctance of the magnetic circuit. This in turn changes the inductance of a coil coupled to the magnetic circuit. The inductance of an N-turn coil is

$$L = \frac{N^2}{R_{\text{mag}}}, \tag{47}$$

where R_{mag} is the reluctance. When the reluctance is essentially all in the air gap, which is usually the case, L becomes

$$L = \frac{N^2 \mu_0 A}{d + \xi}. \tag{48}$$

When ξ is a small displacement, the change in inductance becomes

$$\Delta L = -\frac{L_0 \xi}{d}. \tag{49}$$

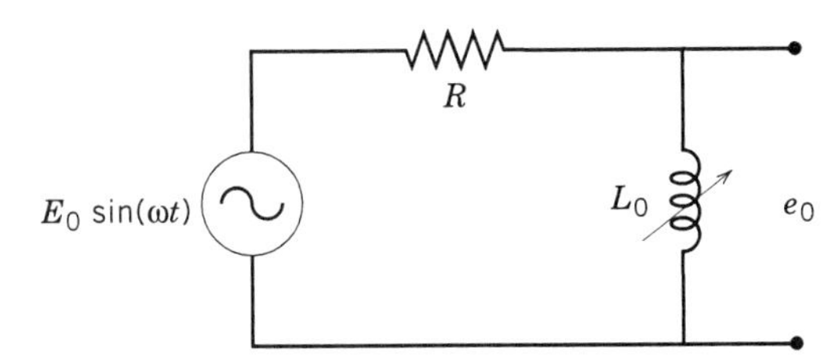

Fig. 17 Circuit for converting an inductance charge to an electrical output.

When the variable inductor is placed in the series circuit of Fig. 17 and $R \gg \omega_0 L$, the output potential is

$$e_0 = -E_0 \cos \omega_0 t \left(\frac{\omega L_0}{R} \right) \left(1 - \frac{\xi}{d} \right). \qquad (50)$$

As in the capacitive case ξ is an amplitude modulation term that can be demodulated.

Again the inductor can be part of a resonant circuit, as in Fig. 16*b*, to produce a frequency shift proportional to the displacement ξ:

$$\omega = \omega_0 + \frac{\omega_0}{2d} \xi. \qquad (51)$$

An FM detector converts the frequency shift to a potential proportional to ξ.

It should be noted that each of the above nonreciprocal transducer principles extend down to zero frequency or static mechanical functions. In the reciprocal transducers it is either impractical or impossible to do this.

5 NEW TECHNOLOGIES

Over the past 20 years there have been rapid advances in acoustical sensors and actuators. These advances have matched advances in physics in general and can be classed into four areas: fiber-optic transducers, solid-state sensors and actuators, new materials, and novel techniques. New materials such as composites, thin films, and magnetostrictive rare earths were discussed in Section 3. Below we describe each of the other areas in some detail.

5.1 Fiber-Optic Transducers

Fiber-optic transducers offer many advantages over conventional nonoptical transducers: higher resolution, higher signal-to-noise ratio, immunity to electromagnetic interference, and compatibility with fiber-optic communication systems. In addition, they are easier to use than most other optical transducers. Hence they are rapidly finding their way into sensors spanning the frequency range from DC to over 1 MHz.

All fiber-optic sensors have four main portions: a light source, a sensor, a detector, and a demodulator. The light source may be incoherent or coherent. The detector includes circuitry designed to process the optical signals.

Fiber-optic sensors are divided into two classes: those in which the light carried in a fiber leaves the fiber, reflects off of a vibrating object, and then returns to the detector via a fiber (extrinsic) and those in which the light never leaves the fiber (intrinsic). Figure 18*a* shows a fiber-optic lever, a simple example of an extrinsic fiber-optic sensor. The main component of this instrument is a fiber-optic probe consisting of multiple fibers, half of

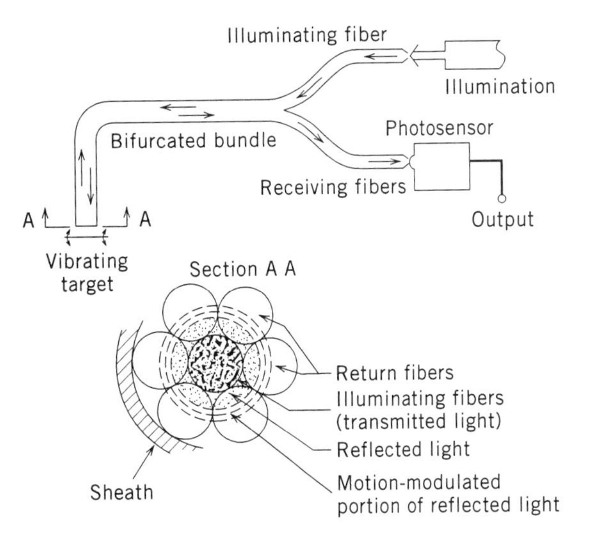

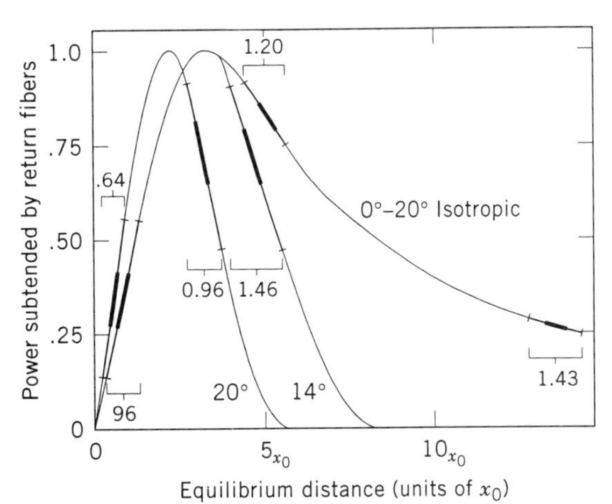

Fig. 18 Fiber-optic lever.

which transmit light and half of which carry the light reflected off of the target back to the detector. The detector generates an output voltage proportional to the intensity of the reflected light. No modulation is required. As shown in Fig. 18*b*, the output voltage is null when the probe touches the target since the light reflects specularly back into the sending fibers rather than the receiving fibers. As the distance between the target and the probe end increases, the output voltage rises rapidly until the entire surface of the probe is illuminated. Increases in separation beyond this distance result in a voltage drop due to square-law reduction in reflected light intensity on the probe surface.

Intrinsic fiber-optic sensors are designed so that the length of a fiber varies with changes in a parameter to be measured. Typically the length variation is determined by comparison to a reference length of fiber. An example of such a sensor is shown in Fig. 19, which illustrates a Mach–Zehnder interferometer. In this sensor a coherent light source is split so that part of the beam travels in a reference fiber and part in a sensing fiber. The reference fiber is wrapped around a piezoelectric crystal, which is used as a phase modulator. By applying a potential across the piezoelectric, the reference fiber length is controlled, thus permitting the modulator to establish a bias point and to correct for any initial mismatch in the length of the two fibers. The sensing fiber is exposed to acoustic pressure that produces a strain in the fiber. This causes a change in the refractive index and thus a change in the effective length of the fiber. By recombining the reference and sensor signals and monitoring the phase shift from the original, the sensor fiber length change is very accurately determined.

Also, the light from the end of a fiber can be shined on a vibrating surface. The light reflected back into the fiber carries a phase change that can be sensed by the interferometer.

Buckman[10] has developed a system based on a 1.5

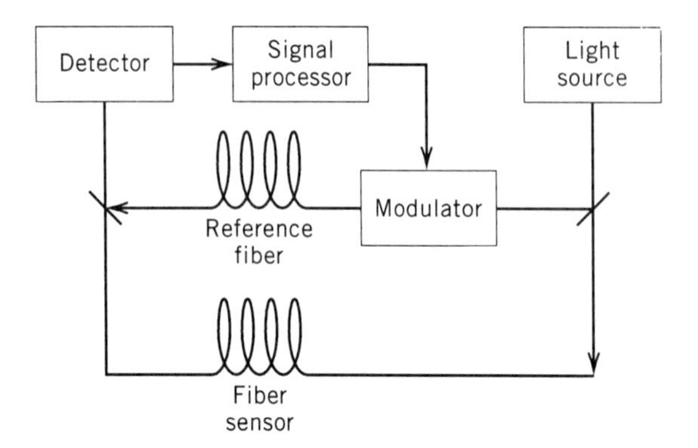

Fig. 19 Interferometer fiber-optic sensor.

Mach–Zehnder interferometer. It uses phase feedback to compensate for environmental effects, control the operating point, and increase the sensitivity.

The fiber-optic devices discussed above are also representative of the common modulation schemes. Most fiber-optic devices are based on amplitude modulation, as in the optic lever, or phase modulation, as in the Mach–Zehnder interferometer. Phase modulation fiber-optic sensors are more sensitive, but the signals are more difficult to modulate.

For more information on fiber optic sensors see Refs. 11–13.

5.2 Solid-State Sensors and Actuators

For the past 15 years there has been substantial work in the area of solid-state sensors, and in the last few years there has been increasing interest in solid-state actuators. Solid-state transducers are generally devices made using the standard materials found in microelectronic devices such as silicon. The advantages of solid-state sensors and actuators are many: There is good fabrication process control, miniaturization is easily achieved, transducers may be integrated with the electronics on a single chip, and costs per item are low. The notable successes in solid-state sensors include devices for the measurement of acoustic pressure.

The construction of solid-state sensors differs substantially from the standard fabrication processes involved in microelectronics. Two items in particular have presented problems: size scales and feature shapes. The sizes of sensor and actuator transducing elements are generally much larger than standard microelectronics features, and there have been problems obtaining consistently high yields of large-feature items. Further, most solid-state transducers require nonstandard feature shapes such as cavities. These nonplanar shapes can be difficult to obtain, although selective chemical etches, which follow preferred crystal orientations, have permitted rapid progress overcoming this stumbling block. Another difficulty with solid-state sensors and actuators has been the desire to integrate the processing electronics with the device. Although such an approach offers advantages from the perspective of signal-to-noise ratio, the yields for electronics alone are still substantially higher than those for the transducer alone. Hence, insisting upon integration usually results in loss of a large amount of viable electronics. In addition, many of the fabrication processes for sensors and actuators seem to be fairly incompatible with electronics processing. For further information about the fabrication of solid-state sensors see Ref. 14.

Acoustic solid-state transducers generally have been limited to pressure sensors (for air and water), although

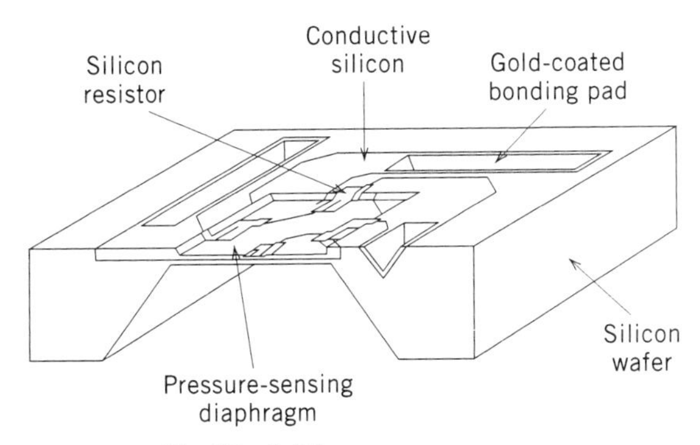

Fig. 20 Solid-state pressure sensor.

attempts to build sound generators have been reported. Solid-state pressure sensors have been designed using a piezoresistive effect, a piezoelectric effect, or the effect of stress on a *p–n* junction. A varying capacitance as described in Section 4.2 is also used. Of these, the varying capacitive and piezoresistive sensors have received the most attention.

In typical capacitive solid-state pressure sensors such as shown in Fig. 20, a moving membrane is coated with metal and a bias voltage is applied between the membrane and a fixed backplate. Devices such as this are miniaturized versions of conventional condensor microphones, except that the stiffness is usually determined by the size of the air cavity rather than the membrane tension. However, the solid-state capacitive transducers may be more sensitive and usually are useful up to much higher frequencies due to the increase in the first resonant frequency that accompanies the miniaturization. Their main disadvantage is a low capacitance of generally 1–3 pF, necessitating close location of processing electronics and causing poorer low-frequency response than their macroscopic counterparts. For further information on capacitive solid-state pressure sensor see Ref. 15.

Piezoresistive solid-state pressure sensors are more numerous than the capacitive type. In these devices, resistors are typically diffused into a membrane and used in a half-bridge or full-bridge form. Membrane motions cause strain and thus result in differential resistances. These devices are easier to fabricate than the capacitive pressure sensors, and thus are less expensive, but are generally more limited in range due to hysteresis and nonlinearities. In addition, the piezoresistive solid-state sensors are susceptible to interference due to stressing the package since they sense differential stresses.

5.3 Novel Techniques

In addition to advances in materials, recent years have seen advances in techniques used in acoustic instruments. Among these are laser-generated sound, magnetohydrodynamic transducers, acoustic heat engines and refrigerators, and acoustic microscopy.

Laser-generated sound is the process of using a laser to generate sound in a medium. Most of the work has involved fluid media, although there has been work on laser-generated sound in solids. Several transduction mechanisms have been identified for laser-generated sound in fluids, varying widely in efficiency and directivity. The two mechanisms that have received the greatest attention are the thermal mechanism and surface ablation.

In the thermal mechanism a laser shined into a fluid medium causes the medium to heat and expand. By modulating the laser, the expansion of the medium is modulated, resulting in sound wave propagation. The advantage of this approach to sound generation in a fluid is that no instruments need to be located in the fluid medium, where they are subject to corrosion. However, the process of sound generation is very inefficient. Attempts to increase the source sensitivity using a moving laser beam have met with moderate success.

In the surface ablation mechanism of laser sound generation the laser is focused onto the surface of a fluid where it induces a localized boiling. This causes a momentum transfer into the medium that radiates away as sound. Although much more efficient than the thermal mechanism of laser sound generation, this blast mechanism is less directive and has proved difficult to control and analyze. Surface ablation using a laser has also been used to create short-duration acoustic pulses in solid materials.

Magnetohydrodynamic (MHD) acoustic instruments are transducers that rely on the coupling between magnetic, electric, and velocity fields in a conducting medium much like the linear electromagnetic method of Section 2.2. A typical realization of an MHD acoustic transducer is shown in Fig. 21 and uses salt water in a duct. Permanent magnets establish a magnetic field in one direction, and an applied electric field in an orthogonal direction results in a mean velocity of

$$U = \frac{V_0}{d\mu H} \tag{52}$$

for the medium in the third orthogonal direction, which is generally taken to be along the axis of the duct. Here V_0 is the mean electric potential, H is the magnetic field established by the permanent magnets, and d is the effective diameter of the duct. In addition to the mean fluid flow (which is nonexistent in the absence of a DC bias voltage), there is an oscillating fluid motion that causes acoustic radiation into the medium surrounding

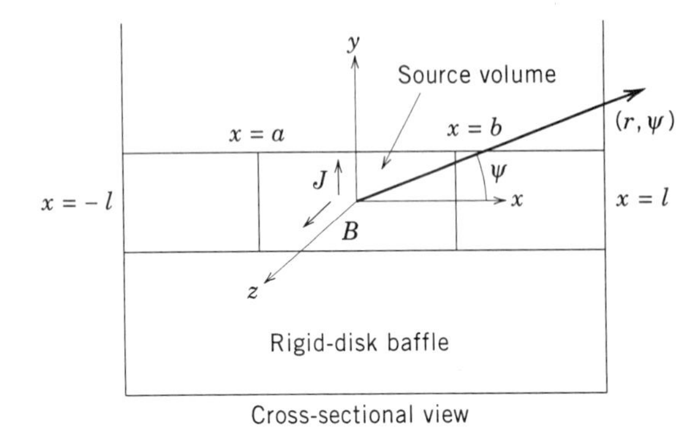

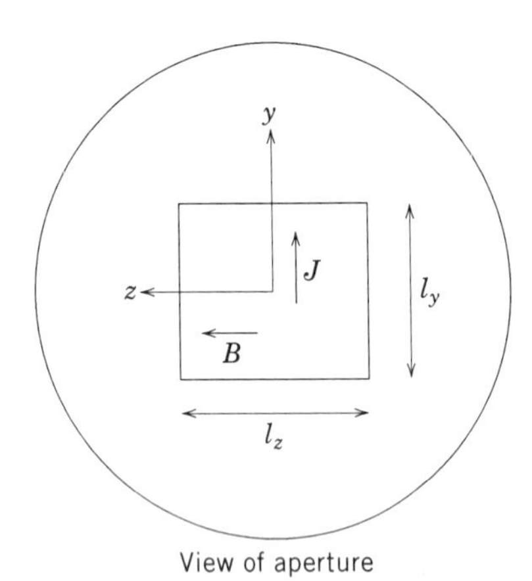

Fig. 21 Magnetohydrodynamic transducer.

the source. Alternatively, the MHD transducer may be used as a receiver, in which case the fluid oscillation caused by incoming sound results in the establishment of an electric field. These MHD transducers offer the advantage of broadband operation, but they are less efficient than piezoelectric counterparts.

Acoustic refrigerators and heat engines based on both MHD and thermally induced sound have been studied. In these devices thermal variations are used to generate sound waves that propagate in a duct. The thermoacoustic sound-generating mechanism is analogous to the thermal mechanism for sound generation by the laser discussed above. The sound in the duct is coupled to electrical power through an MHD mechanism. The result is a compact instrument that has high-energy conversion efficiency. For example, Migliori and Swift[16] have reported on a liquid–sodium acoustic engine in which an imposed temperature difference is used to generate sound; most of the sound energy is recovered in an MHD transduction process.

REFERENCES

1. F. A. Fisher, *Fundamentals of Electroacoustics*, Interscience, New York, 1955.
2. F. V. Hunt, *Electroacoustics*, Harvard University Press, Cambridge, MA, 1954.
3. G. M. Sessler, *Electrets, Topics in Physics*, Vol. 33, Springer Verlag, Berlin, 1987.
4. G. S. Kino, *Acoustic Waves*, Prentice-Hall, Englewood Cliffs, NJ, 1987.
5. W. P. Mason, *Physical Acoustics*, Vol. 1, Part A, Academic, New York, 1964.
6. J. Curie and P. Curie, "Development par compression de l'electricite' polaric dans les cristaux hemiedres a faces inclinces," *Bull. Soc. Mineral. France*, Vol. 3, 1880, pp. 90–93.
7. R. E. Newham, *Ferroelectrics*, Vol. 68, 1986, pp. 1–32.
8. A. E. Clark, in E. P. Wohlfarth (Ed.), *Ferromagnetic Materials*, Vol. I, North-Holland, New York, 1980.
9. O. B. Wilson, *An Introduction to the Theory and Design of Sonar Transducers*, U.S. Government Printing Office, Washington, DC, 1985.
10. A. B. Buckman, D. G. Pritchett, Jr., and K. Park, "Sensitivity Enhancement, Common Mode Coupled Mach Zehnder Fiber Optic Sensor Circuit with Electro Optic Feedback," *Opt. Lett*, Vol. 14, 1989, pp. A86–A88.
11. H. V. Winsor, *SPIE*, Vol. 239, 1980, pp. 252–258.
12. W. O. Grant, *Understanding Lightwave Transmission: Applications of Fiber Optics*, Harcourt Brace Jovanovich, San Diego, 1988.
13. D. A. Krohn, *Fiber Optic Sensors—Fundamentals and Applications*, Instrument Society of America, Research Triangle Park, NC, 1988.
14. R. S. Muller, R. T. Howe, S. P. Sentaria, R. L. Smith, and R. M. White, *Microsensors*, IEEE Press, New York, 1991.
15. W. H. Ko, *Sensors Actuators*, Vol. 10, 1986, pp. 303–320.
16. A. Migliori and G. W. Swift, "Performance of a Liquid-Sodium Thermoacoustic Engine," *J. Acoust. Soc. Am.*, Vol. 84, 1988, p. S37A.

111

LOUDSPEAKER DESIGN

BRADLEY M. STAROBIN

1 INTRODUCTION

A loudspeaker system is a collection of interdependent electromechanical components that together radiate acoustical energy into a space in conformance with the electrical energy supplied to it. As such it is a *transducer*. This chapter explores the design constraints, goals, and solutions that are imposed upon and evidenced by modern-day loudspeaker systems for applications ranging from professional sound reinforcement through consumer home entertainment and automotive products.

Direct-radiating electrodynamic loudspeaker systems—their transduction mechanism, electroacoustic circuit representations, and major components—are the primary focus of this chapter. *Direct radiators* inject sound power into a space without any intermediate impedance-matching devices (such as a horn), instead coupling directly the vibratory energy of their sound-radiating elements with the acoustic medium.[1]

As they pertain to loudspeakers in general and dynamic direct radiators in particular, topics including sound radiation, room effects, directivity, and acoustic circuit analysis are covered in this chapter. The chapter concludes with some remarks about modern measurement techniques.

1.1 Some Definitions[1]

Some of the important terms that appear in this chapter are introduced and defined here. As required, additional ones are presented later in their appropriate context.

Note: References to chapters appearing only in the *Encyclopedia* are preceded by *Enc.*

A loudspeaker *drive element* (or simply, *driver*) is an electro-mechano-acoustic transducer. *Electrodynamic* (or simply, *dynamic*) *drivers* are the familiar fixed-magnet, moving-coil type that incorporate cones and domes as their radiating elements. *Woofers* are low- to midfrequency (20 Hz–6 kHz) cone drivers, while *tweeters* handle the mid-to-high bands (1.0–20 kHz) and normally employ dome-shaped diaphragms.

A driver's *piston band* is the frequency range over which its radiation is nondirectional. Its upper bound is the frequency whose wavelength equals the driver's circumference; over the range $f < 11,000/d$ (where f is frequency in hertz and d is the driver's diameter in centimetres, using 345 m/s for the speed of sound in air), the driver is acoustically small.

An *enclosure* is a cabinet in which a driver is mounted. Depending on the type of loudspeaker system, it may or may not have any additional apertures (besides the driver's). Conventional electrodynamic direct radiators incorporate an enclosure of some description.

A *baffle* is the support structure for the drivers; it is the front face of the enclosure for a conventional direct radiator.

A *crossover* is the electronic dividing network that filters and distributes to the system's drive elements the electrical driving signal.

A loudspeaker *system* is the sum total of its drive element(s), radiation aid(s) (baffle, enclosure, and/or horn), and crossover.

Small-signal analysis concerns behavior within a loudspeaker system's linear range, where input levels are not so large that thermal effects and suspension anomalies appreciably affect performance, while *large-signal* analysis considers ultimate system capabilities, where these nonlinearities are significant.

1.2 Performance Requirements

A program's bandwidth, dynamic range, and intended playback levels impose a set of performance requirements on loudspeaker systems. Faithful reproduction of musical performances requires a three-decade bandwidth (20 Hz–20 kHz, to match the human hearing system's), a dynamic range of 100 dB, and the ability to replicate in situ peak sound pressure levels (SPLs) near the discomfort threshold of 120 dB (re 20 μPa).[2]

Further criteria include sufficiently low distortion products, including timbral (spectral deviations), phase (nonuniform group delay), and harmonic, so as to be inaudible. Another criterion is freedom from operational noises, such as mechanical buzzes, rubs, and air turbulence.

2 SOUND RADIATION

The pertinent attributes of the acoustic medium and space are treated here.

2.1 Radiation Resistance: Implications on System Response[3,4]

The real part of complex radiation impedance, radiation resistance is a measure of an acoustic medium's ability to dissipate energy. It relates radiated sound power (P_a, in watts) to a source's volume velocity according to the formula

$$P_a = |U_0^2| R_{\mathrm{ar}}, \tag{1}$$

where U_0 is root-mean-square volume velocity (in units of cubic metres per second) and R_{ar} is acoustic radiation resistance [in units of mks (metre-kilogram-second) acoustic ohms, or newton-seconds per metre to the fifth power].

Figure 1 plots the acoustic radiation resistance as "seen" by a rigid piston in an infinitely extended baffle versus the quantity ka. The acoustic size (with respect to a wavelength) of a radiator is given by the product of acoustic wavenumber k ($= \omega/c = 2\pi/\lambda$) and a, the radius of the piston (in metres), where ω is the radian frequency ($=2\pi f$, where f is cyclical frequency), c is speed of sound (=345 m/s in air at room temperature), and λ ($=c/f$) is acoustic wavelength (in metres).

The frequency at which $ka = 2$ marks the lower bound

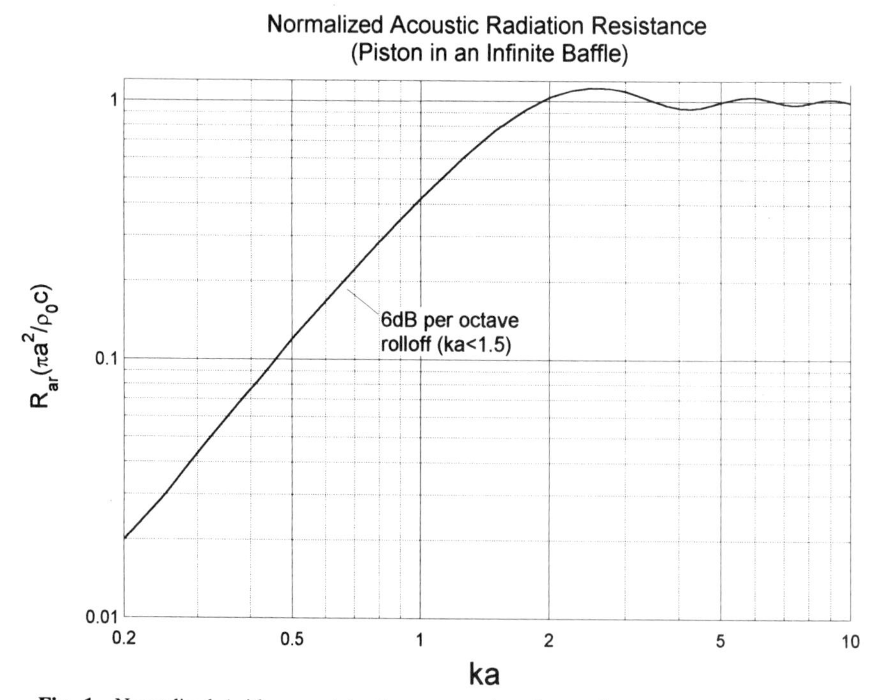

Fig. 1 Normalized (with respect to the asymptotic value as $ka \to \infty$) acoustic radiation resistance [Re(Z_{ar})] vs. ka.

of the range where R_{ar} is approximately constant. Above $ka = 2$, R_{ar} tends toward $\rho_0 c/\pi a^2$, where ρ_0 is density and $\rho_0 c$ characteristic impedance of the medium. Under normal conditions for air, $\rho_0 = 1.18$ kg/m^3 and $\rho_0 c = 407$ mks rayles. Below approximately $ka = 1.5$, the radiation resistance decreases by 6 dB with each lower octave.

For direct-radiating cone loudspeakers, R_{ar}'s 6 dB/octave rolloff implies that cone velocity would have to double with each lower octave in order to maintain flat power response. Generally, this requirement is satisfied within the frequency band bounded by the system resonance and the transition frequency. This is the region of mass-controlled response, where cone velocity is inversely proportional to frequency. But as velocity doubles, cone displacement quadruples. In part, displacement limitations determine the output capability and restrict the passband of practical loudspeakers.

2.2 Boundary Effects

Loudspeaker–room interactions profoundly affect in-room system performance, particularly at low frequencies. Sound power doubling [3 dB PWL (acoustic power level) increase] with each halving of solid angle into which an acoustic source radiates obtains in-room SPL boosts of 6 dB [single rigid boundary (2π steradians)], 12 dB [two boundary junction (π)] or 18 dB [corner ($\frac{1}{2}\pi$)] over free-space (4π) conditions. The inverse proportionality of radiation resistance and solid angle explains this low-frequency ($\lambda \gg$ driver–boundary distance) phenomenon. When driver–boundary distances approach $\frac{1}{4}\lambda$, destructive interference between direct and reflected energy causes response notching, while beyond $\frac{3}{4}\lambda$ boundaries have little effect.[5,6]

A second factor is the spatial variation of system response that results from the characteristic low modal density of small spaces at low frequencies. These effects on the overall speaker–room system response depend on the room's shape (dimensional ratios), the room's size, and both the loudspeakers' and listeners' locations relative to room boundaries (where pressure maxima are located). Small-room acoustics is covered in Chapter 74.

A third loudspeaker–room interaction is "room gain." For a closed-box radiator, the room transfer function within the "pressure response" regime mimics the loudspeaker's velocity response, yielding a 12-dB boost with each lower octave.[7] This phenomenon occurs over the range where wavelengths are at least twice the longest dimensions of the space. In automotive environments, where typical dimensions do not exceed 2 m, the imposed low-frequency boost takes effect below 90 Hz and explains the surprising bass performance of even modest car speaker systems.

A host of other room artifacts are summarized in Ref. 5, including timbral effects beyond the low-frequency regime and some that are manifest only in the perceived spatial representation of acoustic events by multichannel systems (*imaging*), which fall within the domains of stereophony and psychoacoustics.

2.3 Directivity[3]

Fundamentally, the physical size of an acoustic source relative to a wavelength determines its *directivity*, the degree to which sound energy is concentrated about the primary axis of radiation. Directivity index $\mathrm{DI}(f) = 10 \log_{10} Q(f)$, where $Q(f)$ (directivity factor) is the frequency-dependent ratio of a radiator's actual sound intensity to that of an equivalent point source on the primary radiation axis at a specified distance. A loudspeaker is nondirectional (DI $\approx$ 1.0 dB) over its piston band, beyond which its radiation pattern is increasingly concentrated about the primary axis of radiation. For a rigid circular piston in the end of a long tube, DI increases from 1.1 to 9.6 dB as ka increases from 0.5 to 3.0.

Directivity has important implications both on intrasystem crossover design and intersystem array element arrangement (orientation and relative spacing). The latter pertains to sound reinforcement applications. As an example of the former, directivity considerations dictate a low crossover point between a relatively large driver and tweeter in a two-way system. Otherwise, the overly abrupt expansion of system coverage angle (moving up in frequency from the midwoofer's to the tweeter's passband) would generate too much variability in the acoustic frequency response as a function of listener location; the resulting poor acoustic power response gives rise to an unnatural spectral impression for listeners positioned far off of the primary radiation axis or in the reverberant field.[5,8] Furthermore, uniformity of intensity of arriving reflections (with respect to frequency) has been correlated with natural imaging and high intelligibility.[9]

Beamwidth versus ka Figure 2 shows how beamwidth decreases nonlinearly with increasing ka, while Table 1 indicates the upper bounds of the frequency ranges within which drivers of various sizes achieve particular minimum coverage angles. These data assume that the drivers behave as rigid pistons through their passbands. That "cone breakup" (explained in Section 3.1) tends to reduce effective radiating area (reducing directivity effects) is partially offset by the directivity increases associated with conical diaphragm shape.

2.4 Diffraction[10,11]

This self-interference phenomenon occurs when sound waves encounter discontinuities as they propagate along the baffle. The resulting time-delayed radiation from

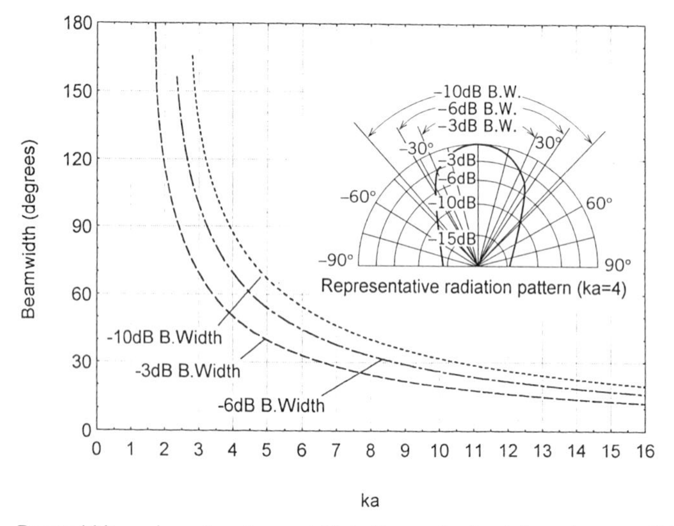

Fig. 2 Beamwidth vs. ka, where beamwidth is the angle about the primary radiation axis included by the radiation axes along which the level is −3, −6, and −10 dB with respect to the on-axis level.

TABLE 1 Upper-Bound Frequency for Selected −6 dB Beamwidths by Driver Size

Nominal Driver Size[a]	Beamwidth 30°	60°	90°	120°
1 in. tweeter	≫20 kHz	≫20 kHz	18.5 kHz	15.1 kHz
4.5 in midrange	14.5	7.5	5.3	4.3
8 in. woofer	7.3	3.8	2.6	2.1
12 in. woofer	4.7	2.4	1.7	1.4

[a]Actual radiating surface may be smaller than implied by nominal driver size [=approximate diaphragm diameter (tweeters) or frame diameter (mid/woofers); see Section 3.1].

these virtual secondary sources combines in a complex manner, interfering both constructively and destructively, with direct-wave radiation. Listening-axis-dependent narrow-band peaks and dips in the sound pressure amplitude response result. The cited references outline their mechanism and some measures for reducing these effects, including the use of dissipative "acoustic blankets" attached to the baffle about the drive units. Finally, both the shape of the baffle and positioning of the drive units on it relative to the edges have an important bearing on diffraction effects; spherically shaped enclosures are the most innocuous, cubical baffles the least, and nonsymmetrical driver locations (with respect to baffle edges) reduce response variations when conventional rectangular baffles are used.

3 LOUDSPEAKER SYSTEM COMPONENTS

The major components of a dynamic direct-radiating system proper are the driver, enclosure, and crossover network.

3.1 Moving-Coil Drivers[12]

These are employed in the vast majority of modern loudspeaker systems.* An exploded view of a typical midwoofer is shown in Fig. 3. The major subcomponents are the motor, diaphragm, and suspension. The latter two, along with the voice coil (also a part of the motor system) and dust cap, make up the moving system.

The Motor[13] Referring to Fig. 3, the permanent magnet, front ring (top plate), pole piece/bottom (back) plate, and the voice-coil/former together comprise the *motor system.*

The operating principle of a moving-coil driver is described in detail in Chapter 110. When electrical current passes through a conductor located in a magnetic

*Alternative transduction mechanisms are employed by *planar* loudspeakers, a class comprising ribbon and electrostatic types. These open-panel systems are distinguished from enclosed direct radiators by their extended diaphragms of extremely low moving mass (accounting for their superior transient performance), the absence of an enclosure proper, and their dipolar radiation patterns. Chapter 112 covers the transduction mechanisms of planar systems as they pertain to microphones.

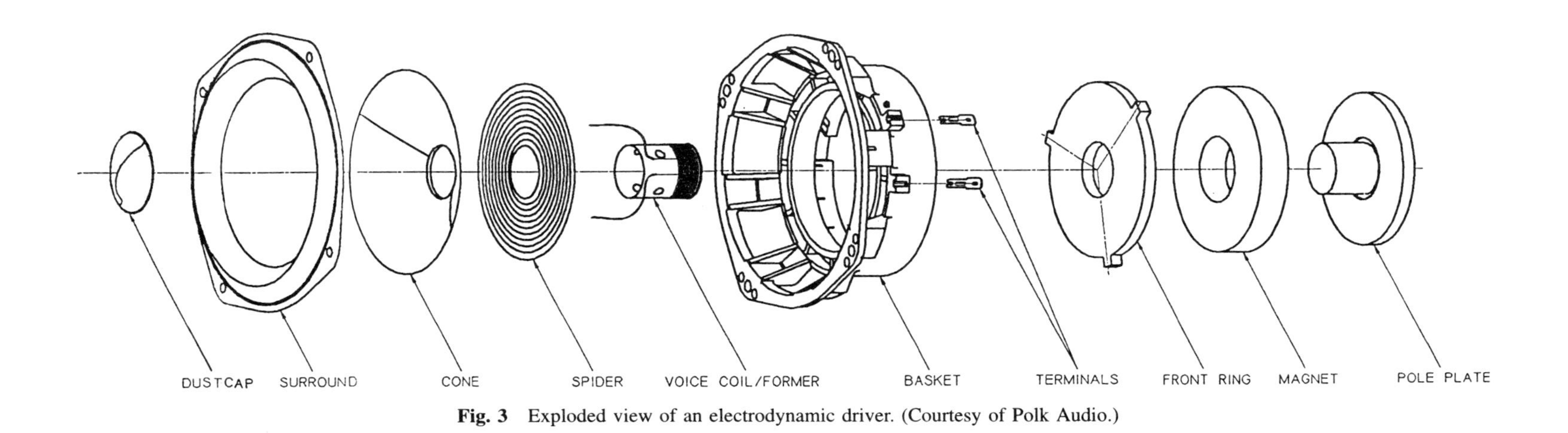

Fig. 3 Exploded view of an electrodynamic driver. (Courtesy of Polk Audio.)

field, a force is exerted on the wire. Its magnitude is the product of the input signal current I (amperes) and motor strength Bl (webers per metre or tesla-metre) where B is the flux density (normally expressed in units of gauss, where 10 kG = 1T = 1 Wb/m^2) in the voice coil gap and l is the length (in metres) of wire in that field. That is, $|F| = (Bl)I$.

This electromotive force acts along the voice coil's axis of symmetry (generally, perpendicular to the axis of the conductor) and reverses during each cycle for AC signals. By rigidly attaching a diaphragm to one end of a lightweight cylinder (the voice coil *former*), around which the electrical conductor is wound (collectively, with the former, the *voice coil*) and suspending this assembly within a permanently magnetized air gap, this transduction mechanism is put to work in dynamic loudspeaker drivers.

The *magnetic circuit* (or *motor structure*) a subset of the motor system, encompasses the magnet, front and back plates, pole piece, and air gap. Ideal geometries efficiently concentrate the flux within the air gap and set up a symmetric field about it so as to ensure a constant electromagnetic force on the voice coil as it travels through the gap. Magnetically permeable mild steel is most commonly used in the composition of the circuit proper, the pole piece, and plates. Most dynamic drivers utilize permanent magnets composed of ceramic ferrites. Rare earth magnets such as neodymium and iron boron have gained favor in applications where physical size and weight are at a premium, such as automotive uses. Their expense and susceptibility to demagnetization at high temperatures have prevented more widespread application.

The most fundamental goal for the transducer engineer is to design a cost-effective motor system that achieves the driver's target motor strength value. For typical consumer drive units, Bl generally does not exceed 15 T-m; corresponding flux density values range from 4–20 kG (4–12 kG for woofers, 10–20 kG for tweeters). The larger motor strength values required of drivers employed in professional loudspeaker systems are achieved by using massive motor structures and large-diameter voice coils (≥50 mm).

Moving System This includes the voice coil, diaphragm, dust cap, and suspension.

Voice Coil.[13] Engineered jointly with the motor structure, its principal design attributes are diameter, height (axial winding distance along the former), electrical impedance, former material (aluminum, Kapton®, or other engineered composites, chosen for low mass, rigidity, and thermal robustness), conductor properties [type (insulated copper or aluminum), *resistivity* (electrical resistance per unit length, determined by material, cross-section, and gauge), and winding arrangement (number of layers)]. These interdependently address the host driver's targets for acoustic efficiency, motor strength, electrical DC resistance (DCR), resonant damping, passband, power handling, and acoustic output capability. Motor systems that employ voice coils of relatively large diameter have advantages in efficiency (larger Bl, due to a longer conductor in the magnetic gap), and power handling (better thermal capability) over otherwise comparable smaller voice coil systems. Partially offsetting the first performance benefit, the second one has implications on voice coil mass as it is achieved by employing a low resistivity (low gauge, high mass) conductor and several winding layers may be required in order to achieve the target DC resistance. Furthermore, voice coil diameter and number of winding layers have implications on the driver's acoustic frequency response; electrical inductance increases with either, thereby limiting the driver's passband by restricting its top end acoustic output (see Section 4.3).

Finally, a driver's maximum required excursion has a bearing on voice coil height and motor system geometry. As shown in Fig. 4, the standard motor-structure/voice-coil geometries are "underhung" (short coil, long gap) and "overhung" (long coil, short gap); while underhung geometries offer better linearity and lower voice coil mass, overhung designs have advantages in efficiency,

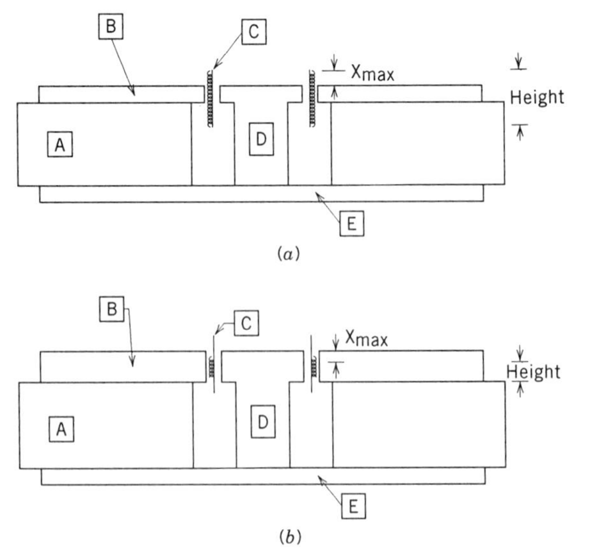

Fig. 4 Cross-sectional view of an electrodynamic driver's motor system, illustrating (*a*) overhung and (*b*) underhung magnetic-gap/voice-coil geometries. Motor system subcomponents are A: magnet, B: top plate, C: voice coil; D: pole piece, and E: bottom plate. Peak linear excursion ($X_{\max}$) and voice coil height are indicated.

excursion capability, and overall cost-effectiveness; they remain the overwhelming choice of loudspeaker designers. For overhung designs, larger height increases peak linear excursion (X_{max}); generally, height has implications on DCR, mass, and L.

Diaphragm.[14,15] The vast majority of modern dynamic drivers use cones composed of paper or various types of plastics (polypropylene composites being very common), the latter often with stiffening agents such as talc, mica, graphite, or glass fiber. Engineered woven fibrous materials such as Kevlar®, carbon, or glass are increasingly common. Other exotic designs include light, thin stamped metals (such as aluminum) alone or in laminar constructions with damping materials.

Mass–stiffness ratio and damping are the most important properties. Generally, a relatively light, stiff cone (small mass–stiffness ratio) is desirable as high cone stiffness facilitates extended midfrequency performance, and low mass permits a smaller motor strength for achieving a target efficiency rating. However, there are special cases when large moving mass (normally accompanied by large motor strength to compensate for the efficiency reduction) is advantageous, particularly when ambitious low-frequency performance objectives are to be achieved subject to unusually severe constraints on maximum enclosure size.

In part, cone material properties affect the severity and frequency at which "breakup" occurs. Cone flexure comprises sound-radiating structural vibrations (generally, both longitudinal and bending waves, which are strongly coupled) within the diaphragm material. Radial waves originate at the cone–voice-coil junction and propagate out toward the compliant surround, where, to some degree, they reflect back toward the dust cap. The result is concentric vibration modes whose resonance frequencies depend on material mass and stiffness. Some representative nodal patterns are shown in Fig. 5. As the mass–stiffness ratio is made smaller, these modes occur at higher frequencies, ideally beyond the passband of the driver. While rather severe irregularities (narrow-band deviations of 8 dB or more) in the pressure response of the driver near its top end can result, the damping afforded by the cone material itself and its terminations presented by the surround and dust cap can reduce these effects appreciably.

Concentrically propagating bending waves give rise to radial modes. Due to the low bending stiffness for these waves, radial modes tend to occur at much lower frequencies (near 100 Hz for an 8-in. cone) than do concentric modes; the effects of radial modes on pressure response are much more innocuous and often negligible.

Dust Cap. As its name implies, the dust cap's function is to seal the gap between the voice coil and the pole piece that would otherwise be exposed to foreign particles. In addition, perforated dust caps provide voice coil cooling, a function otherwise served by a vented pole piece. Material composition (paper, cloth, felt, foam, and various plastics are common) and shape have important, if not readily predictable, effects on frequency response. These and other design considerations are outlined in Ref. 13.

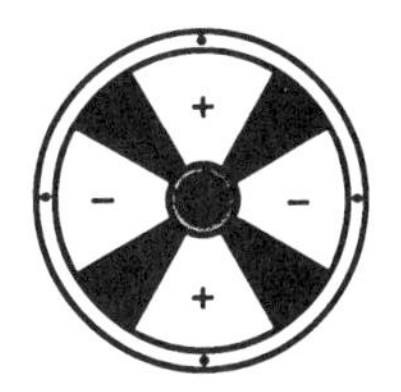

RADIAL CONE MODES

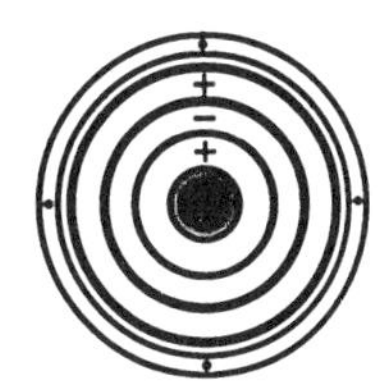

CONCENTRIC CONE MODES

Fig. 5 Cone vibration modes. Radial bending waves form concentric nodal patterns (relatively high frequency) while concentrically propagating waves give rise to radial modes (low frequency). (Reprinted with permission from *The Loudspeaker Design Cookbook*, 5th ed., by Vance Dickason. Copyright © 1995.).

Driver Suspension.[16,17] The *spider* and *surround* together form the driver's suspension, contributing to the restoring force on the cone in proportion to displacement over its linear (small-signal) range. The surround's main tasks are to keep the cone axially centered over the pole plate and provide a damped termination at the cone edge for reducing structural cone vibrations. Common materials are treated cloth, urethane foam, and "rubber" compounds (natural and synthetic). The spider, typically composed of formed cloth (treated cotton blends and Nomex® are common), provides most of the suspension's stiffness and contributes to axial centering.

The Frame The frame serves as both the driver's foundation, supporting and locating its components, and an interface with the enclosure. Cast aluminum, stamped steel, and molded plastic composites are most common. The frame's most important performance attribute is its ability to isolate the enclosure from the driver's moving components, ideally presenting an infinite mechanical impedance to both the enclosure and the driver components with which it interfaces. Soft closed-cell (to prevent air leakage) foam gaskets help to reduce transmission of vibrational energy between the frame and baffle.

Tweeters[18,19] The operating principle and major design attributes of moving-coil tweeters are largely the same as low- to midfrequency cone drivers'. They are contrasted by their small size (12, 19, and 25 mm in diameter are most common), dome-shaped (usually convex) diaphragms (driven at their periphery), and the (common) use of "ferrofluid" (a magnetically conductive, heat dissipative substance) in the voice coil gap for sufficient power handling and damping. Finally, designers place extreme emphasis on reducing a tweeter's moving mass; transient response (so critical to the performance of high-frequency transducers) depends largely on that mechanical attribute. In part, this explains the wide variety of materials that have been employed, such as pure metals (Al, Ti), plastics, woven materials (silk and man made), and composites of these and other materials.

3.2 The Enclosure

The most obvious function of the enclosure is to control transmission of the driver's rear-radiated sound energy, thereby preventing its mutual cancellation with front-radiated energy at low frequencies. The enclosure also serves to acoustically "load" the driver by presenting to it an appropriate acoustic impedance given the characteristics of the driver and the host system's application, particularly with regard to low-frequency and large-signal performance. Particular enclosure arrangements for the major classes of direct radiators are covered in Section 4.

Internal Modes Like any acoustic space, the enclosure volume resonates at particular frequencies that depend on its shape and dimensions. Standing waves can cause irregularities in the frequency response, evidenced by peaks and dips of varying magnitude and bandwidth (depending on modal damping), via both driver re-radiation of their energy and their resonant acoustic impedances presented to the driver. These effects may be minimized by increasing the frequency at which internal cabinet modes occur (by reducing the longest cabinet dimensions), maintaining certain dimensional ratios with the enclosure (such as 5:3:2 for a rectangular enclosure) so as to reduce the individual predominance of modes by forcing them to occur at prescribed frequency intervals, locating the driver internally so that it is near the offending modes' pressure minima, and strategically placing (along the modes' velocity maxima) dissipative material (such as fiberglass, long fiber Dacron, or open-cell foams) inside the enclosure. Generally, this last technique is ineffective below approximately 800 Hz for the listed materials employed in reasonable packing densities. Finally, cabinet shape plays an important role. Sufficiently nonparallel internal surfaces prevent the formation of low-order normal modes, and instead facilitate the generation of a more diffuse acoustic field (larger modal density over the pertinent bandwidth) inside the cabinet.

While standing waves along height and width axes and their effects on loudspeaker system performance are generally amenable to the techniques listed above, a rectangular enclosure's depth modes are more difficult. For enclosures of substantial internal volume and reasonable dimensional ratios they occur within the frequency range over which conventional damping materials are only marginally effective. Furthermore, they are unavoidably excited. This is because the drive units are necessarily mounted to depth modes' antinodal surfaces (pressure maxima, velocity minima)—namely, the baffle itself. A technique applicable to vented systems for mitigating this problem involves active cancellation (destructive interference) of the fundamental (half-wavelength) depth mode's re-radiated energy (typically evidenced by a 1–5 dB peak, approximately $\frac{1}{3}$ of an octave in bandwidth, centered at the depth mode resonance frequency) by means of resonant radiation of the vent itself, thereby reducing the magnitude of the response peak associated with re-radiation and providing improved midrange clarity. This scheme entails choosing the port's length so that its fundamental (axial) mode coincides with the enclosure's fundamental depth mode and appropriately positioning the vent on the baffle near the affected drive unit. Since this technique fixes its length, the vent's cross-sectional area must be chosen to achieve the desired tuning frequency; alternatively, multiple vents may be employed whereby at least one is configured (as described above) to reduce depth mode re-radiation, and the others (collectively with the port of the prescribed length) to achieve the target port tuning frequency.[20]

Cabinet Vibrations[21,22] As a perfectly rigid structure; the ideal enclosure contributes no sound to the total system's sound radiation, but the performance of even high-quality systems may suffer from cabinet vibration problems. Excited either acoustically (airborne) by the internal pressure fluctuations distributed over the inside surfaces of the cabinet walls, or mechanically (structure-borne) via transmission of driver frame reaction forces, enclosure resonances color a loudspeaker system's output. While wood products such as particle board, MDF (medium density fiberboard), and plywood are used in the fabrication of the great majority of loudspeaker enclosures, plastics, honeycomb aluminum and other materials also have been utilized.

With respect to a particular resonance, cabinet radiation is stiffness controlled below, mass controlled above, and damping controlled within the modal bandwidth. Respectively increasing stiffness, mass, and damping decreases vibration over these regimes. Ideally, a panel of small enough extent and composed of a material of a sufficiently low mass–stiffness ratio would have its fundamental resonance either beyond the passband of the system's low-frequency drive element or within the range of effectiveness of dissipative box-stuffing materials. When panel resonances do occur within the woofer's passband, material damping of the panel itself reduces vibration. Sandwich constructions with a constrained layer of a damping material or application of a damping compound to enclosure panels can help in this regard. Finally, limiting unsupported dimensions (to raise resonance frequencies), while varying those dimensions via irregularly spaced internal bracing (for favorable frequency intervals between modes) is a time-honored technique for avoiding vibration problems.

Recent research has detailed a reciprocal means by which the sound radiation associated with cabinet vibration may be estimated. That method involves quantifying the cabinet's vibrational response to both distributed acoustic and mechanical point forces. The enclosure's radiation efficiency is then computed from these transfer functions.[23]

3.3 Crossover Networks[12,24]

The crossover is an electronic dividing network composed of high-pass (HP) and low-pass (LP) sections that respectively favor (pass) signals above and below the characteristic cutoff frequency, comparatively blocking signals (by presenting a high-impedance load to the amplifier) below and above cutoff. Individual sections are characterized by their electrical slope (6 dB/octave per order), resonance frequency, and magnification factor Q (quality factor) at resonance (and attendant transient response). Bandpass filters, series combinations of HP and LP filters, are employed in multiway loudspeaker systems, which by definition employ more than two types of drivers.

Ideally, the crossover provides a smooth transition between the passbands of a multiple-driver loudspeaker system such that (1) the combined acoustic frequency and phase response are not unduly sensitive to listening location and (2) HP rolloff rates are sufficiently high to provide adequate protection from overload (which can cause failure due to overexcursion or excessive temperatures associated with voice coil power dissipation) below the high-passed drive elements' passbands. Appropriate combinations of cutoff frequency (relative to the tweeter's free-air resonance) and filter order satisfy the second criterion for a two-way system. For satisfying these goals, design considerations include the drivers' behavior beyond their nominal passbands (acoustic rolloff, ultimately 12 dB/octave steeper than electrical, can have a critical effect on system response), relative output level (typically, tweeters require attenuation), driver offset (noncoincidence in space of their "acoustic centers"), allowable lobing (upper limit on system directivity through the crossover range), and response tilt (angular deviation of the preferred listening axis from the normal to the baffle).

The major types of networks are minimum-phase, all-pass and non-all-pass. By definition, all-pass filters provide flat amplitude but variable phase response, the audibility of which is debatable. Only first-order networks are minimum phase, but insufficient protection of high-passed drive elements below cutoff, excessive demands on drive units' amplitude response smoothness well beyond nominal passbands, and high variability of system response (due to the comb filtering associated with the drive elements' physical offset and the wide overlap of their output) limit their effective implementation. Higher order all-pass filters are less sensitive to driver offset and provide better sub-passband protection for high-passed drive elements. For these reasons, most high-quality production loudspeakers employ at least second-order filters on their tweeters, often in *nonsymmetric* (high- and low-pass sections of different orders) arrangements with lower order low-pass filters. Finally, the non-all-pass filters can be manipulated to provide flat amplitude response by utilizing noncoincident HP/LP cutoff frequencies (usually by spreading them). In fact, the reality of crossover design is such that the "textbook" filters are rarely practically achievable, but by carefully manipulating element values and employing appropriate topologies, satisfactory results are usually obtainable.

Figure 6 compares the magnitude response of first- through fourth-order Butterworth ($Q = 0.707$) filters. The characteristic attenuation rate of 6 dB/octave per order is plainly evident. Again, increasingly steep rolloff rates provide lower sensitivity to both driver offset and extra-passband response irregularities.

The two types of crossover networks (passive and active) are described briefly here.

***Passive Networks*[25–27]** The location of passive networks relative to the source, power amplifier(s), and drivers is shown in Fig. 7*a*. Standard in consumer loudspeaker systems, passive networks incorporate combinations of resistors, capacitors, and inductors. As energy storage devices, the latter two components are inherently frequency dependent, while resistors are electrical dampers, removing energy. These types of filters require

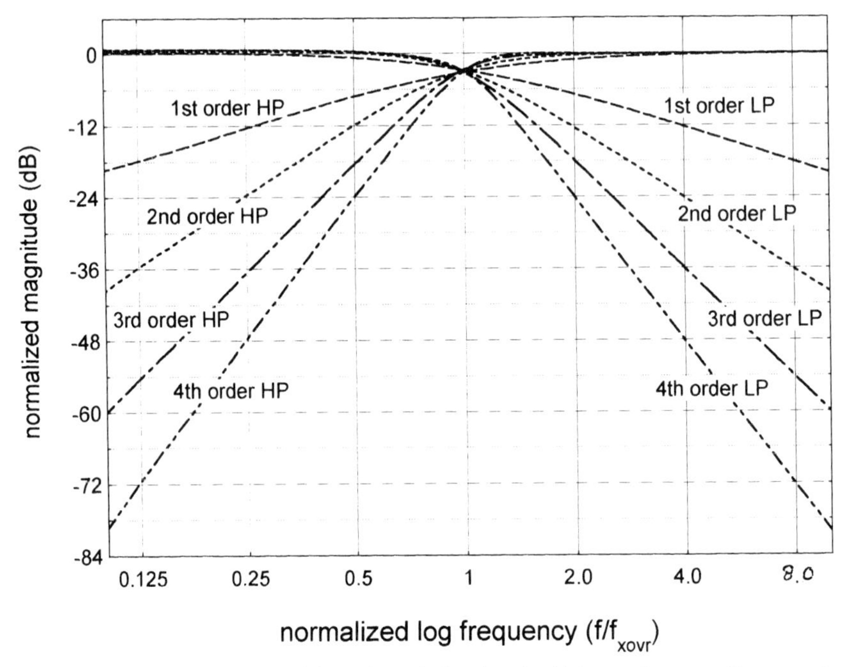

Fig. 6 Magnitude response of first- through fourth-order high-pass/low-pass Butterworth filter pairs.

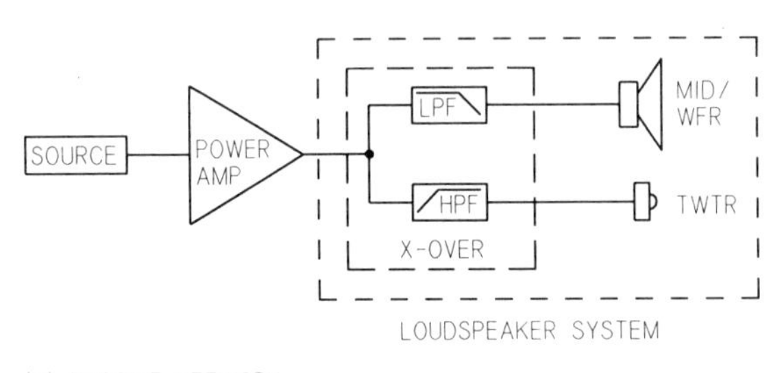

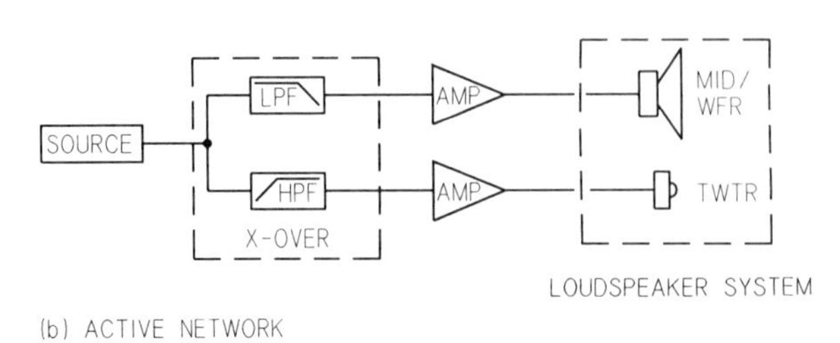

Fig. 7 Schematic representations of (*a*) passive and (*b*) active crossover networks for two-way systems. (From J. R. Ashley and A. L. Kaminsky, "Active and Passive Filters as Loudspeaker Crossover Networks," *J. Audio Eng. Soc.*, June 1971.)

no gain and no external power supply; hence, they are called passive.

Constant voltage (series) and constant resistance (parallel) comprise the major classes; the latter are employed in the vast majority of commercial designs. The standard topologies of first- to fourth-order parallel symmetric (HP, LP sections of the same order) networks are shown in Fig. 8. That the number of reactive elements per section equals that section's electrical order may be noted. In addition to the elements shown, sections for response correction, impedance equalization, and (tweeter) attenuation are often included. The well-documented Butterworth ($Q = 0.707$), Bessel ($Q = 0.58$), Linkwitz–Riley ($Q = 0.49$, squared Butterworth) filters of various orders may be achieved with these topologies. While formulas for calculating particular filters' element values are available from a variety of sources (including the cited references), they assume purely resistive terminations and hence are valuable only for approximating element values. Crossover design lends itself to computer modeling programs, which take into account the complex impedance of actual drivers.

Active Networks Their location in the audio reproduction chain is shown in Fig. 7*b*. Unlike passive filters, active networks require external power, operating at line level to filter and distribute the signals to multiple-power amplifiers, which in turn drive directly the loudspeaker system's transducers. Integrated circuitry, particularly op-amps in combination with resistors and capacitors, can provide almost any desired response shape. A number of low-frequency ("actively assisted") alignments are realizable only by employing a prescribed response augmentation. Typically, these feature shelved or low-Q low-frequency behavior in the absence of active equalization, to be "boosted" appropriately by the filter. Finally, "digital" speakers operate on the program signal in the digital domain, appropriately filtering, delaying, and otherwise tailoring it for the particular drive units before digital-to-analog conversion and subsequent power amplification.

Comparisons: Passive Versus Active The main advantages of passive networks include no need for a power supply and the requirement of only one power amplifier channel per loudspeaker channel, as opposed to one per filter bandwidth. The major disadvantages are the bulk and expense of passive components (particularly, low DC resistance (DCR) series inductors in LP sections and low-dissipation-factor, high-voltage series capacitors in HP sections), the sensitivity losses (1–3 dB SPL) due to inductors' power dissipation (their "parasitic" resistance, reflected by DCR), and finally their thermal sen-

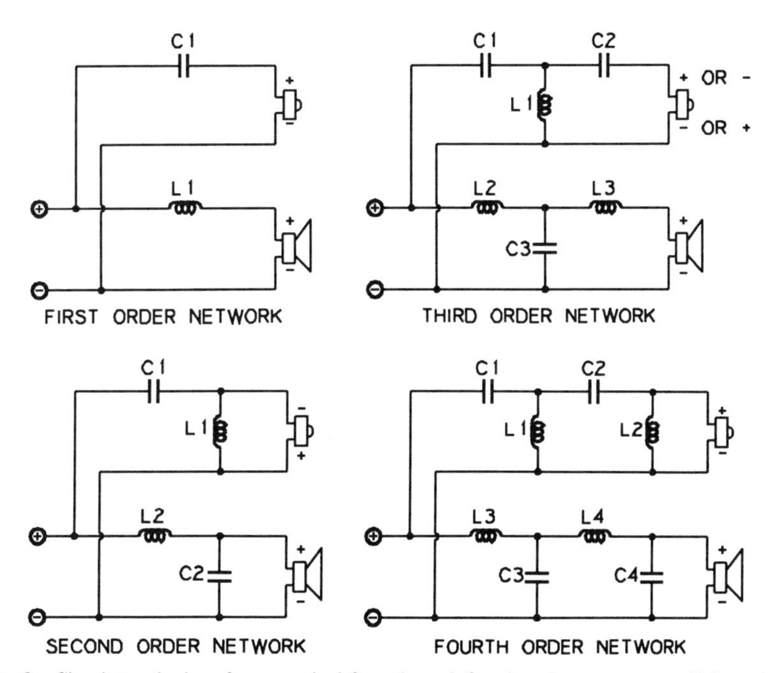

Fig. 8 Circuit topologies of symmetrical first- through fourth-order two-way parallel passive crossover networks. (Reprinted with permission from *The Loudspeaker Design Cookbook, 5th ed.*, by Vance Dickason. Copyright © 1995.)

sitivity, which gives rise to response variability at high drive levels and contributes to "power compression" (see Section 4.4).

Advantages of active networks are the electrical isolation afforded by their high input and the low output impedance; response shapes are independent of (variable) source and load impedances and permit the use of multiple filters serially cascaded. Variable gain and frequency shaping are some other major advantages.

4 DIRECT-RADIATING LOUDSPEAKER SYSTEMS*

The two major classes of dynamic direct radiators are (1) infinite baffle/closed box, whose sole radiation source is the driver diaphragm, and (2) bass reflex, whose enclosures feature tuned resonators that augment the driver's radiation at low frequencies.

4.1 Generalized Model and Circuit[1,3]

A generalized direct radiator is illustrated in Fig. 9. The system includes a driver and a port [or passive radiator (PR)] and allows for enclosure leakage. Respectively, these contribute acoustic volume velocities U_D, U_P, and U_L. At very low frequencies, where the spacing between and dimensions of the sources are much smaller than an acoustic wavelength, the system may be treated as a combination of simple (nondirectional), coincident sources whose net volume velocity is the vector superposition of each source's contribution: $\mathbf{U}_0 = \mathbf{U}_D + \mathbf{U}_P + \mathbf{U}_L$. Equation (1) may be invoked for computation of radiated power from this result.

Figure 10 shows an acoustic circuit of the impedance type for the generalized system depicted in Fig. 8. The circuit's elements and their associated International System (SI) units are as follows:

e_g	Open-circuit output voltage of generator (volts)
B	Magnetic flux density in driver air gap (weber/m^2)
l	Length of voice coil conductor in magnetic field of air gap (metres)
L	Driver voice coil electrical inductance (henrys)
R_g	Output resistance of generator (ohms)
R_e	Driver voice coil resistance (ohms)

*This section is based on R. H. Small's seminal series of papers from the early 1970s (Refs. 1, 28, 30, 32, 33). Readers interested in further study are encouraged to explore them.

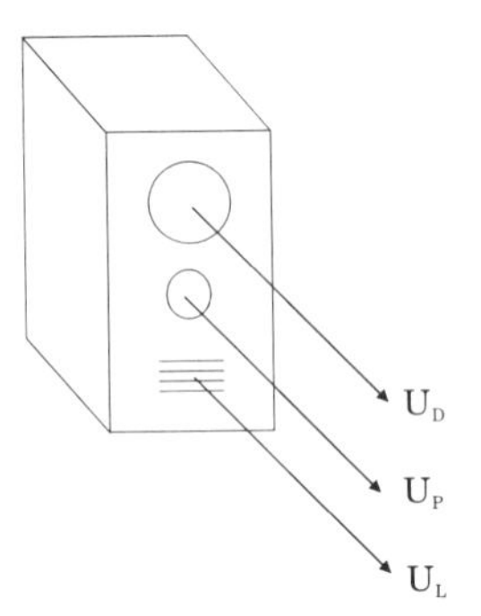

Fig. 9 Generalized model of a direct-radiating loudspeaker system, with volume velocity contributions from the driver (volume velocity U_D), port/passive radiator (U_P), and enclosure leakage (U_L).

S_d	Effective projected surface area of driver diaphragm (m^2)
M_{as}, C_{as}, R_{as}	Acoustic mass (of moving system, including air load, kg·m^4), compliance (m^5/N), and resistance (associated with the driver's suspension, N·s/m^5)
C_{ab}, R_{ab}	Acoustic compliance (of the air inside enclosure, m^5/N) and resistance of enclosure losses due to internal energy absorption (N·s/m^5)
R_{al}	Acoustic resistance of enclosure losses due to leakage (N·s/m^5)
M_{ap}, C_{ap}, R_{ap}	Acoustic mass (including air load), compliance (m^5/N), and resistance (N·s/m^5) associated with port or PR
$R_{ar,\mathrm{DLP}}$	Acoustic radiation resistance pertaining to diaphragm, leakage, or port/PR (N·s/m^5)
$X_{ar,\mathrm{DLP}}$	Acoustic radiation reactance pertaining to diaphragm, leakage, or port/PR (N·s/m^5)

This circuit models the piston-band behavior of direct radiators; the frequency-dependent parameters of the driver's voice coil (see Section 4.2) are omitted here, as their effect is negligible over the low-frequency range of interest. Though they are shown in the circuit for completeness, radiation impedance terms (also frequency dependent) are so small compared to those of the other circuit elements that they may be neglected in the analysis, to be invoked later for computation of acoustic output.

Closed-Box Direct Radiators[28,29] These are depicted in Fig. 11*a*. Acoustic suspension (AS) arrangements are relatively compact systems whose driver's

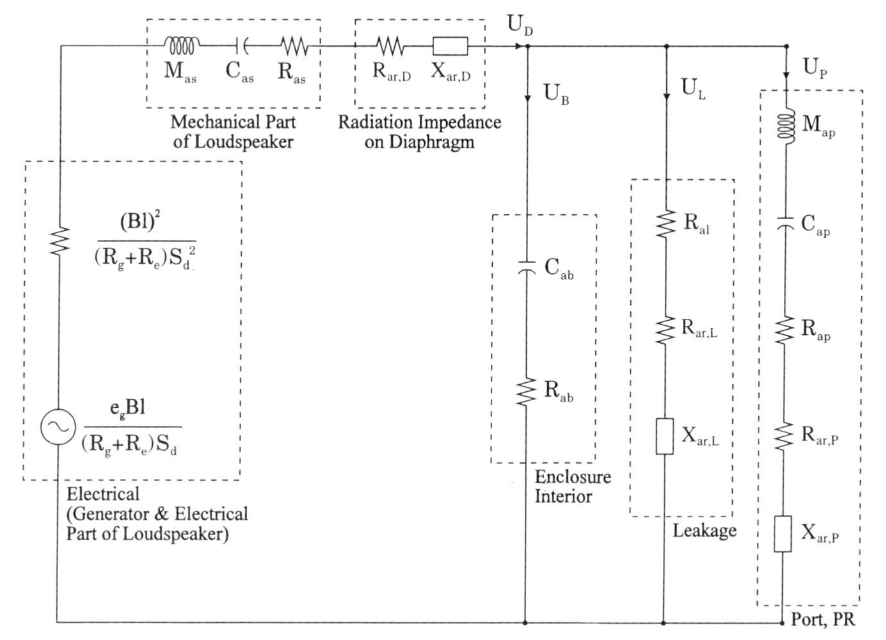

Fig. 10 Acoustic analogous circuit of the impedance type for a generalized direct radiator. Volume velocities through the three parallel branches (box, leakage, and port/PR) are U_B, U_L, and U_P, as indicated. Driver volume velocity is U_D. (From R. H. Small, "Direct-Radiator Loudspeaker System Analysis," *J. Audio Eng. Soc.*, June 1972, with additions.)

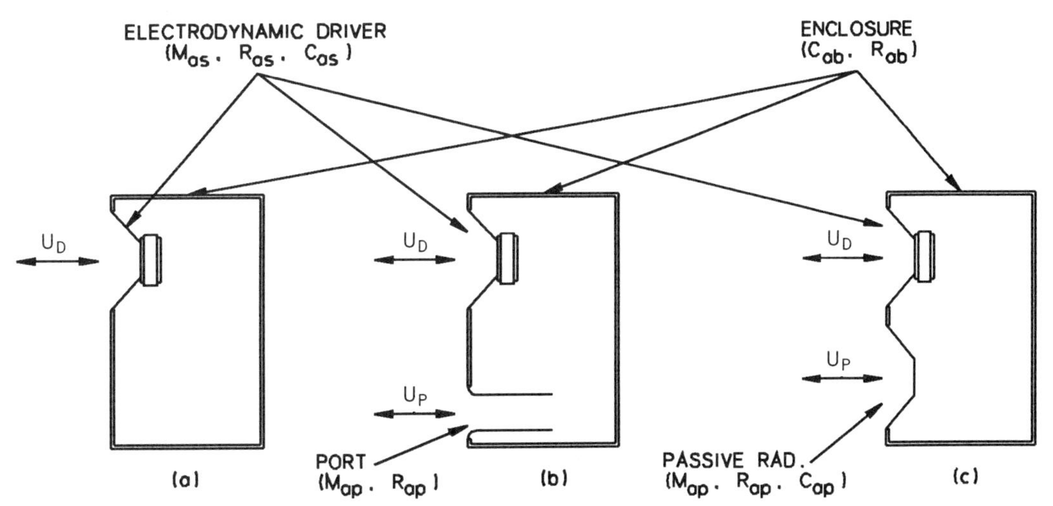

Fig. 11 Direct-radiator loudspeaker systems: (*a*) sealed box; (*b*) vented bass reflex; (*c*) droned bass reflex. The acoustic parameters of their major subsystems (driver, enclosure, port/PR) are indicated.

compliance is at least three times larger than the enclosure's ($3C_{ab} \leq C_{as}$), while infinite baffle (IB) systems have smaller compliance ratios. By design, AS systems have a dramatic enclosure size advantage over IB (or large closed-box) systems; an enclosure of relatively high air stiffness (low C_{ab}, a necessarily small box) is required if it is to provide the restoring force exerted on the cone, while IB systems rely only on the driver's suspension for the restoring force.

Removing the enclosure leakage and the port/PR branches from the circuit yields the circuit for an ideal closed-box loudspeaker. Short circuiting the elements C_{ab} and R_{ab} (which makes their values infinite and zero, respectively) converts this model further into one for an ideal IB system.

Bass-Reflex Systems[30,31] Figures 11*b*,*c* show the major components of the two types of bass-reflex systems. A vented system incorporates a tuned aperture, normally a vent (or port) of a prescribed cross-sectional area and length, whose "confined" air mass (M_{ap}) resonates with the enclosure's air spring (of compliance C_{ab}). The acoustic forces associated with pressure variations inside the cabinet excite this resonance. Thus, the woofer's "back wave" communicates with external acoustic space via a Helmholtz resonator. In PR systems, the vent is replaced with an acoustically driven diaphragm ("drone cone").

As shown in Fig. 10, the port PR is modeled as a series reactance (mass) and resistance. Eliminating the leakage branch and shorting out C_{ap} obtains the circuit for an ideal vented system.

4.2 Small-Signal Parameters[1]

The parameters of a direct radiator's three subsystems (driver, source, and enclosure) together determine small-signal performance.

Driver Parameters The fundamental electromechanical driver parameters are R_e, Bl, S_d, C_{ms} ($=C_{as}/S_d^2$), M_{ms} ($=M_{as}S_d^2$), and R_{ms} ($=R_{as}S_d^2$). The latter three are the mechanical analogs of their acoustic counterparts introduced in Fig. 10.

Four related parameters, convenient for the design and analysis of loudspeaker systems, are the so-called Thiele–Small parameters:

- f_s Free-air (no-baffle) resonance frequency of the driver's moving system, $=(1/2\pi)(1/C_{ms}M_{ms})^{1/2}$
- V_{as} Acoustic compliance of the driver, expressed as an equivalent volume of air, $=S_d^2\rho_0 c^2 C_{ms}$
- Q_{ms} Mechanical Q factor of the driver at f_s, considering nonelectrical resistance only, $=1/2\pi f_s C_{ms}R_{ms}$
- Q_{es} electrical Q of the driver, $=2\pi f_s R_e M_{ms}/(Bl)^2$

A fifth parameter, β [$=(Bl)^2/R_e = 2\pi f_s M_{ms}/Q_{es}$], is convenient for evaluating driver design trade-offs. As a measure of normalized motor strength, β is proportional to acoustic efficiency and inversely proportional to Q_{es} (see Section 4.3).

Driver Electrical Impedance Figure 12*a* shows a dynamic driver's electrical equivalent circuit. The elements L_{ces} [$=C_{ms}(Bl)^2$], C_{mes} [$=M_{ms}/(Bl)^2$], and R_{es} [$=(Bl)^2/R_{ms}$] represent respectively the driver's compliance, mass, and resistance. The parameters L_{evc} and R_{evc} are, respectively, the frequency-dependent inductive and resistive portions of the voice coil's reactive rise while R_e is its DC resistance, normally 1.5–7 Ω (corresponding to nominal impedance ratings of 2–8 Ω) for consumer product applications. Looking into the circuit from its input terminals, the steady-state impedance magnitude features a peak at f_s (free-air resonance) and a reactive rise at high frequencies. Figure 12*b* shows this curve and presents expressions for Q_{ms} and Q_{es}.

Source Parameters In practice, modern power amplifiers have flatter response over a wider bandwidth (compared to a loudspeaker system's) and a much smaller output impedance ($R_g \ll R_e$). Generally, the source's small-signal effects on system performance are negligible.

Enclosure Parameters These appear in the vertical branches of the circuit shown in Fig. 10. Box compliance C_{ab} (proportional to enclosure volume V_b) is critical to the low-frequency performance of any type of enclosed system; larger V_b allows for some combination of more extended bass and higher output capability (see Section 4.6). The other enclosure parameters pertain to the port/PR (M_{ap}, C_{ap}, R_{ap}) and to box losses (R_{ab}, R_{al}).

Bass-Reflex System Box Losses[30] The three sources are leakage (Q_l), internal absorption [Q_a ($=1/C_{ab}R_{ab}$)], and vent resistance [Q_p ($=1/2\pi f_b C_{ab}R_{ap}$]. Normally, these parallel losses are lumped into a representative value for leakage (Q_l', where $1/Q_l' = 1/Q_l + 1/Q_a + 1/Q_p$). For most systems Q_l' (often simply Q_l, but understood to include all of the box losses) falls between 5 and 12.

System Parameters Certain inter-subsystem dimensionless ratios define low-frequency alignments. There

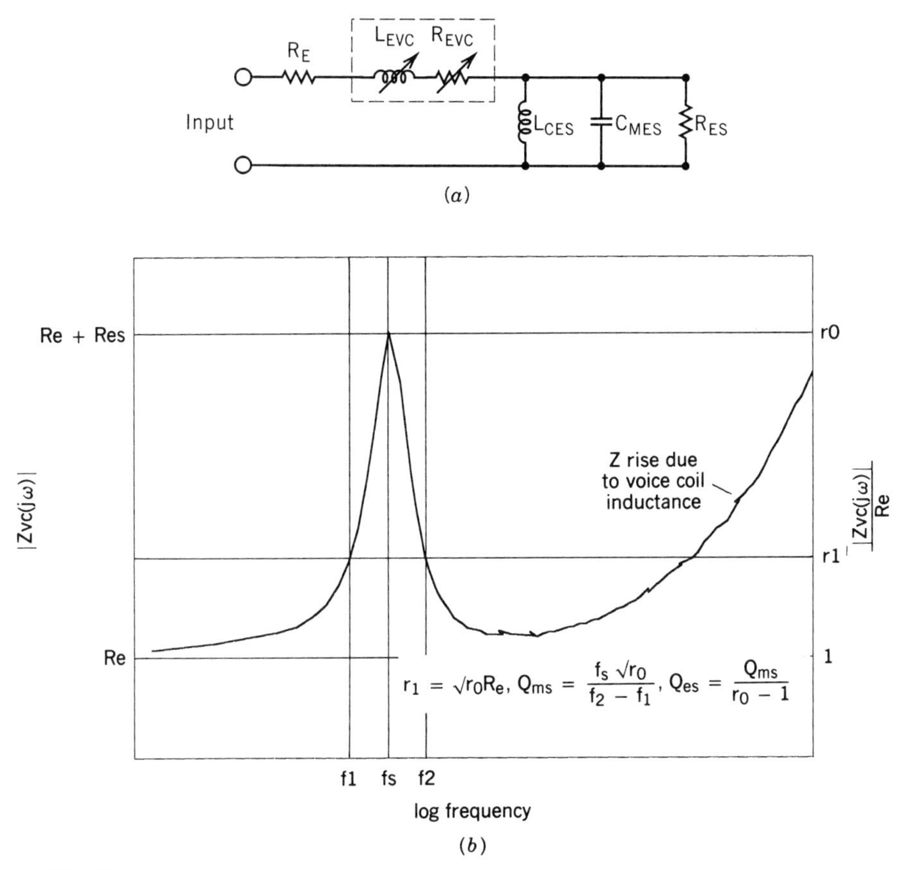

Fig. 12 Electrodynamic driver's equivalent electrical circuit (*a*) and impedance magnitude (*b*); expressions for mechanical and electrical quality factors (Q_{ms} and Q_{es}, respectively) are indicated. (From R. H. Small, "Direct-Radiator Loudspeaker System Analysis," *J. Audio Eng. Soc.*, June 1972, with additions.)

are two system parameters that are relevant to any type of direct radiator:

1. Compliance ratio (α), expressed as either C_{as}/C_{ab} or V_{as}/V_{ab}, which relates the driver's and enclosure's compliances
2. Total system Q's at the various resonances of interest

For closed-box systems, the box resonance ratio is the pertinent system parameter. It is expressed as either f_c/f_s or Q_{tc}/Q_{ts}, where f_c is the closed-box system resonance and Q_{tc} $(=1/2\pi f_c C_{at} R_{atc})$ is the total system Q at f_c. Here, C_{at} $[=C_{ab}C_{as}/(C_{ab} + C_{as})$ and R_{atc} $[=R_{ab} + R_{as} + (Bl)^2/(R_g + R_e)S_d^2]$ are, respectively, the total system compliance and resistance, while Q_{ts} $[=Q_{es}Q_{ms}/(Q_{es} + Q_{ms})]$ is the total driver Q at f_s.

Applicable to vented bass-reflex systems are the following system parameters:

ω_b Radian frequency of the enclosure-vent system resonance, $=2\pi f_b = (1/C_{ab}M_{ap})^{1/2}$

Q_l The Q of the resonance at ω_b associated with leakage losses but whose value may reflect combined box losses, $=\omega_b C_{ab} R_{al}$

h System tuning ratio, $=f_b/f_s$

4.3 Frequency Response

Solving the appropriate equivalent circuit obtains a general response function for radiated sound power that assumes the form of a high-pass filter whose coefficients are functions of the system's masses, compliances, and resistances.

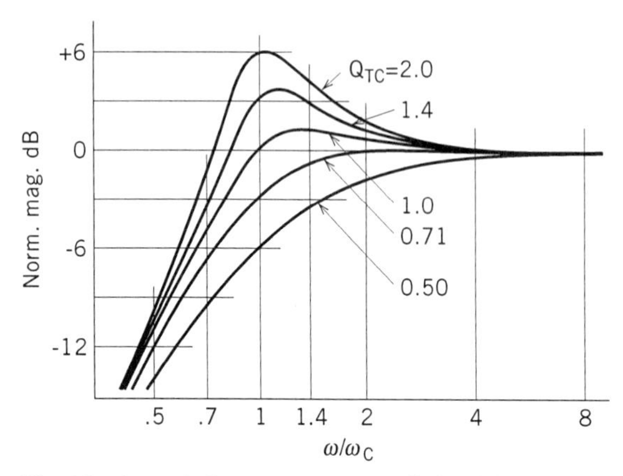

Fig. 13 Acoustic frequency response of closed-box systems for various values of Q_{tc}.

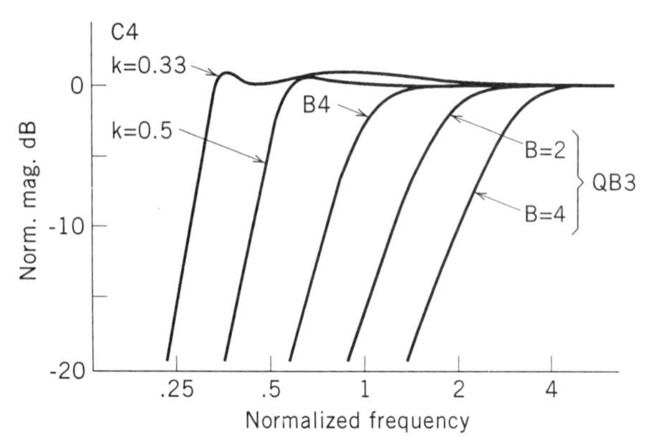

Fig. 14 Normalized acoustic response curves for various bass-reflex vented alignments: B4 ≡ fourth-order Butterworth, maximally flat; C4 ≡ fourth-order Chebyshev, equal ripple (two realizations shown); QB3 ≡ quasi-third-order Butterworth, (two realizations shown). Curves have been displaced along frequency axis for clarity. (From R. H. Small, "Vented-Box Loudspeaker Systems Part I: Small-Signal Analysis," *J. Audio Eng. Soc.*, June 1973.)

Closed-Box Systems: Q_{tc} [28] The characteristic response function for closed-box systems takes the form of a second-order (12-dB/octave) high-pass filter, but higher and lower order rolloffs are achievable. Pressure response shapes for various values of Q_{tc} are shown in Fig. 13. Low-frequency output may be increased by choosing an underdamped response ($Q_{tc} \geq 0.707$), but only at the expense of a peak near f_c ($Q_{tc} \geq 1.0$), which will be audible as bass "overhang" (evident in the transient response as "ringing" near f_c) and a steeper rolloff rate. An overdamped ($Q_{tc} \leq 0.707$) system's output is lower than the reference level near resonance, below which its rolloff approaches a more gentle 6-dB/octave rate.

There exist certain combinations of Q_{ts}, α, and f_c/f_s that yield the "name" filter response shapes [critically damped ($Q_{tc} = 0.5$), Bessel ($Q_{tc} = 0.577$), Butterworth ($Q_{tc} = 0.707$), Chebyshev ($Q_{tc} > 0.707$)]. Some of these appear in Fig. 13. Tabulations of their associated system parameter values are presented in Ref. 12.

Vented Bass-Reflex Systems: Alignments [30,32] The response function of a vented-box system assumes the form of a fourth-order (24-dB/octave) high-pass filter, but its shape can be manipulated to mimic that of higher or lower order filters.

The term "alignment" refers to particular combinations of system parameters that yield, along the continuum of low-frequency response shapes, ones associated with the known filter classes (and their attendant transient behaviors). The pressure magnitude response shapes of several common alignments are shown in Fig. 14. Detailed design protocol, including tables, charts, and formulas for computing vent dimensions and components' small-signal parameters for achieving particular alignments, appear in Refs. 12, 31, and 32.

Passive Radiator Systems The results of the analytical modeling of vented systems are generally valid for PR systems; they are fourth-order systems. While a variety of alignments are achievable (detailed in Ref. 12), PR systems are difficult to tune precisely in practice.

Maximum Power-Available Efficiency [3] This is a measure of acoustic power radiation in proportion to the maximum power that the electrical generator can supply. It is instructive to consider power-available efficiency (PAE) through a direct radiator's entire passband. Ignoring cone breakup, PAE assumes the curve shape of Fig. 15 for closed-box systems, while bass-reflex systems behave identically within regions *C*, *D*, and *E*. There are five distinct regions:

Region *A* Stiffness control [first-order (6-dB/octave) high-pass filter (HPF)] and radiation resistance behavior (another first-order HPF) in this region give rise to the (nominal) +12-dB/octave acoustic power output increase.

Region *B* Near f_c, PAE is damping controlled. Resistance terms $[(Bl)^2/(R_g + R_e), R_{ms}, R_{ar}]$ govern the response.

Region *C* Mass control (first-order low-pass filter) and the still increasing radiation resistance (first-order HPF) negate each other. Cone drivers of unusually low mass and/or electrical voice coil inductance exhibit a rising PAE within this region.

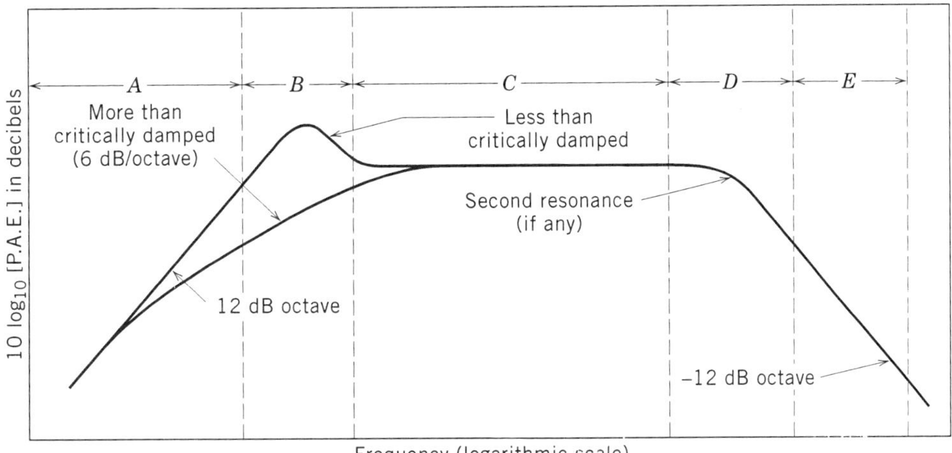

Fig. 15 Normalized PAE for a closed-box loudspeaker system. Acoustic power response mirrors PAE, while on-axis SPL parallels PAE at low frequencies, deviating progressively more severely beyond the piston band. (From L. L. Beranek, *Acoustics*, Acoustical Society of America, New York, 1986, with permission.)

Region D A second resonance occurs here if $R_g + R_e$ is comparable to $L\omega$, where the response is governed by these and other damping terms. This is the transition region between rising and constant radiation resistance, near where $ka = 2$.

Region E The term R_{ar} is approximately constant; mass control and the generator output's inverse proportionality with frequency (due to voice coil inductance) combine to generate the −12-dB/octave behavior, which takes effect beyond approximately $ka = 4$.

Reference Efficiency **(η_0)** Also called *power transfer ratio*, η_0 gives the generated sound power level per input watt within the flat region of response (region C in Fig. 14). As such, it is purely driver dependent. Two equivalent expressions for η_0 are

$$\eta_0 = K_1 \frac{f_s^3 V_{as}}{Q_{es}} = K_2 f_s^2 S_d^2 \frac{C_{ms}}{M_{ms}} \beta, \tag{2}$$

where $K_1 = 9.64 \times 10^{-10}$ (for V_{as} expressed in liters) and $K_2 = 2.15 \times 10^{-5}$ (MKS units for all variables). Typical values of η_0 for dynamic direct radiators fall between 0.5 and 1.5%.

Diaphragm Displacement Determined directly from volume velocity U_d, which can be obtained by solving the system's acoustic circuit, the system displacement function $X(s)$ is a *low*-pass filter for any direct radiator. For closed-box systems, $X(s)$ takes the form of a simple second-order low-pass filter, as shown in Fig. 16.

For bass-reflex systems, $X(s)$ is more complicated. Figure 17 shows diaphragm displacement curves for several common alignments. These curves' characteristic features are the minima at f_b (vent enclosure tuning frequency) and the increased excursion below f_b that typically exceeds that of comparable acoustic suspension

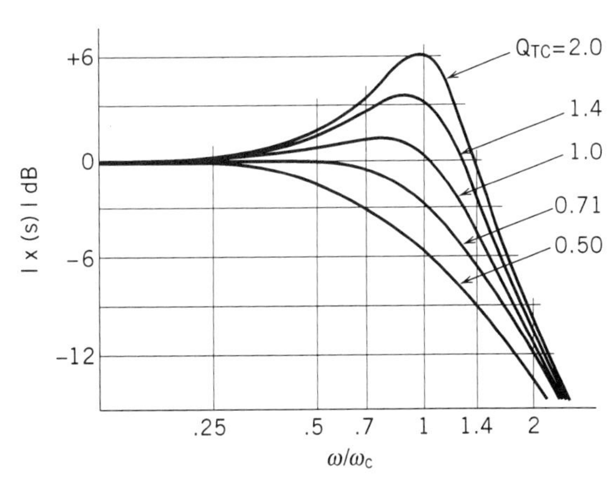

Fig. 16 Normalized diaphragm displacement of closed-box systems for various values of Q_{tc}. (From R. H. Small, "Direct-Radiator Loudspeaker System Analysis," *J. Audio Eng. Soc.*, June 1972.)

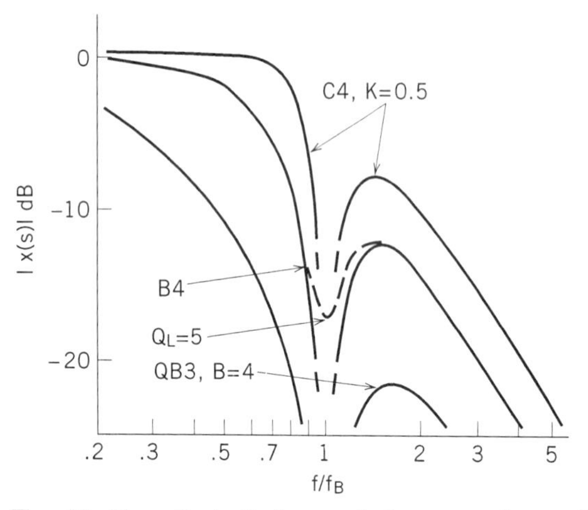

Fig. 17 Normalized diaphragm displacement of vented bass-reflex systems for various alignments: B4 ≡ fourth-order Butterworth maximally flat (shown for $Q_l = 5, \infty$); C4 ≡ fourth-order Chebyshev, equal ripple (one realization shown); QB3 ≡ quasi-third-order Butterworth (one realization shown). (From R. H. Small, "Vented-Box Loudspeaker Systems. Part I: Small-Signal Analysis," *J. Audio Eng. Soc.*, June 1973.)

systems. Since driver and vent radiation are out of phase in this frequency range, this increased excursion does not correlate with system acoustic output, which actually falls at a rate of 24 dB/octave below f_b. In practice, especially for professional applications, active high-pass filters are placed in the signal path in order to limit cone excursion below port tuning.

4.4 Large-Signal Performance[1]

The two pertinent ratings concern power handling (maximum input power) and radiated acoustic power (output limits). These and other large-signal performance attributes are treated presently.

Input Power Rating The input power rating $P_{E(\max)}$ is determined by the voice coil's ability to dissipate heat. Independent of enclosure type, it is the maximum input power rating for the *driver*, depending only on its thermal capability over the middle portion of its passband (region *C* of Fig. 14), where cone excursion is relatively low.

Displacement-Limited Power Ratings These are both driver and loudspeaker system dependent. Determining factors are prevention of damage to the suspension and allowable distortion (AM, FM, and harmonic) levels.

Displacement Limits. The fundamental large-signal parameter is diaphragm peak displacement volume V_d ($=S_d X_{\max}$), where $X_{\max}$ is the peak displacement as determined by the criteria above. Here, $X_{\max}$ (and therefore V_d) may be smaller than its theoretical limit (as predicted from voice-coil/gap geometry), depending on whether distortion exceeds acceptable limits at that excursion level. This parameter also limits a loudspeaker system's frequency-dependent volume velocity capability: $|U_{0,\max}| = (X_{\max} S_d)\omega$.

Electrical Power Capacity (P_{ER}) and Acoustic Power Radiation (P_{AR}). Expressions for P_{ER} and P_{AR} appear in Ref. 1. In practice, their values are determined experimentally. "Blow-up" power tests utilize shaped (rolled-off at the frequency extremes to mirror the spectral content of typical program material), preconditioned (normally 6 dB peak-to-rms voltage) noise to drive the speaker at progressively higher levels until failure. Both acoustic output and the input power level are recorded over the course of the test. "Life" tests employ similarly processed signals to drive the speaker at a constant level over a predetermined time (often 8 hours) or until failure. The electronic standards agencies, Electronic Industries Association (EIA) (RS-426-A) and the International Electrochemical Commission (IEC) (268-5) have documented protocol for these tests.

Power Compression[12] This large-signal phenomenon occurs when acoustic output level does not increase proportionally with drive level. Characteristic of electrically conductive materials are their thermal sensitivity; resistivity (ohms per unit length of conductor) increases nonlinearly with temperature. This means that a driver's voice coil presents a larger electrical impedance magnitude to the amplifier at higher steady-state input levels. Consistent with the inverse proportionality of η_0 and R_e, the result is a lower acoustic output level per input volt. Temperature–resistivity effects also apply to inductors in passive crossovers, causing further increases in the load presented to an amplifier. Ultimately, most drivers fail at temperatures approaching 250°C when adhesive bonds break down, but maximum acoustic output levels are achieved well below the steady-state input levels associated with failure.

The attendant large-signal effects on acoustic frequency response are characterized by decreased damping; for closed-box systems, the response shape can effectively "jump" to curves of progressively higher Q_{tc} as input levels escalate. Proportionality of Q_{tc} and R_e explain this behavior. Since R_e can more than double its small-signal value at very high temperatures, response shapes can change enormously. Similarly, the response shapes of bass-reflex systems are less damped with increasing input power.

Bass-Reflex Considerations For vented designs, a large-signal concern is turbulent flow noise associated with excessive air velocities at the mouth of the port. Choosing a vent design of appropriate cross-sectional size, shape, and termination flare obviates this problem. Using a turbulence threshold for air velocity of 5% of the speed of sound (about 17 m/s), output power ratings impose a minimum vent area (in square metres) of $S_{v,\min} = 0.8 f_b V_d$.[31] While vents even larger than this offer better linearity and less power compression (attributable to nonlinear vent losses), length grows with S_v^2 for a fixed f_b; practical enclosures impose a limit on S_v. Nevertheless, sound reinforcement systems commonly utilize vent areas that are comparable to S_d, expressly for avoiding vent noise problems at their extremely high system operating levels.

Analogously for droned systems, the concern is excursion capability of the PR. For a PR of the same size as the driver, mechanical noises are generally avoidable if the PR's "throw" exceeds that of the driver by at least a factor of 2.

4.5 Comparison of Bass-Reflex and Closed-Box Systems[33]

Bass-reflex systems offer performance advantages involving some realizable combination of lower cone displacement, lower distortion, higher efficiency, lower cutoff, and/or a smaller enclosure size. For example, a bass-reflex system can be made to achieve the same low-frequency extension with lower distortion (due to lower cone displacement), better efficiency (by 3 dB), greater output capability (by about 6 dB, but mitigated somewhat by larger sub-passband displacement sensitivity), and higher power handling than a comparable acoustic suspension system of the same enclosure volume and driver size.

Some advantages of closed-box systems include lower sensitivity to mistuning and better "room coupling" for (arguably) smoother, more extended bass performance. The latter is explained by their less steep low-frequency rolloff.

4.6 Design Trade-Offs for Low-Frequency Extension

A woofer cannot simultaneously provide deep bass extension at high efficiency and operate in a small enclosure. These trade-offs have been quantified by *Hoffman's iron law*, which states that low-frequency efficiency is proportional to enclosure volume and the cube of its low-frequency cutoff. Equation (2) is an expression of these trade-offs. As an example, for a given low-frequency alignment [constant compliance ratio (V_{as}/V_{ab}), box resonance ratio (f_c/f_s, closed-box systems), and system tuning ratio (f_b/f_s, vented systems)] and a fixed Q_{es}, reducing the cutoff frequency by one-third of an octave while maintaining a constant η_0 requires doubling the enclosure volume. Alternately, the lower system cutoff can be achieved by accepting a 50% (3-dB) decrease in efficiency.

4.7 Required Amplifier Power versus System Efficiency versus Room Size[12]

Once both the small-signal efficiency and large-signal power ratings for a loudspeaker system have been determined, the amplifier power required to achieve particular program SPLs in a real room can be considered. Figure 18 plots acoustic power versus room volume for various program SPLs, assuming average absorptive constants and a balanced direct/reverberant field listening position. As an example, a single system of 1% efficiency placed in an 80-m^3 space (IEC listening room,[34] typical domestic size) requires 0.4 W (acoustic power) ÷ 1% (efficiency) = 40 W$_{\text{rms}}$ (electrical power) to achieve an average program SPL of 100 dB (re 20 μPa); each of a pair of such systems requires 20 W$_{\text{rms}}$, assuming incoherent summation (3-dB increase in SPL) of their output. To determine feasibility, the 0.2-W output level should be compared to P_{AR} and the 20-W input level to P_{ER}. If the P_{AR} and P_{ER} ratings were 1 W and 125 W$_{\text{rms}}$, respectively, then a pair of such systems would be capable of generating peak SPLs of about 105 dB [102 dB per system, corresponding to 1 W (acoustic) in an 80-m^3 room]. While these are subjectively high SPLs, they are well below the typical 110–120-dB range of peak acoustic levels of musical performances measured at audience locations.[2]

4.8 Other Types of Direct Radiators[12]

These include bandpass, compound woofer, passively assisted, and transmission lines.

Bandpass (BP) enclosure systems feature "buried" low-frequency drivers that face *into* a bass-reflex type resonator and are further acoustically loaded by a second chamber located behind the driver. Fourth-order (symmetrical second-order HP/LP acoustic filters) systems utilize a sealed rear chamber, while sixth-order systems vent (or drone) the rear chamber. That, by virtue of their dual-chamber design, BP systems are *acoustically* low pass filtered makes them attractive for certain applications where an electronic crossover is impractical, such as the separate woofer module of a three-piece satellite

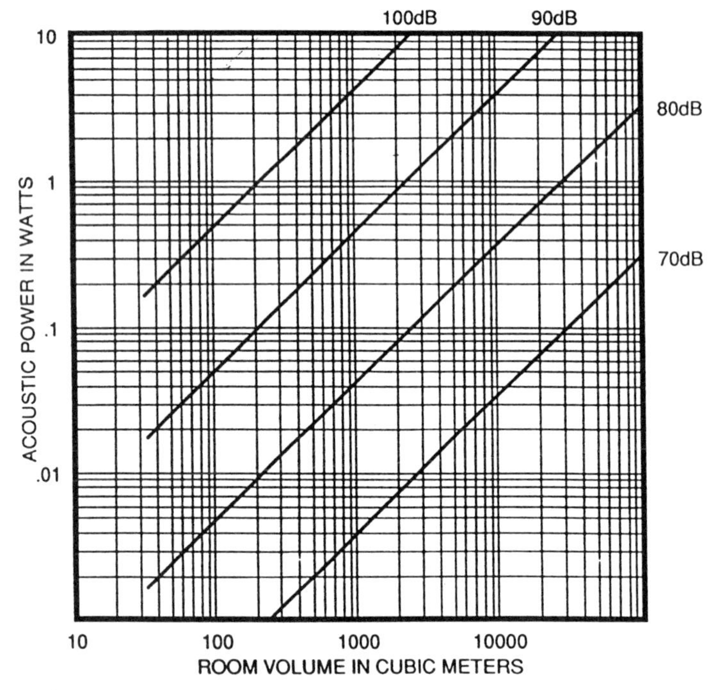

Fig. 18 Acoustic power vs. room size for various program SPLs. (Reprinted, with permission, from *The Loudspeaker Design Cookbook*, 5th ed., by Vance Dickason. Copyright © 1995.)

(mid- to high-frequency module)/woofer stereo system. Finally, efficiency gains of some 3–6 dB with respect to comparable high-pass direct radiators may be realized.[35]

Compound woofer systems utilize multiple drivers in various arrangements to achieve lower distortion, a lower system cutoff, and reduced enclosure size requirements. "Isobarik" (constant-pressure) systems are push–pull arrangements of two woofers that share a small sealed chamber. Loading on one of the woofers is achieved in the usual manner (closed box, bass reflex) while the other couples directly with the room. Greatly reduced distortion products (particularly odd order) and box volume requirements (half that associated with a single woofer) are the major performance advantages of this scheme.

Passively assisted (capacitively coupled) arrangements combine underdamped closed-box systems and a large-value (typically 200–1000 μF) series capacitor to generate extended low-frequency response. These third-order systems (the capacitor contributes an extra order of rolloff) utilize high Q_{ts} drivers and small cabinets [about half the volume required for a sealed-box system (of $Q_{tc} \approx 1.0$) for the same driver] to generate the required underdamped response (in the absence of the capacitor). The capacitor effectively redistributes the energy associated with the response, extending useful response by approximately one-third of an octave in typical cases. Cap-coupled systems trade efficiency for low-frequency extension and small-box operation, since the high Q_{ts} is achieved by some combination of high Q_{ms} (large M_{ms}) and high Q_{es} (small β).

Transmission lines are large, (ideally) nonresonant systems that attempt to absorb the driver's back wave via a dissipative acoustical labyrinth. Their performance features a lower system cutoff than comparable sealed or bass-reflex systems, with a rolloff rate similar to that of overdamped closed-box systems, approaching first order and affording subjectively deep, smooth in-room bass performance. The required length of the duct is at least 0.25λ at the cutoff frequency, necessitating a large enclosure.

5 HORN-LOADED LOUDSPEAKERS[3]

A horn is an impedance-matching device that effectively increases the acoustic power output of a source and gives a directional preference to the radiated power. Compared to 1–2% for direct radiators, horn-loaded loudspeakers can achieve efficiency levels of 10–50% but do so at the expense of physical size (full-bandwidth horn-loaded speakers are necessarily much larger than direct radiators) and frequency response smoothness (resonances along a horn's throat can generate excessive response peaks). However, their marked increase in sensitivity, output capability, and directivity control renders horn-loaded loudspeakers well-suited for sound reinforcement applications. *Enc.* Ch. 161 discusses the details of horn-loaded loudspeakers, including their operating principle, design considerations, performance attributes, and applications.

6 MEASUREMENTS[12,36,37]

Determining a loudspeaker system's performance characteristics (including magnitude and phase response, sensitivity, power output, and distortion) relies on capturing its acoustic pressure response to a known (or captured, for two-channel measurement systems) stimulus with a calibrated microphone and postprocessing those results appropriately.

Modern techniques include time-delay spectrometry (TDS, a form of swept sine wave analysis, whereby the response to a time-varying single-frequency input is captured and appropriately processed) and two-channel spectral analysis, which utilizes random, white noise as a stimulus. Single-channel maximum-length sequence (MLS) processes, whose stimuli are deterministic broad band noise, are a third technique. MLS schemes facilitate computation of true anechoic frequency response (plus a host of other performance characteristics) from a derived impulse response whose room-dependent portion has been "windowed out"; performing an FFT on the remaining portion obtains the anechoic acoustic frequency response.

Assessing low-frequency performance presents its own set of challenges. In accordance with the time–frequency uncertainty principle, capturing complete waveforms over a range bounded at the low end by f_l requires a time window of $T_l \geq 1/f_l$ seconds. Therefore, measurements conducted under either true half-space (outdoor) or anechoic conditions would be required in order to perform valid full-bandwidth frequency response measurements since T_l generally far exceeds the arrival time of initial room reflections. Near-field techniques, whereby the microphone is placed in extreme proximity to the radiating surface (measurement distance $r \ll a$), are a practical alternative. The complex sum of a system's near-field output (driver plus port or PR, weighted in proportion to their diameters) is proportional to far-field sound pressure over the range $ka < 1$ ($f < 1.6f_b$ for ported systems) and may be spliced to a far-field measurement at an appropriately low frequency for generating a full-bandwidth response curve. Two other practical techniques for assessing low-frequency behavior include ground-plane and in-box measurements. These are detailed in the cited references.

Ascertaining values for individual component and system parameters may be achieved by following some well-documented procedures. For direct radiators, most of the parameters are derived from electrical impedance curves whose characteristic features contain a wealth of information regarding system Q factors, system resonance frequencies, and the parameter values that give rise to them. References 1 and 12 outline the protocol of these and other pertinent measurement schemes.

REFERENCES

1. R. H. Small, "Direct-Radiator Loudspeaker System Analysis," *J. Audio Eng. Soc.*, Vol. 20, No. 5.
2. L. D. Fielder, "Dynamic-Range Issues in the Modern Digital Audio Environment," *J. Audio Eng. Soc.*, Vol. 43, No. 5.
3. L. L. Beranek, *Acoustics*, Acoustical Society of America, New York, 1986.
4. E. M. Villchur, "Problems of Bass Production in Loudspeakers," *J. Audio Eng. Soc.*, Vol. 5, No. 3.
5. F. E. Toole, "Loudspeaker and Rooms for Stereophonic Sound Reproduction," *AES 8th International Conference*, 1990.
6. Allison, "The Influence of Room Boundaries on Loudspeaker Power Output," *J. Audio Eng. Soc.*, Vol. 22, No. 5.
7. P. W. Mitchell, "Great Bass: The Acoustic Transfer Function of the Car," *Sound Vision*, Vol. 11, No. 1, 1994,
8. J. Kates, "Optimum Loudspeaker Directional Patterns," *J. Audio Eng. Soc.*, Vol. 28, No. 11.
9. D. Queen, "The Effect of Loudspeaker Radiation Patterns on Stereo Imaging and Clarity," *J. Audio Eng. Soc.*, Vol. 27, No. 5.
10. J. Kates, "Loudspeaker Cabinet Reflection Effects," *J. Audio Eng. Soc.*, Vol. 27, No. 5.
11. H. Olson, "Direct Radiator Loudspeaker Enclosures," *J. Audio Eng. Soc.*, Vol. 17, No. 1.
12. V. Dickason, *The Loudspeaker Design Cookbook*, 5th ed., Audio Amateur Press, Peterborough, NH, 1995.
13. M. R. Gander, "Moving-Coil Loudspeaker Topology as an Indicator of Linear Excursion Capability," *J. Audio Eng. Soc.*, Vol. 29, No. 1.
14. D. Barlow, G. Galletly, and J. Mistry, "The Resonances of Loudspeaker Diaphragms," *J. Audio Eng. Soc.*, Vol. 29, No. 10.

15. F. Frankort, "Vibration Patterns and Radiation Behavior of Loudspeaker Cones," *J. Audio Eng. Soc.*, Vol. 26, No. 9.
16. T. Shindo, O. Yashima, and H. Suzuki, "Effect of Voice-Coil and Surround on Vibration and Sound Pressure Response of Loudspeaker Cones," *J. Audio Eng. Soc.*, Vol. 28, No. 7.
17. M. Klasco, "Santoprene® Rubber—A New Surround Material," *Voice Coil*, Vol. 9, Issue 10.
18. J. Kates, "Radiation from a Dome," *J. Audio Eng. Soc.*, Vol. 24, No. 9.
19. W. Bottenberg, L. Melillo, and K. Raj, "The Dependence of Loudspeaker Design Parameters on the Properties of Magnetic Fluids," *J. Audio Eng. Soc.*, Vol. 28, No. 1.
20. United States Utility Patent, "Improved Bass-Reflex Loudspeaker," Serial No. 08/519365, B. Starobin inventor.
21. P. W. Tappan, "Loudspeaker Enclosure Walls," *J. Audio Eng. Soc.*, Vol. 10, No. 3.
22. J. K. Iverson, "The Theory of Loudspeaker Cabinet Resonances," *J. Audio Eng. Soc.*, Vol. 21, No. 3.
23. B. Starobin and S. Arzoumanian, "Reciprocal Technique for Determining the Sound Radiation Due to Loudspeaker Enclosure Vibrations," *ASA 98th Conference*, New Orleans, LA, May 1992.
24. J. Ashley and A. Kaminsky, "Active and Passive Filters as Loudspeaker Crossover Networks," *J. Audio Eng. Soc.*, Vol. 19, No. 6.
25. R. Bullock, "Loudspeaker Crossover Systems: An Optimal Choice," *J. Audio Eng. Soc.*, Vol. 30, No. 7.
26. R. Bullock, "Satisfying Loudspeaker Crossover Constraints with Conventional Networks—Old and New Designs," *J. Audio Eng. Soc.*, Vol. 31, No. 7.
27. R. H. Small, "Constant-Voltage Crossover Network Design," *J. Audio Eng. Soc.*, Vol. 19, No. 1.
28. R. H. Small, "Closed Box Loudspeaker Systems—Part I: Analysis," *J. Audio Eng. Soc.*, Vol. 20, No. 10.
29. E. M. Villchur, "Revolutionary Loudspeaker and Enclosure," *Audio*, Vol. 38, No. 10, 1954.
30. R. H. Small, "Vented-Box Loudspeaker Systems Part I: Small-Signal Analysis," *J. Audio Eng. Soc.*, Vol. 21, No. 5.
31. A. N. Thiele, "Loudspeakers in Vented Boxes," *Proc. IREE (Australia)*, Vol. 22, Aug., 1961; reprinted in *J. Audio Eng. Soc.*, Vol. 19, Nos. 5 and 6.
32. R. H. Small, "Vented Box Loudspeaker Systems Part III: Synthesis," *J. Audio Eng. Soc.*, Vol. 21, No. 7.
33. R. H. Small, "Vented Box Loudspeaker Systems Part II: Large-Signal Analysis," *J. Audio Eng. Soc.*, Vol. 21, No. 6.
34. IEC Publ. 268-13, "Sound System Equipment, Part 13, Listening Tests on Loudspeakers," International Electrotechnical Committee, Geneva, Switzerland, 1986.
35. E. R. Geddes, "An Introduction to Band-Pass Loudspeaker Systems," *J. Audio Eng. Soc.*, Vol. 37, No. 5.
36. D. B. Keele, Jr., "Low-Frequency Assessment by Nearfield Sound-Pressure Measurement," *J. Audio Eng. Soc.*, Vol. 22, No. 3.
37. R. H. Small, "Simplified Loudspeaker Measurements at Low Frequencies," *J. Audio Eng. Soc.*, Vol. 20, No. 1.

112

TYPES OF MICROPHONES

Ilene J. Busch-Vishniac and Elmer L. Hixson

1 INTRODUCTION

The most common type of acoustic transducer is a microphone, that is, a device designed to produce an electrical signal (usually potential) that mirrors the acoustic pressure (or pressure gradient) present in the air immediately in front of the microphone face. Many types of microphones are available commercially, and the purpose of this chapter is to describe the major types, contrasting them in terms of the physical parameters of interest and their suitability for various applications.

Five types of properties of microphones determine their suitability to serve in a specific acoustic application:

- Electroacoustic performance: the ability of the microphone to perform the task for which it was designed, generally measured in terms of its sensitivity, directivity, frequency response, transient response, linearity, signal-to-noise ratio, and dynamic range. In general terms, a microphone's sensitivity is defined as the ratio of the amplitude of the electrical signal produced to the acoustic pressure (or pressure gradient) amplitude present. The directivity of the microphone is a measure of the extent to which it responds preferentially to sound that arrives from a certain direction.
- Electrical characteristics: the output impedance of the microphone, which determines the amplification methods that are suitable.
- Sensitivity to external influences: the ability of the microphone to operate independently of the air temperature, relative humidity, and wind speed.
- Cost.

In what follows we discuss the microphone types in terms of these performance characteristics. The relative importance of each type of property is dependent upon the application for which the microphone is intended.

While microphones enjoy widespread use, their applications can be put into four main classes:

- Communication microphones: those intended for speech communication between a small number of individuals. Examples include telephone microphones and hearing aids. For this class of applications, high sensitivity, good ruggedness, small size, and low cost are considered essential.
- Sound recording and broadcasting microphones: those intended for high-fidelity reproduction of speech and music. For this class of applications, it is necessary that the microphones have a specified directivity and that they be highly reliable.
- General-purpose microphones: those intended for sound reinforcement systems, public address systems, and home use. The demands on these microphones are quite variable.
- Measurement microphones: those intended for laboratory measurement of acoustic pressure. These microphones must be very accurate and highly stable. Depending on the specific application, a broad dynamic range or frequency range may be demanded.

Note: References to chapters appearing only in the *Encyclopedia* are preceded by *Enc.*

2 SPECIFIC TYPES OF MICROPHONES

There are a number of means to distinguish between the various popular types of microphones. In this section we draw the distinction on the basis of transduction mechanism, paralleling the development in Chapter 110. The condenser and electret microphones are two examples of acoustic sensors in which the transduction mechanism is based upon changes in the energy stored in an electric field (capacitive device). Piezoelectric microphones are capacitive devices that linearly link the mechanical and electrical fields. The moving-coil electrodynamic and ribbon microphones are two examples of sensors in which the transduction is based on interactions of magnetic fields (inductive device). The carbon button microphone is a resistive element that operates through changes in the resistance as a function of pressure. Each of these is described below.

2.1 Condenser Microphones

Conversion of acoustic energy to electrical energy in a condenser microphone occurs because the sound pressure causes small motions of the microphone diaphragm. The diaphragm serves as one plate of a parallel-plate capacitor in which the other plate is stationary, as shown in Fig. 1. Motion of the diaphragm thus serves to change the capacitance, which results in a change in the electrical potential. The application of a static charge allows the pressure-induced change in capacitance to produce a potential proportional to pressure.

The transduction mechanism described above is inherently quadratic in nature (as discussed in Chapter 110). To produce a nominally linear device, the motion of the membrane must be kept small relative to the plate undisturbed separation, and the DC bias voltage between the plates must be kept high relative to the voltage signal level. With these assumptions, the second-harmonic distortion term is small and the sensitivity σ of the microphone is given by

$$\sigma = \frac{e}{p} = \frac{e_0}{\gamma p_0}, \quad (1)$$

where e_0 is the bias voltage, γ is the ratio of specific heats for the medium between the plates ($\gamma = 1.4$ for air), p_0 is the undisturbed pressure of the fluid medium between the plates, and p is the acoustic pressure amplitude. Note that Eq. (1) assumes that adiabatic compression of the fluid trapped between the plates dominates the device stiffness. If the diaphragm stiffness dominates, then the sensitivity is instead given by

$$\sigma = \frac{e_0 a^2}{8T d_0}, \quad (2)$$

where a is the diaphragm radius, d_0 is the undisturbed plate separation, and T is the diaphragm tension.

Because condenser microphones have a capacitive internal impedance C_0, static or very low frequency pressures are impractical to measure. This is because potential sensing instruments have a finite input resistance. In general, the lower frequency limit is given by

$$f = \frac{1}{2\pi R_{in} C_0}, \quad (3)$$

where R_{in} is the instrument input resistance. This low-frequency limit is averted by replacing the DC bias voltage with a high-frequency AC bias voltage. When C_0 is made part of a parallel LC circuit that determines the frequency of a high-frequency oscillator, the change in C with acoustic pressure causes a modulation frequency change that can be monitored using a frequency modulation detector. Condenser microphones have been built to monitor sound at frequencies of fractions of 1 Hz using this method. However, it should be noted that in order to obtain a flat response to such low frequencies it is generally necessary to eliminate the rear pressure equalization port from the transducer because it is acoustically transparent at very low frequencies.

The upper frequency limit of a condenser microphone is the lowest membrane resonant frequency and is determined by the tension and mass of the membrane. The response at this resonance is controlled by acoustic resistance behind the membrane. Precision condenser microphones with flat response up to about 150 kHz are readily available. A flat frequency and a low moving mass couple to give a microphone a very good transient response.

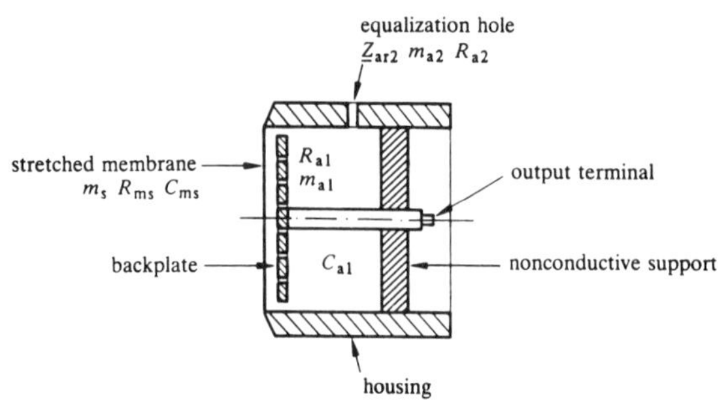

Fig. 1 Condenser microphone using a stretched membrane and a perforated backplate to form the parallel capacitive plates. (From Ref. 1.)

If a condenser microphone is used to monitor sound

at frequencies for which the wavelength is large compared to the diaphragm dimensions, then the microphone can be made nominally omnidirectional; that is, it will respond uniformly to sound incident from any direction. At high frequencies diffraction affects the directional performance. When the diameter of the membrane is approximately equal to the acoustic wavelength, sound arriving at normal incidence experiences a pressure doubling at the membrane, causing a 6-dB error. At grazing incidence pressure averaging across the face can reduce sensitivity. By controlling the acoustic resistance, the normal incidence response can be corrected; however, directivity still remains. When used in a diffuse sound field with waves arriving from random directions, a different compensation is used to account for directivity. Microphones can be compensated to yield a flat frequency response to free field, random incidence, and uniform pressure.

Condenser microphones are generally high-impedance devices requiring amplification to be located near the sensor itself. It is not infrequent for the amplifier to limit the dynamic range of such microphones, but ranges of up to 140 dB are readily available.

A condenser microphone made with a stainless steel membrane and other metal parts is the most stable with temperature and other environmental changes. It holds its calibration for long periods of time and may be used as a secondary standard.

For further information on condenser microphones see Merhaut,[2] Rossi,[1] Beranek,[3] and Wong and Embleton.[4]

2.2 Electret Microphones

Electret microphones are very similar to the condenser microphone described above. The main difference is that the externally applied bias voltage is eliminated through use of a polarized material referred to as an electret. Typical electret microphones use a thin polymer film coated on one side with metal as the moving diaphragm. The film is permanently polarized at a level comparable to that used in biasing a conventional condenser microphone. Such a microphone is shown in Fig. 2. An alternative that is increasing in popularity is to place the electret film onto the stationary plate and use a thin metal foil as the moving diaphragm. This configuration is referred to as a backplate electret and offers the advantage of permitting separate optimization of the diaphragm performance and electrical performance.

Because electret microphones are biased at levels similar to those used in condenser microphones, their sensitivities are roughly the same. Similarly, the linearity, frequency response, and transient response of an electret microphone are close to those of a condenser microphone of similar size. However, due to the use of different amplification strategies, the electret transducer tends to have a broader dynamic range and lower self-noise.

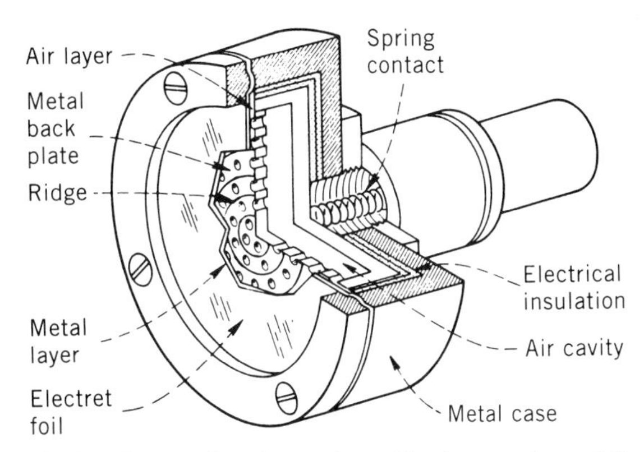

Fig. 2 Electret microphone using a thin electret polymer foil suspended over a perforated backplate. (From Ref. 5.)

The advantage of electret microphones, when compared to conventional condenser microphones, is that they operate without the need for an external power supply, and they are available at very low cost. Electret microphones are now the standard microphone type in consumer products such as telephones and tape recorders. Over 600 million electret microphone units are sold annually worldwide. They are available in a wide variety of sizes and configurations, some of which exhibit strong directivity. For more information on electret microphones, see Ref. 6.

2.3 Piezoelectric Microphones

Piezoelectric microphones rely on piezoelectric materials for the fundamentally linear conversion of mechanical energy to electrical energy. As described in *Enc.* Ch. 52 and Chapter 110, conventional piezoelectric materials are crystals and ceramics that, when distorted by sound pressure, produce voltages proportional to their dimensional strain. Such materials are used in microphones in two fundamentally different ways: A diaphragm responds to sound pressure by moving and the motion is transmitted to a piezoelectric material embedded in the device, or a piezoelectric material itself is used as the diaphragm of a microphone. The advantage of a piezoelectric microphone is that, like the electret microphone, it does not require the attachment of an external power supply.

Figure 3*a* shows a realization of a piezoelectric microphone in which a diaphragm is connected to a pin that drives a piezoelectric bimorph internal to the microphone. A bimorph is two pieces of piezoelectric material connected in opposition so that the voltage output from the desired strain is twice that which would occur from a

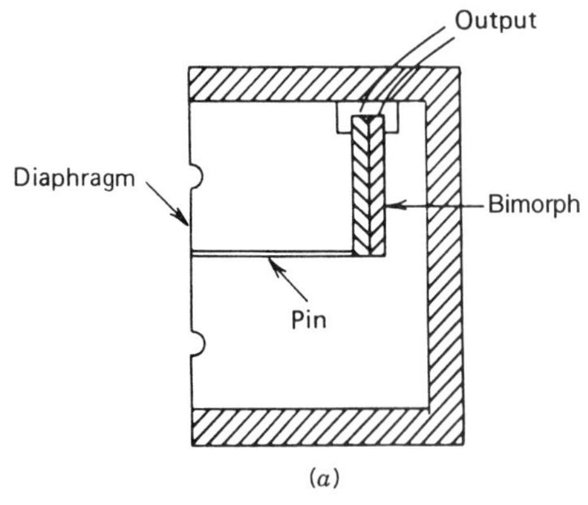

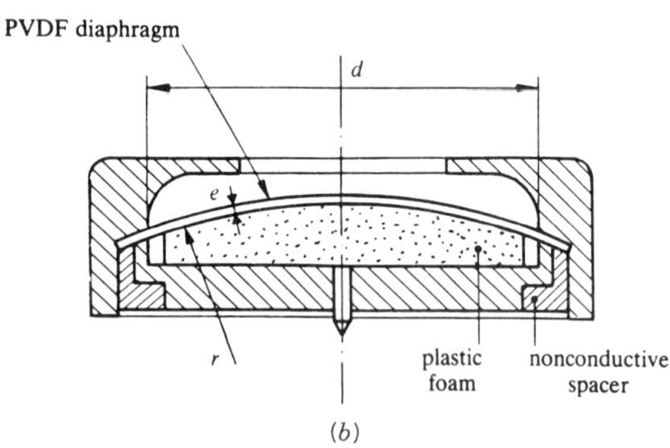

Fig. 3 Example piezoelectric microphones. (*a*) A piezoelectric microphone in which a solid pin converts motion of a diaphragm to motion of a piezoelectric bimorph. (From Ref. 7.) (*b*) Piezoelectric polymer microphone using a curved piezoelectric polymer diaphragm to directly convert sound pressure to a potential. (From Ref. 1.)

single element while the output from orthogonal strains is eliminated. While this microphone has the advantage of good linearity, it tends to be lower in sensitivity than condenser and dynamic microphones.

Figure 3*b* shows a second example of a piezoelectric microphone. In this sensor a piezoelectric polymer film in a dome shape is used as a diaphragm that responds to sound pressure. As the diaphragm deforms, the electrical potential across the piezoelectric material varies. In both cases shown in Fig. 3, the low-frequency limit in performance is set by the RC constant of the device, as in Eq. (3), and the high-frequency limit by the resonance of the diaphragm (or bimorph).

With proper construction, a piezoelectric microphone can exhibit reasonable sensitivity in the range from 20 Hz to 10 kHz. The electrical impedance can be controlled so that it is possible to use fairly long cables to attach the microphone to its amplifier. With the decrease in cost of amplification in the last two decades, piezoelectric microphones have fallen out of favor, since higher fidelity, higher sensitivity counterparts are available without significant cost increase.

For more information on piezoelectric microphones consult Wilson[8] and Kino.[9]

2.4 Moving-Coil Electrodynamic Microphones

Moving-coil electrodynamic microphones, often referred to as dynamic microphones, are the sensor counterpart of conventional electrodynamic loudspeakers. They are based on a fundamentally linear transduction mechanism that involves interaction of magnetic fields. A typical example is shown in Fig. 4. In this dynamic microphone a sound pressure is incident upon a diaphragm to which a coil is rigidly attached. Motion of the diaphragm results in a motion of the coil within the magnetic field generated by a permanent magnet. This gives rise to a potential. The geometry shown in Fig. 4 aligns the magnetic field **B,** the coil current, and the diaphragm velocity in three mutually orthogonal directions. This results in the maximum sensitivity

$$\sigma = \frac{e}{p} = \frac{Bl}{Z_s}, \tag{4}$$

where Z_s, the specific acoustic impedance, is the ratio of the pressure at the microphone face to the linear velocity of the diaphragm and l is the length of the magnetic path. This impedance is a function of the particulars of the microphone construction and generally includes a resistance term associated with flow through capillaries in the silk cloth and a stiffness term associated with the diaphragm suspension. For common microphones, the sen-

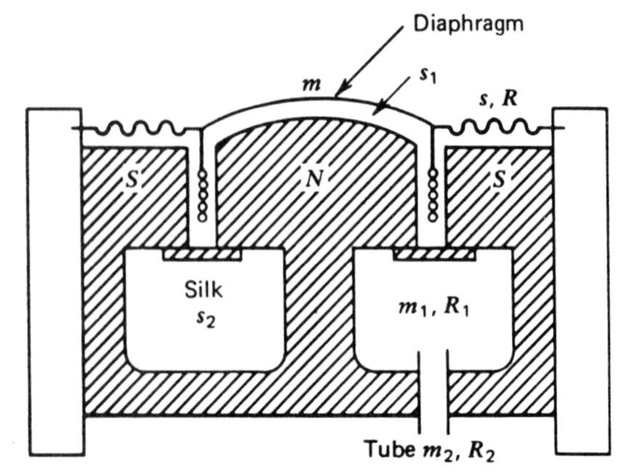

Fig. 4 Moving-coil microphone in which motion of the diaphragm induces a current in a coil. (From Ref. 7.)

sitivity of dynamic microphones is typically about 10 dB less than that for condenser microphones.

The ability of a dynamic microphone to respond to low frequencies is generally limited by resonance of the port supplied for pressure equalization of the outside air and the air in the chamber. This resonance typically lies between 40 and 100 Hz. The upper frequency limit of the microphone is typically determined by the frequency at which the diaphragm has its fundamental resonance. While passive and active damping can be added to the system to minimize the resonant boost at this fundamental frequency, the microphone sensitivity ceases to be flat and displays a decreasing amplitude with increasing frequency above the resonance.

As in the condenser microphone and electret microphone cases, the dynamic microphone is nominally omnidirectional up to a frequency at which the wavelength of sound is comparable to the size of the diaphragm. Since dynamic microphones tend to be larger than the condenser and electret microphones, the frequency range of dynamic microphones is somewhat more restricted at the high-frequency end. Further, owing to their low damping they tend to exhibit greater ringing when exposed to impulsive sound. However, these disadvantages are offset, in some applications, by the fact that the self-noise in dynamic microphones is lower since it stems from noise created by motion of electrons in the coil rather than amplifier noise as in the condenser microphone case. An electret microphone with a field-effect transistor (FET) preamplifier has about the same level of self-noise as a dynamic microphone.

Dynamic microphones are fairly low impedance sensors, which makes them particularly suitable for applications that require long cable lengths. They display a good tolerance to temperature and humidity variations and are very rugged.

For further information on dynamic microphones see Beranek[3] and Kinsler et al.[7]

2.5 Ribbon Microphones

Electrodynamic ribbon microphones are quite similar to moving-coil electrodynamic microphones, but the diaphragm and coil are replaced by a corrugated ribbon suspended in the gap of a magnetic circuit. A typical realization is shown in Fig. 5. In this microphone the ribbon is exposed to sound on both its front and back. It responds to a difference in the sound pressure by deforming primarily in bending. This motion in the presence of a magnetic field induces a potential. Because the ribbon microphone as typically constructed responds to a spatial pressure difference, it is classified as a gradient microphone.

Since the principle of operation of a ribbon microphone is fundamentally the same as that for a dynamic microphone, their sensitivities tend to be roughly the same. However, the ribbon microphone has a lower moving mass than the dynamic microphone. This results in

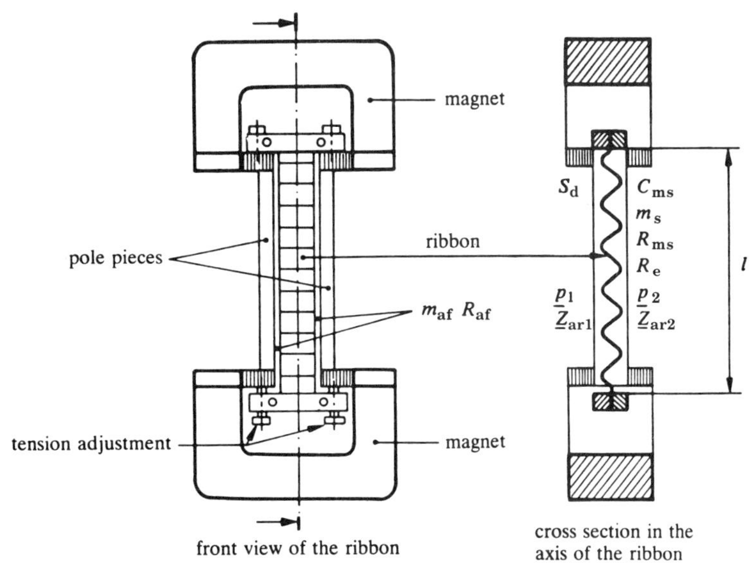

Fig. 5 Ribbon microphone in which a low-mass conductive ribbon moves in response to incident sound. (From Ref. 1.)

a better transient response but greater susceptibility to wind noise problems in the ribbon microphone.

The lower frequency limit of the ribbon microphone is determined by the fundamental resonance of the ribbon. The ribbon is intentionally made very compliant, so that the fundamental bending mode is quite low. For a typical ribbon of aluminum with a length of 20–50 mm, a width of 2–3 mm, and a thickness of several micrometres, the fundamental resonance is between 10 and 30 Hz. The high-frequency limit is determined by diffraction effects in the microphone. Since the ribbon tends to be smaller than the moving dome in a dynamic microphone, the ribbon microphone tends to have a higher high-frequency limit.

Because the ribbon microphone is a pressure gradient sensor, its response is not omnidirectional. In particular, if sound is incident from ±90°, then the path of the sound to the front of the ribbon is identical to that of the back of the ribbon. Under these circumstances there is no pressure gradient and the ribbon will not move. The performance of the microphone is clearly a function of angle, and its sensitivity is described by

$$\sigma = \frac{BlAh}{c_0 m}\cos(\theta) = \frac{Bl}{Z_s}\cos(\theta), \tag{5}$$

where h is the ribbon thickness, c_0 is the speed of sound propagation in air, m is the ribbon mass, l is the length of the ribbon, B is the magnetic flux density, A is the effective area of the ribbon, Z_s is the specific acoustic impedance, and θ is measured from normal to the ribbon face. Figure 6 shows the sensitivity normalized amplitude as a function of angle, that is, the directivity plot. The classic figure-8 structure is known as a bidirectional response.

The pressure gradient or velocity microphone has three primary advantages: First, because of its directivity, it tends to discriminate against reverberant sound and background noise, since these sound contaminants are generally diffuse. This permits the desired sound source to be located farther from the microphone than would be possible in microphones that are omnidirectional. Second, the response is identical for forward-going and backward-going sound, so it is ideal for use in picking up the speech of two individuals facing each other. Third, because of the sharp nulls in the response at ±90°, it is possible to eliminate or greatly reduce the significance of a localized noise source by aiming the null at it.

A peculiarity of the ribbon microphone relates to its performance for sound sources that are located in the near field of the sensor. For such near sources the acoustic wavefront is spherical rather than planar, and for

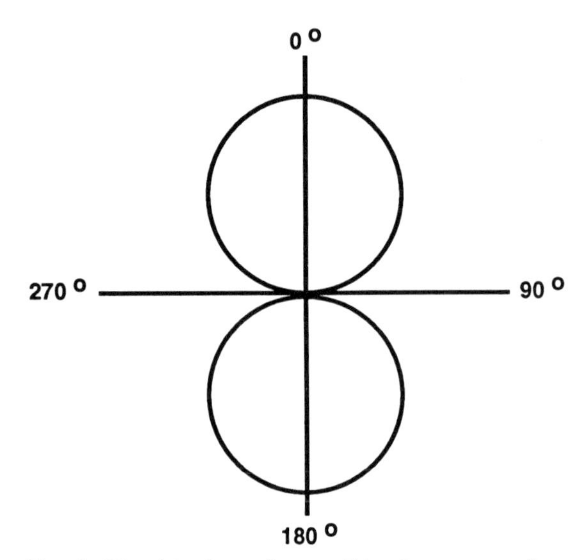

Fig. 6 Directivity factor for a traditional pressure gradient microphone.

spherical waves the ratio of the pressure to the pressure gradient is not $\rho_0 c_0$, as it is for plane waves, but instead is

$$Z = \rho_0 c_0 \frac{kr}{1 + (kr)^2}, \tag{6}$$

where k is the acoustic wavenumber ($k = \omega/c_0, \omega$ being the angular frequency), ρ_0 is the mass density of air, c_0 is the speed of sound in air, and r is the distance from the source. As a result, there is an enhancement of the low-frequency response for close sources.

One recent advance in ribbon microphones is the introduction of a printed ribbon made of a polymer film upon which aluminum is deposited. The resulting ribbon can be made extremely thin (0.025 mm), permitting extension of the frequency response to high ranges (15,000–20,000 Hz), although the corresponding lower limit tends to increase some.

In general, the ribbon microphone is not as rugged as the dynamic microphone. Large-amplitude bursts of pressure can cause the ribbon to rupture and permanently damage the sensor.

For more information on ribbon microphones consult Huber.[10]

2.6 Carbon Microphones

The carbon microphone is a resistive sensor that has been used over the years extensively in situations where it

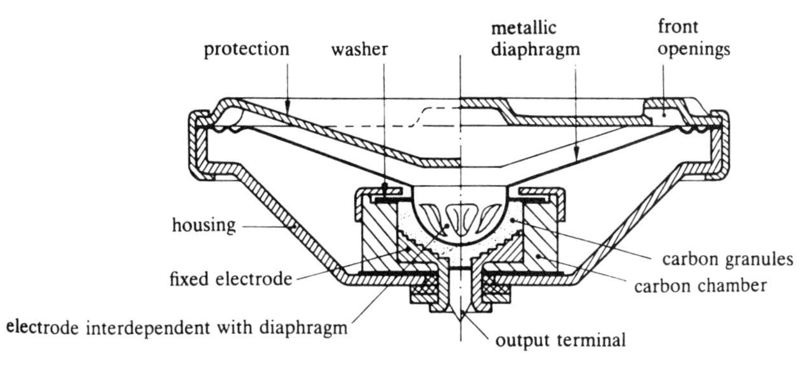

Fig. 7 Carbon microphone in which the electrical resistance changes with motion of a membrane. (From Ref. 1.)

is acceptable to trade a high sensitivity for a poor linearity. In the carbon microphone sound is incident on a diaphragm attached to a plunger that applies a force to carbon granules. An example is shown in Fig. 7. The resistance of the volume of carbon granules, R_c is a function of the force applied and is generally approximated as a linear function:

$$R_c = R_0 + \alpha x, \tag{7}$$

where R_0 is the undisturbed resistance, α is a function of the material properties, and x is the displacement of the plunger. In the presence of a DC voltage bias e_0, the sensitivity of the carbon microphone is given as

$$\sigma = \frac{e_0 \alpha A}{R_0 k_m}, \tag{8}$$

where k_m is the mechanical stiffness of the diaphragm and A is its area.

The lower frequency limit of the carbon microphone is determined by the time constant of the system: the ratio of the moving mass to the resistance. The upper limit of the microphone is set by the diaphragm resonance. Combined, these result in a flat frequency response range from about 70 to 2000 Hz in conventional carbon microphones. Improvements at the high-frequency end can be made by increasing the stiffness or decreasing the size of the diaphragm, but at a cost of microphone sensitivity.

Carbon microphones exhibit poor linearity and poor dynamic range but are very rugged. They have been used exclusively for situations in which speech communication is necessary but high fidelity is not required. They are a poor choice of microphone for high-humidity environments, as the carbon granules coalesce. In recent years, carbon microphones have been replaced as the standard telephone microphones by electret sensors.

For more information on carbon microphones, see Kinsler et al.[7]

3 MICROPHONE DIRECTIVITY

The microphones described above may be generally classed into two types: those that respond to pressure and are nominally omnidirectional and those that respond to a pressure gradient and are directive. While the ribbon microphone is the only microphone described above in the pressure gradient category, any of the pressure microphones discussed above could be made to respond as gradient microphones simply by providing a second acoustic port behind the diaphragm. Such microphones would then exhibit directional characteristics similar to the conventional ribbon microphone and would be useful for elimination of far-field noise in favor of near sounds. Further, the ribbon microphone could be made to operate as a pressure microphone rather than a pressure gradient microphone by simply enclosing one side of the sensor.

The most commonly used measures of angular variation of acoustic transducers are the beamwidth, directivity factor, and directivity index. The beamwidth is typically experimentally determined and is the angular range embracing the peak over which the response is within a specified amount (usually a factor of 0.5) of the peak value. Here the term "response" refers to a quantity, such as sound intensity or mean-squared pressure, which is proportional to energy or power. The directivity factor (DF) is the ratio of the peak response at radius r to the response averaged over a sphere of radius r. The directivity index is a logarithmic measure of the directivity factor: DI = 10 $\log_{10}$ (DF).

There are two primary reasons for seeking microphones that exhibit directive response: They permit source localization in space and they discriminate against unwanted noise sources that tend to be diffuse, such as reverberation and background noise. There are two general causes of directive response: the geometry of individual microphones and the result of summing multiple microphones in arrays. We consider each of these below.

First consider the behavior of a single acoustic transducer. If the transducer is large compared to the wavelength of the sound, it is not well approximated by a point receiver and it will generally exhibit directive response. The precise form of the directivity depends upon the exact geometry of the transducer. For example, many directional transducers expose the energy-converting material on both the front and back faces. The result is an instrument that responds at least in part to pressure differences. Such microphones exhibit a far-field directivity that is generally represented using polar plots of pressure as a function of angle. For conventional pressure elements, there is no rear port, so they are omnidirectional. For first-order gradient elements, there is negligible resistance in the rear port and, as discussed in the case of the ribbon microphone, they display a bidirectional response. (See Fig. 6). Cardioid elements of various types combine pressure and first-order gradient behavior by designing the rear sound port to have a specified resistance, resulting in a broad beam pattern that rejects sound approaching from the rear of the microphone. The precise nature of the pattern is determined by the port resistance. Figure 8 shows a few such cardioid microphone directivity patterns that result and the names given to the microphones generating them. Note that these patterns have broad main lobes with no side lobes and are not frequency dependent. Higher order gradient microphones can be constructed by combining the gradient elements above in a manner such that the differ-

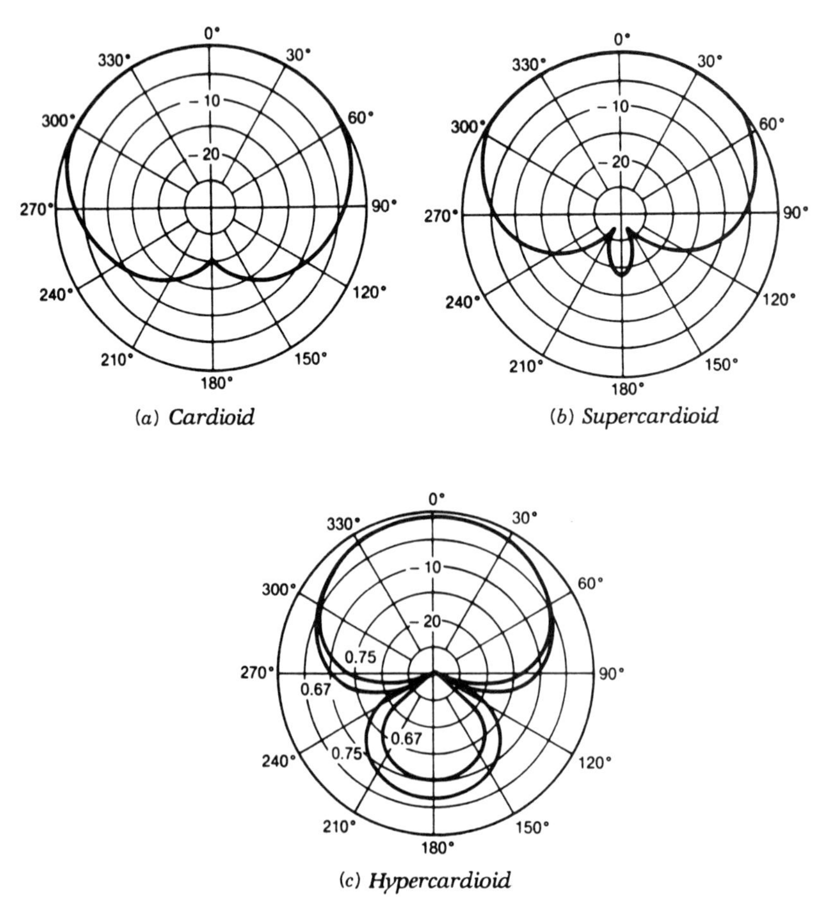

Fig. 8 Various cardioid-type directivity functions: (*a*) standard cardioid; (*b*) supercardioid; (*c*) a hypercardioid. (From Ref. 10.)

ence of the pressure gradients is obtained. The directivity pattern is then given by the product of that for each element and that of the microphone array structure.

Another way to achieve a directive microphone is through the judicious choice of structural geometries. For example, a small microphone surrounded by a parabolic reflector preferentially responds to sound approaching from the front if the transducer is located at the focus, because sound that is not traveling normal to the dish is scattered away from the focus. Such directional microphones are used extensively in sports broadcasting, surveillance, and nature recording.

Now consider the effect of summing the output signals from transducers in arrays. For a sensor array with length large compared to a wavelength, there will be a wide variation in the phases and possibly the amplitudes of sound observed by each element due to path differences to the source. For such a sensor array one can write the response O (either a voltage or a current) as

$$O = \sum \frac{a_i p}{r_i} \exp(jkr_i), \qquad (9)$$

where a_i is the complex amplification gain associated with element i, p is the pressure amplitude of the source at a unit distance, and r_i is the distance from the source to element i. Note that since k is proportional to frequency, the response is frequency dependent. If the sound source is in the far field of the transducer array, then the phase changes dominate over the amplitude variations, and the response can be approximated by

$$O = \frac{p}{r} \sum a_i \exp(jkr_i). \qquad (10)$$

For example, consider a line array of n omnidirectional sensors, as shown in Fig. 9*a*. Assuming the source to be in the far field, the response goes as

$$O = \frac{p}{r} \sum a_i \exp[jk(r \pm x_i \sin\theta)], \qquad (11)$$

where x_i is the distance to element i from the array center and θ is the angle between the normal to the array center and the source. If the amplifier gains and element spacings are uniform, Fig. 9*b* shows the resulting directivity pattern at a number of frequencies. Note that there are a main lobe whose beamwidth decreases with increasing frequency and a number of side lobes. If the amplifier gains are not uniform, the directivity pattern is modified. For example, by introducing phase delays, one can steer the main beam away from the broadside ($\theta = 0°$) angle. Further, using various amplitude shadings, one may narrow or widen the beamwidth and raise or lower the side lobes. The commonly used array shadings are discussed in virtually all texts on antennas and filters.

Using these same principles, one may build two- and three-dimensional arrays of various sorts. Typically, the directive response for arrays is frequency dependent and includes side lobes as well as a main lobe.

It should be noted that it is possible to build acoustical transducer arrays in which the elements are themselves directional. In this case, the product theorem guarantees that the directivity factor of the array is the product of the directivity of the elements and that of the array geometry. (See Ref. 7, for example.)

4 MINIATURE MICROPHONES

Miniature microphones are needed to meet three different types of acoustical requirements:

- Microphones that must fit into small volumes due to physical constraints: This includes hearing aid microphones and microphones mounted in the engine walls of automobiles.
- Microphones for high-frequency measurement of sound pressure: The ideal microphone monitors an acoustic field without affecting it significantly. The extent to which a microphone approaches this ideal is largely dependent upon its size, with smaller sizes being required as the frequency range of interest increases.
- Low-cost, mass-produced microphones: While electret microphones are available at low cost, most microphones are expensive sensors, and this impedes the broader use of microphones.

There are two approaches to the miniaturization of microphones: Decrease the size of conventional microphones and use alternative fabrication techniques, possibly with associated alternative transduction methods. Both approaches have been the subject of great scrutiny. Electret microphones are the microphones that have been most successfully miniaturized using conventional fabrication techniques. As a result, they are the standard components of hearing aids. Techniques normally used in microelectronic manufacturing have been the most successful miniaturization approaches and rely on unconventional construction techniques to make microphones. The solid-state microphones produced using microelectronic fabrication techniques often rely on methods of transduction not used in macroscopic microphones. These are described in Chapter 110.

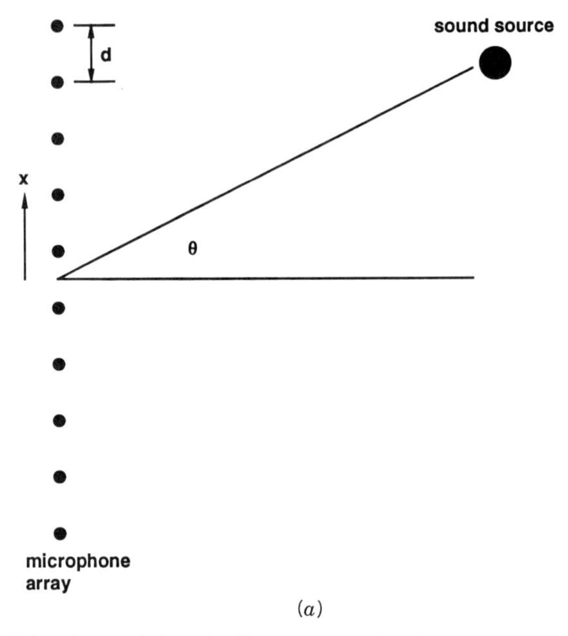

(*a*)

Fig. 9 Directivity characteristics of a line array of microphones: (*a*) schematic and (*b*) directivity pattern as a function of the sound wavelength relative to the element separation. [(*b*) From Ref. 3.]

5 MICROPHONE CALIBRATION

There are generally three calibration methods used in microphones: comparison calibration, absolute direct calibration, and absolute reciprocal calibration. In comparison methods the microphone to be calibrated is compared to a microphone of known characteristics. This method is fast and eliminates the need for specialized equipment but is the least precise of the techniques because errors in calibration propagate.

In absolute direct calibration a specialized piece of equipment is used to provide a known calibration signal, which must be nearly independent of the load seen by the calibrator. A common acoustical example of this type of technique is the use of pistonphones for calibration of microphones. Pistonphones use reciprocating pistons to change the volume of a calibrated cavity in order to generate a well-known sound level. The advantage of absolute direct calibration methods is that they generally are fast and very precise. However, since they require the purchase of dedicated instrumentation, they tend to be expensive. Further, in order to obtain the desired precision, they generally offer absolute calibration at a very limited number of frequencies. For example, most pistonphones operate at a single frequency only. Complete calibration of a microphone then requires the use of the comparison method as well as absolute direct calibration.

The third calibration procedure commonly used for microphones is absolute reciprocal calibration. It is this method that separates the calibration of acoustical instrumentation from calibration methods common to other fields. Reciprocal calibration of a microphone requires three transducers: one to be calibrated, one that is reciprocal (meaning it can operate either as a sensor or as an actuator), and one additional sound source. It is not necessary that any of the transducers have known calibration characteristics.

Although there are now many reciprocal calibration methods, all require a minimum of three measurements. The following is a method for reciprocal calibration of a microphone or hydrophone. First, reciprocal transducer B is placed a distance l from source C. A current i_C is run through C and the open circuit potential of B, e_B, is monitored. Next, transducer B is replaced by the microphone whose calibration we desire, microphone A, and the measurement is repeated, yielding e_A. Finally, sound source C is replaced by B, a current i_B is run through it, and e'_A, the new open circuit voltage from A, is found. The sensitivity of A, defined as the ratio of the open cir-

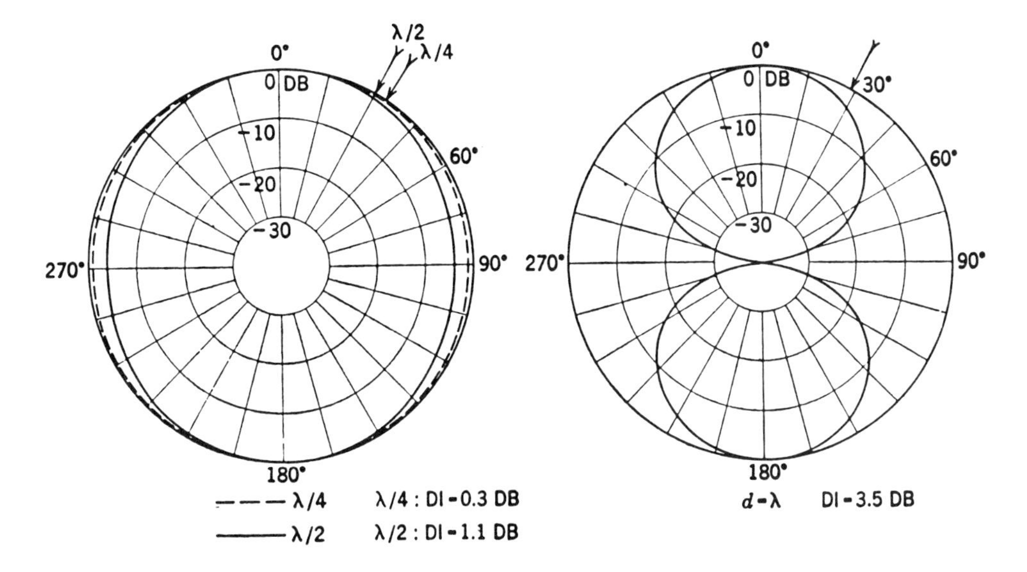

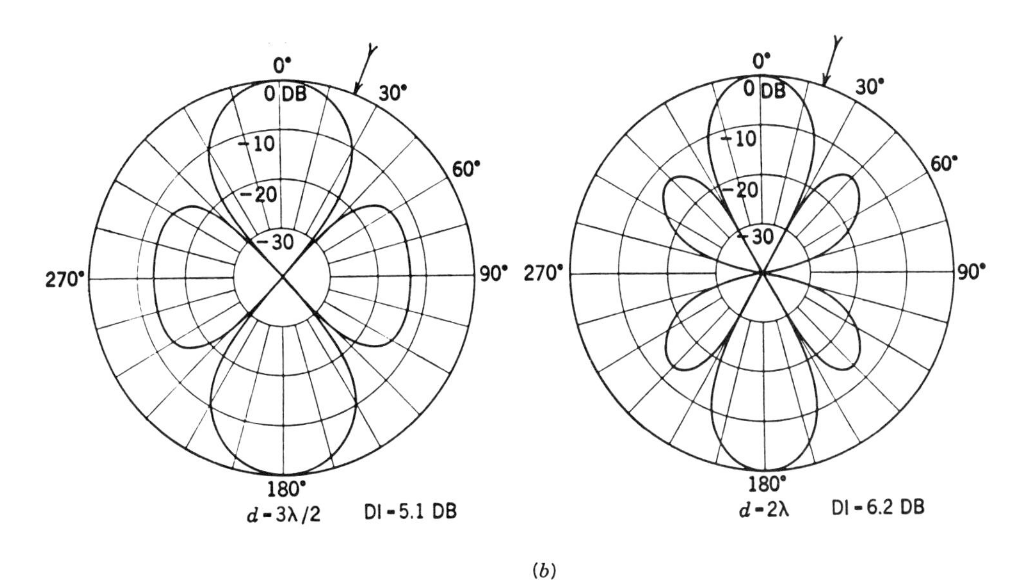

(*b*)

Fig. 9 (*Continued*)

cuit voltage to the sound pressure, is found from

$$\sigma^2 = \frac{e_A e'_A}{e_B i_B} \frac{2l\lambda}{\rho_0 c_0}, \tag{12}$$

where λ is the wavelength of the sound.

The advantages of reciprocal calibration compared to direct absolute and comparison calibration are that it is very accurate, can be applied at many frequencies, and requires only electrical and length calibration equipment. The major disadvantage is that it is more tedious than the other calibration methods.

For additional, more detailed information on microphone calibration, see Chapter 108.

REFERENCES

1. M. Rossi, *Acoustics and Electroacoustics*, P. R. W. Rouse (Trans.), Artech House, Norwood, CT, 1988.
2. J. Merhaut, *Theory of Electroacoustics*, McGraw-Hill, New York, 1981.
3. L. L. Beranek, *Acoustics*, AIP, New York, 1986.
4. G. S. K. Wong and T. F. W. Embleton (Eds.), *AIP Handbook of Condenser Microphones*, AIP, New York, 1995.
5. G. M. Sessler and J. E. West, *J. Acoust. Soc. Am.*, 1973, pp. 1589–1600.
6. G. M. Sessler (Ed.), *Electrets, Topics in Applied Physics*, Vol. 33, Springer-Verlag, Berlin, 1987.
7. L. E. Kinsler, A. R. Frey, A. B. Coppens, and J. V. Sanders, *Fundamentals of Acoustics*, 3rd ed., Wiley, New York, 1982.
8. O. B. Wilson, *An Introduction to the Theory and Design of Sonar Transducers*, Peninsula, Los Altos, CA, 1988.
9. G. S. Kino, *Acoustic Waves*, Prentice-Hall, Englewood Cliffs, 1987.
10. D. M. Huber, *Microphone Manual*, Sams, Indianapolis, 1988.

113

MAGNETIC RECORDING REPRODUCING SYSTEMS

MARVIN CAMRAS

1 INTRODUCTION

A sound recording is information on how the amplitude of a sound wave varies with the passage of time. Such a record may be played back at a later time by causing the information to control a source of sound waves. Of the many conceivable ways of recording and reproducing sound, the common ones are wavy grooves on a disc surface that guide a mechanical stylus (phonograph), photographic tracks that vary in transparency or in reflection of light (sound-on-film and laser disc), and tracks on magnetic materials that retain varying magnetization along the track (tape recording).

Magnetic recording has advantages of erasibility, high density, low cost, immediate playback, and freedom from sophisticated processing. For these reasons magnetic recording has become the most popular of all methods where recording as well as playback are required. It is also a large factor in playback-only applications using compact cassettes.

Figure 1 shows a basic record–play system using magnetic recording. A tape having a magnetizable surface stored on reel A is threaded over a magnetic head B that supplies a magnetic field at recording gap C, the magnetic field strength varying according to the sound to be recorded. The magnetized tape D is pulled past the head by a capstan roller E and pinch roller F and spooled on a takeup reel G. The tape is afterward rewound to reel A while the head is deenergized.

For playback, the above process is repeated except that the winding on head B is connected through an amplifier to a loudspeaker. The magnetized elements on the tape induce a voltage in the head winding, which after equalization is a replica of the gap field variations during recording. The recording may be erased by an additional head H that produces a strong, steady AC field. Usually the erase head is energized while recording, thus removing any previous recorded components. The unique elements in a magnetic recording system are the head and the record medium. The other components are the mechanical drive and the electrical–electronic amplifiers, equalizers, controls, indicators, and power supplies.

Note: References to chapters appearing only in the *Encyclopedia* are preceded by *Enc.*

2 MAGNETIC HEADS

Figure 2 shows the interior construction of a magnetic head, which is often called a ring head because its magnetic circuit is somewhat ring shaped. Other constructions have also been used, especially chisel-shaped (stylus) pole pieces with sharp edges that contact the tape.

The objective is to concentrate the magnetic field in the smallest region of tape that is practical, giving a high recording density of a maximum number of wavelengths per millimetre. To achieve high densities, the head gaps are very small, on the order of a 1 or 2 μm (0.00004–0.00008 in.). The small gaps give low output voltage and require very smooth uniform tape surface and good contact between head and tape.

The head cores are traditionally made of Permalloy (80% Ni, 20% Fe) laminations for high permeability and low magnetic retentivity. In recent years ceramiclike materials such as MnZn–ferrite have been advocated. These are harder than Permalloy, wear longer, and do not have to be laminated. Unfortunately, their brittleness allows crumbling at the gap edges, spoiling the recording density. This fault is remedied by covering the gap faces with sendust, a durable alloy (85% Fe, 9.6% Si, 5.4% Al) that allows the finest gaps to be retained but increases the head cost.

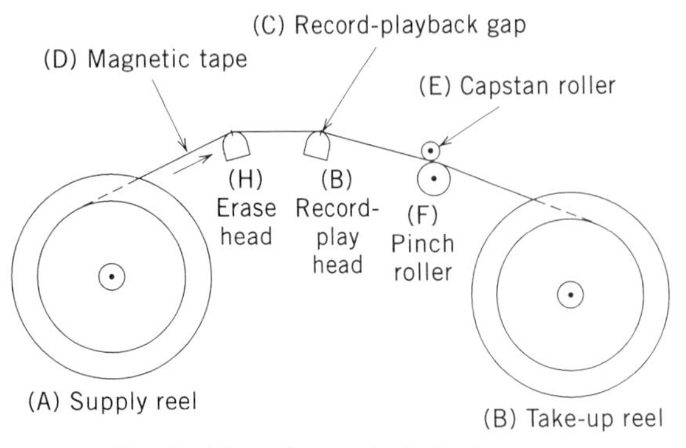

Fig. 1 Magnetic record–playback system.

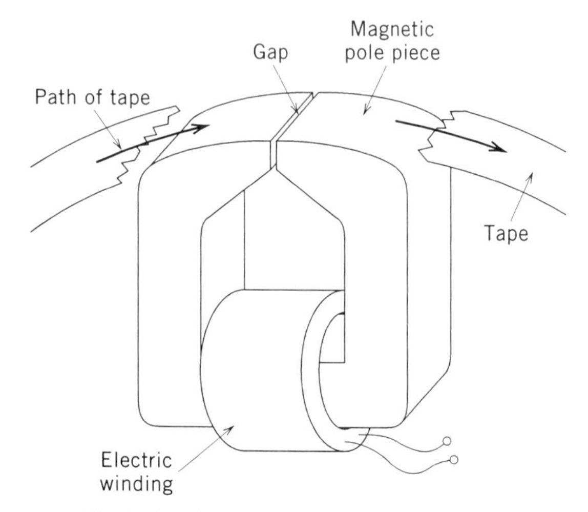

Fig. 2 Interior construction of a magnetic head.

3 MAGNETIC TAPE

Most tapes nowadays consist of a very thin layer of magnetizable material deposited on a smooth, durable, nonmagnetic backing such as polyester (Mylar) for tapes and floppy discs or on aluminum for hard discs. The magnetic layer thickness is about 12 μm or less, down to about 1 μm for high-density tapes.

The most common magnetic layer is a dried suspension of acicular gamma ferrite particles about 0.25 μm long in a lacquer binder based on vinyl or polyutherane polymers. Just after the tape is coated, while the suspension is still fluid, it is subjected to a magnetic field that aligns the magnetic particles. This aligning or orienting step increases recording intensity and decreases noise. Coating is done under the most stringent clean-room conditions, since contaminants result in noise and dropouts. Alternative particles for audio magnetic tape layers are chromium dioxide and cobalt-coated gamma ferrite, both of which have coercivities of about 600. These require "high-bias" settings, provided on many recorders, and give better high-frequency response than ordinary tape.

Continuous layers of evaporated metal containing Fe, Ni, or Co give even higher densities but are mostly used for video recording tapes, as in camcorders.

4 CIRCUITS

Conventional audio amplifiers with a voltage gain of about 80–100 dB are suitable for magnetic tape recorders. For analog recording, equalization is required to compensate for the inductance of electromagnetic heads. In the recording mode the head current is kept nearly constant over the audio frequency range by a resistor in series with the head winding, the resistance being high enough to swamp out the head impedance variations.

Also applied to the recording head is a high-frequency bias, which is a steady AC current whose frequency (say 200 kHz) is about 10 times as high as the highest audio frequency to be recorded and whose amplitude is about 10 times the maximum audio amplitude. The bias frequency is too high to be recorded but acts as a catalyst to give distortion-free low-noise magnetization on the magnetic tape layer, which otherwise would be highly nonlinear. The playback amplifier uses a low-noise preamplifier in the first stage and includes an equalizer *RC* or similar integrating circuit to compensate for the $d\Phi/dt$ rise in playback voltage with rise in recorded frequency for a constant magnetization of the tape. Otherwise, the amplifier is quite ordinary.

5 DESIGN FEATURES AND CONSIDERATIONS

The basic equalization to give flat frequency response is shown in Fig. 3 and has been discussed in connection with the amplifier circuits. Depending on the quality of the recording system (resolution and losses), some additional boost of high frequencies may be added during recording to compensate for head losses and for the fact that the program material may not be rich in high frequencies.

The dynamic range from background noise to maximum signal output may be as low as 40 dB in home recorders or as high as 70 dB in master recorders. It is usually limited by the relative background noise level, which may be reduced by choosing higher tape speeds, wider recorded tracks, and high-energy tapes of homogeneous composition and smooth surface.

Overall background noise can be minimized by equal-

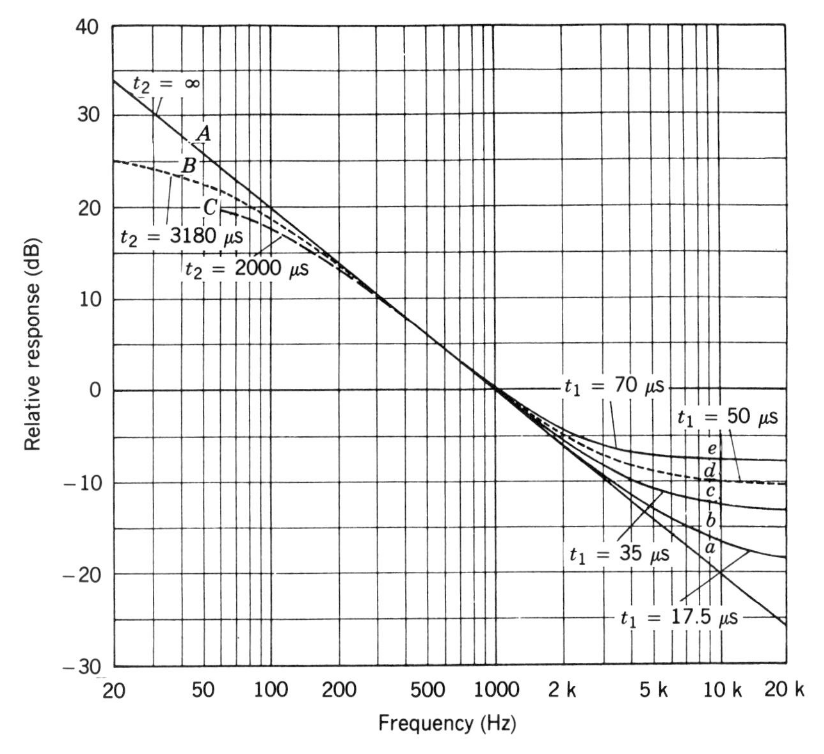

Fig. 3 Playback equalization for professional recorders. The departures from the straight line *Aa* compensate for, e.g., preemphasis during recording.

ization during recording (preequalization) such that the tape is magnetized to high levels, especially at high frequencies. Thus the recorded signal will be well above the tape noise. The playback equalization (postequalization) is then adjusted to deemphasize the preemphasized recording for an overall flat response. This procedure also deemphasizes the playback noise. Even more effective are dynamic noise suppressors such as Dolby and DBX, which adjust the equalization continuously according to the signal being recorded.

6 RESOLUTION

The resolution capability of a recording system has many names, such as bits per millimetre, shortest recorded wavelength, and recording density. The higher the resolution, the more information can be recorded on a given length of tape, usually at the expense of higher background noise. The most important factors in achieving high linear resolution are to use the smallest possible head gap and the smoothest possible tape surface.

Head gap size has decreased over the years, a practical size being 2 μm (0.00008 in.) long, used for professional analog recorders and compact cassettes. In video cassette recorders smaller gaps as low as 0.25 μm are common. The gaps are made by evaporating a SiO or other nonmagnetic film layer on the pole face of a high-permeability core that includes a copper winding for the audio waveform to be recorded or played. Erase heads are of similar construction, except for larger gaps 50–250 μm long.

7 TAPE SPEEDS

Considering the high-fidelity audio spectrum of 20–20,000 Hz or a restricted spectrum of 400–3000 Hz of some telecommunication devices and the head–tape resolution, tape speeds relative to a stationary head have become standardized at 30, 15, 7.5, 3.75, and 1.875 in./s for analog recording. The highest speed is used for master recordings, where cost is no object—the 1.875 speed for compact cassettes. These speeds are too slow for digital recording, which requires bandwidths in the megahertz region. Therefore digital audio recorders resort to video recording techniques with rotating heads, which may scan the tape at 228 in./s while the tape itself moves forward at only 0.66 in./s (VHS-LP II, 4-h format).

8 TAPE TRANSPORTS

The term *tape drive* implies a tape in ribbon form is driven past stationary record, playback, or erase heads for single or multiple tracks. The tape moves between the supply reel and takeup reel as in Fig. 1, the reels typically being 10.5 in. in diameter for professional recorders that are loaded and threaded by hand. Reel diameters of 7.5 and 5.0 in. are also common. The tape is driven by a precision capstan backed by a flexible pinch roller, with great precautions taken to ensure the steadiest possible tape velocity.

For domestic tape recorders used in homes, schools, and telephone answering, the compact cassette is almost universal. It contains both reels and is simply placed into the recorder with no threading or other adjustment.

All of the above tapes must be rewound for replay. Alternatively, the record can be made in the form of a disc or belt, where any part of its surface is available immediately by setting a movable head on that part. There are many other combinations for specialized uses, such as cards with magnetic stripes for automatic bank tellers and single-reel endless tapes (Lear Stereo-eight).

9 DIGITAL RECORDING

Digital recording accomplishes miracles totally impossible with analog recording, such as zero wow and flutter, 90 dB signal-to-noise ratios or higher, and distortion levels as low as anyone wants to specify. All of these are costly, so it does not pay to overspecify beyond the limitations of other audio components such as loudspeakers, microphones, and the human ear.

To enter the digital domain, the analog signal is sampled at regular intervals. The sample amplitude is measured with great accuracy (one part per 100,000,000 for a 96-dB signal-to-noise ratio system) and the amplitude is recorded as a number, say 92,141,735. The sampling must be done more often than twice the highest frequency in the audio signal (Nyquist number), so that 44,000 samples per second is feasible for 20 kHz audio (20,000 × 2 + 10%). The series of digital samples are in the megahertz range and are recorded and played back with videotape recorders. In digital form the audio may be processed with digital filters, error correctors, and other techniques not available in the analog domain, particularly to store successive numbers in buffers and read them out in perfect crystal controlled time intervals, even though the intervals become somewhat irregular due to mechanical fluctuations in the tape drive. Thus wow and flutter are eliminated.

The digital signals are eventually restored to analog by constructing a variable amplitude dictated by the succession of digital numbers digital-to-analog conversion). The results are extremely precise at higher amplitudes and low frequencies, less so at low amplitudes and high frequencies, which are a source of trouble and of weaknesses for these "perfect" digital systems.

10 APPLICATIONS OF MAGNETIC RECORDERS

Magnetic recording has become the most common and most versatile of all recording systems. Its audio realm ranges from studio master recorders with 24 simultaneous high-fidelity tracks to digital recorders using videotapes to broadcast station reel-to-reel machines to the omnipresent compact cassettes that themselves have thousands of uses in entertainment, education, offices, and commerce. In fact, it is almost certain that any reproduced sound at some stage passes through a magnetic record–playback process.

114

DIGITAL AUDIO

JAMES W. BEAUCHAMP AND ROBERT C. MAHER

1 INTRODUCTION

This chapter presents a variety of issues and topics related to the recording, playback, storage, and processing of digitized audio signals. First, the traditional methods for converting continuous waveforms into series of discrete, quantized numbers are discussed, followed by a description of several modern digital-to-analog conversion techniques. Next, common transmission and storage standards for digital audio are presented. The chapter concludes with a brief introduction to digital filters and other digital audio processing applications.

2 BACKGROUND

Experiments employing the digitization of sound were first performed in the late 1950s and early 1960s for computer analysis and synthesis of speech signals, synthesis of musical sounds, and simulation of reverberation, although the method of pulse code modulation (PCM) digital encoding was invented earlier.[1] Most of this work was carried out at Bell Telephone Laboratories, Murray Hill, New Jersey.[2] By the early 1970s this method was being used for high-performance recording,[3] although results were still distributed by analog means. Digital recorders for commercial recording studios became available in the late 1970s and were commonly used for master recordings. A significant breakthrough occurred in the 1980s, with the invention of the compact disc (CD), which soon replaced the analog long-playing (LP) record as the most common medium for playback of prerecorded music. For custom digital recordings, machines for digital recording on cassette video tape and then the rotary-head digital audio tape (R-DAT) became available. Both of these media allowed 2 h or more of stereo full-bandwidth recordings. More recently, the Digital Compact Cassette (DCC) (Philips-Matsushita) and the MiniDisc (MD) (Sony) were introduced.

While the basic theory of time and amplitude quantization had been available for decades, the limited speed and accuracy of available electronic hardware placed severe limitations on the speed and accuracy of digital converters, and various methods had to be found to overcome these limitations. Results from signal and control theory and probability were brought to bear on the problem of achieving high signal-to-noise ratios with less nonlinear and frequency-dependent distortion. Coding theory was used to alleviate problems due to less-than-ideal recording media.

The quality advantages of precision digital recording are immediately apparent to the discerning ear. First, there are dramatic reductions in both distortion (both harmonic and intermodulation) and background noise. Second, there is virtually no measurable time variation distortion (i.e., wow and flutter). Dropouts, due to recording media defects, can occur, although they are not as likely as with most analog tape recording. When substantial dropouts do occur, however, they can be catastrophic, and error correction codes are needed to conceal these errors. Redundancy coding can virtually eliminate these errors.

Another advantage of digital audio is that it can be coupled with digital computers or with video playback systems for educational and entertainment purposes; the current term describing these types of systems is *multimedia*. For the well-endowed recording studio and for use in acoustics research, the advantages of digital over

Note: References to chapters appearing only in the *Encyclopedia* are preceded by *Enc.*

analog technology in terms of quality and flexibility are obvious. The accuracy advantages afforded by the digital approach make its use imperative in most new professional implementations.

3 QUANTIZATION OF AUDIO SIGNALS

Analog audio signals, which can be considered simple single-valued functions of time, can be quantized in a multitude of ways. One obvious method would be to approximate a signal $x(t)$ as a series of discrete points given by coordinates (t_i, x_i), where t_i and x_i represent the time and amplitude of the ith "sample." With suitable scaling, the time and amplitude values could be further quantized as integer values encodable in terms of a finite number of digital bits. While an uneven spacing of samples might actually be desirable for certain types of analog signals, it is generally much more practical for wide-band, bipolar audio signals to space the samples uniformly using a fixed intersample time increment ΔT, so that $t_{i+1} = t_i + \Delta T$. Then the signal is simply represented by a series of sample values $(x_0, x_1, x_2, x_3, \cdots)$, where it is tacitly assumed that the sample increment is known. However, the parameter that is usually referred to is the *sample rate* or *sample frequency* f_s, given by $f_s = 1/\Delta T$. The sample rate, as we shall see, defines the maximum possible bandwidth of an analog signal that can be represented by the digital samples.

3.1 Time Quantization and Bandwidth

An important relationship between the sample frequency necessary for proper uniform sampling of an analog signal and the bandwidth of the signal is simply given as

$$f_s > 2f_{\max}, \tag{1}$$

where $f_{\max}$ is the signal bandwidth, or the maximum frequency of significant energy contained in the signal. In other words, the sampling frequency must be greater than twice the signal's bandwidth. This is known as the *sampling theorem*, and proofs of this theorem are given in many textbooks on communication theory.[4] Put another way, given a particular sampling frequency, the highest frequency contained in the signal should be less than $\frac{1}{2}f_s$, which is referred to as the *Nyquist frequency* or *Nyquist limit*, attributed to Harry Nyquist,[5] a scientist at Bell Telephone Laboratories in New Jersey during the 1920s and 1930s.

More interesting is to see how to characterize the frequency spectrum of a signal after it has been sampled in comparison to its original form, regardless of whether the sampling theorem is satisfied. Using Fourier transform theory, it can be shown that if the spectrum of an original signal $x(t)$ is given by $X(f)$, the spectrum after ideal sampling is given by

$$X'(f) = X(f) + X(f_s - f) + X(f_s + f) + X(2f_s - f) + X(2f_s + f) + \cdots. \tag{2}$$

Each pair of terms $X(nf_s - f)$ and $X(nf_s + f)$ is exactly what would result from amplitude modulation of the signal $x(t)$ with a sinusoidal signal of frequency nf_s and thus can be thought of as *side bands*, which are essentially reflections of the original spectrum on either side of the harmonics of the sampling frequency.

A graph of a typical spectrum $X(f)$ in terms of amplitude versus frequency and the corresponding spectrum $X'(f)$ computed after sampling are shown in Figs. 1*a*, *b*. Note that the bandwidth of $X(f)$ is approximately 3500 Hz. In Fig. 1*b*, the sampling theorem is satisfied, since 3500 is less than $\frac{1}{2}f_s = 5000$ Hz; that is, spectrum $X(f)$ and the nearest side band $X(f_s - f)$ do not overlap, and a low-pass filter can easily pass $X(f)$ and reject $X(f_s - f)$. In Fig. 1*c*, however, there is substantial overlap between the highest frequency portion of $X(f)$ and the lowest frequency portion of $X(f_s - f)$. In this case, no low-pass filter can reconstitute the original spectrum or signal. The side-band reflection about $\frac{1}{2}f_s$ is often referred to as *aliasing* or *foldover*.

Some aliasing or foldover error might be tolerable. From our consideration of spectra, we can define the aliasing error as

$$e_{\text{alias}} = \left(\int_{f_s/2}^{\infty} |X(f)|^2 \, df \right)^{1/2}. \tag{3}$$

Any energy in the original signal above the Nyquist frequency contributes to the error, since it is prone to foldover. If this error is negligible compared to the total amplitude of the signal, we could say that the sampling theorem is satisfied "well enough."

For typical audio recording, a fixed sample rate is employed just sufficient to provide adequate bandwidth. Sample rates as low as 8 kHz, giving a theoretical bandwidth of 4 kHz, can be used for speech applications because most intelligibility is coded in spectra below 4000 Hz. For professional music recording, a 48-kHz sample rate is generally the standard, although rates of 44.1 kHz (used for CD players), 44.056 (used for PCM video), 32 kHz, and 22.05 kHz have been used for particular applications. Actually, the usable audio bandwidth

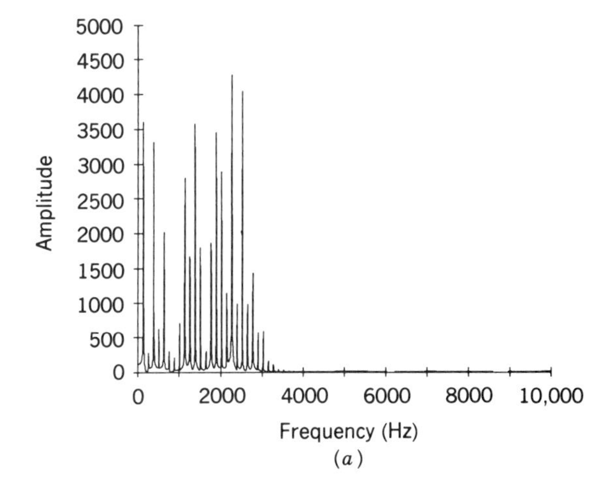

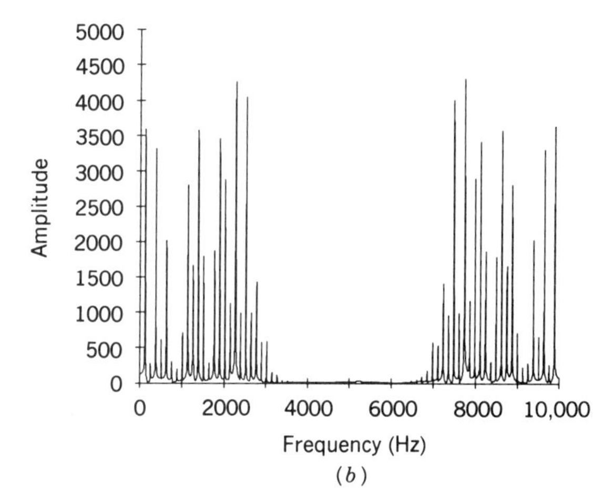

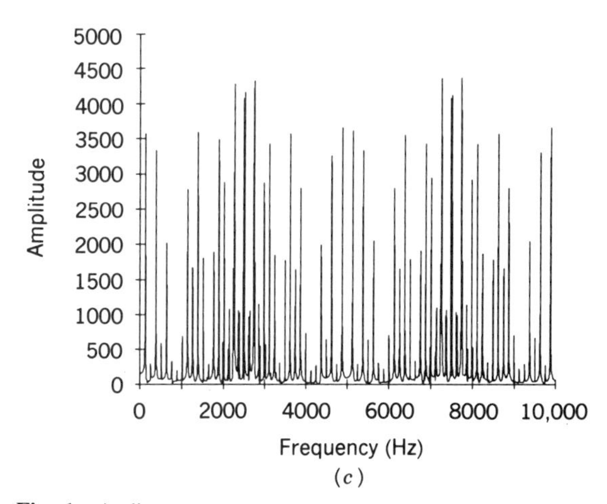

Fig. 1 Audio spectrum: (*a*) original; (*b*) after sampling at 10000 Hz; (*c*) after sampling at 5000 Hz.

is somewhat less than the Nyquist frequency (perhaps by 5–10%) since one must allow for the frequency region, called the "guard band," where the anti-aliasing low-pass filter makes its transition from the passband to the stop-band.

3.2 Amplitude Quantization and Signal-to-Noise Ratio

For storage in digital media, signals must be quantized as signed integer values ranging from $-I_{\max}$ to $I_{\max} - 1$. Generally, the integers are coded as binary values using a fixed number of bits, N. With the sign taking 1 bit, it follows that $I_{\max} = 2^{N-1}$. Therefore, the maximum possible analog signal magnitude corresponds to 2^{N-1} when stored in a digital medium. The process of converting a continuous analog signal $x(t)$ to integer values is called *amplitude quantization* and can be thought of as a nonlinear function $Q[\cdot]$. Thus, the quantized signal can be written $Q[x(t)]$. The *quantization error signal* can then be defined as

$$q(t) = Q[x(t)] - x(t). \tag{4a}$$

With proper rounding, this error is confined to ± 0.5. A primary assumption is that the error signal is spectrally white and random with all values between -0.5 and $+0.5$ equally likely. Since from Eq. (4a) the quantized signal can be expressed as

$$Q[x(t)] = x(t) + q(t), \tag{4b}$$

the quantized signal can be thought of as the original analog signal *plus* the error signal, sometimes referred to as *quantization noise*. Figure 2 depicts the process of quantizing a sine wave to 4 bits and the formation of a noise error signal. In general, the root-mean-square (rms) amplitude of the noise signal $q(t)$ is given by

$$q_{\mathrm{rms}} = \left(\int_{0.5}^{0.5} q^2 \, dq \right)^{1/2} = \frac{1}{\sqrt{12}} \cong 0.2887. \tag{5}$$

Since the largest sine wave that can be encoded has a maximum amplitude of 2^{N-1} and an rms amplitude of $2^{N-1.5}$, the best case signal-to-noise ratio is

$$\mathrm{S/N} = \frac{2^{N-1.5}}{0.2887} = 1.225 \times 2^N. \tag{6a}$$

This can be expressed in terms of decibels as

$$\mathrm{S/N_{dB}} = 20 \log_{10}(1.225 \times 2^N) = 6N + 1.76 \text{ dB}. \tag{6b}$$

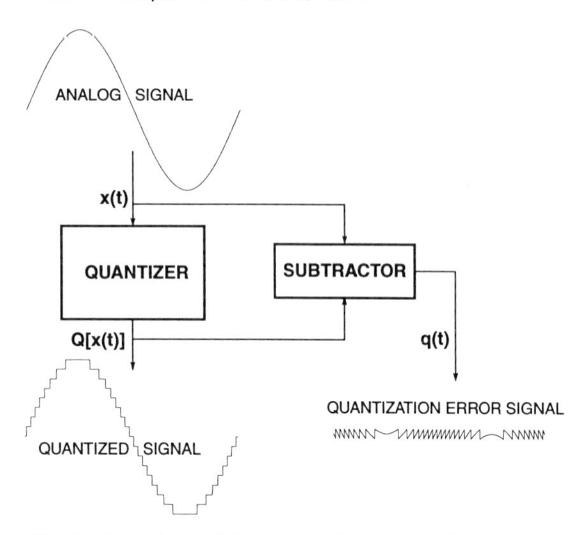

Fig. 2 Quantizing of sine wave and formation of quantization error signal.

For 8 bits we have about 50 dB, 12 bits gives 74 dB, and 16 bits (a common standard) yields 98 dB. Under the best of circumstances, the human ear can hear a 120-dB dynamic range, but this would only be in environments with absolutely quiet backgrounds. Nevertheless, 16 bits is very marginal for professional recording, and efforts have been underway to provide more equivalent bits, up to as many as 24 bits.[6]

3.3 Simultaneous Time and Amplitude Quantization

Of course, in a digital recording system, both time and amplitude quantizations occur. The errors introduced by the two quantizations are different and quite separable. However, during the 1980s there was a growing realization of a trade-off that was possible between the two types. We see in Section 5.2 that for every quadrupling of the sample rate it is possible to reduce the number of bits by 1 for the same audio bandwidth and quantization noise amplitude. Following this logic, it would require a sample rate of 206 GHz to achieve 16-bit equivalent signal-to-noise performance with a Nyquist frequency of 24 kHz! This enormous rate is hardly an advantage, but fortunately there are methods that allow reduction of this figure to a practical level.

4 TRADITIONAL CONVERSION METHODS

As long as one confines oneself to the design of converters having a small number of bits (say less than 10) and low sampling rates (say less than 10 kHz) and is not too concerned about inaccuracies that can occur during transitions between stored samples, the design of digital-to-analog converters (DACs) and analog-to-digital converters (ADCs) is very straightforward. These can be realized with discrete technology consisting of resistors, capacitors, diodes, and transistors for the most part. Or integrated circuits such as operational amplifiers, comparators, analog switches, and various digital logic components can be assembled to do the job. However, when larger numbers of bits and higher sampling rates are required and transitions must be controlled, component speed and accuracies are such that more advanced techniques (e.g., laser trimming of resistor values) may be needed.

4.1 Digital-to-Analog Conversion

The realization for a simple 4-bit all-positive linear DAC is depicted in Fig. 3. It consists of a register to hold the current digital word (representing the current sample), a set of four analog switches that connect a reference voltage to a subset of current-summing resistors, a sample clock generator, a sample-hold circuit, and a low-pass filter. For example, if the digital word to be converted were 1101, switches x_3, x_2, and x_0 would be closed, allowing current to flow through the respective resistors, as determined by Ohm's law. The operational amplifier is essentially a current-summing device whose output voltage is the product of the sum current and the feedback resistor.

In addition, a sample-hold circuit, or "deglitcher," is

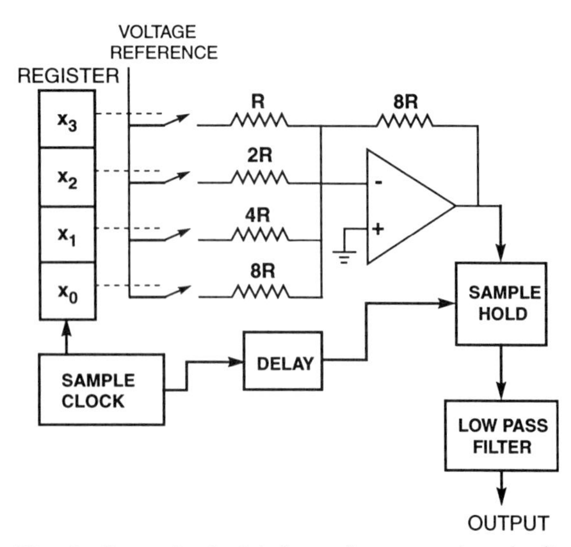

Fig. 3 Conventional digital-to-analog conversion circuit (4-bit example).

needed to capture the amplifier's output, storing it on a capacitor until the next clock pulse arrives. This is because all switches do not close and open simultaneously and instantaneously when a clock pulse occurs, resulting in wild voltage fluctuations, or "glitches," during some transitions (particularly when several bits change, as in the case of the 0111-to-1000 transition). While the sample-hold circuit provides some filtering, it is not sufficient to remove side-band reflection components above $f_s/2$ as described in Section 3.1. Therefore, a sharp-cutoff low-pass analog filter is needed to reject all components for $f > f_s/2$ while passing the signal for $f < f_s/2$.

4.2 Analog-to-Digital Conversion

There are two principal analog-to-digital conversion methods. One, called *flash conversion*, divides the input analog range into an evenly spaced set of discrete values which correspond to the range. As an example, for a 3-bit flash converter, the values $-4, -3, -2, -1, 0, 1, 2, 3$ would be allowed. Any analog value within ±0.5 of one of these values, as determined by a comparator circuit, would be converted to that value in digital form. Since all operations are in parallel, the flash method is very fast; however, it is also very lavish in its use of circuitry. Another method—until recently the method most commonly used for audio converters—is the method of *successive approximation*. Here, bits are determined one at a time, starting with the most significant, by comparing the output of a DAC to the analog input voltage. First, the most significant bit is turned on, corresponding to the center of the analog range; this bit is retained depending on whether the input value is above (retain) or below (do not retain) the DAC value. Then, the next bit is turned on, and a similar process ensues. After N comparisons, the entire digital word is determined.

4.3 Nonlinear Encoding Methods

Thus far, we have discussed the method of linear encoding. However, one often hears that since "the ear hears logarithmically," the quanta needed for digitizing a signal only need to be fixed fractions of the sampled amplitudes. As a result, several attempts have been made to reduce the number of bits per sample needed compared to an equivalent fidelity linear system or to increase the fidelity for the same number of bits.

One attempt is based on the floating-point encoding method, where a fixed number of bits is allocated to a mantissa and a certain number of bits to an exponent.[7] An interesting feature of this method is that most amplitudes can be represented by several different mantissa–exponent combinations. The best accuracy occurs when the mantissa is maximized and the exponent is minimized. If optimized on a sample-by-sample basis, the exponent would be continually changing, a phenomenon that places extremely stringent conditions on the accuracy of digital-to-analog switching hardware. However, if only used to minimize storage, the *block floating-point* method is an effective technique. With this method, a block of N samples is first linearly encoded with a high-quality conventional ADC. Then, the maximum value of the block is determined, the appropriate exponent for this value is computed, and all numbers in the block are converted to floating point with this exponent.

With another nonlinear method, the input signal x is conceptually "predistorted" by a nonlinear distortion function $F(x)$ in such a way that larger values of x are encoded using larger quanta. An inverse nonlinear distortion function $F^{-1}(y)$ is applied when the signal is reconstructed. The entire process is generally implemented with a precalculated table lookup function. One popular theoretical distortion function is the μ-law function.[8]

5 ADVANCED CONVERSION TECHNIQUES

There are several important details concerning the simple analog-to-digital and digital-to-analog concepts described in the last section. First, the practical necessity of limiting the number of binary digits (bits) used to represent the amplitude of each sample means that quantization distortion is unavoidable. Second, because the sample rate of the system cannot be arbitrarily high, the fundamental Nyquist inequality for signal bandwidth requires the presence of frequency-selective filters in the audio path preceding and following the ADC and DAC, respectively. Finally, the subjective performance of a digital audio system may be better if the strengths and weaknesses of the human auditory system are taken into account.

5.1 Quantization Noise and Dither

In a conventional PCM digital audio system (assuming no overload) the quantization error on any given sample is in the range ±0.5, where 1.0 is the step size of the quantizer. If the input signal to the quantizer is a reasonably complex waveform (i.e., not a sinusoid) and is large in amplitude compared to the step size, then it is usually found in practice that (a) the quantization error is uncorrelated with the input signal, (b) the error is independent from sample to sample, (c) the error probability is distributed uniformly over the range ±0.5, and (d) the long-term spectrum of the error is white.[9] However, despite the seemingly random behavior of the quantization error for complex, high-amplitude signals, it is actually *deterministic*; that is, if the quantizer input signal is known exactly, then so is the quantization error. For a sinusoidal input signal spanning only a few quantization levels, the quantization error takes on a tonal or "granular" quality that modulates audibly with changes in the input signal. The level-dependent behavior of the quantization noise is due to the inherently nonlinear character of the quantization process.[10]

The addition of a wide-band noise signal, called *dither*, to the input signal prior to the quantizer can be used to reduce or eliminate the correlation between the input signal and the quantization noise.[11] Properly applied, dither eliminates repetitive sequences of correlated low-level samples. Typical dither signals have either rectangular (±0.5) or triangular (±1.0) probability density functions. Listeners generally judge the broadband, decorrelated noise resulting from dither as less objectionable than the correlated noise of the undithered signals.[12] A further benefit is that dither allows reproduction of signals below the level of 1 bit, whereas without dither nothing is produced below this point. Of course, incorporating dither results in a greater noise level than an undithered signal has. While, because of synchronization problems, it is difficult to subtract the dither signal after quantization to reduce this level, a possible solution has been proposed by Craven and Gerzon.[13] The proposed technique reconstructs the original pseudorandom dither from the least significant bits of the dithered data stream while maintaining compatibility with existing playback systems.

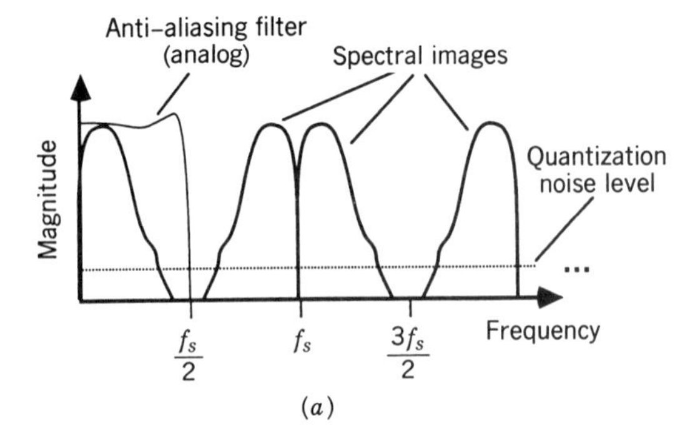

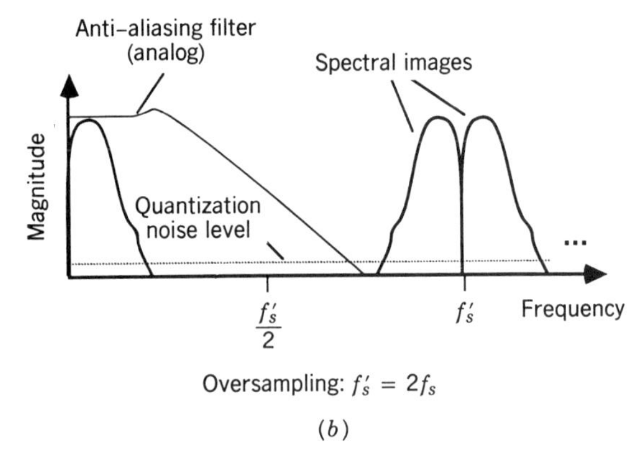

Fig. 4 Illustration of digital-to-analog conversion spectrum in (*a*) a minimal sample rate system and (*b*) an oversampled system.

5.2 Oversampling and Noise Shaping

The requirement that analog input and output signals be strictly band-limited to one-half the sample rate implies that an analog low-pass filter with a very abrupt passband-to-stopband cutoff characteristic must be employed. Sharp-cutoff analog filters are problematic, however, due to their complexity, component sensitivity, and nonlinear phase properties. For analog-to-digital conversion, an approach to avoiding this problem is to use a sample rate that is much higher than the normal sample rate and then to apply a digital low-pass filter to the resulting "oversampled" signal whose cutoff frequency matches the normal Nyquist rate. Since the filter output signal contains essentially no energy above the filter cutoff, it can be safely downsampled (decimated) to regain the normal sample rate.[14]

For digital-to-analog conversion, the digital data stream can be upsampled (interpolated), giving a rate much greater than twice the Nyquist sample rate. To convert the resulting oversampled data stream to an analog signal, only a relatively low-order analog filter is needed to reject the upsampled high-frequency aliased spectral images. Figure 4 illustrates how this method works in the frequency domain.

Oversampling also reduces quantization noise. Since this noise is distributed across a wider bandwidth than for normal sampling, the noise power in any band is reduced by the oversampling factor. As the quantization noise above the signal's baseband is removed by filtering, an improved in-band signal-to-noise ratio results.[15,16] Output quantization noise can be further reduced by *shaping* the quantization noise spectrum to concentrate more of its noise power at frequencies above the audio bandwidth, prior to the digital low-pass stage. As a result, the final signal-to-noise ratio can be improved substantially. This process is called "noise feedback" or "noise shaping." Details are given by Hauser.[17]

An elementary first-order noise-shaping structure is depicted in Fig. 5. Analysis of the structure reveals that the output signal contains a delayed version of the input signal plus a high-pass filtered version of the quantization noise. Hauser's results show that, whereas with no noise shaping oversampling ratios of 4, 16, 64, ... are

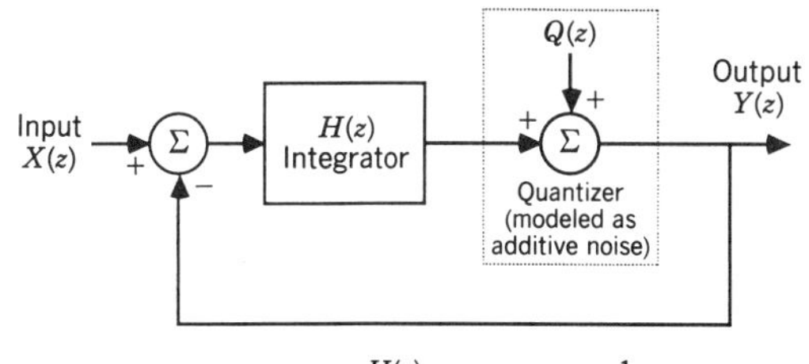

$$Y(z) = X(z)\,\frac{H(z)}{1 + H(z)} + Q(z)\,\frac{1}{1 + H(z)}$$

If $H(z) = \frac{z^{-1}}{1 - z^{-1}}$, then $\frac{H(z)}{1 + H(z)} = z^{-1}$ (simple delay)

and $\frac{1}{1 + H(z)} = 1 - z^{-1}$ (first-order high pass)

(a)

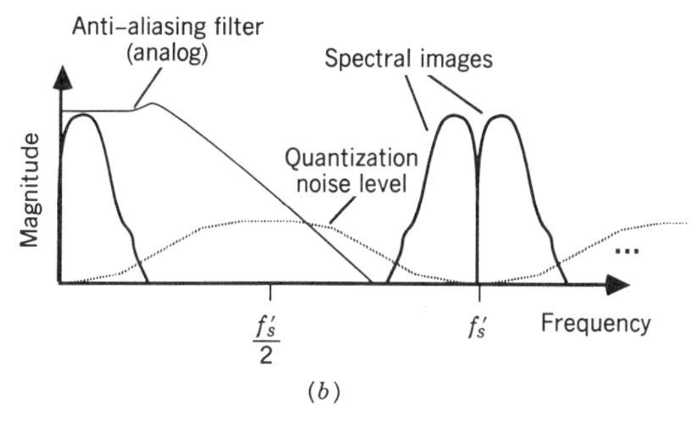

(b)

Fig. 5 Block diagram of a first-order noise-shaping system. With $H(z)$, a simple integrator, the output signal consists of the delayed input signal and high-pass filtered quantization noise.

required to reduce the quantization noise by 1, 2, 3, ... bits, with first-order noise shaping ratios of only 2.4, 3.8, and 5.9 are needed. Even greater improvements result when higher order noise-shaping techniques are used.

5.3 One-Bit Conversion

Conventional high-resolution ADC and DAC structures require many digital bits per sample. Converters with more than 16 bits are difficult to design due to problems with low-level linearity, monotonicity, zero-crossing distortion, and so on. These limitations have encouraged the design of alternative converter structures making use of oversampling and noise-shaping strategies such as those discussed above. Several modern approaches for ADCs and DACs are collectively known as "one-bit" systems, since at some point in the conversion process the signal samples are represented by a single bit.

A widely known oversampling, noise shaping, 1-bit converter technique is called "delta–sigma." The basic first-order structure is shown in Fig. 6. In the forward path, the input analog signal passes through a low-pass filter, $H(z)$, and is quantized with a single-bit quantizer (limiter) at an oversampled rate. The quantized output, a single bit, is fed back through a 1-bit DAC and is then subtracted from the analog input. In this way the signal passing through the filter, $H(z)$, is actually the difference between the input signal and a 1-bit "prediction" of the input signal obtained from the quantized output. This prediction error is obtained at the oversampled rate, meaning that much of the prediction error spectrum lies outside the audio band and can therefore be removed by subsequent decimation filtering.

The internal bit rates required in delta–sigma converters are extremely high. The oversampling factor is typically 48 or more, giving an internal bit rate in excess of 2 MHz. However, the computation load is somewhat less onerous than it might appear, since the internal filtering is done on a single-bit data stream, and fast bitwise hardware multiplier structures can be fabricated quite

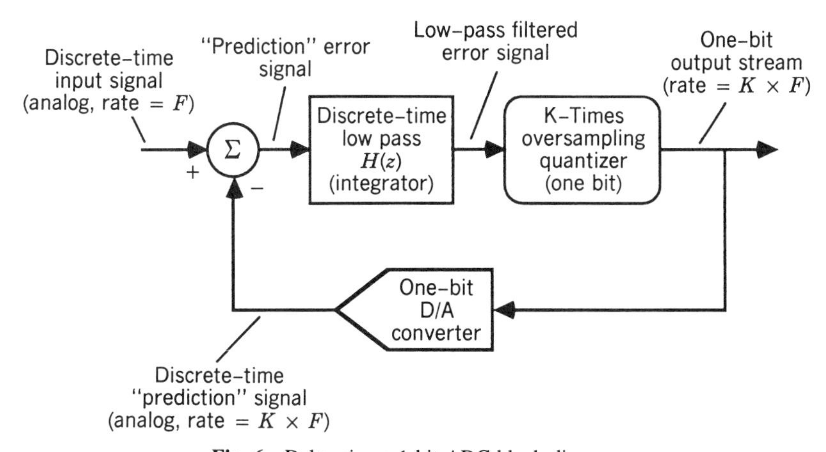

Fig. 6 Delta–sigma 1-bit ADC block diagram.

easily.[16,17] Also, note that most practical implementations of delta–sigma place sophisticated, higher order feedback loops around the basic first-order structure in order to yield more equivalent bits from the structure and reduce the oversampling factor.

6 TRANSMISSION AND STORAGE OF DIGITAL AUDIO DATA

Among the major advantages of digital audio systems over analog systems are the ability to create multiple-generation lossless copies without accumulating noise and the capability to correct transmission errors by including mathematically calculated redundant bits in the digital data stream. These properties mean that a digital audio signal can be conveyed from place to place over a communication channel or storage medium, then reconstructed perfectly for further transmission or for conversion to analog form.

6.1 Consumer and Professional Transmission Standards

It is frequently necessary in digital audio signal processing to send audio data from one device to another. A special interface standard specifically for stereo (two-channel) digital audio was developed by working groups from the Audio Engineering Society (AES) and the European Broadcasting Union (EBU). The professional standard[18] is often referred to as the "AES/EBU" interface.[19] A similar format for consumer applications is the Sony/Philips Digital Interface Format (SPDIF). Both the professional and consumer standards involve serial transmission of stereo digital audio data represented in linear (PCM) form. Up to 24 data bits are provided by the standard (20 data bits and 4 auxilliary bits) plus additional "subcode" bits, resulting in 32 bits of transmitted data per audio sample (per channel). A 48-kHz stereo sample rate results in a transmitted rate of 3.072 million bits per second.[12]

6.2 Compact Disc

By the 1990s the most popular storage medium for prerecorded digital audio was the compact disc (CD).[20] The current CD standard was developed jointly by Philips (Netherlands) and Sony (Japan) and was introduced in the early 1980s. The CD system has been arguably the most successful audio product in history, with disc sales going from essentially zero in 1981 to a 1992 level of more than 200 million discs per year in the United States alone.

The standard CD is 12 cm in diameter and 1.2 mm thick. The primary composition of the CD is a transparent polycarbonate plastic substrate used to help focus the laser readout beam and make the CD structurally rigid. Digital information is stored as a spiral "pit track" formed on the top surface of the substrate, which is covered with a less than 0.1-μm-thick reflective aluminum or gold layer. The top surface of the pit track is protected from external damage by a 20-μm-thick lacquer coating.

Unlike LP records, the CD's pit track spiral starts near the less vulnerable center of the disc and proceeds outward. The pits are only 0.5 μm wide and the track-to-track pitch is 1.6 μm. Digital information is encoded by the transition from pit to land and from land to pit, using an ingenious scheme called eight-to-fourteen modulation (EFM) that does not require every bit in the data stream to be represented by a separate pit transition. Pits vary in length from less than 1 μm to more than 3 μm. A solid-state laser (790 nm) is focused up through the transparent polycarbonate substrate to follow the pit track spiral at a linear rate of about 1.3 m/s. For a 60-min CD, the pit track is approximately 4.7 km (2.9 miles) long. At the standard CD sample rate of 44.1 kHz per channel, with two channels and 16 bits per sample, the audio data bit rate is 1,411,200 bits/s. However, with error correction, synchronization, and the subcode bits added, this increases to 4,321,800 bits/s.

6.3 Digital Audio Tape

The development of recordable/erasable digital audio storage technology has paralleled the development of the read-only compact disc. A variety of recording systems have been made, many of which consist of a digital processor called a PCM adapter that samples the input analog signal and converts the digital audio data stream into a pseudo-video signal suitable for recording on a standard rotating-head video tape recorder. For professional recording purposes requiring editing and multiple tracks, special-purpose digital audio recorders with stationary heads are also available. For consumer and semiprofessional purposes, the R-DAT (rotating-head digital audio tape) record has become quite popular and is standard for CD-quality tape distribution.[15] The R-DAT employs a 2000-rpm helical scan rotating-head design and a cartridge that is smaller than an audio cassette. Although the tape moves through the machine at only about 8 mm/s, it handles a 44.1- or 48-kHz stereo 16-bit sample rate, subcode information, ID coding, and error correction, resulting in a total bit rate of about 3 million bits/s.

6.4 Reduced Bit Rate Digital Audio: Perceptual Coding Systems

In late 1992, two new digital audio formats were unveiled for consumer use: Digital Compact Cassette

(DCC) developed by Philips[21] and MiniDisc (MD) developed by Sony.[22] Both systems were targeted as replacements for the analog cassette. By incorporating perceptual coding techniques to maintain a high degree of perceived audio quality while decreasing the average data rate required to encode a signal, they require much less storage than the CD. Such methods are sometimes referred to as "lossy coding" methods, because the bit-reduced versions of signals they produce cannot be exactly reconstituted to their original forms. DCC uses a multiband polyphase filter bank for analysis, while MD uses a modified discrete cosine transform (MDCT) filter bank. Both schemes use proprietary time–frequency analysis and "bit allocation" strategies to reduce the number of bits required to represent signals. DCC uses a special magnetic tape cassette that is similar in size to a standard analog audio cassette. MD uses a cartridge containing a 64-mm-diameter erasable/recordable optical disk.

Perceptual coding systems for high-fidelity audio exploit the masking, level sensitivity, and critical band properties of the auditory system by distributing the quantization error in frequency bands where it is essentially inaudible due to self-masking by the higher level program material.[23] This method of coding requires a continuous analysis of the temporal and spectral properties of the input audio signal. The Phillips DCC, with its PASC coding format, provides a factor of 4 reduction in the bit rate with virtually no perceived reduction in quality under normal listening conditions.[24]

In addition to the DCC and MD, there are now proposals in Europe to improve radio broadcast fidelity by means of perceptually encoded Digital Audio Broadcasting (DAB). Poor reception and reflected signals are to be overcome by a coding scheme called COFDM (Coded Orthogonal Frequency Division Multiplex), while audio data are to be bit compressed by Masking-pattern Universal Sub-band Integrated Coding and Multiplexing (MUSICAM).[16] In the meantime, another group of coding schemes for both audio and visual image data in television has been internationally developed under the rubric of MPEG (Moving Pictures Expert Group).[25]

7 DIGITAL AUDIO SIGNAL PROCESSING: TOOLS AND APPLICATIONS

The fundamental operations of most signal processing techniques, be they analog or digital, are multiplication and time delay. In analog systems these operations are accomplished with amplifiers and reactive circuit elements. Multiplication in a digital signal processing system is accomplished using bitwise hardware multipliers, while delay requires storing a digital number for some number of sample periods.

7.1 Digital Filters

Digital filters have numerous applications in audio. Example applications are anti-aliasing (for sample rate conversion and oversampling applications), spectrum equalization, speech and musical sound synthesis, acoustic image and space simulation (virtual audio reality), and adaptive-filter noise cancellation.

Digital filters can be conveniently expressed in the form of a difference equation:

$$\sum_{k=0}^{N} a_k y_{n-k} = \sum_{k=0}^{M} b_k x_{n-k}, \qquad n = 0, 1, 2, 3, \ldots, \quad (7)$$

where x_n is a sequence of input samples, y_n is the output sequence, and a_k and b_k represent the coefficients multiplying the output and input samples, respectively. The frequency response of this digital filter can be calculated by substituting $Xe^{j(n-k)\omega t} = x_{n-k}$ and $Ye^{j(n-k)\omega t} = y_{n-k}$, canceling out $e^{in\omega t}$, and solving for $H(j\omega) = Y/X$. For output sample computation, the difference equation can be rearranged to give

$$y_n = \sum_{k=0}^{M} \frac{b_k}{a_0} x_{n-k} - \sum_{k=1}^{N} \frac{a_k}{a_0} y_{n-k}, \quad (8)$$

which expresses the output sequence y_n as the weighted sum of the current input sample, the previous M input samples, and the previous N output samples. Thus, a digital filter expressed in difference-equation form requires memory for several of the past input and output samples as well as the value of the a_k and b_k coefficients. For further details, refer to Ref. 26.

7.2 Digital Delay Devices and Reverberators

Several common techniques in audio signal processing require the use of signal delay lines. A delay line consists of a series of connected registers or its equivalent. A sample that enters the delay line at time nT will transfer from one register to the next at each sample time and will exit the delay line at time $nT + n\Delta T$, where N is the length of the delay line in samples and ΔT is the sample time ($1/f_s$). Figure 7*a* (with the switch open) illustrates this principle.

An *echo generator* or *simple reverberator* can be constructed from a delay line by including a feedback connection with some attenuation to prevent amplitude overflow. Figure 7*a* (with the switch closed) gives a simple realization. Samples fed back to the input are delayed by $\tau = N\Delta T$ and attenuated by factor α, which for stability must be less than unity.

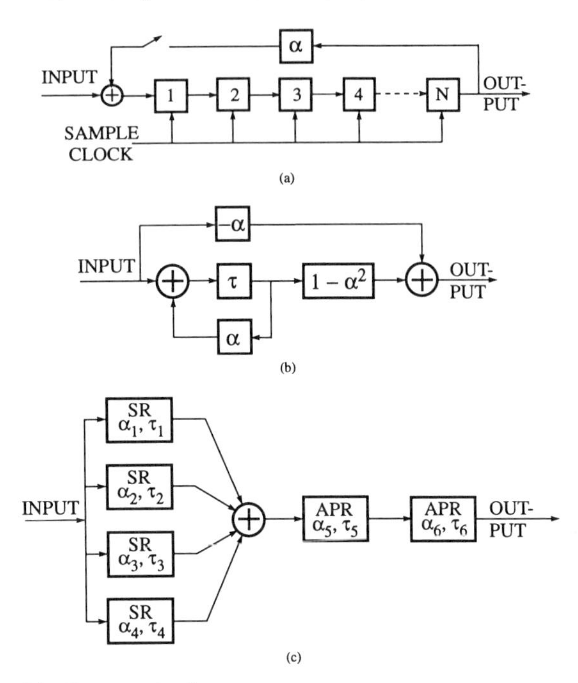

Fig. 7 (*a*) Delay line structure with feedback echo possibility (simple reverberator). (*b*) Schroeder all-pass reverberator with delay τ and attenuation constant α. (*c*) Schroeder natural reverberator consisting of four simple reverberators (SRs) and two all-pass reverberators (APRs).

Practical artificial reverberators combine simple reverberators in an effort to achieve high echo density, imperceptible coloration, and a wide range of reverberation decay times. The first successful digital reverberators were based on two components: the simple reverberator, as described above, and a related reverberator with a unit flat magnitude frequency response, called an *all-pass reverberator*, invented by Schroeder[27] (see Fig. 7*b*). As an example, Fig. 7*c* shows Schroeder's *natural reverberator*, constructed from four simple reverberators (also called *comb filters*, because of their periodic frequency responses) and two all-pass reverberators.

7.3 Sound Synthesis and Analysis

A wide variety of digital synthesis methods are available[28,29] and can be organized into general categories: wavetable, sample playback, additive, subtractive (using digital filters), nonlinear, and physical modeling.

Sound analysis procedures attempt to find the best parameters to resynthesize a given signal. With the time-variant spectrum model any audio signal may be represented by the superposition of sine waves, having time-varying frequencies and amplitudes and possibly a noise signal with a time-varying spectral envelope,[30] and then resynthesized by the method of additive synthesis. The additive synthesis parameters can be found by short-time Fourier transform analysis. Spectral representations can also be used as starting points for development of data-reduced synthesis models, such as nonlinear synthesis[31] and multiple wavetable synthesis.[32]

Analysis techniques can also be used for fundamental frequency detection in speech[33] and music.[34] Generally, frequency detection methods rely on the assumption that the signal is quasi-periodic and consists of a series of harmonic partials. However, many signals are not strictly harmonic, so one may need to take into account psychoacoustic theories of pitch.[35]

7.4 Other Applications

Many other digital audio processes are employed in consumer and professional applications.[15,16,36] A few selected applications are summarized here:

1. *Sample Rate Conversion.* It is frequently necessary to convert digital audio data from one sample rate to another, for example, from the professional 48-kHz rate to the 44.1-kHz rate for CDs. While sample rate conversion can be performed by converting the digital signal to analog and resampling at the new rate, it is more accurate and convenient to resample using methods of digital signal processing.

2. *Noise Reduction and Restoration of Old Recordings.* Techniques have been developed for discriminating between desired signals and undesired noise and other distortions. Methods for removing pops, clicks, and surface noise endured by old gramophone or wax recordings have been developed, and many old recordings have been restored and re-released.[37]

3. *Time Scaling and Frequency Shifting.* Music or speech can be sped up or slowed down without affecting its pitch or basic quality using methods such as granular overlap in the time domain or short-time Fourier transform in the frequency domain. In the latter case, time scaling is simply applied to the amplitude and frequency parameters prior to additive synthesis or inverse transform. Similarly, frequency shifting can be applied to individual components prior to resynthesis without affecting time structure.

4. *Acoustic Space Simulation (Virtual Audio Reality, Auralization).* The complexities of reverberant spaces and the human spatial hearing mechanism demand the precision that digital processing can provide. Recent advances in recording techniques and head-related transfer functions have made it possible to simulate virtual acoustic spaces in real time.[38]

5. *Digital Audio Workstations.* Production of music for commercial release usually involves many individual sound recordings, or tracks, that must be combined in the proper proportion with the proper equalization and processing and at the proper time. Many computer-based digital audio "workstations" have been marketed to perform commonly needed editing, processing, and mixing operations on digital audio material. Sometimes these consist of special software packages with graphical interfaces and hardware peripherals that are added to general-purpose desktop computers. Operations such as time scaling and noise reduction can be standard tools within the workstation environment.

6. *Multimedia.* This refers to end-user software applications on desktop computers that include digital audio. In the 1990s digital audio has rapidly become standard for computers because of its low cost and versatility, thus spawning a plethora of multimedia applications.

8 SUMMARY

A wide range of topics related to digital audio has been presented in this chapter. Because entire books and professional journals relating to digital audio issues are available, this chapter should be viewed as a brief overview of the basic terms and concepts.

To a large extent, digital audio has supplanted many of the audio storage and processing tasks traditionally accomplished with analog components. Digital techniques afford many practical advantages, including stability, noise immunity, error detection/correction, and "perfect" duplication. Furthermore, the majority of research and development work in the professional audio field has been and will continue to be related to digital techniques.

For the forseeable future, any new digital audio system will be judged in terms of cost, convenience, and audio quality compared to the CD. There is currently a trend toward perceptual modeling and data reduction with a goal of obtaining essentially CD subjective quality with substantially reduced bit rate. This trend will continue as enhanced audio quality is incorporated into personal computers and advanced television systems.

REFERENCES

1. H. Reeves, "Electric Signaling System," French Patent No. 852,183, British Patent No. 535,860, U.S. Patent No. 2,272,070, 1938–42.
2. M. V. Mathews, J. E. Miller, and E. E. David, Jr., "Pitch Synchronous Analysis of Voiced Sounds," *J. Acoust. Soc. Am.*, Vol. 33, 1961, pp. 1725–1736.
3. T. Stockham, "A/D and D/A Converters: Their Effect on Digital Signal Fidelity," in L. Rabiner and C. Rader (Eds.), *Digital Signal Processing*, IEEE Press, New York, 1972.
4. R. A. Gabel and R. A. Roberts, *Signals and Linear Systems*, Wiley, New York, 1987.
5. H. Nyquist, "Certain Topics in Telegraph Transmission Theory," *Trans. Am. Inst. Electr. Eng.*, Vol. 47, 1928, pp. 617–644.
6. P. G. Craven, "Toward the 24-Bit DAC: Novel Noise-Shaping Topologies Incorporating Correction for the Nonlinearity in a PWM Output Stage," *J. Audio Eng. Soc.*, Vol. 41, 1993, pp. 291–313.
7. F. F. Lee and D. Lipshutz, "Floating-Point Encoding for Transcription of High-fidelity Audio Signals," *J. Audio Eng. Soc.*, Vol. 25, 1977, pp. 266–272.
8. B. Smith, "Instantaneous Companding of Quantized Signals," *Bell System Tech. J.*, Vol. 36, 1957, pp. 653– 709.
9. B. A. Blesser, "Digitization of Audio: A Comprehensive Examination of Theory, Implementation, and Current Practice," *J. Audio Eng. Soc.*, Vol. 26, 1978, pp. 739–771.
10. R. C. Maher, "On the Nature of Granulation Noise in Uniform Quantization Systems," *J. Audio Eng. Soc.*, Vol. 40, 1992, pp. 12–20.
11. S. P. Lipshitz, R. A. Wannamaker, and J. Vanderkooy, "Quantization and Dither: A Theoretical Survey," *J. Audio Eng. Soc.*, Vol. 40, 1992, pp. 355–375.
12. K. C. Pohlmann, *Principles of Digital Audio*, Sams, Indianapolis, 1989.
13. P. G. Craven and M. A. Gerzon, "Compatible Improvement of 16-bit Systems Using Subtractive Dither," Audio Engineering Society Preprint No. 3356, 1992.
14. R. E. Crochiere and L. R. Rabiner, *Multirate Digital Signal Processing*, Prentice-Hall, Englewood Cliffs, NJ, 1983.
15. J. Watkinson, *The Art of Digital Audio*, Focal, London, 1989..
16. K. C. Pohlmann (Ed.), *Advanced Digital Audio*, Sams, Indianapolis, 1991.
17. M. W. Hauser, "Principles of Oversampling A/D Conversion," *J. Audio Eng. Soc.*, Vol. 39, 1991, pp. 3–26.
18. AES3-1992 (ANSI S4.40-1992), "Recommended Practice for Digital Audio Engineering—Serial Transmission Format for Two-channel Linearly Represented Digital Audio Data," *J. Audio Eng. Soc.*, Vol. 40, 1992, pp. 147–165.
19. R. A. Finger, "AES3-1992: The Revised Two-Channel Digital Audio Interface," *J. Audio Eng. Soc.*, Vol. 40, 1992, pp. 107–116.
20. K. C. Pohlmann, *The Compact Disc: A Handbook of Theory and Use*, A-R Editions, Madison, 1989.
21. G. C. Wirtz, "Digital Compact Cassette: Background and System Description," Audio Engineering Society Preprint No. 3215, 1991.
22. K. Tsutsui, H. Suzuki, O. Shimoyoshi, M. Sonohara, K. Akagari, and R. M. Heddle, "ATRAC: Adaptive Trans-

form Acoustic Coding for MiniDisc," Audio Engineering Society Preprint No. 3456, 1992.

23. E. Zwicker and U. T. Zwicker, "Audio Engineering and Psychoacoustics: Matching Signals to the Final Receiver, the Human Auditory System," *J. Audio Eng. Soc.*, Vol. 39, No. 3, 1991, pp. 115–125.
24. A. J. M. Houstma, "Perception of Bit-Compressed Digitally-Coded Music," *Proceedings of the International Symposium on Musical Acoustics*, Teikyo University of Technology, Tokyo 1992, pp. 143–146.
25. K. Brandenburg and G. Stoll, "ISO-MPEG-1 Audio: A Generic Standard for Coding of High-Quality Digital Audio," *J. Audio Eng. Soc.*, Vol. 42, 1994, pp. 780–792.
26. A. V. Oppenheim and R. W. Schafer, *Discrete-Time Signal Processing*, Prentice-Hall, Englewood Cliffs, NJ, 1989.
27. M. R. Schroeder, "Natural Sounding Artificial Reverberation," *J. Audio Eng. Soc.*, Vol. 10, 1962, pp. 219–223.
28. C. Dodge and T. A. Jerse, *Computer Music: Synthesis, Composition, and Performance*, Schirmer, New York, 1985.
29. F. R. Moore, *Elements of Computer Music*, Prentice-Hall, Englewood Cliffs, NJ, 1990.
30. X. Serra and J. O. Smith, "Spectral Modelling Synthesis: A Sound Analysis/Synthesis System Based on a Deterministic plus Stochastic Decomposition," *Computer Music J.*, Vol. 15, No. 1, 1990, pp. 12–24.
31. J. Beauchamp, "Synthesis by Spectral Amplitude and 'Brightness' Matching of Analyzed Musical Instrument Tones," *J. Audio Eng. Soc.*, Vol. 30, 1982, pp. 396–406.
32. A. Horner, J. Beauchamp, and L. Haken, "Methods for Matching Multiple Wavetable Synthesis of Musical Instrument Tones," *J. Audio Eng. Soc.*, Vol. 41, 1993, pp. 336–356.
33. W. Hess, *Pitch Determination of Speech Signals*, Springer-Verlag, Berlin, 1983.
34. R. C. Maher and J. W. Beauchamp, "Fundamental Frequency Estimation of Musical Signals Using a Two-Way Mismatch Procedure," *J. Acoust. Soc. Am.*, Vol. 95, 1994, pp. 2254– 2263.
35. E. Terhardt, "Pitch, Consonance, and Harmony," *J. Acoust. Soc. Am.*, Vol. 55, 1974, pp. 1061–1069.
36. J. Strawn (Ed.), *Digital Audio Signal Processing*, A-R Editions, Madison, 1985.
37. S. V. Vaseghi and R. Frayling-Cork, "Restoration of Old Gramophone Recordings," *J. Audio Eng. Soc.*, Vol. 40, 1992, pp. 791–801.
38. J.-M. Jot, V. Larcher, and O. Warusfel, "Digital Signal Processing Issues in the Context of Binaural and Transaural Stereophony," Audio Engineering Society Preprint No. 3980, 1995.

INDEX